Infrared and Terahertz Detectors

Infrared and Terahertz Detectors

Third Edition

Antoni Rogalski

CRC Press is an imprint of the
Taylor & Francis Group, an **informa** business

CRC Press
Taylor & Francis Group
6000 Broken Sound Parkway NW, Suite 300
Boca Raton, FL 33487-2742

First issued in paperback 2022

ISBN-13: 978-1-138-19800-5 (hbk)
ISBN-13: 978-1-03-233866-8 (pbk)
DOI: 10.1201/b21951

Visit the Taylor & Francis Web site at
http://www.taylorandfrancis.com

and the CRC Press Web site at
http://www.crcpress.com

"In memory of my daughter Marta"

Contents

Preface to the third edition

Progress in infrared (IR) detector technology has been mainly connected to semiconductor IR detectors, which are included in the class of photon detectors. They exhibit both perfect signal-to-noise performance and a very fast response. But to achieve this, the photon detectors require cryogenic cooling. Cooling requirements are the main obstacles to the more widespread use of IR systems based on semiconductor photodetectors making them bulky, heavy, expensive, and inconvenient to use.

Until the 1990s, despite numerous research initiatives and the appeal of room temperature operation and low-cost potential, thermal detectors have enjoyed limited success compared with cooled photon detectors for thermal imaging applications. Only the pyroelectric vidicon received much attention with the hope that it could be made practical for some applications. Throughout the 1980s and early 1990s, many companies in the United States (especially Texas Instruments and Honeywell's Research Laboratory) developed devices based on various thermal detection principles. In the mid-1990s, this success caused DARPA (Defense Advanced Research Projects Agency) to reduce support for HgCdTe and attempt a major leap with uncooled technology. The desire was to have producible arrays with useful performance, without the burden of fast (*f*/1) long-wavelength infrared optics.

In order to access these new changes in IR detector technology, there was need for a comprehensive introductory account of IR detector physics and operational principles, together with important references. In 2000, the first edition of *Infrared Detectors* was published with the intention of meeting this need. In the second edition published in 2011, about 70% of the contents were revised and updated, and much of the materials were reorganized. The last decade has seen considerable changes with new breakthroughs in detector concepts and performance. It became clear that the book needed substantial revision to continue to serve its purpose.

In this third edition titled *Infrared and Terahertz Detectors*, about 50% of the content is revised and updated. The arrangement of material will be similar to the previous edition, but several new chapters have been introduced and most of the remaining—reorganized.

The monograph is divided into five parts: fundaments of detection, infrared thermal detectors, infrared photon detectors, infrared focal plane arrays (FPAs) and terahertz detectors and focal plane arrays. The first part provides a tutorial introduction to the technical topics that are fundamental to a thorough understanding of different types of IR detectors and systems. The second part presents the theories and technologies of different types of thermal detectors, while the third part—theory and technology of photon detectors. The fourth part concerns IR FPAs where relations between the performance of detector array and IR system quality are considered. The new last part of this monograph constitutes comprehensive review of the present status and trends in the development of different types of terahertz detectors and focal plane arrays.

A short description given below concerns mainly the differences between the third edition and the previously published book (*Infrared Detectors*, second edition, 2011):

- in the fundamentals of IR detection, IR systems are discussed with an emphasis on the difference between night vision systems and thermal imaging systems,
- since heterodyne detection is more widely used in terahertz imaging systems, the chapter devoted to this topic has been considerably broadened,
- in the last decade, considerable progress has been observed in the development of a new class of IR photon detectors, the so-called barrier detectors. As a result, additional topics have been introduced like dedicated barrier and cascade photodetectors,
- in imaging systems with above megapixel formats, the pixel dimension plays a crucial role in determining critical system attributes such as system size, weight, and power consumption (SWaP). The advent of smaller pixels also results in superior spatial and temperature resolutions of imaging systems. The above topic is considered for the development trends of different types of FPAs—both thermal and photon detector arrays,

- it is expected that THz technology is one of the emerging technologies that will change our world. The THz region of the electromagnetic spectrum has proven to be one of the most elusive. Being situated between IR light and microwave radiation, THz radiation is resistant to the techniques commonly employed in these well-established neighboring bands. The last chapter is, in fact, the new chapter devoted to terahertz detectors. It has been completely reorganized with emphasis on status and trends in the development of THz detectors and imaging systems including low-dimensional solids and graphene.

This book is written for those who desire a comprehensive analysis of the latest developments in IR detector technology and a basic insight into the fundamental processes important for evolving detection techniques. Special attention has been given to the physical limits of detector performance and comparisons of performance in different types of detectors. The reader should gain a good understanding of the similarities and contrasts, the strengths and weaknesses of a multitude of approaches that have been developed over a century to improve our ability to sense IR radiation.

The level of presentation is suitable for graduate students in physics and engineering who have received standard preparation in modern solid-state physics and electronic circuits. This book is also of interest to individuals working with aerospace sensors and systems, remote sensing, thermal imaging, military imaging, optical telecommunications, IR spectroscopy, and light detection and ranging. To satisfy the needs of the first group, many chapters discuss the principles underlying each topic and some historical background before bringing to the reader the most recent information available. For those currently in the field, the book can be used as a collection of useful data, as a guide to the literature, and as an overview of topics covering a wide range of applications. The book could also be used as a reference for participants of relevant workshops and short courses.

The new edition of *Infrared and Terahertz Detectors* give, I hope, a comprehensive analysis of the latest developments in IR and THz detector technology and a basic insight into the fundamental processes important for evolving detection techniques. The book covers a broad spectrum of detectors, including theory, types of materials and their physical properties, and detector fabrication.

Antoni Rogalski

Acknowledgements to the third edition

In the course of this writing, many people have assisted me and offered their support. I would like, first, to express my appreciation to the management of the Institute of Applied Physics, Military University of Technology, Warsaw, Poland, for providing the environment in which I worked on the book.

The author has benefited from the kind cooperation of many scientists who are actively working in IR detector technologies. The preparation of this book was aided by many informative and stimulating discussions with the author's colleagues at the Institute of Applied Physics, Military University of Technology in Warsaw. The author thanks the following individuals for providing preprints, unpublished information, and in some cases original figures, which were used in preparing the book: Drs. L. Faraone and J. Antoszewski (University of Western Australia, Perth), Dr. J.L. Tissot (Ulis, Voroize, France), Dr. S.D. Gunapala (California Institute of Technology, Pasadena), Dr. M. Kimata (Ritsumeikan University, Shiga, Japan), Dr. M. Razeghi (Northwestern University, Evanston, Illinois), Drs. M.Z. Tidrow and P. Norton (U.S. Army RDECOM CERDEC NVESD, Fort Belvoir, Virginia), Dr. S. Krishna (University of New Mexico, Albuquerque, New Mexico), Dr. H.C. Liu (National Research Council, Ottawa, Canada), G.U. Perera (Georgia State University, Atlanta, Georgia), Professor J. Piotrowski (Vigo System Ltd., Ożarów Mazowiecki, Poland), Dr. M. Reine (Lockheed Martin IR Imaging Systems, Lexington, Massachusetts), Dr. F.F. Sizov (Institute of Semiconductor Physics, Kiev, Ukraine), and Dr. H. Zogg (AFIF at Swiss Federal Institute of Technology, Zürich). Thanks also to CRC Press, especially Luna Han, who encouraged me to undertake this new edition and for her cooperation and care in publishing this third edition.

Ultimately, it is the encouragement, understanding, and support of my family that provided me the courage to embark on this project and see it to its conclusion.

About the author

Antoni Rogalski is a professor at the Institute of Applied Physics, Military University of Technology in Warsaw, Poland. He is a leading researcher in the field of IR optoelectronics. During the course of his scientific career, he has made pioneering contributions in the areas of theory, design, and technology of different types of IR detectors. In 1997, he received an award from the Foundation for Polish Science (the most prestigious scientific award in Poland) for achievements in the study of ternary alloy systems for IR detectors—mainly an alternative to HgCdTe new ternary alloy detectors such as lead salts, InAsSb, HgZnTe, and HgMnTe. In 2013, he was elected as an ordinary member of the Polish Academy of Sciences (PASs) and a dean of PASs Division Four: Engineering Sciences. In the period 2013–2016, he was a member of the Central Commission for Academic Degrees and Titles.

Professor Rogalski's scientific achievements include determining the fundamental parameters of narrow-gap semiconducting materials estimating the ultimate performance of IR detectors, elaborating on studies of high-quality PbSnTe, HgZnTe, and HgCdTe photodiodes operated in IR spectral ranges; and conducting comparative studies of the performance limitation of HgCdTe photodiodes versus other types of photon detectors (especially QWIP, QDIP, and type II superlattice IR detectors) for third generation IR detectors. His activity is focused on the research of physical processes and phenomena conditioning the best quality detectors' formation. His cooperation with Vigo-System S.A. lasts about 20 years and is, in the Polish reality, a unique example of a cooperation between the academic research team and the High-Tech optoelectronic devices' implementation company.

Professor Rogalski has given over 70 invited plenary talks at international conferences. He is the author and co-author of over 250 indexed papers, cited ca. 7,000 with index h ≈ 40, 13 books, and 30 monographic papers (book chapters). His monumental monograph *Infrared Detectors* published by Taylor and Francis was translated into the Russian and Chinese languages. Another monograph entitles *High-Operating Temperature Infrared Photodetectors* (edited by SPIE Press, 2007), of which he is the co-author, summarizes the globally unique Polish scientific and manufacturing achievements in the field of near room temperature long wavelength IR detectors. Recently he has published the monograph *Antimonide-based infrared detectors – A new perspective* (SPIE Press) as a possible alternative to HgCdTe material system.

Professor Rogalski is a fellow of the International Society for Optical Engineering (SPIE), vice president of the Polish Optoelectronic Committee, editor-in-chief of the journal *Opto-Electronics Review* (1998–2015), deputy editor-in-chief of the *Bulletin of the Polish Academy of Sciences: Technical Sciences*, and a member of the editorial boards of *Journal of Infrared and Millimeter Waves, International Review of Physics, International Journal of Electronics and Telecommunications,* and *Photonics Letters of Poland.* He is an active member of the international technical community—a chair and co-chair, organizer, and member of scientific committees of many national and international conferences on optoelectronic devices and material sciences. For more information see web: http://antonirogalski.com.

Part I

Fundamentals of infrared and terahertz detection

1 Radiometry

This chapter discusses the vocabulary that will be needed in subsequent discussions. The radiometric calculations are a necessary part of the characterization of detectors and the prediction of signal and noise level. The word *radiometry* not only describes the detection and measurement of radiated electromagnetic energy, but is also used to describe the prediction and calculation of the power transferred by radiation from one object or surface to another. The concepts of radiometry are similar to those of ***photometry*** (related to vision and detection by the human eye) and to the transfer of ***photons***, making it convenient to discuss all three together.

The infrared range covers all electromagnetic radiation longer than the visible, but shorter than millimeter waves (Figure 1.1). The divisions between these categories are based on the different source and detector technologies used in each region. Many proposals in the division of IR range have been published and are shown in Table 1.1; these are based on limits of spectral bands of commonly used IR detectors. A wavelength of 1 μm is the sensitivity limit of popular Si detectors. Similarly, 3 μm is a long wavelength sensitivity of PbS and InGaAs detectors; 6-μm wavelength is the sensitivity limit of InSb, PbSe, PtSi detectors, and HgCdTe detectors optimized for a 3–5 μm atmospheric window; and finally, 15 μm is a long wavelength sensitivity limit of HgCdTe detectors optimized for an 8–14 μm atmospheric window.

The IR devices cannot be designed without an understanding of the amount of radiation power that impinges on the detector from the target, and the radiation of the target cannot be understood without a radiometric measurement. This issue is critical to the overall signal-to-noise ratio achieved by the IR system.

Our discussion in this chapter is simplified due to certain provisions and approximations. We specifically consider the radiometry of incoherent sources and ignore the effects of diffraction. In general, we make small-angle assumptions similar to those made for paraxial optics. The sine of an angle is approximated by the angle itself in radians.

This chapter provides some guidance in radiometry. For further details, see [1–7].

1.1 RADIOMETRIC AND PHOTOMETRIC QUANTITIES AND UNITS

Radiometry is the branch of optical physics that deals with the measurement of electromagnetic radiation in the frequency range between 3×10^{13} and 3×10^{16} Hz. This range corresponds to wavelengths between 10 nm and 10 μm and includes the regions commonly called ultraviolet, visible, and infrared. Radiometry deals with the actual energy content of light rather than its perception through a human visual system. Typical radiometric units include watt (radiant flux), watt per steradian (radiant intensity), watt per square meter (irradiance), and watt per square meter per steradian (radiance).

Historically, the power of a light source was obtained by observing brightness of the source. It turns out that brightness, perceived by the human eye, depends upon wavelength and the color of light, and differs from the actual energy contained in the light. The eye is sensitive to radiation over a range of approximately 11 orders of magnitude from bright sunlight to a flash of light containing only a few photons.

The retina of the human eye contains two different types of photoreceptors called rods and cones that produce nerve impulses that are passed on to subsequent stages of the human visual system for processing. The cones are spread over the entire retina, together with a large concentration within a small central area of our vision called the fovea, which results from our high visual acuity at the center of the field of view of

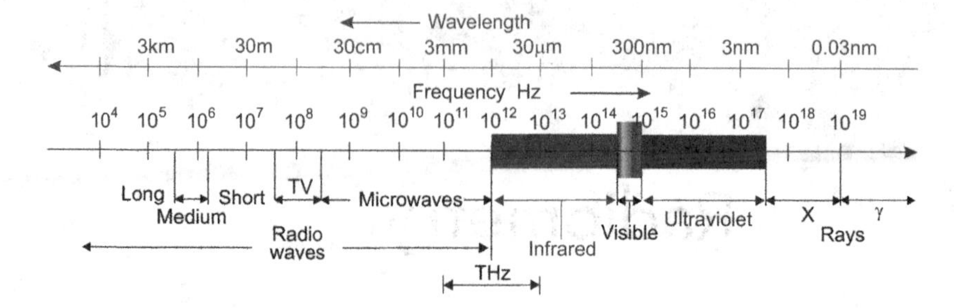

Figure 1.1 Electromagnetic spectrum.

Table 1.1 Division of infrared radiation

REGION (ABBREVIATION)	WAVELENGTH RANGE (μm)
Near-infrared (NIR)	0.78–1
Short wavelength IR (SWIR)	1–3
Medium wavelength IR (MWIR)	3–6
Long wavelength IR (LWIR)	6–15
Very long wavelength IR (VLWIR)	15–30
Far infrared (FIR)	30–100
Submillimeter (SubMM)	100–1000

the eye. The cones are responsible for our daytime color vision. The rods are spread over the entire retina, except the fovea, and are responsible for our nighttime, basically black-and-white, vision.

The eye is most sensitive to the yellow–green light and less sensitive to red and blue lights of the spectrum. To take the difference into account, a new set of physical measures of light is defined for the visible light that parallels the quantities of radiometry, where the power is weighted according to the human response by multiplying the corresponding quantity by a spectral function, called the $V(\lambda)$ function or the spectral luminous efficiency for photopic vision, defined in the domain from 360 to 830 nm, and is normalized to one at its leak, 555 nm (Figure 1.2). The $V(\lambda)$ function tells us the appropriate response of the human eye to various wavelengths. This function was first defined by the Commission Internationale de l'Éclairage (CIE) in 1924 [8] and is an average response of a population of people in a wide range of ages. It should be noted that the $V(\lambda)$ function was defined assuming additivity of sensation and a 2° field of view at relatively high luminance levels (>1 cd/m^2)—the high radiation range is mediated by the cones. The

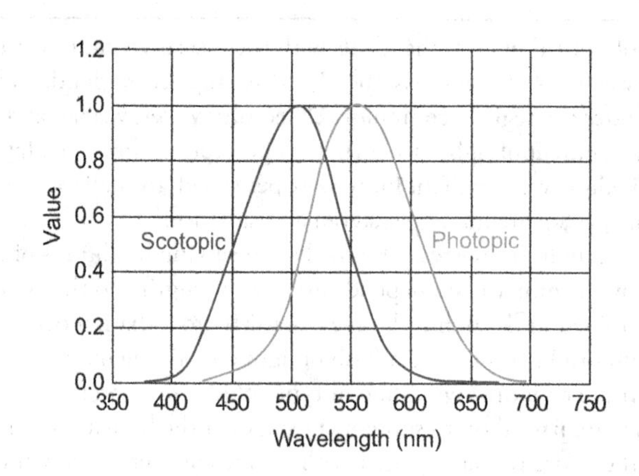

Figure 1.2 CIE spectral luminous efficiency functions.

spectral responsivity of human eyes deviates significantly at very low levels of luminescence (<10^{-3} cd/m^2) when the rods in the eyes are the dominant receptors. This type of vision is called scotopic vision.

Photometry is the measurement of light, which is defined as radiation detectable by the human eye. It is restricted to the visible region and all quantities are weighted by the spectral response of the eye. Typical photometric units include lumen (luminous flux), candela (luminous intensity), lux (illuminance), and candela per square meter (luminance).

The units, as well as the names, of similar properties in photometry differ from those in radiometry. For instance, power is simply called power in radiometry or radiant flux, but it is called the luminous flux in photometry. While the unit of power in radiometry is watt, in photometry it is lumen. A lumen is defined in terms of a superfluous fundamental unit, called candela, which is one of the seven independent quantities of the SI system of units (meter, kilogram, second, ampere, Kelvin, mole, and candela). Candela is the SI unit of the photometric quantity called luminous intensity or luminosity that corresponds to the radiant intensity in radiometry. Table 1.2 lists the radiometric and photometric quantities and units along with translation between both groups of units.

Radiometry is plagued by a confusion of terminology, symbols, definitions, and units. The origin of this confusion is largely because of the parallel or duplicate development of the fundamental radiometric practices by researchers in different disciplines. Consequently, considerable care should be exercised when reading publications. The terminology used in this chapter follows international standards and recommendations [7,9].

1.2 DEFINITIONS OF RADIOMETRIC QUANTITIES

Radiant flux, also called radiant power, is the energy Q (in joules) radiated by a source per unit of time and is defined by

$$\Phi = \frac{dQ}{dt}. \tag{1.1}$$

The unit of radiant flux is the Watt (W = J/s).

Radiant intensity is the radiant flux from a point source emitted per unit solid angle in a given direction and is expressed as

$$I = \frac{d\Phi}{d\Omega} = \frac{\partial^2 Q}{\partial t\, \partial\Omega}, \tag{1.2}$$

Table 1.2 Radiometric and photometric quantities and units

PHOTOMETRIC QUANTITY	UNIT	RADIOMETRIC QUANTITY	SYMBOL	UNIT	UNIT CONVERSION
Luminous flux	lm (lumen)	Radiant flux	ϕ	W (Watt)	1 W = 683 lm
Luminous intensity	cd (candela) = lm/sr	Radiant intensity	I	W/sr	1 W/sr = 683 cd
Illuminance	lx (lux) = lm/m^2	Irradiance	E	W/m^2	1 W/m^2 = 683 lx
Luminance	cd/m^2 = lm/(sr m^2)	Radiance	L	W/(sr m^2)	1 W/(sr m^2) = 683 cd/m^2
Luminous exitance	lm/m^2	Radiant exitance	M	W/m^2	
Luminous exposure	lx s	Radiant exposure		W/(m^2 s)	
Luminous energy	lm s	Radiant energy	Q	J (Joule)	1 J = 683 lm s

where $d\Phi$ is the radiant flux leaving the source and propagating in an element of solid angle $d\Omega$ containing the given direction (see Figure 1.3). The unit of radiant intensity is W/sr.

The solid angle may be expressed in differential form as

$$d\Omega = \frac{dA}{r^2}. \tag{1.3}$$

The unit of solid angle is steradian (sr).

If we use the spherical coordinate system seen in Figure 1.4, and use $dA = r^2 \sin\theta d\theta d\varphi$, we can write an expression for the solid angle subtense of a flat disc of planar half angle θ_{max} as

$$\Omega = \int d\Omega = \int_0^{2\pi} d\varphi \int_0^{\theta_{max}} \sin\theta \, d\theta = 2\pi(1 - \cos\theta_{max}). \tag{1.4}$$

Irradiance is the density of incident radiant flux at a point of surface and is defined as radiant flux per unit area [see Figure 1.5a], as follows

$$E = \frac{\partial\Phi}{\partial A} = \frac{\partial^2 Q}{\partial t \partial A}, \tag{1.5}$$

where $\partial\Phi$ is the radiant flux incident on an element ∂A of the surface containing the point. The unit of irradiance is W/m^2.

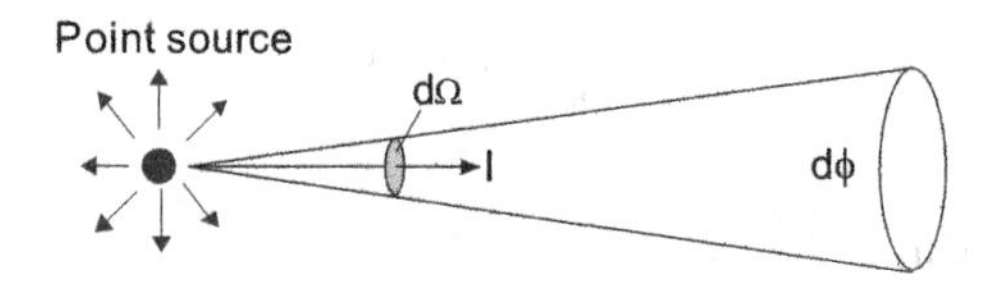

Figure 1.3 Radiant intensity.

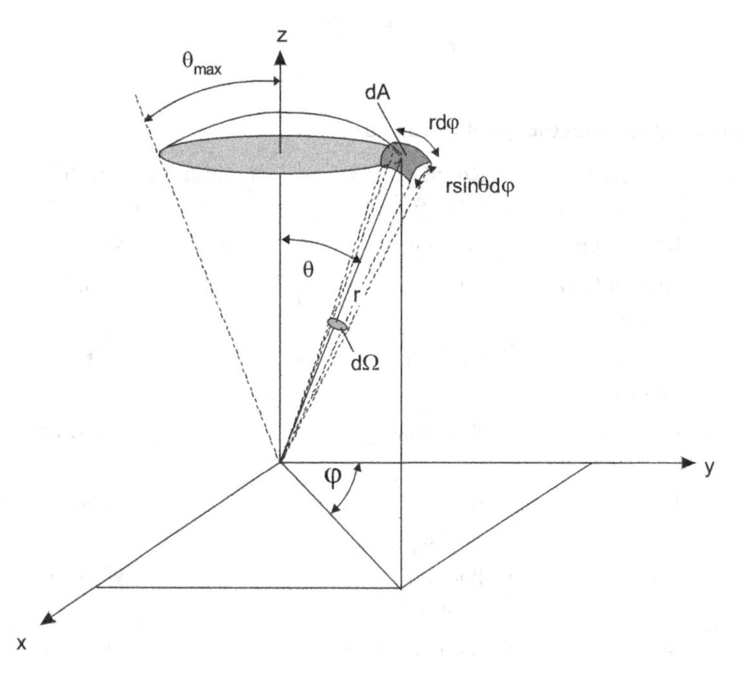

Figure 1.4 Relationship of solid angle to planar angle.

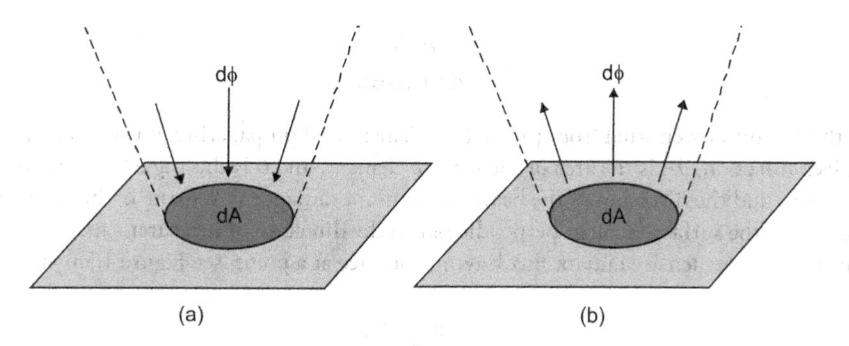

Figure 1.5 (a) Irradiance and (b) radiation exitance.

For strictly point source, a common rule of thumb is the "one-over-r-squared falloff." We consider a receiver of area A, which is placed at various distances from a point source having a uniform radiant intensity I (Figure 1.6). Using Equation 1.2 we get

$$\Phi = I\frac{A}{r^2}, \tag{1.6}$$

and next

$$E = \frac{\Phi}{A} = \frac{I}{r^2}. \tag{1.7}$$

Because the solid angle subtended by the detector falls off as $1/r^2$, the collected flux and the irradiance also decreases proportionally. Note that in the case of extended source, sufficient distance relative to the source is required to assume the relationship (Equation 1.7).

Radiance is the radiant flux per unit solid angle emitted from a surface element in a given direction, per unit projected area of the surface element perpendicular to the direction (see Figure 1.7). This quantity is defined by

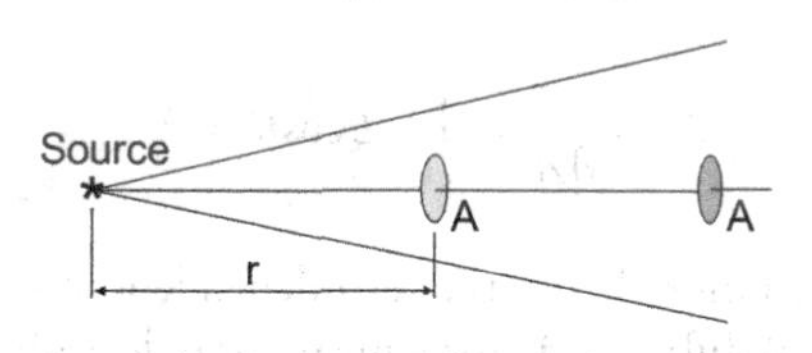

Figure 1.6 Irradiance falloff as a function of r from source.

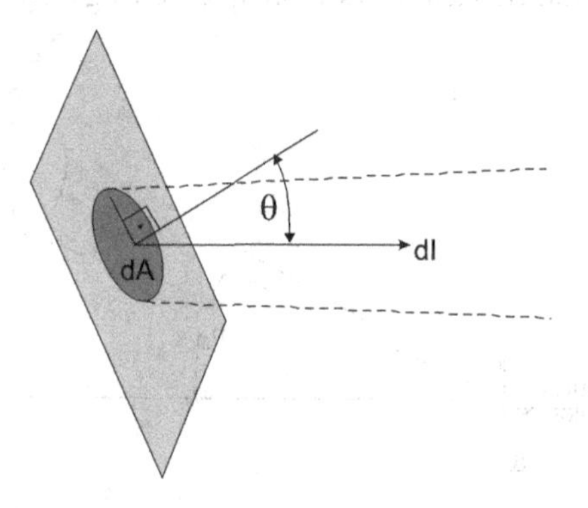

Figure 1.7 Radiance.

$$L = \frac{\partial^2 \Phi}{\partial \Omega \partial A \cos\theta}, \tag{1.8}$$

where $\partial\Phi$ is the radiant flux emitted from the surface element and propagating in the solid angle $\partial\Omega$ containing the given direction, ∂A is the area of the surface element, and θ is the angle between the normal to the surface element and the direction of the beam. The unit of radiance is W/(srm²). The term $\partial A\cos\theta$ gives the projected area of the surface element perpendicular to the direction of measurement.

Radiant exitance is the density radiant flux leaving a surface at a point (see Figure 1.5b) and is defined by

$$M = \frac{\partial \Phi}{\partial A} = \frac{\partial^2 Q}{\partial t \partial A}, \tag{1.9}$$

where $\partial\Phi$ is the radiant flux leaving the surface element. The unit of radiant exitance is W/m².

Irradiance and radiant exitance have the same units but have different interpretations. Irradiance is the amount of power with respect to the unit area that falls on a surface, while radiant exitance is the amount of power per unit area that leaves a surface. Exitance thus characterizes a self-luminous source that is producing energy, while irradiance characterizes a passive receiver surface.

1.3 RADIANCE

Radiance is used to characterize an extended source (see Figure 1.8). Equation 1.8 indicates that the power received by the detector is differential with respect to both the incremental projected area of the source and the incremental solid angle of the detector. Rearranging this equation, we have

$$\partial^2\Phi = L\partial A_s \cos\theta_s \partial\Omega_d, \tag{1.10}$$

and integrating once with respect to source, we obtain intensity

$$I = \frac{\partial \Phi}{\partial \Omega_d} = \int_{A_s} L\cos\theta_s dA_s. \tag{1.11}$$

Similarly, integrating once with respect to detector solid angle, we obtain radiant exitance

$$M = \frac{\partial \Phi}{\partial A_s} = \int_{\Omega_d} L\cos\theta_s d\Omega_d. \tag{1.12}$$

A Lambertian radiator has a constant radiance that is independent of viewing direction. This type of reflector is also referred to as an ideal diffuse radiator (emitter or reflector); see Figure 1.9. In practice, there are no true Lambertian surfaces. Most matte surfaces approximate an ideal diffuse reflector, but typically exhibit semi-specular reflection characteristics at oblique viewing angles. An ideal thermal source

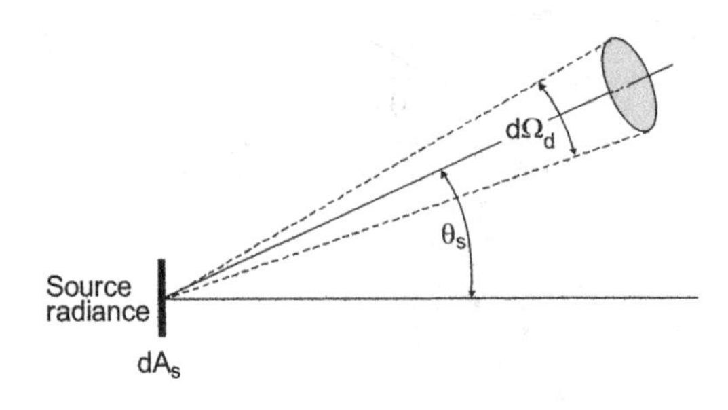

Figure 1.8 Radiance of an extended source.

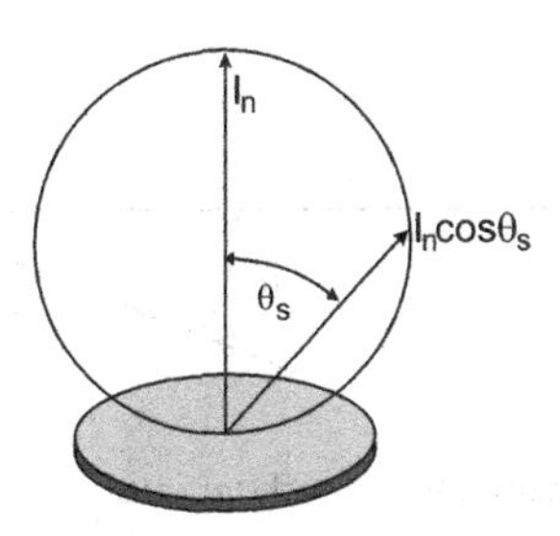

Figure 1.9 Radiant intensity as a function of θ_s for a Lambertian source.

(blackbody) is perfectly Lambertian, while certain special diffusers also closely approximate the condition. An actual source is typically approximately Lambertian within a range of view angles θ_s that is less than 20°.

Even for a Lambertian source, the intensity depends on θ_s. Making the assumption of L, independent of source position, from Equation 1.11 follows

$$I = \frac{\partial \Phi}{\partial \Omega_d} = \int_{A_s} L\cos\theta_s \, dA_s = LA_s\cos\theta_s = I_n\cos\theta_s. \quad (1.13)$$

It is Lambert's cosine law, where I_n is the intensity of the ray leaving in a direction perpendicular to the surface. For non-Lambertian surfaces, the radiance L is a function of the angle itself, and the falloff of I with θ_s is faster than cos θ_s.

To receive relationship between radiation exitance and radiance for a planar Lambertian source, we return to Equation 1.12 and integrate

$$M = \frac{\partial \Phi}{\partial A_s} = \int_{\Omega_d} L\cos\theta_s \, d\Omega_d = \int_0^{2\pi} d\varphi \int_0^{\pi/2} L\cos\theta_s \sin\theta \, d\theta = 2\pi L\frac{1}{2} = \pi L, \quad (1.14)$$

where the Lambertian-source assumption has been used to pull L outside of the angular integrals. For a non-Lambertian source, the integration yields a proportionality constant different from π.

Let us simplify further considerations assuming $\theta_s = 0$. Then, for the geometrical configuration shown in Figure 1.10, the radiant power on the detector can be obtained by multiplying the detector's solid angle by the area of the source and the radiance of the source [10]

$$\Phi_d = LA_s\Omega_d = \frac{LA_sA_d}{r^2} = L\Omega_sA_d. \quad (1.15)$$

From this equation results that the flux on the detector is expressed as the radiance of the source multiplied by an area × solid angle ($A\Omega$) product. To fulfill Equation 1.15, two provisions are required: a small-angle assumption for the approximation of the solid angle of a flat surface by A/r^2 and the flux transfer is unaffected by absorption losses in the system.

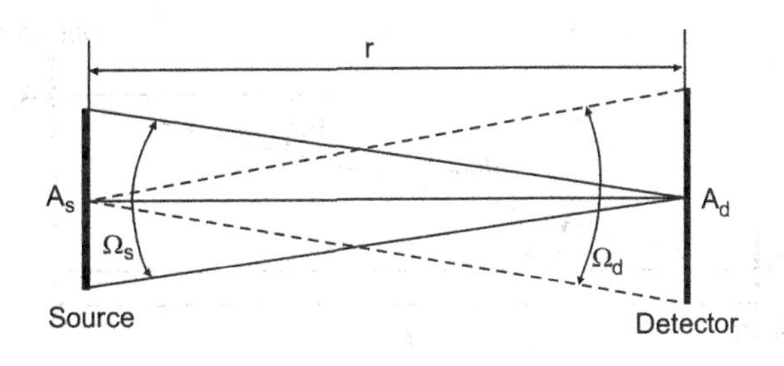

Figure 1.10 Radiant power transfer from source to detector.

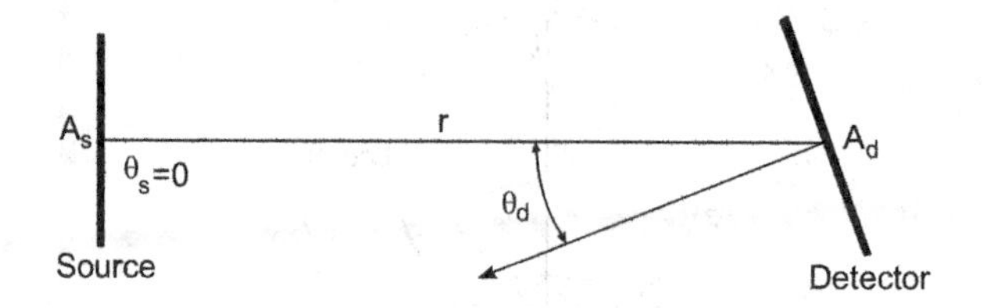

Figure 1.11 Radiant power transfer from source to a titled detector.

Another situation occurs for a tilted receiver shown in Figure 1.11. The source normal is along the line of centers, so $\theta_s = 0$ in this case. The angle θ_d is the angle between the line of centers and the normal to the detector surface. In this situation,

$$\Phi_d = LA_s\Omega_d. \tag{1.16}$$

Assuming that for a titled surface

$$\Omega_d = \frac{A_d \cos\theta_d}{r^2}, \tag{1.17}$$

we find

$$\Phi_d = LA_s \frac{A_d \cos\theta_d}{r^2}. \tag{1.18}$$

Thus, the flux collected and the irradiance (Φ/A_d) are decreased by a factor of $\cos\theta_d$.

We now proceed to calculate the flux on the detector when both θ_s and θ_d are nonzero, assuming a flat Lambertian source (Figure 1.12). A cosine falloff factor arises at both the source and the receiver. In this case,

$$\Phi_d = LA_s \cos\theta_s \frac{A_d \cos\theta_d}{\left(r/\cos\theta_s\right)^2}. \tag{1.19}$$

Assuming that the surfaces of source and detector are parallel and $\theta_s = \theta_d = \theta$, the radiant intensity is proportional to $\cos^4\theta$. On account of this, Equation 1.19 is the so-called cosine to the fourth law.

Finally, we consider the flux transfer in image-forming systems, assuming the limitations of paraxial optics (small angles) as shown in Figure 1.13. Only a certain amount of flux Φ is collected by the optical system, which can be calculated letting the lens aperture (A_{lens}) act as an intermediate receiver. It should be noted that in a more complex system, the entrance pupil is this intermediate receiver and the $A_{lens}\Omega_{obj}$ product is the area–solid angle product of the optical system. At these conditions, the collected radiant flux equals

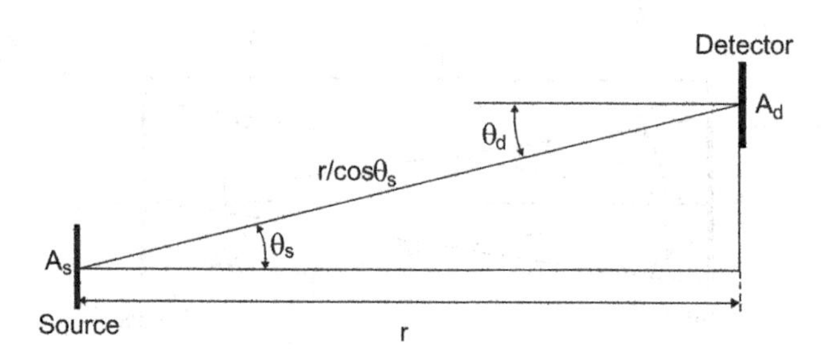

Figure 1.12 Cosine-to-the-fourth ($\cos^4$) law.

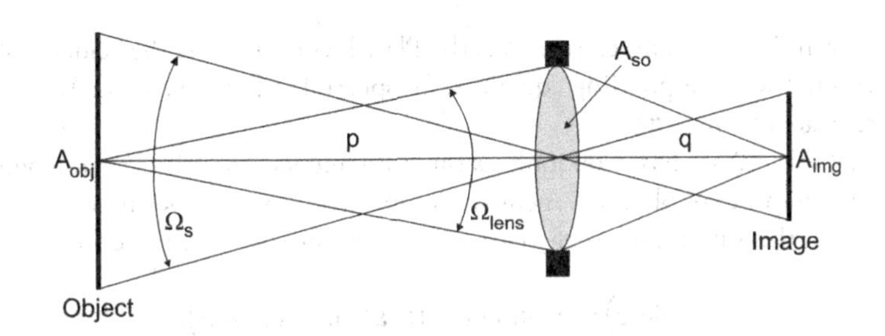

Figure 1.13 Radiant power collected by an optical system. (From Ref. [10])

$$\Phi = LA_{obj}\Omega_{lens} = LA_{lens}\Omega_{obj}, \tag{1.20}$$

and is reformatted by the lens and forms an image of the original object at an appropriate magnification.

The image irradiance can be found simply by dividing the flux collected in Equation 1.20 by the image area:

$$\Phi = LA_{lens}\Omega_{lens} = LA_{lens}\Omega_{obj} = L\frac{A_{lens}A_{obj}}{p^2} = L\frac{A_{lens}A_{img}}{q^2}. \tag{1.21}$$

The last equality was obtained using $A_{img} = A_{obj}(q/p)^2$.

From Equation 1.21, the image irradiance is equal to

$$E_{img} = \frac{\Phi}{A_{imag}} = L\frac{A_{lens}}{q^2}. \tag{1.22}$$

1.4 BLACKBODY RADIATION

All objects are composed of continually vibrating atoms, with higher energy atoms vibrating more frequently. The vibration of all charged particles, including these atoms, generates electromagnetic waves. The higher the temperature of an object, the faster the vibration, and thus the higher the spectral radiant energy. As a result, all objects are continually emitting radiation at a rate with a wavelength distribution that depends upon the temperature of the object and its spectral emissivity, $\varepsilon(\lambda)$.

Radiant emission is usually treated in terms of the concept of a blackbody [4]. A blackbody is an object that absorbs all incident radiation and, conversely according to the Kirchhoff law, is a perfect radiator. The energy emitted by a blackbody is the maximum theoretically possible for a given temperature. A device of this type is a very useful standard source for the calibration and testing of radiometric instruments. Further, most sources of thermal radiation radiate energy in a manner that can be readily described in terms of a blackbody emitting through a filter, making it possible to use the blackbody radiation laws as a starting point for many radiometric calculations.

The blackbody or Planck equation was one of the milestones of physics. The Planck law describes the spectral radiance (spectral radiant exitance) of a perfect blackbody as a function of its temperature and the wavelength of the emitted radiation, in the forms

$$L(\lambda,T) = \frac{2hc^2}{\lambda^5}\left[\exp\left(\frac{hc}{\lambda kT}\right) - 1\right]^{-1} \text{ W/(cm}^2\text{ sr }\mu\text{m)}, \tag{1.23}$$

$$M(\lambda,T) = \frac{2\pi hc^2}{\lambda^5}\left[\exp\left(\frac{hc}{\lambda kT}\right) - 1\right]^{-1} \text{ W/(cm}^2\ \mu\text{m)}, \tag{1.24}$$

where λ is the wavelength, T is the temperature, h is the Planck constant, c is the velocity of light, and k is the Boltzmann constant. The corresponding equations for spectral radiant exitance, $M(\lambda,T)$, and spectral radiance, $L(\lambda,T)$, are related by $M = \pi L$.

The units listed in Table 1.2 are based on joule as the fundamental quantity. An analogous set of quantities can be based on the number of photons. A conversion between two sets of units is easily accomplished using the relationship for the amount of energy carried per photon: $\varepsilon = hc/\lambda$. For example,

$$\phi(\text{joule}/s) = \phi(\text{photon}/s) \times \varepsilon(\text{joule}/\text{photon}). \tag{1.25}$$

In a similar way, the Equations 1.23 and 1.24 may be transformed to the forms

$$L(\lambda,T) = \frac{2c}{\lambda^4}\left[\exp\left(\frac{hc}{\lambda kT}\right)-1\right]^{-1} \text{ photon/(s cm}^2 \text{ sr } \mu\text{m)}, \tag{1.26}$$

$$M(\lambda,T) = \frac{2\pi c}{\lambda^4}\left[\exp\left(\frac{hc}{\lambda kT}\right)-1\right]^{-1} \text{ photon/(s cm}^2 \text{ } \mu\text{m)}. \tag{1.27}$$

Figure 1.14 shows a plot of these curves for a number of blackbody temperatures. As the temperature increases, the amount of energy emitted at any wavelength increases too, and the wavelength of peak emission decreases. The latter is given by Wien's displacement law [11]:

$$\lambda_{mw}T = 2898 \ \mu\text{mK for maximum watts}, \tag{1.28}$$

$$\lambda_{mp}T = 3670 \ \mu\text{mK for maximum Photons}, \tag{1.29}$$

which is derived from the condition for the peak of the exitance function by setting the derivative equal to zero

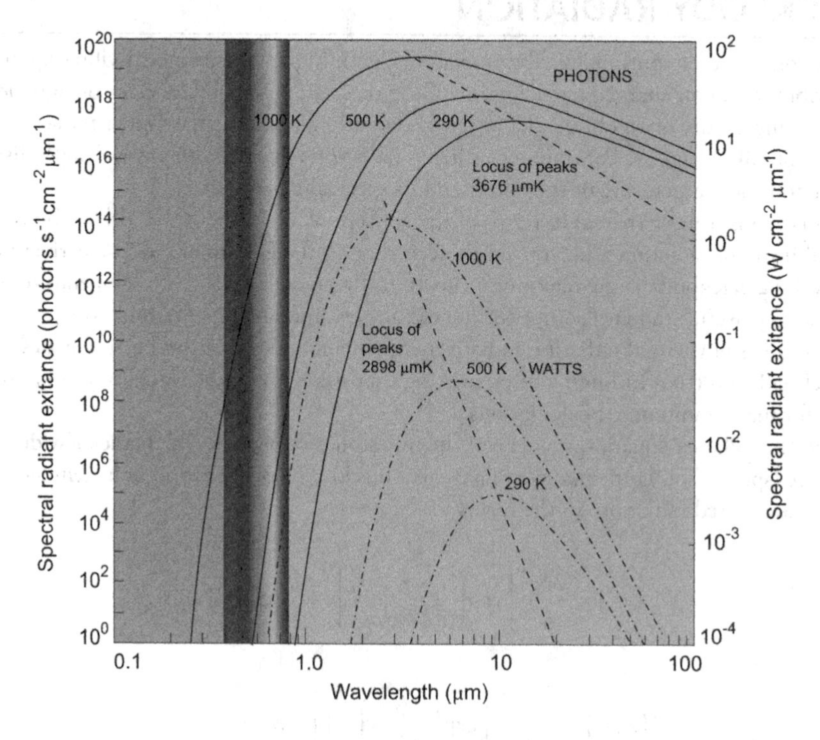

Figure 1.14 Planck's law for spectral radiant exitance. (From Ref. [11])

$$\frac{dM(\lambda,T)}{d\lambda}=0 \tag{1.30}$$

and solving for the wavelength at maximum exitance.

The loci of these maxima are shown in Figure 1.14. Note that for an object at an ambient temperature of 295 K, λ_{mw} and λ_{mp} occur at 10.0 μm and 12.7 μm, respectively. We need detectors operating near 10 μm if we expect to "see" room temperature objects such as people, trees, and vehicles without the aid of reflected light. For hotter objects such as engines, maximum emission occurs at shorter wavelengths. Thus, the waveband 2–15 μm in the infrared or thermal region of the electromagnetic spectrum contains the maximum radiative emission for thermal imaging purposes. It is interesting to note that the λ_{mw} for the sun is near 0.5 μm, very close to the peak of sensitivity of the human eye.

Total radiant exitance from a blackbody at temperature T is the integral of spectral exitance over all wavelengths

$$M(T)=\int_0^\infty M(\lambda,T)d\lambda=\int_0^\infty \frac{2\pi hc^2}{\lambda^5\left[\exp\left(\frac{hc}{\lambda kT}-1\right)\right]}d\lambda=\frac{2\pi^5k^4}{15c^2h^3}T^4=\sigma T^4, \tag{1.31}$$

where $\sigma = 2\pi^5k^4/15c^2h^3$ is called the Stefan–Boltzmann constant and has an approximate value of 5.67×10^{-12} W/cm²K⁴.

The relation determined by Equation 1.31 between the total radiant exitance of a blackbody and its temperature is called the Stefan–Boltzmann law. The total exitance can be interpreted as the area under the spectral exitance curve for a given temperature, as is shown in Figure 1.15.

The radiant exitance of blackbody between λ_a and λ_b is obtained by integrating the Planck law over the integral $[\lambda_a,\lambda_b]$ as shown in Figure 1.15:

$$M_{\Delta\lambda}(T)=\int_{\lambda_a}^{\lambda_b} M(\lambda,T)d\lambda=\int_{\lambda_a}^{\lambda_b}\frac{2\pi hc^2}{\lambda^5\left[\exp\left(\frac{hc}{\lambda kT}-1\right)\right]}d\lambda. \tag{1.32}$$

The exitance of a human body at 300 K is 500 W/m² and for a skin surface of 2 m², the radiative power is of 10^3 W. But the loss of energy is partially compensated by the radiation absorption.

The exitance at 300 K over the interval 8–14 μm is equal to 1.22×10^2 W/m² and is equivalent to the exitance over the interval 3–5 μm at 410 K. If we reported on a curve (the ratio $M_{[8-12]}/M_{[3-5]}$ shown

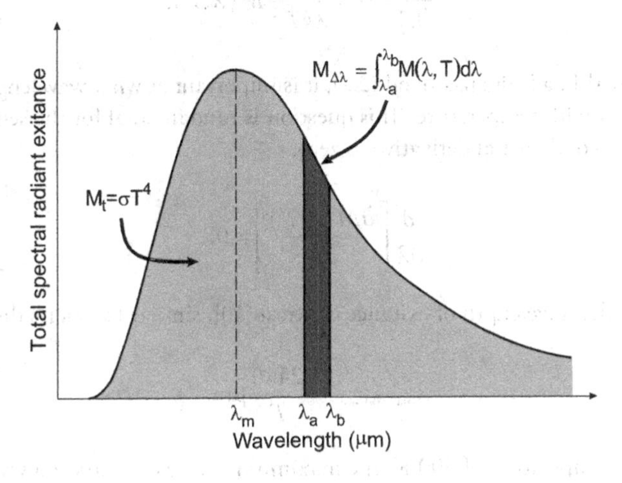

Figure 1.15 Total spectral radiant exitance at temperature T versus wavelength.

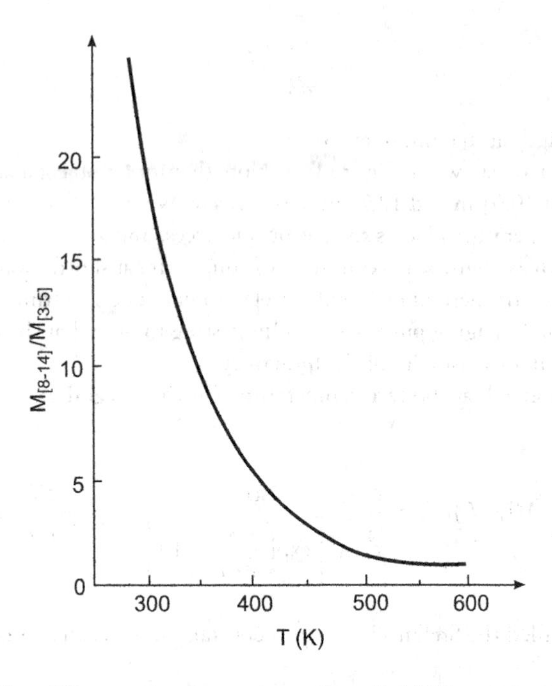

Figure 1.16 Exitance ratio $M_{[8-12]}/M_{[3-5]}$ against temperature. (From Ref. [12])

in Figure 1.16), we notice that the exitance over 8–14 μm is larger than exitance over 3–5 μm up to 600 K [12]. These results show the advantage of using the 8–14 μm region for low-temperature targets.

The temperature variation of the spectral radiant exitance is given by the temperature derivation of Equation 1.24; then

$$\frac{\partial M(\lambda,T)}{\partial T}=\frac{(hc/k)\exp(hc/\lambda kT)}{\lambda T^2\left[\exp(hc/\lambda kT)-1\right]}M(\lambda,T). \tag{1.33}$$

Generally, in thermal imaging, the objects are at temperatures near 300 K for $\lambda_{max} \approx 10\,\mu m$ or near 700 K for $\lambda_{max} \approx 4\,\mu m$. In these two cases, $\lambda << hc/kT$ and then

$$\frac{\partial M(\lambda,T)}{\partial T}=\frac{hc}{\lambda kT^2}M(\lambda,T). \tag{1.34}$$

For a system operating within a finite passband ($\Delta\lambda$), it is important at what wavelength the source (target) exitance changes the most with temperature. This question is fundamental for the sensitivity of an infrared system. Comparing the second partial derivative to zero,

$$\frac{\partial}{\partial\lambda}\left[\frac{\partial M(\lambda,T)}{\partial T}\right]=0, \tag{1.35}$$

produces a constrain on the wavelength of exitance contrast [10], similar to Wien's displacement law

$$\lambda_{\text{max contrast}}=\frac{2410}{T}\,\mu\text{m}. \tag{1.36}$$

For example, at a source temperature of 300 K, the maximum contrast occurs at a wavelength of about 8 μm, which is not the wavelength for maximum exitance.

1.5 EMISSIVITY

As was mentioned previously, the blackbody curve provides the upper limit of the overall spectral exitance of a source for any specific temperature. Most thermal sources are not perfect blackbodies. Many are called graybodies. A graybody is one that emits radiation in exactly the same spectral distribution as a blackbody at the same temperature, but with reduced intensity.

The ratio between the exitance of the actual source and the exitance of a blackbody at the same temperature is defined as emissivity. In general, emissivity depends on λ and T:

$$\varepsilon(\lambda,T)=\frac{M(\lambda,T)_{\text{source}}}{M(\lambda,T)_{\text{blackbody}}}. \tag{1.37}$$

and is a dimensionless number ≤ 1.

For a perfect blackbody $\varepsilon = 1$ for all wavelengths. The emissivity of graybody is independent of λ (see Figure 1.17). A selective source has an emissivity that depends on wavelength.

The total radiant exitance for graybody at all wavelengths is equal

$$M^{gb}(T)=\varepsilon\sigma T^{4}. \tag{1.38}$$

When radiant energy is incident on a surface, fraction α is absorbed, fraction r is reflected, and fraction t is transmitted. Since energy must be conserved, the following relationship can be written:

$$\alpha+r+t=1. \tag{1.39}$$

Kirchhoff observed that at a given temperature, the ratio of the integrated emissivity to the integrated absorptance is a constant for all materials and that it is equal to the radiant exitance of a blackbody at that temperature. Known as Kirchhoff's law, it can be stated as

$$\frac{M(\lambda,T)_{\text{source}}}{\lambda}=M(\lambda,T)_{\text{blackbody}}. \tag{1.40}$$

This law is often paraphrased as "good absorbers are good emitters." Combining Equations 1.31 and 1.37 give

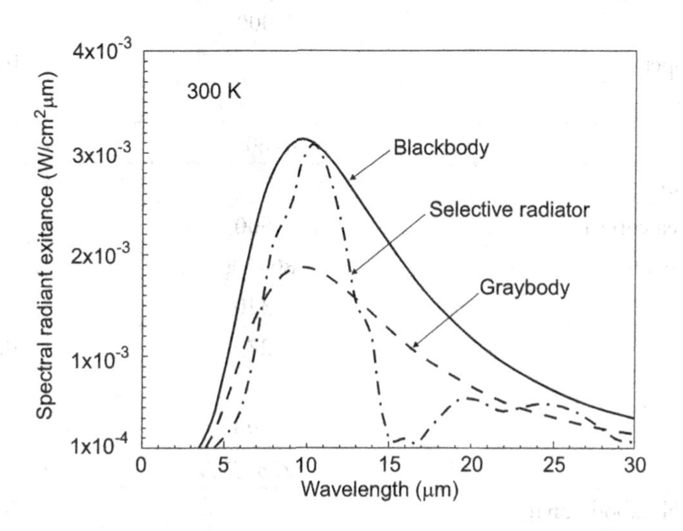

Figure 1.17 Spectral radiant exitance of three different radiators.

$$\frac{\varepsilon\sigma T^4}{\alpha} = \sigma T^4. \tag{1.41}$$

From this follows that

$$\varepsilon = \alpha. \tag{1.42}$$

Thus, the emissivity of any materials at a given temperature is numerically equal to its absorptance at that temperature. Since an opaque material does not transmit energy, $\alpha + r = 1$ and

$$\varepsilon = 1 - r. \tag{1.43}$$

Table 1.3 lists the emissivity of a number of common materials that are frequently a single number, and are seldom given as either a function of λ or T unless it is an essentially well-characterized material [13]. The dependence of emissivity on wavelength results from the fact that many substances (glass, for example) have a negligible absorption and consequent low emissivity at certain wavelengths, while they are almost totally absorbent at other wavelengths. For many materials, emissivity decreases as wavelength increases. For nonmetallic substances, typically $\varepsilon > 0.8$ for room temperature and decreases with increasing temperature. For a metallic substance, the emissivity is very low at room temperature and generally increases in proportion to temperature.

1.6 INFRARED OPTICS

The optical block in an IR system creates an image of observed objects in the plane of the detector (detectors). In the case of a scanning imager, the optical scanning system creates an image with the number of

Table 1.3 Emissivity of a number of materials

MATERIAL	TEMPERATURE (K)	EMISSIVITY
Tungsten	500 1,000 2,000 3,000 3,500	0.05 0.11 0.26 0.33 0.35
Polished silver	650	0.03
Polished aluminium	300 1,000	0.03 0.07
Polished copper		0.02–0.15
Polished iron		0.2
Polished brass	4–600	0.03
Oxidized iron		0.8
Black oxidized copper	500	0.78
Aluminium oxide	80–500	0.75
Water	320	0.94
Ice	273	0.96–0.985
Paper		0.92
Glass	293	0.94
Lampblack	273–373	0.95
Laboratory blackbody cavity		0.98–0.99

Source: After Ref. [13].

pixels much greater than the number of elements of the detector. In addition, optical elements like windows, domes, and filters can be used to protect the system from the environment or to modify the detector spectral response.

There is no essential difference in design rules of optical objectives for visible and IR ranges. The designer of IR optics is only more limited because there are significantly fewer materials suitable for IR optical elements, in comparison with those for the visible range, particularly for wavelengths over 2.5 μm.

There are two types of IR optical element: reflective elements and refractive elements. As the names suggest, the role of reflective elements is to reflect incident radiation, and the role of refractive elements is to refract and transmit incident radiation.

Mirrors used extensively inside IR systems (especially in scanners) are most often met as reflective elements that serve manifold functions in IR systems. Elsewhere they need a protective coating to prevent them from tarnishing. Spherical or aspherical mirrors are employed as imaging elements. Flat mirrors are widely used to fold optical paths, and reflective prisms are often used in scanning systems.

Four materials are most often used for mirror fabrication: optical crown glass, low-expansion borosilicate glass (LEBG), synthetic fused silica, and Zerodur. Less popular in use are metallic substrates (beryllium, copper) and silicon carbide. Optical crown glass is typically applied in non-imaging systems. It has a relatively high thermal expansion coefficient and is employed when thermal stability is not a critical factor. LEBG, known by the Corning brand name Pyrex, is well suited for high-quality front surface mirrors designed for low optical deformation under thermal shock. Synthetic fused silica has a very low thermal expansion coefficient.

Metallic coatings are typically used as reflective coatings of IR mirrors. There are four types of metallic coatings that are used most often: bare aluminium, protected aluminium, silver, and gold. They offer high reflectivity, over about 95%, in the 3–15 μm spectral range. Bare aluminium has a very high reflectance value but oxidizes over time. Protected aluminium is a bare aluminium coating with a dielectric overcoat that arrests the oxidation process. Silver offers better reflectance in the near IR than aluminium and high reflectance across a broad spectrum. Gold is a widely used material and offers consistently very high reflectance (about 99%) in the 0.8–50 μm range. However, gold is soft (it cannot be touched to remove dust) and is most often used in the laboratory.

Most glasses used to manufacture optical elements for visible and near-infrared range transmit light up to about 2.2 μm and can be used for SWIR optics. Thermal imagers use almost exclusively two spectral bands: 3–5 μm or 8–14 μm. Therefore, for infrared optics materials typically considered are those suitable to transmit infrared radiation in the spectral range from 2 to 14 μm.

The list of potential materials that could be used to manufacture infrared refractive optics is quite long: Amorphous Material Transmitting Infrared Radiation (AMTIR-1), barium fluoride (BaF_2), cadmium telluride (CdTe), calcium fluoride (CaF_2), cesium bromide (CsBr), cesium iodide (CsI), fused silica-IR grade, gallium arsenide (GaAs), germanium (Ge), lithium fluoride (LiF), magnesium fluoride (MgF_2), potassium bromide (KBr), potassium chloride (KCl), silicon (Si), sodium chloride (NaCl), thallium bromoiodide (KRS-5), zinc selenide (ZnSe), zinc sulfide (ZnS). However, only the most popular materials used to manufacture refractive optical objectives for thermal imagers will be discussed here. Basic parameters of these materials are presented in Table 1.4 and their IR transmission is shown in Figure 1.18.

Germanium is a silvery metallic-appearing solid of very high refractive index (> 4) that enables the design of high-resolution optical systems using a minimal number of germanium lenses. Its useful transmission range is from 2 to about 15 μm. It is quite brittle and difficult to cut but accepts a very good polish. Germanium is non-hygroscopic and nontoxic, has good thermal conductivity, excellent surface hardness, and good strength. Additionally, due to its very high refractive index, antireflection coatings are essential for any germanium transmitting optical system. Germanium has a low dispersion and is unlikely to need color correction, except in the highest resolution systems. A significant disadvantage of germanium is the serious dependence of its refractive index on temperature, so germanium lenses may need to be athermalized. In spite of high material price and cost of antireflection coatings, germanium is a favorite choice of optical designers of high-performance infrared objectives for thermal imagers.

Infrared chalcogenide classes offer good transmission from about 1 to about 13 μm (from SWIR to LWIR range). Physical properties such as low dn/dT and low dispersion enable optical designers to engineer

Table 1.4 Principal characteristics of some infrared materials

MATERIAL	WAVEBAND (μm)	$n_{4\mu m}$, $n_{10\mu m}$	dn/dT (10^{-6} K^{-1})	DENSITY (g/cm^3)	OTHER CHARACTERISTICS
Ge	2–12	4.0245, 4.0031	424 (4 μm) 404 (10 μm)	5.32	Brittle, semiconductor, can be diamond-turned, visibly opaque, hard
Chalcogenide glasses	3–12	2.5100, 2.4944	55 (10 μm)	4.63	Amorphous IR glass, can be slumped to near-net shape
Si	1.2–7.0	3.4289 (4 μm)	159 (5 μm)	2.329	Brittle, semiconductor, diamond-turned with difficulty, visibly opaque, hard
GaAs	3–12	3.304, 3.274	150	5.32	Brittle, semiconductor, visibly opaque, hard
ZnS	3–13	2.251, 2.200	43 (4 μm) 41 (10 μm)	4.08	Yellowish, moderate hardness and strength, can be diamond-turned, scatters short wavelengths
ZnSe	0.55–20	2.4324, 2.4053	63 (4 μm) 60 (10 μm)	5.27	Yellow–orange, relatively soft and weak, can be diamond-turned, very low internal absorption and scatter
CaF_2	3–5	1.410	−8.1 (3.39 μm)	3.18	Visibly clear, can be diamond-turned, mildly hygroscopic
Sapphire	3–5	1.677(n_o) 1.667(n_e)	6 (o) 12 (e)	3.99	Very hard, difficult to polish due to crystal boundaries
BF7 (Glass)	0.35–2.3		3.4	2.51	Typical optical glass

Source: After Ref. [14].

color-correcting optical systems without thermal defocusing. A moldable feature of these glasses allows a cost-effective manufacture of complex lens geometries in medium to large volumes. Further on, these glasses can be also processed using conventional grinding and polishing techniques, single-point diamond turning if higher performance is to be achieved. Due to these features, infrared chalcogenide glasses made a revolution in the manufacturing of optics for thermal imagers during last decades by enabling mass manufacturing of low cost, good optical performance optical objectives, and now these glasses compete with germanium as the most popular IR optical material. Most popular brands of infrared chalcogenide glasses are AMTIR from Amorphous Materials Inc., GASIR® from Umicore Inc., and IRG glasses from Schott. It should be noted, however, that chalcogenide glasses are more difficult for fabrication of high accuracy lenses compared to germanium.

Physical and chemical properties of silicon are very similar to the properties of germanium. It has a high refractive index (≈3.45), is brittle, does not cleave, takes an excellent polish, and has large dn/dT. Similar to germanium, silicon optics must have antireflection coatings. Silicon offers two transmission ranges: 1–7 and 25–300 μm. Only the first one is used in typical IR systems. The material is significantly cheaper than

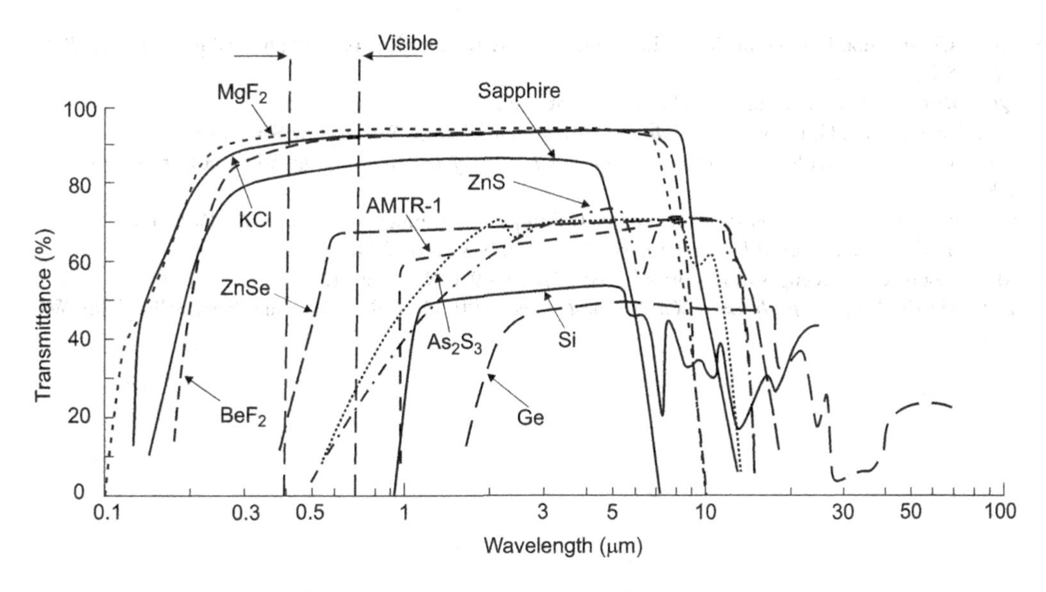

Figure 1.18 Transmission range of infrared materials. (Reproduced from Ref. [14])

germanium, ZnSe, and ZnS. It is used mostly for IR systems operating in the 3–5 μm band. Due to its low density, silicon is a good choice ideal for MWIR objectives with weight constraints.

ZnSe is an optical material of optical properties mostly similar to germanium but of wider transmission range from about 0.55 to about 20 μm, and a refractive index of about 2.4. It is partially translucent when visible and reddish in color. Due to the relatively high refractive index, antireflection coatings are necessary. The chemical resistance of the material is excellent. It is a popular material for lenses for both LWIR and MWIR objectives and for broadband infrared windows.

ZnS offers relatively good transmission in the range from about 3 to 13 μm. It exhibits exceptional high fracture strength, and high hardness, and high chemical resistance. Due to high resistance to rain erosion and high-speed dust abrasion, ZnS is popular for windows or external lenses in thermal imagers used in high-speed airborne applications.

Ordinary glass does not transmit radiation beyond 2.5 μm in the IR region. Fused silica is characterized by a very low thermal expansion coefficient that makes optical systems particularly useful in changing environmental conditions. It offers a transmission range from about 0.3 to 3 μm. Because of low reflection losses due to the low refractive index (≈1.45), antireflection coatings are not needed. However, an antireflection coating is recommended to avoid ghost images. Fused silica is more expensive than BK-7, but still significantly cheaper than Ge, ZnS, and ZnSe, and is a popular material for lenses of IR systems with bands located below 3 mm.

The alkali halides have excellent IR transmission; however, they are either soft or brittle and many of them are attacked by moisture, making them generally unsuitable for industrial applications. For a more detailed discussion of the IR materials, see references [13,15].

REFERENCES

1. F. Grum and R. J. Becherer, *Optical Radiation Measurements*, Vol. 1., Academic Press, San Diego, CA, 1979.
2. W. L. Wolfe and G. J. Zissis, *The Infrared Handbook*, SPIE Optical Engineering Press, Bellingham, WA, 1990.
3. W. L. Wolfe, "Radiation Theory," in *The Infrared and Electro-Optical Systems Handbook*, Vol. 1, ed. G. J. Zissis, 1–48, SPIE Optical Engineering Press, Bellingham, WA, 1993.
4. W. R. McCluney, *Introduction to Radiometry and Photometry*, Artech House, Boston, MA, 1994.
5. W. L. Wolf, *Introduction to Radiometry*, SPIE Optical Engineering Press, Bellingham, WA, 1998.
6. Y. Ohno, "Basic concepts in photometry, radiometry and colorimetry," in *Handbook of Optoelectronics*, Vol. 1, eds. J. P. Dakin and R. G. W. Brown, 287–305, Taylor & Francis, New York, 2006.
7. J. M. Palmer and B. G. Grant, *The Art of Radiometry*, SPIE Press, Bellingham, WA, 2010.

8. CIE – Commission Internationale de l'Eclairage: Compte Rendu, "The basic of physical photometry," *Publ. CIE*, 18(2), 67, 1924.
9. *Quantities and Units*, ISO Standards Handbook, 3rd ed., 1993.
10. E. L. Dereniak and G. D. Boreman, *Infrared Detectors and Systems*, Wiley, New York, 1996.
11. S. G. Burnay, T. L. Williams, and C. H. Jones, *Applications of Thermal Imaging*, Adam Hilger, Bristol, UK, 1988.
12. G. Gaussorgues, *La Thermographie Infrarouge*, Technique et Documentation, Lavoisier, Paris, 1984.
13. W. J. Smith, *Modern Optical Engineering*, McGraw-Hill, New York, 2000.
14. M. E. Couture, "Challenges in IR optics," *Proc. SPIE*, 4369, 649–61, 2001.
15. D. C. Harris, *Materials for Infrared Windows and Domes*, SPIE Optical Engineering Press, Bellingham, WA, 1999.

2 Infrared systems fundamentals

This section concentrates on general aspects of IR systems, which are briefly defined and analyzed to demonstrate their applications. In the work presented here, the emphasis is on the methodology, rather than on the detail description. It is written to clarify and summarize the principles of infrared (IR) technology and combines numerous engineering disciplines necessary for the development of an IR system.

A comprehensive compendium devoted to IR systems was copublished in 1993 by Infrared Information Analysis Center and the International Society for Optical Engineering as *The Infrared and Electro-Optical Systems Handbook* (executive editors: Joseph S. Accetta and David L. Shumaker) [1].

2.1 INFRARED DETECTOR MARKET

Traditionally, IR technologies are connected with controlling functions and night vision problems, with earlier applications connected simply with detection of IR radiation, and later by forming IR images from temperature and emissive differences (systems for recognition and surveillance, tank sight systems, anti-tank missiles, air–air missiles, etc.).

Most of the funding has been provided to fulfill military needs, but peaceful applications have increased continuously, especially since the last decade of the twentieth century (see Figure 2.1). It is predicted currently that the commercial market is about 70% in volume and 40% in value, largely connected with volume production of uncooled imagers [2]. These include medical, industry, earth resources, and energy conservation applications. Medical applications include thermography in which IR scans of the body detect cancers or other trauma, which raise the body surface temperature. Earth resource determinations are done by using IR images from satellites in conjunction with field observation for calibration (in this manner, for example, the area and content of fields and forests can be determined). In some cases, even the health state of a crop is determined from space. Energy conservation in homes and industry has been aided by the use of IR scans to determine the points of maximum heat loss. Demands for these technologies are quickly growing due to their effective applications, for example, in global monitoring of environmental pollution and climate changes, longtime prognoses of agriculture crop yield, chemical process monitoring, Fourier-transform IR spectroscopy, IR astronomy, car driving, IR imaging in medical diagnostics, and others.

Initially developed for the military market by U.S. defense companies, IR uncooled cameras are now widely used in many commercial applications. Currently, microbolometer detectors are produced in larger volumes than all other IR array technologies together. The global IR detectors market was valued at $ 230.3 million in 2014 and is expected to reach an estimated $422.6 million by 2021 and is forecast to grow at a *Compound Annual Growth Rate* (CAGR) of 8.7% from 2016 to 2021—see Figure 2.2 [3]. The major drivers of growth for this market are increased demand for smartphones and tablets, rising concern about safety and security, and increasing automation in the building and industrial sectors. In the year 2014, North America dominated the market and contributed more than a 33% share of the overall market revenue followed by Asia-Pacific. Asia-Pacific is expected to show the highest CAGR during the forecast period.

2.2 NIGHT VISION SYSTEM CONCEPTS

Night vision systems can be divided into two categories: those depending upon the reception and processing of radiation reflected by an object, and those operating with radiation internally generated by an

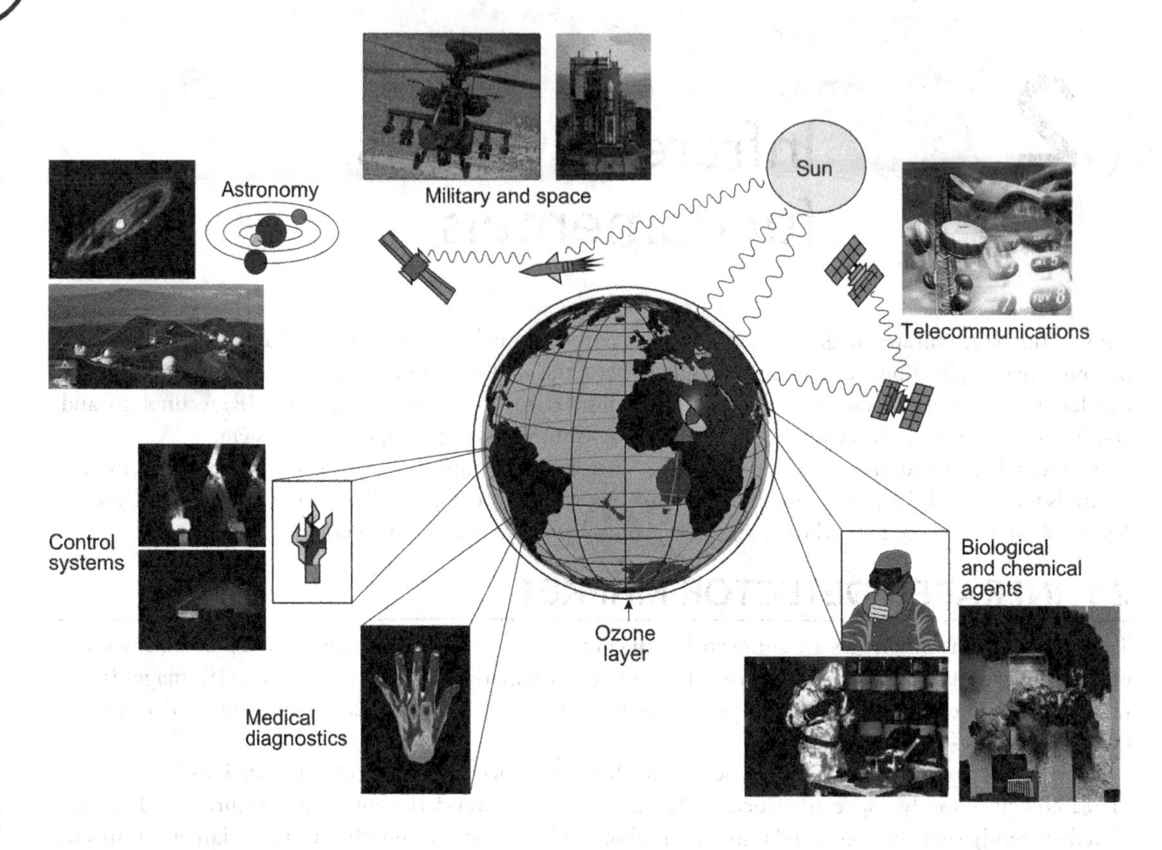

Figure 2.1 Applications of IR detectors.

object. The latter systems are described in the later subsections. These devices gather existing ambient light (starlight, moonlight, or IR light) through the front lens. This light, which is made up of photons goes into a photocathode tube that changes the photons to electrons.

The human visual perception system is optimized to operate in daytime illumination conditions. The visual spectrum extends from about 420 to 700 nm and the region of greatest sensitivity is near the peak wavelength of sunlight, at around 550 nm. However, fewer visible light photons are available at night and only large, high contrast objects are visible.

Surveillance capabilities in the near infrared (NIR) spectral band include passive illumination from the spectral irradiance, caused by nightglow, or with active illumination from eye-safe lasers. Figure 2.3 illustrates the nightglow emission with its maximum intensity in the spectral band 1 and 1.8 μm [4]. It appears that the photon rate in the region from 800 to 900 nm is five to seven times greater than that in a visible region around 500 nm. Moreover, the reflectivity of various materials (e.g., green vegetation, because of its chlorophyll content) is higher, between 800 and 900 nm than at 500 nm. It means that at night, more light is available in the NIR than in the visual region and that against certain backgrounds more contrast is available. In a full moon, the near-IR spectral band is essentially unaffected, because the contribution to the intensity is in the visible range; during full moon, the integrated intensities of moonlight and nightglow are comparable.

A considerable improvement in night vision capability can be achieved with night viewing equipment that consists of an objective lens, image intensifier, and eyepiece (see Figure 2.4). Improved visibility is obtained by gathering more light from the scene with an objective lens than the unaided eye, by use of a photocathode that has higher photosensitivity and broader spectral response than the eye, and by amplification of photo events for visual sensation.

The concepts of image intensification created in the early 1930s were not basically different from those today. However, the early devices suffered from two major deficiencies: poor photocathodes and poor

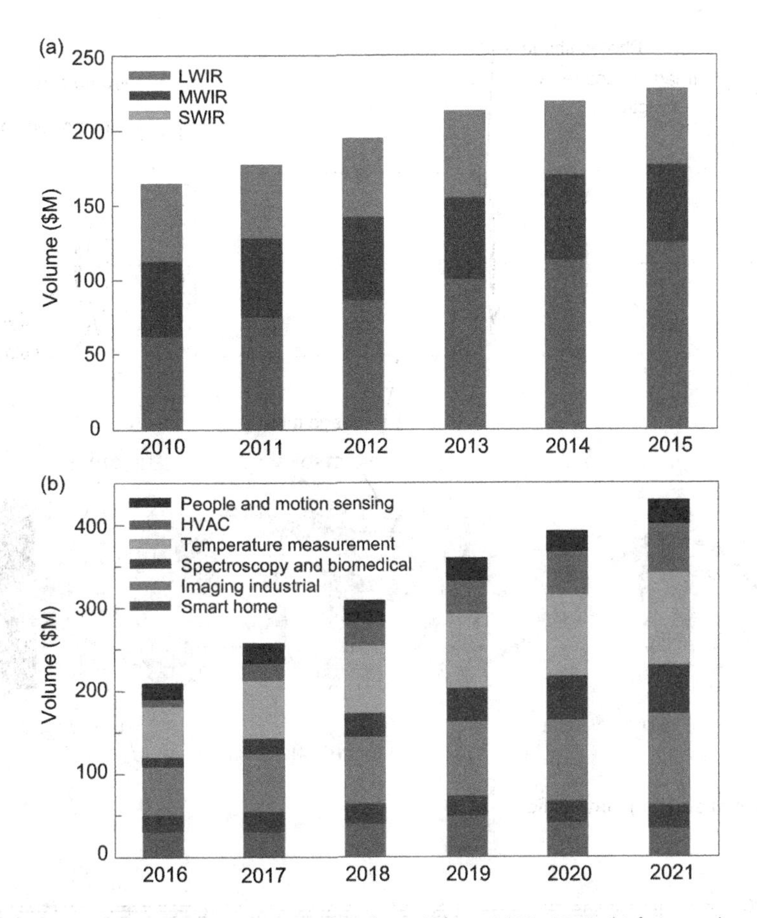

Figure 2.2 Global IR detector market (a) trends by wavelength from 2010–2015; (b) forecast by applications from 2016–2021.

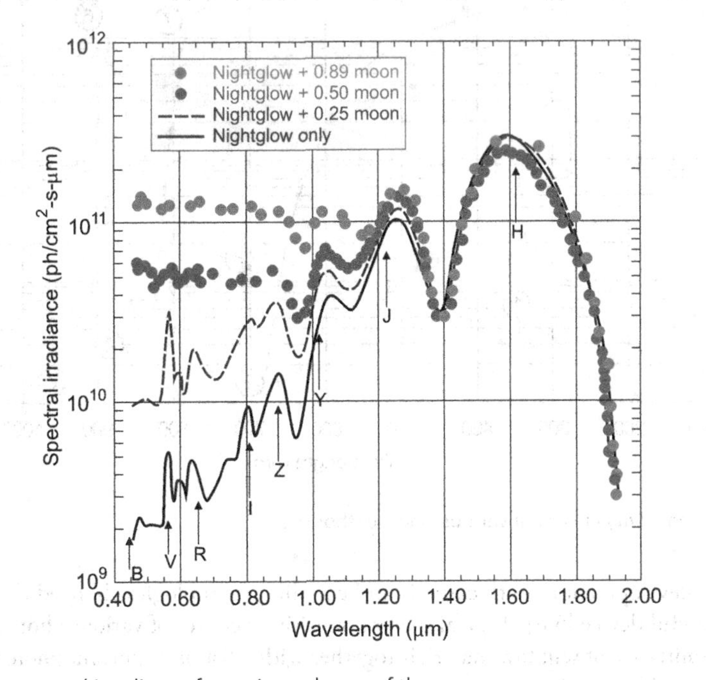

Figure 2.3 Night sky spectral irradiance for various phases of the moon.

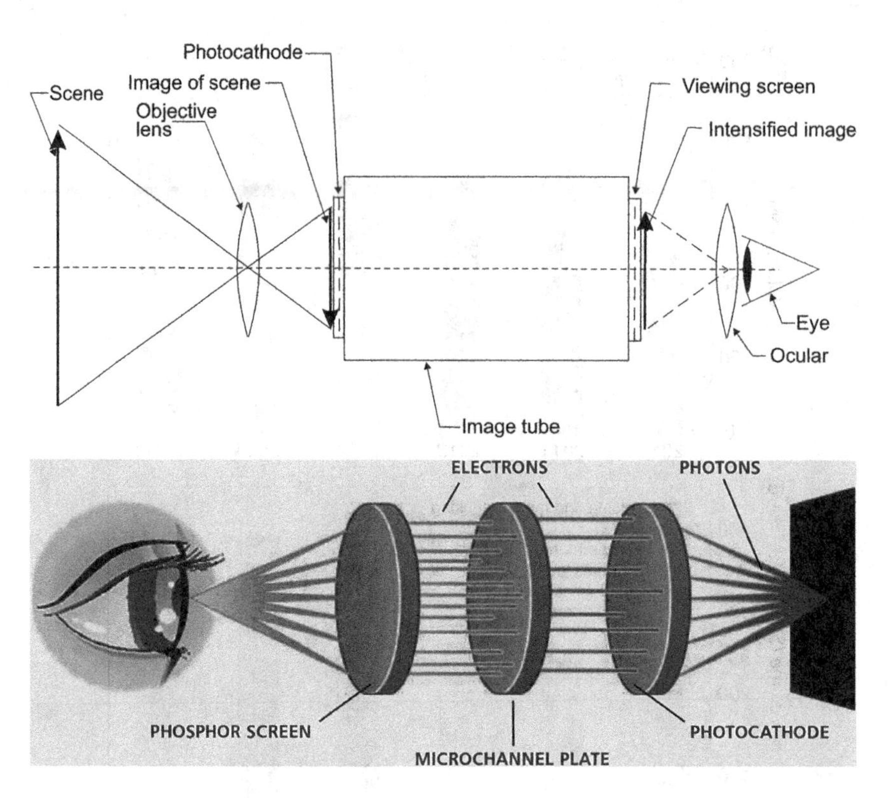

Figure 2.4 Diagram of an image intensifier.

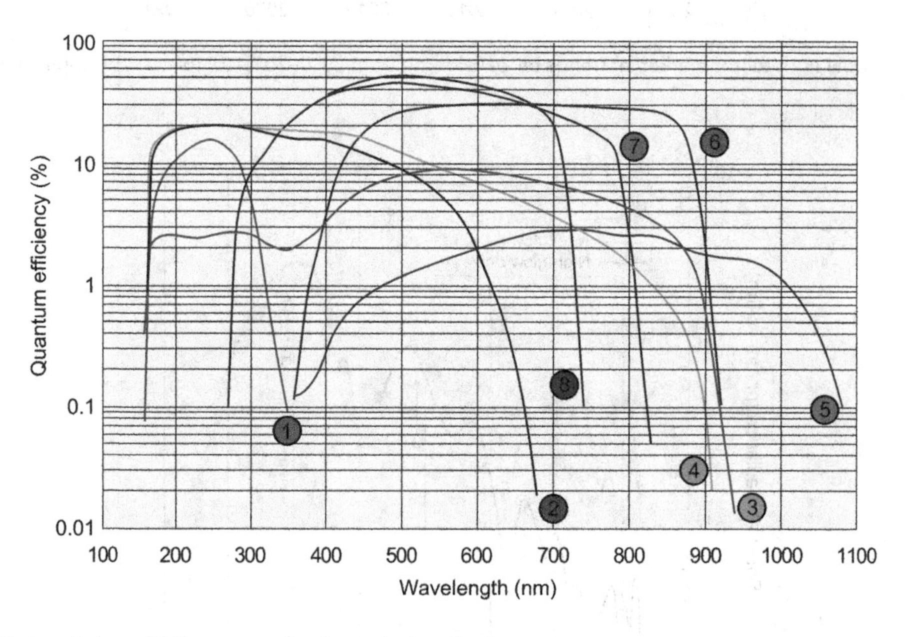

Figure 2.5 Spectral sensitivity curves of various photocathodes.

coupling. Later, the development of both cathode and coupling technologies changed the image intensifier into a much more useful device [5,6]. Typical spectral sensitivity curves of various photocathodes charged with a typical transmittance of window materials together with a list of important photocathodes are shown in Figure 2.5 and below [7].

SUFFIX	PHOTOCATHODE	INPUT WINDOW
1	CsTe	Synthetic silica
2	Bialkali	Synthetic silica
3	Enhanced red multialkali	Synthetic silica
4	Multialkali	Synthetic silica
5	InGaAs	Borosilicate glass
6	GaAs	Borosilicate glass
7	Enhanced red GaAsP	Borosilicate glass
8	GaAsP	Borosilicate glass

As is shown in Figure 2.4, the image intensifier system is built from three main blocks: optical objective, multichannel plate (MCP), and optical ocular. An MCP is a secondary electron multiplier consisting of array of millions of very thin glass channels (of internal diameter $\approx 10\,\mu m$; each capillary works as an independent electron multiplier) bundled in parallel and sliced in the form of disc (see Figure 2.6). Secondary electrons are accelerated by the voltage applied across both ends of the MCP. This process is repeated many times along the channel wall and as a result, a great number of electrons are output from the MCP. Furthermore, the electron flux can be reconverted into an optical image by using a phosphor coating as the rear electrode to provide electroluminescence; this combination provides an image intensifier.

The image intensifiers are classed by generation (Gen) numbers. Gen0 refers to the technology of World War II, employing fragile, vacuum-enveloped photon detectors with poor sensitivity and little gain. Further evolution of image intensifier tubes is presented in Table 2.1. Gen1 represents the technology of the early Vietnam era, the 1960s. In this era, the first passive systems, able to amplify ambient starlight, were introduced. Though sensitive, these devices were large and heavy. Gen1 devices used tri-alkali photocathodes to achieve a gain of about 1,000. By the early 1970s, the MCP amplifier was developed comprising more than two million microscopic conducting channels of hollow glass, each of about 10 mm in diameter, fused into a disc-shaped array. Coupling the MCP with multialkali photocathodes, capable of emitting more electrons per incident photon, produced GenII. GenII devices boasted amplifications of 20,000 and operational lives to 4,000 h. Interim improvements in bias voltage and construction methods produced the GenII$^+$ version. Substantial improvements in gain and bandwidth in the 1980s heralded the advent of GenIII. Gallium

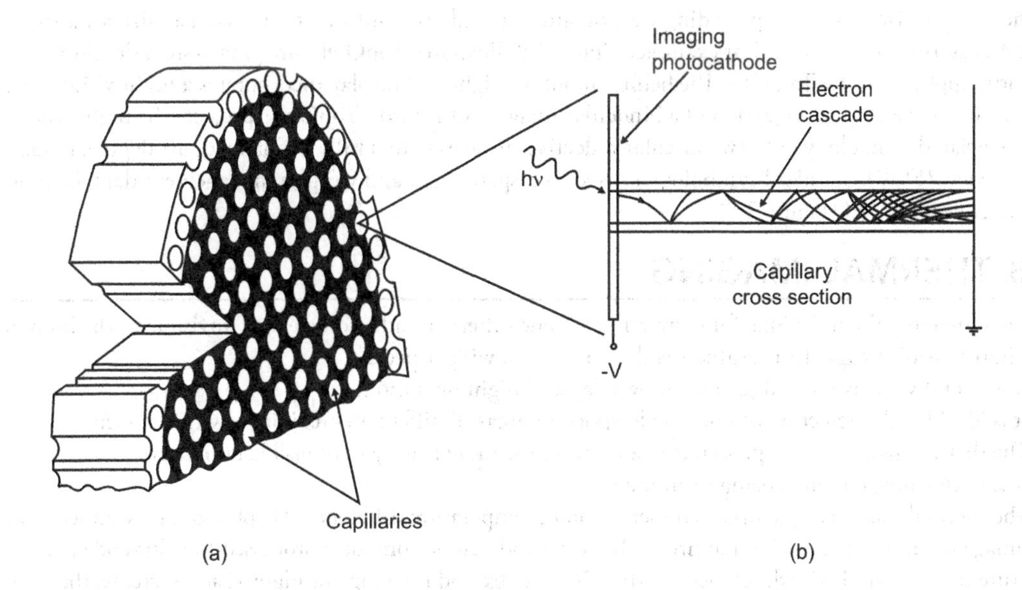

Figure 2.6 Schematic presentation of microchannel plate: (a) cutaway view and (b) a single capillary.

Table 2.1 Image intensifier tubes

GEN 1	GEN 2	GEN 3	GEN 4
• Vietnam War • SbCs i $SbNa_2KCs$ photocathodes (S10,S20) • electrostatic inversion • photosensitivity up to 200 mA/lm	• 1970s • multialkali photocathodes (S25) • MCP • photosensitivity up to 700 mA/lm • operational time to 4,000 h	• early 1980s • GaAs photocathodes • ion barrier to microchannel plate • photosensitivity up to 700 mA/lm • operational live to 10,000 h	• late 1990s • multialkali photocathodes • "filmless" tube • photosensitivity up to 1,800 mA/lm • operational time to 10,000 h

arsenide photocathodes and internal changes in the MCP design resulted in gains ranging from 30,000 to 50,000 and operating lives of 10,000 h.

Image intensifiers were primarily developed for nighttime viewing and surveillance under moonlight or starlight. At present, image intensifier applications have spread from nighttime viewing to various fields including industrial product inspection and scientific research, especially when used with charge-coupled device (CCD) cameras—so-called intensified CCD or ICCD (see Figure 2.7a) [7]. Gate operation models are also useful for observation and motion analysis of high-speed phenomena (high-speed moving objects, fluorescence lifetime, bioluminescence, and chemiluminescence images). Figure 2.7b shows an example of Gen III night vision goggles.

Image intensifiers are widespread in many military applications. The advent of night vision devices and helmet-mounted displays places additional constraints on the helmet, which is now an important element of the cockpit display system, providing weapon aiming, and other information—such as aircraft attitude and status—to the pilot. For example, Figure 2.8 illustrates TopOwl® imaging system developed for airborne applications by Tales [8]. The helmet-mounted sight and display incorporates a night vision system with a 100% overlapped projection of a binocular image on the visor. TopOwl® projects the night scene and associated symbology onto two circular reflective surfaces with a fully overlapped, 40-degree, binocular field of view (FOV). Standard symbology is used to display flight and weapon management data, helping to reduce crew workload (Figure 2.8).

2.3 THERMAL IMAGING

Thermal imaging is a technique for converting a scene's thermal radiation pattern (invisible to the human eye) into a visible image. Its usefulness is due to the following aspects:

- It is a totally passive technique and allows day and night operation;
- It is ideal for the detection of hot or cold spots, or areas of different emissivities, within a scene;
- The thermal radiation can penetrate smoke and mist more readily than visible radiation;
- It is a real-time, remote sensing technique.

The thermal image is a pictorial representation of temperature difference. Displayed on a scanned raster, the image resembles a television picture of the scene and can be computer processed to color-code temperature ranges. Originally developed (in the 1960s) to extend the scope of night vision systems, thermal imagers at first provided an alternative to image intensifiers. As the technology has matured, its range of

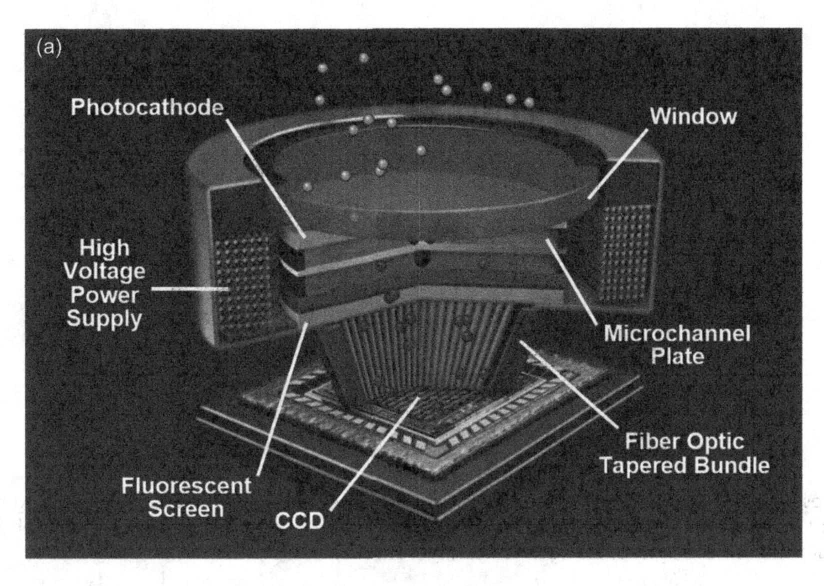

Figure 2.7 Night vision device: (a) proximity-focused image intensifier and (b) Gen III night vision goggles AN/AVS-9 (ITT Night Vision).

application has expanded and now extends into the fields that have little or nothing to do with night vision (e.g., stress analysis, medical diagnostics). In most present-day thermal imagers, an optically focused image is scanned electronically across detectors (many elements or 2D array), the output of which is converted into a visual image. The optics, mode of scanning, and signal processing electronics are closely interrelated. The number of picture points in the scene is governed by the nature of the detector (its performance) or the size of the detector array. The effective number of picture points or resolution elements in the scene is steadily increased.

2.3.1 THERMAL IMAGING SYSTEM CONCEPTS

Due to the existing terminology confusion in literature, we can find at least 11 different terms used as synonyms of the earlier defined thermal imaging systems: thermal imager, thermal camera, thermal imaging camera, forward-looking infrared (FLIR), IR imaging system, thermograph, thermovision, thermal viewer, IR viewer, IR imaging radiometer, thermal viewer, thermal data viewer, and thermal video system. The only

Figure 2.8 Tales TopOwl® helmet incorporates an optical combiner assembly for each eye, allowing the pilot to view the cockpit and the outside world directly with the night imagery superimposed on it. TopOwl® has a 40° FOV and a total headborne weight of 2.2 kg in full configuration.

real difference between the aforementioned terms is that the designations "thermograph," "IR imaging radiometer," and "thermovision" usually refer to thermal cameras used for measurement applications, while the other terms refer to thermal cameras used in observation applications. For example, thermographic imagers supply quantitative temperature (they have built-in specialized software that provides temperature analysis), while radiometers provide quantitative radiometric data on the scene (such as radiance or irradiance) or process these data to yield information about temperatures.

The basic concept of a modern thermal imager system is to form a real image of the IR scene, detect the variation in the imaged radiation, and, by suitable electronic processing, create a visible representation of this variation analogous to conventional television cameras.

IR camera construction is similar to a digital video camera. Detectors are only a part of usable sensor systems. Instead of CCD/complementary metal oxide semiconductor (CMOS) image arrays that video and digital still cameras use, the IR camera detector (see Figure 2.9) is a focal plane array (FPA) of micrometer-size pixels made of various materials sensitive to IR wavelengths. Once a detector is selected, optics (lens) material and filters can be selected to somewhat alter the overall response characteristics of an IR camera system. Figure 2.10 shows the system response for a number of different detectors. The spectral performance of most cameras can be found in their manual or technical specifications.

Thermal imagers have various applications, depending on the platform and the user [9]. Most of them are used in military applications. Military sensor systems include optics, coolers, pointing and tracking

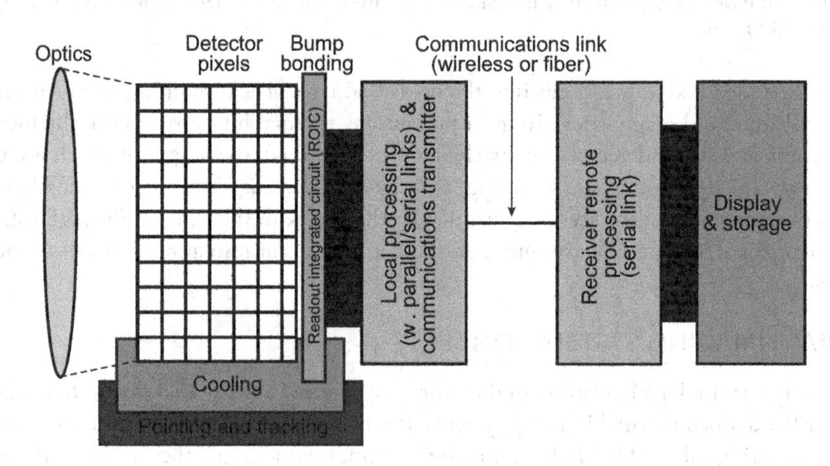

Figure 2.9 Schematic representation of an imaging system showing important subsystems.

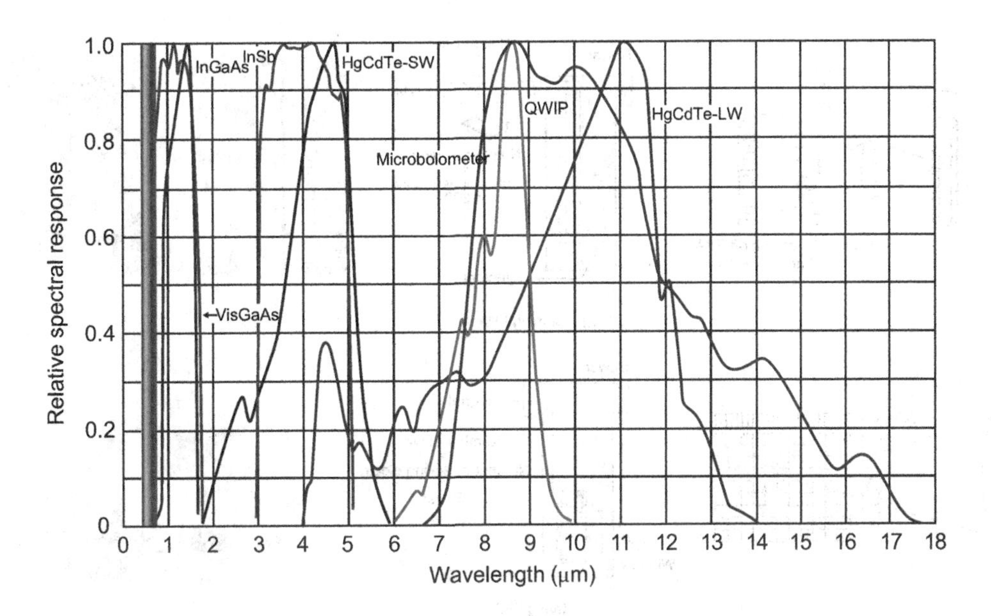

Figure 2.10 Relative response spectra for a number of IR cameras.

systems, electronics, communication, processing together with information extraction subsystems, and displays. So, the process of developing a sensor system is significantly more challenging than fabricating a detector array. They often have multiple fields of view that are user switchable during operation, which gives both a wide, general surveillance mode as well as a high magnification and narrow field for targeting, designating, or detailed intelligence gathering. Many military thermal imagers are integrated with a TV camera and a laser rangefinder. A TV color camera is used during daytime conditions due to its superior image quality. Nonmilitary uses include generic search and track, snow rescue, mountain rescue, illegal border crossing detection, and pilot assistance at night or in bad weather, forest fire detection, firefighting, inspection, and discreet surveillance and evidence gathering. A small but increasing group of thermal imagers enables noncontact temperature measurement and these cameras are used in areas of industry, science, and medicine.

The term "FPA" refers to an assemblage of individual detector picture elements ("pixels") located at the focal plane of an imaging system. Although the definition could include one-dimensional ("linear") arrays as well as two-dimensional (2D) arrays, it is frequently applied to the latter. Usually, the optics part of an optoelectronic images device is limited only to focusing of the image onto the detectors array. These so-called "staring arrays" are scanned electronically usually using circuits integrated with the arrays. The architecture of detector-readout assemblies has assumed a number of forms. The types of readout-integrated circuits (ROICs) include the function of pixel deselecting, antiblooming on each pixel, subframe imaging, output preamplifiers, and may include yet other functions. Infrared imaging systems, which use 2D arrays, belong to so-called "second-generation" systems.

The simplest scanning linear array used in thermal imaging systems consists of a row of detectors (Figure 2.11a). An image is generated by scanning the scene across the strip using, as a rule, a mechanical scanner. At standard video frame rates, at each pixel (detector) a short integration time has been applied and the total charge is accommodated. A staring array is a 2D array of detector pixels (Figure 2.11b) scanned electronically.

The scanning system, which does not include multiplexing functions in the focal plane, belongs to the first-generation systems. A typical example of this kind of detector is a linear photoconductive array (PbS, PbSe, HgCdTe) in which an electrical contact for each element of a multielement array is brought off the cryogenically cooled focal plane to the outside, where one electronic channel is used at ambient temperature for each detector element. The U.S. common module HgCdTe arrays employ 60, 120, or 180 photoconductive elements depending on the application (Figure 2.11 top right).

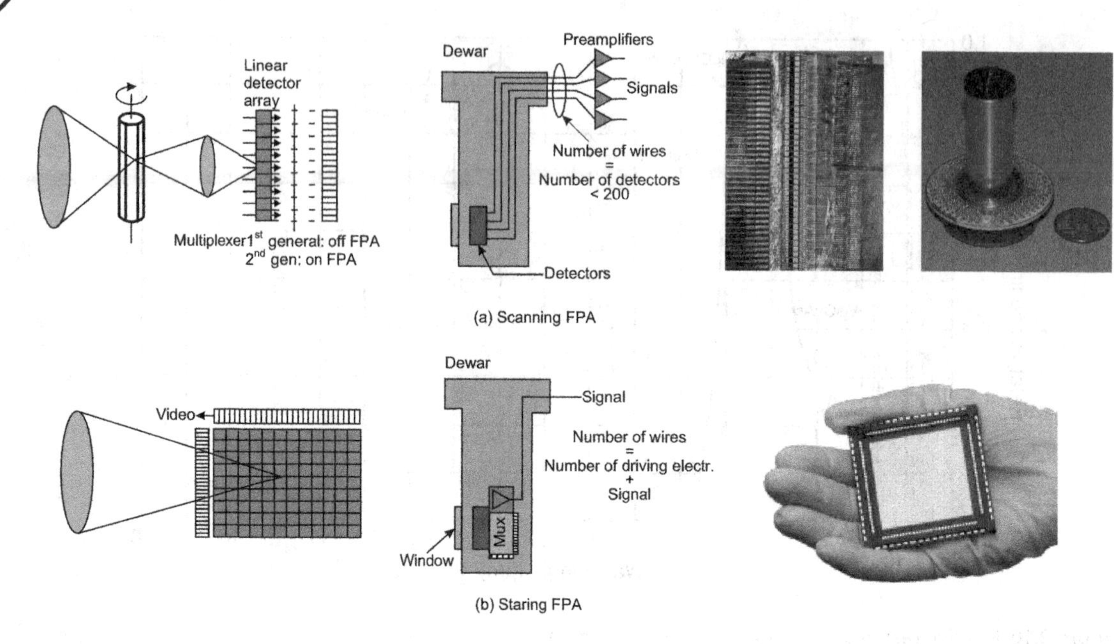

Figure 2.11 Scanning and staring FPAs.

The second-generation systems (full-framing systems), being developed at present, have at least three orders of magnitude more elements (>10^6) on the focal plane than first-generation systems. Intermediary systems are also fabricated with multiplexed scanned photodetector linear arrays in use and with, as a rule, time delay and integration (TDI) functions. Typical examples of these systems are HgCdTe multilinear 288 × 4 arrays fabricated by Sofradir, both for 3–5 and 8–10.5–μm bands, with signal processing in the focal plane (photocurrent integration, skimming, partitioning, TDI function, output preamplification, and some others).

2.3.2 IR CAMERAS VERSUS FLIR SYSTEMS

Historically, a "camera" includes neither the storage medium nor the display, while the "camera system" includes the complete package. At present, the manufacturers offer an optional recording medium (usually CD-ROM), display, and electronics for the display. For example, Figure 2.12 is a photograph of the FLIR P660 IR. This inspection camera has high-resolution 640 × 480 thermal imagery with 30-mK sensitivity, interchangeable lenses, and the flexibility of a tiltable, high-fidelity color LCD. The camera can be used by anyone who does lots of thermal inspections, or who needs to accurately measure small objects from far away.

Figure 2.13a shows representative camera architecture with three distinct hardware pieces: a camera head (which contains optics, including collecting, imaging, zoom, focusing, and spectral filtering assembles), electronics/control processing box, and the display. Electronics and motors to control and drive moving parts must be included. The control electronics usually consist of communication circuits, bias generators, and clocks. Usually, the camera's sensor (FPA) needs cooling and therefore some form of cooler is included, along with its closed-loop cooling control electronics. The signal from FPA is of low voltage and amperage and requires analog preprocessing (including amplification, control, and correction), which is located physically near the FPA and included in the camera head. Often, the A/D is also included here. For user convenience, the camera head often contains the minimum hardware needed to keep volume, weight, and power to a minimum.

Typical costs of cryogenically cooled imagers around $50,000 restrict their installation to critical military applications allowing conducting operations in complete darkness. Moving from cooled to uncooled operation (e.g., using silicon microbolometer) reduces the cost of an imager to below $10,000. Less expensive IR cameras present a major departure from camera architecture presented in Figure 2.13a.

Figure 2.12 FLIR P660 IR camera.

Cameras usually produce high-quality imagers with a temperature sensitivity of 20–50 mK. Details and resolution vary by optics and focal planes. A good camera produces an image akin to that of a black and white television.

"FLIR" is an archaic 1960s jargon for FLIR to distinguish these systems from IR line scanners, which look down rather than looking forward. Conversely, most sensors that do look forward are not considered to be FLIRs (e.g., cameras and astronomical instruments). The term "FLIR" should be eliminated from IR technospeak, but is still used and is likely to remain in the jargon for a while.

It is difficult to explain the difference between a camera and an FLIR system. In general, FLIRs are designed for specific applications and specific platforms, their optics are integrated into the package, and they are used mostly by people. Cameras usually rely on "imaging" of a "target" and are designed for generic purposes, without much consideration for form and fit; they can be used with many different fore-optics and are often used by computers and machines (not just people).

The term "FLIR" usually implies military or paramilitary use, air-based units, and scanners. The FLIR provides automatic search, acquisition, tracking, precision navigation, and weapon delivery functions. A typical FLIR is comprised of several line-replaceable units (LRUs), such as an FLIR optical assembly mounted on a gyro-stabilized platform, an electronics module containing all necessary electronic circuits, and a cryogenically cooled detector array, a power supply unit, and a control and processing assembly.

A representative FLIR architecture with the video signal output (to support LRU for image and higher order processing) is shown in Figure 2.13b. Many systems depart significantly from the architecture of Figure 2.13b.

In the 1960s, the earliest FLIRs were linear scanners. In the 1970s, first-generation common modules (including a Dewar containing 60, 120, or 180 discrete elements of photoconductive HgCdTe) were introduced. The next generation of FLIRs employed a dense linear array of photovoltaic HgCdTe, usually 2(4) × 480 or 2(4) × 960 elements in TDI for each element. At present, these systems are replaced by full-framing FLIRs that employ staring arrays (HgCdTe, InSb, and QWIP). For example, Figure 2.14 shows eLRAS3 system (Long-Range Scout Surveillance System) fabricated by Raytheon. eLRAS3 provides the real-time ability to detect, recognize, identify, and geo-locate distant targets outside the threat acquisition.

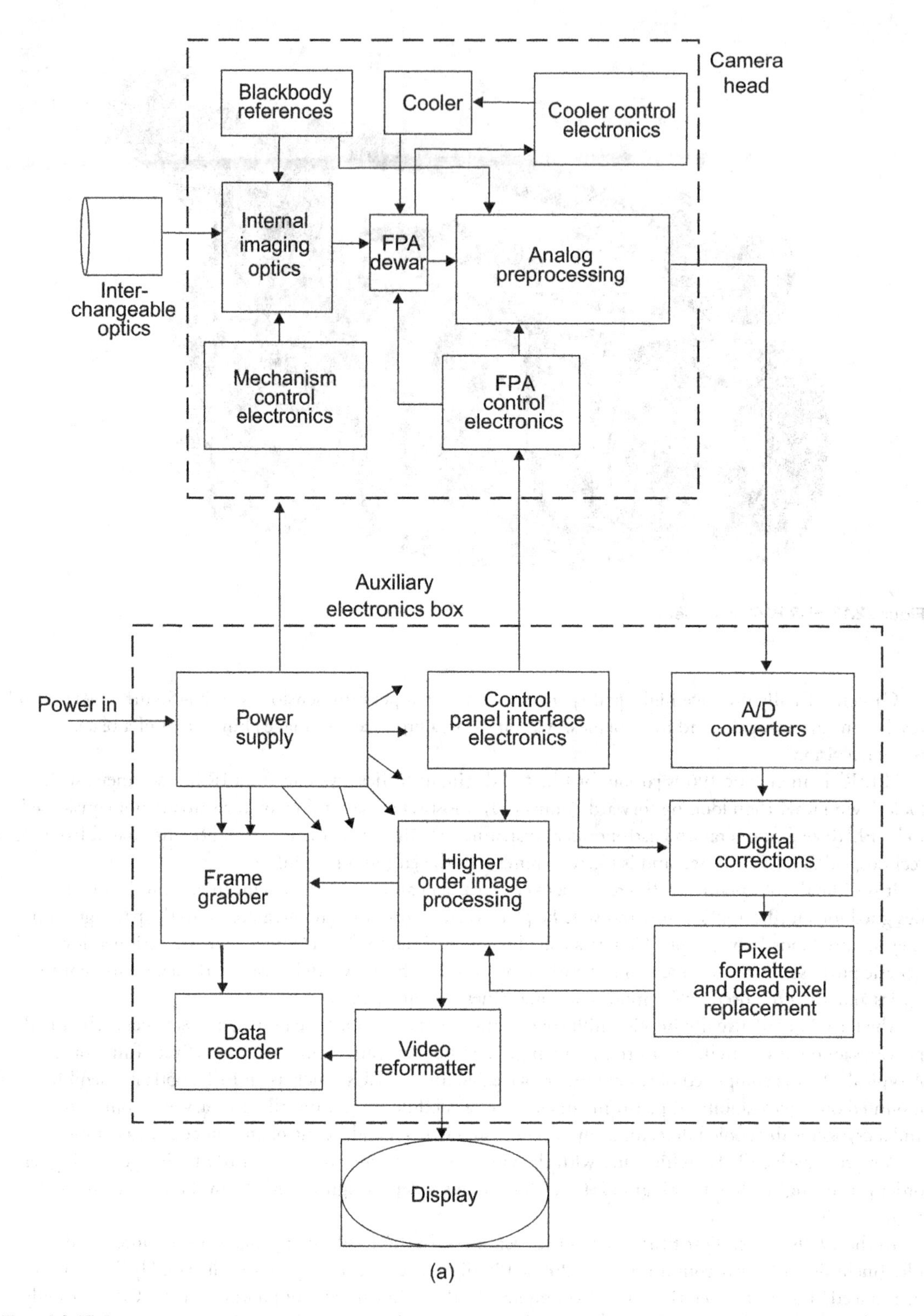

Figure 2.13 Representative IR camera system architecture (a) and FLIR system architecture (b). (Adapted after Ref. [9])

(Continued)

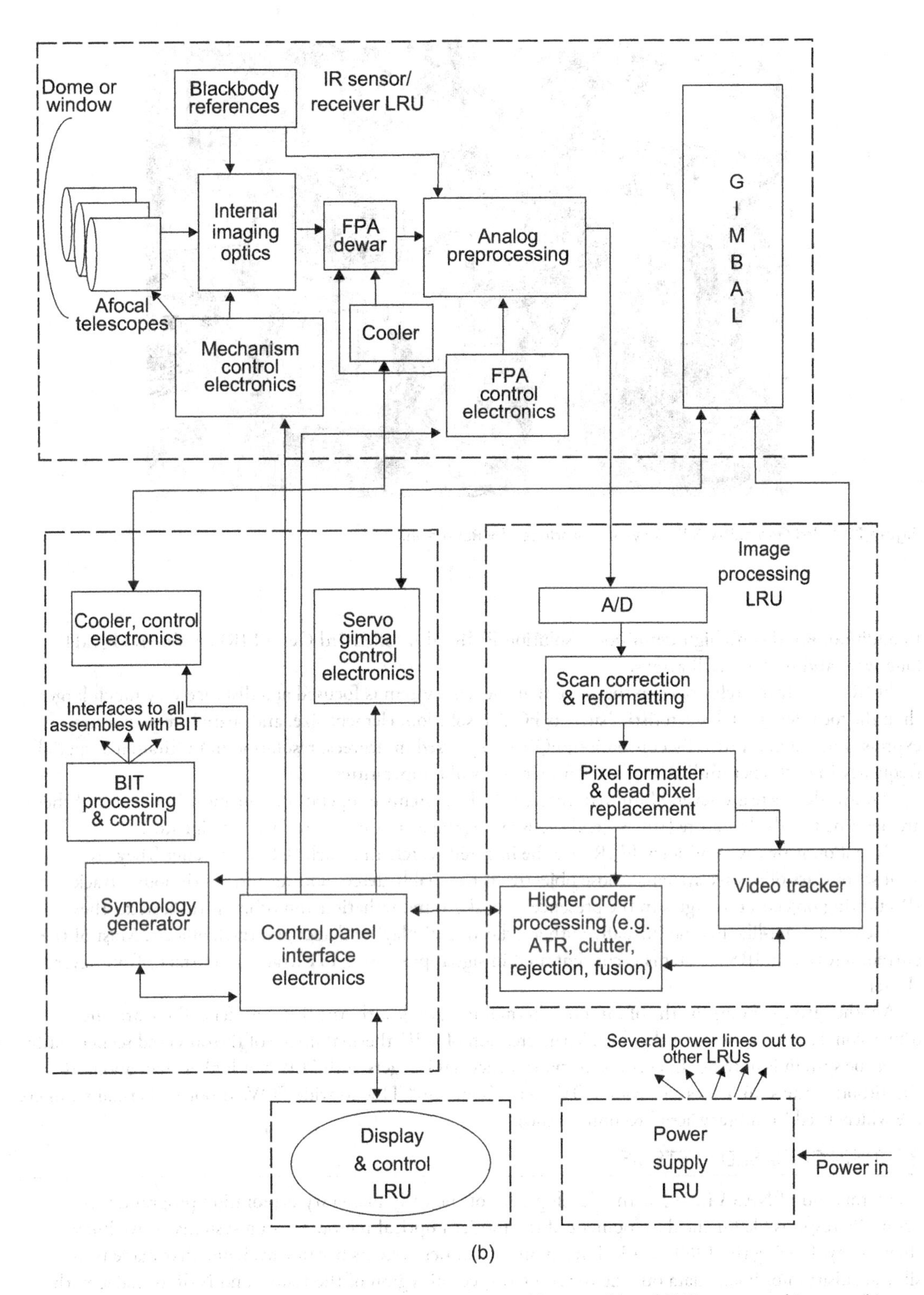

Figure 2.13 (CONTINUED) Representative IR camera system architecture (a) and FLIR system architecture (b). (Adapted after Ref. [9])

Figure 2.14 3rd Gen eLRAS3 FLIR system produced by Raytheon.

In addition, Raytheon's high-definition resolution FLIR (also called 3rd Gen FLIR) combines HgCdTe long-wave and mid-wave IR arrays.

FLIRs usually use telescopes in the sense that the lens system is focused at a distance very much larger than the focal length. Characteristics such as FOV, resolution, element size, and spatial frequency are expressed in angular units. By convention, FOV is expressed in degrees, resolution in milliradians, spatial frequency in cycles per milliradian, and noise in units of temperatures.

Worldwide, there are several hundred different FLIR systems in operation. The most important of them are described in the literature [10]. Several FLIRs integrate a laser ranger or target designator.

Recent outgrowths of military FLIRs are the infrared search and track (IRST) systems. They are a subset or class of passive systems whose objective is to reliably detect, locate, and continuously track IR-emitting objects and targets in the presence of background radiation and other disturbances. They are used in a radar-like manner (usually with a radar-like display) to detect and track objects. Most of the current research in IRST systems is concentrated in signal processing to extract target tracks from severe clutter.

Another group of outgrowths of military thermal imagers is airborne line scanners. These are one-dimensional scanning systems that enable the creation of a 2D thermal image of the observed scenery only when the system is moving. In contrast to typical thermal imagers with FOV not higher than about 40°, the airborne line scanners can provide FOV up to about 180°. Due to wide FOV, airborne thermal scanners are widely used in military aerial reconnaissance.

2.3.3 SPACE-BASED SYSTEMS

The formation of NASA in 1958, and development of the early planetary exploration program, was primarily responsible for the development of the modern optical remote sensing systems, as we know them today. During the 1960s, optical mechanical scanner systems became available that made possible acquisition of image data outside the limited spectral region of the visible and NIR available with film. "Eye in the Sky" was the first successfully flown long-wavelength sensor launched in 1967. A major milestone was the development of the Landsat Multispectral Scanner because it provided the first multispectral synoptic in digital form. The period following the launch of Landsat-1 in 1972 stimulated the development of a new series of airborne and spaceborne sensors. Since that time, hundreds of space-based sensors have been orbited.

The main advantages of space IR sensors are as follows [7]:

- the ability to tune the orbit to cover a ground swath in optimal spatial or temporal way;
- a lack of atmospheric effects on observation;
- global coverage;
- the ability to engage in legal clandestine operations.

Hitherto, anti-satellite weapons do not exist, so satellites are relatively safe from attack. The disadvantages of satellite systems are protracted and excessive costs of fabrication, launch, and maintenance of satellites. Moreover, such operations as repair and upgrade are difficult, expensive, and usually not possible.

The space-based systems installed on space platforms usually perform one of the following functions: military/intelligence gathering, astronomy, earth environmental/resources sensing, or weather monitoring. Hence these functions can be classified as forms of earth remote sensing and astronomy.

Figure 2.15 shows representative space sensor architecture. It should be stressed, however, that many individual space sensors do not have this exact architecture.

Intelligence and military services from wealthy nations have long employed space-based sensors to acquire information. A satellite-borne IR warning receiver, designed to detect intercontinental ballistic missiles, is a strategic system that protects a large area or nation. The U.S. spends about $10 billion per year on space reconnaissance. Although the cold war is over, the long-term strategic monitoring to access military and economic might is still important. Intelligence gathering of crop data and weather trends from space has also been used by hunger relief organizations to more effectively forecast droughts and famines.

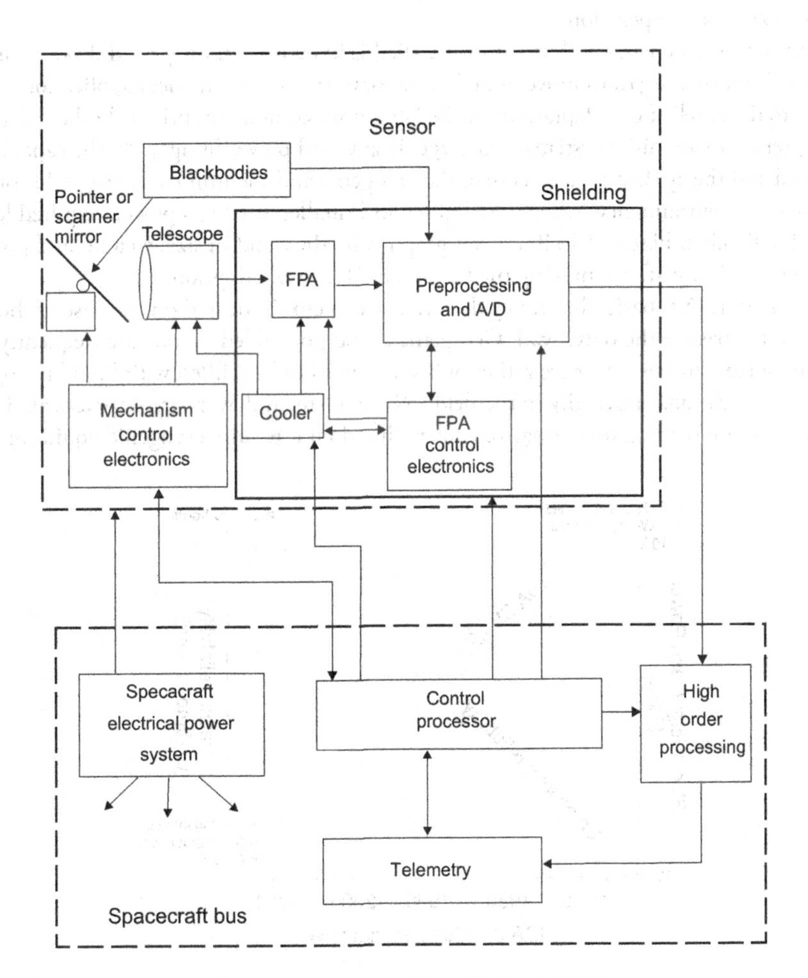

Figure 2.15 Representative space sensor architecture. (Adapted after Ref. [9])

The military also has space-based surveillance for missile launches. Additionally, space basing provides excellent viewing geometries for global events as nuclear explosions and environmental changes that the military is concerned about.

Imaging with IR FPAs provides increasingly detailed and quantitative information about relatively cool objects in the space of our galaxy and beyond. Dwarf stars, for example, or giant Jupiter-like planets in other distant solar systems, do not emit much visible and ultraviolet light, so they are extremely faint at these wavelengths. Also, the longer IR wavelengths can penetrate dusty and optically opaque nebulous molecular clouds in interstellar space where new stars and planetary systems are forming.

There are several unique reasons for conducting astronomy in space [9]:

- To eliminate the influence of absorption, emission, and scattering of IR radiation;
- To answer basic cosmological and astronomical questions (e.g., the formation of stars, protoplanetary discs, extrasolar planets, brown dwarfs, dust and interstellar media, protogalaxies, the cosmic distance scale, and ultraluminous galaxies);
- To observe the earth's environment (detecting the subtle changes indicating environmental stresses and trends).

2.4 COOLER TECHNOLOGIES

Generally, there are two main types of detectors—cooled and uncooled. Cooled detectors require cooling below ambient temperatures. Although uncooled sensors offer significant advantages in terms of cost, lifetime, size, weight, and power, cooled sensors offer significantly enhanced range, resolution, and sensitivity as a result of the lower noise operation.

In the IR industry, the housing with detector installed is known as an integrated detector assembly (or IDA). When IR sensors began to move out of the laboratory and into tactical applications in the 1960s, the first sensors to do so relied on adaptations of the laboratory equipment, primarily the refrigerators and vacuum systems. These initial systems were large, heavy, and power-hungry. As the capability of the systems increased and the applications extended, the cryogenic and vacuum packaging solutions were better suited for tactical environments—they were lighter and smaller, used less power, required less support, and offered high reliability. Figure 2.16 illustrates graphically the trend in size, weight, and power (SWaP) consumption and reliability after removing the barriers to IR sensor adoption.

The housing on an IDA is basically a fancy dewar. The detector is located on the base of the inner wall with a window in the base of the outer wall. Cryogenic liquid pour-filled dewars are frequently used for detector cooling in laboratories. They are rather bulky and need to be refilled with liquid nitrogen every few hours. For many applications, especially in the field LN_2, pour-filled dewars are impractical, due to which many manufacturers are turning to alternative coolers that do not require cryogenic liquids or solids. There

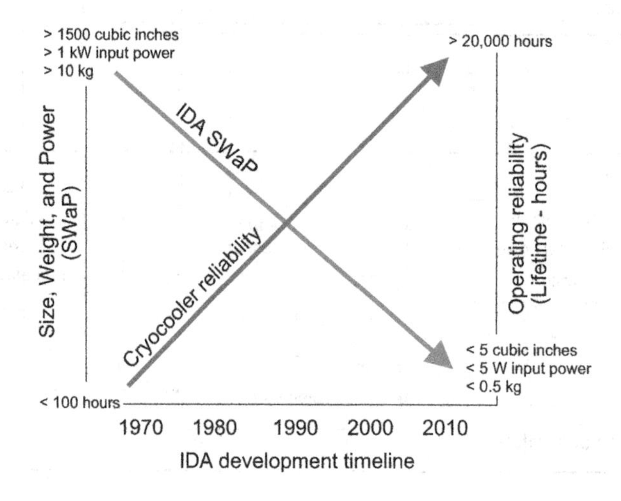

Figure 2.16 Evolution of cryocooler and IR detector packaging.

are many design considerations and challenges that go into developing an IDA. Figure 2.17 illustrates common elements and features of an IR dewar. Inside the dewar are several additional key components:

- Cold shield—to limit the FOV of the detector to the desired optical bundle. The internal features and surfaces of the cold shield are such that stray light coming from outside the desired bundle but coming through the aperture is absorbed and prevented from impinging upon the detector. Typically, cold shields eliminate more than 95% of the stray light.
- Cold filter—to eliminate the energy outside the region of interest. By cooling the filter, the self-emission of the filter element is minimized. Cold filters are fabricated from a variety of substrate materials, most commonly silicon, BK7 glass, sapphire, and germanium.

Figure 2.18 is a chart depicting IR operating temperature and wavelength regions spanned by a variety of available IR detector technologies. Typical operating temperatures range from 4 K to just below

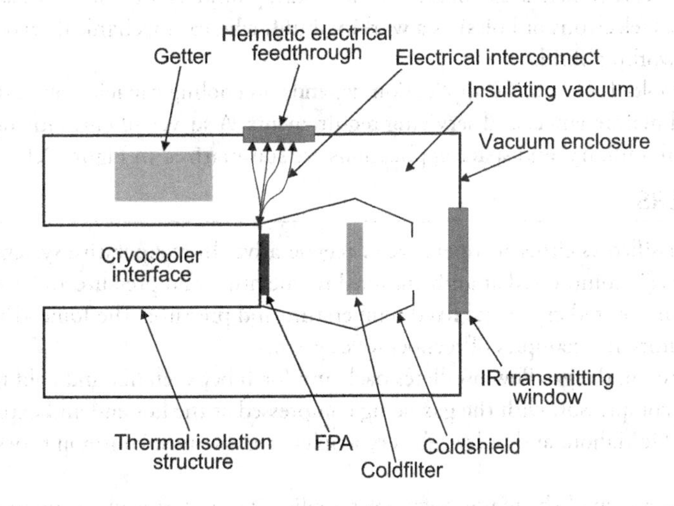

Figure 2.17 Components of a dewar.

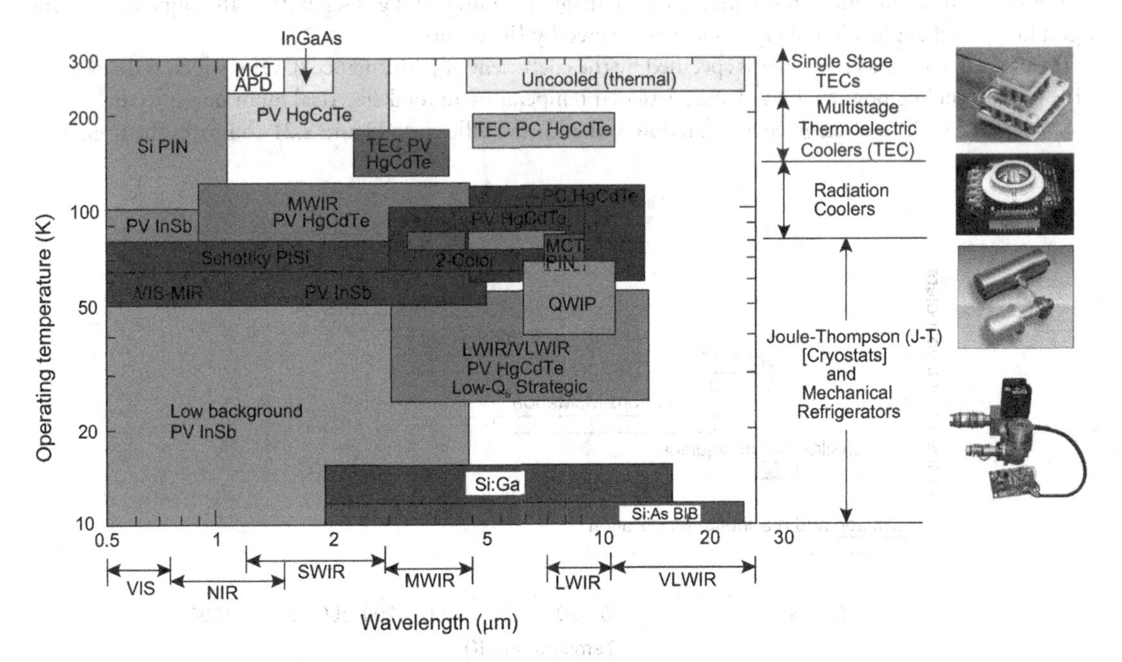

Figure 2.18 The operating temperature and wavelength regions spanned by a variety of available IR detector technologies.

room temperature, depending on the detector technology. Most modern cooled detectors operate in the temperature range from < 10 K to 150 K, depending on the detector type and performance level. 77 K is a very common temperature because this is relatively easily achievable with liquid nitrogen. Uncooled detectors, despite their title, typically incorporate some degree of temperature control near or slightly below room temperature (~250 K–300 K) to minimize noise, optimize resolution, and maintain stable operating temperature.

The method of cooling varies according to the operating temperature and the system's logistical requirements [11,12]. The two technologies currently available for addressing the cooling requirements of IR and visible detectors are closed cycle refrigerators and thermoelectric (TE) coolers. Closed cycle refrigerators can achieve the cryogenic temperatures required for cooled IR sensors, while TE coolers are generally the preferred approach to temperature control for uncooled visible and IR sensors. The major difference between the TE and mechanical cryocoolers is the nature of the working fluid. A TE cooler is a solid-state device that uses charge carriers (electrons or holes) as a working fluid, whereas mechanical cryocoolers use a gas such as helium as the working fluid.

The selection of a cooler for a specific application depends on cooling capacity, operating temperature, procurement, cost and maintenance, and servicing requirements. A survey of currently operating cryogenic systems for commercial, military, and space applications are summarized in Figure 2.19.

2.4.1 CRYOCOOLERS

Cryocoolers can be classified as either recuperative or regenerative. In recuperative systems, gas flows in a single direction. The gas is compressed at ambient fixed temperature and pressure and allowed to expand through an orifice to the desired cryogenic fixed temperature and pressure. The Joule–Thomson (JT) and Brayton cycle refrigerators are examples of recuperative systems.

In a regenerative system, the gas flow oscillates back and forth between hot and cold regions driven by a piston, diaphragm, or compressor, with the gas being compressed at the hot end and expanded at the cold end. Stirling, Gifford-McMahon, and pulse tube cryocoolers are the most common types of regenerative cryocooler systems.

Figure 2.20 presents a map of the major cryocooler applications in terms of the temperature and net refrigeration power required. The major commercial applications include cryopumps for semiconductor fabrication facilities, magnetic resonance imaging magnet cooling, and gas separation and liquefaction. The largest low-power application of cryocoolers is covered by IR sensors.

The performance of a cryocooler is specified by the coefficient of performance (COP), which is defined as the ratio of cooling power achieved at a particular temperature to total electrical input power to the cryocooler. The COP is often given as a fraction of the Carnot efficiency. Figure 2.21 compares the relative

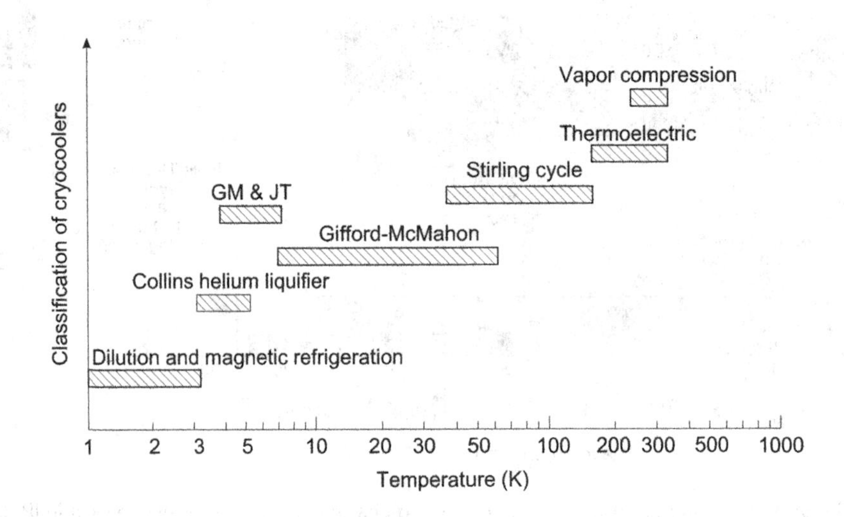

Figure 2.19 Temperature ranges for commercial refrigerators. (Adapted after Ref. [11])

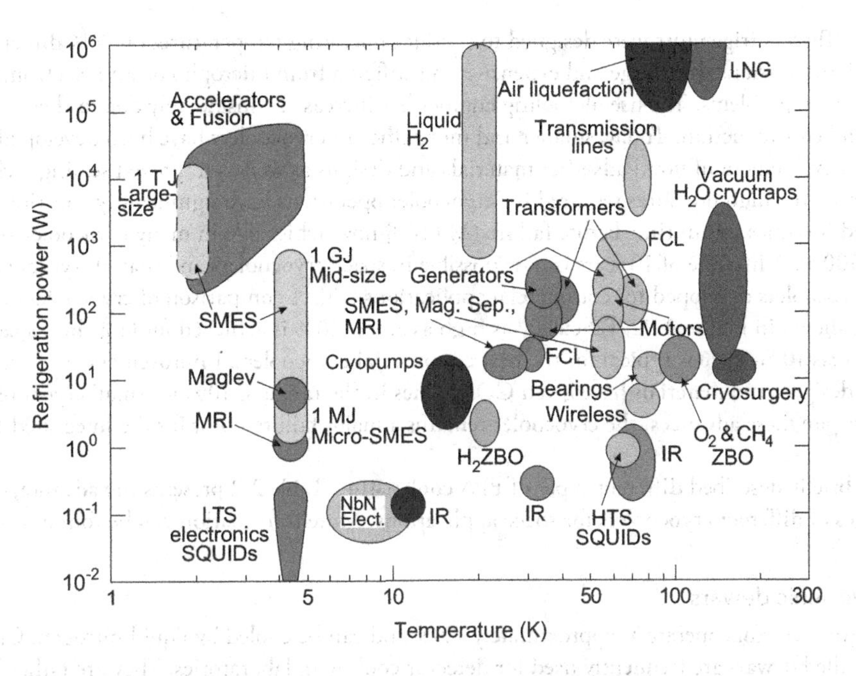

Figure 2.20 Map of cryocooler applications in the plane of refrigeration power versus temperature. (Adapted after Ref. [13]). Note: SMES—Superconducting Magnetic Energy Storage, MRI—Magnetic Resonance Imaging, LTS—Low Temperature Superconductivity, SQIDs—Superconducting Quantum Interference Devices, LNG—Liquid Natural Gas, FCL—Freon Coolant Line, IR—Infrared, ZBO—zero boil-off, and HTC—High-Temperature Superconductivity.

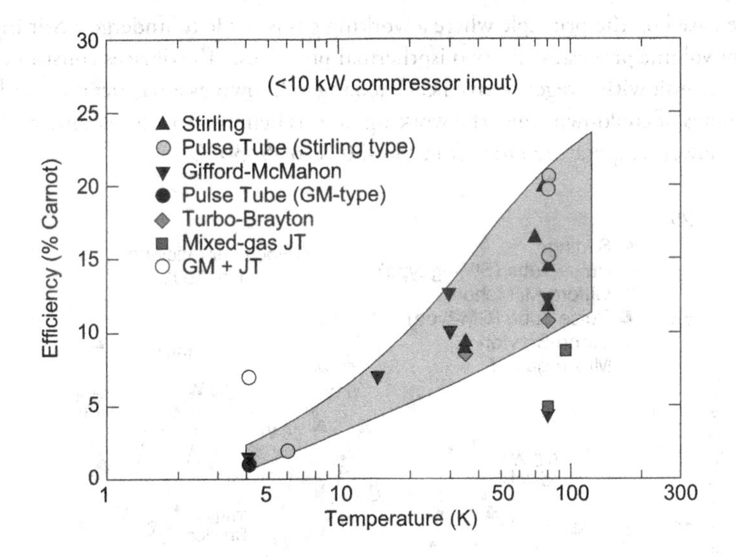

Figure 2.21 The efficiency of small cryocoolers as a function of cold-end temperature. (Adapted after Ref. [13]). Note: GM—Gifford-McMahon, and JT—Joule–Thomson.

performance of the different technologies as a fraction of the limiting ideal efficiency. For lower temperatures, the efficiency drops significantly. At 4 K, typical efficiency is about 1% or less. Generally, recuperative systems have advantages in terms of reduced noise and vibration, whereas the regenerative systems tend to obtain higher efficiencies and greater reliability at the temperatures of interest for many IR detector applications.

Ever since the late 1970s, military systems have overcome the problem of LN_2 operation by utilizing Stirling closed-cycle refrigerators to generate the cryogenic temperatures necessary for critical IR detector components. These refrigerators were designed to produce operating temperatures of 77 K directly from DC power. Early versions were large and expensive and suffered from microphonic and electromagnetic interference noise problems. The use of cooling engines has increased considerably due to their efficiency, reliability, and cost reduction. Today, smaller and more efficient cryocoolers have been developed and refined. The development of novel adsorber materials and designs as well as improved sealing techniques for maintaining the high pressures required for cryocooler operations have significantly contributed to the improved lifetimes [mean time before failure (MTBF)] now achievable in many cryocooler systems (5,000–10,000 h). A lifetime of 10 years is now possible in space cryocoolers and that of 5 years is possible in similar cryocoolers developed for commercial applications [13]. A comparison of cryocooler efficiencies near 80 K is shown in Figure 2.22. Efficiency as high as about 20% is achieved for large best space cryocoolers, whereas 10%–20% is typical for the best commercial cryocoolers. Improvements in heat exchanger, recuperator designs, and materials have given COP values in the range of 10% for small cryocoolers. However, despite these advances, the cryocooler remains a major failure point for the integrated IR sensor system.

We have briefly described different types of FPA coolers [14]. Table 2.2 presents the advantages and disadvantages of different cryocoolers for space applications. More information can be found in Refs [13] and [15].

2.4.1.1 Cryogenic dewars

Most 8–14-μm detectors operate at approximately 77 K and can be cooled by liquid nitrogen. Cryogenic liquid pour-filled dewars are frequently used for detector cooling in laboratories. They are rather bulky and need to be refilled with liquid nitrogen every few hours. For many applications, especially in the field LN_2, pour-filled dewars are impractical, so many manufacturers are turning to alternative coolers that do not require cryogenic liquids or solids.

2.4.1.2 Stirling cycle

Stirling coolers are based on the principle where a working gas is made to undergo a Stirling cycle consisting of two constant volume processes and two isothermal processes. The devices consist of a compressor pump and a displacer unit with a regenerative heat exchanger, known as a regenerator. Stirling engines require several minutes of cooldown time; the working fluid is helium. The development of two-stage devices extends the lower temperature range from 60–80 K to 15–30 K.

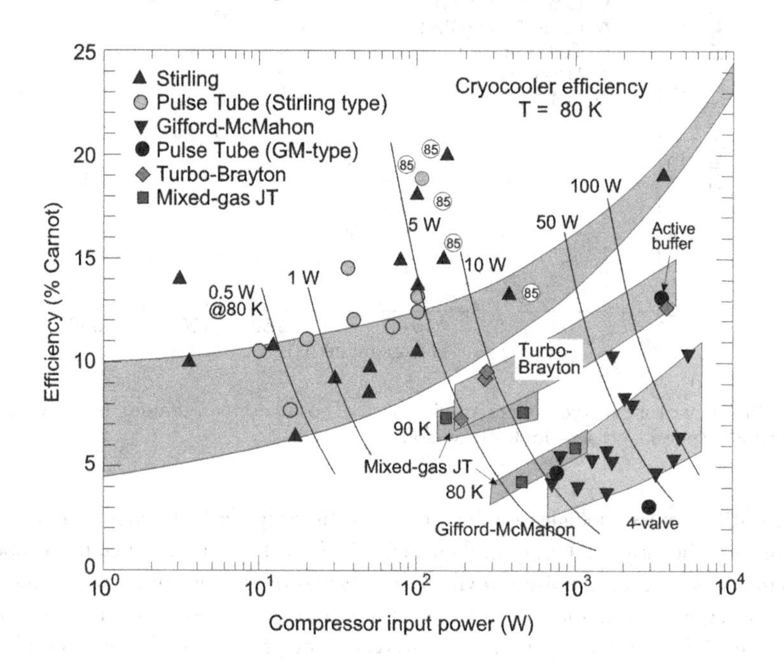

Figure 2.22 Efficiencies of various types of cryocoolers at 80 K. (Adapted after Ref. [13])

Table 2.2 Cryocoolers for space applications

COOLER	TYPICAL TEMPERATURE (K)	TYPICAL HEAT LIFT	ADVANTAGES	DISADVANTAGES
Radiator	80	0.5 W	Reliable, low vibration, long lifetime	Complicates orbit
Stirling—1 stage	80	0.8 W	Efficient, heritage	Vibrations
Stirling—2 stage	20	0.06 W	Intermediate temp	Under development
Pulse tube	80	0.8 W	Lower vibrations, efficiency comparable to Stirling	Difficulties in scaling down to small sizes while maintaining high efficiency, larger diameter cold finger
Joule–Thomson	80	0.01 W	Low vibrations	Requires hybrid design
Sorption	10	0.1 W	Low vibrations	Low efficiency, under development
Brayton	65	8 W	High capacity	Complex
ADR	0.05	0.01 mW	The only way to reach these temps.	Large magnetic field
Peltier	170	1 W	Lightweight	High temp, low efficiency

Source: Adapted after Ref. [14].

Both JT and engine-cooled detectors are housed in precision-bore dewars into which the cooling device is inserted (see Figure 2.23). The detector, mounted in the vacuum space at the end of the inner wall of the dewar and surrounded by a cooled radiation shield compatible with the convergence angle of the optical system, looks out through an IR window. In some dewars, the electrical leads to detector elements are embedded in the inner wall of the dewar to protect them from damage due to vibration. After 3,000–5,000 h of operation, Stirling coolers require factory service to maintain performance. Because the dewar and the cooler is one IDA unit, the entire unit must be serviced together.

Split Stirling coolers are also fabricated. The detector is mounted on the dewar bore, and the cold finger of the cooler is thermally connected to the dewar by a bellows. A fan is necessary to dissipate the heat. The cooler can be easily removed from the detector/dewar for replacement.

Figure 2.24 shows the four sizes of Stirling cryocoolers used for military tactical applications. The refrigeration powers listed for each cooler are for a temperature of about 77 to 80 K, except the 1.75-W system, which is for a temperature of 67 K [13]. Their COPs range from about 3% to 6% of Carnot as the size increases. All of the shown coolers use linear drive motors with a dual-opposed arrangement to reduce vibration. The displacer driven pneumatically with the oscillating pressure in the system gives rise to considerable vibration (because only one displacer is used). Space applications of Stirling cryocoolers require greatly improved reliabilities, and a MTBF of 10 years is now usually specified for these applications.

2.4.1.3 Pulse tube

The moving displacer in the Stirling cryocoolers has several disadvantages. It is a source of vibration, has a limited lifetime, and contributes to axial heat conduction as well as to a shuttle heat loss. Pulse tube coolers

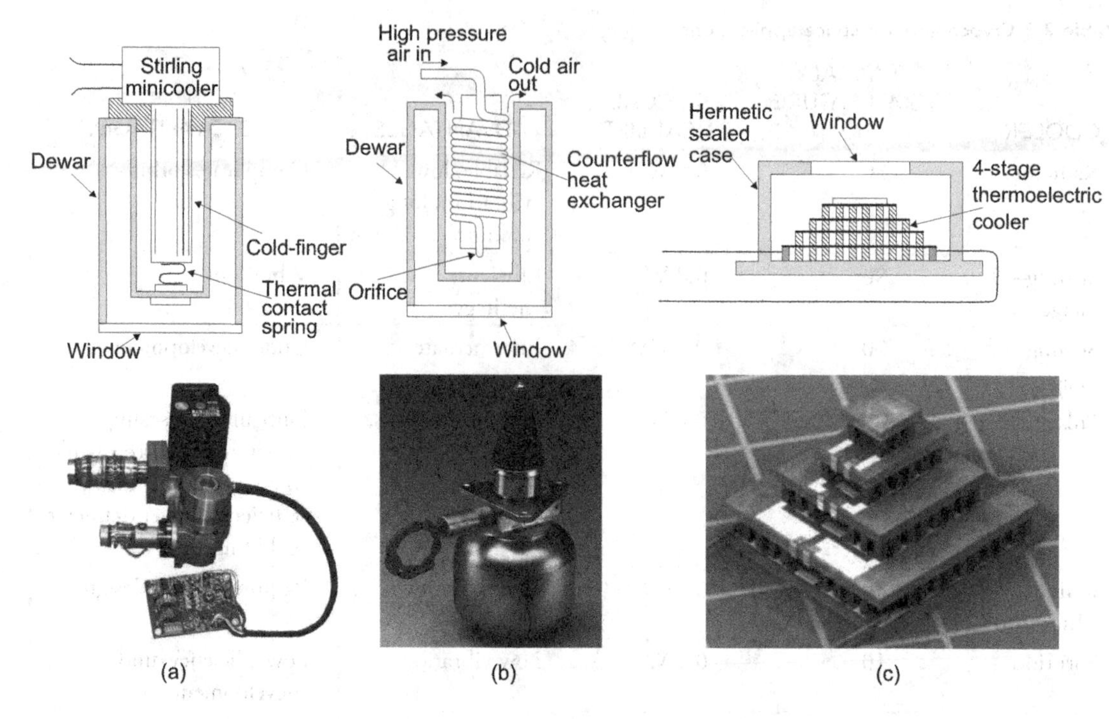

Figure 2.23 Three ways of cooling IR detectors: (a) Stirling cycle engine, (b) JT cooler, and (c) four-stage TE cooler (Peltier cooling).

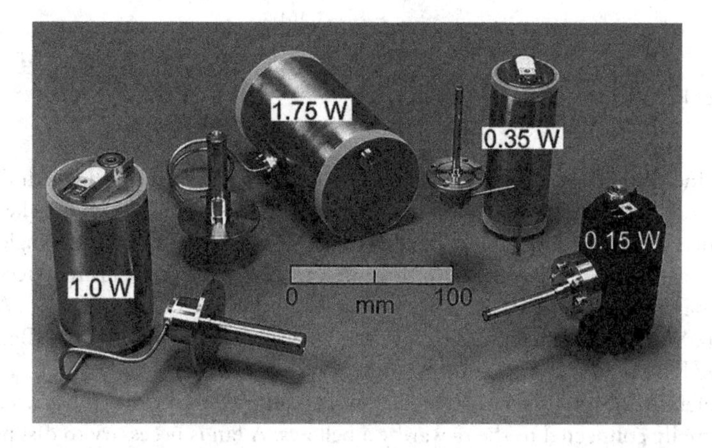

Figure 2.24 Four sizes of Stirling cryocoolers with dual-opposed linear compressors. (After Ref. [13]). The refrigeration powers listed for each cooler are for a temperature of about 77 to 80 K, except the 1.75 W system, which is for a temperature of 67 K.

are similar to Stirling coolers. However, their thermodynamic processes are quite different. The proper gas motion in phase with the pressure is achieved by the use of an orifice, along with a reservoir volume to store the gas during a half cycle. The reservoir volume is large enough that negligible pressure oscillation occurs in it during the oscillating flow. The oscillating flow through the orifice separates the heating and cooling effects just as the displacer does for the Stirling refrigerator.

Since there are no moving parts at the cold end, reliability is theoretically higher than Stirling cycle machines. Efficiencies approaching Stirling cycle coolers can be achieved and several recent missions have demonstrated their usefulness in space.

2.4.1.4 Joule–Thomson coolers

The design of JT coolers is based on the fact that as a high-pressure gas expands upon leaving a throttle valve, it cools and liquefies (leading to isenthalpic cooling). The coolers require a high-pressure gas supply from bottles and compressors. Although this is an irreversible process, with correspondingly low efficiency, JT coolers are simple, reliable, and have low electrical and mechanical noise levels.

Using compressed air, temperatures of the order of 80 K can be achieved in 1 or 2 minutes. The gas used must be purified to remove water vapor and carbon dioxide that could freeze and block the throttle valve. Specially designed JT coolers using argon are suitable for ultrafast cooldown (a few seconds cooling time). Recent advances in JT cryocoolers have been associated with the use of mixed gases as the working fluid rather than pure gases.

Figure 2.23b shows a typical JT cryocooler used for missile guidance, where miniature finned tubing is used for the heat exchanger. An explosive valve is used to start the flow of gas from the high-pressure bottle, and after flowing through the system, the gas is vented to the atmosphere.

2.4.1.5 Sorption

This type of cooler is essentially a JT cooler that uses a thermochemical process to provide gas compression with no moving parts. Powdered sorbent materials (e.g., metal hydrides) are electrically heated and cooled to pressurize, circulate, and adsorb a working fluid such as hydrogen. The disadvantage of sorption coolers is low efficiency, which may be increased by the use of mixed working gases. These coolers are expected to be useful in long-life space missions where very low vibration levels are required.

2.4.1.6 Brayton

In Brayton cryocoolers (sometimes referred to as the reverse-Brayton cycle to distinguish it from a heat engine) cooling occurs as the expanding gas does work. Coolers consist of a rotary compressor, a rotary turbo-alternator (expander), and a counterflow heat exchanger (as opposed to the regenerator found in Stirling or pulse tube coolers). The compressor and expander use high-speed miniature turbines on gas bearings and thus small machines are very difficult to build. Brayton coolers have high efficiencies and are practically vibration-free. This low vibration is often required with sensitive telescopes in satellite applications. These coolers are primarily used for low-temperature experiments (less than 10 K), where a large machine is inevitable, or for large-capacity devices at higher temperatures (although these requirements are quite rare). The expansion engine provides for good efficiency over a wide temperature range, although not as high as some Stirling and pulse tube cryocoolers at temperatures above about 50 K.

2.4.1.7 Adiabatic demagnetization

Adiabatic demagnetization refrigeration (ADR) has been used on the ground for many years to achieve millikelvin temperatures after a first-stage cooling process. The process utilizes the magnetocaloric effect with a paramagnetic salt. These coolers are currently under development for space use.

2.4.1.8 ^{3}He coolers

When a mixture of two isotopes of helium is cooled below approximately 870 mK, the mixture undergoes spontaneous phase separation to form a ^{3}He-rich phase and a ^{3}He-poor phase.

A helium-3 (^{3}He) refrigerator is a simple device used for obtaining temperatures down to about 250 mK. By evaporative cooling of helium-4 (^{4}He) (the more common isotope of helium), a 1-K pot liquefies a small amount of ^{3}He in a small vessel called a ^{3}He pot. Evaporative cooling of the liquid ^{3}He, usually driven by adsorption (an internal charcoal sorption pomp) cools the ^{3}He pot to a fraction of a kelvin.

Using dilution refrigerators, a temperature above 50 mK can be received. A two-stage Gifford-McMahon cryocooler can be used for ^{3}He condensation.

2.4.1.9 Passive coolers

Passive coolers require no input power and have been used for many years in space science applications due to their relatively high reliability and low vibration levels. Radiators and stored cryogens belong to passive coolers.

Radiators are panels radiating heat and are the workhorse of satellite cooling due to their extremely high reliability. They have low mass and a lifetime limited only by surface contamination and degradation. Radiators have severe limitations on the heat load and temperature (typically in the milliwatt range at 70 K). Usually, two and three stages are used to baffle the lowest temperature stage or patch. In this case, the first stage consists of a highly reflective baffle (e.g., a cone) to shield the patch from the spacecraft, earth, or shallow-angle sunlight.

Stored cryogens are dewars containing a cryogen such as liquid helium or solid neon. They are used to achieve temperatures below those offered by radiators (heat is absorbed by either boiling or sublimation) and provide excellent temperature stability with no exported vibrations. However, stored cryogens substantially increase the launch mass of the vehicle and limit the lifetime of the mission to the amount of cryogen stored. They have also proved to be of limited reliability.

2.4.2 PELTIER COOLERS

Thermoelectric cooling of detectors is simpler and less costly than closed cycle cooling. Thermoelectric coolers work by exploiting the Peltier effect that refers to the creation of heat flux at the junction of two dissimilar conductors in the presence of current flow. The optimal performance of the thermoelectrical material in a device is determined by the dimensionless figure of merit

$$ZT = \frac{S^2\sigma}{k}, \tag{2.1}$$

where S is the Seebeck coefficient, σ is the electrical conductivity, k is the thermal conductivity, and T is the average temperature.

It appears that the Peltier cooler's COP is a function of ZT and the overall temperature difference between the hot side and cold side of the cooler. The typical maximum temperature difference between the hot side and the cold side of a single stage TE cooler, referred to as ΔT_{max}, is around 70°C (see Figure 2.25). At maximum temperature difference, the COP goes to zero. Conversely, at zero ΔT, a TE cooler achieves maximum heat pumping capacity. Therefore, TE coolers typically operate at the minimal ΔT that provides acceptable detector performance. To achieve ΔT of 70 K or more, TE coolers are stacked in "stages." The theoretical range for ΔT_{max} achievable using existing commercial materials is plotted in Figure 2.26 as a function of the number of stages [16].

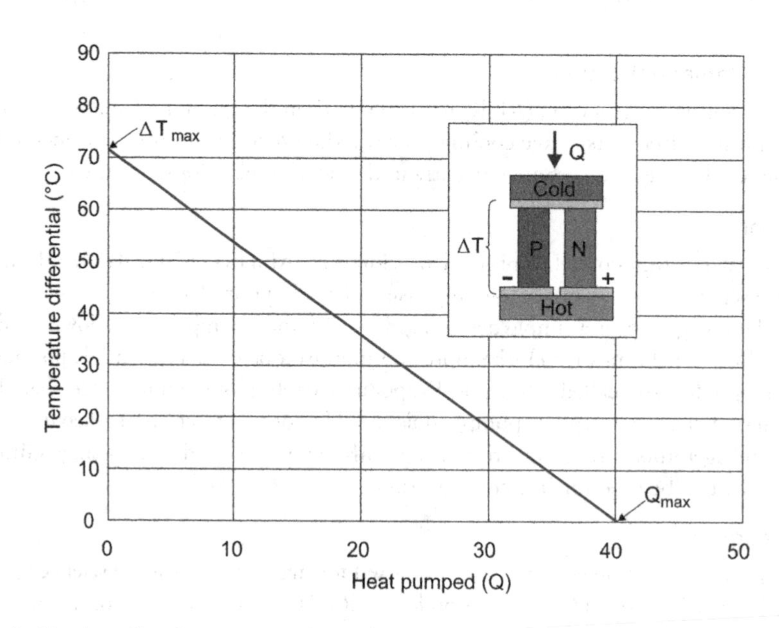

Figure 2.25 Typical load profile of a one-stage TE cooler.

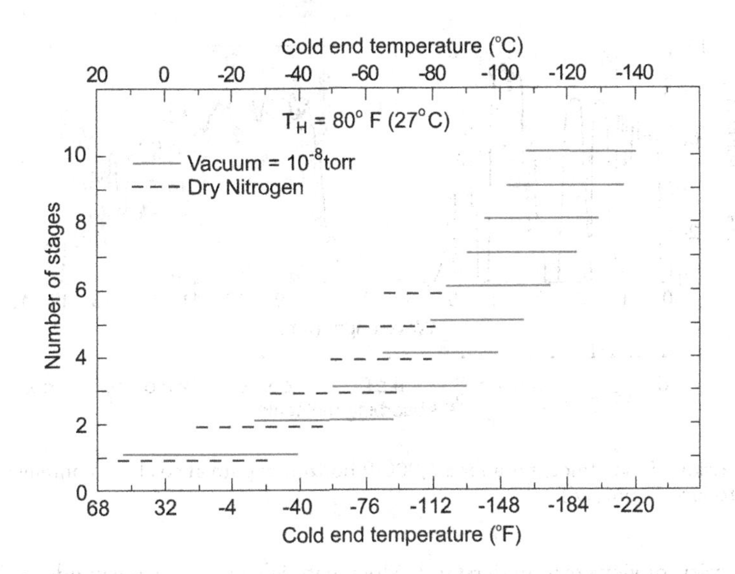

Figure 2.26 Typical performance range of TE modules. (Adapted after Ref. [16])

Commercially available coolers do not go beyond six stages. They are based on alloys of bismuth telluride and antimony telluride materials exhibiting *ZT* values close to 1, but in a device configuration, *ZT* value is closer to 0.7. Recently, the enhancement in *ZT* value has been achieved in low-dimensional solids via nanostructuring in thin film materials. Similar investigations in bulk materials have indicated that bulk TE performance in the range of *ZT* is about 1.5 near room temperature and approaching 2 at higher temperatures [17]. The modest improvements in bulk materials are transitioning to commercial TE products [18].

In the case of Peltier coolers, detectors are usually mounted in a hermetic encapsulation with a base designed to make good contact with a heat sink. TE coolers can achieve temperatures to ≈ 200° K, have about a 20-year operating life, are small and rugged, and have low input power (<1 W for a two-stage device and <3 W for a three-stage device). Their main disadvantage is low efficiency (see Table 2.2).

The TE coolers used for IR FPA operation include one-stage (TE1, down to −20°C or 253K), two-stage (TE2, down to −40°C or 233K), three-stage (TE3, down to −65°C or 208K), and four-stage (TE4, down to −80°C or 193K). Peltier coolers are also the preferred approach to temperature control at the required level; for example, for uncooled visible and IR sensors.

2.5 ATMOSPHERIC TRANSMISSION AND IR BANDS

Most of the aforementioned applications require transmission through air, but the radiation is attenuated by the processes of scattering and absorption. Scattering causes a change in the direction of a radiation beam; it is caused by absorption and subsequent reradiation of energy by suspended particles. For larger particles, scattering is independent of wavelength. However, for small particles, compared with the wavelength of the radiation, the process is known as Rayleigh scattering and exhibits a λ^{-4} dependence. Therefore, scattering by gas molecules is negligibly small for wavelengths longer than 2 μm. Also, smoke and light mist particles are usually small with respect to IR wavelengths, and IR radiation can, therefore, penetrate further through smoke and mists than visible radiation. However, rain, fog particles, and aerosols are larger and consequently scatter IR and visible radiation to a similar degree.

Figure 2.27 is a plot of the transmission through 6,000 feet of air as a function of wavelength [19]. Specific absorption bands of water, carbon dioxide, and oxygen molecules are indicated, which restrict atmospheric transmission to two windows, at 3–5 μm and 8–14 μm. Ozone, nitrous oxide, carbon monoxide, and methane are less important IR absorbing constituents of the atmosphere.

The short wavelength infrared (SWIR) band offers unique imaging advantages over visible and thermal bands. Like visible cameras, the images are primarily created by reflected broadband light sources, so

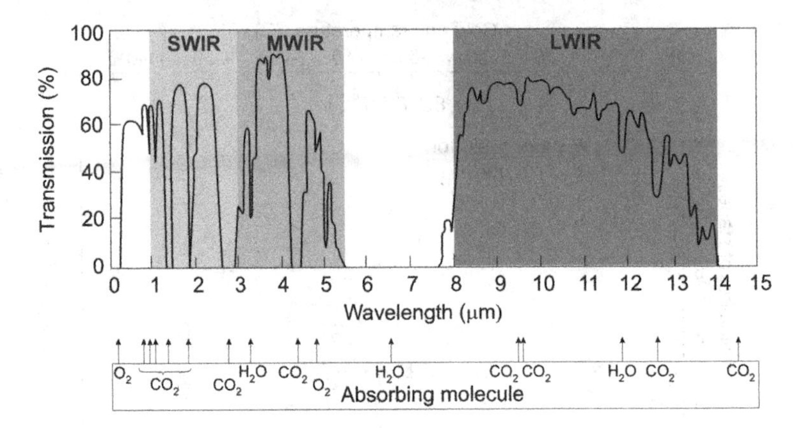

Figure 2.27 Transmission of the atmosphere for a 6,000-ft horizontal path at sea level containing 17 mm of precipitate water. (Reproduced from [19])

SWIR images are easier for viewers to understand. Most materials used to make windows, lenses, and coatings for visible cameras are readily useable for SWIR cameras, keeping costs down. Ordinary glass transmits radiation to about 2.5 μm. SWIR cameras can image many of the same light sources, such as YAG laser wavelengths. Thus, with safety concerns shifting laser operations to the "eye-safe" wavelengths where beams don't focus on the retina (beyond 1.4 μm), SWIR cameras are in a unique position to replace visible cameras for many tasks. Due to the reduced Rayleigh scattering of light at longer wavelengths by particulates in the air, such as dust or fog, SWIR cameras can see through haze better than visible cameras.

In general, the 8–14-μm band is preferred for high-performance thermal imaging because of its higher sensitivity to ambient temperature objects and its better transmission through mist and smoke. However, the 3–5-μm band may be more appropriate for hotter objects, or if sensitivity is less important than contrast. Also, additional differences occur; for example, the advantage of middle wavelength infrared (MWIR) band is a smaller diameter of the optics required to obtain a certain resolution and that some detectors may operate at higher temperatures (TE cooling) than is usual in the long wavelength infrared (LWIR) band where cryogenic cooling is required (about 77 K).

Summarizing, MWIR and LWIR μm spectral bands differ substantially with respect to background flux, scene characteristics, temperature contrast, and atmospheric transmission under diverse weather conditions. Factors that favor MWIR applications are higher contrast, superior clear weather performance (favorable weather conditions, e.g., in most countries of Asia and Africa), higher transmittivity in high humidity, and higher resolution due to ~3 times smaller optical diffraction. Factors that favor LWIR applications are better performance in fog and dust conditions, winter haze (typical weather conditions, e.g., in West Europe, northern United States, Canada), higher immunity to atmospheric turbulence, and reduced sensitivity to solar glints and fire flares. The possibility of achieving higher signal-to-noise (S/N) ratio due to greater radiance levels in LWIR spectral range is not persuasive because the background photon fluxes are higher to the same extent, and also because of readout limitation possibilities. Theoretically, in staring arrays, the charge can be integrated for full frame time, but because of restrictions in the charge-handling capacity of the readout cells, it is much less compared to the frame time, especially for LWIR detectors for which background photon flux exceeds the useful signals by orders of magnitude.

2.6 SCENE RADIATION AND CONTRAST

The total radiation received from any object is the sum of the emitted, reflected, and transmitted radiation. Objects that are not blackbodies emit only the fraction $\varepsilon(\lambda)$ of blackbody radiation, and the remaining fraction, $1 - \varepsilon(\lambda)$, is either transmitted or, for opaque objects, reflected. When the scene is composed of objects and backgrounds of similar temperatures, reflected radiation tends to reduce the available contrast. However, reflections of hotter or colder objects have a significant effect on the appearance of a thermal scene. The powers of 290 K blackbody emission and ground-level solar radiation in MWIR and

Table 2.3 Power available in each MWIR and LWIR imaging bands

IR REGION (μm)	GROUND-LEVEL SOLAR RADIATION (W/m²)	EMISSION FROM 290 K BLACKBODY (W/m²)
3–5	24	4.1
8–13	1.5	127

Source: After Ref. [20].

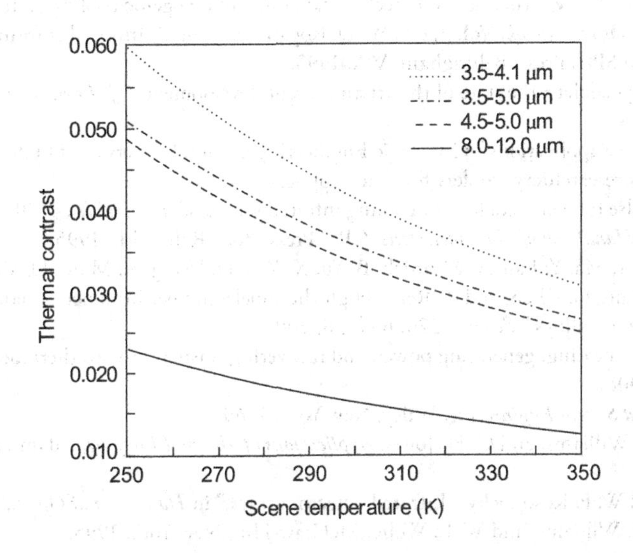

Figure 2.28 Spectral photon contrast in the MWIR and LWIR bands.

LWIR bands are given in Table 2.3 [20]. We can see that while reflected sunlight has a negligible effect on 8–13-μm imaging, it is important in the 3–5-μm band.

A thermal image arises from temperature variations or differences in emissivity within a scene. When the temperatures of a target and its background are nearly the same, detection becomes very difficult. The thermal contrast is one of the important parameters for IR imaging devices. It is the ratio of the derivative of spectral radiant exitance to the spectral radiant exitance

$$C = \frac{\partial M(\lambda,T)/\partial T}{M(\lambda,T)}. \tag{2.2}$$

Figure 2.28 is a plot of C for several MWIR sub-bands and the 8–12 LWIR spectral bands [21]. The contrast in a thermal image is small when compared with visible image contrast due to differences in reflectivity. We can notice the contrast in the MWIR bands at 300 K is 3.5%–4% compared to 1.6% for the LWIR band. This, while the LWIR band may have higher sensitivity for ambient temperature objects, the MWIR band has greater contrast.

REFERENCES

1. J. S. Accetta and D. L. Shumaker (eds.), *The Infrared and Electro-Optical Systems Handbook*, SPIE Press, Bellingham, 1993.
2. P. R. Norton, "Infrared detectors in the next millennium," *Proc. SPIE*, 3698, 652–65, 1999.
3. http://www.lucintel.com/global_infrared_detector_market_2016.aspx.
4. M. L. Vatsia, U. K. Stich, and D. Dunlap, "Night sky radiant sterance from 450 nm to 2000 nanometers", Project IS662709D617, U. S. Army Electronics Command, Fort Belvoir, Virginia, VA, September 1972.

5. K. Chrzanowski, "Review of night vision technology," *Opto-Electron. Rev.*, 21, 153–82, 2013.
6. K. Chrzanowski and A. Rogalski, "Infrared devices and techniques (Revision)," *Metrol. Meas. Syst.*, 21(4), 565–618, 2014.
7. http://www.hamamatsu.com/resources/pdf/etd/II_TII0004E02.pdf.
8. https://customeronline.thalesgroup.com/sites/default/files/asset/document/hel_topowl_en.pdf.
9. J. L. Miller, *Principles of Infrared Technology*, Van Nostrand Reinhold, New York, 1994.
10. S. B. Campana (ed), *The Infrared and Electro-Optical Systems Handbook*, Vol. 5, Passive Electro-Optical Systems, SPIE Press, Bellingham, WA, 1993.
11. A. R. Jha, *Cryogenic Technology and Applications*, Elsevier, Oxford, 2006.
12. P. T. Blotter and J. C. Batty, "Thermal and mechanical design of cryogenic cooling systems," in *The Infrared and Electro-Optical Systems Handbook*, Vol. 3, ed., W. D. Rogatto, 343–433, Infrared Information Analysis Center, Ann Arbor, MI, and SPIE Press, Bellingham, WA, 1993.
13. R. Radebaugh, "Cryocoolers: the state of the art and recent developments," *J. Phys.: Condens. Matter* 21, 164219-1–9, 2009.
14. "Cryocoolers for space applications," Cryogenic Engineering Group, University of Oxford. http://www.eng.ox.ac.uk/cryogenics/research/cryocoolers-for-space-applications.
15. R. Radebaugh, "Pulse tube cryocoolers for cooling infrared sensosrs," *Proc. SPIE*, 4130, 363–79, 2000.
16. D. M. Rowe, *CRC Handbook of Thermoelectrics*, CRC Press, Boca Raton, FL, 1995.
17. B. Poudel, Q. Hao, Y. Ma, Y. Lan, A. Minnich, B. Yu, X. Yan, D. Wang, A. Muto, D. Vashaee, X. Chen, J. Liu, M. S. Dresselhaus, G. Chen, and Z. Ren, "High-thermoelectric performance of nanostructured bismuth antimony telluride bulk alloys," *Science*, 320, 634–38, 2008.
18. L. E. Bell, "Cooling, heating, generating power, and recovering waste heat with thermoelectric systems," *Science*, 321, 1457–1461 (2008).
19. R. Hudson, *Infrared System Engineering*, Wiley, New York, 1969.
20. S. G. Burnay, T. L. Williams, and C. H. Jones, *Applications of Thermal Imaging*, Adam Hilger, Bristol, UK, 1988.
21. L. J. Kozlowski and W. F. Kosonocky, "Infrared detector arrays," in *Handbook of Optics*, eds. M. Bass, E. W. Van Stryland, D. R. Williams, and W. L. Wolfe, McGraw-Hill, New York, 1995.

3 Characterization of infrared detectors

Infrared (IR) radiation itself was unknown until 218 years ago, when Herschel's experiment with the thermometer was first reported. The first detector consisted of a liquid in a glass thermometer with a specially blackened bulb to absorb radiation. Herschel built a crude monochromator that used a thermometer as a detector so that he could measure the distribution of energy in sunlight. In April 1800, he wrote [1]:

> Thermometer No. 1 rose 7 degrees in 10 minutes by an exposure to the full red colored rays. I drew back the stand... thermometer No. 1 rose, in 16 minutes, 8⅜ degrees when its centre was ½ inch out of the visible rays.

The early history of IR was reviewed about 60 years ago in two well-known monographs [2,3]. A lot of historical information can also be found in more recently published papers [4,5].

The most important steps in the development of IR detectors are the following [6,7]:

- In 1821, Seebeck discovered the thermoelectric effect and soon thereafter demonstrated the first thermocouple.
- In 1829, Nobili constructed the first thermopile by connecting a number of thermocouples in series.
- In 1833, Melloni modified the thermocouple and used bismuth and antimony for its design.

Langley's bolometer appeared in 1880 [7]. Langley used two thin ribbons of platinum foil, connected to form two arms of a Wheatstone bridge. Langley continued to develop his bolometer for the next 20 years (400 times more sensitive than his first efforts). His latest bolometer could detect the heat from a cow at a distance of a quarter of a mile. The beginning of the development of IR detectors was connected with thermal detectors.

The photoconductive effect was discovered by Smith in 1873 when he experimented with selenium as an insulator for submarine cables [8]. This discovery provided a fertile field of investigation for several decades, though most of the effort was of doubtful quality. By 1927, over 1,500 articles and 100 patents had been listed on the photosensitive selenium [9]. Work on the IR photovoltaic effect in naturally occurring lead sulfide or galena was announced by Bose in 1904 [10]; however, this effect was not used in a radiation detector for the next several decades.

The photon detectors were developed in the twentieth century. The first IR photoconductor was developed by Case in 1917 [11]. He discovered that a substance composed of thallium and sulfur exhibited photoconductivity. Later, he found that the addition of oxygen greatly enhanced the response [12]. However, instability of resistance in the presence of light or polarizing voltage, loss of responsivity due to overexposure to light, high noise, sluggish response, and lack of reproducibility seemed to be inherent weaknesses.

Since about 1930 the development of IR technology has been dominated by the photon detectors. In about 1930, the appearance of the Cs-O-Ag phototube, with more stable characteristics, to a great extent discouraged further development of photoconductive cells until about 1940. At that time, interest in improved detectors began in Germany [13,14]. In 1933, Kutzscher at the University of Berlin discovered that lead sulfide (from natural galena found in Sardinia) was photoconductive and had a response to about 3 μm. This work was, of course, done under great secrecy and the results were not generally known until after 1945. Lead sulfide was the first practical IR detector deployed in a variety of applications during the war. In 1941, Cashman improved the technology of thallous sulfide detectors, which led to successful production [15]. Cashman, after success with thallous sulfide detectors, concentrated his efforts on lead sulfide, and after World War II found that other semiconductors of the lead salt family (PbSe and PbTe) showed promise as IR detectors [15]. Lead sulfide photoconductors were brought to the manufacturing stage of

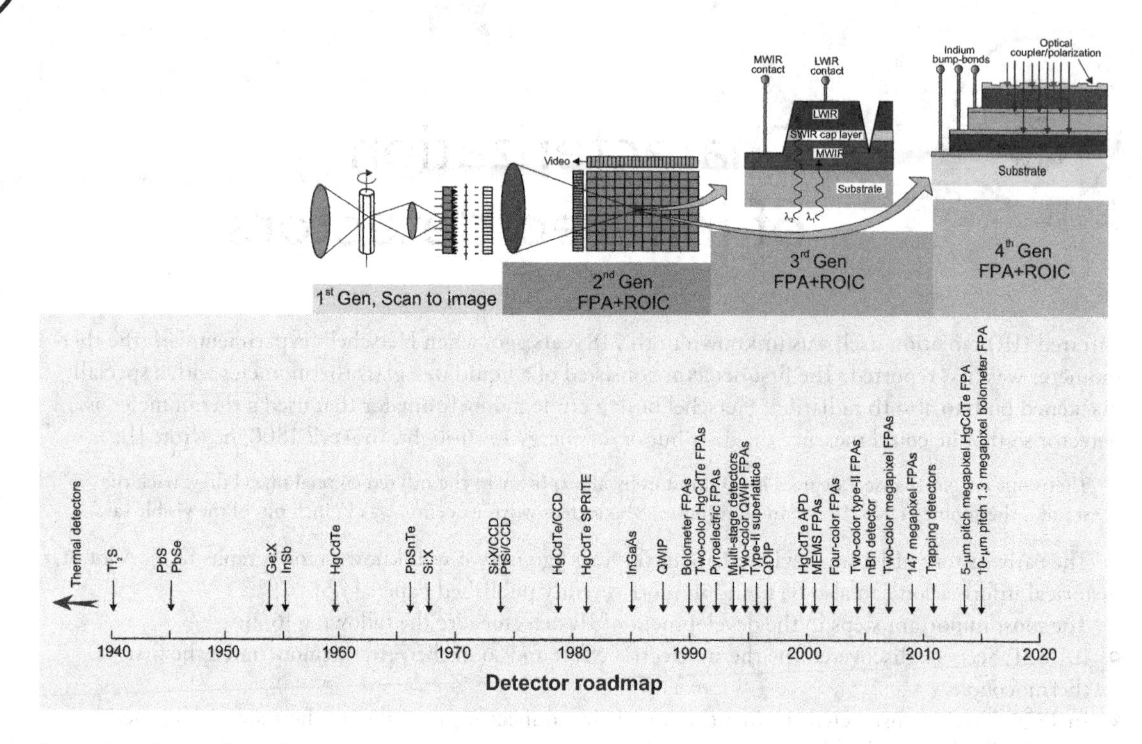

Figure 3.1 History of the development of IR detectors and systems. Four generation systems can be considered for principal military and civilian applications: first-generation (scanning systems), second-generation (staring systems–electronically scanned), third-generation (staring systems with large number of pixels and two-color functionality), and fourth-generation (staring systems with very large number of pixels, multicolor functionality and other on-chip functions; e.g., better radiation/pixel coupling, avalanche multiplication in pixels, polarization/phase sensitivity).

development in Germany in about 1943. They were first produced in the United States at Northwestern University, Evanston, Illinois in 1944 and in 1945, at the Admiralty Research Laboratory in England [16].

Many materials have been investigated in the IR field. Observing a history of the development of the IR detector technology, a simple theorem, after Norton [17], can be stated: *"All physical phenomena in the range of about 0.1–1 eV can be proposed for IR detectors."* Among these effects are thermoelectric power (thermocouples), change in electrical conductivity (bolometers), gas expansion (Golay cell), pyroelectricity (pyroelectric detectors), photon drag, Josephson effect (Josephson junctions, SQUIDs), internal emission (PtSi Schottky barriers), fundamental absorption (intrinsic photodetectors), impurity absorption (extrinsic photodetectors), low-dimensional solids [superlattice (SL), quantum well (QW), and quantum dot (QD) detectors], different type of phase transitions, and so on.

Figure 3.1 gives approximate dates of significant development efforts for the materials mentioned. The years during World War II saw the origins of modern IR detector technology. Recent success in applying IR technology to remote sensing problems has been made possible by the successful development of high-performance IR detectors over the last seven decades. Photon IR technology combined with semiconductor material science, photolithography technology developed for integrated circuits, and the impetus of Cold War military preparedness have propelled extraordinary advances in IR capabilities within a short time period during the last century [18].

3.1 HISTORICAL ASPECTS OF MODERN IR TECHNOLOGY

During the 1950s, IR detectors were built using single-element-cooled lead salt detectors, primarily for anti–air-missile seekers. Lead salt detectors were known to bea polycrystalline and were produced by vacuum evaporation and chemical deposition from a solution, followed by a post-growth sensitization process [16]. The preparation process of lead salt photoconductive detectors was usually not well understood and reproducibility could only be achieved after following well-tried recipes. The first extrinsic photoconductive detectors were reported in the early 1950s [19] after the discovery of the transistor, which stimulated a considerable improvement in the growth and material purification techniques. Since the techniques for controlled impurity introduction became available for germanium at an earlier date, the first high-performance extrinsic detectors were based on germanium. Extrinsic photoconductive response from copper, zinc, and gold impurity levels in germanium gave rise to devices using the 8–14-μm long wavelength IR (LWIR) spectral window and beyond the 14–30-μm very long wavelength IR region. The extrinsic photoconductors were widely used at wavelengths beyond 10 μm prior to the development of the intrinsic detectors. They must be operated at lower temperatures to achieve a performance similar to that of intrinsic detectors, and a sacrifice in quantum efficiency is required to avoid thick detectors.

In 1967, the first comprehensive extrinsic Si detector–oriented paper was published by Soref [20]. However, the state of extrinsic Si was not changed significantly. Although Si has several advantages over Ge (namely, a lower dielectric constant giving shorter dielectric relaxation time and lower capacitance, higher dopant solubility and a larger photoionization cross-section for higher quantum efficiency, and lower refractive index for lower reflectance), these were not sufficient to warrant the necessary development efforts needed to bring it to the level of the highly developed Ge detectors by then. After being dormant for about 10 years, extrinsic Si was reconsidered after the invention of charge-coupled devices by Boyle and Smith [21]. In 1973, Shepherd and Yang [22] proposed the metal-silicide/silicon Schottky-barrier detectors. For the first time it became possible to have much more sophisticated readout schemes—both detection and readout could be implemented in one common silicon chip.

At the same time, rapid advances were being made in narrow bandgap semiconductors that would later prove useful in extending wavelength capabilities and improving sensitivity. The first such material was InSb, a member of the newly discovered III–V compound semiconductor family. The interest in InSb stemmed not only from its small energy gap but also from the fact that it could be prepared in single crystal form using a conventional technique. The end of the 1950s and the beginning of the 1960s saw the introduction of narrow-gap semiconductor alloys in III–V ($InAs_{1-x}Sb_x$), IV–VI ($Pb_{1-x}Sn_xTe$), and II–VI ($Hg_{1-x}Cd_xTe$) material systems. These alloys allowed the bandgap of the semiconductor and hence the spectral response of the detector to be custom tailored for specific applications. In 1959, research by Lawson and coworkers [23] triggered the development of variable bandgap $Hg_{1-x}Cd_xTe$ (HgCdTe) alloys, providing an unprecedented degree of freedom in IR detector design. This first paper reported both photoconductive and photovoltaic response at the wavelength extending out to 12 μm. Soon thereafter, working under a U.S. Air Force contract with the objective of devising an 8–12 μm background-limited semiconductor IR detector that would operate at temperatures as high as 77 K, the group lead by Kruse at the Honeywell Corporate Research Center in Hopkins, Minnesota developed a modified Bridgman crystal growth technique for HgCdTe. They soon reported both photoconductive and photovoltaic detection in rudimentary HgCdTe devices [24].

The fundamental properties of narrow-gap semiconductors (high optical absorption coefficient, high electron mobility, and low thermal generation rate), together with the capability for bandgap engineering, make these alloy systems almost ideal for a wide range of IR detectors. The difficulties in growing HgCdTe material, significantly due to the high vapor pressure of Hg, encouraged the development of alternative detector technologies over the past 40 years. One of these was PbSnTe, which was vigorously pursued in parallel with HgCdTe in the late 1960s and early 1970s [25–27]. PbSnTe was comparatively easy to grow and good quality LWIR photodiodes were readily demonstrated. However, in the late 1970s, two factors led to the abandonment of PbSnTe detector work: high dielectric constant and large temperature coefficient of expansion (TCE) mismatch with Si. The scanned IR imaging systems of the 1970s required relatively fast response times so that the scanned image was not smeared in the scan direction. With the trend today toward staring arrays, this consideration might be less important than it was when first-generation systems were being designed. The second drawback, large TCE, can lead to failure of the indium bonds in hybrid structure (between silicon readout and the detector array) after repeated thermal cycling from room temperature to the cryogenic temperature of operation.

The material technology development was and continues to be primarily for military applications. In the United States, the Vietnam War caused the military services to initiate the development of IR systems that could provide imagery arising from the thermal emission of terrain vehicles, buildings, and people. As photolithography became available in the early 1960s, it was applied to make IR detector arrays. Linear array technology was first applied to PbS, PbSe, and InSb detectors. The discovery in the early 1960s of extrinsic Hg-doped germanium [28] led to the first FLIR systems operating in the LWIR spectral window using linear arrays. Because the detection mechanism was based on an extrinsic excitation, it required a two-stage cooler to operate at 25 K. The cooling requirements of intrinsic narrow bandgap semiconductor detectors are much less stringent. Typically, to obtain the background-limited infrared photodetector (BLIP) performance detectors for the 3–5-μm spectral region are operated at 200 K or less, while those for the 8–14 μm are typically operated at liquid nitrogen temperature. In the late 1960s and early 1970s, first-generation linear arrays of intrinsic HgCdTe photoconductive detectors were developed. These allowed LWIR FLIR systems to operate with a single-stage cryo engine, making the systems much more compact, lighter, and requiring significantly less power consumption.

Exactly, HgCdTe has inspired the development of the three "generations" of detector devices. The first generation, linear arrays of photoconductive detectors, has been produced in large quantities and is in widespread use today. The second generation, two-dimensional (2-D) arrays of photovoltaic detectors, are now in high-rate production. At the present stage of development, staring arrays have about 10^6 elements and are scanned electronically by circuits integrated with the arrays. These 2-D arrays of photodiodes connected with indium bumps to a readout integrated circuit chip as a hybrid structure are often called a sensor chip assembly (SCA). Third-generation devices defined here to encompass the more exotic device structure embodied in two-color detectors and hyperspectral arrays are now in demonstration programs (see Figure 3.1).

Early assessment of the concept of the second-generation system showed that PtSi Schottky barriers, InSb, and HgCdTe photodiodes or high-impedance photoconductors such as PbSe, PbS, and extrinsic silicon detectors were promising candidates because they have impedances suitable for interfacing with the field-effect transistor input of readout multiplexes. Photoconductive HgCdTe detectors were not suitable due to their low impedance and high-power dissipation on the focal plane. A novel British invention, the SPRITE detector [29,30] extended conventional photoconductive HgCdTe detector technology by incorporating signal time delay and integration (TDI) within a single elongated detector element. Such a detector replaces a whole row of discrete elements of a conventional serial-scanned detector and external associated amplifiers and time delay circuitry. Although only used in small arrays of about 10 elements, these devices have been produced in the thousands.

In the late 1970s and through the 1980s, HgCdTe technology efforts focused almost exclusively on photovoltaic device development because of the need for low-power dissipation and high impedance in large arrays to interface to readout input circuits. This effort is finally paying off with the birth of HgCdTe second-generation IR systems that provide large 2-D arrays in both linear formats with TDI for scanning imagers, and in square and rectangular formats for staring arrays. The first megapixel hybrid HgCdTe focal plane arrays (FPAs) were fabricated in mid-1990s. However, present HgCdTe FPAs are limited by the yield of arrays, which increases their cost. In such a situation, alternative alloy systems for IR detectors, such as quantum well infrared photodetectors (QWIPs) and type-II superlattices (T2SLs), are investigated.

Recently, there has been considerable progress towards III-V antimonide-based low-dimensional solids development and device design innovations. Their development results from two primary motivations: the perceived challenges of reproducibly fabricating high-operability HgCdTe FPAs at a reasonable cost and theoretical predictions of lower Auger recombination for T2SL detectors compared to HgCdTe. Lower Auger recombination translates into a fundamental advantage for T2SL over HgCdTe in terms of lower dark current and/or higher operating temperature, provided other parameters such as Shockley–Read–Hall lifetimes are equal.

The trend of increasing pixel numbers is likely to continue in the area of large format arrays. This increase will be continued using close-butted mosaics of several SCAs. Raytheon manufactured a 4×4 mosaics of 2K×2K HgCdTe SCAs and assisted in assembling it into the final focal-plane configuration to survey the entire sky in the Southern Hemisphere at four IR wavelengths [31]. With 67 million pixels, this is currently the world's largest IR focal plane. Although there are currently limitations to reducing the size of the gaps between active detectors on adjacent SCAs, many of these can be overcome. It is predicted that a focal plane of 100 megapixels and larger will be possible, constrained only by budgets but not by technology [32].

A negative aspect of support by defense agencies has been the associated secrecy requirements that inhibit meaningful collaborations among research teams on a national and especially on an international level. In addition, the primary focus has been on FPA demonstration and much less on establishing the knowledge base. Nevertheless, significant progress has been made over four decades. At present, HgCdTe is the most widely used variable gap semiconductor for IR photo-detectors. Over the years it has successfully fought off major challenges from extrinsic silicon and lead–tin telluride devices, but despite that, it has more competitors today than ever before. These include Schottky barriers on silicon, SiGe heterojunctions, AlGaAs multiple QWs, InAs/GaSb T2SLs, high-temperature superconductors, and especially two types of thermal detectors: pyroelectric detectors and silicon bolometers. However, it is interesting that none of these competitors can compete in terms of fundamental properties. They may promise to be more manufacturable, but never to provide higher performance or, with the exception of thermal detectors, to operate at higher or even comparable temperatures.

As is mentioned above, monolithic extrinsic Si detectors were demonstrated first in the mid-1970s [33–35], but were subsequently set aside because the process of integrated circuit fabrication degraded the detector-quality material properties. Historically, Si:Ga and Si:In were the first mosaic FPA photoconductive materials because early monolithic approaches were compatible with these dopants. The photoconductive material was made in either conventional or impurity band conduction (IBC)—or blocked impurity band (BIB)—technologies. Extrinsic photoconductors must be made relatively thick because they have much lower photon capture cross-section than intrinsic detectors. However, BIB detectors have a unique combination of photoconductive and photovoltaic characteristics, including extremely high impedance, reduced recombination noise, linear photoconductive gain, high uniformity, and superb stability. Megapixel detector arrays with cutoff wavelength to 28 μm are now available [36]. Specially doped IBCs operate as solid-state photomultipliers (SSPMs) and visible light photon converters in which photoexcited carriers allow counting of individual photons at low flux levels. Standard SSPMs respond from 0.4 to 28 μm.

As was mentioned previously, the development of IR technology has been dominated by the photon detectors since about 1930. However, the photon detectors require cryogenic cooling. This is necessary to prevent the thermal generation of charge carriers. The thermal transitions compete with the optical ones, making noncooled devices very noisy. The cooled thermal camera usually uses the Sterling cycle cooler, which is the expensive component in the photon detector IR camera, and the cooler's lifetime is only around 10,000 h. Cooling requirements are the main obstacle to the widespread use of IR systems based on semiconductor photon detectors making them bulky, heavy, expensive, and inconvenient to use.

The use of thermal detectors for IR imaging has been the subject of research and development for many decades. However, in comparison with photon detectors, thermal detectors have been considerably less exploited in commercial and military systems. The reason for this disparity is that thermal detectors are popularly believed to be rather slow and insensitive in comparison with photon detectors. As a result, the worldwide effort to develop thermal detectors has been extremely small relative to that of the photon detectors.

It must not be inferred from the preceding outline that work on thermal detectors has not also been actively pursued. Indeed, some interesting and important developments have taken place along this line. In 1947, for example, Golay constructed an improved pneumatic IR detector [37]. This gas thermometer has been used in spectrometers. The thermistor bolometer, originally developed by Bell Telephone Laboratories, has found widespread use in detecting radiation from low-temperature sources [38,39]. The superconducting effect has been used to make extremely sensitive bolometers.

Thermal detectors have also been used for IR imaging. Evaporographs and absorption edge image converters were among the first nonscanned IR imagers. Originally an evaporograph was employed in which the radiation was focused onto a blackened membrane coated with a thin film of oil [40]. The differential rate of evaporation of the oil was proportional to the radiation intensity. The film was then illuminated with visible light to produce an interference pattern corresponding to the thermal picture. The second thermal imaging device was the absorption edge image converter [41]. Operation of this device was based upon utilizing the temperature dependence of the absorption edge of the semiconductor. The performance of both imaging devices was poor because of the very long time constant and the poor spatial resolution. Despite numerous research initiatives and the attractions of ambient temperature operation and low cost potential, thermal detector technology has enjoyed limited success in competition with cooled photon detectors for thermal imaging applications. A notable exception was the pyroelectric vidicon (PEV) [42] that was widely used by firefighting and emergency service organizations. The PEV tube can be considered analogously to the visible television camera tube except that the photoconductive target is replaced by a pyroelectric detector and germanium faceplate. Compact, rugged PEV imagers have been offered for military applications but suffer the disadvantage of low tube life and fragility, particularly the reticulated vidicon tubes required for enhanced spatial resolution. The advent of the staring FPAs, however, marked the development of devices that would someday make uncooled systems practical for many, especially commercial, applications. The defining effort in this field was undertaken by Texas Instruments with contractual support from the Army Night Vision Laboratory [4]. The goal of this program was to build a staring FPA system based on ferroelectric detectors of barium strontium titanate. Throughout the 1980s and early 1990s, many other companies developed devices based on various thermal detection principles.

The second revolution in thermal imaging began at the end of the twentieth century. The development of uncooled IR arrays capable of imaging scenes at room temperature has been an outstanding technical achievement. Much of the technology was developed under classified military contracts in the United States, so the public release of this information in 1992 surprised many in the worldwide IR community [43]. There has been an implicit assumption that only cryogenic photon detectors operating in the 8–12-μm atmospheric window had the necessary sensitivity to image room temperature objects. Although thermal detectors have been little used in scanned imagers because of their slow response, they are currently of considerable interest for 2-D electronically addressed arrays where the bandwidth is low and the ability of thermal devices to integrate over a frame time is an advantage [44–49]. Much recent research has focused on both hybrid and monolithic uncooled arrays and has yielded significant improvements in the detectivity of both bolometric and pyroelectric detector arrays. Honeywell has licensed bolometer technology to several companies for the development and production of uncooled FPAs for commercial and military systems. At present, the compact megapixel microbolometer cameras are produced by Raytheon, DRS, BAE, L-3, and FLIR Systems in the United States. The U.S. government allowed these manufacturers to sell their devices to foreign countries, but not to divulge manufacturing technologies. In recent years, several countries, including the United Kingdom, France, Japan, and Korea, have picked up the ball, determined to develop their own uncooled imaging systems. As a result, although the United States has a significant lead, some of the most exciting and promising developments for low-cost uncooled IR systems may come from non-U.S. companies (e.g., microbolometer FPAs with series p-n junction elaborated by Mitsubishi Electric). This approach is unique, based on an all-silicon version of the microbolometer.

3.2 CLASSIFICATION OF IR DETECTORS

The majority of detectors can be classified into two broad categories: photon detectors and thermal detectors.

Progress in IR detector technology is connected with semiconductor IR detectors, which are included in the class of photon detectors. In this class of detectors, the radiation is absorbed within the material by interaction with electrons either bound to lattice atoms or to impurity atoms or with free electrons. The observed electrical output signal results from the changed electronic energy distribution. The fundamental optical excitation processes in semiconductors are illustrated in Figure 3.2 [50–52]. In QWs [Figure 3.2b], the intersubband absorption takes place between the energy levels of a QW associated with the conduction band (n-doped) or valence band (p-doped). In the case of type-II InAs/GaSb SL [Figure 3.2c], the SL bandgap is determined by the energy difference between the electron miniband E_1 and the first heavy-hole state HH_1 at the Brillouin zone center. A consequence of the type-II band alignment is the spatial separation of electrons and holes.

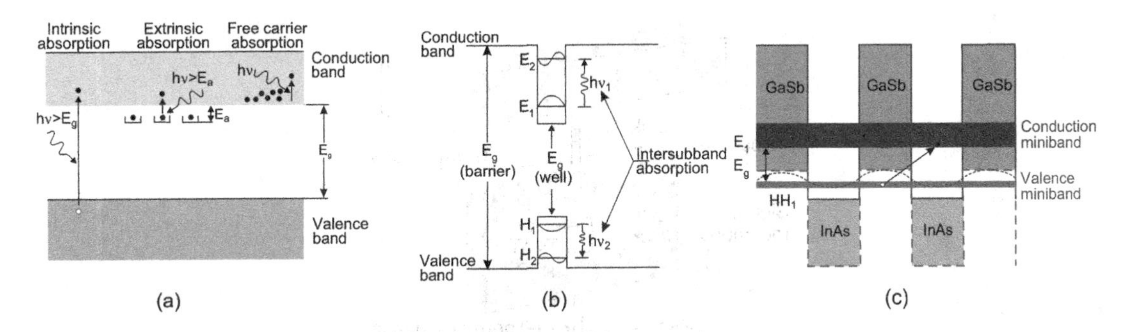

Figure 3.2 Optical excitation processes in: bulk semiconductors (a), QWs (b), and type-II InAs/GaSb SLs (c).

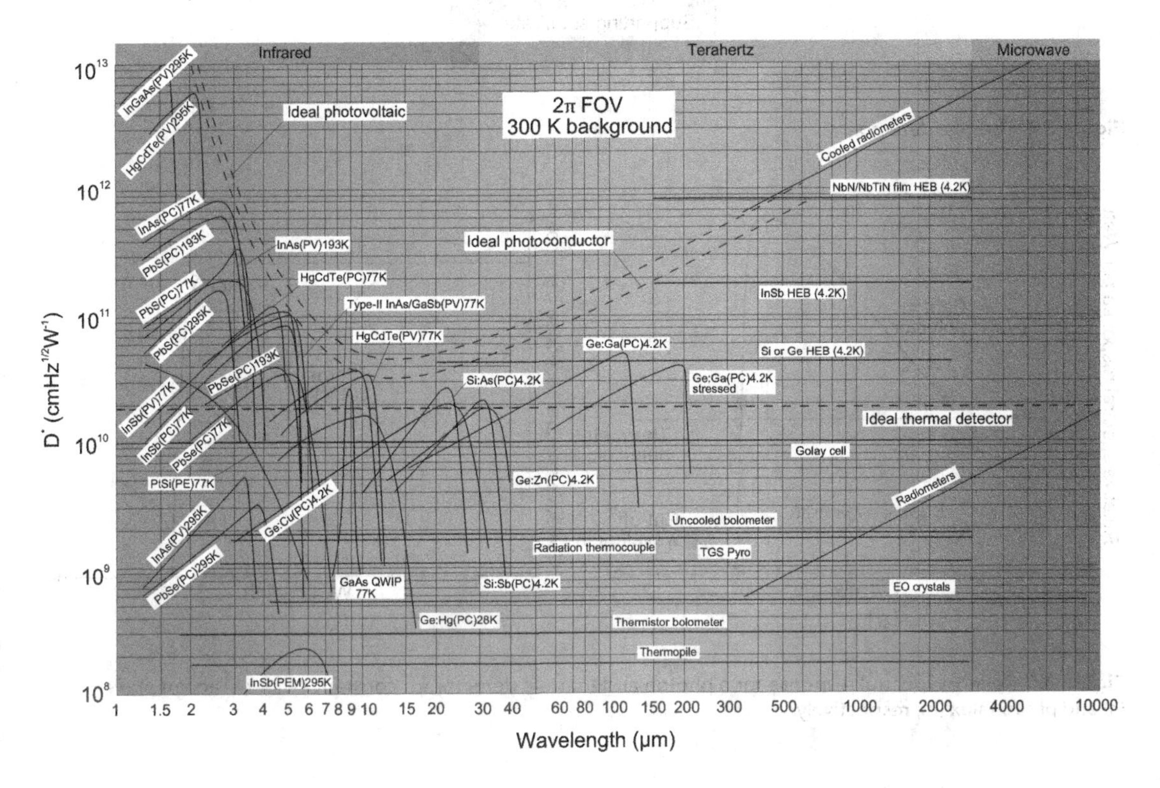

Figure 3.3 Comparison of the D^* of various available detectors when operated at the indicated temperature. Chopping frequency is 1,000 Hz for all detectors except the thermopile (10 Hz), thermocouple (10 Hz), thermistor bolometer (10 Hz), Golay cell (10 Hz), and pyroelectric detector (10 Hz). Each detector is assumed to view a hemispherical surrounding at a temperature of 300 K. Theoretical curves for the background-limited D^* (dashed lines) for ideal photovoltaic and photoconductive detectors and thermal detectors are also shown. *PC*, photoconductive detector, *PV*, photovoltaic detector, *PEM*, photoelectromagnetic detector, and HEB, hot electron bolometer.

Spectral detectivity curves for a number of commercially available IR detectors are shown in Figure 3.3. Interest has centered mainly on the wavelengths of the two atmospheric windows 3–5 μm (middle wavelength (MWIR)) and 8–14 μm, though in recent years there has been increasing interest in longer wavelengths stimulated by space applications. The spectral character of the background is influenced by the transmission of the atmosphere (see Figure 2.27 [50]), which controls the spectral ranges of the IR for which the detector may be used when operating in the atmosphere.

The second class of detectors is composed of thermal detectors. In a thermal detector, the incident radiation is absorbed to change the material temperature, and the resultant change in some physical property is used to generate an electrical output. The detector is suspended on lags, which are connected

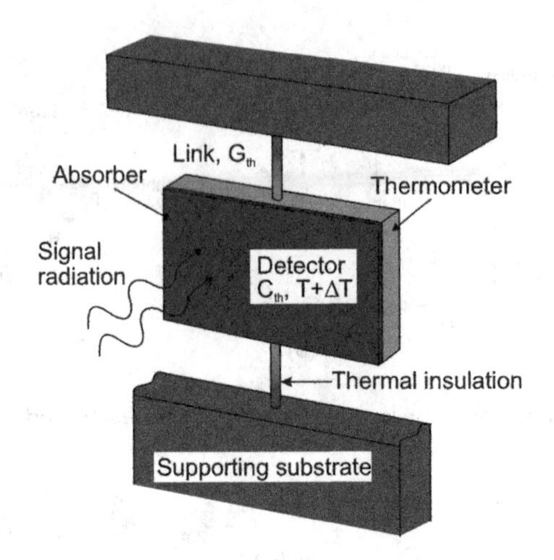

Figure 3.4 Thermal detector.

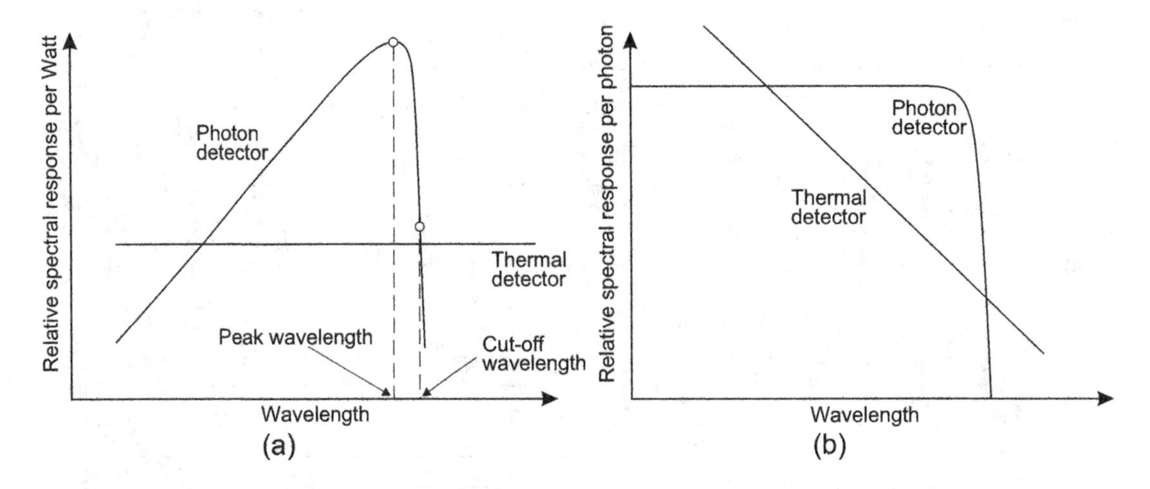

Figure 3.5 Relative spectral response for a photon and thermal detector for constant incident radiant power (a) and photon flux (b), respectively.

to the heat sink (see Figure 3.4). Attention is directed toward three approaches that have found the greatest utility in IR technology; namely, bolometers, pyroelectric, and thermoelectric effects. In pyroelectric detectors, a change in the internal electrical polarization is measured, whereas in the case of thermistor bolometers a change in the electrical resistance is measured.

The relative response of IR detectors is plotted as a function of wavelength with either a vertical scale of W^{-1} or $photon^{-1}$ (see Figure 3.5). The photon detectors show a selective wavelength dependence of response per unit incident radiation power. Their response is proportional to the rate of arrival photons as the energy per photon is inversely proportional to wavelength. In consequence, the spectral response increases linearly with increasing wavelength [see Figure 3.5a] until the cutoff wavelength is reached, which is determined by the detector material. The cutoff wavelength is usually specified as the long wavelength point at which the detector responsivity falls to 50% of the peak responsivity. For thermal detectors, the signal does not depend upon the photonic nature of the incident radiation. Thus, thermal effects are generally wavelength independent; the signal depends upon the radiant power (or its rate of change) but not upon its spectral content.

The photon detectors exhibit both perfect signal-to-noise performance and a very fast response. But to achieve this, the photon detectors require cryogenic cooling. Photon detectors having long wavelength

Table 3.1 Comparison of IR detectors

DETECTOR TYPE			ADVANTAGES	DISADVANTAGES
Thermal (thermopile, bolometers, pyroelectric)			Light, rugged, reliable, & low-cost Room temperature operation	Low detectivity at high frequency Slow response (ms order)
Photon	Intrinsic	IV–VI (PbS, PbSe, PbSnTe)	Easier to prepare More stable materials	Very high thermal expansion coefficient Large permittivity
		II–VI (HgCdTe)	Easy bandgap tailoring Well-developed theory & experience Multicolor detectors	Nonuniformity over large area High cost in growth and processing Surface instability
		III–V (InGaAs, InAs, InSb, InAsSb)	Good material & dopants Advanced technology Possible monolithic integration	Heteroepitaxy with large lattice mismatch Long wavelength cutoff limited to 7 μm (at 77 K)
	Extrinsic (Si:Ga, Si:As, Ge:Cu, Ge:Hg)		Very long wavelength operation Relatively simple technology	High thermal generation Extremely low-temperature operation
	Free carriers (PtSi, Pt_2Si, IrSi)		Low-cost, high yields Large & close packed 2-D arrays	Low quantum efficiency Low-temperature operation
	QWs	Type I (GaAs/AlGaAs, InGaAs/AlGaAs)	Matured material growth Good uniformity over a large area Multicolor detectors	High thermal generation Complicated design and growth
		Type II (InAs/GaSb, InAs/InAsSb)	Low Auger recombination rate Easy wavelength control Multicolor detectors	Complicated design and growth Sensitive to the interfaces
	QDs	InAs/GaAs, InGaAs/InGaP, Ge/Si	Normal incidence of light Low thermal generation	Complicated design and growth

limits above about 3 μm are generally cooled. This is necessary to prevent the thermal generation of charge carriers. The thermal transitions compete with the optical ones, making noncooled devices very noisy.

Depending on the nature of the interaction, the class of photon detectors is further subdivided into different types as shown in Table 3.1. The most important are intrinsic detectors, extrinsic detectors, photoemissive (metal silicide Schottky barriers) detectors, and QW detectors. Depending on how the electric or magnetic fields are developed, there are various modes such as photoconductive, photovoltaic, photoelectromagnetic (PEM), and photoemissive ones. Each material system can be used for different modes of operation.

More recently, this standard classification has been reconsidered by Kinch [51], as discussed briefly below. Photon detectors can be divided into two broad classes; namely, majority and minority carrier devices. The material systems used are:

1. Direct bandgap semiconductors—minority carriers
   - Binary alloys: InSb, InAs
   - Ternary alloys: HgCdTe, InGaAs

Table 3.2 Infrared thermal detectors

DETECTOR	METHOD OF OPERATION
Bolometer Metal Semiconductor Superconductor Ferroelectric Hot electron	Change in electrical conductivity
Thermocouple/Thermopile	Voltage generation, caused by a change in temperature of the junction of two dissimilar materials
Pyroelectric	Changes in spontaneous electrical polarization
Golay cell/Gas microphone	Thermal expansion of a gas
Absorption edge	Optical transmission of a semiconductor
Pyromagnetic	Changes in magnetic properties
Liquid crystal	Changes of optical properties

- Type II, III SLs: InAs/GaSb, HgTe/CdTe

2. Extrinsic semiconductors—majority carriers
   - Si:As, Si:Ga, Si:Sb
   - Ge:Hg, Ge:Ga
3. Type I SLs—majority carriers
   - GaAs/AlGaAs QWIPs
4. Silicon Schottky barriers—majority carriers
   - PtSi, IrSi
5. High-temperature superconductors—minority carriers

All of these material systems have been serious players in the IR marketplace with the exception of type III SLs and high-temperature superconductors.

In contrast to photon detectors, thermal detectors are typically operated at room temperature. They are usually characterized by modest sensitivity and slow response (because heating and cooling of a detector element is a relatively slow process), but they are cheap and easy to use. They have found widespread use in low-cost applications, which do not require high performance and speed. Being unselective, they are frequently used in IR spectrometers. A list of thermal effects is included in Table 3.2.

Up until the 1990s, thermal detectors have been considerably less exploited in commercial and military systems in comparison with photon detectors. In the last decade, however, it has been shown that extremely good imagery can be obtained from large thermal detector arrays operating uncooled at TV frame rates. The speed of thermal detectors is quite adequate for nonscanned imagers with 2-D detectors. The moderate sensitivity of thermal detectors can be compensated by a large number of elements in 2-D electronically scanned arrays. With large arrays of thermal detectors, the best values of temperature sensitivity below 0.05 K could be reached because effective noise bandwidths less than 100 Hz can be achieved.

Finally, the third category of IR detectors, the so-called radiation field detectors, which respond directly to the radiation field should be mentioned. This class of detectors was developed after 1970, but has not had a significant impact [52]. Recently, their importance has increased due to the wider application of harmonic mixers in the far-IR (terahertz) and submillimeter regions [53].

3.3 DETECTOR OPERATING TEMPERATURE

A key difference between intrinsic and extrinsic detectors is that extrinsic detectors require much cooling to achieve high sensitivity at a given spectral response cutoff in comparison with intrinsic detectors. The low-temperature operation is associated with longer wavelength sensitivity to suppress noise due to thermally induced transitions between close-lying energy levels.

There is a fundamental relationship between the temperature of the background viewed by the detector and the lower temperature at which the detector must operate to achieve background-limited performance. HgCdTe photodetectors with a cutoff wavelength of 12.4 μm operate at 77 K. One can scale the results of this example to other temperatures and cutoff wavelengths by noting that for a given level of detector performance, $T\lambda_c \approx$ constant [54]; that is, the longer λ_c the lower is T while their product remains roughly constant. This relation holds because quantities that determine detector performance vary mainly as an exponential of $E_{exc}/kT = hc/kT\lambda_c$, where E_{exc} is the excitation energy, k is the Boltzmann constant, h is the Planck constant, and c is the velocity of light.

The detector temperature of operation can be approximated as

$$T_{\max} = \frac{300K}{\lambda_c[\mu m]}. \quad (3.1)$$

The general trend is illustrated in Figure 3.6 for seven high-performance detector materials suitable for low-background applications: Si, InGaAs, InSb, T2SL, HgCdTe photodiodes, Si:As BIB detectors; and extrinsic Ge:Ga unstressed and stressed detectors. Terahertz photoconductors are operated in extrinsic mode.

The most widely used photovoltaic detector is the p-n junction, where a strong internal electric field exists across the junction even in the absence of radiation. Photons incident on the junction produce free hole electron pairs that are separated by the internal electric field across the junction, causing a change in voltage across the open-circuit cell or a current to flow in the short-circuited case.

Photoconductors that utilize excitation of an electron from the valence to conduction band are called intrinsic detectors. Instead, those that operate by exciting electrons into the conduction band or holes into the valence band from impurity states within the band (impurity-bound states in energy gap, QWs, or QDs), are called extrinsic detectors. A key difference between intrinsic and extrinsic detectors is that

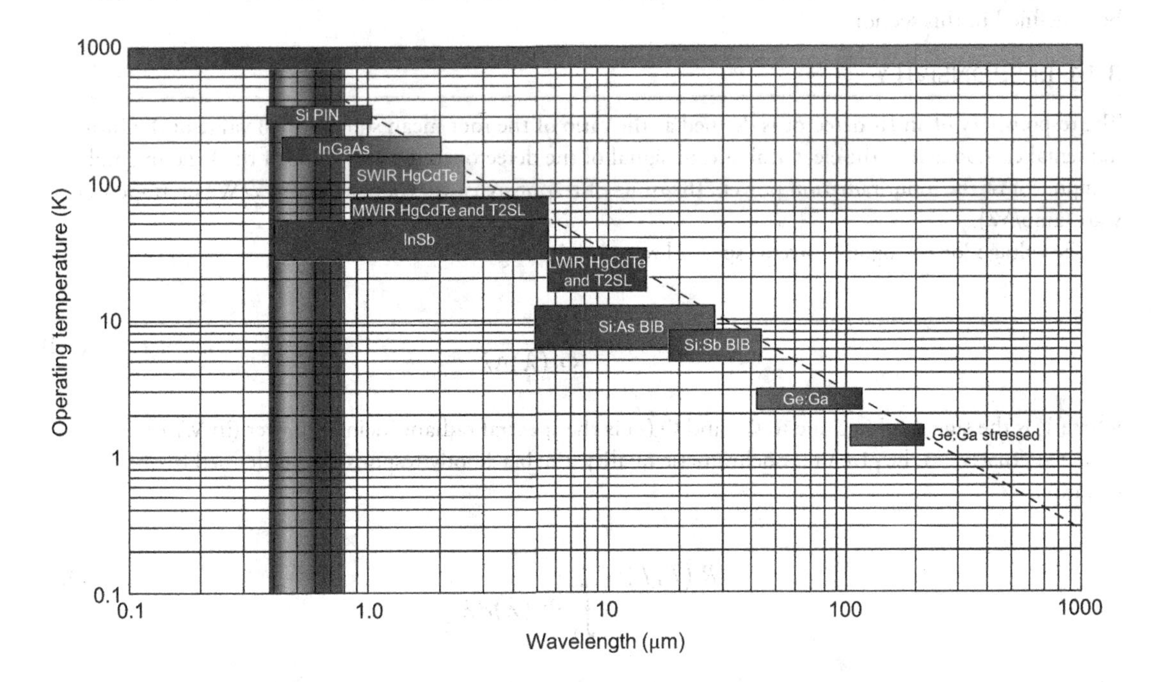

Figure 3.6 Operating temperatures for low-background material systems with their spectral band of greatest sensitivity. The dashed line indicates the trend toward lower operating temperature for longer wavelength detection.

extrinsic detectors require much cooling to achieve high sensitivity at a given spectral response cutoff in comparison with intrinsic detectors. Low-temperature operation is associated with longer wavelength sensitivity to suppress noise due to thermally induced transitions between close-lying energy levels. Intrinsic detectors are most common at the short wavelengths, below 20 μm. In the longer wavelength region, the photoconductors are operated in extrinsic mode. One advantage of photoconductors is their current gain, which is equal to the recombination time divided by the majority carrier transit time. This current gain leads to higher responsivity than is possible with non-avalanching photovoltaic detectors. However, the series problem of photoconductors operated at low temperature is nonuniformity of detector element due to recombination mechanisms at the electrical contacts and its dependence on the electrical bias.

Recently, interfacial workfunction, internal photoemission detectors, QW and QD detectors, which can be included to extrinsic photoconductors, have been proposed especially for IR and THz spectral bands [55]. The very fast time response of QW and QD semiconductor detectors make them attractive for heterodyne detection.

3.4 DETECTOR FIGURES OF MERIT

It is difficult to measure the performance characteristics of IR detectors because of the large number of experimental variables involved. A variety of environmental, electrical, and radiometric parameters must be taken into account and carefully controlled. With the advent of large, two-dimensional detector arrays, detector testing has become even more complex and demanding.

This section is intended to serve as an introductory reference for the testing of IR detectors. Numerous texts and journals cover this issue, including: *Infrared System Engineering* [50], by R. D. Hudson; *The Infrared Handbook* [56], edited by W. L. Wolfe and G. J. Zissis; *The Infrared and Electro-Optical Systems Handbook* [57], edited by W. D. Rogatto; and *Fundamentals of Infrared Detector Operation and Testing* [58] and its second edition [59], by J. D. Vincent . In this volume, we have restricted our consideration to detectors whose output consists of an electrical signal that is proportional to the radiant signal power [58–62].

The measured data described in this text are sufficient to characterize a detector. However, to provide ease of comparison between detectors, certain figures of merit, computed from the measured data, have been defined in this section.

3.4.1 RESPONSIVITY

The responsivity of an IR detector is defined as the ratio of the root mean square (rms) value of the fundamental component of the electrical output signal of the detector to the rms value of the fundamental component of the input radiation power. The units of responsivity are volts per watt (V/W) or amperes per watt (amp/W).

The voltage (or analogous current) spectral responsivity is given by

$$R_v(\lambda, f) = \frac{V_s}{\Phi_e(\lambda)\Delta\lambda}, \tag{3.2}$$

where V_s is the signal voltage due to Φ_e, and $\Phi_e(\lambda)$ is the spectral radiant incident power (in W/m).

An alternative to the given monochromatic quality, the blackbody responsivity, is defined by the equation

$$R_v(T, f) = \frac{V_s}{\int_0^\infty \Phi_e(\lambda)d\lambda}, \tag{3.3}$$

where the incident radiant power is the integral over all wavelengths of the spectral density of power distribution $\Phi_e(\lambda)$ from a blackbody. The responsivity is usually a function of the bias voltage V_b, the operating electrical frequency f, and the wavelength λ.

3.4.2 NOISE EQUIVALENT POWER

The noise equivalent power (NEP) is the incident power on the detector generating a signal output equal to the rms noise output. Stated another way, the NEP is the signal level that produces a signal-to-noise ratio (SNR) of 1. It can be written in terms of responsivity:

$$\mathrm{NEP} = \frac{V_n}{R_v} = \frac{I_n}{R_i}. \tag{3.4}$$

The unit of NEP is watt.

The NEP is also quoted for a fixed reference bandwidth, which is often assumed to be 1 Hz. This "NEP per unit bandwidth" has a unit of watts per square root Hertz ($\mathrm{W/Hz^{1/2}}$).

3.4.3 DETECTIVITY

The detectivity D is the reciprocal of NEP:

$$D = \frac{1}{\mathrm{NEP}}. \tag{3.5}$$

It was found by Jones [63] that for many detectors the NEP is proportional to the square root of the detector signal that is proportional to the detector area, A_d. This means that both NEP and detectivity are functions of electrical bandwidth and detector area, so a normalized detectivity D^* (or D-star) suggested by Jones [63,64] is defined as

$$D^* = D\left(A_d \Delta f\right)^{1/2} = \frac{\left(A_d \Delta f\right)^{1/2}}{\mathrm{NEP}}. \tag{3.6}$$

The importance of D^* is that this figure of merit permits comparison of detectors of the same type, but having different areas. Either a spectral or a blackbody D^* can be defined in terms of a corresponding type of NEP.

Useful equivalent expressions to Equation 3.6 include:

$$D^* = \frac{\left(A_d \Delta f\right)^{1/2}}{V_n} R_v = \frac{\left(A_d \Delta f\right)^{1/2}}{I_n} R_i = \frac{\left(A_d \Delta f\right)^{1/2}}{\Phi_e}(\mathrm{SNR}), \tag{3.7}$$

where D^* is defined as the rms SNR in a 1-Hz bandwidth per unit rms incident radiant power per square root of detector area. D^* is expressed in unit $\mathrm{cmHz^{1/2}W^{-1}}$, which recently is called "Jones."

The blackbody $D^*(T,f)$ may be found from spectral detectivity:

$$D^*(T,f) = \frac{\int_0^\infty D^*(\lambda,f)\Phi_e(T,\lambda)d\lambda}{\int_0^\infty \Phi_e(T,\lambda)d\lambda} = \frac{\int_0^\infty D^*(\lambda,f)E_e(T,\lambda)d\lambda}{\int_0^\infty E_e(T,\lambda)d\lambda}, \tag{3.8}$$

where $\Phi_e(T,\lambda) = E_e(T,\lambda)A_d$ is the incident blackbody radiant flux (in W), and $E_e(T,\lambda)$ is the blackbody irradiance (in $\mathrm{W/cm^2}$).

3.4.4 QUANTUM EFFICIENCY

Photon detectors are based on photon absorption in semiconductor materials. A signal whose photon energy is sufficient to generate photocarriers will continuously lose energy as the optical field propagates through the semiconductor (see Figure 3.7). Inside the semiconductor, the field decays exponentially as

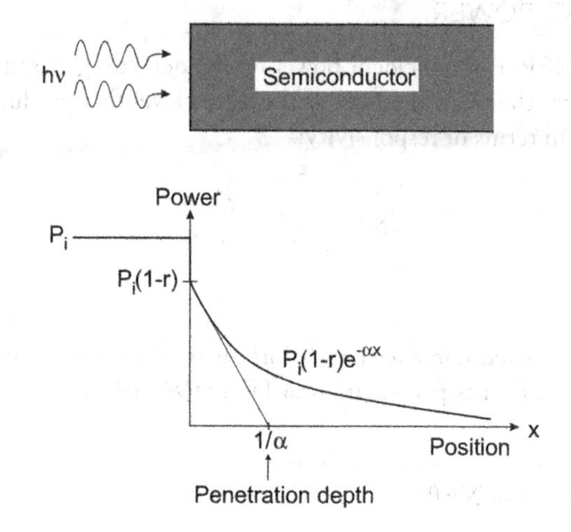

Figure 3.7 Optical absorption in a semiconductor.

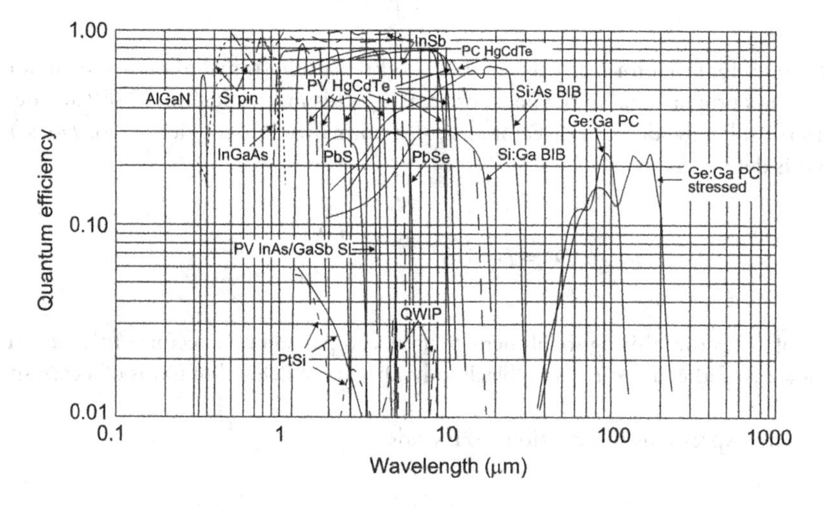

Figure 3.8 Quantum efficiency of different detectors.

energy is transferred to the photocarriers. The material can be characterized by an absorption length, α, and a penetration depth, $1/\alpha$. Penetration depth is the point at which $1/e$ of the optical signal power remains.

The power absorbed in the semiconductor as a function of position within the material is then

$$P_a = P_i(1-r)\left(1-e^{-\alpha x}\right). \tag{3.9}$$

The number of photons absorbed is the power (in Watts) divided by the photon energy ($E=h\nu$). If each absorbed photon generates a photocarrier, the number of photocarriers generated per number of incident photons for a specific semiconductor with reflectivity r is given by

$$\eta(x) = (1-r)\left(1-e^{-\alpha x}\right), \tag{3.10}$$

where $0 \le \eta \le 1$ is a definition for the detector's quantum efficiency as the number of electron hole pairs generated per incident photon.

Figure 3.8 shows the quantum efficiency of some of the detector materials used to fabricate arrays of ultraviolet (UV), visible, and IR detectors. Photocathodes and AlGaN detectors are being developed in the UV region. Silicon p-i-n diodes are shown with and without an antireflection coating. Lead salts (PbS and PbSe) have intermediate quantum efficiencies, while PtSi Schottky barrier types and QWIPs have low values. InSb can respond from the near UV out to 5.5 μm at 80 K. A suitable detector material for near-IR (1.0–1.7 μm) spectral range is InGaAs lattice matched to the InP. Various HgCdTe alloys, in both photovoltaic and photoconductive configurations, cover from 0.7 μm to over 20 μm. InAs/GaSb strained layer SLs have emerged as an alternative to the HgCdTe. Impurity-doped (Sb, As, and Ga) silicon BIB detectors operating at 10 K have a spectral response cutoff in the range of 16 to 30 μm. Impurity-doped Ge detectors can extend the response out to 100–200 μm.

3.5 FUNDAMENTAL DETECTIVITY LIMITS

The ultimate performance of IR detectors is reached when the detector and amplifier noise is low compared to the photon noise. The photon noise is fundamental in the sense that it arises not from any imperfection in the detector or its associated electronics but rather from the detection process itself, as a result of the discrete nature of the radiation field. The radiation falling on the detector is a composite of that from the target and that from the background. The practical operating limit for most IR detectors is not the signal fluctuation limit (SFL) but the background fluctuation limit, also known as the BLIP limit.

The expression for shot noise can be used to derive the BLIP detectivity

$$D^*_{BLIP}(\lambda, f) = \frac{\lambda}{hc}\left(\frac{\eta}{2Q_B}\right)^{1/2}, \tag{3.11}$$

where η is the quantum efficiency and Q_B is the total background photon flux density reaching the detector

$$Q_B = \sin^2(\theta/2)\int_0^{\lambda_c} Q(\lambda, T_B)\, d\lambda. \tag{3.12}$$

The Planck photon emittance (in unit photons $cm^{-2}s^{-1}\mu m^{-1}$) at temperature T_B is given by

$$Q(\lambda, T_B) = \frac{2\pi c}{\lambda^4\left[\exp(hc/\lambda k T_B) - 1\right]} = \frac{1.885\times 10^{23}}{\lambda^4\left[\exp(14.388/\lambda T_B) - 1\right]}. \tag{3.13}$$

Figure 3.9 shows the dependence of the integral background flux density on the wavelength for different blackbody temperatures and 2π field of view (FOV). The values of the integral Equation 3.10 are given in the tables by Lowan and Blanch [65].

From Equation 3.12, we obtain

$$\frac{Q_B(\theta)}{Q_B(2\pi)} = \sin^2(\theta/2), \tag{3.14}$$

and the background-limited D^* relative to 2π FOV becomes

$$\frac{D^*_{BLIP}(\theta)}{D^*_{BLIP}(2\pi)} = \frac{1}{\sin(\theta/2)}. \tag{3.15}$$

The D^* varies with FOV as $[\sin(\theta/2)]^{-1}$. Figure 3.10 is a curve showing how ideal D^* is improved as the cone angle θ is reduced for any given background temperature [66].

Equation 3.11 holds for photovoltaic detectors, which are shot noise–limited. Photoconductive detectors that are generation–recombination noise–limited have a lower D^*_{BLIP} by a factor of $\sqrt{2}$

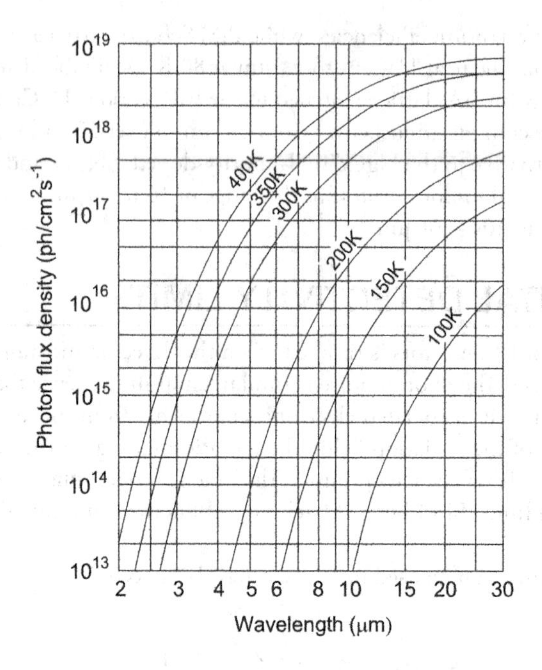

Figure 3.9 Dependence of the integral background flux density on wavelength for different blackbody temperatures and 2π FOV (From Ref. [50])

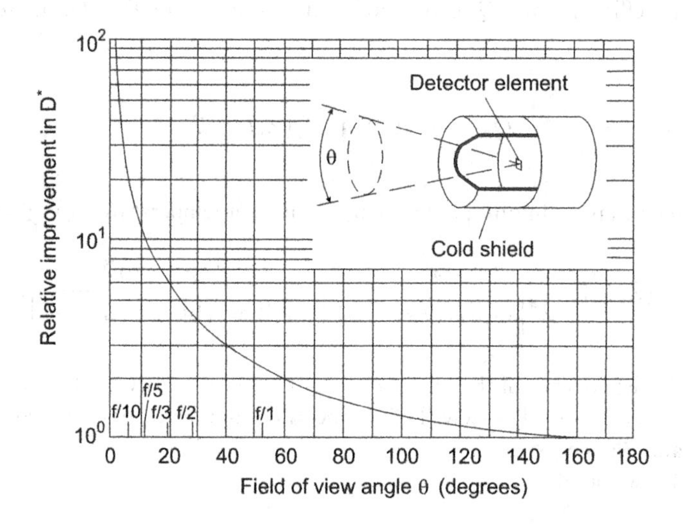

Figure 3.10 Relative improvement factor for detectivity with a reduction in the FOV cone angle for a BLIP detector. (From Ref. [66])

$$D^*_{\mathrm{BLIP}}(\lambda, f) = \frac{\lambda}{2hc}\left(\frac{\eta}{Q_B}\right)^{1/2}. \tag{3.16}$$

The photon noise–limited expressions Equations 3.11 and 3.16 are only for Poisson statistics, when the Bose–Einstein factor $b=[\exp(hc/\lambda kT_B)-1]^{-1}$ is near 1. If the Bose–Einstein factor is included, for example, Equation 3.16 becomes

$$D^*_{\mathrm{BLIP}}(\lambda, f) \frac{\eta\lambda}{2hc\sin(\theta/2)}\left[\int_0^{\lambda_c} \eta(\lambda) Q(\lambda, T_B)(1+b)\,d\lambda\right]^{-1/2}. \tag{3.17}$$

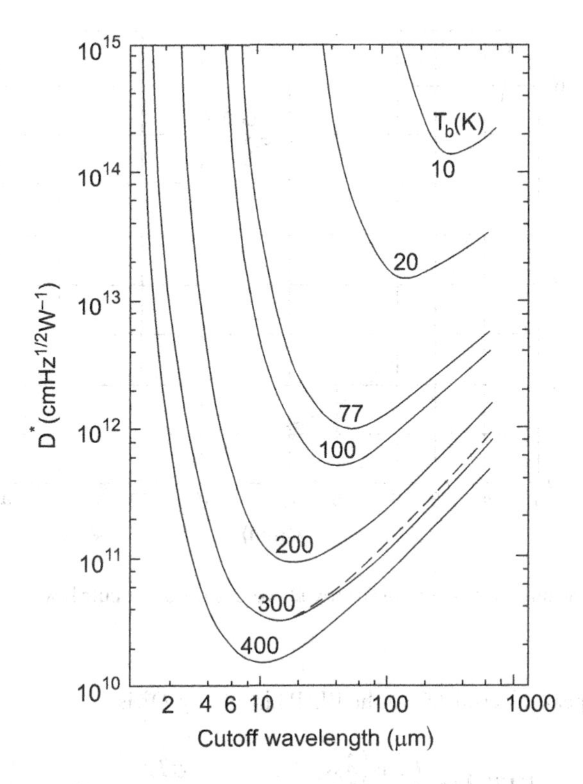

Figure 3.11 Detectivity at λ_c versus λ_c for ideal photoconductive detectors operating at T_b=400, 300, 200, 100, 77, 20, and 10 K for a 2π FOV. The dashed line for 300 K neglects the boson factor. (From Ref. [67])

The highest performance possible will be obtained by the ideal detector with unity quantum efficiency and ideal spectral responsivity [$R(\lambda)$ increases with wavelength to the cutoff wavelength λ_c at which the responsivity drops to zero]. This limiting performance is of interest for comparison with actual detectors. The detectivity of ideal photoconductors at λ_c as a function of λ_c based on numerical integration is shown as a function of background temperature T_B, for a 2π FOV, in Figure 3.11 [67]. The dashed line for $T_B = 300$ K is the detectivity obtained by neglecting the boson factor, which is seen to make a small but increasing effect as the wavelength is extended. As T_B is decreased, the boson factor correction becomes increasingly less significant. Values of D^*_{BLIP} versus λ for various background conditions are given in the literature [3,68–71].

The detectivity of BLIP detectors can be improved by reducing the background photon flux, Φ_b. Practically, there are two ways to do this: a cooled or reflective spectral filter to limit the spectral band or a cooled shield to limit the angular FOV of the detector (as described earlier). The former eliminates background radiation from spectral regions in which the detector need not respond. The best detectors yield background-limited detectivities in quite narrow FOV.

It can be shown that when the signal source is a blackbody at temperature T_s, and the radiation background is a blackbody at temperature T_b, then the background noise–limited blackbody D^*_{BLIP} as a function of the peak spectral D^*_{BLIP} is

$$D^*_{\mathrm{BLIP}}(T_s, f) = D^*_{\mathrm{BLIP}}(\lambda_p, f)\frac{(hc/\lambda_p)}{\sigma T_s^4}\int_0^{\lambda_p} Q(T_s, \lambda)d\lambda, \tag{3.18}$$

where λ_p is the wavelength of peak detectivity, which is also the cutoff wavelength for an ideal photon detector, and σ is the Stefan–Boltzmann constant. All of the D^*_{BLIP} expressions have assumed a Lambertian source subtending a half-angle of $\pi/2$ radians.

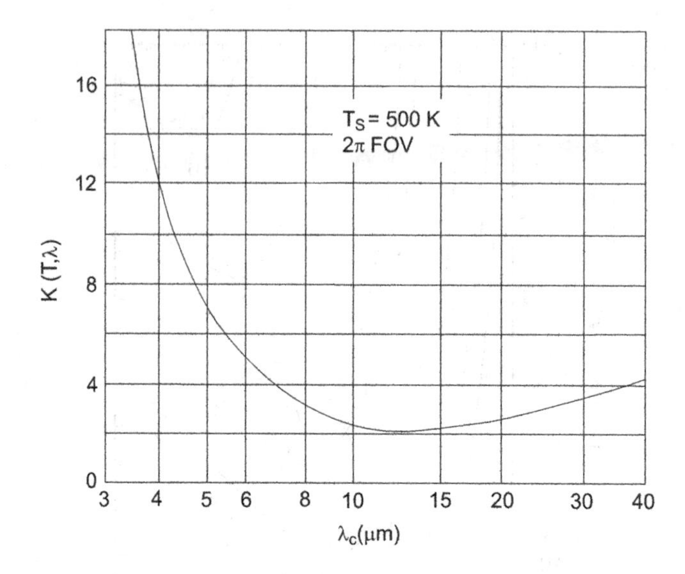

Figure 3.12 The ratio of peak spectral D^* to blackbody D^* versus detector cutoff wavelength for T_s=500 K and 2π FOV. (From Ref. [68])

The ratio of the BLIP peak spectral D^* to the BLIP blackbody D^* is

$$K(T,\lambda)=\frac{D^*_{\mathrm{BLIP}}(\lambda_p,f)}{D^*_{\mathrm{BLIP}}(T_s,f)}=\frac{\sigma T_s^4}{\frac{hc}{\lambda_p}\int_0^{\lambda_p}Q(T_s,\lambda)d\lambda}. \tag{3.19}$$

Figure 3.12 is a plot of $K(\lambda)$ for T_s=500 K and a 2π steradian FOV [67]. The quantity $K(T,\lambda)$ is useful because IR detector testing yields blackbody D^* values. Peak spectral D^* is then calculated using $K(T,\lambda)$.

When detectors are operated in conditions where the background flux is less than the signal flux, the ultimate performance of detectors is determined by the SFL. It is achieved in practice with photomultipliers operating in the visible and UV region, but it is rarely achieved with solid-state devices, which are normally detector noise or electronic noise–limited. This limit is also applicable to longer wavelength detectors when the background temperature is very low. The NEP and detectivity of detectors operating in this limit have been derived by a number of authors (see Kruse et al. [3,68]).

The NEP in the SFL is given as [2,68–71]

$$\mathrm{NEP}=\frac{2hc\Delta f}{\eta\lambda}, \tag{3.20}$$

when Poisson statistics are applicable. This threshold value implies a low number of photons per observation interval. A more meaningful parameter is the probability that a photon will be detected during an observation period. Kruse [68] shows that the minimum signal power to achieve 99% probability that a photon will be detected in an observation period t_o is

$$\mathrm{NEP}_{\min}=\frac{9.22hc\Delta f}{\eta\lambda}, \tag{3.21}$$

where Δf is assumed to be $1/2t_o$. Note that the detector area does not enter into the expression and that $\mathrm{NEP}_{\min}$ depends linearly upon the bandwidth, which differs from the case in which the detection limit is set by internal or background noise.

Seib and Aukerman [72] also have derived an expression for the SFL identical to Equation 3.21 except that the multiplicative constant is not 9.22 but $2^{3/2}$ for an ideal photoemissive or photovoltaic detector and

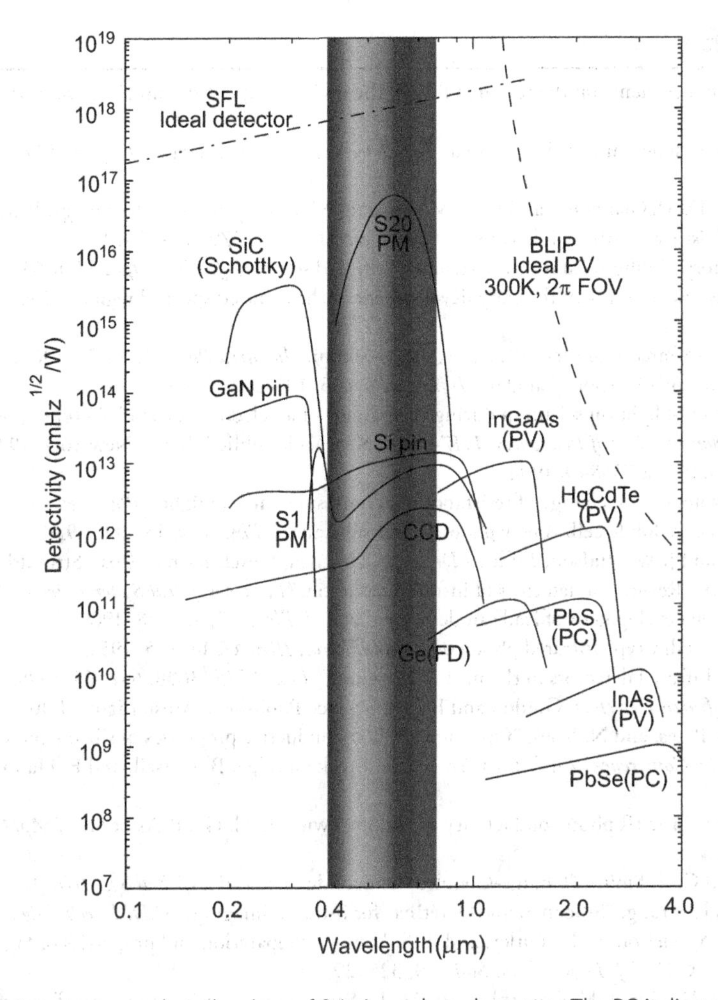

Figure 3.13 Detectivity versus wavelength values of 0.1–4 μm photodetectors. The PC indicates a photoconductive detector; PV, photovoltaic detector; and PM indicates a photomultiplier.

$2^{5/2}$ for a photoconductor. This difference in the constant arises from the differing assumptions as to the manner in which the detector is employed and the minimum detectable SNR.

Assuming Seib and Aukerman approximation for SFL limit of the photovoltaic detector, the corresponding detectivity is

$$D^* = \frac{\eta\lambda}{2^{2/3}hc}\sqrt{\frac{A_d}{\Delta f}}. \tag{3.22}$$

It is interesting to determine the composite signal fluctuation and background fluctuation limits. Figure 3.13 illustrates the spectral detectivities over the wavelength range from 0.1 μm to 4 μm, assuming a background temperature of 290 K and a 2π steradian FOV (applicable only to the background fluctuation limit). Note that the intersections of curves for signal fluctuation and background fluctuation limits lie about 1.2 μm. At wavelengths below 1.2 μm the SFL dominates; the converse is true above 1.2 μm. Below 1.2 μm, the wavelength dependence is small. Above 1.2 μm, it is very large due to the steep dependence of detectivity upon wavelength of the short wavelength end of the 290-K background spectral distribution (Figure 3.13).

It will be seen (Chapter 6) that by employing optical heterodyne detection it is possible to achieve the SFL with IR detectors even in the presence of ambient background temperature.

REFERENCES

1. W. Herschel, "Experiments on the refrangibility of the invisible rays of the sun," *Philos. Trans. R. Soc. Lon.*, 90, 284, 1800.
2. R. A. Smith, F. E. Jones, and R. P. Chasmar, *The Detection and Measurement of Infrared Radiation*, Clarendon, Oxford, 1958.
3. P. W. Kruse, L. D. McGlauchlin, and R. B. McQuistan, *Elements of Infrared Technology*, Wiley, New York, 1962.
4. A. Rogalski, "History of infrared detectors," *Opto-Electr. Rev.*, 14, 279–308, 2012.
5. C. Corsi, "History highlights and future trends of infrared sensors," *J. Mod. Opt.*, 57, 1663–1686, 2009.
6. E. S. Barr, "Historical survey of the early development of the infrared spectral region," *Am. J. Phys.*, 28, 42–54, 1960.
7. E. S. Barr, "The infrared pioneers—II. Macedonio Melloni," *Infrared Phys.*, 2, 67–73, 1962; "The infrared pioneers—III. Samuel Pierpont Langley," *Infrared Phys.*, 3, 195–206, 1963.
8. W. Smith, "Effect of light on selenium during the passage of an electric current," *Nature*, 7, 303, 1873.
9. M. F. Doty, *Selenium, List of References, 1917–1925*, New York Public Library, New York, 1927.
10. J. C. Bose, U. S. Patent 755840, 1904.
11. T. W. Case, "Notes on the change of resistance of certain substrates in light," *Phys. Rev.*, 9, 305–10, 1917.
12. T. W. Case, "The thalofide cell: A new photoelectric substance," *Phys. Rev.* 15, 289, 1920.
13. R. D. Hudson and J. W. Hudson, *Infrared Detectors*, Dowden, Hutchinson & Ross, Stroudsburg, PA, 1975.
14. E. W. Kutzscher, "Review on detectors of infrared radiation," *Electro-Optical Systems Design*, 5, 30, June 1973.
15. D. J. Lovell, "The development of lead salt detectors," *Am. J. Phys.*, 37, 467–78, 1969.
16. R. J. Cushman, "Film-type infrared photoconductors," *Proc. IRE*, 47, 1471–5, 1959.
17. P. R. Norton, "Infrared detectors in the next millennium," *Proc. SPIE*, 3698, 652–65, 1999.
18. A. Rogalski, *Infrared Detectors*, Gordon and Breach Science Publishers, Amsterdam, 2000.
19. E. Burstein, G. Pines, and N. Sclar, "Optical and photoconductive properties of silicon and Germanium," in *Photoconductivity Conference at Atlantic City*, eds. R. Breckenbridge, B. Russell, and E. Hahn, 353–413, Wiley, New York, 1956.
20. R. A. Soref, "Extrinsic IR photoconductivity of Si doped with B, Al, Ga, P, As or Sb," *J. Appl. Phys.*, 38, 5201–9, 1967.
21. W. S. Boyle and G. E. Smith, "Charge-Coupled Semiconductor Devices," *Bell Sys. Tech. J.*, 49, 587–93, 1970.
22. F. Shepherd and A. Yang, "Silicon Schottky retinas for infrared imaging," *IEDM Tech. Dig.*, 310–3, 1973.
23. W. D. Lawson, S. Nielson, E. H. Putley, and A. S. Young, "Preparation and properties of HgTe and mixed crystals of HgTe-CdTe," *J. Phys. Chem. Solids*, 9, 325–29, 1959.
24. P. W. Kruse, M. D. Blue, J. H. Garfunkel, and W. D. Saur, "Long wavelength photoeffects in mercury selenide, mercury telluride and mercury telluride-cadmium telluride," *Infrared Phys.*, 2, 53–60, 1962.
25. J. Melngailis and T. C. Harman, "Single-crystal lead-tin chalcogenides," in *Semiconductors and Semimetals*, Vol. 5, eds. R. K. Willardson and A. C. Beer, 111–74, Academic Press, New York, 1970.
26. T. C. Harman and J. Melngailis, "Narrow Gap Semiconductors," in *Applied Solid State Science*, Vol. 4, ed. R. Wolfe, 1–94, Academic Press, New York, 1974.
27. A. Rogalski and J. Piotrowski, "Intrinsic infrared detectors," *Prog Quant Electron.*, 12, 87–289, 1988.
28. S. Borrello and H. Levinstein, "Preparation and properties of mercury doped infrared detectors," *J. Appl. Phys.*, 33, 2947–50, 1962.
29. C. T. Elliott, D. Day, and B. J. Wilson, "An integrating detector for serial scan thermal imaging," *Infrared Phys.*, 22, 31–42, 1982.
30. A. Blackburn, M. V. Blackman, D. E. Charlton, W. A. E. Dunn, M. D. Jenner, K. J. Oliver, and J. T. M. Wotherspoon, "The practical realisation and performance of SPRITE detectors," *Infrared Phys.*, 22, 57–64, 1982.
31. A. Hoffman, "Semiconductor processing technology improves resolution of infrared arrays," *Laser Focus World*, 81–4, February 2006.
32. A. W. Hoffman, P. L. Love, and J. P. Rosbeck, "Mega-pixel detector arrays: Visible to 28 μm," *Proc. SPIE*, 5167, 194–203, 2004.
33. J. C. Fraser, D. H. Alexander, R. M. Finnila, and S. C. Su, "An extrinsic Si CCD for detecting infrared radiation," in *Digest of Technical Papers*, 442–5, IEEE, New York, 1974.
34. K. Nummendal, J. C. Fraser, S. C. Su, R. Baron, and R. M. Finnila, "Extrinsic silicon monolithic focal plane array technology and applications," in *Proceedings of CCD Applications International Conference*, 19–30, Noval Ocean Systems Center, San Diego, CA, 1976.

35. N. Sclar, R. L. Maddox, and R. A. Florence, "Silicon monolithic infrared detector array," *Appl. Opt.*, 16, 1525–32, 1977.
36. E. Beuville, D. Acton, E. Corrales, J. Drab, A. Levy, M. Merrill, R. Peralta, and W. Ritchie, "High performance large infrared and visible astronomy arrays for low background applications: Instruments performance data and future developments at raytheon," *Proc. SPIE.*, 6660, 66600B, 2007.
37. M. J. E. Golay, "A pneumatic infrared detector," *Rev. Sci. Instrum.*, 18, 357–62, 1947.
38. E. M. Wormser, "Properties of thermistor infrared detectors," *J. Opt. Soc. Am.*, 43, 15–21, 1953.
39. R. W. Astheimer, "Thermistor infrared detectors," *Proc. SPIE.*, 443, 95–109, 1983.
40. G. W. McDaniel and D. Z. Robinson, "Thermal imaging by means of the evaporograph," *Appl. Opt.*, 1, 311–24, 1962.
41. C. Hilsum and W. R. Harding, "The theory of thermal imaging, and its application to the absorption-edge image tube," *Infrared Phys.*, 1, 67–93, 1961.
42. A. J. Goss, "The pyroelectric vidicon: A review," *Proc. SPIE.*, 807, 25–32, 1987.
43. R. A. Wood and N. A. Foss, "Micromachined bolometer arrays achieve low-cost imaging," *Laser Focus World*, 101–6, June, 1993.
44. R. A. Wood, "Monolithic silicon microbolometer arrays," in *Semiconductors and Semimetals*, Vol. 47, eds. P. W. Kruse and D. D. Skatrud, 45–121, Academic Press, San Diego, CA, 1997.
45. C. M. Hanson, "Hybrid pyroelectric–ferroelectric bolometer arrays," in *Semiconductors and Semimetals*, Vol. 47, eds. P. W. Kruse and D. D. Skatrud, 123–74, Academic Press, San Diego, CA, 1997.
46. P. W. Kruse, "Uncooled IR focal plane arrays," *Opto-Electron. Rev.*, 7, 253–58, 1999.
47. R. A. Wood, "Uncooled microbolometer infrared sensor arrays," in *Infrared Detectors and Emitters: Materials and Devices*, eds. P. Capper and C. T. Elliott, 149–74, Kluwer Academic Publishers, Boston, MA, 2000.
48. R. W. Whatmore and R. Watton, "Pyroelectric materials and devices," in *Infrared Detectors and Emitters: Materials and Devices*, eds. P. Capper and C. T. Elliott, 99–147, Kluwer Academic Publishers, Boston, MA, 2000.
49. P. W. Kruse, *Uncooled Thermal Imaging. Arrays, Systems, and Applications*, SPIE Press, Bellingham, WA, 2001.
50. R. D. Hudson, *Infrared System Engineering*, Wiley, New York, 1969.
51. M. A. Kinch, "Fundamental physics of infrared detector materials," *J. Electron. Mater.*, 29, 809–17, 2000.
52. C. T. Elliott and N. T. Gordon, "Infrared detectors," in *Handbook on Semiconductors*, Vol. 4, ed. C. Hilsum, 841–936, Elsevier, Amsterdam, 1993.
53. H.-W. Hübers, "Terahertz heterodyne receivers," *IEEE J Sel. Top. Quantum Electron.*, 14, 378–91, 2008.
54. D. Long, "Photovoltaic and photoconductive infrared detectors," in *Optical and Infrared Detectors*, ed., R. J. Keyes, 101–47, Springer, Berlin, 1980.
55. A. Rogalski, *Infrared Detectors*, 2nd ed., CRC Press, Boca Raton, FL, 2010.
56. W. I. Wolfe and G. J. Zissis, eds., *The Infrared Handbook*, Office of Naval Research, Washington, DC, 1985.
57. W. D. Rogatto, ed., *The Infrared and Electro-Optical Systems Handbook*, Infrared Information Analysis Center, Ann Arbor, MI, and SPIE Optical Engineering Press, Bellingham, WA, 1993.
58. J. D. Vincent, *Fundamentals of Infrared Detector Operation and Testing*, Wiley, New York, 1990.
59. J. D. Vincent, S.E. Hodges, J. Vampola, M. Stegall, and G. Pierce, *Fundamentals of Infrared and Visible Detector Operation and Testing*, Wiley, Hoboken, NJ, 2016.
60. W. L. Eisenman, J. D. Merriam, and R. F. Potter, "Operational characteristics of infrared photodetectors," in *Semiconductors and Semimetals*, Vol. 12, eds. R. K. Willardson and A. C. Beer, 1–38, Academic Press, New York, 1977.
61. T. Limperis and J. Mudar, "Detectors," in *The Infrared Handbook*, eds. W. L. Wolfe and G. J. Zissis, 11.1–11.104, Environmental Research Institute of Michigan, Office of Naval Research, Washington, DC, 1989.
62. D. G. Crove, P. R. Norton, T. Limperis, and J. Mudar, "Detectors," in *The Infrared and Electro-Optical Systems Handbook*, Vol. 3, ed. W. D. Rogatto, 175–283, Infrared Information Analysis Center, Ann Arbor, MI, and SPIE Optical Engineering Press, Bellingham, WA, 1993.
63. R. C. Jones, "Performance of detectors for visible and infrared radiation," in *Advances in Electronics*, Vol. 5, ed. L. Morton, 27–30, Academic Press, New York, 1952.
64. R. C. Jones, "Phenomenological description of the response and detecting ability of radiation detectors," *Proc. IRE.*, 47, 1495–502, 1959.
65. A. N. Lowan and G. Blanch, "Tables of Planck's radiation and photon functions," *J. Opt. Soc. Am.* 30, 70–81, 1940.
66. P. R. Bratt, "Impurity germanium and silicon infrared detectors," in *Semiconductors and Semimetals*, Vol. 12, eds. R. K. Willardson and A. C. Beer, 39–141, Academic Press, New York, 1977.

67. N. Sclar, "Properties of doped silicon and germanium in infrared detectors," *Progress in Quantum Electronics* 9, 149–257, 1984.
68. P. W. Kruse, "The photon detection process," in *Optical and Infrared Detectors*, ed. R. J. Keyes, 5–69, Springer, Berlin, 1977.
69. R. W. Boyd, *Radiometry and the Detection of Optical Radiation*, Wiley, New York, 1983.
70. R. H. Kingston, *Detection of Optical and Infrared Radiation*, Wiley, New York, 1983.
71. E. L. Dereniak and G. D. Boremen, *Infrared Detectors and Systems*, Wiley, New York, 1996.
72. D. H. Seib and L. W. Aukerman, "Photodetectors for the 0.1 to 1.0 μm spectral region," in *Advances in Electronics and Electron Physics*, Vol. 34, ed. L. Morton, 95–221, Academic Press, New York, 1973.

Fundamental performance limitations of infrared detectors

As noted in Chapter 3, infrared detectors fall into two broad categories: photon and thermal. Although thermal detectors have been available commercially in single element form for many decades, their exploitation in imaging arrays started in the last decade of the twentieth century.

This chapter discusses the fundamental limitations to IR detector performance imposed by the statistical nature of the generation, recombination processes, and radiometric considerations. We will try to establish the ultimate theoretical sensitivity limit that can be expected for a detector operating at a given temperature. The models presented here are applicable to any of the detector classes mentioned in Chapter 3. The nonfundamental limitations will be addressed later in this book.

Photon detectors are fundamentally limited by generation–recombination noise arising from photon exchange with radiation background. Thermal detectors are fundamentally limited by temperature fluctuation noise arising from radiant power exchange with a radiating background. Due to fundamentally different types of noises, these two classes of detectors have different dependencies of detectivities on wavelength and temperature. The photon detectors are favored at a long wavelength infrared (LWIR) and lower operating temperatures. The thermal detectors are favored at a very long wavelength spectral range.

In this chapter, we first examine fundamental infrared detection processes for both categories of detectors. Next, the comparative studies of thermal and photon detectors are carried out. Different types of thermal as well as photon detectors are discussed in detail in Part II and Part III of the book, respectively. However, some elementary detection process concepts must be understood to fully appreciate the limitation of sensitivity imposed by noise processes within these devices.

4.1 THERMAL DETECTORS

Thermal detectors are classified according to the operating schemes: thermopile scheme, bolometer scheme, and pyroelectric scheme. In the present section, the general principles of thermal detectors are described.

4.1.1 PRINCIPLE OF OPERATION

The performance of a thermal detector is calculated in two stages. First, by consideration of the thermal characteristics of the system, the temperature rise produced by the incident radiation is determined. Second, this temperature rise is used to determine the change in the property that is being used to indicate the signal. The first stage of the calculations is common to all thermal detectors, but the details of the second stage differ for the different types of thermal detectors.

Thermal detectors operate on a simple principle that when heated by incoming IR radiation their temperature increases and the temperature changes are measured by any temperature-dependent mechanism, such as thermoelectric voltage, resistance, or pyroelectric voltage.

The simplest representation of the thermal detector is shown in Chapter 3, Figure 3.4. The detector is represented by a thermal capacitance C_{th} coupled via a thermal conductance G_{th} to a heat sink at a constant temperature T. In the absence of a radiation input, the average temperature of the detector will also be T, although it will exhibit a fluctuation near this value. When a radiation input is received by the detector, the rise in temperature is found by solving the heat balance equation [1–3]:

$$C_{th}\frac{d\Delta T}{dt}+G_{th}\Delta T=\varepsilon\Phi, \tag{4.1}$$

where ΔT is the temperature difference due to optical signal Φ, between the detector and its surroundings, and ε is the emissivity of the detector. The analogies between thermal and electrical circuits are given in Table 4.1. The thermal circuit (Figure 3.4) corresponds to an electric circuit shown in Figure 4.1.

Assuming the radiant power to be a periodic function,

$$\Phi = \Phi_o e^{i\omega t}, \tag{4.2}$$

where Φ_o is the amplitude of sinusoidal radiation, the solution of differential heat radiation is

$$\Delta T = \Delta T_o e^{-(G_{th}/C_{th})t} + \frac{\varepsilon \Phi_o e^{i\omega t}}{G_{th} + i\omega C_{th}}. \tag{4.3}$$

The first term is the transient part and as time increases, this term exponentially decreases to zero, so it can be dropped with no loss of generality for the change in temperature. Therefore, the change in temperature of any thermal detector due to incident radiative flux is

$$\Delta T = \frac{\varepsilon \Phi_o}{(G_{th}^2 + \omega^2 C_{th}^2)^{1/2}}. \tag{4.4}$$

Equation 4.4 illustrates several features of the thermal detector. Clearly, it is advantageous to make ΔT as large as possible. To do this, the thermal capacity of the detector (C_{th}) and its thermal coupling to its surroundings (G_{th}) want to be as small as possible. The interaction of the thermal detector with the incident radiation need to be optimized while reducing as far as possible all other thermal contacts with its surroundings. This means that a small detector mass and fine connecting wires to the heat sink are desirable.

Equation 4.4 shows that as ω is increased, the term $\omega^2 C_{th}^2$ will eventually exceed G_{th}^2 and then ΔT will fall inversely as ω. A characteristic thermal response time for the detector can therefore be defined as

$$\tau_{th} = \frac{C_{th}}{G_{th}} = C_{th} R_{th}, \tag{4.5}$$

where $R_{th} = 1/G_{th}$ is the thermal resistance. Then Equation 4.4 can be written as

Table 4.1 Thermal-electric analogies

THERMAL		ELECTRIC	
VALUE	UNIT	VALUE	UNIT
Heat energy	J	Charge	C
Heat flow	W	Current	A
Temperature	K	Voltage	V
Thermal impedance	K/W	Impedance	Ω
Heat capacitance	J/K	Capacitance	F

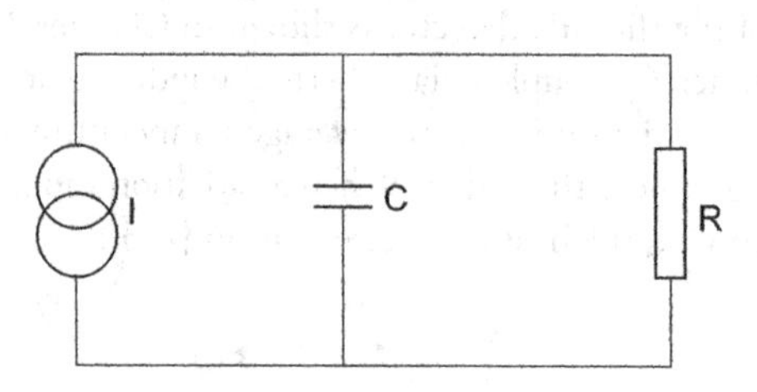

Figure 4.1 Electrical analog of thermal detector.

$$\Delta T = \frac{\varepsilon \Phi_o R_{th}}{(1+\omega^2\tau_{th}^2)^{1/2}}. \tag{4.6}$$

The typical value of thermal time constant is in the millisecond range. This is much longer than the typical time of a photon detector. There is a trade-off between sensitivity, ΔT, and frequency response. If one wants a high sensitivity, then a low-frequency response is forced upon the detector.

For further discussion, we introduce the coefficient K, which reflects how good the temperature changes translates into the electrical output voltage of detector [4]

$$K = \frac{\Delta V}{\Delta T}. \tag{4.7}$$

Then, the corresponding rms voltage signal due to temperature changes ΔT is

$$\Delta V = K\Delta T = \frac{K\varepsilon \Phi_o R_{th}}{(1+\omega^2\tau_{th}^2)^{1/2}}. \tag{4.8}$$

The voltage responsivity R_v of the detector is the ratio of the output signal voltage ΔV to the input radiation power and is given by

$$R_v = \frac{K\varepsilon R_{th}}{(1+\omega^2\tau_{th}^2)^{1/2}}. \tag{4.9}$$

As the last expression shows, the low-frequency voltage responsivity ($\omega \ll 1/\tau_{th}$) is proportional to the thermal resistance and does not depend on the heat capacitance. The opposite is true for high frequencies ($\omega \gg 1/\tau_{th}$). In this case, R_v is not dependent on R_{th} and is inversely proportional to the heat capacitance.

As stated previously, the thermal conductance (thermal resistance) from the detector to the outside world should be small (high). The smallest possible thermal conductance would occur when the detector is completely isolated from the environment under vacuum with only radiative heat exchange between it and its heat-sink enclosure. Such an ideal model can give us the ultimate performance limit of a thermal detector. This limiting value can be estimated from the Stefan–Boltzmann total radiation law.

If the thermal detector has a receiving area A of emissivity ε, when it is in thermal equilibrium with its surroundings it will radiate a total flux $A\varepsilon\sigma T^4$ where σ is the Stefan–Boltzmann constant. Now if the temperature of the detector is increased by a small amount dT, the flux radiated is increased by $4A\varepsilon\sigma T^3 dT$. Hence, the radiative component of the thermal conductance is

$$G_R = \frac{1}{(R_{th})_R} = \frac{d}{dT}(A\varepsilon\sigma T^4) = 4A\varepsilon\sigma T^3. \tag{4.10}$$

In this case

$$R_v = \frac{K}{4\sigma T^3 A(1+\omega^2\tau_{th}^2)^{1/2}}. \tag{4.11}$$

When the detector is in thermal equilibrium with the heat sink, the fluctuation in the power flowing through the thermal conductance into the detector is [5,6]

$$\Delta P_{th} = \left(4KT^2G\right)^{1/2}, \tag{4.12}$$

which will be the smallest when G assumes its minimum value (i.e., G_R). Then ΔP_{th} will be a minimum and its value gives the minimum detectable power for an ideal thermal detector.

The minimum detectable signal power—or noise equivalent power (NEP)—is defined as the rms signal power incident upon the detector required to equal the rms thermal noise power. Hence, if the temperature fluctuation associated with G_R is the only source of noise,

$$\varepsilon\text{NEP} = \Delta P_{th} = \left(16A\varepsilon\sigma kT^5\right)^{1/2}, \tag{4.13}$$

or

$$\text{NEP} = \left(\frac{16A\sigma kT^5}{\varepsilon}\right)^{1/2}. \tag{4.14}$$

If all the incident radiation is absorbed by the detector, $\varepsilon = 1$, and then

$$\text{NEP} = \left(16A\varepsilon\sigma kT^5\right)^{1/2} = 5.0\times10^{-11}\,\text{W}, \tag{4.15}$$

for $A = 1\ \text{cm}^2$, $T = 290$ K, and $\Delta f = 1$ Hz.

4.1.2 NOISE MECHANISMS

To determine the NEP and detectivity (D^*) of a detector, it is necessary to define a noise mechanism. For any detector, there are a number of noise sources that impose fundamental limits to the detection sensitivity.

One major noise is the Johnson noise. This noise in a Δf bandwidth for a resistor of resistance R is

$$V_J^2 = 4kTR\Delta f, \tag{4.16}$$

where k is the Boltzmann constant and Δf is the frequency band. This noise has a white character.

Two other fundamental noise sources are important for assessing the ultimate performance of a detector: thermal fluctuation noise and background fluctuation noise.

Thermal fluctuation noise arises from temperature fluctuations in the detector. These fluctuations are caused by heat conductance variations between the detector and the surrounding substrate with which the detector element is in thermal contact.

The variance in temperature ("temperature" noise) can be shown to be [2,5,6]

$$\overline{\Delta T^2} = \frac{4kT^2\Delta f}{1+\omega^2\tau_{th}^2}R_{th}. \tag{4.17}$$

From this equation it is evident that thermal conductance, $G_{th} = 1/R_{th}$, as the principal heat loss mechanism, is the key design parameter that affects the temperature fluctuation noise. Figure 4.2 shows exemplary temperature fluctuation noise (rms value of temperature fluctuation) for a typical IR sensitive micromechanical detector [7]. Note that the signal follows the same roll-off at higher frequencies as the temperature fluctuation noise does.

The spectral noise voltage due to temperature fluctuations is

$$V_{th}^2 = K^2\overline{\Delta T^2} = \frac{4kT^2\Delta f}{1+\omega^2\tau_{th}^2}K^2R_{th}. \tag{4.18}$$

A third noise source is background noise resulting from radiative heat exchange being observed between the detector at temperature T_d and the surrounding environment at temperature T_b. It is the ultimate limit of a detector's performance capability and is given for a 2π field of view (FOV) by [2,5,6]

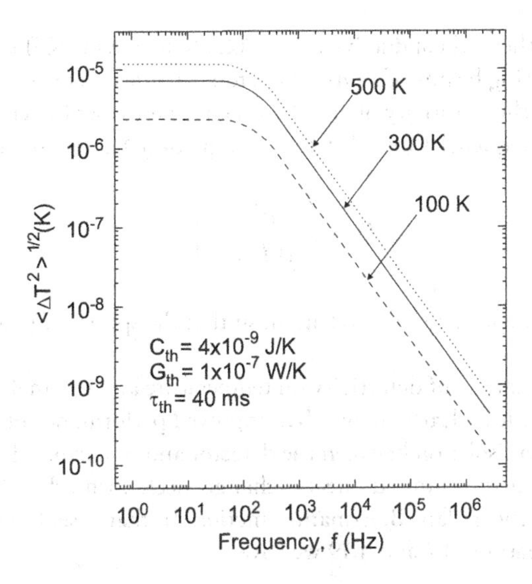

Figure 4.2 Spectral density of temperature fluctuation noise calculated for a typical thermal infrared detector. (From Ref. [7])

$$V_b^2 = \frac{8k\varepsilon\sigma A(T_d^2 + T_b^2)}{1+\omega^2\tau_{th}^2} K^2 R_{th}^2, \tag{4.19}$$

where σ is the Stefan–Boltzmann constant.

In addition to the fundamental noise sources mentioned earlier, $1/f$ is an additional noise source that is often found in the thermal detector and can affect detector performance. It can be described by the empirical form

$$V_{1/f}^2 = k_{1/f} \frac{I^\delta}{f^\beta} \Delta f, \tag{4.20}$$

where the coefficient $k_{1/f}$ is a proportionality factor, δ and β are coefficients whose values are about 1. The $1/f$ power law noise is difficult to characterize analytically because the parameters $k_{1/f}$, δ, and β are very much dependent upon material preparation and processing, including contacts and surfaces.

The square of total noise voltage is

$$V_n^2 = V_{th}^2 + V_b^2 + V_{1/f}^2. \tag{4.21}$$

4.1.3 DETECTIVITY AND FUNDAMENTAL LIMITS

According to Equations 3.7, 4.16, 4.18, and 4.21, the detectivity of a thermal detector is given by

$$D^* = \frac{K\varepsilon R_{th} A^{1/2}}{\left(1+\omega^2\tau_{th}^2\right)^{1/2}\left(\frac{4kT_d^2K^2R_{th}}{1+\omega\tau_{th}^2} + 4kTR + V_{1/f}^2\right)^{1/2}}. \tag{4.22}$$

In the case of a typical operation condition of the thermal detector, while it operates in a vacuum or a gas environment at reduced pressures, heat conduction through the supporting microstructure of the device is dominant heat loss mechanism. However, in the case of an extremely good thermal isolation, the principal heat loss mechanism can be reduced to only radiative heat exchange between the detector and its surroundings. In the atmospheric environment, heat conduction through air is likely to be a dominant heat dissipation mechanism. The thermal conductivity of air (2.4×10^{-2} Wm^{-1}K^{-1}) is larger than the thermal conductance through supporting beams of a typical micromechanical detector.

The fundamental limit to the sensitivity of any thermal detector is set by temperature fluctuation noise. Under this condition at low frequencies ($\omega \ll 1/\tau_{th}$), from Equation 4.22 we obtain

$$D_{th}^{*} = \left(\frac{\varepsilon^2 A}{4kT_d^2 G_{th}} \right)^{1/2}. \tag{4.23}$$

It is assumed here that ε is independent of wavelength, so that the spectral D_{λ}^{*} and blackbody $D^{*}(T)$ values are identical.

Figure 4.3 shows the dependence of detectivity on temperature and thermal conductance plotted for different detector active areas. It is clearly shown that improved performance of thermal detectors can be achieved by increasing thermal isolation between the detector and its surrounding.

If radiant power exchange is the dominant heat exchange mechanism, then G is the first derivative with respect to the temperature of the Stefan–Boltzmann function. In that case, known as the background fluctuation noise limit, from Equations 3.7 and 4.19 we have

$$D_{b}^{*} = \left[\frac{\varepsilon}{8k\sigma (T_d^5 + T_b^5)} \right]^{1/2}. \tag{4.24}$$

Note that D_b^{*} is independent of A, as is to be expected.

In many practical instances the temperature of the background, T_b, is room temperature, 290 K. Figure 4.4 shows the photon noise–limited detectivity for an ideal thermal detector having an emissivity of unity, operated at 290 K and lower, as a function of background temperature [6].

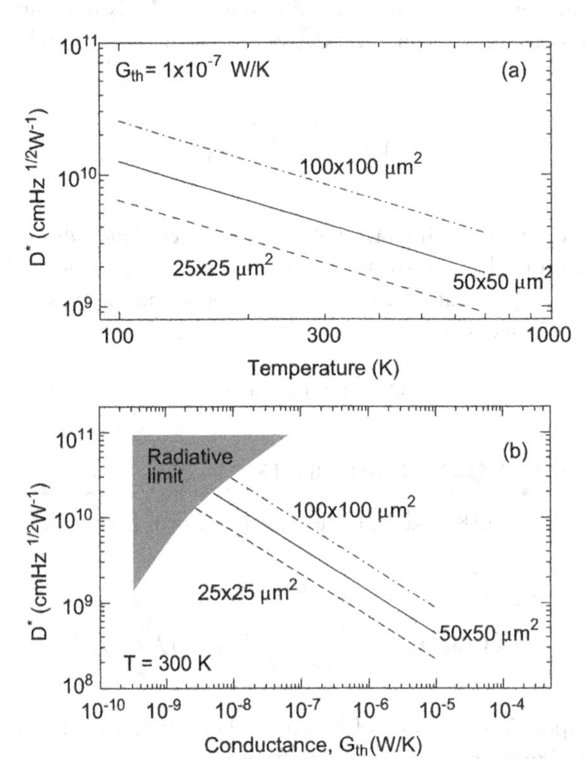

Figure 4.3 Temperature fluctuation noise–limited detectivity for thermal infrared detectors of different areas plotted (a) as a function of the detector temperature and (b) as a function of the total thermal conductance between the detector and its surroundings. (From Ref. [7])

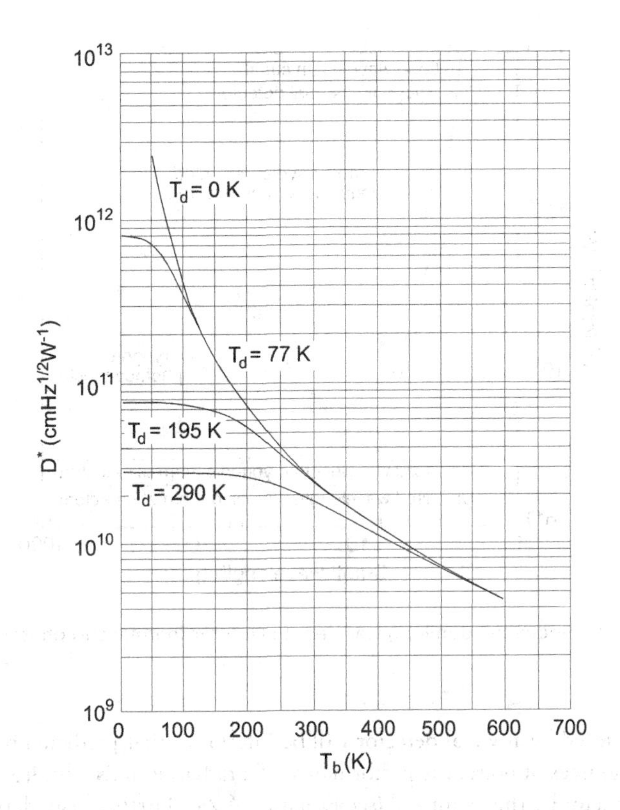

Figure 4.4 Temperature fluctuation noise-limited detectivity of thermal detectors as a function of detector temperatures T_d and background temperature T_b for 2π FOV and $\varepsilon = 1$. (From Ref. [6])

Equations 4.23 and 4.24 and Figure 4.4 assume that background radiation falls upon the detector from all directions when the detector and background temperature are equal, and from the forward hemisphere only when the detector is at cryogenic temperatures. We see that the highest possible D^* to be expected for a thermal detector operated at room temperature and viewing a background at room temperature is 1.98×10^{10} cmHz$^{1/2}$W^{-1}. Even if the detector or background (not both) were cooled to absolute zero, the detectivity would improve only by the square root of two. This is the basic limitation of all thermal detectors. The background noise–limited photon detectors have higher detectivities as a result of their limited spectral responses (what is shown in Figure 3.3).

Up to now we have only considered a flat spectral response of the thermal detector. In practice, it is sometimes necessary to limit the spectral responsivity of a detector by means of cooled filters. Assuming ideal filters, we can calculate the variation of detectivity as a function of both short and long wavelength cutoffs, λ_{c1} and λ_{c2}, respectively. If detector emissivity $\varepsilon = 0$ for wavelengths except between λ_{c1} and λ_{c2}, and if ε is independent of wavelength between λ_{c1} and λ_{c2}, then Equation 4.24 is replaced by [8]

$$D_b^* = \left[\frac{\varepsilon}{8k\sigma T_d^5 + F(\lambda_{c1},\lambda_{c2})}\right]^{1/2}, \tag{4.25}$$

where

$$F(\lambda_{c1},\lambda_{c1}) = 2\int_{\lambda_{c1}}^{\lambda_{c2}} \frac{h^2c^3}{\lambda^6}\frac{\exp(hc/\lambda kT_b)}{[\exp(hc/\lambda kT_b)-1]^2}d\lambda. \tag{4.26}$$

Figure 4.5 illustrates Equation 4.25 as a function of λ for the case of a long wavelength cutoff λ_{c2} (i.e., $\varepsilon = 1$ for $\lambda < \lambda_{c2}$ and $\varepsilon = 0$ for $\lambda > \lambda_{c2}$), and for the case of a short wavelength cutoff λ_{c1} (i.e., $\varepsilon = 0$ for $\lambda < \lambda_{c1}$ and $\varepsilon = 1$ for $\lambda > \lambda_{c1}$). The background temperature is 300 K.

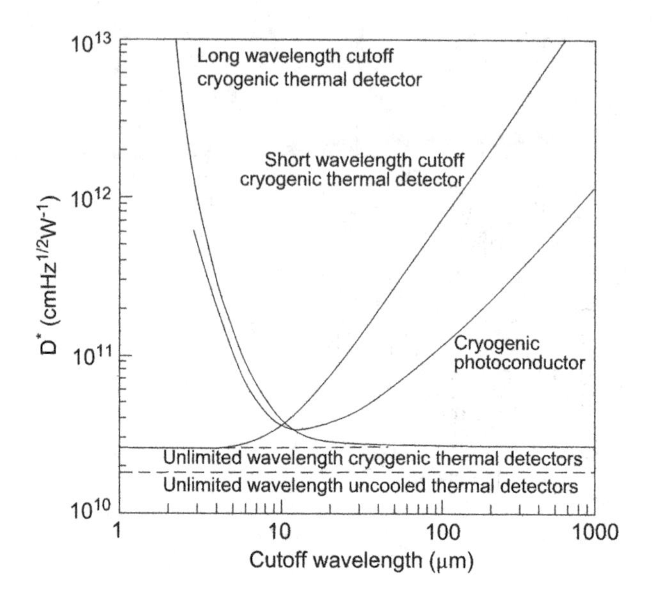

Figure 4.5 Dependence of detectivity upon long wavelength limit for thermal and photon detectors. (From Ref. [8])

The performance achieved by any real detector will be inferior to that predicted by Equation 4.23. Even in the absence of other sources of noise, the performance of a radiation noise–limited detector will be worse than that of an ideal detector by the factor $\varepsilon^{1/2}$ (see Equation 4.24). Further degradation of performance will arise from the following:

- Encapsulation of detector (reflection and absorption losses at the window)
- The effects of excess thermal conductance (influence of electrical contacts, conduction through the supports, influence of any gas—conduction and convection)
- The additional noise sources

Figure 4.6 shows the performance of a number of thermal detectors operating at room temperature [9]. Typical values of detectivities of thermal detectors at 10 Hz change in the range between 10^8 and 10^9 cm Hz$^{1/2}$W^{-1}.

Most thermal detectors can be tailored to have somewhat different properties and the user should contact the manufacturer for detailed information. Table 4.2 gives the general flavor of the performance of different thermal detectors.

4.2 PHOTON DETECTORS

4.2.1 PHOTON DETECTION PROCESS

Modern photodetectors operate on the basis of the internal photoeffects, in which the photoexcited carriers remain within the sample. The most important of the internal photoeffects are photoconductivity and photovoltaic effects. As is shown in Figure 3.7, the field decays exponentially as energy is transferred to the photocarriers. The semiconductor material is characterized by an absorption length, α, and a penetration depth, $1/\alpha$. Penetration depth is the point at which $1/e$ of the optical signal power remains.

Figure 4.7 shows the measured intrinsic absorption coefficients for various narrow gap photodetector materials. The absorption coefficient and corresponding penetration depth vary among the different materials. It is well-known that the absorption curve for direct transitions between parabolic bands at photon energy greater than energy gap, E_g, obeys a square root law

$$\alpha(h\nu) = \beta\left(h\nu - E_g\right)^{1/2}, \tag{4.27}$$

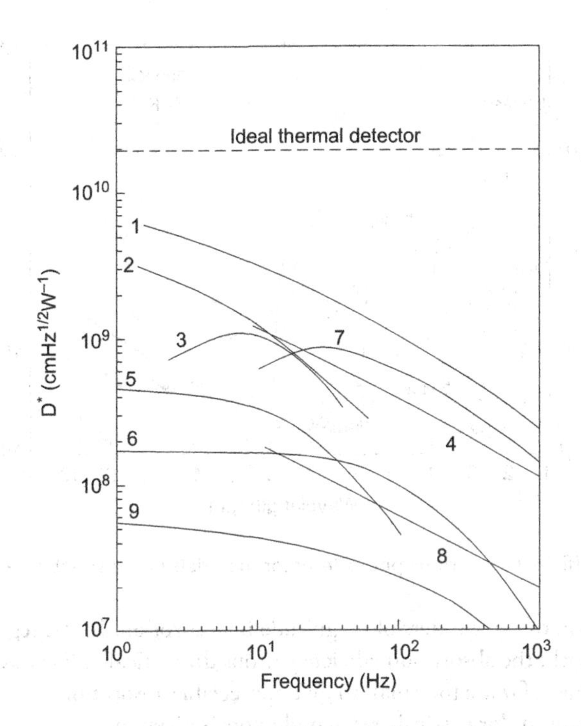

Figure 4.6 Performance of uncooled thermal detectors: (1) alaine-doped triglycine sulfate (TGS) pyroelectric detector (A = 1.5 × 1.5 mm²); (2) spectroscopic thermopile (A = 0.4 mm², τ_{th} = 40 ms); (3) Golay cell; (4) TGS pyroelectric detector in ruggedized encapsulation (0.5 × 0.5 mm²); (5) Sb–Bi evaporated film thermopile (A = 0.12 × 0.12 mm², τ_{th} = 13 ms); (6) immersed thermistor (A = 0.1 × 0.1 mm², τ_{th} = 2 ms); (7) $LiTaO_3$ pyroelectric detector; (8) Plessey lead zirconate titanate (PZT) ceramic pyroelectric detector; and (9) thin film bolometer. (From Ref. [9])

Table 4.2 General Properties of Thermal Detectors

TYPE	TEMPERATURE (K)	D* (cmHz$^{1/2}$/W)	NEP (WHz$^{1/2}$)	τ_{th} (ms)	SIZE (mm²)
Silicon bolometer	1.6		3×10^{-15}	8	0.25–0.70
Metal bolometer	2–4	1×10^{8}		10	
Thermistor bolometer	300	$(1–6) \times 10^{8}$		1–8	0.01–10
Germanium bolometer	2–4		5×10^{-13}	0.4	1.5
Carbon bolometer	2–4		3×10^{-12}	10	20
Superconducting bolometer (NbN)	15		2×10^{-11}	0.5	5 × 0.25
Thermocouples	300		$(2–10) \times 10^{-10}$	10–40	0.1 × 1 to 0.3 × 3
Thermopiles	300			3.3–10	1–100
Pyroelectrics	300	$(2–5) \times 10^{8}$		10–100*	2x2
Golay cell	300	1×10^{9}	6×10^{-11}	10–30	10

* Shorter values can be obtained at the expense of NEP.

where β is a constant. As can be readily seen in Figure 4.7, in an middle wavelength infrared (MWIR) spectral region, the absorption edge value changes between 2×10^3 cm^{-1} and 3×10^3 cm^{-1}; in the LWIR region, it is about 10^3 cm^{-1}.

Since α is a strong function of the wavelength, for a given semiconductor the wavelength range in which appreciate photocurrent can be generated is limited. Near the material's bandgap, there is tremendous

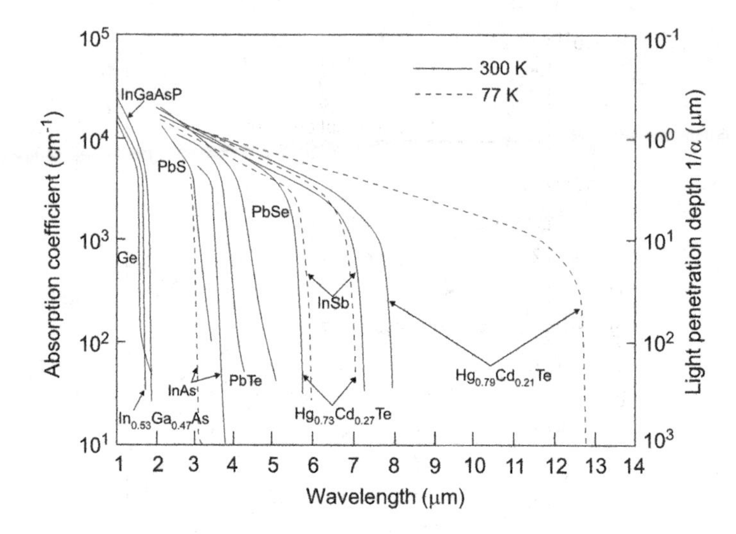

Figure 4.7 Absorption coefficient for various photodetector materials in the spectral range of 1–14 μm.

variation causing a variation of three orders of magnitude in absorption. In the region of the material's maximum usable wavelength, the absorption efficiency drops dramatically. For wavelengths longer than cutoff wavelength, the values of α are too small to give appreciable absorption.

The absorption coefficient, α, for extrinsic semiconductors is given by

$$\alpha = \sigma_p N_i. \tag{4.28}$$

This is the product of the photoionization cross-section, σ_p, and the neutral impurity concentration, N_i. It is desirable to make α as large as possible. The upper limit of N_i is set by either "hopping" or "impurity band" conduction. Practical values of α for optimized doped photoconductors are in the range 1–10 cm^{-1} for Ge and 10–50 cm^{-1} for Si. Thus, to maximize quantum efficiency, the thickness of the detector crystal should not be less than about 0.5 cm for doped Ge and about 0.1 cm for doped Si. Fortunately, for the most extrinsic detectors, the drift length of photocarriers is sufficiently long that quantum efficiencies approaching 50% can be obtained.

The absorption coefficient is considerably modified for low-dimensional solids. Figure 4.8 shows the infrared absorption spectra for different n-doped, 50-period GaAs/Al$_x$Ga$_{1-x}$As quantum well infrared

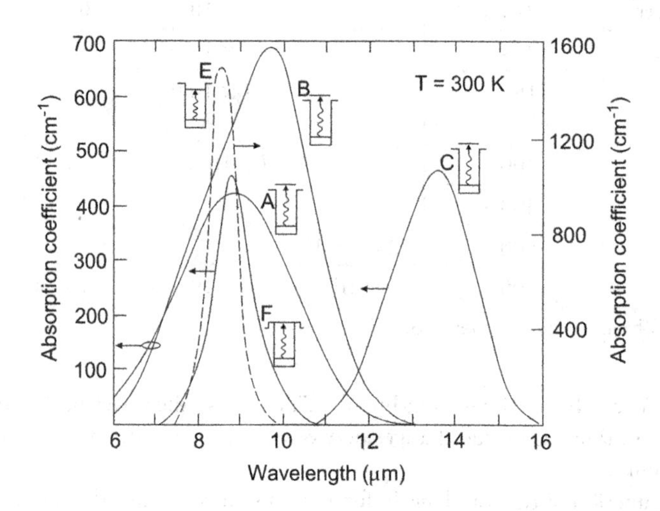

Figure 4.8 Absorption coefficient spectra measured at $T = 300$ K for different QWIP samples. (From Ref. [10])

photodetector (QWIP) structures measured at room temperatures using a 45° multipass waveguide geometry [10]. The spectra of the bound-to-bound continuum (B-C) QWIP (samples A, B, and C) are much broader than the bound-to-bound (B-B; sample F) or bound-to-quasibound (B-QB) QWIP (sample F). Correspondingly, the value of the absorption coefficient for the B-C QWIP is significantly lower than that of the B-B QWIP, due to the conservation of oscillator strength. It appears that the low-temperature absorption coefficient $\alpha_p(77\text{ K}) \approx 1.3\alpha_p(300\text{ K})$ and $\alpha_p(\Delta\lambda/\lambda)/N_D$ is a constant ($\Delta\lambda$ is the full width at half-α_p, N_D is the well's doping) [10]. The typical value of absorption coefficient in 77 K in LWIR region is between 600 and 800 cm^{-1}. Comparing Figures 4.7 and 4.8 we can notice that the absorption coefficients for direct band-to-band absorption are higher than that for intersubband transitions.

For an ensemble of quantum dots (QDs), the absorption spectra can be modeled using a Gaussian line shape in the form [11]

$$\alpha(E) = \alpha_o \frac{n_1}{\delta} \frac{\sigma_{QD}}{\sigma_{ens}} \exp\left[-\frac{\left(E - E_g\right)^2}{\sigma_{ens}^2}\right], \tag{4.29}$$

where α_o is the maximum absorption coefficient, n_1 is the areal density of electrons in the QD ground state, δ is the QD density, and $E_g = E_2 - E_1$ is the energy of the optical transition between ground and excited states in the QDs. The expressions σ_{QD} and σ_{ens} are the standard deviations in the Gaussian line shape for intraband absorption in a single QD and for the distribution in energies for the QD ensemble, respectively. Thus, the terms n_1/δ and σ_{QD}/σ_{ens} describe a decrease in absorption due to the absence of available electrons in the QD ground state and inhomogeneous broadening, respectively.

The optical absorption between the ground and excited levels is found to have a value [12]

$$\alpha_o \approx \frac{3.5\times10^5}{\sigma}, \text{ in cm}^{-1} \tag{4.30}$$

where σ is the linewidth of the transition in meV. Equation 4.30 indicates the trade-off between the absorption coefficient and the absorption linewidth, σ. For very uniform QDs, the theoretically predicted absorption coefficient by Equation 4.29 can be considerably higher in comparison with those measured for narrow gap intrinsic materials.

4.2.2 MODEL OF PHOTON DETECTOR

Let us consider a generalized model of a photodetector, which by its optical area A_o is coupled to a beam of IR radiation [13–19]. The detector is a slab of homogeneous semiconductor with actual "electrical" area, A_e, and thickness t (see Figure 4.9). Usually, the optical and electrical areas of the device are the same or are similar. However, the use of some kind of optical concentrator can increase the A_o/A_e ratio by a large factor.

The current responsivity of the photodetector is determined by the quantum efficiency, η, and by the photoelectric gain, g. The quantum efficiency describes how well the detector is coupled to the impinging radiation. It is defined here as the number of electron hole pairs generated per incident photon in an intrinsic detector, the number of generated free unipolar charge carriers in an extrinsic detector, or the number of charge carriers with energy sufficient to cross the potential barrier in a photoemissive detector. The photoelectric gain is the number of carriers passing contacts per one generated pair in an intrinsic detector, or the number of charge carriers in other types of detectors. This value shows how well the generated charge carriers are used to generate the current response of a photodetector. Both values are assumed here as constant over the volume of the device.

The spectral current responsivity is equal to

$$R_i = \frac{\lambda\eta}{hc} qg, \tag{4.31}$$

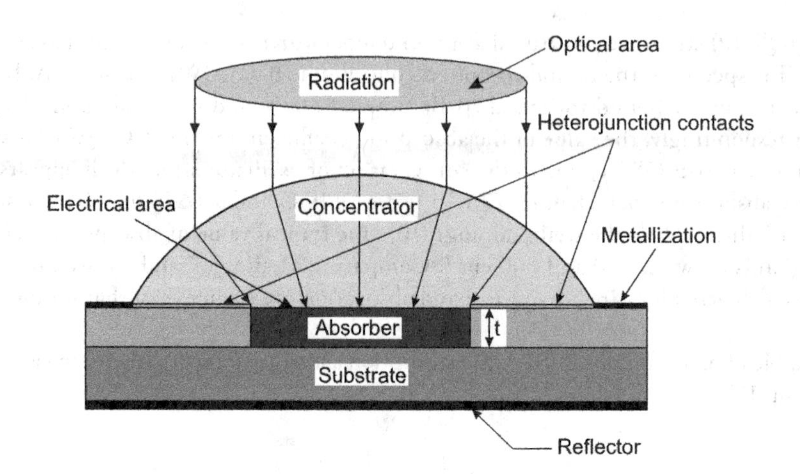

Figure 4.9 Model of a photodetector.

where λ is the wavelength, h is the Planck constant, c is the velocity of light, and q is the electron charge. The current that flows through the contacts of the device is noisy due to the statistical nature of the generation and recombination processes: fluctuation of optical generation, thermal generation, and radiative and nonradiative recombination rates. Assuming that the current gain for the photocurrent and the noise current are the same, the noise current is

$$I_n^2 = 2q^2g^2(G_{op} + G_{th} + R)\Delta f, \tag{4.32}$$

where G_{op} is the optical generation rate, G_{th} is the thermal generation rate, R is the resulting recombination rate, and Δf is the frequency band.

It should be noted that the effects of a fluctuating recombination frequently can be avoided by arranging for the recombination process to take place in a region of the device where it has little effect due to a low photoelectric gain, for example, at the contacts in sweep-out photoconductors, at the backside surface of a photoelectromagnetic detector, or in the neutral regions of the diodes. The generation processes with their associated fluctuations, however, cannot be avoided by any means [20,21].

Detectivity, D^*, is the main parameter to characterize normalized signal-to-noise performance of detectors and can be defined as

$$D^* = \frac{R_i(A_o\Delta f)^{1/2}}{I_n}. \tag{4.33}$$

4.2.2.1 Optical generation noise

Optical generation noise is photon noise due to fluctuation of the incident flux. The optical generation of the charge carriers may result from three different sources:

- Signal radiation generation
- Background radiation generation
- Thermal self-radiation of the detector itself at a finite temperature.

The optical signal generation rate (photons/s) is

$$G_{op} = \Phi_s A_o \eta, \tag{4.34}$$

where Φ_s is the signal photon flux density.

If recombination does not contribute to the noise,

$$I_n^2 = 2\Phi_s A_o \eta q^2 g^2 \Delta f, \tag{4.35}$$

and

$$D^* = \frac{\lambda}{hc}\left(\frac{\eta}{2\Phi_s}\right)^{1/2}. \tag{4.36}$$

This is the ideal situation, when the noise of the detector is determined entirely by the noise of the signal photons. Usually, the noise due to the optical signal flux is small compared to the contributions from background radiation or thermal generation–recombination processes. An exception is heterodyne detection, when the noise due to the powerful local oscillator radiation may dominate.

Background radiation frequently is the main source of noise in a detector. Assuming no contribution due to recombination,

$$I_n^2 = 2\Phi_B A_o \eta q^2 g^2 \Delta f, \tag{4.37}$$

where Φ_B is the background photon flux density. Therefore,

$$D^*_{\mathrm{BLIP}} = \frac{\lambda}{hc}\left(\frac{\eta}{2\Phi_B}\right)^{1/2}. \tag{4.38}$$

Once the background-limited performance is reached, quantum efficiency, η, is the only detector parameter that can influence a detector's performance.

Figure 4.10 shows the peak spectral detectivity of a background-limited photodetector operating at 300, 230, and 200 K, versus the wavelengths calculated for 300-K background radiation and hemispherical FOV ($\theta = 90°$). The minimum D^*_{BLIP} (300 K) occurs at 14 μm and is equal to 4.6×10^{10} cmHz$^{1/2}$/W. For photodetectors that operate at near-equilibrium conditions, such as non–sweep-out photoconductors, the recombination rate is equal to the generation rate. For these detectors, the contribution of recombination to the noise will reduce D^*_{BLIP} by a factor of $2^{1/2}$. Note that D^*_{BLIP} does not depend on the area and the A_o/A_e ratio. As a consequence, the background-limited performance cannot be improved by making A_o/A_e large.

The highest performance possible will be obtained by the ideal detector with unity quantum efficiency and ideal spectral responsivity [$R(\lambda)$ increases with wavelength to the cutoff wavelength λ_c at which the responsivity drops to zero]. This limiting performance is of interest for comparison with actual detectors.

The detectivity of background-limited infrared photodetector (BLIP) detectors can be improved by reducing the background photon flux, Φ_B. Practically, there are two ways to do this: a cooled or reflective

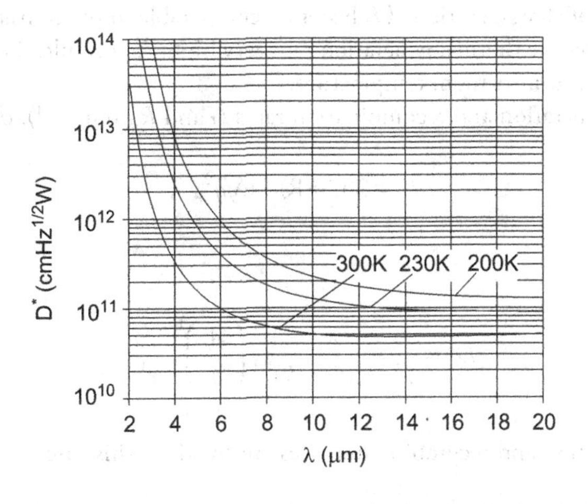

Figure 4.10 Calculated spectral detectivities of a photodetector limited by the hemispherical FOV background radiation of 300 K as a function of the peak wavelength for detector operating temperatures of 300, 230, and 200 K. (From Ref. [22])

spectral filter to limit the spectral band or a cooled shield to limit the angular FOV of the detector (as described earlier). The former eliminates background radiation from spectral regions in which the detector need not respond. The best detectors yield background-limited detectivities in quite narrow FOVs.

In contrast to the signal and background-related processes, optical generation is connected with the detector itself and may be of importance for detectors operating at near–room temperatures. The related ultimate performance is usually calculated, assuming blackbody radiation, and taking into account the reduced speed of light and wavelength due to a >1 refractive index of the detector material and full absorption of photons with energy larger than the bandgap [23]. The carrier generation rate per unit area is

$$g_a = 8\pi c n^2 \int_0^{\infty} \frac{d\lambda}{\lambda^4 \left(e^{hc/\lambda T} - 1\right)}, \tag{4.39}$$

where n is the refractive index. Note that the generation rate is a factor of $4n^2$ larger compared to the 180° FOV background generation for $\eta = 1$. Therefore, the resulting detectivity

$$D^* = \frac{\lambda \eta A_o}{hc(2g_a A_e)^{1/2}}. \tag{4.40}$$

will be a factor of $2n$ lower when compared to BLIP detectivity for a background at detector temperature ($A_e = A_o$). The internal thermal radiation limited D^* can be improved by making A_o/A_e large, in contrast to D^*_{BLIP}.

When Humpreys reexamined the existing theories of radiative recombination and internal optical generation [24,25], he indicated that most photons emitted as a result of radiative recombination are immediately reabsorbed inside the detector, generating charge carriers. Due to reabsorption, the radiative lifetime is highly extended, which means that the internal optical generation–recombination processes could be practically noiseless in optimized devices. Therefore, the ultimate limit of performance is set by the signal or background photon noise.

4.2.2.2 Thermal generation and recombination noise

Infrared photodetectors operating at near–room temperature and low-temperature devices operated at low background irradiances are generally limited by thermal generation and recombination mechanisms rather than by photon noise. For effective absorption of IR radiation in a semiconductor, we must use materials with a low energy of optical transitions compared to the energy of photons to be detected, for example, semiconductors with a narrower bandgap. A direct consequence of this fact is that at near–room temperatures, the thermal energy of charge carriers, kT, becomes comparable to the transition energy. This enables thermal transitions, making the thermal generation rate very high. As a result, the long wavelength detector is very noisy when operated at near-room temperature.

For uniform volume generation and recombination rates G and R (in $m^{-6}s^{-1}$), the noise current is

$$I_n^2 = 2(G+R)A_e t \Delta f q^2 g^2, \tag{4.41}$$

therefore

$$D^* = \frac{\lambda}{2^{1/2} hc(G+R)^{1/2}} \left(\frac{A_o}{A_e}\right)^{1/2} \frac{\eta}{t^{1/2}}. \tag{4.42}$$

At equilibrium, the generation and recombination rates are equal. In this case

$$D^* = \frac{\lambda \eta}{2hc(Gt)^{1/2}} \left(\frac{A_o}{A_e}\right)^{1/2}. \tag{4.43}$$

4.2.3 OPTIMUM THICKNESS OF PHOTODETECTOR

For a given wavelength and operating temperature, the highest performance can be obtained by maximizing $\eta/[(G+R)t]^{1/2}$. This is the condition for the highest ratio of the quantum efficiency to the square root of the sum of the sheet thermal generation and recombination rates. This means that high quantum efficiency must be obtained with a thin device.

In further calculations, we will assume $A_e = A_o$, perpendicular incidence of radiation, and negligible front and backside reflection coefficients. In this case,

$$\eta = 1 - e^{-\alpha_t}. \tag{4.44}$$

where α is the absorption coefficient. Then

$$D^* = \frac{\lambda}{2^{1/2}hc}\left(\frac{\alpha}{G+R}\right)^{1/2} F(\alpha t), \tag{4.45}$$

where

$$F(\alpha t) = \frac{1-e^{-\alpha t}}{(\alpha t)^{1/2}}. \tag{4.46}$$

Function $F(\alpha t)$ achieves its maximum 0.638 for $t = 1.26/\alpha$. In this case, $\eta = 0.716$ and the highest detectivity is

$$D^* = 0.45\frac{\lambda}{hc}\left(\frac{\alpha}{G+R}\right)^{1/2}. \tag{4.47}$$

The detectivity can also be increased by a factor of $2^{1/2}$ for double pass of radiation. This can be achieved by the use of a backside reflector. A simple calculation shows that the optimum thickness in this case is half that of the single pass case, while the quantum efficiency remains equal to 0.716.

At equilibrium, the generation and recombination rates are equal. Therefore

$$D^* = \frac{\lambda}{2hc}\eta(Gt)^{-1/2} \tag{4.48}$$

If the recombination process is uncorrelated with the generation process that contributes the detector noise,

$$D^* = \frac{\lambda}{2^{1/2}hc}\eta(Gt)^{-1/2}. \tag{4.49}$$

4.2.4 DETECTOR MATERIALS FIGURE OF MERIT

To summarize the aforementioned discussion, the detectivity of an optimized infrared photodetector of any type can be expressed as

$$D^* = 0.31\frac{\lambda}{hc}k\left(\frac{\alpha}{G}\right)^{1/2}, \tag{4.50}$$

where $1 \le k \le 2$ and dependent on the contribution of recombination and backside reflection [22]. The k-coefficient can be modified by using more sophisticated coupling of the detector with IR radiation; for example, using photonic crystals or surface plasmon polaritons.

As we can see, the ratio of the absorption coefficient to the thermal generation rates, α/G, is the main figure of merit of any materials for infrared detectors. This figure of merit can be utilized to predict ultimate performance of any infrared detector and to select possible material candidates for use as detectors [17,26].

The α/G ratio versus temperature for various material systems capable of band gap tuning is shown in Figure 4.11 for a hypothetical energy gap equal to 0.25 eV ($\lambda = 5\,\mu$m) [Figure 4.11a] and 0.124 eV ($\lambda = 10\,\mu$m) [Figure 4.11b]. Procedures used in calculations of α/G for different material systems are given in Ref. 27. Analysis shows that the narrow gap semiconductors are more suitable for high-temperature photodetectors in comparison to competing technologies such as extrinsic devices, QWIP and quantum dot IR photodetector (QDIP) devices. The main reason for high performance of intrinsic photodetectors is the high density of states in the valence and conduction bands, which results in strong absorption of infrared radiation. Figure 4.11b predicts that recently emerging competing IR material, type-II SL, is the most efficient material technology for IR detection in the long wavelength region, theoretically perhaps even better than HgCdTe, if the influence of SRH lifetime is not considered. It is characterized by a high absorption coefficient and relatively low fundamental (band-to-band) thermal generation rate. However, this theoretical prediction has not been confirmed by experimental data.

The calculation of the figure of merit involves the determination of the absorption coefficient and thermal generation rate, taking into account various processes of fundamental and less fundamental nature.

It should be noted that the importance of thermal generation rate as a material figure of merit was recognized for the first time by Long [28]. It was used in many papers of English workers [29,30] related to high-operating temperature detectors. More recently, Kinch [31] introduced the thermal generation rate within $1/\alpha$ depth per unit of area as the figure of merit. This is actually the inversed α/G figure of merit originally proposed by Piotrowski and Rogalski [17].

In the Kinch criterion, the BLIP condition can be described as [31]

$$\frac{\eta\Phi_B\tau}{t} > n_{th}, \tag{4.51}$$

where n_{th} is the density of thermal carriers at the temperature T, τ is the carrier lifetime, Φ_B is the total background photon flux density (unit cm^{-2}s^{-1}) reaching the detector, and t is the detector's thickness. Rearranging, for the BLIP requirements we have

$$\frac{\eta\Phi_B}{t} > \frac{n_{th}}{\tau}; \tag{4.52}$$

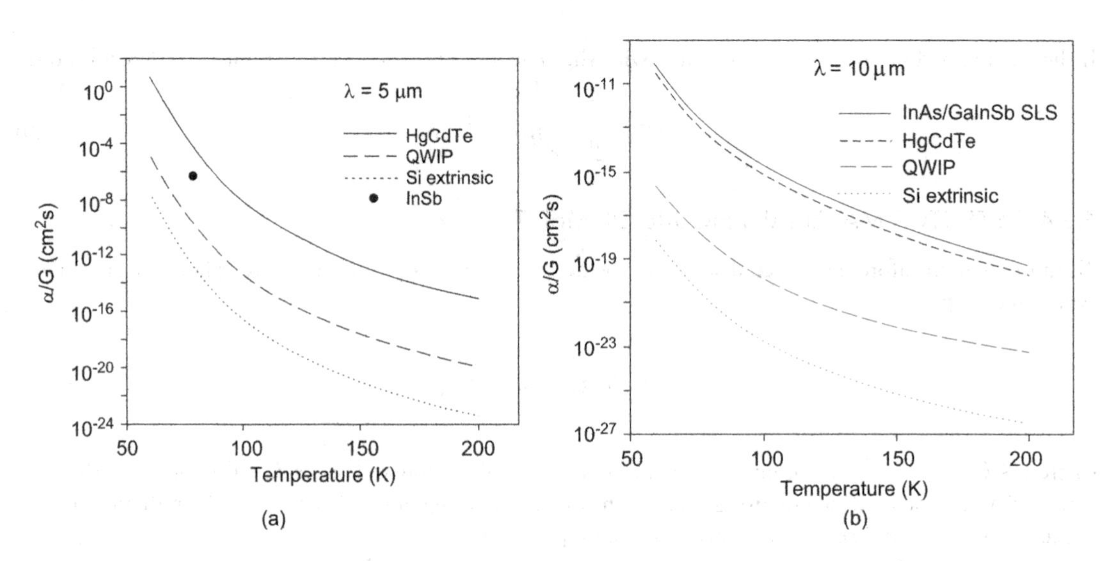

Figure 4.11 α/G ratio versus temperature for (a) MWIR – $\lambda = 5\,\mu$m, and (b) LWIR – $\lambda = 10\,\mu$m photodetectors based on HgCdTe, InSb (for MWIR only), QWIP, Si extrinsic and type-II superlattice (for LWIR only) material technology.

that is, the photon generation rate per unit volume needs to be greater than the thermal generation rate per unit volume. The carriers can be either majority or minority in nature. Using $\eta = \alpha t$ where α is the absorption coefficient in the material, we obtain

$$\Phi_B > \frac{n_{th}}{\alpha\tau} = G_{th}. \tag{4.53}$$

The normalized thermal generation, $G_{th} = n_{th}/(\alpha\tau)$, predicts the ultimate performance of any infrared material and can be used to compare the relative performance of different materials as a function of temperature and energy gap (cutoff wavelength).

4.3 COMPARISON OF FUNDAMENTAL LIMITS OF PHOTON AND THERMAL DETECTORS

In further considerations, we follow Kinch [31], assuming that the thermal generation rate of the IR material is the key parameters that enable comparison of different material systems.

The normalized dark current

$$J_{\text{dark}} = G_{th}q \tag{4.54}$$

directly determines thermal detectivity (see Equation 4.49)

$$D^* = \frac{\eta\lambda}{hc}(2G_{th})^{-1/2}. \tag{4.55}$$

The normalized dark current densities for the various materials used in infrared detector technologies in LWIR spectral region ($E_g = 0.124$ eV, $\lambda_c = 10$ μm) are shown in Figure 4.12 [32]. In addition, the *f*/2 background flux current density is also shown. The extrinsic silicon, the high-temperature superconductors (HTSC), and the photoemissive (silicon Schottky barrier) detectors are hypothetical, but are included for comparison. In the calculations carried out for different material systems we have followed the procedures used in Kinch's paper [31], except QDIPs, where the Phillips model is used [11]. The parameters representative for self-assembled InAs/GaAs QDs reported in the literature are as follows [11,33]: $\alpha_o = 5 \times 10^4$ cm^{-2},

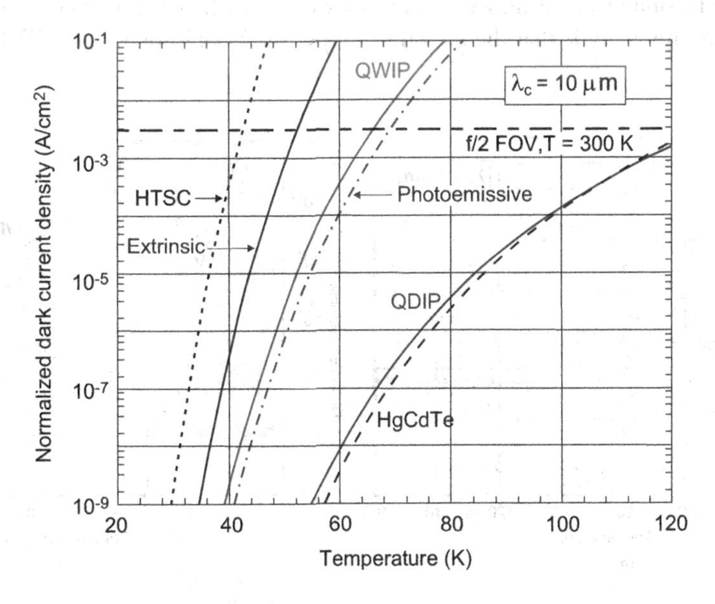

Figure 4.12 Temperature dependence of the normalized dark current of various LWIR ($\lambda_c = 10$ μm) material technologies. The *f*/2 background flux current density is also shown. (From Ref. [32])

$V = 5.3 \times 10^{-19}$ cm^{-3}, $\delta = 5 \times 10^{10}$ cm^{-2}, $\tau = 1$ ns, $N_d = 1 \times 10^{11}$ cm^{-2} and the detector thickness $t = 1/\alpha_o$. The calculations are described in detail in Martyniuk and Rogalski [33].

In Phillips's paper, an ideal QD structure is assumed with two electron energy levels (the excited state coincides with the conduction band minimum of the barrier material). Also, inhomogeneous broadening of the dot ensemble is neglected ($\sigma_{QD}/\sigma_{ens} = 1$; see Equation 4.29). Above assumptions determine the high performance of QDIPs.

In the MWIR and LWIR regions, the dominant positions have HgCdTe photodiodes. QWIPs are mainly used in LWIR tactical systems operating at lower temperature, typically 65–70 K, where cooling is not an issue. Beyond 15 μm, good performance is achieved using extrinsic silicon detectors. These detectors are termed impurity band conduction detectors and have found a niche market for the astronomy and civil space communities because HgCdTe has not yet realized its potential at low temperatures and reduced background.

Figure 4.12 displays that tunable bandgap alloy, HgCdTe, demonstrates the highest performance (the lowest dark current/thermal generation and the highest BLIP operating temperature). These estimations are confirmed by experimental data [34,35]. For very uniform QD ensembles, the QDIP performance can be close to the HgCdTe one and potentially can exceed that of HgCdTe in the region of high operation temperatures.

Figure 4.13 compares the thermal detectivities of various photodetectors with a cutoff wavelength in MWIR ($\lambda_c = 5$ μm) and LWIR ($\lambda_c = 10$ μm) regions. The assumed typical quantum efficiencies are indicated in the figure. Theoretical estimations for QDIPs are carried out assuming low quantum efficiency ≈ 2% often measured in practice. The value of 67% is typical for HgCdTe photodiodes (without antireflection coating). It should be noticed, however, that rapid progress has been made in the performance of QDIP devices, especially at near–room temperature. Lim et al. have announced a quantum efficiency of 35% for detectors with a peak detection wavelength around 4.1 μm [36].

Estimation of detectivity for InAs/GaInSb strained-layer superlattices (SLSs) are based on several theoretical papers [37–39]. Early calculations showed that an LWIR type-II InAs/GaSb SLS should have an absorption coefficient comparable to an HgCdTe alloy with the same cutoff wavelength [37]. Figure 4.13b predicts that type-II superlattices are the most efficient detectors of IR radiation in long wavelength regions. It is an even better material than HgCdTe; it is characterized by the high absorption coefficient and relatively low thermal generation rate. However, hitherto, this theoretical prediction has been not confirmed by experimental data, the main reason being the influence of the Shockley–Read generation–recombination mechanism, which causes lower carrier lifetime (higher thermal generation rate). It is clear from this analysis that the fundamental performance limitation of QWIPs is unlikely

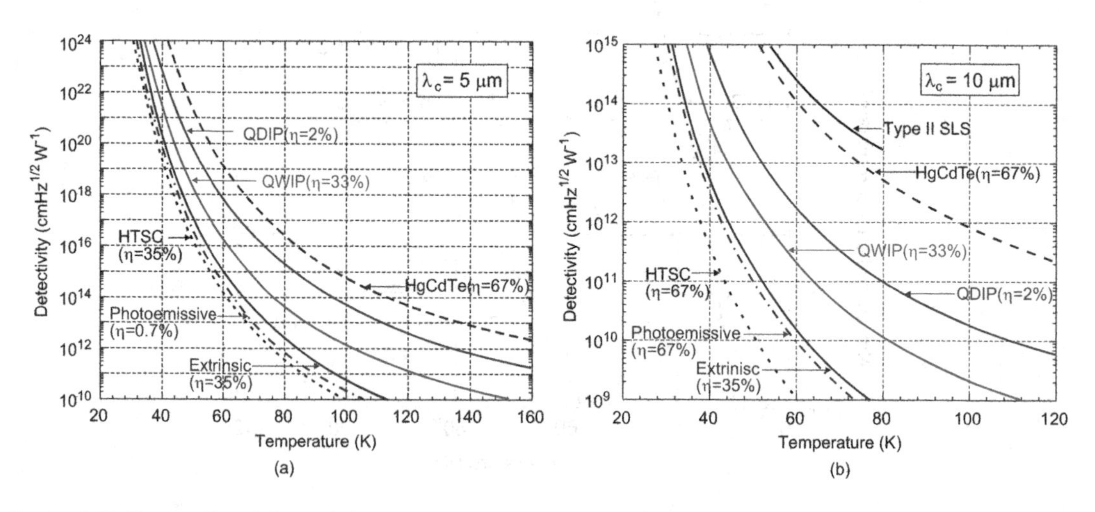

Figure 4.13 The predicted thermal detectivity versus temperature for various (a) MWIR ($\lambda_c = 5$ μm) and (b) LWIR ($\lambda_c = 10$ μm) photodetectors. The assumed quantum efficiencies are indicated. (From Ref. [32])

to rival HgCdTe photodetectors. However, the performance of very uniform QDIP [when $\sigma_{QD}/\sigma_{ens} = 1$] is predicted to rival HgCdTe. We can also notice from Figure 4.13 that AlGaAs/GaAs QWIP is better material than extrinsic silicon.

The BLIP temperature is defined as the device operating at a temperature at which the dark current is equal to the background photocurrent, given an FOV and a background temperature.

In Figure 4.14, plots of the calculated temperature required for BLIP operation in *f*/2 FOV are shown as a function of cutoff wavelength for various types of detectors. HgCdTe detectors with background-limited performance operate in practice with thermoelectric coolers in the MWIR range, but the LWIR detectors ($8 \leq \lambda_c \leq 12$ μm) operate at ≈100 K. HgCdTe photodiodes exhibit higher operating temperatures compared to extrinsic detectors, silicide Schottky barriers, QWIPs, QDIPs, and HTSCs. Type II SLSs are omitted in our considerations. The cooling requirements for QWIPs with cutoff wavelengths below 10 μm are less stringent in comparison with extrinsic detectors, Schottky-barrier devices, and HTSCs.

It has been shown by Phillips [11] that the QD detector performance may be degraded by orders of magnitude for the values of $\sigma_{ens}/\sigma_{QD} = 100$, which are indicative of the current state of QD fabrication technology. It is well-known that reduced optical absorption in QDs due to size nonuniformity results in an increase in the normalized dark current and a reduction in detectivity. The nonuniformity also has strong influence on the BLIP temperature. Increase of σ_{ens}/σ_{QD} ratio from 1 to 100 causes decrease of T_{BLIP} by several tens of degrees [33].

Due to fundamentally different types of noise, thermal and photon detectors have different dependencies of detectivities on wavelength and temperature. The temperature dependence of the fundamental limits of D^* of photon and thermal detectors for different levels of background are shown in Figure 4.15 [40]. In comparison with Kruse's paper [41], these studies are reexamined taking into account updated theories of different types of detectors.

It is evident from Figure 4.15a that in the LWIR spectral range, the performance of intrinsic IR detectors (HgCdTe photodiodes) is higher than for other types of photon detectors. HgCdTe photodiodes with background-limited performance operate at temperatures below ≈80 K. HgCdTe is characterized by high optical absorption coefficient and quantum efficiency and relatively low thermal generation rate compared to extrinsic detectors and QWIPs. The extrinsic photon detectors require more cooling than intrinsic photon detectors having the same long wavelength limit.

The theoretical detectivity value for the thermal detectors is much less temperature dependent than for the photon detectors. At temperatures below 50 K and zero background, LWIR thermal detectors are

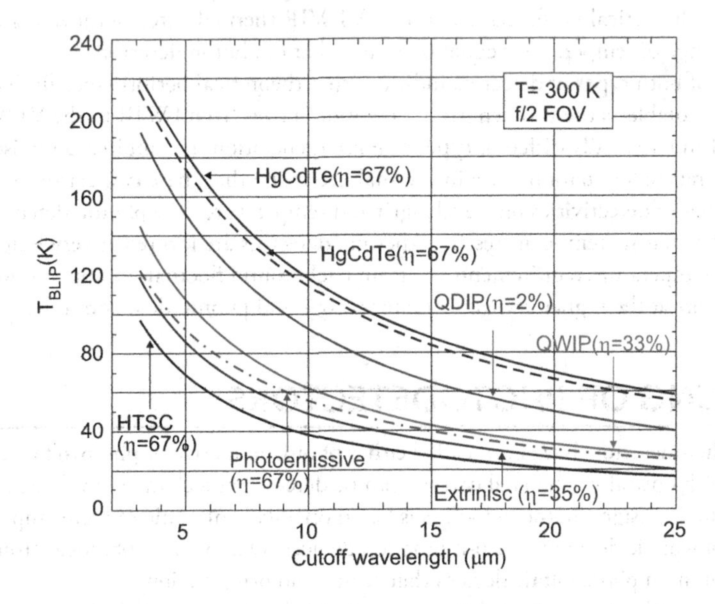

Figure 4.14 Estimation of the temperature required for background-limited operation of various types of photodetectors. (From Ref. [32])

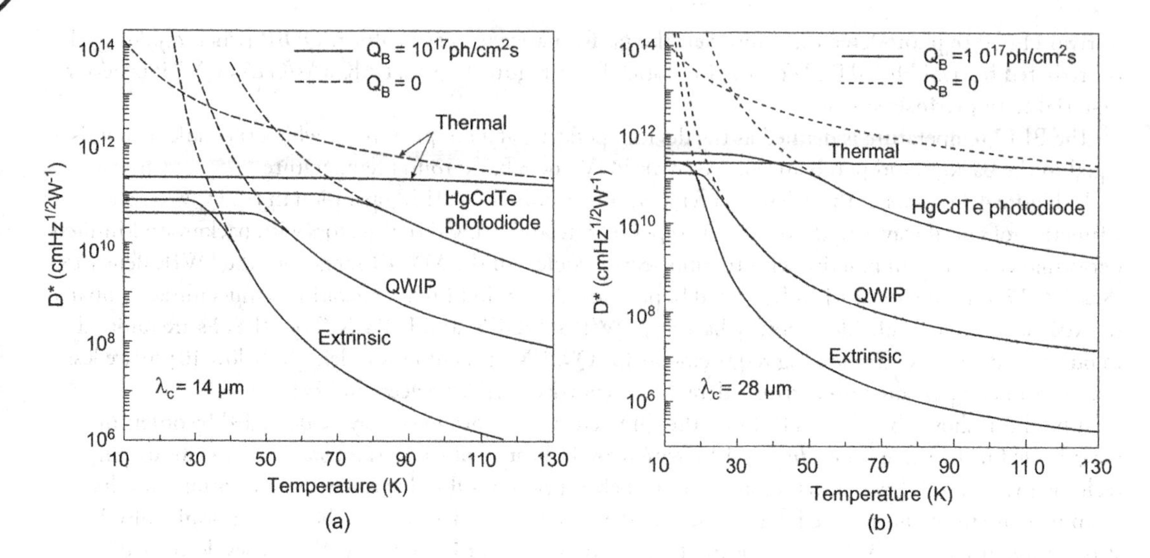

Figure 4.15 Theoretical performance limits of photon and thermal detectors as a function of detector temperature at wavelength 14 μm (a) and 28 μm (b), zero background and background of 10^{17} photons/cm²s. (From Ref. [40])

characterized by D^* values lower than those of LWIR photon detectors. However, at temperatures above 60 K, the limits favor the thermal detectors. At room temperature, the performance of thermal detectors is much better than LWIR photon detectors. The above relations are modified by the influence of background; this is shown in Figure 4.15a for a background of 10^{17} photons/cm²s. It is interesting to notice that the theoretical curves of D^* for photon and thermal detectors show similar fundamental limits at low temperatures.

Similar considerations have been carried out for VLWIR detectors operated in the 14–50 μm spectral range. The calculation results are presented in Figure 4.15b. Detectors operating within this range are cryogenic Si and Ge extrinsic photoconductors and cryogenic thermal detectors, usually bolometers. Nevertheless, theoretical prediction for intrinsic detectors (HgCdTe photodiodes) is also included. Figure 4.15b shows that the theoretical performance limit of VLWIR thermal detectors at zero and high backgrounds in a wide range of temperatures equal or exceed that of photon detectors.

The comparison of both types of detectors indicates that theoretical performance limits for thermal detectors are more favorable as the wavelength of operation moves from LWIR to the VLWIR. It is due to the influence of fundamentally different types of noise (generation–recombination noise in photon detectors and temperature fluctuation noise in thermal detectors) that these two classes of detectors have different dependencies of detectivities on wavelength and temperature. The photon detectors are favored at LWIR and lower operating temperatures. The thermal detectors are favored at very long wavelength spectral range. The temperature requirements to attain background fluctuation noise performance generally favor thermal detectors at the higher cryogenic temperatures and photon detectors at the lower cryogenic temperatures.

4.4 MODELING OF PHOTODETECTORS

Traditionally, IR photodetectors have been called either photoconductive or photovoltaic detectors based on the principle by which optically generated carriers can be detected as a change in voltage or current across the element. The simple design photoconductors is based on a flake of semiconductor supplied with ohmic contacts, while photovoltaic detectors are p-n junction devices. Dember and photoelectromagnetic effect detectors are less common photovoltaic devices that require no p-n junction.

However, advances in heterostructure devices such as the development of the heterojunction photoconductors, double-layer heterostructure photodiodes, and introduction of nonequilibrium mode of operation make this distinction unclear. Moreover, the photovoltaic structures are frequently biased, exhibiting signal due to both photovoltages at junctions and from the photoconductivity of some regions.

An optimized photodetector (see Figure 4.9) of any type may be a 3-D monolithic heterostructure that consists of the following regions [18,22,42]:

- Concentrator of IR radiation that directs incident radiation onto absorber (an example is an immersion lens made of a wide gap semiconductor).
- Absorber of IR radiation where the generation of free carriers occurs (this is a narrow gap semiconductor with bandgap, doping, and geometry selected for the highest ratio of the optical-to-thermal generation rates).
- Contacts to the absorber, which sense optically generated charge carriers (contacts should not contribute to the dark current of the device; examples are wide gap heterojunction contacts used in the modern devices).
- Passivation of the absorber (the surfaces of the absorber must be insulated from the ambient by a material that also doesn't contribute to carrier generation); in addition, passivation repels the carriers optically generated in absorber, keeping them away from surfaces where recombination can reduce the quantum efficiency.
- Retroreflector to enhance absorption (examples are metal or dielectric mirrors; optical resonant cavity structures can also be used).

The above conditions can be fulfilled using heterojunctions like N^+-p-p^+ and P^+-n-n^+ with heavily doped contact regions (symbol "+" denotes strong doping, capital letter denotes wider gap). Homojunction devices (like n-p, n^+-p, p^+-n) suffer from surface problems; excess thermal generation results in increased dark current and recombination, which reduces photocurrent.

Modeling of the photodetectors is a strategically important task necessary to understand photodetector properties and optimize their design. Analytical models were developed for specific types of IR devices based on idealized structures, operating in equilibrium and nonequilibrium modes. These models make some features of the devices' operation easy to understand and analyze.

In general, however, the operation of the advanced devices can no longer be described by analytical models. Omitting specific features of the narrow gap materials, such as degeneracy and nonparabolic conduction band, may result in enormous errors. The nonequilibrium mode of operation of infrared photodetectors brings further complications. The devices are based on an absorber that is near intrinsic or just extrinsic at the operating temperature. The properties of the device differ from those with extrinsic absorber [43]. First, drift and diffusion are dominated by ambipolar effects due to space charge coupling between electrons and holes. Second, the concentration of charge carriers in near-intrinsic materials can be driven to levels considerably below intrinsic concentrations. As a result, the perturbation can be described only in terms of large signal theory. Third, carrier concentration in low bandgap materials is dominated by Auger generation and recombination.

An accurate description of more and more complex device architectures including doping and bandgap grading, heterojunctions, 2-D and 3-D effects, ambipolar effects, nonequilibrium operation, and surface, interface, and contact effects can be achieved only with a solution of the fundamental equations that describe the electrical behavior of semiconductor devices. These partial differential equations include continuity equations for electrons and holes and Poisson equation

$$\frac{dn}{dt} = \frac{1}{q}\vec{\nabla} \times \vec{J}_n + G_n - R_n, \tag{4.56}$$

$$\frac{dp}{dt} = \frac{1}{q}\vec{\nabla} \times \vec{J}_p + G_p - R_p, \tag{4.57}$$

$$\varepsilon_0 \varepsilon_r \nabla^2 \psi = -q\left(N_d^+ - N_a^- + p + n\right) - \rho_s, \tag{4.58}$$

where Ψ is the electrostatic potential defined as the intrinsic Fermi potential, ρ_s is the surface charge density, N_d^+ and N_a^- are concentrations of ionized donors and acceptors.

The solution of the Equations set 4.56 through 4.58 makes it possible to analyze stationary and transient phenomena in semiconductor devices. The main problem with the solution of these equations is their non-linearity and complex dependences of their parameters. In many cases, some simplifications are possible. From Boltzmann transport theory, the current densities $\vec{J}_n$ and $\vec{J}_p$ can be written as functions of the carrier concentrations and the quasi-Fermi potentials for electrons and holes, Φ_n and Φ_p

$$\vec{J}_n = -q\mu_n n\vec{\nabla}_n, \tag{4.59}$$

$$\vec{J}_p = -q\mu_p n\vec{\nabla}_p. \tag{4.60}$$

Alternatively, $\vec{J}_n$ and $\vec{J}_p$ can be written as functions of Ψ, n, and p, consisting of drift and diffusion components

$$\vec{J}_n = q\mu_n \vec{E}_e + qD_n\vec{\nabla}n, \tag{4.61}$$

and

$$\vec{J}_p = q\mu_p \vec{E}_h + qD_p\vec{\nabla}p, \tag{4.62}$$

where D_n and D_p are the electron and hole diffusion coefficients.

If the effects of bandgap narrowing are neglected and the Boltzmann carrier statistic is assumed

$$\vec{E}_n = \vec{E}_p = \vec{E} = -\vec{\nabla}\Psi. \tag{4.63}$$

The steady-state behavior of 1-D devices can be described by the set of five differential equations with suitable boundary conditions: two transport equations for electrons and holes, two continuity equations for electrons and holes, and the Poisson equation, which are all related to Van Roosbroeck [44]:

$$J_n - qD_n\frac{dn}{dx} - q\mu_n n\frac{d\Psi}{dx}, \quad \text{current transport for electrons} \tag{4.64}$$

$$J_p = qD_p\frac{dp}{dx} - q\mu_p p\frac{d\Psi}{dx}, \quad \text{current transport for holes} \tag{4.65}$$

$$\frac{1}{q}\frac{dJ_n}{dx} + (G - R) = 0, \quad \text{continuity equation for electrons} \tag{4.66}$$

$$\frac{1}{q}\frac{dJ_p}{dx} - (G - R) = 0, \quad \text{continuity equation for holes} \tag{4.67}$$

$$\frac{d^2\Psi}{dx^2} = -\frac{q}{\varepsilon_o\varepsilon_r}\left(N_d^+ - N_a^- + p - n\right) \quad \text{Poisson's equation.} \tag{4.68}$$

Many papers devoted to the solution of these equations have been published, from papers of Gummel [45] and de Mari [46] to recent commercially available numerical programs. The fundamental equations cannot be solved analytically without the approximations, even for the 1-D steady-state case. Therefore, the numerical solutions must be applied. The numerical solution is composed of three steps: (i) grid generation step, (ii) discretization to transform the differential equations into the linear algebraic equations, and

(iii) the solution. The Newton direct method is usually used to solve the matrix equation [47]. Other methods are also used to improve the convergence speed and reduce the number of iterations [48].

Since experiments with complex device structures are complicated, costly, and time-consuming, numerical simulations have become a critical tool to develop advanced detectors [49]. Some laboratories have developed suitable software; for example, Stanford University (USA), Military Technical University (Poland) [50], Honyang University (South Korea) [48], and others. Commercial simulators are available from several sources, including Medici (Technology Modeling Associates), Semicad (Dawn Technologies), Atlas/Blaze/Luminouse (Silvaco International, Inc.), APSYS (Crosslight Software, Inc.), and others. The APSYS, for example, is a full 2-D/3-D simulator that solves not only the Poisson equation and the current continuity equations (including such features as field-dependent mobilities and avalanche multiplication), but also the scalar wave equation for photonic waveguiding devices (such as waveguide photodetectors) and heat transfer equations with flexible thermal boundary conditions and arbitrary temperature-dependent parameters.

Although existing simulators still do not fully account for all semiconductor properties important for photodetectors, they are already invaluable tools for analysis and development of improved infrared IR photodetectors. In addition to device simulators, process simulators are being developed that facilitate advanced device growth technologies development [51,52].

REFERENCES

1. A. Rogalski, *Infrared Detectors*, Gordon and Breach, Amsterdam, 2000.
2. J. T. Houghton and S. D. Smith, *Infra-Red Physics*, Oxford University Press, Oxford, UK, 1966.
3. E. L. Dereniak and G. D. Boreman, *Infrared Detectors and Systems*, Wiley, New York, 1996.
4. J. Piotrowski, "Breakthrough in infrared technology: The micromachined thermal detector arrays," *Opto-Electronics Review* 3, 3–8, 1995.
5. A. Smith, F. E. Jones, and R. P. Chasmar, *The Detection and Measurement of Infrared Radiation*, Clarendon, Oxford, 1968.
6. W. Kruse, L. D. McGlauchlin, and R. B. McQuistan, *Elements of Infrared Technology*, Wiley, New York, 1962.
7. P. G. Datskos, "Detectors: Figures of Merit," in *Encyclopedia of Optical Engineering*, ed. R. Driggers, 349–57, Marcel Dekker, New York, 2003.
8. F. J. Low and A. R. Hoffman, "The detectivity of cryogenic bolometers," *Applied Optics* 2, 649–50, 1963.
9. E. H. Putley, "Thermal detectors," in *Optical and Infrared Detectors*, ed. R. J. Keyes, 71–100, Springer, Berlin, 1977.
10. B. F. Levine, "Quantum-well infrared photodetectors," *J. Appl. Phys.*, 74, R1–R81, 1993.
11. J. Phillips, "Evaluation of the fundamental properties of quantum dot infrared detectors," *Journal of Applied Physics* 91, 4590–94, 2002.
12. J. Singh, *Electronic and Optoelectronic Properties of Semiconductor Structures*, Cambridge University Press, New York, 2003.
13. A. Rogalski, "Infrared detectors: Status and trends," *Prog. Quant. Electron.* 27, 59–210, 2003.
14. A Rogalski, "Photon detectors," in *Encyclopedia of Optical Engineering*, ed. R. Driggers, 1985–2036, Marcel Dekker Inc., New York, 2003.
15. P. Norton, "Detector focal plane array technology," in *Encyclopedia of Optical Engineering*, ed. R. Driggers, 320–8, Marcel Dekker Inc., New York, 2003.
16. J. Piotrowski, "$Hg_{1-x}Cd_xTe$ Infrared photodetectors," in *Infrared Photon Detectors*, Vol. PM20, 391–494, SPIE Press, Bellingham, WA, 1995.
17. J. Piotrowski and A. Rogalski, Comment on "Temperature limits on infrared detectivities of $InAs/In_xGa_{1-x}Sb$ superlattices and bulk $Hg_{1-x}Cd_xTe$" *J. Appl. Phys.*, 74, 4774, 1993, *J. Appl. Phys.*, 80(4), 2542–4, 1996.
18. J. Piotrowski and A. Rogalski, "New generation of infrared photodetectors," *Sens. Actuators*, A67, 146–52, 1998.
19. J. Piotrowski, "Uncooled operation of IR photodetectors," *Opto-Electron. Rev.*, 12, 111–22, 2004.
20. T. Ashley and C. T. Elliott, "Non-equilibrium mode of operation for infrared detection," *Electron. Lett.*, 21, 451–2, 1985.
21. T. Ashley, T. C. Elliott, and A. M. White, "Non-equilibrium devices for infrared detection," *Proc. SPIE.*, 572, 123–32, 1985.
22. J. Piotrowski and A. Rogalski, *High-Operating Temperature Infrared Photodetectors*, SPIE Press, Bellingham, WA, 2007.

23. S. Jensen, "Temperature limitations to infrared detectors," *Proc. SPIE*, 1308, 284–92, 1990.
24. R. G. Humpreys, "Radiative lifetime in semiconductors for infrared detectors," *Infrared Phys*., 23, 171–5, 1983.
25. R. G. Humpreys, "Radiative lifetime in semiconductors for infrared detectors," *Infrared Phys*., 26, 337–42, 1986.
26. A. Rogalski, "HgCdTe Infrared detector material: History, status, and outlook," *Rep. Prog. Phys*., 68, 2267–336, 2005.
27. A. Rogalski, "Quantum well photoconductors in infrared detectors technology," *J. Appl. Phys*. 93, 4355–91, 2003.
28. D. Long, "Photovoltaic and photoconductive infrared detectors," in *Optical and Infrared Detectors*, ed. R. J. Keyes, 101–47, Springer-Verlag, Berlin, 1977.
29. C. T. Elliott and N. T. Gordon, "Infrared detectors," in *Handbook on Semiconductors*, Vol. 4, ed. C. Hilsum, 841–936, North-Holland, Amsterdam, 1993.
30. C. T. Elliott, "Photoconductive and non-equilibrium devices in HgCdTe and related alloys," in *Infrared Detectors and Emitters: Materials and Devices*, eds. P. Capper and C. T. Elliott, 279–312, Kluwer Academic Publishers, Boston, MA, 2001.
31. M. A. Kinch, "Fundamental physics of infrared detector materials," *J. Electron. Mater*., 29, 809–17, 2000.
32. P. Martyniuk and A. Rogalski, "Quantum-dot infrared photodetectors: Status and outlook," *Prog. Quant. Electron*., 32, 89–120, 2008.
33. P. Martyniuk and A. Rogalski, "Insight into performance of quantum dot infrared photodetectors," *Bull. Pol. Acad. Sci.: Tech. Sci*., 57, 103–16, 2009.
34. A. Rogalski, K. Adamiec, and J. Rutkowski, *Narrow-Gap Semiconductor Photodiodes*, SPIE Press, Bellingham, WA, 2000.
35. M. A. Kinch, *Fundamentals of Infrared Detector Materials*, SPIE Press, Bellingham, WA, 2007.
36. H. Lim, S. Tsao, W. Zhang, and M. Razeghi, "High-performance InAs quantum-dot infrared photoconductors grown on InP substrate operating at room temperature," *Appl. Phys. Lett*., 90, 131112, 2007.
37. D. L. Smith and C. Mailhiot, "Proposal for strained type II superlattice infrared detectors," *J. Appl. Phys*., 62, 2545–8, 1987.
38. C. H. Grein, H. Cruz, M. E. Flatte, and H. Ehrenreich, "Theoretical performance of very long wavelength InAs/$In_xGa_{1-x}Sb$ superlattice based infrared detectors," *Appl. Phys. Lett*., 65, 2530–2, 1994.
39. C. H. Grein, P. M. Young, M. E. Flatté, and H. Ehrenreich, "Long wavelength InAs/InGaSb infrared detectors: Optimization of carrier lifetimes," *J. Appl. Phys*., 78, 7143–52, 1995.
40. R. Ciupa and A. Rogalski, "Performance limitations of photon and thermal infrared detectors," *Opto-Electron. Rev*., 5, 257–66, 1997.
41. P. W. Kruse, "A comparison of the limits to the performance of thermal and photon detector imaging arrays," *Infrared Phys. Technol*., 36, 869–82, 1995.
42. J. Piotrowski and A. Rogalski, "Uncooled long wavelength infrared photon detectors," *Infrared Phys. Technol*., 46, 115–31, 2004.
43. M. White, "Auger suppression and negative resistance in low gap diode structures," *Infrared Phys*., 26, 317–24, 1986.
44. W. Van Roosbroeck, "Theory of the electrons and holes in germanium and other semiconductors," *Bell Sys. Tech. J*., 29, 560–607, 1950.
45. H. K. Gummel, "A self-consistent iterative scheme for one-dimensional steady state transistor calculations," *IEEE Trans. Electron Devices*, ED 11, 455–65, 1964.
46. A. De Mari, "An accurate numerical steady-state one-dimensional solution of the p-n junction," *Solid State Electron*., 11, 33–58, 1968.
47. M. Kurata, *Numerical Analysis of Semiconductor Devices*, Lexington Books, DC Heath, Lexington, MA, 1982.
48. S. D. Yoo, N. H. Jo, B. G. Ko, J. Chang, J. G. Park, and K. D. Kwack, "Numerical simulations for HgCdTe related detectors," *Opto-Electron. Rev*. 7, 347–56, 1999.
49. K. Kosai, "Status and application of HgCdTe device modeling," *J Electron. Mater*. 24, 635–40, 1995.
50. K. Jóźwikowski, "Numerical modeling of fluctuation phenomena in semiconductor devices," *J. Appl. Phys*. 90, 1318–27, 2001.
51. J. L. Meléndez and C. R. Helms, "Process modeling and simulation of $Hg_{1-x}Cd_xTe$. Part I: Status of Stanford university mercury cadmium telluride process simulator," *Journal of Electronic Materials* 24, 565–71, 1995.
52. J. L. Meléndez and C. R. Helms, "Process modeling and simulation for $Hg_{1-x}Cd_xTe$. Part II: Self-diffusion, interdiffusion, and fundamental mechanisms of point-defect interactions in $Hg_{1-x}Cd_xTe$," *J. Electron. Mater*., 24, 573–9, 1995.

5 Coupling of infrared radiation with detectors

There are different methods of light coupling in a photodetector to enhance quantum efficiency [1]. A notable example of methods described for thin-film solar cells [2,3] can be applied for infrared photodetectors. In general, they can be divided into four groups, as is shown in Figure 5.1, using optical concentration, antireflection structures, optical path increase, and light localization.

5.1 STANDARD COUPLING

Semiconductor materials used for photodetectors have large values of refractive index and thus large values of reflection coefficient at the device surface. This reflection is minimized using antireflection structures. The simplest way to enhance absorption is the use of retroreflector for the double pass of infrared radiation. In thin devices, the quantum efficiency can be significantly enhanced using interference phenomena to set up a resonant cavity within the photodetector [1,4]. Various optical resonator structures are used. In the simplest method, interference occurs between the waves reflected at the rear, highly reflective surface and at the front surface of the semiconductor. The thickness of the semiconductor is selected to set up the standing waves in the structure with peaks at the front and nodes at the back surface. The quantum efficiency oscillates with thickness of the structure, with the peaks at a thickness corresponding to an odd multiple of $\lambda/4n$, where n is the refractive index of the semiconductor. The gain in quantum efficiency increases with n. With the use of interference effects, a strong and highly nonuniform absorption can be achieved even for long wavelength radiation with a low absorption coefficient.

Another possible way of improving the performance of IR photodetector is to increase the apparent "optical" size of the detector in comparison with the actual physical size using a suitable concentrator that compresses impinging IR radiation. The concentration efficiency can be then defined as the ratio between the optical and the electrical area, minus absorption and scattering losses (see Section 4.2.2).

This must be achieved without the reduction of acceptance angle, or at least, with limited reduction to angles required for fast optics of IR systems. Various types of suitable optical concentrators can be used including optical cones, conical fibers, and other types of reflective, diffractive, and refractive optical concentrators [5].

An efficient way to achieve an effective concentration of radiation is utilizing an immersion lens. There are a number of different structures to serve the same purpose. Roughly, one could divide them into refractive, reflective, and diffractive elements, although hybrid solutions are also possible. The examples of solutions are shown in Figure 5.2. Microlenses monolithically integrated with detectors, typically used in charge-coupled device (CCD) and complementary metal oxide semiconductor (CMOS) active pixel imagers for visible application, concentrate the incoming light into the photosensitive region when they are accurately deposited over each pixel (see Figure 5.2a). When the fill factor is low and microlenses are not used, the light falling elsewhere is either lost or, in some cases, creates artifacts in the imagery by generating electrical currents in the active circuitry. An example of the concept of uncooled infrared array sensor with microlenses is shown in Figure 5.2b [6].

The principle of operation of a hemispherical immersion lens is shown in Figure 5.3 [5]. The detector is located at the center of curvature of the immersion lens. The lens produces an image of the detector. No spherical or coma aberration exists (aplanatic imaging). Due to immersion, the apparent linear size of the detector increases by a factor of n. The image is located at the detector plane. The use of a hemispheric immersion lens in combination with an objective lens of an imaging optical system is shown in Figure 5.3b. The immersion lens plays the role of a field lens, which increases the FOV of the optical system.

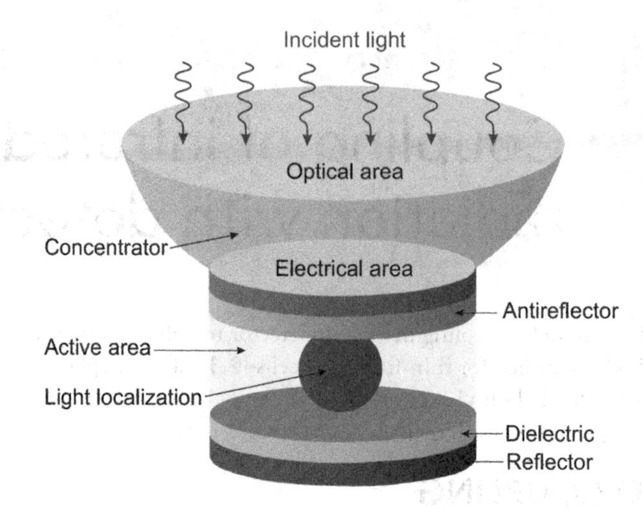

Figure 5.1 Methods of absorption enhancement in a photodetector using optical concentrator, antireflection structure, structures for optical path increase (cavity enhancement), and light localization structures.

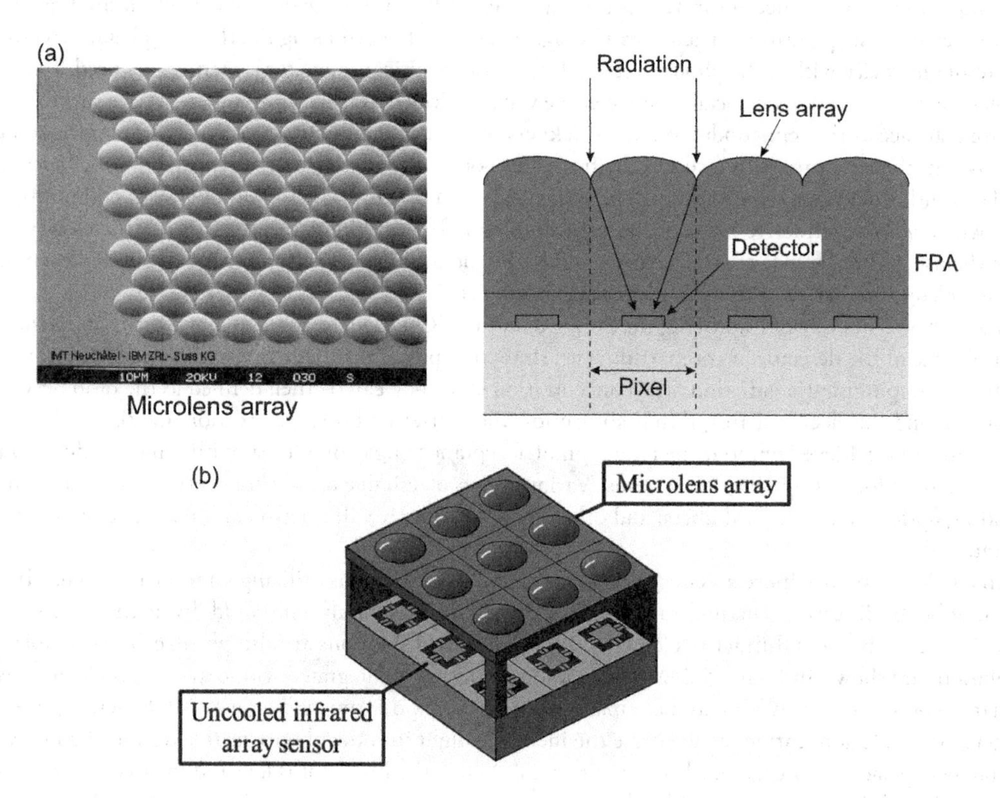

Figure 5.2 Microlenses for infrared arrays sensors: (a) micrograph of and cross-sectional drawing of microlensed FPA, (b) concept of uncooled infrared array sensor with microlenses.

The limit to the compression is determined by the Lagrange invariant ($A\,\Omega$ product) and the sin condition for an aplanatic system [5]. In air, the physical and the apparent size of the detector are related by the equation

$$n^2 A_e \sin\theta' = A_o \sin\theta, \tag{5.1}$$

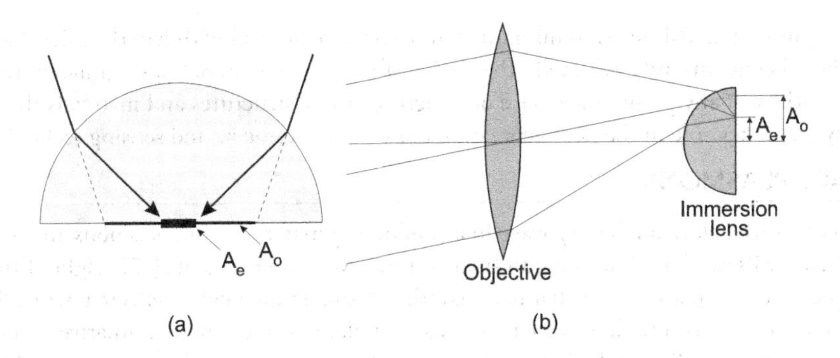

Figure 5.3 Principle of optical immersion (a), and ray tracing for an optical system with a combination of an objective lens and an immersion lens (b).

where n is the lens refractive index; A_e and A_o are the physical and apparent size of the detector, respectively; and θ and θ' are the marginal ray angles before refraction at the lens and at the image, respectively. Therefore,

$$\frac{A_o}{A_e} = n^2 \frac{\sin\theta'}{\sin\theta}. \tag{5.2}$$

For the hemispherical lens, the marginal ray angles are 90°. Therefore, the area gain is n^2. Larger gain can be obtained for a hyper-hemisphere used as an aplanatic lens [5]. This results in an apparent increase of linear detector size by a factor of n^2. In this case, the image plane is shifted.

An alternative approach is the use of a compound parabolic concentrator (also called a Winston collector or Winston cone) [7,8]. QinetiQ has developed a micromachining technique involving dry etching to fabricate the cone concentrators for the detector and luminescent devices [9,10].

As mentioned earlier, the simplest way of radiation coupling with the detector-active region is to place a mirror at the backside of the detector, thus doubling the optical path through the active region. At present, however, much more sophisticated methods are used including different cavities with reflective walls, as well as surface structuring. These methods belong to the optical trapping approaches. The advent of nanophotonics enabled optical trapping such as the subwavelength optical localization utilizing plasmonic nanocomposites. Some metal-dielectric structures ensure the possibility of light localization on a level much smaller than the operating wavelength.

Advances in optoelectronics-related materials science, such as metamaterials and nanostructures, have opened doors for new nonclassical approaches to device design methodologies, which are expected to offer enhanced performance along with reduced product cost for a wide range of applications. Surface plasmons (SPs) are widely recognized in the field of surface science following the pioneering work of Ritchie in 1950s [11]. The relative ease of manipulating SPs opens an opportunity for their applications to photonics and optoelectronics for scaling down optical and electronic devices to nanometric dimensions. For the first time, it is possible to reliably control light at the nanoscale. Additionally, plasmonics takes advantage of the very large (and negative) dielectric constant of metals, to compress the wavelength and enhance electromagnetic (EM) fields in the vicinity of metal conductors. Coupling light into semiconductor materials remains a challenging and active research topic. Micro- and nanostructured surfaces have become a widely used design tool to increase light absorption and enhance the performance of broadband detectors without employing antireflection coatings.

The methods of photodetector enhancement are presented in the next section.

5.2 PLASMONIC COUPLING

New solutions arising from the use of plasmonic structures open novel avenues for photodetectors development [12–16]. The goal of infrared plasmonics is to increase the absorption in a given volume of detector's material. As mentioned in Section 4.2, smaller volumes provide lower noise, while higher absorption results in a stronger output signal. This leads to miniaturized detector structures with length scales that are much smaller than those being currently achieved. The choice of plasmonic material has significant implications for the ultimate utility of any plasmonic device or structure. These structures and materials that support SP excitations may play a key role in the next-generation optical interconnects and sensing technologies.

5.2.1 SURFACE PLASMONS

A plasmon is a quantized electron density wave in a conducting material. Bulk plasmons are longitudinal excitations, whereas SPs can have both longitudinal and transverse components [12]. Light of frequency below the frequency of the plasmon for that material (the plasma frequency) is reflected, while light above the plasma frequency is transmitted. Surface plasmons on a plane surface are nonradiative EM modes, that is, they cannot be generated directly by light nor can they decay spontaneously into photons. However, if the surface is rough, or has a grating on it, or is patterned in some way, light around the plasma frequency couples strongly with the SPs, creating what is called a polariton, or a surface plasmon polariton (SPP)—a transverse-magnetic optical surface wave that may propagate along the surface of a metal until energy is lost either via absorption in the metal or radiation into free space.

In its simplest form, an SPP is an EM excitation (coupled EM field/charge-density oscillation) that propagates in a wave-like fashion along the planar interface between a metal and a dielectric medium and whose amplitude decays exponentially with increasing distance into each medium from the interface. Thus, the SPP is a surface EM wave, whose field is confined to the near vicinity of the dielectric–metal interface. This confinement leads to an enhancement of the field at the interface, resulting in an extraordinary sensitivity of the SPP to surface conditions. The intrinsically two-dimensional (2D) nature of SPPs prohibits them from directly coupling to light. Usually, a surface metal grating is required for the excitation of SPPs by normally incident light. Moreover, since the EM field of an SPP decays exponentially with distance from the surface, it cannot be observed in conventional (far-field) experiments unless the SPP is transformed into light by its interaction with a surface grating.

Schematic representation of an electron density wave propagating along a metal-dielectric interface is shown in Figure 5.4. The charge-density oscillations and associated EM fields comprise SPP waves. The local electric field component is enhanced near the surface and decays exponentially with distance in a direction normal to the interface.

The interaction between the surface charge density and the EM field results in the momentum of the SP mode, $\hbar k_{SP}$, being greater than that of a free-space photon of the same frequency, $\hbar k_o$ ($k_o = \omega/c$ is the free space wavevector)—see Figure 5.4(c). Solving Maxwell's equations under the appropriate boundary conditions yields the SP dispersion relation, that is the frequency-dependent SP wavevector [17]

$$k_{SP} = k_o\left(\frac{\varepsilon_d\varepsilon_m}{\varepsilon_d + \varepsilon_m}\right)^{1/2}. \tag{5.3}$$

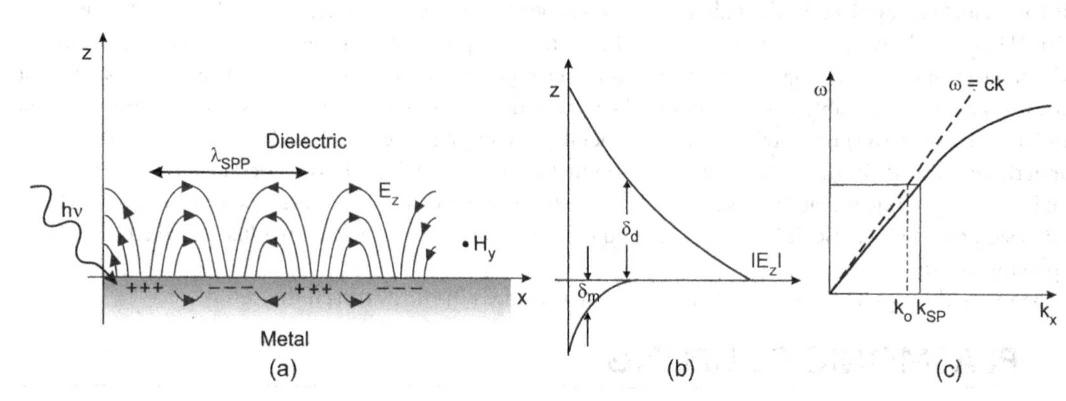

Figure 5.4 (a) Schematic illustration of electromagnetic wave and surface charges at the interface between the metal and the dielectric material; (b) the local electric field component is enhanced near the surface and decays exponentially with distance in a direction normal to the interface, and (c) dispersion curve for an SP mode.

The frequency-dependent permittivity of the metal, ε_d, and the dielectric material, ε_m, must have opposite signs if SPs are to be possible at such an interface. This condition is satisfied for metals because ε_m is both negative and complex (the latter corresponding to absorption in the metal). The increase in $\hbar k_{SP}$ momentum is associated with the binding of the SP to the surface (the resulting momentum mismatch between light and SPs of the same frequency must be bridged if light is to be used to generate SPs).

In contrast to the propagating nature of SPs along the surface, the field perpendicular to the surface decays exponentially with distance from the surface. This field is evanescent in nature, as a consequence of the bound, nonradiative SPs, which prevents power from propagating away from the surface.

SP mode propagates on a flat metal surface with gradual attenuation owing to losses arising from absorption in the metal. The propagation length, δ_{SP}, can be found from the SP dispersion equation as [18]

$$\delta_{SP} = \frac{1}{2k''_{SP}} = \frac{\lambda}{2\pi}\left(\frac{\varepsilon'_m + \varepsilon_d}{\varepsilon'_m \varepsilon_d}\right)^{3/2} \frac{(\varepsilon'_m)^2}{\varepsilon''_m}, \tag{5.4}$$

where k''_{SP} is the imaging part of the complex SP wavevector, $k_{SP} = k'_{SP} + ik''_{SP}$; ε'_m and ε''_m are the real and imaging parts of the dielectric function of the metal, $\varepsilon_m = \varepsilon'_m + i\varepsilon''_m$. The propagation length is dependent on the dielectric constant of the metal and the incident wavelength.

SPs were first studied in the visible region. The vast majority of plasmonics research has focused on the shorter wavelength side of the optical frequency range. In the visible spectrum, silver is the metal with lowest losses, where propagation distances are typically in the range of 10–100 μm. In addition, for a longer incident wavelength, such as the near-infrared telecommunication wavelength of 1.55 μm, the propagation length of silver increases toward 1 mm. For a relatively absorbing metal such as aluminium, the propagation length is 2 μm at a wavelength of 500 nm [14].

It appears that common metals such as gold or silver have plasmon resonances in blue or deep ultraviolet wavelength ranges. An increased research effort has recently been directed to the infrared. However, when moving from the visible to the infrared range, metal films with arrays of holes that ordinarily show optical transmission are quite opaque. There are no metals available whose plasmon resonances are in the IR range under 10 microns in wavelength. Moreover, the integration of plasmonic structure with active detector region is intrinsically incompatible due to the low-quality metal deposition techniques in comparison with the high-quality epitaxial growth of semiconductors or dielectrics. As a result, many intrinsic plasmonic properties can be masked by the poor metal quality or poor semiconductor-metal interfaces. In addition, wavelength tuneability is difficult to realize since the plasmonic resonance frequency is fixed for a given metal. Thus, other alternatives to metals such as highly doped semiconductors have been proposed; for example, InAs/GaSb bilayer structure [16,19].

Figure 5.5 shows schematically the difference in surface-enhanced absorption between visible and infrared wavelengths. As shown in Figure 5.5(a), the optical field of an SPP is closely bound to the material surface at visible wavelengths but weakly bound at mid-IR wavelengths. Metal surfaces with indentations or holes can give rise to leaky waveguides (spoof plasmonics) or can be used to couple light into dielectric waveguides, as is shown in Figure 5.5(b). This effective approach is based on a deep subwavelength pitch grating in the surface of the metal [20]. This design leads not only to resonance in the grating but also to extremely tight confinement of light and is used for improving the performance of quantum well infrared photodetector (QWIP) and quantum dot infrared photodetectors (QDIP) detectors [21,22].

Current plasmonic devices at telecommunication and optical frequencies face significant challenges due to losses encountered in the constituent plasmonic materials. These large losses seriously limit the practicality of these metals for many novel applications. Apart from traditional plasmonic materials (the noble metals), the newer material systems with infrared plasma wavelengths [transition metal nitrides, transparent conducting oxides (TCOs), and silicides] are considered [16,23,24]. TCOs have been shown to be effective plasmonic materials in the infrared region while transition metal nitrides extend into the visible spectrum. Figure 5.6 shows various classes of materials that are grouped using two important parameters that determine the optical properties of conducting materials: the carrier concentration and carrier mobility. In plasmonics, the carrier concentration has to be high enough to provide a negative real part of the dielectric

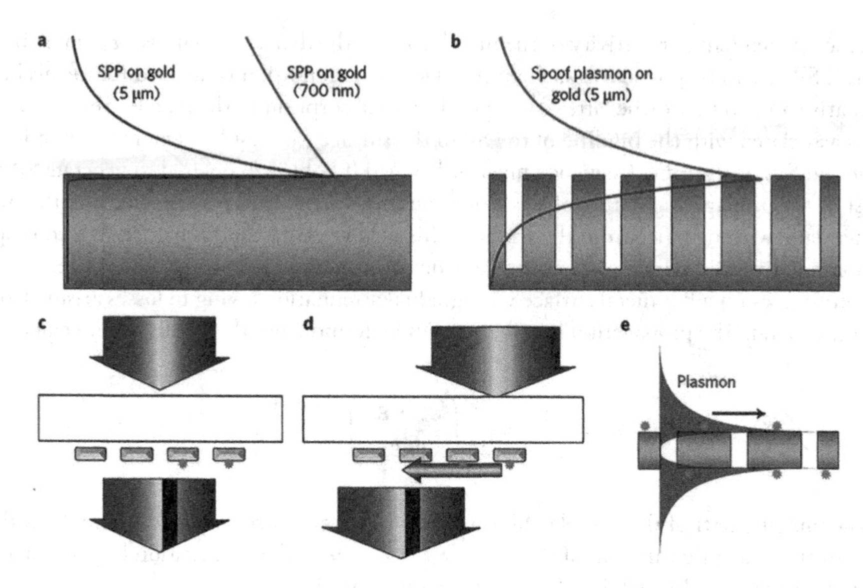

Figure 5.5 Optical field of SPs and surface-enhanced infrared absorption. (a) The optical field of an SPP is closely bound to the material surface at visible wavelengths (700 nm) but weakly bound at mid-IR wavelengths (5 μm). (b) Optical field of spoof plasmons on gold at a wavelength of 5 μm, showing strong confinement. (c) Chemicals (stars) on gold islands exhibit better absorption than an unstructured substrate. (d) Using SPs to increase the absorption enhancement. (e) SP-enhanced infrared absorption in a hole array. Plasmons bound to the surface interact with the molecules deposited on and inside the hole array. (Adapted after Ref. [13])

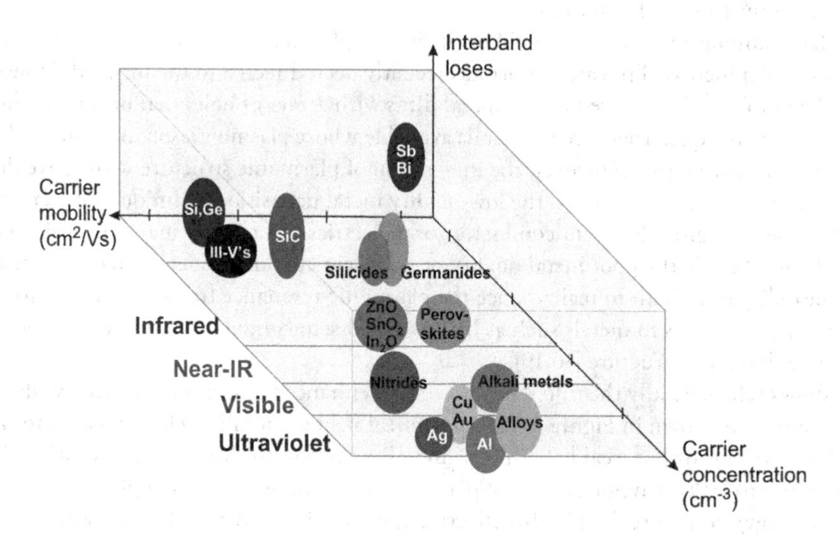

Figure 5.6 Material space for plasmonics and metamaterial applications. Important material parameters such as carrier concentration (maximum doping concentration for semiconductors), carrier mobility, and interband losses form the optimization phase space for various applications. While spherical bubbles represent materials with low interband losses, elliptical bubbles represent those with larger interband losses in the corresponding part of the electromagnetic spectrum. (Adapted after Ref. [24])

permittivity. In addition, tunability of dielectric permittivity values with a change in carrier concentration is desirable. Lower carrier mobilities could indicate higher damping losses and thus higher material losses.

5.2.2 PLASMONIC COUPLING OF INFRARED DETECTORS

Different architectures are used to support SPPs on metalo-dielectric structures involving planar metal waveguides; metal gratings; nanoparticles such as islands, spheres, rods, and antennas; or optical transmission through one or many subwavelength holes in a metal film. However, great challenges still remain to fully realize many promised potentials.

Figure 5.7 presents three popular geometries to enhance the detector's photoresponse: (a) grating couplers to convert incident light to SPPs, which are focused inside a small-scale detector; (b) particle antenna on a small-scale detector; and (c) metallic photonic crystal (PC) structures to enhance the photoresponse. The inclusion of an antenna or resonator enhances the photoresponse or makes the detector wavelength and polarization-specific.

In the first architecture, a nanoscale semiconductor photodetector is discussed. Small area photodetectors benefit from low noise levels, a low-junction capacitance, and a possible high-speed operation. However, due to the decrease of the active area of the semiconductor detector under the same optical power density, a lower output is obtained. A nanoantenna in close proximity to the active material of a photodetector allows us to take advantage of the concentrated plasmonic fields [25]. Its role is to convert free-space plane waves into SPs bound to a patterned metal surface without reflection. The antenna-like structure to couple incident radiation to SPs is a technique very popular for THz detectors [26].

In Figure 5.7(b), we have shown an integrating detector nanoscale structure whose photoresponse is enhanced by a local plasmon resonance. Resonant antennas can confine strong optical fields inside a subwavelength volume. By designing the structure in such a way that the region with highly confined optical fields overlaps with the active region of the photodetector, a strong enhancement of the photocurrent can be achieved using both SPP and localized SPP (LSPP) resonances.

LSSPs are charge oscillations that are bound to a small metal particle or nanostructure. These oscillations can be represented by the displacement of charge in the sphere. For example, for a metal sphere in a dielectric, the field inside the metal is given by electrostatic approximation as [27]

$$E_{in} = \frac{3\varepsilon_d E_o}{\varepsilon_m + 2\varepsilon_d}, \tag{5.5}$$

where E_o is the electric field away from the sphere. Ignoring the imaginary contributions to the relative permittivities in Eq. (5.5), it is clear that the field inside the sphere diverges when $\varepsilon_m = 2\varepsilon_d$, which leads to a strong enhancement of the field on the outer surface of the sphere (which is limited in practice by the imaginary part of ε_m). The quality of the resonance is limited by the dispersion of the metal and dielectric, as is clear from the field enhancement denominator in Eq. (5.5).

The third way of enhancing the photoresponse of a photodetector is shown in Figure 5.7(c). The photoresponse enhancing is achieved by the inclusion of a metallic PC on the detector area or arranging the detector structures in a periodic way, forming a PC structure. Integration of a resonant structure with the detector increases the interaction length between the incoming light and the active semiconductor region. This design is interesting for thin-film semiconductor detectors with a large absorption length.

Since the absorption coefficient is a strong function of the wavelength, the wavelength range in which an appreciable photocurrent can be generated is limited for a given detector material. So, broadband absorption is usually inadequate due to quantum efficiency roll-off. Research on PC structures with a periodic

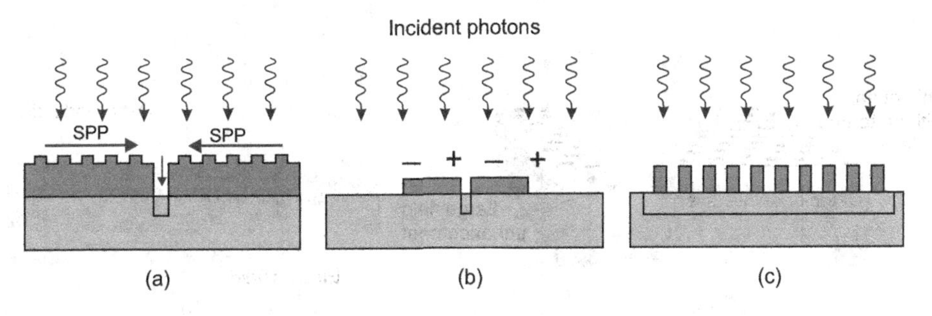

Figure 5.7 Different architectures for plasmon-enhanced detectors: (a) grating couplers to focus the generated SPPs inside a small-scale detector, (b) particle antenna on a small-scale detector, and (c) metallic PC structures.

refractive index modulation have opened up several ways for the control of light. Most existing devices are realized as 2D PC structures, as they are compatible with standard semiconductor processing [28–30].

PC crystal represents a regular array of holes (defects) that is used to modify the local refractive index to provide localized modes in the "photonic" band structure (see Figure 5.8). By removing a single hole, an energy well for photons is formed similar to that of electrons in a quantum wire structure. The periodic variation in the refractive index gives rise to Bragg scattering of photons, which opens up forbidden energy gaps in the in-plane photon dispersion relation. The PC has a grating effect that "diffracts" the normally incident radiation to the in-plane direction. In addition, a $\lambda/2$ high-index slab is used to trap photons in the vertical direction by internal reflection at the air-slab interface. As a result, the combination of Bragg reflection from the 2D PC and internal reflection results in a three-dimensionally (3D) confined optical mode.

Figure 5.9 shows the schematic design of infrared detector, which provides an opportunity for near-field detection of enhanced evanescent waves transmitted through a structured surface by using a nearby buried quantum detector (with a distance much smaller than the wavelength of the incident EM field).

An example of a metal PC-integrated detector design is shown in Figure 5.10 along with a schematic cross-sectional view of the sample structure [30]. The PC is a 100-nm thick Au film perforated with a 3.6-μm period square array of circular holes having a diameter of 1.65 ± 0.05 μm. This array of circular holes couples to surface plasma waves at 11.3 and 8.1 μm for reverse and forward bias, respectively, where InAs QDIP exhibit the strongest detectivity (up to 30-fold enhancement).

By confining the quantum dots in a waveguide structure and using a metallic grating coupler, a considerable increase in absorption is observed. Figure 5.11 shows the low-temperature photoresponse (10 K) of the metal PC QDIP and reference devices at −3.0 V and 3.4 V [30]. The arrows in the figure indicate the reference devices which exhibit two, rather broad color responses with indistinct peaks for both −3.0 V and 3.4 V cases. The peak shift with applied voltage has been interpreted in terms of the quantum-confined

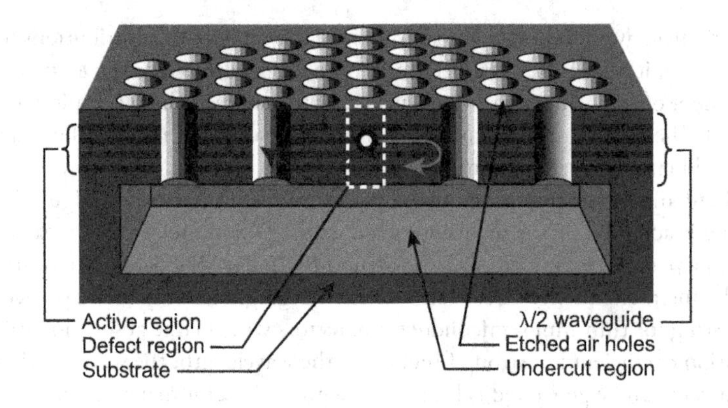

Figure 5.8 Cross section through the PC microcavity.

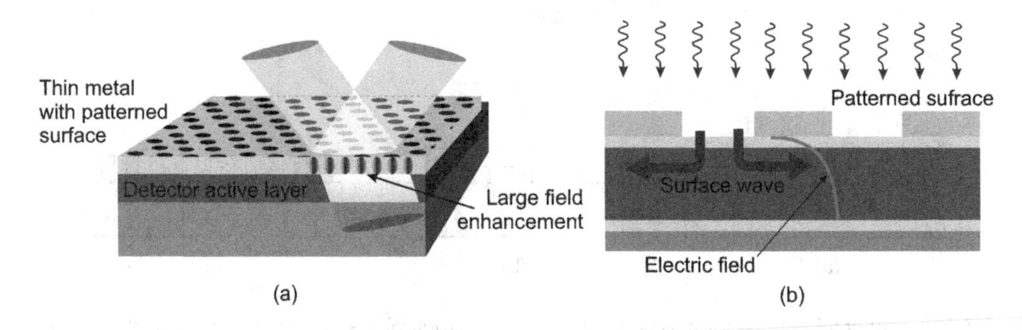

Figure 5.9 Conceptual design of infrared detector enhanced by SPPs: (a) general view and (b) cross section.

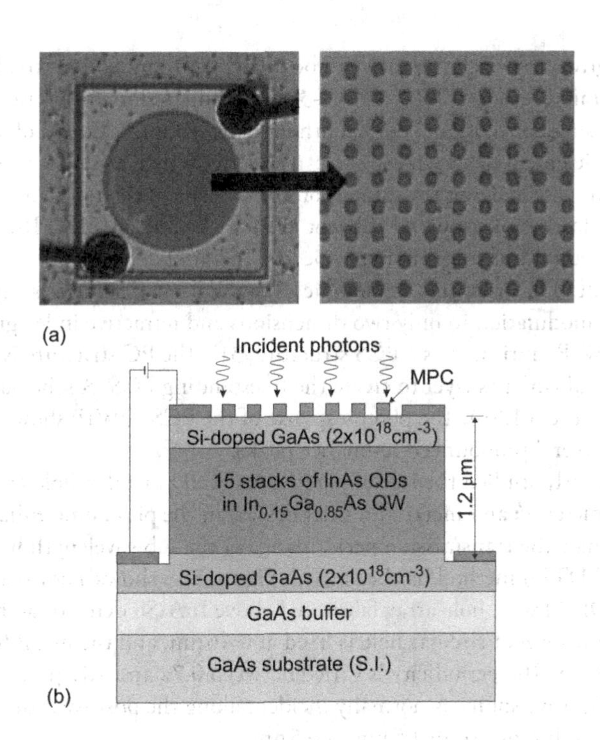

Figure 5.10 (a) Optical microscope images of the metal PC device with 16 times magnification revealing the details of the metal PC. The period of the circular holes is 3.6 µm; (b) a schematic cross-sectional structure of the metal PC device. (Adapted after Ref. [30])

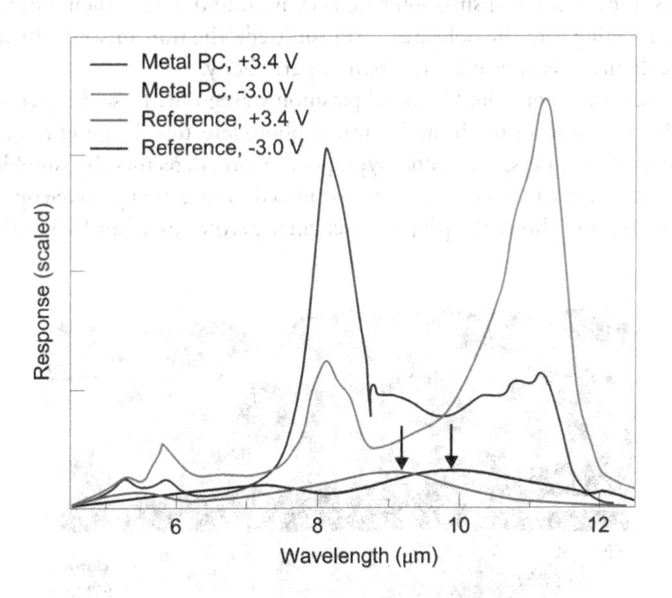

Figure 5.11 Spectral response curves of the reference device (two spectra at the bottom with the arrows indicating the highest peak in each spectrum) and the metal PC device (other two spectra with higher responsivity) for –3.0V and 3.4V at 10 K. (Adapted after Ref. [30])

Stark effect. On the other hand, the metal PC devices have totally different voltage-dependent spectral responsivities in both the peak wavelengths and especially the response intensity. They show four peaks at identical wavelengths, but varying amplitude for both biases. The peak at 11.3 µm, which is much stronger than that of the reference device, is dominant for reverse bias, while the peak at 8.1 µm is more intense than any other peak for forward bias. The two remaining peaks at 5.8 and 5.4 µm are relatively weak.

The advantage of the given approach is that it can be easily incorporated into the FPA fabrication process of present-day infrared sensors. Holes with 2–3–μm diameter for a response wavelength range of 8–10 μm can be defined using conventional optical lithography. An introduction of straightforward modification of a single or multielement defect in the PC can selectively increase the response of photons with a specific energy. Therefore, by changing the dimensions of the defect, the resonance wavelength can be altered, leading to the fabrication of a spectral element in each pixel of the FPA. This would have a revolutionary impact on multispectral imaging and hyperspectral imaging detectors.

An important class of 2D PC structures is photonic crystal slabs (PCSs) consisting of a dielectric structure, with a periodic modulation in only two dimensions and refractive index guiding in the third. Figure 5.12(a) shows a QWIP fabricated as a PCS structure [31]. The PC structure is underetched by selective removal of the sacrificial AlGaAs layer to create the freestanding PCS. A schematic illustration of the final device is shown in Figure 5.12(b). The photoresponse of the PCS-QWIP shows a wider response peak but additionally displays several pronounced resonance peaks.

Qiu et al. [32] have recently studied the role of parameters of 2D metallic hole arrays (the periodicity of hole arrays, p; hole diameter, d; and metal film thickness, t) in the plasmonic enhancement of InAsSb infrared detector. To evaluate the transmission performance of the subwavelength hole array, the 3D finite-difference time-domain (FDTD) method has been used. Figure 5.13 shows a cross-sectional view and top view of 2D hole array (2DHA) with hole array fabricated above InAsSb detector-active region.

In estimations, the diameter d of circular hole is fixed at 0.46 μm, and the metal film is made of gold, with a thickness of $t = 20$ nm. The periodicity is varied between 0.72 and 1.12 μm, with other parameters remaining unchanged. The wave source is normally incident along the positive z-direction, polarized in the x-direction, with a wavelength range from 1.5 μm to 6.5 μm.

The transmission efficiency shown in Figure 5.14 has been calculated for the main peak when either hole diameter is fixed at $d = 0.46$ μm or periodicity is fixed at $p = 0.92$ μm. As shown in this figure, at the resonance wavelengths, the highest transmission efficiency is around 3.85, which indicates much more light than that is directly impinging into the hole area is transmitted. The transmission efficiency reaches a maximum value when hole diameter is approximately half of periodicity.

Utilizing either a single-metal or a double-metal plasmon waveguide, Rosenberg et al. [21] have considered a plasmonic PC resonator for use in mid-infrared photodetectors. Its good frequency and polarization selectivity can be used in hyperspectral and hyperpolarization detectors. By suitable scaling of the PC holes and waveguide width, such a resonator can be optimized for use at any wavelength from the terahertz to the visible bands. Figure 5.15 shows the proposed schematic structure of an FPA with double-metal

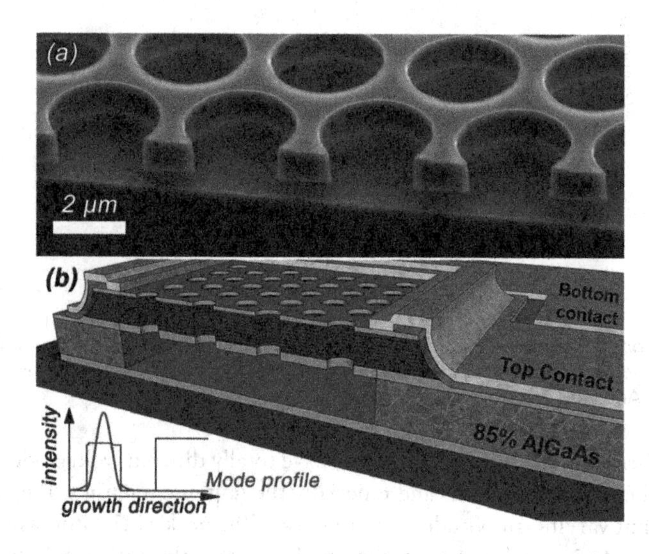

Figure 5.12 PCS-QWIP design: (a) scanning electron microscope (SEM) image of a cleaved PCS, (b) cross section through the PCS-QWIP structure. (After Ref. [31])

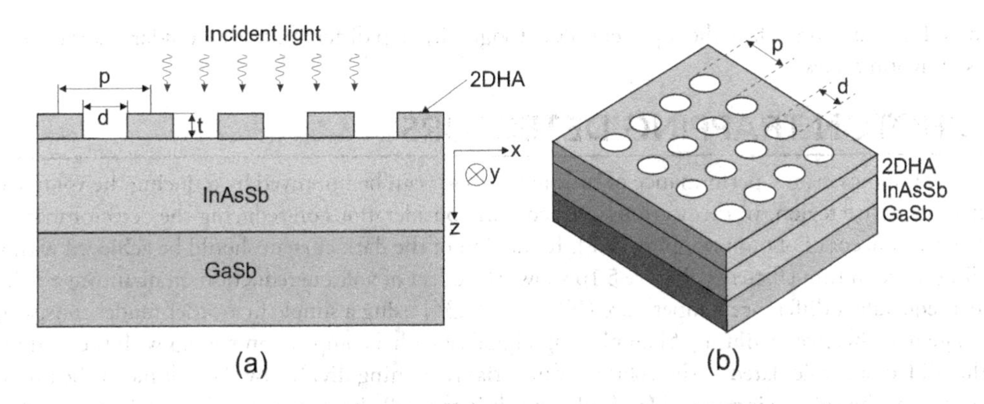

Figure 5.13 A cross-sectional view (a) and top view (b) of 2DHA. (Adapted after Ref. [32])

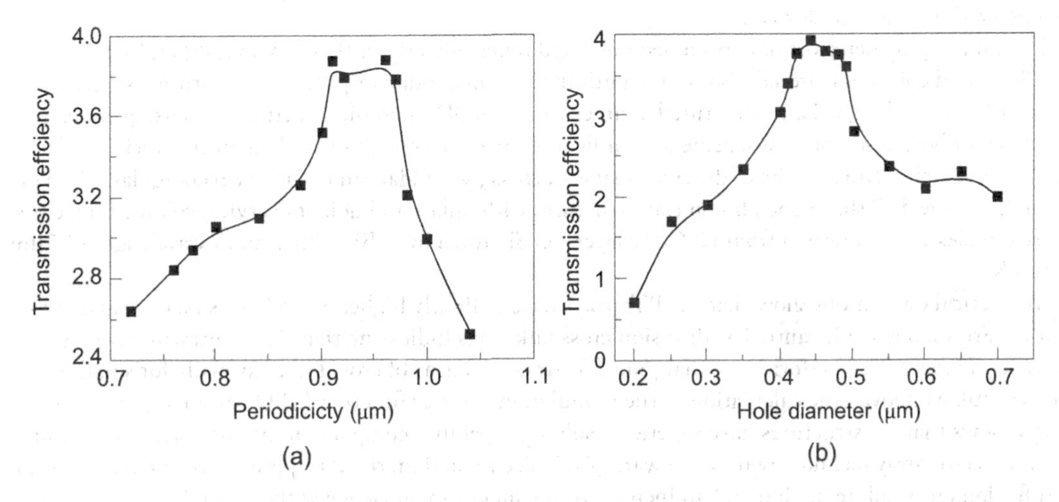

Figure 5.14 A plot of transmission efficiency as a function of (a) the periodicity p ($d = 0.46\,\mu m$) and (b) hole diameter d ($p = 0.92\,\mu m$). (Adapted after Ref. [32])

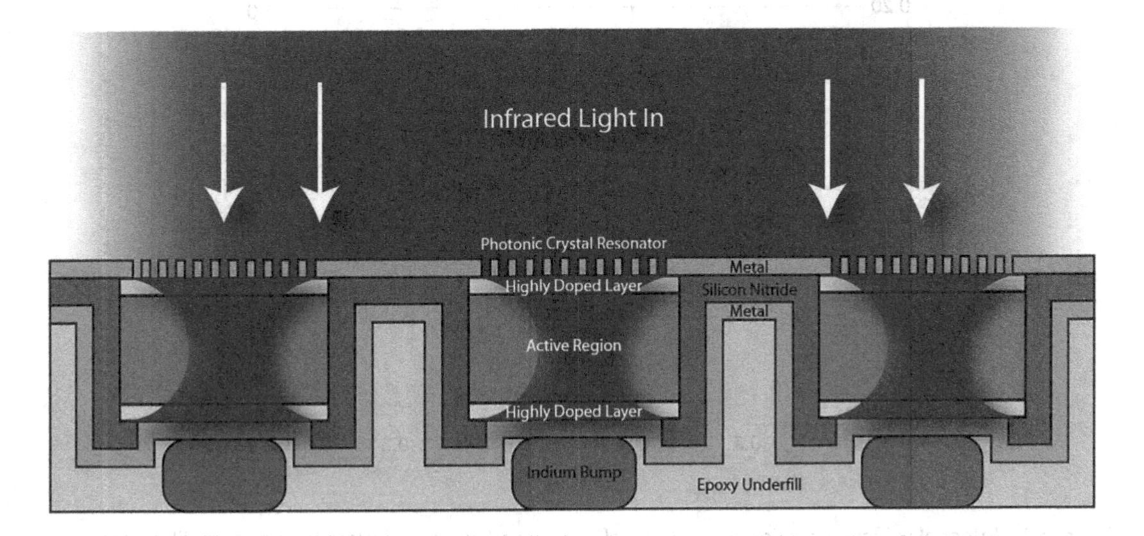

Figure 5.15 A design schematic for a resonant double-metal plasmonic PC FPA. (Adapted after Ref. [21])

plasmonic PC resonators. Only the top metal PC lithography step differs from the standard fabrication process of hybrid arrays.

5.3 PHOTON TRAPPING DETECTORS

As Eq. (4.43) indicates, the performance of infrared detector can be improved by reducing the volume of the detector's active region. In this section, we focus our considerations on reducing the detector material volume via a concept of photon trapping (PT). Reduction of the dark current should be achieved without degrading the quantum efficiency. Figure 5.16 shows the effect of volume reduction on quantum efficiency and noise equivalent difference temperature (NEDT) [33,34] using a simple first-order model consisting of the Bruggeman effective medium [35] combining HgCdTe with a composition x ≈ 0.3 with the void material. The fill factor is calculated as the volume of material remaining divided by the volume of the unit cell. As expected, as the volume is reduced (and fill factor is increased) the quantum efficiency is increased and the *NEDT* value generally decreases improving the performance until a critical point when photon collection begins to decrease faster than noise, and hence the overall performance degrades. The modeled trends are observed in measured devices.

Photon trapping detectors have been demonstrated independently in II–VI [33,34,36] and III–V [36–39]-based epitaxial materials. Subwavelength in size semiconductor pillar arrays within a single detector has been designed and structured as an ensemble of 3D photonic structure units using either a top-down or bottom-up process scheme to significantly increase absorption and quantum efficiency. The sub-element architecture can be of different shapes such as pyramidal, sinusoidal, or rectangular [37]. For example, Figure 5.17 shows the photon trap structures with pillars and holes of varying volume fill factors. These samples were fabricated from HgCdTe layers on Si grown by MBE with a cutoff wavelength of 5 μm at 300 K.

Theoretical estimations show that the PT arrays have a slightly higher optical cross talk compared to non-PT arrays, but significantly less diffusion cross talk, thus indicating that PT arrays will have significantly better device performance than non-PT arrays in terms of cross talk, especially for small pixel pitches [40]. Moreover, the calculation of the modulation transfer function (MTF) from a spot scan of the arrays shows that PT structures have superior resolving capability compared to non-PT structures. Thus, as the detector array technology moves toward pixels size reduction, the PT approach is an effective means of diffusion cross talk reduction and an increase of the quantum efficiency without employing antireflection coatings.

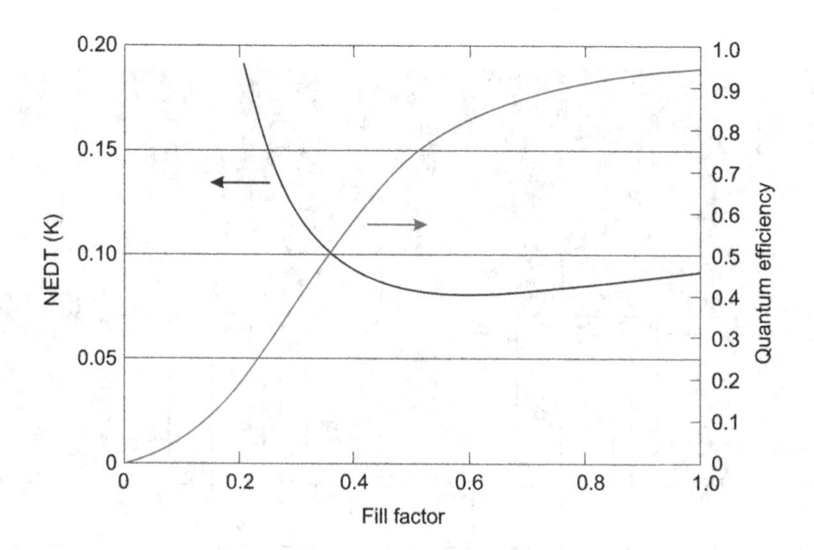

Figure 5.16 Effect of volume reduction on quantum efficiency and noise equivalent temperature difference. (Adapted after Ref. [34])

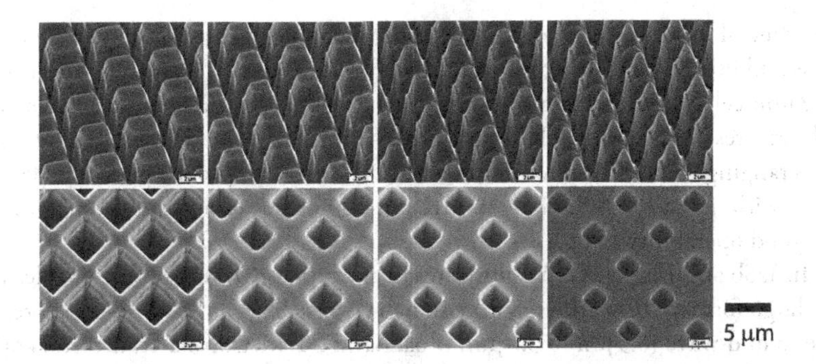

Figure 5.17 Examples of PT HgCdTe microstructures with test PC fields with varying fill factor for FTIR demonstration. (After Ref. [33])

The finite-difference time-domain simulation of pillar structures indicates resonance between them and confirms that the PT process via total internal reflection, effectively serving as a waveguide to direct incident energy away from the removed regions and into the remaining absorber material. For example, Figure 5.18 presents the optical generation rate as a function of wavelength from 0.5 μm to 5.0 μm for a single HgCdTe pillar and the uppermost part of the absorber layer beneath the pillars [41]. At a wavelength of 0.5 μm, the optical generation is concentrated at the edge of the pillar, and as the wavelength is increased, the optical generation region gradually extends deeper into the pillars. At 5.0 μm, there is very little of optical generation at the tip of the pillar, but instead, there is a significant optical generation further into the pillar and into the absorber layer. It has been found that the absorption enhancement is weakly dependent on the pillar lattice type, but the lattice period does have a significant impact on the enhancement [42].

It has been confirmed experimentally, that volume reduction leads to improved device performance and consequently to a higher operating temperature of detector arrays. Process improvements, such as unique

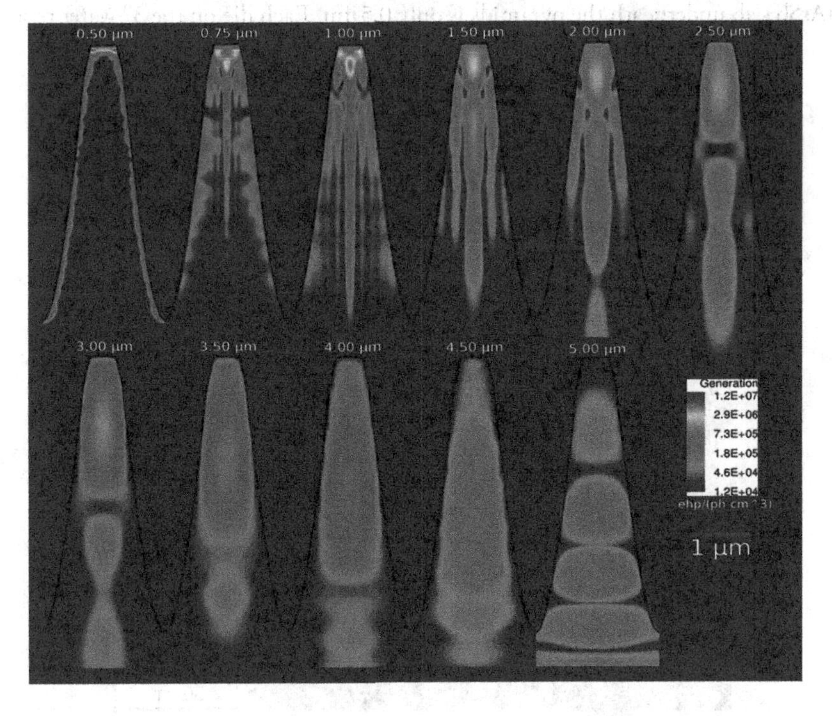

Figure 5.18 Calculated optical generation profile inside the pillar of HgCdTe photon trapping structure back-illuminated with plane waves within the wavelength range of 0.5 – 5.0 μm. This pillar is one of a 2D array of HgCdTe pillars like those shown in figure 5.17. (Adapted after Ref. [41])

self-aligned contact metal processes and advanced stepper technologies have been developed to achieve the critical dimensions and lithography alignments required for advanced PT detector designs with a smaller feature within the unit cell. Figure 5.19 shows advanced hexagonal PC designs with holes in a standard 30-μm mesa with features on a 5-μm pitch. Large format MBE HgCdTe/Si epitaxial wafer arrays with cutoff wavelengths ranging from 4.3 μm to 5.1 μm at 200 K exhibit improved performance compared to unpatterned mesas, with measured NEDT of 40 mK and 100 mK at temperatures of 180 K and 200 K, respectively, with good operability [34].

Utilization of InAsSb absorber on GaAs substrates instead of HgCdTe absorber enables fabrication of MWIR low-cost, large-format HOT FPAs. Souza and coworkers [37–39] have described research efforts in developing visible to mid-wave (0.5 μm to 5.0 μm) broadband PT InAsSb-based detectors operating at high temperature (150–200K) with low dark current and high quantum efficiency.

The $InAs_{0.82}Sb_{0.18}$ ternary alloy with a 5.25-μm cutoff wavelength at 200 K was grown on a lattice-mismatched GaAs substrate. To compare the detector performance, both 128 × 128/60-μm and 1024 × 1024/18-μm detector arrays consisting of bulk absorber structures as well as photon-trapped pyramid structures were fabricated. This novel detector design was based on pyramidal PT InAsSb structures in conjunction with barrier-based device architecture to suppress both the generation-recombination dark current, as well as the diffusion current through absorber-reduced volume. Absence of depletion regions in the narrow-gap absorption layer is immunity of barrier detector to dislocations and other defects, which may allow growth on lattice-mismatched substrates with a reduced penalty of excess dark current generated by misfit dislocations. The pixel arrays were defined very simply by etching through the contact layer up to the barrier. Figure 5.20(a) shows a 5-μm cutoff wavelength nBn detector structure operated at 200 K with AlAsSb barrier and pyramid-shaped absorbers fabricated in the n-type InAsSb absorber. On the basis of optical simulation, the engineered pyramidal structures minimize the reflection and provide >90 % absorption over the entire 0.5 μm to 5.0 μm spectral range [see Figure 5.20(b)], while providing up to 3× reduction in absorber volume.

Figure 5.21 shows an SEM top and side views of 4.5-μm height pyramid structures of InAsSb nBn barrier on GaAs substrate. The spacing between adjacent pyramids is smaller than 0.5 μm. The thickness of the InAsSb slab underneath the pyramids is only 0.5 μm. Each die on the 3" wafer represents a

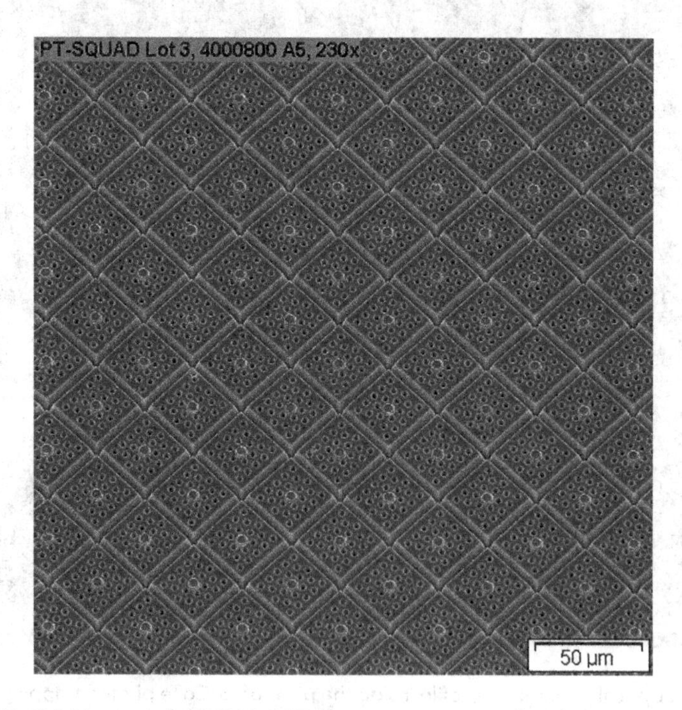

Figure 5.19 MWIR 512×512 30-μm pitch MBE HgCdTe/Si array consisting of PC holes on a 5-μm pitch in a standard mesa. (After Ref. [34])

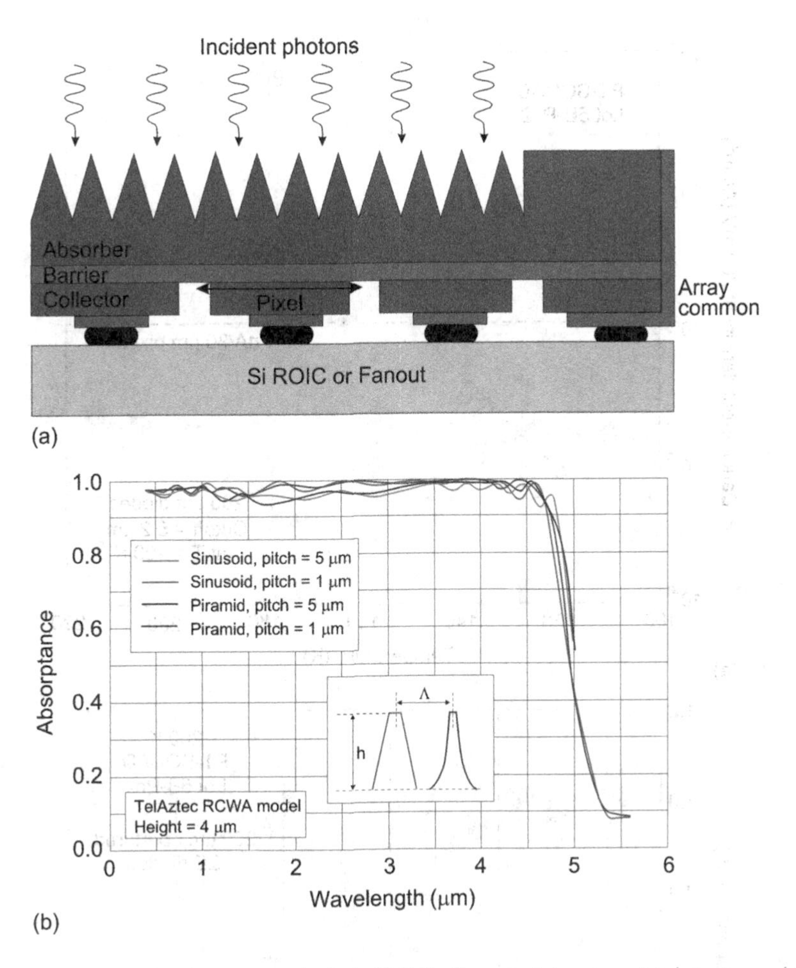

Figure 5.20 Photon trapping nBn detector in the InAsSb/AlAsSb material system: (a) detector architecture with pyramid-shaped absorber layer. (After Ref. [37]), (b) optical simulation of broadband detector response. (After Ref. [36])

Figure 5.21 Snapshot photo of fabricated staggered pyramids on a 3″ substrate (a), SEM pictures of staggered pyramids (b). (After Ref. [38])

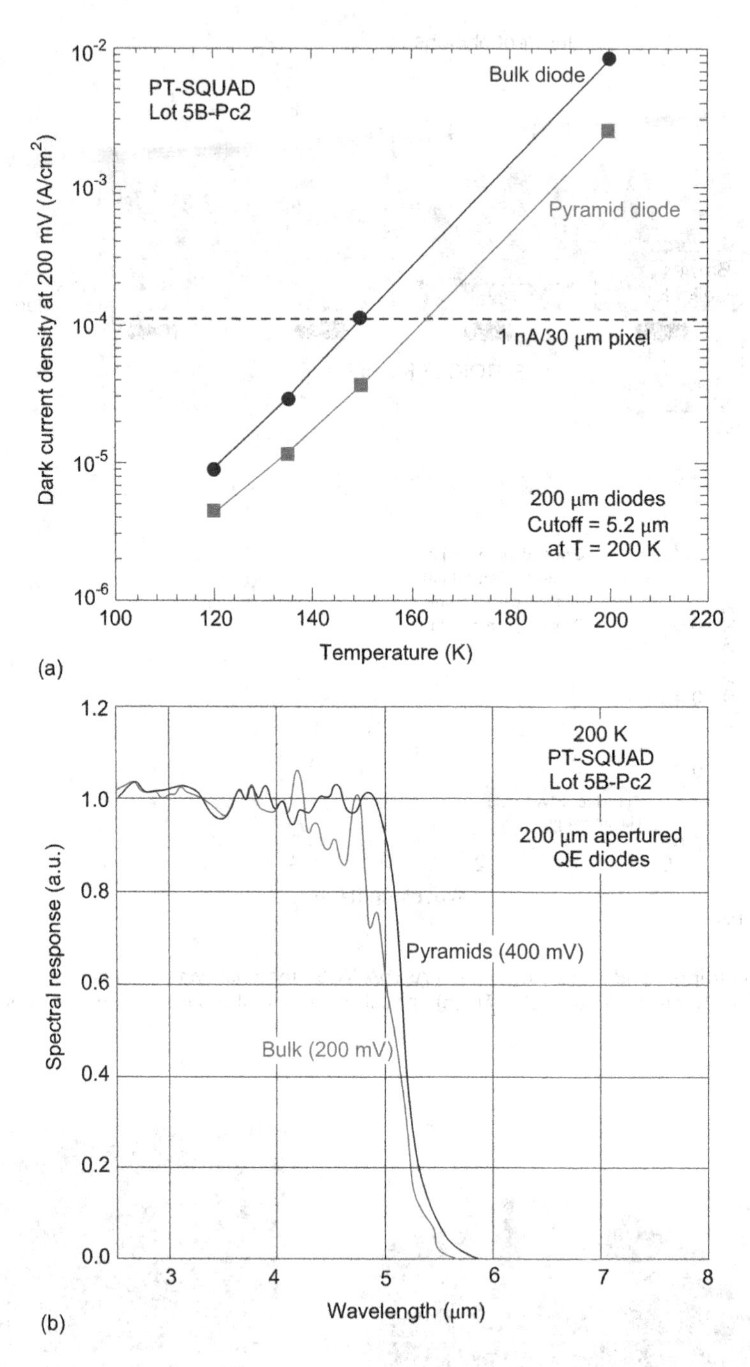

Figure 5.22 Comparison of dark current density between nBn InAsSb bulk detectors and pyramid detectors at different temperature (a), and spectral response comparison between bulk- and pyramid-based detectors (b). (After Ref. [38])

1024 × 1024 pixel 18-μm pitch FPA. After pyramid etching, the wafer was flipped over and bonded temporarily to another 3" handling substrate using an epoxy. The entire 600-μm thick GaAs growth substrate was subsequently removed using a high etch rate and high selectivity ICP dry etch process.

The measured dark current density in the pyramidal structured diodes is reduced by a factor of 3 in comparison with conventional diodes with the bulk absorber [see Figure 5.22(a)], which is consistent with volume reduction due to the creation of the absorber topology. High detectivity (> 1.0×10^{10} cmHz$^{1/2}$W^{-1}) and high internal quantum efficiency (> 90%) has been achieved. The spectral response measurements reveal that staggered pyramids suppress etalon effects and make the spectral response more flat [see Figure 5.22(b)], as expected from simulation results. Despite the small absorber volume of pyramidal detectors, the internal quantum efficiency is higher than 80% and detectivity above 1×10^{10} cm $\sqrt{\text{Hz}}$ / W at 200 K has been estimated over the entire 0.5-μm to 5.0-μm spectral range.

REFERENCES

1. Z. Jakšić, *Micro and Nanophotonics for Semiconductor Infrared Detectors*, Springer, Heidelberg, 2014.
2. S. J. Fonash, *Solar Cell Device Physics*, Elsevier, Amsterdam, 2010.
3. G. Li, R. Zhu, Y. Yang, "Polymer solar cells," *Nat. Photonics*, 6(3), 153–61, 2012.
4. M. S. Ünlü and S. Strite, "Resonant cavity enhanced photonic devices," *J. Appl. Phys.*, 78, 607–639, 1995.
5. J. Piotrowski and A. Rogalski, *High-Operating-Temperature Infrared Photodetectors*, SPIE Press, Bellingham, WA, 2007.
6. R. Yamazaki, A. Obana, and M. Kimata, "Microlens for uncooled infrared array sensor," *Electron. Commun. Jpn.*, 96(2), 1–8, 2013.
7. R. Winston, "Principles of solar concentrators of a novel design," *Sol. Energy*, 16(2), 89–95, 1974.
8. R. Winston, "Dielectric compound parabolic concentrators," *Appl. Opt.*, 15(2), 291–3, 1976.
9. M. K. Haigh, G. R. Nash, N. T. Gordon, J. Edwards, A. J. Hydes, D. J. Hall, A. Graham, J. Giess, J. E. Hails, and T. Ashley, "Progress in negative luminescent $Hg_{1-x}Cd_xTe$ diode arrays," *Proc. SPIE*, 5783, 376–83, 2005.
10. G. J. Bowen, I.D. Blenkinsop, R. Catchpole, N.T. Gordon, M.A.C. Harper, P.C. Haynes, L. Hipwood, C.J. Hollier, C. Jones, D.J. Lees, C.D. Maxey, D. Milner, M. Ordish, T.S. Philips, R.W. Price, C. Shaw, and P. Southern, "HOTEYE: A novel thermal camera using higher operating temperature infrared detectors," *Proc. SPIE*, 5783, 392–400, 2005.
11. R. H. Ritchie, "Plasma losses by fast electrons in thin films", *Phys. Rev.*, 106, 874–81, 1957.
12. S. A. Maier, *Plasmonic: Fundamentals and Applications*, Springer, New York, 2007.
13. R. Stanley, "Plasmonics in the mid-infrared," *Nat. Photonics*, 6, 409–11, 2012.
14. J. Zhang, L. Zhang, and W. Xu, "Surface plasmon polaritons: physics and applications," *J. Phys. D: Appl. Phys.*, 45, 113001, 2012.
15. P. Berini, "Surface plasmon photodetectors and their applications," *Laser Photonics Rev.*, 8, 197–220, 2013.
16. Y. Zhong, S.D. Malagari, T. Hamilton, and D. Wasserman, "Review of mid-infrared plasmonic materials," *J. Nanophotonics*, 9, 093791-1–21, 2015.
17. J.R. Sambles, G.W. Bradbery, and F.Z. Yang, "Optical-excitation of surface-plasmons: An introduction," *Contemp. Phys.*, 32, 173–183, 1991.
18. H. Raether, *Surface Plasmons*, ed. G. Hohler, Springer, Berlin, 1988.
19. D. Li and C. Z. Ning, "All-semiconductor active plasmonic system in mid-infrared wavelengths," *Opt. Express*, 19(15), #147367, 2011.
20. P. Bouchon, F. Pardo, B. Portier, L. Ferlazzo, P. Ghenuche, G. Dagher, C. Dupuis, N. Bardou, R. Haidar, and J.-L. Pelouard, "Total funneling of light aspect ratio plasmonic nanoresonators," *Appl. Phys. Lett.*, 98, 191109, 2011.
21. J. Rosenberg, R.V. Shenoi, S. Krishna, and O. Painter, "Design of plasmonic photonic crystal resonant cavities for polarization sensitive infrared photodetectors," *Opt. Exp.*, 18(4), 3672–3686, 2010.
22. C.-C. Chang, Y.D. Sharma, Y.-S. Kim, J.A. Bur, R.V. Shenoi, S. Krishna, D. Huang, and S.-Yu Lin, "A surface plasmon enhanced infrared photodetector based on InAs quantum dots," *Nano Lett.*, 10, 1704–1709, 2010.
23. G.V. Naik, V.M. Shalaev, and A. Boltasseva, "Alternative plasmonic materials: Beyond gold and silver," *Adv. Mater.*, 25, 3264–3294, 2013.
24. J. B. Khurgin and A. Boltasseva, "Reflecting upon the losses in plasmonics and metamaterials," *MRS Bullet.*, 37, 768–779, 2012.
25. P. Biagioni, J.-S.Huang, and B. Hecht, "Nanoantennas for visible and infrared radiation," *Rep. Prog. Phys.*, 75, 024402, 2012.
26. K. Ishihara, K. Ohashi, T. Ikari, H. Minamide, H. Yokoyama, J.-I. Shikata, and H. Ito, "Therahertz-wave near field imaging with subwavelength resolution using surface-wave-assisted bow-tie aperture," *Appl. Phys. Lett.*, 89, 201120, 2006.
27. U. Kreibig and M. Vollmer, *Optical Properties of Metal Clusters*, Springer, Berlin, 1995.
28. K. T. Posani, V. Tripathi, S. Annamalai, N.R. Weisse-Bernstein, S. Krishna, R. Perahia, O. Crisafulli, and O. J. Painter, "Nanoscale quantum dot infrared sensors with photonic crystal cavity," *Appl. Phys. Lett.*, 88, 151104, 2006.

29. K. T. Posani, V. Tripathi, S. Annamalai, S. Krishna, R. Perahia, O. Crisafulli, and O. Painter, "Quantum dot photonic crystal detectors," *Proc. SPIE*, 6129, 612906-1–8, 2006.
30. S. C. Lee, S. Krishna, and S. R. J. Brueck, "Quantum dot infrared photodetector enhanced by surface plasma wave excitation," *Opt. Exp.*, 17(25) 23160–23168, 2009.
31. S. Kalchmair, H. Detz, G. D. Cole, A. M. Andrews, P. Klang, M. Nobile, R. Gansch, C. Ostermaier, W. Schrenk, and G. Strasser, "Photonic crystal slab quantum well infrared photodetector," *Appl. Phys. Lett.*, 98, 011105, 2011.
32. S. Qiu, L.Y.M. Tobing, Z. Xu, J. Tong, P. Ni, and D.-H. Zhang, "Surface plasmon enhancement on infrared photodetection," *Procedia Eng.*, 140, 152–158, 2016.
33. J. G. A. Wehner, E. P. G. Smith, G. M. Venzor, K. D. Smith, A. M. Ramirez, B. P. Kolasa, K. R. Olsson, and M.F Vilela, "HgCdTe photon trapping structure for broadband mid-wavelength infrared absorption," *J. Electron. Mater.*, 40, 1840–1846, 2011.
34. K. D. Smith, J.G.A. Wehner, R. W. Graham, J. E. Randolph, A. M. Ramirez, G. M. Venzor, K. Olsson, M. F. Vilela, and E.P.G Smith, "High operating temperature mid-wavelength infrared HgCdTe photon trapping focal plane arrays, *Proc. SPIE.*, 8353, 83532R, 2012.
35. D. A. G. Bruggeman, "Berechnung verschiedener physikalischer Konstanten von heterogenen Substanzen," *Ann. Phys.* (Leipzig), 24, 636679, 1935.
36. N. K. Dhar and R. Dat, "Advanced imaging research and development at DARPA," *Proc. SPIE*, 8353, 835302, 2012.
37. A. I. D'Souza, E. Robinson, A. C. Ionescu, D. Okerlund, T. J. de Lyon, R. D. Rajavel, H. Sharifi, D. Yap, N. Dhar, P. S. Wijewarnasuriya, and C. Grein, "MWIR $InAs_{1-x}Sb_x$ nCBn detectors data and analysis," *Proc. SPIE*, 8353, 835333, 2012.
38. H. Sharifi, M. Roebuck, T. De Lyon, H. Nguyen, M. Cline, D. Chang, D. Yap, S. Mehta, R. Rajavel, A. Ionescu, A. D'Souza, E. Robinson, D. Okerlund, and N. Dhar, "Fabrication of high operating temperature (HOT), visible to MWIR, nCBn photon-trap detector arrays," *Proc. SPIE*, 8704, 87041U, 2013.
39. A. I. D'Souza, E. Robinson, A. C. Ionescu, D. Okerlund, T. J. de Lyon, R. D. Rajavel, H. Sharifi, N. K. Dhar, P. S. Wijewarnasuriya, and C. Grein, "MWIR InAsSb barrier detector data and analysis," *Proc. SPIE*, 8704, 87041V, 2013.
40. J. Schuster and E. Bellotti, "Numerical simulation of crosstalk in reduced pitch HgCdTe photon-trapping structure pixel arrays," *Opt. Exp.*, 21(12) 14712, 2013.
41. C. A. Keasler, "Advanced numerical modeling and characterization of infrared focal lane arrays," PhD Thesis, Boston University, 2012.
42. C. A. Keasler and E. Bellotti, "A numerical study of broadband absorbers for visible to infrared detectors," *Appl. Phys. Lett.*, 99, 091109-1–3, 2011.

6 Heterodyne detection

Most current IR detectors that have been considered in the previous chapters are used in the direct detector mode. In direct detection, the output electrical signal is linearly proportional to the signal power. But in terms of the amplitude of the electric field carried by radiation, photodetectors are quadratic. As such, no information regarding the phase of the signal is retained in the electrical output. When the radiation field at the detector is only that of the signal, we talk of direct detection (Figure 6.1a), by far the most usual case in applications.

A heterodyne detection, in contrast, produces an output signal proportional to the electric field strength of the signal and thus the phase of the optical field is preserved in the phase of the electrical signal. In the 1960s, the laser made it possible to generate intense coherent beams of light for the first time, similar to those developed at radio frequencies many years earlier. It was first demonstrated that the heterodyne technique worked in the optical region as it did in the radio region. The main virtues of this method detection are higher sensitivity, higher and more easily obtained selectivity, plus the possibility of detection of all types of modulation and easier tuning over a wide range. Very weak signals can be detected because mixing the local field with the incoming photons allows the signal to be amplified. However, the main advantage of the technique is that the signals are downconverted to frequencies where extremely low-noise electronics can be used to amplify them. Heterodyne receivers are the only detection systems that can offer high spectral resolution ($\nu/\Delta\nu > 10^6$, where v is the frequency) combined with high sensitivity. Coherent optical detection has been developed since 1962, but compact and stable production of this system is more difficult and the system is more expensive and troublesome than its radiotechnique equivalent. At present,

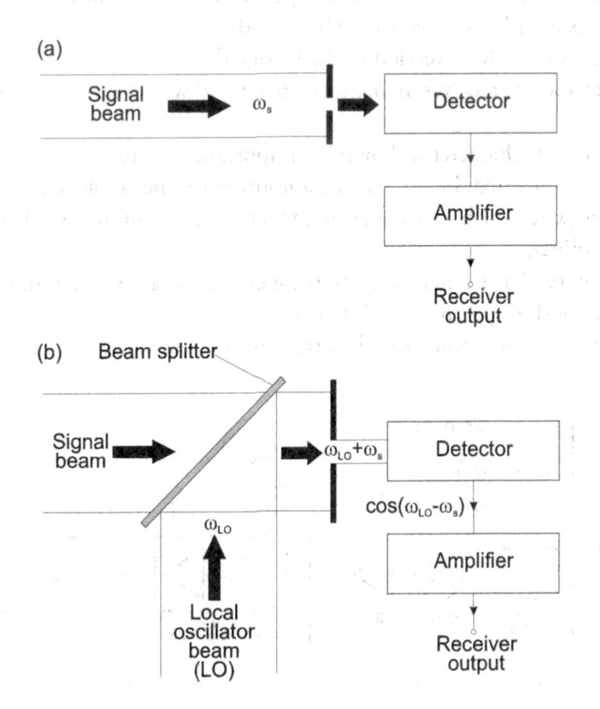

Figure 6.1 The generalized system of (a) direct and (b) coherent detection.

coherent receivers monopolize radio applications; they are not used as widely at infrared or optical frequencies because of their narrow spectral bandwidths, small fields of view, and their inability to be constructed in simple large-format arrays.

Infrared heterodyne detection has been commonly used for many years in commercial and domestic radio receivers and also for the microwave range of an electromagnetic spectrum. This technique is used for the construction of Doppler velocimeters, laser range finders, spectroscopy [particular light detection and ranging (LIDAR) systems], and telecommunication systems. Today, the heterodyne principle is the basis of nearly all radio and television receivers as well as wireless telecommunications. It can be dated back to more than a hundred years when in 1901, R. A. Fessenden, a Canadian-born radio engineer, filed the first patent regarding the heterodyne principle, which was granted in 1902 [1]. During World War I, work in several countries led to the first practical heterodyne receivers. For a review on the history of heterodyne technology, see Ref. [2]. The development of faster and more sensitive detectors has been a major reason for the opening up of the terahertz (THz) spectral region. The THz spectral range is taken to be 0.1–10 THz, which roughly coincides with the somewhat older definitions of the submillimeter and far-infrared spectral range.

The past 20 years have seen a revolution in THz systems, as advanced materials research provided new and high-power sources, and the potential of THz for advanced physics research and commercial applications was demonstrated. High-resolution heterodyne spectroscopy in the THz region is an important technique for the investigation of the chemical composition, evolution, and dynamic behavior of astronomical objects, and in the investigation of the earth's atmosphere. More recently, THz heterodyne receivers have been applied to imaging in biomedicine and security.

A heterodyne receiver consists of two basic subsystems: the front-end and the back-end (see Figure 6.2). The major components of the front-end deals with electromagnetic radiation are as follows:

- A local oscillator (LO) that delivers the reference frequency to the mixer.
- A mixer onto which the signal radiation and the radiation from the LO impinge [the mixer's output signal is at the difference of the signal and the LO frequencies, the so-called intermediate frequency (IF)].
- Optical elements that couple the signal radiation and the LO radiation onto the mixer.

The components of the back-end deals with the downconverted IF signal are as follows:

- An IF processor that amplifies the difference signal generated by the mixer (note that the first amplifier that follows the mixer is usually assigned to the front-end).
- A spectrometer or detector for the detection of the IF signal.

In this chapter, only a heterodyne receiver, namely, the front-end with the mixer as its core element, is considered.

The heterodyne technique is characterized by three important features:

- Since the signal is downconverted, low-frequency amplifiers can be employed. This allows the use of heterodyne receivers at extremely high frequencies, where direct amplification is not possible due to the lack of high-speed amplifiers.
- High-frequency selectivity, that is, many transmitting channels can be fitted into a given frequency band (high spectral resolution spectroscopy is possible).
- The narrow bandwidth detection process with corresponding low noise.

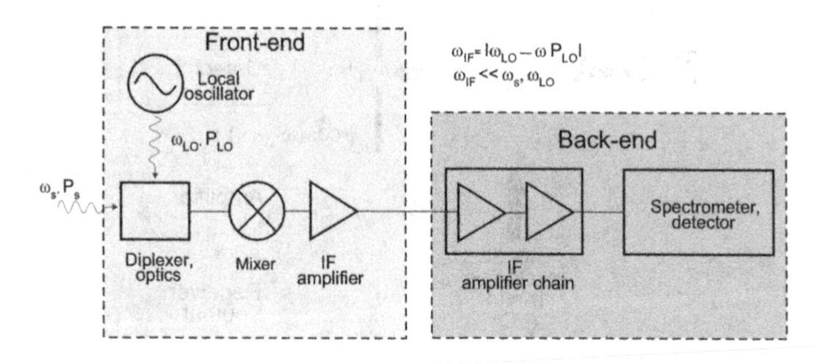

Figure 6.2 Scheme of a heterodyne receiver.

6.1 HETERODYNE DETECTION THEORY

Mixing takes place in a device with a nonlinear characteristic such as a nonlinear current-voltage (*I-V*) curve. A mixer with a nonlinear *I-V* curve can be expanded into a power series as

$$I(V) = k_0 + k_1 V + k_2 V^2 + \ldots = \sum_{i=0}^{\infty} k_i V^i. \tag{6.1}$$

The voltage change is induced by the electric field of the LO and the signal, as follows:

$$V = V_{LO}\sin(\omega_{LO}t) + V_s\sin(\omega_s t). \tag{6.2}$$

Combining the last two equations, and after some algebraic and trigonometric manipulation, this leads to

$$\begin{aligned} I &= k_0 + k_1\ V_{LO}\sin(\omega_{LO}t) + V_s\sin(\omega_s t)\ + \frac{k_2}{2}\left(V_{LO}^2 + V_s^2\right) \\ &- \frac{k_2}{2}\left(V_{LO}^2 - V_s^2\right)\cos(2\omega_{LO}t) + k_2 V_{LO} V_s \cos\ (\omega_{LO} - \omega_s)t\ - k_2 V_{LO} V_s \cos\ (\omega_{LO} + \omega_s)t\ + \ldots \end{aligned} \tag{6.3}$$

From the last equation we obtain that the mixer produces a frequency spectrum of the form

$$\omega_k = |l\omega_{LO} \pm m\omega_s| \qquad l,\ m = \ 0,\ 1,\ 2, \ldots, \tag{6.4}$$

and since the power of the signal is usually much smaller than the power from the LO, and because the power in the higher harmonics is approximately proportional to $1/l^2$, only three frequency components are important:

$$\begin{array}{ll} \omega_{LO} + \omega_s & \text{sum frequency,} \\ |\omega_{LO} - \omega_s| = \omega_{IF} & \text{intermediate frequency,} \\ 2\omega_{LO} - \omega_s & \text{image frequency.} \end{array} \tag{6.5}$$

Figure 6.3 illustrates the frequency conversion process in a mixer. As is shown, two frequencies are transferred to the intermediate frequencies, *IF*, $\omega_{LO}-\omega_{IF}$ and $\omega_{LO}+\omega_{IF}$, called the lower sideband (LSB) and upper sideband (USB), respectively. A mixer can be operated in a way that both sidebands are present at the IF—it is called as a double sideband (DSB). On the other side, if one sideband is suppressed

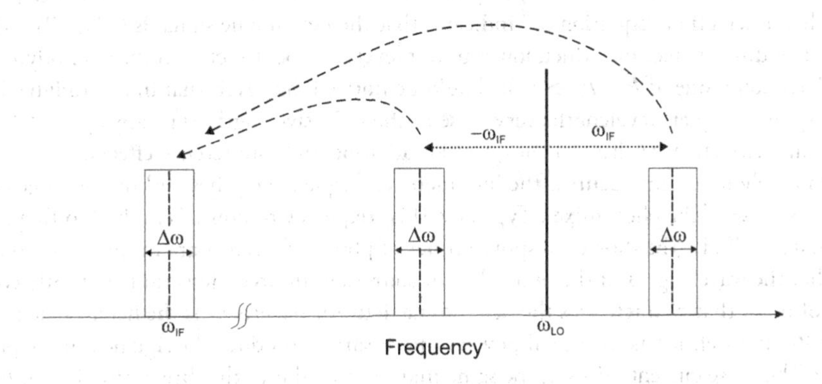

Figure 6.3 Illustration of the frequency conversion process in a mixer.

(e.g., by a filter in the signal path), only one sideband is downconverted and it then becomes a single-sideband (SSB) mixer.

In the SSB operation, the receiver is configured so that, at the image sideband, the mixer is connected to a termination within the receiver. There is no external connection to the image frequency, and the complete receiver is functionally equivalent to an amplifier followed by a frequency converter.

In the DSB operation, on the other hand, the mixer is connected to the same input port at both USB and LSBs. The DSB receivers can be operated in two modes:

- In SSB operation to measure narrow-band signals contained entirely within one sideband—for detection of such narrow-band signals, power collected in the image band of a DSB receiver degrades the measurement sensitivity.
- In DSB operation to measure broadband (or continuum) sources whose spectrum covers both sidebands—for continuum radiometry, the additional signal power collected in the image band of a DSB receiver improves the measurement sensitivity.

Infrared heterodyne detection is analogous to millimeter wave techniques. In heterodyne detection, a coherent optical signal beam of photon flux Φ_s is mixed with a laser LO beam of photon flux Φ_{LO} at the input of a detector, as shown schematically in Figure 6.1b. Two optical beams are well collimated and are aligned so that their wave fronts are parallel. Thus, the weaker coherent signal, which is collinear with the LO and absorbed within the detector, is mixed with the LO, producing the photocurrent generated at the intermediate or difference frequency $\omega_{IF} = \omega_{LO} - \omega_s|$ [3–6]

$$I_{ph} = I_{LO} + I_s + 2\eta(\omega_{IF})q(\Phi_{LO}\Phi_s)^{1/2} A\cos(\omega_{IF}t), \tag{6.6}$$

where I_{LO} and I_s are the direct current (DC) photocurrents due to Φ_{LO} and Φ_s, respectively. For photodiode:

$$I_{LO} = \eta(0)q\Phi_{LO}A$$

$$I_S = \eta(0)q\Phi_S A$$

The DC quantum efficiency $\eta(0)$ and the alternating current (AC) quantum efficiency $\eta(\omega_{IF})$ govern the DC photocurrent ($I_{LO}+I_s$) and the modulated photocurrent, respectively. The values of $\eta(0)$ and $\eta(\omega_{IF})$ are different when the frequency response is limited by carrier diffusion to the junction [6].

The rms value of the heterodyne signal current $I_H(\omega_{IF})$ obtained from Equation 6.6 is equal to

$$I_H(\omega_{IF}) = \eta(\omega_{IF})q(2\Phi_{LO}\Phi_s)^{1/2} A = \frac{\eta(\omega_{IF})q}{h\nu}(2P_{LO}P_s)^{1/2} = R_i\left(\frac{2P_{LO}}{P_s}\right)^{1/2}, \tag{6.7}$$

where P_{LO} is the LO radiation power, P_s is the signal radiation power, and R_i is the current responsivity as measured for direct detection. Equation 6.7 indicates that the heterodyne signal is $(2P_{LO}/P_s)^{1/2}$ times greater than in the case of direct detection. Much lower power levels can be detected in the heterodyne mode than in the direct detection mode, if $P_{LO}/P_s >> 1$. It should be noticed, however, that in heterodyne detection, the detector responds only at wavelengths very close to the LO wavelength (typically $\lambda_{LO} \pm < 0.003\,\mu m$) [7]. This makes them relatively insensitive to background radiation and interference effects.

The modulated photocurrent occurs if the intermediate frequency ω_{IF} lies within the range covered by the frequency response of the photomixer. Typically, this frequency response is limited to frequencies less than approximately 1 THz (the shortest response times of photon detectors are in the picosecond range). This implies that the wavelengths of the two coherent sources in indirect detection are nearly equal.

The figure of merit that characterizes the heterodyne detector sensitivity is the heterodyne noise equivalent power NEP_H. It is defined as the signal power P_s necessary to produce the signal-to-noise power ratio $(S/P)_P$ of unity. The noise current arises in the same manner as it did in the direct case. For sufficient LO power ($I_{LO} > I_s$) and reverse-biased photodiode $I_n^2 = 2qI_{LO}\Delta f$, where Δf is the bandwidth of the IF channel following the detector. This is the case of a well-designed heterodyne receiver in which the noise is dominated by LO-induced shot or generation–recombination noise. In this case, a final $(S/P)_P$ is given by

$$\left(\frac{S}{N}\right)_P = \frac{I_H^2}{I_n^2} = \frac{\eta^2(\omega_{IF})}{\eta(0)} \frac{P_s}{h\nu\Delta f}$$

and then

$$NEP_H = \frac{\eta(0)}{\eta^2(\omega_{IF})} h\nu\Delta f. \tag{6.8}$$

We can see that in heterodyne detection, the sensitivity of the photodiode is reduced by a factor $\eta(\omega_{IF})/\eta(0)$ from the value $\eta(\omega_{IF})P_s/h\nu\Delta f$ obtained when carrier diffusion to the junction is ignored. It should be noticed that the sensitivity degradation factor $\eta(\omega_{IF})/\eta(0)$ is common to both heterodyne detection and background-limited infrared photodetector (BLIP) direct detection [6].

For the case of a photoconductor heterodyne receiver, the results described here are not directly applicable. In the limit of large LO power, photoconductors display generation–recombination noise induced by LO signal fluctuations. For the photoconductor detectors [3],

$$NEP_H = \frac{2h\nu}{\eta(\omega_{IF})} \Delta f. \tag{6.9}$$

These devices are less sensitive by a factor of $2/\eta$ than the perfect quantum counter.

A simple and more general expression for the NEP_H can be obtained assuming that the responsivity and internal noise of the detector are unaffected by the presence of the LO. Then

$$NEP_H = \left[\frac{(NEP_D)^2}{2P_{LO}} + \frac{h\nu}{\eta(\omega_{IF})}\right]\Delta f, \tag{6.10}$$

where NEP_D is the noise equivalent power for direct detection.

Practical application of heterodyne detection is most successful on the infrared region due to such factors as sensitivity improvement due to reduced quantum noise, easier alignment of signal and LO, and greater diffraction-limited acceptance angle for a given aperture size [3,7,8]. In this wavelength region, strong and stable 10.6-μm line LOs exist. Moreover, in contrast to thermal imaging, extremely low dark detector current is not needed for good coherent detection. The main requirement for good heterodyne detection sensitivity is high quantum efficiency at IF and at the LO power level needed for shot-noise-limited operation. Achievement of shot-noise-limited operation and high frequency at either THz frequencies or at high operating temperatures have been the challenges for high-performance photomixers.

The basic block diagram of the heterodyne optical receiver is shown in Figure 6.4. Signal radiation containing information is, after being passed through an input optical filter and beam splitter, arranged to coherently combine or "mix" with a light beam of an LO at the detector surface. A beam-splitter can be made in many ways, the simplest being a glass plate with adequate refraction coefficient. In a general case, a device fulfilling such a role is called a direction coupler, as an analogy to microwave or radio devices. A detector used for signal mixing has to have a square-law characteristic for detecting the electronic field of the light, but this is conveniently typical of most optical detectors (photodiode, photoconductor, photomultiplier, avalanche photodiode, etc.). This signal is next amplified. An electrical filter of IF extracts the desired difference component of the signal which next undergoes a demodulation process. The design and operation principle of the subsequent electrical detector depends on the nature of the modulation of the signal. The signal from a load resistance passes through an output filter to a receiver output and by means of an LO frequency controller, it controls a laser. A frequency control loop is used for the LO laser to maintain a constant frequency difference $\omega_{LO} - \omega_s = \omega_{IF}$ with the input signal. An indispensable condition for

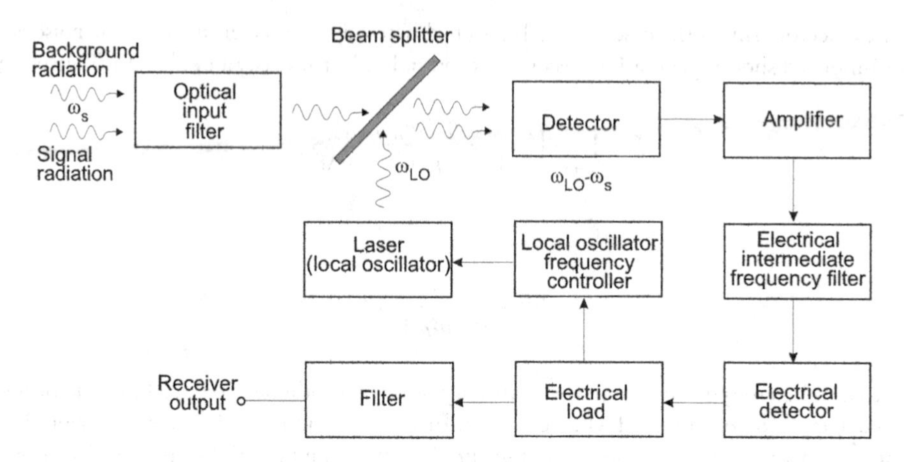

Figure 6.4 Block diagram of a heterodyne detection optical receiver.

efficient coherent detection is to match the polarization, and to the shape of both waveforms of both beams to match the profile of the detector surface.

6.2 INFRARED HETERODYNE TECHNOLOGY

Early work on 10.6-μm photomixers centered on the liquid-helium-cooled copper-doped Ge photoconductors, HgCdTe, and lead salt photodiodes [3,7–11]. However, as HgCdTe photodiodes were developed, they quickly replaced the Ge photoconductors, which have very high LO power requirements and a factor-of-2 lower sensitivity, and lead salt photodiodes, which are slow due to a very large dielectric constant of the material [12,13].

Heterodyne detection systems with 1-GHz base bandwidth and sensitivities that approach the ideal limits were reported during the 1960s [3,9]. Figure 6.5 shows the experimental results of the heterodyne

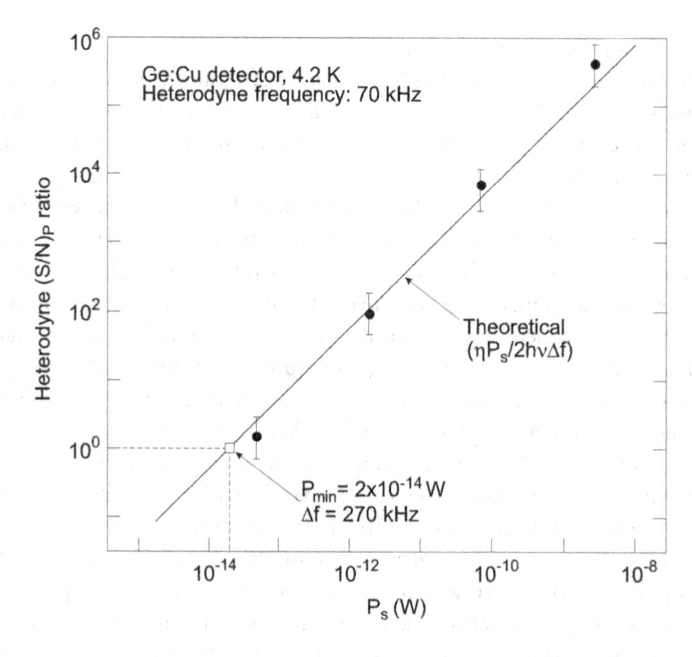

Figure 6.5 The heterodyne signal-to-noise ratio for Ge:Cu doped photoconductor at 4.2 K. The filled circles represent the observed signal-to-noise power ratio data points; the solid line presents the theoretical result. (From Ref. [4])

signal-to-noise ratio for Ge:Cu doped operated at 4.2 K [4]. The field circles present the observed signal-to-noise power ratio data points, $(S/N)_p$, as a function of the signal beam radiation power P_s. Only noise arising from the presence of the LO beam (which was the dominant contribution to the noise) is considered. A plot of the theoretically expected result $(S/N)_P = \eta P_s/2h\nu\Delta f$ is also shown. Using an estimated quantum efficiency $\eta = 0.5$, it is seen to be in good agreement with the experimental data. With a heterodyne signal centered at about 70 kHz, and an amplifier bandwidth of 270 kHz, the experimentally observed NEP_H was seen to be 7×10^{-20} W, which is to be compared with the expected value $(2/\eta)h\nu\Delta f \approx 7.6 \times 10^{-20}$ W. Usually, detectors have higher noise and require more LO power that can overpower the cooler or heat the detector, resulting in an NEP_H less than theoretical.

The frequency spectrum of the shot noise reveals the roll-off into the external circuit (e.g., RC) but is not a good indicator of the bandwidth of the photodiode, because in addition to circuit effects, the frequency response of a photodiode depends on the time required for carriers to diffuse and transit across the space charge region. It is easy to design an optimum heterostructure with transparent wider energy-gap layers to minimize carrier diffusion, but the p-i-n diode structure needed for low capacitance has been feasible with HgCdTe epitaxial layer. Most 10-μm HgCdTe photomixers have been the n-n⁻-p homojunction photodiodes illuminated from the n-type side. Since the active area of a photomixer is defined by the incident LO pattern, the detector area involves a trade-off between accommodating the LO and junction capacitance. Low-capacitance n-n⁻- p etched mesa and planar HgCdTe photodiodes have been fabricated by Hg diffusion into the p-type material. Impurity diffusion and ion implantation have not been as successful in making gigahertz-bandwidth photomixers. Wideband HgCdTe photomixers are operated at reverse biases in the range of 0.5–2 V. Operation in this mode results in lower junction capacitance, which facilitates good impedance matching and good coupling to the first stage preamplifier (preferably cooled together with the detector). The diffused junction with diameters of 100–200 μm typically have capacitances in the range of 1–5 pF and RC roll-off frequencies of 0.5–3 GHz [10,13]. The largest wideband arrays developed for target tracking were 12-element arrays of 1.5-GHz photomixers in the configuration of a central quadrant array surrounded by eight additional detectors.

Figure 6.6 shows the dependence of heterodyne NEP as a function of LO power [14]. The solid lines are calculated from Equation 6.7 for two different values of NEP_D and quantum efficiency $\eta = 50\%$. It can be seen that a small value of NEP_D allows the use of a low P_{LO} to approach the quantum limit.

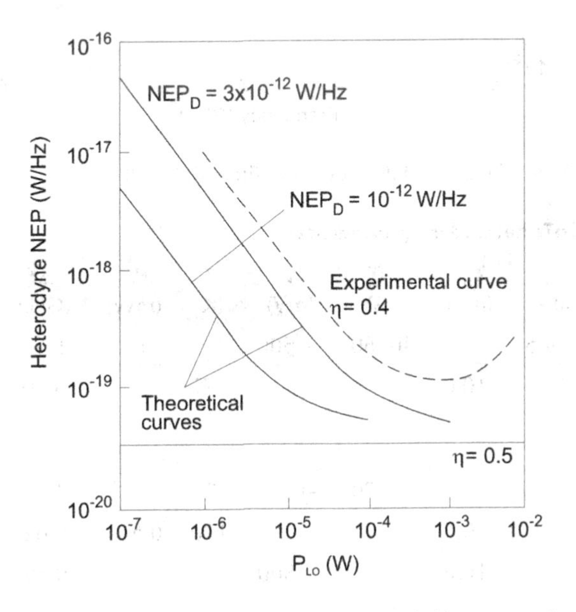

Figure 6.6 The heterodyne NEP plotted against the LO power. The solid lines are calculated from Equation 6.7 for two different values of NEP in direct detection and quantum efficiency, $\eta = 0.5$. The dashed line is a typical curve for the experimental performance of HgCdTe performance. (From Ref. [14])

With a detector exhibiting poor NEP_D, good performance can still be achieved, provided the necessary increase in P_{LO} does not affect the internal responsivity and noise processes through increased carrier density or heating effects.

Figure 6.7 shows the best values of NEP_H at 10.6 µm obtained at Lincoln Laboratory and at Honeywell as a function of frequency for HgCdTe photodiodes at 77 K [15]. At 1 GHz, NEP_H is only a factor of 2 above the theoretical quantum limit that for $\lambda = 10.6$ µm has the value of 1.9×10^{-20} W/Hz. The average sensitivities of many 12-element photodiode arrays, 4.3×10^{-20} W/Hz, have been close to this value [15].

Table 6.1 gives a summary of some of the heterodyne HgCdTe photodiodes detailed in the literature [16].

Most high frequency, 10-µm HgCdTe photomixers were operated at about 77 K, although they can work well at somewhat higher temperatures. At elevated temperatures, p-type photoconductors have higher sensitivities than the more common n-type photoconductors in both heterodyne and direct detection [14–19]. Moreover, for wide bandwidth photoconductors, the LO power requirement can be considerably larger than that of a photodiode, due to low photoconductive gain. The dependence of HgCdTe photoconductor NEP on LO power (and bias power) has been successfully modeled by taking into account the effect of heating and bandfilling on carrier lifetime and optical absorption [15]. The best performance is obtained with small (50–100 µm square) devices, because the LO power requirement

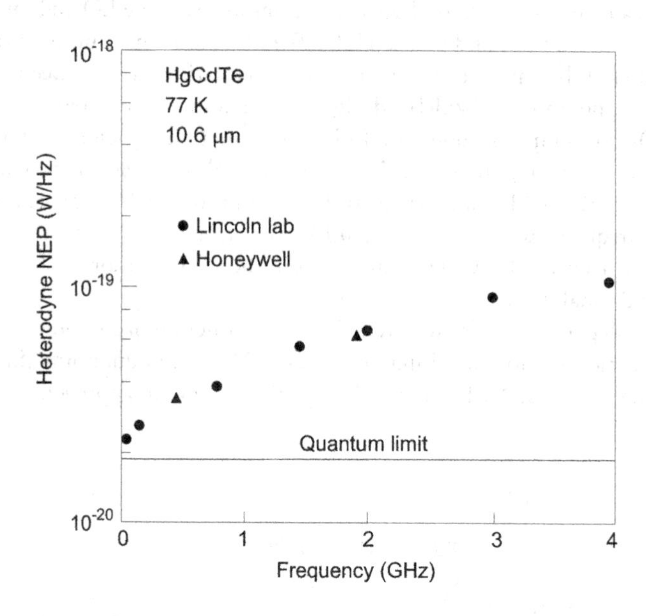

Figure 6.7 Heterodyne NEP_H as a function of IF frequency. (From Ref. [15])

Table 6.1 Examples of HgCdTe heterodyne photodiodes

x	A (10^{-4} cm^2)	λ_c (µm)	λ_{LO} (µm)	$\eta(0)$ (%)	V (mV)	T (K)	P_{LO} (mW)	Δf (GHz)	NEP (10^{-19} W/Hz)
0.19	1	12.5–14.5		40–60	−500		0.5	1.4	
0.2		10.7–12.5	10.6			77		>2.0	0.43 (1 GHz) 0.62 (1.8 GHz) 1.1 (4 GHz)
0.2	0.12	12		70	−1100	77		3–4	
	1.8		10.6	21		170	0.5–1	0.023	8.0 (10 GHz)
			10.6		−800	77		0.85	1.0 (20 MHz) 1.65 (1.5 GHz) 3.0 (1.5 GHz, 130 K)

Source: After Ref. [16].

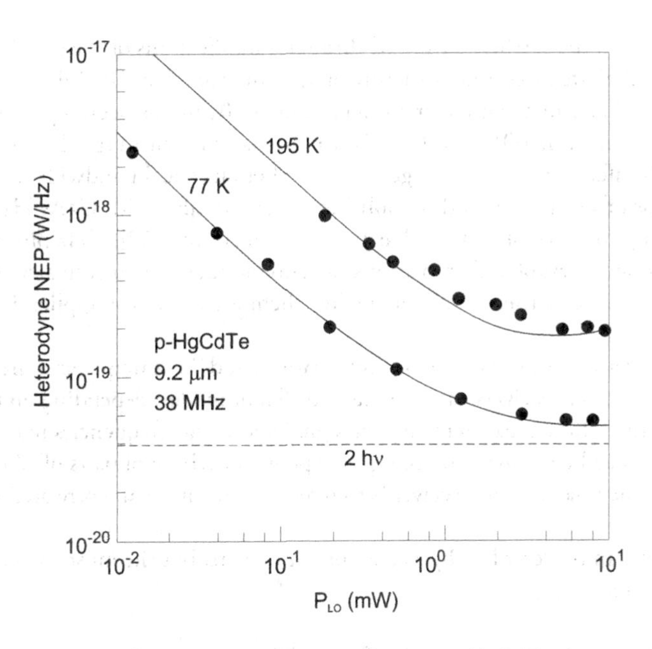

Figure 6.8 Calculated and measured NEP_H versus LO power for p-type HgCdTe photoconductor at 77 and 195 K. (From Ref. [15])

is proportional to the volume of the photoconductor, and heat sinking improves with reduced size. Figure 6.8 shows the calculated and measured NEP_H as a function of LO power for a $100 \times 100\,\mu m^2$ p-type photomixer at 77 and 195 K [15]. At 77 K, LO-noise-limited operation is easily achieved, but at 195 K, maximum LO-induced noise is comparable to or less than amplifier and dark current g-r noise. At this high LO power level, the quantum efficiency has dropped considerably below the low power value of over 70%. Table 6.2 summarizes the relevant information concerning HgCdTe photoconductive heterodyne detector operated at 193 K [16].

Table 6.2 HgCdTe photoconductive heterodyne mixer

MATERIAL	x = 0.18–0.19
Type	p-type, $N_a \approx 2 \times 10^{17}$ cm^{-3}
Surface	Native oxide passivated, ZnS AR coated
Detector temperature	193 K
Sensitive area	$100 \times 100\,\mu m^2$; $A_{opt} = 10^{-4}$ cm^2
Substrate	Sapphire heat sink
Response time	Down to a few ns
Bandwidths	Up to 100 MHz
Responsivity at 20 kHz	67 V/W
Detectivity at 20 kHz	2.7×10^8 cmHz$^{1/2}$W^{-1}
Minimum NEP at 193 K	2×10^{-19} W/Hz
Theoretical limit	4×10^{-20} W/Hz
P_{LO}	7 mW
λ_{LO}	10.6 μm

Source: After Ref. [16].

Infrared photomixers can be further optimized through modifications of the standard photoconductor geometry using an interdigitated electrode structure or an immersion lens [19,20].

It should be noticed that wideband coherent detection at 10 μm has been reported for GaAs/AlGaAs quantum well IR photodetectors (QWIPs) [21,22]. In comparison with HgCdTe-based technology, GaAs/AlGaAs QWIPs offers several advantages—the higher electrical bandwidth, more robust and tolerant it is to high levels of LO power, and monolithically compatible with GaAs HEMT amplifiers. The heterodyne detection up to an IF of 82 GHz has been demonstrated [22]. THz photon detectors have been demonstrated using intersubband transitions in semiconductor quantum structure. The potential of the very fast time response of these detectors makes them attractive for applications in THz heterodyne detection [23].

The most sensitive receivers at microwave, millimeterwave, and THz frequencies are based on the heterodyne principle. These mixer receivers can operate in different modes, depending on the configuration of the receiver and the nature of the measurement. The signal and image frequencies may be separated in the correlator, or the image may be removed by appropriate phase switching of pairs of LOs. The function of separating or dumping the image in the receiver is to remove some of the uncorrelated noise to improve the system sensitivity.

Heterodyne receivers can be described by a series of parameters, but the most encountered one is the receiver noise temperature,

$$T = T_{\text{mixer}} + LT_{IF}. \tag{6.11}$$

Here, the insides indicate the noise contribution of the mixer and the intermediate first amplifier stage, respectively, and L is the mixer conversion loss. More information about noise and its measurement in radio frequency equipment can be found in Hewlett Packard [24].

Generally, in THz receivers, the noise of a mixer is quoted in terms of an SSB, T^{SSB}, or DSB, T^{DSB}, mixer noise temperature. The quantum-noise limit for an SSB system noise temperature is [23,24]

$$T^{\text{SSB}} = \frac{hv}{k}. \tag{6.12}$$

This is the system noise temperature of a broad mixer receiver when performing narrowband measurements (within a single sideband). If we instead perform broadband (continuum) measurements, the desired signal will be twice as large and the ideal system noise temperature will be [25,26]

$$T^{\text{DSB}} = \frac{hv}{2k}. \tag{6.13}$$

The noise for a direct detector is normally quoted in terms of an NEP. To convert between an NEP and a T^{SSB}, one can use the relationship [27]

$$T^{\text{SSB}} = \frac{\text{NEP}^2}{2\alpha k P_{\text{LO}}}, \tag{6.14}$$

where P_{LO} is the incident LO power and a is the coupling factor of the radiation to the mixer. A similar relationship can be found for T^{DSB}

$$T^{\text{DSB}} = \frac{\text{NEP}^2}{4\alpha k P_{\text{LO}}}. \tag{6.15}$$

The technology that has traditionally been available for THz receivers utilizes Schottky-barrier diode mixers pumped by gas laser LOs. Recently, the first integrated transceivers have been fabricated–see Figure 6.9.

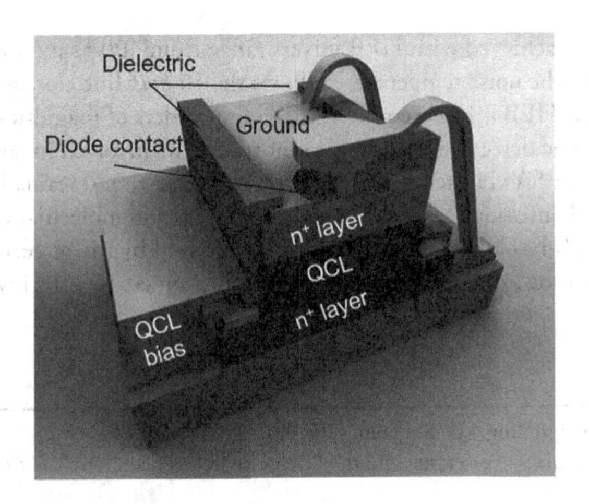

Figure 6.9 Structure of the integrated transceiver.

The Schottky diode is on top of the quantum cascade laser waveguide ridge. The integrated transceiver replaces the discrete LO and mixer units as well as the coupling optics.

The DSB noise temperatures achieved with Schottky diode mixers are presented in Figure 6.10 [28]. The noise temperature of such receivers has essentially reached a limit of about 50 $h\nu/k$ in the frequency range below 3 THz. Above 3 THz, there occurs a steep increase, mainly due to increasing losses of the antenna and reduced performance of the diode itself.

In the last two decades, an impressive improvement in receiver sensitivities has been achieved using superconducting mixers, with both superconductor–insulator–superconductor (SIS) and hot electron bolometer (HEB) mixers. In Figure 6.10, selected receiver noise temperatures are plotted. Nb-based SIS mixers yield almost quantum limited performance up to a gap frequency of 0.7 THz. The SIS mixers offer the best sensitivities, closely approaching the quantum limit.

Unlike Schottky diodes and SIS mixers, the HEB mixer is a thermal detector. The overall time constant of the process involved in the mixing is about few tens of picoseconds maximum, so the bolometric process is fast to follow the IF, but slow to respond directly to the incident LO or signal field. Figure 6.10 shows

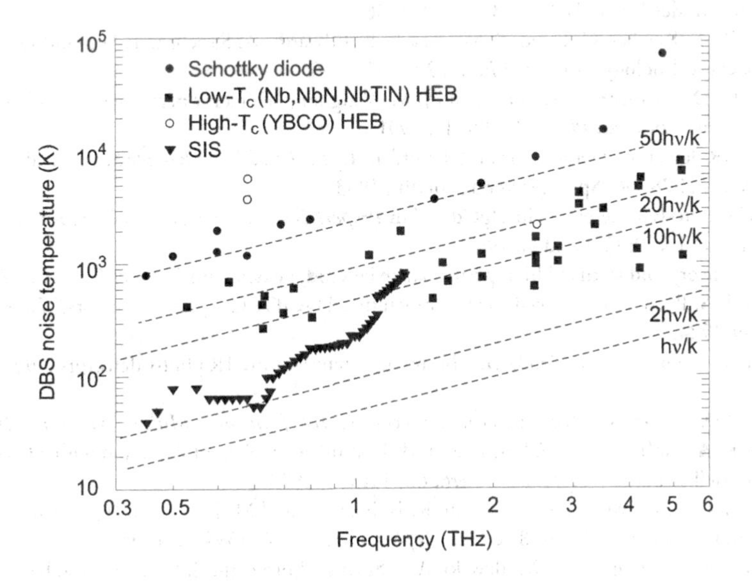

Figure 6.10 DSB noise temperature of Schottky diode mixers, SIS mixers, and HEB mixers operated in the THz spectral band. (From Ref. [28])

that DSB noise temperatures achieved with HEB mixers range from 400 K at 600 GHz up to 6800 K at 5.2 THz. Up to 2.5 THz, the noise temperature follows the 10 $h\nu/k$ line closely. In comparison with Schottky-barrier technology, HEB mixers require three to four orders of magnitude less LO power.

High-resolution heterodyne detection is an important technique involved in large astronomy and atmospheric projects such as ESA's Herschel Space Observation and the Atacama Large Millimeter Array to measure the astronomical emissions from various states of stars and molecular clouds [29]. The fine-line frequencies identified by high-resolution detection are characterized by the molecular species and their rotational transition. Due to the same reason, heterodyne detection enables the investigation of the Earth's atmosphere.

REFERENCES

1. R. Fessenden, "Wireless signaling," U. S. Patent 706,740, (1902).
2. J. S. Belrose, "Reginald Aubrey Fessenden and the birth of wireless telephony", *IEEE Antennas Propag. Mag.*, 44, 38, 2002.
3. M. C. Teich, "Coherent detection in the infrared," in *Semiconductors and Semimetals*, Vol. 5, ed. R. K. Willardson and A. C. Beer, 361–407, Academic Press, New York, 1970.
4. M. C. Teich, "Infrared heterodyne detection," *Proc. IEEE* 56, 37–46, 1968.
5. R. H. Kingston, *Detection of Optical and Infrared Radiation*, Springer-Verlag, Berlin, 1979.
6. D. L. Spears and R. H. Kingston, "Anomalous noise behavior in wide-bandwidth photodiodes in heterodyne and background-limited operation," *Appl. Phys. Lett.*, 34, 589–90, 1979.
7. R. J. Keyes and T. M. Quist, "Low-level coherent and incoherent detection in the infrared," in *Semiconductors and Semimetals*, Vol. 5, ed. R. K. Willardson and A. C. Beer, 321–59, Academic Press, New York, 1970.
8. F. R. Arams, E. W. Sard, B. J. Peyton, and F. P. Pace, "Infrared Heterodyne Detection with Gigahertz IF Response," in *Semiconductors and Semimetals*, Vol. 5, eds. R. K. Willardson and A. C. Beer, 409–34, Academic Press, New York, 1970.
9. F. R. Arams, E. W. Sard, B. J. Peyton, and F. P. Pace, "Infrared 10.6-micron heterodyne detection with gigahertz IF capability," *IEEE J. Quant. Electron.*, QE-3, 484–92, 1967.
10. C. Verie and M. Sirieix, "Gigahertz cutoff frequency capabilities of CdHgTe photovoltaic detectors at 10.6 μm," *IEEE J. Quant. Electron.*, QE-8, 180–91, 1972.
11. A. M. Andrews, J. A. Higgins, J. T. Longo, E. R. Gertner, and J. G. Pasko, "High-speed $Pb_{1-x}Sn_xTe$ photodiodes," *Appl. Phys. Lett.*, 21, 285–7, 1972.
12. D. J. Wilson, R. Foord, and G. D. J. Constant, "Operation of an intermediate temperature detector in a 10.6 μm heterodyne rangefinder," *Proc. SPIE*, 663, 155–8, 1986.
13. I. Melngailis, W. E. Keicher, C. Freed, S. Marcus, B. E. Edwards, A. Sanchez, T. Yee, and D. L. Spears, "Laser radar component technology," *Proc. IEEE*, 84, 227–67, 1996.
14. D. J. Wilson, G. D. J. Constant, R. Foord, and J. M. Vaughan, "Detector performance studies for CO_2 laser heterodyne systems," *Infrared Phys.*, 31, 109–15, 1991.
15. D. L. Spears, "IR detectors: Heterodyne and direct," in *Optical and Laser Remote Sensing*, eds. D. K. Killinger and A. Mooradian, 278–86, Springer-Verlag, Berlin, 1983.
16. G. Galeczki, "Heterodyne detectors in HgCdTe," in *Properties of Narrow Gap Cadmium-Based Compounds*, ed. P. Capper, 347–58, INSPEC, London, 1994.
17. D. L. Spears, "Theory and status of high performance heterodyne detectors," *Proceedings of SPIE*, 300, 174, 1981.
18. W. Galus and F. S. Perry, "High-speed room-temperature HgCdTe CO_2-laser detectors," *Laser Focus/Electro-Opt.*, 11, 76–9, 1984.
19. J. Piotrowski, W. Galus, and M. Grudzień, "Near room-temperature IR photo-detectors," *Infrared Phys.*, 31, 1–48, 1991.
20. T. Kostiuk and D. L. Spears, "30 μm heterodyne receiver," *Int. J. Infrared Millim. Waves*, 8, 1269–79, 1987.
21. E. R. Brown, K. A. McIntosh, F. W. Smith, and M. J. Manfra, "Coherent detection with a GaAs/AlGaAs multiple quantum well structure," *Appl. Phys. Lett.*, 62, 1513–5, 1993.
22. H. C. Liu, J. Li, E. R. Brown, K. A. McIntosh, K. B. Nichols, and M. J. Manfra, "Quantum well intersubband heterodyne infrared detection Up to 82 GHz," *Appl. Phys. Lett.*, 67, 1594–6, 1995.
23. H. C. Liu, H. Luo, C. Song, Z. R. Wasilewski, A. J. Spring Thorpe, and J. C. Cao, "Terahertz quantum well photodetectors," *IEEE J. Sel. Top. Quantum Electron.* 14, 374–7, 2008.
24. *Fundamentals of RF and Microwave Noise Figure Measurements*, Application Note 57-1, Hewlett Packard, July 1983.

25. A. R. Kerr, M. J. Feldman, and S.-K. Pan, "Receiver noise temperature, the quantum noise limit, and the role of the zero-point fluctuations," in *Proceedings of the 8th International Space Terahertz Technology Symposium* March 25–27, pp. 101–11, 1997. http://colobus.aoc.nrao.edu/memos, as MMA Memo 161.
26. E. L. Kollberg and K. S. Yngvesson, "Quantum-noise theory for terahertz hot electron bolometer mixers," *IEEE Trans. Microwave Theory Technol.*, 54, 2077–89, 2006.
27. B. S. Karasik and A. I. Elantiev, "Noise temperature limit of a superconducting hot-electron bolometer mixer," *Appl. Phys. Lett.*, 68, 853–5, 1996.
28. H.-W. Hübers, "Terahertz heterodyne receivers," *IEEE J. Sel. Topics Quantum Electron.*, 14, 378–91, 2008.
29. C. Kulesa, "Terahertz spectroscopy for astronomy: From comets to cosmology", *IEEE Trans. Terahertz Sci. Technol.*, 1, 232–40, 2011.

Infrared thermal detectors

7 Thermopiles

The thermocouple was discovered in 1821 by the Russian-born German physicist, J. Seebeck [1]. He discovered that at the junction of two dissimilar conductors, a voltage could be generated by a change in temperature (see Figure 7.1). Using this effect, Melloni produced the first bismuth–copper thermocouple detector in 1833 [2], to investigate the infrared spectrum. The small output voltage of thermocouples, of the order of some μV/K for metal thermocouples, prevented the measurements of very small temperature differences. Connecting several thermocouples in series, first carried out by Nobili in 1829, generated a higher and therefore measurable voltage.

The thermopile is one of the oldest IR detectors and is a collection of thermocouples connected in series to achieve better temperature sensitivity. For a long time, thermopiles were slow, insensitive, bulky, and costly devices. But with developments in semiconductor technology, thermopiles can be optimized for specific applications. Recently, thanks to conventional complementary metal-oxide-semiconductor (CMOS) processes, the thermopile's on-chip circuitry technology has opened the doors to mass production. Although thermopiles are not as sensitive as bolometers and pyroelectric detectors, they will replace these in many applications due to their reliable characteristics and good cost/performance ratio.

7.1 BASIC PRINCIPLE AND OPERATION OF THERMOPILES

The internal voltage responsible for current flow in a thermocouple is directly proportional to the temperature difference between the two junctions,

$$\Delta V = \alpha_s \Delta T, \tag{7.1}$$

where α_s is the Seebeck coefficient commonly expressed in μV/K.

The coefficient α_s is the effective or relative Seebeck coefficient of the thermocouple composed of two dissimilar conductors "a" and "b" by electrically joining one set of their ends. Consequently, a thermovoltage is equal to:

$$\Delta V = \alpha_s \Delta T = (\alpha_a - \alpha_b)\Delta T, \tag{7.2}$$

where α_a and α_b are the absolute Seebeck coefficients of the materials a and b. It should be noted that both relative and absolute Seebeck coefficients are temperature-dependent and the proportionality between the generated potential difference and the temperature gradient is valid only within the limit of a small temperature difference.

The output voltage of a single thermocouple is usually not sufficient; therefore, a number of thermocouples are connected in series to form a so-called thermopile. Figure 7.2 shows a thermopile that is constructed by a series connection of three thermocouples. The thermopile can be used as an infrared detector if the thermocouples are placed on a suspended dielectric layer and if an absorber layer is placed close to or on the top hot contacts of the thermopile. An important factor for obtaining a large output voltage from a thermopile is to obtain a high thermal isolation to maximize the temperature difference between hot and cold junctions, ΔT, for a specific absorber power. Corresponding to Equation 7.2, the totalized output voltage will be N times as high as for a single element:

$$\Delta V = N(\alpha_a - \alpha_b)\Delta T, \tag{7.3}$$

with N as the number of joined thermocouples.

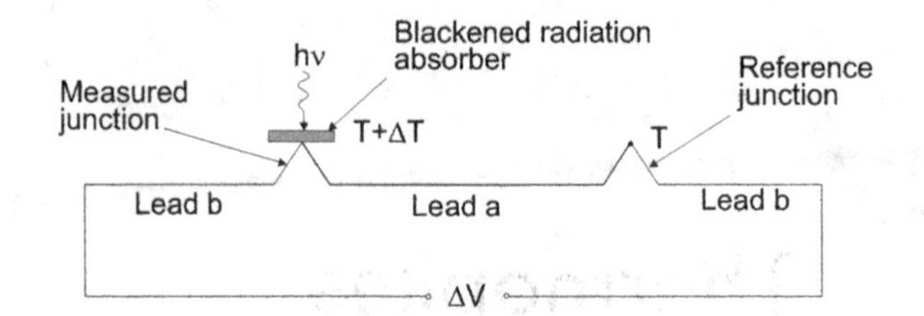

Figure 7.1 Two dissimilar leads connected in series.

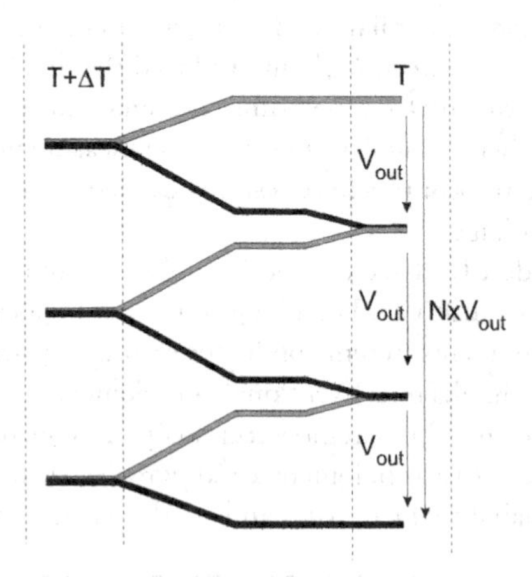

Figure 7.2 Schematic drawing of a three-thermocouple thermopile.

Apart from the Seebeck effect, yet another important thermoelectric offsets exist—the Peltier or Thomson effects. The latter two effects are present only when current flows in a closed thermoelectric circuit. The Peltier effect may give rise to considerable asymmetries in thermoelectric effect and special care due to this effect should be taken when designing thermal devices using heating resistors. This effect is reversible, since heat is absorbed or released depending on the direction of the current. The Thomson effect deals with similar heat exchange from a wire instead of a junction; when a wire has a temperature gradient along its length, a Thomson electromotive force is developed.

The Peltier coefficient Π (in V), which quantifies the ratio of heat absorption to electrical current, is equal to the Seebeck coefficient times the absolute temperature, called the first Kelvin relation,

$$\Pi = \alpha_s T. \tag{7.4}$$

A common value for the Peltier coefficient is 100–300 mV, which is significant if the heating voltage of, for instance, 1–3 V is used. In such cases, a Peltier heat flow of 10% of the generated (irreversible) Joule heat will flow from one contact to the other.

The literature describes a number of theoretical expressions for the Seebeck coefficient. Generally, and without going into thermodynamics, Pollock [3] states that the Seebeck effect does not arise as a result of the junction of dissimilar materials, nor is it directly affected by the Thomson or Peltier effects. Ashcroft and Mermin describe the Seebeck effect as a consequence of the mean electron velocity [4]. This assumption finally implicates a gradient of the electrochemical potential $\Phi(T)$ [in V] with respect to the temperature. On the basis of this, the output voltage of a thermocouple can be expressed as: [5]

$$\Delta V = \left[\frac{d\Phi_a(T)}{dT} - \frac{d\Phi_b(T)}{dT}\right]\Delta T. \tag{7.5}$$

A more descriptive explanation of the absolute Seebeck effect in an extrinsic nondegenerate semiconductor has been given by van Herwaarden [6]. He identifies two contributions to the Seebeck effect:

- Due to a change in the Fermi–Dirac distribution as the result of the temperature gradient.
- Due to change of the absolute value of the band edges being the result of an electric field induced by a net diffusion current and by phonon-drag currents.

A simplified expression for heavily doped silicon (>10^{19} cm^{-3}) at room temperature is: [7]

$$\alpha_n = -\frac{k}{q}\left(\ln\frac{N_c}{N_d} + 4\right) \quad \text{for n-type silicon,} \tag{7.6}$$

$$\alpha_p = \frac{k}{q}\left(\ln\frac{N_v}{N_a} + 4\right) \quad \text{for p-type silicon.,} \tag{7.7}$$

where N_c and N_v are the density of states in the conduction and valence bands, respectively; N_d is the donor concentration in n-type silicon; and N_a is the acceptor concentration in p-type silicon. Here, α_s has a positive sign for p-type silicon, whereas a negative sign is selected for n-type silicon. Further considerations of the Seebeck effect in semiconductors can be found in Graf and colleagues [8].

The Seebeck effect as it has been treated here is valid for bulk materials. Influence of other effects, such as grain size and grain boundary, can be found in the thin-film structure. A description of the effects in very thin films has been given by Salvadori et al. [9].

Table 7.1 lists the parameters for selected thermoelectric materials [8,10]. The bismuth/antimony (Bi/Sb) thermocouple is the most classical material pair in conventional thermocouples, not just from the historical point of view [11]. The Bi/Sb also has the highest Seebeck coefficient and the lowest thermal conductivity of all metal thermocouples. Despite the long tradition of metal thermocouples, new advantages can be found by using semiconducting materials, such as silicon (crystalline, polycrystalline) for thermoelectric materials due to the possibility of using standard integrated circuit processes. The Seebeck coefficient of semiconductor materials depends on the variation of the Fermi level of the semiconductor with respect to temperature; therefore, for semiconductor thermopiles, the magnitude and sign of the Seebeck coefficient and resistivity can be adjusted in the doping type and doping level.

For practical silicon sensor design purposes, it is very convenient to approximate the Seebeck coefficient as a function of electrical resistivity [12]:

$$\alpha_s = \frac{mk}{q}\ln\left(\frac{\rho}{\rho_o}\right), \tag{7.8}$$

with $\rho_o \cong 5\times10^{-6}$ Ωm and m ≅ 2.6 as constants. Typical values of the Seebeck coefficient of silicon are 500–700 μV/K for the optimum compromise between low resistance and high Seebeck coefficient. Equation 7.8 suggests that the Seebeck coefficient of a semiconductor increases in magnitude with increased resistivity, and therefore, with decreasing doping level. However, a thermopile material with very low electrical resistivity is not necessarily the best choice for a particular infrared detector, as the Seebeck coefficient is only one of the parameters influencing its overall performance.

For a wide variety of surface micromachined devices, polysilicon has rapidly become the most important material. The popularity of polysilicon in this area is a direct result of its mechanical properties and its relatively well-developed deposition and processing technologies. These characteristics along with the capability of utilizing established IC processing techniques make it a natural selection.

The measured Seebeck coefficient for polysilicon is shown in Figure 7.3 [13]. The Seebeck coefficient for p-type polysilicon and n-type polysilicon are almost the same, but the signs are opposite. The value of this coefficient greatly depends on impurity concentration. The used impurity concentrations are between 10^{19} and 10^{20} cm^{-3}.

Table 7.1 Parameters of selected thermoelectric materials at near-room temperature

SAMPLE	α_a (μV/K)	REFERENCE ELECTRODE	ρ (μΩm)	G_{th} (W/mK)
p-Si	100–1000		10–500	≈150
p-poly-Si	100–500		10–1000	≈20–30
p-Ge	420	Pt		
Sb	48.9	Pt	18.5	0.39
Cr	21.8			
Fe	15		0.086	72.4
Ca	10.3			
Mo	5.6			
Au	1.94		0.023	314
Cu	1.83		0.0172	398
In	1.68			
Ag	1.51		0.016	418
W	0.9			
Pb	−1.0			
Al	−1.66		0.028	238
Pt	−5.28		0.0981	71
Pd	−10.7			
K	−13.7			
Co	−13.3	Pt	0.0557	69
Ni	−19.5		0.0614	60.5
Constantan	−37.25	Pt		
Bi	−73.4	Pt	1.1	8.1
n-Si	−450	Pt	10–500	≈150
n-poly-Si	−100 to −500		10–1000	≈20–30
n-Ge	−548	Pt		

Source: After Refs. [8,10].

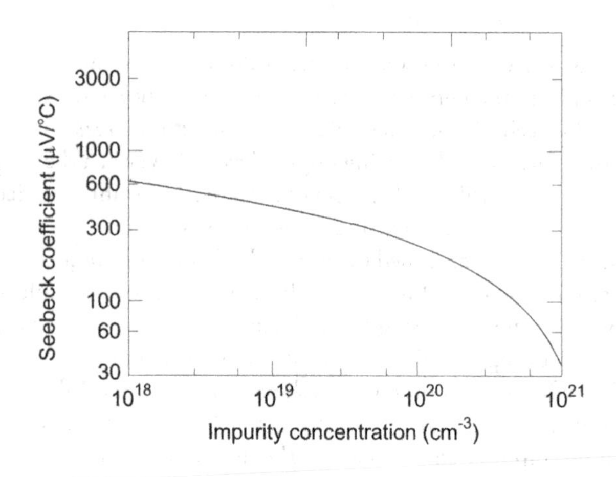

Figure 7.3 Seebeck coefficient for polysilicon. (From Ref. [13])

7.2 FIGURES OF MERIT

In further discussions, we will follow the considerations carried out in Chapter 4. Taking into account Equations 4.7 and 7.1, we notice that $K = \alpha_s$.

A consequence of Equation 4.9 is that the voltage responsivity is equal:

$$R_v = \frac{\alpha R_{th} \varepsilon}{\left(1 + \omega^2 \tau_{th}^2\right)^{1/2}}. \tag{7.9}$$

At very low frequencies, $\omega^2 \tau_{th}^2 << 1$, and then

$$R_v = \frac{\alpha \varepsilon}{G_{th}}. \tag{7.10}$$

Usually, the thermocouple rms noise voltage is dominated in the frequency range 0.1–1000 Hz by the thermal noise of the thermocouple resistance R. Then, according to Equation 4.22,

$$D^* = \frac{\alpha_s \varepsilon A_d^{1/2}}{G_{th} (4kTR)^{1/2}}. \tag{7.11}$$

If N thermocouples are placed in series, the responsivity is increased by N:

$$R_v = \frac{N \alpha \varepsilon}{G_{th}\left(1 + \omega^2 \tau_{th}^2\right)^{1/2}}, \tag{7.12}$$

and then for the thermopile with the dominant contribution of thermal noise, the detectivity can be expressed as:

$$D^* = \frac{\alpha_s \varepsilon (N A_d)^{1/2}}{G_{th} (4kTR_e)^{1/2}}, \tag{7.13}$$

where R_e is the electrical resistance of each thermocouple in the thermopile.

To produce an efficient device, the junction thermal capacity C_{th} must be minimized, to give as short a response as possible (see Equation 4.5), and the absorption coefficient optimized, which is often achieved by blackening the sensor. By careful design, it is possible for a thermocouple to be 99% efficient, with a spectrally flat response from visible to beyond 40 μm. For further discussion of black absorbers, see Blevin and Geist [14]. The spectral response is also determined by the material of the encapsulation window.

The junction should be fabricated from two materials with:

- a large Seebeck coefficient α_s,
- low thermal conductivity G_{th} (to minimize the heat transfer between the hot and cold junctions),
- low volume resistivity (to reduce the noise and heat developed by the flow of current).

Keep in mind that by scaling down a device by a certain percentage, the surface area only decreases with the square root (see Equation 7.11); the miniaturization of the thermopile is an appropriate way to increase the overall detectivity.

Unfortunately, these requirements are incompatible in view of the Wiedemann–Franz law [15] relating the thermal conductivity G_{th} and the electrical resistivity ρ:

$$\frac{G_{th} \rho}{T} = L. \tag{7.14}$$

L is known as the Lorentz number and has very nearly a constant value for most materials, especially metals, except at very low temperatures. This leads naturally to the well-known criterion of the figure of merit for thermoelectric materials in which a maximum value of

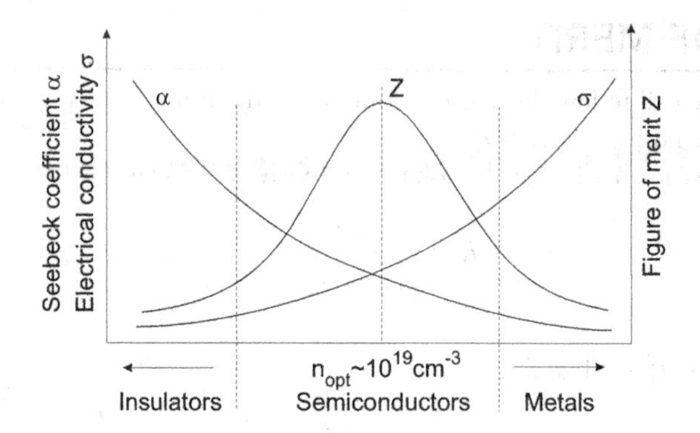

Figure 7.4 Thermoelectric properties of metals, semiconductors, and insulators. (From Ref. [20])

$$Z = \frac{\alpha_s^2}{\rho G_{th}} \tag{7.15}$$

is sought [16]. It is important to note that this thermoelectric figure of merit is defined in terms of output power delivered into an optimum load resistance, rather than the open-circuit voltage that enters directly into the definition of the responsivity, Equation 7.9.

The criteria of thermoelectric materials have been considered in detail in numerous papers and publications. Ioffe's development was based upon the influence of carrier concentration [17] and did not consider the influence of a change in effective mass or mobility. These aspects are reviewed and summarized by Egli [18], Cadoff and Miller [16], and in many more recently published papers [19–22].

Equation 7.11 indicates that the thermocouple materials should be chosen from low-resistivity materials. However, a lower ρ value also gives a lower Seebeck coefficient. Therefore, an optimum point needs to be determined considering all these parameters based on the figure of merit. Figure 7.4 shows a graph for the visual interpretation of the thermoelectric properties of metals, semiconductors, and insulators [20]. An optimum value for the figure of merit for semiconductors is achieved at a doping value of about 10^{19} cm^{-3} for single crystal silicon and polysilicon materials. A similar conclusion was predicted by Ioffe more than 50 years ago [17].

It should be noted that when the number of thermocouples is increased to obtain a high output voltage, it also increases the thermal conduction between the hot and cold junctions and the series electrical resistance and thermal noise. This means that care should be taken to optimize the number of thermocouples for a thermopile. Increasing the number of thermocouples does not necessarily increase the performance.

Equation 7.15 is valid for a single thermocouple material. For a thermocouple constructed with two different materials, *a* and *b*, the figure of merit is defined as follows: [16]

$$Z = \frac{(\alpha_a - \alpha_b)^2}{\left(\sqrt{\rho_a G_a} + \sqrt{\rho_b G_b}\right)^2}, \tag{7.16}$$

where ρ_a and ρ_b are electrical resistivities of the materials, and G_a and G_b are the thermal conductivities of the materials, respectively.

The total thermal conductivity G_{th} contains all contributions of thermal conductivity between absorber and heat sink (including thermal conductivity of the surrounding gas G_g, the support and thermoelectric conductors G_s, as well as the radiation losses G_R):

$$G_{th} = G_g + G_s + G_R + N(G_a - G_b). \tag{7.17}$$

The Z values of selected thermoelectric material pairs are listed in Table 7.2 [23].

Table 7.2 Z values of thermoelectric junction pairs at room temperature

JUNCTION PAIR	Z (K^{-1})
Chromel/Constantan	1.0×10^{-4}
Al/n-polySi or p-polySi	1.1×10^{-5}
n-polySi/p-polySi	1.4×10^{-5}
Bi/Sb	1.8×10^{-4}
$Bi_{0.87}Sb_{0.13}$/Sb	7×10^{-4}
n-PbTe/p-PbTe	1.3×10^{-3}
n-Bi_2Te_3/p-Bi_2Te_3	2×10^{-3}

Source: After Ref. [23].

7.3 THERMOELECTRIC MATERIALS

The first thermopiles were constructed from fine metallic wires, the most popular combinations being bismuth-silver, copper-constantan, and bismuth-bismuth/tin alloy [24]. The two wires are joined to form the thermoelectric junction and a blackened receiver, usually a thin gold foil that defines the sensitive area attached directly to the junction.

The development of semiconductors produced materials with much larger Seebeck coefficients, and hence the possibility of constructing thermopiles with increased sensitivities. However, the production of fine wire was impracticable. To make contacts, a new technique was developed in which the gold foil receiver was used as a constructing link between two active elements. The alloys recommended by Schwartz for this construction were (33% Te, 32% Ag, 27% Cu, 7% Se, 1% S) for the positive electrode and (50% Ag_2Se, 50% Ag_2S) for the negative material [15]. The responsivity of these devices was increased by about an order of magnitude (3×10^9 cmHz$^{1/2}$W^{-1} had been achieved) if they were mounted in a vacuum filled with a gas that had a low thermal conductivity; for example, xenon. The response time of these devices, usually about 30 min, was reduced when the thickness of the deposited films was reduced. However, due to the resistance increase of the device, the Johnson noise increased.

Although the sensitivity of the older thermopiles using metals is much lower than those using semiconductor elements, the metal elements can be made more robust and stable so that they are still widely used where a high degree of reliability and long-term stability is required. They have been successfully used in a number of space instruments, ground-based meteorological instruments, and in industrial radiation pyrometers [25,26].

Better-quality thermopile infrared detectors have been generally realized using vacuum evaporation (bismuth and antimony) and shadow masking of the thermocouple materials on thin plastic or alumina substrates [15,24]. This approach resulted in relatively large structures that lack the batch fabrication and process flexibility typical of devices employing the highly developed silicon IC technology. To profit from this technology, thermopile detectors that did use silicon, but only as a supporting structure, were realized [27,28].

Good electrical conductors (e.g., gold, copper, and silver) have very poor thermoelectric power (see Table 7.1). Metals with higher resistivity (especially antimony and bismuth), however, possess high thermoelectric power in combination with low thermal conductivity—they became the "classical" thermoelectrical materials. By doping these materials in combination with Se or Te, the thermoelectric coefficient has been improved up to 230 μV/K [11]. Fote and coworkers have improved the performance of thermopile linear arrays by combining Bi–Te and Bi–Sb–Te thermoelectric materials [29,30]. Compared with most other thermoelectric materials, their D^* values are highest (what is shown in Figure 7.5). However, Bi–Sb–Te materials are not readily available in a CMOS technology.

Silicon, as a promising material for thermoelectric devices due to its high Seebeck coefficient, was recognized as early as the 1950s [31,32]. However, early attempts to implement silicon into practical thermopile devices lacked definitive success primarily because the large number of couples required to generate a

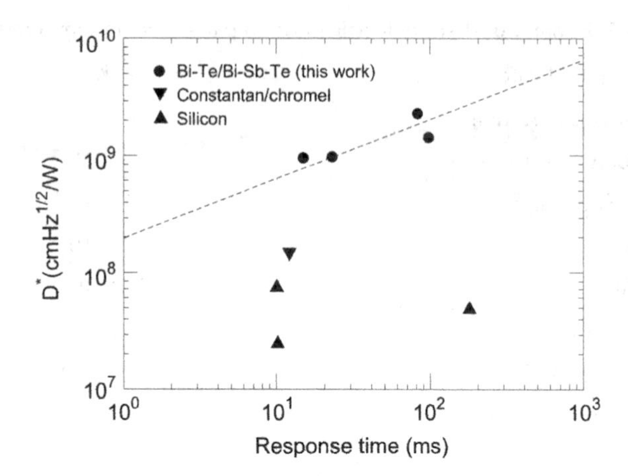

Figure 7.5 Representative data from literature showing reported D^* values as a function of response time for thin-film thermopile linear arrays. The dashed line represents the Fote and Jones results. Its slope indicates D^* proportional to the square root of response time, which is typical for thermopiles or bolometers with different geometries and the same material system. (From Ref. [29])

meaningful output made the devices excessively large. Silicon is a very good heat conductor, and the silicon substrate spoils the sensitivity with a thermal short-circuit. It turns out to be very rewarding to remove this silicon using planar technology [27], with its batch fabrication features and micromachining.

The last 20 years have seen significant advances in the development of integrated silicon thermopiles. The introduction of microfabrication facilitated practical solutions to miniaturization and consequently, a number of microfabricated thermopile embodiments have been presented in the literature, with varying degrees of success. Early efforts included silicon in their design merely as a substrate, using thin film metals for the thermoelectric materials. For example, Lahiji and Wise fabricated a thermopile in which Bi/Sb pairs were evaporated onto a thin silicon membrane insulated with a chemical vapor deposition (CVD) oxide [28]. Later designs incorporated silicon in the active thermopile junctions. The influence of aluminum interconnections on chips commonly used in ICs is negligible compared to the Seebeck coefficient for silicon.

The advantages of semiconductors in thermopile fabrication result from the following reasons:

- Semiconductors offer significantly higher Seebeck coefficients than metals.
- Semiconductor micromachining offers considerable miniaturization of devices that effectively reduce their thermal capacity.
- Production of high-performance thermopiles is compatible with standard IC processes like CMOS.

The significant progress made in recent years with microsensors is primarily due to micromachining technology. Micromachining is the fabrication of small, robust structures with submicron precision using a combination of photolithography and selective etching. Micromachining is possible in many materials, but silicon is favored because many micromachining techniques are similar to silicon-processing techniques. Silicon also enables the incorporation of electronics monolithically linked with the microstructures. The most widely used approach is n-polySi/Al thermopiles. Although Al has a very low Seebeck coefficient, this approach is used widely, as it is easy to implement with post-CMOS processes. The p-polySi/n-polySi approach is very attractive, as it provides relatively high Seebeck coefficient. The basic concepts of bulk and surface micromachining technology are discussed, for example, in Ristic's monograph [33].

Table 7.3 presents some representative parameters of different micromachined CMOS thermopiles [8]. It should be mentioned that different alternative materials not included in Table 7.3 have been proposed. Dehe et al. [34] proposed AlGaAs/GaAs thermopile with high Seebeck coefficient in GaAs of about −670 μV/K, which results from the comparably high charge carrier mobility of 470 cm^2/Vs (for comparison, polysilicon has a carrier mobility of 24 cm^2/Vs). However, high thermal conductivity avoids the common usage of AlGaAs thermopiles. Also, a concept for the realization of InGaAs/InP micromachined

Table 7.3 Characteristics data of different micromachined thermopiles

SORT	AREA (mm^2)	D^* (10^7 JONES)	R (μV/W)	MATERIAL SYSTEM	τ (ms)	α_s (μV/K)	COUPLES	ATM.	DATA
CB	0.013	0.68	10	Al/poly		58	20		Sim
CB	0.77	1.5	25	Al/poly		58	200		Sim
CB	15.2	5		p-Si/Al	300	700	44	Air	Meas
MB	15.2	10	>10	p-Si/Al		700	44	Vac	Meas
MB	0.12	1.7	12	Al/poly	10	−63	4 × 10		Meas
MB	0.3	2	44	n,p poly	18	200	4 × 12		Meas
MB	0.15	2.4	72	n,p poly	10	200	4 × 12		Meas
MB	0.15	2.4	150	n,p poly	22	200	4 × 12	Kr	Meas
MB	0.12	1.74	12	Al/poly AMS	10	65	10	Air	Meas
MB	0.12	1.78	28	Al/poly AMS	20	65	2 × 24	Air	Meas
MB	0.42	4.4	11	InGaAs/InP				Air	Meas
M	0.42	71	184	InGaAs/InP				Vac	Meas
M	4	6	6	Bi/Sb	15	100	60		Meas
M	4	3.5	7	n-poly/Au	15		60		Sim
M	4	4.8	9.6	p-poly/Au	15		60		Sim
M	0.25	9.3	48	p-poly/Al	20		40		
M	3.28	13	12	p-poly/Al	50		68		
M	0.2	55	180	Bi/Sb	19	100	72	Air	Meas
M	0.2	88	290	Bi/Sb	35	100	72	Kr	Meas
M	0.2	52	340	$Bi_{0.50}Sb_{0.15}Te_{0.35}$/$Bi_{0.87}Sb_{0.13}$	25	330	72	Air	Meas
C	0.2	77	500	$Bi_{0.50}Sb_{0.15}Te_{0.35}$/$Bi_{0.87}Sb_{0.13}$	44	330	72	Kr	Meas
C	9	26	14.8	Bi/Sb	100		72	Ar	Meas
C	0.785	29	23.5	Bi/Sb	32		15	Ar	Meas
C	0.06	25	194	Si	12		20	Ar	Meas
C	0.37	5.6	36	CMOS	<6				Meas
C	1.44	8.7	27	CMOS	<6				Meas
C	0.37	5.6	36	CMOS	<6				Meas
C	1.44	4.6	12	CMOS	30				
C	0.2	45	200		20		72	N_2	Meas
C	1.44	35	100		30		200	N_2	Meas
C	0.49	21	110	BiSb/NiCr	40		100		Meas
C	0.49	6	35	CMOS	25				Meas
C	1.44	8	20	CMOS	35				Meas
C	0.6	24	80	CMOS	<40			Vac	Meas

Source: After Ref. [8].

CB: cantilever beam thermopile, MB: micro bridge thermopile, M: membrane thermopile, C: commercial thermopile, Sim: simulated data, Meas: measured data.

thermoelectric sensors has been presented [35]. Key futures of this material system are high thermal resistivity (0.09 Km/W) and high carrier mobility. This is combined with high Seebeck coefficient of 790 μV/K for p-type InGaAs and –450 μV/K for n-InGaAs.

7.4 MICROMACHINED THERMOPILES

Thermopile infrared detectors can easily be integrated on standard CMOS chips in several structures by adaptational post-CMOS surface or the bulk micromachining process. The increase in production capacity in recent years is due to a very good compromise between cost-efficiency and performance. Baltes et al. [36] have made a systematic assessment of the compatibility of the thermopile fabrication with the CMOS technology, including the issue of materials of the thermopile absorbers.

A schematic drawing of a modern thermopile structure is shown in Figure 7.6a [8]. It consists of series thermocouples supported by a micromachined isolating membrane. From a technological and economical point of view, the common industrial device today is the CMOS thermopile, which consists of Al/Si thermocouples. The hot junctions of the thermopile located on a membrane is covered with an absorbing layer. In contrast, the cold junctions lie on the substrate rim, which acts as a heat sink. Usually, the link between the hot and cold region is formed by a very thin (several micron thick) silicon membrane (Figure 7.6b) or cantilever beam, which contains the thermopile. The heat generated in the hot region flows through the silicon membrane (or cantilever beam) to the cold region, usually at an ambient temperature, causing a temperature difference across the thermal resistance of the membrane (or cantilever beam).

7.4.1 DESIGN OPTIMIZATION

An accurate analysis of the micromachined thermopile is complicated due to its complex device architecture and the influence of 3-D heat flow effects on device performance. For example, to achieve a maximum of detectivity for a thermopile on a membrane, two conditions should be fulfilled [23]. The first is:

$$\frac{A_a}{A_b} = \left(\frac{G_{tha}\rho_a}{G_{thb}\rho_b}\right)^{1/2}, \tag{7.18}$$

where A_a is the cross-section area of material a, which is deposited upon a leg connecting the membrane to the substrate, and A_b is the corresponding value for material b. The G_{tha} and G_{thb} are the thermal conductivities of materials a and b, respectively.

The second condition is:

$$N\left(G_{tha}A_a + G_{thb}A_b\right) = 2G_{thm}tw, \tag{7.19}$$

where G_{thm} is the thermal conductivity of the membrane material, and t and w are the thickness and width of the legs; it is assumed that there are two identical legs. Equation 7.19 states that the thermal conductance to the substrate through the thermoelectric lines is equal to that through the membrane material.

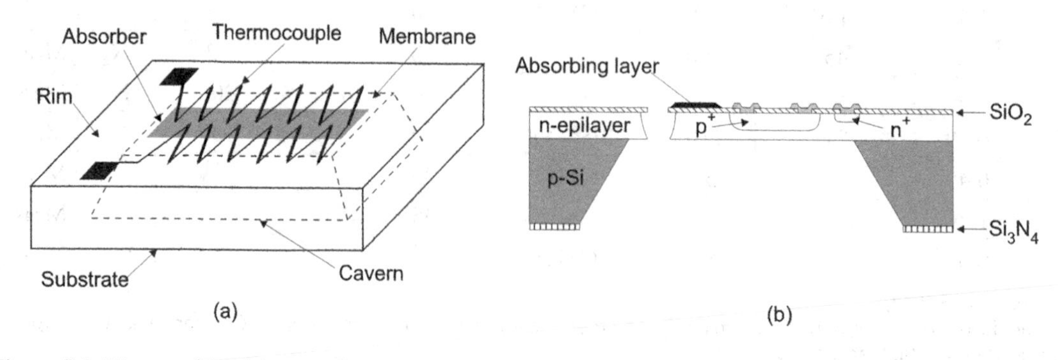

Figure 7.6 Micromachined thermopile: (a) general drawing and (b) schematic cross section.

Assuming the given conditions, detectivity reaches its maximum value given by: [23]

$$D^* = \frac{Z^{1/2}\tau_{th}^{1/2}\beta\kappa}{2^{3/2}\left(C_m t + C_{abs}\right)^{1/2}\left(kT\right)^{1/2}}, \tag{7.20}$$

where C_m is the heat capacity per unit volume of the membrane material, and C_{abs} is the heat capacity per unit surface area of the absorbing material. Also, β is a detector fill factor, and κ is an optical absorption coefficient defined as the fraction of the radiant power falling on the sensitive area, which is absorbed by that area.

The thermal response time of the optimized configuration becomes

$$\tau_{th} = \frac{\left(C_m t + C_{abs}\right)Al}{4G_{thm}tw}. \tag{7.21}$$

Here l is the length of a leg.

Equation 7.20 indicates that thermopile optimization concerns optimization through the choice of thermoelectric material and optimization through the design of the structure. In practice, however, silicon compatibility technological steps in fabrication monolithic arrays considerably limit the choices.

7.4.2 THERMOPILE CONFIGURATIONS

Generally, two thermopile configurations are reported in the literature [8]; deposited either in a one-layer [12,37,38] or a multilayer configuration [11,39].

The most common thermopile structure is the one-layer thermopile where both thermocouple leads are deposited using conventional lithography. Thermocouple leads lie in a single plane one beside the other. The one-layer structures allow simple and fast processing and provide good thermal isolation due to its flatness. An example of this structure is shown in Figure 7.7 [40]. The structure consists of p-type silicon strips in an n-type epilayer or well, interconnected by aluminum strips. The device contains a 10-μm thick cantilever beam with one half covered by an absorbing layer and the other half by a 44-strip thermopile. Shallow p- and n-type diffusions were carried out successively. During the n-type diffusion, n^+ islands were created to achieve a good contact to the epilayer for the electrochemically controlled etching step. The doping of the strips is optimized so that the Seebeck coefficient is high (700 μV/K). A 750 Å low pressure chemical vapor deposition (LPCVD) Si_3N_4 layer was grown on the back side of the wafer. Next, a two-stage etch process was undertaken—first, electrochemically controlled etching was used to stop etching at the epilayer/substrate junction and to form a membrane; second, reactive plasma etching in CF_4+6% O_2 was used to form the cantilever beam structure. Lastly, the absorbing area of each beam

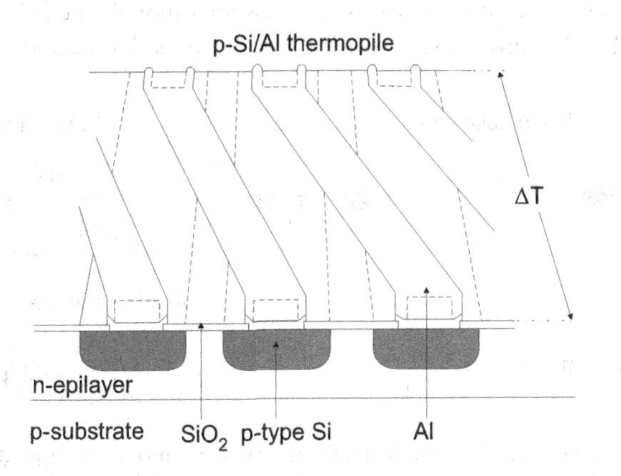

Figure 7.7 Schematic representation of the integrated Al/p-Si thermopile. (From Ref. [40])

was coated with a layer of an absorbent material. The final structure is similar to that shown in Figure 7.6b. The detectors had a responsivity in air of approximately 6 V/W and detectivity measured in air for a 500-K blackbody source of approximately 5×10^7 cmHz$^{1/2}$W^{-1}.

In the case of multilayer configuration, one thermocouple lead lies upon the other separated by a thin insulating layer (e.g., photoresist). The insulating layer is removed only at the hot and cold ends of the first patterned thermoelectric film, forming a small contact window [11]. However, designing a multilayer thermopile requires a careful balance between higher integration density, higher internal electrical resistance, and lower thermal resistance (the last two effects are due to the thicker layer stack of a multilayer thermopile).

7.4.3 MICROMACHINED THERMOPILE TECHNOLOGY

From the point of view of micromachining techniques, thermopiles can be generally devoted into bulk micromachined and surface-bulk micromachined devices [36]. The first type of structure is made by accurate micromachining of the silicon substrate—the membrane etch process is done from the back side to remove the entire bulk silicon under the membrane. On the contrary, in surface-bulk micromachined thermopile, the micromachined process is provided from the front side through windows in the deposited thin film stacks. Then, only a part of the silicon substrate is removed, forming a cavity under the membrane.

The fabrication process for bulk micromachined thermopiles utilizing microjoinery is shown in Figure 7.8. A comprehensive description of this process is given by Allison and colleagues [41]. The strips of alternating p- and n-type silicon are fabricated on separate n- and p-type, single crystal wafers, bonded together, and electrically connected in series to form the p/n couple junctions, as shown in the cross-section diagram of Figure 7.8c. The basic n- and p-type elements with mating grooves commonly fabricated by anisotropic KOH etching are bonded using either organic adhesives, frit glass, or high-temperature bonding. The etching process terminates on (111) crystal planes that provide flat mating surfaces for bonding. Before the wafers can be bonded, an electrically insulating thermal SiO_2 oxide film is grown on the surface of the grooves. Finally, the series connection is carried out by lift-off patterning of e-beam–evaporated aluminum. In this way, the thermopile, as a linear array of thermocouples with a chip size of 0.5 × 3.5 cm^2 and approximately 100-μm thickness, is fabricated.

In spite of the simple fabrication of bulk micromachined thermopile, this kind of device is characterized by high thermal conductivity between cold and hot junctions, which results in poor performance. To improve the thermal isolation, different freestanding micromachined structures have been developed. One such solution is a closed micromachined structure shown in Figure 7.9 in which the heat from cold contacts is isolated. The cold region containing cold junctions is formed by a wafer-thin rim around the etched membrane. The rim serves both as a heat sink as well as suspension of the etched structure and mechanical protection [42].

The closed-membranes micromachining process has been reported in Lahiji and Wise [28] and Elbel [37]. Lahiji and Wise, to develop a fabrication process for a monolithic Bi/Sb detector, began using (100)-oriented Si wafers [28]. The process starts with the thermal oxidation of the wafer to a thickness of

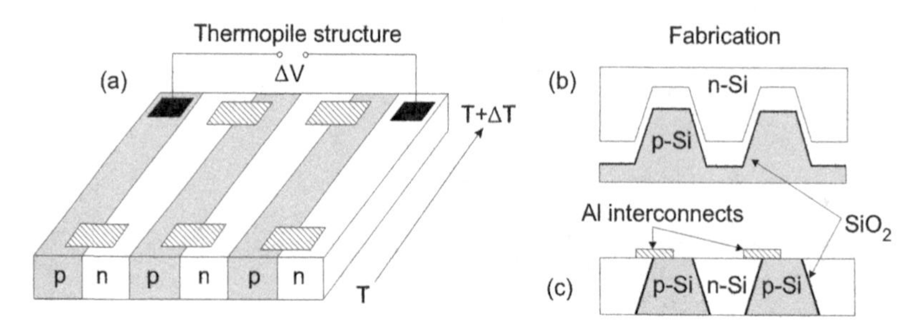

Figure 7.8 Bulk micromachined silicon thermopile: (a) Diagram of thermopile design, (b) Fabrication sequence of joints (grooves) formed in both p- and n-type Si by KOH etching, and (c) Final device structure (annealed, diced, and packaged).

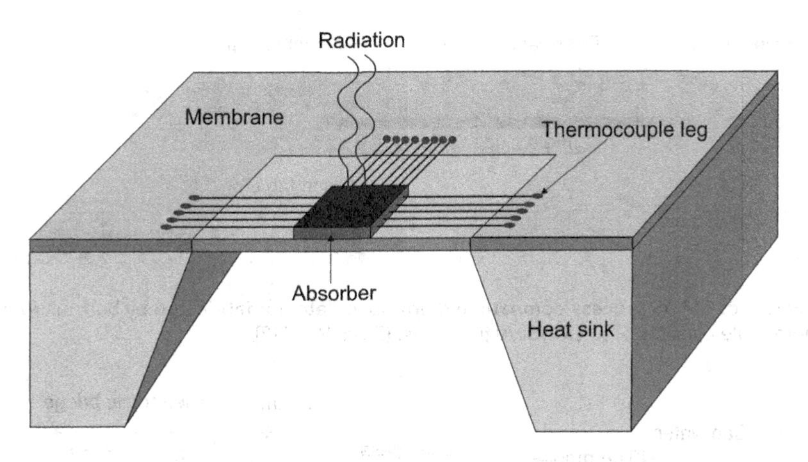

Figure 7.9 Cross section of a bulk micromachined thermopile.

about 0.8 μm. Next, the desired patterns are defined on the front and back side of the wafer. After that, a shallow boron diffusion layer over the membrane area is used as an etch stop for the anisotropic etch process. Because of the high electrical conductivity of this layer, a thin dielectric layer is next deposited on the front side, isolating the thermopile. The best results have been achieved using a very thin thermal oxide layer followed by thin CVD layers of silicon dioxide and silicon nitride. Finally, the thermocouple materials and interconnecting metallization are deposited and patterned using conventional lithography.

A simpler fabrication process that is widely used today has been proposed by Elbel [37]. In the first step, on the front side of a (100)Si wafer, a thin stress-optimized CVD $Si_3N_4/SiO_2/Si_3N_4$ stack is deposited. As a pattern mask for the etch process on the back side of the wafer, a single Si_3N_4 layer is deposited. The membrane thickness is defined by the etch-stop bottom layer of the dielectric stack on the front. Using this describing process, a 1-μm thick membrane with low thermal conductivity (2.4 W/mK) compared to bulk Si (140 W/mK) can be realized.

A further increase of the thermal resistance can be found by reducing the contact between the membrane and silicon rim using cantilever beam membranes. In this way, the sensitivity of the device considerably increases, but unfortunately, the mechanical stability reduces. The thermal beam resistance can be determined by the thermal sheet resistance times the length-to-width (l/w) ratio of the beam. In comparison with circular membranes, the l/w ratio of cantilever thermocouple is of the order of five, thus increasing the thermal resistance by a factor of 10 [8]. Moreover, because of increasing the absorbing area without reducing the thermal resistance, the responsivity increases considerably compared to the circular membrane. It should be noticed that the overall performance reduces considering the increased time constant and a mechanically more fragile structure.

Usually, bulk-micromachined beam thermopiles are fabricated by an anisotropic etch process, which consists of two phases [43,44]. In the first phase, the entire bulk silicon under the membrane is removed from the back side using electrochemically controlled etching to stop the etching at the epilayer/substrate junction. Next, the front side of the wafer is prepared for a plasma etch step, realizing the membrane on three sides. Figure 7.10 shows an example of the pixel structure on n^+-polysilicon/Al thermopile fabricated with a silicon bulk micromachining technology [45].

Surface-bulk micromachining is a combination of bulk and surface micromachining [46]. Fabrication of structures in surface micromachining occurs from surface-deposited thin films being released by etching away sacrificial layers beneath functional layers. However, in contrast to conventional surface micromachined devices, bulk silicon is removed through small openings in the surface layers, which reduces the complexity of the production process. From a technological point of view, a surface-bulk micromachined CMOS thermopile with closed membrane and cap wafer is the best device configuration today [8].

An example of a closed CMOS surface-bulk micromachined thermopile is shown in Figure 7.11. In the first step of fabrication of a Bosch thermopile chip (Figure 7.11a), multiple oxide and epilayers forming membrane and thermocouples on the substrate surface have been provided. An anisotropic etching was

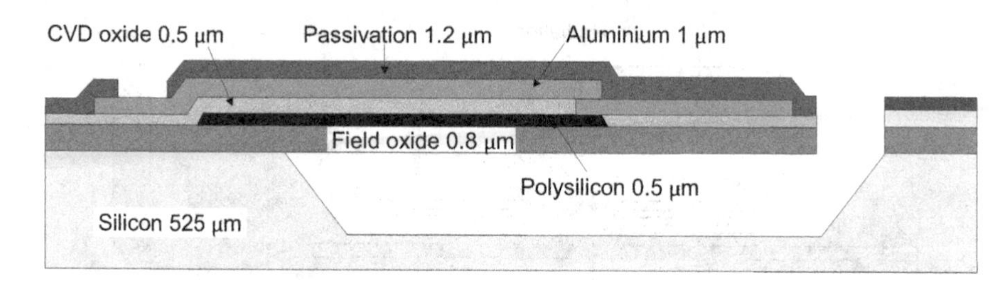

Figure 7.10 Structure of CMOS process-compatible thermopile detector fabricated by bulk micromachining process. The thermopile consists of n⁺-polySi/Al junctions. (From Ref. [45])

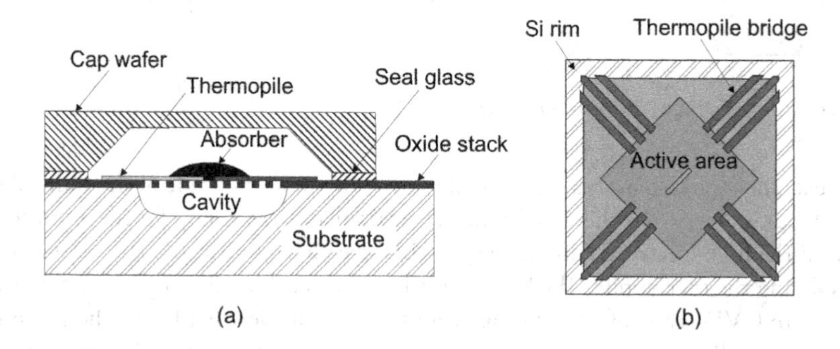

Figure 7.11 Surface-bulk micromachined thermopile detectors: (a) Bosch thermopile chip. (From Ref. [8]) (b) Bridge thermopile structure.

applied to perforate surface layers to access the substrate and the substrate under the surface was removed through these etch windows. In the final step, the thermopile was hermetically sealed. This thermopile with a closed membrane is mechanically more stable than that with a beam-type membrane.

To improve mechanical stability, different kinds of bridge-type structures have been proposed. For example, Figure 7.11b shows a sensor configuration with a 4-bridge structure and absorber on the central membrane [47].

To achieve high sensitivity, the detector's absorption efficiency must be high, but the thermal mass of the absorption layer must be small. In general, absorbers can be divided into three groups: metallic films, porous metal blacks, and thin-film stacks.

The metallic films are characterized by a narrow spectral bandwidth and their absorption strongly depends on film thickness. A maximum absorption of 50% for a 17-µm thick Au layer has been reported [48]. Porous metal blacks are deposited by electroplating (e.g., platinum) and evaporation (e.g., gold). Another way of enhancing absorption is the implementation of interference phenomena to set up a resonant cavity within the detector. With this kind of absorber, 90% of the radiation in the 8–14-µm wavelength range can be absorbed. More information about absorbers can be found in References 48–51.

REFERENCES

1. J. T. Seebeck, "Magnetische polarisation der metalle und erze durch temperatur-differenz," *Abh. Dtsch. Akad. Wiss. Berlin*, 265–373, 1822.
2. M. Melloni, "Ueber den durchgang der wärmestrahlen durch verschiedene körper," *Ann. Phys.*, 28, 371–8, 1833.
3. D. D. Pollock, "Thermoelectric phenomena," in *CRC Handbook of Thermoelectrics*, ed. D. M. Rowe, 7–17, CRC Press, Boca Raton, FL, 1995.
4. N. W. Ashcroft and N. D. Mermin, *Solid State Physics*, Saunders College, Philadelphia, PA, 1976.
5. S. M. Sze, *Semiconductor Devices*, Wiley, New York, 2002.
6. A. W. Van Herwaarden, "The Seebeck effect in silicon ICs," *Sens. Actuators*, 6, 245–54, 1984.
7. J. H. Kiely, D. V. Morgan, and D. M. Rowe, "The design and fabrication of a miniature thermoelectric generator using MOS process techniques," *Meas. Sci. Technol.*, 5, 182–9, 1994.

8. A. Graf, M. Arndt, M. Sauer, and G. Gerlach, "Review of micromachined thermopiles for infrared detection," *Meas. Sci. Technol.*, 18, R59–5, 2007.
9. M. C. Salvadori, A. R. Vaz, F. S. Teixeira, M. Cattani, and I. G. Brown, "Thermoelectric effect in very thin film Pt/Au thermocouples," *Appl. Phys. Lett.*, 88, 133106, 2006.
10. J. Schieferdecker, R. Quad, E. Holzenkämpfer, and M. Schulze, "Infrared thermopile sensors with high sensitivity and very low temperature coefficient," *Sens. Actuators, A*, 46–47, 422–7, 1995.
11. F. Völklein, A. Wiegand, and V. Baier, "High-sensitive radiation thermopiles made of Bi-Sb-Te films," *Sens. Actuators*, 29, 87–91, 1991.
12. A. W. Van Herwaarden and P. M. Sarro, "Thermal sensors based on the Seebeck effect," *Sens. Actuators*, 10, 321–46, 1986.
13. T. Kanno, M. Saga, S. Matsumoto, M. Uchida, N. Tsukamoto, A. Tanaka, S. Itoh, et al., "Uncooled infrared focal plane array having 128×128 thermopile detector elements," *Proc. SPIE*, 2269, 450–9, 1994.
14. W. R. Blevin and J. Geist, "Influence of black coatings on pyroelectric detectors," *Appl. Opt.*, 13, 1171–8, 1974.
15. A. Smith, F. E. Jones, and R. P. Chasmar, *The Detection and Measurement of Infrared Radiation*, Clarendon, Oxford, UK, 1968.
16. I. B. Cadoff and E. Miller, *Thermoelectric Materials and Devices*, Reinhold, New York, 1960.
17. A. F. Ioffe, *Semiconductor Thermoelements and Thermoelectric Cooling*, Infosearch Ltd., London, 1957.
18. P. H. Egli, *Thermoelectricity*, Wiley, New York, 1958.
19. H. J. Goldsmid, "Conversion efficiency and figure-of-merit," in *CRC Handbook of Thermoelectrics*, ed. D. M. Rowe, 19–26, CRC Press, Boca Raton, FL, 1995.
20. F. Voelklein, "Review of the thermoelectric efficiency of bulk and thin-film materials," *Sens. Mater.*, 8, 389–408, 1996.
21. D. M. Rowe, G. Min, V. Kuznietsov, and A. Kaliazin, "Effect of a Limit to the Figure-of-Merit on Thermoelectric Generation," *Energy Conversion Engineering Conference and Exhibition, (IECEC)*, pp. 123–34, 35th Intersociety, Las Vegas, NV, 2000.
22. T. Akin, "CMOS-based thermal sensors," in *Advanced Micro and Nanosystems*, Vol. 2, eds. H. Baltes, O. Brand, G. K. Fedder, C. Hierold, J. Korvink, and O. Tabata, 479–511, Wiley, Weinheim, Germany, 2005.
23. P. W. Kruse, *Uncooled Thermal Imaging. Arrays, Systems, and Applications*, SPIE Press, Bellingham, WA, 2001.
24. B. Stevens, "Radiation thermopiles," in *Semiconductors and Semimetals*, Vol. 5, eds. R. K. Willardson and A. C. Beer, 287–317, Academic Press, New York, 1970.
25. A. J. Drummond, "Precision radiometry and its significance in atmospheric and space physics," in *Advances in Geophysics*, Vol. 14, 1–52, Academic Press, New York, 1970.
26. R. W. Astheimer and S. Weiner, "Solid-backed evaporated thermopile radiation detectors," *Appl. Opt.*, 3, 493–500, 1964.
27. C. Shibata, C. Kimura, and K. Mikami, "Far infrared sensor with thermopile structure," *Proceedings of the 1st Sensor Symposium* 221–25, Japan, 1981.
28. G. R. Lahiji and K. D. Wise, "A batch-fabricated silicon thermopile infrared detector," *IEEE Trans. Electron Devices*, ED-29, 14–22, 1982.
29. M. C. Fote, E. W. Jones, and T. Caillat, "Uncooled thermopile infrared detector linear arrays with detectivity greater than 10^9 cmHz$^{1/2}$/W," *IEEE Trans. Electron Devices*, 45, 1896–1902, 1998.
30. M. C. Fote and E. W. Jones, "High performance micromachined thermopile linear arrays," *Proc. SPIE*, 3379, 192–7, 1998.
31. C. Herring, "Theory of the thermoelectric power of semiconductors," *Phys. Rev.*, 96, 1163–87, 1954.
32. T. H. Geballe and G. W. Hull, "Seebeck effect in silicon," *Phys. Rev.*, 98, 940–7, 1955.
33. L. Ristic, ed., *Sensor Technology and Devices*, Artech House, Boston, MA, 1994.
34. A. Dehé, K. Fricke, and H. L. Hartnagel, "Infrared thermopile sensor based on AlGaAs-GaAs micromachining," *Sens. Actuators A*, 46–47, 432–6, 1995.
35. A. Dehé, D. Pavlidids, K. Hong, and H. L. Hartnagel, "InGaAs/InP thermoelectric infrared sensors utilizing surface bulk micromachining technology," *IEEE Trans. Electron. Devices*, 44, 1052–8, 1997.
36. H. Baltes, O. Paul, and O. Brand, "Micromachined thermally based CMOS microsensors," *Proc. IEEE*. 86, 1660–78, 1998.
37. T. Elbel, "Miniaturized thermoelectric radiation sensor," *Sens. Mater.*, A3, 97–109, 1991.
38. A. Mzerd, F. Tchelibou, A. Sackda, and A. Boyer, "Improvement of thermal sensors based on Bi_2Te_3; Sb_2Te_3, and $Bi_{0.1}Sb_{1.9}Te_3$," *Sens. Actuators*, A47, 387–90, 1995.
39. T. Elbel, S. Poser, and H. Fischer, "Thermoelectric radiation microsensors," *Sens. Actuators*, A42, 493–6, 1994.

40. P. M. Sarro and A. W. Van Herwaarden, "Infrared detector based on an integrated silicon thermopile," *Proc. SPIE*, 807, 113–8, 1987.
41. S. C. Allison, R. L. Smith, D. W. Howard, C. Gonzalez, and S. D. Collins, "A bulk micromachined silicon thermopile with high sensitivity," *Sens. Actuators*, A102, 32–9, 2003.
42. I. Simon and M. Arndt, "Thermal and gas sensing properties of a micromachined thermal conductivity sensor for the detection of hydrogen in automotive applications," *Sens. Actuators*, A98–98, 104–8, 2002.
43. A. W. Van Herwaarden, P. M. Sarro, and H. C. Meijer, "Integrated vacuum sensor," *Sens. Actuators*, 8, 187–96, 1985.
44. P. M. Sarro and A. W. Van Herwaarden, "Silicon cantilever beams fabricated by electrochemically controlled etching for sensor applications," *J. Electrochem. Soc.*, 133, 1724–9, 1986.
45. R. Lenggenhager, H. Baltes, J. Peer, and M. Forster, "Thermoelectric infrared sensors by CMOS technology," *IEEE Electron Devices*, 13, 454–6, 1992.
46. J. M. Bustillo, R. T. Howe, and R. S. Muller, "Surface micromachining for microelectromechanical systems," *Proc. IEEE*, 86, 1552–74, 1998.
47. C.-H. Du and C. Lee, "Investigation of thermopile using CMOS compatible process and front-side Si bulk etching," *Proc. SPIE*, 4176, 168–78, 2000.
48. W. Lang, K. Kühl, and H. Sandmaier, "Absorbing layers for thermal infrared detectors," *Sens. Actuators*, A34, 243–8, 1992.
49. N. Nelms and J. Dowson, "Goldblack coating for thermal infrared detectors," *Sens. Actuators*, A120, 403–7, 2005.
50. A. Hadni and X. Gerbaux, "Infrared and millimeter wale absorber structured for thermal detectors," *Infrared Phys.*, 30, 465–78, 1990.
51. A. D. Parsons and D. J. Fedder, "Thin-film infrared absorber structures for advanced thermal detectors," *J. Vac. Sci. Technol.*, A6, 1686–9, 1988.

8 Bolometers

Another widely used thermal detector is the bolometer. The bolometer is a resistive element constructed from a material with a very small thermal capacity and a large temperature coefficient so that the absorbed radiation produces a large change in resistance. In contrast to the thermocouple, the device is operated by passing an accurately controlled bias current through the detector and monitoring the output voltage. The change in resistance is like to the photoconductor; however, the basic detection mechanisms are different. In the case of a bolometer, radiant power produces heat within the material, which in turn produces the resistance change. There is no direct photon–electron interaction.

The first bolometer, designed in 1880 by American astronomer S. P. Langley for solar observations [1], used a blackened platinum absorber element and a simple Wheatstone bridge sensing circuit. Langley was able to make bolometers that were more sensitive than the thermocouples available at that time. Although other thermal devices have been developed since that time, the bolometer remains one of the most used infrared (IR) detectors.

Modern bolometer technology development started in the early 1980s with the work of Honeywell on vanadium oxide (VO_x) and Texas Instruments on amorphous silicon (a-Si). Most of the technology was developed under classified military contacts in the United States and so the public release of this information in 1992 surprised many in the worldwide infrared community. Figure 8.1 shows the cross section of a thin-film bolometer prepared by silicon micromachining compatible with integrated circuit processing technology that enables the development of very large, low-cost, monolithic two-dimensional arrays.

Development efforts in uncooled FPAs are now basically going in two directions:

- Arrays for military and high-end commercial applications with the highest possible performance.
- Arrays for commercial applications with the lowest possible cost.

The key factor is to find a high-performance sensor together with high thermal isolation in the smallest possible area.

8.1 BASIC PRINCIPLE AND OPERATION OF BOLOMETERS

The relative temperature coefficient of resistance (TCR) is defined as

$$\alpha = \frac{1}{R}\frac{dR}{dT}. \tag{8.1}$$

The change of voltage of a constant current-biased bolometer is

$$\Delta V = I\Delta R = IR\alpha\Delta T.$$

So, in this case, $K = IR\alpha$ (see Equation 4.7), and according to Equation 4.9, the voltage responsivity is:

$$R_v = \frac{IR\alpha R_{th}\varepsilon}{\left(1+\omega^2\tau_{th}^2\right)^{1/2}}. \tag{8.2}$$

The expressions for voltage responsivity of a bolometer and a thermocouple are similar, with $n\alpha$ replaced by $IR\alpha_s$. The responsivity is inversely proportional to the thermal conductance ($G_{th} = 1/R_{th}$), which is also true for the thermocouple.

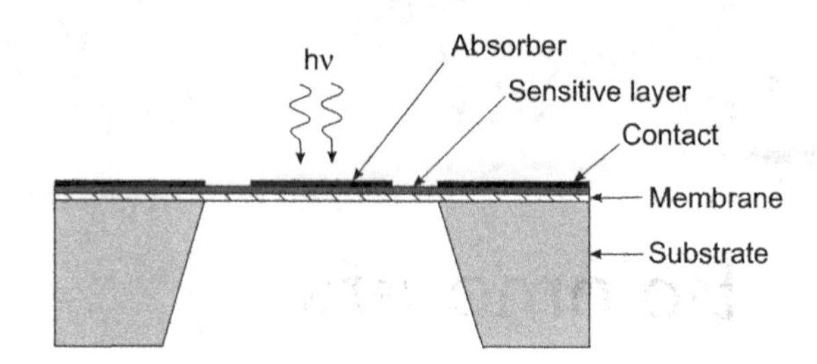

Figure 8.1 Schematic cross section of a thin-film bolometer.

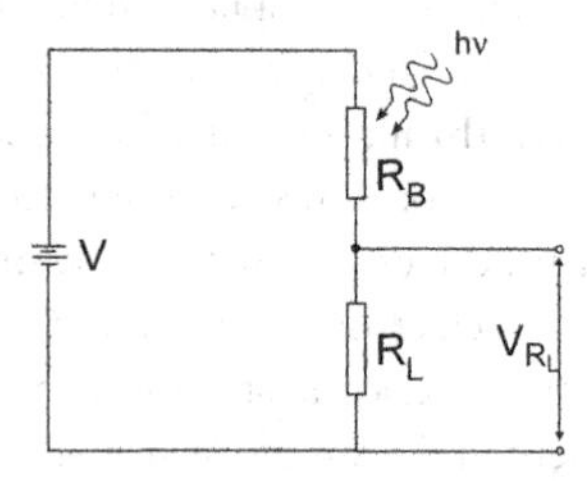

Figure 8.2 Bolometer circuit.

The maximum bias current is limited by the maximum allowed element temperature T_{max}. Therefore,

$$I^2 R = G_{th}\left(T_{max} - T\right), \tag{8.3}$$

and

$$R_v = \alpha\varepsilon\left[\frac{RR_{th}\left(T_{max} - T\right)}{1+\omega^2\tau_{th}^2}\right]^{1/2}. \tag{8.4}$$

The value of R_v is controlled in part by $R_{th} = 1/G_{th}$; bolometers with a high thermal conductance are fast (see Equation 4.5), but their responsivity is low. The key to developing highly sensitive bolometers is having a high temperature coefficient α, a very low thermal mass C_{th}, and excellent thermal isolation (low thermal conductance G_{th}).

These considerations are widely used for the simplest model, which omits the Joulean heating of the bias current and assumes a constant electrical bias. An accurate representation of a bolometer is a complex and difficult task. It has been analyzed in detail in Refs [2–4].

More complexity is introduced when the heat flow equation (like Equation 4.1) includes the Joulean heating due to electrical bias, and a load resistor, R_L, is introduced into a bolometer circuit to distinguish between voltage source operation ($R_L >> R_B$, where R_B is the bolometer resistance) and current source operation ($R_L << R_B$) (see Figure 8.2). If the circuit is opened and if no signal radiation is present, the bolometer is at ambient temperature T_o. Closing the circuit causes current flow and Joulean heating in the bolometer resistance, R_B. As a result, its temperature increases to T_1. If radiation now falls upon the bolometer, its temperature changes by ΔT to the new value T. This results in a resistance change in the bolometer, causing a change in the voltage across R_L.

Kruse carried out analysis of the bolometer operation, including the Joulean heating and constant electrical bias [4]. It appears that the behavior of the detector seriously depends on the temperature dependence of bolometer resistance upon temperature.

The resistance of a piece of semiconductor can be shown to be the form

$$R = R_o T^{-3/2}\exp\left(\frac{b}{T}\right), \tag{8.5}$$

where R_o and b are constants. For a semiconductor at room temperature,

$$\alpha = -\frac{b}{T^2}. \tag{8.6}$$

For a metal that has a linear dependence of resistance on temperature, that is,

$$R = R_o\left[1+\gamma(T-T_o)\right], \tag{8.7}$$

and thus

$$\alpha = \frac{\gamma}{1+\gamma(T-T_o)}. \tag{8.8}$$

Here γ is the temperature coefficient of the detector material.

In the case of bolometer operation under constant electrical bias and Joulean heating, the solution of a heat balance equation is similar to that described by Equation 4.3 [4] and has a form

$$\Delta T = \Delta T_o e^{-(G_e/C_{th})t} + \frac{\varepsilon\Phi_o e^{i\omega t}}{G_e + i\omega C_{th}}. \tag{8.9}$$

The first term of Equation 8.9 represents a transient, whereas the second term is a periodic function. Here G_e is the "effective" thermal conductance defined as

$$G_e = G - G_o(T_1 - T_o)\alpha\left(\frac{R_L - R_B}{R_L + R_B}\right), \tag{8.10}$$

where G_o is the average thermal conductance through a detector medium in the temperature range between T_1 and T_0, and G is the thermal conductance when the bolometer is at temperature T. Equation 8.10 indicates that G_e is the difference in two terms. The G_e is positive if

$$G > G_o(T_1 - T_o)\alpha\left(\frac{R_L - R_B}{R_L + R_B}\right), \tag{8.11}$$

and then the transient term goes to zero with time and only the periodic function remains. However, in the case when

$$G < G_o(T_1 - T_o)\alpha\left(\frac{R_L - R_B}{R_L + R_B}\right), \tag{8.12}$$

where G_e is negative, it means that the bolometer temperature will increase exponentially with time (see Equation 8.9) reaching burnout. This can happen with semiconductors but not with metals [2].

Assuming that $R_L >> R_B$, it can be shown that the responsivity is given by

$$R_v = \frac{\alpha I_b R_B \varepsilon}{G_e(1+\omega\tau_e)^{1/2}}, \tag{8.13}$$

where τ_e is defined as

$$\tau_e = \frac{C_{th}}{G_e}. \tag{8.14}$$

Here, τ_e is known as the "effective thermal response time." The dependence of thermal capacity and τ on temperature due to bias current heating is termed the "electrothermal effect."

Usually, when large focal plane arrays (FPAs) are implied, the electrical bias is pulsed rather than continuous, and the heating is due to the electrical bias (Joulean), and the incident absorbing radiant flux. In this situation, the heat transfer equation is nonlinear and numerical solutions must be obtained [5].

In addition to radiation noise and temperature noise associated with the thermal impedance of the element, Johnson noise associated with resistance R is one of the most important noise sources. With room temperature bolometers, amplifier noise should not be important but with cryogenic devices, it is usually the dominant noise source. With some types of bolometers, low-frequency current noise is important and is the principal factor limiting current.

8.2 TYPES OF BOLOMETERS

Usually, a bolometer is a thin, blackened flake or slab, whose impedance is highly temperature dependent. Bolometers may be divided into several types. The most commonly used ones are the metal, the thermistor, and the semiconductor bolometers. A fourth type is the superconducting bolometer. This bolometer operates on a conductivity transition in which the resistance changes dramatically over the transition temperature range. Figure 8.3 shows schematically the temperature dependence of resistance of different types of bolometers. General bolometer properties are given in Table 4.2.

8.2.1 METAL BOLOMETERS

Typical materials used for metal bolometers are nickel, bismuth, platinum, or antimony. These metals are in use where their high, long-term stability is an essential requirement. Being made from metal, these bolometers need to be small so that the heat capacity is low enough to allow reasonable sensitivity. Most metal bolometers are formed as film strips, about 100–500 Å thick, via vacuum evaporation or sputtering. They are often coated with a black absorber such as evaporated gold or platinum black.

A typical value of the temperature coefficient is positive for metals and is equal to about 0.3%/°C. The metal bolometers operate at room temperature and have detectivities of the order 1×10^8 cmHz$^{1/2}$W^{-1}, with response times of approximately 10 ms. Unfortunately, they are generally rather fragile, thus limiting their use in certain applications. Nevertheless, metal film bolometer arrays have been fabricated in various linear and 2-D formats, and have been successfully employed in non-imaging IR sensors designed for remote surveillance applications [6]. The technology of these devices is generally limited to small arrays due to power consumption and amplifier design constraints associated with matching to the low detector impedance.

Between metals, titanium films are more frequently used in the bolometers due to the following reasons: titanium can be used in a standard silicon process line, low thermal conductivity (0.22 W/Kcm in bulk material—far lower than that of most other metals), and low $1/f$ noise [7,8]. However, metal in thin-film form has a temperature coefficient resistance of 0.004%/K, considerably lower than that for competitor materials, which causes it to be of little use in uncooled bolometer arrays.

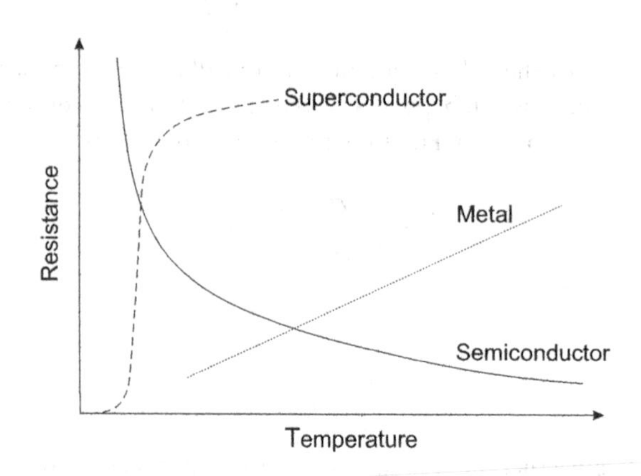

Figure 8.3 Temperature dependence of resistance of three material types.

Antenna-coupled metal microbolometers at room temperature (usually with bismuth or niobium) are also operated in the very long wavelength IR (LWIR) region (10–100 µm). The noise equivalent power (NEP) values of about 10^{-12} W/Hz$^{1/2}$ can be obtained with suspended microbridges or low thermal conductance buffer layers with the silicon substrate. See Figure 8.21 for some experimental data for these type of detectors.

8.2.2 THERMISTORS

Thermistor materials were first developed at Bell Laboratories during World War II. Single pixel thermistor bolometers have been commercially available for about 60 years. They have found wide applications, ranging from burglar alarms, to fire detection systems, industrial temperature measurement, space-borne horizon sensors, and radiometers. They are also useful in radiometric applications where a uniform spectral response is desired. Thermistors have a long life, good stability when biased properly, and are highly resistant to nuclear radiation.

Thermistor bolometers are constructed from a sintered mixture of various semiconducting oxides that have a higher TCR than metals (2%/°C–4%/°C) and are generally much more rugged. They crystallize in the spinel structure. In their final form, they are semiconductor flakes, 10-µm thick. The negative temperature coefficient depends on the band gap, the impurity states, and the dominant conduction mechanism. This coefficient is not constant but varies as T^{-2}, which is a result of the exponential dependence of semiconductor resistivity [9]. The sensitive material in a thermistor bolometer is typically made of wafers of manganese, cobalt, and nickel oxides, sintered together and mounted on electrically insulating but thermally conducting materials such as sapphire [10,11]. The sapphire is then mounted on a metallic heat sink to control the time constant of the device. The sensitive area is blackened to improve its radiation absorption characteristics. Typical room temperature resistivity of a thermistor changes between 250 and 2500 Ωcm. They can be made in sizes ranging from 0.05 to 5 mm square. The principal material investigated is $(MnNiCO)_3O_4$, which has a negative TCR of about 0.04/°C at room temperature.

The usual construction uses a matched pair of devices for a single unit (Figure 8.4). One of the pairs of thermistors is shielded from radiation and fitted into a bridge such that it acts as the load resistor. This arrangement allows one to optimize the signal from the active element by compensating for any ambient temperature changes. The result can be a dynamic range of a million to one. The device sensitivity and response time cannot both be optimized as improved heat sinking, to reduce the time constant, prevents the detector from reaching its maximum temperature, thus reducing the responsivity. Considerations carried out by Astheimer [12] indicate that Johnson noise–limited detectivity of this type of bolometer at room temperature can be described by the equation

$$D^* = 3\times 10^9 \tau^{1/2} \quad \mathrm{cmHz^{1/2}W^{-1}} \tag{8.15}$$

with τ in seconds. The time constant changes typically from 1 to 10 ms. Their sensitivity closely approaches that of the thermopile for frequencies higher than 25 Hz. At lower frequencies, there may be excess or $1/f$ noise.

Since the thermistor is Johnson noise–limited, an improvement in detectivity can be realized by placing a hemispherical or hyper-hemispherical lens over its surface [13]. This procedure will not improve

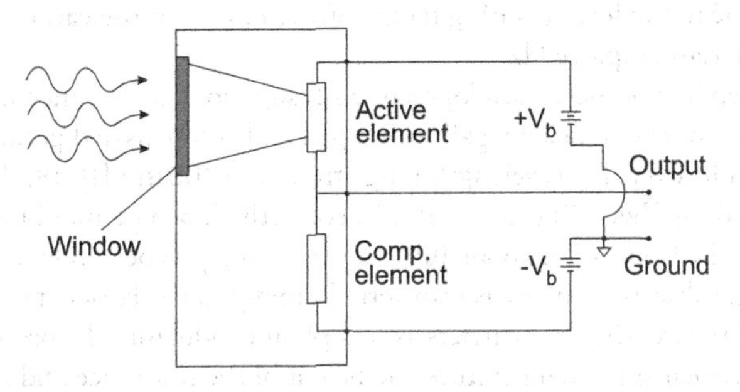

Figure 8.4 Typical bias circuit of a thermistor bolometer. The compensating element is shielded from incident radiation.

the signal-to-noise ratio if the detector is photon noise–limited. The detector must be optically coupled with the lens, which can be accomplished by depositing the detector directly on the plane of the lens (see Figure 5.3). The rays directed to the edge of the detector are refracted by the lens, giving the detector an appearance of being larger by a factor of n (or n^2 in the case of hyper-hemispherical lenses), where n is the index of refraction of the lens. Since the detector is a two-dimensional device, the virtual area increase is n^2. As a result, the signal-to-noise ratio is increased by n^2 (or n^4). Immersion also reduces the bias power dissipation by factors n^2 (or n^4), thereby reducing the heat load of coolers and enabling higher bias power dissipation densities to be achieved. The requirements on the lens material are that it should have as large an index of refraction as possible, and should be electrically insulating so as not to short the thermistor film. Germanium, silicon, and arsenic triselenide are among the most useful materials. Thermistor detectors are particularly adaptable to immersion because the low thermal impedance inherent in immersion permits higher bias voltage to be applied.

The thermistor detectors are mainly fabricated by bonding thin flakes of the thermistor material to a substrate. Device performance parameters such as responsivity, noise, and response time are very dependent on operator skill and experience and processing conditions. To overcome these drawbacks and reduce the cost of thermistors, the possibility of evaporated films has been investigated [14,15]. The most popular are sputtering techniques.

The thermistor spectral responsivity is essentially flat with an upper wavelength determined by the transmission of the window that encapsulates the chip. Because they exhibit a large amount of $1/f$ noise arising at grain boundaries, thermistor thin films have not been found to be useful in uncooled bolometer thermal imaging arrays.

8.2.3 SEMICONDUCTOR BOLOMETERS

A significant improvement in the performance of bolometer detectors is achieved if the devices are cooled as the resistance changes are much greater than at room temperature, and thicker devices can be fabricated to improve IR absorption without increasing the thermal capacity due to the reduced specific heat in the cooled material. The ultimate sensitivity can be orders of magnitude higher than that for a room temperature device. In practice, for most applications, it is necessary to have an aperture in the enclosure admitting some room temperature background radiation.

Semiconductor bolometers are the most highly developed form of thermal detectors for low light levels and are the detectors of choice for many applications, especially in the infrared and submillimeter spectral range. They must be constructed carefully to ensure that they are isolated from the thermal surroundings, and the techniques typically used to construct them do not lend themselves to the efficient development of large arrays.

The typical bias circuit of the bolometer is similar to that shown in Figure 8.2. A large value of load resistance is almost always used to minimize the Johnson noise. The general noise considerations can be found in Mather [16].

The modern history of infrared bolometers begins with the introduction of the carbon resistance bolometer by Boyle and Rogers [17]. At this time, carbon radio resistors were widely used by low temperature physicists as thermometers at liquid helium temperatures. They were easy to build, inexpensive, and were of moderate heat capacity due to the low operating temperature. However, the carbon-specific heat was not as low as the crystalline materials employed later.

The next important step in bolometer development with superior performance was the invention of a low-temperature bolometer based on heavily gallium-doped and compensated germanium with sensitivities close to the theoretical limits over the wavelength range from 5 to 100 μm [18,19]. Its mode of operation has been discussed in detail by Zwerdling et al. [20]. In correctly doped germanium (typical concentrations of gallium are about 10^{16} cm^{-3} with about 10^{15} cm^{-3} in giving p-type conductivity with a compensation ratio of about 0.1), the absorbed energy is transferred rapidly to the lattice, raising the temperature of the sample rather than of the existing free carriers as in a photoconductor. The optimum level of doping is determined by two requirements: the temperature coefficient of the resistance and a resistance that allows for an efficient coupling to a low noise amplifier. Since the resistance is large at low temperatures, the semiconductor must be heavily doped close to the metal–insulator transition, with a majority donor impurity and a compensating minority acceptor impurity. In these conditions, the dominant conduction mechanism is hopping of carriers from one dopant atom to another. The temperature coefficient and the resistance depend strongly on the degree of compensation, because the hopping conduction mechanism is determined by the distance between dopant atoms. Typical doping concentrations are 10^{16} cm^{-3} for Ge and 10^{-18} cm^{-3} for Si.

The temperature dependence of semiconductor resistance is described by Equation 8.5. However, at very low temperatures (<10 K), the semiconductor material must be doped more heavily than assumed in Equation 8.5 so that the dominant conductivity mode is hopping. This mechanism freezes out relatively slowly; the resistance is given by an empirical expression of the form

$$R(T) = R_0 \exp\left(\sqrt{\frac{T_0}{T}}\right). \tag{8.16}$$

The constants T_0 and R_0 are of the order of 2–10 K and 0.1–0.5 Ω. We then have

$$\alpha(T) = \frac{1}{R}\frac{dR}{dT} = -\frac{1}{2}\sqrt{\frac{T_0}{T^3}}. \tag{8.17}$$

Note that $\alpha < 0$ and has strong temperature dependence. This is in contrast to superconducting bolometers, which have positive α, allowing for electrothermal feedback. An excellent agreement between Equation 8.16 and experimental data was found for Ge bolometers [21,22].

The absorption efficiency can be increased by mounting the device in an integrating cavity, which is discussed by Putley [18]. For small apertures at 4.2 K, the inherent Ge bolometer noise was divided equally between Johnson noise and photon noise [23]. At larger apertures, the photon noise may predominate over the inherent detector noise, depending on the performance of the background. The advantages of a well-known material with reproducible properties, high stability, and low noise led to its application to infrared astronomy and medium and long wavelengths as well as to laboratory infrared spectroscopy. Over most of the far infrared (FIR) spectrum, the performance of a germanium bolometer is comparable to that of the best photon detectors, with the added advantage of being a broadband device. Further improvement in device performance was possible. Draine and Sievers have obtained an *NEP* of 3×10^{-16} WHz$^{-1/2}$ for a device operating at 0.5 K [24]. However, this very low value could probably be exploited in an experiment conducted entirely at cryogenic temperature and requires a cryogenically operated low noise amplifier.

More attention has been recently given to the use of Si as an alternative to Ge. In comparison with Ge, Si has a lower specific heat (by a factor of 5), easier materials preparation, and more advanced device fabrication technology. Silicon bolometers with NEPs of 2.5×10^{-14} WHz$^{1/2}$ that compare favorably with germanium have been reported by Kinch [25].

Details of the fabrication and performance of modern bolometers are described in review papers [21,22,26]. It appears that Ge and Si submillimeter bolometers with improved performance can be fabricated using more uniform materials obtained by neutron transmutation doping (NTD) and by ion implantation.

The weak absorption of neutrons by Ge leads to a uniform doping concentration. Natural Ge contains five stable isotopes: ^{70}Ge, ^{72}Ge, ^{73}Ge, ^{74}Ge, and ^{76}Ge. When a nucleus captures a neutron, it may become a stable isotope, as in the case of ^{72}Ge and ^{73}Ge, transmutating into ^{73}Ge and ^{74}Ge respectively, which do not contribute to the doping process. Otherwise, the nucleus may undergo a beta decay or a K capture. Hence, the doping process produces both acceptors and donors. For Ge with natural isotopic abundances, NTD-Ge has a compensation of 0.32. Therefore, R_0 and T_0 can be changed independently only by changing the dimensions of the crystal or by isotopically enriching Ge. Special care has to be taken with respect to the contacts. Soldered In contacts, which were often used for Ge bolometers in the early days, introduce excess $1/f$ noise. Metallized ion-implanted and annealed contacts give better results.

The chip bolometers described earlier combine the functions of radiation absorption and thermometry. These two functions are difficult to comply, especially for chip bolometers designed for millimeter and

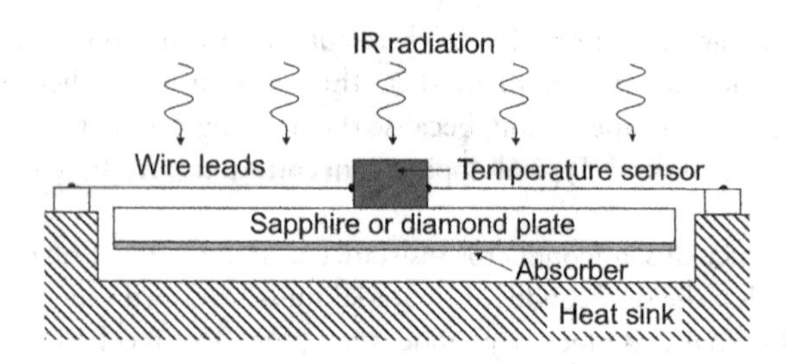

Figure 8.5 Composite bolometer. (From Ref. [28])

submillimeter wavelengths. The bulk absorption coefficient of Ge and Si material with a useful resistivity decreases at low frequencies. Consequently, bolometers must typically be one or more millimeters thick and the resulting heat capacity is a significant limitation. To overcome these limitations, composite bolometers are fabricated to lower heat capacity and consistently to reduce the time constant [22,27]. They look promising for low-noise and low-temperature detection.

The composite bolometer is made up of three parts: the radiation-absorbing material, the substrate that determines its active area, and the temperature sensor as shown in Figure 8.5 [28]. The absorber is made of a thin film whose thickness and composition are adjusted so that the emissivity is very high—into the hundreds of microns wavelength region. Usually, black bismuth and nichrome absorbers have been used. The temperature sensor (e.g., germanium) is bonded both mechanically and thermally to the substrate via an epoxy or varnish. Thus, the substrate and film combination act as an efficient absorbing element of a large effective area and low heat capacity for a temperature sensor that can be made very small.

Early composite bolometers used sapphire substrates that have about 1/60th the heat capacity of germanium and have negligible absorptivity up to ~300 cm^{-1} at low temperatures. This means that a larger detector area can be made with no corresponding frequency response loss. Diamond is transparent well beyond 1000 cm^{-1} and has a heat capacity of approximately 1/600th that of germanium; thus, much larger active areas can be achieved. It is now widely used for low background bolometers at $T \sim 1$ K. Silicon substrates have larger lattice specific heat, but are useful for temperature $\ll 1$ K because the lattice heat capacity is small enough and they have smaller impurity heat capacity than the diamond. Examples of composite bolometer fabrications are described by Richards [22].

Modern well-developed semiconductor bolometer technologies exist to produce arrays of hundreds of pixels that are operated at temperatures of 100–300 mK (see Section 28.10). They are typically fabricated by lithography and micromachining techniques of Si or Si_3N_4 and use thermistors of ion-implanted silicon or neutron transmutation–doped Ge.

8.3 MICROMACHINED ROOM-TEMPERATURE BOLOMETERS

A new generation of monolithic Si bolometers has been introduced by Downey et al. [29]. They presented a bolometer concept in which a thin Si substrate supported by narrow Si legs is micromachined from a Si wafer using the techniques of optical lithography. A conventional Bi film absorber was used on the back of the substrate, but the thermometer was created directly in the Si substrate by implanting P and B ions to achieve a suitable donor density and compensation ratio. However, the performance of the bolometer was not good. Further progress in monolithic Si bolometer technology is fascinating. One of the most widely used approaches is microbolometers implemented using surface micromachined bridges on complementary metal-oxide–semiconductor (CMOS)–processed wafers. The surface micromachining technique allows the deposition of temperature-sensitive layers with very small thickness, very low mass, and very good thermal isolation on top of bridges over the readout circuit chips. After removal of the sacrificial layers between the bridge structures and readout circuit chips, thermally isolated and suspended detector structures packaged in vacuum are obtained. Honeywell Sensor and System Development Center in Minneapolis, Minnesota, began developing silicon micro-machined IR sensors in the early 1980s. The goal of the work that continued under classified programs sponsored by DARPA and U.S. Army Night Vision and Electronic Sensor Directorate was aimed at producing low-cost night vision systems amenable for use throughout the military

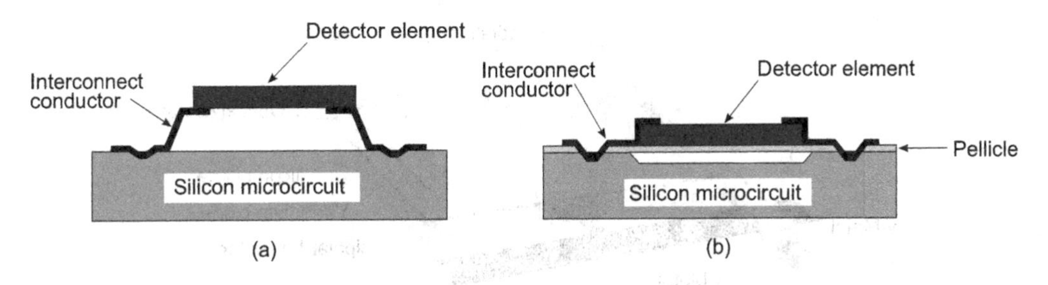

Figure 8.6 Thermal detector element design: (a) microbridge detector element, and (b) pellicle-supported detector element. (From Ref. [34])

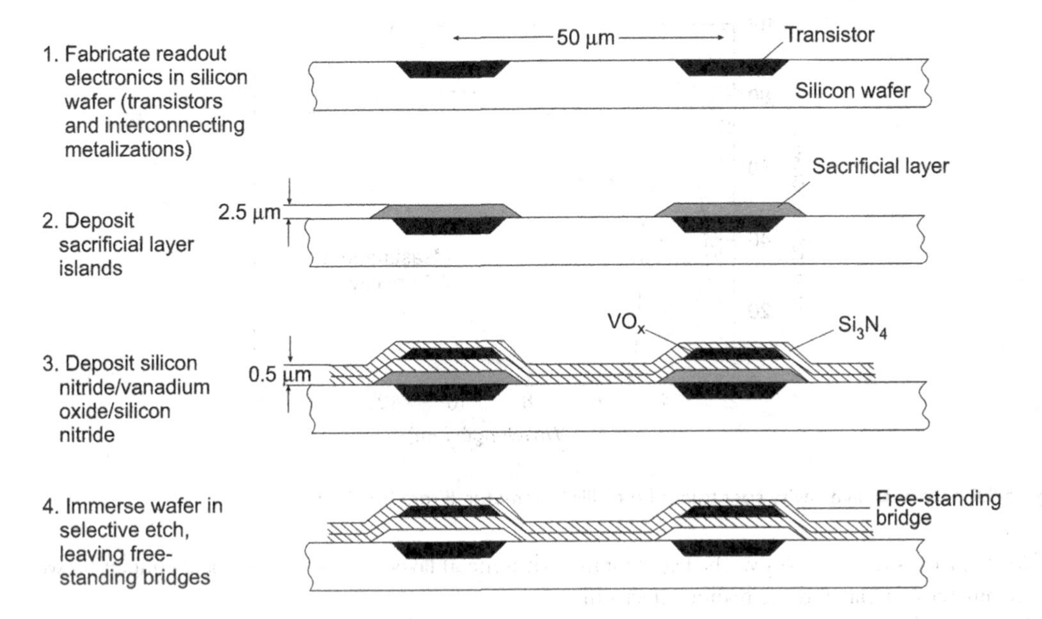

Figure 8.7 Simplified process of two-level Honeywell microbolometer fabrication. (From Ref. [36])

with noise-equivalent temperature difference of 0.1°C using *f*/1 optics. Both Si bolometer arrays and pyroelectric arrays of Texas Instruments have exceeded that goal [30,31].

Considerations carried out in Chapter 25 suggest that FPAs of thermal detectors can resolve very small temperature differences, comparable to those attained with the current scanned cryogenic imagers.

The two options used for the bolometers structure are microbridge- and pellicle-supported designs (see Figure 8.6). The former comprises detector elements that are supported on legs above the plane of the microcircuit. The legs are designed to have a high thermal resistance and carry electrical conductors from the detector to the microcircuit. This approach is implied in the Honeywell microbolometer design [31–33]. The second concept consists of detector elements deposited onto a thin dielectric pellicle that is coplanar with the surface of the wafer and is the basis of the original Australian monolithic detector technology [34,35].

The basic fabrication process used to fabricate Honeywell silicon microbolometers together with a brief explanation of the micromachining process steps is shown in Figure 8.7 [36]. This process consists of depositing and patterning a sacrificial layer on a substrate containing the electronics. The detecting area is defined by a thin membrane upon which a thin film of the detecting material is deposited. The final microbolometer pixel structure is shown in Figure 8.8 [31]. The microbolometer consists of a 0.5-μm thick bridge of Si_3N_4 suspended about 2 μm above the underlying silicon substrate. The bridge is supported by two narrow legs of Si_3N_4. The Si_3N_4 legs provide the thermal isolation between the microbolometer and the heat-sink readout substrate. A bipolar input amplifier is normally required, and this can be obtained with

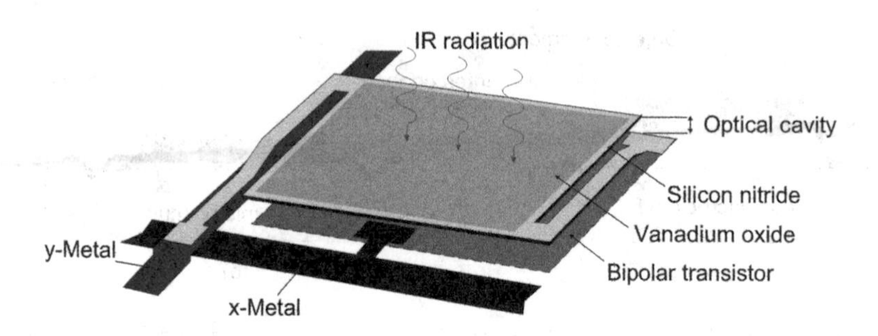

Figure 8.8 Bridge structure of the Honeywell microbolometer.

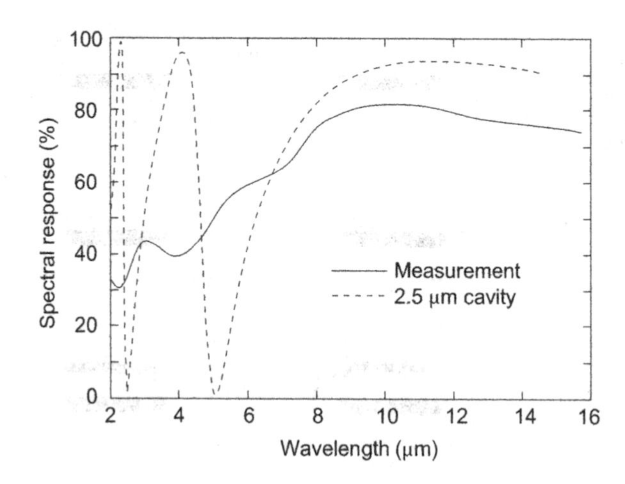

Figure 8.9 Quarter-wave cavity spectrum of the Ulis bolometer. (From Ref. [37])

biCMOS processing technology. The legs contain a thin metal layer to provide electrical contact between the bolometer material and the readout electronics.

A reflective layer on the substrate below the membrane (typically aluminum) causes incidental infrared radiation that is completely absorbed by the detecting material to be reflected back through the material, thereby increasing the amount absorbed. This is most effective when the spacing between the absorbing layer and the reflecting layer is one-fourth of the wavelength of the incident radiation. The peak absorption wavelength λ_p dictated by the optical cavity results from the following equation

$$nt = \frac{(2k+1)\lambda_p}{4}, \tag{8.18}$$

where n is the index of refraction of the transmission media in the cavity (vacuum), t is the cavity thickness, and k is the resonant order. For $n = 1$ (vacuum), $t = 2.5\,\mu m$, and $k = 0$, $\lambda_p = 10\,\mu m$ that corresponds to the standard microbolometer array designated to be efficient in the LWIR region. However, for the next order ($k = 1$), a large absorption in 3–5-μm wavelength band ($\lambda_p = 3.33\,\mu m$) is predicted. For example, Figure 8.9 compares the spectral response of the Ulis microbolometer to the theoretical quarter wave cavity spectrum [37]. It is the most commonly used resonant cavity design. In the second type of design, the resonant optical cavity is part of the bolometer membrane. This solution is rarely used.

The Si_3N_4 is used because of its excellent processing characteristics. This allowed microbolometers to be fabricated with thermal isolation close to the attainable physical limit, which is about 1×10^8 K/W for a 50-μm square detector. It was demonstrated that, with a microbolometer having a thermal isolation of 1×10^7 K/W [38], a typical incident IR signal of 10 nW was sufficient to change the microbolometer temperature by 0.1 K [32]. The measured thermal capacity was about 10^{-9} J/K, which corresponds to a thermal time constant of 10 ms. Honeywell has determined that the microbridges are robust structures that can tolerate shocks of several thousand g-forces. Encapsulated in the center of the Si_3N_4 bridge is a thin layer (500 Å) of polycrystalline VO_x as the active bolometer's material. The VO_x assures good combination of high TCR, electrical resistivity, and fabrication capability, which has resulted in pixels with a responsivity of 250,000 V/W in response to 300 K blackbody radiation [33].

Conventional bolometers are operated in a vacuum package to minimize the thermal conduction between the bolometers and their surroundings through the surrounding gas. The vacuum atmosphere in which the bolometers are operated is typically on the order of 1 Pa.

8.3.1 MICROBOLOMETER SENSING MATERIALS

Today, the most common bolometer temperature sensing materials are VO_x, amorphous silicon (a-Si), and silicon diodes. Figure 8.10 shows the layer systems of VO_x and a-Si microbolometer bridges. The performance of bolometers is typically limited by the $1/f$ noise [38,39] that can vary several orders of magnitude for different materials and even small variations of the material composition can dramatically change the $1/f$ noise value. Monocrystalline materials can have significantly lower $1/f$ noise constant as compared to amorphous or polycrystalline materials [38,40]. However, for most materials, $1/f$ noise constant is not well documented in the literature.

8.3.1.1 Vanadium oxide

The most popular thermistor material used in the fabrication of the micromachined silicon bolometers is vanadium oxide, VO_x. A thin film of the mixed oxides sputtered on a Si_3N_4 microbridge substrate was originally developed at Honeywell. Vanadium is a metal with a variable valence forming a large number of oxides. Preparation of these materials in both bulk and thin-film forms is very difficult given the narrowness of the stability range of any oxide. Some of the vanadium oxides, among them the best known being VO_2, V_2O_3, and V_2O_5, show a temperature-induced crystallographic transformation that is accomplished by reversible semiconductor (low-temperature phase) to metal (high-temperature phase) phase transition, with a significant change in electrical and optical properties (Figure 8.11) [41–44]. Vanadium dioxide undergoes their transition in the temperature range from about 50°C to 70°C. The V_2O_5 can be formed by the ion beam from a vanadium metal target in high O_2 partial pressure, but its resistance at room temperature is very high. The V_2O_3 has low formation energy and undergoes a transition from semiconductor to metal phase at low temperatures, so its resistance is very low at room temperature. This oxide is important to the fabrication of low-noise microbolometers.

VO_x films are prepared by a variety of techniques including reactive radio frequency (RF) sputtering, pulsed laser deposition, annealing, and oxidation of evaporated vanadium under controlled conditions [45]. Several groups are looking to find a fabrication procedure to obtain high TCR in combination with sufficient low sheet resistance (e.g., [46,47]).

Figure 8.12 shows the TCR in dependence on the resistivity for thin films of mixed vanadium oxides [3]. From the point of view of infrared imaging application, the most important property of VO_x is its

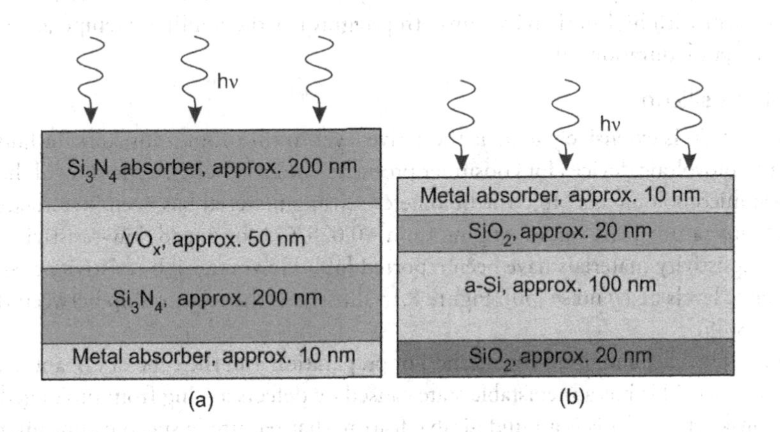

Figure 8.10 Layer systems of a microbolometer bridge: (a) VO_x and (b) a-Si.

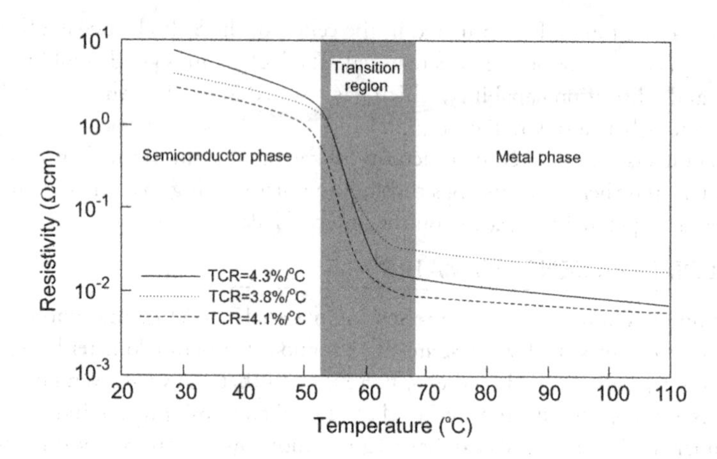

Figure 8.11 Resistivity versus temperature characteristics of three VO_2 films. (From Ref. [41])

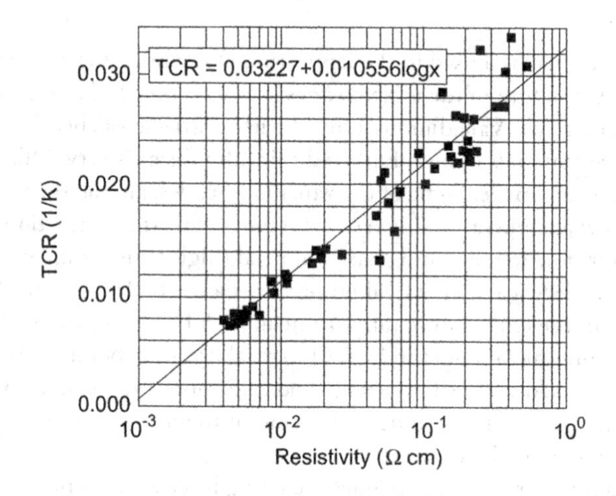

Figure 8.12 The TCR versus resistivity for thin films of mixed vanadium oxides. (From Ref. [3])

high negative TCR at ambient temperature, which exceeds 3%/°. However, there are two reasons for not using VO_x material with higher x-values and substantially higher TCR. In the region of higher x-value, the reproducibility of oxide property suffers due to scattering of experimental data. In addition, a Joulean heating becomes a problem with high-resistivity films. This aggravates the nonlinear temperature versus time problem during the pulse duration [3].

8.3.1.2 Amorphous silicon

Amorphous silicon (a-Si) is extensively used as the active layer in thin-film transistors for liquid crystal displays, small-area photovoltaic devices for consumer products, and solar cells. Syllaios et al. have described technology of a-Si microbolometers [48]. The bolometer-sensing material has been investigated for at least 20 years. Room-temperature TCR values ranging from $-0.025°C^{-1}$ for doped, low-resistivity films to $0.06°C^{-1}$ for high-resistivity materials have been reported [49]. However, high-resistivity a-Si is characterized by unacceptable levels of $1/f$ noise [50]. Figure 8.13 illustrates the relationship between the TCR and the resistivity of a-Si [51].

Properties of the films depend upon the method of preparation and the type of dopant. Amorphous hydrogenated silicon (a-Si:H) has a metastable state caused by defects arising from prolonged illumination (Staebler and Wronski effect). This is an undesirable feature that requires a specific annealing cycle during preparation (methodology for reliability enhancement is described in [52]). If not removed, it adversely

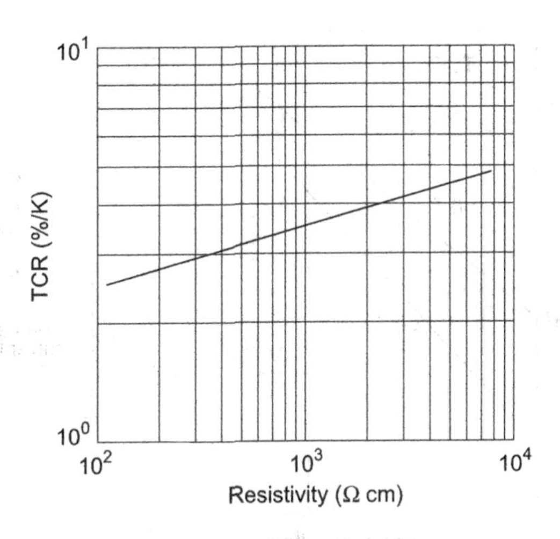

Figure 8.13 TCR as a function of electrical resistivity of a-Si. (From Ref. [51])

affects long-term reliability. Hydrogenated a-Si can only be produced by a nonequilibrium process, such as plasma-enhanced chemical vapor deposition (PECVD) or sputtering. The TCR and sheet resistance are therefore a direct consequence of the deposition procedure: doping concentration, deposition temperature, and annealing [53]. The a-Si can be deposited at a very low temperature, as low as 75°C.

Because the typical resistivity of a-Si films is several orders of magnitude higher than that of VO_x, a-Si finds application in uncooled arrays in which the bias is continuous rather than pulsed. This choice is dictated by the fact that the bolometer signal depends on $I_b\alpha R$, whereas the power dissipation, which causes the rise in detector temperature, depends on $I_b^2 R$ (here I_b is the bias current).

8.3.1.3 Silicon diodes

For temperature sensing, the forward-biased p-n junction [54–57] or Schottky-barrier junction [58] can be also used. In the case of an ideal diffusion-limited characteristic and a sufficiency large forward-bias voltage, the current flowing in the p-n junction is given by the following equation:

$$I = AJ_s \exp\left(\frac{qV}{kT}\right), \tag{8.19}$$

$$J_s = KT^{(3+m/2)} \exp\left(-\frac{E_g}{kT}\right),$$

where A is the junction area, J_s is the saturation current, q is the electron charge, E_g is the bandgap energy, m is a constant determined by the temperature dependence of the diffusion constant and carrier lifetime, and K is the temperature-independent constant.

When the diode is driven in a constant-current mode, the temperature sensitivity of V is expressed as follows:

$$\left.\frac{dV}{dT}\right|_{I=\text{const}} = \frac{V}{T} - \left(3 + \frac{m}{2}\right)\frac{k}{q} - \frac{E_g}{qT} \approx \frac{qV - E_g}{qT}. \tag{8.20}$$

dV/dT is insensitive to the process fluctuations because m, a process-sensitive parameter, is included only in the second term of Equation (8.20) and this term is negligible when compared with the other two terms. At typical forward bias condition of 0.6V for a silicon diode at room temperature, the current approximately doubles every 10°C, and the voltage decreases linearly with temperature with a coefficient of approximately 2 mV/°C and gives a temperature coefficient of about 0.2%/°C, an order of magnitude lower than the best resistive devices. For example, the diode forward voltage of OMEGA Engineering's temperature sensing

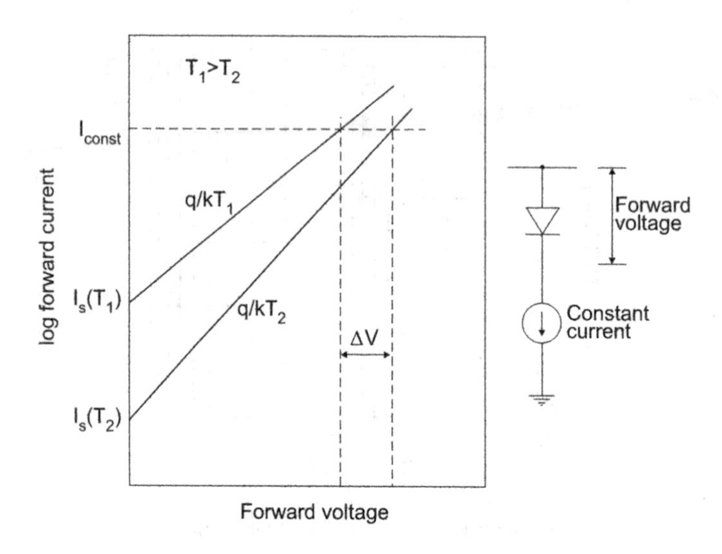

Figure 8.14 Operation of forward-biased p-n junction sensor.

diode biased with an accurate 10 μA current in the temperature range above 100 K decreases nearly linearly with temperature as indicated by the equation $V = 1.245 - 0.0024(\text{V/kT})T$, where V is expressed in volts [60].

To explain the operation of a diode sensor, Figure 8.14 shows forward I–V characteristics of a diode for two different temperatures [61]. When the p-n junction is operated in a constant-current mode, the forward voltage measured across the junction reflects the temperature of the device. As is shown in Figure 8.14, the forward voltage is determined by J_s, the intersection with the current axis and the gradient q/kT. The forward voltage becomes smaller as the temperature rises.

If the detector is constructed with diodes serially connected by metal straps, the forward current is given by

$$I = AJ_s \exp\left(\frac{qV}{nkT}\right), \tag{8.21}$$

and the temperature coefficient of the forward current can be expressed by

$$\left.\frac{dV}{dT}\right|_{I=\text{const}} \approx n\frac{q(V/n) - E_g}{qT}, \tag{8.22}$$

where n is the number of series. From the last equation we know that the temperature coefficient of forward voltage is proportional to the number of series.

The main noise in p-n junction is shot noise, which is a result of the statistical process of carriers flowing over a potential barrier. It can be given by the following equation:

$$V_n = \sqrt{2qI}\,\frac{dV}{dI}. \tag{8.23}$$

In addition to the shot noise, two additional types of noise should be taken into account: the Johnson noise associated with the resistance and the $1/f$ noise. The diodes with low $1/f$ power law noise can be manufactured in a standard CMOS technique. An advantage of diodes is the fabrication possibility of smaller pixel size in comparison to resistors. Also, the first realization and characterization of monolithic uncooled infrared sensor arrays based on a-Si thin-film transistors have been demonstrated [62,63].

Uncooled FPAs with p-n junction sensors have been successfully developed and IR cameras with some of these devices are commercially available [64,65]. For example, Figure 8.15 shows a pixel structure with a

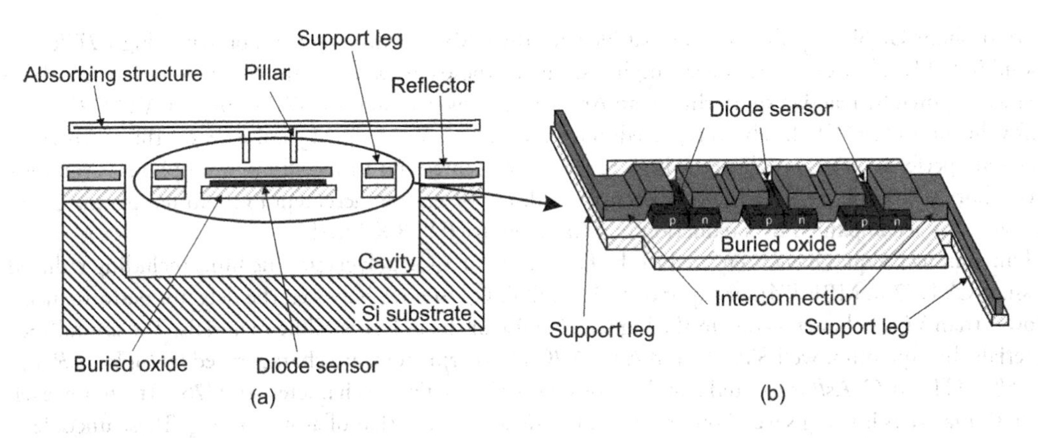

Figure 8.15 Pixel structure of SOI diode uncooled infrared FPA, showing cross section (a) and p-n junction temperature sensor (b). (After Ref. [61])

Table 8.1 Materials employed as thermo-sensing films in microbolometers

MATERIALS	*TCR* (K^{-1})	E_a (eV)	σ_{300K} ($\Omega^{-1}cm^{-1}$)
VO_x	0.021	0.16	2×10^{-1}
a-Si:H (PECVD)	0.1–0.13	0.8–1	$\sim 1 \times 10^{-9}$
a-Si:H,B (PECVD)	0.028	0.22	5×10^{-3}
a-Ge_xSi_y:H (PECVD)	0.043	0.34	1.6×10^{-6}
Poly-SiGe	0.024	0.18	9×10^{-2}
$Ge_xSi_{1-x}O_y$	0.042	0.32	2.6×10^{-2}
YBaCuO	0.033	0.026	1×10^{-3}

p-n junction uncooled sensor on an SOI wafer, which was proposed by Ishikawa et al. [66]. Using silicon-on-insulator (SOI) technology it was possible to fabricate a freestanding structure with diode sensor, support legs, horozontal addressing line, and vertical signal line placed on the same level. The IR absorbing structure consists of a thin absorbing metal and a dielectric film for maintaining the shape of this metal. An interference absorber is formed with this absorbing structure and the reflector deposited on the lower level. A 90% fill factoe is feasible with this SOI diode pixel.

8.3.1.4 Other materials

Table 8.1 shows the most employed materials as thermo-sensing films in microbolometers [67]. Since the temperature coefficient of resistance $TCR \approx E_a/kT^2$, both *TCR* and the activation energy, E_a, are directly related.

As was mentioned previously, VO_x is one of the most employed materials in large microbolometer arrays; however, this material is not compatible with Si CMOS standard technology and its *TCR* is not very large, around 0.02 K^{-1}. a-Si:H and boron-doped a-Si:H are also widely employed in microbolometer arrays. Intrinsic a-Si:H is compatible with CMOS technology and has a very high *TCR*, around 0.1–0.13 K^{-1}; however, it is a highly resistive material, resulting in high resistive microbolometers that present a mismatch impedance with the readout circuits. Boron-doped a-Si:H not only has moderated resistivity, but also a reduced *TCR* of 0.028 K^{-1}. Therefore, none of those materials can be considered the optimum one as sensing material in microbolometers.

At present, several research programs are focused toward enhancement of performance levels in excess of 10^9 $cmHz^{1/2}W^{-1}$. Both amorphous as well as polycrystalline compound materials are being studied. It is anticipated that new materials (e.g., SiGe, SiGeO, SiC) will be the basis of the next generation of semiconductor film bolometers [68].

Amorphous $Ge_xSi_{1-x}O_y$ films are compatible with the CMOS technology and present a high *TCR*, around 0.042 K^{-1}; however, they have a high resistance. These compounds with Ge content in the order of 85% were grown by reactive sputtering in an Ar or Ar:O_2 environment [69–71] or by PECVD [72]. The *TCR* value up to 5.1%/°C has been reported; however, the relatively high $1/f$ noise lowers the potential bolometer performance. To reduce the high resistivity presented in intrinsic films, amorphous germanium-silicon-boron alloys, a-$Ge_xSi_yB_z$:H, have been studied. However, the increment in room-temperature conductivity was accompanied by a reduction in *TCR*, to above 0.028 K^{-1} [67].

Different techniques have been used in the fabrication of SiGe polycrystalline films including reduced-pressure CVD [73], MBE [74], or vapor deposition [75]. Up to now, however, the detector performance is lower than VO_x bolometers due to the higher $1/f$ noise of the polycrystalline materials. The crystalline materials, like quantum well Si/GeSi and AlGaAs/GaAs thermistors, are characterized by high *TCR* (up to 4 5%/°C for AlGaAs/GaAs) and simultaneously very low $1/f$ noise characteristics [76]. The noise level for both materials is being several orders of magnitude lower than that of a-Si and VO_x. These uncooled thermistor materials are hybridized with readout circuits by using conventional flip-chip assembly or wafer-level adhesion bonding [45].

Another type of thermistor material is $YBa_2Cu_3O_{6+x}$ (YBaCuO), which belongs to a class of cooper oxides and is best known as a high-temperature superconductor (HTSC). Its conduction properties can be changed from metallic ($0.5 \le x \le 1$) to insulating ($0 \le x \le 0.5$) by suitably decreasing the oxygen content. For $x \approx 1$, YBaCuO possesses an orthorhombic crystal structure, exhibits metallic conductivity, and becomes superconductive upon cooling below its critical temperature. As x decreases to 0.5, the crystal structure undergoes a phase transition to a tetragonal structure and it exhibits semiconducting conductivity characteristics because it exists in a Fermi glass state. As x is decreased further below 0.3, YBaCuO becomes a Hubbard insulator with a well-defined energy gap on the order of 1.5 eV [77].

In the semiconducting state, YBaCuO exhibits a relatively large TCR (34%/°C–4%/°C) over a 60-K temperature range near room temperature. The large TCR, combined with the ease of thin-film fabrication that is compatible with CMOS processing, makes YBaCuO attractive to microbolometer applications [78–82]. These devices have demonstrated voltage responsivities over 10^3 V/W, detectivities above 10^8 cmHz$^{1/2}$/W, and thermal time constant less than 15 ms using typical air-gap microbolometer structures [82].

It should be mentioned that perovskite metal-oxide manganites with colossal magnetoresistance effect and high room temperature TCR of 4.4%/°C [83] and thin-film carbon [84] have been also proposed for thermal imaging.

Most of the reported uncooled bolometers are fabricated by building up an FPA consisting of pixels directly on top of a pre-manufactured readout integrated circuit (ROIC) on standard CMOS process using thin thermistor layer deposition and patterning process. The advantages of this monolithic integration is a simpler process, but it is limited to materials and processes that are compatible with the ROIC. Since processing temperatures above 400°C–450°C are prohibited, the choices for the IR material are limited to materials that can be obtained by processing at such temperatures.

The bolometer performance is determined by its TCR, but also by the noise in the material. It is well-known that the high-quality single crystalline materials present very low $1/f$ noise. Transfer bonding allows the use of high-quality crystalline materials for temperature sensing. One such material is monocrystalline Si/GeSi quantum wells [85]. The thermistor material is optimized and deposited on a carrier wafer, then transferred to the ROIC by low-temperature direct wafer bonding and subsequent removal of the carrier [86].

8.4 SUPERCONDUCTING BOLOMETERS

The concept of a superconducting transition edge bolometer is not new [87–93]. Conventional-type superconducting bolometers are large area structures where the detector film acts as the radiation absorber. These have been built using tin [92] and niobium nitride [89]. The first bolometer to actually utilize the superconducting transition was a composite structure using a blackened aluminum foil absorber in conjunction with tantalum temperature sensor [89]. Another such composite structure makes use of aluminum as the temperature sensor and bismuth as the absorber [93].

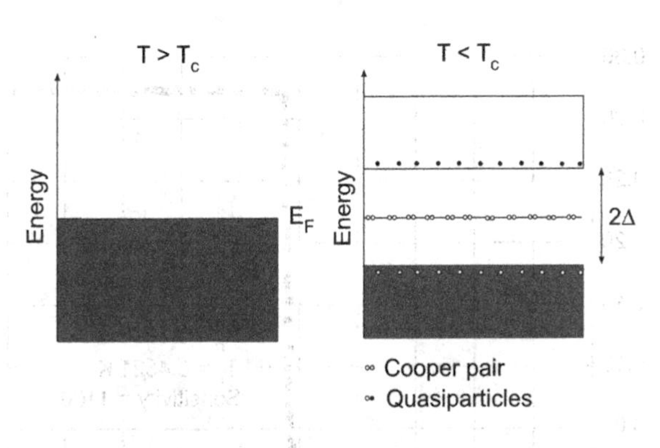

Figure 8.16 Energy diagram (a) above and (b) below the superconducting transition.

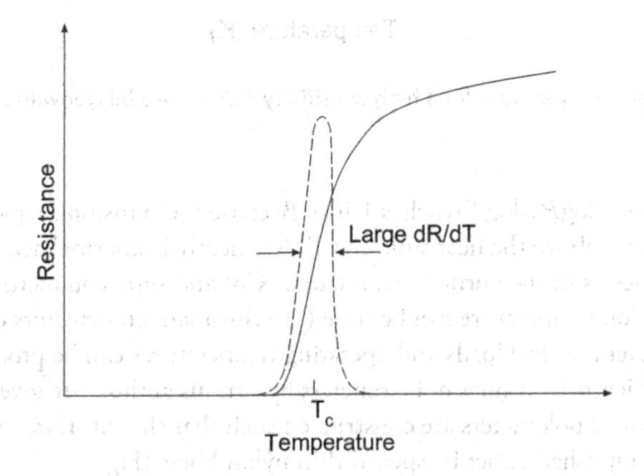

Figure 8.17 Typical resistance versus temperature plot with the derivative *dR/dT* superimposed on the transition edge.

Below a certain temperature, defined as the critical temperature T_c, the electrons at the Fermi energy condense into a coherent state made up of Cooper pairs, as shown in Figure 8.16. The parameter 2Δ is the superconducting energy gap. According to Bardeen–Cooper–Schrieffer theory, the value of 2Δ is given by $3.53kT_c$, which is in reasonably good agreement with the measured value of $(3.2–4.6)kT_c$ for superconducting elements [94]. For a transition temperature of 90 K (typical for YBaCuO), the energy gap predicted from this relationship is 27 meV. Often given the scatter in the experimental data for different materials, the precise value of the energy gap cannot be determined. As in an ordinary bolometer, there is not a direct photon–lattice interaction, so the response is slow, but independent of wavelength.

The theory, construction principles, and performance of superconductor bolometers are considered in several reviews [25,93–102]. The general theory of bolometers described in Section 8.1 is also relevant for superconducting bolometers.

Figure 8.17 illustrates the operation of a bolometer in which the resistive element is a superconductor maintained near the midpoint of the superconducting-to-normal transition edge. This detector is usually called a transition edge sensor (TES). Because of the steepness of the edge near the critical temperature T_c, a small change in temperature causes a large change in resistance; so *dR/dT* is large and the width can be < 0.001 K. It was shown [96] that the value of D^* depends on the time constant of the detector: $D^* = \text{const} \times \tau^{1/2}$ (for comparison, see Equation 8.15). For bolometers especially operated in FIR regions so-called impulse detectivity as a figure of merit have been introduced. It is defined as the ratio between the detectivity and the square root of the response time, $D^*/\tau^{1/2}$, in units cm/J [97].

The transition between the superconducting and the normal stage can be used as an extremely sensitive detector. For example, Figure 8.18 shows a superconducting transition of a bilayer with 400 Å of molybdenum and 750 Å of gold, yielding a normal resistance of 330 mΩ [103]. Near its transition temperature of

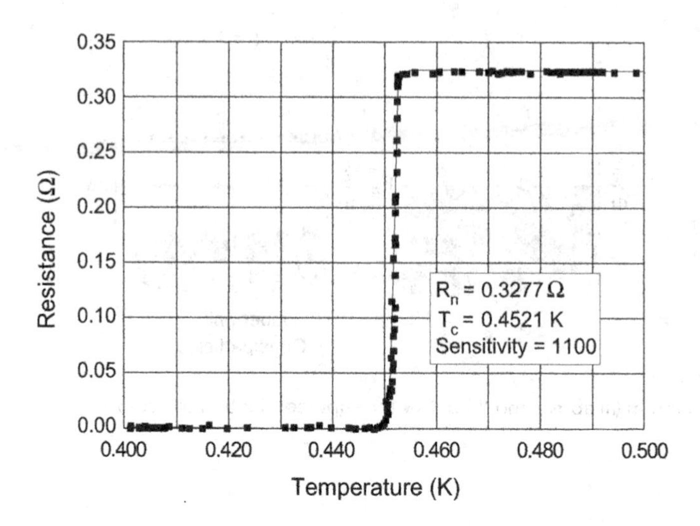

Figure 8.18 Resistance versus temperature for a high-sensitivity TES Mo/Au bilayer with superconducting transition at 444 mK. (From Ref. [103])

440 mK, the sensitivity $\alpha \equiv d\log R/d\log T$ reaches 1,100. Because the transition region is narrow (≈1 mK) compared to the temperature above the heat sink, the TES is nearly isothermal across the transition. By varying the relative thicknesses of the normal metal (Au or Cu) and superconducting metal (Mo) in the bilayer process, the transition temperature can be tuned. In this manner, detectors optimized for performance in a variety of different optical loads and operating temperatures can be produced.

The experimentally achieved D^* values of bolometers by various authors are given in Table 8.2 [97,99]. The most sensitive and slowest bolometers are constructed such that the substrate makes a poor contact with the base. This is accomplished either by special thin nylon fibers (Figure 8.19a [97]; materials 1 and 2 in Table 8.2), or the heat is drawn off through the thin substrate, the ends of which are connected to the base (Figure 8.19b; samples 3–5 in Table 8.2). Faster response is realized in bolometer constructions on a solid substrate (see Figure 8.19c). In some cases, the superconducting film makes a direct contact with the massive substrate made from material having a high heat conductivity, such as sapphire (sample 7 in Table 8.2), makes contact with liquid helium (samples 8 and 9 in Table 8.2), or sandwiches a heat-insulating layer between the base and the superconducting sensitive element (sample 10 in Table 8.2). The last design has made it possible to build bolometers for the microsecond region having a threshold flux close to the limiting value set by fluctuations of the background radiation power.

A major difficulty to couple incident radiation to a superconducting bolometer resides in the high reflectivity of superconducting materials, especially the long wavelength and FIR. For example, YBaCuO exhibits > 98% reflectivity at $\lambda > 20\,\mu$m [104]. To overcome this problem, porous and granular black metals (usually silver and gold) offer both a large absorption and a low specific heat to give a reasonable compromise between the absorbing layer and satisfactory time response. However, this type of device is characterized by a rather slow response time. The impulse detectivity ranges in the 10^{10}–10^{11} cm/J interval and is one or two orders of magnitude lower than the realizable values with antenna-coupled devices.

The antenna-coupled design (Figure 8.19d) gives an effective way of increasing the sensitivity of the thermal radiation detector while retaining a fast response. In this case, a thin-film antenna deposited onto a substrate receives the radiation, which induces displacement currents in it having a frequency corresponding to the radiation wavelength. The high-frequency currents heat the thin-film bolometer, which fulfills the role of converting thermal power into an electrical signal.

In 1977, Schwarz and Ulrich had the first paper devoted to room temperature antenna-coupled metallic film infrared detectors [105]. In contrast to absorbing layer coupling, antenna coupling gives selective responsivity to both the spatial mode and the polarization of the incoming radiation. Since the effective detector area is an order of λ^2, this leads to a large absorbing area (also to a large thermal mass and slow response time) in the case of FIR radiation. However, fabrication of microbolometers (see Figure 8.19e),

Table 8.2 Parameters of superconductor thermal radiation detectors

MATERIAL	ELEMENT SIZE (mm^2)	TEMPERATURE (K)	SENSITIVITY (V/W)	TIME CONSTANT (s)	D^*($cmHz^{1/2}W^{-1}$)	*NEP* ($W/Hz^{1/2}$)	REMARKS (SUBSTRATE/ ANTENNA)
1. Sn	3×2	3.05	850	10^{-2}	3.6×10^{11}	7×10^{-13}	
2. Al	4×4	1.27	3.5×10^4	8×10^{-2}	1.2×10^{14}	7×10^{-13}	
3. Ni + Sn	1×1	0.4	2.2×10^6	10^{-3}	2.2×10^{13}	3.4×10^{-15}	
4. Pb + Sn	–	4.8	10^4	6×10^{-3}	–	4.5×10^{-15}	
5. NbN	0.1×0.1	6.5	5×10^5	10^{-4}	–	–	
6. Sn	0.15×0.15	3.7	10^4	6×10^{-3}	10^{10}	1.6×10^{-12}	
6. Pb + Sn	1×1	3.9	24	7×10^{-9}	1.2×10^9	8.4×10^{-11}	
7. Sn	10×10	3.63	1	2×10^{-8}	10^9	10^{-9}	
8. Ag + Sn	2.3×2.3	2.1	2.2	5×10^{-9}	2.6×10^9	9×10^{-10}	
9. Sn	1×1	3.3	4200	2×10^{-6}	5×10^{10}	2×10^{-12}	
10. Pb + Sn	0.02×0.00225	4.7	5700	2×10^{-8}	–	3×10^{-13}	
11. Mo:Ge	–	0.1	10^9	10^{-6}	1×10^{16}	1×10^{-18}	
12. Pb	–	3.7	10^5	10^{-8}	5×10^{11}	2×10^{-14}	Sapphire
13. Au + Pb + Sn	–	3.7	6000	2×10^{-8}	2×10^{11}	5×10^{-14}	Quartz/V antenna
14. YBaCuO	1×1	20	0.1	4×10^{-7}	2.5×10^6	4×10^{-13}	
15. YBaCuO	1×1	86	40	1.3×10^{-2}	6.7×10^7	1.5×10^{-9}	
16. YBaCuO	0.01×0.09	40	4×10^3	10^{-3}	10^8	2.5×10^{-11}	
17. YBaCuO	0.1×0.1	86	15	1.6×10^{-4}	3.3×10^7	3×10^{-10}	
18. YBaCuO	$0.1 \times$	80	10^3 (A/W)	6×10^{-2}	3×10^8	10^{-10}	
19. YBaCuO	–	90	2000	10^{-6}	2×10^9	5×10^{-12}	YSZ
20. YBaCuO	–	91	480	2×10^{-5}	2.2×10^9	4.5×10^{-12}	YSZ/Log-periodic
21. YBaCuO	–	90	4000	2×10^{-7}	4×10^9	2.5×10^{-12}	Si_3N_4/Suspended bridge
22. YBaCuO	–	88	2180	1×10^{-5}	1.1×10^9	9×10^{-12}	Si_3N_4/Log-periodic
23. YBaCuO	–	85	240	3×10^{-7}	8.3×10^8	1.2×10^{-11}	$NdGaO_3$/Bow-tie

Source: After Refs. [97,99].

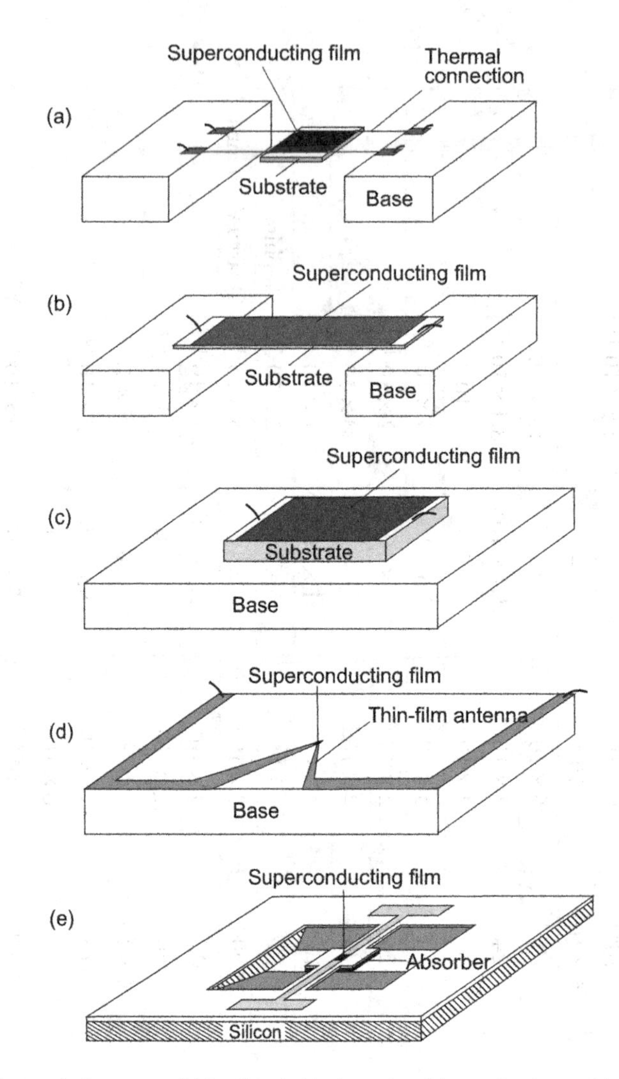

Figure 8.19 Superconductor bolometers: (a) isothermal bolometer, (b) non-isothermal bolometer, (c) bolometer on solid substrate, (d) antenna-coupled bolometer, and (e) micromachined bolometer. (From Ref. [97])

using conventional lithography and micromachining techniques, reaches time constants in the μs range and good detectivity. In this case, an antenna with a rather large effective area can feed a superconducting microbridge of a much smaller area (a few μm²).

Two frequency-independent antenna families, shown in Figure 8.20, can be built. In the first one, the antenna geometry is defined by angles (not by geometrical lengths) [99]; bow-tie and spiral antennas fall into so-called equiangular antennas. In the second family (see Figure 8.20c), the antenna is built up from coupled elements (e.g., dipoles). The finite dimensions of the structures restrict the antenna bandwidth roughly from $2r_{min}$ to $2r_{max}$ in terms of wavelength. Perturbations in the radiation bandwidth limits are overcome with the log-periodic structure.

Another class of antenna structures belongs to end-fire antennas, which are derived from long-wire traveling wave antennas. As opposed to previously described structures whose radiation direction lies in a plane orthogonal to the antenna plane, it lies along the antenna plane for end-fire geometries. These structures offer the possibility of making compact detector arrays with pixels containing V-antennas like those shown in Figure 8.19d.

Table 8.2 contains performance characteristics and Figure 8.21 presents detectivity as a function of response time for antenna-coupled superconductor bolometers operated in the FIR spectral region. For

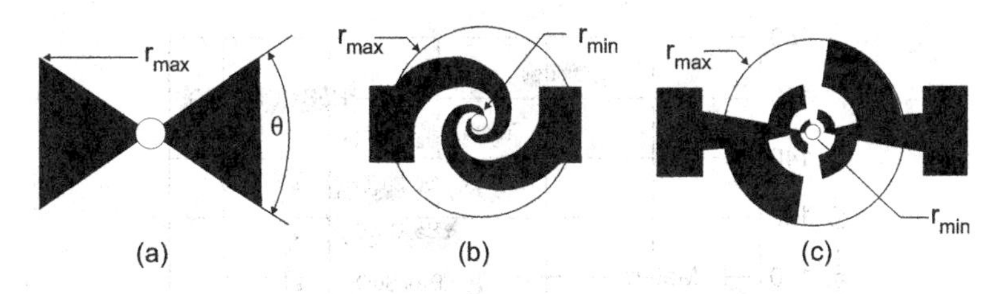

Figure 8.20 Frequency-independent planar antenna geometries: (a) bow-tie equiangular, (b) spiral equiangular, and (c) circular log-periodic. The sensing element is located at the center of the structure. (From Ref. [99])

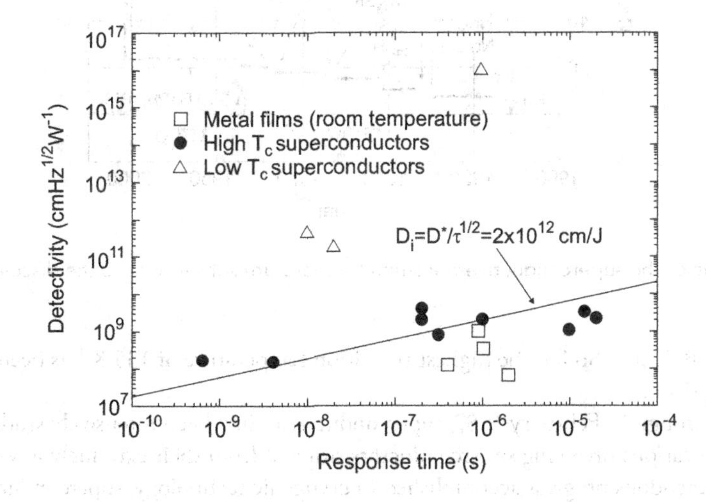

Figure 8.21 Detectivity as a function of response time for antenna-coupled FIR bolometers. (From Ref. [99])

the best detectors, *NEP* values close to the 2×10^{-12} W/Hz$^{1/2}$ phonon noise limited prediction have been obtained [99]. For liquid nitrogen (LN)-cooled bolometers based on YBaCuO bolometers, a simple relation between detectivity and response time is observed: $D^* = 2 \times 10^{12}\,\tau^{1/2}$, where D^* in cmHz$^{1/2}$/W and τ in s.

The experimental data of room temperature detectors presented in Figure 8.21 concern bismuth or niobium microbridge sensors operated in the 10–100-µm wavelength range. The best *NEP* values have been obtained with suspended microbridges or low thermal conductance buffer layers with silicon substrates.

At present, conventional low critical temperature superconductors are used rarely. They offer unsurpassed performance in terms of both voltage responsivity and *NEP*. Their low operating temperature leads to high impulse detectivity, the order of 10^{15} cm/J, due to lower specific heat. It should be also noted that very low resistance of these sensor causes impedance matching between antenna and sensor.

8.5 HIGH-TEMPERATURE SUPERCONDUCTING BOLOMETERS

The important discovery by Müller and Bednorz of a new class of superconducting materials [106,107], so called HTSCs, is undoubtedly one of the major breakthroughs in material science at the end of the twentieth century. Figure 8.22 shows the progression of the superconducting transition temperatures from the discovery of the phenomenon in mercury by Onnes in 1911. Between 1911 and 1974, the critical temperatures of metallic superconductors steadily increased from 4.2 K in mercury up to 23.2 K in sputtered Nb_3Ge films. Nb_3Ge held the record for the critical temperature in metallic superconductors until the unexpected discovery of superconductivity at 39 K in the intermetallic MgB_2 [108]. The first superconducting oxide $SrTiO_3$, characterized by transition temperature as low as 0.25 K, was discovered in 1964. Discovery of high-temperature superconductivity in the cuprate $(La,Ba)_2CuO_4$ ($T_c \approx 30$ K) opened a new field of research. Within less than a year, a critical temperature well above 77 K could be achieved

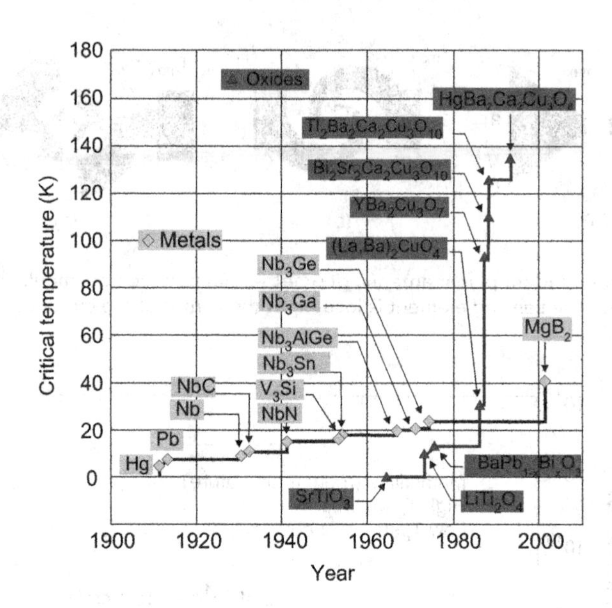

Figure 8.22 Evolution of the superconductive transition temperature subsequent to the discovery of the phenomenon.

in $YbBa_2Cu_3O_{7-x}$ (YBaCuO). So far, the highest transition temperature of 135 K has been found in $HgBa_2Ca_2Cu_3O_{8+x}$ [109].

Prior to a breakthrough in February 1987, superconductivity had been intensively studied, but the development of applications involving superconductors suffered from their extremely low working temperatures. In spite of tremendous progress accomplished in cryogenic technology, superconductivity has been used either when no classical alternative was available or when the required performances were impossible to achieve using a non-superconducting solution. Moreover, in the case of superconducting infrared detectors, their usefulness was limited due to the stringent temperature control required, their poor radiation absorption characteristics due to being thin, and their fragility. The performance was generally limited by amplifier noise rather than by radiation fluctuations.

All HTSC materials are so-called oxygen-deficient perovskites and their basic crystalline structure is similar to that of $CaTiO_3$ (parent mineral for the perovskite family). Although several HTSCs are known, attention is directed toward YBaCuO since it has received far more attention. Enomoto and Murakami made the earliest photoresponse measurements using granular $BaPb_{0.7}Bi_{0.3}O_3$ and reported encouraging results [110]. The parameters of YBaCuO appropriate to the device design are reviewed by Kruse [111]. The interest in using BiSrCaCuO has decreased because of the difficulties in establishing one superconducting phase. There have been doubts about using TlBaCaCuO because of the poisonous and volatile element Tl, even though the critical temperature is 125 K [107].

For HTSC, several authors have theoretically considered YBaCuO films from a theoretical viewpoint [25,111–115]. The performance of cryogenic superconducting detectors should be one to two orders of magnitude better than uncooled thermal detectors. As the transition temperature of good YBaCuO films is about 90 K, LN is a convenient cryogen for YBaCuO films. Richards and coworkers estimated that NEP in the range of $(1–20) \times 10^{-12}$ W/Hz$^{1/2}$ [114], depending on a substrate, should be available. Such performance could exceed that of any other detectors operating at or above LN temperatures for wavelengths greater than 20 μm.

For a properly designed microbolometer detector, the value of the thermal conductance G_{th} is dominated by the support structure, not by YBaCuO. Values as small as 2×10^{-7} W/K are possible [111]. Since the density of YBaCuO is 6.3 g/cm^3, the specific heat at 90 K is 195 mJg^{-1}K^{-1}, a 75×75 μm^2 pixel 0.30-μm thick has a thermal capacity C_{th} of 2.1×10^{-9} J/K; thus, the thermal time constant given by Equation 4.5 is 1.0×10^{-2} s. Assuming an absorptance ε of 0.8, the value of the temperature fluctuation noise–limited detectivity given by Equation 4.23 is 2.1×10^{10} cmHz$^{1/2}$W^{-1}. To attain this value, the bias current must equal or exceed 3.5 μA. At lower values of bias current, the detector will be Johnson noise–limited. As the bias is raised to 3.5 μA, the microbolometer will become temperature fluctuation noise–limited. Assuming that the TCR is 0.33 K^{-1}, the low-frequency responsivity (see Equation 8.2) at a bias current of 3.5 μm will be 6.1×10^3 V/W.

The performance of YBaCuO HTSC bolometers in comparison with typical photon detectors in 2D FPAs operated at 77 K has been presented by Verghese et al. [115] and is shown in Figure 8.23. The calculations are carried out using values of measured properties of YBaCuO films on yttria-stabilized zirconia (YSZ) buffer layers on Si and Si_3O_4 (on Si) with $\tau = 10$ ms, diffraction limited throughput, and *f*/6 optics. The D^* can be as high as 3×10^{10} $cmHz^{1/2}W^{-1}$. Also shown, for comparison, are the photon noise–limited D^* and estimates of the performance of detectors now used in imaging arrays. We can see that D^* falls at short wavelengths because the membrane technology is not able to provide a small enough G_{th} for a small detector area, and τ becomes shorter than 10 ms. It also falls at long wavelengths because the resistance fluctuation noise becomes important in large area detectors. The D^* of a Si_3O_4 membrane bolometer has a potentially useful peak near 10 μm, which is a very important wavelength for thermal imaging. Because of higher bulk thermal conductivity of silicon, the region of photon noise–limited D^* appears at longer wavelengths than for Si_3N_4 membrane bolometer. The limit to D^* from bolometer noise is higher than that of bolometers on Si_3O_4 because of the lower NEP of YBaCuO on silicon.

The given analysis clearly indicates that HTSC bolometers are especially useful in the FIR ($\lambda > 20$ μm), where it is difficult to find sensitive detectors that operate at relatively high temperatures (e.g., $T > 77$ K). Moreover, it is estimated that the pixel production cost of HTSC bolometers is potentially several orders of magnitude less than that of HgCdTe and InSb.

A review of substrate materials and deposition techniques suitable for the fabrication of high-quality HTSC films for electronic and optoelectronic applications is given by Sobolewski [112]. The SEM micrographs of HTSC thin films reveal two types of structures—random or granular structure and oriented or epitaxial structure. The random structure consists of many small (≈1 μm) superconducting grains embedded in a non-superconducting matrix. Point contacts between grains may act as Josephson junctions. Films made of this type of structure have broad transitions and relatively low T_c and the critical current I_c. The grain boundaries result also in excessive 1/*f* noise.

The oriented structure is a crystal growth with c-axis perpendicular to the plane of the substrate. Films made of this structure have sharp transitions and relatively high T_c and I_c. For detector applications, the

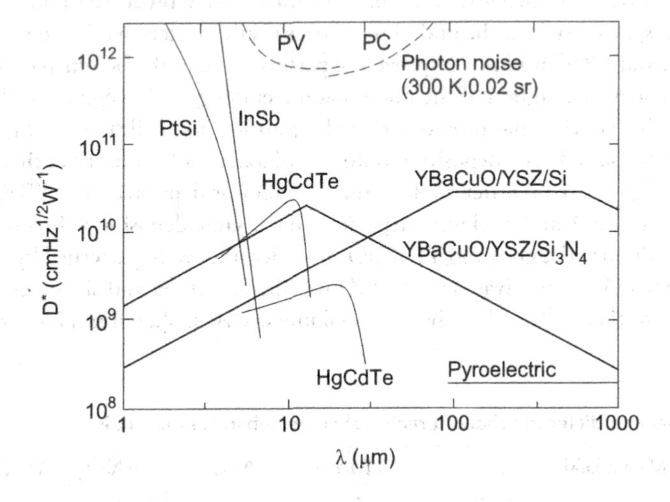

Figure 8.23 Detectivity as a function of wavelength for diffraction-limited pixels with field of view, FOV = 0.02 sr (*f*/6 optics) and $\tau = 10$ ms. The thick lines show the predicted D^* for HTSC bolometers on silicon and Si_3N_4 membranes using YBaCuO films. These lines were calculated using estimates for the minimum achievable heat capacity and thermal conductance and using measurements of voltage noise in HTSC bolometers. Typical values of D^* for InSb, PtSi, and HgCdTe detectors in 2D FPAs operated at 77 K are shown for comparison. Also shown are the photon noise limits for photovoltaic and photoconductive detectors which view 300 K radiation in a 0.02 sr FOV. (From Ref. [115])

film thickness must be less than the optical penetration depth of the material, ≈0.15 μm. For granular structures, the film should be patterned into a microbridge to reduce the number of Josephson junctions biased. Ideally, a linear chain of junctions is preferable so that there is only a single conduction path. For epitaxial structures, we must artificially create weak links in the film to form the Josephson junction.

It is generally accepted that good-quality HTSC films require high-quality dielectric substrates, which combine desired dielectric properties with a good lattice match, enabling epitaxial growth of the films. Except for diamond, most suitable substrate materials have similar volume-specific heat at 77–90 K. In all cases, it is very much larger than is seen at liquid helium temperatures. Consequently, the thermal time tends to be long. Therefore, one important requirement for a substrate material is strength, so that it can be made very thin. Some substrates that are favorable for film growth, such as $SrTiO_3$ and $LaAlO_3$, are too weak to produce thin layers of millimeter dimensions. However, good-quality bolometers have been fabricated using these substrates [22,26,97–100,114–119]. Also, such substrates as silicon, sapphire, ZrO_2, or SiN are used [98–100]. Much attention has been given to silicon substrates because of their compatibility with on-chip electronics implementation in semiconductor technology.

In general, however, the substrate should fulfill additional issues. To minimize the phonon escape time [e.g., for phonon-cooled, hot electron bolometers (HEB)], the substrate should have a high thermal conductivity and offer a low thermal interface resistance R_b to the superconducting film. Second, it should have good properties to propagate the radiation signal when the readout circuitry is implemented (i.e., in the GHz range); within this respect, the dielectric loss tangent should be low and the dielectric constant should be fitted to the propagation line and also be comparable with the antenna size. Finally, the substrate material should be transparent to signal radiation; for example, in FIR sensors the receiving antenna is usually illuminated from the substrate back side by means of a focusing lens. See Table 8.3 for gathered substrate parameters [120].

The main efforts in HTSC bolometers technology was directed toward improving the performance of microbolometer FPAs fabricated by micromachining on silicon substrates. At the beginning, the YBaCuO films in these devices were sandwiched between two layers of silicon nitride with thin YSZ layers to buffer YBaCuO from the silicon nitride [121,122]. These 125 × 125 μm² devices were estimated to have an *NEP* of 1.1×10^{-12} W/Hz$^{1/2}$ near 5 Hz with a 5 μA bias (neglecting contact noise). A drawback of this design was that the YBaCuO was grown on an a-Si nitride underlayer, which precludes the possibility of epitaxial YBaCuO growth. The YBaCuO, therefore, was polycrystalline with a broad resistance transition, which limits the bolometer responsivity, and the grain boundaries result in excessive $1/f$ noise.

Incorporating epitaxial YBaCuO films improve the performance of the bolometers [123–127]. Figure 8.24 shows a schematic diagram of the microbolometer design using epitaxial YBaCuO films. The fabrication process of these devices has been described by Johnson and colleagues [124]. The superconductor film was deposited by pulsed laser deposition onto an epitaxial YSZ buffer layer that had been deposited onto a bare, unoxidized 3-in silicon wafer. Gold contact metal was deposited onto YBaCuO film by RF sputtering. The YSZ, YBaCuO, and gold were deposited in the same deposition chamber, without breaking vacuum between depositions. The gold and YBaCuO meander lines were patterned by conventional photolithography, the YBaCuO was passivated with YSZ and silicon nitride, and silicon etch pits were formed by anisotropic etching to thermally isolate the microbolometers. As is shown in Figure 8.24, the membrane

Table 8.3 Thermal and dielectric characteristics of some substrate materials

MATERIAL	MgO	Al_2O_3	$LaAlO_3$	$YAlO_3$	YSZ
Thermal conductivity at 90 K(W/Kcm)	3.4	6.4	0.35	0.3	0.015
R_b with YBaCuO at 90 K (KW^{-1}cm^2)	5×10^{-4}	10^{-3}	10^{-3}	–	10^{-3}
$\tan\delta$ at 10 GHz, 77 K	7×10^{-6}	8×10^{-6}	5×10^{-6}	10^{-5}	4×10^{-4}
ε_r at 10 GHz, 77 K	10	10	23	16	32

Source: After Ref. [120].

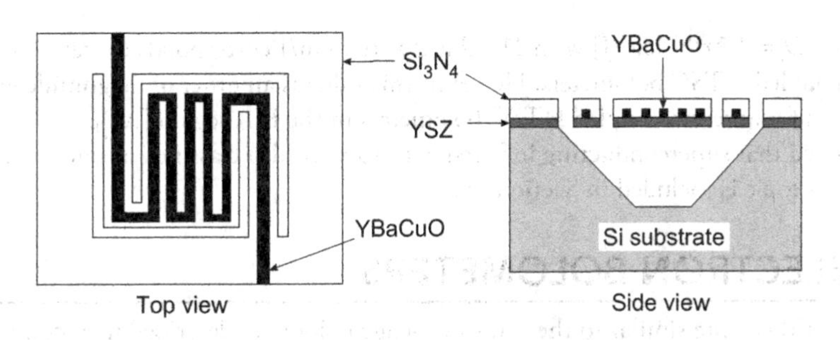

Figure 8.24 Schematic diagram of YBaCuO microbolometer using epitaxial YBaCuO on an epitaxial YSZ buffer layer on a silicon substrate (From Ref. [123])

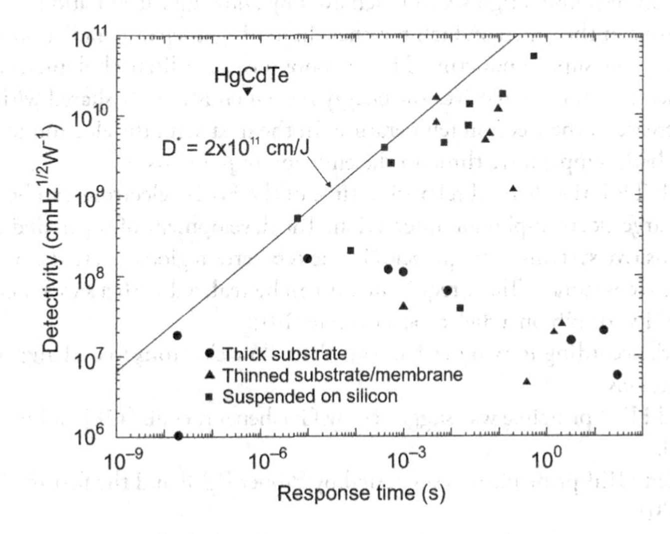

Figure 8.25 The detectivity as a function of the response time for thin-film HTSC bolometers (λ = 0.8–20 μm). (From Ref. [100])

is suspended over an etch pit on a silicon wafer, and supported only by lateral silicon nitride legs approximately 8 μm wide.

For the aforementioned devices with the size $140 \times 105\,\mu m^2$, detectivity of $(8\pm2) \times 10^9$ cmHz$^{1/2}$W^{-1} has been measured on a single element at a bias current of 2 μA. It is one of the highest D^* reported on any semiconducting microbolometer operating at temperatures higher than 70 K. The NEP was 1.5×10^{-12} W/Hz$^{1/2}$ at 2 Hz, at a temperature of 80.7 K. The thermal constant was 105 ms. The noise power spectral density was found to scale with frequency as $1/f^{3/2}$. Linear arrays of microbolometers have also been fabricated. The measured responsivity of detectors varied by less than 20% over the 6-mm length of the 64-element linear array.

Further progress occurred with the deposition of YBaCuO on very thin sapphire. Broadly speaking, sapphire is commercially available as thin as 25 μm, a thickness obtainable by mechanical polishing. Further thinning, necessary for reduced time constants and operating at frequencies above the $1/f$ noise region, is possible by chemical polishing. By using chemical–mechanical polishing techniques to thin sapphire down to 7 μm, a bolometer with 1.2×10^{10} cmHz$^{1/2}$W^{-1} near 4 Hz has been achieved [126].

The progress attained toward realizing HTSC photon detectors is described in many papers (e.g., see [97–100,127]). Since the appearance 30 years ago of the first reports demonstrating a thermal detection action with HTSC sensors, progress in their technology has benefited mainly from newly developed superconducting nanostructures, which are especially promising candidates for FIR detection [98,100]. In the mid-infrared range, their detectivities are similar to those of LN-cooled photon detectors. The latter, however, is still leading in terms of response time. Figure 8.25 compares the performance of HTSC bolometers, where a plot of detectivity as a function of response time is shown. The performance of a typical cooled photoconductive HgCdTe (λ = 3–20 μm) is also included in Figure 8.25. As pointed out previously, an impulse detectivity $D_i = D^*\tau^{-1/2}$ [cm/J] with D_i value 2×10^{11} cm/J corresponds to state-of-the-art results for the absorber-coupled HTSC bolometers. However, this value is an order of magnitude lower than the average D_i observed for antenna-coupled HTSC bolometers in the FIR region [100].

It should be noted that superconducting infrared detectors are also classified as photon detectors. More information on this topic is included in Section 28.12.

8.6 HOT ELECTRON BOLOMETERS

In principle, the HEB is quite similar to the transition-edge bolometer described in Section 8.4, where small temperature changes caused by the absorption of incident radiation strongly influence resistance of biased sensor near its superconducting transition. The main difference between HEBs and ordinary bolometers is the speed of their response. High speed is achieved by allowing the radiation power to be directly absorbed by the electrons in the superconductor, rather than using a separate radiation absorber and allowing the energy to flow to the superconducting TES via phonons, as ordinary bolometers do. After photon absorption, a single electron initially receives the energy $h\nu$, which is rapidly shared with other electrons, producing a slight increase in the electron temperature. In the next step, the electron temperature subsequently relaxes to the bath temperature through the emission of phonons.

In comparison with TES, the thermal relaxation time of the HEB's electrons can be made fast by choosing a material with a large electron-phonon interaction. The development of superconducting HEB mixers has led to the most sensitive systems at frequencies in the terahertz region, where the overall time constant has to be a few tens of picoseconds. These requirements can be realized with a superconducting microbridge made from NbN, NbTiN, or Nb on a dielectric substrate [101].

The HEBs can work according to two mechanisms that allow electrons to exchange their energy faster than they heat the phonons:

- The phonon-cooled HEB principle was suggested by Gershenzon et al. [128] and first realized by Karasik et al. [129].
- The diffusion-cooled HEB principle was suggested by Prober [130] and the first realization was reported by Skalare et al. [131].

A synthetic presentation of both mechanisms has been given by McGrath [132].

Figure 8.26a shows the basic operation of the phonon-cooled bolometer. In this type of device, hot electrons transfer their energy to the phonons with the time τ_{eph}. In the next step, the excess of phonon energy escapes toward the substrate with the time τ_{esc}. Several conditions should be fulfilled to make phonon-cooled mechanism effective: (i) the electron–electron interaction time (τ_{ee}) must be much shorter than τ_{eph}; (ii) the superconducting film must be very thin (a few nm) and the film to substrate thermal conductance must be very high ($\tau_{esc} << \tau_{eph}$) to obtain an efficient phonon escape from superconductor to substrate; and (iii) the substrate thermal conductivity must be very high and very good thermal contact between the substrate and a cold finger must be insured.

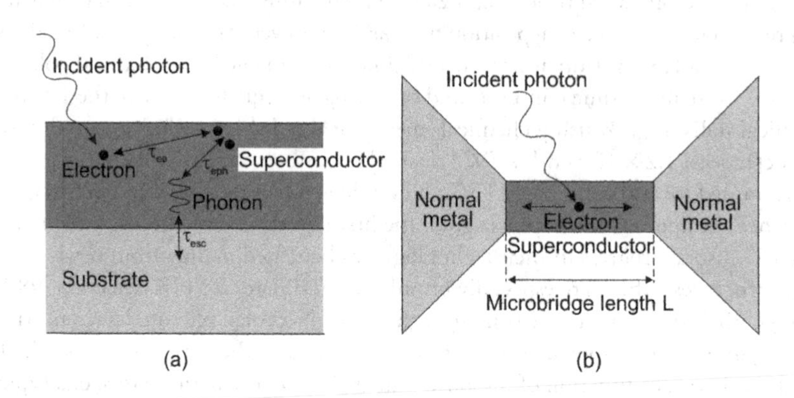

Figure 8.26 HEB mechanisms: (a) phonon-cooled and (b) diffusion-cooled principles.

In diffusion-cooled bolometer—the principle of mechanism is shown in Figure 8.26b—hot electrons transfer their energy by diffusion to a normal metal, which forms the electrical contacts to the external detector readout circuitry and/or the arms of a planar antenna. In this case, the length of the superconducting microbridge must be very short, with maximum value $L_{max} = 2(D_e\tau_{ee})^{1/2}$, where D_e is the electron diffusivity. As was shown by Burke et al. [133], the bolometer bandwidth is inversely proportional to the squared microbridge length, which lies in the submicronic range (see Figure 8.27). The bandwidth of diffusion-cooled bolometers is not limited by such parameter as τ_{eph}. As a result, in comparison with phonon-cooled bolometers, larger intermediate frequency (IF) values are obtained. For diffusion cooling, the interface between the contact pads and the superconducting film is crucial, while for phonon cooling, the interface between the film and the substrate is crucial. It should be pointed out that the distinction is, to some extent, arbitrary since phonon cooling also exists in diffusion-cooled bolometers and vice versa.

Typically, phonon-cooled HEBs are made from ultrathin films of NbN, whereas diffusion-cooled devices use Nb or Al. Current state-of-the-art NbN technology is capable of routinely delivering 3-nm thick devices that are 500 nm^2 in size with transition temperature T_c above 9 K and the transition width of 0.5 K. NbN films are deposited on a dielectric (typically high resistivity >10 kΩcm silicon) by dc reactive magnetron sputtering. The superconducting bridge is defined by means of electron beam lithography. Its length varies between 0.1 and 0.4 μm and the width between 1 and 4 μm. For example, Figure 8.28 shows an example of cross-section of a central part for a planar antenna with the NbN hot electron microbridge.

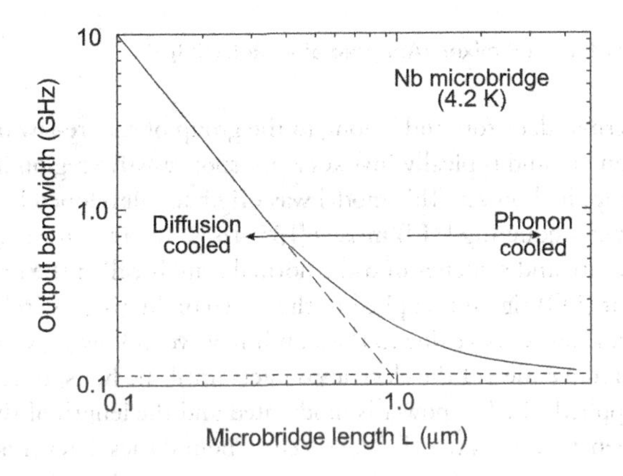

Figure 8.27 Output bandwidth as a function of microbridge length for niobium HEB mixers. For L shorter than 1 μm, the cooling mechanism is by electron diffusion to a normal metal; for larger L, the phonon-cooling mechanism is the dominant one. (From Ref. [133])

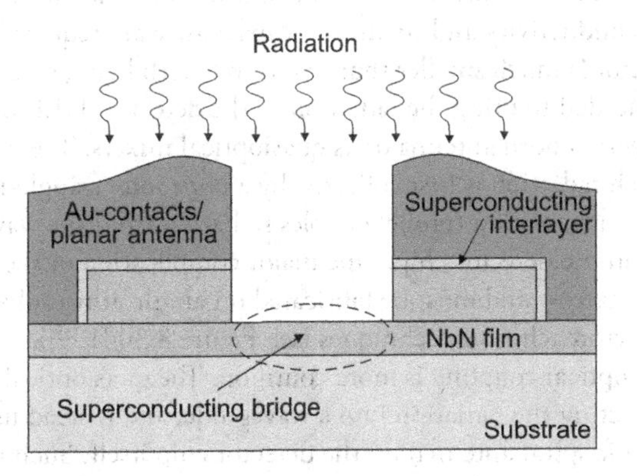

Figure 8.28 Structure of a phonon-cooled HEB. The superconducting microbridge is indicated by a dashed ellipse. The electrical contacts between the microbridge and the Au contacts of the planar antenna are made by a superconducting interlayer.

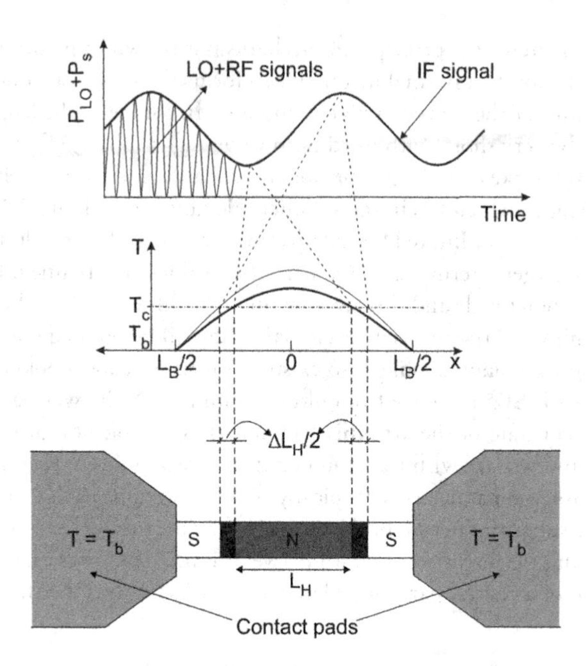

Figure 8.29 Hot-spot model of an HEB mixer. (Adapted after Ref. [138])

The HEB mixers are thermal detectors and belong to the group of square-law mixers. Theory of HEBs is still under development and typically invokes a hot-spot resistive region in the center, whose size responds to changes in the applied power. This model was originally developed by Skocpol et al. [134], and was later applied to superconducting HEB mixers [135–137]. The region where the actual temperature exceeds the critical temperature and switches into the normal state is called the hot-spot (see Figure 8.29 [138]). When local oscillator (LO) signal is applied to the superconducting microbridge, the central part of the microbridge becomes normal-resistive due to the standing wave, following which the hotspot is formed and its boundaries begin to move toward the electric contacts until the hot spot reaches thermal equilibrium. When the RF is applied, the LO power is modulated and the length of the hotspot is oscillated periodically, with the frequency same as the IF. The speed of boundaries determines the response time, but other effects such as the interaction of radiation with magnetic vortices play a role. When comparing diffusion-cooled and phonon-cooled HEBs, the latter provide a smaller noise temperature, and are therefore, preferred.

Superconducting HEB detectors are finding a key role in FIR and terahertz wavelengths. If the bolometer is to be used as a mixer, a bandwidth must be optimized, which means that for a given IF signal, $\tau \ll 1/\omega_{IF}$. In other words, conductivity and small heat capacitance are required [see Equation (4.4)]. Moreover, because the detector is much smaller than the wavelength being received, an antenna and associated coupling circuitry are needed to bring the radiation to the detector. HEB mixers can be made either in a waveguide configuration with a horn antenna or as quasioptical mixers. The more traditional approach is waveguide coupling, in which radiation is first collected by a horn into a single-mode waveguide (typically a rectangular guide), and then, a transition (probe) couples radiation from the waveguide onto a lithographed thin-film transmission line on the detector chip. One major complication of the waveguide approach is that the mixer chip must be very narrow and must be fabricated on an ultrathin substrate. These requirements are helpful using modern micromachining techniques (see Figure 8.30 [139]).

Above ~1 THz, the quasioptical coupling is more common. The quasioptical coupling approach omits the intermediate step of collecting the radiation into a waveguide, and instead uses a lithographed antenna (e.g., twin-slot or a logarithmic spiral antenna) on the detector chip itself. Such mixers are substantially simpler to fabricate and may be produced using a thick substrate (see Figure 8.31 [140]). The substrate with the feed antenna and microbridge is mounted to the reserve side of a hyper-hemispherical or elliptical lens. The reflection loss at the lens surface can be minimized with a quarter-wavelength antireflection coating.

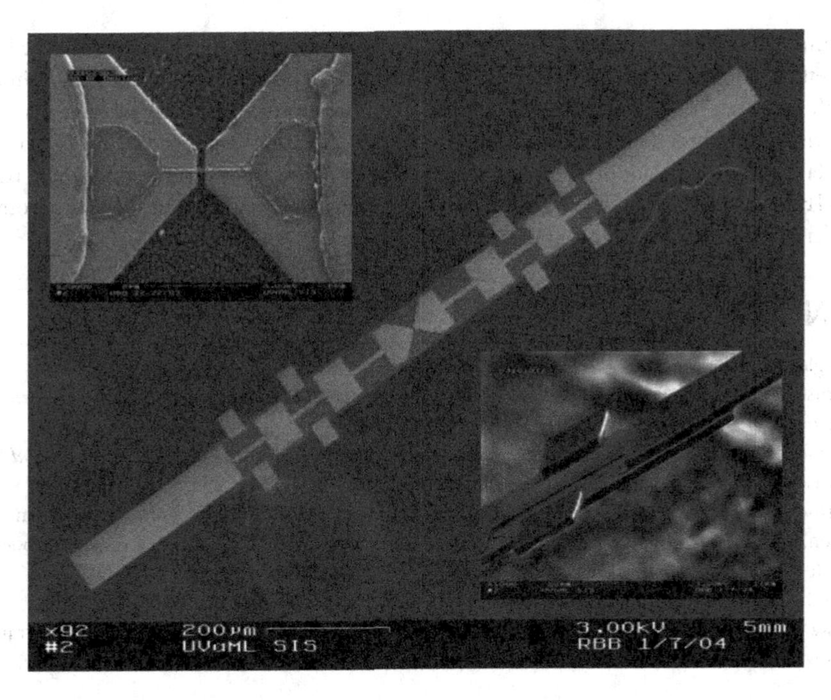

Figure 8.30 Images of a 585-GHz diffusion-cooled HEB mixer chip for a waveguide mount, fabricated using an ultrathin silicon substrate. The dimensions of the HEB bridge are 150 nm long by 75 nm wide; the chip itself is 800 µm long and 3 µm thick. Protruding from the sides and ends of the chip are 2-µm thick gold leads, which provide electrical and thermal contact to the waveguide block, as well as mechanical support for the chip. (From Ref. [139])

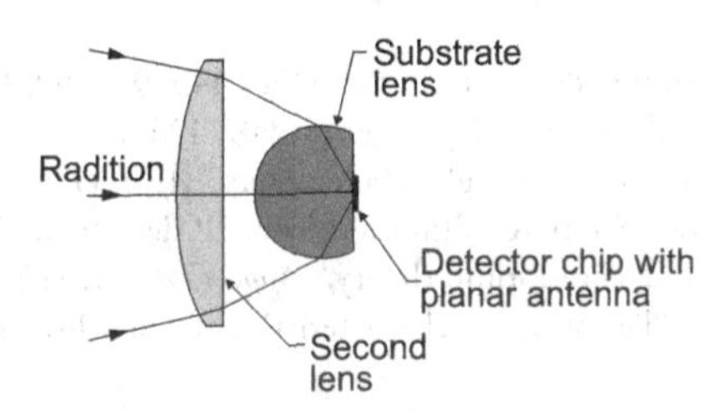

Figure 8.31 Schematic diagram of the "reverse-microscope" quasioptical coupling approach. (From Ref. [140])

The performance of selected HEBs is shown in Figure 6.10. At all frequencies, phonon-cooled HEBs have a lower noise temperature than that of diffusion-cooled devices, although improvements to the latter type may be possible [101]. More information about HEB mixers is given in Section 28.11.1.

Other HEB materials such as a normal metal (usually copper), high T_c superconductors, and n-type InSb should be mentioned in this brief discussion. HTSCs exhibit a very short electron-phonon interaction time (typically between 1 and 2 ps at 80–90 K in YBaCuO), so only phonon-cooled devices have been realized to date (electron diffusion mechanism is negligible). Moreover, the analysis of HTSC HEBs is rather different from low-temperature counterparts due to the high operating temperature [141]. Limited results have been obtained with YBaCuO HEBs [99,100].

InSb HEBs have found practical applications. Since their bandwidth is about 4 MHz, their use is limited. The parameters of these bolometers are largely set by the InSb [142,143]. Typical voltage responsivities are 100 to 1,000 V/W, the thermal conductance is about 5×10^{-5} W/K, and the thermal capacity of the electron sea $C_{th} \approx (3/2)nkV$, where n is the carrier concentration and V is the volume of the detector. Assuming detector volume as 10^{-2} cm^3 and $n \approx 5 \times 10^{13}$ cm^{-3}, we can estimate thermal capacity as $C_{th} \approx 10^{-11}$ J/K and the detector time constant as 2×10^{-7} s (see Equation 4.5). For this n-type InSb sample operating at 4 K, the thermally limited NEP that equals 2×10^{-13} W/Hz$^{1/2}$ has been estimated.

The quantum efficiency of the IR detector depends on the absorption coefficient. Since the free carrier absorption coefficient increases as λ^2, the performance of a device utilizing this effect should improve as the wavelength increases. The value for $\alpha \approx 22$ cm^{-1} at 1-mm wavelength is comparable with that found in extrinsic germanium photoconductive detectors [142], but the value found at 100 μm is too small ($\alpha \approx$ 0.30 cm^{-1}) to fabricate an efficient detector at this wavelength. From these estimations, it can be noted that n-type InSb HEBs are useful at wavelengths of 1 mm or somewhat less, but these devices become ineffective at a wavelength shorter than about 300 μm.

REFERENCES

1. S. P. Langley, "The bolometer," *Nature*, 25, 14–16, 1881.
2. P. W. Kruse, L. D. McGlauchlin, and R. B. McQuistan, *Elements of Infrared Technology: Generation, Transmission, and Detection*, Wiley, New York, 1962.
3. R. A. Wood, "Monolithic silicon microbolometer arrays," in *Uncooled Infrared Imaging Arrays and Systems*, eds. P. W. Kruse and D. D. Skatrud, 43–121, Academic Press, San Diego, CA, 1997.
4. P. W. Kruse, *Uncooled Thermal Imaging. Arrays, Systems, and Applications*, SPIE Press, Bellingham, WA, 2001.
5. C. Jansson, U. Ringh, and K. Liddiard, "Theoretical analysis of pulse bias heating of resistance bolometer infrared detectors and effectiveness of bias compensation," *Proc. SPIE*, 2552, 644–52, 1995.
6. K. C. Liddiard, "Thin-film resistance bolometer IR detectors," *Infrared Phys.*, 24, 57–64, 1984.
7. A. Tanaka, S. Matsumoto, N. Tsukamoto, S. Itoh, K. Chiba, T. Endoh, A. Nakazato, et al., "Infrared focal plane array incorporating silicon IC process compatible bolometer," *IEEE Trans. Electron Devices*, 43, 1844–80, 1996.
8. S.-B. Ju, Y.-J. Yong, and S.-G. Kim, "Design and fabrication of high fill-factor microbolometer using double sacrificial layers," *Proc. SPIE*, 3698, 180–89, 1999.
9. J. M. Shive, *Semiconductor Detectors*, Van Nostrand, New York, 1959.
10. E. M. Wormser, "Properties of thermistor infrared detectors," *J. Optic. Soc. Am.*, 43, 15–21, 1953.
11. R. De Waars and E. M. Wormser, "Description and properties of various thermal detectors," *Proc. IRE*, 47, 1508–13, 1959.
12. R. W. Astheimer, "Thermistor infrared detectors," *Proc. SPIE*, 443, 95–109, 1984.
13. R. C. Jones, "Immersed radiation detectors," *Appl. Opt.*, 1, 607, 1962.
14. S. G. Bishop and W. J. Moore, "Chalcogenide glass bolometers," *Appl. Opt.*, 12, 80–83, 1973.
15. S. Baliga, A. Doctor, and M. Rost, "Sputtered film thermistor IR detectors," *Proc. SPIE*, 2225, 72–78, 1994.
16. J. C. Mather, "Bolometer noise: Nonequilibrium theory," *Appl. Opt.*, 21, 1125–29, 1982.
17. W. S. Boyle and K. F. Rogers, Jr., "Performance characteristics of a new low-temperature bolometer," *J. Opt. Soc. Am.*, 49, 66–69, 1959.
18. E. H. Putley, "Thermal detectors," in *Optical and Infrared Detectors*, ed. R. J. Keyes, 7–1100, Springer, Berlin, 1977.
19. N. Coron, "Infrared helium cooled bolometers in the presence of background radiation: optimal parameters and ultimate performances," *Infrared Phys.*, 16, 411–19, 1976.
20. S. Zwerdling, R. A. Smith, and J. P. Theriault, "A fast, high-responsivity bolometer detector for the very far infrared," *Infrared Phys.*, 8, 271–336, 1968.
21. E. E. Haller, "Physics and design of advanced IR bolometers and photoconductors," *Infrared Phy.*, 25, 257–66, 1985.
22. P. L. Richards, "Bolometers for infrared and millimeter waves," *J. Appl. Phys.*, 76, 1–24, 1994.
23. F. J. Low, "Low-temperature germanium bolometer," *J. Opt. Soc. Am.*, 51, 1300–04, 1961.
24. B. T. Draine and A. J. Sievers, "A high responsivity, low-noise germanium bolometer for the far infrared," *Opt. Comm.*, 16, 425–28, 1976.
25. M. A. Kinch, "Compensated silicon-impurity conduction bolometer," *J. Appl. Phys.*, 42, 5861–63, 1971.
26. E. E. Haller, "Advanced far-infrared detectors," *Infrared Phys. Technol.*, 35, 127–46, 1994.
27. N. S. Nishioka, P. L. Richards, and D. P. Woody, "Composite bolometers for submillimeter wavelengths," *Appl. Opt.*, 17, 1562–67, 1978.
28. E. L. Dereniak and D. G. Crowe, *Optical Radiation Detectors*, Wiley, New York, 1984.
29. P. M. Downey, A. D. Jeffries, S. S. Meyer, R. Weiss, F. J. Bachner, J. P. Donnelly, W. T. Lindley, R. W. Mountain, and D. J. S. Silversmith, "Monolithic silicon bolometers," *Appl. Opt.*, 23, 910–14, 1984.
30. R. E. Flannery and J. E. Miller, "Status of uncooled infrared imagers," *Proc. SPIE*, 1689, 379–95, 1992.

31. R. A. Wood, C. J. Han, and P. W. Kruse, "Integrated uncooled IR detector imaging arrays," *Proceedings of IEEE Solid State Sensor and Actuator Workshop*, 132–35, Hilton Head Island, SC, June 1992.
32. R. A. Wood, "Micromachined bolometer arrays achieve low-cost imaging," *Laser Focus World*, 101–6, June 1993.
33. R. A. Wood, "Uncooled thermal imaging with monolithic silicon focal planes," *Proc. SPIE*, 2020, 322–29, 1993.
34. K. C. Liddiard, "Thin film monolithic arrays for uncooled thermal imaging," *Proc. SPIE*, 1969, 206–16, 1993.
35. M. H. Unewisse, S. J. Passmore, K. C. Liddiard, and R. J. Watson, "Performance of uncooled semiconductor film bolometer infrared detectors," *Proc. SPIE*, 2269, 43–52, 1994.
36. R. A. Wood, "High-performance infrared thermal imaging with monolithic silicon focal planes operating at room temperature," *Electron Devices Meeting, IEDM'93, Technical Digest, International*, 175–77, 1993.
37. B. Fieque, J. L. Tissot, C. Trouilleanu, A. Crates, and O. Legras, "Uncooled microbolometer detector: Recent developments at Ulis," *Infrared Phys. Technol.*, 49, 187–91, 2007.
38. M. Kohin and N. Buttler, "Performance limits of uncooled VO_x microbolometer focal plane arrays," *Proc. SPIE*, 5406, 447–53, 2004.
39. P. W. Kruse, "Can the 300 K radiating background noise limit be attained by uncooled thermal imagers?" *Proc. SPIE*, 5406, 437–46, 2004.
40. F. Niklaus, C. Jansson, A. Decharat, J.-E. Källhammer, H. Pettersson, and G. Stemme, "Uncooled infrared bolometer arrays operating in a low to medium vacuum atmosphere: performance model and tradeoffs," *Proc. SPIE*, 6542, 65421M, 2007.
41. H. Jerominek, F. Picard, and D. Vincent, "Vanadium oxide films for optical switching and detection," *Opt. Eng.*, 32, 2092–99, 1993.
42. E. Kuźma, "Contribution to the technology of critical temperature resistors," *Electron Technol.*, 26(2/3), 129–42, 1993.
43. H. Jerominek, T. D. Pope, M. Renaud, N. R. Swart, F. Picard, M. Lehoux, S. Savard, et al., "64 × 64, 128 × 128 and 240 × 320 Pixel uncooled IR bolometric detector arrays," *Proc. SPIE*, 3061, 236–47, 1997.
44. C. Chen, X. Yi, J. Zhang, and B. Xiong, "Micromachined uncooled IR bolometer linear array using VO_2 thin films," *Int. J. Infrared Millim. Waves*, 22, 53–58, 2001.
45. F. Niklaus, C. Vieider, and H. Jakobsen, "MEMS-based uncooled infrared bolometer arrays: A review," *Proc. SPIE*, 6836, 68360D, 2007.
46. M. Soltani, M. Chaker, E. Haddad, R. V. Kruzelecky, and J. Margot, "Effects of Ti-W codoping on the optical and electrical switching of vanadium dioxide thin films grown by a reactive pulsed laser deposition," *Appl. Phys. Lett.*, 85, 1958–60, 2004.
47. Y.-H. Han, K.-T. Kim, H.-J. Shin, and S. Moon, "Enhanced characteristics of an uncooled microbolometer using vanadium-tungsten oxide as a thermoelectric material," *Appl. Phys. Lett.*, 86, 254101–3, 2005.
48. A. J. Syllaios, T. R. Schimert, R. W. Gooch, W. L. McCardel, B. A. Ritchey, and J. H. Tregilgas, "Amorphous silicon microbolometer technology," *MRS Proc.*, 609, A14.4 1–6, 2000.
49. K. C. Liddiard, U. Ringh, C. Jansson, and O. Reinhold, "Progress of Swedish-Australian research collaboration on uncooled smart IR sensors," *Proc. SPIE*, 3436, 578–84, 1998.
50. B. I. Craig, R. J. Watson, and M. H. Unewisse, "Anisotropic excess noise within a-Si:H," *Solid-State Electron.*, 39, 807–12, 1996.
51. J. L. Tissot, F. Rothan, C. Vedel, M. Vilain, and J.-J. Yon, "LETI/LIR's amorphous silicon uncooled IR systems," *Proc. SPIE*, 3379, 139–44, 1998.
52. J. L. Tissot, J. L. Martin, E. Mottin, M. Vilain, J. J. Yon, and J. P. Chatard, "320 × 240 microbolometer uncooled IRFPA development," *Proc. SPIE*, 4130, 473–79, 2000.
53. J. F. Brady, T. S. Schimert, D. D. Ratcliff, R. W. Gooch, B. Ritchey, P. McCardel, K. Rachels, et al., "Advances in amorphous silicon uncooled IR systems," *Proc. SPIE*, 3698, 161–67, 1999.
54. A. Tanaka, M. Suzuki, R. Asahi, O. Tabata, and S. Sugiyama, "Infrared linear image sensor using a poly-Si pn junction diode array," *Infrared Phys.*, 33, 229–36, 1992.
55. Y. P. Xu, R. S. Huang, and G. A. Rigby, "A silicon-diode-based infrared thermal detector array," *Sens. Actuators A*, 37–38, 226–30, 1993.
56. M. Ueno, O. Kaneda, T. Ishikawa, K. Yamada, A. Yamada, M. Kimata, and M. Nunoshita, "Monolithic uncooled infrared image sensor with 160 × 120 pixels," *Proc. SPIE*, 2552, 636–43, 1995.
57. J.-K. Kim and C.-H. Han, "A new uncooled thermal infrared detector using silicon diode," *Sens. Actuators*, A89, 22–27, 2001.
58. J. E. Murguia, P. K. Tedrow, F. D. Shepherd, D. Leahy, and M. M. Weeks, "Performance analysis of a thermionic thermal detector at 400 K, 300 K, and 200 K," *Proc. SPIE*, 3698, 361–75, 1999.

59. S. M. Sze, *Physics of Semiconductor Devices*, 2nd ed., John Wiley and Sons, New York, 1982.
60. *Omega Complete Temperature Measurement Handbook and Encyclopedia*, Vol. 26, Omega Engineering Inc., Stamford, CT, U-1–24, 1989.
61. M. Kimata, H. Yagi, U. Ueno, J. Nakanishi, T. Ishikawa, Y. Nakaki, M. Kawai, E. Endo, Y. Kosasayama, Y. Ohota, T. Shugino, and T. Sone, "Silicon infrared focal plane arrays," *Proc. SPIE*, 4288, 286–297, 2001.
62. L. Dong, R. F. Yue, and L. T. Liu, "A high performance single-chip uncooled a-Si TFT infrared sensor," *Proc. Transducers 2003*, Vol. 1, 312–15, 2003.
63. L. Dong, R. Yue, and L. Liu, "Fabrication and characterization of integrated uncooled infrared sensor arrays using a-si thin-film transistors as active elements," *J. Microelectromechan. Sys.*, 14, 1667–77, 2005.
64. M. Kimata, M. Ueno, M. Takeda, and T. Seto, "SOI diode uncooled infrared focal plane arrays", *Proc. SPIE*, 6127, 6127-1–11, 2006.
65. M. Kimata, "IR imaging" in *Comprehensive Microsystems*, Vol. 3, eds. Y. B. Gianchandani, O. Tabata, and H. Zappe, 113–62, Elsevier, Amsterdam, 2008.
66. T. Ishikawa, M. Ueno, K. Endo, Y. Nakaki, H. Hata, T. Sone, and M. Kimata, "Low-cost 320×240 uncooled IRFPA using conventional silicon IC process," *Opto-Electr. Rev.*, 7, 297–303, 1999.
67. M. Moreno, A. Torres, R. Ambrosio, and A. Kosarev, "Un-cooled microbolometers with amorphous germanium-silicon (a-GexSiy:H) thermo-sensing films," in *Bolometers*, ed. A.G.U. Perera, 23–50, InTech, Rijeka, 2012.
68. M. H. Unewisse, B. I. Craig, R. J. Watson, O. Reinhold, and K. C. Liddiard, "The growth and properties of semiconductor bolometers for infrared detection," *Proc. SPIE*, 2554, 43–54, 1995.
69. E. Iborra, M. Clement, and L. Herrero, "Sangrador IR uncooled bolometers based on amorphous GeSiO on silicon micromachined structures," *J. Microelectromechan. Sys.*, 11, 322–29, 2002.
70. D. Butler and M. Rana, "Radio frequency sputtered SiGe and SiGeO thin films for uncooled infrared detectors," *Thin Solid Films*, 514, 355–60, 2006.
71. A. Ahmed and R. Tait, "Noise behavior of amorphous GeSiO for microbolometer applications," *Infrared Phys. Technol.*, 46, 468–72, 2005.
72. M. Moreno, A. Kosarev, A. Torres, and R. Ambrosio, "Fabrication and performance comparison of planar and sandwich structures of micro-bolometers with Ge thermo-sensing layer," *Thin Solid Films*, 515, 7607–10, 2007.
73. V. Leonov, N. Perova, P. De Moor, B. Du Bois, C. Goessens, B. Grietens, A. Verbist, C. Van Hoof, and J. Vermeiren, "Micromachined Poly-SiGe bolometer arrays for infrared imaging and spectroscopy," *Proc. SPIE*, 4945, 54–63, 2003.
74. I. Chistokhin, I. Michailovsky, B. Fomin, and E. Cherepov, "Polycrystalline layers of silicon-germanium alloy for uncooled IR bolometers," *Proc. SPIE*, 5126, 407–14, 2003.
75. R. Yue, L. Dong, and L. Liu, "Monolithic uncooled 8 × 8 bolometer arrays based on Poly-SiGe thermistor," *Int. J. Infrared Millim. Waves*, 27, 995–1003, 2006.
76. S. Wissmar, L. Hoglund, J. Andersson, C. Vieider, S. Susan, and P. Ericsson, "High signal to noise ratio quantum well bolometer material," *Proc. SPIE*, 6401, 64010N, 2006.
77. G. Yu and A. J. Heeger, "Photoinduced charge carriers in insulating cuprates: Fermi glass insulator, metal–insulator transition and superconductivity," *Int. J. Modern Phys.*, B7, 3751, 1993.
78. P. C. Shan, Z. Celik-Butler, D. P. Butler, A. Jahanzeb, C. M. Travers, W. Kula, and R. Sobolewski, "Investigation of semiconducting YBaCuO thin films: A new room temperature bolometer," *J. Appl. Phys.*, 80, 7118–23, 1996.
79. L. Mechin, J. C. Villegier, and D. Bloyet, "Suspended epitaxial YbaCuO microbolometers fabricated by silicon micromachining: Modeling and measurements," *J. Appl. Phys.*, 81, 7039–47, 1997.
80. A Jahanzeb, C. M. Travers, Z. Celik-Butler, D. P. Butler, and S. G. Tan, "A semiconductor YBaCuO microbolometer for room temperature ir imaging," *IEEE Trans. Electron Devices*, 44, 1795–801, 1997.
81. L. Phong and S. Qiu, "Room temperature YBaCuO microbolometers," *J. Vac. Sci. Technol.*, A18, 635–38, 2000.
82. M. Almasri, Z. Çelik-Butler, D. P. Butler, A. Yarafanakul, and A. Yildiz, "Semiconducting YBaCuO microbolometers for uncooled broad-band IR sensing," *Proc. SPIE*, 4369, 264–73, 2001.
83. J. Kim and A. Grishin, "Free-standing epitaxial $La_{1-x}(Sr,Ca)_xMnO_3$ membrane on Si for uncooled infrared microbolometer," *Appl. Phys. Lett.*, 87, 033502, 2005.
84. M. Liger and Y.-C. Tal, "A 32 × 32 Parylene-pyrolyzed carbon bolometer imager," *Proc. MEMS 2006*, 106–9, 2006.
85. S. Wissmar, L. Höglund, J. Andersson, C. Vieider, S. Savage, and P. Ericsson, "SiGe quantum wells for uncooled long wavelength infrared radiation (LWIR) sensors," *J. Phys.: Conf. Ser.*, 100 042029, 1–5, 2008.
86. A. Roer, A. Laladatu, M. Bring, E. Wolla, E. Hohler, and G. Kittilsland, "High performance LWIR microbolometer with Si/SiGe quantum well thermistor and wafer level packaging," *Proc. SPIE*, 8185, 818507-1–8, 2011.

87. D. H. Andrews, W. F. Brucksch, Jr., W. T. Ziegler, and E. R. Blanchard, "Attenuated superconductors: I. For measuring infra-red radiation," *Rev. Sci. Instrum.*, 13, 281–92, 1942.
88. R. M. Milton, "A superconducting bolometer for infrared measurements," *Chem. Rev.*, 39, 419–22, 1946.
89. D. H. Andrews, R. M. Milton, and W. DeSorbo, "A fast superconducting bolometer," *J. Opt. Soc. Am.*, 36, 518–24, 1946.
90. N. Fuson, "The Infra-red sensitivity of superconducting bolometers," *J. Opt. Soc. Am.*, 38, 845–53, 1948.
91. H. D. Martin and D. Bloor, "The applications of superconductivity to the detection of radiant energy," *Cryogenics*, 1, 159, 1961.
92. C. L. Bertin and K. Rose, "Radiant-energy detection by superconducting films," *J. Appl. Phys.*, 39, 2561–68, 1968.
93. J. Clarke, G. I. Hoffer, P. L. Richards, and N. H. Yeh, "Superconductive bolometers for submillimeter wavelengths," *J. Appl. Phys.*, 48, 4865–79, 1977.
94. C. P. Poole, H. A. Farach, and R. J. Creswick, *Superconductivity*, Academic Press, San Diego, CA, 1995.
95. K. Rose, C. L. Bertin, and R. M. Katz, "Radiation detectors," in *Applied Superconductivity*, Vol. 1, ed. V. L. Newhouse, 268–308, Academic Press, New York, 1975.
96. K. Rose, "Superconductive FIR detectors," *IEEE Trans. Electron Devices*, ED-27, 118–25, 1980.
97. I. A. Khrebtov, "Superconductor infrared and submillimeter radiation receivers," *Sov. J. Opt. Technol.*, 58, 261–70, 1991.
98. H. Kraus, "Superconductive bolometers and calorimeters," *Superconductor Sci. Technol.*, 9, 827–42, 1996.
99. A. J. Kreisler and A. Gaugue, "Recent progress in HTSC bolometric detectors at terahertz frequencies," *Proc. SPIE*, 3481, 457–68, 1998.
100. A. J. Kreisler and A. Gaugue, "Recent progress in high-temperature superconductor bolometric detectors: From the mid-infrared to the far-infrared (THz) range," *Superconductor Sci. Technol.*, 13, 1235–45, 2000.
101. J. Zmuidzinas and P. L. Richards, "Superconducting detectors and mixers for millimeter and submillimeter astrophysics," *Proc. IEEE*, 92, 1597–616, 2004.
102. G. H. Rieke, "Infrared detector arrays for astronomy," *Annual Rev. Astron. Astrophys.*, 45, 77–115, 2007.
103. D. J. Benford and S. H. Moseley, "Superconducting transition edge sensor bolometer arrays for submillimeter astronomy," *Proceedings of the International Symposium on Space and THz Technology*. Available: http://www.eecs.umich.edu/~jeast/benford_2000_4_1.pdf.
104. Z. Zhang, T. Le, M. Flik, and E. Carvalho, "Infrared optical-constant of the HTc superconductor $YBa_2Cu_3O_7$," *J. Heat Transfer*, 116, 253–56, 1994.
105. S. E. Schwarz and B. T. Ulrich, "Antenna coupled thermal detectors," *J. Appl. Phys.*, 85, 1870–73, 1977.
106. K. A. Müller and J. G. Bednorz, "The discovery of a class of high-temperature superconductors," *Science*, 237, 1133–39, 1987.
107. J. G. Bednorz and K. A. Müller, "Possible High T_c Superconductivity in the Ba-La-Cu-O system," *Zeitschrift fur Physik B-Condensed Matter*, 64, 189–93, 1986; "Perovskite-type oxides-the new approach to high-T_c superconductivity," *Rev. Modern Phys.*, 60, 585–600, 1988.
108. J. Nagamatsu, N. Nakagawa, T. Muranaka, Y. Zenitani, and J. Akimitsu, "Superconductivity at 39°K in magnesium diboride," *Nature*, 410, 63–64, 2001.
109. A. Schilling, M. Cantoni, J. D. Guo, and H. R. Ott, "Superconductivity above 130 K in the Hg-Ba-Ca-Cu-O system," *Nature*, 363, 56–58, 1993.
110. Y. Enomoto and T. Murakami, "Optical detector using superconducting $BaPb_{0.7}Bi_{0.3}O_3$," *J. Appl. Phys.*, 59, 3807–14, 1986.
111. P. W. Kruse, "Physics and applications of high-T_c superconductors for infrared detectors," *Semiconductor Sci. Technol.*, 5, S229–S329, 1990.
112. R. Sobolewski, "Applications of high-T_c superconductors in optoelectronics," *Proc. SPIE*, 1512, 14–27, 1991.
113. Q. Hu and P. L. Richards, "Design analysis of high T_c superconducting microbolometer," *Appl. Phys. Lett.*, 55, 2444–46, 1989.
114. P. L. Richards, J. Clarke, R. Leoni, Ph. Lerch, S. Verghese, M. B. Beasley, T. H. Geballe, R. H. Hammond, P. Rosenthal, and S. R. Spielman, "Feasibility of the high T_c superconducting bolometer," *Appl. Phys. Lett.*, 54, 283–85, 1989.
115. S. Verghese, P. L. Richards, K. Char, D. K. Fork, and T. H. Geballe, "Feasibility of infrared imaging arrays using high-T_c superconducting bolometers," *J. Appl. Phys.*, 71, 2491–98, 1992.
116. J. C. Brasunas, S. H. Moseley, B. Lakew, R. H. Ono, D. G. McDonald, J. A. Beall, and J. E. Sauvageau, "Construction and performance of a high-temperature-superconductor composite bolometer," *J. Appl. Phys.*, 66, 4551–54, 1989.

117. J. Brasunas and B. Lakew, "High T_c bolometer developments for planetary missions," *Proc. SPIE*, 1477, 166–73, 1991.
118. S. Verghese, P. L. Richards, S. A. Sachtjen, and K. Char, "Sensitive bolometers using high-T_c superconducting thermometers for wavelengths 20–300 µm," *J. Appl. Phys.*, 24, 4251–53, 1993.
119. J. Brasunas and B. Lakew, "High T_c bolometer with record performance," *Appl. Phys. Lett.*, 64, 777–78, 1994.
120. M. J. Burns, A. W. Kleinsasser, K. A. Delin, R. P. Vasquez, B. S. Karasik, W. R. McGraph, and M. C. Gaidis, "Fabrication of High-Tc hot-electron bolometric mixers for terahertz applications," *IEEE Trans. Appl. Superconductivity*, 7, 3564–67, 1997.
121. B. R. Johnson, T. Ohnstein, H. Marsh, S. B. Dunham, and P. W. Kruse, "$YBa_2Cu_3O_7$ superconducting microbolometer linear arrays," *Proc. SPIE*, 1685, 139–45, 1992.
122. B. R. Johnson and P. W. Kruse, "Silicon microstructure superconducting microbolometer infrared arrays," *Proc. SPIE*, 2020, 2–7, 1993.
123. M. C. Foote, B. R. Johnson, and B. D. Hunt, "Transition edge $Yba_2Cu_3O_{7-x}$ microbolometers for infrared staring arrays," *Proc. SPIE*, 2159, 2–9, 1994.
124. B. R. Johnson, M. C. Foote, H. A. Marsh, and B. D. Hunt, "Epitaxial $YBa_2Cu_3O_7$ superconducting infrared microbolometers on silicon," *Proc. SPIE*, 2267, 24–30, 1994.
125. B. R. Johnson, M. C. Foote, and H. A. Marsh, "High performance linear arrays of $YBa_2Cu_3O_7$ superconducting infrared microbolometers on silicon," *Proc. SPIE*, 2475, 56–61, 1995.
126. B. Lakew, J. C. Brasunas, A. Pique, R. Fetting, B. Mott, S. Babu, G. M. Cushman, "High-T_c superconducting bolometer on chemically-etched 7 µm thick sapphire," *Physics C*, 229, 69–74, 2000.
127. J. Brasunas, "High-temperature superconducting IR detectors," in *Handbook of High-Temperature Superconductor Electronics*, chapter 11, ed. N. Khare, Marcel Dekker, New York, 2003.
128. E. M. Gershenzon, G. N. Gol'tsman, I. G. Gogidze, Y. P. Gusev, A. J. Elant'ev, B. S. Karasik, and A. D. Semenov, "Millimeter and submillimeter range mixer based on electronic heating of superconducting films in the resistive state," *Superconductivity*, 3, 1582–97, 1990.
129. B. Karasik, G. N. Gol'tsman, B. M. Voronov, S. I. Svechnikov, E. M. Gershenzon, H. Ekström, S. Jacobsson, E. Kollberg, and K. S. Yngvesson, "Hot electron quasioptical NbN superconducting mixer," *IEEE Trans. Appl. Superconductivity*, 5, 2232–35, 1995.
130. D. E. Prober, "Superconducting terahertz mixer using a transition-edge microbolometer," *Appl. Phys. Lett.*, 62, 2119–21, 1993.
131. A. Skalare, W. R. McGrath, B. Bumble, H. G. LeDuc, P. J. Burke, A. A. Vereijen, R. J. Schoelkopf, and D. E. Prober, "Large bandwidth and low noise in a diffusion-cooled hot-electron bolometer mixer," *Appl. Phys. Lett.*, 68, 1558–60, 1996.
132. W. R. McGrath, "Novel hot-electron bolometer mixers for submillimeter applications: An overview of recent developments," *Proc. URSI Int. Sympos. Signals Sys. Electron.*, 147–52, 1995.
133. P. J. Burke, R. J. Schoelkopf, D. E. Prober, A. Skalare, W. R. McGrath, B. Bumble, and H. G. LeDuc, "Length scaling of bandwidth and noise in hot-electron superconducting mixers," *Appl. Phys. Lett.*, 68, 3344–46, 1996.
134. W. J. Skocpol, M. R. Beasly, and M. Tinkham, "Self-heating hotspots in superconducting thin-film microbridges," *J. Appl. Phys.*, 45, 4054–66, 1974.
135. A. D. Semenov, G. N. Gol'tsman, and R. Sobolewski, "Hot-electron effect in superconductors and its applications for radiation sensors," *Superconductor Sci. Technol.*, 15, R1–R16, 2002.
136. A. D. Semenov and H.-W. Hübers, "Frequency bandwidth of a hot-electron mixer according to the hot-spot model," *IEEE Trans. Appl. Superconductivity*, 11, 196–99, 2001.
137. H.-W. Hübers, "Terahertz heterodyne receivers," *IEEE J. Select. Topics Quant. Electron.*, 14, 378–91, 2008.
138. http://www.sron.nl/index.php?option=com_content&task=view&id=44&Itemid=111.
139. R. B. Bass, "Hot electron bolometers on ultra-thin silicon chips with beam leads for a 585 GHz receiver." PhD dissertation. Available: http://www.ece.virginia.edu/sis/Papers/rbbpapers/diss.pdf.
140. D. Rutledge and M. Muha, "Imaging antenna arrays," *IEEE Trans. Antennas Propag.*, AP-30, 535–40, 1982.
141. B. S. Karasik, W. R. McGrath, and M. C. Gaidis, "Analysis of a high-T_c hot-electron superconducting mixer for terahertz applications," *J. Appl. Phys.*, 83, 1581–89, 1997.
142. M. A. Kinch and B. V. Rollin, "Detection of millimetre and submillimetre wave radiation by free carrier absorption in a semiconductor," *Br. J. Appl. Phys.*, 14, 672–76, 1963.
143. E. H. Putley, "InSb submillimeter photoconductive devices," in *Semiconductors and Semimetals*, Vol. 12, eds. R. K. Willardson and A. C. Beer, 143–68, Academic Press, New York, 1977.

9 Pyroelectric detectors

Whenever a pyroelectric crystal undergoes a change of temperature, surface charge is produced in a particular direction as a result of the change in its spontaneous polarization with temperature. This effect has been known as a physically observable phenomenon for many centuries, being described by Theophrastus in 315 BC [1]. Its name "pyroelectricity" was introduced by Brewster [2]. The concept of using the pyroelectric effect for detecting radiation was proposed very early by Ta [3]; however, in practice, little progress was made due to the lack of suitable materials. The importance of the pyroelectric effect in infrared detection became obvious about 60 years ago, due to scientific activity from such authors as Chynoweth [4], Cooper [5,6], Hadni et al. [7], and others [8–13]. A widely acclaimed review of work up to 1969 has been published by Putley [14] and further developments have been reported by Baker et al. [15], Putley [16], Liu and Long [17], Marshall [18], Porter [19], Joshi and Dawar [20], Whatmore [21,22], Ravich [23], Watton [24], and Robin et al. [25]. More recently published papers have shown that the pyroelectric micromachining version of uncooled thermal detectors reaches fundamental limits [26–34].

9.1 BASIC PRINCIPLE AND OPERATION OF PYROELECTRIC DETECTORS

Pyroelectricity has been known for the last twenty-four centuries, but in 1938, a chemist at the Sorbonne in Paris, Ta, proposed that tourmaline crystals could be used as IR sensors in spectroscopy [3,32]. Some research on pyroelectric detectors was conducted in the next decade in the United Kingdom, United States, and Germany, but the results appeared only in classified documents. In 1962, Cooper presented the first theory of the pyroelectric detector and conducted experiments using barium titanate [5,6]. Also, that year, Lang proposed the use of pyroelectric devices for measuring temperature changes as small as 0.2 μK. Later on, an explosive growth of papers in pyroelectric infrared detectors had begun [32].

Pyroelectric materials are those with a temperature-dependent spontaneous electrical polarization. There are 32 known crystal classes—twenty-one are noncentrosymmetric and ten of these exhibit temperature-dependent spontaneous polarization. Under equilibrium conditions, the electrical asymmetry is compensated by the presence of free charges. If, however, the temperature of the material is changed at a rate faster than these compensating charges can redistribute themselves, an electrical signal can be observed. This means that the pyroelectric detector is an AC device, unlike other thermal detectors that detect temperature levels rather than temperature changes. This generally limits the low-frequency operation, and for a maximum output signal, the rate of charge of the input radiation should be comparable to the electrical time constant of the element.

Most pyroelectric are also ferroelectric, which means that the direction of their polarization can be reversed at the application of a suitable electric field, and the polarization reduces to zero at some temperature known as the Curie temperature T_C. Generally, the pyroelectric materials considered for thermal detector arrays are the lead-based perovskite oxides such as lead titanate [$PbTiO_3$: PT]. These materials have structural similarities with the mineral perovskite ($CaTiO_3$). The basic formula is ABO_3; where *A* is lead, *O* is oxygen, and *B* may be one, or a mixture, of cations; for example, lead zirconate titanate [$Pb(ZrTi)O_3$: PZT], barium strontium titanate [$BaSrTiO_3$: BST], lead scandium tantalate [$Pb(Sc_{0.5}Ta_{0.5})O_3$: PST], and lead magnesium niobate [$Pb(Mg_{1/3}Nb_{2/3})O_3$: PMN]. Often dopants are added to these basic formulations to enhance or tune the material properties. Above Curie temperature, T_C, these materials form a symmetric nonpolar, cubic structure (Figure 9.1). Above this temperature, the material is paraelectric and it has no

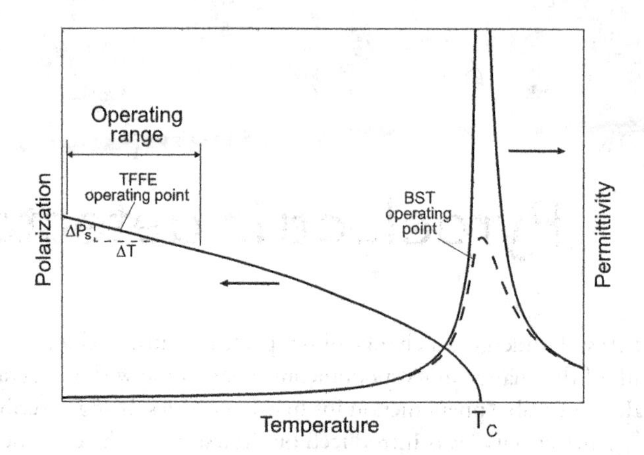

Figure 9.1 Thermal behavior of ferroelectric material. The dashed line shows the effect of an applied field on permittivity.

pyroelectric activity. On cooling, they undergo a structural phase transition to form a polar, ferroelectric phase.

These materials can be further subdivided into two groups. The conventional pyroelectric materials, such as PT and PZT, operate at room temperature well below their Curie temperature without the need for an applied field. Requirements for detector temperature stabilization are minimal or can be eliminated since there is little variation in detector performance over quite a large temperature range.

It is, however, possible to operate ferroelectrics at or above T_C, with an applied bias field, in the mode of a dielectric bolometer. Its operation is associated with the change of permittivity with a temperature in the region of the transition. The permittivity is strongly temperature dependent, but less so with an applied field (see dashed line in Figure 9.1). The applied bias charges the element, and the heating due to the incident radiation results in an increment of permittivity and hence a signal voltage. This second group of materials (including BST, PST, and PMN) has T_C slightly below the detector operating temperature, resulting in minimal pyroelectricity.

In general, the case for ferroelectric materials, the electrical displacement, D, is a sum of contribution from the spontaneous (zero field) polarization, P_s, and the field-induced polarization (i.e., $\varepsilon_o\varepsilon_r E$). It is also important to realize that the permittivity around the transition is nonlinear and hence an integral is required:

$$D = P_s(T) + \varepsilon_o \int_0^E \varepsilon_r (E', T) dE', \tag{9.1}$$

where ε_0 is the permittivity of free space and ε_r is the relative permittivity of the pyroelectric material.

The pyroelectric coefficient is the change in displacement with temperature,

$$p = \left.\frac{dD}{dT}\right|_E = \frac{dP_s}{dT} + \varepsilon_0 \int_0^E \frac{d\varepsilon_r}{dT} dE'. \tag{9.2}$$

To obtain high pyroelectric coefficient from dielectric bolometer materials, it is desirable to have a large variation in permittivity with temperature and/or high bias fields should be applied. As mentioned previously, the bias field generally reduces the permittivity variation and even introduces positive slopes; hence, there is a limit to the benefits gained from simply applying high fields.

9.1.1 RESPONSIVITY

The pyroelectric detector can be considered as a small capacitor with two conducting electrodes mounted perpendicularly to the direction of spontaneous polarization, as shown in Figure 9.2 with its equivalent electrical circuit. If the temperature of the material is varied rapidly, the internal dipole moment will

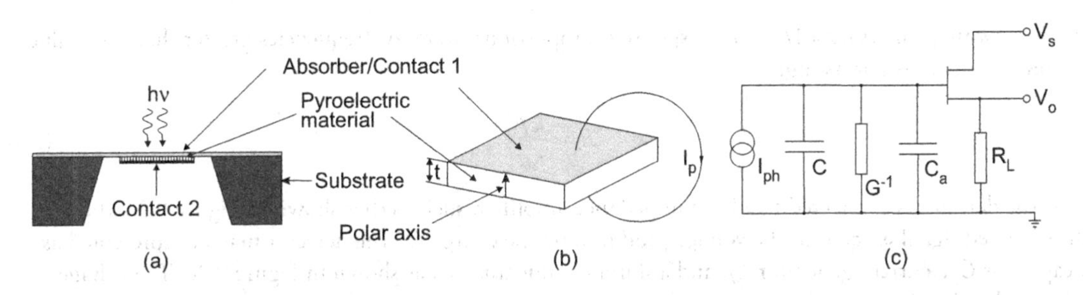

Figure 9.2 Pyroelectric detector: (a) schematic cross section, (b) pyroelectric element, and (c) equivalent electrical circuit.

change, producing a transient voltage. This pyroelectric effect is exploited to produce sensitive detectors of modulated infrared radiation, operating at ambient temperatures. When the detector is operated, the change in polarization will appear as a charge on the capacitor and a current will be generated, the magnitude of which depends on the temperature rise and the pyroelectrical coefficient, p, of the material.

The polarization change due to a change in temperature ΔT is described by

$$P = p\Delta T, \tag{9.3}$$

The pyroelectric charge generated is given by

$$Q = pA\Delta T, \tag{9.4}$$

so, the effect of temperature change on a pyroelectric material is to cause a current $I_{ph} = dQ/dT$ to flow in an external circuit (see Figure 9.2), such that:

$$I_{ph} = Ap\frac{dT}{dt}, \tag{9.5}$$

where A is the detector area, p is the component of the pyroelectric coefficient normal to the electrodes, and dT/dt is the rate of change of temperature with time. Taking into account Equation 4.4, the photocurrent is equal to:

$$I_{ph} = \frac{\varepsilon pA\Phi_o\omega}{G_{th}\left(1+\omega^2\tau_{th}^2\right)^{1/2}}. \tag{9.6}$$

To make a pyroelectric device work, it is necessary to modulate the source of energy. This can be achieved by mechanical chopping or by moving the detector relative to the source of radiation. The element with an area A and thickness t gives a thermal capacitance $C_{th} = c_{th}At$, (where c_{th} is the volume-specific heat) connected via a thermal conductance G_{th} to a heat sink. This gives a thermal time constant $\tau_{th} = C_{th}/G_{th}$.

The detector possesses an electrical capacitance C and presents an electrical conductance G ($G^{-1} = R$ is a parallel resistance) to a low noise–high input impedance buffer amplifier such as a MOSFET, with input capacitance C_a. In practice, the amplifier resistance is large compared with the shunt resistor G^{-1}, and can be ignored; but C_a is not always small compared with the detector capacitance C. This produces an electrical time constant $\tau_e = (C_a + C)/G$. The τ_{th} and τ_e are the fundamental parameters that determine the frequency response.

The current responsivity is:

$$R_i = \frac{I_{ph}}{\Phi_o} = \frac{\varepsilon pA\omega}{G_{th}\left(1+\omega^2\tau_{th}^2\right)^{1/2}}, \tag{9.7}$$

and for low frequencies ($\omega \ll 1/\tau_{th}$) the response is proportional to ω. At frequencies greater than this value, the response is constant, being:

$$R_i = \frac{\varepsilon p}{c_{th} t}. \tag{9.8}$$

If the detector is connected to a high impedance amplifier, such as that shown in Figure 9.2c, then the observed signal is equal to the voltage produced by the charge Q. The detector may be represented as a capacitor C, a current generator I_{ph}, and a shunt conductance G, as shown in Figure 9.2c. The voltage generated is therefore given by:

$$V = \frac{I_{ph}}{\left(G^2 + \omega^2 C^2\right)^{1/2}}, \tag{9.9}$$

and then the voltage responsivity is equal to:

$$R_v = \frac{V}{\Phi_o} = \frac{R\varepsilon pA\omega}{G_{th}\left(1+\omega^2\tau_{th}^2\right)^{1/2}\left(1+\omega^2\tau_e^2\right)^{1/2}}, \tag{9.10}$$

where $\tau_e = C/G$ is the electrical time constant. The last equations simplify at frequencies that are high compared with $(\tau_{th})^{-1}$ and $(\tau_e)^{-1}$ to give

$$R_v = \frac{\varepsilon p}{\varepsilon_o \varepsilon_r c_{th} A\omega}. \tag{9.11}$$

Equation 9.11 shows that, at high frequencies, the voltage responsivity of a pyroelectric detector is inversely proportional to frequency. At low frequencies, this is modified by the electrical and thermal time constants, as in Equation 9.10, so that the true frequency response is of the form shown in Figure 9.3. The maximum value occurs at a frequency of $(\tau_e \tau_{th})^{-1/2}$ with a value of:

$$R_{v\,\max} = \frac{\varepsilon pAR}{G_{th}\left(\tau_e + \tau_{th}\right)}. \tag{9.12}$$

From Equation 9.12 it is easy to show that the responsivity is maximized by minimizing G_{th}. The thermal capacity should also be reduced, within the constraint of maintaining an appropriate thermal time constant τ_{th}.

At the frequencies, $\omega = (\tau_e)^{-1}$ and $\omega = (\tau_{th})^{-1}$:

$$R_v = \frac{R_{v\,\max}}{\sqrt{2}}. \tag{9.13}$$

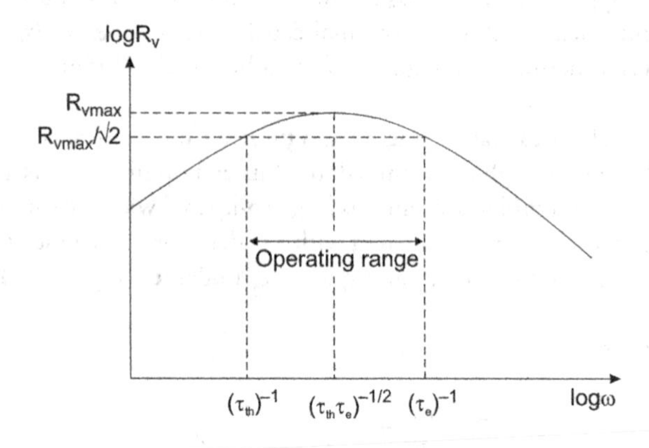

Figure 9.3 Frequency dependence of voltage responsivity of a pyroelectric detector.

It is not possible to distinguish between τ_e and τ_{th} from responsivity measurements alone. Putley has discussed in detail the analysis of the performance from responsivity and noise measurements together [35].

The selection of τ_e and τ_{th} is determined by a number of factors. For a low-frequency, high-sensitivity operation, the device is mounted with freely suspended active elements to minimize conduction of heat to the surroundings. The thermal capacity of the element is adjusted to maximize the response at the frequency of interest. To realize this device, a thin, low-thermal-capacity, high-electrical-capacity element can be used. Typically, τ_{th} is within the range of 0.01–10 s. The τ_e, however, can be anywhere between 10^{-12} and 100 s, depending on the sizes of the detector capacitance and the shunt resistor.

For high-frequency operation, one of the time constants (usually τ_e) is reduced so that its inverse is greater than the maximum frequency of interest. This can be done by minimizing the electrical capacitance of the element (using an edge-electrode structure) and feeding the output into a 50-Ω line. Because the speed of the pyroelectric response is limited only by the frequency of the vibrational polarization of the crystal lattice (about 10^{12} Hz), these detectors have the potential to be extremely fast. Austan and Glass have experimentally verified response time of 9 ns [36], while Roundy et al. have demonstrated practical detectors with a response time of 170 ps [37].

These considerations of detector response do not take into account the input resistance of the amplifier (R_a) that will appear in parallel with the resistor R. For low-frequency detectors, $R_a >> R$ and R_a can be ignored in this case. For fast detectors, $R_a << R$ and thus R_a determines the electrical time constant and the device responsivity.

More rigorous analyses of pyroelectric detectors have been performed by many authors, taking into account the effects of mounting techniques and black coatings [36–42]. The aforementioned treatment is, however, adequate for the majority of applications.

Generally, in bulk material devices, $\tau_e < \tau_{th}$. However, considerations carried out by Putley [16] and Porter [19] pointed out that τ_e can also be larger than τ_{th}, depending on materials and electrical elements. This situation is in typical thin-film structures. The major consequences of decreasing thickness of the pyroelectric material, t, are an increase of the electrical capacity and a decrease of the heat capacity. In addition, it is difficult to improve thermal insulation to the surroundings by the same amount as the thickness, because bulk pyroelectric materials are good thermal insulators [28]. Since the ratio τ_e/τ_{th} scales roughly as

$$\frac{RC(t)}{C_{th}(t)/G_{th}} \propto \frac{1}{t^2}, \tag{9.14}$$

the ratio switches from <1 to >1 when scaling down from a single crystal to a thin film. The frequency behavior for thin films is similar to those shown in Figure 9.3; however, the time constants have the opposite order (i.e., $\tau_e > \tau_{th}$). As a consequence, the voltage responsivity in the intermediate-frequency region (operating region; see Figure 9.3) is determined by other parameters:

- For bulk devices

$$R_v \cong \frac{\varepsilon pAR}{C_{th}}. \tag{9.15}$$

- For thin-film devices

$$R_v \cong \frac{\varepsilon pA}{CG_{th}}. \tag{9.16}$$

In bulk devices, a value of parallel resistance of 10 GΩ is typically applied (R should not exceed the gate impedance of the amplifier). The last equation indicates that in thin-film detectors, the parallel resistance is not directly involved in the voltage responsivity. It may be avoided because thin-film capacitors exhibit larger currents than bulk capacitors.

Assuming that $C = \varepsilon_o\varepsilon_r A/t$, Equation 9.16 can be further modified to

$$R_v \cong \frac{\varepsilon p t}{\varepsilon_o \varepsilon_r G_{th}}, \tag{9.17}$$

which shows that the responsivity is independent of the detector area A.

9.1.2 NOISE AND DETECTIVITY

There are three major noise sources in a pyroelectric detector with a shunt resistor [14,17,19,21,28]:

- Thermal fluctuation noise
- Johnson noise
- Amplifier noise

The first two types of noises are described in Section 4.1. The Johnson noise connected with the shunt resistor R is described by Equation 4.16. However, in most devices at moderate frequencies of operation (1 Hz–1 kHz), the noise is dominated by the AC electrical conductance of the detector element. The AC conductance of the device has two components: a frequency-independent component R^{-1} and a frequency-dependent component G_d:

$$G_d = \omega C \tan\delta, \tag{9.18}$$

where $\tan\delta$ is the loss tangent of the detector material. For frequencies much less than $\omega = (RC\tan\delta)^{-1}$, the Johnson noise is simply given by:

$$V_{Jr}^2 = \frac{4kTR\Delta f}{1+\omega^2\tau_e^2}, \tag{9.19}$$

which leads to an ω^{-1} dependence at frequencies $\omega \gg \tau_e^{-1}$.

For frequencies much greater than $\omega = (RC\tan\delta)^{-1}$, the noise generated by the AC conductance of the detector element will dominate, so that:

$$V_{Jd}^2 = 4kT\Delta f \frac{\tan\delta}{C}\frac{1}{\omega} \text{ for } C \gg C_a \tag{9.20}$$

This type of noise, also called dielectric noise, dominates at high frequencies.

It is interesting to compare relative magnitudes of these various noise sources for a typical detector. These are shown plotted as a function of frequency in Figure 9.4 [21]. It has been assumed that both the thermal and electrical time constants are longer than 1 second. In nearly all practical detectors the thermal

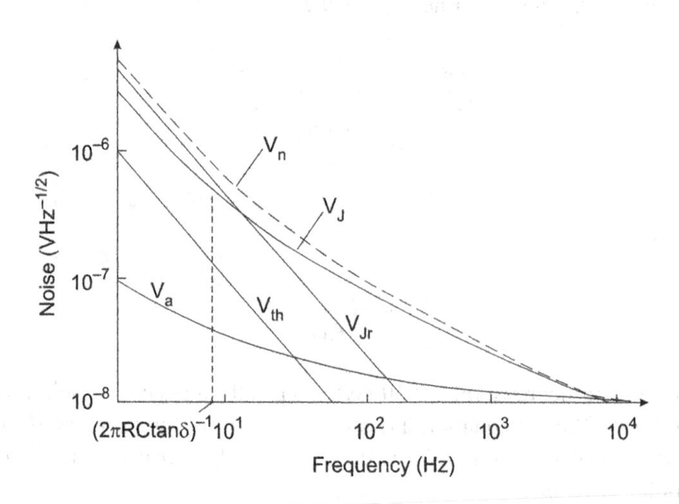

Figure 9.4 Relative magnitudes of noise voltages in a typical pyroelectric detector. (From Ref. [21]. With permission.)

noise is insignificant and is often ignored in calculations. It can be seen that the loss-controlled Johnson noise dominates above 20 Hz, while below this frequency, the resistor-controlled Johnson noise and the amplifier current noise (V_{ai}) contribute almost equally significantly to the total noise. At very high frequencies, the amplifier voltage noise (V_{av}) dominates.

At high frequencies [greater than τ_e^{-1} and $(RC\tan\delta)^{-1}$], the detectivity (from Equations 3.6, 9.11, and 9.20) is given by the equation

$$D^* = \frac{\varepsilon t}{(4kT)^{1/2}} \frac{p}{c_{th}(\varepsilon_o \varepsilon_r \tan\delta)^{1/2}} \frac{1}{\omega^{1/2}}. \quad (9.21)$$

The fall in detectivity with a frequency of $\omega^{-1/2}$ means that the D^* will be at a maximum of a rather higher frequency than R_v (see Equation 9.11) and falls more slowly (as $\omega^{-1/2}$) than R_v (as ω^{-1}) above this maximum. For most detectors, D^* maximizes in the 1–100 Hz range and a reasonably flat D^* can be achieved in the range of a few Hz to a several hundred Hz.

There are a number of other sources of unwanted signals in pyroelectric detectors, mostly associated with the environment. Environmental temperature fluctuations can give rise to spurious signals at low frequencies, or, if the rate of external temperature change is very large, the detector amplifier can be saturated.

A major limitation to the usefulness of pyroelectric detectors is that they are microphonic; electrical outputs are produced by mechanical vibration or acoustic noise. This microphonic signal may dominate all other noise sources if the detector is in a high vibration environment. The basic course of microphony is the piezoelectric nature of pyroelectric materials, meaning that a change in polarization is produced by a mechanical strain as well as by a change in temperature. In general, lower microphony is obtained by making the mounting of the pyroelectric less rigid. Further reduction in microphony has been achieved by using compensated detectors or by selection of a material with low piezoelectric coupling to the dominant strain components. Shorrocks et al. have discussed methods by which the microphony of pyroelectrical arrays may be reduced to a very low level [43].

Two other sources of environment noise affect pyroelectric detector operation. If a pyroelectric detector is subjected to changes in ambient temperature, fast pulses are sometimes observed, superimposed on the normal pyroelectric response. These pulses occur in a random fashion, but their number and amplitude increase with the rate of increasing temperature. It is suggested that these spurious noise signals are due to ferroelectric domain wall movements. These can be minimized by good materials selection and appear to be lower in ceramics than some single crystal materials such as $LiTaO_3$. Finally, electromagnetic interference is a source of unwanted signals. Detectors operated at low frequencies, having a very high input impedance preamplifier, require careful screening. This is generally achieved by using electrically conducting windows of germanium or silicon that are connected to the earthed metal can.

9.2 PYROELECTRIC MATERIAL SELECTION

Many pyroelectric materials have been investigated for detector applications. However, the choice is not an obvious one as it depends on many factors including the size of the detector required, the operating temperature, and the frequency of operation.

It is possible to formulate a number of figures of merit (FoM) that describe the contribution of the physical properties of a material to the performance of a device. For example, the current responsivity (see Equation 9.8) is proportional to

$$F_i = \frac{p}{c_{th}}, \quad (9.22)$$

Instead, the voltage responsivity (see Equation 9.11) is proportional to

$$F_v = \frac{p}{\varepsilon_o \varepsilon_r c_{th}}. \quad (9.23)$$

For thin-film pyroelectric detectors, a voltage responsivity FoM can be introduced as (see Equation 9.17):

$$F_v^* = \frac{p}{\varepsilon_o \varepsilon_r}. \tag{9.24}$$

The conventional sensitivity FoM, detectivity D^*, is of little practical use because of frequency dependencies and filter factors. However, its analytical expression is useful for examining the relative importance of various parameters. In the case of a detector dominated by the AC Johnson noise (see Equation 9.21), the detectivity is proportional to

$$F_d = \frac{p}{c_{th}(\varepsilon_o \varepsilon_r \tan\delta)^{1/2}}, \tag{9.25}$$

which forms the FoM for pyroelectric detectors.

A useful FoM that includes the effect of input capacitance of the circuit with which the detector is used is

$$F = \frac{1}{C_d + C_L} \frac{p}{c_{th}}. \tag{9.26}$$

This equation reduces to F_i or F_v when C_L is comparatively small or large, respectively.

The relevant FoM for the materials used in pyroelectric vidicons is F_{vid}:

$$F_{vid} = \frac{F_v}{G_{th}}, \tag{9.27}$$

where G_{th} is the thermal conductivity of the pyroelectric. The dependence of F_{vid} on G_{th} can be eliminated by dicing a thermal imaging target into individual islands (using the reticulation process).

A responsivity FoM is valuable in selecting a material with responsivity sufficiently high that preamplifier noise is small compared to temperature fluctuation noise. A Johnson noise–sensitivity FoM is valuable in selecting a material whose Johnson noise is small compared with the temperature fluctuation noise. Thus, both FoM must be large to ensure temperature fluctuation noise–limited performance.

An ideal material should have a large pyroelectric coefficient, low dielectric constant, low dielectric loss, and low volume-specific heat. The possibility of satisfying these requirements in a single material is not promising. While it is generally true that a large pyroelectric coefficient and a small dielectric constant are desirable, it is also true that these two parameters are not independently adjustable. Thus, we find that materials having a high pyroelectric coefficient also have a high dielectric constant, and materials having a low dielectric constant also have a low pyroelectric coefficient. This means that different detector–preamplifier sizes and configurations will be optimized with different materials [21]. Thus, Equation 9.25 is a better responsivity FoM, assuming one knows the pixel geometry and the circuit with which the detector material will be used. Table 9.1 shows the parameter values and traditional FoM for typical materials. The traditional FoM indicate, for example, that triglycine sulfate (TGS) and lithium tantalate ($LiTaO_3$) should be much better than BST and PST; however, sensor system results indicate the contrary.

The state of the art in pyroelectric materials and assessments of their relative merits for different applications have been reviewed by Whatmore [21,22], Watton [24], Muralt [28], and others [32,34,44–47]. Characteristics of pyroelectric detector materials are given in Tables 9.1–9.3 [28,31,46].

The pyroelectric materials can broadly be classified into three categories: single crystals, ceramics (polycrystalline), and polymers. Batra and Aggarwal monograph [34] reviews the relevant techniques for processing key materials.

9.2.1 SINGLE CRYSTALS

Among the single crystals, the most notable success has been achieved with TGS $(NH_2CH_2COOH)_3H_2SO_4]$. It possesses attractive properties, a high pyroelectric coefficient, a reasonably low dielectric constant and thermal conductivity (high value of F_v). However, this material is rather hygroscopic, difficult to handle, and shows poor long-term stability, both chemically and electrically. Its low Curie temperature is a major

Table 9.1 Properties of bulk and polymer pyroelectric materials

MATERIAL	STRUCTURE	p ($\mu Cm^{-2}K^{-1}$)	ε	$\tan\delta$	c_{th} ($10^6 Jm^{-3}K^{-1}$)	F_v^* ($kVm^{-1}K^{-1}$)	F_v (m^2C^{-1})	F_d ($10^{-5}Pa^{-1/2}$)	T_C (°C)
$NaNO_2$	Single crystal	40	4	0.02		1,130			164
$LiTaO_3$	Single crystal	230	47	<0.01	3.2	553	0.17	5–35	620
TGS	Single crystal	280	38	0.01	2.3	832	0.36	6.6	49
DTGS	Single crystal	550	43	0.02	2.6		0.53	8.3	61
ATGSAs	Single crystal	70	32	<0.01	2.6		0.99	>16	51
SBN-50	Single crystal	550	400	0.003	2.3	155	0.07	7.3	121
$(Pb,Ba)_5Ge_3O_{11}$	Single crystal	320	81	0.001	2.0	446	0.22	18.9	70
$PbZrTiO_3$ PZFNTU	Ceramic	380	290	0.003	2.5	148	0.06	5.5	230
$PbTiO_3$	Ceramic	180	190	0.01	3.0	107	0.04	1.5	490
$PbTiO_3$ PCWT 4–24	Ceramic	380	220	0.01	2.5	195	0.08	3.4	255
$BaSrTiO_3$67/33	Ceramic, field-induced	1500	8800	0.004	2.6			12.4	25
$PbSc_{0.5}Ta_{0.5}O_3$	Ceramic, field-induced	3000–6000	Up to 15000		2.7			14–16	25
P(VDF/TrFE)50/50	Co-polymer film	40	18	0.03	2.3	251	0.11	0.8	49
P(VDF/TrFE)80/20	Co-polymer film	31	7	0.015	2.3	500	0.22	1.4	135

Source: After Refs. 28 and 46.

Table 9.2 Properties of thin-film pyroelectric materials on silicon substrates

MATERIAL/TEXTURE/ ELECTRODE	DEPOSITION METHOD/ SUBSTRATE	p (μCm^{-2}K^{-1})	ε	$\tan\delta$	c_{th} (10^6Jm^{-3}K^{-1})	F_v^* (kVm^{-1}K^{-1})	F_v (m^{2}C^{-1})	F_d (10^{-5}Pa$^{-1/2}$)
$PbTiO_3$/(001)+(100)Pt	Sol-gel and sputter	130–145	180–260	0.014–0.035	2.7	57–88	0.02–0.03	0.7–1.1
PZT15/85/(111)Pt	Sol-gel	160–220	200–230	0.01–0.015	2.7	78–113	0.03–0.04	1.3–1.5
PZT25/75(111)Pt	Sputter	200	300	0.01	2.7	75	0.028	1.4
PZT30/70(111)Pt	Sol-gel	200	340	0.011	2.7	66	0.025	1.3
Mod. PZT (Mn-doped)	Ceramic	356	218	0.007	2.6		0.07	5.1
PTL10–20Pt and Si	Ion beam, sputter, sol-gel, MOD	200–576	153–550	0.01–0.024	2.7	41–425	0.02–0.15	0.7–4.1
Porous PCT15/(11)Pt	Sol-gel	220	90	0.01	2.0	276	0.14	3.9
$LiNbO_3$/(006)Pt	Sputter	71	30	0.01	3.2	267	0.08	1.4
YBaCuO/Nb	Sputter	4000						3.2
Epitaxial films								
$PbTiO_3$/(001)Pt	Sputter/MgO	250	97	0.006	3.2	291	0.09	3.4
PZT45/55(001)Pt	Sputter/MgO	420	400	0.013	3.1	119	0.04	2.0
PZT52/48(100)	YBaCuO/$LaAlO_3$ PLD	500	100	0.02	3.1	57	0.02	1.2
PZT90/10(111)Pt	Sapphire/sputter	450	350	0.02	3.2	145	0.05	1.7
PLT5–15/(001)Pt	Sputter/MgO	400–1300	100–350	0.006–0.01	3.2	196–565	0.06–0.17	2.6–8.9
PLZT7.5/8/92–20/80/(001)Pt	Sputter/MgO	360–820	193–260	0.013–0.017	2.6	160–480	0.06–0.18	2.2–6.7
PCT30/(001)Pt	Sputter/MgO	520	290	0.02	3.0	202	0.06	2.4

Source: After Refs. 28 and 46.

Table 9.3 Properties of pyroelectric thin film suitable for induced pyroelectricity

MATERIAL/TEXTURE/ ELECTRODE	DEPOSITION METHOD/ SUBSTRATE	INDUCED p ($\mu Cm^{-2}K^{-1}$)	ε	$tan\delta$	c_{th} ($10^6 Jm^{-3}K^{-1}$)	F_v^* ($kVm^{-1}K^{-1}$)	F_v (m^2C^{-1})	F_d ($10^{-5}Pa^{-1/2}$)
$PbSc_{0.5}Ta_{0.5}0_3$/sapphire	RF-sputter, 900°C	6,000 (25°C –30°C)	6,500	0.03	2.5	104	0.04	6–9
$PbSc_{0.5}Ta_{0.5}0_3$/CdGa-garner	Sol-gel, 900°C	3,800	9,000	0.002	2.7	50	0.02	11
$PbSc_{0.5}Ta_{0.5}0_3$/Si/Pt	Sol-gel, 700°C	200–450	900	0.02	2.7	25–57	0.02	0.6–1.3
$PbSc_{0.5}Ta_{0.5}0_3$/Si/Pt	Sol-gel, 630°C	490	700	0.008	2.7	60	0.02	2.6
PbMgZn-NbO/(100)Pt/MgO	Sol-gel, 900°C	14,000 (15°C)	1,600	0.004	2.85	989	0.34	20–40
$K_{0.89}Na_{0.11}Ta_{0.55}Nb_{0.45}O_3/KTO_3$	LPE, 930°C	5,200 (66°C)	1,200 (66°C)	0.02	2.9	50	0.02	3.9

Source: After Ref. 28.

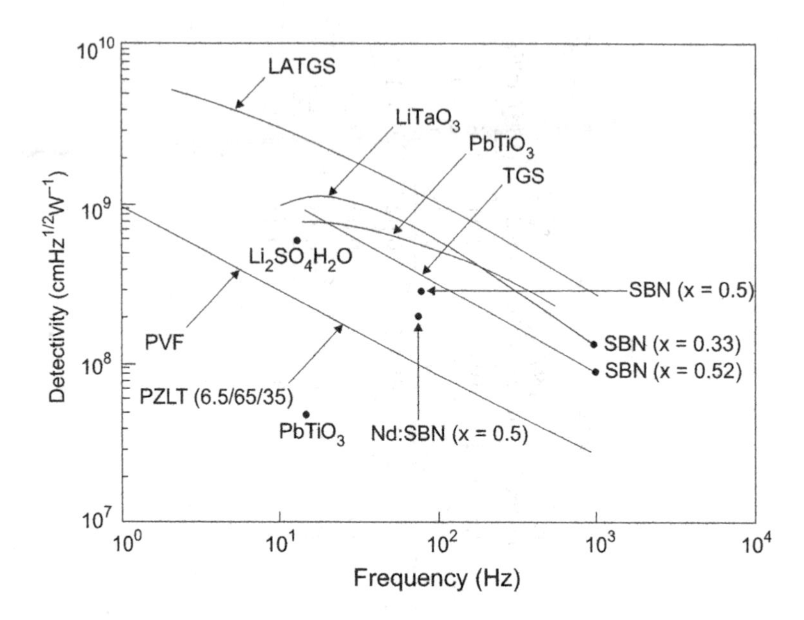

Figure 9.5 Discrete pyroelectric detector performance. (From Ref. [23])

disadvantage, particularly for detectors that are required to meet military specifications. In spite of these problems, TGS is frequently used for high-performance single element detectors and it is the preferred material for vidicon targets. Several variants on pure TGS have been developed to overcome the problem of the low Curie temperature. The alanine and arsenic acid-doped materials (ATGSAs) are particularly interesting because of their low dielectric losses and high pyroelectric coefficients (see Table 9.1). Detectors with D^* values of 2×10^9 cmHz$^{1/2}$W^{-1} have been obtained at 10 Hz (see Figure 9.5 [23]).

Lithium tantalate, $LiTaO_3$, gives an inferior performance to TGS, due to its lower pyroelectric coefficient and slightly higher relative permittivity (lower value of F_v). It has the following advantages: high chemical stability, very low loss (so F_d is favorable), very high Curie temperature, and insolubility in water. The material is widely used for single element detectors, although there can be problems associated with thermally induced transient noise spikes from this material when used in very low-frequency devices. It is not particularly favorable to use for thermal imaging arrays because of its low permittivity. Its thermal conductivity is quite high, so it is not a good material for pyroelectric vidicons. Good single crystals of $LiTaO_3$ can be produced by the Czochralski technique. It is readily available commercially.

Strontium barium niobate (SBN) is the next single crystal pyroelectric material. In fact, it is a family name for a range of solid solutions defined by the formula $Sr_{1-x}Ba_xNb_2O_6$, in which x can be varied from 0.25 to 0.75. SBN-50 ($x=0.50$) has a favorable F_d FoM. Depending on the composition, the ferroelectric transition can be tuned between 40 and 200°C. A high field-induced effect has been applied in uncooled thermal imaging, based on ferroelectric materials with a near–room temperature phase transition. Its high dielectric constant makes it a good candidate for thermal imaging arrays. The SBN is produced by the Czochralski technique, but good-quality large single crystals are relatively difficult to grow.

9.2.2 PYROELECTRIC POLYMERS

Ferroelectric polymers based on polyvinylidene fluoride (PVDF) and copolymers with trifluoroethylene (PVDF-TrFE) possess relatively low pyroelectric coefficient and low dielectric constants with high losses, so their FoM are inferior to the other materials. This class of materials is attractive for use in pyroelectric vidicon primarily because of their superior mechanical properties, in the case of fabrication in their samples (< 6 μm), low permittivities, and low thermal conductivity. Their low permittivities make them well-suited to large area detectors, but they are rather poorly suited for large area arrays. They are, however, obvious candidates for very low-cost detectors since they are readily available in large thin sheets that do not require expensive lapping and polishing processes necessary for other materials [48,49]. Their low heat conductivity and dielectric constant reduce the cross talk between neighboring elements in multielement detectors. The performance of PVDF detectors is inferior to the other categories of materials, except for very large detectors operating at high frequencies. Their low glass temperatures are severe obstacles for many applications.

The PVDF is commercially available as poled and electroded polymer sheets of varying thicknesses and needs to be mechanically stretched before poling to develop its ferroelectric properties. However, the PVDF–TrFE copolymers can be cast from the melt or methyl ethyl ketone solution directly into the ferroelectric phase and, therefore, they are particularly interesting for direct deposition onto a silicon substrate for making arrays.

9.2.3 PYROELECTRIC CERAMICS

Another class of materials, polycrystalline ferroelectric ceramics show promise for use in pyroelectric detectors. They offer a number of advantages over the materials listed earlier:

- They are relatively cheap to manufacture in large areas using standard mixed-oxide processes
- They are both mechanically and chemically robust (they can be processed into thin wafers)
- They possess high Curie temperature
- They do not suffer from thermally induced noise spikes
- They can be modified by the inclusion of selected dopant elements into the lattice to control such parameters as: p, ε_r, tanδ, Curie temperature, electrical impedance, and mechanical properties (controlling grain size of material)

There is a vast range of ceramic materials that consist of solid solutions of PZ (lead zirconate, $PbZrO_3$) and PT (lead titanate, $PbTiO_3$), and very similar oxides. These have been developed over a period of many years to satisfy a variety of ferroelectric, piezoelectric, electro-optic, and pyroelectric requirements. An example of modification of the electrical properties of a pyroelectric ceramic has been given by Whatmore [21]. The ceramic can be poled in any desired direction by the application of a suitable electric field. The morphotropic phase boundary compositions of the PZT system are generally avoided for pyroelectric applications because these have high permittivities, which are detrimental to the FoM.

The conventional pyroelectric ceramics are still favored for most practical applications because of the stability of their properties over the normal operating temperature range (the Currie temperature is typically above 200°C) and because they do not need an applied DC bias field to operate them.

There have been various experimental studies to improve the FoM of modified PZ compositions [46]. One prospect is through the exploitation of the step in the spontaneous polarization at phase transition [50].

Ceramic devices have D^* values in the 10^8 cmHz$^{1/2}$W^{-1} range. The performance of these devices is comparable or better than lithium tantalate, except in the case of large detectors.

Resistivities of modified ceramics cover the range of 10^9–10^{11} Ωcm^2. This means that the gate bias resistor in Figure 9.2c, which is generally around 10^{11}–10^{12} Ω, an expensive item, can be eliminated as a separate component by adjusting the resistivity of the material to suit the electrical time constant required. This is particularly important where large numbers of elements are involved in an array.

Using bulk pyroelectrics in the fabrication of infrared detectors leads to several drawbacks; the material must be cut, lapped, and polished to make a thin, well-insulated and sensitive layer. In addition, the array fabrication requires metallization on both faces and bonding to a silicon readout circuit to yield a complete hybrid array. On account of this, in the last decade, there has been a growth of interest in the integration of pyroelectric thin films directly onto silicon substrates as a means for both reducing array fabrication costs and increasing performance through reduced thermal mass and improved thermal isolation [31].

Properties of thin-film materials differ from those of bulk materials in as much as microstructure and substrate influence, which are of importance [28]. In contrast to bulk ceramics, thin films can be grown textured or even completely oriented in the case of epitaxy (see Table 9.2). The performance similar to the single crystal materials is obtained for the optimal texture when the polar axis stays perpendicular to the electrodes everywhere in the film. Also, considerable improvement of thin-film properties is possible in the case of materials that only exist as polycrystalline ceramics in bulk form (e.g., PZT, PLT). For example, a good demonstration of this case is epitaxial $PbTiO_3$, whose FoM F_v^* was measured as 291 kV/mK for thin films, whereas only 107 kV/mK is reached in bulk ceramics [28].

There is a trade-off between temperature stability of materials and size of the pyroelectric effect. Materials with high critical temperatures such as $LiTaO_3$ and $PbTiO_3$ are more adequate for simple and reliable devices. The relevant properties of materials gathered in Table 7.2 indicate that $PbTiO_3$-derived compounds with PZT (15%–30%) Zr films are favorite materials, but they can be replaced by PLT or PCT. Pure $PbTiO_3$ has been mostly abandoned because of too-high dielectric losses and difficulties to pole. We also notice that $LiTaO_3$ thin-film pyroelectric applications are far from being as advanced as its applications in bulk detectors.

The oxide materials (modified lead zirconate titanates or the dielectric bolometer materials) possess the right properties (high ε and high F_d) as ceramics sintered around 1200°C. However, for integrating ferroelectric thin films directly on silicon places causes a very important constraint on the temperature at which the ferroelectric can be grown. The interconnect metallization on the chips should not be taken above 500°C for any length of time, and this places an upper limit on the ferroelectric layer process temperature. Fortunately, many techniques have been researched for ferroelectric thin-film deposition. These include chemical solution deposition—particularly sol-gel or metalorganic deposition (MOD) and metalorganic chemical vapor deposition [28,31]. It appears that sol-gel deposition provides an excellent technique for thin-film growth of Mn-doped PZT films at 560°C, with a FoM F_d exceeding those of many bulk materials (p of 3.52×10^{-4} C/Km^2 and F_d of 3.85×10^{-5} Pa$^{-1/2}$).

9.2.4 DIELECTRIC BOLOMETERS

The conventional materials discussed earlier are ferroelectrics operated well below T_C, where the polarization is not permanently affected by changes in ambient temperature. It is, however, possible to operate ferroelectrics at or above T_C, with an applied bias field, in the mode of a dielectric bolometer [51]. Current developments in the area of pyroelectric materials include the use of dielectric bolometers.

With the application of an external electric field, the total polarization is described by Equation 9.1. Below T_C, P_s is large compared to the second term; thus D and P_s are often used interchangeably. However, it is clear from Figure 9.6a that the maximum pyroelectric effect (i.e., the maximum slope of P versus T) occurs near T_C, and therefore it seems desirable to operate there.

The field-enhanced pyroelectric coefficient is described by Equation 9.2. From this equation, it can be noted that the induced part of the pyroelectric coefficient depends not only upon the temperature rate of change of the permittivity, but also upon the field dependence of the rate change. Because of that, it is not a simple matter to calculate the field effect. At all temperatures, the dielectric behavior is nonlinear; that is, the gradient of permittivity varies with the applied field and the dielectric peak and $d\varepsilon/dT$ are both depressed with increasing field (see Figure 9.6b). Note that the pyroelectric coefficient maximum is somewhat lower in temperature than the peak capacitance value (see Figure 9.6b). The capacitance data represents a biased sample, and both the dielectric constant and pyroelectric coefficient maxima occur at

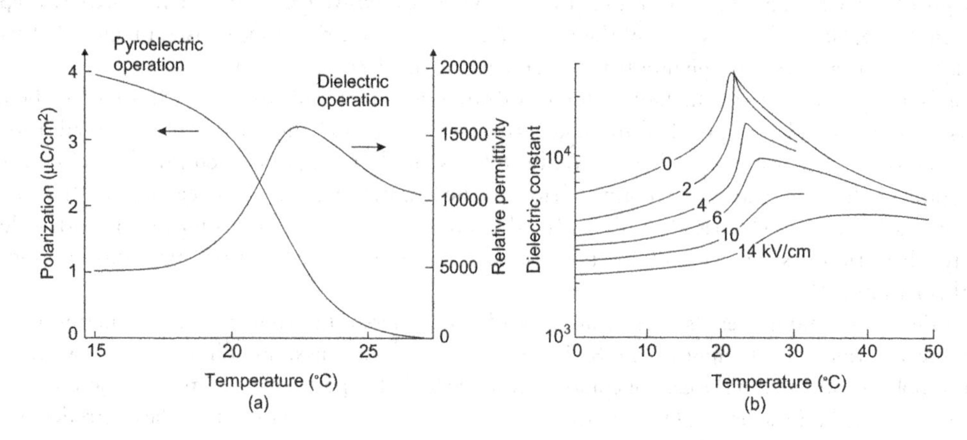

Figure 9.6 Barium strontium titanate ceramic: operating mode for (a) ferroelectric ceramic and (b) dielectric constant. (From Ref. [26]. With permission.)

temperatures above the Curie temperature. As the operating point continues to diverge from the Curie temperature, dielectric contributions to polarization become dominant.

Thus, the application of an electric field gives several benefits to the detector performance [52]:

- It adds an induced polarization to the spontaneous polarization
- It suppresses the dielectric permittivity, especially as it peaks near the transition
- It broadens the response peak, easing temperature control limits
- It suppresses dielectric loss, reducing noise
- It stabilizes polarization near the transitions, providing predictable performance

Several materials have been examined in dielectric bolometer mode, including potassium thallium niobate, $KTa_xNb_{1-x}O_3$ (KTN), lead zinc niobate, $Pb(Zn_{1/3}Nb_{2/3})O_3$ (PZN), barium strontium titanate, $Ba_{1-x}Sr_xTiO_3$ (BST), lead magnesium niobate, $Pb(Mg_{1/3}Nb_{2/3})O_3$, (PMN), and more recently, lead scandium tantalate, $Pb(Sc_{1/2}Ta_{1/2})O_3$ (PST) [22]. Dielectric bolometers require stringent bias and temperature stabilization. Properties of pyroelectric materials with transitions near ambient and operating with an electric field are shown in Tables 9.1 and 9.3.

The BST ceramic is a relatively well-behaved material with a very high permittivity. When Sr moves from 40% to 0% in the compound, T_C moves from 0°C to 120°C. Typical values of dielectric constant higher than 30,000 were noticed in the BST 67/33 material used in the high-density array development program [52]; the 17-µm thickness of the BST ceramic reached with difficulty appears as the lowest limit. The peak F_d (10.5×10^{-5} $Pa^{-1/2}$) achieved by BST65/35 is over twice that for the modified PZ or PT ceramics [22].

All the oxide materials for dielectric bolometers have very high dielectric constants (>1,000) under the operational temperatures and fields coupled with very high pyroelectric coefficients. These make them well-suited for small-area detectors in general, and very large arrays of small elements in particular.

Operating in the dielectric mode, single crystal BST should present little if any advantage over ceramic BST. The pyroelectric coefficient of ceramic BST more than doubles that of single crystal material of ostensibly the same composition measured under similar conditions. Likewise, the ceramic dielectric constant exceeds the single crystal value. Other attributes that make the ceramic BST more desirable than the single crystal are ease and cost of fabrication, material uniformity, superior performance, electrical resistance, resistance to aging, and amenable to doping.

The BST technology is a cumbersome bulk ceramic technology that requires grounding and polishing of ceramic wafers sliced from a boule, laser reticulation of pixels, multiple thinning, and planarization steps. The arrays are connected to the silicon readout circuit by compression bonds. The process suffers from a thermal isolation problem due to the thick mesa structure and BST surface degradation due to the thinning procedure.

Next-generation uncooled pyroelectric detectors are required to operate in a normal pyroelectric mode without bias and temperature stabilization. In addition, it is desired to use a thin-film pyroelectric detector technology to utilize the state-of-the-art micromachining technology for fabrication focal plane arrays (FPAs).

9.2.5 CHOICE OF MATERIAL

To obtain a direct comparison between the various pyroelectric materials is very difficult as the detector area and operating frequency will affect the performance, and account must also be taken of the environmental operating conditions. Porter has compared devices operating under different conditions for detector areas ranging from 100 to 0.01 mm² [19]. If a maximum detectivity is required, then for a given field-effect transistor (FET), the element area is an important consideration as it will affect the matching between the element capacitance and the amplifier capacitance. For large-area elements, the low dielectric constant materials dominate; TGS and lithium tantalate appear to be the best devices for all frequencies, except at very high frequencies (>10 kHz) when the polymer film devices begin to dominate. A more complicated situation is if the element area is reduced to 1 mm², the most commonly used order of magnitude. No one material is best at all frequencies. However, for small devices, the high dielectric constant materials are better (e.g., SBN because of the better capacitative match between the element and the amplifier). For intermediate-area devices, the performance of all devices is comparable.

It should be emphasized that the aforementioned discussion only shows the trends, as varying the detector parameters or the FET amplifier could alter this situation. Other factors such as environmental stability, availability, cost, and manufacturing considerations are also very important.

Stringent requirements are put on the materials used in the fabrication of FPAs, where very thin ferroelectric films are used [27,29,53–56]. Most ferroelectrics tend to lose their interesting properties as the thickness is reduced. However, some ferroelectric materials seem to maintain their properties better than others. This seems particularly true for PT and related materials, whereas for BST, the material does not hold its properties well in the thin-film form.

9.3 DETECTOR DESIGNS

Uncooled pyroelectric-IR detectors are used in many applications, including air quality monitoring, gas analyzers, atmospheric temperature measurement, biomedical imaging, industry control systems, facial recognition, fire alarm systems, infrared imaging, IR spectrometers and interferometers, laser detection, meteorology, pyroelectric vidicons, pyrometers, remote sensing, UV to FIR (THz) detection, and x-ray detectors. Thermal imaging is dominated by microbolometer arrays although smaller lower cost thermal imagers using pyroelectric detector arrays have been developed and are now creating a new market for the so-called "visual thermometers."

Figure 9.7 shows the typical design for a single-element pyroelectric detector. Most commercial detectors consist of about 2×2 mm of lithium tantalate and wafers of a few microns in thickness mounted on a shallow cylindrical substrate. To increase absorption, a metallic thin film, metal black paste, and evaporated or electroplated metal black are commonly used as absorbers. Due to the very high impedance of the sensor, a simple FET [junction-gate FET or metal–oxide–semiconductor (MOSFET)] with an appropriate load resistance, or an operational amplifier with feedback resistance in the detector package, is utilized to convert impedance to low output and to detect small charges.

Pyroelectric materials are available in bulk or in thin-film forms for sensor applications. These materials can be integrated using appropriate micromachining techniques into micromechanical structures. Rapid advances in thin-film technology have produced detectors directly coupled to integrated circuits. Critical to the realization of a micromachined sensor is the deposition of the pyroelectric material in the form of thin films from 0.1 μm, with pyroelectric properties approaching those of the corresponding bulk materials. Typically, the permittivity of thin films is lower than that of bulk materials. The details of the techniques

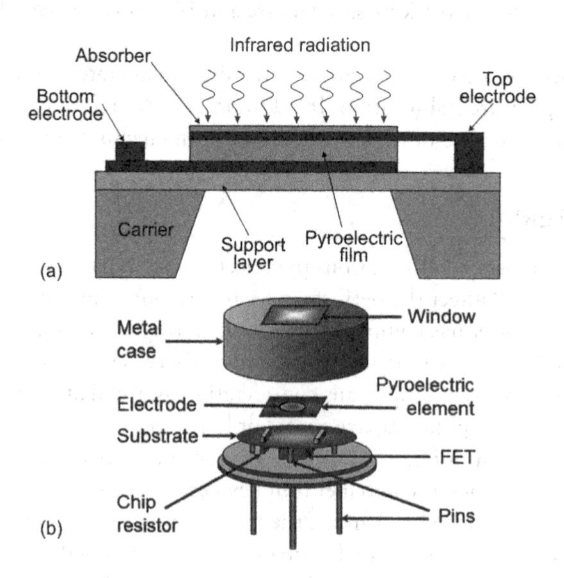

Figure 9.7 Single-element pyroelectric detector: (a) a typical design and (b) a cross section of thin-film pyroelectric sensor.

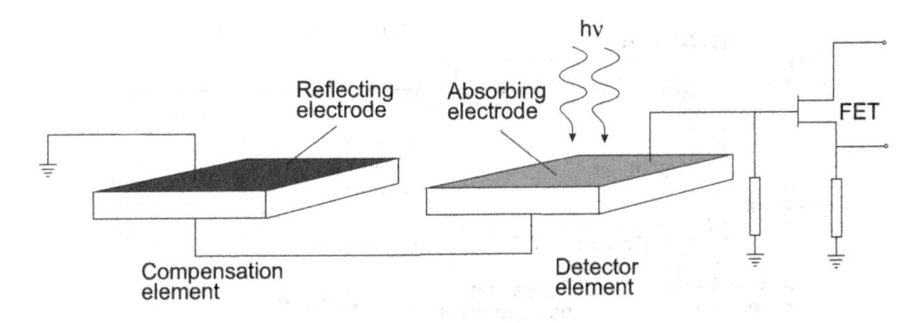

Figure 9.8 Compensated pyroelectric detectors. (From Ref. [21]. With permission.)

and the issues relating to the deposition and quality control of the films are too large for this short review; however, several excellent articles are available, for example, Refs. 28, 47, and 57.

A disadvantage of the single element is its sensitivity to a drift in ambient temperature and piezoelectric microphonic noise. To reduce this type of interference, two oppositely polarized detector elements are connected either in series or in parallel. The compensating element (see Figure 9.8 [21]) is coated with a reflecting electrode and/or mechanically screened so that it is not subject to the input radiation flux. The compensating element should be placed in a position that makes it thermally and mechanically similar to the detector element so that a signal due to temperature changes or mechanical stresses is canceled.

The temperature sensitivity along with the low noise performance can be also achieved in quartz oscillators in the principle of operation of several thermal sensors based on acoustic waves [58]. However, this principle of operations is a relatively unexplored technology for uncooled IR sensors. The major practical challenges to this approach were associated with the difficulties of producing large arrays of identical mechanical elements [59]. This problem, however, is specific to quartz since high-density micromachining on quartz substrates is difficult. With the current state of technology, it is possible to fabricate large, densely packed arrays of resonators using other thin-film piezoelectric materials.

More recently, Gokhale et al. have developed a novel uncooled IR detector using a combination of piezoelectric, pyroelectric, and resonant effects with a sensitivity that is orders of magnitude higher than that of quartz-based resonators [60,61]. Figure 9.9 shows the basic principle of transduction for a resonant IR detector—the change of its mechanical resonant frequency caused by the temperature rise induced by the absorbed infrared radiation. The mechanical resonator is driven in an open loop or used in a self-sustaining oscillator by using simple and common feedback circuits. The shift in resonance frequency and amplitude is compared to the reference resonator. The detector architecture consists of a high-Q GaN micromechanical resonator—GaN possesses high piezoelectric and pyroelectric coefficients and can be grown on a silicon substrate via low-cost batch fabrication.

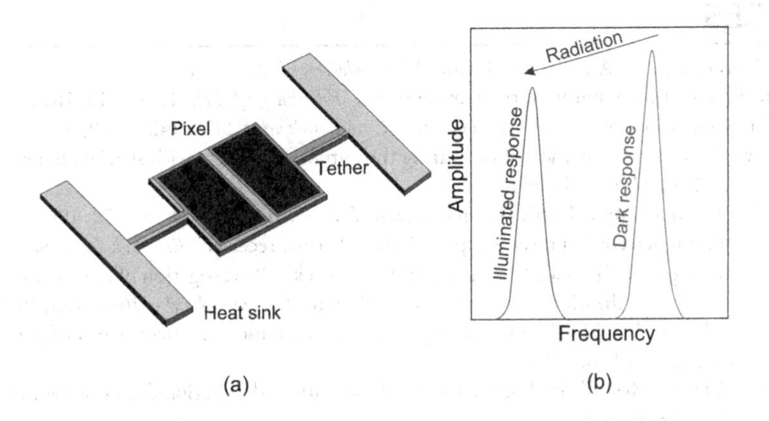

Figure 9.9 Resonator-mode infrared detector: (a) a thin-film resonator mechanically suspended by thin tethers, (b) generated heat shifts in frequency characteristics of resonator because of pyroelectric/piezoelectric effects.

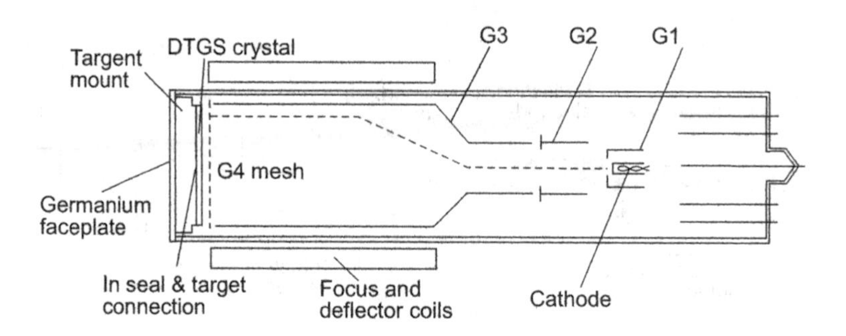

Figure 9.10 Schematic representation of a pyroelectric vidicon tube.

9.4 PYROELECTRIC VIDICON

Pyroelectric devices only respond to changes in incident radiation and are ideally suited for the detection of very small changes in flux whilst operating with a large background level of incident energy. The second characteristic of note is their broad spectral response from microwaves to x-rays.

The most important application of pyroelectric detectors is thermal imaging. The concept of the pyroelectric vidicon tube was first proposed by Hadni [62], and commercial devices were demonstrated as early as 1970 [63]. A schematic of a tube is shown in Figure 9.10. This device can be considered analogous to the visible television camera tube except that the photoconductive target is replaced by the pyroelectric detector and germanium faceplate. The target consists of a disc of pyroelectric material (20-μm thick, 2-cm diameter), with a transparent electrode on the front surface. An infrared lens produces a thermal image on this target and the resulting charge distribution is read off the back surface by the scanning electron beam. The original tubes were fabricated with TGS but better results have been achieved with deuterated TGS and TGFB material [64,65].

The major factor limiting the resolution achievable with pyroelectric vidicons is thermal diffusion within the target. This causes the thermal resolution to degrade rapidly as the spatial frequency increases. For this reason, reticulated targets are being developed [66]. The vidicon tubes achieve 0.2°C resolution in an image consisting of 100 TV lines, and with reticulation, the spatial resolution increases to 400 TV lines or more [67]. Of the tube programmers in the United States, United Kingdom, and France, only the UK technology at EEV Co. Ltd. went on to manufacture, with applications in fire service cameras and industrial maintenance.

Although good-quality images have been obtained from the pyroelectric vidicon, recent work is mainly directed toward fabrication of large 2-D FPAs. This allows improvement on the temperature resolution of the systems and produces more robust and lighter imagers.

REFERENCES

1. S. B. Lang, "Pyroelectricity: A 2300-year history," *Ferroelectrics* 7, 231–4, 1974.
2. D. Brewster, "Observation of pyroelectricity of minerals," *Edinburgh J. Sci.*, 1, 208–14, 1824.
3. Y. Ta, "Action of radiations on pyroelectric crystals," *Comptes Rendus* 207, 1042–4, 1938.
4. A. G. Chynoweth, "Dynamic method of measuring the pyroelectric effect with special reference to barium titanate," *J. Appl. Phys.*, 27, 78–84, 1956.
5. J. Cooper, "A fast-response pyroelectric thermal detector," *J. Sci. Instrum.*, 39, 467–72, 1962.
6. J. Cooper, "Minimum detectable power of a pyroelectric thermal receiver," *Rev. Sci. Instrum.*, 33, 92–5, 1962.
7. A. Hadni, Y. Henninger, R. Thomas, P. Vergnat, and B. Wyncke, "Investigation of pyroelectric properties of certain crystals and their utilization for detection of radiation," *Comptes Rendus* 260, 4186, 1965.
8. G. A. Burdick and R. T. Arnold, "Theoretical expression for the noise equivalent power of pyroelectric detectors," *J. Appl. Phys.*, 37, 3223–6, 1966.
9. J. H. Ludlow, W. H. Mitchell, E. H. Putley, and N. Shaw, "Infrared radiation detection by pyroelectric effect," *J. Sci. Instrum.*, 44, 694–6, 1967.
10. H. P. Beerman, "Pyroelectric infrared radiation detector," *Am. Ceram. Soc. Bull.*, 46, 737, 1967.

11. A. M. Glass, "Ferroelectric strontium-barium-niobate as a fast and sensitive detector of infrared radiation," *Appl. Phys. Lett.*, 13, 147–9, 1968.
12. R. W. Astheimer and F. Schwarz, "Thermal imaging using pyroelectric detectors: Mylar supported TGS," *Appl. Opt.*, 7, 1687–95, 1968.
13. R. J. Phelan, Jr., R. J. Mahler, and A. R. Cook, "High D* pyroelectric polyvinylfluoride detectors," *Appl. Phys. Lett.*, 19, 337–8, 1971.
14. E. H. Putley, "The Pyroelectric Detector," in *Semiconductors and Semimetals*, Vol. 5, eds. R. K. Willardson and A. C. Beer, 259–85, Academic Press, New York, 1970.
15. G. Baker, D. E. Charlton, and P. J. Lock, "High performance pyroelectric detectors," *Radio Electron. Eng.*, 42, 260–4, 1972.
16. E. H. Putley, "Thermal Detectors," in *Optical and Infrared Detectors*, ed. R. J. Keyes, 71–100, Springer, Berlin, 1977.
17. S. T. Liu and D. Long, "Pyroelectric detectors and materials," *Proc. IEEE* 66, 14–26, 1978.
18. D. E. Marshall, "A review of pyroelectric detector technology," *Proc. SPIE* 132, 110–17, 1978.
19. S. G. Porter, "A brief guide to pyroelectric detectors," *Ferroelectrics* 33, 193–206, 1981.
20. J. C. Joshi and A. L. Dawar, "Pyroelectric materials, their properties and applications," *Phys. Status Solidi A* 70, 353–69, 1982.
21. R. W. Whatmore, "Pyroelectric devices and materials," *Rep. Prog. Phys.*, 49, 1335–86, 1986.
22. R. W. Whatmore, "Pyroelectric ceramics and devices for thermal infra-red detection and imaging," *Ferroelectrics* 118, 241–59, 1991.
23. L. E. Ravich, "Pyroelectric detectors and imaging," *Laser Focus/Electro-Opt.*, 22, 104–15, July 1986.
24. R. Watton, "Ferroelectric materials and design in infrared detection and imaging," *Ferroelectrics* 91, 87–108, 1989.
25. P. Robin, H. Facoetti, D. Broussoux, G. Vieux, and J. L. Ricaud, "Performances of advanced infrared pyroelectric detectors," *Rev. Tech. Thomson.*, 22(1), 143–86, 1990.
26. H. Betatan, C. Hanson, and E. G. Meissner, "Low cost uncooled ferroelectric detector," *Proc. SPIE* 2274, 147–56, 1994.
27. M. A. Todd, P. A. Manning, O. D. Donohue, A. G. Brown, and R. Watton, "Thin film ferroelectric materials for microbolometer arrays," *Proc. SPIE* 4130, 128–39, 2000.
28. P. Muralt, "Micromachined infrared detectors based on pyroelectric thin films," *Rep. Prog. Phys.*, 64, 1339–88, 2001.
29. C. M. Hanson, H. R. Beratan, and J. F. Belcher, "Uncooled infrared imaging using thin-film ferroelectrics," *Proc. SPIE* 4288, 298–303, 2001.
30. P. W. Kruse, *Uncooled Thermal Imaging. Arrays, Systems, and Applications*, SPIE Press, Bellingham, WA, 2001.
31. R. W. Whatmore and R. Watton, "Pyroelectric materials and devices," in *Infrared Detectors and Emitters: Materials and Devices*, eds. P. Capper and C. T. Elliott, 99–147, Kluwer Academic Publishers, Boston, MA, 2000.
32. S. B. Lang, "Pyroelectricity: From ancient curiosity to modern imaging tool," *Phys. Today* 58(8), 31–6, 2005.
33. R. W. Whatmore, Q. Zhang, C. P. Shaw, R. A. Dorey, and J. R. Alock, "Pyroelectric ceramics and thin films for applications in uncooled infra-red sensor arrays," *Phys. Scr., T* 129, 6–11, 2007.
34. A.K. Batra, and M.D. Aggarwal, *Pyroelectric Materials: Infrared Detectors, Particle Accelerators and Energy Harvesters*, SPIE Press, Bellingham, 2013.
35. E. H. Putley, "A method for evaluating the performance of pyroelectric detectors," *Infrared Phys.*, 20, 139–47, 1980.
36. D. H. Austan and A. M. Glass, "Optical generation of intense picosecond electrical pulses," *Appl. Phys. Lett.*, 20, 398–9, 1972.
37. C. B. Roundy, R. L. Byer, D. W. Phillion, and D. J. Kuizenga, "A 170 psec pyroelectric detector," *Opt. Commun.*, 10, 374–7, 1974.
38. W. R. Blevin and J. Geist, "Influence of black coatings on pyroelectric detectors," *Appl. Opt.*, 13, 1171–8, 1974.
39. A. Van der Ziel, "Pyroelectric response and D* of thin pyroelectric films on a substrate," *J. Appl. Phys.*, 44, 546–9, 1973.
40. R. M. Logan and K. More, "Calculation of temperature distribution and temperature noise in a pyroelectric detector: I. gas-filled tube," *Infrared Physics* 13, 37–47, 1973.
41. R. M. Logan, "Calculation of temperature distribution and temperature noise in a pyroelectrical detector: II. Evacuated tube," *Infrared Phys.*, 13, 91–8, 1973.
42. R. L. Peterson, G. W. Day, P. M. Gruzensky, and R. J. Phelan, Jr., "Analysis of response of pyroelectric optical detectors," *J. Appl. Phys.*, 45, 3296–303, 1974.
43. N. M. Shorrocks, R. W. Whatmore, M. K. Robinson, and S. G. Parker, "Low microphony pyroelectric arrays," *Proc. SPIE* 588, 44–51, 1985.
44. A. Mansingh and A. K. Arora, "Pyroelectric films for infrared applications," *Indian J. Pure Appl. Phys.*, 29, 657–64, 1991.

45. A. Sosnin, "Image infrared converters based on ferroelectric-semiconductor thin-layer systems," *Semicond. Phys., Quantum Electron. Optoelectron.*, 3, 489–95, 2000.
46. R. W. Whatmore, "Pyroelectric arrays: Ceramics and thin films," *J. Electroceram.*, 13, 139–47, 2004.
47. S. Tadigadapa and K. Mateti, "Piezoelectric MEMS sensors: State-of-the-art and perspectives", *Meas. Sci. Technol.*, 20, 092001, 2009.
48. S. B. Lang and S. Muensit, "Review of some lesser-known applications of piezoelectric and pyroelectric polymers," *Appl. Phys. A* 85, 125–34, 2006.
49. J. L. Coutures, R. Lemaitre, E. Pourquier, G. Boucharlat, and P. Tribolet, "Uncooled infrared monolithic imaging sensor using pyroelectric polymer," *Proc. SPIE* 2552, 748–54, 1995.
50. R. Clarke, A. M. Glazer, F. W. Ainger, D. Appleby, N. J. Poole, and S. G. Porter, "Phase transitions in lead zirconate-titanate and their applications in thermal detectors," *Ferroelectrics* 11, 359–64, 1976.
51. R. A. Hanel, "Dielectric bolometer: A new type of thermal radiation detector," *J. Opt. Soc. Am.*, 51, 220–25, 1961.
52. C. Hanson, H. Beratan, R. Owen, M. Corbin, and S. McKenney, "Uncooled thermal imaging at Texas Instruments," *Proc. SPIE* 1735, 17–26, 1992.
53. R. Watton, "IR bolometers and thermal imaging: The role of ferroelectric materials," *Ferroelectrics* 133, 5–10, 1992.
54. R. Watton and P. Manning, "Ferroelectrics in uncooled thermal imaging," *Proc. SPIE* 3436, 541–54, 1998.
55. R. K. McEwen and P. A. Manning, "European uncooled thermal imaging sensors," *Proc. SPIE* 3698, 322–37, 1999.
56. C. M. Hanson, H. R. Beratan, and D. L. Arbuthnot, "Uncooled thermal imaging with thin-film ferroelectric detectors," *Proc. SPIE* 6940, 694025, 2008.
57. P. Muralt, "Ferroelectric thin films for micro-sensors and actuators: A review," *J. Micromech. Microeng.*, 10, 136–146, 2000.
58. J. R. Vig, R. L. Filler, and K. Yoonkee, "Uncooled IR imaging array based on quartz microresonators," *J. Microelectromech. Syst.*, 5, 131–7, 1996.
59. P. Kao and S. Tadigadapa, "Micromachined quartz resonator based infrared detector array," *Sens. Actuators A, Phys.*, 149, 189–192, 2009.
60. V. J. Gokhale, Y. Sui, and M. Rais-Zadeh, "Novel uncooled detector based on gallium nitride micromechanical resonators," *Proc. SPIE* 8353, 835319, 2012.
61. V. J. Gokhale and M. Rais-Zadeh, "Uncooled infrared detectors using gallium nitride on silicon micromechanical resonators," *J. Microelectromech. Syst.*, 23, 803–810, 2014.
62. A. Hadni, "Possibilities actualles de detection du rayonnement infrarouge," *J. Phys.*, 24, 694–702, 1963.
63. E. H. Putley, R. Watton, W. M. Wreathall, and S. D. Savage, "Thermal imaging with pyroelectric television tubes," *Adv. Electron. Electron Phys.*, 33A, 285–292, 1972.
64. R. Watton, "Pyroelectric materials: Operation and performance in thermal imaging camera tubes and detector arrays," *Ferroelectrics* 10, 91–8, 1976.
65. E. H. Stupp, "Pyroelectric vidicon thermal imager," *Proc. SPIE* 78, 23–7, 1976.
66. S. E. Stokowski, J. D. Venables, N. E. Byer, and T. C. Ensign, "Ion-beam milled, high-detectivity pyroelectric detectors," *Infrared Phys.*, 16, 331–4, 1976.
67. A. J. Goss, "The pyroelectric vidicon: A review," *Proc. SPIE* 807, 25–32, 1987.

10 Pneumatic detectors

The Golay cell is a sensitive photo-acoustic or pneumatic infrared (IR) thermal detector. Invented by Marcel Golay in the late 1940s, it offers the best performance among thermal infrared detectors [1,2]. Despite some disadvantages, such as high cost (about $5,000) and relatively large size, the Golay cell is still commercially available and is used where high performance is essential. New versions of miniaturized micromachined Golay cells that utilize capacitive as well as tunneling displacement transducers have been developed.

10.1 GOLAY DETECTOR

The Golay cell (Figure 10.1) is a detector consisting of a hermetically sealed container filled with gas (usually xenon for its low thermal conductivity) and arranged such that expansion of the gas under heating by a photon signal distorts a flexible membrane on which a mirror is mounted. Within the cell there is a thin absorbing metallic film with an impedance approximately matching that of free space—a resistance of 270 Ω per square produced optimum performance. Light from the source is condensed by the lens system through a grid and is then reflected by the flexible mirror back through the grid onto the detector. The movement of the mirror is used to deflect a beam of light shining on a photocell and thus producing a change in the photocell current as the output. In modern Golay cells, the photocell is replaced by a solid-state photodiode and a light-emitting diode is used for illumination [3]. The reliability and stability of this arrangement are significantly better than that of the earlier Golay cells that used a tungsten filament lamp and a vacuum photocell.

To compensate for changes in the ambient temperature, and to make the device into an AC detector, there is a small leak from the gas cell into a reservoir. This leak leads to a time constant of a few Hz. As a result, since the detector noise rises rapidly below 10 Hz, the optimum modulation speed for incoming radiation is in the range of 10–20 Hz. Currently, Golay detectors are manufactured in-house. Delivery includes a detector head and a power supply unit. Their responsivity can be calibrated individually very precisely with blackbody sources at frequencies corresponding to the near-IR band. This calibration remains accurate to frequencies down to at least 0.3 THz.

Golay cells have a relatively high sensitivity at room temperature and a flat response over a very broad range of frequencies from 0.1 to 20 THz. The responsivity and the noise equivalent power (NEP) are about 1×10^5 V/W and 1×10^{-10} W/$\sqrt{\text{Hz}}$, respectively. The speed of a Golay cell depends on the heat capacity of the cavity and the heat conductance. Golay cells are slow detectors with a response time of the order of a few 10 ms (typical 15 ms), faster than that of the bolometer, thermistor, or thermocouple. The detector is very fragile and subject to mechanical failure (susceptible to mechanical vibrations). The performance specifications typical for the Golay cell are given in Table 10.1.

During the recent years, Golay cell detectors have been used for detecting terahertz (THz) radiation. In the long process of signal detection, the energy accumulation in the detector can cause a DC drift of the detected signal and reduce the accuracy of the detector. To avoid these circumstances, a chopper is used to reduce the noise. The materials used for the entrance window of a detector are high-density polyethylene (HDPE), polymethylpentene, and diamond. Figure 10.2 shows spectral responsivities of THz Golay cells manufactured by Microtech Instruments Inc.

(a)

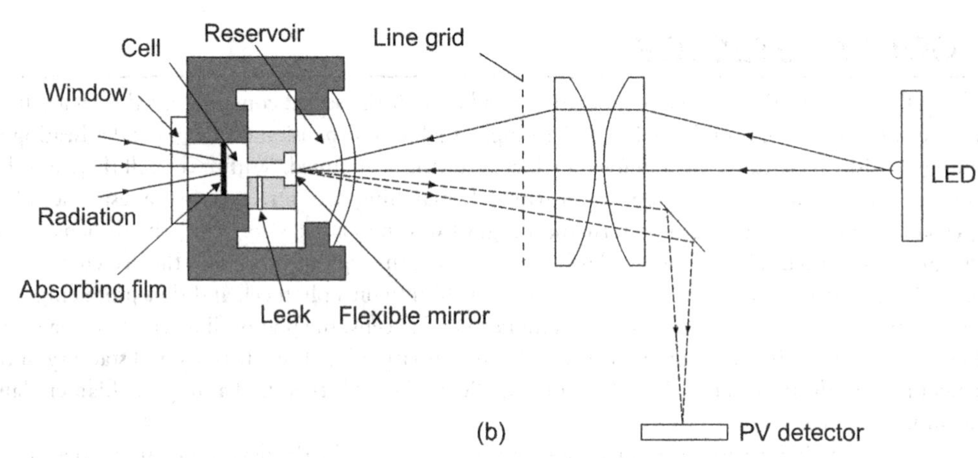

(b)

Figure 10.1 The photo (a) and schematic diagram (b) of the Golay cell detector.

Table 10.1. Performance specifications for the golay cell (after Ref. 4)

Detector size	6 mmØ
Window material	HDPE or Diamond
Wavelength range	7 to 8000 μm
Maximum power	10 μW
Optimum chopping frequency	20 Hz
NEP	1.2×10^{10} W/Hz$^{1/2}$
Responsivity	1.5×10^{5} V/W @ 20 Hz
Detectivity	7×10^{9} cmHz$^{1/2}$/W
Power requirement	VAC
Operating temperature range	5 to 40°C
Package size	126×45×87 mm
Response time	25 ms

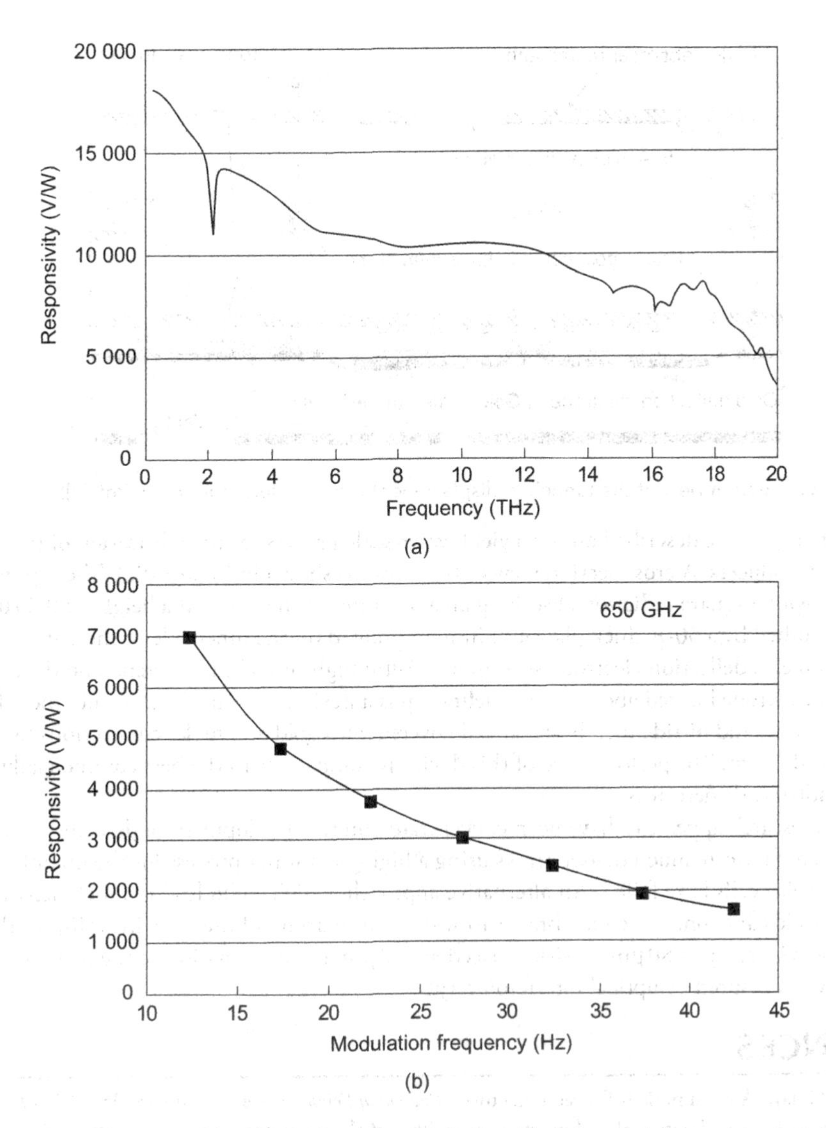

Figure 10.2 Spectral responsivities of THz Golay cells manufactured by Microtech Instruments Inc. Golay cell with polyethylene (a) and diamond windows (b), respectively. (Adapted after Ref. [5])

10.2 MICROMACHINED GOLAY-TYPE SENSORS

The Golay cell is known to be the most sensitive thermal detector, that is, of the order of subnanowatts. However, high cost, large size equipped with a power supply, and requiring a window in the size of a few millimeters can be considered as some of the drawbacks for Golay cells. Some efforts have been demonstrated recently based on the MEMS (microelectromechanical systems) technology to miniaturize devices. However, the miniaturization failed to be developed satisfactorily due to having a large gas chamber.

New miniature Golay cells fabricated by silicon micromachining techniques have been described utilizing capacitive [6] as well as tunneling displacement transducers [7,8]. Initial work on micromachined tunneling Golay cells were made at the Jet Propulsion Laboratory (Pasadena, California). The devices were made in a low yield process that involved the hand assembly and gluing together of the sensor parts. This mode of fabrication produced devices with large variations in key operating parameters. Prototype devices have been operated with NEP better than 3×10^{-10} $\mathrm{WHz^{-1/2}}$ at 25 Hz [8].

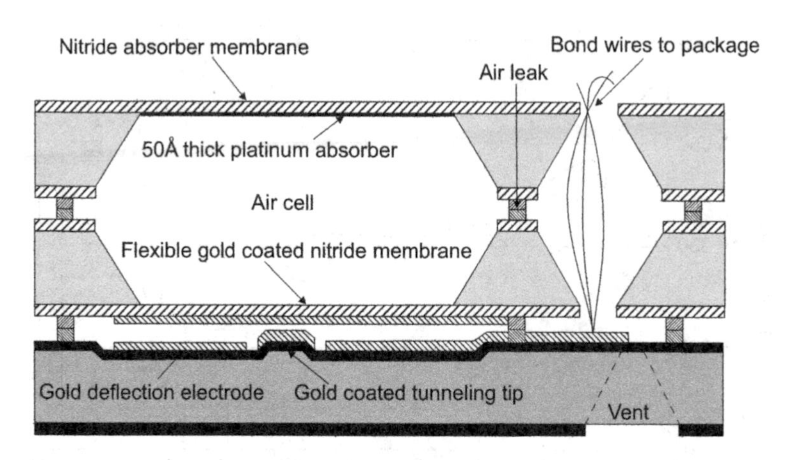

Figure 10.3 Cross-section view of the tunneling displacement infrared detector. (From Ref. [9])

Ajakaiye et al. [9] have described an 80% yield wafer-scale process for the fabrication of tunneling displacement transducers. A cross-section view of the sensor is shown in Figure 10.3. The top two parts form gas cells, with a square radiation absorbing area of 2 mm on the side and a height of 0.85 μm. Infrared radiation is absorbed by a 50-Å thick platinum film evaporated on the inner side of the 1-μm thick upper nitride membrane. A deflection electrode surrounds a 7-μm high tunneling tip etched on the bottom wafer. The deflection electrode located above the tunneling tip is a flexible 0.5-μm thick membrane. The tip, deflection electrode, and nitride membrane are all covered with gold and make connections to three pads located next to the vent. The performance of this device is comparable to the best commercially available uncooled broadband IR detectors.

In practice, research appears to have been concentrated on the development of detectors of a few mm^2 area. Attempts to fabricate much denser arrays using a high-yield wafer process for the manufacture of miniaturized Golay cells have failed. An alternative approach, which could lead to much denser arrays, uses robust and flexible nanocomposite membranes to seal uniform arrays of microcavities [10]. In this experiment with a 64×64 array of 80 μm^2 cavities spaced at 150 μm, the deformation of the membrane with temperature was measured by optical interferometry.

REFERENCES

1. M. J. E. Golay, "A pneumatic infra-red detectors," *Review of Scientific Instruments* 18, 357–62, 1947.
2. M. J. E. Golay, "The theoretical and practical sensitivity of the pneumatic infrared detector," *Review of Scientific Instruments* 20, 816, 1949.
3. J. R. Hickley and D. B. Daniels, "Modified optical system for the Golay detector," *Review of Scientific Instruments* 40, 732–33, 1969.
4. https://gentec-eo.com/Content/downloads/application-note/AN_201924_THz_R1.pdf
5. http://www.mtinstruments.com/downloads/Golay%20Cell%20Datasheet%20Revised.pdf
6. J. B. Chevrier, K. Baert, T. Slater, and A. Verbist, "Micromachined infrared pneumatic detector for gas sensors," *Microsystem Technologies* 1, 71–74, 1995.
7. T. W. Kenny, J. K. Reynolds, J. A. Podosek, E. C. Vote, L. M. Miller, H. K. Rockstad, and W. J. Kaiser, "Micromachined infrared sensors using tunneling displacement transducers," *Review of Scientific Instruments* 67, 112–28, 1996.
8. T. W. Kenny, "Tunneling infrared sensors," in *Uncooled Infrared Imaging Arrays and Systems*, eds. P. W. Kruse and D. D. Skatrud, 227–67, Academic Press, San Diego, CA, 1997.
9. O. Ajakaiye, J. Grade, C. Shin, and T. Kenny, "Wafer-scale fabrication of infrared detectors based on tunneling displacement transducers," *Sensors & Actuators* A134, 575–81, 2007.
10. C. Jiang, M. E. McConney, S. Singamaneni, E. Merrick, Y. Chen, J. Zhao, L. Zhang, and V. V. Tsukruk, "Thermo-optical arrays of flexible nanoscale nanomembranes freely suspended over microfabricated cavities as IR microimagers," *Chem. Mater.* 18, 2632 (2006).

11 Novel thermal detectors

At present, vanadium oxide and amorphous silicon (a-Si) microbolometers are technologies of choice for uncooled thermal imaging. The military continues to explore imaging systems where performance criteria are the most important factors: pixel count, sensitivity, and response time. However, their sensitivity limitations [1] and the still-significant prices encouraged many research teams to explore other infrared (IR) sensing techniques with the potential for improved performance with reduced detector costs. Lower-cost technologies are necessary to pursue consumer applications for more widespread "mass-consumable" applications, such as private property alarm surveillance, household fire detectors in addition to smoke detectors, occupancy detectors for intelligent building automation, pedestrian collision prevention on pedestrian crossings, road traffic monitoring for smart traffic management, simple thermal imaging for automotive to prevent pedestrian or animal collision, wildfire monitoring systems, and so on. Recently, thermal imaging modules for less than $1,000 have been produced. It means a tenfold reduction in costs, compared with the approximate price for current IR imaging systems (see Table 11.1).

One of the most promising IR sensing methods is the use of thermally actuated microelectromechanical systems (MEMS), with a reported detectivity of 10^8 cmHz$^{1/2}$W^{-1}. These novel uncooled detectors are also considered in this chapter. It is expected that novel uncooled detector arrays could become very attractive for a number of applications due to their inherent simplicity, high sensitivity, and rapid response to radiation.

11.1 NOVEL UNCOOLED DETECTORS

Despite successful commercialization of uncooled microbolometers suitable for thermal imaging, the community is still searching for a platform for thermal imagers that combines affordability, convenience of operation, and excellent performance. Recent advances in MEMS systems have led to the development of uncooled IR detectors operating as micromechanical thermal detectors as well as micromechanical photon detectors. Between them, the most important are bimaterial microcantilevers that mechanically respond to the absorption of the radiation [2]. These sensing structures were originally invented at the Oak Ridge National Laboratory (ORNL) in the mid-1990s [3–7], and subsequently developed by ORNL [8–12], the Sarnoff Corporation [13,14], Sarcon Microsystems [15–17], and others for imaging [2,18–25] and photo spectroscopic applications [26,27].

The thermomechanical detector approach was pioneered by Barnes et al. when they coated microcantilevers with a metal as the sensing active layer to form the bimaterial [28]. Figure 11.1 shows a schematic diagram of the capacitively sensed microcantilever structure. The microcantilever is attached mechanically and electrically to the substrate at the end by an anchor, and the second end is free to bend under the influence of any changes in stress along the arm. The IR radiation is absorbed by the microcantilever paddle materials along with a tuned resonant absorption cavity. The absorbed radiation is converted to heat in the microcantilever structure thermally isolated from the substrate by thermal isolation arms like those used in bolometers.

The cantilever structure contains a bimaterial region that is fabricated from two layers that possess significantly different thermal expansion coefficients; for example, a low thermal expansion coefficient SiO_2 substrate layer ($\alpha = 0.5\times10^{-6}$ K^{-1}), overlying an Al layer with a thermal expansion of $\alpha = 23\times10^{-6}$ K^{-1} [16]. When the incident radiation heats the structure, the bimaterial region of the cantilever bends up out of the plane due to the dissimilar thermal expansion of the bimaterials (approximately 0.1 μm for every degree of temperature change). Table 11.2 shows the generally used bimaterial combination. Usually, SiN_x and SiO_2 are used as IR absorbing materials, whereas Au and Al are used as contacts and reflectors.

Table 11.1 Performance and cost comparison

FEATURES	MICROBOLOMETERS	PYROMETERS	THERMOMECHANICAL
Conversion sensitivity (%K^{-1})	2–4	2–3	20–100
Response time (ms)	15–20	15–20	5–10
Dynamic range	10^4	10^3	$>10^5$
Optics	Large, expensive	Large, expensive	Small, cheap
Power requirements	Low	Low	Low
Ease of fabrication	Difficult	Difficult	Standard IC fabrication
Size	Moderately small	Moderately small	Small
Cost of camera	\$10–50K	\$5–20K	\$1–10K

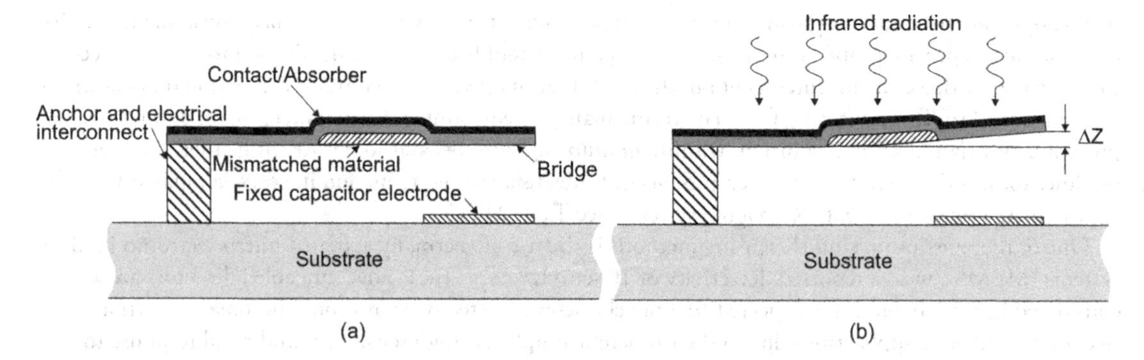

Figure 11.1 Schematic diagram of the operating principle of the bimaterial microcantilever IR detector: (a) nominal position, and (b) with increased radiation.

Table 11.2 Properties of several materials used in the design of cantilevers

	YOUNG'S MODULE (GPa)	THERMAL CONDUCTIVE COEFFICIENT ($Wm^{-1}K^{-1}$)	THERMAL EXPANSION COEFFICIENT ($10^{-6}K^{-1}$)	HEAT CAPACITY ($Jkg^{-1}K^{-1}$)	DENSITY (10^3 kgm^{-3})	EMISSIVITY (8–14 μm)
SiN_x	180	5.5±0.5	0.8	691	2.40	0.8
Au	73	296.0	14.2	129	19.3	
SiO_2	46–92	1.1	0.05–12.3	–	2.20	
Al	80	237.0	23.6	908	2.70	0.01
Si	100	135	2.6	700		

Silicon nitride does have some native absorption due to the stretching frequencies of chemical bonds present in the materials. To increase the speed response of the pixel, the mass and therefore the thickness of SiN are kept thin (300 nm) and thus the total native absorption is small. To improve the absorption, a thin metal film is used to get both good IR absorption and good visible reflection for the optical readout.

The primary fundamental limits to the performance of microcantilevers are related to the properties of the thermal detectors themselves: the background fluctuation limit and the temperature fluctuation limit (see Chapter 4).

In the case of thermomechanical IR detectors, there is an additional fundamental limitation that is related to spontaneous microscopic mechanical motion (oscillation) of any suspended microstructure due to its thermal energy. It appears that for the majority of the readout means, these oscillations are indistinguishable from temperature-induced bending and, therefore, directly contribute to the detector noise [11,29]. According to Sarid [30], thermomechanical noise is frequency independent at low frequencies and the cantilever tip displacement noise is equal to:

$$\left\langle \delta z_{TM}^{2} \right\rangle^{1/2} = \left(\frac{4kT\Delta f}{Qk_s\omega_o} \right)^{1/2}, \quad (11.1)$$

where Q (quality factor) is the ratio of the resonance frequency, ω_o, to the resonance peak width, and k_s is the spring constant, defined as the ratio of the force applied to the microcantilever divided by the displacement of the tip. An alternative model predicts that when the damping is due to intrinsic friction process rather than due to viscous damping of the medium, the density of thermomechanical noise follows a $1/f^{1/2}$ trend below the mechanical resonance [31].

Using Equation 11.1, the predicted limits to noise equivalent power (NEP) and detectivity due to thermomechanical noise are:

$$\mathrm{NEP} = \frac{1}{R_z}\left(\frac{4kT\Delta f}{Qk_s\omega_o} \right)^{1/2} \quad \text{and} \quad D^* = \frac{1}{R_z}\left(\frac{4kT}{AQk_s\omega_o} \right)^{1/2}, \quad (11.2)$$

where R_z is the responsivity of the detector.

An important advantage of thermomechanical detectors is that they are essentially free of intrinsic electronic noise and can be combined with a number of different readout techniques with high sensitivity. Depending on readout techniques, the novel uncooled detectors can be devoted to:

- Capacitative [11,13–17]
- Optical [5,7,9,11,12,18–20,22–25]
- Piezoresistive [3,4]
- Electron tunneling [32]

11.1.1 ELECTRICALLY COUPLED CANTILEVERS

In electrically coupled thermal transducers, bending of the cantilever changes its capacitance. This capacitance change is converted into an electrical signal that is proportional to the amount of absorbed IR light. All unwanted external vibrations are damped using an actively tuned resonant RC-circuit. For example, Figure 11.2 shows a schematic diagram of the operation of a capacitive-coupled detector, with the variable plate microcantilever capacitive sensor forming one arm of the bridge circuit [16]. Symmetric and

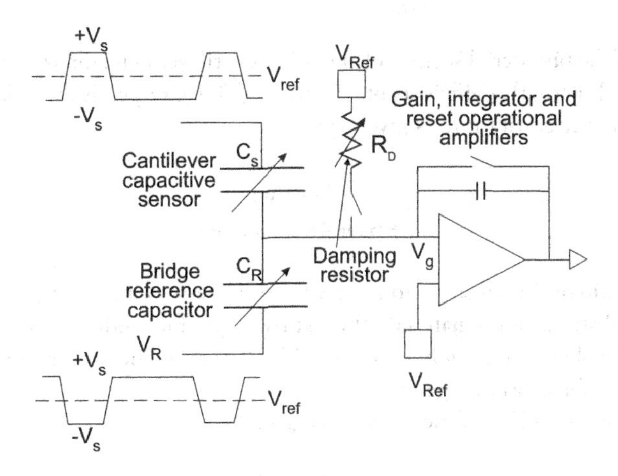

Figure 11.2 A circuit diagram showing the microcantilever bridge circuit, damping resistor, and signal gain amplifier. (From Ref. [16])

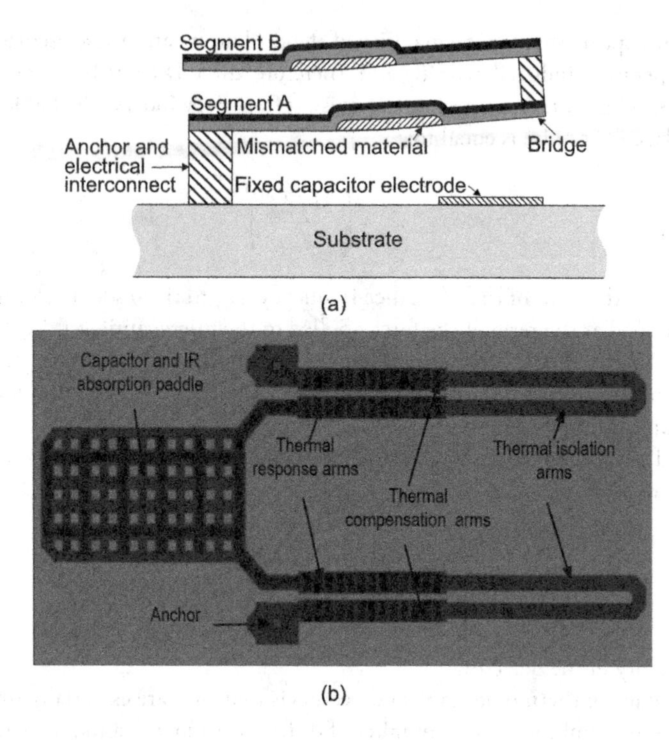

Figure 11.3 Ambient temperature compensation: (a) schematic diagram showing the operating principle, and (b) thermally compensated 25-μm pixel pitch structure elaborated at Multispectral Imaging. (From Ref. [33])

oppositely phased voltage pulses, $\pm V_s$, are applied to the cantilever and bridge reference capacitors, C_s and C_R, respectively, around a reference voltage, V_{Ref}. If the C_s and C_R are the same, the voltage value at the common node between the capacitors is zero. When IR radiation falls on the microcantilever, the paddle moves up, increasing the capacitor gap, thereby decreasing the detector capacitance and generating an offset voltage, V_g, at the input to the gain and integrator circuits.

When the detector temperature increases from T_o to T, the deflection of the microcantilever tip (see Figure 11.3b) is given by [16]:

$$\Delta Z = \frac{3L_p^2}{8t_{bi}}(\alpha_{bi} - \alpha_{\text{subs}})(T - T_o)K_o, \tag{11.3}$$

where L_p is the length of the bimaterial section of the microcantilever detector, α_{bi} and α_{subs} are the bimaterial and substrate material thermal coefficients of expansion (TCE), respectively, t_{bi} is the thickness of the high TCE bimaterial, and the constant K_o is given by:

$$K_o = \frac{8(1+x)}{4+6x+4x^2+nx^3+1/nx}, \tag{11.4}$$

where $x = t_{\text{subs}}/t_{\text{bi}}$ is the ratio of the substrate to bimaterial thicknesses and $n = E_{\text{subs}}/E_{\text{bi}}$ is the ratio of the Young's moduli of the substrate and bimaterial. The last two equations indicate that the maximum microcantilever banding can be obtained using bimaterials with large differences in their thermal coefficients of expansion and optimizing the cantilever geometry.

The voltage responsivity (in VK^{-1}) of the detector is given by

$$R_v = \frac{V_s C_s}{C_T Z_{\text{gap}}} \frac{\Delta Z}{\Delta T}, \tag{11.5}$$

Table 11.3 Modeled NEDT noise values (mK) for different microcantilever pixel structures

PIXEL DIMENSIONS	50 (μm)	25 (μm)	17 (μm)
Background thermal noise	1.2	2.1	3.5
Temperature fluctuation noise	5.2	7.3	10.4
Thermomechanical noise	0.7	0.7	1.0
ROIC noise sources			
1/f+white noise	9.7	7.1	7.4
kTC^{-1} noise	7.0	8.7	15.1
Switching and other correlated noise	Small?	Small?	Small?
Total NEDT	13.1	13.7	19.8

Source: After Ref. [33].

where Z_{gap} is the effective vacuum gap in the sensor and C_T is the sum of total capacitances appearing at the input to the operational amplifier.

For the electrically coupled cantilevers, a parameter called the temperature coefficient of capacitance (TCC) is defined in a way analogous to the temperature coefficient of resistance of the bolometer:

$$\mathrm{TCC} = \frac{1}{C_s}\frac{\Delta C}{\Delta T} = \frac{1}{Z_{gap}}\frac{\Delta Z}{\Delta T}. \tag{11.6}$$

For capacitive sensors, TCC > 30% K^{-1} has been measured [17]. The modeled performance can be much greater, up to 100% K^{-1} depending on the required dynamic range. Thermomechanical noise is comparable to or less than the background thermal conductance noise for a properly tuned and damped sensor array. The major noise contribution for the present devices is in the readout integrated circuit (ROIC) (kTC^{-1} noise, preamplifier, and switching noise). Table 11.3 summarizes the noise equivalent difference temperature (NEDT) values modeled for various noise sources with different pixel structures [33].

A typical difference in thermal expansion coefficients of metal–ceramic bimaterial designs is inherently limited to $\Delta a < 20\times10^{-6}$ K^{-1}. It has recently been suggested that the polymer–ceramic bimaterial cantilevers dramatically enhance thermally induced bending due to much more efficient actuation of readily expandable polymer nanolayers with $\Delta a < 200\times10^{-6}$ K^{-1}, combined with low thermal conductivity [34]. These new composite structures have been introduced with a combination of polymer brush layer, silver nanoparticles, and carbon nanotubes to enhance IR absorption and reinforce nanocomposite coating. This new cantilever design allows the achievement of nearly fourfold improvement in thermal sensitivity compared to the metal-coated counterparts. The serious drawback of the polymer–ceramic cantilevers at the present stage of development is their noncompatibility with traditional microfabrication technology.

Microcantilever IR detectors can have NEDT as small as 5 mK for 50 μm square pixels using silicon nitride for thermal isolation [13]. However, several important issues need to be addressed before the full potential can be realized. Between them we can distinguish: (i) mechanical noise inherent in micromechanical systems, (ii) nonuniformity of microcantilevers in large arrays, and (iii) high sensitivity of thermal IR detectors to the ambient temperature changes.

The effects of mechanical noise can be largely eliminated by tailoring the resonance frequencies and the stiffness of microcantilevers. Figure 11.3b shows the ribbed single pixel structure to enhance the stiffness of the paddle and thermal isolation arms. The transverse corrugated structure on the bimaterial arm is used to reduce delamination issues and increase the bimaterial responsivity. This figure also shows a design of the pixel that is immune to ambient temperature changes and other sources of interfering mechanical stress [35]. The sensor contains a second bimaterial and thermal isolation structure (see also Figure 11.3a). The operation can be described as follows:

- When the ambient temperature rises, Segment A cantilever bends upward
- Segment B cantilever support point follows the movement of Segment A cantilever

- Segment B cantilever tracks lower temperature (except for radiation loading)
- Segment B cantilever bends and keeps capacitor plate at the same distance from the substrate

Both the IR sensing (Segment A) and additional (Segment B) structures respond identically but in opposite sense to the changes in ambient temperature, thus nulling out any substrate temperature-induced motion in the cantilever paddle. In practice, leg Segment B is behind Segment A, rather than above it (see Figure 11.3b [33]). The compensation structure has the additional advantage since it also nulls out any mechanical bending of the bimaterial and isolation arms due to residual stresses created during the detector fabrication process.

The world's first uncooled infrared camera based on silicon MEMS technology was presented in the United States by Sarcon Microsystems, Inc. of Knoxville, Tennessee, in 2003. A schematic plan view of 320×240 pixel arrays is shown in Figure 11.4a. Each cantilever, microengineered with silicon carbide (SiC) on a silicon substrate with integrated electronics, acts as the top plate of the air-gap capacitor and is coated with a thin metal absorber. It is supported by two U-shaped arms, which are anchored at their ends. The inner part of the arms is coated with a bimaterial film, and the outer part provides the electrical interconnect and also thermal isolation between the sensing element and the readout electronics in the substrate. Figure 11.4b shows an evaluation unit for use in a thermal imager.

Sarcon Microsystems encountered technical problems and founded later that year, but in 2005, Multispectral Imaging, Inc. was granted an exclusive, worldwide license from ORNL to commercialize

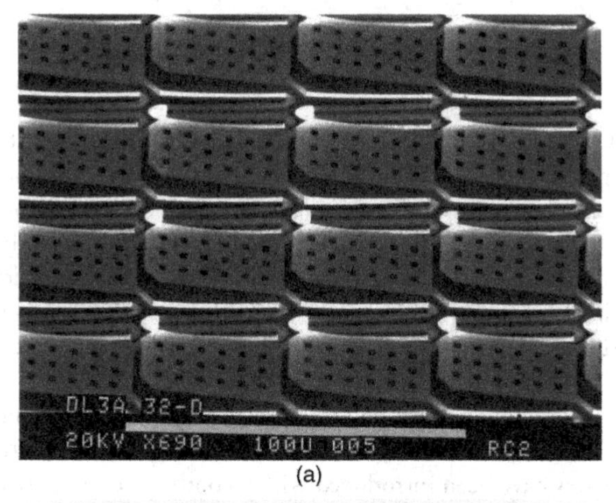

(a)

(b)

Figure 11.4 First 320×240 MEMS-based uncooled IR detector array (Sarcon Microsystems, Inc.): (a) schematic of microcantilevers; (b) customer evaluation unit with MEMS sensor array.

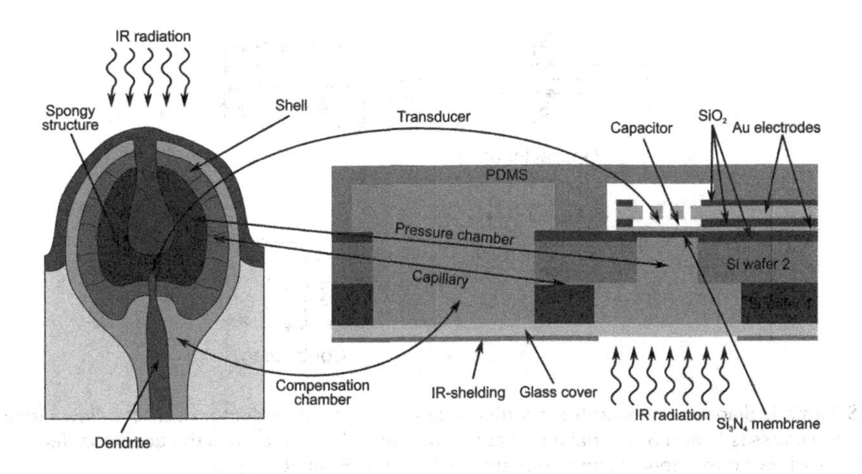

Figure 11.5 Transfer of the biological sensor to the technological sensor. Left: Simplified cross-section of the beetle's sensillum. Right: Cross-section of the technological IR sensor (not to scale) (Adapted after Ref. [36])

the technology. Whilst cantilever-based sensors appear to offer considerable potential, they are unlikely to bring to an end to the search for a high-performance uncooled thermal sensing array.

New ideas are bioinspired. For example, the beetle *Melanophila acuminata* detects forest fires from a distance as far as 50 km away because burned trees provide the environment for their larvae to develop and hatch into adults. The beetle detects IR radiation via a specialized structure known as the pit organ, which contains 60–70 sensors, termed "sensilla." Each sensillum consists of a minute sphere made from the same material as the beetle's outer shell and is connected by nerves to a highly sensitive mechanoreceptive sensory cell. The outer sphere absorbs incoming IR radiation, and absorption causes the sphere to expand. This micromechanical event is measured by the mechanoreceptor and a neuronal signal is generated.

Recently, Siebke et al. [36] have presented the biomimetic approach of an uncooled IR sensor based on the beetle receptors (see Figure 11.5), where a fluid-filled pressure cell expends upon absorbing IR radiation. The expending fluid deflects one electrode of a plate capacitor, causing a change in capacitance. The complete detector system has been incorporated into a silicon chip with an area of a few mm^2 using MEMS technology.

Detailed modeling of the thermomechanical response of the cantilever pixels gives thermal response times in the 5–10 ms range. This theoretical prediction has been confirmed by experimental data [25]. These thermal detectors have slow response times in comparison with photon detectors. However, the micromechanical structures can also be used as photon detectors with faster response times and higher performance than that of micromechanical thermal detectors [37–41].

The absorption of photons by a solid result in temperature changes and lattice thermal expansion that in turn gives rise to acoustic waves at frequencies corresponding to the amplitude modulation of the incident photon radiation. When a silicon microcantilever is exposed to photons, the excess charge carriers generated induce an electronic stress that causes the semiconductor microcantilever to deflect (see Figure 11.6 [38]). Generation of electrons and holes in semiconductors results in the development of a local mechanical strain. The surface stresses S_1 and S_2 are balanced at equilibrium, generating the radial force F_r along the medial plane of the microcantilevers. These stresses become unequal upon exposure to photons, producing the bending force, F_z, which displaces the tip of the microcantilever. The extent of bending is proportional to the radiation intensity.

Results of works published by Datskos et al. [39–41] demonstrated that microstructures represent an important development in MEMS photon detector technology and can be expected to provide the basis for further development. Up until now, however, progress in their development has been small.

11.1.2 OPTICALLY COUPLED CANTILEVERS

The infrared radiation detection and subsequent reconstruction of an image can be also based on the deflection of individual microcantilever pixels using an optical technique, which was adapted from standard atomic force microscope (AFM) imaging systems [5]. With this approach, the array does not require

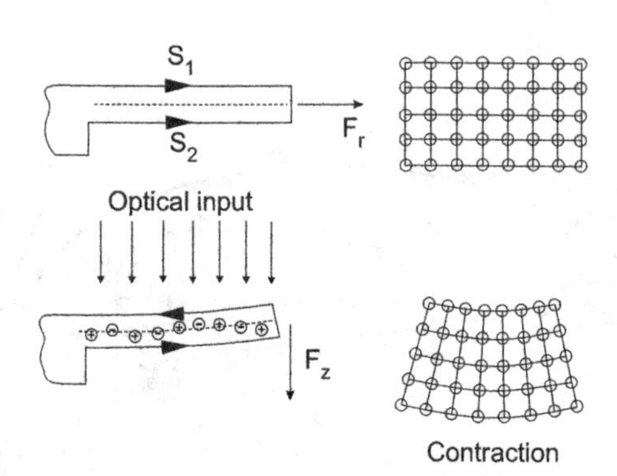

Figure 11.6 Schematic diagram showing the bending process of a semiconductor microcantilever exposed to radiation. Surface stresses S_1 and S_2 are balanced at equilibrium. Also depicted is the accompanied contraction of the silicon lattice following the generation of electron-hole pairs. (From Ref. [38])

metallization to individually address each pixel. In comparison to electrically coupled cantilevers, the optical readout has a number of important advantages [22]:

- The array is simpler to fabricate, enabling reduced cost.
- The need for an integrated ROIC is eliminated.
- The layout complexity of matrix addressing is not required.
- Parasitic heat from ROIC is eliminated.
- The absence of electrical contacts between pixels and substrate eliminates a thermal leakage path.

The most important practical implication of the above approach is, however, related to their straightforward scalability to much larger (>2000×2000) arrays [12].

The responsivities of individual pixels for a particular array of microcantilevers can have slight variations, and in addition, part of pixels can be slightly stressed. As a consequence, the deformations of some of them will not be detectable by readout. Fortunately, recently developed computational algorithms restore images or videos that contain missing or degraded pixel information [12].

Figure 11.7 demonstrates a schematic diagram and components of the optomechanical IR imaging system [42]. It consists of an IR imaging lens, a microcantilever focal plane array (FPA), and an optical readout. Visible light that comes from the light-emitting diode (LED)/laser becomes parallel via the collimating lens. Subsequently, the parallel light is reflected by the pixels of the FPA and then passes through a transforming lens. The reflected diffracting rays synthesize the spectra of the cantilever array on the rear focal plane of the transforming lens. When the incident IR flux is absorbed by the pixels, their temperature rises and then causes a small deflection of the cantilevers. Consequently, the changes in the reflected distribution of visible light are collected and analyzed by a conventional charge-coupled device (CCD) or a complementary metal–oxide–semiconductor (CMOS) camera. The small aperture of the lens mounted on a camera makes it possible to achieve the required angle-to-intensity conversion. This simple optical readout uses 1-mW power of the light beam, while the power per FPA pixel is a few nanowatts. The dynamic range, intrinsic noise, and resolution of the camera largely determine the performance of the system.

In the optomechanical imaging system, the infrared sensor array is physically separated from the readout. The modularity of the optical readout architecture allows for an extra design freedom that is not possible in bolometers, negating fundamental trade-offs, such as NETD versus thermal time constant. Agiltron has reported photmechanical imagers operating with fast frame rates up to 1,000 fps [43].

To minimize fabrication complexity, Datskos and coworkers [12,42] have devised a process flow for the FPA fabrication that involves only three photolithographic steps and relies on well-established methods of surface micromachining. The process starts with double-sided polished Si wafers. To pattern 5-μm tall anchoring posts for suspending the structures, a photoresist mask is etched using SF_6 reactive ion etching. Then, a 6.5-μm thick sacrificial layer of silicon oxide is deposited on the Si surface with posts using plasma-enhanced chemical vapor deposition (PECVD) at 250°C. This step is followed by chemo-mechanical polishing for surface planarization until a 4.5-μm thick oxide layer flush with the posts remains. This sacrificial layer thickness is chosen to form an optimal resonant cavity between the Si substrate and the pixel

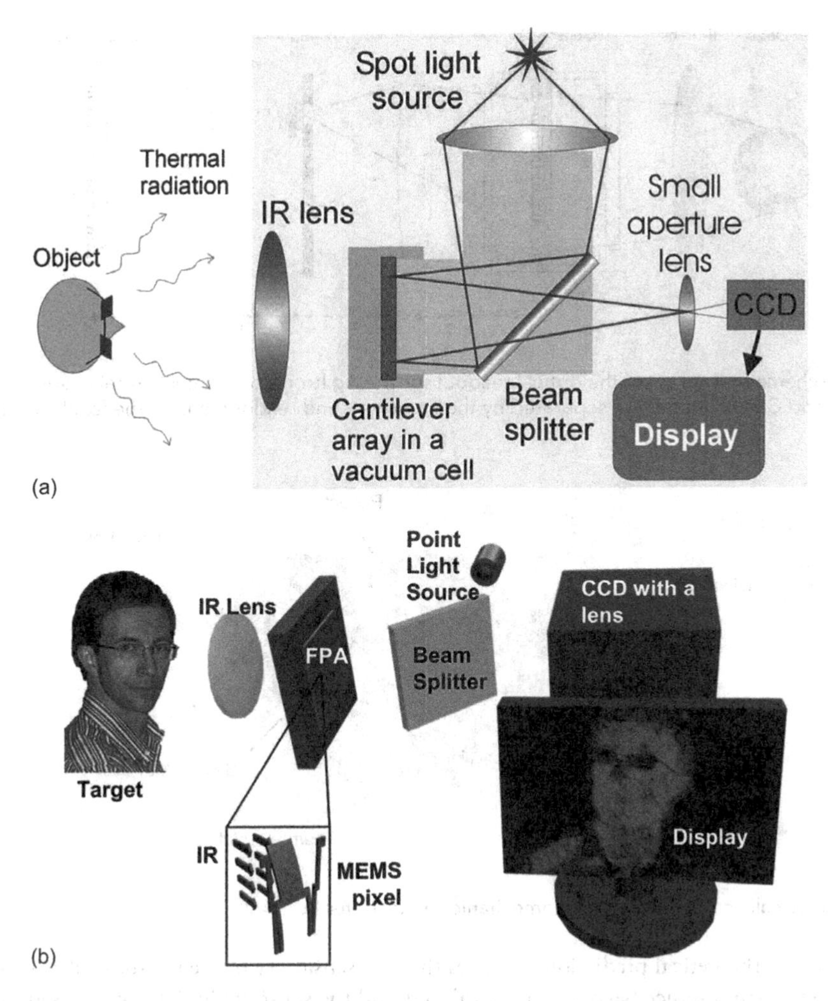

Figure 11.7 Uncooled optical-readable IR imaging system: (a) schematic diagram, and (b) components of the thermal imager. (From Ref. [42])

layer. Next, a 600-nm thick SiN_x layer is deposited on the planarized oxide layer. After this step, the Au metallization is e-beam evaporated on the previously deposited 5-nm thick Cr adhesion layer. The second photolithography involves lift-off patterning of a 120-nm Au layer evaporated on SiN_x, which corresponds to the superposition of the bimaterial leg sections and reflective regions of the pixel heads. Definition of the detector geometry in the SiN_x layer is carried out in the third photolithography. Finally, a sacrificial layer is removed by wet etching in HF followed by rinsing and CO_2 critical point drying.

In the photomechanical imager fabricated by Agiltron, Inc., readout light supplied by an LED illuminates the spatially varying deflected pixels, which are simultaneously projected onto a CMOS imager using a $4f$ optical readout system—see Figure 11.8 [44]. Agiltron has announced the commercialization of this imager for THz spectral region [45].

A photomechanical pixel is illustrated in Figure 11.9. It contains a dielectric absorber layer and metal reflector layer separated by an optical absorption cavity to enhance infrared absorption. Two sets of arms—compensation and sensor arms—connect and thermally isolate the paddle from the substrate. The sensor arm deflects at an angle proportional to the paddle temperature, which equals the substrate temperature plus any heat absorbed from the scene. Instead, the compensation arm deflects at an angle proportional to only the substrate temperature. Because the arms are situated so that the rotation of the compensation arm opposes the rotation of the sensor arm, the net deflection of the paddle is thus proportional to the scene temperature only.

Cantilever-based IR imaging devices demonstrated to date are less sensitive than the theoretical predictions. There are many ways of improving sensitivity including the design and improving the processes and

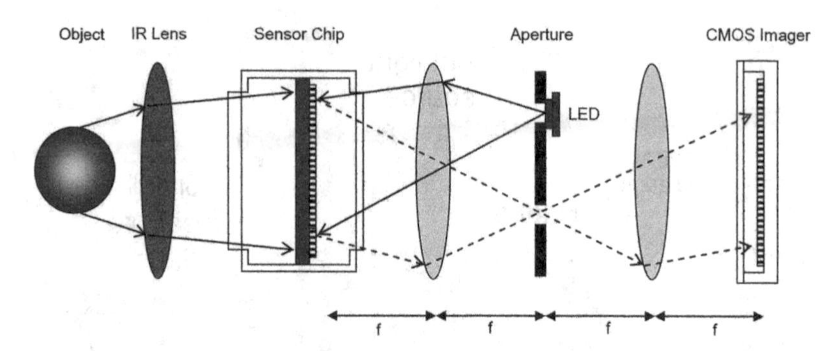

Figure 11.8 A schematic diagram of the optical readout for the Agiltron's photomechanical imager. The sensor chip, aperture, and CMOS imager are separated by the first or second readout lens by the focal length f (after Ref. [44])

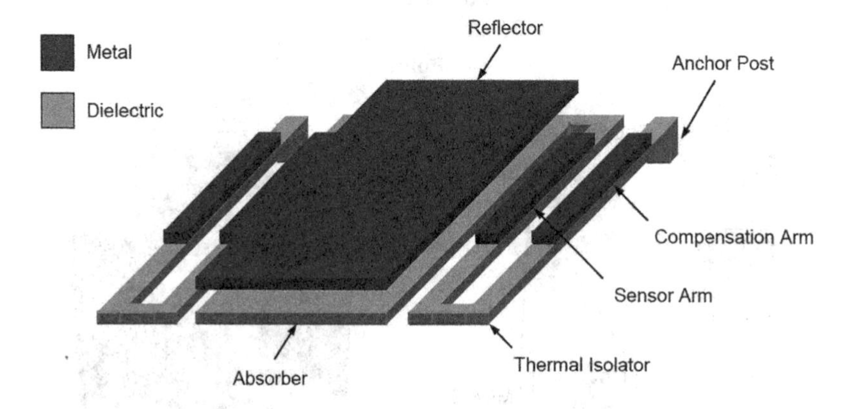

Figure 11.9 A thermally compensated photomechanical pixel (after Ref. [46])

readout system. The theoretical prediction indicates that the sensitivity of the microcantilevers is inversely proportional to the gap distance between the cantilevers and the substrate [47]. Cantilevers are usually anchored to a silicon substrate with a 2–3 μm spacing between them. Small gaps result in high performance; however, small gaps also lead to problems caused by stiction and the sacrificial layer remaining in the released structure. Moreover, the IR flux must transmit through the silicon substrate and only 54% of the incident radiation can reach the cantilevers. As a result of this, new kinds of design structures are explored.

One of the novel designs is a substrate-free uncooled IR detector based on an optical-readable method [25]. The detector is composed of a bimaterial cantilever array, without a silicon substrate, which is eliminated in the fabrication process. An example of this structure is shown in Figure 11.10. The cantilever with a 1-μm thick SiN_x main structure layer incorporates an IR absorber/reflector, two bimaterial arms, and two thermal isolation arms. A thin Au reflection layer and thick Au bimaterial layers were deposited on the IR absorber and the bimaterial arms, respectively. A bulk silicon process that includes Si-glass anodic bonding and deep reactive ion etching was developed to remove the substrate silicon and form frames for every FPA pixel. Spectroscopy analysis has shown that approximately 50% of the incoming IR flux is lost due to the double-side polished substrate in the backside illumination mode [2]. The thermomechanical sensitivity of the cantilever pixel was measured as 0.11 $\mu m K^{-1}$.

The simplicity of the detector's micromachining process is a key factor for a competitive commercial IR detector. Steffanson et al. [48] have presented different designs and different micromachining technologies consisting of only four lithographic steps. The arch-type sensor was produced using low-pressure chemical vapor deposition oxide as a sacrificial layer and nitride as the arch, enabling highly cost-efficient batch operation. Figure 11.11 shows micrograph images of freestanding microcantilever arrays on silicon substrate rails.

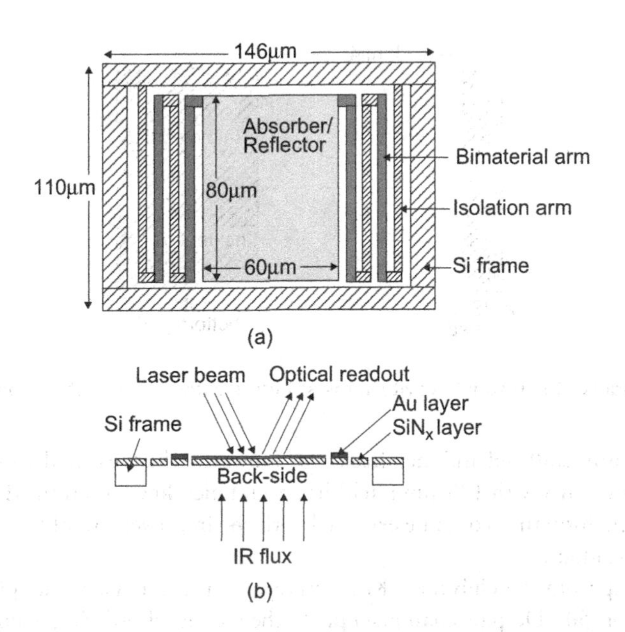

Figure 11.10 Schematic diagram of a cantilever pixel: (a) the top view, (b) the cross-section. (From Ref. [25])

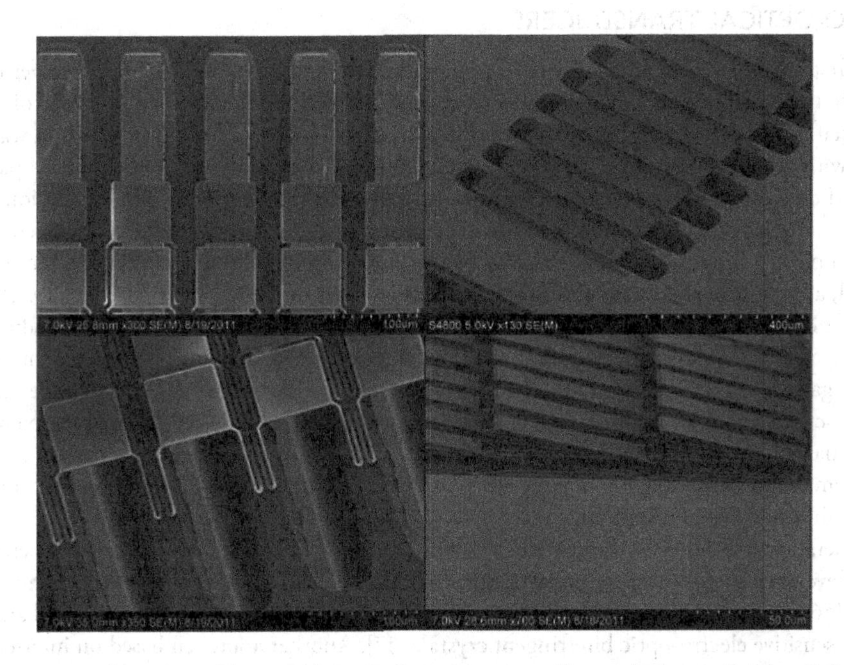

Figure 11.11 Micrograph images of freestanding cantilever arrays on silicon substrate rails (after Ref. [2])

Another modification of substrate-free optical-readable FPA is the introduction of double bimaterial-layer cantilever pixels [49]. The top layer of the cantilever pixels consists of two materials with large mismatching TCE: SiN_x and Au, which convert IR radiation into mechanical deflection (see Figure 11.12). The bottom layer is also SiN_x cantilever, which partially serves as the thermal isolation legs. Such geometry forms the resonant cavity, which together with substrate-free design considerably enhances the absorption of incident IR radiation (IR radiation passes through the bottom SiN_x absorption pad). The theoretical analysis suggests that the temperature resolution of the imaging system can reach 7 mK.

A multifold interval metalized leg configuration is a novel design that has proposed to increase the sensitivity of optical-readable bimaterial microcantilever arrays [24]. This multifold configuration consists of

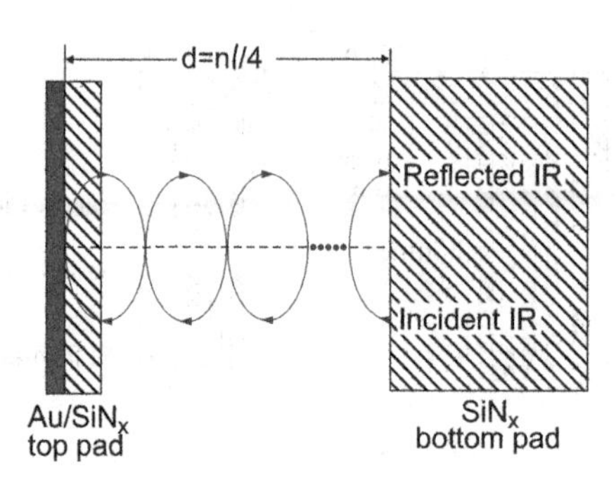

Figure 11.12 Optical model of the resonate cavity in the structure without substrate. (From Ref. [49])

alternatively connected unmetallized and metalized legs. However, the measured sensitivity (NEDT about 400 mK for 160×160 elements with 120-μm pitch) has been much lower than the theoretical value (28.1 mK). Therefore, better performance could be expected with the improvement of fabrication techniques and using less noise optical readout.

More recently, a group from the University Koç, Turkey, has presented a 35-μm pixel pitch thermomechanical MEMS detector [50]. Despite small pixel pitch, the measured NEDT is comparable to the state-of-the-art sensors that are larger in size.

11.1.3 PYRO-OPTICAL TRANSDUCERS

Among the first nonscanned IR imagers were evaporographs and absorption-edge image converters. In an evaporograph, the radiation was focused onto a blackened membrane coated with a thin film of oil [51,52]. The differential rate of evaporation of the oil was proportional to radiation intensity. The film was then illuminated with visible light to produce an interference pattern corresponding to the thermal picture.

The second thermal imaging device was the absorption-edge image converter [53]. Operation of the device was based upon utilizing the temperature dependence of the absorption edge of a semiconductor. When the sample of a suitable material is viewed by transmitted monochromatic light at a wavelength near the threshold, any variations in temperature appear as differences in the transmitted intensity. The image converter built by Harding, Hilsum, and Northrop used amorphous selenium as the semiconductor, with absorption edges of 580–660 nm [53]. The absorption edge shifted by 0.27 nm/°C, equivalent to a change in the energy gap of 9.7×10^{-4} eV/°C [54]. The theory and applications of the absorption-edge image converter were discussed by Hilsum and Harding [55]. They showed that bodies 10°C above ambient temperature can be imaged.

The performance of both imaging devices was poor because of the very long time constant (up to several seconds) and the poor spatial resolution.

A new generation of solid-state thermal devices with optical readout has been proposed recently. Carr and Setiadi developed pyro-optical pixels with phase-change materials whose absorption changed with temperature [56]. Secundo, Lubianiker, and Granat proposed waveguide-based sensitive pixel arrays based on thermally sensitive electro-optic birefringent crystals [57]. Another approach based on interference on a polymer membrane with high TCE has been described by Flusberg et al. [58,59].

Figure 11.13 shows the basic concept of the pyro-optical IR-visible transducer. IR radiation from the scene is imaged on the IR-visible transducer. Each pixel of thermally tunable transducer acts as a wavelength translator, converting IR radiation into visible signals that can be detected by visible sensors (eye, CCD, or CMOS image cameras). Conversion of IR to visible radiation is received by probing the temperature change of the pixels by visible light. To combine both IR and visible radiation, a high-transmitted splitter is used. The transducer shown in Figure 11.13 as reflective in the visible may instead be transmissive. It is analogous to a bolometer, but with optical bias.

An example of a newly designed, optically readable pyro-optical direct-view imager is the RedShift's Thermal Light Valve™ (TLV) [60,61]. The TLV chip is based on a class of optical active thin films developed by Aegis Semiconductor Inc. for telecom applications [62] and is composed of a thermally tunable

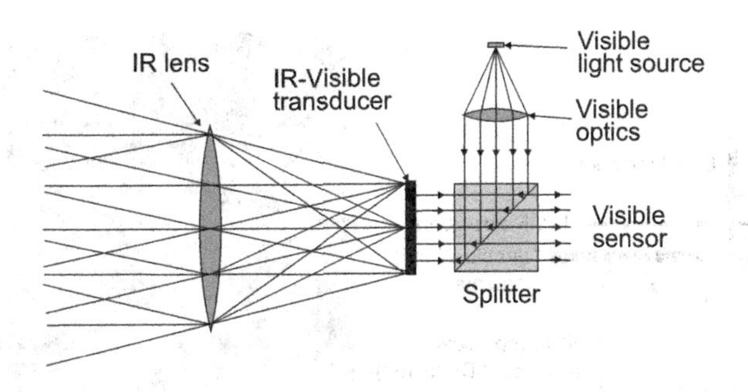

Figure 11.13 Concept of the pyro-optical transducer.

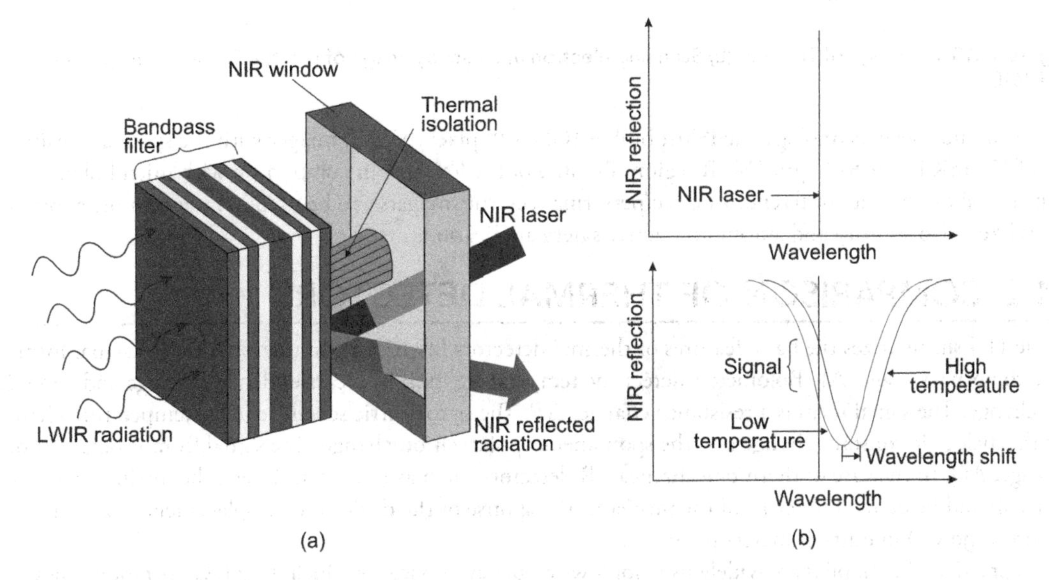

Figure 11.14 RedShift's TLV pixel: (a) filter pixel standing on a post on a substrate and the light paths of IR from scene and NIR probing light, and (b) the "wavelength converter" principle.

Fabry–Perot bandpass filters (pixels) standing on thermally resistive posts on an optically reflective and thermally conductive substrate (see Figure 11.14a). A thermal image is obtained by measuring the pixel-to-pixel variation in reflection of the near-infrared (NIR) probe signal (an 850-nm vertical-cavity surface-emitting laser) using CMOS imagers. Because the NIR probe signal reflected from the TLV depends on the incident IR radiation, the intensity of light received by the CMOS imager is effectively modulated by the IR signature of the observed scene.

The Fabry–Perot structure is based on high-index a-Si and low-index SiN_x thin films, which have been used extensively for many years in solar cells and flat panel displays. These materials are deposited using PECVD. This technique produces uniform, dense materials in high-volume manufacturing environments. The spectral tunability of the filter can be achieved by changing its cavity—a product of thickness and index of refraction. RedShift achieves tunability by changing the index of refraction. The a-Si layers have a refractive index that changes with temperature at a rate of $6\times10^{-5}K^{-1}$ at 300 K (normalized by index).

Figure 11.15 shows the design of TLV and scanning electron microscopy image of tunable 2-D filter array [61]. The array consists of sensor filters that absorb long wavelength IR (LWIR) radiation from the scene and reference filters, both connected to the substrate through thermal isolating anchors. In its base state, the filter array forms a mirror that reflects the NIR probe beam. Due to IR radiation heating, the temperature of a sensor filter diverges from that of the substrate and local reference filters. The phase of the reflected NIR probe beam is then shifted and reflection amplitude over the NIR readout wavelength range increases.

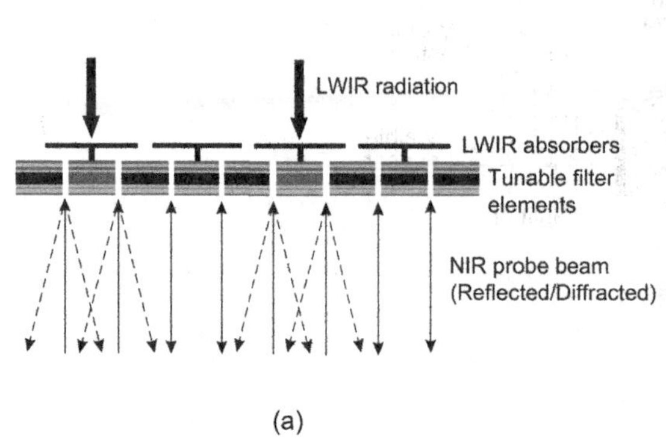

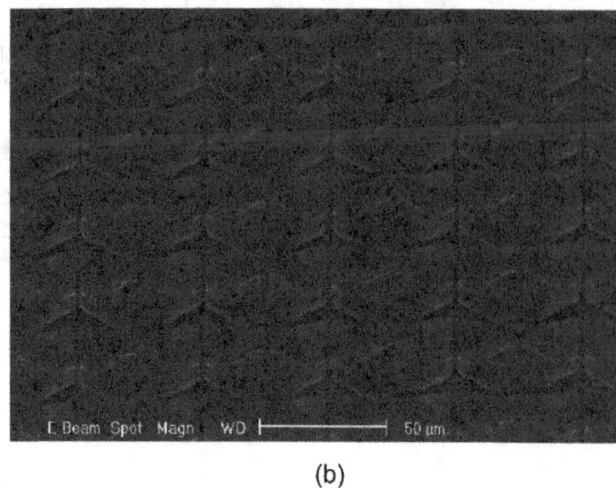

Figure 11.15 (a) Design of TLV, and (b) Scanning electron microscopy image of tunable 2-D filter array. (From Ref. [61])

Using the above technology, RedShift built a 160×120 pixel thermal imager with the spectral sensitivity of 150 mK in the 8–12 μm LWIR region. Because of the low cost in comparison with microbolometer cameras (about 5 times difference in price/performance), the imagers can be used in large-volume applications like video security and automotive active safety applications.

11.2 COMPARISON OF THERMAL DETECTORS

Table 11.4 summarizes the basic features of thermal detectors [29,63,64]. In thermopiles, the signal form is electromotive voltage, ΔV. Bolometers detect the temperature itself by a carrier density change and a mobility change. The signal form is a resistance change, ΔR. The pyroelectric scheme detects temperature change by the dielectric constant change and the spontaneous polarization change. The signal form is polarization change, ΔQ. In the case of thermomechanical IR detectors, such as microcantilevers, the intrinsic responsivity should be defined in terms of the mechanical response of the device (i.e., displacement, ΔZ, per absorbed power) in units of meters per watts.

In practice, thermopiles are widely used for low-frequency applications including DC operation. They experience serious competition from pyroelectric detectors and bolometers that offer better performance at

Table 11.4 Basic parameters of thermal detectors

TYPE OF DETECTOR	SIGNAL VS TEMPERATURE	CHARACTERISTIC PARAMETER	K	ELECTRICAL NOISE POWER DENSITY	BIAS POWER
Thermocouple	$\propto \Delta T$	$\alpha_s = \frac{dV}{dT}$	$\alpha_s = \frac{dV}{dT}$	$4kTR$	no
Bolometer	$\propto T$	$\alpha = \frac{1}{R}\frac{dR}{dT}$	IRa	$4kTR$	I^2R
Pyroelectric	$\sim\propto \frac{dT}{dt}$	$p = \frac{dP}{dT}$	$\frac{pA\omega R}{\left(1+\omega^2\tau^2\right)^{1/2}}$	$\frac{4kTR}{1+\omega^2\tau^2}$	no
Forward-bias diode	$\propto T$	$\alpha = \frac{dV}{dT}$	α	$\frac{4(kT)^2}{2I}$	IV
Microcantilever	$\propto T$	$\mathrm{TCC} = \frac{1}{Z_{gap}}\frac{\Delta Z}{\Delta T}$			

high frequencies. Bolometers can be used in conjunction with optical immersion, which enables a very good performance and a response time of ≈1 ms. Since thermopiles detect the temperature difference between the hot junction and the cold junctions, and since the cold junctions are located on the heat reservoir, the cold junction plays an important role of temperature reference. Therefore, thermopiles do not need an operation stabilizer, while the bolometer does. Since the temperature change at the absorber by the incident IR is much smaller than the operation temperature change, and since it is difficult for the preamplifier to sense the resistance change according to the range for the whole operation temperature change, operation temperature stabilization is often needed for the bolometer. To increase the temperature dependence of resistance and dielectric constant and to realize large responsivity, bolometer and pyroelectric detectors often use a thermoelectric material with a transition point, and the operation temperature should be set near the transition temperature. In this case, operation temperature control is necessary. However, it should be marked that the improvements of readout circuits can eliminate the thermoelectric stabilization of the bolometer.

The thermopile detector has a temperature reference inside so that chopping is not needed. The bolometer also does not need a chopper, because it detects the temperature itself. On the other hand, the pyroelectric detector needs a chopper, because it detects the temperature change. Moreover, the pyroelectric detector cannot be used under a circumstance where the vibration is large, because it is adversely affected by the microphonic noise.

REFERENCES

1. C. M. Hanson, "Barriers to background-limited performance for uncooled IR sensors," *Proc. SPIE* 5406, 454–64, 2004.
2. M. Steffanson and I. W. Rangelow, "Microthermomechanical infrared sensors," *Opto-Electron. Rev.*, 22, 1–15, 2014.
3. P. I. Oden, P. G. Datskos, T. Thundat, and R. J. Warmack, "Uncooled thermal imaging using a piezoresistive microcantilevers," *Appl. Phys. Lett.*, 69, 3277–9, 1996.
4. P. G. Datskos, P. I. Oden, T. Thundat, E. A. Wachter, R. J. Warmack, and S. R. Hunter, "Remote infrared detection using piezoresistive microcantilevers," *Appl. Phys. Lett.*, 69, 2986–8, 1996.
5. E. A. Wachter, T. Thundat, P. I. Oden, R. J. Warmack, P. D. Datskos, and S. L. Sharp, "Remote optical detection using microcantilevers," *Rev. Sci. Instrum.*, 67, 3434–9, 1996.
6. P. I. Oden, E. A. Wachter, P. G. Datskos, T. Thundat, and R. J. Warmack, "Optical and infrared detection using microcantilevers," *Proc. SPIE* 2744, 345–54, 1996.
7. P. G. Datskos, S. Rajic, and I. Datskou, "Photoinduced and thermal stress in silicon microcantilevers," *Appl. Phys. Lett.*, 73, 2319–21, 1998.
8. P. G. Datskos, S. Rajic, and I. Datskou, "Detection of infrared photons using the electronic stress in metal/semiconductor cantilever interfaces," *Ultramicroscopy* 82, 49–56, 2000.
9. L. R. Senesac, J. L. Corbeil, S. Rajic, N. V. Lavrik, and P. G. Datskos, "IR imaging using uncooled microcantilever detectors," *Ultramicroscopy* 97, 451–8, 2003.
10. P. G. Datskos, N. V. Lavrik, and S. Rajic, "Performance of uncooled microcantilever thermal detectors," *Rev. Sci. Instrum.*, 75, 1134–48, 2004.
11. P. Datskos and N. Lavrik, "Uncooled infrared MEMS detectors," in *Smart Sensors and MEMS*, eds. S. Y. Yurish and M. T. Gomes, 381–419, Kluwer Academic, Dordrecht, The Netherlands, 2005.
12. N. Lavrik, R. Archibald, D. Grbovic, S. Rajic, and P. Datskos, "Uncooled MEMS IR imagers with optical readout and image processing," *Proc. SPIE* 6542, 65421E, Dordrecht, The Netherlands, 2007.
13. R. Amantea, C. M. Knoedler, F. P. Pantuso, V. K. Patel, D. J. Sauer, and J. R. Tower, "An uncooled IR imager with 5 mK NEDT," *Proc. SPIE* 3061, 210–22, 1997.
14. R. Amantea, L. A. Goodman, F. Pantuso, D. J. Sauer, M. Varghese, T. S. Villani, and L. K. White, "Progress towards an uncooled IR imager with 5 mK NEDT," *Proc. SPIE* 3436, 647–59, 1998.
15. S. R. Hunter, R. A. Amantea, L. A. Goodman, D. B. Kharas, S. Gershtein, J. R. Matey, S. N. Perna, Y. Yu, N. Maley, and L. K. White, "High sensitivity uncooled microcantilever infrared imaging arrays," *Proc. SPIE* 5074, 469–80, 2003.
16. S. R. Hunter, G. Maurer, L. Jiang, and G. Simelgor, "High sensitivity uncooled microcantilever infrared imaging arrays," *Proc. SPIE* 6206, 62061J, 2006.
17. S. R. Hunter, G. Maurer, G. Simelgor, S. Radhakrishnan, J. Gray, K. Bachir, T. Pennell, M. Bauer, and U. Jagadish, "Development and optimization of microcantilever based IR imaging arrays," *Proc. SPIE* 6940, 694013, 2008.

18. T. Ishizuya, J. Suzuki, K. Akagawa, and T. Kazama, "Optically readable Bi-material infrared detector," *Proc. SPIE* 4369, 342–9, 1998.
19. T. Perazzo, M. Mao, O. Kwon, A. Majumdar, J. B. Varesi, and P. Norton, "Infrared vision using uncooled micro-optomechanical camera," *Appl. Phys. Lett.*, 74, 3567–9, 1999.
20. P. Norton, M. Mao, T. Perazzo, Y. Zhao, O. Kwon, A. Majumdar, and J. Varesi, "Micro-optomechanical infrared receiver with optical readout—MIRROR," *Proc. SPIE* 4028, 72–8, 2000.
21. J. E. Choi, "Design and control of a thermal stabilizing system for a MEMS optomechanical uncooled infrared imaging camera," *Sens. Actuators* A104, 132–42, 2003.
22. J. Zhao, "High sensitivity photomechanical MW-LWIR imaging using an uncooled MEMS microcantilever array and optical readout," *Proc. SPIE* 5783, 506–13, 2005.
23. B. Jiao, C. Li, D. Chen, T. Ye, S. Shi, Y. Qu, L. Dong, et al., "A novel opto-mechanical uncooled infrared detector," *Infrared Phys. Technol.*, 51, 66–72, 2007.
24. F. Dong, Q. Zhang, D. Chen, Z. Miao, Z. Xiong, Z. Guo, C. Li, B. Jiao, and X. Wu, "Uncooled infrared imaging device based on optimized optomechanical micro-cantilever array," *Ultramicroscopy* 108, 579–88, 2008.
25. X. Yu, Y. Yi, S. Ma, M. Liu, X. Liu, L. Dong, and Y. Zhao, "Design and fabrication of a high sensitivity focal plane array for uncooled IR imaging," *J. Micromech. Microeng.*, 18, 057001, 2008.
26. J. R. Barnes, R. J. Stephenson, C. N. Woodburn, S. J. O'Shea, M. E. Welland, J. R. Barnes, R. J. Stephenson, et al., "A femtojoule calorimeter using micromechanical sensors," *Rev. Sci. Instrum.*, 65, 3793–8, 1994.
27. J. Varesi, J. Lai, T. Perazzo, Z. Shi, and A. Majumdar, "Photothermal measurements at picowatt resolution using uncooled micro-optomechanical sensors," *Appl. Phys. Lett.*, 71, 306–8, 1997.
28. J. R. Barnes, R. J. Stephenson, C. N. Woodburn, S. J. O'Shea, M. E. Welland, T. Rayment, J. K. Gimzewski, et al., "A femtojoule calorimeter using micromechanical sensors," *Rev. Sci. Instrum.*, 65, 3793–8, 1994.
29. P. G. Datskos, "Detectors: Figures of Merit," in *Encyclopedia of Optical Engineering*, ed. R. Driggers, 349–57, Marcel Dekker, New York, 2003.
30. D. Sarid, *Scanning Force Microscopy*, Oxford University Press, New York, 1991.
31. E. Majorana and Y. Ogawa, "Mechanical noise in coupled oscillators," *Phys. Lett.*, A233, 162–8, 1997.
32. T. W. Kenny, J. K. Reynolds, J. A. Podosek, E. C. Vote, L. M. Miller, H. K. Rockstad, and W. J. Kaiser, "Micromachined infrared sensors using tunneling displacement transducers," *Rev. Sci. Instrum.*, 67, 112–8, 1996.
33. S. R. Hunter, G. S. Maurer, G. Simelgor, S. Radhakrishnan, and J. Gray, "High sensitivity 25μm and 50μm pitch microcantilever IR imaging arrays," *Proc. SPIE* 6542, 65421F, 2007.
34. Y. H. Lin, M. E. McConney, M. C. LeMieux, S. Peleshanko, C. Jiang, S. Singamaneni, and V. V. Tsukurk, "Trilayered ceramic-metal-polymer microcantilevers with dramatically enhanced thermal sensitivity," *Adv. Mater.*, 18, 1157–61, 2006.
35. J. L. Corbeil, N. V. Lavrik, S. Rajic, and P. G. Datskos, "'Self-leveling' uncooled microcantilever thermal detector," *Appl. Phys. Lett.*, 81, 1306–8, 2002.
36. G. Siebke, K. Gerngroβ, P. Holik, S. Schmitz, M. Rohloff, S. Tätzner, and S. Steltenkamp, "An uncooled capacitive sensor for IR detection," *Proc. SPIE* 9070, 90701W-1–10, 2014.
37. P. G. Datskos, S. Rajic, and I. Datskou, "Photo-induced stress in silicon microcantilevers," *Appl. Phys. Lett.*, 73, 2319–21, 1998.
38. P. D. Datskos, S. Rajic, I. Datskos, and C. M. Eger, "Novel photon detection based electronically-induced stress in silicon," *Proc. SPIE* 3379, 173–81, 1998.
39. P. G. Datskos, "Micromechanical uncooled photon detectors," *Proc. SPIE* 3948, 80–93, 2000.
40. P. G. Datskos, S. Rajic, and I. Datskou, "Detection of infrared photons using the electronic stress in metal-semiconductor cantilever interfaces," *Ultramicroscopy* 82, 49–56, 2000.
41. P. G. Datskos, S. Rajic, L. R. Senesac, and I. Datskou, "Fabrication of quantum well microcantilever photon detectors," *Ultramicroscopy* 86, 191–206, 2001.
42. P. Datskos and N. Lavrik, "Simple thermal imagers use scalable micromechanical arrays," *SPIE Newsroom* 10.1117/2.1200608.036, 2006.
43. L. Zhang, F. P. Pantuso, G. Jin, A. Mazurenko, M. Erdtmann, S. Radhakrishnan, and J. Salerno, "High-speed uncooled MWIR hostile fire indicator," *Proc. SPIE* 8012, 801219, 2011.
44. M. Erdtmann, L. Zhang, G. Jin, S. Radhakrishnan, G. Simelgor, and J. Salerno, "Optical readout photomechanical imager: from design to implementation", *Proc. SPIE* 7298, 72980I, 2009.
45. M. Erdtmann, L. Zhang, S. Radhakrishnan, S. Wu, T. M. Goyette, and A. J. Gatesman, "Uncooled photomechanical terahertz imagers", *Proc. SPIE* 8363, 83630C, 2012.
46. M. Erdtmann, G. Simelgor, S. Radhakrishnan, L. Zhang, Y. Liu, P. Y. Emelie, and J. Salerno, "Photomechanical imager FPA design for manufacturability", *Proc. SPIE* 7660, 766017, 2010.

47. B. Li, "Design and simulation of an uncooled double-cantilever microbolometer with the potential for ~mK NETD," *Sens. Actuators* A112, 351–9, 2004.
48. M. Steffanson, T. Ivanov, and I. W. Rangelow, "Methodology for micro-fabricating free standing micro-mechanical structures for infrared detection", *Proc. IR2*, 105–9, 2013.
49. S. Shi, D. Chen, B. Jiao, C. Li, Y. Qu, Y. Jing, T. Ye, et al., "Design of a novel substrate-free double-layer-cantilever FPA applied for uncooled optical-readable infrared imaging system," *IEEE Sens. J.*, 7, 1703–10, 2007.
50. U. Adiyan, F. Civitci, O. Ferhanoglu, H. Torun, and H. Urey, "A 35-μm pitch IR thermo-mechanical MEMS sensor with AC-coupled optical readout", *IEEE J. Selected Topics in Quantum Electronics*, 21(4), 2701306, 2015.
51. P. W. Kruse, L. D. McGlauchlin, and R. B. McQuistan, *Elements of Infrared Technology*, Wiley, New York, 1962.
52. G. W. McDaniel and D. Z. Robinson, "Thermal imaging by means of the evaporograph," *Appl. Opt.*, 1, 311–24, 1962.
53. W. R. Harding, C. Hilsum, and D. C. Northrop, "A new thermal image-converter," *Nature* 181, 691–2, 1958.
54. C. Hilsum, "The absorption edge of amorphous selenium and its change with temperature," *Proc. Phys. Soc.*, B69, 506–12, 1956.
55. C. Hilsum and W. R. Harding, "The theory of thermal imaging, and its application to the absorption-edge image tube," *Infrared Phys.*, 1, 67–93, 1961.
56. W. Carr and D. Setiadi, "Micromachined pyro-optical structure," U.S. Patent No. 6,770,882.
57. L. Secundo, Y. Lubianiker, and A. J. Granat, "Uncooled FPA with optical reading: Reaching the theoretical limit," *Proc. SPIE* 5783, 483–95, 2005.
58. A. Flusberg and S. Deliwala, "Highly sensitive infrared imager with direct optical readout," *Proc. SPIE* 6206, 62061E, 2006.
59. A. Flusberg, S. Swartz, M. Huff, and S. Gross, "Thermal-to-visible transducer (TVT) for thermal-IR imaging," *Proc. SPIE* 6940, 694015, 2008.
60. M. Wagner, E. Ma, J. Heanue, and S. Wu, "Solid state optical thermal imagers," *Proc. SPIE* 6542, 65421P, 2007.
61. M. Wagner, "Solid state optical thermal imaging: Performance update," *Proc. SPIE* 6940, 694016, 2008.
62. M. Wu, J. Cook, R. DeVito, J. Li, E. Ma, R. Murano, N. Nemchuk, M. Tabasky, and M. Wagner, "Novel low-cost uncooled infrared camera," *Proc. SPIE* 5783, 496–505, 2005.
63. J. T. Houghton and S. D. Smith, *Infra-Red Physics*, Oxford University Press, Oxford, UK, 1966.
64. J. Piotrowski, "Breakthrough in infrared technology—The micromachined thermal detector arrays," *Opto-Electron. Rev.*, 3, 3–8, 1995.

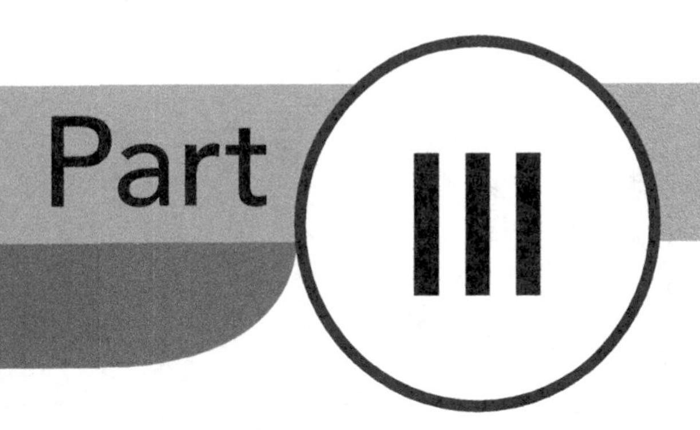

Infrared photon detectors

12 Theory of photon detectors

The interaction of infrared (IR) radiation with electrons results in several photoeffects such as photoconductive, photovoltaic, photoelectromagnetic (PEM), Dember, and photon drag. On the basis of these photoeffects, different types of detectors have been of interest, but only photoconductive and photovoltaic (p-n junction and Schottky barrier) detectors have been widely exploited.

Photoeffects, which occur in structures with built-in potential barriers, are essentially photovoltaic and result when excess carriers are injected optically into the vicinity of such barriers. The role of the built-in electric field is to cause the charge carriers of opposite sign to move in opposite directions depending upon the external circuit. Several structures are possible for observing the photovoltaic effect. These include p-n junctions, heterojunctions, Schottky barriers, and metal–insulator–semiconductor (MIS) photocapacitors. Each of these different types of devices has certain advantages for IR detection, depending on the particular applications. Recently, more interest has been focused on p-n junction photodiodes for use with silicon hybrid focal plane arrays (FPAs) for direct detection in the 3–5 and 8–14 μm spectral regions. In this application, photodiodes are preferred over photoconductors because of their relatively high impedance, matched directly into the input stage of a silicon readout, and lower power dissipation. Furthermore, the photodiodes have a faster response than photoconductors because the strong field in the depletion region imparts a large velocity to the photogenerated carriers. Also, photodiodes are not affected by many of the trapping effects associated with photoconductors.

The basic theory of different types of photon detectors will be presented in this chapter in a uniform structure convenient for the various detector materials.

12.1 PHOTOCONDUCTIVE DETECTORS

A number of excellent treatises and papers have been published on photoconductive detectors [1–13]. Many of them considered HgCdTe photoconductors, because in the last five decades the work in this area has been devoted almost exclusively to these detectors. Our purpose is to present an up-to-date description of the theory and principles of photoconductors in a form most suitable for design and applications.

12.1.1 INTRINSIC PHOTOCONDUCTIVITY THEORY

The photoconductive detector is essentially a radiation-sensitive resistor. The operation of a photoconductor is shown in Figure 12.1. A photon of energy $h\nu$ greater than the band-gap energy E_g is absorbed to produce electron hole pairs, thereby changing the electrical conductivity of the semiconductor. For direct narrow gap semiconductors, the optical absorption is very much higher than that in extrinsic detectors.

In almost all cases, the change in conductivity is measured by means of electrodes attached to the sample. For low resistance material, where the sample resistance is typically 100 Ω, the photoconductor is usually operated in a constant current circuit as shown in Figure 12.1. The series load resistance is large compared to the sample resistance, and the signal is detected as a change in voltage developed across the sample. For high-resistance photoconductors, a constant voltage circuit is preferred and the signal is detected as a change in current in the bias circuit.

We assume that the signal photon flux density $\Phi_s(\lambda)$ is incident on the detector area $A = wl$ and that the detector is operated under constant current conditions (i.e., $R_L >> R$). We suppose further that the illumination and the bias field are weak, and the excess carrier lifetime τ is the same for majority and minority carriers. To derive an expression for voltage responsivity, we take a one-dimensional approach for simplicity. This is justified for a detector thickness t that is small with respect to minority carrier diffusion length.

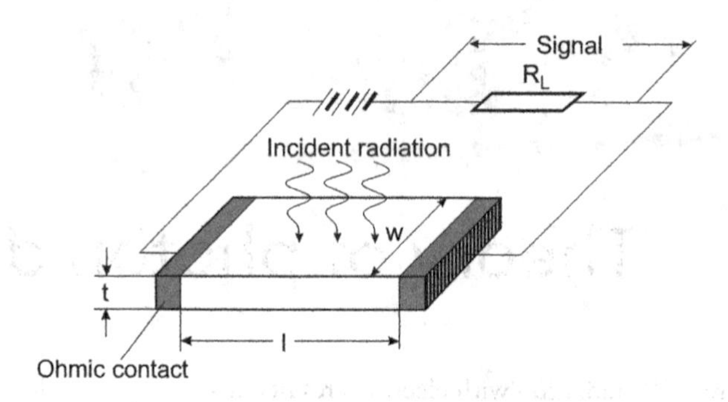

Figure 12.1 Geometry and bias of a photoconductor.

We also neglect the effect of recombination at the front and rear surfaces. Initially, we will consider simple photoconductivity effects due to the influence of bulk material properties.

The basic expression describing either intrinsic or extrinsic photoconductivity in semiconductors under equilibrium excitation (i.e., steady state) is

$$I_{ph} = q\eta A\Phi_s g, \tag{12.1}$$

where I_{ph} is the short circuit photocurrent at zero frequency (DC); that is, the increase in current above the dark current accompanying irradiation. The photoconductive gain, g, is determined by the properties of the detector (i.e., by which detection effect is used and the material and configuration of the detector).

In general, photoconductivity is a two-carrier phenomenon and the total photocurrent of electrons and holes is

$$I_{ph} = \frac{qwt\left(\Delta n\mu_e + \Delta p\mu_h\right)V_b}{l}, \tag{12.2}$$

where μ_e is the electron mobility, μ_h is the hole mobility; V_b is the bias voltage, and

$$n = n_o + \Delta n; \quad p = p_o + \Delta p, \tag{12.3}$$

n_o and p_o are the average thermal equilibrium carrier densities, and Δn and Δp are the excess carrier concentrations.

Taking the conductivity to be dominated by electrons (this is found to be the case in all known high-sensitivity photoconductors) and assuming uniform and complete absorption of the light in the detector, the rate equation for the excess electron concentration in the sample is [14]

$$\frac{d\Delta n}{dt} = \frac{\Phi_s\eta}{t} - \frac{\Delta n}{\tau}, \tag{12.4}$$

where τ is the excess carrier lifetime. In the steady condition, the excess carrier lifetime is given by the equation

$$\tau = \frac{\Delta nt}{\eta\Phi_s}. \tag{12.5}$$

Equating Equations 12.1 to 12.2 gives

$$g = \frac{tV_b\mu_e\Delta n}{l^2\eta\Phi_s}, \tag{12.6}$$

and invoking Equation 12.5, we get for the photoconductive gain

$$g = \frac{\tau\mu_e V_b}{l^2} = \frac{\tau}{l^2/\mu_e V_b}. \tag{12.7}$$

So, the photoconductive gain can be defined as

$$g = \frac{\tau}{t_t}, \tag{12.8}$$

where t_t is the transit time of electrons between ohmic contacts. This means that the photoconductive gain is given by the ratio of free carrier lifetime, τ, to transit time, t_t, between the sample electrodes. The photoconductive gain can be less than or greater than unity depending upon whether the drift length, $L_d = v_d\tau$, is less than or greater than interelectrode spacing, l. The value of $L_d > l$ implies that a free charge carrier swept out at one electrode is immediately replaced by injection of an equivalent free charge carrier at the opposite electrode. Thus, a free charge carrier will continue to circulate until recombination takes place.

When $R_L >> R$, a signal voltage across the load resistor is essentially the open circuit voltage

$$V_s = I_{ph}R_d = I_{ph}\frac{l}{qwtn\mu_e}, \tag{12.9}$$

where R_d is the detector resistance. Assuming that the change in conductivity upon irradiation is small compared to the dark conductivity, the voltage responsivity is expressed as

$$R_v = \frac{V_s}{P_\lambda} = \frac{\eta}{lwt}\frac{\lambda\tau}{hc}\frac{V_b}{n_o}, \tag{12.10}$$

where the absorbed monochromatic power $P_\lambda = \Phi_s Ahv$.

The expression (Equation 12.10) shows clearly the basic requirements for high photoconductive responsivity at a given wavelength λ: one must have high quantum efficiency η, long excess-carrier lifetime τ, the smallest possible piece of crystal, low thermal equilibrium carrier concentrations n_o, and the highest possible bias voltage V_b.

The frequency-dependent responsivity can be determined by the equation

$$R_v = \frac{\eta}{lwt}\frac{\lambda\tau_{ef}}{hc}\frac{V_b}{n_o}\frac{1}{\left(1+\omega^2\tau_{ef}^2\right)^{1/2}}, \tag{12.11}$$

where τ_{ef} is the effective carrier lifetime.

The given simple model takes no account of additional limitations related to the practical conditions of photoconductor operations such as sweep-out effects or surface recombination. These are specified in the next sections.

12.1.1.1 Sweep-out effects

Equation 12.11 shows that voltage responsivity increases monotonically with the increase of bias voltage. However, there are two limits on applied bias voltage, namely, thermal conditions (Joule heating of the detector element) and sweep-out of minority carriers. The thermal conductance of the detector depends on the device fabrication procedure. The trend to smaller element dimensions (typically, e.g., 50 × 50 μm²) is conditioned by the extension of photoconductor technology to two-dimensional close-packed arrays for thermal imaging. If the excess carrier lifetime is long (usually exceeds 1 μs in 8–14 μm devices at 77 K and 10 μs in 3–5 μm devices at higher temperatures), we cannot ignore the effects of contacts and of drift and diffusion on the device performance. At moderate bias fields, minority carriers can drift to the ohmic contacts in a short time compared to the recombination time in the material. Removal of carriers at an ohmic contact in this way is referred to as sweep-out [15,16]. Minority carrier sweep-out limits the maximum applied voltage of V_b. The effective carrier lifetime can be reduced considerably in detectors where the minority carrier diffusion length exceeds the detector length (even at very low bias voltages) [17–21]. At low bias, the average drift length of the minority carriers is very much less than the detector length l, and the minority carrier lifetime is determined by the bulk recombination modified by diffusion to surface and contacts. The carrier densities are uniform along the length of the detector. At higher values of the applied field, the drift length of the minority carriers is comparable to or greater than l. Some of the excess minority carriers are lost at an electrode, and to maintain space charge equilibrium, a drop in excess majority carrier density is necessary. This way, the majority carrier lifetime is reduced. It should be pointed out that the loss of the majority carriers at one ohmic contact is replenished by injection at the other, but minority carriers are not replaced. At high bias, the excess carrier density is nonuniformly distributed along the length of the sample.

We follow Rittner and derive the optically generated excess minority carrier concentration under sweep-out conditions [15]. The excess carrier concentration $\Delta p\ (x, t) = p(x, t) - p_o$ within the semiconductor is governed by ambipolar transport. The ambipolar continuity equation for a one-dimensional case under steady state and electrical neutrality conditions may be written as

$$\frac{\partial^2(\Delta p)}{\partial x^2} + \frac{L_d}{L_D^2}\frac{\partial(\Delta p)}{\partial x} + \frac{\Delta p}{L_D^2} + G_s = 0, \quad (12.12)$$

where

$L_d = \tau\mu_a E$ drift length,

$L_D = (D_D\tau)^{1/2}$ diffusion length,

$\mu_a = \dfrac{(n_o - p_o)\mu_e\mu_h}{n_o\mu_e + p_o\mu_h}$ ambipolar drift mobility,

$D_D = \dfrac{D_e p_o\mu_h + D_h n_o\mu_e}{n_o\mu_e + p_o\mu_h}$ ambipolar diffusion coefficient.

Other marks have their usual meanings: $D_{e,h} = (kT/q)\mu_{e,h}$ are the respective carrier diffusion coefficients, G_s is the signal generation rate, $E = V_b/l$ is the bias electrical field, and k is the Boltzmann constant.

The major assumption in the Rittner model relates to the boundary conditions at the metal–semiconductor interface, at $x = 0$ and $x = l$. In this model, one assumes that this interface is characterized by infinite recombination velocity, which means that the photoconductor contacts are completely ohmic. The appropriate boundary conditions are

$$\Delta p(0) = \Delta p(l) = 0. \quad (12.13)$$

The solution to Equation 12.12 is

$$\Delta p = G_s\tau\left[1 - C_1\exp(\alpha_1 x) + C_2\exp(\alpha_2 x)\right], \quad (12.14)$$

where

$$\alpha_{1,2} = \frac{1}{2L_D^2}\left[-L_d \pm \left(L_d^2 + 4L_D^2\right)^{1/2}\right]. \quad (12.15)$$

Taking into account the boundary conditions (Equation 12.13), we have

$$C_1 = \frac{1 - \exp(\alpha_2 l)}{\exp(\alpha_2 l) - \exp(\alpha_1 l)};\ C_2 = \frac{1 - \exp(\alpha_1 l)}{\exp(\alpha_2 l) - \exp(\alpha_1 l)}. \quad (12.16)$$

The total number of carriers contributing to photoconductivity is obtained by integrating Equation 12.14 over the length of the samples:

$$\Delta P = wt\int_0^l \Delta p(x)\,dx.$$

Note that the signal generation rate G_s is related to the total signal flux Φ_s and the quantum efficiency η, by the expression $G_s = \eta\Phi_s/t$. Then

$$\Delta P = \eta\Phi_s\tau w\int_0^l \left[1 + C_1\exp(\alpha_1 x) + C_2\exp(\alpha_2 x)\right]dx. \tag{12.17}$$

Alternatively, we can write

$$\Delta P = \eta\Phi_s\tau_{ef} \quad \text{where} \quad \tau_{ef} = \gamma\tau \tag{12.18}$$

It can be proved that

$$\gamma = 1 + \frac{(\alpha_1 - \alpha_2)th(\alpha_1 l/2)th(\alpha_2/2)}{\alpha_1\alpha_2(l/2)\left[th(\alpha_2 l/2) - th(\alpha_1 l/2)\right]}. \tag{12.19}$$

In this situation, the voltage responsivity is [3]

$$R_v = \frac{\eta}{lwt}\frac{\lambda\tau_{ef}}{hc}\frac{V_b(b+1)}{bn+p}\frac{1}{\left(1+\omega^2\tau_{ef}^2\right)^{1/2}}, \tag{12.20}$$

where $b = \mu_e/\mu_h$, and at the low-frequency modulation ($\omega\tau_{ef} \ll 1$)

$$R_v = \frac{\eta}{lwt}\frac{\lambda\tau_{ef}}{hc}\frac{V_b(b+1)}{bn+p}. \tag{12.21}$$

We obtain similar formulas as previously seen (Equation 12.11), except that the carrier lifetime τ is replaced by τ_{ef}. Because $\gamma \le 1$, so always $\tau_{ef} \le \tau$. The lifetime degradation problem associated with ohmic contacts can be eliminated by the use of an overlap structure [17,18], heterojunction contact [19,22,23], or highly doped contact [18,19,23].

Practically, the contacts are characterized by a recombination velocity that can be varied from infinity (ohmic contacts) to zero (perfectly blocking contacts). In the latter case, a more intensely doped region at the contact (e.g., n^+ for n-type devices) causes a built-in electric field that repels minority carriers, thereby reducing recombination and increasing the effective lifetime and responsivity. More sophisticated blocking contacts and their influence on the performance of intrinsic photoconductors have been considered by Kumar et al. [24,25]. The experimental results show that contact recombination velocities as low as a few hundred cm/s can be achieved [19,23,26].

In general, the electric field distribution in photoconductors is not homogeneous. In this instance, these structures cannot be adequately described by analytical methods and require a numerical solution. Numerical techniques have been used to solve the carrier transport equations for several device configurations [21]. Usually, the Van Roosbroeck model is used [27] (see Section 4.4).

Analysis of the influence of the sweep-out effect on the photoconductor performance has been carried out by Elliott et al. [13,20]. It appears that the formulas (Equations 12.20 and 12.21) given are quite generally applicable, provided that n and p are replaced by somewhat different values n' and p' under high bias conditions. The values of n' and p' depend on the source of the minority carriers and on the nature of the minority carrier injecting contact. For general cases, the responsivity can be written as:

$$R_v = \frac{\eta}{lwt}\frac{\lambda\tau_{ef}}{hc}\frac{V_b(b+1)}{bn'+p'}, \tag{12.22}$$

where

$$\tau_{ef} = \tau\left[1-\frac{\tau}{\tau_a}\left\{1-\exp\left(-\frac{\tau_a}{\tau}\right)\right\}\right],$$

and $\tau_a = l/\mu_a E$, the time for a minority carrier to drift the sample length.

Under very high bias conditions, the voltage responsivity saturates to the value [20]

$$R_v = \frac{\eta q\lambda}{2hc}(1+b)\frac{\mu_h}{\mu_a}R', \tag{12.23}$$

where R' is the device resistance; $\mu_a = \mu_h$ in the n-type material, and $\mu_a = \mu_e$ in the p-type material.

12.1.1.2 Noise mechanisms in photoconductors

All detectors are limited in the minimum radiant power that they can detect by some form of noise that may arise in the detector itself, in the radiant energy to which the detector responds, or in the electronic system following the detector. Careful electronic design including that of low noise amplification can reduce system noise below that in the output of the detector. That topic will not be included here.

We can distinguish two groups of noise—the radiation noise and the noise internal to the detector. The radiation noise includes signal fluctuation noise and background fluctuation noise [28,29]. Under most operating conditions, the background fluctuation limit discussed in Section 3.5 is operative for IR detectors, whereas the signal fluctuation limit is operative for ultraviolet and visible detectors.

The random processes occurring in semiconductors give rise to internal noise in detectors even in the absence of illumination. There are two fundamental processes responsible for the noise: fluctuations in the velocities of free carriers due to their random thermal motion, and fluctuations in the densities of free carriers due to randomness in the rates of thermal generation and recombination [30].

The photon noise voltage can be calculated according to the theory presented in Van der Ziel's monograph [30]:

$$V_{ph} = \frac{2\pi^{1/2}V_b}{(lw)^{1/2}t}\frac{1+b}{bn+p}\int_{\nu_o}^{\infty}\frac{\eta(\nu)\nu^2\exp(h\nu/kT_B)d\nu}{c^2\left[\exp(h\nu/kT_B)-1\right]^2}\frac{\tau(\Delta f)^{1/2}}{(1+\omega^2\tau^2)^{1/2}}, \tag{12.24}$$

where T_B is the temperature of the background and ν_o is the frequency corresponding to the long-wavelength limit of the detector, λ_c.

A number of internal noise sources are usually operative in photoconductive detectors. The fundamental types are Johnson–Nyquist (sometimes called thermal) noise and generation–recombination (g–r) noise. The third form of noise, not amenable to exact analysis, is called $1/f$ noise because it exhibits a $1/f$ power law spectrum to a close approximation.

The total noise voltage of a photoconductor is

$$V_n^2 = V_{gr}^2 + V_J^2 + V_{1/f}^2. \tag{12.25}$$

Johnson–Nyquist noise is associated with the finite resistance R of the device. This type of noise is due to the random thermal motion of charge carriers in the crystal and not due to fluctuations in the total number of these charge carriers. It occurs in the absence of external bias as a fluctuating voltage or current depending upon the method of measurement. Small changes in the voltage or current at the terminals of the device are due to the random arrival of charge at the terminals. The root mean square of Johnson–Nyquist noise voltage in the bandwidth Δf is given by Equation 4.16. This type of noise has a "white" frequency distribution.

At finite bias currents, the carrier density fluctuations cause resistance variations, which are observed as noise exceeding Johnson–Nyquist noise. This type of excess noise in photoconductive detectors is referred to as g–r noise. The g–r noise is due to the random generation of free charge carriers by the crystal vibrations and their subsequent random recombination. Because of the randomness of the generation and recombination processes, it is unlikely that there will be exactly the same number of charge carriers in the free state at succeeding instances of time. This leads to conductivity changes that will be reflected as fluctuations in current flow through the crystal.

The (g–r) noise voltage for equilibrium conditions is equal

$$V_{gr}^2 = 2(G+R)lwt(Rqg)^2\Delta f, \tag{12.26}$$

where G and R in the first bracket are the volume generation and recombination rates.

Many forms of g–r noise expression exist, depending upon the internal properties of the semiconductors. The expression for noise in a near-intrinsic photoconductor has been given by Long [31]:

$$V_{gr} = \frac{2V_b}{(lwt)^{1/2}}\frac{1+b}{bn+p}\left(\frac{np}{n+p}\right)^{1/2}\left(\frac{\tau\Delta f}{1+\omega^2\tau^2}\right)^{1/2}. \tag{12.27}$$

Generation–recombination noise usually dominates the noise spectrum of photoconductors at intermediate frequencies. It should be noted that in the high-bias regime, the expressions for *g–r* noise are different from those at low bias [20].

The rms *g–r* noise current for an extrinsic n-type photoconductor with carrier lifetime τ can be written as [30]

$$I_{gr}^2 = \frac{4I^2\overline{\Delta N^2}\tau\Delta f}{N^2\left(1+\omega^2\tau^2\right)}, \tag{12.28}$$

where N is the number of carriers in the detector. Usually, in an extrinsic semiconductor, there will be some counter doping (i.e., electrons trapped at deep-lying levels). If the number of deep traps is small compared to the number of electrons (electrons being the majority carriers), then the variance ΔN^2 is equal to N [30]. The current flowing in the device is $I = Nqg/\tau$; hence

$$I_{gr}^2 = \frac{4qIg\Delta f}{1+\omega^2\tau^2}. \tag{12.29}$$

The $1/f$ noise is characterized by a spectrum in which the noise power depends approximately inversely upon frequency. Infrared detectors usually exhibit $1/f$ noise at low frequency. At higher frequencies, the amplitude drops below that of one of the other types of noise: the generation–recombination noise and Johnson noise.

The general expression for the noise current is

$$I_{1/f} = \left(\frac{KI_b^{\alpha}\Delta f}{f^{\beta}}\right)^{1/2}, \tag{12.30}$$

where K is the proportionality factor, I_b is the bias current, α is a constant whose value is about 2, and β is a constant whose value is about unity.

In general, $1/f$ noise appears to be associated with the presence of potential barriers at the contacts, interior, or surface of the semiconductor. Reduction of $1/f$ noise to an acceptable level is an art that depends greatly on the processes employed in preparing the contacts and surfaces. Up until now, no fully satisfactory general theory has been formulated. The two most current models for the explanation of $1/f$ noise were considered [32]: Hooge's model [33], which assumes fluctuations in the mobility of free charge carriers, and McWhorter's model [30], based on the idea that the free carrier density fluctuates.

The low-frequency noise voltage described by the Hooge expression is

$$V_{1/f}^2 = \alpha_H \frac{V^2}{Nf} \Delta f, \tag{12.31}$$

where α_H is the Hooge constant and N is the number of charge carriers. Frequently, the low frequency is characterized by the $1/f$ noise knee frequency $f_{1/f}$:

$$V_{1/f}^2 = V_{gr}^2 \frac{f_{1/f}}{f}. \tag{12.32}$$

The value of the Hooge constant and $f_{1/f}$ is usually considered the technology-related property of the device. There are, however, quantum $1/f$ noise theories describing the $1/f$ noise as the fundamental material property [7]. The Hooge constant in the range of $5 \times 10^{-3} - 3.4 \times 10^{-5}$ has been measured frequently below the lower limit calculated according to existing theories [34].

12.1.1.3 Quantum efficiency

In most photoconductor materials, the internal quantum efficiency η_o is nearly unity; that is, almost all photons absorbed contribute to the photoconductive phenomenon. For a detector, as a slab of material, shown in Figure 12.1, with surface reflection coefficients r_1 and r_2 (on the top and bottom surfaces, respectively) and absorption coefficient α, the internal photogenerated charge profile in the y-direction is [35]

$$S(y) = \frac{\eta_o (1-r_1)\alpha}{1 - r_1 r_2 \exp(-2\alpha t)} \left[\exp(-\alpha y) + r_2 \exp(-2\alpha t)\exp(-\alpha y)\right]. \tag{12.33}$$

The external quantum efficiency is simply the integral of this function over the detector thickness:

$$\eta = \int_0^t S(y)\,dy = \frac{\eta_o (1-r_1)\left[1 + r_2 \exp(-\alpha t)\right]\left[1 - \exp(\alpha t)\right]}{1 - r_1 r_2 \exp(-\alpha t)}. \tag{12.34}$$

When r_1 and $r_2 = r$, the quantum efficiency is reduced to

$$\eta = \frac{\eta_o (1-r)\left[1 - \exp(\alpha t)\right]}{1 - r\exp(-\alpha t)}. \tag{12.35}$$

Intrinsic detector materials tend to be highly absorptive; hence, in a practical well-designed detector assembly, only the top surface reflection term is significant, and then

$$\eta \approx \eta_o (1-r) \approx 1 - r. \tag{12.36}$$

By antireflection coating of the front surface of the detector, this quantity can be made greater than 0.9.

12.1.1.4 Ultimate performance of photoconductors

Usually intrinsic or lightly doped n-type materials are used for fabrication of IR photoconductors. However, if band-to-band recombination mechanisms are dominant, ultimate photoconductor performances are expected in lightly doped p-type materials. This situation usually occurs in the case of long-wavelength near–room temperature HgCdTe photoconductors [36,37].

The classical long-wavelength near–room temperature photoconductors operated at weak optical excitation and at steady-state conditions can be satisfactorily described by a simple model, in which such phenomena as sweep-out, surface recombination, interference within the device, edge effects, and the influence of background radiation are neglected. In the optimum case, when reflection coefficients from the front and rear surfaces are $r_1 = 0$ and $r_2 = 1$, the quantum efficiency is given by

$$\eta = \eta_o\left[1-\exp(-2\alpha t)\right] \approx 1-\exp(-2\alpha t). \tag{12.37}$$

Under the above conditions, the expression for voltage responsivity specified by Equation 12.21 becomes:

$$R_v = \frac{V_b}{hc}\frac{\mu_e+\mu_h}{n_o\mu_e+p_o\mu_h}\frac{\tau\left[1-\exp(-2\alpha t)\right]}{lwt}. \tag{12.38}$$

Taking into account only Johnson–Nyquist and g–r noise (the $1/f$ noise can be minimized by appropriate fabrication techniques and can be neglected), the detectivity is equal to:

$$D^* = \frac{R_v\left(lw\Delta f\right)^{1/2}}{\left(V_J^2+V_{gr}^2\right)^{1/2}}. \tag{12.39}$$

We can distinguish two cases: the first, when V_{gr} saturates at a level above V_J—the g–r noise limited case; and the second, when the saturated level of V_{gr} is less than V_J—the Johnson noise/sweep-out limited case.

The g–r noise limited case always applies to background-limited detectors. The g–r noise limited detectivity is obtained from Equations 12.21, 12.27, and 12.39 as

$$D_{gr}^* = \frac{\lambda}{2hc}\frac{\eta}{t^{1/2}}\left(\frac{n+p}{np}\right)^{1/2}\tau^{1/2}. \tag{12.40}$$

This can be written as Equation 4.48, where G is the sum of all the generation processes per unit volume. The $(n+p)\tau/(np)$ can be used as a generalized, doping dependent figure of merit of the semiconductor that determines the ultimate performance of the photoconductor. Since $\alpha \approx 1/t$, Equation 12.40 can be written as

$$D_{gr}^* = \frac{\lambda\eta}{2hc}\left(\frac{n+p}{n_i}\right)^{1/2}\left(\frac{\alpha\tau}{n_i}\right)^{1/2}, \tag{12.41}$$

where the $\alpha\tau/n_i$ can be treated as the material figure of merit for photoconductors [38], which actually is the α/G figure of merit (see Section 4.2.4).

If, for an idealized detector structure, we ignore the nonfundamental generation processes that occur at surfaces and electrodes, the total generation rate can be expressed as the sum of the rates due to three types of bulk processes: Auger, radiative, and Shockley–Read. The radiative term is due to photons absorbed in the detector that have been emitted from the detector enclosed or have been received through a lens from the ambient temperature scene. The fundamental limit to detector performance is reached when the detector is cooled sufficiently for the radiative generation to dominate, provided that this term is principally caused by photons from the scene (see next section).

In most practical applications, photoconductive detectors are operated at reduced temperatures to eliminate thermally generated transitions and noise due to power dissipation. Joule heating due to bias current produces a rise in detector temperature, and as a consequence, an interface between the detector and the cooling receiver is necessary. Johnson noise–limited detectivity due to power dissipation is observed in large detectors operating in the short wavelength range under reduced background.

12.1.1.5 Influence of background

Under a condition of excess carrier density generation by a background radiation flux density Φ_b, the carrier densities are given by

$$n = n_o + \frac{\eta\Phi_b\tau}{t}, \quad p = p_o + \frac{\eta\Phi_b\tau}{t}.$$

As Φ_b increases, the influence of the background appears initially as an increase of minority carrier density. In normal operation, the detector is cooled sufficiently so that the thermally excited minority carriers are negligible compared to the photon excited excess carriers. Then, the g–r noise is entirely due to the background photon flux density. For the photoconductors operating in the background flux density, background-limited performance requires that two conditions be satisfied $\eta\Phi_b\tau/t > p_o$ for an n-type sample ($\eta\Phi_b\tau/t > n_o$ for a p-type sample) and $V_{gr}^2 > V_j^2$. The second condition states that the applied bias voltage across the detector must be large enough that the g–r noise dominates the Johnson–Nyquist noise contributions. If these conditions are satisfied, the detectivity is given by

$$D_b^* = \frac{\lambda}{2hc}\left[\frac{\eta(n+p)}{\Phi_b n}\right]^{1/2}. \tag{12.42}$$

At moderate background influence ($p_o < \Delta n = \Delta p < n_o$), we obtain the detectivity defined by the equation:

$$D_b^* = \frac{\lambda}{2hc}\left[\frac{\eta}{\Phi_b}\right]^{1/2}. \tag{12.43}$$

However, with high background fluxes and high-purity material, fulfillment of the condition $\Delta n = \Delta p >> n_o$, p_o is possible, and "photovoltaic" background limited infrared photodetector (BLIP) detectivity is achieved:

$$D_b^* = \frac{\lambda}{hc}\left(\frac{\eta}{2\Phi_b}\right)^{1/2}. \tag{12.44}$$

It should be noted that the photovoltaic BLIP detectivity is difficult to achieve in practice because it is a function of carrier densities and decreases at high background levels. This type of lifetime behavior has been observed experimentally [39].

12.1.1.6 Influence of surface recombination

The photoconductive lifetime in general provides a lower limit to the bulk lifetime, due to the possibility of enhanced recombination at the surface. Surface recombination reduces the total number of steady-state excess carriers by reducing the recombination time. It can be shown that τ_{ef} is related to the bulk lifetime by the expression [40]

$$\frac{\tau_{ef}}{\tau} = \frac{A_1}{\alpha^2 L_D^2 - 1}, \tag{12.45}$$

where

$$A_1 = L_D\alpha\left[\frac{(\alpha D_D + s_1)\left\{s_2\left[ch(t/L_D) - 1\right] + (D_D/L_D)sh(t/L_D)\right\}}{(D_D/L_D)(s_1+s_2)ch(t/L_D) + (D_D^2/L_D^2 + s_1s_2)sh(t/L_D)}\right.$$

$$\left. - \frac{(\alpha D_D - s_2)sh\left\{s_1\left[ch(t/L_D) - 1\right] + (D_D/L_D)sh(t/L_D)\right\}\exp(\alpha t)}{(D_D/L_D)(s_1+s_2)ch(t/L_D) + (D_D^2/L_D^2 + s_1s_2)sh(t/L_D)} - \left[1 - \exp(-\alpha t)\right]\right].$$

D_D is the ambipolar diffusion coefficient, s_1 and s_2 are the surface recombination velocities at the front and back surfaces of the photoconductor, and $L_D = (D_D\tau)^{1/2}$.

If the absorption coefficient α is large, $\exp(-\alpha t) \approx 0$ and $s_1 << \alpha D_D$, Equation 12.45 is simplified to the well-known expression [15,20,28]

$$\frac{\tau_{ef}}{\tau} = \frac{D_D}{L_D} \frac{s_2\left[ch(t/L_D)-1\right]+(D_D/L_D)sh(t/L_D)}{L_D(D_D/L_D)(s_1+s_2)ch(t/L_D)+(D_D^2/L_D^2+s_1+s_2)sh(t/L_D)}. \quad (12.46)$$

Further simplification for $s_1 = s_2 = s$ leads to

$$\frac{1}{\tau_{ef}} = \frac{1}{\tau} + \frac{2s}{t}. \quad (12.47)$$

Considerations carried out by Gopal indicate that for accurate modeling of photoconductors, surface recombination effects should be considered as directly influencing the quantum efficiency rather than the carrier lifetime [40]. According to this in the case $r_1 = r_2 = r$,

$$\eta = \frac{(1-r)A_1}{\left[1-r\exp(-\alpha t)\right](\alpha^2 L_D^2 - 1)}. \quad (12.48)$$

If $s_1 = s_2 = 0$, this equation reduces to Equation 12.35.

For low temperatures, the diffusion length is so large that the typical photoconductor is invariably operated in the mode $t/L < 1$, and if $s << 1$, $\tau_{ef} = [1/\tau + 2s/t]^{-1} \approx t/2s$, the detectivity becomes [41]

$$D^* = \frac{\eta\lambda}{2hc}\left(\frac{n_o + p_o}{2n_i^2 s}\right)^{1/2}. \quad (12.49)$$

The essential point of the given discussion is that a finite value of surface recombination velocity may have a strong effect on the attainable detectivity.

12.1.2 EXTRINSIC PHOTOCONDUCTIVITY THEORY

A number of extrinsic photoconductor reviews have previously been published, the first of which, "Optical and Photoconductive Properties of Silicon and Germanium," by Burstein, Picus, and Sclar [42], appeared in 1956. This was followed by "Photoconductivity of Germanium" in 1959 by Newman and Tyler [43], "Far Infrared Photoconductivity" in 1964 by Putley [44], "Impurity Germanium and Silicon Infrared Detectors" by Bratt [5] in 1977, and "Properties of Doped Silicon and Germanium Infrared Detectors" by Sclar [8] in 1984. The last two reviews are still timely and very comprehensive and are ones to which we will make numerous references. In a more recently published review [45], Kocherov et al. have considered certain peculiarities of the operation of extrinsic detectors under a low background.

In the beginning, major emphasis was directed to the Ge detectors. At present, however, there is considerable interest in Si devices because of their potential for the fabrication of very large for thermal imaging [46,47]. The attraction of extrinsic silicon lies in the highly developed MOS technology and the possibility of integrating the detectors with charge transfer devices (CTD) for the readout and signal processing.

There are two simple configurations used in biasing extrinsic photoconductors: transverse bias and parallel bias. These are illustrated in Figure 12.2 [8]. In the transverse case, the electric field and the resulting current flow are transverse to the incident photon flux; the photocarrier generation profile is independent of distance in the direction of current flow. In the longitudinal case, the electric field is parallel to the photon flux, and the photocarrier generation profile varies exponentially in the direction of current flow. The distinction between bias configurations becomes important for large absorptance ($\alpha l > 1$). Analysis carried out by Nelson indicates (Figure 12.3) that for the optimum condition of longitudinal geometry, the responsivity peak is about 87% of the normalized value at $\alpha l \cong 1.5$ and then declines with a further increase

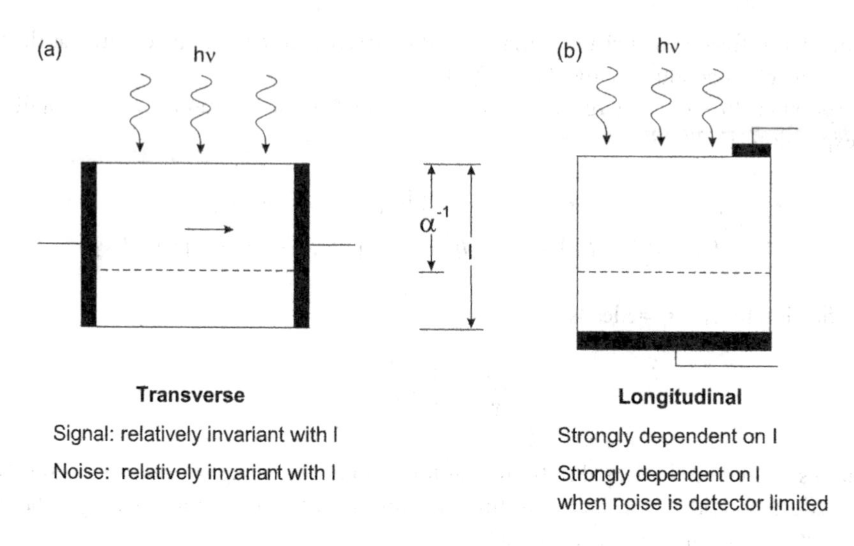

Figure 12.2 Activation geometry comparison for (a) transverse and (b) longitudinal detectors. (From Ref. [8])

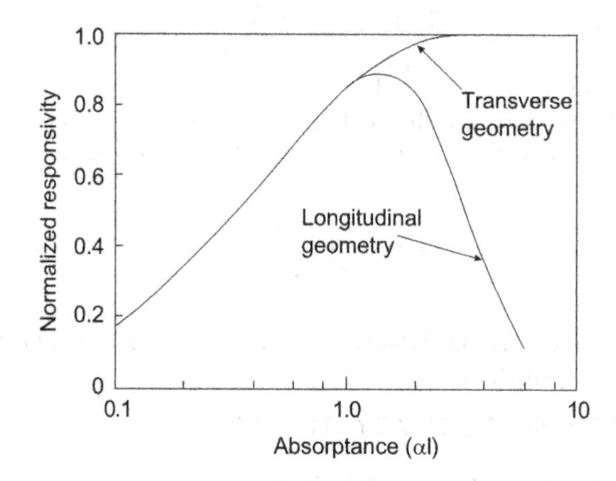

Figure 12.3 Normalized responsivity for longitudinal and transverse detector geometry versus absorptance for a detector with ideal surface coatings ($r_1 = 0$; $r_2 = 1$) and a photoconductive gain of unity. (From Ref. [35])

in αl [35]. Adherence to the condition $\alpha l \cong 1.5$, accordingly, represents an optimum design criterion for detectors employing longitudinal geometries. The reason for the inferiority of the longitudinal geometry is illustrated in Figure 9.2. For the transverse geometry, the unactivated detector depth merely represents a high resistance shunt that has little influence on signal or noise. For the longitudinal geometry, this unactivated depth is electrically in series with the activated depth. Consequently, it acts to quench the signal level and to possibly increase the noise. Transverse bias has historically been used in arrays of discrete detectors while parallel bias is now used in monolithic arrays. The longitudinal geometry detector exhibits a far more uniform sensitivity to a spot scan compared to the transverse detector, which can be very significant in scanned arrays. Because the latter promises better economy and performance, the subsequent analysis will assume parallel bias.

For the following discussion, we will analyze the geometrical model of Figure 12.2b and assume a simple energy level model of an n-type extrinsic semiconductor consisting of a photoionizable donor level and a compensating acceptor level; properties of a corresponding p-type model would be analogous. We assume that the photoconductor crystal contains N_d majority shallow donor impurities and N_a minority shallow acceptor impurities (i.e., $N_d > N_a$). At very low temperatures ($kT \ll E_i$, where E_i is the binding energy of the electrons to the donors) and in dark ($N_d - N_a$) donors will bind an electron and are therefore neutral while N_a donors will have given up their electron to the compensating acceptors. The number of electrons in the conduction band will be extremely small, resulting in high resistivity. The semiconductor material is further characterized by the lifetime τ, the mobility μ, and the quantum efficiency η.

When the signal photon flux density Φ_s with $h\nu \geq E_d$ enters the crystal and is absorbed by neutral donors, bound electrons will be excited into the conduction band. The free electrons will travel in the externally applied electric field E with a velocity $v = \mu E$. The photocurrent is then given by Equation 12.1, where the photoconductive gain g is

$$g = \frac{\mu\tau E}{l}. \tag{12.50}$$

Then,

$$I_{ph} = \frac{q\eta\mu\tau}{l} E\Phi_s A. \tag{12.51}$$

It is clear from Equation 12.51 that a high photocurrent requires high mobility, long lifetime, and as short a detector as is consistent with a high quantum efficiency, which will be discussed later.

The photoconductive gain of extrinsic photoconductors depends on frequency due to carrier sweep-out and dielectric relaxation. Sweep-out effects are more difficult to understand [45,48–50] but are not generally as important in practice as in intrinsic photoconductors.

Let us consider a detector subjected to a short light pulse. The pulse will produce n_{op} electrons and an equal density of positively charged donors. The electrons are swept out of the detector in a transit time, leaving behind a uniform distribution of ionized donors. It is assumed here that the drift length $L_d = \mu\tau E$ is larger than the detector length l. The detector relaxes back to its neutral state within the dielectric relaxation time

$$\tau_\rho = \varepsilon\varepsilon_o\rho, \tag{12.52}$$

where ε is the dielectric constant, ε_o is the permittivity of space, and ρ is the resistivity of the detector. Assuming that $\rho = (qn_{op}\mu)^{-1}$ and $n_{op} = \eta\Phi_s\tau/l$, the dielectric relaxation frequency is given by

$$f_\rho = \frac{q\eta\mu\tau\Phi_s}{2\pi\varepsilon\varepsilon_o}. \tag{12.53}$$

For typical parameters of Si, $\eta = 0.3$, $\mu = 8 \times 10^3$ cm^2/Vs, $\tau = 10^{-8}$ s, and $l = 0.05$ cm, the last equation gives $f_\rho \cong 1.2 \cong 10^{-11}\ \Phi_s$ Hz. For low background applications, where $\Phi_b \approx 10^{12}$ photons/cm^{2}s, f_ρ is only 12 Hz, while for conventional 300 K terrestrial imaging, f_ρ is only in the low-kHz range.

Dielectric relaxation time effects are observed when holes are swept out of the detector without replenishment from contacts. It means that the photoconductive gain should be frequency dependent. There are several models that describe this frequency dependence. The first model [48] predicts a gain drop at f_ρ, while the second model [49] predicts a corner frequency of $f_\rho/2g_o$, where g_o is the low-frequency gain given by Equation 12.50.

More recently published papers refer to nonlinear phenomena and anomalous transient response of cooled extrinsic photoconductors (see e.g., [45,51–55]). The commonly observed behavior in these photoconductors is investigated by performing the dynamic response analysis of the space charge to illumination with full account of the regions near the injecting electrical contacts. Detector anomalies in the transient response, spiking, and noise are currently attributed to electric field effects at the injecting contacts. A high local electric field value creates, for instance, a hot carrier distribution with changes substantially in the mobility, changes drastically in the capture cross section, the impact ionization coefficient, and consequently, the dynamic state of the carriers.

Excess carriers generated in response to an increase in photon illumination can either drift or diffuse to a contact region, where they recombine. This limits the initial gain of the device. Since changes in injection require local changes in the space charge electric field in the region adjacent to the contact, the charge that is lost to the contact cannot be immediately replaced in the bulk by increased injection. As a result, the transient response consists of a slow and a fast component, with their relative magnitudes dependent on the ratio of diffusion and drift lengths to the device length. The slow transient response is controlled by out-diffusion and sweep-out and the establishment of a counteracting electric field barrier, but the fast component is determined by the carrier lifetime (see also Fig. 14.3).

Photoconductors made from doped Ge and Si exhibit values of g to 10, but values between 0.1 and 1 are more typical, because of the low lifetime achieved thus far. Hence, using a frequency-independent gain is reasonable. However, with material improvements, lifetime improvements can be expected with resultant gain increases, and then the frequency dependence of the gain will need to be considered.

Since photoconductivity from extrinsic detectors arises from the photoionization of impurities, it is necessary that the detector be operated under circumstances that permit the free charge carriers to be trapped at the impurities. The major competing process is that of thermal ionization, which dictates a cooling requirement for the detector to suppress this contribution. In the absence of background, the thermal equilibrium concentration of the electrons, n_{th}, is determined by a balance between the rate of thermal ionization of the neutral impurity centers and the rate of recombination at the ionized centers. The general model is rather complicated (e.g., see the discussion carried out in [5] and [8]). For steady state conditions, at the low temperature of operation of impurity photoconductors (when $kT \ll E_i$ and $n \ll N_d, N_a$), the thermal equilibrium free charge carrier is equal to:

$$n_{th} = \frac{N_c}{\delta}\left(\frac{N_d - N_a}{N_a}\right)\exp\left(-\frac{E_d}{kT}\right). \tag{12.54}$$

Here N_c is the density of states in the conduction band and δ is the degeneracy factor, which is four for p-type and two for n-type impurities. High n_{th} would make the detector useless, and there are two options to reduce n_{th}: reduce the temperature to freeze the electrons, or add compensating acceptors. The former is clearly undesirable, and hence the latter method is used. For example, the effects of residual boron impurities in the Si:In detector are compensated by donor concentrations to achieve moderate cooling requirements (50–60 K). For Czochralski-grown Si, where $N_B = (5\text{–}10) \times 10^{13}$ cm^{-3}, it is obviously very difficult to achieve the desired compensation. For the float zone, where the boron concentration is lower by a factor of 10–50, precise compensation is easier to obtain, provided a compensating impurity like phosphorus can be introduced at such low levels. A promising method is the use of neutron transmutation doping where the thermal neutrons in a nuclear reactor interact with the Si lattice transmuting a small fraction of the silicon atoms into a known concentration of phosphorus donors [56]. Figure 12.4 demonstrates the power of the neutron transmutation doping technique for very precise compensation. The Czochralski-grown sample had a fairly high N_B of 1.5×10^{14} cm^{-3}, with a residual phosphorus concentration of 5.9×10^{13} cm^{-3}. After neutron irradiation, $N_P - N_B = 1.9 \times 10^{14}$ cm^{-3}.

Additional charge carriers may be added to the semiconductor by absorption of external radiation or by impact ionization. Theoretical and experimental results indicate that phonon-assisted cascade recombination process is the dominant mechanism for free electron or hole recombination at ionized impurities in Ge and Si [28]. Thus,

$$\tau = \frac{1}{B(N_a + n)}. \tag{12.55}$$

In most practical cases, $n \ll N_a$, so that Equation 12.55 becomes

$$\tau = \frac{1}{BN_a}. \tag{12.56}$$

The recombination coefficient B is given by

$$B = \langle v \rangle \sigma_c, \tag{12.57}$$

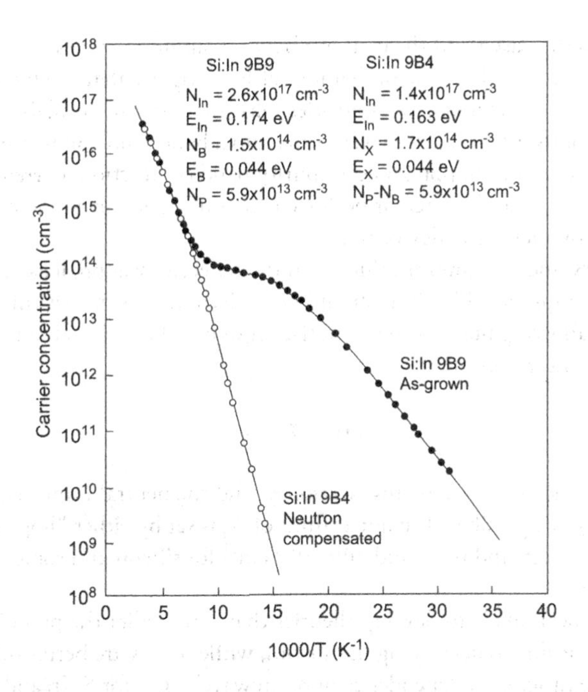

Figure 12.4 Carrier concentration versus reciprocal temperature for an uncompensated as-grown sample (9B9) and a neutron-compensated sample (9B4). The concentrations on the figure refer to In, X (0.11 eV level), B, and P-B (net concentration). (From Ref. [56])

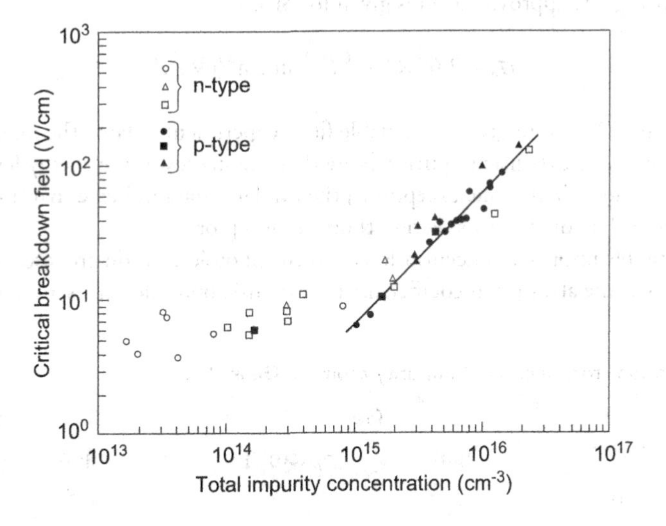

Figure 12.5 Critical impact ionization breakdown field for shallow level impurities in Ge at 4–5 K. (From Ref. [5])

where $< v > = (8kT/\pi m^*)^{1/2}$ is the average free carrier velocity and σ_c is the capture cross section of the recombination center.

Impact ionization is caused by free carriers gaining sufficient energy from the applied electric field to ionize neutral impurity atoms. This effect is manifested by a sharp increase in current through the crystal at some critical field strength E_c. Impact ionization not only creates additional free carriers but also produces excessive electrical noise due to the sporadic nature of the breakdown in different regions of the crystal. The critical field increases with an increasing majority impurity concentration, because higher concentrations reduce the carrier mobility through neutral impurity scattering. Figure 12.5 shows representative experimental data [5].

As the concentration is increased and the distance between atoms becomes sufficiently small, carriers can hop from one impurity to another. The probability of hopping is enhanced by compensating impurities, which by ionizing some of the majority impurities makes empty sites available for carriers to hop into. For still higher concentrations, the impurity level forms into a band, and conduction takes place by carriers flowing within this band. For both hopping and impurity band conduction, current flows without the need to excite holes into the valence band. Detector performance is degraded by reducing the ratio of photoconductive/dark current and by increasing device noise.

Quantum efficiency assumes maximum value when the reflectivity approachs zero at the front surface and unity at the back (see Equation 12.37). It should be noticed, however, that this case can introduce optical cross talk in FPAs by allowing nonabsorbed irradiation to be reflected back into the device.

The absorption coefficient α is given by

$$\alpha = \sigma_p N_i. \tag{12.58}$$

This is the product of the photoionization cross section σ_p and the neutral impurity concentration. It is desirable to make α as large as possible. The upper limit of N_i is set by either "hopping" or "impurity band" conduction, as discussed earlier, and is around 10^{15}–10^{16} cm^{-3} for silicon and somewhat lower for germanium (see Table 12.1) [5,8].

Various attempts have been made to develop theories that can predict the photoionization cross section [8]. Some of these are applicable to deep-lying impurities, while others are better suited to impurities with shallow energy levels. The functional dependence of σ_p on wavelength for Si:In and Si:Ga detector materials is shown in Figure 12.6 [57]. It rises from zero wavelength to a maximum at $\lambda_c/2$ and then decreases. Although it is not constant, it has a rather broad maximum and the absorption coefficient is reasonably constant over a useful wavelength range. The dependence of the maximum value for photoionization cross section σ_o on E_i for hydrogenic approximation is given for Si as

$$\sigma_o = 2.65\times10^{-18} E_i^{-2} \text{ in cm}^2(\text{eV})^2, \tag{12.59}$$

which is shown in Figure 12.7 [8] to give a reasonable fit to experimental data. The maximum value varies with the energy level of the extrinsic impurity. Note that the shallower the energy level, the larger the photoionization cross section. With some exceptions, the available data indicate that, for a given energy, the donors achieve a higher value for the cross section than the acceptors.

Using typical acceptable impurity concentrations and the photoionization cross sections, it can be seen from Equation 12.58 that the absorption coefficients for extrinsic photodetectors are some three orders

Table 12.1 Photoionization cross section of impurity atoms in Ge and Si

		Ge		Si	
IMPURITY	TYPE	λ_c (μm)	σ_p (cm^{-2})	λ_c (μm)	σ_p (cm^{-2})
Al	p			18.5	8×10^{-16}
B	p	119	1.0×10^{-14}	28	1.4×10^{-15}
Be	p	52		8.3	5×10^{-18}
Ga	p	115	1.0×10^{-14}	17.2	5×10^{-16}
In	p	111		7.9	3.3×10^{-17}
As	n	98	1.1×10^{-14}	23	2.2×10^{-15}
Cu	p	31	1.0×10^{-15}	5.2	5×10^{-18}
P	n	103	1.5×10^{-14}	27	1.7×10^{-15}
Sb	n	129	1.6×10^{-14}	29	6.2×10^{-15}

Source: After Refs. [5,8].

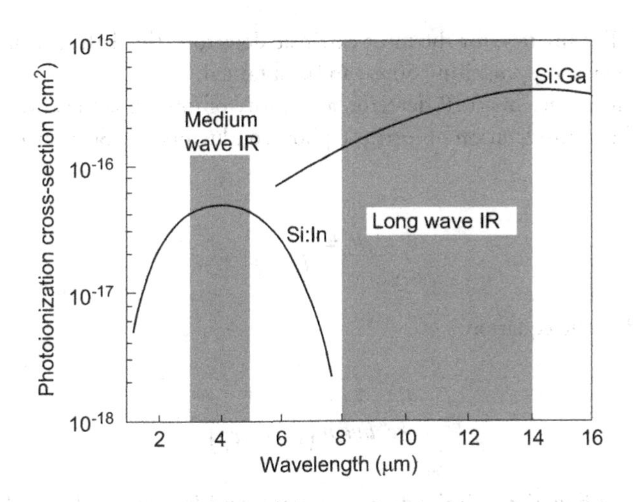

Figure 12.6 Photoionization cross section versus wavelength for Si:In and Si:Ga IR detector material. (From Ref. [57])

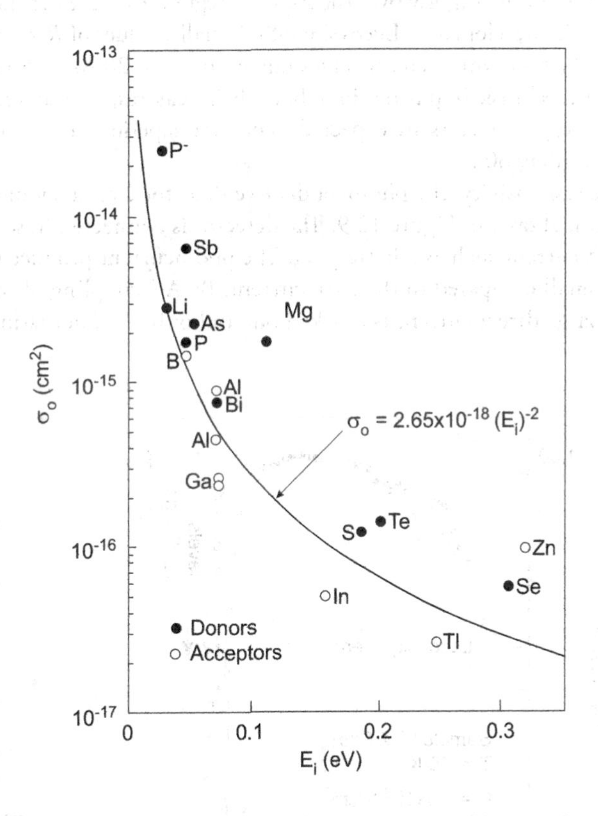

Figure 12.7 Impurity photoionization cross section at peak response versus impurity binding energy in Si. (From Ref. [8])

of magnitude less than those for direct absorption in intrinsic photoconductors. Practical values of α for optimized photoconductors are in the range of 1–10 cm^{-1} for Ge and 10–50 cm^{-1} for Si. Thus, to maximize quantum efficiency, the thickness of the detector crystal should be not less than about 0.5 cm for doped Ge and about 0.1 cm for doped Si. There is a limit in the thickness of extrinsic detectors because photocarriers generated beyond the drift length $L_d = \mu\tau E$ recombine before being collected (photoconductive gain $g = L_d/l$

decreases as l increases). Fortunately, for the most extrinsic detectors, the drift length is sufficiently long such that quantum efficiencies approaching 50% can be obtained.

In describing the performance of an IR detector, a quantity of interest is the current (or voltage) responsivity. Analogical to consideration of intrinsic photoconductors (see Section 12.1.1), the short-circuit current responsivity is

$$R_i = \frac{I_{ph}}{P_\lambda}, \tag{12.60}$$

which can be converted to the equation:

$$R_i = \frac{\eta\lambda}{hc}\frac{\tau}{lwt}\frac{I}{n}\frac{1}{\left(1+\omega^2\tau^2\right)^{1/2}}, \tag{12.61}$$

where I is the dark current flow through the detector circuit, and lwt is the volume of the detector element. It can be shown that for $\alpha l < 1$, $R_i \propto \alpha\lambda \propto \sigma_o$, the responsivity is proportional to σ_o.

Figure 12.8 shows the relative spectral response of neutron-compensated Si:In detector at 10 K [56]. The measured response differs only slightly over the 2–8 μm region from the generally accepted theoretical model for deep impurities developed by Lucovsky [58]. Usually, values of R_i reach up to 100 A/W for the very best Si photoconductors, while typical values range from 1 to 20 A/W. It has been found that for a given energy level, n-type extrinsic impurities in Si have their peak response at longer wavelengths than do p-type impurities, so n-type detectors are expected to provide superior temperature characteristics for a given wavelength response [8,59,60].

To determine voltage responsivity, the photoconductive detector circuit should be considered. The practical detector circuit is shown in Figure 12.9. The detector is connected in series with a load resistor R_L and a source of direct current such as a battery V_b. The photocurrent produced by incoming signal photons is usually very small compared to the dark current. By AC coupling of the detector circuit to the preamplifier, the large direct current is blocked out and only the fluctuating signal current is measured.

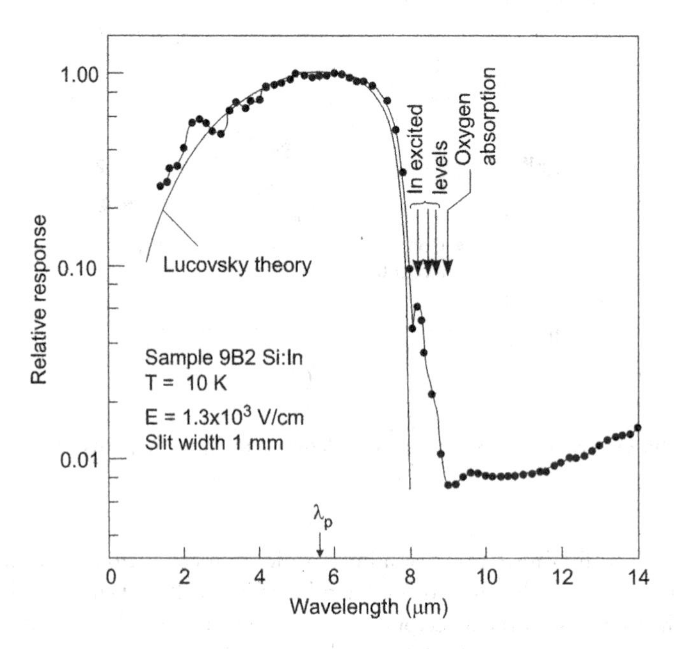

Figure 12.8 Experimental and theoretical relative spectral response for neutron-compensated Czochralski Si:In sample with $N_{In} = 2.5 \times 10^{17}$ cm^{-3}, $N_p - N_B = 1.6 \times 10^{14}$ cm^{-3}, and $N_B = 1.3 \times 10^{14}$ cm^{-3}. (From Ref. [56])

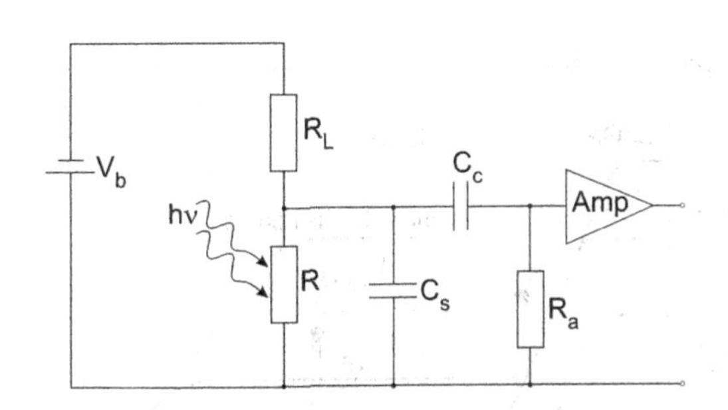

Figure 12.9 Practical detector circuit.

It can be shown that

$$\Delta V = I\frac{\Delta R}{R}\frac{RR_L}{R+R_L}, \tag{12.62}$$

where ΔV is the signal voltage, and where ΔR is a small change in detector resistance R due to signal radiation. This equation is valid for an "ohmic" photoconductor provided that the amplifier input resistance R_a is much greater than the detector resistance. Most impurity detectors are decidedly "nonohmic," and in this case, Equation 12.62 should be replaced by:

$$\Delta V = I\frac{\Delta R}{R_{dc}}\frac{R_{ac}R_L}{R_{ac}+R_L}, \tag{12.63}$$

R_{ac}, given by dV/dI, is the AC resistance of the detector; R_{dc}, given by V/I, is the DC resistance. Voltage responsivity may now be easily obtained because:

$$R_v = R_i\frac{R_{ac}R_L}{R_{ac}+R_L}. \tag{12.64}$$

The open circuit voltage responsivity is sometimes required, and this can be obtained from Equation 12.64 by letting $R_L >> R_{ac}$, so that

$$R_{vo} = R_i R_{ac}. \tag{12.65}$$

The ultimate sensitivity of the detector is determined by the signal-to-noise ratio. Maximum performance can be achieved only when the noise in the device is of the g–r type. For the usual case at low temperatures, $n << N_a$, N_d and g–r noise current are given by [8]:

$$I_{gr} = 2I\left(\frac{\tau\Delta f}{nlwt}\right)^{1/2}\frac{1}{\left(1+\omega^2\tau^2\right)^{1/2}}. \tag{12.66}$$

Then, since the detectivity

$$D^* = \frac{R_i\left(A\Delta f\right)^{1/2}}{I_n}, \tag{12.67}$$

inserting the g–r noise (Equation 12.66) and the current responsivity (Equation 12.61) into Equation 12.67 gives

$$D^* = \frac{\eta\lambda}{2hc}\left(\frac{\tau}{nl}\right)^{1/2}. \tag{12.68}$$

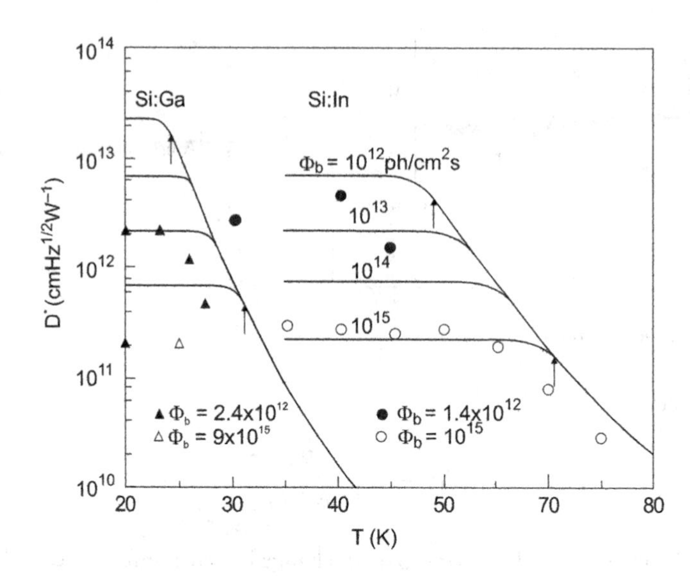

Figure 12.10 Measured and calculated detectivity D^* as a function of temperature for a neutron-compensated Si:In photoconductor and a doped Si:Ga photoconductor. The solid lines are calculated for $\eta = 50\%$ and $l = 0.05$ cm at $\lambda = 12$ μm (Si:Ga) and 4 μm (Si:In). The data points are measured values. (From Ref. [50])

When both thermal and photon generation are important, the free carrier density may be written as a sum of two terms, $n = n_{th} + n_{op}$, with n_{th} given by Equation 12.54 and

$$n_{op} = \frac{\eta\Phi\tau}{l} = \frac{\eta\Phi}{l}\frac{1}{BN_a}, \tag{12.69}$$

where Equation 12.56 has been used to express the lifetime in terms of B, which contains the temperature dependence. The detectivity can thus be written with explicit temperature dependence as:

$$D^* = \frac{\eta\lambda}{2hc}\left[\eta\Phi + \frac{lB}{\delta}(N_d - N_a)N_c \exp\left(-\frac{E_i}{kT}\right)\right]^{-1/2}. \tag{12.70}$$

Maximum D^* is achieved when $n_{th} \ll n_{op}$; that is, the thermal carrier concentration should be reduced for the detector to be dominated by optically generated carriers. This is shown graphically in Figure 12.10 for Si:In and Si:Ga photoconductors [50]. For temperature $T < T_{BLIP} \approx 60$ K, the Si:In detector is background-limited for background radiation flux density $\Phi_b = 10^{15}$ photons/cm^{2}s. T_{BLIP} is a function of Φ_b to which the detectors are exposed. We can see from Figure 12.10 that the Si:In data lie 5–10 K below theory, mainly as a result of the 0.11-eV level contaminant, while the Si:Ga data are about 3–5 K below. As expected, the temperature decreases for reduced backgrounds, as shown by the 3 dB arrows.

12.1.3 OPERATING TEMPERATURE OF INTRINSIC AND EXTRINSIC IR DETECTORS

In this section we compare detectivity, as a function of operating temperature, for intrinsic and extrinsic photoconductors based on previously established relations and measured data. The effects of such parameters as impurity concentration, free carrier lifetime, and capture cross section are evaluated.

We begin with the expression for D^* at peak when the g–r noise is dominant, as this sets the upper D^* limit, independent of bias and detector area. Combining Equations 12.21, 12.27, and 12.39, D^* for intrinsic photoconductors with low excess impurity density can be expressed as follows:

$$D_{in}^* = \frac{\eta\lambda}{2hc}\left[\frac{\tau_{in}}{t_{in}\left(n_{ph} + n_i\right)}\right]^{1/2}, \tag{12.71}$$

where n_{ph} is the optically generated carrier density, and n_i is the intrinsic carrier concentration. Equation 12.71 gives an upper limit of D^* not achievable in measurements because of such effects as trapping centers in the forbidden band, excess impurities, temperature-dependent excess noise, or preamplifier requirements.

In the case of extrinsic photoconductors with the same assumptions of thermal and optical g–r noise dominant, D^* for an n-type detector can be expressed as:

$$D^*_{ex} = \frac{\eta\lambda}{2hc}\left[\frac{\tau_{ex}}{t_{ex}(n_{ph}+n_{th})}\right]^{1/2}, \tag{12.72}$$

according to Equation 12.68.

Both for intrinsic and extrinsic detectors, the BLIP conditions are fulfilled when $n_{ph} > (n_{th}$ or $n_i)$ and D^* is determined by Equation 12.43. The temperature corresponding to the transition from thermal to background-limited noise is found by equating the thermal and background generated free carrier densities. For extrinsic photoconductors, by equating Equations 12.54 and 12.69 we can obtain:

$$T_{\mathrm{BLIP}} = \frac{E_d}{k}\left\{\ln\left[\left(\frac{tN_d}{\eta}\right)\frac{BN_c}{\delta\Phi_b}\right]\right\}^{-1}. \tag{12.73}$$

This equation shows that T_{BLIP} is, for given field of view (FOV), a function of the impurity parameters E_d, σ_c ($B \propto \sigma_c$), σ_p (which determines the absorption coefficient and quantum efficiency), and the background flux Φ_b. The temperature limitation of extrinsic Si detectors has been considered by Bryan [61].

One of the principal factors that determine the stringent cooling requirements for extrinsic detectors is the large value of σ_c associated with the commonly used dopants, since very low values for the recombination time result even when a small fraction of the dopants are thermally ionized. The capture cross section σ_{ex} of extrinsic photoconductors is larger than the corresponding σ_{in} of intrinsic photoconductors. The shallow-level impurities (B and As) show typically $\sigma_c \approx 10^{-11}$ cm^2, while the deep-level impurities (In, Au, Zn) show $\sigma_c = 10^{-13}$ cm^2 (see Table 12.2) [8]. By comparison, Figure 12.11 shows the σ_c of several intrinsic photoconductors to be $\sigma_{in} = 1.2 \times 10^{-17}$ cm^2 [62].

Milnes indicates that conventional dopants, with an attractive charge for the recapture of photocarriers, have values for σ_c between 10^{-15} and 10^{-12} cm^2; neutral impurities have σ_c about 10^{-17} cm^2 and 10^{-15} cm^2, and repulsive centers have σ_c less than 10^{-22} cm^2 (see Figure 12.12) [63]. The σ_c of an impurity atom depends on its recombination potential, which is smaller for neutral or repulsive centers than present attractive coulomb centers. Elliott et al. have discussed the possibility of obtaining higher temperature operations using neutral or repulsive centers [64]. It is suggested that this might be achieved by using very

Table 12.2 Capture cross section of impurity atoms in Ge and Si

IMPURITY	GERMANIUM		SILICON	
ATOM	T(K)	σ_c (cm^2)	T(K)	σ_c (cm^2)
B			4.2	8×10^{-12}
Al	10	2×10^{-12}		
In			77	10^{-13}
As	10	10^{-12}	10	10^{-11}
Cu	10	5×10^{-12}		
Au	80	1×10^{-13}	77	10^{-13}
Zn			80–200	10^{-13}
Cd	8	1×10^{-11}		
Hg	20	3.6×10^{-12}		

Source: After Ref. [8].

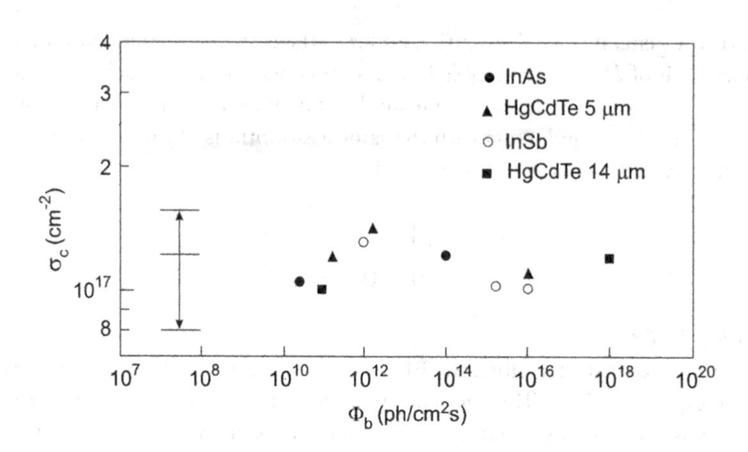

Figure 12.11 σ_c of some intrinsic photoconductors, calculated from response-time data. (From Ref. [62])

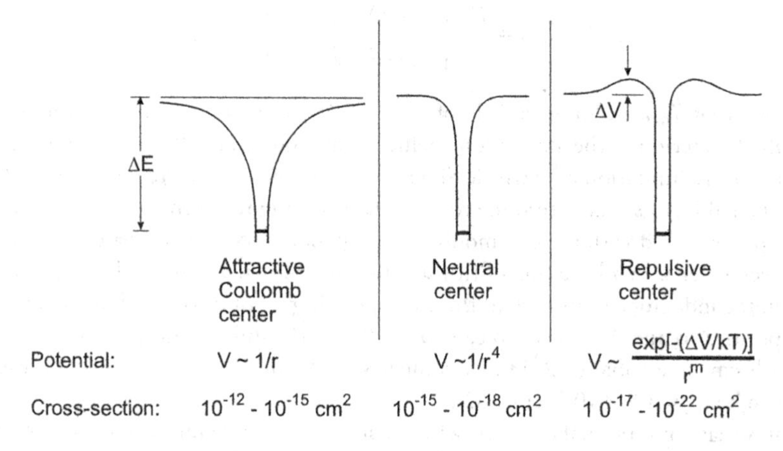

Figure 12.12 Potential distributions and estimated cross sections for recombination for attractive coulomb, neutral, and repulsive charge centers. (From Ref. [8])

deep levels, for example, an acceptor level at the appropriate ionization energy from the conduction band, counter doped with a shallow donor impurity; compensation with impurities of the opposite type produces recombination sites that are neutral or repulsive, respectively. Although some increase in operating temperature has been obtained from counter doping, the benefits in operating temperatures are not as large as were predicted. The probable reason for this is that the capture–cross sections for neutral and repulsive centers increase with temperature and the values given by Milnes [63] have generally been measured at very low temperatures.

Figure 12.13 illustrates the direct effect of smaller σ_c on higher operating temperatures for extrinsic Si photoconductors [20]. In calculations, the photoionization cross section, σ_p, is assumed to have the wavelength dependence predicted by Lucovsky [58] and the following numerical values were used for the p-type material: refractive index = 3.44, $N_v = 1.7 \times 10^{15}\ T^{3/2}$ cm^{-3}, $\delta = 4$, $v_{th} = 9.5 \times 10^5\ T^{1/2}$ cm/s, and $E_{ef}/E_o = 2.5$; where E_{ef} and E_o are electric fields as defined in Lucovsky's paper. The scene temperature is 295 K and the FOV is 30°. The experimental results for Si IR detectors for use at higher background fluxes are taken from Sclar [8]. The plots are consistent with the observed behavior of the most studied impurities, assuming $\sigma_c = 10^{-12}$ cm^2. They also indicate that T_{BLIP} greater than 50 K should be achievable in the 8–14 μm band and greater than 80 K in the 3–5 μm band, using dopants with a similar value of σ_c.

In Figure 4.14, plots of the calculated temperature required for background-limited operation in *f*/2 FOV are shown as a function of cutoff wavelength [65,66]. The calculation for silicon extrinsic detectors has been carried out in terms of a dimensionless parameter $a(\eta)$ [8], which is of the order of unity, and the value of this parameter $Q = [a(\eta)(m^*/m)^{3/2}/\delta](B/\sigma_p)$ of 10^{10}. We can see that the operating temperature

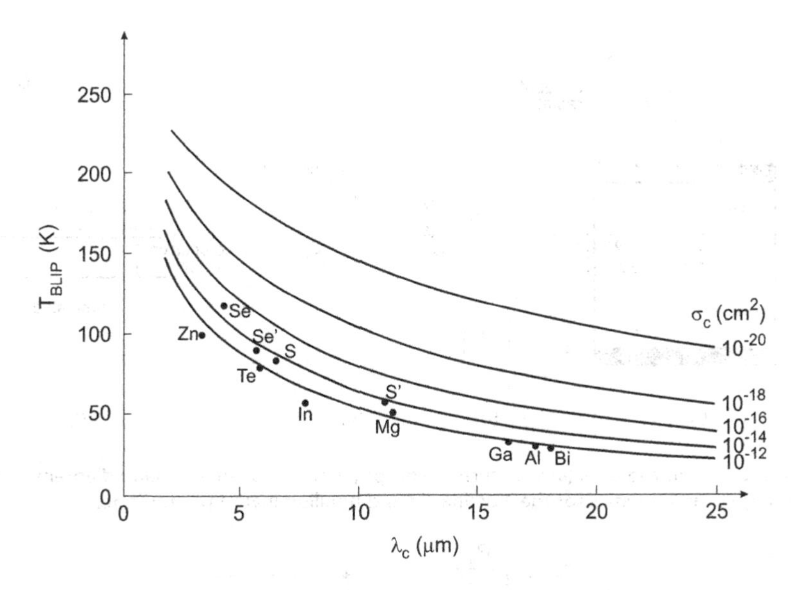

Figure 12.13 T_{BLIP} versus cutoff wavelength for extrinsic Si photoconductors as a function of the thermal capture cross section. (Experimental data are taken from *Progress in Quantum Electronics*, 9, 149–257, 1984.) 295 K background, 30° FOV. (From Ref. [20])

of "bulk" intrinsic IR detectors (HgCdTe) is higher than for other types of photon detectors. Intrinsic materials are characterized by high optical absorption coefficient and quantum efficiency and relatively low thermal generation rate compared to extrinsic detectors, silicide Schottky barriers, and quantum well IR photodetectors.

Extrinsic photoconductors are made thicker than intrinsic photoconductors since the absorption cross section of extrinsic photoconductors is smaller than that of the intrinsic photoconductors. For Si detectors, typically $t_{ex} = 0.1$ cm, whereas for intrinsic ones $t_{in} = 10^{-3}$ cm. As is pointed out, t_{ex} is a direct consequence of σ_p, which is a function of wavelength (see Figure 12.6) and is independent of temperature, irradiance, and impurity concentration [67]. Therefore, σ_p is not subject to control and is a fixed parameter for each impurity.

12.2 P-N JUNCTION PHOTODIODES

The most common example of a photovoltaic detector is the abrupt p-n junction prepared in the semiconductor, which is often referred to simply as a photodiode. The operation of the p-n junction photodiode is illustrated in Figure 12.14. Photons with energy greater than the energy gap, incident on the front surface of the device, create electron hole pairs in the material on both sides of the junction. By diffusion, the electrons and holes generated within a diffusion length from the junction reach the space charge region. The electron hole pairs are then separated by the strong electric field; minority carriers are readily accelerated to become majority carriers on the other side. This way, a photocurrent is generated that shifts the current-voltage characteristic in the direction of negative or reverse current, as shown in Figure 12.14d.

The equivalent circuit of a photodiode is shown in Figure 12.15. The photodiode has a small series resistance R_s, a total capacitance C_d consisting of junction and packaging capacitances, and a bias (or load) resistor R_L. The amplifier following the photodiode has an input capacitance C_a and a resistance R_a. For practical purposes, R_s is much smaller than the load resistance R_L and can be neglected.

The total current density in the p-n junction is usually written as:

$$J(V,\Phi) = J_d(V) - J_{ph}(\Phi), \tag{12.74}$$

where the dark current density, J_d, depends only on V and the photocurrent depends only on the photon flux density Φ.

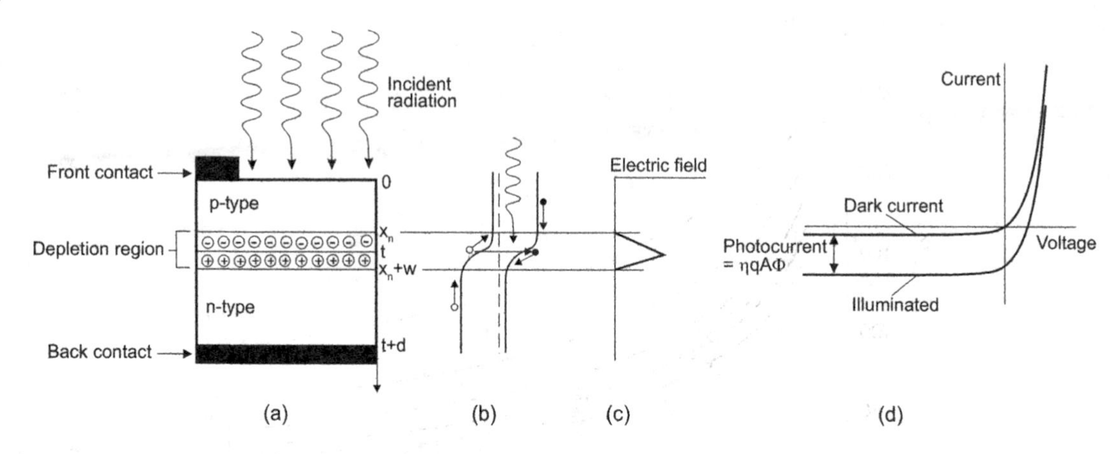

Figure 12.14 p-n junction photodiode: (a) structure of abrupt junction, (b) energy band diagram, (c) electric field, and (d) current-voltage characteristics for the illuminated and nonilluminated photodiode.

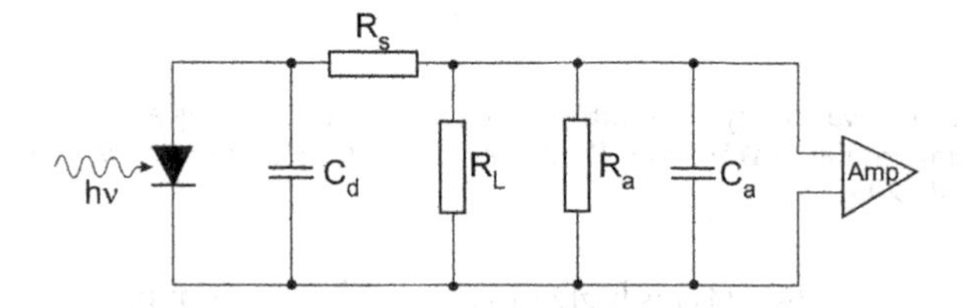

Figure 12.15 Equivalent circuit of an illuminated photodiode (the series resistance includes the contact resistance as well as the bulk p- and n-regions).

Generally, the current gain in a simple photovoltaic detector [e.g., not an avalanche photodiode (APD)] is equal to 1, and then according to Equation 12.1, the magnitude of photocurrent is equal to:

$$I_{ph} = \eta q A \Phi. \tag{12.75}$$

However, the electric field that is responsible for the removal of excess majority carriers created when a photodiode is illuminated induces an additional minority carrier flow to the junction. As a result, for photodiodes where mixed conduction is significant, there is a gain associated with the collection photocurrent, which depends on the mobility ratio and that can be increased or decreased by application of bias [68]. The gain applies to excitation close to the junction. This effect can be the cause of anomalously low junction resistance in photodiodes. The theory of a conventional photodiode with a gain equal to 1 has been considered in the later section.

The dark current and photocurrent are linearly independent (which occurs even when these currents are significant) and the quantum efficiency can be calculated in a straightforward manner [69–72].

If the p-n diode is open-circuited, the accumulation of electrons and holes on the two sides of the junction produces an open-circuit voltage (Figure 12.14d). If a load is connected to the diode, a current will be conducted in the circuit. The maximum current is realized when an electrical short is placed across the diode terminals and this is called the short-circuit current.

The open-circuit voltage can be obtained by multiplying the short-circuit current by the incremental diode resistance $R = (\partial I/\partial V)^{-1}$ at $V = V_b$:

$$V_{ph} = \eta q A \Phi R, \tag{12.76}$$

where V_b is the bias voltage and $I = f(V)$ is the current-voltage characteristic of the diode.

In many direct applications the photodiode is operated at zero-bias voltage:

$$R_o = \left(\frac{\partial I}{\partial V}\right)^{-1}_{|V_b=0}. \tag{12.77}$$

A frequently encountered figure of merit for a photodiode is the R_oA product

$$R_oA = \left(\frac{\partial J}{\partial V}\right)^{-1}_{|V_b=0}, \tag{12.78}$$

where $J = I/A$ is the current density.

In the detection of radiation, the photodiode is operated at any point of the *I-V* characteristic. Reverse bias operation is usually used for very high-frequency applications to reduce the *RC* time constant of the devices.

12.2.1 IDEAL DIFFUSION-LIMITED P-N JUNCTIONS

12.2.1.1 Diffusion current

Diffusion current is the fundamental current mechanism in a p-n junction photodiode. Figure 12.14a shows a one-dimensional photodiode model with an abrupt junction where the spatial charge of width w surrounds the metallographic junction boundary $x = t$, and two quasineutral regions $(0, x_n)$ and $(x_n + w, t + d)$ are homogeneously doped. The dark current density consists of electrons injected from the n-side over the potential barrier into the p-side and an analogous current due to holes injected from the p-side into the n-side. The current-voltage characteristic for an ideal diffusion-limited diode is given by:

$$I_D = AJ_s\left[\exp\left(\frac{qV}{kT}\right) - 1\right], \tag{12.79}$$

where [70,71]

$$J_s = \frac{qD_h p_{no}}{L_h}\frac{\gamma_1 ch(x_n/L_h) + sh(x_n/L_h)}{\gamma_1 sh(x_n/L_h) + ch(x_n/L_h)} + \frac{qD_e n_{po}}{L_e}\frac{\gamma_2 ch[(t+d-x_n-w)/L_e] + sh[(t+d-x_n-w)/L_e]}{\gamma_2 sh[(t+d-x_n-w)/L_e] + ch[(t+d-x_n-w)/L_e]}, \tag{12.80}$$

in which case $\gamma_1 = s_1L_h/D_h$, $\gamma_2 = s_2L_e/D_e$, p_{no} and n_{po} are the concentrations of minority carriers on both sides of the junction, s_1 and s_2 are the surface recombination velocities at the illuminated (for holes in n-type material) and back photodiode surface (for electrons in p-type material), respectively. The value of the saturation current density, J_s, depends on minority carrier diffusion lengths (L_e, L_h), minority carrier diffusion coefficients (D_e, D_h), surface recombination velocities (s_1, s_2), minority carrier concentrations (p_{no}, n_{po}), and junction design (x_n, t, w, d).

For a junction with thick quasineutral regions $[x_n >> L_h, (t + d - x_n - w) >> L_e]$, the saturation current density is equal to:

$$J_s = \frac{qD_h p_{no}}{L_h} + \frac{qD_e n_{po}}{L_e}, \tag{12.81}$$

and when the Boltzmann statistic is valid, $n_o p_o = n_i^2, D = (kT/q)\mu$, and $L = (D\tau)^{1/2}$, then

$$J_s = (kT)^{1/2} n_i^2 q^{1/2}\left[\frac{1}{p_{po}}\left(\frac{\mu_e}{\tau_e}\right)^{1/2} + \frac{1}{n_{no}}\left(\frac{\mu_h}{\tau_h}\right)^{1/2}\right], \tag{12.82}$$

where p_{po} and n_{no} are the hole and electron majority carrier concentrations, and τ_e and τ_h the electron and hole lifetimes in the p- and n-type regions, respectively. Diffusion current varies with temperature as n_i^2.

The resistance at zero bias can be obtained from Equation 12.79 by differentiation of *I-V* characteristics,

$$R_o = \frac{kT}{qI_s}, \tag{12.83}$$

and then the R_oA product determined by diffusion current is:

$$(R_oA)_D = \left(\frac{dJ_D}{dV}\right)^{-1}_{|V_b=0} = \frac{kT}{qJ_s}. \tag{12.84}$$

If $\gamma_1 = \gamma_2 = 1$ and in the case of a diode with thick regions on both sides of the junction, the R_oA is given by the equation:

$$(R_oA)_D = \frac{(kT)^{1/2}}{q^{3/2}n_i^2}\left[\frac{1}{n_{no}}\left(\frac{\mu_h}{\tau_h}\right)^{1/2} + \frac{1}{p_{po}}\left(\frac{\mu_e}{\tau_e}\right)^{1/2}\right]^{-1}. \tag{12.85}$$

The photodiodes with thick regions on both sides of the junction are not realized in practice. Analysis of the effect of the structure of a classical photodiode (of thick p-type region) has shown that the R_oA product for junction depths $0 < t < 0.2L_h$ and surface recombination velocities $0 < s_1 < 10^6$ cm/s differs from product $(R_oA)_o$ for a photodiode with thick regions on both sides of the junction by a factor of 0.3–2 [72]. This shows that the R_oA product calculated for photodiodes with thick p-type and n-type regions is a good approximation of the product for photodiodes of optimum construction. For n-p$^+$ type junctions, the $R_oA/(R_oA)_o$ ratio has the form

$$\frac{R_oA}{(R_oA)_o} = \frac{\gamma_1 ch(x_n/L_h) + sh(x_n/L_h)}{\gamma_1 sh(x_n/L_h) + ch(x_n/L_h)}. \tag{12.86}$$

In Figure 12.16, the dependence of the $R_oA/(R_oA)_o$ on the n-p$^+$ (n$^+$ -p) junction depth at various γ_1 values is presented [72]. For $\gamma_1 < 1$ (blocking contact [69]) we get $R_oA > (R_oA)_o$, the increase of R_oA being particularly high for small values of γ_1 and $x_n/L_h \to 0$. Fabrication of such structures may involve great technological difficulties connected with the need to fulfill the condition $s_2 = 0$. In that case, it is advantageous to use n-p-p$^+$ (p-n-n$^+$) structures, since the potential barrier between the p- (n-) and p$^+$ -type (n$^+$ -type) regions limits the flow of minority carriers to the region with more impurities.

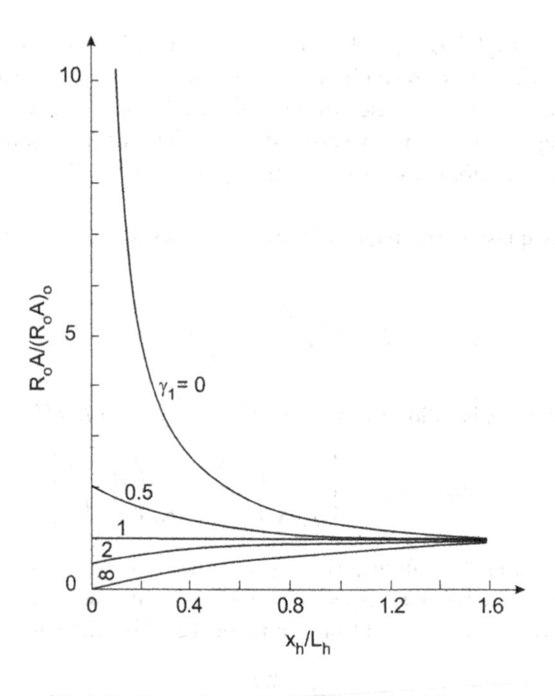

Figure 12.16 The dependence of $R_oA/(R_oA)_o$ on the normalized depth of the n-p$^+$ (n$^+$ -p) junction at different surface recombination velocities γ_1 = 0; 0.5; 1; 2; and ∞. (From Ref. [72])

For the n⁺ -p diode structure, the junction resistance is limited by diffusion of minority carriers from the p side into the depletion region. In the case of conventional bulk diodes, where $d >> L_e$:

$$(R_oA)_D = \frac{(kT)^{1/2}}{q^{3/2} n_i^2} N_a \left(\frac{\tau_e}{\mu_e}\right)^{1/2}. \tag{12.87}$$

By thinning the substrate to a thickness smaller than the minority carrier diffusion length (thus reducing the volume in which diffusion current is generated), the corresponding R_oA product increases, provided that the back surface is properly passivated to reduce surface recombination. In the case of the n⁺ -p junction, if the thickness of the p-type region is such that $d << L_e$, we obtain:

$$(R_oA)_D = \frac{kT}{q^2} \frac{N_a}{n_i^2} \frac{\tau_e}{d}. \tag{12.88}$$

As result, R_aA can increase by a factor of L_e/d. Of course, analogical formulas can be obtained for p⁺-n junctions.

The thickness of the illuminated p⁺ -n (n⁺ -p) junction from the p⁺ (n⁺) side must be small to eliminate absorption of radiation by the free carriers. In n-p⁺ (p-n⁺) structures illuminated from the n(p)-type side, the major contribution to the quantum efficiency comes from the region with fewer impurities of the n(p)-type. That is why the thickness of that region, and hence the depth of the junction, should be greater (i.e., $0.2L_h < t < 0.4L_h$ [72]). On the other hand, at lower t values and $0 < \gamma_1 < 1$, we can obtain a significant increase in the R_oA product (see Figure 12.16). It follows that the optimum depth of the junction is shifted toward smaller t values.

In conclusion, it should be noted that the influence of design on the R_oA product is determined by the diffusion component of the current density. In the case when the R_oA product is determined by another mechanism, the aforementioned considerations are not justifiable. The consideration concerning the quantum efficiency still remains valid.

12.2.1.2 Quantum efficiency

Three regions contribute to photodiode quantum efficiency: two neutral regions of different types of conductivity and the spatial charge region (see Figure 12.14). Thus [70,71]:

$$\eta = \eta_n + \eta_{DR} + \eta_p, \tag{12.89}$$

where

$$\eta_n = \frac{(1-r)\alpha L_h}{\alpha^2 L_h^2 - 1}\left\{\frac{\alpha L_h + \gamma_1 - e^{-\alpha x_n}\left[\gamma_1 ch(x_n/L_h) + sh(x_n/L_h)\right]}{\gamma_1 sh(x_n/L_h) + ch(x_n/L_h)} - \alpha L_h e^{-\alpha x_n}\right\}, \tag{12.90}$$

$$\eta_p = \frac{(1-r)\alpha L_e}{\alpha^2 L_e^2 - 1} e^{-\alpha(x_n+w)} \times\left\{\frac{(\gamma_2 - \alpha L_e)e^{-\alpha(t+d-x_n-w)} - sh\left[(t+d-x_n-w)/L_e\right] - \gamma_2 ch\left[(t+d-x_n-w)/L_e\right]}{ch\left[(t+d-x_n-w)/L_e\right] - \gamma_2 sh\left[(t+d-x_n-w)/L_e\right]} + \alpha L_e\right\}, \tag{12.91}$$

$$\eta_{DR} = (1-r)\left[e^{-\alpha x_n} - e^{-\alpha(x_n+w)}\right]. \tag{12.92}$$

In the following section, we shall consider the internal quantum efficiency, neglecting the losses due to reflection of the radiation from the illuminated photodiode surface. Obtaining high photodiode quantum efficiency requires that the illuminated region of the junction be sufficiently thin so that the generated carriers may reach the junction potential barrier by diffusion.

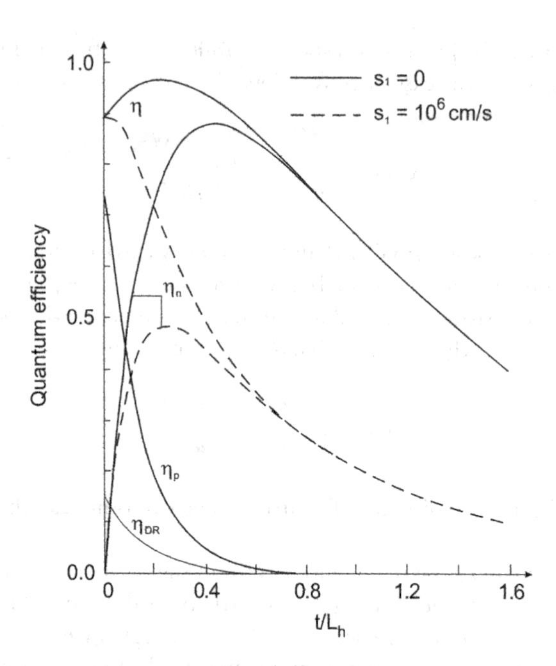

Figure 12.17 The dependence of quantum efficiency on the normalized thickness of the junction-illuminated region at $s_1 = 0$ ($\gamma_1 = 0$) and $s_1 = 10^6$ cm/s ($\gamma_1 = 7$). In the calculations it was assumed that $d = \infty$, $w = 0.3$ μm, $r = 0$, and $\alpha = 5 \times 10^3$ cm^{-1}. (From Ref. [72])

In Figure 12.17, the relationship is presented between the components of the photodiode quantum efficiency and the normalized thickness of the junction-illuminated region t/L_h at an infinite thickness of the p-type region [72]. The calculations are carried out for the typical absorption of 5×10^3 cm^{-1} at a wavelength close to the intrinsic absorption edge in narrow-gap semiconductors and for $L_e = L_h = 15$ μm, $w = 0.3$ μm. The quantum efficiency of the depletion layer gradually decreases with increasing t, but it is small and plays no major role. The total quantum efficiency attains its maximum at $t \approx 0.2L_h$ for $s_1 = 0$. This maximum is shifted toward smaller t values as the surface recombination velocity s_1 increases. The position of the total quantum efficiency maximum also depends on the absorption coefficient. When the absorption coefficient increases, the depth of the junction at which the total efficiency attains maximum decreases.

The surface recombination velocity significantly affects η in the range of high absorption coefficient values (small wavelengths) so the depth of radiation penetration $1/\alpha$ is very small. The quantum efficiency is constant for all values of the absorption coefficient when the surface recombination velocity is much smaller than a certain characteristic value s_o and then decreases to a smaller but also constant value in the range $s_1 \gg s_o$. In Van De Wiele [71], it was found that the value s_o can be determined from the formula $s_o = (D_h/L_h)\, cth(x_n/L_h)$ and that it is independent of the absorption coefficient.

Normally, the photodiode is designed so that most of the radiation is absorbed in one side of the junction, for example, in the p-type side in Figure 12.14a. This could be achieved in practice either by making the n-type region very thin or by using a heterojunction in which the bandgap in the n-region is larger than the photon energy so that most of the incident radiation can reach the junction without being absorbed. If the back contact is several minority carrier diffusion lengths, L_e, away from the junction, the quantum efficiency is given by:

$$\eta(\lambda) = (1-r)\frac{\alpha(\lambda)L_e}{1+\alpha(\lambda)L_e}. \tag{12.93}$$

If the back contact is less than a diffusion length away from the junction, the quantum efficiency tends to:

$$\eta(\lambda) = (1-r)\left[1-e^{-\alpha(\lambda)d}\right], \tag{12.94}$$

where d is the thickness of the p-type region. It has been assumed that the back contact has zero surface recombination velocity and that no radiation is reflected from the back surface. Thus, if the given conditions hold, a high quantum efficiency can be achieved using an antireflection coating to minimize the reflectance of the front surface, ensuring that the device is thicker than the absorption length.

It should be noticed that many works (e.g., [73–75]) based on numerical and analytical approaches presented computer solutions for the two-dimensional and three-dimensional cases of photodiodes.

12.2.1.3 Noise

In comparison with photoconductive detectors, the two fundamental processes responsible for thermal noise mechanisms (fluctuations in the velocities of free carriers due to their random motion, and due to randomness in the rates of thermal generation and recombination) are less readily distinguishable in the case of junction devices, giving rise jointly to shot noise on the minority carrier components, which make up the net junction current. The random thermal motion is responsible for fluctuations in the diffusion rates in the neutral regions of junction devices and g–r fluctuations both in the neutral regions and in the depletion region. We will show later that for a junction device at zero bias (i.e., when the net junction current is zero), the resulting noise is identical to Johnson noise associated with the incremental slope of the device.

A general theory of noise in photodiodes that is applicable at arbitrary bias and to all sources of leakage current has not been developed [76]. The intrinsic noise mechanism of a photodiode is shot noise in the current passing through the diode. It is generally accepted that the noise in an ideal diode is given by:

$$I_n^2 = \left[2q\left(I_D + 2I_s\right) + 4kT\left(G_J - G_o\right)\right]\Delta f, \tag{12.95}$$

where $I_D = I_s[\exp(qV/kT) - 1]$, G_J is the conductance of the junction, and G_o is the low-frequency value of G_J. In the low-frequency region, the second term on the right-hand side is zero. For a diode in thermal equilibrium (i.e., without applied bias voltage and external photon flux), the mean square noise current is just the Johnson–Nyquist noise of the photodiode zero bias resistance $\left(R_o^{-1} = qI_s/kT\right)$:

$$I_n^2 = \frac{4kT}{R_o}\Delta f, \tag{12.96}$$

and

$$V_n = 4kTR_o\Delta f. \tag{12.97}$$

Note that the mean-square shot noise in reverse bias is half the mean-square Johnson–Nyquist noise at zero bias.

For a diode exposed to background flux density Φ_b, an additional current $I_{ph} = q\eta A\Phi_b$ constitutes a statistically independent contribution to the mean square noise current. Then [3]

$$I_n^2 = 2q\left[q\eta A\Phi_b + \frac{kT}{qR_o}\exp\left(\frac{qV}{\beta kT}\right) + \frac{kT}{qR_o}\right]\Delta f, \tag{12.98}$$

where

$$R_o = \left(\frac{dI}{dV}\right)^{-1}_{|V=0} = \frac{\beta kT}{qI_s}, \tag{12.99}$$

is the dark resistance of the diode at zero bias voltage. In the case of zero bias voltage, Equation 12.96 becomes

$$I_n^2 = \frac{4kT\Delta f}{R} + 2q^2\eta A\Phi_b\Delta f, \tag{12.100}$$

and then

$$V_n = \left(4kT + 2q^2\eta A\Phi_b R_o\right)R_o\Delta f \tag{12.101}$$

These forms of the noise equation at zero bias voltage are generally assumed to be applicable independent of the other sources of current.

Equation 12.98 predicts that the shot noise is decreased by operating the diode under a reverse bias. In the absence of a background-generated current, the current noise is equal to the Johnson noise ($4kT\Delta f/R_o$) at zero bias, and it tends to the usual expression for shot noise ($2qI_D\Delta f$) for voltages greater than a few kT in either direction. However, this improved performance under reverse bias is quite difficult to achieve in practice. In real devices, the current noise often increases under a reverse bias, due to presumably $1/f$ noise in the leakage current. In the earlier analysis, $1/f$ noise has been ignored.

12.2.1.4 Detectivity

In the case of the photodiode, the photoelectric gain is usually equal to 1; so according to Equation 4.31, the current responsivity is

$$R_i = \frac{q\lambda}{hc}\eta, \tag{12.102}$$

and the detectivity (see Equation 4.33) can be determined as

$$D^* = \frac{\eta\lambda q}{hc}\left[\frac{4kT}{R_oA} + 2q^2\eta\Phi_b\right]^{-1/2}, \tag{12.103}$$

and is obtained from Equations 12.100 and 12.102.

For the last formula we may distinguish two important cases:

- Background-limited performance; if $4kT/R_oA \ll 2q^2\eta\Phi_b$, then we obtain Equation 4.38
- Thermal noise–limited performance; if $4kT/R_oA \gg 2q^2\eta\Phi_b$, then

$$D^* = \frac{\eta\lambda q}{2hc}\left(\frac{R_oA}{kT}\right)^{1/2}. \tag{12.104}$$

Figure 12.18 shows detectivity against R_oA for diodes of 12 μm and 5 μm cutoff with $\eta = 50\%$ operated in 300 K, $f/1$ background. The minimum requirements for near–background-limited operation are 1.0 and 160 Ωcm^2, at 77 and 195 K, respectively, if the criterion $4kT/R_oA = 2q^2\eta\Phi_b$ is used [77].

The detectivity in the absence of background photon flux can also be expressed as:

$$D^* = \frac{\eta\lambda q}{hc}\left[\frac{A}{2q(I_D + 2I_s)}\right]^{1/2}. \tag{12.105}$$

In reverse biases, I_d tends to $-I_s$ and the expression in brackets tends to I_s.

From the discussion, it is shown that the performance of an ideal diffusion-limited photodiode can be optimized by maximizing the quantum efficiency and minimizing the reverse saturation current, I_s. For a diffusion-limited photodiode, the general expression for the saturation current of the electron from the p-type side is (see Equation 12.80):

$$I_s^p = A\frac{qD_en_{po}}{L_e}\frac{sh(d/L_e) + (s_2L_e/D_e)ch(d/L_e)}{ch(d/L_e) + (s_2L_e/D_e)sh(d/L_e)}. \tag{12.106}$$

It is normally possible to minimize the leakage current from the side that does not contribute to the photosignal. The minority carrier generation rate and hence the diffusion current can be greatly reduced, in theory at least, by increasing doping or the bandgap on the inactive side of the junction.

If the back contact is several diffusion lengths away from the junction, then Equation 12.106 tends to

$$I_s = \frac{qD_en_{po}}{L_e}A. \tag{12.107}$$

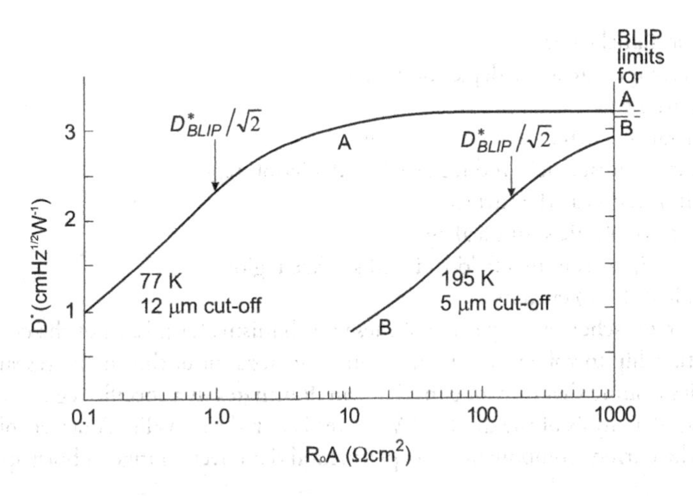

Figure 12.18 Detectivity as a function of R_oA for zero bias photodiodes with η = 50% operated in a 300 K, f/1 background. (From Ref. [77])

As the back contact is brought closer to the junction, the leakage current can either increase or decrease, depending on whether the surface recombination velocity is greater than the diffusion velocity D_e/L_e. In the limiting case where $d \ll L_e$, the saturation current is reduced by a factor d/L_e relative to Equation 12.107 for $s = 0$ and increased by a factor of L_e/d for $s = \infty$. If the surface recombination velocity is small, then Equation 12.107 can usually be written in the form:

$$I_s = qGV_{\text{diff}}, \tag{12.108}$$

where G is the bulk minority carrier generation rate per unit volume and V_{diff} is the effective volume of material from which the minority carriers diffuse to the junction. The effective volume is AL_e for $L_e \ll d$ and tends to Ad for $L_e \gg d$. For p-type material, the generation rate is given by:

$$G = \frac{n_{po}}{\tau_e} = \frac{n_i^2}{N_d \tau}. \tag{12.109}$$

The discussion indicates that the performance of the device is strongly dependent on the properties of the back contact. The most common solution to this problem is to move the back contact many diffusion lengths away to one side and to ensure that all the surfaces are properly passivated. Alternatively, the back contact itself can sometimes be designed to have a low surface recombination velocity by introducing a barrier for minority carriers between the metal contact and the rest of the device. This barrier can be made by increasing the doping or the bandgap near the contact, which effectively isolates the minority carriers from the high recombination rate at the contact.

12.2.2 REAL P-N JUNCTIONS

In the previous section, photodiodes were analyzed in which the dark current was limited by diffusion. However, this behavior is not always observed in practice, especially for wide-gap semiconductor p-n junctions. Several additional excess mechanisms are involved in determining the dark current–voltage characteristics of the photodiode. The dark current is the superposition of current contributions from three diode regions: bulk, the depletion region, and surface. Between them we can distinguish:

1. Thermally generated current in the bulk and depletion region
    - Diffusion current in the bulk p and n regions
    - Generation–recombination current in the depletion region
    - Band-to-band tunneling
    - Intertrap and trap-to-band tunneling

- Anomalous avalanche current
- The ohmic leakage across the depletion region

2. Surface leakage current
    - Surface generation current from surface states
    - Generation current in a field-induced surface depletion region
    - Tunneling induced near the surface
    - The ohmic or nonohmic shunt leakage
    - Avalanche multiplication in a field-induced surface region
3. Space charge–limited (SCL) current

Figure 12.19 illustrates schematically some of these mechanisms [20]. Each of the components has its own individual relationship to voltage and temperature. On account of this, many researchers analyzing the I-V characteristics assume that only one mechanism dominates in a specific region of the diode bias voltage. This method of analysis of the diode's I-V curves is not always valid. A better solution is to numerically fit the sum of the current components to experimental data over a range of both applied voltage and temperature.

In a later section, we will be concerned with the current contribution of high-quality photodiodes with high R_oA products limited by:

- Generation–recombination within the depletion region
- Tunneling through the depletion region
- Surface effects
- Impact ionization
- SCL current

12.2.2.1 Generation–recombination current

The importance of this current mechanism was first pointed out by Sah et al. [78], and was later extended by Choo [79]. It appears that one space charge region g–r current could be more important than its diffusion current, especially at low temperatures, although the width of the space charge region is much less than the minority carrier diffusion length. The generation rate in the depletion region can be very much greater than in the bulk of the material. In reverse bias, the current can be given by an equation similar to Equation 12.108.

$$I = qG_{dep}V_{dep}, \tag{12.110}$$

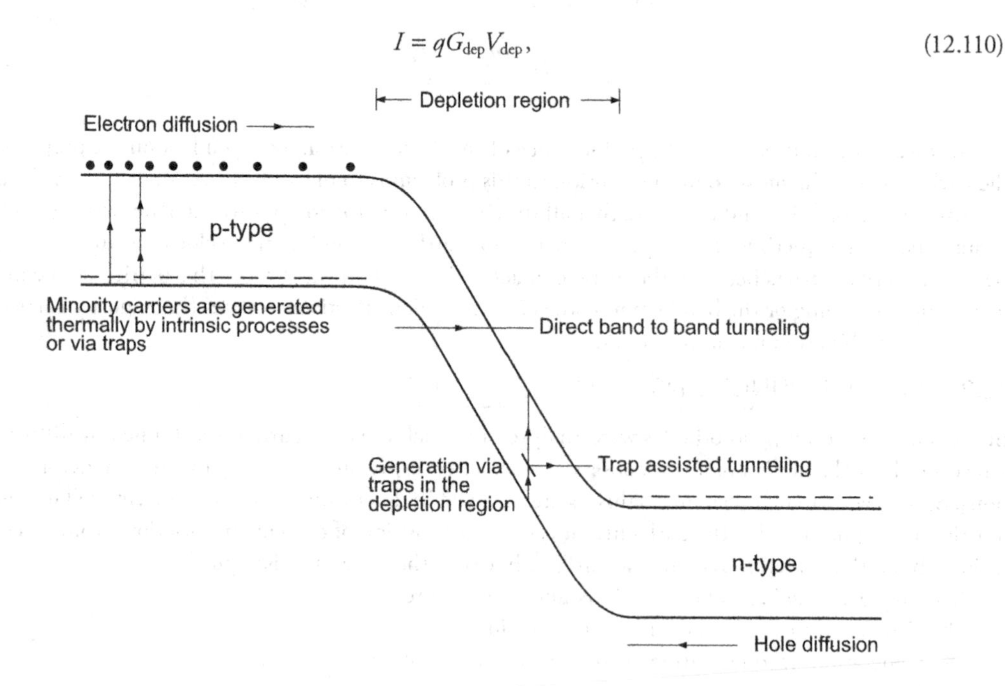

Figure 12.19 Schematic representation of some of the mechanisms by which dark current is generated in a reverse-biased p-n junction. (From Ref. [20])

where G_{dep} is the generation rate and V_{dep} is the volume of the depletion region. In particular, the generation rate from traps in the depletion region is given by the Shockley–Read–Hall (SRH) formula

$$G_{dep} = \frac{n_i^2}{n_1\tau_{eo} + p_1\tau_{ho}}, \tag{12.111}$$

where n_1 and p_1 are the electron and hole concentrations that would be obtained if the Fermi energy was at the trap energy, and τ_{eo} and τ_{ho} are the lifetimes in strong n-type and p-type materials. Normally, one of the terms in the denominator of Equation 12.111 dominate and for the case of a trap at the intrinsic level, $n_i = p_1 = n_i$, giving

$$G_{dep} = \frac{n_i}{2\tau_o}, \tag{12.112}$$

and the g–r current of the depletion region is equal to:

$$J_{GR} = \frac{qwn_i}{2\tau_o}. \tag{12.113}$$

The comparison of Equations 12.109 and 12.112 indicates that the generation rate in the bulk of the material is proportional to n_i^2, whereas the generation rate in the depletion region is proportional to n_i for a mid-gap state.

The width of the depletion region, and therefore its volume, increases with reverse voltage. For an abrupt junction,

$$w = \sqrt{\frac{2\varepsilon_o\varepsilon_r\left(V_{bi} \pm V\right)}{qN_aN_d\left(N_a + N_d\right)}}, \tag{12.114}$$

where N_a and N_d are the acceptor and donor concentrations, respectively; $V_{bi} = \left(kT/q\right)\ln\left(N_aN_d/n_i^2\right)$ is the built-in voltage; and V is the applied voltage. For a linearly graded junction, the width of the depletion region depends on $V^{1/3}$.

The g–r current varies roughly as the square root of the applied voltage ($w \sim V^{1/2}$) for the case of an abrupt junction, or as the cube root ($w \sim V^{1/3}$) for a linearly graded junction. This behavior, in which the current increases with reverse bias, can be contrasted with a diffusion-limited diode, where the reverse current is independent of voltage above a few kT/q.

The space charge region g–r current varies with temperature as n_i; that is, less rapidly than the diffusion current that varies as n_i^2; so that a temperature is finally reached at which the two currents are comparable and below this temperature, the g–r current dominates. Usually, at the lowest temperatures, current density may saturate due to the existence of a weakly temperature-dependent shunt resistance. The effect of space charge recombination is to be seen most readily in wide-bandgap semiconductors.

In the Sah–Noyce–Shockley theory [78], the doping levels were assumed to be the same on the two sides of the junction, and a single recombination center located in the vicinity of the gap was also assumed. The g–r current density under reverse-bias voltage and for forward-bias voltage values that are less than V_{bi} by several kT/q was derived as

$$J_{GR} = \frac{qn_iw}{\left(\tau_{eo}\tau_{ho}\right)^{1/2}}\frac{2sh\left(qV/2kT\right)}{q\left(V_{bi} - V\right)/kT}f(b), \tag{12.115}$$

where τ_{eo}, τ_{ho} are the carrier lifetimes with the depletion region. The function $f(b)$ is a complicated expression involving the trap level E_t, the two lifetimes, and applied voltage V:

$$f(b) = \int_0^\infty \frac{dx}{x^2 + 2bx + 1},$$

$$b = \exp\left(-\frac{qV}{2kT}\right)ch\left[\frac{E_t - E_i}{kT} + \frac{1}{2}\ln\left(\frac{\tau_{eo}}{\tau_{ho}}\right)\right],$$

where E_i is the intrinsic Fermi level. The function $f(b)$ has a maximum value of $\pi/2$, which occurs at small values of b (forward biases > $2kT/q$); $f(b)$ decreases as b increases. When $E_t = E_i$ and $\tau_{eo} = \tau_{ho} = \tau_o$, the recombination center has its maximum effect and then for symmetrical junction parameters $f(b) \approx 1$, and J_{GR} is determined by Equation 12.113.

For small applied bias, the function $f(b)$ may be taken as independent of V. We can also neglect the bias dependence of the depletion width w for small bias. The zero bias resistance is then found by differentiating Equation 12.115 and setting $V = 0$:

$$(R_oA)_{GR} = \left(\frac{dJ_{GR}}{dV}\right)^{-1}_{V=0} = \frac{V_b(\tau_{eo}\tau_{ho})^{1/2}}{qn_iwf(b)}. \tag{12.116}$$

For simplicity, we further assume $\tau_{ho} = \tau_{eo} = \tau_o$, $E_t = E_i$ and $f(b) = 1$. Then, Equation 12.116 becomes

$$(R_oA)_{GR} = \frac{V_b\tau_o}{qn_iw}. \tag{12.117}$$

In evaluating Equation 12.117, the term of greatest uncertainty is τ_o.

12.2.2.2 Tunneling current

The third type of dark current component that can exist is a tunneling current caused by electrons directly tunneling across the junction from the valence band to the conduction band (direct tunneling) or by electrons indirectly tunneling across the junction by way of intermediate trap sites in the junction region (direct tunneling or trap-assisted tunneling [TAT]; see Figure 12.19).

The usual direct tunneling calculations assume a particle of constant effective mass incident on a triangular or parabolic potential barrier. For the triangular potential barrier, [80]

$$J_T = \frac{q^2EV_b}{4\pi^2\hbar^2}\left(\frac{2m^*}{E_g}\right)^{1/2}\exp\left[-\frac{4(2m^*)^{1/2}E_g^{3/2}}{3q\hbar E}\right]. \tag{12.118}$$

For an abrupt p-n junction, the electric field can be approximated by

$$E = \left[\frac{2q}{\varepsilon_o\varepsilon_s}\left(\frac{E_g}{q} \pm V_b\right)\frac{np}{n+p}\right]^{1/2}. \tag{12.119}$$

The tunnel current is seen to have an extremely strong dependence on energy gap, applied voltage, and an effective doping concentration $N_{ef} = np/(n+p)$. It is relatively insensitive to temperature variation and the shape of the junction barrier. For the parabolic barrier, [80]

$$J_T \propto \exp\left[-\frac{(\pi m^*)^{1/2}E_g^{3/2}}{2^{3/2}q\hbar E}\right]. \tag{12.120}$$

Anderson [81], based on the Wentzel-Kramers-Brillouin (WKB) approximation and Kane **k·p** theory, developed expressions for direct interband tunneling in narrow-gap semiconductors for asymmetrically abrupt p-n homo-junctions. Anderson's expressions are convenient to use as a first estimation, especially for bias voltage near zero. But due to the extreme sensitivity of tunneling to the electric field, it may lead to orders of magnitudes of error when applied to the general device structure. This was demonstrated by Beck and Byer in their tunneling calculations of linearly graded junctions with various slopes [82]. Next, Adar has presented a new technique for direct band-to-band tunneling (BBT) in narrow-gap semiconductors including spatial integration throughout the depletion region at each bias point [83].

Apart from direct BBT, tunneling is possible by means of indirect transitions in which impurities or defects within the space charge region act as intermediate states [84]. This is a two-step process in which one step is a thermal transition between one of the bands and the trap and the other is a tunneling between the trap and the other band. The tunneling process occurs at lower fields than direct BBT because the electrons have a shorter distance to tunnel (see Figure 12.19). This type of tunneling has been reported in HgCdTe p-n junctions at low temperatures [85–93]. Small but finite acceptor activation energies are generally observed in p-type HgCdTe.

Trap-assisted tunneling can occur via the following:

- A thermal transition with rate $\gamma_p p_1$ to trap center in the bandgap located at energy E_t from the valence band [where γ_p is the hole recombination coefficient for the trap centers of density N_t, and $p_t = N_v \exp(-E_t/kT)$]
- A tunnel transition with rate $\omega_v N_v$ (where ω_v represents the carrier tunneling probability between the center and the band)
- Followed by tunnel transition to the conduction band with rate $\omega_c N_c$

The total trap-assisted current is then given by [86]:

$$J_T = qN_t w\left(\frac{1}{\gamma_p p_1 + \omega_v N_v} + \frac{1}{\omega_c N_c}\right)^{-1}. \quad (12.121)$$

On account of the low density of states associated with the small value of the conduction band electron mass, thermal transitions from the conduction band are ignored. For the limiting case $\omega_c N_c < \gamma_p p_1$ and $\omega_c N_c \cong \omega_v N_v$, then

$$J_T = qN_t \omega_c N_c w. \quad (12.122)$$

Assuming a parabolic barrier and uniform electric field, the tunneling rate from a neutral center and to the conduction band is given by

$$\omega_c N_c = \frac{\pi^2 q m^* E M^2}{h^3 (E_g - E_t)} \exp\left[-\frac{(m^*/2)^{1/2} E_g^{3/2} F(a)}{2qE\hbar}\right], \quad (12.123)$$

where $a = 2(E_t/E_g) - 1$, $F(a) = (\pi/2) - a(1 - a^{1/2})^{1/2} - (1/\sin a)$. The matrix element M is associated with the trap potential. The experimentally determined value for the quantity $M^2(m^*/m)$ for silicon is found to be 10^{23} Vcm^3 [84]. A similar value is assumed for HgCdTe [86,91,92]. Tunneling increases exponentially with decreasing effective mass; thus the light hole mass dominates tunnel processes between the valence band and trap centers. In comparison with direct tunneling, indirect tunneling is critically dependent not only on the electric field (doping concentration) but also on the density of recombination centers and their location in the bandgap (via the geometrical factor $0 < F(a) < \pi$). Most of the tunnel current will pass through the trap level with the highest transition probability (i.e., the carriers choose the path of least resistance). If the conduction band and light hole masses are approximately equal, then the maximum tunneling probability occurs for midgap states for which $\omega_c N_c \cong \omega_v N_v$. Often in the absence of detailed information on the location of trap states in the material, the theoretical treatment assumes a single midgap SRH center. Assuming another trap level changes the overall magnitude of the calculated tunnel current, the general behavior of the tunnel current with electric field and temperature will be similar. In general, tunnel current varies exponentially with an electric field and has a relatively weak temperature dependence in comparison with diffusion and depletion currents.

The influence of different junction current components on the R_oA product for various types of junctions fabricated in narrow-gap semiconductors has been considered in many papers. For example, Figure 12.20 shows the dependence of the R_oA product components on the dopant concentrations for n^+-p $Hg_{0.78}Cd_{0.22}Te$ abrupt junctions at 77 K [94]. We can see that the performance of the junctions at zero bias is limited by the tunneling current if the substrate doping is too high. The doping concentration of 10^{16} cm^{-3} or less is required to produce high R_oA products. Below this concentration, the R_oA product is determined by diffusion current with minority carrier lifetime conditioned by the Auger 7 process. However, to obtain the highest possible value of the R_oA product and avoid the effect of a fixed insulator charge of a passivation junction layer, the technological process of photodiode preparation should be performed in a manner to obtain a dopant concentration above 10^{15} cm^{-3} [94]. From this, it follows that the optimum concentration

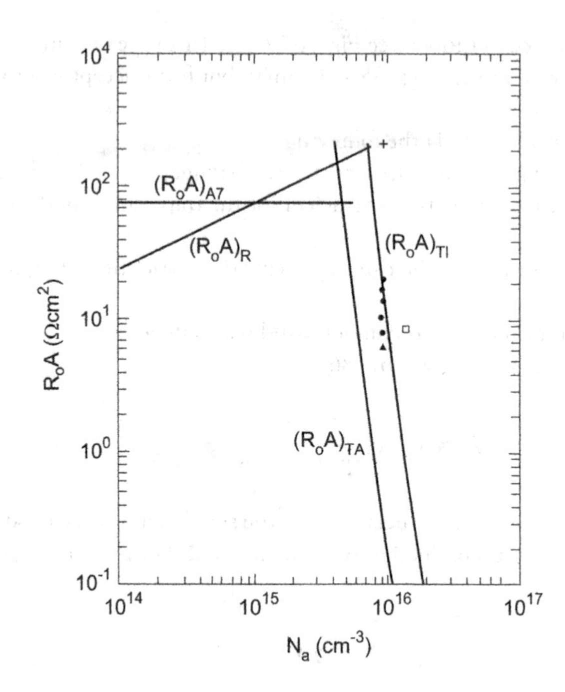

Figure 12.20 The dependence of the R_oA product on acceptor concentration for n⁺ -p $Hg_{0.78}Cd_{0.22}Te$ photodiodes at 77 K. The experimental values (•, +, ▲, □) are taken from different papers. (From Ref. [94])

range on the lower doping side of the junction is $10^{15} < N_a < 10^{16}$ cm^{-3}. From the comparison of the theoretical curves with the experimental data, it may be seen that a satisfactory consistency has not been achieved. The main reason for these discrepancies is probably due to the inconsistency of the abrupt junction model with the experimental junction profiles. The Anderson tunneling theory assumes a uniform charge model for the potential barrier and nonparabolicity of the band structure. Smaller values of tunnel current (higher experimental R_oA products for acceptor concentration $10^{16} < N_a < 2 \times 10^{16}$ cm^{-3}; see Figure 12.20) can be associated with a decrease of electric field away from the metallographic junction boundary. Furthermore, the tunnel current calculations are strictly valid only for the approximately empty well condition. As the well fills, the tunnel current tends to decrease due to the collapse of the well.

12.2.2.3 Surface leakage current

In a real p-n junction, particularly in wide-gap semiconductors and at low temperatures, additional dark current related to the surface occurs. Surface phenomena play an important part in determining photovoltaic detector performance. The surface provides a discontinuity that can result in a large density of interface states. These generate minority carriers by the SRH mechanism, and can increase both the diffusion and depletion region–generated currents. The surface can also have a net charge, which affects the position of the depletion region at the surface.

The surface of actual devices is passivated so as to stabilize the surface against chemical and heat-induced changes as well as to control surface recombination, leakage, and related noise. Native oxides and overlying insulators are commonly employed in p-n junction fabrication that introduces fixed charge states, which then induce accumulation or depletion at the semiconductor–insulator surface. We can distinguish three main types of states on the semiconductor–insulator interface, namely, fixed insulator charge, low surface states, and fast surface states. The fixed charge in the insulator modifies the surface potential of the junction. A positively charged surface pushes the depletion region further into the p-type side and a negatively charged surface pushes the depletion region toward the n-type side. If the depletion region is moved toward the more highly doped side, the field will increase, and tunneling becomes more likely. If it is moved toward the more lightly doped side, the depletion region can extend along the surface, greatly increasing the depletion region–generated currents. When sufficient fixed charge is present, accumulated

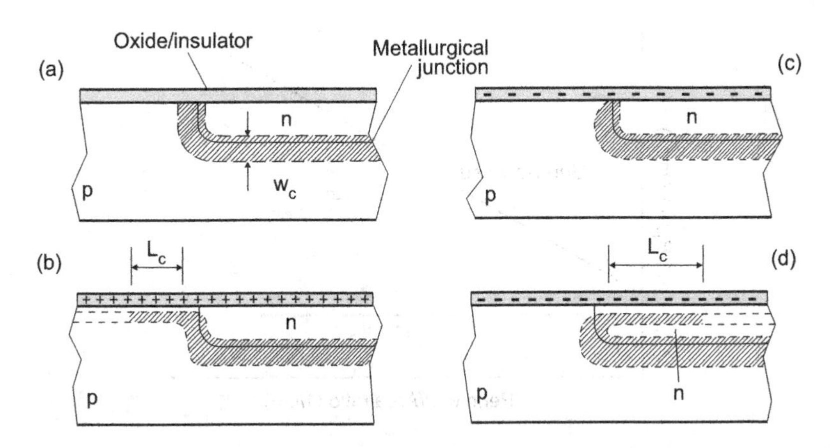

Figure 12.21 Effect of fixed insulator charge on the effective junction space charge region: (a) flat band condition, (b) positive fixed charge (inversion of the p side, formation of an n-type surface channel), (c) negative fixed charge (accumulation of the p side, field-induced junction at the surface), and (d) large negative fixed charge (inversion of the n side, formation of p-type surface channel). (From Ref. [95])

and inverted regions as well as n-type and p-type surface channels are formed (see Figure 12.21) [95]. An ideal surface would be electrically neutral and would have a very low density of surface states. The ideal passivation would be a wide gap insulator grown with no fixed charge at the interface.

Fast interface states, which act as g–r centers, and fixed charge in the insulator cause a variety of surface-related current mechanisms. The kinetics of g–r through fast surface states is identical to that through bulk SRH centers. The current in a surface channel is given by

$$I_{\mathrm{GRS}} = \frac{qn_i w_c A_c}{\tau_o}, \tag{12.124}$$

where w_c is the channel width and A_c is the channel area.

Apart from g–r processes occurring at the surface and within surface channels, there are other surface-related current mechanisms, termed surface leakage, with ohmic or breakdown-like current-voltage characteristics. They are nearly temperature independent. Surface breakdown occurs in a region of high electric field (Figure 12.21c, where the depletion layer intersects the surface; Figure 9.21d, very narrow depletion layer).

To eliminate the influence of leakage current, p-n junctions are usually passivated. The effectiveness of passivation is commonly evaluated using the variable area diode array (VADA) method. The dark current density can be expressed as the summation of the bulk component of dark current and the surface leakage current. The surface dependence of inverse of the dynamic resistance-area product at zero bias R_0A of passivated diode can be approximated as

$$\frac{1}{R_0A} = \frac{1}{R_0A_{\mathrm{bulk}}} + \frac{1}{r_{\mathrm{surface}}}\frac{P}{A}, \tag{12.125}$$

where $(R_0A)_{\mathrm{bulk}}$ is the bulk R_0A contribution ($\Omega\mathrm{cm}^2$), r_{surface} is the surface resistivity (Ωcm), P is the diodes perimeter, and A is the diodes area. Note that the bulk and surface components of the R_0A product have different dependencies of junction geometry.

The slope of the function given by Equation 12.125 is directly proportional to the surface-dependent leakage current of the diode. Figure 12.22 schematically illustrates the $1/R_0A$ vs (P/A) dependence for the diode with ideal and non-ideal passivations. If the bulk current dominates the detector performance, then the curve has a slope close to zero. If the surface leakage is significant, then an increase in the dark current density is observed for smaller devices—surface resistivity indicates weaker dependence of the diode's characteristics on the surface effects.

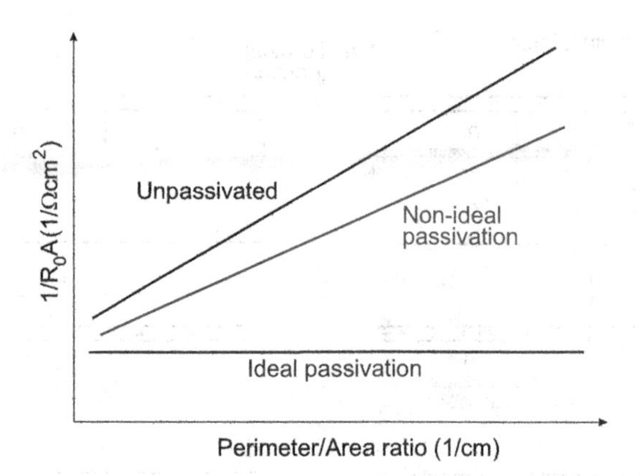

Figure 12.22 Dependence of the dynamic resistance-area product at zero bias vs. perimeter-to-area ratio for VADA diodes with ideal and nonideal passivations.

The zero bias resistance area product for the case of uniformly distributed bulk and surface g–r centers is given by

$$\left(\frac{1}{R_oA}\right)_{GR} = \frac{en_iw}{V_b}\left(\frac{1}{\tau_o} + \frac{S_oP}{A}\right), \tag{12.126}$$

where V_b is the built-in voltage of the p-n junction. The second term in parentheses is for g–r centers localized at the surface, and the parameter S_o is referred to as the g–r surface recombination velocity, which is proportional to the density of g–r defects.

Band bending at the p-n junction surface can be controlled by a gate electrode overlaid around the junction perimeter on an insulating film.

The dark current as a sum of several independent contributions can also be written as follows:

$$I = I_s\left[\exp\frac{q(V - IR_s)}{\beta kT} - 1\right] + \frac{V - IR_s}{R_{sh}} + I_T, \tag{12.127}$$

where R_s is the series resistance and R_{sh} is the shunt resistance of the diode. In the case of the predominant diffusion current, the β coefficient approaches unity, but when the g–r current is mainly responsible for carrier transport, $\beta = 2$.

12.2.2.4 Space charge–limited current

In the case of wide bandgap p-n junctions, the forward current-voltage characteristics are often described by the equation:

$$J \propto \exp\left(\frac{qV}{\beta kT}\right), \tag{12.128}$$

where the diode ideality factor $\beta > 2$. This value of β does not fall within the range that results when diffusion current ($\beta = 1$) or depletion layer current ($\beta = 2$) dominate the forward-bias current conduction. This behavior is typical for an insulator with shallow and/or deep traps and thermally generated carriers.

Space charge–limited current flow in solids has been considered in detail by Rose [96], Lampert [97], and Lampert and Mark [98]. They loosely defined materials with $E_g \leq 2$ eV as semiconductors and those with $E_g \geq 2$ eV as insulators.

When a sufficiently large field is applied to an insulator with ohmic contacts, electrons will be injected into the bulk of the material to form a current that is limited by space charge effects. When trapping centers are present, they capture many of the injected carriers, thus reducing the density of free carriers.

At low voltages where the injection of carriers into the semi-insulating material is negligible, Ohm's law is obeyed and the slope of the J-V characteristics defines the resistivity ρ of the material. At some applied voltage V_{TH}, the current begins to increase more rapidly than linearly with applied voltages. Here V_{TH} is given by

$$V_{TH} = 4\pi \times 10^{12} q p_t \frac{t^2}{\varepsilon}, \tag{12.129}$$

where t is the thickness of the semi-insulating material, and p_t is the density of trapped holes, if one carrier (holes) SCL current is considered. As the voltage is continuously increased beyond V_{TH}, additional excess holes are injected into the material and the current density is given by:

$$J = 10^{-13} \mu_h \varepsilon \theta \frac{V^2}{t^2}, \tag{12.130}$$

where θ is the probability of trap occupation determined as a ratio of the densities of free to trapped holes. Here θ is given by

$$\theta = \frac{p}{p_t} = \frac{N_v}{N_t} \exp\left(-\frac{E_t}{kT}\right), \tag{12.131}$$

where N_v is the effective density of states in the valence band, N_t is the density of traps, and E_t is the depth of traps from the top of the valence band. When the applied voltage further increases, the square-law region of Equation 12.130 will terminate in a steeply rising current that increases until it becomes the trap-free SCL currents given by

$$J = 10^{-13} \mu_h \varepsilon \frac{V^2}{t^2}. \tag{12.132}$$

The density of traps N_t can be determined from the voltage V_{TFL} at which the traps are filled and the currents rise sharply:

$$V_{TFL} = 4\pi \times 10^{12} q N_t \frac{t^2}{\varepsilon}. \tag{12.133}$$

The trap depth E_t can also be calculated from Equation 12.133 using the value of N_t. But if Equation 12.130 is measured as the function of temperature and a plot of $\ln(\theta T^{-3/2})$ versus $1/T$ is possible, E_t and N_t can be obtained directly from these data without referring to Equation 12.133.

For the purpose of illustration of the above phenomenon in semi-insulating materials, consider the four $\log J$ versus $\log V$ graphs in Figure 12.23 [99]. Figure 12.23a represents an ideal insulator where $I \propto V^2$,

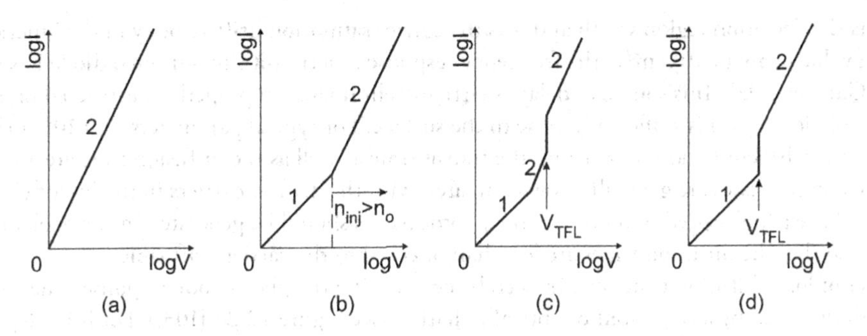

Figure 12.23 Schematic drawings of the logarithmic dependence of current versus voltage for (a) an ideal insulator, (b) a trap insulator with thermally generated free carriers, (c) an insulator with shallow traps and thermal free carriers, and (d) an insulator with deep traps and thermal free carriers. (From Ref. [99])

indicating SCL current flow. In other words, there are no thermally generated carriers resulting from impurity-band or band-to-band transitions; the conduction is only within the conduction band as a result of carrier injection. As shown in Figure 12.23b, ohmic conduction is obtained in trap-free insulators in the presence of thermally generated free carriers, n_o. When the injected carrier density n_{inj} exceeds n_o ($n_{inj} > n_o$), ideal insulator characteristics are observed ($I \propto V^2$). Shallow traps contribute to an $I \propto V^2$ regime at a lower voltage followed by a sharp transition to an ideal insulator, square-law regime as shown in Figure 12.23c. The sharp transition corresponds to an applied voltage; V_{TFL} is required to fill a discrete set of traps that are initially unoccupied. Figure 12.23d illustrates the case of a material with deep traps that have become filled when n_{inj} becomes comparable to n_o. The voltage at which this occurs is V_{TFL}. Therefore, at $V < V_{TFL}$, ohmic conduction is observed, and at $V > V_{TFL}$, SCL current flow dominates.

12.2.3 RESPONSE TIME

Wide-bandwidth photodiodes are used both for direct and heterodyne detection. The considerable interest in high-frequency photodiodes has been mostly due to their application at 10.6 μm (involving CO_2 laser heterodyne detection) for lidar systems and applications for optical fiber communications.

The upper-frequency response of a photodiode may be determined by basically three effects: the time of carrier diffusion to the junction depletion region, τ_d; the transit time of carrier drift across the depletion region, τ_t; and the RC time constant associated with circuit parameters including the junction capacitance C and the parallel combination of diode resistance and external load (the series resistance is neglected).

The photodiode parameters responsible for these three factors are the absorption coefficient α, the depletion region width w_{dep}, the photodiode junction and package capacitances C_d, the amplifier capacitance C_a, the detector load resistance R_L, the amplifier input resistance R_a, and the photodiode series resistance R_s (see Figure 12.15).

Photodiodes designed for fast response are generally constructed so that the absorption of radiation occurs in the p-type region. This ensures that most of the photocurrent is carried by electrons that are more mobile than holes (whether by diffusion or drift). The frequency response for the diffusion process in a backside illuminated diode has been calculated by Sawyer and Rediker as a function of diode thickness, diffusion constant, absorption depth, minority carrier lifetime, and surface-recombination velocity [100]. Assuming that the diffusion length is greater than both the diode thickness and the absorption depth, the cutoff frequency where the response drops by $\sqrt{2}$ is given by [20,101]

$$f_{\text{diff}} = \frac{2.43D}{2\pi t^2}, \tag{12.134}$$

where D is the diffusion constant and t is the diode thickness.

The depletion region transit time is equal to

$$\tau_t = \frac{w_{\text{dep}}}{v_s}, \tag{12.135}$$

where w_{dep} is the depletion region width and v_s is the carrier saturation drift velocity in the junction field, which has a value of about 10^7 cm/s. The frequency response of a transit time–limited diode has been derived by Gartner [102]. This source of delay is virtually eliminated in properly constructed photodiodes where the depletion region is sufficiently close to the surface. For typical parameters $\mu_e = 10^4$ cm²/Vs, $v_s = 10^7$ cm/s, $\alpha = 5 \times 10^3$ cm^{-1} and $w_{\text{dep}} = 1$ μm, the transit time as well as the diffusion time are about 10^{-11} s.

The diffusion processes are generally slow compared with the drift of carriers in the high-field region. Therefore, to have a high-speed photodiode, the photocarriers should be generated in the depletion region or close to it so that the diffusion times are less than or equal to the carrier drift times.

The effect of long diffusion times can be seen by considering the photodiode response time when the detector is illuminated by a step input of optical radiation (see Figure 12.24 [103]). For fully depleted photodiodes, when $w_{\text{dep}} >> 1/\alpha$, the rise and fall time are generally the same. The rise and fall times of the photodiode follow the input pulse quite well (Figure 12.24b). If the photodiode capacitance is larger, the

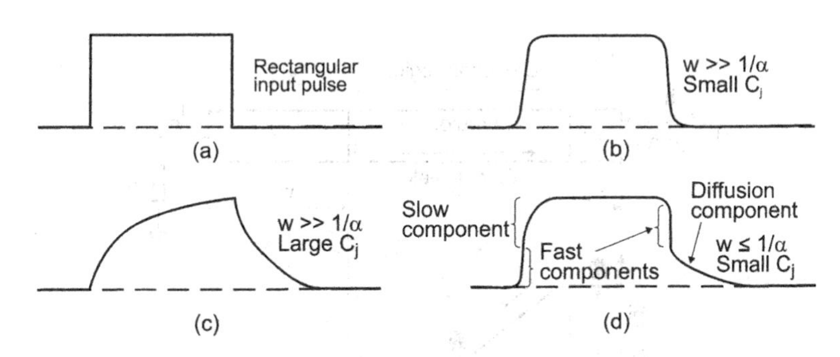

Figure 12.24 Photodiode pulse responses under various detector parameters. (From Ref. [103])

response time becomes limited by the *RC* time constant of the load resistor R_L and the photodiode capacitance (Figure 12.24c):

$$\tau_{RC} = \frac{AR_T}{2}\left(\frac{q\varepsilon_o\varepsilon_s N}{V}\right)^{1/2}, \tag{12.136}$$

$R_T = R_d(R_s + R_L)/(R_s + R_d + R_L)$ where R_s, R_d and R_L are the series, diode, and load resistances. The detector behaves approximately like a simple *RC* low-pass filter with a pass band given by

$$\Delta f = \frac{1}{2\pi R_T C_T}. \tag{12.137}$$

In a general case, R_T is the combination of the load and amplifier resistance and C_T is the sum of the photodiode and amplifier capacitance (see Figure 12.15).

To reduce the *RC* time constant, we can decrease the majority carrier concentration adjacent to the junction, increase *V* by means of a reverse bias, decrease the junction area, or lower either the diode resistance or the load resistance. Except for the application of a reverse bias, all of these changes reduce the detectivity. The trade-off between response time and detectivity is then apparent.

If the depletion layer is too narrow, electron hole pairs generated in the n and p regions would have to diffuse back to the depletion region before they could be collected. Devices with very thin depletion regions thus tend to show distinct slow- and fast-response components, as shown in Figure 12.24d; the fast component is due to carriers generated in the depletion region, whereas the slow component arises from the diffusion carriers.

Generally, a reasonable compromise between high-frequency response and high quantum efficiency is required, and the thickness of the absorption region is usually between $1/\alpha$ and $2/\alpha$.

12.3 P-I-N PHOTODIODES

The p-i-n photodiode is a popular alternative to the simple p-n photodiodes especially for ultra-fast photodetection in optical communication, measurement, and sampling systems. In p-i-n photodiodes, an undoped i-region (p^- or n^-, depending on the method of junction formation) is sandwiched between p^+ and n^+ regions. Figure 12.25 shows the schematic representation of a p-i-n diode, an energy band diagram under reverse-bias conditions together with optical absorption characteristics. Because of the very low density of free carriers in the i-region and its high resistivity, any applied bias drops entirely across the i-region, which is fully depleted at zero bias or very low value of reverse bias. Typically, for a doping concentration of ~10^{14}–10^{15} cm^{-3} in the intrinsic region, a bias voltage of 5–10 V is sufficient to deplete several micrometers, and the electron velocity also reaches the saturation value.

The p-i-n photodiode has a "controlled" depletion layer width, which can be tailored to meet the requirements of photoresponse and bandwidth. A trade-off is necessary between response speed and quantum efficiency. For high response speed, the depletion layer width should be small but for high quantum

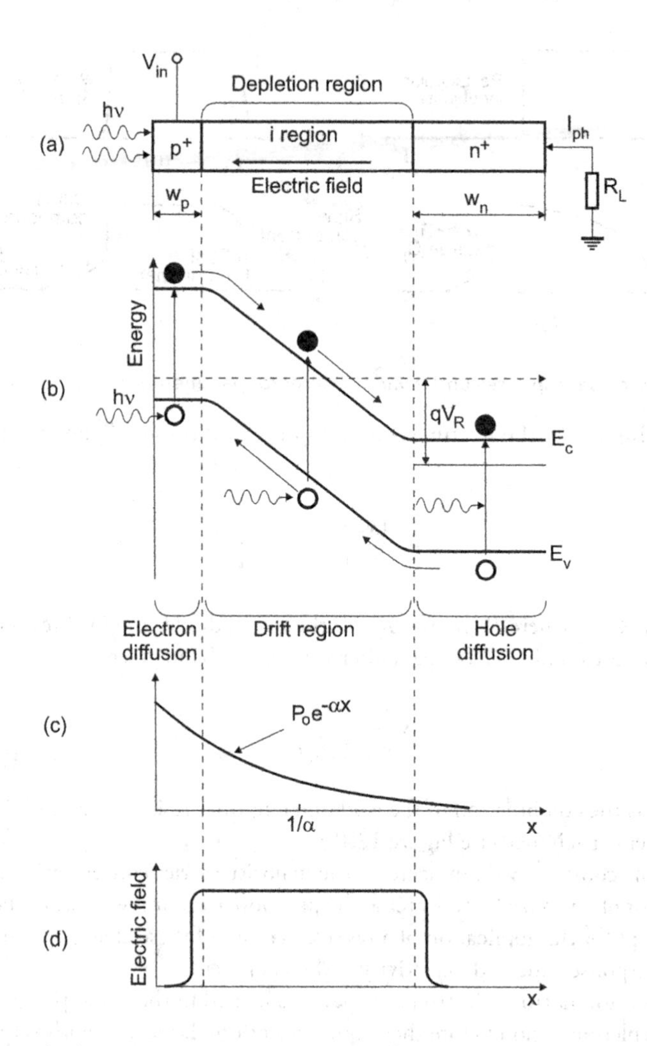

Figure 12.25 p-i-n photodiode: (a) Structure, (b) Energy band diagram, (c) Carrier generation characteristic, and (d) Electric field profile.

efficiency (or responsivity) the width should be large. An external resonant microcavity approach has been proposed to enhance quantum efficiency in such a situation [104,105]. In this approach, the absorption region is placed inside a cavity so that a large portion of the photons can be absorbed even with small detection volume.

The response speed of a p-i-n photodiode is ultimately limited either by transit time or by circuit parameters. The transit time of carriers across the i-layer depends on its width and the carrier velocity. Usually, even for moderate reverse biases, the carriers drift across the i-layer with saturation velocity. The transit time can be reduced by reducing the i-layer thickness. Fabricating the junction close to the illuminated surface can minimize the effect of diffusion of carrier created outside the i-layer.

The transit time of the p-i-n photodiode is shorter than that obtained in a p-n photodiode even though the depletion region is longer than in the p-n photodiode case due to carriers traveling at near their saturation velocity virtually the entire time they are in the depletion region (in p-n junction, the electric field peaked at the p-n interface and then rapidly diminished). For p and n regions less than one diffusion length in thickness, the response time to diffusion alone is typically 1 ns/μm in p-type silicon and about 100 ps/μm in p-type III–V materials. The corresponding values for n-type III–V materials is several nanoseconds per micrometer due to the lower mobility of holes.

Two generic p-i-n photodiodes, front-illuminated and back-illuminated are commonly used (see Figure 12.26). For practical applications, photoexcitation is provided through either an etched opening in the

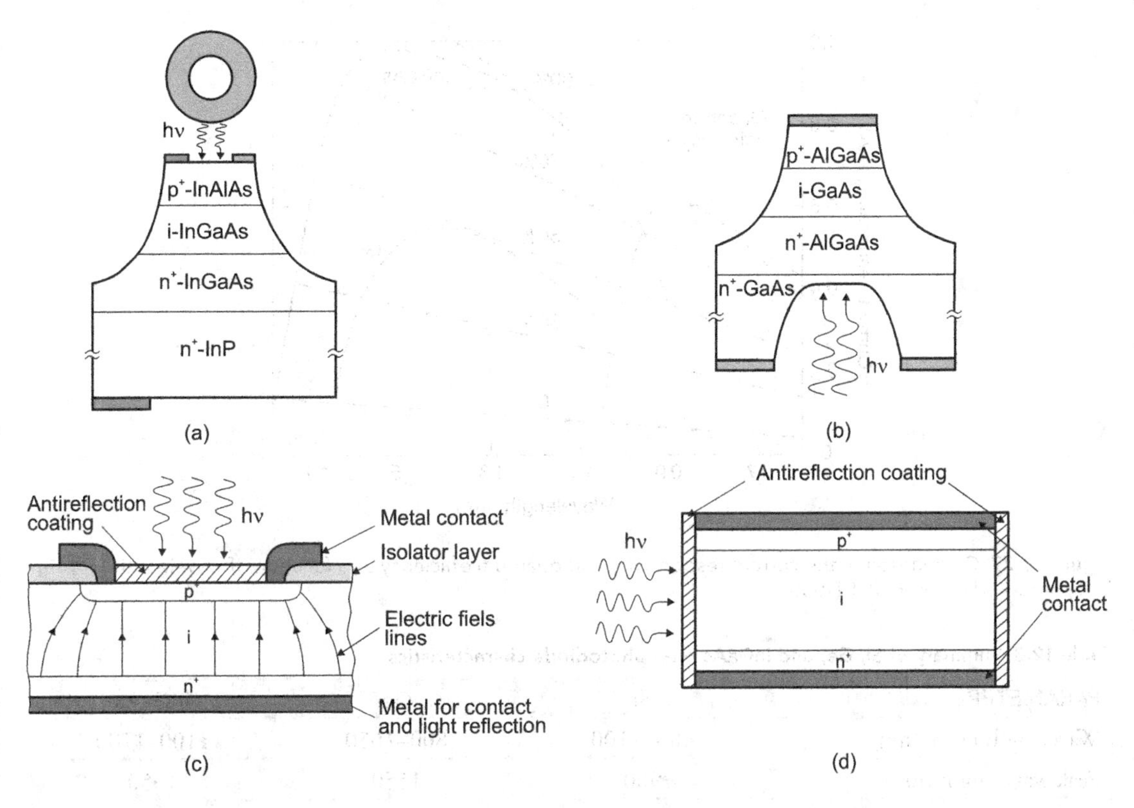

Figure 12.26 Device configurations of p-i-n photodiodes: (a) front-illuminated mesa, (b) back-illuminated mesa, (c) front-illuminated planar, and (d) parallel-illuminated planar.

top contact or an etched hole in the substrate. The latter reduces the active area of the diode to the size of the incident light beam. Sidewalls of mesas are covered using passivation materials such as polyimide. A compromise between quantum efficiency and response can be reached if the light is incident from the side, parallel to the junction as shown in Figure 12.26d. The light can also be allowed to strike at an angle that creates multiple reflections inside the device, substantially increasing the effective absorption and quantum efficiency. This solution is often used to couple a detector with one-mode fiber.

For the 1.0–1.55-μm wavelength range suitable for optical communications, Ge and a few III-V compound semiconductor alloys are materials for p-i-n photodiodes, primarily because of their large absorption coefficients (see Figure 4.7). Typical p-i-n photodiode responsivities as a function of wavelength in the near infrared (NIR) spectral region are shown in Figure 12.27 [103]. Representative values are 0.65 A/W for silicon at 900 nm and 0.45 A/W for germanium at 1.3 μm. For InGaAs, typical values are 0.9 A/W at 1.3 μm and 1.0 A/W at 1.55 μm. At present, III-V semiconductors have largely replaced germanium as materials for fiber optical compatible detectors. Due to small bandgap, the dark current of Ge photodiodes degrades the signal-to-noise ratio. Also, the passivation technique for Ge photodiodes is not satisfactory. Table 12.3 compares the performance values for p-i-n photodiodes derived from various vendor data sheets.

The principal source of noise in a p-i-n photodiode is g–r noise; it is larger than the Johnson noise since the dark current in a reverse-biased junction is very low.

12.4 AVALANCHE PHOTODIODES

When the electric field in a semiconductor is increased above a certain value, the carriers gain enough energy (greater than the bandgap) so that they can excite electron hole pairs by impact ionization. An APD operates by converting each detected photon into a cascade of moving carrier pairs [80,106–112]. The device is a strongly reverse-biased photodiode in which the junction electric field is large; the charge carriers therefore accelerate, acquiring enough energy to excite new carriers by the process of impact ionization.

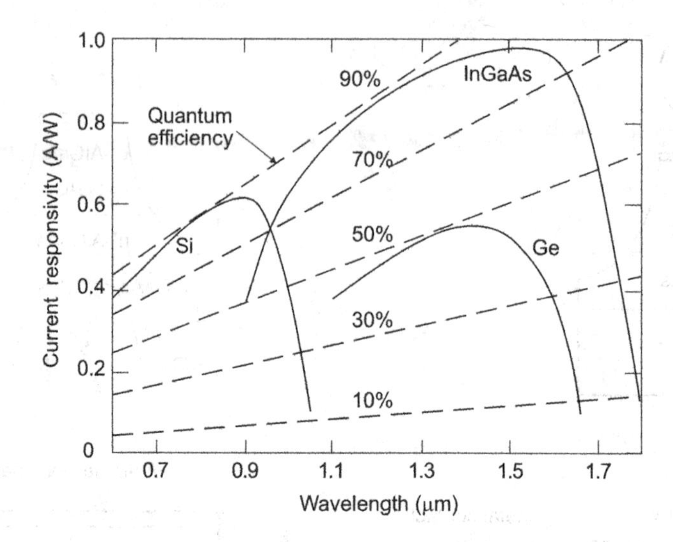

Figure 12.27 Comparison of the current responsivity and quantum efficiency as a function of wavelength for p-i-n NIR photodiodes. (From Ref. [103])

Table 12.3 Summary of Si, Ge, and InGaAs p-i-n photodiode characteristics

PARAMETER	Si	Ge	InGaAs
Wavelength range (nm)	400–1100	800–1650	1100–1700
Peak wavelength (nm)	900	1550	1550
Current responsivity (A/W)	0.4–0.6	0.4–0.5	0.75–0.95
Quantum efficiency (%)	65–90	50–55	60–70
Dark current (nA)	1–10	50–500	0.5–2.0
Rise time (ns)	0.5–1	0.1–0.5	0.05–0.5
Bandwidth (GHz)	0.3–0.7	0.5–3	1–2
Bias voltage (–V)	5	5–10	5
Capacity (pF)	1.2–3	2–5	0.5–2

In an optimally designed photodiode, the geometry of the APD should maximize photon absorption (for example, by assuming the form of a p-i-n structure) and the multiplication region should be thin to minimize the possibility of localized uncontrolled avalanches (instabilities or microplasmas) being produced by strong electric field. Greater electric field uniformity can be achieved in a thin region. These two conflicting requirements call for an APD design in which the absorption and multiplication regions are separate. For example, Figure 12.28 shows a reach-through p^+-π-p-n^+ APD structure that accomplishes this. Photon absorption occurs in the wide π-region (very lightly doped p region). Electrons drift through the π-region into a thin p-n^+ junction, where they experience a sufficiently strong electric field to cause avalanching. The reverse bias applied across the device is large enough for the depletion layer to reach through the p and π regions into the p^+ contact layer.

The avalanche multiplication process is illustrated in Figure 12.28d. A photon absorbed at point 1 creates an electron hole pair. The electron accelerates under the effect of the strong electric field. The acceleration process is constantly interrupted by random collisions with the lattice in which the electron loses some of its acquired energy. These competing processes cause the electron to reach an average saturation velocity. The electron can gain enough kinetic energy that, upon collision with an atom, can break the lattice bonds, creating a second electron hole pair. This is called impact ionization (at point 2). The newly created electron and hole both acquire kinetic energy from the field and create additional electron hole pairs (e.g., at point 3). These in turn continue the process, creating other electron hole pairs. The microscopic manner in which

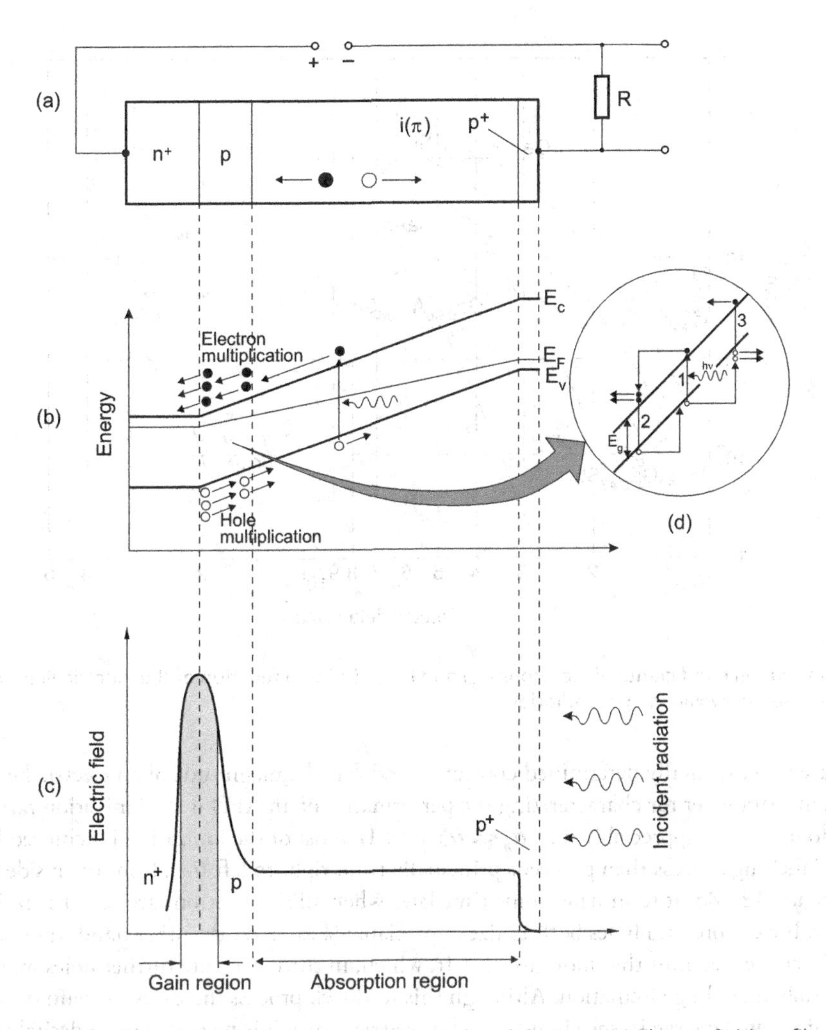

Figure 12.28 Avalanche photodiode: (a) Structure, (b) Energy band diagram, (c) Electric field profile, and (d) Schematic representation of the multiplication process.

a carrier gains energy from an electric field and undergoes impact ionization depends on the semiconductor band structure and the scattering environment (mainly optical phonons) it finds itself. A good review of the theory of the impact ionization process in the semiconductor at low fields is given by Capasso [107].

The abilities of electrons and holes to impact ionize are characterized by the ionization coefficients α_e and α_h. These quantities represent ionization probabilities per unit length. The ionization coefficients increase with the depletion layer electric field and decrease with increasing device temperature.

Figure 12.29 shows the curves of α_e and α_h as a function of the electric field for several semiconductors important in APDs [113]. As is shown, starting from a few 10^5 V/cm, the ionization coefficients steeply increase with a small increase of the electric field, while for field < 10^5 V/cm, ionization is negligible in all of the semiconductor compounds. In some semiconductors, electrons ionize more efficiently than holes (Si, GaAsSb, InGaAs, for which $\alpha_e > \alpha_h$), while in others the reverse is true (Ge, GaAs, where $\alpha_h > \alpha_e$).

The ionization coefficients increase with the applied electric field and decrease with increasing device temperature. The increase with the field is due to additional carrier velocity, while the decrease with temperature is due to an increase in nonionizing collisions with thermally excited atoms. For a given temperature, the ionization coefficients are exponentially dependent on the electric field and have a functional form of

$$\alpha = a\exp\left(-\left[\frac{b}{E}\right]^c\right), \tag{12.138}$$

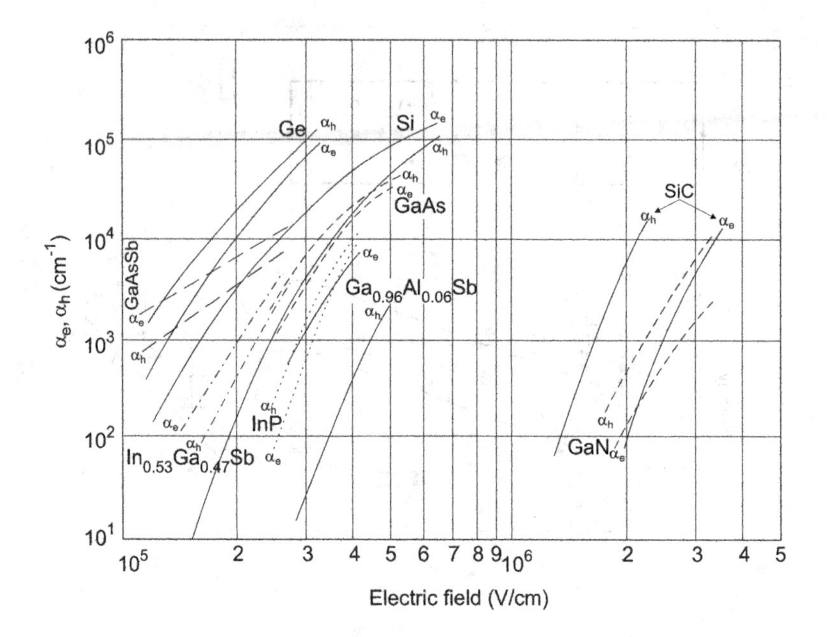

Figure 12.29 Ionization coefficients of electrons (α_e) and holes (α_h) as a function of the electric field for some semiconductors used in avalanche photodiodes.

where a, b, c are experimentally determined constants, and E is the magnitude of an electric field.

An important parameter for characterizing the performance of an APD is the ionization ratio $k = \alpha_h/\alpha_e$. When holes do not ionize appreciably (i.e., $\alpha_h \ll \alpha_e$; $k \ll 1$), most of the ionization is achieved by electrons. The avalanching process then proceeds principally from right to left (i.e., from the p side to n side) as shown in Figure 12.28d. It terminates some time later when all the electrons arrive at the n side of the depletion layer. If electrons and holes both ionize appreciably ($k \approx 1$), on the other hand, those holes moving to the right create electrons that move to the left, which, in turn, generate further holes moving to the right, in a possibly unending circulation. Although this feedback process increases the gain of the device (i.e., the total generated charge in the circuit per photocarrier pair), it is nevertheless undesirable for several reasons: it is time-consuming and therefore reduces the device bandwidth; it is random and therefore increases the device noise; and it can be unstable, thereby causing avalanche breakdown. It is therefore desirable to fabricate APDs from materials that permit only one type of carrier (either electrons or holes) for impact ionization. If electrons have the higher ionization coefficient, for example, optimal behavior is achieved by injecting the electron of a photocarrier pair at the p edge of the depletion layer and by using a material whose value of k is as low as possible. If holes are injected, the hole of a photocarrier pair should be injected at the n edge of the depletion layer and k should be as large as possible. The ideal case of single-carrier multiplication is achieved when $k = 0$ or ∞.

A comprehensive theory of avalanche noise in APDs was developed by McIntyre [114,115]. The noise of an APD per unit bandwidth can be described by the formula

$$\langle I_n^2 \rangle = 2qI_{ph} \langle M \rangle^2 F, \tag{12.139}$$

where I_{ph} is the unmultiplied photocurrent (signal), $< M >$ is the average avalanche gain, and F is the excess noise factor associated with M that arises from the stochastic nature of the ionization process.

McIntyre showed that:

$$F_e(M_e) = k\langle M_e \rangle + (1-k)\left(2 - \frac{1}{\langle M_e \rangle}\right), \tag{12.140}$$

and for the case when holes initiate multiplication,

$$F_h(M_h) = \frac{1}{k}\langle M_h \rangle + \left(1 - \frac{1}{k}\right)\left(2 - \frac{1}{\langle M_h \rangle}\right). \tag{12.141}$$

In p-n and p-i-n reverse-biased photodiodes without gain, $< M > = 1$, $F = 1$ and the well-known shot noise formula indicate the noise performance of the device. In the avalanche process, if every injected photocarrier underwent the same gain M, the factor would be unity, and the resulting noise power would only be the input shot noise due to the random arrival of signal photons, multiplied by the gain squared. The avalanche process is, instead, intrinsically statistical, so that individual carriers generally have different avalanche gains characterized by a distribution with an average $< M >$. This causes additional noise called avalanche excess noise, which is conveniently expressed by the F factor in Equation 12.139. As was mentioned earlier, to achieve a low F, not only must α_e and α_h be as different as possible, but also the avalanche process must be initiated by carriers with the higher ionization coefficient. According to McIntyre's rule, the noise performance of ADP can be improved by more than a factor of 10 when the ionization ratio is increased to 5. Most of III–V semiconductors have $0.4 \leq k \leq 2$.

The gain mechanisms are very temperature-sensitive because of the temperature dependence of the electron and hole ionization rates.

Equations 12.140 and 12.141 have been derived under the condition that the ionization coefficients are in local equilibrium with the electric field and hence the designation "local field" model. In most semiconductor materials, this local approximation provides an accurate prediction of the excess noise factors for thick avalanche regions (>1 μm). From Equation 12.140 we can see that the lowest excess noise is obtained when k is minimized for electron initiated multiplication, as shown in Figure 12.30. It is well-known, however, that the impact ionization is nonlocal and carriers injected into the high-field are "cool" and require a certain distance to attain a sufficient energy to ionize [116]. The distance in which no impact ionization occurs is referred to as the "dead space" $d_{e(h)}$ for electrons (holes). If the multiplication region is thick, the dead space can be neglected and the local field model provides an accurate description of the APD characteristics. Figure 12.31 shows a schematic of the ionization path length probability distribution function (PDF) for electrons, $h_e(x)$, including both local model and hard dead space model [116]. The value of d for electrons and holes is approximately given by E_{th}/qE, where E_{th} is the threshold energy for ionization and depends on the semiconductor band structure and the electric field, E. The dead-space effect can

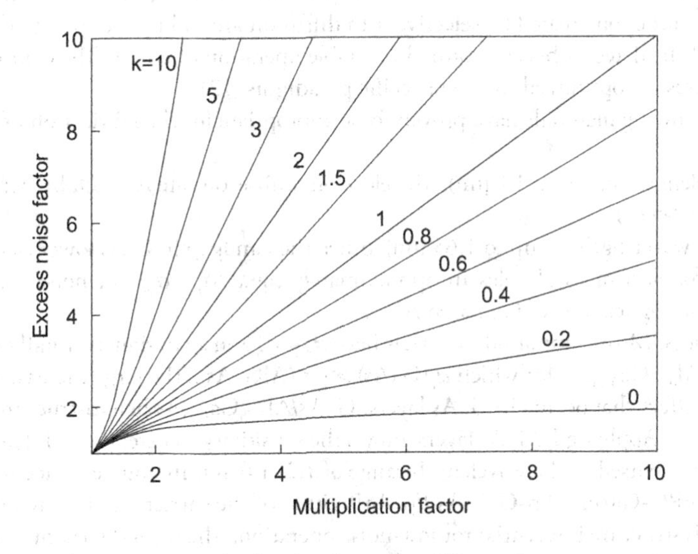

Figure 12.30 Excess noise factor versus multiplication factor for different k.

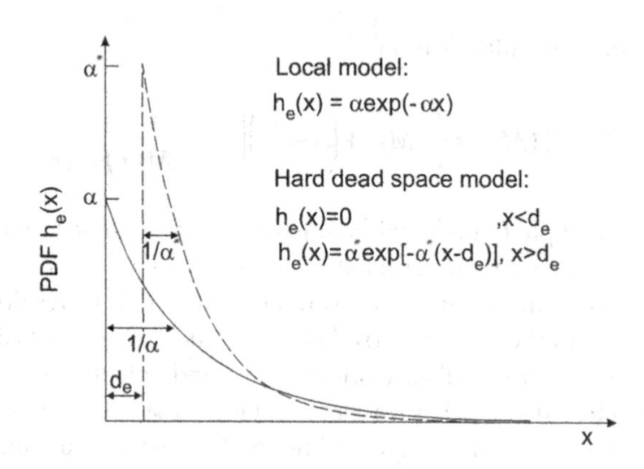

Figure 12.31 PDFs of ionization path lengths in local and hard dead space models. (From Ref. [117])

be significant and a large reduction in the excess noise can occur as a result of a much narrow PDF than the local model. In consequence, APDs with low excess noise can be obtained even with $k \sim 1$ [117–119].

Reducing the avalanche region length has other unexpected benefits—increased speed. The photodiode gain-bandwidth product results from the time required for the avalanche process to build up or decay; the higher the gain, the higher the associated time constant and thus lower the bandwidth. However, Emmons showed that the bandwidth limitation is removed when either α_e or $\alpha_h = 0$ [120]. For nonzero ionization coefficients, the frequency dependence of electron-initiated mean gain is approximately given by

$$M(\omega) = \frac{M_o}{\sqrt{1+(\omega M_o k\tau)^2}}, \tag{12.142}$$

where M_o is the DC gain, and τ is approximately (within a factor of ~2) the carrier transit time across the multiplication region.

Two of the most important objectives in APD design are the reduction of dark current and the enhancement of device speed. To obtain the best performance, several structural and materials requirements must be met. First and foremost, it is important to ensure uniformity of carrier multiplication over the entire photosensitive area of the diode. The device material in which avalanching occurs must be defect free, and great care must be taken in device fabrication. An essential problem is the excessive leakage current at the junction edges. In Si APDs, the common technique used to alleviate this problem is to incorporate a guard ring, which is an n-p junction created by selective-area diffusion around the periphery of the diode. Very careful regulation of the detector bias is required for stable operation of APDs. The commonly available silicon APD structures are optimized to meet specific paradigms [121].

At present the following materials have proved to be appropriate for the fabrication of high-performance APDs:

- Silicon (for wavelengths of 0.4 to 1.1 μm). The electron ionization rate is much higher than the hole ionization rate ($\alpha_e >> \alpha_h$);
- Germanium (for wavelengths of up to 1.65 μm). Since the bandgap in Ge is lower than in Si and the ionization rates for electrons and holes are approximately equal ($\alpha_e \approx \alpha_h$), the noise is considerably higher and this limits the applications of Ge-based APDs;
- GaAs-based devices. Most compound materials have $\alpha_e \approx \alpha_h$ and so designers usually use heterostructures like GaAs/$Al_{0.45}Ga_{0.55}As$, for which α_e(GaAs) >> α_e(AlGaAs). The large increase in gain occurs due to the avalanche effect that occurs in GaAs layers. GaAs/$Al_{0.45}Ga_{0.55}As$ heterostructures are in spectral range below 0.9μm. Applying InGaAs layers allows the sensitivity to extend to ≈1.4 μm;
- InP-based devices are used in the wavelength range of 1.2–1.6 μm. In double lattice–matched heterostructure n^+-InP/n-GaInAsP/p-GaInAsP/p^+-InP, either of the carriers are injected into the high field region—this structure is essential for low-noise operation. The second structure, p^+-InP/n-InP/n-InGaAsP/n^+-InP, is similar to the Si reach through devices. The absorption occurs in the relatively wide InGaAsP layers and avalanche multiplication of the minority carriers proceeds in the n-InP layer.

Table 12.4 Summary of Si, Ge, and InGaAs APD characteristics

PARAMETER	Si	Ge	InGaAs
Wavelength range (nm)	400–1100	800–1650	1100–1700
Peak wavelength (nm)	830	1300	1550
Current responsivity (A/W)	50–120	2.5–25	–
Quantum efficiency (%)	77	55–75	60–70
Avalanche gain	20–400	50–200	10–40
Dark current (nA)	0.1–1	50–500	10–50 ($M = 10$)
Rise time (ns)	0.1–2	0.5–0.8	0.1–0.5
Gain × Bandwidth (GHz)	100–400	2–10	20–250
Bias voltage (V)	150–400	20–40	20–30
Capacity (pF)	1.3–2	2–5	0.1–0.5

- $Hg_{1-x}Cd_xTe$ APDs. These devices are electron-initiated ones and their operation has been demonstrated for a broad range of compositions from $x = 0.7$ to 0.21 corresponding to cut-off wavelengths from 1.3 μm to 11 μm. Thus, HgCdTe APD at gain = 100 provides 10 to 20 times less noise than InGaAs or InAlAs APDs and 4 times less noise than Si APDs.

The choice of materials for an APD is dictated by the applications; the most popular ones are laser range finding, high-speed receivers, and single-photon counting. InGaAs used for fiber optic communication is more expensive than germanium, but provide lower noise and higher frequency response for a given active area. Germanium APDs are recommended for applications in which amplifier noise is high or cost is a prime consideration. Table 12.4 lists the parameters of Si, Ge, and InGaAs APDs. They are given as guidelines for comparison purposes.

Much of the recent work on APDs has focused on developing new structures and incorporating alternative materials that will yield lower noise and higher speed while maintaining optimal gain levels [112,118]. For example, low excess noise, due to dead space effect, can be achieved using submicron InAlAs or InP avalanche regions with InGaAs absorption regions. Both InAlAs and InP are lattice matched to InGaAs. The α_e/α_h ratio has been found to be significantly larger than α_h/α_e ratio in InP at low electric fields. The excess noise factor at a given gain is significantly lower in InAlAs than that in InP due to a large α_e/α_h ratio in the former and the beneficial effect of the dead space in the latter.

It should be mentioned that the APD can also be biased at voltage larger than the infinity-gain voltage in such a way that the arrival of a single photon precipitates avalanche breakdown, thereby creating a large current pulse that signifies a subsequent photon. This may be carried out either by passive or active means. This working regime is called counting mode or single photon avalanche detector (SPAD), also known as a Geiger-mode avalanche detector, after the work pioneered by Cova and coworkers [122]. SPAD is potentially very sensitive, comparable to that of photomultipliers. However, it has to be noted that once the avalanche at infinite gain is initiated, further photons eventually detected during pulse duration and circuit recovery time are ignored. From this point of view, the SPAD is similar to a Geiger counter than a photomultiplier.

12.5 SCHOTTKY-BARRIER PHOTODIODES

Schottky-barrier photodiodes have been studied quite extensively and have also found application as ultraviolet, visible, and IR detectors [123–132]. These devices reveal some advantages over p-n junction photodiodes: fabrication simplicity, absence of high-temperature diffusion processes, and high-speed response.

12.5.1 SCHOTTKY–MOTT THEORY AND ITS MODIFICATIONS

According to a simple Schottky–Mott model, the rectifying property of the metal–semiconductor contact arises from the presence of an electrostatic barrier between the metal and the semiconductor, which is due to the difference in work functions ϕ_m and ϕ_s of the metal and semiconductor, respectively. For example, for a metal contact with an n-type semiconductor, ϕ_m should be greater, while for a p-type semiconductor it should be less than ϕ_s. The barrier heights in both these cases, shown in Figure 12.32a and 12.32b, are given by

$$\phi_{bn} = \phi_m - \chi_s \tag{12.143}$$

and

$$\phi_{bp} = \chi_s + E_g - \phi_m, \tag{12.144}$$

respectively, where χ_s is the electron affinity of the semiconductor. Values of work functions, electron affinities, and bandgap energies for some metals and semiconductors are given in Table 12.5.

The potential barrier between the interior of the semiconductor and the interface, known as band bending, is given by

$$\psi_s = \phi_m - \phi_s \tag{12.145}$$

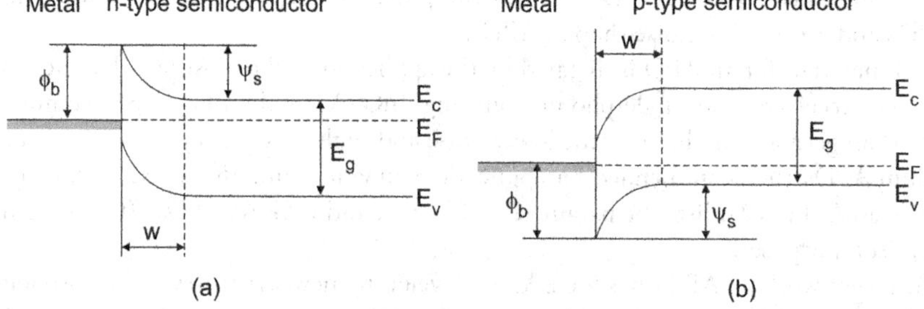

Figure 12.32 Equilibrium energy band diagram of Schottky-barrier junctions: (a) metal–(n-type) semiconductor, and (b) metal–(p-type) semiconductor.

Table 12.5 Work functions, electron affinities, and bandgap energies for some metals and semiconductors

METAL	Φ_m [eV]	SEMICONDUCTOR	χ [eV]	E_g [eV]
Mg	3.68	Ge	4.0	0.661
Al	4.08	Si	4.05	1.12
Zn	4.3	GaAs	4.07	1.424
Ti	4.33	4H-SiC	3.2	2.36
Hg	4.5	6H-SiC	3.45	3.23
Cr	4.5	3C-SiC	4.0	2.86
Mo	4.6	AlN	0.6	6.026
Cu	4.7	GaP	3.8	2.26
Ag	4.73	GaN	4.1	3.2
Co	5.0	ZnO	4.35	3.37
Ni	5.01	InP	4.38	1.344
Au	5.1	InSb	4.59	0.17
Pd	5.12	CdS	4.8	2.42
Pt	6.35	InAs	4.9	0.354

in both cases. If $\phi_b > E_g$, the layer of the p-type semiconductor adjacent to the surface is inverted in type and we have a p-n junction within the material. However, in practice, the built-in barrier does not follow such a simple relationship with ϕ_m and is effectively reduced due to interface states originating either from surface states or from metal-induced gap states and/or due to interface chemical reactions of metal and semiconductor atoms.

In the literature, there are numerous and considerable variations among experimental data on ϕ_m [128,130]. Their analysis indicates an empirical relationship of the type

$$\phi_b = \gamma_1 \phi_m + \gamma_2, \tag{12.146}$$

where γ_1 and γ_2 are constants characteristic of the semiconductors. Two limit cases, namely, $\gamma_1 = 0$ and $\gamma_1 = 1$ indicative of Bardeen barrier (when the influence of localized surface states is decisive) and ideal Schottky barrier, respectively, can be visualized from such an empirical relation. It has also been pointed out by various workers that the slope parameter $\gamma_1 = \partial\phi_b/\partial\phi_m$ can be used for describing the extent of Fermi level stabilization or pinning for a given semiconductor. The parameters γ_1 and γ_2 have been used by some workers for estimating the interface state density.

It was shown by Cowley and Sze that, according to the Bardeen model, the barrier height in the case of an n-type semiconductor is given approximately by [133]

$$\phi_{bn} = g(\phi_m - \chi_s) + (1-g)(E_g - \phi_o) - \Delta\phi, \tag{12.147}$$

where $\gamma = \varepsilon_i/(\varepsilon_i + q\delta D_s)$. The ϕ_o is the position of the neutral level of the interface states measured from the top of the valence band, $\Delta\phi$ is the barrier lowering due to image forces, δ is the thickness of the interfacial layer, and ε_i its total permittivity. The surface states are assumed to be uniformly distributed in energy within the bandgap, with a density D_s per electron-volt per unit area. If there are not surface states, $D_s = 0$, and neglecting $\Delta\phi$, Equation 12.147 reduces to Equation 12.143. If the density of states is very high, γ becomes very small and ϕ_{bn} approaches the value $E_g - \phi_o$. This is because a very small deviation of the Fermi level from the neutral level can produce a large dipole moment, which stabilizes the barrier height by a sort of negative feedback effect. The Fermi level is pinned relative to the band edges by the surface states.

A similar analysis for the case of a p-type semiconductor shows that ϕ_{bp} is approximately given by

$$\phi_{bp} = \gamma(E_g - \phi_m + \chi_s) + (1-\gamma)\phi_o \tag{12.148}$$

Hence, if ϕ_{bn} and ϕ_{bp} refer to the same metal on n- and p-type specimens of the same semiconductor, we should have

$$\phi_{bn} + \phi_{bp} \cong E_g, \tag{12.149}$$

if the semiconductor surface is prepared in the same way in both cases, so that δ, ε_j, D_s, and ϕ_o are the same. This relationship holds quite well in practice. It is usually true that $\phi_{bn} > E_g/2$, and $\phi_{bn} > \phi_{bp}$.

12.5.2 CURRENT TRANSPORT PROCESSES

The current transport in metal–semiconductor contacts is mainly due to majority carriers, in contrast to p–n junctions, where it is mainly due to minority carriers. The current can be transported in various ways under forward-bias conditions as shown in Figure 12.33. The four processes are [129]:

1. Emission of electrons from the semiconductor over the top of the barrier into the metal
2. Quantum mechanical tunneling through the barrier
3. Recombination in the space charge region
4. Recombination in the neutral region (equivalent hole injection from the metal to the semiconductor)

The inverse processes occur under reverse bias. In addition, we may have edge leakage current due to a high electric field at the contact periphery or interface current due to traps at the metal–semiconductor interface.

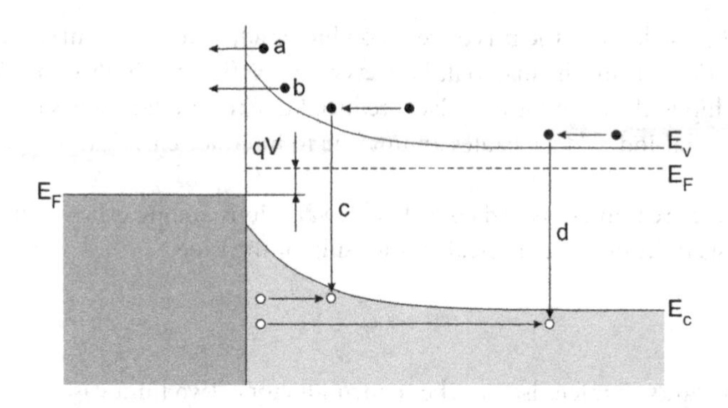

Figure 12.33 Four basic transport processes in the forward-biased Schottky barrier on an n-type semiconductor. (From Ref. [129])

The transport of electrons over the potential barrier has been described by various theories, namely, diffusion [134,135], thermionic emission [136], and unified thermionic emission–diffusion [133]. It is now widely accepted that, for high-mobility semiconductors with impurity concentrations of practical interest, the thermionic emission theory appears to explain qualitatively the experimentally observed *I-V* characteristics [137]. Some workers [138–140] have also included in the simple thermionic theory the quantum effects (i.e., quantum-mechanical reflection and tunneling of carriers through the barrier) and have tried to obtain modified analytical expressions for the current-voltage relation. This, however, has essentially led to a lowering of the barrier height and a rounding-off of the top.

The thermionic emission theory by Bethe [136] is derived from the assumptions that the barrier height is much larger than kT, thermal equilibrium is established at the plane that determines emission, and the existence of a net current flow does not affect this equilibrium. Bethe's criterion for the slope of the barrier is that the barrier must decrease by more than kT over a distance equal to the scattering length. The resulting current flow will depend only on the barrier height and not on the width, and the saturation current is not dependent on the applied bias. Then the current density of majority carriers from the semiconductor over the potential barrier into the metal is expressed as

$$J_{MSt} = J_{st}\left[\exp\left(\frac{qV}{\beta kT}\right)-1\right], \tag{12.150}$$

where saturation current density

$$J_{st} = A^*T^2\exp\left(-\frac{\phi_b}{kT}\right) \tag{12.151}$$

and $A^* = 4\pi qk^2m^*/h^3$ equals $120(m^*/m)$ Acm^{-2}K^{-2}, is the Richardson constant, m^* is the effective electron mass, and β is an empirical constant close to unity. Equation 12.150 is similar to the transport equation for p-n junctions. However, the expressions for the saturation current densities are quite different.

It appears that the diffusion theory is applicable to low-mobility semiconductors and the current density expressions of the diffusion and thermionic emission theories are basically very similar. However, the saturation current density for the diffusion theory

$$J_{sd} = \frac{q^2D_eN_c}{kT}\left[\frac{q(V_{bi}-V)2N_d}{\varepsilon_o\varepsilon_s}\right] \tag{12.152}$$

varies more rapidly with the voltage but is less sensitive to temperature compared with the saturation current density J_{st} of the thermionic emission theory.

A synthesis of the thermionic emission and diffusion approaches described has been proposed by Crowell and Sze [137]. They assumed Bethe's criterion for the validity of the thermionic emission (the mean free path should exceed the distance within which the barrier falls by kT/q from its maximum value) and also took into account the effects of optical phonon scattering in the region between the top of the barrier and the metal and of the quantum mechanical reflection of electrons that have sufficient energy to surmount the barrier. Their combined effect is to replace the Richardson constant A^* with $A^{**} = f_p f_q q A^*$, where f_p is the probability of an electron reaching the metal without being scattered by an optical phonon after having passed the top of the barrier, and f_q is the average transmission coefficient. The f_p and f_q depend on the maximum electric field in the barrier, the temperature, and the effective mass. Generally speaking, the product $f_p f_q$ is on the order of 0.5.

With respect to the band diagram shown in Figure 12.32, at smaller values of photon energy, such that $q\phi_b < h\nu < E_g$, electrons photoexcited in the metal can surmount the barrier by thermionic emission, transit across the semiconductor, and be collected at the contact. This process extends the spectral response of the diode to photon energies lower than the bandgap. However, since the thermionic emission process can be slow, it is not a very desirable mode of operation. The most efficient mode of operation is when photon energy is greater than the energy gap ($h\nu > E_g$). If the metal layer is semitransparent, photons absorbed in the semiconductor create the electron hole pairs, which are moving in opposite directions with their respective saturation and are collected.

Knowing J_{st} with Equation 12.151, R_oA can be calculated from

$$(R_oA)_{MS} = \left(\frac{dJ_{MSt}}{dV}\right)^{-1}_{|V=0} = \frac{kT}{qJ_{st}} = \frac{k}{qA^*T}\exp\left(\frac{\phi_b}{kT}\right). \tag{12.153}$$

The current responsivity may be written in the form

$$R_i = \frac{q\lambda}{hc}\eta, \tag{12.154}$$

and the voltage responsivity

$$R_v = \frac{q\lambda}{hc}\eta R, \tag{12.155}$$

where $R = (dI/dV)^{-1}$ is the differential resistance of the photodiode.

At this point, it is important to discuss some significant differences between photoconductive, p-n junction, and Schottky-barrier detectors.

The photoconductive detectors exhibit the important advantage of the internal photoelectric gain, which relaxes requirements to a low noise preamplifier. The advantages of p-n junction detectors relative to photoconductors are: low or zero bias currents; high impedance, which aids coupling to readout circuits in FPAs; capability for high-frequency operation and the compatibility of the fabrication technology with planar-processing techniques. In comparison with Schottky barriers, the p-n junction photodiodes also indicate some important advantages. The thermionic emission process in a Schottky barrier is much more efficient than the diffusion process and therefore for a given built-in voltage, the saturation current in a Schottky diode is several orders of magnitude higher than in the p-n junction. In addition, the built-in voltage of a Schottky diode is smaller than that of a p-n junction with the same semiconductor. However, the high-frequency operation of p-n junction photodiodes is limited by the minority-carrier storage problem. The Schottky-barrier structures are majority-carrier devices and therefore have inherently fast responses and large operating bandwidths. In other words, the minimum time required to dissipate the carriers injected by the forward bias is dictated by the recombination lifetime. In a Schottky barrier, electrons are injected from the semiconductor into the metal under forward bias if the semiconductor is n-type. Next they thermalize very rapidly ($\approx 10^{-14}$ s) by carrier–carrier collisions, and this time is negligible compared to the minority carrier–recombination lifetime. There are examples of photodiodes with bandwidths in excess of 100 GHz. The diode is usually operated under a reverse bias.

Table 12.6 Properties of Silicides

SILICIDE	RESISTIVITY (μΩcm)	FORMATION TEMPERATURE (°C)	Å OF Si PER Å OF METAL	Å OF SILICIDE PER Å OF METAL
$CoSi_2$	18–25	>550	3.64	3.52
$MoSi_2$	80–250	>600	2.56	2.59
$NiSi_2$	≈ 50	750	3.65	3.63
Pd_2Si	30–35	>400	0.68	≈ 1.69
PtSi	28–35	600–800	1.32	1.97
$TaSi_2$	30–45	>600	2.21	2.40
TiS_2	14–18	>700	2.27	2.51
WSi_2	30–70	>600	2.53	2.58

12.5.3 SILICIDES

Most contacts used in semiconductor devices are subjected to heat treatment. This may be deliberate, to promote adhesion of the metal to the semiconductor, or unavoidable, because high temperatures are needed for other processing stages that occur after the metal is deposited. It is important to avoid the melting of rectifying contacts, because if this happens the interface may become markedly nonplanar, with sharp metallic spikes projecting into the semiconductor that can cause tunneling through the high-field region at the tip of the spike and may severely degrade the electrical characteristics.

The effect of heat on metal–silicon contacts is particularly important if the metal is capable of forming a silicide, which is a stoichiometric compound. Most metals, including all the transition metals, form silicides after appropriate heat treatment (see Table 12.6). These silicides may form as a result of solid state reactions at temperatures of about one-third to one-half the melting point of the silicide in degrees Kelvin [139]. Studies on a number of transition metal silicides-Si systems have revealed that ϕ_b decreases almost linearly with the eutectic temperature [140]. The vast majority of silicides exhibit metallic conductivity so that, if a metallic silicide is formed as a result of heat treatment of a metal–silicon contact, the silicide–silicon junction behaves like a metal–semiconductor contact and may exhibit rectifying properties. Moreover, because the silicide–silicon interface is formed some distance below the original surface of the silicon, it is free from contamination and is very stable at room temperature. It also exhibits very good mechanical adhesion. Contacts formed in this way generally show stable electrical characteristics, which are very close to ideal [141,142]. The unique feature of silicide–silicon devices is compatible with the silicon planar processing technology.

In 1973, Shepherd and Yang, of the Rome Air Development Center, Hanscom AFB, Massachusetts, proposed the concept of silicide Schottky-barrier detector FPAs as a much more reproducible alternative to HgCdTe FPAs for IR imaging [143]. Since then, the development of silicide Schottky-barrier FPA technology has progressed from the demonstration of the initial concepts in the 1970s to the development of high-resolution scanning and staring FPAs that are being considered for many applications for IR imaging in the 1–3 and 3–5 μm spectral bands [144]. PtSi/p–Si detectors are well-situated for the 3–5 μm wavelength range. Alternative silicides of interest are Pd_2Si and IrSi; Pd_2Si Schottky barriers exhibit a cutoff wavelength of $\lambda = 3.7$ μm that does not match with the IR transparent atmospheric window. IrSi Schottky barriers exhibit a low barrier energy and cutoff wavelengths reported in the range from $\lambda = 7.3$ to 10.0 mm [145].

More information about properties and technology of silicide Schottky-barrier detectors and FPAs can be found in Chapter 15.

12.6 METAL–SEMICONDUCTOR–METAL PHOTODIODES

Another form of metal–semiconductor photodiode is the metal–semiconductor–metal or MSM photodiode illustrated in Figure 12.34 [80,110,111]. This structure is physically similar to the interdigitated

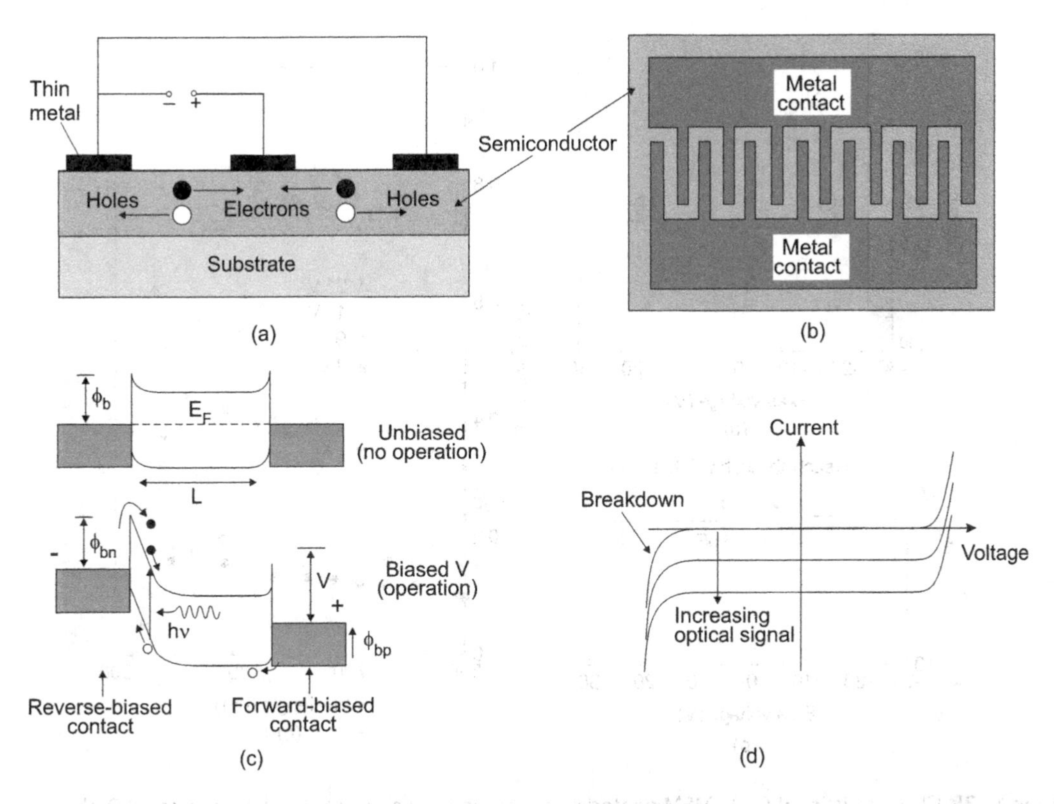

Figure 12.34 Metal–semiconductor–metal photodiode: (a) Cross section structure, (b) Top view, (c) Unbiased (nonoperational) and biased (operational) energy band diagrams, and (d) Current-voltage characteristics.

photoconductor that is illustrated in Figure 12.34b except that the metal–semiconductor and semiconductor–metal junctions are fabricated as Schottky barriers instead of ohmic contacts. Being a planar structure, the MSM photodiode lends itself to monolithic integration and can be fabricated using processing steps nearly identical to those required for making field-effect transistors [146].

The MSM photodiode is essentially a pair of Schottky diodes connected back-to-back. Absorbed photons generate electron hole pairs in the semiconductor. The holes drift with the applied electric field to the negative contacts while electrons drift to the positive contacts. The quantum efficiency of the MSM photodiode is dependent on the shadowing caused by the metal electrodes. Dual Schottky barriers provide a lower dark current than a Schottky diode alone, and the capacitance of an MSM diode is smaller than that of a p-i-n photodiode, which enhances the response speed. The MSM diode operates under a bias, which forward-biases one contact while reverse-biasing the other. Diagrams of the biased and unbiased energy band states are shown in Figure 12.34c. At low biases, electron injection at the reverse-biased contact dominates the conduction mechanism. At higher biases, this conduction is supplemented by the injection of holes at the forward-biased contact. Figure 12.34d illustrates the I-V characteristics for an MSM photodiode for three levels of illumination. The depletion region at the reverse-biased contact is much larger than the depleted region near the forward-biased contact, but when they meet, the device is said to achieve the "reach through" condition. Although internal gain has been observed for these structures, it has not been fully understood and modeled [110].

For example, Figure 12.35 shows the current-voltage characteristics of a GaAs MSM device with WSi_x contacts [147]. The dark current is of order 1 nA, which is comparable to a p-i-n photodiode. For both polarities, the capacitance decreases with bias up to the reach-through voltage, after which it remains nearly unchanged. The increase of the responsivity with bias (see Figure 12.35c) indicates an internal gain, even at low bias values. The latter precludes the avalanche multiplication effect. It is suggested that the gain at low biases can be operative due to traps or surface defects having a long lifetime. The barrier for hole transit can be lowered by electrons accumulated at the conduction band minimum.

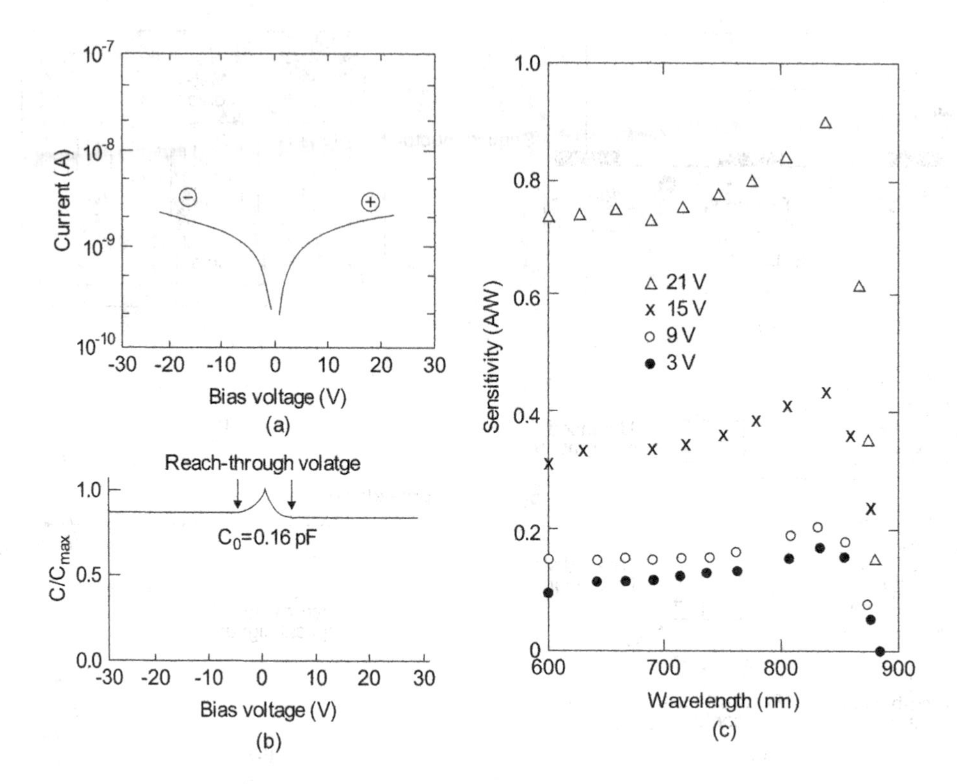

Figure 12.35 Characteristics of GaAs MSM photodiode with tungsten silicide Schottky contacts: (a) Dark current characteristics, (b) Capacitance-voltage characteristics, and (c) Spectral responsivity as a function of bias. (From Ref. [147])

12.7 MIS PHOTODIODES

The MIS structure is the most useful device in the study of the semiconductor surface. This structure has been extensively studied because it is directly related to most planar devices and integrated circuits. Before the 1970s, MIS devices as IR detectors had no importance whatever. In 1970, Boyle and Smith [148] published a paper reporting the charge-coupling principle, which is a simple, extremely powerful concept based upon the transfer of charge packets in an MIS structure. In the area of IR technique, the use of charge transfer device (CTD) holds the key to substantial improvements in thermal imaging. The general theory of MIS capacitors is reviewed in many excellent papers (e.g., [86,149–151]).

In this section, only p-channel MIS with the n-type substrate will be considered, although all of the arguments can be extended to n-channel MIS with appropriate changes.

The simple MIS devices consist of a metal gate separated from a semiconductor surface by an insulator of thickness t_i and dielectric constant ε_i (Figure 12.36). By applying a negative voltage V_b to the metal electrode, electrons are repelled from the I-S interface, creating a depletion region (Figure 12.37). In effect, a potential well for minority carriers (holes) is created. The surface potential Φ_s is related to the gate voltage and other parameters by [152–154]

$$\Phi_s = V_G - V_{FB} + \frac{qN}{C_i} - \frac{qN_d\varepsilon_0\varepsilon_s}{C_i^2} + \frac{1}{C_i}\left[-2q\varepsilon_0\varepsilon_s N_d\left(V_G - V_{FB} + \frac{qN}{C_i}\right) + \left(\frac{qN_d\varepsilon_0\varepsilon_s}{C_i}\right)^2\right]^{1/2}, \quad (12.156)$$

where V_{FB} is the flatband voltage, $C_i = \varepsilon_o\varepsilon_i/t$ is the insulator capacitance per unit area, N is the number of mobile electrons in the inversion layer per unit area, $N_d = n_o$ is the substrate doping density, and ε_s is the dielectric constant of the semiconductor. From Equation 12.156 we can see that the surface potential can be controlled by the proper choice of gate voltage, doping density, and insulator thickness.

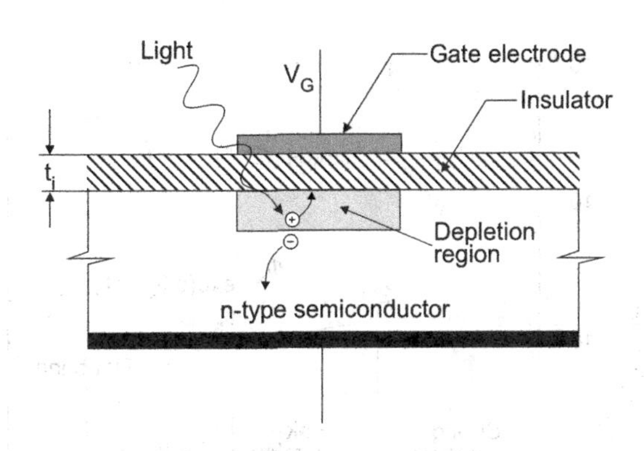

Figure 12.36 MIS structure.

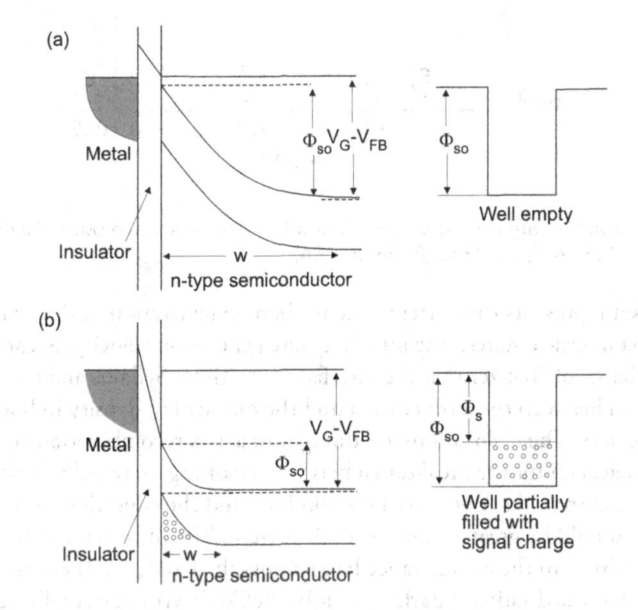

Figure 12.37 Energy band diagram for a p-channel MIS structure: (a) band bending at deep depletion and the empty potential well representation, and (b) band bending with mobile charge at the I-S interface and the partially filled potential well representation.

Initially, no charge is present in the potential well ($N = 0$ in Equation 12.156), resulting in a relatively large surface potential Φ_{so}. However, carriers are collected in the potential well as a result of photon absorption, injection from an input diffusion, thermal generation, or tunneling, and the potential across the insulator and the semiconductor will be redistributed as shown in Figure 12.37b. In steady state, the potential well is completely filled up and the surface potential assumes its final value [80]

$$\Phi_{sf} \approx 2\Phi_F = -\frac{2kT}{q}\ln\left(\frac{n_o}{n_i}\right) = -\frac{kT}{q}\ln\left(\frac{n_o}{p_o}\right). \tag{12.157}$$

Φ_F is the potential difference between the bulk Fermi level and the intrinsic Fermi level. It should be noticed that at this surface potential strong inversion begins (Figure 12.38 [86]), and surface concentration of holes becomes greater than the bulk majority carrier concentration.

The thermal current-generation mechanisms for MIS structure are similar to those previously discussed for p-n junction (see Section 12.2). For n-type material [86],

$$J = qn_i\left(\frac{n_i L_h}{N_d \tau_h} + \frac{w_d(t)}{\tau} + \frac{1}{2}S\right) + \eta q \Phi_b + J_t, \tag{12.158}$$

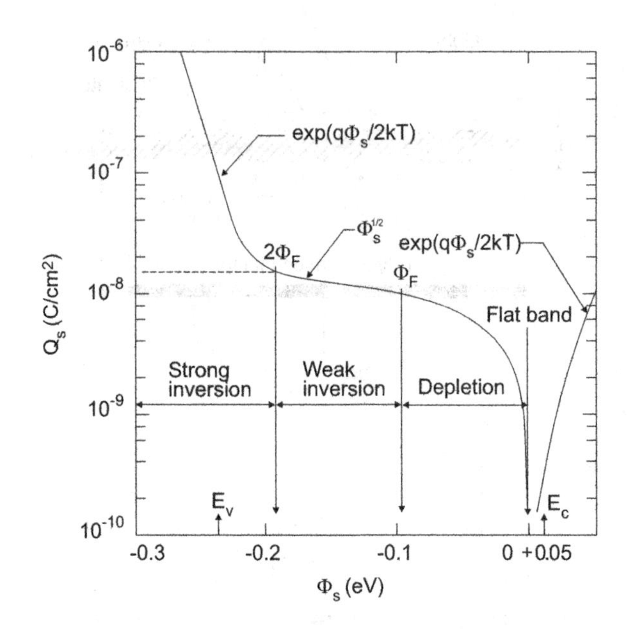

Figure 12.38 Variation of space charge density $Q_s = qN$ as a function of surface potential Φ_s for 0.25 eV p-channel MIS HgCdTe with $N_d = 2 \times 10^{15}$ cm^{-3}, $T = 77$ K. (From Ref. [86])

where the term in bracket represents the currents due to thermal generation in the neutral bulk, the depletion region, and via interface states. The interface state generation velocity, S, can be greatly reduced by maintaining a bias charge of "flat zero" at the interface at all time. Syllaios and Colombo [155] have found a strong correlation between the dark current and the dislocation density in both MWIR and LWIR HgCdTe MIS devices. The dislocations are thought to act as recombination centers and degrade the lifetime, and these defects limit the breakdown bias voltage of HgCdTe MIS devices. The second term in Equation 12.158 is the current due to the background flux, and the condition for background-limited performance is that this should be greater than the dark current. The third term is due to tunneling of carriers from the valence band to the conductance band across the bandgap. The considerations carried out by Kinch [86] indicate that the dominant dark current in HgCdTe MIS devices, for reasonable surface recombination velocities ($s < 10^2$ cm/s), is due to generation in the depletion region.

Equation 12.157 can be used to calculate the maximum charge density that can be stored on an MIS capacitor. Solving Equation 12.156 for $N = N_{max}$ gives

$$N_{max} = \frac{C_i}{q}\left(V_G - V_{FB} - 2\Phi_{sf} - V_B\right), \tag{12.159}$$

where $V_B = (4q\varepsilon_o\varepsilon_s\Phi)^{1/2}/C_i$. Typically, V_{FB}, $2\Phi_{sf}$, and V_B are smaller than V_b; therefore, $N_{max} \approx C_iV_b/q$. If we assume similar values for C_i and V_b to those achieved in the Si/SiO$_2$ structure, then the maximum stored charge is $\approx 10^{12}$ electrons/cm^2. In practice, the maximum storage capacity is taken to about 0.5 N_{max}, to prevent the stored charge from diffusing away and is around 0.4×10^{12} electrons/cm^2 at 4.5 µm and 0.15×10^{12} electrons/cm^2 at 9.5 µm [20].

Equation 12.159 shows that the storage capacity can be increased by increasing the insulator capacitance, or the gate voltage, or by reducing the doping in the semiconductor. However, the electric field in the semiconductor must be kept below breakdown. A common breakdown mechanism in narrow bandgap semiconductors is tunneling. Calculations carried out by Anderson for MIS devices of HgCdTe [156], InSb, and PbSnTe indicate that tunneling current increases rapidly as the bandgap is reduced and this current imposes a severe limitation for cutoff wavelengths greater than 10 µm. Unlike thermal processes, the tunneling current cannot be reduced by cooling. Goodwin et al. [157] have shown that the tunneling currents can be greatly reduced in HgCdTe by growing a heterojunction, which is arranged such that the highest fields occur in the wide bandgap material.

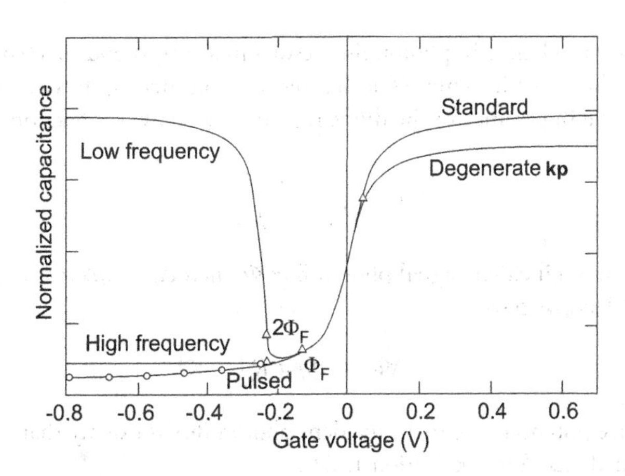

Figure 12.39 Capacitance versus gate voltage for p-channel MIS HgCdTe at 77 K, $E_g = 0.25$ eV, $N_d = 10^{15}$ cm^{-3}, $\Phi_F = 0.092$ eV, $C_i = 2.1 \times 10^{-7}$ F/cm^2. (From Ref. [86])

The maximum storage time for an unilluminated device can be approximated by

$$\tau_c = \frac{Q_{\max}}{J_{\text{dark}}} = \frac{qN_{\max}}{J_{\text{dark}}}, \tag{12.160}$$

and its typical value at 77 K varies from 100 s at 4.5 μm to 100 μs at 10 μm [86]. The storage time is a critical parameter in the operation of a CTD because it establishes the minimum frequency of operation. A long storage time is obtained by reducing the number of bulk generation centers, reducing the number of surface states, and by reducing the temperature. A serious limitation of the storage time in narrow-gap semiconductors is caused by tunneling.

The total capacitance of the idealized MIS device (with $V_{FB} = 0$) is a series combination of the insulator capacitance C_i and the semiconductor depletion layer capacitance C_d

$$C = \frac{C_i C_d}{C_i + C_d}, \tag{12.161}$$

where

$$C_d = \left(\frac{\varepsilon_o \varepsilon_s q^2}{2kT}\right)^{1/2} \frac{n_{no}\left[\exp(q\Phi_s/kT) - 1\right] - p_{no}\left[\exp(-q\Phi_s/kT) - 1\right]}{\left\{n_{no}\left[\exp(q\Phi_s/kT) - q\Phi_s/kT - 1\right] - p_{no}\left[\exp(-q\Phi_s/kT) + q\Phi_s/kT - 1\right]\right\}^{1/2}}. \tag{12.162}$$

The theoretical variation of C versus gate voltage V_b for an idealized p-channel MIS HgCdTe with $E_g = 0.25$ eV and $n_o = 10^{15}$ cm^{-3} is shown in Figure 12.39 [86]. At positive gate voltage we have an accumulation of electrons and therefore C_d is large. As the voltage is reduced to the negative value, a depletion region is formed near the I–S interface and total capacitance decreases. The capacitance achieves a minimum value and then increases until it again approximates C_i in the strong inversion range. Note that the increase of the capacitance in the negative voltage region depends on the ability of minority carriers to follow the applied AC signal. MIS capacitance-voltage curves measured at higher frequencies do not show the increase of capacitance in this voltage region. Figure 12.39 also shows the capacitance under deep depletion, sufficiently fast pulse conditions, which is directly related to the operation of a CCD.

For the case of an incident photon flux, surface inversion occurs for lower values of Φ_s than the thermal equilibrium case. The associated depletion width will also be narrower for the incident photon flux case. The relationship between p_o and Φ_s is given by Equation 12.157 to be

$$\Delta\Phi_{sf} = \frac{kT}{q}\frac{\Delta p_o}{p_o}. \tag{12.163}$$

Thus, a change of Δp_o due to a change in photon flux results in a corresponding change in $\Delta\Phi_{sf}$.

The last expression can be recast into more familiar terms. Considering the case of the MIS device limited by diffusion current, the impedance of the diode region is given by (see Section 12.2.1)

$$RA = \frac{kT}{qJ_D} = \frac{kT\tau_h}{q^2 p_o L_h}. \quad (12.164)$$

Assuming that Δp_o is due to an incident signal photon flux Φ_s then $\Delta p_o = \eta\Delta\Phi_s\tau_h/L_h$, and substituting in Equations 12.163 and 12.164, we have

$$\Delta\Phi_{sf} = \eta q\Phi_s RA. \quad (12.165)$$

Thus, the change in surface potential due to an incident photon flux is exactly that expected for an open-circuit photodiode of impedance R (see Equation 12.76).

12.8 NONEQUILIBRIUM PHOTODIODES

The major drawback of IR detectors is the need for cooling to suppress thermal generation of free carriers, which results in noise. Elliott and other British scientists [158,159] have proposed a new approach to reduce the detector cooling requirements, which is based on the nonequilibrium mode of operation. Their concept relies on the suppression of the Auger processes by decreasing the free carrier concentration below its equilibrium values. This can be achieved, for example, in a biased l-h or heterojunction contacts. The above-mentioned possibilities have been demonstrated in the case of HgCdTe n-type photoconductors [159,160] and photodiodes [161], and InSb photodiodes [162].

The nonequilibrium devices are based on a near-intrinsic, narrow gap, epitaxial layer that is contained between two wider gap layers or between one wider gap layer and one very heavily doped layer. Examples are P-π-N, P-π-N$^+$, P-ν-N$^+$, where the capital letter means wide gap, the + symbol indicates high doping in excess of 10^{17} cm^{-3}, π is near-intrinsic p, and ν is near-intrinsic n. These devices contain one p-n junction, which is operated in reverse bias producing extraction of minority carriers. The other, isotype junction, is excluding to minority carriers preventing their injection into the π or ν layer. For example, let us consider a heterostructure P-π-N shown in Figure 12.40 [163]. Both P and N regions are transparent to photons with energy close to or just above the bandgap of the π region so that both can act as windows. In equilibrium, the electron and hole concentrations in the π region, n_o and p_o are close to the intrinsic value, n_i, which is typically 10^{16}–10^{17} cm^{-3} as illustrated in Figure 12.40c.

Even at zero bias, the device structure shown in Figure 12.40 has two important advantages over homojunction for IR detector:

- The noise generation is confined to the active volume (the wide-gap regions have very low thermal generation rates and isolate the active region from carrier generation at the contacts).
- The doping level and type in the active region can be chosen to maximize carrier lifetime and minimize noise.

However, larger improvements of detector performance are expected under reverse bias as a result of the following phenomena:

- Minority carrier exclusion and extraction occur at the P-π and π-P junctions, respectively; as a result, both carrier types in active region decrease (minority electron density by several orders, while the majority hole concentration falls to the net doping level) as shown in Figure 12.40c.
- As a consequence of these processes, the thermal generation involving Auger processes falls, so that the saturation current (I_s) is less than would be expected from the zero bias resistance (R_o); that is, $I_s < kT/qR_o$ (see Equation 12.83) and a region of negative conductance is predicted to occur [164].

At the present stage of device technology, the Auger suppression nonequilibrium photodiodes suffer from large $1/f$ noise, and hence the improvement in the detectivity resulting from the reduced leakage currents can only be realized at high frequencies. However, this is not a problem for heterodyne systems operated at higher intermediate frequencies.

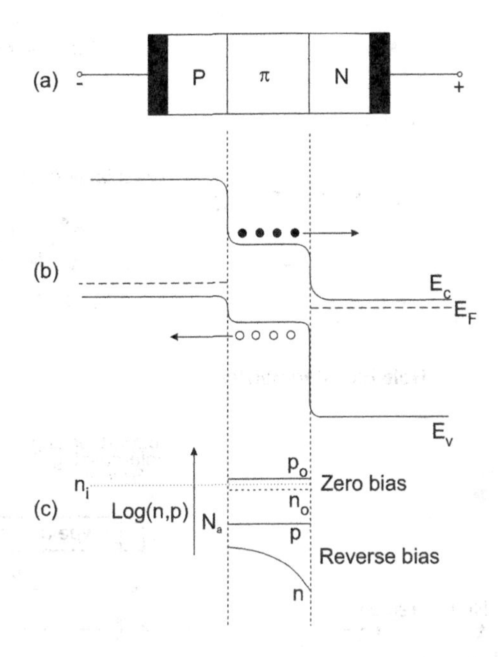

Figure 12.40 Schematic drawing of an extracting P-π-N heterostructure photodiode: (a) Multilayer structure, (b) Band edges under reverse bias, and (c) Current densities in the π region. (From Ref. [163])

12.9 BARRIER PHOTODETECTORS

Historically, the first barrier detector was proposed by A.M. White in 1983 [165] as a high impedance photoconductor. It postulates an n-type heterostructure with a narrow-gap absorber region coupled to a thin wide bandgap layer, followed by a narrow bandgap contact region. A.M. White in his prescient patent also proposed a bias-selectable two-color detector realized and exploited currently in HgCdTe and in the type-II superlattice (T2SL) material systems.

The barrier detector concept assumes almost zero one-band offset approximation throughout the heterostructure, allowing the flow of only minority carriers in a photoconductor. Little or no valence band offset (VBO) was difficult to realize using standard IR detector materials such as InSb and HgCdTe. The situation has changed dramatically in the middle of the first decade of the twenty-first century after the introduction of 6.1 Å III-V material detector family and when the first high-performance detectors (in InAs and InAsSb) [166,167] and FPAs were demonstrated. Introduction of unipolar barriers in various designs based on T2SLs drastically changed the architecture of IR detectors [168]. In general, unipolar barriers are used to implement the barrier detector architecture for increasing the collection efficiency of photogenerated carriers and reducing dark current generation without inhibiting photocurrent flow. The ability to tune the positions of the conduction and valence band edges independently in a broken-gap T2SL is especially helpful in the design of unipolar barriers.

12.9.1 PRINCIPLE OF OPERATION

The term "unipolar barrier" was coined to describe a barrier that can block one carrier type (electron or hole) but allows an unimpeded flow of the other. Between different types of barrier detectors, the most popular is the nBn detector, shown in Figure 12.41. The n-type semiconductor on one side of the barrier constitutes a contact layer for biasing the device, while the n-type narrow-bandgap semiconductor on the other side of the barrier is a photon-absorbing layer whose thickness should be comparable to the absorption length of light in the device, typically several microns. The same doping type in the barrier and active layers is key to maintaining a low, diffusion-limited dark current. The barrier needs to be carefully engineered. It must be nearly lattice matched to the surrounding material and have zero offset in the one band

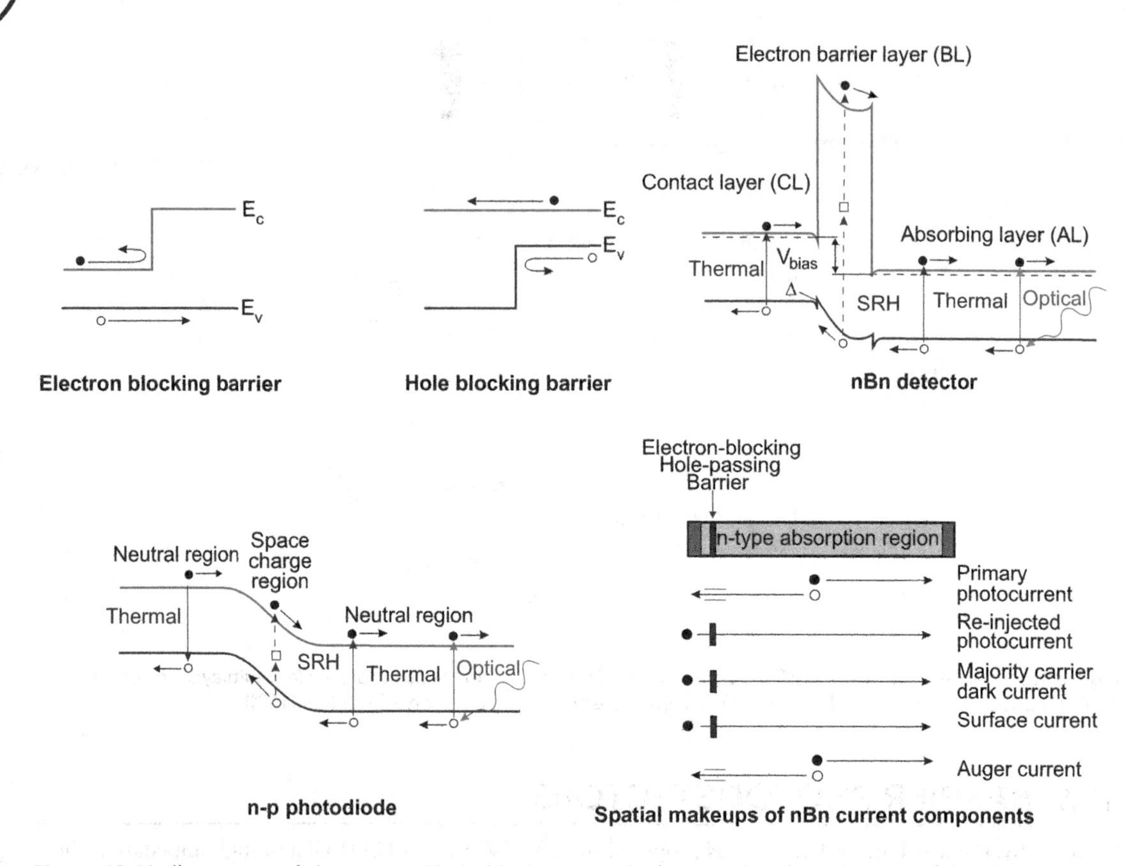

Figure 12.41 Illustrations of electron- and hole-blocking unipolar barriers, bandgap diagram of the nBn barrier detector (the valence-band offset, Δ, is shown explicitly) and p-n photodiode. The bottom right side of the nBn barrier detector shows spatial makeups of the various current components and barrier blocking.

and a large offset in the other. It should be located near the minority carrier collector and away from the region of optical absorption. Such barrier arrangement allows photogenerated holes to flow to the contact (cathode) while majority carrier dark current, re-injected photocurrent, and surface current are blocked. Effectively, the nBn detector is designed to reduce the dark current (associated with SRH processes) and noise without impeding the photocurrent (signal). In particular, the barrier serves to reduce the surface leakage current—the benefit of the nBn architecture is self-passivation. Spatial makeups of the various current components and barrier blocking in the nBn detector are shown in the bottom right side of Figure 12.41 [169].

Other key benefits connected with the absence of depletion regions in the narrow-gap absorption layer is the immunity of nBn detector to dislocations and other defects, which may allow growth on lattice-mismatched substrates such as GaAs with a reduced penalty of excess dark current generated by misfit dislocations.

A new insight on barrier detectors helped M. Reine and co-workers develop numerical simulations and analytical models to better understand the physics and operation of simple, ideal, defect-free nBn devices with p-type [170] and n-type barriers [171]. For detectors with p-type barrier, the approximation model is analogous to the well-known depletion approximation for the conventional p-n junction with new boundary conditions for ideal back-to-back photodiodes.

The papers [172,173] established a criterion for the combination of bias voltage and barrier concentration that allows operation with no depletion region in the narrow-gap absorption layer. A valence band barrier is present for an n-type barrier (see Figure 12.41) that can significantly impede hole current transport between the absorption layer and the contact layer, which can require large bias voltages to overcome. In contrast, a p-type barrier has no barrier, but rather a potential well for holes in the valence band that does not impede hole transport between the absorption layer and the contact layer. However, in the last case, a

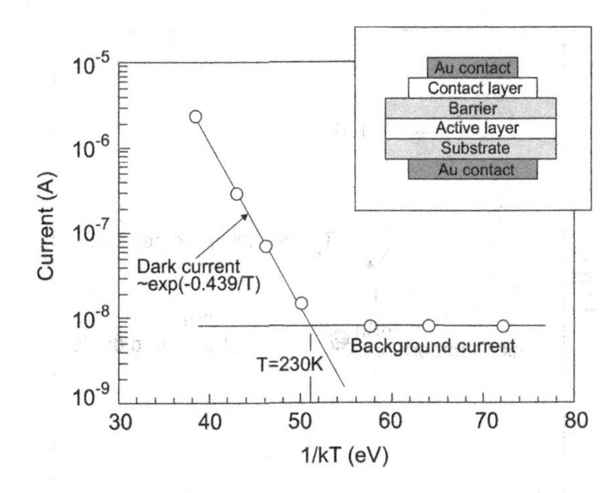

Figure 12.42 Arrhenius plot of the current of an InAs nBn exposed to room temperature background radiation via 2π steradians. Inset: schematic of an nBn device. (From Ref. [167])

p-type barrier layer inherently causes depletion regions to form in the narrow-gap absorption layer for all bias voltages, which should be avoided (depletion regions cause excessive g–r dark currents).

The nBn detector is essentially a "minority carrier photoconductor" with unity gain, due to the absence of majority carrier flow, and in this respect is similar to a photodiode—the junction (space charge region) is replaced by an electron blocking unipolar barrier (B), and the p-contact is replaced by an n-contact. It can be stated, that the nBn design is a hybrid between photoconductor and photodiode.

The first InAs nBn structure described by Maimon and Wicks [167] consists of an n-type narrow-bandgap thin contact layer, a 50–100–nm thick wide-bandgap layer with a barrier for electrons and no barrier for holes, and a thick n-type narrow-bandgap absorbing layer. The high barrier layer is thick enough so that there is negligible electron tunneling through it (i.e., 5–100 nm thick and height over 1 eV). Due to a new heterostructure device design and processing, the nBn detectors demonstrate promising results for suppression of surface leakage currents. The insert of Figure 12.42 shows the nBn InAs structure after standard processing [167]. The detector is defined by etching the contact layer with a selective etchant that stops at the barrier. Gold contact is deposited on the contact layer and on the substrate, and the active layer is covered with the barrier layer. As a result, an additional surface passivation can be eliminated. This is a major advantage compared to InAs-InAsSb-GaSb material system photodiodes, as there is no good passivation.

The InAs device consists of three molecular beam epitaxy- (MBE) grown layers: 3-μm thick InAs (N_d ~ 2×10^{16} cm^{-3}), a 100-nm thick AlAsSb barrier, and n-type InAs contact layer (N_d ~ 1×10^{18} cm^{-3}). The growth was made on InAs substrate lattice matched to GaSb. As Figure 12.42 shows, at higher temperatures, the dark device current exhibits a thermal activation energy of 0.439 eV close to InAs bandgap. At temperatures below 230 K, the device operates in BLIP conditions. The BLIP temperature is at least 100 K higher than that of commercial InAs photodiodes [167].

Figure 12.43 shows a typical Arrhenius plot of the dark current in a conventional diode and in an nBn detector [172]. The diffusion current typically varies as $T^3\exp(-E_{g0}/kT)$, where E_{g0} is the band gap extrapolated to zero temperature, T is the temperature, and k is the Boltzmann constant. The generation–recombination current varies as $T^{3/2}\exp(-E_{g0}/2kT)$ and is dominated by the generation of electrons and holes by SRH traps in the depletion region. Because in an nBn detector, there is no depletion region, the generation–recombination contribution to the dark current from the photon-absorbing layer is totally suppressed. The lower portion of Arrhenius plot for the standard photodiode has a slope that is roughly half that of the upper portion. The solid line (nBn) is an extension of the high-temperature diffusion-limited region to temperatures below T_c, which is defined as the crossover temperature at which the diffusion and generation–recombination currents are equal. In the low-temperature region, the nBn detector offers two important advantages. First, it exhibits a higher signal-to-noise ratio than a conventional diode operating at the same temperature. Second, it operates at a higher temperature than a conventional diode with the same dark current. The latter is depicted by the horizontal dashed line in Figure 12.43.

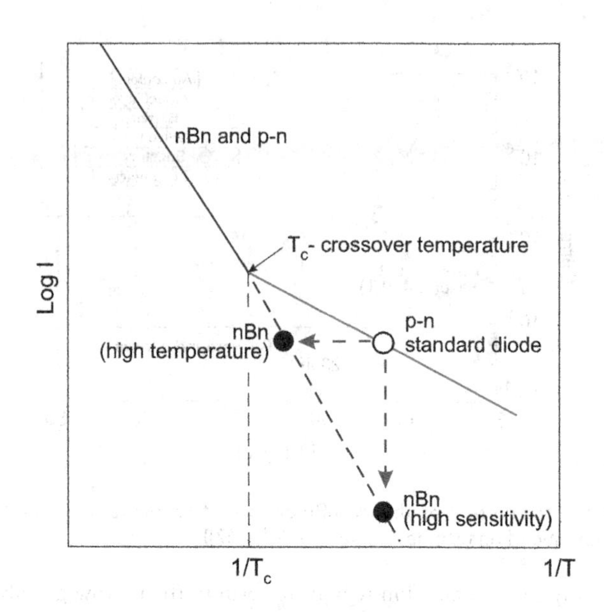

Figure 12.43 A schematic Arrhenius plot of the dark current in a standard diode and in an nBn device.

The absence of a depletion region offers a way for materials with relatively poor SRH lifetimes, such as all III-V compounds, to overcome the disadvantage of large depletion dark currents.

The operating principles of the nBn and related detectors have been described in detail in the literature [167–175]. While the idea of nBn design has originated with bulk InAs materials [167], its demonstration using T2SL-based materials facilitates the experimental realization of the barrier detector concept with better control of band edge alignments [176].

Klipstein et al. [177] have considered a wide family of barrier detectors, which they divide into two groups: XB_nn and XB_pp detectors (see Figure 12.44). In the case of the former group, all designs have the same n-type B_nn structural unit but use different contact layers (X), in which either the doping, material, or both are varied. If we consider, for example, C_pB_nn and nB_nn devices, C_p is the p-type contact made from a different material than the active layer, whereas n is the n-type contact made from the same material as the active layer. In the case of a pB_nn structure, the p-n junction can be located at the interface between the heavily doped p-type material and the lower doped barrier, or within the lower doped barrier itself. Our barrier detector family also has p-type members, designated as XB_pp, which are polarity-reversed versions of the n-type detectors. The pBp architecture should be employed when the surface conduction of the materials is p-type and must be used with the p-type absorbing layer. This structure can be realized using, for example, a p-type InAs/GaSb T2SLs as the absorbing layer [175,178,179]. In addition, the so-called pMp device consists of two p-doped superlattice active regions and a thin M-structure with a higher energy barrier. The bandgap difference between the superlattice M structures falls in the valence band, creating a valence band barrier for majority of holes in the p-type semiconductor.

Unipolar barriers can also be inserted into a conventional p-n photodiode architecture [169,180]. There are two possible locations into which a unipolar barrier can be implemented: (i) outside of depletion layer in the p-type layer, or (ii) near the junction, but at the edge of the n-type absorbing layer (see Figure 12.45). Depending on the barrier placement, different dark current components are filtered. For example, placing the barrier in the p-type layer blocks surface current, but currents due to diffusion, generation–recombination, TAT, and BBT cannot be blocked. If the barrier is placed in the n-type region, the junction-generated currents and surface currents are effectively filtered out. The photocurrent shares the same spatial makeup as the diffusion current, which is shown in Figure 12.46.

Unipolar barriers can significantly improve the performance of IR photodiodes, as is shown in Figure 12.47 for InAs material system. For InAs, $AlAs_{0.18}Sb_{0.82}$ is an ideal electron blocking unipolar barrier material. Theoretical predictions suggest that the VBO should be less than kT for $AlAs_ySb_{1-y}$ barrier composition in the range of $0.14 < y < 0.18$. Figure 12.47 compares the temperature-dependent R_0A product data for

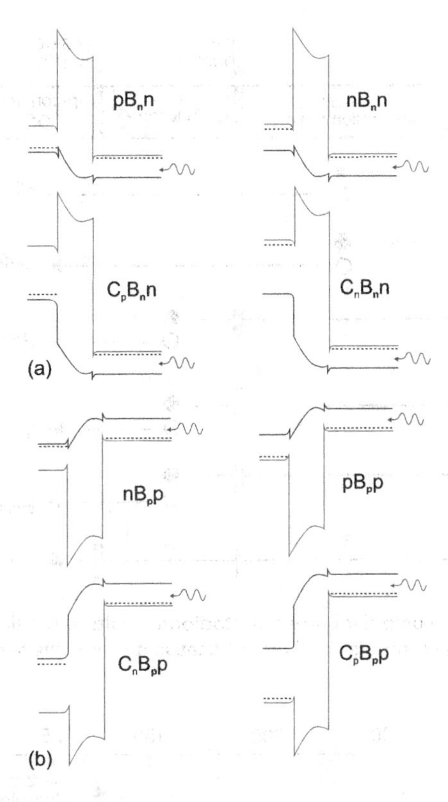

Figure 12.44 Schematic band profile configurations under operating bias for XB$_n$n (a) and XB$_p$p (b) barrier detector families. In each case, the contact layer (X) is on the left, and IR radiation is incident onto the active layer on the right. When X is composed of the same material as the active layer, both layers have the same symbol (denoting the doping type), otherwise it is denoted as C (with the doping type as a subscript). (After Ref. [177])

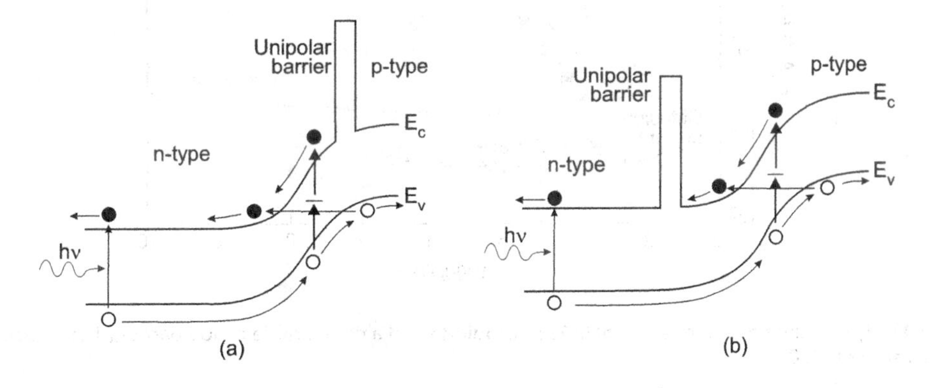

Figure 12.45 Band diagrams of a p-side (a) and an n-side (b) unipolar photodiode under bias. (Adapted after Ref. [169])

an n-side unipolar barrier photodiode with that of a conventional p-n photodiode. The unipolar barrier photodiode shows performance near "Rule 07" with activation energy near the bandgap of InAs indicating diffusion-limited performance and six orders of magnitude higher R_0A value in low-temperature range than that of the conventional p-n junction.

The "Rule 07" criterion manifests the performance of p-on-n HgCdTe photodiode architecture, which is limited by Auger 1 diffusion current from 10^{15} cm^{-3} n-type material, and is the popular mark of reference to compare the performance of any type of detector with HgCdTe state of the art. Any detector architecture that is limited by Auger 7 p-type diffusion, or by depletion currents will not behave according to "Rule 07." In fact, the appropriate criterion to be used for comparative studies is the detector dark current relative to system flux current.

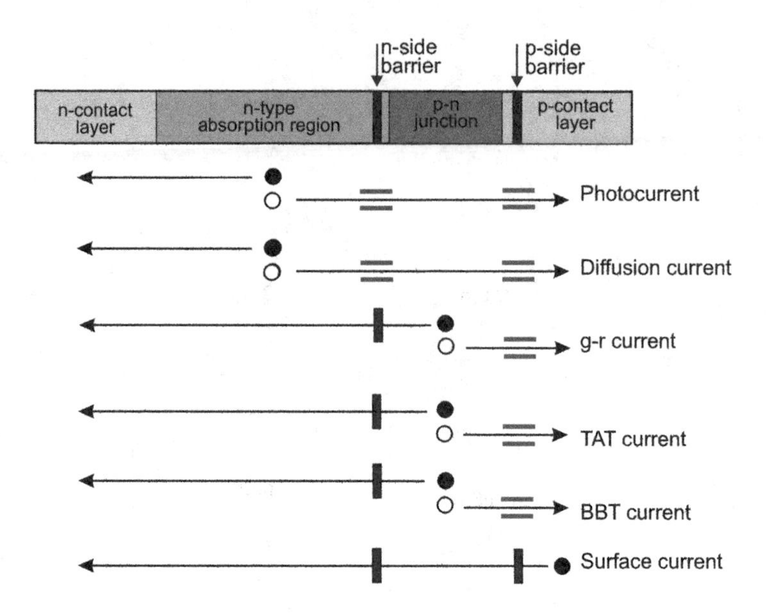

Figure 12.46 Placing the barrier in a unipolar barrier photodiode results in the filtering of surface currents and junction-related currents. Diffusion current is not filtered because it shares the same spatial makeup as the photocurrent. (Adapted after Ref. [180])

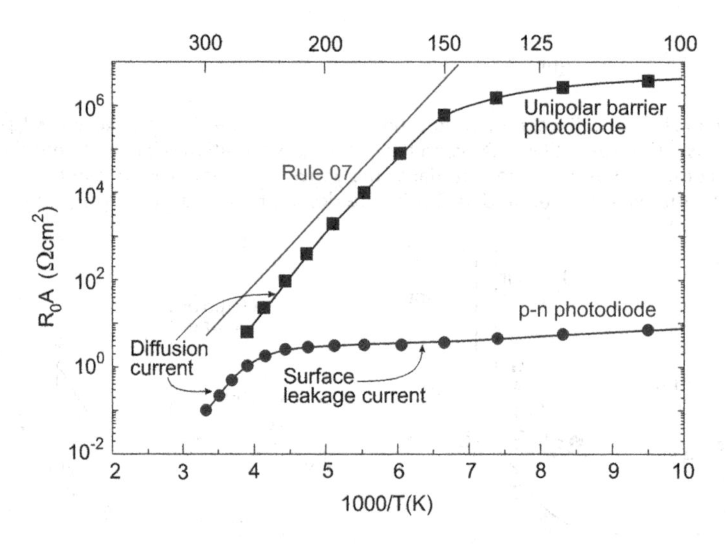

Figure 12.47 R_0A product of a conventional InAs photodiode and a comparable n-side barrier photodiode. (Adapted after Ref. [180])

12.10 PHOTOELECTROMAGNETIC, MAGNETOCONCENTRATION, AND DEMBER DETECTORS

Aside from photoconductive detectors, photodiodes, and barrier detectors, three other junctionless devices have been used mainly for uncooled IR photodetectors: PEM detectors, magnetoconcentration detectors, and Dember effect detectors. Certainly they are niche devices, but they are still in production and used with important applications, including very fast uncooled detectors with long-wavelength IR radiation. The devices have been reviewed by Piotrowski and Rogalski [38,181].

12.10.1 PHOTOELECTROMAGNETIC DETECTORS

The first experiments on the PEM effect were performed with Cu_2O by Kikoin and Noskov in 1934 [182]. Nowak's monographic paper summarizes the results of his investigations on the PEM effect [183], which has been replicated worldwide during the last 50 years. For a long time, the PEM effect has been used mostly for InSb room temperature detectors in the middle- and far-IR band [184]. However, the uncooled InSb devices with a cutoff wavelength at ≈7 μm exhibit no response in the 8–14 μm atmospheric window and relatively modest performance in the 3–5 μm window. $Hg_{1-x}Cd_xTe$ and closely related $Hg_{1-x}Zn_xTe$ and $Hg_{1-x}Mn_xTe$ alloys made it possible to optimize the performance of PEM detectors at any specific wavelength [185].

12.10.1.1 PEM effect

The PEM effect is caused by diffusion of photogenerated carriers due to the photo-induced carrier in-depth concentration gradient and by deflection of electron and hole trajectories in opposite directions by the magnetic field (Figure 12.48). If the sample ends are open-circuit in the *x*-direction, a space charge builds up that gives rise to an electric field along the *x*-axis (open-circuit voltage). If the sample ends are short-circuited in the *x*-direction, a current flows through the shorting circuit (short-circuit current). In contrast to photoconductors and PV devices, the generation of PEM photovoltage (or photocurrent) requires not simply optical generation, but instead, the formation of an in-depth gradient of photogenerated carriers. Typically, this is accomplished by nonhomogeneous optical generation due to radiation absorption in the near surface region of the device.

Generally, the carrier transport in PEM devices cannot be adequately described by analytical methods and therefore require a numerical solution. Consider the assumptions that make it possible to obtain analytic solutions: homogeneity of the semiconductor, nondegenerate statistics, negligible interface and edge effects, independence of material properties on the magnetic and electric fields, and equal Hall and drift mobilities.

The transport equations for electrons and holes both in *x* and *y* take the form:

$$J_{hx} = qp\mu_h E_x + \mu_h B J_{hy}, \tag{12.166}$$

$$J_{hy} = qp\mu_h E_y - \mu_h B J_{hx} - qD_h \frac{dp}{dy}, \tag{12.167}$$

$$J_{ex} = qn\mu_e E_x - \mu_e B J_{ey}, \text{ and} \tag{12.168}$$

$$J_{ey} = qn\mu_e E_y + \mu_e B J_{ex} + qD_e \frac{dn}{dy}, \tag{12.169}$$

where B is the magnetic field in the *z*-direction, E_x and E_y are the *x* and *y* components of the electric field, J_{ex} and J_{ey} are the *x* and *y* components of the electron current density, and J_{hx} and J_{hy} are the analogous components of the hole current density.

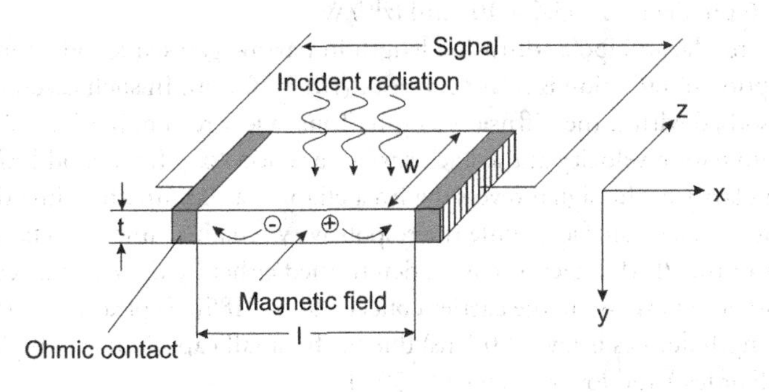

Figure 12.48 Schematic of PEM effect.

The E_y can be eliminated in Equations 12.166 through 12.169 with the condition

$$J_y = J_{ey} + J_{hy} = 0. \tag{12.170}$$

The other equation to be used is the continuity equation for y-direction currents:

$$\frac{dJ_{hy}}{dy} = -\frac{dJ_{ey}}{dy} = q(G - R), \tag{12.171}$$

where G and R denote the carrier generation and recombination rates, respectively.

As a result, a nonlinear second order differential equation for p can be obtained from the set of transport equations for electrons and holes,

$$A_2\frac{d^2p}{dy^2} + A_1\left(\frac{dp}{dy}\right)^2 + A_0\frac{dp}{dy} - (G - R) = 0, \tag{12.172}$$

where A_2, A_1, and A_0 are coefficients dependent on semiconductor parameters, and electric and magnetic fields. Equation 12.172, with boundary conditions for the front and back side surfaces, determines the hole distribution in the y-direction. The electron concentration can be calculated from electric quasineutrality equations. In consequence, x-direction currents and electric fields can be calculated.

Lile reported an analytical solution for the small-signal steady-state PEM photovoltage. The voltage responsivity of the PEM detector that can be derived from the Lile solution is [186,187]

$$R_v = \frac{\lambda}{hc}\frac{B}{wt}\frac{\alpha z(b+1)}{n_i(b+z^2)}\frac{Z(1-r_1)}{Y(a^2+\alpha^2)}, \tag{12.173}$$

where $b = \mu_e/\mu_h$, $z = p/n_i$, w and t are the width and thickness of the detector, r_1 is the front reflectance, and a is the reciprocal diffusion length in the magnetic field.

The PEM photovoltage is generated along the length of the detector, so the signal linearly increases with the length of the detector and is independent on the device widths for the same photon flux density. This results in a good responsivity for large area devices, in contrast to conventional junction photovoltaic devices.

Analysis of Equation 12.173 indicates that the maximum voltage responsivities can be reached in strong magnetic fields ($B \approx 1/\mu_e$) for samples with high resistance. In the case $\mu_e/\mu_h >> 1$, the resistance of the detector reaches its maximum value at the point $p/n_i \approx (\mu_e/\mu_h)^{1/2}$ and the highest value for R_v is reached for lightly doped p-type material [186,187]. The acceptor concentration in narrow gap semiconductors is adjusted to a level of about (2–3) × 10^{17} cm^{-3} to achieve detectivity-optimized detectors. Due to the high mobility of minority carriers in p-type devices, a magnetic field of ≈2 T is sufficient for good performance. The voltage responsivity of the detectivity-optimized device is ≈0.6 V/W. The maximum theoretical detectivity of uncooled 10.6 μm device is ≈3.4 × 10^7 cmHz$^{1/2}$/W.

At room temperature, the ambipolar diffusion length in narrow-gap semiconductors is small (several μm), while the absorption of radiation is relatively weak ($1/\alpha \approx 10$ μm). In such cases, the radiation is almost uniformly absorbed within the diffusion length. Thus, a low recombination velocity at the front surface and a high recombination velocity at the back surface are necessary for a good PEM detector response. In such devices, the polarity of the signal reverses with a change of illumination direction from the low to the high recombination velocity surface, while the responsivity remains almost unchanged.

The response time of the PEM detector may be determined either by the RC time constant or by the decay time of the gradient of excess charge carrier concentration [185]. Typically, the RC time constant of uncooled long-wavelength devices is low (< 0.1 ns) due to the small capacitance of high-frequency optimized devices (≈ 1 pF or less) and low resistance (≈ 50 Ω).

The decay of gradient of carrier's concentration may be caused by volume recombination or ambipolar diffusion. While the first mechanism reduces excess concentration, the second one tends to make the excess concentration uniform. Therefore, the response time can be significantly shorter than the recombination time if the thickness is shorter than the diffusion length. The resulting response time is

$$\frac{1}{\tau_{ef}} = \frac{1}{\tau} + \frac{2D_a}{t^2}. \tag{12.174}$$

For a p-type HgCdTe device with a thickness t of 2 μm, the response time is ≈5×10^{-11} s. Even shorter response times are achieved with thinner layers, but at the cost of radiation absorption, quantum efficiency, and detectivity.

12.10.1.2 Fabrication and performance

Only epitaxial devices are manufactured at present [188,189]. The preparation of PEM detectors is essentially very similar to that of photoconductive ones, with the exception of the back surface preparation, which in PEM devices is subjected to a special mechanochemical treatment to achieve a high recombination velocity. This procedure is not mandatory when using graded gap structures. Electrical contacts are usually made by Au/Cr deposition and gold wires are then attached. Figure 12.49 shows the cross section of a sensitive element of a PEM detector.

Figure 12.50 schematically shows the housings of PEM detectors, which are based on standard TO-5 or, for larger elements, TO-8 transistor cans. For high-frequency operation, the active elements of PEM detectors are accommodated in housing, which incorporates a miniature two-element permanent magnet and pole pieces. Magnetic fields approaching 2 T are achievable with the use of modern rare-earth magnetic materials for the permanent magnet and cobalt steel for the pole pieces.

The best PEM devices exhibit measured voltage responsivity exceeding 0.15 V/W (width of 1 mm) and detectivities of 1.8×10^7 cmHz$^{1/2}$/W, a factor of ≈2 below the predicted ultimate value [187]. The reason for this is a lower (than optimum) magnetic field and faults in detector construction, which are probably nonoptimum surface processing and material composition/doping profile.

The fast response of PEM detectors has been confirmed by observations of CO_2 laser self-mode-locking and free-electron laser experiments [38]. When detecting ordinary or low repetition rate short pulses,

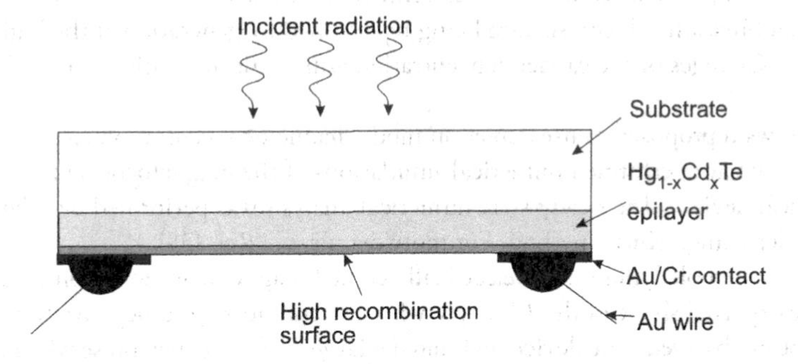

Figure 12.49 Cross section of a back-side illuminated sensitive element of PEM detector.

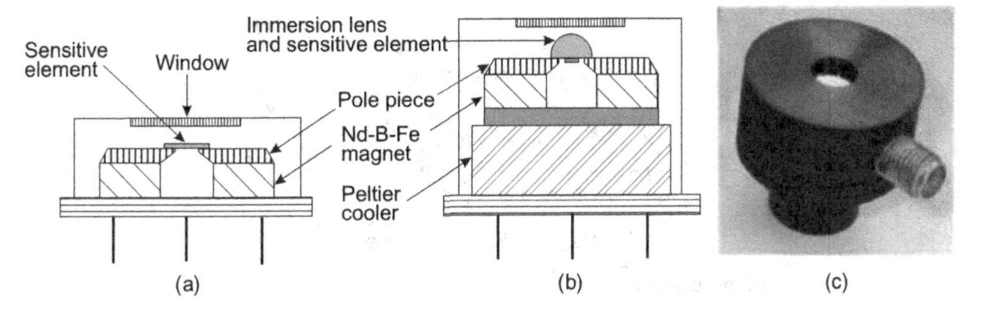

Figure 12.50 Schematic of the (a) housing of an ambient temperature and (b) a thermoelectrically cooled, optically immersed PEM detector, and (c) a picture of high-frequency optimized specialized housing. (From Ref. [38])

signal voltages up to ≈1 V are obtained with 1-mm long detectors being limited by strong optical excitation effects. The maximum signal voltages for chopped CO_2 radiation are much lower due to radiation heating. In good heat dissipation–design devices, they exceed 30 mV per mm.

Both theoretical and measured performances of PEM detectors are inferior to those of photoconductor detectors. PEM detectors have, however, additional important advantages, which make them useful in many applications. In contrast to photoconductors, they do not require electric bias. The frequency characteristics of the PEM detector are flat over a wide frequency range, starting from zero frequency. This is due to the lack of the low-frequency noise and very short response time. The resistance of PEM detectors does not decrease with increasing size, which makes it possible to achieve the same performance for small and large area devices. With the resistance typically close to 50 Ω, the devices are conveniently coupled directly to wideband amplifiers.

PEM detectors have been also fabricated with other HgTe-based ternary alloys; $Hg_{1-x}Zn_xTe$ and $Hg_{1-x}Mn_xTe$ [38]. It seems that the use of $Hg_{1-x}Zn_xTe$ and $Hg_{1-x}Mn_xTe$ offers no advantages when compared to $Hg_{1-x}Cd_xTe$ in terms of performance and speed of response.

12.10.2 MAGNETOCONCENTRATION DETECTORS

If a semiconductor plate is electrically biased and placed in a crossed magnetic field, then the spatial distribution of electron hole pairs along the crystal section deviates from the equilibrium value. It is hence called magnetoconcentration effect. Such redistribution is efficient in the plate whose thickness is comparable with the ambipolar diffusion length of charge carriers. The Lorentz force deflects electrons and holes drifting in the electric field in the same direction, resulting in an increase in the concentration of carriers at one surface and decrease at the opposite one, depending on the direction of the magnetic and electric field.

The sample shown in Figure 12.51 with small recombination velocity at the front-side surface and a high at the back side is especially interesting for IR detector and source application. When Lorentz force moves carriers toward the high recombination velocity surface (depletion mode), the carriers recombine there, resulting in depletion in the volume with exception of the region close to the high-surface recombination surface. With opposite direction of the Lorentz force (enrichment mode), the carriers are moved toward the low recombination velocity surface being replenished by generation at the high recombination velocity surface. Changes of the carrier concentration will result in positive or negative luminescence [190–192].

Djuric and Piotrowski proposed to use depletion mode magnetoconcentration effect to suppress the Auger [193–195]. They have performed numerical simulations of the magnetoconcentration devices and reported first practical devices. The steady-state numerical analysis was performed by solving the equations using the fourth-order Runge-Kutta method. For more details, see Ref. [38].

Practical, 10.6-μm uncooled and thermoelectrically cooled magnetoconcentration detectors have been reported [196]. The expected shape of the *I-V* curve with current saturation, negative resistance, and oscillation regions has been observed. The devices exhibited a large low-frequency noise when biased to achieve a sufficient depletion of the semiconductor. Anomalously high noise has been generated in the region of

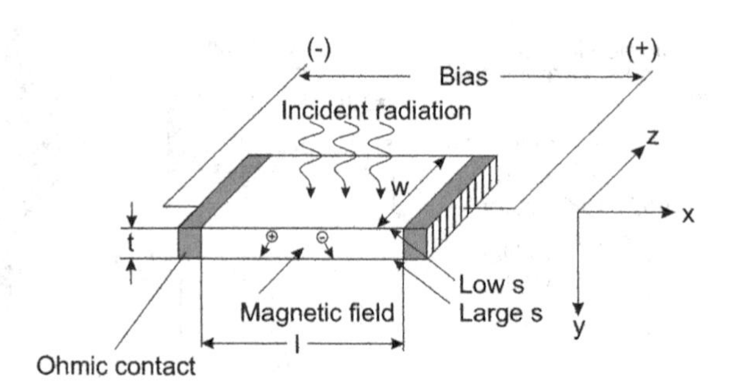

Figure 12.51 Magnetoconcentration effect (depletion mode). Electrons and holes are pushed by the action of the Lorentz force toward the back-side surface with a high recombination velocity.

negative resistance. Significant reduction of noise and improvement of performance was observed at high frequencies (>100 kHz), with careful selection of bias current. With such properties, the devices are suitable only for some broadband applications and further efforts are necessary to make them useful in typical applications.

12.10.3 DEMBER DETECTORS

Dember detectors are a type of photovoltaic devices, based on bulk photodiffusion voltage in a simple structure with only one type of semiconductor doping supplied with two contacts [197]. When radiation is incident on the surface of a semiconductor generating electron hole pairs, a potential difference is usually developed in the direction of the radiation (see Figure 12.52), as a result of the difference in diffusion of electrons and holes. The Dember effect electrical field restrains the electrons with higher mobility, while holes are accelerated, thus making both fluxes equal.

The Dember effect device can be analyzed solving the transport and continuity equations, assuming zero total currents in x and y directions. The steady-state photovoltage under conditions of weak optical excitation and assuming electroneutrality can be expressed as [198]

$$V_d = \int_0^t E_z(z)\,dz = \frac{kT}{q}\frac{\mu_e - \mu_h}{n_o\mu_e + p_o\mu_h}\left[\Delta n(0) - \Delta n(t)\right], \tag{12.175}$$

where E_z is an electric field in the z-direction and $\Delta n(z)$ is the excess electron concentration. As this expression shows, two conditions are required for generation of the photovoltage—the distribution of photogenerated carriers should be nonuniform and the diffusion coefficients of electrons and holes must be different. The gradient may result from a nonuniform optical generation or/and from different recombination velocities at the front and back surfaces of the device.

Djuric and Piotrowski have estimated the theoretical voltage responsivity and detectivity of Dember detector for boundary conditions at the top and bottom surface

$$D_a\frac{\partial \Delta n}{\partial z} = s_1\Delta n(0) \quad \text{at } z = 0, \tag{12.176}$$

$$D_a\frac{\partial \Delta n}{\partial z} = s_2\Delta n(t) \quad \text{at } z = t \tag{12.177}$$

The maximum voltage responsivity can be achieved for:

- Optimized p-type doping
- A low surface recombination velocity and a low reflection coefficient at illuminated side contact
- A large recombination velocity and a large reflection coefficient at nonilluminated side contact

Since the Dember device is not biased, the detectivity is determined by the Johnson–Nyquist thermal noise.

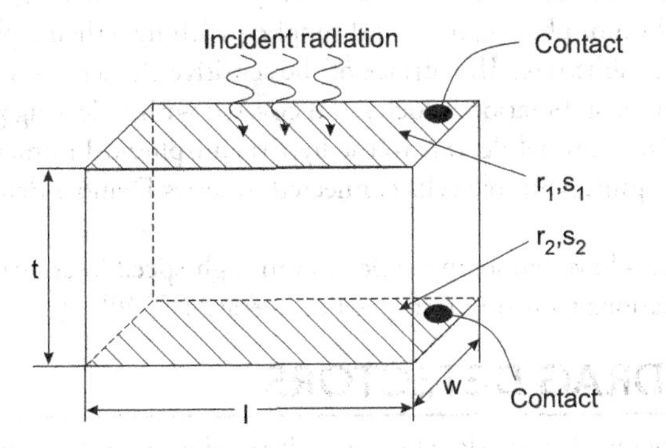

Figure 12.52 Schematic of Dember detector.

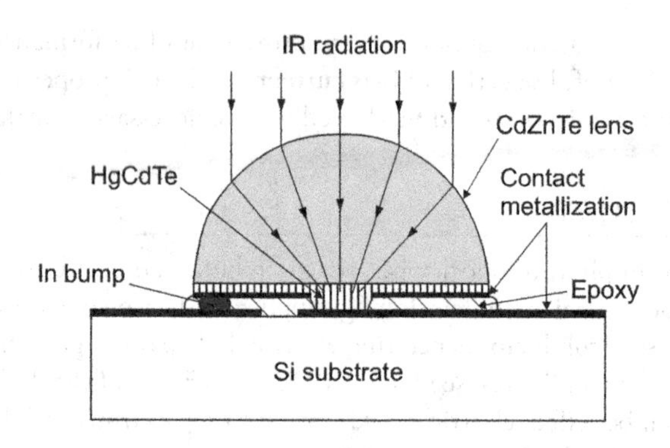

Figure 12.53 Schematic cross section of an experimental monolithic optically immersed Dember detector. (From Ref. [38])

The theoretical design of $Hg_{1-x}Cd_xTe$ and practical Dember effect detectors have been reported [185,196,199]. The best performance is achievable for a device thickness slightly larger than the ambipolar diffusion length. More thin devices exhibit low voltage responsivity while more thick have excessive resistance and large related Johnson noise. As in the case of PEM detectors, the best performance is achieved in the p-type material. The calculated detectivity of Dember detectors is comparable to that of photoconductors operated under the same conditions. Detectivities as high as $\approx 2.4 \times 10^8$ cmHz$^{1/2}$/W and of $\approx 2.2 \times 10^9$ cmHz$^{1/2}$/W are predicted for optimized 10.6-μm devices at 300 and 200 K, respectively.

The interesting feature of the Dember device is a significant photoelectric gain larger at zero bias condition. For optimum doping the gain is about 1.7 and increases in decreasing thickness. The gain is caused by the ambipolar effects. At zero bias and shortened device, the photo-generated electrons may travel several times between contacts before the holes will recombine or diffuse to the back side contact.

Very low resistance, low voltage responsivities, and noise voltages well below the noise level of the best amplifiers pose serious problems in achieving the potential performance, however. For example, the optimized uncooled 10.6-μm devices with only 7×7 μm^2 size will have a resistance of $\approx 7\ \Omega$, voltage responsivity of ≈ 82 V/W and noise voltage of ≈ 0.35 nV/Hz$^{1/2}$. The low resistance and the low voltage responsivity is the reason why the simple Dember detectors cannot compete with other types of photodetectors.

There are some potential ways to overcome this difficulty. One is the connection of small area Dember detectors in series. The other possibility is to use optical immersion [38].

Due to ambipolar diffusion of excess carriers, the response time of a Dember detector is shorter than the bulk recombination time and can be very short for devices with a thickness much less than the diffusion length. The response time of detectivity-optimized devices is about 1 ns, shortening with p-type doping, with some expense in performance. The high-frequency optimized Dember detectors should be prepared from heavily doped materials using optical resonant cavity.

Figure 12.53 shows a schematic of a small size, optically immersed Dember effect detector manufactured by Vigo Systems [38]. The sensitive element has been prepared from a $Hg_{1-x}Cd_xTe$-graded gap epilayers grown by the isothermal vapor phase epitaxy and supplied with hyperhemispherical lens formed directly in the transparent (Cd,Zn)Te substrate. The surface of the sensitive element was subjected to a special treatment to produce a high recombination velocity and covered with reflecting gold. The "electrical" size of the sensitive element is 7×7 μm^2 while, due to the hyperhemispherical immersion, the apparent optical size is increased to $\approx 50 \times 50$ μm^2. Multiple cells connected in series Dember detectors have been used for large area devices.

HgCdTe Dember detectors have found an application in high-speed laser beam diagnostics of industrial CO_2 lasers and other applications requiring fast speed of operation [200].

12.11 PHOTON-DRAG DETECTORS

In heavily doped semiconductors, free carrier absorption is the dominant absorption mechanism for wavelengths longer than the absorption edge. An incident stream of photons having wavelengths consistent with

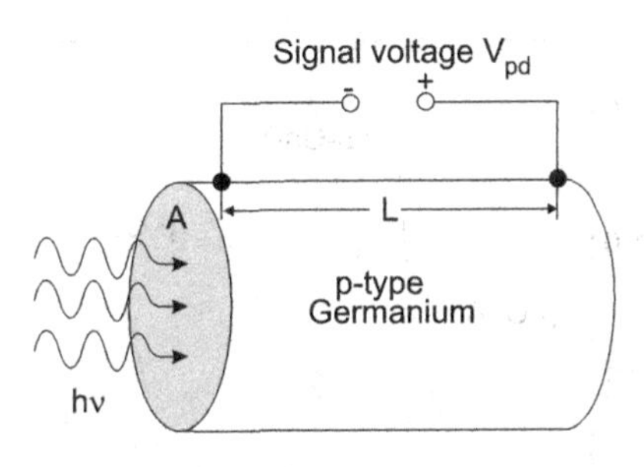

Figure 12.54 Photon drag detector structure.

free carrier absorption will transfer momentum to free carriers, effectively pushing them in the direction of the Poynting vector. Thus, a longitudinal electric field is established within the semiconductor that can be detected through electrodes attached to the sample. The transfer of momentum from photons to free carriers in semiconductors is called the photon drag effect. The radiation pressure builds up a voltage difference between the front and back surfaces. When the device is operated with a high-impedance amplifier, the net current is zero, an electric field is set up to oppose the photon drag force, and the potential change along the bar provides the output signal.

The photon-drag detector structure is shown in Figure 12.54. A typical photon drag device consists of a cylindrical or rectanguloid rod of semiconductor material, with sufficient doping to absorb radiation over the length of the rod at the wavelength of interest. Normally, the detector has a length to diameter ratio sufficient to ensure a uniform field, regardless of where the light beam strikes the end of the rod.

In the classical model, when the photon energy $h\nu \ll kT$ and absorption are determined by free carriers, the photon drag voltage can be estimated by considering the momentum p possessed by each photon, $p = E/c$, where E is photon energy and c is the speed of light. The rate of change of momentum in x-direction per unit volume is given by

$$\frac{dp(x)}{dt} = \frac{n_r P\alpha \exp(-\alpha x)}{Ac}, \tag{12.178}$$

where P is the power of the light flux, which falls on the cross sectional A. In open-circuit condition, the average rate of change of momentum to each carrier must be balanced by an electromotive force acting on the carrier (taking n-type as an example)

$$qE(x) = \frac{n_r P\alpha \exp(-\alpha x)}{Acn}. \tag{12.179}$$

Integrating this longitudinal electric field over the entire length, L, the photon drag voltage is equal to:

$$V_{pd} = \frac{n_r P[1-\exp(-\alpha L)]}{qAcn}. \tag{12.180}$$

The last equation is found to obey only for wavelengths longer than a few 100 μm. For shorter wavelengths, responsivity is wavelength dependent and V_{pd} value is less than that given by Equation 12.180.

The microscopic theory of the classical photon drag effect was developed by Gurevich and Rumyantsev in 1967 [201]. The theory of the photon drag given by Gibson and Montasser is valid in the wavelength range 1–10 μm [202]. A photon drag effect on quantum well structures, arising from intersubband transition rather than interband transition, was suggested by Grinberg and Luryi [203]. The photon drag detectors are described in more detail in Gibson and Kimmitt [204].

In general, the mechanism controlling the photon drag effect is complicated, especially in the case when light absorption is determined by interband optical transitions. Due to the participation of the lattice in the momentum conservation law, the drag of carriers can be opposite to the light propagation direction. As a result, the sign of the photon drag voltage changes with the wavelength in the range of 1–10 μm [205].

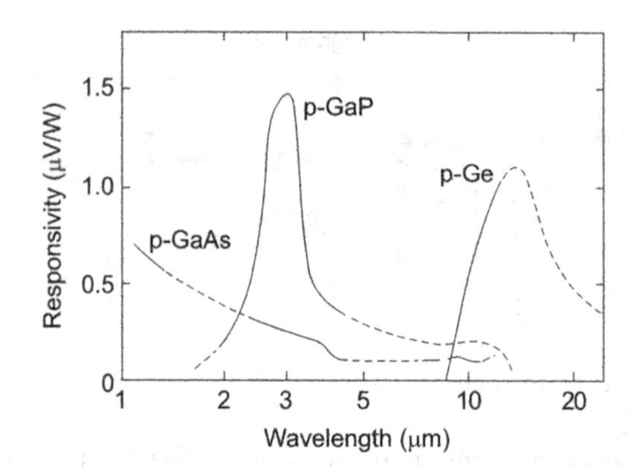

Figure 12.55 Responsivity of 3.2 Ωcm p-Ge, 5 Ωcm n-GaP, and 2.5 Ωcm p-Ge longitudinal detectors. All the detectors have an active area of 4 × 4 mm, a resistance of 50 Ω, and are oriented in a 111 direction. The responsivity of 1 μV/W is equivalent to an NEP of ≈ 10^{-3} W/Hz$^{1/2}$. (From Ref. [204])

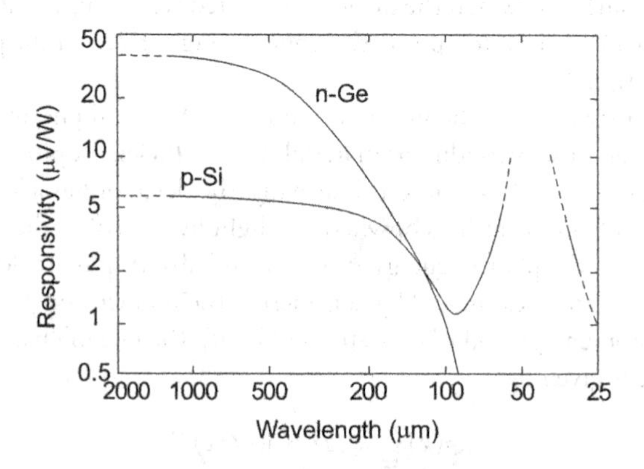

Figure 12.56 Responsivity of 30 Ωcm n-Ge and p-Si, longitudinal detectors oriented in a 100 direction. The active area is 4 × 4 mm and the detector resistance are n-Ge, 250 Ω; and p-Si, 350 Ω. The responsivity of 10 μV/W is equivalent to an NEP of ≈ 2 × 10^{-4} W/Hz$^{1/2}$ for n-Ge and 2.6 × 10^{-4} W/Hz$^{1/2}$ for p-Si. (From Ref. [204])

The absorption of p-type germanium at $\lambda = 10.6$ μm is due to electron transitions from the light hole band to vacant states in the heavy hole band. Photoconductivity due to these intraband transitions was first observed in the heavily doped sample by Feldman and Hergenrother [206]. The photon drag effect was used for the detection of short CO_2-laser pulses by Gibson et al. [207] and Danishevskii et al. [208]. Gibson et al. showed that a Q-switched CO_2 laser working at 10.6-μm wavelength can transfer sufficient momentum to produce a longitudinal electromagnetic force in a rectangular germanium rod of 4-cm length [207]. Similar results have been achieved using p-type tellurium [209].

The responsivity measured experimentally is very low, typically in the range of 1–40 μV/W, but the devices are extremely fast (less than 1 ns) and they operate at room temperature. When using laser sources, there is normally quite sufficient power for a good S/N-ratio.

At the shortest wavelengths, p-GaAs has the best sensitivity (see Figure 12.55) [204]. For 2–11 μm, n-GaP is a good choice. The peak response around 3 μm is due to optical rectification and severe absorption sets extend to around 12 μm. At longer wavelengths (Figure 12.56), p-type silicon appears to be a good available detector. The nature of the direct valence band transitions leads to high photon drag coefficients.

Detectors exploiting the photon drag effect are useful as laser detectors because of their rapid response and ability to absorb large amounts of power without damage because their absorption is small (the radiation is absorbed over a large volume of material). Signal linearity with power is good up to a power density of ≈50 MW/cm^2. Damage of the detector starts to occur around 100 MW/cm^2. Their lack of sensitivity makes them of little use as IR detectors for most applications.

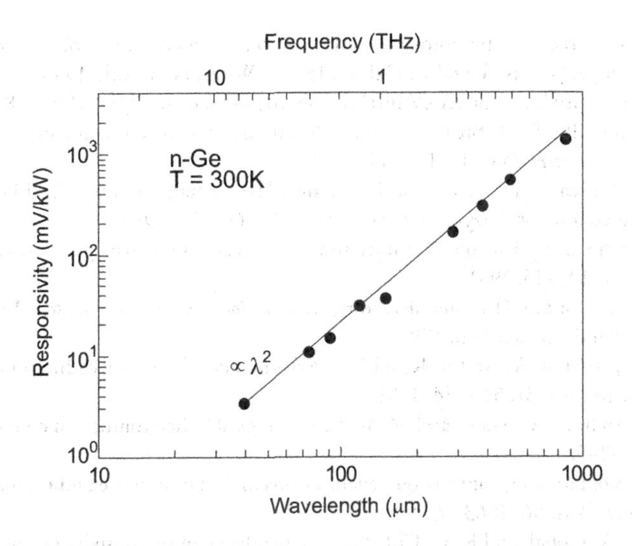

Figure 12.57 Responsivity of an n-type Ge:Sb photon drag detector after 20-dB amplification as a function of wavelength. (From Ref. [212])

Recently, interest in photon drag detectors has increased and is stimulated by impressive progress in the development of THz detector technology [210,211]. For THz detection, p-type Ge detectors are not well suited. The reason is that due to direct intersubband transitions in the valence band and Drude absorption of free carriers, the spectrum of responsivity is sharply structured with several zeros and sign inversions. In Figure 12.57, the responsivity after 20-dB amplification of an n-type germanium detector is plotted as a function of wavelength [212]. Due to rapid free carrier momentum relaxation at room temperature, the response time is very short and may be less than 1 ps.

REFERENCES

1. A. Rose, *Concepts in Photoconductivity and Allied Problems*, Interscience, New York, 1963.
2. D. Long and J. Schmit, "Mercury-cadmium telluride and closely related alloys," in *Semiconductors and Semimetals*, Vol. 5, eds. R. K. Willardson and A. C. Beer, 175–255, Academic Press, New York, 1970.
3. D. Long, "Photovoltaic and photoconductive infrared detectors," in *Optical and Infrared Detectors*, ed. R. J. Keyes, 101–47, Springer-Verlag, Berlin, 1977.
4. W. L. Eisenman, J. D. Merriam, and J. F. Potter, "Operational characteristics of infrared photodetectors," in *Semiconductors and Semimetals*, Vol. 12, eds. R. K. Willardson and A. C. Beer, 1–38, Academic Press, New York, 1977.
5. P. R. Bratt, "Impurity germanium and silicon infrared detectors," in *Semiconductors and Semimetals*, Vol. 12, eds. R. K. Willardson and A. C. Beer, 39–142, Academic Press, New York, 1977.
6. R. H. Kingston, *Detection of Optical and Infrared Radiation*, Springer-Verlag, Berlin, 1979.
7. R. M. Broudy and V. J. Mazurczyck, "(HgCd)Te photoconductive detectors," in *Semiconductors and Semimetals*, Vol. 18, eds. R. K. Willardson and A. C. Beer, 157–99, Academic Press, New York, 1981.
8. N. Sclar, "Properties of doped silicon and germanium infrared detectors," *Prog. Quant. Electron.* 9, 149–257, 1984.
9. A. Rogalski and J. Piotrowski, "Intrinsic infrared detectors," *Prog. Quant. Electron.* 12, 87–289, 1988.
10. A. Rogalski, "Photoconductive detectors," in *Infrared Photon Detectors*, ed. A. Rogalski, 13–49, SPIE Optical Engineering Press, Bellingham, WA, 1995.
11. E. L. Dereniak and G. D. Boreman, *Infrared Detectors and Systems*, Wiley, New York, 1996.
12. C. T. Elliott, "Photoconductive and non-equilibrium devices in HgCdTe and related alloys," in *Infrared Detectors and Emitters: Materials and Devices*, eds. P. Capper and C. T. Elliott, 279–312, Kluwer Academic Publishers, Boston, 2001.
13. A. Rogalski, *Infrared Detectors*, 2nd ed., CRC Press, Boca Raton, FL, 2010.
14. R. A. Smith, *Semiconductors*, Cambridge University Press, Cambridge, UK, 1978.

15. E. S. Rittner, "Electron processes in photoconductors," in *Photoconductivity Conference at Atlantic City 1954*, eds. R. G. Breckenbridge, B. Russel, and E. Hahn, 215–68, Wiley, New York, 1956.
16. R. L. Williams, "Sensitivity limits of 0.1 eV intrinsic photoconductors," *Infrared Phys.* 8, 337–43, 1968.
17. M. A. Kinch, S. R. Borrello, B. H. Breazeale, and A. Simmons, "Geometrical enhancement of HgCdTe photoconductive detectors," *Infrared Phys.* 17, 137–45, 1977.
18. J. F. Siliquini, C. A. Musca, B. D. Nener, and L. Faraone, "Performance of optimized $Hg_{1-x}Cd_xTe$ long wavelength infrared photoconductors," *Infrared Phys. Technol.* 35, 661–71, 1994.
19. Y. J. Shacham-Diamand and I. Kidron, "Contact and bulk effects in intrinsic photoconductive infrared detectors," *Infrared Phys.* 21, 105–15, 1981.
20. C. T. Elliott and N. T. Gordon, "Infrared detectors," in *Handbook on Semiconductors*, Vol. 4, ed. C. Hilsum, 841–936, North-Holland, Amsterdam, 1993.
21. A. Jóźwikowska, K. Jóźwikowski, and A. Rogalski, "Performance of mercury cadmium telluride photoconductive detectors," *Infrared Phys.* 31, 543–54, 1991.
22. D. L. Smith, D. K. Arch, R. A. Wood, and M. W. Scott, "HgCdTe heterojunction contact photoconductor," *Appl. Phys. Lett.* 45, 83–5, 1984.
23. D. L. Smith, "Effects of blocking contacts on generation-recombination noise and responsivity in intrinsic photoconductors," *J. Appl. Phys.* 56, 1663–9, 1984.
24. R. Kumar, S. Gupta, V. Gopal, and K. C. Chhabra, "Dependence of responsivity on the structure of a blocking contact in an intrinsic HgCdTe photoconductor," *Infrared Phys.* 31, 101–7, 1991.
25. V. Gopal, R. Kumar, and K. C. Chhabra, "A n^{2+} -n^+ -n blocking contact structure for an intrinsic photoconductor," *Infrared Phys.* 31, 435–40, 1991.
26. T. Ashley and C. T. Elliott, "Accumulation effects at contacts to n-type cadmium-mercury-telluride photoconductors," *Infrared Phys.* 22, 367–76, 1982.
27. W. Van Roosbroeck, "Theory of the electrons and holes in germanium and other semiconductors," *Bell Sys. Tech. J.* 29, 560–607, 1950.
28. P. W. Kruse, L. D. McGlauchlin, and R. B. McQuistan, *Elements of Infrared Technology*, Wiley, New York, 1962.
29. P. W. Kruse, "The photon detection process," in *Optical and Infrared Detectors*, ed. R. J. Keyes, 5–69, Springer-Verlag, Berlin, 1977.
30. A. Van der Ziel, *Fluctuation Phenomena in Semiconductors*, Butterworths, London, 1959.
31. D. Long, "On generation-recombination noise in infrared detector materials," *Infrared Phys.* 7, 169–70, 1967.
32. T. D. Kleinpenning, "1/f noise in p-n junction diodes," *J. Vac. Sci. Technol. A* 3, 176–82, 1985.
33. F. N. Hooge, "1/f noise is no surface effect," *Phys. Lett.* 29A, 123–40, 1969.
34. Z. Wei-jiann and Z. Xin-Chen, "Experimental studies on low frequency noise of photoconductors," *Infrared Phys.* 33, 27–31, 1992.
35. R. D. Nelson, "Infrared charge transfer devices: The silicon approach," *Opt. Eng.* 16, 275–83, 1977.
36. K. Jóźwikowski and J. Piotrowski, "Ultimate performance of $Cd_xHg_{1-x}Te$ photoresistors as a function of doping," *Infrared Phys.* 25, 723–7, 1985.
37. J. Piotrowski, W. Galus, and M. Grudzień, "Near room-temperature IR photo-detectors," *Infrared Phys.* 31, 1–48, 1991.
38. J. Piotrowski and A. Rogalski, *High-Operating Temperature Infrared Photodetectors*, SPIE Press, Bellingham, WA, 2007.
39. S. Borrello, M. Kinch, and D. LaMont, "Photoconductive HgCdTe detector performance with background variations," *Infrared Phy.* 17, 121–5, 1977.
40. V. Gopal, "Surface recombination in photoconductors," *Infrared Phy.* 25, 615–8, 1985.
41. A. Kinch and S. R. Borrello, "0.1 eV HgCdTe photodetectors," *Infrared Phy.* 15, 111–24, 1975.
42. E. Burstein, G. Picus, and N. Sclar, "Optical and photoconductive properties of silicon and germanium," in *Photoconductivity Conference at Atlantic City*, eds. R. Breckenbridge, B. Russell, and E. Hahn, 353–413, Wiley, New York, 1956.
43. R. Newman and W. W. Tyler, "Photoconductivity in germanium," in *Solid State Physics*, Vol. 8, eds. F. Steitz and D. Turnbill, 49–107, Academic Press, New York, 1959.
44. E. H. Putley, "Far infrared photoconductivity," *Phys. Status Solidi* 6, 571–614, 1964.
45. V. F. Kocherov, I. I. Taubkin, and N. B. Zaletaev, "Extrinsic silicon and germanium detectors," in *Infrared Photon Detectors*, ed. A. Rogalski, 189–297, SPIE Optical Engineering Press, Bellingham, WA, 1995.
46. A. W. Hoffman, P. J. Love, and J. P. Rosbeck, "Mega-pixel detector arrays: Visible to 28 μm," *Proc. SPIE* 5167, 194–203, 2004.
47. G. H. Rieke, "Infrared detector arrays for astronomy," *Ann. Rev. Astron. Astrophys.* 45, 77–115, 2007.

48. M. M. Blouke, E. E. Harp, C. R. Jeffus, and R. L. Williams, "Gain saturation in extrinsic photoconductors operating at low temperatures," *J. Appl. Phys.* 43, 188–94, 1972.
49. A. F. Milton and M. M. Blouke, "Sweepout and dielectric relaxation in compensated extrinsic photoconductors," *Phys. Rev.* 3B, 4312–30, 1971.
50. D. K. Schroder, "Extrinsic silicon focal plane arrays," in *Charge-Coupled Devices*, ed. D. F. Barbe, 57–90, Springer-Verlag, Heidelberg, Germany, 1980.
51. R. W. Westervelt and S. W. Teitsworth, "Nonlinear transient response of extrinsic Ge far-infrared photoconductors," *J. Appl. Phys.* 57, 5457–69, 1985.
52. N. M. Haegel, C. A. Latasa, and A. M. White, "Transient response of infrared photoconductors: The role of contacts and space charge," *Appl. Phys.* A56, 15–21, 1993.
53. N. M. Haegel, C. B. Brennan, and A. M. White, "Transport in extrinsic photoconductors: A comprehensive model for transient response," *J. Appl. Phys.* 80, 1510–4, 1996.
54. N. M. Haegel, S. A. Sampei, and A. M. White, "Electric field and responsivity modeling for far-infrared blocked impurity band detectors," *J. Appl. Phys.* 93, 1305–10, 2003.
55. N. M. Haegel, W. R. Schwartz, J. Zinter, A. M. White, and J. W. Beeman, "Origin of the hook effect in extrinsic photoconductors," *Appl. Opt.* 40, 5748–54, 2001.
56. R. N. Thomas, T. T. Braggins, H. M. Hobgood, and W. J. Takei, "Compensation of residual boron impurities in extrinsic indium-doped silicon by neutron transmutation of silicon," *J. Appl. Phys.* 49, 2811–20, 1978.
57. H. M. Hobgood, T. T. Braggins, J. C. Swartz, and R. N. Thomas, "Role of neutron transmutation in the development of high sensitivity extrinsic silicon IR detector material," in *Neutron Transmutation in Semiconductors*, ed. J. M. Meese, 65–90, Plenum Press, New York, 1979.
58. G. Lucovsky, "On the photoionization of deep impurity centers in semiconductors," *Solid State Commun.* 3, 299–302, 1965.
59. N. Sclar, "Extrinsic silicon detectors for 3–5 and 8–14 μm," *Infrared Phys.* 16, 435–48, 1976.
60. N. Sclar, "Survey of dopants in silicon for 2–2.7 and 3–5 μm infrared detector application," *Infrared Phys.* 17, 71–82, 1977.
61. E. Bryan, "Operation temperature of extrinsic Si photoconductive detectors," *Infrared Phys.* 23, 341–8, 1983.
62. S. R. Borrello, C. G. Roberts, B. H. Breazeale, and G. R. Pruett, "Cooling requirements for BLIP performance of intrinsic photoconductors," *Infrared Phys.* 11, 225–32, 1971.
63. A. G. Milnes, *Deep Impurities in Semiconductors*, John Wiley, New York, 1973.
64. C. T. Elliott, P. Migliorato, and A. W. Vere, "Counterdoped extrinsic silicon infrared detectors," *Infrared Phys.* 18, 65–71, 1978.
65. A. Rogalski, "Quantum well infrared photodetectors among the other types of semiconductor infrared detectors," *Infrared Phys. Technol.* 38, 295–310, 1997.
66. P. Martyniuk and A. Rogalski, "Quantum-dot infrared photodetectors: Status and outlook," *Prog. Quant. Electron.* 32, 89–120, 2008.
67. N. Sclar, "Temperature limitation for IR extrinsic and intrinsic photodetectors," *IEEE Trans. Electron Devices* ED-27, 109–18, 1980.
68. A. P. Davis, C. T. Elliott, and A. M. White, "Current gain in photodiode structures," *Infrared Phys.* 31, 575–7, 1991.
69. R. W. Dutton and R. J. Whittier, "Forward current-voltage and switching characteristics of p^+ -n-n+ (Epitaxial) diodes," *IEEE Trans. Electron. Devices* ED-16, 458–67, 1969.
70. H. J. Hovel, *Semiconductors and Semimetals*, Vol. 11, eds. R. K. Willardson and A. C. Berr, Academic Press, New York, 1975.
71. F. Van De Wiele, "Quantum efficiency of photodiode," in *Solid State Imaging*, eds. P. G. Jespers, F. Van De Wiele, and M. H. White, 41–76, Noordhoff, Leyden, The Netherlands, 1976.
72. A. Rogalski and J. Rutkowski, "Effect of structure on the quantum efficiency and R_oA product of lead-tin chalcogenide photodiodes," *Infrared Phys.* 22, 199–208, 1982.
73. Z. Djuric and Z. Jaksic, "Back side reflection influence on quantum efficiency of photovoltaic devices," *Electron. Lett.* 24, 1100–1, 1988.
74. J. Shappir and A. Kolodny, "The response of small photovoltaic detectors to uniform radiation," *IEEE Trans. Electron. Devices* ED-24, 1093–8, 1977.
75. D. Levy, S. E. Schacham, and I. Kidron, "Three-dimensional analytical simulation of self- and cross-responsivities of photovoltaic detector arrays," *IEEE Transact. Electron. Devices* ED-34, 2059–70, 1987.
76. M. J. Buckingham and E. A. Faulkner, "The theory of inherent noise in p-n junction diodes and bipolar transistors," *Radio Electron. Eng.* 44, 125–40, 1974.

77. P. Knowles, "Mercury cadmium telluride detectors for thermal imaging," *GEC J. Res.* 2, 141–56, 1984.
78. C. T. Sah, R. N. Noyce, and W. Shockley, "Carrier generation and recombination in p–n junctions and p–n junction characteristics," *Proc. IRE* 45, 1228–43, 1957.
79. S. C. Choo, "Carrier generation-recombination in the space-charge region of an asymmetrical p–n junction," *Solid-State Electron.* 11, 1069–77, 1968.
80. S. M. Sze, *Physics of Semiconductor Devices*, Wiley, New York, 1981.
81. W. W. Anderson, "Tunnel contribution to $Hg_{1-x}Cd_xTe$ and $Pb_{1-x}Sn_xTe$ p–n junction diode characteristics," *Infrared Phys.* 20, 353–61, 1980.
82. W. A. Beck and N. Y. Byer, "Calculation of tunneling currents in (Hg,Cd)Te photodiodes using a two-sided junction potential," *IEEE Trans. Electron. Devices* ED-31, 292–7, 1984.
83. R. Adar, "Spatial integration of direct band-to-band tunneling currents in general device structures," *IEEE Trans. Electron. Devices* 39, 976–81, 1992.
84. C. T. Sah, "Electronic processes and excess currents in gold-doped narrow silicon junctions," *Phys. Rev.* 123, 1594–612, 1961.
85. J. Y. Wong, "Effect of trap tunneling on the performance of long-wavelength $Hg_{1-x}Cd_xTe$ photodiodes," *IEEE Trans. Electron. Devices* ED-27, 48–57, 1980.
86. M. A. Kinch, "Metal-insulator-semiconductor infrared detectors," in *Semiconductors and Semimetals*, Vol. 18, eds. R. K. Willardson and A. C. Beer, 313–78, Academic Press, New York, 1981.
87. H. J. Hoffman and W. W. Anderson, "Impurity-to-band tunneling in $Hg_{1-x}Cd_xTe$," *J. Vac. Sci. Technol.* 21, 247–50, 1982.
88. D. K. Blanks, J. D. Beck, M. A. Kinch, and L. Colombo, "Band-to-band processes in HgCdTe: Comparison of experimental and theoretical studies," *J. Vac. Sci. Technol.* A6, 2790–4, 1988.
89. Y. Nemirovsky, D. Rosenfeld, A. Adar, and A. Kornfeld, "Tunneling and dark currents in HgCdTe photodiodes," *J. Vac. Sci. Technol.* A7, 528–35, 1989.
90. Y. Nemirovsky, R. Fastow, M. Meyassed, and A. Unikovsky, "Trapping effects in HgCdTe," *J. Vac. Sci. Technol.* B9, 1829–39, 1991.
91. D. Rosenfeld and G. Bahir, "A model for the trap-assisted tunneling mechanism in diffused n–p and implanted n^+ –p HgCdTe photodiodes," *IEEE Trans. Electron. Devices* 39, 1638–45, 1992.
92. Y. Nemirovsky and A. Unikovsky, "Tunneling and 1/f noise currents in HgCdTe photodiodes," *J. Vac. Sci. Technol.* B10, 1602–10, 1992.
93. G. Sarusi, A. Zemel, A. Sher, and D. Eger, "Forward tunneling current in HgCdTe photodiodes," *J. Appl. Phys.* 76, 4420–5, 1994.
94. A. Rogalski, "Analysis of the R_oA product in n^+ -p $Hg_{1-x}Cd_xTe$ photodiodes," *Infrared Phys.* 28, 139–53, 1988.
95. M. B. Reine, A. K. Sood, and T. J. Tredwell, "Photovoltaic infrared detectors," in *Semiconductors and Semimetals*, Vol. 18, eds. R. K. Willardson and A. C. Beer, 201–311, Academic Press, New York, 1981.
96. A. Rose, "Space-charge-limited currents in solids," *Phys. Rev.* 97, 1538–44, 1955.
97. M. A. Lampert, "Simplified theory of space-charge-limited currents in an insulator with traps," *Phys. Rev.* 103, 1648–56, 1956.
98. M. A. Lampert and P. Mark, *Current Injections in Solids*, eds. H. G. Booker and N. DeClaris, Academic Press, New York, 1970.
99. J. A. Edmond, K. Das, and R. F. Davis, "Electrical properties of ion-implanted p-n junction diodes in β-SiC," *J. Appl. Phys.* 63, 922–9, 1988.
100. D. E. Sawyer and R. H. Rediker, "Narrow base germanium photodiodes," *Proc. IRE* 46, 1122–30, 1958.
101. G. Lucovsky and R. B. Emmons, "High frequency photodiodes," *Appl. Opt.* 4, 697–702, 1965.
102. W. W. Gartner, "Depletion-layer photoeffects in semiconductors," *Phys. Rev.* 116, 84–7, 1959.
103. G. Kreiser, *Optical Fiber Communications*, McGraw-Hill Book Co., Boston, 2000.
104. A. G. Dentai, R. Kuchibhotla, J. C. Campbell, C. Tasi, and C. Lei, "High quantum efficiency, long-wavelength InP/InGaAs microcavity photodiode," *Electron. Lett.* 27, 2125–7, 1991.
105. M. S. Ünlü and M. S. Strite, "Resonant cavity enhanced photonic devices," *J. Appl. Phys.* 78, 607–39, 1995.
106. G. E. Stillman and C. M. Wolfe, "Avalanche photodiodes," in *Semiconductors and Semimetals*, Vol. 12, eds. R. K. Willardson and A. C. Beer, 291–393, Academic Press, New York, 1977.
107. F. Capasso, "Physics of avalanche diodes," in *Semiconductors and Semimetals*, Vol. 22D, ed. W. T. Tang, 2–172, Academic Press, Orlando, FL, 1985.
108. T. Kaneda, "Silicon and germanium avalanche photodiodes," in *Semiconductors and Semimetals*, Vol. 22D, ed. W. T. Tang, 247–328, Academic Press, Orlando, FL, 1985.

109. T. P. Pearsall and M. A. Pollack, "Compound semiconductor photodiodes," in *Semiconductors and Semimetals*, Vol. 22D, ed. W. T. Tang, 173–245, Academic Press, Orlando, FL, 1985.
110. P. Bhattacharya, *Semiconductor Optoelectronics Devices*, Prentice Hall, Upper Saddle River, New Jersey, 1993.
111. S. B. Alexander, *Optical Communication Receiver Design*, SPIE Optical Engineering Press, Bellingham, WA, 1997.
112. J. C. Campbell, S. Demiguel, F. Ma, A. Beck, X. Guo, S. Wang, X. Zheng, et al., "Recent advances in avalanche photodiodes," *IEEE J. Select. Topics Quant. Electron.* 10, 777–87, 2004.
113. S. Donati, *Photodetectors. Devices, Circuits, and Applications*, Prentice Hall, New York, 2000.
114. R. J. McIntryre, "Multiplication noise in uniform avalanche diodes," *IEEE Trans. Electron. Devices* ED-13, 164–68, 1966.
115. R. J. McIntyre, "The distribution of gains in uniformly multiplying avalanche photodiodes: Theory," *IEEE Trans. Electron. Devices* ED-19, 703–13, 1972.
116. Y. Okuto and C. R. Crowell, "Ionization coefficients in semiconductors: A nonlocalized property," *Phys. Rev.* B10, 4284–96, 1974.
117. J. P. R. David and C. H. Tan, "Material considerations for avalanche photodiodes," *IEEE J. Select. Topics Quant. Electron.* 14, 998–1009, 2008.
118. J. C. Campbell, "Recent advances in telecommunications avalanche photodiodes," *J. Lightwave Technol.* 25, 109–21, 2007.
119. G. J. Rees and J. P. R. David, "Why small avalanche photodiodes are beautiful," *Proc. SPIE* 4999, 349–62, 2003.
120. R. B. Emmons, "Avalanche-photodiode frequency response," *J. Appl. Phys.* 38, 3705–14, 1967.
121. S. Melle and A. MacGregor, "How to choice avalanche photodiodes," *Laser Focus World*, 145–56, October 1995.
122. S. Cova, A. Lacaita, M. Ghioni, G. Ripamonti, and T. A. Louis, "20 ps timing resolution with single-photon avalanche diodes," *Rev. Sci. Instrum.* 60, 1104–10, 1989.
123. H. K. Henish, *Rectifying Semiconductor Contacts*, Clarendon Press, Oxford, UK, 1957.
124. M. M. Atalla, "Metal-semiconductor Schottky barriers, devices and applications," in *Proceedings of 1966 Microelectronics Symposium* 123–57, Munich-Oldenberg, 1966.
125. F. A. Padovani, "The voltage-current characteristics of metal-semiconductor contacts," in *Semiconductors and Semimetals*, Vol. 7A, eds. R. K. Willardson and A. C. Beer, 75–146, Academic Press, New York, 1971.
126. A. G. Milnes and D. L. Feught, *Heterojunctions and Metal-Semiconductor Junctions*, Academic Press, New York, 1972.
127. V. L. Rideout, "A review of the theory, technology and applications of metal-semiconductor rectifiers," *Thin Solid Films* 48, 261–91, 1978.
128. E. H. Rhoderick, *Metal-Semiconductor Contacts*, Clarendon Press, Oxford, UK, 1978.
129. E. R. Rhoderick, "Metal-semiconductor contacts," *IEE Proc.* 129, 1–14, 1982.
130. S. C. Gupta and H. Preier, "Schottky barrier photodiodes," in *Metal-Semiconductor Schottky Barrier Junctions and Their Applications*, ed. B. L. Sharma, 191–218, Plenum, New York, 1984.
131. W. Monch, "On the physics of metal-semiconductor interfaces," *Reports Prog. Phys.* 53, 221–78, 1990.
132. R. T. Tung, "Electron transport at metal-semiconductor interfaces: General theory," *Phys. Rev.* B45, 13509–23, 1992.
133. A. M. Cowley and S. M. Sze, "Surface states and barrier height of metal-semiconductor systems," *J. Appl. Phys.* 36, 3212–20, 1965.
134. W. Schottky and E. Spenke, "Quantitative treatment of the space-charge boundary-layer theory of the crystal rectifiers," *Wiss. Veroff. Siemens-Werken.* 18, 225–91, 1939.
135. E. Spenke, *Electronic Semiconductors*, McGraw-Hill, New York, 1958.
136. H. A. Bethe, "Theory of the boundary layer of crystal rectifiers," *MIT Radiation Laboratory Report* 43-12, 1942.
137. C. R. Crowell and S. M. Sze, "Current transport in metal-semiconductor barriers," *Solid-State Electron.* 9, 1035–48, 1966.
138. S. Y. Wang and D. M. Bloom, "100 GHz bandwidth planar GaAs Schottky photodiode," *Electron. Lett.* 19, 554–5, 1983.
139. J. W. Mayer and K. N. Tu, "Analysis of thin-films structures with nuclear backscattering and x-ray diffraction," *J. Vac. Sci. Technol.* 11, 86–93, 1974.
140. G. Ottaviani, K. N. Tu, and J. M. Mayer, "Interfacial reaction and Schottky barrier in metal-silicon systems," *Phys. Rev. Lett.* 44, 284–87, 1980.

141. M. P. Lepselter and S. M. Sze, "Silicon Schottky barrier diode with near-ideal I–V characteristics," *Bell Sys. Technol. J.* 47, 195–208, 1968.
142. J. M. Andrews and M. P. Lepselter, "Reverse current-voltage characteristics of metal-silicide Schottky Diodes," *Solid-State Electron.* 13, 1011–23, 1970.
143. F. D. Shepherd and A. C. Yang, "Silicon Schottky retinas for infrared imaging," *IEDM Technical Digest*, 310–13, Washington, DC, 1973.
144. W. F. Kosonocky, "Review of infrared image sensors with Schottky-barrier detectors," *Optoelectron.—Devices Technol.* 6, 173–203, 1991.
145. W. A. Cabanski and M. J. Schulz, "Electronic and IR-optical properties of silicide/silicon interfaces," *Infrared Phys.* 32, 29–44, 1991.
146. J. S. Wang, C. G. Shih, W. H. Chang, J. R. Middleton, P. J. Apostolakis, and M. Feng, "11 GHz bandwidth optical integrated receivers using GaAs MESFET and MSM technology," *IEEE Photon. Technol. Lett.* 5, 316–8, 1993.
147. M. Ito and O. Wada, "Low dark current GaAs metal-semiconductor-metal (MSM) photodiodes using WSi_x contacts," *IEEE J. Quant. Electron.* QE-62, 1073–7, 1986.
148. W. S. Boyle and G. E. Smith, "Charge-coupled semiconductor devices," *Bell Sys. Technol. J.* 49, 587–93, 1970.
149. A. S. Grove, *Physics and Technology of Semiconductor Devices*, Wiley, New York, 1967.
150. E. R. Nicollian and J. R. Brews, *MOS Physics and Technology*, Wiley, New York, 1982.
151. D. G. Ong, *Modern MOS Technology: Process, Devices and Design*, McGraw-Hill Book Company, New York, 1984.
152. D. F. Barbe, "Imaging devices using the charge-coupled concept," *Proc. IEEE* 63, 38–67, 1975.
153. E. S. Young, *Fundamentals of Semiconductor Devices*, McGraw-Hill Book Company, New York, 1978.
154. W. D. Baker, "Intrinsic focal plane arrays," in *Charge-Coupled Devices*, ed. D. F. Barbe, 25–56, Springer, Berlin, 1980.
155. A. J. Syllaios and L. Colombo, "The influence of microstructure on the impedance characteristics on HgCdTe MIS devices," *IEDM Tech. Digest* 137–48, 1982.
156. W. W. Anderson, "Tunnel current limitation of narrow bandgap infrared charge coupled devices," *Infrared Phys.* 17, 147–64, 1977.
157. M. W. Goodwin, M. A. Kinch, and R. J. Koestner, "Metal-insulator-semiconductor properties of molecular-beam epitaxy grown HgCdTe heterostructure," *J. Vac. Sci. Technol.* A8, 1226–32, 1990.
158. T. Ashley and C. T. Elliott, "Non-equilibrium devices for infrared detection," *Electron. Lett.* 21, 451–2, 1985.
159. T. Ashley, C. T. Elliott, and A. T. Harker, "Non-equilibrium modes of operation for infrared detectors," *Infrared Phys.* 26, 303–15, 1986.
160. T. Ashley, C. T. Elliott, and A. M. White, "Non-equilibrium devices for infrared detection," *Proc. SPIE* 572, 123–33, 1985.
161. C. T. Elliott, "Non-equilibrium modes of operation of narrow-gap semiconductor devices," *Semicond. Sci. Technol.* 5, S30–7, 1990.
162. T. Ashley, A. B. Dean, C. T. Elliott, M. R. Houlton, C. F. McConville, H. A. Tarry, and C. R. Whitehouse, "Multilayer InSb diodes grown by molecular beam epitaxy for near ambient temperature operation," *Proc. SPIE* 1361, 238–44, 1990.
163. C. T. Elliott, "Advanced heterostructures for $In_{1-x}Al_xSb$ and $Hg_{1-x}Cd_xTe$ detectors and emiters," *Proc. SPIE* 2744, 452–62, 1996.
164. A. M. White, "Auger suppression and negative resistance in low gap pin diode structure," *Infrared Phys.* 26, 317–24, 1986.
165. A. M. White, "Infrared detectors," U.S. Patent 4,679,063 (22 September 1983).
166. P. C. Klipstein, "Depletionless photodiode with suppressed dark current and method for producing the same," U.S. Patent 7,795,640 (2 July 2003).
167. S. Maimon and G. Wicks, "nBn detector, an infrared detector with reduced dark current and higher operating temperature," *Appl. Phys. Lett.* 89, 151109-1–3, 2006.
168. D. Z.-Y. Ting, A. Soibel, L. Höglund, J. Nguyen, C. J. Hill, A. Khoshakhlagh, and S. D. Gunapala, "Type-II superlattice infrared detectors," in *Semiconductors and Semimetals*, Vol. 84, eds. S. D. Gunapala, D. R. Rhiger, and C. Jagadish, 1–57, Elsevier, Amsterdam, 2011.
169. G.R. Savich, J.R. Pedrazzani, D.E. Sidor, and G.W. Wicks, "Benefits and limitations of unipolar barriers in infrared photodetectors," *Infrared Phys. Technol.* 59, 152–5, 2013.
170. M. Reine, J. Schuster, B. Pinkie, and E. Bellotti, "Numerical simulation and analytical modeling of InAs nBn infrared detectors with p-type barriers," *J. Electron. Mater.* 42 (11), 3015–33, 2013.

171. M. Reine, J. Schuster, B. Pinkie, and E. Bellotti, "Numerical simulation and analytical modeling of InAs nBn infrared detectors with n-type barrier layers," *J. Electron. Mater.* 43 (8), 2915–34, 2014.
172. P. Klipstein, "XBn barrier photodetectors for high sensitivity operating temperature infrared sensors," *Proc. SPIE*. 6940, 69402U-1–11, 2008.
173. D. Z. Ting, C. J. Hill, A. Soibel, J. Nguyen, S. A. Keo, M. C. Lee, J. M. Mumolo, J. K. Liu, and S. D. Gunapala, "Antimonide-based barrier infrared detectors," *Proc. SPIE* 7660, 76601R-1–12, 2010.
174. P. Martyniuk and A. Rogalski, "HOT infrared photodetectors," *Opto-Electron. Rev.* 21, 240–58, 2013.
175. P. C. Klipstein, "XB_nn and XB_pp infrared detectors", *J. Cryst. Growth* 425, 351–256, 2015.
176. J. B. Rodriguez, E. Plis, G. Bishop, Y. D. Sharma, H. Kim, L. R. Dawson, and S. Krishna, "nBn structure based on InAs/GaSb type-II strained layer superlattices," *Appl. Phys. Lett.* 91, 043514-1–2, 2007.
177. P. Klipstein, D. Aronov, E. Berkowicz, R. Fraenkel, A. Glozman, S. Grossman, O. Klin, I. Lukomsky, I. Shtrichman, N. Snapi, M. Yassem, and E. Weiss, "Reducing the cooling requirements of mid-wave IR detector arrys," *SPIE Newsroom*, 2011, doi:10.1117/2.1201111.003919.
178. M. Razeghi, S. P. Abdollahi, E.K. Huang, G. Chen, A. Haddadi, and B.M. Nquyen, "Type-II InAs/GaSb photodiodes and focal plane arrays aimed at high operating temperatures," *Opto-Electr. Rev.* 19, 261–9, 2011.
179. M. Razeghi, "Type II superlattice enables high operating temperature," *SPIE Newsroom*, 2011, doi:10.1117/2.1201110.003870.
180. G.R. Savich, J. R. Pedrazzani, D. E. Sidor, S. Maimon, and G. W. Wicks, "Dark current filtering in unipolar barrier infrared detectors," *Appl. Phys. Lett.* 99, 121112, 2011.
181. J. Piotrowski and A. Rogalski, "Photoelectromagnetic, magnetoconcentration and Dember infrared detectors," in *Narrow-Gap II-VI Compounds for Optoelectronic and Electromagnetic Applications*, ed. P. Capper, 507–25, Chapman & Hall, London, 1997.
182. I. K. Kikoin and M. M. Noskov, "A new photoelectric effect in copper oxide," *Physik Zeit Der Soviet Union*, 5, 586, 1934.
183. M. Nowak, "Photoelectromagnetic effect in semiconductors and its applications," *Prog. Quant. Electron.* 11, 205–346, 1987.
184. P. W. Kruse, "Indium antimonide photoelectromagnetic infrared detector," *J. Appl. Phys.* 30, 770–8, 1959.
185. J. Piotrowski, W. Galus, and M. Grudzień, "Near room-temperature IR photodetectors," *Infrared Phys.* 31, 1–48, 1991.
186. D. L. Lile, "Generalized photoelectromagnetic effect in semiconductors," *Phys. Rev.* B8, 4708–22, 1973.
187. D. Genzow, M. Grudzień, and J. Piotrowski, "On the performance of noncooled CdHgTe photoelectromagnetic detectors for 10,6 μm radiation," *Infrared Phys.* 20, 133–8, 1980.
188. J. Piotrowski, "HgCdTe detectors," in *Infrared Photon Detectors*, ed. A. Rogalski, 391–3, SPIE Optical Engineering Press, Bellingham, WA, 1995.
189. J. Piotrowski, "Uncooled operation of IR photodetectors," *Opto-Electron. Rev.* 12, 11–122, 2004.
190. P. Berdahl, V. Malutenko, and T. Marimoto, "Negative luminescence of semiconductors," *Infrared Phys.* 29, 667–72, 1989.
191. V. Malyutenko, A. Pigida, and E. Yablonovsky, "Noncooled infrared magnetoinjection emitters based on $Hg_{1-x}Cd_xTe$," *Optoelectron. Devices Technol.* 7, 321–8, 1992.
192. T. Ashley, C. T. Elliott, N. T. Gordon, R. S. Hall, A. D. Johnson, and G. J. Pryce, "Negative luminescence from $In_{1-x}Al_xSb$ and $Cd_xHg_{1-x}Te$ diodes," *Infrared Phys. Technol.* 36, 1037–44, 1995.
193. Z. Djuric and J. Piotrowski, "Room temperature IR photodetector with electromagnetic carrier depletion," *Electron. Lett.* 26, 1689–91, 1990.
194. Z. Djuric and J. Piotrowski, "Infrared photodetector with electromagnetic carrier depletion," *Opt. Eng.* 31, 1955–60, 1992.
195. Z. Djuric, Z. Jaksic, A. Vujanic, and J. Piotrowski, "Auger generation suppression in narrow-gap semiconductors using the magnetoconcentration effect," *J. Appl. Phys.* 71, 5706–8, 1992.
196. J. Piotrowski, W. Gawron, and Z. Djuric, "New generation of near room-temperature photodetectors," *Opt. Eng.* 33, 1413–21, 1994.
197. H. Dember, "Uber die vorwartsbewegung von elektronen durch licht," *Physik Z.* 32, 554, 856, 1931.
198. J. Auth, D. Genzow, and K. H. Herrmann, *Photoelectrische Erscheinungen*, Akademie Verlag, Berlin, 1977.
199. Z. Djuric and J. Piotrowski, "Dember IR photodetectors," *Solid-State Electron.* 34, 265–9, 1991.
200. H. Heyn, I. Decker, D. Martinen, and H. Wohlfahrt, "Application of room-temperature infrared photo detectors in high-speed laser beam diagnostics of industrial CO_2 lasers," *Proc. SPIE* 2375, 142–53, 1995.
201. L. E. Gurevich and A. A. Rumyantsev, "Theory of the photoelectric effect in finite crystals at high frequencies and in the presence of an external magnetic field," *Soviet Phys. Solid State* 9, 55, 1967.

202. A. F. Gibson and S. Montasser, "A theoretical description of the photon-drag spectrum of p-type germanium," *J. Phys. C: Solid State Phys.* 8, 3147–57, 1975.
203. A. A. Grinberg and S. Luryi, "Theory of the photon-drag effect in a two-dimensional electron gas," *Phys. Rev. B* 38, 87, 1987.
204. A. F. Gibson and M. F. Kimmitt, "Photon drag detection," in *Infrared and Millimeter Waves*, Vol. 3, ed. K. J. Button, 182–219, Academic Press, New York, 1980.
205. A. F. Gibson and A. C. Walker, "Sign reversal of the photon drag effect in p-type germanium," *Journal of Physics C* 4, 2209–19, 1971.
206. J. M. Feldman and K. M. Hergenrother, "Direct observation of the excess light hole population in optically pumped p-type germanium," *Appl. Phys. Lett.* 9, 186, 1966.
207. A. F. Gibson, M. F. Kimmitt, and A. C. Walker, "Photon drag in germanium," *Appl. Phys. Lett.* 17, 75–7, 1970.
208. A. M. Danishevskii, A. A. Kastalskii, S. M. Ryvkin, and I. D. Yaroshetskii, "Dragging of free carriers by photons in direct interband transitions," *Soviet Phys. JETP* 31, 292, 1970.
209. S. Panyakeow, J. Shirafuji, and Y. Inuishi, "High-performance photon drag detector for a CO_2 laser using p-type tellurium," *Appl. Phys. Lett.* 21, 314–6, 1972.
210. S. D. Ganichev and W. Prettl, *Intense Terahertz Excitation of Semiconductors*, Clarendon Press, Oxford, 2005.
211. E. Bründermann, H.-W. Hübers, and M.F. Kimmitt, *Terahertz Techniques*, Springer-Verlag, Berlin, 2012.
212. S. D. Ganichev, Ya. V. Terent'ev, and I. D. Yaroshetskii, "Photon-drag photodetectors for the far-IR and submillimeter regions," *Soviet Tech. Phys. Lett.* 11, 20–1, 1985.

13 Intrinsic silicon and germanium detectors

Silicon is the semiconductor that has dominated the electronics industry for over 50 years. While the first transistor fabricated in Ge and III-V semiconductor material compounds may have higher mobilities, higher saturation velocities, or larger bandgaps, silicon devices account for over 97% of all microelectronics [1]. The main reason is that silicon is the cheapest microelectronic technology for integrated circuits. The dominance of silicon can be traced to a number of natural properties of silicon, but more importantly, two insulators of silicon, SiO_2 and Si_3N_4, allow the deposition and selective etching processes to be developed with exceptionally high uniformity and yield.

Photodetectors are perhaps the oldest and the best understood silicon photonic devices [2,3]. Recently, the interest in utilizing Si-based optical components to realize a fully monolithic solution for high-performance optical interconnects is on the rise [4]. Silicon, being an indirect bandgap semiconductor with a centrosymmetric crystalline structure, is not directly suited for optoelectronics. Moreover, a Si bandgap of 1.16 eV prevents its use in the second (1.3 μm) and third (1.55 μm) window of optical fiber communications. Despite these facts, the rather unique success of Si as a semiconductor for electronics has motivated a large amount of research toward the development of silicon-based optoelectronic devices. The cost of silicon-integrated circuits has remained constant around 1 U.S. cent per square mm for a number of decades, but the number of their elements (transistors, passive, and other components) has been increasing at an exponential rate with time [1]. In addition, in complementary metal–oxide–semiconductor (CMOS) architectures now dominated, apart from leakage currents, power is only dissipated when gates are switched. The low leakage currents achievable with silicon insulators and p-n implanted isolation, combined with a higher thermal conductivity than many other semiconductors, have allowed higher densities in silicon technology than any other technology driving the integrated circuit developments. While microelectronics has dominated the twentieth-century technologies, some of the authors have predicted that silicon photonics will be a major technology in the twenty-first century [2,3]. Currently, silicon is becoming an important candidate for optical functionalities.

In this chapter, we review the recent achievements of silicon and germanium technologies for the fabrication of near-infrared (NIR) photodetectors. To probe beyond this chapter, readers should consult the excellent monographs [4–7] and reviews [8–11]. Table 13.1 lists the properties of silicon and germanium in room temperatures [4,5].

13.1 SILICON PHOTODIODES

Silicon photodiodes are widely applied in the spectral range below 1.1 μm and are even used for X-ray and gamma-ray detectors. The main types are as follows:

- p-n junctions generally formed by diffusion (ion implantation is also used),
- p-i-n junctions (because of the thicker active region, they have enhanced near-IR spectral response),
- UV-enhanced and blue-enhanced photodiodes, and
- avalanche photodiodes (APDs).

In the planar photodiode structure (diffused or implanted)—the cross section is shown in Figure 13.1—the highly doped p^+-region is very thin (typically about 1 μm) and is coated with the thin dielectric film (SiO_2 or Si_3N_4) that serves as an antireflection layer. The diffused junction can be formed either by a p-type impurity such as born into an n-type bulk silicon wafer, or the n-type impurity, such as phosphorous, into a p-type bulk silicon wafer. To form an ohmic contact, another impurity diffusion (often coupled with implanting technique) into the back side of the wafer is necessary. The contact pads are deposited on the

Table 13.1 Properties of Si and Ge at 300 K

PROPERTIES	Si	Ge
Atoms (cm^{-3})	5.02×10^{22}	4.42×10^{22}
Atomic weight	28.09	72.60
Breakdown field (Vcm^{-3})	$\sim3\times10^{5}$	$\sim10^{5}$
Crystal structure	Diamond	Diamond
Density (g cm^{-3})	2.329	5.3267
Dielectric constant	11.9	16.0
Effective density of states in the conduction band (cm^{-3})	2.86×10^{19}	1.04×10^{19}
Effective density of states in the valence band (cm^{-3})	2.66×10^{19}	6.0×10^{19}
Effective mass (conductivity)		
Electrons (m_e/m_o)	0.26	0.39
Holes (m_h/m_o)	0.19	0.12
Electron affinity (V)	4.05	4.0
Energy gap (eV)	1.12	0.67
Index of refraction	3.42	4.0
Intrinsic carrier concentration (cm^{-3})	9.65×10^{9}	2.4×10^{13}
Intrinsic resistivity (Ωcm)	3.3×10^{5}	
Lattice constant (Å)	5.43102	5.64613
Linear coefficient of thermal expansion (°C^{-1})	2.59×10^{-6}	5.8×10^{-6}
Melting point (°C)	1,412	937
Minority carrier lifetime (μs)		
Electrons (p-type)	800	1,000
Holes (n-type)	1,000	1,000
Mobility ($cm^2\ Vs^{-1}$)		
μ_e (electrons)	1,450	3,900
μ_h (holes)	505	1,900
Optical phonon energy (eV)	0.063	0.037
Specific heat (J/g°C)	0.7	
Thermal conductivity (W cmK^{-1})	1.31	0.31

front defined active area, and on the back side, completely covering the surface. An antireflection coating reduces the reflection of the light for specific predefined wavelength. The nonactive area on the top is covered with a thick layer of SiO_2. Depending on the photodiode application, different design structures are used. By controlling the thickness of the bulk substrate, both speed response and sensitivity of the photodiode can be controlled (See Section 12.2).

Note that the photodiodes can be operated as unbiased (photovoltaic) or reverse-biased (photoconductive) modes (Figure 13.2). The amplifier's function is a simple current-to-voltage conversion (photodiode operates in a short-circuit mode). Mode selection depends upon the speed requirements of the application, and the amount of dark current that is tolerable. The unbiased mode of operation is preferred when a photodiode is used in low-frequency applications (up to 350 kHz) as well as ultra–low light applications. Application of a reverse bias can greatly improve the speed of response and linearity of the devices. This is due to an increase in the depletion region width and consequently a decrease in junction capacitance. The drawback of applying a reverse bias is an increase in the dark and noise currents.

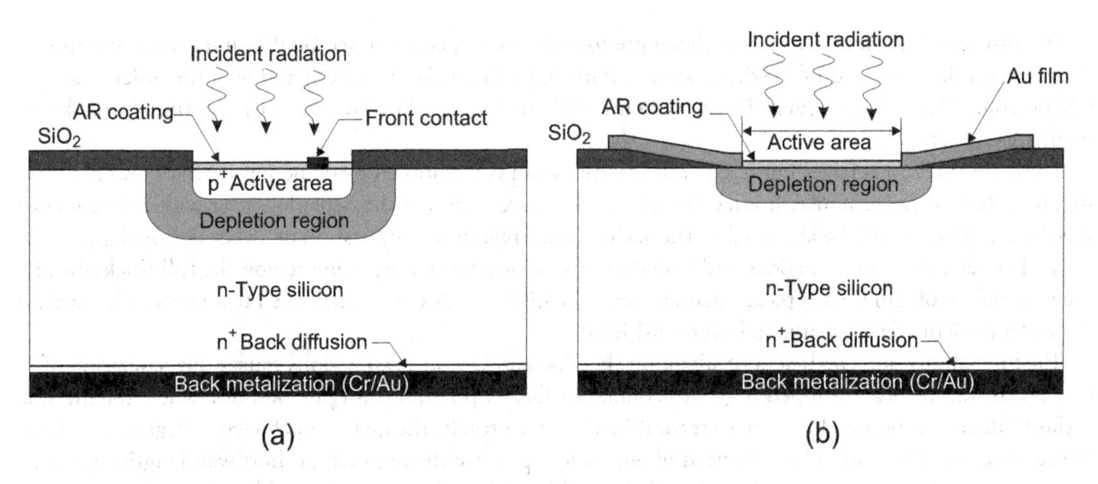

Figure 13.1 Cross section of silicon photodiodes: (a) p-n junction, and (b) Schottky barrier.

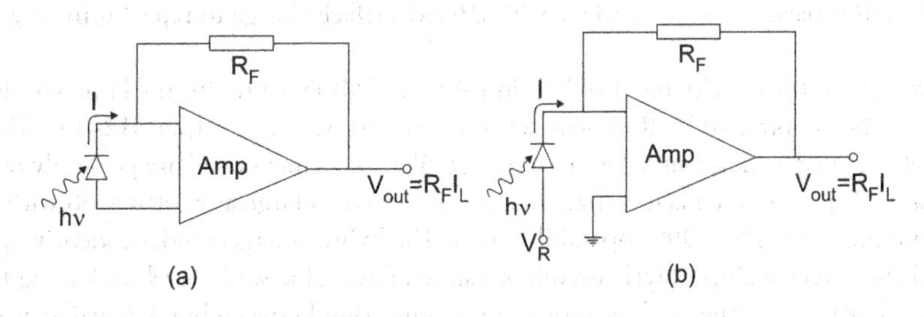

Figure 13.2 Modes of photodiode operation: (a) photovoltaic mode, and (b) photoconductive mode.

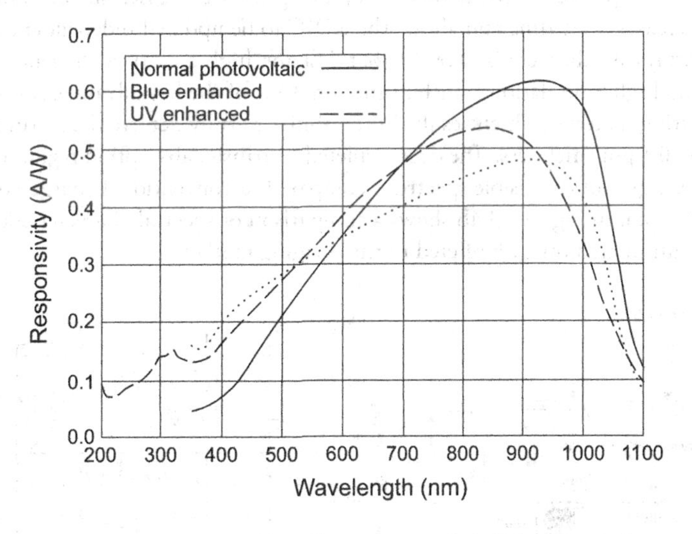

Figure 13.3 Typical current responsivity of several different types of planar diffused silicon photodiodes (After UDT Sensors, Inc., Catalog; https://www.physics.utoronto.ca/~astummer/Archives/2008%20X-ray%20PD%20TIAs/Docs/Photodiode_UDT_catalog.pdf)

Typical spectral characteristics of planar diffusion photodiodes are shown in Figure 13.3. Due to indirect photon absorption, they do not have sharp absorption edges. Si photodiodes usually have quantum efficiencies close to 100% at selected intervals within the spectral range from ≈400 to 1000 nm. Special front surface treatment allows the high-efficiency region of Si photodiodes to be extended to below 400 nm. The usual approach used to reduce dark current is to provide special structure like guard rings and surface-leakage stoppers.

The time constant of p-n junction silicon photodiodes is generally limited by RC time constant rather than by the inherent speed of the detection mechanism (drift and/or diffusion) and is in the order of a microsecond. Detectivity is typically between mid-10^{12} and 10^{13} cmHz$^{1/2}$W^{-1}, usually amplifier-limited for small area detectors.

The p-i-n detector is faster but is also less sensitive than conventional p-n junction detector and has slightly extended red response. It is a consequence of the extension of the depletion layer width since longer wavelength photons will be absorbed in the active device region. Incorporation of a very lightly doped region between the *p* and *n* regions and a modest reverse bias form a depletion region the full thickness of the material (≈500 μm for a typical silicon wafer). The higher dark current collected from generation within the wider depletion layer results in lower sensitivity.

The high absorption coefficient of silicon in the blue and UV spectral regions causes the generation of carriers within the heavily doped p^+ (or n^+) contact surface of p-n and p-i-n photodiodes, where the lifetime is short due to the high and/or surface recombination. As a result, the quantum efficiency degrades rapidly in these regions. Blue- and UV-enhanced photodiodes optimize the response at short wavelengths by minimizing near-surface carrier recombination. This is achieved by using very thin and highly graded p^+ (or n^+ or metal Schottky) contacts, by using lateral collection to minimize the percentage of the surface area that is heavily doped, and/or passivating the surface with a fixed surface charge to repeal minority carriers from the surface.

Impressive progress in the development of hybrid p-i-n Si-CMOS arrays for the large visible and near-IR imaging market has been obtained by Raytheon Vision Systems (See Figure 13.4) [12–16]. The large format imagers are loosely defined as detector arrays that exceed 5k×5k elements and are generally over 2×2 cm^2 in area. At the same time, the pitch is less than 10 μm. The arrays as large as 8160×8160 with an 8-μm pitch are being produced with >99.99% operability [17]. The hybrid imagers independently optimize the readout chip and the detector chip. Raytheon offers a suite of available wafer level packaging technologies and processes in this area with fine pitch electrical interconnection between bonded wafers using direct bond hybridization (3-μm interconnect on 6-μm pitch). Alignment is verified using an infrared microscope and is approximately 1–1.5 μm across the 200-mm bond length for processed silicon wafer pairs.

This flexibility decreases cycle time and allows the ROIC to be updated independent of the detector and vice versa. The arrays are characterized by 100% fill factor, high quantum efficiency in 400–900–nm wavelength range, and high modulation transfer function. One of the key advantages is the use of high-resistivity silicon starting material, which results in the ability to fully deplete the intrinsic region of the detectors with up to 200-μm thickness. The deep depleted (intrinsic) absorption region is more sensitive to the longer (red) wavelengths of the visible spectrum compared to conventional charge-coupled device imagers (Figure 13.4b). The plot in Figure 13.4b shows a comparison of spectral characteristics of the various front and back side illuminated imagers offered in the imaging market.

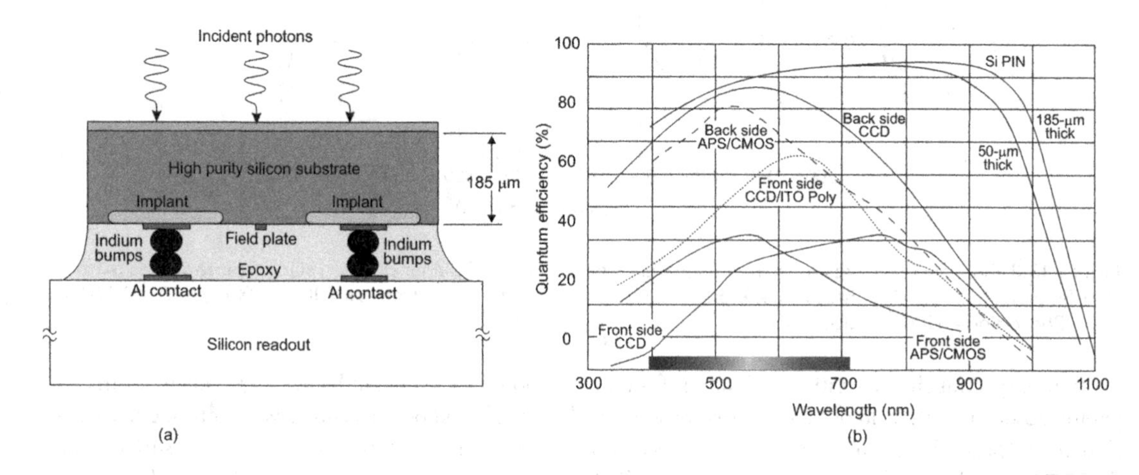

Figure 13.4 (a) Hybrid p-i-n Si-CMOS imager unit cells, and (b) quantum efficiency comparison of various technologies. (From Ref. [14])

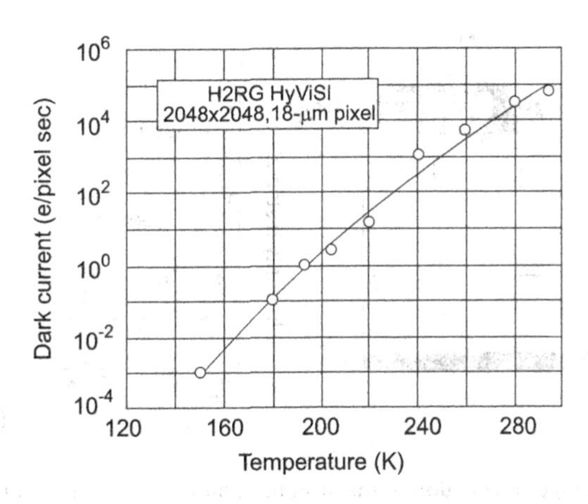

Figure 13.5 Dark current of 18-μm pixel p-i-n Si detector (H2RG-18 FPA fabricated by Teledyne Imaging Sensors). (From Ref. [13])

Silicon p-i-n detector dark current is due to thermal generation in the depletion region and at the surface of the detector. At the present stage of development, the dark current is 5–10 nA cm^{-2} at room temperature for 18-μm pixel, which corresponds to one electron per pixel per second at ~195 K (See Figure 13.5 [13]). Further reduction to ~1 nA cm^{-2} is possible. With a thick detector layer, the p-i-n photodiode must have a strong electric field within the detector layer to push the photocharge to the p-n junction to minimize charge diffusion and obtain good point spread function. A bias up to 50 volts is applied to the back surface contact.

The APDs are especially useful where both fast response and high sensitivity are required. They are most commonly used for communication and active sensing applications. In addition to APD operation in the linear mode, beyond the breakdown voltage Geiger-mode operation is realized (typically much higher than biases of linear-mode detectors) and forms the basis for Geiger-mode APDs (GM-APDs), a technology widely used today for active sensing. At this high voltage, a metastable state is reached where carriers generated through photon absorption (or dark) entering the depletion region begin avalanche multiplication and cause a sustained chain of impact ionizations, resulting in a measurable current pulse. Materials with high probabilities of electron or hole ionization are usually good for Geiger-mode detectors. GM-APDs are commonly based on Si, InGaAs, and InGaAsP [18,19]. They are being used as photon-counting detectors in ladar systems and have been highly successful in 3-D imaging applications at 1.06 and 1.55 μm.

3-D imaging systems operate with either a single detector element or a linear array of detectors, with scanned laser systems. 3-D flash imaging can be performed with 2-D arrays of APD detectors or focal plane arrays (FPAs). The FPAs require special integrated circuit readouts that provide circuitry for each pixel; they are used to detect the laser return pulse for signal peak amplitude and time delay (range). Because the laser needs to illuminate a large area, laser power is critical.

An n$^+$-p-π-p$^+$ epiplanar avalanche Si photodiode configuration developed at the Institute of Electron Technology is shown in Figure 13.6 [20]. An initial material is Si wafer with π-type epilayer (ρ_π=200–300 Ωcm, x_π=30–35 μm; see Figure 13.6b) on p$^+$ Si(111) substrate. The choice of π-type highly resistive layer ensures higher participation of electrons than holes in the detection process. The n-type guard ring is provided by prediffusion of phosphorous followed by rediffusion that takes place during a thermal treatment of the active region. The p$^+$-type channel stopper is made by implanting and then rediffusing boron. The 150-nm thick SiO_2 antireflection layer covers the photodiode active region. The active (photosensitive, avalanche) region constitutes the central region of the n$^+$-p abrupt junction obtained by arsenic diffusion from amorphous silicon to the p-type area previously formed by boron implantation followed by boron rediffusion.

The basic parameters of the photodiodes developed at Excelitas Technologies are listed in Table 13.2. The diameter of the active area varies from 0.5 to 3 mm to accommodate a large variety of applications. The "S" series of the C30902 family of APDs can be used in either their normal linear mode or for photon counter in the Geiger mode. The temperature control is achieved with a thermoelectric cooler that improves noise and responsivity or maintains constant responsivity over a wide range of ambient temperatures.

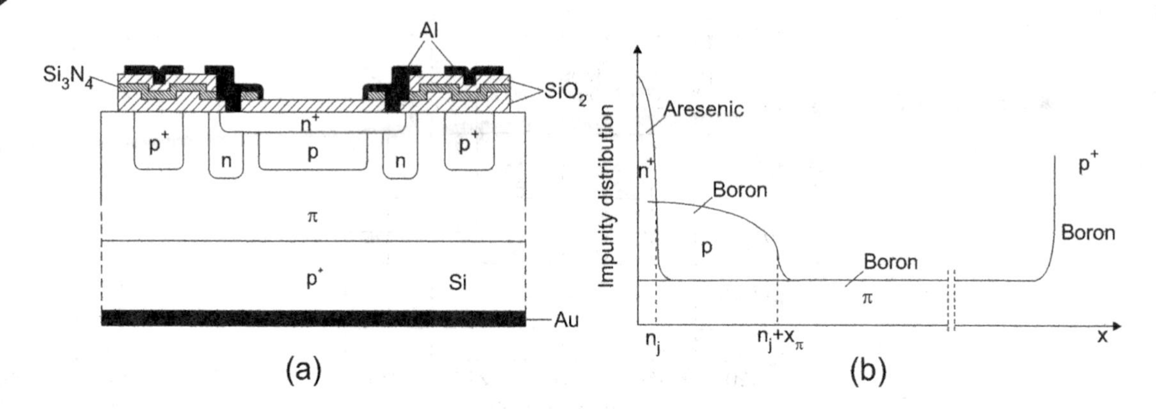

Figure 13.6 n^+-p-π-p^+ Si APD: (a) cross section of the structure, and (b) distribution of dopants in active region. (From Ref. [20])

While normal photodiodes become Johnson or thermal noise–limited when used with low-impedance load resistor for fast response, APDs make use of internal multiplication and keep the detector noise above the Johnson noise level. An optimum gain exists below which the system is limited by receiver noise and above which the shot noise dominates receiver noise and the overall noise increases faster than the signal (see Figure 13.7). A very careful bias control is essential for stable performance. Noise is a function of the detector area and increases as gain increases. Signal-to-noise ratio improvements of one to two orders of magnitude over a non-avalanche detector can be achieved. Typical detectivity is $(3–5)\times 10^{14}$ cmHz$^{1/2}$W^{-1}.

Table 13.3 lists the characteristics of commonly available silicon APD structures optimized to meet specific paradigms. The excessive leakage current along the junction edges due to junction curvature effect or high-field concentration is eliminated by using a guard-ring or surface-beveled structure.

The Schottky-barrier diode can be used as an efficient photodiode. Since it is a majority carrier device, minority carrier storage and removal problems do not exist and therefore higher bandwidths can be expected. In comparison with p-i-n photodiode, Schottky photodiode has narrower active regions and hence transit times are very short. This type of device also offers lower parasitic resistance and capacitance and has the capability to operate at frequencies >100 GHz. However, narrow active regions also cause lower quantum efficiency. Surface traps and recombination cause substantial loss of generated carriers at the surface.

The general structure of silicon Schottky photodiode is shown in Figure 13.1b. The silicon devices are usually fabricated by evaporating a thin Au layer (~150 Å) on high resistivity n-type materials. Due to the deposited gold layer, the reflection coefficient in the wavelength range $\lambda > 800$ nm is above 30%. This results in decreasing the responsivity in this spectral range. Response time is in the picosecond region, corresponding to a bandwidth of about 100 GHz.

The absorption in silicon is generally poor because of the indirect bandgap. The absorption length in Si is almost 20 μm compared with 1.1 μm for GaAs and 0.27 μm for Ge. Consequently, it is difficult to design a Si photodetector with high efficiency in silicon CMOS processes.

Due to the long absorption length of silicon, the traditional vertical photodiode would have to be many microns thick to attain reasonable quantum efficiency. In bulk Si detectors, a majority of the electron hole pairs are generated deep within the substrate, far from the high-field drift region or generated by surface electrodes. Second, the operating voltage would have to be high to deplete the absorption region. As a result, a 3-dB bandwidth is seriously limited. Moreover, attempting to incorporate a thick p-i-n structure into the silicon CMOS process is simply impractical.

A photodiode structure that is more compatible with CMOS is the lateral p-i-n structure shown in Figure 13.8a. The design consists of alternating p-type and n-type interdigitated fingers separated by the absorption region, similar to an MSM photodetector layout. This structure features low capacitance per unit area; however, slow drift of the carrier to the electrodes from a deep region severely limits the bandwidth. Therefore, it is beneficial to block the deep carriers at the expense of quantum efficiency. One method of solution involves placing an insulating layer, such as SiO_2, a couple of microns below the surface.

Table 13.2 Typical parameters of silicon APDs developed at Excelitas Technologies

UNIT	ACTIVE DIAMETER (mm)	CAPACITANCE (pF)	RISE/FALL TIME (ns)	DARK CURRENT (nA)	BREAKDOWN VOLTAGE MIN/MAX (V)	TEMP. COEFFICIENT ($V°C^{-1}$)	TYPICAL GAIN	RESPONSIVITY 900 nm (A/W)	NEP (fW/√Hz)	PACKAGE
C30817EH	0.8	2	2	50	300/475	2.2	120	75	1	TO-5
C30884E	0.8	4	1	100	190/290	1.1	100	63	13	TO-
C30902BH	0.5	1.6	0.5	15	185/265	0.7	150	60	3	Ball lens TO-18
C30902BSTH	0.5	1.6	0.5	15	185/265	0.7	150	60	3	ST receptacle
C30902EH	0.5	1.6	0.5	15	185/265	0.7	150	60	3	TO-18, flat window
C30902EH-2	0.5	1.6	0.5	15	185/265	0.7	150	60	3	TO-18, 905 nm filter
C30902SH	0.5	1.6	0.5	15	185/265	0.7	250	108	0.9	TO-18, flat window
C30902SH-2	0.5	1.6	0.5	15	185/265	0.7	250	108	0.9	TO-18, 905 nm filter
C30916EH	1.5	3	3	100	315/490	2.2	80	50	20	TO-5
C30921EH	0.25	1.6	0.5	15	185/265	0.7	150	60	3	TO-18, flat window
C30921SH	0.25	1.6	0.5	15	185/265	0.7	250	108	0.9	TO-18, light pipe
C30954EH	0.8	2	2	50	300/475	2.4	120	75	13	TO-5
C30955EH	1.5	3	2	100	315/490	2.4	100	70	14	TO-5
C30956EH	3	10	2	100	325/500	2.4	75	45	25	TO-8

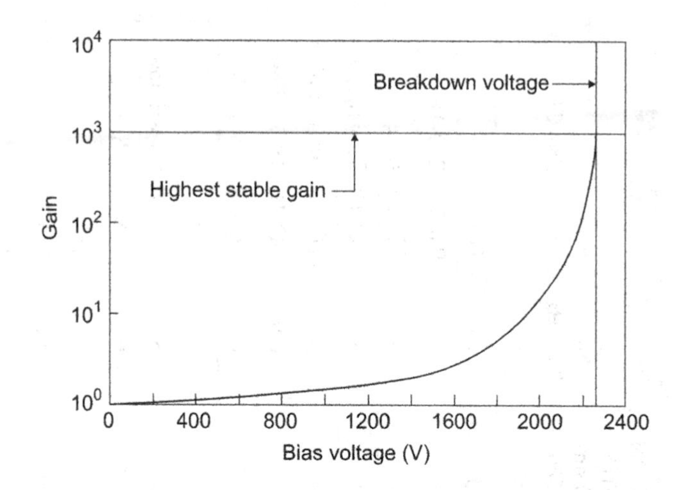

Figure 13.7 Gain as a function of reverse bias can reach 1,000. This operating point is very close to breakdown and requires careful bias control (After Advanced Photonix Inc. Avalanche Catalog.) www.advance photonix.com/ap_products/.

Table 13.3 Properties of commonly available silicon APD structures

	BEVELED EDGE	EPITAXIAL	REACH-THROUGH
Structure	M A	M	M A
Absorption region	Wide	Narrow	Medium to wide
Multiplication region	Wide	Narrow	Narrow
Typical size (diameter, mm)	up to 15 mm	up to 5 mm	up to 5 mm
Gain	50–1,000	1–200	13–300
Excess noise factor	Excellent (k ≈ 0.0015)	Good (k ≈ 0.03)	Good to excellent (k ≈ 0.0015)
Operating voltage (V)	500–2,000 V	80–300 V	150–500 V
Response time	Slow	Fast	Fast
Capacitance	Low	High	Low
Blue response	Good	Good	Poor
Response	Excellent	Poor	Good

Source: After Ref. [21].

The thickness of the oxide is adjusted to maximize the reflectivity at the desired wavelength. An alternative way to block the slow carriers is to use a p-n junction as a screening terminal. The active area is placed inside the n-well surrounded by substrate contacts.

A novel approach to improve bulk detectors was described by Yang et al. [22], who demonstrated a lateral trench detector (LTD) that consisted of a lateral p-i-n detector with 7-μm deep trench electrodes (see Figure 13.8b). These detectors were fabricated on a p-type (100) silicon with a resistivity of 11–16 Ωcm. The trench width at the top surface is approximately 0.35 μm. To fill the trenches with n^+ and p^+ amorphous silicon (a-Si) sequentially, borosilicate glass was used as a sacrificial material. The a-Si inside the trenches was in situ doped with a concentration of 1×10^{20} cm^{-3} for phosphorous and 6×10^{20} cm^{-3} for boron. In the final step, devices were annealed at a high temperature to crystallize a-Si into polysilicon, activate the dopants, and drive the dopants from trenches into the silicon substrate, forming junctions away from the trench sidewalls.

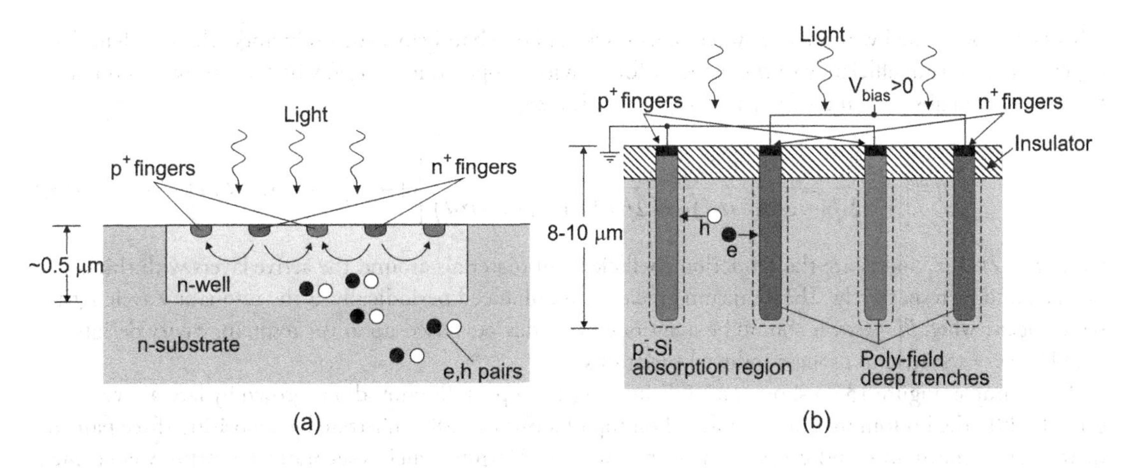

Figure 13.8 Si p-i-n lateral photodiodes: (a) interdigitated and (b) trenched.

LTDs presented in Yang and colleagues [22] were able to simultaneously show high quantum efficiency (68%) and high-speed operation (3 GHz) at 670 nm. However, at longer wavelengths, the devices suffered from degraded bandwidth due to carrier generation and collection below electrodes.

To improve the performance of Si-based photodetectors, the use of silicon-on-insulator (SOI) has been investigated extensively [23]. This technology is particularly attractive when taken into account the widespread acceptance of the SOI technology as a platform for high-performance CMOS devices [24]. The main benefit of using SOI technology is the buried insulator that prevents carriers generated in the substrate, below the oxide, from reaching the surface electrodes above the oxide. In addition, the refractive index contrast of the buried oxide causes the reflection of a portion of the incident light back into the absorbing layer, thus improving quantum efficiency.

Liu et al. have demonstrated MSM detectors on thin SOI substrates with bandwidth as high as 140 GHz [25]. However, due to the thin absorbing layer of 200 nm, these devices have very low external quantum efficiency, below 2%. The SOI detectors with thicker absorbing layers were not only characterized by improved quantum efficiency (to 24% at 840 nm) but they also resulted in a reduced bandwidth of 3.4 GHz [26].

The resonance cavity enhanced (RCE) detector design is another technique that is applied to Si detectors [27,28]. Figure 13.9a shows a schematic representation of a general RCE active layer structure that can be applied to different device configurations, such as Schottky diodes, MSM photodetectors, and APDs. The top and bottom distributed Bragg reflection mirrors consist of alternating layers of non-absorbing larger bandgap materials. The active layer of thickness d is a small bandgap semiconductor positioned between two mirror structures at respective distances l_1 and l_2 from the top and bottom mirrors. The two end

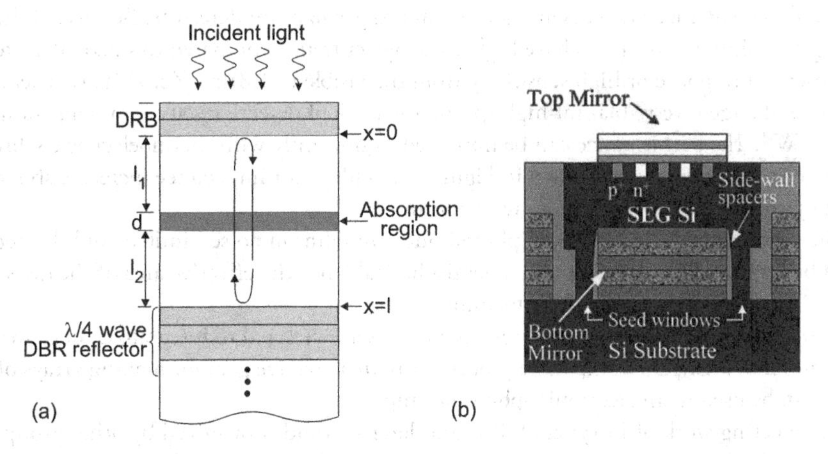

Figure 13.9 Resonant cavity enhanced photodetector: (a) a general structure, and (b) schematic cross section of the interdigitated p-i-n photodiode. (From Ref. [29])

mirrors can be formed with quarter-wave ($\lambda/4$) stacks of large bandgap semiconductors. The wavelength-dependent quantum efficiency of the detector for a cavity of optical length βl, which an active region of thickness d and absorption coefficient α, is given as [27,28]

$$\eta = \left[\frac{1 + r_2 \exp(-\alpha d)}{1 - 2\sqrt{r_1 r_2}\exp(-\alpha d)\cos(2\beta L) + r_1 r_2 \exp(-\alpha d)}\right](1 - r_1)\left[1 - \exp(-\alpha d)\right], \tag{13.1}$$

where $\beta = 2\pi/\lambda$, r_1 and r_2 are the refraction coefficients of materials around the active layers with thicknesses l_1 and l_2, respectively. The quantum efficiency is enhanced periodically at the resonant wavelengths, which occur when $\beta l = m\pi$. It should be mentioned that devices based upon the resonant cavity design could be very sensitive to process-induced variations.

For example, Figure 13.9b shows an RCE-interdigitated p-i-n Si photodiode grown by lateral overgrowth [29]. The bottom mirror was formed on top of a p-type (100) substrate by depositing three pairs of quarter-wavelength SiO_2 and polysilicon layers. Two 20×160 μm trenches separated by 40 μm were etched in the mirror to serve as seed windows for the subsequent selective epitaxial growth. SiO_2 sidewall spacers on the mirror were formed to prevent the nucleation of defects at the edges of the polysilicon during Si epitaxial process. Next, interdigitated p-i-n photodiodes were formed in the epitaxial silicon by sequential As and BF_2 implants and anneals. Following metallization, two dielectric mirror pairs (ZnS–MgF) were evaporated to form the Fabry–Perot cavity.

13.2 GERMANIUM PHOTODIODES

Germanium photodiodes are usually fabricated by diffusion of arsenide into a p-type germanium (gallium doped to a 10^{15} cm^{-3} concentration and resistivity of 0.8 Ωcm). After creation of a 1-μm thick n-type region, an oxide passivation film is deposited to reduce surface conductivity in the vicinity of the p-n junction. Finally, an antireflection coating is deposited (germanium has a high refractive index, $n \sim 4$, and a useful transmission range of 2–23 μm). Germanium is an excellent candidate material for the immersion lens because of its high refractive index.

Germanium does not form a stable oxide. GeO_2 is soluble in water, which leads to two process challenges: device passivation and stability. The lack of a high-quality passivation layer makes it difficult to achieve a low dark current. Interestingly, by scaling the device to smaller dimensions, a higher dark current can be tolerated.

Three germanium photodiode types are available: p-n junction, p-i-n junction, and APD [30]. They are readily made in areas ranging from 0.05 to 3 mm^2, and capability exists to make the area as small as 10×10 μm^2 or as large as 500 mm^2. The upper limit is imposed by raw material uniformity.

The previous discussion on silicon detectors applies in general to germanium ones, with the exception that the blue and UV-enhanced devices are not relevant to germanium detectors. Because of the narrower bandgap, the germanium photodiodes have higher leakage currents, compared to silicon detectors. They offer submicrosecond response or high sensitivity from the visible to 1.8 μm. Zero bias is generally used for high sensitivity and large reverse bias for high speed. The peak of detectivity at room temperature is above 2×10^{11} cmHz$^{1/2}$W^{-1}. The performance can be improved significantly with thermoelectric cooling or cooling to liquid nitrogen temperature, as is shown in Figure 13.10 (detector impedance increases about the order of magnitude by cooling 20°C below room temperature).

Usually, the performance of germanium photodiodes are Johnson noise–limited, and the detector performance can be improved by immersion in a hemispherical lens. The effective area of the detector increases by n^2, where n is the refractive index of the medium.

At the present stage of development, the research efforts are directed to integrate G_e detectors on silicon (Ge/Si) substrates into a CMOS-compatible process. It is an attractive goal for making arrays of on-chip detectors that can be used in an electronic–photonic chip.

Since the pioneering work of Luryi et al. [31] and later methods optimized by other groups (involved using graded SiGe buffer layers to reduce the density of threading dislocations arising in the Ge/GeSi layers), a number of different approaches have been proposed for growing Ge (or GeSi) films on silicon

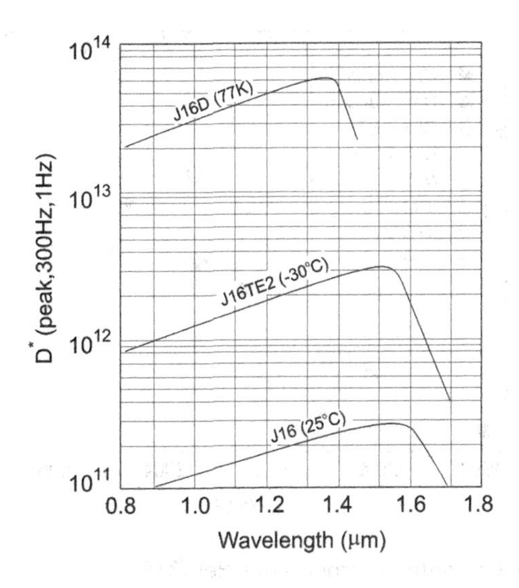

Figure 13.10 Detectivity as a function of wavelength for germanium photodiode at three temperatures (After Teledyne Judson Technologies http://www.teledynejudson.com/#

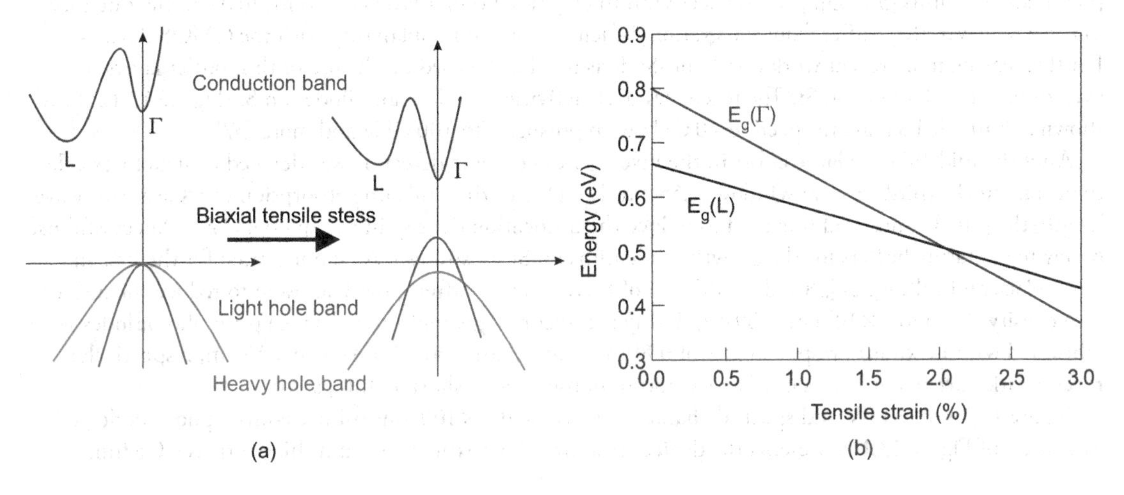

Figure 13.11 Effect of tensile strain on the band structure of germanium: (a) changes of the band diagram, (b) plot of the bandgap energies for the Γ and L bands.

substrates aimed at both optimizing the electronic quality of the films while preserving compatibility with the standard silicon technology. In recent years, a two-step epitaxial growth technique (where the growth temperature is ramped up between the growth steps) has also been developed to directly deposit Ge/SiGe on Si without using the buffer layer and a similar dislocation density ~10^6–10^7 cm^{-2} has been obtained [11]. In addition, since the lattice constant of Ge is 4.17% larger than that of Si, this two-step method can introduce tensile strain to the Ge layer during fabrication, which could transform Ge from an indirect to a direct bandgap material (See Figure 13.11), thus greatly improving the optoelectronic properties of Ge. Other techniques, such as H_2 annealing, are also adopted to improve the Ge-on-Si quality [11,32].

The strain resulting from lattice mismatch modifies the band structure and causes dislocation defects that increase the leakage current of photodiodes. Strain limits the thickness of Ge layers that can be epitaxially grown on silicon. Taking into account the bandwidth point of view, thin Ge films are preferred since it minimizes the carrier transit time. However, it comes at the expense of reduced absorption and diminished responsivity. Nevertheless, excellent progress is being made toward Ge on silicon detectors.

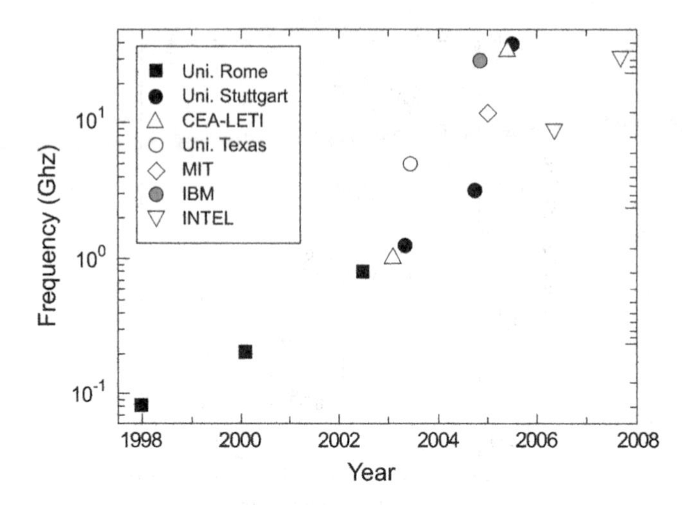

Figure 13.12 Speed evolution of Ge photodetectors. (From Ref. [37])

To overcome the above limitations, the following approaches have been pursued: low Ge content alloys, incorporating of carbon, graded buffers, low-temperature thin buffers, and, more recently, the growth of polycrystalline films [33–35]. p-i-n devices with thick graded buffer layers demonstrated excellent quality [35,36]; however, they suffer from integration difficulties due to nonplanarity with the CMOS devices. For this reason, more recent work on Ge-on-Si detectors has focused on the use of thin buffer layers or even direct growth of Ge on Si. The recent research activities on Ge photodiodes on Si (Figure 13.12) have shown a dramatic increase in speed to 40 GHz with prospects to 100 GHz and more [37].

After the mid-1990s, a large effort in the investigation of SiGe detectors was devoted to structures able to operate in the third spectral window (1.55 μm) [7]. Due to the vanishing absorption of SiGe at this wavelength, the pure Ge material was used to reduce the dislocation density in the epitaxial layer. Successful use of low-temperature buffers for the growth of Ge films on Si substrates was demonstrated for the first time (see Colace and colleagues [38]). The addition of thermal cycles after growth allowed to reduce the dislocation density down to 2×10^{7} cm^{-2} [39,40]. Using this technology, high-performance p-i-n photodiodes were fabricated with maximum responsivities of 0.89 A/W at 1.3 μm and 0.75 A/W at 1.55 μm, respectively; reverse dark currents of 15 mA cm^{-2} at 1 V and response time as short as 180 ps [40].

Figure 13.13 shows I-V and spectral characteristics of a 100×100 μm^2 p-i-n Ge-on-Si photodiode [41]. The insert of Figure 13.13a presents the device structure. Approximately 2-μm thick intrinsic Ge film was

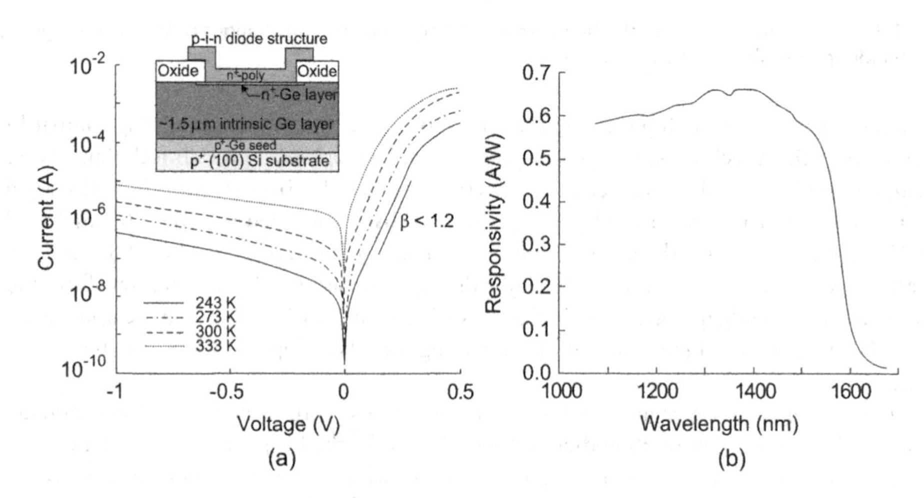

Figure 13.13 Characteristics of a 100×100 μm^2 p-i-n Ge-on-Si photodiode: (a) I-V characteristics as a function of temperature, and (b) responsivity versus wavelength (without AR coating). (From Ref. [41])

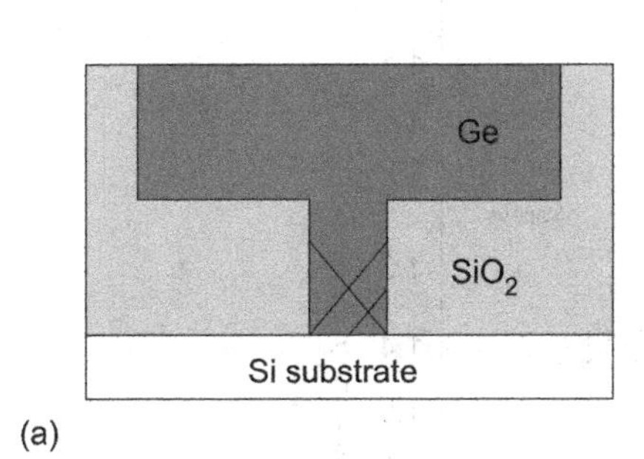

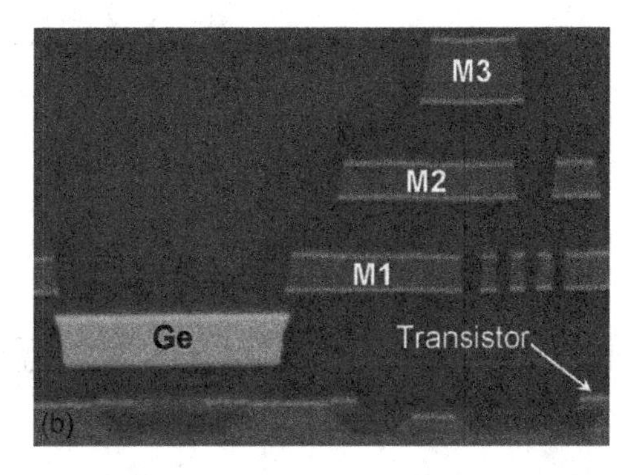

Figure 13.14 Growth of germanium island: (a) dislocation trapping, and (b) scanning electron microscope (SEM) of germanium photodiode embedded in CMOS stack. (From Ref. [42])

deposited on p⁺-(100)Si substrate by a two-step deposition process (by ultra-high vacuum chemical vapor deposition (CVD) and by low-pressure CVD), and capped with a 0.2-μm n⁺-polysilicon layer. After Ge growth, standard CMOS processes were used to deposit and pattern a dielectric (SiON) film to open up windows to the Ge surface. Then poly Si was deposited and implanted with phosphorus into the underlying Ge to form a vertical p-i-n junction in germanium. The diode has an ideality factor less than 1.2 at 300 K with a perimeter-dominated reverse leakage current of ~40 mA cm^{-2} at −1 V bias. The leakage current is believed to be due to surface states introduced during the Ge passivation process. The responsivity without AR coating is 0.5 A/W at 1.55 μm.

More recently, NoblePeal Vision developed monolithic germanium imaging arrays in which Ge islands are integrated with the silicon transistors and metal layers of a CMOS process [42]. The key to growing germanium islands is illustrated schematically in Figure 13.14a. Selective germanium epitaxial growth is carried out in an aperture formed in the dielectric layer that overlies the transistors. Dislocations at the Si–Ge interface propagate at an angle of 60° to the interface and terminate on the sidewalls. After formation of the germanium island, the conventional CMOS process is proceeded to connect the germanium photodiode to the circuitry (See Figure 13.14b). An ion implantation was used to form lateral photodiodes. Using this innovative growth technique, a prototype 128×128 imager at a 10-μm pitch was designed in a high-volume 0.18-μm CMOS technology.

13.3 SiGe PHOTODIODES

There are several technologies such as InGaAs, PbS, and HgCdTe, which cover NIR spectrum. SiGe offers a low-cost alternative approach for NIR sensors that can cover spectral band up to 1.6 microns. The attractive features of SiGe-based IR FPAs take advantage of silicon-based technology, which can promise very small feature size and compatibility with the silicon CMOS circuit for signal processing.

The $Si_{1-x}Ge_x$ alloy grown on Si is an ideal material because, owing to its 100% complete miscibility, the SiGe can be continuously tuned from the Si (1.1 eV) down to the Ge bandgap (0.66 eV). The lattice mismatch between Si and Ge leads to strained heteroepitaxy due to which for a given Ge-content x, a $Si_{1-x}Ge_x$ alloy can only be deposited on Si up to a critical thickness without the formation of dislocations. The strain also leads to a change in the SiGe bandgap structure (See Figure 13.15 [36]). One can see the big influence of the strain on the bandgap when one compares the unstrained curve (dashed) with the strained line (solid lines). The SiGe technology has been introduced to capture the increased carrier mobility advantages of strained-layer epitaxy, which is 46% for n-channel devices and 60%–80% for p-channel devices. It is decisive in the capture of SiGe technology into high-speed CMOS technology. The sharp drop of the bandgap for $x > 0.85$ for the unstrained alloy indicates the transition point to the Ge-like band structure for this high Ge-content where the conduction band minimum changes from the Δ-point to the L-point <111> of the Brillouin zone [43,44].

A number of epitaxial systems have been demonstrated including molecular beam epitaxy (MBE), plasma-enhanced CVD, sputtering, and laser-assisted growth. For production lines used in CMOS and Si

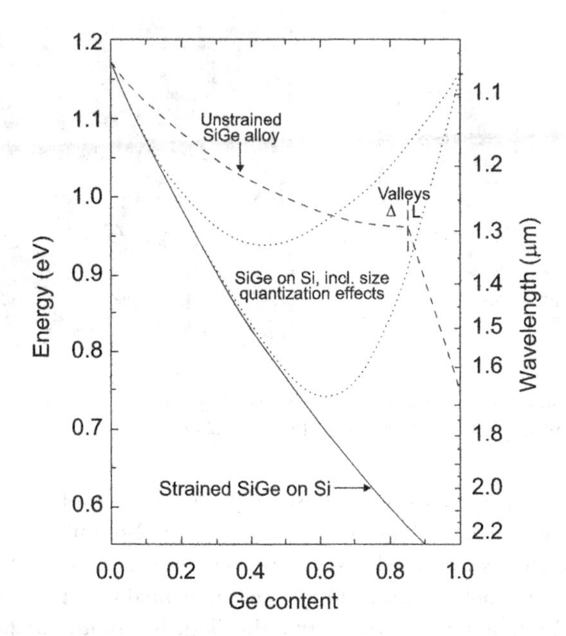

Figure 13.15 The fundamental bandgap of SiGe alloys versus Ge-content. The dashed line corresponds to unstrained SiGe bulk samples and the solid line to strained SiGe with the lateral constant of Si. The dotted lines refer to the effect of critical layer thickness and size quantization. (From Ref. [43])

bipolar factories, only conventional CVD at high temperatures (>1000°C) has, to date, demonstrated the quality, uniformity, and through-put required by companies for production.

In one of the first NIR p-i-n Ge on silicon detectors, a graded SiGe buffer was used to separate the active intrinsic layer of the diode from the highly dislocated Si–Ge interface region [31]. The large dislocation density, however, gave rise to reverse dark currents in excess of 50 mA cm^{-2} at 1 V. To reduce the dark current, an active layer made of a GeSi/Si superlattice strained-layer superlattice (SLS) was introduced by Temkin et al. [45]. The bandgap of SLS is smaller than that of unstrained layers, and therefore, it is interesting to increase the optical absorption coefficient at a small detector thickness.

The SLS is preferred to SiGe alloys of similar average composition because they allowed the deposition of layers well above the critical thickness (strain symmetrization). When strained SiGe is grown on a relaxed Si, the band alignment at the heterointerface of Si and SiGe is of type I, which means that the offset lies predominantly within the valence band (See Figure 13.16b). On the other hand, when strained Si is grown on relaxed SiGe, a staggered band alignment of type II results at the heterointerface. In this

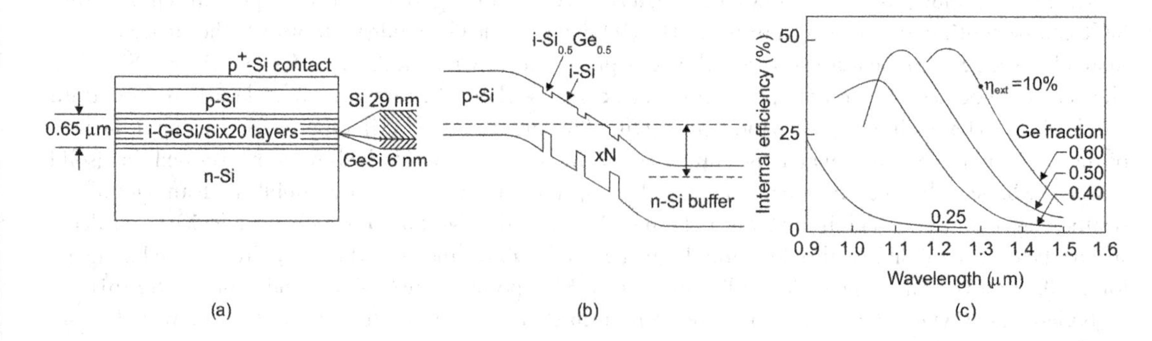

Figure 13.16 p-i-n SiGe superlattice photodiode: (a) schematic diagram of the device structure, (b) energy band diagram, and (c) room temperature spectral response as a function of Ge content of the active layer. (From Ref. [1])

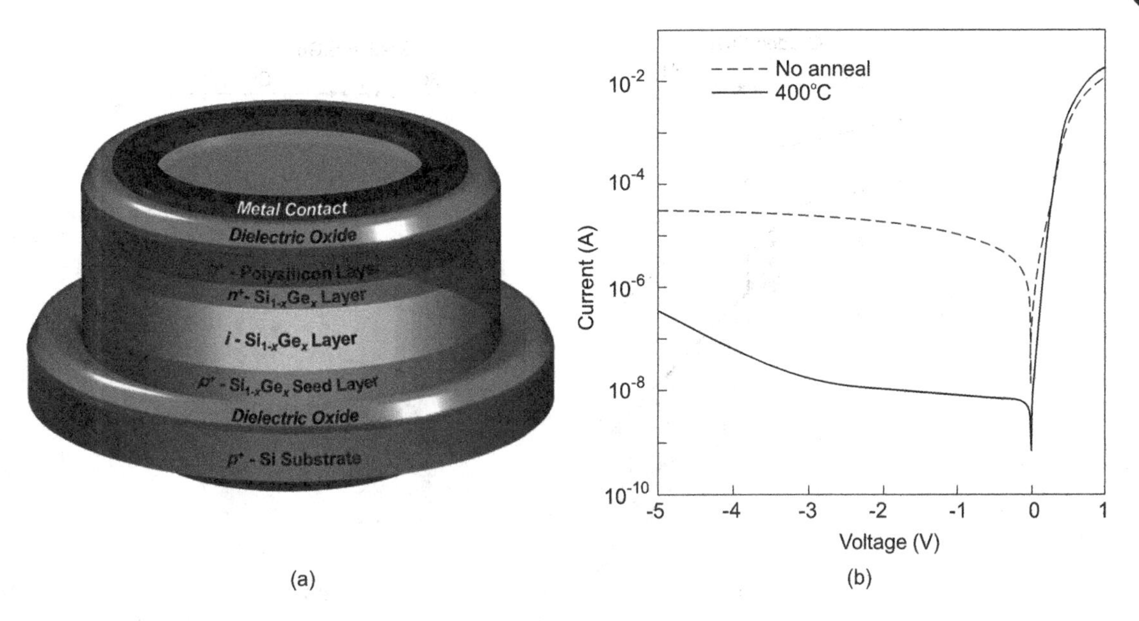

Figure 13.17 p-i-n SiGe photodiode: (a) schematic structure of photodiode after fabrication, (b) I-V characteristics for $10 \times 10\,\mu m^2$ photodiode with and without 400°C post-metallization anneal in N_2 for 45 min (Adapted after Ref. [46])

case, the conduction band of SiGe is higher than that of Si and the valence band in SiGe is lower than that of Si.

Commonly, the photodetector is grown layer by layer on a substrate and the light hits the detector perpendicular to the surface. To overcome the poor sensitivity due to weak absorption in the thin layers of vertical diodes, the so-called waveguide photodetectors were built. Here the absorber is shaped as a thin, narrow but very long rectangular waveguide. The light travels parallel to the surface through the detector structure and the absorption length can be up to some millimeters depending on the length of the waveguide.

Figure 13.16 presents p-i-n SLS SiGe waveguide photodiode characteristics [44]. The p-i-n detector consists of n-type and p-type Si layers on either side of an undoped 20-period superlattice active region grown by MBE on n-type Si substrate. Serious change in spectra responsivity occurs for composition $x > 0.40$, which clearly shows an efficient response at wavelengths of 1.3 μm. As the Ge fraction is increased from $x = 40$ in 10% increments, the photocurrent response peaks at 1.08, 1.12, and 1.23 μm, respectively. Using waveguide geometry, the internal quantum efficiency in the order of 40% at 1.3 μm in SL with the Ge fraction of $x = 0.6$ was measured.

Figure 13.17 shows the mesa structure p-i-n SiGe photodiode with improved performance [46]. Following the deposition of the polysilicon layer, an activation anneal is performed, which serves to outdiffuse dopant atoms from the polysilicon layer into the underlying Ge/SiGe to form a vertical p-i-n junction. Next, the SiO_2 passivation layer is deposited using a low-temperature CVD-based process. Following the metallization process, the samples have been annealed in nitrogen, as shown in Figure 13.17b, which reduces the dark current by up to three orders of magnitude for small area photodiodes.

The first avalanche SiGe photodiodes were demonstrated at Bell Labs in 1986 with an overall thickness ranging from 0.5 to 2 μm [47]. Among them, the most effective was based on 3.3-nm thick $Ge_{0.6}Si_{0.4}$ layers with 39-nm Si-spacers with a quantum efficiency of about 10% at 1.3 μm and a maximum responsivity of 1.1 A/W. More stable characteristics were obtained at 30 V reverse bias with a responsivity of 4 A/W.

Ishikawa has presented recently a new low-noise and low-voltage Ge/graded-SiGe heterojunction APDs shown schematically in Figure 13.18. In Figure 13.18a, electron hole pairs are generated by the direct optical absorption in the Ge layer, followed by the injection of generated electrons into the graded SiGe layer without potential barriers. Then, the electrons are launched into the Si multiplication layer. Since the kinetic energy for the injected electron is suddenly increased by the conduction band discontinuity ΔE_c,

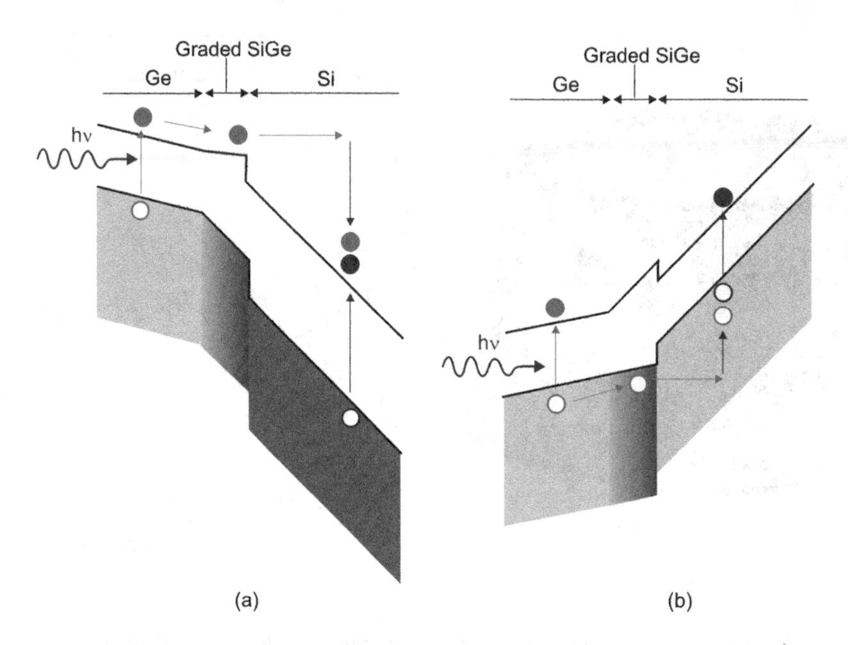

Figure 13.18 Schematic band diagram for n-Ge/i-SiGe/i-Ge heterostructure APDs (Adapted after Ref. [49])

the impact ionization is enhanced across the abrupt SiGe/Si interface, which effectively realized low-voltage Ge/Si APDs. Similarly, the multiplication layer of SiGe/Ge in Figure 13.18b should be effective for the hole-injection APDs.

It should be mentioned that short-period SiGe superlattices and SiGe quantum well structures have been grown by MBE epitaxy on Si substrates for NIR, middle, and long-wavelength infrared detection [50,51]. More information about quantum well infrared detectors can be found in Chapter 19.

REFERENCES

1. D. J. Paul, "Si/SiGe heterostructures: From material and physics to devices and circuits," *Semicond. Sci. Technol.*, 19, R75–108, 2004.
2. L. Pavesi, "Will silicon be the photonic material of the third millennium?" *J. Phys.: Condens. Matter* 15, R1169–96, 2003.
3. B. Jajali and S. Fathpour, "Silicon photonics," *J. Lightwave Technol.*, 24, 4600–15, 2006.
4. S. M. Sze, *Physics of Semiconductor Devices*, Wiley, New York, 1981.
5. S. Adachi, *Properties of Group-IV, III-V and II-VI Semiconductors*, John Wiley & Sons, Chichester, 2005.
6. D. Wood, *Optoelectronic Semiconductor Devices*, Prentice-Hall, Trowbridge, UK, 1994.
7. H. Zimmermann, *Integrated Silicon Optoelectronics*, Springer, New York, 2000.
8. T. P. Pearsall, "Silicon-germanium alloys and heterostructures: Optical and electronic properties," *CRS Crit. Rev. Solid State Mater. Sci.*, 15, 551–600, 1989.
9. J. C. Bean, "Silicon-based semiconductor heterostructures: Column IV bandgap engineering," *Proc. IEEE* 80, 571–87, 1992.
10. R. A. Soref, "Silicon-based optoelectronics," *Proc. IEEE* 81, 1687–1706, 1993.
11. A. K. Sood, J. W. Zeller, R. A. Richwine, Y. R. Puri, H. Efstathiadis, P. Haldar, N. K. Dhar, and D. L. Polla, "SiGe Based Visible-NIR Photodetector Technology for Optoelectronic Applications", Chapter 10 in *Advances in Optical Fiber Technology: Fundamental Optical Phenomena and Applications*, eds. M. Yasin, H. Arof and S. W. Harun, InTech, 2015, doi:10.5772/58517.
12. T. Chuh, "Recent developments in infrared and visible imaging for astronomy, defense and homeland security," *Proc. SPIE* 5563, 19–34, 2004.
13. Y. Bai, J. Bajaj, J. W. Beletic, and M. C. Farris, "Teledyne imaging sensors: Silicon CMOS imaging technologies for X-ray, UV, visible and near infrared," *Proc. SPIE* 7021, 702102, 2008.
14. S. Kilcoyne, N. Malone, M. Harris, J. Vampola, and D. Lindsay, "Silicon p-i-n focal plane arrays at Raytheon," *Proc. SPIE* 7082, 70820J, 2008.

15. S. Kilcoyne, N. Malone, B. Kean, J. Cantrell, J. Fierro, L. Meier, S. DeWalt, C. Hewitt, J. Wyles, J. Drab, G. Grama, G. Paloczi, J. Vampola, and K. Brown, "Advancements in large-format SiPIN hybrid focal plane technology", *Proc. SPIE* 9219, 921906-111, 2014.
16. B. Starr, L. Mears, C. Fulk, J. Getty, E. Beuville, R. Boe, C. Tracy, E. Corrales, S. Kilcoyne, J. Vampola, J. Drab, R. Peralta, and C. Doyle, "RVS large format arrays for astronomy", *Proc. SPIE* 9915, 99152X-1–14, 2016.
17. J. Drab, "Multilevel wafer stacking for 3D circuit integration," *Raytheon Technol. Today* 1, 30–31, 2015.
18. B. F. Aull, A. H. Loomis, D. J. Young, R. M. Heinrichs, B. J. Felton, P. J. Daniels, and D. J. Landers, "Geiger-mode avalanche photodiodes for three-dimensional imaging," *Lincoln Lab. J.*, 13(2), 33–550, 2002.
19. M. Entwistle, M. A. Itzler, J. Chen, M. Owens, K. Patel, X. Jiang, K. Slomkowski, and S. Rangwala, "Geiger-mode APD camera system for single photon 3-D LADAR imaging," *Proc. SPIE* 8375, 83750D-1–12, 2012.
20. I. Węgrzecka and M. Węgrzecki, "Silicon photodetectors: The state of the art," *Opto-Electron. Rev.*, 5, 137–46, 1997.
21. S. Melle and A. MacGregor, "How to choice avalanche photodiodes," *Laser Focus World* 145–56, October 1995.
22. M. Yang, K. Rim, D. L. Rogers, J. D. Schaub, J. J. Welser, D. M. Kuchta, D. C. Boyd, et al., "A high-speed, high-sensitivity silicon lateral trench photodetector," *IEEE Electron Device Lett.*, 23, 395–7, 2002.
23. S. J. Koester, J. D. Schaub, G. Dehlinger, and J. O. Chu, "Germanium-on-SOI infrared detectors for integrated photonic applications," *IEEE J. Sel. Top. Quantum Electron.*, 12, 1489–1502, 2006.
24. G. G. Shahidi, "SOI technology for the GHz era," *IBM J. Res. Dev.*, 46, 121–31, 2002.
25. M. Y. Liu, E. Chen, and S. Y. Chou, "140-GHz metal-semiconductor-metal photodetectors on silicon-on-insulator substrate with a scaled active layer," *Appl. Phys. Lett.*, 65, 887–8, 1994.
26. C. L. Schow, R. Li, J. D. Schaub, and J. C. Campbell, "Design and implementation of high-speed planar Si photodiodes fabricated on SOI substrates," *IEEE J. Quantum Electron.*, 35, 1478–82, 1999.
27. M. S. Ünlü and M. S. Strite, "Resonant cavity enhanced photonic devices," *J. Appl. Phys.*, 78, 607–39, 1995.
28. M. S. Ünlü, G. Ulu, and M. Gökkavas, "Resonant cavity enhanced photodetectors," in *Photodetectors and Fiber Optics*, ed. H. S. Nalwa, 97–201, Academic Press, San Diego, CA, 2001.
29. J. D. Schaub, R. Li, C. L. Schow, L. C. Campbell, G. W. Neudeck, and J. Denton, "Resonant-cavity-enhanced high-speed Si photodiode grown by epitaxial lateral overgrowth," *IEEE Photonics Technol. Lett.*, 11, 1647–9, 1999.
30. A. Bandyopadhyay and M. J. Deen, "Photodetectors for optical fiber communications," in *Photodetectors and Fiber Optics*, ed. H. S. Nalwa, 307–68, Academic Press, San Diego, CA, 2001.
31. S. Luryi, A. Kastalsky, and J. C. Bean, "New infrared detector on a silicon chip," *IEEE Trans. Electron Devices* ED-31, 1135–9, 1984.
32. P. C. Eng, S. Song, and B. Ping, "State-of-the-art photodetectors for optoelectronic integration at telecommunication wavelength," *Nanophotonics* 4, 277–302, 2015.
33. G. Masini, L. Colace, and G. Assanto, "Poly-Ge near-infrared photodetectors for silicon based optoelectronics," *ICTON* Th.C 4, 207–10, 2003.
34. G. Masini, L. Colace, F. Petulla, G. Assanto, V. Cencelli, and F. DeNotaristefani, "Monolithic and hybrid near infrared detection and imaging based on poly-Ge photodiode arrays," *Opt. Mater.*, 27, 1079–83, 2005.
35. S. B. Samavedam, M. T. Currie, T. A. Langdo, and E. A. Fitzgerald, "High-quality germanium photodiodes integrated on silicon substrates using optimized relaxed buffers," *Appl. Phys. Lett.*, 73, 2125–7, 1998.
36. J. Oh, J. C. Campbell, S. G. Thomas, S. Bharatan, R. Thoma, C. Jasper, R. E. Jones, and T. E. Zirkle, "Interdigitated Ge p-i-n photodetectors fabricated on a Si substrate using graded SiGe buffer layers," *IEEE J. Quantum Electron.*, 38, 1238–41, 2002.
37. E. Kasper and M. Oehme, "High speed germanium detectors on Si," *Phys. Status Solidi C* 5, 3144–9, 2008.
38. L. Colace, G. Masini, G. Assanto, G. Capellini, L. Di Gaspare, E. Palange, and F. Evangelisti, "Metal-semiconductor-metal near-infrared light detector based on epitaxial Ge/Si," *Appl. Phys. Lett.*, 72, 3175, 1998.
39. H.-C. Luan, D. R. Lim, K. K. Lee, K. M. Chen, J. G. Sandland, K. Wada, and L. C. Kimerling, "High-quality Ge epilayers on Si with low threading-dislocation densities," *Appl. Phys. Lett.*, 75, 2909–11, 1999.
40. S. Farma, L. Colace, G. Masini, G. Assanto, and H.-C. Luan, "High performance germanium-on-silicon detectors for optical communications," *Appl. Phys. Lett.*, 81, 586–8, 2002.
41. F. X. Kärtner, S. Akiyama, G. Barbastathis, T. Barwicz, H. Byun, D. T. Danielson, F. Gan, et al., "Electronic photonic integrated circuits for high speed, high resolution, analog to digital conversion," *Proc. SPIE* 6125, 612503, 2006.
42. C. S. Rafferty, C. A. King, B. D. Ackland, I. Aberg, T. S. Sriram, and J. H. O'Neill, "Monolithic germanium SWIR imaging array," *Proc. SPIE* 6940, 69400N, 2008.
43. H. Presting, "Infrared silicon/germanium detectors," in *Handbook of Infrared Detection Technologies*, eds. M. Henini and M. Razeghi, 393–448, Elsevier, Kidlington, UK, 2002.
44. C. W. Liu, and L. J. Chen, "SiGe/Si heterostructures," in *Encyclopedia of Nanoscience and Nanotechnology*, ed. H. S. Nalwa Vol. 9, 1–18, American Scientific Publisher, 2004.

45. H. Temkin, T. P. Pearsall, J. C. Bean, R. A. Logan, and S. Luryi, "Ge_xSi_{1-x} strained-layer superlattice waveguide photodetectors operating near 1.3 μm," *Appl. Phys. Lett.*, 48, 963–5, 1986.
46. A. K. Sood, R. A. Richwine, A. W. Sood, Y. R. Puri, N. DiLello, J. L. Hoyt, T. I. Akinwande, N. Dhar, R. S. Balcerak, and T. G. Bramhall, "Characterization of SiGe-detector arrays for visible-NIR imaging sensor applications", *Proc. SPIE* 8012, 801240-1–10, 2011.
47. T. P. Pearsall, H. Temkin, J. C. Bean, and S. Luryi, "Avalanche gain in Ge_xSi_{1-x}/Si infrared waveguide," *IEEE Electron Device Lett.*, 7, 330–2, 1986.
48. N. Izhaky, M. T. Morse, S. Koehl, O. Cohen, D. Rubin, A. Barkai, G. Sarid, R. Cohen, and M. J. Paniccia, "Development of CMOS-compatible integrated silicon photonics devices," *IEEE J. Sel. Top. Quantum Electron.*, 12, 1688–98, 2006.
49. Y. Ishikawa, "Ge/SiGe for silicon photonics", *Proc. SPIE* 10131, 101310C-1–9, 2016.
50. H. Presting, "Near and mid infrared silicon/germanium based photodetection," *Thin Solid Films* 321, 186–95, 1998.
51. H. Presting, J. Konle, M. Hepp, H. Kibbel, K. Thonke, R. Sauer, W. Cabanski, and M. Jaros, "Mid-infrared silicon/germanium focal plane arrays," *Proc. SPIE* 3630, 73–89, 1999.

Extrinsic silicon and germanium detectors

Historically, an extrinsic photoconductor detector based on germanium was the first extrinsic photodetector. After that, photodetectors based on silicon and other semiconductor materials, such as GaAs or GaP, have appeared.

Extrinsic photodetectors are used in a wide range of the IR spectrum extending from a few μm to approximately 300 μm. They are the principal detectors operating in the range of $\lambda > 20$ μm. The spectral range of particular photodetectors is determined by the doping impurity and by the material into which it is introduced. For the shallowest impurities in GaAs, the long wavelength cutoff of photoresponse is around 300 μm. Detectors based on silicon and germanium have found the widest application as compared with extrinsic photodetectors on other materials and will be considered in this chapter.

The research and development of extrinsic IR photodetectors has been ongoing for more than 60 years [1–3]. In the 1950s and 1960s, germanium could be made purer than silicon; doped Si then needed more compensation than doped Ge and was characterized by shorter carrier lifetimes than extrinsic germanium. Today, the problems with producing pure Si have been largely solved, with the exception of boron contamination. Si has several advantages over Ge; for example, three orders of magnitude higher impurity solubilities are attainable, hence thinner detectors with better spatial resolution can be fabricated from silicon. Si has lower dielectric constant than Ge, and the related device technology of Si has now been more thoroughly developed, including contacting methods, surface passivation, and mature metal oxide semiconductor (MOS) and charge-coupled device technologies. Moreover, Si detectors are characterized by a superior hardness in nuclear radiation environments.

The availability of a highly developed silicon MOS technology facilitates the integration of large detector arrays with charge-transfer devices for readout and signal processing. The well-established technology also helps in the manufacturing of uniform detector arrays and the formation of low-noise contacts. Although the potential of large extrinsic silicon FPAs for terrestrial applications has been examined, interest has declined in favor of HgCdTe and InSb with their more convenient operating temperatures. Strong interest in doped silicon continues for space applications, particularly in low background flux and for wavelengths from 13 to 30 μm, where compositional control is difficult for HgCdTe. Success in the technology of photodetectors, creation of deep cooled, low-noise semiconductor preamplifiers and multiplexers, as well as unique designs of photodetector devices and equipments for deep cooling have ensured the achievement of a record-breaking detectivity close to the radiation limit even under exceedingly low space backgrounds, 8–10 orders of magnitude lower than that of the room background [4,5]. This is unattainable for the other types of photodetectors at the present time.

14.1 EXTRINSIC PHOTOCONDUCTIVITY

The development and manufacture of extrinsic photodetectors are mainly concentrated in the United States. The programs on the use of off-atmospheric astronomy have spread especially intensively after the outstanding success of the Infrared Astronomical Satellite (IRAS) [6,7], which used 62 discrete photodetectors arranged in the focal plane. A number of National Aeronautics and Space Administration (NASA)- and National Science Foundation (NSF)-supported programs have been initiated by various research centers and universities. The array manufacturers have taken a strong interest and have done far more for astronomers than they might have anticipated [8,9]. They are Raytheon Vision Systems, DRS Technologies, and Teledyne Imaging Sensors [10,11]. A deep cooled space telescope located in vacuum can cover the entire IR range from 1 to 1,000 μm. Far-infrared astronomy provides key information about the formation

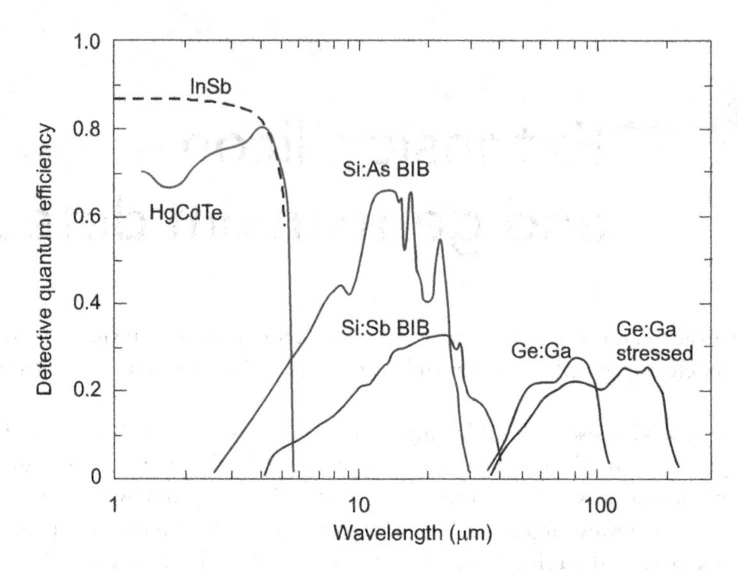

Figure 14.1 Typical quantum efficiency of semiconductors used in the wavelength range between 1 and 300 μm.

and evolution of galaxies, stars, and planets. The low level of background irradiation makes it possible to improve the sensitivity of such systems by increasing the integration time (to hundreds of seconds).

The fundamentals of the classical theory of extrinsic photoconductor detectors have been disclosed earlier in this book (See Sections 12.1.2). In comparison with intrinsic photoconductivity, extrinsic photoconductivity is far less efficient because of limits in the amount of impurity that can be introduced into a semiconductor without altering the nature of the impurity states (See Figure 14.1). Intrinsic detectors are most common at the short wavelengths, below 20 μm. Terahertz photoconductors, with a cutoff wavelength above 30 μm, are operated in the extrinsic mode. It is interesting to note the presence of a marked gap in the wavelength coverage of around 40 μm.

As Figure 14.1 shows, Si-based detectors (so called bocked impurity band (BIB) devices), doped with arsenic and antimony, are the materials of choice for operation at wavelengths from 5 μm to 40 μm. Several attempts have been made to provide a similar technology for operation in the far-IR by switching from silicon to semiconductor materials that provide a shallower impurity band. Both Ge-based and GaAs-based systems have been attempted, with greater success achieved in Ge.

A key difference between intrinsic and extrinsic detectors is that extrinsic detectors require much cooling to achieve high sensitivity at a given spectral response cutoff in comparison with intrinsic detectors. The low-temperature operation is associated with longer wavelength sensitivity to suppress noise due to thermally induced transitions between close-lying energy levels. The long wavelength cutoff can be approximated by Equation 3.1 and is illustrated in Figure 3.6.

This chapter gives a brief survey of extrinsic detector technology, comments on certain peculiarities of the operation of these devices, and describes different types of extrinsic photodetectors that have been developed—extrinsic photoconductors, BIB devices and solid-state photomultipliers (SSPM). The main versions of extrinsic photodetector FPAs are presented in Section 26.2.

14.2 TECHNOLOGY OF EXTRINSIC PHOTOCONDUCTORS

Silicon and germanium extrinsic photoconductors are usually made using techniques drawn from the semiconductor industry [4,12]. In the Czochralski method, a crystal is pulled out of the melt without touching the walls of the vessel. This technique avoids stresses in the crystal that can develop if it solidifies in contact with the crucible walls, and also prevents the escape of impurities from the walls into the crystal. The material can be doped by adding impurities to the melt from which the crystal is grown. Alternatively, the purity of the material can be improved after growth by zone refining the crystal. The dopant impurities can be added in a late stage of zone refining by placing the impurity in the melted zone. Doping of grown crystals can also be accomplished by exposing a crystal to impurity vapors or by bringing its surface into contact with a sample of the impurity material; in both of these cases, the impurity atoms are distributed into the crystal by diffusion at high temperature. Floating zone growth combines features of Czochralski growth and zone refining. The crystal is grown onto a seed from a melted portion of an ingot that is suspended to avoid contact with the walls of the crucible.

To counteract the problem of unwanted thermal conductivity due to impurities with small values of excitation energy E_i, these dopants are compensated by adding carefully controlled amounts of the opposite kind of impurity, for example, n-type if the unwanted impurity is p-type (See Section 12.1.2). Although some degree of compensation is frequently critical in obtaining good detector performance, heavily compensated material is usually inferior to material that starts as pure as possible and needs to be only lightly compensated. A high concentration of compensating impurities usually results in the rapid capture of photoexcited charge carriers and reduces the photoconductive gain. Problems with control of the compensation process arise because of nonuniformities in the sample or because of imprecision in the exact amount of compensating material. Finally, hopping conductivity may be enhanced because the atoms of the compensating impurity ionize some of the majority impurity atoms, making available empty sites into which carriers can hop. The effect of proper compensation is to fill up the contaminating impurity while only partially filling the desired impurity; the material then freezes out at a temperature determined by the energy level of the desired impurity. During compensation, the operating temperature is raised at the expense of the decreased carrier lifetime. Undercompensation leads to thermal generation from shallow impurities, which results in excess noise, unless the operating temperature is lowered. Overcompensation degrades carrier lifetime and causes poor detector responsivity, but does not further increase the operating temperature.

Achieving uniformity in both the dopant and the compensating impurities is of paramount importance in creating high-quality Si:X (Ge:X) material, especially that used for an array. Factors that influence uniformity include the level of dopant concentration, the segregation coefficient and vapor pressure of the dopant at the solidification of the host crystal, pickup from the atmosphere or the crucible during crystal growth, pulling and rotation rates during growth, defects in the form of precipitation and dislocations, and possible high temperature–induced effects from processing material into an array. High-quality Si:X material is produced by using the float zone growth technique. This technique gives material with the lowest density of contamination elements. Uniformity of the major dopant must be grown-in by using high-speed rotation rates (6 rpm) and low pulling rates (4 mm min^{-1}). In the case of Si:Ga, annealing for 16 days at 1,300°C is necessary to achieve maximum uniformity. This annealing process is successful only for crystals low in oxygen and carbon that must necessarily be produced by float zone growth. By comparison, Czochralski-grown crystals are full of contaminating impurities including oxygen, which can introduce unwanted donors, and carbon and can alter the impurity activation energy level.

Both phosphorus and boron are found in polycrystalline silicon. Phosphorus can be removed by a succession of vacuum float zone operations, whereas boron cannot. Phosphorus is usually removed, then reintroduced to compensate for the boron. A typical compensated impurity concentration is $N_d - N_a = 10^{13}$ cm^{-3}. A method of compensating the boron in extrinsic Si is called neutron transmuting doping and involves the transmuting of a fraction of the Si atoms into the P donor via controlled neutron irradiation.

To make the detector, a thin wafer of material is cut from the crystal with a precision saw and then polished. Two sides of this wafer form the contacts of the detector. To smooth the variations in the electric fields near the contacts, heavy doping is used in the semiconductor where it joins the actual contact. The doping can be applied by implantation of impurity atoms of a similar type to the majority impurity (for example, boron for a p-type material and phosphorus for n-type material) and then annealing the crystal damage by heating [13]. Thermal annealing restores single crystallinity and activates the implanted atoms. Metallization with a thin 200 Å Pd adhesion layer followed by a few thousand Å of Au completes the contact formation process. A transparent contact can be made by adding a second layer of ion-implanted impurities at a lower energy than the first, such that it does not penetrate so far into the crystal and at a higher impurity density, so it has large electrical conductivity. For an opaque contact, metal is evaporated over the ion-implanted layers. Individual detectors or arrays of detectors are cut from the wafer with a precision saw. The surface damage left by the saw can be removed by etching the detectors. Finally, naturally-grown silicon oxide (at elevated temperatures) is an excellent protective layer for silicon photodetectors. In the case of germanium and many other semiconductors, protective layers must be added by more complex processing steps.

Different additional techniques, including standard photolithography, are used to produce the semiconductor detectors. Epitaxy is a particularly useful technique for growing a thin crystalline layer on a preexisting crystal structure. In comparison with silicon, germanium and many other semiconductors require more complex processing steps especially in fabrication protection or insulating layers.

14.3 PECULIARITIES OF THE OPERATION OF EXTRINSIC PHOTOCONDUCTORS

The studies on the photoconductivity in high-resistivity crystals, to which extrinsic photoconductors belong at low temperatures and backgrounds, began simultaneously with the works or the studies of the extrinsic photoconductivity as a phenomenon since the early 1950s. The most interesting works carried out at that time included the studies of the mechanism of photocurrent gain, estimation of the space charge–limited currents (SCLC) [14–16], ascertainment of the peculiarities of the sweep-out of photogenerated carriers in strong electric fields [17–21], screening of electric fields in compensated semiconductors [22], and the study of thermal and generation–recombination noises in photoconductors [18]. Conventional extrinsic photoconductors have only one type of mobile carrier, unlike intrinsic devices. The low impedance of the intrinsic photoconductive devices results in space charge neutrality being relatively easy to maintain, and hence the excess distribution of both electrons and holes moves in one direction under an applied bias. Space charge neutrality can be violated in high-impedance photoconductors, and this can lead to unusual effects as noted later.

The earliest experimental studies have revealed a strong dependence of the parameters of extrinsic photoconductor detectors on the methods of fabrication and characteristics of the contacts, especially at the frequencies that are higher than the inverse of the dielectric relaxation time. The necessity of detailed studies of the mechanisms of this influence including the peculiarities of the behavior of the contacts with compensated semiconductors at low temperatures and backgrounds was emphasized in Sclar's review papers published in the 1980s [4,23–27]. It should be noted that these review papers have been of great importance in the systematization of the data on extrinsic photoconductor detectors, as well as in the popularization and incorporation of these devices into IR optoelectronic equipment.

Considerable progress in understanding the operation of extrinsic photoconductors at low temperatures and backgrounds has been achieved in the last four decades [4,8,28–35]. An interesting historical overview of the theory of nonstationary behavior of low background extrinsic detectors, developed in Russia and ignored elsewhere, is given by Fouks [31]. More comprehensive theoretical discussion is presented by Kocherov and colleagues [33]. It has been shown that the heavy-doped contacts with the compensated semiconductors, being ohmic under the influence of the constant or low frequency–modulated electric fields, become the effective injectors at the frequencies that are higher than the inverse of the trap recharging time. In this case, the injection of charge carriers is controlled by the electric field near the contact and observed with the essentially smaller bias than for SCLC.

The screening length in extrinsic semiconductors is dependent on the frequency. At low frequencies, the screening of electric fields takes place over the distance of the order of the Debye length, which is much smaller than the length of the detector. The screening length greatly increases and can exceed the detector length at the frequencies that are higher than the dielectric relaxation frequency, equal to the inverse of the dielectric relaxation time. In this case, the sweep-out of the photogenerated carriers and photoconductive gain saturation with increase in the electric field takes place.

The frequency dependence of the impedance and photoresponse, transient behavior on an abrupt change of photosignal or voltage, as well as frequency dependence of the spectral density of noise for an extrinsic photoconductive detector were obtained using the new concepts of contact properties [29,31–37]. In addition, it has been shown that the trap recharging resulting from injection of charge carriers leads to unstable behavior of the monopolar plasma of carriers in a compensated semiconductor even with the linear current–voltage characteristic. The leading role of the excessive concentration of the majority carriers in non–steady-state recombination processes has been established, and the mechanism of suppression of the shot noise of the current of the injecting contact at the intermediate frequencies has been found.

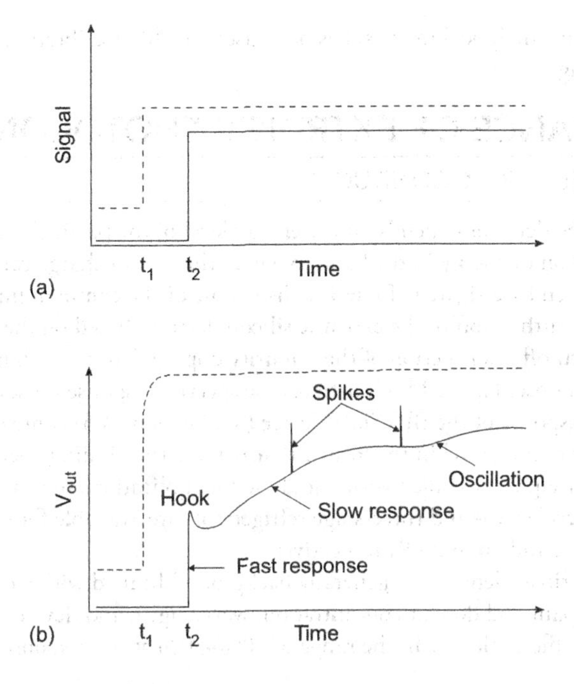

Figure 14.2 Response of an extrinsic photoconductor to two-step input signals: (a) one from a moderate background (dashed line) and the other from a nearly zero one (solid line). (b) The response of the detector. (From Ref. [38])

In the traditional discussion of photoconductivity, the initial rapid response is due to the generation and recombination of charges in the active region of the detector (See Figure 14.2) [38]. For the moderate background, the detector follows the signal reasonably (dashed line). In low background conditions, typical for extrinsic detectors, the necessity to emit a charge carrier to maintain equilibrium is communicated across the detector at roughly the dielectric relaxation time constant, and the detector space charge adjusts to a new configuration at this relatively slow rate. The result of this slow injection of new carriers is a slow adjustment of the electric field near the injecting contact to a new equilibrium under a new level of illumination. Due to this phenomenon, the response shows multiple components, including fast and slow response, a hook anomaly, voltage spikes, and oscillations (solid line).

The hook response artifact (so named because of the appearance of the electric output waveform) can result from the nonuniform illumination of the detector volume in transverse contact detectors; due to shading by the contacts, the portion of light under them is reduced. When illumination is increased, the resistance of the rest of the detector volume is driven down, leaving a high-resistance layer near the contacts for the carriers in most of the electric field and where the recovery to an equilibrium state occurs only over the dielectric relaxation time. As a result, after the initial fast response, the photoconductive gain in the detector area is driven down, and the overall response decreases slowly. Hook response is associated with the quantum mechanical tunneling of charge carriers through the voltage barrier at the contact. Spikes are produced when charges are accelerated in the electric fields near contacts and acquire sufficient energy to impact-ionize impurities in the material, creating a mini-avalanche of charge carriers. Care in the construction of the electrical contacts of detectors can minimize hook response and spiking. In transparent Ge detectors, where the entire detector volume including the areas under the contacts is fully illuminated, the hook behavior is reduced or eliminated [37]. Spiking and hook response are also reduced at a low photoconductive gain (for example, by operating the detector at reduced bias voltage).

Extrinsic photoconductors are relatively well-behaved in a high-background condition that produces a steady concentration of free carriers. This reduces the dielectric relaxation time constant and hence allows the rapid approach to a new equilibrium appropriate for the new signal level. At low backgrounds, however, the signal must be extracted while the detector is still in a nonequilibrium state and the input signal must be deduced from the partial response of the detector. A variety of approaches, including empirically fitted corrections and approximate analytical models, has been used [39,40]. Calibration of data under low backgrounds can be challenging.

14.4 PERFORMANCE OF EXTRINSIC PHOTOCONDUCTORS

14.4.1 SILICON-DOPED PHOTOCONDUCTORS

The spectral response of the detector depends on the energy level of the particular impurity state and the density of states as a function of energy in the band to which the bound charge carrier is excited. A number of other impurities have been investigated. Table 14.1 lists some of the common impurity levels and the corresponding long wavelength cutoff of the extrinsic silicon detector based on them. Note that the exact long wavelength spectral cutoff is a function of the impurity doping density, with higher densities giving slightly longer spectral response. Figure 14.3 illustrates the spectral response for several extrinsic detectors [41]. The longer spectral response of the BIB Si:As device (See Section 14.5) compared with the bulk Si:As device is due to the higher doping level in the former that reduces the binding energy of an electron.

Extrinsic detectors are frequently cooled with liquid He for applications such as ground- and space-based astronomy. Closed-cycle two- and three-stage refrigerators are available for use with these detectors for cooling from 20 to 60 K and 10 to 20 K, respectively.

The performance of extrinsic detectors is generally background limited with a quantum efficiency that varies with the specific dopant and dopant concentration, wavelength, and device thickness (See Section 12.1.2). Typical quantum efficiencies are in the range of 10%–50% at the response peak.

Table 14.1 Common impurity levels used in extrinsic Si IR detectors. Operating temperature depends upon background flux level

IMPURITY	ENERGY (meV)	CUTOFF (μm)	TEMPERATURE (K)
Indium	155	8	40–60
Bismuth	69	18	20–30
Gallium	65	19	20–30
Arsenic	54	23	13
Antimony	39	32	10

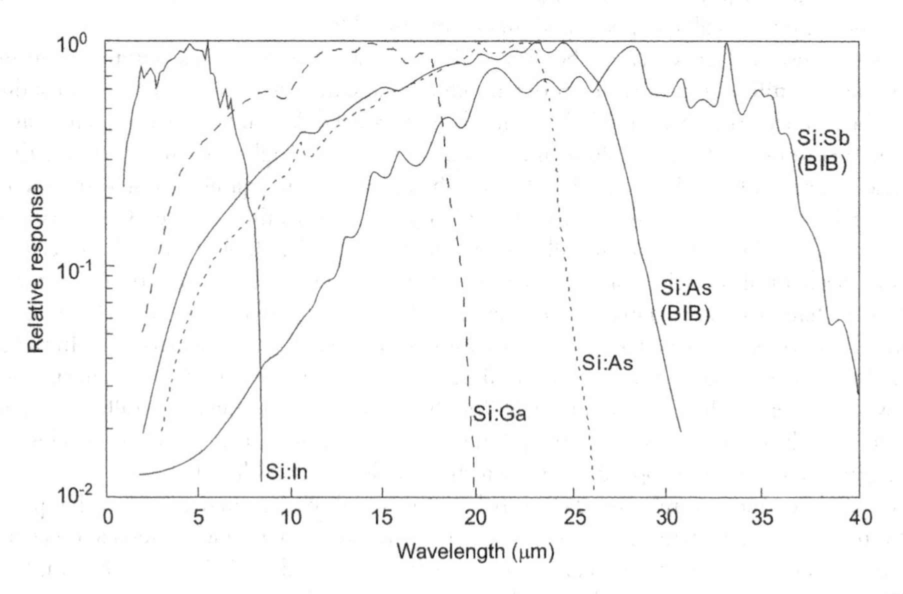

Figure 14.3 Examples of extrinsic silicon detector spectral response. Shown are Si:In, Si:Ga, and Si:As bulk detectors and Si:As and Si:Sb BIB.

For the 3–5 μm range, Si doped with the acceptor In (Si:In) offers an ideal choice. The In hole ground state lies at E_v + 156 meV (E_v is energy at the valence band top) leading to a photoconductive onset at $\lambda \approx 8$ μm, with a peak response at 7.4 μm. Every Si crystal unavoidably contains some residual B acceptors with a hole binding energy of E_v + 45 meV. A further level called In-X that has been linked to an In-C center is located at E_v + 111 meV. Besides these three acceptor levels, there are always P donor levels present. Alexander et al. have evaluated the influence of the temperature and compensation of the shallow levels on detector responsivity [42]. At a very low-temperature operation, when thermal generation from even the shallowest residual acceptor (E_B = 45 meV) is negligible, highest responsivity is always obtained by minimizing the concentration of residual donors. Very high responsivities may be obtained when the shallow acceptors are closely compensated because the effective cross section for capture in these centers is small due to the high probability of thermal reemission, leading to long free carrier lifetimes (as high as 200 ns). Values of 100 AW^{-1} have been obtained in Si:In. The close-to-perfect compensation has been achieved by a doping process using the transmutation of ^{30}Si nuclei into ^{31}P nuclei by thermal-neutron capture followed by β-decay [43,44]. The performance achieved with Si:In detectors is shown in Figure 12.10 and are compared with the theoretical curves.

For the longer wavelength detectors, Ga is chiefly used. The detector material is usually prepared by doping the melt during vertical float zone crystal growth. However, low activation energy (0.074 eV) of Ga is not optimum for use in the 8–14 μm window and unnecessarily contributes to its low operational temperature. Magnesium (Mg) has a more suitable energy level, but according to Sclar, appears to have a shallow 0.044 eV level associated with it, requiring low temperatures or extra compensation [23]. If this level can be eliminated, then perhaps Mg is the ideal 8–14 μm level. Of other detectors, Si:Al and Si:Bi reveal various disadvantages [23]. During heat treatments of Si:Al wafers, Al precipitation and the formation of interstitial Al_2C_3 is frequently experienced, which is detrimental for preparing uniform detector arrays. In the case of Si:Bi, growing uniform crystals encompassing suitable doping concentrations by the float zone method has proved to be difficult because of the high vapor pressure of Bi at growth temperatures. Si:Bi grown by the Czochralski method exhibits higher B concentrations that decrease detector performance.

The comprehensive reviews of Si and Ge detectors operated in various spectral ranges are presented in Sclar's papers [4,23–27]. Table 14.2 summarizes the status of Si detectors for use in the atmospheric windows of 2–2.5 μm, 3–5 μm, and 8–14 μm. The next table (Table 14.3) presents a summary of the properties of doped silicon and doped germanium devices for low-background space applications.

Table 14.2 Performance of Si IR detectors

DETECTOR	$(E_i)_{op}$ (meV)	$(E_i)_{th}$ (meV)	λ_p (μm)	λ_c(T) [μm;(K)]	$\eta(\lambda_p)$ (%)	T_{BLIP} (30FOV;λ_p) [K;(μm)]
Si:Zn(p)		316	2.3	3.2(50–110)	20	103(32)
Si:Tl(p)	246	240	3.5	4.3(78)	>1	
Si:Se(n)	306.7	300	3.5	4.1(78)	24	122
Si:In(p)	156.9	153	5.0	7.4(78)	48	60
Si:Te(n)	198.8	202	4.6	6.3(78)	25	77
Si:S(n)	186.42	174	5.5	6.8(78)	13	78
Si:Se'(n)	205	200	5.5	6.2(78)		85
Si:Ga(p)	74.05	74	15.0	17.8(27)	30(13.5)	32(13.5)
Si:Al(p)	70.18	67	15.0	18.4(29)	6(13.5)	32(13.5)
Si:Bi(n)	70.98	69	17.5	18.7(29)	35(13.5)	32(13.5)
Si:Mg(n)	107.5	108	11.5	12.1–12.4(29)	2(11)	50(11)
Si:S'(n)	109	102	11.0	12.1(5)	<1	55

Source: After Ref. [4].

Table 14.3 Status summary of some Si and Ge IR detectors for low-background applications

DETECTOR	$(\Delta E)_{opt}$ (meV)	λ_p (μm)	λ_c(T) μm(K)	$\eta(\lambda_p)$ (%)	Φ_B ($phcm^{-2}s^{-1}$)	NEP($\lambda;T;f$) ($WHz^{-1/2}$)	λ(μm); T(K); f(Hz)
Si:As	53.76	23	24–24.5(5)	50(T) 20(L)	9×10^6 6.4×10^7	0.88×10^{-17} 4.0×10^{-17}	(19;6;1.6) (23;5;5)
Si:P	45.59	24/26.5	28/29(5)	~30(T)	2.5×10^8	7.5×10^{-17}	(28;4.2;10)
Si:Sb	42.74	28.8	31(5)	58(T) 13(L)	1.2×10^8 1.2×10^8	5.6×10^{-17} 5.5×10^{-17}	(28.8;5;5) (28.8;5;5)
Si:Ga	74.05	15.0	18.4(5)	47(T)	6.6×10^8	1.4×10^{-17}	(15;5;5)
Si:Bi	70.98	17.5	18.5(27)	34(L)	$<1.7 \times 10^8$	3×10^{-17}	(13;11;–)[a]
Ge:Li	9.98	125 (calc)			8×10^8	1.2×10^{-16}	(120;2;13)
Ge:Cu	43.21	23	29.5(4.2)	50	5×10^{10}	1.0×10^{-15}	(12;4.2;1)
Ge:Be[b]	24.81	39	50.5(4.2)	100[b]	1.9×10^{10}	1.8×10^{-16}	(43;3.8;20)
Ge:Ga	11.32	94	114(3)	34	6.1×10^9	5.0×10^{-17}	(94;3;150)
Ge:Ga[b]	11.32	94	114(3)	~100[b]	5.1×10^9	2.4×10^{-17}	(94;3;150)
Ge:Ga[b](s)[c]	~6	150	193(2)	73[b]	2.2×10^{10}	5.7×10^{-17}	(150;2;150)

Source: After Ref. [4].

T and L indicate transverse and longitudinal geometry detectors, respectively.

[a] Signal integrated for 1 s.

[b] Results obtained with an integrating cavity.

[c] (s): stress = 6.6×10^3 $kgcm^{-2}$.

14.4.2 GERMANIUM-DOPED PHOTOCONDUCTORS

Germanium extrinsic detectors have largely been supplanted by silicon detectors, as discussed previously, for both high and low background applications where the comparable spectral response can be obtained, but germanium devices are still of interest for very long wavelengths. Germanium photoconductors have been used in a variety of IR astronomical experiments, both airborne and space-based. An example of the space application is ISOPHOT, the photometer for ESA's infrared space observatory, which uses extrinsic photoconductors at wavelengths ranging from 3 to more than 200 μm [45]. The highly successful IRAS mission marks the beginning of modern, far-infrared photoconductor research and development [46]. Very shallow donors, such as Sb, and acceptors, such as B, In, or Ga, provide cutoff wavelengths in the region of 100 μm. Figure 14.4 shows the spectral response of the extrinsic germanium photoconductors doped with Zn, Be, Ga, and of stressed gallium-doped germanium [30]. Despite a large amount of effort recently in the development of very sensitive thermal detectors, germanium photoconductors remain the most sensitive detectors for wavelengths shorter than 240 μm.

The achievement of low noise equivalent power (NEP) values in the range of a few parts 10^{-17} $WHz^{-1/2}$ (See Table 14.3) was made possible by advances in crystal growth development and by controlling the residual minority impurities down to 10^{10} cm^{-3} in a doped crystal [13,47]. As a result, a high lifetime and mobility value and thus a higher photoconductive gain have been obtained.

Ge:Be photoconductors cover the spectral range from ≈30 to 50 μm. Beryllium, a double acceptor in Ge with energy levels at $E_v + 24.5$ meV and $E_v + 58$ meV, poses special doping problems because of its strong oxygen affinity. The Be doping concentrations of $5 \times 10^{14} - 1 \times 10^{15}$ cm^{-3} give significant photon absorption in 0.5–1 mm thick detectors, while at the same time keeping dark currents caused by hopping conduction at levels as low as a few tens of electrons per second. Ge:Be detectors with responsivities >10 AW^{-1} at $\lambda = 42$ μm and quantum efficiency 46% have been reported at low background [13].

Ge:Ga photoconductors are the best low background photon detectors for the wavelength range from 40 to 120 μm. Since the absorption coefficient for a material is given by the product of the photoionization cross section and the doping concentration (See Equation 12.58), it is generally desirable to maximize this concentration. The practical limit occurs when the concentration is so high that impurity band conduction results in an excessive dark current. For Ge:Ga, the onset of impurity banding occurs at approximately 2×10^{14} cm^{-3}, resulting in an absorption coefficient of only 2 cm^{-1} and typical values of quantum efficiency range from 10% to 20% [48]. Consequently, the detectors must either have long physical absorption path lengths or be mounted in an integrating cavity. Table 14.4 gives characteristic parameters for the Ge:Ga detectors [49].

Application of uniaxial stress along the (100) axis of Ge:Ga crystals reduces the Ga acceptor binding energy, extending the cutoff wavelength to ≈240 μm [50,51]. In making practical use of this effect, it is essential to apply and maintain very uniform and controlled pressure to the detector so that the entire

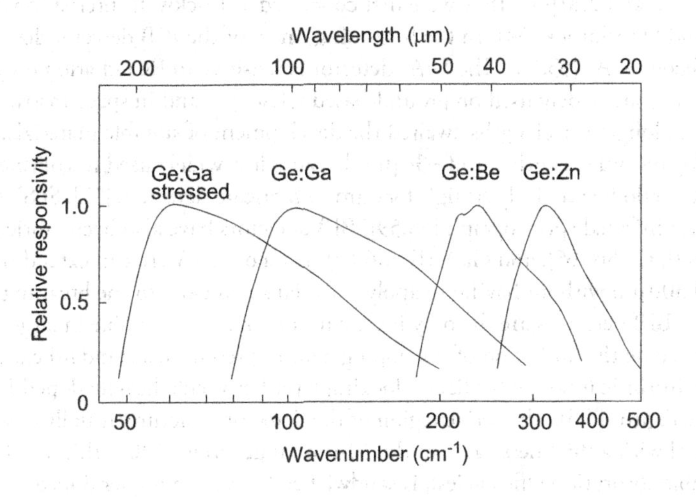

Figure 14.4 Relative spectral response of some germanium extrinsic photoconductors. (From Ref. [30])

Table 14.4 Typical Ge:Ga detector parameters

PARAMETER	VALUE
Acceptor concentration	$2 \times 10^{14} cm^{-3}$
Donor concentration	$<1 \times 10^{11} cm^{-3}$
Typical bias voltage	$50 mV mm^{-1}$
Operating temperature	<1.8 K
Responsivity	$7 AW^{-1}$
Quantum efficiency	20%
Dark current	$<180 es^{-1}$

Source: After Ref. [49].

detector volume is placed under stress without exceeding its breaking strength at any point. A number of mechanical stress modules have been developed. The stressed Ge:Ga photoconductor systems have found a wide range of astronomical and astrophysical applications [45,49,52,53].

14.5 BLOCKED IMPURITY BAND DEVICES

From Section 12.1.2 it is evident that so as to maximize the quantum efficiency and detectivity of extrinsic photoconductors, the doping level should be as high as possible. This is particularly important when the devices are required to be radiation hard and are made as thin as possible to minimize the absorbing volume for ionizing radiation. The limit to the useful doping that is possible in conventional extrinsic detectors is set by the onset of impurity banding. This occurs when the doping level is sufficiently high that the wave functions of neighboring impurities overlap and their energy level is broadened to a band that can support hopping conduction. When this occurs, it limits the detector resistance and photoconductive gain and also increases the dark current and noise. In Si:As, for example, these effects become important for doping levels above $7 \times 10^{16} cm^{-3}$. To overcome the impurity banding effect and, in addition, to improve radiation hardness and reduce the optical cross-talk between adjacent elements of an array, the BIB device was proposed. The BIB detectors, also called the IBC detectors, have demonstrated other significant advantages, such as freedom from the irregular behavior typical of photoconductive detectors (spiking, anomalous transient response), increased frequency range for constant responsivity, and superior uniformity of response over the detector area and from detector to detector.

The BIB devices made from either doped silicon and doped germanium are sensitive to IR wavelength range located between 2 and 220 μm. They were first conceived at Rockwell International Science Center in 1977 by Petroff and Stapelbroek [54]. In the beginning, most of the BIB detector development centered on arsenic-doped silicon, Si:As [55,56]. The Si:As detector is sensitive to IR radiation only in the 2–30 μm wavelength range—they are widely used on ground-based telescopes and in space instruments. Extension of BIB performance to longer wavelengths awaited the development of suitable materials. Phosphorus is another attractive dopant (wavelength cutoff ~34 μm) because it is widely used in commercial integrated circuits and hence it would be relatively straightforward to fabricate detectors [57]. Si:Sb together with Si:As are installed in Spitzer infrared spectrograph [58,59]. BIB detectors have also been fabricated with Ge:Ga [60–62], Ge:B [63,64], Ge:Sb [65], and GaAs:Te [66,67]. Data on GaAs:Te can extend the detector spectral approach to beyond 300 μm without having to apply uniaxial stress close to the breaking limit of detector's pixel. However, Si:As BIB detectors are the only form currently readily available in large format arrays.

BIB detectors overcome the limitation of the doping density present in a standard extrinsic photoconductor by placing a thin intrinsic (undoped) silicon blocking layer between a heavily doped IR active layer and a planar contact (See Figure 14.5). The active region of the detector structure, usually based on epitaxially grown n-type material with a thickness value in the 10-μm range (about 100 × thinner than bulk photoconductors for comparable absorption efficiencies), is sandwiched between a higher doped degenerate substrate electrode and an undoped blocking layer. Doping of the active layer is high enough (typically 100 × greater

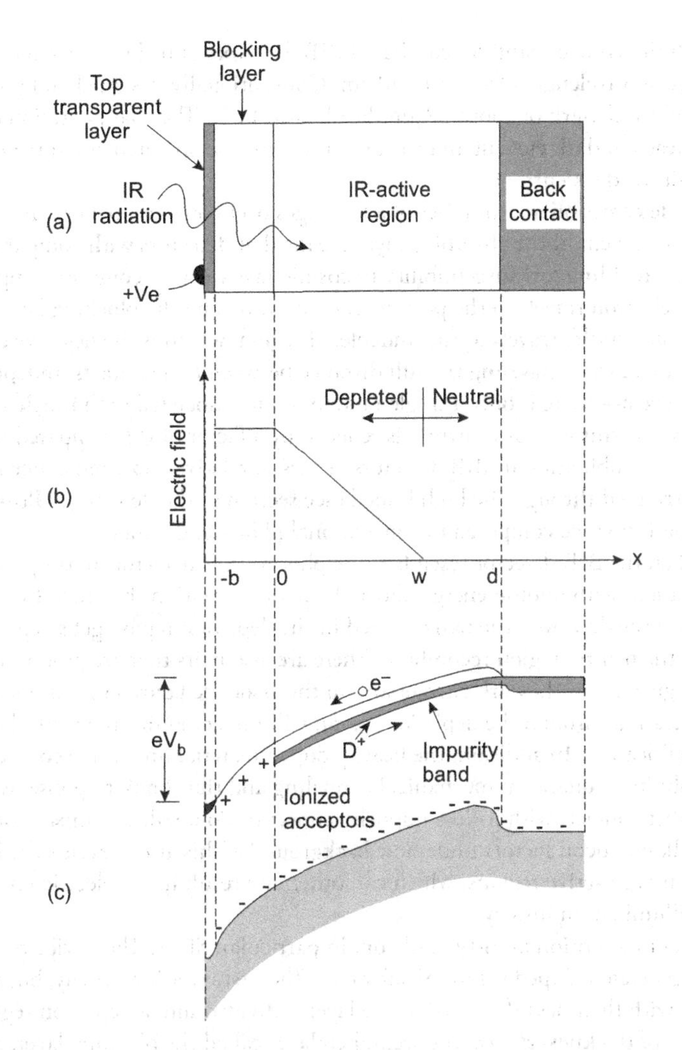

Figure 14.5 Blocked impurity band detector: (a) cross section, (b) electric field, and (c) energy band diagram of a positively biased detector.

than possible for bulk photoconductors) for the onset of an impurity band to display a high quantum efficiency for impurity ionization (in the case of Si:As BIB, the active layer is doped to ≈ 5×10^{17}cm^{-3}). The heavily doped n-type IR-active layer has a small concentration of negatively charged compensating acceptor impurities ($N_a \approx 10^{13}$cm^{-3}). In the absence of an applied bias, charge neutrality requires an equal concentration of ionized donors. Whereas the negative charges are fixed at acceptor sites, the positive charges associated with ionized donor sites (D^+ charges) are mobile and can propagate through the IR-active layer via the mechanism of hopping between occupied (D^0) and vacant (D^+) neighboring sites. A positive bias to the transparent contact creates a field that drives the preexisting D^+ charges toward the substrate, while the undoped blocking layer prevents the injection of new D^+ charges. A region depleted of D^+ charges is therefore created, with a width depending on the applied bias and on the compensating acceptor concentration.

BIB detectors effectively use the hopping conductivity associated with "impurity banding" in relatively heavily doped semiconductors. Because of the presence of the blocking layer, BIB detectors do not follow the usual photoconductor model. The behavior of BIB detectors is closer to that of a reverse-biased photodiode except that the photoexcitation of electrons occurs from the donor impurity band to the conduction band. Due to the heavy doping of the active layer, the impurity band increases in width, therefore effectively decreasing the energy gap between the impurity band and the conduction band. As a result, BIB detectors typically offer spectral responses extended toward longer wavelengths compared to bulk-type photoconductors with the same dopant. Silicon-based BIB detectors doped with arsenic and antimony have the materials of choice at wavelengths from 5 to 40 μm. Conventionally designed and processed Si:As BIB detectors have a cutoff wavelength of about 28 μm [See Figure 14.3]. Thermal excitation of electrons across the narrow bandgap leads to dark current and the detectors must be operated at temperatures sufficiently low ($T < 13$ K) to limit the dark current.

The design of BIB detectors offers a number of advantages over conventional extrinsic photoconductors. The high absorption coefficient of the absorbing layer means that detectors with comparatively small active volumes can be made, providing low susceptibility to cosmic rays without compromising quantum efficiency. Moreover, the electron travels to the positive contact traversing the blocking layer while the positive donor state (not the donor itself) travels to the undepleted region, which is the negative contact. The net result is equivalent to one carrier traveling the full distance between two contacts, independent of the location of the ionization event—there is only a single random event associated with the detection of a single photon. Consequently, the rms g-r noise current is reduced by a factor of $\sqrt{2}$ compared with conventional photoconductors (the recombination in BIB detectors occurs in relatively low resistance material and is not a random process distributed through the high impedance section of the detector). Ultimately, BIB devices provide better noise performance compared to conventional photoconductors.

As mentioned earlier, the BIB detector resembles the photovoltaic detector in its operation. Pairs are created when IR radiation, with photon energy above the photoionization threshold for the chosen donor impurity, is incident on the detector. The pairs created in the depletion region get swept out by its electric field; pairs created in the neutral region recombine. There are also pairs that are produced thermally in the depletion region that give rise to the dark current and to the associated dark current noise. For large electric fields, electrons accelerating through the depletion region will impact ionize the neutral donors producing internal gain amplification. In addition, the heavily doped material under one contact of BIB detectors reduces the possibility of quantum mechanical tunneling and thus hook response, while the intrinsic material under the other contact, with its low impurity concentration, reduces impact ionization and thus spiking relative to bulk photoconductors under low backgrounds. This also prevents the build of large space charge regions in the n-type active regions, which can otherwise result in a dielectric relaxation response that depends on the illumination history.

We will now focus our attention on n-type silicon, in particular, Si:As. The device consists of two layers deposited on a degenerately doped n-type Si substrate. The first layer is a heavily, but not degenerately, doped IR active layer with thickness d. The IR active layer is divided into a depletion region of thickness w and a neutral region of thickness $d - w$. The second epilayer called the blocking layer, of thickness b, is intrinsic or, at most, lightly doped. Finally, a shallow n^+ implant transparent contact for incoming IR radiation is put on top of the blocking layer.

In the IR active layer, there is a small concentration of acceptors (boron), which are assumed to be totally compensated and hence all ionized, $N_a^- = N_a\left(N_a \approx 10^{13}\,\text{cm}^{-3}\right)$. In the absence of an applied bias, charge neutrality requires an equal concentration of ionized donors, $N_d^+ = N_a$. The negative charges associated with the fixed acceptor sites are immobile, whereas the positive charges associated with ionized donor sites are mobile. Because of the heavy doping concentration of donors, there is a high probability of donor charge hopping from one site to another (called the impurity band effect). Electrons associated with the N_d^0 neutral sites hop to the vacant N_d^+ sites, which can be viewed as "holes" in the impurity band moving in the opposite direction. Applying a positive bias to the transparent contact, the N_d^+ charges are swept out through the IR active layer away from the interface with the blocking layer, while the undoped blocking layer prevents the injection of new N_d^+ charges. A region depleted of N_d^+ charges is therefore created. Since the ionized acceptor charges are not mobile, a negative space charge is left in the depletion region. The electric field is largest in the blocking layer and decreases linearly with the distance in the IR-active layer (See Figure 14.5b) in accordance with the Poisson equation. If the blocking layer is taken ideally as intrinsic (is devoid of all impurities), the width, w, of the depletion region is given by

$$w = \left[\frac{2\varepsilon_0\varepsilon_s}{qN_a}V_b + b^2\right]^{1/2} - b, \qquad (14.1)$$

where V_b is the applied bias voltage. The width of the depletion region defines the active region of the device since an appropriate electric field exists only in this region.

Assuming $N_a = 10^{13}\,\text{cm}^{-3}$, $V_b = 4\,\text{V}$, and $b = 4\,\mu\text{m}$, we derive $w = 19.2\,\mu\text{m}$. For a typical donor concentration $N_d = 5 \times 10^{17}\,\text{cm}^{-3}$, it can be seen that $\sigma_i N_d w = 2.12$, and the absorptive quantum efficiency will be about 88% for a single pass, or as high as 98% for a double pass in a detector with the reflective back surface. The quantum efficiency will increase with increasing V_b until $w \geq d$, where d is the thickness of the IR-active layer. Assuming that $w = d$ and typical values of other parameters, the electric field near the blocking layer is large, ≈2,800 Vcm^{-1}, far larger than the bulk conventional photoconductors [38].

The mean free path of the electrons in this large region is ≈0.2 µm. An electron that is accelerated across this path by a field of 2,800 Vcm^{-1} will acquire an energy of 0.056 eV. The excitation energy for Si:As is 0.054 eV (corresponding λ_c of 23 µm); therefore, 0.054 eV of energy will be adequate to ionize the neutral arsenic impurity atoms, resulting in two conduction band electrons. Repeated collisions can lead to significant gain and, because this is a statistical process, additional noise due to gain dispersion that can be characterized by an excess noise factor, β. This factor exceeds unity at higher bias voltages, and the detector therefore operates with increased noise. The product of the quantum efficiency η and the gain g gives the quantum yield that is proportional to the responsivity $R = (\lambda\eta/hc)qg$ (See Equation 4.31). The detective quantum efficiency is defined as the ratio of the quantum efficiency to the excess noise factor η/β. Under background-limited conditions, the signal-to-noise ratio is proportional to the square root of η/β. For a more detailed analysis of the BIB detector, see Szmulowicz and Madarsz [68].

DRS Technologies has demonstrated a longer cutoff wavelength of the BIB detectors [See Figure 14.6]. The far-infrared extended BIB detectors are fabricated by a further increase in donor doping. Average overlap of donor wave functions becomes large, resulting in the impurity conduction band becoming wider and the gap between the impurity band and the conduction band becoming narrower. The narrowed bandgap results in longer wavelength response. The points according to the two models are represented in Figure 14.6. The proposed broadband detector in the 10–50 -µm (goal 3–100 -µm) is based on the As-doped Si BIB detectors to operate at 10–12 K.

The details of the arrangement of steps in detector preparation depend on whether the detector is to be front illuminated through the first contact and blocking layer or back illuminated through the second contact. These two geometries are illustrated in Figure 14.7 [38]. Fabrication of a back-illuminated BIB detector begins with an ion-implanted buried electrical contact. The relatively highly doped IR-active layer and undoped blocking layer are grown epitaxially over the buried contact. Detector elements (pixels in

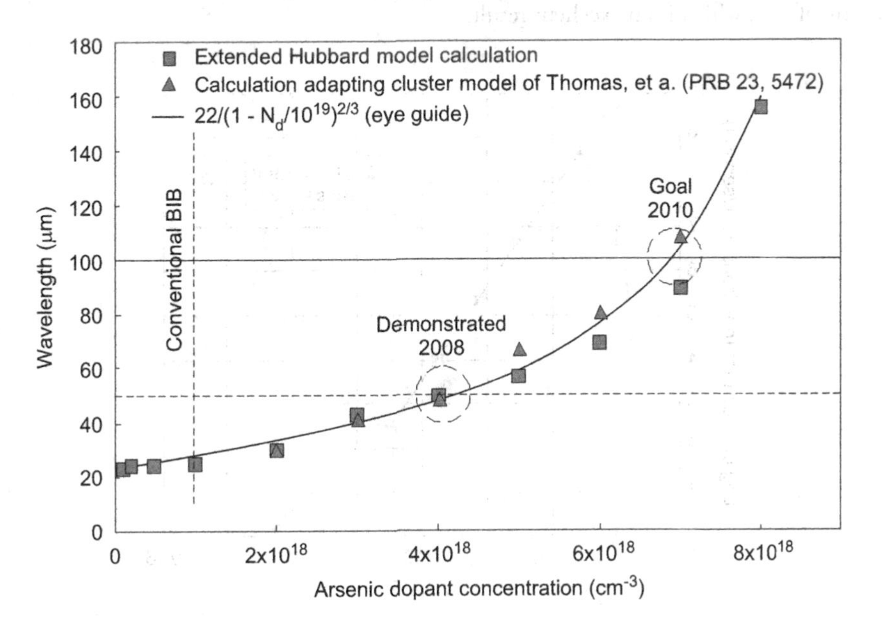

Figure 14.6 Cutoff wavelength vs. donor doping for Si:As BIB detectors (after Ref. [69])

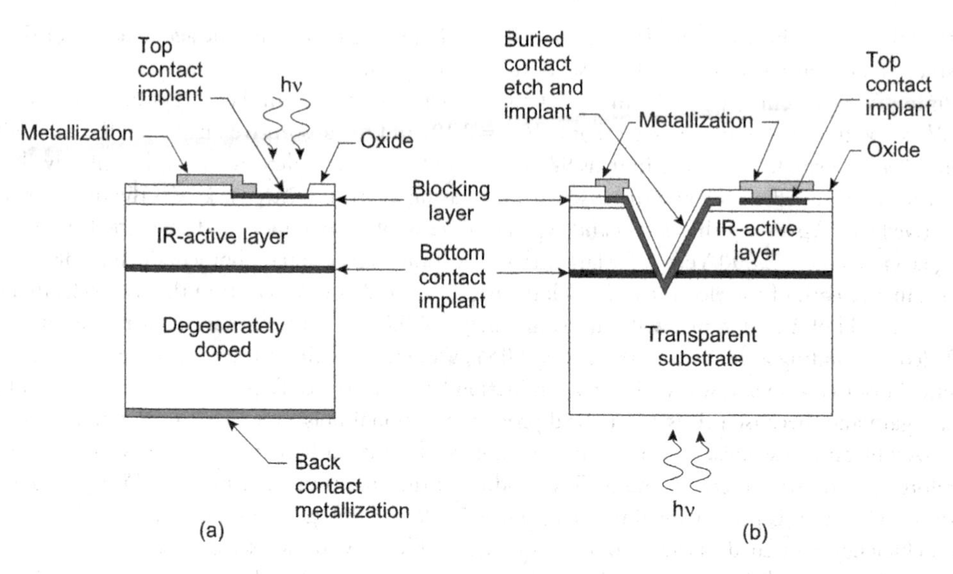

Figure 14.7 (a) Front-illuminated BIB detector compared with (b) a back-illuminated detector. (From Ref. [38])

array) are patterned and defined by another ion implant to form the second contact. The detector structures are completed by using standard silicon microlithographic processing to delineate metal lines, provide for electrical connection to the buried contact, and form indium bumps. In the front-illuminated detector, a transparent contact is implanted into the blocking layer and the second contact is made by growing the detector on an extremely heavily doped, electrically conducting (degenerate) substrate. In the latter case, a thin degenerate but transparent contact layer is grown underneath the active layer on a high purity, transparent substrate. The bias voltage on the buried contact is established through a V-shaped etched trough, metal-coated to make it conductive, and placed to one side of the detector.

Typically, the required operating temperature of Si:As BIB detectors is somewhere between 6–10 K. Figure 14.8 gives the Arrhenius plot for two reverse-bias voltages showing the exponential dependence of the thermal activation. The dark current is shown to be approximately 1 electron/pixel/second at an operating temperature of 7 K, which is an excellent result.

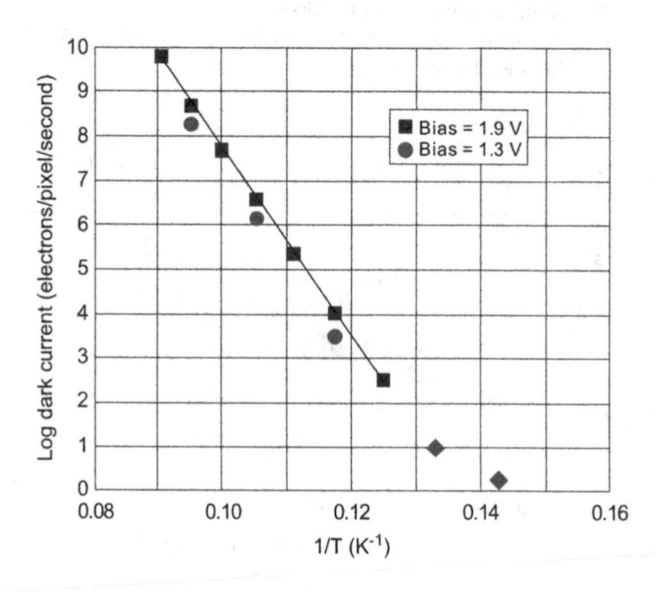

Figure 14.8 Typical Arrhenius plot of dark current for two different reverse bias voltages (after Ref. [70])

The great success achieved with Si:As BIB arrays has so far not been carried over successfully to the Ge BIB devices. The art and the science of Ge epitaxy were not nearly as far developed as for Si and it was also very different in many aspects. Several technological hurdles exist, including the very limited experience with Ge epitaxy and the non-existence of high purity Ge epitaxial growth techniques. Structural defects and impurities continue to affect detector performance.

The fabrication of a Ge BIB detector requires the use of a high-purity epitaxial technique. A different approach to fabricate these detectors that are based on the use of ultrapure Ge doped with ion implantation has led to functional devices that show low leakage currents but exhibit low responsivity because of the rather thin implanted layer [63]. It appears that liquid phase epitaxy has many advantages over MBE and CVD techniques (lower growth temperature, lower dopant level) for the growth of epitaxial Ge layers for BIB detectors [71,72]. It can be expected that continued development efforts together with some new ideas will lead to functional Ge BIB detectors that can be used in low background astronomical and astrophysical observations. Recently, Wada et al. [62] have successfully demonstrated Ge BIB detectors fabricated by the surface-activated bond technology whose cutoff wavelengths are longer than 160 μm without any stressing mechanism. Table 14.5 shows the characteristics of these devices fabricated.

For optimal GaAs BIB operation, unintentional doping of the blocking layer must be below 10^{13} cm^{-3}, but the requirements for the absorbing layer are more challenging. Compared to elemental Ge and Si semiconductors, one encounters difficulties that are due to the nature of GaAs. Any deviation from perfect stoichiometry leads to large concentrations of native defects. These defects are typically charged and form deep centers. In the interest of the highest possible purity, liquid phase epitaxy (LPE) technique is preferred. A reproducible growth of GaAs LPE layers with residual doping below 10^{12} cm^{-3} has been demonstrated [73].

14.6 SOLID-STATE PHOTOMULTIPLIERS

The SSPM is closely related to the BIB detector and is illustrated schematically in Figure 14.9 [74]. Successful operation of SSPM requires optimization of a number of parameters, including not only the construction of the detector but also its bias voltage and operating temperature. The internal structure of the detector is similar to a Si:As BIB except that a well-defined gain region is grown between the blocking layer and the IR-absorbing layer with higher compensating acceptor level ($5 \times 10^{13} - 1 \times 10^{14}$ cm^{-3}) and a thickness of about 4 μm. Typically, the IR-absorbing region has a lower acceptor concentration. When the detector is properly biased, the gain region goes into depletion and a strong electric field is developed across it as with the BIB detector. A higher electric field falls from about 8,000 Vcm^{-1} in the blocking layer over

Table 14.5 Characteristics of the Ge BIB devices fabricated by the surface-activated bond technology

Device width and length	1×1 mm^2
IR absorption layer	
Thickness	0.5 mm
Dopant concentration	10^{16} cm^{-3} (Ga)
Blocking layer	
Thickness	0.5 mm
Carrier concentration	$<4 \times 10^{12}$ cm^{-3} (p-type)
Device performance	
Dark current	<5 fA
Responsivity	2 AW^{-1}
Cutoff wavelength	160 μm
NEP	$<2 \times 10^{-17}$ WHz$^{-1/2}$

Source: After Ref. [62].

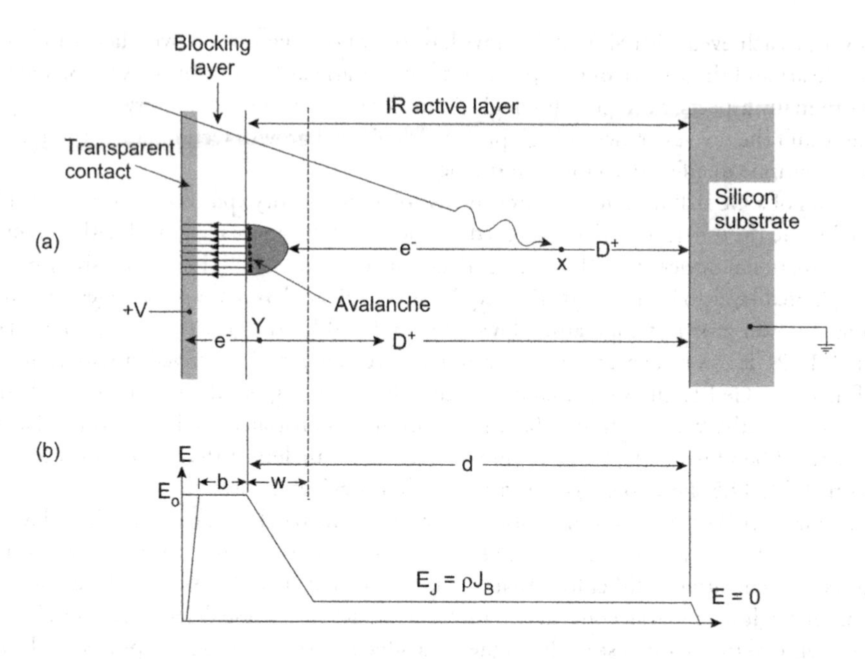

Figure 14.9 (a) Solid-state photomultiplier configuration with a schematic representation of the generation of bias current and effects following absorption of a photon, and (b) electric field profile. (From Ref. [74])

a depletion width in the IR-active layer. Over part of this depletion layer (4-μm thick), the electric field exceeds the critical field for impact ionization of neutral donors (≈ 2,500 Vcm^{-1}). A larger field than with a normal BIB detector increases the amount of avalanching. To the right of the depletion region is a uniform-field (≈ 1,000 Vcm^{-1}) drift region, ≈ 25 μm thick (See Figure 14.9b).

As in the case of BIB detectors, the positive charges associated with ionized donor sites (D^+ charges) are mobile. An electron-D^+ pair created by an IR photon (or by thermal generation) at point x in Figure 14.10a is separated by the electrical field $E_J = \rho J_B$ (where J_B is the bias current, and ρ is the resistivity of the IR-active layer). The electron drifts rapidly to the left, with a negligible probability for impact ionization while in the low-field region, and the D^+-charge drifts more slowly to the right. An avalanche of well-defined gain ($M \approx 4 \times 10^4$ at V = 7 V and T = 77 K) occurs for each electron entering the steeply increasing electric field region adjacent to the blocking layer [74].

The SSPM devices are capable of detecting continuously the individual photons in the wavelength range from 0.4 to 28 μm [74], indicating the applicability of these devices to astronomical applications. The output pulses have submicrosecond rise times and amplitudes well above the readout noise. A counting quantum efficiency of over 30% was observed at 20-μm wavelength and over 50% in the visible light region. Optimum photon-counting performance occurs for temperatures between 6 and 10 K for count rates less than 10^{10} counts/cm^{2}s of detector area.

The performance of the SSPM is described in detail by Hays et al. [75].

REFERENCES

1. B. V. Rollin and E. L. Simmons, "Long wavelength infrared photoconductivity of silicon at low temperatures," *Proc. Phys. Soc.*, B65, 995–6, 1952.
2. E. Burstein, J. J. Oberly, and J. W. Davisson, "Infrared photoconductivity due to neutral impurities in silicon," *Phys. Rev.*, 89(1), 331–2, 1953.
3. B. V. Rollin and E. L. Simmons, "Long wavelength infrared photoconductivity of silicon at low temperatures," *Proc. Phys. Soc.*, B66, 162–68, 1953.
4. N. Sclar, "Properties of doped silicon and germanium infrared detectors," *Prog. Quantum Electron.*, 9, 149–257, 1984.

5. C. R. McCreight, M. E. McKelvey, J. H. Goebel, G. M. Anderson, and J. H. Lee, "Detector arrays for low-background space infrared astronomy," *Laser Focus/Electro-Opt.*, 22, 128–33, November 1986.
6. F. J. Low, C. A. Beichman, F. C. Gillett, J. R. Houck, G. Neugebauer, D. E. Langford, R. G. Walker, and R. H. White, "Cryogenic telescope on the infrared astronomical satellite (IRAS)," *Proc. SPIE* 430, 288–96, 1983.
7. N. W. Boggess, "NASA space programs in infrared astronomy," *Laser Focus/Electro-Opt.*, 116–27, June 1984.
8. G. H. Rieke, "Infrared detector arrays for astronomy," *Ann. Rev. Astron. Astrophys.*, 45, 77–115, 2007.
9. T. Sprafke and J. W. Beletic, "High-performance infrared focal plane arrays for space applications," *OPN* 22–7, June 2008.
10. B. Starr, L. Mears, C. Fulk, J. Getty, E. Beuville, R. Boe, C. Tracy, E. Corrales, S. Kilcoyne, J. Vampola, J. Drab, R. Peralta, C. Doyle, "RVS large format arrays for astronomy," *Proc. SPIE* 9915, 99152X-1–14, 2016.
11. O. Diazovski, "Focal plane arrays for optical payloads," in *Optical Payloads for Space Missions*, ed. S.-E. Qian, 793–837, John Wiley & Sons, West Sussex, UK, 2016.
12. P. R. Bratt, "Impurity germanium and silicon infrared detectors," in *Semiconductors and Semimetals*, Vol. 12, eds. R. K. Willardson and A. C. Beer, 39–142, Academic Press, New York, 1977.
13. E. E. Haller, "Advanced far-infrared detectors," *Infrared Phys. Technol.*, 35, 127–46, 1994.
14. A. Rose, "Space-charge-limited currents in solids," *Phys. Rev.*, 97, 1538–44, 1955.
15. M. A. Lampert, "Simplified theory of space-charge-limited currents in an insulator with traps," *Phys. Rev.*, 103, 1648–56, 1956.
16. M. A. Lampert and P. Mark, *Current Injections in Solids*, Academic Press, New York, 1970.
17. A. Rose, "Performance of photoconductors," *Proc. IRE* 43, 1850–69, 1955.
18. A. Rose, *Concepts in Photoconductivity and Allied Problems*, Interscience, New York, 1963.
19. R. L. Williams, "Response characteristics of extrinsic photoconductors," *J. Appl. Phys.*, 40, 184–90, 1969.
20. A. F. Milton and M. M. Blouke, "Sweepout and dielectric relaxation in compensated extrinsic photoconductors," *Phys. Rev.*, 3B, 4312–30, 1971.
21. M. M. Blouke, E. E. Harp, C. R. Jeffus, and R. L. Williams, "Gain saturation in extrinsic photoconductors operating at low temperatures," *J. Appl. Phys.*, 43, 188–94, 1972.
22. S. M. Ryvkin, *Photoelectric Effects in Semiconductors*, Consultants Bureau, New York, 1964.
23. N. Sclar, "Extrinsic silicon detectors for 3–5 and 8–14 μm," *Infrared Phys. Technol.*, 16, 435–48, 1976.
24. N. Sclar, "Survey of dopants in silicon for 2–2.7 and 3–5 μm infrared detector application," *Infrared Phys. Technol.*, 17, 71–82, 1977.
25. N. Sclar, "Temperature limitation for IR extrinsic and intrinsic photodetectors," *IEEE Trans. Electron Devices* ED-27, 109–18, 1980.
26. N. Sclar, "Development status of silicon extrinsic IR detectors," *Proc. SPIE* 409, 53–61, 1983.
27. N. Sclar, "Development status of silicon extrinsic IR detectors," *Proc. SPIE* 443, 11–41, 1984.
28. D. K. Schroder, "Extrinsic Silicon Focal Plane Arrays," in *Charge-Coupled Devices*, ed. D. F. Barbe, 57–90, Springer-Verlag, Heidelberg, Germany, 1980.
29. R. W. Westervelt and S. W. Teitsworth, "Nonlinear transient response of extrinsic Ge far-infrared photoconductors," *J. Appl. Phys.*, 57, 5457–69, 1985.
30. J. Leotin, "Far infrared photoconductive detectors," *Proc. SPIE* 666, 81–100, 1986.
31. B. I. Fouks, "Nonstationary Behaviour of Low Background Photon Detectors," in *Proceedings of ESA Symposium on Photon Detectors for Space Instruments*, pp. 167–74, 1992.
32. N. M. Haegel, C. A. Latasa, and A. M. White, "Transient response of infrared photoconductors: The roles of contacts and space charge," *Appl. Phys.*, A 56, 15–21, 1993.
33. V. F. Kocherov, I. I. Taubkin, and N. B. Zaletaev, "Extrinsic Silicon and Germanium detectors," in *Infrared Photon Detectors*, ed. A. Rogalski, 189–297, SPIE Optical Engineering Press, Bellingham, WA, 1995.
34. N. M. Haegel, C. B. Brennan, and A. M. White, "Transport in extrinsic photoconductors: A comprehensive model for transient response," *J. Appl. Phys.*, 80, 1510–14, 1996.
35. N. M. Haegel, J. C. Simoes, A. M. White, and J. W. Beeman, "Transient behavior of infrared photoconductors: Application of a numerical model," *Appl. Opt.*, 38, 1910–19, 1999.
36. A. Abergel, M. A. Miville-Deschenes, F. X. Desert, M. Perault, H. Aussel, and M. Sauvage, "The transient behaviour of the long wavelength channel of ISOCAM," *Exp. Astron.*, 10, 353–68, 2000.
37. N. M. Haegel, W. R. Schwartz, J. Zinter, A. M. White, and J. W. Beeman, "Origin of the hook effect in extrinsic photoconductor," *Appl. Opt.*, 34, 5748–54, 2001.
38. G. H. Rieke, *Detection of Light: From the Ultraviolet to the Submillimeter*, 2nd ed., Cambridge University Press, Cambridge, 2003.

39. A. Coulais, B. I. Fouks, J.-F. Giovanelli, A. Abergel, and J. See, "Transient response of IR detectors used in space astronomy: What we have learned from the ISO satellite," *Proc. SPIE* 4131, 205–17, 2000.
40. J. Schubert, B. I. Fouks, D. Lemke, and J. Wolf, "Transient response of ISOPHOT Si:Ga infrared photodetectors: Experimental results and application of the theory of nonstationary processes," *Proc. SPIE* 2553, 461–9, 1995.
41. P. R. Norton, "Infrared image sensors," *Opt. Eng.*, 30, 1649–63, 1991.
42. D. H. Alexander, R. Baron, and O. M. Stafsudd, "Temperature dependence of responsivity in closely compensated extrinsic infrared detectors," *IEEE Trans. Electron Devices* ED-27, 71–77, 1980.
43. R. N. Thomas, T. T. Braggins, H. M. Hobgood, and W. J. Takei, "Compensation of residual boron impurities in extrinsic indium-doped silicon by neutron transmutation of silicon," *J. Appl. Phys.*, 49, 2811–20, 1978.
44. H. M. Hobgood, T. T. Braggins, J. C. Swartz, and R. N. Thomas, "Role of Neutron Transmutation in the Development of High Sensitivity Extrinsic Silicon IR Detector Material," in *Neutron Transmutation in Semiconductors*, ed. J. M. Meese, 65–90, Plenum Press, New York, 1979.
45. J. Wolf, C. Gabriel, U. Grözinger, I. Heinrichsen, G. Hirth, S. Kirches, D. Lemke, et al., "Calibration facility and preflight characterization of the photometer in the infrared space observatory," *Opt. Eng.*, 33, 26–36, 1994.
46. G. H. Rieke, M. W. Werner, R. I. Thompson, E. E. Becklin, W. F. Hoffmann, J. R. Houck, F. J. Low, W. A. Stein, and F. C. Witteborn, "Infrared astronomy after IRAS," *Science* 231, 807–14, 1986.
47. N. M. Haegel and E. E. Haller, "Extrinsic germanium photoconductor material: Crystal growth and characterization," *Proc. SPIE* 659, 188–94, 1986.
48. J.-Q. Wang, P. I. Richards, J. W. Beeman, J. W. Haegel, and E. E. Haller, "Optical efficiency of far-infrared photoconductors," *Appl. Opt.*, 25, 4127–34, 1986.
49. E. Young, J. Stansberry, K. Gordon, and J. Cadien, "Properties of germanium photoconductor detectors," in *Proceedings of the Conference of ESA SP-481*, eds. L. Metcalfe, A. Salama, S. B. Peschke, and M. F. Kessler, 231–5, VilSpa, European Space Agency, 2001.
50. A. G. Kazanskii, P. L. L. Richards, and E. E. Haller, "Far-infrared photoconductivity of uniaxially stressed germanium," *Appl. Phys. Lett.*, 31, 496–7, 1977.
51. E. E. Haller, M. R. Hueschen, and P. L. Richards, "Ge:Ga photoconductors in low infrared backgrounds," *Appl. Phys. Lett.*, 34, 495–7, 1979.
52. N. Hiromoto, M. Fujiwara, H. Shibai, and H. Okuda, "Ge:Ga far-infrared photoconductors for space applications," *Jpn. J. Appl. Phys.*, 35, 1676–80, 1996.
53. Y. Doi, S. Hirooka, A. Sato, M. Kawada, H. Shibai, Y. Okamura, S. Makiuti, T. Nakagawa, N. Hiromoto, and M. Fujiwara, "Large-format and compact stressed Ge:Ga array for the astro-F (IRIS) mission," *Adv. Space Res.*, 30, 2099–104, 2002.
54. M. D. Petroff and M. G. Stapelbroek, "Blocked Impurity Band Detectors," *U.S. Patent*, No. 4 568 960, filed October 23, 1980, granted February 4, 1986.
55. S. B. Stetson, D. B. Reynolds, M. G. Stapelbroek, and R. L. Stermer, "Design and performance of blocked-impurity-band detector focal plane arrays," *Proc. SPIE* 686, 48–65, 1986.
56. D. B. Reynolds, D. H. Seib, S. B. Stetson, T. L. Herter, N. Rowlands, and J. Schoenwald, "Blocked impurity band hybrid infrared focal plane arrays for astronomy," *IEEE Trans. Nucl. Sci.*, 36, 857–62, 1989.
57. H. H. Hogue, M. L. Guptill, D. Reynolds, E. W. Atkins, and M. G. Stapelbroek, "Space Mid-IR detectors from DRS," *Proc. SPIE* 4850, 880–9, 2003.
58. J. E. Huffman, A. G. Crouse, B. L. Halleck, T. V. Downes, and T. L. Herter, "Si:Sb blocked impurity band detectors for infrared astronomy," *J. Appl. Phys.*, 72, 273–5, 1992.
59. J. E. Van Cleve, T. L. Herter, R. Butturini, G. E. Gull, J. R. Houck, B. Pirger, and J. Schoenwald, "Evaluation of Si:As and Si:Sb blocked-impurity-band detectors for SIRTF and WIRE," *Proc. SPIE* 2553, 502–13, 1995.
60. D. M. Watson and J. E. Huffman, "Germanium blocked impurity band far infrared detectors," *Appl. Phys. Lett.*, 52, 1602–4, 1988.
61. D. M. Watson, M. T. Guptill, J. E. Huffman, T. N. Krabach, S. N. Raines, and S. Satyapal, "Germanium blocked impurity band detector arrays: Unpassivated devices with bulk substrates," *J. Appl. Phys.*, 74, 4199–206, 1993.
62. T. Wada, Y. Arai, S. Baba, M. Hanaoka, Y. Hattori, H. Ikeda, H. Kaneda, C. Koch, A. Miyachi, K. Nagase, H. Nakaya, M. Ohno, S. Oyabu, T. Suzuki, S. Ukai, K. Watanabe, and K. Yamamoto, "Development for germanium blocked impurity band far-infrared image sensors with fully-depleted silicon-on-insulator CMOS readout integrated circuit," *J. Low Temp. Phys.*, 184, 217–24, 2016.
63. I. C. Wu, J. W. Beeman, P. N. Luke, W. L. Hansen, and E. E. Haller, "Ion-implanted extrinsic Ge photodetectors with extended cutoff wavelength," *Appl. Phys. Lett.*, 58, 1431–3, 1991.

64. J. W. Beeman, S. Goyal, L. R. Reichetz, and E. E. Haller, "Ion-implanted Ge:B far-infrared blocked impurity-band detectors," *Infrared Phys. Technol.*, 51, 60–5, 2007.
65. J. Bandaru, J. W. Beeman, and E. E. Haller, "Growth and performance of Ge:Sb blocked impurity band (BIB) detectors," *Proc. SPIE* 4486, 193–9, 2002.
66. L. A. Reichertz, J. W. Beeman, B. L. Cardozo, N. M. Haegel, E. E. Haller, G. Jakob, and R. Katterloher, "GaAs BIB photodetector development for far-infrared astronomy," *Proc. SPIE* 5543, 231–8, 2004.
67. L. A. Reichertz, J. W. Beeman, B. L. Cardozo, G. Jakob, R. Katterloher, N. M. Haegel, and E.E. Haller, "Development of a GaAs-based BIB detector for sub-mm wavelengths," *Proc. SPIE* 6275, 62751S, 2006.
68. F. Szmulowicz and F. L. Madarsz, "Blocked impurity band detectors: An analytical model: Figures of merit," *J. Appl. Phys.*, 62, 2533–40, 1987.
69. H. Hogue, E. Atkins, D. Reynolds, M. Salcido, L. Dawson, D. Molyneux, and M. Muzilla, "Update on blocked impurity band detector technology from DRS", *Proc. SPIE* 7780, 778004-1–10, 2011.
70. D. Ives, G. Finger, G. Jakob, S. Eschbaumer, L. Mehrgan, M. Meyer, and J. Stegmeier, "AQUARIUS: The next generation mid-IR detector for ground based astronomy", *Proc. SPIE* 8453, 8453-1–13, 2012.
71. J. E. Huffman, "Infrared detectors for 2 to 220 μm astronomy," *Proc. SPIE* 2274, 157–69, 1994.
72. N. M. Haegel, "BIB detector development for the far infrared: From Ge to GaAs," *Proc. SPIE* 4999, 182–94, 2003.
73. E. E. Haller and J. W. Beeman, "Far infrared Photoconductors: Recent Advances and Future Prospects," in *Far-IR Sub-mm&MM Detectors Technology Workshop*, Monterey, April, 1–3, pp. 2–06, 2002.
74. M. D. Petroff, M. G. Stapelbroek, and W. A. Kleinhaus, "Detection of individual 0.4–28 μm wavelength photons via impurity-impact ionization in a solid-state photomultiplier," *Appl. Phys. Lett.*, 51, 406–8, 1987.
75. K. M. Hays, R. A. La Violette, M. G. Stapelbroek, and M. D. Petroff, "The Solid State Photomultiplier-Status of Photon Counting Beyond the Near-Infrared," in *Proceedings of the Third Infrared Detector Technology Workshop*, ed. C. R. McCreight, 59–80, NASA Technical Memorandum 102209, 1989. http://adsabs.harvard.edu/abs/1989tidt.work...59H.

15 Photoemissive detectors

In 1973, Shepherd and Yang of Rome Air Development Center (Rome, New York) proposed the concept of silicide Schottky-barrier detector FPAs as a much more reproducible alternative to HgCdTe FPAs for infrared (IR) thermal imaging [1]. For the first time, it became possible to have much more sophisticated readout schemes—both detection and readout could be implemented in one common silicon chip. Since then, the development of the Schottky-barrier technology has progressed continuously and currently offers large IR image sensor formats. Despite lower quantum efficiency (QE) than other types of IR detectors, the PtSi Schottky-barrier detector technology has exhibited remarkable advances. Such attributes as monolithic construction, uniformity in responsivity and signal to noise (the performance of an IR system ultimately depends on the ability to compensate for the nonuniformity of an FPA using external electronics and a variety of temperature references), and the absence of discernible $1/f$ noise make Schottky-barrier devices a formidable contender to the mainstream IR systems and applications [2–9]. While PtSi Schottky-barrier detectors are operated in the short and middle-wavelength IR spectral bands, long-wavelength IR detectors have already been demonstrated with Si-based heterojunction IR photoemission detectors. The photodetection mechanism in the later detectors is the same as that of the Schottky-barrier detectors.

In this chapter, the Si-based photoemissive detectors are reviewed.

15.1 INTERNAL PHOTOEMISSION PROCESS

The original electron photoemission model from metals into vacuum was described by Fowler in the 1930s [10]. In the 1960s, Fowler's photoyield model was modified based on the studies of internal photoemission of hot electrons from metal films into a semiconductor [11,12]. Cohen et al. modified the Fowler emission theory to account for emission into semiconductors [13].

As illustrated in Figure 15.1, internal photoemission resembles electron emission from metal to vacuum by photon irradiation. The incident photons are absorbed in the metal and generate electron hole pairs. The excited electrons randomly walk in the metal films until they reach the interface between the metal and the semiconductor. Finally, the electrons surmount the barrier and are emitted into the semiconductor. The internal photoemission process involves three steps:

- Photoabsorption in the electrode that gives rise to a hot carrier gas
- Hot carrier transport in the electrode and in the semiconductor prior to barrier emission
- Emission over the Schottky barrier

Unlike intrinsic detectors, the QE of the Schottky-barrier detector depends on the photon energy because of the strong dependence of the emission probability on the energy of the excited electron.

Assuming that the probability of electron excitation is independent of the energy of the initial and final states and that there is an abrupt transition from filled to empty states at the Fermi level (see Figure 15.1), the total number of possible excited states is given by

$$N_T \int_{E_F}^{E_F+h\nu} \frac{dN}{dE} dE, \tag{15.1}$$

where dN/dE is the density of states of the metal, E_F is the Fermi energy, $h\nu$ is the incident photon energy, and E is the electron energy with respect to the metal conduction band edge. The photoemission occurs when an electron is excited to a state for which the component of the momentum normal to the interface

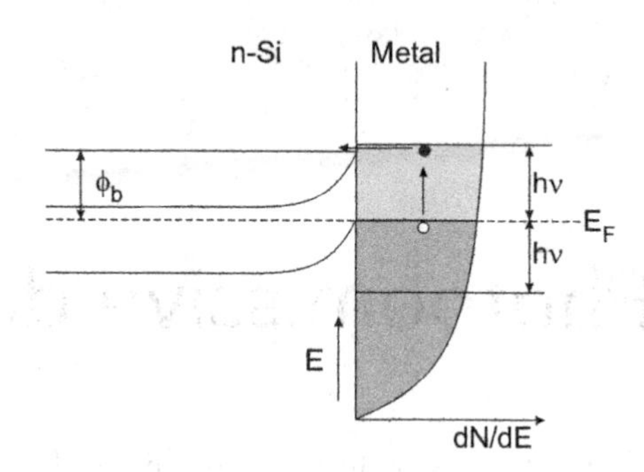

Figure 15.1 Internal photoemission in the Schottky-barrier detector.

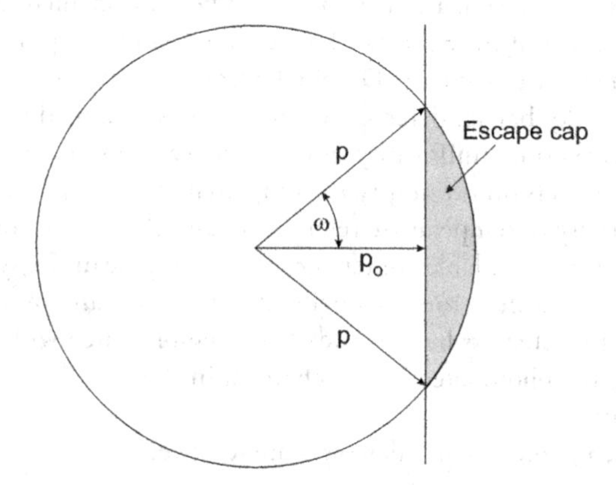

Figure 15.2 Momentum criterion for internal photoemission. p_o is the momentum corresponding to the barrier height. The excited electrons with momentum included within the escape cap are emitted into the semiconductor.

corresponds to a kinetic energy equal to or greater than the barrier. Therefore, the number of states that meet the momentum criterion is

$$N_E = \int_{E_F+\phi_b}^{E_F+h\nu} \frac{dN}{dE} P(E)\,dE, \tag{15.2}$$

where $P(E)$ is the photoemission probability for the electron with an energy E and ϕ_b is the height of the barrier. If the momentum distribution of the electrons is isotropic, $P(E)$ can be calculated as shown in Figure 15.2. In this figure, p is the momentum of the excited electron and p_o is the momentum corresponding to the barrier height:

$$p = \sqrt{2m^* E}, \tag{15.3}$$

$$p_o = \sqrt{2m^*(E_F+\phi_b)}, \tag{15.4}$$

where m^* is the effective mass of the electron. $P(E)$ is the ratio of the surface area of the sphere included within the escape cap to the total surface area of the sphere and is equal to

$$P(E) = \frac{1}{2}(1-\cos\omega) = \frac{1}{2}\left(1-\sqrt{\frac{E_F+\phi_b}{E}}\right). \tag{15.5}$$

For further discussion, we assume that dN/dE can be considered to be independent of the energy over the energy range of interest because the Fermi energy is much greater than the photon energy. Equations 15.1 and 15.2 then become

$$N_T = \frac{dN}{dE} h\upsilon. \tag{15.6}$$

$$N_E = \frac{dN}{dE} \frac{(h\upsilon - \phi_b)^2}{8E_F}. \tag{15.7}$$

Here we assume that there are no collisions of electrons or energy losses before the excited electron reaches the interface. Thus, the internal QE, which is the ratio of N_E to N_T, is given by

$$\eta_i = \frac{1}{8E_F} \frac{(h\upsilon - \phi_b)^2}{h\upsilon}. \tag{15.8}$$

This simple theory described by Cohen et al. [13] was later extended by Dalal [14], Vickers [15], and Mooney and Silverman [16]. Following these authors, the general form for the QE for internal photoemission is given by

$$\eta = C_f \frac{(h\upsilon - \phi_b)^2}{h\upsilon}, \tag{15.9}$$

where C_f is the Fowler emission coefficient. The Fowler coefficient provides an energy-independent measure of the efficiency of the internal photoemission. Its value may be approximated by

$$C_f = \frac{H}{8E_F}, \tag{15.10}$$

where H is a device and a voltage dependent factor.

Equation 15.9 converted to wavelength variables is given by

$$\eta = 1.24C_f \frac{(1 - \lambda/\lambda_c)^2}{\lambda}. \tag{15.11}$$

C_f depends upon the physical and geometric parameters of the Schottky electrode. Values of λ_c and C_f as high as 6 μm and 0.5 $(eV)^{-1}$, respectively, have been obtained in PtSi-Si [3]. Schottky photoemission is independent of such factors as semiconductor doping, minority carrier lifetime, and alloy composition, and as a result of this, has spatial uniformity characteristics that are far superior to those of the other detector technologies. Uniformity is limited only by the geometric definition of the detectors.

Figure 15.3 compares the spectral QE of the typical photon detectors. From this figure, it seems reasonable to choose high-QE intrinsic photodetectors. The effective QE of Schottky-barrier detectors in the 3–5 μm atmospheric window is very low, of the order of 1%, but useful sensitivity is obtained by means of near–full frame integration in area arrays. An extension of this technology to the long wavelength band is possible using IrSi (see Figure 15.3), but this will require cooling below 77 K [4].

The current responsivity (see Equation 4.31 with $g = 1$) can be expressed as

$$R_i = qC_f\left(1 - \frac{\lambda}{\lambda_c}\right)^2. \tag{15.12}$$

Two specific properties of photoemissive detectors follow from the last two equations. The photoresponse decreases with wavelength and the QE is low compared to that of bulk detectors. Both of these properties are a direct result of conservation of momentum during carrier emission over the potential barrier.

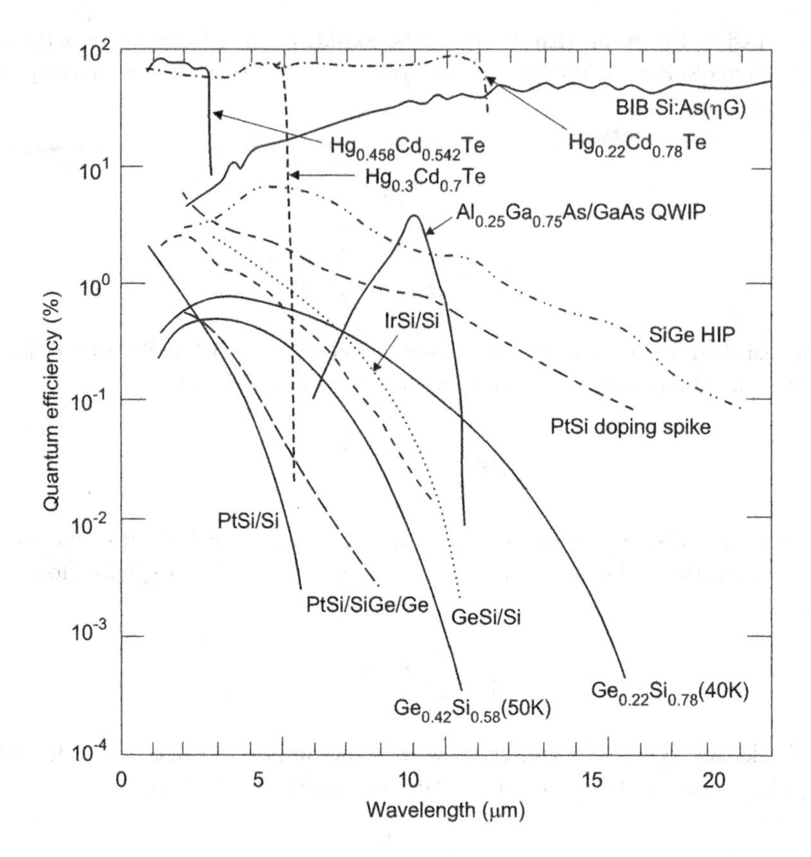

Figure 15.3 Quantum efficiency versus wavelength for several detector materials. The detectors include HgCdTe intrinsic photodiodes, BIB extrinsic detectors, GaAs-based quantum well infrared photodetectors, and Si-based photoemissive detectors (PtSi, IrSi, PtSi/SiGe, PtSi doping spike, and SiGe heterojunction internal photoemission).

The majority of excited carriers, which do not have enough momentum normal to the barrier, are reflected and not emitted. Figure 15.4 shows typical spectral responses of Pd_2Si, PtSi, and IrSi Schottky-barrier detectors [7].

Figures 12.32 and 15.1, which are often given as the energy band diagram for a Schottky barrier, are misleading because they give the impression that the peak of the Schottky barrier potential occurs at the semiconductor–electrode interface. The electric field near the Schottky barrier has an influence on the barrier height. When the carrier is injected into the semiconductor, it feels an attractive force called the image force. As a result, the effective barrier height is reduced. This lowering is called the Schottky effect. As a result of this effect, the peak potential always occurs in the semiconductor, typically at a depth of 5–50 nm (see Figure 15.5 [17]).

The magnitude of the barrier lowering $\Delta\phi_b$ is given by [17]

$$\Delta\phi_b = \sqrt{\frac{qE}{4\pi\varepsilon_o\varepsilon_s}}, \tag{15.13}$$

where q is the electron charge, E is the electric field near the barrier, ε_o is the permittivity of free space, and ε_s is the dielectric constant of silicon. The electric field is given by

$$E = \sqrt{\frac{2qN_d}{\varepsilon_o\varepsilon_s}\left(V + V_{bi} - \frac{kT}{q}\right)}, \tag{15.14}$$

where N_d is the impurity concentration of silicon, V is the applied voltage, and V_{bi} is the built-in potential. The last equation indicates that the barrier height can be controlled by the reverse-bias voltage and

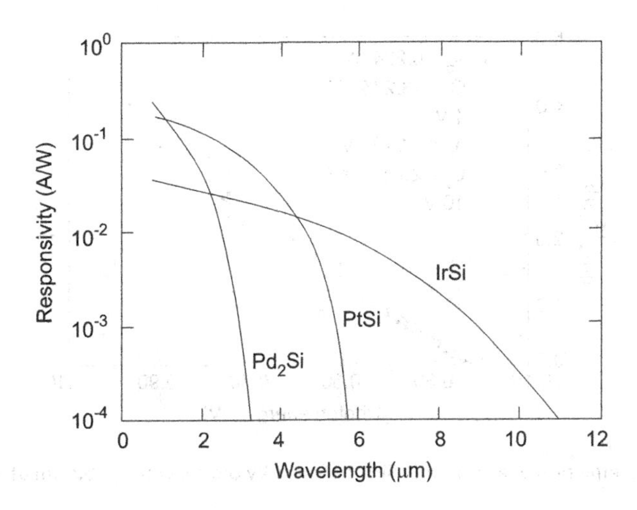

Figure 15.4 Spectral response of Pd_2Si, PtSi, and IrSi Schottky-barrier detectors. (From Ref. [7])

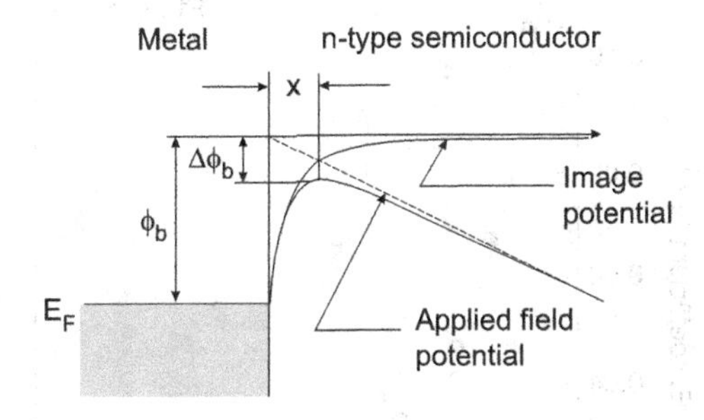

Figure 15.5 Barrier lowering by Schottky effect. The attraction force between the emitted electron and the induced positive charge reduces the barrier height by $\Delta\phi_b$. (From Ref. [17])

the impurity concentration of the substrate. The distance between the interface and the potential maximum becomes shorter for larger electric fields, and this shift of the potential maximum enhances the QE coefficient.

From Equation 15.9 we find that the QE of the Schottky-barrier detector is expressed by two parameters—the barrier height and the Fowler emission coefficient. We can determine those two parameters from the plot of $\sqrt{\eta \times h\nu}$ versus $h\nu$. This type of analysis is known as a Fowler plot. The Fowler coefficient is determined from the square of the slope and the barrier potential from the intercept of the plot. Figure 15.6 shows the Fowler plot based on spectral responsivity data from a PtSi/p-Si detector fabricated at the Rome Laboratory [18]. Image lowering improves the emission efficiency and extends the spectral response to longer wavelengths with increased voltage.

15.1.1 SCATTERING EFFECTS

Hot carrier transport in the electrode includes elastic scattering at surfaces and grain boundaries, as well as inelastic scattering with phonons and Fermi electrons [16]. Elastic scattering redirects carriers, thereby increasing the emission probability. Also, the phonon scattering redirects carriers and may improve the emission probability. However, the loss of energy to the lattice during phonon emission lowers the probability for the transit of the potential barrier. In the case of multiple phonon scattering events, the carrier energy drops below that of the potential peak and the photoemission does not occur. The carrier is "thermalized" to the Fermi level.

The choice of electrode thickness is a compromise between keeping it thin enough for hot carriers to reach the silicon interface without loss of energy and making it thick enough to absorb the radiation.

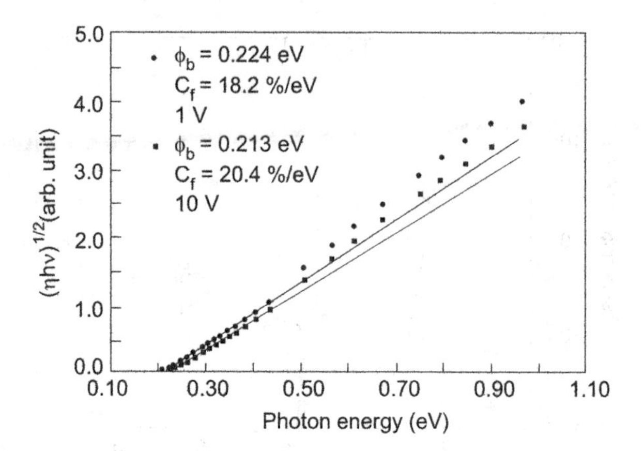

Figure 15.6 Fowler photoemission analysis for PtSi/p-Si Schottky diode, with λ_c=5.5 μm at 1 V bias and λ_c=5.8 μm at 10 V. (From Ref. [18])

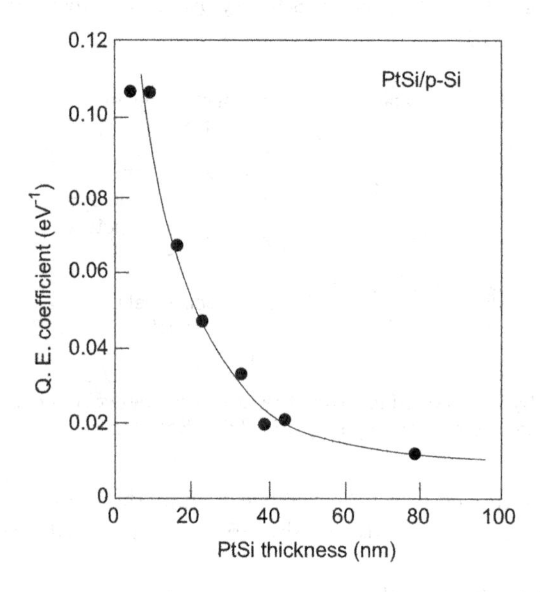

Figure 15.7 Dependence of QE coefficient of PtSi Schottky-barrier detector on PtSi thickness. (From Ref. [22])

To optimize emission efficiency, the photoemission electrode must be designed to maximize the ratio of elastic to inelastic scattering events. Many papers reported a great improvement in the QE of PtSi/p-Si Schottky-barrier detectors by thinning the metal films as shown in Figure 15.7 [19–22]. This observation has been explained assuming that the interface scattering (at the metal–silicon and metal–dielectric interfaces) redirects the momentum without energy loss and the direction of momentum is independent of its previous orientation [13–15]. When the thickness of the metal film becomes much less than the attenuation length of the electron, multiple interface scattering occurs in the metal film and the QE is improved because the redirection of the momentum by the interface scattering increases the probability of emission.

Mooney et al. [16,23] extended the Vickers model [15] to take into account the effect of carriers removed by emission and the energy loss by electron–phonon scattering. This model explains the fine structure of the spectral response, including roll-off from a linear fit of the modified Fowler plot for higher photon energy and a finite response for photon energies below the intercept of the linear region with the photon energy axis. The roll-off for large photon energy is caused by the reduction in the number of available carriers by prior emission events in the high QE region. The finite response below the energy of the extrapolated barrier height is related to the energy loss by the electron–phonon scattering. While only a few phonon collisions suffice to thermalize the hot carrier for low excitation energy, at higher energy, the carrier is less easily thermalized and more probably redirected into the escape direction and hence phonon scattering tends to increase the QE. This effect makes the apparent extrapolated barrier height higher than the actual barrier height. This is the reason why the barrier height measured by electrical methods is always lower than that measured by optical methods.

15.1.2 DARK CURRENT

The current flowing through the barrier in silicide Schottky-barrier diodes is dominated by thermionic emission current. The thermionic emission theory gives the current–voltage characteristic expressed by Equation 12.150. The effective Richardson constant for holes in silicon, A^*, is about 30 A cm^{-2}K^2 in a moderate electric field range [17].

The Schottky-barrier diode is operated under reverse bias in the IR focal plane arrays (FPAs). In the reverse-biased condition, barrier lowering due to the Schottky effect has to be taken into consideration. For a reverse bias greater than $3kT/q$, from Equation 12.151 we get

$$J_{st} = A^* T^2 \exp\left[-\frac{q(\phi_b - \Delta\phi_b)}{kT}\right], \tag{15.15}$$

where $\Delta\phi_b$ is the magnitude of the barrier lowering calculated by Equation 15.13. By Equation 15.15, we can determine the effective barrier height at a certain reverse bias from the plot of J_{st}/T^2 versus $1/T$. Figure 15.8 is a Richardson analysis for the PtSi diode of Figure 15.6, biased at 1 V [18]. The Richardson constant is determined with less accuracy from the vertical intercept. Richardson analysis is often linear over more than five orders of magnitude. The presence of any leakage current or excess series resistance causes this plot to saturate. Thus, the Richardson analysis allows assessment of data quality. It is important to note that the barrier height obtained from the electrical measurement (ϕ_{bt}) is lower than that obtained from the optical measurement (ϕ_{bo}), as discussed previously:

$$\phi_{bt} = \phi_{bo} - nh\nu, \tag{15.16}$$

where $nh\nu$ is the average energy loss by electron–phonon scattering. The typical measured value of this energy loss is from 20–50 meV. The lower values are observed in the most efficient devices where elastic scattering is dominant.

Using Equations 15.11 and 15.15, we can estimate the temperature required for the background-limited operation of Schottky-barrier detectors as a function of cutoff wavelength, assuming that a range of barrier heights could be made. The temperature T_{BLIP} is calculated at which J_{st} is equal to the background current.

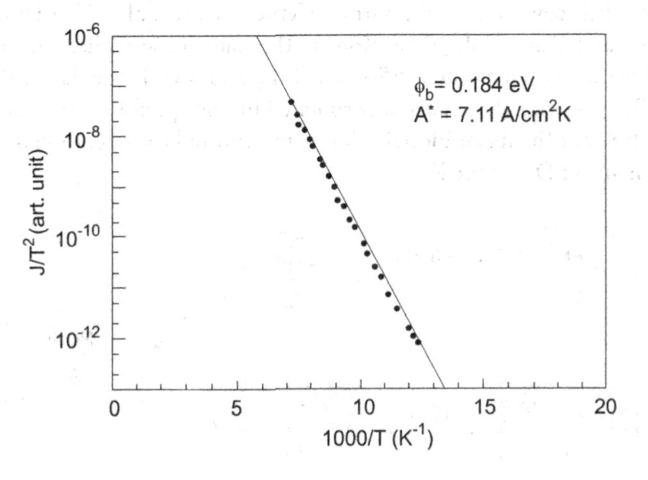

Figure 15.8 Richardson thermionic emission analysis PtSi/p-Si diode, at 1 V bias. Note the difference in barrier potential compared to Fowler analysis in Figure 15.6. (From Ref. [18])

The results are shown in Figure 4.14. The cooling requirements are more stringent in comparison with intrinsic detectors but are comparable with extrinsic silicon devices. For operation in the 8–12 μm band, cooling below 80 K is required—typical temperature is about 45 K.

15.1.3 METAL ELECTRODES

There are five silicides used for Schottky-barrier IR detectors: palladium silicide (Pd_2Si), platinum silicide (PtSi), iridium silicide (IrSi), cobalt silicide (Co_2Si), and nickel silicide (NiSi).

The solid–solid chemical reaction between the Pt and Si substrate is well-defined and controllable. The main steps in the formation of PtSi are illustrated in Figure 15.9 [24]. The initial phase of a Pt_2Si layer begins to form at a temperature of 300°C. The thickness of the Pt_2Si layer grows with the square root of the anneal time until all of the Pt is consumed, yielding a Pt_2Si–Si interface. Further, absence of source elemental Pt leads to a change in the composition of the interface in favor of the formation of an Si-rich phase, namely, PtSi. Thickness of PtSi layer is also proportional to the square root of the annealing time. Finally, additional annealing causes consumption of the Pt_2Si and the formation of the PtSi/Si junction interface. Postannealing oxidation forms an SiO_2 layer on the outer surface of the PtSi layer. This procedure is suitable for forming the resonant cavity in the Schottky-barrier structure.

The first thermal image was achieved at the General Electric Company using a high landing velocity vidicon, as described by Spratt and Schwarz [25], and a palladium silicide diode focal plane target, which was fabricated at Rome Laboratory. The vidicon sensitivity, angular resolution, and dynamic range were limited. As a result, tube sensor development was discontinued.

In the same year, Kohn et al. [26] at the David Sarnoff Research Center (then RCA Laboratory, Princeton, New Jersey) demonstrated the first solid-state IR imaging using a monolithic charge-coupled devices (CCD) array with Pd_2Si/p-Si Schottky photodiodes, fabricated using the process developed at Rome Laboratory. The CCD sensor had excellent dynamic range, good sensitivity from 1 to 3 μm, and angular resolution equal to the detector pitch. The barrier of the Pd_2Si detector was 0.34 eV and the cutoff wavelength was 3.5 μm. As a result, these devices had limited thermal sensitivity and research efforts were directed toward the development of PtSi, which had the potential for photoresponse beyond 5 μm. There was continuing interest in the use of Pd_2Si to measure reflected energy in the spectral band between 1 and 3 μm, for satellite-based Earth resources assessment (e.g., [27–31]). The operating temperature was around 130 K, which is compatible with the current satellite passive cooling technology.

The most popular Schottky-barrier detector is the PtSi detector on p-type silicon, which can be used for detection in the 3–5 μm spectral range. In 1979, Capone et al. [32] described a 25×50 element PtSi monolithic IR-CCD array with λ_c=4.6 μm (barrier height of 0.27 eV), QE coefficient of 0.036 eV^{-1}, 16% fill factor, 2% photoresponse uniformity, and an NEDT of 0.8°C. There have been steady advances in PtSi device processing and FPA development ever since. Significant advances were reported by Kosonocky [33] and Sauer [34] and their coworkers at the Sarnoff Center, Gates et al. [35] at Hughes Aircraft, Kimata et al. [36] at Mitsubishi, and Clark et al. [37] at Kodak. The state of the art has advanced to where several vendors in the United States and Japan offer PtSi monolithic arrays with sizes larger than 480×640 and NEDTs below 0.1 K. These devices have been incorporated in high-performance IR cameras. In 1991 Yutani and coworkers [38] at Mitsubishi Electric Company reported the operation of a 1,040×1,040 element PtSi array with an NEDT of 0.1 K.

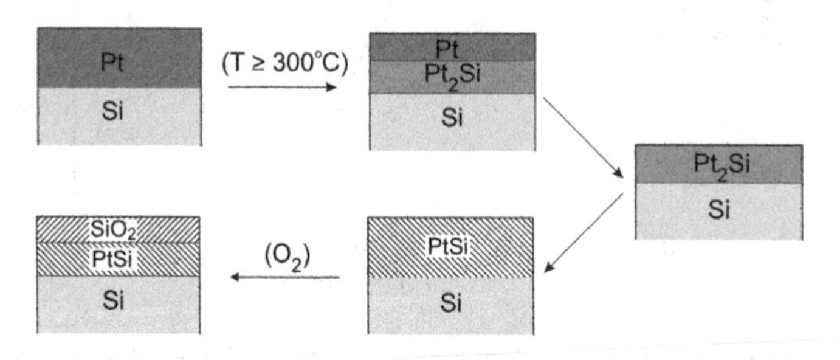

Figure 15.9 Formation of PtSi and Pt_2Si films. Arrows indicate increasing time or temperature. (From Ref. [24])

The IrSi Schottky-barrier on p-type silicon is expected to have the lowest barrier. In 1982, Pellegrini et al. [39] measured photoresponse beyond 8 μm in iridium silicide detectors. They also discussed the difficulty of IrSi formation compared with PtSi formation [40]. Unlike PtSi formation, Si is the major diffusing species during the whole reaction process, and the contaminations at the original Si surface remain on the IrSi/Si interface and thus degrade diode characteristics. The other difficulty is related to the phase control. IrSi process repeatability is poor and array uniformities are substantially inferior to that of either Pd_2Si or PtSi. It is very difficult to obtain a clean interface with the IrSi detector because the diffusing atom is not Ir but Si. Despite these problems, IrSi arrays have been demonstrated [41].

$CoSi_2$ and NiSi detectors have also been developed for remote sensing applications in short-wavelength IR bands. Optical barrier heights of $CoSi_2$ and NiSi detectors on p-type silicon were reported to be 0.44 and 0.40 eV, respectively [42,43].

15.2 CONTROL OF SCHOTTKY-BARRIER DETECTOR CUTOFF WAVELENGTH

The cutoff wavelength or operating temperature of Schottky-barrier detectors limits specific applications. The barrier potential may be raised to reduce dark current, or to allow an increase in the operating temperature. Lowering the barrier of the PtSi detector improves cold night thermal imaging performance but requires operation at a lower temperature.

Enhancement of the electric field near the barrier reduces the effective barrier height. Tsubouchi et al. have observed the barrier lowering of PtSi Schottky-barrier detectors by the Schottky effect [44]. The effective barrier height is also reduced by narrowing the potential profile to the onset of tunneling. Shannon demonstrated raising and lowering of Ni Schottky barrier potential by very shallow ion implantation [45]. This technique was employed by Pellegrini et al. [46] and Wei et al. [47] to adjust the cutoff wavelength of PtSi Schottky-barrier detectors. Tsaur et al. applied this technique to an IrSi Schottky-barrier detector, and obtained a cutoff wavelength longer than 12 μm [48].

Fathauer et al. controlled the barrier potential of both p-type and n-type $CoSi_2$/Si Schottky diodes by molecular-beam epitaxy (MBE) growth of thin n^+ Ga-doped Si layer at the metal–semiconductor interface [49]. They reported that the barrier height was reduced from 0.35 to 0.25 eV. Lin et al. demonstrated a PtSi detector with a 22-μm cutoff wavelength by introducing a 1-nm thick doping spike at the interface [50–52]. A thin doping spike is essential to reduce the optical barrier height. To make a very thin doping spike with high impurity concentrations, they developed a low-temperature Si-MBE technology using elemental boron as a dopant source.

The barrier height may also be controlled by the use of an alloy of two metals. Tsaur et al. fabricated a Schottky-barrier detector with a combination of PtSi and IrSi [53]. A Pt-Ir silicide Schottky barrier equal to 0.135 eV was obtained by sequentially depositing 1.5 nm Ir and 0.5-nm Pt films onto p-type silicon. They reported that better diode characteristics can be obtained using this technology compared with IrSi detectors alone.

Schottky barrier detectors are generally fabricated on (100) substrates. However, it appears that the use of an (111) orientation raises the barrier potential nearly 0.1–0.313 eV. This reduces the detector cutoff wavelength to 4.0 μm and raises the operating temperature above 100 K.

The existence of surface states also affects the barrier height. The barrier height of PtSi barriers decreases with the reduction of the surface states [54].

15.3 OPTIMIZED STRUCTURE AND FABRICATION OF SCHOTTKY-BARRIER DETECTORS

The Schottky-barrier detector is typically operated in the back-side illumination mode. This mode of operation is possible because silicon is transparent in the IR spectral range of interest. The silicon substrate serves as a refraction index–matching layer for the silicide film, causing the back-side illumination mode to have a reduced reflection loss than the front-side illumination mode. The difference in the QE between the back-side and front-side illumination modes was reported to be a factor of 3 [55].

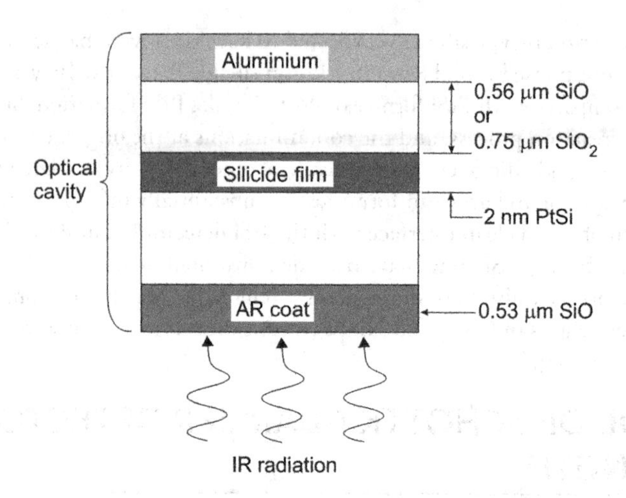

Figure 15.10 Cross-sectional view of the PtSi Schottky-barrier detector, with optical cavity tuned for maximum response at a wavelength of 4.3 μm. (From Ref. [6])

To achieve maximum responsivity, the Schottky-barrier detector is usually constructed with an "optical cavity." The optical cavity structure consists of the metal reflector and the dielectric film between the reflector and the metal electrode of the Schottky-barrier diode. According to fundamental optical theory, the effect of the optical cavity depends on the thickness and refractive index of the dielectric films and the wavelength. The conventional 1/4 wavelength design for the optical cavity thickness is a good first approximation for optimizing the responsivity [56]. For example, Figure 15.10 shows the cross section of a PtSi Schottky-barrier detector with an optical cavity tuned for maximum response at 4.3 μm [6].

It should be added that the front-side illuminated Schottky-barrier detectors were reported with the IR signal introduced on the silicide side of the detector to achieve a broad spectral response including IR, visible, and ultraviolet [55,57], and in the case of direct Schottky injection for achieving 100% fill factor [58]. Figure 15.11 shows the spectral response of a front-side illuminated PtSi Schottky-barrier detector in the wavelength range between 0.4 and 5.2 μm [59]. The QE in the visible and near-infrared spectral regions is

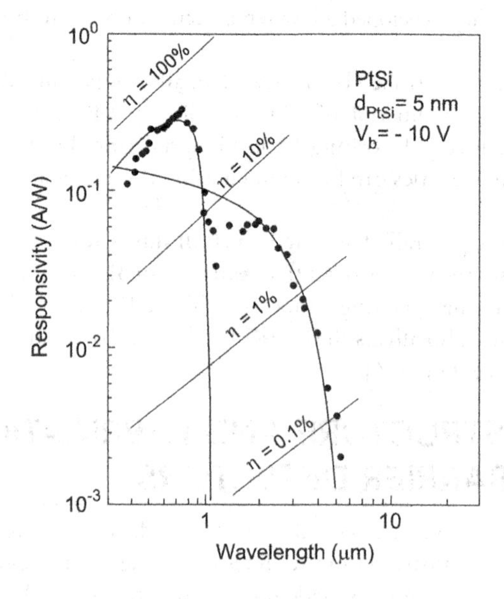

Figure 15.11 Spectral response of a front-side illuminated PtSi Schottky-barrier detector. The fitting curves in the IR and visible spectral regions are calculated based on internal photoemission and intrinsic mechanisms, respectively. (From Ref. [59])

higher than in the IR spectral range due to electron hole creation in the silicon substrate for a wavelength range shorter than the fundamental absorption edge of silicon.

The main advantage of the Schottky-barrier detectors is that they can be fabricated as monolithic arrays in a standard silicon VLSI process. Typically, the silicon arrays are completed up to the Al metallizational step. A Schottky-contact mask is used to open the SiO_2 surface to p-type (100) silicon (with resistivity 30–50 Ωcm) at the Schottky-barrier detector location. In the case of PtSi detectors, a very thin layer of Pt (1–2 nm) is deposited and sintered (annealed at a temperature in the range of 300°C–600°C) to form PtSi, and the unreacted Pt on the SiO_2 surfaces is removed by dip-etching in hot aqua regia [60]. The Schottky-barrier structure is then completed by a deposition of a suitable dielectric (usually SiO_2) for forming the resonant cavity, removing this dielectric outside the Schottky-barrier regions, and depositing and defining Al for the detector reflector and the interconnects of the Si readout multiplexer. In the case of the 10-μm IrSi Schottky-barrier detectors, the IrSi was formed by in situ vacuum annealing and the unreacted Ir was removed by reactive ion etching [61].

15.4 NOVEL INTERNAL PHOTOEMISSIVE DETECTORS

15.4.1 HETEROJUNCTION INTERNAL PHOTOEMISSIVE DETECTORS

In 1971, Shepherd et al. recommended replacing the metal electrode of the Schottky-barrier detector with degenerate semiconductors [62]. The free carrier density of a degenerate semiconductor is at least two orders of magnitude less than that of metal silicides. Because of this, the carrier absorption coefficient is lower and degenerate semiconductor electrodes must be made thicker, to match the metal silicide absorption efficiencies. In 1970, degenerate electrode devices were fabricated at Rome Laboratory using alloying and liquid phase epitaxy; however, due to higher phonon scattering losses during transport in the thicker electrode, the emission efficiencies of these devices were not competitive with those of metal silicide devices, and work was discontinued [62].

The development in the MBE technology made it possible to fabricate high-quality Ge_xSi_{1-x} (GeSi) thin films onto silicon substrates, and several works concerning the realization of ideas utilizing the internal photoemission of GeSi/Si heterojunction diodes for IR detection have been reported [63–74]. Lin et al. demonstrated the first GeSi/Si heterojunction detector in 1990 [63,64]. Figure 15.12 shows the structure and energy band diagram of this detector [68]. An energy barrier ϕ_b is established between GeSi and Si. The bandgap energy of the strained SiGe is smaller than that of Si, and by varying Ge composition and the doping concentration of the Ge_xSi_{1-x} layer, the cutoff wavelength of the device can be varied from 2 to 25 μm [65,68]. Figure 15.3 compares the quantum efficiencies of several technologies, between them a conventional PtSi/p-Si detector, IrSi/p-Si detector, PtSi/p-$Si_{0.85}Ge_{0.15}$/p-Si detector, PtSi/p-Si detector with a 1-nm doping spike of 2×10^{20} cm^{-3} boron, and p-$Si_{0.7}Ge_{0.3}$/p-Si heterojunction detector with a boron doping of 5×10^{20} cm^{-3} in the SiGe layer.

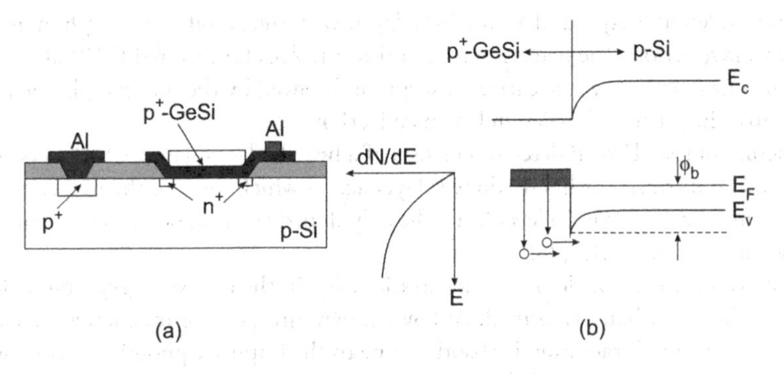

Figure 15.12 Construction (a) and operation (b) of Ge_xSi_{1-x} heterojunction photoemissive detector. (From Ref. [68])

Absorption mechanisms in the $Ge_{1-x}Si_x/Si$ detector are free carrier absorption and intravalence band transitions. Assuming that the density of state of the valence band is proportional to the square root of the energy, the QE can be expressed as [70]:

$$\eta = \frac{A}{8E_F^{1/2}\left(E_F+\phi_b\right)^{1/2}} \frac{\left(h\nu-\phi_b\right)^2}{h\nu}, \tag{15.17}$$

where A is the absorptance, relatively wavelength-independent in the long-wavelength region. This expression is similar for silicide-barrier detectors (see Equation 15.9). Since E_F is usually much smaller for semiconductors than for metals, η is expected to be significantly greater for heterojunction detectors than for silicide Schottky-barrier detectors. On the other hand, the absorptance A of $Ge_{1-x}Si_x$ is expected to be smaller than that of the silicide Schottky-barrier detectors because of the lower density of free carriers in $Ge_{1-x}Si_x$.

The optimum thickness of $Ge_{1-x}Si_x$ is reported to be about 20 nm [68], which is an order of magnitude thicker than that of the metal electrode for silicide-barrier detectors. This thickness is a compromise between achieving a possible high absorptance of GeSi layer and an influence of elastic and inelastic scattering. (See discussion of scattering effects in Section 15.1.1.) Tsaur et al. discussed the improvement of the QE of the heterojunction detector with a metal overlayer and double-heterojunction structure [68]. Park et al. proposed a stacked GeSi/Si heterojunction detector, with which both the optical absorption and internal QE can be optimized [71–73]. The stacked $Si_{0.7}Ge_{0.3}/Si$ detector consisted of several periods of degenerated boron-doped thin (≤5 nm) $Si_{0.7}Ge_{0.3}$ layers, and undoped thick (≈30 nm) Si layers have exhibited photoresponse at wavelengths ranging from 2 to 20 μm with QE of about 4% and 1.5% at 10 and 15 μm wavelengths, respectively. The detectors showed near-ideal thermionic emission–limited dark current characteristics.

Recently, Lao and Perera [75] have focused on the internal photoemission phenomena in the IR wavelength range that corresponds to the energy of photons less than the value of the bandgap. This review paper discusses the generic processes based on heterojunction structures, characterizing p-type band structure and the band offset at the heterointerface, and IR photodetection including a novel concept of IR photoresponse extension based on an energy transfer mechanism between hot and cold carriers including split-off band heterojunction detectors, wavelength-extended photovoltaic photodetectors, and hot hole photodetectors with a response beyond the bandgap spectral limit.

15.4.2 HOMOJUNCTION INTERNAL PHOTOEMISSIVE DETECTORS

The concept of a homojunction internal photoemissive detector for IR was first realized in 1988 by Tohyama et al. [76]. This detector had a useful spectral response from 1 to 7 μm, although the reported QE was low. Next, O'Neil reported that the QE of the homojunction detector had been improved to a practical level (several percent). His group has developed a 128×128 element array using this technology [77].

More recently, various detector approaches based on a high-low Si and GaAs homojunction interfacial workfunction internal photoemission (HIWIP) junctions have been discussed by Perera et al. [78–85]. The operation of HIWIP detectors is based on the internal photoemission occurring at the interface between a heavily doped absorber/emitter layer and an intrinsic layer, with the cutoff wavelength mainly determined by the interfacial workfunction. The detection mechanism involves far infrared (FIR) absorption in the highly doped thin emitter layers by free carrier absorption followed by the internal photoemission of photoexcited carriers across the junction barrier and then collection.

The basic structure of the HIWIP detector consists of a heavily doped layer, which acts as the IR absorber region, and an intrinsic (or lightly doped) layer across which most of the bias is dropped. According to the doping concentration level in the heavily doped layer, these detectors can be divided into three types as shown in Figure 15.13 [78].

In the Type I detector, when the doping concentration, N_d, in the n^+-layer is high but below the Mott critical value, N_c, an impurity band is formed. At low temperatures, the Fermi energy is located in the impurity band. The incident IR radiation is absorbed due to the impurity photoionization, with a workfunction given by $\Delta = E_c^{n+} - E_F$, where E_c^{n+} is the conduction band edge in the n^+-layer. An electric field is formed in the i-layer by an external bias to collect photoexcited electrons generated in the n^+-layer.

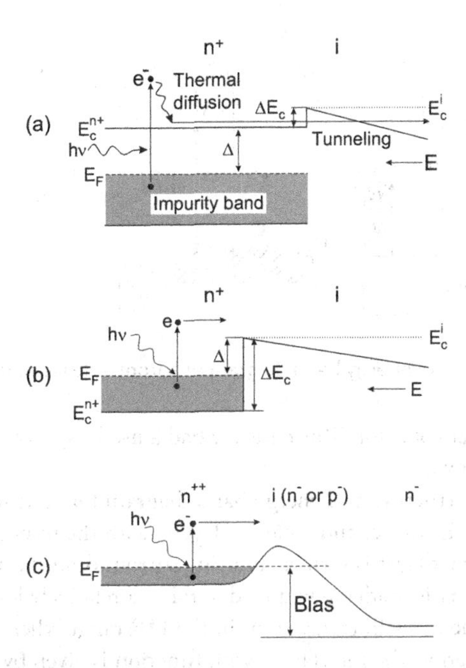

Figure 15.13 Energy band diagram of three different types of homojunction internal photoemissive detectors. (a) Type I: $N_d < N_c\left(E_F < E_c^{n+}\right)$, (b) Type II: $N_c < N_d < N_o\left(E_c^{n+} < E_F < E_c^{i}\right)$ (should be: N_c<N_d.....), and (c) Type III: $N_d > N_o\left(E_F > E_c^{i}\right)$. Here N_c is the Mott critical concentration and N_o is the critical concentration corresponding to $\Delta = 0$. In (a) and (b), the conduction band edge of the i-layer is represented by a dotted line for $V_b = 0$ and by solid line for $V_b > V_o$. (From Ref. [78])

Type I detectors are analogous to semiconductor photoemissive detectors in their operation, which can be described by a three-step process:

- Photoexcitation of electrons from filled impurity band states below Fermi level into empty states in the conduction band
- Rapid thermalization of photoexcited electrons into the bottom of the conduction band by phonon relaxation and next their diffusion to the emitting interface
- Tunneling of the electrons through an interfacial barrier, ΔE_c, which is due to the offset of the conduction band edge caused by the bandgap narrowing effect

Type II detectors are analogous to Schottky-barrier IR detectors in their operation. When the doping concentration is above the Mott transition, the impurity band is linked with the conduction band edge, and n⁺-layer becomes metallic. Even in this case, the Fermi level can still be below the conduction band edge of the i-layer $\left(E_F < E_c^i\right)$ due to the bandgap narrowing effect, giving rise to a work function $\Delta = E_c^i - E_F$, unless N_d exceeds a critical concentration N_o at which $\Delta = 0$. One of the unique features of Type II detectors is the fact that there is no restriction on the cutoff wavelength, because the work function can become arbitrarily small with increasing doping concentration. High-performance Si detectors with $\lambda_c > 40\,\mu m$ can be realized [78].

In the Type III detector, the doping concentration is so high that the Fermi level is above the conduction band edge of the i-layer, the n^+-layer becomes degenerate, and a barrier associated with space charge region is formed at the n^+-i interface due to electron diffusion. The barrier height depends on the doping concentration and the applied voltage, giving rise to an electrically tunable λ_c. As the barrier voltage is increased, the barrier height is reduced, the spectral response shifts toward a longer wavelength and the signal increases at a given wavelength. This change in response is caused by image lowering, leveraged by the p-n junction at the interface. Type III device was first demonstrated by Tohyama et al. [76] using a structure composed of a degenerate n^{++} hot-carrier, a depleted barrier layer (lightly doped p, n or i), and

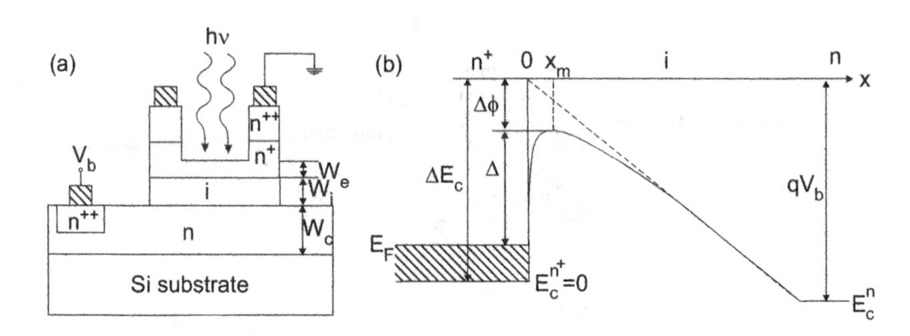

Figure 15.14 (a) Basic structure, and (b) energy band diagram of a front-side illuminated n⁺-i HIWIR. (From Ref. [80])

a lightly doped n-type hot-carrier collector. This detector had a useful spectral response from 1 to 7 µm, although the reported QE was low.

Figure 15.14 shows the basic structure and energy band diagram for an n^+-i HIWIP detector [80]. The structure consists of the emitter, intrinsic, and collector layers, with the respective thicknesses represented by W_e, W_i, and W_c. The top contact layer is formed as a ring surrounding the active area to minimize absorption loss. The collector layer is moderately doped and has a relatively low resistance due to the impurity band conduction, while it is still transparent in the FIR range when the photon energy is smaller than the impurity ionization energy. The interfacial work function is given by $\Delta = \Delta E_c - E_F - \Delta\phi$, where ΔE_c is the conduction band edge offset due to bandgap narrowing in the heavily doped emitter layer, E_F is the Fermi energy, and $\Delta\phi$ is the lowering of interfacial barrier height due to image force effect. To get a high internal QE, the emitter layer thickness should be thin enough, in spite of the reduction of the photon absorption efficiency. Thus, the optimal thickness is a tradeoff of photon absorption and hot electron scattering.

Si HIWIP FIR detectors could have a performance comparable to that of conventional Ge FIR photoconductors [86] or Ge blocked impurity band (BIB) detectors [87]. In addition to Si, significant bandgap shrinkage has been observed for heavily doped p-GaAs. Better carrier transport properties of GaAs may produce an improved performance for this type of device. The p-GaAs HIWIP FIR detectors show great potential in becoming a strong competitor in FIR applications—responsivity of 3.10 ± 0.05 AW^{-1}, QE of 12.5%, and detectivity of 5.9×10^{10} $cmHz^{1/2}W^{-1}$ at 4.2 K, for λ_c from 80 to 100 µm.

REFERENCES

1. F. D. Shepherd and A. C. Yang, "Silicon Schottky retinas for infrared imaging," *Tech. Dig. IEDM* 310–3, 1973.
2. W. F. Kosonocky, "Infrared image sensors with Schottky-barrier detectors," *Proc. SPIE* 869, 90–106, 1987.
3. F. D. Shepherd, "Schottky diode based infrared sensors," *Proc. SPIE* 443, 42–9, 1984.
4. F. D. Shepherd, "Silicide infrared staring sensors," *Proc. SPIE* 930, 2–10, 1988.
5. W. F. Kosonocky, "Review of Schottky-barrier imager technology," *Proc. SPIE* 1308, 2–26, 1990.
6. W. F. Kosonocky, "Review of infrared image sensors with Schottky-barrier detectors," *Optoelectron.: Devices Technol.*, 6, 173–203, 1991.
7. M. Kimata and N. Tsubouchi, "Schottky Barrier Photoemissive Detectors," in *Infrared Photon Detectors*, ed. A. Rogalski, 299–349, SPIE Optical Engineering Press, Bellingham, WA, 1995.
8. M. Kimata, "Metal Silicide Schottky Infrared Detector Arrays," in *Infrared Detectors and Emitters: Materials and Devices*, eds. P. Capper and C. T. Elliott, 77–98, Kluwer Academic Publishers, Boston, MA, 2000.
9. M. Kimata, "Silicon Infrared Focal Plane Arrays," in *Handbook of Infrared Detection Technologies*, eds. M. Henini and M. Razeghi, 353–92, Elsevier, Oxford, 2002.
10. R. H. Fowler, "The analysis of photoelectric sensitivity curves for clean metals at various temperatures," *Phys. Rev.*, 38, 45–57, 1931.
11. C. R. Crowell, W. G. Spitzer, L. E. Howarth, and E. E. Labate, "Attenuation length measurements of hot electrons in metal films," *Phys. Rev.*, 127, 2006.
12. R. Stuart, F. Wooten, and W. E. Spicer, "Monte-Carlo calculations pertaining to the transport of hot electrons in metals," *Phys. Rev.*, 135(2A), 495–504, 1964.

13. J. Cohen, J. Vilms, and R. J. Archer, "Investigation of Semiconductor Schottky Barriers for Optical Detection and Cathodic Emission," *Air Force Cambridge Research Labs.* Report No. 68–0651 (1968) and No. 69–0287 (1969).
14. V. L. Dalal, "Simple model for internal photoemission," *J. Appl. Phys.*, 42, 2274–9, 1971.
15. V. E. Vickers, "Model of Schottky-barrier hot electron mode photodetection," *Appl. Opt.*, 10, 2190–2, 1971.
16. J. M. Mooney and J. Silverman, "The theory of hot-electron photoemission in Schottky barrier detectors," *IEEE Trans. Electron Devices* ED-32, 33–9, 1985.
17. S. M. Sze, *Physics of Semiconductor Devices*, Wiley, New York, 1981.
18. F. D. Shepherd, "Infrared internal emission detectors," *Proc. SPIE* 1735, 250–61, 1992.
19. R. Taylor, L. Skolnik, B. Capone, W. Ewing, F. Shepherd, S. Roosild, B. Cochrun, M. Cantella, J. Klein, and W. Kosonocky, "Improved platinum silicide IRCCD focal plane," *Proc. SPIE* 217, 103–10, 1980.
20. H. Elabd and W. F. Kosonocky, "The photoresponse of thin-film PtSi Schottky barrier detector with optical cavity," *RCA Rev.*, 43, 543–7, 1982.
21. H. Elabd and W. F. Kosonocky, "Theory and measurements of photoresponse for thin film Pd_2Si and PtSi infrared Schottky-barrier detectors with optical cavity," *RCA Rev.*, 43, 569–89, 1982.
22. M. Kimata, M. Denda, T. Fukumoto, N. Tsubouchi, S. Uematsu, H. Shibata, T. Higuchi, T. Saeki, R. Tsunoda, and T. Kanno, "Platinum silicide Schottky-barrier IR-CCD image sensor," *Jpn. J. Appl. Phys.*, 21(Suppl. 21-1), 231–5, 1982.
23. J. M. Mooney, J. Silverman, and M. M. Weeks, "PtSi internal photoemission; theory and experiment," *Proc. SPIE* 782, 99–107, 1987.
24. E. L. Dereniak and G. D. Boreman, *Infrared Detectors and Systems*, Wiley, New York, 1996.
25. J. P. Spratt and R. F. Schwarz, "Metal-silicon Schottky diode arrays as infrared retinae," *Tech. Dig. IEDM*, 306–9, 1973.
26. E. Kohn, S. Roosild, F. Shepherd, and A. Young, "Infrared imaging with monolithic CCD-addressed Schottky-barrier detector arrays," *International Conference on Application of CCD's*, 59–69, 1975.
27. H. Elabd, T. S. Villani, and J. R. Tower, "High density Schottky-barrier infrared charge-coupled device (IRCCD) sensors for short wavelength infrared (SWIR) application at intermediate temperature," *Proc. SPIE* 345, 161–71, 1982.
28. H. Elabd, T. Villani, and W. Kosonocky, "Palladium-silicide Schottky-barrier IR-CCD for SWIR applications at intermediate temperatures," *IEEE Electron Device Lett.*, EDL-3, 89–90, 1982.
29. J. R. Tower, A. D. Cope, L. E. Pellon, B. M. McCarthy, R. T. Strong, K. F. Kinnard, A. G. Moldovan, et al., "Development of multispectral detector technology," *Proc. SPIE* 570, 172–83, 1985.
30. J. R. Tower, L. E. Pellon, B. M. McCarthy, H. Elabd, A. G. Moldovan, W. F. Kosonocky, J. E. Kakshoven, and D. Tom, "Shortwave infrared 512×2 line sensor for earth resources applications," *IEEE Trans. Electron Devices* ED-32, 1574–83, 1985.
31. J. R. Tower, A. D. Cope, L. E. Pellon, B. M. McCarthy, R. T. Strong, K. F. Kinnard, A. G. Moldovan, et al., "Design and performance of 4×5120-element visible and 5×2560-element shortwave infrared multispectral focal planes," *RCA Rev.*, 47, 226–55, 1986.
32. B. Capone, L. Skolnik, R. Taylor, F. Shepherd, S. Roosild, and W. Ewing, "Evolution of a Schottky infrared charge-coupled (IRCCD) staring mosaic focal plane," *Opt. Eng.*, 18, 535–41, 1979.
33. W. F. Kosonocky, F. V. Shallcross, T. S. Villani, and J. V. Groppe, "160×244 element PtSi Schottky-barrier IR-CCD image sensor," *IEEE Trans. Electron Devices* ED-32, 1564–73, 1985.
34. D. J. Sauer, F. V. Shallcross, F. L. Hsueh, G. M. Meray, P. A. Levine, H. R. Gilmartin, T. S. Villani, B. J. Esposito, and J. R. Tower, "640×480 MOS PtSi IR sensor," *Proc. SPIE* 1540, 285–96, 1991.
35. J. L. Gates, W. G. Connelly, T. D. Franklin, R. E. Mills, F. W. Price, and T. Y. Wittwer, "488×640-element platinum silicide Schottky focal plane array," *Proc. SPIE* 1540, 262–73, 1991.
36. M. Kimata, M. Denda, N. Yutani, S. Iwade, and N. Tsubouchi, "512×512 element PtSi Schottky-barrier infrared image sensor," *IEEE J. Solid-State Circuits* SC-22, 1124–9, 1987.
37. D. L. Clark, J. R. Berry, G. L. Compagna, M. A. Cosgrove, G. G. Furman, J. R. Heydweiller, H. Honickman, R. A. Rehberg, P. H. Solie, and E. T. Nelson, "Design and performance of a 486–640 pixel platinum silicide IR imaging system," *Proc. SPIE* 1540, 303–11, 1991.
38. N. Yutani, H. Yagi, M. Kimata, J. Nakanishi, S. Nagayoshi, and N. Tsubouchi, "1040×1040 element PtSi Schottky-barrier IR image sensor," *IEDM Tech. Dig.*, 175–8, 1991.
39. P. W. Pellegrini, A. Golubovic, C. E. Ludington, and M. M. Weeks, "IrSi Schottky barrier diodes for infrared detection," *IEDM Tech. Dig.*, 157–60, 1982.
40. P. W. Pellegrini, A. Golubovic, and C. E. Ludington, "A comparison of iridium silicide and platinum silicide photodiodes," *Proc. SPIE* 782, 93–8, 1987.

41. B-Y. Tsaur, M. J. McNutt, R. A. Bredthauer, and B. R. Mattson, "128×128-element IrSi Schottky-barrier focal plane arrays for long wavelength infrared imaging," *IEEE Electron Devices Lett.*, 10, 361–3, 1989.
42. J. Kuriański, J. Vermeiren, C. Claeys, W. Stessens, K. Maex, and R. De Keersmaecker, "Development and evaluation of $CoSi_2$ Schottky barrier infrared detectors," *Proc. SPIE* 1157, 145–52, 1989.
43. J. Kuriański, J. Van Dammer, J. Vermeiren, M. Maex, and C. Claeys, "Nickel silicide Schottky barrier detectors for short wavelength infrared applications," *Proc. SPIE* 1308, 27–34, 1990.
44. N. Tsubouchi, M. Kimata, M. Denda, M. Yamawaki, N. Yutani, and S. Uematsu, "Photoresponse improvement of PtSi-Si Schottky-barrier infrared detectors by ion implantation," *Tech. Dig. 12th ESSDERC* 169–71, 1982.
45. J. M. Shannon, "Reducing the effective height of a Schottky barrier using low-energy ion implantation," *Appl. Phys. Lett.*, 24, 369–71, 1974; "Control of Schottky barrier height using highly doped surface layers," *Solid-State Electron.*, 19, 537–43, 1976.
46. P. Pellegrini, M. Weeks, and C. E. Ludington, "New 6.5 μm photodiode for Schottky barrier array applications," *Proc. SPIE* 311, 24–9, 1981.
47. C-Y. Wei, W. Tantraporn, W. Katz, and G. Smith, "Reduction of effective barrier height in PtSi-p-Si Schottky diode using low-energy ion implantation," *Thin Solid Films* 93, 407–12, 1982.
48. B-Y. Tsaur, C. K. Chen, and B. A. Nechay, "IrSi Schottky-barrier infrared detectors with wavelength response beyond 12 μm," *IEEE Electron Device Lett.*, 11, 415–7, 1990.
49. R. W. Fathauer, T. L. Lin, P. J. Grunthaner, P. O. Andersson, and J. Maserijian, "Modification of the Schottky barrier height of MBE-grown CoSi2/Si(111) diodes by the use of selective Ga doping," in *Proceedings of the 2nd International Symposium on MBE (Honolulu)*, pp. 228–34, 1987.
50. T. L. Lin, J. S. Park, T. George, E. W. Jones, R. W. Fathauer, and J. Maserijian, "Long-wavelength PtSi infrared detectors fabricated by incorporating a p^+ doping spike grown by molecular beam epitaxy," *Appl. Phys. Lett.*, 62, 3318–20, 1993.
51. T. L. Lin, J. P. Park, T. George, E. W. Jones, R. W. Fathauer, and J. Maserijian, "Long-wavelength infrared doping-spike PtSi detectors fabricated by molecular beam epitaxy," *Proc. SPIE* 2020, 30–5, 1993.
52. T. L. Lin, J. S. Park, S. D. Gunapala, E. W. Jones, and H. M. Del Castillo, "Doping-spike PtSi Schottky infrared detectors with extended cutoff wavelengths," *IEEE Trans. Electron Devices* 42, 1216–20, 1995.
53. B-Y. Tsaur, M. M. Weeks, and P. W. Pellegrini, "Pt-Ir silicide Schottky-barrier IR detectors," *IEEE Electron Device Lett.*, 9, 100–102, 1988.
54. B-Y. Tsaur, J. P. Mattia, and C. K. Chen, "Hydrogen annealing of PtSi-Si Schottky barrier contacts," *Appl. Phys. Lett.*, 57, 1111–3, 1990.
55. B-Y. Tsaur, C. K. Chen, and J. P. Mattia, "PtSi Schottky-barrier focal plane arrays for multispectral imaging in ultraviolet, visible, and infrared spectral bands," *IEEE Electron Device Lett.*, 11, 162–4, 1990.
56. J. M. Kurianski, S. T. Shanahan, U. Theden, M. A. Green, and J. W. V. Storey, "Optimization of the cavity for silicide Schottky infrared detectors," *Solid-State Electron.*, 32, 97–101, 1989.
57. C. K. Chen, B. Nechay, and B.-Y. Tsaur, "Ultraviolet, visible, and infrared response of PtSi Schottky-barrier detectors operated in the front illumination mode," *IEEE Trans. Electron Devices* 38, 1094–1103, 1991.
58. W. F. Kosonocky, T. S. Villani, F. V. Shallcross, G. M. Meray, and J. J. O'Neil, "A Schottky-barrier image sensor with 100% fill factor," *Proc. SPIE* 1308, 70–80, 1990.
59. M. Kimata, M. Denda, S. Iwade, N. Yutani, and N. Tsubouchi, "A wide spectral band detector with PtSi/p-Si Schottky-barrier," *Inter. J. Infrared MM Waves* 6, 1031–41, 1985.
60. W. F. Kosonocky, F. V. Shallcross, T. S. Villani, and J. V. Groppe, "160×244 element PtSi Schottky-barrier IR-CCD image sensor," *IEEE Trans. Electron Devices* ED-32, 1564–72, 1995.
61. B.-Y. Tsaur, M. M. Weeks, R. Trubiano, P. W. Pellegrini, and T.-R. Yew, "IrSi Schottky-barrier infrared detectors with 10-μm cutoff wavelength," *IEEE Trans. Electron Device Lett.*, 9, 650–3, 1988.
62. F. D. Shepherd, V. E. Vickers, and A. C. Yang, "Schottky-barrier photodiode with degenerate semiconductor active region," U. S. Patent No. 3,603,847, September 7, 1971.
63. T. L. Lin, A. Ksendzov, S. M. Dajewski, E. W. Jones, R. W. Fathauer, T. N. Krabach, and J. Maserjian, "A novel Si-based LWIR detector: The SiGe/Si heterojunction internal photoemission detector," *IEDM Tech. Dig.*, 641–4, 1990.
64. T. L. Lin and J. Maserjian, "Novel $Si_{1-x}Ge_x$/Si heterojunction internal photoemission long-wavelength infrared detectors," *Appl. Phys. Lett.*, 57, 1422–4, 1990.
65. T. L. Lin, A. Ksendzov, S. M. Dejewski, E. W. Jones, R. W. Fathauer, T. N. Krabach, and J. Maserjian, "SiGe/Si heterojunction internal photoemission long-wavelength infrared detectors fabricated by molecular beam epitaxy," *IEEE Trans. Electron Devices* 38, 1141–4, 1991.

66. T. L. Lin, E. W. Jones, T. George, A. Ksendzov, and M. L. Huberman, "Advanced Si IR detectors using molecular beam epitaxy," *Proc. SPIE* 1540, 135–9, 1991.
67. B.-Y. Tsaur, C. K. Chen, and S. A. Marino, "Long-wavelength GeSi/Si heterojunction infrared detectors and 400×400-element imager arrays," *IEEE Electron Device Lett.*, 12, 293–6, 1991.
68. B.-Y. Tsaur, C. K. Chen, and S. A. Marino, "Long-wavelength $Ge_{1-x}Si_x$/Si heterojunction infrared detectors and focal plane arrays," *Proc. SPIE* 1540, 580–95, 1991.
69. B.-Y. Tsaur, C. K. Chen, and S. A. Marino, "Heterojunction $Ge_{1-x}Si_x$/Si infrared detectors and focal plane arrays," *Opt. Eng.*, 33, 72–8, 1994.
70. T. L. Lin, J. S. Park, S. D. Gunapal, E. W. Jones, and H. M. Del Castillo, "Photoresponse model for $Si_{1-x}Ge_x$/Si heterojunction internal photoemission infrared detector," *IEEE Electron Device Lett.*, 15, 103–5, 1994.
71. T. L. Lin, J. S. Park, S. D. Gunapala, E. W. Jones, and H. M. Del Castilo, "Si_{1-x}Gex/Si heterojunction internal photoemission long wavelength infrared detector," *Proc. SPIE* 2474, 17–23, 1994.
72. J. S. Park, T. L. Lin, E. W. Jones, H. M. Del Castillo, T. George, and S. D. Gunapala, "Long-wavelength stacked $Si_{1-x}Ge_x$/Si heterojunction internal photoemission infrared detectors," *Proc. SPIE* 2020, 12–21, 1993.
73. J. S. Park, T. L. Lin, E. W. Jones, H. M. Del Castillo, and S. D. Gunapala, "Long-wavelength stacked SiGe/Si heterojunction internal photoemission infrared detectors using multiple SiGe/Si layers," *Appl. Phys. Lett.*, 64, 2370–2, 1994.
74. H. Wada, M. Nagashima, K. Hayashi, J. Nakanishi, M. Kimata, N. Kumada, and S. Ito, "512×512 element GeSi/Si heterojunction infrared focal plane array," *Opto-Electron. Rev.*, 7, 305–11, 1999.
75. Y. F. Lao and A. G. U. Perera, "Physics of internal photoemission and its infrared applications in the low-energy limit," *Adv. OptoElectron.*, Article ID 1832097, 1–18, 2016, http://dx.doi.org/10.1155/2016/1832097.
76. S. Tohyama, N. Teranishi, K. Kunoma, M. Nishimura, K. Arai, and E. Oda, "A new concept silicon homojunction infrared sensor," *IEDM Tech. Dig.*, 82–5, 1988.
77. W. F. O'Neil, "Nonuniformity corrections for spectrally agile sensor," *Proc. SPIE* 1762, 327–39, 1992.
78. A. G. U. Perera, H. X. Yuan, and M. H. Francombe, "Homojunction internal photoemission far-infrared detectors: Photoresponse performance analysis," *J. Appl. Phys.*, 77, 915–24, 1995.
79. A. G. U. Perera, "Physics and Novel Device Applications of Semiconductor Homojunctions," in *Thin Solid Films*, Vol. 21, eds. M. H. Francombe and J. L. Vossen, 1–75, Academic Press, New York, 1995.
80. H. X. Yuan and A. G. H. Perera, "Dark current analysis of Si homojunction interfacial work function internal photoemission far-infrared detectors," *Appl. Phys. Lett.*, 66, 2262–4, 1995.
81. A. G. U. Perera, H. X. Yuan, J. W. Choe, and M. H. Francombe, "Novel homojunction interfacial workfunction internal photoemission (HIWIP) tunable far-infrared detectors for astronomy," *Proc. SPIE* 2475, 76–87, 1995.
82. W. Shen, A. G. U. Perera, M. H. Francombe, H. C. Liu, M. Buchanan, and W. J. Schaff, "Effect of emitter layer concentration on the performance of GaAs p^+–i homojunction far-infrared detectors: A comparison of theory and experiment," *IEEE Trans. Electron Devices* 45, 1671–7, 1998.
83. A. G. U. Perera and W. Z. Shen, "GaAs homojunction interfacial workfunction internal photoemission (HIWIP) far-infrared detectors," *Opto-Electron. Rev.*, 7, 153–80, 1999.
84. A. G. H. Perera, "Semiconductor Photoemissive Structures for Far Infrared Detection," in *Handbook of Thin Devices*, Vol. 2, ed. M. H. Francombe, 135–70, Academic Press, San Diego, CA, 2000.
85. A. G. H. Perera, "Silicon and GaAs as Far-Infrared Detector Material," in *Photodetectors and Fiber Optics*, ed. H. S. Nalwa, 203–37, Academic Press, San Diego, CA, 2001.
86. E. E. Haller, "Advanced far-infrared detectors," *Infrared Phys. Technol.*, 35, 127, 1994.
87. D. W. Watson, M. T. Guptill, J. E. Huffman, T. N. Krabach, S. N. Raines, and S. Satyapal, "Germanium blocked-impurity-band detector arrays: Unpassivated devices with bulk substrates," *J. Appl. Phys.*, 74, 4199, 1993.

16

III–V Detectors

In the middle and late 1950s it was discovered that InSb had the smallest energy gap of any semiconductor known at that time and its applications as a middle wavelength infrared (MWIR) detector became obvious [1,2]. The energy gap of InSb is less well matched to the 3–5 μm band at higher operating temperatures, and better performance can be obtained from $Hg_{1-x}Cd_xTe$. InAs is a similar compound to InSb but has a larger energy gap [3], so that the threshold wavelength is 3–4 μm, and both photoconductive and photovoltaic detectors have been fabricated. The photoconductive process in InSb has been studied extensively, and more details can be found in Morten and King [4], Kruse [5], and Elliott and Gordon [6].

Indium antimonide detectors have been extensively used in high-quality detection systems and have found numerous applications in the defense and space industry for more than 40 years. Perhaps the best known (and most successful) of these systems has been the Sidewinder air-to-air antiaircraft missile. Manufacturing techniques for InSb are well-established, and the invention of charge-coupled devices (CCD) and complementary metal oxide semiconductor (CMOS) hybrid devices has increased the interest in this semiconductor.

At the beginning of the 1990s, several national agencies (e.g., in U.S., Germany, and France) switched its research emphasis to III–V low-dimensional solid materials (quantum wells and superlattices), as an alternative technology option to HgCdTe. New emerging strategies include type-II superlattice photodiodes and barrier structures such as nBn detectors [7].

16.1 SOME PHYSICAL PROPERTIES OF III–V NARROW GAP SEMICONDUCTORS

Development of crystal growth techniques in the early 1950s led to InSb and InAs bulk single crystal detectors. Since then, the quality of single crystal growth has improved immensely. Several methods have been developed, among which the Czochralski, Bridgman, and vertical gradient freeze (VGF) methods are the most popularly used. Every crystal growth method has advantages and disadvantages. For commercial production, trade-offs need to be made to maintain balance between crystal quality, desirable electrical and optical properties, infrastructure investments and operational costs, and so on. Whilst the quantity of III–V compound semiconductor materials produced by the Czochralski method remains low when compared with more volume-oriented crystal growth methods such as VGF, the substrate industry has met the challenge to develop more volume production-oriented approaches to crystal growth and expended road map of advanced III–V devices technologies that are reliant on Czochralski-grown materials. For small-scale fundamental research, much wider crystal growth methods can be suitably adopted, based on available infrastructure.

Single crystals are grown with relatively high purity, low dislocation density, and ingot sizes that permit wafer diameters up to 6 inches, suitable for convenient handling and photolithography [7]. The growth of InSb single crystals is reviewed by several authors [8–12]. A wide spectrum of topics in materials that today's engineers, material scientists, and physicists need is included in a comprehensive treatise on III–V electronic and photonic materials gathered in the two *Springer Handbooks* [13,14].

For commercial production of InSb, the majority of ingots are grown in the (211) direction. Production of (100) InSb is possible, but difficulties are encountered in the growth of this orientation type and applications of (100) InSb remain limited. At present, the ultrahigh pure InSb bulk crystal's carrier concentration can be less than 10^{-13} cm^3. Typical etch pit densities below 10^2 cm^{-2} are found for InSb, which is considered one of the lowest defect type compound semiconductor materials available commercially.

Unlike InSb, the reaction of the elements to form an InAs compound is not a simple matter. To keep the arsenic from disappearing because of its high vapor pressure near the melting point, it is necessary to let the constituents react in a sealed quartz ampoule. Purification of InAs is also more difficult than InSb. InAs single crystals up to 100-mm diameter are being commercially grown using high-pressure liquid-encapsulated Czochralski (LEC) technique. pBN crucibles and ultralow water content B_2O_3 encapsulation are being used. For research purposes, InAs single crystals can be grown in vacuum-sealed crucibles using vertical Bridgman/VGF method. In the last case, an appropriately placed As reservoir inside the crucible and a pressure vessel for providing counteracting pressure outside the crucible are required during crystal growth.

Also, in the case of growth of GaSb single crystals from a melt, several challenges must be overcome that require modifications of the conventional Czochralski equipment. Problems include controlling oxide scum on the melt surface and evaporation of Sb from the melt surface. Initially, researches used the LEC growth, using molten B_2O_3 to prevent water from reaching the surface for oxidation scum and preventing Sb vapor from escaping [11]. This method is still in use today but presents its own set of complications (the molten B_2O_3 causes the melt contamination, changes in surface tension and viscosity, significant alteration of the growth process, and modifies the heat and energy flow at the melt meniscus). In addition, since B_2O_3 is hygroscopic, extra care must be taken to keep it dry by vacuum baking or bubbling with dry N_2. Growth without encapsulant has been revisited a number of times. GaSb crystals grown without encapsulant in a hydrogen environment had reduced oxide formation, higher crystal quality, and less twinning probability than encapsulated runs. Historical challenges with Czochralski GaSb crystal growth are described in more details in Ref. [15].

The majority of GaSb detector production relies upon epitaxial growth using 3" diameter substrates, with a small volume of 4" material consumed too. However, to bring GaSb substrate production technologies to the same level of maturity as InSb, interests in 6" GaSb substrates for very large area detector applications has emerged recently.

GaSb is intrinsically p-type, with hole carrier concentrations of $\sim 10^{17}\,cm^{-3}$ at room temperature. The residual hole concentration of GaSb bulk crystals is about $2\times 10^{16}\,cm^{-3}$. The intrinsic defect is mostly Ga antisite defects or complexes of $V_{Ga}Ga_{Sb}$ providing acceptor sites. High resistivity or n-type GaSb can be obtained by compensation doping with group VI elements such as Te—they are characterized by lower absorption coefficient in comparison with the undoped ones.

Compared with other antimonide-based III–V compounds, little work has been done on the investigation of the AlSb. The large, high-quality AlSb single crystals are rarely fabricated and their surfaces react rapidly with air. AlSb single crystals are not commercially available; in their Czochralski process, Al_2O_3 crucibles are the most suitable.

Due to an immiscible gap of the multielement antimonides, their growth process is very immature and the ternary and quaternary antimonide bulk crystal materials are rarely used.

The technology of semi-insulating InP substrates is becoming of great interest to an increasing number of applications in high-speed devices, such as MOSFETs, optoelectronics communications, and solar cells. The high-pressure LEC growth is normally used for InP. The problems involved are analogous to those for GaAs growth, with the major addition of an increased tendency for twinning. Loss of phosphorus (P_4) is still a problem and leads to a deterioration in crystal quality. Dislocation densities are still seen to be high. Some of these problems have been alleviated by modifications to the basic LEC process by vapor pressure–controlled Czochralski process [16,17]. Lower dislocation densities are produced by the VGF method at a given crystal diameter. InP single crystals 100 mm in diameter and 80 mm in length can be obtained.

The $In_xGa_{1-x}As$ ternary alloy has been of great interest for the short wavelength infrared (SWIR), low-cost detector applications. The $x=0.53$ alloy is lattice-matched to InP substrates, has a bandgap of 0.73 eV, and covers the wavelength range from 0.9 to 1.7 μm. The $In_xGa_{1-x}As$ with 53% InAs is often called "standard InGaAs" without bothering to note the values of "x" or "$1-x$". This is mature material, driven by the mass production of fiber optic receivers at 1.3 and 1.55 μm. At present, InGaAs is also becoming the choice for high-temperature operations in the 1–3 μm spectrum. By increasing the indium content to $x=0.82$, the wavelength response of $In_xGa_{1-x}As$ can be extended out to 2.6 μm. Single element InGaAs detectors have been made with up to 2.6 μm cutoffs, while linear arrays and cameras have been demonstrated to 2.2 μm.

The energy bandgap of $In_xGa_{1-x}As_yP_{1-y}$ quaternary system (with the values of x and y varying independently) range from 0.35 eV (InAs) to 2.25 eV (GaP), with InP (1.29 eV) and GaAs (1.43 eV) falling in between [18,19]. Adding either element changes the lattice constant, so successfully growing the compound on an InP substrate requires balancing the composition to match the lattice constant to the substrate. The room temperature bandgap of $In_xGa_{1-x}As_yP_{1-y}$ lattice-matched to InP is represented by the following expression:

$$E_g\left(\text{in eV}\right) = 1.35 - 0.72y + 0.12x^2. \tag{16.1}$$

InGaAsP alloys have been epitaxially grown by hydride and chloride vapor phase epitaxy (VPE), liquid phase epitaxy (LPE), molecular beam epitaxy (MBE), and metalorganic chemical vapor deposition (MOCVD) [20]. A brief comparison of the four techniques is given by Olsen and Ban [21]. While each of these techniques has certain advantages, hydride VPE is well-suited for InGaAsP/InP optoelectronic devices.

The substitution of a fraction of antimony sites in InSb with isovalent arsenic reduces the energy gap of InSb-InAs ($InAs_{1-x}Sb_x$) to a value lower than the energy gap of either of the parent binary compounds. Consequently, $InAs_{1-x}Sb_x$ ternary alloy has the lowest energy gap among the III–V semiconductors. A room temperature energy gap in both 3–5 and 8–14-μm atmospheric wavelength windows can be achieved. However, progress in this ternary system has been limited by crystal synthesis problems. The large separation between the solidus and liquidus [see Figure 16.1(a)] and the lattice mismatch (6.9% between InAs and InSb) place stringent demands upon the method of crystal growth. These difficulties are being overcome systematically using MBE and MOCVD growth methods.

Considering the $Ga_{1-x}In_xSb$ (GaInSb) pseudobinary phase diagram shown in Figure 16.1(b), the separation between the solidus and liquidus curves leads to alloy segregation. The vertical Bridgman or vertical gradient freezing process is the most suitable method for growing large-diameter GaInSb bulk crystals. The established process has been successfully demonstrated in laboratory-scale experiments for growing GaInSb crystals (up to 50-mm diameter) with a wide range of alloy compositions.

Table 16.1 presents some physical properties of semiconducting families, including narrow-gap semiconductors, used in the fabrication of infrared photodetectors. All compounds have a diamond (D) or a zincblende (ZB) crystal structure. Moving across the table from left to right, there is a trend in change of chemical bond from the covalent group IV semiconductors to the more ionic II–VI semiconductors with increasing of the lattice constant. The chemical bonds become weaker and the materials become softer, which is reflected by the values of the bulk. The materials with larger contribution of covalent bound are more mechanically robust, which leads to better manufacturability. This is evidenced in the dominant

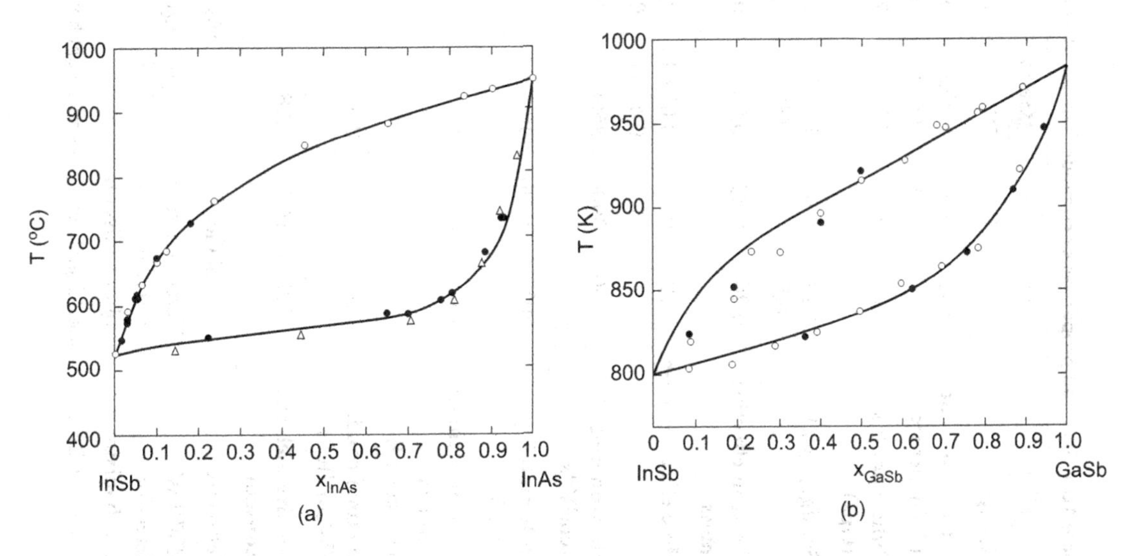

Figure 16.1 Pseudobinary phase diagram for the InAs-InSb (a) and InSb-GaSb systems (b).

Table 16.1 Selected properties of common families of semiconductors used in the fabrication of infrared photodetectors

	Si	Ge	GaAs	AlAs	InP	InGaAs	AlInAs	InAs	GaSb	AlSb	InSb	HgTe	CdTe
Group	IV	IV	III-V	III-V	III-V	III-V	III-V	III-V	III-V	III-V	III-V	II-VI	II-VI
Lattice constant (Å)/structure	5.431 (D)	5.658 (D)	5.653 (ZB)	5.661 (ZB)	5.870 (ZB)	5.870 (ZB)	5.870 (ZB)	6.058 (ZB)	6.096 (ZB)	6.136 (ZB)	6.479 (ZB)	6.453 (ZB)	6.476 (ZB)
Bulk modulus (Gpa)	98	75	75	74	71	69	66	58	56	55	47	43	42
Bandgap (eV)	1.124 (id)	0.660 (id)	1.426 (d)	2.153 (id)	1.350 (d)	0.735 (d)		0.354 (d)	0.730 (d)	1.615 (id)	0.175 (d)	−0.141 (d)	1.475 (d)
Electron effective mass	0.26	0.39	0.067	0.29	0.077	0.041		0.024	0.042	0.14	0.014	0.028	0.090
Hole effective mass	0.19	0.12	0.082(L) 0.45(H)	0.11(L) 0.40(H)	0.12(L) 0.55(H)	0.05(L) 0.60(H)		0.025(L) 0.37(H)	0.4	0.98	0.018(L) 0.4(H)	0.40	0.66
Electron mobility ($cm^2V^{-1}s^{-1}$)	1450	3900	8500	294	5400	13800		3×10^4	5000	200	8×10^4	26500	1050
Hole mobility ($cm^2V^{-1}s^{-1}$)	505	1900	400	105	180			500	880	420	800	320	104
Electron saturation velocity (10^7 cm s^{-1})	1.0	0.70	1.0	0.85	1.0			4.0			4.0		
Thermal cond. ($Wcm^{-1}K^{-1}$)	1.31	0.31	0.5		0.7			0.27	0.4	0.7	0.15		0.06
Relative dielectric constant	11.9	16.0	12.8	10.0	12.5			15.1	15.7	12.0	17.9	21	10.2
Substrate	Si,Ge		GaAs		InP			InAs,GaSb			InSb	CdZnTe,GaAs,Si	
MW/LW detection mechanism	Heterojunction internal photoemission		QWIP,QDIP		QWIP			Bulk (MW) Superlattice (MW/LW) Band-to-band (B-B)			Bulk B-to-B	Bulk Band-to-band	

D, diamond; ZB, zincblende; id, indirect; d, direct; L, light hole; H, heavy hole position of silicon in electronic materials and GaAs in optoelectronics ones. On the other hand, the bandgap energy of semiconductors on the right side of the table tends to have smaller values. Due to their direct bandgap structure, strong band-to-band absorption leading to high quantum efficiency is observed (e.g., in InSb and HgCdTe).

Table 16.2 gives a detailed description of physical parameters of narrow-gap III–V semiconductors: InAs, InSb, GaSb, $InAs_{0.35}Sb_{0.65}$, and $In_{0.53}Ga_{0.47}As$. Between them, InSb and InGaAs have been the most broadly investigated.

The temperature independent portions of the Hall curves indicate that most of the electrically active impurity atoms in InSb have shallow activation energies and above 77 K are thermally ionized. The Hall coefficient for p-type samples is positive in the low-temperature extrinsic range and reverses sign to become negative in the intrinsic range because of the higher mobility of the electrons (the mobility ratio $b=\mu_e/\mu_h$ of the order of 10^2 is observed). The transition temperature for the p-type samples, at which R_H changes sign, depends on the purity. The samples become intrinsic above a certain temperature (above 150 K for pure n-type samples), and below these temperatures (below 100 K for pure n-type samples), there is little variation of Hall coefficients.

There are various carrier scattering mechanisms in semiconductors, as shown in Figure 16.2 for InSb [23]. Reasonably pure n-type and p-type samples exhibit an increase in mobility up to approximately 20–60 K after which the mobility decreases due to polar and electron hole scattering. Carrier mobility systematically increases with a decrease in impurity concentration both in temperatures 77 as well as 300 K.

In alloy semiconductors, the charged carriers see potential fluctuations as a result of the composition disorder. This kind of scattering mechanisms, so-called alloy scattering, is important in some III–V ternaries and quaternaries. Let us simply express the total carrier mobility μ_{tot} in alloy $A_xB_{1-x}C$ as [13]

$$\frac{1}{\mu_{tot}(x)} = \frac{1}{x\mu_{tot}(AC)+(1-x)\mu_{tot}(BC)} + \frac{1}{\mu_{al,0}/[x/(1-x)]}. \tag{16.2}$$

The first term in Equation 16.2 comes from the linear interpolation scheme and the second term accounts for the effects of alloying. For example, Figure 16.3 plots the electron Hall mobility in $Ga_xIn_{1-x}P_yAs_{1-y}$/InP quaternary [13]. The quaternary is an alloy of the constituents $In_{0.53}Ga_{0.47}As$ ($y=0$) and InP ($y=1.0$) and the values $\mu_{tot}(In_{0.53}Ga_{0.47}As)=13{,}000$ and $\mu_{al,0}=3{,}000$ cm^2 Vs^{-1} have been used. The experimental data correspond to those for relatively pure samples.

Optical properties of InSb have been reviewed by Kruse [5]. Because of the very small effective mass of electrons, the conduction band density of states is small and it is possible to fill the available band states by doping, thereby appreciably shifting the absorption edge to shorter wavelengths. This has been referred to as the Burstein–Moss effect (Figure 16.4 [24]).

The physical properties of InAs are similar to those of InSb. An InAs material with a cutoff wavelength of 3.5 μm makes it of limited utility for the middle wavelength band even though it could theoretically be operated near 190 K. Development work has been limited by growth and passivation problems.

The III–V detector materials have a ZB structure and direct energy gap at the Brillouin zone center. The shape of the electron band and the light mass hole band is determined by the **k·p** theory. The momentum matrix element varies only slightly for different materials and has an approximate value of 9.0×10^{-8}eV cm. Then, the electron effective masses and conduction band densities of states are similar for materials with the same energy gap.

These materials have a conventional negative temperature coefficient of the energy gap that is well described by the Varshni relation [25]

$$E_g(T) = E_o - \frac{\alpha T^2}{T+\beta}, \tag{16.3}$$

where α and β are fitting parameter characteristics of a given material.

Figure 16.5 shows the composition dependence of the energy gap and the electron effective mass at the Γ-conduction bands of $Ga_xIn_{1-x}As$, $InAs_xSb_{1-x}$, and $Ga_xIn_{1-x}Sb$ ternaries.

Table 16.2 Physical properties of narrow-gap III–V compounds

	T(K)	InAs	InSb	GaSb	$InAs_{0.35}Sb_{0.65}$	$In_{0.53}Ga_{0.47}As$
Lattice structure		cub.(ZnS)	cub.(ZnS)	cub. (ZnS)	cub.(ZnS)	cub.(ZnS)
Lattice constant a (nm)	300	0.60584	0.647877	0.6094	0.636	0.58438
Thermal expansion coefficient α ($10^{-6}K^{-1}$)	300 80	5.02	5.04 6.50	6.02		
Density ρ (g cm^{-3})	300	5.68	5.7751	5.61		5.498
Melting point T_m (K)		1210	803	985		
Energy gap E_g (eV)	4.2	0.42	0.2357	0.822	0.138	0.627
	80	0.414	0.228	0.725	0.136	0.75
	300	0.359	0.180		0.100	
Thermal coefficient of E_g	100–300	-2.8×10^{-4}	-2.8×10^{-4}			-3.0×10^{-4}
Effective masses:						
m_e^*/m	4.2	0.023	0.0145	0.042		0.041
	300	0.022	0.0116		0.0101	0.0503
m_{lh}^*/m						
m_{hh}^*/m	4.2	0.026	0.0149			
	4.2	0.43	0.41	0.28	0.41	0.60
Momentum matrix element P (eV cm)		9.2×10^{-8}	9.4×10^{-8}			
Mobilities:						
μ_e ($cm^2V^{-1}s^{-1}$)	77	8×10^4	10^6		5×10^5	70,000
	300	3×10^4	8×10^4	5×10^3	5×10^4	13,800
μ_h ($cm^2V^{-1}s^{-1}$)	77		1×10^4	2.4×10^3		
	300	500	800	880		
Intrinsic carrier concentration n_i (cm^{-3})	77	6.5×10^3	2.6×10^9		2.0×10^{12}	
	200	7.8×10^{12}	9.1×10^{14}		8.6×10^{15}	5.4×10^{11}
	300	9.3×10^{14}	1.9×10^{16}		4.1×10^{16}	
Refractive index n_r		3.44	3.96	3.8		
Static dielectric constant ε_s		14.5	17.9	15.7		14.6
High-frequency dielectric constant ε_∞		11.6	16.8	14.4		
Optical phonons:						
LO (cm^{-1})		242	193		≈ 210	
TO (cm^{-1})		220	185		≈ 200	

Source: After Refs. [18],[19], and [22].

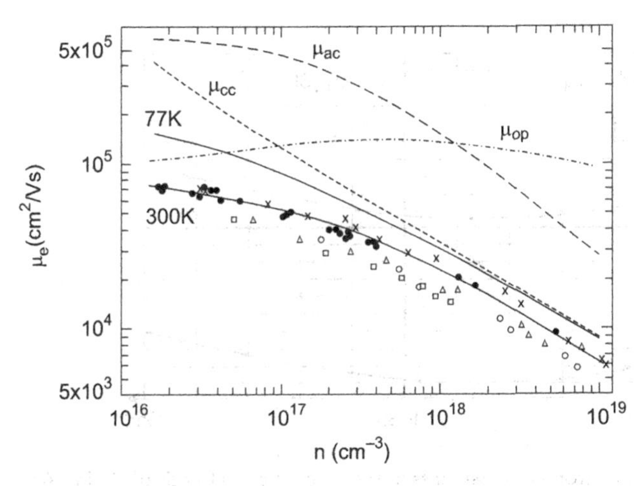

Figure 16.2 Electron mobility in n-type InSb at 300 and 77 K versus free electron concentration—dashed lines denote the theoretical mobilities at 300 K for charged center, polar optical, and acoustic scattering modes. The experimental data are taken at 300 K. (From Ref. [23])

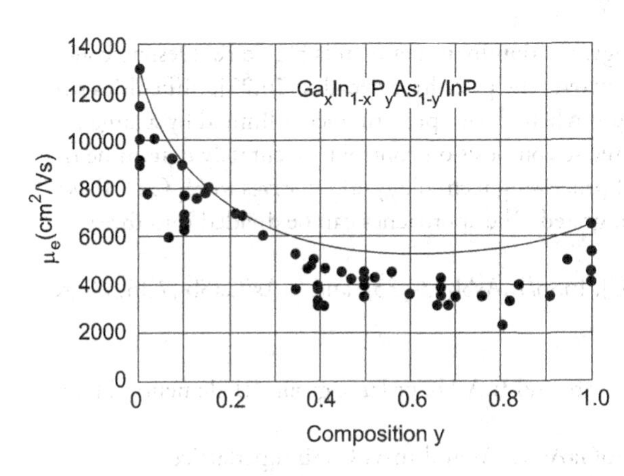

Figure 16.3 Electron Hall mobility in the $Ga_xIn_{1-x}P_yAs_{1-y}$/InP quaternary. The experimental data correspond to those for relatively pure samples. The solid line represents the results calculated using Equation 16.2 with the values of $\mu_{tot}(In_{0.53}Ga_{0.47}As) = 13,000$ and $\mu_{al,0} = 3,000$ cm^2 Vs^{-1}. (From Ref. [13])

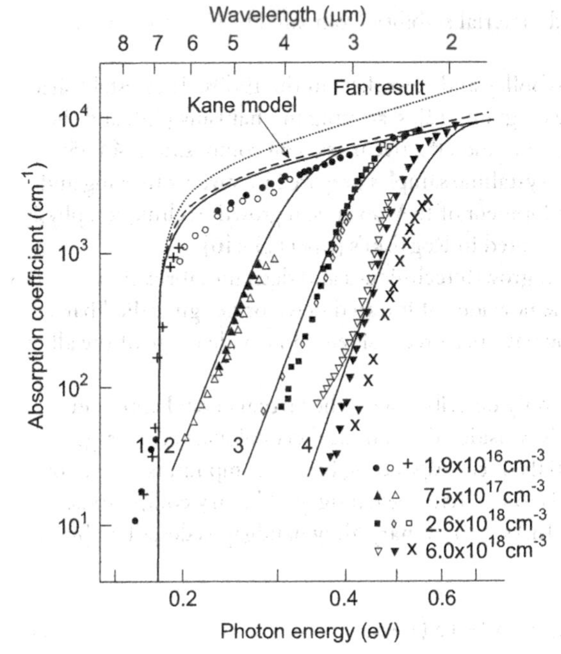

Figure 16.4 Dependence of the optical absorption coefficient of InSb upon photon energy at 300 K. Carrier concentration: 1.9×10^{16}cm^{-3} (1), 7.5×10^{17}cm^{-3} (2), 2.6×10^{18}cm^{-3} (3), and 6.0×10^{18}cm^{-3} (4). (From Ref. [24])

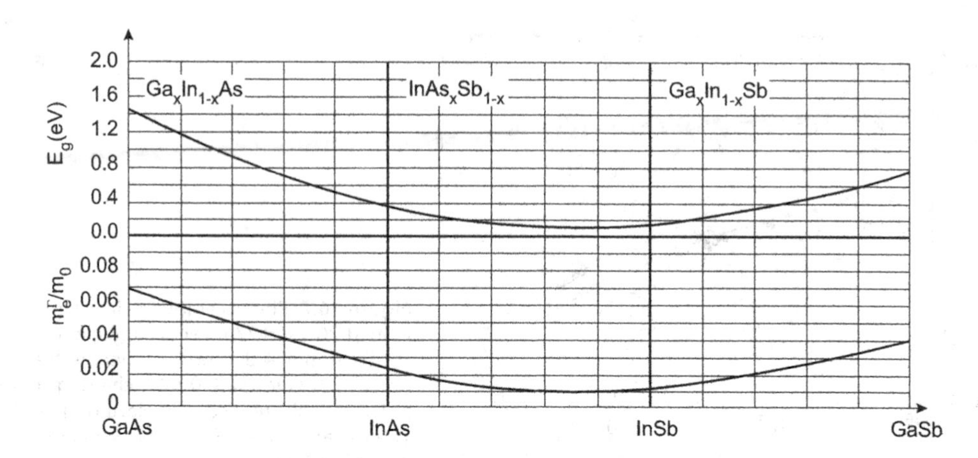

Figure 16.5 Variation of the bandgap energy and electron effective mass at the Γ-conduction bands of $Ga_xIn_{1-x}As$, $InAs_xSb_{1-x}$, and $Ga_xIn_{1-x}Sb$ ternary alloys at room temperature.

For long wavelength (8–12 μm) detector technology, the dominant material is HgCdTe. Despite considerable progress in HgCdTe photovoltaic technology over the past three decades [26,27], difficulties still remain—particularly for wavelengths exceeding 10 μm where device performance is limited by a large tunneling dark current and a sensitive dependence on precise composition control to accurately determine the energy gap. Given the maturity of III–V growth and processing technology, alternatives to HgCdTe-based III–V semiconductors have been investigated and suggested. The approaches can be divided into three categories:

- Use of superlattices such as AlGaAs/GaAs [28–33], InSb/InAsSb [34,35], and InAs/GaSb [7,36,37] (see Chapters 19 and 20)
- Use of quantum dots (see Chapter 21)
- Addition of large group V elements, Bi, in InAs, InSb, and InAsSb, or large group III elements, Tl, in InAs, InP, and InSb [38,39]

Hitherto, the best results have been obtained using AlGaAs/GaAs and InAs/GaSb superlattices.

The development of $InAs_{1-x}Sb_x$ ternary alloys has a long history. InAsSb ternary alloy as an alternative to HgCdTe for IR applications has been demonstrated in the middle of 1970s [40]. A III–V detector technology would benefit from superior bond strengths and material stability (compared to HgCdTe), well-behaved dopants, and high-quality III–V substrates.

The properties of InAsSb were first investigated by Woolley and coworkers in the 1960s. They established the InAs–InSb miscibility [41], the pseudobinary phase diagram [42], scattering mechanisms [43], and the dependence of fundamental properties such as bandgap [44] and effective masses on composition [44,45]. All of the above measurements were performed on polycrystalline samples prepared by various freezing and annealing techniques. The review of the early stage development of InAsSb crystal growth techniques, physical properties, and detector fabrication procedures is presented in Rogalski's papers [35,46].

Recently, the rapid development of $InAs_{1-x}Sb_x$ crystal growth techniques and detector fabrication procedure has been observed due to the discovery of the new idea of infrared detector design, called barrier detector. Also, electronic properties of this ternary alloy have been reconsidered in a wide range of the alloy composition [38,47–50].

The electronic properties of ternary alloys are commonly described with the virtual crystal approximation approach [51]. In this model, the disordered alloy is considered by an ideal crystal with an average potential modeled by linear interpolation of the potentials of corresponding binary compounds. The alloy bandgap depends nonlinearly on the composition and is lower than the bandgap of binary compounds.

The nonlinearity of the composition dependence of $InAs_{1-x}Sb_x$ ternary alloy bandgap is described by the bowing parameter C as

$$E_g(x) = E_{gInSb} + E_{gInAs}(1-x) - Cx(1-x). \qquad (16.4)$$

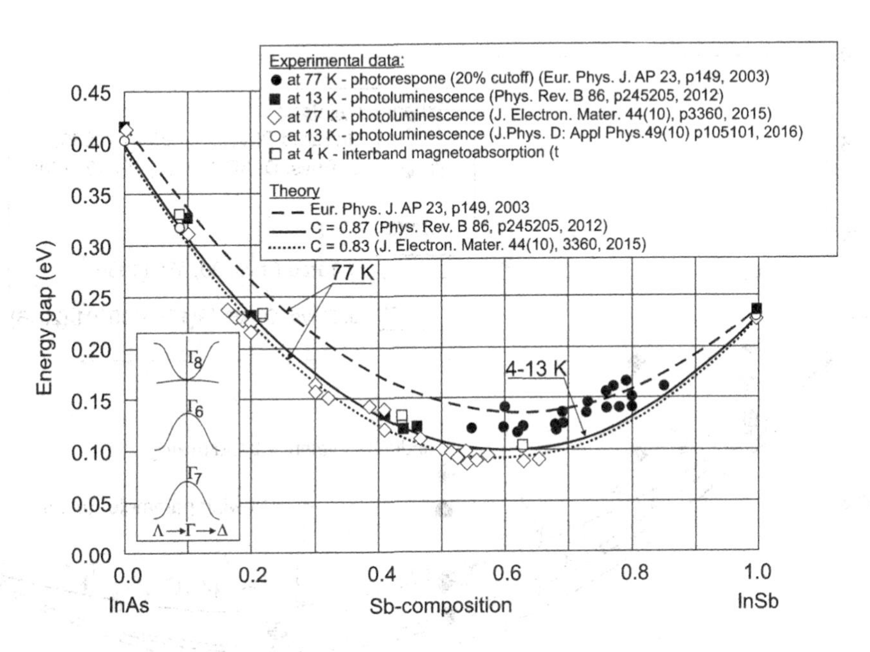

Figure 16.6 Bandgap energy of $InAs_{1-x}Sb_x$ as a function of the Sb composition. The experimental data are taken with different papers as is marked.

Initial reports based on experimental data at temperatures above or near 100 K put the direct-gap bowing parameter in InAsSb at 0.58–0.6 eV [18]. Theoretical considerations led to a higher projected bowing parameter of 0.7 eV, which was recommended by Rogalski and Jozwikowski [52]. More recent photoluminescence studies on unrelaxed MBE-grown $InAs_{1-x}Sb_x$ in a wide range of composition gave the bowing parameter from 0.83 to 0.87 [48,49]. Figure 16.6 summarizes the experimental data and theoretical predictions of energy gap in the temperature range of 4–77 K published in different papers. Discrepancies in the $E_g(x,T)$ dependence can be caused by several reasons including structural quality of samples and CuPt-type ordering effect. It is possible that the low energy gap data in earlier reports were masked by electron filling of the conduction band, which would be due to background doping as these samples were grown with various degrees of residual strain and relaxation. High-quality unstrained unrelaxed InAsSb epilayers have been developed by using a special graded buffer layer that accommodates the large difference between the lattice constant of the substrate and alloy. Electron diffraction patterns of unstrained InAsSb alloys, described in Refs. [48] and [49], showed an ordering-free distribution of group V elements, which indicates that the observed energy gaps of ternary alloys are inherent (both ordering and residual strain effects were eliminated).

The described $E_g(x,T)$ dependency indicates that conventional InAsSb has a sufficiently small gap at 77 K for operation in the 8–14-μm wavelength range and differs from that previously described by Wieder and Clawson [53]

$$E_g(x,T) = 0.411 - \frac{3.4\times10^{-4}T^2}{210+T} - 0.876x + 0.70x^2 + 3.4\times10^{-4}xT(1-x). \quad (16.5)$$

It has been found that the bandgap energy of InAsSb ternary compound is generally a square function of the composition and indicates a fairly weak dependence of the band edge on composition in comparison with HgCdTe (see Fig. 19.7). The minimum of E_g appears at composition $x \approx 0.63$.

To obtain a good agreement between experimental room temperature–effective masses and calculations, Rogalski and Jozwikowski [52] have taken into account the conduction–valence band mixing theory [54] (see Figure 16.7). More recently published low-temperature data, especially for middle composition range, refer to lower electron effective masses. Estimated negative bowing parameter for the electron effective mass is $C_m = 0.038$ and is slightly less than expected from the Kane model ($C_m = 0.045$), reaching the lowest

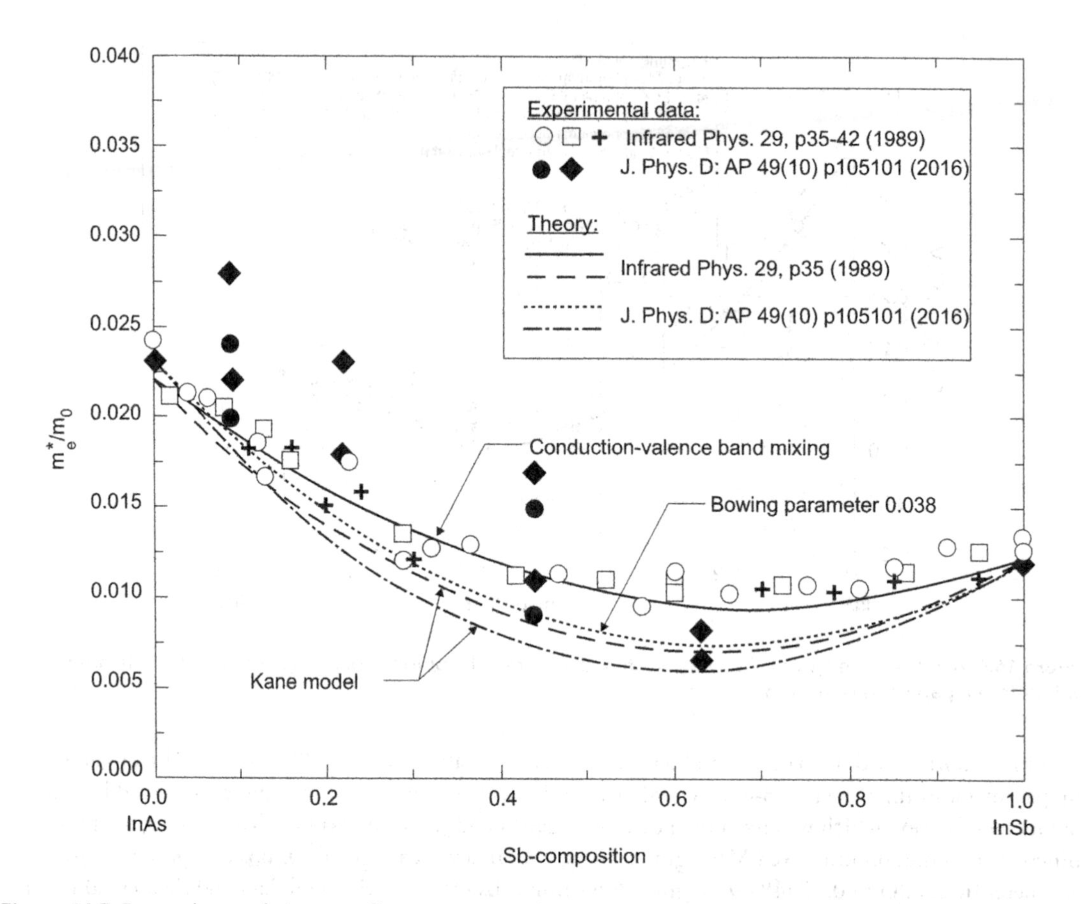

Figure 16.7 Dependence of electron effective mass on the composition for $InAs_{1-x}Sb_x$ alloy system. The experimental data are taken from Refs. [50] and [52].

effective mass ($0.0082m_0$ at $x=0.63$ and 4 K) ever reported for III–V semiconductors. A possible reason for the effective mass value discrepancy, shown in Figure 16.7, is the mixing of the conduction and valence band states caused by the random potential due to alloy disorder.

The measured results of spin–orbit splitting energy of $InAs_{1-x}Sb_x$ ternary alloy, Δ, by Cripps et al. [55] are in strong disagreement with the paper of Van Vechten et al. published in 1972 [56]. Almost no bowing for the Δ parameter as a function of x (Sb fraction) was observed. A good approximation gives

$$\Delta(x) = 0.81x + 0.373(1-x) + 0.165x(1-x) \text{ in eV.} \tag{16.6}$$

This spin–orbit splitting bandgap energy much more satisfactorily agrees with measurements and it is independent of temperature. However, the recently measured composition and temperature dependence of energy gap are consistent with that predicted from the literature.

The intrinsic carrier concentration in InAsSb as a function of composition x for various temperatures can be approximated by the following relation [46]:

$$n_i = \left(1.35 + 8.50x + 4.22\times10^{-3}T - 1.53\times10^{-3}xT - 6.73x^2\right)\times10^{14}T^{3/2}E_g^{3/4}\exp\left(-\frac{E_g}{2kT}\right). \tag{16.7}$$

For a given temperature, the maximum of n_i appears at $x\approx0.63$, which corresponds to the minimum energy gap.

The first measurements of transport properties of n-type InAsSb alloys were performed on samples prepared in the late 1960s by various freezing and annealing techniques [43,57]. The properties of high-quality InAsSb epitaxial layers with x < 0.35 manufactured by LPE are similar to those of pure InAs (when $n = 2\times10^{16}$ cm^{-3}, typical mobilities are 30,000 cm^2 Vs^{-1} at 300 K and 50,000 cm^2 Vs^{-1} at 77 K). For InSb-rich alloys with $x \geq 0.90$, typical mobilities are 60,000 cm^2 Vs^{-1} at 300. When As was added to $InAs_{1-x}Sb_x$ alloys, the background carrier concentration increased to a low 10^{17} cm^{-3}; instead, the mobility first increased and then decreased by a factor of 1.5 to 2 for temperatures decreasing from 300 to 77 K. At the present stage of MBE-growth development, the background electron concentration in $InAs_{1-x}Sb_x$ alloys with 40% Sb at 77 K is as low as 1.5×10^{15} cm^{-3} [49].

Chin et al. have calculated the electron mobility of InAsSb by considering all the possible scattering mechanisms such as impurities, acoustic phonons, optical phonons, alloy scattering, and dislocations [58,59]. Comparison with the experiment confirms that dislocation scattering has a strong effect on transport, while alloy scattering limits mobility in ternary samples grown with a minimum of defects.

The ternary alloy $Ga_xIn_{1-x}Sb$ is an important material for the fabrication of detectors designed for middle wavelength IR applications. The long wavelength limit of $Ga_xIn_{1-x}Sb$ detectors has been tuned compositionally from 1.52 μm ($x = 1.0$) at 77 K to 6.8 μm ($x = 0.0$) at room temperature. The bandgap energy of $In_xGa_{1-x}As_ySb_{1-y}$ at room temperature can be fitted by the relationship [60]:

$$E_g(x) = 0.726 - 0.961x - 0.501y + 0.08xy + 0.451x^2 - 1.2y^2 + 0.021x^2y + 0.62xy^2. \tag{16.8}$$

The lattice matching condition to GaSb imposes the additional constraint that x and y are related as $y = 0.867/(1-0.048x)$.

16.2 InGaAs PHOTODIODES

The need for high-speed, low-noise $In_xGa_{1-x}As$ (InGaAs) photodetectors for use in lightwave communication systems operating in the 1–1.7 μm wavelength region is well-established [61–70]. Having lower dark current and noise than indirect-bandgap germanium, the competing near-IR material, the material addresses both entrenched in a variety of thermal-imaging applications not practical with cryogenically cooled detectors [71–78]. The applications now include low-cost industrial thermal imaging, eye-safe surveillance, online process control, and subsurface inspection of fine art.

The SWIR wavelength band offers unique imaging advantages over visible and thermal bands. Like visible cameras, the images are primarily created by reflected broadband light sources, making SWIR images easier for viewers to understand. Most materials used to make windows, lenses, and coatings for visible cameras are readily usable for SWIR cameras, keeping costs down. Ordinary glass transmits radiation to about 2.5 μm. SWIR cameras can image many of the same light sources, such as YAG laser wavelengths. Thus, with safety concerns shifting laser operations to the "eye-safe" wavelengths where beams won't focus on the retina (beyond 1.4 μm), SWIR cameras are in a unique position to replace visible cameras for many tasks. Due to the reduced Rayleigh scattering of light at longer wavelengths, particulate in the air, such as dust or fog, SWIR cameras can see through haze better than visible cameras.

The InAs/GaAs ternary system bandgaps span 0.35 eV (3.5 μm) for InAs to 1.43 eV (0.87 μm) for GaAs. By changing the alloy composition of the InGaAs absorption layer, the photodetector responsivity can be maximized at the desired wavelength of the end user to enhance the signal-to-noise ratio. Figure 16.8 shows the spectral response of three such InGaAs detectors at room temperature, whose cutoff wavelength is optimized at 1.7, 2.2, and 2.5 μm, respectively. The spectral response of an $In_{0.53}Ga_{0.47}As$ focal plane array (FPA) to the night spectrum makes it a better choice for use in a night vision camera in comparison with the current state-of-the-art technology for enhancing night vision—GaAs Gen III image-intensifier tubes. Figure 16.8 also marks the key laser wavelengths.

The fundamental device parameters (energy bandgap, absorption coefficient, and background carrier concentration) distinguish InGaAs from germanium [21]. Low background doping levels ($n = 1\times10^{14}$ cm^{-3}) and high mobilities (11,500 cm^2 Vs^{-1}) for InGaAs at room temperature were achieved [79].

InGaAs-detector processing technology is similar to that used with silicon, but the detector fabrication is different. The InGaAs detector's active material is deposited onto a substrate using chloride VPE [21,80] or MOCVD [81,82] techniques adjusted for thickness, background doping, and other requirements. Planar

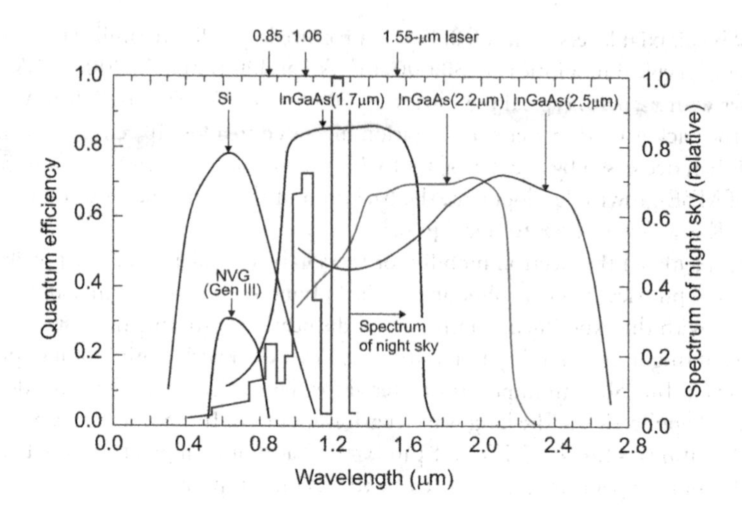

Figure 16.8 Quantum efficiency of silicon, InGaAs, and night vision tube detectors in the visible and SWIR regions. Key laser wavelengths are also noted. $In_{0.53}Ga_{0.47}As$ photodiode has nearly three times higher quantum efficiency than GaAs Gen III photocathodes; InGaAs also overlaps the illumination spectrum of the night sky more.

technology evolved from the older mesa technology and at present is widely used due to its simple structure and processing as well as high reliability and low cost.

Both p-i-n and avalanche InGaAs photodiode structures with total layer thicknesses of several microns are fabricated [83]. The absorption coefficient of $In_{0.53}Ga_{0.47}As$ at 1.55 μm of 7000 cm^{-1} is more than an order of magnitude larger than that of Ge, so that a thickness of 1.5 μm can absorb >70% of the incident photons. This enables high-speed p-i-n photodiodes to operate at up to ~10 GHz with good quantum efficiency.

It is well-known that the internal gain of APDs provides a higher sensitivity in optical receivers than p-i-n photodiodes [84–86]; however, at the cost of more complex epitaxial wafer structures and bias circuits. In comparison to APDs operating in the same wavelength region, p-i-n photodiodes offer the advantages of lower dark current, larger frequency bandwidth, and simpler driving circuitry. Thus, although p-i-n diodes do not have internal gain, an optimal combination of a p-i-n diode with a low-noise, large-bandwidth transistor has led to high sensitivity optical receivers operating up to several Gbit/s.

16.2.1 P–I–N InGaAs PHOTODIODES

Figure 16.9 shows the structure of an InGaAs back-side illuminated (BSI) p-i-n photodiode [87]. The starting substrate is n$^+$-InP on which is deposited approximately 1 μm of n$^+$-InP as a buffer layer. The 3–4 μm of the n$^-$-InGaAs active layer is then deposited followed by a 1 μm n$^-$-InP cap layer. The structure is covered with Si_3N_4. The p-i-n photodiodes are formed by the diffusion of zinc through the InP cap into the active layer. Ohmic contacts were formed by the sintering of an Au/Zn alloy. At this point, the substrate is thinned to approximately 100 μm and a sintered Au/Ge alloy is used as the back ohmic contact. The last step is the deposition of 20-μm columns of indium on the front contacts.

When the indium content of the alloy is increased, the long wavelength cut off extends to cover the entire traditional near-IR band [80–82]. The lattice constant of $In_{0.8}Ga_{0.2}As$ ($\lambda_c \approx 2.5$ μm) is about 2% larger than that of the InP substrate and the $InAs_yP_{1-y}$ gradient layer structure is used to accommodate the difference [71]. Published literature describes a structure having 15 layers with y increasing from 0.0 to 0.68 [88]. By selective removing of various layers, the p-n junctions with different wavelength response could be formed. However, due to the smaller bandgap and interface defects resulting from the lattice mismatch, longer wavelength InGaAs photodiodes have considerably higher dark currents than those fabricated from the lattice-matched alloy, especially at lower voltages. The mismatch between the absorption layers and the InP substrate provide midgap generation–recombination (g–r) centers, increasing the

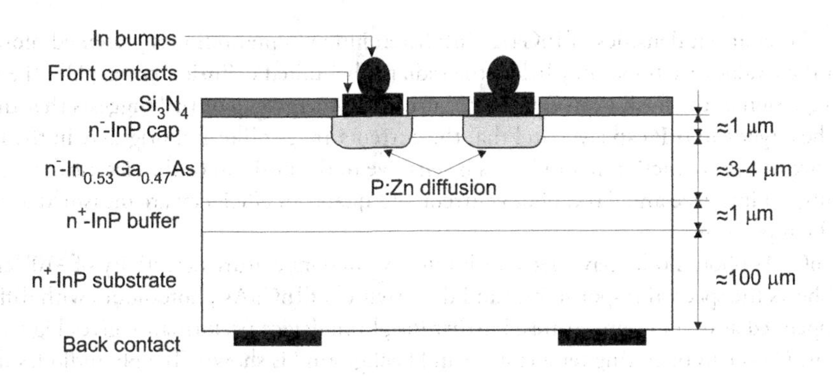

Figure 16.9 Cross section of a planar BSI p-i-n InGaAs photodiode. (From Ref. [87])

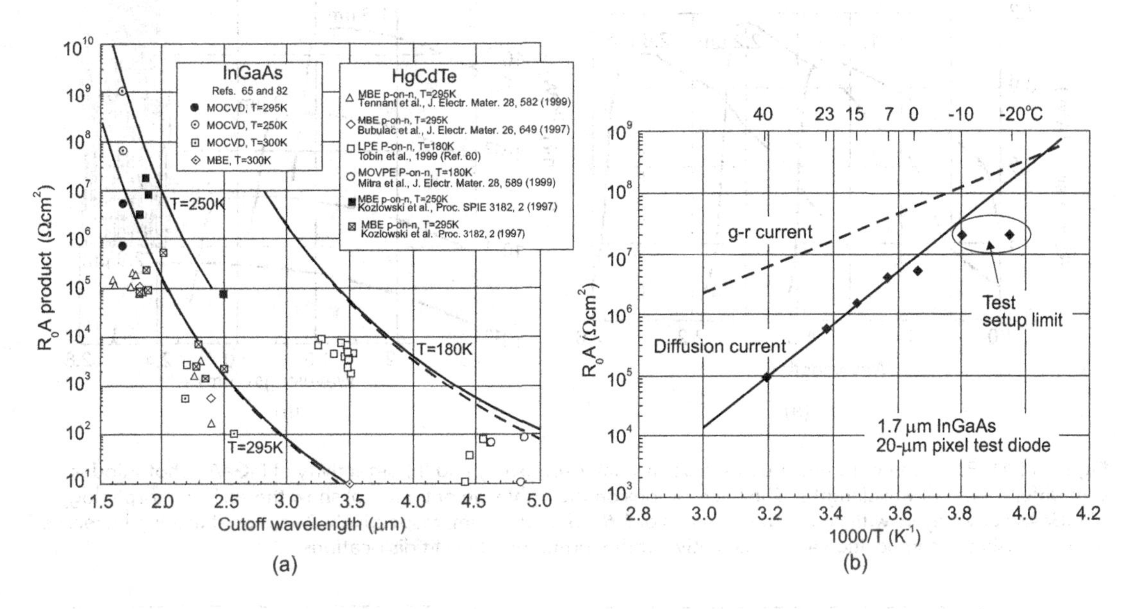

Figure 16.10 R_oA product of InGaAs photodiodes: (a) versus wavelength at 295 and 250 K. (From Refs. [73,92].) (b) Temperature dependence for 1.7-μm InGaAs 20-μm pixel photodiode. (From Ref. [76])

g–r current. Depending on wavelength and illumination direction, the quantum efficiency between 15% and 95% has been measured.

Joshi et al. have reported "Four Rules" for reducing leakage currents in the lattice-mismatched InGaAs photodiodes [80]:

- Grow epitaxial layers with compositionally abrupt interfaces
- Keep lattice mismatch between the adjacent layers below 0.12%
- Dope the active InGaAs layer to 1 to 5×10^{17} cm^{-3}
- Thermally cycle the wafer after growth

A useful photodiode parameter is the R_oA product. Figure 16.10 shows that the highest quality InGaAs photodiodes have been grown by MOCVD [73,89–91]. Their performance agrees with the radiative limit. Due to the similar band structure of InGaAs and HgCdTe ternary alloys, the ultimate fundamental performance of both types of photodiodes are similar in the wavelength range $1.5<\lambda<3.7$ [92]. Figure 16.10b shows the temperature dependence of zero-bias resistance in the region from –20°C to 40°C for 1.7-μm InGaAs 20-μm pixel diode [76]. It can be seen that R_oA is diffusion limited throughout the temperature range tested. At –100 mV, the current is diffusion limited at T≥7°C and becomes g–r limited at lower temperatures. At this bias voltage, mean dark current is about 70 fA at room temperature and about 25 fA at 4°C.

Analysis of dark current densities of InGaAs/InP heterojunction photodiodes presented more recently indicates that their values are noticeably below the radiatively limited diffusion curve [93]. The data fitted an empirical equation of the form $\exp(-E_g(T)/2kT)$ using the energy gap of InP suggests that the junction is located in the larger gap InP cap layer and that the currents are g–r like and originate in the InP [94]. The performance of heterojunction photodiodes is sensitive to the position of the electrical junction relative to the metallurgical interface and dark, photo current and quantum efficiency are measured as a function of voltage and temperature.

Standard InGaAs photodiodes have detector-limited room temperature detectivity of ~10^{13} cmHz$^{1/2}$W^{-1}. Figure 16.11 shows the spectral responsivities and detectivities of InGaAs photodiodes with different cutoff wavelengths operated at room temperatures. Further insight in device performance gives Figure 16.12 [73], where the mean D^* versus operating temperature and background is shown. The photodiodes fabricated

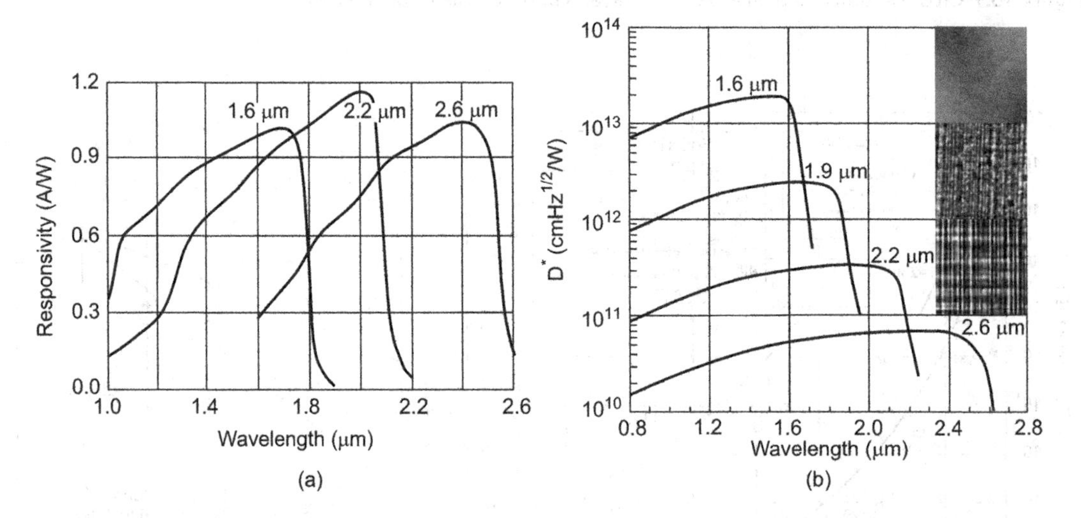

Figure 16.11 Room temperature spectral response (a) responsivity and (b) detectivity of InGaAs photodiodes with different cutoff wavelengths. The micrographs on the right side of figure (b) show the surface morphology of InGaAs active layers with cutoff wavelengths of 1.6, 1.9, and 2.2 μm, respectively. The cross-hatching increases with increasing In composition and is indicative of the formation of misfit dislocations.

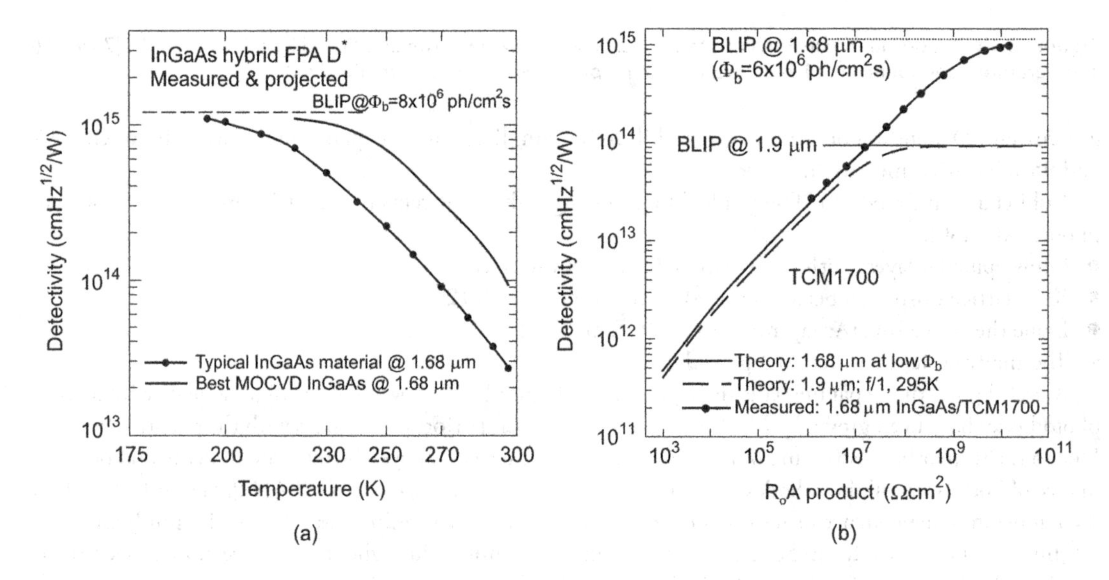

Figure 16.12 Measured and theoretical detectivity of InGaAs photodiodes: (a) versus temperature for 1.7-μm InGaAs; projected trendline shows MOCVD InGaAs capability, and (b) versus R_oA product. (From Ref. [73])

with the radiatively limited MOCVD-fabricated material would exhibit D^* of about 8×10^{13}cm Hz$^{1/2}$ W^{-1} at 295 K. The dependence of detector D^* on R_oA product at the very low short wavelength IR backgrounds is shown in Figure 16.12b. The highest mean D^* exceeding 10^{15}cm Hz$^{1/2}$ W^{-1} implies R_oA product $>10^{10}$ Ωcm^2.

16.2.2 InGaAs AVALANCHE PHOTODIODES

The earliest APDs for telecommunication application were based on silicon utilizing a reach-through structure, with a relatively thin high electric field multiplication region and a much thicker carrier drift region with lower field [95]. This structure with a relatively low operating voltage gives high quantum efficiency, a large α_e/α_h ratio (typically 10–100), modest speed, high gain, and very low noise. As fiber-based optical communication systems evolved, there was a requirement for APDs capable of detecting light at 1.3 and 1.55–1.65 µm. Germanium APDs capable of detecting photons in this wavelength range have α_h/α_e ratio of 1.5, so low excess noise could not be obtained [96–98]. Moreover, the quantum efficiency drops off rapidly beyond 1.5 µm and photodiodes suffer from a high thermal generation rate.

For high-speed receivers, one has to produce a short response time of the photogenerated carriers and a high-bandwidth product. These parameters are limited mainly by the shape of the heterojunction between absorption and multiplication zone and the doping profile throughout the device. In the case of InGaAs, attempts to obtain significant avalanche gain by increasing the electric field is not possible due to the onset of a tunneling mechanism that causes very high leakage current. The small electron effective mass results in rapidly increasing the tunneling current at low fields above 150 kV cm^{-1} [64,90,99]. These problems were overcome by the combination of an InGaAs region capable of absorbing photons under a low electric field, and a lattice-matched wider bandgap InP region producing avalanche multiplication. The resulting structure is known as a separate absorption multiplication APD (SAM-APD).

An example of SAM-APD is shown in Figure 16.13, together with the band diagram of the entire heterostructure. Light is absorbed in the InGaAs and holes (with a higher impact ionization coefficient than electrons; this ensures low-noise operation) are swept to an InP junction, where the avalanche multiplication takes place. This structure combines low leakage, due to the junction being placed in the high bandgap material (InP), with sensitivity at longer wavelength provided by the lower-bandgap InGaAs absorption region. Dark currents of the order pA can be obtained. However, there is a potential problem with the operation of a SAM-APD. Holes can accumulate at the valence band discontinuity at the InGaAs/InP heterojunction and thereby increase the response time. To alleviate this problem, a graded bandgap InGaAsP layer can be inserted between the InP and InGaAs. This modified structure is known as a separate absorption graded multiplication APD (SAGM-APD). The APD can achieve 5–10 dB better sensitivity than p-i-n, provided that the multiplication noise is low and the gain bandwidth product is sufficiently high.

In practical devices, the 1–2 µm thick absorption region is undoped. The graded layer (0.1–0.3 µm) and the avalanching layer (1–2 µm) are doped to 1×10^{16}cm^{-3}. The p$^+$-layer can be thin and doped to

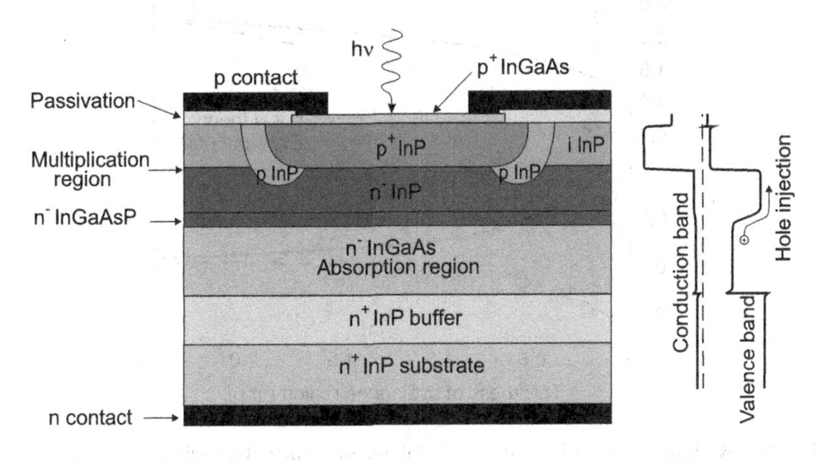

Figure 16.13 Cross section of SAGM-APD InP-based photodiode.

10^{17}–10^{18}cm^{-3}. The junction is usually fabricated by zinc p$^+$-type diffusion in the InP for multiplication and Cd diffusion (or implantation) for guard-ring into the top InP layer through structures SiO_2 masks in closed ampoules. The excellent stability of the device was demonstrated by long-term aging measurements (after 30,000 hours at 150°C and a 100-μA reverse current) [79].

The critical part of InGaAs SAM-APD is the thickness and doping in the field control layer between InGaAs absorption layer and InP multiplication layer. The avalanche gain and excess noise are predominantly determined by the ionization coefficients of InP [100]. In InP, the hole ionization coefficient, α_h, is greater than electron ionization coefficient, α_e, and the α_h/α_e ratio varies from 4 at low fields to 1.3 at the highest fields. Then, according to the McIntyre local theory, this gives rise to APDs with noise corresponding to $1/k \sim 0.4$ in 1-μm thick avalanching structures using hole-initiated multiplication (see Figure 16.14).

In the past decade, the performance of APDs for optical fiber communication systems has improved as a result of improvements in materials and the development of advanced device structures [69,70,101]. Improvements concern introduction as continual or step grading in the bandgap of the absorption/multiplication regions to prevent carrier trapping and attempts to introduce a field control layer. These novel structures are described in more detail in a review of telecommunication APDs by Campbell [70]. The width of the avalanching region in these devices has continued to shrink due to a requirement for higher operating speed. The McIntyre local model is incapable of predicting noise performance of devices with

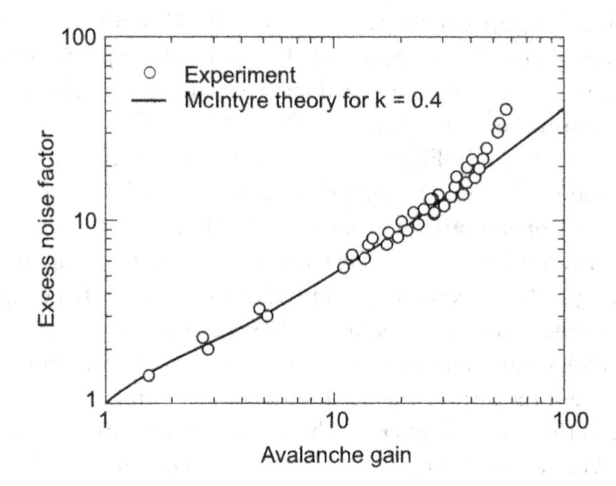

Figure 16.14 Excess noise factor versus avalanche gain for SAGM-APD InP-based photodiode.

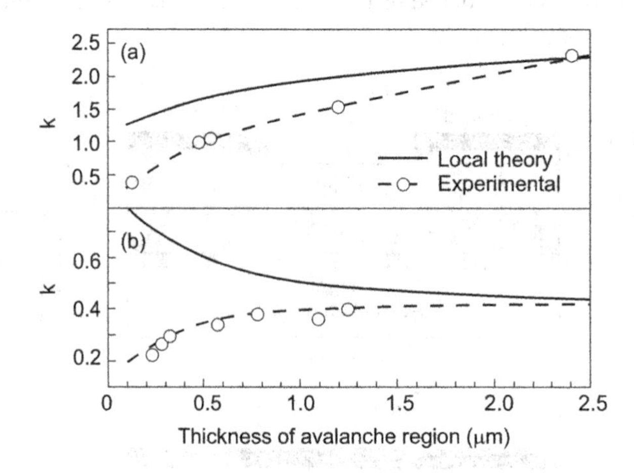

Figure 16.15 The effective ionization coefficient ratio in InP versus avalanche region thickness for (a) electron and (b) hole-initiated multiplication. (From Ref. [101])

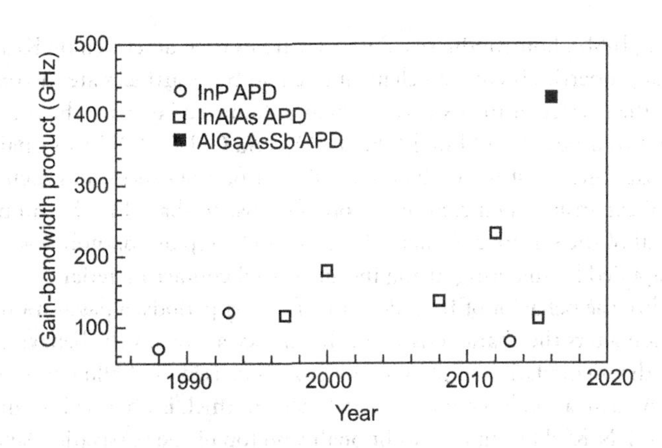

Figure 16.16 Comparison of typical experimental gain–bandwidth products of APDs with InP and InAlAs avalanche layers.

thin avalanching widths (see Figure 16.15 [101]). An InP diode with an avalanching width of 0.25 μm would exhibit a noise performance equivalent of $1/k\sim0.25$ for hole-initiated multiplication, although the actual ionization coefficient ratio at such electric fields would yield a much larger value of ~0.7 using the McIntyre model. The significant reduction of excess noise is due to the dead space effect observed in submicron avalanche regions. Further improvement in device performance can be achieved by introducing InAlAs, also lattice matched to InGaAs and InP, as a replacement for the InP as the multiplication region. The α_e/α_h ratio has been found to be significantly larger than the α_h/α_e ratio in InP at a low electric field. Also, the excess noise factor at a given gain is significantly lower in InAlAs than that in InP due to the large α_e/α_h ratio in the former and the beneficial effect of the dead space in the latter.

Experimental reports of the Gain-Bandwidth Product (GBP) are limited to 80–120 GHz and 105–160 GHz for the InGaAs/InP APDs and InGaAs/$In_{0.52}Al_{0.48}As$ APDs, respectively, as illustrated in Figure 16.16. Recently, Xie et al. have demonstrated the first InGaAs/AlGaAsSb APD with a high GBP over 400 GHz that is compatible to InP technology, with clear exploitation in telecom/data communication networks at 10 Gb/s or above [102]. This is a step change in performance compared to previous reports of GBP values from 1550-nm wavelength APDs. The APD exhibits insignificant band-to-band tunneling current, the maximum responsivity of 5.33 A W^{-1} at 1550 nm, a maximum −3 dB bandwidth of 14 GHz.

16.3 BINARY III–V PHOTODETECTORS

16.3.1 InSb PHOTOCONDUCTIVE DETECTORS

The basic requirements for a photoconductive detector is a material of very high purity into which small and carefully controlled amounts of a particular impurity can be introduced. InSb photoconductors are usually fabricated of p-type material (a dopant such as germanium is often used) with a free carrier concentration of less than 10^{14} cm^{-3} at 77 K, hole mobility of 7000 cm^2 Vs^{-1}, and dislocation density of less than 10^2 cm^{-2}. The higher resistance of p-type photoconductors and the absence of sweep-out allow higher responsivities than are possible in an n-type device [6]. In p-type material at low temperatures, the lifetime for the two types of carriers are very different ($\tau_e<10^{-9}$ s, $\tau_h\approx10^9/p$) [103]. The effect of the strong minority carrier trapping reduces their drift length in the device, thereby increasing the bias field at which sweep-out onset occurs. It allows achievement of higher D^* values than in small detectors made of trap-free, p-type material. High-quality photoconductors are also made of n-type materials with carrier concentration below 10^{14} cm^{-3} and electron mobility of 7.5×10^5 cm^2 Vs^{-1} at 77 K [104,105].

The photosensitive area of the detectors is made from several tens of μm^2 to several mm^2, depending upon their usage. A small photosensitive area is used for quick response and a large photosensitive area for high responsivity. For the 300 K operating devices, the optimum thickness of elements is 5–10 μm, while for 195 and 77 K photoconductive elements, the thickness is about 25 μm [5].

The steps in preparing InSb photoconductive detectors are exactly described by Kruse [5]. A wafer that has the desired electrical properties is cut into element size and both surfaces are mirror-lapped using fine alumina powder. Next, the surface of the oxidized film and damaged layer are chemically removed by etching in a mixture of equal volumes of undiluted HF, HNO_3, and CH_3COOH to a final thickness. Elements are then mounted on a sapphire substrate with epoxy resin. Also, other substrates such as Ge and Irtran 2, for which the thermal expansion coefficient is reasonably close to that of InSb, can be used. An alternative approach for removal of the sensitive elements from the slab employs photolithographic techniques. Electrical contacts are applied by soldering, using indium-based contact material.

To control and stabilize the behavior of InSb detectors for long periods, a passivation layer on the surface is needed. This passivation alters the characteristics of the surface as well as the behavior of the detector [104]. Sunde has found that degradation of photoconductors is caused by shallow traps in the oxide layer on the InSb surface [106]. A passivation layer approximately 50-nm thick is obtained by anodization, which is usually carried out in a 0.1-N KOH solution. Additionally, on top of the passivation layer, a layer of SiO_x or ZnS about 0.5-μm thick is evaporated to provide a more stable surface. This layer is also the antireflection coating.

Detectors for different applications are usually supplied with self-contained, thermally insulated encapsulations complete with window, radiation shield, lead outputs, and provision for cooling.

The detectivity of p-type InSb photoconductive detectors at 77 K over a wide range of doping levels are g–r noise limited at bias voltages well below the sweep-out onset or the power dissipation limit. The g–r noise expression in the case of excitation and recombination for a single type of center is given by [107]

$$V_{gr}(0) = \frac{2V_b\tau_h^{1/2}}{(lwtp)^{1/2}}, \tag{16.9}$$

and the responsivity [5,6]

$$R_v = \frac{\eta\lambda V_b\tau_h}{hclwtp}. \tag{16.10}$$

Thus, from these equations and Equation 4.33,

$$D^* = \frac{\eta\lambda\tau_h^{1/2}}{2hct^{1/2}p^{1/2}}. \tag{16.11}$$

A more detailed theory of the photoconductive effect in InSb in which expressions are derived for the spectral responsivity and detectivity for three operating temperatures (77, 195, and 300 K) is given by Kruse [5].

Figure 16.17 illustrates the detectivities of InSb photoconductors as functions of wavelength, with modulation frequency as an independent parameter. The detectivity of available detectors operating at 77 K is usually in the range of 5×10^{10}–10^{11} cm Hz$^{1/2}$ W^{-1} at a sensitivity peak of 5.3 μm. Devices approaching the background-limited detectivity can be made. Responsivities in excess of 10^5 VW^{-1} for a 0.5×0.5 mm^2 element at optimum signal-to-noise ratio are obtained, and the resistance is typically 2 kΩ per square. The noise characteristics do not always vary linearly to the bias current, and over a certain critical value, noise increases rapidly. Up to 100–150 Hz, the noise is proportional to $1/f$. From 150 Hz to several kHz, g–r noise is predominant. At higher frequency ranges, noise is limited by Johnson noise.

InSb photoconductors are also used without cooling or with thermoelectric cooling in the temperature range from 190 to 300 K. As the operating temperature increases, the long wavelength response increases and the detectivity decreases considerably. At room temperature, detectivities at peak wavelength $\lambda_p \approx 6$ μm of 2.5×10^8 cm Hz$^{1/2}$ W^{-1} may be achieved. The response of a 1×1 mm^2 detector may be 0.5 VW^{-1}. The low time constant, about 0.05 μs, makes these detectors suitable for high-speed operation applications. For 77 K photoconductive detectors, the time constant is normally in the range of 5–10 μs.

Pines and Stafsudd have reported photoconductive data on high-quality n-type InSb photoconductors [105]. Their studies show that the parameters of surface-passivated detectors are limited by bulk

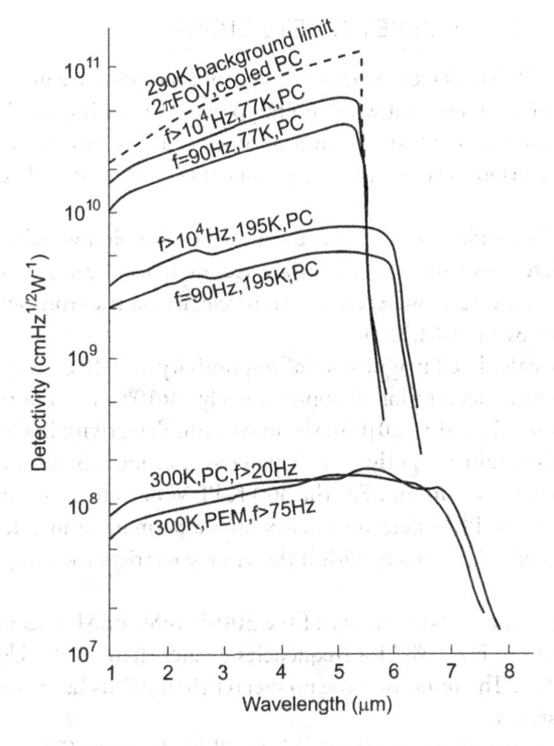

Figure 16.17 Typical spectral variation of detectivity for InSb PC and PEM detectors operating at 77, 195, and 300 K. (From Ref. [5])

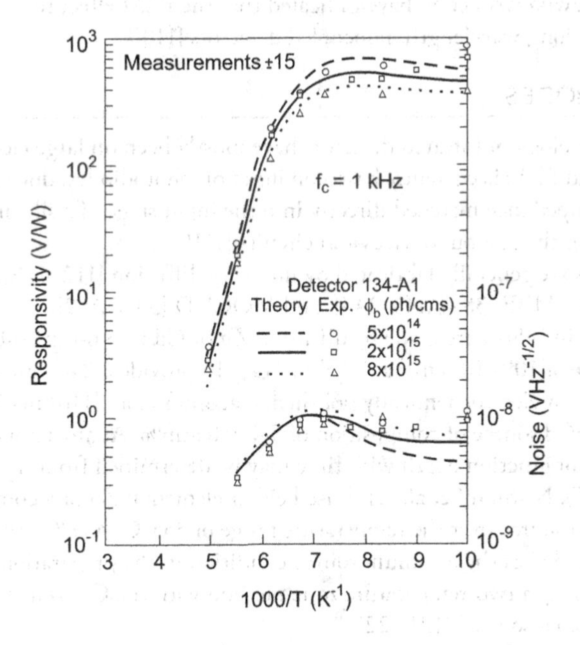

Figure 16.18 Responsivity and noise as a function of temperature in n-type InSb photoconductors. (From Ref. [105])

material properties. Figure 16.18 shows the responsivity and noise as a function of temperature in an n-type photoconductive detector. At higher temperatures, the responsivity and noise depend slightly on the background photon flux density. The roll-off in responsivity and noise at higher temperatures can be correlated with the decrease in the carrier lifetime and the increase in the majority carrier concentrations.

16.3.2 InSb PHOTOELECTROMAGNETIC DETECTORS

InSb photoelectromagnetic (PEM) detectors operated at room temperature offer good performance in the 5–7 μm interval. Difficulties associated with cooling prevent operating the detectors at 195 or 77 K. Moreover, when cooled they offer no advantage over photoconductive or photovoltaic detectors. A comprehensive review of the operation, technology, and properties of InSb PEM detectors has been done by Kruse [5].

The preparation of the PEM InSb detectors is the same as for the photoconductive devices. However, whereas photoconductive detectors require a low surface recombination velocity on both the front and the back surface, PEM detectors require a low recombination velocity on the front surface and high recombination velocity on the back (see Section 12.10.1).

The highest theoretically calculated magnitudes of responsivity and detectivity have been found in p-type materials having a hole concentration of approximately 7×10^{16} cm^{-3}. For the assumed values of the parameters (λ_c=6.6 μm, B=0.7 T, and t=20 μm), the maximum detectivity is about 6×10^{8} cm Hz$^{1/2}$ W^{-1} and responsivity 5 VW^{-1}. To obtain a slightly p-type material, a residual donor in InSb single crystals is usually compensated with zinc or cadmium. For the 300 K PEM detectors, the optimum thickness is about 25 μm. The housing design of the PEM detector incorporates a permanent magnet (see Figure 12.50). The requirement for a high magnetic flux density within the sample restricts the sample width to a value of about 1 mm or less.

Figure 16.17 illustrates the spectral detectivity of the 300-K InSb PEM detector. The detectivity has a peak of approximately 2×10^{8} cm Hz$^{1/2}$ W^{-1} for frequencies greater than 75 Hz. Usually, the peak value of D^* occurs at 6.2 μm [108,109]. The response time no greater than 0.2 μs has been found. It is an attractive feature in wideband applications.

The region of maximum responsivity of uncooled InSb PEM detectors (5.5–6.5 μm) is outside the "atmospheric windows." The realization of noncooled detectors for radiation with wavelengths longer than 8 μm is a real problem. Jóźwikowski et al. have indicated that the PEM effect in $InAs_{0.35}Sb_{0.65}$ can be used as a basic principle for the long-wavelength, noncooled detectors [110].

16.3.3 InSb PHOTODIODES

Recent efforts in the technology of infrared detectors have mostly been on large electronically scanned FPAs. The design of hybrid FPAs is currently based on junction photodiodes, due to reduced electrical power dissipation, high impedance matched directly into the input stage of a silicon readout, and less stringent noise requirements for the readout devices and circuits [111].

The III–V photodiodes are generally fabricated by impurity diffusion [112–122], ion implantation [123–135], LPE [136–144], MBE [38,47,145–148], and MOCVD [38,47,149].

Initially, p-n junctions in InSb were made by diffusing Zn or Cd into n-type substrates with net donor concentration in the range of 10^{14}–10^{15} cm^{-3} at 77 K [3,112]. To provide a flat, damage-free surface, the substrates were chemically or electrochemically polished. Catagnus et al. [116] made p-n junctions in closed ampoules using the In:Sb:Cd source of composition of 5:45:50 atm%. At any temperature between 250°C and 400°C, the variation of junction depth with time may be determined from the formula x=40.5 cm hr$^{-1/2}$(t)$^{1/2}$exp(−0.80 eV/kT). Nishitani et al. [117] used either elemental Zn or a combination of elemental Zn and Sb as the diffusion source over the temperature range of 355°C–455°C. The diffusion source of $N_{Sb}/N_{Zn}\geq5$ was recommended as the optimum source condition in the preparation of a p-n junction. A modified technology utilizing a two-temperature zone method with the Cd source at 380°C and the InSb substrate at 440°C has been also used [121,122].

Figure 16.19 illustrates 10 major processing steps for the diffusion fabrication of InSb mesa photodiodes [150]. The active surface area of the detector is coated with an anodic layer for stabilizing surface states and an antireflection layer of SiO.

It was found, however, by several researchers [125] that Zn and Cd produced a porous surface that is difficult to remove. To circumvent this surface problem and to make high-quality p-n junctions, implantation of the light ions Be and Mg have been used [123–135]. Implantation of sulfur [124] and protons [123] to form n-p diodes and zinc and cadmium to form p-n diodes have been reported. It appears that InSb can

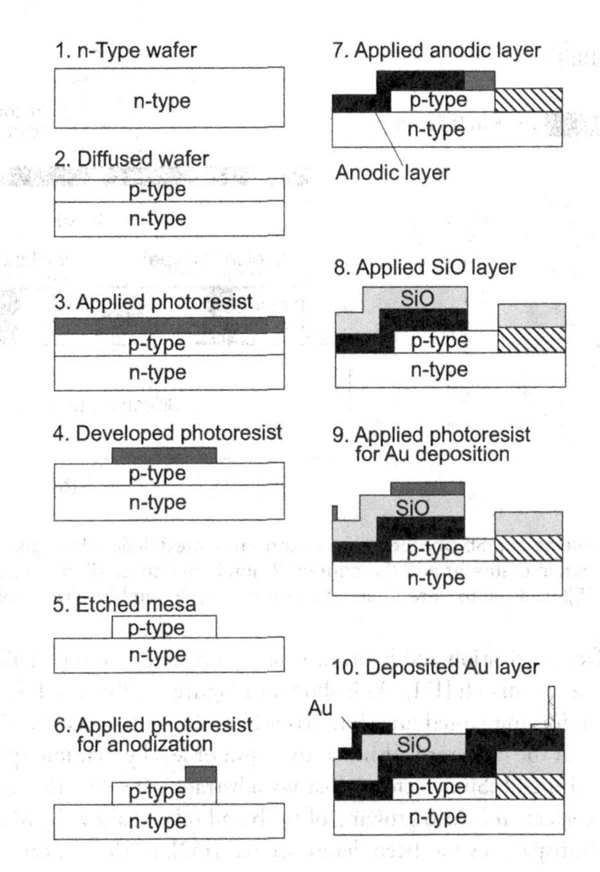

Figure 16.19 Processing steps for the diffusion fabrication of InSb mesa photodiodes. (From Ref. [150])

be made amorphous by implanting it at room temperature with heavy ions [129]. Zn and Cd may be too heavy to implant in InSb.

InSb photodiodes produced by Be implantation have been reported by Hurwitz and Donnelly [125]. This developed diode process has been used for monolithic InSb CCD arrays, for flat-zero input structures and simple extraction of charge at the output or in charge-skimming circuits [126]. Beryllium implantation is also used for source and drain formation for InSb MOSFETs. Diodes were fabricated on (100)-oriented slices with donor concentrations of 10^{14}–10^{15} cm^{-3}. Before implantation, each wafer was polished and anodized in a 0.1 N solution of NH_4OH at 30 V for 30 s to protect the surface during subsequent processing. The 5-μm thick photoresist was then applied to each sample and an array of 0.5-mm diameter holes opened in the photoresist that formed a mask against ions. The best photodiodes were made with ion implantation energies of 100–200 keV, with doses of (2–5) × 10^{14} cm^{-2}. Samples were tilted 7° from the normal to reduce channeling. After removal of the photoresist, the samples were coated with 100 nm of pyrolytic SiO_2 deposited at 350°C for 2–3 minutes and were annealed at 350°C for 15 minutes. This process acts effectively to anneal out the radiation damage caused by the room temperature Be ion implantation. Next, approximately 0.4 μm of InSb was removed together with the surface inversion layer formed during annealing. It was found to be vital for the fabrication of good diodes. Immediately after etching and rinsing, the wafers were coated with 150 nm of silicon oxynitride and with 100 nm of SiO_2 to provide a stable surface. To permit adjustment of the surface potential for optimum diode performance, a field guard ring was employed. Contact to the p-type region was made by plating of 0.5-μm-thick Au covered by 1.5 μm of In. The substrate was contacted with In. A cross-sectioned wafer of a finished diode is shown in Figure 16.20a [125].

At present in InSb photodiode fabrication, epitaxy is usually not used; instead, the standard manufacturing technique begins with bulk n-type single crystal wafers with the donor concentration of about 10^{15} cm^{-3}. Relatively large bulk grown crystals with 4–6-inch diameters are available on the market. An array hybrid with a size of up to 4,096×4,096 is possible because the InSb detector material is thinned to

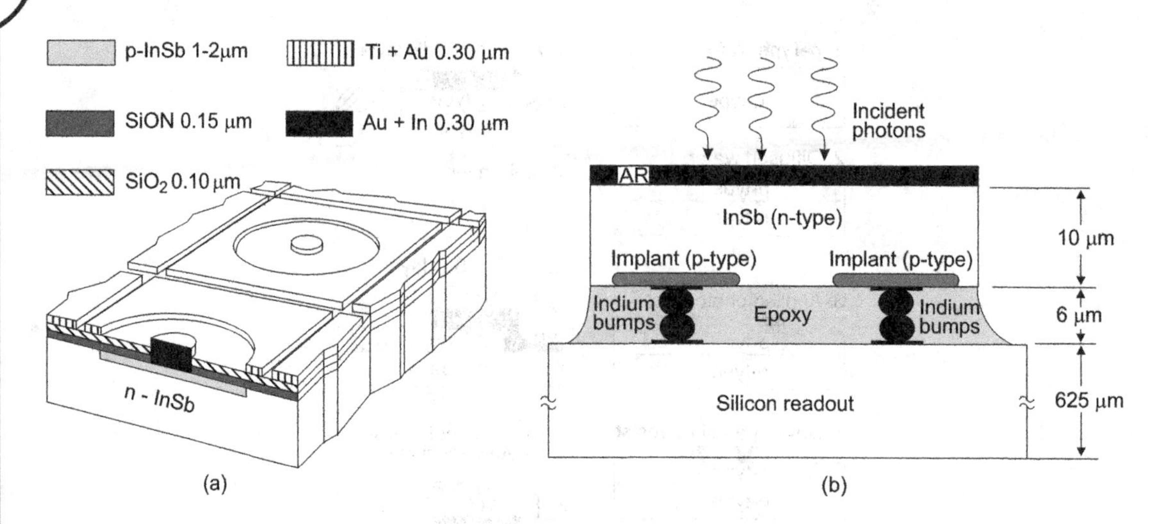

Figure 16.20 Cross-section views of InSb photodiodes: (a) ion-implanted, field-plate-guarded InSb photodiode (the implanted region is 0.5 mm in diameter and the contact 75 μm in diameter, distance between adjacent field plates is 5 μm). (From Ref. [125]) (b) Architecture of an InSb sensor chip assembly. (From Ref. [151])

less than 10 μm (after surface passivation and hybridization to a readout chip) that allows it to accommodate the InSb/silicon thermal mismatch [151]. As is shown in Figure 16.20b, the BSI InSb p-on-n detector is a planar structure with an ion-implanted junction. After hybridization, epoxy is wicked between the detector and the Si ROIC and the detector is thinned to 10 μm or less by diamond-point-turning. One important advantage of a thinned InSb detector is that no substrate is needed; these detectors also respond in the visible portion of the spectrum. Also growing of InSb and related alloys by MBE together with doping of substrate to induce transparency has been demonstrated [152]. In the last case, the thinning of the detector material is not required.

The forward I-V characteristics are given by $J=J_o\exp(qV/\beta kT)$ with $\beta\approx 1.7$. The diffusion current component $J_D=J_s[\exp(qV/kT)-1]$ was determined from the forward I-V curves. J_s was 5–7×10^{-10} A cm^{-2} at 77 K and was negligibly small compared to the generation–recombination current density in reverse bias. The measured values of R_oA were above 10^6 Ωcm^2 at 77 K. Figure 16.21a shows measured current density versus reverse voltage for planar p$^+$-n junctions in InSb formed by Be-ion implantation [126]. The base material was obtained using either Czochralski or LPE growth techniques. The current in the small bias region (A) is clearly g–r limited. The solid curve labeled J_{gr} is calculated using Equation 12.113 and the parameters $N_d=6.6\times10^{14}$cm^{-3}, $V_{bi}=0.209$ V, $\tau_o=10^{-8}$s, $E_t=E_i$ and $n_i=2.7\times10^9$cm^{-3}. As V increases, the photodiodes enter a breakdown region where J increases superlinearly with V. The current in the breakdown region is a strong function of doping and is believed to be due to interband tunneling. Solid curves (B) and (C) are the tunnel currents calculated using an equation similar to Equation 12.118 for $N_d=6.6\times10^{14}$cm^{-3} and 1.4×10^{14}cm^{-3}. The broken curves (*D*) and (*E*) are calculated for liquid helium temperature assuming an interband tunneling model. We can see that tunneling is dominating for reverse voltage exceeding ≈2 V. The current densities of the LPE diodes are about a factor of five lower than for Czochralski material. The longer minority carrier lifetime observed in the epitaxial material translates directly into reduced dark current density in the photodiodes for a given operating temperature [153].

Figure 16.21b compares the dependence of dark current on temperature between InSb and HgCdTe photodiodes used in the current high-performance large FPAs [154]. This comparison suggests that MWIR HgCdTe photodiodes have a significantly higher performance in the 30–120 K temperature range. The InSb devices are dominated by g–r currents in the 60–120 K temperature range because of a defect center in the energy gap, whereas MWIR HgCdTe detectors do not exhibit g–r currents in this temperature range and are limited by diffusion currents. In addition, wavelength tunability has made HgCdTe the preferred material.

Capacitance–voltage measurements gave a functional dependence of C versus V as C^{-2}, confirming that the Be diodes are one-sided abrupt junctions. A zero-bias RC time constant is about 0.4 ms.

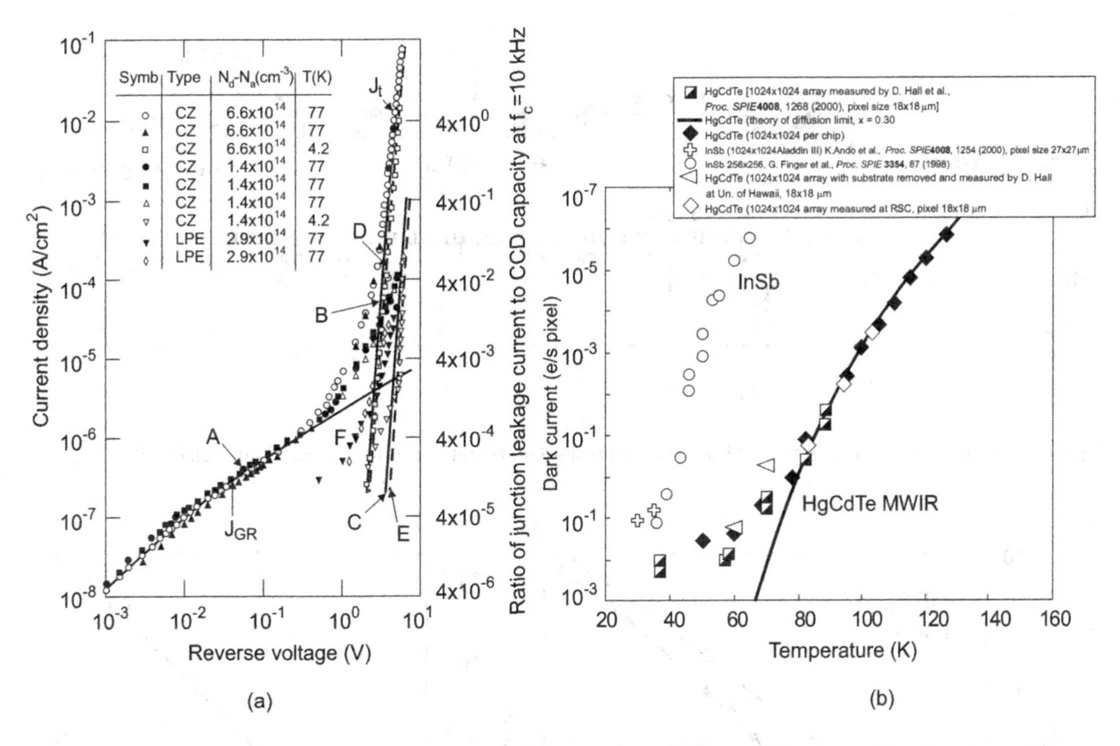

Figure 16.21 Current density versus reverse voltage for InSb photodiodes: (a) planar p⁺-n junctions formed by Be-ion implantation; solid curves are the theoretical fit to the data. (From Ref. [126]) (b) The comparison of dependence of dark current on temperature between highest reported value for InSb arrays and MBE-grown HgCdTe MWIR FPAs (with 18×18 μm pixels). (From Ref. [154])

The simplest possible InSb photodiode structures, requiring only an insulating layer and surface metallization for contacts, are typically used for large, single element detectors. However, there are several disadvantages of this scheme—the bonding pad capacitance becomes a significant portion of the total device capacitance (especially for small area detectors); in closely spaced, multiple element arrays, the bonding pads can no longer be used to completely blind all active area between elements. In multiple element imaging and spectroscopy systems, the resulting exposed "semiactive" areas between photodiodes may result in a loss of resolution (the minority carrier diffusion length in the n-region of InSb diodes is on the order of 20–30 μm). To maximize the available resolution and response in InSb photodiodes, the bulk material is thinned to about 10 μm. For high-performance FPAs in the 2–5 μm range, the photodiode parameters must be well-defined to design a complete detector/preamplifier package. InSb material is highly uniform and, combined with diffusion or an implanted process in which the device geometry is precisely controlled, the resulting detector array responsivity is excellent.

Bloom and Nemirovsky have presented an improved processing technology of BSI, planar, gate-controlled (evaporated titanium) InSb photodiodes with an active n-type region with carrier concentration of about 10^{15} cm^{-3} [155,156]. The photodiode design is shown in Figure 16.22. The p⁺-n junctions were formed by implanting beryllium on (111) wafers with a dose of 5×10^{14} cm^{-2} and energy of 100 keV, followed by annealing at 350°C for half an hour in a nitrogen ambient. For both front- and back-side surfaces they used improved surface passivation utilizing a modified UV photoassisted SiO_x deposition (PHOTOX). In this process, the photolysis of the reactant gases (SiH_4 and N_2O) occurs at 50°C by collisions with excited mercury atoms that absorb the ultraviolet radiation at a wavelength of 2,537 Å. The strongly accumulated interface with reduced surface recombination velocity was formed on the In face. The resultant band-banding produces an electric field that retards the flow of minority carriers to the surface recombination sites, sweeps the generated holes to the junctions, and increases the quantum efficiency.

Wimmers et al. have presented the status of InSb photodiode technology at Cincinnati Electronics (Mason, Ohio) for a wide variety of linear arrays and FPAs [118,120]. Fabrication techniques for InSb

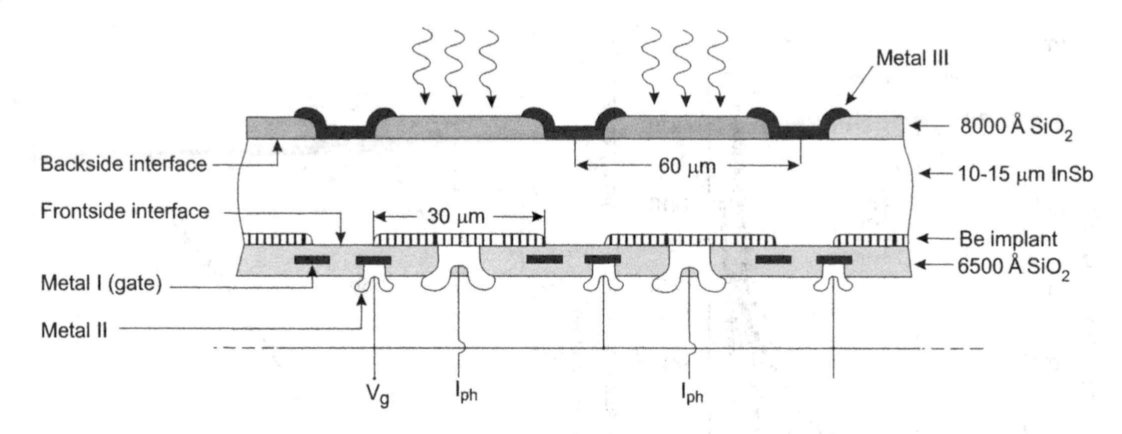

Figure 16.22 Detailed cross section of BSI, gate-controlled InSb matrix photodiodes. (From Ref. [156])

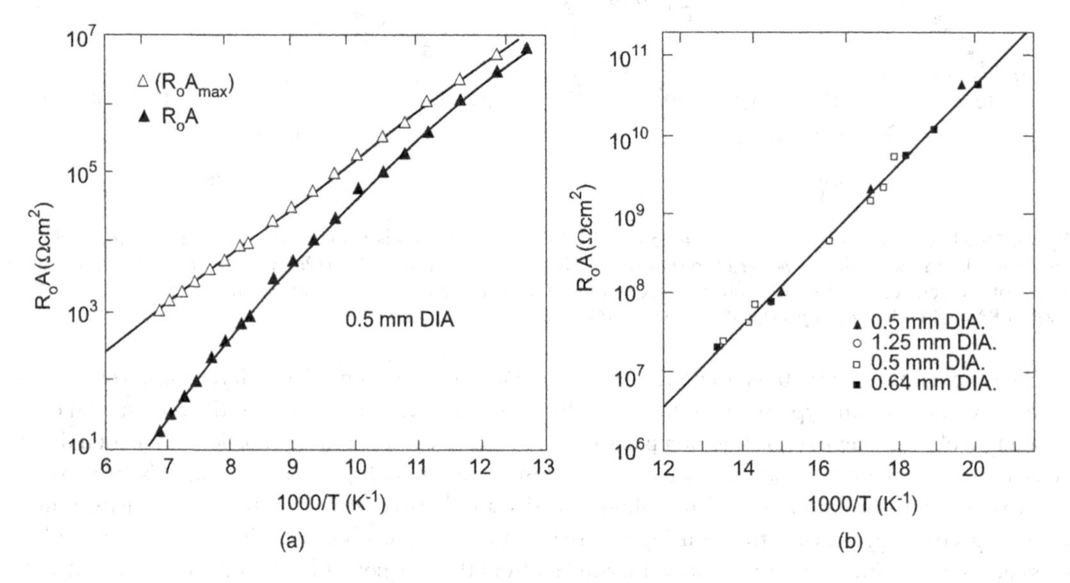

Figure 16.23 R_oA and $(RA)_{max}$ versus temperature for InSb photodiodes: (a) 150–80 K. (From Ref. [120]) (b) 80–50 K. (From Ref. [118])

photodiodes use gaseous diffusion and a subsequent etch results in a p-type mesa on n-type substrate with donor concentration about 10^{15} cm^{-3}. A highly controlled diffusion process allows p-layer diffusion to occur with little surface damage, eliminating the need for deep diffusion and subsequent etch-back. This permits total "mesa" heights of only a few microns. A grounded "buried-metallization" process, independent of bond-pad metallization, was developed to render the surface of the InSb opaque, with the exception of the active area and the contact area. The accuracy of the photolithography along with the controlled diffusion process provides excellent uniformity of response.

Typical InSb photodiode resistance area (RA) product at 77 K is 2×10^6 Ωcm^2 at zero bias and 5×10^6 Ωcm^2 at slight reverse biases of approximately 100 mV (see Figures 16.23 and 16.24 [118,120]). This characteristic is beneficial when the detector is used in the capacitive discharge mode. As element size decreases below 10^{-4} cm^{-2}, the ratio of the circumference to area is increased and some slight degradation in resistance due to surface leakage occurs.

The performance of FPAs depends upon the capacitance of the detector element. InSb photodiode capacitance can be effectively modeled as the sum of a voltage-dependent junction capacitance (which agrees well with the abrupt junction model) and a bond pad capacitance [120]. For diodes with a relatively

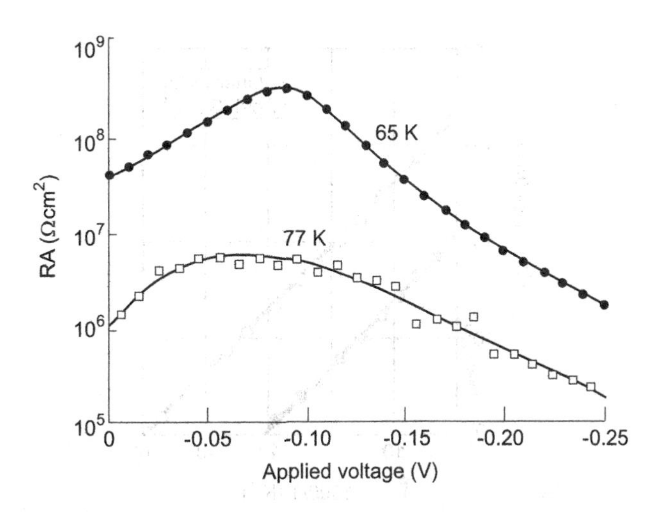

Figure 16.24 Dependence of the RA product on applied reverse voltage for 3.2 × 3.2 mil InSb photodiode at 77 and 65 K. (From Ref. [120])

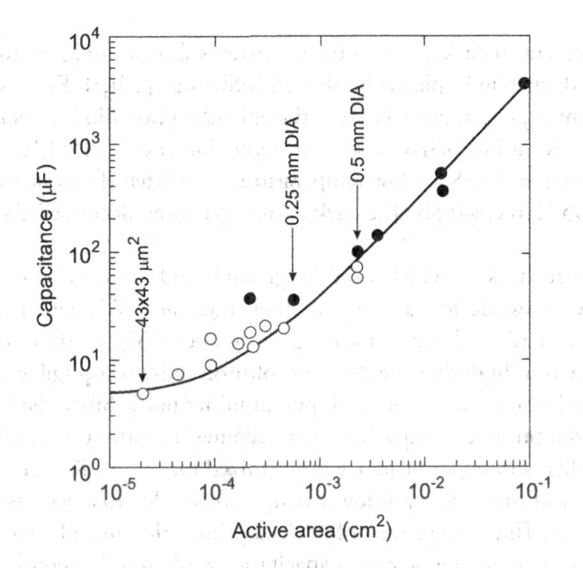

Figure 16.25 Zero-bias capacitance as a function of active area for InSb photodiodes at 77 K. The solid line is theoretically calculated. (From Ref. [118])

large active area, neither the bond pad capacitance nor the capacitance associated with the contact area represents a significant fraction of the total diode capacitance. However, as the active area decreases, the total diode capacitance becomes more strongly dependent on these two terms. Figure 16.25 shows zero-bias capacitance as a function of the active area for InSb photodiodes at 77 K [118].

As is mentioned previously, the best-quality InSb photodiodes are generation–recombination limited. In this limit, Shockley–Read–Hall (SRH) traps created by imperfections in the semiconductor crystal lattice provide energy states located in the semiconductor bandgap. In a standard planar technology, p-on-n junctions are created by ion implantation into n-type substrates. A new approach involving MBE growth has been adopted for reducing the dark current. Because in in situ MBE epilayer growth of p-n structures, an implantation damage is avoided, the diodes have a much lower concentration of G-R centrs than in standard planar p-n junctions [148]. The dark current is thus reduced according to the ratio of concentrations of G-R centers in the standard and MBE grown structures. After the growth of high-quality epilayer homojunctions on an InSb substrate, the diodes are isolated by etching of mesa structures through the p-n junctions.

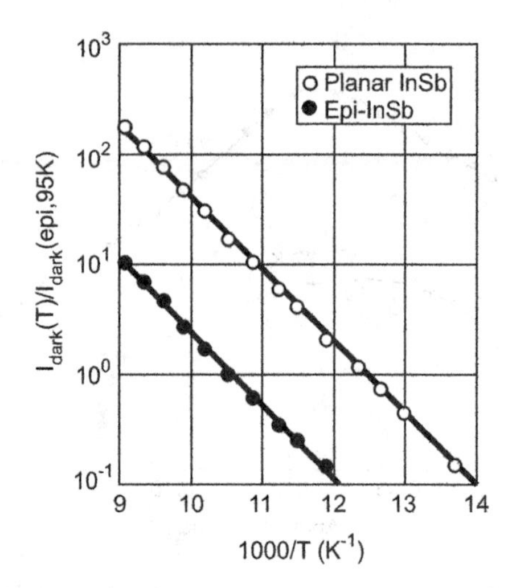

Figure 16.26 Temperature dependence of the dark currents of 15-μm pitch InSb diodes fabricated using planar (open points) and MBE technology. Fitting lines are calculated assuming G-R formula (after Ref. [148])

An example of improvement in dark current characteristic is shown in Figure 16.26, where the temperature dependence of the dark current in planar implanted InSb and epi-InSb FPAs with a 15-μm pitch is compared. The dark current is normalized to that at the epi-InSb photodiode operated at 95 K. The solid lines are fitted assuming G-R limited behavior with an activation energy of 0.12 eV, which corresponds to approximately half the bandgap of InSb at low temperatures. Note that the same dark current is achieved at 80 K in planar InSb and 95 K in epi-InSb. The dark current has been decreased about 17× by going to the MBE-based technology.

The InSb photovoltaic detectors are widely used for ground- and space-based infrared astronomy. For applications in astrophysics, these devices are very often operated at 4–7 K with a resistive or capacitive transimpedance amplifier to achieve the lowest noise performance [157]. At these low temperatures, the InSb photodiode resistance is so high that the detector Johnson noise is negligible, and the dominant noise sources are either the feedback resistance or input amplifier noise. Since the latter scales directly with the combined detector and input circuit capacitance, it becomes important to minimize these. The InSb photodiodes described earlier, whose performance is optimized for 60–80 K operation, have been shown to lose long wavelength quantum efficiency at lower temperatures, due to a decrease in the minority carrier lifetime in the n-region [158]. Thus, the process of device optimization must be redone for low-temperature applications, with emphasis on reducing detector capacitance while simultaneously maximizing the quantum efficiency [159]. A reduction in doping density in the n-type region to $\approx 10^{14} cm^{-3}$, along with other minor process modifications, minimizes the decrease of the carrier lifetime, providing an added benefit of decrease in capacitance. This approach also reduces the *RA* product slightly, but the *RA* product still increases exponentially with decreasing temperature, until detector resistance once again is not a significant noise contribution.

The detectivity versus wavelength for a typical InSb photodiode is shown in Figure 16.27. Detectivity increases with reduced background flux (narrow FOV and/or cold filtering), as illustrated in the figure. InSb photodiodes can also be operated in the temperature range above 77 K. Of course, the RA products degrade in this region. At 120 K, RA products of $10^4\ \Omega cm^2$ are still achieved with slight reverse bias, making background limited infrared photodetector (BLIP) operation possible. The quantum efficiency in InSb photodiodes optimized for this temperature range remains unaffected up to 160 K [120]). Process modifications such as increases in doping density allow responsivity to remain unchanged by increasing temperature.

Some of the design constraints associated with bulk devices with implanted or diffusion junctions can be relaxed using epitaxial methods. Epitaxial, BSI InSb photodiodes have been grown on Te-doped InSb

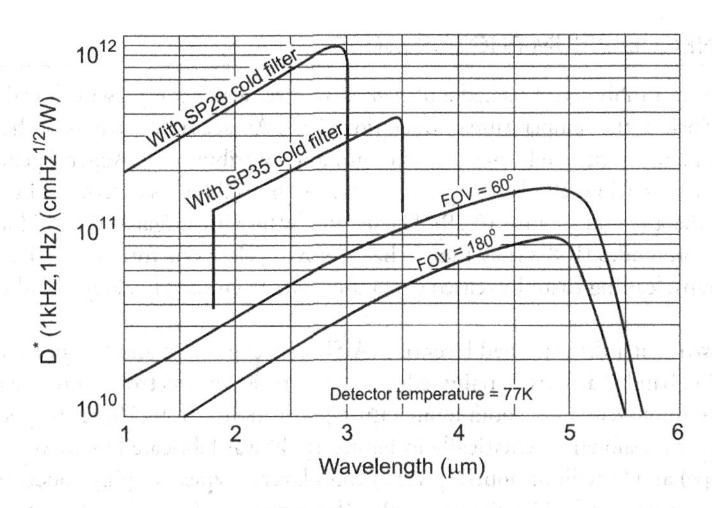

Figure 16.27 Detectivity as a function of wavelength for an InSb photodiode operating at 77 K. (From Judson Catalog, *Infrared Detectors*, http://www.judsontechnologies.com.)

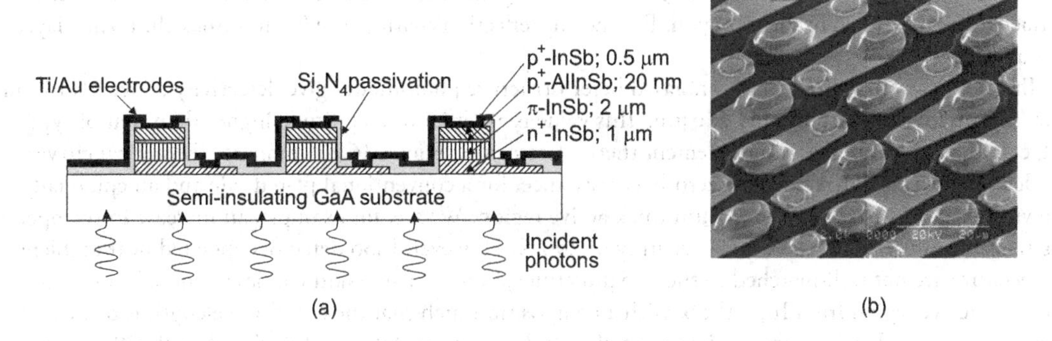

Figure 16.28 InSb photovoltaic infrared sensor: (a) schematic structure of multiple photodiodes connected in series, and (b) scanning electron microscope (SEM) photograph. (From Ref. [165])

substrates. Degenerate n-type doping of the substrates allows them to be made transparent through the Burstein–Moss effect. A doping of 2×10^{18}cm^{-3} is sufficient to obtain transparency at 80 K over the majority of the 3–5 μm wavelength range [160].

The InSb photodiodes grown heteroepitaxially on Si and GaAs substrates by MBE have also been reported [38,161–166]. More recently, Kuze et al. [165,166] have developed a novel microchip-sized InSb photodiode sensor, on semi-insulating GaAs(100) substrate, operating at detection. The sensor consists of 910 photodiodes connected in series (see Figure 16.28). Each photodiode consists of MBE-grown 1-μm thick n$^+$-InSb layer, followed by a 2-μm thick π-InSb absorber layer. To reduce the diffusion of photoexcited electrons, a 20-nm thick p$^+$-$Al_{0.17}In_{0.83}Sb$ barrier layer was grown on the π-InSb layer. Finally, a 0.5-μm thick π-InSb layer was grown as the top contact. The n- and p-type dopants Sn and Zn were used, respectively, with concentrations of 7×10^{18}cm^{-3} for the n$^+$-layer, 6×10^{16}cm^{-3} for the π layer, and 2×10^{18}cm^{-3} for the p$^+$ layer. To insulate mesa structures, a 300-nm thick plasma CVD passivation Si_3N_4 layer was deposited. Finally, after Ti/Au lift-off metallization, a 300-nm thick SiO_2 passivation also grown by plasma CVD was made. The length of a single InSb photodiode was 20 μm. The final external dimensions of the photovoltaic infrared sensor were $1.9\times2.7\times0.4$ mm^3. The sensitivity and the noise equivalent difference temperature were 127 μV K^{-1} and 1 mK Hz$^{-1/2}$, respectively.

16.3.4 InSb NONEQUILIBRIUM PHOTODIODES

The first nonequilibrium InSb detectors had a p^+-π-n^+ structure, where π represents low doped p-type material that is intrinsic at the temperature of operation [167]. An accurate analysis of the source of current in the diode at room temperature indicates the predominant contribution of Auger 7 generation in the p^+ material. At temperatures below 200 K, the performance of photodiodes is determined by the Shockley–Read generation in the π region (Figure 16.29). Davies and White investigated the residual currents in Auger suppressed photodiodes [168]. They found that removing electrons from the active region alters the occupancy of the traps, leading to an increased generation rate from the Shockley–Read traps in the active region.

Next, it was shown that a thin strained layer of InAlSb between the p^+ and π regions produces a barrier in the conduction band that substantially reduces the diffusion of electrons from the p^+ layer to the π region, leading to an improvement in room temperature performance [169,170]. This type of p^+-P^+-π-n^+ InSb/$In_{1-x}Al_xSb$ structure shown schematically in Figure 16.30 was fabricated by MBE at 420°C using silicon-doped (n-type) and beryllium-doped (p-type) InSb layers. Typical doping concentrations and thickness of individual layers are marked in Figure 16.30a. The composition, x, of the $In_{1-x}Al_xSb$ layer, is 0.15 giving a conduction band barrier height, which is estimated as 0.26 eV. The central region is typically 3-μm thick and not deliberately doped. Circular diodes of 300-μm diameter were fabricated by mesa etching to the p^+ region and were passivated with an anodic oxide. Sputtered chromium/gold contacts were applied to the top of each mesa, with an annual geometry having an internal diameter of 180 μm and an external diameter of 240 μm and to the p^+ region. For the antireflection coating, the 0.7-μm thick thin oxide layer was used.

The p^+-P^+-π-n^+ InSb/$In_{1-x}Al_xSb$ unbiased heterostructure photodiodes give detectivity above 2×10^9 cm $Hz^{1/2}$ W^{-1}, with peak responsivity at 6 μm. This value is an order of magnitude higher than that of typical, commercially available, single element thermal detectors. Figure 16.31 compares theoretical curves for detectivity, calculated from the zero-bias resistance, for a conventional p^+-n diode and an epitaxially grown p^+-P^+-v-n^+ structure with a 3-μm thick active region. We see, for example, an increase in the operating temperature of about 40 K in the vicinity of 200 K. However, InSb detectors operated near ambient temperature are not well-matched to the 3–5 μm atmospheric transmission window. One solution is to form the active region from $In_{1-x}Al_xSb$, with a composition such that the cutoff wavelength is decreased to the optimum value. To obtain a 5-μm cutoff would require $x\approx0.023$ at 200 K and $x\approx0.039$ at room temperature. The predicted detectivity for material with a constant 5-μm cutoff wavelength is plotted by the dotted line in Figure 16.31 [170]. The total increase in D^* compared with a conventional bulk device is more than a factor of 10. Background-limited detectivity in 2π FOV can be achieved at 200 K, which is

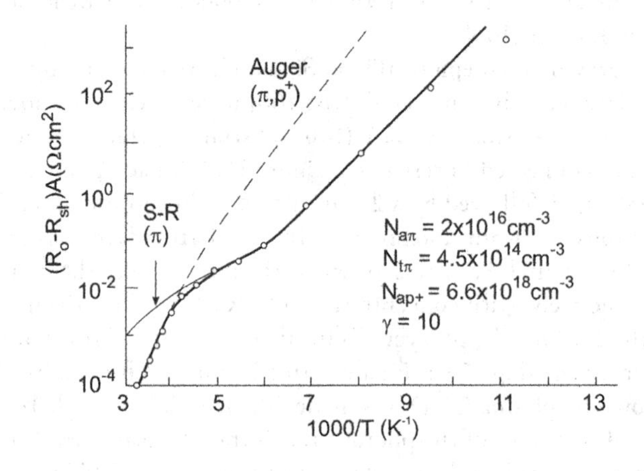

Figure 16.29 Temperature dependence of the R_oA product for p^+-π-n+InSb photodiode. Circles indicate experimental data; the solid lines show calculated values based on both Shockley–Read and Auger generation mechanisms in the π and p^+ regions. (From Ref. [167])

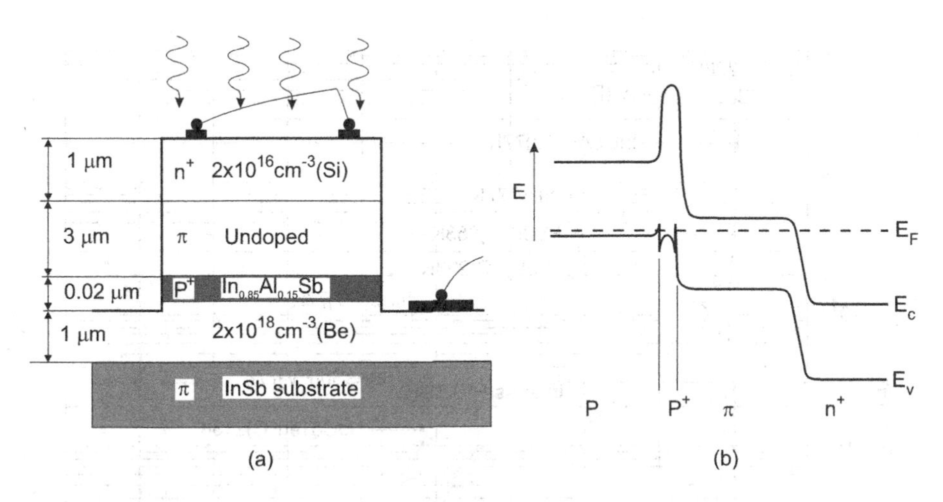

Figure 16.30 Schematic cross section of the (a) p⁺-P⁺-π-n⁺ InSb/$In_{1-x}Al_xSb$ heterostructure photodiode, and (b) its energy band diagram. (From Ref. [170])

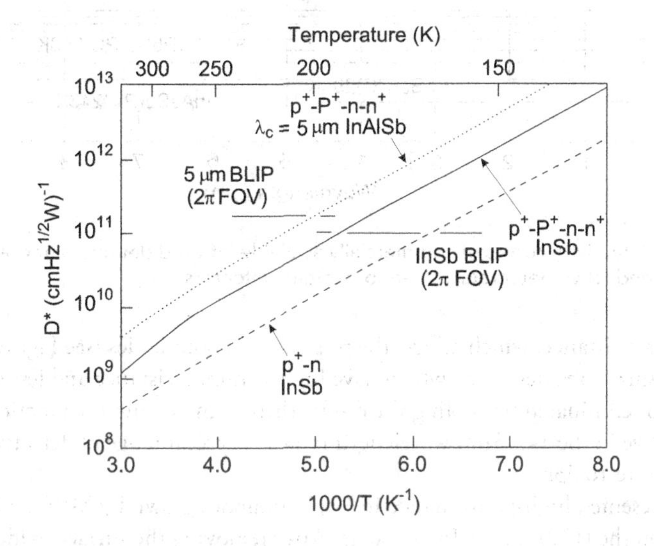

Figure 16.31 Calculated detectivity versus temperature for InSb photodiodes comparing bulk InSb (dashed), epitaxial InSb (solid), and epitaxial InAlSb with 5-μm cutoff wavelength (dotted). (From Ref. [170])

achievable with Peltier coolers leading to the possibility of high performance, compact, and comparatively inexpensive imaging systems.

16.3.5 InAs PHOTODIODES

InAs detectors have been made to operate in the photoconductive, photovoltaic, and PEM modes. Recently, however, wider applications (laser warning receivers, process control monitors, temperature sensors, pulsed laser monitors, and infrared spectroscopy) have found InAs photodiodes operated at near room temperature. The photodiodes are mainly fabricated by ion implantation [124,135], the diffusion method [3,114]. More recently, high-performance InAs photodiodes are fabricated using epitaxial techniques; LPE [143,144,171], MBE [148,172–176], and MOCVD [174,175]. A significant breakthrough in growing and investigation of narrow band-gap $A^{III}B^{V}$ heterostructures in the system InAs–InSb–GaSb was obtained in the Ioffe Physico-Technical Institute, St. Petersburg, Russia. Their performance can be compared with commercially available Hamamatsu's detectors—see Figure 16.32.

The diode sensitivity, the speed of response, impedance, and peak wavelength can be optimized by operation at the proper temperature. At room temperature, the shunt resistance of the InAs photodiode is

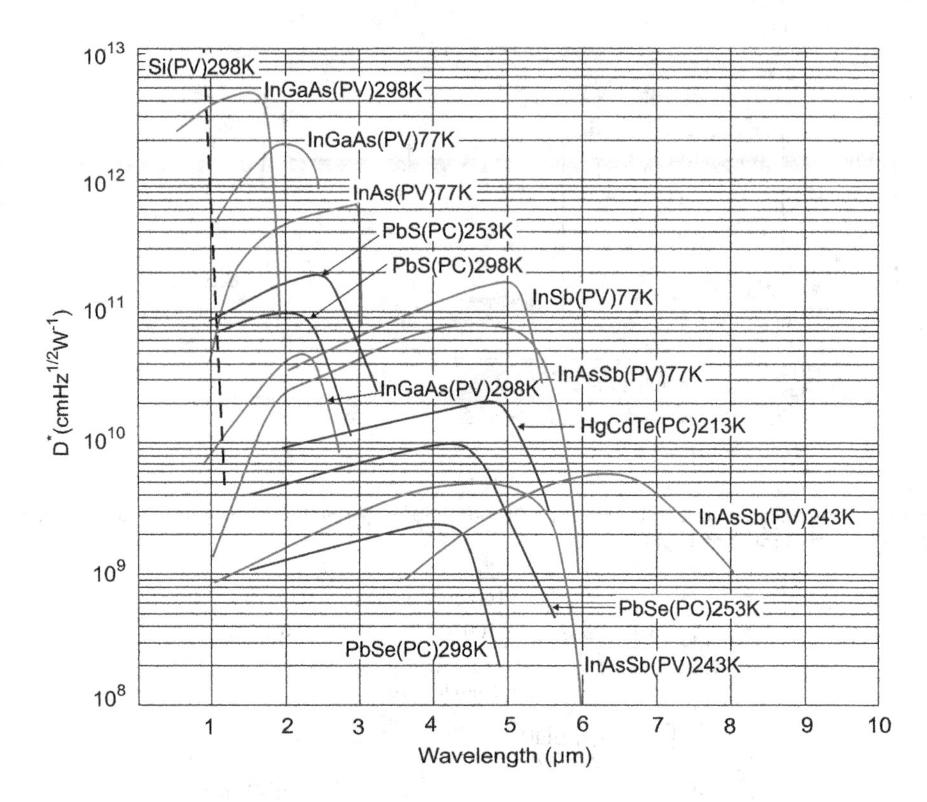

Figure 16.32 Spectral detectivity curves of commercially available infrared detectors operated at different temperatures. PC—photoconductive detectors, PV—photovoltaic detectors.

comparable with series resistance, which affects the response of photodiodes (see Figure 16.33a). This effect is less pronounced in small area detectors, which have higher shunt resistance and less surface area. The effect is also reduced or eliminated by cooling the diode, thereby increasing the junction resistance. InAs photodiodes are sensitive in the 1–3.6 μm wavelength range. A typical range of detectivity for InAs photodiodes is shown in Figure 16.33b.

Kuan et al. have presented high-performance InAs photodiodes grown by MBE [147,173]. The diode structure was grown on the (100) n-type InAs wafers. After removing the surface oxides (by slowly heating to 500°C), the InAs epilayers were grown at 500°C under optimized growth conditions. The p-i-n photodiode structure consisted of a 0.2-μm thick n-type buffer layer (Si-doped to 1×10^{18} cm^{-3}), followed by a 1-μm thick n-type InAs active layer (Si-doped to 5×10^{16} cm^{-3}). Then, a 0.72-μm thick undoped InAs layer was grown, followed by a 0.1-μm thick p-type InAs layer (Be doped to 1×10^{18} cm^{-3}), and finally a 0.1-μm thick InAs contact layer (exponentially graded doping from 1×10^{18} cm^{-3} to 1×10^{19} cm^{-3}) was deposited. The same structure was also grown except for the undoped InAs layer for the p-n diode. The schematic diagram and device structure of the InAs gate-controlled photodiode are shown in Figure 16.34 [173].

Before fabricating unpassivated and passivated InAs diodes, a special chemical treatment and two-step photolithographic procedures were used [173]. In the case of gate-controlled photodiodes, the epilayers were first mesa etched into circular dots measuring 200 μm in diameter and then the photo-CVD technique was applied to deposit the 300-nm thick SiO_2. A second photolithographic step was used to remove the SiO_2 from the defined area of 10×4 μm^2 on the p-type layer so as to make electrical contacts. A double layer of 100-nm thick Au-Be and 300-nm thick Au was evaporated sequentially and lifted off to form a p-type ohmic contact. A third photolithographic step was used to define the pad with a dot having a 40-μm diameter and the gate electrode that covered the junction perimeter. Then, a double layer of 12 nm of Cr and 300 nm of Au were evaporated and lifted off.

Figure 16.35a shows the typical 77 K and room temperature I-V characteristics of unpassivated p-i-n and p-n photodiodes, respectively [173]. At 77 K, the dark current of unpassivated p-i-n photodiode is

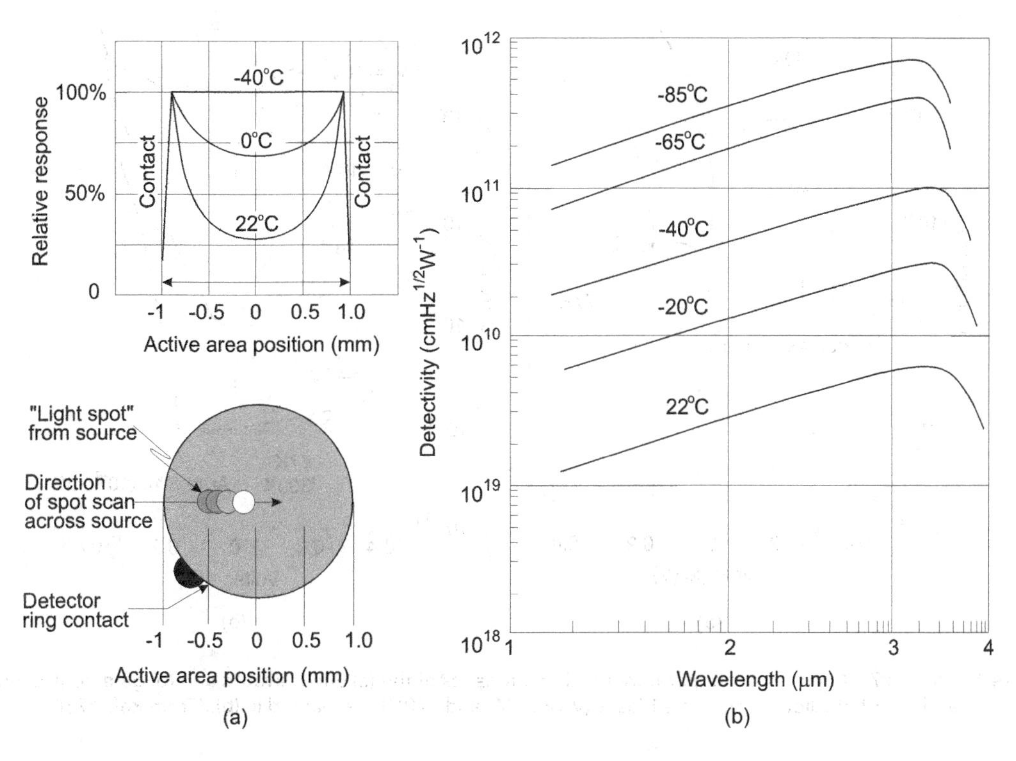

Figure 16.33 InAs photodiodes: (a) response variation across the 2-mm active area of InAs photodiodes, and (b) detectivity versus wavelength at different temperatures (After product brochure of Judson Inc., http://www.judsontechnologies.com).

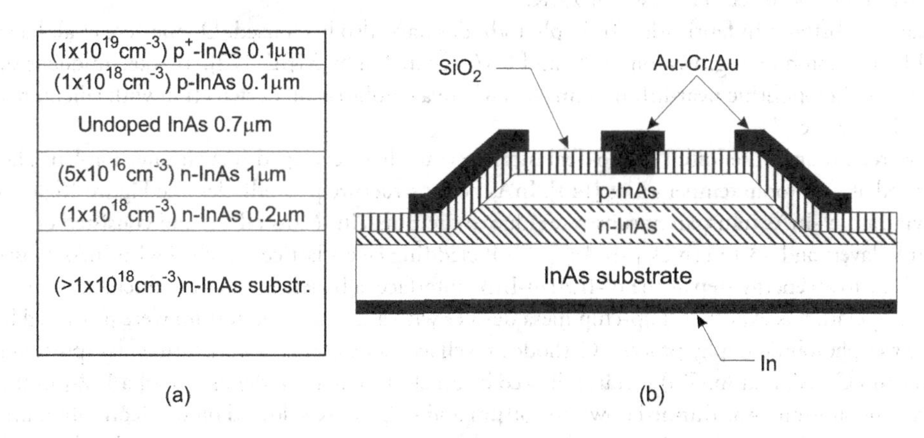

Figure 16.34 InAs gate-controlled photodiode: (a) schematic diagram, and (b) device structure. (From Ref. [173])

disturbed by the background thermal radiation, proof of which is the existence of a photovoltage. We can also see that the reverse dark current of unpassivated photodiode depends on the diode reverse bias both at 77 and 300 K indicating the existence of a shunt leakage current. Figure 16.35b shows the typical 77 and 300 K I-V characteristics of gate-controlled p-i-n photodiodes under different gate bias at 0, –16, and –40 V. The strong dependence of the diode reverse dark current on the gate voltage indicates that the reverse leakage current is flowing through the surface region. When the gate bias V_g approaches –40 V, the reverse dark current becomes nearly independent of reverse biases, which indicates that the diode is leakage free. It is obvious that the I-V characteristics of unpassivated p-i-n photodiode are similar to and even better than the p-i-n gate-controlled photodiode under gate bias V_g=–40 V. This indicates that passivation

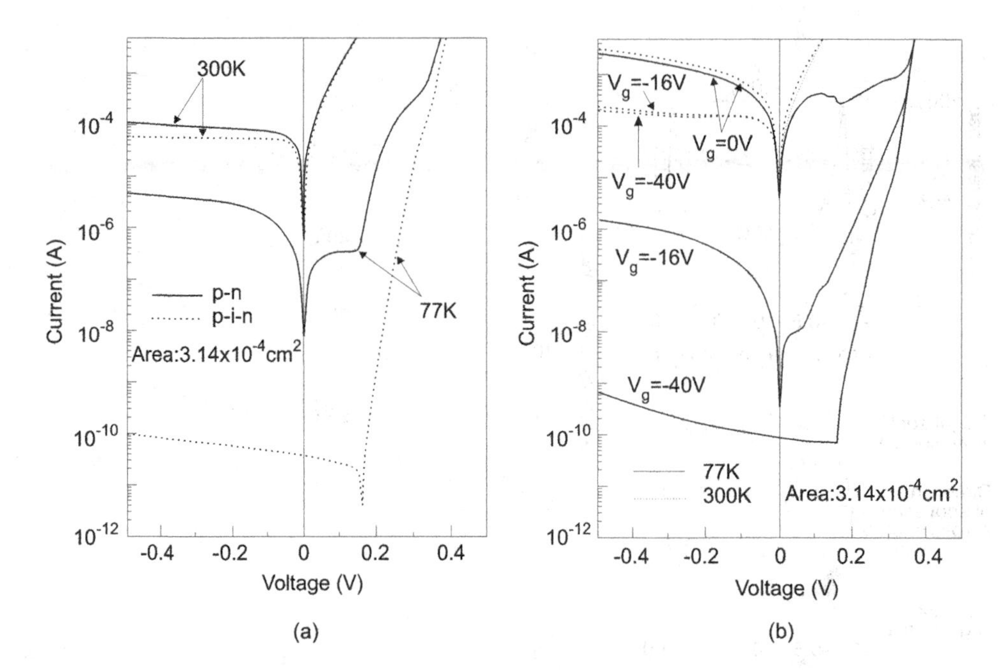

Figure 16.35 The 77 K and room temperature I-V characteristics of the (a) unpassivated and (b) gate-controlled InAs p-i-n and p-n photodiodes. The gate biases (V_g=0, –16, and –40V) are marked in (b). (From Ref. [173])

of InAs p-i-n photodiode degrades the device performance. The unpassivated p-i-n photodiode exhibits R_oA product of 8.1 Ωcm² at room temperature and 1.3 M Ωcm² at 77 K. When illuminated under a 500 K blackbody source, the photodiode detectivity limited by Johnson noise is 1.2×10^{10} cm Hz$^{1/2}$ W^{-1} at room temperature and 8.1×10^{11} cm Hz$^{1/2}$ W^{-1} at 77 K.

Alternative substrates in fabrication InAs photodiodes have also been used. Dobbelaere et al. have produced InAs photodiodes grown on GaAs and GaAs-coated Si by MBE [172]. This technique is suitable for fabrication of monolithic near-infrared imagers where a combination of detection with silicon readout electronics is possible.

The research group at the Ioffe Physico-Technical Institute has developed InAs immersion-lens photodiodes operated at near-room temperatures [144]. InAs heterostructure photodiodes (see Figure 16.36) were LPE grown onto n⁺-InAs transparent substrates (due to the Burstein–Moss effect) and consisted of ~3-μm thick n-InAs layers and ~3-μm thick p-$InAs_{1-x-y}Sb_xP_y$ cladding layers lattice-matched with InAs substrate ($y\sim2.2x$). Due to an energy step at the n⁺-InAs/n-InAs interface, a beneficial hole confinement for the photodiode operation is expected. Flip-chip mesa devices with a diameter of 280 μm were processed by a multistage wet photolithography process. Cathode as well as anode contacts were formed by sputtering of Cr, Ni, Au(Te) and Cr, Ni, and Au(Zn) metals followed by an electrochemical deposition of a 1–2-μm thick gold layer. Next, the substrate was thinned down to 150 μm and chips were soldered onto silicon submounts with Pb-Sn contact pads. Finally, the 3.5-mm wide silicon lens was attached to the substrate side of a chip by a chalcogenide glass with high refractive index (n=2.4)—see Figure 16.37(a). It is obvious, that field of view of immersion photodiodes is considerably lower than that for the uncoated device (decrease down to 15 deg).

Figure 16.37(b) shows the detectivity spectra of InAs heterostructure immersion photodiode. Superior detectivity of these photodiodes, in comparison with z commercial Hamamatsu and Judson devices, reflects improvements associated with board mirror contact, asymmetric doping, immersion effect, and radiation collection by inclined mesa walls. The narrow spectral responses are a result of filtering in the substrate and intermediate layers. Peak wavelengths shift to long wavelengths as the temperature is increased due to bandgap narrowing at higher temperatures. However, the short wavelength spectra are more sensitive to temperature than the long wavelength ones, probably due to the progressively poor transparency of n⁺-InAs near the absorption edge at elevated temperatures due to the elimination of the conduction band electron degeneration [144].

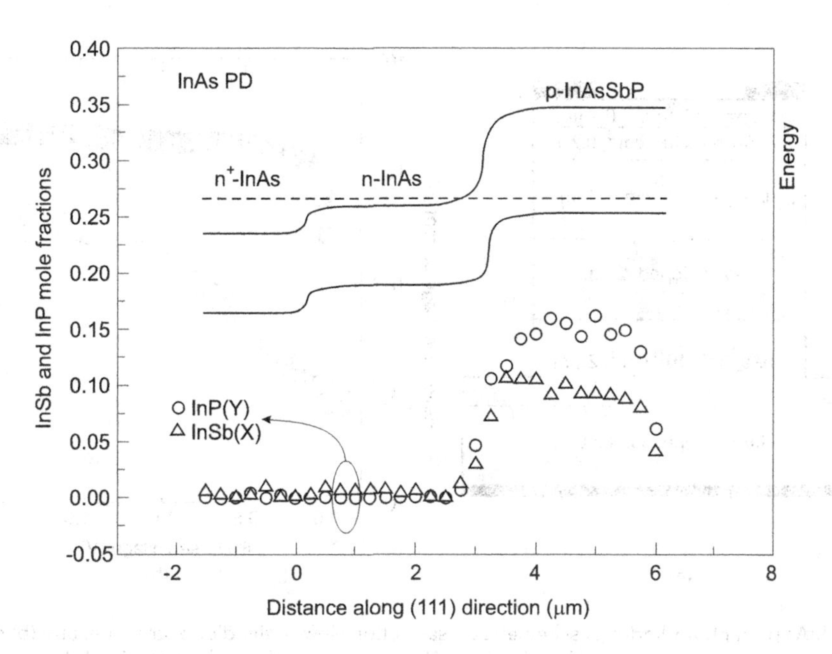

Figure 16.36 Alloy composition in InAs photodiode structure (composition profile) together with the band diagram. (Adapted after Ref. [144])

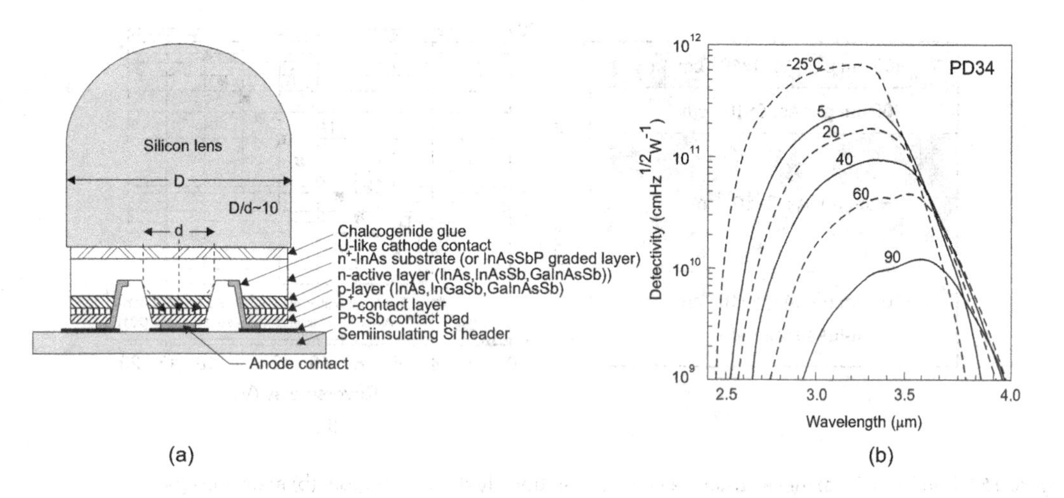

Figure 16.37 InAs heterostructure immersion photodiode: (a) construction of the immersion photodiode, and (b) detectivity spectral at near-room temperatures. (From Ref. [144])

Considerable progress in the development of high-quality InAs photodiodes has been achieved by the research group at the University of Shieffield [175,176]. The MBE-grown InAs p-i-n wafers were fabricated into circular mesa diodes—see Figure 16.38(a)—with 0.2-μm AlAsSb blocking layer placed at the top of the p-type cap layer. This photodiode design is similar to that proposed by Ashley et al. [169] (see also Figure 16.30), where the wider bandgap blocking layer at the interface to the intrinsic region "excludes" electrons generated in the p-type cap from diffusing to the junction. Further reduction in the reverse leakage current was achieved through studies of wet etching using a range of etchants. Room temperature reverse dark leakage current measurements on devices with and without the blocking layer are shown in Figure 16.38(b). It can be seen that the addition of the blocking layer reduces the low bias leakage current by more than an order of magnitude.

For high-sensitivity applications, semiconductor APD detectors are most useful because they provide an internal gain on the detectors. They are most commonly used for communication and active sensing applications.

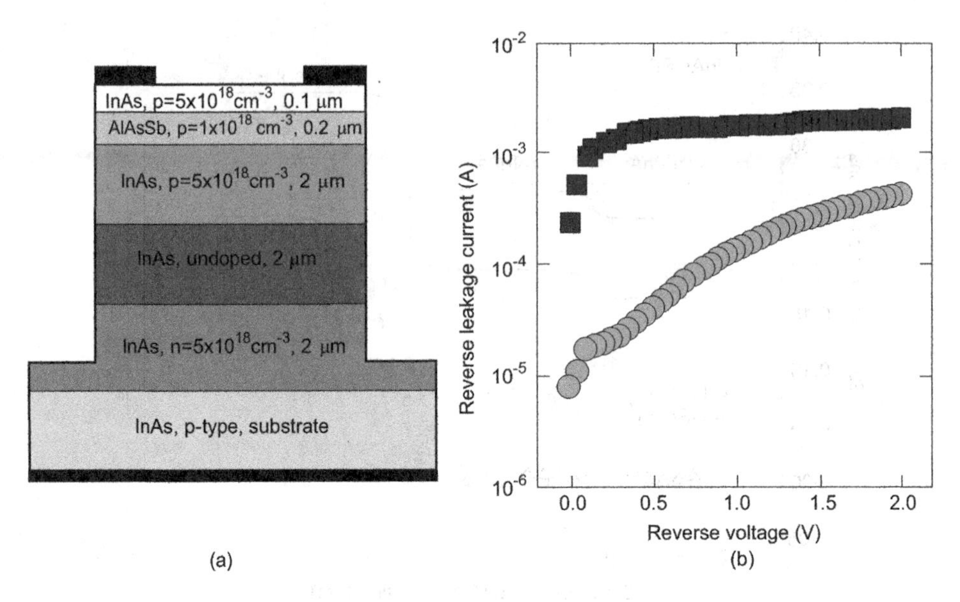

Figure 16.38 InAs p-i-n photodiode: (a) schematic cross-section view of the diode structure, and (b) room temperature reverse leakage current comparison for 3.14×10^{-4}cm^2 area diodes with and without the blocking layer. (Adapted after Ref. [175])

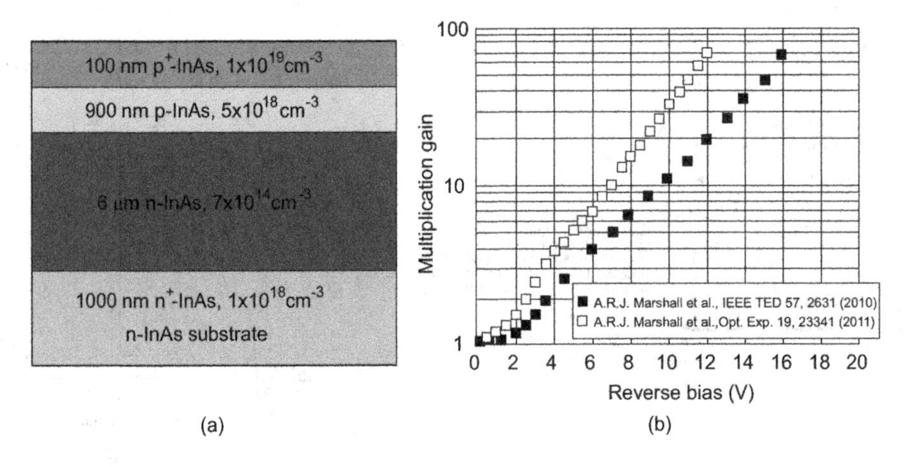

Figure 16.39 InAs APD: (a) mesa structure with unintentionally doped i-region, (b) measured gain.

The local field model (McIntyre's model) for APD is described in Section 12.4. Beck et al. were the first to report APD characteristics consistent with $k=0$ in 2001 [177], when they reported results from $Hg_{0.7}Cd_{0.3}Te$ APDs. They have since shown that for a number of compositions the impact ionization for holes remains essentially zero in $Hg_{1-x}Cd_xTe$ APDs detecting in the short, mid, and long-wave infrared [178]. They coined the phrase electron-APD (e-APD) to describe such APDs where only electrons undergo impact ionization. In this case, the excess noise factor is <2 and independent of gain. However, the low bandgap HgCdTe also results in relatively high dark current at room temperature, which necessitates operation at low temperature.

New impact on the development of InAs APDs has been obtained recently [175,179]. Similarly to HgCdTe devices, InAs APDs have also demonstrated $k\approx0$, with moderately low dark current at room temperature. It has been shown that in InAs p-i-n diodes, significant electron-initiated multiplication can be achieved whilst hole-initiated multiplication in InAs n-i-p diodes remains negligible across the same electric field range [180].

Figure 16.39(a) shows an InAs APD MBE-grown mesa structure with i-region as thick as 6 μm. Beryllium and silicon were used as acceptors and donors, respectively. The n-type background doping

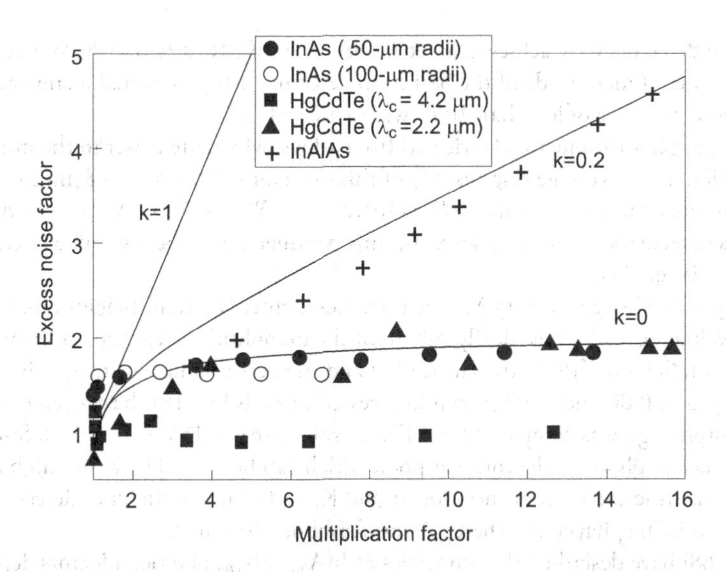

Figure 16.40 A comparison between the F_e reported on APDs of different materials including InAs diodes with a 3.5-μm intrinsic width and radii of 50 and 100 μm, HgCdTe photodiodes with cutoff wavelengths of 4.2 and 2.2 μm, and an InAlAs diode. (Adapted after Ref. [176])

concentration in the i-region was below 1×10^{15} cm^{-3}. To suppress surface leakage current, devices with diameters up to 500 μm were wet etched using 1:1:1 (phosphoric acid, hydrogen peroxide, deionized water), followed by 30 seconds etching in 1:8:80 (sulfuric acid, hydrogen peroxide, deionized water). The etched mesa sidewalls were additionally covered with SU-8 passivation. Good ohmic contact is easily formed by depositing Ti/Au (20/150 nm) without annealing.

Figure 16.39(b) shows that the measured gain in InAs APDs increases exponentially with reverse bias, showing no sign of breakdown, a signature of $k \approx 0$ [181,182]. Devices yielded room temperature multiplication gains >300 [179]. The bandwidth is transit time–limited in the range of 2–3 GHz independent of gain.

Figure 16.40 shows the excess noise factor versus the multiplication factor in InAs APDs and compares with data for different materials. The F_e measured on InAs p-i-n diodes falls slightly below the local model prediction for $k=0$ as given by Equation 12.140. This is comparable to that reported for SWIR HgCdTe e-ADPs, although it is somewhat higher than that reported for MWIR HgCdTe e-APDs. To explain such excess noise below the lower limit case of the local model, it is necessary to consider the influence of "dead space," which is neglected from the local model. The distance in which no impact ionization occurs is referred to as the "dead space" for electrons. If the multiplication region is thick, the "dead space" can be neglected and the local field model provides an accurate description of the APD characteristics. The excess noise in the InAlAs APD rises with increasing multiplication as is for all conventional APDs in which both carriers undergo impact ionization.

The realization of InAs e-APDs has brought the ideal avalanche multiplication and excess noise characteristics into the readily available III–V material system for more widespread applications previously only achievable in the less readily available HgCdTe system. These properties make InAs APDs attractive for a number of near- and mid-infrared sensing applications including remote gas sensing, light detection and ranging, and both active and passive imaging.

16.4 InAsSb PHOTODETECTORS

A summary of works for the fabrication of InAsSb photodetectors before 2010 is given in Rogalski's monograph [183].

Only a few papers are devoted to some aspects of theory and technology of $InAs_{1-x}Sb_x$ (InAsSb) photoconductors. Bethea et al. [184,185] have fabricated and characterized photoconductive detectors obtained by MBE growth of InAsSb on semi-insulating GaAs substrate. In spite of large lattice mismatch (about 14%) between the epitaxial layer and substrate, good quality $InAs_{0.02}Sb_{0.98}$ photoconductors have been obtained [185]. Such detectors have achieved a detectivity value of 3×10^{10}cm Hz$^{1/2}$ W^{-1} at $\lambda=5.4$ µm, which is within an order of magnitude of the best InSb detectors; a high internal quantum efficiency of 47% and a high-speed response of less than 10 ns were achieved.

The performance of photoconductors fabricated from alloys, which are closer to the minimum bandgap composition, are inferior. The voltage responsivity of this detector is 1.5 VW^{-1} at 8 µm for a bias voltage 9 V, which corresponds to a detectivity of $D^*=10^8$cm Hz$^{1/2}$ W^{-1} at 77 K. A photoconductive lifetime determined for this detector is low and equals 9 ns. This parameter and the low value of electron mobility indicate a poor material quality.

A research group at the Interuniversity Microelectronics Center (Leuven, Belgium) has demonstrated the first coplanar technology, which is ideally suited for the monolithic integration of InAsSb-based infrared detectors with Si CCD [186]. The fabrication and the performance of $InAs_{1-x}Sb_x$ photoconductive detectors (with x equal to 0.80 and 0.95) grown in recessed Si wells by MBE have been demonstrated. The InAsSb epilayer morphology was compared for different substrate conditions—Si-well, Si-mesa, and GaAs [187]. The performance of photoconductors was poor, which can be caused by a very high doping level (of the order of the intrinsic carrier concentration at 300 K) and a large number of defects in the interface region between the InAsSb epilayer and the GaAs buffer (about 10^8cm^{-2}).

Podlecki et al. [188] have described the properties of $InAs_{0.91}Sb_{0.09}$ photoconductors deposited by MOCVD on GaAs substrates. The 2-µm thick detectors with an active area of 200 µm^2 and sputtered Au-Sn contacts have been fabricated. The excess noise led to a rather low 3.65 µm detectivity $D^*=5.3\times10^{10}$cm Hz$^{1/2}$ W^{-1} ($V_b=3.75$ V, $f=10^5$ Hz, $T=80$ K). Also, Kim et al. [189] have reported the room temperature–operating photoconductors based on p-type $InAs_{0.23}Sb_{0.77}$ grown on GaAs substrates by MOCVD. The photoconductor structure composed of two epitaxial p-$InAs_{0.23}Sb_{0.77}$/p-InSb layers. InSb layer was used as a buffer layer with 2% lattice mismatch to $InAs_{0.23}Sb_{0.77}$ and also as a confinement layer for the electron in the active layer. P-type doping in the InSb-like band structures ensures a better compromise between optical and thermal generation than n-type doping at near room temperatures. A cutoff wavelength of around 14 µm at room temperature was observed, indicating a smaller bandgap than the expected value. The decrease may have been caused by structural ordering.

Recently, the main effort in InAsSb detector technology has been shifted to the development of photodiodes as useful devices for the second-generation thermal imaging systems and the next generation very-low-loss fiber communication systems. During the last three decades, high-quality InAsSb photodiodes for the 3–5-µm spectral region have been developed. The long wavelength limit of $InAs_{1-x}Sb_x$ detectors has been tuned compositionally from 3.1 µm ($x=0.0$) to 7.0 µm ($x\approx0.6$) at room temperature. To realize IR detectors in all potentially capable operating regions, lattice-matched substrates are necessary. This problem seems to be resolved by using $Ga_{1-x}In_xSb$ substrates. In this case, the lattice parameter can be tuned between 6.095 Å (GaSb) and 6.479 Å (InSb). Several research groups have succeeded in growing GaInSb single crystals [12,190]. One composition worth noting is $Ga_{0.38}In_{0.62}Sb$ that is lattice matched to $InAs_{0.35}Sb_{0.65}$, which has a bandgap minimum, corresponding to ~12 µm.

A variety of InAsSb photodiode configurations have been proposed including mesa and planar, n-p, n-p$^+$, p$^+$-n, and p-i-n structures. The techniques used to form p-n junctions have included diffusion of Zn, Be ion implantation, and the creation of p-type layers on n-type material by LPE, MBE, and MOCVD. The photodiode technology relies essentially on n-type material with concentrations generally about 10^{16}cm^{-3}.

At first, $InAs_{0.85}Sb_{0.15}$ photodiodes grown by a step-graded LPE technique on InAs substrates were reported in 1977 [40]. A series of InAsSb compositionally step-graded buffer layers were introduced between active layers and the InAs substrate to relieve strain caused by lattice mismatch. The device was operated in the BSI mode. In this case, photons enter through the InAs substrate and through sufficiently thick buffer layers and on a filter layer to absorb most incoming photons with energy greater than the energy bandgap of the filter layer. The cutoff wavelength is mainly determined by the energy bandgap of the active layer and spectral characteristics can be controlled by the Sb composition of the filter layer and the active layer. The carrier concentration across the junction is $\approx10^{15}$cm^{-3} for n-type and $\approx10^{16}$cm^{-3} for p-type. Mesa photodiodes fabricated in this way exhibit excellent characteristics as narrow-band BSI infrared detectors. Half-widths of spectral response as narrow as 176 nm at 77 K with a peak internal quantum

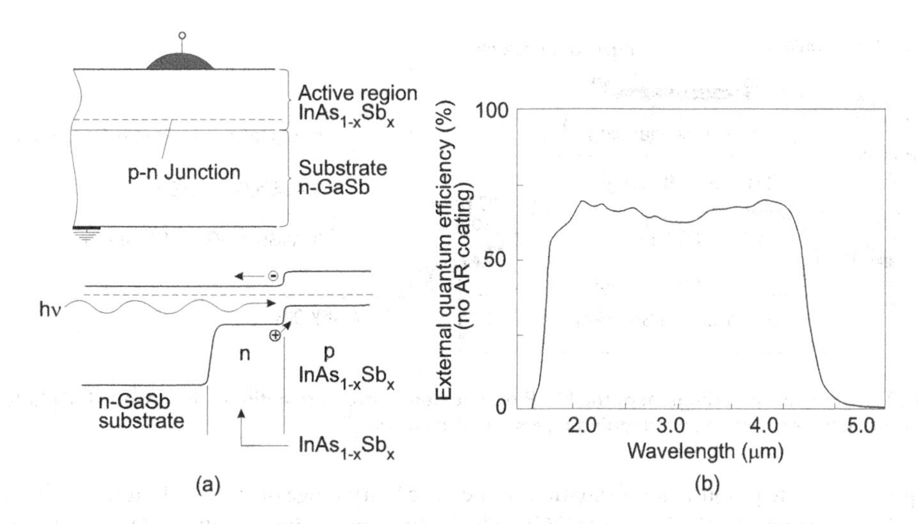

Figure 16.41 Back-side illuminated $InAs_{0.86}Sb_{0.14}$/GaSb photodiode: (a) device structure and energy band diagram of the structure, and (b) spectral response at 77 K. (From Ref. [191])

efficiency of 70% has been achieved. The zero-bias resistance area, R_oA, products are in the 10^5 Ωcm² range, with the best ones achieving 2×10^7 Ωcm².

The best performances of InAsSb photodiodes have been obtained when lattice-matched $InAs_{1-x}Sb_x$/GaSb ($0.09\leq x\leq 0.15$) device structures were used [191]. Lattice mismatch up to 0.25% for the $InAs_{0.86}Sb_{0.14}$ epitaxial layer can be accommodated in terms of low etch-pit density ($\approx10^4$ cm^{-2}). The structure of a BSI $InAs_{1-x}Sb_x$/GaSb photodiode is shown in Figure 16.41(a). The photons enter through the GaSb transparent substrate and reach the $InAs_{1-x}Sb_x$ active layer where they are absorbed. The GaSb substrate determines the short-wavelength cut-on value, which is 1.7 μm at 77 K; instead, the active region establishes the long-wavelength cutoff value [see Figure 16.41(b)]. The p-n junctions were obtained as homojunctions using the LPE technique. The carrier concentrations, both in the undoped n-type layer and in the Zn-doped p-type layer, were approximately 10^{16} cm^{-3}. The high quality of $InAs_{0.86}Sb_{0.14}$ photodiodes was demonstrated by a high R_oA product in excess of 10^9 Ωcm² at 77 K.

High-performance $InAs_{0.89}Sb_{0.11}$ photodiodes have also been obtained by Be ion implantation [192]. The as-grown LPE layers on (100) GaSb substrates were n-type with a typical carrier concentration of 10^{16} cm^{-3}. The implantation mask was formed by 100 nm of CVD SiO_2 deposited at 200°C and next covered with about 5 μm of photoresist or about 700 nm of aluminum. The Be ion implantation was performed using a 100 keV beam and a total dose of 5×10^{15} cm^{-2}. Following the implantation, annealing was carried out at 550°C for about 1 hour. The EBIC analysis of the $InAs_{0.89}Sb_{0.11}$ planar junction and C-V data confirmed the junction formation by the thermodiffusion mechanism.

Attempts to MBE grow $InAs_{0.85}Sb_{0.15}$ p-i-n junctions on lattice-mismatched substrates—InAs (lattice mismatch 1%), GaAs (8.4%), and Si (12.8%)—have not given good results [193]. The performance of these photodiodes was inferior in comparison with the ones fabricated using LPE. Their R_oA product was almost three orders lower than photodiodes obtained by LPE [40,191,192]—below 50 Ωcm² at 77 K for diodes on InAs. The diodes exhibited significantly larger reverse leakage currents. The presence of defects reduces the carrier lifetime so that the g–r currents become increasingly important.

Rogalski has performed an analysis of a resistance-area product (R_oA) of n-p⁺ abrupt $InAs_{0.85}Sb_{0.15}$ junctions [46]. In the temperature range above 160 K, the R_oA product follows the diffusion model, whereas in the temperature range $80\leq T\leq 160$ K, R_oA fits a g–r model. The characteristic lifetime in the depletion region τ_o determined from the theoretical fit to the g–r model is found to be 0.03–0.5 μs. In the best photodiodes, it was determined to be 0.55 μs [191]. The theoretical estimates yield for the radiative $(R_oA)_R$ and Auger recombination $(R_oA)_{A1}$ values of the R_oA product of several orders of magnitude larger. Tunneling current produces an abrupt lowering of R_oA at a concentration a little above 10^{16} cm^{-3}.

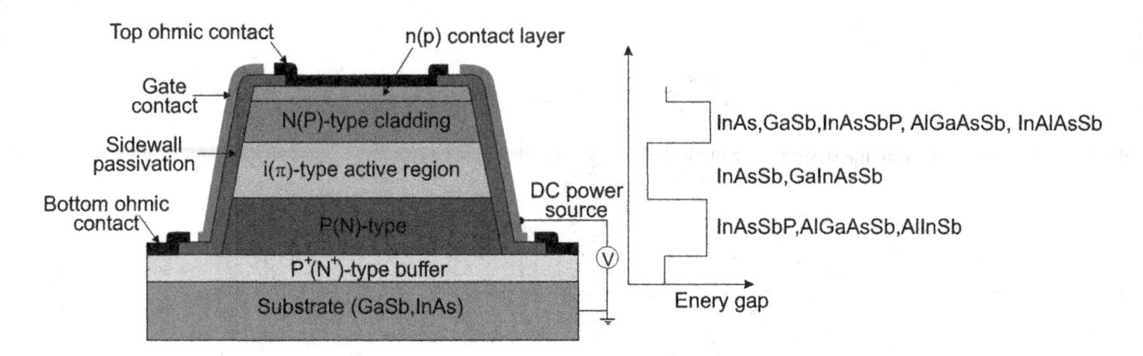

Figure 16.42 Schematic band diagram of the N-i-P double heterostructure antimonide–based III–V photodiodes. Different combinations of active and cladding layers are also shown.

Attempts to fabricate p-n junction formation in the miscibility range of $InAs_{1-x}Sb_x$ ternary alloy over the composition range $0.4 \leq x \leq 0.7$ using MOCVD have not given positive results [194]. The p⁺-n junctions were formed by Zn diffusion into the undoped n-type epitaxial layer with carrier concentration in the range of 10^{16} cm^{-3}. The forward and reverse characteristics were affected by the g–r current of the depletion region and by surface leakage current. It is believed that recombination centers in the depletion layer were caused by diffusion-induced damages and by lattice-mismatch dislocations between the $InAs_{0.60}Sb_{0.40}$ epilayer and the InSb substrate.

To improve device performance (lower dark current and higher detectivity), several groups have developed P-i-N heterostructure devices of an unintentionally doped InAsSb active layer sandwiched between P and N layers of larger bandgap materials. The lower minority carrier concentration in the high bandgap layers resulted in a lower diffusion dark and higher R_oA product and detectivity. Figure 16.42 shows a schematic band diagram of the N-i-P double heterostructure antimonide–based III-V photodiodes together with the different combinations of active and cladding layers in the device structure. Depending on the contact configurations and transparency of substrates, both back-side and front-side illumination can be used. Usually, p-type GaSb and n-type InAs are used. Despite the relatively low absorption coefficients, substrates required thinning to small thicknesses, even less than 10 μm. InAs is fragile and many fabrication processes are not possible. This obstacle can be overcome by using heavily doped n^+-InAs substrates where strong degeneracy of the electrons in the conduction band occurs at relatively low electron concentration (>10^{17} cm^{-3}). For example, the Burstein–Moss shift in heavily doped n^+-InAs ($n = 6 \times 10^{18}$ cm^{-3}) makes the corresponding substrates transparent to the 3.3 μm [144].

Antimonide-based ternary and quaternary alloys are well-established as materials for developing of MWIR photodiodes for near–room temperature operation. Many articles have discussed the properties of mid-IR photodiodes and many of the investigations were made in the Ioffe Institute; see, for example, Refs. [144,171,195]. These LPE-grown heterostructure devices consist of n-type InAs(100) substrates with $n = 2 \times 10^{16}$ cm^{-3} (for undoped) or $n^+ = 2 \times 10^{18}$ cm^{-3} (for the Sn-doped) substrates, ≈ 10-μm thick undoped n-$InAs_{1-x}Sb_x$ active layers and finally p-$InAs_{1-x-y}Sb_xP_y$(Zn) claddings (contact layer) (see Figure 16.43). The narrow-gap InAsSb active layer was surrounded by semiconductors with wider energy gap.

In device processing, the standard optical photolithography and wet chemical etching processes have been implemented to obtain 26-μm high circular mesas ($\varnothing_m = 190$ μm) and 55-μm deep grooves for separation of the 580 × 430 μm rectangular chips. Next, circular Au- or Ag-based reflective anode ($\varnothing_a = 170$ μm) and cathode contacts were formed on the same chip side by sputtering and thermal evaporation in vacuum followed by 3-μm thick gold-plating deposition. Finally, the flip-chip bonding/packaging procedure has been implemented using the 1,800 × 900 μm submount made from the semi-insulating Si wafer with Pb-Sn bonding pads. Photodiode chips were mounted upside down, the n^+-InAs side being an "entrance window" for the incoming radiation as shown in the insert in Figure 16.43. Some chips were equipped with aplanatic hyperhemispherical Si immersion lenses ($\varnothing = 3.5$ mm) with antireflection coating using a chalcogenide glass as an optical glue between Si and n^+-InAs (or n-InAs). The final construction of immersion photodiode is similar to that shown in Figure 16.37(a).

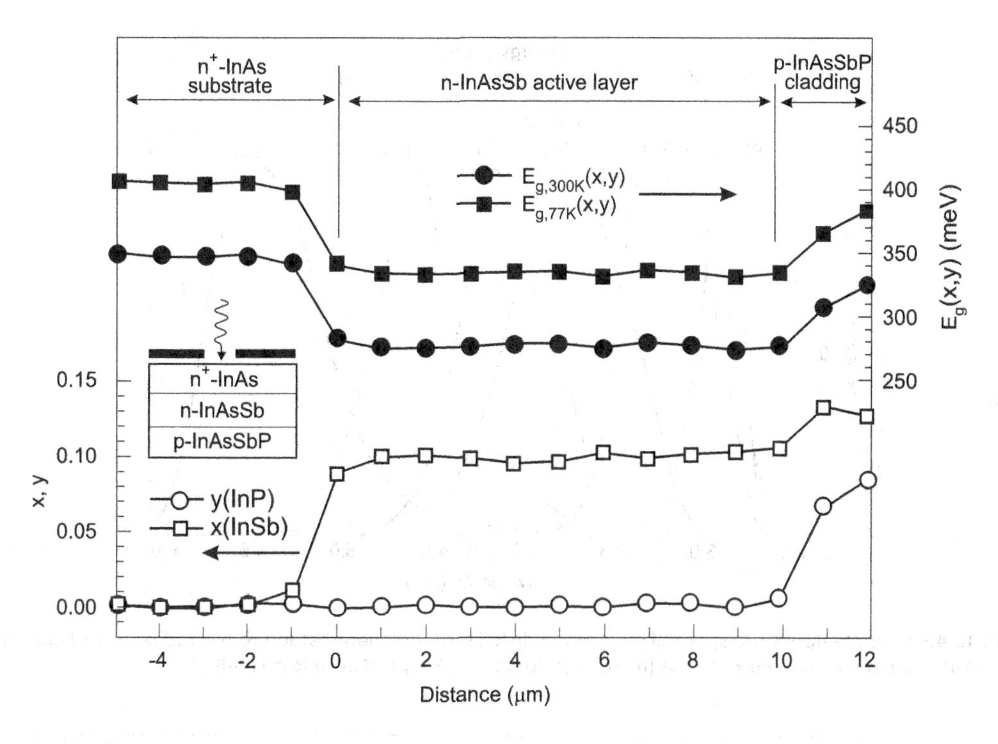

Figure 16.43 Alloy composition and energy bandgap structure versus distance for n⁺-InAs/n-InAsSb/p-InAsSbP double heterostructure photodiode (Adapted after Ref. [171])

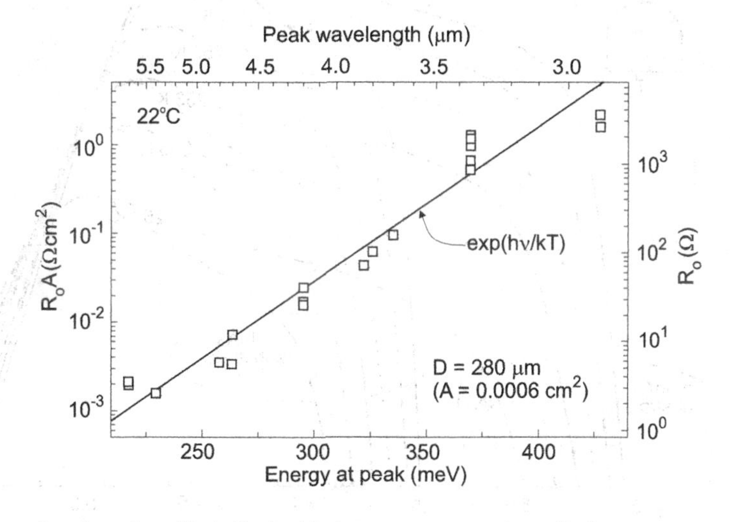

Figure 16.44 R_0A product in series of InAsSb double heterostructure photodiodes at room temperature (Adapted after Ref. [144])

Figure 16.44 summarizes experimental data of zero-bias resistivity and R_0A product vs. photon energy for n⁺-InAs/n-InAsSb/p-InAsSbP double heterostructure photodiodes. An exponential dependence of R_0A product, approximated by $\exp(E_g/kT)$, indicates that the diffusion current determines the transport properties of the heterojunctions with negligible leakage current flow mechanisms at $T > 190$ K. At lower temperatures, generation–recombination at high bias and tunneling at low bias prevail.

Figure 16.45 shows normalized spectral responsivity curves for InAsSb double heterostructure photodiodes with different cutoff wavelengths at room temperature. The photodiodes are characterized by the narrow spectral response [with a full width at half maximum (FWHM) of about 0.3–0.8 μm] resulting

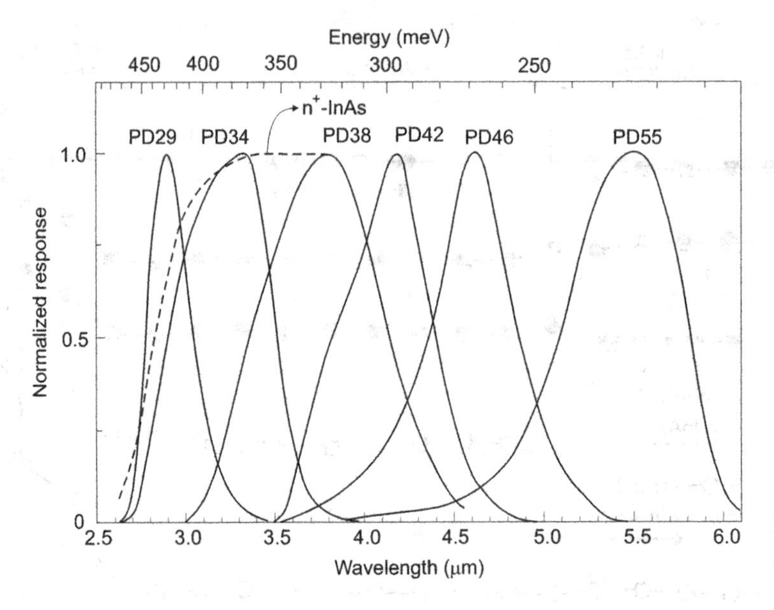

Figure 16.45 Room temperature spectral response of InAsSb double heterostucture photodiodes and normalized transmission of 175-μm thick n^+-InAs [$n^+=(3–6)\times10^{18}cm^{-3}$] (Adapted after Ref. [144])

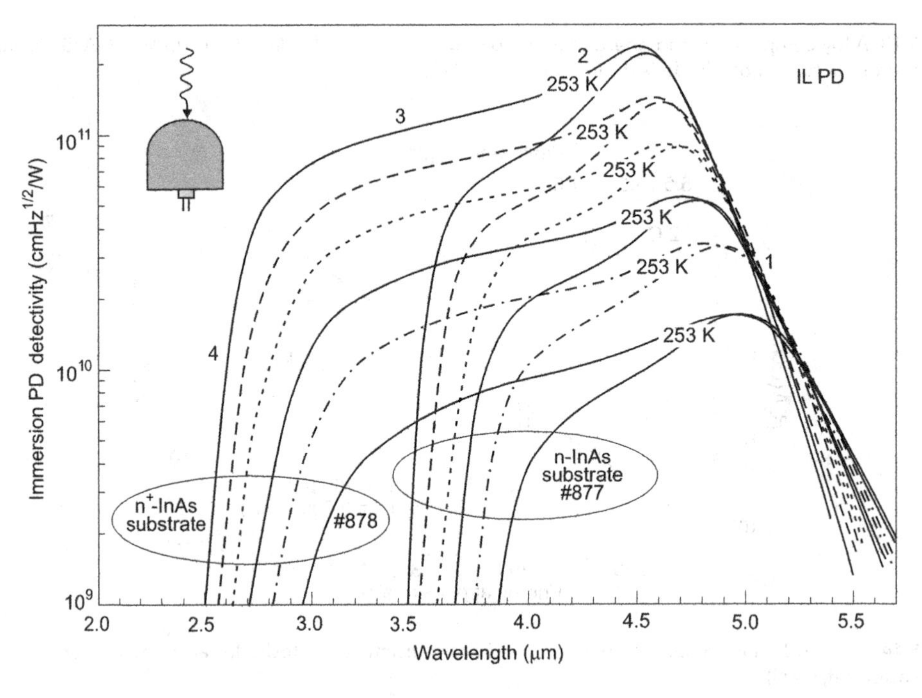

Figure 16.46 Detectivity spectra at different temperatures for InAsSb double heterostucture photodiodes with n-InAs and n^+-InAs substrates (Adapted after Ref. [171])

from spectral filtering in the substrate and intermediate layers. The narrowest response of PD29 device is due to poor transmission of n^+-InAs at short wavelengths. On the other hand, the spectral response of the long-wave photodiodes exhibits broad shoulder at short wavelengths that originates from diffusion of carriers created in the high-energy InAsSbP regions toward the narrow-gap p-n junction.

Figure 16.46 shows the Johnson-limited detectivity spectra of InAsSb double heterostucture photodiodes at different temperatures. As seen from this figure, the responsivity spectra bear four distinct

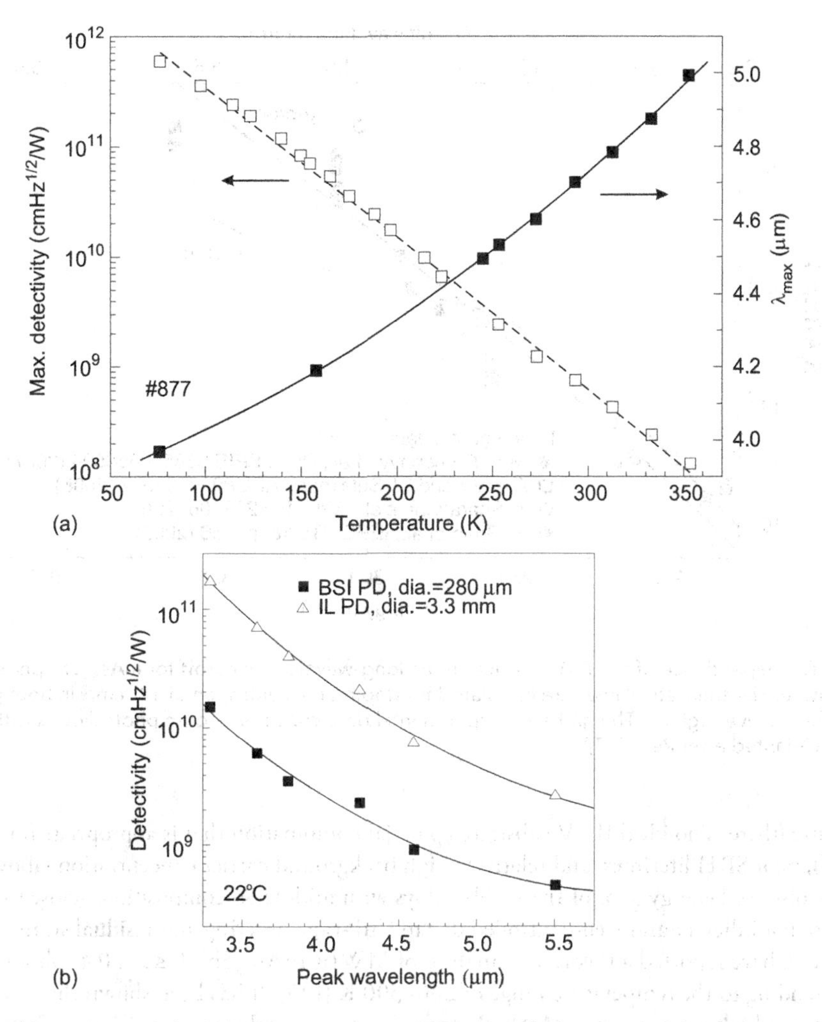

Figure 16.47 Detectivity of InAsSb double heterostucture photodiodes: (a) maximum detectivity and peak wavelength vs. temperature (after Ref. [171]); (b) peak detectivity of photodiodes without (back-side illuminated – BSI) and with Si lenses (immersion illuminated – IL) (Adapted after Ref. [144])

regions—(1) the cutoff region (4.7 < λ < 5.5 μm), (2) sharp longwave response decline region, (3) smooth response decline region, and finally, (4) fast shortwave response decline region. The latter region is due to transmission degradation in heavily doped n^+-InAs substrate. In this case, the Moss–Burstein effect-associated absorption edge shift in n^+-InAs is as large as 1 μm (see difference in short wavelength spectral responsivity between heavily doped #878 sample, and undoped #877 sample).

Figure 16.47(a) summarizes the peak detectivity of photodiodes depending on temperature and wavelength. Next figure, Figure 16.47(b), presents the detectivities depending on peak wavelength for BSI and coated photodiodes (with immersion lens, IL). For the IL photodiodes, the peak detectivities are generally about a decade higher than those for bare chip photodiodes—see Figure 16.47(b). The photodiodes developed at the Ioffe Physical-Technical Institute are superior to the majority published in literature. At the same time, the achievable R_0A products are lower than that given in Ref. [144].

Despite the promise of the III–V–based detectors, HgCdTe remains the highest performing IR material technology for a number of applications. The main obstacles in the rapid development of InAsSb photodiodes are difficulties in the preparation of single crystals and epitaxial layers. During the last 20 years, high-quality InAsSb photodiodes for the 3–5-μm spectral region have been developed. However, a longstanding hope that the InAsSb might become a useful material in the 8–12-μm spectral band has not realized to date. In comparison with HgCdTe, the main obstacles in achieving this goal are poor quality

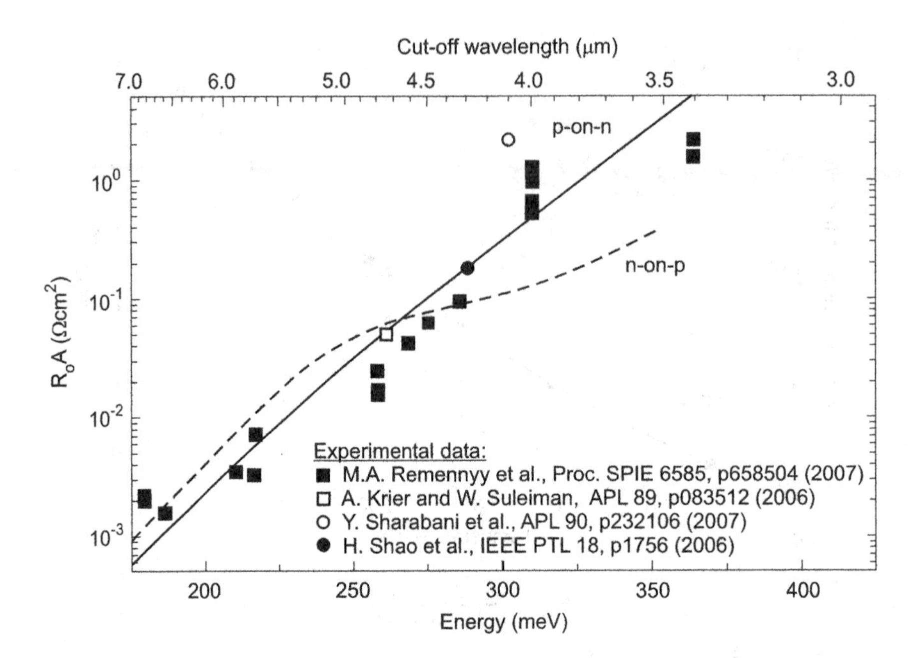

Figure 16.48 The dependence of the R_0A product on the long-wavelength cutoff for $InAs_{1-x}Sb_x$ photodiodes at room temperature. The theoretical lines are calculated for doping concentration of 10^{16}cm^{-3} in both p-on-n and n-on-p 5-μm thick active regions. The gathered experimental data concerns p-on-n photodiodes with n-type active region. (Adapted after Ref. [197])

crystal structure (there is no ideal III–V substrate/epitaxial combination that is appropriate for the LWIR spectral band), poor SRH lifetimes, and relatively high background carrier concentration (above 10^{15}cm^{-3}). Moreover, the observed energy gaps of $InAs_{1-x}Sb_x$ alloys with middle Sb compositions (close to minimum energy gap) are not inherent and well-controlled due to CuP-type ordering and residual strain effects.

Rogalski et al. have reported a theoretical analysis of MWIR $InAs_{1-x}Sb_x$ ($0 \leq x \leq 0.4$) photodiodes, with operations extending to the temperature range of 200–300 K [196]. It has been shown that the theoretical performance of high-temperature InAsSb photodiodes is comparable to that of HgCdTe photodiodes. More recently, Wróbel et al. have considered the effects of doping profiles on room temperature MWIR InAsSb photodiode parameters (R_0A product and detectivity) [197]. In theoretical estimations, a new insight into composition dependence of spin–orbit-splitting bandgap energy is taken into account [198].

Figure 16.48 shows the dependence of the R_0A product on the long-wavelength cutoff for $InAs_{1-x}Sb_x$ photodiodes at 300 K. The theoretical lines are calculated for doping concentration of 10^{16}cm^{-3} in both p-on-n and n-on-p 5-μm thick active regions. It is clearly shown that the influence of Auger S mechanism on n-on-p photodiodes with composition active region close to InAs ($0 \leq x \leq 0.15$; λ_c<4.5 μm) considerably decreases R_0A product in comparison with p-on-n devices, where influence of Auger S mechanisms is eliminated. However, if the composition of the active region $x \geq 0.15$ (λ_c>4.5 μm), the structure based on p-type material is more optimal than that of the p-on-n one.

The theoretically predicted performance is comparable with the experimental values of p-on-n photodiodes with n-type active region. The agreement between both types of data is good. Some of experimental data are located above the theoretical line. It is probably observed that if the thickness of the photodiode active region is smaller than the minority carrier diffusion length, reducing the volume, in which diffusion current is generated, causes a decrease in the corresponding dark current and an increase in R_0A.

The next figure (Figure 16.49) presents spectral dependence of thermal noise–limited detectivity of p-on-n InAsSb photodiodes operated at room temperature assuming quantum efficiency η=0.7 and doping concentration in active region of 10^{16}cm^{-3}. It shows that the upper detectivity experimental data coincide well with theoretical prediction. Discrepancy between both types of results increases with cutoff wavelength increasing, which is mainly caused by decreasing experimentally measured quantum efficiency.

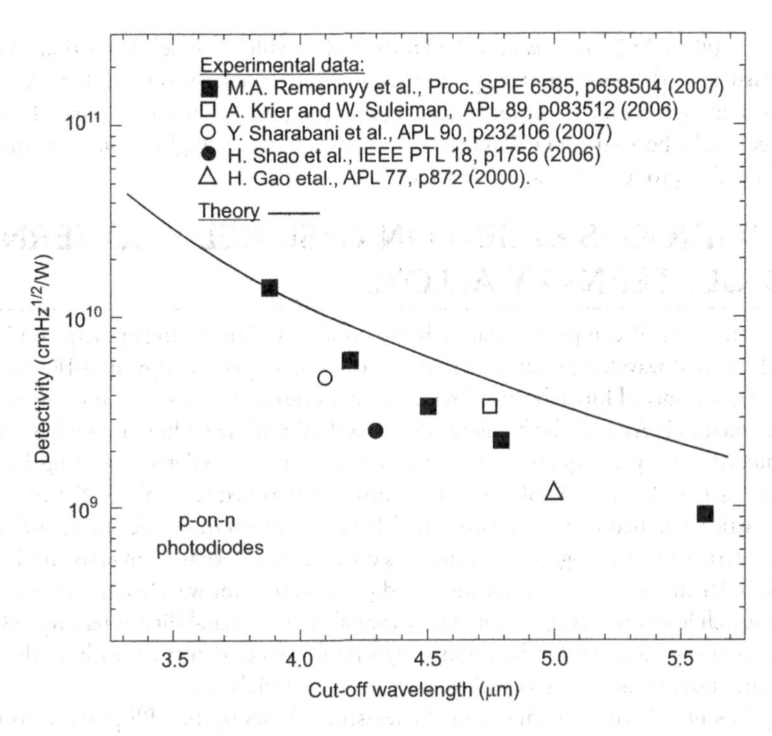

Figure 16.49 The dependence of the detectivity on the long-wavelength cutoff for p-on-n $InAs_{1-x}Sb_x$ photodiodes at room temperature. The theoretical line is calculated for 5-μm thick active region with doping concentration of $10^{16}cm^{-3}$. The gathered experimental data are taken from literature. (Adapted after Ref. [197])

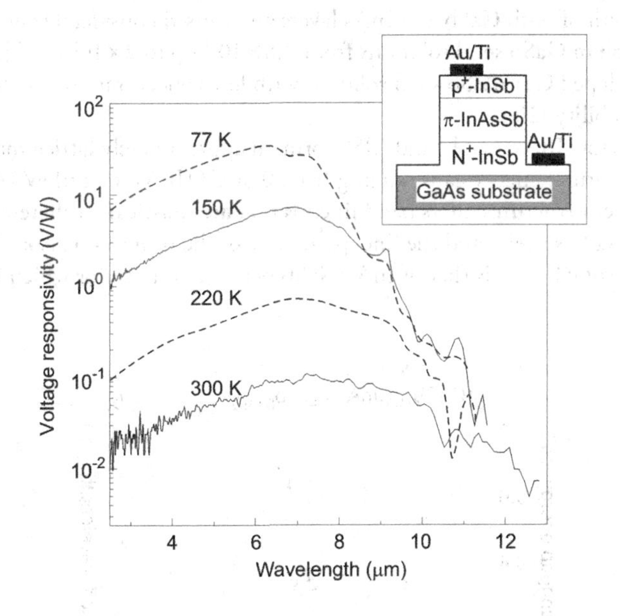

Figure 16.50 Spectral response of the p^+-InSb/π-$nAs_{0.15}Sb_{0.85}$/n^+-InSb heterojunction device at various temperatures. Inside, a schematic device structure is shown. (From Ref. [200])

The first InAsSb-based long wavelength (8–14 μm) photodiode operating at room temperature has been described by Kim et al. [199]. The structure, grown by low-pressure MOCVD, is designed for the operation of BSI (GaAs substrate side). Figure 16.50 shows the voltage responsivity at various temperatures and inside the figure—schematic of device structure [200]. Photoresponse up to 13 μm has been obtained at 300 K in an p^+-InSb/π-$nAs_{0.15}Sb_{0.85}$/n^+ -InSb heterojunction device. The peak voltage responsivity is

9.13×10^{-2} VW^{-1} at 300 K. At 77 K, it is only 2.85×10^{1} VW^{-1}, which is much lower than the expected value. Possible reasons are the poor interface properties due to the lattice mismatch between the absorber and contact layers and high dark current due to the high doping level in the active layer. Introducing the AlInSb buffer layer as the bottom contact layer blocks carriers from the highly dislocated interface resulting in an increase of the R_oA product and detectivity [38].

16.5 PHOTODIODES BASED ON GaSb-RELATED TERNARY AND QUATERNARY ALLOYS

Ternary and quaternary III–V compound materials are suitable for fabricating optoelectronic devices in the near and mid-infrared wavelength range including laser-diode spectroscopy, mid-IR fiber optics, laser rangefinding, free space optical links for high-frequency communications, and so on. The availability of binary substrates, such as InAs and GaSb, allows the growth of multilayer homo- and heterostructures, where lattice-matched ternary and quaternary layers could be tailored to detect wavelengths in the range of 0.8–4 µm. The bandgap of $Ga_xIn_{1-x}As_ySb_{1-y}$ can be continuously tuned from about 475 to 730 meV while the remaining is lattice matched to a GaSb substrate [18,19], as shown in Figure 16.51, and in contrast to leading ternary materials in this range such as InGaAs on InP. Both ternary (InGaSb and InAsSb) and quaternary (InGaAsSb and AlGaAsSb) indicated good performance for wavelength range ≥2 µm, but they are still on the research level not being commercially available. The availability of ternary InGaSb virtual substrates has a promising potential for developing high-performance detectors, without the influence of the binary substrates usually used for processing the ternary materials [201].

The knowledge about a thermodynamic data of coexisting phases in the LPE process has been achieved by the development of the original method based on excess thermodynamic functions and linear combinations of chemical potentials [202]. In this way, the phase diagrams for the Ga-In-As-Sb, In-As-Sb-P, Ga-In-As-Pb, and Ga-Al-As-Sb systems have been calculated. An idea was introduced using Pb as a neutral solvent during LPE growth of both GaSb and InAs layers and caused considerable decrease of the structural defect concentration in GaSb solid solutions from 2.8×10^{17} up to 2×10^{15} cm^{-3} [203]. Moreover, the use of Pb introduces undoped GaInAsSb solid solution with low concentration of defects and impurities and with high carrier mobility [204].

The three semiconductors InAs, GaSb, and AlSb form an approximately lattice-matched set around 6.1 Å, with (room temperature) energy gaps ranging from 0.36 eV (InAs) to 1.61 eV (AlSb) [205]. The combination of their heterostructures offers band lineups that are drastically different from those of the more widely studied AlGaAs system, and the lineups are one of the principal reasons for interest in the 6.1 Å family. The most exotic lineup is that of InAs/GaSb heterojunctions, for which it was found in 1977

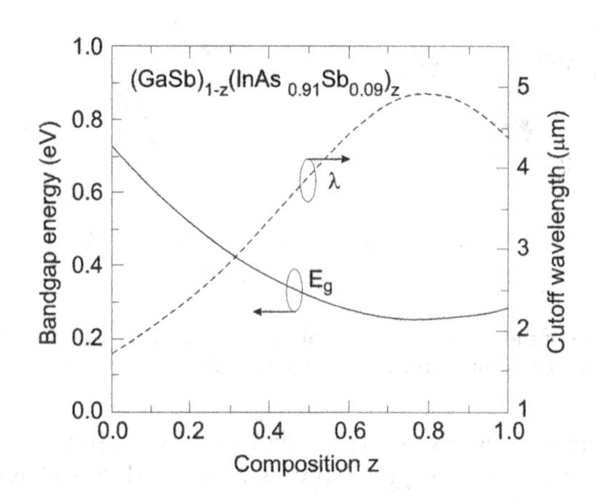

Figure 16.51 The bandgap of $Ga_xIn_{1-x}As_ySb_{1-y}$ with x and y concentrations chosen in the ratio $(GaSb)_{1-z}(InAs_{0.91}Sb_{0.09})_z$ can be tuned continuously from about 475 to 730 meV while remaining lattice matched to a GaSb substrate.

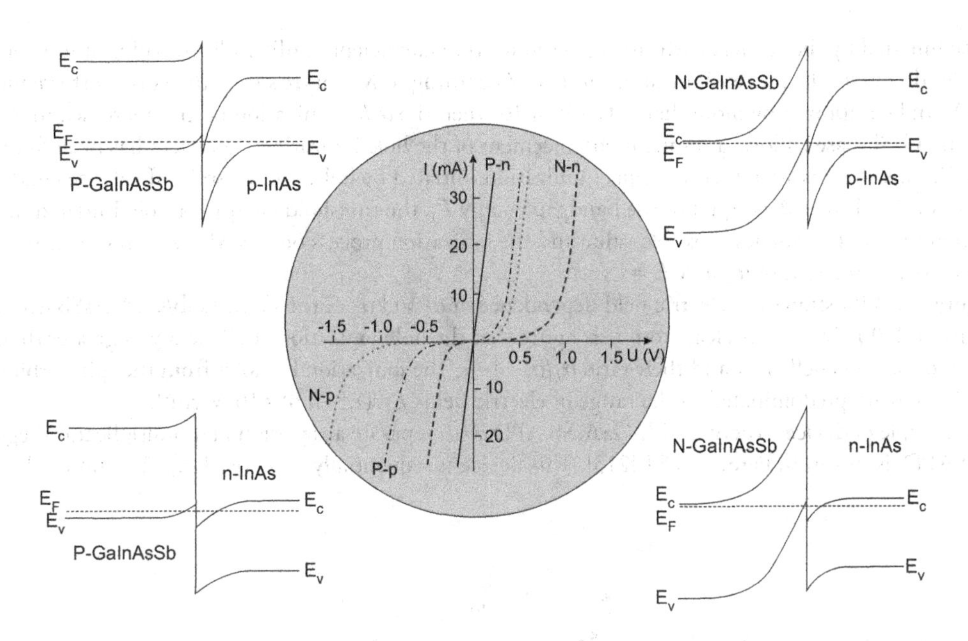

Figure 16.52 Energy band diagrams for four types of single broken-gap GaInAsSb/InAs heterojunctions and their I–V characteristics at 77 K. (From Ref. [207])

by Sakaki et al. [206] and that they exhibit a broken gap lineup: at the interface, the bottom of conduction band of InAs lines up below the top of the valence band of GaSb with a break in the gap of about 150 meV. In such a heterostructure, with partial overlapping of the InAs conduction band with the GaSb-rich solid solution valence band, electrons and holes are spatially separated and localized in self-consistent quantum wells formed on both sides of the heterointerface. This leads to unusual tunneling-assisted radiative recombination transitions and novel transport properties. For example, Figure 16.52 shows the simulation of approximate energy band diagrams for the four types of GaInAsSb/InAs heterojunction (N-n, N-p, P-p, and P-n) [207]. As one can see in the figure, all of the rectifying heterostructures (N-n, N-p, and P-p) demonstrate a large space charge region in the heterojunction. The significant overlap owing to a large bending of the conduction band in GaInAsSb and the valence band in InAs at the heterointerface leads to a strong confinement of the carriers in the self-consistent potential wells on both sides of the heteroboundary. When this overlap disappears, an unconfined movement of carriers on both sides of the junction results in the ohmic (metallic) behavior of the P-GaInAsSb/n-InAs structure. For the N-n heterojunction, the barrier height is close to the bandgap value of the GaInAsSb solid solution and for the case of the P-p structure to the InAs bandgap.

Similar to N-GaInAsSb/n-InAs izotype heterojunction shown in Figure 16.52, also N-GaSb/n-InAsSb rectifying heterostructure has the unique type II broken gap interface [208,209]. A large barrier for electrons is formed in the GaSb side of the interface. Due to differences in electron affinity between two materials, electrons are transferred from the GaSb side to the InAsSb side across the interface. The resultant band banding leads to the formation of a barrier for electrons in the GaSb side and a two-dimensional electron gas in the InAsSb side. The barrier of the N-n interface is comparable to the energy gap of the wider bandgap material (GaSb). Sharabani et al. have shown that N-GaSb/n-$In_{0.91}As_{0.09}Sb$ heterostructure is a promising material for high–operating temperature MWIR detectors [210]. The BLIP temperature was found to be 180 K, and the R_oA products of 2.5 and 180 Ωcm^2 were measured at 300 and 180 K, respectively.

Narrow-gap III–V semiconductors and their alloys are also promising materials for developing high-speed, low-noise APDs. Mikhailova and Andreev have published a comprehensive review paper devoted to 2–5-µm APDs [195].

It is a well-known fact that the excess avalanche noise factor and thus the signal-to-noise ratio of an APD depends on the ratio of electron and hole impact ionization coefficients (α_e and α_h, respectively). To achieve a low noise factor, not only must α_e and α_h be as different as possible, but also the avalanche process must be initiated by the carriers with the higher ionization coefficient. Unlike silicon APDs, it was found that holes dominate the impact ionization process. According to McIntyre's rule, the noise performance of APD can be improved by more than a factor of 10 when the α_h/α_e ionization ratio is increased to 5. For InAs- and GaSb-based alloys, a resonant enhancement of the hole ionization coefficient has been found [195,211]. This effect is attributed to impact ionization initiated by holes from a split-off valence band: if the spin–orbit splitting Δ is equal to the bandgap energy E_g, the threshold energy for hole-initiated impact ionization reaches the smallest possible value and the ionization process occurs with zero momentum. This leads to a strong increase of α_h at $\Delta/E_g=1$.

Figure 16.53 illustrates the electric field dependence of α_e and α_h in the GaInAsSb/GaAlAsSb heterostructure at 230 K [212]. It is clear from this figure that the hole ionization coefficient was greater than the electron ionization coefficient and their ratio α_h/α_e~4–5. The ionization by holes from the spin–orbit splitting valence band predominated in the range of electric fields $E=(1.5–2.3)\times10^5$ V cm^{-1}.

An example of device structure of InGaAsSb APD with separate absorption and multiplication region, (SAM) APD, is shown in Figure 16.54 [213]. This device is sequentially composed of a 2.2-μm thick

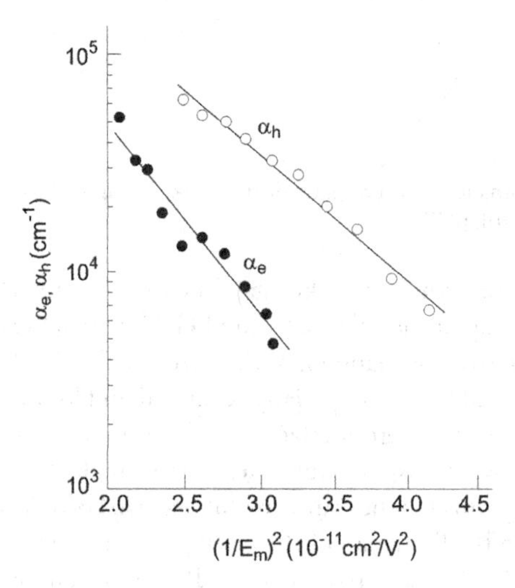

Figure 16.53 Dependence of the hole and electron ionization coefficients on the square of the reciprocal maximum electric field in the $Ga_{0.80}In_{0.20}As_{0.17}Sb_{0.83}$ solid solution at 230 K. (From Ref. [212])

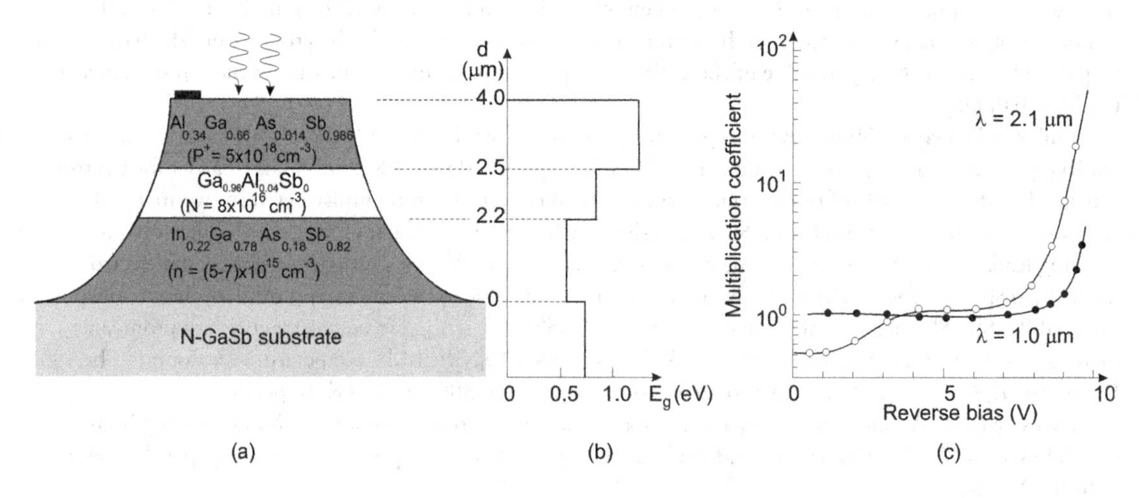

Figure 16.54 SAM APD $Ga_{0.80}In_{0.20}As_{0.17}Sb_{0.83}$ /$Ga_{0.96}Al_{0.04}Sb$ with "resonant" composition in the avalanche region: (a) schematic device structure, (b) bandgap structure, and (c) multiplication coefficient versus reverse bias. (From Andreev, I. A., Afrailov, M. A., Baranov, A. N., Marinskaya, N. N., Mirsagatov, M. A., Mikhailova, M. P., Yakovlev, Yu. P., *Soviet Technical Physics Letters*, 15, 692–96, 1989.) (From Ref. [213])

Te-compensated $Ga_{0.78}In_{0.22}As_{0.18}Sb_{0.82}$ layer with electron concentration $(5–7)\times10^{15}cm^{-3}$; a 0.3-μm thick n-$Ga_{0.96}Al_{0.04}Sb$ "resonant" composition layer with electron concentration of $8\times10^{16}cm^{-3}$; and a 1.5-μm thick $Al_{0.34}Ga_{0.66}As_{0.014}Sb_{0.986}$ window layer with hole concentration of $5\times10^{18}cm^{-3}$. The location of the p-n junction coincides with the heterointerface between two wide-gap materials. The space charge region lies in the n-$Ga_{0.96}Al_{0.04}Sb$/p-$Al_{0.34}Ga_{0.66}As_{0.014}Sb_{0.986}$ heterointerface and results in predominant multiplication of holes in the n-$Ga_{0.96}Al_{0.04}Sb$ multiplication region. The maximum values of the multiplication factor were measured to be M=30–40 at room temperature. The breakdown voltage determined by the wide-gap material was about 10–12V. As a band resonance condition takes place in $Ga_{0.96}Al_{0.04}Sb$ at 0.76eV, high values of α_h/α_e ratio up to 60 are achieved. Thus, an essentially unipolar multiplication by holes is provided that reduces the excess noise problem in these APDs.

16.6 NOVEL Sb-BASED III–V NARROW GAP PHOTODETECTORS

$In_{1-x}Tl_xSb$ (InTlSb) was proposed as a potential IR material in the LWIR region [214,215]. The TlSb is predicted as a semimetal. By alloying TlSb with InSb, the bandgap of InTlSb could be varied from −1.5 to 0.26eV. Assuming a linear dependence of the bandgap on alloy composition, $In_{1-x}Tl_xSb$ can then be expected to reach a bandgap of 0.1 eV at x=0.08, while exhibiting a similar lattice constant as InSb since the radius of Tl atom is very similar to In. At this gap, InTlSb and HgCdTe have very similar band structure. This implies that InTlSb has comparable optical and electrical properties to HgCdTe. In the structural aspect, InTlSb is expected to be more robust due to stronger bonding. The estimated miscibility limit of Tl in ZB InTlSb was estimated to be approximately 15%, which is sufficient to obtain an energy gap down to 0.1 eV. Figure 16.55 shows the expected relationships between bandgap energy and lattice constant

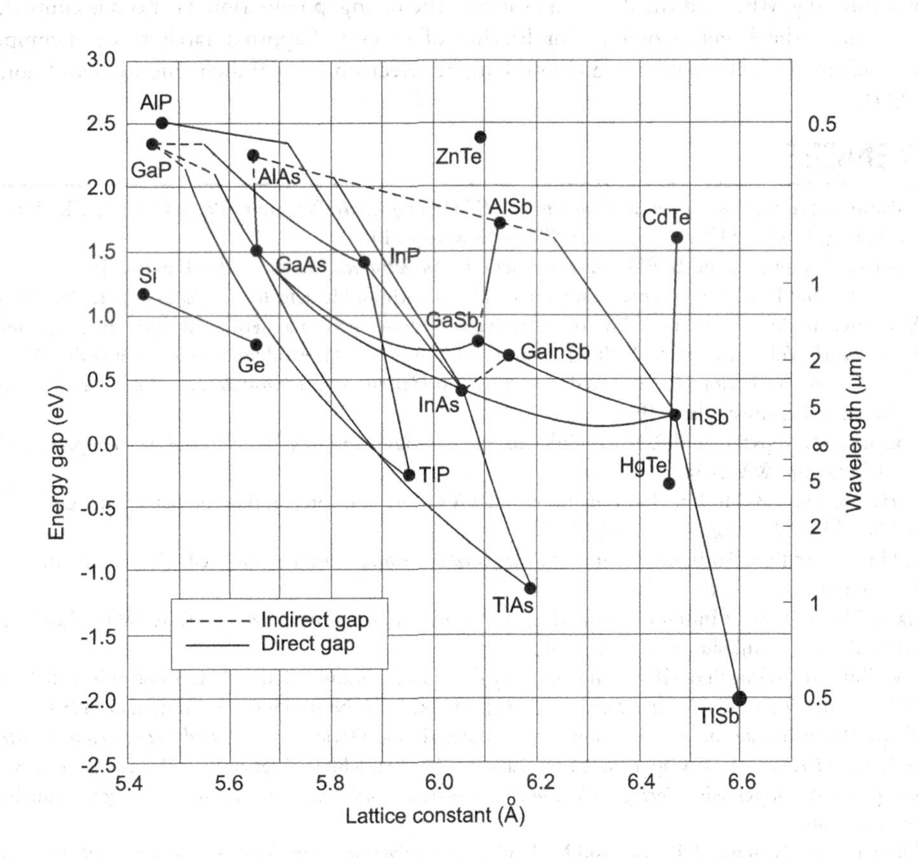

Figure 16.55 Composition and wavelength diagrams of a number of semiconductors with D and ZB structures versus their lattice constants. Tl-based III–Vs are also included.

for Tl-based III–V ZB alloys [39]. The room temperature operation of InTlSb photodetectors has been demonstrated with a cutoff wavelength of approximately 11 μm [163,216].

Van Schilfgaarde et al. showed that another ternary alloy, $In_{1-x}Tl_xP$ (InTlP), is a promising material for IR detectors [217]. It was shown that this material can cover the bandgap from 1.42 eV (InP at 0 K) to 0 eV using a small lattice mismatch with InP. Optical measurements verified the reduction of the bandgap by the addition of Tl into the InP [218].

As another alternative to the HgCdTe material system, $InSb_{1-x}Bi_x$ (InSbBi) has been considered because the incorporation of Bi into InSb produces a rapid reduction in the bandgap of 36 meV/%Bi. Thus, only a few percentages of Bi are required to reduce the bandgap energy.

The growth of an InSbBi epitaxial layer is difficult due to the large solid-phase miscibility gap between InSb and InBi. The successful growth of an InSbBi epitaxial layer on InSb and GaAs (100) substrates with a substantial amount of Bi (~5%) was demonstrated using low-pressure MOCVD [219,220]. The responsivity of the $InSb_{0.95}Bi_{0.05}$ photoconductor at 10.6 μm was 1.9×10^{-3} VW^{-1} at room temperature, and the corresponding Johnson noise–limited detectivity was 1.2×10^6 cm Hz$^{1/2}$ W^{-1}. The effective carrier lifetime estimated from bias voltage-dependent responsivity was approximately 0.7 ns at 300 K.

In diluted nitride alloys of III–V semiconductors, a strong negative bandgap bowing effects is observed [221,222]. Most papers concentrated on the alloys GaAsN and GaInAsN owing to their technological importance for fiber communications at wavelengths of 1.3 and 1.55 μm.

Preliminary estimates of the band structure of $InSb_{1-x}N_x$ were made using a semiempirical **k·p** model. The theoretically predicted variation in bandgap structure indicates a decrease in bandgap of 110 meV (fractional change of 63%) at 1% N, which clearly offers potential for long-wavelength applications [223]. These theoretical predictions were experimentally confirmed with measurements of response wavelengths of light-emitting diodes. The $InSb_{1-x}N_x$ samples with up to 10% of the *N* were grown by combining MBE and the *N* plasma source. The bandgap reduction has been accompanied by an enhancement in the Auger recombination lifetime of a factor of approximately three in comparison with an equivalent HgCdTe bandgap, due to the higher electron mass and conduction band nonparabolicity [224].

REFERENCES

1. C. Hilsum and A. C. Rose-Innes, *Semiconducting III-V Compounds*, Pergamon Press, Oxford, UK, 1961.
2. O. Madelung, *Physics of III-V Compounds*, Wiley, New York, 1964.
3. T. S. Moss, G. J. Burrel, and B. Ellis, *Semiconductor Optoelectronics*, Butterworths, London, 1973.
4. F. D. Morten and R. E. King, “Photoconductive indium antimonide detectors,” *Appl. Opt.*, 4, 659–63, 1965.
5. P. W. Kruse, “Indium Antimonide Photoconductive and Photoelectromagnetic Detectors,” in *Semiconductors and Semimetals*, Vol. 5, eds. R. K. Willardson and A. C. Beer, 15–83, Academic Press, New York, 1970.
6. C. T. Elliott and N. T. Gordon, “Infrared Detectors,” in *Handbook of Semiconductors*, Vol. 4, ed. C. Hilsum, 841–936, Elsevier, Amsterdam, 1993.
7. A. Rogalski, M. Kopytko, and P. Martyniuk, *Antimonide-Based Infrared Detectors – A new Perspective*, SPIE Press, Bellingham, WA, 2018.
8. K. F. Hulme and J. B. Mullin, “Indium antimonide: A review of its preparation, properties and device applications,” *Solid-State Electron.*, 5, 211–47, 1962.
9. K. F. Hulme, “Indium Antimonide,” in *Materials Used in Semiconductor Devices*, ed. C. A. Hogarth, 115–62, Wiley-Interscience, New York, 1965.
10. S. Liang, “Preparation of Indium Antimonide,” in *Compound Semiconductors*, eds. R. K. Willardson and H. L. Georing, 227–37, Reinhold, New York, 1966.
11. J. B. Mullin, “Melt-Growth of III-V Compounds by the Liquid Encapsulation and Horizontal Growth Techniques,” in *III-V Semiconductor Materials and Devices*, ed. R. J. Malik, 1–72, North Holland, Amsterdam, 1989.
12. W. F. M. Micklethwaite and A. J. Johnson, “InSb: Materials and Devices,” in *Infrared Detectors and Emitters: Materials and Devices*, eds. P. Capper and C. T. Elliott, 178–204, Kluwer Academic Publishers, Boston, MA, 2001.
13. S. Kasap and P. Capper, eds., *Springer Handbook of Electronic and Photonic Materials*, Springer, Heidelberg, Germany, 2006.
14. G. Dhanaraj, K. Byrappa, V. Prasad, and M. Dudley, eds., *Springer Handbook of Crystal Growth*, Springer, Heidelberg, Germany, 2010.

15. M. J. Furlong, B. Martinez, M. Tybjerg, B. Smith, and A. Mowbray, "Growth and characterization of ≥6" epitaxy-ready GaSb substrates for use in large area infrared imaging applications", *Proc. SPIE* 9451, 94510S-1–7, 2015.
16. T. Asahi, K. Kainosho, K. Kohiro, A. Noda, K. Sato, and O. Oda, "Growth of III-V and II-VI Single Crystals by the Vertical Gradient-freeze Method" in *The Technology of Crystal Growth and Epitaxy*, eds. H. J. Scheel and T. Fukuda, 323–348, Wiley, Chichester, UK, 2003.
17. O. Oda, *Compound Semiconductor Bulk Materials and Characterizations*, World Scientific Publishing, Sinapore, 2007.
18. S. Adachi, *Physical Properties of III-V Semiconducting Compounds: InP, InAs, GaAs, GaP, InGaAs, and InGaAsP*, Wiley-Interscience, New York, 1992; Properties of Group-IV, III-V and II-VI Semiconductors, John Wiley & Sons, Ltd., Chichester, UK, 2005.
19. I. Vurgaftman, J. R. Meyer, and L. R. Ram-Mohan, "Band parameters for III–V compound semiconductors and their alloys," *J. Appl. Phys.*, 89, 5815–75, 2001.
20. M. Ilegems, "InP-based Lattice-matched Heterostructures," in *Properties of Lattice-Matched and Strained Indium Gallium Arsenide*, ed. P. Bhattacharya, 16–25, IEE, London, 1993.
21. G. H. Olsen and V. S. Ban, "InGaAsP: The next generation in photonics materials," *Solid State Technol.*, 99–105, February 1987.
22. A. Rogalski, K. Adamiec, and J. Rutkowski, *Narrow-Gap Semiconductor Photodiodes*, SPIE Press, Bellingham, WA, 2000.
23. W. Zawadzki, "Electron transport phenomena in small-gap semiconductors," *Adv. Phys.*, 23, 435–522, 1974.
24. F. P. Kesamanli, Y. V. Malcev, D. N. Nasledov, Y. I. Uhanov, and A. S. Filipczenko, "Magnetooptical investigations of InSb conduction band," *Fiz. Tverd. Tela* 8, 1176–81, 1966.
25. Y. P. Varshni, "Temperature dependence of the energy gap in semiconductors," *Physica* 34, 149–54, 1967.
26. A. Rogalski, "Infrared detectors: Status and trends," *Prog. Quantum Electron.*, 27, 59–210, 2003.
27. A. Rogalski, "HgCdTe infrared detector material: History, status, and outlook," *Rep. Prog. Phys.*, 68, 2267–336, 2005.
28. B. F. Levine, "Quantum well infrared photodetectors," *J. Appl. Phys.*, 74, R1–81, 1993.
29. S. D. Gunapala and K. M. S. V. Bandara, "Recent Developments in Quantum-Well Infrared Photodetectors," in *Physics of Thin Films*, Vol. 21, eds. M. H. Francombe and J. L. Vossen, 113–237, Academic Press, New York, 1995.
30. S. D. Gunapala and S. V. Bandara, "Quantum Well Infrared Photodetector (QWIP)," in *Handbook of Thin Film Devices*, Vol. 2, eds. M. H. Francombe, 63–99, Academic Press, San Diego, CA, 2000.
31. H. C. Liu, "An Introduction to the Physics of Quantum Well Infrared Photodetectors and Other Related New Devices," in *Handbook of Thin Film Devices*, Vol. 2, ed. M. H. Francombe, 101–34, Academic Press, San Diego, CA, 2000.
32. H. Schneider and H. C. Liu, *Quantum Well Infrared Photodetectors. Physics and Applications*, Springer, Berlin, 2007.
33. *Advances in Infrared Photodetectors*, eds. S. D. Gunapala, D. R. Rhiger, and C. Jagadish, Elsevier, Amsterdam, 2011.
34. F. F. Sizov and A. Rogalski, "Semiconductor superlattices and quantum wells for infrared optoelectronics," *Prog. Quantum Electron.*, 17, 93–164, 1993.
35. A. Rogalski, *New Ternary Alloy Systems for Infrared Detectors*, SPIE Optical Engineering Press, Bellingham, WA, 1994.
36. D. L. Smith and C. Mailhiot, "Proposal for strained type II superlattice infrared detectors," *J. Appl. Phys.*, 62, 2545–8, 1987.
37. L. Bürkle and F. Fuchs, "InAs/(GaIn)Sb Superlattices: A Promising Material System for Infrared Detection," in *Handbook of Infrared Detection and Technologies*, eds. M. Henini and M. Razeghi, 159–89, Elsevier, Oxford, UK, 2002.
38. M. Razeghi, "Overview of antimonide based III-V semiconductor epitaxial layers and their applications at the Center for Quantum Devices," *Eur. Phys. J. Appl. Phys.*, 23, 149–205, 2003.
39. H. Asahi, "Tl-Based III-V Alloy Semiconductors," in *Infrared Detectors and Emitters: Materials and Devices*, eds. P. Capper and C. T. Elliott, 233–49, Kluwer Academic Publishers, Boston, MA, 2001.
40. D. T. Cheung, A. M. Andrews, E. R. Gertner, G. M. Williams, J. E. Clarke, J. L. Pasko, and J. T. Longo, "Backside-illuminated $InAs_{1-x}Sb_x$-InAs narrow-band photodetectors," *Appl. Phys. Lett.*, 30, 587–98, 1977.
41. J. C. Woolley and B. A. Smith, "Solid solution in III-V compounds," *Proc. Phys. Soc.*, 72, 214–23, 1958.
42. J. C. Woolley and J. Warner, "Preparation of InAs-InSb alloys," *J. Electrochem. Soc.*, 111, 1142–5, 1964.
43. M. J. Aubin and J. C. Woolley, "Electron scattering in InAsSb alloys," *Can. J. Phys.*, 46, 1191–8, 1968.
44. J. C. Woolley and J. Warner, "Optical energy-gap variation in InAs-InSb alloys," *Can. J. Phys.*, 42, 1879–85, 1964.
45. E. H. Van Tongerloo and J. C. Woolley, "Free-carrier Faraday rotation in $InAs_{1-x}Sb_x$ alloys," *Can. J. Phys.*, 46, 1199–206, 1968.

46. A. Rogalski, "InAsSb infrared detectors," *Prog. Quantum Electron.*, 13, 191–231, 1989.
47. W. Zhang and M. Razeghi, "Antimony-Based Materials for Electro-Optics", in *Semiconductor Nanostructures for Optoelectronic Applications*, ed. T. Steiner, 229–88, Artech House, Norwood, Boston, MA 02062, 2004.
48. S. P. Svensson, W. L. Sarney, H. Hier, Y. Lin, D. Wang, D. Donetsky, L. Shterengas, G. Kipshidze, and G. Belenky, "Band gap of $InAs_{1-x}Sb_x$ with native lattice constant," *Phys. Rev. B* 86, 245205-1–6, 2012.
49. Y. Lin, D. Donetsky, D. Wang, D. Westerfeld, G. Kipshidze, L. Shterengas, W. L. Sarney, S. P. Svensson, and G. Belenky, "Development of bulk InAsSb alloys and barrier heterostructures for long-wavelength infrared detectors," *J. Electron. Mater.*, 44(10), 3360–6, 2015.
50. S. Suchalkin, L. Ludwig, G. Belenky, B. Laikhtman, G. Kipshidze, Y. Lin L. Shterengas, D. Smirnov, S. Luryi, W.L. Sarney, and S.P. Svensson, "Electronic properties of unstrained narrow gap $InAs_{1-x}Sb_x$ alloys," *J. Phys. D: Appl. Phys.*, 49(10), 105101, 2016.
51. J. M. Schoen, "Augmented-plane-wave virtual-crystal approximation, " *Phys. Rev.*, 184(3), 858–63, 1969.
52. A. Rogalski and K. Jóźwikowski, "Intrinsic carrier concentration and effective masses in $InAs_{1-x}Sb_x$," *Infrared Phys.*, 29, 35–42, 1989.
53. H. H. Wieder and A. R. Clawson, "Photo-electronic properties of $InAs_{0.07}Sb_{0.93}$ films," *Thin Solid Films* 15, 217–21, 1973.
54. O. Berolo, J. C. Woolley, and J. A. Van Vechten, "Effect of disorder on the conduction-band effective mass, valence-band spin-orbit splitting, and the direct band gap in III-V alloys," *Phys. Rev. B* 8, 3794, 1973.
55. S. A. Cripps, T. J. C. Hosea, A. Krier, V. Smirnov, P. J. Batty, Q. D. Zhuang, H. H. Lin, P. W. Liu, and G. Tsai, "Determination of the fundamental and spin-orbit-splitting band gap energies of InAsSb-based ternary and pentenary alloys using mid-infrared photoreflectance," *Thin Solid Films* 516, 8049–58, 2008.
56. J. A. Van Vechten, O. Borolo, and J. C. Woolley, "Spin-orbit splitting in compositionally disordered semiconductors," *Phys. Rev. Lett.*, 29, 1400, 1972.
57. W. M. Coderre and J. C. Woolley, "Electrical properties of electron effective mass in III-V alloys", *Can. J. Phys.*, 46, 1207, 1968.
58. V. W. L. Chin, R. J. Egan, and T. L. Tansley, "Electron mobility in $InAs_{1-x}Sb_x$ and the effect of alloy scattering," *J. Appl. Phys.*, 69, 3571, 1991.
59. R. J. Egan, V. W. L. Chin, and T. L. Tansley, "Dislocation scattering effects on electron mobility in InAsSb," *J. Appl. Phys.*, 69, 2473–8, 1994.
60. A. Joullie, F. Jia Hua, F. Karouta, H. Mani, and C. Alibert, "III-V alloys based on GaSb for optical communications at 2.0–4.5 μm," *Proc. SPIE* 587, 46–57, 1985.
61. T. P. Pearsall and M. Papuchon, "The $Ga_{0.47}In_{0.53}As$ homojunction photodiode: A new avalanche photodetector in the near infrared between 1.0 and 1.6 μm," *Appl. Phys. Lett.*, 33, 640–2, 1978.
62. S. R. Forrest, R. F. Leheny, R. E. Nahory, and M. A. Pollack, "$In_{0.53}Ga_{0.47}As$ photodiodes with dark current limited by generation-recombination and tunneling," *Appl. Phys. Lett.*, 37, 322–5, 1980.
63. N. Susa, H. Nakagome, O. Mikami, H. Ando, and H. Kanbe, "New InGaAs/InP avalanche photodiode structure for the 1–1.6 μm wavelength region," *IEEE J. Quantum Electron.*, QE-16, 864–70, 1980.
64. S. R. Forrest, R. G. Smith, and O. K. Kim, "Performance of $In_{0.53}Ga_{0.47}As$/InP avalanche photodiodes," *IEEE J. Quantum Electron.*, QE-18, 2040–8, 1980.
65. G. E. Stillman, L. W. Cook, G. E. Bulman, N. Tabatabaie, R. Chin, and P. D. Dapkus, "Long-wavelength (1.3- to 1.6-μm) detectors for fiber-optical communications," *IEEE Trans. Electron Devices* ED-29, 1365–71, 1982.
66. G. E. Stillman, L. W. Cook, N. Tabatabaie, G. E. Bulman, and V. M. Robbins, "InGaAsP photodiodes," *IEEE Trans. Electron Devices* ED-30, 364–81, 1983.
67. T. P. Pearsall and M. A. Pollack, "Compound Semiconductor Photodiodes," in *Semiconductors and Semimetals*, Vol. 22D, ed. W. T. Tang, 173–245, Academic Press, Orlando, FL, 1985.
68. D. E. Ackley, J. Hladky, M. J. Lange, S. Mason, S. R. Forrest, and C. Staller, "Linear arrays of InGaAs/InP avalanche photodiodes for 1.0–1.7 μm," *Proc. SPIE* 1308, 261–72, 1990.
69. J. C. Campbell, S. Demiguel, F. Ma, A. Beck, X. Guo, S. Wang, X. Zheng, et al., "Recent advances in avalanche photodiodes," *IEEE J. Sel. Top. Quantum Electron.*, 10, 777–87, 2004.
70. J. C. Campbell, "Recent advances in telecommunications avalanche photodiodes," *J. Lightwave Technol.*, 25(1), 109–21, 2007.
71. G. H. Olsen, A. M. Joshi, S. M. Mason, K. M. Woodruff, E. Mykietyn, V. S. Ban, M. J. Lange, J. Hladky, G. C. Erickson, and G. A. Gasparian, "Room-temperature InGaAs detector arrays for 2.5 μm," *Proc. SPIE* 1157, 276–82, 1989.
72. L. J. Kozlowski, W. E. Tennant, M. Zandian, J. M. Arias, and J. G. Pasko, "SWIR staring FPA performance at room temperature," *Proc. SPIE* 2746, 93–100, 1996.

73. L. J. Kozlowski, K. Vural, J. M. Arias, W. E. Tennant, and R. E. DeWames, "Performance of HgCdTe, InGaAs and quantum well GaAs/AlGaAs staring infrared focal plane arrays," *Proc. SPIE* 3182, 2–13, 1997.
74. M. H. Ettenberg, M. J. Cohen, G. H. Olsen, and J. J. Kennedy, "InGaAs focal plane arrays and cameras for man-portable near-infrared imaging," *Proc. SPIE* 3701, 225–32, 1999.
75. A. Hoffman, T. Sessler, J. Rosbeck, D. Acton, and M. Ettenberg, "Megapixel InGaAs for low background applications," *Proc. SPIE* 5783, 32–8, 2005.
76. H. Yuan, G. Apgar, J. Kim, J. Laquindanum, V. Nalavade, P. Beer, J. Kimchi, and T. Wong, "FPA development: From InGaAs, InSb, to HgCdTe," *Proc. SPIE* 6940, 69403C, 2008.
77. G. Martin, "High Performance SWIR Imaging Cameras", http://www.raptorphotonics.com/wp-content/uploads/2015/10/Ninox-White-Paper-Final.pdf.
78. "Introduction to Scientific InGaAs FPA Cameras", Technical Note, http://www.princetoninstruments.com/userfiles/files/ technotes/Introduction- to-scientific-InGaAs-FPA-cameras.pdf.
79. I. Gyuro, "MOVPE for InP-based optoelectronic device application," *III-Vs Rev.*, 9(2), 30–5, 1996.
80. A. M. Joshi, G. H. Olsen, S. Mason, M. J. Lange, and V. S. Ban, "Near-infrared (1–3 μm) InGaAs detectors and arrays: Crystal growth, leakage current and reliability," *Proc. SPIE* 1715, 585–93, 1992.
81. R. U. Martinelli, T. J. Zamerowski, and P. A. Longeway, "2.6 μm InGaAs photodiodes," *Appl. Phys. Lett.*, 53, 989–91, 1988.
82. G. H. Olsen, M. J. Lange, M. J. Cohen, D. S. Kim, and S. R. Forrest, "Three-band 1.0–2.5 μm near-infrared InGaAs detector array," *Proc. SPIE* 2235, 151–9, 1994.
83. D. E. Ackley, J. Hladky, M. J. Lange, S. Mason, S. R. Forrest, and C. Staller, "Linear arrays of InGaAs/InP avalanche photodiodes for 1.0–1.7 μm," *Proc. SPIE* 1308, 261–72, 1990.
84. S. D. Personick, "Receiver design for digital fiber-optic communication systems, parts I and II," *Bell Syst. Tech. J.*, 52, 843–86, 1973.
85. R. G. Smith and S. D. Personick, "Receiver Design for Optical Fiber Communications Systems," in *Semiconductor Devices for Optical Communication*, pp. 89–160, ed. H. Kressel, Springer-Verlag, New York, 1980.
86. S. R. Forrest, "Sensitivity of Avalanche Photodetector Receivers for Highbit-Rate Long-Wavelength Optical Communication Systems," in *Semiconductors and Semimetals*, Vol. 22, Chapter 4, ed. W. T. Tang, Academic Press, Orlando, FL, 1985.
87. M. J. Cohen and G. H. Olsen, "Room-temperature InGaAs camera for NIR imaging", *Proc. SPIE* 1946, 436–443, 1993.
88. V. S. Ban, G. H. Olsen, and A. M. Joshi, "High performance InGaAs detectors and arrays for near-infrared spectroscopy," *Spectroscopy* 6(3), 49–52, 1991.
89. M. Gallant, N. Puetz, A. Zemel, and F. R. Shepherd, "Metalorganic chemical vapor deposition InGaAs p-i-n photodiodes with extremely low dark current," *Appl. Phys. Lett.*, 52, 733–5, 1988.
90. A. Zemel and M. Gallant, "Current-voltage characteristics of metalorganic chemical vapor deposition InP/InGaAs p-i-n photodiodes: The influence of finite dimensions and heterointerfaces," *J. Appl. Phys.*, 64, 6552–61, 1988.
91. V. Tetyorkin, A. Sukach and A. Tkachuk, "InAs infrared photodiodes", in *Advances in Photodiodes*, ed. G.F.D. Betta, InTech, 2011, http://www.intechopen.com/books/advances-in-photodiodes/inas-infrared-photodiodes.
92. A. Rogalski and R. Ciupa, "Performance limitation of short wavelength infrared InGaAs and HgCdTe photodiodes," *J. Electron. Mater.*, 28, 630–6, 1999.
93. E. Dewames, R. Littleton, K. Witte, A. Wichman, E. Belloti, and J. Pellegrono, "Electro-optical characteristics of P⁺n $In_{0.53}Ga_{0.47}As$ hetero-junction photodiodes in large format dense focal plane arrays," *J. Electron. Mater.*, 44(8), 2813–22, 2015.
94. J. A. Trezza, N. Masaum, and M. Ettenberg, "Analytic modeling and explanation of ultra-low noise in dense SWIR detector arrays", *Proc. SPIE* 8012, 80121Y-1–12, 2011.
95. H. W. Ruegg, "An optimized avalanche photodiode," *IEEE Trans. Electron Devices* ED-14, 239–51, 1967.
96. H. Melchior and W. T. Lynch, "Signal and noise response of high speed germanium avalanche photodiodes," *IEEE Trans. Electron Devices* ED-13, 820–38, 1966.
97. H. Ando, H. Kanbe, T. Kimura, T. Yamaoka, and T. Kaneda, "Characteristics of germanium avalanche photodiodes in the wavelength region of 1–1.6 μm," *IEEE J. Quantum Electron.*, QE-14, 804–9, 1978.
98. T. Mikawa, S. Kagawa, T. Kaneda, Y. Toyama, and O. Mikami, "Crystal orientation dependence of ionization rates in germanium," *Appl. Phys. Lett.*, 37, 387–9, 1980.
99. J. S. Ng, J. P. R. David, G. J. Rees, and J. Allam, "Avalanche breakdown voltage of $In_{0.53}Ga_{0.47}As$," *J. Appl. Phys.*, 91, 5200–2, 2002.
100. L. W. Cook, G. E. Bulman, and G. E. Stillman, "Electron and hole ionization coefficients in InP determined by photomultiplication measurements," *Appl. Phys. Lett.*, 40, 589–91, 1982.

101. J. P. R. David and C. H. Tan, "Material considerations for avalanche photodiodes," *IEEE J. Sel. Top. Quantum Electron.*, 14, 998–1009, 2008.
102. S. Xie, X. Zhou, S. Zhang, D. J. Thomson, X. Chen, G. T. Reed, J. S. Ng, and C. H. Tan, "InGaAs/AlGaAsSb avalanche photodiode with high gain-bandwidth product", *Opt. Express* 24(21), 24242, 2016.
103. R. W. Zitter, A. J. Strauss, and A. E. Attard, "Recombination processes in p-type indium antimonide," *Phys. Rev.*, 115, 266–73, 1969.
104. M. Y. Pines and O. M. Stafsudd, "Surface effect in InSb photoconductors," *Infrared Phys.*, 19, 559–61, 1979.
105. M. Y. Pines and O. M. Stafsudd, "Characteristics of n-type InSb," *Infrared Phys.*, 19, 563–9, 1979.
106. E. Sunde, "Impact of surface treatment on the performance of cooled photoconductive indium antimonide detectors," *Phys Scr.*, 25, 768–71, 1982.
107. K. M. Van Vliet, "Noise limitations in solid state photodetectors," *Appl. Opt.*, 6, 1145–69, 1967.
108. P. W. Kruse, "Indium antimonide photoelectromagnetic infrared detector," *J. Appl. Phys.*, 30, 770–8, 1959.
109. P. W. Kruse, L. D. McGlauchlin, and R. B. McQuistan, *Elements of Infrared Technology: Generation, Transmission, and Detection*, Wiley, New York, 1962.
110. K. Jóźwikowski, Z. Orman, and A. Rogalski, "On the performance of non-cooled (In,As)Sb photoelectromagnetic detectors for 10.6 μm radiation," *Phys. Status Solidi (a)* 91, 745–51, 1985.
111. D. A. Scribner, M. R. Kruer, and J. M. Killiany, "Infrared focal plane array technology," *Proc. IEEE* 79, 66–85, 1991.
112. M. Talley and D. P. Enright, "Photovoltaic effect in InAs," *Phys. Rev.*, 95, 1092–4, 1954.
113. L. L. Chang, "Junction delineation by anodic oxidation in InSb(As,P)," *Solid Sate Electron.*, 10, 539–44, 1967.
114. R. L. Mozzi and J. M. Lavine, "Zn-diffusion damage in InSb diodes," *J. Appl. Phys.*, 41, 280–5, 1970.
115. C. W. Kim and W. E. Davern, "InAs charge-storage photodiode infrared vidicon targets," *IEEE Trans. Electron Devices* ED-18, 1062–9, 1971.
116. P. C. Catagnus, C. Polansky, and J. P. Spratt, "Diffusion of cadmium into InSb," *Solid State Electron.*, 16, 633–5, 1973.
117. K. Nishitani, K. Nagahama, and T. Murotani, "Extremely reproducible zinc diffusion into InSb and its application to infrared detector array," *J. Electron. Mater.*, 12, 125–41, 1983.
118. J. T. Wimmers and D. S. Smith, "Characteristics of InSb photovoltaic detectors at 77 K and below," *Proc. SPIE* 364, 123–31, 1983.
119. R. Adar, Y. Nemirovsky, and I. Kidron, "Bulk tunneling contribution to the reverse breakdown characteristics of InSb gate controlled diodes," *Solid-State Electron.*, 30, 1289–93, 1987.
120. J. T. Wimmers, R. M. Davis, C. A. Niblack, and D. S. Smith, "Indium antimonide detector technology at cincinnati electronics corporation," *Proc. SPIE* 930, 125–38, 1988.
121. T. P. Sun, S. C. Lee, and S. J. Yang, "The current leakage mechanism in InSb p^+ -n diodes," *J. Appl. Phys.*, 67, 7092–7, 1990.
122. W. H. Lan, S. L. Tu. Y. T. Cherng, Y. M. Pang, S. J. Yang, and K. F. Huang, "Field-induced junction in InSb gate-controlled diodes," *J. Appl. Phys.*, 30, L1–3, 1991.
123. A. G. Foyt, W. T. Lindley, and J. P. Donnelly, "N-p junction photodetectors in InSb fabricated by proton bombardment," *J. Appl. Phys.*, 16, 335–7, 1970.
124. P. J. McNally, "Ion implantation in InAs and InSb," *Radiat. Eff. Defects Solids* 6, 149–53, 1970.
125. C. E. Hurwitz and J. P. Donnelly, "Planar InSb photodiodes fabricated by Be and Mg ion implantation," *Solid State Electron.*, 18, 753–6, 1975.
126. R. D. Thom, T. L. Koch, J. D. Langan, and W. L. Parrish, "A fully monolithic InSb infrared CCD array," *IEEE Trans. Electron Devices* ED–27, 160–70, 1980.
127. C. Y. Wei, K. L. Wang, E. A. Taft, J. M. Swab, M. D. Gibbons, W. E. Davern, and D. M. Brown, "Technology developments for InSb infrared images," *IEEE Trans. Electron Devices* ED–27, 170–5, 1980.
128. J. P. Rosbeck, I. Kassi, R. M. Hoendervoog, and T. Lanir, "High performance be implanted InSb photodiodes," *IEEE IEDM* 81, 161–4, 1981.
129. J. P. Donnelly, "The electrical characteristics of ion implanted semiconductors," *Nucl. Instrum. Methods Phys. Res.*, 182/183, 553–71, 1981.
130. M. Fujisada and T. Sasase, "Effects of insulated gate on ion implanted InSb p^+-n junctions," *J. Appl. Phys.*, 23, L162–4, 1984.
131. H. Fujisada and M. Kawada, "Temperature dependence of reverse current in Be Ion implanted InSb p^+-n junctions," *J. Appl. Phys.*, 24, L76–8, 1985.
132. S. Shirouzu, T. Tsuji, N. Harada, T. Sado, S. Aihara, R. Tsunoda, and T. Kanno, "64×64 InSb focal plane array with improved two layer structure," *Proc. SPIE* 661, 419–25, 1986.

133. J. P. Donnelly, "Ion Implantation in III-V Semiconductors," in *III-V Semiconductor Materials and Devices*, ed. R. J. Malik, 331–428, North-Holland, Amsterdam, 1989.
134. I. Bloom and Y. Nemirovsky, "Quantum efficiency and crosstalk of an improved backside-illuminated indium antimonide focal plane array," *IEEE Trans. Electron Devices* 38, 1792–6, 1991.
135. V. P. Astachov, Y. A. Danilov, V. F. Dutkin, V. P. Lesnikov, G. Y. Sidorova, L. A. Suslov, I. I. Taukin, and Y. M. Eskin, "Planar InAs photodiodes," *Pisma Zh. Tekh. Fiz.*, 18(3), 1–5, 1992.
136. O. V. Kosogov and L. S. Perevyaskin, "Electrical properties of epitaxial p^+-n junctions in indium antimonide," *Fiz. Tekh. Poluprovodn.*, 8, 1611–4, 1970.
137. K. Kazaki, A. Yahata, and W. Miyao, "Properties of InSb photodiodes fabricated by liquid phase epitaxy," *J. Appl. Phys.*, 15, 1329–34, 1976.
138. N. N. Smirnova, S. V. Svobodchikov, and G. N. Talalakin, "Reverse characteristic and breakdown of InAs photodiodes," *Fiz. Tekh. Poluprovodn.*, 16, 2116–20, 1982.
139. Z. Shellenbarger, M. Mauk, J. Cox, J. South, J. Lesko, P. Sims, M. Jhabvala, and M. K. Fortin, "Recent progress in GaInAsSb and InAsSbP photodetectors for mid-infrared wavelengths," *Proc. SPIE* 3287, 138–45, 1998.
140. N. D. Stoyanov, M. P. Mikhailova, O. V. Andreichuk, K. D. Moiseev, I. A. Andreev, M. A. Afrailov, and Y. P. Yakovlev, "Type II heterojunction photodiodes in a GaSb/InGaAsSb system for 1.5–4.8 μm spectral range," *Fiz. Tekh. Poluprovodn.*, 35, 467–73, 2001.
141. T. N. Danilova, B. E. Zhurtanov, A. N. Imenkov, and Y. P. Yuakovlev, "Light emitting diodes on GaSb alloys for mid-infrared 1.4–4.4 μm spectral range," *Fiz. Tekh. Poluprovodn.*, 39, 1281–1311, 2005.
142. A. Krier, X. L. Huang, and V. V. Sherstnev, "Mid-Infrared Electroluminescence in LEDs Based on InAs and Related Alloys," in *Mid-Infrared Semiconductor Optoelectronics*, ed. A. Krier, 359–94, Springer-Verlag, London, 2006.
143. B. A. Matveev, "LED-Photodiode Opto-Pairs," in *Mid-Infrared Semiconductor Optoelectronics*, ed. A. Krier, 395–428, Springer-Verlag, London, 2006.
144. M. A. Remennyy, B. A. Matveev, N. V. Zotova, S. A. Karandashev, N. M. Stus, and N. D. Ilinskaya, "InAs and InAs(Sb)(P) (3–5 μm) immersion lens photodiodes for potable optic sensors," *Proc. SPIE* 6585, 658504, 2007.
145. T. Ashley, A. B. Dean, C. T. Elliott, C. F. McConville, and C. R. Whitehouse, "Molecular-beam growth of homoepitaxial InSb photovoltaic detectors," *Electron. Lett.*, 24, 1270–2, 1988.
146. G. S. Lee, P. E. Thompson, J. L. Davis, J. P. Omaggio, and W. A. Schmidt, "Characterization of molecular beam epitaxially grown InSb layers and diode structures," *Solid-State Electron.*, 36, 387–9, 1993.
147. C. H. Kuan, R. M. Lin, S. F. Tang, and T. P. Sun, "Analysis of the dark current in the bulk of InAs diode detectors," *J. Appl. Phys.*, 80, 5454–8, 1996.
148. I. Shtrichman, D. Aronov, M. Ben Ezra, I. Barkai, E. Berkowicz, M. Brumer, R. Fraenkel, A. Glozman, S. Grossman, E. Jacobsohn, O. Klin, P. Klipstein, I. Lukomsky, L. Shkedy, N. Snapi, M. Yassen, and E. Weiss, "High operating temperature epi-InSb and XBn-InAsSb photodetectors," *Proc. SPIE* 8353, 83532Y, 2012.
149. R. M. Biefeld and S. R. Kurtz, "Growth, Properties and Infrared Device Characteristics of Strained InAsSb-Based Materials," in *Infrared Detectors and Emitters: Materials and Devices*, eds. P. Capper and C. T. Elliott, 205–32, Kluwer Academic Publishers, Boston, MA, 2001.
150. W. S. Chan and J. T. Wan, "Auger analysis of InSb IR detector arrays," *J. Vac. Sci. Technol.*, 14, 718–22, 1977.
151. P. J. Love, K. J. Ando, R. E. Bornfreund, E. Corrales, R. E. Mills, J. R. Cripe, N. A. Lum, J. P. Rosbeck, and M. S. Smith, "Large-format infrared arrays for future space and ground-based astronomy applications," *Proc. SPIE* 4486, 373–84, 2002.
152. T. Ashley, R. A. Ballingall, J. E. P. Beale, I. D. Blenkinsop, T. M. Burke, J. H. Firkins, D. J. Hall, et al., "Large format MWIR focal plane arrays," *Proc. SPIE* 4820, 400–5, 2003.
153. S. R. Jost, V. F. Meikleham, and T. H. Myers, "InSb: A key material for IR detector applications," *Mater. Res. Soc. Symp. Proc.*, 90, 429–35, 1987.
154. M. Zandian, J. D. Garnett, R. E. DeWames, M. Carmody, J. G. Pasko, M. Farris, C. A. Cabelli, et al., "Mid-wavelength infrared p-on-on $Hg_{1-x}Cd_xTe$ heterostructure detectors: 30–120 kelvin state-of-the-art performance," *J. Electron. Mater.*, 32, 803–9, 2003.
155. I. Bloom and Y. Nemirovsky, "Bulk lifetime determination of etch-thinned InSb wafers for two-dimensional infrared focal plane arrays," *IEEE Trans. Electron Devices* 39, 809–12, 1992.
156. I. Bloom and Y. Nemirovsky, "Surface passivation of backside-illuminated indium antimonide focal plane array," *IEEE Trans. Electron Devices* 40, 309–14, 1993.
157. D. N. B. Hall, R. S. Aikens, R. R. Joyse, and T. W. McCurnin, "Johnson-noise limited operation of photovoltaic InSb detectors," *Appl. Opt.*, 14, 450–3, 1975.
158. R. Schoolar and E. Tenescu, "Analysis of InSb photodiode low temperature characteristics," *Proc. SPIE* 686, 2–11, 1986.

159. J. T. Wimmers and D. S. Smith, "Optimization of InSb detectors for use at liquid helium temperatures," *Proc. SPIE* 510, 21, 1984.
160. T. Ashley and N. T. Gordon, "Epitaxial structures for reduced cooling of high performance infrared detectors," *Proc. SPIE* 3287, 236–43, 1998.
161. A. Tevke, C. Besikci, C. Van Hoof, and G. Borghs, "InSb infrared p–i–n photodetectors grown on GaAs coated Si substrates by molecular beam epitaxy", *Solid-State Electronics* 42(6), 1039–1044, 1998.
162. E. Michel, J. Xu, J. D. Kim, I. Ferguson, and M. Razeghi, "InSb infrared photodetectors on Si substrates grown by molecular beam epitaxy," *IEEE Photonics Technol. Lett.*, 8, 673–5, 1996.
163. E. Michel and M. Razeghi, "Recent advances in Sb-based materials for uncooled infrared photodetectors," *Opto-Electron. Rev.*, 6, 11–23, 1998.
164. I. Kimukin, N. Biyikli, T. Kartaloglu, O. Aytür, and E. Ozbay, "High-speed InSb photodetectors on GaAs for mid-IR applications," *IEEE J. Sel. Top. Quantum Electron.*, 10, 766–70, 2004.
165. E. G. Camargo, K. Ueno, T. Morishita, M. Sato, H. Endo, M. Kurihara, K. Ishibashi, and M. Kuze, "High-sensitivity temperature measurement with miniaturized InSb Mid-IR sensor," *IEEE Sen. J.*, 7, 1335–9, 2007.
166. M. Kuze, T. Morishita, E. G. Camargo, K. Ueno, A. Yokoyama, M. Sato, H. Endo, Y. Yanagita, S. Toktuo, and H. Goto, "Development of uncooled miniaturized InSb photovoltaic infrared sensors for temperature measurements," *J. Cryst. Growth* 311, 1889–92, 2009.
167. T. Ashley, A. B. Dean, C. T. Elliott, M. R. Houlton, C. F. McConville, H. A. Tarry, and C. R. Whitehouse, "Multilayer InSb diodes grown by molecular beam epitaxy for near ambient temperature operation," *Proc. SPIE* 1361, 238–44, 1990.
168. A. P. Davis and A. M. White, "Residual noise in Auger suppressed photodiodes," *Infrared Phys.*, 31, 73–9, 1991.
169. T. Ashley, A. B. Dean, C. T. Elliott, A. D. Johnson, G. J. Pryce, A. M. White and C. R. Whitehouse, "A heterojunction minority carrier barrier for InSb devices," *Semicond. Sci. Technol.*, 8, S386–9, 1993.
170. C. T. Elliott, "Advanced heterostructures for $In_{1-x}Al_xSb$ and $Hg_{1-x}Cd_xTe$ detectors and emiters," *Proc. SPIE* 2744, 452–62, 1996.
171. P. N. Brunkov, N. D. Il'inskaya, S. A. Karandashev, A. A. Lavrov, B. A. Matveev, M. A. Remennyi, N. M. Stus', and A. A. Usikov, "InAsSbP/$InAs_{0.9}Sb_{0.1}$/InAs DH photodiodes ($\lambda_{0.1}$=5.2 μm, 300 K) operating in the 77–353 K temperature range," *Infrared Phys. Technol.*, 73, 232–7, 2015.
172. W. Dobbelaere, J. De Boeck, P. Heremens, R. Mertens, and G. Borghs, "InAs p-n diodes grown on GaAs and GaAs-coated Si by molecular beam epitaxy," *Appl. Phys. Lett.*, 60, 868–70, 1992.
173. R. M. Lin, S. F. Tang, S. C. Lee, C. H. Kuan, G. S. Chen, T. P. Sun, and J. C. Wu, "Room temperature unpassivated InAs p-i-n photodetectors grown by molecular beam epitaxy," *IEEE Trans. Electron Devices* 44, 209–13, 1997.
174. V. Tetyorkin, A. Sukach and A. Tkachuk, "Infrared photodiodes on II-VI and III-V narrow-gap semiconductors", in *Photodiodes. From Fundamentals to Applications,* ed. I. Yun, IntechOpen, 2012, http://dx.doi.org/10.5772/52930, https://www.intechopen.com/books/photodiodes-from-fundamentals-to-applications/infrared-photodiodes-on-ii-vi-and-iii-v-narrow-gap-semiconductors.
175. A. R. J. Marshall, C. H. Ting, J. P. R. David, J. S. Ng, and M. Hopkinson, "Fabrication of InAs photodiodes with reduced surface leakage current," *Proc. SPIE* 6740, 67400H-1–9, 2007.
176. A. R. J. Marshall, "The InAs Electron Avalanche Photodiode", in *Advances in Photodiodes,* eds. G. F. D. Betta, InTech, London, 2011, http://www.intechopen.com/books/advances-in-photodiodes/the-inas-electron-avalanche-photodiode.
177. J. D. Beck, C.-F. Wan, M. A. Kinch, and J. E. Robinson, "MWIR HgCdTe avalanche photodiodes," *Proc. SPIE* 4454, 188–97, 2001.
178. J. Beck, C. Wan, M. Kinch, J. Robinson, P. Mitra, R. Scritchfield, F. Ma, and J. Campbell, "The HgCdTe electron avalanche photodiode," *J. Electron. Mater.*, 35, 1166–73, 2006.
179. S. Bank, S. J. Maddox, W. Sun, H. P. Nair, and J. C. Campbell, "Recent progress in high gain InAs avalanche photodiodes," *Proc. SPIE* 9555, 955509, 2015.
180. A. R. J. Marshall, C. H. Tan, and J. P. R. David, "Impact ionization in InAs electron avalanche photodiodes," *IEEE Trans. Electron Devices* 57(10), 2631–8, 2010.
181. A. R. J. Marshall, P. J. Ker, A. Krysa, J. P. R. David, and C. H. Tan, "High speed InAs electron avalanche photodiodes overcome the conventional gain-bandwidth product limit," *Opt. Exp.*, 19(23), 23341–9, 2011.
182. W. Sun, Z. Lu, X. Zheng, J. C. Campbell, S. J. Maddox, H. P. Nair, and S. R. Bank, "High-gain InAs avalanche photodiodes," *IEEE J. Quant. Electron.*, 49(2), 154–61, 2013.
183. A. Rogalski, *Infrared Detectors*, 2nd ed., CRC Press, Boca Raton, FL, 2010.
184. C. G. Bethea, M. Y. Yen, B. F. Levine, K. K. Choi, and A. Y. Cho, "Long wavelength $InAs_{1-x}Sb_x$/GaAs detectors prepared by molecular beam epitaxy," *Appl. Phys. Lett.*, 51, 1431–2, 1987.

185. C. G. Bethea, B. F. Levine, M. Y. Yen, and A. Y. Cho, "Photoconductance measurements on $InAs_{0.22}Sb_{0.78}$/GaAs grown using molecular beam epitaxy," *Appl. Phys. Lett.*, 53, 291–2, 1988.
186. W. Dobbelaere, J. De Boeck, M. Van Hove, K. Deneffe, W. De Raedt, R. Mertens, and G. Borghs, "Long wavelength infrared photoconductive InAsSb detectors grown in Si wells by molecular beam epitaxy," *Electron. Lett.*, 26, 259–61, 1990.
187. J. De Boeck, W. Dobbelaere, J. Vanhellemont, R. Mertens, and G. Borghs, "Growth and structural characterization of embedded InAsSb on GaAs-coated patterned silicon by molecular beam epitaxy," *Appl. Phys. Lett.*, 58, 928–30, 1991.
188. J. Podlecki, L. Gouskov, F. Pascal, F. Pascal-Delannoy, and A. Giani, "Photodetection at 3.65 µm in the atmospheric window using $InAs_{0.91}Sb_{0.09}$/GaAs heteroepitaxy," *Semicond. Sci. Technol.*, 11, 1127–30, 1996.
189. J. D. Kim, D. Wu, J. Wojkowski, J. Piotrowski, J. Xu, and M. Razeghi, "Long-wavelength InAsSb photoconductors operated at near room temperatures (200–300 K)," *Appl. Phys. Lett.*, 68, 99–101, 1996.
190. P. S. Dutta, "III-V ternary bulk substrate growth technology: A review," *J. Cryst. Growth* 275, 106–12, 2005.
191. L. O. Bubulac, A. M. Andrews, E. R. Gertner, and D. T. Cheung, "Backside-illuminated InAsSb/GaSb broadband detectors," *Appl. Phys. Lett.*, 36, 734–6, 1980.
192. L. O. Bubulac, E. E. Barrowcliff, W. E. Tennant, J. P. Pasko, G. Williams, A. M. Andrews, D. T. Cheung, and E. R. Gertner, "Be ion implantation in InAsSb and GaInSb," *Institute of Physics Conference Series No. 45*, 519–29, 1979.
193. W. Dobbelaere, J. De Boeck, P. Heremans, R. Mertens, and G. Borghs, "$InAs_{0.85}Sb_{0.15}$ infrared photodiodes grown on GaAs and GaAs-coated Si by molecular beam epitaxy," *Appl. Phys. Lett.*, 60, 3256–8, 1992.
194. P. K. Chiang and S. M. Bedair, "p-n junction formation in InSb and $InAs_{1-x}Sb_x$ by metalorganic chemical vapor deposition," *Appl. Phys. Lett.*, 46, 383–5, 1985.
195. M. P. Mikhailova and I. A. Andreev, "High-Speed Avalanche Photodiodes for the 2–5 µm Spectral Range," in *Mid-Infrared Semiconductor Optoelectronics*, ed. A. Krier, 547–92, Springer-Verlag, London, 2006.
196. A. Rogalski, R. Ciupa, and W. Larkowski, "Near room-temperature InAsSb photodiodes: Theoretical predictions and experimental data," *Solid-State Electron.*, 39, 1593–600, 1996.
197. J. Wróbel, R. Ciupa, and A. Rogalski, "Performance limits of room-temperature InAsSb photodiodes," *Proc. SPIE* 7660, 766033-1–9, 2010.
198. S. A. Cripps, T. J. C. Hosea, A. Krier, V. Smirnov, P. J. Batty, Q. D. Zhuang, H. H. Lin, P. W. Liu, and G. Tsai, "Determination of the fundamental and spin-orbit-splitting band gap energies of InAsSb-based ternary and pentenary alloys using mid-infrared photoreflectance," *Thin Solid Films* 516, 8049–58, 2008.
199. J. D. Kim, S. Kim, D. Wu, J. Wojkowski, J. Xu, J. Piotrowski, E. Bigan, and M. Razeghi, "8–13 µm InAsSb heterojunction photodiode operating at near room temperature," *Appl. Phys. Lett.*, 67, 2645–7, 1995.
200. J. D. Kim and M. Razeghi, "Investigation of InAsSb infrared photodetectors for near-room temperature operation," *Opto-Electron. Rev.*, 6, 217–30, 1998.
201. T. Refaat, N. Abedin, V. Bhagwat, I. Bhat, P. Dutta, and U. Singh, "InGaSb photodetectors using an InGaSb substrate for 2-µm applications," *Appl. Phys. Lett.*, 85, 1874–6, 2004.
202. A. M. Litvak and N. A. Charykov, "New thermodynamic method for calculation of the diagrams of double and triple systems including In, Ga, As, and Sb," *Zh. Inorg. Mater.*, 27, 225–30, 1990.
203. Y. P. Yakovlev, I. A. Andreev, S. Kizhayev, E. V. Kunitsyna, and P. Mikhailova, "High-speed photothodes for 2.0-4.0 µm spectral range," *Proc. SPIE* 6636, 66360D, 2007.
204. E. V. Kunitsyna, I. A. Andreev, N. A. Charykov, Y. V. Soloviev, and Y. P. Yakovlev, "Growth of $Ga_{1-x}In_xAs_ySb_{1-y}$ solid solutions from the five-component Ga-In-As-Sb-Pb melt by liquid phase epitaxy," *J. Appl. Surf. Sci.*, 142, 371–4, 1999.
205. H. Kroemer, "The 6.1 Å family (InAs, GaSb, AlSb) and its heterostructures: A selective review," *Phys. E* 20, 196–203, 2004.
206. H. Sakaki, L. L. Chang, R. Ludeke, C. A. Chang, G. A. Sai-Halasz, and L. Esaki, "$In_{1-x}Ga_xAs$-$GaSb_{1-y}As_y$ heterojunctions by molecular beam epitaxy," *Appl. Phys. Lett.*, 31, 211–3, 1977.
207. M. P. Mikhailova, K. D. Moiseev, and Y. P. Yakovlev, "Interface-induced optical and transport phenomena in type II broken-gap single heterojunctions," *Semicond. Sci. Technol.*, 19, R109–28, 2004.
208. A. K. Srivastava, J. L. Zyskind, R. M. Lum, B. V. Dutt, and J. K. Klingert, "Electrical characteristics of InAsSb/GaSb heterojunctions," *Appl. Phys. Lett.*, 49, 41–3, 1986.
209. M. Mebarki, A. Kadri, and H. Mani, "Electrical characteristics and energy-band offsets in n-$InAs_{0.89}Sb_{0.11}$/n-GaSb heterojunctions grown by the liquid phase epitaxy technique," *Solid State Commun.*, 72, 795–8, 1989.
210. Y. Sharabani, Y. Paltiel, A. Sher, A. Raizman, and A. Zussman, "InAsSb/GaSb heterostructure based mid-wavelength-infrared detector for high temperature operation," *Appl. Phys. Lett.*, 90, 232106, 2007.

211. O. Hildebrand, W. Kuebart, K. W. Benz, and M. H. Pilkuhn, "$Ga_{1-x}Al_xSb$ avalanche photodiodes: Resonant impact ionization with very high ratio of ionization coefficients," *IEEE J. Quantum Electron.*, 17, 284–8, 1981.
212. I. A. Andreev, M. P. Mikhailava, S. V. Mel'nikov, Y. P. Smorchkova, and Y. P. Yakovlev, "Avalanche multiplication and ionization coefficients of GaInAsSb," *Sov. Phys. Semicond.*, 25, 861–5, 1991.
213. I.A. Andreev, M.A. Afrailov, A.N. Baranov, N.N. Marinskaya, M.A. Mirsagatov, M.P. Mikhailova, Yu.P. Yakovlev, "Low-noise avalanche photodiodes with separated absorption and multiplication regions for the spectral interval 1.6–2.4 μm," *Sov. Tech. Phys. Lett.* 15, 692–696, 1989.
214. M. Van Schilfgaarde, A. Sher, and A. B. Chen, "InTlSb: An infrared detector material?" *Appl. Phys. Lett.*, 62, 1857–9, 1993.
215. A. B. Chen, M. Van Schilfgaarde, and A. Sher, "Comparison of $In_{1-x}Tl_xSb$ and $Hg_{1-x}Cd_xTe$ as long wavelength infrared materials," *J. Electron. Mater.*, 22, 843–6, 1993.
216. Y. S. Park, "Current status of infrared detectors and focal plane arrays", *Journal of the Korean Physical Society* 32(3), 443–451, 1998.
217. M. Van Schilfgaarde, A. B. Chen, S. Krishnamurthy, and A. Sher, "InTlP: A proposed infrared detector material," *Appl. Phys. Lett.*, 65, 2714–6, 1994.
218. M. Razeghi, "Current status and future trends of infrared detectors," *Opto-Electron. Rev.*, 6, 155–94, 1998.
219. J. J. Lee and M. Razeghi, "Novel Sb-based materials for uncooled infrared photodetector applications," *J. Cryst. Growth* 221, 444–9, 2000.
220. J. L. Lee and M. Razeghi, "Exploration of InSbBi for uncooled long-wavelength infrared photodetectors," *Opto-Electron. Rev.*, 6, 25–36, 1998.
221. J. N. Baillargeon, P. J. Pearah, K. Y. Cheng, G. E. Hofler, and K. C. Hsieh, "Growth and luminescence properties of GaPN," *J. Vac. Sci. Technol. B* 10, 829–31, 1992.
222. M. Weyers, M. Sato, and H. Ando, "Red shift of photoluminescence and absorption in dilute GaAsN alloy layers," *Jpn. J. Appl. Phys.*, 31(7A), L853–5, 1992.
223. T. Ashley, T. M. Burke, G. J. Pryce, A. R. Adams, A. Andreev, B. N. Murdin, E. P. O'Reilly, and C. R. Pidgeon, "$InSb_{1-x}N_x$ growth and devices," *Solid-State Electron.*, 47, 387–94, 2003.
224. B. N. Murdin, M. Kamal-Saadi, A. Lindsay, E. P. O'Reilly, A. R. Adams, G. J. Nott, J. G. Crowder, et al., "Auger recombination in long-wavelength infrared InN_xSb_{1-x} alloys," *Appl. Phys. Lett.*, 78, 1568–70, 2001.

17 HgCdTe Detectors

The 1959 publication by Lawson and coworkers triggered the development of variable bandgap $Hg_{1-x}Cd_xTe$ (HgCdTe) alloys providing an unprecedented degree of freedom in the infrared (IR) detector design [1]. During the conference on Infrared Technology and Applications held in Orlando, Florida, in April 2009, a special session was organized to celebrate the 50th anniversary of this first publication [2]. This session brought together most of the research centers and industrial companies that have participated in the subsequent development of HgCdTe. Figure 17.1 shows the three Royal Radar Establishment inventors of HgCdTe (W. D. Lawson, S. Nielson, and A. S. Young) who disclosed the compound ternary alloy in a 1957 patent [3]. They were joined by E. H. Putley in the first publication dated in 1959 [1].

HgCdTe is a pseudobinary alloy semiconductor that crystallizes in the zinc blende structure. Because of its bandgap tunability with x, $Hg_{1-x}Cd_xTe$ has evolved to become the most important/versatile material for detector applications over the entire IR range. As the Cd composition increases, the energy gap for $Hg_{1-x}Cd_xTe$ gradually increases from the negative value for HgTe to the positive value for CdTe. The bandgap energy tunability results in IR detector applications that span the short wavelength (SWIR: 1–3 μm), middle wavelength (MWIR: 3–5 μm), long wavelength (LWIR: 8–14 μm), and very long wavelength (VLWIR: 14–30 μm) IR ranges.

HgCdTe technology development was and continues to be primarily for military applications. A negative aspect in respect to defense agencies has been the associated secrecy requirements that inhibit meaningful collaborations among research teams on a national and, especially, on an international level. In addition, the primary focus has been on focal plane array (FPA) demonstration and much less on establishing the knowledge base. Nevertheless, significant progress has been made over five decades. At present, HgCdTe is the most widely used variable gap semiconductor for IR photodetectors.

17.1 HgCdTe HISTORICAL PERSPECTIVE

The first paper published by Lawson et al. [1] reported both photoconductive and photovoltaic response at wavelengths extending till 12 μm and made the understated observation that this material showed promise for intrinsic IR detectors. At that time, the importance of the 8–12 μm atmospheric transmission window was well-known for thermal imaging, which enabled night vision by imaging the emitted IR radiation from the scene. Since 1954, Cu-doped Ge extrinsic photoconductive detector was known [4], but its spectral response extended to 30 μm (far longer than required for the 8–12 μm window) and to achieve background-limited performance, it was necessary for the Ge:Cu detector to cool down to liquid helium temperature. In 1962, it was discovered that the Hg acceptor level in Ge has an activation energy of about 0.1 eV [5], and the detector arrays were soon made from this material; however, the Ge:Hg detectors were cooled to 30 K to achieve maximum sensitivity. It was also clear from theory that the intrinsic HgCdTe detector (where the optical transitions were direct transitions between the valence band and the conduction band) could achieve the same sensitivity at much higher operating temperatures (as high as 77 K). Early recognition of the significance of this fact led to intensive development of HgCdTe detectors in a number of countries including England, France, Germany, Poland, the former Soviet Union, and the United States [6]. However, little has been written about the early development years, for example, existence of the work going on in the U.S. was classified until the late 1960s.

Figure 17.2 gives the approximate dates of significant development efforts for HgCdTe IR detectors; Figure 17.3 gives additional insight into the time line of the evolution of detectors and key developments in process technology [7]. Photoconductive devices were built in the U.S. as early as 1964 at Texas

Figure 17.1 The discoverers of HgCdTe ternary alloy. (From Ref. [3])

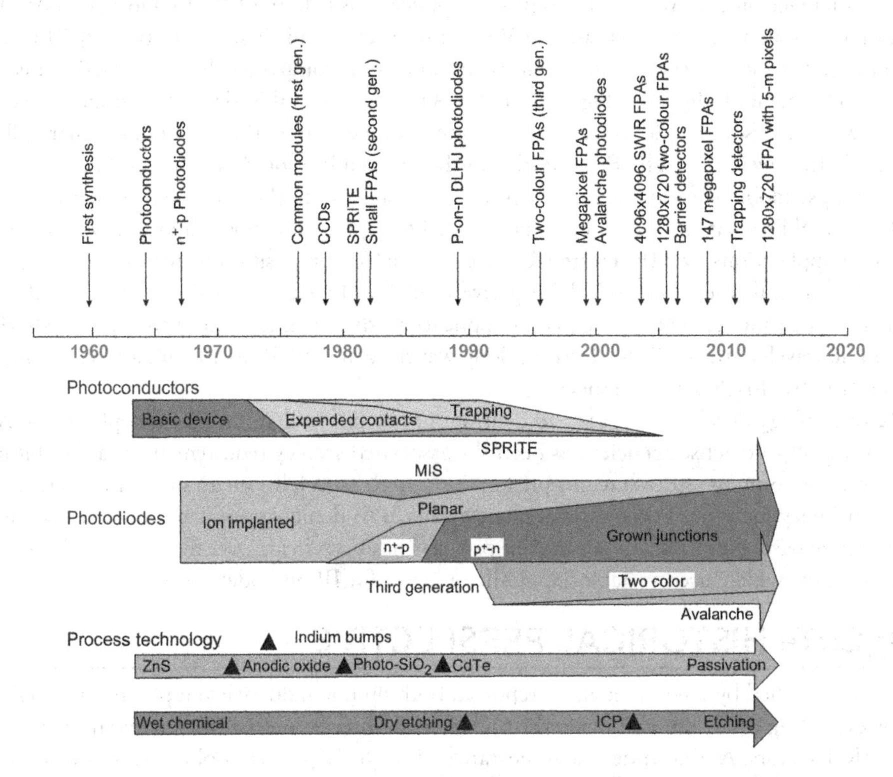

Figure 17.2 A time line of the evolution of HgCdTe IR detectors and the key developments in process technology that made them possible. (From Ref. [7])

Instruments, after the development of a modified Bridgman crystal growth technique. The first report of a junction intentionally formed to make an HgCdTe photodiode was by Verie and Granger [8], who used Hg in-diffusion into p-type material doped with Hg vacancies. The first important application of HgCdTe photodiodes was as high-speed detectors for CO_2 laser radiation [9]. The French pavilion at the 1967 Montreal Expo illustrated a CO_2 laser system with HgCdTe photodiode. However, the high performance MWIR and LWIR linear arrays developed and manufactured in the 1970s were n-type photoconductors used in the first generation scanning systems. In 1969, Bartlett et al. [10] reported the background-limited performance of photoconductors operated at 77 K in the LWIR spectral region. The advantage in material preparation and detector technology has led to devices approaching theoretical limits of responsivity and detectivity over wide ranges of temperature and background [11].

A novel variation of the standard photoconductive device, the SPRITE detector (Signal Processing in the Element), was invented in England [12,13]. A family of thermal imaging systems has utilized this device, however its usage has declined. The SPRITE detector provides signal averaging of a scanned image spot accomplished by synchronization between the drift velocity of minority carriers along the length of photoconductive bar of a material and the scan velocity of the imaging system. Then the image signal builds up a bundle of minority charges that is collected at the end of the photoconductive bar, effectively integrating the signal for a significant length of time and thereby improving the signal-to-noise ratio (SNR).

The scanning system, which does not include multiplexing functions in the focal plane, belongs to the first generation systems. The U.S. common module HgCdTe arrays employ 60, 120, or 180 photoconductive elements depending on the application.

After the invention of charge coupled devices (CCDs) by Boyle and Smith [14], the idea of an all solid-state electronically scanned two-dimensional (2D) IR detector array caused attention to be directed toward HgCdTe photodiodes. These include p-n junctions, heterojunctions, and MIS (metal-insulator-semiconductor) photocapacitors. Each of these different types of devices has certain advantages for IR detection, depending on the particular application. More interest has been focused on the first two structures, so further considerations are restricted to p-n junctions and heterostructures. Photodiodes with their very low power dissipation, inherently high impedance, negligible $1/f$ noise, and easy multiplexing on focal plane silicon chip, can be assembled into 2D arrays containing a very large number of elements, limited only by existing technologies. They can be reverse-biased for even higher impedance and can therefore match electrically with compact low-noise silicon readout preamplifier circuits. The response of photodiodes remains linear to significantly higher photon flux levels than that of photoconductors (because of higher doping levels in the photodiode absorber layer and because the photogenerated carriers are collected rapidly by the junction). At the end of the 1970s, the emphasis was directed toward large photovoltaic HgCdTe arrays in the MW and LW spectral bands for thermal imaging. Recent efforts have been extended to SW (e.g., for starlight imaging in the SW range) as well as to VLWIR space borne remote sensing beyond 15 μm.

The second generation systems (full-framing systems) have, typically, three orders of magnitude more elements ($>10^6$) on the focal plane than the first generation systems, and the detector elements are configured in a 2D array. These staring arrays are scanned electronically by circuits integrated with the arrays. These readout integrated circuits (ROICs) include, for example, pixel deselecting, antiblooming on each pixel, subframe imaging, output preamplifiers, and other functions. The second generation HgCdTe devices are 2D arrays of photodiodes. This technology began in the late 1970s and it took the next decade to reach volume production. The hybrid architecture was first demonstrated in the mid-1970s [15]; indium bump bonding of readout electronics provides for multiplexing the signals from thousands of pixels onto a few output lines, greatly simplifying the interface between the vacuum-enclosed cryogenic sensor and the system electronics. The detector material and multiplexer are optimized independently. Other advantages of the hybrid FPAs are near 100% fill factors and an increased signal-processing area on the multiplexer chip.

The MIS photocapacitor is usually formed on the n-type absorber layer with a thin semitransparent metal film as a gate electrode. The insulator of choice is a thin native oxide. The only motivation for developing HgCdTe MIS detectors was the allure of realizing a monolithic IR CCD, with both detection and multiplexing taking place in the same material. However, because of the nonequilibrium operation of the MIS detector (usually a bias voltage pulse of several volts is applied across the capacitor to drive the surface into deep depletion), much larger electric fields are set up in the depletion region of the MIS device than in the p-n junction, resulting in a defect-related tunneling current that is several orders of magnitude larger than the fundamental dark current. The MIS detector requires much higher material quality than the photodiode, which still has not been achieved. For this reason, the development of HgCdTe MIS detector was abandoned around 1987 [16,17].

In the last decade of the twentieth century, the third generation of HgCdTe detectors emerged because of a tremendous impetus in the detector developments (see Chapter 27). This generation of detectors has emerged from technological achievements in the growth of heterostructure devices used in the production of second generation systems [18].

17.2 HgCdTe: TECHNOLOGY AND PROPERTIES

High-quality semiconductor material is essential for the production of high performance and affordable IR photodetectors. The material must have a low defect density, a large size of wafers, uniformity, and reproducibility of intrinsic and extrinsic properties. To achieve these characteristics, HgCdTe materials evolved from high-temperature, melt-grown, bulk crystals to low-temperature, liquid and vapor phase epitaxy (VPE). However, the cost and availability of large-area and high-quality $Hg_{1-x}Cd_xTe$ are still the main considerations for producing affordable devices.

17.2.1 PHASE DIAGRAMS

A solid understanding of phase diagrams is essential for the proper design of the growth process. The phase diagrams and their implications for $Hg_{1-x}Cd_xTe$ crystal growth have been discussed extensively [19–21]. The ternary Hg-Cd-Te phase diagrams have been established both theoretically and experimentally throughout the Gibbs triangle [22–24]. Brice summarized the works of more than 100 researchers on the Hg-Cd-Te phase diagram as a numerical description [23], which is convenient for the design of growth processes [25].

The generalized associated solution model has proven to be successful in explaining experimental data and in predicting the phase diagram of the entire Hg-Cd-Te system. It was assumed that the liquid phase was a mixture of Hg, Te, Cd, HgTe, and CdTe. The gas phase of the material contained Hg, Cd atoms, and Te_2 molecules. The composition of the solid material can be described by a generalized formula $(Hg_{1-x}Cd_x)_{1-y}Te_y$. The familiar $Hg_{1-x}Cd_xTe$ formula corresponds to the pseudobinary CdTe and HgTe alloy ($y = 0.5$) with complete mutual solubilities. At present, it is believed that the sphalerite pseudobinary phase region in $Hg_{1-x}Cd_xTe$ is extended in Te-rich material with a width in the order of 1%. The width narrows at lower temperatures. The consequence of such a form of diagram is a tendency for Te precipitation. An excess of Te is due to vacancies in the metal-sublattice, which results in p-type conductivity of pure materials. Low-temperature annealing at 200 K–300 K reduces the native defect (predominantly acceptor) concentration and reveals an uncontrolled (predominantly donor) impurity background. Weak Hg-Te bonding results in low activation energy for defect formation and Hg migration in the matrix. This can cause bulk and surface instabilities.

Most of the problems with crystal growth are due to the marked difference between the solidus and liquidus curves (see Figure 17.3) resulting in the segregation of binaries during crystallization from melts. The segregation coefficient for growth from melts depends on Hg pressure. Serious problems also arise from high Hg pressures over pseudobinary and Hg-rich melts. A full appreciation of the P_{Hg}-T diagram shown in Figure 17.4a is therefore essential [26]. Curves are the partial pressures of Hg along boundaries for solid solutions of composition x where the solid solution is in equilibrium with another condensed phase as well as with the vapor phase. We can see that for $x = 0.1$ and $1{,}000/T = 1.3$ K^{-1} HgCdTe exists at Hg pressure of 0.1 (Te-saturated) and 7 (Hg-saturated) atm. The atomic fraction of Te decreases, as the Hg partial pressure increases, over a small but nonzero range near 0.5 atomic fraction. Even at $x = 0.95$ and Te-saturated conditions, Hg is the predominant

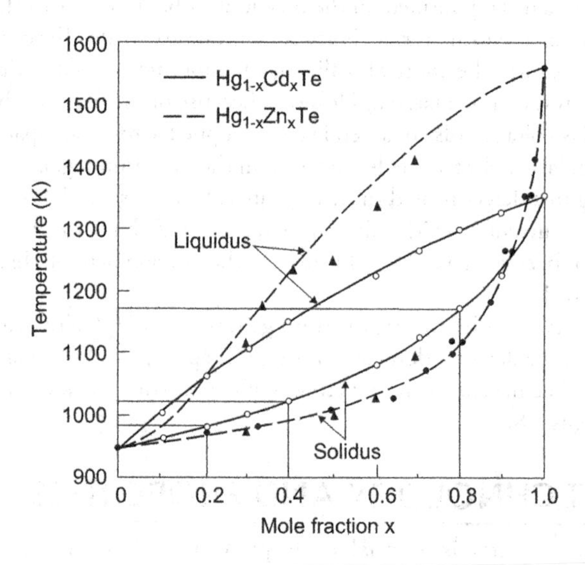

Figure 17.3 Liquidus and solidus lines in the HgTe-CdTe and HgTe-ZnTe pseudobinary systems.

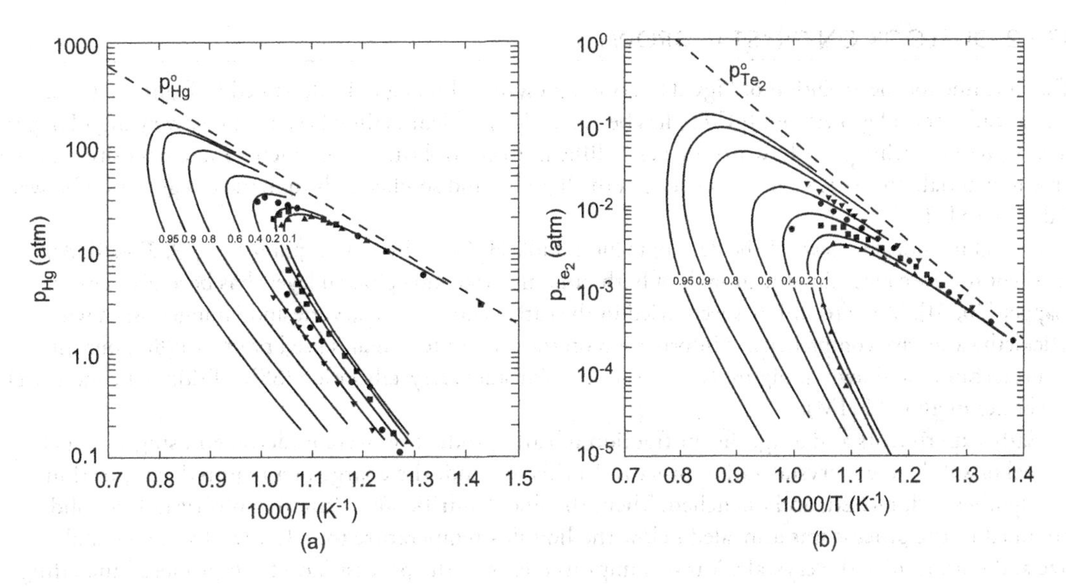

Figure 17.4 Partial pressures of (a) Hg and (b) Te_2 along the three-phase curves for various solid solutions. The labels are the values of x in the formula $Hg_{1-x}Cd_xTe$. (From Ref. [26])

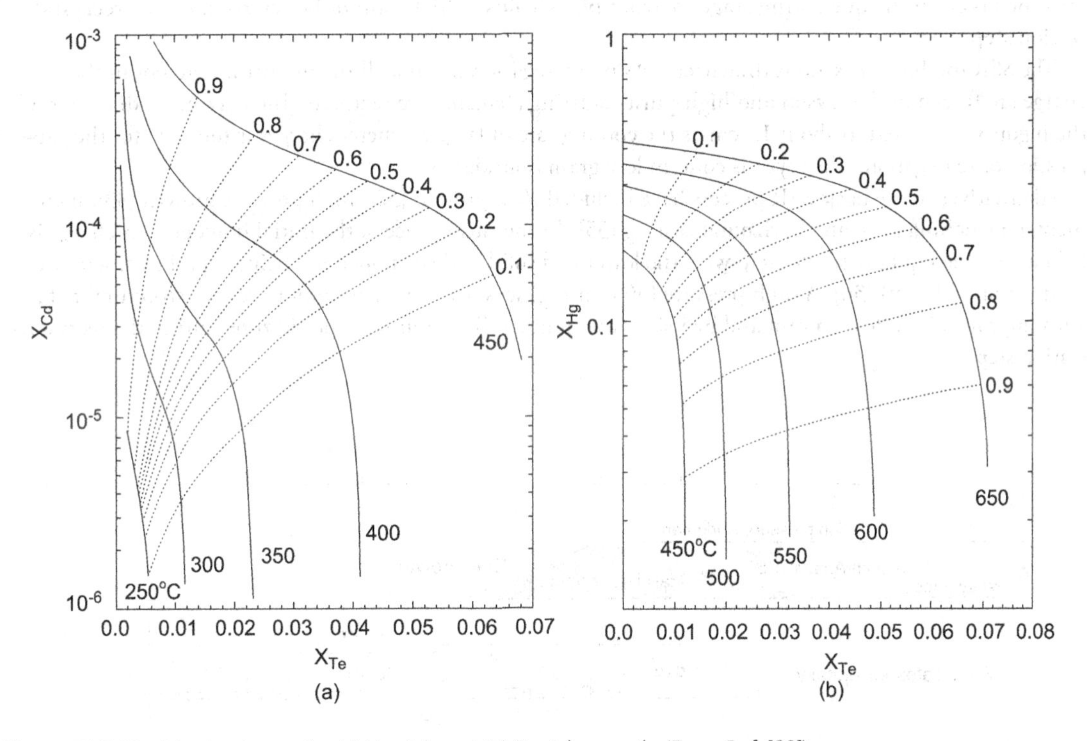

Figure 17.5 Liquidus isotherms for (a) Hg-rich and (b) Te-rich growth. (From Ref. [22])

vapor species and no solid solution contains exactly 0.5 atomic fraction Te. These features are highly significant in controlling the native defect concentrations and hence the electrical properties of HgCdTe. Figure 17.4b shows that in comparison with Hg pressure the partial pressure of Te_2 is several orders of magnitude lower. Figure 17.5 shows the low-temperature liquidus and solid-solution isoconcentration lines in the Hg-rich and Te-rich corners [22]. Figure 17.5a shows that, for example, at 450°C, a solid solution containing 0.90 mole fraction CdTe is in equilibrium with a liquid containing 7×10^{-4} atomic fraction Cd and 0.014 atomic fraction Te. One sees that almost pure CdTe(s) crystallizes from very Hg-rich liquids.

17.2.2 OUTLOOK ON CRYSTAL GROWTH

The time line for the evolution of HgCdTe crystal growth technologies is illustrated in Figure 17.6 [7]. Historically, crystal growth of HgCdTe has been a major problem mainly because a relatively high Hg pressure is present during growth, which makes it difficult to control the stoichiometry and composition of the grown material. The wide separation between the liquidus and solidus leads to marked segregation between CdTe and HgTe.

Several historical reviews of the development of bulk HgCdTe have been published [19, 27–29]. An excellent review concerning the growth of both bulk material and epitaxial layers has been given by Capper [30, 31]. Many techniques were tried in the early years (see, e.g., Verie and Granger [8] in which Micklethwaite gave comprehensive information on the growth techniques used prior to 1980), but three prime techniques, shown in Figure 17.6, survived: solid state recrystallization (SSR), Bridgman, and traveling heater method (THM).

Early experiments and a significant fraction of early production were undertaken using a quench-anneal or solid-state recrystalization process. In this method, the charge of a required composition was synthesized, melted, and quenched. Then, the fine dendritic mass (highly polycrystalline solid) obtained in the process was annealed below the liquidus temperature for a few weeks to recrystallize and homogenize the crystals. Various improvements of the process have been proposed including temperature gradient annealing and slow cooling down to prevent precipitation of Te. The material usually requires low-temperature annealing for adjusting the concentration of native defects. Care must be taken in the quenching stage to avoid pipes/voids, which cannot be removed by the recrystalization step.

The SSR method has serious drawbacks. As no segregation occurs, all the impurities present in the charge are frozen in the crystal and high-purity starting elements are required. The maximum diameter of the ingots was limited to about 1.5 cm as the cooling rate of large diameter charges is too slow for the suppression of segregation. The crystals contain low grain boundaries.

Alternatives to the basic SSR process have included *slush* growth [32], high pressure growth [33], incremental quenching [34], and horizontal casting [35]. In the slush process, the initial homogenous charge is held across the liquidus-solidus gap with the lower end solid and the upper end being liquid (temperature gradient of 10 K/cm) [30]. A high pressure (30 atm Hg) was used in an attempt to reduce structural defects with improved heat flow control and by using intergranual Te as a moving liquid zone during the recrystalization step.

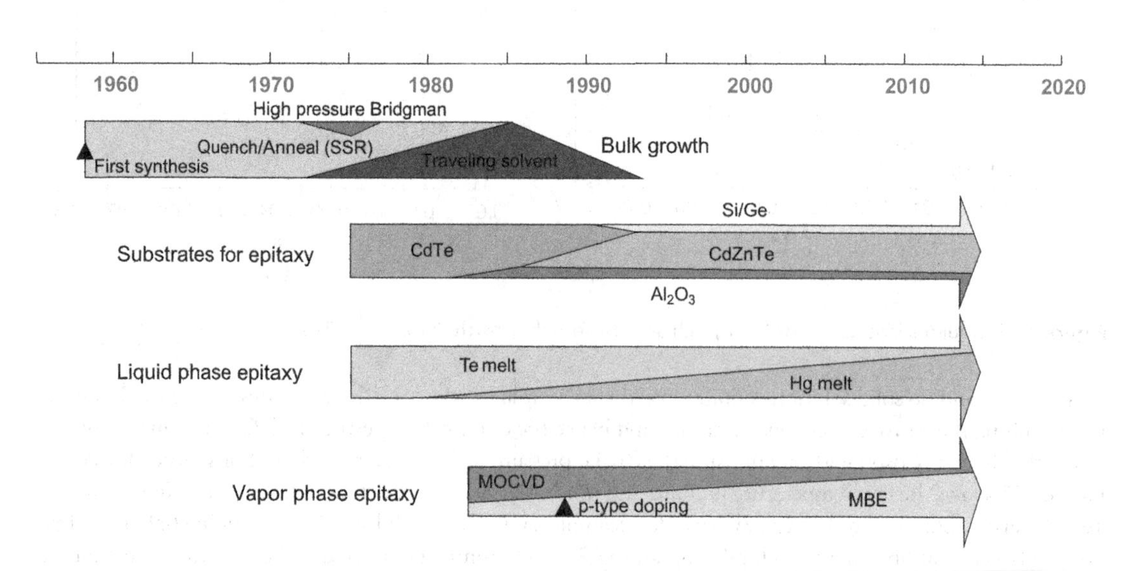

Figure 17.6 Evolution of the HgCdTe crystal growth technology from 1958 to present. (From Ref. [7])

Bridgman growth was attempted for several years near the mid-1970s. Limits on controlling melt mixing in the Bridgman process necessitated a means of stirring melts contained in sealed, pressurized ampoules. The addition of accelerated crucible rotation technique (ACRT) in which the melt is subjected to periodic acceleration/deceleration at rotation rates of up to 60 rpm causes a marked improvement over normal Bridgman growth, particularly for high x material, that is, better reproducibility of the process, relatively flat interfaces, and decrease in the number of major grains (typically 10 to 1 in the $x = 0.2$ region) [36–39]. Crystals of up to 20 mm in diameter were produced, with x value up to 0.6 in the tip regions of some crystals.

At the same time, solvent growth methods from Te-rich melts were initiated to reduce the growth temperature. One successful implementation was the THM that resulted in crystals up to 5 cm diameter [40]. The perfect quality of crystals grown by this method is achieved at the cost of a low growth rate [41].

Bulk HgCdTe crystals were initially used for all types of IR photodetectors. At present, they are still used for some IR applications such as n-type single element photoconductors, SPRITE detectors, and linear arrays. Bulk growth produced thin rods, generally up to 50 mm in diameter, about 60 cm in length, and with a nonuniform distribution of composition. Large 2D arrays could not be realized with bulk crystals. Another drawback of bulk material was the need to thin the bulk wafers, usually cut to about 500 μm, and finally down to a final device thickness of about 10 μm. Also, further fabrication steps (polishing the wafers, mounting them on suitable substrates, and polishing to the final device thickness) was very labor intensive.

In comparison to bulk growth techniques, epitaxial techniques offer the possibility of growing large-area (≈100 cm^2) epilayers and fabricating sophisticated device structures with good lateral homogeneity, abrupt and complex composition, and doping profiles that can be configured to improve the performance of photodetectors. The growth is performed at low temperatures (see Figure 17.7) [42], which makes it possible to reduce the native defect density.

Among the various epitaxial techniques, liquid phase epitaxy (LPE) is the most technologically mature method. The LPE is a single crystal growth process in which growth from a cooling solution occurs onto a substrate. Another technique, VPE growth of HgCdTe is typically carried out with nonequilibrium methods that can also be applied to metalorganic chemical vapor deposition (MOCVD), molecular beam

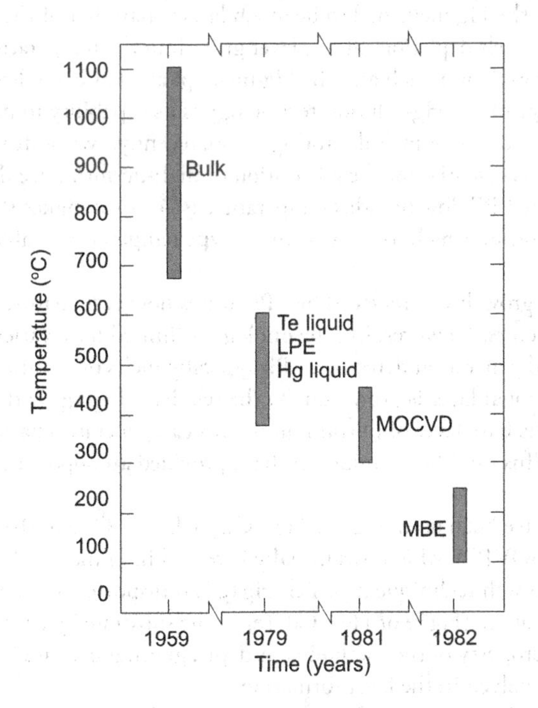

Figure 17.7 Temperature ranges for the growth of HgCdTe by various growth techniques versus date of first reported attempt. (From Ref. [42])

epitaxy (MBE), and their derivatives. The great benefit of MBE and MOCVD over equilibrium methods is the ability to modify the growth conditions dynamically during growth to tailor bandgaps, add and remove dopants, prepare surfaces and interfaces, add passivations, perform anneals, and even grow on selected areas of a substrate. The growth control is exercised with great precision to obtain basic material properties comparable to those routinely obtained from equilibrium growth.

Epitaxial growth of HgCdTe layers requires a suitable substrate. CdTe was used initially, since it was available from commercial sources in reasonably large sizes. The main drawback to CdTe is that it has a few percentage lattice mismatches with LWIR and MWIR HgCdTe. By the mid-1980s, it was demonstrated that the addition of a few percentage of ZnTe to CdTe (typically 4%) could create a lattice-matched substrate. CdTe and closely lattice-matched CdZnTe substrates are typically grown with the modified vertical or horizontal unseeded Bridgman technique. Most commonly the (111) and (100) orientations have been used, although others have been tried. Twinning, which occurs in (111) layers, can be prevented by a suitable misorientation of the substrate. Growth conditions found to be nearly optimal for the (112)B orientation were selected. The limited size, purity problems, Te precipitates, dislocation density (routinely in the low $10^4 cm^{-2}$ range), nonuniformity of lattice match, and high price ($50–$500 per cm^2, polished) are remaining problems to be solved. It is believed that these substrates will continue to be important for a long time, particularly for the highest performance devices.

The LPE growth of a thin layer of HgCdTe on CdTe substrates began in the early to mid-1970s. Both Te-solution growth (420°C–500°C) and Hg-solution growth (360°C–500°C) have been used with equal success in a variety of configurations. The pioneering work on Te corner of the phase diagram of the Hg-Cd-Te system was published in 1980 [43] and the associated LPE growth equipment provided the necessary groundwork that led to several variations of open-tube LPE from Te solutions (see, e.g., [44–47]). Initially, Te solutions with dissolved Cd (Cd has a high solubility in Te) and saturated with Hg vapor were used to efficiently grow HgCdTe in the temperature range of 420°C–600°C. This allowed small volume melts to be used with slider techniques that did not appreciably deplete during the growth run. Experiments with Hg-solvent LPE began in the late 1970s. The Santa Barbara Research Center (SBRC) pioneered the phase diagram study of the Hg corner of the Hg-Cd-Te system [48] and, after several years of experimenting, a reproducible Hg-solution technology was developed [49]. Because of the limited solubility of Cd in Hg, the volume of the Hg melts had to be much larger than that of the Te melts (typically about 20 kg) in order to minimize melt depletion during layer growth in the temperature range of 380°C–500°C. This precluded the slider growth approach and the Hg-melt epitaxy has been developed using large dipping vessels. One major advantage of the Hg-solution technology is its capability to produce layers of excellent surface morphology due to the ease of melt decanting. More recently, two additional unique characteristics have been recognized as essential for the fabrication of high-performance double-layer heterojunction (DLHJ) photodiodes by LPE: low liquidus temperature (< 400°C) makes the cap-layer growth step feasible; and the ease of incorporating both p-type and n-type temperature-stable impurities, such as As, Sb, and In, during growth.

In the early 1990s, bulk growth was replaced by LPE and is now very mature for the production of first and second generation detectors. However, LPE technology is limited for a variety of advanced HgCdTe structures required for third generation detectors. LPE typically melts off a thin layer of the underlying material each time an additional layer is grown due to the relatively high growth temperature. Additionally, the gradient in x-value in the base layer of p^+ -on-n junctions can generate a barrier to carrier transport in certain cases due to interdiffusion. These limitations have provided an opportunity for VPE, especially MBE and MOCVD.

Various VPE methods have been used to grow $Hg_{1-x}Cd_xTe$ layers. One of the oldest is the isothermal vapor phase epitaxy (ISOVPE), which was initially invented in France [50]. ISOVPE is a relatively simple, quasiequilibrium growth technique in which HgTe is transported at a relatively high temperature (400°C–600°C) from the source (HgTe or $Hg_{1-x}Cd_xTe$) to the substrate by evaporation-condensation mechanisms. An inherent property of the method is in-depth grading because interdiffusion of deposited and substrate materials is involved in the layer formation.

The era of MBE and MOCVD growth of HgCdTe began in the early 1980s by adopting both methods that had been well established in the III–V semiconductor materials. Through the following decade, a variety of metalorganic compounds were developed along with a number of reaction chamber designs [51,52]. The MOCVD growth of HgCdTe can be achieved by one of two alternative processes, direct alloy growth (DAG) and the interdiffused multilayer process (IMP). The DAG is faced with several severe problems mainly because of the great difference in stability between HgTe and CdTe together with the higher reactivity of Te precursors with the Cd alkyl than with metallic Hg. The IMP technique overcomes these problems [53,54]. The successive layers of CdTe and HgTe have the combined thickness of two consecutive layers that is approximately 0.1 μm. These layers are deposited and interdiffused during growth or during a short annealing in situ at the end of the growth. The activation of As, Sb, and In dopants is easier to achieve with IMP when introduced during the CdTe growth cycle and with a Cd/Te flux ratio above 1 [55]. In comparison with In, I is a more stable donor dopant that can be used to control local doping within a 3×10^{14}–2×10^{18} cm^{-3} range with 100% activation when following a standard stoichiometric anneal [56].

The MBE growth of HgCdTe is carried out with effusive sources that contain Hg, Te_2, and CdTe. A specially designed Hg-source oven was successfully designed to overcome the low sticking coefficient of Hg at the growth temperature [57–60]. Surface growth temperature in MBE HgCdTe plays a critical role in the introduction of extended defects. The optimized growth temperature is in the range of 185°C–190°C. At lower temperatures, an excess of Hg is obtained at the surface because the sticking coefficient of Hg increases as the temperature is reduced. The excess Hg produces microtwin defects. These defects are detrimental to the electrical properties of the epilayer and devices. The etch pit density (EPD) values of material grown under these conditions are high (10^6–10^7 cm^{-2} range). If growth temperatures under the same conditions are raised above 190°C, then a deficiency of Hg is obtained at the surface and void defects are formed. Currently, under the best optimized Hg/Te_2 flux growth conditions, the lowest concentration of void defects that has been observed is around 100 cm^{-2}. Dust particles and/or substrate related surface imperfections may account for this. The EPD values for epilayers grown under these conditions are low (10^4–10^5 cm^{-2}). Significant efforts are being spent on As and Sb-doping to improve incorporation during the MBE process and to reduce the temperature required for activation. The metal saturation conditions cannot be reached at the temperatures required for high-quality MBE growth. The necessity to activate acceptor dopants at high temperatures eliminates the benefits of low-temperature growth. Nearly 100% activation was achieved for a 2×10^{18} cm^{-3} As concentration with a 300°C activation anneal followed by a 250°C stoichiometric anneal [61].

At present, MBE is the dominant vapor phase method for the growth of HgCdTe. It offers low temperature growth under an ultrahigh vacuum environment, in situ n-type and p-type doping, and control of composition, doping, and interfacial profiles. MBE is now the preferred method for growing complex layer structures for multicolor detectors and for avalanche photodiodes. Although the quality of MBE material is not yet on a par with LPE, it has made tremendous progress in the past decade. A key to its success has been the doping ability and the reduction of EPDs to below 10^5 cm^{-2}.

The growth temperature is less than 200°C for MBE, but around 350°C for MOCVD, making it more difficult to control p-type doping in MOCVD due to the formation of Hg vacancies at higher growth temperatures. As is a preferable dopant for p-type layers, while In is preferable for n-type layers. Several laboratories consistently report undoped MOCVD and MBE grown layers with impurity levels below 10^{14} cm^{-3} indicating that source material purity from commercial vendors now appears adequate, although it could still use some improvements. The remaining problems of the two methods are twin formations, requirement of a very good surface preparation prior to growth, uncontrolled doping, dislocation density, and composition inhomogenities.

The lowest reported carrier concentrations and the longest lifetimes in MOCVD and MBE grown layers have been achieved in HgCdTe films grown onto CdZnTe substrates. The substrates are typically grown by the modified vertical and horizontal unseeded Bridgman technique [62,63]. Near lattice-matched CdZnTe substrates have severe drawbacks such as lack of large area, high production cost and, more importantly, a difference in thermal expansion coefficient (TEC) between the CdZnTe substrates and the silicon readout integrated circuit. Furthermore, interest in large-area 2D IR FPAs (larger than megapixel arrays) have resulted in limited applications of CdZnTe substrates. Currently, readily producible CdZnTe substrates are limited to areas of approximately 60 cm^2. Large semiconductor wafers for IR detector fabrication enable lower cost through the printing of more die per wafer or enable larger single piece detector arrays.

A viable approach to cheap substrates is the use of hybrid substrates, which consist of laminated structures with wafers of bulk crystal and are covered with buffer lattice-matched layers. Four issues dominate alternative substrates: lattice mismatch, nucleation phenomena, thermal expansion mismatch, and majority species contamination [64,65]. Bulk Si, GaAs, and sapphire are some of the high-quality, low-cost, and readily available crystals that have been shown to be useful substrates for $Hg_{1-x}Cd_xTe$. The buffer layers are a few micrometers thick CdTe or (Cd, Zn) Te, obtained in situ or ex situ with a nonequilibrium growth, typically from the vapor phase. The feasibility of growing high-quality $Hg_{1-x}Cd_xTe$ on hybrid substrates was demonstrated first by Rockwell International. This technology is referred to as a producible alternative to CdTe for epitaxy (PACE) [66,67]. The substrates are CdTe/sapphire (PACE 1), CdTe/GaAs (PACE 2), and Si/GaAs/ CdZnTe (PACE 3).

Sapphire has been widely used as a substrate for HgCdTe epitaxy. In this case, a CdTe (CdZnTe) film is deposited on the sapphire prior to the growth of HgCdTe. This substrate has excellent physical properties and can be purchased in large wafer sizes. The large lattice mismatch with HgCdTe is accommodated by a CdTe buffer layer. Sapphire is transparent with the UV to about 6 μm in wavelength and has been used in back side illuminated SWIR and MWIR detectors (it is not acceptable for back side illuminated LWIR arrays because of its opacity beyond 6 μm).

For the 8–12 μm LWIR band, the CdTe/GaAs (PACE 2) has been developed with detectors fabricated on GaAs substrates [68]. Because GaAs has a TEC comparable to CdZnTe, these PACE 2 FPAs will have the same size limitations as CdZnTe-based hybrids, unless GaAs readout circuits are used. Moreover, the GaAs initiation layer can result in Ga contamination of the II-VI films along with undesirable additional cost and process complexity. More recently, published papers have indicated that by using CdZnTe buffer layers the problem of Ga contamination can be resolved [69].

The use of Si substrates is very attractive in IR FPA technology not only because it is less expensive and available in large area wafers but also because the coupling of the Si substrates with Si readout circuitry in an FPA structure allows the fabrication of very large arrays exhibiting long-term thermal cycle reliability. The $8 \times 8\,cm^2$ bulk CdZnTe substrate is the largest commercially available material, and it is unlikely to become much larger than its present size. With the cost of 6-inch Si substrates being ≈ $100 versus $10,000 for the $8 \times 8\,cm^2$ CdZnTe, significant advantages of HgCdTe/Si are evident [70]. Despite the large lattice mismatch (≈19%) between CdTe and Si, MBE has been successfully used for the heteroepitaxial growth of CdTe on Si. Using optimized growth condition for Si (211)B substrates and a CdTe/ZnTe buffer system, epitaxial layers with EPD in the range of $10^6\,cm^{-2}$ have been obtained. This value of EPD has little effect on both MWIR and LWIR HgCdTe/Si detectors [70,71]. By comparison, HgCdTe epitaxial layers grown by MBE or LPE on bulk CdZnTe have typical EPD values in the 10^4–mid-$10^5\,cm^{-2}$ range where there is a negligible effect of dislocation density on the detector performance. MBE has been successfully used for the heteroepitaxial growth of MWIR HgCdTe photodiodes on composite CdTe/Si substrates. However, it has proved difficult to attain the best LW photodiode performance when grown on Si by MBE, which is observed on lattice-matched CdZnTe substrates.

17.2.3 DEFECTS AND IMPURITIES

Native defect properties and impurity incorporation still constitute a field of intensive research. Various aspects of defects in bulk crystals and epilayers such as electrical activity, segregation, ionization energies, diffusivity, and carrier lifetimes have been summarized in many reviews [72–81].

17.2.3.1 Native defects

The defect structures of undoped and doped $Hg_{1-x}Cd_xTe$ can be explained with the quasichemical approach [82–87]. The dominant native defect in $Hg_{1-x}Cd_xTe$ is a double ionizable acceptor associated with metal lattice vacancies. Some direct measurements show much larger vacancy concentrations than those that follow from Hall measurements, indicating that most vacancies are neutral [88].

In contrast to numerous early findings, now it seems to be established that the native donor defect concentration is negligible. As-grown undoped and pure $Hg_{1-x}Cd_xTe$, including that grown in Hg-rich LPE, always exhibits p-type conductivity with the hole concentration depending on composition, growth temperature, and Hg pressure during growth, reflecting correspondence to the concentration of vacancies.

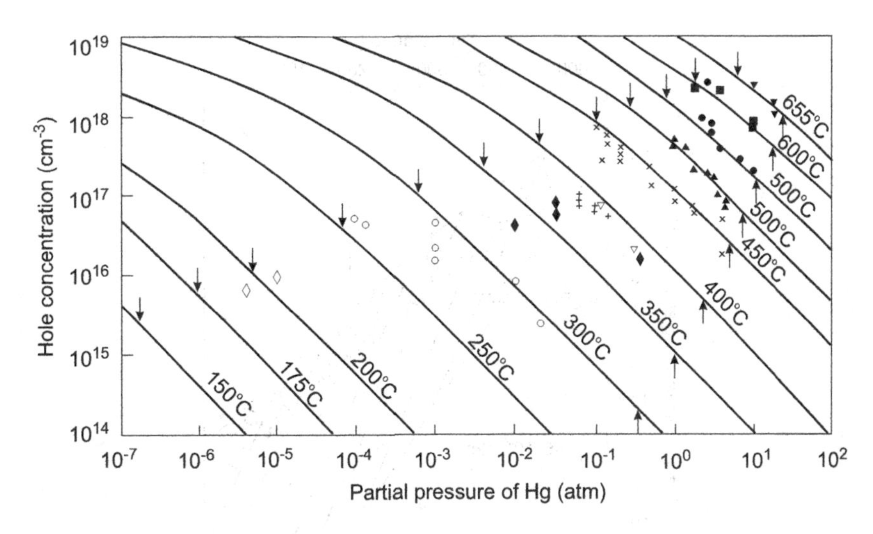

Figure 17.8 The 77 K hole concentration in $Hg_{0.80}Cd_{0.20}Te$ calculated according to the quasichemical approach as a function of the partial pressure of Hg and annealing temperature (150°C–655°C). Arrows define the material existence region. (From Ref. [74])

The equilibrium concentration of vacancies and Hg pressures over Te-saturated $Hg_{1-x}Cd_xTe$ are:

$$c_V\left[\mathrm{cm}^{-3}\right]=\left(5.08\times10^{27}+1.1\times10^{28}x\right)P_{Hg}^{-1}\exp\left(\frac{-\left(1.29+1.36x-1.8x^2+1.375x^3\right)\mathrm{eV}}{kT}\right), \quad (17.1)$$

$$p_{Hg}[\mathrm{atm}]=1.32\times10^{5}\exp\left(-\frac{0.635\mathrm{eV}}{kT}\right). \quad (17.2)$$

The Hg pressure over Hg-saturated $Hg_{1-x}Cd_xTe$ is close to the saturated Hg pressure

$$p_{Hg}[\mathrm{atm}]=\left(5.0\times10^{6}+5.0\times10^{6}x\right)\times\exp\left(\frac{-0.99+0.25x}{kT}\mathrm{eV}\right). \quad (17.3)$$

Figure 17.8 shows the hole concentration as a function of the partial Hg pressure, showing $1/p_{Hg}$ dependence of the native acceptor concentration [74], which is in agreement with the predictions of the quasichemical approach for narrow-gap $Hg_{1-x}Cd_xTe$. Annealing in Hg vapors reduces the hole concentration by filling the vacancies. Low-temperature (<300°C) annealing in Hg vapors reveals the background impurity level, causing the p-to-n conversion in some crystals. For example, Figure 17.9 shows the iso-hole concentration plot for $Hg_{0.80}Cd_{0.20}Te$, indicating the possibility of obtaining identical hole concentrations for anneals at two different temperatures and partial pressures of Hg [75]. HgCdTe crystals with a vacancy concentration of about 10^{15} cm^{-3} can be obtained either with an anneal at $T = 300$°C and $p_{Hg} = 7\times10^{-2}$ atm or with an anneal at $T = 200$°C and $p_{Hg} = 6\times10^{-5}$ atm, the concentration of Hg interstitial for the lower temperature anneal being lower. If Hg interstitials are the Shockley-Read centers, the minority carrier lifetimes in samples prepared at lower temperatures should be higher than that in samples prepared at the highest temperature, even though the Hg vacancy concentrations in both samples are the same.

Samples with higher residual donor concentration turn n-type at higher temperatures and show higher electron concentration. Unexpected effects may arise from Te precipitates [89]. Hg diffusing into material dissolves precipitates and drives the major impurities ahead of Hg, leaving the core p-type. On further annealing, these impurities may redistribute throughout the slice, turning the whole sample p-type. A variety of effects may cause unexpected n-type behavior contamination: surface layers formed during cool down, strain, dislocations, twins, grain boundaries, substrate orientation, oxidation, and perhaps other parameters.

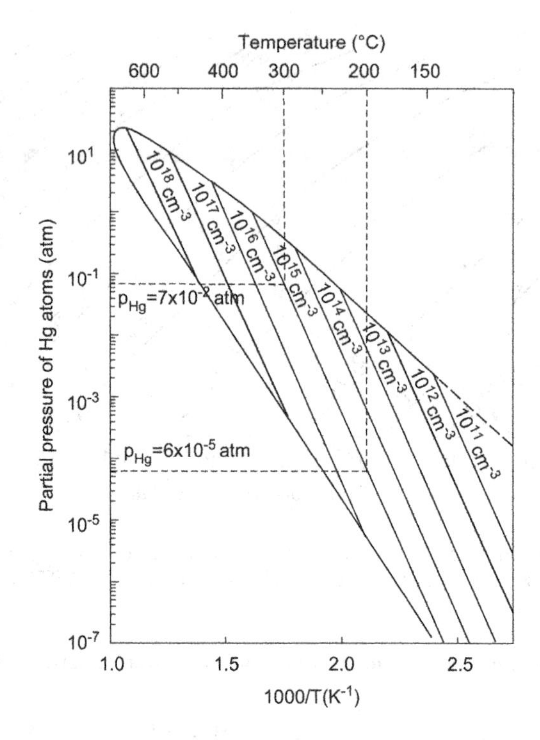

Figure 17.9 Iso-hole concentration plot for $Hg_{0.80}Cd_{0.20}Te$ indicating the possibility of obtaining identical hole concentrations for anneals at two different temperatures and partial pressures of Hg. (From Ref. [75])

Native defects play a dominant role in the diffusion behavior [90]. Vacancies have very high diffusivities even at low temperatures. For example, to form a junction a few micrometers deep in 10^{16}cm^{-3} material, it requires only about 15 minutes at a temperature of 150°C–200°C. This corresponds to diffusion constants in the order of 10^{-10}cm^2/s. The presence of dislocations can enhance the vacation mobility even further, while the presence of Te precipitates may retard the motion of Hg into lattice.

17.2.3.2 Dopants

The electrical behavior of dopants has been extensively reviewed by Capper [76]. Donor behavior is expected for the elements from group IIIB on the metal lattice site, and for group VIIB elements on the Te site. In is most frequently used as a well-controlled dopant for n-type doping due to its high solubility and moderately high diffusion. The experimental data can be explained, assuming that at low (<10^{18}cm^{-3}) concentration In incorporates as a single ionizable donor occupying a metal lattice site. At high In concentration, In incorporates as a neutral complex corresponding to In_2Te_3. The bulk materials are typically doped by direct addition to melts. In is frequently introduced during epitaxy and by diffusion; it has been used for many years as a contact material for the n-type photoconductors and the n-type side of photodiodes.

Among the group VIIB elements, only I that was occupying the Te sites proved to be a well-behaved donor with concentrations in the range of 10^{15}–10^{18}cm^{-3} [78,91]. The electron concentration was found to increase with Hg pressure. Acceptor behavior is expected from elements in the I group (Ag, Cu, and Au) substituting for metal lattice sites, and from elements in the V group (P, As, Sb, Bi) substituting for Te sites.

Ag, Cu, and Au are shallow single acceptors [72,76]. They are very fast diffusers that limit the applications for devices. Significant diffusion of Ag and especially Cu occurs at room temperature [92]. Hole concentrations have been obtained that are roughly equal to Cu concentration: up to 10^{19}cm^{-3}. But the behavior of Au is more complex. Au does not seem to be very useful as a controllable acceptor, though it has proven to be useful for contacts.

The amphoteric behavior of the VB group elements (P, As, Sb) has been established [75,78]. They are acceptors substituting for Te sites and donors at metal sites; therefore, metal-rich conditions are necessary to introduce dopants at Te sites. As proved to be the most successful p-type dopant, to date, for the formation of stable junctions [93–96]. The main advantages are very low diffusity, stability in lattice, low activation energy, and the possibility of controlling concentration over a wide (10^{15}–10^{18}cm^{-3}) range. Intensive efforts are currently underway to reduce the high temperature (400°C) and high Hg pressures required to activate As as an acceptor.

17.3 FUNDAMENTAL HgCdTe PROPERTIES

HgCdTe ternary alloy is a nearly ideal IR detector material system. Its position is conditioned by three key features [97]:

- Tailorable energy bandgap over the 1–30 μm range.
- Large optical coefficients that enable high quantum efficiency.
- Favorable inherent recombination mechanisms that lead to high operating temperature (HOT).

These properties are direct consequences of the energy band structure of this zinc-blende semiconductor. Moreover, the specific advantages of HgCdTe are the ability to obtain both low and high carrier concentrations, high mobility of electrons, and low dielectric constant. The extremely small change in lattice constant with composition makes it possible to grow high-quality layered and graded gap structures. As a result, HgCdTe can be used for the detectors operated at various modes (photoconductor, photodiode, or MIS detector).

Table 17.1 summarizes the various material properties of $Hg_{1-x}Cd_xTe$ [97]; Table 17.2 compares the important parameters of HgCdTe with other narrow-gap semiconductors used in IR detector fabrication.

Table 17.1 Summary of the material properties for the $Hg_{1-x}Cd_xTe$ ternary alloy, listed for the binary components HgTe and CdTe, and for several technologically important alloy compositions

PROPERTY	HgTe	$Hg_{1-x}Cd_xTe$						CdTe
X	0	0.194	0.205	0.225	0.31	0.44	0.62	1.0
a (Å)	6.461	6.464	6.464	6.464	6.465	6.468	6.472	6.481
T (K)	77	77	77	77	140	200	250	300
E_g (eV)	−0.261	0.073	0.091	0.123	0.272	0.474	0.749	1.490
λ_c (μm)	–	16.9	13.6	10.1	4.6	2.6	1.7	0.8
n_i (cm^{-3})	–	1.9×10^{14}	5.8×10^{13}	6.3×10^{12}	3.7×10^{12}	7.1×10^{11}	3.1×10^{10}	4.1×10^{5}
m_c/m_o	–	0.006	0.007	0.010	0.021	0.035	0.053	0.102
g_c	–	−150	−118	−84	−33	−15	−7	−1.2
$\varepsilon_s/\varepsilon_o$	20.0	18.2	18.1	17.9	17.1	15.9	14.2	10.6
$\varepsilon_\infty/\varepsilon_o$	14.4	12.8	12.7	12.5	11.9	10.8	9.3	6.2
n_r	3.79	3.58	3.57	3.54	3.44	3.29	3.06	2.50
μ_e (cm^2/Vs)	–	4.5×10^5	3.0×10^5	1.0×10^5	–	–	–	–
μ_{hh} (cm^2/Vs	–	450	450	450	–	–	–	–
$b=\mu_e/\mu_\eta$	–	1000	667	222	–	–	–	–
τ_R (μs)	–	16.5	13.9	10.4	11.3	11.2	10.6	2
τ_{A1} (μs)	–	0.45	0.85	1.8	39.6	453	4.75×10^3	
$\tau_{typical}$ (μs)	–	0.4	0.8	1	7	–	–	–
E_p (eV)	19							
Δ (eV)	0.93							
m_{hh}/m_o	0.40–0.53							
ΔE_v (eV)	0.35–0.55							

Source: After Ref. [97].

τ_R and τ_{A1} calculated for n-type HgCdTe with $N_d = 1\times10^{15}$cm^{-3}, the last four material properties are independent of or relatively insensitive to alloy composition.

Table 17.2 Some physical properties of narrow-gap semiconductors

	E_g (eV)		n_i (cm^{-3})			μ_e(10^{4}cm^2/Vs)		μ_h(10^{4}cm^2/Vs)	
MATERIAL	77 K	300 K	77 K	300 K	ε	77 K	300 K	77 K	300 K
InAs	0.414	0.359	6.5×10^3	9.3×10^{14}	14.5	8	3	0.07	0.02
InSb	0.228	0.18	2.6×10^9	1.9×10^{16}	17.9	100	8	1	0.08
$In_{0.53}Ga_{0.47}As$	0.66	0.75		5.4×10^{11}	14.6	7	1.38		0.05
PbS	0.31	0.42	3×10^7	1.0×10^{15}	172	1.5	0.05	1.5	0.06
PbSe	0.17	0.28	6×10^{11}	2.0×10^{16}	227	3	0.10	3	0.10
PbTe	0.22	0.31	1.5×10^{10}	1.5×10^{16}	428	3	0.17	2	0.08
$Pb_{1-x}Sn_xTe$	0.1	0.1	3.0×10^{13}	2.0×10^{16}	400	3	0.12	2	0.08
$Hg_{1-x}Cd_xTe$	0.1	0.1	3.2×10^{13}	2.3×10^{16}	18.0	20	1	0.044	0.01
$Hg_{1-x}Cd_xTe$	0.25	0.25	7.2×10^8	2.3×10^{15}	16.7	8	0.6	0.044	0.01

17.3.1 ENERGY BANDGAP

The electrical and optical properties of $Hg_{1-x}Cd_xTe$ are determined by the energy gap structure in the vicinity of the Γ-point of the Brillouin zone, essentially in the same way as for InSb. The shape of the electron band and the light-mass hole band are determined by the **k·p** interaction and, hence, by the energy gap and the momentum matrix element. The energy gap of this compound at 4.2 K ranges from −0.300 eV for semimetallic HgTe, goes through zero at about $x = 0.15$, and extends up to 1.648 eV for CdTe.

Figure 17.10 plots the energy bandgap $E_g(x, T)$ for $Hg_{1-x}Cd_xTe$ versus alloy composition parameter x at temperatures of 77 K and 300 K. Also plotted is the cutoff wavelength $\lambda_c(x, T)$, defined as that wavelength at which the response has dropped to 50% of its peak value.

A number of expressions approximating $E_g(x, T)$ are available at present. The most widely used expression is due to Hansen et al. [98].

$$E_g = -0.302 + 1.93x - 0.81x^2 + 0.832x^3 + 5.35 \times 10^{-4}(1 - 2x)T, \tag{17.4}$$

where E_g is in eV and T is in K.

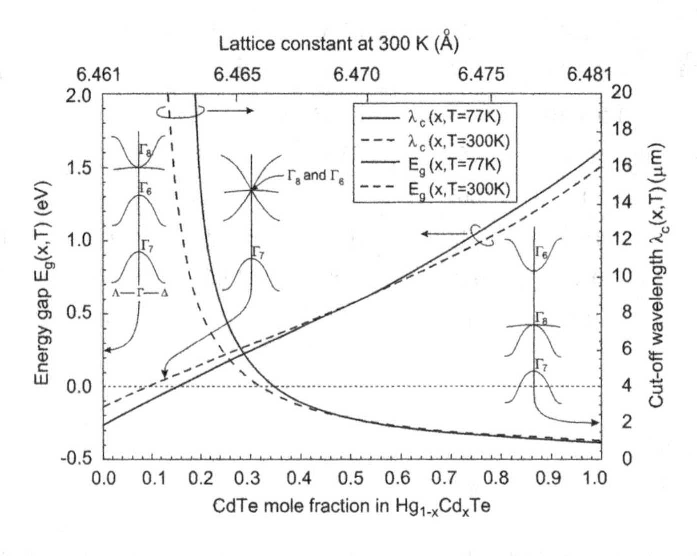

Figure 17.10 The bandgap structure of $Hg_{1-x}Cd_xTe$ near the Γ-point for three different values of the forbidden energy gap. The energy bandgap is defined as the difference between the Γ_6 and Γ_8 band extrema at $\Gamma = 0$.

The expression that has become the most widely used for intrinsic carrier concentration is that by Hansen and Schmit [99] who used their own $E_g(x, T)$ relationship of Equation 17.4, the **k·p** method, and a value of 0.443 m_o for heavy hole effective mass ratio

$$n_i = (5.585 - 3.82x + 0.001753T - 0.001364xT) \times 10^{14} E_g^{3/4} T^{3/2} \exp\left(-\frac{E_g}{2kT}\right). \tag{17.5}$$

The electron m_e^* and light hole m_{lh}^* effective masses in the narrow-gap Hg compounds are close and they can be established according to the Kane band model. Here, we used Weiler's expression [100].

$$\frac{m_o}{m_e^*} = 1 + 2F + \frac{E_p}{3}\left(\frac{2}{E_g} + \frac{1}{E_g + \Delta}\right), \tag{17.6}$$

where E_p = 19 eV, Δ = 1 eV, and $F = -0.8$. This relationship can be approximated by $m_e^*/m \approx 0.0071E_g$, where E_g is in eV. The effective mass of heavy hole m_{hh}^* is high; the measured values range between 0.3–0.7 m_o. The value of m_{hh}^* = 0.55 m_o is frequently used in the modeling of IR detectors.

17.3.2 MOBILITIES

Due to small effective masses the electron mobilities in HgCdTe are remarkably high, while heavy-hole mobilities are two orders of magnitude lower. A number of scattering mechanisms dominate the electron mobility [100–104]. The x-dependence of the mobility results primarily from the x-dependence of the bandgap, and the temperature dependence primarily from the competition among various scattering mechanisms that are temperature dependent.

The electron mobilities in HgCdTe are primarily determined by ionized impurity scattering (CC) in the low-temperature region and by polar longitudinal-optical phonon (LO) scattering above the low-temperature region as shown in Figure 17.11a [105]. Figure 17.11b depicts the composition dependence of mobility in undoped and doped samples at 77 K and 300 K. Extremely high values of mobility for high-purity samples are observed near the semiconductor-semimetal transition, where the electron

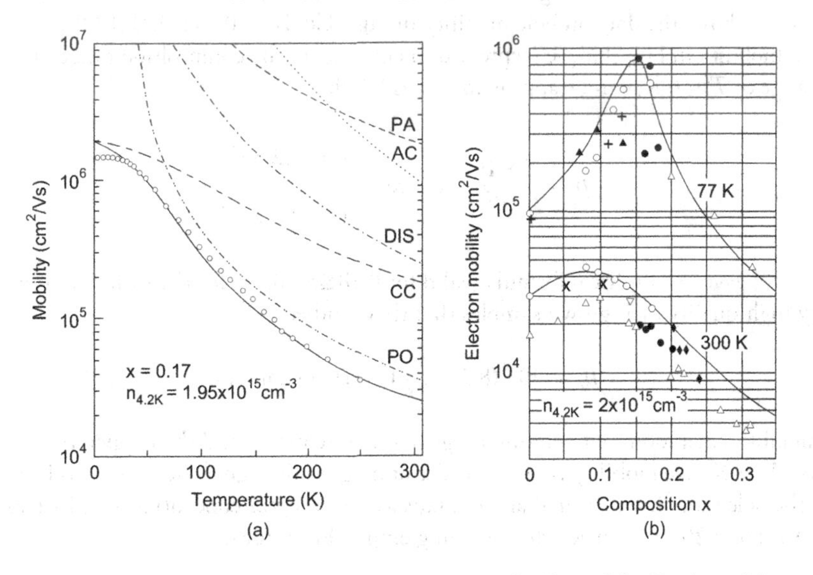

Figure 17.11 Electron mobility in $Hg_{1-x}Cd_xTe$: (a) versus temperature for $Hg_{0.83}Cd_{0.17}Te$; the solid curve is theoretically assumed for mixed scattering modes: by charged center (CC), polar (PO), disorder (DIS), acoustic (AC), and piezoacoustic (PA) scattering modes; (b) versus composition at 77 K and 300 K; curves are calculated with electron concentration of 2 × 10^{15}cm^{-3} at 4.2 K, experimental points for approximately corresponding concentrations are taken from various works. (From Ref. [105])

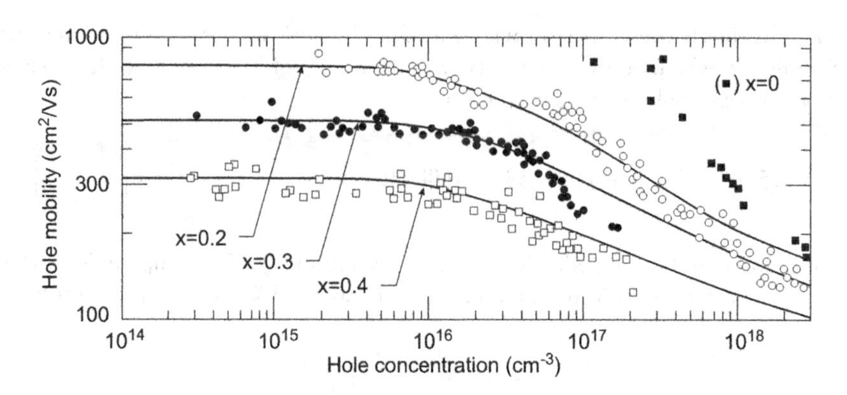

Figure 17.12 Hole mobility as a function of hole concentration for $Hg_{1-x}Cd_xTe$ at 77 K. The solid curves represent calculated data concerning combined lattice and ionized impurity scattering. (From Ref. [108])

effective mass has its minimum value. It seems that the theory correctly describes the highest mobilities. For $Hg_{0.78}Cd_{0.22}Te$ LPE layer, the threshold carrier concentration above which the CC begins to dominate is about $1 \times 10^{16} cm^{-3}$ for n-type and about $1 \times 10^{17} cm^{-3}$ for p-type materials [106]. The electron mobility data, when plotted versus temperature, often exhibit a broad peak at $T < 100$ K, particularly for LPE samples. On the other hand, mobility data obtained from high-quality bulk samples do not exhibit these peaks. It is believed that these peaks are associated with scattering by charged centers or are related to the anomalous electrical behavior that are closely associated with sample inhomogeneities [104].

The transport properties of holes are less studied than those of electrons mainly because the contribution of holes to the electrical conduction is relatively small due to their low mobility. Very often in transport measurements, the electron contribution predominates even in p-type materials unless the electron density is sufficiently low. Comprehensive analysis of different hole scattering mechanisms in $Hg_{1-x}Cd_xTe$ (x = 0.2–0.4) has been carried out by Yadava et al. [107]. They concluded that the heavy-hole mobility is largely governed by the CC, unless the strain field or dislocation scattering below 50 K or the polar scattering above 200 K becomes dominant. The light-hole mobility is mainly governed by the acoustic phonon scattering. Figure 17.12 shows the data on hole mobility in $Hg_{1-x}Cd_xTe$ ($0.0 \le x \le 0.4$) [108].

The electron mobility in $Hg_{1-x}Cd_xTe$ (expressed in cm^2/Vs), in the composition range of $0.2 \le x \le 0.6$ and temperature range of $T > 50$ K, can be approximated as [109]:

$$\mu_e = \frac{9 \times 10^8 s}{T^{2r}} \text{ where } \begin{array}{l} r = (0.2/x)^{0.6} \\ s = (0.2/x)^{7.5} \end{array} \tag{17.7}$$

Higgins et al. [110] gave an empirical formula (valid for $0.18 \le x \le 0.25$) for the variation of μ_e with x at 300 K for very high-quality melt grown samples that they studied:

$$\mu_e = 10^{-4}(8.754x - 1.044)^{-1} \text{in cm}^2/\text{Vs}. \tag{17.8}$$

The hole mobilities at a room temperature range of 40–80 cm^2/Vs, and the temperature dependence is relatively weak. A 77 K hole mobility is by one order of magnitude higher. According to Dennis and colleagues [111], the hole mobility measured at 77 K falls as the acceptor concentration is increased and in the composition range of 0.20–0.30 yields the following empirical expression:

$$\mu_h = \mu_o \left[1 + \left(\frac{p}{1.8 \times 10^{17}}\right)^2\right]^{-1/4}, \tag{17.9}$$

where $\mu_o = 440\ cm^2/Vs$.

For modeling IR photodetectors, the hole mobility is usually calculated assuming that the electron-to-hole mobility ratio $b = \mu_e/\mu_h$ is constant and equal to 100.

The minority carrier mobility is one of the fundamental material properties affecting the performance of HgCdTe along with carrier concentration, composition, and minority carrier lifetime. For materials having acceptor concentrations $<10^{15}\,\text{cm}^{-3}$, literature results give comparable electron mobilities to those found in n-type HgCdTe [$\mu_e(n)$]. As the acceptor concentration increases, the deviation from n-type electron mobilities increases, resulting in lower electron mobilities for p-type material [$\mu_e(p)$]. Typically, for $x = 0.2$ and $N_a = 10^{16}\,\text{cm}^{-3}$, $\mu_e(p)/\mu_e(n) = 0.5–0.7$, while for $x = 0.2$ and $N_a = 10^{17}\,\text{cm}^{-3}$, $\mu_e(p)/\mu_e(n) = 0.25–0.33$. For $x = 0.3$, however, $\mu_e(p)/\mu_e(n)$ ranges from 0.8 for $N_a = 10^{16}\,\text{cm}^{-3}$ to 0.9 for $N_a = 10^{17}\,\text{cm}^{-3}$ [112]. It was found that at temperatures above 200 K there was very little difference between the electron mobility in epitaxial p-type HgCdTe layers and that measured directly in n-type layers.

17.3.3 OPTICAL PROPERTIES

Optical properties of HgCdTe have been investigated mainly at energies near the bandgap [25,80,113,114]. There still appears to be considerable disagreement among the reported results concerning absorption coefficients. This is caused by different concentrations of native defects and impurities, nonuniform composition and doping, thickness inhomogeneities of samples, mechanical strains, and different surface treatments.

In most compound semiconductors, the band structure closely resembles the parabolic energy versus the momentum dispersion relation. The optical absorption coefficient would then have a square root dependence on energy that follows the electronic density of states, often referred to as the Kane model [115]. The above bandgap absorption coefficient can be calculated for InSb-like band structure semiconductors, such as $Hg_{1-x}Cd_xTe$, including the Moss-Burstein shift effect. Corresponding expressions were derived by Anderson [116]. Beattie and White proposed an analytic approximation with a wide range of applicability for band-to-band radiative transition rates in direct, narrow bandgap semiconductors [117].

In high-quality samples, the measured absorption in the SW region is in good agreement with the Kane model calculation, while the situation appears to be complicated in the LW edge because of the appearance of an absorption tail extending at energies lower than the energy gap.

This tail has been attributed to the composition-induced disorder. According to Finkman and Schacham [118], the absorption tail obeys a modified Urbach's rule:

$$\alpha = \alpha_o \exp\left[\frac{\sigma(E - E_o)}{T + T_o}\right] \text{in cm}^{-1}, \tag{17.10}$$

where T is in K, E is in eV, and $\alpha_o = \exp(53.61x - 18.88)$, $E_o = -0.3424 + 1.838x + 0.148x^2$ (in eV), $T_o = 81.9$ (in K), $\sigma = 3.267 \times 10^4(1 + x)$ (in K/eV) are fitting parameters that vary smoothly with composition. The fit was performed with data at $x = 0.215$ and $x = 1$ and for temperatures between 80 K and 300 K.

Assuming that the absorber coefficient for large energies can be expressed as:

$$\alpha(h\nu) = \beta(h\nu - E_g)^{1/2}, \tag{17.11}$$

and many researchers assume that this rule can be applied to HgCdTe. For example, Schacham and Finkman used the following fitting parameter $\beta = 2.109 \times 10^5[(1 + x)/(81.9 + T)]^{1/2}$, which is a function of composition and temperature [119]. The conventional procedure used to locate the energy gap is to use the point inflection, that is, to exploit the large change in the slope of $\alpha(h\nu)$ that is expected when the band-to-band transition overtakes the weaker Urbach contribution. To overcome the difficulty in locating the onset of the band-to-band transition, the bandgap was defined as that energy value where $\alpha(h\nu) = 500\,\text{cm}^{-1}$ [118]. Schacham and Finkman analyzed the crossover point and suggested that $\alpha = 800\,\text{cm}^{-1}$ was a better choice [119]. Hougen analyzed the absorption data of n-type LPE layers and suggested that the best formula was $\alpha = 100 + 5{,}000x$ [120].

Chu et al. [121] have reported similar empirical formulas for absorption coefficient at the Kane and Urbach tail regions. They received the following modified Urbach rule of the form:

$$\alpha = \alpha_o \exp\left[\frac{\delta(E - E_o)}{kT}\right], \tag{17.12}$$

where $\ln \alpha_0 = -18.5 + 45.68x$

$E_o = -0.355 + 1.77x$

$\delta/kT = (\ln \alpha_g - \ln \alpha_o)/(E_g - E_o)$

$\alpha_g = -65 + 1.88T + (8694 - 10.314T)\, x$

$E_g\,(x, T) = -0.295 + 1.87x - 0.28x^2 + 10^{-4}\,(6 - 14x + 3x^2)T + 0.35x^4$

The meaning of the parameter α_g is that $\alpha = \alpha_g$ when $E = E_g$, the absorption coefficient at the bandgap energy. When $E < E_g$ and $\alpha < \alpha_g$, the absorption coefficient obeys the Urbach rule in Equation 17.12.

Chu et al. [122] have also found an empirical formula for the calculation of the intrinsic optical absorption coefficient at the Kane region.

$$\alpha = \alpha_g \exp\left[\beta\left(E - E_g\right)\right]^{1/2}, \tag{17.13}$$

where the parameter β depends on the alloy composition and temperature, $\beta(x, T) = -1 + 0.083T + (21 - 0\ 13T)x$. Expanding Equation 17.13, one finds a linear term, $(E - E_g)^{1/2}$, which fits the square root law between α and E proper for parabolic bands (see Equation 17.11).

Figure 17.13 shows the intrinsic absorption spectrum for $Hg_{1-x}Cd_xTe$ with $x = 0.170–0.443$ at temperatures 300 K and 77 K. The absorption strength generally decreases as the gap becomes smaller due to the decrease in the conduction band effective mass and the $\lambda^{-1/2}$ dependence of the absorption coefficient on wavelength λ. It can be seen that the calculated Kane plateaus according to Sharma and colleagues [123] and Equation 17.13 link closely to the calculated Urbach absorption tail from Equation 17.12 at the turning point α_g. Since the tail effect is not included in the Anderson model [116], the curves calculated according to this model fall down sharply at energies adjacent to E_g. At 300 K, the line shapes derived for the absorption coefficient above α_g have almost the same tendency; however, the Chu et al. expression (Equation 17.12) shows better agreement with the experimental data. At 77 K, the curves for the expressions of Anderson and Chu et al. are in agreement with the measurements, but discrepancies occur for the empirical parabolic rule of Sharma et al. [123] and these deviations increase with decreasing x. The degree of band nonparabolicity increases as the temperature or x decreases, resulting in increasing discrepancy between the experimental result and the square root law. In general, Chu et al.'s empirical rule and the Anderson model agree well with the experimental data for $Hg_{1-x}Cd_xTe$, with x in the range of 0.170–0.443 and at temperatures 4.2 K–300 K, but the Anderson model fails to explain the absorption near E_g [124].

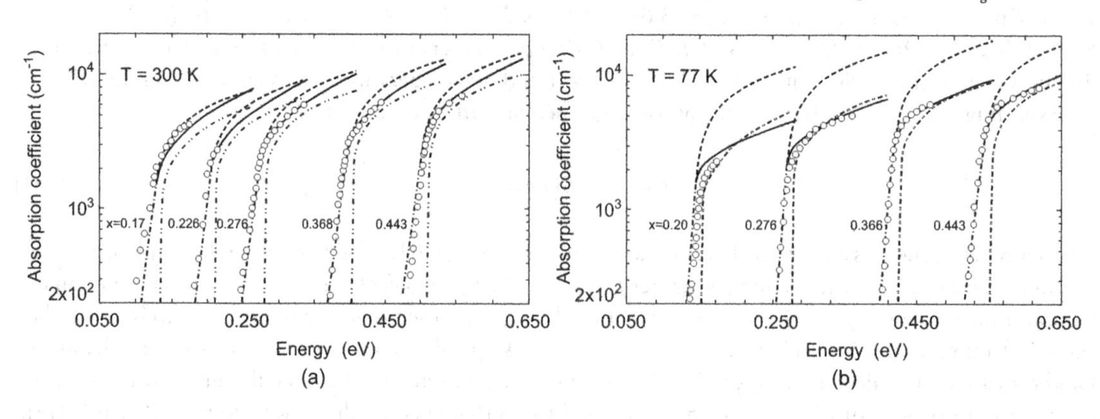

Figure 17.13 Intrinsic absorption spectrum of $Hg_{1-x}Cd_xTe$ samples with $x = 0.170–0.443$ at (a) 300 K and (b) 77 K. Symbols indicate experimental data (From Refs. [121,122]), the dash-double-dotted curves are according to Anderson's model, medium-dashed curves (From Ref. [123]), the solid curves are from Equation 17.13 of Chu et al., and the dash-dotted lines below E_g are from Equation 17.12. (From Ref. [124])

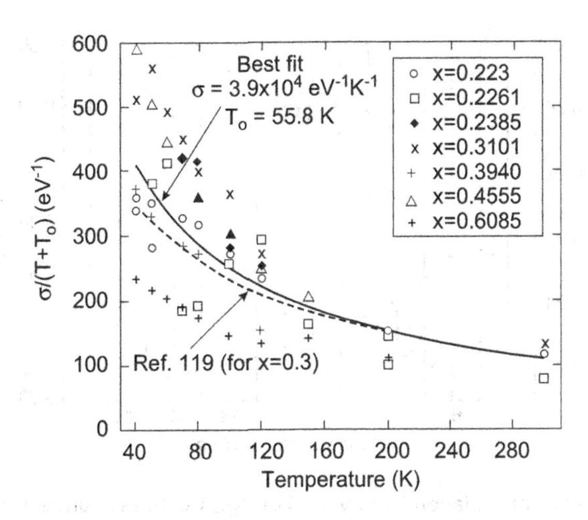

Figure 17.14 Band-tail parameter $\sigma/(T+T_o)$ versus temperature for different compositions and a proposed model based on the best overall fit along with values (From Ref. [118]) for $x = 0.3$. (From Ref. [127])

More recently, it has been suggested [125,126] that narrow bandgap semiconductors, such as HgCdTe, more closely resemble a hyperbolic band structure relationship with an absorption coefficient given by:

$$\alpha = \frac{K\sqrt{(E-E_g+c)^2-c^2}\left(E-E_g+c\right)}{E}, \tag{17.14}$$

where c is the parameter defining hyperbolic curvature of the band structure and K is the parameter defining the absolute value of absorption coefficient. This theoretical prediction has been confirmed by the experimental measurements of optical properties of MBE HgCdTe grown samples with uniform compositions [127,128]. The fitting parameters for bandgap tail and hyperbolic regions of the absorption coefficient defined in Equations 17.10 and 17.14 have been extracted by determination of the transition point between both the regions. It has been realized that the derivative of the absorption coefficient has a maximum between the Urbach and hyperbolic regions. Figure 17.14 shows the measured exponential slope parameter values $\sigma/(T+T_o)$ versus temperature and compares them to values given by Finkman and Schacham [118] for arbitrarily chosen composition of $x = 0.3$. This choice of composition does not have a significant effect on the values obtained, where the values given by Finkman and Schacham have small compositional dependence in the region of interest ($0.2 < x < 0.6$), where the parameter $\sigma/(T+T_o)$ is proportional to $(1+x)^3$. This parameter shows no clear correlation with composition when there is significant scattering in the data at cryogenic temperatures. The trend of decreasing values with increasing temperature is in agreement with an increase in thermally excited absorption processes, where the values obtained at lower temperatures are more indicative of the quality of the layers grown.

It should be noted that the above discussed expressions do not take into account the influence of doping on the absorption coefficient, so they are not very useful in modeling the LW uncooled devices.

$Hg_{1-x}Cd_xTe$ and the closely related alloys exhibit significant absorption below the absorption edges, which can be related to intraband transitions in both the conduction and valence bands and intervalence band transitions. This absorption does not contribute to the optical generation of charge carriers.

The Kramers and Kronig interrelations are usually used to estimate the dependence of the refractive index on temperature [129–131]. For $Hg_{1-x}Cd_xTe$ with x 0.276–0.540 and temperatures 4.2 K–300 K, the following empirical formula can be used [130]:

$$n^2(\lambda,T) = A + \frac{B}{1-(C/\lambda)^2} + D\lambda^2, \tag{17.15}$$

where A, B, C, and D are fitting parameters, which vary with composition x and temperature T. Equation 17.15 can also be used for $Hg_{1-x}Cd_xTe$ with x from 0.205 to 1 at room temperature.

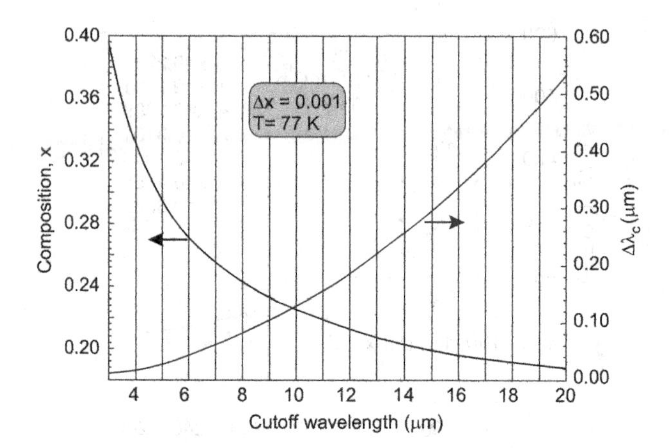

Figure 17.15 The cutoff wavelength variation (right y-axis) of $Hg_{1-x}Cd_xTe$ as a function of cutoff wavelengths (x-axis) with a fixed composition fluctuation of x = 0.001 during the growth.

The high frequency dielectric constant, ε_∞, and the static constant, ε_o, are usually derived from reflectivity data in evaluating the real and imaginary parts of ε. The dielectric constants are not linear functions of x and temperature dependence is not observed within the experimental resolution [101]. These dependences can be described by the following relations:

$$\varepsilon_\infty = 15.2 - 15.6x + 8.2x^2, \tag{17.16}$$

$$\varepsilon_o = 20.5 - 15.6x + 5.7x^2. \tag{17.17}$$

One of the main problems with $Hg_{1-x}Cd_xTe$ detectors is producing a homogeneous material. Recall that the variation of x can be related to the cutoff wavelength by $\lambda_c[\mu m] = 1.24/E_g(x)[eV]$, where E_g is given by Equation 17.4. Substitution and rearrangement of the terms yield:

$$\lambda_c[\mu m] = \frac{1}{-0.244 + 1.556x + \left(4.31\times 10^{-4}\right)T(1-2x) - 0.65x^2 + 0.671x^3}. \tag{17.18}$$

The differential of Equation 17.17 relates x variation in manufacturing to variation in the cutoff wavelength:

$$d\lambda_c = \lambda_c^2\left(1.556 - 8.62\times 10 - 4T - 1.3x + 2.013x^2\right)dx. \tag{17.19}$$

Figure 17.15 shows some uncertainty in the cutoff wavelength for x variation of 0.1%. This variation in x value is for a very good material. For SW (≈3 μm) and MW (≈5 μm) materials, the variation in cutoff wavelength is not large. However, for the LW materials (≈20 μm), the uncertainty in the cutoff wavelength is large, above 0.5 μm, and cannot be neglected. This response variation causes radiometric calibration problems in that the radiation is detected over a different spectral region than expected.

Absorption measurements are possibly the most common routine methods to determine and map the compositions of bulk crystals and epitaxial layers. Typically, 50% or 1% cut-on wavelengths are used for thick (>0.1 mm) samples [80,120,132,133]. Various methods have been used for thinner samples. According to Higgins et al. [110], for thick samples:

$$x = \frac{w_n(300K) + 923.3}{10683.98}, \tag{17.20}$$

where w_n is the 1% absolute transmission cut-on wave number. Compositions of epitaxial layers are usually determined from a wavelength corresponding to half of the maximum transmission $0.5T_{max}$ [80]. The determination can be complicated by the presence of composition grading.

The UV and visible reflectance measurements are also useful in composition determination, particularly for characterization of the surface (10–30 nm penetration depth) region [133]. Usually, the position of the peak reflectivity at the E_1 bandgap location is measured and the composition is calculated from the experimental expression:

$$E_1 = 2.087 + 0.7109x + 0.1421x^2 + 0.3623x^3. \tag{17.21}$$

17.3.4 THERMAL GENERATION-RECOMBINATION PROCESSES

The generation processes that compete against the recombination processes directly affect the performance of photodetectors, setting up a steady-state concentration of carriers in semiconductors subjected to the thermal and optical excitation and, frequently, determining the kinetics of photogenerated signals. Generation-recombination (GR) processes in semiconductors are widely discussed in literature (see, for example, [79,81,134–139]. Here, we reproduce only some dependencies directly related to the performance of photodetectors. Assuming bulk processes only, there are three main thermal GR processes to be considered in the narrow bandgap semiconductors, namely, Shockley-Read (SR), radiative, and Auger.

The GR mechanisms are generally well established in HgCdTe ternary alloys. The carrier lifetime measurements of MWIR, with cutoff wavelength $\lambda_c = 5\,\mu\text{m}$ and LWIR, with $\lambda_c = 10\,\mu\text{m}$, for both n-type and p-type HgCdTe at 77 K are summarized in Figures 17.16 and 17.17. The trend lines of carrier lifetimes are given according to Kinch et al. [139]. The experimental data are taken from many sources.

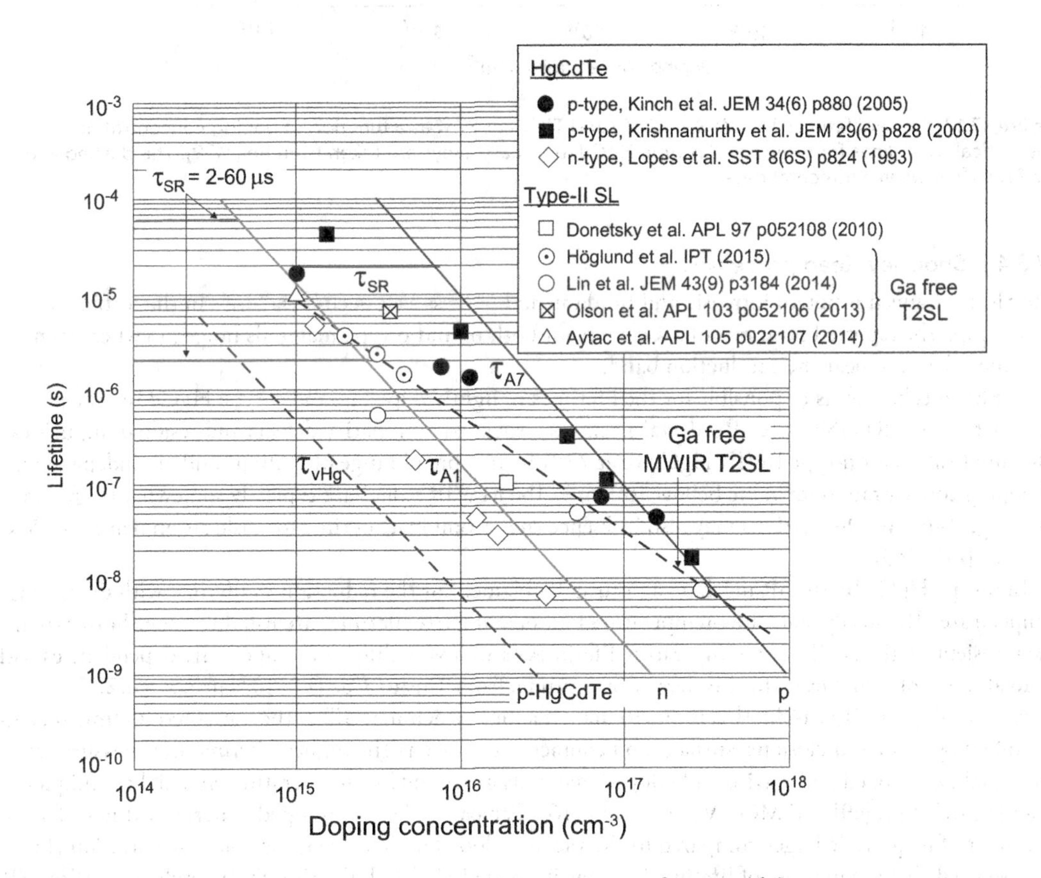

Figure 17.16 Carrier lifetimes for MWIR HgCdTe and T2SLs at 77 K as a function of doping concentration. Theoretical trend lines for n-type and p-type HgCdTe ternary alloys are taken from Ref. [139]. The dashed line for Ga-free T2SLs follows experimental data.

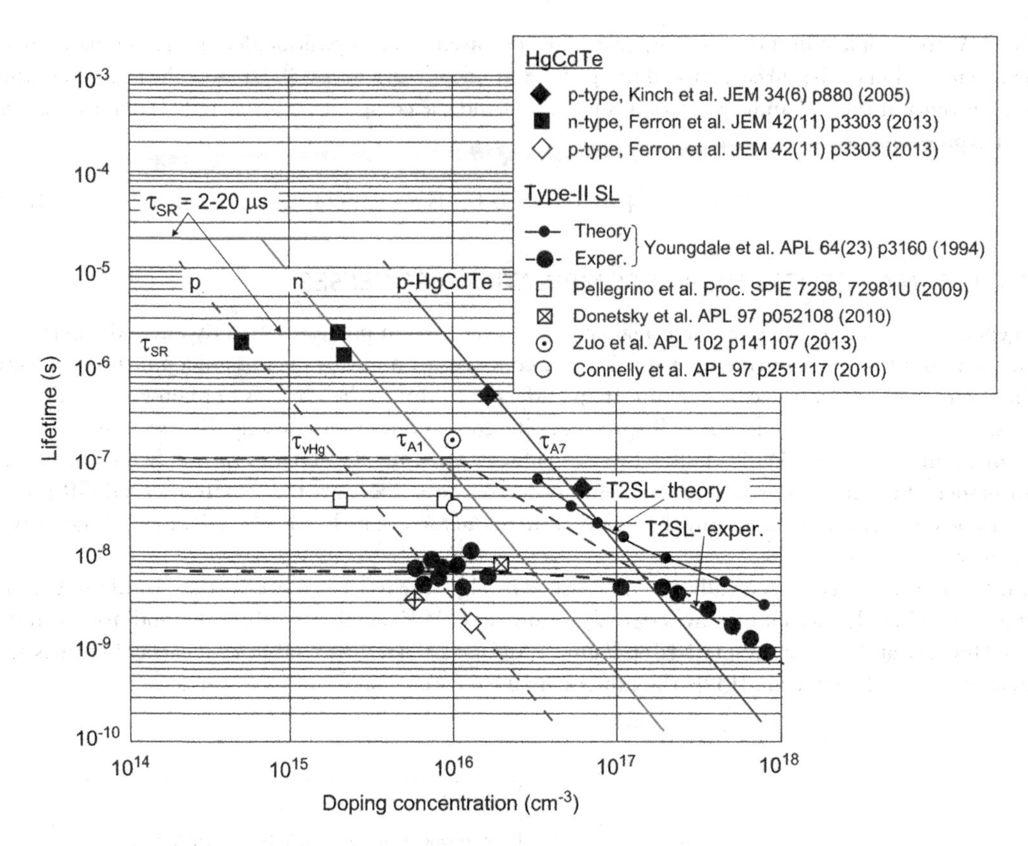

Figure 17.17 Carrier lifetimes for LWIR HgCdTe and T2SLs at 77 K as a function of doping concentration. Theoretical trend lines for n-type and p-type HgCdTe ternary alloys are taken from Ref. [139]. The dashed line for T2SLs follows experimental data.

17.3.4.1 Shockley-Read processes

The SR mechanism is not an intrinsic and fundamental process as it occurs via levels in the forbidden energy gap. The reported positions of SR centers for both n- and p-type materials range anywhere from near the valence to near the conduction band.

The SR mechanism is responsible for the lifetimes in lightly doped n- and p-type HgCdTe. The possible factors are SRH (Shockley-Read-Hall) centers associated with native defects and residual impurities. Measured values for n-type LWIR HgCdTe at 77 K lie in a broad range of 2–20 μs and are independent of doping concentration for value bellow 10^{15} cm^{-3}. The MWIR values are typically somewhat longer, in the range 2–60 μs. Dislocations may also influence the recombination time for dislocation densities >5 × 10^5 cm^{-2} [140–143].

In p-type HgCdTe, the SR mechanism is usually blamed for the reduction in lifetime with decreasing temperature. The steady-state, low-temperature photoconductive lifetimes are usually much shorter than the transient lifetimes. The low-temperature lifetimes exhibit very different temperature dependencies with a broad range of values over three orders of magnitude, from 1 ns to 1 μs ($p \approx 10^{16}$ cm^{-3}, $x \approx 0.2$, $T \approx 77$ K, vacancy doping) [135,143]. This is due to many factors, which may affect the measured lifetime including inhomogeneities, inclusions, surface, and contact phenomena. The highest lifetime was measured in high-quality undoped and extrinsically doped materials grown by low temperature epitaxial techniques from Hg-rich LPE [49] and MOCVD [91,144–146]. Typically, Cu or Au-doped materials exhibit lifetimes one order of magnitude larger compared to the vacancy-doped ones of the same hole concentration [143]. It is believed that the increase of lifetime in impurity-doped $Hg_{1-x}Cd_xTe$ arises from a reduction of the SR centers. This may be due to the low-temperature growth of doped layers or due to low-temperature annealing of doped samples.

The origin of the SR centers in vacancy-doped, p-type material is not clear at present. These centers seem not to be the vacancies themselves and thus may be removable [147]. Vacancy-doped materials with the same carrier concentration, but created under different annealing temperatures, may produce different lifetimes. One possible candidate for recombination centers is Hg interstitials [148]. Vacancy-doped $Hg_{1-x}Cd_xTe$ exhibits SR recombination center densities roughly proportional to the vacancy concentration.

Measurements at Leonardo DRS [138] give lifetime values for extrinsic p-type material.

$$\tau_{ext} = 9 \times 10^9 \frac{p_1 + p}{pN_a}, \tag{17.22}$$

where

$$p_1 = N_v \exp\left(\frac{q(E_r - E_g)}{kT}\right), \tag{17.23}$$

and E_r is the SR center energy relative to the conduction band. Experimentally, E_r was found to lie at the intrinsic level for As, Cu, and Au dopants, giving $p_1 = n_i$.

For vacancy-doped p-type $Hg_{1-x}Cd_xTe$:

$$\tau_{vac} = 5 \times 10^9 \frac{n_1}{pN_{vac}}, \tag{17.24}$$

where

$$n_1 = N_c \exp\left(\frac{qE_r}{kT}\right). \tag{17.25}$$

E_r is ≈ 30 mV from the conduction band ($x = 0.22–0.30$).

As follows from these expressions and Figures 17.16 and 17.17 [139], doping with the foreign impurities (Au, Cu, and As for p-type material) gives lifetimes significantly increased compared to native doping of the same level.

Although a considerable research effort is still necessary, the SR process does not represent a fundamental limit to the performance of the photodetectors. The values of τ_{SR}, more recently gathered by Kinch and given in Table 17.3, are considerably larger in the lower temperature range, and are >200 μs to 50 ms depending on the cutoff wavelength.

17.3.4.2 Radiative processes

Radiative generation of charge carriers is a result of absorption of internally generated photons. The radiative recombination is an inversed process of annihilation of electron-hole pairs with emission of photons. The radiative recombination rates were calculated for conduction-to-heavy-hole-band and conduction-to-light-hole-band transitions using an accurate analytical form [117].

For a long time, internal radiative processes have been considered as the main fundamental limit of detector performance and the performance of practical devices has been compared to this limit. The role of

Table 17.3 Values of τ_{SR} in $Hg_{1-x}Cd_xTe$ deduced from reported I–V and FPA characteristics

	X COMPOSITION	τ_{SRH} (μs)
LWIR	0.225	>100 at 60 K
MWIR	0.30	>1,000 at 110 K
MWIR	0.30	~50,000 at 89 K
SWIR	0.455	>3,000 at 180 K

Source: After Ref. [137]

radiative mechanism in the detection of IR radiation has been critically reexamined [149–151]. Humpreys [150] indicated that most of the photons emitted in photodetectors as a result of radiative decay are immediately reabsorbed, so that the observed radiative lifetime is only a measure of how well photons can escape from the body of the detector. Due to reabsorption the radiative lifetime is highly extended and dependent on the semiconductor geometry. Therefore, internal combined GR processes in one detector are essentially noiseless. In contrast, the recombination act with cognate escape of a photon from the detector or the generation of photons by thermal radiation from outside the active body of the detector are noise producing processes. This may readily happen for a case of detector array where an element may absorb photons emitted by other detectors or a passive part of the structure [151,152]. Deposition of reflective layers (mirrors) on the back and the side of the detector may significantly improve optical insulation preventing noisy emission and absorption of thermal photons.

It should be noted that internal radiative generation could be suppressed in detectors operated under reverse bias where the electron density in the active layer is reduced to well below its equilibrium level [153,154].

As follows from the above considerations, the internal radiative processes, although of fundamental nature, do not limit the ultimate performance of IR detectors.

17.3.4.3 Auger processes

Auger mechanisms dominate the generation and recombination processes in high-quality, narrow-gap semiconductors, such as $Hg_{1-x}Cd_xTe$ and InSb, at near room temperatures [155,156]. The Auger generation is essentially the impact ionization by electrons of holes in the high-energy tail of Fermi-Dirac distribution. The band-to-band Auger mechanisms in InSb-like band structure semiconductors are classified into ten photonless mechanisms. Two of them have the smallest threshold energies ($E_T \approx E_g$) and are denoted as Auger 1 (A1) and Auger 7 (A7; see Figure 17.18). In some wider-bandgap materials (e.g., InAs and low x $InAs_{1-x}Sb_x$) in which the split-off band energy Δ is comparable to E_g, the Auger process involving split-off band (AS process) may also play an important role.

The A1 generation is the impact ionization by an electron, generating an electron-hole pair, so this process involves two electrons and one heavy hole. It is well known that the A1 process is an important recombination mechanism in n-type $Hg_{1-x}Cd_xTe$, particularly for x around 0.2 and at higher temperatures [79,81,135,157,158]. In Figure 17.19, experimental results from Kinch et al. [157] are compared with theoretical data of the intrinsic A1 carrier lifetime τ^i_{A1} and intrinsic radiative carrier lifetime τ^i_R. There is excellent agreement between the experimental and numerical data. Even in the extrinsic range at temperatures below 140 K, the measured lifetime seems to be governed by the A1 effect where the following relation

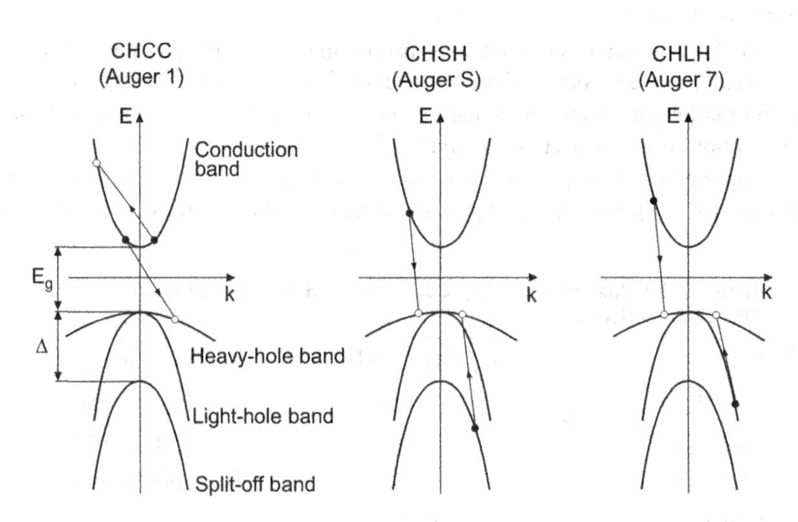

Figure 17.18 The three band-to-band Auger recombination processes. (Arrows indicate electron transitions; •, occupied state; o, unoccupied state.)

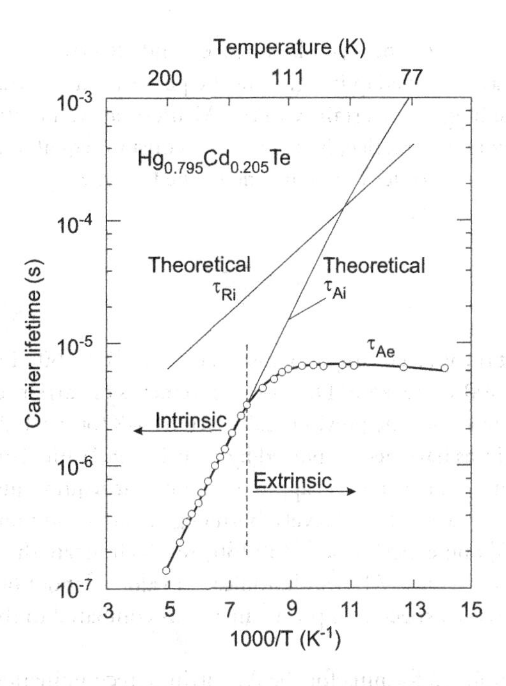

Figure 17.19 Theoretical and experimental lifetime data versus temperature for n-type $Hg_{0.795}Cd_{0.205}Te$ ($n_{77K} = 1.7 \times 10^{14} cm^{-3}$, $\mu_{77K} = 1.42 \times 10^5 cm^2/Vs$); solid lines represent the theoretical values for radiative and A1 recombination, respectively. (From Ref. [157])

should hold: $\tau_{A1} \approx 2\tau^i_{A1}(n_i/n_o)^2$. An interesting feature is the behavior of A1 generation and recombination with degenerate n-type doping. Due to the low density, the Fermi level moves high into the conduction band with n-type doping, so the concentration of minority holes is strongly reduced and the threshold energy required for the Auger transition is increased. This results in suppression of the A1 processes in heavy doped n-type material.

A7 generation is the impact generation of electron-hole pair by a light hole, involving one heavy hole, one light hole, and one electron [159–161]. This process may dominate in p-type material. Heavy p-type doping has no dramatic effect on the A7 generation and recombination rates due to the much higher density of states. The corresponding Auger recombination mechanisms are inverse processes of electron-hole recombination with energy transferred to an electron or a hole. Strong temperature and bandgap dependence is expected, since lowered temperature and increased bandgap strongly reduces the probability of these heat-stimulated transitions.

The net generation rate due to the A1 and A7 processes can be described as [162]:

$$G_A - R_A = \frac{n_i^2 - np}{2n_i^2}\left[\frac{n}{(1+an)\tau^i_{A1}} + \frac{p}{\tau^i_{A7}}\right], \tag{17.26}$$

where τ^i_{A1} and τ^i_{A7} are the intrinsic A1 and A7 recombination times, and n_i is the intrinsic concentration. The last equation is valid for a wide range of concentrations, including degeneration, which easily occurs in n-type materials. This is expressed by the finite value of a. According to White [162], $a = 5.26 \times 10^{-18} cm^3$. Due to the shape of the valence band, the degeneracy in p-type material occurs only at very high doping levels, which is not achievable in practice.

The A1 intrinsic recombination time is equal [155].

$$\tau^i_{A1} = \frac{h^3\varepsilon_s^2}{2^{3/2}\pi^{1/2}q^4 m_o}\frac{\varepsilon_s^2(1+\mu)^{1/2}(1+2\mu)\exp\left[\left(\frac{1+2\mu}{1+\mu}\right)\frac{E_g}{kT}\right]}{(m_e^*/m)|F_1F_2|^2(kT/E_g)^{3/2}}, \tag{17.27}$$

where μ is the ratio of conduction to the heavy-hole, valence-band effective mass, ε_s is the static frequency dielectric constant, and $|F_1F_2|$ are the overlap integrals of the periodic part of the electron wave functions. The overlap integrals cause the biggest uncertainty in the A1 lifetime. Values in the range of 0.1–0.3 have been obtained by various authors. In practice, it is taken as a constant equal to anywhere between 0.1 and 0.3, leading to changes by almost an order of magnitude in the lifetime.

The ratio of A7-to-A1 intrinsic times

$$\gamma = \frac{\tau_{A7}^i}{\tau_{A1}^i} \tag{17.28}$$

is another term of high uncertainty. According to Casselman et al. [159,160], for $Hg_{1-x}Cd_xTe$ over the range $0.16 \le x \le 0.40$ and $50\text{ K} \le T \le 300$ K, $3 \le \gamma \le 6$. Direct measurements of carrier recombination show the ratio γ to be larger than expected from the previous calculations (≈ 8 for $x \approx 0.2$ at 295 K) [163]. Accurate calculations of the Auger lifetimes have been reported by Beattie and White [164]. The flat valence band model has been used to obtain a simple analytic approximation that requires just two parameters to cover a wide range of temperature and carrier Fermi levels, both degenerated and nondegenerate. More recently published theoretical [165,166] and experimental [139,166] results indicate that this ratio is even higher; the data presented in Figures 17.16 and 17.17 would indicate a value of about 60. As γ is higher than unity, higher recombination lifetimes are expected in p-type materials compared to the n-type materials of the same doping.

Kinch [139] delivered a simplified formula for the A1 intrinsic recombination time.

$$\tau_{A1}^i = 8.3 \times 10^{-13} E_g^{1/2} \left(\frac{q}{kT}\right)^{3/2} \exp\left(\frac{qE_g}{kT}\right), \tag{17.29}$$

where E_g is in eV.

As the Equations 17.26 and 17.28 show, the Auger generation and recombination rates are strongly dependent on temperature via dependence of carrier concentration and intrinsic time on temperature. Therefore, cooling is a natural and a very effective way to suppress Auger processes.

Until recently, the n-type A1 lifetime was deemed to be well established. Krishnamurthy et al. [166] have shown that the full band calculations indicate that the radiative and Auger recombination rates are much slower than those predicted by expressions used in the literature (theory of Beattie and Landsberg [167]). It appears that a trap state tracking the conduction band edge with very small activation energy can explain the lifetimes in n-doped MBE samples.

The p-type A7 lifetime has long been subject to controversy. Detailed calculations of the Auger lifetime in p-type HgCdTe reported by Krishnamurthy and Casselman [165] suggest significant deviation from the classic $\tau_{A7} \sim p^{-2}$ relation. The decrease of τ_{A7} with doping is much weaker, resulting in significantly longer lifetimes in highly doped p-type low x materials (factor of ≈ 20 for $p = 1 \times 10^{17}\text{ cm}^{-3}$, $x = 0.226$, and $T = 77$–300 K).

17.4 AUGER-DOMINATED PHOTODETECTOR PERFORMANCE

17.4.1 EQUILIBRIUM DEVICES

Let us consider the Auger-limited detectivity of photodetectors. At equilibrium, the generation and recombination rates are equal. Assuming that both rates contribute to noise (see Equation 4.43),

$$D^* = \frac{\lambda\eta}{2hc(G_A t)^{1/2}} \left(\frac{A_o}{A_e}\right)^{1/2}. \tag{17.30}$$

If nondegenerate statistics is assumed, then

$$G_A = \frac{n}{2\tau_{A1}^1} + \frac{p}{2\tau_{A7}^1} = \frac{1}{2\tau_{A1}^1}\left(n + \frac{p}{\gamma}\right). \tag{17.31}$$

The resulting Auger generation achieves its minimum in just extrinsic p-type materials with $p = \gamma^{1/2}n_i$. It leads to an important conclusion about optimum doping for the best performance. In practice, the required p-type doping level would be difficult to achieve for liquid nitrogen (LN)-cooled and SW devices. Moreover, the p-type devices are more vulnerable to nonfundamental limitations (contacts, surface, SR processes) than the n-type ones. This is the reason why low-temperature and the SW photodetectors are typically manufactured from lightly doped n-type materials. In contrast, just p-type doping is clearly advantageous for the near room temperature and LW photodetectors.

The Auger dominated detectivity is

$$D^* = \frac{\lambda}{2^{1/2}hc}\left(\frac{A_o}{A_e}\right)^{1/2}\frac{\eta}{t^{1/2}}\left(\frac{\tau_{A1}^i}{n+p/\gamma}\right)^{1/2}. \quad (17.32)$$

As follows from Equation 4.50, for the optimum thickness devices,

$$D^* = 0.31k\frac{\lambda\alpha^{1/2}}{hc}\frac{\left(2\tau_{A1}^i\right)^{1/2}}{\left(n+p/\gamma\right)}. \quad (17.33)$$

This expression can be used for determination of the optimum detectivity of A1/A7 as a function of wavelength, material bandgap, and doping.

To estimate the wavelength and temperature dependence of D^*, let us assume a constant absorption for photons with energy equal to the bandgap. For extrinsic materials ($p = N_a$ or $n = N_d$),

$$D^* \sim \left(\tau_{A1}^i\right)^{1/2} \sim n_i^{-1} \sim \exp\left(\frac{E_g}{2kT}\right) = \exp\left(\frac{hc/2\lambda_{co}}{kT}\right). \quad (17.34)$$

In this case, the ultimate detectivity will be inversely proportional to the intrinsic concentration. This behavior should be expected at shorter wavelengths and lower temperatures when the intrinsic concentration is low.

For intrinsic materials and for materials doped for the minimum thermal generation, where $p = \gamma^{1/2}n_i$, $n = n_i/\gamma^{1/2}$, and $n + p = 2\gamma^{-1/2}n_i$, stronger $D^* \sim n_i^{-2}$ dependence can be expected.

$$D^* \sim \frac{\left(\tau_{A1}^i\right)^{1/2}}{n_i} \sim n_i^{-2} \sim \exp\left(\frac{E_g}{kT}\right) = \exp\left(\frac{hc/\lambda_{co}}{kT}\right). \quad (17.35)$$

Figure 17.20a shows the calculated detectivity of Auger GR-limited $Hg_{1-x}Cd_xTe$ photodetectors as a function of wavelength and temperature of operation [168]. The calculations have been performed for 10^{14}cm^{-3} doping, which is the lowest donor doping level that, at present, is achievable in a controllable manner, in practice. Values as low as $\approx 1 \times 10^{13}$cm^{-3} are at present achievable in labs, while values of 3×10^{14}cm^{-3} are more typical in industry. LN cooling potentially makes it possible to achieve background-limited infrared photodetector (BLIP; 300 K) performance over the entire 2–20 μm range. The 200 K cooling, which is achievable with Peltier coolers, would be sufficient for the BLIP operation in the middle and SW regions (<5 μm).

The detector performance can be improved further by the use of optical immersion. However, the theoretical limit of detectivity for uncooled detectors remains about one or almost two orders of magnitude below D^*_{BLIP}(300 K, 2π) for ≈5 μm and 10 μm wavelengths, respectively. Improvement by a factor of ≈2 is still possible with optimum p-type doping.

17.4.2 NONEQUILIBRIUM DEVICES

Auger generation appeared to be a fundamental limitation to the performance of IR photodetectors. However, British workers [169,170] proposed a new approach to reducing the photodetector cooling requirements, which is based on the nonequilibrium mode of operation. This is one of the most exciting

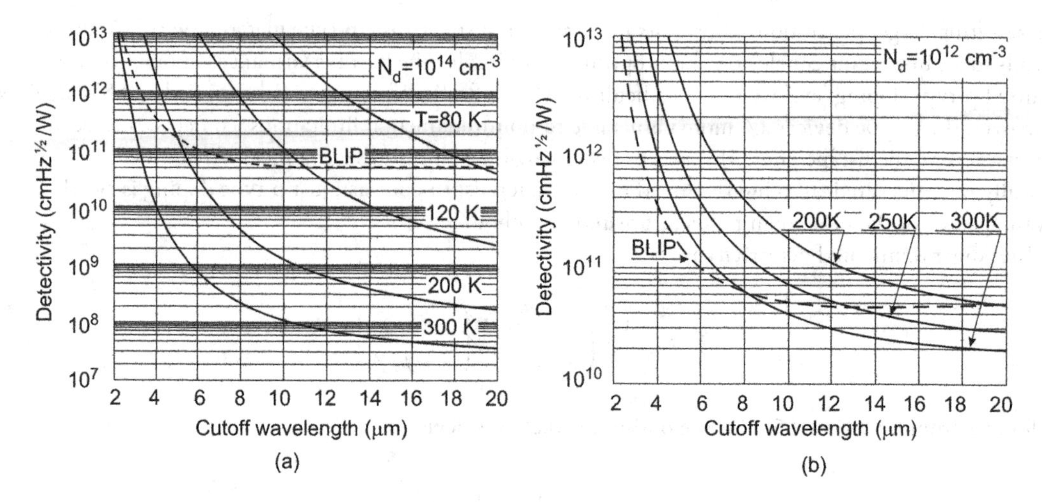

Figure 17.20 Calculated performance of (a) equilibrium and (b) nonequilibrium Auger GR-limited $Hg_{1-x}Cd_xTe$ photodetectors as a function of wavelength and temperature of operation. BLIP detectivity has been calculated for 2π field of view (FOV), $T_B = 300$ K, η = 1.

events in the field of IR photodetectors operating without cryogenic cooling. Their concept relies on dependence of the Auger processes on the concentration of free carriers. The suppression of Auger processes may be achieved by decreasing the free carrier concentration below equilibrium values. Nonequilibrium depletion of semiconductors can be applied to reduce concentration of the majority and minority carrier concentration. This can be achieved in some devices based on lightly doped narrow-gap semiconductors, for example, in biased low-high (l-h) doped or heterojunction contact structures, MIS structures, or using the magnetoconcentration effect. Under strong depletion, the concentration of both majority and minority carriers can be reduced below the intrinsic concentration. The majority carrier concentration saturates at the extrinsic level, while the concentration of minority carriers is reduced below the extrinsic level. Therefore, the necessary condition for deep depletion is a very light doping of the semiconductor, below the intrinsic concentration.

Let us discuss the fundamental limits of performance of the Auger-suppressed devices. At first, consider a detector based on ν-type material. At strong depletion, $n = N_d$, and the Auger generation rate is

$$G_A = \frac{N_d}{2\tau_{A1}^i}. \tag{17.36}$$

As follows from Equation 17.36, deep depletion makes the recombination rate negligible compared to the generation rate, so it is readily eliminated as a noise source in depleted materials. Therefore,

$$D^* = \frac{\lambda}{hc}\frac{\eta}{t^{1/2}}\left(\frac{\tau_{A1}^i}{N_d}\right)^{1/2}. \tag{17.37}$$

Similarly, for π-type materials,

$$G_A = \frac{N_a}{2\gamma\tau_{A1}^i}, \tag{17.38}$$

and

$$D^* = \frac{\lambda}{hc}\frac{\eta}{t^{1/2}}\left(\frac{\tau_{A1}^i}{N_a/\gamma}\right)^{1/2}. \tag{17.39}$$

Again, as in the equilibrium case, the use of π-type material is advantageous, improving the detectivity by a factor of $\gamma^{1/2}$ as compared to ν-type material of the same doping ($\gamma > 1$).

Comparing the corresponding equations for equilibrium and nonequilibrium modes, we can find that the use of the nonequilibrium mode of operation may reduce the Auger generation rate by a factor n_i/N_d in lightly doped material, with a corresponding improvement of detectivity by $(2n_i/N_d)^{1/2}$. The additional gain factor of $2^{1/2}$ is due to the negligible recombination rate in the depleted semiconductor. The gain for p-type material is even larger—a factor of $[2(\gamma + 1)n_i/N_a]^{1/2}$, taking into account the elimination of A1 and A7 recombinations. Additional depletion-related improvements can also be expected from the increased absorption due to reduced band-filling effect.

The resulting improvement may be quite large, particularly for LWIR devices operating at near room temperatures, as shown in Figure 17.20b for very low doping, that is, 10^{12} cm^{-3}. Potentially, the BLIP performance can be obtained without cooling at all. The BLIP limit can be achieved by [168]:

- using materials with controlled doping at very low levels ($\approx 10^{12}$ cm^{-3})
- using extremely high-quality materials with very low concentrations of SR centers
- proper design of a device that prevents thermal generation at surfaces, interfaces, and contacts
- using a thermal dissipation device whose design makes it possible to achieve a state of strong depletion.

The requirements for the BLIP performance, particularly doping concentration, can be significantly eased by the use of optical immersion.

17.5 PHOTOCONDUCTIVE DETECTORS

The first results on photoconductivity in $Hg_{1-x}Cd_xTe$ were reported by Lawson et al. in 1959 [1]. Ten years later, in 1969, Bartlett et al. reported the background-limited performance of photoconductors operated at 77 K in the 8–14 μm LWIR spectral region [10]. The advances in material preparation and detector technology have led to devices approaching theoretical limits of responsivity over wide ranges of temperature and background [11,171–173]. The largest market was for the 60-, 120-, and 180-element units in the Common Module military thermal imaging viewers. Photoconductivity was the most common mode of operation for 3–5 um and 8–14 μm $Hg_{1-x}Cd_xTe$ n-type photodetectors for many years.

In 1974, Elliott reported a major advance in IR detectors in which the detection, time delay, and integration functions in serial scan thermal imaging systems were performed within a simple three-lead filament photoconductor, SPRITE [134].

The further development of photoconductors is connected with the elimination of deleterious effect of sweep-out [174,175] by the application of accumulated [176–178] or heterojunction [179,180] contacts. Heterojunction passivation has been used to improve stability [181,182]. The operation of 8–14 μm photodetectors has been extended to ambient temperatures [168,183–186]. Means applied to improve the performance of photoconductors operated without cryogenic cooling include the optimized p-type doping, the use of optical immersion, and optical resonant cavities. Elliott and other British scientists introduced Auger-suppressed excluded photoconductors [169,170].

The research activity on photoconductors has been significantly reduced in the last three decades, reflecting maturity of the devices. At the same time, $Hg_{1-x}Cd_xTe$ photodetectors are still manufactured and used in many important applications.

The physics and principle of operation of intrinsic photoconductors are summarized in Section 12.1.1. HgCdTe photoconductive detectors have been reviewed by a number of authors [134,153,168,183,187–191].

17.5.1 TECHNOLOGY

The photoconductors can be prepared either from $Hg_{1-x}Cd_xTe$ bulk crystals or epilayers. Figure 17.21 shows a typical structure of photoconductor. The main part of all the structures is a 3–20 μm flake of $Hg_{1-x}Cd_xTe$, supplied with electrodes. The optimum thickness of the active elements (few μm) depends upon the wavelength and temperature of operation and is smaller in uncooled, LW devices. The front side surface is usually covered with a passivation layer and AR coating. The back-side surface of the device is also being passivated. In contrast, the back side surface of the epitaxial layer grown on the CdZnTe substrate does not require any passivation since increasing bandgap prevents reflection of minority carriers. The devices are bonded to heat conductive substrates.

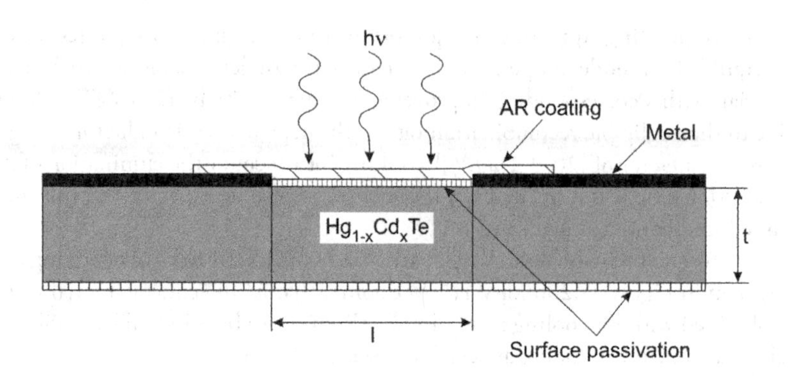

Figure 17.21 Typical structure of HgCdTe photoconductor.

To increase the absorption of radiation, the detectors are frequently supplied with a gold back reflector [183,187] insulated from the photoconductor with a ZnS layer or substrate. The thicknesses of the semiconductor and the two dielectric layers are frequently selected to establish the optical resonant cavities with standing waves in the structure with peaks at the front and nodes at the back surfaces. For effective interference, the two surfaces must be sufficiently flat.

Numerous fabrication procedures are being used by various manufacturers [181,183,185,192–194]. Technological methods of modern microelectronics are used in $Hg_{1-x}Cd_xTe$ photoconductor manufacturing, however extreme care is necessary to prevent any mechanical and thermal damage of the material. Fabrication usually starts from selection of the starting materials, which are $Hg_{1-x}Cd_xTe$ wafers or epilayers. For selection, they are typically characterized by composition, doping, and minority lifetime mapping.

The key processing steps for manufacturing $Hg_{1-x}Cd_xTe$ photoconductors from bulk $Hg_{1-x}Cd_xTe$ are:

- Preparation of back-side surface of the $Hg_{1-x}Cd_xTe$ wafer. The procedure involves careful polishing of one side of the $Hg_{1-x}Cd_xTe$ slab with fine (0.3–1 μm) alumina powder, cleaning in organic solvents, followed by etching in 1%–10% bromine in methanol solution for several minutes and washing in methanol. Alternatively, various chemical-mechanical polishing procedures can be used. The back-side preparation is completed with passivation, which differs for n- and p-type materials.
- Bonding to a substrate. Sapphire, Ge, Irtran 2, Si, and alumina ceramics are the most common substrates for bulk-type photoconductors. Epoxy resin is generally used for the bonding of $Hg_{1-x}Cd_xTe$ wafers to substrates. The thickness of the epoxy layer should be kept below 1 μm for good heat dissipation.
- Thinning the slab to its final thickness and preparation of the front side surface. This is done by lapping, polishing, and etching, followed by surface passivation and an antireflection coating (usually ZnS). The individual elements are then delineated with wet or dry etching with the use of photolithography. The side walls of active elements are also often passivated.
- Electric contact preparation. Vacuum evaporation, sputtering, electroplating, and galvanic or chemical metallization after further photolithography is used. External contacts are supplied by gold wire ultrasonic bonding, conductive epoxy bonding, or by soldering with indium. Expanded contact pads are sometimes used to prevent damage of the semiconductor.

Preparation of a photoconductor from epilayers is more straightforward, as laborious thinning to a very low thickness and back-side preparation is not required. The use of ISOVPE [183,195], LPE [196–199], MOCVD [184,185,200–203], and MBE [204] photoconductors have been reported. The CdZnTe, which is typically used for substrates, has relatively poor thermal conductivity. Therefore, for the best heat dissipation, the substrates must be thinned below 30 μm and the photoconductors must be fixed to a good heat sinking support. The heat dissipation is much easier in small size devices ($<50 \times 50\,\mu m^2$) in which 3D dissipation is significant and no thinning of substrates is necessary. Another solution is to use the epilayers deposited on good heat conducting materials, such as sapphire, Si, and GaAs.

The low-temperature epitaxial techniques make it possible to grow complex multilayer photoconductor structures that can be used as multicolor devices or devices with shaped spectral response [184,203].

Passivation is one of the most critical steps in the preparation of photoconductors. The passivation must seal the semiconductor, stabilizing it chemically, and often it also acts as an antireflection coating. An excellent review of $Hg_{1-x}Cd_xTe$ native and deposited insulator layers has been published by Nemirovsky and Bahir [205]. Passivation of n-type materials is commonly performed with the use of anodic oxidation in 0.1 N KOH in a 90% water solution of ethyl glycol [206–208]. Typically, 100 nm thick oxide layers are grown. Good interface properties of the n-$Hg_{1-x}Cd_xTe$-oxide interface are due to accumulation (10^{11}–10^{12} electrons per cm^{-2}) of the semiconductor surface during oxidation. Passivation by pure chemical oxidation in an aqueous solution of $K_3Fe(CN)_6$ and of KOH is also used [209]. Dry methods of growing native oxide have been attempted, such as plasma [210] and photochemical oxidation [211]. Passivation can be improved by overcoating with ZnS or SiO_x layers [212]. Another approach to passivation is based on direct accumulation of the surface to repel minority holes by a shallow ion milling [213,214].

Passivation of p-type materials is of strategic importance for near room temperature devices based on p-type absorbers. It should be admitted that the passivation still presents practical difficulties. Oxidation is not useful for p-type $Hg_{1-x}Cd_xTe$ because it causes inversion of the surface. In practice, sputtered or electron beam evaporated ZnS, with an option of a second layer coating [215], is usually used for the passivation of p-type materials. Native sulfides [216] and fluorides [217] have also been proposed.

The use of CdTe for passivation is very promising since it has high resistivity, is lattice matched, and is chemically compatible to $Hg_{1-x}Cd_xTe$ [218–220]. Excellent passivation can be obtained with a graded CdTe-$Hg_{1-x}Cd_xTe$ interface [215]. Barriers can be found in both the conduction and valence band. The best heterojunction passivation can be obtained during epitaxial growth [221]. Directly grown in situ CdTe layers lead to low fixed interface charge. The indirectly grown CdTe passivation layer is not as good as the directly grown, but it is acceptable in some applications. Low thickness (10 nm) of CdTe is recommended in some papers to prevent $Hg_{1-x}Cd_xTe$ lattice stress [205].

Contact preparation is another critical step. Evaporated In has been used for a long time for contact metallization to n-type material [187,192]. Multilayer metallization, Cr-Au, Ti-Au, Mo-Au, is more frequently used at present. Metallization is often preceded by a suitable surface treatment. Ion milling was found very useful to accumulate n-type surfaces and it seems to be the most preferable surface treatment prior to metallization of n-type material. Chemical and dry etch are also used. Preparation of good contacts to p-type material is more difficult. Evaporated, sputtered, or electroless deposited Au and Cr-Au are the most frequently used for contacts to p-type materials.

17.5.2 PERFORMANCE OF PHOTOCONDUCTIVE DETECTORS

17.5.2.1 Devices for operation at 77 K

HgCdTe photoconductive detectors operating at 77 K in the 8–14 μm range are widely used in the first generation thermal imaging systems in linear arrays of up to 200 elements, although custom 2D arrays up to 10 × 10 have been made for unique applications. The production processes of these devices are well established. The material used is n-type with an extrinsic carrier density of about $1 \times 10^{14} cm^{-3}$. The low hole diffusion coefficient makes n-type devices less vulnerable to contact and surface recombination. In addition, n-type materials exhibit a lower concentration of SR centers and there are good methods of surface passivation.

Commercially available HgCdTe photoconductive detectors are typically manufactured in a square configuration with active size from 25 μm to 4 mm. The length of the photoconductors used in high-resolution thermal imaging systems (≈50 μm) is typically less than the minority carrier diffusion and drift length in cooled HgCdTe, resulting in the reduction of photoelectric gain due to diffusion and drift of photogenerated carriers to the contact regions, called sweep-out effect [134,174,175,177,187,188,222]. This causes the saturation of response with increasing electric field. The behavior of a typical device, showing the saturation in responsivity (at about 10^5 V/W) is shown in Figure 17.22 [134].

The n-type HgCdTe photoconductive detectors (with $E_g \approx 0.1$ eV at 77 K) approaching theoretical limits of performance have been described by Kinch et al. [11,171,172], Borello et al. [173], and Siliquini et al. [181,199,223]. Their generation and recombination carrier mechanisms are clearly dominated by the A1 mechanism. Background radiation has a decisive influence on performance since the concentration of

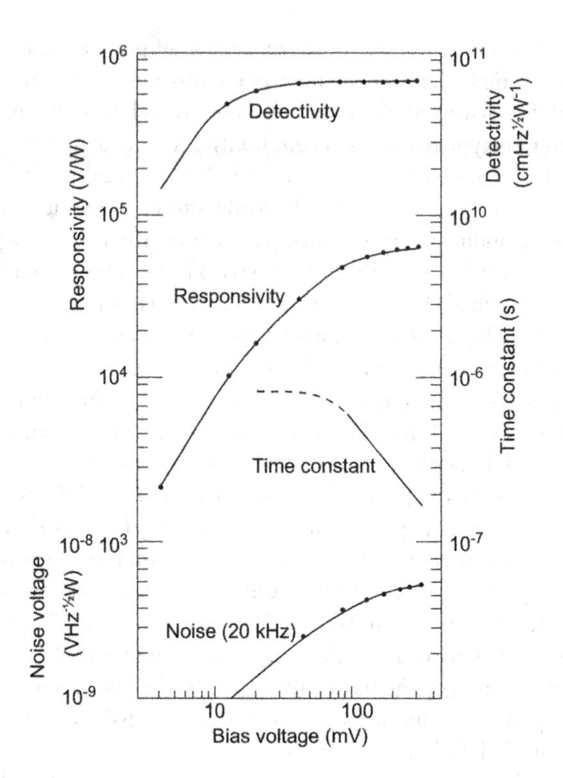

Figure 17.22 Characteristics of a 50 μm HgCdTe photoconductive detector operated at 80 K as a function of voltage. The measurements were made in 30° FOV and the responsivity values refer to the peak wavelength response at 12 μm. (From Ref. [134])

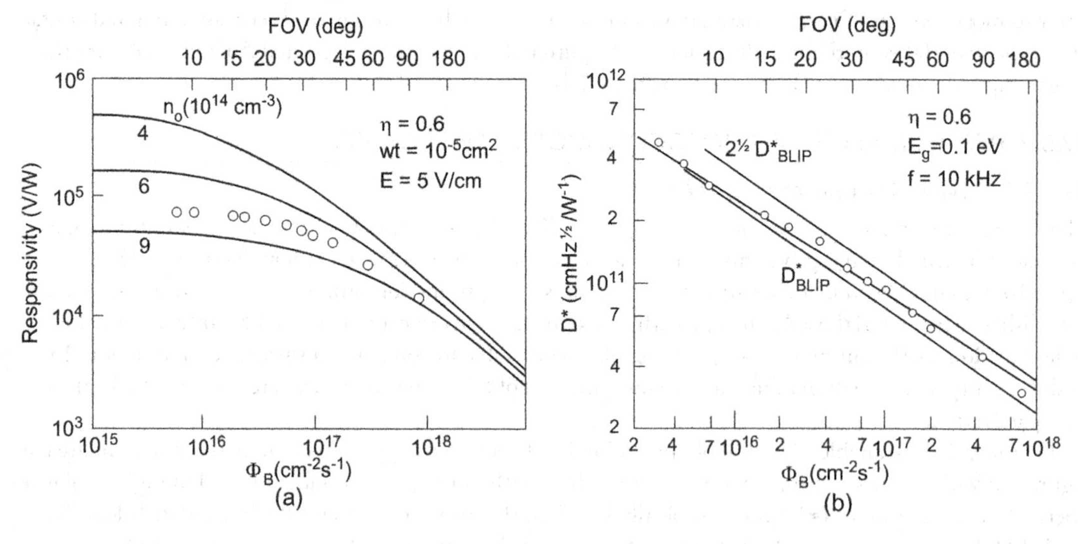

Figure 17.23 Dependence of voltage (a) responsivity and (b) detectivity on background photon flux for a 0.1 eV HgCdTe photoconductive detector. (From Ref. [176])

both majority and minority carriers in 77 K, 8–14 μm devices and the concentration of minority carriers in 3–5 μm devices are typically determined by background flux. Near-BLIP performance can also be achieved at elevated temperatures up to about 200 K [172,224]. Figure 17.23 shows the influence of 300 K background photon flux on photoconductor parameters [173]. The density of background-generated holes and, for high fluxes, also electrons may dominate the thermally-generated carriers, decreasing the recombination

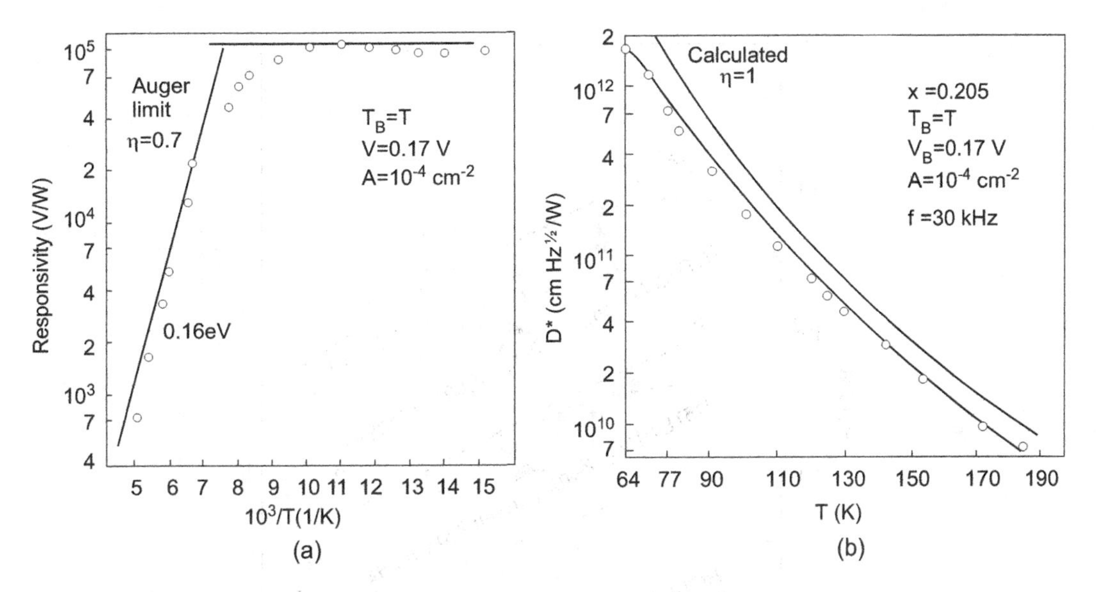

Figure 17.24 Measured and calculated (a) responsivity and (b) detectivity versus temperature for a $Hg_{0.795}Cd_{0.205}Te$ photoconductor. (From Ref. [11])

time. The effects of background radiation tend to override any nonuniformity that might be present in the bulk material with regard to element resistance, responsivity, and noise.

Figure 17.24 shows the calculated and measured low-background responsivity and detectivity of a photoconductor as a function of temperature [11]. The generation and recombination rates are clearly dominated by the A1 mechanism. The 77 K detectivity achieves a value of about 10^{12} cmHz$^{1/2}$/W, closely approaching the limits predicted by theory.

The dependence of the voltage responsivity on detector length for 8–14 μm HgCdTe photoconductors operated at 77 K, 200 K, and 300 K is shown in Figure 17.25 [222]. The marked region inside the figure indicates the ranges of voltage responsivity for the detector series produced by Judson Infrared, Infrared Associates, and Vigo-System. Improved performance of devices operated at 200 K and 300 K are achieved assuming p-type doping materials with hole concentrations $p \approx \gamma^{1/2} n_i$. In case of ohmic contacts (Rittner model), sweep-out effect significantly reduces responsivity of the detector with short active sizes operated at 77 K. This effect is negligible at 300 K. In the range of detector lengths below 100 μm, the experimental results exceed theoretical calculations based on the Rittner model. It may be conditioned by intentional or coincidental processing procedures in which contacts deviate from ohmicity. Application of a high-low doping contact barrier leads to enhancement in responsivity of photoconductive detectors. Ashley and Elliott [177] have shown that by means of a n^+-n ion milled contact the responsivity can be enhanced by a factor of five.

The positive feature of sweep-out is an improvement of the high-frequency characteristics. In the high-bias condition, the response time is determined by the transit time of the minority carriers between the electrodes rather than by the excess carrier lifetime. Since the recombination processes are partially arranged to take place in the contact region, they do not contribute to recombination noise. As a result, the GR noise decreases by a factor of $2^{1/2}$ and the sweep-out/GR-limited devices may exhibit improved detectivity by the same factor. Under high bias, D^*_{BLIP} is identical to the photovoltaic case.

The reduced gain may, however, cause the Johnson-Nyquist noise to dominate with deleterious effect to detectivity. Sweep-out has been recognized as a major limitation of the 8–14 μm photoconductor performance, when they are short and are operated at low temperatures with low background radiation. The influence of sweep-out is even stronger for shorter wavelength devices. For $\lambda < 5$ μm photoconductors with low GR rates, these effects become significant at low and elevated temperatures, even for relatively long devices, making the devices Johnson-Nyquist/sweep-out limited.

Various approaches to reduce the undesirable sweep-out effects in devices with small sensitive areas have been suggested. Kinch et al. [172] proposed an overlap structure geometry in which the device

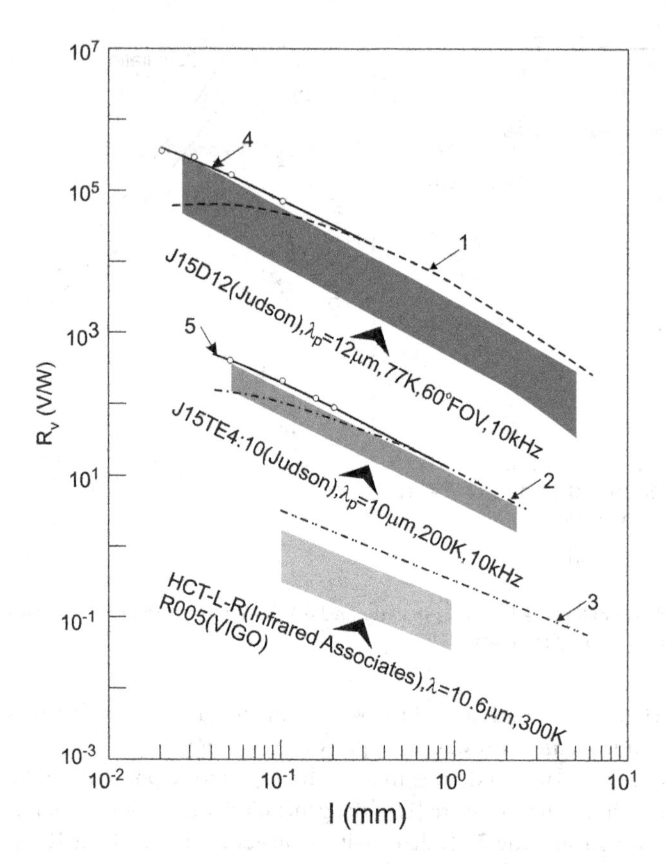

Figure 17.25 Dependence of the voltage responsivity on detector length for 8–14 μm HgCdTe photoconductive detectors operated at 77 K, 200 K, and 300 K. The marked regions indicate ranges of voltage responsivity for detector series produced by different manufacturers. The theoretical curves 1–3 are calculated using the Rittner model (see Section 12.1.1); the curves 4 and 5 are calculated assuming a high-low contact regions. (From Ref. [222])

length is greater than the required sensitive length. The end regions are covered with an opaque layer. By this means the effects of carrier diffusion to the contacts, at low bias, or minority-hole sweep-out at the cathode, at high bias, are reduced. The theory of the overlap structure was reported by Smith [225]. According to Shacham-Diamond and Kidron [176], partial blocking of excess carriers occurs at the cathode of n-type photoconductors, thus enhancing the current responsivity and reducing the time response. The n^+-n contact can be used to isolate regions containing minority holes from high recombination rate regions [177–179]. Due to degeneracy in the n^+ region, the potential barrier to holes is much larger than the simple Boltzmann factor $(kT/q)\ln(n/n_o)$. The effective recombination velocity at the blocking contact below 100 cm/s can be achieved. For this, the electron concentration must be degenerate within the distance of 1 μm or less.

The carrier sweep-out can also be virtually eliminated with the heterojunction contact photoconductor proposed by Smith et al. [180,226]. His calculations predict elimination of the saturation of responsivity and a very large increase in responsivity. The responsivity increase of an order of magnitude using a heterojunction contact to the HgCdTe with $x = 0.20$ has been obtained [226].

Also, combination of different approaches, such as the combined overlap/blocking contact device structure [199] or heterostructure photoconductor with blocking contacts [223], has been used to improve the device performance. The second type of devices is insensitive to the condition of the semiconductor/passivant interface.

The accumulation and heterojunction contacts for the reflection of minority carriers are equally applicable to reduce the recombination at the active and back side surface of photoconductors. Excessive accumulation, on the other hand, can lead to a large shunt conductance, which can also degrade detectivity [227].

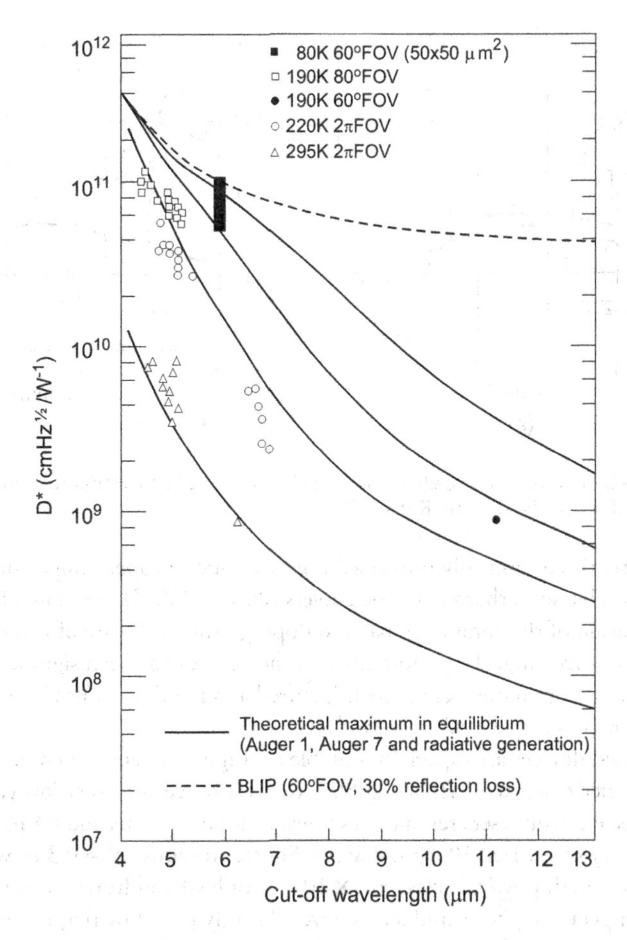

Figure 17.26 Detectivity versus cutoff wavelength for n-type HgCdTe photoconductive detectors. The theoretical curves are calculated including Auger generation and radiative generation only. The experimental points are for 230 μm square n-type detectors, except where indicated. (From Ref. [134])

17.5.2.2 Devices for operation above 77 K

The performance of HgCdTe photoconductors is reduced at higher temperatures. For many applications, however, there are significant advantages in accepting this fact; for example, the input power for cooling engines can be reduced and their life can be extended, the operation temperature above 180 K can be achieved with thermoelectric coolers.

The carrier lifetimes at higher temperatures are short, being fundamentally limited by Auger processes, and the GR noise-limited performance is obtained [134,136]. Since $\gamma > 1$ (see Equation 17.32), there is in principle an advantage in using p-type material. In practice, however, p-type photoconductors are difficult to passivate and low-$1/f$ noise contacts are difficult to form. For these reasons, the majority of devices for the higher temperature operation are n-type. Figure 17.26 shows examples of the detectivity as a function of cutoff wavelength, obtained from 230 μm square n-type devices operated at different temperatures. For comparison, theoretical limiting detectivity is shown assuming an extrinsic concentration of 5×10^{14} cm^{-3}, thickness of 7 μm, reflection coefficient at front and back surface of 30%, and $f/1$ optics [134].

The p-type HgCdTe photoconductors are used as laser receivers, where the bandwidth is usually high and $1/f$ noise is unimportant. Intermediate temperature operation of p-type devices in the LWIR region has been reported by several authors [134,183,191,194,228–232]. The measured detectivity at 193 K at relatively high modulation frequency of 20 kHz was 7×10^8 cmHz$^{1/2}$W^{-1}. The minimum value of the NEP (noise equivalent power) for heterodyne detection was observed to be about 1×10^{-19} WHz^{-1} with a bandwidth of 100 MHz [232].

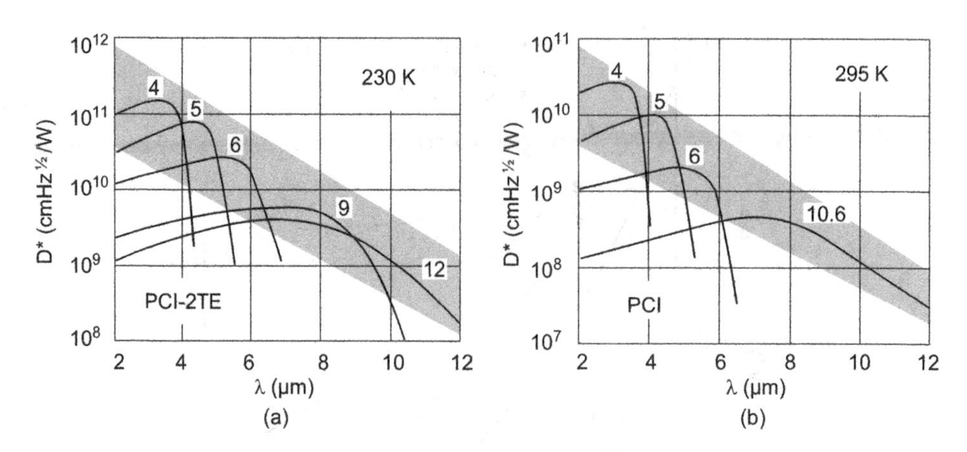

Figure 17.27 Spectral detectivities of optically immersed $Hg_{1-x}Cd_xTe$ photoconductors: (a) uncooled and (b) cooled with two-stage Peltier coolers. (From Ref. [185])

The measured detectivities of optically immersed photoconductors operating at ambient temperature or at temperatures achievable with thermoelectric coolers (200 K–250 K) are shown in Figure 17.27 [185]. Due to careful optimization of the compositional and doping profile, the use of metal back reflectors, better surfaces, and contact processing, the performance of the devices has been significantly improved. The operation of two-stage cooled photodetectors manufactured at Vigo-System has been extended to ≈16 μm with detectivities of ≈2×10^9 cmHz$^{1/2}$/W at 12 μm [168].

The optically immersed devices are especially suitable for high-frequency operation due to a short lifetime, absorber resistance close to 50 Ω, negligible series resistance, and very low capacitance. The measurements using free electron lasers revealed response times of ≈0.6 ns and ≈4 ns at 300 K and 230 K, respectively, for detectivity-optimized 10.6 μm devices. Shorter response of ≈ 0.3 ns was observed in specially designed devices of small physical sizes (≈10×10 μm^2 or less) and heavier p-type doping [168].

The present high-temperature photoconductors have relatively poor low-frequency properties. Typically, the $1/f$ knee frequencies of 10 kHz have been observed in uncooled 10.6 μm detectors at electrical fields of ≈40 V/cm (typical large size Vigo R005 photoconductors). The Hooge constant of about ≈10^{-4} has been deduced from the low-field measurements. Rapid, nonlinear increase of the Hooge constant has been observed for stronger electric fields. Cooling to ≈ 200 K reduces the Hooge constant by a factor of ≈2, which in conjunction with much lower electrical fields requirements makes it possible to achieve the GR-limited operation at frequencies of ≈1000 Hz and higher. Proportionality between GR and $1/f$ noise in $Hg_{0.8}Cd_{0.2}Te$ photoconductors have been observed in a wide temperature range 77 K–250 K [233]. The reason for the poor low-frequency performance is fully understood at present; inadequate surface passivation and contact technology is usually blamed. No one existing theory can explain qualitatively the observed low-frequency noise [234].

It should be noted that the measured detectivity of uncooled ≈10 μm photoconductors is many orders of magnitude higher compared to other ambient-temperature 10.6 μm detectors with subnanosecond response time, such as photon drag detectors, fast thermocouples, bolometers, and pyroelectric ones.

17.5.3 OTHER MODES OF PHOTOCONDUCTOR OPERATIONS

17.5.3.1 Trapping-mode photoconductors

If the minority carrier (hole) is somehow trapped before it can be swept out at the negative electrode, then the electron current keeps flowing for a longer time, which increases the photoconductive gain. Significant improvements in the gain of photoconductors have been realized by the development of trapping-mode HgCdTe detectors in the 1980s [235,236]. The device structure and its energy bandgap profile are shown in Figure 17.28. LPE was used to grow these structures on CdTe substrates. The post grown low-temperature annealing forms n-type lightly doped detector active layer ($n \approx 10^{14}$ cm^{-3}), while retaining the p-type layer with the p-type trapping region within a compositionally graded interface between the

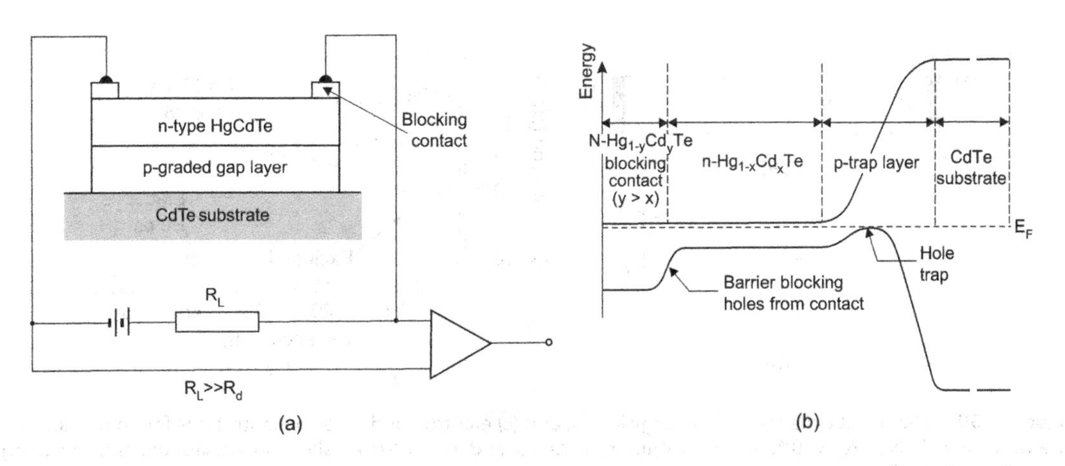

Figure 17.28 Trapping-mode HgCdTe photoconductive detectors: detector structure and its biasing (a) and band diagram (b). (From Ref. [235])

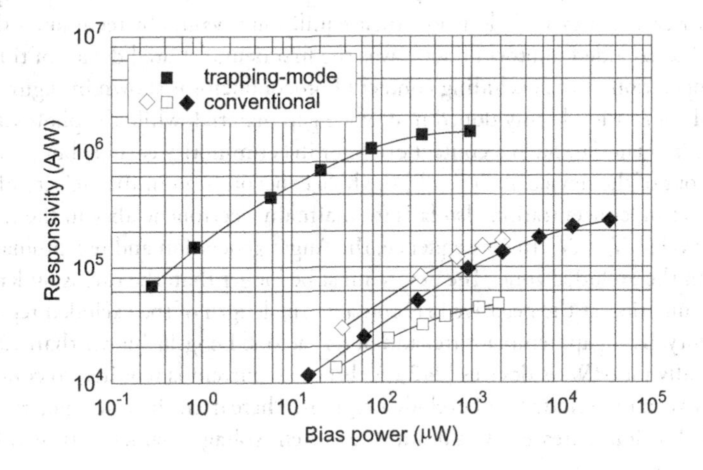

Figure 17.29 Comparison of the bias dependence of responsivity of trapping-mode HgCdTe devices with conventional photoconductive detectors at 80 K. The presented data are for devices with dimensions of about 50 × 50 μm² and 12 μm cutoff wavelengths. The impedance of all devices is of the order of 100 Ω. (From Ref. [236])

epitaxial layer and the substrate, underlining the entire n-type layer. Minority-carrier holes are separated from majority-carrier electrons by the junction, producing a reduction in the depletion layer width. Due to low electron concentration, the depletion layer width at the p-n junction is substantial and the tunneling leakage current is minimal. Large photoconductive gain (of the order of 1,000 to 2,000) results both from the existing hole trap region and the blocking N-n HgCdTe interface near the contacts.

Figure 17.29 compares the responsivity of 12 μm cutoff conventional and trapping-mode HgCdTe photoconductors at 80 K [236]. The presented data are for devices with dimensions of about 50 × 50 μm². The impedance of all the devices is of the order of 100 Ω. Trapping-mode devices have at least two orders of magnitude lower bias requirements to achieve 10^5 V/W responsivity—of the order of 0.12 W/cm² compared with 12 W/cm² [237]. This two orders of magnitude lower bias power greatly reduces the bias heat load for large multielement arrays.

A further benefit of trapping-mode devices is significantly lower $1/f$ noise [238]. In conventional HgCdTe photoconductive detectors, the $1/f$ noise knees are typically 1 kHz, but in their high frequency counterparts at 80 K and for $f/2$ background flux conditions, they are only of the order of a few hundred hertz.

17.5.3.2 Excluded photoconductors

Elliott and other English workers [134,153,169,170,188,239–244] proposed and developed a new approach to reducing the photodetector cooling requirements, which is based on the nonequilibrium mode of

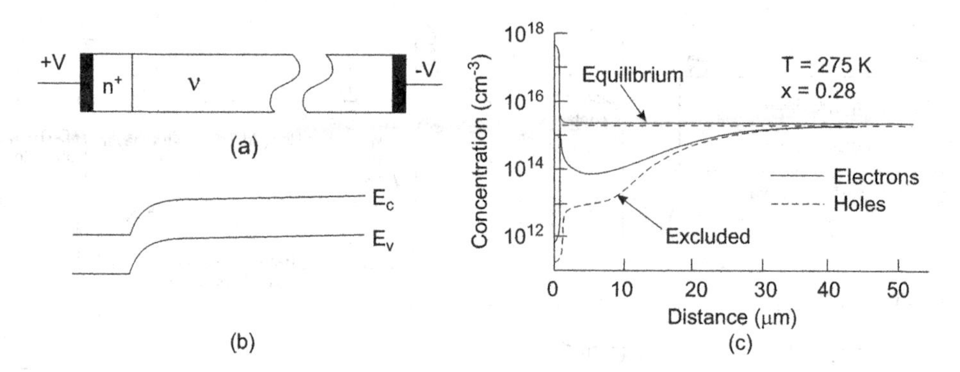

Figure 17.30 (a) Schematic diagram, (b) energy levels, and (c) electron and hole concentrations for an n⁺-v structure with $x = 0.28$, $N_d - N_a = 10^{14}$cm^{-3}, $\tau_{SR} = 4\,\mu$m, $\tau_{Ai} = 2.4\,\mu$s, and $J = 48$ A/cm^2, showing equilibrium and excluding levels. (From Ref. [170])

operation. It appears that the Auger generation process with its associated noise can be suppressed in devices where the carrier densities are held below their equilibrium values by techniques that do not involve large electric fields. The excluded photoconductor was the first demonstrated device of this type.

The principle of operation of the excluding contact photoconductor is shown in Figure 17.30 [170]. The positively biased contact is a highly doped n⁺ or wide gap material, while the photosensitive area is a near-intrinsic n-type (v) material. Such a contact does not inject minority carriers but permits the majority electrons to flow out of the device. As a result, the hole concentration in the vicinity of the contact is decreased and the electron concentration also falls (to maintain electroneutrality in the region) to a value close to the extrinsic value $N_d - N_a$. In consequence, the Auger generation and recombination processes become suppressed in the excluded zone. The device must be longer than the exclusion length to avoid the effect of carrier accumulation at the negative bias contact. The length of the excluded region depends on the bias current density, bandgap, temperature, and other factors. Lengths higher than 100 μm have been observed experimentally for MWIR devices [242]. A threshold current is required to counter the back diffusion current between unexcluded and excluded regions. Thereafter, the resistance rises rapidly as the length of the excluded region increases. As a result, the current-voltage characteristic exhibits saturation for above-threshold currents.

The accurate analysis of nonequilibrium devices can be done only by numerical solution of the full continuity equation for electrons and holes (see Section 4.4). The usual approximations break down in the case of a nonequilibrium mode of operation due to large departures from equilibrium.

The practical realization of nonequilibrium devices depends on several important limitations. The electric field in an excluded region must be sufficiently low to avoid heating a device as a whole and the electrons above the lattice temperature. The heating of the structure can be prevented by a proper heat sink design, and it seems not to be a serious limitation—at least for a single element and low-area devices. Electron heating sets a maximum field that has been estimated as 1,000 V/cm in materials for uncooled 5 μm devices and a few hundred V/cm in materials for ≈10 μm devices operated at 180 K. Electron heating is not an important constraint in the 3–5 μm band, but it seems to restrict the usefulness of exclusion at 10.6 μm and longer wavelengths in the 8–14 μm band. Very low doping ($<10^{14}$cm^{-3}) is required for effective exclusion; however, the values 3×10^{14}cm^{-3} are typical in industry. Exclusion may be inhibited by any non-Auger generation such as the SR or surface generation. A large electrical field may result in flicker noise.

Practical excluded HgCdTe photoconductors have been fabricated from low-concentration, bulk-grown material with the n⁺-regions formed by ion milling [170,186]. In contrast to the equilibrium-mode photoconductors that are usually based on extrinsically p-type doped material, the excluded devices are fabricated from very low-concentration, n-type bulk HgCdTe with the n⁺ regions formed by ion milling or degenerate extrinsic doping. The device is schematically shown in Figure 17.31 [170]. In order to avoid the effect of accumulation at the negative contact, the devices are three-lead structures with the sensitive area defined by an opaque mask, and a side-arm potential probe is used for a readout contact. Such device geometry limits the size of the active area to the highly depleted region, which prevents thermal generation

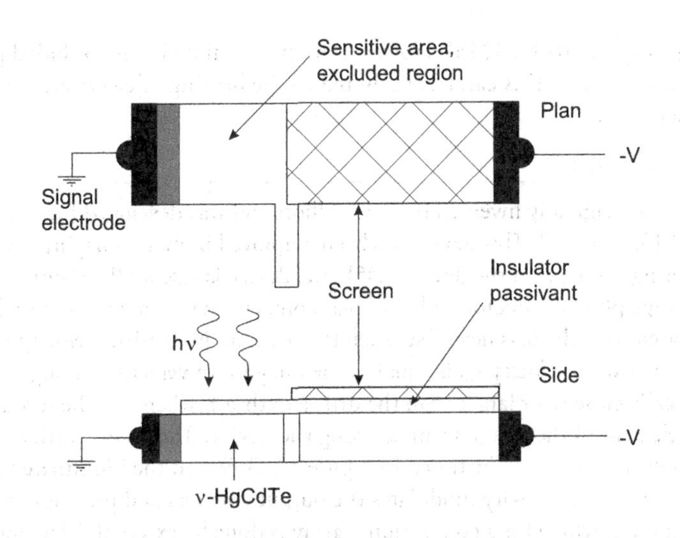

Figure 17.31 Schematic of a three-lead excluding HgCdTe photoconductive detector. (From Ref. [170])

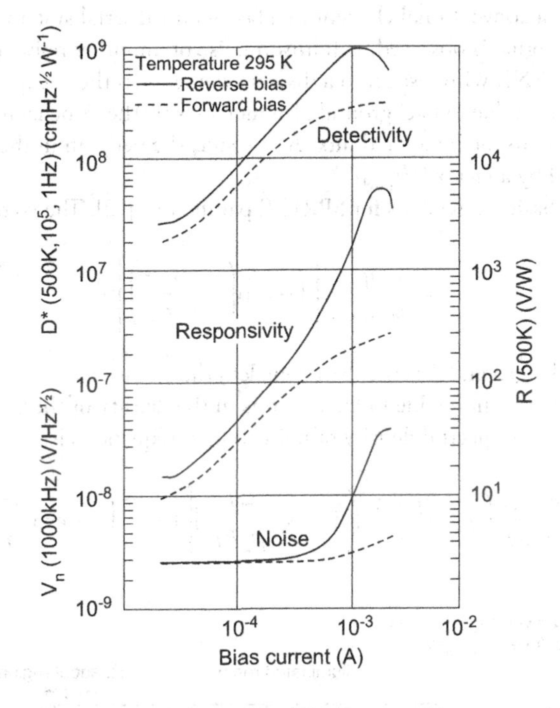

Figure 17.32 Noise, responsivity (500 K), and detectivity (500 K) versus bias current for $Hg_{0.72}Cd_{0.28}Te$, excluding photoconductor, measured at 295 K. (From Ref. [134])

in the nondepleted part and at the negative contact by contributing to the noise measured at the readout electrode. ZnS is used for passivation as the usual native oxide passivation produces an accumulated surface, which shunts the excluded region.

The detector parameters, noise, responsivity, and detectivity, are shown for both directions of bias current in Figure 17.32 [134]. The responsivity and noise increase to large values in the excluded direction of bias due to two effects: the increased impedance in the excluded region and the increase in the effective carrier lifetime-to-transit time. The improvement of detectivity is more modest due to high flicker noise levels at reverse bias. An uncooled 10 μm × 10 μm photoconductor with 4.2 μm cutoff has exhibited a 500 K blackbody voltage responsivity of 10^6 V/W and a detectivity of 1.5×10^9 cm $Hz^{1/2}$/W at a modulation frequency of 20 kHz [243]. However, no practical 8–14 μm excluded photoconductors have been demonstrated to date. This can be attributed to the heating of electrons at the high electric fields required for exclusion.

17.5.4 SPRITE DETECTORS

The SPRITE detector was originally invented by T. C. Elliott and was developed further almost exclusively by British workers [12,13,245–257]. This device has been employed in many imaging systems [255]. Figure 17.33 shows the operating principle of the device [245]. The device is essentially ≈1-mm-long, 62.5-μm-wide, and a 10-μm-thick n-type photoconductor with two bias contacts and a readout potential probe. The device is constant current biased with the bias field E set such that the ambipolar drift velocity v_a, which approximates to the minority hole drift velocity v_d, is equal to the image scan velocity v_s along the device. The length of the device L is typically close to or larger than the drift length $v_d\tau$, where τ is the recombination time.

Consider now an element of the image scanned along the device. The excess carrier concentration in the material increases during scan, as illustrated in Figure 17.33. When the illuminated region enters the readout zone, the increased conductivity modulates the output contacts and provides an output signal. Thus, the signal integration, which for a conventional array is done by external delay line and summation circuitry, is done in the SPRITE detector in the element itself.

The integration time approximates the recombination time τ for long devices. It becomes much longer than the dwell time τ_{pixel} on a conventional element in a fast-scanned serial system. Thus, a proportionally larger ($\propto\tau/\tau_{pixel}$) output signal is observed. If Johnson noise or amplifier noise dominates, it leads to a proportional increase in the SNR with respect to a discrete element. In the background-limited detector, the excess carrier concentration due to background also increases by the same factor, but the corresponding noise is proportional only to the integrated flux. As a result, the net gain in the SNR with respect to a discrete element is increased by a factor $(\tau/\tau_{pixel})^{1/2}$.

Elliott et al. derived the basic expressions for SPRITE parameters [12]. The voltage responsivity is:

$$R_v = \frac{\lambda}{hc}\frac{\eta\tau El}{nw^2t}\left[1-\exp\left(-\frac{L}{\mu_a E\tau}\right)\right], \tag{17.40}$$

where l is the readout zone length and L is the drift zone length.

The dominant noise is the GR noise due to fluctuations in the density of thermal and background-radiation generated carriers. The spectral density of noise at low frequencies is:

$$V_n^2 = \frac{4E^2 l\tau}{n^2 wt}\left(p_o+\frac{\eta Q_B\tau}{t}\right)\left(1-\exp\frac{-L}{\mu_a E\tau}\right)\left[1-\frac{\tau}{\tau_a}\left(1-\exp\frac{-\tau_a}{\tau}\right)\right]. \tag{17.41}$$

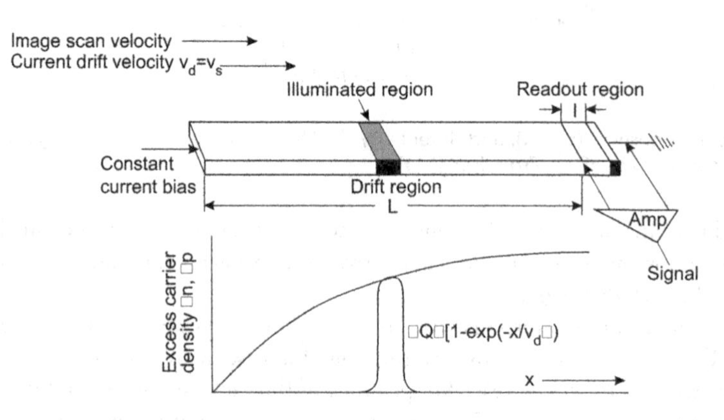

Figure 17.33 The operating principle of a SPRITE detector. The upper part of the figure shows a HgCdTe filament with three ohmic contacts. The lower part shows the buildup of excess carrier density in the device as a point in the image that is scanned along it. (From Ref. [245])

Figure 17.34 Schematic of an eight-row SPRITE array with bifurcated readout zones. (From Ref. [245])

For a long, background-limited device in which $L >> \mu_a E\tau$ and $\eta Q_B \tau / t >>, p_o$,

$$D^* = \frac{\lambda\eta^{1/2}}{2hc}\left(\frac{l}{Q_B w}\right)^{1/2}\left[1-\frac{\tau}{\tau_a}\left(1-\exp\frac{-\tau_a}{\tau}\right)\right]^{-1/2}, \tag{17.42}$$

and at sufficiently high speeds such that $\tau_a << \tau$, and

$$D^* = (2\eta)^{1/2} D^*_{BLIP}\left(\frac{l}{w}\right)^{1/2}\left(\frac{\tau}{\tau_a}\right)^{1/2}. \tag{17.43}$$

For a nominal resolution size $w \times w$, the pixel rate is v_a/w and

$$D^* = (2\eta)^{1/2} D^*_{BLIP}(s\tau)^{1/2}. \tag{17.44}$$

Based on the above considerations, long lifetimes are required to achieve large gains in the SNR. Useful improvement in detectivity relative to a BLIP-limited discrete device can be achieved when the value of $s\tau$ exceeds unity. The performance of the device can be described in terms of the number of BLIP-limited elements in a serial array giving the same SNR,

$$N_{eq}(BLIP) = 2s\tau. \tag{17.45}$$

For example, a 60 μm wide element scanned at a speed of 2×10^4 cm/s and with $\tau = 2$ μs gives N_{eq}(BLIP) = 13.

Figure 17.34 shows a photograph of an eight-row bifurcated SPRITE detector [245]. The bifurcated readout allows close-packed arrays without stagger. Readouts are shown at both ends, but only one set is used.

As shown in Table 17.4, to achieve usable performance in the 8–14 μm band, the SPRITE devices require LN cooling, while three- or four-stage Peltier coolers are sufficient for effective operation in the 3–5 μm range. The performance achieved in the 8–12 μm band is illustrated in Figure 17.35 [13]. The detectivity increases with the square root of the bias field, except at the higher fields where the element temperature is raised by Joule heating. In addition, this parameter increases with increasing cold shield effective *f*/# to about *f*/4. To avoid reductions of the carrier lifetime resulting from increased carrier density due to the background flux, efficient cold shielding with *f*/# of two or larger is used.

Table 17.4 Performance of SPRITE detectors

MATERIAL	MERCURY CADMIUM TELLURIDE	
Number of elements	8	
Filament length (μm)	700	
Nominal sensitive area (μm^2)	62.5 × 62.5	
Operating band (μm)	8–14	3–5
Operating temperatures (K)	77	190
Cooling method	Joule-Thompson or heat engine	Thermoelectric
Bias field (V/cm)	30	30
Field of view	*f*/2.5	*f*/2.0
Ambipolar mobility (cm^2/Vs)	390	140
Pixel rate per element (pixel/s)	1.8×10^6	7×10^5
Typical element resistance (Ω)	500	4.5×10^3
Power dissipation (mW per element) total	9 <80	1 <10
Mean D* (500 K, 20 kHz, 1 Hz) 62.5 × 62.5 μm (10^{10} $cmHz^{1/2}/W$)	>11	4–7
Responsivity (500 K), 62.5 × 62.5 μm (10^4 V/W)	6	1

Source: After Ref. [13]

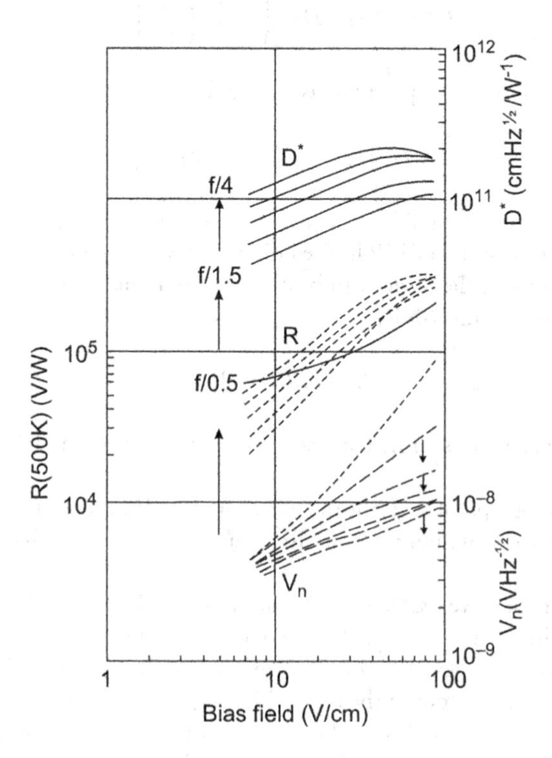

Figure 17.35 Variation of the performance of a SPRITE detector with bias field and field of view: $\lambda_c = 11.5\mu m$, $T = 77$ K. (From Ref. [13])

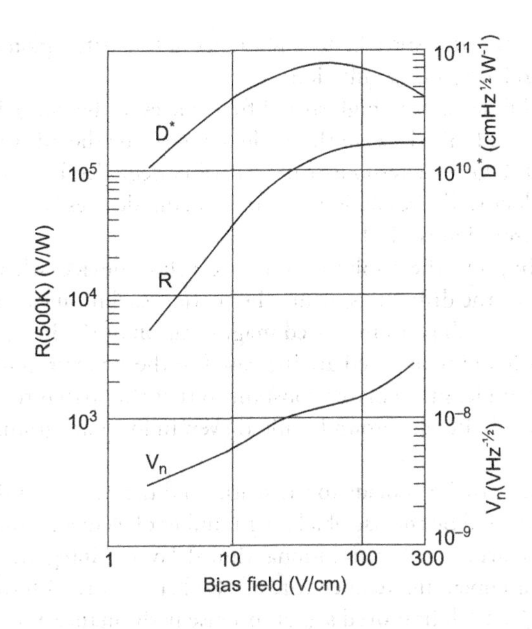

Figure 17.36 Performance of a 3–5 μm SPRITE operating at 190 K. (From Ref. [13])

An example of results obtained from a SPRITE operating in the 3–5 μm band is shown in Figure 17.36 [13]. Useful performance in this band can be obtained at temperatures up to about 240 K.

The spatial resolution of the SPRITE detector, when the scan velocity and the carrier velocity are matched throughout the device length, is determined by the diffusive spread of the photogenerated carriers and the spatial averaging in the readout zone. This can be expressed through the modulation transfer function (MTF) [245].

$$\mathrm{MTF} = \left(\frac{1}{1+k_s^2 L_d^2}\right)\left[\frac{2\sin(k_s l/2)}{k_s l}\right], \tag{17.46}$$

where k_s is spatial frequency and L_d is diffusion length.

The SPRITE detectors are fabricated from lightly doped ($\approx 5 \times 10^{14}$cm^{-3}) n-type HgCdTe. Both bulk material and epilayers are being used [255]. Single and two, four, eight, 16, and 24 element arrays have been demonstrated; the eight-element arrays are the most common (Figure 17.34). In order to manufacture the devices in line, it is necessary to reduce the width of the readout zone and the corresponding contacts to bring them out parallel to the length of the element within the width of the element as shown in Figure 17.34. Various modifications of the device geometry (Figure 17.37 [252]) have been proposed to improve both detectivity and spatial resolution. The modifications have included horn geometry of the

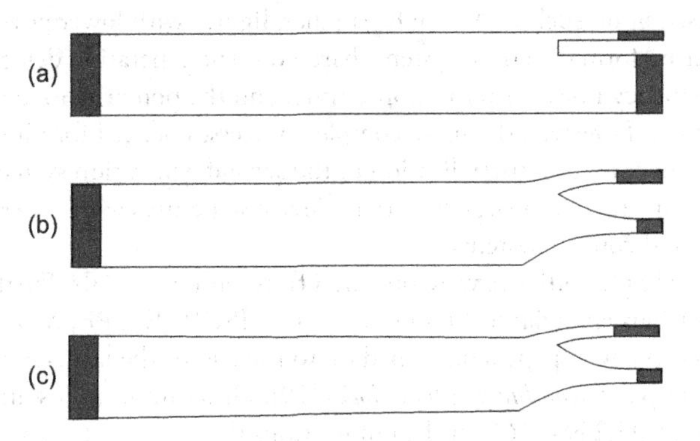

Figure 17.37 The evolution in the geometry of SPRITE devices. (From Ref. [256])

readout zone to reduce the transit time spread and slight taper of the drift region to compensate a slight change of drift velocity due to background radiation.

The response of the SPRITE detector to high spatial frequencies in the image is limited fundamentally either by spatial averaging due to limited size of the readout zone or by the diffusive spread of photogenerated carriers in the filament [245]. The response can be further degraded by imperfect matching of carrier drift velocity to the image velocity. The resolution size for 8–14 μm devices is ≈55 μm. At 200 K, for the 3–5 μm devices, the spatial resolution is ≈140 μm.

One possible method to improve the resolution is to use a short device, where the transit time is less than the lifetime to reduce the diffusive spread. The spatial resolution can also be improved by the use of anamorphic optics, which gives increased magnification of the image in the scan direction [249,252]. The detector length and scan speed are increased in the same proportion as the increase in magnification, but the diffusion length remains constant so that the spatial resolution is improved. Since the SPRITE detectors remain background limited even in low background flux, the SNR remains unaffected.

A number of improvements have been made to the spatial and thermal resolutions [254,255]. System thermal sensitivity can be increased by the use of a larger number of elements. In addition to parallel arrays, 2D 8 × 4 parallel/serial arrays with conventional time delay and integration along a row have been demonstrated. Moving from an operating temperature of ≈80 K to ≈70 K with the cutoff wavelength shift from 12 μm to 12.5 μm can give both improved signal-to-noise performance and spatial resolution. In the 3–5 μm band, the main improvement has been achieved by the use of more effective five- and six-stage Peltier coolers; 500 K blackbody detectivities as high as 5×10^{10} cmHz$^{1/2}$/W have been achieved with eight-element arrays.

Despite remarkable successes, SPRITE detectors have important limitations, such as limited size, stringent cooling requirements, and the necessity to use fast mechanical scanning. The ultimate size of a SPRITE array is limited by the significant heat load imposed by Joule heating. This means that SPRITE detectors are transition-stage devices to the staring 2D arrays area.

17.6 PHOTOVOLTAIC DETECTORS

The development of HgCdTe photodiodes was stimulated initially by their applications as high-speed detectors, mostly for direct and heterodyne detection of 10.6 μm CO_2 laser radiation [9,258]. Operations of such photodiodes at 77 K in the heterodyne mode, at frequencies up to several GHz, is possible because of the low junction capacitance, which in turn is a result of its relatively low static dielectric constant. In the mid-1970s, attention turned to the photodiodes for passive IR imaging applications in the two commonly used atmospheric windows at 3–5 μm and 8–14 μm. At that time, it was seen that many IR applications in the future would be needed for higher radiometric performance and/or higher spatial resolution than could be achieved with the first generation photoconductive detectors. The main limitation of photoconductive detectors is that they cannot easily be multiplexed on the focal plane. In contrast to photoconductors, photodiodes can be assembled in 2D arrays containing more than megapixel elements, limited only by existing technologies. Systems based upon such FPAs can be smaller, lighter with lower power consumption, and can result in much higher performance than systems based on first generation detectors. Photodiodes can also have less low-frequency noise, faster response time, and the potential for a more uniform spatial response across each element. However, the more complex processes needed for photovoltaic detectors have influenced slower development and industrialization of the second generation systems. Another point is that, unlike photoconductors, there is a large variety of device structures with different passivations, junction forming techniques, and contact systems.

Initially, the first HgCdTe photodiodes were prepared from bulk materials. Further development was, however, dominated by various epitaxial techniques including ISOVPE, LPE, MBE, and MOCVD. A more detailed historical review of papers published up to the end of the last century is presented in the monograph: *Narrow-Gap Semiconductor Photodiodes* [79]. Other important sources of information are included (see, e.g., [194,259–266]) in the publications that came out each year since 1981, in the workshop-style technical meetings devoted exclusively to the physics and chemistry of HgCdTe and related semiconductor and IR materials (published initially in *Journal of Vacuum Science and Technology* and next in *Journal of Electronic Materials*), and in numerous *Proceedings of SPIE*.

This chapter will concentrate mainly on the present status of HgCdTe photodiode physics and technology important for the fabrication of large FPAs. The success in HgCdTe photodiode technology has stimulated programs on third generation IR detector technology being performed in centers around the world (see Chapter 27) [18].

17.6.1 JUNCTION FORMATION

The p-n HgCdTe junctions have been formed by numerous techniques including Hg in- and out-diffusion, impurity diffusion, ion implantation, electron bombardment, plasma-induced type conversion, doping during growth from vapor or liquid phase, and other methods [79]. To avoid citing hundreds of related references, some recently published monographs and reviews are being mentioned here [31,79,262,263,266].

The low binding energies and ionic bond nature of HgCdTe give rise to two important effects, which are influential in most junction forming processes. The first is the role of Hg, which is liberated readily by processes such as ion implantation and ion beam milling. This creates a much deeper junction than would be expected from the implantation range. A second effect is the role of dislocations, which may play a part in annihilating vacancies. The role of Hg interstitials, dislocations, and ion bombardment in the junction forming process is complex and not well understood in detail. Despite the complex physics involved, manufacturers have received good phenomenological control of the junction depth and dopant profiles with a variety of processes. Recently, epitaxial techniques with doping during growth are most often used for preparing p-on-n junctions. MBE and MOCVD have been successfully accomplished with As doping during growth.

17.6.1.1 Hg in-diffusion

It is relatively straightforward to achieve local type conversion of the material by neutralizing the vacancies by the in-diffusion of Hg. The n-type conductivity is originated from a background donor impurity. The knowledge of Hg in-diffusion process has been summarized by Dutton et al. [267]. The Hg diffusion process at 200°C is accompanied by vacancy out-diffusion, which creates a graded distribution of vacancies near the metallurgical junction. To form junctions requires only 10–15 minutes at temperatures of 200°C–250°C in material with low vacancy concentrations of 10^{16} cm^{-3}. This corresponds to diffusion constants of the order of 10^{-10} cm^{2}s^{-1}. The presence of dislocations can enhance the vacancy mobility even further, while the presence of Te microprecipitates may retard the motion of Hg into the lattice. Crystal defects, such as dislocations and microprecipitates, can also harbor background impurities, which will further affect the location and quality of the junction.

Hg in-diffusion into vacancy-doped (10^{16}–10^{17} cm^{-3}) HgCdTe, originally proposed by Verie et al. [9,258] at the beginning of the 1970s, has been the most widely used for very fast photodiodes in which a low concentration (10^{14}–10^{15} cm^{-3}) n-type region is necessary for large depletion width and low junction capacitance. Initially, a mesa configuration was used for this type of device. Spears and Freed [268,269] reported an improvement of the technique using n^+-n-p planar structure. A 0.5 µm ZnS was sputtered onto the surface of vacancy-doped HgCdTe. After etching openings in the ZnS through a photoresist mask and before removing the photoresist, a ≈10 nm In layer was sputtered onto the surface. A low-concentration n-type layer about 5 µm deep with thin n^+ skin was produced within 30 minutes, during 240°C Hg-vapor diffusion in a sealed ampoule. Photoresist liftoff technique was used to define the sputtered In-Au bonding pads. A ZnS mask provides passivation around the junction perimeter. Fabrication of similar n^+-n-p structures for heterodyne applications was described by Shanley et al. [270].

Further modification in preparing n-p planar junctions was presented by Parat et al. [271]. Following a Hg-saturated anneal at 220°C for 25 hours, the MWIR HgCdTe layers grown by MOCVD become n-type with carrier concentrations around 5×10^{14} cm^{-3}. However, the presence of a 0.5–0.8-µm-thick CdTe cap layer acts as an effective barrier for Hg diffusion and provides excellent junction passivation. By opening windows in this cap, the underlying HgCdTe layer can be annealed and converted to n-type in a selective manner.

Jenner and Blackman [272] have reported a variation of the Hg in-diffusion method, which uses anodic oxide to act as a source of free Hg. This technique is particularly suitable for high-speed devices in which low and uniform doping n-type regions are necessary. Anodic oxidation produces a Hg-rich layer at the interface between oxide and semiconductor. During anneal, Hg diffuses into a material background giving rise to n-type characteristics of the material. The anodic oxide layer acts as an out-diffusion mask, preventing the loss of Hg into the vacuum on gas ambient. Brogowski and Piotrowski have calibrated this method [273]. The square root dependence of the junction depth on anneal time for short times is a clear indication of the diffusion nature of the p-to-n type conversion. The existence of maximum depth of the junctions shows that the source of free Hg is a finite one. After a prolonged annealing the Hg source becomes exhausted, Hg diffuses deeply into the bulk, resulting in a reconversion of conductivity type at the surface of the material.

17.6.1.2 Ion milling

Conversion of vacancy-doped p-type $Hg_{1-x}Cd_xTe$ to n-type during low-energy ion bombardment became another important technique of junction fabrication [213,274–279]. Neither donor ions nor post-annealing is required. The ion beam injects a small proportion of the Hg atoms (approximately 0.02% of the gas ions) into the lattice. These then neutralize acceptor-like Hg vacancies and leave the lattice weakly n-type by background donor atoms.

The ion energy is usually less than 1 keV and the dose normally varies between 10^{16} and 10^{19} cm^{-2}. But even at the lower dose, when the surface is etched very gently, ion milling results in substantial changes of the electrical properties of HgCdTe over a large depth. Blackman et al. have shown that the depth of the p-n junction depends on dose and can extend a few hundred μm from the surface [213]. The electrical properties of ion-etched HgCdTe measured by differential Hall effect have shown that after ion milling a thin n-type degraded layer, of approximately 1 μm in thickness with low mobility and high concentration of electrons, occurs close to the surface [277,280]. Below this damaged region, an n-type doping profile, which decreases exponentially away from the surface, and a low doped n-region of controllable width and with high electron mobility are created. The n^+ grade leads to a very effective reflecting contact for minority carriers, leading to a highly sensitive n-type region, as revealed by the electron beam induced current analysis. A deeper junction is created by a higher beam current, longer milling time, lower beam voltage, and higher ion mass. The whole process is carried out at a low temperature, preventing the original material and passivation quality, which is an advantage of this technique. The type conversion of p-type material using an ion beam has been used commercially by GEC-Marconi Infrared Ltd for HgCdTe FPAs since the late 1970s [281].

The diffusion of Hg during ion milling is very quick even in comparison with annealing experiments at 500°C. To explain this phenomenon, diffusion mechanisms via fast ways (dislocations, grain boundaries, stacking faults) were discussed. A model of the diffusion of Hg in HgCdTe, taking into account the recombination of Hg interstitials with Hg vacancies, was presented [282,283]. Based on this model it was shown that there is in principle no discrepancy between the high value of the diffusion constant of Hg interstitials and that of the radioactive Hg self-diffusion constant established from equilibrium annealing experiments.

The local damage due to the impinging ions is restricted to a distance of the order of the ion range, while the depth of the converted zone is much larger and remains roughly proportional to the thickness of the layer removed by ion bombardment.

17.6.1.3 Ion implantation

The ion implantation in HgCdTe is a well-established approach for fabricating HgCdTe photovoltaic devices with n-on-p type junctions [284,285]. It is a common method of HgCdTe photodiode fabrication since it avoids heating of this metallurgically sensitive material and permits a precise control of the junction depth. Many manufacturers obtain the desired p-type level by controlling the density of acceptor-like Hg vacancies within a carrier concentration range of 10^{16}–10^{17} cm^{-3}. The n^+-p structures are produced by Al, Be, In, and B ions implantation into vacancy-doped p-type material, but the technique typically uses ion implantation of light species (usually B and Be) to form n-region. B is possibly most frequently used, perhaps due to the fact that B is also a standard implant for Si. Regardless of the nature of the implanted species, an n-type electrical activity is associated with implantation damage as observed first by Foyt et al. [286].

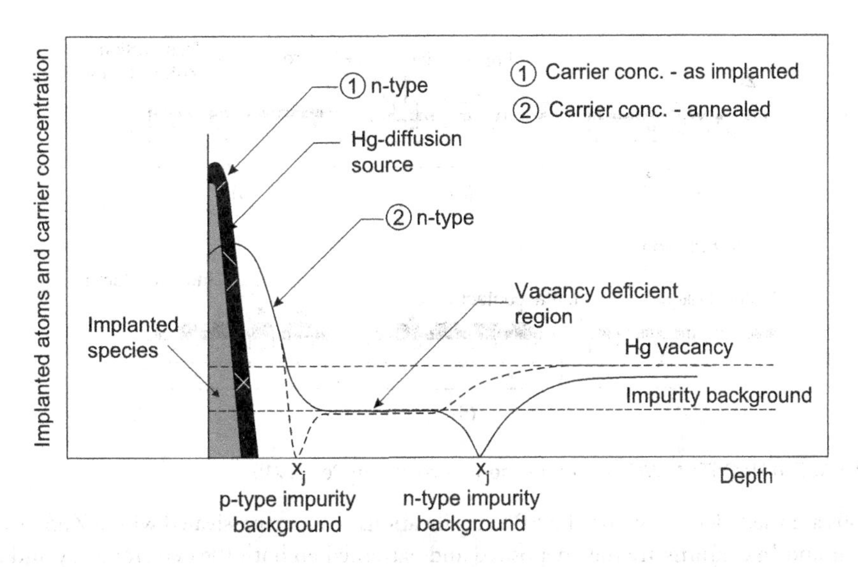

Figure 17.38 Qualitative model of displaced role of Hg on junction formation in ion-implanted HgCdTe. (From Ref. [287])

A summary of the complex phenomena associated with ion implantation in HgCdTe is shown in Figure 17.38. A fundamental concept developed for light species implantation in HgCdTe is the generation of free Hg atoms by the irradiation process of ion implantation (knockout Hg atoms) and their diffusion from the implanted source [287]. This concept is the basis of n-on-p junction formation in certain conditions of background type and concentration of carriers by the diffusion of Hg from the implant-induced Hg source, as illustrated in Figure 17.38. In this figure, the typical profile implanted species atom concentration and the corresponding carrier concentrations before (curve 1) and after (curve 2) postimplant anneal are sketched. A fraction of Hg interstitials are the diffusing elements primarily responsible for the changes in p-type behavior of the starting material when it is doped by Hg vacancies (i.e., the interstitials annihilate Hg vacancies that are encountered), revealing the net doping due to impurity background that can either be n- or p-type. The junctions form at the end of the annihilated region. Typically, the p-n junctions are located at a depth of 1–3 μm, which is large compared with the submicron range of implanted ions (<300 Å). If the net background doping is p-type, the junctions are of n^+-p type (dashed line). When the net background doping is n-type, the annihilated region is n-type. The junctions thus formed are n^+ -n^--p type (full line). This is the most desirable case since the junction is located away from the implant-defect region in a material, which is free of irradiation-induced defects.

The mesa diodes are made by implantation on a passivated surface with subsequent etching for element separation. For planar devices, prior to implantation, the substrates are covered with dielectric layers (photoresists, ZnS, CdTe) with the openings acting as a mask for impinging ions and thus defining the junction area. Implantation is typically performed at room temperature; the substrates, if oriented, are inclined to the beam axis to avoid ion channeling. Doses of 10^{12}–10^{15} cm^{-2} and energies of 30–200 keV are applied. No postimplant annealing was found necessary to achieve high performance, particularly at lower doses and for SWIR and MWIR devices. Some workers concluded that LWIR photodiode performance can be improved if postimplant annealing is used [288,289]. The postimplant anneals remove the radiation damage with a temperature dependence that varies with implanted species and with implant/anneal conditions. Examples are shown for In, B, and Be for which anneals at temperatures of ≥300°C annihilated the electrically-active defects. Post-annealing also modifies the characteristics of the p-type base layer when doped by Hg vacancies. The electrically-active ion implantation has been observed by Bahir et al. after rapid thermal annealing (0.3 seconds) of P and B implants with a continuous wave (cw) CO_2 laser [290].

An example of a processing sequence for PACE 1 HgCdTe MWIR FPAs is shown in Figure 17.39. At the beginning, a layer of CdTe is grown via MOCVD on the sapphire [291]. HgCdTe is then grown on the CdTe buffer via LPE. The junctions are formed by B ion implantation and thermal annealing. Planar and

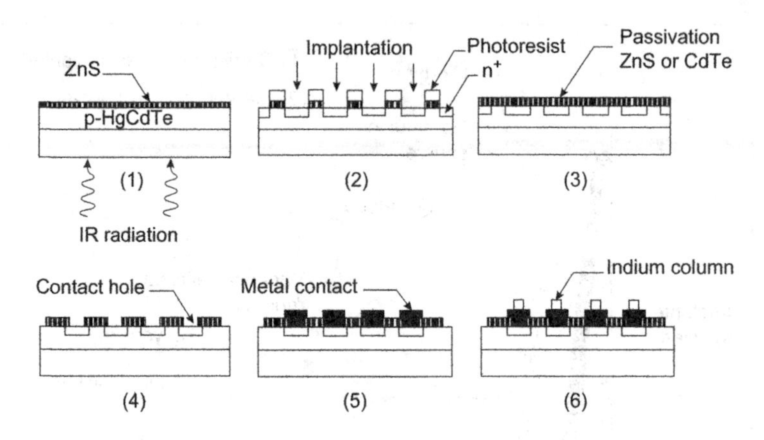

Figure 17.39 PACE 1I HgCdTe MWIR processing sequence. (From Ref. [291])

mesa junctions are used, depending on the FPA specifications, and are passivated with a ZnS or CdTe film. Metal contacts and In columns are then deposited and patterned on both the detector array and the multiplexing readout, and the hybrid is fabricated by mating the detector array to the readout via In column interconnects.

Until the 1990s, less attention had been paid to the implanted p-on-n photodiodes [284,285]. The junctions were formed by Au, Ag, Cu, P, and As implantation in n-type HgCdTe followed by annealing. Establishing p-on-n junctions in HgCdTe using extrinsic dopants has taken an increasingly critical role in HgCdTe photodiode technology, especially in LW spectral range. In the published results of p-on-n devices, the electrical junctions are controlled by the tail components in the As diffusion profile rather than by a classical component, which is representative of a volume diffusion mechanism.

The As redistribution from the ion-implanted source as well as from a grown source (As incorporated in a part of the growth cycle) typically exhibits a multicomponent behavior in which the tail component varies from sample to sample. A model that explains the nature of these components has been proposed by Bubulac et al. [292,293]. In this model, the component (1) (see Figure 17.40) extends from the surface to its

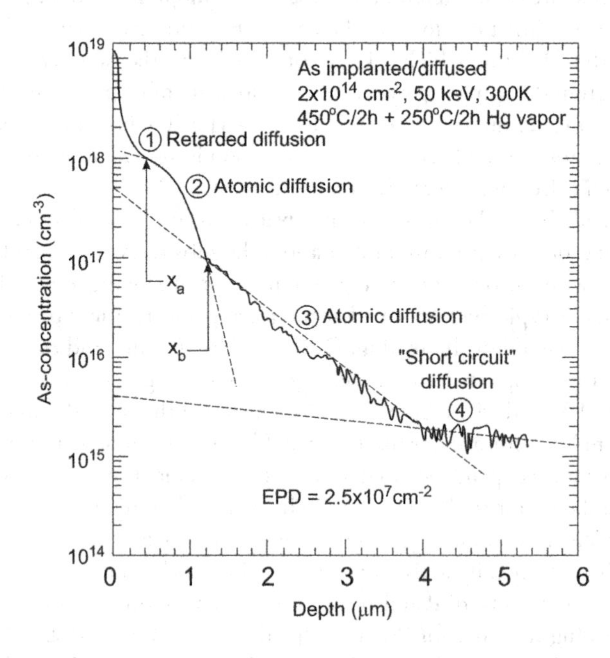

Figure 17.40 Typical example of As redistribution from the SIMS and the proposed model for As diffusion from a SIMS: (1) retarded diffusion; (2) atomic diffusion, As starts on Te-sublattice; (3) atomic diffusion, As starts on metal-sublattice; and (4) short-circuit diffusion. (From Ref. [292])

intersection with the deeper Gaussian-like component (2). As distribution in the near-surface region occurs due to a retarded diffusion mechanism operating in the radiation damage region. It was shown that the depth of component (1) is related to the dislocation density (EPD) in the starting material and on implant-anneal conditions. The diffusion mechanism proposed for deeper component (2) is atomic in nature and vacancy-based in which As starts on the Te-sublattice in the surface damage region. Because of a large number of Hg interstitials generated in this region as a result of Hg knockout, the conditions under which As is introduced in the lattice approximates a Hg-rich condition, resulting in a larger fraction of As incorporation on the Te-sublattice versus the metal-sublattice. The As diffusion of this component is described by Fick's law with a constant diffusion coefficient of $D = 2.5 \times 10^{-14}$ cm²/s at 400°C and $D = 2.0 \times 10^{-13}$ cm²/s at 450°C. It was also determined that the diffusion coefficient in this component (2) is independent of EPD in the material. A linear dependence of the diffusion length of As as a function of the square root of time confirms the Gaussian redistribution of As in component (2) from the As ion implanted source. The excellent reproducibility of the diffusion coefficient on samples grown by various growth techniques on various substrates (e.g., MOCVD/GaAs/Si, LPE/CdTe, and LPE/sapphire) is evidence that this diffusion mechanism is representative of the bulk diffusion in HgCdTe. The diffusion mechanism in this component is the most desirable mechanism for a controlled junction formation process.

The tailing component (3) is generated by a fraction of As atoms, which is introduced on the metal-sublattice sites and diffused by an enhanced vacancy-based mechanism. The enhancement is a consequence of departure from the point defect equilibrium within the atomic diffusion zone (in the damage region, a continuous metal vacancy flux to the diffusion zone is maintained). Since vacancies enhance the substantial diffusion, the atomic diffusion in component (3) is affected by the gradient of vacancy distribution and thus the resulting diffusion coefficient is position dependent. The component (3) of As diffusion controls the location of the electrical junction. This component is electrically complex, consisting of n- and p-type active As, and perhaps also neutral As. A site transfer with the associated change in electrical activity is possible, depending on subsequent thermal treatment and phase equilibria. The principal consideration in minimizing the As tailing component is to reduce the dislocation density of the starting material. An increase of about one order of magnitude in EPD ($2.6 \times 10^{6} - 2.7 \times 10^{7}$ cm^{-2}) causes an increase in surface damage depth by a factor of 9 (0.05–0.45 μm) and an increase of about two orders of magnitude in the As concentration ($10^{15} - 10^{17}$ cm^{-3}).

In the presence of nested dislocations in the starting material, a further enhancement of As diffusion was observed (i.e., component 4). In Si, it has been shown that impurities can move in dislocations or in clustered defects by short circuit mechanism. The experiments showed that the most effective way to minimize component (4) was to reduce the EPD of the material.

17.6.1.4 Reactive ion etching

Reactive ion etching (RIE) is an effective anisotropic etching technology that is widely used in Si and GaAs semiconductors allowing delineation of a high density of active device elements incorporating small features. This plasma-induced technique is an alternative to the ion implantation junction formation technology [294,295]. The postimplant annealing, necessary in implantation technology to produce high-quality photodiodes, is not needed in plasma technology.

Figure 17.41 shows the H_2/CH_4 RIE etch depth observed in two vacancy-doped p-type HgCdTe samples as a function of the partial pressures of H_2 and CH_4 [282,294,298]. The etch depth increases from zero methane level, reaches a maximum at about (H_2/CH_4) 0.8, and decreases to the zero-hydrogen level. The p-n junction depth decreases with the increasing methane fraction in the mixture.

In plasma-induced type conversion, the accelerated plasma ions sputter the HgCdTe surface, liberate Hg atoms from ordinary lattice positions, and create a source of Hg interstitials under the etched surface. Some of these atoms diffuse quickly into the material where they decrease the concentration of acceptors by the interaction of point defects, mainly Hg vacancies. Residual or native donor impurities then start to dominate the conductivity and cause the p-n conversion. In the extrinsically doped material, conversion extends beyond Hg vacancies. White et al. [296] have proposed a p-n conversion model where the junction formation mechanism is thought to be a mixture of RIE-induced damage, Hg interstitial formation to which hydrogen forms strong bonds, and hydrogen-induced neutralization of acceptors.

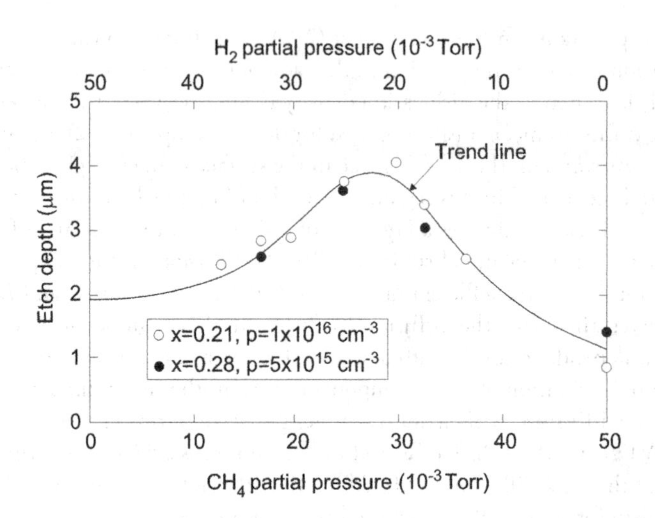

Figure 17.41 Etch depth in $Hg_{1-x}Cd_xTe$ as a function of H_2 and CH_4 partial pressures (H_2/CH_4 RIE): (o) $x = 0.21$, $p = 1 \times 10^{16} cm^{-3}$; (•) $x = 0.28$, $p = 5 \times 10^{15} cm^{-3}$. The etch time is 10 minutes, and the RF power is 180 W. (From Ref. [294])

Exposure of p-type HgCdTe to CH_4/H_2 RIE plasma has been to fabricate n-p junctions [294,297,298]. The plasma-induced type of conversion may also be used with P-on-n heterostructures to insulate the junction of high-performance photodiodes [299,300].

17.6.1.5 Doping during growth

Doping during epitaxial growth has become the preferred technique at present, with inherent advantage of integration of the material growth and device processing. High-performance photodiodes can be obtained by successive growth of doped layers using LPE from Te-rich solutions or Hg-rich solutions [46,47], MBE [61], and MOCVD [54]. Mynbaev and Ivanov-Omski have reviewed available publications concerned with doping of HgCdTe epitaxial layers and heterostructures [301].

Epitaxial techniques make it possible to grow in situ multilayer structures. Stable and readily achievable at temperatures below 300°C dopants are required to prevent interdiffusion processes. In and I are preferred n-dopants and As is the preferred p-type dopant for in situ doping.

As was mentioned above, the As dopants must reside on the Te site to accomplish p-doping. This requires either growth or annealing at relatively high temperatures under cation-rich conditions. LPE growth from a Hg melt (at temperatures around 400°C) satisfies this condition automatically, although some postgrowth annealing may be required. MOCVD and MBE have successfully accomplished As doping during growth. Mitra et al. [56] have reviewed the progress made in the development of IMP MOCVD for the in situ growth of HgCdTe p-on-n junction devices for FPAs. It is shown that MOCVD-IMP has progressed to the point that sophisticated bandgap-engineered multilayer HgCdTe device structures can be grown in situ on a repeatable and dependable basis. Good run-to-run repeatability and control have been demonstrated for a series of identical grown runs of a MW/LW p-n-N-p double-heterojunction dual-band device structure.

Due to low sticking coefficient of As, in situ p-type doping has been a challenging issue. Incorporation of As into the layer under Te-rich conditions (i.e., high concentration of Hg vacancies) facilitates the activation of As [302]. MBE growth under Hg-rich conditions results in twin defect formation, and the As dopant may incorporate at the surface boundaries of twin defects. As a result, As-doped samples do not activate even at the highest annealing temperatures. Effective, near 100% activation is achieved for samples annealed under isothermal conditions at temperatures higher than 300°C.

Figure 17.42 illustrates cross section schematics of two HgCdTe photodiode architectures fabricated using In and As doping during MBE growth [303]. The two-color detector is an n-p^+ -n HgCdTe triple-layer heterojunction (TLHJ) back-to-back p^+ -n photodiode structure grown on a (211) oriented CdZnTe substrate, and the single-color detector is a p^+ -n DLHJ detector design on a (211) Si substrate with ZnTe and CdTe buffer layers [303,304].

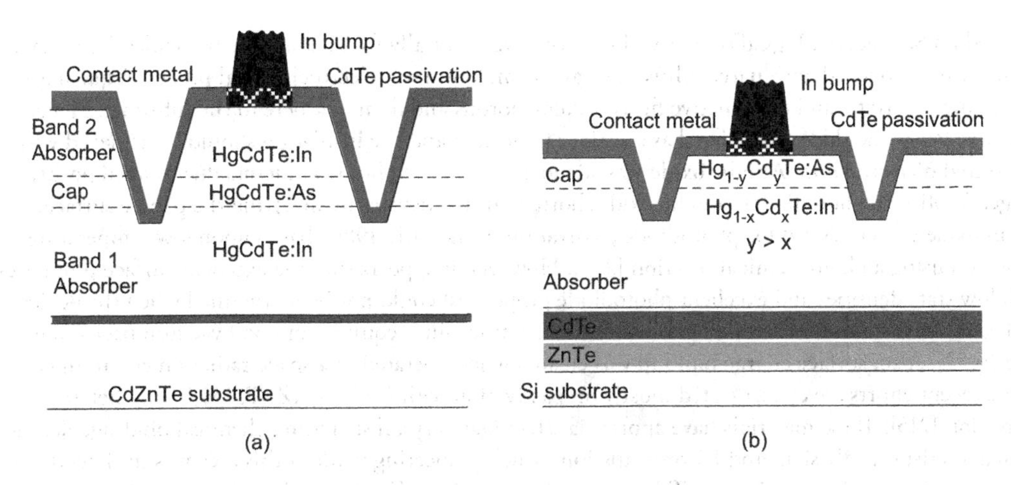

Figure 17.42 Cross section schematic illustrations of MBE-grown HgCdTe detector architectures: (a) two-color TLHJ HgCdTe/CdZnTe and (b) single-color DLHJ HgCdTe/Si. (From Ref. [303])

One important advantage of the p-on-n device (see Figure 17.42b) is that the n-type $Hg_{1-x}Cd_xTe$ carrier concentration is easy to control in the $10^{15}\,cm^{-3}$ range using extrinsic doping—usually In or I (for the n-on-p device, the p-type carrier concentration at this low level is difficult). The wider bandgap capping $Hg_{1-y}Cd_yTe$ ($y \approx x + 0.04$) is 0.5–1 μm thick when not intentionally doped. The structures are terminated with a thin (500 Å) CdTe layer for protection of the surface.

The formation of planar p-on-n photodiodes is achieved by first selectively implanting As through windows made on a mask of photoresist/ZnS and then diffusing the As through the cap layer into the narrow gap layer [305]. The ion-implanted diffusion source is formed by a shallow implant with As for a dose of about $1 \times 10^{14}\,cm^{-2}$, at energies of 50–350 keV [305,306]. Usually, after implantation, the structures are annealed under Hg overpressure. The samples undergo two consecutive annealings; one at high temperatures ($T \geq 350°C$ for a short time, e.g., 20 minutes [307]) and the other at 250°C for 24 hours immediately after. The first annealing is performed to anneal out the radiation-induced damage and to diffuse and electrically-activate As by substituting As atoms on the Te-sublattice, while the second one is to annihilate Hg vacancies formed in the HgCdTe lattice during growth and the diffusion of As to restore the sample background to n-type. The concentration of As dopant in p-type cap layer is at a level of about $10^{18}\,cm^{-3}$.

17.6.1.6 Passivation

In general, there are two different passivation approaches depending on whether the defects are located at the semiconductor surface or are located in the bulk: one is surface passivation and the other is bulk passivation. The key technology needed to make detectors possible is surface passivation, which is targeted at reducing surface/interface defect recombination centers and interface trapped charges. The bulk passivation is targeted at passivating bulk defect recombination centers.

Passivation is a critical step in the HgCdTe photodiode technology that greatly affects surface leakage current and device thermal stability, so it is treated as a proprietary process by most manufacturers. A surface potential of less than 100 meV can accumulate, deplete, or invert the surface significantly, thus drastically affecting device performance. In comparison with photoconductive detectors, passivation of photodiodes is more difficult, since the same coating must stabilize simultaneously regions of n- and p-type conductivities. The most difficult is the passivation of p-type material due to its tendency for inversion.

Passivation of HgCdTe has been done by several techniques where a comprehensive review was given by Nemirovsky et al. [205,218,308,309]. Passivation technologies can be classified into three categories: native films (oxides [205,308,309], sulfides [205,216,310], fluorides [217,311]), deposited dielectrics (ZnS [312,313], SiO_x [212,314], Si_3N_4 [315], polymers) and in situ grown heterostructures where a wider bandgap material is the passivant. A two-layer combination of a thick deposited dielectric film upon a thin native film of heterostructure is often the preferred passivation.

Based on Si's success, HgCdTe passivation efforts were initially focused mainly on oxides. Two major problems are associated with native films: they are formed by a wet electrochemical process requiring a conducting substrate, and thick native films become porous and do not adhere to the substrate. Hence, the native layers should be considered as a surface treatment and the insulation should be achieved with a deposited dielectric film. Anodic oxide was adequate for n-type photoconductors due to fixed positive charge. Applied to photodiodes, anodic oxide shorted out the devices by inverting the p-type surface. Silicon oxide was employed for photodiode passivation in the early 1980s based upon low-temperature deposition using a photochemical reaction [212]. However, it appears that the excellent surface properties (with low state densities and excellent photodiode properties) could not be maintained when the device was heated in vacuum for extended periods of time, a procedure required for good vacuum packaging integrity [7]. Also, surface charge buildup was created when operated in a space-radiation environment.

The recent efforts are concentrated mostly on passivation with CdTe, CdZnTe [309], and heterojunction passivation [215]. These materials have appropriate bandgap, crystal structure, chemical binding, electrical characteristics, adhesion, and IR transmission. Much pioneering work in this area was initially done in France at Societe Anonymique de Telecommunicacion (SAT) [316]. It is desirable to use intrinsic CdTe with the layer-stoichiometry, minimal stress, and a low (below 10^{11} cm^{-2}) interface fixed charge density. The layers are sputtered, e-beam is evaporated, and grown mainly by MOCVD and MBE.

The CdTe passivation can be obtained during epitaxial growth [221,317,318] or by the postgrowth deposition [220,319] or anneal of the grown heterostructures in Hg/Cd vapors [320]. Directly grown in situ CdTe layers lead to low fixed interface charge, the indirectly grown ones are acceptable, but are not as good as the directly grown. Procedures for surface preparation prior to indirect CdTe deposition has been proposed [298,319]. Encouraging results have been achieved in in situ grown MOCVD and MBE CdTe layers, but additional surface pre- and posttreatments are necessary to achieve near-flat band conditions. Annealing at 350°C for 1 hour exhibited the lowest density of fixed surface states [317]. With low CdTe doping, however, the cap layer will be fully depleted and this may affect the interface charge. Bubulac et al. [321] have evaluated e-beam and MBE grown by SIMS and atomic force microscopy to determine the usefulness of the material for passivation of HgCdTe. It has been suggested that CdTe grown by MBE at 90°C is more thermally stable and denser as compared to the e-beam grown material. The wide material is not necessarily single crystalline and, in many cases, a polycrystalline layer provides the passivation. CdTe passivation is stable during vacuum packaging bake cycles and shows little effect from the radiation found in space applications. Diodes do not show a variation in R_oA product with diode size, indicating that surface perimeter effects can be neglected.

Bahir et al. [221] have discussed the results of TiAu/ZnS/CdTe/HgCdTe metal insulator semiconductor heterostructures. Samples with indirectly grown CdTe showed hysteresis, which can be attributed to the slow interface traps. A fixed charge density of $(5\pm2)\times10^{10}$ cm^{-2} in p-type $Hg_{0.78}Cd_{0.22}Te$ has been reported. There is also evidence of the presence of slow interface traps in directly grown CdTe. The asymmetry of the CdTe/HgCdTe interface (see Figure 17.43) leads to large band bending and to carrier inversion in p-type HgCdTe. In the presence of ZnS layer at the top, the band bending in CdTe is eliminated, resulting in flat band conditions in the CdTe/ HgCdTe interface. Sarusi et al. [219] have investigated the passivation properties of CdTe/p-$Hg_{0.77}Cd_{0.23}Te$ heterostructure grown by MOCVD. The interface recombination velocity has been determined to be 5000 cm/s, which is a lower value than for ZnS-passivated surfaces. The results demonstrated that CdTe is an ideal candidate for passivation of photodiodes based on p-type HgCdTe.

It is well recognized that some degree of interface grading in CdTe/HgCdTe is beneficial for surface passivation [298]. This grading may shift the original HgCdTe defective surface to the wider-gap CdTe region, which results in more thermally stable devices and in improvements in surface passivation, irrespective of the CdTe growth method.

The passivation with ZnS, CdTe, or wider-gap $Hg_{1-x}Cd_xTe$ is often overcoated with SiO_x, SiN_x, and ZnS [298]. Thick deposited dielectric films are required to achieve the protection of the interface from environmental conditions and to insulate the metallization pattern of the contacts and the bonding pads from the substrate. In addition, in front illuminated detectors, optical properties of the dielectric films are of great importance: the film must exhibit excellent transmission in the relevant wavelength region and must possess an appropriate index of refraction for achieving a well-matched, antireflection coating at the

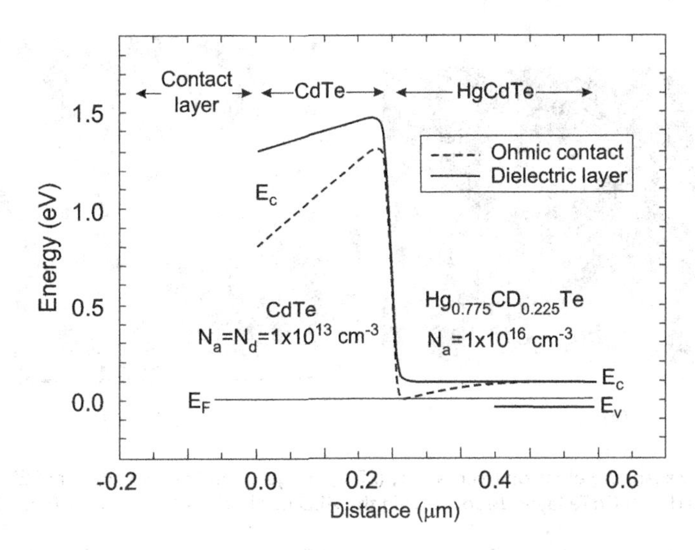

Figure 17.43 Effect of the contact layer on the energy band diagram of CdTe/HgCdTe. Ohmic contact to CdTe causes carrier inversion in p-type HgCdTe, while the dielectric layer leads to nearly flat band conditions. (From Ref. [221])

required wavelength. The most commonly used deposited dielectrics are ZnS (either evaporated or magnetron sputtered), CVD-SiO_2 (either thermal or low-temperature photo-assisted), and Si_3N_4 (deposited using electron cyclotron resonance or plasma enhanced).

Mestechkin et al. [220] have examined the bake stability for HgCdTe wafers and photodiodes with CdTe surface passivation deposited by thermal evaporation. It was found that the bakeout process (a ten-day vacuum bakeout at 80°C) generated additional defects at the CdTe/HgCdTe interface and degraded the photodiode parameters. Annealing at 220°C under an Hg vapor pressure following the CdTe deposition suppressed the interface defect generation process during bakeout and stabilized the HgCdTe photodiode performance.

Bulk passivation is especially important for HgCdTe detectors grown on alternative substrates. For epitaxial layers grown on Si, there are a number of threading dislocations due to the large lattice and the coefficient of thermal expansion mismatch between HgCdTe and the substrates. Hydrogen is reported to effectively passivate the defect states in HgCdTe epilayers. It is expected that hydrogen, introduced into HgCdTe using electron cyclotron resonance plasma hydrogenation, will eliminate the gap states by involving several hydrogen atoms and host-lattice structural rearrangements in defect-hydrogen complexes [322]. In another work [323], HgCdTe/CdTe/Si layers passivated with hydrogen demonstrate lifetimes comparable to those grown on CdZnTe, even though their dislocation densities were higher. This is due to the fact that the incorporated hydrogen passivates both the scattering and recombination centers and thus increases the carrier lifetime and mobility [324]. Defect-related leakage current still represents an ongoing issue for HgCdTe photodiodes, especially in the LWIR range. One potential approach for the improvement of passivation is hybrid passivation, for example, combination of CdTe and ZnS layer depositions with hydrogen passivation [325].

17.6.1.7 Device processing

The critical issue in the development of high-performance HgCdTe photodiodes is a reliable fabrication process. The fabrication of modern mesa-isolated IR FPAs with pixel size below 10 μm is far more challenging than single element detectors and it requires the fabrication of well-isolated pixels with high aspect ratios, smooth sidewalls, and high uniformity. Now pixel etching plays a more critical role in comparison with the early days of IR technology, since back then the arrays were smaller and the relatively large pixels (typically 50 μm) were fabricated mainly via wet chemical etching generally based on alcohol solutions of Br_2 and Br_2-HBr, including Br_2-lactic acid, Br_2-ethylene glycol, and HBr-H_2O_2 [295]. Wet isotropic etching commonly results in nonuniform etch rates that is not suitable for fabricating large format HgCdTe arrays.

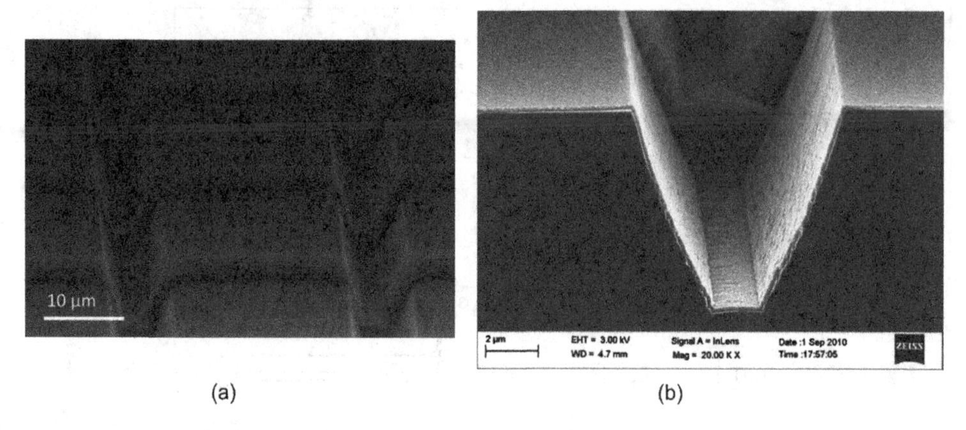

Figure 17.44 Top-view scanning electron microscopy (SEM) image (a) and cross-sectional SEM image (b) of an HgCdTe FPA passivated with CdTe layer deposited via the ALD method. (Adapted after Ref. [327])

Currently, dry etching technology, which commenced for HgCdTe from the late 1980s, is dominated by plasma etching using a hydrocarbon/H_2-based methyl radical etching chemistry [295,326]. There are two main plasma generation systems that are widely used: electron cyclotron resonance (ECR) and inductively coupled plasma (ICP) processes. Various processing parameters of plasma etching have been studied and optimized to achieve a clean and smooth surface as well as to reduce the wafer damage and etching lag, such as substrate temperature, plasma gases used, radio frequency (RF) power, direct current (DC) bias, chamber pressure, Ar/H_2 gas ratio, photoresist parameters, and ion angular distribution.

For CdTe surface passivation of HgCdTe detectors, low temperature is required for the passivation layer deposition to prevent Hg depletion from the surface. Hence, conventional methods such as vacuum evaporation at room temperature are typically used for CdTe deposition on planar detectors. With the increase in complexity of the detector structures, other methods of deposition such as chemical vapor deposition and atomic layer deposition (ALD) have attracted increased attention for their potential to obtain improved conformal coverage for high-aspect-ratio structures. For example, Figure 17.44 shows the conformal ALD deposition of CdTe passivation films on the sidewalls of high-aspect-ratio, small-area mesa diode structures [327].

The issues connected with contacts in photodiodes are contact resistance, contact surface recombination, contact $1/f$ noise, and long-term and thermal stability of devices. The contacts determine the performance and reliability of a device [309]. Ideal ohmic contacts are obtained when the work function of the metal is less than the electron affinity of the n-type semiconductor (the barrier is negative), while for a p-type semiconductor it is obtained when the barrier is positive. The various mechanisms proposed for metal/semiconductor barrier formation are reviewed in Section 12.5. Current favored theories are based on the presence of surface or interface states, which when present on a high enough density at a particular energy in the bandgap can pin the Fermi level producing a barrier height that is approximately metal independent [328,329]. It was found that for $Hg_{1-x}Cd_xTe$ with $x<0.4$, ohmic contacts are expected on n-type and Schottky barriers are expected on p-type HgCdTe. For $x>0.4$, Schottky barriers are expected on both n- and p-type material. The common method of preparing ohmic contacts is by highly doping the surface region so that the space charge layer width is substantially reduced and the electrons tunnel through yielding low-resistance contacts. In practice, however, ohmic contacts are generally prepared according to empirically derived recipes and the underlying science of the surface is often poorly understood. Practical contacts are often composed of several layers of different metals, which are required in order to promote adhesion and reduce solid state reactions.

Interfacial reactions between HgCdTe and various metal overlayers can be classified into four groups: ultrareactive, reactive, intermediate reactive, and unreactive. It depends on the relative heat of formation of HgTe and the overlayer metal telluride [330], and on the heat of formation of intermetallic compounds, Cd and Hg, with the overlayer metal [331]. Deposition of an ultrareactive Ti or reactive metal (Al, In, Cr) results in the formation of a metal telluride and induces a loss of Hg from the interfacial regions.

Conversely, deposition of an unreactive metal, such as Au, results in a stoichiometric interface with little loss of Hg.

The most popular metal for n-$Hg_{1-x}Cd_xTe$ for many years is In [17,192,260], which has a low work function. Leech and Reeves examined In/n-HgCdTe contacts that displayed an ohmic nature over the range of composition $x = 0.30–0.68$ [332]. Carrier transport in these contacts has been attributed to a process of thermionic field emission. This behavior was attributed to the rapid in-diffusion of In, a substitutional donor in HgCdTe, to form a n^+ region below the contact. The value of specific contact resistance ranged from 2.6×10^{-5} Ωcm^2 at $x = 0.68$ through to 2.0×10^{-5} Ωcm^2 at $x = 0.30$, correlating with changes in the sheet resistivity of the HgCdTe.

In general, ohmic contacts to p-type HgCdTe is more difficult to realize since a larger work function of the metal contact is required. Au, Cr/Au, and Ti/Au have been most frequently used for both p-type HgCdTe. Investigations carried out by Beck et al. [333] have shown that Au and Al contacts to p-type $Hg_{0.79}Cd_{0.21}Te$ exhibited ohmic characteristics where the specific contact resistance varies from 9×10^{-4} Ωcm^2 to 3×10^{-3} Ωcm^2 at room temperature. The diameter dependence of $1/f$ noise implied that noise in the Au contacts originated at or near the Au/HgCdTe interface, while the noise in the Al contacts originated from a surface conduction layer near the contact.

For lightly doped p-type $Hg_{1-x}Cd_xTe$, no good contacts exist and all the metals tend to form Schottky barriers. The problem is especially difficult for high x composition material. This problem could be solved by heavy doping of semiconductor in the region close to the metallization to increase tunnel current, but the required level of doping is difficult to achieve in practice. One practical solution is to use rapid narrowing of bandgap at the $Hg_{1-x}Cd_xTe$-metal interface [334].

17.6.2 FUNDAMENTAL LIMITATION TO HgCdTe PHOTODIODE PERFORMANCE

From consideration carried out previously [see Section 17.4], the Auger mechanisms impose fundamental limitations to the HgCdTe photodiode performance. Assuming that the saturation dark current is only due to thermal generation in the base layer and that its thickness is low compared to the diffusion length,

$$J_s = Gtq, \tag{17.47}$$

where G is the generation rate in the base layer. Then the zero-bias resistance-area product is (see Equation 12.83)

$$R_oA = \frac{kT}{q^2Gt}. \tag{17.48}$$

Taking into account the A7 mechanism in extrinsic p-type region of n^+ -on-p photodiode, we receive

$$R_oA = \frac{2kT\tau^i_{A7}}{q^2N_at}, \tag{17.49}$$

and the same equation for p-on-n photodiode is

$$R_oA = \frac{2kT\tau^i_{A1}}{q^2N_dt}, \tag{17.50}$$

where N_a and N_d are the acceptor and donor concentrations in the base regions, respectively.

As Equations 17.49 and 17.50 show, the R_oA product can be decreased by reduction in the thickness of the base layer. Since $\gamma = \tau^i_{A7}/\tau^i_{A1} > 1$, a higher R_oA value can be achieved in the p-type base devices as compared to the n-type devices of the same doping level. Detailed analysis shows that the absolute maximum of R_oA is achievable with base layer doping, producing $p = \gamma^{1/2}n_i$, which corresponds to the minimum of thermal generation. The required p-type doping is difficult to achieve in practice for low-temperature photodiodes (the control of hole concentration below 5×10^{15} cm^{-3} level is difficult) and the p-type material suffers from some nonfundamental limitations, such as contacts, surface, and SR processes.

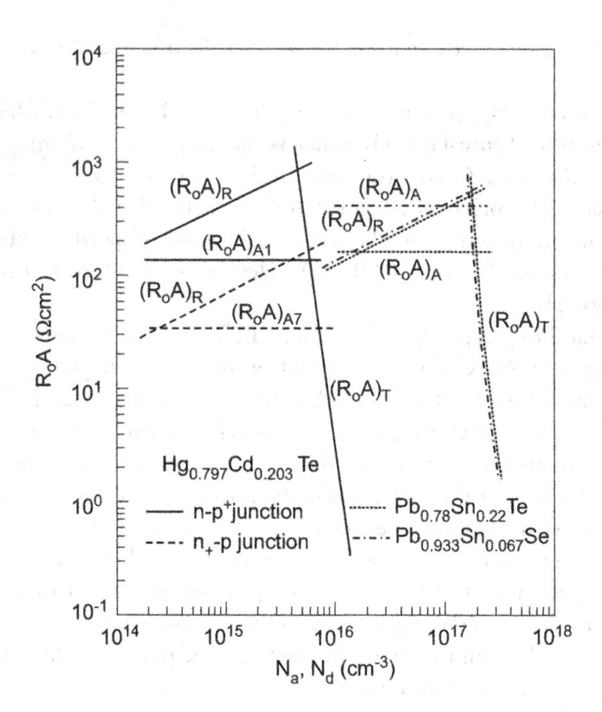

Figure 17.45 The dependence of the R_oA product components on the dopant concentrations for the one-sided abrupt junctions produced in $Hg_{0.797}Cd_{0.203}Te$, $Pb_{0.78}Sn_{0.22}Te$, and $Pb_{0.933}Sn_{0.067}Se$. (From Ref. [335])

In 1985, Rogalski and Larkowski indicated that due to the lower minority carrier diffusion length (lower mobility of holes) in the n-type region of p$^+$-on-n junctions with thick n-type active region, the diffusion-limited R_oA product of such junctions is larger than that for the n$^+$-on-p ones (see Figure 17.45) [335]. These theoretical predictions were next confirmed by experimental results obtained for p$^+$-on-n HgCdTe junctions.

The thickness of the base region should be optimized for near-unity quantum efficiency and a low dark current. This is achieved with a base thickness slightly higher than the inverse absorption coefficient for single pass devices; $t = 1/\alpha$ (which is ≈ 10 µm) or half of $1/\alpha$ for double pass devices (devices supplied with a retroreflector). Low doping is beneficial for a low thermal generation and high quantum efficiency. Since the diffusion length in the absorbing region is typically longer than its thickness, any carrier generated in the base region can be collected, giving rise to the photocurrent.

Different HgCdTe photodiode architectures have been fabricated that are compatible with back side and front side illuminated hybrid FPA technology. The eight most important architectures are included in Table 17.5, which summarizes the applications of HgCdTe photodiode designs by the major FPA manufacturers today.

Figures of configurations II and III show cross sections of the two most important n-on-p HgCdTe junction structures adapted in fabrication of multicolor detectors. The structure III pioneered by SAT has been the most widely developed and used by Sofradir [337]. The first type of structure (configuration I) is a vertically integrated photodiode (VIP™) developed by DRS Infrared Technologies [342]. This structure currently referred to as a high-density vertically integrated photodiode (HDVIP) is similar to the British-developed loophole photodiode (configuration II) [281]. This n$^+$-n$^-$-p architecture is formed around the via, both by the etching process itself and by a subsequent ion implant step. Low-background In n$^-$-doping levels of 1.5–5 × 10^{14} cm^{-3} are routinely used. The p-type dopant is usually Cu (acceptor concentration about 4 × 10^{16} cm^{-3}). The p$^+$-p noninjection contacts are formed in each cell of the FPA and are joined electrically by a top surface metal grid as shown in Figure 17.46.

The cross sections of the second type of structure used in fabrication of multicolor detectors, heterojunction p-on-n HgCdTe photodiode are illustrated as configuration VI (see also Figure 17.42a). In these so-called DLHJ structures, an absorber layer about 10 µm thick is doped with In at 1 × 10^{15} cm^{-3} or less, and is

Table 17.5 HgCdTe photodiode architectures used for hybrid FPAs

CONFIGURATION		ARCHITECTURE	JUNCTION FORMATION	COMPANY	REFERENCES
I	n-on-p VIP	hν; n-HgCdTe; p-HgCdTe; Epoxy; Silicon ROIC	Ion implantation forms n-on-p diode in p-type HgCdTe, grown by Te-solution LPE on CdZnTe and epoxied to Si ROIC wafer; over the edge contact.	DRS Infrared Technologies (formerly Texas Instruments)	336
II	n-p loophole	hν; p-HgCdTe; n; n; p-HgCdTe; Epoxy; Silicon ROIC	Ion beam milling forms n-type islands in p-type Hg vacancy-doped layer grown by Te-solution LPE on CdZnTe, and epoxied onto Si ROIC wafer; cylindrical lateral collection diodes	GEC-Marconi Infrared (GMIRL)	262,281
III	n⁺-on-p planar	n^+-HgCdTe; p-HgCdTe; CdZnTe substrate	Ion implant into acceptor-doped p-type LPE film grown by Te-solution slider	Sofradir (Societe Francaise de Detecteurs Infrarouge)	337
IV	n^+-n^--p planar homojunctions	n-HgCdTe; p-HgCdTe; CdZnTe or sapphire substrate	B implant into Hg-vacancy p-type, grown by Hg-solution tipper on 3″ dia. sapphire with MOCVD CdTe buffer; ZnS passivation	Rockwell/Boeing	338
V	n^+-n^--p planar homojunctions	Si_3N_4; SiO_2; Widegap graded layer; n^+-HgCdTe; p-HgCdTe; CdZnTe buffer; GaAs substrate	MBE-grown n-type layer on GaAs substrate with CdZnTe buffer layer. B implant into converted p-type layer. SiO_2/Si_3N_4 passivation.	Institute of Semiconductor Physics, Novosibirsk	69

(continued)

Table 17.5 (*Continued*) HgCdTe photodiode architectures used for hybrid FPAs

CONFIGURATION		ARCHITECTURE	JUNCTION FORMATION	COMPANY	REFERENCES
VI	P-on-n mesa	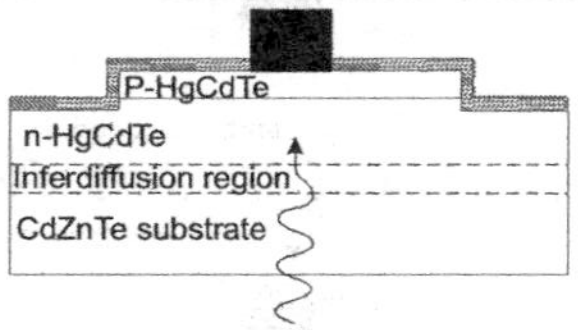 P-HgCdTe n-HgCdTe Interdiffusion region CdZnTe substrate	1. Two-layer LPE on CdZnTe: Base: Te-solution slider, In-doped Cap: Hg-solution dipper, As-doped 2. MOCVD in situ on CdZnTe I-doped base, As-doped cap	IR Imaging Systems, Sanders – A Lockheed Martin Company (LMIRIS)	339,56
VII	P-on-n mesa	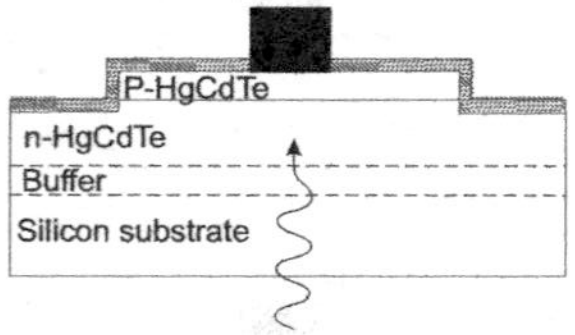 P-HgCdTe n-HgCdTe Buffer Silicon substrate	1.Two-layer LPE on CdZnTe or Si:Base: Hg-solution dipper, In-dopedCap: Hg-solution dipper, As-doped 2. MBE in situ on CdZnTe or SiIn-doped base, As-doped cap	Raytheon Infrared Center of Excellence (RIRCoE, formerly SBRC) and Hughes Research Laboratories (HRLs)	340, 341
VIII	P-on-n planar buried heterostructure	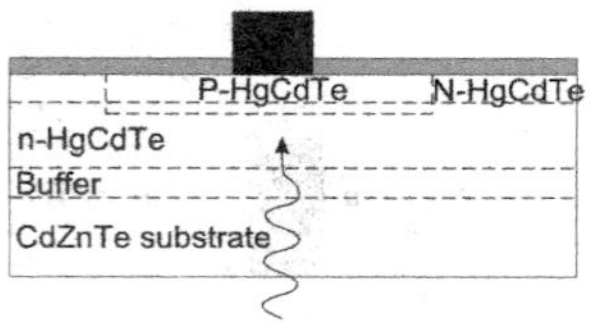 P-HgCdTe N-HgCdTe n-HgCdTe Buffer CdZnTe substrate	As implant into In-doped N-n or N-n-N film grown by MBE on CdZnTe	Rockwell/Boeing	338
IX	P-on-n mesa heterjunction	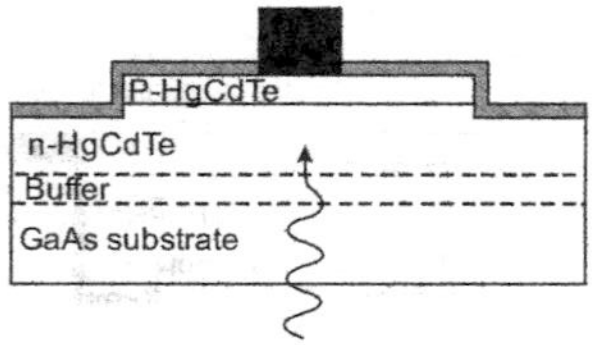 P-HgCdTe n-HgCdTe Buffer GaAs substrate	MOCVD *in situ* on GaAs Iodine-doped base, arsenic-doped cap	Selex Galileo	99

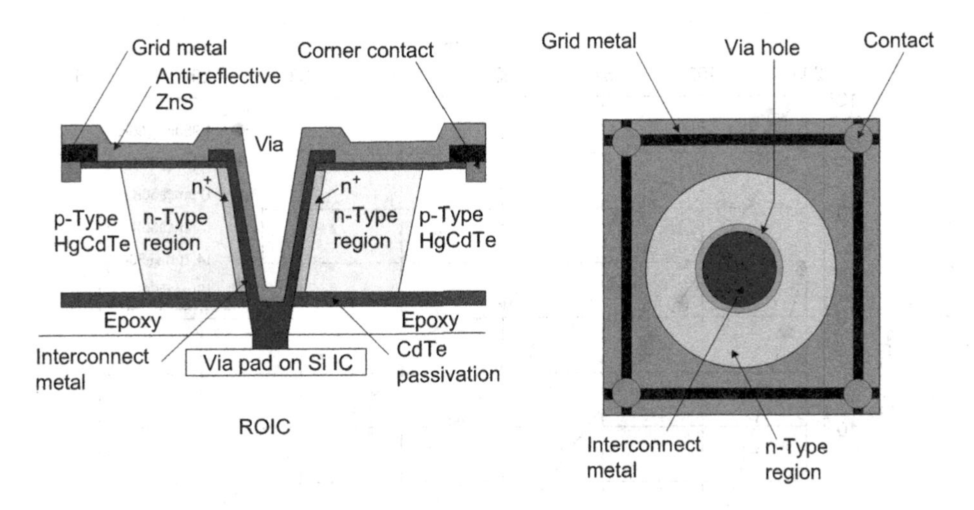

Figure 17.46 DRS's HDVIP™ n^+-n^--p HgCdTe photodiode. (From Ref. [342])

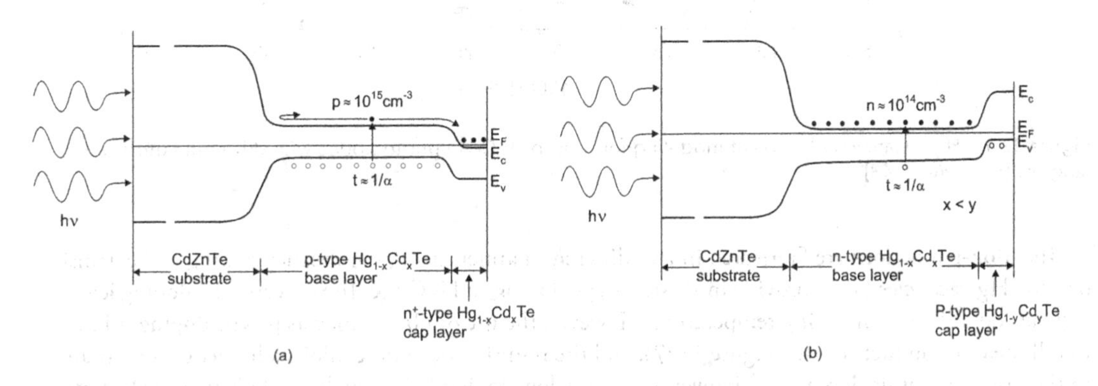

Figure 17.47 Schematic band diagrams of (a) n^+-on-p and (b) p-on-n heterojunction photodiodes.

sandwiched between the CdZnTe substrate and the highly As-doped, wider-gap region. Contacts are made to the p^+-layer in each pixel and to the common n-type layer at the edge of the array (not shown). IR flux is incident through the IR-transparent substrate.

The formation of planar p-on-n photodiodes is achieved by selective area As ion implantation through the cap layer into the narrow-gap base layer [307]. The dopant activation step is achieved by a two-step thermal anneal under Hg overpressure; the first step, at high temperature, activates the dopant by substituting As atoms on the Te-sublattice, and the second, at lower temperature, annihilates the Hg vacancies formed in the HgCdTe lattice during growth and high-temperature annealing step.

Figure 17.47 shows the schematic band profiles of the most commonly used unbiased homo-(n^+-on-p) and heterojunction (P-on-n) photodiodes. To avoid contribution of the tunneling current, a doping concentration, in the base region, below 10^{16} cm^{-3} is required. In both photodiodes, the lightly doped narrow-gap absorbing region [base of the photodiode: p(n)-type carrier concentration of about 1×10^{15} cm^{-3} (1×10^{14} cm^{-3})] determines the dark current and photocurrent. The internal electric fields at interfaces are blocking for minority carriers and the influence of surface recombination is eliminated. Also, suitable passivation prevents the influence of surface recombination. In is most frequently used as a well-controlled dopant for n-type doping due to its high solubility and moderately high diffusion. Elements of the VB group are acceptors substituting the Te sites. They are very useful for fabrication of stable junctions due to very low diffusivity. As proved to be the most successful p-type dopant to date.

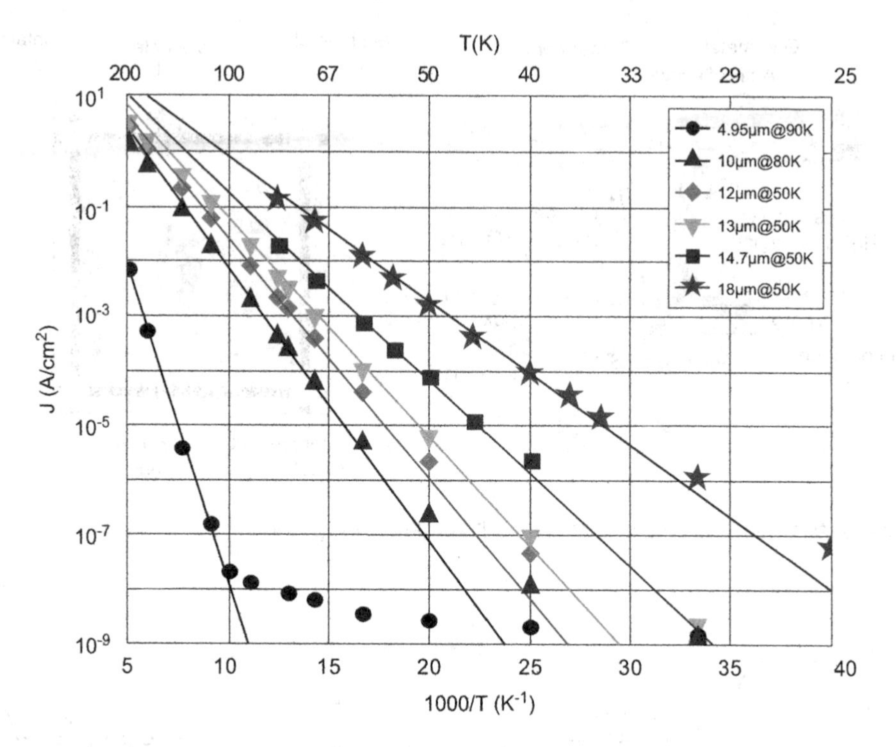

Figure 17.48 Hg-vacancy dark current modeling for n-on-p HgCdTe photodiodes with different cutoff wavelengths. (From Ref. [343])

The n-on-p junctions are fabricated in two different manners using Hg vacancy doping and extrinsic doping. Hg vacancies (V_{Hg}) provide intrinsic p-type doping in HgCdTe. In this case, the doping level depends on only one annealing temperature. However, the use of Hg vacancy as p-type doping is known to kill the electron lifetime (see Figure 17.17), and the resulting detector exhibits a higher current than in the case of extrinsic doping use. However, for very low doping ($<10^{15}$ cm^{-3}), the hole lifetime becomes SR-limited and does not depend on doping anymore [139]. V_{Hg} technology leads to low minority diffusion length of the order of 10–15 μm, depending on doping level. Generally, n-on-p vacancy-doped diodes give rather high diffusion currents but lead to a robust technology as its performance weakly depends on doping level and absorbing layer thickness. Simple modeling manages to describe dark current behavior of V_{Hg}-doped n-on-p junctions over a range of at least eight orders of magnitude (see Figure 17.48 [343]). In case of extrinsic doping, Cu, Au, and As are often used. Due to higher minority carrier lifetime, extrinsic doping is used for low dark current (low flux) applications [147]. The extrinsic doping usually leads to larger diffusion length and allows lower diffusion current but might exhibit performance fluctuations, thus affecting yield and uniformity.

In case of p-on-n configuration, the typical diffusion length is up to 30–50 μm for low doping levels, such as 10^{15} cm^{-3}, typically reached with In doping and the dark carrier generation is volume-limited by the absorbing layer volume itself [343]. Different behaviors of saturation current depending on diffusion lengths is explained by Equations 12.87 and 12.88. If $t >> L$, the saturation current is inversely proportional to $N\tau^{1/2}$. On the contrary, if $t << L$, the saturation dark current is inversely proportional to $N\tau$.

Although n-on-p technology (Hg vacancy doping) offers excellent yield and production capability, p-on-n technology is necessary to achieve low dark current and thus allows good SNR performance. Compared to n-on-p technology, the detector temperature can be higher 10 K–20 K while keeping the same performance.

Tennant et al. [344] have developed a simple empirical relationship that describes the dark current behavior with temperature and wavelengths for the better Teledyne HgCdTe diodes and arrays (primarily double-layer planar heterojunction (DLPH) structure devices). It is called Rule 07 and predicts the

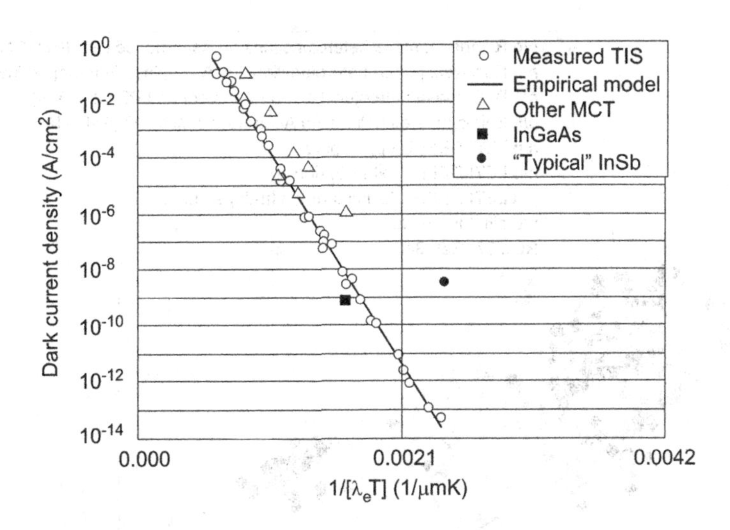

Figure 17.49 The data fit leading to Rule 07. (From Ref. [344])

dark current density within a factor of 2.5 over a 13 order of magnitude range. The formula for Rule 07 is approximately (exact formula given in the reference):

$$J_{\text{dark}} = 8367 \exp\left(-\frac{1.44212q}{k\lambda_c T}\right) \quad \text{for} \quad \lambda_c \geq 4.635\,\mu\text{m}, \tag{17.51}$$

and

$$J_{\text{dark}} = 8367 \exp\left\{-\frac{1.44212q}{k\lambda_c T}\left[1-0.2008\left(\frac{4.635-\lambda_c}{4.635\lambda_c}\right)^{0.544}\right]\right\} \quad \text{for} \quad \lambda_c < 4.635\,\mu\text{m}, \tag{17.52}$$

where λ_c is the cutoff wavelength in μm, T is the operating temperature in K, q is the electron charge, and k is the Boltzmann's constant (the last two in SI units). Rule 07 was developed for operating temperature cut-off wavelength products between 400 μmK and ~1,700 μmK, and for operating temperatures above 77 K.

It should be marked, however, that the Rule 07 criterion is merely a manifestation of a detector architecture that is limited by A1 diffusion currents from 10^{15} cm^{-3} n-type material. Any detector architecture that is limited by A7 p-type diffusion or by depletion currents will not behave according to Rule 07.

Rule 07 is also an excellent tool for a quick comparison of R_oA product of other material systems with HgCdTe. However, caution should be taken when expanding it to other parameters, such as detectivity and lower operating temperatures.

Figure 17.49 shows the fit data, both with Teledyne data and some of the better data from other vendors for comparison, including representative data from both InSb (~5.3 μm) and InGaAs (~1.7 μm). InGaAs shows lower dark current density than DLPH HgCdTe, but InSb has much higher dark current density. Note that λ_e equals the cutoff wavelength for wavelengths above ~5 μm, but it is somewhat longer than the cutoff for wavelengths below 5 μm.

Figure 17.50 is an accumulation of R_oA data with different cutoff wavelengths [343] taken from the Laboratoire Infrarouge (LIR) of the Electronics and Information Technology Laboratory (LETI) using both V_{Hg} and extrinsic doping n-on-p HgCdTe photodiodes compared with P-on-n data from other laboratories, extracted from various recent literature reports using various techniques and different diode structures [345–348]. It is clearly shown that extrinsic doping of n-on-p photodiodes leads to higher R_oA product (lower dark current), whereas Hg vacancy doping remains on the bottom part of the plot. P-on-n structures are characterized by the lowest dark current (highest R_oA product); see the trend line calculated with Teledyne (formerly Rockwell Scientific) empirical model [344].

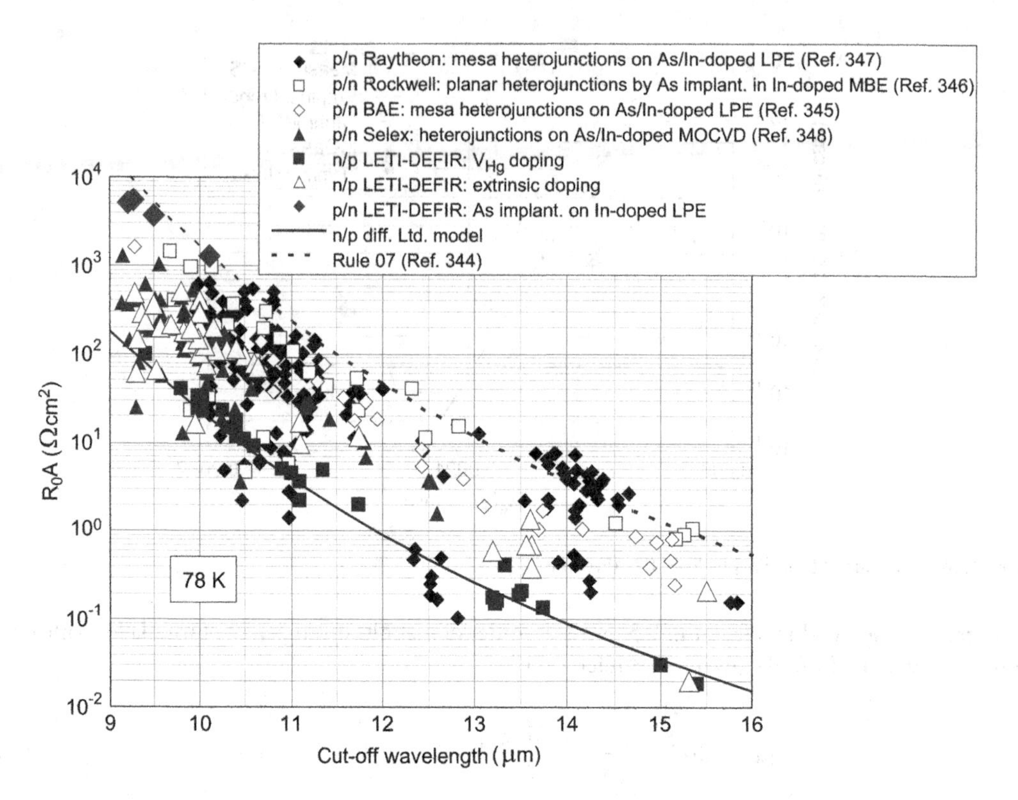

Figure 17.50 R_oA product versus cutoff wavelength at 78 K, summarized with bibliographic data. (From Ref. [343])

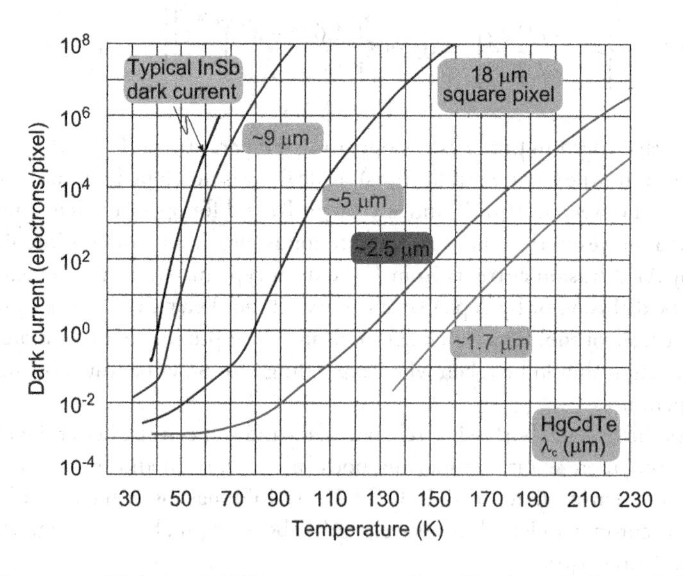

Figure 17.51 The dark current of Teledyne's MBE-grown p-on-n HgCdTe photodiodes scaled for an 18 μm square pixel. (From Ref. [349])

An additional insight on dark current achieved by Teledyne gives Figure 17.51 [349] where the dark current scaled for an 18 μm square pixel of MBE-grown P-on-n HgCdTe photodiodes is shown. The dark current less than 0.01 electrons per pixel per second for 2.5 and 5 μm cutoff wavelengths can be achieved. It should be marked that the cutoff wavelength is shown with the approximation symbol, since λ_c is a function of temperature and there will be a slight variation in the cutoff wavelength of a HgCdTe photodiode as it cools.

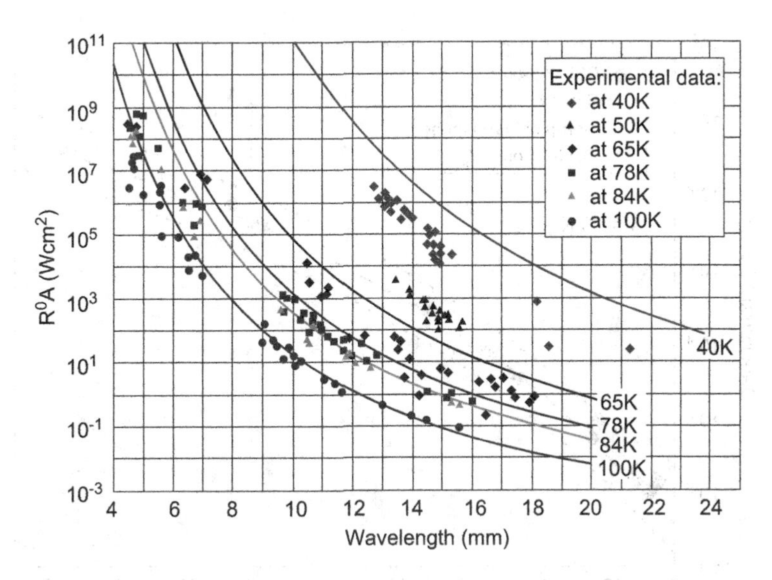

Figure 17.52 R_0A product versus cutoff wavelength for Teledyne's (formerly Rockwell Scientific) p-on-n HgCdTe photodiode data at various temperatures compared to the theoretical 1D diffusion model. (From Ref. [346])

The quality of HgCdTe photodiodes has improved steadily over the past 30 years as materials and device-processing science evolved and has progressed to the point where there is usually no clear indication of GR current. The plots of R_oA data versus temperature generally follow a diffusion current dependence at higher temperatures, and transition into a comparatively temperature-independent tunneling-like regime at lower temperatures. Figure 17.52 illustrates the highest measurable R_oA values of P-on-n HgCdTe photodiodes versus cutoff wavelength at different temperatures [346]. The solid lines are theoretically calculated using a 1D model, which assumes that the diffusion current from a narrower bandgap n-side is dominant, and the minority carrier recombination via Auger and radiative process. The R_oA of Teledyne HgCdTe photodiodes exhibit near-theoretical performance for various growth material cutoffs at various temperatures. The average value of the R_oA product at 77 K for a 10 μm cutoff HgCdTe photodiode is around 1000 Ωcm² and drops to 200 Ωcm² at 12 μm. At 40 K, the R_oA product varies between 10^6 and 10^8 Ωcm² at 12 μm.

Several major advantages in the development of VLWIR HgCdTe technology have been presented by several research groups [17,346,350–352]. Continued improvements in defect reduction have enabled the R_oA products to follow the diffusion current limit down to temperatures of 30 K–40 K for cutoff wavelengths out to 20 μm. Usually, development of such VLWIR photodiodes is based on LPE technology. The different values of dark current obtained at 78 K are gathered onto Figure 17.53. As can be seen, the experimental data for p-on-n devices perfectly follows the diffusion line represented by Rule 07. In case of n-on-p devices, the dark current is above one order of magnitude larger.

The A1 and A7 recombination mechanisms are relatively insignificant in the wide-gap HgCdTe alloys needed for SWIR and MWIR applications, thus the only fundamental recombination mechanism to be considered is radiative recombination. This is illustrated in Figure 16.10a for three groups of p-on-n devices fabricated using LPE, MBE, and MOCVD. For photodiodes at temperature 180 K, the R_oA data fall generally about a factor of ten below the theoretical curves, indicating that a lifetime mechanism other than traditional radiative recombination is lowering the lifetime. According to DeWames et al. [353], a shallow SR recombination center, possibly process-induced, is responsible for the reduced lifetime. It appears that for photodiodes operated at room temperature with cutoff wavelengths less than 3.5 μm, the R_oA product fall short of the limits calculated with traditional radiative recombination equations. However, the highest quality SW photodiodes fabricated with HgCdTe (and also with InGaAs alloys), have performance levels in agreement with the radiative limit.

The performance of conventional p-n junction LW HgCdTe photovoltaic detectors operating at near-room temperature is very poor due to low quantum efficiency (low diffusion length and weak absorption of radiation) and a low dynamic resistance [168]. Even more serious problems are caused by the series

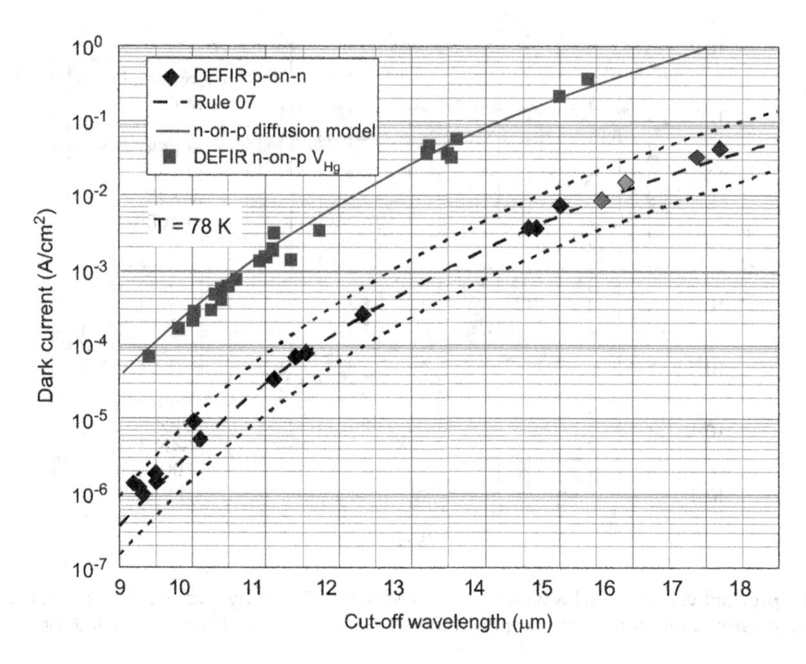

Figure 17.53 Summary of dark currents measured at 78 K for both n-on-p and p-on-n HgCdTe photodiodes from IWIR to VLWIR spectral ranges. (Adapted after Ref. [352])

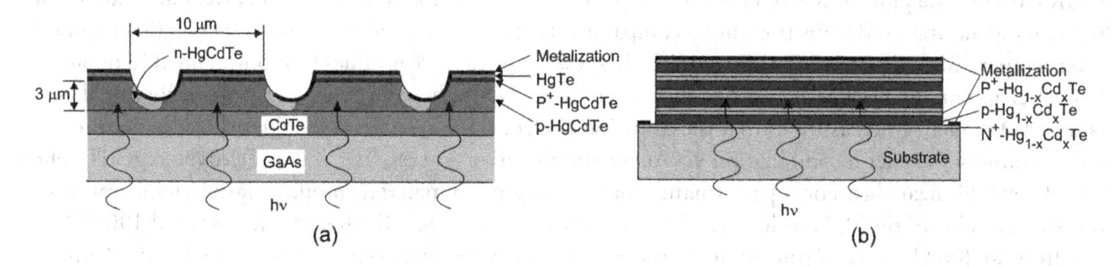

Figure 17.54 Back side illuminated multiple heterojunction devices: (a) junction's planes perpendicular to the substrate and (b) four-cells stacked multiple detector. (From Ref. [164])

resistance and extremely low junction resistance and voltage responsivity. This can be overcome with the development of multiple heterojunction photovoltaic devices in which short elements are connected in series. An example is a device with the junction's planes perpendicular to the substrate (Figure 17.54a). The multiheterojunction device consisted of a structure based on back side illuminated n$^+$-p-P photodiodes. This device was the first commercially available uncooled and unbiased LW photovoltaic detector introduced in 1995. Such devices are characterized by large voltage responsivity, fast response time, but they suffer from nonuniform response across the active area and dependence of response on the polarization of incident radiation.

More promising are the stacked photovoltaic cells monolithically connected in series shown in Figure 17.54b. They are capable of achieving both good quantum efficiency and a large differential resistance. Each cell is composed of p-type doped narrow-gap absorber and heavily doped N$^+$ and P$^+$ heterojunction contacts. The incoming radiation is absorbed only in the absorber regions, while the heterojunction contacts collect the photogenerated charge carriers. Such devices are capable of achieving high quantum efficiency, large differential resistance, and fast response. The practical problem is the shortage of adjacent N$^+$ and P$^+$ regions. This can be achieved by employing tunnel currents at the N$^+$ and P$^+$ interface.

Figures 17.55 and 17.56 show the performance of the HOT HgCdTe devices [354]. Without optical immersion, MWIR photovoltaic detectors are sub-BLIP devices with performance close to the GR limit, but well-designed, optically immersed devices approach BLIP limit when thermoelectrically cooled with

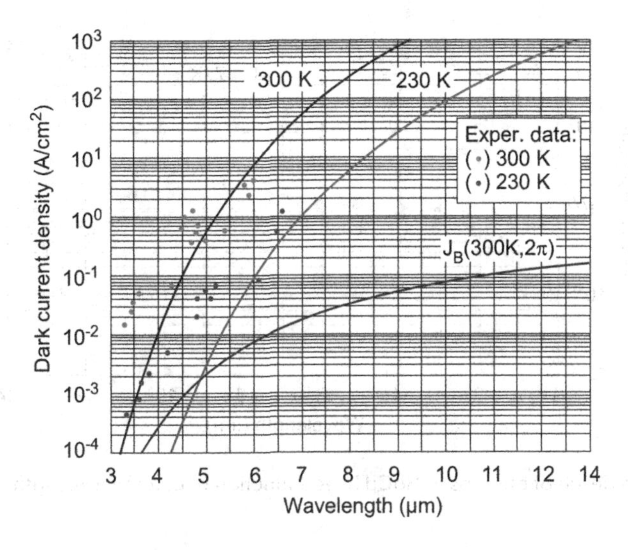

Figure 17.55 Dark current of optically immersed HOT HgCdTe devices at 300 and 230 K. (From Ref. [354])

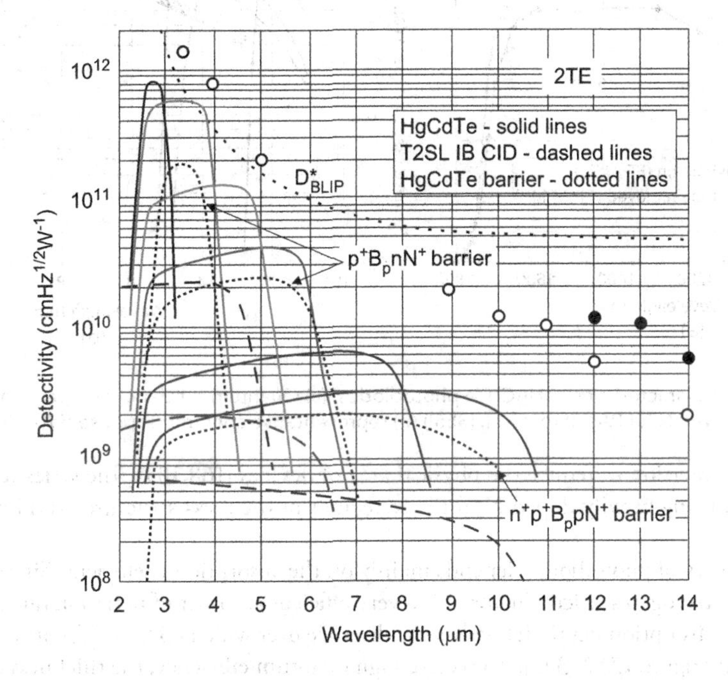

Figure 17.56 Typical spectral detectivity curves of HgCdTe immersed detectors and with two-stage thermoelectric (TE) coolers (solid lines). The best experimental data (white dots) are measured for detectors with FOV equal to 36 degrees. BLIP detectivity is calculated for FOV = 2π. Black dots are measured for detectors with four-stage TE coolers (courtesy Vigo-Systems). Also, spectral detectivity curves for three T2SL interband (IB) cascade IR detectors (CIDs) (without immersion) and barrier detectors (without immersion) are shown for comparison (dashed and dotted lines, respectively).

two-stage Peltier coolers. The situation is less favorable for >8 μm LWIR photovoltaic detectors; they show detectivities below the BLIP limit by an order of magnitude. Typically, the devices are used at zero bias. The attempts to use Auger-suppressed nonequilibrium devices were not successful due to large $1/f$ noise extending to ≈100 MHz in extracted photodiode.

The HOT devices are characterized by a very fast response. The uncooled ≈10 μm photodetectors show ≈1 ns or less response time. The *RC* time constant of photovoltaic devices can be shortened by

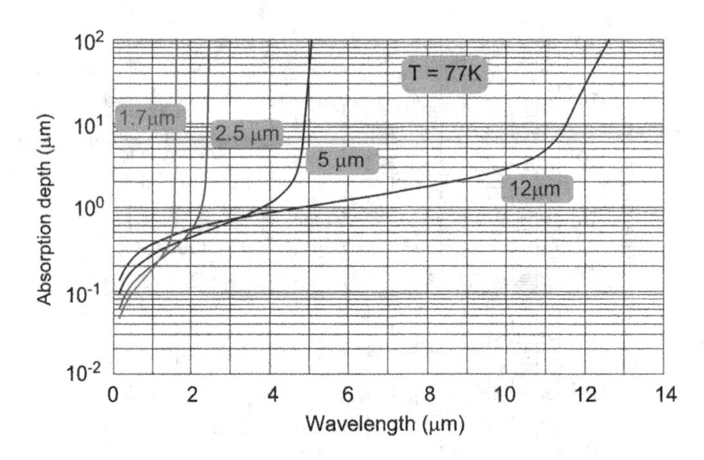

Figure 17.57 Absorption depth of photons in HgCdTe as a function of cutoff wavelength at 77 K. (From Ref. [349])

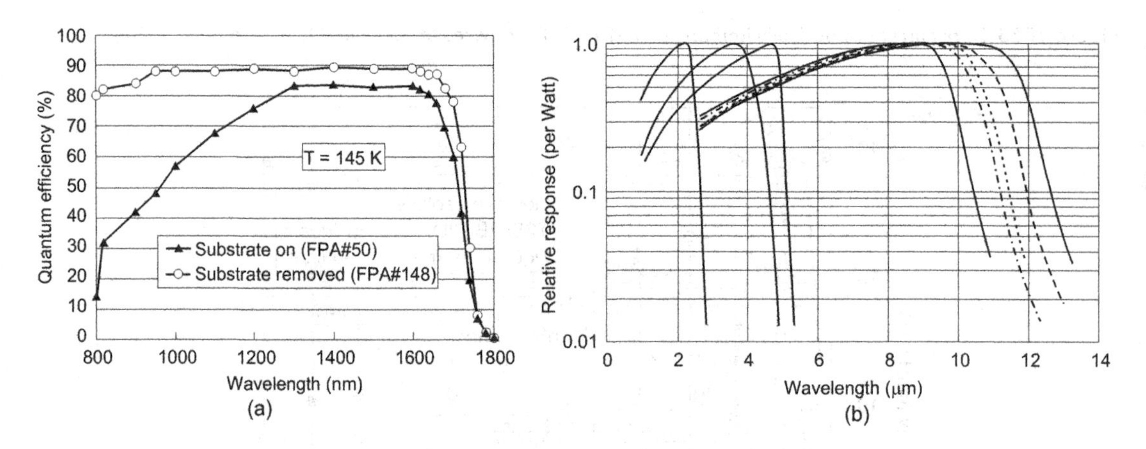

Figure 17.58 Spectral characteristics of HgCdTe photodiodes: (a) Quantum efficiency comparison for substrate-on versus substrate-removed NIR FPAs. (From Ref. [355]) (b) representative spectral response data. (From Ref. [236])

the use of optical immersion to reduce the physical area of devices [168,185]. The series resistance was minimized to ≈1 Ω using heavily doped N^+ for base regions of the mesa structures and improved anode contact.

Spectral responsivity of photodiodes depends mainly on the absorption coefficient. Since HgCdTe ternary alloy is a direct bandgap semiconductor, it is a very efficient absorber of radiation due to high absorption coefficient. The absorption depth defined as the distance over which 63% (1–1/e) of the photon flux is absorbed is shown in Figure 17.57 [349]. To receive high quantum efficiency, the thickness of active detector layer should be equal, at least to the cutoff wavelength. Typical quantum efficiency with antireflection coating is about 90%. Substrate removing of the back side illuminated photodiode results in an increase of quantum efficiency in the SW spectral range. This effect is shown in Figure 17.58a after removal of the CdZnTe substrate after the hybridization process [355]. HgCdTe photodiodes are available to cover the spectral range from 1–20 μm. Figure 17.58b illustrates the representative spectral response from photodiodes. Spectral cutoff can be tailored by adjusting the HgCdTe alloy composition [356].

17.6.3 NONFUNDAMENTAL LIMITATION TO HgCdTe PHOTODIODE PERFORMANCE

Recently, Gravrand et al. have summarized DEFIR (CEA-LETI and Sofradir joint laboratory) achievements in the fabrication of HgCdTe photodiodes showing the dark current evolution with respect to E_g/kT (see Figure 17.59) [357]. Diffusion-limited photodiodes are usually obtained at intermediate temperature following proportionality to the intrinsic carrier concentration squared $n_i^2\left[\propto \exp\left(E_g/kT\right)\right]$. When the

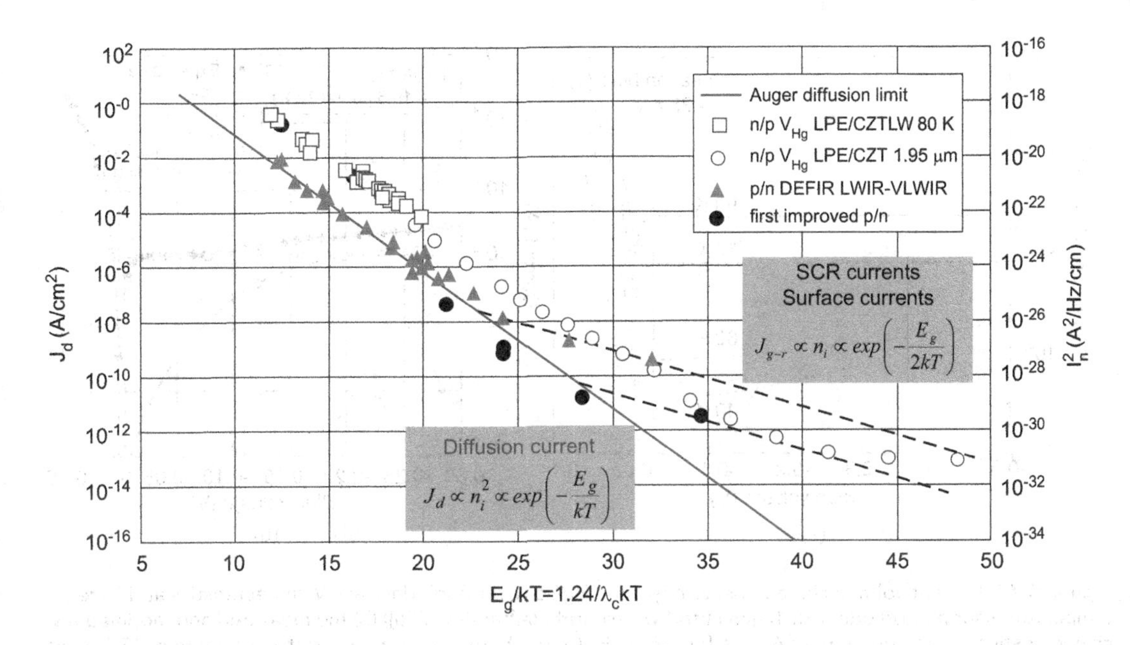

Figure 17.59 Dark current and noise in HgCdTe development. (Adapted after Ref. [357])

noise is limited by shot noise, this dark current plot also represents the dark noise power spectral density $i_n^2 = 2qJ_{dark}$, which is indicated on the right-hand side secondary axis. The dark current lies on a different line depending on the device design (Hg-vacancy-doped n-on-p or p-on-n). At lower temperature, the diffusion current becomes so low that other dark current sources are dominated for which the thermal evolution is different. These currents are related to defect physics rather than bulk physics and are not a fundamental limitation of the photodiode. For example, the space charge region (SCR) GR current is proportionally the intrinsic carrier concentration n_i [$\propto exp(E_g/2kT)$]. As is shown in Figure 17.59, the improved p-on-n device has been optimized to reduce the depletion region current. The low-frequency noise is expected to evolve proportionally to the current squared—represented by dashed lines in the figure. Some detectors switch from the shot noise limit to low-frequency limit and, in consequence, result in performance degradation.

17.6.3.1 Current-voltage characteristics

Many additional excess mechanisms affecting the dark current of HgCdTe photodiodes arise from nonfundamental sources located in the base and cap layer, the depletion region, and the surface [79]. As the operating temperature is lowered, the thermal dark current mechanisms become weaker and allow other mechanisms to prevail. The main leakage mechanisms of HgCdTe photodiodes are: generation in the depletion region, interband tunneling, trap-assisted tunneling (TAT), and impact ionization. Some of them are caused by structural defects in the p-n junction. These mechanisms receive much attention now, particularly because they ultimately determine the array uniformity, yield, and cost for some applications, especially those with lower operating temperatures.

Many authors have successfully modeled the key reverse bias leakage mechanism in LWIR HgCdTe photodiodes at ≲77 K in terms of TAT[16,358–368]. An alternative model to explain the reverse bias leakage current in LWIR HgCdTe loophole photodiodes has been proposed by Elliott et al. [369] This model is based upon an impact ionization effect within the depletion layer and, in fairly low doped devices, this gives a good fit to practical observations. The reverse bias characteristics arising from impact ionization have two main features: first, the leakage current is fairly insensitive to temperature over a wide range and, secondly, the current increases much more slowly with reverse bias than the currents due to tunneling.

In high-quality $Hg_{1-x}Cd_xTe$ photodiodes with $x \approx 0.20$, the diffusion current in the zero-bias and the low-bias region is usually the dominant current down to 40 K [342,370–372]. At medium values of reverse bias, the dark current is mostly due to TAT. TAT dominates the dark current also at zero bias and very low

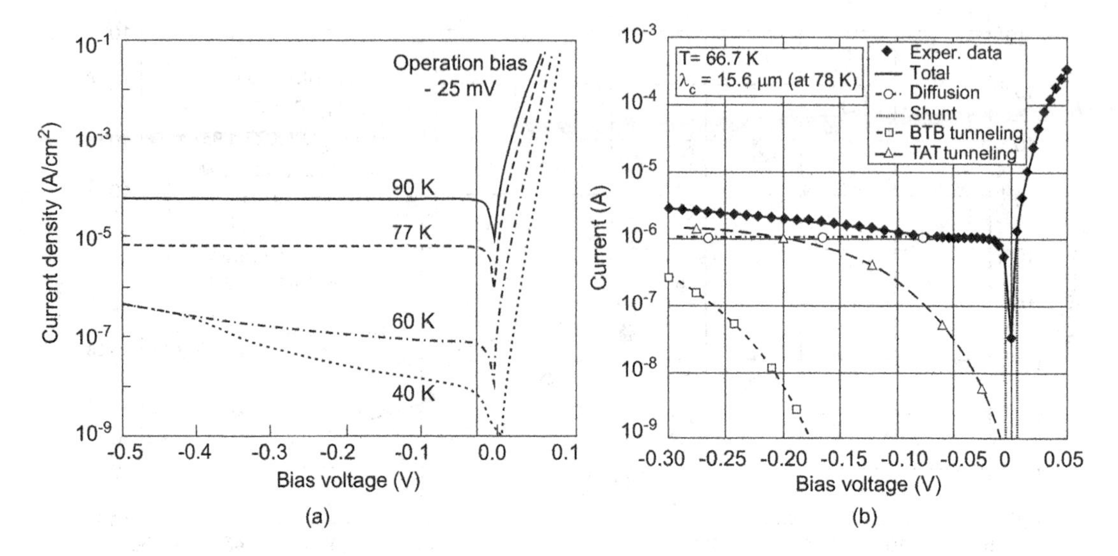

Figure 17.60 Current-voltage characteristics of p-on-n HgCdTe photodiodes: (a) I–V characteristics at different temperatures for photodiode with 12-μm cutoff wavelength. (After Ref. [356]) (b) the measured and modeled I–V characteristics at a temperature of 66.7 K for p-on-n HgCdTe photodiode with a cutoff wavelength of 15.6 μm at 78 K. (Adapted after Ref. [371])

temperatures (below 30 K). At high values of reverse bias, bulk band-to-band tunneling (BTB) dominates. At very low temperatures, below 30 K, significant spreads in the R_oA product distributions are typically observed due to the onset of tunneling currents associated with localized defects. Moreover, HgCdTe photodiodes often have an additional surface-related component of the dark current, particularly at low temperatures.

In ideal photodiodes, the diffusion current is dominant, therefore their leakage current is very low and insensitive to the detector bias. Leakage current is the primary contribution of unwanted noise. Figure 17.60 shows typical current-voltage characteristics of a p-on-n HgCdTe photodiode at temperatures between 40 K and 90 K for a 12 μm cutoff detector at 40 K [356]. The leakage current is less than 10^{-5} A/cm^2 at 77 K [see Figure 17.60a]. The bias-independent leakage current makes it easier to achieve better FPA uniformity as well as to reduce the detector bias-control requirements during changes in photocurrent. Figure 17.60b shows that the measured and modeled data at a temperature of 66.7 K for this type of photodiode, with a cutoff wavelength of 15.6 μm at 78 K, equates with 18.7 μm at 40 K and 20.0 μm at 28 K [371]. The I-V characteristics were found to be diffusion-limited by the ideal diffusion currents over much of the temperature and bias ranges analyzed. BTB was found to limit the current at the largest reverse biases (>200 mV) and lowest temperatures. The TAT mechanisms limit the current under low temperatures and moderate bias (50 mV < V < 200 mV).

Figure 17.61 presents the *I-V* curves measured at different temperatures for LWIR p-on-n and VLWIR n-on-p photodiodes fabricated by DEFIR [343]. It is shown that thermal current strongly decreases with temperature, but the high bias current exhibits the opposite thermal behavior, which is the usual signature of tunnel contributions. In order to minimize tunneling currents, low doping levels are preferable on at least one side of the junction. Note, however, that the p-on-n photodiode active region is doped to $N_d \leq 10^{15}$ cm^{-3}, whereas the n-on-p doping level is higher, $N_a \approx 10^{16}$ cm^{-3}. We can also see that the diffusion-limited performance extends to larger reverse biases at higher temperatures.

Chen et al. [373] carried out a detailed analysis of the wide distribution of the R_o values of HgCdTe photodiodes operating at 40 K. Figure 17.62 shows the cumulative distribution function, R_o, obtained in devices with a cutoff wavelength of 9.4–10.5 μm. It is clear that while some devices exhibit a fair operability with R_o values spanning only two orders of magnitude, other devices show a poor operability with R_o values spanning more than 5–6 orders of magnitude. Lower performance, with R_o values below 7×10^6 Ω at 40 K, is usually due to gross metallurgical defects, such as dislocation clusters and loops, pin holes,

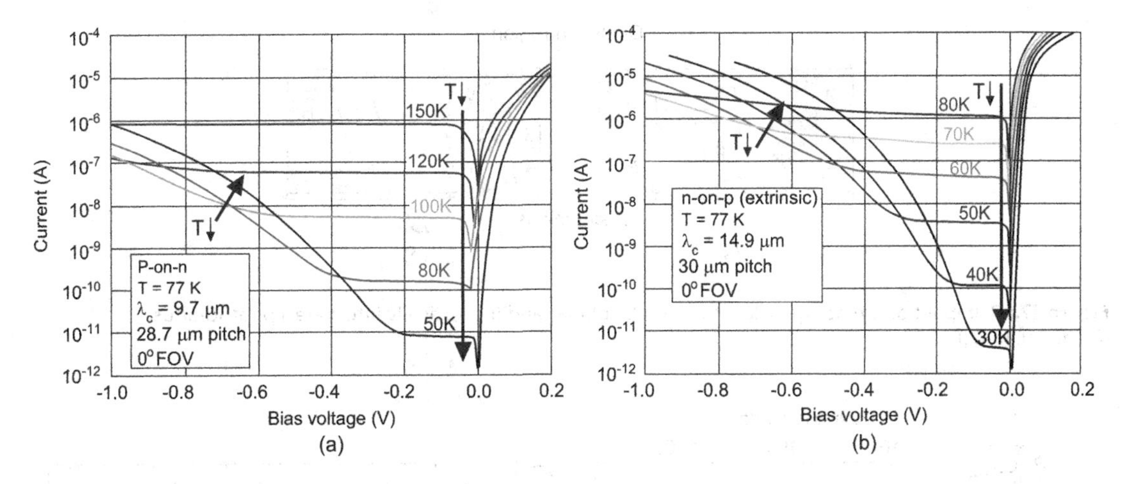

Figure 17.61 Typical dark current-voltage characteristics of (a) a p-on-n and (b) n-on-p HgCdTe photodiodes at different temperatures. (From Ref. [343])

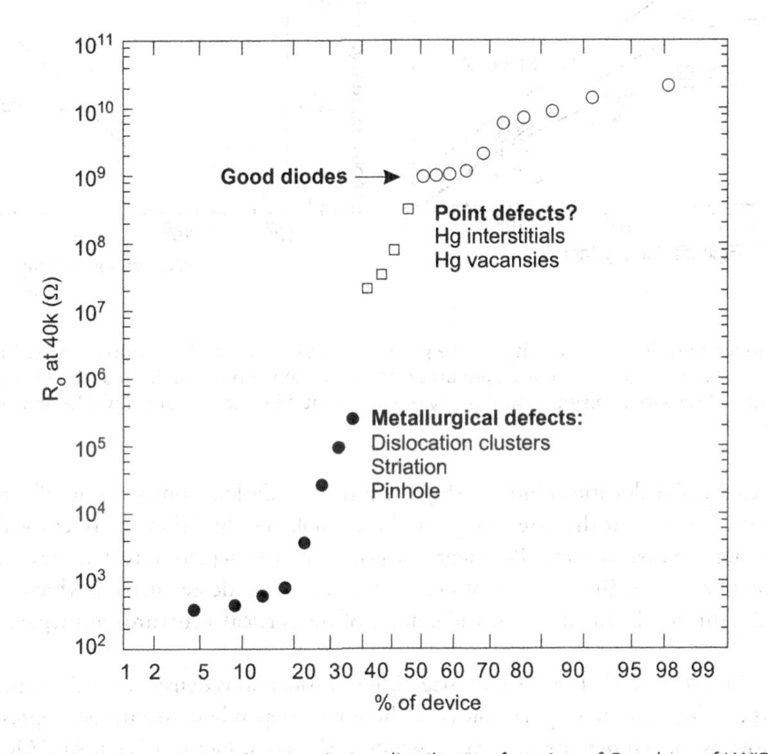

Figure 17.62 Detailed analysis separates the cumulative distribution function of R_o values of LWIR p-on-n HgCdTe photodiodes (fabricated by LPE) into three regions: good diodes, diodes affected by point defects, and diodes affected by metallurgical defects. (From Ref. [373])

striations, Te inclusions, and heavy terracing. However, diodes with R_o values 7×10^6–1×10^9 Ω at 40 K contained no visible defects (Hg interstitials and vacancies).

17.6.3.2 Dislocations and 1/f noise

Dislocations are known to increase the dark current and the $1/f$ noise current [140–142]. At 77 K, the requirement on dislocation density for LWIR material is $<2 \times 10^5$ cm^{-2}. MWIR, on the other hand, can tolerate higher densities of dislocations at 77 K, but evidence is mounting for the fact that this is no longer true at higher operating temperatures [342]. The reverse bias characteristics of HgCdTe diodes depend

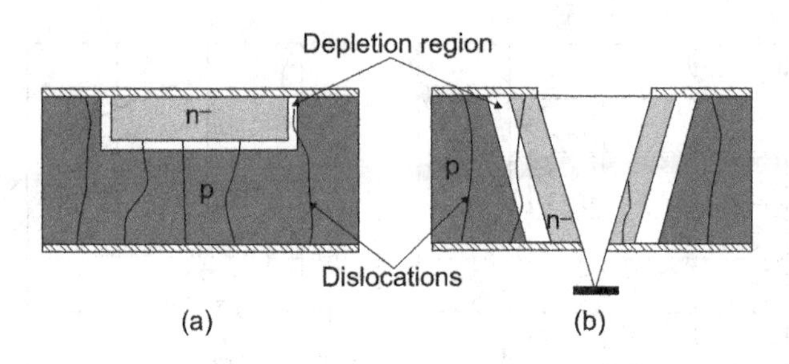

Figure 17.63 Impact of threading dislocations on (a) planar and (b) vertically integrated photodiodes. (From Ref. [342])

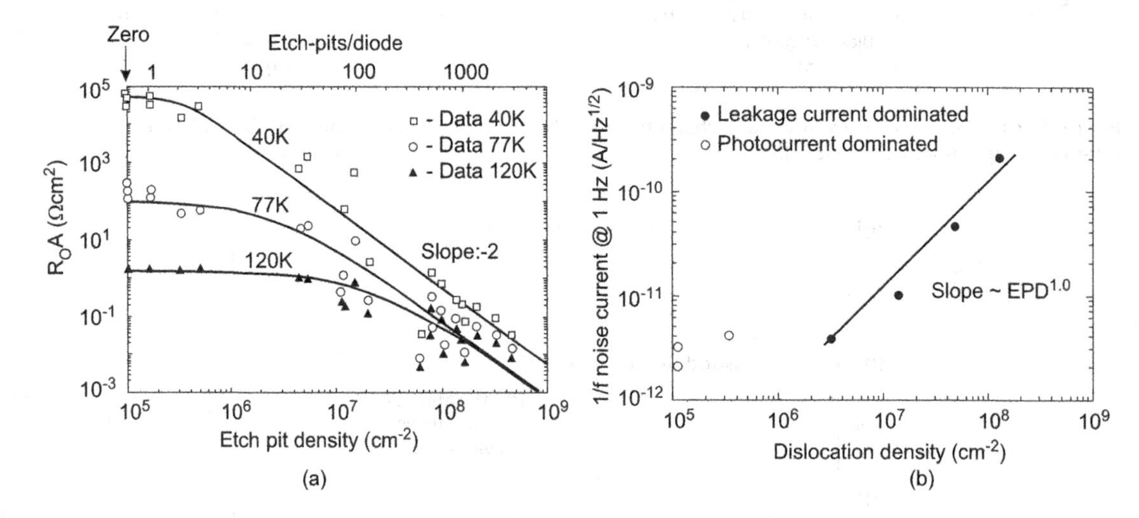

Figure 17.64 Influence of dislocation density on the parameters of HgCdTe photodiodes: (a) R_oA product versus EPD, showing the fit of model to data for a 9.5 µm array (at 78 K), measured at 120, 77, and 40 K at zero FOV and (b) 1/*f* noise current at 1 Hz versus dislocation density measured at 78 K for 10.3 µm HgCdTe photodiode array (*f*/2 FOV). (From Ref. [141])

strongly on the density of dislocations intercepting the junction. Dislocations are a significant source of tunnel current and 1/*f* noise, but they are thought to be a problem only if they intersect the depletion region of p-n junction. In case of vertically integrated geometry, the depletion region presents an extremely small cross section to the threading dislocations relative to planar diode geometries, which is shown in Figure 17.63 [342]. This results in an obvious advantage of the vertical structure with regard to potential detector defects.

Johnson et al. [141] showed that in the presence of high dislocations densities the R_oA product decreases as the square of the dislocation density; the onset of the square dependence occurs at progressively lower dislocation densities as the temperature decreases, which is shown in Figure 17.64a. At 77 K, R_oA begins to decrease at an EPD of approximately 10^6 cm^{-2}, while at 40 K, R_oA is immediately affected by the presence of one or more dislocations in the diode. The scatter in the R_oA data at large EPD may be associated with the presence of an increased number of pairs of interacting dislocations in some of those diodes; these pairs are more effective in reducing the R_oA than the individual dislocations. To describe the dependence of the R_oA product with dislocation density, a phenomenological model was developed that was based on the conductances of individual and interacting dislocations that shunt the p-n junctions. As Figure 17.64a shows, this model was found to give a reasonable fit to the experimental data.

In general, the 1/*f* noise appears to be associated with the presence of potential barriers at the contacts, interior, or surface of the semiconductor. Reduction of 1/*f* noise to an acceptable level is an art, which depends greatly on the processes employed in preparing the contacts and surfaces. Up until now, no fully satisfactory general theory has been formulated. The two most current models for the explanation of 1/*f* noise are [374]: Hooge's model [375], which assumes fluctuations in the mobility of free charge carriers, and McWhorter's model [376], which is based on the idea that the free carrier density fluctuates.

Tobin et al. [377] have reported the following relation for the $1/f$ noise of implanted n$^+$ -p MWIR HgCdTe photodiodes:

$$I_{1/f} = \alpha I_l^{\beta} f^{-1/2}, \tag{17.53}$$

where α and β are empirical constants with a value of 1×10^{-3} and 1, respectively, and I_l is the leakage current. It was found that $1/f$ noise is independent of photocurrent and diffusion current, but it is linearly related to the surface generation current. It was proposed that $1/f$ noise in reverse-biased HgCdTe photodiodes is a result of the modulation of the surface generation current by fluctuations in the surface potential. Work by Chung et al. also support this link [378]. Hoffman and Anderson [359] have developed a model involving TAT across a pinched-off depletion region and have shown that surface potential fluctuation model can explain the empirical relationship. Bajaj et al. [379] have found a similar relationship with GR currents originating from junction defects, such as dislocations. The correlation between $1/f$ noise and tunneling process (in particular TAT) has been supported in other papers [361,365,380,381]. More recently, a relevant model for HgCdTe has been developed by Schiebel [382], which explains the experimental data well, including the modulation of the surface generation current by the fluctuations of the surface potential and the influence of TAT across a pinched-off depletion region.

Johnson et al. [141] presented the effects of dislocations on the $1/f$ noise. Figure 17.64b shows that, at low EPD, the noise current is dominated by the photocurrent, while at higher EPD the noise current varies linearly with EPD. It appears that dislocations are not the direct source of the $1/f$ noise, but rather increase this noise only through their effect on the leakage current. The $1/f$ noise current varies as $I^{0.76}$ (where I is the total diode current); similar to the fit of data taken on undamaged diodes. A similar variation in $1/f$ noise was reported for leakage currents of MWIR PACE 1 HgCdTe photodiodes where the leakage current varied with changes in temperature, bias voltage, and electron irradiation damage [383].

Measurements of $1/f$ noise taken at DRS for CdTe passivated vertically integrated n$^+$ -n$^-$-p HgCdTe photodiodes also indicate a dependence of noise on the device dark current density, as shown in Figure 17.65, for a material with dislocation density below 2×10^5cm^{-2} [342]. The noise depends on the absolute value of dark current, regardless of the x composition or operating temperature [384]. These observations are in good agreement with Schiebel's theory for realistic surface trap densities in the 10^{12}cm^{-2} range.

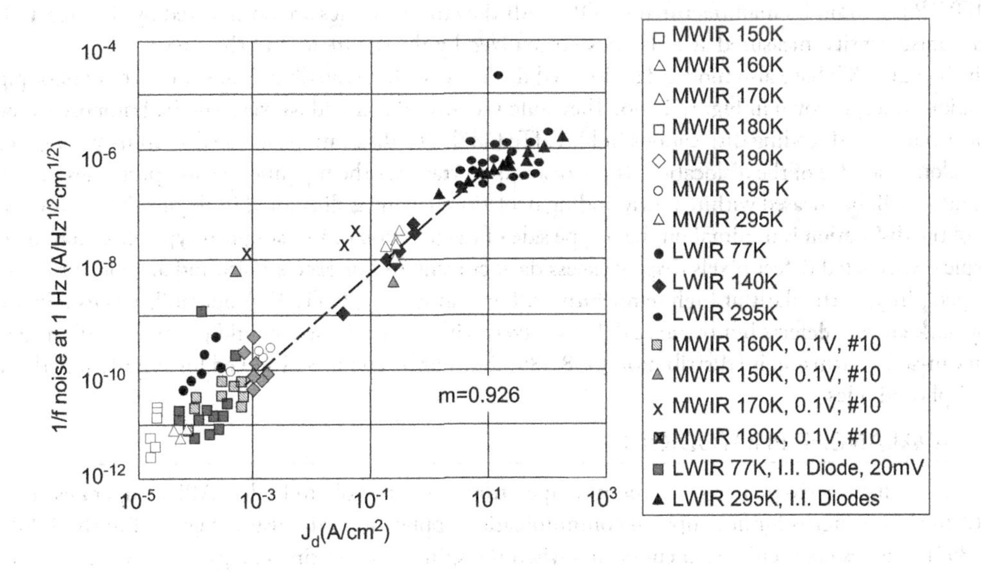

Figure 17.65 1/f noise figure versus dark current density for various HgCdTe photodiodes. (From Ref. [342])

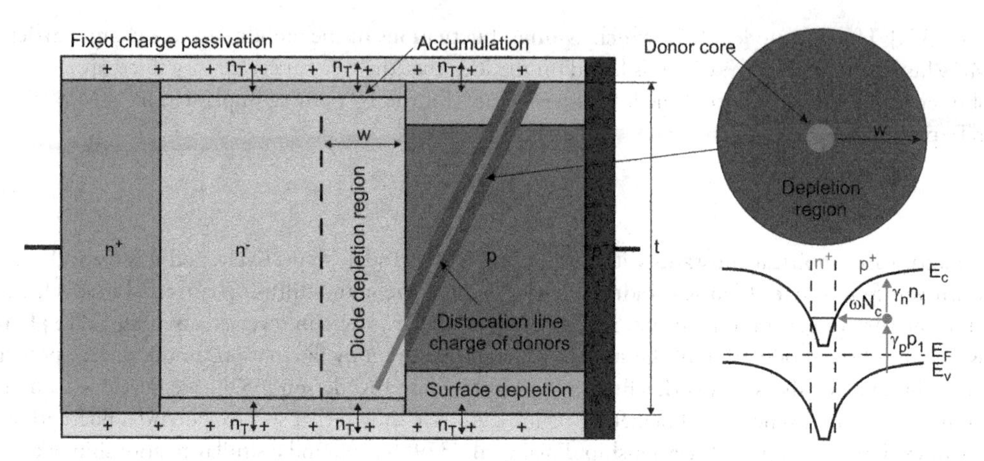

Figure 17.66 n⁺-n⁻-p-p⁺ diode architecture with fixed positively charged passivated surface together with donor-pipe dislocation concept in p-semiconductor volume. (Adapted after Ref. [385])

A new insight on 1/*f* noise was given in a paper published by Kinch et al. [385] assuming McWhorter's fluctuating surface trap model. A simple n^+-n^--p-p^+ diode geometry shown in Figure 17.66 has been considered. The following assumptions are made: the fixed charge in the passivation is positive and generates an accumulation layer on the n-side and a depletion layer at the surface on the p-side; the donor concentration $n^- \ll p$, the acceptor concentration (the main depletion region is formed completely on the n-side); the surface on the p-side may or not may not support an inversion layer depending on the magnitude of the fixed positive charge in the passivation and the reverse bias on the diode.

From Ref. [385] results that if depletion current dominates, then the systemic 1/*f* noise will vary as n_i, whereas for diffusion current it will vary as n_i^2. Typically, at lower temperatures, where depletion currents prevail, the diode 1/*f* performance will be dominated by the side with the lower doping. However, at higher temperatures, where Auger generation dominates, the 1/*f* noise will be independent of the doping concentration on both sides of the junction. These general conclusions are supported by experimental data for both MWIR and LWIR HgCdTe photodiodes [385].

Similar correlation between 1/f noise current density and dark current has been achieved recently by Hassis et al. [386]. Figure 16.67 compares the noise performance of DEFIR photodiodes with data reported for HDVIP photodiodes manufactured by DRS. All the current values are normalized by the area and the current noise density, measured at 1 Hz, was normalized by the square root of the area.

Kinch et al. [385] have also modeled an isolated defect noise that is attributed to the effect of donor-pipe dislocation concept shown in Figure 17.66. They followed after Baker and Maxey, who had proposed a model for the behavior of threading dislocations in HgCdTe [387]. The dislocation is treated as an n-type pipe of donors along the edge of the dislocation. If the dislocation traverses the n-p junction and protrudes into the p-volume, it will be encased within a surrounding depletion region, as illustrated in Figure 17.66. However, the effect of the dislocation is minimal on the n-type side of the junction or in a simple n-type photoconductor.

Typically, isolated defect pixels display excess dark current and/or excess noise and are a leading cause of FPA operability, particularly at high temperatures. For example, HgCdTe FPA operability is usually limited not by dark current defects but by noise defects. Pixels with high 1/*f* noise should produce a tail in the root-mean-square (rms) noise distribution [388,389]. Similar behavior is observed for type-II superlattice (T2SL) photodiodes.

17.6.4 AVALANCHE PHOTODIODES

HgCdTe, as an attractive material for room temperature avalanche photodiodes (APDs), operates at 1.3–1.6 μm wavelengths for fiber optical communication applications and was recognized in the 1980s [390–392]. The resonant enhancement occurs when the spin-orbit splitting energy in the valence band, Δ, is equal to the fundamental energy gap, E_g. This has the beneficial effect, first pointed out by Verie et al., of

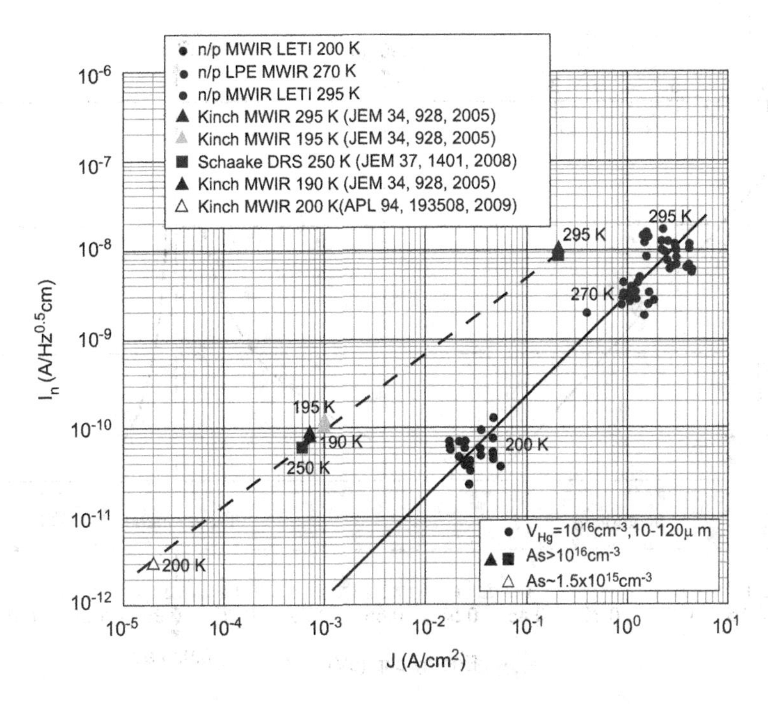

Figure 17.67 Noise current density versus dark current for HgCdTe diffusion-limited diodes. (Adapted after Ref. [386])

making the electron and hole impact ionization rates quite different, which is highly desirable for low-noise APDs. Early work by Alabedra et al. [391] reported on HgCdTe APDs with $x \approx 0.73$ ($E_g = 0.92\,\text{eV}$). The band structure of HgCdTe gives k-values close to 0—a highly favorable ratio of hole-to-electron multiplication during avalanche conditions, resulting in very little noise gain. These properties give HgCdTe APDs a figure of merit better than in the InGaAs APDs, where k is about 0.45. Si, in comparison, has an ionization ratio of 0.02, and therefore much lower excess noise, however Si is not sensitive to wavelengths greater than 1.1 μm.

Another APD application is LADAR (laser radar), where a pulsed laser system is obtained by 3D imagery. Three-dimensional imagery has been traditionally obtained with scanned laser systems that operate in the visible and near-infrared (NIR) out to 1 μm. However, eye-safe lasers are required if there is a chance that humans will be in the scene. The eye-safe range is around 1.55 μm in the NIR region. In addition, the advantages of a single pulse, flood-illuminated, laser imaging system caused researchers to consider the 2D arrays of APD sensors. The gated active/passive system promises target detection and identification at longer ranges compared to the conventional passive only imaging systems. Because of the desired operating ranges (up to 10 km, typically) and the fact that laser power is limited, there is a need for a NIR solid-state detector with a sensitivity approaching single photon [393].

As shown in Figure 17.68 [394], the avalanche properties of $Hg_{1-x}Cd_xTe$ vary dramatically with bandgap. Leveque et al. [395] described two regimes in which the ratio $k = \alpha_h/\alpha_e$ of the hole ionization coefficient-to-electron ionization coefficient is either much greater or much lesser than unity. For cutoff wavelengths shorter than approximately 1.9 μm ($x = 0.65$ at 300 K), they predict $\alpha_h >> \alpha_e$ because of resonant enhancement of the hole ionization coefficient when $E_g \cong \Delta = 0.938\,\text{eV}$. The situation, when $k = \alpha_h/\alpha_e >> 1$, is favorable for low-noise APDs with hole-initiated avalanche. Using this regime, deLyon et al. [396] demonstrated back-illuminated multilayer separate absorption and multiplication avalanche photodiode (SAM-APD) grown in situ by MBE on CdZnTe with a cutoff wavelength of 1.6 μm and a multiplication cutoff wavelength of 1.3 μm. Avalanche gains in the range of 30–40 were demonstrated at reverse-bias voltages of 80–90 V in 25-element mini arrays.

The unique crystal lattice properties of HgCdTe allow two types of noise-free linear avalanche in quite distinct modes—pure electron initiated (e-APD) for bandgaps < 0.65 eV (λ_c >1.9 μm) and pure hole initiated

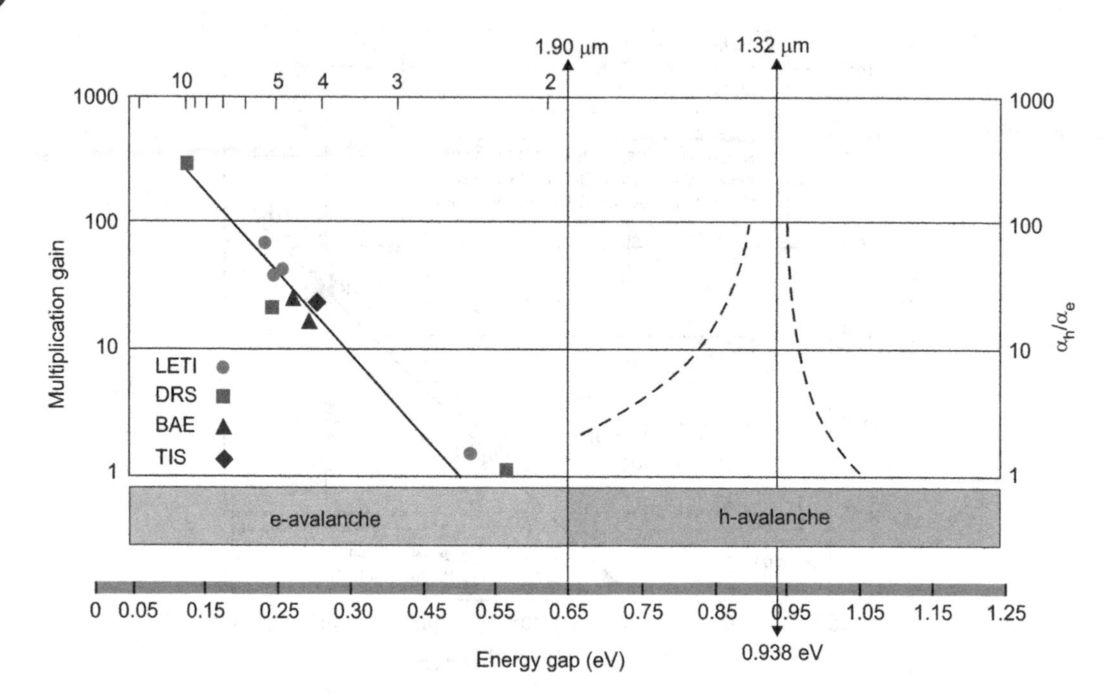

Figure 17.68 The distinct e-APD and h-APD regimes of $Hg_{1-x}Cd_xTe$ crossover at $E_g \approx 0.65$ eV ($\lambda_c = 1.9\,\mu m$). At lower bandgaps, the e-APD gain increases exponentially (material for four manufacturers shows remarkably consistent results). (From Ref. [394])

(h-APD) centered on a bandgap of 0.938 eV ($\lambda_c = 1.32\,\mu m$) corresponding to a resonance with spin-orbit splitting. Both utilize very similar architectures consisting of a SAM layer graded into a photodetection layer of a lower bandgap.

Initially, several isolated experimental reports were published to verify the predicted small values of $k = \alpha_h/\alpha_e$ (<0.1) in $Hg_{1-x}Cd_xTe$ with λ_c longer than 1.9 μm [397,398]. One of the earliest studies of the electron-initiated multiplication on a LWIR HgCdTe ($\lambda_c = 11\,\mu m$) showed that reasonable gains could be obtained at low voltages (5.9 at −1.4 V) [369]. However, the clear and compelling advantages of the electron-initiated avalanche process in MWIR lateral-collection n^+-n^--p (with p-type absorber regions) were first reported in 2001 by Beck et al. [399]. Soon, thereafter, a theory by Kinch et al. [400] substantiated by Monte Carlo simulations by the University of Texas group [401] has been used to develop an empirical model to fit the experimental data obtained at DRS Infrared Technologies. The large inequity between α_e and α_h results from three key features of the HgCdTe energy band structure: (i) the electron effective mass is much smaller than the heavy hole effective mass (electrons have a much higher mobility), (ii) a much lower scattering rate by optical phonons, and (iii) a factor-of-two lower ionization threshold energy (there are no subsidiary minima in the conduction band to which energetic electrons can scatter; and the light holes are not important).

In 1999, DRS researches proposed an APD based on their cylindrical p-around-n HDVIP. This architecture is shown in Figure 17.46 and is also used in the production of FPAs. It is a front side illuminated photodiode with high quantum efficiency response from the visible region to the IR cutoff (see Figure 17.69 [402]). The device geometry and operation are illustrated in the next figure (Figure 17.70). If the reverse bias increases from a typical value of 50 mV to several volts, the centralized n-region becomes fully depleted and produces the high-field region in which multiplication occurs. The hole-electron pairs are optically generated in the surrounding p-type absorption region and next diffuse to the multiplication region and thus comprise the injection species.

The Monte Carlo theory modeling predicts the bandwidth of the multiplication process typically to above 2 GHz [403]. Record gain bandwidth G × BW > 16 THz has also been measured (see Figure 17.71) where the measured gain versus bias voltage is shown [404]. Large pixels with high bandwidth are obtained

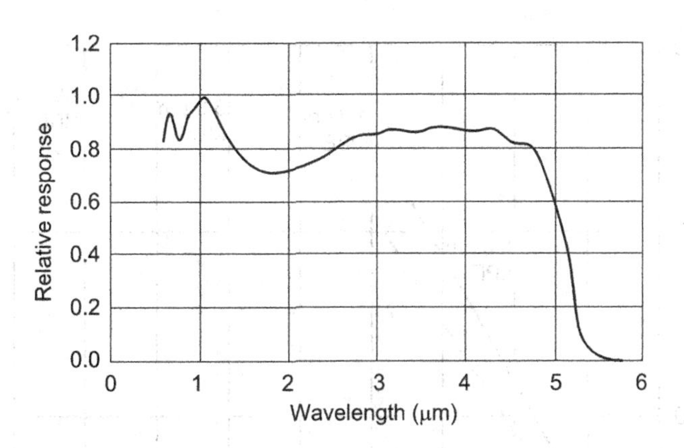

Figure 17.69 Relative spectral response of 5.1 μm HgCdTe HDVIP at 80 K. (From Ref. [402])

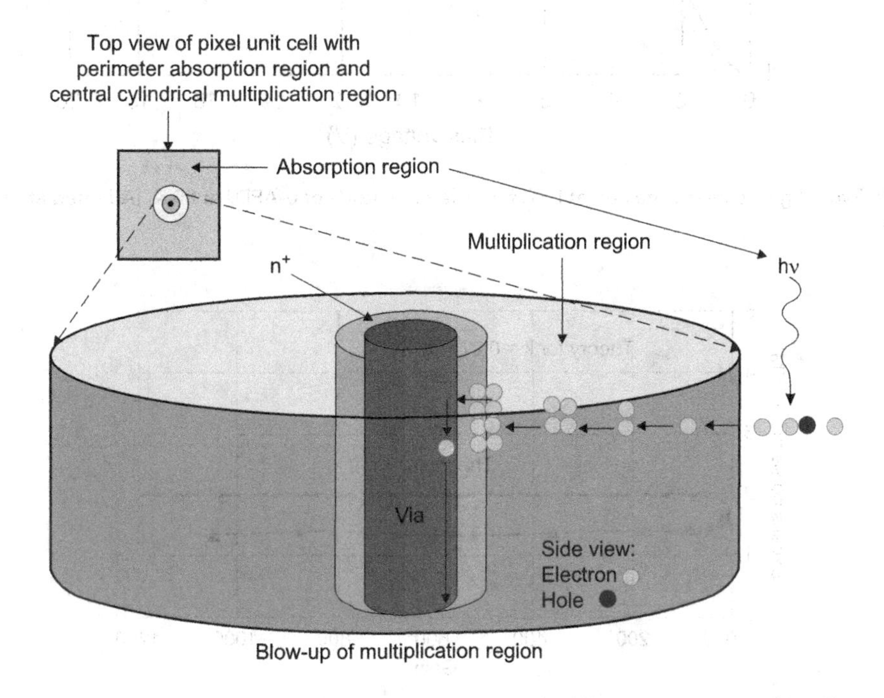

Figure 17.70 Cross section view of electron avalanche process in HgCdTe HDVIP. (From Ref. [402])

by connecting the APDs together in parallel in an N × N configuration with small capacitance due to the cylindrical junction geometry.

It appears that the performance of MWIR HgCdTe APDs can be extended to SW and LW photodiodes [405]. Wider bandgap HgCdTe of 0.56 eV (2.2 μm cutoff) also showed similar exponential gain behavior. The operating voltages were, however, higher and the excess noise factor was about 2 suggesting that while $k \approx 0$, electron-phonon scattering is present.

Experimental data of HDVIPs reveal almost ideal APD; the device is characterized by uniform, exponential gain voltage characteristics that is consistent with a hole-to-electron ionization coefficient ratio $k = \alpha_h/\alpha_e = 0$. Excess noise data on 4.3-μm cutoff photodiodes indicate a gain independent excess noise factor of 1.3 out to gains of greater than 1,000 (see Figure 17.72), suggesting that the electron ionization process is ballistic [401,405].

More recently, other configurations and architectures have been reported that corroborate the essential features of the electron-initiated avalanche process. Figure 17.73a shows schematic illustration of the LETI's

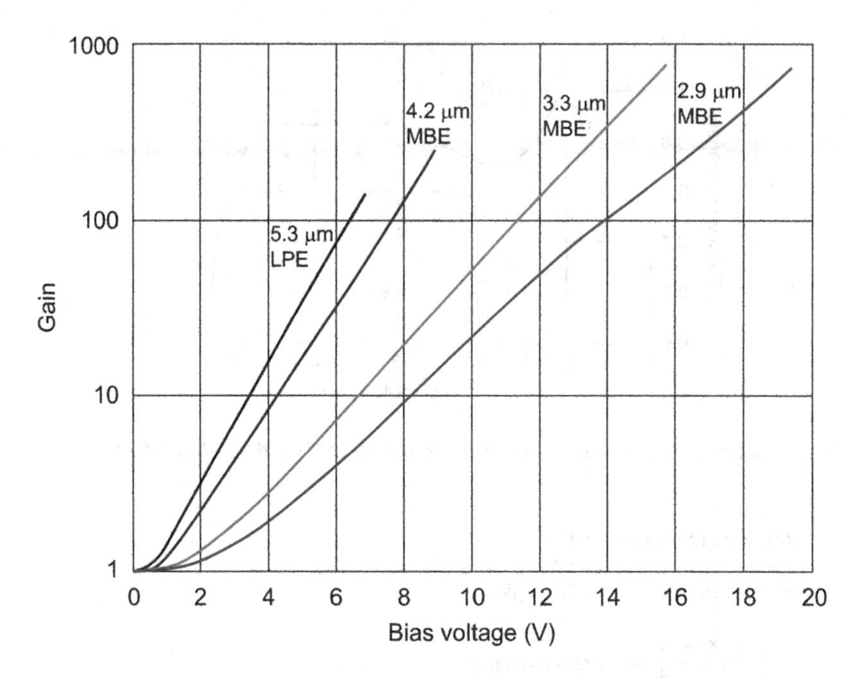

Figure 17.71 Typical gain curves obtained at LETI for different cutoffs of e-APDs at 80 K. (Adapted after Ref. [404])

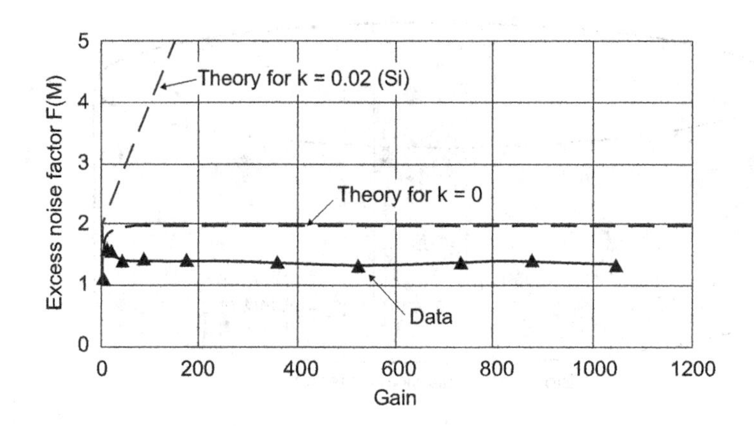

Figure 17.72 Excess noise factor versus gain data at 80 K on a 4.3-μm APD in an 8 × 8 array compared to McIntyre's original theory. (From Ref. [403])

p-i-n homojunction formed by transforming a narrow region close to the surface in a vacancy-doped p-type layer ($N_a = 3 \times 10^{16}$ cm^{-3}) into a n$^+$ region with a doping level of $N_d = 1 \times 10^{18}$ cm^{-3} [406]. During formation of the n$^+$ region, n$^-$ region is generated by the suppression of Hg vacancies to the residual doping of the epitaxy, typically $N_d = 3 \times 10^{14}$ cm^{-3}. The extension of the n$^-$ layer is correlated to the depth of the n$^+$ layer. The quantum efficiency of the diodes was typically about 50% due to a reduced optical fill factor and the use of a nonoptimized antireflecting coating. The width of the n$^-$ layer is of the order of 1–3 μm.

Selex in Southampton has developed the MOCVD growth mesa heterojunctions of HgCdTe on 2″ GaAs substrates (see Figure 17.73b) [407]. Mesa heterojunctions allow the doping and bandgap to be varied freely through the device structure, and the depletion region is a tightly controlled width determined at the material growth stage. For APDs, it allows the absorber, p-n junction region, and multiplication region to be independently optimized. Each pixel of FPA is electrically isolated by a mesa slot that extends though the absorber to eliminate lateral collection and blooming.

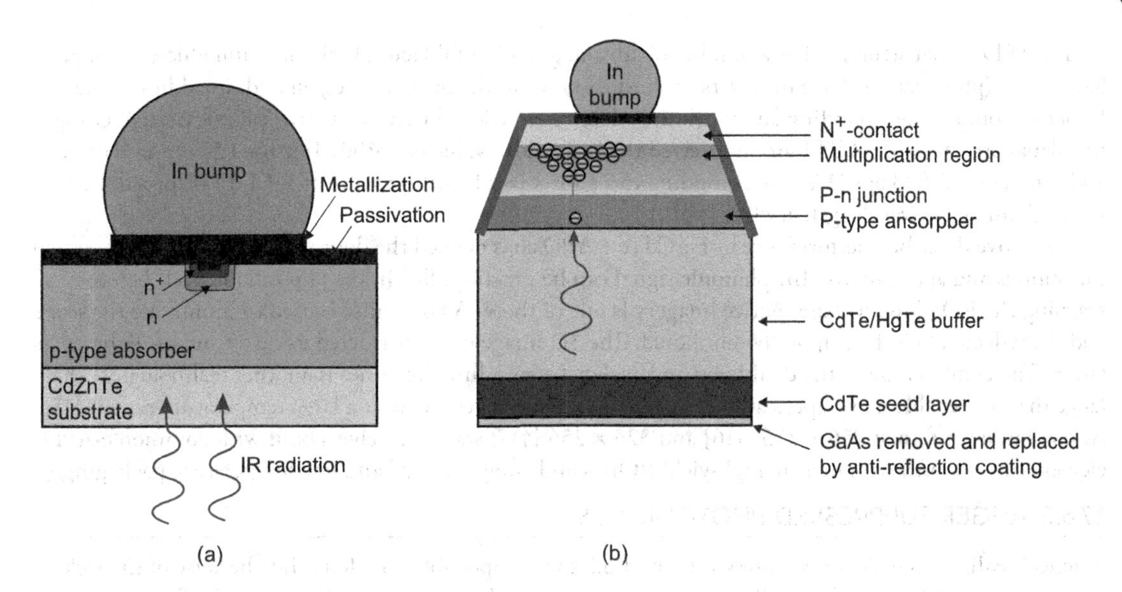

Figure 17.73 Cross section of back side illuminated HgCdTe e-APD: (a) LETI (After Ref. [406]) and (b) Selex (After Ref. [407]) architectures.

Table 17.6 Typical performance of SWIR and MWIR HgCdTe APDs at T = 80 K

PARAMETER	SWIR	MWIR
Quantum efficiency	60%–80%	
Max gain	2,000	13,000
Bias at $M = 100$	12–14	7–10
Excess noise factor F	1.1–1.4	
Quantum efficiency to F ratio	40%–70%	
Typical response time τ_{90-10}	0.5–20 ns	
Maximum Gain × BW product	2.1 THz	

Source: After Ref. [409].

The highest gain of 5,300 at −12.5 V has been reported by Perrais et al. [408] for a back-illuminated planar p-i-n 30-μm pitch photodiode with a cutoff wavelength of 5.0 μm, formed in a HgCdTe layer grown by MBE on a CdZnTe substrate. Similar to SW e-APD behavior, the highest gain is generally consistent with exponential increase with both cutoff wavelength and reverse bias voltage.

Typical performances of such APDs at $T = 80$ K are compared in Table 17.6. The maximum gain in this table corresponds to the highest stable gain values and ranges from 2,000 for SWIR APDs to higher than 10,000 in MWIR APDs. The usefulness of such high gains depends on the dark-current noise of the APD, the observation time, and the noise of the detection electronics. Stable gain associated with low noise has been observed up to room temperature in SWIR APDs.

Baker et al. [410,411] have reported the first laser gated imaging at 90 K in the eye-safe region at 1.57 μm using a lateral-collection loophole HgCdTe e-ARD 320 × 256 FRA having 24 × 24 μm unit cells, cutoff wavelengths of 4.2–4.6 μm, and gains of over 100 at −7 V.

The HgCdTe APD design, based on front-illuminated lateral-collection p-around-n HDVIP architecture developed by DRS, was used to demonstrate the feasibility of a MWIR active/passive 128 × 128 gated imaging system composed of 40 μm-pitch with cutoff wavelengths of 4.2to 5 μm and a custom readout integrated circuit [412,413]. Median gains as high as 946 at −11 V bias and a sensitivity of less than one photon for a 1 μs gate width were measured.

The APD planar structure has a number of advantages: high fill factor, high quantum efficiency, high bandwidth (photocarriers have to traverse a short distance to the depletion region and, in addition, the favorable compositional grading layer with the effective electric field accelerates the photocarriers). Using the planar structure, the LETI group received the bandwidth value of ≈ 400 MHz for 1.55 μm radiation and a gain of ≈ 2,800 at −11 V, corresponding to a gain × bandwidth product of ≈ 1.1 THz for a device with 5.3 μm cutoff wavelength at 80 K [414].

The above described performance of HgCdTe e-APDs has opened the door to new passive/active system capabilities and applications. The photonic signal can be pre-amplified in the photodiode itself before reaching the ROIC input stage. Active imagery is one of those. A laser pulse is used to illuminate the scene and the reflected light is temporally monitored. The 3D image is reconstructed using a time-of-flight calculation. The combination of the dual-band and avalanche gain functionalities is another technological challenge that will enable many applications, such as dual band detection over a large temperature range [415]. At present, large format 256 × 256 [416] and 320 × 256 [417] staring receivers built with common e-APD elements are coupled with mature high-yield ROICs and compact TEC and low-temperature packaging.

17.6.5 AUGER-SUPPRESSED PHOTODIODES

Practical realization of Auger-suppressed photodiodes was impossible for a long time because of the lack of technology to obtain wide-gap P contacts to $Hg_{1-x}Cd_xTe$ absorber region. Therefore, the first Auger-suppressed devices were III-V heterostructures with InSb absorbers [243,244,418,419]. The first reported $Hg_{1-x}Cd_xTe$ Auger extracting diodes were so-called proximity-extracting diode structures (see Figure 17.74) in which additional guard reverse-biased n^+-n^- junctions were placed in the current path between the p^+ and n^+ regions to intercept the electron injected from the p^+ region. Philips Components Ltd fabricated practical proximity-extracting devices in both linear and cylindrical geometries [243]. The devices have been fabricated from bulk-grown $Hg_{1-x}Cd_xTe$ with a cutoff wavelength of 9.3 μm at 200 K ($x = 0.2$). The acceptor concentration of 8×10^{15} cm^{-3} was due to native doping. The n^+ -regions were fabricated by ion milling. The I-V characteristics of such structures are complex and difficult to interpret because of the occurrence of bipolar transistor action and impact ionization, but the general features were predictable when the standard transistor modeling was applied. A current reduction of a factor of 48 has been obtained by biasing the guard junction, but the extracted current was much greater than that predicted for an Auger-suppressed, SR-limited case. This is possibly due to a surface-generated current. The measured 500 K blackbody detectivity of a 320 μm^2 optical area device at a modulation frequency of 20 kHz was 1×10^9 cmHz$^{1/2}$/W.

The use of wide-gap P contact is a straightforward way to eliminate harmful thermal generation in the p-type region. The practical realization of such devices would require a well-established multilayer epitaxial technology capable of growing high-quality heterostructures with complex gap and doping profiles. This technology became available in the early 1990s and three-layer n^+-π-P^+, N^+-π-P^+ heterostructural photodiodes have been demonstrated [420] and gradually improved [419–424]. The arrangement of the photodiode in the optical resonant cavity makes it possible to use a thin extracted zone without loss in quantum

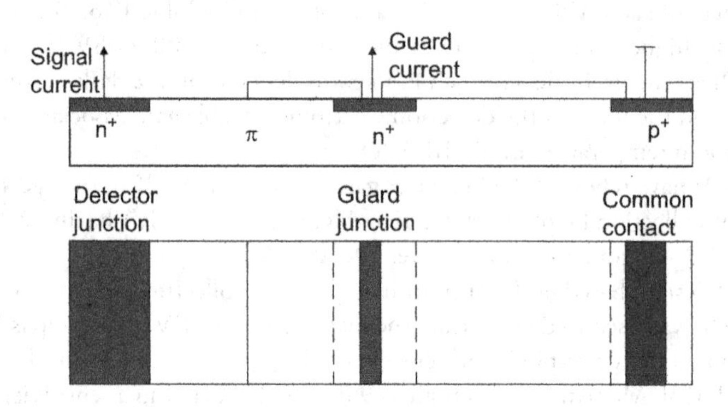

Figure 17.74 Schematic structure of proximity-extracting photodiode. (From Ref. [243])

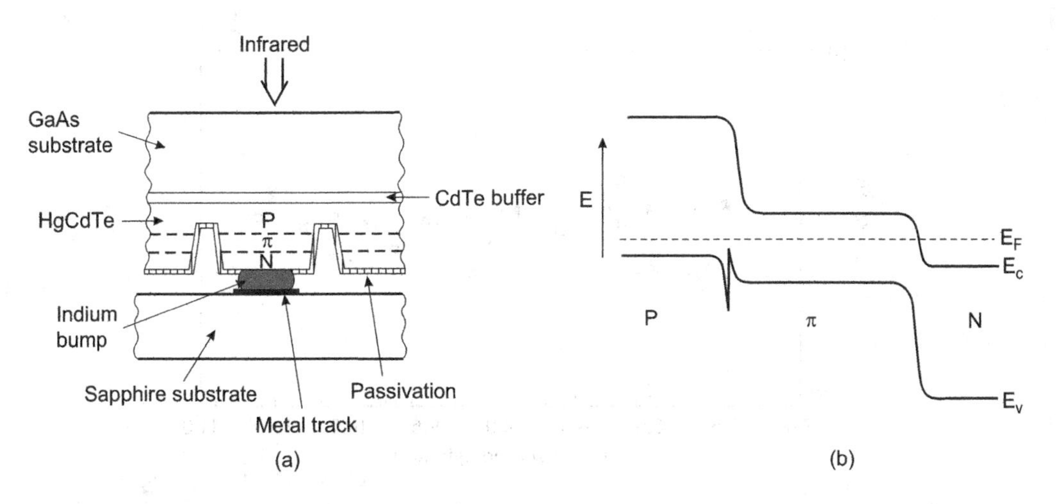

Figure 17.75 Schematic cross section of (a) the P^+-π-N^+ HgCdTe heterostructure photodiode, and (b) its energy band diagram. (From Ref. [419])

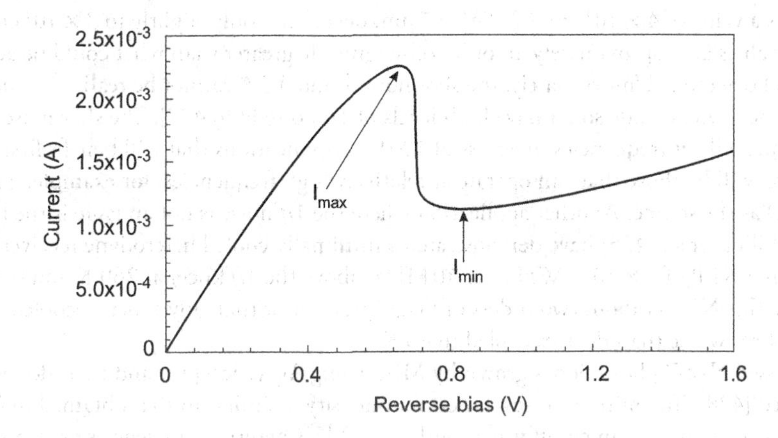

Figure 17.76 An example of current-voltage characteristics of a P-p-N heterostructure showing the positions of I_{max} and I_{min}. (From Ref. [425])

efficiency, which is also favorable for the reduction of saturation current, and it results in minimized noise and bias power dissipation.

Figure 17.75 shows the device structure of N^+-π-P^+ HgCdTe heterostructure, intended for operation at temperatures ≥ 145 K, and its appropriate energy band diagram [419,422]. Typical parameters for the LW devices would be $x = 0.184$ in the active π region, $x = 0.35$ in the P^+ region, and $x = 0.23$ in the N^+ region. The structures were grown on CdZnTe and GaAs substrates using IMP MOCVD. The n and P^+ regions were doped with As to typical levels of 7×10^{15} cm^{-3} and 1×10^{17} cm^{-3}, respectively, and the N^+ region was doped with I to a concentration of 3×10^{17} cm^{-3}. Diodes were defined by etching circular trenches to produce 64-element linear arrays with common contact to the P^+ region at each end. These slotted mesa devices were passivated with 0.3-μm-thick ZnS and metalized with Cr/Au. Finally, the electrical contact to the mesas is achieved by In bump bonding the array onto a Au lead-out pattern on a sapphire carrier.

For low bias, the device behaves as a linear resistor until the electric field exceeds the critical value for exclusion or extraction (Figure 17.76 [425]). Then the current drops sharply from its maximum value I_{max}, a further increase of voltage gradually decreases the current to minimum I_{min}. At high voltages, the current increase occurred due to diode breakdown. As a result, the dynamic resistance increases to high values in the regions close to the transition range. The devices have shown negative resistances for temperatures above 190 K.

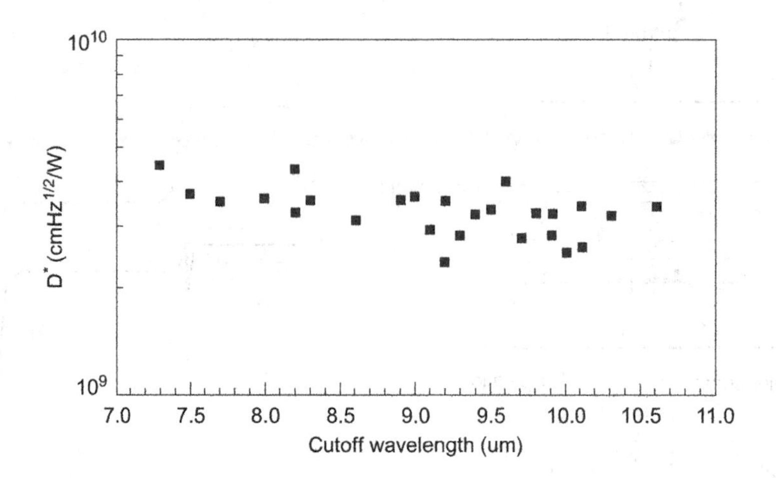

Figure 17.77 Shot-noise detectivity of HgCdTe nonequilibrium detectors at 300 K. (From Ref. [419])

The shot noise-limited detectivity as a function of cutoff wavelength is shown in Figure 17.77 [419]. This figure shows a value of 4×10^9 cmHz$^{1/2}$/W at 7 µm, decreasing only slightly to 3×10^9 cmHz$^{1/2}$/W at 11 µm. These values are approximately an order of magnitude greater than what could be achieved with uncooled thermal detectors. Unfortunately, the shot noise-limited D^* cannot be realized in imaging applications because the devices made so far have high levels of $1/f$ noise [426,427]. The shot noise level is only observed experimentally at frequencies in excess of 1 MHz. Applications that will benefit first from these devices, therefore, will be those that can operate at relatively high frequencies; for example, gas detection using an IR LED as the source. Another application where the $1/f$ noise is not an issue is the laser heterodyne detection. Elliott et al. [425] have demonstrated a minimally cooled heterodyne receiver for CO_2 laser radiation that has a NEP of 2×10^{-19} WHz^{-1} at 40 MHz (above the $1/f$ knee) at 260 K and 0.3 mW local oscillator power. This NEP is about two orders of magnitude better than any other uncooled device and only a factor of three worse than devices cooled to 80 K.

Auger-suppressed N-π-P photodiodes grown by MBE using Ag as acceptor and In as donor dopant were also demonstrated [428]. The minimum reverse current density is similar to that obtained in MOVPE-grown material at the same ≈9 µm cutoff wavelength at 300 K. Quantum efficiencies exceeding 100% have been measured and attributed to carrier multiplication due to the relatively high bias across the sample or to mixed conduction effects.

More recently, improved N^+-N^--π-P^--P^+ $Hg_{1-x}Cd_xTe$ heterostructures with refined bandgap and doping profiles were reported [422,427]. The N and P layers were used to minimize generation between the regions of different compositions. The doping of π-region was $\approx 2 \times 10^{15}$ cm^{-3}. The thickness of π-region was typically 3 µm wide that ensured good quantum efficiency with mesa contact acting as a reflector. All the layers have been annealed for 60 hours at 220°C in Hg-rich nitrogen with the same temperature of Hg reservoir. Mesa structures passivated with CdTe and ZnS have been received.

The Auger-suppressed devices exhibit a high-low frequency noise with $1/f$ knee frequencies from 100 to a few MHz for ≈ 10 µm devices at room temperature. This reduces their SNR at frequencies of ≈1 kHz to a level below that for equilibrium devices. The $1/f$ noise remains the main obstacle to achieving background-limited noise equivalent difference temperature (NEDT) in 2D arrays at near-room temperatures.

The $1/f$ noise level is much lower in MWIR devices and they look capable of achieving useful D^* values for imaging applications in very high frame rate or high-speed choppers [429]. Typically, the low frequency noise current is proportional to the bias current with $a \approx 2 \times 10^{-4}$. The HOTEYE thermal imaging camera based on a 320 × 256 FPA with cutoff wavelength 4 µm at 210 K has been demonstrated. The pixel structures were P^+-p-N^+ starting from the GaAs substrate. The histogram peaks at an NEDT of around 60 mK with $f/2$ optics were measured.

The reason for the large $1/f$ noise is not clear [427,430]. Traps in depletion region, high-field areas, hot electrons, and background optical generation are among the possible reasons. For MWIR devices, the origin of the noise was suspected to be the depletion region [430], exponentially dependent on bandgap; noise ∝ exp($E_g/2kT$).

17.6.6 MIS PHOTODIODES

The interest in HgCdTe MIS structures was mostly connected with the possibility of using them in monolithic FPAs [16,17,431]. These structures enable not only the detection of IR radiation but also advanced signal management. For nearly two decades, technologists have sought to develop fully monolithic CCD imagers in HgCdTe for detection of IR radiation. Initial works concentrated on p-channel CCDs due to the maturity of growth and the doping control on n-type material [16,431–433]. However, due to the difficulty of forming stable p^+-n junctions in HgCdTe, readout structures could not be incorporated in the devices. The necessity of utilizing off-chip readout circuitry increased the parasitic capacitance at the sense node and decreased the charge-to-voltage conversion efficiency, resulting in a limited dynamic range. To alleviate these difficulties, it became necessary to develop n-channel CCDs. The use of p-type material allowed for stable diode formation by ion implantation and provided a means for the development of HgCdTe metal oxide semiconductor field effect transistors (MISFETs) [434–436]. With the demonstration of MISFET-based amplifiers in HgCdTe [437], the path was cleared for creating a fully monolithic CCD in HgCdTe [438–440]. The MIS structures are also used as a tool for investigation of surface and interface properties of HgCdTe.

The operation and properties of HgCdTe MIS structures have been extensively reviewed by Kinch [431], and the general theory of MIS capacitor is described in Section 12.7. The minority carriers generated by absorbed radiation in MIS photodetectors are trapped in the well, while the majority carriers are forced into the neutral bulk by the surface potential. Although MIS detectors are essentially capacitors, their dark current can be compared with the dark current in conventional p-n junction photodiodes, allowing a reference to the R_oA product. The sources of dark current are essentially the same as in the p-n photodiodes. The dark current limits the maximum storage or integration time at low background and is a source of noise. The sources of noise are similar to those in a reverse-biased p-n junction [431,441]. However, this problem is more serious for MIS structures as, in contrast to the p-n junction photodiode, they have to operate in strong depletion to achieve adequate storage capacity. Another reason for the high dark current is the use of weakly doped material for the high breakdown voltage, resulting in high GR current from the depletion region. Practically, the dark current should be reduced below the background-generated current. This condition sets the maximum operating temperature of the devices. The optimization of the thickness of the base layer can either increase the operating temperature or a longer integration time can be achieved.

Trapping the charge in fast interface states at the edge of the CCD well is the primary limitation of the charge transfer efficiency at low frequencies [431]. Since the transfer loss is proportional to the density of the fast states, the latter should be minimized. At high frequencies, transfer efficiency sharply decreases due to limited time, which is required for charge transport between adjacent wells. The corner frequency, equal to about 1 MHz for typical p-type channel devices, can be increased by reducing the gate length and increasing the maximum employable gate voltages. A much higher corner frequency would be expected for n-channel devices as a result of the high electron-to-hole mobility ratio.

For the maximum charge storage capability, the MIS structures are operated at maximum effective gate voltages, which are limited by breakdown due to tunnel current. BLIP performance requires that the tunnel current be lower than the current due to incident photon flux, and this criterion is used to define an upper limit on applied electric field, the breakdown field E_{bd} for the material in question. The breakdown voltage decreases with decreasing gap and increasing doping level. This limitation is especially severe for LWIR devices, which require a large capacity to store background-generated currents. Tactical background flux levels are shown in Figure 17.78a for a typical $f/2$ system, and values for E_{bd} are 3×10^4 V/cm for 5 μm cutoff wavelength and 8×10^3 V/cm for 11.5 μm [16]. These values are in agreement with experimental data for good-quality p-type HgCdTe with doping concentration. E_{bd} values for n-type HgCdTe are typically lower by a factor of 30% due to lower density of states in the conduction band (which is by a factor of 10^2–10^3 lower than the density of states in the valence band) and the effects of inversion layer quantization for the p-type case [442].

The dependence of available well capacity on doping concentration is shown in Figure 17.78b, assuming an insulator capacity per area unit of 4×10^{-8} F/cm^2. Integrating MIS FPAs require well capacities in

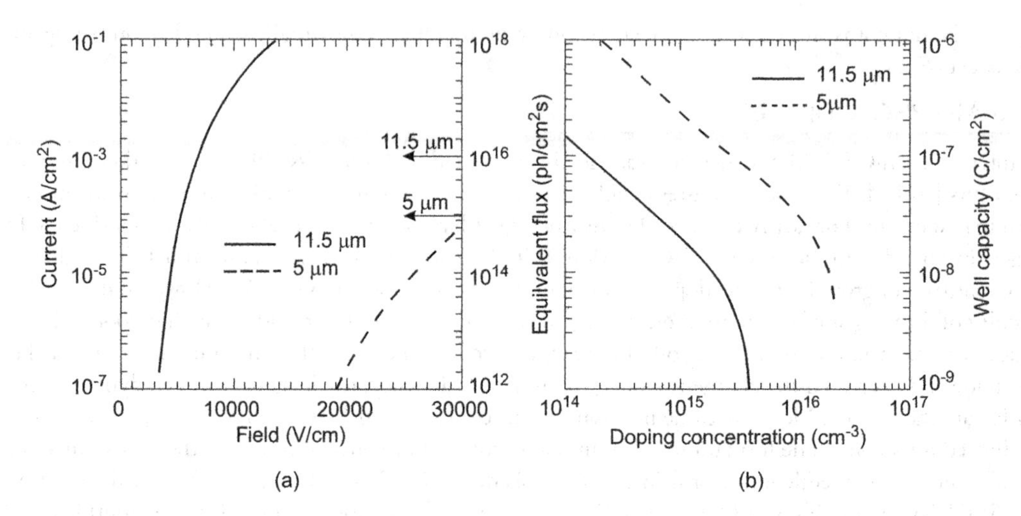

Figure 17.78 (a) Tunnel current versus electric field and (b) well capacity versus hole concentration, for 5 μm and 11.5 μm p-type HgCdTe. (From Ref. [16])

the 10^{-8}–10^{-7} C/cm^2 range for pixel noise to dominate. Such well capacities are readily available for 5 μm HgCdTe at doping concentrations in the >10^{15} cm^{-3} range. However, the LWIR device operation imposes much more stringent requirements due to the narrow bandgap involved and the relatively low value of $E_{bd} = 8 \times 10^3$ V/cm. The n-type HgCdTe with low doping concentrations (10^{14} cm^{-3}–10^{15} cm^{-3}) is available, but its performance does not improve at doping concentrations below 8×10^{14} cm^{-3} as predicted by Figure 17.78b, but rather it reaches a plateau at ≈2×10^{-8} C/cm^2 (equivalent to an applied voltage $V = 0.5$ V). The p-type HgCdTe of this quality is not reproducibly available. An example of the device is the 10 μm bulk n-type HgCdTe-based MIS detector reported by Borrello et al. [439], which exhibits at 80 K quantum efficiency above 50% and peak $D^* = 8 \times 10^{10}$ cmHz$^{1/2}$/W at *f*/1.3 shielding.

MIS devices were fabricated using LPE HgCdTe epilayers with a thickness lower than the minority carrier diffusion length to reduce the dark diffusion current [440]. MWIR devices with 5 μm cutoff wavelength at 77 K have produced average detectivity values exceeding 3×10^{13} cmHz$^{1/2}$W^{-1} for a background flux level of 6×10^{12} photon/cm^{2}s. An increase in the device operating temperature produced additional dark current, lowering the effective signal storage capacity of the CCD, and lowering the D^* value. For 5.25 μm cutoff devices, the maximum practical operating temperature was 100 K.

Unlike the photoconductors and p-n junction photodiodes, the MIS device operates under strong non-equilibrium conditions with large electric fields in the deep depletion regions. This makes the MIS device much more sensitive to material defects than the photoconductors and photodiodes. This sensitivity caused the monolithic approach to be abandoned in favor of various hybrid approaches.

17.6.7 SCHOTTKY BARRIER PHOTODIODES

Photoreceivers based on the Schottky barriers exhibit high bandwidth and simpler fabrication compared to p-n photodiodes. However, the performance of HgCdTe Schottky barrier photodiodes is not useful for the detection of IR radiation.

The traditional models of metal-semiconductor (M-S) interface are described in Section 12.5. However, the physical picture of M-S has been modified and a variety of phenomena have been observed at the microscopic M-S interfaces, which form interfacial regions with new electronic and chemical structures. Several new models have been suggested to explain the Fermi level position in the junction interface, which includes the modified work function model of Freeouf and Woodall [443], metal-induced gap states (MIGS) [444], and Spicer's native defects model [445]. According to the MIGS model, when a metal is in intimate contact with a semiconductor surface, the tails of the metal wave functions can tunnel into the bandgap of the semiconductor, leading to MIGs that are capable of strong Fermi level pinning.

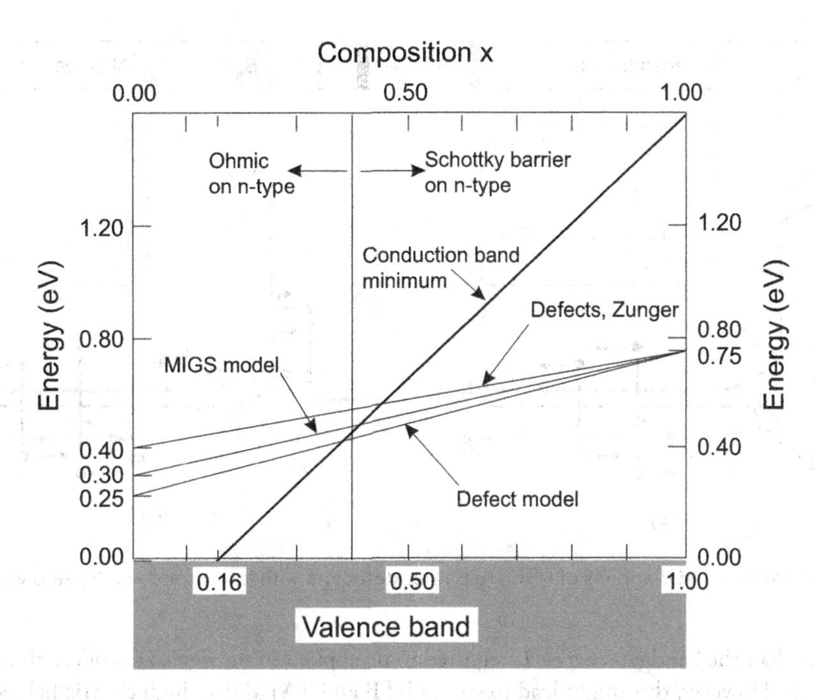

Figure 17.79 Lower limit of the Fermi level pinning position as a function of composition. Two models, MIGS (Ref. [444]) and defect (Refs. [446,447]) were used for the extrapolation. Near x = 0.4, the Fermi level moves into the conduction band providing ohmic contacts on n-type material. (From Ref. [328])

The Schottky barrier height and ohmic contacts on $Hg_{1-x}Cd_xTe$ over the whole composition have been discussed by Spicer et al. [328] in the framework of the current theories. Their predictions are summarized in Figure 17.79. The lower limit of the pinning positions due to the various mechanisms is represented by the solid lines across the whole composition range. For the defect-based model, two similar lines depending on the theoretical calculation of either Kobayoshi et al. [446] or Zunger [447] was used. The MIGS-derived pinning positions were based on the work of Tersoff [444]. It is interesting to note that all models predict a pinning position situated in the conduction band below a x value around 0.4–0.5.

Polla and Sood [448,449] reported Schottky barrier photodiodes on narrow-gap bulk p-type $Hg_{1-x}Cd_xTe$. Barrier metals (Al, Cr, and Mn) were thermally evaporated and ZnS was sputtered as a passivation layer. Schottky diodes on bulk and MOCVD n-type $Hg_{1-x}Cd_xTe$ ($x = 0.6–0.7$) have been reported by Leech and Kibel [450]. All of the metal contacts (Ag, Au, Cu, Pd, Pt, Sb) formed rectifying contacts on the etched surfaces with Schottky barrier heights in the range 0.7–0.8 eV; only Ti gave near-ohmic behavior. The ideality factors for the rectifying contacts were greater than unity, indicating the presence of transport mechanisms other than thermionic emission. The diode characteristics are strongly dependent on the surface preparation.

17.7 BARRIER PHOTODETECTORS

Contrary to the III-V semiconductor-based heterostructures, HgCdTe material does not exhibit a near-zero valance band offset, which is the key item limiting the performance of nBn detectors [451–456]. Devices exhibit poor responsivity and detectivity, especially at low temperatures, where the low-energy minority carriers generated by optical absorption are not able to overcome the valence band energy barrier. Reduction of valence band offsets to a reasonably low value, by the adjustment of the Cd mole fraction in the barrier, results in a corresponding reduction of the barrier in the conduction band below a critical level, thus increasing the majority carrier dark current and reducing the responsivity (detectivity) at high temperatures.

At low bias, the valence band barrier (ΔE_v) inhibits the minority carrier hole flow between the absorber and contact cap layer [Figure 17.80a]. Depending on the wavelength of operation, a relatively high bias,

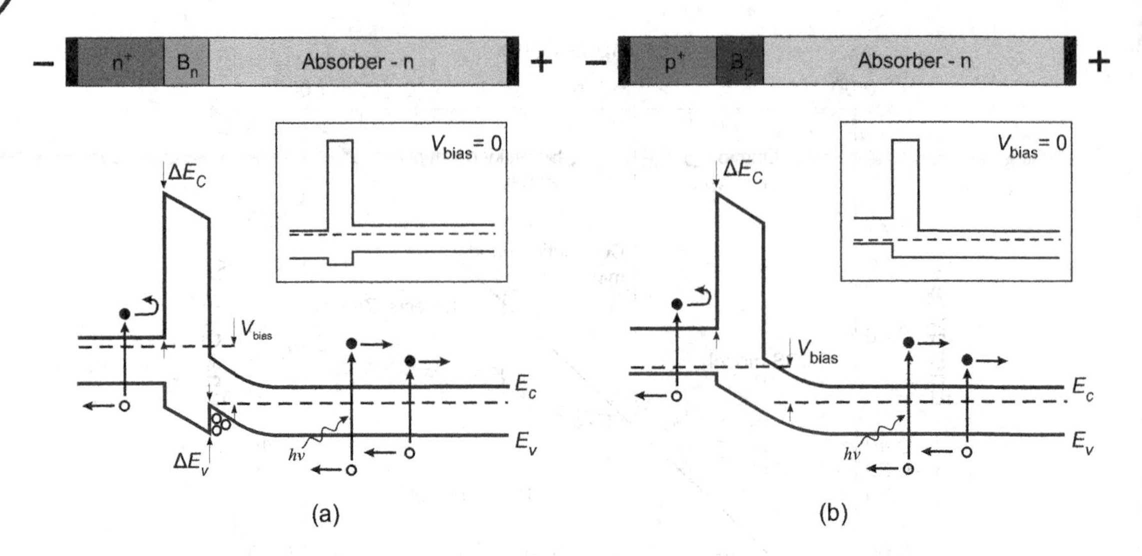

Figure 17.80 Schematic band diagrams of HgCdTe barrier detectors with (a) nonzero and (b) zero valence band offset.

typically greater than the bandgap energy, is required to be applied to the device to collect all the photogenerated carriers. However, this might lead to strong BTB and TAT due to high electric field within the depletion layer.

In HgCdTe material, proper p-type doping of the barrier reduces the valence band offset and increases the offset in the conduction band [457]. The device with the barrier only in the conduction band is similar to that proposed in Ref. [458] in which a p-type barrier is interposed between two narrow-gap n-type regions. Furthermore, due to the presence of the barrier, it is possible to replace the n-type cap contact by the p-type layer [Figure 17.80b] without affecting the dark current. Then, the p-n junction is located at the interface between the heavily doped p-type barrier and lightly doped absorber and its depletion region extends into the absorbing layer. It is similar to one of the realization of III-V semiconductor-based MWIR devices and is named as XBn structure [459] in which X stands for the n- or p-type contact layer. The advantage of a XBn detector over a conventional photodiode appears to be significant at low temperatures.

To overcome HgCdTe band offset issues, the band gap discontinuity should be efficiency eliminated by grading of the barrier composition and doping density profiles. Practical implementation of barrier detector architecture requires well-established epitaxial technology, such as a MBE or MOCVD, with the ability to control composition, thickness, and dopants during growth facilitates. Both methods have advantages and disadvantages in the fabrication of heterostructure detector designs [460].

First, MWIR HgCdTe nB_nn (n-type barrier) devices grown by MBE on a bulk CdZnTe substrate were reported by Itsuno et al. [451–453]. The epitaxial nB_nn structure consists of three In-doped n-type layers: a narrow-bandgap cap contact layer, a wide-bandgap barrier, and a narrow-bandgap absorber. Detectors were plasma etched in planar-mesa and mesa geometry. The schematic cross section of front side illuminated nB_nn planar-mesa and mesa devices is illustrated in Figure 17.81.

The existence of the valence band barrier is the main issue limiting the performance of HgCdTe nB_nn detectors. Figure 18.82a shows the measured current-voltage characteristic at 77 K of plasma etched mesa device. The turn-on voltage (required to align the valence band and enabling the collection of minority carrier (holes) from the absorber layer) is at a reverse bias value around 0.2 V. Shown in Figure 17.82b, the wavelength-dependent relative response measurement at 77 K and −0.2 V bias exhibits a cutoff wavelength of 5.2 μm, consistent with the band edge of the $Hg_{0.71}Cd_{0.29}Te$ absorber layer.

Recently published papers by a research group at the Institute of Applied Physics, Military University of Technology have shown that MOCVD technology is an excellent tool for the HgCdTe barrier architecture growth with a wide range of composition, donor/acceptor doping, and without postgrown annealing [461–463]. The device concept of a specific barrier bandgap architecture integrated with Auger-suppression is a good solution for HOT IR detectors.

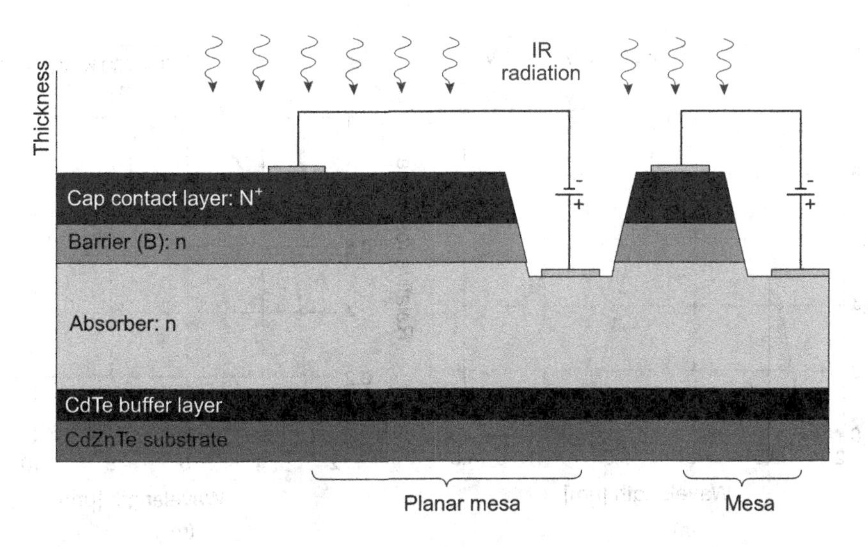

Figure 17.81 Schematic cross section of MBE-grown HgCdTe nB_nn detector with planar-mesa and mesa geometry.

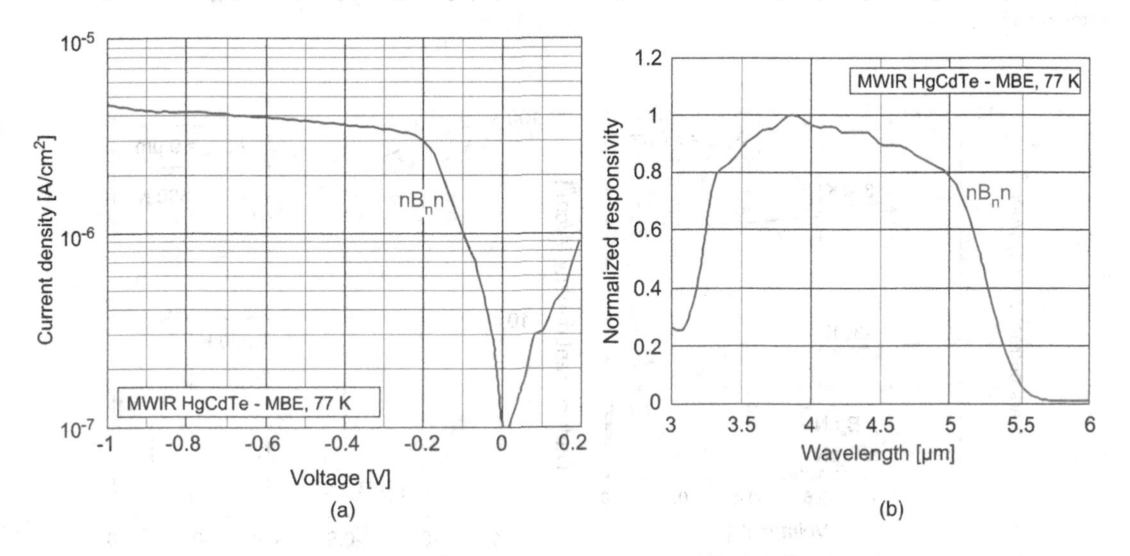

Figure 17.82 Experimental characteristics of second iteration mesa MWIR HgCdTe nB_nn detector grown by MBE: (a) current-voltage characteristic at 77 K and (b) relative spectral response at 77 K and 0.2V reverse bias. (Adapted after Ref. [452])

The barrier structures optimized at 50% cutoff wavelengths up to 3.6, 6, and 9 μm at 230 K were grown in Aixtron AIX-200 MOCVD system on GaAs substrates after the CdTe buffer layer. Devices have p^+-B_p cap-barrier structural unit, intentionally undoped (due to donor background concentration with n-type conductivity), or a low p-type doped absorption layer and wide-bandgap highly doped N^+ bottom contact layer. The relative spectra response of the devices is presented in Figure 17.83. The devices are bottom illuminated through the N^+ layer that plays the role of an IR transmitting window for photons with energies below the bandgap.

Figure 17.84 shows the measured, at 230 K and 300 K, current-voltage characteristics for MOCVD-grown HgCdTe barrier detectors optimized for different spectral ranges. Reverse-biased detectors (both p^+BpnN^+ and p^+BppN^+) with a 3.6 μm cutoff wavelength exhibited very low dark currents in the range of (2–3) × 10^{-4} A/cm² at 230 K. The data indicate that the dark current is mostly due to diffusion current. Very low threshold voltages (–0.1 V) of these detectors indicate that there is no valence band barrier.

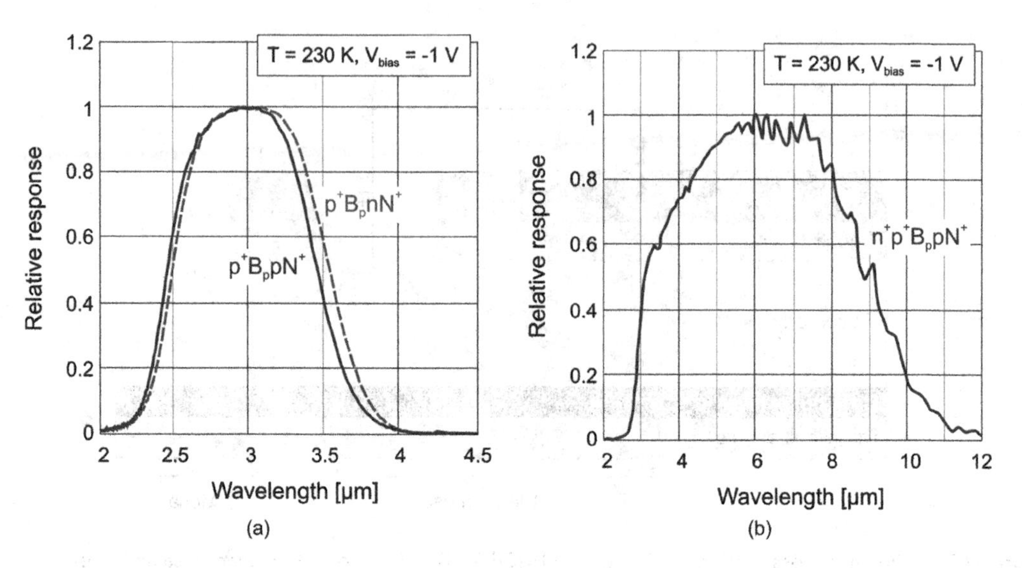

Figure 17.83 Relative spectral response for a back side illuminated, thermoelectrically-cooled (230 K) MOCVD-grown HgCdTe barrier detector optimized at 3.6 µm (a) and 9 µm (b) cutoff wavelengths. Relative spectral response was calculated on the basis of measured values of current responsivity at bias voltage of 1 V. (Adapted after Ref. [463])

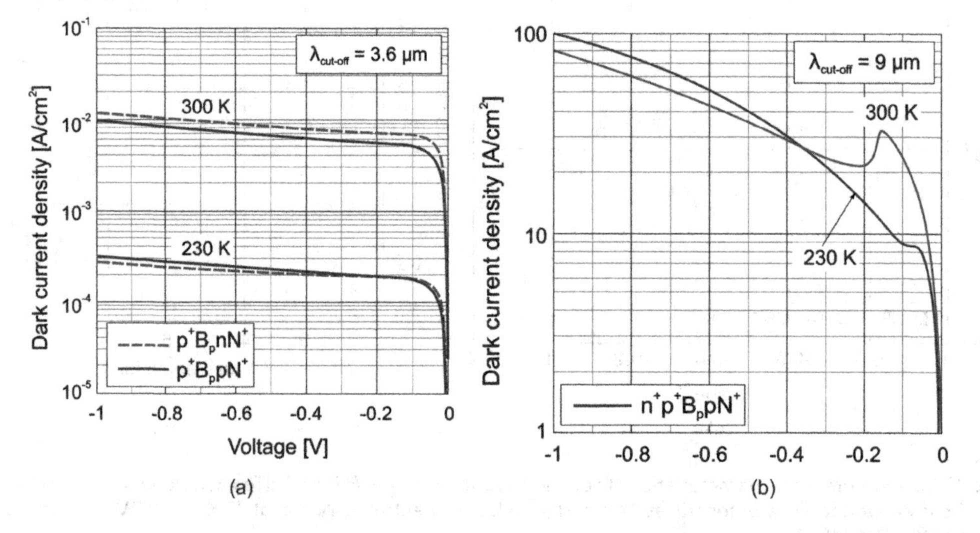

Figure 17.84 Current-voltage characteristics for an MOCVD-grown HgCdTe barrier detector operated at 230 K and 300 K and optimized at 3.6 µm (a) and 9 µm (b) cutoff wavelength. (Adapted after Ref. [463])

$p^+B_ppN^+$ detector with a 6 µm cutoff wavelength and $n^+p^+B_ppN^+$ detector with a 9 µm cutoff wavelength show a suppression of Auger generation that is especially evident at 300 K where the negative dynamic resistance area occurred. Under reverse bias, the electrons are extracted from the absorber region by a positive electrode connected to the bottom N^+-layer. The electrons are also excluded from the absorber near the B_p-p junction because they cannot be injected through the barrier. However, devices with a p-type absorbing layer indicate tunneling, which dominates the leakage current for higher biases. These tunneling effects are especially due to TAT at a decisive heterojunction.

Figure 17.85 compares the minimum dark current density of the analyzed structures to the values given by the Rule 07. The detectors optimized at a 3.6 µm cutoff wavelength show an order of magnitude lower dark current densities than those determined by Rule 07.

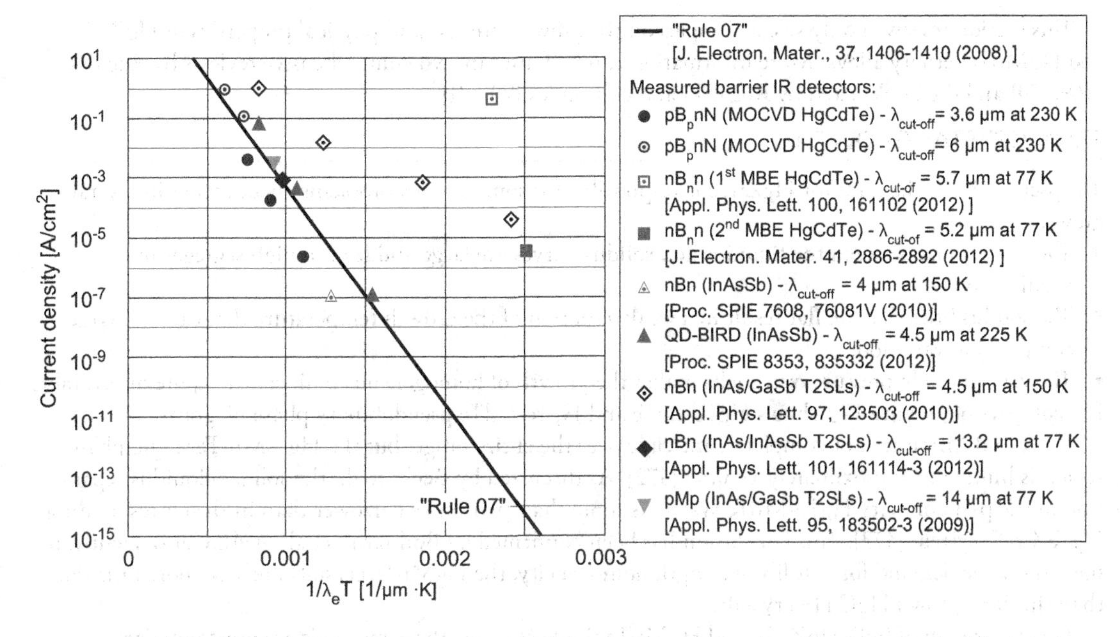

Figure 17.85 Comparison of different barrier IR detectors to the *Rule 07*. Devices fabricated in our laboratory are marked with a darker field on the legend. (Adapted after Ref. [462])

Figure 17.56 compares the detectivity of nonimmersed HgCdTe barrier detectors (dotted lines) with the optically immersed HgCdTe photodiodes manufactured by Vigo-System S.A. The detectivity of HgCdTe detectors with p-type barriers is comparable to the value-marked HgCdTe photodiodes. Implementing the optical immersion for HgCdTe barrier detectors might increase the detectivities by an order of magnitude using GaAs lenses.

17.8 Hg-BASED ALTERNATIVE DETECTORS

Among the small-gap II-VI semiconductors for IR detectors, only Cd, Zn, Mn, and Mg have been shown to open the bandgap of the Hg-based binary semimetals, HgTe and HgSe, to match the IR wavelength range. It appears that the amount of Mg to be introduced in HgTe to match the 10 μm range is insufficient to reinforce the Hg-Te bond [464]. The main obstacles in the technological development of $Hg_{1-x}Cd_xSe$ are the difficulties in obtaining type conversion. In the above alloy systems, $Hg_{1-x}Zn_xTe$ (HgZnTe) and $Hg_{1-x}Mn_xTe$ (HgMnTe) occupy a privileged position.

Neither HgZnTe nor HgMnTe have ever been systematically explored in the device context. There are several reasons for this. First, preliminary investigations of these alloy systems came into the scene after HgCdTe detector development was well on its way. Secondly, the HgZnTe alloy presents a more serious technological challenge than HgCdTe. In case of HgMnTe, Mn is not a group II element and, as a result, HgMnTe is not truly a II-VI alloy. This ternary compound was viewed with some suspicion by those not directly familiar with its crystallographic, electrical, and optical behavior. In such a situation, proponents of the parallel development of HgZnTe and HgMnTe for IR detector fabrication encountered considerable difficulties in selling the idea to industries and funding agencies.

In 1985, Sher et al. [465] showed from theoretical consideration that the weak Hg-Te bond is destabilized by alloying it with CdTe but stabilized by ZnTe. Many groups worldwide have become very interested in this prediction and, more specifically, in the growth and properties of the HgZnTe alloy system as the material for photodetection application in the IR spectral region. But the question of lattice stability in case of HgMnTe compounds is rather ambiguous. According to Wall et al. [466], the Hg-Te bond stability of this alloy is similar to that observed in the binary narrow-gap parent compound. This conclusion contradicts other published results [467]. It has been established that the incorporation of Mn in CdTe destabilizes its lattice because of the Mn 3d orbitals hybridizing into the tetrahedral bonds [468].

This section reviewed only selected topics on the growth process and physical properties of HgZnTe and HgMnTe ternary alloys. More information can be found in two comprehensive reviews by Rogalski [469,470] and the books cited by Rogalski and colleagues [79,471].

17.8.1 CRYSTAL GROWTH

The pseudobinary diagram for HgZnTe is responsible for some serious problems encountered in crystal growth including:

- The separations between the liquidus and solidus curves are large and lead to high segregation coefficients.
- The solidus lines that are flat result in a weak variation of the growth temperature that causes a large composition variation.
- The very high Hg pressure over melts makes the growth of homogeneous bulk crystals quite unfavorable.

For comparison, Figure 17.3 shows HgTe-ZnTe and HgTe-CdTe pseudobinary phase diagrams.

HgTe and MnTe are not completely miscible over the entire range, but the $Hg_{1-x}Mn_xTe$ single-phase region is limited to approximately $x<0.35$ [472]. As discussed by Becla et al., the solidus-liquidus separation in the pseudobinary HgTe-MnTe system is more than two times narrower than in the corresponding HgTe-CdTe system [473]. This conclusion has been confirmed by Bodnaruk et al. [474]. Consequently, to meet the same demand for cutoff wavelength homogeneity, the HgMnTe crystals must be more uniform than similarly grown HgCdTe crystals.

For the growth of bulk HgZnTe and HgMnTe single crystals, three methods are most popular: Bridgman-Stockbarger, SSR, and THM. The best quality HgZnTe crystals have been produced by THM. Using this method, Triboulet et al. [475] produced $Hg_{1-x}Zn_xTe$ crystals ($x \approx 0.15$) with a longitudinal homogeneity of ± 0.01 mol and radial homogeneity of ± 0.01 mol. The source material was a cylinder composed of two cylindrical segments—one HgTe, the other ZnTe—the cross section of which was in the ratio corresponding to the desired composition.

To improve the crystalline quality of HgMnTe single crystals, different modified techniques have been used. Gille et al. [476] demonstrated $Hg_{1-x}Mn_xTe$ ($x \approx 0.10$) single crystals grown by THM with standard deviation $\Delta x = \pm 0.003$ along a 16 mm diameter slice of crystals. Becla et al. [473] decreased the radial macrosegregation and eliminated the small-scale compositional undulations in the vertical Bridgman-grown material by applying a 30 kG magnetic field. Takeyama and Narita [477] developed an advanced crystal growth method, called the modified two-phase mixture method, to produce highly homogeneous, large single crystals of ternary and quaternary alloys.

The best performance of modern devices, however, requires more sophisticated structures. These structures are only achieved by using the epitaxial growth techniques. Additionally, when compared to bulk growth techniques, epitaxial techniques offer important advantages including lower temperatures and Hg vapor pressures, shorter growth times, and reduced precipitation problems that enable the growth of large-area samples with good lateral homogeneity. The above advantages have prompted research in a variety of thin-film growth techniques, such as VPE, LPE, MBE, and MOCVD. The first studies of LPE-crystal growth of HgCdZnTe and HgCdMnTe from Hg-rich solutions have demonstrated that the homogeneity of epilayers can be improved by incorporating Zn or Mn during the crystal growth [478]. More recently, considerable progress has been achieved in HgMnTe film fabrication by MOCVD using an IMP [479]. Depending on growth conditions, both n- and p-type layers may be produced with extrinsic electron and hole concentrations of the order of 10^{15} cm^{-3} and 10^{14} cm^{-3}, respectively.

All the epitaxial growth processes depend on the identification of suitable substrates. They require large-area, single-crystal substrates [480]. The large difference in the lattice parameters of HgTe and ZnTe induces strong interactions between cations. Vegard's law appears to be obeyed relatively better in HgZnTe than in HgCdTe and, for $Hg_{1-x}Zn_xTe$ at 300 K, $a(x) = 6.461–0.361x$ (Å) [471]. In comparison with HgCdTe, the lattice parameter of $Hg_{1-x}Mn_xTe$, $a(x) = 6.461–0.121x$ (Å), varies with x much more rapidly, which is a disadvantage from the point of view of the epitaxial growth of multilayer heterostructures that is required for advanced IR devices. The lattice parameter of the zinc blende compound $Cd_{1-x}Zn_xTe$ ($Cd_{1-x}Mn_xTe$) indicates a simple matter: to find suitable substrates for the epitaxial growth of $Hg_{1-x}Zn_xTe$ ($Hg_{1-x}Mn_xTe$). However, Bridgman-grown CdMnTe crystals are highly twinned and thus unusable as epitaxial substrates.

17.8.2 PHYSICAL PROPERTIES

The physical properties of both the ternary alloys are determined by the energy gap structure near the Γ-point of the Brillouin zone. The shape of the electron band and the light-mass hole band is determined by the **k·p** theory. The bandgap structure of HgZnTe near the Γ-point is similar to that of the HgCdTe ternary illustrated in Figure 17.10. The bandgap energy of HgZnTe varies approximately 1.4 times (two times for HgMnTe) as fast with the composition parameter x as it does for HgCdTe.

Both HgZnTe and HgMnTe exhibit compositional-dependent optical and transport properties, similar to HgCdTe materials with the same energy gap. Some physical properties of alternative alloys indicate a structural advantage in comparison with HgCdTe. Introducing ZnTe in HgTe decreases statistically the ionicity of the bond, improving the stability of the alloy. Moreover, because the bond length of ZnTe (2.406 Å) is 14% shorter than that of HgTe (2.797 Å) or CdTe (2.804 Å), the dislocation energy per unit length and the hardness of the HgZnTe alloy are higher than that of HgCdTe. The maximum degree of microhardness for HgZnTe is more than twice that for HgCdTe [480]. HgZnTe is a material that is more resistant to dislocation formation and plastic deformation than HgCdTe.

The as-grown $Hg_{1-x}Zn_xTe$ material is highly p-type in the 10^{17} cm^{-3} range with mobilities in the hundreds of cm^2/Vs. These values indicate that its conduction is dominated by holes arising from Hg vacancies. After a low-temperature anneal ($T \leq 300$°C) accomplished with an excess of Hg (which annihilates the Hg vacancies), the material is converted to low n-type in the mid-10^{14} cm^{-3} to low-10^{15} cm^{-3} range, which has mobilities ranging from 10^4 to 4×10^5 cm^2/Vs. A study by Rolland et al. [481] showed that the n-type conversion occurs only for crystals with composition $x \leq 0.15$. The Hg diffusion rate is slower in HgZnTe than in HgCdTe. Interdiffusion studies between HgTe and ZnTe indicate that the interdiffusion coefficient is approximately ten times lower in HgZnTe than in HgCdTe [475].

Berding et al. [482] and Granger et al. [483] gave the theoretical description of the scattering mechanisms in HgZnTe. To obtain a good fit to experimental data for $Hg_{0.866}Zn_{0.134}Te$, they considered phonon dispersion plus ionized impurity scattering plus core dispersion without compensation in their mobility calculations. Theoretical calculations of electron mobilities indicate that the disorder scattering is negligible for HgZnTe alloy [475]. In contrast, the hole mobilities are likely to be limited by alloy scattering and the predicted alloy hole mobility of HgZnTe is approximately a factor of two less than what was found for HgCdTe. Additionally, Abdelhakiem et al. [484] confirmed that the electron mobilities are very close to the HgCdTe ones for the same energy gap and the same donor and acceptor concentrations.

The HgMnTe alloy is a semimagnetic narrow-gap semiconductor. The exchange interaction between band electrons and Mn^{2+} electrons modifies their band structure, making it dependent on the magnetic field at very low temperature. In the range of temperatures typical for IR detector operation (≥77 K), the spin-independent properties of HgMnTe are practically identical to the properties of HgCdTe, which are discussed exhaustively in the literature. The studies carried out by Kremer et al. [485] confirmed that the annealing of samples in Hg vapor eliminates the Hg vacancies, with the resulting material being n-type due to some unknown native donor. The diffusion rate of Hg into HgMnTe is the same as into HgCdTe. Measurements of the transport properties of $Hg_{1-x}Mn_xTe$ ($0.095 \leq x \leq 0.15$) indicate deep donor and acceptor levels into the energy gap, which influence not only the temperature dependencies of the Hall coefficient, conductivity, and Hall mobility, but also the minority carrier lifetimes [486,487]. Theoretical considerations of the electron mobilities in HgCdTe and HgMnTe indicate that at room temperature the mobilities are nearly the same. But at 77 K, the electron mobilities are approximately 30% less for HgMnTe when compared with the same concentration of defects [488].

The measured carrier lifetime in both ternary alloy systems is a sensitive characteristic of semiconductors that depends on material composition, temperature, doping, and defects. The Auger mechanism governs the high-temperature lifetime, and the SR mechanism is mainly responsible for low-temperature lifetimes. The reported positions of SR centers for both n- and p-type materials range anywhere from near the valence to near the conduction band. Comprehensive reviews of GR mechanisms and the carrier lifetime experimental data for both ternary alloys are given by Rogalski et al. [79,471].

Tables 17.7 and 17.8 contain lists of standard approximate relationships for material properties of HgZnTe and HgMnTe, respectively. Most of these relationships have been taken from Rogalski [469,470].

Table 17.7 Standard relationships for $Hg_{1-x}Zn_xTe$ (0.10 ≤ x ≤ 0.40)

PARAMETER	RELATIONSHIP
Lattice constant $a(x)$ (nm) at 300 K	$0.6461–0.0361x$
Density γ (g/cm^3) at 300 K	$8.05–2.41x$
Energy gap E_g (eV)	$-0.3+0.0324x^{1/2}+2.731x-0.629x^2+0.533x^3$ $+5.3\times10^{-4}T\left(1-0.76x^{1/2}-1.29x\right)$
Intrinsic carrier concentration n_i (cm^{-3})	$\left(3.607+11.370x+6.584\times10^{-3}T-3.633\times10^{-2}xT\right)$ $\times10^{14}E_g^{3/4}T^{3/2}\exp\left(-5802E_g/T\right)$
Momentum matrix element P (eVcm)	8.5×10^{-8}
Spin-orbit splitting energy Δ (eV)	1.0
Effective masses: m_e^*/m m_h^*/m	$5.7\times10^{-16}E_g/P^2$ E_g in eV; P in eVcm 0.6
Mobilities: μ_e (cm^2/Vs) μ_h (cm^2/Vs)	$9\times10^8 b/T^{2a}$ $a=(0.14/x)^{0.6}$; $b=(0.14/x)^{7.5}$ $\mu_e(x,T)/100$
Static dielectric constant ε_s	$20.206-15.153x+6.5909x^2-0.951826x^3$
High-frequency dielectric constant ε_∞	$13.2+19.1916x+19.496x^2-6.458x^3$

Source: Data was obtained from Ref. [469].

Some of these parameters, for example, the intrinsic carrier concentration, have since been reexamined. For example, Sha et al. [489] concluded that their improved calculations of intrinsic carrier concentration were approximately 10%–30% higher than those obtained earlier by Jóźwikowski and Rogalski [490]. However, the new calculations also should be treated as approximations since the dependence of the energy gap on composition and temperature ($E_g(x, T)$ is necessary in the calculations of n_i) is still under serious discussion [491].

17.8.3 HgZnTe PHOTODETECTORS

The technology for HgZnTe IR detectors has benefited greatly from the HgCdTe device technology base [469,471,492]. In comparison with HgCdTe, HgZnTe detectors are easier to prepare due to their relatively higher hardness. The development of device technology requires reproducible high-quality, electronically-stable interfaces with a low interface state density. It was found that the tendency to form surface inversion layers on HgZnTe by anodization is considerably lower than that of HgCdTe [493]. Also, fixed charges at the anodic oxide-HgZnTe interface (2×10^{10} cm^{-2} at 90 K) are lower [494]. Additionally, it has been observed that the anodic oxide-HgZnTe interface is more stable under thermal treatment than the anodic-HgCdTe interface [495].

The first HgZnTe photoconductive detectors were fabricated by Nowak in the early 1970s [496]. Because of their early stage of development, the performances of these devices were inferior to that of HgCdTe. Then Piotrowski et al. demonstrated that p-type $Hg_{0.885}Zn_{0.115}Te$ can be used as a material for high-quality ambient 10.6 μm photoconductors [497,498]. These photoconductors, working at 300 K, can achieve 10^8 cmHz$^{1/2}$W^{-1} detectivity with optimized composition, doping, and geometry. Aspects of theoretical performance for both photoconductive and photovoltaic detectors are discussed by Rogalski and colleagues [469–471,499–502].

Table 17.8 Standard relationships for $Hg_{1-x}Mn_xTe$ (0.08 ≤ x ≤ 0.30)

PARAMETER	RELATIONSHIP
Lattice constant $a(x)$ (nm) at 300 K	$0.6461–0.0121x$
Density γ (g/cm³) at 300 K	$8.12–3.37x$
Energy gap E_g (eV)	$-0.253+3.446x+4.9\times10^{-4}xT-2.55\times10^{-3}T$
Intrinsic carrier concentration n_i (cm⁻³)	$\left(4.615-1.59x+2.64\times10^{-3}T-1.70\times10^{-2}xT+34.15x^2\right)$ $\times10^{14}E_g^{3/4}T^{3/2}\exp\left(-5802E_g/T\right)$
Momentum matrix element P (eVcm)	$(8.35-7.94x)\times10^{-8}$
Spin-orbit splitting energy Δ (eV)	1.08
Effective masses: m_e^*/m m_h^*/m	$5.7\times10^{-16}\,Eg/P^2$ E_g in eV; P in eVcm 0.5
Mobilities: μ_e (cm²/Vs) μ_h (cm²/Vs)	$9\times10^8 b/T^{2a}$ $a=(0.095/x)^{0.6}$; $b=(0.095/x)^{7.5}$ $\mu_e(x,T)/100$
Static dielectric constant ε_s	$20.5-32.6x+25.1x^2$
High frequency dielectric constant ε_∞	$15.2–28.8x+28.2x^2$

Source: Data was obtained from Ref. [470].

Several different techniques have yielded p-n HgZnTe junctions, including Hg in-diffusion [469,492,503], Au diffusion [469,492,504], ion implantation [469,470,492,505,506], and ion etching [214,469,492,500]. To date, the ion implantation method gives the best-quality n⁺-p HgZnTe photodiodes.

The HgZnTe photodiode characteristics at 77 K were similar to those of HgCdTe photodiodes in 1980s [492]. Encouraging results were also achieved using a HgCdZnTe quaternary alloy system [507].

17.8.4 HgMnTe PHOTODETECTORS

Among the different types of HgMnTe IR detectors, primarily p-n junction photodiodes have been developed [492,508,509]. Also, the quaternary HgCdMnTe alloy system is an interesting material for IR applications. The presence of Cd, a third cation in this system, makes it possible to use composition to tune not only the bandgap but also other energy levels, in particular the spin-orbit-split band Γ_7 [510,511]. Because of this flexibility, the system appears advantageous, especially for APDs.

Becla produced good-quality p-n HgMnTe and HgCdMnTe junctions by annealing as-grown, p-type samples in Hg-saturated atmospheres [508]. These junctions were made in HgMnTe or HgCdMnTe bulk samples grown by THM, and epitaxial layers grown isothermally on the CdMnTe substrate.

More recently, high-quality planar and mesa HgMnTe photodiodes have been fabricated by Kosyachenko et al. [512,513] using ion etching in a system generating an Ar beam of 500–1,000 eV energy and 0.5–1 mA/cm² current density. The hole concentration of annealed as-grown Bridgman wafers selected for photodiode preparation was (2–5) × 10¹⁶ cm⁻³. For $Hg_{1-x}Mn_xTe$ photodiodes operated at 80 K with cutoff wavelength of 10–11 μm, the R_oA product is equal to 20–30 Ωcm², whereas for photodiodes with λ_c of 7–8 μm, $R_oA \approx 500$ Ωcm² has been obtained.

The potential advantage of the HgCdMnTe system is connected with the bandgap spin-orbit-splitting resonance ($E_g = \Delta$) effects in the impact ionization phenomena in APDs. At room temperature, HgCdTe and HgMnTe systems provide an $E_g = \Delta$ resonance at 1.3 μm and 1.8 μm, respectively. To demonstrate the above possibility of obtaining high-performance HgCdMnTe APDs, Shin et al. used the B-implantation method to fabricate the mesa-type structures for this quaternary alloy [514]. The R_oA product was 2.62 × 10² Ωcm², which is equivalent to a detectivity value of 1.9 × 10^{11} cmHz$^{1/2}$W^{-1} at 300 K. The breakdown voltage defined at the dark current 10 μA was over 110 V.

Becla et al. [515] have developed HgMnTe APDs, which have increased speed and performance compared to its standard line of photodiodes. Several p-n and n-p mesa type structures were fabricated, which permitted the injection of minority carriers from both n- and p-type regions and led to hole-initiated and electron-initiated avalanche gain. Avalanche gain for 7 μm devices was more than 40, while 10.6 μm detectors showed gains better than 10.

Other types of HgMnTe detectors have also been fabricated. Becla et al. described $Hg_{0.92}Mn_{0.08}Te$ photoelectromagnetic (PEM) detectors with an acceptor concentration of approximately 2 × 10^{17} cm^{-3} [516]. The best performance of PEM detectors was achieved using $Hg_{1-x}Mn_xTe$ with composition x of approximately 0.08–0.09. At that composition range, the peak detectivity of the detector was in the region of 7–8 μm.

REFERENCES

1. W. D. Lawson, S. Nielson, E. H. Putley, and A. S. Young, "Preparation and properties of HgTe and mixed crystals of HgTe-CdTe," *J. Phys. Chem. Solids* 9, 325–29, 1959.
2. B. J. Andresen, G. F. Fulop, and P. R. Norton, "Infrared technology and applications XXXV," *Proceedings of SPIE* 7298, Bellingham, WA, 2009.
3. T. Elliot, "Recollections of MCT work in the UK at Malvern and Southampton," *Proc, SPIE* 7298, 72982M, 2009.
4. E. Burstein, J. W. Davisson, E. E. Bell, W. J. Turner, and H. G. Lipson, "Infrared photoconductivity due to neutral impurities in germanium," *Phys. Rev.* 93, 65–8, 1954.
5. S. Borrello and H. Levinstein, "Preparation and properties of mercury-doped germanium," *J. Appl. Phys.* 33, 2947–50, 1962.
6. D. Long and J. L. Schmit, "Mercury-Cadmium Telluride and Closely Related Alloys," in *Semiconductors and Semimetals*, Vol. 5, eds. R. K. Willardson and A. C. Beer, 175–255, Academic Press, New York, 1970.
7. P. Norton, "HgCdTe infrared detectors," *Opto-Electron. Rev.* 10, 159–74, 2002.
8. C. Verie and R. Granger, "Propriétés de junctions p-n d'alliages $Cd_xHg_{1-x}Te$," *Comptes Rendus de l'Académie des Sciences* 261, 3349–52, 1965.
9. G. C. Verie and M. Sirieix, "Gigahertz cutoff frequency capabilities of CdHgTe photovoltaic detectors at 10.6 μm," *IEEE J. Quant. Electron.* 8, 180–4, 1972.
10. B. E. Bartlett, D. E. Charlton, W. E. Dunn, P. C. Ellen, M. D. Jenner, and M. H. Jervis, "Background limited photoconductive detectors for use in the 8–14 micron atmospheric window," *Infrared Phys. Technol.* 9, 35–6, 1969.
11. M. A. Kinch, S. R. Borrello, and A. Simmons, "0.1 eV HgCdTe photoconductive detector performance," *Infrared Phys.* 17, 127–35, 1977.
12. C. T. Elliott, D. Day, and B. J. Wilson, "An integrating detector for serial scan thermal imaging," *Infrared Phys.* 22, 31–42, 1982.
13. A. Blackburn, M. V. Blackman, D. E. Charlton, W. A. E. Dunn, M. D. Jenner, K. J. Oliver, and J. T. M. Wotherspoon, "The practical realization and performance of SPRITE detectors," *Infrared Phys.* 22, 57–64, 1982.
14. W. S. Boyle and G. E. Smith, "Charge-coupled semiconductor devices," *Bell Sys. Technol. J.* 49, 587–93, 1970.
15. R. Thom, "High density infrared detector arrays," U. S. Patent No. 4,039,833, 8/2/77.
16. M. A. Kinch, "MIS Devices in HgCdTe," in *Properties of Narrow Gap Cadmium-Based Compounds*, EMIS Datareviews Series No. 10, ed. P. Capper, 359–63, IEE, London, 1994.
17. M. B. Reine, "Photovoltaic Detectors in HgCdTe," in *Infrared Detectors and Emitters: Materials and Devices*, eds. P. Capper and C. T. Elliott, 313–76, Chapman and Hall, London, 2000.
18. A. Rogalski, J. Antoszewski, and L. Faraone, "Third-generation infrared photodetector arrays," *J. Appl. Phys.* 105, 091101–44, 2009.
19. W. F. H. Micklethweite, "The Crystal Growth of Mercury Cadmium Telluride," in *Semiconductors and Semimetals*, Vol. 18, eds. R. K. Willardson and A. C. Beer, 48–119, Academic Press, New York, 1981.
20. S. Sher, M. A. Berding, M. van Schlifgaarde, and A.-B. Chen, "HgCdTe status review with emphasis on correlations, native defects and diffusion," *Semicond. Sci. Technol.* 6, C59–70, 1991.
21. H. R. Vydyanath, "Status of Te-Rich and Hg-rich liquid phase epitaxial technologies for the growth of (Hg,Cd) Te alloys," *J. Electron. Mater.* 24, 1275–85, 1995.
22. R. F. Brebrick, "Thermodynamic modeling of the Hg-Cd-Te systems," *J. Cryst. Growth* 86, 39–48, 1988.
23. J. C. Brice, "A numerical description of the Cd-Hg-Te phase diagram," *Prog. Cryst. Growth Charact. Mater.* 13, 39–61, 1986.

24. T. C. Yu and R. F. Brebrick, "Phase Diagrams for HgCdTe," in *Properties of Narrow Gap Cadmium-Based Compounds*, EMIS Datareviews Series No. 10, ed. P. Capper, 55–63, IEE, London, 1994.
25. *Properties of Narrow Gap Cadmium-Based Compounds*, EMIS Datareviews Series No. 10, ed. P. P. Capper, IEE, London, 1994.
26. T. Tung, C. H. Su, P. K. Liao, and R. F. Brebrick, "Measurement and analysis of the phase diagram and thermodynamic properties in the Hg-Cd-Te system," *J. Vac. Sci. Technol.* 21, 117–24, 1982.
27. H. Maier and J. Hesse, "Growth, Properties and Applications of Narrow-Gap Semiconductors," in *Crystal Growth, Properties and Applications*, ed. H. C. Freyhardt, 145–219, Springer Verlag, Berlin, 1980.
28. J. H. Tregilgas, "Developments in recrystallized bulk HgCdTe," *Prog. Cryst. Growth Character. Mater.* 28, 57–83, 1994.
29. P. Capper, "Bulk Growth Techniques," in *Narrow-gap II-VI Compounds for Optoelectronic and Electromagnetic Applications*, ed. P. Capper, 3–29, Chapman & Hall, London, 1997.
30. P. Capper, "Narrow-Bandgap II–VI Semiconductors: Growth," in *Springer Handbook of Electronic and Photonic Materials*, ed. S. Kasap and P. Capper, 303–24, Springer, Heidelberg, Germany, 2006.
31. P. Capper and J. Garland, *Mercury Cadmium Telluride: Growth, Properties, and Applications*, John Wiley & Sons, Chichester, UK, 2011.
32. T. C. Harman, "Single crystal growth of $Hg_{1-x}Cd_xTe$," *J. Electron. Mater.* 1, 230–42, 1972.
33. A. W. Vere, B. W. Straughan, D. J. Williams, N. Shaw, A. Royle, J. S. Gough, and J. B. Mullin, "Growth of $Cd_xHg_{1-x}Te$ by a pressurised cast-recrystallise-anneal technique," *J. Cryst. Growth* 59, 121–29, 1982.
34. L. Colombo, A. J. Syllaios, R. W. Perlaky, and M. J. Brau, "Growth of large diameter (Hg,Cd) Te crystals by incremental quenching," *J. Vac. Sci. Technol. A* 3, 100–4, 1985.
35. R. K. Sharma, V. K. Singh, N. K. Nayyar, S. R. Gupta, and B. B. Sharma, "Horizontal casting for the growth of $Hg_{1-x}Cd_xTe$ by solid state recrystalization," *J. Cryst. Growth* 131, 565–73, 1993.
36. P. Capper, "Bridgman growth of $Cd_xHg_{1-x}Te$: A. Review," *Prog. Cryst. Growth Character. Mater.* 19, 259–93, 1989.
37. P. Capper, "The role of accelerated crucible rotation in the growth of $Hg_{1-x}Cd_xTe$ and CdTe/CdZnTe," *Prog. Cryst. Growth Character. Mater.* 28, 1–55, 1989.
38. P. Capper, J. Gosney, J. E. Harris, E. O'Keefe, and C. D. Maxey, "Infra-red materials activities at GEC-Marconi Infra-red Limited: Part I: Bulk growth techniques," *GEC J. Res.* 13, 164–74, 1996.
39. P. Capper, "Bulk Growth of Mercury Cadmium Telluride (MCT)," in *Mercury Cadmium Telluride: Growth, Properties, and Applications*, eds. P. Capper and J. Garland, 320, John Wiley & Sons, Chichester, UK, 2011.
40. L. Colombo, R. R. Chang, C. J. Chang, and B. A. Baird, "Growth of Hg-based alloys by the travelling heater method," *J. Vac. Sci. Technol. A* 6, 2795–99, 1988.
41. R. Triboulet, "The travelling heater method (THM) for $Hg_{1-x}Cd_xTe$ and related materials," *Prog. Cryst. Growth Character. Mater.* 28, 85–114, 1994.
42. E. R. Gertner, "Epitaxial mercury cadmium telluride," *Ann. Rev. Mater. Sci.* 15, 303–28, 1985.
43. T. C. Harman, "Liquidus isotherms, solidus lines and LPE growth in the Te-Rich corner of the Hg-Cd-Te system," *J. Electron. Mater.* 9, 945–61, 1980.
44. B. Pelliciari, "State of the art of LPE HgCdTe at LIR," *J. Cryst. Growth* 86, 146–60, 1988.
45. L. Colombo and G. H. Westphal, "Large volume production of HgCdTe by dipping liquid phase epitaxy," *Proc. SPIE* 2228, 66–72, 1994.
46. P. Capper, T. Tung, and L. Colombo, "Liquid phase epitaxy," in *Narrow-gap II-VI Compounds for Optoelectronic and Electromagnetic Applications*, ed. P. Capper, 30–70, Chapman & Hall, London, 1997.
47. P. Capper, "Liquid Phase Epitaxy of MCT," in *Mercury Cadmium Telluride: Growth, Properties, and Applications*, eds. P. Capper and J. Garland, 95112, John Wiley & Sons, Chichester, UK, 2011.
48. P. E. Herning, "Experimental determination of the mercury-rich corner of the Hg-Cd-Te phase diagram," *J. Electron. Mater.* 13, 1–14, 1984.
49. T. Tung, L. V. DeArmond, R. F. Herald, P. E. Herning, M. H. Kalisher, D. A. Olson, R. F. Risser, A. P. Stevens, and S. J. Tighe, "State of the art of Hg-melt LPE HgCdTe at Santa Barbara Research Center," *Proc. SPIE* 1735, 109–31, 1992.
50. G. Cohen-Solal and Y. Marfaing, "Transport of photocarriers in $Cd_xHg_{1-x}Te$ graded-gap structures," *Solid State Electron.* 11, 1131–47, 1968.
51. R. F. Hicks, "The chemistry of the organometallic vapor-phase epitaxy of mercury cadmium telluride," *Proc. IEEE* 80, 1625–40, 1992.
52. S. J. C. Irvine, "Metal-organic Vapour Phase Epitaxy," in *Narrow-gap II-VI Compounds for Optoelectronic and Electromagnetic Applications*, ed. P. Capper, 71–96, Chapman & Hall, London, 1997.

53. J. Tunnicliffe, S. J. C. Irvine, O. D. Dosser, and J. B. Mullin, "A new MOCVD technique for the growth of highly uniform CMT," *J. Cryst. Growth* 68, 245–53, 1984.
54. C. D. Maxey, "Metal-organic Vapor Phase Epitaxy (MOVPE) Growth," in *Mercury Cadmium Telluride: Growth, Properties, and Applications*, edited by P. Capper and J. Garland, 113–29, John Wiley & Sons, Chichester, UK, 2011.
55. J. C. Irvine, "Recent development in MOCVD of $Hg_{1-x}Cd_xTe$," *Proc. SPIE* 1735, 92–9, 1992.
56. P. Mitra, F. C. Case, and M. B. Reine "Progress in MOVPE of HgCdTe for advanced infrared detectors," *J. Electron. Mater.* 27, 510–20, 1998.
57. J. M. Arias-Cortes, "MBE of HgCdTe for Electro-Optical Infrared Applications," in *II-VI Semiconductor Compounds*, ed. M. Pain, 509–36, World Scientific Publishing, Singapore, 1993.
58. O. K. Wu, T. J. deLyon, R. D. Rajavel, and J. E. Jensen, "Molecular Beam Epitaxy of HgCdTe," in *Narrow-gap II-VI Compounds for Optoelectronic and Electromagnetic Applications*, ed. P. Capper, 97–130, Chapman & Hall, London, 1997.
59. T. J. de Lyon, R. D. Rajavel, J. A. Roth, and J. E. Jensen, "Status of HgCdTe MBE Technology," in *Handbook of Infrared Detection and Technologies*, eds. M. Henini and M. Razeghi, 309–52, Elsevier, Oxford, UK, 2002.
60. J. Garland, "MBE Growth of Mercury Cadmium Telluride," in *Mercury Cadmium Telluride: Growth, Properties, and Applications*, eds. P. Capper and J. Garland, 131–49, John Wiley & Sons, Chichester, UK, 2011.
61. T. S. Lee, J. Garland, C. H. Grein, M. Sumstine, A. Jandeska, Y. Selamet, and S. Sivananthan, "Correlation of arsenic incorporation and its electrical activation in MBE HgCdTe," *J. Electron. Mater.* 29, 869–72, 2000.
62. A. Noda, H. Kurita and R. Hirano, "Bulk Growth of CdZnTe/CdTe Crystals," in *Mercury Cadmium Telluride: Growth, Properties, and Applications*, eds. P. Capper and J. Garland, 21–50, John Wiley & Sons, Chichester, UK, 2011.
63. J. Garland and R. Sporken, "Substrates for the Epitaxial Growth of MCT," in *Mercury Cadmium Telluride: Growth, Properties, and Applications*, eds. P. Capper and J. Garland, 75–93, John Wiley & Sons, Chichester, UK, 2011.
64. W. E. Tennant, C. A. Cockrum, J. B. Gilpin, M. A. Kinch, M. B. Reine, and R. P. Ruth, "Key issue in HgCdTe-based focal plane arrays: An industry perspective," *J. Vac. Sci. Technol. B* 10, 1359–69, 1992.
65. R. Triboulet, A. Tromson-Carli, D. Lorans, and T. Nguyen Duy, "Substrate issues for the growth of mercury cadmium telluride," *J. Electron. Mater.* 22, 827–34, 1993.
66. E. R. Gertner, W. E. Tennant, J. D. Blackwell, and J. P. Rode, "HgCdTe on sapphire: A new approach to infrared detector arrays," *J. Cryst. Growth* 72, 462–67, 1987.
67. L. J. Kozlowski, S. L. Johnston, W. V. McLevige, A. H. B. Vandervyck, D. E. Cooper, S. A. Cabelli, E. R. Blazejewski, K. Vural, and W. E. Tennant, "128 × 128 PACE 1 HgCdTe hybrid FPAs for thermoelectrically cooled applications," *Proc. SPIE* 1685, 193–203, 1992.
68. D. D. Edwall, J. S. Chen, J. Bajaj, and E. R. Gertner, "MOCVD HgCdTe/GaAs for IR detectors," *Semicond. Sci. Technol.* 5, S221–4, 1990.
69. V. S. Varavin, V. V. Vasiliev, S. A. Dvoretsky, N. N. Mikhailov, V. N. Ovsyuk, Yu. G. Sidorov, A. O. Suslyakov, M. V. Yakushev, and A. L. Aseev, "HgCdTe epilayers on GaAs: Growth and devices," *Opto-Electron. Rev.* 11, 99–111, 2003.
70. J. M. Peterson, J. A. Franklin, M. Readdy, S. M. Johnson, E. Smith, W. A. Radford, and I. Kasai, "High-quality large-area MBE HgCdTe/Si," *J. Electron. Mater.* 36, 1283–86, 2006.
71. R. Bornfreund, J. P. Rosbeck, Y. N. Thai, E. P. Smith, D. D. Lofgreen, M. F. Vilela, A. A. Buell, et al., "High-performance LWIR MBE-grown HgCdTe/Si focal plane arrays," *J. Electron. Mater.* 37, 1085–91, 2007.
72. P. Capper, "A review of impurity behavior in bulk and epitaxial $Hg_{1-x}Cd_xTe$," *J. Vac. Sci. Technol. B* 9, 1667–86, 1991.
73. S. Sher, M. A. Berding, M. van Schlifgaarde, and A. B. Chen, "HgCdTe status review with emphasis on correlations, native defects and diffusion," *Semicond. Sci. Technol.* 6, C59–70, 1991.
74. H. R. Vydyanath, "Mechanisms of incorporation of donor and acceptor dopants in $Hg_{1-x}Cd_xTe$ alloys," *J. Vac. Sci. Technol. B* 9, 1716–23, 1991.
75. H. R. Vydyanath, "Incorporation of dopants and native defects in bulk $Hg_{1-x}Cd_xTe$ crystals and epitaxial layers," *J. Cryst. Growth* 161, 64–72, 1996.
76. P. Capper, "Intrinsic and Extrinsic Doping," in *Narrow-Gap II-VI Compounds for Optoelectronic and Electromagnetic Applications*, ed. P. Capper, 211–37, Chapman & Hall, London, 1997.
77. Y. Marfaing, "Point Defects in Narrow Gap II–VI Compounds," in *Narrow-Gap II-VI Compounds for Optoelectronic and Electromagnetic Applications*, ed. P. Capper, 238–67, Chapman & Hall, London, 1997.
78. H. R. Vydyanath, V. Nathan, L. S. Becker, and G. Chambers, "Materials and process issues in the fabrication of high-performance HgCdTe infrared detectors," *Proc. SPIE* 3629, 81–7, 1999.
79. A. Rogalski, K. Adamiec, and J. Rutkowski, *Narrow-Gap Semiconductor Photodiodes*, SPIE Press, Bellingham, WA, 2000.
80. J. Chu and A. Sher, *Physics and Properties of Narrow Gap Semiconductors*, Springer, New York, 2008.

81. J. Chu and A. Sher, *Device Physics of Narrow Gap Semiconductors*, Springer, New York, 2010.
82. M. A. Berding, M. van Schilfgaarde, and A. Sher, "$Hg_{0.8}Cd_{0.2}Te$ native defect: Densities and dopant properties," *J. Electron. Mater.* 22, 1005–10, 1993.
83. J. L. Melendez and C. R. Helms, "Process modeling of point defect effects in $Hg_{1-x}Cd_xTe$," *J. Electron. Mater.* 22, 999–1004, 1993.
84. S. Holander-Gleixner, H. G. Robinson, and C. R. Helms, "Simulation of HgTe/CdTe interdiffusion using fundamental point defect mechanisms," *J. Electron. Mater.* 27, 672–79, 1998.
85. P. Capper, C. D. Maxey, C. L. Jones, J. E. Gower, E. S. O'Keefe, and D. Shaw, "Low temperature thermal annealing effects in bulk and epitaxial $Cd_xHg_{1-x}Te$," *J. Electron. Mater.* 28, 637–48, 1999.
86. M. A. Berding "Equilibrium properties of indium and iodine in LWIR HgCdTe," *J. Electron. Mater.* 29, 664–8, 2000.
87. D. Chandra, H. F. Schaake, J. H. Tregilgas, F. Aqariden, M. A. Kinch, and A. Syllaios, "Vacancies in $Hg_{1-x}Cd_xTe$," *J. Electron. Mater.* 29, 729–31, 2000.
88. H. Wiedemayer and Y. G. Sha, "The direct determination of the vacancy concentration and p-T phase diagram of $Hg_{0.8}Cd_{0.2}Te$ and $Hg_{0.6}Cd_{0.4}Te$ by dynamic mass-loss measurements," *J. Electron. Mater.* 19, 761–71, 1990.
89. H. F. Schaake, J. H. Tregilgas, J. D. Beck, M. A. Kinch, and B. E. Gnade, "The effect of low temperature annealing on defects, impurities and electrical properties," *J. Vac. Sci. Technol. A* 3, 143–49, 1985.
90. D. A. Stevenson and M. F. S. Tang, "Diffusion mechanisms in mercury cadmium telluride," *J. Vac. Sci. Technol. B* 9, 1615–24, 1991.
91. P. Mitra, Y. L. Tyan, F. C. Case, R. Starr, and M. B. Reine, "Improved arsenic doping in metalorganic chemical vapor deposition of HgCdTe and *in situ* growth of high performance long wavelength infrared photodiodes," *J. Electron. Mater.* 25, 1328–35, 1996.
92. M. Tanaka, K. Ozaki, H. Nishino, H. Ebe, and Y. Miyamoto, "Electrical properties of HgCdTe epilayers doped with silver using an $AgNO_3$ solution," *J. Electron. Mater.* 27, 579–82, 1998.
93. S. P. Tobin, G. N. Pultz, E. E. Krueger, M. Kestigian, K. K. Wong, and P. W. Norton, "Hall effect characterization of LPE HgCdTe p/n heterojunctions," *J. Electron. Mater.* 22, 907–14, 1993.
94. D. Chandra, M. W. Goodwin, M. C. Chen, and L. K. Magel, "Variation of arsenic diffusion coefficients in HgCdTe alloys with temperature and hg pressure: Tuning of p on n double layer heterojunction diode properties," *J. Electron. Mater.* 24, 599–608, 1995.
95. D. Shaw, "Diffusion in mercury cadmium telluride—An update," *J. Electron. Mater.* 24, 587–97, 1995.
96. L. O. Bubulac, D. D. Edwall, S. J. C. Irvine, E. R. Gertner, and S. H. Shin, "p-Type doping of double layer mercury cadmium telluride for junction formation," *J. Electron. Mater.* 24, 617–24, 1995.
97. M. B. Reine, "Fundamental Properties of Mercury Cadmium Telluride," in *Encyclopedia of Modern Optics*, ed. B. Guenther and D. Steel, Academic Press, London, 2004.
98. G. L. Hansen, J. L. Schmit, and T. N. Casselman, "Energy gap versus alloy composition and temperature in $Hg_{1-x}Cd_xTe$," *J. Appl. Phys.* 53, 7099–101, 1982.
99. G. L. Hansen and J. L. Schmit, "Calculation of intrinsic carrier concentration in $Hg_{1-x}Cd_xTe$," *J. Appl. Phys.* 54, 1639–40, 1983.
100. M. H. Weiler, "Magnetooptical Properties of $Hg_{1-x}Cd_xTe$ Alloys," in *Semiconductors and Semimetals*, Vol. 16, eds. R. K. Willardson and A. C. Beer, 119–91, Academic Press, New York, 1981.
101. R. Dornhaus, G. Nimtz, and B. Schlicht, *Narrow-Gap Semiconductors*, Springer Verlag, Berlin, 1983.
102. J. R. Meyer, C. A. Hoffman, F. J. Bartoli, D. A. Arnold, S. Sivanathan, and J. P. Faurie, "Methods for magnetotransport characterization of IR detector materials," *Semicond. Sci. Technol.* 8, 805–23, 1991.
103. R. W. Miles, "Electron and hole mobilities in HgCdTe," in *Properties of Narrow Gap Cadmium-Based Compounds*, EMIS Datareviews Series No. 10, ed. P. P. Capper, 221–6, IEE, London, 1994.
104. J. S. Kim, J. R. Lowney, and W. R. Thurber, "Transport Properties of Narrow-Gap II–VI Compound Semiconductors," in *Narrow-gap II-VI Compounds for Optoelectronic and Electromagnetic Applications*, ed. P. Capper, 180–210, Chapman & Hall, London, 1997.
105. J. J. Dubowski, T. Dietl, W. Szymańska, and R. R. Gałązka, "Electron scattering in $Cd_xHg_{1-x}Te$," *J. Phys. Chem. Solids* 42, 351–63, 1981.
106. M. C. Chen and L. Colombo, "The majority carrier mobility of n-type and p-type $Hg_{0.78}Cd_{0.22}Te$ liquid phase epitaxial films at 77 K," *J. Appl. Phys.* 73, 2916–20, 1993.
107. R. D. S. Yadava, A. K. Gupta, and A. V. R. Warrier, "Hole scattering mechanisms in $Hg_{1-x}Cd_xTe$," *J. Electron. Mater.* 23, 1359–78, 1994.
108. D. A. Nelson, W. M. Higgins, R. A. Lancaster, R. Roy, and H. R. Vydyanath, *Extended Abstracts of U.S. Workshop on the Physics and Chemistry of Mercury Cadmium Telluride*, p. 175, Minneapolis, MN, October 28–30, 1981.

109. J. P. Rosbeck, R. E. Star, S. L. Price, and K. J. Riley, "Background and temperature dependent current-voltage characteristics of $Hg_{1-x}Cd_xTe$ photodiodes," *J. Appl. Phys.* 53, 6430–40, 1982.
110. W. M. Higgins, G. N. Seiler, R. G. Roy, and R. A. Lancaster, "Standard relationships in the properties of $Hg_{1-x}Cd_xTe$," *J. Vac. Sci. Technol. A* 7, 271–5, 1989.
111. P. N. J. Dennis, C. T. Elliott, and C. L. Jones, "A method for routine characterization of the hole concentration in p-type cadmium mercury telluride," *Infrared Phys.* 22, 167–9, 1982.
112. S. Barton, P. Capper, C. L. Jones, N. Metcalfe, and N. T. Gordon, "Electron mobility in p-type grown $Cd_xHg_{1-x}Te$," *Semicond. Sci. Technol.* 10, 56–60, 1995.
113. P. M. Amirtharaj and J. H. Burnett, "Optical Properties of MCT," in *Narrow-gap II-VI Compounds for Optoelectronic and Electromagnetic Applications*, ed. P. Capper, 133–79, Chapman & Hall, London, 1997.
114. J. Chu and Y. Chang, "Optical Properties of MCT," in *Mercury Cadmium Telluride: Growth, Properties, and Applications*, eds. P. Capper and J. Garland, 205–38, John Wiley & Sons, Chichester, UK, 2011.
115. E. O. Kane, "Band structure of InSb," *J. Phys. Chem. Solids* 1, 249–61, 1957.
116. W. W. Anderson, "Absorption constant of $Pb_{1-x}Sn_xTe$ and $Hg_{1-x}Cd_xTe$ Alloys," *Infrared Phys.* 20, 363–72, 1980.
117. R. Beattie and A. M. White, "An analytic approximation with a wide range of applicability for band-to-band radiative transition rates in direct, narrow-gap semiconductors," *Semicond. Sci. Technol.* 12, 359–68, 1997.
118. E. Finkman and S. E. Schacham, "The exponential optical absorption band tail of $Hg_{1-x}Cd_xTe$," *J. Appl. Phys.* 56, 2896–900, 1984.
119. S. E. Schacham and E. Finkman, "Recombination mechanisms in p-type HgCdTe: freezout and background flux effects," *J. Appl. Phys.* 57, 2001–9, 1985.
120. C. A. Hougen, "Model for infrared absorption and transmission of liquid-phase epitaxy HgCdTe," *J. Appl. Phys.* 66, 3763–6, 1989.
121. J. Chu, Z. Mi, and D. Tang, "Band-to-band optical absorption in narrow-gap $Hg_{1-x}Cd_xTe$ semiconductors," *J. Appl. Phys.* 71, 3955–61, 1992.
122. J. Chu, B. Li, K. Liu, and D. Tang, "Empirical rule of intrinsic absorption spectroscopy in $Hg_{1-x}Cd_xTe$," *J. Appl. Phys.* 75, 1234–5, 1994.
123. R. K. Sharma, D. Verma, and B. B. Sharma, "Observation of below band gap photoconductivity in mercury cadmium telluride," *Infrared Phys. Technol.* 35, 673–80, 1994.
124. B. Li, J. H. Chu, Y. Chang, Y. S. Gui, and D. Y. Tang, "Optical absorption above the energy band gap in $Hg_{1-x}Cd_xTe$," *Infrared Phys. Technol.* 37, 525–31, 1996.
125. S. Krishnamurthy, A. B. Chen, and A. Sher, "Near band edge absorption spectra of narrow-gap III–V semiconductor alloys," *J. Appl. Phys.* 80, 4045–48, 1996.
126. S. Krishnamurthy, A. B. Chen, and A. Sher, "Electronic structure, absorption coefficient, and Auger rate in HgCdTe and thallium-based alloys," *J. Electron. Mater.* 26, 571–7, 1997.
127. K. Moazzami, D. Liao, J. Phillips, D. L. Lee, M. Carmody, M. Zandian, and D. Edwall, "Optical absorption properties of HgCdTe epilayers with uniform composition," *J. Electron. Mater.* 32, 646–50, 2003.
128. K. Moazzami, J. Phillips, D. Lee, D. Edwall, M. Carmody, E. Piquette, M. Zandian, and J. Arias, "Optical-absorption model for molecular-beam epitaxy HgCdTe and application to infrared detector photoresponse," *J. Electron. Mater.* 33, 701–8, 2004.
129. F. Tong and N. M. Ravindra, "Optical properties of $Hg_{1-x}Cd_xTe$," *Infrared Phys.* 34, 207–12, 1993.
130. K. Liu, J. H. Chu, and D. Y. Tang, "Composition and temperature dependence of the refractive index in $Hg_{1-x}Cd_xTe$," *J. Appl. Phys.* 75, 4176–9, 1994.
131. S. Rolland, "Dielectric Constant and Refractive Index of HgCdTe," in *Properties of Narrow Gap Cadmium-Based Compounds*, EMIS Datareviews Series No. 10, Chapter 4A, ed. P. P. Capper, 80–5, IEE, London, 1994.
132. J. L. Pautrat and N. Magnea, "Optical Absorption in HgCdTe and Related Heterostructures," in *Properties of Narrow Gap Cadmium-Based Compounds*, EMIS Data reviews Series No. 10, ed. P. Capper, 75–9, IEE, London, 1994.
133. S. L. Price and P. R. Boyd, "Overview of compositional measurements techniques for $Hg_{1-x}Cd_xTe$ with emphasis on IR transmission, energy dispersive X–ray analysis and optical reflectance," *Semicond. Sci. Technol.* 8, 842–59, 1993.
134. C. T. Elliott and N. T. Gordon, "Infrared Detectors," in *Handbook on Semiconductors*, Vol. 4, ed. C. Hilsum, 841–936, North-Holland, Amsterdam, 1993.
135. V. C. Lopes, A. J. Syllaios, and M. C. Chen, "Minority carrier lifetime in mercury cadmium telluride," *Semicond. Sci. Technol.* 8, 824–41, 1993.
136. M.A. Kinch, *Fundamentals of Infrared Detectors*, SPIE Press, Bellingham, WA, 2007.
137. M. A. Kinch, *State-of-the-Art Infrared Detector Technology*, SPIE Press, Bellingham, WA, 2014.
138. M. A. Kinch, "Fundamental physics of infrared detector materials," *J. Electron. Mater.* 29, 809–17, 2000.

139. M. A. Kinch, F. Aqariden, D. Chandra, P.-K. Liao, H. F. Schaake, and H. D. Shih, "Minority carrier lifetime in p-HgCdTe," *J. Electron. Mater.* 34, 880–4, 2005.
140. Y. Yamamoto, Y. Miyamoto, and K. Tanikawa, "Minority carrier lifetime in the region close to the interface between the anodic oxide and CdHgTe," *J. Cryst. Growth* 72, 270–4, 1985.
141. S. M. Johnson, D. R. Rhiger, J. P. Rosbeck, J. M. Peterson, S. M. Taylor, and M. E. Boyd, "Effect of dislocations on the electrical and optical properties of long-wavelength infrared HgCdTe photovoltaic detectors," *J. Vac. Sci. Technol. B* 10, 1499–506, 1992.
142. K. Jóźwikowski and A. Rogalski, "Effect of dislocations on performance of LWIR HgCdTe photodiodes," *J. Electron. Mater.* 29, 736–41, 2000.
143. M. C. Chen, L. Colombo, J. A. Dodge, and J. H. Tregilgas, "The minority carrier lifetime in doped and undoped p-type $Hg_{0.78}Cd_{0.22}Te$ liquid phase epitaxy films," *J. Electron. Mater.* 24, 539–44, 1995.
144. P. Mitra, T. R. Schimert, F. C. Case, R. Starr, M. H. Weiler, M. Kestigian, and M. B. Reine, "Metalorganic chemical vapor deposition of HgCdTe for photodiode applications," *J. Electron. Mater.* 24, 661–8, 1995.
145. M. J. Bevan, M. C. Chen, and H. D. Shih, "High-quality p-type $Hg_{1-x}Cd_xTe$ prepared by metalorganic chemical vapor deposition," *Appl. Phys. Lett.* 67, 3450–2, 1996.
146. R. DeWames, and J. Pellegrino, "Electrical characteristics of MOVPE grown MWIR N^+p(As) HgCdTe heterostructure photodiodes build on GaAs substrates," *Proc. SPIE* 8353, 83532P (2012).
147. G Destefanis and J. P. Chamonal, "Large improvement in HgCdTe photovoltaic detector performances at LETI," *J. Electron. Mater.* 22, 1027–32, 1993.
148. C. L. Littler, E. Maldonado, X. N. Song, Z. You, J. L. Elkind, D. G. Seiler, and J. R. Lowney, "Investigation of mercury interstitials in $Hg_{1-x}Cd_xTe$ alloys using resonant impact-ionization spectroscopy," *J. Vac. Sci. Technol. B* 9, 1466–70, 1991.
149. R. G. Humpreys, "Radiative lifetime in semiconductors for infrared detectors," *Infrared Phys.* 23, 171–5, 1983.
150. R. G. Humpreys, "Radiative lifetime in semiconductors for infrared detectors," *Infrared Phys.* 26, 337–42, 1986.
151. T. Elliott, N. T. Gordon, and A. M. White, "Towards background-limited, room-temperature, infrared photon detectors in the 3–13 µm wavelength range," *Appl. Phys. Lett.* 74, 2881–3, 1999.
152. N. T. Gordon, C. D. Maxey, C. L. Jones, R. Catchpole, and L. Hipwood, "Suppression of radiatively generated currents in infrared detectors," *J. Appl. Phys.* 91, 565–8, 2002.
153. C. T. Elliott and C. L. Jones, "Non-equilibrium Devices in HgCdTe," in *Narrow-Gap II-VI Compounds for Optoelectronic and Electromagnetic Applications*, ed. P. Capper, 474–85, Chapman & Hall, London, 1997.
154. T. Ashley, N. T. Gordon, G. R. Nash, C. L. Jones, C. D. Maxey, and R. A. Catchpole, "Long- wavelength HgCdTe negative luminescent devices," *Appl. Phys. Lett.* 79, 1136–8, 2001.
155. A. R. Beattie and P. T. Landsberg, "Auger effect in semiconductors," *Proc. Royal Soc. London A* 249, 16–29, 1959.
156. J. S. Blakemore, *Semiconductor Statistics*, Pergamon Press, Oxford, UK, 1962.
157. M. A. Kinch, M. J. Brau, and A. Simmons, "Recombination mechanisms in 8–14 µm HgCdTe," *J. Appl. Phys.* 44, 1649–63, 1973.
158. G. Nimtz, "Recombination in narrow-gap semiconductors," *Phys. Rep.* 63, 265–300, 1980.
159. T. N. Casselman and P. E. Petersen, "A comparison of the dominant Auger transitions in p-type (Hg,Cd)Te," *Solid State Commun.* 33, 615–9, 1980.
160. T. N. Casselman, "Calculation of the Auger lifetime in p-type $Hg_{1-x}Cd_xTe$," *J. Appl. Phys.* 52, 848–54, 1981.
161. P. E. Petersen, "Auger Recombination in Mercury Cadmium Telluride," in *Semiconductors and Semimetals*, Vol. 18, eds. R. K. Willardson and A. C. Beer, 121–55, Academic Press, New York, 1981.
162. A. M. White, "The characteristics of minority-carrier exclusion in narrow direct gap semiconductors," *Infrared Phys.* 25, 729–41, 1985.
163. C. M. Ciesla, B. N. Murdin, T. J. Phillips, A. M. White, A. R. Beattie, C. J. G. M. Langerak, C. T. Elliott, C. R. Pidgeon, and S. Sivananthan, "Auger recombination dynamics of $Hg_{0.795}Cd_{0.205}Te$ in the high excitation regime," *Appl. Phys. Lett.* 71, 491–3, 1997.
164. R. Beattie and A. M. White, "An analytic approximation with a wide range of applicability for electron initiated Auger transitions in narrow-gap semiconductors," *J. Appl. Phys.* 79, 802–13, 1996.
165. S. Krishnamurthy and T. N. Casselman, "A detailed calculation of the Auger lifetime in p-type HgCdTe," *J. Electron. Mater.* 29, 828–31, 2000.
166. S. Krishnamurthy, M. A. Berding, Z. G. Yu, C. H. Swartz, T. H. Myers, D. D. Edwall, and R. DeWames, "Model of minority carrier lifetimes in doped HgCdTe," *J. Electron. Mater.* 34, 873–9, 2005.
167. P. T. Landsberg, *Recombination in Semiconductors*, Cambridge University Press, Cambridge, UK, 2003.
168. J. Piotrowski and A. Rogalski, *High-Operating Temperature Infrared Photodetectors*, SPIE Press, Bellingham, WA, 2007.

169. T. Ashley and C. T. Elliott, "Non-equilibrium mode of operation for infrared detection," *Electron. Lett.* 21, 451–2, 1985.
170. T. Ashley, T. C. Elliott, and A. M. White, "Non-equilibrium devices for infrared detection," *Proc. SPIE* 572, 123–32, 1985.
171. M. A. Kinch and S. R. Borrello, "0.1 eV HgCdTe photodetectors," *Infrared Phys.* 15, 111–24, 1975.
172. M. A. Kinch, S. R. Borrello, D. H. Breazeale, and A. Simmons, "Geometrical enhancement of HgCdTe photoconductive detectors," *Infrared Phys.* 17, 137–45, 1977.
173. S. R. Borrello, M. Kinch, and D. Lamont, "Photoconductive HgCdTe detector performance with background variations," *Infrared Phys.* 17, 121–5, 1977.
174. S. P. Emmons and K. L. Ashley, "Minority-carrier sweepout in 0.09-eV HgCdTe," *Appl. Phys. Lett.* 20, 241–2, 1972.
175. M. R. Johnson, "Sweep-out effects in $Hg_{1-x}Cd_xTe$ photoconductors," *J. Appl. Phys.* 43, 3090–3, 1972.
176. Y. J. Shacham-Diamond and I. Kidron, "Contact and bulk effects in intrinsic photoconductive infrared detectors," *Infrared Phys.* 21, 105–15, 1981.
177. T. Ashley and C. T. Elliott, "Accumulation effects at contacts to n-type cadmium-mercury-telluride photoconductors," *Infrared Phys.* 22, 367–76, 1982.
178. M. White, "Recombination in a graded n-n contact region in a narrow-gap semiconductor," *J. Phys. C: Solid State Phys.* 17, 4889–96, 1984.
179. D. L. Smith, "Effect of blocking contacts on generation-recombination noise and responsivity in intrinsic photoconductors," *J. Appl. Phys.* 56, 1663–9, 1984.
180. D. K. Arch, R. A. Wood, and D. L. Smith, "High responsivity HgCdTe heterojunction photoconductor," *J. Appl. Phys.* 58, 2360–70, 1985.
181. J. F. Siliquini, K. A. Fynn, B. D. Nener, L. Faraone, and R. H. Hartley, "Improved device technology for epitaxial $Hg_{1-x}Cd_xTe$ infrared photoconductor arrays," *Semicond. Sci Technol.* 9, 1515–22, 1994.
182. C. A. Musca, J. F. Siliquini, B. D. Nener, L. Faraone, and R. H. Hartley, "Enhanced responsivity of HgCdTe infrared photoconductors using MBE grown heterostructures," *Infrared Phys. Technol.* 38, 163–7, 1997.
183. J. Piotrowski, W. Galus, and M. Grudzień, "Near room-temperature IR photodetectors," *Infrared Phys.* 31, 1–48, 1990.
184. J. Piotrowski, "Uncooled operation of IR photodetectors," *Opto-Electron. Rev.* 12, 111–22, 2004.
185. J. Piotrowski and A. Rogalski, "Uncooled long wavelength infrared photon detectors," *Infrared Phys. Technol.* 46, 115–31, 2004.
186. T. Ashley, C. T. Elliott, and A. M. White, "Infra-red detector using minority carrier exclusion," *Proc. SPIE* 588, 62–8, 1985.
187. R. M. Broudy and V. J. Mazurczyck, "(HgCd)Te Photoconductive Detectors," in *Semiconductors and Semimetals*, Vol. 18, ed. R. K. Willardson and A. C. Beer, 157–99, Academic Press, New York, 1981.
188. C. T. Elliott, "Photoconductive and Non-equilibrium Devices in HgCdTe and Related Alloys," in *Infrared Detectors and Emitters: Materials and Devices*, ed. P. Capper and C. T. Elliott, 279–312, Kluwer Academic Publishers, Boston, MA, 2001.
189. A. Rogalski, *Infrared Detectors*, CRC Press, Boca Raton, FL, 2011.
190. I. M. Baker, "MCT Photoconductuve Infrared Detectors," in *Mercury Cadmium Telluride: Growth, Properties, and Applications*, eds. P. Capper and J. Garland, 429–46, John Wiley & Sons, Chichester, UK, 2011.
191. J. Piotrowski and A. Piotrowski, "Room temperature IR photodetectors," in *Mercury Cadmium Telluride: Growth, Properties, and Applications*, eds. P. Capper and J. Garland, 513–37, John Wiley & Sons, Chichester, UK, 2011.
192. D. Long and J. L. Schmit, "Mercury-Cadmium Telluride and Closely Related Alloys," in *Semiconductors and Semimetals*, Vol. 5, ed. R. K. Willardson and A. C. Beer, 175–255, Academic Press, New York, 1970.
193. D. L. Spears, "Heterodyne and direct detection at 10.6 μm with high temperature p-type photoconductors," *Proceedings of IRIS Active Systems*, 1–15, San Diego, 1982.
194. J. Piotrowski, "$Hg_{1-x}Cd_xTe$ Infrared Photodetectors," in *Infrared Photon Detectors*, ed. A. Rogalski, 391–494, SPIE Optical Engineering Press, Bellingham, WA, 1995.
195. Z. Djuric, "Isothermal vapour-phase epitaxy of mercury-cadmium telluride (Hg,Cd)Te," *J. Mater. Sci.: Mater. Electron.* 5, 187–218, 1995.
196. K. Nagahama, R. Ohkata, and T. Murotani, "Preparation of high quality n-$Hg_{0.8}Cd_{0.2}Te$ epitaxial layer and its application to infrared detector (λ = 8–14 μm)," *Jap. J. Appl. Phys.* 21, L764–6, 1982.
197. T. Nguyen-Duy and D. Lorans, "Highlights of recent results on HgCdTe thin film photoconductors," *Semicond. Sci. Technol.* 6, 93–5, 1991.

198. B. Doll, M. Bruder, J. Wendler, J. Ziegler, and H. Maier, "3–5 μm Photoconductive Detectors on Liquid-Phase-Epitaxial-MCT," in *Fourth International Conference on Advanced Infrared Detectors and Systems*, 120–4, IEE, London, 1990.
199. J. F. Siliquini, C. A. Musca, B. D. Nener, and L. Faraone, "Performance of optimized $Hg_{1-x}Cd_xTe$ long wavelength infrared photoconductors," *Infrared Phys. Technol.* 35, 661–71, 1994.
200. L. T. Specht, W. E. Hoke, S. Oguz, P. J. Lemonias, V. G. Kreismanis, and R. Korenstein, "High performance HgCdTe photoconductive devices grown by metalorganic chemical vapor deposition," *Appl. Phys. Lett.* 48, 417–8, 1986.
201. C. G. Bethea, B. F. Levine, P. Y. Lu, L. M. Williams, and M. H. Ross, "Photoconductive $Hg_{1-x}Cd_xTe$ detectors grown by low-temperature metalorganic chemical vapor deposition," *Appl. Phys. Lett.* 53, 1629–31, 1988.
202. R. Druilhe, A. Katty, and R. Triboulet, "MOVPE Grown (Hg,Cd)Te Layers for Room Temperature Operating 3–5 μm Photoconductive Detectors," in *Fourth International Conference on Advanced Infrared Detectors and Systems*, 20–4, IEE, London, 1990.
203. M. C. Chen and M. J. Bevan, "Room-temperature midwavelength two-color infrared detectors with HgCdTe multilayer structures by metal-organic chemical-vapor deposition," *J. Appl. Phys.* 78, 4787–9, 1995.
204. S. Yuan, J. Li He, J. Yu, M. Yu, Y. Qiao, and J. Zhu, "Infrared photoconductor fabricated with a molecular beam epitaxially grown CdTe/HgCdTe heterostructure," *Appl. Phys. Lett.* 58, 914–6, 1991.
205. Y. Nemirovsky and G. Bahir, "Passivation of mercury cadmium telluride surfaces," *J. Vac. Sci. Technol. A* 7, 450–9, 1989.
206. P. C. Catagnus and C. T. Baker, "Passivation of mercury cadmium telluride," U. S. Patent 3,977,018, 1976.
207. Y. Nemirovsky and E. Finkman, "Anodic oxide films on $Hg_{1-x}Cd_xTe$," *J. Electrochem. Soc.* 126, 768–70, 1979.
208. E. Bertagnolli, "Improvement of anodically grown native oxides on n-(Cd,Hg)Te," *Thin Solid Films* 135, 267–75, 1986.
209. A. Gauthier, "Process for passivation of photoconductive detectors made of HgCdTe," U. S. Patent 4,624,715, 1986.
210. Y. Nemirovsky and R. Goshen, "Plasma anodization of $Hg_{1-x}Cd_xTe$," *Appl. Phys. Lett.* 37, 813–4, 1980.
211. S. P. Buchner, G. D. Davis, and N. E. Byer, "Summary abstract: Photochemical oxidation of (Hg,Cd)Te," *J. Vac. Sci. Technol.* 21, 446–7, 1982.
212. Y. Shacham-Diamand, T. Chuh, and W. G. Oldham, "The electrical properties of Hg-sensitized 'photox'-oxide layers deposited at 80°C," *Solid-State Electron.* 30, 227–33, 1987.
213. M. V. Blackman, D. E. Charlton, M. D. Jenner, D. R. Purdy, and J. T. M. Wotherspoon, "Type conversion in $Hg_{1-x}Cd_xTe$ by ion beam treatment," *Electron. Lett.* 23, 978–9, 1987.
214. P. Brogowski, H. Mucha, and J. Piotrowski, "Modification of mercury cadmium telluride, mercury manganese tellurium, and mercury zinc telluride by ion etching," *Phys. Status Solidi A* 114, K37–40, 1989.
215. P. H. Zimmermann, M. B. Reine, K. Spignese, K. Maschhoff, and J. Schirripa, "Surface passivation of HgCdTe photodiodes," *J. Vac. Sci. Technol. A* 8, 1182–4, 1990.
216. Y. Nemirovsky, L. Burstein, and I. Kidron, "Interface of p-type $Hg_{1-x}Cd_xTe$ passivated with native sulfides," *J. Appl. Phys.* 58, 366–73, 1985.
217. E. Weiss and C. R. Helms, "Composition, growth mechanism, and stability of anodic fluoride films on $Hg_{1-x}Cd_xTe$," *J. Vac. Sci. Technol. B* 9, 1879–85, 1991.
218. Y. Nemirovsky, "Passivation with II–VI compounds," *J. Vac. Sci. Technol. A* 8, 1185–7, 1990.
219. G. Sarusi, G. Cinader, A. Zemel, and D. Eger, "Application of CdTe epitaxial layers for passivation of p-Type $Hg_{0.77}Cd_{0.23}Te$," *J. Appl. Phys.* 71, 5070–6, 1992.
220. A. Mestechkin, D. L. Lee, B. T. Cunningham, and B. D. MacLeod, "Bake stability of long-wavelength infrared HgCdTe photodiodes," *J. Electron. Mater.* 24, 1183–7, 1995.
221. G. Bahir, V. Ariel, V. Garber, D. Rosenfeld, and A. Sher, "Electrical properties of epitaxially grown CdTe passivation for long-wavelength HgCdTe photodiodes," *Appl. Phys. Lett.* 65, 2725–7, 1994.
222. A. Jóźwikowska, K. Jóźwikowski, and A. Rogalski, "Performance of mercury cadmium telluride photoconductive detectors," *Infrared Phys.* 31, 543–54, 1991.
223. C. A. Musca, J. F. Siliquini, K. A. Fynn, B. D. Nener, L. Faraone, and S. J. C. Irvine, "MOCVD-grown wider-bandgap capping layers in $Hg_{1-x}Cd_xTe$ long-wavelength infrared photoconductors," *Semicond. Sci. Technol.* 11, 1912–22, 1996.
224. J. F. Siliquini, C. A. Musca, B. D. Nener, and L. Faraone, "Temperature dependence of $Hg_{0.68}Cd_{0.32}Te$ infrared photoconductor performance," *IEEE Trans. Electron Dev.* 42, 1441–8, 1995.
225. D. L. Smith, "Theory of generation-recombination noise and responsivity in overlap structure photoconductors," *J. Appl. Phys.* 54, 5441–8, 1983.

226. D. L. Smith, D. K. Arch, R. A. Wood, and M. W. Scott, "HgCdTe heterojunction contact photoconductor," *J. Appl. Phys.* 45, 83–5, 1984.
227. J. R. Lowney, D. G. Seiler, W. R. Thurber, Z. Yu, X. N. Song, and C. L. Littler, "Heavily accumulated surfaces of mercury cadmium telluride detectors: Theory and experiment," *J. Electron. Mater.* 22, 985–91, 1993.
228. M. C. Wilson and D. J. Dinsdale, "CO_2 detection with cadmium mercury telluride," in *Third International Conference on Advanced Infrared Detectors and Systems*, 139–45, IEE, London, 1986.
229. D. L. Spears, "IR Detectors: Heterodyne and Direct," in *Optical and Laser Remote Sensing*, eds. D. K. Killinger and A. Mooradian, 278–86, Springer-Verlag, Berlin, 1983.
230. Z. Djuric, Z. Jaksic, Z. Djinovic, M. Matic, and Z. Lazic, "Some theoretical and technological aspects of uncooled HgCdTe detectors: A review," *Microelectron. J.* 25, 99–114, 1994.
231. F. A. Capocci, A. T. Harker, M. C. Wilson, D. E. Lacklison, and F. E. Wray, "Thermoelectrically-cooled cadmium mercury telluride detectors for CO_2 laser radiation," in *Second International Conference on Advanced Infrared Detectors and Systems*, 40–4, IEE, London, 1983.
232. D. J. Wilson, R. Foord, and G. D. J. Constant, "Operation of an intermediate temperature detector in a 10.6 μm heterodyne rangefinder," *Proc. SPIE* 663, 155–8, 1986.
233. Y. Li and D. J. Adams, "Experimental studies on the performance of $Hg_{0.8}Cd_{0.2}Te$ photoconductors operating near 200 K," *Infrared Phys.* 35, 593–5, 1994.
234. Z. Wei-jiann and Z. Xin-Chen, "Experimental studies on low frequency noise of photoconductors," *Infrared Phys.* 33, 27–31, 1992.
235. P. R. Norton, "Structure and method of fabricating a trapping-mode photodetector," *International Application Published under The Patent Cooperation Treaty* PCT/US86/002516, International Publication Number WO87/03743, 18 June 1987.
236. P. R. Norton, "Infrared image sensors," *Optic. Eng.* 30, 1649–63, 1991.
237. P. Norton, "HgCdTe for NASA EOS missions and detector uniformity benchmarks," *Proceedings of the Innovative Long Wavelength Infrared Detector Workshop*, 93–4, Pasadena, Jet Propulsion Laboratory, April 24–26, 1990.
238. D. G. Crove, P. R. Norton, T. Limperis, and J. Mudar, "Detectors," in *The Infrared and Electro-Optical Systems Handbook*, Vol. 3, ed. W. D. Rogatto, 175–283, Infrared Information Analysis Center, Ann Arbor, MI, and SPIE Optical Engineering Press, Bellingham, WA, 1993.
239. T. Ashley, C. T. Elliott, and A. T. Harker, "Non-equilibrium mode of operation for infrared detectors," *Infrared Phys.* 26, 303–15, 1986.
240. A. M. White, "Generation-recombination processes and Auger suppression in small-bandgap detectors," *J. Cryst. Growth* 86, 840–8, 1988.
241. A. P. Davis and A. M. White, "Effects of residual Shockley-Read traps on efficiency of auger-suppressed IR detector diodes," *Semicond. Sci. Technol.* 5, S38–40, 1990.
242. C. T. Elliott, "Future infrared detector technologies," in *Fourth International Conference on Advanced Infrared Detectors and Systems*, 61–6, IEE, London, 1990.
243. C. T. Elliott, "Non-equilibrium mode of operation of narrow-gap semiconductor devices," *Semicond. Sci. Technol.* 5, S30–7, 1990.
244. T. Ashley and T. C. Elliott, "Operation and properties of narrow-gap semiconductor devices near room temperature using non-equilibrium techniques," *Semicond. Sci. Technol.* 8, C99–105, 1991.
245. C. T. Elliott, "Infrared Detectors with Integrated Signal Processing," in *Solid State Devices*, eds. A. Goetzberger and M. Zerbst, 175–201, Verlag Chemie, Weinheim, Germany, 1983.
246. T. Ashley, C. T. Elliott, A. M. White, J. T. M. Wotherspoon, and M. D. Johns, "Optimization of spatial resolution in SPRITE detectors," *Infrared Phys.* 24, 25–33, 1984.
247. J. T. M. Wotherspoon, R. J. Dean, M. D. Johns, T. Ashley, C. T. Elliott, and M. A. White, "Developments in SPRITE infra-red detectors," *Proc. SPIE* 810, 102–12, 1985.
248. R. F. Leftwich and R. Ward, "Latest developments in SPRITE detector technology," *Proc. SPIE* 930, 76–86, 1988.
249. A. Campbell, C. T. Elliott, and A. M. White, "Optimisation of SPRITE detectors in anamorphic imaging systems," *Infrared Phys.* 27, 125–33, 1987.
250. A. B. Dean, P. N. J. Dennis, C. T. Elliott, D. Hibbert, and J. T. M. Wotherspoon, "The serial addition of SPRITE infrared detectors," *Infrared Phys.* 28, 271–8, 1988.
251. C. M. Dyson, "Thermal-radiation imaging devices and systems," U. K. Patent Application 2,199,986 A, 1988.
252. C. T. Elliott, "SPRITE detectors and staring arrays in $Hg_{1-x}Cd_xTe$," *Proc. SPIE* 1038, 2–8, 1988.
253. G. D. Boreman and A. E. Plogstedt, "Modulation transfer function and number of equivalent elements for SPRITE detectors," *Appl. Opt.* 27, 4331–5, 1988.

254. S. P. Braim, A. Foord, and M. W. Thomas, "System implementation of a serial array of SPRITE infrared detectors," *Infrared Phys.* 29, 907–14, 1989.
255. J. Severn, D. A. Hibbert, R. Mistry, C. T. Elliott, and A. P. Davis, "The design and performance options for SPRITE arrays," in *Fourth International Conference on Advanced Infrared Detectors and Systems*, 9–14, IEE, London, 1990.
256. A. P. Davis, "Effect of high signal photon fluxes on the responsivity of SPRITE detectors," *Infrared Phys.* 33, 301–5, 1992.
257. K. J. Barnard, G. D. Boreman, A. E. Plogstedt, and B. K. Anderson, "Modulation-transfer function measurement of SPRITE detectors: Sine-wave response," *Appl. Opt.* 31, 144–7, 1992.
258. M. Rodot, C. Verie, Y. Marfaing, J. Besson, and H. Lebloch, "Semiconductor lasers and fast detectors in the infrared (3 to 15 Microns)," *IEEE J. Quant. Electron.* 2, 586–93, 1966.
259. M. B. Reine, A. K. Sood, and T. J. Tredwell, "Photovoltaic Infrared Detectors," in *Semiconductors and Semimetals*, Vol. 18, eds. R. K. Willardson and A. C. Beer, 201–311, Academic Press, New York, 1981.
260. A. Rogalski and J. Piotrowski, "Intrinsic infrared detectors," *Prog. Quant. Electron.* 12, 87–289, 1988.
261. A. I. D'Souza, P. S. Wijewarnasuriya, and J. G. Poksheva, "HgCdTe Infrared Detectors," in *Thin Films*, Vol. 28, 193–226, 2001.
262. I. M. Baker, "Photovoltaic IR Detectors," in *Narrow-gap II-VI Compounds for Optoelectronic and Electromagnetic Applications*, ed. P. Capper, 450–73, Chapman & Hall, London, 1997.
263. M. B. Reine, "Photovoltaic Detectors in MCT," in *Infrared Detectors and Emitters: Materials and Devices*, eds. P. Capper and C. T. Elliott, 313–76, Kluwer Academic Publishers, Boston, MA, 2001.
264. I. M. Baker, "HgCdTe 2D Arrays: Technology and Performance Limits," in *Handbook of Infrared Technologies*, eds. M. Henini and M. Razeghi, 269–308, Elsevier, Oxford, UK, 2002.
265. A. Rogalski, "Infrared detectors: Status and trends," *Progress in Quantum Electronics* 27, 59–210, 2003.
266. I. M. Baker, "II-VI Narrow-Bandgap Semiconductors for Optoelectronics," in *Springer Handbook of Electronic and Photonic Materials*, eds. S. Kasap and P. Capper, 855–85, Springer, Heidelberg, Germany, 2006.
267. D. T. Dutton, E. O'Keefe, P. Capper, C. L. Jones, S. Mugford, and C. Ard, "Type conversion of $Cd_xHg_{1-x}Te$ grown by liquid phase epitaxy," *Semicond. Sci. Technol.* 8, S266–9, 1993.
268. D. L. Spears and C. Freed, "HgCdTe varactor photodiode detection of cw CO_2 laser beats beyond 60 GHz," *Appl. Phys. Lett.* 23, 445–7, 1973.
269. D. L. Spears, "Planar HgCdTe quadrantal heterodyne arrays with GHz response at 10.6 μm," *Infrared Phys.* 17, 5–8, 1977.
270. J. F. Shanley, C. T. Flanagan, and M. B. Reine, "Elevated temperature n^+ -p $Hg_{0.8}Cd_{0.2}Te$ photodiodes for moderate bandwidth infrared heterodyne applications," *Proc. SPIE* 227, 117–22, 1980.
271. K. K. Parat, H. Ehsani, I. B. Bhat, and S. K. Ghandhi, "Selective annealing for the planar processing of HgCdTe devices," *J. Vac. Sci. Technol. B* 9, 1625–9, 1991.
272. M. D. Jenner and M. V. Blackman, "Method of manufacturing an infrared detector device," U. S. Patent 4,318,217, 1982.
273. P. Brogowski and J. Piotrowski, "The p-to-n conversion of HgCdTe, HgZnTe and HgMnTe by anodic oxidation and subsequent heat treatment," *Semicond. Sci. Technol.* 5, 530–2, 1990.
274. J. M. T. Wotherspoon, U. S. Patent 4.411.732, 1983.
275. M. A. Lunn and P. S. Dobson, "Ion beam milling of $Cd_{0.2}Hg_{0.8}Te$," *J. Cryst. Growth* 73, 379–84, 1985.
276. J. L. Elkind, "Ion mill damage in n-HgCdTe," *J. Vac. Sci. Technol. B* 10, 1460–5, 1992.
277. V. I. Ivanov-Omskii, K. E. Mironov, and K. D. Mynbaev, "$Hg_{1-x}Cd_xTe$ doping by ion-beam treatment," *Semicond. Sci. Technol.* 8, 634–7, 1993.
278. P. Hlidek, E. Belas, J. Franc, and V. Koubele, "Photovoltaic spectra of HgCdTe diodes fabricated by ion beam milling," *Semicond. Sci. Technol.* 8, 2069–71, 1993.
279. K. D. Mynbaev and V. I. Ivanov-Omski, "Modification of $Hg_{1-x}Cd_xTe$ properties by low-energy ions," *Semiconductors* 37, 1127–50, 2003.
280. G. Bahir and E. Finkman, "Ion beam milling effect on electrical properties of $Hg_{1-x}Cd_xTe$," *J. Vac. Sci. Technol. A* 7, 348–53, 1989.
281. M. Baker and R. A. Ballingall, "Photovoltaic CdHgTe: Silicon hybrid focal planes," *Proc. SPIE* 510, 121–9, 1984.
282. E. Belas, P. Höschl, R. Grill, J. Franc, P. Moravec, K. Lischka, H. Sitter, and A. Toth, "Ultrafast diffusion of Hg in $Hg_{1-x}Cd_xTe$ ($x \approx 0.21$)," *J. Cryst. Growth* 138, 940–3, 1994.
283. E. Belas, R. Grill, J. Franc, A. Toth, P. Höschl, H. Sitter, and P. Moravec, "Determination of the migration energy of Hg interstitials in (HgCd)Te from ion milling experiments," *J. Cryst. Growth* 159, 1117–22, 1996.

284. G. Destefanis, "Electrical doping of HgCdTe by ion implantation and heat treatment," *J. Cryst. Growth* 86, 700–22, 1988.
285. L. O. Bubulac, "Defects, diffusion and activation in ion implanted HgCdTe," *J. Cryst. Growth* 86, 723–34, 1988.
286. A. G. Foyt, T. C. Harman, and J. P. Donnelly, "Type conversion and n–p junction formation in $Cd_xHg_{1-x}Te$ produced by proton bombardment," *Appl. Phys. Lett.* 18, 321–3, 1971.
287. L. O. Bubulac and W. E. Tennant, "Role of Hg in junction formation in ion-implanted HgCdTe," *Appl. Phys. Lett.* 51, 355–7, 1987.
288. P. G. Pitcher, P. L. F. Hemment, and Q. V. Davis, "Formation of shallow photodiodes by implantation of boron into mercury cadmium telluride," *Electron. Lett.* 18, 1090–2, 1982.
289. L. J. Kozlowski, R. B. Bailey, S. C. Cabelli, D. E. Cooper, G. McComas, K. Vural, and W. E. Tennant, "640 × 480 PACE HgCdTe FPA," *Proc. SPIE* 1735, 163–73, 1992.
290. G. Bahir, R. Kalish, and Y. Nemirovsky, "Electrical properties of donor and acceptor implanted $Hg_{1-x}Cd_xTe$ following CW CO_2 laser annealing," *Appl. Phys. Lett.* 41, 1057–9, 1982.
291. R. B. Bailey, L. J. Kozlowski, J. Chen, D. Q. Bui, K. Vural, D. D. Edwall, R. V. Gil, A. B. Vanderwyck, E. R. Gertner, and M. B. Gubala, "256 × 256 Hybrid HgCdTe infrared focal plane arrays," *IEEE Trans. Electron Devices* 38, 1104–9, 1991.
292. L. O. Bubulac, D. D. Edwall, and C. R. Viswanathan, "Dynamics of arsenic diffusion in metalorganic chemical vapor deposited HgCdTe on GaAs/Si substrates," *J. Vac. Sci. Technol. B* 9, 1695–704, 1991.
293. L. O. Bubulac, S. J. Ivine, E. R. Gertner, J. Bajaj, W. P. Lin, and R. Zucca, "As diffusion in $Hg_{1-x}Cd_xTe$ for junction formation," *Semicond. Sci. Technol.* 8, S270–5, 1993.
294. O. P. Agnihorti, H. C. Lee, and K. Yang, "Plasma induced type conversion in mercury cadmium telluride," *Semicond. Sci. Technol.* 17, R11–9, 2002.
295. V. Srivastav, R. Pal, and H. P. Vyas, "Overview of etching technologies used for HgCdTe," *Opto-Electron. Rev.* 13(3), 197–211 (2005).
296. J. K. White, R. Pal, J. M. Dell, C. A. Musca, J. Antoszewski, L. Faraone, and P. Burke, "p-to-n type-conversion mechanisms for HgCdTe exposed to H_2/CH_4 plasma," *J. Electron. Mater.* 30, 762–7, 2001.
297. J. M. Dell, J. Antoszewski, M. H. Rais, C. Musca, J. K. White, B. D. Nener, and L. Faraone, "HgCdTe mid-wavelength IR photovoltaic detectors fabricated using plasma induced junction technology," *J. Electron. Mater.* 29, 841–8, 2000.
298. J. K. White, J. Antoszewski, P. Ravinder, C. A. Musca, J. M. Dell, L. Faraone, and J. Piotrowski, "Passivation effects on reactive ion etch formed n-on-p junctions in HgCdTe," *J. Electron. Mater.* 31, 743–8, 2002.
299. T. Nguyen, C. A. Musca, J. M. Dell, R. H. Sewell, J. Antoszewski, J. K. White, and L. Faraone, "HgCdTe long-wavelength IR photovoltaic detectors fabricated using plasma induced junction formation technology," *J. Electron. Mater.* 32, 615–21, 2003.
300. E. P. G. Smith, G. M. Venzor, M. D. Newton, M. V. Liguori, J. K. Gleason, R. E. Bornfreund, S. M. Johnson, et al., "Inductively coupled plasma etching got large format HgCdTe focal plane array fabrication," *J. Electron. Mater.* 34, 746–53, 2005.
301. K. D. Mynbaev and V. I. Ivanov-Omski, "Doping of epitaxial layers and heterostructures based on HgCdTe," *Semiconductors* 40, 1–21, 2006.
302. H. R. Vydyanath, J. A. Ellsworth, and C. M. Devaney, "Electrical activity, mode of incorporation and distribution coefficient of group V. Elements in $Hg_{1-x}Cd_xTe$ grown from tellurium rich liquid phase epitaxial growth solutions," *J. Electron. Mater.* 16, 13–25, 1987.
303. E. P. G. Smith, R. E. Bornfreund, I. Kasai, L. T. Pham, E. A. Patten, J. M. Peterson, J. A. Roth, et al., "Status of two-color and large format HgCdTe FPA technology at Raytheon Vision Systems," *Proc. SPIE* 6127, 61261F, 2006.
304. E. P. G. Smith, G. M. Venzor, Y. Petraitis, M. V. Liguori, A. R. Levy, C. K. Rabkin, J. M. Peterson, M. Reddy, S. M. Johnson, and J. W. Bangs, "Fabrication and characterization of small unit-cell molecular beam epitaxy grown HgCdTe-on-Si mid-wavelength infrared detectors," *J. Electron. Mater.* 36, 1045–51, 2007.
305. J. M. Arias, J. G. Pasko, M. Zandian, L. J. Kozlowski, and R. E. DeWames, "Molecular beam epitaxy HgCdTe infrared photovoltaic detectors," *Optic. Eng.* 33, 1422–28, 1994.
306. L. O. Bubulac and C. R. Viswanathan, "Diffusion of As and Sb in HgCdTe," *J. Cryst. Growth* 123, 555–66, 1992.
307. M. Arias, J. G. Pasko, M. Zandian, S. H. Shin, G. M. Williams, L. O. Bubulac, R. E. DeWames, and W. E. Tennant, "Planar p-on-n HgCdTe heterostructure photovoltaic detectors," *Appl. Phys. Lett.* 62, 976–8, 1993.
308. Y. Nemirovsky, N. Mainzer, and E. Weiss, "Passivation of HgCdTe," in *Properties of Narrow Gap Cadmium-Based Compounds*, EMIS Datareviews Series No. 10, ed. P. P. Capper, 284–90, IEE, London, 1994.

309. Y. Nemirovsky and N. Amir, "Surfaces/Interfaces of Narrow-Gap II–VI Compounds," in *Narrow-gap II-VI Compounds for Optoelectronic and Electromagnetic Applications*, ed. P. Capper, 291–326, Chapman & Hall, London, 1997.
310. Y. Nemirovsky and L. Burstein, "Anodic sulfide films on $Hg_{1-x}Cd_xTe$," *Appl. Phys. Lett.* 44, 443–4, 1984.
311. E. Weiss and C. R. Helms, "Composition, growth mechanism, and oxidation of anodic fluoride films on $Hg_{1-x}Cd_xTe$ ($x \approx 0.2$)," *J. Electrochem. Soc.* 138, 993–9, 1991.
312. A. Kolodny and I. Kidron, "Properties of ion implanted junctions in mercury cadmium-telluride," *IEEE Trans. Electron Devices* ED-27, 37–43, 1980.
313. P. Migliorato, R. F. C. Farrow, A. B. Dean, and G. M. Williams, "CdTe/HgCdTe indium-diffused photodiodes," *Infrared Phys.* 22, 331–6, 1982.
314. M. Lanir and K. J. Riley, "Performance of PV HgCdTe arrays for 1–14-μm applications," *IEEE Trans. Electron. Devices* ED-29, 274–9, 1982.
315. N. Kajihara, G. Sudo, Y. Miyamoto, and K. Tonikawa, "Silicon nitride passivant for HgCdTe n^+–p diodes," *J. Electrochem. Soc.* 135, 1252–5, 1988.
316. J. F. Ameurlaire and G. D. Cohen-Solal, U. S. Patent No. 3,845,494, 1974; G. D. Cohen-Solal and A. G. Lussereau, U. S. Patent No. 3,988,774, 1976; J. H. Maille and A. Salaville, U. S. Patent No. 4,132,999, 1979.
317. D. J. Hall, L. Buckle, N. T. Gordon, J. Giess, J. E. Hails, J. W. Cairns, R. M. Lawrence, et al., "High-performance long-wavelength HgCdTe infrared detectors grown on silicon substrates," *Appl. Phys. Lett.* 84, 2113–5, 2004.
318. Y. Nemirovsky, N. Amir, and L. Djaloshinski, "Metalorganic chemical vapor deposition CdTe passivation of HgCdTe," *J. Electron. Mater.* 25, 647–54, 1995.
319. Y. Nemirovsky, N. Amir, D. Goren, G. Asa, N. Mainzer, and E. Weiss, "The interface of metalorganic chemical vapor deposition-CdTe/HgCdTe," *J. Electron. Mater.* 24, 1161–7, 1995.
320. S. Y. An, J. S. Kim, D. W. Seo, and S. H. Suh, "Passivation of HgCdTe p-n diode junction by compositionally graded HgCdTe formed by annealing in a Cd/Hg atmosphere," *J. Electron. Mater.* 31, 683–7, 2002.
321. L. O. Bubulac, W. E. Tennant, J. Bajaj, J. Sheng, R. Brigham, A. H. B. Vanderwyck, M. Zandian, and W. V. McLevige, "Characterization of CdTe for HgCdTe surface passivation," *J. Electron. Mater.* 24, 1175–82, 1995.
322. H. Jung, H. Lee, and C. Kim, "Enhancement of the steady state minority carrier lifetime in HgCdTe photodiode using ECR plasma hydrogenation," *J. Electron. Mater.* 25, 1266, 1996.
323. M. Carmody, J. G. Pasko, P. D. Edwall, M. Daraselia, L. A. Almeida, J. Molstad, J. H. Dinan, J. K. Markunas, Y. Chen, G. Brill, and N. K. Dhar, "Long wavelength infrared, molecular beam epitaxy, HgCdTe-on-Si diode performance," *J. Electron. Mater.* 33, 531, 2004.
324. P. Boieriu, C. H. Grein, J. Garland, S. Velicu, C. Fulk, A. Stoltz, L. Bubulac, J. H. Dinan, and S. Sivananthan, "Effects of hydrogen on majority carrier transport and minority carrier lifetimes in long-wavelength infrared HgCdTe on Si," *J. Electron. Mater.* 35, 1385, 2006.
325. W. D. Hu, X. S. Chen, Z. H. Ye, and W. Lu, "A hybrid surface passivation on HgCdTe long wave infrared detector with in-situ CdTe deposition and high-density hydrogen plasma modification," *Appl. Phys. Lett.* 99, 091101, 2011.
326. C. R. Eddy, Jr., E. A. Dobisz, J. R. Meyer, and C. A. Hoffman, "Electron cyclotron resonance reactive ion etching of fine features in $Hg_{1-x}Cd_xTe$ using CH_4/H_2 plasmas," *J. Vac. Sci. Technol. A* 11, 1763, 1986.
327. N. LiCausi, S. Rao, and I. B. Bhat, "Low-pressure chemical vapor deposition of CdS and atomic layer deposition of CdTe films for HgCdTe surface passivation," *J. Electron. Mater.* 40, 1668, 2011.
328. W. E. Spicer, D. J. Friedman, and G. P. Carey, "The electrical properties of metallic contacts on $Hg_{1-x}Cd_xTe$," *J. Vac. Sci. Technol. A* 6, 2746–51, 1988.
329. S. P. Wilks and R. H. Williams, "Contacts to HgCdTe," in *Properties of Narrow Gap Cadmium-Based Compounds*, EMIS Datareviews Series No. 10, ed. P. P. Capper, 297–9, IEE, London, 1994.
330. G. D. Davis, "Overlayer interactions with (HgCd)Te," *J. Vac. Sci. Technol. A* 6, 1939–45, 1988.
331. J. F. McGilp and I. T. McGovern, "A simple semiquantitative model for classifying metal-compound semiconductor interface reactivity," *J. Vac. Sci. Technol. B* 3, 1641–4, 1985.
332. P. W. Leech and G. K. Reeves, "Specific contact resistance of indium ohmic contacts to n-type $Hg_{1-x}Cd_xTe$," *J. Vac. Sci. Technol. A* 10, 105–9, 1992.
333. W. A. Beck, G. D. Davis, and A. C. Goldberg, "Resistance and 1/f noise of Au, Al, and Ge contacts to (Hg,Cd)Te," *J. Appl. Phys.* 67, 6340–6, 1990.
334. A. Piotrowski, P. Madejczyk, W. Gawron, K. Kłos, M. Romanis, M. Grudzień, A. Rogalski, and J. Piotrowski, "MOCVD growth of $Hg_{1-x}Cd_xTe$ heterostructures for uncooled infrared photodetectors," *Opto-Electron. Rev.* 12, 453–8, 2004.

335. A. Rogalski and W. Larkowski, "Comparison of photodiodes for the 3–5.5 μm and 8–12 μm spectral regions," *Electron Technol.* 18(3/4), 55–69, 1985.
336. A. Turner, T. Teherani, J. Ehmke, C. Pettitt, P. Conlon, J. Beck, K. McCormack, et al., "Producibility of VIP™ scanning focal plane arrays," *Proc. SPIE* 2228, 237–48, 1994.
337. P. Tribolet, J. P. Chatard, P. Costa, and A. Manissadjian, "Progress in HgCdTe homojunction infrared detectors," *J. Cryst. Growth* 184/185, 1262–71, 1998.
338. J. Bajaj, "State-of-the-Art HgCdTe materials and devices for infrared imaging," in *Physics of Semiconductor Devices*, eds. V. Kumar and S. K. Agarwal, 1297–309, Narosa Publishing House, New Delhi, 1998.
339. G. N. Pultz, P. W. Norton, E. E. Krueger, and M. B. Reine, "Growth and characterization of p-on-n HgCdTe liquid-phase epitaxy heterojunction material for 11–18 μm applications," *J. Vac. Sci. Technol. B* 9, 1724–30, 1991.
340. T. Tung, M. H. Kalisher, A. P. Stevens, and P. E. Herning, "Liquid-phase epitaxy of $Hg_{1-x}Cd_xTe$ from Hg solution: A route to infrared detector structures," *Mater. Res. Soc. Sympos. Proc.* 90, 321–56, 1987.
341. T. J. DeLyon, J. E. Jensen, M. D. Gorwitz, C. A. Cockrum, S. M. Johnson, and G. M. Venzor, "MBE growth of HgCdTe on silicon substrates for large-area infrared focal plane arrays: A review of recent progress," *J. Electron. Mater.* 28, 705–11, 1999.
342. M. Kinch, "HDVIP™ FPA technology at DRS," *Proc. SPIE* 4369, 566–78, 2001.
343. O. Gravrand, Ph. Chorier, and H. Geoffray, "Status of very long infrared wave focal plane array development at DEFIR," *Proc. SPIE* 7298, 729821, 2009.
344. W. E. Tennant, D. Lee, M. Zandian, E. Piquette, and M. Carmody, "MBE HgCdTe technology: A very general solution to IR detection, described by "Rule 07," a very convenient heuristic," *J. Electron. Mater.* 37, 1407–10, 2008.
345. J. A. Stobie, S. P. Tobin, P. Norton, M. Hutchins, K.-K. Wong, R. J Huppi, and R. Huppi, "Update on the imaging sensor for GIFTS," *Proc. SPIE* 5543, 293–303, 2004.
346. T. Chuh, "Recent developments in infrared and visible imaging for astronomy, defense and homeland security," *Proc. SPIE* 5563, 19–34, 2004.
347. A. S. Gilmore, J. Bangs, A. Gerrish, A. Stevens, and B. Starr, "Advancements in HgCdTe VLWIR materials," *Proc. SPIE* 5783, 223–30, 2005.
348. C. L. Jones, L. G. Hipwood, C. J. Shaw, J. P. Price, R. A. Catchpole, M. Ordish, C. D. Maxey, et al., "High performance MW and LW IRFPAs made from HgCdTe grown by MOVPE," *Proc. SPIE* 6206, 620610, 2006.
349. J. W. Baletic, R. Blank, D. Gulbransen, D. Lee, M. Loose, E. C. Piquette, T. Sprafke, W. E. Tennant, M. Zandian, and J. Zino, "Teledyne imaging sensors: Infrared imaging technologies for astronomy & civil space," *Proc. SPIE* 7021, 70210H, 2008.
350. M. B. Reine, S. P. Tobin, P. W. Norton, and P. LoVecchio, "Very long wavelength (>15 μm) HgCdTe photodiodes by liquid phase epitaxy," *Proc. SPIE* 5564, 54–64, 2004.
351. L. Mollard, G. Bourgeois, C. Lobre, S. Gout, S. Viollet-Bosson, N. Baier, G. Destefanis, O. Gravrand, J. P. Barnes, F. Milesi, A. Kerlain, L. Rubaldo, and A. Manissadjian, "p-on-n HgCdTe infrared focal-plane arrays: From short-wave to very-long-wave infrared," *J. Electron. Mater.* 43, 802–7, 2014.
352. O. Gravrand, J. Rothman, C. Cervera, N. Baier, C. Lobre, J. P. Zanatta, O. Boulade, V. Moreau, and B. Fieque," HgCdTe detectors for space and science imaging: General issues and latest achievements," *J. Electron. Mater.* 45, 4532–41, 2016.
353. R. E. DeWames, D. D. Edwall, M. Zandian, L. O. Bubulac, J. G. Pasko, W. E. Tennant, J. M. Arias, and A. D'Souza, "Dark current generating mechanisms in short wavelength infrared photovoltaic detectors," *J. Electron. Mater.* 27, 722–6, 1998.
354. A. Rogalski, "HgTe-based photodetectors in Poland," *Proc. SPIE* 7298, 72982Q, 2009.
355. E. C. Piquette, D. D. Edwall, H. Arnold, A. Chen, and J. Auyeung, "Substrate-removed HgCdTe-based focal-plane arrays for short-wavelength infrared astronomy," *J. Electron. Mater.* 37, 1396–400, 2008.
356. P. Norton, "Status of infrared detectors," *Proc. SPIE* 3379, 102–14, 1998.
357. O. Gravrand, G. Destefanis, S. Bisotto, N. Baier, J. Rothman, L. Mollard, D. Brellier, L. Rubaldo, A. Kerlain, V. Destefanis, and V. Vuillermet, "Issues in HgCdTe research and expected progress in infrared detector fabrication," *J. Electron. Mater.* 42, 3349–58, 2013.
358. J. Y. Wong, "Effect of trap tunneling on the performance of long-wavelength $Hg_{1-x}Cd_xTe$ photodiodes," *IEEE Trans. Electron Devices* ED-27, 48–57, 1980.
359. H. J. Hoffman and W. W. Anderson, "Impurity-to-band tunneling in $Hg_{1-x}Cd_xTe$," *J. Vac. Sci. Technol.* 21, 247–50, 1982.
360. D. K. Blanks, J. D. Beck, M. A. Kinch, and L. Colombo, "Band-to-band processes in HgCdTe: Comparison of experimental and theoretical studies," *J. Vac. Sci. Technol. A* 6, 2790–4, 1988.

361. R. E. DeWames, G. M. Williams, J. G. Pasko, and A. H. B. Vanderwyck, "Current generation mechanisms in small band gap HgCdTe p-n junctions fabricated by ion implantation," *J. Cryst. Growth* 86, 849–58, 1988.
362. Y. Nemirovsky, D. Rosenfeld, R. Adar, and A. Kornfeld, "Tunneling and dark currents in HgCdTe photodiodes," *J. Vac. Sci. Technol. A* 7, 528–35, 1989.
363. Y. Nemirovsky, R. Fastow, M. Meyassed, and A. Unikovsky, "Trapping effects in HgCdTe," *J. Vac. Sci. Technol. B* 9, 1829–39, 1991.
364. D. Rosenfeld and G. Bahir, "A model for the trap-assisted tunneling mechanism in diffused n-p and implanted n^+-p HgCdTe photodiodes," *IEEE Trans. Electron. Devices* 39, 1638–45, 1992.
365. Y. Nemirovsky and A. Unikovsky, "Tunneling and 1/f noise currents in HgCdTe photodiodes," *J. Vac. Sci. Technol. B* 10, 1602–10, 1992.
366. K. Jóźwikowski, M. Kopytko, A. Rogalski, and A. Jóźwikowska, "Enhanced numerical analysis of current-voltage characteristics of long wavelength infrared n-on-p HgCdTe photodiodes," *J. Appl. Phys.* 108, 074519-1–11, 2010.
367. K. Jóźwikowski, M. Kopytko, and A. Rogalski, "Enhanced numerical analysis of current-voltage characteristics of long wavelength infrared p-on-n HgCdTe photodiodes," *Bull. Pol. Acad. Sci.: Tech. Sci.* 58 (4), 52333, 2010.
368. R. E. DeWames, "Review of an assortment of IR materials-devices technologies used for imaging in spectral bands ranging from the visible to very long wavelength," *Proc. SPIE* 9819, 981924-1–28, 2016.
369. T. Elliott, N. T. Gordon, and R. S. Hall, "Reverse breakdown in long wavelength lateral collection $Cd_xHg_{1-x}Te$ diodes," *J. Vac. Sci. Technol. A* 8, 1251–3, 1990.
370. G. M. Williams and R. E. DeWames, "Numerical simulation of HgCdTe detector characteristics," *J. Electron. Mater.* 24, 1239–48, 1995.
371. A. S. Gilmore, J. Bangs, and A. Gerrishi, "VLWIR HgCdTe detector current-voltage analysis," *J. Electron. Mater.* 35, 1403–10, 2006.
372. O. Gravrand, E. De Borniol, S. Bisotto, L. Mollard, and G. Destefanis, "From long infrared to very long wavelength focal plane arrays made with HgCdTe $n^+ n^-/p$ ion implantation technology," *J. Electron. Mater.* 36, 981–7, 2007.
373. M. C. Chen, R. S. List, D. Chandra, M. J. Bevan, L. Colombo, and H. F. Schaake, "Key performance-limiting defects in p-on-n HgCdTe heterojunction infrared photodiodes," *J. Electron. Mater.* 25, 1375–82, 1996.
374. T. D. Kleinpenning, "1/f noise in p-n junction diodes," *J. Vac. Sci. Technol. A* 3, 176–82, 1985.
375. F. N. Hooge, "1/f noise is no surface effect," *Phys. Lett.* 29A, 123–40, 1969.
376. A.L. McWhorter, "1/f Noise and Germanium Surface Properties," in *Semiconductor Surface Physics*, ed. R.H. Kingston, 207–28, Pennsylvania University Press, Philadelphia, 1957.
377. S. P. Tobin, S. Iwasa, and T. J. Tredwell, "1/f noise in (Hg,Cd)Te photodiodes," *IEEE Trans. Electron Devices* ED-27, 43–8, 1980.
378. H. K. Chung, M. A. Rosenberg, and P. H. Zimmermann, "Origin of 1/f noise observed in $Hg_{1-x}Cd_xTe$ variable area photodiode arrays," *J. Vac. Sci. Technol. A* 3, 189–91, 1985.
379. J. Bajaj, G. M. Williams, N. H. Sheng, M. Hinnrichs, D. T. Cheung, J. P. Rode, and W. E. Tennant, "Excess (1/f) noise in $Hg_{0.7}Cd_{0.3}Te$ p-n junctions," *J. Vac. Sci. Technol. A* 3, 192–4, 1985.
380. R. E. DeWames, J. G. Pasko, E. S. Yao, A. H. B. Vanderwyck, and G. M. Williams, "Dark current generation mechanisms and spectral noise current in long-wavelength infrared photodiodes," *J. Vac. Sci. Technol. A* 6, 2655–63, 1988.
381. Y. Nemirovsky and D. Rosenfeld, "Surface passivation and 1/f noise phenomena in HgCdTe photodiodes," *J. Vac. Sci. Technol. A* 8, 1159–66, 1990.
382. R. A. Schiebel, "A model for 1/f noise in diffusion current based on surface recombination velocity fluctuations and insulator trapping," *IEEE Trans. Electron Devices* ED-41, 768–78, 1994.
383. J. Bajaj, E. R. Blazejewski, G. M. Williams, R. E. DeWames, and M. Brawn, "Noise (1/f) and dark currents in midwavelength infrared PACE-I HgCdTe photodiodes," *J. Vac. Sci. Technol. B* 10, 1617–25, 1992.
384. M. A. Kinch, C.-F. Wan, and J. D. Beck, "Noise in HgCdTe photodiodes," *J. Electron. Mater.* 34, 928–32, 2005.
385. M.A. Kinch, R.L. Strong, and C.A. Schaake, "1/f noise in HgCdTe focal-plane arrays," *J. Electron. Mater.* 42(11), 324351, 2013.
386. W. Hassis, O. Gravrand, J. Rothman, and S. Benahmed, "Low-frequency noise characteristics of HgCdTe infrared photodiodes operating at high temperatures," *J. Electron. Mater.* 42, 3288–96, 2013.
387. I.M. Baker and C.D. Maxey, "Summary of HgCdTe 2D array technology in the U.K," *J. Electron. Mater.* 30(6), 2829, 2000.
388. L.O. Bubulac, J.D. Benson, R.N. Jacobs, A.J. Stoltz, M. Jaime-Vasquez, L.A. Almeida, A. Wang, L. Wang, R. Hellmer, T. Golding, J.H. Dinan, M. Carmody, P.S. Wijewarnasuriya, M.F. Lee, M.F. Vilela, J. Peterson, S.M. Johnson, D.F. Lofgreen, and D. Rhiger, "The distribution tail of LWIR HgCdTe-on-Si FPAs: A hypothetical physical mechanism," *J. Electron. Mater.* 40(3), 2808, 2011.

389. R.L. Strong and M.A. Kinch, "Quantification and modeling of RMS noise distributions in HDVIP® infrared focal plane arrays," *J. Electron. Mater.* 43(8), 2824–30, 2014.
390. C. Verie, F. Raymond, J. Besson, and T. Nquyen Duy, "Bandgap spin-orbit splitting resonance effects in $Hg_{1-x}Cd_xTe$ Alloys," *J. Cryst. Growth* 59, 342–46, 1982.
391. R. Alabedra, B. Orsal, G. Lecoy, G. Pichard, J. Meslage, and P. Fragnon, "An $Hg_{0.3}Cd_{0.7}Te$ avalanche photodiode for optical-fiber transmission systems at $\lambda = 1.3\,\mu m$," *IEEE Trans. Electron. Devices* ED-32, 1302–6, 1985.
392. B. Orsal, R. Alabedra, M. Valenza, G. Lecoy, J. Meslage, and C. Y. Boisrobert, "$Hg_{0.4}Cd_{0.6}Te$ 1.55-μm avalanche photodiode noise analysis in the vicinity of resonant impact ionization connected with the spin-orbit split-off band," *IEEE Trans. Electron. Devices* ED-35, 101–7, 1988.
393. J. Vallegra, J. McPhate, L. Dawson, and M. Stapelbroeck, "Mid-IR couting array using HgCdTe APDs and the Medipix 2 ROIC," *Proc. SPIE* 6660, 666000, 2007.
394. D. N. B. Hall, B. S. Rauscher, J. L. Pipher, K. W. Hodapp, and G. Luppino, "HgCdTe optical and infrared focal plane array development in the next decade," *Astro2010: The Astronomy and Astrophysics Decadal Survey*, Technology Development Papers, no. 28.
395. G. Leveque, M. Nasser, D. Bertho, B. Orsal, and R. Alabedra, "Ionization energies in $Cd_xHg_{1-x}Te$ avalanche photodiodes," *Semicond. Sci. Technol.* 8, 1317–23, 1993.
396. T. J. deLyon, B. A. Baumgratz, G. R. Chapman, E. Gordon, M. D. Gorwitz, A. T. Hunter, M. D. Jack, et al., "Epitaxial growth of HgCdTe 1.55 μm avalanche photodiodes by MBE," *Proc. SPIE* 3629, 256–67, 1999.
397. N. Duy, A. Durand, and J. L. Lyot, "Bulk crystal growth of $Hg_{1-x}Cd_xTe$ for avalanche photodiode applications," *Mater. Res. Soc. Sympos. Proc.* 90, 81–90, 1987.
398. R. E. DeWames, J. G. Pasko, D. L. McConnell, J. S. Chen, J. Bajaj, L. O. Bubulac, E. S. Yao, et al., *Extended Abstracts, 1989 Workshop on the Physics and Chemistry of II-VI Materials*, San Diego, October 3–5, 1989.
399. J. D. Beck, C.-F. Wan, M. A. Kinch, and J. E. Robinson, "MWIR HgCdTe avalanche photodiodes," *Proc. SPIE* 4454, 188–97, 2001.
400. M. A. Kinch, J. D. Beck, C. F. Wan, F. Ma, and J. Campbell, "HgCdTe electron avalanche photodiodes," *J. Electron. Mater.* 33, 630–9, 2004.
401. F. Ma, X. Li, J. C. Campbell, J. D. Beck, C.-F. Wan, and M. A. Kinch, "Monte Carlo simulations of $Hg_{0.7}Cd_{0.3}Te$ avalanche photodiodes and resonance phenomenon in the multiplication noise," *Appl. Phys. Lett.* 83, 785, 2003.
402. J. Beck, C. Wan, M. Kinch, J. Robinson, P. Mitra, R. Scritchfield, F. Ma, and J. Campbell, "The HgCdTe electron avalanche photodiode," *IEEE LEOS Newslett.*, October 8–12, 2006.
403. J. Beck, C. Wan, M. Kinch, J. Robinson, P. Mitra, R. Scritchfield, F. Ma, and J. Campbell, "The HgCdTe electron avalanche photodiode," *J. Electron. Mater.* 35, 1166–73, 2006.
404. O. Gravrand and G. Destefanis, "Recent progress for HgCdTe quantum detection in France," *Infrared Phys. Technol.* 59, 163–71, 2013.
405. M. K. Kinch, "A theoretical model for the HgCdTe electron avalanche photodiode," *J. Electron. Mater.* 37, 1453–59, 2008.
406. J. Rothman, G. Perrais, P. Ballet, L. Mollard, S. Gout, and J.-P. Chamonal, "Latest developments of HgCdTe e-APDs at CEA LETI-Minatec," *J. Electron. Mater.* 37, 1303–10, 2008.
407. I. Baker, C. Maxey, L. Hipwood and K. Barnes, "Leonardo (formerly Selex ES) infrared sensors for astronomy: Present and future," *Proc. SPIE* 9915, 991505-1–8, 2016.
408. G. Perrais, O. Gravrand, J. Baylet, G. Destefanis, and J. Rothman, "Gain and dark current characteristics of planar HgCdTe avalanche photodiodes," *J. Electron. Mater.* 36, 963–70, 2007.
409. J. Rothman, E. de Borniol, O. Gravrand, P. Kern, P. Feautrier, J.-B. Lebouquin, and O. Boulade, "MCT APD focal plane arrays for astronomy at CEA-LETI," *Proc. SPIE* 9915, 9915-1–12, 2016.
410. I. Baker, S. Duncan, and J. Copley, "Low noise laser gated imaging system for long range target identification," *Proc. SPIE* 5406, 133–44, 2004.
411. I. Baker, P. Thorne, J. Henderson, J. Copley, D. Humphreys, and A. Millar, "Advanced multifunctional detectors for laser-gated imaging applications," *Proc. SPIE* 6206, 620608, 2006.
412. J. Beck, M. Woodall, R. Scritchfield, M. Ohlson, L. Wood, P. Mitra, and J. Robinson, "Gated IR imaging with 128 × 128 HgCdTe electron avalanche photodiode FPA," *Proc. SPIE* 6542, 654217, 2007.
413. J. Beck, M. Woodall, R. Scritchfield, M. Ohlson, L. Wood, P. Mitra, and J. Robinson, "Gated IR imaging with 128 × 128 HgCdTe electron avalanche photodiode FPA," *J. Electron. Mater.* 37, 1334–43, 2008.
414. G. Perrais, J. Rothman, G. Destefanis, and J.-P. Chamonal, "Impulse response time measurements in $Hg_{0.7}Cd_{0.3}Te$ MWIR avalanche photodiodes," *J. Electron. Mater.* 37, 1261–73, 2008.

415. G. Perrais, J. Rothman, G. Destefanis, J. Baylet, P. Castelein, J.-P. Chamonal, and P. Tribolet, "Demonstration of multifunctional bi-colour-avalanche gain detection in HgCdTe FPA," *Proc. SPIE* 6395, 63950H, 2006.
416. M. Jack, G, Chapman, J. Edwards, W. Mc Keag, T. Veeder, J. Wehner, T. Roberts, T. Robinson, J. Neisz, C. Andressen, R. Rinker, D. N. B. Hall, S. M. Jacobson, F. Amzajerdian, and T. D. Cook, "Advances in LADAR components and subsystems at Raytheon," *Proc. SPIE* 8353, 83532F-1–17, 2012.
417. P. Feautrier, J.-L. Gach and P. Wizinowich, "State of the art IR cameras for wavefront sensing using e-APD MCT arrays," *Proceedings of the Conference* held 25–30 October, 2015 at UCLA, https://escholarship.org/uc/item/4p4339x0.
418. T. Ashley, A. B. Dean, C. T. Elliott, M. R. Houlton, C. F. McConville, H. A. Tarry, and C. R. Whitehouse, "Multilayer InSb diodes grown by molecular beam epitaxy for near ambient temperature operation," *Proc. SPIE* 1361, 238–44, 1990.
419. C. T. Elliott, "Advanced heterostructures for $In_{1-x}Al_xSb$ and $Hg_{1-x}Cd_xTe$ detectors and emiters," *Proc. SPIE* 2744, 452–62, 1996.
420. C. T. Elliott, "Non-Equilibrium Devices in HgCdTe," in *Properties of Narrow Gap Cadmium-Based Compounds*, EMIS Datareviews Series No. 10, ed. P. Capper, 339–46, IEE, London, 1994.
421. C. T. Elliott, N. T. Gordon, R. S. Hall, T. J. Phillips, A. M. White, C. L. Jones, C. D. Maxey, and N. E. Metcalfe, "Recent results on metalorganic vapor phase epitaxially grown HgCdTe heterostructure devices," *J. Electron. Mater.* 25, 1139–45, 1996.
422. D. Maxey, C. L. Jones, N. E. Metcalfe, R. Catchpole, M. R. Houlton, A. M. White, N. T. Gordon, and C. T. Elliott, "Growth of fully doped $Hg_{1-x}Cd_xTe$ heterostructures using a novel iodine doping source to achieve improved device performance at elevated temperatures," *J. Electron. Mater.* 25, 1276–85, 1996.
423. T. Ashley, C. T. Elliott, N. T. Gordon, R. S. Hall, A. D. Johnson, and G. J. Pryce, "Room temperature narrow gap semiconductor diodes as sources and detectors in the 5–10 μm wavelength region," *J. Cryst. Growth* 159, 1100–3, 1996.
424. M. K. Haigh, G. R. Nash, N. T. Gordon, J. Edwards, A. J. Hydes, D. J. Hall, A. Graham, J. Giess, J. E. Hails, and T. Ashley, "Progress in negative luminescent $Hg_{1-x}Cd_xTe$ diode arrays," *Proc. SPIE* 5783, 376–83, 2005.
425. C. T. Elliott, N. T. Gordon, T. J. Phillips, H. Steen, A. M. White, D. J. Wilson, C. L. Jones, C. D. Maxey, and N. E. Metcalfe, "Minimally cooled heterojunction laser heterodyne detectors in metalorganic vapor phase epitaxially grown $Hg_{1-x}Cd_xTe$," *J. Electron. Mater.* 25, 1146–50, 1996.
426. C. T. Elliott, N. T. Gordon, R. S. Hall, T. J. Phillips, C. L. Jones, and A. Best, "1/f noise studies in uncooled narrow gap $Hg_{1-x}Cd_xTe$ non-equilibrium diodes," *J. Electron. Mater.* 26, 643–8, 1997.
427. C. L. Jones, N. E. Metcalfe, A. Best, R. Catchpole, C. D. Maxey, N. T. Gordon, R. S. Hall, T. Colin, and T. Skauli, "Effect of device processing on 1/f noise in uncooled Auger-suppressed CdHgTe diodes," *J. Electron. Mater.* 27, 733–9, 1998.
428. T. Skauli, H. Steen, T. Colin, P. Helgesen, S. Lovold, C. T. Elliott, N. T. Gordon, T. J. Phillips, and A. M. White, "Auger suppression in CdHgTe heterostructure diodes grown by molecular beam epitaxy using silver as acceptor dopant," *Appl. Phys. Lett.* 68, 1235–7, 1996.
429. G. J. Bowen, I. D. Blenkinsop, R. Catchpole, N. T. Gordon, M. A. Harper, P. C. Haynes, L. Hipwood, et al., "HOTEYE: A novel thermal camera using higher operating temperature infrared detectors," *Proc. SPIE* 5783, 392–400, 2005.
430. M. K. Ashby, N. T. Gordon, C. T. Elliott, C. L. Jones, C. D. Maxey, L. Hipwood, and R. Catchpole, "Novel $Hg_{1-x}Cd_xTe$ device structure for higher operating temperature detectors," *J. Electron. Mater.* 32, 667–71, 2003.
431. M. A. Kinch, "Metal-Insulator-Semiconductor Infrared Detectors," in *Semiconductors and Semimetals*, Vol. 18, eds. R. K. Willardson and A. C. Beer, 313–78, Academic Press, New York, 1981.
432. R. A. Chapman, S. R. Borrello, A. Simmons, J. D. Beck, A. J. Lewis, M. A. Kinch, J. Hynecek, and C. G. Roberts, "Monolithic HgCdTe charge transfer device infrared imaging arrays," *IEEE Trans. Electron Devices* ED-27, 134–46, 1980.
433. A. F. Milton, "Charge Transfer Devices for Infrared Imaging," in *Optical and Infrared Detectors*, ed. R. J. Keyes, 197–228, Springer-Verlag, Berlin, 1980.
434. G. M. Williams and E. R. Gertner, "n-Channel M.I.S.F.E.T. in epitaxial HgCdTe/CdTe," *Electron Lett.* 16, 839, 1980.
435. A. Kolodny, Y. T. Shacham-Diamond, and I. Kidron, "n-Channel MOS transistors in mercury-cadmium telluride," *IEEE Trans. Electron Devices* 27, 591–5, 1980.
436. Y. Nemirovsky, S. Margalit, and I. Kidron, "n-channel insulated-gate field-effect transistors in $Hg_{1-x}Cd_xTe$ with $x = 0.215$," *Appl. Phys. Lett.* 36, 466–8, 1980.
437. R. A. Schiebel, "Enhancement mode HgCdTe MISFETs and circuits for focal plane applications," *IEDM Tech. Digest* 132–5, 1987.

438. T. L. Koch, J. H. De Loo, M. H. Kalisher, and J. D. Phillips, "Monolithic n-channel HgCdTe linear imaging arrays," *IEEE Trans. Electron Devices* ED-32, 1592–607, 1985.
439. S. R. Borrello, C. G. Roberts, M. A. Kinch, C. E. Tew, and J. D. Beck, "HgCdTe MIS photocapacitor detectors," in *Fourth International Conference on Advanced Infrared Detectors and Systems*, 41–7, IEE, London, 1990.
440. M. V. Wadsworth, S. R. Borrello, J. Dodge, R. Gooh, W. McCardel, G. Nado, and M. D. Shilhanek, "Monolithic CCD imagers in HgCdTe," *IEEE Trans. Electron Devices* 42, 244–50, 1995.
441. J. P. Omaggio, "Analysis of dark current in IR detectors on thinned p-type HgCdTe," *IEEE Trans. Electron Devices* 37, 141–52, 1990.
442. R. A. Chapman, M. A. Kinch, A. Simmons, S. R. Borrello, H. B. Morris, J. S. Wrobel, and D. D. Buss, "$Hg_{0.7}Cd_{0.3}Te$ charge-coupled device shift registers," *Appl. Phys. Lett.* 32, 434–6, 1978.
443. J. L. Freeouf and J. M. Woodall, "Schottky barriers: An effective work function model," *Appl. Phys. Lett.* 39, 727–9, 1981.
444. J. Tersoff, "Schottky barriers and semiconductor band structures," *Phys. Rev.* 32, 6968–71, 1985.
445. W. E. Spicer, I. Lindau, P. Skeath, C. Y. Su, and P. Chye, "Unified mechanism for Schottky-barrier formation and III–V oxide interface states," *Phys. Rev. Lett.* 44, 420–3, 1980.
446. A. Kobayoshi, O. F. Shankey, and J. D. Dow, "Chemical trends for defect energy levels in $Hg_{1-x}Cd_xTe$," *Phys. Rev.* 25, 6367–79, 1982.
447. A. Zunger, "Composition-dependence of deep impurity levels in alloys," *Phys. Rev. Lett.* 54, 849–50, 1985.
448. D. L. Polla and A. K. Sood, "Schottky barrier photodiodes in $Hg_{1-x}Cd_xTe$," *IEDM Tech. Digest*, 419–20, 1978.
449. D. L. Polla and A. K. Sood, "Schottky barrier photodiodes in p $Hg_{1-x}Cd_xTe$," *J. Appl. Phys.* 51, 4908–12, 1980.
450. P. W. Leech and M. H. Kibel, "Properties of Schottky diodes on n-type $Hg_{1-x}Cd_xTe$," *J. Vac. Sci. Technol. B* 9, 1770–6, 1991.
451. A.M. Itsuno, J.D. Phillips, and S. Velicu, "Mid–wave infrared HgCdTe nBn photodetector", *Appl. Phys. Lett.* 100, 161102 (2012).
452. A.M. Itsuno, J.D. Phillips, and S. Velicu, "Design of an Auger-suppressed unipolar HgCdTe NBnN photodetector", *J. Electron. Mater.* 41, 2886–92 (2012).
453. S. Velicu, J. Zhao, M. Morley, A.M. Itsuno, and J.D. Philips, "Theoretical and experimental investigation of MWIR HgCdTe nBn detectors", *Proc. SPIE* 8268, 82682X-1–13 (2012).
454. M. Kopytko, J. Wróbel, K. Jóźwikowski, A. Rogalski, J. Antoszewski, N. D. Akhavan, G. A. Umana-Membreno, L. Faraone, and C. R. Becker, "Engineering the bandgap of unipolar HgCdTe-based nBn infrared photodetectors," *J. Electron. Mater.* 44, 158–66 (2014).
455. M. Kopytko, A. Kębłowski, W. Gawron, P. Madejczyk, A. Kowalewski, and K. Jóźwikowski, "High-operating temperature MWIR nBn HgCdTe detector grown by MOCVD," *Opto-Electron. Rev.* 21, 402–5 (2013).
456. N.D. Akhavn, G. Jolley, G.A. Umma-Membreno, J. Antoszewski, and L. Faraone, "Design of band engineered HgCdTe nBn detectors for MWIR and LWIR applications," *IEEE Trans. Electron Devices* 62, 722–8 (2015).
457. M. Kopytko, "Design and modelling of high-operating temperature MWIR HgCdTe nBn detector with *n*- and *p*-type barriers," *Infared Phys. Technol.* 64, 47–55 (2014).
458. A. White, "Infrared detectors," U.S. Patent 4,679,063 (22 September 1983).
459. P. Klipstein, "XBn barrier photodetectors for high sensitivity operating temeprature infrared sensors", *Proc. SPIE.* 6940, 69402U-1–11 (2008).
460. R. Pelzel, "A comparison of MOVPE and MBE growth technologies for III-V epitaxial structures," *The International Conference on Compound Semiconductor Manufacturing Technology 2013, CS Mantech.*
461. M. Kopytko, A. Kębłowski, W. Gawron, A. Kowalewski, and A. Rogalski, "MOCVD grown HgCdTe barrier structures for HOT conditions," *IEEE Trans. Electron Devices* 61(11), 3803–7, 2014.
462. M. Kopytko and A. Rogalski, "HgCdTe barrier infrared detectors," *Prog. Quant. Electron.* 47, 1–18, 2016.
463. M. Kopytko, K. Jóźwikowski, P. Martyniuk, W. Gawron, P. Madejczyk, A. Kowalewski, O. Markowska, A. Rogalski, and J. Rutkowski, "Status of HgCdTe barrier infrared detectors grown by MOCVD in Military University of Technology," *J. Electron. Mater. 45*, 4563–73, 2016.
464. R. Triboulet, "Alternative small gap materials for IR detection," *Semicond. Sci. Technol.* 5, 1073–9, 1990.
465. A. Sher, A. B. Chen, W. E. Spicer, and C. K. Shih, "Effects influencing the structural integrity of semiconductors and their alloys," *J. Vac. Sci. Technol. A* 3, 105–11, 1985.
466. A. Wall, C. Caprile, A. Franciosi, R. Reifenberger, and U. Debska, "New ternary semiconductors for infrared applications: $Hg_{1-x}Mn_xTe$," *J. Vac. Sci. Technol. A* 4, 818–22, 1986.
467. K. Guergouri, R. Troboulet, A. Tromson-Carli, and Y. Marfaing, "Solution hardening and dislocation density reduction in CdTe crystals by Zn addition," *J. Cryst. Growth* 86, 61–5, 1988.

468. P. Maheswaranathan, R. J. Sladek, and U. Debska, "Elastic constants and their pressure dependences in $Cd_{1-x}Mn_xTe$ with $0 \leq x \leq 0.52$ and in $Cd_{0.52}Zn_{0.48}Te$," *Phys. Rev. B* 31, 5212–6, 1985.
469. A. Rogalski, "$Hg_{1-x}Zn_xTe$ as a potential infrared detector material," *Prog. Quant. Electron.* 13, 299–353, 1989.
470. A. Rogalski, "$Hg_{1-x}Mn_xTe$ as a new infrared detector material," *Infrared Phys.* 31, 117–66, 1991.
471. A. Rogalski, *New Ternary Alloy Systems for Infrared Detectors*, SPIE Press, Bellingham, WA, 1994.
472. R. T. Dalves and B. Lewis, "Zinc blende type HgTe-MnTe solid solutions: I," *J. Phys. Chem. Solids* 24, 549–56, 1963.
473. P. Becla, J. C. Han, and S. Matakef, "Application of strong vertical magnetic fields to growth of II–VI pseudo-binary alloys: HgMnTe," *J. Cryst. Growth* 121, 394–8, 1992.
474. O. A. Bodnaruk, I. N. Gorbatiuk, V. I. Kalenik, O. D. Pustylnik, I. M. Rarenko, and B. P. Schafraniuk, "Crystalline structure and electro-physical parameters of $Hg_{1-x}Mn_xTe$ crystals," *Nieorganicheskie Mater.* 28, 335–9, 1992 (in Russian).
475. R. Triboulet, "(Hg,Zn)Te: a new material for IR detection," *J. Cryst. Growth* 86, 79–86, 1988.
476. P. Gille, U. Rössner, N. Puhlmann, H. Niebsch, and T. Piotrowski, "Growth of $Hg_{1-x}Mn_xTe$ crystals by the travelling heater method," *Semicond. Sci. Technol.* 10, 353–7, 1995.
477. S. Takeyama and S. Narita, "New techniques for growing highly-homogeneous quaternary $Hg_{1-x}Cd_xMn_yTe$ single crystals," *Jpn. J. Appl. Phys.* 24, 1270–3, 1985.
478. T. Uchino and K. Takita, "Liquid phase epitaxial growth of $Hg_{1-x-y}Cn_xZn_yTe$ and $Hg_{1-x}Cd_xMn_yTe$ from Hg-rich solutions," *J. Vac. Sci. Technol. A* 14, 2871–4, 1996.
479. A. B. Horsfall, S. Oktik, I. Terry, and A. W. Brinkman, "Electrical measurements of $Hg_{1-x}Mn_xTe$ films grown by metalorganic vapour phase epitaxy," *J. Cryst. Growth* 159, 1085–9, 1996.
480. R. Triboulet, A. Lasbley, B. Toulouse, and R. Granger, "Growth and characterization of bulk HgZnTe crystals," *J. Cryst. Growth* 76, 695–700, 1986.
481. S. Rolland, K. Karrari, R. Granger, and R. Triboulet, "P-to-n conversion in $Hg_{1-x}Zn_xTe$," *Semicond. Sci. Technol.* 14, 335–40, 1999.
482. M. A. Berding, S. Krishnamurthy, A. Sher, and A. B. Chen, "Electronic and transport properties of HgCdTe and HgZnTe," *J. Vac. Sci. Technol. A* 5, 3014–8, 1987.
483. R. Granger, A. Lasbley, S. Rolland, C. M. Pelletier, and R. Triboulet, "Carrier concentration and transport in $Hg_{1-x}Zn_xTe$ for x near 0.15," *J. Cryst. Growth* 86, 682–8, 1988.
484. W. Abdelhakiem, J. D. Patterson, and S. L. Lehoczky, "A comparison between electron mobility in n-type $Hg_{1-x}Cd_xTe$ and $Hg_{1-x}Zn_xTe$," *Mater. Lett.* 11, 47–51, 1991.
485. R. E. Kremer, Y. Tang, and F. G. Moore, "Thermal annealing of narrow-gap HgTe-based alloys," *J. Cryst. Growth* 86, 797–803, 1988.
486. P. I. Baranski, A. E. Bielaiev, O. A. Bodnaruk, I. N. Gorbatiuk, S. M. Kimirenko, I. M. Rarenko, and N. V. Shevchenko, "Transport properties and recombination mechanisms in $Hg_{1-x}Mn_xTe$ alloys ($x \sim 0.1$)," *Fizyka i Technika Poluprovodnikov* 24, 1490–3, 1990.
487. M. M. Trifonova, N. S. Baryshev, and M. P. Mezenceva, "Electrical properties of n-type $Hg_{1-x}Mn_xTe$ alloys," *Fizyka i Technika Poluprovodnikov* 25, 1014–7, 1991.
488. W. A. Gobba, J. D. Patterson, and S. L. Lehoczky, "A comparison between electron mobilities in $Hg_{1-x}Mn_xTe$ and $Hg_{1-x}Cd_xTe$," *Infrared Phys.* 34, 311–21, 1993.
489. Y. Sha, C. Su, and S. L. Lehoczky, "Intrinsic carrier concentration and electron effective mass in $Hg_{1-x}Zn_xTe$," *J. Appl. Phys.* 81, 2245–9, 1997.
490. K. Jóźwikowski and A. Rogalski, "Intrinsic carrier concentrations and effective masses in the potential infrared detector material, $Hg_{1-x}Zn_xTe$," *Infrared Phys.* 28, 101–7, 1988.
491. C. Wu, D. Chu, C. Sun, and T. Yang, "Infrared spectroscopy of $Hg_{1-x}Zn_xTe$ alloys," *Jpn. J. Appl. Phys.* 34, 4687–93, 1995.
492. A. Rogalski, "Hg-based Alternatives to MCT," in *Infrared Detectors and Emitters: Materials and Devices*, eds. P. Capper and C. T. Elliott, 377–400, Kluwer Academic Publishers, Boston, MA, 2001.
493. D. Eger and A. Zigelman, "Anodic oxides on HgZnTe," *Proc. SPIE* 1484, 48–54, 1991.
494. K. H. Khelland, D. Lemoine, S. Rolland, R. Granger, and R. Triboulet, "Interface properties of passivated HgZnTe," *Semicond. Sci. Technol.* 8, 56–82, 1993.
495. Yu. V. Medvedev and N. N. Berchenko, "Thermodynamic properties of the native oxide-$Hg_{1-x}Zn_xTe$ interface," *Semicond. Sci. Technol.* 9, 2253–7, 1994.
496. Z. Nowak, Doctoral Thesis, Military University of Technology, Warsaw, Poland, 1974 (in Polish).
497. J. Piotrowski, K. Adamiec, A. Maciak, and Z. Nowak, "ZnHgTe as a material for ambient temperature 10.6 μm photodetectors," *Appl. Phys. Lett.* 54, 143–4, 1989.

498. J. Piotrowski, K. Adamiec, and A. Maciak. "High-temperature 10,6 μm HgZnTe photodetectors," *Infrared Phys.* 29, 267–70, 1989.
499. J. Piotrowski and T. Niedziela, "Mercury zinc telluride longwavelength high temperature photoconductors," *Infrared Phys.* 30, 113–9, 1990.
500. A. Rogalski, J. Rutkowski, K. Józwikowski, J. Piotrowski, and Z. Nowak, "The performance of $Hg_{1-x}Zn_xTe$ photodiodes," *Appl. Phys. A* 50, 379–84, 1990.
501. K. Jóźwikowski, A. Rogalski, and J. Piotrowski, "On the performance of $Hg_{1-x}Zn_xTe$ photoresistors," *Acta Phys. Polon. A* 77, 359–62, 1990.
502. J. Piotrowski, T. Niedziela, and W. Galus, "High-temperature long-wavelength photoconductors," *Semicond. Sci. Technol.* 5, S53–6, 1990.
503. R. Triboulet, T. Le Floch, and J. Saulnier, "First (Hg,Zn)Te infrared detectors," *Proc. SPIE* 659, 150–2, 1988.
504. Z. Nowak, J. Piotrowski, and J. Rutkowski, "Growth of HgZnTe by cast-recrystallization," *J. Cryst. Growth* 89, 237–41, 1988.
505. J. Ameurlaine, A. Rousseau, T. Nguyen-Duy, and R. Triboulet, "(HgZn)Te infrared photovoltaic detectors," *Proc. SPIE* 929, 14–20, 1988.
506. R. Triboulet, M. Bourdillot, A. Durand, and T. Nguyen Duy, "(Hg,Zn)Te among the other materials for IR detection," *Proc. SPIE* 1106, 40–7, 1989.
507. D. L. Kaiser and P. Becla, "$Hg_{1-x-y}Cd_xZn_yTe$: Growth, properties and potential for infrared detector applications," *Mater. Res. Soc. Sympos. Proc.* 90, 397–404, 1987.
508. P. Becla, "Infrared photovoltaic detectors utilizing $Hg_{1-x}Mn_xTe$ and $Hg_{1-x-y}Cd_xMn_yTe$ alloys," *J. Vac. Sci. Technol. A* 4, 2014–8, 1986.
509. P. Becla, "Advanced infrared photonic devices based on HgMnTe," *Proc. SPIE* 2021, 22–34, 1993.
510. S. Takeyama and S. Narita, "The band structure parameters determination of the quaternary semimagnetic semiconductor alloy $Hg_{1-x-y}Cd_xMn_yTe$," *J. Phys. Soc. Jpn.* 55, 274–83, 1986.
511. S. Manhas, K. C. Khulbe, D. J. S. Beckett, G. Lamarche, and J. C. Woolley, "Lattice parameters, energy gap, and magnetic properties of the $Cd_xHg_yMn_xTe$ alloy system," *Phys. Status Solidi B* 143, 267–74, 1987.
512. L. A. Kosyachenko, I. M. Rarenko, S. Weiguo, and L. Zheng Xiong, "Charge transport mechanisms in HgMnTe photodiodes with ion etched p-n junctions," *Solid-State Electron.* 44, 1197–202, 2000.
513. L. A. Kosyachenko, I. M. Rarenko, S. Weiguo, L. Zheng Xiong, and G. Qibing, "Photoelectric properties of HgMnTe photodiodes with ion etched p-n junctions," *Opto-Electron. Rev.* 8, 251–62, 2000.
514. S. H. Shin, J. G. Pasko, D. S. Lo, W. E. Tennant, J. R. Anderson, M. Górska, M. Fotouhi, and C. R. Lu, "$Hg_{1-x-y}Cd_xMn_yTe$ alloys for 1.3–1.8 μm photodiode applications," *Mater. Res. Soc. Sympos. Proc.* 89, 267–74, 1987.
515. P. Becla, S. Motakef, and T. Koehler, "Long wavelength HgMnTe avalanche photodiodes," *J. Vac. Sci. Technol. B* 10, 1599–601, 1992.
516. P. Becla, M. Grudzień, and J. Piotrowski, "Uncooled 10.6 μm mercury manganise telluride photoelectromagnetic infrared detectors," *J. Vac. Sci. Technol. B* 9, 1777–80, 1991.

18 IV–VI detectors

Around 1920, Case investigated the thallium sulfide photoconductor—one of the first photoconductors to give a response in the near-infrared (NIR) region of approximately 1.1 μm [1]. The next group of materials to be studied was the Pb salts (PbS, PbSe, and PbTe), which extended the wavelength response to 7 μm. The PbS photoconductors from natural galena found in Sardinia were originally fabricated by Kutzscher at the University of Berlin in the 1930s [2]. However, for any practical application, it was necessary to develop a technique for producing synthetic crystals. PbS thin-film photoconductors were first produced in Germany, next in the United States at Northwestern University in 1944, and then in England at the Admiralty Research Laboratory in 1945 [3]. During World War II, the Germans produced systems that used PbS detectors to detect hot aircraft engines. Immediately after the war, communications, fire control, and search systems began to stimulate a strong development effort that has extended to the present day. The *Sidewinder* heat-seeking, IR-guided missiles received a great deal of public attention. After 60 years, low-cost, versatile PbS and PbSe polycrystalline thin films remain the photoconductive detectors of choice for many applications in the 1–3 μm and 3–5 μm spectral ranges. Current developments with Pb salts are in the focal plane arrays (FPAs) configuration.

The study of the IV–VI semiconductors received fresh impetus in the mid-1960s with a discovery at Lincoln Laboratories [4,5]: PbTe, SnTe, PbSe, and SnSe form solid solutions in which the energy gap varies continuously through zero so that it is possible to obtain any required small energy gap by selecting the appropriate composition. For 10 years, during the late 1960s to mid-1970s, HgCdTe alloy detectors were in serious competition with IV–VI alloy devices (mainly PbSnTe) for developing photodiodes because of the latter's production and storage problems [6,7]. The PbSnTe alloy seemed easier to prepare and appeared more stable. However, the development of PbSnTe photodiodes was discontinued because the chalcogenides suffered from two significant drawbacks. The first drawback was a high permittivity that resulted in a high diode capacitance, and therefore a limited frequency response. For scanning systems under development at that time, this was a serious limitation. However, for the staring imaging systems that use 2D arrays (which are currently under development), this would not be such a significant issue. The second drawback to IV–VI compounds was their very high thermal coefficient of expansion [8] (a factor of seven higher than Si). This limited their applicability in hybrid configurations with Si multiplexers. Development of ternary Pb salt FPAs was nearly entirely stopped in the early 1980s in favor of HgCdTe. Today, with the ability to grow these materials on alternative substrates, such as Si, this would not be a fundamental limitation either. Moreover, IV–VI materials were the only choice to fabricate MWIR (medium wavelength IR) laser diodes before the invention of the quantum cascade lasers and they continue to be of importance today [9–12].

In this chapter, we begin with a survey of fundamental properties of Pb salt chalcogenides and go on to describe detailed technology and properties of IV–VI photoconductive and photovoltaic IR detectors.

18.1 MATERIAL PREPARATION AND PROPERTIES

18.1.1 CRYSTAL GROWTH

The properties of the Pb salt binary and ternary alloys have been extensively reviewed [6,7,13–23]. Therefore, only some of their most important properties will be mentioned here.

The development of pseudobinary alloy systems, especially $Pb_{1-x}Sn_xTe$ (PbSnTe) and $Pb_{1-x}Sn_xSe$ (PbSnSe), has brought about major advances in the 8–14 μm wavelength region. Their energy gap varies

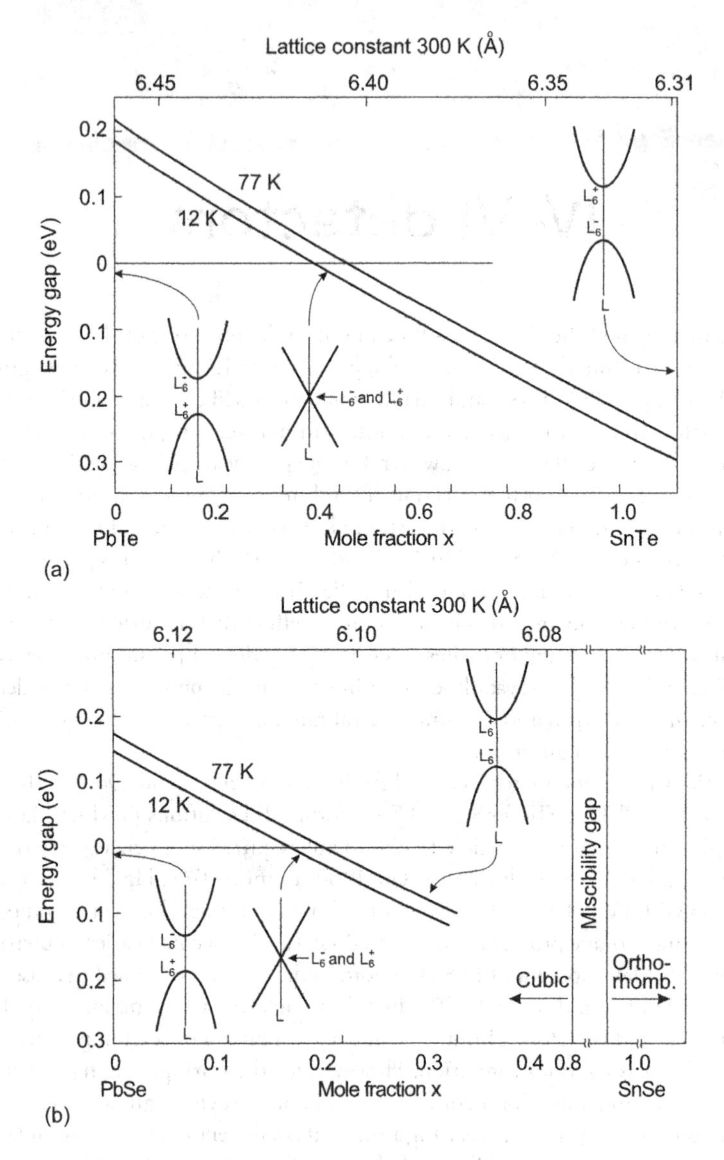

Figure 18.1 Energy gap versus mole fraction x and lattice constant for (a) $Pb_{1-x}Sn_xTe$ and (b) $Pb_{1-x}Sn_xSe$. Schematic representation of the valence and conduction bands.

continuously through zero so that it is possible, by selecting the appropriate composition, to obtain any required small energy gap (Figure 18.1). In comparison with $Hg_{1-x}Cd_xTe$ material system, the cutoff wavelength of IV–VI materials is less sensitive to composition.

It should be noted that besides PbSnTe and PbSnSe, a number of other Pb salts, such as PbS_xSe_{1-x} (PbSSe) and $PbTe_{1-x}Se_x$ (PbTeSe), are of interest for detection. Moreover, choosing $Pb_{1-x}Y_xZ$ materials with Y=Sr or Eu (Z=Te or Se), wider bandgap compounds can be obtained characterized by lower refractive indices (see Figure 18.2). This allows a high freedom to design more elaborate device structures including epitaxial Bragg mirrors, which can be realized by molecular-beam epitaxy (MBE) in different IV–VI optoelectronic devices [24].

The lead chalcogenide semiconductors have the face centered cubic (rock salt) crystal structure and hence obtain the name lead salts. Thus, they have (100) cleavage planes and tend to grow in the (100) orientation, although they can also be grown in the (111) orientation. Only SnSe possesses the orthorhombic-B29 structure. Therefore, the ternary compounds PbSnTe and PbSSe exhibit complete solid solubility, while the existence range of PbSnSe with rock-salt structure is restricted to the Pb-rich side ($x < 0.4$).

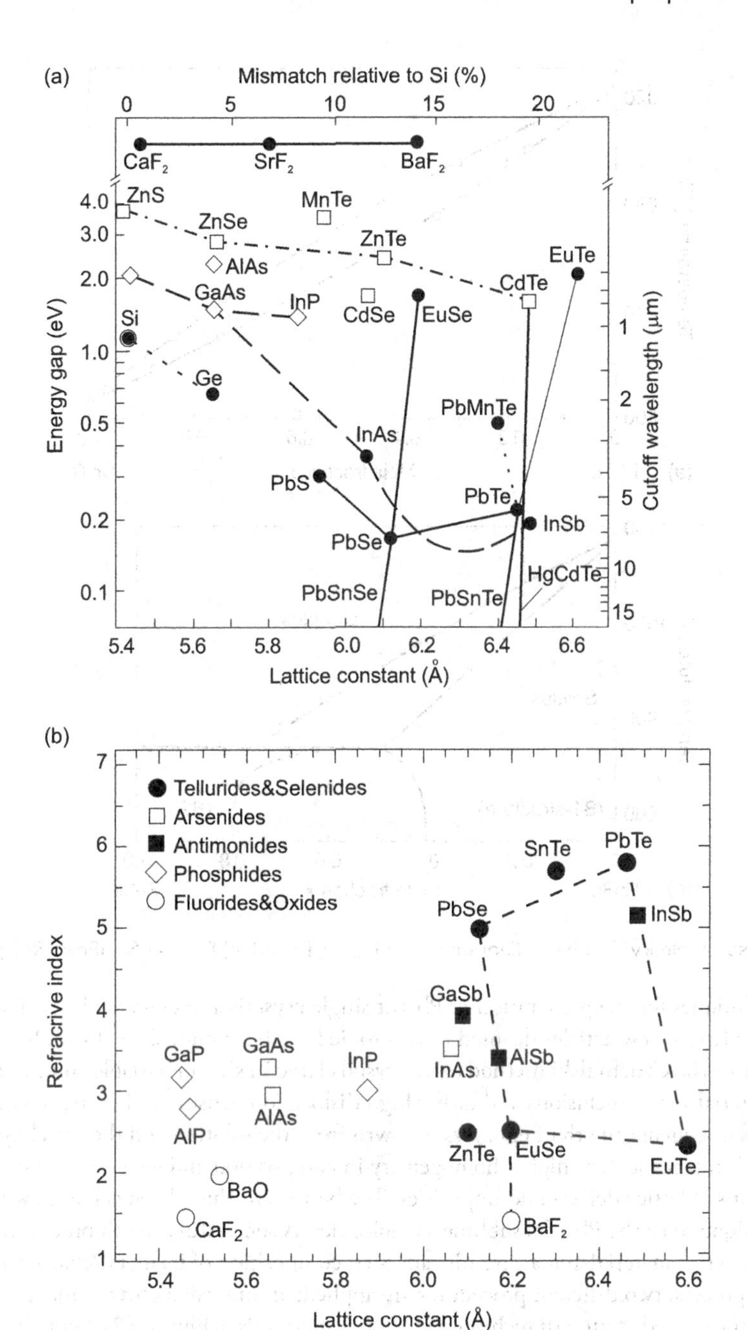

Figure 18.2 (a) Bandgap energy and corresponding emission wavelength (right-hand scale) and (b) bandgap refractive index of various III–V, II–VI, IV–VI, group IV semiconductors and selected fluorides and oxides plotted versus lattice constant. (From Ref. [24])

The crystalline properties of ternary alloys (PbSnTe, PbSnSe, and PbSSe) are comparable to those of the binary compounds (PbTe, PbSe, PbS, and SnTe). Despite the less fundamental physical properties relative to HgCdTe, PbSnTe and PbSnSe have received a good deal of attention as materials for photodiodes; the main reason being the much easier material technology. The separation between the liquidus and solidus curves is much smaller in ternary IV–VI alloys (see Figure 18.3) [25]. As a result, it has been relatively easy to grow PbSnTe and PbSnSe crystals that are homogeneous in composition. A second difference is in the vapor pressures of the elements, which are similar in magnitude for all the three elements in the ternary IV–VI alloys. Thus, vapor growth techniques have been successfully used in growing Pb salts.

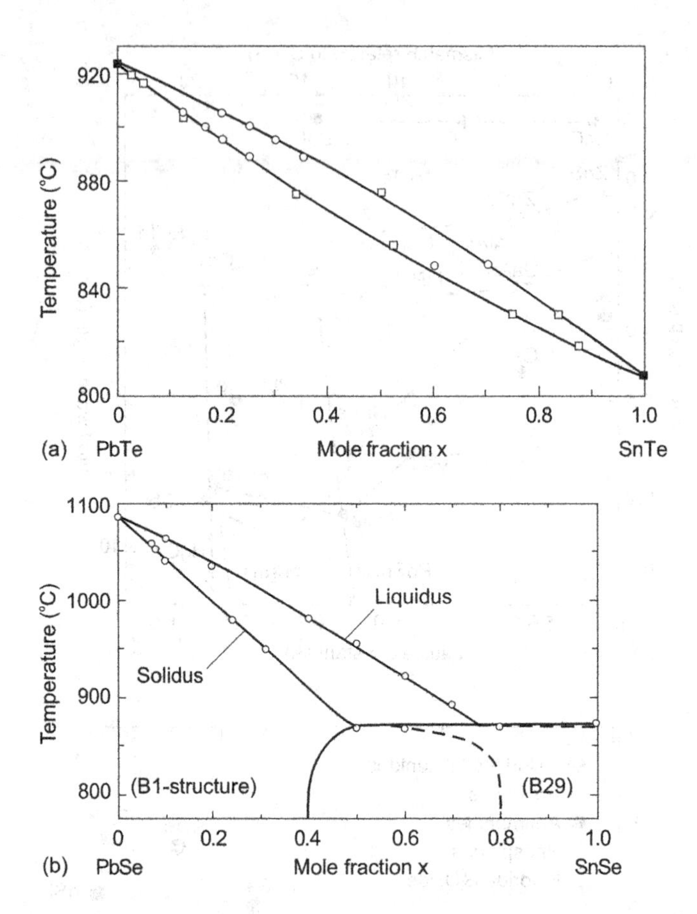

Figure 18.3 The pseudobinary T-x phase diagram of (a) $Pb_{1-x}Sn_xTe$ and (b) $Pb_{1-x}Sn_xSe$. (From Ref. [25])

Numerous techniques for the preparation of Pb salt single crystals and epitaxial layers have been investigated. Several excellent review articles devoted to this topic have been published [15,16,26,27].

Bridgman-type or the Czochralski methods give crystals large in size, of variable composition, and frequently the material shows inclusions and rather high dislocation densities. They are mainly used as substrates for the subsequent growth of epilayers. Growth from the solution and the traveling solvent method offer interesting advantages such as higher homogeneity in composition and lower temperatures leading to lower concentrations of lattice defects and impurities. The best results have been achieved with the sublimation growth technique since the Pb salts sublime as molecules. A successful growth process requires a very high purity of the source material and a carefully adjusted composition of its metal/chalcogen ratio [16]. For the vapor growth process, two different procedures are applied: an unseeded growth and a seeded growth technique. Using the unseeded growth technique, the largest crystals exhibiting 2–3 cm^2 (100) facets can be obtained with stoichiometric or slightly metal-rich source material.

A minimal contact with the quartz walls of the ampoule leads to an excellent metallurgical structure. Disadvantage of this growth procedure is the nonreproducible shape of the crystals that implies problems in large scale device production; but the main advantage is the growth of crystals with controlled carrier concentrations down to the 10^{17} cm^{-3} range without the need for lengthy annealing procedures [16]. Large crystals with excellent homogeneity and structural quality have been grown with seeded growth technique. As a seed, (111)-oriented single crystal Pb salt slices or (111)-oriented BaF_2 crystals are used (BaF_2 provides a lattice match and a good thermal expansion match comparable to Pb salts). In other methods developed by Tamari and Shtrikman [28], a quartz tip is used and the first deposit of material acts as a nonoriented seed in the growth. Good results have also been obtained by using the method developed by Markov and Davydov [29] for lead salts [30]. In the latter method, the crystals do not touch the ampoule wall and the dislocation density is low, typically 10^2–10^3 cm^{-2}.

Thin single crystal films of IV–VI compounds have found broad applications in fundamental research and application. The epitaxial layers are usually grown by either VPE (vapor phase epitaxy) or LPE (liquid phase epitaxy) techniques. Recently, the best quality devices have been obtained using MBE [24,27,31].

In the late 1970s, excellent results were obtained with LPE technique in the fabrication of PbTe/PbSnTe heterojunctions. A great deal of experimental research has been performed on solid-liquid equilibria in the Pb–Sn–Te system. Szapiro et al. [32] calculated the phase diagram of the Pb–Sn–Te system in the (Pb+Sn)-rich region using the modified model of regular associated solutions and received reasonable agreement with the calculated lines at higher concentrations of Sn in liquid ($x > 0.3$) and poor agreement for $x < 0.2$. Surface morphology of the epilayers was found to be primarily related to the substrate orientation and their surface preparation. Generally, (100)-oriented PbTe or PbSnTe substrates are employed. It should be noticed that the LPE technique leads to some difficulties for other ternary Pb salts [33].

Due to the fact that the IV–VI compounds evaporate predominantly in the form of binary molecules, which means that nearly congruent evaporation occurs, in MBE growth the main constituents are supplied from compound effusion sources loaded with PbTe, PbS, PbS, SnTe, SnSe, or GeTe. While the degree of dissociation is only a few percentage, it notably increases for the Sn and Ge chalcogenides. By changing the total group IV to group VI flux ratio, the background carrier concentration and the type of carriers can be controlled. Excess group IV flux leads to n-type and excess group VI flux to p-type conductivity in the layers.

Concerning the substrate materials, the lattice mismatch of Pb salts to common semiconductor substrates such as Si or GaAs is rather large (10% and more, see Figure 18.2a). In addition, the thermal expansion coefficient of the IV–VI compounds of around 20×10^{-6}/K differ strongly from that of Si as well as that of the zinc blende-type III–V or II–VI compounds (typically less than 6×10^{-6}/K). As a consequence, large thermal strains are induced in the epitaxial layers during cooling of the samples to room temperature and below after sample growth. A best compromise in these respects is achieved for BaF_2 substrates, in spite of its different crystal structure (calcium fluoride structure). As shown in Figure 18.2a, BaF_2 shows only a moderate lattice mismatch to PbSe or PbTe (–1.2% and +4.2%, respectively) and, moreover, the thermal expansion coefficient is almost exactly matched to that of the Pb salt compounds. Furthermore, BaF_2 is highly insulating and optically transparent in the mid-IR region. Due to its high ionic character, good BaF_2 surfaces can be obtained easily only for the (111) surface orientation. In conclusion, for Pb salt MBE growth, BaF_2 has been the most widely used substrate material.

Renewed interest in the growth of IV–VI epitaxial layers started in the mid-1980s with the growth of high-quality MBE layers on Si(111) substrates employing a very thin CaF_2 buffer layer. It appears that the thermal expansion mismatch between the IV–VI layers and Si relaxes through the glide of dislocations on (100) planes, which are inclined with respect to the (111) surface plane. The threading ends of the misfit dislocations glide to remove the mechanical thermal strain built up and therefore increase the structural layer quality [34].

18.1.2 DEFECTS AND IMPURITIES

The Pb salts can exist with very large deviations from stoichiometry and it is difficult to prepare material with carrier concentrations below about 10^{17} cm^{-3} [16,35–37]. For PbSnTe alloys, the solidus field shifts considerably toward the Te-rich side of the stoichiometric composition with increasing SnTe content, and very high hole concentrations are obtained at the Te-rich solidus lines. The solidus lines for several $Pb_{1-x}Sn_xTe$ alloys shown in Figure 18.4 have been determined by means of an isothermal annealing technique that is also useful for reducing the carrier concentration and converting the carrier type of crystals [16]. Low electron and hole concentrations in the range 10^{15} cm^{-3} have been obtained by isothermal annealing or by LPE at low temperature. In $Pb_{0.80}Sn_{0.20}Te$, which is of particular importance in device applications, the conversion of the carrier type occurs at a temperature of 530°C.

The width of the solidus field is large in IV–VI compounds (≈0.1%) making the doping by native defects very efficient. Deviations from stoichiometry create n- or p-type conduction. It is generally accepted that vacancies and interstitials are formed and that they control the conductivity. From the absence of any observable freeze-out it appears unlikely that native defects in Pb salts form hydrogen-like states. Native defects associated with excess metal (nonmetal vacancies or possibly metal interstitials) yield acceptor levels,

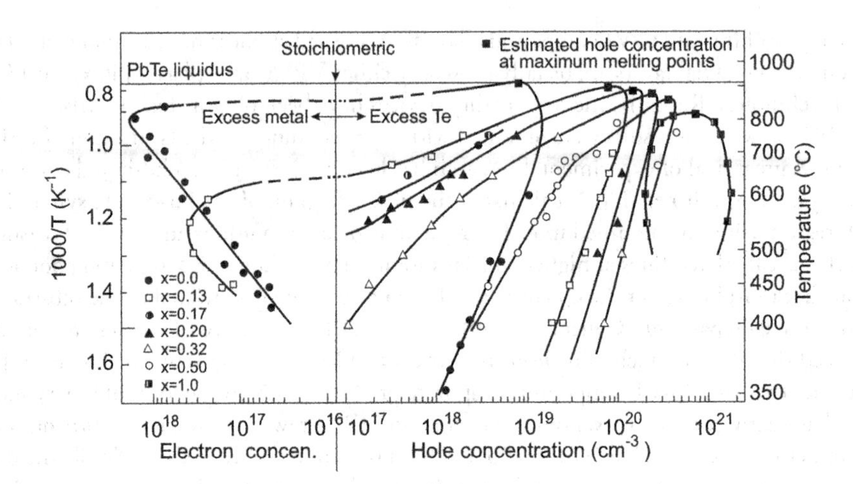

Figure 18.4 Carrier concentrations at 77 K (300 K for SnTe) versus isothermal annealing temperature for several $Pb_{1-x}Sn_xTe$ compositions. (From Ref. [16])

while those which result from excess nonmetal (metal vacancies or possibly nonmetal interstitials) yield donor levels. Parada and Pratt [38,39] first pointed out that the strong perturbations around the defects in PbTe cause valence band states to shift to the conduction band. As a result, a Te vacancy provides two electrons for the conduction band and a Pb vacancy two holes for the valence bands. The Pb interstitial yields a single electron, whereas the Te interstitial was found to be neutral. Similar results were obtained by Hemstreet [40] on the basis of scattered-wave cluster calculations for PbS, PbTe, and SnTe. Also Lent et al. [41] have used a simple chemical theory of s- and p-bonded substitutional point defect in PbTe and PbSnTe, which corroborates the experimental data [42] and the predictions of Parada and Pratt.

In crystals grown from high-purity elements, the effects of foreign impurities are usually negligible when the carrier concentration due to lattice defects is above 10^{17} cm^{-3}. Below this concentration, foreign impurities can play a role by compensating the lattice defects and other foreign impurities. A compilation of the results on impurity doping of IV–VI semiconductors is given by Dornhaus, Nimtz, and Richter [43]. Most of the impurities can be assumed to have shallow or even resonant levels [44]. However, deep donor levels of unidentified and identified defects were also found in Pb salts [45]. For example, it was found that In has an energy level in $Pb_{1-x}Sn_xTe$, which is resonant with the conduction band for small x values and is situated in the bandgap for higher mole fraction [46]. Such behavior of In level was theoretically explained by Lent et al. [41].

Cd is a compensating impurity in $Pb_{1-x}Sn_xTe$ and has the very important property of reducing carrier concentration and changing conductivity type in as-grown p-type materials. Silberg and Zemel [47] carried out Cd diffusion in a two-temperature zone furnace in which the samples were held at a fixed temperature of 400°C, while the Cd source temperature was varied between 150°C and 310°C. Figure 18.5 shows the dependence of electron concentration of the Cd-diffused $Pb_{1-x}Sn_xTe$ samples measured at 77 K on the Cd concentrations N_{Cd}. The electron concentration in the saturation region decreases appreciably with increasing x values. It appears that after vacancy sites become fully compensated by the Cd atoms most of Cd distributes uniformly in the crystal lattice as electrically inactive impurity.

18.1.3 SOME PHYSICAL PROPERTIES

The Pb salts have direct energy gaps, which occur at the Brillouin zone edge at the L point. The effective masses are therefore higher and the mobilities lower than for a zinc blende structure with the same energy gap at the Γ point (the zone center). Due to band inversion in PbSnTe and PbSnSe, the energy gaps approach zero at certain compositions (see Figure 18.1). Therefore, with these ternary compounds, very long wavelength cutoff of IR detectors can be achieved. The constant energy surfaces are ellipsoids characterized by the longitudinal and transverse effective masses m_l^* and m_t^*, respectively. The anisotropy factor for PbSnTe is of the order of ten and increases with decreasing bandgap. It is much less, about two, for PbSnSe and PbSSe.

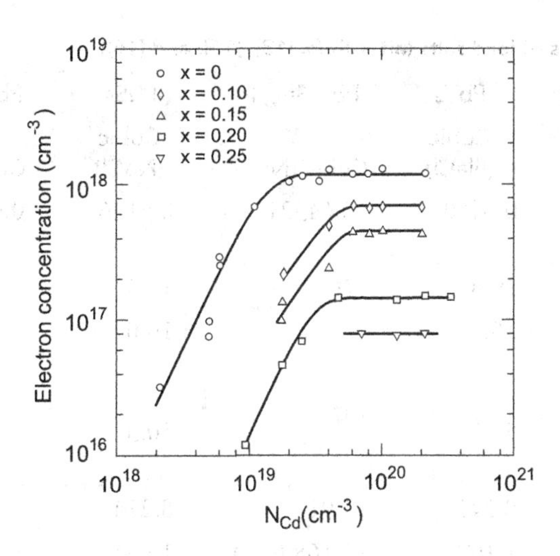

Figure 18.5 Electron concentration as a function of Cd concentration measured at 77 K on $Pb_{1-x}Sn_xTe$ crystals with different Sn mole fractions. (From Ref. [47])

Table 18.1 contains a list of material parameters for different types of binary and ternary Pb salts.

In the vicinity of the energy gap in Pb salts, a system of three conduction bands and three valence bands has been observed. The temperature and x-dependence of the effective masses at the band edges may be expressed by [15]

$$\frac{1}{m^*(x,T)} = \frac{1}{m^*_{cv}} \frac{E_g(0,0)}{E_g(x,T))} + \frac{1}{m^*_F}, \tag{18.1}$$

where m^*_{cv} determines a contribution due to the interactions between the nearest extremes of the valence band and conduction band and m^*_F is the far band contribution. The functions in Equation 18.1 for the four effective masses m^*_{et} (conduction band, transverse), m^*_{eh} (valence band, transverse), m^*_{el} (conduction band, longitudinal), and m^*_{hl} (valence band, longitudinal) are described by Preier [15] and Dornhaus et al. [17].

Exact knowledge of $E_g(x,T)$ dependences for the ternary compounds is required for reliable calculations over a wide range of composition. On the basis of the study of experimental results for the ternary compounds, good agreement with experimental data was obtained using formulas of the Grisar type $E_g(x,T) = E_1 + [E^2 + \alpha(T+\theta)^2]^{1/2}$ [15]:

$$E_g(x,T) = 171.5 - 535x + \sqrt{(12.8)^2 + 0.19(T+20)^2} \text{ in meV}, \tag{18.2}$$

$$E_g(x,T) = 125 - 1021x + \sqrt{400 + 0.256T^2} \text{ in meV}, \tag{18.3}$$

$$E_g(x,T) = 263 - 138x + \sqrt{400 + 0.265T^2} \text{ in meV}, \tag{18.4}$$

for $Pb_{1-x}Sn_xTe$, $Pb_{1-x}Sn_xSe$, and $PbS_{1-x}Se_x$, respectively.

Rogalski and Jóźwikowski [48] have calculated the intrinsic carrier concentrations in $Pb_{1-x}Sn_xTe$, $Pb_{1-x}Sn_xSe$, and $PbS_{1-x}Se_x$ in terms of the six-band **k·p** model of Dimmock. By fitting the calculated nonparabolic n_i values to the expression for parabolic bands, the following approximations were obtained:

for $Pb_{1-x}Sn_xTe$ ($0 \leq x \leq 0.40$)

$$n_i = \left(8.92 - 34.46x + 2.25\times10^{-3}T + 4.12\times10^{-2}xT + 97.00x^2\right)\times10^{14} E_g^{3/4}T^{3/2}\exp\left(-\frac{E_g}{2kT}\right), \tag{18.5}$$

Table 18.1 Physical properties of lead salts (after Refs. [13], [15], and [16])

	T (K)	PbTe	$Pb_{0.8}Sn_{0.2}Te$	PbSe	$Pb_{0.93}Sn_{0.7}Se$	PbS
LATTICE STRUCTURE		Cubic (NaCl)	Cubic (NaCl)	Cubic (NaCl)	Cubic (NaCl)	Cubic (NaCl)
Lattice constant a (nm)	300	0.6460	0.64321	0.61265	0.6118	0.59356
Thermal expansion coefficient α (10^{-6} K^{-1})	300	19.8	20	19.4		20.3
	77	15.9		16.0		
Heat capacity C_p (J mol^{-1} K^{-1})	300	50.7		50.3		47.8
Density γ (g/cm^3)	300	8.242	7.91	8.274		7.596
Melting point T_m (K)		1,197	1,168 (sol.)	1,354	1,325 (sol.)	1,400
			1,178 (liq.)		1,340 (liq.)	
Bandgap E_g (eV)	300	0.31	0.20	0.28	0.21	0.42
	77	0.22	0.11	0.17	0.10	0.31
	4.2	0.19	0.08	0.15	0.08	0.29
Thermal coefficient of E_g (10^{-4} eVK^{-1})	80–300	4.2	4.2	4.5	4.5	4.5
Effective masses						
m^*_{et}/m	4.2	0.022	0.011	0.040	0.037	0.080
m^*_{ht}/m		0.025	0.012	0.034	0.021	0.075
m^*_{el}/m		0.19	0.11	0.070	0.041	0.105
m^*_{hl}/m		0.24	0.13	0.068	0.040	0.105
Mobilities						
μ_e (cm^2/Vs)	77	3 × 104	3 × 104	3 × 104	3 × 104	1.5 × 104
μ_h (cm^2/Vs)		2 × 104	2 × 104	3 × 104	2 × 104	1.5 × 104
Intrinsic carrier concentration n_i (cm^{-3})	77	1.5×10^{10}	3×10^{13}	6×10^{11}	8×10^{13}	3×10^{7}
Static dielectric constant ε_s	300	380		206		172
	77	428		227		184
High frequency dielectric constant ε_∞	300	32.8	38	22.9	26.0	17.2
	77	36.9	42	25.2	30.9	18.4
Optical phonons						
LO (cm^{-1})	300	114	120	133		212
TO (cm^{-1})	77	32		44		67

for $Pb_{1-x}Sn_xSe$ $(0 \le x \le 0.12)$

$$n_i = \left(1.73 - 3.68x + 3.77\times10^{-4}T + 1.60\times10^{-2}xT + 8.92x^2\right)\times10^{15} E_g^{3/4} T^{3/2} \exp\left(-\frac{E_g}{2kT}\right), \quad (18.6)$$

and for $PbS_{1-x}Se_x$ $(0 \le x \le 1)$

$$n_i = \left(2.14 - 8.85\times10^{-1}x + 6.12\times10^{-4}T + 6.47\times10^{-4}xT + 3.32\times10^{-1}x^2\right)\times10^{15} E_g^{3/4} T^{3/2} \exp\left(-\frac{E_g}{2kT}\right). \quad (18.7)$$

A large number of experimental and theoretical works have been directed toward the elucidation of the dominant scattering mechanisms in Pb salts [17,49–52]. As a consequence of the similar valence and conduction bands of Pb salts, the electron and hole mobilities are approximately equal for the same temperatures and doping concentrations. Room temperature mobilities in Pb salts are 500–2,000 cm^2/Vs [17]. In many high-quality single crystal samples, the mobility due to lattice scattering varies as $T^{-5/2}$ [14]. This behavior has been ascribed to a combination of polar-optical and acoustical lattice scattering and achieves the limiting values in the range of 10^5–10^6 cm^2/Vs due to defect scattering (see Figure 18.6) [53]. The scattering mechanism considerations for the binary alloys are also applicable to the mixed crystals. However, the importance of the acoustical phonon scattering is diminished for the small-gap materials, since the mobility due to this type of mechanism is proportional to the reciprocal of the density of states. In these materials, scattering by impurity atoms and vacancies is important even at room temperature. Additionally, the lack of chemical order in the mixed crystals results in relatively strong disorder scattering of electrons in samples with comparatively low carrier concentrations at low temperatures. This type of scattering is dominant in PbSnTe with carrier concentrations below approximately 10^{18} cm^{-3} at 4.2 K. Figure 18.7 shows the carrier concentration dependence of the electron mobility for $Pb_{1-x}Sn_xTe$ $(0.17 \le x \le 0.20)$ at 77 K [52]. In the concentration range $\ge 10^{18}$ cm^{-3}, the nonelastic scattering on the impurity and vacancy potentials is decisive.

The interband absorption of the Pb salts is more complicated as compared with the standard case due to anisotropic multivalley structure of both conduction and valence bands, nonparabolic Kane-type energy

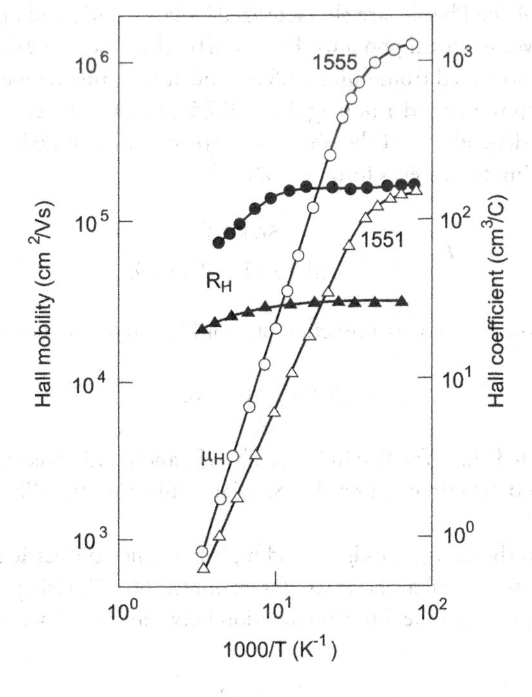

Figure 18.6 Temperature dependence of the Hall mobility and the Hall coefficient for two PbTe epitaxial layers on BaF_2 substrates. (From Ref. [53])

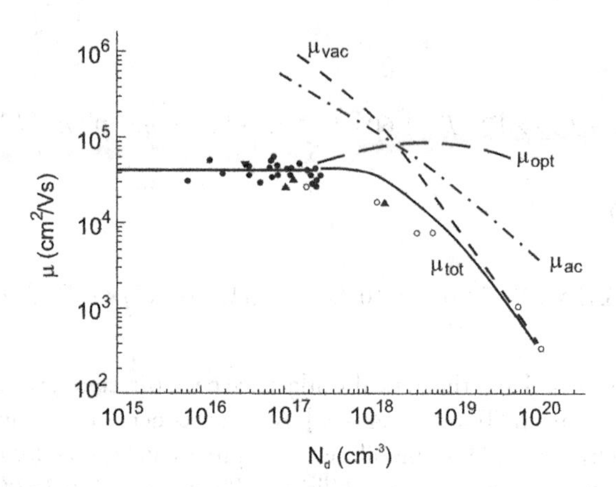

Figure 18.7 Carrier concentration dependence of the electron mobility for $Pb_{1-x}Sn_xTe$ ($0.17 \leq x \leq 0.20$) at 77 K. The curves are calculated for scattering of carriers by the longitudinal optical phonons (μ_{opt}), the acoustical phonons (μ_{ac}), and the vacancy potential (μ_{vac}). (From Ref. [52])

dispersion, and **k**-dependent matrix elements. Analytical expressions for the absorption coefficients of energies near the absorption edge have been given by several workers [54–58]. The temperature and composition dependence of the absorption coefficient near the absorption edge are well described by the two-band model for which the following expression has been obtained [55]:

$$\alpha(z) = \frac{2q^2\left(m_t^{*2}m_l^*\right)^{1/2}}{\varepsilon_o n_r c\pi\hbar^4 E_g^{1/2}} \frac{2P_t^2 + P_l^2}{3} \frac{(z-1)^{1/2}}{z} \frac{(z+1)^{1/2}}{\sqrt{2}} \frac{\left(1+2z^2\right)}{3z}\left(f_v - f_c\right), \tag{18.8}$$

where $z = h\nu/E_g$, P_t and P_l are the transverse and longitudinal momentum matrix elements, n_r is refractive index, and the factor $(f_v - f_c)$ describes the band filling and nearly equals unity in the nondegenerate case. The formula for the Burstein–Moss factor is given by Anderson [54].

The dielectric properties of the Pb salts are characterized by large static and optical dielectric constants and low frequencies of transverse optical phonons. For PbSnTe, the observed values of the static dielectric constant have been widely distributed from 400 to 5,800, and at the same temperature these values have been scattered in the range up to one order of magnitude [17,59,60]. More recently, Butenko et al. [61] have determined the temperature dependence of the static dielectric constant of PbTe in a wide temperature range of 10 K–300 K according to Barrett's formula [62]

$$\varepsilon_o = \frac{1.356\times 10^5}{36.14\coth(36.14/T) + 49.15}. \tag{18.9}$$

The high frequency dielectric constant as a function of x for $Pb_{1-x}Sn_xTe$ can be described by the relation [59]:

$$\varepsilon_\infty = 27.4 + 22.0x - 6.4x^2. \tag{18.10}$$

Shani et al. [63] have calculated the refractive index of PbSnTe and good agreements have been obtained between their calculations and experimental results. Similar results for other Pb salts have been presented by Jensen and Tarabi [64,65].

The interrelations between the energy bandgaps and high-frequency dielectric constant and the refractive index have been summarized by several authors; see, for example, [66,67]. Using a classical oscillator theory, Herve and Vandamme have proposed the following relation between the refractive index and energy gap:

$$n^2 = 1 + \left(\frac{13.6}{E_g + 3.4}\right)^2. \tag{18.11}$$

This relation is accurate for most of the compounds used in optoelectronics structures and for wide-gap semiconductors, but it cannot properly describe the behavior of the IV–VI group. For Pb salts, better agreement between experimental and theoretical results can be obtained using the model of Wemlple and DiDomenico [66].

18.1.4 GENERATION-RECOMBINATION PROCESSES

Although direct one-phonon recombination as well as plasmon recombination have been proposed theoretically [68,69] and identified experimentally [70,71], for very small energy gaps of IV–VI compounds, in samples with gaps of 0.1 eV or more, SRH (Shockley-Read-Hall), radiative, and Auger recombinations are dominant [72–75].

Ziep et al. [68] have calculated the lifetime determined by radiative recombination in terms of the Kane-type two-band model case and Boltzmann statistics. However, it appears that in case of mirror-symmetric band structure of Pb salts $\left(m_e^* \approx m_h^*\right)$, a good approximation for the recombination rate is the following equation [6]:

$$G_R = \frac{10^{-15} n_r E_g^2 n_i^2}{(kT)^{3/2} K^{1/2} (2+1/K)^{3/2} (m^*/m)^{5/2}} \text{ in cm}^3/\text{s}, \quad (18.12)$$

where $K = m_l^*/m_t^*$ is the effective mass anisotropy coefficient. The mass m^* can be determined if we know the longitudinal m_l^* and transverse m_t^* components of the effective mass, since $m^* = \left[1/3\left(2/m_t^* + 1/m_l^*\right)\right]^{-1}$. In Equation 18.12, the values of kT and E_g should be expressed in electron volts.

For a long time, the Auger process was considered to be a low-efficiency channel for nonradiative recombination in semiconductors of the IV–VI type. The valence band and the conduction band with mirror-reflection symmetry occur at point L of the Brillouin zone (number of valley w=4). In such a case, the energy and momentum conservation laws are difficult to fulfill for impact recombination, especially for carriers near the band edges when only single-valley interaction is taken into account. Since the pioneering Emtage's paper was published in 1976 [76] many theoretical and experimental works have been published [77–85] in which the intervalley interaction of carriers is considered, and it has been found that even at lower temperatures the lifetime of carriers is determined by impact recombination. According to the intervalley carrier interaction model (Figure 18.8), an electron and hole from valley (a), characterized by heavy mass m_l^* and a third carrier from valley (b) with a light mass m_t^* (in PbSnTe mass anisotropy coefficient $K > 10$) participate in impact recombination in the given direction. As a result of this interaction, the heavy electron and hole carriers recombine, and the liberated energy and momentum are transferred to the light carrier.

Two cases should be considered:

- All scattered carriers are at a definite point of the Brillouin zone.
- The initial carriers are in different valleys of the band.

In case of Boltzmann statistics [69],

$$\left(\tau_A^j\right)^{-1} = C_A^j\left(n_o^2 + 2n_o p_o\right), \quad (18.13)$$

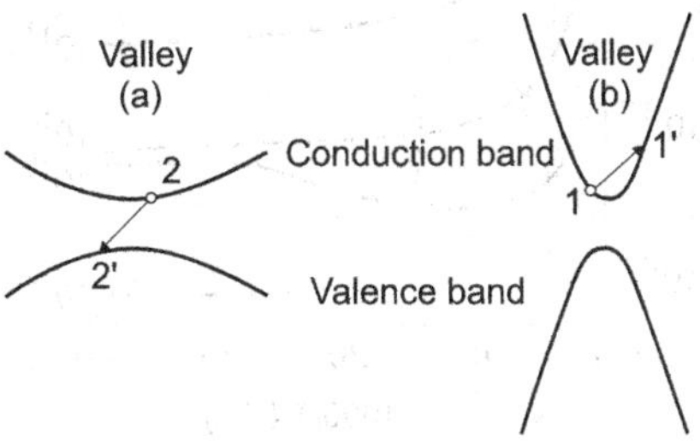

Figure 18.8 Intervalley Auger recombination. (a) An electron and hole from valley, characterized by heavy mass m_l^* and (b) a third carrier from valley with a light mass m_t^*.

where C_A^j is the Auger coefficient for j-th recombination mechanism. Then,

$$\tau_A^{-1} = \left(\tau_A^a\right)^{-1} + \left(\tau_A^b\right)^{-1}. \tag{18.14}$$

Different approximations for C_A, especially for the process (b), can be found in the literature [76,81,82,86–88]. For the intervalley process proposed by Emtage within the parabolic band model [76],

$$C_A = (2\pi)^{5/2}\frac{w-1}{w^2}\frac{q^2}{\left(4\pi\varepsilon_o\varepsilon_\infty\right)^2}(kT)^{1/2}E_g^{-7/2}\frac{\hbar^3}{m_l^{*1/3}m_t^{*2/3}}\exp\left(-\frac{E_gK^{-1}}{2kT}\right). \tag{18.15}$$

Nonparabolic bands of the Kane type are expected to reduce the Auger transition rate [80,82]. However, the Emtage expression is a good approximation of more exactly calculated Auger coefficients [81].

Ziep et al. [86] carried out a comprehensive examination of nonradiative and radiative recombination mechanisms in the mixed crystals $Pb_{0.78}Sn_{0.22}Te$ and $Pb_{0.91}Sn_{0.09}Se$ for a temperature interval $20 < T <$ 400 K and in a doping range $0 < N_d < 10^{19}$ cm^{-3}. In the calculations, the degeneracy of the carrier gas, the anisotropy, and to some extent the nonparabolicity of the band structure were taken into account. Their calculated data for $Pb_{0.78}Sn_{0.22}Te$ are shown in Figure 18.9. The radiative lifetime as well as the Auger lifetime have a maximum in the transition from the extrinsic to the intrinsic region. At low temperatures in the range of low dopant concentrations, the lifetime is determined by radiative recombination. As the temperature increases, the Auger recombination comes to the fore and determines the carrier lifetimes at room temperature. At comparable energy gaps, the radiative lifetimes differ only slightly in PbSnTe and PbSnSe compounds. However, due to the smaller anisotropy of the isoenergetic surface in PbSnSe as compared with PbSnTe, the Auger lifetime is higher in PbSnSe than in PbSnTe. At carrier concentrations above 10^{19} cm^{-3}, the plasmon recombination dominates over radiative and Auger recombinations.

Dmitriev [87] has calculated the Auger carrier lifetime in PbSnTe, taking into account the carrier degeneracy and the exact expressions for the overlap integrals [88]. The calculated lifetimes are greater than those obtained previously.

Experimental investigations of Pb salts confirm that the carrier lifetime is determined by band-to-band recombination as well as by any SRH recombination. The results of the first extensive investigation of $Pb_{1-x}Sn_xTe$ ($0.17 \le x \le 0.20$) revealed an extremely large variation of the lifetime ($10^{-12} < \tau < 10^{-8}$ s at 77 K)

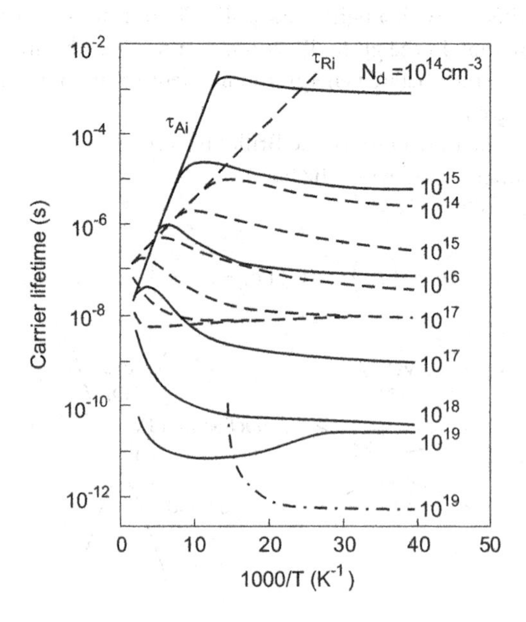

Figure 18.9 Calculated lifetime versus temperature for Auger (———), radiative (– – –), and plasmon (– · – · –) recombinations in $Pb_{0.78}Sn_{0.22}Te$. (From Ref. [86])

measured from photoconductivity and photoelectromagnetic effect (PEM) effects (see Figure 18.10) [74,79,89]. The highest observed lifetime fits quite well to the straight line calculated according to Emtage's theory of Auger recombination [86]. All Cd- and In-doped samples are characterized by considerably lower lifetimes in spite of their lower free-carrier concentrations. Donor levels with ionization energies 12–25 meV are presumed to work as recombination centers.

In case of PbTe epitaxial layers on BaF_2 substrates, a good agreement with the Emtage theory has been obtained by Lischka and Huber [78]. However, these authors observed a second lifetime branch with a much longer lifetime than that for Auger recombination (Figure 18.11). This branch was attributed to deep levels acting as minority carrier traps at intermediate temperatures [73]. Zogg et al. [83] observed up to four lifetime branches determined by the transient photoconductivity method in n- and p-type PbSe epitaxial layers. The shortest lifetimes were in agreement with calculated values for direct (Auger and radiative) recombination below ≈250 K for samples with carrier concentrations $\leq 2\times10^{17}$ cm^{-3}. From the longer observed lifetime branches, three impurity levels (separated between 20 and 50 meV from the nearer band edge), acting principally as minority carrier traps at intermediate temperature, were calculated. Also, Shahar et al. [84] confirmed that the carrier lifetime in undoped $Pb_{0.8}Sn_{0.2}Te$ layers grown by LPE is determined by band-to-band radiative and Auger recombination mechanisms in the temperature range 10 K–110 K, while in In-doped PbTe layers, recombination takes place via nonradiative centers.

Two other groups, Schlicht et al. [90] and Weiser et al. [91], have not obtained agreement between experimental data on PbSnTe epitaxial layers on (111) BaF_2 substrates and the calculated Auger lifetime. Measurements of the photoconductive decay time yield values that at 77 K are longer than predicted by the Emtage theory. Also, the dependence of lifetime on carrier density in the extrinsic range of conductivity was not observed. Weiser et al. [91] suggest that the lifetime enhancement by about two orders of magnitude at $T\approx 14$ K is the result of photon recycling and explains the lifetime enhancement by about one order of magnitude at 77 K by overestimation of the recombination rate within the Emtage theory or by the fact that strains in the sample reduce the importance of the multivalley Auger process. Genzow et al. [92] theoretically investigated the influence of uniaxial strain on the Auger and radiative recombinations in PbSnTe and obtained quite good agreement of their theoretical results with the experimental data given by Weiser et al. [91].

We conclude that the mechanism of Auger recombination in small-gap IV–VI semiconductors is expected to be important but is not fully clarified. Experimental evidence for this mechanism is not yet unambiguously presented. Some disagreement between the experimental results reported by the various

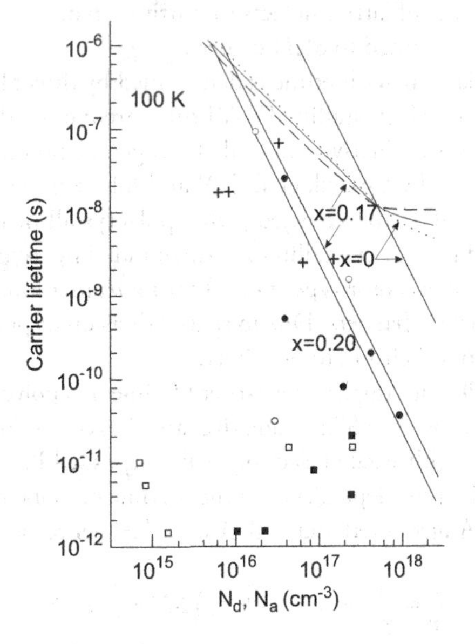

Figure 18.10 PEM lifetime (T=100 K) versus doping level for $Pb_{1-x}Sn_xTe$ ($0.17\leq x\leq 0.20$) samples of different origins: Bridgman-annealed (•), THM (+), vapor-phase (○), uncompensated (□), and Cd-compensated (■). The full lines are calculated curves for Auger recombination, the dash-dotted lines for radiative recombination. (From Ref. [74])

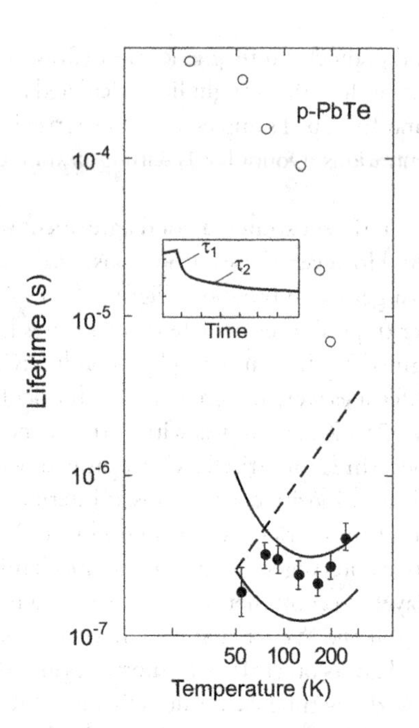

Figure 18.11 Carrier lifetime versus temperature for p-PbTe: experimental values of lifetime τ_1 (•; $p=9.5\times10^{16}$ cm^{-3}), solid curves are calculated for Auger recombination assuming m_l^*/m_t^* (the upper curve)=10 and $m_l^*/m_t^* = 14$ (the lower curve); experimental data of a second time constant τ_2 occurring in the decay of the photocurrent pulse (O). (From Ref. [78])

authors may be related to differences in sample preparation as well as to differences in the experimental methods used to measure the lifetime. Also, these discrepancies are assumed to be due to the lack of accurate band parameters for Pb salts and to the theoretical description of screening in the Auger process [72]. In spite of these differences some conclusions may be drawn:

- With decreasing energy gaps and increasing temperature, impact recombination becomes more important.
- In samples having a large number of lattice defects (impurities, native defects, misfit dislocations), the recombination mechanism is attributed to SRH centers.
- In defect-free undoped material, carrier lifetime is determined by direct band-to-band recombination.

The recombination processes in polycrystalline IV–VI films are modified by grain boundary potential barriers. Different theories of photoconductivity in lead chalcogenide polycrystalline layers have been proposed. Their brief review is given by Espevik et al. [93] and Johnson [94]. Neustroev and Osipov [95] developed a theoretical model based on the knowledge that polycrystalline materials are a two-phase system. Low-ohmic, n-type conductance crystallites are surrounded by oxygen-saturated p-type inversion layers with a large concentration of acceptor-type states. During irradiation, electrons and holes generating in crystallites are separated by surface barriers. Due to spatial division of photocarriers, their lifetime rises abruptly and, consequently, so does their photosensitivity.

Similarly, as in single-crystal Pb salt samples, the carrier lifetime in a polycrystalline material is determined by three principal recombination processes: SRH, radiative, and Auger. Vaitkus et al. [96] have investigated picosecond photoconductivity in highly excited electrolytically deposited PbS and vacuum-evaporated PbTe polycrystalline films. They described the dependence of the lifetime on nonequilibrium concentrations, taking into account linear, trap-assisted Auger, interband radiative, and Auger recombinations [97].

$$\frac{1}{\tau} = \frac{1}{\tau_R} + \left(\gamma_{At} N_t + \gamma_R\right)\Delta N + \gamma_A \Delta N^2, \tag{18.16}$$

where γ_{At} is the trap Auger coefficient, N_t is the trap density, γ_R is the radiative recombination coefficient, and γ_A is the interband Auger recombination coefficient. Good agreement of the theoretical curves, based on Equation 18.16, with experimental points in all the investigated films (freshly made and annealed) has been obtained, assuming $\gamma_{At}N_t+\gamma_R=2.2\times10^{-11}$ cm^3/s and $\gamma_A=5.3\times10^{-29}$ cm^6/s. Interband Auger recombination in the grain bulk is the main mechanism of carriers. The structure and the film preparation process influence only the slower recombination.

18.2 POLYCRYSTALLINE PHOTOCONDUCTIVE DETECTORS

A number of Pb salt photoconductive reviews have been published [22,23,94,98–108]. One of the best reviews of development efforts in Pb salt detectors was published by Johnson [94].

18.2.1 DEPOSITION OF POLYCRYSTALLINE LEAD SALTS

Although the fabrication methods developed for these photoconductors are not completely understood, their properties are well established. Unlike most other semiconductor IR detectors, Pb salt photoconductive materials are used in the form of polycrystalline films approximately 1 μm thick and with individual crystallites ranging in size from approximately 0.1 μm to 1.0 μm. They are usually prepared by chemical deposition using empirical recipes, which generally yields better uniformity of response and more stable results than the evaporative methods [101–105].

The PbSe and PbS films used in commercial IR detectors are made by chemical bath deposition (CBD); the oldest and most-studied PbSe and PbS thin-film deposition method. It was used to deposit PbS in 1910 [108]. The basis of CBD is a precipitation reaction between a slowly produced anion (S^{2-} or Se^{2-}) and a complexed metal cation. The commonly used precursors are Pb salts, $Pb(CH_3COO)_2$ or $Pb(NO_3)_2$, thiourea [$(NH_2)_2CS$] for PbS, and selenourea [$(NH_2)_2CSe$] for PbSe, all in alkaline solutions. Pb may be complexed with citrate, ammonia, triethanolnamine, or with selenosulfate itself. Most often, however, the deposition is carried out in a highly alkaline solution where OH^- acts as the complexing agent for Pb^{2+}.

In CBD, the film is formed when the product of the concentrations of the free ions is larger than the solubility product of the compound. Thus, CBD demands very strict control over the reaction temperature, pH, and precursor concentrations. In addition, the thickness of the film is limited, the terminal thickness usually being 300–500 nm. Therefore, in order to get a film with a sufficient thickness (approximately 1 μm in IR detectors, for example), several successive depositions must be done. The benefit of CBD compared to gas phase techniques is that CBD is a low-cost temperature method and the substrate may be temperature sensitive with the various shapes.

As-deposited PbS films exhibit significant photoconductivity. However, a postdeposition baking process is used to achieve final sensitization. In order to obtain high-performance detectors, lead chalcogenide films need to be sensitized by oxidation. The oxidation may be carried out by using additives in the deposition bath by postdeposition heat treatment in the presence of oxygen or by chemical oxidation of the film. The effect of the oxidant is to introduce sensitizing centers and additional states into the bandgap and thereby increase the lifetime of the photoexcited holes in the p-type material.

The baking process changes the initial n-type films to p-type films and optimizes performance through the manipulation of resistance. The best material is obtained using a specific level of oxygen and a specific bake time. Only a small percentage (3%–9%) of oxygen influences the absorption properties and response of the detector. Temperatures ranging from 100°C to 120°C and time periods from a few hours to in excess of 24 hours are commonly employed to achieve final detector performance optimized for a particular application. Other impurities added to the chemical deposition solution for PbS have a considerable effect on the photosensitivity characteristics of the films [105]. The $SbCl_2$, $SbCl_3$, and As_2O_3 prolong the induction period and increase the photosensitivity by up to ten times that of films prepared without these impurities. The increase is thought to be caused by the increased absorption of CO_2 during the prolonged induction period. This increases $PbCO_3$ formation and thus photosensitivity. Arsine sulfide also changes the oxidation states on the surface. Moreover, it has been found that essentially the same performance characteristics can be achieved by baking in an air or a nitrogen atmosphere. Therefore, all of the constituents necessary for sensitization are contained in the raw PbS films as deposited.

The preparation of PbSe photoconductors is similar to the PbS ones. The postdeposition baking process for PbSe detectors operating at 77 K is carried out at a higher temperature (> 400°C) in an oxygen atmosphere. However, for detectors to be used at ambient and/or intermediate temperatures, the oxygen or air bake is immediately followed by baking in a halogen gas atmosphere at temperatures in the range of 300°C–400°C [94]. According to Torquemada and colleagues [109], I plays a key role in the sensitization of the PbSe layers obtained by thermal evaporation in a vacuum on thermally oxidized Si. The halogen behaves as a transport agent during the PbSe recrystallization process, and promotes the fast growth of PbSe microcrystals. Oxygen is trapped in the PbSe lattice during the recrystallization process, as it happens in chemically deposited PbSe films. The introduction of halogens in the PbSe sensitization procedure is a highly efficient technique for the incorporation of oxygen to the semiconductor lattice in electrically active positions. If halogens are not introduced during the PbSe sensitization, the oxygen is incorporated inside the lattice of microcrystals only by diffusion, which is a less efficient way.

A variety of materials can be used as substrates, but the best detector performance is achieved using single-crystal quartz material. PbSe detectors are often matched with Si to obtain higher collection efficiency.

Photoconductors also have been fabricated from epitaxial layers without baking that resulted in devices with uniform sensitivity, uniform response time, and no aging effects. However, these devices do not offset the increased difficulty and cost of fabrication.

18.2.2 FABRICATION

The films are deposited either over or under plated Au electrodes and on fused quartz, crystal quartz, single crystal sapphire, glass, various ceramics, single-crystal strontium titanate, Irtran II (ZnS), Si, and Ge. The most commonly used substrate materials are fused quartz for ambient operation and single-crystal sapphire for detectors used at temperatures below 230 K. The very low thermal expansion coefficient of fused quartz relative to PbS films results in poorer detector performance at lower operating temperatures. Different shapes of substrates are used: flat, cylindrical, or spherical. To obtain higher collection efficiency, detectors may be deposited directly by immersion into optical materials with high indices of refraction (e.g., into strontium titanate). Pb salts cannot be immersed directly; special optical cements must be used between the film and the optical element.

As was mentioned above, in order to obtain high-performance detectors, lead chalcogenide films must be sensitized by oxidation, which may be carried out by using additives in the deposition bath, by postdeposition heat treatment in the presence of oxygen, or by chemical oxidation of the film. Unfortunately, in the older literature, the additives are seldom identified and are often referred to only as an ocidant [93,103]. A more recent paper deals with the effects of H_2O_2 and $K_2S_2O_8$, both in the deposition bath and in the postdeposition treatment [110]. It was found that both treatments increase the resistivity of PbS films. Although the resistivity usually increases during the oxidation, a different behavior also has been observed [111]. A sensitized PbS film may significantly degrade in air without an overcoating. Possible overcoating materials are As_2S_3, CdTe, ZnSe, Al_2O_3, MgF_2, and SiO_2. Vacuum-deposited As_2S_3 has been found to have the best optical, thermal, and mechanical properties, and it has improved the detector performance. The drawback of the As_2S_3 coating is the toxicity of As and its precursors. Overall, however, the electric properties (resistivity, and in particular detectivity) of Pb salt thin films are rather poorly reported, although there are many papers on PbS and PbSe film growth. The effects of annealing and oxidation treatments on detectivity have not been reported accurately either.

To explain the photoconductivity process in thin-film Pb salt detectors, three theories have been proposed [104]. The first consists of increases in carrier density during illumination; oxygen is assumed to introduce a trapping state that inhibits recombination. The second mechanism is based primarily on the increase in the mobility of free carriers in the barrier model. It is assumed that potential barriers are formed during sensitization by heating in oxygen, either between the crystallites of the film or between n- and p-regions in nonuniform films. The third model (commonly referred to as the generalized theory of photoconductivity in semiconducting films in general and for the Pb salts in particular) was proposed by Petritz. In the review paper [104], Bode concluded that the Petritz theory provided a reasonable framework for general use, even though the complex mechanisms in Pb salts were still unsolved. More recently, Espevik et al. [93] provided significant additional support for the Petritz theory.

The PbS and PbSe materials are peculiar because they have relatively long response times that affect the significant photoconductive gain. It has been suggested that during the sensitization process the films are oxidized, converting the outer surfaces of exposed PbS and PbSe films to PbO or a mixture of $PbO_xS(Se)_{1-x}$, and forming a heterojunction at the surface (which is shown schematically in Figure 18.12) [112]. Oxide heterointerfaces create conditions for trapping minority carriers or separating majority carriers and thereby extends the lifetime of the material. As was mentioned above, without the sensitization (oxidation) step, Pb salt materials have very short lifetimes and low responses.

The formation of p-n junction during the sensitization process is preferred in Ref. [113], where the charge-separation-junction (CSJ) model has been introduced. According to this model, during the sensitization process of p-type polycrystalline PbSe, a thin outer layer of crystallite is converted into n-type due to iodination. Due to the small sizes of the crystallites in the range of 0.2–0.5 μm, all p-type cores and n-type outer layers are charged due to majority carrier diffusion and the photon-induced electrons and holes are spatially separated due to the built-in potential in such charged region. When bias voltage is applied, holes mainly travel in the p-type cores through all the connection channels and electrons travel in the n-type outer layer. Because of such spatial carrier separation, the minority photon-induced carrier concentration is significantly reduced and, in consequence, the recombination mechanisms are significantly inhibited, leading to much enhanced lifetime and thus photoresponse.

The basic detector fabrication steps are electrode deposition and delineation, active-layer deposition and delineation, passivation overcoating, mounting (cover/window), and Pb wire attachment. Figure 18.13 shows a cross section of a typical PbS detector structure [94]. The entire device is overcoated for environmental protection and sensitivity improvement. Standard packing consists of electrode deposition and placement in a metal case with a window over the top. Wire leads are placed in a groove in a substrate and Au electrodes are usually vacuum-evaporated onto the film using a mask.

Photolithographic delineation methods are used for complex, high-density patterns of small element sizes. The outer electrodes are produced with vacuum deposition of bimetallic films such as TiAu. To passivate and optimize transmission of radiation into the detector, usually a quarter-wavelength thick

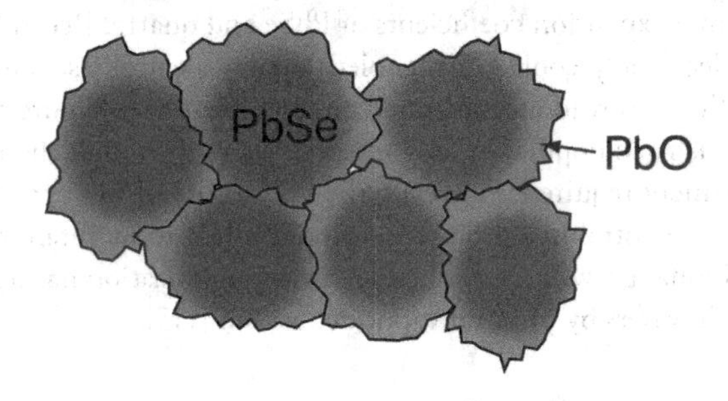

Figure 18.12 PbSe polycrystalline films after the sensitization process. The films are coated with PbO, forming a heterojunction on the surface. (From Ref. [112])

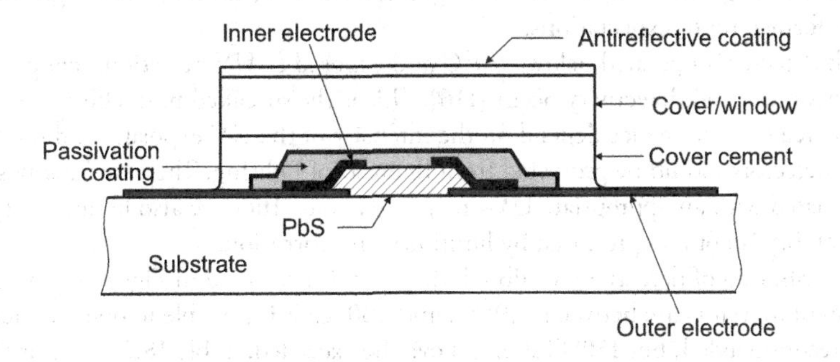

Figure 18.13 Typical structural configuration of PbS detector. (Reprinted from Ref. [94])

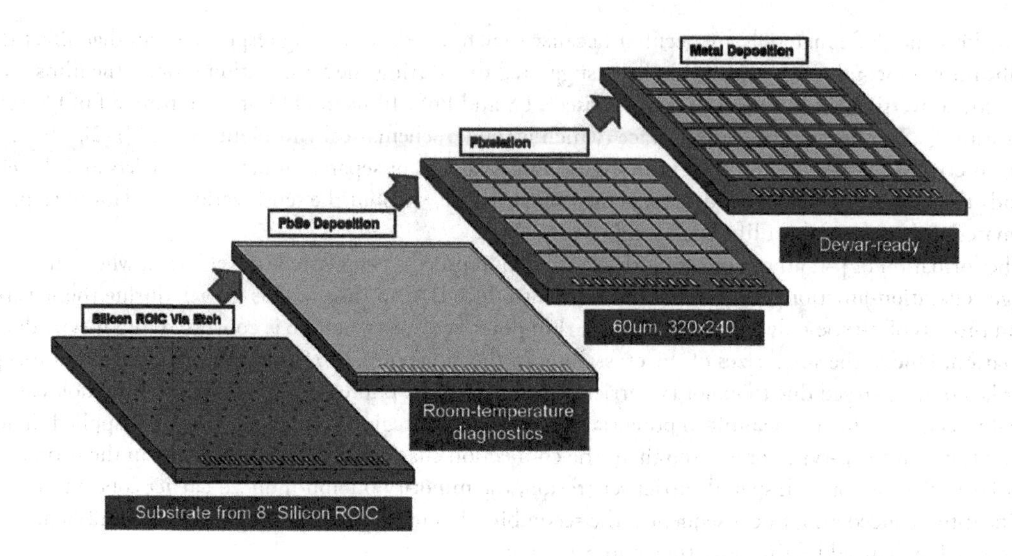

Figure 18.14 High-level conceptual drawing of Pb salt direct deposition FPA fabrication. (after Ref. [114])

overcoating of As_2S_3 is used. Normally, detectors are sealed between a cover plate and the substrate with epoxy cement. The cover plate material is ordinarily quartz, but other materials such as sapphire may be used to transmit longer wavelengths. This technique seals the detector reasonably well against humid environments. The top surface of the cover is normally made antireflective with a material such as MgF_2.

As was mentioned previously, traditionally, high-sensitivity PbSe is fabricated on a quartz substrate, and most commercially available detectors are fabricated on quartz substrates. This is primarily due to the relatively well-matched thermal expansion coefficients of PbSe and quartz. Recently, progress in the development of lead-salt thermoelectrically-cooled (TE-cooled) imaging sensors is connected with their architecture integrated monolithically on Si readout integrated circuit (ROIC; see Figure 18.14) [114]. The method of wet chemical deposition has been popular among research groups through the 1980s due to the low start-up infrastructure investment required. However, the uniformity of wet chemically-deposited material has required additional process control in a production environment to maintain run-to-run reproducibility. Aside from this traditional growth method, at least one organization has demonstrated high-purity PbSe material deposited on Si wafers by physical vapor deposition [115].

18.2.3 PERFORMANCE

Standard active area sizes are typically 1, 2, or 3 mm squares. However, most manufacturers offer sizes ranging from 0.08×0.08 to 10×10 mm^2 for PbSe detectors, and 0.025×0.025 to 10×10 mm^2 for PbS detectors. Active areas are generally square or rectangular; detectors with more exotic geometries have sometimes not performed up to expectations.

When a Pb salt detector is operated below –20°C and exposed to UV radiation, semipermanent changes in responsivity, resistance, and detectivity occur [116]. This is the so-called flash effect. The amount of change and the degree of permanence depend on the intensity of the UV exposure and the length of exposure time. Pb salt detectors should be protected from fluorescent lighting. They are usually stored in a dark enclosure or overcoated with an appropriate UV-opaque material. They are also hermetically sealed so that their long-term stability is not compromised by humidity and corrosion.

The spectral distribution of detectivity of Pb salt detectors is presented in Figure 18.15. Usually the operating temperatures of detectors are between –196°C and 100°C; it is possible to operate them at a temperature higher than recommended, but 150°C should never be exceeded. Table 18.2 contains the performance range of detectors fabricated by various manufacturers [106].

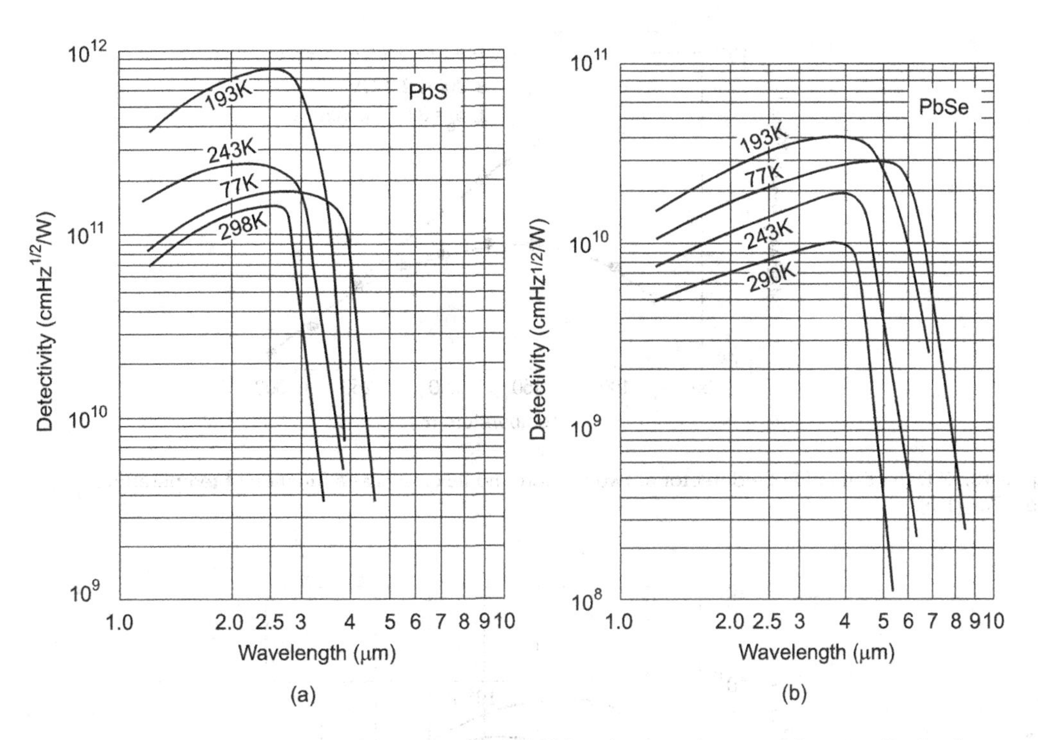

Figure 18.15 Typical spectral detectivity for (a) PbS and (b) PbSe photoconductors. (After New England Photoconductor data sheet, www.nepcorp.com)

Table 18.2 Performance of lead salt detectors (2π FOV, 300 K background) (after Ref. [106])

	T (K)	SPECTRAL RESPONSE (μm)	λ_p (μm)	D^* (λ_p, 1,000 Hz, 1) (cmHz$^{1/2}$ W^{-1})	$R/\square$ (MΩ)	τ (μs)
PbS	298	1–3	2.5	(0.1–1.5) × 10^{11}	0.1–10	30–1,000
	243	1–3.2	2.7	(0.3–3) × 10^{11}	0.2–35	75–3,000
	195	1–4	2.9	(1–3.5) × 10^{11}	0.4–100	100–10,000
	77	1–4.5	3.4	(0.5–2.5) × 10^{11}	1–1,000	500–50,000
PbSe	298	1–4.8	4.3	(0.05–0.8) × 10^{10}	0.05–20	0.5–10
	243	1–5	4.5	(0.15–3) × 10^{10}	0.25–120	5–60
	195	1–5.6	4.7	(0.8–6) × 10^{10}	0.4–150	10–100
	77	1–7	5.2	(0.7–5) × 10^{10}	0.5–200	15–150

Below 230 K, background radiation begins to limit the detectivity of PbS detectors. This effect becomes more pronounced at 77 K, and peak detectivity is no greater than the value obtained at 193 K. Figure 18.16 shows typical peak detectivity values for the PbSe detector [116]. Depending on operating temperature, background flux, and chemical additives, the detector impedance per square can be adjusted in the range of 10^6–10^9 Ω/□. Quantum efficiency of approximately 30% is limited by incomplete absorption of the incident flux in the relatively thin (1–2 μm) detector material. The responsivity uniformity of PbS and PbSe detectors is generally of the order of 3%–10%.

The typical frequency response of PbS detectors is shown in Figure 18.17 [117]. An optimum operating frequency arises from the combined effects of the $1/f$ noise at low frequencies and the high-frequency roll-off in responsivity.

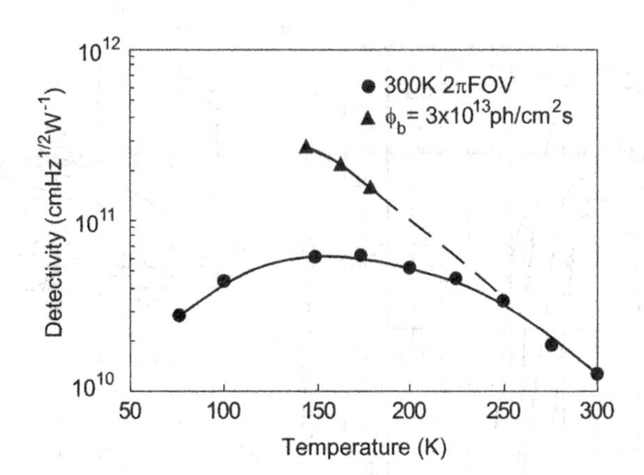

Figure 18.16 Detectivity of PbSe detector at two background flux levels as a function of temperature. (From Ref. [116])

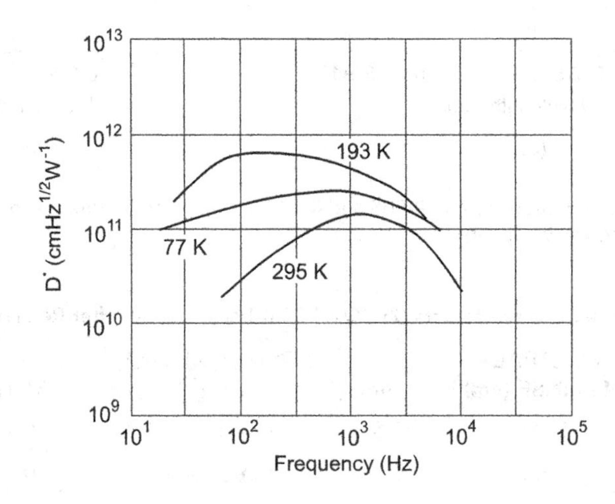

Figure 18.17 Peak spectral detectivity as a function of frequency for a PbS photoconductive detector at 295 K, 193 K, and 77 K operating temperatures. (From Ref. [117])

18.3 P-N JUNCTION PHOTODIODES

In comparison with photoconductive detectors, Pb salt photodiodes have not found wide commercial applications. As was mentioned previously, development of PbSnTe and PbSnSe FPAs ternary was nearly entirely stopped in the early 1980s in favor of HgCdTe. However, through advancements in growth techniques, especially using MBE growth, vast improvements in device performance are realizable. One such innovation is microelectromechanical system (MEMS)-based tunable IR detector to deliver voltage-tunable multiband IR FPA [118,119], also called adaptive FPA.

In this section, we begin with a survey of fundamental performance limits of lead salt chalcogenide photovoltaic detectors and go on to describe in more detail the technologies and properties of different types of these photodiodes.

18.3.1 PERFORMANCE LIMIT

Considerations carried out in this section are proper for an one-sided abrupt junction model. Surface leakage effects will not be considered since they may be minimized by appropriate surface treatment or the use of a guard ring structure. No distinction needs to be made between n^+-on-p and p^+-on-n structures owing to the mirror symmetry of valence and conduction bands.

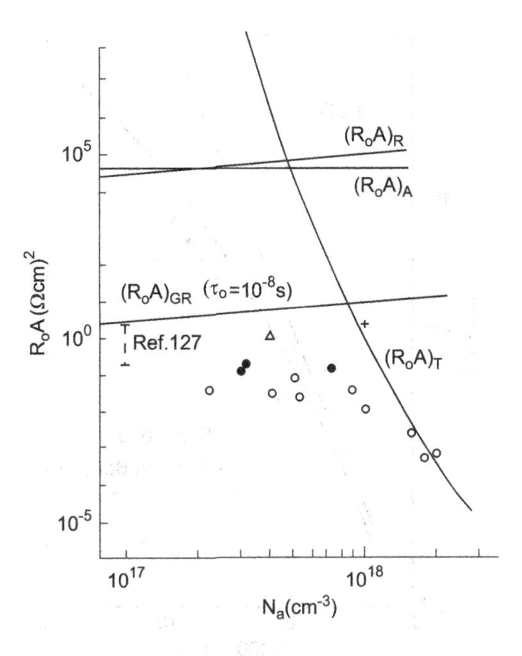

Figure 18.18 The dependence of the R_oA product on the doping concentration for one-sided abrupt PbTe junctions at 77 K. The experimental points are from Refs. [125](•), [123](○), [126](+), [127] and [128](Δ). (From Ref. [123])

The R_oA product determined by the diffusion current for the p⁺-n junction is (see Chapter 12, Equation 12.87):

$$(R_oA)_D = \frac{(kT)^{1/2}}{q^{3/2} n_i^2} N_d \left(\frac{\tau_h}{\mu_h}\right)^{1/2}. \qquad (18.17)$$

The R_oA product controlled by depletion layer current is given by Equation 12.116. Relating the width w of the abrupt junction depletion layer to the concentration N_d and assuming that $V_b = E_g/q$, we get

$$(R_oA)_{GR} = \frac{E_g^{1/2} \tau_o N_d^{1/2}}{q n_i (2\varepsilon_o \varepsilon_s)^{1/2}}. \qquad (18.18)$$

The ultimate values of the R_oA product for Pb salts, PbS, PbSe, and PbTe, abrupt p-n junctions and ideal Schottky junctions within the temperature range between 77 K and 300 K were calculated by Rogalski and coworkers [120–124]. For example, the dependence of the R_oA product on the concentration of dopants for PbTe photodiodes at 77 K is shown in Figure 18.18 [123]. At 77 K, the R_oA product is determined by the generation current of the junction depletion layer. The theoretical estimate yield for the radiative and Auger recombination values of the R_oA product are several orders of magnitude larger. Tunneling current produces an abrupt lowering of the R_oA at a concentration of about 10^{18} cm^{-3}.

Figure 18.19 shows the temperature dependence of the R_oA product for PbTe n⁺-p junctions [124]. For comparison in this figure, experimental data are also included. From the comparison of the theoretical curves with the experimental data, it may be seen that a satisfactory consistence has been achieved in a high temperature range of operation, where the dependence of $R_oA(T)$ follows a diffusion-limited behavior as revealed by the approximately $\exp(E_g/kT)$ slope. With lowering temperature the discrepancies increase. These discrepancies seem to be controlled by the state of junction production technology. The dashed line shown in Figure 18.19 is calculated assuming only band-to-band, generation-recombination mechanisms (the SRH mechanism is omitted). We can see that in the high temperature region of operation, the R_oA product of best quality photodiodes is determined by band-to-band, generation-recombination

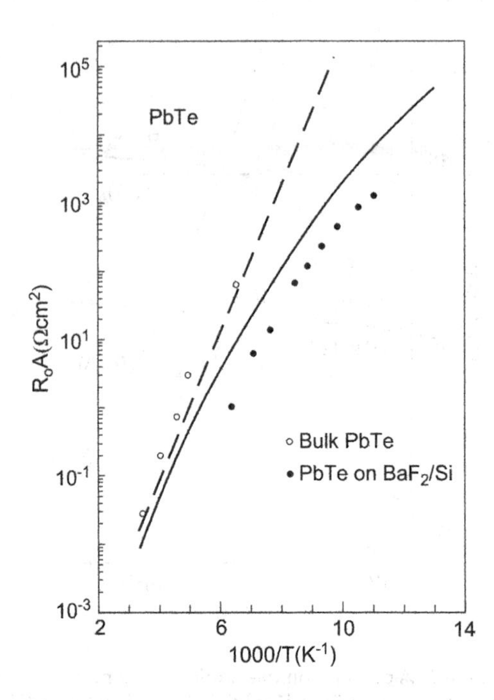

Figure 18.19 The temperature dependence of the R_oA product for n⁺-p PbTe photodiodes. The solid lines are calculated assuming SR, radiative, and Auger processes. The dashed lines are calculated regardless SR mechanism. The experimental data are taken from Refs. [128](○) and [129](•). (From Ref. [124])

mechanisms. Some experimental data are situated above the dashed lines, which is probably caused by the influence of series resistance of photodiodes. It is clearly shown that the R_oA product of photodiodes fabricated in PbTe epitaxial layers on BaF_2/Si substrates is lower than that for photodiodes made in bulk material.

The performance of Pb salt photodiodes is inferior to HgCdTe photodiodes, and is below theoretical limits. Considerable improvements are possible by improving the material quality (to reduce trap concentration) and optimizing the device fabrication technique. Better results should be obtained using buried p-n junctions with a thin wider-bandgap cap layer. This technique is successfully used in the fabrication of double-layer heterostructure HgCdTe photodiodes (see Section 17.6.1). The wider-bandgap cap layer contributes a negligible amount of thermally generated diffusion current compared with that by the bulk opposite type absorber layer.

During the early 1970s, Pb salt ternary alloy (mainly PbSnTe) photodiode technology was advancing rapidly [6,7,19]. The performance of PbSnTe photodiodes was better than that of the HgCdTe ones at that time. The dependence of the R_oA product on the long wavelength cutoff for long wavelength infrared (LWIR) PbSnTe photodiodes at 77 K is shown in Figure 18.20 [130]. In this figure, a selection of experimental data is also observed. The Auger recombination contribution to R_oA increases with the increase of composition x (λ_c increasing) in the base region of the photodiode. For the composition range $x > 0.22$, the R_oA product is determined by the Auger recombination [136]. A satisfactory agreement between the theoretical curves and the experimental data has been achieved for n^+-p-p^+ homojunction structures. In the short wavelength region, the discrepancy between the theoretical curve and experimental data increases, which is due to additional currents in the junctions (such as the generation-recombination current of the depletion region or the surface leakage current) that are not considered. The theoretically calculated curve for double layer heterojunction (DLHJ) structures (n^+-PbSeTe/p-PbSnTe/n-PbSeTe) is situated above the experimentally measured values of the R_oA product, which indicates the potential possibilities of constructing higher-quality PbSnTe photodiodes. Up until now, however, this type of PbSnTe photodiode structure has not been fabricated.

In Figure 18.21, the R_oA product versus temperature is presented under a 0° field of view (FOV) for $Pb_{0.80}Sb_{0.20}Te$ photodiode with a cutoff wavelength of 11.8 µm at 77 K [130]. Good agreement between

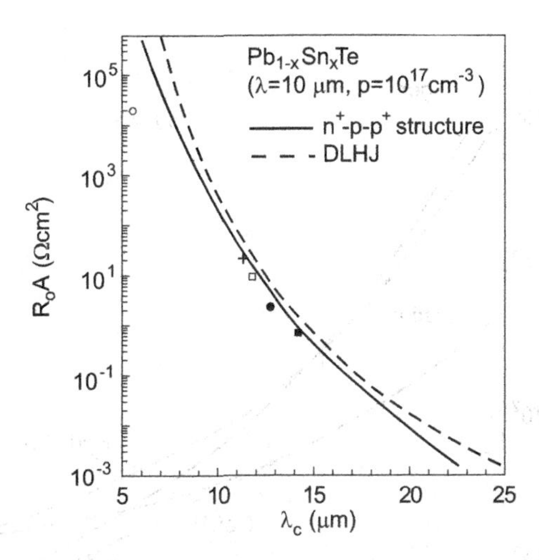

Figure 18.20 Dependence of the R_oA product on the long wavelength cutoff for PbSnTe photodiodes at 77 K. The experimental data are taken from Refs. [128](○), [131](+), [132](•), [133,134](□), and [135](■). The solid line is calculated for n⁺-p-p⁺ PbSnTe homojunction photodiodes, instead the dashed line is calculated for DLHJ PbSnTe photodiode structures. (From Ref. [130])

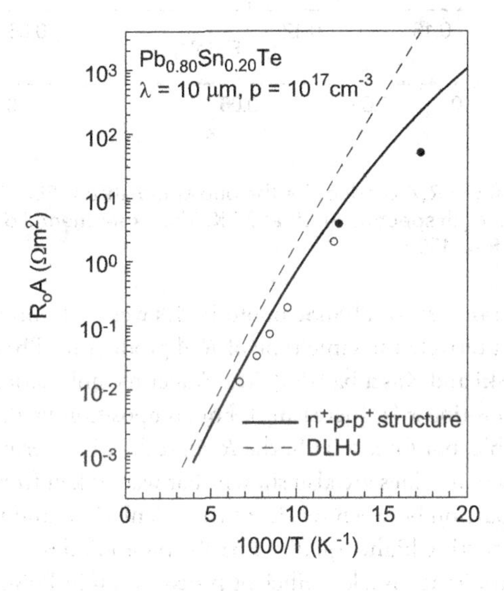

Figure 18.21 Temperature dependence of the R_oA product of $Pb_{0.80}Sn_{0.20}Te$ photodiode. The experimental data are taken from Refs. [131](○) and [137](•). The solid line is calculated for n⁺-p-p⁺ $Pb_{0.80}Sn_{0.20}Te$ homojunction photodiode, instead the dashed line is calculated for DLHJ $Pb_{0.80}Sn_{0.20}Te$ photodiode structure. (From Ref. [130])

experimental data and theoretical calculations (solid line) has been achieved. However, it should be noticed that for more optimized DLHJ structure, theoretically predicted values of the R_oA product are higher (see dashed line). The increase of R_oA product for DLHJ photodiodes will be more emphasized in case of higher quality p-type base PbSnTe layer, when the contribution of Shockley-Read (SR) generation will be suppressed (for higher values of τ_{no} and τ_{po}; in our calculations we assumed $\tau_{no}=\tau_{po}=10^{-8}$ seconds). It should be noted that due to inherently higher Auger generation rate in PbSnTe in comparison with HgCdTe, the enhancement of R_oA product of PbSnTe photodiodes is more limited in comparison with HgCdTe photodiodes.

The PbSnTe photodiodes were preferred over the PbSnSe photodiodes because the high-quality single crystals and epitaxial layers could be fabricated more easily from PbSnTe. However, more recently, rapid advancement in the technology of fabricating monolithic PbSnSe Schottky barriers on Si substrates has been achieved by the research group at the Swiss Federal Institute of Technology [129,138–148].

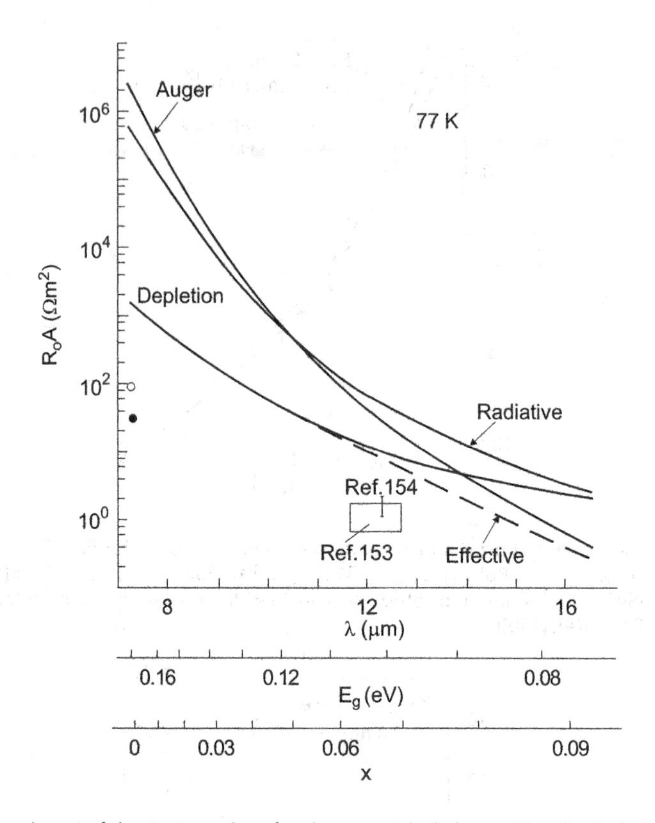

Figure 18.22 The dependence of the R_oA product for the one-sided abrupt $Pb_{1-x}Sn_xSe$ junction (for optimum concentration) on the long wavelength spectral cutoff at 77 K. The experimental data are taken from Refs. [153], [154], [155](O) and [156](•). (From Ref. [150])

The theoretically limited parameters of PbSnSe photodiodes were determined in several papers [149–154]. Figure 18.22 shows quantitatively the same type of R_oA product for PbSnSe junctions as for PbSnTe junctions, calculated by Rogalski and Kaszuba [150]. The Auger recombination contribution to R_oA increases with increasing composition x (λ_c increasing). For composition $x \approx 0.08$, $(R_oA)_{GR}$ ($\tau_o = 10^{-8}$ seconds) and $(R_oA)_A$ values are comparable, but for $x > 0.08$, the R_oA product is determined by Auger recombination. In Figure 18.22, experimental values are also shown that were taken from the literature concerning the Schottky diodes. The comparison between results of the calculations and the experimental data shows potential possibilities for constructing higher quality PbSnSe photodiodes.

The Auger recombination in PbSnSe is a less efficient process than in PbSnTe and in this connection $(R_oA)_{APbSnSe} > (R_oA)_{APbSnTe}$ at comparable E_g. It was first noticed by Preier [151] who compared the theoretically determined dependence of R_oA product components on the doping concentrations in these compounds, exhibiting an energy gap of 0.1 eV at 77 K. The results of calculations are plotted in Figure 17.45 and seem to justify the efforts to use PbSnSe as a detector material. So far, photodiodes with better performance have been prepared from PbSnTe material.

$PbS_{1-x}Se_x$ ternary alloys have become of some interest as a material for laser diodes with emission wavelengths between 4 and 8 μm [15,31]. This mixed semiconductor also shows promise as a photodetector in the 3–5 μm region at temperatures above 77 K. The R_oA product analysis was carried out by Rogalski and Kaszuba [152]. At 77 K, the main contribution to the current flow through the junctions comes from the generation-recombination current in the depletion layer. A satisfactory consistency between experimental and theoretical results has been achieved for the effective lifetime in the depletion layer that is equal to 10^{-10} seconds. As the temperature increases, the influence of the depletion layer decreases, and instead the influence of Auger recombination increases, especially with increasing composition x (λ_c increasing). At room temperature, the Auger process is decisive in a wide range of composition. Only for $x=0$ are contributions of the diffusion and the depletion currents comparable.

18.3.2 TECHNOLOGY AND PROPERTIES

A wide variety of techniques have been used to form p-n junctions in Pb salts. They have included interdiffusion, diffusion of donors, ion implantation, proton bombardment, and creation of n-type layers on p-type material by vapor epitaxy or LPE. A summary of works for the fabrication of high-quality p-n junction and Schottky-barrier photodiodes is given in Refs. [21 and 22].

PbSnTe photodiodes are the most developed of the Pb salt devices, particularly for the 8–14 μm spectral region. Mesa and planar photodiodes are fabricated using standard photolithographic techniques. Performance and stability of the device is especially limited by surface preparation and by the passivation technique that is usually kept proprietary by the producer. It appears that the presence of Pb, Sn, and Te oxides on PbSnTe surfaces almost always produces high leakage currents [157]. The native oxide was found to be an insufficient passivant because the oxidized surfaces contain an unstable TeO_2 [158]. An anodic oxidation is often used for device passivation. The anodic oxide was grown electrolytically from a glycerol-rich solution of water, ethanol, and potassium hydroxide through an anodization/dissolution process [159]. Another type of electrolyte was used by Jimbo et al. [160].

The surrounding atmosphere has a considerable influence on the electrical properties of Pb salt materials [161,162]. According to Sun et al. [163] this process is due to the adsorption of oxygen, the diffusion of Sn ions from the bulk to the surface, and the oxidation of Pb, Sn, and Te, resulting in the formation of a depletion layer in the n-type samples and an accumulation layer in p-type samples. Photolithographically formed SiO_2 diffusion masks are usually used in the planar PbSnTe photodiodes. About 100 nm of SiO_2 is deposited at a temperature between 340°C and 400°C using a silane-oxygen reaction. However, according to Jakobus et al. [164], the PbSnTe surface becomes strongly p-type after coating with pyrolytic SiO_2, therefore radio frequency (RF)-sputtered Si_3N_4 was used by them. The junction areas of Schottky-barrier photodiodes are often delineated by windows in a vacuum-deposited layer of BaF_2 [165,166]. The BaF_2 provided a comparable lattice match and the best thermal expansion match to PbSnTe.

Ohmic contacts to n-type regions are usually realized by In evaporation and to p-type regions by chemical or vacuum deposition of Au. Possible passivating and antireflection (AR) surface coatings such as ZnS, Al_2O_3, MgF_2, Al_2S_3, and Al_2Se_3 have been tried [19]. Samples overcoated with As_2S_3 are completely insulated from the effects of oxygen [161]. It was identified that one of the most important factors that limits the performance of small area devices is damage introduced during the bonding of the leads [167]. Therefore, the leads are attached remotely from the sensitive area to obtain high-quality photodiodes.

18.3.2.1 Diffused photodiodes

Historically, interdiffusion was the first technique used in the fabrication of p-n junction in Pb salts [5,6], which produced a change in type due to a change in the stoichiometric defects. In PbSnTe, an n-type region was formed in a p-type substrate by diffusion using a metal-rich PbSnTe source at temperatures of 400°C –500°C for alloy compositions of about x=0.20. Although high-performance devices and detector arrays were produced ($D^* > 10^{10}$ cmHz$^{1/2}$ W^{-1} in the 8–14 μm region at 77 K), there were difficulties in obtaining reproducible results. Therefore, detectors have also been made by the diffusion of foreign impurities into bulk p-type PbSnTe to form a junction. It was observed that the position or depth of the p-n junction obtained by Al, In, and Cd diffusion varies from slice to slice and tends to drift at temperatures over 100°C [123]. Moreover, the junction diffusion rate increased with decreasing hole concentration in the substrate. Despite the above technological problems, high quality photodiodes formed by diffusing Cd [168,169] or In [131,133,170] into p-type PbSnTe crystals have been fabricated.

According to Wrobel [170], n-p$^+$ junctions should be formed by the diffusion of Cd into as-grown $Pb_{1-x}Sn_xTe$ ($x \approx 0.20$) single crystals at a temperature of 400°C for 1.5 hour. Cd diffusion has also been carried out in a two-temperature zone furnace (specimen at 400°C, source up to 250°C with 2% Cd in In alloy as a diffusion source) [169,171]. During this process, the p$^+$-type material is converted to n-type with a carrier concentration of about 10^{17} cm^{-3}, which is optimum for photodiodes with cutoff wavelengths at about 12 μm. The junction depth is smaller than 10 μm after an 1-hour diffusion time. Upon comparing the measured R_oA product with calculated values, approximate values for the lifetimes within depletion layers and for the minority carrier lifetimes within n-type regions have been established. The carrier lifetime at

77 K is determined by Schockley-Read centers and is lower than 10^{-9} s [169]. This value is consistent with the carrier lifetimes that were determined from photoconductivity and PEM measurements in homogeneously Cd-doped samples [74,172].

In diffusion into p-type PbTe [128] and p-type $Pb_{1-x}Sn_xTe$ ($x \approx 0.20$) [167] has been used to make high-quality photodiodes. Suitable PbSnTe material with an optimum hole concentration of about 10^{16}–10^{17} cm^{-3} can be made only by low temperature growing methods, for instance, by LPE [133] or by annealing in controlled atmospheres [170] in case of other growing methods. The p-type PbTe material with a hole concentration of 4×10^{17} cm^{-3} can be grown by LPE if the growth melt is doped with As [128]. A procedure for In deposition and diffusion is described by Lo Vecchio et al. [167].

In 1975, Chia et al. [133] and DeVaux et al. [173] described results from In-diffused planar arrays of $Pb_{1-x}Sn_xTe$ ($x \approx 0.20$) homojunctions. The diodes of lower base carrier concentrations had a larger R_oA product. From measurements of the forward-bias I-V characteristics and from measurements of the R_oA temperature dependence, they concluded that bulk diffusion currents limit the R_oA product at temperatures above about 70 K. The measured average R_oA product of a ten-element array equal to 3.8 Ωcm^2 at 78 K is consistent with a minority carrier lifetime of 1.7×10^{-8} seconds. This array had a $Pb_{0.785}Sn_{0.215}Te$ substrate with 1.7×10^{16} cm^{-3} base hole concentration. At lower temperatures, the R_oA product tends to saturate, which is attributed to surface leakage. Figure 18.23 shows the spectral dependence of detectivity of two photodiodes of a ten-element array. Except for some scatter at low wavelength, the curves are indistinguishable. The oscillatory nature of the spectral response is due to the AR coating action of the insulator. At 80 K, with 10 mV reverse bias, an average D^* of 1.1×10^{11} cmHz$^{1/2}$ W^{-1} (*f*/5 FOV) was measured (η=79%).

Wang and Lorenzo [134] have also fabricated high-quality planar devices by impurity diffusion into low-concentration ($p=4\times10^{16}$ cm^{-3} or less) LPE layers. As a diffusion mask they used an oxide layer. At 77 K, the average R_oA product was 1.85 Ωcm^2 for an array of 124 elements each with an area 2.5×10^{-5} cm^2. The average quantum efficiency without AR coating was 40% and the typical peak detectivity of 2.6×10^{10} cmHz$^{1/2}$ W^{-1} was measured at 11 μm in a 2π FOV. A striking feature of the arrays was the extreme uniformity in spectral response, the λ_c values varied by less than 1%. The above technique is suitable for the routine fabrication of high-density planar PbSnTe arrays of high quality.

More recently, John and Zogg [174] manufactured p-n PbTe junction photodiode by overgrowth of a Bi-doped n$^+$ cap layer onto p-type base layer. Generally, the p-n junction works well with tellurides, while selenides exhibit a too high diffusion in order to obtain reliable devices.

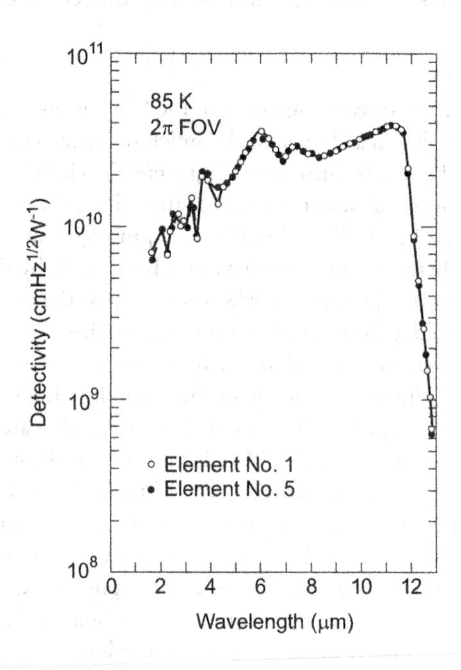

Figure 18.23 Detectivity as a function of wavelength for two photodiodes of a $Pb_{0.785}Sn_{0.215}Te$ array. (From Ref. [133])

18.3.2.2 Ion implantation

Most of the implant work in IV–VI semiconductors has been devoted to the formation of n-on-p junction photodiodes, mainly covering the 3–5 μm spectral region.

In early works, proton bombardment was used to convert p-type PbTe and PbSnTe to n-type [175]. The electron concentration in the n-type layer was about 10^{18} cm^{-3}. Wang et al. [176,177] studied the electrical and annealing properties of proton bombardment of the $Pb_{0.76}Sn_{0.24}Te$ layer. They found that the defects were annealed out at temperatures around 100°C, changing the conductivity back to p-type. Good devices were obtained without knowing the details about doping, lattice damage, and annealing. Next, Donnelly [178] and Palmetshofer [179] have reviewed the investigation and understanding of ion implantation in IV–VI semiconductors.

The best diodes in some of the Pb salts were reported by Donnelly et al. [126,180–182]. These diodes were fabricated by Sb^+ implantation. It was found that Sb becomes fully activated at an annealing temperature of about 300°C. The Sb^+ ion implantation was carried out using a 400 keV beam and a total dose of $(1–2)\times10^{14}$ cm^{-2}. Following implantation, the photoresist implantation mask was removed and each sample was coated with a 150 nm layer of pyrolitic SiO_2 for 2–5 minutes at a temperature between 340°C and 400°C. This annealing step is sufficient to effectively anneal out the radiation damage caused by room temperature implantation. Holes were then opened in the oxide and the Au contacts were electroplated on the sample. The β parameter in the forward-bias I-V characteristics had a value of about 1.6, while the reverse characteristics showed a soft power-law breakdown due to tunneling. PbTe photodiodes had a R_oA product at 77 K as high as 2.1×10^4 Ωcm^2 and a detectivity limited by background [126]. With a reduced background (77 K shield) at 50 Hz, the measured detectivity was 1.6×10^{12} cmHz$^{1/2}$ W^{-1}, which is slightly below the theoretical amplifier plus thermal noise limit. The peak detectivity $\lambda_p=4$ μm and the cutoff wavelength $\lambda_c=5.1$ μm are shifted toward a shorter wavelength region, due to band filling in the highly doped p-type substrate ($p>10^{18}$ cm^{-3}).

Implantation of In seems to be less appropriate for p-n junction fabrication, because of the low electrical activity and a saturation of the carrier concentration at higher doses [179].

Implantation of the constituent elements for the photodiode preparation gave good results in PbSSe diode fabrication [180].

18.3.2.3 Heterojunctions

An alternative technology adopted for the preparation of long wavelength photodiodes is the use of heterojunctions of n-type PbTe (PbSeTe) deposited onto p-type PbSnTe substrates by LPE [8,133,135,137,183–186], VPE [187,188], MBE [189], and hot wall epitaxy (HWE) [190].

The best results have been obtained using the LPE method. Either type of material with carrier concentrations of 10^{15}–10^{17} cm^{-3} can be grown and used without annealing. Using the LPE method an advanced concept in PbSnTe photodiode design with the fabrication of back side illuminated diodes has been successfully realized (see inset in Figure 18.24 [185]). It permits complete optical utilization of the electrical area of the photodiode, significantly reduces the optical dead area of an array, and increases the optically sensitive area of the diode because of the refractive index mismatch between PbTe and air. Moreover, wider-energy gap material on one side of the junction results in reduction of the saturation current. Figure 18.24 shows the spectral response characteristics of the back side illuminated n-PbTe/p-PbSnTe heterojunction at 77 K. Filtering to 6 μm is caused by absorption in the PbTe substrate.

It should be noticed that the spectral response of LPE-grown PbTe/PbSnTe heterojunctions depends on the PbTe growth temperature. Migration of the p-n junction away from the PbTe-PbSnTe interface is possible during epitaxial growth due to the quite large interdiffusion coefficients of the native defects in PbTe and PbSnTe. The p-n junctions shifts into the PbTe layer during PbTe growth at $T>480$°C [191,192]. A pure PbTe spectral response of n-PbTe/p-PbSnTe heterostructures obtained by HWE technique has been observed elsewhere [193,194]. More recently published papers have indicated that the spectral response of heterojunctions depends on an electric field in the PbSnTe layer [195] and on the relation between the n and p concentrations [196].

Lattice mismatch between PbTe and $Pb_{1-x}Sn_xTe$ (0.4% for $x=0.2$) introduces strain-relieving misfit dislocations that may be important in determining the performances of photodiodes [197]. Kesemset and

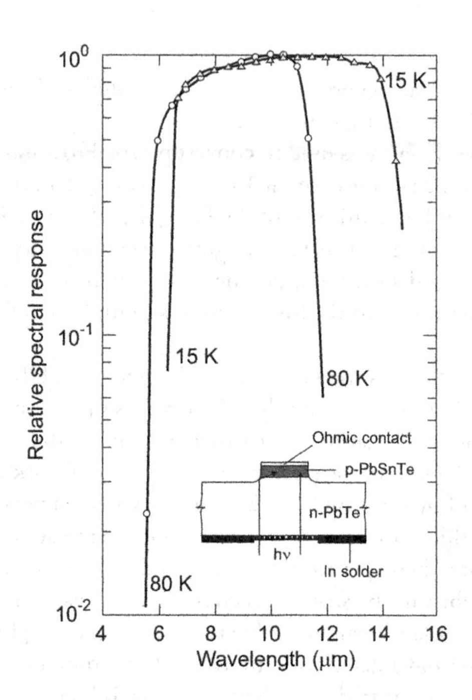

Figure 18.24 Spectral response of n-PbTe/p-PbSnTe back side illuminated mesa photodiode at 77 K. (From Ref. [185])

Fonstad [198] have found the relatively high interface recombination velocity (≈10^5 cm/s) in the active region of $Pb_{0.86}Sn_{0.14}Te$/PbTe double-heterostructure laser diodes. With decreasing lattice mismatch, the interface recombination also decreases. The lattice-matched PbSnTe/PbTeSe configuration offers a promising solution to the problem of mismatch arising in this system [199]. It was confirmed for n-PbTeSe/p-PbSnTe photodiodes operated at temperatures below 77 K [137].

Wang and Hampton [132] reported high-quality n-PbTe/p-$Pb_{0.80}Sn_{0.20}Te$ mesa arrays produced by using an LPE wiper technique. The p-type $Pb_{0.80}Sn_{0.20}Te$ were first grown on the (100) $Pb_{0.80}Sn_{0.20}Te$ single crystals followed by an n-type PbTe growth. The $Pb_{0.80}Sn_{0.20}Te$ epilayer was grown at temperatures from 540°C to about 500°C with a cooling rate from 0.1 to 0.25°C/min (typical thickness was 20 μm after about 3 hour of growth). The PbTe layer was started at the temperature where the $Pb_{0.80}Sn_{0.20}Te$ layer growth stopped. The thickness of the PbTe layer was about 5 μm after 40 minutes of growth. At 80 K, the average R_oA product for an 18-element array was 3.3 Ωcm² and the mean D^* at $\lambda_p = 10$ μm was 2.3×10^{10} cmHz$^{1/2}$ W^{-1} for 2π FOV and 300 K background.

Studies of the electrical properties of p-$Pb_{0.80}Sn_{0.20}Te$/n-PbTe inverted heterostructure diodes have been reported by Wang et al. [185]. Typical forward and reverse I–V characteristics at various temperatures are given in Figure 18.25. In this figure, continuous curves represent the measured values and the dashed curves represent the theoretical values. Diffusion current is predominant at temperatures above 80 K. The activation energy in the diffusion-dominated region is 0.082 eV. As the temperature decreases, GR current becomes important and the activation energy reduces to 0.044 eV. The values of activation energy are roughly equal to E_g and $E_g/2$, respectively, at $T = 15$ K. The GR current is predominant in the temperature range $30 < T < 80$ K. Below 30 K, the surface-related leakage current becomes evident for the small area diodes ($A < 10^{-4}$ cm²). This current increases as the reverse-bias voltage increases. The excess leakage current of large area diodes is probably conditioned by bulk defects.

The LPE technique can be used in the fabrication of two-color PbTe/PbSnTe detectors. PbTe/PbSnTe heterostructures were proposed that cover both (3–5 μm and 8–14 μm) atmospheric windows [184]. Planar p-n junctions have been made using an In diffusion technique. The PbTe diodes had an average D^* of 10^{11} cmHz$^{1/2}$ W^{-1} at $\lambda_p = 4.6$ μm, and the PbSnTe diodes had an average of 2.2×10^{10} cmHz$^{1/2}$ W^{-1} at their 9.8 μm peak.

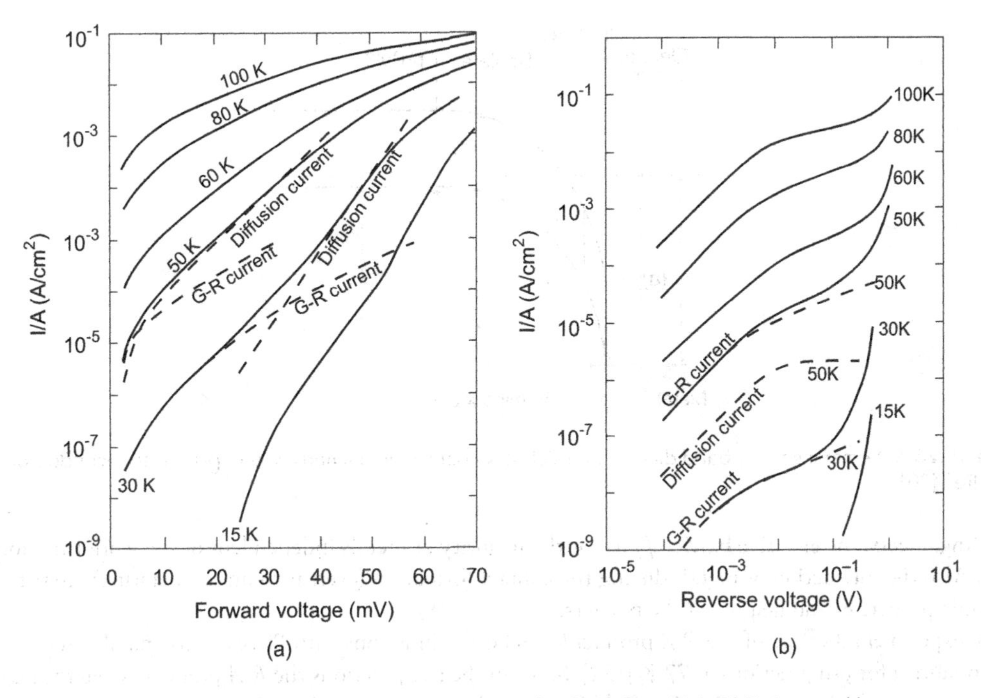

Figure 18.25 (a) Forward and (b) reverse I–V characteristics of p-$Pb_{0.80}Sn_{0.20}Te$/n-PbTe inverted heterostructure diode. (From Ref. [185])

18.4 SCHOTTKY-BARRIER PHOTODIODES

It has been pointed out [125,200–202] that in case of Schottky-barrier photodiodes with p-type Pb salt semiconductors, the effective barrier height ϕ_b is independent of metal work function. In case of moderate carrier concentrations ($\approx 10^{17}$ cm^{-3}), ϕ_b does not appreciably exceed the energy gap. Assuming $\phi_b = E_g$, the expression (Equation 12.153) takes the form:

$$(R_oA)_{MS} = \frac{h^3}{4\pi q^2 kT}\frac{1}{m^*}\exp\left(\frac{E_g}{kT}\right). \tag{18.19}$$

18.4.1 SCHOTTKY BARRIER CONTROVERSIAL ISSUE

The experimental values of the R_oA product for the Schottky junctions are consistent with the theoretical curves within the higher temperature range, while at 77 K deviations from the values calculated are very large [122]. This divergence seems to be due to the use of Equation 18.19, which does not take account of additional processes occurring in Schottky junctions with p-type narrow-gap semiconductors. Despite the fact that Gupta and colleagues [149], using the above-mentioned relationship, achieved good agreement with experimental data for ternary compounds PbSnTe and PbSnSe, their results seem to be accidental for the energy gap equal to 0.1 eV. The scheme of energy bands for such a junction has been proposed by Walpole and Nill [201] and is shown in Figure 18.26, where three regions may be distinguished: inverted, depleted, and bulk. In the ideal junction model, only processes (a), that is, the hole emission from the Fermi level in metal to the valence band for $h\nu = \phi_b$ are considered. No account is taken of the excitation of hole-electron pairs in the inverted region (processes (b)) or of the band-to-band excitation of hole-electron pairs in the depleted region (processes (c)). The last excitations are of particular importance, because the depleted region is wide due to a high dielectric constant. According to Nill and colleagues [200], the barrier height ϕ_b for holes is considerably lowered to take the value ϕ_{be} slightly exceeding the energy gap E_g. Since the kinetic energy of the hole slightly exceeds E_g, the narrow top of the barrier is transparent due to

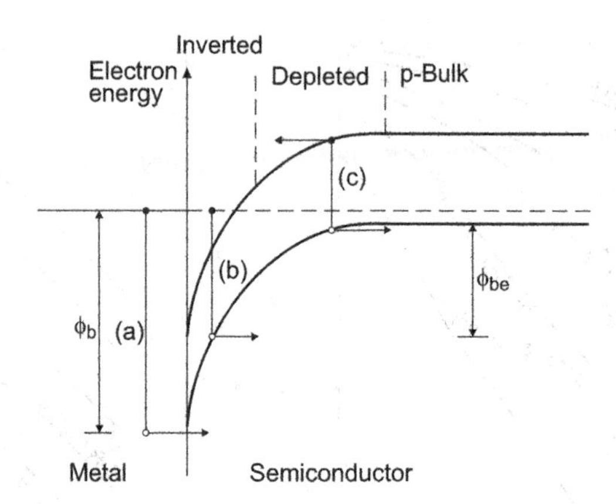

Figure 18.26 Schematic energy band diagram for Schottky barrier with a narrow-gap p-type semiconductor. (From Ref. [201])

tunneling effects, the effective barrier ϕ_{be} is for the majority of metals independent of the work function of metal. Also the interaction of metals during the contact formation (even at room temperature) alters the electronic properties and heights of the barriers.

The experimental values of the R_oA product for Schottky junctions with PbTe are comparable with those obtained for p-n junctions at 77 K [122]. Since in the last junctions the R_oA product is determined by generation-recombination processes it should be concluded that for metal-semiconductor (M-S) junctions an essential contribution to the R_oA comes from the depletion region. Several researchers [138,165,203,204] observed two saturation current generation mechanisms for Pb-barrier Pb salt photodiodes: diffusion-limited and depletion-limited behavior. In these papers, a tendency for R_oA saturation at lower temperatures has been explained by a weak temperature-dependent shunt resistance. Instead, Maurer [205] explained the temperature dependence of the R_oA product of Pb-PbTe Schottky barriers, assuming simple thermionic emission theory and band-to-band tunneling.

Up to the present, knowledge of the M-S interface in IV–VI materials is quite fragmentary, which leads to uncertainty about the nature of the p-n junction in metal-barrier photodiodes. However, we cannot exclude the possibility that a shallow-diffused n^+-p junction is formed owing to changes in the stoichiometry of the IV–VI layer surface caused by the metal layer. Likewise, Chang et al. [206] and Grishina et al. [207] showed experimentally that evaporated films of In on PbTe and PbSnTe chemically react at room temperature to give a structure with an intermediate layer between the metal and semiconductor.

Sizov et al. [202] evaluated the activity of the metal with PbTe using the bulk data on standard enthalpies and entropies of the tellurides formation:

Ga, Zn, Mn, Ti, Cd, In, **PbTe**, Mo, Sn, Ge, Cu, Tl, Pt, Ag, Hg, Bi, Sb, As, Au.

The metals to the left of PbTe, during contact formation on the PbTe surface, must interact with the semiconductor even at room temperature, forming a new species that must alter the electronic properties and the heights of barriers. The metals to the right of PbTe will not interact with the semiconductor near room temperature.

On account of the above divergence, different methods of Schottky barrier preparation on IV–VI compounds have been elaborated. Initially specimens were cooled (nominally to 77 K) during Pb deposition [201]. According to a U.S. patent [208], the metal layer (In, Pb) should be evaporated in high vacuum immediately after the epitaxial layer has been cooled to room temperature without breaking the vacuum. A layer of SiO_2 is then deposited by thermal in situ evaporation to protect the M-S contact from the atmosphere. A contrary method of Schottky barrier preparation has been described by Buchner et al. [209]. They showed that deposition of Pb onto a $Pb_{0.8}Sn_{0.2}Te$ surface gave a rectifying barrier only if the surface had been previously exposed to the atmosphere. It was interpreted by the presence of oxygen at the surface of PbSnTe that prevents the migration of Sn across the interface. Grishina et al. [210] noticed that improvement of the electrical characteristics of Pb(In)-$Pb_{0.77}Sn_{0.23}Te$ Schottky barriers is caused by the influence of native oxide that prevents interaction between the metal and semiconductor. More recently, it has been demonstrated that the usage of a thin chemical oxide as an intermediate layer between metal and semiconductor improves the Schottky barrier properties, while anodic oxide passivation raises the thermal and temporal stability as well as lowers the scattering of parameters from element to element in multielement linear arrays [211].

Pb-PbTe rectifying contacts were obtained with both clean and air-exposed surfaces [212]. In contrast to $Pb_{0.8}Sn_{0.2}Te$, the oxidation of PbTe tends to saturate at about monolayer coverage (continuing oxidation of PbSnTe is accompanied by the diffusion of Sn from the bulk to the surface). Thermal stability of Pb-PbTe devices has been obtained by washing with pure water and then vacuum baking at 150°C for periods of up to 12 hours [165]. Schoolar et al. [154,156] prepared Schottky-barrier PbSnSe and PbSSe photodiodes using vacuum-annealed (at 170°C for 30 minutes) epitaxial films to desorb a surface oxide layer and cooling the films to 25°C prior to depositing the Pb or In barriers. It was also found that the presence of Cl in the interface vastly improved I–V characteristics of Schottky junctions [213–215]. Further studies of interfaces are necessary to resolve the present controversial issues concerning Schottky barrier formation.

A new insight into the theory of Schottky barrier Pb salt photodiodes has been presented by Paglino et al. [144]. By using the Schottky barrier fluctuation model introduced by Werner and Guttler [216], it is assumed that the Schottky barrier height ϕ_b has a continuous Gaussian distribution σ around a mean value ϕ. Due to the exponential dependence of saturation current (see Equation 12.151) on ϕ_b, it follows that the effective barrier responsible for current is given by:

$$\phi_b = \phi - \frac{q\sigma^2}{2kT}. \tag{18.20}$$

This barrier ϕ_b is smaller than the mean value ϕ that is derived from capacitance-voltage characteristics. Therefore, since ϕ_b depends on temperature, no straight line is obtained in the Richardson plot. In order to get consistent results with I-V characteristics at $V \neq 0$, Werner and Güttler showed that the changes of the barrier $\Delta\phi$ and square of the barrier fluctuation $\Delta\sigma^2$ with the applied bias voltage V are proportional to this voltage,

$$\Delta\phi_b(V) = \phi_b(V) - \phi_b(0) = \rho_2 V \quad \Delta\sigma^2(V) = \sigma^2(V) - \sigma^2(0) = \rho_3 V, \tag{18.21}$$

where ρ_2 and ρ_3 are negative constants. These settings lead to a temperature dependence of the ideality factors β,

$$\frac{1}{\beta} - 1 = -\rho_2 + \rho_3 \frac{q}{2kT}, \tag{18.22}$$

which has been observed with numerous Schottky barriers fabricated with different materials over a large temperature range [216].

A plot of the R_oA product versus temperature for Pb-PbSe Schottky barrier photodiode on Si substrate is shown in Figure 18.27 [144]. A near-perfect fit is obtained over the whole temperature range. The fluctuation σ leads to the saturation of the J_{st} or R_oA product at low temperature. For Pb-PbSnSe Schottky barriers, these fluctuations have an assumed Gaussian distribution with a width σ of up to 35 meV. The values depend on the structural quality; higher quality devices show lower σ. The ideality factors β are correctly described with the model even if $\beta > 2$. It is expected that the barrier fluctuation is caused by the threading dislocations; lower densities of these dislocations lead to lower σ and higher saturation R_oA values at lower temperatures. The dislocation densities for PbSe were in the 2×10^7–5×10^8 cm^{-2} range for the 3–4-μm-thick as-grown layers. Higher R_oA products are obtainable by lowering these densities by thermal annealing [145], which sweeps the threading ends of the misfit dislocations over appreciable distances (in the cm range) to the edges of the sample [34].

18.4.2 TECHNOLOGY AND PROPERTIES

In comparison with p-n junction preparation, a considerably simpler technique for Schottky barrier photodiode fabrication involves evaporation of a thin metal layer onto the semiconductor surface. It is a planar technology with the potential for cheap fabrication of large arrays. This technique has been successfully applied to IV–VI compounds. An excellent review of Schottky barrier IV–VI photodiodes with emphasis on thin-film devices has been given by Holloway [165].

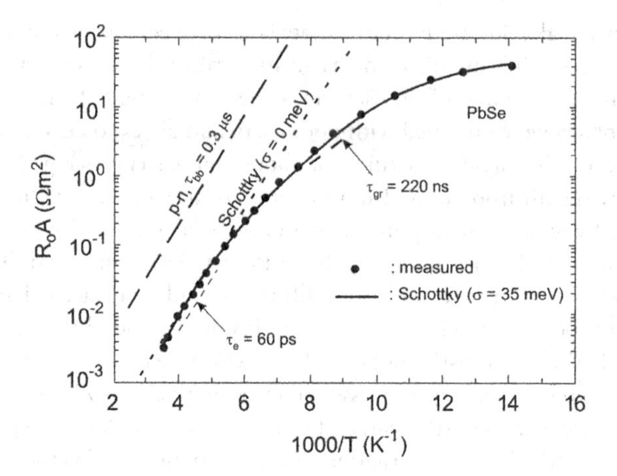

Figure 18.27 Resistance area product R_oA versus inverse temperature for Pb–PbSe Schottky barrier photodiodes on Si substrates. The values are fitted with the barrier fluctuation model (solid line). For comparison, the values calculated for the ideal Schottky barrier photodiode are plotted as well as the values from the p-n theory in the diffusion case for band-to-band recombination limited lifetime τ_{bb}=0.3 μs as well as for a recombination lifetime τ_e=60 ps and GR limited lifetime τ_{gr}=220 ns. (From Ref. [144])

In most cases, the Schottky barriers of IV–VI semiconductors are made by the evaporation of Pb or In metals. If bulk crystal semiconductor material is used, the photodiodes are front side illuminated through semitransparent electrodes [164,217]. Since 1971, when the first high-performance, thin-film photodiodes were reported by the Ford group [218], the devices have been back side illuminated through BaF_2 substrates. Development of these devices is presented by Holloway [165]. The thin-film IV–VI photodiodes were made of epitaxial layers of Pb salts using HWE or MBE. These films are p-type with hole concentrations of about 10^{17} cm^{-3}. Development of the MBE technique in the next two decades have offered the possibility of growing Pb salt epitaxial layers on Si substrates with an appropriate buffer layer, which is stacked CaF_2-BaF_2 or CaF_2-SrF_2-BaF_2 of only ≈200 nm total thickness [129] and of a quality sufficient for IR device integration. This opens the way to fully monolithic, heteroepitaxial FPAs. Figure 18.28 gives four main configurations in which Schottky barrier photodiodes were fabricated.

The high-performance, thin-film IV–VI photodiodes were developed in the 1970s by two U.S. research groups at the Ford Motor Company and at the Naval Surface Weapons Center. Pb-(p)PbTe photodiodes prepared in stringent conditions of cleanliness had a R_oA product of 3×10^4 Ωcm² at 80 K–85 K, which together with the quantum efficiency of 90%, influenced by interference, corresponds to a Johnson-noise-limited $D^*=10^{13}$ cmHz$^{1/2}$ W^{-1} for a peak wavelength near 5 μm [165]. The best photodiodes had the region of contact between the semiconductor and the Pb layer, defined by a window in an evaporated layer of BaF_2. This configuration of photodiode has been used in the fabrication of 100-element arrays. The photolithographic methods, including the BaF_2 window and delineation techniques for the semiconductor and the Pb and Pt metallization, are briefly described by Holloway [165].

In the next two decades, the research group at the Swiss Federal Institute of Technology has continued to pursue thin-film IV–VI photodiode technology and has made significant progress [9,118,129,143–148,174]. In 1985, Zogg and Huppi [219] proposed a new monolithic heteroepitaxial PbSe/Si integrated circuit obtained by epitaxial growth of PbSe films onto Si using a (Ca,Ba)F_2 buffer layer. The lattice mismatch of ≈14% between Si and BaF_2 (see Figure 18.2) can be overcome using graded layers consisting of CaF_2 at the Si and BaF_2 at the IV–VI alloy interface. Thermal expansion coefficient mismatch between the IV–VI alloys and the fluorides (about 20×10^{-6} K^{-1} at 300 K) and Si (2.6×10^{-6} K^{-1} at 300 K) [220] was not found to be detrimental. It appears that this thermal expansion mismatch relaxes through a glide of dislocations, and the layers survive more than 800 thermal cycles (between 300 and 80 K) without problems [143]. By the development of this technology, considerable improvement in fully monolithic Pb salt FPAs has been achieved using different binary and ternary IV–VI alloys (PbS [221,222], PbSe [140,223,224], PbTe [174,204,223], PbEuSe [221], and PbSnSe [138,141,143]). An epitaxial stacked CaF_2-BaF_2 buffer layer of ≈200 nm thickness was usually used.

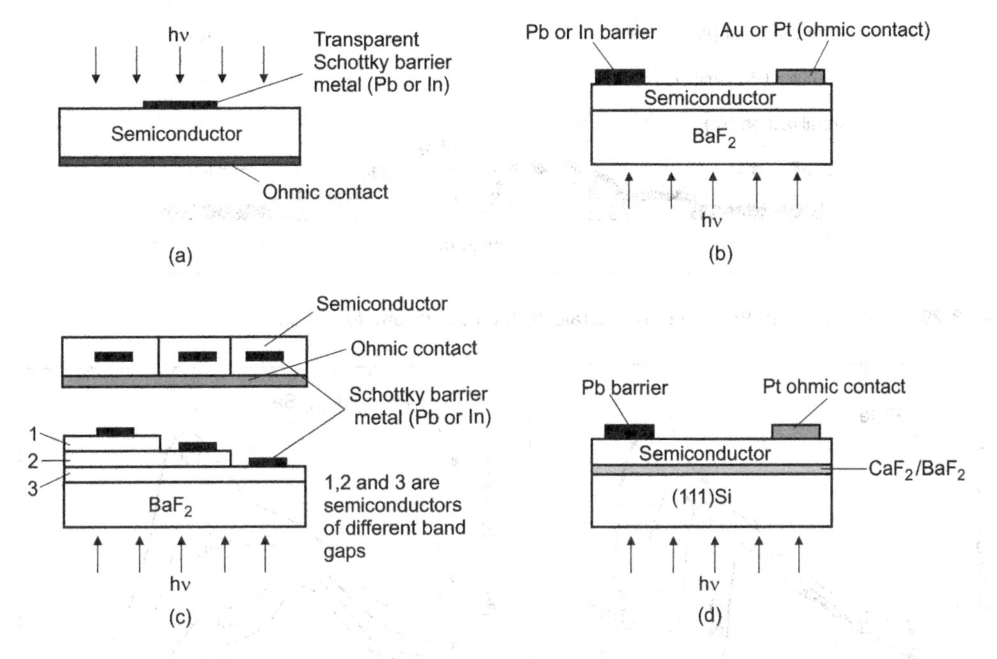

Figure 18.28 Various configurations of Schottky barrier photodiodes: (a) conventional photodiode front side illuminated, (b) back side illuminated photodiode, (c) multispectral (three-color) detector, and (d) heteroepitaxial photodiode on Si substrate.

The fabrication process of PbSnSe photodiodes starts with a layer deposition in the second growth chamber at 350°C–400°C onto a CaF_2 buffer on a 3-inch Si(111)-wafer. Typical carrier concentration in p-type layers is $(2–5)\times 10^{17}$ cm^{-3}; thickness of the layers is 3–4 µm. Growth with (111) orientation is preferred, since thermal mismatch strain relaxation occurs by dislocation glide in the main {100} < 110 > glide system with the glide planes inclined to the surface. When grown on (100)-oriented substrates, thermal mismatch strain relaxation has to occur via higher glide systems, which normally leads to cracking of the layers as soon their thickness exceeds about 0.5 µm [146,225]. Dislocation densities down to 10^6 cm^{-2} have been obtained in unprocessed layers of a few µm thickness, while for layers used to fabricate detectors, dislocation densities range from the low 10^7 to low 10^8 cm^{-2} range. The further basic steps in the device fabrication are as follows:

- Vacuum deposition of ≈200 nm Pb ex situ or in situ in the MBE growth chamber with the sample held at room temperature.
- Overgrowth with ≈100 nm vacuum-deposited Ti for protection of the Pb.
- Vacuum deposition of Pt for better adhesion of the following Au layer and delineation of the Pt.
- Definition of sensitive areas: etching of Ti with a selective Ti-etchant, etching of Pb with a selective Pb-etchant.
- Electroplating of Au for ohmic contacts and on top of the Pb/Ti/Pt blocking contacts.
- Etching of the PbSnSe layer.
- Spin-on, photolithographic patterning and curve of a polymide for insulation and planarization.
- Vacuum deposition of Al and delineation of the fan-out by wet etching.

The active areas were 30–70 µm in diameter. Illumination is from the back side through the IR transparent Si substrate. Figure 18.29 shows a schematic cross section of a device [145]. No surface passivation was needed, surface effects were rather negligible as deduced from the R_oA-value diodes with different sizes.

The spectral response curves of PbTe and $Pb_{0.935}Sn_{0.065}Se$ photodiodes on Si substrates are shown in Figure 18.30 [129]. Typical quantum efficiencies are around 50% without AR coating. Note the pronounced interference effects in case of $Pb_{0.935}Sn_{0.065}Se$ photodiode, which help to increase considerably the response near the peak wavelength at 50 K, but which lead to some decrease near the cutoff at 77 K. In order to optimize the peak response at a specific temperature, one has to optimize the thickness of PbSnSe as well as that of the buffer layer in order to obtain constructive interference in the active region.

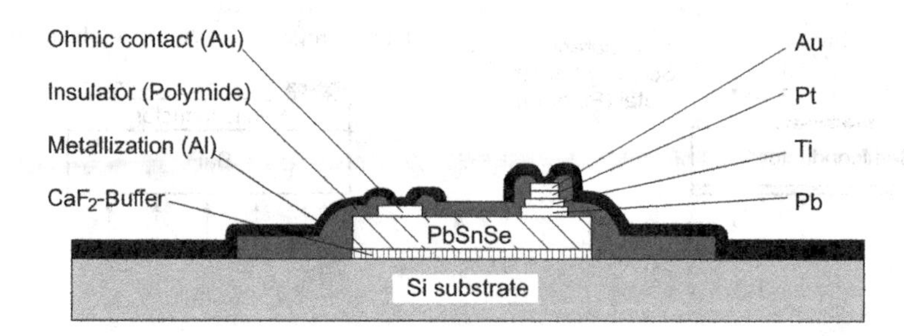

Figure 18.29 Cross section of PbSnSe photovoltaic IR detector. (From Ref. [146])

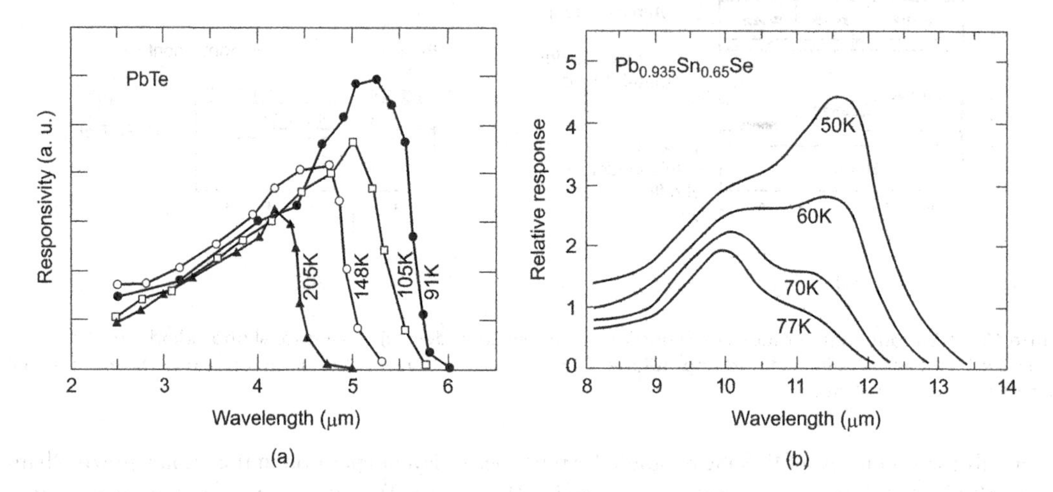

Figure 18.30 Spectral responses of Pb salt photodiodes on Si substrates at different temperatures: (a) PbTe photodiode, and (b) $Pb_{0.935}Sn_{0.065}Se$ photodiode. (From Ref. [129])

Figure 15.31 shows R_oA products at 77 K as a function of cutoff wavelength for different Pb salt photodiodes on Si with stacked BaF_2/CaF_2 and CaF_2 buffer layers. Although these values are considerably above the background limited infrared photodetectors (BLIP) limit (for 300 K, 2π FOV and η=50%), they are still significantly below the theoretical limit given by Auger recombination (see Figure 18.22). The performance of IV–VI photodiodes is inferior to that of the HgCdTe photodiodes; their R_oA products are two orders of magnitude below the values for p$^+$-on-n HgCdTe. R_oA products of PbSnSe photodiodes with a cutoff wavelength of 10.5 μm are about 1 Ωcm^2 at 77 K.

Figure 18.32 shows the temperature dependence of the R_oA product for n$^+$-p $Pb_{0.935}Sn_{0.065}Se$ photodiode [124]. For comparison in this figure, also included are experimental data taken from Zogg and colleagues [138]. For temperatures down to 100 K, the increase in R_oA follows a diffusion-limited behavior as revealed by the approximately $\exp(E_g/kT)$ slope. In the range between 100 K and 77 K, the behavior may be attributed to depletion-layer generation-recombination mechanism. Theoretical prediction for a back side ohmic contact of the photodiode (dashed curve in Figure 18.32) indicates considerable decrease of the R_oA product. This structure of photodiode is inferior since the minority carrier diffusion length in base p-type region is larger than the thickness of the photodiode. Results of calculations presented in Figure 18.32 indicate the potential possibilities of constructing higher quality PbSnSe photodiodes. In the region of higher temperature, where the experimental dependence of $R_oA(T)$ follows a diffusion-limited behavior, the dominant generation-recombination mechanism is not fundamental band-to-band mechanism (radiative or Auger processes). It means that diffusion-limited behavior of photodiode is determined by SR generation-recombination mechanism with values of τ_{no} and τ_{po} below 10^{-8} seconds (which was assumed in calculations). It is probably caused by relatively high trap concentrations, which dominate the excess carrier lifetime in p-type material by the SR generation-recombination mechanism.

It was found that the maximal R_oA products of PbSnSe photodiodes are determined by the dislocation densities. It is shown in Figure 18.33 [148], where the saturation values of the R_oA products for various

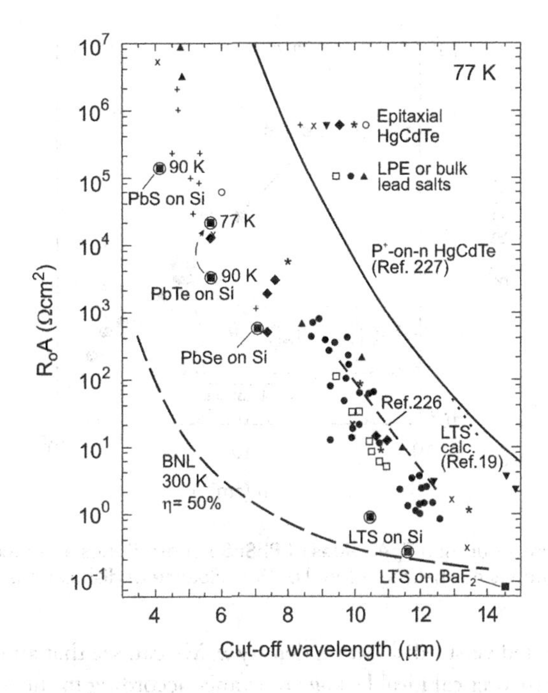

Figure 18.31 Experimental R_oA products at 77 K versus cutoff wavelength for different Pb salt photodiodes in comparison with HgCdTe photodiodes. The 300 K background noise limit (BNL) for 180° FOV and 50% quantum efficiency is included. The dotted line labelled LTS is the calculated ultimate value for PbSnSe. (From Ref. [19]) The broken line is the upper limit of numerous experimental HgCdTe data points. (From Ref. [226]) The solid line represents calculated data for p-on-n HgCdTe photodiodes. (From Refs. [140 and 227])

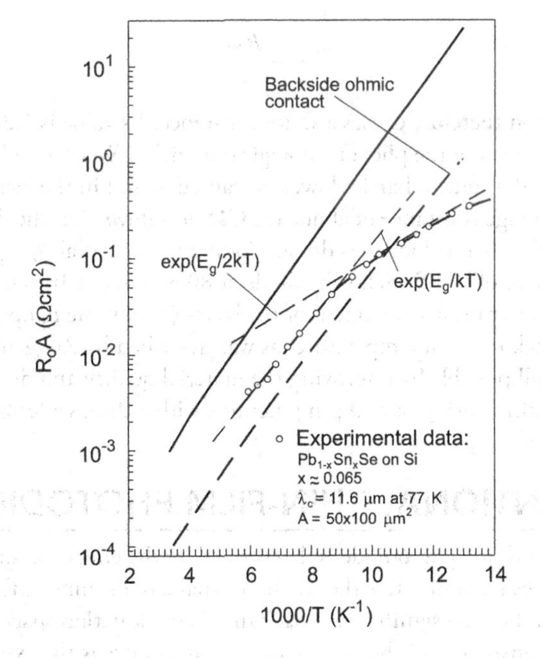

Figure 18.32 The temperature dependence of the R_oA product for n⁺-p $Pb_{0.935}Sn_{0.065}Se$ photodiode. The solid line is calculated for back side illuminated photodiode structure with 3-μm-thick p-type base layer with carrier concentration 10^{17} cm^{-3} and 0.5-μm-thick n⁺-cap layer with electron concentration 10^{18} cm^{-3}. Instead, the dashed line is calculated assuming a back side ohmic contact of the photodiode. The experimental data (O) are taken from Zogg [138]; the diffusion and depletion regimes are indicated by thin dashed lines. (From Ref. [124])

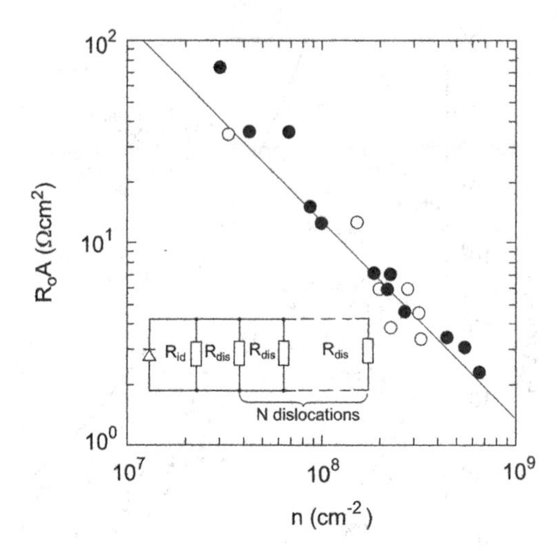

Figure 18.33 Low-temperature saturation R_oA values of PbSnSe photodiodes as a function of dislocation density. The influence of the composition is normalized. Model of the influence of dislocations on the leakage resistance in the insert. (From Ref. [148])

PbSnSe photodiodes are plotted versus dislocation density n. We can see that an inverse linear dependence is observed: $R_oA \sim 1/n$. The theoretical ideal leakage resistance according to the Schottky theory without barrier fluctuation is R_{id}. If N dislocations end in the active area A of the device, and if each dislocation gives rise to an additional leakage resistance R_{dis}, the measured differential resistance is [145]:

$$\frac{1}{R_{ef}} = \frac{1}{R_{id}} + \frac{1}{R_{dis}}, \tag{18.23}$$

and for $R_{id} \gg R_{dis}/N$

$$R_oA = \frac{R_{disl}}{n}, \tag{18.24}$$

since $N = nA$. Each dislocation therefore causes a shunt resistance. Its value is 1.2 GΩ for a PbSe photodiode at 80 K. The fluctuations σ in the phenomenological model of Werner and Güttler (see Equation 18.20) are therefore explained as due to barrier lowering caused at and in the vicinity of dislocations. Since each dislocation causes a leakage resistance of about 1.2 GΩ, it follows that the dislocation density should be below 2×10^6 cm^{-2} in order that dislocations do not dominate the actual R_oA product of PbSe photodiodes (the theoretical R_oA value of a PbSe Schottky diode at 80 K is about 10^3 Ωcm^2). Dislocation densities in range are obtainable after proper treatment of the layers (after some temperature cycles from room temperature to 77 K and back to room temperature) as was described by Zogg and colleagues [34,148].

Much improvement is still possible by improving the material quality and device fabrication technique. Better results should be obtained using buried p-n junctions with a thin, wider-bandgap cap layer rather than blocking Pb contacts.

18.5 UNCONVENTIONAL THIN-FILM PHOTODIODES

In practice, the response speed of a photodiode is determined by the effects of junction capacitance, dynamic resistance, and series resistance together with external circuit impedance (see Section 12.2.3). In case of Pb salt photodiodes, the only significant capacitance is the junction space-charge region capacitance C (due to a high dielectric constant) and the only significant resistance is the external resistance R_L; then the upper-frequency limit f_c is [228]:

$$f_c = \frac{1}{2\pi R_L C}. \tag{18.25}$$

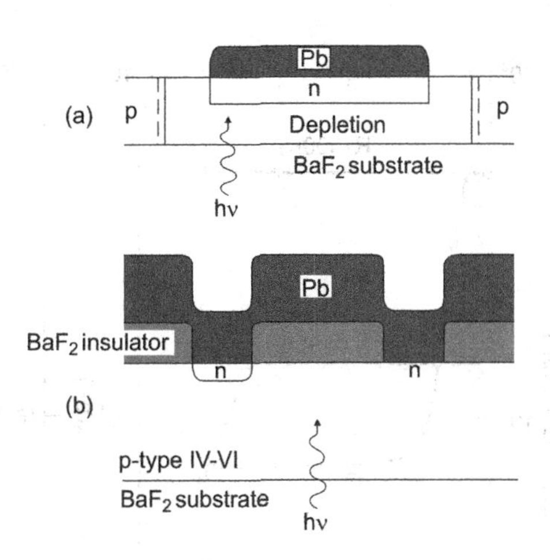

Figure 18.34 The concept of (a) the pinched-off photodiode and (b) the lateral collection photodiode. (From Ref. [230])

For the abrupt junction, we can take $C=\varepsilon_s\varepsilon_o A/w$ and $w=[2\varepsilon_s\varepsilon_o(V_b-V)/qN_{ef}]^{1/2}$ where $N_{ef}=N_aN_d/(N_a+N_d)$ is the effective doping concentration in the space-charge region, and V_b is the built-in junction voltage.

Holloway and coworkers [127,229–231] developed two unconventional photodiodes that gave reduced junction capacitance (Figure 18.34). One approach uses the pinched-off photodiode (Figure 18.34a), which is operated with its depletion region extended from a n-type region through a semiconductor to an insulating substrate. If the bias is changed, the change in the depletion layer width is confined to the periphery of the diode. Some properties of a typical three-quarter wave PbTe pinched-off photodiode are shown in Figure 18.35. This structure was capable of withstanding 150°C for at least 8 hours without degradation of its 80 K performance and required back bias for pinch-off (quarter-wave structures were unusable). The capacitance of the device decreases from a zero-bias value of 700 pF to about 70 pF after application of a back bias greater than about 150 mV. The value of 70 pF is dominated by the lead-out capacitance. The noise current at 1 kHz remains constant for back biases <0.3 V and is close to that calculated for the 300 K background; further bias causes significant contribution of $1/f$ noise. Three-quarter wave devices with a larger area (5×10^{-3} cm^2) showed almost two orders of magnitude reduction in their junction back bias capacitance. The reduction in capacitance of the pinch-off photodiodes significantly decreases the limitation on the operating frequency of IV–VI photodiodes.

The second approach to capacitance reduction is the lateral collection photodiode (Figure 18.34b), which involves a matrix of small p-n junctions that collect photogenerated minority electrons from the intervening nonjunction p-type region. Detailed analysis carried out by Holloway [231] shows that useful collection efficiencies can be obtained with collector separations of up to two diffusion lengths. On the other hand, a reduction in junction capacitance can be obtained if the collector diameters are smaller than this dimension. For IV–VI semiconductors, the diffusion length is of the order of 10–20 µm.

Thin-film lateral collection photodiodes made up of layers of PbTe and PbSeTe on BaF_2 substrates have been described by Holloway et al. [229]. With these devices, a capacitance reduction by a factor of 20 and an increase in the Johnson noise-limited D^* by a factor of three relative to conventional photodiodes have been demonstrated.

It should be noticed that the lateral diffusion of photogenerated carriers significantly enhanced the photoresponse of bulk p-n junctions with dimensions reduced to values that are comparable to the minority carrier diffusion lengths in the semiconductors with which they are made [232–234].

Thin-film IV–VI narrow-band photodiodes have been demonstrated by the research group at the Naval Surface Weapons Center [154,156,213]. Epitaxial layers with slightly different energy gaps were deposited (using HWE technique) on either side of a cleaved BaF_2 substrate (see insert in Figure 18.36). The Schottky barrier (Pb or In) was made on the layer with the smaller energy gap and the layer with the larger energy gap acted as a cooled absorption filter. The spectral response of narrow-band PbSSe photodiodes is shown in Figure 18.36. These photodiodes give half-bandwidths of 0.1–0.2 µm and peak quantum efficiencies of 40%–50%.

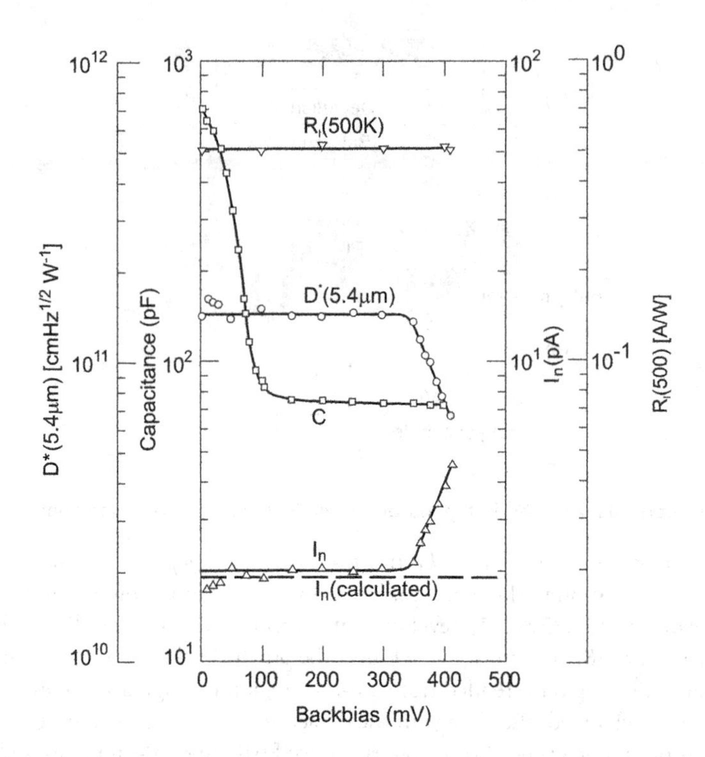

Figure 18.35 Bias-dependent properties of a three-quarter wave pinched-off PbTe photodiode at 80 K and 2π FOV. The blackbody and noise measurements were made at 1 kHz with 10 Hz bandwidths. Thickness of the PbTe film is 0.54 μm. The photodiode area is 6×10^{-4} cm^2. (From Ref. [127])

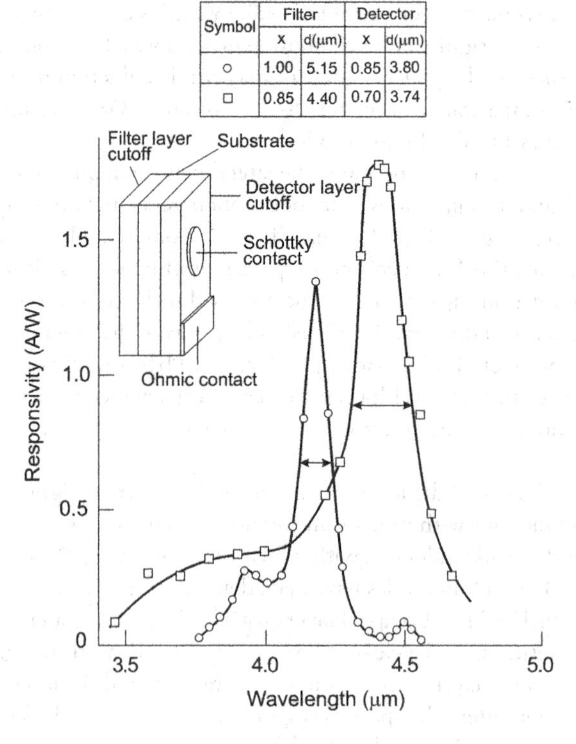

Figure 18.36 Narrow-band configuration and spectral response of PbSSe Schottky barrier photodiodes. (From Ref. [156])

The extension of narrow-band detection technique to longer wavelengths with photodiodes prepared from PbSnSe epitaxial films has been described by Schoolar and Jensen [154]. The spectral half-bandwidth of 0.6 μm at 10.6 μm has been achieved for a 6.67-μm-thick $Pb_{0.935}Sn_{0.065}Se$ photodiode with a 4.12-μm-thick $Pb_{0.095}Sn_{0.05}Se$ filter at 77 K (Johnson-noise-limited $D^* = 7\times10^{10}$ $cmHz^{1/2}$ W^{-1} with cold filter and 20° FOV).

An original narrow-band device containing a lateral collection photodiode on a BaF_2 substrate and a combination of a dielectric stack and a metal reflector has been proposed by Holloway [235]. The peak position of this device may be tuned by changing the angle of incidence of radiation. This suggests that such a structure may find an application as a simple spectrometer.

Multispectral PbSSe and PbSnSe photovoltaic detectors have been also developed at the Naval Surface Weapons Center Laboratories [213,236]. Two-, three-, and four-color detectors were prepared by depositing Pb Schottky barrier contacts onto stair-stepped structures obtained by subsequent HWE deposition of the Pb salt alloy semiconductors on BaF_2 substrates (see Figure 18.28c). The spectral responsivities of these back side illuminated devices cover spectral bands separated by the bandgaps of the underlying layers.

18.6 TUNABLE RESONANT CAVITY ENHANCED DETECTORS

By use of MEMS fabrication techniques, arrays of devices, such as etalons, can be fabricated on an IR detector array that permits tuning of the incident radiation on the detector. If the etalons can be programmed to change the distance from the detector surface by the order of IR wavelengths, the detector responds to all the wavelengths in a waveband sequentially [237].

Zogg and coworkers have elaborated tunable resonant cavity enhanced (RCE) IV–VI detectors operated in the mid-wavelength band [238,239]. Figure 18.37 shows a schematic cross section of a tunable resonant cavity PbTe detector on Si substrate [238]. The top mirror is fabricated with piezo-actuators or MEMS technology and hybridized to the detector chip. The Bragg mirror is realized by MBE with IV–VI materials and consist of quarter wavelength pairs with alternating high- and low-refractive index layers. As is shown in Figure 18.2b, their refractive index is large ($n \approx 5$ to 6) for the narrow-band compositions, while, for example, for EuSe $n = 2.4$ or for BaF_2 $n = 1.4$. A few quarter wavelength pairs are sufficient to obtain near 100% reflectivity across a broad spectral band. It is due to the high index contrast between the constituents of Bragg mirror.

The detector active layer with an area 7.5×10^{-4} cm^2, shown in Figure 18.37, forms a 0.3-μm-thick n^+-p structure grown on a 1.5 pair Bragg mirror. In fact, it is a heterojunction with a top Bi-doped n^+-$Pb_{1-x}Sr_xTe$

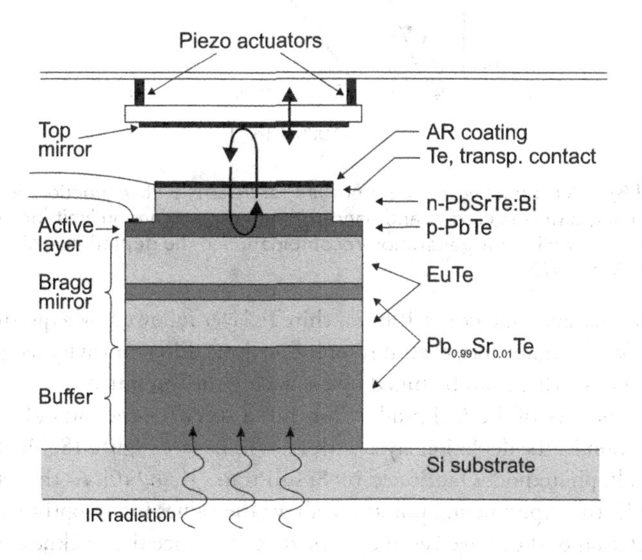

Figure 18.37 Schematic cross section of a PbTe-on-Si RCE detector. (From Ref. [138])

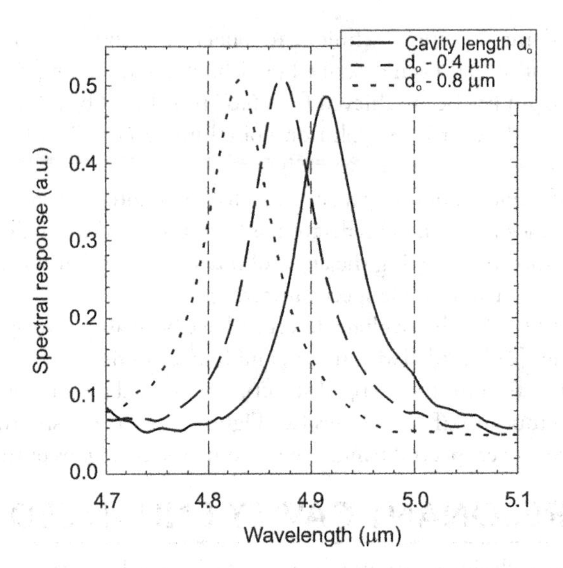

Figure 18.38 Three spectral response with different cavity lengths obtained with the piezo-actuated mirror at 100 K. (From Ref. [239])

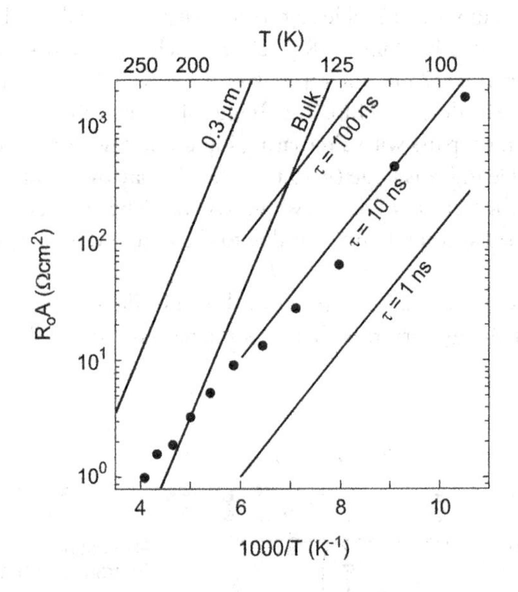

Figure 18.39 Measured (•) R_oA product versus temperature for RCE n⁺-p PbTe junction on Si substrate. Solid lines are calculated assuming diffusion current and band-to-band recombination limit for thick (bulk) and thin (0.3-μm-thick) photodiodes as well as for generation-recombination in the depletion region for three SR lifetimes (100, 10, and 1 ns). (From Ref. [239])

window. The top IR transparent contact is a 100 nm thin Te layer followed by a quarter wavelength TiO_2 AR coating. Figure 18.38 illustrates the spectral response at three different cavity lengths [239]. The spectrum exhibits only one peak, which can be tuned by displacing the top mirror.

The temperature dependence of the R_oA product for thin n^+-p PbTe junction indicates performance limitation by SR recombination in the depletion layer with $\tau_{SR} \approx 10$ ns (see Figure 18.39). This value is higher as compared to PbTe bulk photodiodes fabricated on Si substrates [174,240]. At the temperature range between 250 K to 200 K, the experimental points are above the values for an optimized bulk photodiode. It results from the limitation of the active volume of the detector, since the thickness of the absorbing layer is smaller than the electron diffusion length (see Equation 12.88). However, the experimental values are still much below the theoretical thin limit since for a 0.3-μm-thick absorber layer, the R_oA product should increase 40 times.

Another solution of tunable cavity enhanced detectors is a device with the movable MEMS mirror that consists of Au-coated square mirror plate attached to four symmetrically arranged suspension legs [118]. The counterelectrodes are placed on a glass support wafer at a distance of 10 μm. A movement of more than 3 μm is achieved by applying a voltage of 30 V between the silicon mirror membrane and the electrodes on the glass wafer.

18.7 LEAD SALTS VERSUS HgCdTe

Theoretically, in comparison with HgCdTe, Pb salt ternary alloys have

- stronger chemical bonds,
- more resilience to defects (see Figure 18.40, where dependence of the R_oA product on dislocation density is shown for photodiodes with $\lambda_c \approx 10$ μm at 77 K),
- cutoff wavelength is less sensitive to composition,
- less sensitive to passivation requirement,

which should lead to large and highly uniform FPAs operated at higher temperatures. The consequence of the above properties is potentially significant, particularly for very LWIR spectral region. The challenges are soft material, high dielectric constant (low speed), and higher thermal mismatch with Si.

Pb salt ternary alloys (PbSnTe and PbSnSe) seemed easier to prepare and appeared more stable. Development of IV–VI alloy photodiodes was discontinued because the chalcogenides suffered two significant drawbacks: high dielectric constant and very high thermal coefficients of expansion (TCE).

High dielectric constant influences the space-charge region capacitance and the response speed of photodiode. Figure 18.41 shows plots of the cutoff frequency versus applied reverse-bias voltage for one-side abrupt n-p$^+$ $Pb_{0.78}Sn_{0.22}Te$ and $Hg_{0.797}Cd_{0.203}Te$ photodiodes at 77 K ($\lambda_c \approx 12$ μm) [228]. For reverse-bias voltage above 1 V and doping concentrations in the space-charge region above 10^{15} cm^{-3}, an avalanche breakdown can be observed. The cutoff frequency has been calculated for a load resistance of 50 Ω and for a junction area of 10^{-4} cm^2. Various values of donor concentrations are marked in Figure 18.41. In this figure, the space-charge region width and the capacitance per unit C/A are also shown. We can see that cutoff frequencies of 2 GHz can be realized with HgCdTe photodiodes when n-side doping concentration is not larger than 10^{14} cm^{-3} at reverse bias voltages. The cutoff frequency for PbSnTe photodiodes is almost an order of magnitude smaller. It should be noted that the cutoff frequency is decreased by any junction series resistance and stray capacitance.

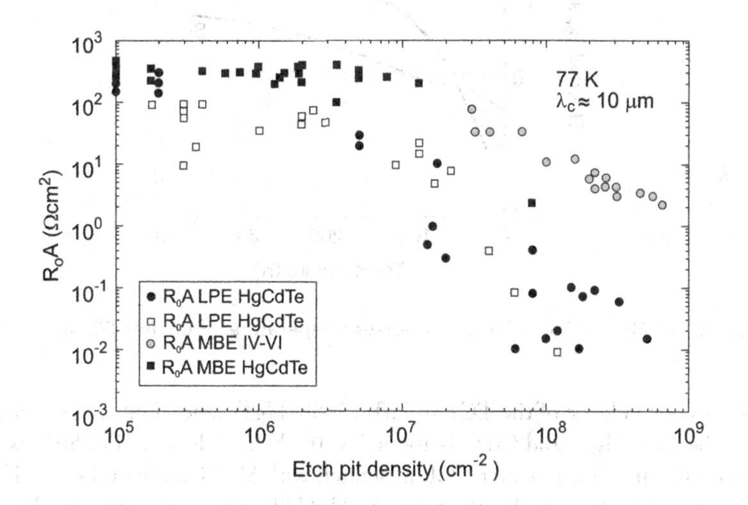

Figure 18.40 Dependence of the R_oA product on dislocation density for HgCdTe and IV–VI photodiodes with cutoff wavelength ≈10 μm at 77 K. (From Tidrow, M., private communication, 2007)

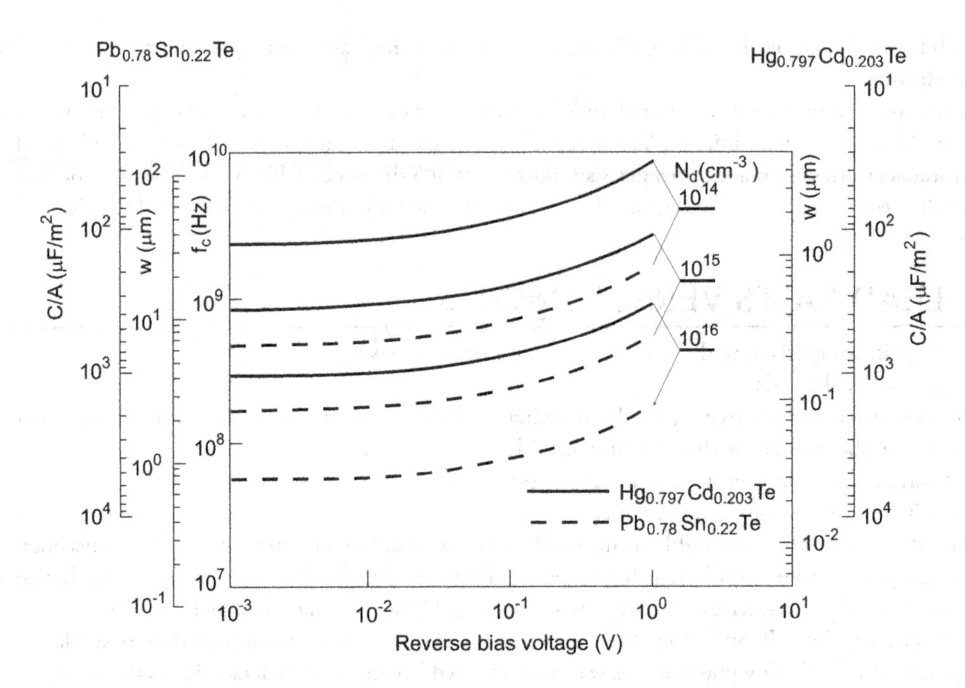

Figure 18.41 Cutoff frequency for one-side abrupt n-p⁺ $Pb_{0.78}Sn_{0.22}Te$ and $Hg_{0.797}Cd_{0.203}Te$ photodiodes ($\lambda_c \approx 12\,\mu m$) at 77 K with an active area of 10^{-4} cm². The additional scales correspond to the depletion width and the junction capacitance per unit area. (From Ref. [228])

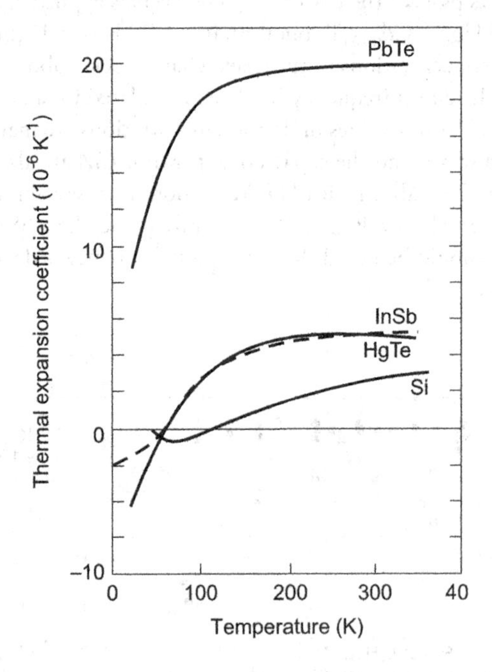

Figure 18.42 Linear TCE of PbTe, InSb, HgTe, and Si versus temperature. (From Ref. [220])

Figure 18.42 shows dependence of the TCE of PbTe, InSb, HgTe ,and Si on temperature [220]. At room temperature, the TCE HgTe and CdTe is about 5×10^{-6} K^{-1}, while that of PbSnTe is in the range of 20×10^{-6} K^{-1}. This results in a much greater TCE mismatch with Si (TCE about 3×10^{-6} K^{-1}). It should be noted that both Ge and GaAs have TCE values close to HgCdTe, giving detectors on those materials that have no significant advantage in this respect.

The doping concentrations in the base regions of HgCdTe and PbSnTe photodiodes are different. For both types of photodiodes, the tunneling current (and R_oA product) is critically dependent on doping concentration. To produce high R_oA products for HgCdTe and Pb salt photodiodes, the doping concentration of 10^{16} and 10^{17} cm^{-3} (or less) are required, respectively (see Figure 17.45). The maximum available doping levels due to the onset of tunneling are more than an order of magnitude higher with IV–VIs than with HgCdTe photodiodes. This is due to their high permittivities ε_s because the tunneling contribution of the R_oA product contains factors $\exp[\mathrm{const}(m^*\varepsilon_s/N)^{1/2}E_g]$ (see Equation 12.120). The maximum allowable concentrations above 10^{17} cm^{-3} are easily controllable in IV–VIs grown by MBE. Tunneling current is not a limiting issue in Pb salt photodiodes.

In comparison with HgCdTe detectors, the main performance limit of Pb salt is connected with SR centers. It is not clear whether it is due to residual or native defects in material. However, it is obvious that these SR centers exert a dominating influence on the magnitude of the depletion current. Domination of depletion current, unlike the case of HgCdTe with the same bandgap, requires a lower temperature of operation for BLIP performance than for HgCdTe.

REFERENCES

1. T. W. Case, "Notes on the change of resistance of certain substrates in light," *Phys. Rev.* 9, 305–310, 1917; "The thalofide cell: a new photoelectric substance," *Phys. Rev.* 15, 289, 1920.
2. E. W. Kutzscher, "Letter to the editor," *Electro-Opt. Syst. Des.* 5, 62, 1973.
3. R. J. Cashman, "Film-type infrared photoconductors," *Proc. IRE* 47, 1471–75, 1959.
4. J. O. Dimmock, I. Melngailis, and A. J. Strauss, "Band structure and laser action in $Pb_{1-x}Sn_xTe$," *Phys. Rev. Lett.* 16, 1193–1196, 1966.
5. A. J. Strauss, "Inversion of conduction and valence bands in $Pb_{1-x}Sn_xSe$ alloys," *Phys. Rev.* 157, 608–611, 1967.
6. J. Melngailis and T. C. Harman, "Single-crystal lead-tin chalcogenides," in *Semiconductors and Semimetals*, Vol. 5, eds. R. K. Willardson and A. C. Beer, 111–174, Academic Press, New York, 1970.
7. T. C. Harman and J. Melngailis, "Narrow gap semiconductors," in *Applied Solid State Science*, Vol. 4, ed. R. Wolfe, 1–94, Academic Press, New York, 1974.
8. J. T. Longo, D. T. Cheung, A. M. Andrews, C. C. Wang, and J. M. Tracy, "Infrared focal planes in intrinsic semiconductors," *IEEE Trans. Electron Devices* ED-25, 213–232, 1978.
9. H. Zogg and A. Ishida, "IV–VI (lead chalcogenide) infrared sensors and lasers," in *Infrared Detectors and Emitters: Materials and Devices*, eds. P. Capper and C. T. Elliott, 43–75, Kluwer, Boston, 2001.
10. G. Springholz, "Molecular beam epitaxy of IV–VI heterostructures and superlattices," in *Lead Chalcogenides: Physics and Applications*, ed. D. Khoklov, 123–207, Taylor & Francis Inc., New York, 2003.
11. H. Zogg, "Lead chalcogenide infrared detectors grown on silicon substrates," in *Lead Chalcogenides: Physics and Applications*, ed. D. Khoklov, 587–615, Taylor & Francis Inc., New York, 2003.
12. G. Springholz, T. Schwarzl, and W. Heiss, "Mid-infrared vertical cavity surface emitting lasers based on the lead salt compounds," in *Mid-Infrared Semiconductor Optoelectronics*, ed. A. Krier, 265–301, Springer Verlag, London, 2007.
13. R. Dalven, "A review of the semiconductor properties of PbTe, PbSe, PbS and PbO," *Infrared Phys.* 9, 141–184, 1969.
14. Yu. I. Ravich, B. A. Efimova, and I. A. Smirnov, *Semiconducting Lead Chalcogenides*, Plenum Press, New York, 1970.
15. H. Preier, "Recent advances in lead-chalcogenide diode lasers," *Appl. Phys.* 10, 189–206, 1979.
16. H. Maier and J. Hesse, "Growth, properties and applications of narrow-gap semiconductors," in *Crystal Growth, Properties and Applications*, ed. H. C. Freyhardt, 145–219, Springer Verlag, Berlin, 1980.
17. R. Dornhaus, G. Nimtz, and B. Schlicht, *Narrow-Gap Semiconductors*, Springer Verlag, Berlin, 1983.
18. A. V. Ljubchenko, E. A. Salkov, and F. F. Sizov, *Physical Fundamentals of Semiconductor Infrared Photoelectronics*, Naukova Dumka, Kiev, 1984.
19. A. Rogalski and J. Piotrowski, "Intrinsic infrared detectors," *Prog. Quantum Electron.* 12, 87–289, 1988.
20. A. Rogalski, "IV–VI detectors," in *Infrared Photon Detectors*, ed. E. Rogalski, 513–559, SPIE Optical Engineering Press, Bellingham, WA, 1995.
21. A. Rogalski, K. Adamiec, and J. Rutkowski, *Narrow-Gap Semiconductor Photodiodes*, SPIE Press, Bellingham, WA, 2000.
22. A. Rogalski, *Infrared Detectors*, CRC Press, Boca Raton, FL, 2011.

23. A. Rogalski, “Narrow-gap semiconductors for infrared detectors,” in *Handbook of Luminescent Semiconductor Materials*, eds. L. Bergman and J. L. McHale, 191–253, CRC Press, Boca Raton, FL, 2012.
24. G. Springholz and G. Bauer, “Molecular beam epitaxy of IV–VI semiconductor hetero- and nano-structures,” *Phys. Status Solidi B* 244, 2752–2767, 2007.
25. T. C. Harman, “Control of imperfections in crystals of $Pb_{1-x}Sn_xTe$, $Pb_{1-x}Sn_xSe$, and $PbS_{1-x}Se_x$,” *J. Nonmet.* 1, 183–194, 1973.
26. S. G. Parker and R. E. Johnson, “Preparation and properties of (Pb,Sn)Te,” in *Preparation and Properties of Solid State Materials*, ed. W. R. Wilcox, 1–65, Marcel Dekker, Inc., New York, 1981.
27. D. L. Partin and J. Heremans, “Growth of narrow bandgap semiconductors,” in *Handbook on Semiconductors*, Vol. 3, ed. S. Mahajan, 369–450, Elsevier Science B.V., Amsterdam, 1994.
28. N. Tamari and H. Shtrikman, “Growth study of large nonseeded $Pb_{1-x}Sn_xTe$ single crystals,” *J. Electron. Mater.* 8, 269–288, 1979.
29. E. V. Markov and A. A. Davydov, “Vapour phase growth of cadmium sulphide single crystals,” *Neorg. Mater.* 11, 1755–258, 1975.
30. K. Grasza, “Estimation of the optimal conditions for directional crystal growth from the vapour phase with no contact between crystal and ampoule wall,” *J. Cryst. Growth* 128, 609–612, 1993.
31. D. L. Partin, “Molecular-beam epitaxy of IV–VI compound heterojunctions and superlattices,” in *Semiconductors and Semimetals*, eds. R. K. Willardson and A. C. Beer, 311–336, Academic Press, Boston, 1991.
32. S. Szapiro, N. Tamari, and H. Shtrikman, “Calculation of the phase diagram of the Pb–Sn–Te system in the (Pb+Sn)-rich region,” *J. Electron. Mater.* 10, 501–516, 1981.
33. J. Kasai and W. Bassett, “Liquid phase epitaxial growth of $Pb_{1-x}Sn_xSe$,” *J. Cryst. Growth* 27, 215–220, 1974.
34. P. Müller, H. Zogg, A. Fach, J. John, C. Paglino, A. N. Tiwari, M. Krejci, and G. Kostorz, “Reduction of threading dislocation densities in heavily lattice mismatched PbSe on Si(111) by glide,” *Phys. Rev. Lett.* 78, 3007–3010, 1997.
35. Y. Huand and R. F. Brebrick, “Partial pressures and thermodynamic properties for lead telluride,” *J. Electrochem. Soc.* 135, 486–496, 1988.
36. Y. Huand and R. F. Brebrick, “Partial pressures and thermodynamic properties of PbTe–SnTe solid and liquid solutions with 13, 20, and 100 mole percent SnTe,” *J. Electrochem. Soc.* 135, 1547–1559, 1988.
37. Y. G. Sha and R. F. Brebrick, “Explicit incorporation of the energy-band structure into an analysis of the defect chemistry of PbTe and SnTe,” *J. Electron. Mater.* 18, 421–443, 1989.
38. N. J. Parada and G. W. Pratt, “New model for vacancy states in PbTe,” *Phys. Rev. Lett.* 22, 180–182, 1969.
39. G. W. Pratt, “Vacancy and interstitial states in the lead salts,” *J. Nonmet.* 1, 103–109, 1973.
40. L. A. Hemstreet, “Cluster calculations of the effects of lattice vacancies in PbTe and SnTe,” *Phys. Rev.* B12, 1213–1216, 1975.
41. C. S. Lent, M. A. Bowen, R. A. Allgaier, J. D. Dow, O. F. Sankey, and E. S. Ho, “Impurity levels in PbTe and $Pb_{1-x}Sn_xTe$,” *Solid State Commun.* 61, 83–87, 1987.
42. L. Palmetshofer, “Ion implantation in IV–VI semiconductors,” *Appl. Phys.* A34, 139–153, 1984.
43. R. Dornhaus, G. Nimtz, and W. Richter, *Solid-State Physics*, Springer Verlag, Berlin. 1976.
44. K. Lischka, “Bound defect states in IV–VI semiconductors,” *Appl. Phys.* A29, 177–189, 1982.
45. V. I. Kaidanov and Yu. I. Ravich, “Deep and resonant states in $A^{IV}B^{VI}$ semiconductors,” *Usp. Fiz. Nauk* 145, 51–86, 1985.
46. B. A. Akimov, L. I. Ryabova, O. B. Yatsenko, and S. M. Chudinov, “Rebuilding of energy spectrum in $Pb_{1-x}Sn_xTe$ alloys doped with in under the pressure and the variation of their composition,” *Fiz. Tekh. Poluprovodn* 13, 752–759, 1979; B. A. Akimov, A. V. Dmitriev, D. R. Khokhlov, and L. I. Ryabova, “Carrier transport and non-equillibrium phenomena in doped PbTe and related materials,” *Phys. Status Solidi A* 137, 9–55, 1993.
47. E. Silberg and A. Zemel, “Cadmium diffusion studies of PbTe and $Pb_{1-x}Sn_xTe$ crystals,” *J. Electron. Mater.* 8, 99–109, 1979.
48. A. Rogalski and K. Jóźwikowski, “The intrinsic carrier concentration in $Pb_{1-x}Sn_xTe$, $Pb_{1-x}Sn_xSe$ and $PbS_{1-x}Se_x$,” *Phys. Status Solidi A* 111, 559–565, 1989.
49. Yu. Ravich, B. A. Efimova, and V. I. Tamarchenko, “Scattering of current carriers and transport phenomena in lead chalcogenides. I. Theory,” *Phys. Status Solidi B* 43, 11–33, 1971.
50. Yu. Ravich, B. A. Efimova, and V. I. Tamarchenko, “Scattering of current carriers and transport phenomena in lead chalcogenides. II. Experiment,” *Phys. Status Solidi B* 43, 453–469, 1971.
51. W. Zawadzki, “Mechanisms of electron scattering in semiconductors,” in *Handbook on Semiconductors*, Vol. 1, ed. W. Paul, 423, North Holland, Amsterdam, 1980.

52. F. F. Sizov, G. V. Lashkarev, M. V. Radchenko, V. B. Orletzki, and E. T. Grigorovich, "Peculiarities of carrier scattering mechanisms in narrow gap semiconductors," *Fiz. Tekh. Poluprovodn.* 10, 1801–1808, 1976.
53. A. Lopez-Otero, "Hot wall epitaxy," *Thin Solid Films* 49, 3–57, 1978.
54. W. W. Anderson, "Absorption constant of $Pb_{1-x}Sn_xTe$ and $Hg_{1-x}Cd_xTe$ alloys," *Infrared Phys.* 20, 363–372, 1980.
55. D. Genzow, K. H. Herrmann, H. Kostial, I. Rechenberg, and A. E. Yunovich, "Interband absorption edge in $Pb_{1-x}Sn_xTe$," *Phys. Status Solidi B* 86, K21–K25, 1978.
56. D. Genzow, A. G. Mironow, and O. Ziep, "On the interband absorption in lead chalcogenides," *Phys. Status Solidi B* 90, 535–542, 1978.
57. O. Ziep and D. Genzow, "Calculation of the interband absorption in lead chalcogenides using a multiband model," *Phys. Status Solidi B* 96, 359–368, 1979.
58. B. L. Gelmont, T. R. Globus, and A. V. Matveenko, "Optical absorption and band structure of PbTe," *Solid State Commun.* 38, 931–934, 1981.
59. N. Suzuki and S. Adachi, "Optical constants of $Pb_{1-x}Sn_xTe$ Alloys," *J. Appl. Phys.* 79, 2065–2069, 1996.
60. S. Adachi, *Properties of Group-IV, III–V and II–VI Semiconductors*, John Wiley & Sons, Ltd, Chichester, UK, 2005.
61. A. V. Butenko, R. Kahatabi, E. Mogilko, R. Strul, V. Sandomirsky, Y. Schlesinger, Z. Dashevsky, V. Kasiyan, and S. Genikhov, "Characterization of high-temperature PbTe p-n junctions prepared by thermal diffusion and by ion implantation," *J. Appl. Phys.* 103, 024506, 2008.
62. J. H. Barrett, "Dielectric constant in perovskite type crystals," *Phys. Rev.* 86, 118–120, 1952.
63. Y. Shani, R. Rosman, and A. Katzir, "Calculation of the refractive index of lead-tin-telluride for infrared devices," *IEEE J. Quantum Electron.* QE-20, 1110–1114, 1984.
64. B. Jensen and A. Tarabi, "Dispersion of the refractive index of ternary compound $Pb_{1-x}Sn_xTe$," *Appl. Phys. Lett.* 45, 266–268, 1984.
65. B. Jensen and A. Tarabi, "The refractive index of compounds PbTe, PbSe, and PbS," *IEEE J. Quantum Electron.* QE-20, 618–21, 1984.
66. K. H. Harrmann, V. Melzer, and U. Muller, "Interband and intraband contributions to refractive index and dispersion in narrow-gap semiconductors," *Infrared Phys.* 34, 117–136, 1993.
67. P. Herve and L. K. J. Vandamme, "General relation between refractive index and energy gap in semiconductors," *Infrared Phys. Technol.* 35, 609–615, 1994.
68. O. Ziep, D. Genzow, M. Mocker, and K. H. Herrmann, "Nonradiative and radiative recombination in lead chalcogenides," *Phys. Status Solidi B* 99, 129–138, 1980.
69. O. Ziep, "Estimation of second-order Auger recombination in lead chalcogenides," *Phys. Status Solidi B* 115, 161–170, 1983.
70. D. A. Cammack, A. V. Nurmikko, G. W. Pratt, and J. R. Lowney, "Enhanced interband recombination in $Pb_{1-x}Sn_xTe$," *J. Appl. Phys.* 46, 3965–3969, 1975.
71. D. M. Gureev, O. I. Davarashvili, I. I. Zasavitskii, B. N. Matsonashvili, and A. R. Shotov, "Radiative recombination in epitaxial $Pb_{1-x}Sn_xSe$ ($0 \leq x \leq 0.4$) layers," *Fiz. Tekh. Poluprovodn.* 13, 1752–1755, 1979.
72. G. Nimtz, "Recombination in narrow-gap semiconductors," *Phys. Rep.* 63, 265–300, 1980.
73. K. Lischka and W. Huber, "Carrier recombination and deep levels in PbTe," *Solid State Electron.* 21, 1509–1512, 1978.
74. K. H. Harrmann, "Recombination in small-gap $Pb_{1-x}Sn_xTe$," *Solid State Electron.* 21, 1487–1491, 1978.
75. D. Khokhlov, ed. *Lead Chalcogenides: Physics and Applications*, Taylor & Francis Inc., New York, 2003.
76. P. R. Emtage, "Auger recombination and junction resistance in lead-tin telluride," *J. Appl. Phys.* 47, 2565–2568, 1976.
77. K. Lischka, W. Huber, and H. Heinrich, "Experimental determination of the minority carrier lifetime in PbTe p-n junctions," *Solid State Commun.* 20, 929–931, 1976.
78. K. Lischka and W. Huber, "Auger recombination in PbTe," *J. Appl. Phys.* 48, 2632–2633, 1977.
79. T. X. Hoai and K. H. Herrmann, "Recombination in $Pb_{0.82}Sn_{0.18}Te$ at high levels of optical excitation," *Phys. Status Solidi B* 83, 465–470, 1977.
80. O. Ziep, M. Mocker, D. Genzow, and K. H. Herrmann, "Auger recombination in PbSnTe-like semiconductors," *Phys. Status Solidi B* 90, 197–205, 1978.
81. O. Ziep and M. Mocker, "A new approach to Auger recombination," *Phys. Status Solidi B* 98, 133–142, 1980.
82. R. Rosman and A. Katzir, "Lifetime calculations for Auger recombination in lead-tin-telluride," *IEEE J. Quantum Electron.* QE18, 814–817, 1982.
83. H. Zogg, W. Vogt, and W. Baumgartner, "Carrier recombination in single crystal PbSe," *Solid State Electron.* 25, 1147–1155, 1982.

84. A. Shahar, M. Oron, and A. Zussman, "Minority-carrier diffusion-length measurement and lifetime in $Pb_{0.8}Sn_{0.2}Te$ and indium-doped PbTe liquid phase layers," *J. Appl. Phys.* 54, 2477–2482, 1983.
85. M. Mocker and O. Ziep, "Intrinsic recombination in dependence on doping concentration and excitation level," *Phys. Status Solidi B* 115, 415–425, 1983.
86. O. Ziep, M. Mocker, and D. Genzow, "Theorie der band-band-rekombination in bleichalkogeniden," *Wissensch. Zeitschr. Humboldt-Univ. Berlin, Math.-Nataruiss. Reihe* XXX, 81–97, 1981.
87. A. V. Dmitriev, "Calculation of the Auger lifetime of degenerate carriers in the many-valley narrow gap semiconductors," *Solid State Commun.* 74, 577–581, 1990.
88. S. D. Beneslavskii and A. V. Dmitriev, "Auger transitions in narrow-gap semiconductors with Lax band structures," *Solid State Commun.* 39, 811–814, 1981.
89. P. Berndt, D. Genzow, and K. H. Herrmann, "Recombination analysis in 10 µm $Pb_{1-x}Sn_xTe$," *Phys. Status Solidi A* 38, 497–503, 1976.
90. A. Schlicht, R. Dornhaus, G. Nimtz, L. D. Haas, and T. Jakobus, "Lifetime measurements in PbTe and PbSnTe," *Solid State Electron.* 21, 1481–1485, 1981.
91. K. Weiser, E. Ribak, A. Klein, and M. Ainhorn, "Recombination of photocarriers in lead-tin telluride," *Infrared Phys.* 21, 149–154, 1981.
92. A. Genzow, M. Mocker, and E. Normantas, "The influence of uniaxial strain on the band to band recombination in $Pb_{1-x}Sn_xTe$," *Phys. Status Solidi B* 135, 261–270, 1986.
93. S. Espevik, C. Wu, and R. H. Bube, "Mechanism of photoconductivity in chemically deposited lead sulfide layers," *J. Appl. Phys.* 42, 3513–3529, 1971.
94. T. H. Johnson, "Lead salt detectors and arrays. PbS and PbSe," *Proc. SPIE* 443, 60–94, 1984.
95. L. N. Neustroev and V. V. Osipov, "Physical properties of photosensitive polycrystalline lead chalcogenide films," *Mikroelektronika* 17, 399–416, 1988.
96. J. Vaitkus, M. Petrauskas, R. Tomasiunas, and R. Masteika, "Photoconductivity of highly excited $A^{IV}B^{VI}$ thin films," *Appl. Phys.* A54, 553–555, 1992.
97. J. Vaitkus, R. Tomasiunas, K. Tumkevicius, M. Petrauskas, and R. Masteika, "Picosecond photoconductivity of polycrystalline PbTe films," *Fiz. Tekh. Poluprovodn.* 24, 1919–1922, 1990.
98. T. S. Moss, "Lead salt photoconductors," *Proc. IRE* 43, 1869–1881, 1955.
99. R. J. Cashman, "Film-type infrared photoconductors," *Proc. IRE* 47, 1471–1475, 1959.
100. P. W. Kruse, L. D. McGlauchlin, and R. B. McQuistan, *Elements of Infrared Technology*, Wiley, New York, 1962.
101. D. E. Bode, T. H. Johnson, and B. N. McLean, "Lead selenide detectors for intermediate temperature operation," *Appl. Opt.* 4, 327–331, 1965.
102. J. N. Humphrey, "Optimization of lead sulfide infrared detectors under diverse operating conditions," *Appl. Opt.* 4, 665–675, 1965.
103. T. H. Johnson, H. T. Cozine, and B. N. McLean, "Lead selenide detectors for ambient temperature operation," *Appl. Opt.* 4, 693–696, 1965.
104. D. E. Bode, "Lead salt detectors," in *Physics of Thin Films*, Vol. 3, eds. G. Hass and R. E. Thun, 275–301, Academic Press, New York, 1966.
105. T. S. Moss, G. J. Burrel, and B. Ellis, *Semiconductor Optoelectronics*, Butterworths, London, 1973.
106. R. E. Harris, "PbS… Mr. versatility of the detector world," *Electro-Opt. Syst. Des.* 47–50, 1976.
107. R. E. Harris, "PbSe... Mr super sleuth of the detector world," *Electro-Opt. Syst. Des.* 42–44, 1977.
108. R. H. Harris, "Lead-salt detectors," *Laser Focus/Electro-Optics*, 87–96, 1983.
109. M. C. Torquemada, M. T. Rodrigo, G. Vergara, F. J. Sanchez, R. Almazan, M. Verdu, P. Rodriguez, et al., "Role of halogens in the mechanism of sensitization of uncooled PbSe infrared photodetectors," *J. Appl. Phys.* 93, 1778–1784, 2003.
110. C. Nascu, V. Vomir, I. Pop, V. Ionescu, and R. Grecu, "The study of PbS films: VI. Influence of oxidants on the chemically deposited PbS thin films," *Mater. Sci. Eng.* B41, 235–240, 1996.
111. I. Grozdanov, M. Najdoski, and S. K. Dey, "A simple solution growth technique for PbSe thin films," *Mat. Lett.* 38, 28, 1999.
112. S. Horn, D. Lohrmann, P. Norton, K. McCormack, and A. Hutchinson, "Reaching for the sensitivity limits of uncooled and minimaly-cooled thermal and photon infrared detectors," *Proc. SPIE* 5783, 401–411, 2005.
113. B. Weng, J. Qiu, L. Zhaoa, Z. Yuana, C. Changa, and Z. Shi, "Recent development on the uncooled mid-infrared PbSe detectors with high detectivity", *Proc. SPIE* 8993, 899311, 2014
114. K. Green, S.-S. Yoo, C. Kauffman, "Lead salt TE-cooled imaging sensor development", *Proc. SPIE* 9070, 90701G, 2014.

115. R. L. Herrero, M. T. Montojo Supervielle and A. B. Ramirez, "VPD PbSe technology: the road towards the industrial maturity", *Proc. SPIE* 7660, 766034, 2010.
116. P. R. Norton, "Infrared image sensors," *Opt. Eng.* 30, 1649–1663, 1991.
117. E. L. Dereniak and G. D. Boreman, *Infrared Detectors and Systems*, Wiley, New York, 1996.
118. N. Quack, S. Blunier, J. Dual, F. Felder, M. Arnold, and H. Zogg, "Mid-infrared tunable resonant cavity enhanced detectors," *Sensors* 8, 5466–5478, 2008.
119. A. Rogalski, J. Antoszewski, and L. Faraone, "Third-generation infrared photodetector arrays," *J. Appl. Phys.* 105, 091101–091144, 2009.
120. A. Rogalski, "Detectivity limits for PbTe photovoltaic detectors," *Infrared Phys.* 20, 223–229, 1980.
121. A. Rogalski and J. Rutkowski, "R_oA Product for PbS and PbSe abrupt p-n junctions," *Opt. Appl.* 11, 365–370, 1981.
122. A. Rogalski and J. Rutkowski, "Temperature dependence of the R_oA product for lead chalcogenides photovoltaic detectors," *Infrared Phys.* 21, 191–199, 1981.
123. A. Rogalski, W. Kaszuba, and W. Larkowski, "PbTe photodiodes prepared by the hot-wall evaporation technique," *Thin Solid Films* 103, 343–353, 1983.
124. A. Rogalski, R. Ciupa, and H. Zogg, "Computer modeling of carrier transport in binary salt photodiodes," *Proc. SPIE* 2373, 172–181, 1995.
125. J. Baars, D. Bassett, and M. Schulz, "Metal-semiconductor barrier studies of PbTe," *Phys. Status Solidi A* 49, 483–488, 1978.
126. J. P. Donnelly, T. C. Harman, A. G. Foyt, and W. T. Lindley, "PbTe photodiodes fabricated by Sb^+ ion implantation," *J. Nonmet.* 1, 123–128, 1973.
127. H. Holloway and K. F. Yeung, "Low-capacitance PbTe photodiodes," *Appl. Phys. Lett.* 30, 210–212, 1977.
128. C. H. Gooch, H. A. Tarry, R. C. Bottomley, and B. Waldock, "Planar indium-diffused lead telluride detector arrays," *Electronics Letters* 14, 209–210, 1978.
129. H. Zogg, S. Blunier, T. Hoshino, C. Maissen, J. Masek, and A. N. Tiwari, "Infrared sensor arrays with 3–12 μm cutoff wavelengths in heteroepitaxial narrow-gap semiconductors on silicon substrates," *IEEE Trans. Electron Devices* 38, 1110–1117, 1991.
130. A. Rogalski and R. Ciupa, "PbSnTe photodiodes: theoretical predictions and experimental data," *Opto-Electron. Rev.* 5, 21–29, 1997.
131. C. A. Kennedy, K. J. Linden, and D. A. Soderman, "High-performance 8–14 μm $Pb_{1-x}Sn_xTe$ photodiodes," *Proc. IEEE* 63, 27–32, 1976.
132. C. C. Wang and S. R. Hampton, "Lead telluride-lead tin telluride heterojunction diode array," *Solid State Electron.* 18, 121–125, 1975.
133. P. S. Chia, J. R. Balon, A. H. Lockwood, D. M. Randall, and F. J. Renda, "Performance of PbSnTe diodes at moderately backgrounds," *Infrared Phys.* 15, 279–285, 1975.
134. C. C. Wang and J. S. Lorenzo, "High-performance, high-density, planar PbSnTe detector arrays," *Infrared Phys.* 17, 83–88, 1977.
135. C. C. Wang and M. E. Kim, "Long-wavelength PbSnTe/PbTe inverted heterostructure mosaics," *J. Appl. Phys.* 50, 3733–3737, 1979.
136. M. Grudzień and A. Rogalski, "Photovoltaic detectors $Pb_{1-x}Sn_xTe$ ($0 \leq x \leq 0.25$). Minority carrier lifetimes. Resistance-area product," *Infrared Phys.* 21, 1–8, 1981.
137. V. F. Chishko, V. T. Hryapov, I. L. Kasatkin, V. V. Osipov, and O. V. Smolin, "High detectivity $Pb_{0.8}Sn_{0.2}Te$–$PbSe_{0.08}Te_{0.92}$ heterostructure photodiodes," *Infrared Phys.* 33, 275–280, 1992.
138. H. Zogg, C. Maissen, J. Masek, S. Blunier, A. Lambrecht, and M. Tacke, "Heteroepitaxial $Pb_{1-x}Sn_xSe$ on Si infrared sensor array with 12 μm cutoff wavelength," *Appl. Phys. Lett.* 55, 969–971, 1989.
139. H. Zogg, C. Maissen, J. Masek, S. Blunier, A. Lambrecht, and M. Tacke, "Epitaxial lead chalcogenide IR sensors on Si for 3–5 and 8–12 μm," *Semicond. Sci. Technol.* 5, S49–S52, 1990.
140. H. Zogg, C. Maissen, J. Masek, T. Hoshino, S. Blunier, and A. N. Tiwari, "Photovotaic infrared sensor arrays in monolithic lead chalcogenides on silicon," *Semicond. Sci. Technol.* 6, C36–C41, 1991.
141. T. Hoshino, C. Maissen, H. Zogg, J. Masek, S. Blunier, A. N. Tiwari, S. Teodoropol, and W. J. Bober, "Monolithic $Pb_{1-x}Sn_xSe$ infrared sensor arrays on Si prepared by low-temperature process," *Infrared Phys.* 32, 169–175, 1991.
142. J. Masek, T. Hoshino, C. Maissen, H. Zogg, S. Blunier, J. Vermeiren, and C. Claeys, "Monolithic lead-chalcogenide IR-Arrays on silicon: fabrication and use in thermal imaging applications," *Proc. SPIE* 1735, 54–61, 1992.
143. H. Zogg, A. Fach, C. Maissen, J. Masek, and S. Blunier, "Photovoltaic lead-chalcogenide on silicon infrared sensor arrays," *Opt. Eng.* 33, 1440–1449, 1994.

144. C. Paglino, A. Fach, J. John, P. Müller, H. Zogg, and D. Pescia, "Schottky-barrier fluctuations in $Pb_{1-x}Sn_xSe$ infrared sensors," *J. Appl. Phys.* 80, 7138–7143, 1996.
145. A. Fach, J. John, P. Müller, C. Paglino, and H. Zogg, "Material properties of $Pb_{1-x}Sn_xSe$ epilayers on Si and their correlation with the performance of infrared photodiodes," *J. Electron. Mater.* 26, 873–877, 1997.
146. H. Zogg, A. Fach, J. John, P. Müller, C. Paglino, and A. N. Tiwari, "PbSnSe-on-Si: Material and IR-device properties," *Proc. SPIE* 3182, 26–29, 1998.
147. H. Zogg, "Lead chalcogenide on silicon infrared sensor arrays," *Opto-Electron. Rev.* 6, 37–46, 1998.
148. H. Zogg, "Photovoltaic IV–VI on silicon infrared devices for thermal imaging applications," *Proc. SPIE* 3629, 52–62, 1999.
149. S. C. Gupta, B. L. Sharma, and V. V. Agashe, "Comparison of Schottky barrier and diffused junction infrared detectors," *Infrared Phys.* 19, 545–548, 1979.
150. A. Rogalski and W. Kaszuba, "Photovoltaic detectors $Pb_{1-x}Sn_xSe$ ($0 \le x \le 0.12$). Minority carrier lifetimes. Resistance-area product," *Infrared Phys.* 21, 251–259, 1981.
151. H. Preier, "Comparison of the junction resistance of (Pb,Sn)Te and (Pb,Sn)Se infrared detector diodes," *Infrared Phys.* 18, 43–46, 1979.
152. A. Rogalski and W. Kaszuba, "$PbS_{1-x}Se_x$ ($0 \le x \le 1$) photovoltaic detectors: carrier lifetimes and resistance-area product," *Infrared Phys.* 23, 23–32, 1983.
153. D. K. Hohnke, H. Holloway, K. F. Yeung, and M. Hurley, "Thin-film (Pb,Sn)Se photodiodes for 8–12 µm operation," *Appl. Phys. Lett.* 29, 98–100, 1976.
154. R. B. Schoolar and J. D. Jensen, "Narrowband detection at long wavelengths with epitaxial $Pb_{1-x}Sn_xSe$ films," *Appl. Phys. Lett.* 31, 536–538, 1977.
155. D. K. Hohnke and H. Holloway, "Epitaxial PbSe Schottky-barrier diodes for infrared detection," *Appl. Phys. Lett.* 24, 633–635, 1974.
156. R. B. Schoolar, J. D. Jensen, and G. M. Black, "Composition-tuned PbS_xSe_{1-x} Schottky-barrier infrared detectors," *Appl. Phys. Lett.* 31, 620–622, 1977.
157. R. W. Grant, J. G. Pasko, J. T. Longo, and A. M. Andrews, "ESCA surface studies of $Pb_{1-x}Sn_xTe$ devices," *J. Vacuum Sci. Technol.* 13, 940–947, 1976.
158. M. Bettini and H. J. Richter, "Oxidation in air and thermal desorption on PbTe, SnTe and $Pb_{0.8}Sn_{0.2}Te$ surface," *Surf. Sci.* 80, 334–343, 1979.
159. D. L. Partin and C. M. Thrush, "Anodic oxidation of lead telluride and its alloys," *J. Electrochem. Soc.* 133, 1337–1340, 1986.
160. T. Jimbo, M. Umeno, H. Shimizu, and Y. Amemiya, "Optical properties of native oxide film anodically grown on PbSnTe," *Surf. Sci.* 86, 389–397, 1979.
161. J. D. Jensen and R. B. Schoolar, "Surface charge transport in PbS_xSe_{1-x} and $Pb_{1-x}Sn_xSe$ epitaxial films," *J. Vacuum Sci. Technol.* 13, 920–925, 1976.
162. A. Rogalski, "Effect of air on electrical properties of $Pb_{1-x}Sn_xTe$ layers on a mica substrate," *Thin Solid Films* 74, 59–68, 1980.
163. S. Sun, S. P. Buchner, N. E. Byer, and J. M. Chen, "Oxygen uptake on an epitaxial PbSnTe (111) surface," *J. Vacuum Sci. Technol.* 15, 1292–1297, 1978.
164. T. Jakobus, W. Rothemund, A. Hurrle, and J. Baars, "$Pb_{0.8}Sn_{0.2}Te$ infrared photodiodes by indium implantation," *Revue de Physique Applquee* 13, 753–756, 1978.
165. H. Holloway, "Thin-film IV–VI semiconductor photodiodes," in *Physics of Thin Films*, Vol. 11, eds. G. Haas, M. H. Francombe, and P. W. Hoffman, 105–203, Academic Press, New York, 1980.
166. W. Larkowski and A. Rogalski, "High-performance 8–14 µm PbSnTe Schottky barrier photodiodes," *Opt. Appl.* 16, 221–229, 1986.
167. P. Lo Vecchio, M. Jasper, J. T. Cox, and M. B. Barber, "Planar $Pb_{0.8}Sn_{0.2}Te$ photodiode array development at the Night Vision Laboratory," *Infrared Phys.* 15, 295–301, 1975.
168. M. R. Johnson, R. A. Chapman, and J. S. Wrobel, "Detectivity limits for diffused junction PbSnTe detectors," *Infrared Phys.* 15, 317–329, 1975.
169. J. Rutkowski, A. Rogalski, and W. Larkowski, "Mesa Cd-diffused $Pb_{0.80}Sn_{0.20}Te$ photodiodes," *Acta Phys. Pol.* A67, 195–198, 1985.
170. J. S. Wrobel, "Method of forming p-n junction in PbSnTe and photovoltaic infrared detector provided thereby," *Patent USA 3,911,469*, 1975.
171. R. Behrendt and R. Wendlandt, "A study of planar Cd-diffused $Pb_{1-x}Sn_xTe$ photodiodes," *Phys. Status Solidi A* 61, 373–380, 1980.
172. P. T. Landsberg, *Recombination in Semiconductors*, Cambridge University Press, Cambridge, UK, 2003.

173. L. H. DeVaux, H. Kimura, M. J. Sheets, F. J. Renda, J. R. Balon, P. S. Chia, and A. H. Lockwood, "Thermal limitations in PbSnTe detectors," *Infrared Phys.* 15, 271–277, 1975.
174. J. John and H. Zogg, "Infrared p-n-junction diodes in epitaxial narrow gap PbTe layers on Si substrates," *Appl. Phys. Lett.* 85, 3364–3366, 1999.
175. E. M. Logothetis, H. Holloway, A. J. Varga, and W. J. Johnson, "N-p junction IR detectors made by proton bombardment of epitaxial PbTe," *Appl. Phys. Lett.* 21, 411–413, 1972.
176. T. F. Tao, C. C. Wang, and J. W. Sunier, "Effect of proton bombardment on $Pb_{0.76}Sn_{0.24}Te$," *Appl. Phys. Lett.* 20, 235–237, 1972.
177. C. C. Wang, T. F. Tao, and J. W. Sunier, "Proton bombardment and isochronal annealing of p-type $Pb_{0.76}Sn_{0.24}Te$," *J. Appl. Phys.* 45, 3981–3987, 1974.
178. J. P. Donnelly, "The electrical characteristics of ion implanted compound semiconductors," *Nucl. Instrum. Methods* 182/183, 553–571, 1981.
179. L. Palmetshofer, "Ion implantation in IV–VI semiconductors," *Appl. Phys.* A34, 139–153, 1984.
180. J. P. Donnelly and T. C. Harman, "P-n Junction PbS_xSe_{1-x} photodiodes fabricated by Se^+ ion implantation," *Solid State Electron.* 18, 288–290, 1975.
181. J. P. Donnelly, T. C. Harman, A. G. Foyt, and W. T. Lindley, "PbS photodiodes fabricated by Sb^+ ion implantation," *Solid State Electron.* 16, 529–534, 1973.
182. J. P. Donnelly and H. Holloway, "Photodiodes fabricated in epitaxial PbTe by Sb^+ ion implantation," *Appl. Phys. Lett.* 23, 682–683, 1973.
183. A. M. Andrews, J. T. Longo, J. E. Clarke, and E. R. Gertner, "Backside-illuminated $Pb_{1-x}Sn_xTe$ heterojunction photodiode," *Appl. Phys. Lett.* 26, 438–441, 1975.
184. A. H. Lockwood, J. R. Balon, P. S. Chia, and F. J. Renda, "Two-color detector arrays by PbTe/$Pb_{0.8}Sn_{0.2}Te$ liquid phase epitaxy," *Infrared Phys.* 16, 509–514, 1976.
185. C. C. Wang, M. H. Kalisher, J. M. Tracy, J. E. Clarke, and J. T. Longo, "Investigation on leakage characteristics of PbSnTe/PbTe inverted heterostructure diodes," *Solid State Electron.* 21, 625–632, 1978.
186. Nugraha, W. Tamura, O. Itoh, K. Suto, and J. Nishizawa, "Te vapor pressure dependence of the p-n junction properties of PbTe liquid phase epitaxial layers," *J. Electron. Mater.* 27, 438–441, 1999.
187. W. Rolls and D. V. Eddolls, "High detectivity $Pb_{1-x}Sn_xTe$ photovoltaic diodes," *Infrared Phys.* 13, 143–147, 1972.
188. R. W. Bicknell, "Electrical and metallurgical examination of $Pb_{1-x}Sn_xTe$/PbTe heterojunctions," *J. Vacuum Sci. Technol.* 14, 1012–1015, 1977.
189. V. V. Tetyorkin, V. B. Alenberg, F. F. Sizov, E. V. Susov, Yu. G. Troyan, A. V. Gusarov, V. Yu. Chopik, and K. S. Medvedev, "Carrier transport mechanisms and photoelectrical properties of PbSnTe/PbTeSe heterojunctions," *Infrared Phys.* 30, 499–504, 1990.
190. A. Rogalski, "$Pb_{1-x}Sn_xTe$ photovoltaic detectors for the range of atmospheric window 8–14 μm prepared by a modified hot-wall evaporation technique," *Electron Technology* 12(4), 99–107, 1978.
191. D. Eger, A. Zemel, S. Rotter, N. Tamari, M. Oron, and A. Zussman, "Junction migration in PbTe-PbSnTe heterostructures," *J. Appl. Phys.* 52, 490–495, 1981.
192. D. Yakimchuk, M. S. Davydov, V. F. Chishko, I. J. Tsveibak, V. V. Krapukhin, and I. A. Sokolov, "Current-voltage characteristics of p-$Pb_{0.8}Sn_{0.2}Te$/n-$PbTe_{0.92}Se_{0.08}$ heterojunctions," *Fizika i Tekhnika Poluprovodnikov* 22, 1474–1478, 1988.
193. I. Kasai, D. W. Bassett, and J. Hornung, "PbTe and $Pb_{0.8}Sn_{0.2}Te$ epitaxial films on cleaved BaF_2 substrates prepared by a modified hot-wall technique," *J. Appl. Phys.* 47, 3167–3171, 1976.
194. A. Rogalski, "n-PbTe/p$^+$-$Pb_{1-x}Sn_xTe$ heterojunctions prepared by a modified hot wall technique," *Thin Solid Films* 67, 179–186, 1980.
195. D. Eger, M. Oron, A. Zussman, and A. Zemel, "The spectral response of PbTe/PbSnTe heterostructure diodes at low temperatures," *Infrared Phys.* 23, 69–76, 1983.
196. E. Abramof, S. O. Ferreira, C. Boschetti, and I. N. Bendeira, "Influence of interdiffusion on N-PbTe/P-PbSnTe heterojunction diodes," *Infrared Phys.* 30, 85–91, 1990.
197. N. Tamari and H. Shtrikman, "Dislocation etch pits in LPE-grown $Pb_{1-x}Sn_xTe$ (LTT) heterostructures," *J. Appl. Phys.* 50, 5736–5742, 1979.
198. D. Kasemset and G. Fonstad, "Reduction of interface recombination velocity with decreasing lattice parameter Mismatch in PbSnTe heterojunctions," *J. Appl. Phys.* 50, 5028–5029, 1979.
199. D. Kasemset, S. Rotter, and C. G. Fonstad, "Liquid phase epitaxy of PbTeSe lattice-matched to PbSnTe," *J. Electron. Mater.* 10, 863–878, 1981.
200. K. W. Nill, A. R. Calawa, and T. C. Harman, "Laser emission from metal-semiconductor barriers on PbTe and $Pb_{0.8}Sn_{0.2}Te$," *Appl. Phys. Lett.* 16, 375–377, 1970.

201. J. N. Walpole and K. W. Nill, "Capacitance-voltage characteristic of the metal barrier on p PbTe and n InAs: effect of inversion layer," *J. Appl. Phys.* 42, 5609–5617, 1971.
202. F. F. Sizov, V. V. Tetyorkin, Yu. G. Troyan, and V. Yu. Chopick, "Properties of the Schottky barriers on compensated PbTe < Ga >," *Infrared Phys.* 29, 271–277, 1989.
203. W. Vogt, H. Zogg, and H. Melchior, "Preparation and properties of epitaxial PbSe/BaF_2/PbSe structures," *Infrared Phys.* 25, 611–614, 1985.
204. C. Maissen, J. Masek, H. Zogg, and S. Blunier, "Photovoltaic infrared sensors in heteroepitaxial PbTe on Si," *Appl. Phys. Lett.* 53, 1608–1610, 1988.
205. W. Maurer, "Temperature dependence of the R_oA product of PbTe Schottky diodes," *Infrared Phys.* 23, 257–260, 1983.
206. B. Chang, K. E. Singer, and D. C. Northrop, "Indium contacts to lead telluride," *J. Phys. D: Appl. Phys.* 13, 715–723, 1980.
207. T. A. Grishina, I. A. Drabkin, Yu. P. Kostikov, A. V. Matveenko, N. G. Protasova, and D. A. Sakseev, "Study of in-$Pb_{1-x}Sn_xTe$ interface by Auger spectroscopy," *Izv. Akad. Nauk SSSR, Neorg. Mater.* 18, 1709–1713, 1982.
208. K. P. Scharnhorst, R. F. Bis, J. R. Dixon, B. B. Houston, and H. R. Riedl, "Vacuum deposited method for fabricating an epitaxial PbSnTe rectifying metal semiconductor contact photodetector," U.S Patent 3,961,998, 1976.
209. S. Buchner, T. S. Sun, W. A. Beck, N. E. Dyer, and J. M. Chen, "Schottky barrier formation on (Pb,Sn)Te," *J. Vacuum Sci. Technol.* 16, 1171–1173, 1979.
210. T. A. Grishina, N. N. Berchenko, G. I. Goderdzishvili, I. A. Drabkin, A. V. Matveenko, T. D. Mcheidze, D. A. Sakseev, and E. A. Tremiakova, "Surface-barrier $Pb_{0.77}Sn_{0.23}Te$ structures with intermediate layer," *J. Techn. Phys.* 57, 2355–2360, 1987.
211. N N Berchenko, A. I. Vinnikova, A. Yu. Nikiforov, E. A. Tretyakova, and S. V. Fadyeev, "Growth and properties of native oxides for IV–VI optoelectronic devices," *Proc. SPIE* 3182, 404–407, 1997.
212. T. S. Sun, N. E. Byer, and J. M. Chen, "Oxygen uptake on epitaxial PbTe(111) surfaces," *J. Vacuum Sci. Technol.* 15, 585–589, 1978.
213. A. C. Bouley, T. K. Chu, and G. M. Black, "Epitaxial thin film IV–VI detectors: device performance and basic material properties," *Proc. SPIE* 285, 26–32, 1981.
214. A. C. Chu, A. C. Bouley, and G. M. Black, "Preparation of epitaxial thin film lead salt infrared detectors," *Proc. SPIE* 285, 33–35, 1981.
215. M. Drinkwine, J. Rozenbergs, S. Jost, and A. Amith, "The Pb/PbSSe interface and performance of Pb/PbSSe photodiodes," *Proc. SPIE* 285, 36–43, 1981.
216. J. H. Werner and H. H. Güttler, "Barrier inhomogeneities at Schottky contacts," *J. Appl. Phys.* 69, 1522–1533, 1991.
217. D. W. Bellavance and M. R. Johnson, "Open tube vapor transport growth of $Pb_{1-x}Sn_xTe$ epitaxial films for infrared detectors," *J. Electron. Mater.* 5, 363–380, 1976.
218. E. M. Logothetis, H. Holloway, A. J. Varga, and E. Wilkes, "Infrared detection by Schottky barrier in epitaxial PbTe," *Appl. Phys. Lett.* 19, 318–320, 1971.
219. H. Zogg and M. Huppi, "Growth of high quality epitaxial PbSe onto Si using a $(Ca,Ba)F_2$ buffer layer," *Appl. Phys. Lett.* 47, 133–135, 1985.
220. J. Baars, "New aspects of the material and device technology of intrinsic infrared photodetectors," in *Physics of Narrow Gap Semiconductors*, eds. E. Gornik, H. Heinrich, and L. Palmetshofer, 280–182, Springer, Berlin, 1982.
221. J. Masek, C. M. Maissen, H. Zogg, S. Blunier, H. Weibel, A. Lambrecht, B. Spanger, H. Bottner, and M. Tacke, "Photovoltaic infrared sensor arrays in heteroepitaxial narrow gap lead-chalcogenides on silicon," *J. Phys., Coll.* C4, 587–590, 1988.
222. J. Masek, A. Ishida, H. Zogg, C. Maissen, and S. Blunier, "Monolithic photovoltaic PbS-on-Si infrared-sensor array," *IEEE Electron Dev. Lett.* 11, 12–14, 1990.
223. H. Zogg and P. Norton, "Heteroepitaxial PbTe-Si and (Pb,Sn)Se-Si structures for monolithic 3–5 μm and 8–12 μm infrared sensor arrays," *IEDM Techn. Dig.* 121–124, 1985.
224. P. Collot, F. Nguyen-Van-Dau, and V. Mathet, "Monolithic integration of PbSe IR photodiodes on Si substrates for near ambient temperature operation," *Semicond. Sci. Technol.* 9, 1133–1137, 1994.
225. H. Zogg, A. Fach, J. John, J. Masek, P. Müller, C. Paglino, and W. Buttler, "PbSnSe-on-Si LWIR sensor arrays and thermal imaging with JFET/CMOS read-out," *J. Electron. Mater.* 25, 1366–1370, 1996.
226. J. Ameurlaine, A. Rousseau, T. Nguyen-Duy, and R. Triboulet, "(HgZn)Te infrared photovoltaic detectors," *Proc. SPIE* 929, 14–20, 1988

227. A. Rogalski and R. Ciupa, "Long wavelength HgCdTe photodiodes: n⁺-on-p versus p-on-n structures," *J. Appl. Phys.* 77, 3505–3512, 1995.
228. A. Rogalski and W. Larkowski, "Comparison of photodiodes for the 3–5.5 μm and 8–14 μm spectral regions," *Electron Technol.* 18(3/4), 55–69, 1985.
229. H. Holloway, M. D. Hurley, and E. B. Schermer, "IV–VI semiconductor lateral-collection photodiodes," *Appl. Phys. Lett.* 32, 65–67, 1978.
230. H. Holloway, "Unconventional thin film IV–VI photodiode structures," *Thin Solid Films* 58, 73–78, 1979.
231. H. Holloway, "Theory of lateral-collection photodiodes," *J. Appl. Phys.* 49, 4264–4269, 1978.
232. H. Holloway and A. D. Brailsford, "Peripheral photoresponse of a p-n junction," *J. Appl. Phys.* 54, 4641–4656, 1983.
233. H. Holloway and A. D. Brailsford, "Diffusion-limited saturation current of a finite p-n junction," *J. Appl. Phys.* 55, 446–453, 1984.
234. H. Holloway, "Peripheral electron-beam induced current response of a shallow p-n junction," *J. Appl. Phys.* 55, 3669–3675, 1984.
235. H. Holloway, "Quantum efficiencies of thin-film IV–VI semiconductor photodiodes," *J. Appl. Phys.* 50, 1386–1398, 1979.
236. R. B. Schoolar, J. D. Jensen, G. M. Black, S. Foti, and A. C. Bouley, "Multispectral PbS_xSe_{1-x} and $Pb_ySn_{1-y}Se$ photovoltaic infrared detectors," *Infrared Phys.* 20, 271–275, 1980.
237. J. Carrano, J. Brown, P. Perconti, and K. Barnard, "Tuning in to detection," *SPIE's OEmagazine*, 20–22, 2004.
238. F. Felder, M. Arnold, M. Rahim, C. Ebneter, and H. Zogg, "Tunable lead-chalcogenide on Si resonant cavity enhanced midinfrared detector," *Appl. Phys. Lett.* 91, 101102, 2007.
239. H. Zogg, M. Arnold, F. Felder, M. Rahim, C. Ebneter, I. Zasavitskiy, N. Quack, S. Blunier, and J. Dual, "Epitaxial lead chalcogenides on Si got Mid-IR detectors and emitters including cavities," *J. Electron. Mater.* 37, 1497–1503, 2008.
240. D. Zimin, K. Alchalabi, and H. Zogg, "Heteroepitaxial PbTe-on-Si pn-junction IR-sensors: correlations between material and device properties," *Phys. E* 13, 1220–1223, 2002.

19 Quantum well infrared photodetectors

Since the initial proposal by Esaki and Tsu [1] and the advent of MBE, the interest in semiconductor superlattices (SLs) and quantum well (QW) structures has increased continuously over the years, driven by technological challenges, new physical concepts and phenomena as well as promising applications. A new class of materials and heterojunctions with unique electronic and optical properties has been developed. Here we focus on devices that involve infrared (IR) excitation of carriers in low-dimensional solids (QWs, quantum dots (QDs), and SLs). A distinguishing feature of these IR detectors is that they can be implemented in chemically stable wide-bandgap materials as a result of the use of intraband processes. On account of this, it is possible to use such material systems as GaAs/Al$_x$Ga$_{1-x}$As (GaAs/AlGaAs), In$_x$Ga$_{1-x}$As/In$_x$Al$_{1-x}$As (InGaAs/InAlAs), InSb/InAs$_{1-x}$Sb$_x$ (InSb/InAsSb), InAs/Ga$_{1-x}$In$_x$Sb (InAs/GaInSb), and Si$_{1-x}$Ge$_x$/Si (SiGe/Si) as well as other systems, although most of the experimental works have been carried out with AlGaAs. Some devices are sufficiently advanced that there exists the possibility of incorporating them in high-performance integrated circuits. High uniformity of epitaxial growth over large areas shows promise for the production of large-area two-dimensional arrays. In addition, flexibility associated with control over composition during epitaxial growth can be used to tailor the response of QW IR detectors to particular IR bands or multiple bands.

Among the different types of quantum well infrared photodetectors (QWIPs), technology of the GaAs/AlGaAs multiple QW detector is the most mature. Rapid progress has recently been made in the performance of these detectors [2–15]. Detectivities have improved dramatically and are now high enough so that megapixel focal plane arrays (FPAs) with long wavelength infrared (LWIR) imaging performance comparable to state of the art HgCdTe are fabricated [16,17]. Despite huge research and development efforts, large photovoltaic HgCdTe FPAs remain expensive, primarily because of the low yield of operable arrays. The low yield is due to the sensitivity of LWIR HgCdTe devices to defects and surface leakage, which is a consequence of basic material properties. With respect to HgCdTe detectors, GaAs/AlGaAs QW devices have a number of potential advantages including the use of standard manufacturing techniques based on mature GaAs growth and processing technologies, highly uniform and well-controlled molecular beam epitaxy (MBE) growth on greater than 6-inch GaAs wafers, high yield and thus low cost, more thermal stability, and extrinsic radiation hardness.

This chapter is devoted to the properties and applications of QW structures for IR detection. It is difficult to cover such topics since the technology of the above devices is being rapidly developed and new concepts for these devices are currently being proposed. It is assumed that the phenomena, materials, and optical and electrical properties of QWs and SLs are well known to the readers. Several introductory textbooks on QW physics have been written by Bastard [18], Weisbuch and Vinter [19], Shik [20], Harrison [21], Bimberg, Grundmann, and Ledentsov [22], and Singh [23]. Only the elementary properties of QWs are described below. Because Chapters 20 and 21 are devoted to SL and QD photodetectors, we will concentrate on different types of low-dimensional solids in the next section.

19.1 LOW DIMENSIONAL SOLIDS: BACKGROUND

Rapid progress in the development of epitaxial growth techniques has made it possible to grow semiconductor structures at one-monolayer accuracy. The device structure dimensions can be compared to wavelengths of the relevant electron or hole wave functions, at least in the growth direction. This means that one can do electrical engineering at the quantum mechanical level. The electron confinement within a sufficiently narrow region of semiconductor material can significantly change the carrier energy spectrum, and novel physical properties are expected to emerge. These novel properties will give rise to new semiconductor devices as well as drastically improved device characteristics [24,25]. Most expected improvements in electronic and optoelectronic device performance originate from the change in the density of states.

In addition to QW case, where energy barriers for electron motion exist in one direction of propagation, one can also imagine electron confinement in two directions and, as the ultimate case, in all three directions. The structures of these kinds are now known as quantum wires and QDs. Thus, the family of dimensionalities of the device structures involves bulky semiconductor epilayer (three-dimensional (3D)), thin epitaxial layer of QW (two-dimensional (2D)), elongated tube or quantum wire (one-dimensional (1D)) and, finally, isolated island of QD (zero-dimensional (0D)). These three cases are shown in Figure 19.1.

In a crystalline semiconductor, the electrons and holes that determine the transport and optical properties are considered as quasi free with the effective mass m^* taking account of the periodic crystal potential. The dimensions of a bulk semiconductor crystal are macroscopic on the scale of the de Broglie wavelength

$$\lambda = \frac{h}{(2m^*E)^{1/2}} \tag{19.1}$$

and thus the so-called size quantization is negligible. The quasi-free electron wave functions are given by the Bloch functions:

$$\Psi_{j\vec{k}}^{3D}(r) = \frac{1}{V} u_{j\vec{k}}(\vec{r}) \exp\left(i\vec{k}\vec{r}\right), \tag{19.2}$$

where V is the macroscopic volume, $k = 2\pi/\lambda$ is the electron wave vector, and j is the band index. The Bloch functions have the free particle wave function $\exp\left(i\vec{k}\vec{r}\right)$ as the envelope, instead allowed k-values, as defined by the crystal boundaries, are quasi continuous. The energy dispersion of quasi-free electrons (holes) at the bottom of the conduction band (at the top of the valence band) is given by

$$E^{3D} = \frac{\hbar^2 k}{2m^*}, \tag{19.3}$$

where $\hbar$ is Planck's constant. This quadratic dispersion results in a parabolic density of state function.

$$\rho^{3D} = \frac{1}{2\pi^2}\left(\frac{2m^*}{\hbar^2}\right)^{3/2} E^{1/2}. \tag{19.4}$$

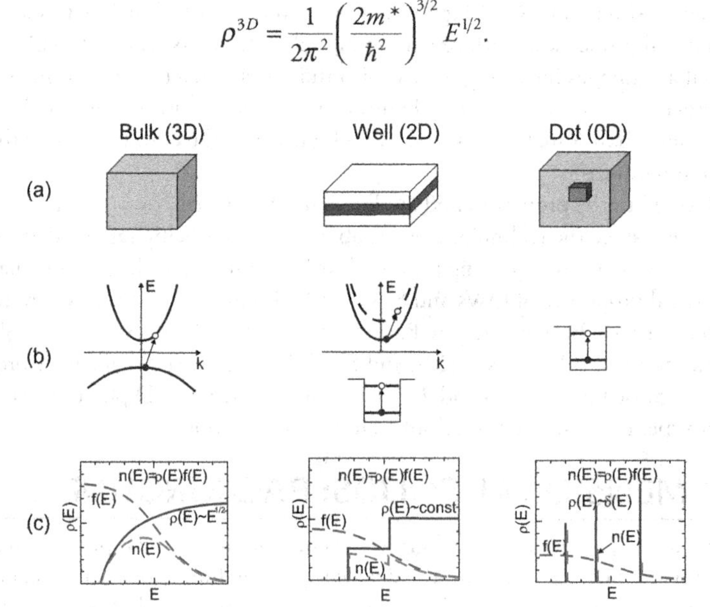

Figure 19.1 Illustrations of (a) quantum nanostructures, (b) energy versus wave vector, (c) density of states and energy spreading (where electrons have the Boltzmann distribution) in bulk, QW and QD.

While the density of states characterizes the energy distribution of allowed states, Fermi function f, gives the probability of electron occupation at a certain energy even if no allowed state exists at all at that energy. Thus, the product of ρ and Fermi function describes the total concentration of the charge carriers of the given type in a crystal.

$$n = \int \rho(E) f(E) dE. \tag{19.5}$$

The transport and optical properties of semiconductors are basically determined by the uppermost valence band and lowest conduction band, which are separated in energy by the bandgap E_g. The bandgap structure of GaAs consists of a heavy hole, light hole, split-off valence band, and the lowest conduction band. GaAs is a direct gap semiconductor with the maximum of the valence band and minimum of the conduction band at the same position in the Brillouin zone at $k = 0$ (Γ-point). AlAs is an indirect gap semiconductor with the lowest conduction band minimum close to the boundary of the Brillouin zone in (100) direction at the X-point. In addition, the conduction band minimum in AlAs is highly unisotrop with a longitudinal and transverse electron mass equal to 1.1 m_l^* and $m_t^* = 0.2m$, respectively.

Confinement of electrons in one or more dimensions modifies the wave functions, dispersion, and density of states. The effective potential associated with spatial variation of the conduction and valence band edges is spatially modulated in so-called compositional SLs that consist of alternating layers of two different semiconductors. Figure 19.2 illustrates the electronic states associated with planar QWs and SLs [26]. Widths of minibands in SLs are related by the uncertainty principle to well-to-well tunneling times. In a typical case, the barrier regions would be AlGaAs layers and the wells would be GaAs. The typical distance scales (≈50 Å) over which composition is varied is small compared with distances over which ballistic electron propagation has been observed. This supports the notion that wave functions can be coherent throughout the structures that have spatially modulated composition. The wave functions just referred to are the envelope wave functions of effective mass theory. Spatial variation of the host composition introduces a number of sublattices, but the basic picture of potentials and wave functions guides much of the effort in the field of quantum devices.

Each QW can be considered to be a three-dimensional rectangular potential well. When the thickness of the well is much lesser than the transverse dimensions ($L_z << L_x, L_y$), and the thickness is comparable to the de Broglie wavelength of the carriers in the well, quantization of the carrier motion in the z-direction must be taken into account in the dispersion carrier dynamics. Motion in the x and y directions is not quantized, so that each state of the system corresponds to a subband. Electrons (or holes) in such a well can be regarded as a two-dimensional electron (or hole) gas. When the well is infinitely deep, the Schrödinger equation energy eigenvalues are:

$$E^{2-D} = E_{n_z} + \frac{\hbar^2 \left(k_x^2 + k_y^2\right)}{2m^*} \tag{19.6}$$

with the confinement energy

$$E_{n_z} = \frac{\hbar^2}{2m}\left(\frac{n_z \pi}{L_z}\right)^2, \tag{19.7}$$

where k_x and k_y are momentum vectors along the x and y axes, and n_z is the quantum number ($n_z = 1, 2,...$). The electron wave functions are represented by plane waves in the x and y directions and by even or odd harmonic functions in the z direction:

$$\psi_{n_z}^{2-D} = \left(\frac{2}{L}\right) \exp(ik_x x) \exp(ik_y y) \left(\frac{2}{L_z}\right)^{1/2} \sin(k_{n_z}). \tag{19.8}$$

Confinement of electrons by potential wells with finite height (see Figure 19.2a) does not affect the principal features of size quantization as described above; however, it modifies the results in three important respects:

- The confinement energy for a given quantum state characterized by the quantum number n_z is lower for a finite barrier height.
- Only a finite number of quantized states is bound in a well with finite barrier height (for an infinitely high barrier an infinite number of quantized states exists); when the width of a single QW is decreased, the first excited state merges from the well and becomes a virtual state (Figure 19.3); for example, Figure 19.4 illustrates the link between bound states and virtual states as the parameters of a single QW are varied.
- The electron wave functions do not vanish at the boundary but penetrate into the barrier where the amplitude drops exponentially.

The latter effect actually provides the base for the formation of SLs. If the overlap occurs, the confined energy levels split into a manifold of levels given by the number of coupled potential wells. For a sufficiently large number of coupled wells, these split levels form a quasi continuous energy band as illustrated in Figure 19.2b.

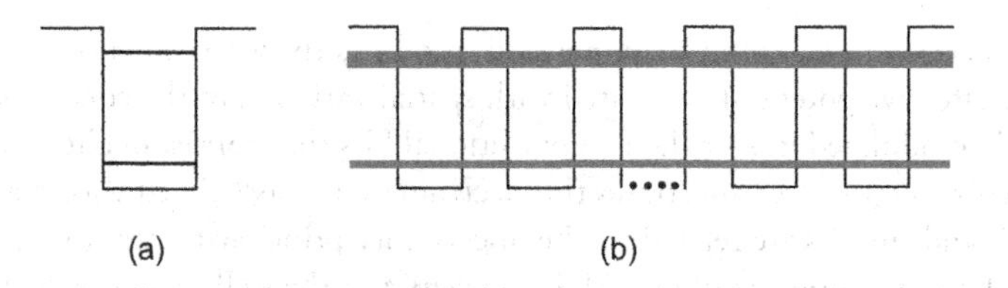

Figure 19.2 (a) Electron bound states in a QW and (b) the formation of minibands in a SL. The confining potentials are associated with the conduction band edge. (From Ref. [26])

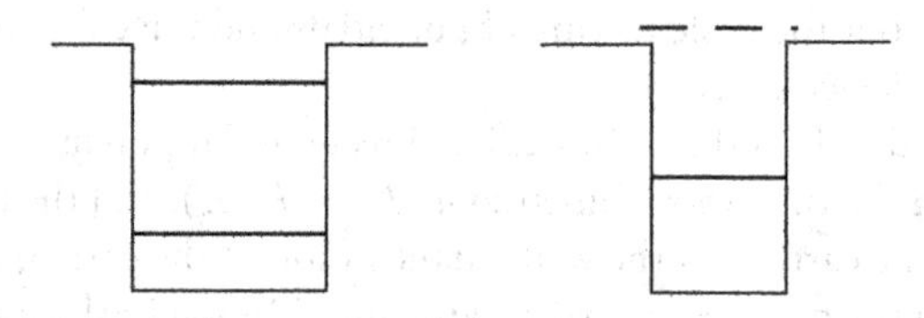

Figure 19.3 Illustration of the formation of a virtual state as the width of the QW is decreased. (From Ref. [26])

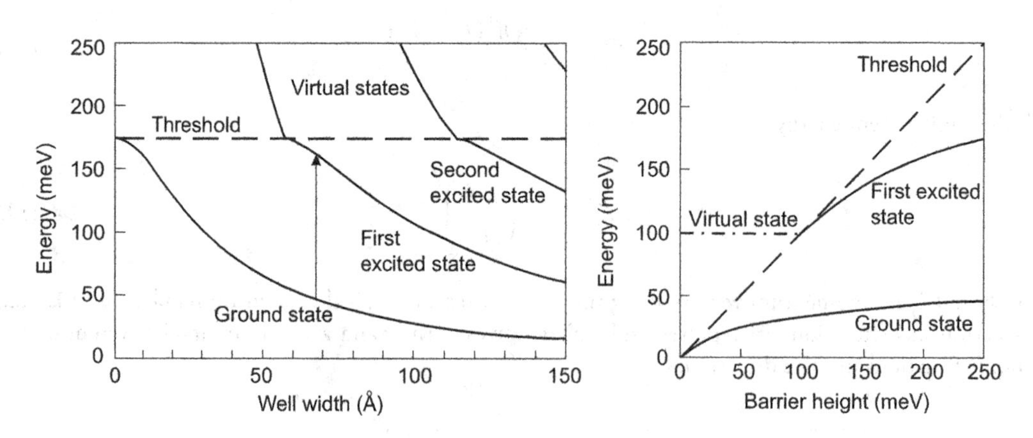

Figure 19.4 Dependence of energy levels on well width in a $GaAs/Al_{0.25}Ga_{0.75}As$ QW. The zero of the energy scale is at the bottom of the well. The barrier height dependence for a 75 Å QW is shown on the right. (From Ref. [26])

In general, E_n and ψ_n are eigenvalues and eigenfunctions of a finite one-dimensional potential well. The energy levels in the k_x and k_y directions form a continuum and, for discrete values of E_n (i.e., for each bound state), a two-dimensional in the $k_x - k_y$ plane will form. Each of these two-dimensional energy bands gives rise to a band density of states that is energy independent. The density of state function changes from the smooth parabolic shape to a function

$$\rho^{2-D} = \frac{m^*}{\pi\hbar^2}\sum_{n_z}\delta\left(E - E_{n_z}\right), \tag{19.9}$$

where $\delta(E)$ is the Heaviside step function with $\delta(E \geq E_{nz}) = 1$ and $\delta(E < E_{nz}) = 0$. The cumulative density of states is steplike in character up to the energy at which the discrete, bound-state spectrum gives way to the continuum of free (unbound) states.

Farther confinement in two or, finally, three dimensions results in size quantization in the corresponding directions and stronger discretization of the energy spectrum and the density of state distribution approaching atomic behavior for three-dimensional confinement.

An ideal QD, also known as a quantum box, is a structure capable of confining electrons in all three directions, thus allowing zero dimension in their degrees of freedom. The energy spectrum is completely discrete, similar to that in an atom. The total energy is the sum of three discrete components:

$$E^{0-D} = E_{nx} + E_{my} + E_{lz} = \frac{\hbar^2 k_{nx}^2}{2m^*} + \frac{\hbar^2 k_{my}^2}{2m^*} + \frac{\hbar^2 k_{lz}^2}{2m^*}, \tag{19.10}$$

where n, m, and l are integers (1, 2,...) used to index the quantized energy levels and quantized wave numbers, which result from the confinement of the electron motion in the x, y, and z-directions, respectively.

As for the bulk material, the most important characteristic of a QD is its electron density of states in the conduction band given by:

$$\rho^{0-D}(E) = 2\sum_{n,m,l}\delta\left[E_{nx} + E_{my} + E_{lz} - E\right]. \tag{19.11}$$

Each QD level can accommodate two electrons with different spin orientations.

The density of states of zero-dimensional electrons consists of Dirac functions, occurring at the discrete energy levels $E(n,m,l)$, as shown in Figure 19.1. The divergences in the density of states shown in Figure 19.1 are for ideal electrons in a QD and are smeared out in reality by a finite electron lifetime ($\Delta E \geq \hbar/\tau$). Since QDs have a discrete, atom-like energy spectrum, they can be visualized and described as artificial atoms. This discreteness is expected to render the carrier dynamics very different from that in higher-dimensional structures where the density of states is continuous over a range of energy values.

The energy position of QD (also QW) level essentially depends on the geometrical sizes and even one monolayer variation of the size can significantly affect the energy of optical transition. Fluctuations of the geometrical parameters result in corresponding fluctuations of the quantum levels over the array of dots. Random fluctuations also affect the density of states on nonuniform array of QDs.

The self-assembling method for fabricating QDs has been recognized as one of the most promising methods for forming QDs that can be practically incorporated into IR photodetectors. In the crystal growth of highly lattice-mismatched materials system, self-assembling formation of nanometer-scale 3D islands has been reported [27]. The lattice mismatch between a QD and the matrix is the fundamental driving force of self assembling. In(Ga)As on GaAs is the most commonly used material system because lattice mismatch can be controlled by the In alloy ratio up to about 7%.

Both QW and QD structures are used in the fabrication of IR detectors. In general, quantum dot infrared photodetectors (QDIPs) are similar to QWIPs, but with the QWs replaced by QDs, which have size confinement in all spatial directions.

Figure 19.5 shows the schematic layers of a QWIP and a QDIP [28]. In both cases, the detection mechanism is based on the intraband photoexcitation of electrons from confined states in the conduction band wells or dots into the continuum. The emitted electrons drift toward the collector in the electric field provided by the applied bias, and the photocurrent is created. It is assumed that the potential profile at the conduction band edge along the growth direction for both structures have a similar shape as shown in Figure 19.5b.

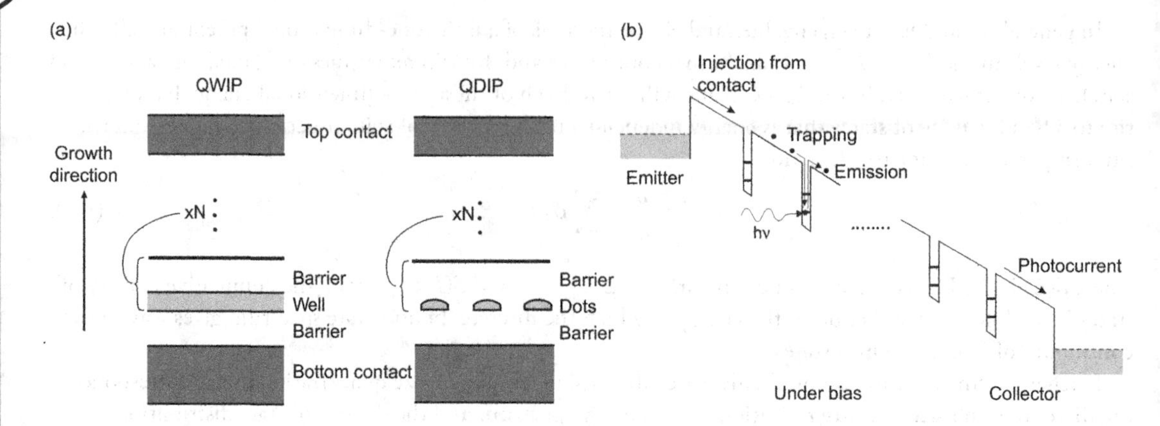

Figure 19.5 (a) Schematic layers of QWIP and QDIP and (b) potential profile for both structures under bias. For QDIP, influence of wetting layer is neglected. (From Ref. [28])

The self-assembled QDs in QDIPs are wide in the in-plane direction and narrow in the growth direction. The strong confinement is therefore in the growth direction, while the in-plane confinement is weak, resulting in several levels in the dots (see Figure 19.6). In this situation, the transitions between the in-plane confinement levels give rise to the normal incidence response.

19.2 MULTIPLE QUANTUM WELLS AND SUPERLATTICES

All the above considerations are devoted to mainly compositional SLs that consist of alternating layers of two different semiconductors. The second kind of SL that was also originated by Esaki and Tsu is the doping SL [1]. These types of SLs consist of alternating n- and p-type layers of a single semiconductor. Electrical fields generated by the charged dopants modulate the electronic potential. SLs in which both the composition and the doping are modulated have been also considered [29].

19.2.1 COMPOSITIONAL SUPERLATTICES

Heteroepitaxy of materials with different bandgaps is of fundamental interest for the composition SL growth. An important requirement for high-quality compositional QW and SLs is the matching of the lattice constants of the constituents with different bandgaps, except for the so-called pseudomorphic systems, where internal stress or strain due to lattice mismatch is used to tailor the electronic band structure in

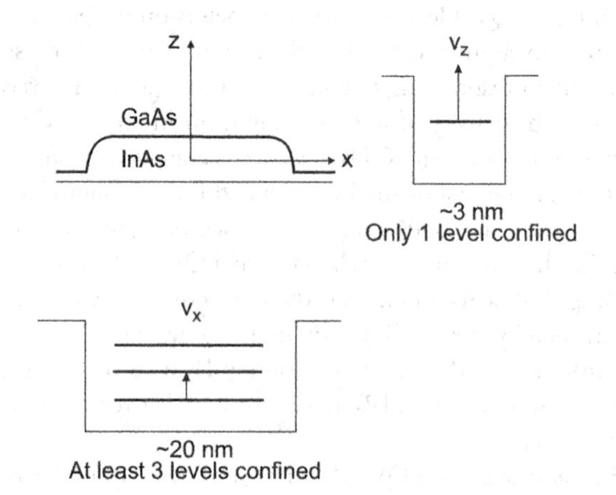

Figure 19.6 Illustration of transitions under polarized light in the growth direction (*z*) or in the in-plane directions (*x* or *y*). The strong confinement in the growth direction is represented by a narrow well; whereas the in-plane wide potential well leads to several states. The upward arrows indicate the strongest transitions for *z* and *x* polarized lights. (From Ref. [28])

addition to the effects due to confinement. The status of strained-layer SLs has been reviewed by Mailhiot and Smith [30].

The condition for lattice match is a severe restriction for the materials that can be considered for heteroepitaxy to form QW structures; however, nature still provides enough freedom. Figure 19.7 shows the plot of energy gaps at 4.2 K versus lattice constants for zinc-blende semiconductors together with Si and Ge. The joining lines represent the ternary alloys except for Si-Ge, GaAs-Ge, and InAs-GaSb. MnSe and MnTe are not shown here because their stable crystal structures are not zinc blende. The most extensively studied structures are based on the GaAs/AlGaAs material system that is conditioned by almost perfect natural lattice match between GaAs and $Al_xGa_{1-x}As$ for all values of x. From Figure 19.7 it results that the energy gap generally decreases with an increase in the lattice constant or the atomic number [31]. It should also be noted that all the binary compounds fall into five distinct columns shown by the shaded areas, suggesting that the lattice constants are alike as long as the mean atomic numbers of the binary constituents are the same.

The physical properties of the respective QW structures are strongly determined by the band discontinuities at the interface (i.e., the band alignment). An abrupt discontinuity in the local band structure is usually associated with a gradual band bending in its neighborhood, which reflects space-charge effects. The conduction- and valence-band discontinuities will determine the character of carrier transport across the interfaces, and so they are the most important quantities, which determine the suitability of present SLs or QWs for IR detector purposes. The presence of an additional SL periodic potential changes the electronic spectrum of a semiconductor in such a manner that the Brillouin zone is divided into a series of minizones giving rise to narrow subbands separated by minigaps (see, e.g., Bastard [18]). Thus, the SLs will possess new properties not exhibited by homogeneous semiconductors. Surprisingly, the corresponding values for the band discontinuities of the conduction band (ΔE_c) and the valence band (ΔE_v) cannot be obtained by simple considerations. Band lineups based on electron affinity do not work in most cases when two semiconductors form a heterostructure. This is because of subtle charge sharing effects that occur across atoms on the interface. There have been a number of theoretical studies that can predict general trends in bands lineup (e.g., [32–36]). However, the techniques are quite complex and the heterostructure design usually depends on experiments to provide lineup information [37–39]. It must be taken into account that electrical and optical methods do not measure the band offsets themselves but instead the quantities associated with the electronic structure of the heterostructure. The band offset determinations from such experiments require an appropriate theoretical model. The reported fundamental parameters ΔE_c and ΔE_v are slightly different, even for the most widely studied GaAs/$Al_xGa_{1-x}As$ material system. For the latter system, values of ΔE_c:ΔE_v = 6:4 are widely accepted for the composition range $0.1 \leq x \leq 0.4$.

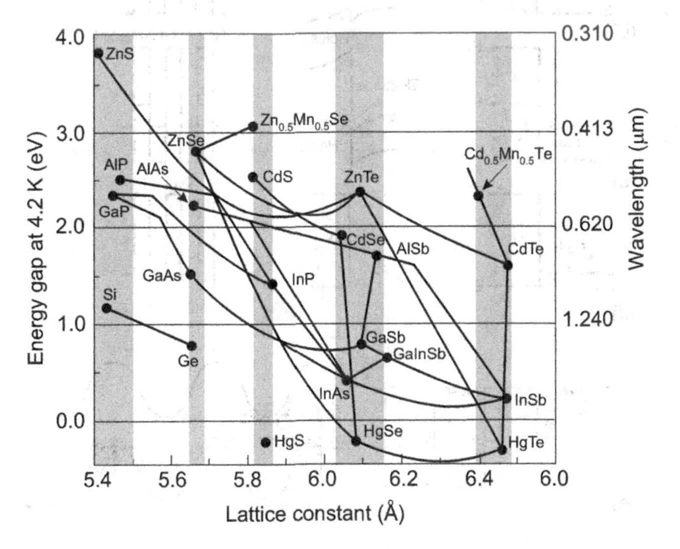

Figure 19.7 A plot of the low-temperature energy bandgaps for a number of semiconductors with the diamond and zinc blende structures versus their lattice constants. The shaded regions highlight several families of semiconductors with similar lattice constants. (From Ref. [31])

Depending on the values for band discontinuities, known heterointerfaces can be classified into four groups: type I, type II-staggered, type II-misaligned, and type III as shown in Figure 19.8.

Type I occurs for systems such as GaAs/AlAs, GaSb/AlSb, strained-layer structure GaAs/GaP, and most II-VI and IV-VI semiconductor structures with a nonzero bandgap. The sum of ΔE_c and ΔE_v is seen to be equal to the bandgap difference $E_{g2} - E_{g1}$ of the two semiconductors. The electrons and holes are confined in one of the semiconductors that are in contact. Such types of SLs and multiple quantum wells (MQWs) are preferentially used as effective injection lasers in which the threshold currents can be made much lower than those of heterolasers.

Type II structures can be divided into two groups: *staggered* (Figure 19.8b) and *misaligned* (Figure 19.8c) structures. Here it is seen that $\Delta E_c - \Delta E_v$ equals the bandgap difference $E_{g2} - E_{g1}$. The type II- staggered structure is found in certain SLs of ternary and quaternary III-Vs where the bottom of the conduction band and the top of the valence band of one of the semiconductors are below the corresponding values of the other (e.g., as in the case of $InAs_xSb_{1-x}$/InSb, $In_{1-x}Ga_xAs/GaSb_{1-y}As_y$ structures). As a consequence, the bottom of the conduction bands and the top of the valence bands are located in opposite layers of SLs or MQWs, so the spatial separation of confined electrons and holes takes place. Such structures potentially can be used as photodetectors since photoinduced nonequilibrium carriers are spatially separated. The type II-misaligned structure is an extension of this in which the conduction band states of semiconductor 1 overlap the valence band states of semiconductor 2. This has been established as occurring, for example, to InAs/GaSb, PbTe/PbS, and PbTe/SnTe systems. Electrons from GaSb valence band enter the InAs conduction band and produce a dipole layer of electron and hole gas, shown in Figure 19.8c. With smaller periods of SLs or MQWs, it is possible to observe the semimetalic-to-semiconductor transition and to use such systems as photosensitive structures, in which the spectral detectivity range can be changed by the thickness of the components.

Type III structures are formed from one semiconductor with a positive bandgap (e.g., $E_g = E_{\Gamma 6} - E_{\Gamma 8} > 0$, such as CdTe or ZnTe) and one semiconductor with a negative bandgap $E_g = E_{\Gamma 6} - E_{\Gamma 8} < 0$ (e.g., HgTe-type

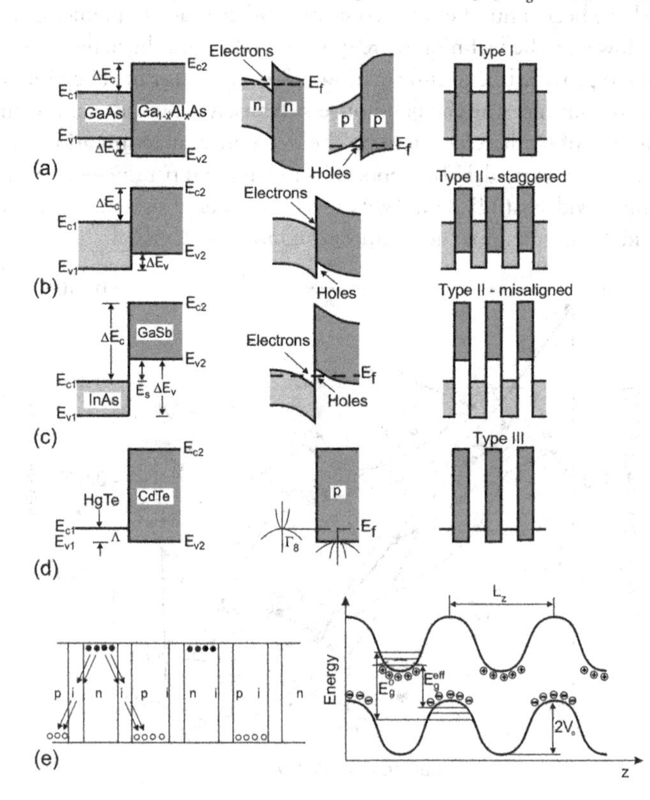

Figure 19.8 Various types of semiconductor SL and MQW structures: (a) type I structure, (b) type II-*staggered* structure, (c) type II-*misaligned* structure, (d) type III structure, and (e) n-i-p-i-structure. L_z is the period of the structures, $2V_o$ is the modulation potential, and E_g^{eff} is the effective bandgap in the n-i-p-i-structures.

semiconductors). At all temperatures, the HgTe-type semiconductors behave like semimetals since there are no activation energies between the light- and heavy-hole states in the Γ_8 band (see Figure 19.8d). This type of SL cannot be formed with the III-V compounds.

19.2.2 DOPING SUPERLATTICES

A spatial modulation of the doping in an otherwise homogeneous lattice can produce a SL effect; that is, a spatial modulation of the band structure that induces a reduction in the Brillouin zone of electrons and new energy bands in the SL direction. The realization of such structures is achieved using periodic n-doped, undoped, p-doped, undoped, n-doped,..., multilayer structure. So far, nearly all the experimental investigations and most of the theoretical studies on doping SLs have dealt with GaAs-doping SL structures not containing intrinsic regions. The term *n-i-p-i crystals*, however, became popular for the whole class of doping SLs. A doping superlattice was first considered in the original proposal of Esaki and Tsu [1] and next pursued especially by Ploog and Döhler [40–42].

The basic concept of the n-i-p-i SL is explained with the help of Figure 19.8e. The doping SL causes the potential to oscillate between the n and p layers (in the same semiconductor), creating a reduced energy gap E_g^{eff} that separates the electron potential valley in the conduction band from the hole potential valley in the valence band. Charged particles are subject to a self-consistent potential [43]. The values of the effective potential $2V_o$ and the effective bandgap E_g^{eff} depend on the doping levels N_d, the dielectric constant values ε_r, and the thicknesses of the layers. At equilibrium,

$$E_g^{eff} = E_g^o - \frac{q^2 N_d}{\varepsilon_o \varepsilon_r}\left(\frac{a}{4} + b\right), \quad (19.12)$$

where a is the layer thickness of n-type and p-type regions, b is the thickness of the i-type layers, $E_g^0 = E_c - E_v$ is the intrinsic bandgap, and the donor and acceptor concentrations are assumed to be equal. For equal, uniform doping levels N_d and zero thickness undoped layers, the periodic potential arcs and has an amplitude:

$$V_o = \frac{q^2}{8\varepsilon_o \varepsilon_r} N_d a^2. \quad (19.13)$$

For GaAs with $N_d = 10^{18}\,\mathrm{cm}^{-3}$ and $a = 500\,\text{Å}$, $V_o = 400$ meV.

Equation 19.12 neglects the additional terms due to quantization of energy in the potential valleys of the n- and p-layers when a is very small. The quantized energy levels in the potential wells are approximately the harmonic oscillator levels:

$$E_{e,h} = \hbar\left(\frac{q^2 N_d}{\varepsilon_o \varepsilon_r m_{e,h}^*}\right)\left(n + \frac{1}{2}\right). \quad (19.14)$$

For electrons in GaAs, for instance, the subband separation is 40.2 meV for the above parameters.

The Equation 19.12 gives the equilibrium value of E_g^{eff} with no applied voltage between the layers. If, however, the n- and p-layers can be contacted separately, using methods described by Döhler [41], then the very interesting possibility arises of controlling the energy gap as a function of the applied voltage.

The n-i-p-i structures are qualitatively similar to the type-II SLs in that the separation in free space of electrons and holes reduces the overlap of the electron and hole wave functions and, consequently, the absorption coefficient. This effect is at least partially compensated by the increased carrier lifetime, which is also a consequence of the spatial separation. These structures can also be considered as potential photodetectors due to the spatial separation of photoinduced carriers [44].

19.2.3 INTERSUBBAND OPTICAL TRANSITIONS

The description of the electron confinement is simplest for a one-dimensional rectangular potential of a well with infinitely high barriers (see Section 19.1). For this model, all the results for description of the performance of the optoelectronic devices can be obtained analytically and, though they are not applicable quantitatively for real structures, the insight gained from such a model is transferable to the finite-barrier case.

In the finite well case, the position of the energy levels changes considerably compared to the infinite well case, even for a parabolic dispersion law. The nonparabolicity of the dispersion law, the multivalley band structure (e.g., the case of n-Si and n-Ge), and the finite heights of the barriers change the description considerably. The electron wave functions already do not vanish at the boundaries of the well but penetrate into the barriers (the amplitudes drop exponentially in the barriers), which is the basis for the formation of SLs. The amplitudes of the envelope wave functions (together with Bloch functions) both in the wells and in the barriers determine the intensity of interband and intersubband (intraband) optical transitions (see Figure 19.9). For more familiar acquaintance with the analysis of optical transitions in SLs and QWs, one can see, e.g., [18,23,45].

To obtain the absorption coefficient values one needs to calculate the dipole matrix element. From theoretical consideration results (see, e.g., [20]) it can be seen that the allowed dipole optical transitions are split into two classes:

- Interband transitions that take place between QW subbands originating from different band extrema i, j and defined by atomic-like dipole matrix elements.
- Intersubband (intraband, $i = j$) optical transitions that are defined by dipole matrix elements between the envelope functions of the same band.

The optical dipole moment can be expressed as:

$$M \sim \int \phi_F(z)\vec{\varepsilon}\cdot\vec{r}\phi_I(z)dr, \tag{19.15}$$

where ϕ_I and ϕ_F are initial and final envelope wave functions, $\vec{\varepsilon}$ is the polarization vector of the incident photons, and z is the growth direction of the QW. These yield a dipole matrix element in the order of the size of the QW for intersubband transitions compared to the value of the atomic size for interband transitions. For an infinite well, the value of the dipole matrix element $\langle z \rangle$ between the ground and first excited state is $16L/9\pi^2$ (~0.18 L_w; L_w is the width of the QW). Since the envelope wave functions in Equation 19.15 are orthogonal, M is nonzero due to the component of $\vec{\varepsilon}\cdot\vec{r}$ perpendicular to the QW (along the growth direction). Therefore, the optical electric field must also have a component along this direction in order to induce an intersubband transition; thus, normal incidence radiation will not be absorbed.

The intensity of intersubband optical transitions is proportional to $\cos^2\phi$, where ϕ is the angle between the plane of the QW and the electromagnetic electric field vector. Levine et al. [46] have shown that the polarization selection rule $\alpha \propto \cos^2\phi$ is experimentally confirmed as shown in Figure 19.10 [2]. The IR intersubband absorption was measured at 8.2 μm in a doped GaAs/AlGaAs QW SL using a multipass waveguide geometry. The multipass waveguide geometry increased the net intersubband absorption by approximately two orders of magnitude, hence allowing the accurate measurements of the oscillator strength, the polarization selection rule, and the line shape.

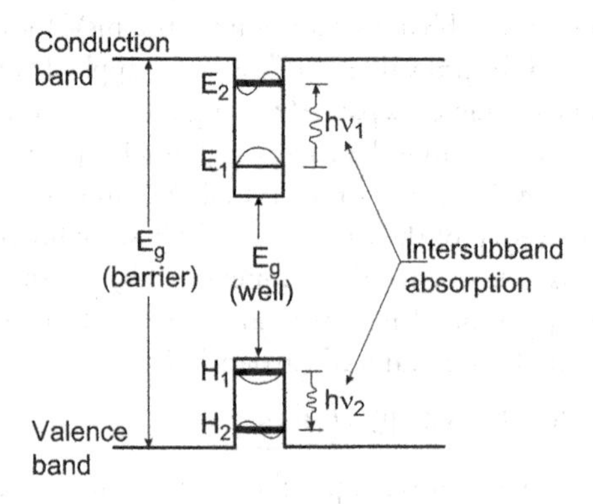

Figure 19.9 Schematic band diagram of a QW. Intersubband absorption can take place between the energy levels of a QW associated with the conduction band (n-doped) or the valence band (p-doped).

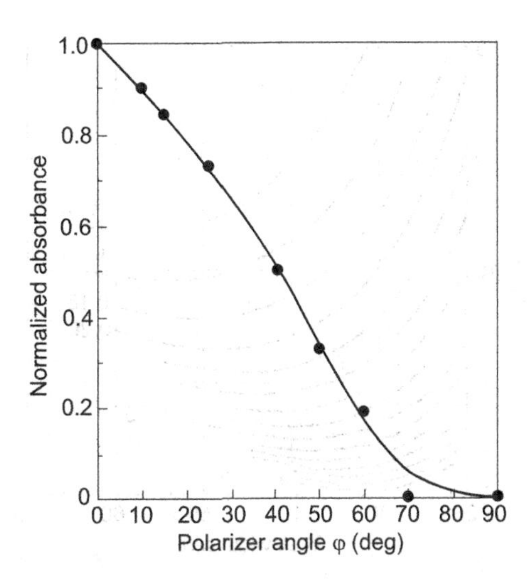

Figure 19.10 Measured intersubband absorbance (normalized to $\varphi = 0$) versus polarizer angle at Brewster's angle $\theta_B = 73°$ for a doped 8.2 μm GaAs/AlGaAs QW SL. The solid line is drawn through the points as a guide to the eye. (From Ref. [2])

The absorption coefficient $\alpha(h\nu)$ associated with an optical transition of an electron promoted from the ground state E_1 to excited state E_2 by absorbing a photon $h\nu$ can be expressed as [47]:

$$\alpha(h\nu) - \frac{\rho^{2-D}}{L}\frac{\pi q^2 \hbar}{2n_r \varepsilon_o m^* c}\sin^2 \varphi f(E_2) g(E_2), \tag{19.16}$$

where L is the period length of the MQW structure, m^* is the effective mass of the electrons in the well, n_r is the index of refraction, $f(E_2)$ is the oscillator strength, and $g(E_2)$ is the one-dimensional density of the final state. When the scattering effects are neglected, the density of final state $g(E_2)$ in the continuum is given simply by [47]:

$$g(E_2) = \frac{L}{2\pi\hbar}\left(\frac{m_b^*}{2}\right)^{1/2}\frac{1}{\sqrt{E_2 - H}}, \tag{19.17}$$

where m_b^* is the effective mass of the electrons in the barrier and H is the barrier height.

According to Equation 19.16, the photon energy at the absorption peak is determined by the product of $f(E_2)$ and $g(E_2)$; however, due to the singularity of $g(E_2)$ at $E_2 = H$, the absorption strongly peaks close to the barrier height. In reality, the density of states in the continuum is strongly modified by the presence of the well and is broadened by impurity scattering and both will tend to smooth out the singularity. As a result, it is safe to assume that the local density of states in the continuum varies relatively slowly compared to $f(E_2)$ when E_2 is close to H [47]. Under this assumption, the absorption peak is approximately determined by the energy dependence of the oscillator strength f. Hence, the peak absorption wavelength λ_p can be obtained by computing the value of $E_2 = E_m$ where f is maximized:

$$\lambda_p = \frac{2\pi\hbar c}{E_m - E_1}. \tag{19.18}$$

Choi [47] has performed calculations on the detector wavelength, the absorption linewidth, and the oscillator strength of a typical GaAs/Al$_x$Ga$_{1-x}$As MQW photodetector with aluminum molar ratio in the barrier ranging from 0.14 to 0.42 and the QW width ranging from 20 to 70 Å. Figure 19.11 shows the absorption peak wavelength λ_p as a function of well width. Within the detector parameters shown, λ_p can be varied from 5 μm to over 25 μm.

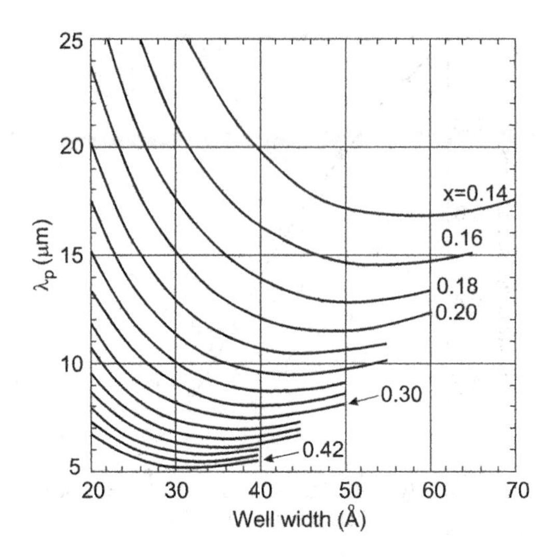

Figure 19.11 The position of the absorption peak λ_p for a given Al molar ratio x as a function of well width. The x changes 0.02 in steps between each curve. (From Ref. [47])

The intersubband absorption has been investigated theoretically and experimentally by a lot of authors as a function of QW width, barrier height, temperature, and doping density in the well [2,5,6]. Bandara et al. [48] have shown that, for high doping ($N_d > 10^{18}\,\mathrm{cm}^{-3}$), exchange interaction can significantly lower the ground-state subband energy and that the direct Coulomb shift can increase the excited-state subband energy; consequently, the peak absorption wavelength shifts to higher energy. In addition to absorption peak shift at high doping densities, the absorption linewidth broadens and the oscillation strength increases linearly with the doping density. Furthermore, there are shifts in the peak absorption wavelength and absorption linewidth with the temperature. The experimentally observed linewidths are within $\Delta v = 50$–$120\,\mathrm{cm}^{-1} \approx 6$–$15$ meV and are governed by the process of longitudinal optical (LO) phonon scattering, with the intersubband relaxation times between the second excited and the first ground states within $\tau_{21} \approx (0.2$–$0.9)$ ps [4,49]. With temperature lowering, there is a small decrease in the position of the peak absorption wavelength and the absorption linewidth. Hasnain et al. [50] have observed an increase in the peak absorption by approximately 30% in comparison with typical maximum room temperature value of $\alpha \approx 700\,\mathrm{cm}^{-1}$. Manasreh et al. [51] have explained this temperature shift by including the collective plasma, exciton-like, Coulomb and exchange interactions, nonparabolicity, and the temperature dependence of the bandgaps and effective mass.

To demonstrate the difference between the absorption spectra line shapes for different MQW structures, Figure 4.8 shows the normalized absorption spectra at $T = 300$ K. The very large difference in spectral width is apparent with the bound excited-state transitions ($\Delta\lambda/\lambda = 9\%$–$11\%$) being three to four times narrower than for the continuum excited state [52]. Much broader spectra in the case of bound-to-continuum transitions are caused by the broadening of the extended continuum excited states.

The major drawback of using n-type III-V Γ-point extrema MQWs is that the intersubband transitions are forbidden for normally incident radiation. Considerable efforts were applied to observe the normal incidence IR absorption both in n-type and p-type III-V and Si/SiGe MQWs (e.g., see [53–63]).

In case of n-type structures, the normal incidence absorption is possible due to the multivalley band structure when the principal axes of the effective mass tensor in the ellipsoids of equal energy are tilted with respect to the growth direction [53]. In such a situation, the normal incidence intersubband absorption may be strong enough and reach in n-Si QWs the values of about 10^3–$10^4\,\mathrm{cm}^{-1}$ for (110) or (111)-growth directions and the free-carrier concentrations in the ground state of about $10^{19}\,\mathrm{cm}^{-3}$. Heavily doped Si is easier to achieve than n-GaAs and corresponding Fermi levels are lower in Si layers due to the greater density of states.

For p-type Si/SiGe normal incidence MQW photodetectors, the great advantage lies in an entirely Si-based technology. But there exist some metallurgical problems in SiGe layer growth with large Ge content. Also, the p-type MQW devices are restricted in transport and responsivity characteristics due to much lower carrier mobilities compared to n-type III-V compound MQW intersubband devices.

As was earlier shown [54] in p-type QWs, the nonorthogonal nature of the hole envelope function is the reason for allowed hole-intersubband transitions with any polarization of light since, for finite in-plane wave vectors, the eigenfunctions of the multiband effective mass Hamiltonian are weighted hole-envelope function linear combinations. In addition to intersubband transitions, transitions between different hole bands were also observed (see Figure 19.12). For these so-called intervalence-subband transitions, the same selection rule is applicable [60–62]. Such transitions occur not only in the 8–12 μm spectral range but also can be extended to the 3–5 μm IR radiation range [61].

The p-type Si/SiGe QWIPs can have broadband photoresponse (8–14 μm), attributed to strain and quantum confinement-induced mixing of heavy, light, and split-off hole bands [63].

It must be taken into account that the intersubband transitions in SLs or QWs are possible only if their initial states are occupied by carriers. So, the intersubband photodetectors are extrinsic in nature.

19.2.4 INTERSUBBAND RELAXATION TIME

There is a considerable interest in the direct determination of the intersubband relaxation processes as these processes determine the operating frequencies of MQW devices, their total quantum efficiency (QE), and the photoconductive gain. The relevant interactions leading to carrier capture are electron-phonon, electron-impurity, and electron-electron scattering. Usually, theoretical considerations are restricted to electrons in n-type GaAs/AlGaAs QWIPs, since the band structure is known reasonably well, and since coupling between heavy and light hole states in the valence band introduces additional complications [13].

The most relevant interaction is the Fröhlich interaction between electrons and LO phonons. Because the confined carriers are free to move within the plane, there is no energy gap separating the confined from the unconfined states, and the density of empty final states satisfying the energy-momentum conservation for phonon emission is high. Consequently, transitions from the extended states (above the barriers) to confined states are very fast, resulting in extremely short excited-carrier lifetime of the order of picoseconds. At large carrier densities (above 10^{18} cm^{-3}), electron-impurity and electron-electron scattering play an important role, whereas radiative relaxation is several orders of magnitude less efficient than the electron-phonon interaction and can be neglected in this context.

If LO phonons play a decisive role in the intersubband relaxation processes, then the intersubband relaxation time will depend on whether the energy separation between the subbands is greater or lower than the LO-phonon energy E_{LO} (e.g., in GaAs E_{LO} = 36.7 meV). If separation between the ground and excited states in the well is lower than the LO energy, then these phonons would not play any role in relaxation processes and the lifetime may be rather long (and so the escape probability of the carriers out of the well under the exterior electric field will be large); if not, then the intersubband relaxation time should be less than 1 ps [64]. The situation discussed was proved in the experiments on picosecond time-resolved Raman spectroscopy. In broad GaAs QWs (L_w = 21.5 nm), in which the energy separation ΔE_{12} between the ground and excited

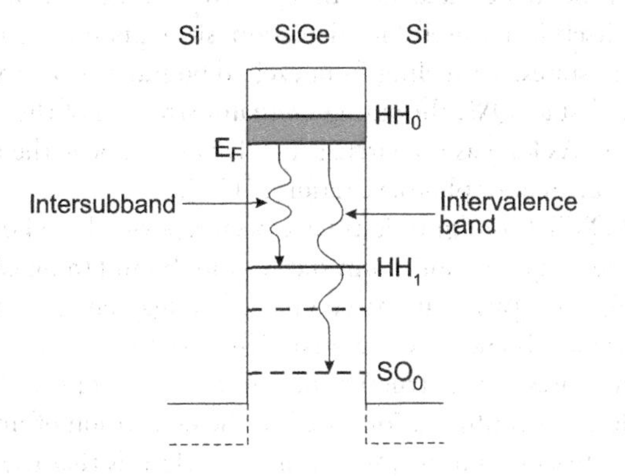

Figure 19.12 Band diagram of the QW structures showing possible subband transitions. (From Ref. [62])

states is less than the LO-phonon energy ($\Delta E_{12} = 26.8$ meV$<E_{LO} = 36.7$ meV) and thus the carrier scattering via LO-phonon excitation cannot take place, the intersubband relaxation times of several hundred picoseconds were observed [65]. These long lifetimes were explained by longitudinal acoustical phonon scattering of the carriers in the upper excited states of the QWs. In another paper [66], the intersubband relaxation times in excess of 500 ps have been observed at low temperatures. For narrower wells ($L_w = 116$ Å, $\Delta E_{12} = 64.2$ meV$>E_{LO}$), where LO phonons play a major role in relaxation processes, the intersubband relaxation time was too short to be measured with the resolution of about 8 ps of the setup used.

Other experiments on the intersubband relaxation time determination within the conditions $\Delta E_{12} > h\nu_1$ showed that under such conditions the times of intersubband relaxation processes of about $\tau_{12} \approx (1\text{–}10)$ ps were observed in GaAs and related MQWs [4]. The experiments carried out on intersubband relaxation time determination differ substantially on the τ_{12} values. The results obtained depend much on the value of photoexcited carrier densities, the degree of confinement of the particle in the excited state, and so on.

Simple estimations carried out for intersubband energies when $E_2 - E_1 >> E_{LO}$, in case of bound-to-continuum QWIPs, give [67]

$$\frac{1}{\tau_{LO}} = \frac{q^2 \lambda_c E_{LO} I_1}{4h^2 c L_p}\left(\frac{1}{\varepsilon_\infty} - \frac{1}{\varepsilon_s}\right), \tag{19.19}$$

where λ_c is the cutoff wavelength, L_p is the QWIP period, and $I_1 \approx 2$ a dimensionless integral. This equation gives the capture time of about 5 ps for typical QWIP parameters.

The lifetime of excited carriers from the ground to continuum states depends on the energy of the states above the well as the capture probability depends on the energy position of the particle above the well. For AlGaAs MQWs, the excited carriers are captured into the well by the emission of polar optical phonons and, for excited states slightly above the well, the lifetime may be less than 20 ps [68]. Thus, taking into account the thin layered structure of intersubband QW photodetectors, high speed operation (>1 GHz) can be provided for a variety of IR applications such as picosecond CO_2-laser pulse investigations, high frequency heterodyne experiments, new telecommunication requirements with new IR fiber materials, and so forth.

19.3 PHOTOCONDUCTIVE QWIP

The concept of using IR photoexcitation out of quantum wells as a means of IR detection was suggested by Smith et al. [69,70]. Coon and Karunasiri [71] made a similar suggestion and pointed out that the optimum response should occur when the first excited state lies near the classical threshold for photoemission from the quantum well. West and Eglash [72] were the first to demonstrate large intersubband absorption between the confined states in a 50 GaAs quantum well. In 1987, Levine and coworkers [73] fabricated the first QWIP operating at 10 μm. This detector design was based on a transition between two confined states in the QW and subsequent tunneling out of the well by an applied electric field. It appears that transitions between the ground state and the first excited state have relatively large oscillator strengths and absorption coefficient. However, this by itself is not useful for detection, since photoexcited carriers may not readily escape from the excited bound states. Tunneling from excited bound states is exponentially suppressed. By decreasing the size of a dual-state QW, the strong oscillator strength of the excited bound state can be pushed up into the continuum. As long as the virtual state is not far above the threshold, the excited state remains effective in the enhancement of photoexcitation [74,75].

Up to the present, several QWIP configurations have been reported based on transitions from bound to extended states, from bound to quasi-continuum states, from bound to quasi-bound states, and from bound to miniband states [13]. All QWIPs are based on the bandgap engineering of layered structures of wide-bandgap (relative to thermal IR energies) materials. The structure is designed such that the energy separation between two of the states in the structure match the energy of the IR photons to be detected.

Figure 19.13 shows two detector configurations used in the fabrication of multicolor QWIP FPAs. The major advantage of the bound-to-continuum QWIP (Figure 19.13a) is that the photoelectron can escape from the QW to the continuum transport states without being tunneled through the barrier. As a result,

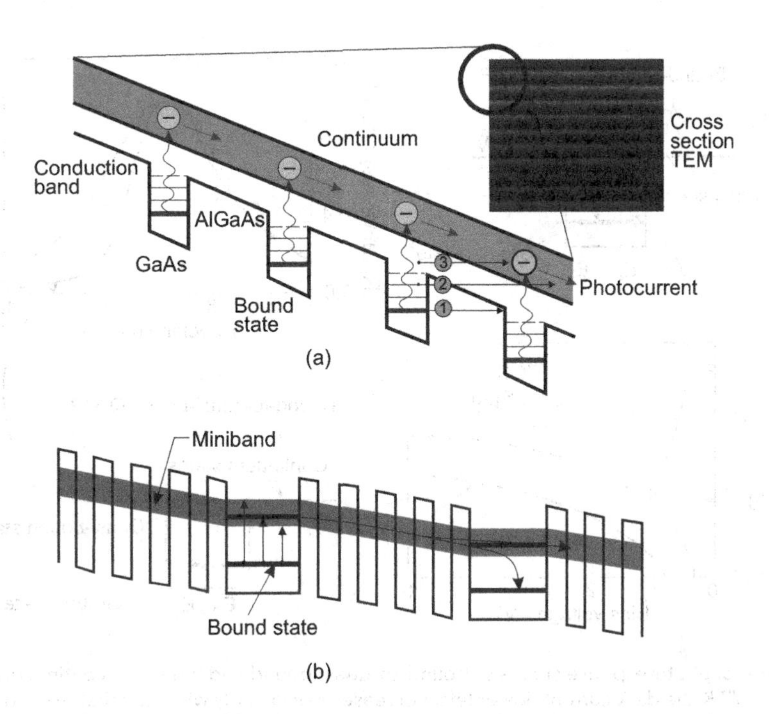

Figure 19.13 Band diagram of demonstrated QWIP structures: (a) bound-to-continuum (after Ref. [8]) and (b) bound-to-miniband. Three mechanisms creating dark current are also shown in (a): ground-state sequential tunneling (1), intermediate thermally-assisted tunneling (2), and thermionic emission (3).

the voltage bias required to efficiently collect the photoelectrons can be reduced dramatically, thereby lowering the dark current. Furthermore, since the photoelectrons are collected without having to tunnel through a barrier, the AlGaAs barriers can be made thicker without reducing the photoelectron collection efficiency. The multilayer structure consists of a periodic array of Si-doped ($N_d \approx 10^{18}$cm^{-3}) GaAs QWs of thickness L_w separated by undoped-Al$_x$Ga$_{1-x}$As barriers of thickness L_b. The heavy n-type doping in the wells is required to ensure that freezeout does occur at low temperatures and that a sufficient number of electrons are available to absorb the IR radiation. For operation at $\lambda = 7$–11 µm, typically $L_w = 40$ Å, $L_b = 500$ Å, $x = 0.25$–0.30, and 50 periods are grown. In order to shift the intersubband absorption to longer wavelength, the x value is decreased to $x = 0.15$ and, in addition, in order to maintain the strong optical absorption and reasonably sharp cutoff line shape, the QW width is increased from 50 Å to 60 Å. This optimization allows the same bound state to excited continuum state optical absorption and efficient hot electron transport and collection. It appears that the dark current decreases significantly when the first excited state is decreased in energy from the continuum to the well top in a bound-to-quasi-bound QWIP (Figure 19.14 [74]), without sacrificing responsivity. In comparison with narrow response of bound-to-bound transitions, the bound-to-continuum transitions are characterized by a broader response. The simple QWIP structures shown in Figures 19.5a and 19.13a are based on the photoemission of electrons from the QWs. They are unipolar devices with contacts on both sides, which require typically 50 wells for sufficient absorption (although 10–100 have been used).

A miniband transport QWIP contains two bound states, the higher-energy one being in resonance with the ground-state miniband in the SL barrier (see Figure 19.13b). In this approach, IR radiation is absorbed in the doped QWs, exciting an electron into the miniband that provides the transport mechanism, until it is collected or recaptured into another QW. Thus, the operation of this miniband QWIP is analogous to that of a weakly-coupled MQW bound-to-continuum QWIP. In this device structure, the continuum states above the barriers are replaced by the miniband of the SL barriers. The miniband QWIPs have lower photoconductive gain than bound-to-continuum QWIPs, because the photoexcited electron transport occurs in the miniband where electrons have to pass through many thin heterobarriers resulting in a lower mobility.

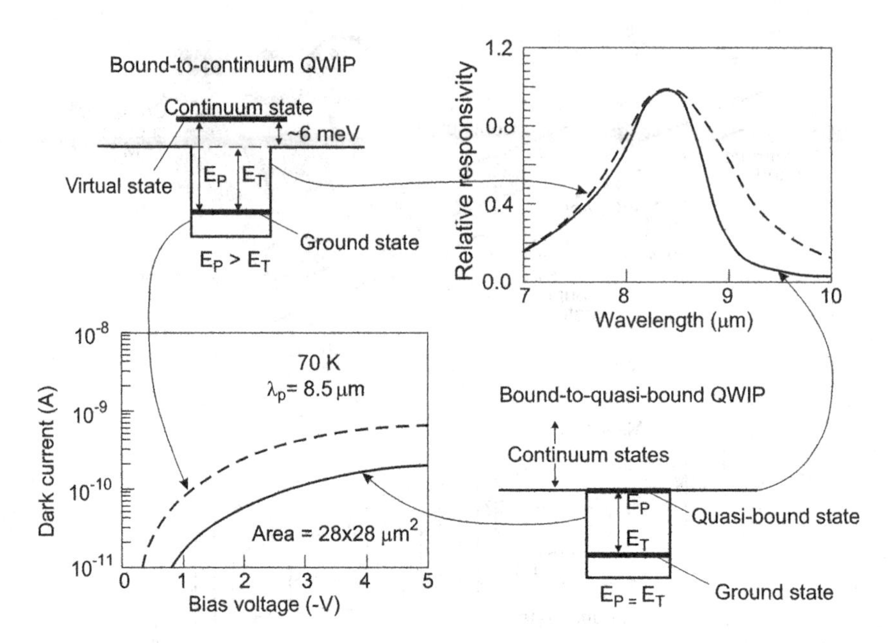

Figure 19.14 In typical photoresponse curves of bound-to-quasi-bound and bound-to-continuum 8.5 μm QWIPs at a temperature of 77 K the dark current (lower left) decreases significantly when the first excited state is dropped from the continuum to the well top, bound-to-quasi-bound QWIP, without sacrificing the responsivity (upper right). The first excited state now resonating with the barrier top produces sharper absorption and photoresponse. (From Ref. [74])

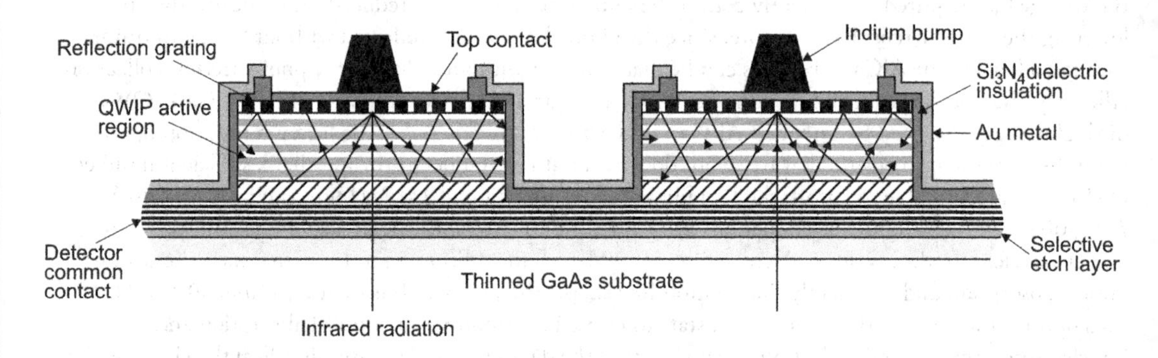

Figure 19.15 Cross section of a detector element in a QWIP array.

19.3.1 FABRICATION

The QW AlGaAs/GaAs structures are mainly grown by MBE on semi-insulating GaAs substrate, although excellent SL structures can also be grown using MOCVD [76,77]. At present, GaAs substrates are available with a diameter up to 8 inches, however, in QWIP, processing lines are typically 4-inch substrates. Process technology starts with the epitaxial growth of structure with a periodic array of Si-doped ($N_d \approx 10^{18}\,cm^{-3}$) GaAs QWs and the sequential growth of an etch-stop layer (usually AlGaAs) used for substrate removal. The QWIP active region is sandwiched between two n-type GaAs contact layers about 1 μm thick (also heavily doped to $N_d \approx 10^{18}\,cm^{-3}$) and an etch stop followed by a sacrificial layer for the grating. For optical coupling, usually 2D reflective diffraction gratings are fabricated (see Section 19.6). Further, process technology includes etching mesas through the SL to the bottom contact layer, followed by ohmic contacts to the n^+- oped GaAs contact layers. These steps can be accomplished using wet chemical or dry etching techniques. In selective etching, usually, ion beam etching is practical for pattering the grating coupler into each pixel.

The same processes are used for the fabrication of both middle wavelength infrared (MWIR) and LWIR devices. Difference concerns the mesa definitions. This modification is necessary since the MWIR QWIPs are based on InGaAs/AlGaAs system whereas the LWIR devices contain In-free GaAs/AlGaAs epilayers. For these reasons, the mesas are defined by reactive ion beam etching for MWIR QWIPs and by chemically assisted ion beam etching for LWIR QWIPs [13].

Ohmic contacts are evaporated (e.g., AuGe/Ni/Au) and alloyed by rapid thermal annealing (e.g., at 425°C for 20 seconds [78]). Usually, the gratings on each pixel is covered by metallization (Au), which is advantageous in comparison with using the ohmic contact metal in order to increase IR absorption in an active detector's region. The surface of the detector array is passivated with silicon nitride and, to provide electrical contact to each detector element, openings in the nitride are formed. Finally, in order to facilitate the hybridization to Si readout, a separate metallization is evaporated. Figure 19.15 shows the cross section of a pixel in a QWIP array.

After dicing the wafers into single chips and hybridization to Si readout, the GaAs substrate is removed in order to reduce the mechanical stress between two chips and prevent optical crosstalk arising from light propagation between pixels. The process of substrate removal is accomplished using a sequence of mechanical lapping, wet chemical polishing, and a selective wet chemical etching. The last process is stopped at a dedicated etch-stop layer previously deposited.

19.3.2 DARK CURRENT

A good understanding of the dark current is crucial for the design and optimization of a QWIP detector because dark current contributes to the detector noise and dictates the operating temperature.

Initial efforts with multiple QW photoconductive detectors showed appreciable dark currents in which are involved three relevant mechanisms: tunneling, phonon-assisted tunneling, and thermionic emission out of QWs. In Figure 19.16 [79], individual current contributions for devices with areas $A = 2 \times 10^{-5}\,\text{cm}^2$ at $V_b = 0.05\,\text{V}$ are shown as functions of temperature. We can see that tunneling is the dominant dark current mechanism in the low-temperature regime, while thermionic emission limits the performance at high operating temperature.

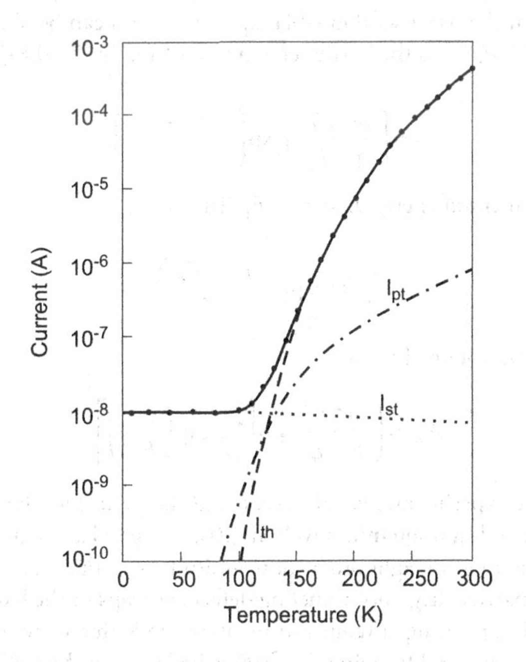

Figure 19.16 Temperature dependence of the dark current at low bias for bound-to-bound multiple QW $Al_{0.36}Ga_{0.64}As/GaAs$ device (50 wells with 70 Å wells and 140 Å barriers). The subscripts *th*, *st*, and *pt* refer to thermionic, tunneling, and phonon-assisted tunneling mechanisms, respectively. (From Ref. [79])

Further, rapid progress in multiple QW IR detectors was conditioned by the use of thinner quantum wells and pushing the excited state into the continuum. The linear relation between responsivity and the bias voltage for these devices is completely different from the highly nonlinear photoresponse obtained from devices containing two bound states in the QWs. In particular, the bound-to-bound state device requires a substantial bias $V_b > 0.5\,V$ before any photosignal is observed, whereas the bound-to-extended detectors generate a photocurrent at a very low bias. The reason for this difference is that the bound-to-bound state detector requires a large electric field to assist the tunneling escape of the photoexcited carriers out of the well, and thus at low bias the excited carriers cannot be collected resulting in a negligible photoresponse. The barrier thickness could be greatly increased (e.g., $L_b \approx 500$ Å), thereby dramatically lowering the undesirable dark current.

In the further considerations we follow after Levine et al. They shown that thermionic-assisted tunneling is a major source of the dark current [2,5,8,80]. To calculate the dark current I_d, we first determine the effective number of electrons n that are thermally excited out of the well into the continuum transport states as a function of bias voltage V:

$$n = \left(\frac{m^*}{\pi \hbar^2 L_p}\right) \int_{E_o}^{\infty} f(E) T(E,V)\, dE, \tag{19.20}$$

where the first factor containing the effective mass m^* is obtained by dividing the two-dimensional density of states by the SL period L_p (to convert it into an average three-dimensional density), and where $f(E)$ is the Fermi factor $f(E) = \{1 + \exp[(E - E_o - E_F)/kT]\}^{-1}$, E_o is the ground state energy, E_F is the two-dimensional Fermi level, and $T(E,V)$ is the bias-dependent tunneling current transmission factor for a single barrier. Equation 19.20 accounts for both thermionic emission above the energy barrier E_b (for $E > E_b$) and thermionically-assisted tunneling (for $E < E_b$). The bias-dependent dark current is

$$I_d(V) = qn(V)v(V)A, \tag{19.21}$$

where q is the electronic charge, A is the device area, and v is the average transport velocity (drift velocity) given by $v = \mu F[1 + (\mu F/v_s)^2]^{-1/2}$, where μ is the mobility, F is the average field, and v_s is the saturated drift velocity.

A much simpler expression that is a useful low-bias approximation can be obtained by setting $T(E) = 0$ for $E < E_b$ and $T(E) = 1$ for $E > E_b$ (E_b is the barrier energy), resulting in [2,80,81]

$$n = \left(\frac{m^* kT}{\pi \hbar^2 L_p}\right) \exp\left(-\frac{E_c - E_F}{k_T}\right), \tag{19.22}$$

where we have set the spectral cutoff energy $E_c = E_b - E_1$. Therefore,

$$\frac{I_d}{T} \propto \exp\left(-\frac{E_c - E_F}{kT}\right), \tag{19.23}$$

where the Fermi energy can be obtained from

$$N_d = \left(\frac{m^* kT}{\pi \hbar^2 L_w}\right) \ln\left[1 + \exp\left(\frac{E_F}{kT}\right)\right]. \tag{19.24}$$

Figure 19.17 compares the experimental (solid curves) and theoretical (dashed) dark I-V curves at various temperatures for a 50-period multiquantum well SL [81]. The good agreement between theory and experiment is achieved over a range of eight orders of magnitude of dark current and demonstrates the high quality of the AlGaAs barriers (e.g., no tunneling defects or traps in the barriers).

For AlGaAs/GaAs QWIPs operating at temperatures above 45 K (for 15 μm devices), the thermionic emission dominates the dark current. Dropping the first excited state to the well top (bound-to-quasibound QWIPs; see Figure 19.14) theoretically causes the dark current to drop by a factor of ~6 at temperature 70 K for 9 μm devices [82]. This compares well with the four-fold drop experimentally observed.

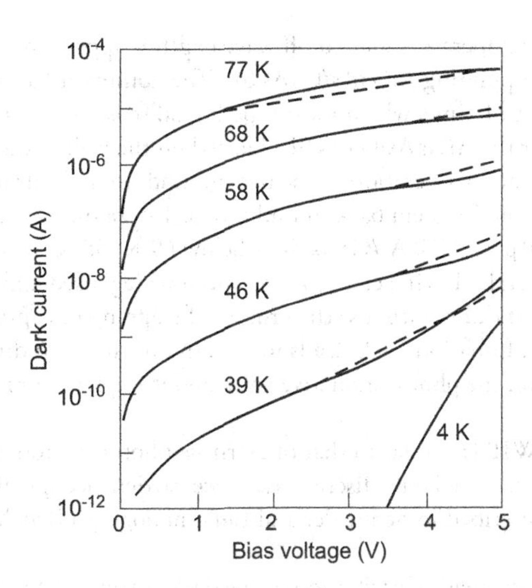

Figure 19.17 Comparison of experimental (solid curves) and theoretical (dashed) dark current-voltage curves at various temperatures for a 50-period, 200-μm-diameter mesa $Al_{0.25}Ga_{0.75}As$/GaAs detector having a doping density of 1.2×10^{18} cm^{-3} ($L_w = 40$ Å, $L_b = 480$ Å, $\lambda_c = 10.7$ μm). (From Ref. [81])

The bound-to-quasi-bound QWIP still preserves the photocurrent [75,82]. The first excited state could be pushed deeper into the well to increase the barrier to thermionic emission, but this would drop the photocurrent to unacceptably low levels. Dark current was also reduced by cutting the well-doping density to decrease the ground-state electrons available for thermionic emission and by increasing each barrier thickness in the QW stack.

Using a similar consideration, Kinch and Yariv [83] have presented an investigation of the fundamental physical limitations of individual MQW IR detectors as compared to ideal HgCdTe detectors. Figure 19.18 compares the thermal generation current versus temperature for AlGaAs/GaAs MQW SLs and HgCdTe alloys at $\lambda_c = 8.3$ μm and 10 μm. Calculations were carried out for a specific set of device parameters ($\tau = 8.5$ ps, $t = 1.7$ μm, $L_w = 40$ Å, $L_p = 340$ μm, and $N_d = 2\times10^{18}$ cm^{-3}) chosen to agree with the already published data for detectors with $\lambda_c = 8.3$ μm [84]. For $\lambda_c = 10$ μm, the QW width is changed to $L_w = 30$ Å and the remaining parameters assumed the same. It is apparent from Figure 19.18 that for HgCdTe the thermal

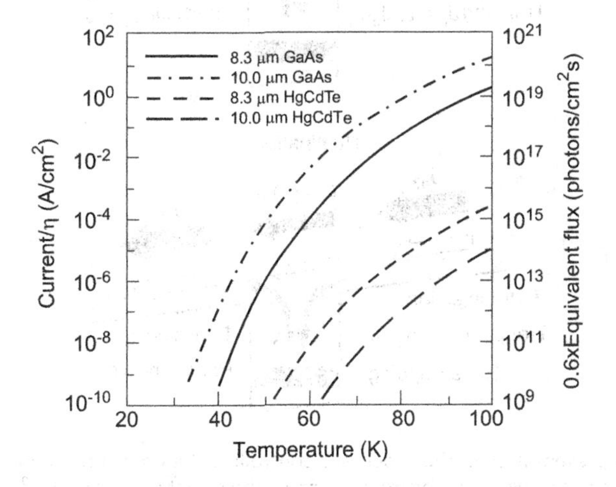

Figure 19.18 Thermal generation current versus temperature for GaAs/AlGaAs MQWs and HgCdTe alloy detectors at $\lambda_c = 8.3$ μm and 10 μm. The assumed effective QEs are $\eta = 12.5$ and 70% for GaAs/AlGaAs and HgCdTe detectors, respectively. (From Ref. [83])

generation rate at any specific temperature and cutoff wavelength is approximately five orders of magnitude smaller than for the corresponding AlGaAs/GaAs SL. The dominant factor favoring HgCdTe in this comparison is the excess carrier lifetime, which for n-type HgCdTe is above 10^{-6} seconds at 80 K, compared to 8.5×10^{-12} seconds for the AlGaAs/GaAs SL. Plotted on the right-hand axis of Figure 19.18 is the equivalent minimum temperature of operation in the background limited infrared photodetectors (BLIP) condition. For example, at a typical system background flux of 10^{16} photons/cm^{2}s, the required temperature of operation for the 8.3 μm (10 μm) AlGaAs/GaAs SL is below 69 K (58 K) to achieve the BLIP condition.

Even though the model given by Levine et al. [80] has been widely used and is in good agreement with many experimental data, it did not discuss the process of trapping or capture to balance the electron emission or escape. As a result, Equation 19.22 leads to certain misunderstandings. For example, it ignores the implicit dependence of J_d on the photoconductive gain, and it implies an unrealistic proportionality between J_d and $1/L_p$.

The device operation of QWIP is similar to that of extrinsic photodetectors, but in contrast with conventional detectors its distinct feature is the discreteness since carriers occupy discrete QWs. Details of the carrier behavior in QWs are described in Schneider and Liu's monograph [13]. We further follow after this monograph.

Figure 19.19 (top) shows a schematic distribution of the dark current paths. In the barrier region, the current flows as a 3D flux and the current density J_{3D} is equal to the dark current J_d. In the vicinity of each well, the capture (current density J_c) and emission (current density J_e) of electrons from the well must be balanced by the trapping or capture of electrons into the well under a steady state condition, so $J_c = J_e$. If we define a trapping or capture probability p_c, we must have $J_c = p_c J_{3D}$, and the sum of the captured and uncaptured fractions must equal the current in the barrier region [13]:

$$J_{3D} = J_c + (1 - p_c) J_{3D} = J_e + (1 - p_c) J_{3D}. \tag{19.25}$$

The dark current can be determined by calculating either J_{3D} directly or by calculating J_e, and in the latter case $J_d = J_e/p_c$.

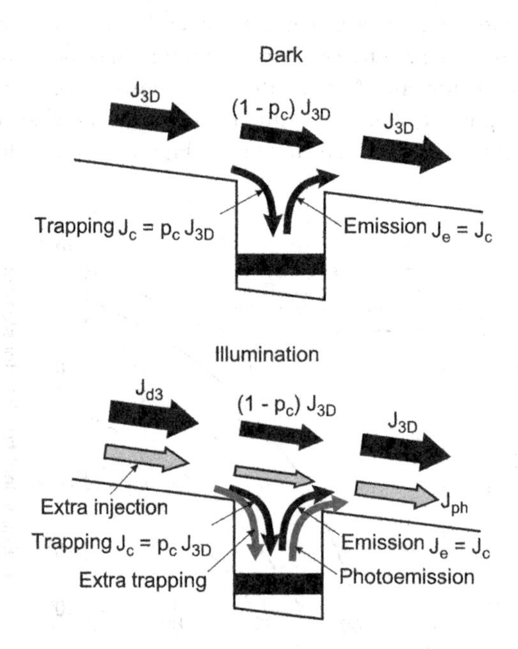

Figure 19.19 Schematic representation of the processes controlling the dark current and photocurrent. The top part shows the dark current paths, while the bottom indicates the direct photoemission and the extra current injection from the contact to balance the loss of electrons from the well. The dark current paths remain the same under illumination. The collected total photocurrent is the sum of the direct photoexcited and the extra injection contributions. (From Ref. [13])

Liu, in review papers [7,85] and monograph [13], together with critical comments, presented several established physical models of the QWIP dark current. Between them we can distinguish:

- carrier drift model,
- emission capture model, and
- several self-consistent and numerical models.

In the carrier drift model, first presented by Kane et al. [86], only drift carrier contribution is taken into account (diffusion is neglected). The dark current is given by, for example, Equation 19.21, $J_d = qn_{3D}v(F)$, where $n_{3\text{-}D}$ is a 3D electron density on top of the barrier. In this way, the SL barriers are treated as a bulk semiconductor, which is justified because the barriers are thick (much thicker than the wells). The only 2D QW effect comes for the evaluation of the Fermi level. Assuming a complete ionization (wells are degenerately doped), the 2D doping density N_d equals the electron density within a given well. Then assuming the relation between N_d and the Fermi energy E_f as $N_d = (m/\pi\hbar^2)E_f$, a simple calculation yields:

$$n_{3D} = 2\left(\frac{m_b kT}{2\pi\hbar^2}\right)^{3/2} \exp\left(-\frac{E_a}{kT}\right), \tag{19.26}$$

if $E_a/kT >> 1$, which is appropriate for most practical cases. Here, m_b is the barrier effective mass and E_a is the thermal activation energy equal to the energy difference between the top of the barrier and the top of the Fermi level in the well.

The simple carrier drift model compares well with experiments in the regime of a low applied electric field. This is a consequence of the key step in the evaluation of n_{3D} by taking the equilibrium value at zero bias with the Fermi level determined by the well doping.

The second emission capture model, gives adequate results for a large range of applied fields. It was originally presented by Liu et al. [87]. This model calculates first J_e and then dark current as $J_d = J_e/p_c$.

The escape current density can be written as:

$$J_e = \frac{qn_{2D}}{\tau_{sc}}, \tag{19.27}$$

where n_{2D} is a 2D electron density on the upper part of the ground state subband and τ_{sc} is the scattering time to transfer these electrons from the 2D subband to the nonconfined continuum on top of the barrier.

The capture probability is related to the relevant time constants by:

$$p_c = \frac{\tau_t}{\tau_c + \tau_t}, \tag{19.28}$$

where τ_c is the capture time (lifetime) for an excited electron back into the well and τ_t is the transit time for an electron across one QW region. For actual devices at operating electric fields, $p_c << 1$ ($\tau_c >> \tau_t$) and the dark current becomes:

$$J_d = \frac{J_e}{p_c} = q\frac{n_{2D}}{\tau_{sc}}\frac{1}{p_c} = q\frac{n_{2D}\tau_c}{\tau_{sc}\tau_t} = q\frac{n_{2D}}{L_p}\frac{\tau_c}{\tau_{sc}}v, \tag{19.29}$$

where $L_p = L_w + L_b$. n_{2D}/τ_{sc} represents the thermal escape or generation of electrons from the QW, and $1/p_c$, as shown later, is proportional to the photoconductive gain that implies the dependence of dark current on the photoconductive gain.

The final expression for dark current in Liu's model is given by [13,87]:

$$J_d = \frac{qv\tau_e}{\tau_{sc}}\int_{E_1}^{\infty}\frac{m}{\pi\hbar^2 L_p}T(E,F)\left[1+\exp\left(\frac{E-E_F}{kT}\right)\right]^{-1} dE. \tag{19.30}$$

For pure thermionic emission regime, when the transmission coefficient $T(E,F) = 0$ for E below the barrier, the last equation becomes

$$J_d = \frac{qv\tau_c}{\tau_{sc}}\frac{m}{\pi\hbar^2 L_p}kT\exp\left(-\frac{E_a}{kT}\right), \tag{19.31}$$

and closely resembles Equation 19.21 together with Equation 19.26.

Analysis of several established models of dark current in QWIP structures, with varying degrees of complexity, give good agreement between them and convergence with experiments [13]. However, realistic calculations of scattering or trapping rates are extremely complicated and have not performed so far.

The magnitude of QWIP dark current can be modified using different device structures, doping densities, and bias conditions. Figure 19.20 shows the QWIP I-V characteristics for temperatures ranging from 35 K to 77 K, measured in a device at the 9.6 μm spectral peak [88]. It shows typical operation at 2 V applied bias in the region where the current varies slowly with bias, between the initial rise in current at low voltage and the later rise at high bias. Typical LWIR QWIP dark current is about 10^{-4} A cm^{-2} at 77 K. Thus, a 9.6- μm QWIP must be cooled to 60 K to have a leakage current comparable to that of a 12 μm HgCdTe photodiode operating at a temperature that is 25°C higher.

19.3.3 PHOTOCURRENT

The bottom part of Figure 19.19 shows the additional processes that occur in QW as a result of the incidence of IR radiation. The direct photoemission of electrons from the well contributes to the observed photocurrent in the collector. All dark current paths remain unchanged.

Photoconductive gain is an important parameter that affects the spectral responsivity and detectivity of detector (see Section 4.2.2). This parameter is defined as the number of electrons flowing through the external circuit for each photon absorbed and is a result of the extra current injection from the contact necessary to balance the loss of electrons from the well due to photoemission. As Figure 19.19 shows, the total photocurrent consists of contributions from the direct photoemission and the extra current injection.

The magnitude of photocurrent is independent of the number of wells if the absorption for each well is the same. Considering two neighboring wells, the processes of photoemission and refilling are identical for both wells. The same arguments can be made for any subsequent well. This means that the photocurrent is unaffected by adding more wells as long as the magnitude of absorption and hence photoemissions from all the wells remain the same.

Assuming n_{ex} as the number of the excited electrons from one well and taking into account a rate equation, we have

$$\frac{dn_{ex}}{dt} = \Phi\eta^{(1)} - \frac{n_{ex}}{\tau_{esc}} - \frac{n_{ex}}{\tau_{relax}}. \tag{19.32}$$

Next, solving Equation 19.32 for n_{ex} with regard to $i_{ph}^{(1)} = qn_{ex}/\tau_{esc}$ for one well under steady state ($dn_{ex}/dt = 0$), gives

$$i_{ph}^{(1)} = q\Phi\eta^{(1)}\frac{\tau_{relax}}{\tau_{relax} + \tau_{esc}} = q\Phi\eta\frac{p_e}{N}, \tag{19.33}$$

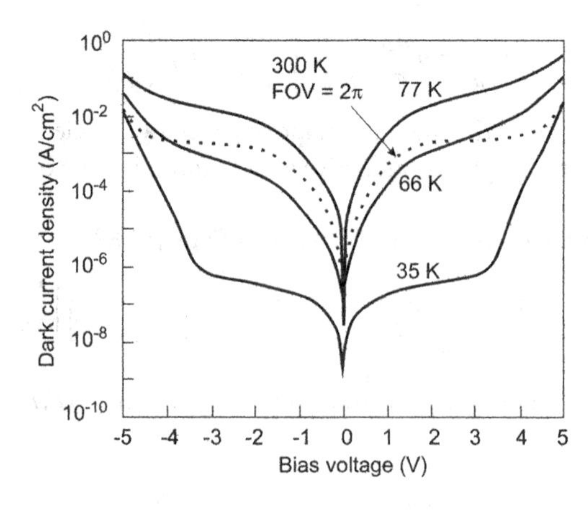

Figure 19.20 Current-voltage characteristics of a QWIP detector having a peak response of 9.6 μm at various temperatures, along with the 300 K background window current measured at 30 K with an 180° FOV. (From Ref. [88])

where Φ is the incident photon number per unit time, the superscript (1) indicates the quantities for one well, τ_{esc} is the escape time, τ_{relax} is the intersubband relaxation time, $\eta = N\eta^{(1)}$ is the total absorption QE (we have assumed that the amount of absorption is the same for all the wells), and N is the number of wells. The escape probability for an excited electron from the well is given by

$$p_e = \frac{\tau_{relax}}{\tau_{relax} + \tau_{esc}}. \tag{19.34}$$

The injection current, $i_{ph}^{(1)} = i_{ph}^{(1)}/p_c$, which refills the well to balance the loss due to emission, equals the photocurrent

$$I_{ph} = \frac{i_{ph}^{(1)}}{p_c} = q\Phi\eta\frac{p_e}{Np_c} \tag{19.35}$$

and

$$g_{ph} = \frac{p_e}{Np_c} \tag{19.36}$$

is the photoconductive gain.

We comment on several aspects of Liu's model. In the conventional theory of photoconductivity, $g_{ph} = \tau_c/\tau_{t,tot}$ (see Equation 12.8 and Rose [89]) where $\tau_{t,tot} = (N+1)\,\tau_t$ is the total transit time across the detector active region. Under the approximation $p_e \approx 1$, $p_c \approx \tau/\tau_c << 1$ and $N >> 1$, the gain expression given by Equation 19.36 and the conventional theory become the same:

$$g_{ph} \approx \frac{1}{Np_c} \approx \frac{\tau_c}{\tau_{t,tot}} = \frac{\tau_c v}{NL_p}. \tag{19.37}$$

The capture time, also called lifetime of carriers, is associated with the scattering of electrons (trapping) into the ground state subband. The condition $p_e \approx 1$ is fulfilled for a bound-to-continuum case, while for a bound-to-bound case this is no longer true. If the absorption is proportional to N, the photocurrent is independent of N since g_{ph} is inversely proportional to N. Photocurrent independence of N is equivalent to its independence of device length in the conventional theory. It should also be mentioned that this independence does not mean that the detector performance is independent of the number of wells because of noise considerations.

The dependence of photoconductive gain on the number of wells for different values of p_c and $p_e = 1$ together with some existing experimental data is shown in Figure 19.21 [13,90]. Most of the reported detector samples have 50 QWs, and a range of gain values from about 0.25 to 0.80 have been observed.

Concerning estimation of the time scales involved in QWIPs, it can be assumed that τ_c is approximately 5 ps. The transit time is mainly determined by the high-field drift velocity of an excited electron in the barrier region. For typical parameters of $v = 10^7$ cm s^{-1} and $L_p = 30$–50 nm, $\tau_t \approx L_p/v$ is estimated to be in the range of 0.3–0.5 ps. One therefore expects a capture probability ($p_c = \tau_t/(\tau_c + \tau_t) \approx \tau_t/\tau_c$) to be in the range of 0.06–0.10, consistent with the experiments.

19.3.4 DETECTOR PERFORMANCE

The conventional theory of photoconductivity (see Section 12.1.1) is normally employed to describe MQW photoconductors where the electron recirculates through the SL for the time $\tau_{t,tot}$, and thus the hot-electron mean free path can be much larger than the SL length. However, because the hot-electron lifetime is very short, so $g_{ph} > 1$ is received only in the low-period SLs (see Figure 19.21).

The current responsivity is given by

$$R_i = \frac{\lambda\eta}{hc} q g_{ph} \tag{19.38}$$

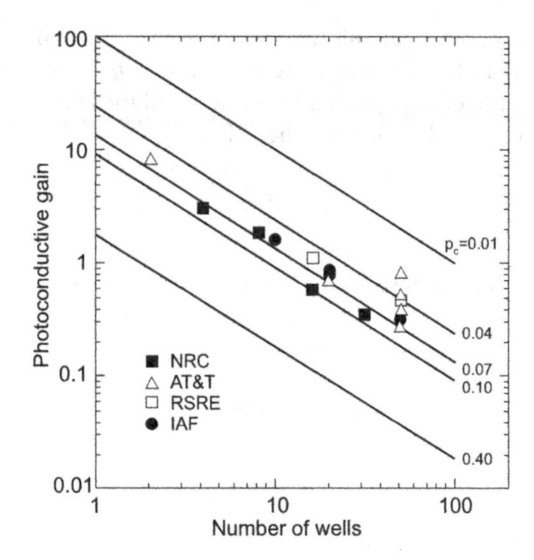

Figure 19.21 Calculated photoconductive gain versus the number of wells for different capture probabilities. Experimental data are taken from Liu et al. (■), Levine et al. (Δ), Kane et al. (□), and Schneider et al. (•). (From Ref. [13])

and depends on both QE and photoconductive gain. A high absorption does not necessarily give rise to a large photocurrent. It was shown that the optimum occurs when the excited state is in close resonance with the top of the barrier [91,92]. Then, the photoexcited electrons effectively escape from the wells to give rise to a large photocurrent.

In Section 4.2.1, it is marked that the QWIP absorption spectra, its value and shape depends on the design of the active region. As is shown in Figure 4.8, the spectra of the bound-to-continuum (samples A, B, and C) are much broader than the bound-to-bound or bound-to-quasi-bound (samples E and F). Table 19.1 gives the absorption values and the corresponding spectral parameters for different n-doped GaAs/AlGaAs QWIP structures. Also included are the values of QE.

Since $\eta \propto N$ and $g_{ph} \propto 1/N$, there is nothing that can be done about the number of wells to improve responsivity. Analysis carried out by Liu [85] indicates that the escape probability must be made close to unity, which is fulfilled for the bound-to-continuum case. If a bound-to-bound design is employed, one must have the excited state close to the top of the barrier. For a typical 10 GaAs/AlGaAs QWIP under a typical field of 10 kV cm^{-1}, this dictates that the excited state should be lower than about 10 meV below the top of the barrier.

The normalized spectra of responsivity are given in Figure 19.22 for the same samples A–F [2,8]. Again, we see that the bound and quasi-bound excited-state QWIP samples are much narrower ($\Delta\lambda/\lambda$ = 10%–12%) than the continuum structure ($\Delta\lambda/\lambda$ = 19%–28%; $\Delta\lambda$ is the spectral width for which responsivity drops to half value).

The detectivity can be determined using

$$D^* = R_i \frac{(A\Delta f)^{1/2}}{I_n}, \qquad (19.39)$$

where A is the detector area and Δf is the noise bandwidth (taken as Δf = 1 Hz).

In general, a photoconductive detector has several sources of noise. The most important are: Johnson noise, the generation-recombination noise (dark noise), and photon noise (connected with current induced by incident photons). For QWIPs, the $1/f$ noise seldom limits the detector performance. Its nature is complicated and is still an ongoing research topic [13].

In a conventional photoconductor, the noise gain equals the photoconductive gain $g_n = g_{ph}$. However, in a QWIP detector, g_n is different from g_{ph}, which was explained by Liu [93]. The standard generation-recombination noise can be written as [89]:

$$I_n^2 = 4qg_nI_d\Delta f. \qquad (19.40)$$

Table 19.1 Parameters for different n-doped, 50-period $Al_xGa_{1-x}As$ QWIP structures

SAMPLE	WELL WIDTH (Å)	BARRIER WIDTH (Å)	COMPOSITION (X)	DOPING DENSITY (10^{18} cm^{-3})	INTERSUBBAND TRANSITION*	λ_p (μm)	λ_c (μm)	$\Delta\lambda$ (μm)	$\Delta\lambda/\lambda$ (%)	α_p(77 K) (cm^{-1})	η (77 K) (%)
A	40	500	0.26	1.0	B-C	9.0	10.3	3.0	33	410	13
B	40	500	0.25	1.6	B-C	9.7	10.9	2.9	30	670	19
C	60	500	0.15	0.5	B-C	13.5	14.5	2.1	16	450	14
E	50	500	0.26	0.42	B-B	8.6	9.0	0.75	9	1,820	20
F	45	500	0.30	0.5	B-QB	7.75	8.15	0.85	11	875	14

Source: After Ref. [2].

* Type of transition: bound-to-continuum (B-C), bound-to-bound (B-B), bound-to-quasi-bound (B-QB).

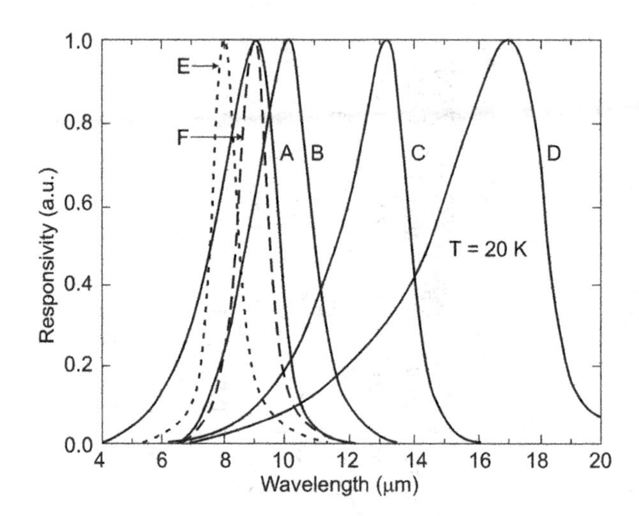

Figure 19.22 Normalized responsivity spectra versus wavelength measured at $T = 20$ K for samples A–F. (From Refs. [2,8])

This equation takes into account influences of both generation and recombination processes. In case of a QWIP detector, the generation-recombination noise should consist of contributions connected with carrier emission and capture (fluctuations in i_e and i_c). Seeing that

$$I_d = \frac{i_e^{(1)}}{p_c} = \frac{i_e}{Np_c}, \tag{19.41}$$

where $i_e = Ni_e^{(1)}$ is the total emission current from all N wells and, equivalently,

$$I_d = \frac{i_c^{(1)}}{p_c} = \frac{i_c}{Np_c}, \tag{19.42}$$

we have

$$I_n^2 = 2q\left(\frac{1}{Np_c}\right)^2 (i_e + i_c)\Delta f = 4q\left(\frac{1}{Np_c}\right)^2 i_e\Delta f = 4q\frac{1}{Np_c}I_d\Delta f. \tag{19.43}$$

Comparing the last equation with Equation 19.40, the noise gain is defined by

$$g_n = \frac{1}{Np_c}, \tag{19.44}$$

and is different from the photoconductive gain (see Equation 19.36). It appears that this expression is valid for small QW capture probabilities (i.e., $p_c \ll 1$). QWIPs satisfy this condition at the usual operating bias (i.e., 2–3 V).

The last equation does not give a correct explanation in the limit of high capture probability $p_c \approx 1$ (or equivalent low noise gain). The appropriate model for this case was first given by Beck [94] using stochastic considerations and taking into account that a high capture probability is not necessarily connected with a low escape probability. A more general expression has a form

$$I_n^2 = 4qg_nI_d\left(1 - \frac{p_c}{2}\right)\Delta f, \tag{19.45}$$

which can apply even in low-bias conditions where capture probabilities for carriers traversing the wells are high. For the case of $p_c \approx 1$, this expression equals the shot-noise expression for N series-connected detectors.

Since $I_d \propto \exp[-(E_c-E_F)/kT]$ (see Equations 19.21 and 19.22) and $D^* \propto (R_i/I_n)$, we have

$$D^* = D_o \exp\left(\frac{E_c}{2kT}\right). \tag{19.46}$$

Based on this relationship, Levine et al. [2,52] have reported useful empirical D^* values by the best fit $T = 77$ K detectivity for the n-type

$$D^* = 1.1\times10^6 \exp\left(\frac{hc}{2kT\lambda_c}\right)\text{cmHz}^{1/2}\text{W}^{-1} \tag{19.47}$$

and for p-type GaAs/AlGaAs QWIPs

$$D^* = 2\times10^5 \exp\left(\frac{hc}{2kT\lambda_c}\right)\text{cmHz}^{1/2}\text{W}^{-1}. \tag{19.48}$$

Figure 19.23 shows the detectivity versus cutoff energy for both n-type and p-type GaAs/AlGaAs QWIPs. It should be noted that the experimental results are for a 45° polished input facet and that optimized gratings and optical cavities can be expected to improve the performance. Note that although Equations 19.47 and 19.48 are fitted to data taken at 77 K, they are expected to be valid over a wide range of temperatures.

Rogalski [95] used simple analytical expressions for detector parameters described by Andersson [67]. Figure 19.24 shows the dependence of detectivity on the long wavelength cutoff for GaAs/AlGaAs QWIPs at different temperatures. The satisfactory agreement with experimental data in a wide range of cutoff wavelengths $8 \le \lambda_c \le 19$ μm and temperatures $35 \le T \le 77$ K has been obtained, considering the samples have different doping, different methods of crystal growth (MBE, MOCVD, and gas-source MBE), different spectral widths, different excited states (continuum, bound, and quasi continuum), and even in one case a different material system (InGaAs). As a matter of fact, Equation 19.47 is in nice agreement with the results presented in Figure 19.24.

19.3.5 QWIP VERSUS HgCdTe

Rogalski [95] has also compared the detectivity of GaAs/AlGaAs QWIPs with the theoretical ultimate performance of n$^+$-p HgCdTe photodiodes limited by Auger mechanism in the base region. In the range of cutoff wavelength $8 \le \lambda_c \le 24$ μm and operating temperature ≤ 77 K, the detectivity of HgCdTe photodiodes is considerably higher. All the QWIP detectivity data for devices with cutoff wavelength near 9 μm is clustered between 10^{10} and 10^{11} cm Hz$^{1/2}$ W^{-1} at an operating temperature close to 77 K. However, the advantage of HgCdTe is less distinct in temperature range below 50 K due to the problems associated with HgCdTe material (p-type doping, Shockley-Read-Hall recombination, trap-assisted tunneling, surface and interface instabilities).

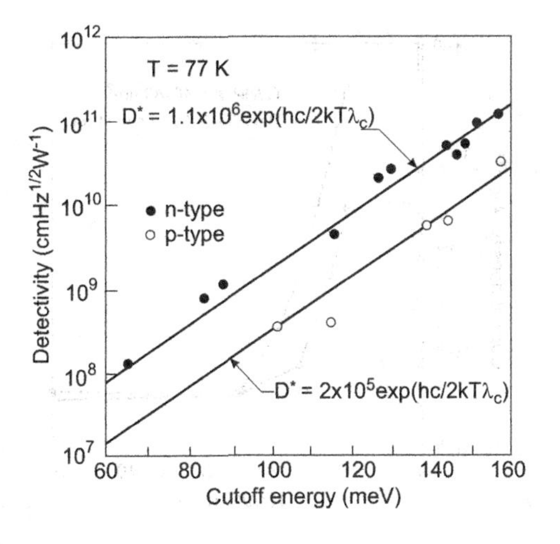

Figure 19.23 Detectivity at 77 K versus cutoff energy for n-doped GaAs/AlGaAs QWIPs (solid circles) and p-doped GaAs/AlGaAs QWIPs (open circles). The straight lines are best fits to the measured data. (From Ref. [52])

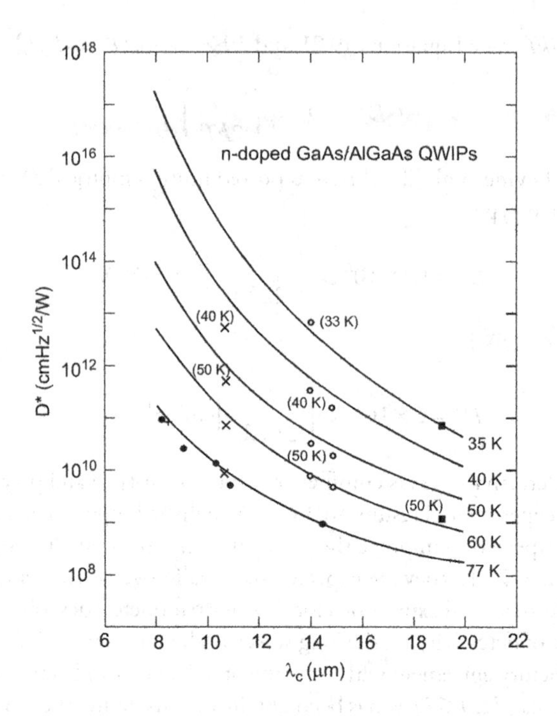

Figure 19.24 Detectivity versus cutoff wavelength for n-doped GaAs/AlGaAs QWIPs at temperatures≤77 K. The solid lines are theoretically calculated. The experimental data are taken from Refs. [52] (•), [81] (×), [96] (+), [97] (○), and [98] (■). (From Ref. [95])

Additional insight into the difference in the temperature dependence of the dark currents is given by Figure 19.25 [99], where the current density versus inverse temperature for a GaAs/AlGaAs QWIP and an HgCdTe photodiode, both with $\lambda_c = 10\,\mu m$, is shown. The current density of both the detectors at temperatures lower than 40 K is similar and is limited by tunneling, which is temperature independent. The thermionic emission regime for the QWIP (≥40 K) is highly temperature dependent, and *cuts on* very rapidly. At 77 K, the QWIP has a dark current that is approximately two orders of magnitude higher than that of the HgCdTe photodiode.

LWIR QWIPs cannot compete with the HgCdTe photodiodes as single devices, especially for higher temperature operations (>70 K) due to the fundamental limitations associated with the intersubband

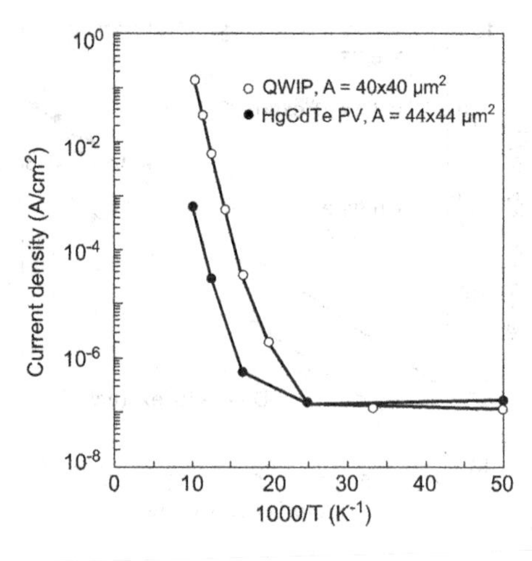

Figure 19.25 Current density versus temperature for a HgCdTe photodiode and a GaAs/AlGaAs QWIP with $\lambda_c = 10\,\mu m$. (From Ref. [99])

transitions. In addition, QWIP detectors have relatively low QEs, typically less than 10%. Figure 19.26 compares the spectral η of an HgCdTe photodiode to that of a QWIP. A higher bias voltage can be used to boost η in the QWIP. However, an increase in the reverse-bias voltage also causes an increase of the leakage current and associated noise, which limits any potential improvement in the system performance. HgCdTe has high optical absorption and a wide absorption band irrespective of the polarization of the radiation, which greatly simplifies the detector array design. The QE of HgCdTe photodiodes is routinely around 70% without an antireflection (AR) coating and is in excess of 90% with an AR coating. Moreover, it is independent of the wavelength over the range from less than 1 μm to near the cutoff of the detector. The wide-band spectral sensitivity with a near-perfect η enables greater system collection efficiency and allows a smaller aperture to be used. This makes HgCdTe FPAs useful for imaging, spectral radiometry, and long-range target acquisition. However, it should be noted that because of high photon fluxes, current LWIR staring array performance is mostly limited by the charge handling capacity of the readout integrated circuit (ROIC) and the background (warm optics). Thus, the spectral response band of QWIP detectors, with a full-width, half-maximum of about 15%, is not a major drawback at LWIR wavelengths.

At the present stage of technology development, QWIP devices are not suitable for space-based remote sensing applications due to dielectric relaxation effects and flux memory effects. In low irradiance environments and associated low-temperature operation, the responsivity of QWIPs depends on frequency and the frequency response depends on the operating conditions (temperature, photon irradiance, bias voltage, and the dynamic resistance of the detector). The typical frequency response is empirically similar to the dielectric relaxation effects observed in bulk extrinsic Si and Ge photoconductors under similar operational conditions. The frequency response has flat regions at both low and high frequencies and the response rolls off between these two levels at a frequency point that is proportional to the inverse of the dynamic resistance of the detector [100] (see Figure 19.27). The dynamic resistance is set by a combination of detector

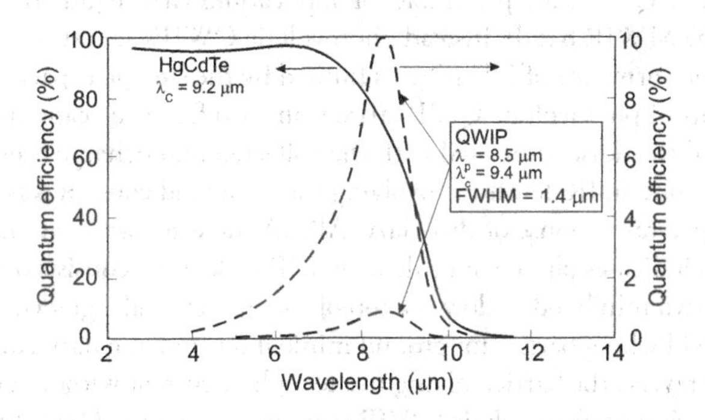

Figure 19.26 QE versus wavelength for a HgCdTe photodiode and GaAs/AlGaAs QWIP detector with similar cutoffs.

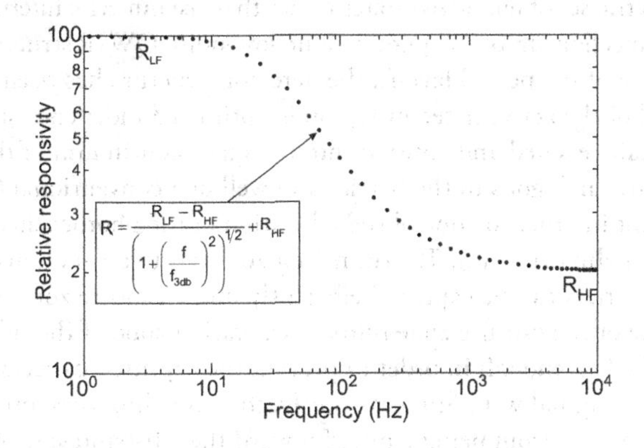

Figure 19.27 Generalized frequency response of QWIP detector. (From Ref. [100])

bias, photon irradiance, and operating temperature. Under typical ambient background conditions, the dynamic resistance is low and the roll-off, which takes place at frequencies in the range of 100 kHz, is not normally evident.

Even though QWIPs are photoconductive devices, several of its properties such as high impedance, fast response time, and low power consumption are well matched with the requirements for large FPA fabrication. The main drawbacks of LWIR QWIP FPA technology are the performance limitations for applications requiring short integration time and the requirement to operate at a lower temperature than HgCdTe of comparable wavelengths. The main advantages of QWIPs are linked to pixel performance uniformity and to the availability of large size arrays. The large established industrial infrastructure in III-V materials/device growth, processing, and packaging brought about by the application of GaAs-based devices in the telecommunication industry gives QWIPs a potential advantage in producibility and cost, whereas the only major use of HgCdTe, to this date, is for IR detectors.

A more detailed comparison of the two technologies has been given by Tidrow et al. [101] and Rogalski et al. [12,14,102,103].

19.4 PHOTOVOLTAIC QWIP

The standard QWIP structure, pioneered by Levine and his group [2] and discussed in the previous sections, is a photoconductive detector, where the photoexcited carriers are swept out of nominally symmetric QWs by an external electric field. A key result of photovoltaic QWIP structures is the application of internal electric fields. These devices in principle can be operated without external bias voltage and vanishing of the dark current and suppression of the recombination noise should be expected [13]. However, their photocurrent is associated with a much smaller gain by comparison to photoconductive QWIPs. The reduced photocurrent and the reduced noise give rise to detectivities similar to photoconductive devices [104]. In conclusion, photoconductive QWIPs are preferable for applications that require high responsivity (e.g., for sensors operating in the MWIR band), instead photovoltaic QWIPs are attractive in camera systems operating in LWIR. The performance of LW FPAs is limited by the storage capacity of the readout circuit. In this context, the benefits of photovoltaic QWIP arise from two facts: the capacitor is less effective when loaded by dark current and the noise associated with the collected photocharge is extremely small [13].

The first experimental work on IR detectors involving the miniband concept was carried out by Kastalsky et al. in 1988 [105]. The spectral response of this GaAs/AlGaAs detector with extremely small QE was in the range 3.6–6.3 μm and indicates photovoltaic detection. This detector consists of a bound-to-bound miniband transition (i.e., two minibands below the top of the barrier) and a graded barrier between the SL and the collector layer as a blocking barrier for ground miniband tunneling dark current. Electrons excited into the upper miniband traverse the barrier, giving rise to a photocurrent without external bias voltage. Further evolution in the design of photovoltaic QWIP structures is analyzed by Schneider and Liu in their monograph [13]. Here we concentrate on the final development of photovoltaic structure at Fraunhofer IAF, so-called the four-zone QWIP [106–109]. The photovoltaic effect in this structure arises from the carrier transfer among an asymmetric set of quantized states rather than asymmetric internal electric fields.

The photoconduction mechanism of the photovoltaic low-noise QWIP structure is explained in Figure 19.28 [107]. Because of the period layout, the detector structure has been called a four-zone QWIP [109]. Each period of the active detector region is optimized independently. In the excitation zone (1), carriers are optically excited and emitted into the quasi continuum of the drift zone (2). The first two zones (1 and 2) are analogous to the barrier and well of a conventional QWIP. Moreover, two additional zones are present in order to control the relaxation of the photoexcited carriers, namely a capture zone (3) and a tunneling zone (4). The tunneling zone has two functions; it blocks the carriers in the quasi continuum (carriers can be captured efficiently into a capture zone) and transmits the carriers from the ground state of the capture zone into the excitation zone of the subsequent period. This tunneling process has to be fast enough in order to prevent the captured carriers from being reemitted thermoelectrically into the original well. Simultaneously, the tunneling zone provides a large barrier to prevent the photoexcited carriers from being emitted toward the left-hand side of the excitation zone. In this way, the noise associated with the carrier capture is suppressed.

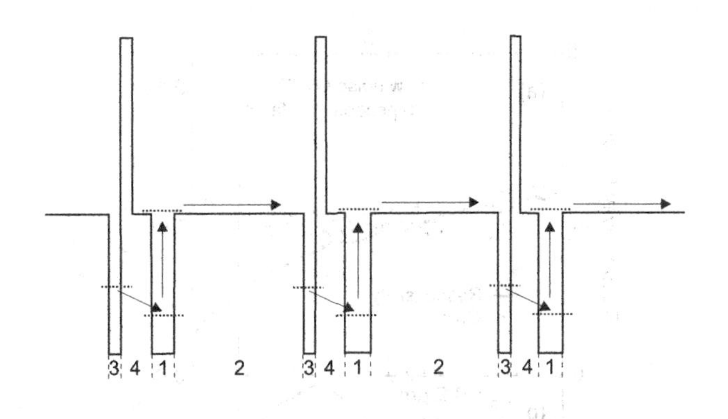

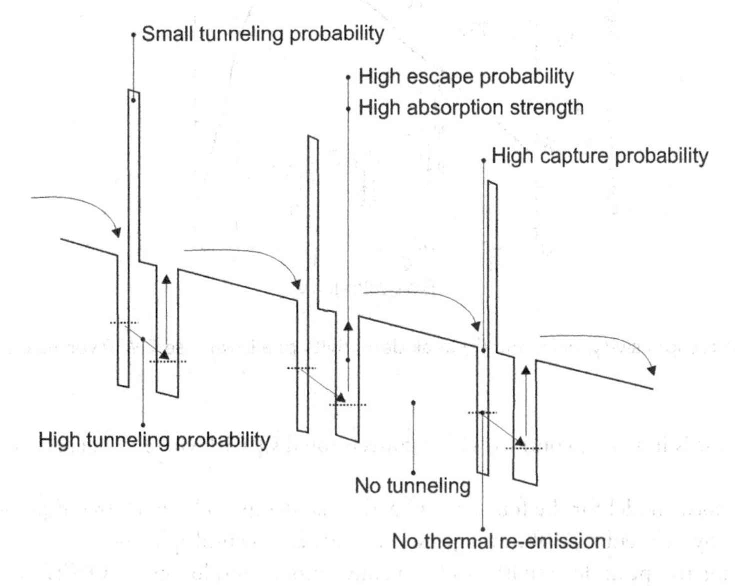

Figure 19.28 Schematic (a) band edge distribution and (b) transport mechanism of the four-zone QWIP. Potential distribution (1) emission zone, (2) drift zone (3) capture zone, and (4) tunneling zone. (From Ref. [107])

Figure 19.28b depicts several requirements for the carrier transport under a finite applied electric field that determines efficient implementation of the four-zone structure. As is shown, the tunnel barrier has to exhibit low probability for tunneling at high energies to receive shorter capture time into the narrow QW than the tunneling escape time. In addition, the time constant for tunneling has to be shorter than that for thermionic reemission from the narrow QW back into the wide QW. An important detail of the tunneling zone is the steplike barrier. The separation between the emission zone and the high-energy part of the tunneling zone is required to reduce the absorption line to a value comparable to that of the conventional QWIP.

The experimentally demonstrated (MBE grown on (100)-oriented semi-insulating GaAs substrate) four-zone structure contains (in the growth direction) an active region with 20 periods of nominally 3.6-nm GaAs (capture zone), 45-nm $Al_{0.24}Ga_{0.76}As$ (drift zone), 4.8-nm GaAs (excitation zone), and a sequence of 3.6-nm $Al_{0.24}Ga_{0.76}As$, 0.6-nm AlAs, 1.8-nm $Al_{0.24}Ga_{0.76}As$, and 0.6-nm AlAs (tunneling zone). The 4.8-nm GaAs wells are n-doped to a sheet concentration of $4\times10^{11}\,cm^{-2}$ per well. The active region is sandwiched between Si-doped ($1.0\times10^{18}\,cm^{-3}$) n-type contact layers.

Figure 19.29 summarizes the performance of a typical 20-period, low-noise QWIP with a cutoff wavelength of 9.2 μm [108]. The peak responsivity is 11 mVW^{-1} at zero bias (photovoltaic operation) and about 22 mAW^{-1} in the range between −2 and −3 V. Between −1 and −2 V, a gain of about 0.05 is observed. The detectivity has its maximum around −0.8 V and about 70% of this value is obtained at zero bias. Due to the asymmetric nature of the transport process, the detectivity strongly depends on the sign of the bias

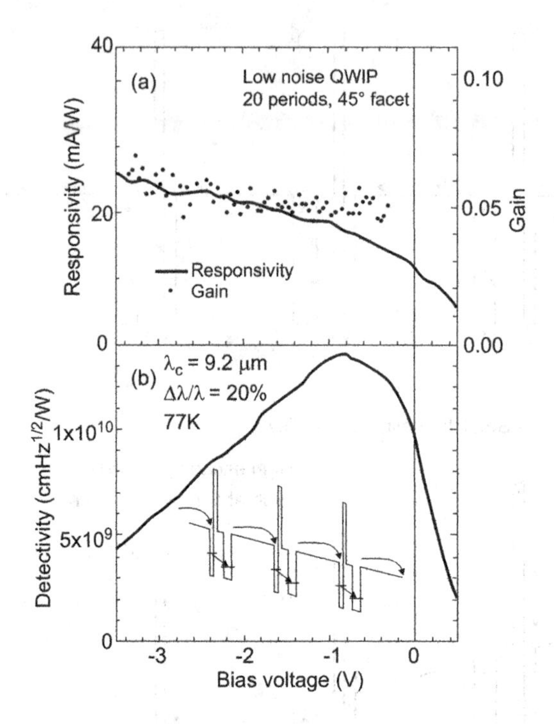

Figure 19.29 (a) Peak responsivity, gain, and (b) peak detectivity of a low-noise QWIP versus bias voltage. (From Ref. [108])

voltage. This behavior is in strong contrast with a conventional QWIP where the detectivity vanishes at zero bias.

An appropriate noise model for the four-zone QWIP was first given by Beck (see Equation 19.45) and was next developed by Schneider [110] in the presence of avalanche multiplication.

Figure 19.30 compares peak detectivities of both conventional and low-noise QWIP structures as functions of the cutoff wavelength [108]. The low-noise QWIPs show similar detectivities as the conventional ones, which are in good agreement with a thermionic emission model.

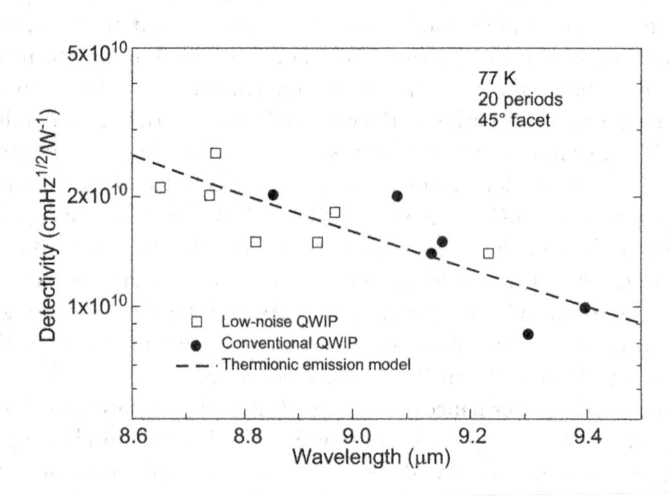

Figure 19.30 Peak detectivities of low-noise QWIPs and photoconductive QWIPs at 77 K versus cutoff wavelength. (From Ref. [108])

19.5 SUPERLATTICE MINIBAND QWIP

In addition to QW, the SL is another promising structure for an IR photodetector but has drawn less attention. The SL intersubband photodetector with a graded barrier was fabricated for photovoltaic detection in the range of 3.6–6.3 μm and 8–10.5 μm in 1988 [105] and 1990 [111], respectively.

The SLs alone were applied in the detection of the wavelength range of 5–10 μm in 1991 [112]. The structures consisted of 100 periods of GaAs QWs of either $L_b = 30$ or 45 Å barriers of $Al_{0.28}Ga_{0.72}As$ and $L_w = 40$ Å GaAs wells (doped $N_d = 1 \times 10^{18}$ cm^{-3}) sandwiched between doped-GaAs contact layers. The absolute values for the peak absorption coefficients were $\alpha = 3100$ and $\alpha = 1800$ cm^{-1} for the $L_b = 30$ and 45 Å structures, respectively, and the detectivities were about 2.5×10^9 cm Hz$^{1/2}$ W^{-1} for $T = 77$ K. Also, the SLs with a blocking layer operated in a low-bias region were demonstrated [113,114]. More recently, a modified voltage tunable SL IR photodetector (SLIP) has been implemented to fabricate two-color FPAs [115,116]. The works on the SLIPs indicate that their advantages include a broader absorption spectrum, lower operation voltage, and more flexible miniband engineering than the conventional QWIP.

The SL miniband detectors use a concept of IR photoexcitation between minibands (ground state and first excited state) and transport of these photoexcited electrons along the excited state miniband. An energy miniband is formed when the carrier de Broglie wavelength becomes comparable to the barrier thickness of the SL. Thus, the wave functions of the individual wells tend to overlap due to tunneling.

Figure 19.31 shows the schematic conduction band diagrams for different miniband structures. Depending on where the upper excited states are located and the barrier layer structure, the intersubband transitions can be based on the bound-to-continuum miniband, bound-to-miniband, and step bound-to-miniband. Among them, the GaAs/AlGaAs QWIP structures using the bound-to-miniband transitions are the most widely used material systems for the fabrication of large FPAs.

Placing the excited state in the continuum increases the thermionic dark current because of the lower barrier height. This fact is more critical for LWIR detectors because the photoexcitation energy becomes even smaller. To improve the detector performance, a new class of MQW IR detectors has received much interest because of their potential for large, uniform FPAs with high sensitivity. Studies by Yu et al. [117–119] revealed that by replacing the bulk AlGaAs barrier in the QWIP with a short-period SL-barrier layer structure (see Figure 19.31b), a significant improvement of the intersubband absorption and thermionic emission property can be obtained. The physical parameters are chosen so that the first excited state in the enlarged wells is merged and lined up with the ground state of the miniband in the SL barrier layer to achieve a large oscillation strength and intersubband absorption. The electron transport in these MQWs

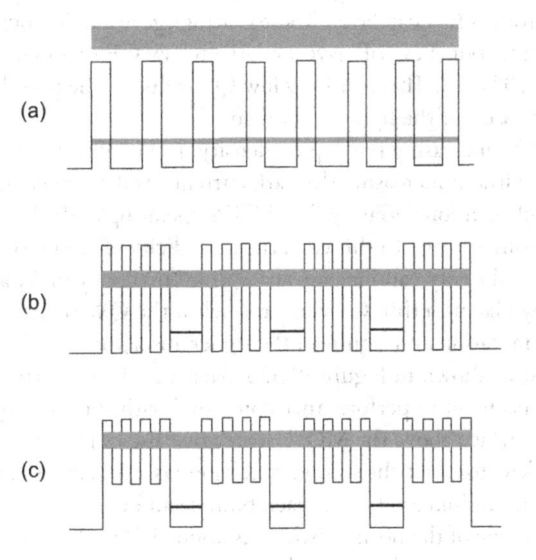

Figure 19.31 Schematic drawing of miniband structures: (a) bound-to-continuum miniband, (b) bound-to-miniband, and (c) step bound-to-miniband.

is based on the bound-to-miniband transition, SL miniband resonant tunneling, and coherent transport mechanism. Thus, the operation of this miniband QWIP is analogous to that of a weakly-coupled MQW bound-to-continuum QWIP. In this device structure, the continuum states above the barriers are replaced by the miniband of the SL barriers. The use of two bound states in the enlarged QW removes the requirement imposed by the bound-to-continuum transition design for a unique solution of the well width and barrier height for a given wavelength (i.e., it is possible to obtain the same operating wavelength with a continuous range of well widths and barrier heights). These miniband QWIPs show lower photoconductive gain than bound-to-continuum QWIPs because the photoexcited electron transport occurs in the miniband where electrons have to transport through many thin heterobarriers resulting in a lower mobility.

In the first GaAs/AlGaAs MQW detector with enlarged wells [117], a 40-period GaAs QW with a well width of 88 Å and a dopant density 2.0×10^{18} cm^{-3} was used. The barrier layer on each side of the GaAs QW consists of five-period undoped AlGaAs (58 Å)/GaAs (29 Å) SL layers that were grown alternatively with the GaAs QWs. The active structure was sandwiched between a 1-μm-thick GaAs buffer layer grown on semi-insulating GaAs and 0.45-μm-thick GaAs cap layer with a dopant density 2.0×10^{18} cm^{-3} to facilitate ohmic contacts. To enhance the light coupling efficiency, a planar transmission metal grating coupler (consisting of regularly spaced metal grating fingers) was developed. The dark current in this type of miniband transport QWIPs is dominated by the thermionic-assisted tunneling conduction via the miniband for $T \geq$ 60 K, whereas resonant tunneling conduction prevails for $T \leq 40$ K. For bias voltage 0.2 V, detectivity $D^* = 1.6 \times 10^{10}$ cm Hz$^{1/2}$ W^{-1} at $\lambda = 8.9$ μm and $T = 77$ K was found. Beck et al. [120,121] adopted this bound-to-miniband approach and demonstrated excellent IR imagers using FPAs in sizes from 256×256 to 640×480.

In order to further reduce the undesirable electron tunneling from the doped QWs and improve the performance, a step bound-to-miniband QWIP was designed and measured (shown in Figure 19.31c). This QWIP consists of GaAs/AlGaAs SL barriers, but with a strained QW of $In_{0.07}Ga_{0.93}As$ [119,122].

New ideas of SL miniband detectors are still presented; for more details see, for example, Li [123].

19.6 LIGHT COUPLING

A key factor in QWIP FPA performance is the light coupling scheme. Illumination of the detector at 45° restricts detector geometries to single elements and one-dimensional arrays. The majority of existing gratings are designed for 2D FPAs. The array illumination is through the substrate back side.

Goossen et al. [124,125] and Hasnain et al. [126] developed a method to couple light efficiency into two-dimensional arrays. They placed gratings on top of the detector that deflect the incoming light away from the direction normal to the surface. The gratings were made by either depositing fine metal strips on top of the QW or etching grooves in a cap layer. These gratings gave a light coupling efficiency comparable to the 45° illumination scheme, but they still gave a relatively low QE of about 10%–20% for QWIPs having 50 periods and $N_d = 1 \times 10^{18}$ cm^{-3}. This relatively low QE is due to the poor light coupling efficiency and the fact that only one polarization of the light is absorbed.

The QE can be improved by increasing the doping density in the QWs, but this leads to a higher dark current. To increase the QE without increasing the dark current, Andersson et al. [76,127] and Sarusi et al. [128] developed a two-dimensional grating for QWIPs operating at the 8–10 μm spectral range, which absorbed both polarization components. In this case, the periodicity of the grating is repeated in both directions. The addition of an optical cavity can increase absorption further by making the radiation pass through the MQW structure twice by placing a thin GaAs mirror below the QW structure (see Figure 19.32a).

Many more passes of IR radiation and significantly higher absorption can be achieved with a randomly roughened reflecting surface, as shown in Figure 19.32b. Sarusi et al. [129] have demonstrated almost an order of magnitude enhancement in performance compared with the 45° scheme by using carefully designed random reflecting surface above the MQW structure. The randomness prevents the light from being diffracted out of the detector after the second reflection (as happens in Figure 19.32a). Instead, the light is scattered at a different random angle after each bounce and can only escape if it is reflected toward the surface within a critical angle of the normal (which is about 17° for GaAs/air interface). The random surfaces are made from GaAs using standard photolithography and selective dry etching, which allows the feature sizes in the pattern to be controlled accurately and the pixel-to-pixel uniformity needed for

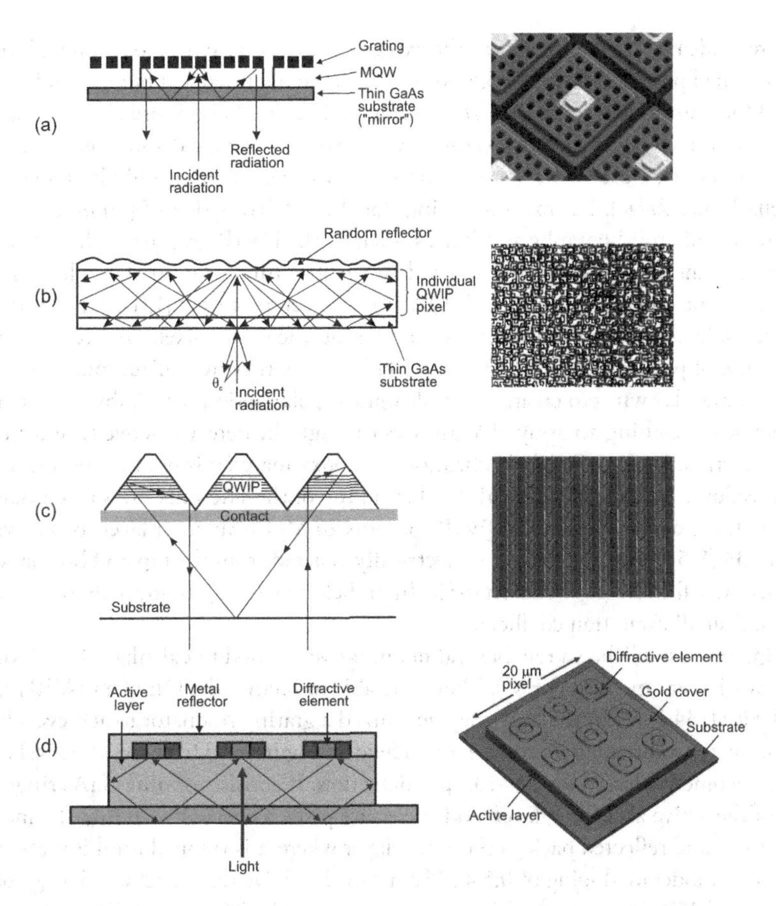

Figure 19.32 Gratings light coupling mechanisms used in QWIPs: (a) gratings with optical cavity, (b) random scatterer reflector, (c) corrugated quantum wells, and (d) resonator detector.

high sensitivity imaging arrays to be preserved. To reduce probability of the light escaping, the surface has three distinct scattering surfaces (see details in [2,5,128]). Experiments show that the maximum response is obtained when the unit cell is equal to the wavelength of the QWIP maximum response. If the unit is larger than this, the number of scatterings on the detector surface decrease and the light is scattered less efficiently. On the other hand, if the unit cell size is smaller, the scattering surface becomes smoother and the efficiency again decreases. Naturally, thinning down the substrate enables more bounces of light and therefore higher responsivity. This also reduces the amount of light that bounces from a pixel to its neighbor (which is bad for the performance of large detector arrays). The thinning or complete removing of the substrate also allows the GaAs/AlGaAs detector array to stretch and accommodate the thermal expansion mismatch with the Si ROIC.

It should be noticed that one of the main differences between the effect of the cross grating and the random reflector is the shape of the responsivity curve; unlike the cross grating, the random reflector has little impact on the bandwidth of the response curve since the scattering efficiency of the random reflector is significantly less wavelength-dependent than for the regular grating. Therefore, for the QWIPs with random reflectors, the integrated responsivity is enhanced by nearly the same amount as the peak responsivity.

The light coupling, such as diffraction gratings and random gratings achieve high QE only when the detector size is large. In addition, because of its wavelength dependence, each grating design is only suitable for a specific wavelength. A size and wavelength independent coupling scheme is much needed.

More recently, Schimert and coworkers [130] reported QWIPs in which the diffraction gratings are etched into the QW stack itself, thereby forming a dielectric grating. This design reduces the conducting detector area (and the leakage current) by a factor of about four while maintaining a QE of 15%. The reported peak D^* of 7.7×10^{10} cm Hz$^{1/2}$ W^{-1} in a QWIP with peak wavelength of 8.5 μm is the highest reported at 77 K.

The gratings are made by etching them into the extra layer grown after the top contact layer and are in the form of either etched pits (see Figure 16.32a) or trenches leaving unetched bumps, and the Au metal is then evaporated for near perfect reflection. The grating period should be approximately the wavelength inside the material, that is, $d = \lambda / n_r$, where λ is the wavelength to be detected and n_r is the refractive index. In practice, λ is chosen to be close to the cutoff wavelength. The etch depth should be about one-fourth of the inside wavelength (i.e., $\lambda/4n_r$). Diffraction gratings for the MWIR with 1.65 μm period were successfully fabricated with contact photolithography and RIE, similar to the LWIR gratings with 2.95 μm period [13].

In order to simplify the array production, a new detector structure for normal incidence light coupling, which is referred to as the corrugated QWIP (C-QWIP) has been proposed [131,132]. The device structure is shown in Figure 19.32c. In large FPAs fabricated at present, the entire pixels are occupied by a single corrugation (see top view of pixels in Figure 19.32c) [115,116]. This structure utilizes total internal reflection at the sidewalls of triangular wires to create favorable optical polarization for IR absorption. These wires are created by chemically etching an array of V grooves through the detector active region along a specific crystallographic direction. During FPA hybridization, an epoxy material is used to connect the detector array with the Si readout. This epoxy material that lies on top of the sidewalls can substantially reduce the internal reflection. Therefore, at present C-QWIP contains an MgF_2/Au cover layer for sidewall protection (see Figure 19.33 [115]). This cover layer is electrically isolated from the top and bottom contacts of the pixels. The dielectric film MgF_2 is chosen for its high dielectric strength, low conducting current, low refractive index, and small excitation coefficient.

Choi et al. [133] have established a reliable and quantitative method to calculate the QE of any detector by adopting a finite-element method (FEM). They were able to predict the QE of a QWIP regardless of its structural complexity [134,135]. In this way they optimized a grating resonator to achieve a high QE and designed a ring resonator to broaden its coupling bandwidth. Figure 19.32d shows the pixel of resonator QWIP (R-QWIP) geometry design for 7.5–10.5 μm detection. It consists of nine GaAs rings as diffractive elements on top of the active absorbing layer, and the entire pixel is covered with Au. The incident light is diffracted by the rings and reflected back to the active layer where it is trapped until it is absorbed. 25-μm pitch detectors with a moderate doping of 0.5×10^{18} cm^{-3} in the active region achieved a QE of 37% and a conversion efficiency of 15% in a 1.3-μm-thick active material and 35% QE and 21% conversion efficiency in a 0.6-μm-thick active material [136].

In an isotropic optical coupling scheme, 2D grating is used to eliminate polarization sensitivity. For the polarimetric QWIPs, linear instead of 2D gratings are used. The use of a microscanner makes it possible to design a camera that resolves the polarimetric components of the scene radiation. Such a discriminating imager, added without significant loss of sensitivity or increased cost, may be beneficial in locating difficult targets. Thales formed four linear gratings rotated by 45° to each other on a set of four detector elements. This pattern is then replicated across the whole array. The layout and a scanning electron microscope (SEM) picture from an actual array are shown in Figure 16.34 [137].

Although gratings are successfully incorporated in commercially available QWIP FPAs, they can be further improved. Development of microfabrication technology and techniques from new fields, such as computer-generated holograms and photonic crystals, could be explored for more efficient optical couplers [13].

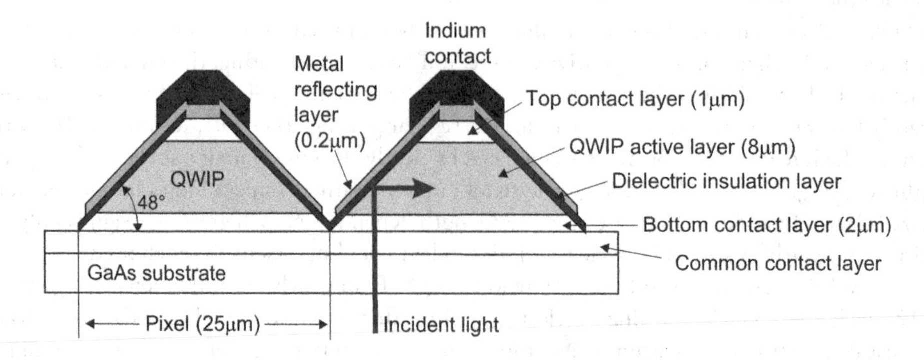

Figure 19.33 The side view of the C-QWIP pixels with 25 μm pitch. (From Ref. [115])

Figure 19.34 SEM picture of a polarimetric QWIP array. (From Ref. [137])

19.7 RELATED DEVICES

19.7.1 P-DOPED GaAs/AlGaAs QWIPs

Up until now, most of the studies have been centered on the n-type GaAs/AlGaAs QWIPs. However, for n-type QWIPs, due to quantum mechanical selection rules, normal incidence absorption is forbidden without use of metal or dielectric grating couplers. The original impetus for the study of p-QWIPs was their ability to absorb light at normal incidence. In p-type QWIPs, the normal incidence absorption is allowed due to the mixing between the off zone center ($\mathbf{k} \neq 0$) heavy-hole and light-hole states [54,138]. Because of the larger effective mass (hence lower optical absorption coefficient) and the lower hole mobilities, the performance of p-QWIPs are in general lower than that of n-QWIPs [2,5,139–142]. However, if the biaxial compressive strain is introduced into the QW layers of a p-QWIP, then the effective mass of the heavy holes will be reduced, which in turn can improve the overall device performance [143].

In the type I MQWs, the wells for holes, as those for electrons, are in GaAs layers. For moderate levels of doping ($\lesssim 5\times10^{18}$ cm^{-3}) and well thicknesses not exceeding 50 Å, the lowest heavy-hole-like subband (HH_1) is filled only partly and all the other energy subbands (HH_2 and light-hole like LH_1) are empty at 77 K. Only these three subbands are found to be underneath the $Ga_{1-x}Al_xAs$ ($x \approx 0.3$) barriers. Holes from the HH_1 subband can be photoexcited to these subbands or to other energy subbands in the continuum HH_{ext} and LH_{ext} (see Figure 19.35a). The theoretical analysis of the experimental results on the normal incidence absorption and responsivity of p-type $GaAs/Ga_{0.77}Al_{0.3}As$ QW structures showed the $HH_1 \rightarrow LH_{ext}$ transitions to be the dominant mechanism for IR absorption in the $\lambda \approx 7$ μm region [144]. The $HH_1 \rightarrow LH_1$ transitions are out of the observed spectral range.

Levine et al. [57] experimentally demonstrated the first QWIP that uses hole intersubband absorption in the GaAs valence band. The samples were grown on a (100) semi-insulating substrate, using gas-source MBE, and consisted of 50 periods of $L_w = 30$ Å (or $L_w = 40$ Å) QWs (doped $N_a = 4\times10^{18}$ cm^{-3} with Be) separated by $L_b = 300$ Å barriers of $Al_{0.3}Ga_{0.7}As$, and capped by $N_a = 4\times10^{18}$ cm^{-3} contact layers. The experimental values of the photoconductive gain for these structures are $g = 0.024$ and $g = 0.034$, respectively. These values are more than an order of magnitude smaller than for n-type QWIPs. The values of QE $\eta \geq 15\%$ and escape probability $p_e \geq 50\%$ were comparable to those of n-type GaAs/AlGaAs QWIPs in spite of the fact that heavy hole effective mass ($m_{hh} \approx 0.5\, m_o$) is much larger than that of the electrons ($m_e \approx 0.073\, m_o$). The

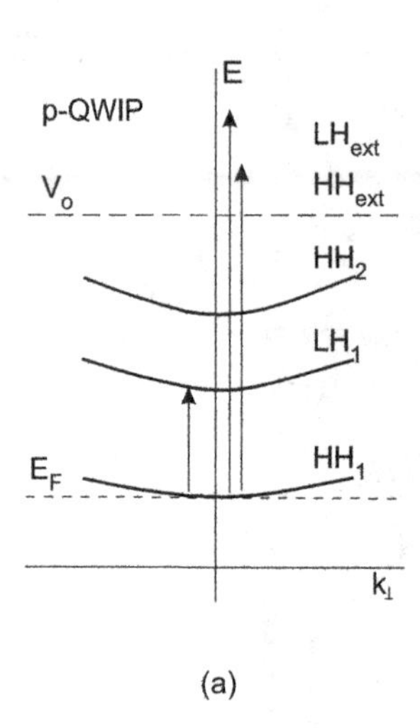

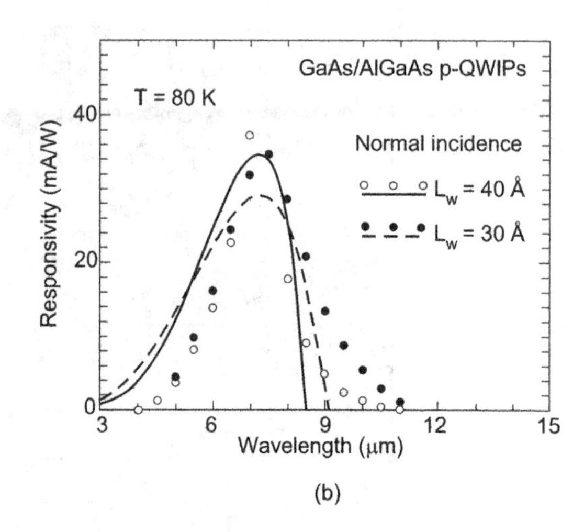

Figure 19.35 p-type GaAs/Ga$_{1-x}$Al$_x$As QW IR photodetector: (a) schematic energy band (the energy of holes is taken to be positive), (b) responsivities of the p-GaAs/Ga$_{0.7}$Al$_{0.3}$As QWIPs. The experimental data (circles) are taken from Levine et al. [57] for structures with L_w = 40 Å and L_w = 30 Å. The solid line and dashed line are calculated. (From Ref. [144])

calculated and observed responsivities for these structures are shown in Figure 19.35b. Good agreement is obtained in the region of $\lambda<\lambda_c$. The small value of responsivity at $\lambda>\lambda_c$ is due to $LH_1 \rightarrow LH_{ex}$.

The performance of p-type GaAs/GaAlAs normal incidence QWIP intersubband photodetectors is below that of the corresponding n-type intersubband detectors for the same wavelength. The detectivities of p-type GaAs/GaAlAs QWIPs are by a factor of 5.5 lower than the detectivities for n-type photodetectors (see Equations 19.47 and 19.48). At present, p-type GaAs/AlGaAs QWIPs are less explored for IR imaging.

19.7.2 HOT-ELECTRON TRANSISTOR DETECTORS

The essential feature of all the designs of QWIPs is that electrons in the lower (ground) state cannot flow in response to the applied electric field, but electrons in the upper (excited) state do flow, thereby yielding photocurrent. The operating temperature of III-V MQW IRdetectors is still needed to kept lower than 77 K due to the large dark current at high temperatures. To increase the operating temperature, it is desirable to reduce the dark current of the detector while maintaining a high detectivity. To reduce the dark current, Choi et al. [145–147] have proposed a new device structure, the IR hot-electron transistor (IHET). Its physics was discussed in detail by Choi [148]. In this device, an energy filter is added to the QW stack that requires a third terminal, but which preferentially removes the leakage current over photocurrent. Under some conditions, the resultant IHET has significantly improved the signal-to-noise ratio as compared to the standard two-terminal QWIP. However, it has not been possible to implement the three-terminal detector into an FPA. The band structure of IHET is shown in Figure 19.36a [147].

The improvement IHET was grown on a (100)-semi-insulating substrate [147]. The first layer was a 0.6-μm-thick n$^+$-GaAs layer doped to 1.2×10^{18} cm^{-3} as the emitter layer. Next, an IR-sensitive 50-period Al$_{0.25}$Ga$_{0.75}$As/GaAs SL structure nominally identical to that reported by Levine et al. [80] was deposited, except that their barrier width was 480 Å instead of 200 Å. On top of the SL structure, thin 300 Å In$_{0.15}$Ga$_{0.85}$As base layer was grown, followed by a 0.2-μm-thick Al$_{0.25}$Ga$_{0.75}$As electron energy high pass filter and a 0.1-μm-thick n$^+$-GaAs ($n = 1.2\times10^{18}$ cm^{-3}) as the collector layer. The emitter and collector areas of the detector were 7.92×10^{-4} cm^2 and 2.25×10^{-4} cm^2, respectively. The detector configuration together with emitter I_E and collector I_C dark currents are shown in Figure 19.36b. Because the thin In$_{0.15}$Ga$_{0.85}$As base layer with a large Γ-L valley separation improves the photocurrent transfer ratio, the detectivity of the transistor increase to 1.4×10^{10} cm Hz$^{1/2}$ W^{-1} at 77 K with a cutoff wavelength of 9.5 μm, two times as large as the companion state of the art GaAs MQW detector. With further optimization of the device parameters, a broadband 10 μm IHET with a detectivity close to 10^{11} cm Hz$^{1/2}$ W^{-1} should be achievable at 77 K [147].

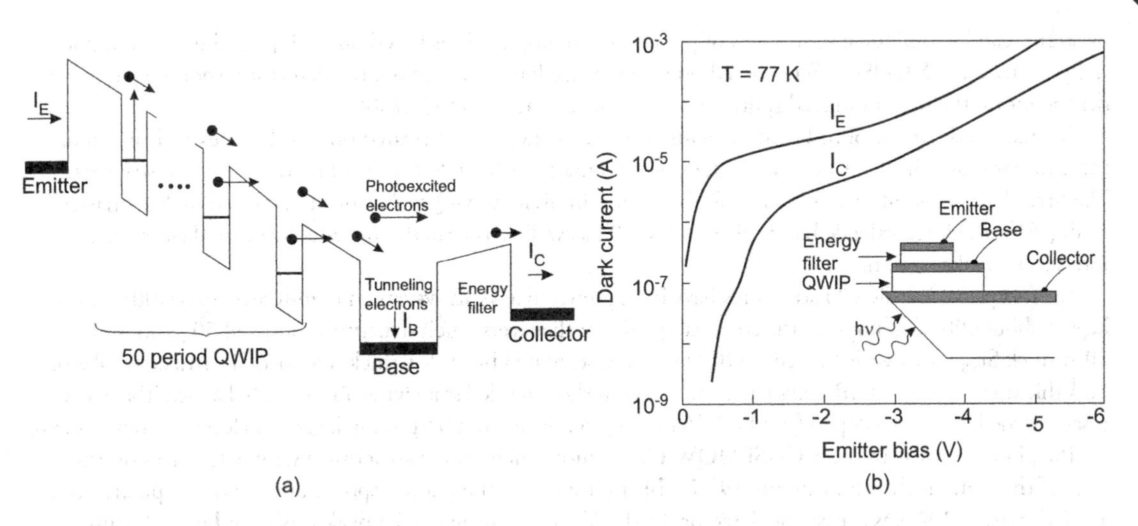

Figure 19.36 IR hot-electron transistor: (a) conduction band diagram and (b) emitter dark current I_E and the collector dark current I_C as a function of the emitter bias at 77 K. The insert shows the schematic device configuration. (From Ref. [147])

Recently, an IHET 5×8 array with a common base configuration that allows two-terminal readout integration was investigated and fabricated for the first time [148]. The study that the IHET structure is compatible with existing electronic readout circuits for photoconductors in producing sensitive focal plane arrays.

Usually, the dark current of IHETs is two to four orders of magnitude lower than that of QWIPs [149]. This fact is especially important in the wavelength range required in many space applications (3–18 µm) where the background photon flux is very low. Also, for thermal imaging purposes, it is desirable to incorporate the extended wavelength QWIPs into FPAs. Progress in the development of very LWIR IHETs has been achieved [150–152].

Potentially, GaAs-based QWIPs can be monolithically integrated with GaAs circuits and a concept of QWIP integration with a high electron mobility transistor was demonstrated [153]. Also, other transistor ideas have been proposed to improve the QWIP performance [154] or to achieve monolithic integration [155]. Hitherto however, the GaAs technology has not been developed in the area of readout circuits.

19.7.3 SiGe/Si QWIPs

Depending on the composition, the bandgap of $Si_{1-x}Ge_x$ alloys varies from 1.1 to 0.7 eV and thus, they are suitable for detector operation in 0.5–1.8 µm wavelength region. However, the large lattice mismatch ($\Delta a_o/a_o \approx 4.2\%$ at room temperature) between Ge (lattice constant $a_o = 5.657$ Å) and Si ($a_o = 0.5431$ Å) hampers the fabrication of integrated photoelectronic devices on Si substrates because of large misfit dislocations at the interface that are not desirable for obtaining good device performance. Still, it was reported [156] that high-quality lattice-mismatched structures can be grown pseudomorphically by MBE technique at the low temperature regime ($T \approx 400°C–500°C$) without misfit dislocations, provided the layers are thinner than the critical thickness h_c. In this case, the lattice mismatch is accommodated by a distortion in the layers, giving a built-in coherent strain in them, and photodetectors sensitive in the mid- and long IR wavelength regions can be fabricated on the basis of Si/SiGe SLs and MQWs.

The critical thickness of a strained SiGe layer on Si substrate is strongly dependent on the growth parameters, especially the substrate temperature. For a typical growth temperature $T \approx 500°C$ for a $Si_{0.5}Ge_{0.5}$, a critical thickness is about 100 Å. In case of multiple layer growth, the critical thickness is obtained using the average Ge composition $x_{Ge} = (x_1 d_1 + x_2 d_2)/(d_1 + d_2)$, where the x-s and d-s are the Ge content and thickness of each constituent layer, respectively.

The strain not only changes the bandgaps of the constituents, but it splits the degeneracy of heavy- and light-hole bands of Si and $Si_{1-x}Ge_x$ layers and also removes the degeneracy of conduction band (multivalley band structure) [157,158]. The changes in the band structure of $Si_{1-x}Ge_x$/Si strained-layer devices, introduced by strain, can be used for several types of photodetector applications based on both p- and n-type conductivity devices. For SiGe/Si QWIPs based on intervalence band absorption, the detector response is strongly dependent on the strain-induced splitting of the valence bands [62,159,160].

The major advantage of Si-based detectors is the fact they are fabricated on Si substrates and thus monolithic integration with Si electronic readout devices makes it feasible to manufacture very large-scale arrays. The first observation of intersubband IR absorption in SiGe/Si MQWs has been described by Karunasiri et al. [161]. The large valence band offset of SiGe/Si as well as the small hole effective mass favors the hole intersubband absorption.

The p-type MQW SiGe/Si structures have been grown in a MBE system on high-resistivity Si(100) wafers kept at about 600°C to improve the epitaxial quality of the layers. Such structures consist of 50 periods of 30-Å-thick $Si_{0.85}Ge_{0.15}$ wells (doped $p \approx 10^{19}$ cm^{-3}) and separated by 500-Å-thick undoped Si barriers. A QW of 30 Å thickness can absorb IR energy near 10 μm with the extended state lying above the Si barrier. The entire SL is sandwiched between a doped ($p = 1 \times 10^{19}$ cm^{-3}) 1 μm bottom and 0.5 μm top layers for electrical contacts [59].

The photoresponse of the SiGe/Si MQW (200-μm-diameter mesa structures) with a 45° facet on the edge of the wafer is shown in Figure 19.37. This figure shows the photoresponse at 0° and 90° polarizations at 77 K, with a 2-V bias across the detector. In the 0° polarization case, a peak was found near 8.6 μm. For the 90° polarization, a peak at 7.2 μm was observed. The responsivities for both cases were the same, 0.3 A W^{-1}. The responsivity for an unpolarized beam with a peak near 7.5 μm is about 0.6 A W^{-1} and is approximately the sum of two polarization cases. The responsivity measured by illuminating the light normally on the back side of the devices is shown by the dashed curve in Figure 19.37. The photoresponse at normal light incidence seemed to be due to the internal photoemission caused by free-carrier absorption. More detailed discussion of absorption mechanisms in SiGe/Si heterostructures is carried out by Park et al. [60] The estimated detectivity for the above nonoptimized SiGe/Si MQW IR detectors is about 1×10^9 cm Hz$^{1/2}$ W^{-1} at 9.5 μm and 77 K.

The first intervalence-subband IR detector for the 3–5 μm wavelength region was demonstrated with relatively high detectivity $D^*(3\,\mu m) = 4 \times 10^{10}$ cm Hz$^{1/2}$ W^{-1} [61]. As the Ge composition is increased, the peak photoresponse moves toward a shorter wavelength, in agreement with transmission coefficient data. The photoresponse shows several peaks, not observed in room temperature absorption spectra, which is connected with a different type of transition revealed under the bias, when carriers can tunnel from the excited states through the barrier to reach the contact (except the carriers excited to the continuum states).

Normal incidence hole intersubband QWIPs are also possible in pseudomorphic GeSi/Si QWs [162]. People et al. [63,163] have described the fabrication and performance characteristics of pseudomorphic p-type $Si_{0.75}Ge_{0.25}$/Si QWIP on (001) Si substrate. The 40 Å $Si_{0.75}Ge_{0.25}$ QWs were B doped by ion

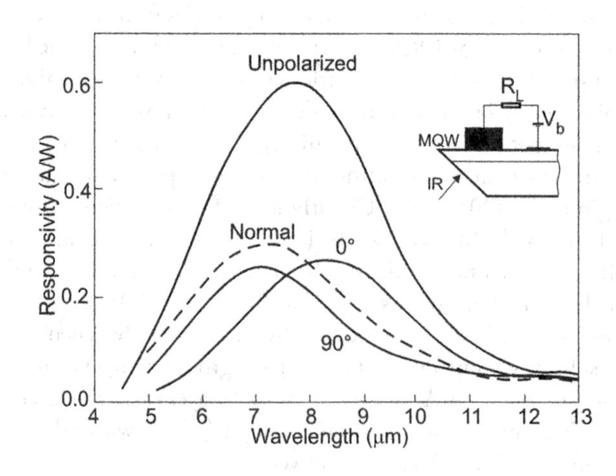

Figure 19.37 Responsivity of $Si_{0.85}Ge_{0.15}$/Si QWIP at 77 K for two polarization angles with a 2 V bias. IR radiation is illuminated on the facet at the normal such that the incidence angle on the MQW structure is 45°, as shown in the inset. Dashed curve shows the responsivity for normal back side illumination. (From Ref. [59])

implantation of 2 keV and had carrier concentrations $\approx 4\times 10^{18}$ cm^{-3}. The 300 Å Si barrier layers were undoped. These devices show broadband response (8–14 µm) that is attributed to the strain and quantum confinement-induced mixing of heavy, light, and split-off hole bands. The 200-µm-diameter devices show detectivities 3.3×10^9 cm Hz$^{1/2}$ W^{-1}, responsivities 0.04 A W^{-1}, and differential resistance 10^6 Ω at operating bias −2.4 V and $T = 77$ K with no cold shields (field of view (FOV) 2π, 300 K).

Strong intersubband IR optical absorption at normal incidence in Si/Si$_{1-x}$Ge$_x$ strained-layer MQWs is possible not only in p-type structures but also in n-type structures [53], which was first demonstrated for (110) Si/Si$_{1-x}$Ge$_x$ MQW samples with different Ge compositions and doping concentrations, with the position of absorption peaks ranging from 4.9 to 5.8 µm [55]. It was earlier shown [53] that for (110) and (111) growth directions, the normal intersubband absorption coefficients in Si QWs of the order of 10^4 cm^{-1} can be achieved in Si QWs doped up to 10^{19} cm^{-3}, which is possible in MBE procedure.

The demonstration of the normal incidence of radiation on SiGe/Si SLs indicates the possible realization of IR FPAs without the use of the grating couplers normally required for AlGaAs/GaAs intersubband detectors. Also, very encouraging are the theoretically predicted performance of SiGe/Si QWIPs [53,164,165]. However, SiGe/Si QWIP detectors have not received detectivities comparable with n-type GaAs/AlGaAs material system; they are about one order lower at 77 K in the LWIR region. For this reason, the development of SiGe/Si QWIPs was abandoned in the mid-1990s.

19.7.4 QWIPs WITH OTHER MATERIAL SYSTEMS

The GaAs/AlGaAs MQW detector response can also be designed to operate in shorter wavelength spectral ranges. However, the short wavelength limit in the AlGaAs/GaAs material system imposed by keeping the AlGaAs barriers direct is $\lambda = 5.6$ µm. If the Al concentration x increased beyond $x = 0.45$, the indirect X valley becomes the lowest bandgap. Since Γ–X scattering together with GaAs X-barrier trapping in such structures can result in inefficient carrier collection, thus leading to a poor responsivity, this has been thought to be highly undesirable. The limited conduction band discontinuity of AlGaAs/GaAs system (with acceptable Al mole fractions) makes it impossible to grow epilayer structures that are sensitive in the 3–5 µm MWIR window. For this reason, initially, Levine et al. [166] investigated the promise of the AlInAs/InGaAs system for MWIR QWIPs using a 50-well AlInAs-InGaAs epilayer structure with 50 Å InGaAs wells and 100-Å-thick AlInAs barriers resulting in a bound-to-bound QWIP with an absorption peak at $\lambda_p = 4.4$ µm and $\Delta\lambda/\lambda_p \approx 7\%$. Latter, Hasnain et al. [167] investigated the direct gap system $In_{0.53}Ga_{0.47}As/In_{0.52}Al_{0.48}As$ and demonstrated the MQW IR detectors operating at $\lambda_p = 4.2$ µm with a detectivity $D^* = 2\times 10^{10}$ cmHz$^{1/2}$W^{-1} at 77 K and a background-limited detectivity 2.3×10^{12} cm Hz$^{1/2}$ W^{-1} at temperatures up to 120 K.

A larger conduction band discontinuity can be achieved with the AlGaAs/InGaAs material system, which has been the standard material system for MWIR QWIPs in spite of the degrading effects and limitations of lattice-mismatched epitaxy.

An AlInAs/InGaAs lattice-matched structure with a sufficiently large conduction band discontinuity is an alternative to the strained AlGaAs/InGaAs material system for both single-band MWIR and stacked multi-band QWIP FPAs. When combined with LWIR InP/InGaAs or InP/InGaAsP QW stacks, this material system offers a completely lattice-matched dual- or multi-band QWIP structure on an InP substrate benefiting from the advantages of InP-based QWIPs as well as avoiding the limitations of strained layer epitaxy. More recently reported performance of large format 640×512 AlInAs/InGaAs FPA [168,169] with a cutoff wavelength of 4.6 µm are comparable to the best results reported for MWIR AlGaAs/InGaAs QWIPs [16].

Intersubband absorption and hot electron transport are not limited to the $Al_xGa_{1-x}As$/GaAs material system. The long wavelength SL detectors have been demonstrated using InP-based material system such as lattice-matched GaAs/$Ga_{0.5}In_{0.5}P$ ($\lambda_p = 8$ µm), n-doped (p-doped) 1.3 µm $In_{0.53}Ga_{0.47}As$/InP (7–8 µm and 2.7 µm, respectively), 1.3 µm InGaAsP/InP ($\lambda_c = 13.2$ µm), 1.55 InGaAsP/InP ($\lambda_c = 9.4$ µm) heterosystems. The responsivities of n-doped $In_{0.53}Ga_{0.47}As$/ InP MQW IR photoconductors were in fact somewhat larger than those obtained in equivalent AlGaAs/GaAs ones; detectivity D^* of 9×10^{10} cm Hz$^{1/2}$ W^{-1} was measured for a detector operating at 77 K at a wavelength of 7.5 μm [96,170]. The first short wavelength ($\lambda_c = 2.7$ μm) detector with p-type $Ga_{0.47}In_{0.53}As/InP$ material system has been demonstrated by Gunapala et al. [171]. In the last case, the detector works at the normal incidence of IR radiation. The extensive literature on QWIPs using materials other than GaAs/AlGaAs is reviewed by Levine [2], Gunapala and Bandara [5], and Li [123].

For most of the GaAs-based QWIPs demonstrated thus far, GaAs is the low-bandgap well material and the barriers are lattice-matched AlGaAs, GaInP, or AlInP. However, it is interesting to consider GaAs as the barrier material since the transport in binary GaAs is expected to be superior to that of a ternary alloy. To achieve this, Gunapala et al. [172,173] have used the lower-bandgap nonlattice-matched alloy $In_xGa_{1-x}As$ as well material together with the GaAs barriers. It has been demonstrated that strain-layer heterostructures can be grown for lower In composition ($x<0.15$), which results in lower barrier heights. Therefore, this heterobarrier system is very suitable for very long wavelength ($\lambda > 14$ μm) QWIPs. Excellent hot electron transport and high detectivity $D^* = 1.8 \times 10^{10}$ cm Hz$^{1/2}$ W^{-1} at $\lambda_p = 16.7$ μm were achieved at temperature $T = 40$ K [173]. The large responsivity and detectivity values are comparable to those achieved with the usual lattice-matched GaAs/AlGaAs material system [98].

19.7.5 MULTICOLOR DETECTORS

One of the distinct advantages of the QW approach is the ability to easily produce multicolor (multispectral) detectors, which is desirable for future high-performance IR systems. In general, a multicolor detector is a device where the spectral response varies with parameters like applied bias voltage. Three basic approaches to achieving multicolor detection have been proposed: multiple leads, voltage switched, and voltage tuned.

The first two-color GaAs/AlGaAs QWIP has been realized by Kock et al. [174] by stacking on the same GaAs substrate two series of QWIPs (see Figure 19.38a) with different wavelength selectivities. This approach involves contacting each intermediate conducting layer separating the one-color QWIPs. This leads to a separately addressable multicolor QWIP with multiple electrical terminals. The advantage of this approach is its simplicity in design and its negligible electrical crosstalk between colors. Moreover, each QWIP detector can be optimized independently for a desired detection wavelength. The drawback of this approach is the difficulty of fabricating a many color version. A technical solution for these difficulties is described in many papers, see, for example, [175–177]. Gunapala et al. [178] demonstrated a four-color imager by separating a large array into four stripes, each responding to a different color (see Section 27.3).

A novel concept of the multicolor detector structure was made by Liu et al. [179] by stacking conventional (one-color) QWIPs, separated by thin, heavily doped layers (≈100 nm in test structure). The tunability is achieved by relying on the highly nonlinear and expotential nature of the device dark current voltage characteristics. This implies that an applied voltage across the entire multistack would be distributed among the one-color QWIPs according to their DC resistance values. When the applied voltage is increased from zero, most of the voltage will be dropped across the one-color QWIP with the highest resistance. As the voltage is further increased, an increasing fraction of it will be dropped across the next highest resistance one-color stack of QWs, and so on. The band edge profiles of a three-color version under the highest bias condition is shown schematically in Figure 19.38b. The above structure is similar to that of Grave et al. [180], but with an important difference: the voltage division in Grave et al. is accomplished by high-low field domain formation, which is not quantitatively understood for MQWs [181].

In a practically demonstrated three-color version of such a multicolor QWIP concept, the GaAs well widths for the three stacks were 55, 61, and 66 Å, respectively (32 wells in each stack). The $Al_xGa_{1-x}As$ barriers were all 468 Å thick with alloy fractions of 0.26, 0.22, and 0.19, respectively. The separation between the one-color QWIPs was a 934-Å-thick GaAs layer doped with Si to 1.5×10^{18} cm^{-3}. Three well-resolved peaks at different biases were observed at 7.0, 8.5, and 9.8 μm. The calculated detectivity for the 8.5 μm response at bias $V_b = -3$ V was 5×10^9 cm Hz$^{1/2}$ W^{-1} for unpolarized radiation and 45° facet geometry. It can be improved up to $D^* = 3 \times 10^{10}$ cm Hz$^{1/2}$ W^{-1} for light coupling geometry with 100% absorption.

The advantage of a voltage tunable approach is the simplicity in fabrication (as it requires only two terminals) and the implementation of many colors. The drawback is the difficulty to achieve a negligible electrical crosstalk between colors.

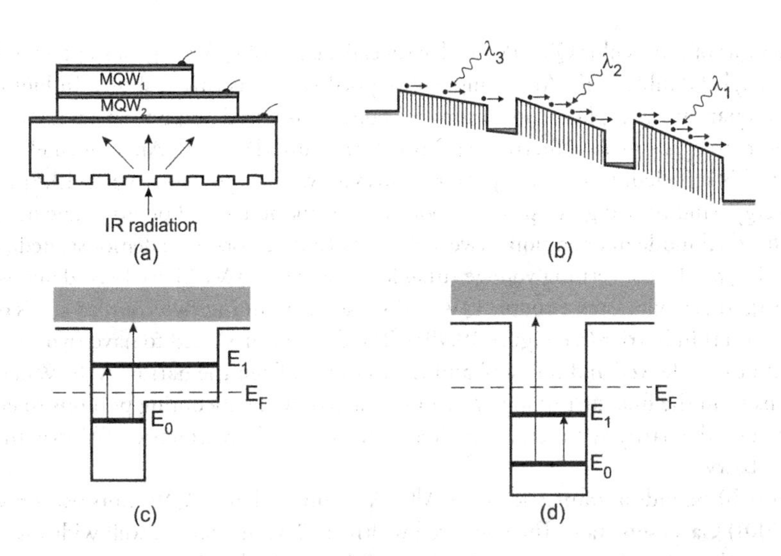

Figure 19.38 Multicolor QW IR detectors based on: (a) intersubband transitions in two series of QWIPs stacked on the same substrate, (b) three-color voltage tunable detector at the highest bias voltage, (c) bound-to-bound and bound-to-extended transitions in asymmetric step multiquantum well structure, and (d) bound-to-continuum states transition mechanism.

Another example of voltage-switched two-color detection is schematically shown in Figure 19.39 [115]. In this case, the unit cell consists of two SLs of QWs with bound-to-miniband transition mechanism. This idea was suggested for the first time by Wang et al. [182]. One SL is tuned to the MWIR band and the other to the LWIR band. Between the SLs is a graded barrier. Under negative bias, photoelectrons generated in the second SL lose energy in the relaxation barrier and are blocked by the first SL. The LW photoelectrons generated in the SL2 pass into the highly conducting energy relaxation layer, resulting in no impedance changes. Under positive bias, the reverse situation occurs and only the SL2 LW photoelectrons pass through the graded barrier and produce an impedance change. This design was used for the fabrication of two-color C-QWIPs [115,116].

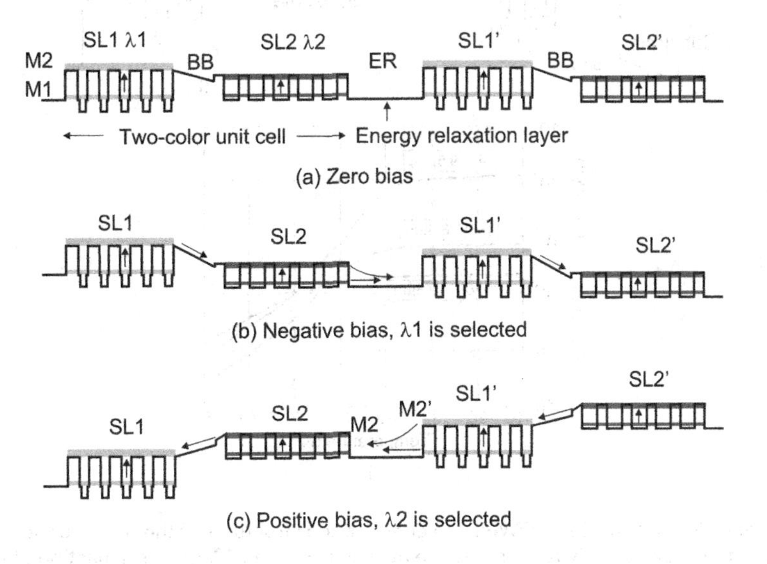

Figure 19.39 Voltage-switched two-color detection mechanism in SL QWs. (From Ref. [115])

Alternative designs of multicolor QWIPs involve special shapes of QWs (e.g., a stepped well or an asymmetrically-coupled double well). An example of stepped-well structure is shown in Figure 19.38c, where a two-color quantum well photoconductor uses bound-to-bound and bound-to-extended transitions of similar oscillator strengths in asymmetric step MQW structure [183,184]. An applied electric field excursion of ±40 kV cm^{-1} is sufficient to shift the peak responsivity wavelength from 8.5 to 13.5 μm [183]. An asymmetrical energy band bending can promote both photoconductive and photovoltaic modes of operation and using this dual mode of operations, two-color QWIP detectors were demonstrated [185].

Very encouraging results concerning voltage tunable three-color QWIP have been demonstrated by Tidrow et al. [186]. This device uses a couple QW units, each containing two coupled QWs of different widths separated by a thin barrier (see Figure 19.40a). The device is designed to have two subbands E_1 and E_2 originated from the wide well and one subband E_2' originated from the narrow well. When the wide well is doped, electrons from the first energy state E_1 can be excited by the incoming photons to either E_2 or E_2' energy states. Because the parity symmetry is broken in the coupled asymmetric QW structure, more than one color can be observed.

The device with 30-periods asymmetric GaAs/AlGaAs coupled double QW units was grown on semi-insulating (001) GaAs substrate. The wide well width is 72 Å, the narrow well width is 20 Å, the $Al_{0.31}Ga_{0.69}As$ barrier between the two coupled wells is 40 Å, and the barrier between the coupled well units is 500 Å. The bottom contact layer is 1,000 nm-doped GaAs and the top contact layer is 500-Å-doped $In_{0.08}Ga_{0.92}As$. The doping in the wide QWs and the contact layer is $n^+ = 1.0 \times 10^{18}$ cm^{-3}. The narrow wells are undoped.

The peak detectivity as a function of temperature is shown in Figure 19.40c. We can see that the detectivity is around 10^{10} cm Hz$^{1/2}$ W^{-1} for the 9.6 and 10.3 μm peaks at 60 K. The responsivity for the 8.4 μm peak is smaller, with the $D^* = 4 \times 10^9$ cm Hz$^{1/2}$ W^{-1} at 60 K. The detection peak is independently selectable among these three wavelengths by tuning the bias voltage. Generally, however, it is difficult to ensure a good QWIP performance for all voltages. To provide a large intersubband transition strength and, at the same time, an easy escape for the excited carriers, the transition final state should be close to the top of the barrier. These two conditions are difficult to fulfill for all voltages [13]. Also, relatively wide wells in

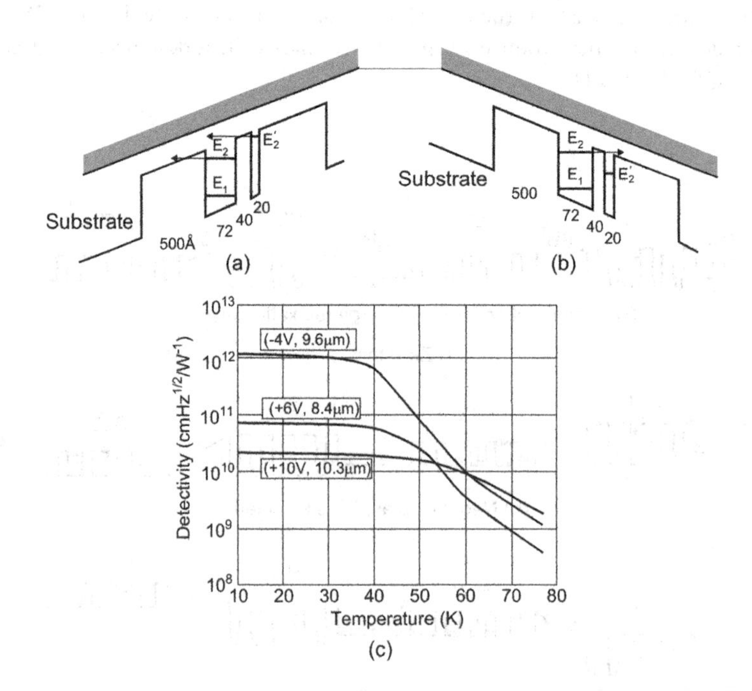

Figure 19.40 Three-color GaAs/AlGaAs QWIP: The energy band structures of the device under (a) positive and (b) negative biases; the peak detectivity of the device at (c) 8.4, 9.6, and 10.3 μm as a function of temperature under bias +6 V, −4 V, and +10 V, respectively. (From Ref. [186])

the case of a stepped well, may lead to an enhanced trapping probability, hence a shorter carrier lifetime. Finally, field-induced changes of the subband structure usually require relatively high external fields (large voltages: see Figure 16.40c), which increase the dark current and noise.

Also, for detectors with symmetric wells, two-color operation is possible [174,187]. This proposal is connected with large filling, when two states in the wells are occupied and optical transitions occur between the different states in the wells (like those shown in Figure 19.38d) or with wells of different thicknesses in which only the ground states are occupied by the carriers. Such MQW photodetectors aim to cover both the 8–12 and 3–5 μm spectral regions.

19.7.6 INTEGRATED QWIP-LED

The innovative concept of frequency-up conversion based on the integration of QWIP with light emitting diodes (QWIP-LED) offers an alternative to the standard hybrid technique in making an imaging device. This approach may lead to devices that are difficult to realize by the standard one, such as ultra-large size sensors [188].

The integrated QWIP-LED concept was independently proposed by Liu et al. [189] and by Ryzhii et al. [190] and was first experimentally demonstrated by Liu and colleagues [189]. The basic idea is shown in Figure 19.41. Under a forward bias, photocurrent electrons from the QWIP recombine with injected holes in LED, giving rise to an increase in LED emission. The QWIP is a photoconductor so that under IR light illumination its resistance decreases, which leads to an increase in the voltage drop across the LED and therefore an increase in the amount of emission. This device is therefore an IR converter. Photocarriers in QWIP have a strong lateral locality. The resulting emission around 0.9 μm can be easily imaged using the well-developed Si charge-coupled device (CCD) array.

The electron well charges capacitance in present commercial CCD (typically 4×10^5 electrons) is almost two orders smaller than that of the readout circuit used in LWIR FPAs. Then CCD is usually needed to operate at full well working point mode using QWIP-LED for IR imaging to get a relatively higher thermal image gray level.

The advantage of the integrated QWIP-LED is technologically important since in this scheme, one can make 2D large format imaging devices without the need of making any circuit readouts. The QWIP-LED device still operates under low temperatures. The up-conversion approach can be easily implemented in multicolor imaging devices in a pixelless geometry [191].

The initial demonstration of the pixelless QWIP-LED used a p-type material to simplify fabrication (avoiding gratings). The next efforts were concentrated on n-type QWIPs and steady improvements were made [192–194]. Due to low performance, further developments failed.

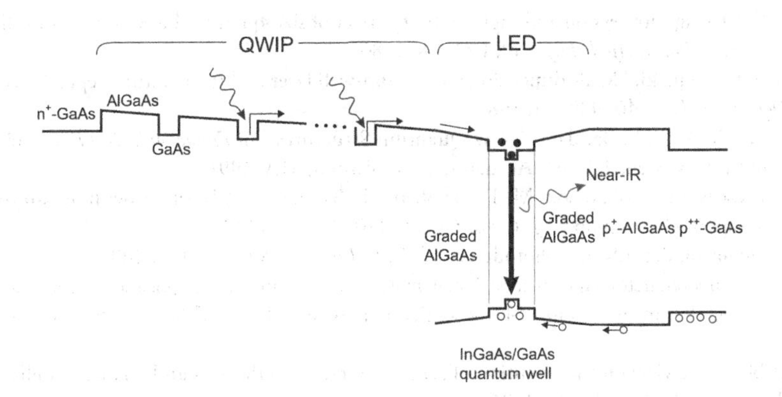

Figure 19.41 Bend edge profile of an integrated QWIP-LED. Under forward bias, the photocurrent generated in QWIP leads to an emission in LED, achieving the up-conversion of QWIP-detected IR signal to LED emission near IR or visible light. (From Ref. [188])

REFERENCES

1. L. Esaki and R. Tsu, "Superlattice and negative conductivity in semiconductors," *IBM J. Res. Dev.*, 14, 61–5, 1970.
2. B. F. Levine, "Quantum-well infrared photodetectors," *J. Appl. Phys.*, 74, R1–81, 1993.
3. M. O. Manasreh, ed., *Semiconductor Quantum Wells and Superlattices for Long-Wavelength Infrared Detectors*, Artech House, Norwood, MA, 1993.
4. F. F. Sizov and A. Rogalski, "Semiconductor superlattices and quantum wells for infrared optoelectronics," *Prog. Quantum Electron.*, 17, 93–164, 1993.
5. S. D. Gunapala and K. M. S. V. Bandara, "Recent Development in Quantum-Well Infrared Photodetectors," in *Thin Films*, Vol. 21, eds. M. Francombe and J. Vossen, 113–237 Academic Press, New York, 1995.
6. K. K. Choi, *The Physics of Quantum Well Infrared Photodetectors*, World Scientific, Singapore, 1997.
7. H. C. Liu, "Quantum Well Infrared Photodetector Physics and Novel Devices," in *Semiconductors and Semimetals*, Vol. 62, eds. H. C. Liu and F. Capasso, 129–96, Academic Press, San Diego, 2000.
8. S. D. Gunapala and S. V. Bandara, "Quantum Well Infrared Photodetectors (QWIP)," in *Handbook of Thin Devices*, Vol. 2, ed. M. H. Francombe, 63–99, Academic Press, San Diego, 2000.
9. J. L. Pan and C. G. Fonstad, "Theory, fabrication and characterization of quantum well infrared photodetectors," *Mater. Sci. Eng.*, R28, 65–147, 2000.
10. S. D. Gunapala and S. V. Bandara, "GaAs/AlGaAs Based Quantum Well Infrared Photodetector Focal Plane Arrays," in *Handbook of Infrared Detection Technologies*, ed. M. Henini and M. Razeghi, 83–119, Elsevier, Oxford, UK, 2002.
11. V. Ryzhi, ed., *Intersubband Infrared Photodetectors*, World Scientific, River Edge, New Jersey, 2003.
12. A. Rogalski, "Quantum well photoconductors in infrared detectors technology," *J. Appl. Phys.*, 93, 4355–91, 2003.
13. H. Schneider and H. C. Liu, *Quantum Well Infrared Photodetectors*, Springer, Berlin, 2007.
14. A. Rogalski, J. Antoszewski, and L. Faraone, "Third-generation infrared photodetector arrays," *J. Appl. Phys.*, 105, 091101-44, 2009.
15. S. D. Gunapala, D. R. Rhiger, and C. Jagadish, *Advances in Photodetectors*, Elsevier, Amsterdam, 2011.
16. S. D. Gunapala, S. V. Bandara, J. K. Liu, C. J. Hill, B. Rafol, J. M. Mumolo, J. T. Trinh, M. Z. Tidrow, and P. D. LeVan, "1024×1024 pixel mid-wavelength and long-wavelength infrared QWIP focal plane arrays for imaging applications," *Semicond. Sci. Technol.*, 20, 473–80, 2005.
17. M. Jhabvala, K. K. Choi, C. Monroy, and A. La, "Development of a 1K×1K, 8–12 μm QWIP array," *Infrared Phys. Technol.*, 50, 234–9, 2007.
18. G. Bastard, *Wave Mechanics Applied to Semiconductor Heterostructures*, Les Editions de Physique, Les Ulis Cedex, Halsted Press, New York, 1988.
19. C. Weisbuch and B. Vinter, *Quantum Semiconductor Structures*, Academic Press, New York, 1991.
20. A. Shik, *Quantum Wells*, World Scientific, Singapore, 1997.
21. P. Harrison, *Quantum Wells, Wires and Dots: Theoretical and Computational Physics*, Wiley, New York, 1999.
22. D. Bimberg, M. Grundmann, and N. N. Ledentsov, *Quantum Dot Heterostructures*, Wiley, Chichester, UK, 2001.
23. J. Singh, *Electronic and Optoelectronic Properties of Semiconductor Structures*, Cambridge University Press, Cambridge, UK, 2003.
24. H. Sakaki, "Scattering suppression and high-mobility effect of size-quantized electrons in ultrafine semiconductor wire structures," *Jpn. J. Appl. Phys.*, 19, L735–8, 1980.
25. Y. Arakawa and H. Sakaki, "Multidimensional quantum-well laser and temperature dependence of its threshold current," *Appl. Phys. Lett.*, 40, 939–41, 1982.
26. D. D. Coon and K. M. S. V. Bandara, "New Quantum Structures," in *Physics of Thin Films*, Vol. 15, eds. M. H. Francombe and J. L. Vossen, 219–64, Academic Press, Boston, MA, 1991.
27. W. Seifert, N. Carlsson, J. Johansson, M.-E. Pistol, and L. Samuelson, " In situ growth of nano-structures by metal-organic vapour chase epitaxy," *J. Cryst. Growth* 170, 39–46, 1997.
28. H. C. Liu, "Quantum dot infrared photodetector," *Opto-Electron. Rev.*, 11, 1–5, 2003.
29. D. H. Döhler, "Semiconductor superlattices: A new material for research and applications," *Phys. Scr.*, 24, 430, 1981.
30. C. Mailhiot and D. L. Smith, "Strained-layer semiconductor superlattices," *Solid State & Mater. Sci.*, 16, 131–60, 1990.
31. L. Esaki, "A bird's-eye view on the evolution of semiconductor superlattices and quantum wells," *IEEE J. Quantum Electron.*, QE-22, 1611–24, 1986.
32. H. Krömer, "Barrier control and measurements: Abrupt semiconductor heterojunctions," *J. Vac. Sci. Technol.*, B2, 433, 1984.

33. J. Tersoff, "Theory of semiconductor heterojunctions: The role of quantum dipoles," *Phys. Rev.*, B30, 4874, 1984.
34. W. A. Harrison, "Elementary Tight-Binding Theory of Schottky-Barrier and Heterojunction Band Line-Ups," in *Two Dimensional Systems: Physics and New Devices, Springer Series in Solid State Sciences*, Vol. 67, eds. G. Bauer, F. Kuchar, and H. Heinrich, 62, Springer, Berlin, 1986.
35. W. Pollard, "Valence-band discontinuities in semiconductor heterojunctions," *J. Appl. Phys.*, 69, 3154–8, 1991.
36. H. Krömer, "Band offsets and chemical bonding: The basis for heterostructure applications," *Phys. Scr.*, 68, 10–16, 1996.
37. G. Duggan, "A critical review of semiconductor heterojunction band offsets," *J. Vac. Sci. Technol.*, B3, 1224, 1985.
38. T. W. Hickmott, "Electrical Measurements of Band Discontinuits at Heterostructure Interfaces," in *Two Dimensional Systems: Physics and New Devices, Springer Series in Solid State Sciences*, Vol. 67, eds. G. Bauer, F. Kuchar, and H. Heinrich, 72, Springer, Berlin, 1986.
39. J. Menendez and A. Pinczuk, "Light scattering determinations of band offsets in semiconductor heterojunctions," *IEEE J. Quantum Electron.*, 24, 1698–711, 1988.
40. K. Ploog and G. H. Döhler, "Compositional and doping superlattices in III–V semiconductors," *Adv. Phys.*, 32, 285–359, 1983.
41. G. H. Döhler, "Doping superlattices ("n-i-p-i crystals")," *IEEE J. Quantum Electron.*, QE-22, 1682–95, 1986.
42. G. H. Döhler, "The physics and applications of n-i-p-i doping superlattices," *CRC Crit. Rev. Solid State Mater. Sci.*, 13, 97–141, 1987.
43. C. Weisbuch, "Fundamental Properties of III–V Semiconductor Two-Dimensional Quantized Structures: The Basis for Optical and Electronic Device Applications," in *Semiconductors and Semimetals*, Vol. 24, eds. R. Willardson and A. C. Beer; *Applications of Multiquantum Wells, Selective Doping, and Superlattices*, ed. R. Dingle, 1–133, Academic Press, New York, 1987.
44. A. Rogalski, *New Ternary Alloy Systems for Infrared Detectors*, SPIE Optical Engineering Press, Bellingham, WA, 1994.
45. D. L. Smith and C. Mailhiot, "Theory of semiconductor superlattice electronic structure," *Rev. Mod. Phys.*, 62, 173–234, 1990.
46. B. F. Levine, R. J. Malik, J. Walker, K. K. Choi, C. G. Bethea, D. A. Kleinman, and J. M. Wandenberg, "Strong 8.2 μm infrared intersubband absorption in doped GaAs/AlAs quantum well waveguides," *Appl. Phys. Lett.*, 50, 273–5, 1987.
47. K. K. Choi, "Detection wavelength of quantum-well infrared photodetectors," *J. Appl. Phys.*, 73, 5230–6, 1993.
48. K. M. S. V. Bandara, D. D. Coon, O. Byungsung, Y. F. Lin, and M. H. Francombe, "Exchange interactions in quantum well subbands," *Appl. Phys. Lett.*, 53, 1931–3, 1988.
49. M. C. Tatham, J. R. Ryan, and C. T. Foxon, "Time-resolved Raman scattering measurement of electron-optical phonon intersubband relaxation in GaAs quantum wells," *Solid State Electron.*, 32, 1497–501, 1989.
50. G. Hasnain, B. F. Levine, C. G. Bethea, R. R. Abbott, and S. J. Hsieh, "Measurement of intersubband absorption in multiquantum well structures with monolithically integrated photodetectors," *J. Appl. Phys.*, 67, 4361–3, 1990.
51. M. O. Manasreh, F. F. Szmulowicz, D. W. Fischer, K. R. Evans, and C. E. Stutz, "Intersubband infrared absorption in a GaAs/$Al_{0.3}Ga_{0.7}As$ quantum well structure," *Appl. Phys. Lett.*, 57, 1790–2, 1990.
52. B. F. Levine, A. Zussman, S. D. Gunapala, M. T. Asom, J. M. Kuo, and W. S. Hobson, "Photoexcited escape probability, optical gain, and noise in quantum well infrared photodetectors," *J. Appl. Phys.*, 72, 4429–43, 1992.
53. C. I. Yang and D. S. Pan, "Intersubband absorption of silicon-based quantum wells for infrared imaging," *J. Appl. Phys.*, 64, 1573–5, 1988.
54. Y. C. Chang and R. B. James, "Saturation of intersubband transitions in p-type semiconductor quantum wells," *Phys. Rev.*, B39, 12672–81, 1989.
55. C. H. Lee and K. L. Wang, "Intersubband absorption in Sb δ-doped $Si/Si_{1-x}Ge_x$ quantum well structures grown on Si (110)," *Appl. Phys. Lett.*, 60, 2264–6, 1992.
56. J. Katz, Y. Zhang, and W. I. Wang, "Normal incidence infrared absorption in AlAs/AlGaAs X-valley multi-quantum wells," *Appl. Phys. Lett.*, 61, 1697–99, 1992.
57. B. F. Levine, S. D. Gunapala, J. M. Kuo, and S. S. Pei, "Normal incidence hole intersubband absorption long wavelength GaAs/$Al_xGa_{1-x}As$ quantum well infrared photodetector," *Appl. Phys. Lett.*, 59, 1864–6, 1991.
58. W. S. Hobson, A. Zussman, B. F. Levine, J. De Long, M. Gera, and L. C. Luther, "Carbon doped GaAs/$Al_xGa_{1-x}As$ quantum well infrared photodetectors grown by organometallic vapor phase epitaxy," *J. Appl. Phys.*, 71, 3642–4, 1992.
59. J. S. Park, R. P. G. Karunasiri, and K. L. Wang, "Normal incidence detector using p-type SiGe/Si multiple quantum wells," *Appl. Phys. Lett.*, 60, 103–5, 1992.

60. J. S. Park, R. P. G. Karunasiri, and K. L. Wang, "Intervalence-subband transition in SiGe/Si multiple quantum wells: Normal incidence detection," *Appl. Phys. Lett.*, 61, 681–3, 1992.
61. R. P. G. Karunasiri, J. S. Park, and K. L. Wang, "Normal incidence infrared detector using intervalence-subband transitions in $Si_{1-x}Ge_x/Si$ quantum wells," *Appl. Phys. Lett.*, 61, 2434–6, 1992.
62. G. Karunasiri, "Intersubband transition in Si-based quantum wells and application for infrared photodetectors," *Jpn. J. Appl. Phys.*, 33, 2401–11, 1994.
63. R. People, J. C. Bean, C. G. Bethea, S. K. Sputz, and L. J. Peticolas, "Broadband (8–14 μm), normal incidence, pseudomorphic Ge_xSi_{1-x}/Si strained-layer infrared photodetector operating between 20 and 77 K," *Appl. Phys. Lett.*, 61, 1122–4, 1992.
64. B. K. Ridley, "Electron scattering by confined LO polar phonons in a quantum well," *Phys. Rev.*, B39, 5282–6, 1989.
65. D. Y. Oberli, D. R. Wake, M. V. Klein, T. Henderson, and H. Morkoc, "Intersubband relaxation of photoexcited hot carriers in quantum wells," *Solid State Electron.*, 31, 413–8, 1988.
66. B. N. Murdin, W. Heiss, C. J. G. M. Langerak, S. -C. Lee, I. Galbraith, G. Strasser, E. Gornik, M. Helm, and C. R. Pidgeon, "Direct observation of the LO phonon bottleneck in wide $GaAs/Al_xGa_{1-x}$ as quantum wells," *Phys. Rev.*, B55, 5171–6, 1997.
67. J. Y. Andersson, "Dark current mechanisms and conditions of background radiation limitation of n-doped AlGaAs/GaAs quantum-well infrared detectors," *J. Appl. Phys.*, 78, 6298–304, 1995.
68. V. D. Shadrin and F. L. Serzhenko, "The theory of multiple quantum-well GaAs/AlGaAs infrared detectors," *Infrared Phys.*, 33, 345–57, 1992.
69. J. C. Smith, L. C. Chiu, S. Margalit, A. Yariv, and A. Y. Cho, "A new infrared detector using electron emission from multiple quantum wells," *J. Vac. Sci. Technol.*, B1, 376–8, 1983.
70. L. C. Chiu, J. S. Smith, S. Margalit, A. Yariv, and A. Y. Cho, "Application of internal photoemission from quantum-well and heterojunction superlattices to infrared photodetectors," *Infrared Phys.*, 23, 93–7, 1983.
71. D. D. Coon and P. G. Karunasiri, "New mode of IR detection using quantum wells," *Appl. Phys. Lett.*, 45, 649–51, 1984.
72. L. C. West and S. J. Eglash, "First observation of an extremely large-dipole infrared transition within the conduction band of a GaAs quantum well," *Appl. Phys. Lett.*, 46, 1156–8, 1985.
73. B. F. Levine, K. K. Choi, C. G. Bethea, J. Walker, and R. J. Malik, "New 10 μm infrared detector using intersubband absorption in resonant tunneling GaAlAs superlattices," *Appl. Phys. Lett.*, 50, 1092–4, 1987.
74. S. Gunapala, M. Sundaram, and S. Bandara, "Quantum wells stare at long-wave IR scenes," *Laser Focus World* 233–40, June 1996.
75. S. D. Gunapala, J. K. Liu, J. S. Park, M. Sundaram, C. A. Shott, T. Hoelter, T. L. Lin, et al., "9-μm cutoff 256×256 $GaAs/Al_xGa_{1-x}As$ quantum well infrared photodetector hand-held camera," *IEEE Trans. Electron Devices* 44, 51–7, 1997.
76. J. Y. Andersson, L. Lundqvist, and Z. F. Paska, "Quantum efficiency enhancement of AlGaAs/GaAs quantum well infrared detectors using a waveguide with a grating coupler," *Appl. Phys. Lett.*, 58, 2264–6, 1991.
77. Z. F. Paska, J. Y. Andersson, L. Lundqvist, and C. O. A. Olsson, "Growth and characterization of AlGaAs/GaAs quantum well structures for the fabrication of long wavelength infrared detectors," *J. Cryst. Growth* 107, 845–9, 1991.
78. W. Bloss, M. O'Loughlin, and M. Rosenbluth, "Advances in multiple quantum well IR detectors," *Proc. SPIE* 1541, 2–10, 1991.
79. K. K. Choi, B. F. Levine, C. G. Bethea, J. Walker, and R. J. Malik, "Multiple quantum well 10 μm GaAs/$Al_xGa_{1-x}As$ infrared detector with improved responsivity," *Appl. Phys. Lett.*, 50, 1814–6, 1987.
80. B. F. Levine, C. G. Bethea, G. Hasnain, V. O. Shen, E. Pelve, R. R. Abbot, and S. J. Hsieh, "High sensitivity low dark current 10 μm GaAs quantum well infrared photodetectors," *Appl. Phys. Lett.*, 56, 851–3, 1990.
81. S. D. Gunapala, B. F. Levine, L. Pfeifer, and K. West, "Dependence of the performance of GaAs/AlGaAs quantum well infrared photodetectors on doping and bias," *J. Appl. Phys.*, 69, 6517–20, 1991.
82. S. D. Gunapala, J. S. Park, G. Sarusi, T. L. Lin, J. K. Liu, P. D. Maker, R. E. Muller, C. A. Shott, and T. Hoelter, "15-μm 128×128 $GaAs/Al_xGa_{1-x}As$ quantum well infrared photodetector focal plane array camera," *IEEE Trans. Electron Devices* 44, 45–50, 1997.
83. M. A. Kinch and A. Yariv, "Performance limitations of GaAs/AlGaAs infrared superlattices," *Appl. Phys. Lett.*, 55, 2093–5, 1989.
84. B. F. Levine, C. G. Bethea, G. Hasnain, J. Walker, and R. J. Malik, "High-detectivity $D^* = 1.0\times10^{10}$ cm $Hz^{1/2}/W$ GaAs/AlGaAs multiquantum well $\lambda = 8.3$ μm infrared detector," *Appl. Phys. Lett.*, 53, 296–8, 1988.
85. H. C. Liu, "An Introduction to the Physics of Quantum Well Infrared Photodetectors and Other Related Devices," in *Handbook of Thin Film Devices*, Vol. 2, ed. M. H. Francombe, 101–34, Academic Press, San Diego, 2000.

86. M. J. Kane, S. Millidge, M. T. Emeny, D. Lee, D. R. P. Guy, and C. R. Whitehouse, "Performance Trade Offs in the Quantum Well Infra-Red Detector," in *Intersubband Transitions in Quantum Wells*, eds. E. Rosencher, B. Vinter, and B. Levine, 31–42, Plenum Press, New York, 1992.
87. H. C. Liu, A. G. Steele, M. Buchanan, and Z. R. Wasilewski, "Dark current in quantum well infrared photodetectors," *J. Appl. Phys.*, 73, 2029–31, 1993.
88. M. Z. Tidrow, J. C. Chiang, S. S. Li, and K. Bacher, "A high strain two-stack two-color quantum well infrared photodetector," *Appl. Phys. Lett.*, 70, 859–61, 1997.
89. A. Rose, *Concepts in Photoconductivity and Allied Problems*, Interscience, New York, 1963.
90. H. C. Liu, "Photoconductive gain mechanism of quantum-well intersubband infrared detectors," *Appl. Phys. Lett.*, 60, 1507–9, 1992.
91. A. G. Steele, H. C. Liu, M. Buchanan, and Z. R. Wasilewski, "Importance of the upper state position in the performance of quantum well intersubband infrared detectors," *Appl. Phys. Lett.*, 59, 3625–7, 1991.
92. H. C. Liu, "Dependence of absorption spectrum and responsivity on the upper state position in quantum well intersubband photodetectors," *Appl. Phys. Lett.*, 73, 3062–7, 1993.
93. H. C. Liu, "Noise gain and operating temperature of quantum well infrared photodetectors," *Appl. Phys. Lett.*, 61, 2703–5, 1992.
94. W. A. Beck, "Photoconductive gain and generation-recombination noise in multiple-quantum-well infrared detectors," *Appl. Phys. Lett.*, 63, 3589–91, 1993.
95. A. Rogalski, "Comparison of the performance of quantum well and conventional bulk infrared photodetectors," *Infrared Phys. Technol.*, 38, 295–310, 1997.
96. S. D. Gunapala, B. F. Levine, D. Ritter, R. A. Hamm, and M. B. Panish, "InGaAs/InP long wavelength quantum well infrared photodetectors," *Appl. Phys. Lett.*, 58, 2024–6, 1991.
97. A. Zussman, B. F. Levine, J. M. Kuo, and J. De Jong, "Extended long-wavelength $\lambda = 11–15\ \mu m$ GaAs/$Al_xGa_{1-x}As$ quantum well infrared photodetectors," *J. Appl. Phys.*, 70, 5101–7, 1991.
98. B. F. Levine, A. Zussman, J. M. Kuo, and J. De Jong, "19 μm cutoff long-wavelength GaAs/ $Al_xGa_{1-x}As$ quantum well infrared photodetectors," *J. Appl. Phys.*, 71, 5130–5, 1992.
99. A. Singh and M. O. Manasreh, "Quantum well and superlattice heterostructures for space-based long wavelength infrared photodetectors," *Proc. SPIE* 2397, 193–209, 1995.
100. D. C. Arrington, J. E. Hubbs, M. E. Gramer, and G. A. Dole, "Nonlinear response of QWIP detectors: Summary of data from four manufactures," *Proc. SPIE* 4028, 289–99, 2000.
101. M. Z. Tidrow, W. A. Beck, W. W. Clark, H. K. Pollehn, J. W. Little, N. K. Dhar, P. R. Leavitt, et al., "Device physics and focal plane applications of QWIP and MCT," *Opto-Electron. Rev.*, 7, 283–96, 1999.
102. A. Rogalski, "Third generation photon detectors," *Opt. Eng.*, 42, 3498–516, 2003.
103. A. Rogalski, "Competitive technologies of third generation infrared photon detectors," *Opto-Electron. Rev.*, 14, 87–101, 2006.
104. C. Schönbein, H. Schneider, R. Rehm, and M. Walther, "Noise gain and detectivity of n-type GaAs/AlAs/AlGaAs quantum well infrared photodetectors," *Appl. Phys. Lett.*, 73, 1251–4, 1998.
105. A. Kastalsky, T. Duffield, S. J. Allen, and J. Harbison, "Photovoltaic detection of infrared light in a GaAs/AlGaAs superlattice," *Appl. Phys. Lett.*, 52, 1320–2, 1988.
106. H. Schneider, M. Walther, J. Fleissner, R. Rehm, E. Diwo, K. Schwarz, P. Koidl, et al., "Low-noise QWIPs for FPA sensors with high thermal resolution," *Proc. SPIE* 4130, 353–62, 2000.
107. H. Schneider, P. Koidl, M. Walther, J. Fleissner, R. Rehm, E. Diwo, K. Schwarz, and G. Weimann, "Ten years of QWIP development at Fraunhofer," *Infrared Phys. Technol.*, 42, 283–9, 2001.
108. H. Schneider, M. Walther, C. Schönbein, R. Rehm, J. Fleissner, W. Pletschen, J. Braunstein, et al., "QWIP FPAs for high-performance thermal imaging," *Phys. E* 7, 101–7, 2000.
109. H. Schneider, C. Schönbein, M. Walther, K. Schwarz, J. Fleissner, and P. Koidl, "Photovoltaic quantum well infrared photodetectors: The four-zone scheme," *Appl. Phys. Lett.*, 71, 246–8, 1997.
110. H. Schneider, "Theory of avalanche multiplication and excess noise in quantum-well infrared photodetectors," *Appl. Phys. Lett.*, 82, 4376–8, 2003.
111. O. Byungsung, J. W. Choe, M. H. Francombe, K. M. S. V. Bandara, D. D. Coon, Y. F. Lin, and W. J. Takei, "Long-wavelength infrared detection in a Kastalsky-type superlattice structure," *Appl. Phys. Lett.*, 57, 503–5, 1990.
112. S. D. Gunapala, B. F. Levine, and N. Chand, "Band to continuum superlattice miniband long wavelength GaAs/$Al_xGa_{1-x}As$ infrared detectors," *J. Appl. Phys.*, 70, 305–8, 1991.
113. K. M. S. V. Bandara, J. W. Choe, M. H. Francombe, A. G. U. Perera, and Y. F. Lin, "GaAs/AlGaAs superlattice miniband detector with 14.5 μm peak response," *Appl. Phys. Lett.*, 60, 3022–4, 1992.

114. C. C. Chen, H. C. Chen, C. H. Kuan, S. D. Lin, and C. P. Lee, "Multicolor infrared detection realized with two distinct superlattices separated by a blocking barrier," *Appl. Phys. Lett.*, 80, 2251–3, 2002.
115. K-K. Choi, C. Monroy, V. Swaminathan, T. Tamir, M. Leung, J. Devitt, D. Forrai, and D. Endres, "Optimization of corrungated-QWIP for large format, high quantum efficiency, and multi-color FPAs," *Infrared Phys. Technol.*, 50, 124–35, 2007.
116. K.-K. Choi, M. D. Jhabvala, and R. J. Peralta, "Voltage-tunable two-color corrugated-QWIP focal plane arrays," *IEEE Electron Device Lett.*, 29, 1011–3, 2008.
117. L. S. Yu and S. S. Lu, "A metal grating coupled bound-to-miniband transition GaAs multiquantum well/superlattice infrared detector," *Appl. Phys. Lett.*, 59, 1332–4, 1991.
118. L. S. Yu, S. S. Li, and P. Ho, "Largely enhanced bound-to-miniband absorption in an InGaAs multiple quantum well with short-period superlattice InAlAs/InGaAs barrier," *Appl. Phys. Lett.*, 59, 2712–4, 1991.
119. L. S. Yu, Y. H. Wang, S. S. Li, and P. Ho, "Low dark current step-bound-to-miniband transition InGaAs/GaAs/AlGaAs multiquantum-well infrared detector," *Appl. Phys. Lett.*, 60, 992–4, 1992.
120. W. A. Beck, J. W. Little, A. C. Goldberg, and T. S. Faska, "Imaging Performance of LWIR Miniband Transport Multiple Quantum Well Infrared Focal Plane Arrays," in *Quantum Well Intersubband Transition Physics and Devices*, eds. H. C. Liu, B. F. Levine, and J. Y. Anderson, 55–68, Kluwer Academic Publishers, Dordrecht, The Netherlands, 1994.
121. W. A. Beck and T. S. Faska, "Current status of quantum well focal plane arrays," *Proc. SPIE* 2744, 193–206 1996.
122. J. Chu and S. S. Li, "The effect of compressive strain on the performance of p-type quantum-well infrared photodetectors," *IEEE J. Quantum Electron* 33, 1104–13, 1997.
123. S. S. Li, "Multi-Color, Broadband Quantum Well Infrared Photodetectors for Mid-, Long-, and Very Long-Wavelength Infrared Applications," in *Intersubband Infrared Photodetectors*, ed. V. Ryzhii, 169–209, World Scientific, Singapore, 2003.
124. K. W. Goossen and S. A. Lyon, "Grating enhanced quantum well detector," *Appl. Phys. Lett.*, 47, 1257–1529, 1985.
125. K. W. Goossen, S. A. Lyon, and K. Alavi, "Grating enhancement of quantum well detector response," *Appl. Phys. Lett.*, 53, 1027–9, 1988.
126. G. Hasnain, B. F. Levine, C. G. Bethea, R. A. Logan, L. Walker, and R. J. Malik, "GaAs/AlGaAs multiquantum well infrared detector arrays using etched gratings," *Appl. Phys. Lett.*, 54, 2515–7, 1989.
127. J. Y. Andersson, L. Lundqvist, and Z. F. Paska, "Grating-coupled quantum-well infrared detectors: Theory and performance," *J. Appl. Phys.*, 71, 3600–10, 1992.
128. G. Sarusi, B. F. Levine, S. J. Pearton, K. M. S. V. Bandara, and R. E. Leibenguth, "Optimization of two dimensional gratings for very long wavelength quantum well infrared photodetectors," *J. Appl. Phys.*, 76, 4989–94, 1994.
129. G. Sarusi, B. F. Levine, S. J. Pearton, K. M. S. V. Bandara, and R. E. Leibenguth, "Improved performance of quantum well infrared photodetectors using random scattering optical coupling," *Appl. Phys. Lett.*, 64, 960–2, 1994.
130. T. R. Schimert, S. L. Barnes, A. J. Brouns, F. C. Case, P. Mitra, and L. T. Claiborne, "Enhanced quantum well infared photodetector with novel multiple quantum well grating structure," *Appl. Phys. Lett.*, 68, 2846–8, 1996.
131. C. J. Chen, K. K. Choi, M. Z. Tidrow, and D. C. Tsui, "Corrugated quantum well infrared photodetectors for normal incident light coupling," *Appl. Phys. Lett.*, 68, 1446–8, 1996.
132. C. J. Chen, K. K. Choi, W. H. Chang, and D. C. Tsui, "Performance of corrugated quantum well infrared photodetectors," *Appl. Phys. Lett.*, 71, 3045–7, 1997.
133. K. K. Choi, M. D. Jhabvala, D. P. Forrai, A. Waczynski, J. Sun, and R. Jones, "Electromagnetic modeling and design of quantum well infrared photodetectors", *IEEE J. Sel. Top. Quantum. Electron.*, 19, 3800310, 2013.
134. K. K. Choi, M. D. Jhabvala, D. P. Forrai, A. Waczynski, J. Sun, and R. Jones, "Electromagnetic design of resonator-QWIPs", *Proc. SPIE* 8268 826820-1–9, 2012.
135. K. K. Choi, M. D. Jhabvala, J. Sun, C. A. Jhabvala, A. Waczynski, and K. Olver, "Resonator-quantum well infrared photodetectors," *Appl. Phys. Lett.*, 103, 201113, 2013.
136. K. K. Choi, S. C. Allen, Y. Wei, J. G. Sun, K. A. Olver, and R. X. Fu, "Long wavelength resonator-QWIPs", *Proc. SPIE* 9817, 981917-1–12, 2016.
137. J. A. Robo, E. Costard, J. P. Truffer, A. Nedelcu, X. Marcadet, and P. Bois, "QWIP focal plane arrays performances from MWIR to VLWIR," *Proc. SPIE* 7298, 7298–15, 2009.
138. S. S. Li and Y.-K. Su (Eds.) *Intersubband Transitions in Quantum Wells: Physics and Devices*, Springer, New York, 1998.
139. H. Xie, J. Katz, and W. I. Wang, "Infrared absorption enhancement in light- and heavy-hole inverted $Ga_{1-x}In_xAs/Al_{1-y}In_yAs$ quantum wells," *Appl. Phys. Lett.*, 59, 3601–3, 1991.
140. H. Xie, J. Katz, W. I. Wang, and Y. C. Chang, "Normal incidence infrared photoabsorption in p-type GaSb/$Ga_xAl_{1-x}Sb$ quantum wells," *J. Appl. Phys.*, 71, 2844–47, 1992.

141. F. Szmulowicz, G. J. Brown, H. C. Liu, A. Shen, Z. R. Wasilewski, and M. Buchanan, "GaAs/AlGaAs p-type multiple-quantum wells for infrared detection at normal incidence: model and experiment," *Opto-Electron. Rev.*, 9, 164–72, 2001.
142. F. Szmulowicz and G. J. Brown, "Whither p-type GaAs/AlGaAs QWIP?" *Proc. SPIE* 4650, 158–66, 2002.
143. K. Hirose, T. Mizutani, and K. Nishi, "Electron and hole mobility in modulation doped GaInAs-AlInAs strained layer superlattice," *J. Cryst. Growth* 81, 130–5, 1987.
144. P. Man and D. S. Pan, "Analysis of normal-incident absorption in p-type quantum-well infrared photodetectors," *Appl. Phys. Lett.*, 61, 2799–801, 1992.
145. K. K. Choi, M. Dutta, P. G. Newman, and M. L. Saunders, "10 μm infrared hot-electron transistors," *Appl. Phys. Lett.*, 57, 1348–50, 1990.
146. K. K. Choi, M. Dutta, R. P. Moekirk, C. H. Kuan, and G. J. Iafrate, "Application of superlattice bandpass filters in 10 μm infrared detection," *Appl. Phys. Lett.*, 58, 1533–5, 1991.
147. K. K. Choi, L. Fotiadis, M. Taysing-Lara, W. Chang, and G. J. Iafrate, "High detectivity InGaAs base infrared hot-electron transistor," *Appl. Phys. Lett.*, 59, 3303–5, 1991.
148. R. Fu, "Structure and process of infrared hot electron transistor arrays", *Sensors* (Basel) 12(5), 6508–6519, 2012
149. K. K. Choi, M. Z. Tidrow, M. Taysing-Lara, W. H. Chang, C. H. Kuan, C. W. Farley, and F. Chang, "Low dark current infrared hot-electron transistor for 77 K operation," *Appl. Phys. Lett.*, 63, 908–10, 1993.
150. C. Y. Lee, M. Z. Tidrow, K. K. Choi, W. H. Chang, and L. F. Eastman, "Long-wavelength $\lambda_c = 18\,\mu m$ infrared hot-electron transistor," *J. Appl. Phys.*, 75, 4731–6, 1994.
151. C. Y. Lee, M. Z. Tidrow, K. K. Choi, W. H. Chang, and L. F. Eastman, "Activation characteristics of a long wavelength infrared hot-electron transistor," *Appl. Phys. Lett.*, 65, 442–4, 1994.
152. S. D. Gunapala, J. S. Park, T. L. Lin, J. K. Liu, and K. M. S. V. Bandara, "Very long-wavelength GaAs/$Al_xGa_{1-x}As$ infrared hot electron transistor," *Appl. Phys. Lett.*, 64, 3003–5, 1994.
153. D. Mandelik, M. Schniederman, V. Umansky, and I. Bar-Joseph, "Monolithic integration of a quantum-well infrared photodetector array with a read-out circuit," *Appl. Phys. Lett.*, 78, 472–4, 2001.
154. V. Ryzhii, "Unipolar darlington infrared phototransistor," *Jpn. J. Appl. Phys.*, 36, L415–7, 1997.
155. K. Nanaka, Semiconductor Devices, Japan Patent 4-364072, 1992.
156. S. C. Jain, J. R. Willis, and R. Bullough, "A review of theoretical and experimental work on the structure of Ge_xSi_{1-x} strained layers and superlattices, with extensive bibliography," *Adv. Phys.*, 39, 127–90, 1990.
157. R. P. G. Karunasiri and K. L. Wang, "Quantum devices using SiGe/Si heterostructures," *J. Vac. Sci. Technol.*, B9, 2064–71, 1991.
158. K. L. Wang and R. P. G. Karunasiri, "SiGe/Si electronics and optoelectronics," *J. Vac. Sci. Technol.*, B11, 1159–67, 1993.
159. R. P. G. Karunasiri, J. S. Park, K. L. Wang, and S. K. Chun, "Infrared photodetectors with SiGe/Si multiple quantum wells," *Opt. Eng.*, 33, 1468–76, 1994.
160. D. J. Robbins, M. B. Stanaway, W. Y. Leong, J. L. Glasper, and C. Pickering, "$Si_{1-x}Ge_x$/Si quantum well infrared photodetectors," *J. Mater. Sci.: Mater. Electron.*, 6, 363–7, 1995.
161. R. P. G. Karunasiri, J. S. Park, Y. J. Mii, and K. L. Wang, "Intersubband absorption in $Si_{1-x}Ge_x$/Si multiple quantum wells," *Appl. Phys. Lett.*, 57, 2585–7, 1990.
162. M. Francombe and J. Vossen (Eds.) *Homojunction and Quantum-Well Infrared Detectors*, Academic Press, New York, 1995.
163. R. People, J. C. Bean, S. K. Sputz, C. G. Bethea, and L. J. Peticolas, "Normal incidence hole intersubband quantum well infrared photodetectors in pseudomorphic $Si_{1-x}Ge_x$/Si," *Thin Solid Films* 222, 120–5, 1992.
164. V. D. Shadrin, V. T. Coon, and F. L. Serzhenko, "Photoabsorption in n–type Si-SiGe quantum-well infrared photodetectors," *Appl. Phys. Lett.*, 62, 2679–81, 1993.
165. V. D. Shadrin, "Background limited infrared performance of n–type Si–SiGe (111) quantum well infrared photodetectors," *Appl. Phys. Lett.*, 65, 70–72, 1994.
166. B. F. Levine, A. Y. Cho, J. Walker, R. J. Malik, D. A. Kleinmen, and D. L. Sivco, "InGaAs/InAlAs multiquantum well intersubband absorption at a wavelength of $\lambda = 4.4\,\mu m$," *Appl. Phys. Lett.*, 52, 1481–3, 1988.
167. G. Hasnain, B. F. Levine, D. L. Sivco, and A. Y. Cho, "Mid-infrared detectors in the 3–5 μm band using bound to continuum state absorption in InGaAs/InAlAs multiquantum well structures," *Appl. Phys. Lett.*, 56, 770–2, 1990.
168. S. Ozer, U. Tumkaya, and C. Besikci, "Large format AlInAs-InGaAs quantum-well infrared photodetector focal plane array for midwavelength infrared thermal imaging," *IEEE Photonics Technol. Lett.*, 19, 1371–3, 2007.
169. M. Kaldirim, Y. Arslan, S. U. Eker, and C. Besikci, "Lattice-matched AlInAs-InGaAs mid-wavelength infrared QWIPs: Characteristics and focal plane array performance," *Semicond. Sci. Technol.*, 23, 085007, 2008.

170. S. D. Gunapala, B. F. Levine, D. Ritter, R. A. Hamm, and M. B. Panish, "Lattice-matched InGaAsP/InP long-wavelength quantum well infrared photoconductors," *Appl. Phys. Lett.*, 60, 636–8, 1992.
171. S. D. Gunapala, B. F. Levine, D. Ritter, R. Hamm, and M. B. Panish, "InP based quantum well infrared photodetectors," *Proc. SPIE* 1541, 11–23, 1991.
172. S. D. Gunapala, K. M. S. V. Bandara, B. F. Levine, G. Sarusi, D. L. Sivco, and A. Y. Cho, "Very long wavelength $In_xGa_{1-x}As/GaAs$ quantum well infrared photodetectors," *Appl. Phys. Lett.*, 64, 2288–90, 1994.
173. S. D. Gunapala, K. M. S. V. Bandara, B. F. Levine, G. Sarusi, J. S. Park, T. L. Lin, W. T. Pike, and J. K. Liu, "High performance InGaAs/GaAs quantum well infrared photodetectors," *Appl. Phys. Lett.*, 64, 3431–3, 1994.
174. A. Kock, E. Gornik, G. Abstreiter, G. Bohm, M. Walther, and G. Weimann, "Double wavelength selective GaAs/AlGaAs infrared detector device," *Appl. Phys. Lett.*, 60, 2011–3, 1992.
175. P. H. Bois, E. Costard, J. Y. Duboz, and J. Nagle, "Technology of multiquantum well infrared detectors," *Proc. SPIE* 3061, 764–71, 1997.
176. E. Costard, P. H. Bois, F. Audier, and E. Herniou, "Latest improvements in QWIP technology at Thomson-CSF/LCR," *Proc. SPIE* 3436, 228–39, 1998.
177. S. D. Gunapala, S. V. Bandara, J. K. Liu, J. M. Mumolo, C. J. Hill, S. B. Rafol, D. Salazar, J. Woollaway, P. D. LeVan, and M. Z. Tidrow, "Towards dualband megapixel QWIP focal plane arrays," *Infrared Phys. Technol.*, 50, 217–26, 2007.
178. S. D. Gunapala, S. V. Bandara, J. K. Liu, S. B. Rafol, J. M. Mumolo, C. A. Shott, R. Jones, et al., "640×512 pixel narrow-band, four-band, and broad-band quantum well infrared photodetector focal plane arrays," *Infrared Phys. Technol.*, 44, 411–25, 2003.
179. H. C. Liu, J. Li, J. R. Thompson, Z. R. Wasilewski, M. Buchanan, and J. G. Simmons, "Multicolor voltage tunable quantum well infrared photodetector," *IEEE Electron Device Lett.*, 14, 566–8, 1993.
180. I. Grave, A. Shakouri, N. Kuze, and A. Yariv, "Voltage-controlled tunable GaAs/AlGaAs multistack quantum well infrared detector," *Appl. Phys. Lett.*, 60, 2362–4, 1992.
181. H. C. Liu, "Recent progress on GaAs quantum well intersubband infrared photodetectors," *Opt. Eng.*, 33, 1461–7, 1994.
182. Y. H. Wang, S. S. Li, and P. Ho, "Voltage-tunable dual-mode operation in InAlAs/InGaAs quantum well infared photodetector for narrow- and broadband detection at 10 μm," *Appl. Phys. Lett.*, 62, 621–3, 1993.
183. E. Martinet, F. Luc, E. Rosencher, P. H. Bois, and S. Delaitre, "Electrical tunability of infrared detectors using compositionally asymmetric GaAs/AlGaAs multiquantum wells," *Appl. Phys. Lett.*, 60, 895–97, 1992.
184. E. Martinet, E. Rosencher, F. Luc, P. H. Bois, E. Costard, and S. Delaitre, "Switchable bicolor (5.5–9.0 μm) infrared detector using asymmetric GaAs/AlGaAs multiquantum well," *Appl. Phys. Lett.*, 61, 246–8, 1992.
185. Y. H. Wang, S. H. S. Li, and P. Ho, "Photovoltaic and photoconductive dual-mode operation GaAs quantum well infrared photodetector for two-band detection," *Appl. Phys. Lett.*, 62, 93–5, 1993.
186. M. Z. Tidrow, K. K. Choi, C. Y. Lee, W. H. Chang, F. J. Towner, and J. S. Ahearn, "Voltage tunable three-color quantum well infrared photodetector," *Appl. Phys. Lett.*, 64, 1268–70, 1994.
187. K. Kheng, M. Ramsteiner, H. Schneider, J. D. Ralston, F. Fuchs, and P. Koidl, "Two-color GaAs/AlGaAs quantum well infrared detector with voltage-tunable spectral sensitivity at 3–5 and 8–12 μm," *Appl. Phys. Lett.*, 61, 666–8, 1992.
188. H. C. Liu, E. Dupont, M. Byloos, M. Buchanan, C. -Y. Song, and Z. R. Wasilewski, "QWIP-LED Pixelless Thermal Imaging Device," in *Intersubband Infrared Photodetectors*, ed. V. Ryzhii, 299–313, World Scientific, Singapore, 2003.
189. H. C. Liu, J. Li, Z. R. Wasilewski, and M. Buchanan, "Integrated quantum well intersub-band photodetector and light emitting diode," *Electron. Lett.*, 31, 832–3, 1995.
190. V. Ryzhii, M. Ershov, M. Ryzhii, and I. Khmyrova, "Quantum well infrared photodetector with optical output," *Jpn. J. Appl. Phys.*, 34, L38–40, 1995.
191. E. Dupont, M. Gao, Z. R. Wasilewski, and H. C. Liu, "Integration of n-type and p-type quantum-well infrared photodetectors for sequential multicolor operation," *Appl. Phys. Lett.*, 78, 2067–9, 2001.
192. E. Dupont, H. C. Liu, M. Buchanan, Z. R. Wasilweski, D. St-Germain, and P. Chevrette, "Pixelless infrared imaging based on the integration of a n-type quantum-well infrared photodetector with a light emitting diode," *Appl. Phys. Lett.*, 75, 563–65, 1999.
193. E. Dupont, M. Byloos, M. Gao, M. Buchanan, C.-Y. Song, Z. R. Wasilewski, and H. C. Liu, "Pixel-less thermal imaging with integrated quantum-well infrared photo detector and light-emitting diode," *IEEE Photonics Technol. Lett.*, 14, 182–4, 2002.
194. E. Dupont, M. Byloos, T. Oogarah, M. Buchanan, and H. C. Liu, "Optimization of quantum-well infrared detectors integrated with light-emitting diodes," *Infrared Phys. Technol.*, 47, 132–43, 2005.

Superlattice detectors

Many types of optoelectronic devices can be enhanced significantly through the introduction of quantum confinement in reduced dimensionality heterostructures. This was the main motivation for the study of superlattices (SLs) as an alternative to infrared (IR) detector materials. The HgTe/CdTe SL system was proposed in 1979, only a few years after the first GaAs/AlGaAs quantum heterostructures were fabricated with molecular-beam epitaxy (MBE). It was anticipated that SL IR materials would have several advantages over bulk HgCdTe (the current industry standard) for this application:

- a higher degree of uniformity, which is importance for detector arrays;
- smaller leakage current due to the suppression of tunneling (larger effective masses) available in SLs; and
- lower Auger recombination rates due to substantial splitting of the light- and heavy-hole bands and increased electron effective masses.

Early attempts to realize SLs with properties suitable for IR detection were unsuccessful, largely because of the difficulties associated with epitaxial deposition of HgTe/CdTe SLs. More recently, significant interest has been shown in multiple quantum well (MQW) AlGaAs/GaAs photoconductors. However, these detectors are extrinsic in nature, and have been predicted to be limited to performance inferior to that of intrinsic HgCdTe detectors [1–6]. On account of this—in addition to the use of intersubband absorption [7–15] and absorption in doping SL [16]—two additional intrinsic intersubband transitions are utilized to directly shift bandgaps into the IR spectral range:

- SL quantum confinement without strain: HgTe/HgCdTe,
- strained type II SLs: InAs/GaSb and InAs/InAsSb.

Figure 20.1 shows schematically the bandgap diagrams of three basic types of SLs used in IR detector fabrications.

Type I SLs consist of alternating thin wider-bandgap layers of AlGaAs and GaAs (see Chapter 19). Their bandgaps are approximately aligned—the valence band (VB) (with symmetry of Γ_8) of one does not overlap with the conduction band (CB) (with symmetry of Γ_6) of the other. Various forms exist, but generally the device is a majority photoconductor with IR absorption achieved by the transitions between the energy levels induced in the CB by dimensional quantization. The AlGaAs layers are very thick barriers that inhibit excess current, such as tunneling through the SL. Absorption coefficients of quantum well infrared photodetectors (QWIPs) are typically very small, and ingenious tricks must be employed to couple efficiently incident normal radiation into the structure due to selection rules for the optical transitions. The advantages of AlGaAs/GaAs QWIP architecture are: foundry material technology, design complexity capability, and low $1/f$ noise. The disadvantages are: high dark current, low quantum efficiency, and low operating temperatures.

Type II SL, representative of which is InAs/GaSb SL, is similar to type I with the exception of the overlapping conduction and VBs in adjacent bands. It utilizes the quantized levels associated with the CB of one layer and the VB of the adjacent layer. The electron and hole levels are separated in real space and transitions only occur in spatial regions in which the wave functions of the carriers overlap. To provide a suitable absorption, extremely thin layers are used. The enhancement of absorption can be achieved by the additional introduction of lattice misfit and strain between alternating layers.

In a type III SL, the alternating layers are of different conduction and VB symmetries. The architecture is essentially that of the type I, except for the use of a semimetal instead of a semiconductor alternated with a semiconductor barrier layer. However, in this case, the thickness of the semimetal layer determines a system of 2D quantized levels in both the conduction and VBs. Conduction of electrons and holes occurs via tunneling through the thin barrier layers of the SL.

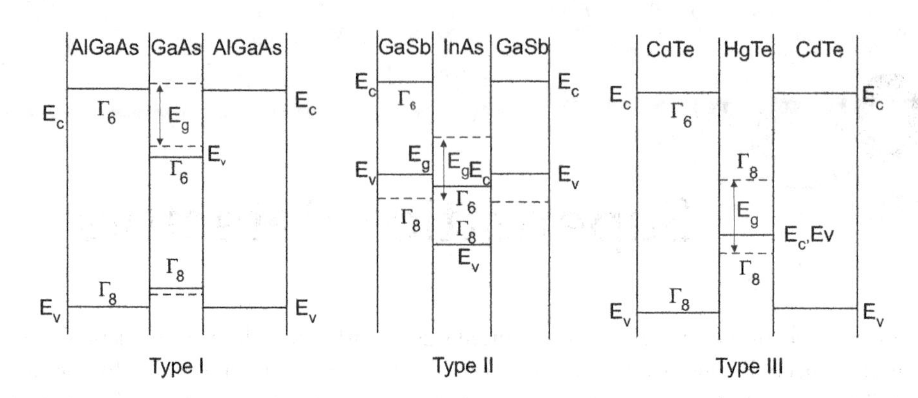

Figure 20.1 Bandgap diagrams of three basic types of SLs designated for IR detector applications. (Adapted after Ref. [15])

Type II and type III SLs are essentially minority carrier intrinsic semiconductor materials. Their absorption coefficients of IR radiation are similar to direct-bandgap alloys and the effective masses are larger than those associated with a direct-bandgap alloy of the same bandgap.

Summarizing, the advantages of type II and type III SLs are: direct-bandgap absorption, large effective masses, and reduced Auger generation (especially to type II InAs/GaSb SLs due to space-charge separation). The disadvantages of type III are: surface passivation and layer interdiffusion at typical processing temperatures. Also, for type II SLs, the surface passivation is a serious issue. However, the main drawback of type II SLs is short Shockley-Read (SR) lifetimes.

20.1 HgTe/HgCdTe SUPERLATTICES

The HgTe/CdTe SL system was the first from a new class of quantum-size structures for IR photoelectronics, which was proposed as a promising new alternate structure for the construction of long-wave (LW) IR detectors to replace those of HgCdTe alloys [17]. Since that time significant theoretical and experimental attention has been given to the study of this new SL system [18–25]. To date, however, attempts to realize HgTe/CdTe SLs with properties suitable for IR detectors with parameters comparable with HgCdTe alloy photodetectors have been unsuccessful in spite of the considerable amount of fundamental research in this field. It seems to be determined by the interface instabilities of SLs due to weak Hg chemical bonding in the material. Significant intermixing even at the very low temperatures (185°C) used in the HgTe/CdTe SL preparation have serious implications on device performance. An interdiffusion coefficient of 3.1×10^{-18}cm^{2}s^{-1} has been found at 185°C. Appreciable intermixing of the HgTe and CdTe layers at temperatures as low as 110°C have been observed [26], which prevents the realization of a low-dimensional solid system in a stable form. The situation is worse due to certain aspects of device processing, such as impurity activation, native defect reduction, and surface passivation. On account of this, the topic concerning HgTe/CdTe SLs will be treated shortly and only more recently published data will be included.

20.1.1 MATERIAL PROPERTIES

The HgTe/CdTe SL appears to belong to type III SLs (see Figures 19.8d and 20.1). This is due to the inverted band structure (Γ_6 and Γ_8) in a zero-gap semiconductor HgTe as compared to those of CdTe, which is a normal semiconductor. Thus, the Γ_8 light-hole band in CdTe becomes the CB in HgTe. When bulk states made of atomic orbitals of the same symmetry but with effective masses of opposite signs are used, the matching up of bulk states belonging to these bands has, as a consequence, the existence of a quasi-interface state that could contribute significantly to optical and transport properties.

As was shown in many theoretical calculations (see, e.g., [23,25]), the VB discontinuity between HgTe and CdTe has a crucial influence on the HgTe/CdTe SL band structure. In earlier publications, the assumptions of significant advantages of these SLs for LWIR and very LWIR detectors were based on a small valence band offset [17], $\Delta E_v \geq 40$ meV, which seemed to be in agreement with the common anion rule of lattice-matched heterojunction interfaces. Of important practical interest were theoretical predictions of a

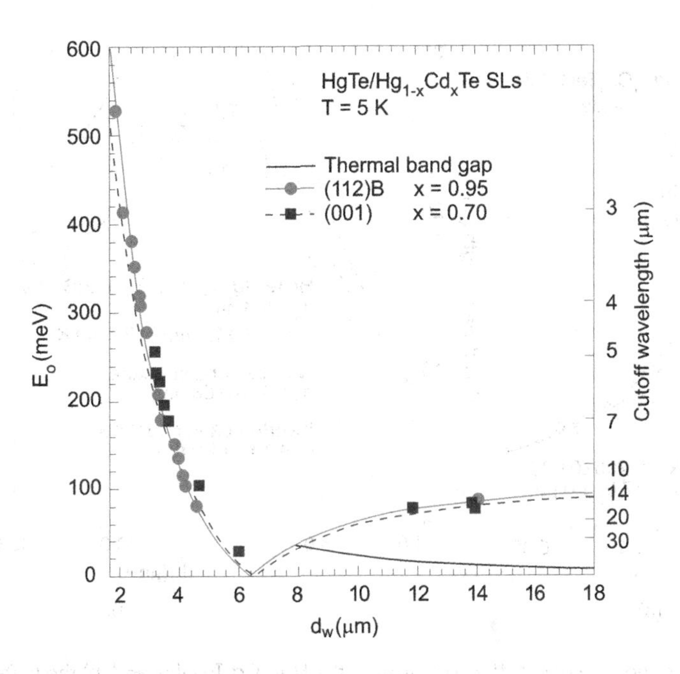

Figure 20.2 Experimental (symbols) and theoretical calculations (solid and dashed lines) bandgap: H1→E1 inter-subband transition of HgTe-based MQWs and the thermal bandgap (dotted line) in the inverted band structure regime. (From Ref. [29])

better control of the bandgap in HgTe/CdTe SL compared to that in HgCdTe alloy and a sufficient reduction of the tunneling currents due to larger effective masses in the SL growth direction. More recently, it has been recognized that numerous aspects of the HgTe/CdTe and related SL properties can only be explained in terms of large valence band offset $\Delta E_v \geq 350$ meV [25]. According to Becker et al. [27], the VB offset between HgTe and CdTe

$$\Delta E_v = \Delta E_{vo} + \frac{d(\Delta E_c)}{dT}T, \tag{20.1}$$

where $\Delta E_{vo} = 570$ meV and $d(\Delta E_c)/dT = -0.40$ meV K^{-1} [27] and ΔE_v is assumed to vary linearly with x in $Hg_{1-x}Cd_xTe$ [28].

The experimental and calculated optical bandgaps of $HgTe/Hg_{1-x}Cd_xTe$ SLs for the (001) and (112) B orientations are shown in Figure 20.2 [29]. If a HgTe thickness, d_w, is less than about 6.2 nm, then the band structure is normal and if $d_w > 6.2$ nm then the band structure is inverted. The SL structure is primarily determined by that of the quantum well (QW) and is influenced to a much lesser degree by the band structure of the barrier ($Hg_{1-x}Cd_xTe$). Only the energy gap changes significantly with alloy composition and temperature. Figure 20.2 shows that experimental and theoretical data are in excellent agreement.

As is mentioned above, the advantages of HgTe/HgCdTe SLs were judged to be insignificant, however, in the very LWIR band they gain sufficient importance to become one of the best detector materials. The required precision of the desired bandgap or cutoff wavelength is less for a SL in comparison with bulk material. It is demonstrated in Figure 20.3. If $\lambda_c = 17.0 \pm 1.0$ μm at 40 K is wished, the required precision for the alloy is ±1.0% and that of the SL is ±2% and ±8% for the SL with a normal and inverted band structure, respectively.

If we take the electron and hole dispersion relations in HgTe/HgCdTe QW to have the approximate form given by Equation 19.6, the density of state has the well-known staircase dependence. The j-th steep occurs at the photon energy:

$$h\nu \approx E_g^w + \frac{\hbar^2 j^2 \pi^2}{2d_w^2}\left(\frac{1}{m_e^w} + \frac{1}{m_h^w}\right), \tag{20.2}$$

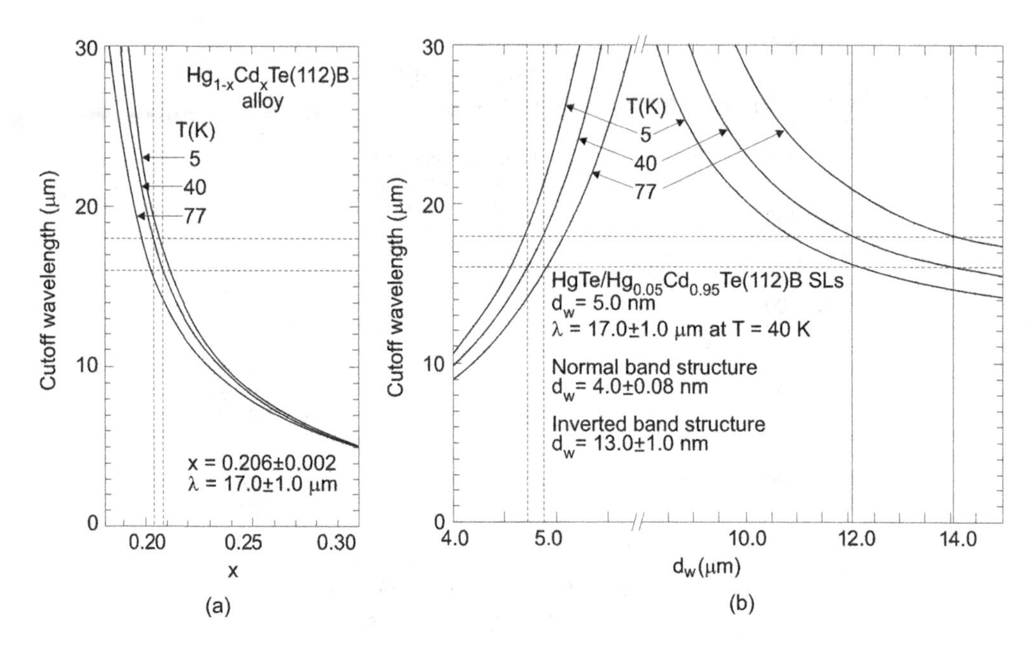

Figure 20.3 (a) The dependence of cutoff wavelength of the $Hg_{1-x}Cd_xTe$ alloy and (b) the $HgTe/Hg_{1-x}Cd_xTe$ SL with normal and inverted band structures at 5 K, 40 K, and 77 K. The precision in x and d_w for these three cases required to produce materials with cutoff wavelengths of 17.0±1.0 μm at 40 K is indicated. (From Ref. [29])

where E_g^w is the bulk energy gap of the well material, m_e^w and m_h^w are the effective masses of electron and hole in the bulk well material before confinement is imposed, and the kinetic energy due to motion in the x, y plane unaffected by the confinement along z. The above equation contrasts with the more gradual $\left(h\nu - E_g^w\right)^{1/2}$ dependence for the density of states in a 3D bulk material such as HgCdTe.

Figure 20.4 shows, for example, the experimental and theoretical absorption coefficients for a SL and an alloy material with similar bandgaps near 60 meV (λ_c=20 μm). Agreement between experiment and theory is very good. Furthermore, the absorption edge for the SL is much steeper and therefore the absorption of the SL is up to a factor of five larger. The large absorption coefficient in a SL represents a distinct advantage over the HgCdTe bulk material, in which α tends to be 1,000–2,000 cm^{-1} at the same proximity to the band edge. This means that the active detector layers can be significantly thinner in the equivalent alloy, of the order of a few microns. This advantage is more pronounced for the LW region, but even for a middle wavelength (MW) band the absorption is appreciably larger. It appears also that a Burstein-Moss effect produces a negligible shift of the absorption coefficient in comparison with the bulk material. This is due to the flatter dispersion in a direction perpendicular to the 2D plane and the corresponding larger density of states.

One of the main advantages concerning HgTe/HgCdTe SLs for IR photoelectronics lies in a variation of the bandgap not with chemical composition in ternary or quaternary alloys, but in a variation of layer thickness of much more stable binary compounds. As the bandgap, the growth-direction effective mass values of electrons and holes, and thus the carrier mobilities, can be tuned over a wide range by varying the barrier thickness (also to be taken into account is a mass-broadening effect, which causes the in-plane effective mass to depend strongly on the growth-direction wave vector) [25]. Whereas the mass in the alloy is fixed by its proportionality to the energy gap; in the SL, the effective mass can be varied independently of E_g simply by adjusting the barrier thickness. A large growth-direction effective mass is desired because the tunneling current scales exponentially with $m^{1/2}$ (see Section 12.2.2). But to achieve high quantum efficiency HgTe/HgCdTe, SLs should be grown with thin barriers (less than 30 Å [30]) to be beyond the region of hopping mobility from one well to the next where the miniband conduction model breaks down.

The electronic band of HgTe/HgCdTe SL can be engineered to suppress Auger recombination relative to that in comparable bulk detectors. SRH lifetimes of up to 20 μs have been reported for HgTe/CdTe SLs grown with MBE [31]. Calculated carrier lifetimes at 80 K for hole concentration of 5×10^{15} cm^{-3} are shown in Figure 20.5 as a solid line [32]. The experimental results for photoconductive decay are in reasonable

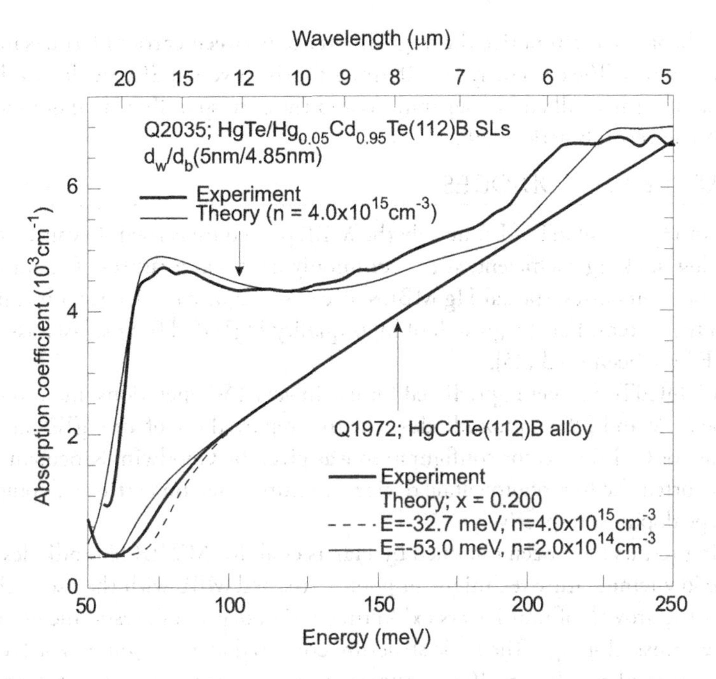

Figure 20.4 The experimental (thick lines) and theoretical (thin lines) absorption coefficients for $HgTe/Hg_{1-x}Cd_xTe$ SL and a $Hg_{1-x}Cd_xTe$ alloy with a bandgap near 60 meV ($\lambda_c \approx 20\,\mu m$) at 40 K. The two theoretical spectra for the alloy are for two different electron concentrations and the corresponding Fermi energies. (From Ref. [29])

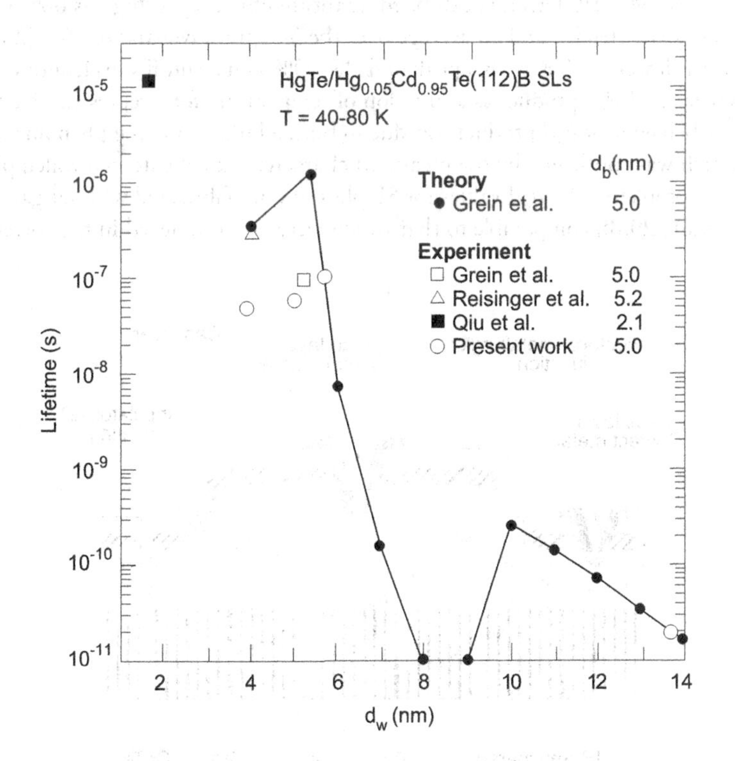

Figure 20.5 The experimental (symbols) and theoretical (solid line) lifetimes for $HgTe/Hg_{1-x}Cd_xTe$ SLs at temperatures between 40 K and 80 K. (From Ref. [29]) The theoretical line is calculated for $HgTe/Hg_{0.05}Cd_{0.95}Te$ SL at 40 K with $d_w = d_b = 5$ nm and an acceptor concentration of $5 \times 10^{15} cm^{-3}$. (From Ref. [32])

agreement with the theory. We can notice the large difference between carrier lifetimes in normal and inverted SL band structures. The extremely fast lifetimes for the inverted SLs are due to the presence of many valence subbands with small energy separations and the corresponding large number of occupied states for efficient Auger recombination [29].

20.1.2 SUPERLATTICE PHOTODIODES

For the fabrication of HgTe/HgCdTe SLs, mainly the MBE procedure is used. Because Hg has both a high vapor pressure and low sticking coefficient, at the commonly used temperatures of about 180°C or less to minimize interdiffusion processes, special Hg MBE sources are required to let a significant amount of Hg vapor pass through the system. For the growth of high-quality HgTe/CdTe SLs, laser-assisted MBE and photo-assisted MBE have been used [25].

In spite of HgTe/HgCdTe SLs being predicted for use in very LW operations, main research efforts were concentrated on the MW and LWIR photodiodes. The first reported use of an HgTe/CdTe SL in a metal insulator semiconductor (MIS) detector configuration was given by Goodwin, Kinch, and Koestner [33]. Wroge et al. [34] reported the first photovoltaic device structure based on extrinsic doping using In and Ag for n-type and p-type dopings, respectively.

More encouraging results have been obtained by Harris et al. for MWIR photodiodes [35]. It appears that combining the low-temperature technique of photon-assisted MBE with the use of the (211)B orientation has allowed for the growth of multilayers exhibiting a high degree of crystalline perfection and in situ n-type and p-type extrinsic dopings. The basic structure consisted of a 3–5 µm n-type layer followed by a 1–2-µm-thick p-type cap layer. Figure 20.6 illustrates a schematic cross section of the mesa structure. To ensure p-type and n-type behaviors in the layers, As at the 10^{17} cm^{-3} and In at the 10^{16} cm^{-3} levels were used, respectively (typical doping levels in an optimized device structure would be about a factor of ten less than this).

Both MWIR and LWIR SL photodiodes exhibiting high quantum efficiency and uniform response have been fabricated. The MWIR detectors exhibited quantum efficiency as high as 66% (at 140 K; see Figure 20.7) at the peak wavelength and an average over the 3–5 µm waveband of 55% [23]. Measured quantum efficiency was lower at 78 K with a peak of 45%–50% and a cutoff wavelength of 4.9 µm. Figure 20.8 shows measured R_oA product as a function of temperature for a representative SL photodiode. The low-temperature behavior of R_oA product was due to both a bulk tunneling phenomenon as well as surface currents, which was confirmed by the measured characteristics of gate-controlled photodiodes. Even with passivation problems, the R_oA values for SL photodiodes fabricated without gates were typically 5×10^5 Ωcm^2 (see Figure 20.8), comparable to that which were often achieved in the corresponding alloy.

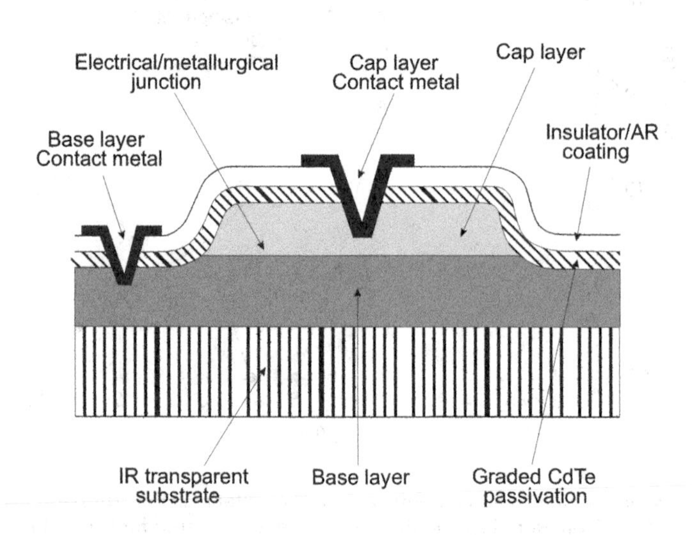

Figure 20.6 Schematic of a processed SL mesa structure. (From Ref. [35])

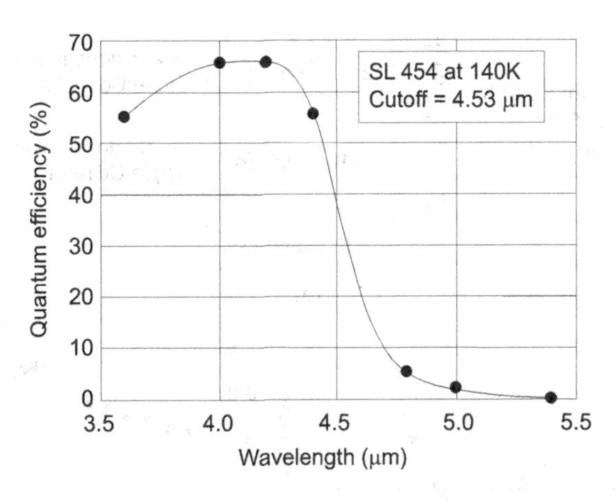

Figure 20.7 Representative spectral response of an HgTe/CdTe SL photodiode with λ_c=4.53 μm at 140 K. The peak response corresponds to 66% quantum efficiency. (From Ref. [23])

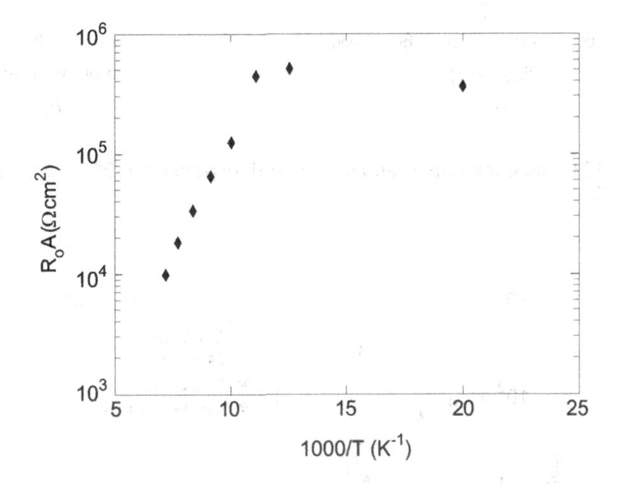

Figure 20.8 Measured R_oA product as a function of temperature for a representative SL photodiode. The low-temperature characteristics are indicative of tunneling processes limiting R_oA (nonoptimal surface passivation). (From Ref. [23])

The potential advantages of the HgTe/CdTe material system have been also demonstrated in the LWIR spectral range. Figure 20.9 shows doping configurations, spectral response, and I-V characteristics for a typical p-on-n LWIR SL photodiode with a cutoff wavelength of 9.0 μm [25]. The measured quantum efficiency was 62% and the R_oA was 60 Ωcm^2. Preliminary results confirm that the growth quality of SL photodiodes has become sufficiently advanced so that a high-performance SL photodiode technology appears feasible. Figure 20.10 shows values of R_oA product at 80 K as a function of wavelength for four HgTe/CdTe SL photodiodes, compared with a number of production n-on-p photodiodes employing HgCdTe bulk materials (Bridgman and traveling heater methods), illustrating comparable performance.

At present, the performance of HgTe/HgCdTe SL photodiodes is inferior in comparison with high-quality HgCdTe photodiodes with a comparable cutoff wavelength. As a result, a lack of research funding has led to an industry-wide suspension of further efforts to develop HgTe/HgCdTe SL IR detectors.

20.2 TYPE II SUPERLATTICES

In case when the constituent materials are rather closely lattice matched, it is possible to design the electronic type II SL or MQW band structure by controlling the layer thickness and the height of the barriers. But it is also possible to grow high-quality III-V type II SL (T2SL) devices with reduced

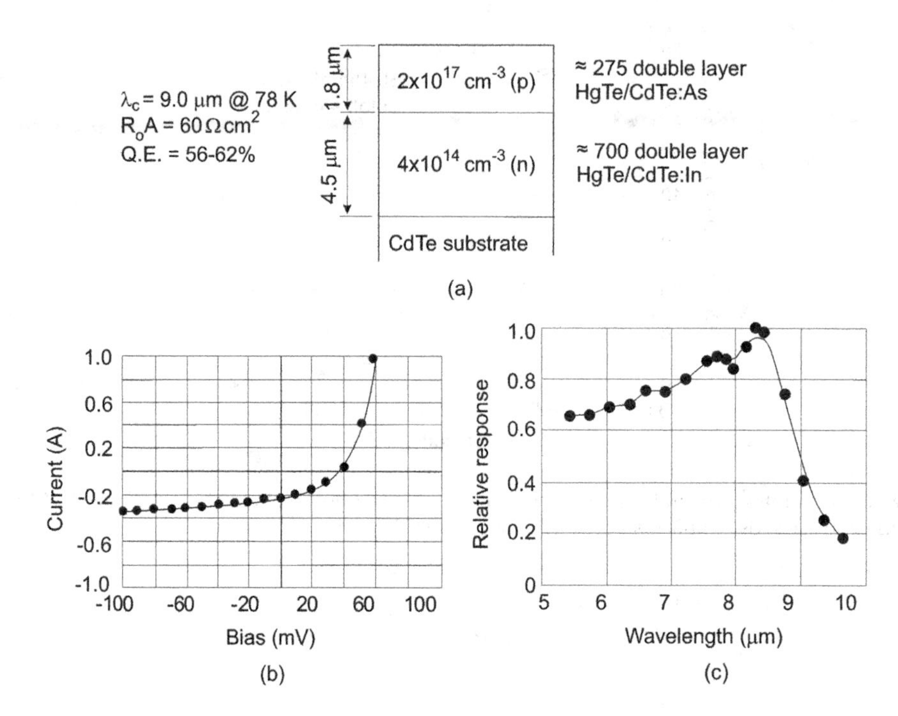

Figure 20.9 (a) Structure, (b) I-V characteristics, and (c) spectral response for (211) HgTe/$Hg_{0.1}Cd_{0.9}$Te SL LWIR photodiode. (From Ref. [25])

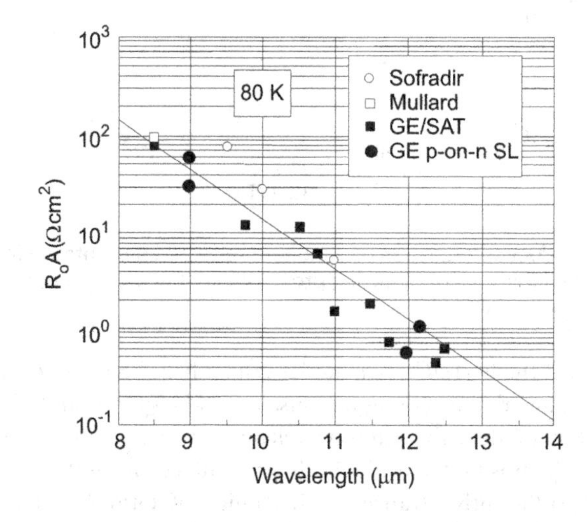

Figure 20.10 R_oA product versus wavelength for (211) HgTe/$Hg_{0.1}Cd_{0.9}$Te SL photodiodes (dark circles). Also shown for comparison are analogous results for production of n-on-p HgCdTe photodiodes. (From Ref. [25])

conduction-valence bandgap for IR detector applications in which a QW layer can be controlled on the atomic scale too, but with a significantly different lattice constant of the well material as compared to that of the barrier material, which gives additional opportunities to design the electronic band structure with deformation potential effects [36–38]. A schematic of the typical strained-layer SL (SLS) structure is given in Figure 20.11a. The thin SLS layers are alternatively in compression and tension so that the in-plane lattice constants of the individual strained layers are equal. The entire lattice mismatch is accommodated by layer strains without the generation of misfit dislocations if the individual layers are below the critical thickness for dislocation generation. Since misfit defects are not generated in SLS structures, the SLS layer

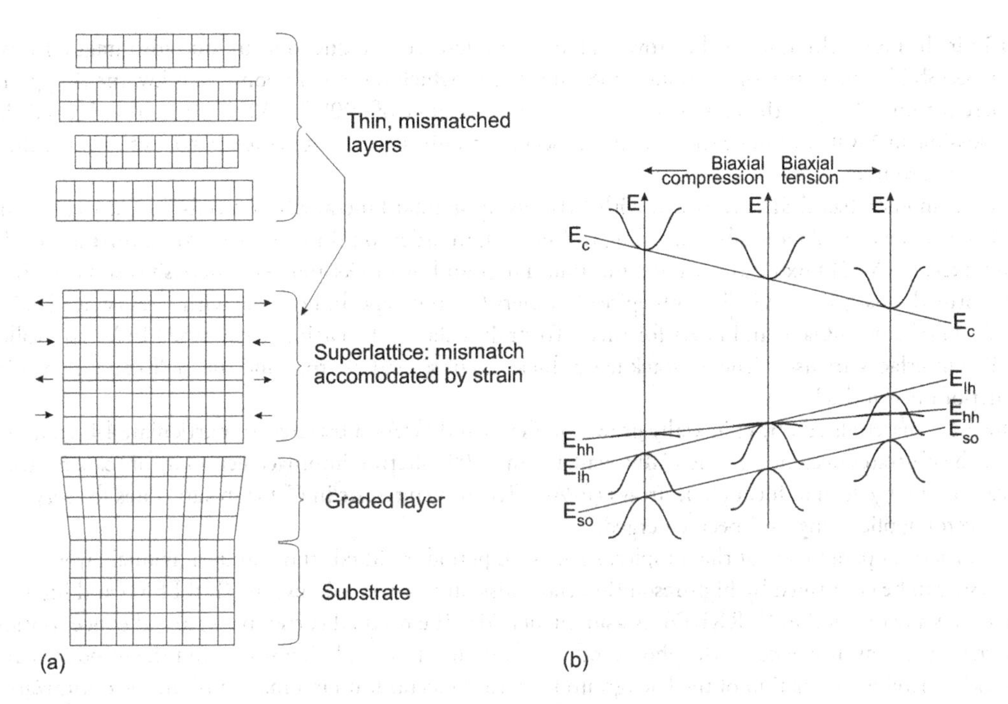

Figure 20.11 Strained-layer SL: (a) Schematic of the fabrication and (b) the biaxial strain-induced shifts of the out-of-growth-plane energies of the VB and CB.

can be of sufficiently high crystalline quality for a variety of scientific and device applications. Strain can change the bandgaps of the constituents and split the degeneracy of heavy- and light-hole bands in such a way that these changes and band splitting can lead not only to energy level reversals in the SL electronic band structure but also to appreciable suppression of the recombination rates of photoexcited carriers [39]. In such systems, the conduction-valence bandgap can be made much smaller than that in any III-V alloy bulk crystals [40]. For example, Figure 20.11b shows the effect of biaxial strain in a tetrahedrally-coordinated, direct-bandgap semiconductor. The out-of-plane conduction, light-hole, heavy-hole, and split-off band energies are shown for different biaxial strain components. As a result, for example, in InAsSb/InSb SLS system, the bandgap of small-bandgap component, InAsSb, is decreased, and the bandgap of the InSb layers is increased. Therefore, from the effects of strain alone, InAsSb can potentially absorb at longer wavelengths than the InAsSb alloys.

Type II band alignment and some of its interesting physical behaviors were originally suggested by Sai-Halasz, Tsu, and Esaki in 1977 [41]. Soon after that they reported the optical absorption by type II SLs [42] and their semimetal behavior [43], and the potential use of this system in IR optoelectronics was recognized.

Progress in the growth of InAs/InAsSb SLSs by both MBE and MOCVD has been observed since 1984, when Osbourn in his theoretical work showed that strain effects were sufficient to achieve wavelength cutoffs above 12 μm in 77 K [36]. The first decade efforts in the development of epitaxial layers are presented in Rogalski's monograph [44]. Difficulties have been encountered in finding the proper growth conditions, especially for SLSs in the middle region of composition [45–47]. This ternary alloy tends to be unstable at low temperatures, exhibiting miscibility gaps, and this can generate phase separation or clustering. Control of alloy composition has been problematic, especially for MBE. Due to the spontaneous nature of CuPt-orderings, which result in substantial bandgap shrinkage, it is difficult to accurately and reproducibly control the desired bandgap for optoelectronic device applications [48].

Although InAsSb T2SL structures were successfully demonstrated in the 1990s [49], they were set aside as potential IR detector materials in favor of the InAs/InGaSb SL. Recently, new impacts in their development are being observed [50] due to the limits of InAs/InGaSb detector performance (short carrier lifetime and reduced quantum efficiency).

MBE is the best technique for the growth of antimonide-based SL due to its unique advantages. In case of InAs/GaSb SL, these structures contain InSb interfaces, which have weak bonds and low melting points. For these reasons, the growth is restricted to a temperature range of 390°C–450°C. This growth condition is not possible in MOCVD since the substrate susceptor requires much higher temperatures to crack the metalorganic sources.

The antimonide-based SLs are grown with MBE using standard metal effusion cells for Ga and In, and valved cracker cells for As and Sb. According to Ref. [51], in order to minimize cross contamination of the anion fluxes, the V/III flux ratio is set at a minimum of 3 for both InAs and GaSb depositions. Growths are performed mostly on 2" GaSb (100) epiready wafers (both n-type and p-type) with rates typically about 1 Å/s for GaSb and InGaSb and lower for InAs. To strain balance the lattice-mismatched InAs, controlled InSb-like interfaces are used. The SL stack has a thickness of several microns and the GaSb buffer layer is an order of micron thick.

The 6.1 Å materials can be epitaxially grown on GaSb and GaAs substrates. In particular, 4-inch diameter GaSb substrates became commercially available in 2009, offering improved economy of scale for the fabrication of large format focal plane arrays (FPAs). Recently, interests in 6" GaSb substrates for very large area detector applications have been emerged.

The structural parameters of the samples, such as SL period, residual strain, and individual layer thickness, can be confirmed by high-resolution transmission electron microscopy (HRTEM) and high-resolution X-ray diffraction (HRXRD) measurements [51]. The residual carrier background concentration of SL structures has influence on the photodiode performance (the depletion width and the minority carrier response). Therefore, reduction of the background carrier concentration is a major task in the optimization of the growth conditions.

InAs/InAsSb SLs have been grown both by MBE and MOCVD on GaSb substrates. The MOCVD growth of these SLs is more suitable than that for the InAs/GaSb ones. It appears that the InAsSb-based SLs is strain balanced simply by the layer thicknesses without any interfacial control [50]. The substrate growth temperature is similar to that used in the fabrication of InAs/GaSb SLs.

20.2.1 PHYSICAL PROPERTIES

6.1 Å III–V semiconductor family plays a decisive role in offering new concepts of high performance IR detectors connected with high design flexibility, direct energy gaps, and strong optical absorption. This family is formed by three semiconductors of an approximately matched lattice constant of around 6.1 Å: InAs, GaSb, and AlSb forms with the low-temperature energy gaps ranging from 0.417 eV (InAs) to 1.696 eV (AlSb) [52]. Like other semiconductor alloys, they are of interest principally for their heterostructures, especially the ones combining InAs with the two antimonides (GaSb and AlSb) and their alloys. This combination offers a band alignment that is drastically different from that of the more widely studied AlGaAs system, and it is the flexibility in the band alignment that forms one of the principal reasons for interest in the 6.1 Å family. The most exotic band alignment is that of InAs/GaSb heterojunctions, which is identified as the broken gap alignment. At the interface, the bottom of the CB of InAs is located below the top of the VB of GaSb by about 150 meV. In such a heterostructure, with partial overlapping of the InAs CB with the GaSb-rich solid solution VB, electrons and holes are spatially separated and localized in self-consistent QWs formed on both sides of the heterointerface. This leads to unusual tunneling-assisted radiative recombination transitions and novel transport properties. As illustrated in Figure 20.12, with the availability of type I (nested or straddling), type II-staggered, and type II-broken gap (misaligned) band offsets between the GaSb/AlSb, InAs/AlSb, and InAs/GaSb material pairs, respectively, there is considerable flexibility in forming a rich variety of alloys and SLs.

The band alignment of T2SL shown in Figure 20.13a creates a situation in which the energy bandgap of the SL can be adjusted to form either a semimetal (for wide InAs and GaInSb layers) or a narrow-bandgap (for narrow layers) semiconductor material. The resulting energy gaps depend upon the layer thicknesses and interface compositions. In the reciprocal **k**-space, the SL is a direct-bandgap material that enables optical coupling as shown in Figure 20.13b. Unlike AlGaAs/GaAs SL, the electrons in an InAs/GaSb T2SL are confined in InAs, while the holes are confined in GaSb. Since the layers are thin, the overlap of electron and hole wave functions in InAs and GaSb, respectively, forms electron and hole minibands. The bandgap

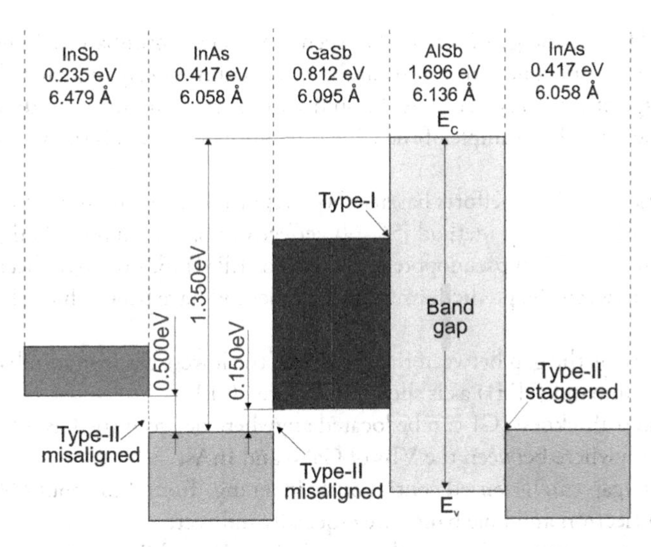

Figure 20.12 Schematic illustration of the low-temperature energy band alignment in the nearly 6.1 Å lattice-matched InAs/GaSb/AlSb material system. Three types of band alignment are available in this material system: Type I (nested) band alignment between GaSb and AlSb, type II-staggered alignment between InAs and AlSb, and type II-misaligned (or broken gap) alignment between InAs and GaSb. The approximate values of band offsets are marked.

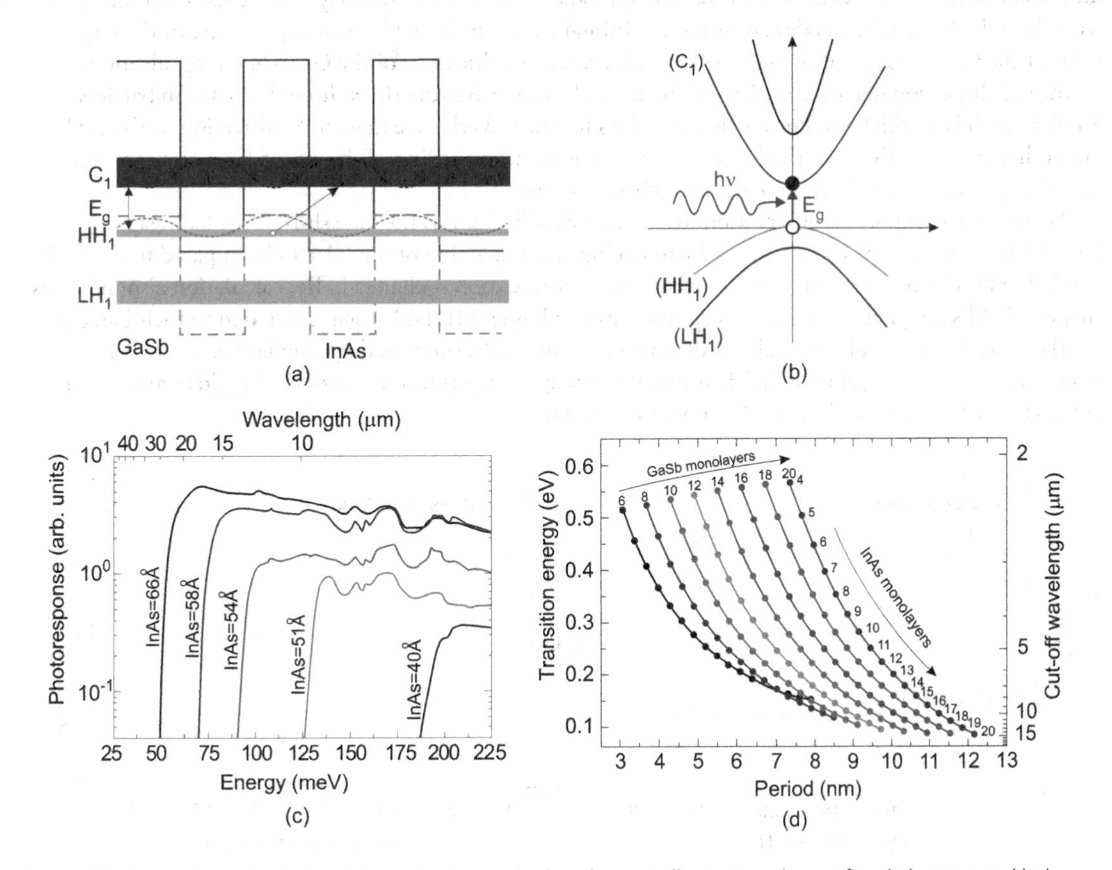

Figure 20.13 InAs/InGaSb strained-layer SL: (a) band edge diagram illustrating the confined electron and hole minibands, which form the energy bandgap; (b) band structure with direct bandgap and absorption process in k-space; (c) experimental data of type II InAs/GaSb SLS cutoff wavelengths change with the InAs thickness while GaSb is fixed at 40 Å; and (d) change in cutoff wavelength with change of InAs and GaSb MLs for InAs/GaSb SLSs.

of the SL is determined by the energy difference between the electron miniband C1 and the first heavy-hole state HH1 at the Brillouin zone center and can be varied continuously in a range between 0 and about 400 meV. One advantage of using type II SL is the ability to fix one component of the material and vary the other to tune wavelength. An example of the wide tunability of the SLs is shown in Figures 20.13c and 20.13d [53,54].

Remarkable theoretical modeling efforts has also been reported in calculating T2SL band structure [55]. Several methods such as the **k·p** method [56–60], effective-bond-orbital method [61], empirical tight binding method [53], and empirical pseudopotential method (EPM) [62–65] have been considered and reasonable agreements between the predictions from each method have been achieved. From theoretical modeling it results that:

- the bandgap is defined as the gap between the bottom of the lowest electron miniband (C1) and the top of the highest hole miniband (HH1) as is shown in Figure 20.13a,
- depending on the layer thickness, C1 can be located anywhere between the CBs of InAs and GaSb, while HH1 can be anywhere between the VBs of GaSb and InAs,
- theoretically the bandgap can be varied continuously in a range from 0 to about 400 meV,
- transitions between electron and hole bands are especially indirect,
- the hole effective mass is extremely high in the growth direction, while the electron effective mass is slightly heavier than that of InAs and weakly dependent on the T2SL design, and
- the hole mobility in the growth direction is extremely low.

Figure 20.14 shows the EPM-calculated variation of the C1 and HH1 band positions of $(InAs)_NGaSb_N$ T2SLs, where N is the layer thickness in monolayers (MLs). As the reference level, the InAs CB set at 0 meV is considered. It is clearly shown that the C1 band is more sensitive to layer thickness than the HH1 band. The thickness of the GaSb layers has a minimal effect on the T2SL bandgap because of the large value of the GaSb heavy-hole mass (~0.41 m_o); however, the thickness of the GaSb has a significant effect on the CB dispersion due to tunneling of the InAs electron wave functions through the GaSb barriers. Similar GaSb layer thickness creates similar values for the CB effective mass in SL direction. It should be noted that changing the layer thickness requires a good understanding of the effects from strain on the material quality since InAs is not lattice matched to GaSb.

An additional important observation for the InAs/GaSb T2SL is the blue shift of the bandgap [56,63,64], which can be explained by the much higher broadening of the C1 band compared to the HH1 band. The HH1 band shifts very slowly when the layer thickness is changed. The calculation also indicates that the T2SLs designated for LW absorption exhibit a higher HH1-LH1 gap compared to their bangaps.

The basic properties of artificial InAs/GaSb T2SL materials, supported by simple theoretical considerations are given by Ting et al. [67]. Their properties may be superior to those of the HgCdTe alloys and are completely different from those of the constituent layers.

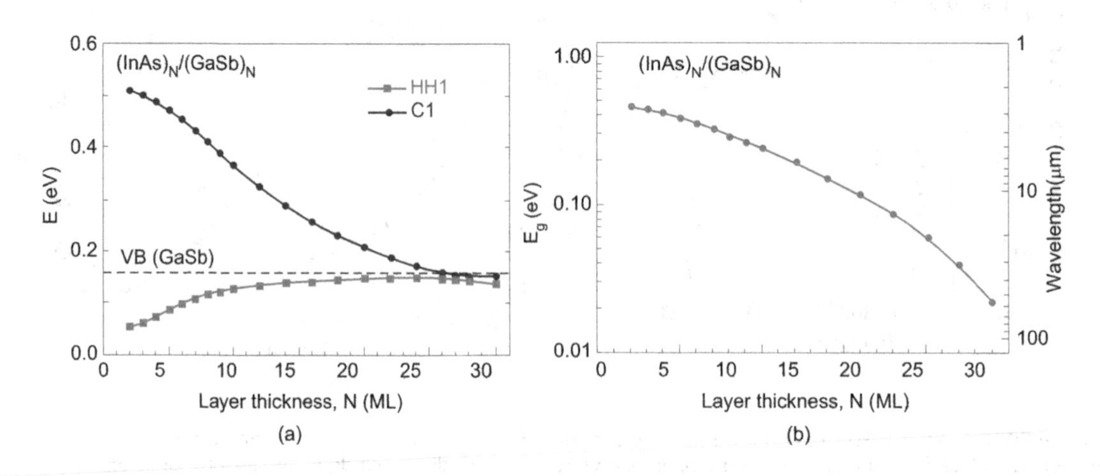

Figure 20.14 (a) Calculated HH1 and C1 band positions for InAs/GaSb SLs with equally thick InAs and GaSb layers. (b) Variation of the bandgap with layer thickness. (Adapted after Ref. [66])

The SL band structure reveals important information about the carrier transport properties. The C1 band shows strong dispersion along both the growth (z) and in-plane direction (x), whereas the HH_1 band is highly anisotropic and appears nearly dispersionless along the growth (transport) direction. The electron effective mass along the growth direction is quite small and even slightly smaller than the in-plaen electron effective mass. The values estimated by Ting et al. [67] for the LWIR SL material (22 ML InAs/6 ML GaSb) are as follows: $m_e^{x*} = 0.023\,m_0$, $m_e^{z*} = 0.022\,m_0$, $m_{hh1}^{x*} = 0.04\,m_0$, and $m_{hh1}^{z*} = 1{,}055\,m_0$. The SL CB structure near the zone center is approximately isotropic in contrary to the highly anisotropic VB structure. For these reasons, we would expect very low hole mobility along the growth direction that is unfavorable for detector design for LWIR FPAs [67]. The estimation of effective masses for MWIR SL material (6 ML InAs/34 ML GaSb) gives [66]: $m_e^{x*} = 0.173\,m_0$, $m_e^{z*} = 0.179\,m_0$, $m_{hh1}^{x*} = 0.0462\,m_0$, and $m_{hh1}^{z*} = 6.8\,m_0$.

The effective masses are not directly dependent on the bandgap energy, as is the case in a bulk semiconductor. The electron effective mass of InAs/GaSb SL is larger compared to $m_e^* = 0.009\,m_0$ in HgCdTe alloy with the same bandgap ($E_g \approx 0.1$ eV). Thus, diode tunneling currents in the SL can be reduced compared to the HgCdTe alloy.

Electronic transport properties of T2SLs are anisotropic. Although in-plane mobilities drop precipitously for thin wells, electron mobilities approaching $10^4 cm^2Vs^{-1}$ have been observed in the InAs/GaSb SLs, with the layers less than 40 Å thick. Among the possible low-temperature scattering mechanisms, two are particularly important for SLs: interface roughness scattering and alloy scattering. Alloy scattering of carriers is shown not to be a factor in electronic transport—it can be concluded that there are no disadvantages in incorporating In into the GaSb layers. Theoretical modeling indicates that the electron mobility perpendicular to the SL layer is nearly equal to the in-plane mobility in low temperature range and decreases strongly with temperature increase (see Figure 20.15) [51]. It should be mentioned that the vertical transport in the photodiode structure has an important influence on the IR detector performance [68].

Since the mobility of holes is much lesser than the electron mobility, hence by keeping the electrons as minority carriers, higher photodetector performance can be expected compared to the performance when holes are minority carriers.

$InAs/InAs_{1-x}Sb_x$ SLs are a viable alternative to the well-studied $InAs/Ga_{1-x}In_xSb$ SLSs; however, the $InAs/InAs_{1-x}Sb_x$ SLs are less studied. Steenbergen et al. [69] have reviewed the band edge alignment models for $InAs\text{-}InAs_{1-x}Sb_x$ system and considered different types of heterojunctions. Figure 20.16 shows three possible band alignments between InAs and $InAs_{1-x}Sb_x$ including two type II band alignments: with the InAs CB higher in energy than the InAsSb CB and with the InAsSb CB higher in energy than the InAs CB.

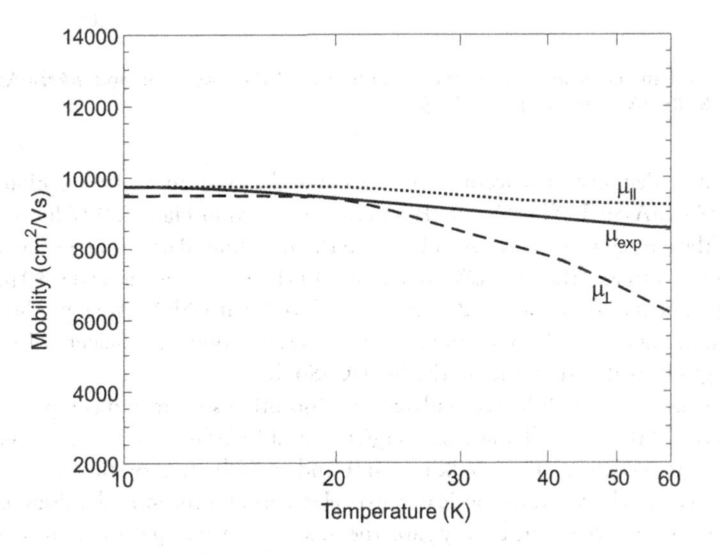

Figure 20.15 Comparison of the calculated horizontal and vertical electron mobilities with the measured horizontal mobility as a function of temperature for the 48.1 Å InAs/20.4 Å GaSb SL. (Adapted after Ref. [51])

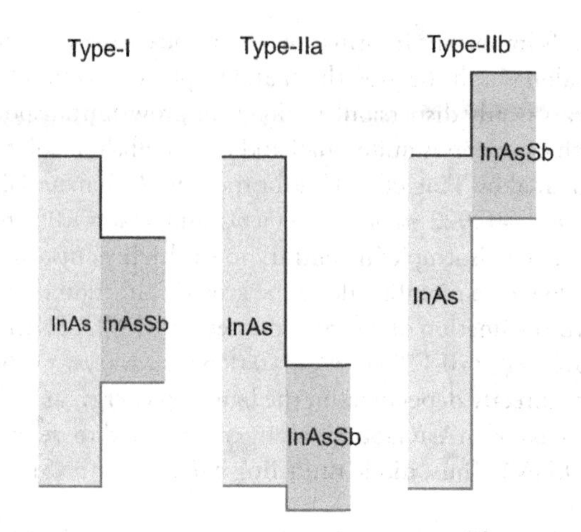

Figure 20.16 Three possible band alignments between InAs and $InAs_{1-x}Sb_x$.

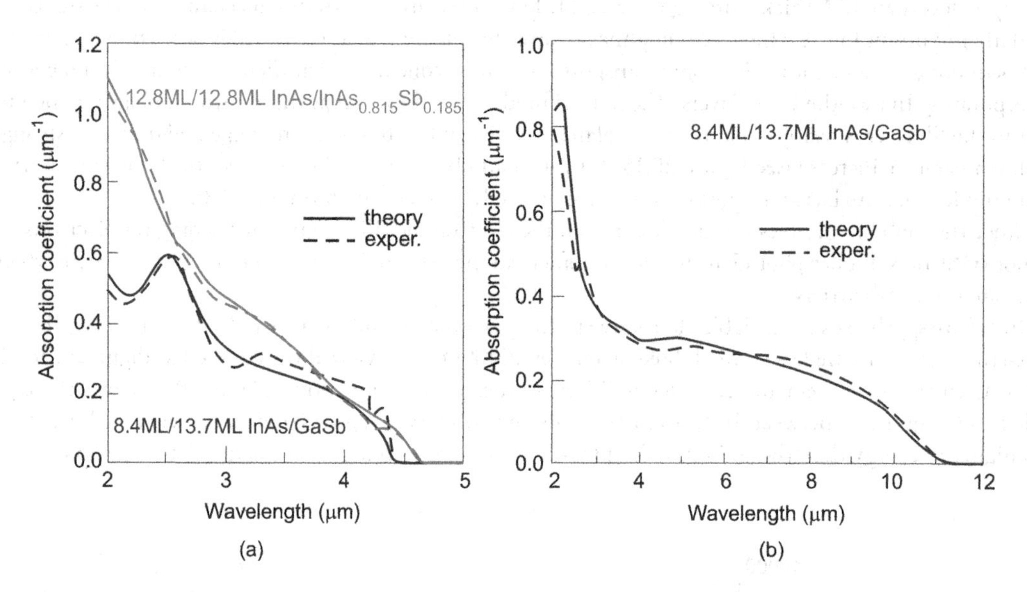

Figure 20.17 Measured and calculated absorption spectra for MWIR InAs/GaSb and InAs/InAsSb T2SLs (a) and LWIR InAs/GaSb T2SL (b). (Adapted after Ref. [59])

Klipstein et al. have demonstrated accurate simulation of the experimental absorption spectra for both InAs/GaSb and InAs/InAsSb T2SLs [57–59]. Results are gathered in Figure 20.17 for both MW and LW material systems. The strong peak below 3 μm is due to the zone boundary HH2 → C1 transitions. Note that this peak is much stronger than the LW zone center LH1 → C1 transition (at ~ 3.4 μm). The main features of the experimental spectrum of 12.8ML/12.8ML InAs/InAsSb SL are reproduced by theoretical calculations shown in Figure 20.17b. We can notice that the absorption coefficient of InAs/InAsSb SL near the cutoff wavelength is weaker than that in the InAs/GaSb SL.

Recently, Vurgaftman et al. [70] have calculated the absorption spectra and compared it to the measured data for LWIR T2SLs and bulk materials (HgCdTe and InAsSb) with the same energy gap—see Figure 20.18. Absorption coefficients for bulk HgCdTe and InAsSb are very similar, which reflects relatively minor differences between the optical matrix elements and the joint densities of states existing in bulk semiconductors with the same energy gap. The SLs with an average lattice constant matched to GaSb have significantly lower absorption. This behavior is explained by the electron-hole overlap in the strain-balanced T2SLs, which occurs primarily in the hole well having a relatively small fraction of the

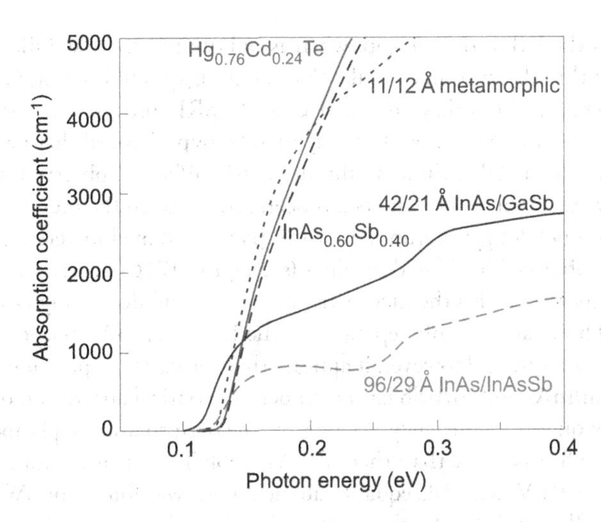

Figure 20.18 Calculated interband absorption coefficients as a function of photon energy at 80 K for bulk $InAs_{0.60}Sb_{0.4}$ and $Hg_{0.76}Cd_{0.24}Te$ and T2SLs: 42 Å InAs/21 Å GaSb, 96 Å InAs/29 Å $InAs_{0.61}Sb_{0.39}$, and 11 Å $InAs_{0.66}Sb_{0.34}$/12 Å $InAs_{0.36}Sb_{0.64}$ metamorphic.

total thickness. It has been shown, however, that the absorption strength in small-period metamorphic $InAs_{1-x}Sb_x/InAs_{1-y}S_y$ SLs is similar to that in the bulk materials.

In comparison with HgCdTe ternary alloys, III-V semiconductor bulk materials exhibit more active SRH centers resulting in lower lifetime. The situation is more complex in case of T2SLs.

In the type II InAs/GaSb SL, separation of electrons (mainly located in the InAs layers) and holes (confined in the GaSb layers) suppresses the Auger recombination mechanisms and thereby enhances carrier lifetime. Optical transitions occur spatially indirectly and thus the optical matrix element for such transitions is relatively small. Theoretical analysis of band-to-band Auger and radiative recombination lifetimes for InAs/GaInSb SLSs showed that in these objects the p-type Auger recombination rates are suppressed by several orders, compared to those of bulk HgCdTe with similar bandgap [71,72], but n-type materials are less advantageous. In p-type SL, Auger rates are suppressed due to lattice-mismatch-induced strain that splits the highest two VBs (the highest light band lies significantly below the heavy-hole band and thus limits the available phase space for Auger transitions). In n-type SL, Auger rates are suppressed by increasing the InGaSb layer widths, thereby flattening the lowest CB and thus limiting the available phase space for Auger transition.

Comparison of theoretically calculated and experimentally observed lifetimes at 77 K for 10 μm InAs/GaInSb SLS and 10 μm HgCdTe is presented in Figure 17.17. The agreement between theory and experiment for carrier densities above 2×10^{17} cm^{-3} is good. The discrepancy between both types of results for lower carrier densities is due to Shockley-Read recombination processes having $\tau \approx 6\times10^{-9}$ second, which has been not taken into account in the calculations. For higher carrier densities, the SL carrier lifetime is two orders of magnitude longer than in HgCdTe, however in a low doping region (below 10^{15} cm^{-3}, necessary in fabricating high performance p-on-n HgCdTe photodiodes), the experimentally measured carrier lifetime in HgCdTe is more than two orders of magnitude longer than in SL.

The promise of Auger suppression is not yet to be observed in practical device material. At present time, the measured carrier lifetime is typically below 100 ns and is limited by SRH mechanism in both MWIR and LWIR compositions. Increase in the minority carrier lifetime to 157 ns has been observed due to the incorporation of InSb interfacial layer in InAs/GaSb T2SL [73]. There is no clear understanding why the minority carrier lifetime varies within the device structure [74].

According to the statistical theory of the SRH process, the SRH rate approaches a maximum as the energy level of the trap center approaches midgap. Analysis of the defect formation energy of native defects is dependent on the location of the Fermi level stabilization energy. In bulk GaSb, the stabilized Fermi level is located near either the VB or the midgap, whereas in bulk InAs, the stabilized Fermi level is located above the CB edge. From this observation it results that the midgap trap levels in GaSb are available for SRH recombination, whereas in InAs they are inactive for the SRH process, suggesting a longer carrier lifetime in bulk InAs than in bulk GaSb material. It may then be hypothesized that native defects associated with GaSb are responsible for the SRH-limited minority carrier lifetimes observed in InAs/GaSb T2SLs.

The origin of the above recombination centers has been attributed to the presence of Ga as the Ga-free InAs/InAsSb SLs possess much longer lifetimes, up to 10 μs for undoped material in the MWIR region [75], comparable to those obtained for HgCdTe alloys (see Figure 17.16). It has been observed that the minority carrier lifetime increases with the increasing Sb content and decreasing layer thickness. In this system, the electrons and holes are spatially separated in the InAs and InAsSb layers so that the recombination process is drastically reduced. However, it appears that the carrier separation (due to electrons and holes spatially separated in InAs and InAsSb layers) reduces the optical absorption of the material, driving toward considerably weak quantum efficiency. As a result, the performance of photodiodes manufactured from Ga-free InAs/InAsSb SLs is worse than that of InAs/GaSb T2SL competitors.

The longest τ_{SR} value for III-V materials equal to about 200 μs was found for SWIR lattice-matched epitaxial InGaAs ternary alloy on InP substrate, with a cutoff wavelength of 1.7 μm [76]. For the most investigated MWIR InSb alloy, the best τ_{SR} value was estimated as ~400 ns for LPE-grown materials [77]. It is interesting to note that bulk-grown InSb is typically inferior and has had a similar SR lifetime issue since its inspection in 1950s. Through the last 50 years the τ_{SR} value has not been improved. Similar values of ~400 ns have been reported for MBE-grown InAsSb bulk alloy and InAs/InAsSb SLs [78]. As mentioned above, τ_{SR} for T2SLs containing Ga is typically an order of magnitude lower.

It is expected that SRH recombination mechanism is presumably associated with some departure of the semiconductors from perfect crystallinity. Since the ionic bond in II-VI alloys is stronger than in the corresponding III-V materials, the electron wave function around the lattice sites has a much stronger confinement rendering the II-VI lattice more immune to the formation of bandgap states due to deviations from crystalline perfection [79].

InAs/GaInSb SLSs and QWs are also employed as the active regions of MWIR lasers operated in the 2.5–6 μm spectral region. Meyer et al. [79,80] have experimentally determined Auger coefficients for the InAs/GaInSb QW W interband cascade laser structures with energy gaps corresponding to 2.5–6.5 μm and have compared their values with that for typical III-V and II-VI type I SLs. The Auger coefficient is defined by the expression $\gamma_3 \equiv 1/\tau_A n^2$. Figure 20.19 summarizes the Auger coefficients at ≈ 300 K for different material systems: a wide variety of type I materials, including bulk and QW III-V semiconductors as

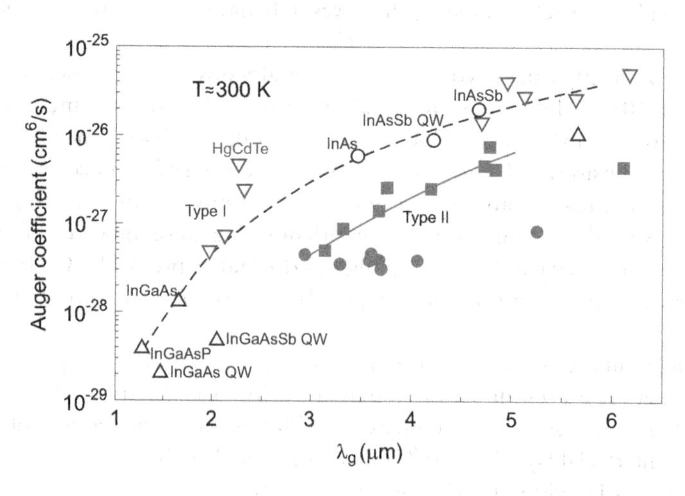

Figure 20.19 Experimental Auger coefficient vs. gap wavelength for type II W interband cascade laser structures (solid circles—Ref. [80]) and various T2SL QWs taken from Ref. [79] (solid squares) along with typical data for a variety of conventional III-V and HgCdTe type I materials (also from Ref. [79]). The solid and dashed lines are guides to the eye.

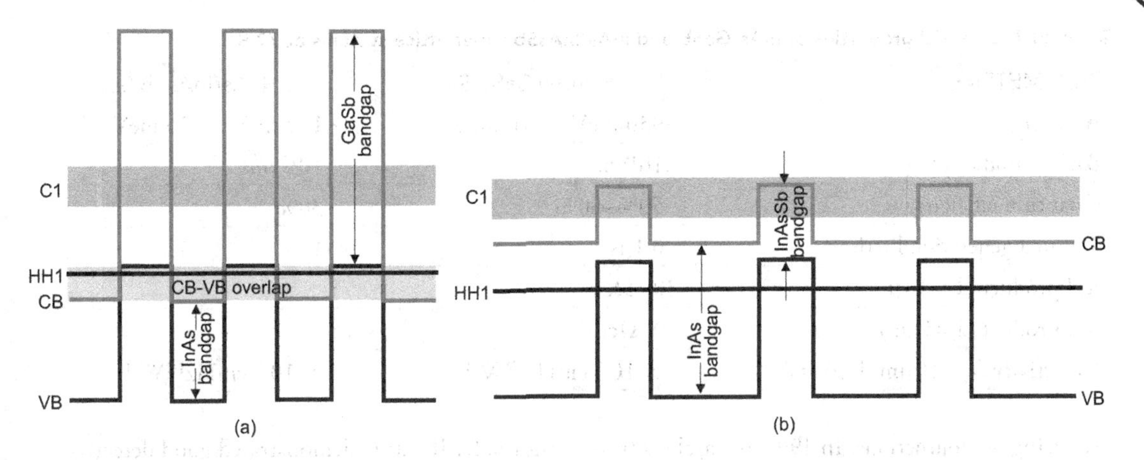

Figure 20.20 Bandgap diagrams for (a) InAs/GaSb and (b) InAs/InAsSb T2SLs. (Adapted after Ref. [81])

well as HgCdTe. We can see that the room temperature Auger coefficients for seven different InAs/GaInSb SLs are found to be nearly an order of magnitude lower than the typical type I results for the same wavelength, indicating a significant suppression of Auger losses in the antimonide T2SLs. Note that all the more recently fabricated interband cascade lasers with $\lambda > 3\,\mu m$ [80] exhibit a significant further reduction in γ_3 in comparison with those described earlier [79]. The data imply that at this temperature the Auger rate is relatively insensitive to details of the band structure. In contrast to MWIR devices, the promise of Auger suppression is yet to be observed in practical LWIR type II device materials.

IR detectors made of InAs/InAsSb SLs on GaSb substrates are in the early stage of development and are less studied than the InAs/GaSb SL counterparts. Due to only two common elements (In and As) in SL layers and a relatively simple interface structure with Sb-changing elements, the InAs/InAsSb SL growth follows with a better controllability and simpler manufacturability.

The interest in InAs/InAsSb SLs stemmed mainly to overcome the carrier lifetime limitations imposed by the GaSb layer in InAs/GaSb SLs. A significantly longer minority carrier lifetime has been obtained in an InAs/InAsSb SL system as compared to that in InAs/GaSb operating at the same wavelength range (for MWIR material at 77 K, ~1 μs and ~100 ns, respectively). Increase in minority carrier lifetimes suggests lower dark currents for InAs/InAsSb SL photodiodes in comparison with InAs/GaSb SL detectors. In practice, however, the dark currents are not low as expected and are higher than that for the InAs/GaSb SL photodiodes.

As was mentioned previously, the InAs/GaSb T2SL band alignment results in an overlap energy (estimated as 140–170 meV) between the CB minimum and the VB maximum of the two materials. For InAs/InAsSb T2SLs, the band offset is defined in terms of the VB offset between InAs and InSb (considered to be about 620 meV). The primary difference in profiles of CB and VB in the InAs/GaSb and InAs/InAsSb T2SLs is illustrated in Figure 20.20.

The band offsets in conduction (ΔE_c) and valence (ΔE_v) bands in the InAsSb SL (ΔE_c ~ 142 meV, ΔE_v ~ 226 meV) are much smaller as compared to that in the InAs/GaSb SL (ΔE_c ~ 930 meV, ΔE_v ~ 510 meV) [25]. This situation suggests higher contribution of tunneling currents in the dark current of InAs/InAsSb SL photodiodes operated at higher temperature. Also, experimental data and theoretical estimation of absorption coefficients in both SLs given by Klipstein et al. indicate a lower absorption coefficient of InAs/InAsSb SL in comparison with the InAs/GaSb one [58]. Table 20.1 compares the essential properties of both the SL systems.

20.3 InAs/GaSb SUPERLATTICE PHOTODIODES

The first InAs/InGaSb SLS photodiodes with photoresponses out to 10.6 μm have been presented by Johnson et al. [82]. The detectors consist of double heterojunctions (DH) of the SL with n-type and p-type GaSb grown on GaSb substrates. The use of heterojunctions in photodiodes offers several advantages

Table 20.1 Essential properties of InAs/GaSb and InAs/InAsSb superlattice systems at 77 K

PARAMETER	InAs/GaSb SL	InAs/InAsSb SL
ΔE_c; ΔE_v	~930 meV; ~510 meV	~142 meV; ~226 meV
Background doping	<10^{15}cm^{-3}	>10^{15}cm^{-3}
Quantum efficiency	~50%–60%	~30%
Thermal generation lifetime	~0.1 μs	~1
R_oA product (λ_c=10 μm)	300 Ωcm^2	?
R_oA product (λ_c=5 μm)	10^7 Ωcm^2	?
Detectivity (λ_c=10 μm, FOV=0)	1×10^{12} cmHz$^{1/2}$W^{-1}	1×10^{11} cmHz$^{1/2}$W^{-1}

over using homojunctions. In 1997, researchers from Franunhofer Institute demonstrated good detectivity (approaching HgCdTe, 8 μm cutoff, 77 K) on individual devices, initiating renewed interest in LWIR detection with type II SLs [83]. While theoretical predictions of detector performance seem to favor the InAs/InGaSb system due to the additional strain provided by the InGaSb layer, the majority of research in the past ten years has been focused on the binary InAs/GaSb system. This is attributed to the complexity of structures grown with large mole fractions of In.

As shown in Figure 20.13, the InAs/GaSb T2SL consists of alternating layers of nanoscale materials. Typically, their thicknesses vary from six to 20 MLs. The overlap of electron (hole) wave functions between adjacent InAs (GaSb) layers results in the formation of electron (hole) minibands in the conduction (valence) band. Optical transitions in two mid-bandgap semiconductors, between holes localized in GaSb layers and electrons confined in InAs layers, are employed in the wide spectral IR detection process between 3 and 30 microns.

Taalat et al. have shown [84] strong influence of the SL composition on both material properties and MWIR photodetector performance, such as the background doping concentration, shape of the spectral responsivity, and the value of dark current. As is shown in Figure 20.21, a bandgap energy of around

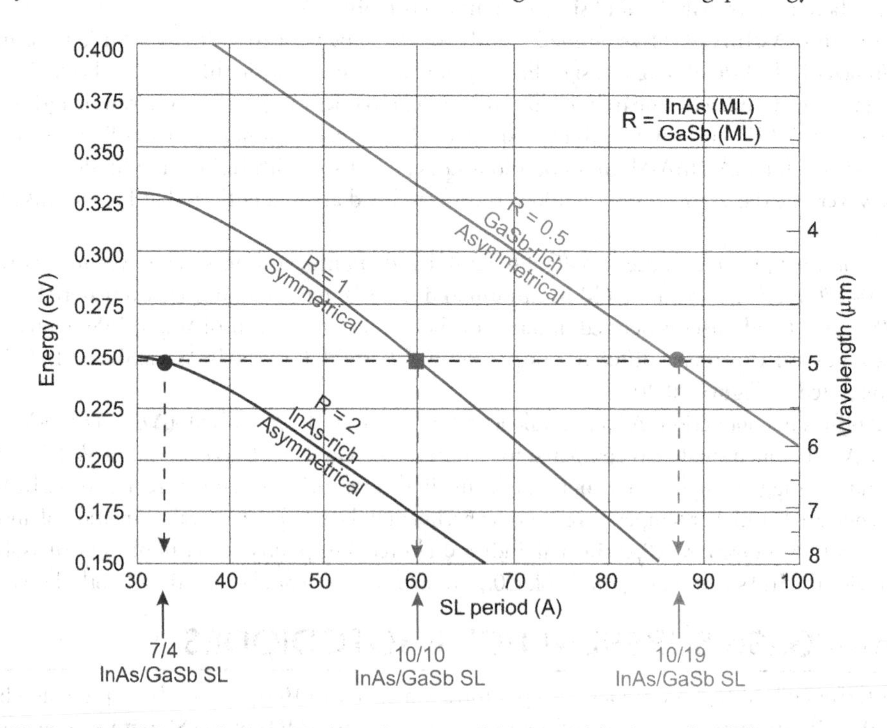

Figure 20.21 Calculated SL bandgap at 77 K as a function of the period thickness for different ratio R=InAs/GaSb. (Adapted after Ref. [84])

248 meV (λ_c=5 μm at 77 K) is achieved at three different SL periods: with a GaSb-rich composition (10 MLs InAs/19 MLs GaSb per period), symmetric with the same InAs and GaSb thicknesses (10 MLs), and InAs-rich composition (7 MLs InAs/4 MLs GaSb).

20.3.1 MWIR PHOTODIODES

At present, SL photodiodes are typically based on p-i-n double heterostructures with an unintentionally doped, intrinsic region between the heavily doped contact portions of the device (see Figure 16.42). The sample presented in this work is the InAs/GaSb T2SL p-i-n detector with the SU-8 passivation fabricated at the Center for High Technology Materials, University of New Mexico, Albuquerque, New Mexico [85]. The device structure was grown on Te-doped epiready (100) GaSb substrates. It consists of 100 periods of ten MLs of InAs:Si ($n=4\times10^{18}\,cm^{-3}$)/ten MLs of GaSb as the bottom contact layer. This was followed by 50 periods of graded, n-doped ten MLs of InAs:Si/ten MLs of GaSb, 350 periods of absorber, 25 periods of ten MLs of InAs:Be ($p=1\times10^{18}\,cm^{-3}$)/ten MLs of GaSb, and finally 17 periods of ten MLs of InAs:Be ($p=4\times10^{18}\,cm^{-3}$)/ten MLs of GaSb which formed a p-type contact layer. 25 periods of the SL structure with graded doping layers were added between the absorber and the contact layers in order to improve the transport of minority carriers in the detector structure. By varying the Be concentration in the InAs layer of the active region, the residually n-type SL is compensated to become slightly p-type. As a consequence, increases in the *RA* product and quantum efficiency were observed in similar structures [86].

Figure 20.22 shows the schematic photodiode structure and its design. A normal-incidence single-pixel mesa photodiode, with a $450\times450\,\mu m^2$ electrical area, was fabricated by photolithography and inductively coupled plasma (ICP) etching. The rest of the fabricated devices were dipped in a phosphoric acid-based solution to remove the native oxide film on the etched mesa sidewalls, then covered with SU-8 (~1.5-μm thickness) to act as the passivation layer. Ohmic contacts were made by depositing Ti/Pt/Au on the contact layers. For details, see Refs. [87] and [88].

The cutoff wavelength is increasing with the increasing temperature—assuming 5.6 μm at 120 K and 6.2 μm at 230 K. This shift can be attributed to the dependence of the bandgap on temperature according to the Varshni formula. In order to explain current-voltage characteristics of the MWIR type II SL photodiodes, a bulk-based model with an effective bandgap of SL material is used.

It is well recognized that the photodiode dark current can be found as a superposition of several mechanisms (see Figure 20.23),

$$I_{dark} = I_{diff} + I_{gr} + I_{btb} + I_{tat} + I_{Rshunt}, \tag{20.3}$$

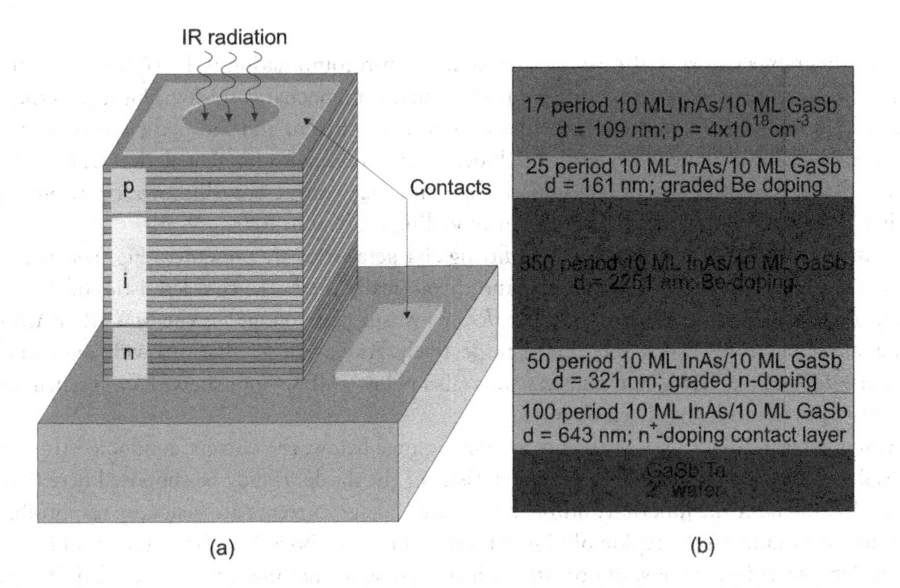

Figure 20.22 MWIR InAs/GaSb type II SL photodiode: (a) schematic device structure and (b) photodiode design.

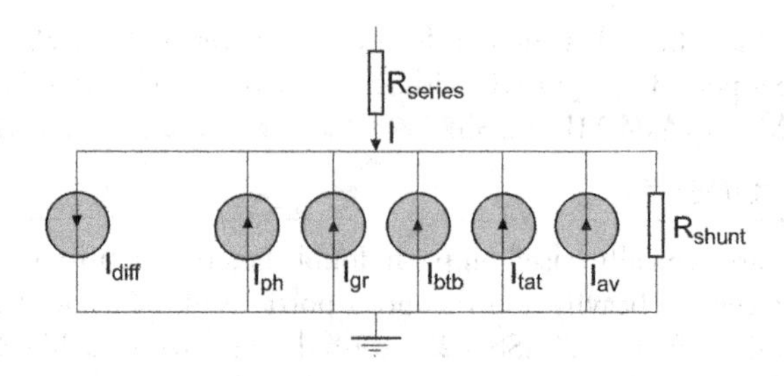

Figure 20.23 Possible currents operating in the photodiode. I_{diff} is ideal diffusion current, I_{ph} is the photocurrent, I_{gr} is due to GR mechanism, I_{btb} is due to band-to-band tunneling, I_{tat} is due to trap-to-band tunneling, and R_{shunt} is due to surface and bulk leakage shunt resistance. Limiting currents act in opposition to diffusion current.

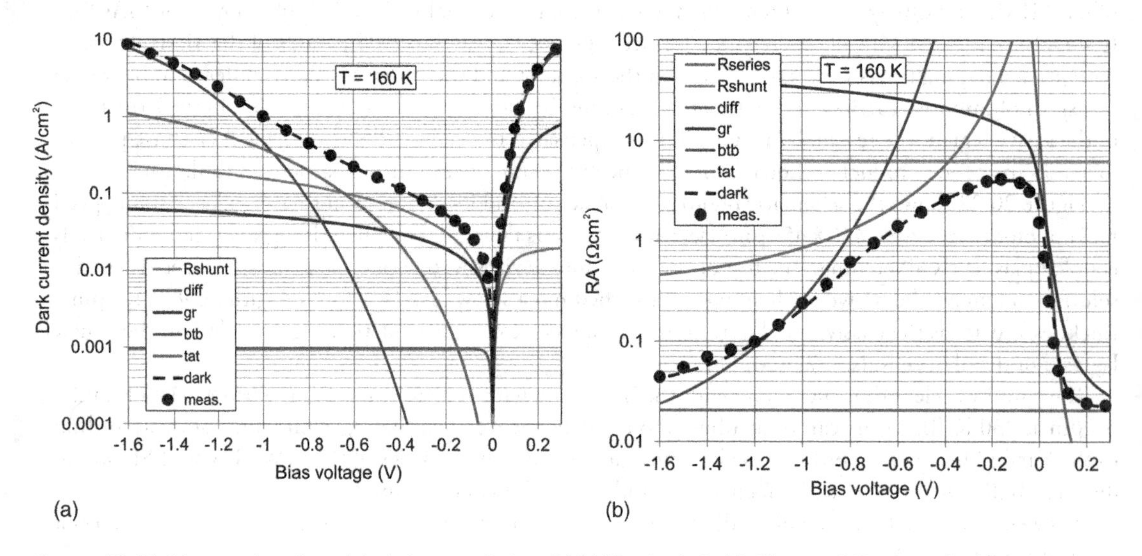

Figure 20.24 Measured and modeled characteristics of MWIR p-i-n InAs/GaSb type II SL photodiode at a temperature 160 K: The dark current density (a) and the resistance area product (b) versus bias voltage. (Adapted after Ref. [90])

including four main mechanisms: diffusion (I_{diff}), generation-recombination (GR) (I_{gr}), band-to-band tunneling (I_{btb}), and trap-assisted tunneling (I_{tat}). The remaining mechanism is current due to the shunt resistance (I_{Rshunt}, originates from the surface and bulk leakage current and shows the presence in the reverse-bias region). The avalanche current, which occurs in diodes with large depletion widths and a high reverse bias voltage, is omitted in our considerations. A summary of these well-known contributing dark currents that have been taken into account in our modeling is given in Refs. [89–91].

Below several examples of the measured and fitting characteristics are presented. Figures 20.24 and 20.25 present the comparisons of experimental and theoretically predicted characteristics of the dark current density, *J-V*, and the resistance area product, *RA*(*V*) versus bias voltage for the MWIR InAs/GaSb SL photodiodes at temperatures 160 K and 230 K, respectively. As we can see that in a wide region of bias voltages between +0.1 V up to −1.6 V and temperatures (also below 160 K—not shown) an excellent agreement between both types of results has been obtained.

At a low temperature (≤ 120 K), in the reverse bias voltage below, the current conducted through the shunt resistance dominates the reverse characteristic of the diode. It may be supposed here that the dislocations that intersect the junction and/or the surface leakage currents are generally responsible for shunt currents in a diode. In the region of higher reverse voltages, above 1 V, the influence of band-to-band tunneling is decisive. Influence of shunt resistance is negligible in thermoelectrically cooled (TE cooled) photodiodes (in the temperature range above 170 K).

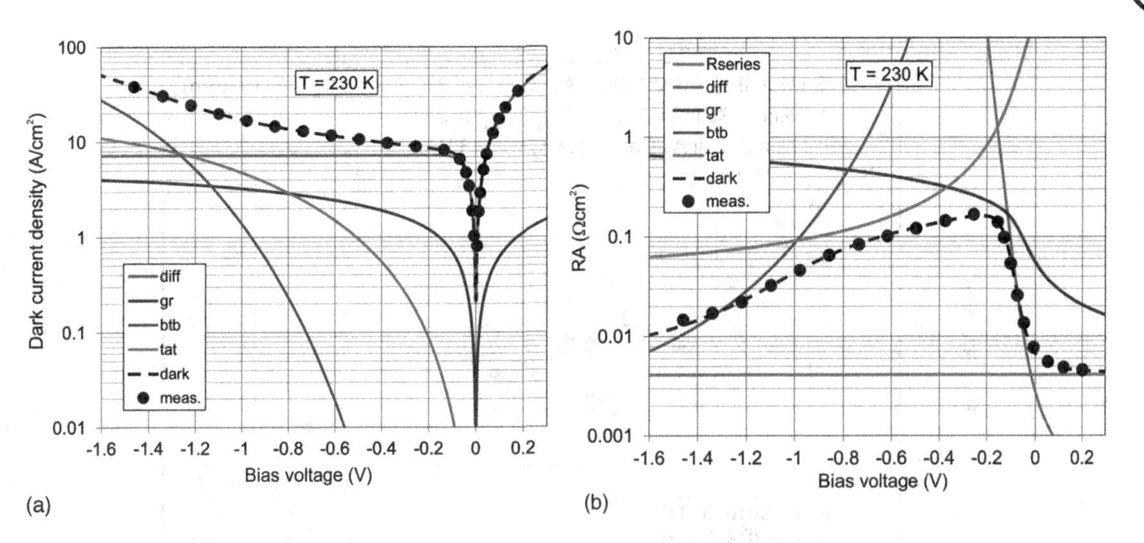

Figure 20.25 Measured and modeled characteristics of MWIR p-i-n InAs/GaSb type II SL photodiode at a temperature 230 K: the dark current density (a) and the resistance area product (b) versus bias voltage. (Adapted after Ref. [90])

With a temperature increase, the contributions of diffusion and GR currents increase in the zero-bias and the low-bias regions and they are dominated at 230 K. At medium values of reverse bias (between 0.6 and 1.0 V at 160 K), the dark current is mostly due to trap-assisted tunneling. At high values of reverse bias (above 1 V at 160 K), a bulk band-to-band tunneling dominates. In the region of forward bias voltage above 0.1 V, the influence of series resistance is decisive.

Wróbel et al. [92] have examined a near midgap trap energy level in $InAs_{10\,ML}/GaSb_{10\,ML}$ type II SLs using thermal analysis of dark current, Fourier transform photoluminescence, and low-frequency noise spectroscopy. Several wafers and diodes with similar period designs and the same macroscopic constructions have been investigated. All the characterization techniques gave nearly the same value of about 140 meV, independent of the substrate type. Additionally, photoluminescence spectra show that the transition related to the trap center is temperature independent.

Figure 20.26 presents a comparison of the R_0A product versus the cutoff wavelength for the HgCdTe photodiodes fabricated in laboratory at the Institute of Applied Physics, Military University of Technology, Warsaw and the type II InAs/GaSb SL photodiodes operating at 230 K [93]. It is clearly seen that the performance of SL devices has reached a comparable level with the state of the art HgCdTe detectors. In fact, the *RA* products of 6.2-μm SL devices are even higher, but they were measured at 0.3 V reverse bias. The dashed lines are trend lines of the HgCdTe device experimental data. The HgCdTe photodiodes still generally offer a better performance, especially at a lower temperature, for example, at 77 K [94], mainly due to a larger quantum efficiency. Typically, the quantum efficiency of our type II photodiodes is about 30% (at 0.3 V reverse bias) in comparison with 70% for the HgCdTe devices. The active region of the SL devices is thinner in comparison with the HgCdTe photodiodes. As a result, the dark current and the R_0A product of devices with high-bandgap contacts are comparable with those of the HgCdTe photodiodes at 230 K.

The research group of Montpellier University [84,95,96] has elaborated the MWIR T2SL p-i-n device structures similar to those shown in Figure 16.42. As marked previously, several InAs/GaSb SL material properties depend strongly on the chosen SL period, such as the temperature bandgap energy, the absorption coefficient, residual doping level, and carrier lifetime. The SL period influences the carrier localization that results in different miniband widths and thus different joint densities of states and shapes of absorption coefficients. For example, Figure 20.27 reports photoresponse spectra of symmetrical InAs/GaSb SL structures having emission in the MWIR wavelength range.

The important observation is the independence of minority carrier lifetime on the interface density. The recombination centers limiting the SL lifetime are located in the binary materials rather than at the interfaces [84,97]. The worst material properties are obtained for the largest GaSb content. In SL samples at 77

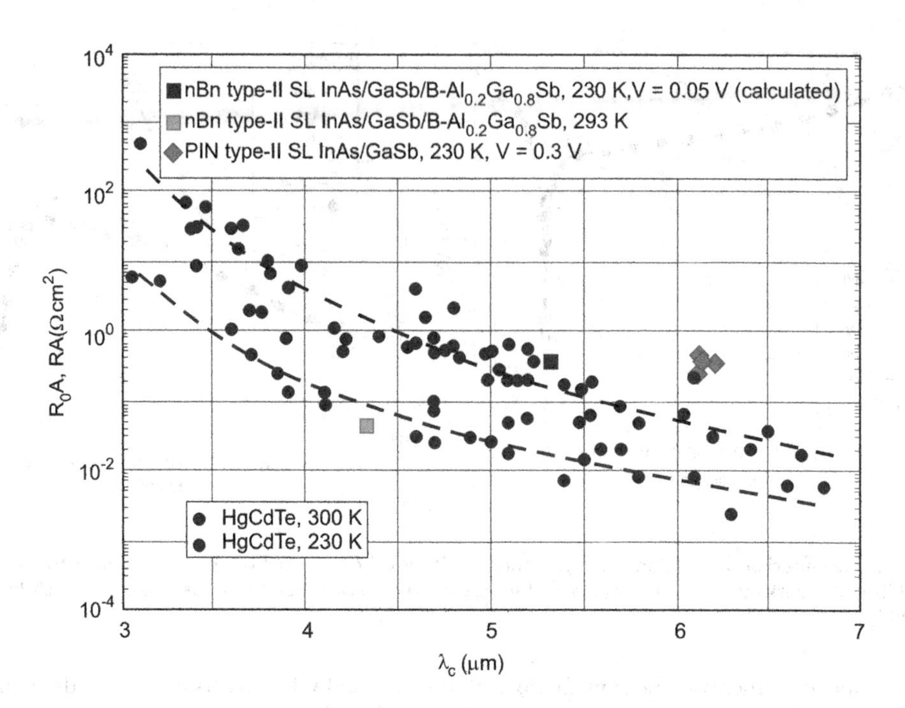

Figure 20.26 Dependence of the *RA* and R_0A products on cutoff wavelength for MWIR InAs/GaSb/B-Al0.2Ga0.8Sb type II SL nBn detector, HgCdTe bulk diodes, and InAs/GaSb type II SL p-i-n diodes operated at near-room temperature.

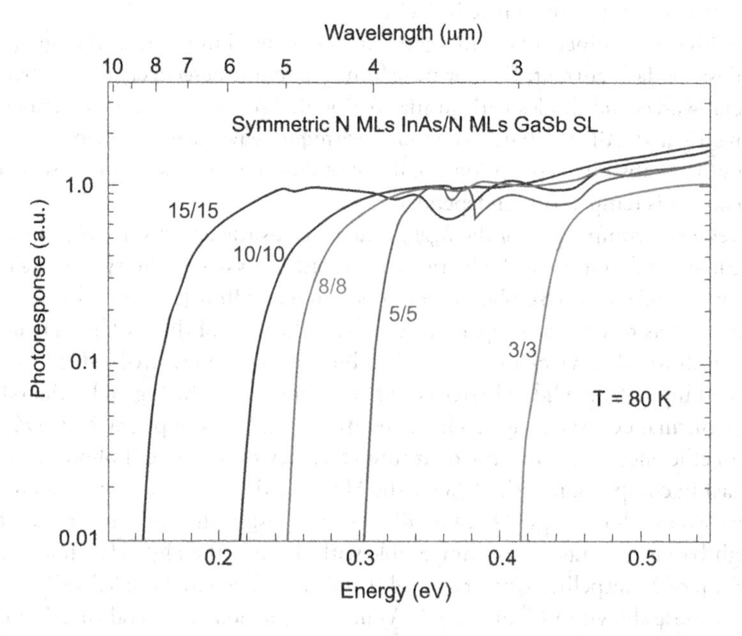

Figure 20.27 Normalized photoresponse spectra of InAs (N)/GaSb (N) symmetrical MWIR SL detector structures with N=3, 5, 8, 10 and 15 MLs. The spectra are recorded at 80 K. (Adapted after Ref. [96])

K with cutoff wavelength close to 5 μm, the residual doping level was found to increase from 6×10^{14}cm^{-3} to 5.5×10^{15}cm^{-3} when the GaSb content in each SL period increased from 36% to 65%. Since the GR-limited current dominating at 77 K is proportional to $n_i/\tau n^{1/2}$ (n is the majority carrier density), an increase of τ correspondingly contributes to a decrease by the same amount of I_{gr}. Figure 20.28 gathers the experimental data of dark current density for GaSb-rich, symmetric, and InAs-rich SL compositions. In case of an

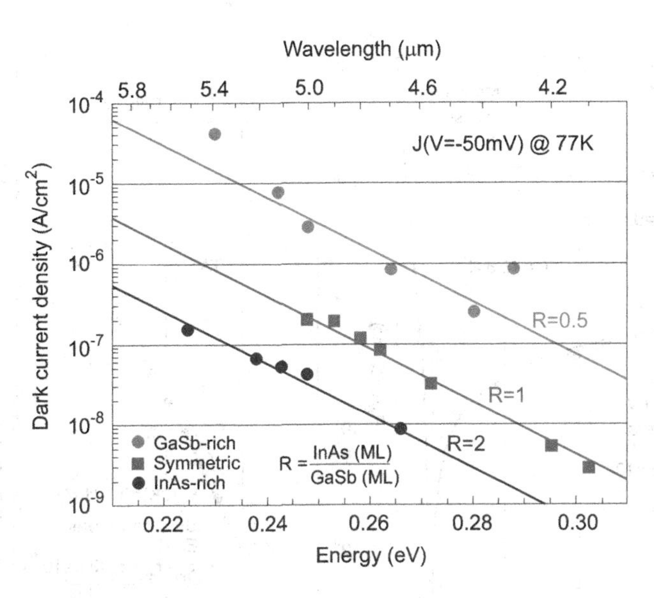

Figure 20.28 Dark current density at 77 K as a function of the period thickness of p-i-n InAs/GaSb T2SL photodiodes for different ratio R=InAs/GaSb. (Adapted after Ref. [96])

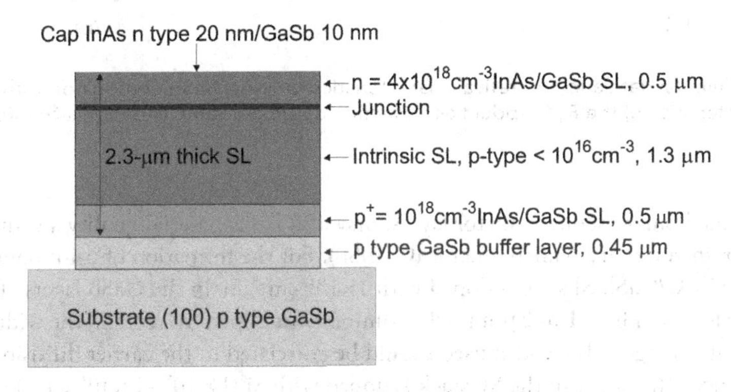

Figure 20.29 Schematic design of p-i-n double heterojunction InAs/GaSb photodiode with 10.5-μm cutoff wavelength.

asymmetric (7/4) InAs-rich InAs/GaSb T2SL structure (R=1.75) with a cutoff wavelength of 5.5 μm at 77 K, the R_0A product as high as 7×10^6 Ωcm^2 at 77 K was reported [96].

20.3.2 LWIR PHOTODIODES

Figure 20.29 shows a cross section scheme of a completely processed mesa detector and design of a 10.5 μm InAs/GaSb SL photodiode. The layers are usually grown with MBE at substrate temperatures around 400°C on undoped (001)-oriented GaSb substrates. With the addition of cracker cells for the group V sources, the SL quality become significantly improved. Despite the relatively low absorption coefficients, GaSb substrates require thinning the thickness below 25 μm in order to transmit appreciable IR radiation [98]. Since the GaSb substrates and buffer layers are intrinsically p-type, the p-type contact layer, intentionally doped with Be at an acceptor concentration of 1×10^{18} atoms/cm^3 is grown first.

Sensors for the LWIR spectral ranges are based on binary InAs/GaSb short-period SLs [99,100]. The layers needed are already so thin that there is no benefit in using GaInSb alloys. The oscillator strength of the InAs/GaSb SL is weaker than the InAs/GaInSb; however, the InAs/GaSb SL, which uses unstrained

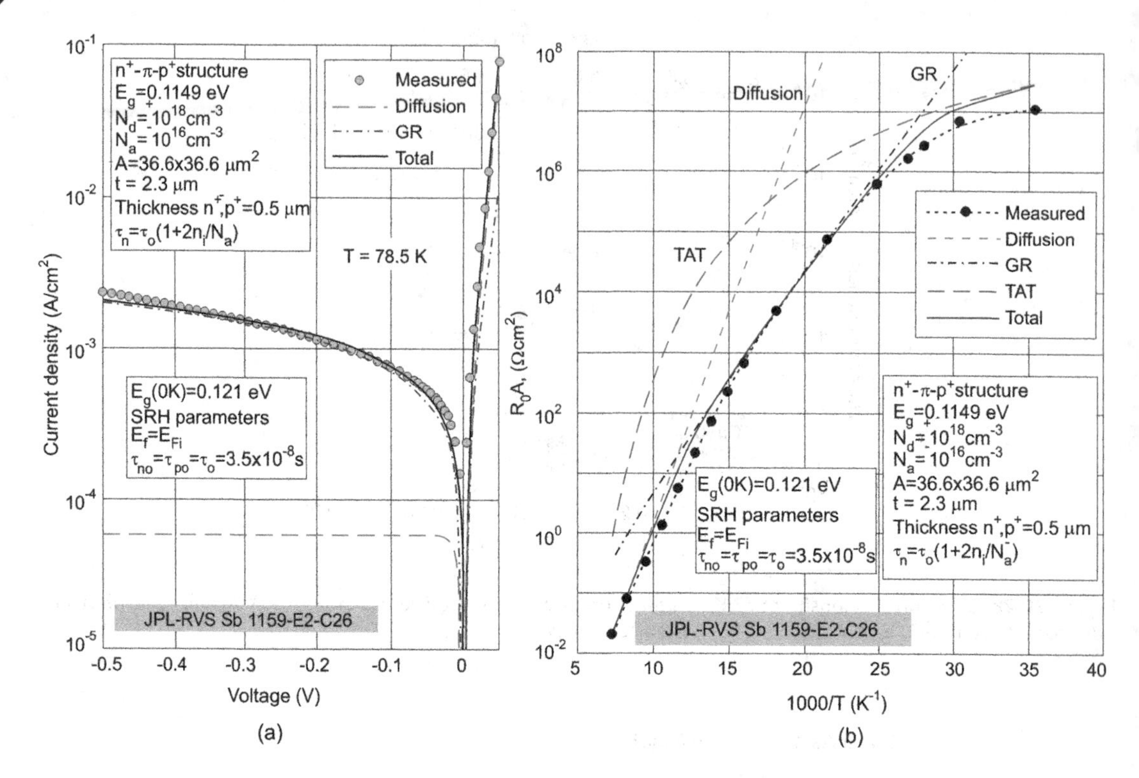

Figure 20.30 Experimental data and theoretical characteristics for InAs/GaSb photodiode with $\lambda_c = 10.5\,\mu$m at 78.5 K: (a) *J-V* characteristic, (b) the R_0A product as a function of temperature. (Adapted after Ref. [101])

and minimally strained binary semiconductor layers, may also have material quality advantages over the SL, which uses a strained ternary semiconductor (GaInSb). For the formation of p-i-n photodiodes the lower periods of the InAs/GaSb SLs are p-doped with $1\times10^{17}\,\text{cm}^{-3}$ Be in the GaSb layers. The acceptor-doped SL layers are followed by a 1 to 2 μm thick, nominally undoped SL region. The width of the intrinsic region does vary in the designs. The width used should be correlated to the carrier diffusion lengths for improved performance. The upper of the SL stack is doped with Si (1×10^{17}–$1\times10^{18}\,\text{cm}^{-3}$) in the InAs layers and is typically 0.5 μm thick. The top of the SL stack is then capped with an InAs:Si ($n \approx 10^{18}\,\text{cm}^{-3}$) layer to provide good ohmic contact.

The main technological challenge for the fabrication of photodiodes is the growth of thick SL structures without degrading the material quality. High-quality SLS materials thick enough to achieve acceptable quantum efficiency is crucial for the success of the technology.

Figure 20.30 shows the experimental data and theoretical prediction of the R_0A product as a function of temperature for InAs/GaSb photodiodes with 10.5-μm cutoff wavelength at 78 K. The photodiodes are depletion region (GR)-limited in the temperature range below 100 K. Space-charge recombination currents dominate the reverse bias at 78 K and taking the dominant recombination centers to be located at the intrinsic Fermi level that is shown in Figure 20.30(a). The trap-assisted tunneling is dominant at $T \leq 40$ K. The performance of LWIR photodiodes in the high temperature range is limited by the diffusion process. At a low temperature and near the zero-bias voltage, the currents are diffusion limited. At larger biases, trap-assisted tunneling currents dominate.

An additional insight in dark current mechanisms in LWIR InAs/GaSb SL photodiodes has been given by Rehm et al. [102,103]. By assuming that the device is affected by sidewall leakage, bulk and sidewall contributions can be described by the well-known relation for the total dark current density:

$$I_{\text{dark}} = I_{\text{dark,bulk}} + \sigma \times P / A, \tag{20.4}$$

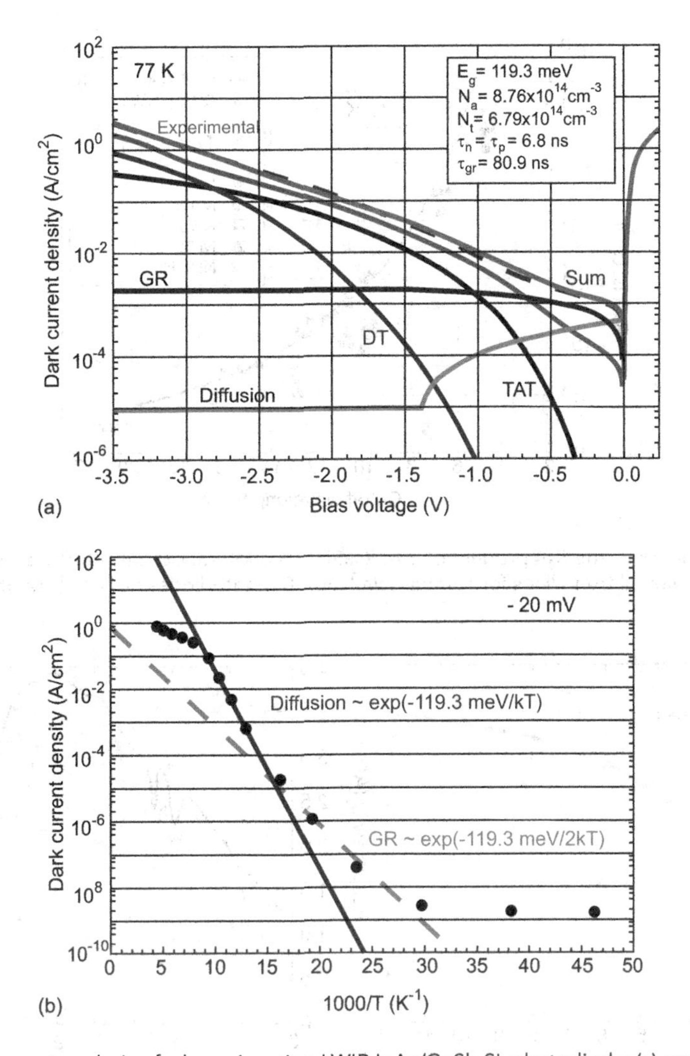

Figure 20.31 Dark current analysis of a homojunction LWIR InAs/GaSb SL photodiode: (a) versus bias voltage at 77 K and (b) versus temperature. (Adapted after Ref. [103])

where σ is the sidewall current per unit length of the mesa sidewall and P/A is the perimeter-to-area ratio of the device. In general, $\sigma(V,T)$ is a function of the applied bias voltage V and temperature T. The bulk dark current $I_{dark,bulk}$ contains components described by the Equation 20.3.

Figure 20.31a shows that the sidewall current does not dominate the behavior of $I(V)$ characteristic at low reverse bias at 77 K. The temperature dependence of $I(V)$ characteristic at 77 K at low reverse bias is diffusion limited [see Figure 20.31b].

The heterojunction device concepts help to considerably reduce the dark current [103]. It was shown by the design of a p$^+$-InAs/GaSb SL absorber combined with N-high gap part realized with a second CB-matched InAs/GaSb SL with higher bandgap.

Figure 20.32 compares the R_0A values of InAs/GaSb SL and HgCdTe photodiodes in the LW spectral range. The solid line denotes the theoretical diffusion-limited performance of p-type HgCdTe material. As can be seen in the figure, the most recent photodiode results for SL devices rival that of the practical HgCdTe devices, indicating that substantial improvement has been achieved in SL detector development.

The quantum efficiency of p-i-n photodiode structure shown in Figure 20.33 critically depends on the thickness of the i(π)-region. By fitting the quantum efficiencies of a series of photodiodes with i-region thicknesses varying from 1 to 4 μm, Aifer et al. [105] determined that the minority carrier electron

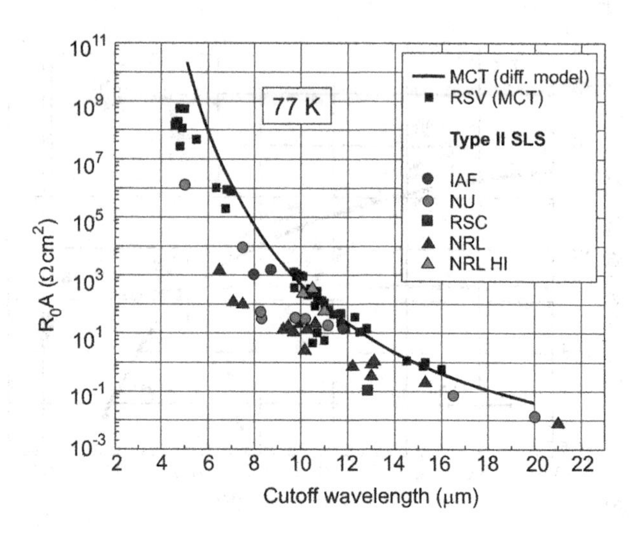

Figure 20.32 Dependence of the R_0A product of InAs/GaSb SLS photodiodes on cutoff wavelength compared to theoretical and experimental trend lines for comparable HgCdTe photodiodes at 77 K. (Adapted after Ref. [104])

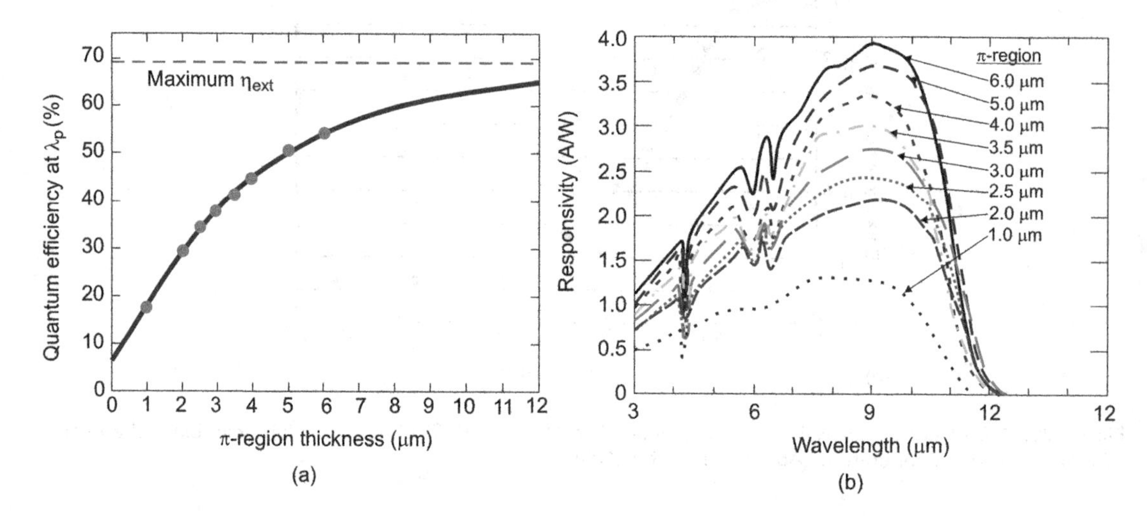

Figure 20.33 Spectral characteristics of InAs/GaInSb SL photodiodes at 77 K: (a) quantum efficiency as a function of π-region thickness; the dashed line represents the maximum possible quantum efficiency without antireflective coating and (b) measured current responsivity of photodiodes with π-region thickness ranging from 1 to 6 μm, the CO_2 absorption at 4.2 μm is visible as well as the water vapor absorption between 5 and 8 μm. (Adapted after Ref. [106])

diffusion length in LWIR is 3.5 μm. This value is considerably lower in comparison with the typical one for high-quality HgCdTe photodiodes. More recently, an external quantum efficiency of 54% has been obtained for a 12-μm cutoff wavelength photodiode by extending the thickness of the π-region to 6 μm. Figure 20.33a shows the dependence of quantum efficiency on the thickness of the π-region and Figure 20.33b presents the spectral current responsivity of eight of the structures with different thicknesses of the π-regions [106].

Figure 20.34 compares the calculated detectivity of type II and P-on-n HgCdTe photodiodes as a function of wavelength and temperature of operation with the experimental data of type II detectors operated at 78 K [107]. The solid lines are theoretical thermal limited detectivities for HgCdTe photodiodes, calculated using the one-dimensional (1D) model that assumes that the diffusion current from the narrower bandgap n-side is dominant, and minority carrier recombination via Auger and radiative process. In calculations,

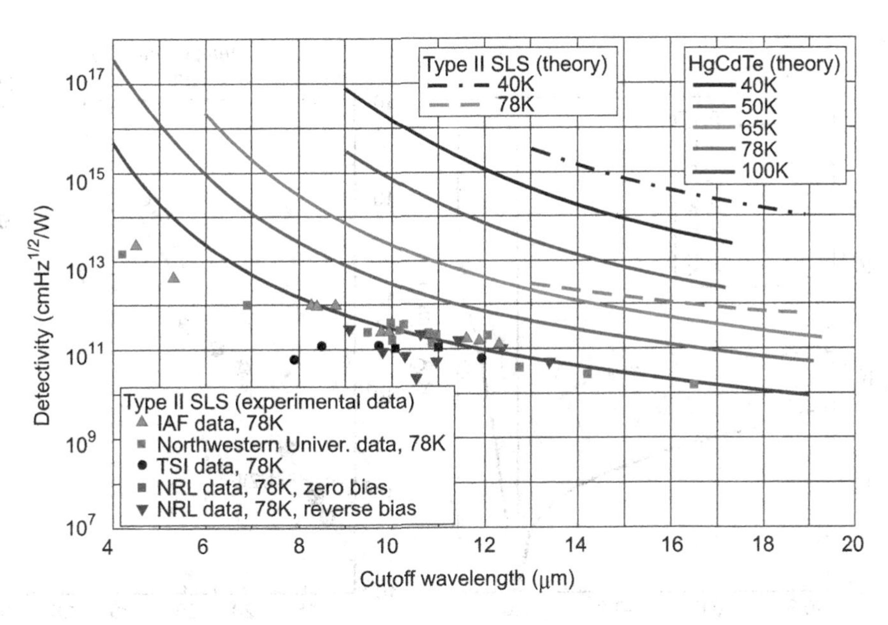

Figure 20.34 The predicted detectivity of type II and P-on-n HgCdTe photodiodes as a function of wavelength and temperature. The experimental data are taken with several sources. (Adapted after Ref. [107])

typical values for the n-side donor concentration ($N_d = 1\times10^{15}\,\text{cm}^{-3}$), the narrow-bandgap active layer thickness (10 μm), and quantum efficiency (60%) have been used. The predicted thermally limited detectivities of the T2SLs are larger than those for HgCdTe [72,108].

From Figure 20.34 it results that the measured thermally limited detectivities of T2SL photodiodes are as yet inferior to the current HgCdTe photodiode performance. Their performance has not achieved theoretical values. This limitation appears to be due to two main factors: relatively high background concentrations (about $5\times10^{15}\,\text{cm}^{-3}$, although values below $10^{15}\,\text{cm}^{-3}$ have been reported [109]) and a short minority carrier lifetime (typically tens of nanoseconds in lightly doped p-type material). Up till now, non-optimized carrier lifetimes have been observed and at desirably low carrier concentrations is limited by the SRH recombination mechanism. The minority carrier diffusion length is in the range of several micrometers. Improving these fundamental parameters is essential to realize the predicted performance of type II photodiodes.

20.4 InAs/InAsSb SUPERLATTICE PHOTODIODES

Interest in InAs/InAsSb SL materials for IR detection is motivated by the carrier lifetime limitations imposed by the GaSb layer in InAs/GaSb SLs. As mentioned in Section 20.2.1, a significantly longer minority carrier lifetime has been obtained in an InAs/InAsSb SL system as compared to an InAs/GaSb SL operating in the same wavelength range and temperature. It is expected that such increase in minority carrier lifetime results in lower dark current for InAs/InAsSb SL detectors in comparison with their InAs/GaSb SL counterparts [110]. In addition, with two common elements (In and As) in SL layers, the InAs/InAsSb SL has a relatively simple interface structure with only one changing element (Sb) that promises a better controllability in epitaxial growth and simpler manufacturability.

Both MWIR [111] and LWIR [112] InAs/InAsSb SL photodiodes have been demonstrated. The experimentally measured dark current density of MWIR photodiodes with a cutoff wavelength 5.4 μm at 77 K was larger than that of the conventional InAs/GaSb SL detectors. This was attributed to an increased probability of carrier tunneling due to reduced VB and CB offsets in InAs/InAsSb SL system.

The better quality LWIR $InAs/InAs_{1-x}Sb_x$ SL photodiodes have been demonstrated by Hoang et al. [112]. Despite the introduction of large amount of Sb ($x=0.43$), the material quality is still good and leads to high performance photodetectors. The cutoff wavelength of the active region is mainly determined by the VB level in the $InAs_{1-x}Sb_x$ layer, which is directly related to the amount of Sb. The samples were MBE grown

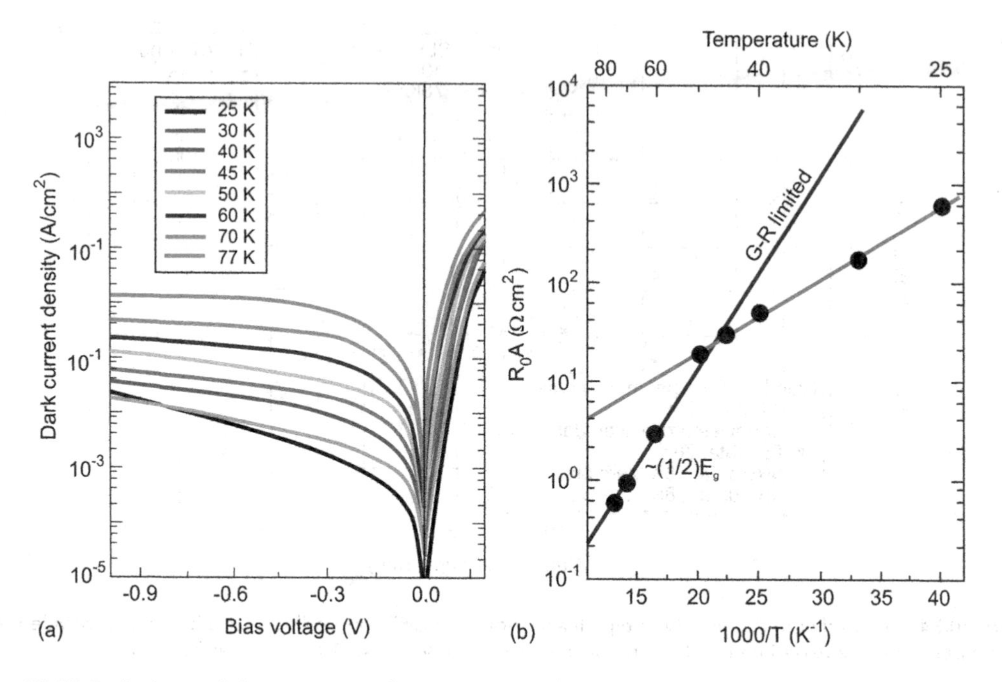

Figure 20.35 Dark electrical characteristics of LWIR InAs/InAsSb SL photodiodes: (a) current-voltage characteristics as functions of temperature; (b) R_0A vs. temperature. (Adapted after Ref. [112])

on Te-doped (001) GaSb substrate. The device structure consisted of 0.5-μm-thick InAsSb buffer layer, followed by a 0.5-μm-thick bottom n-contact (n ~ 10^{18} cm^{-3}), a 0.5-μm-thick slightly n-doped barrier, a 2.3-μm-thick slightly p-doped active region (~ 10^{15} cm^{-3}) and a 0.5-μm-thick top p-contact (p ~ 10^{18} cm^{-3}). Finally, the structure was capped with a 200-nm-thick p-doped GaSb layer. The n-type and p-type dopants are Si and Be, respectively.

The effective passivation of InAs/InAsSb T2SL photodiodes is in very early stages of development. Usually, photodiodes are not passivated. The simplest passivation is based on the common dielectric insulators deposited onto the exposed surface of the device, utilized in the Si industry (such as an oxide or nitride of Si).

Figure 20.35 shows the electrical characteristics of the LWIR InAs/InAsSb SL photodiode in the temperature range from 25 to 77 K. At 77 K, the R_0A product is 0.84 Ωcm² indicating a lower value in comparison with its InAs/GaSb counterpart (see Figure 20.32). Above 50 K, the diode exhibits an Arrhenius-type behavior with associated activation energy of 39 meV, which is approximately half of the active region's bandgap (~80 meV for 15 μm). It indicates the GR current from the active region as the limiting dark current mechanism. Below 50 K, the R_0A deviates from the trend and becomes less sensitive with the temperature variation. This behavior suggests that the dark current is limited by other mechanisms, either the tunneling current or surface leakage, at this range of temperature.

The spectral characteristics of photodiode are shown in Figure 20.36. At 77 K, the sample exhibited a 100% cutoff wavelength at 17 μm and 50% cutoff at 14.6 μm. High quantum efficiency is achieved with reverse bias above 150 mV and saturation at 300 mV. At the saturation, the current responsivity reached a peak of 4.8 AW^{-1}, corresponding to a quantum efficiency of 46% for a 2.3-μm-thick active region. The calculated shot noise and Johnson noise-limited detectivity of the device at 77 K based on the measured quantum efficiency, the dark current, and the RA product is shown in Figure 20.36(b).

20.5 DEVICE PASSIVATION

Despite numerous efforts of various research groups in T2SL device fabrication, the development of effective device passivation scheme is not well established. The mesa sidewalls are a source of excess current.

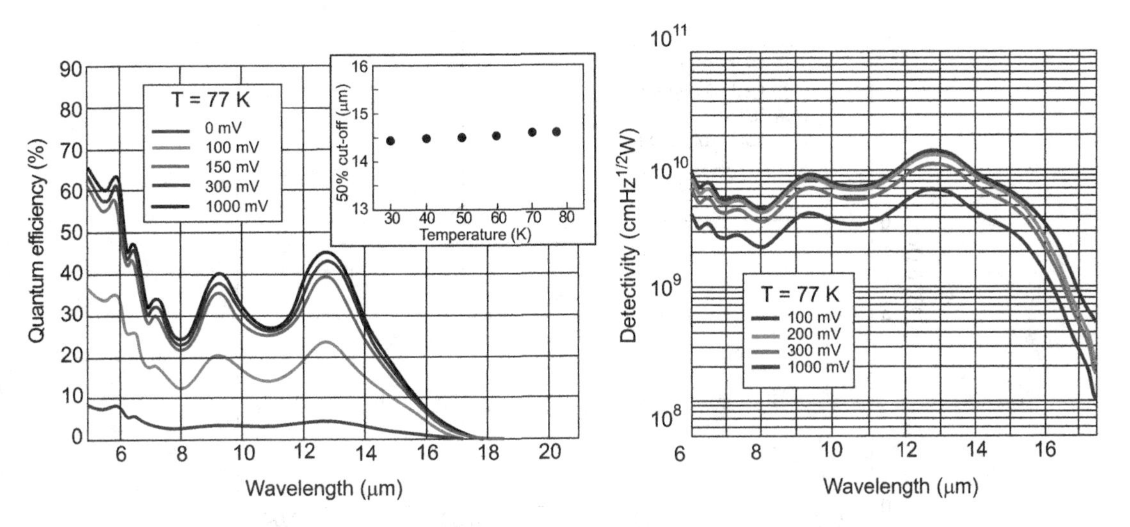

Figure 20.36 Spectral characteristics of LWIR InAs/InAsSb SL photodiode: (a) quantum efficiency spectrum with different applied bias at 77 K. Inset: 50% cutoff wavelength as function of temperature; (b) the calculated shot noise and Johnson noise-limited detectivity of the device at 77 K with different applied bias. (Adapted after Ref. [112])

Besides the efficient suppression of surface leakage currents, a passivation layer suitable for production purposes must withstand various treatments occurring during the subsequent processing of the device. Considerable surface leakage is attributed to the discontinuity in the periodic crystal structure caused by mesa delineation. Scaling of the pixel dimensions makes FPA performance strongly dependent on surface effects due to a large pixel surface/volume ratio. Thus, methods for the elimination of surface currents need to be developed.

The native oxide of most III-V compounds is not beneficial for the natural passivation. The oxidation of GaSb at relatively low temperature follows accordance with the reaction

$$2GaSb+3O_2 \rightarrow Ga_2O_3+Sb_2O_3. \tag{20.5}$$

In the first step of chemical passivation, the solution removes the native oxides at the surface, and then new atoms occupy the dangling bonds to prevent reoxidation of the material, surface contamination, and to minimize band bending.

A review of the passivation techniques for T2SLS detectors may be found in the work of Plis et al. [114], where they are categorized into two directions:

- encapsulation of the etched detector sidewalls with thick layers of dielectrics, organic materials (polyimide and various photoresists), or wider-bandgap III-V material, and
- chalcogenide passivation, that is, saturation of unsatisfied bonds on the semiconductor surfaces by sulfur atoms.

Here, we briefly describe the available passivation techniques and their limitations.

The effectiveness of passivation is commonly evaluated using variable area diode array method. The dark current density can be expressed as the summation of bulk components of dark current and the surface leakage current—see Equation 12.125. The slope of the function $(R_0A)^{-1}=f(P/A)$ is directly proportional to the diode's surface-dependent leakage current ($1/r_{Surface}$). If the bulk current dominates the detector performance, then the dependence $(R_0A)^{-1}=f(P/A)$ has a slope close to zero. If the surface leakage is significant, then an increase in the dark current density is observed for smaller devices.

Figure 20.37 shows exemplary dependences of $(R_0A)^{-1}$ as a function of P/A for LWIR T2SL photodiodes passivated in different ways and operated at 77 K [115,116]. Detectors with mesa sizes ranging within 100–400 μm etched by ICP and electron cyclotron resonance (ECR) with sidewalls encapsulated

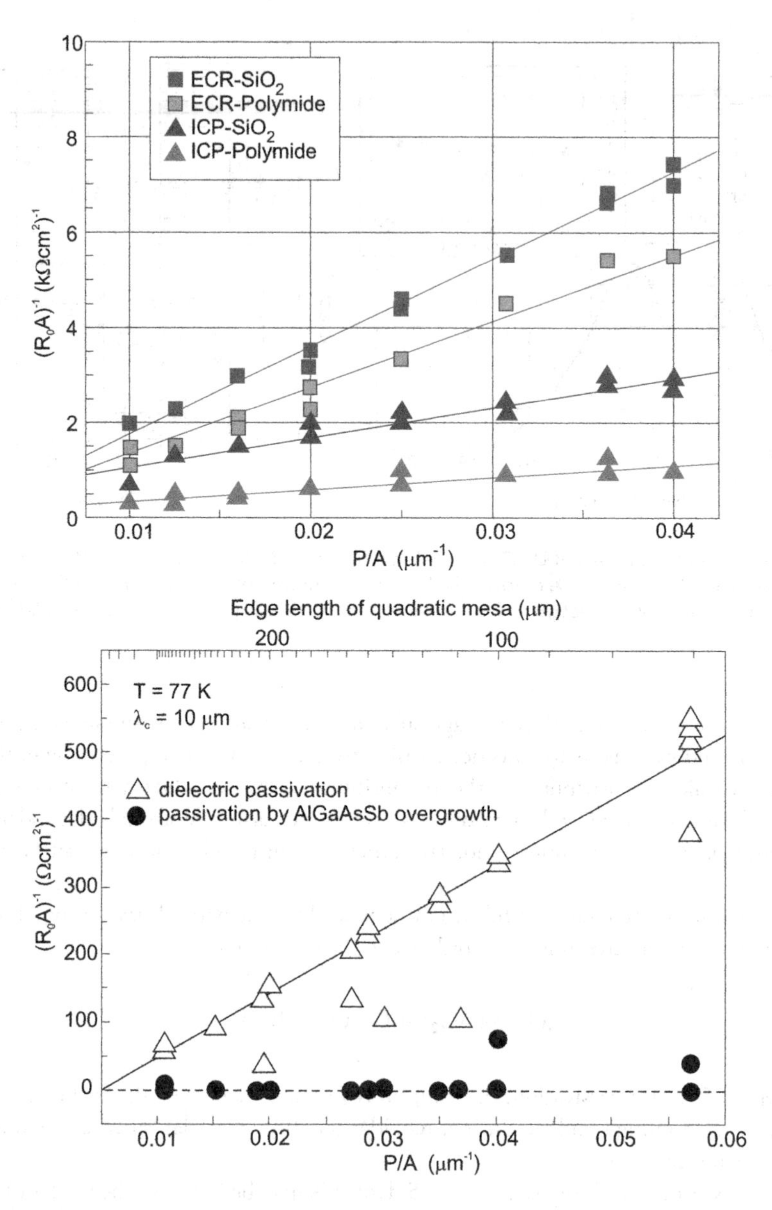

Figure 20.37 Dependence of $(R_0A)^{-1}$ as a function of *P/A* for LWIR T2SL photodiodes at 77 K: (a) detectors fabricated with ICP and ECR etching techniques and protected by SiO_2 and polymide, (b) detectors passivated by $Al_xGa_{1-x}As_ySb_{1-y}$ overgrowth and by the conventional dielectric layer passivation. (Adapted after Refs. [115] and [116])

by polyimide has demonstrated the highest surface resistivity (6.7×10^4 Ωcm) among four treatments (see Figure 20.37a). Comparison of the electrical performance of detectors with the same post-etch encapsulation method (polyimide) revealed an order of magnitude lower dark current density for an ICP-polyimide sample. Dielectric passivation with low fixed and interfacial charge densities at process temperatures substantially lower than the T2LS growth temperature (to prevent T2SL period mixing) presents the challenge of developing high-quality passivation layers. It appears that the band bending at the mesa sidewalls caused by the abrupt termination of the periodic crystal structure induces accumulation or type inversion of charge, which results in surface tunneling currents along the sidewalls. As was demonstrated by Delaunay et al. [117], native fixed charges present in the dielectric passivation layer (e.g., SiO_2) can either improve or deteriorate the narrow-bandgap device's performance. A good method of band-bending control of the

SiO_2-T2SL to establish the flat-band condition and suppress the leakage current is applying a negative bias voltage along the device sidewalls [118].

An effective way of T2SL device passivation is the encapsulation of etched sidewalls with wide-bandgap material or shallow etch technique that isolates the neighboring devices but terminates within a wider-bandgap layer. Rehm et al. have used the MBE overgrowth of lattice-matched, large-bandgap AlGaAsSb layer over etched mesa sidewalls [116]. In order to prevent Al-containing passivation layer from oxidation, a thin silicon nitride layer was deposited after the regrowth process. Figure 20.37b shows the dependence of $(R_0A)^{-1}$ vs. perimeter-to-area ratio for two similar detector structures, passivated by the overgrowth and the conventional dielectric layer passivation. No surface leakage is observed for overgrowth passivation (a slope close to zero). Also, Szmulowicz and Brown [119] proposed mesa sidewall encapsulation with GaSb to eliminate the surface currents. The GaSb encapsulant acts as a barrier to electrons at both the n- and p-type sides of the SL.

It appears that the reproducibility and long-term stability achieved by the SiO_2 passivation layer is more critical for photodiodes in the LWIR range. In general, the inversion potentials are bigger for the higher-bandgap materials, and therefore SiO_2 can passivate high-bandgap materials (MWIR photodiodes) but not the low-bandgap material (LWIR photodiodes). Using this property, a double heterostructure that prevents the inversion of the high-bandgap p-type and n-type SL contact regions has been proposed—see Figure 16.42 [117]. For such a structure, the surface leakage channel at the interface between the active region and the p- or n- contacts is considerably decreased.

Several additional design modifications that dramatically improve the LWIR photodiode dark current and R_0A product have been described. The very shallow slope of the shallow-etched samples demonstrate that it is possible to reduce excess currents due to the sidewalls [119].

An alternate method of eliminating the excess currents due to sidewalls is shallow-etch mesa isolation with band-graded junction [105,120]. The primary effect of the grading is to suppress tunneling and GR currents in the depletion region at low temperatures. Since both processes depend exponentially on the bandgap, it is highly advantageous to substitute a wide gap into the depletion region. In this approach, the mesa etch terminates at just past the junction and exposes only a very thin (300 nm), wider-bandgap region of the diode. Subsequent passivation is therefore in wider-gap materials. As a result, it reduces the electrical junction area, increases the optical fill factor, and eliminates the deep trenches within the detector array. However, if the lateral diffusion lengths are larger than the distance between neighboring pixels in the FPAs (typically several pixels), crosstalk between the FPA elements can be encountered, degrading the image resolution.

Different organic materials, polyimide and photoresists, are also attractive for passivants due to the simplicity of its integration into the T2SL detector fabrication procedure. Organic passivants are usually spin-coated onto a detector at room temperature and with thicknesses varying from 0.2 to 100 µm. A more popular option is SU-8, a high-contrast, epoxy-based negative photoresist developed by IBM [121]. Photo-polymerized SU-8 is mechanically and chemically stable after a hard bake. In several papers, the passivation of MWIR and LWIR T2SL detectors with SU-8 photoresist [122–124], polyimide [125], and AZ-1518 photoresist [126] have been reported. Polyimides are polymers of imide monomers, characterized by good thermal stability and chemical resistance and excellent mechanical properties. For LWIR InAs/GaSb SL photodiodes ($\lambda_c = 11.0\,\mu m$ at 77 K, with the side ranging in size from 25 to 50 µm), passivated with polyimide layers, no surface dependence was observed for the diodes with R_0A values within the range of 6–13 Ωcm^2 [125].

Stimulated by the successful application of sulfide passivation for GaAs surfaces, T2SL devices have been also passivated by alkaline sulfides, including Na_2S and $(NH_4)_2S$ in aqueous solutions. It was found that chalcogenide passivation through an immersion in a sulfur-containing solution or deposition of a sulfur-based layer effectively removes the native oxide with minimal surface etching and creates a covalently bonded sulfur layer. However, the chalcogen-based passivation does not provide physical protection and encapsulation of the device and there are some reports on the temporal instability of such a passivation layer [114].

The most effective technique to reduce surface leakage current in T2SL devices is currently the gating technique shown schematically in Figure 16.42. By creating a metal gate electrode on top of a dielectric passivation layer, the surface leakage can be tailored by an applied voltage [118].

Summarizing the technological passivation problems, there are various passivation techniques developed for T2SL detectors. Up till now, however, there is no universal approach that would treat equally efficiently the SL detectors with different cutoff wavelengths. Especially, more studies have to be conducted on the long-term stabilities of proposed passivation schemes to successfully integrate into the FPA fabrication procedure.

20.6 NOISE MECHANISMS IN TYPE II SUPERLATTICE PHOTODETECTORS

A proper understanding of the noise behavior of T2SL photodetectors is still lacking. The observed noise behavior is very complex and, depending on the circumstances in the diode under test, several mechanisms seem to be present. Detailed data on the noise properties is still sparse in the published literature [127–135]. The last general remark concerns different types of T2SL photodetectors: p-i-n photodiodes, nBn detectors, and interband cascade (IBC) detectors described in Chapters 12 and 22.

According to the classical theory, the fundamental noise current in diffusion- and GR-limited photodiodes is given by:

$$I_n^2 = 2q\left(I_{\text{dark}} + 2I_s\right)\Delta f, \tag{20.6}$$

where q denotes the elementary charge, I_{dark} the total dark current, I_s the reverse bias saturation value of the diffusion current, and Δf the measurement bandwidth. At zero-bias voltage, the Equation 20.6 reduces to the well-known Johnson noise expression. In case of GR-limited photodiode at high bias, $2I_s$ is small compared to I_d and Equation 20.6 tends to the familiar shot noise expression.

For the high-quality T2SL photodiodes limited by the SRH processes, the noise current follows Equation 20.6. An example of such behavior is shown in Figure 20.38 for InAs-rich MWIR InAs/GaSb p-i-n photodiode measured at 7 kHz bandwidth. At low frequency, 1/f noise is the most important. The structure consists of a 200-nm Be-doped (p^+-type doping ~1×10^{18} cm^3) GaSb buffer layer on lattice-matched GaSb substrate, several periods of p^+-doped SL, a non-intentionally doped InAs/GaSb SL active region, several periods of n^+-doped SL, and a 20-nm-thick Te-doped (n^+-type doping ~1×10^{18} cm^3) InAs cap layer. The mid-InAs-rich SL active region is composed of 300 periods of 7.5 InAs MLs and 3.5 GaSb MLs (7.5/3.5 SL structure) for a total thickness of 1 μm. Metallizations were ensured by Cr/Au on top of the mesa and on the back of the substrate.

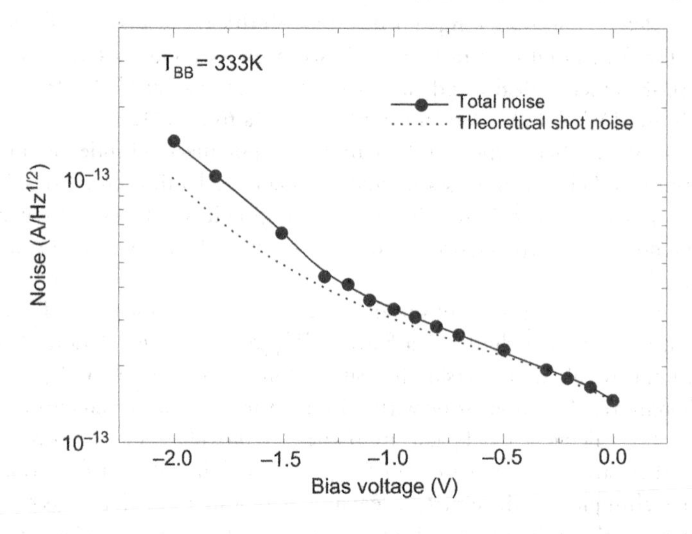

Figure 20.38 Experimental noise versus bias voltage for a 60°C blackbody and 77 K operating temperature and a 7 kHz bandwidth. (Adapted after Ref. [134])

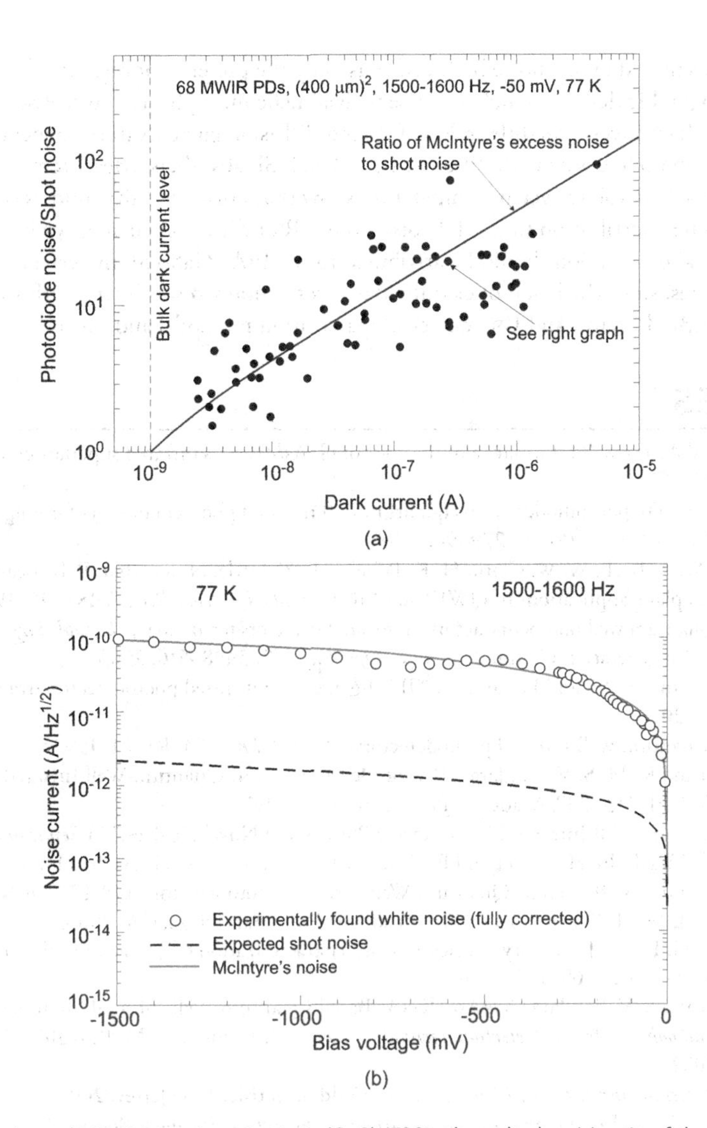

Figure 20.39 Noise data of MWIR homojunction InAs/GaSb T2SL photodiodes: (a) ratio of the experimentally found photodiode white noise to the expected shot noise versus dark current at 77 K and ~ 50 mV reverse bias for a set of 68 photodiodes with a size of 400×400 μm². The dashed line marks the GR-limited bulk dark current level found in small sized diodes; (b) white photodiode noise at 77 K versus bias voltage for the device marked by an arrow in the graph (a). (Adapted after Ref. [133])

In several papers it was shown that the 1/*f* noise is not intrinsically present in the T2SL structure. However, large extraneous frequently-dependent noise is generated by the sidewall leakage currents [127,130]. An effective way to eliminate the sidewall leakage current is the development of reliable passivation.

Authors from Fraunhofer Institut in Freiburg [131–133] have measured the noise of a number of MW and LW p-i-n InAs/GaSb photodiodes with large junction areas of 400×400 μm² and compared it with smaller reference diodes in a regime of low-frequency white noise. The dark current had varied about four orders of magnitude due to the presence of macroscopic defects in those large-area diodes and it was strongly increased compared to the GR-limited value of the bulk. The simple shot noise model completely failed. The shot noise model only explained the noise of devices with a dark current close to the GR-limited bulk level and the deviation of the experimentally observed noise from the expected shot noise increased with increasing dark current—see Figure 20.39. To explain the experimental data, McIntyre's excess noise model for electron-initiated avalanche multiplication was successfully proposed. It was suggested that the increase in dark current and excess noise has been caused by the presence of high electric field domains at the sites of crystallographic defects, which give rise to avalanche multiplication processes.

Ciura et al. [135] have investigated the role of GR and diffusion currents in the generation of $1/f$ noise in different MWIR photodetectors with different InAs/GaSb SL absorbers. Measurements of $1/f$ noise at constant small reverse bias voltage versus temperature show that noise intensity follows the squared leakage current and there is no contribution to the $1/f$ noise from GR or diffusion currents or it is too small to be observed. This general observation should be attributed to the InAs/GaSb SL material rather than to the device-specific features, since the batch of examined devices contained specimens with various architectures (p-i-n photodiode, nBn detector, and IBC detector), passivation methods, and substrates.

REFERENCES

1. M. A. Kinch and A. Yariv, "Performance limitations of GaAs/AlGaAs infrared superlattices," *Appl. Phys. Lett.*, 55, 2093–5, 1989.
2. A. Rogalski, "HgCdTe photodiodes versus quantum well infrared photoconductors for long wavelength focal plane arrays," *Opto-Electron. Rev.*, 6, 279–94, 1998.
3. M. Z. Tidrow, W. A. Beck, W. W. Clark, H. K. Pollehn, J. W. Little, N. K. Dhar, P. R. Leavitt, et al., "Device physics and focal plane applications of QWIP and MCT," *Opto-Electron. Rev.*, 7, 283–96, 1999.
4. A. Rogalski, "Quantum well photoconductors in infrared detectors technology," *J. Appl. Phys.*, 93, 4355–91, 2003.
5. A. Rogalski, "Third generation photon detectors," *Opt. Eng.*, 42, 3498–516, 2003.
6. A. Rogalski, J. Antoszewski, and L. Faraone, "Third-generation infrared photodetector arrays," *J. Appl. Phys.*, 105, 091101–44, 2009.
7. B. F. Levine, "Quantum-well infrared photodetectors," *J. Appl. Phys.*, 74, R1–81, 1993.
8. S. D. Gunapala and K. M. S. V. Bandara, "Recent Development in Quantum-Well Infrared Photodetectors," in *Thin Films*, Vol. 21, 113–237, Academic Press, New York, 1995.
9. H. C. Liu, "Quantum Well Infrared Photodetector Physics and Novel Devices," in *Semiconductors and Semimetals*, Vol. 62, eds. by H. C. Liu and F. Capasso, 129–96, Academic Press, San Diego, CA, 2000.
10. S. D. Gunapala and S. V. Bandara, "Quantum Well Infrared Photodetectors (QWIP)," in *Handbook of Thin Devices*, Vol. 2, ed. M. H. Francombe, 63–99, Academic Press, San Diego, CA, 2000.
11. J. L. Pan and C. G. Fonstad, "Theory, fabrication and characterization of quantum well infrared photodetectors," *Mater. Sci. Eng.*, R28, 65–147, 2000.
12. S. D. Gunapala and S. V. Bandara, "GaAs/AlGaAs Based Quantum Well Infrared Photodetector Focal Plane Arrays," in *Handbook of Infrared Detection Technologies*, eds. M. Henini and M. Razeghi, 83–119, Elsevier, Oxford, UK, 2002.
13. V. Ryzhi, ed., *Intersubband Infrared Photodetectors*, World Scientific, New Jersey, 2003.
14. H. Schneider and H. C. Liu, *Quantum Well Infrared Photodetectors*, Springer, Berlin, 2007.
15. M. A. Kinch, *State-of-the-Art Infrared Detector Technology*, SPIE Press, Bellingham, WA, 2014.
16. G. H. Döhler, "Doping SLs ("n-i-p-i Crystals")," *IEEE J. Quantum Electron.*, QE-22, 1682–95, 1986.
17. J. N. Schulman and T. C. McGill, "The CdTe/HgTe superlattice: Proposal for a new infrared material," *Appl. Phys. Lett.*, 34, 663–5, 1979.
18. D. L. Smith, T. C. McGill, and J. N. Schulman, "Advantages of the HgTe-CdTe superlattice as an infrared detector material," *Appl. Phys. Lett.*, 43, 180–2, 1983.
19. J. P. Faurie, "Growth and properties of HgTe-CdTe and other Hg-based superlattices," *IEEE J. Quantum Electron.*, QE-22, 1656–65, 1986.
20. J. M. Berroir, Y. Guldner, and M. Voos, "HgTe-CdTe superlattices: Magnetooptics and band structure," *IEEE J. Quantum Electron.*, QE-22, 1793–98, 1986.
21. J. R. Meyer, C. A. Hoffman, and R. J. Bartoli, "Narrow-gap II–VI superlattices: Correlation of theory with experiment," *Semicond. Sci. Technol.*, 5, S90–99, 1990.
22. T. H. Myers, J. R. Meyer, C. A. Hoffman, and L. R. Ram-Mohan, "HgTe/CdTe superlattices for IR detection revisited," *Appl. Phys. Lett.*, 61, 1814–6, 1992.
23. T. H. Myers, J. R. Meyer, and C. A. Hoffman, "The III Superlattices for Long-Wavelength Infrared Detectors: The HgTe/CdTe System," in *Semiconductor Quantum Wells and Superlattices for Long-Wavelength Infrared Detectors*, ed. M. O. Manasreh, 207–59, Artech House, Boston, MA, 1993.
24. F. F. Sizov and A. Rogalski, "Semiconductor superlattices and quantum wells for infrared optoelectronics," *Prog. Quantum Electron.*, 17, 93–164, 1993.

25. J. R. Meyer, C. A. Hoffman, and F. J. Bartoli, "Quantum Wells and Superlattices," in *Narrow-Gap II-VI Compounds for Optoelectronic and Electromagnetic Applications*, ed. P. Capper, 363–400, Chapman & Hall, London, 1997.
26. D. K. Arch, J. P. Faurie, J. L. Staudenmann, M. Hibbs-Brenner, and P. Chow, "Interdiffusion in HgTe-CdTe superlattices," *J. Vac. Sci. Technol.*, A4, 2101–5, 1986.
27. C. R. Becker, V. Latussek, A. Pfeuffer-Jeschke, G. Landwehr, and L. W. Molenkamp, "Band structure and its temperature dependence for type-III HgTe/$Hg_{1-x}Cd_xTe$ superlattices and their semimetal constituent," *Phy. Rev.*, B62, 10353–63, 2000.
28. C. K. Shih and W. E. Spicer, "Determination of a natural valence-band offset: The case of HgTe-CdTe," *Phys. Rev. Lett.*, 58, 2594–7, 1987.
29. C. R. Becker, K. Ortner, X. C. Zhang, S. Oehling, A. Pfeyffer-Jeschke, and V. Latussek, "Far Infrared Detectors Based on HgTe/$Hg_{1-x}Cd_xTe$ Superlattices, and In Situ p Type Doping," in *Advanced Infrared Technology and Applications 2007*, ed. M. Strojnik, 79–89, Leon, Mexico, 2008.
30. J. M. Arias, S. H. Shin, D. E. Cooper, M. Zandian, J. G. Pasko, E. R. Gertner, R. E. DeWames, and J. Singh, "P-type arsenic doping of CdTe and HgTe/CdTe superlattices grown by photoassisted and conventional molecular-beam epitaxy," *J. Vac. Sci. Technol.*, A8, 1025–33, 1990.
31. K. A. Harris, R. W. Yanka, L. M. Mohnkern, A. R. Reisinger, T. H. Myers, Z. Yang, Z. Yu, S. Hwang, and J. F. Schetzina, "Properties of (211)B HgTe–CdTe superlattices grown by photon assisted molecular-beam epitaxy," *J. Vac. Sci. Technol.*, B10, 1574–81, 1992.
32. C. H. Grein, H. Jung, R. Singh, and M. F. Flatte, "Comparison of normal and inverted band structure HgTe/CdTe superlattices for very long wavelength infrared detectors," *J. Electron. Mater.*, 34, 905–8, 2005.
33. M. W. Goodwin, M. A. Kinch, and R. J. Koestner, "Metal-insulator-semiconductor properties of HgTe/CdTe superlattices," *J. Vac. Sci. Technol.*, A6, 2685–92, 1988.
34. M. L. Wroge, D. J. Peterman, B. J. Feldman, B. J. Morris, D. J. Leopold, and J. G. Broerman, "Impurity doping of HgTe-CdTe superlattices during growth by molecular-beam epitaxy," *J. Vac. Sci. Technol.*, A7, 435–9, 1989.
35. K. A. Harris, T. H. Myers, R. W. Yanka, L. M. Mohnkern, and N. Otsuka, "A high quantum efficiency *in situ* doped mid-wavelength infrared p-on-n homojunction superlattice detector grown by photoassisted molecular-beam epitaxy," *J. Vac. Sci. Technol.*, B9, 1752–8, 1991.
36. G. C. Osbourn, "InAsSb strained-layer superlattices for long wavelength detector applications," *J. Vac. Sci. Technol.*, B2, 176–8, 1984.
37. D. L. Smith and C. Mailhiot, "Proposal for strained type II superlattice infrared detectors," *J. Appl. Phys.*, 62, 2545–8, 1987.
38. C. Mailhiot and D. L. Smith, "Long-wavelength infrared detectors based on strained InAs-GaInSb type-II superlattices," *J. Vac. Sci. Technol.*, A7, 445–9, 1989.
39. C. H. Grein, P. M. Young, and H. Ehrenreich, "Minority carrier lifetimes in ideal InGaSb/InAs superlattice," *Appl. Phys. Lett.*, 61, 2905–7, 1992.
40. G. C. Osbourn, "Design of III-V quantum well structures for long-wavelength detector applications," *Semicond. Sci. Technol.*, 5, S5–11, 1990.
41. G. A. Sai-Halasz, R. Tsu, and L. Esaki, "A new semiconductor superlattice," *Appl. Phys. Lett.*, 30, 651–3, 1977.
42. G. A. Sai-Halasz, L. L. Chang, J. M. Welter, and L. Esaki, "Optical absorption of $In_{1-x}Ga_xAs$-$GaSb_{1-y}As_y$ superlattices," *Solid State Commun.*, 27, 935–7, 1978.
43. L. L. Chang, N. J. Kawai, G. A. Sai-Halasz, P. Ludeke, and L. Esaki, "Observation of semiconductor-semimetal transition in InAs/GaSb superlattices," *Appl. Phys. Lett.*, 35, 939–42, 1979.
44. A. Rogalski, *New Ternary Alloy Systems for Infrared Detectors*, SPIE Optical Engineering Press, Bellingham, WA, 1994.
45. H. R. Jen, K. Y. Ma, and G. B. Stringfellow, "Long-range order in InAsSb," *Appl. Phys. Lett.*, 54, 1154–6, 1989.
46. S. R. Kurtz, L. R. Dawson, R. M. Biefeld, D. M. Follstaedt, and B. L. Doyle, "Ordering-induced band-gap reduction in $InAs_{1-x}Sb_x$ ($x \approx 0.4$) alloys and superlattices," *Phys. Rev.*, B46, 1909–12, 1992.
47. R. A. Stradling, S. J. Chung, C. M. Ciesla, C. J. M. Langerak, Y. B. Li, T. A. Malik, B. N. Murdin, A. G. Norman, C. C. Philips, C. R. Pidgeon, M. J. Pullin, P. J. P. Tang, and W. T. Yuen, "The evaluation and control of quantum wells and superlattices of III-V narrow gap semiconductors," *Mater. Sci. Eng.*, B44, 260–5, 1997.
48. Y.-H. Zhang, A. Lew, E. Yu, and Y. Chen, "Microstructural properties of InAs/$InAs_xSb_{1-x}$ ordered alloys grown by modulated molecular beam epitaxy," *J. Crys. Growth* 175/176, 833–7, 1997.
49. Y.-H. Zhang, "InAs/$InAs_xSb_{1-x}$ type-II superlattice midwave infrared lasers," in *Optoelectronic Properties of Semiconductors and Superlattices*, Vol. 3: Antimonide-Related Strained-Layer Heterostructures, ed. M. O. Manasreh, 461–500, Gordon and Breach Science Publishers, Amsterdam, 1997.

50. E. H. Steenbergen, O. O. Cellek, H. Li, S. Liu, X. Shen, D. J. Smith, and Y.-H. Zhang, "$InAs/InAs_xSb_{1-x}$ superlattices on GaSb substrates: A promising material system for mid- and long-wavelength infrared detectors", in *The Wonder of Nanotechnology: Quantum Optoelectronic Devices and Applications*, eds. M. Razeghi, L. Esaki, K. Von Klitzing, pp. 59–83, SPIE Press, Bellingham, WA, 2013.
51. G. J. Brown, S. Elhamri, W. C. Mitchel, H. J. Haugan, K. Mahalingam, M. J. Kim, and F. Szmulowicz, "Electrical, Optical, and Structural Studies of InAs/InGaSb VLWIR Superlattices", in *The Wonder of Nanotechnology: Quantum Optoelectronic Devices and Applications*, eds. M. Razeghi, L. Esaki, K. Von Klitzing, pp. 41–58, Bellingham, WA, 2013.
52. H. Kröemer, "The 6.1 Å family (InAs, GaSb, AlSb) and its heterostructures: A selective review", *Physica E* 20, 196–203, 2004.
53. Y. Wei and M. Razeghi, "Modelling of type-II InAs/GaSb superlattices using an empirical tight-binding method and interface engineering", *Phys. Rev.* B69, 085316–7, 2004.
54. F. Rutz, R. Rehm, J. Schmitz, M. Wauro, J. Niemasz, J.-M. Masur, A. Wörl, M. Walther, R. Scheibner, J. Wendler, and J. Ziegler, "InAs/GaSb superlattices for high-performance infrared detection", *Sensor + Test Conferences* 2011, *IRS² Proceedings*, 16–20.
55. E. Machowska-Podsiadlo and M. Bugajski, "Superlattices: Design of InAs/GaSb Superlattices for Optoelectronic Applications—Basic Theory and Numerical Methods", in *CRC Concise Encyclopedia of Nanotechnology*, eds. B. I. Kharisov, O. V. Kharissova, and U. Ortiz-Mendez, 1008–24, CRC Press, Boca Raton, FL 2015
56. F. Szmulowicz, H. Haugan, and G. J. Brown, "Effect of interfaces and the spin-orbit band on the band gaps of InAs/GaSb superlattices beyond the standard envelope-function approximation", *Phys. Rev.*, B 69(15), 155321, 2004.
57. Y. Livneh, P. C. Klipstein, O. Klin, N. Snapi, S. Grossman, A. Glozman, and E. Weiss, "**k·p** model for the energy dispersions and absorption spectra of InAs/GaSb type-II superlattices", *Phys. Rev.*, B86, 235311, 2012.
58. P. C. Klipstein, Y. Livneh, A. Glozman, S. Grossman, O. Klin, N. Snapi, and E. Weiss, "Modeling InAs/GaSb and InAs/InAsSb superlattice infrared detectors", *J. Electron. Mater.*, 43(8), 2984–90, 2014.
59. P. C. Klipstein, E. Avnon, Y. Benny, R. Fraenkel, A. Glozman, S. Grossman, O. Klin, L. Langoff, Y. Livneh, I. Lukomsky, M. Nitzani, L. Shkedy, I. Shtrichman, N. Snapi, A. Tuito, and E. Weiss, "InAs/GaSb type II superlattice barrier devices with a low dark current and a high-quantum efficiency", *Proc. SPIE* 9070, 9070 0U-1–10, 2014.
60. L. W. Wang, S. H. Wei, T. Mattila, A. Zunger, I. Vurgaftman, and J. R. Meyer, "Multiband coupling and electronic structure of $(InAs)_n/(GaSb)_n$ superlattices", *Phys. Rev.*, B60(8), 5590, 1999.
61. X. Cartoixà, D. Z. Y. Ting, and T. C. McGill, "Description of bulk inversion asymmetry in the effective bond-orbital model", *Phys. Rev.*, B68(23), 235319, 2003.
62. A. J. Williamson and A. Zunger, "InAs quantum dots: Predicted electronic structure of free-standing versus GaAs-embedded structures", *Phys. Rev.*, B59(24), 15819, 1999.
63. D. C. Dente, and M. L. Tilton, "Comparing pseudopotential predictions for InAs/GaSb superlattices", *Phys. Rev.*, B66(16), 165307, 2002.
64. P. Piquini, A. Zunger, and R. Magri, "Pseudopotential calculations of band gaps and band edges of shortperiod $(InAs)_n/(GaSb)_m$ superlattices with different substrates, layer orientations, and interfacial bonds", *Phys. Rev.*, B77(11), 115314, 2008.
65. P. Harrison, *Quantum Wells, Wires and Dots*, Wiley, Chichester, UK, 2009.
66. G. Ariyawansa, J. M. Duran, M. Grupen, J. E. Scheihing, T. R. Nelson, and M. T. Eismann, "Multispectral imaging with type II superlattice detectors", *Proc. SPIE* 8353, 83530E-1–14, 2012.
67. D. Z.-Y. Ting, A. Soibel, L. Höglund, J. Nguyen, C. J. Hill, A. Khoshakhlagh, and S. D. Gunapala, "Type-II superlattice infrared detectors," in *Semiconductors and Semimetals*, Vol. 84, eds. S. D. Gunapala, D. R. Rhiger, and C. Jagadish, 1–57, Elsevier, Amsterdam, 2011.
68. G. A. Umana-Membreno, B. Klein, H. Kala, J. Antoszewski, N. Gautam, M. N. Kutty, E. Plis, S. Krishna, and L. Faraone, "Vertical minority carrier electron transport in p-type InAs/GaSb type-II superlattices", *Appl. Phys. Lett.*, 101, 253515, 2012.
69. E. H. Steenbergen, O. O. Cellek, D. Labushev, Y. Qiu, J. M. Fastenau, A. W. K. Liu, and Y.-H. Zhang, "Study of the valence band offsets between InAs and $InAs_{1-x}Sb_x$ alloys," *Proc. SPIE* 8268, 82680K-1–9, 2012.
70. I. Vurgaftman, G. Belenky, Y. Lin, D. Donetsky, L. Shterengas, G. Kipshidze, W. L. Sarney, and S. P. Svensson, "Interband absorption strength in long-wave infrared type-II superlattices with small and large superlattice periods compared to bulk materials", *Appl. Phys. Lett.*, 108, 222101-1–5, 2016.
71. E. R. Youngdale, J. R. Meyer, C. A. Hoffman, F. J. Bartoli, C. H. Grein, P. M. Young, H. Ehrenreich, R. H. Miles, and D. H. Chow, "Auger lifetime enhancement in $InAs\text{-}Ga_{1-x}In_xSb$ superlattices," *Appl. Phys. Lett.*, 64, 3160–2, 1994.

72. C. H. Grein, P. M. Young, M. E. Flatté, and H. Ehrenreich, "Long wavelength InAs/InGaSb infrared detectors: Optimization of carrier lifetimes," *J. Appl. Phys.*, 78, 7143–52, 1995.
73. D. Zuo, P. Qiao, D. Wasserman, and S. L. Chuang, "Direct observation of minority carrier lifetime improvement in InAs/GaSb type-II superlattice photodiodes via interfacial layer control", *Appl. Phys. Lett.*, 102, 141107, 2013.
74. M. Z. Tidrow, L. Zheng, and H. Barcikowski, "Recent success on SLS FPAs and MDA's new direction for development," *Proc. SPIE* 7298, 7298–61, 2009.
75. Y. Aytac, B. V. Olson, J. K. Kim, E. A. Shaner, S. D. Hawkins, J. F. Klem, M. E. Flatte, and T. F. Boggess, "Effects of layer thickness and alloy composition on carrier lifetimes in mid-wave infrared InAs/InAsSb superlattices", *Appl. Phys. Lett.*, 105, 022107, 2014.
76. J. G. Pellegrino, R. DeWames, P. Perconti, C. Billman, and P. Maloney, "HOT MWIR HgCdTe performance on CZT and alternative substrates," *Proc. SPIE* 8353, 83532X, 2012.
77. S. R. Jost, V. F. Meikleham, and T. H. Myers, "InSb: A key material for IR detector applications," *Mat. Res. Symp. Proc.*, 90, 429–36, 1987.
78. E. H. Steenbergen, B. C. Connelly, G. D. Metcalfe, H. Shen, M. Wraback, D. Lubyshev, Y. Qiu, J. M. Fastenau, A. W. K. Liu, S. Elhamri, O. O. Cellek, and Y.-H. Zhang, "Significantly improved minority carrier lifetime observed in a long-wavelength infrared III-V type-II superlattice comprised of InAs/InAsSb," *Appl. Phys. Lett.*, 99, 251110, 2011.
79. J. R. Meyer, C. L. Felix, W. W. Bewley, I. Vurgaftman, E. H. Aifer, L. J. Olafsen, J. R. Lindle, C. A. Hoffman, M. J. Yang, B. R. Bennett, B. V. Shanabrook, H. Lee, C. H. Lin, S. S. Pei, and R. H. Miles, "Auger coefficients in type-II InAs/$Ga_{1-x}In_xSb$ quantum wells," *Appl. Phys. Lett.*, 73, 2857–9, 1998.
80. W. W. Bewley, J. R. Lindle, C. S. Kim, M. Kim, C. L. Canedy, I. Vurgaftman, and J. R. Meyer, "Lifetimes and Auger coefficients in type-II W interband cascade lasers", *Appl. Phys. Lett.*, **93**, 041118-1–3, 2008.
81. G. Ariyawansa, E. Steenbergen, L. J. Bissell, J. M. Duran, J. E. Scheihing, and M. T. Eismann, "Absorption characteristics of mid-wave infrared type-II superlattices", *Proc. SPIE* 9070, 90701J-1–15, 2014.
82. J. L. Johnson, L. A. Samoska, A. C. Gossard, J. L. Merz, M. D. Jack, G. H. Chapman, B. A. Baumgratz, K. Kosai, and S. M. Johnson, "Electrical and optical properties of infrared photodiodes using the InAs/$Ga_{1-x}In_xSb$ superlattice in heterojunctions with GaSb," *J. Appl. Phys.*, 80, 1116–27, 1996.
83. F. Fuchs, U. Weimer, W. Pletschen, J. Schmitz, E. Ahlswede, M. Walther, J. Wagner, and P. Koidl, "High performance InAs/$Ga_{1-x}In_xSb$ superlattice infrared photodiodes," *Appl. Phys. Lett.*, 71, 3251–3, 1997.
84. R. Taalat, J.-B. Rodriguez, M. Delmas and P. Christol, "Influence of the period thickness and composition on the electro-optical properties of type-II InAs/GaSb midwave infrared superlattice photodetectors", *J. Phys. D: Appl. Phys.*, 47, 015101, 2014.
85. H. S. Kim, E. Plis, A. Khoshakhlagh, S. Myers, N. Gautam, Y. D. Sharma, L. R. Dawson, S. Krishna, S. J. Lee, S. K. Noh, "Performance improvement of InAs/GaSb strained layer superlattice detectors by reducing surface leakage currents with SU-8 passivation", *Appl. Phys. Lett.*, 96, 033502-1–3, 2010.
86. D. Hoffman, B. M. Nguyen, P. Y. Delaunay, A. Hood, and M. Razeghi, "Beryllium compensation doping of InAs/GaSb infrared superlattice photodiodes", *Appl. Phys. Lett.*, 91, 143507-1–3, 2007.
87. B. Klein, E. Plis, M. N. Kutty, N. Gautam, A. Albrecht, S. Myers, and S. Krishna," Varshni parameters for InAs/GaSb strained layer superlattice infrared photodetectors", *J. Phys. D: Appl. Phys.*, 44, 075102-1–5, 2011.
88. H. S. Kim, E. Plis, N. Gautam, A. Khoshakhlagh, S. Myers, M. N. Kutty, Y. Sharma, L. R. Dawson, and S. Krishna, "SU-8 passivation of type-II InAs/GaSb strained layer superlattice detectors", *Proc. SPIE* 7660, 76601U-1–9, 2010.
89. A. Rogalski, *Infrared Detectors*, 2nd ed., CRC Press, Boca Raton, FL, 2010.
90. J. Wrobel, P. Martyniuk, E. Plis, P. Madejczyk, W. Gawron, S. Krishna, and A. Rogalski, "Dark current modeling of MWIR type-II superlattice detectors", *Proc. SPIE* 8353, 835316-1–10, 2012.
91. J. Wróbel, E. Plis, W. Gawron, M. Motyka, P. Martyniuk, P. Madejczyk, A. Kowalewski, M. Dyksik, J. Misiewicz, S. Krishna, and A. Rogalski, "Analysis of temperature dependence of dark current mechanisms in mid-wavelength infrared pin type-II superlattice photodiodes", *Sens. Mater.*, 26(4), 235–44, 2014.
92. J. Wróbel, Ł. Ciura, M. Motyka, F. Szmulowicz, A. Kolek, A. Kowalewski, P. Moszczyński, M. Dyksik, P. Madejczyk, S. Krishna, and A. Rogalski, "Investigation of a near mid-gap trap energy level in mid-wavelength infrared InAs/GaSb type-II superlattices", *Semicond. Sci. Technol.*, 30, 115004-1–10, 2015.
93. P. Martyniuk, J. Wróbel, E. Plis, P. Madejczyk, A. Kowalewski, W. Gawron, S. Krishna, and A. Rogalski, "Performance modeling of MWIR InAs/GaSb/B-$Al_{0.2}Ga_{0.8}Sb$ type-II superlattice nBn detector", *Semicond. Sci. Technol.*, 27, 055002-1–10, 2012.
94. D. R. Rhiger, "Performance comparison of long-wavelength infrared type II superlattice devices with HgCdTe," *J. Electron. Mater.*, 40, 1815–22, 2011.

95. P. Christol and J. B. Rodriguez, "Progress on type-II InAs/GaSb superlattice infrared photodetectors: from MWIR to VLWIR spectral domains", *ICSO* 2014 *International Conference on Space Optics*, Tenerife, Canary Islands, 7–10 October 2014.
96. P. Christol, M. Delmas, R. Rossignol, and J. B. Rodriguez, "Influence of the InAs/GaSb super lattice period composition on the electro-optical performances of T2SL infrared photodiode", *3rd International Conference and Exhibition on Lasers, Optics and Photonics*, Valencia, Spain, 1–3 September 2015.
97. S. P. Svensson, D. Donetsky, D. Wang, H. Hier, F. J. Crowne, and G. Belenky, "Growth of type II strained layer superlattice, bulk InAs and GaSb materials for minority lifetime characterization", *J. Cryst. Growth* 334, 103–7, 2011.
98. J. L. Johnson, "The InAs/GaInSb strained layer superlattice as an infrared detector material: An overview", *Proc. SPIE* 3948, 118–32, 2000.
99. G. J. Brown, "Type-II InAs/GaInSb superlattices for infrared detection: An overview," *Proc. SPIE* 5783, 65–77, 2005.
100. M. Razeghi, Y. Wei, A. Gin, A. Hood, V. Yazdanpanah, M. Z. Tidrow, and V. Nathan, "High performance type II InAs/GaSb superlattices for mid, long, and very long wavelength infrared focal plane arrays," *Proc. SPIE* 5783, 86–97, 2005.
101. J. Pellegrini and R. DeWames, "Minority carrier lifetime characteristics in type II InAs/GaSb LWIR superlattice $n^+\pi p^+$ photodiodes," *Proc. SPIE* 7298, 7298–67, 2009.
102. R. Rehm, F. Lemke, J. Schmitz, M. Wauro, and M. Walther, "Limiting dark current mechanisms in antimony-based superlattice infrared detectors for the long-wavelength infrared regime", *Proc. SPIE* 9451, 94510N, 2015.
103. R. Rehm, F. Lemke, M. Masur, J. Schmitz, T. Stadelman, M. Wauro, A. Wörl, and M. Walther, "InAs/GaSb superlattice infrared detectors", *Infrared Phys. Technol.*, 70, 87–92, 2015.
104. C. L. Canedy, H. Aifer, I. Vurgaftman, J. G. Tischler, J. R. Meyer, J. H. Warner, and E. M. Jackson, "Antimonide type-II W photodiodes with long-wave infrared R_oA comparable to HgCdTe," *J. Electron. Mater.*, 36, 852–6, 2007.
105. E. H. Aifer, J. G. Tischler, J. H. Warner, I. Vurgaftman, W. W. Bewley, J. R. Meyer, C. L. Canedy, and E. M. Jackson, "W-structured type-II superlattice long-wave infrared photodiodes with high quantum efficiency," *Appl. Phys. Lett.*, 89, 053519, 2006.
106. B.-M. Nguyen, D. Hoffman, Y. Wei, P.-Y. Delaunay, A. Hood, and M. Razeghi, "Very high quantum efficiency in type-II InAs/GaSb superlattice photodiode with cutoff of 12 μm," *Appl. Phys. Lett.*, 90, 231108, 2007.
107. J. Bajaj, G. Sullivan, D. Lee, E. Aifer, and M. Razeghi, "Comparison of type-II superlattice and HgCdTe infrared detector technologies," *Proc. SPIE* 6542, 65420B, 2007.
108. C. H. Grein, H. Cruz, M. E. Flatte, and H. Ehrenreich, "Theoretical performance of very long wavelength $InAs/In_xGa_{1-x}Sb$ superlattice based infrared detectors," *Appl. Phys. Lett.*, 65, 2530–2, 1994.
109. A. Hood, D. Hoffman, Y. Wei, F. Fuchs, and M. Razeghi, "Capacitance-voltage investigation of high-purity InAs/GaSb superlattice photodiodes," *Appl. Phys. Lett.*, 88, 052112, 2006.
110. D. Lackner, M. Steger, M. L. W. Thewalt, O. J. Pitts, Y. T. Cherng, S. P. Watkins, E. Plis, and S. Krishna, "InAs/InAsSb strain balanced superlattices for optical detectors: Material properties and energy band simulations," *J. Appl. Phys.*, 111, 034507–10, 2012.
111. T. Schuler-Sandy, S. Myers, B. Klein, N. Gautam, P. Ahirwar, Z.-B. Tian, T. Rotter, G. Balakrishnan, E. Plis, and S. Krishna, "Gallium free type II InAs/InAsSb superlattice photodetectors", *Appl. Phys. Lett.*, 101, 071111-1–3, 2012.
112. A. M. Hoang, G. Chen, R. Chevallier, A. Haddadi, and M. Razeghi, "High performance photodiodes based on InAs/InAsSb type-II superlattices for very long wavelength infrared detection", *Appl. Phys. Lett.*, 104, 251105-1–4, 2014.
113. A. Rogalski, P. Martyniuk, and M. Kopytko, "Challenges of small-pixel infrared detectors: a review", *Rep. Prog. Phys.*, 79, 046501-1–42, 2016.
114. E. Plis, M. N. Kutty, and S. Krishna, "Passivation techniques for InAs/GaSb strained layer superlattice detectors", *Laser Photonics Rev.*, 7(1), 45–59, 2013.
115. E. K. Huang, D. Hoffman, B.-M. Nguyen, P.-Y. Delaunay, and M. Razeghi, "Surface leakage reduction in narrow band gap type-II antimonide-based superlattice photodiodes", *Appl. Phys. Lett.*, 94, 053506-1–3, 2009.
116. R. Rehm, M. Walther, F. Fuchs, J. Schmitz, and J. Fleissner, "Passivation of InAs/(GaIn)Sb short-period superlattice photodiodes with 10 μm cutoff wavelength by epitaxial overgrowth with $Al_xGa_{1-x}As_ySb_{1-y}$", *Appl. Phys. Lett.*, 86, 173501-1–3, 2005.
117. P. Y. Delaunay, A. Hood, B. M. Nguyen, D. Hoffman, Y. Wei, and M. Razeghi, "Passivation of type-II InAs/GaSb double heterostructure," *Appl. Phys. Lett.*, 91, 091112-1–3, 2007.

118. G. Chen, B.-M. Nguyen, A. M. Hoang, E. K. Huang, S. R. Darvish, and M. Razeghi, “Elimination of surface leakage in gate controlled type-II InAs/GaSb mid-infrared photodetectors,” *Appl. Phys. Lett.*, 99, 183503, 2011.
119. F. Szmulowicz and G. J. Brown, “GaSb for passivating type-II InAs/GaSb superlattice mesas,” *Infrared Phys. Technol.*, 53, 305–7, 2011.
120. E. H. Aifer, H. Warner, C. L. Canedy, I. Vurgaftman, J. M. Jackson, J. G. Tischler, J. R. Meyer, S. P. Powell, K. Oliver, and W. E. Tennant, “Shallow-etch mesa isolation of graded-bandgap W-structured type II superlattice photodiodes,” *J. Electron. Mater.*, 39, 1070–9, 2010.
121. U.S. Patent No. 4882245, 1989.
122. H. S. Kim, E. Plis, A. Khoshakhlagh, S. Myers, N. Gautam, Y. D. Sharma, L. R. Dawson, S. Krishna, S. J. Lee, and S. K. Noh, “Performance improvement of InAs/GaSb strained layer superlattice detectors by reducing surface leakage currents with SU-8 passivation,” *Appl. Phys. Lett.*, 96, 033502–4, 2010.
123. E. A. DeCuir, Jr., J. W. Little, and N. Baril, “Addressing surface leakage in type-II InAs/GaSb superlattice materials using novel approaches to surface passivation,” *Proc. SPIE* 8155, 815508, 2011.
124. H. S. Kim, E. Plis, N. Gautam, S. Myers, Y. Sharma, L. R. Dawson, and S. Krishna, “Reduction of surface leakage current in InAs/GaSb strained layer long wavelength superlattice detectors using SU-8 passivation,” *Appl Phys. Lett.*, 97, 14351-2–4, 2010.
125. A. Hood, P. Y. Delaunay, D. Hoffman, B. M. Nguyen, Y. Wei, and M. Razeghi, “Near bulk-limited R_0A of long-wavelength infrared type-II InAs/GaSb superlattice photodiodes with polyimide surface passivation,” *Appl. Phys. Lett.*, 90, 233513, 2007.
126. R. Chaghi, C. Cervera, H. Ait-Kaci, P. Grech, J. B. Rodriguez, and P. Christol, “Wet etching and chemical polishing of InAs/GaSb superlattice photodiodes,” *Semicond. Sci. Technol.*, 24, 065010, 2009.
127. A. Soibel, D. Z.-Y. Ting, C. J. Hill, M. Lee, J. Nguyen, S. A. Keo, J. M. Mumolo, and S. D. Gunapala, “Gain and noise of high-performance long wavelength superlattice infrared detector”, *Appl. Phys. Lett.*, 96, 111102-1–3, 2010.
128. V. M. Cowan, C. P. Morath, S. Myers, N. Gautam, and S. Krishna, “Low temperature noise measurement of an InAs/GaSb-based nBn MWIR detector”, *Proc. SPIE* 8012, 801210, 2011.
129. C. Cervera, I. Ribet-Mohamed, R. Taalat, J. P. Perez, P. Christol, and J. B. Rodriguez, “Dark current and noise measurements of an InAs/GaSb superlattice photodiode operating in the midwave infrared domain”, *J. Electron. Mater.*, 41(10), 2714–8, 2012.
130. T. Tansel, K. Kutluer, A. Muti, O. Salihoglu, A. Aydinli, and R. Turan, “Surface recombination noise in InAs/GaSb superlattice photodiodes”, *Appl. Phys. Express* 6 032202-1–4, 2013.
131. A. Wörl, R. Rehm, and M. Walther, “Excess noise in long-wavelength infrared InAs/GaSb type-II superlattice pin-photodiodes”, *22nd International Conference on Noise and Fluctuations* (*ICNF*), 24–28 June 2013.
132. R. Rehm, A. Wörl, and M. Walther, “Noise in InAs/GaSb type-II superlattice photodiodes”, *Proc. SPIE* 8631, 86311M-1–9, 2013.
133. M. Walther, A. Wörll, V. Daumer, R. Rehm, L. Kirste, F. Rutz, and J. Schmitz, “Defects and noise in type-II superlattice infrared detectors”, *Proc. SPIE* 8704, 87040U-1–9, 2013.
134. E. Giard, R. Taalat, M. Delmas, J.-B. Rodriguez, P. Christol, and I. Ribet-Mohamed, “Radiometric and noise characteristics of InAs-rich T2SL MWIR pin photodiodes”, *J. Euro. Opt. Soc. Rap. Public* 9, 14022-1–6, 2014.
135. Ł. Ciura, A. Kołek, J. Wróbel, W. Gawron, and A. Rogalski, “1/f noise in mod-wavelength infrared detectors with InAs/GaSb superlattice absorber”, *IEEE Trans. Electron Devices* 62(6), 2022–6, 2015.

21 Quantum dot infrared photodetectors

The success of quantum well (QW) structures for infrared (IR) detection applications has stimulated the development of quantum dot infrared photodetectors (QDIPs). In general, QDIPs are similar to quantum well infrared photodetectors (QWIPs) but with the QWs replaced by quantum dots (QDs), which have size confinement in all spatial directions.

Recent advances in the epitaxial growth of strained heterostructures, such as InGaAs on GaAs, have led to the realization of coherent islands through the process of self-organization. These islands behave electronically as quantum boxes or QDs. Zero-dimensional quantum confined semiconductor heterostructures have been investigated theoretically and experimentally for some time [1–3]. At present, nearly defect-free QD devices can be fabricated reliably and reproducibly. Also, new types of IR photodetectors taking advantage of the quantum confinement obtained in semiconductor heterostructures have emerged. Like the QWIPs, the QDIPs are based on optical transitions between bound states in the conduction (valence) band in QDs. Also, like the QWIPs, they benefit from a mature technology with large-bandgap semiconductors.

First observations of intersublevel transitions in the far IR region were reported in the early 1990s, either in InSb-based electrostatically defined QDs [4] or in structured two-dimensional (2D) electron gas [5]. The first QDIP was demonstrated in 1998 [6]. Great progress has been made in their development and performance characteristics [7–9] and in their applications to thermal imaging focal plane arrays (FPAs) [10].

Interest in QD research can be traced back to a suggestion by Arakawa and Sakaki in 1982 [1] that the performance of semiconductor lasers could be improved by reducing the dimensionality of the active regions of these devices. Initial efforts at reducing the dimensionality of the active regions focused on using ultrafine lithography coupled with wet or dry chemical etching to form 3D structures. However, it was soon realized that this approach introduced defects (high density of surface states) that greatly limited the performance of such QDs. Initial efforts were mainly focused on the growth of InGaAs nanometer-sized islands on GaAs substrates. In 1993, the first epitaxial growth of defect-free QD nanostructures was achieved by using MBE [11]. Most of the practical QD structures today are synthesized both by MBE and MOCVD.

21.1 QDIP PREPARATION AND PRINCIPLE OF OPERATION

Under certain growth conditions, when the film with the larger lattice constant exceeds a certain critical thickness, the compressive strain within the film is relieved by the formation of a coherent island. Figure 21.1 qualitatively shows the changes in the total energy of a mismatched system versus time [12]. The plot can be divided into three sections: period *a* (2D deposition), period *b* (2D to 3D transition), and period *c* (ripening of islands). In the beginning, the deposition of a 2D layer by layer mechanism leads to a perfect wetting of the substrate. At the point t_{cw} (the critical wetting layer thickness) the stable 2D growth enters into an area of the metastable growth. A supercritically thick wetting layer builds up and the epilayer is potentially ready to undergo a transition toward a Stranski-Krastanow morphology [13]. This transition starts around point X in Figure 21.1 and its dynamic depends primarily on the height of the transition barrier E_a. It is presumed that further growth continues without material supply, simply by consuming the excess material accumulated in the supercritically thick wetting layer. Between the points Y and Z (ripening: period *c*), the process loses most of the excess energy; the mobile material is consumed as a result of the potential differences between smaller and larger islands. These islands may be QDs.

Coherent QD islands are generally formed only when the growth proceeds as the Stranski-Krastanow growth model [13]. The onset of the transformation of the growth process from a 2D layer-by-layer growth

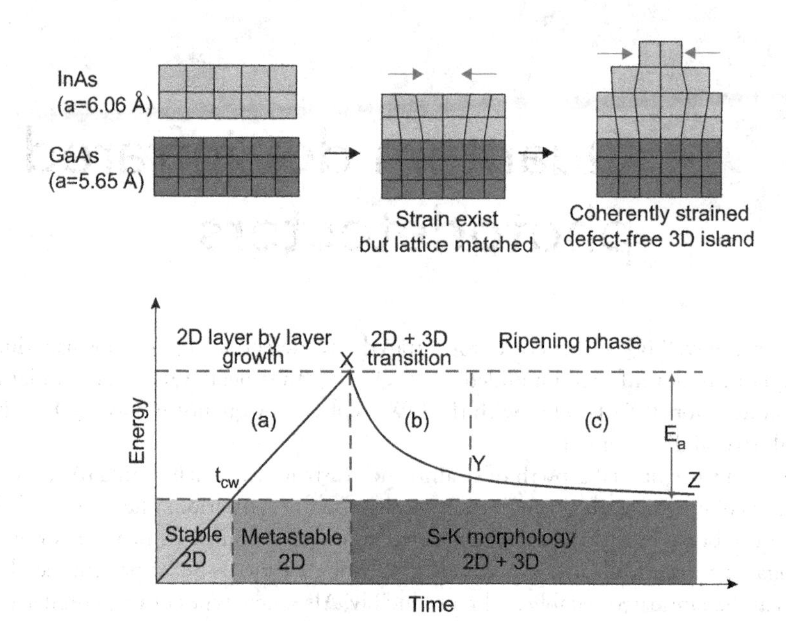

Figure 21.1 Schematics of total energy versus time for the 2D–3D morphology transition. t_{cw} is the critical wetting layer thickness, E_a is the barrier for formation of 3D islands, *X* is the point where a pure strain-induced transition occurs. Between *Y* and *Z*, a slow ripening process continues. (From Ref. [12])

mode to a 3D island growth mode results in a spotty RHEED pattern. This is, in contrast to the conventional streaky pattern, generally observed for the layer-by-layer growth mode. The transition typically occurs after the deposition of a certain number of monolayers. For InAs on GaAs, this transition occurs after about 1.7 monolayers of InAs have been grown; this is the onset of islanding and, hence, QD formation. Noncoherent islands are typically produced by very high material supply and contain misfit dislocations at the interface.

The detection mechanism of QDIP is based on the intraband photoexcitation of electrons from confined states in the conduction band wells or dots into the continuum. The emitted electrons drift toward the collector in the electric field provided by the applied bias and a photocurrent is created. It is assumed that the potential profile at the conduction band edge along the growth direction have a shape similar to QWIP as shown in Figure 19.5. In practice, since the dots are spontaneously self-assembled during growth, they are not correlated between multilayers in the active region.

Two types of QDIP structures have been proposed: conventional structure (vertical, see Figure 21.2) and lateral structure (Figure 21.3). In a vertical QDIP, the photocurrent is collected through the vertical transport of carriers between the top and bottom contacts. The device heterostructure comprises repeated InAs QD layers buried between GaAs barriers with the top and bottom contact layers at active region boundaries. The mesa height can vary from 1 to 4 μm depending on the device heterostructure. The QDs are directly doped (usually with Si) in order to provide free carriers during photoexcitation and an AlGaAs barrier can be included in the vertical device heterostructure in order to block dark current created by thermionic emission [14,15].

The lateral QDIP (Figure 21.3a) collects photocurrent through transport of carriers across a high-mobility channel between the two top contacts, operating much like a field effect transistor. As previously, again AlGaAs barriers are present, but instead of blocking the dark current, these barriers are used to both modulation dope the QDs and to provide the high-mobility channel (see Figure 21.3b). The normally incident IR radiation promotes the carriers from the dots to the continuum from where they are quickly transferred to the high mobility 2D channel on either side due to the favorable band bending. Lateral QDIPs have demonstrated lower dark currents and higher operating temperatures than the vertical QDIPs since the major components of the dark current arise from interdot tunneling and hopping conduction [16]. However, these devices will be difficult to incorporate into a FPA hybrid bump bonded to a Si readout

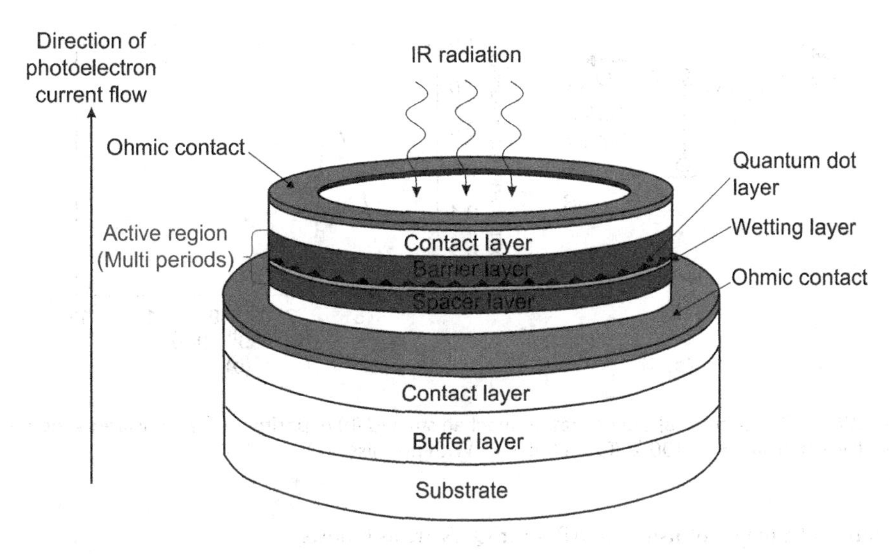

Figure 21.2 Schematic diagram of conventional QD detector structure.

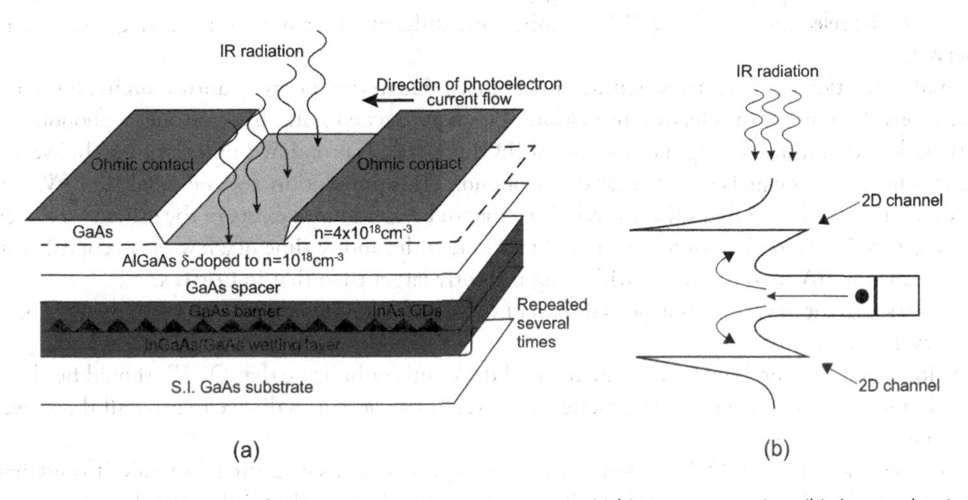

Figure 21.3 Schematic diagram of (a) lateral QD detector structure with a representation; (b) the conduction band profile and photoresponse mechanism.

circuit. Because of this, more effort is directed toward improving the performance of vertical QDIPs, which are more compatible with commercially available readout circuits.

In addition to the standard InAs/GaAs QDIP, several other heterostructure designs have been investigated for use as IR photodetectors [7,8]. An example is InAs QDs embedded in a strain-relieving InGaAs QW, which are known as dot-in-a-well (DWELL) heterostructures (see Figure 21.4) [10,17,18]. This device offers two advantages: challenges in wavelength tuning through dot-size control can be compensated in part by engineering the QW sizes, which can be controlled precisely and the QWs can trap electrons and aid in carrier capture by QDs, thereby facilitating ground state refilling. Figure 21.4b shows DWELL spectral tuning by varying well geometry.

21.2 ANTICIPATED ADVANTAGES OF QDIPs

The quantum mechanical nature of QDIPs leads to several advantages over QWIPs and other types of IR detectors that are available. As in HgCdTe, QWIP, and type II superlattice technologies, QDIPs provide multiwavelength detection. However, QDs provide many additional parameters for tuning the energy spacing between energy levels, such as QD size and shape, strain, and material composition.

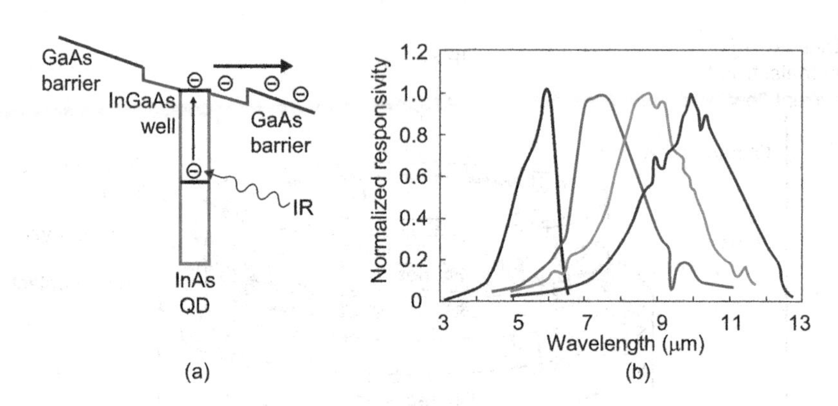

Figure 21.4 DWELL IR detector: (a) the operation mechanism and (b) experimentally measured spectral tunability by varying well width from 55 to 100 Å. (From Ref. [18]. With permission.)

The potential advantages of using QDIPs over QWs are as follows:

- Intersubband absorption may be allowed at normal incidence (for n-type material). In QWIPs, only transitions polarized perpendicularly to the growth direction are allowed due to absorption selection rules. The selection rules in QDIPs are inherently different and a normal incidence absorption is observed.
- Thermal generation of electrons is significantly reduced due to the energy quantization in all three dimensions. As a result, the electron relaxation times from excited states increase due to phonon bottleneck. Generation by longitudinal optical (LO) phonons is prohibited unless the gap between the discrete energy levels equals exactly that of the phonon. This prohibition does not apply to QWs, since the levels are quantized only in the growth direction and a continuum exists in the other two directions (hence generation-recombination (g-r) by LO phonons with capture time of a few picoseconds). Thus, it is expected that *S/N* ratio in QDIPs will be significantly larger than that in QWIPs.
- Lower dark current of QDIPs is expected than of QWIPs due to 3D quantum confinement of the electron wavefunction.

Both the increased electron lifetime and the reduced dark current indicate that QDIPs should be able to provide high temperature operation. In practice, however, it has been a challenge to meet all the above expectations.

Carrier relaxation times in QDs are longer than the typical 1–10 ps measured for QWs. It is predicted that the carrier relaxation time in QDs is limited by electron-hole scattering [19] rather than phonon scattering. For QDIPs, the lifetime is expected to be even larger, greater than 1 ns, since the QDIPs are majority carrier devices due to the absence of holes.

The main disadvantage of QDIP is the large inhomogeneous linewidth of the QD ensemble variation of dot size in the Stranski-Krastanow growth mode [20,21]. As a result, the absorption coefficient is reduced since it is inversely proportional to the ensemble linewidth. Large, inhomogeneously broadened linewidth has a deleterious effect on QDIP performance. Subsequently, the quantum efficiency of QD devices tend to be lower than what is predicted theoretically. Vertical coupling of QD layers also reduces the inhomogeneous linewidth of the QD ensemble; however, it may also increase the dark current of the device since carriers can tunnel through adjacent dot layers more easily. As in other types of detectors, nonuniform dopant incorporation adversely affects the performance of the QDIP. Therefore, improving the QD uniformity is a key issue in increasing absorption coefficient and improving the performance. Thus, the growth and design of a unique QD heterostructure is one of the most important issues related to the achievement of state of the art QDIP performance.

21.3 QDIP MODEL

In further considerations, a QDIP model developed by Ryzhii et al. is adapted [22,23]. The QDIP consists of a stack of QD layers separated by wide-gap material layers (see Figure 21.5). Each QD layer includes

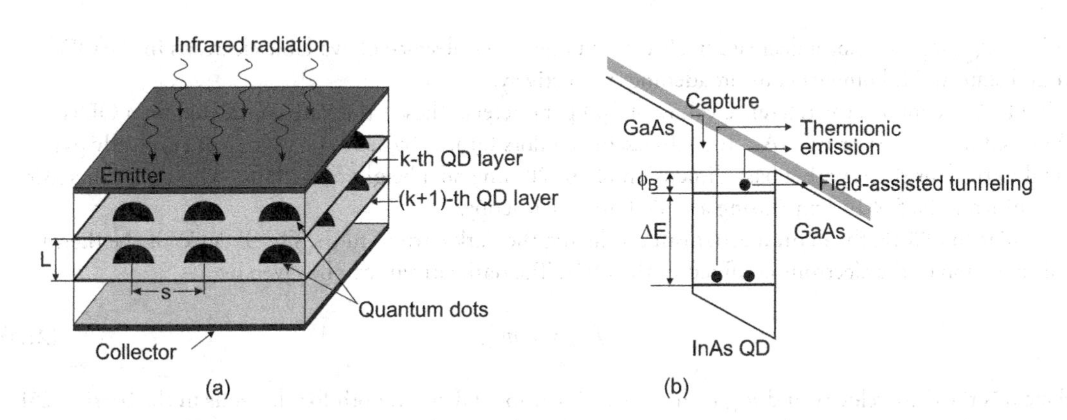

Figure 21.5 Schematic view of (a) the QD structure and (b) conduction band structure of the dot.

periodically distributed identical QDs with a density Σ_{QD} and a sheet density of doping donors equal to Σ_D. In the realistic QDIPs, the lateral size of QDs, a_{QD}, is sufficiently large in comparison with the transverse size, h_{QD}. Consequently, only two energy levels associated with quantization in the transverse direction exist. Relatively sufficiently large lateral size, l_{QD}, causes a large number of bound states in dots and, consequently, is capable of accepting a large number of electrons. Whereas, the transverse size is small in comparison with the spacing between the QD layers, L. The lateral spacing between QDs is equal to $L_{QD} = \sqrt{\Sigma_{QD}}$. The average number of electrons in a QD belonging to the k-th QD layer, $<N_k>$, can be indicated by a solitary QD layer index ($k = 1, 2,\ldots, K$, where K is the total number of the QD layers). The QDIP active region (the stack of QD arrays) is sandwiched between two heavily doped regions that serve as the emitter and collector contacts.

Due to the discrete nature of QDs, the fill factor F should be included for optical absorption in QDs. This factor can be estimated in a simple way as:

$$F = \frac{\sqrt[3]{V}}{s}, \tag{21.1}$$

where V is the quantum dot volume.

For self-assembled QDs, a Gaussian distribution has been observed for the electronic and optical spectra. Phillips modeled the absorption spectra for an ensemble of QDs using a Gaussian line shape in the shape [20]:

$$\alpha(E) = \alpha_0 \frac{n_1}{\delta} \frac{\sigma_{QD}}{\sigma_{ens}} \exp\left[-\frac{(E - E_g)^2}{\sigma_{ens}^2} \right], \tag{21.2}$$

where α_0 is the maximum absorption coefficient, n_1 is the areal density of electrons in the QD ground state, δ is the QD density, and $E_g = E_2 - E_1$ is the energy of the optical transition between ground and excited states in the QDs. It should be noticed that Equation 21.2 estimates the absorption coefficient for the necessary presence of electrons in the QD ground state.

The optical absorption between the ground and excited levels is found to have a value [24]

$$\alpha_0 \approx \frac{3.5 \times 10^5}{\sigma}, \tag{21.3}$$

where σ is the linewidth of the transition in meV. Equation 21.3 indicates the trade-off between the absorption coefficient and the absorption linewidth, σ.

The expressions σ_{QD} and σ_{ens} are the standard deviations in the Gaussian line shape for intraband absorption in a single QD and for the distribution of energies for the QD ensemble, respectively. The terms

n_1/δ and σ_{QD}/σ_{ens} describe a decrease in absorption due to the absence of available electrons in the QD ground state and inhomogeneous broadening, respectively.

Table 21.1 contains the reference values of QD parameters. These values are considered for a QDIP fabricated from GaAs or InGaAs. The self-assembled dots formed by epitaxial growth are typically pyramidal to the lens shape with a base dimension of 10–20 nm and a height of 4–8 nm, with an areal density determined to be 5×10^{10}cm^{-2} using atomic force microscopy.

Similar to QWIP, the main mechanism producing the dark current in the QDIP device is the thermionic emission of the electrons confined in the QDs. The dark current can be given by

$$J_{dark} = evn_{3D}, \tag{21.4}$$

where v is the drift velocity and n_{3D} is the three-dimensional density, both for electrons in the barrier [25]. Equation 21.4 neglects the diffusion contribution. The electron density can be estimated by

$$n_{3D} = 2\left(\frac{m_b kT}{2\pi\hbar^2}\right)^{3/2} \exp\left(-\frac{E_a}{kT}\right), \tag{21.5}$$

where m_b is the barrier effective mass and E_a is the activation energy, which equals the energy difference between the top of the barrier and the Fermi level in the dot. At a higher operating temperature and larger bias voltage, the contribution of field-assisted tunneling through a triangular potential barrier is considerable [26,27].

Figure 21.6 shows, for example, the normalized dark current versus bias for temperature range 20 K–300 K for QDIP with AlGaAs confinement layers below the QD layer and on top of the GaAs cap layers [8]. In such a case, we have the InAs islands into QWs, and the AlGaAs blocking layers effectively improve the dark current and detectivity. As it is shown, at a low temperature (e.g., 20 K), the dark current increases

Table 21.1 Typical parameter values of QDIP fabricated from GaAs or InGaAs

a_{QD}	h	Σ_{QD}	Σ_D	L	K	N_{QD}
10–40 nm	4–8 nm	$(1–10) \times 10^{10}$cm^{-2}	(0.3–0.6) Σ_{QD}	40–100 nm	10–70	8

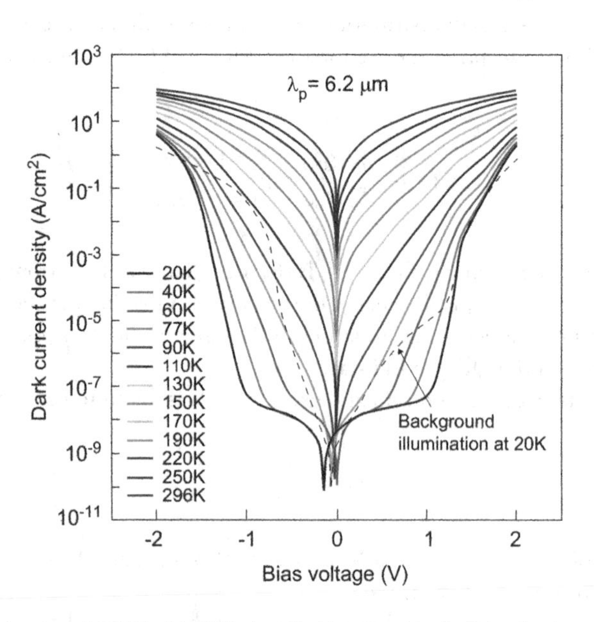

Figure 21.6 Dark current density of QDIP with AlGaAs blocking layer including photocurrent induced by room-temperature background. (From Ref. [8]. With permission.)

rapidly as the bias is increased, which is attributed to electron tunneling between the QDs. For higher bias $|0.2| \leq V_{bias} \leq |1.0|$, the dark current increases slowly. With further increase in bias, the dark current strongly increases, which is largely due to lowering of the potential barriers. Figure 21.6 also shows the photocurrent induced by the room-temperature background. It is clear that background limited infrared photodetector (BLIP) temperature varies with bias.

As Equation 12.1 describes, the photocurrent is determined by the quantum efficiency and gain, g. The photoconductive gain is defined as the ratio of total collected carriers to total excited carriers, whether these carriers are thermally generated or photogenerated. Usually in photoconductors, the gain is greater than one since the carrier lifetime, τ_e, exceeds the carrier transit time, τ_t, through the device between contacts

$$g_{ph} = \frac{\tau_e}{\tau_t}. \tag{21.6}$$

In InAs/GaAs QDIPs, the gain has typical values in the range 1–5. However, the gain strongly depends on the QDIP design and detector polarization. Much higher values, up to several thousands, have been observed [8,21]. The higher gain of the QDIPs in comparison with QWIPs (typically in the range 0.1–50 for similar electric field intensities) is the result of longer carrier lifetimes. The larger photoconductive gain has an influence on higher current responsivity (see Equation 4.31).

The photoconductive gain and the noise gain in the conventional photoconductive detector are equal to each other. It is not the same in QDIPs since these devices are neither homogeneous nor are they bipolar devices. The photoconductive gain in QWIPs is expressed in terms of the capture probability, p_c, as [28,29]:

$$g_{ph} = \frac{1-p_c/2}{Np_c}, \tag{21.7}$$

where $p_c \ll 1$ and N is the number of QW layers. This equation is approximately correct for QDs after including the fill factor, F, in the denominator that takes into account the surface density of discrete dots across the single layer [30]. Then

$$g_{ph} = \frac{1-p_c/2}{Np_cF}. \tag{21.8}$$

Ye et al. [31] have estimated an average value of F as equal to 0.35. The published paper by Lu and colleagues indicates [32] that temperature-dependent photoresponsivity is attributed to temperature-dependent electron capture probability. The capture probability can be changed in a wide region, from below 0.01 to above 0.1, with dependence on bias voltage and temperature.

The noise current of QDIP contains both g-r noise current and thermal noise (Johnson noise) current.

$$I_n^2 = I_{nGR}^2 + I_{nJ}^2 = 4qg_nI_d\Delta f + \frac{4kT}{R}\Delta f, \tag{21.9}$$

where R is the differential resistance of the QDIP, which can be extracted from the slope of the dark current.

It can be shown that the noise gain is related to the electron capture probability, p_c, as

$$g_n = \frac{1}{Np_cF}. \tag{21.10}$$

In a typical QDIP, the thermal noise is significant in the very low bias region. For example, Figure 21.7 shows the bias dependence of the noise current at 77, 90, 105, 120, and 150 K and a measurement frequency of 140 Hz for InAs/GaAs QDIP [31]. The calculated thermal noise current is also shown at 77 K.

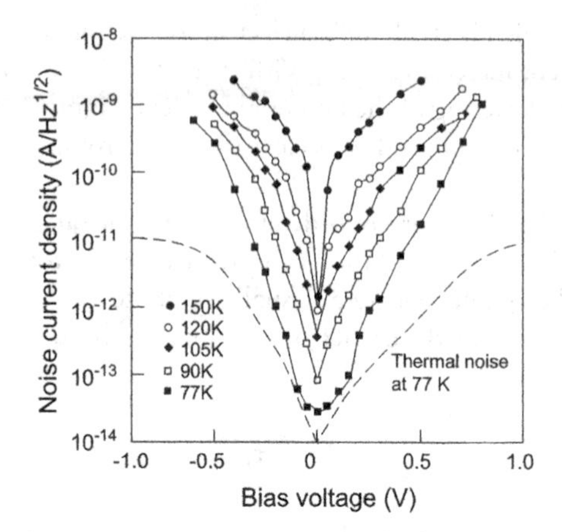

Figure 21.7 Noise current density versus bias voltage at 77, 90, 105, 120, and 150 K. The symbols are measured data. The dashed line is calculated thermal noise current at 77 K. (From Ref. [31]. With permission.)

Thermal noise is significant in the very low bias region $|V_{bias}| \leq 0.1$ V. As the bias increases, the detector noise current increases much faster than the thermal noise and it is primarily g-r noise.

The larger photoconductive gain has influence on the higher current responsivity.

$$R_i = \frac{q\lambda}{hc}\eta g_{ph}. \tag{21.11}$$

Detectivity is defined as the rms signal-to-noise ratio in a 1Hz bandwidth per unit rms incident radiant power per square root of detector area, A_d, and can be determined as:

$$D^* = \frac{(A_d\Delta f)^{1/2}}{I_n}R_i = \frac{q\lambda}{hc}\frac{\eta g_{ph}}{\left(I_{nGR}^2 + I_{nJ}^2\right)^{1/2}}(A_d\Delta f)^{1/2}. \tag{21.12}$$

The quantum efficiency often measured in practice is low, typically ≈2%. It should be noticed, however, that rapid progress has recently been made in the performance of QDIP devices, especially at near room temperature. Lim et al. [33] have announced a quantum efficiency of 35% for detectors with peak detection wavelength around 4.1 μm.

Figure 21.8 shows a typical photoconductivity spectrum of an InAs/GaAs vertical QDIP [25]. The wedge coupling geometry is shown in the inset. The S polarization corresponds to an electric field in the layer plane, while the electric field of the P-polarized excitation has a component along the *z* growth axis and in the layer plane. Clearly the P-polarized response is much stronger than the S-polarized response and the results reported in Figure 21.8 indicate that the photoresponse of a QDIP can be easily measured for an in-plane polarized excitation (i.e., operation at normal incidence is feasible).

The photoconductivity spectra are believed to be due to the fact that self-assembled QDs grown so far for QDIPs are wide in the plane direction (~20 nm) and narrow in the growth direction (~3 nm) [25]. The strong confinement is therefore in the growth direction, while the in-plane confinement is weak, resulting in several levels in the dots (see Figure 19.6). The transitions between the in-plane confined levels give rise to the normal incidence response.

Analysis of the fundamental performance limitation of QDIPs carried out by Phillips indicates that the predicted performance of very uniform QDIP [when $\sigma_{ens}/\sigma_{QD} = 1$] rivals that of HgCdTe (see Figure 21.9 [20]). However, as is mentioned previously, poor QDIP performance is generally linked to two sources: nonoptimal band structure and nonuniformity in QD size. In the Phillips analysis, two electron energy levels are assumed, where the excited state coincides with the conduction band minimum of the barrier material. If the excited state is below the barrier conduction band, photocurrent is difficult to extract,

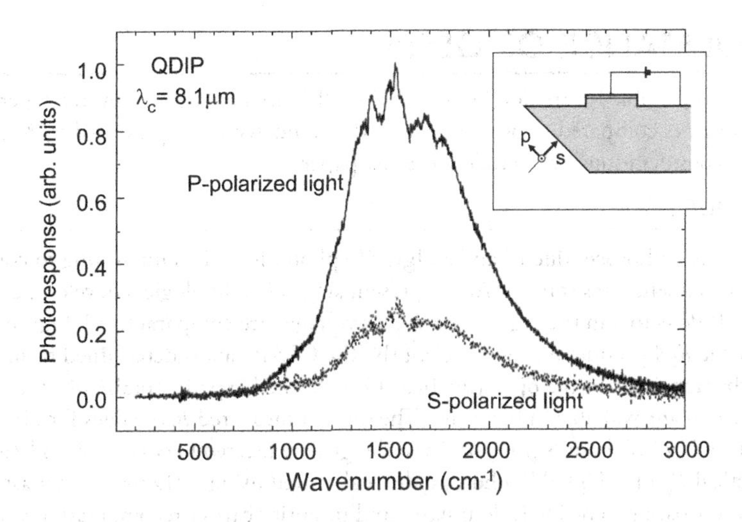

Figure 21.8 P- and S-polarized spectral response curves in the 45° facet detector geometry. The QDIPs have 50 layers of InAs dots separated by 30 nm GaAs barriers. The dot density is about $5 \times 10^9 cm^{-2}$. The number of electrons is estimated to be one per dot. (From Ref. [25]. With permission.)

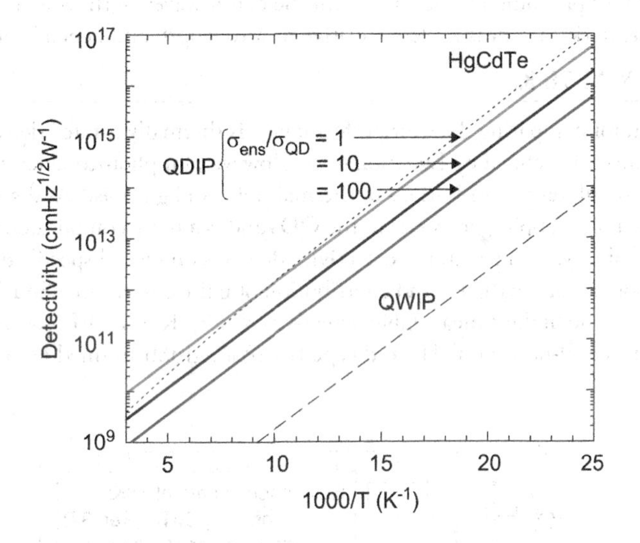

Figure 21.9 Detectivities of HgCdTe, QWIP, and QDIP detectors as functions of temperature with a bandgap energy corresponding to the wavelength 10 μm. (From Ref. [20]. With permission.)

which is reflected in the low responsivity and detectivity. Also, usually, QDs contain additional energy levels between the excited and ground state transitions. If these states are similar to the thermal excitations or permit phonon scattering between levels, carrier lifetime will be dramatically reduced. In consequence, a large increase in dark current and a reduction in detectivity are observed. In case of Stranski-Krastanow growth mode, some degradation of self-assembled QDs occurs due to a coupling 2D wetting layer.

Fabrication of the QDs also strongly affect the absorption properties in the material. If lateral quantum confinement in QDs is small and they resemble QWs more closely, sensitivity to normal incidence radiation is decreased, resulting in an increase in dark current and a reduced detectivity. At the current state of QD fabrication technology, the inhomogeneous broadening of QD energy levels modeled by values of σ_{ens}/σ_{QD} is about 100. Moreover, the experimentally measured quantum efficiency is low by several percentages, and it has not come to equal that of QWIP. Thus, size uniformity of QDs needs to be improved together with the ability to increase the QD density to improve the QDIP's performance.

21.4 PERFORMANCE OF QDIPs

The present status and possible future developments of QDIPs are reviewed in several papers [34–36]. The performance of QDIPs as compared to other types of IR photodetectors is presented by Martyniuk and Rogalski [34]. Our considerations below follow the last paper.

21.4.1 R_0A PRODUCT

In spite of QDIP being a photoconductor and a HgCdTe photodiode, it is interesting to compare their dark currents and incremental resistances. At the present stage of technological development, the dark currents of both MWIR detectors in the region of low bias voltages are comparable [9]. Figure 21.10 displays the dependence of the R_0A product on the wavelength. The QDIP data is determined from the dynamic resistance in I-V characteristics at the operating bias. Only limited experimental R_0A values for the QDIPs marked in Figure 21.10 are available in literature. The highest measured R_0A values for HgCdTe photodiodes operated at 78 K with about 5-μm cutoff wavelength are located between 10^8 and 10^9 Ωcm^2. The solid line is theoretical R_0A for HgCdTe photodiodes, calculated using a 1D model that assumes diffusion current from narrower-bandgap n-side is dominant, and minority carrier recombination via Auger and radiative processes. Theoretical calculations used typical values for the p-side donor concentration ($N_d = 1 \times 10^{15}$ cm^{-3}) and the narrow-bandgap active layer thickness (10 μm).

The R_0A product is an inherent property of the HgCdTe ternary alloy and depends on the cutoff wavelength. The dark current of photodiodes increases with the cutoff wavelength, which is an important difference from QDIPs where dark current is far less sensitive to wavelength and depends on device geometry.

21.4.2 DETECTIVITY AT 78 K

A useful figure of merit, for comparing detector performance, is thermally limited detectivity. In case of photodiodes, this parameter is defined by Equation 4.55. However, for photoconductors, the situation is more complicated due to different contributions of thermal noise and g-r noise. As discussed above, the noise in QDIP originates from the trapping processes in the QDs and is a more complicated function of detector design and capture probability. As a result, the detectivity depends on several specific quantities, such as the quantum efficiency, photoconductive gain, and contribution of noise current (see Equation 21.12).

Figure 21.11 compares the highest measurable detectivities at 77 K of QDIPs found in literature with the predicted detectivities of P-on-n HgCdTe and type II InAs/GaInSb strained-layer superlattice (SLS)

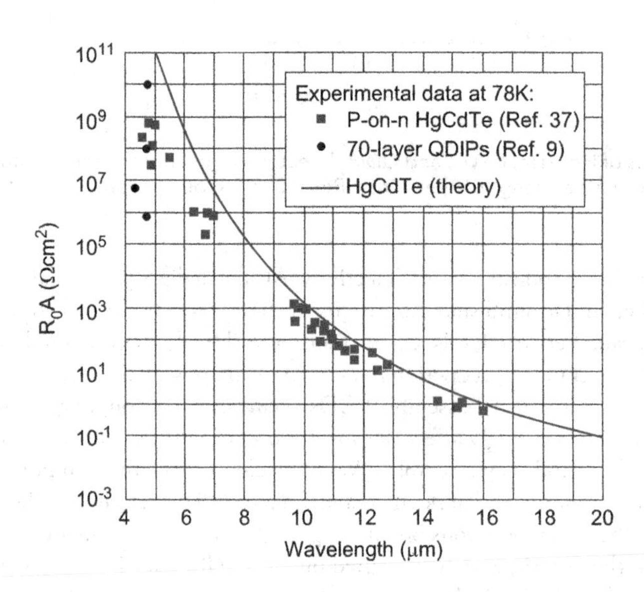

Figure 21.10 R_0A versus wavelength for P-on-n HgCdTe photodiodes and QDIPs at 78 K. Solid line is calculated theoretically assuming 1D n-side diffusion model.

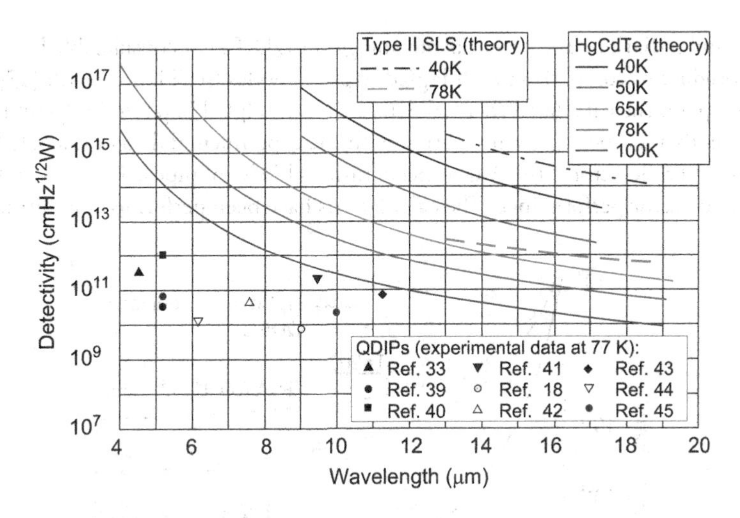

Figure 21.11 The predicted detectivity of P-on-n HgCdTe and type II InAs/GaInSb SLS photodiodes compared with measured QDIP detectivities at 77 K.

photodiodes. It should be insisted that for HgCdTe photodiodes theoretically predicted curves for a temperature range between 50 K and 100 K coincide very well with the experimental data (not shown in Figure 21.11). The predicted thermally limited detectivities for the type II SLS are larger than those for HgCdTe [38].

The measured values of QDIP detectivities at 77 K gathered in Figure 21.11 indicate that QD device detectivities are as yet considerably inferior to current HgCdTe detector performance. In the LWIR region, the upper experimental QDIP data at 77 K coincide with the HgCdTe ones at temperature 100 K.

21.4.3 PERFORMANCE AT HIGHER TEMPERATURE

One of the main potential advantages of QDIPs is low dark current. In particular, the lower dark currents enable higher operating temperatures. Up until now, however, most of the QDIP devices reported in the literature have been working in the temperature range of 77 K–200 K. On account of this fact, it is an interesting insight on achievable QDIP performance in temperatures above 200 K in comparison with other types of detectors.

Most modern IR FPAs are fabricated from the hybrid devices—a detector array made from compound semiconductor materials and a Si signal processing chip called a readout integrated circuit (ROIC). To receive high injection efficiency, the input impedance of the metal oxide semiconductor field effect transistor (MOSFET) must be much lower than the internal dynamic resistance of the detector at its operating point, and the condition $IR_d >> \beta kT/q$ should be fulfilled (R_d is the dynamic impedance of the detector and β is an ideality factor usually in the range 1-2). Generally, it is not a problem to fulfil this inequality for short wavelength infrared (SWIR) and middle wavelength infrared (MWIR) FPAs where the dynamic resistance of detector, R_d, is large, but it is very important for long wavelength infrared (LWIR) designs where R_d is low. There are more complex injection circuits that effectively reduce the input impedance and allow lower detector resistance to be used.

The above requirement is especially critical for near room temperature HgCdTe photodetectors operating in the LWIR region. Their resistance is very low due to a high thermal generation. In materials with a high electron-to-hole ratio, such as HgCdTe, the resistance is additionally reduced by ambipolar effects. Small-sized, uncooled 10.6 μm photodiodes (50 × 50 μm²) exhibit less than 1 Ω zero-bias junction resistances that are well below the series resistance of a diode [46]. As a result, the performance of conventional devices is very poor, so they are not usable for practical applications. The saturation current for a 10 μm photodiode achieves 1,000 A cm^{-2} and it is by four orders of magnitude larger than the photocurrent due to the 300 K background radiation. The potential advantages of QDIPs is a considerably lower dark current and a higher R_0A product in comparison with HgCdTe photodiodes (see Figure 21.12) [47].

Figure 21.13 compares the calculated thermal detectivities of HgCdTe photodiodes and QDIPs as functions of wavelength and operating temperature with the experimental data of uncooled HgCdTe and type II InAs/GaInSb SLS detectors. The Auger mechanism is likely to impose fundamental limitations on the LWIR HgCdTe detector performance. The calculations have been performed for optimized doping

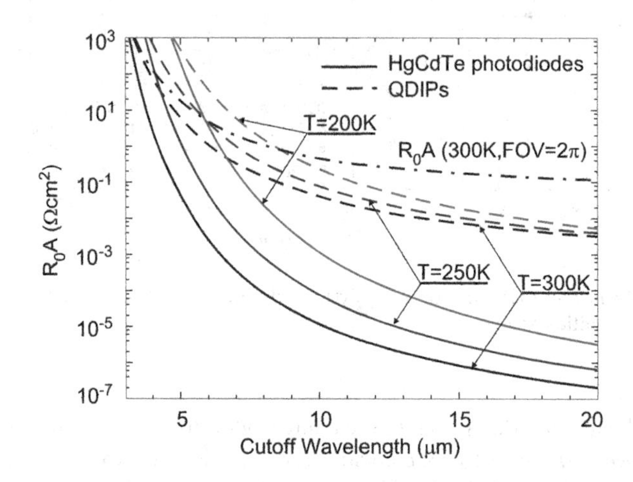

Figure 21.12 R_0A product of HgCdTe photodiodes and QDIPs as a function of wavelength. The calculations for HgCdTe photodiodes have been performed for the optimized doping concentration $p = \gamma^{1/2}n_i$.

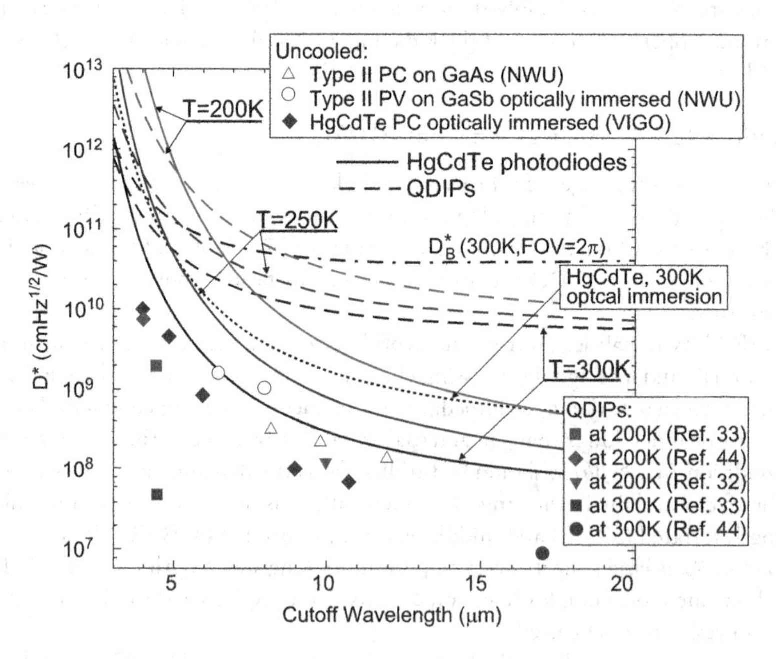

Figure 21.13 Calculated performance of Auger g-r-limited HgCdTe photodetectors as a function of wavelength and operating temperature. BLIP detectivity has been calculated for 2π field of view (FOV), the background temperature is $T_{BLIP} = 300$ K, and the quantum efficiency $\eta = 1$. The calculations for HgCdTe photodiodes have been performed for the optimized doping concentration $p = \gamma^{1/2}n_i$. The experimental data is taken for commercially available uncooled HgCdTe photoconductors (produced by Vigo System) and uncooled type II detectors at the Center for Quantum Devices, Northwestern University (Evanston, Illinois). The experimental data for QDIPs are gathered from the market literature for detectors operated at 200 K and 300 K.

concentration, $p = \gamma^{1/2} n_i$. The experimental data for QDIPs are gathered from the literature for the detectors operated at 200 K and 300 K.

Uncooled LWIR HgCdTe photodetectors are commercially available and manufactured in significant quantities, mostly as single-element devices [46]. They have found important applications in IR systems that require fast response. The results presented in Figure 21.13 confirm that the type II superlattice is a good candidate for IR detectors operating in the spectral range from the mid-wavelength to the very long-wavelength IR. However, comparison of QDIP performance both with HgCdTe and type II superlattice detectors gives evidence that the QDIP is suitable for high temperature. Especially, encouraging results have been achieved for very long-wavelength QDIP devices with a double-barrier resonant tunneling filter with each QD layer in the absorption region [48]. In this type of device, photoelectrons are selectively collected from the QDs by resonant tunneling, while the same tunnel barriers block electrons of dark current due to their broad energy distribution. For the 17 μm detector, a peak detectivity of 8.5×10^6 cmHz$^{1/2}$ W^{-1} has been measured. Up until now, this novel device has demonstrated the highest performance of room-temperature photodetectors. Further improvement in technology and design can result in the application of QDIPs in room temperature FPAs with the advantages of larger operating speeds (shorter frame times) in comparison with thermal detectors (bolometers and pyroelectric devices).

Thermal detectors seem to be unsuitable for the next generation of IR thermal imaging systems, which are moving toward faster frame rates and multispectral operation. A response time much shorter than that achievable with thermal detectors is required for many nonimaging applications. Improvements in technology and design of QDIP detectors make it possible to achieve both high sensitivity and fast response at room temperature.

21.5 COLLOIDAL QDIPs

A fundamentally different approach to QDIPs is to use an active region with 3D quantum confined semiconductor nanoparticles synthesized by inorganic chemistry [49]. This relatively new approach based on colloidal quantum dots (CQDs) offers a promising alternative on several aspects. These nanoparticles can improve QDIP performance compared to epitaxial QDs due to [35]:

- control over CQD synthesis and ability to conduct size filtering, leading to highly uniform ensembles,
- spherical shape of CQDs simplifying the calculations for device modeling and design, and
- greater selection of active region materials since strain considerations that dominate the growth of epitaxial QDs are eliminated.

Moreover, CQDs offer several attractive features for imagery [50]:

- cost reduction of fabrication compared to epitaxial growth (such solution are used as spin coated or inject printed),
- the material can be deposited on any material (there is no need of epitaxial matching),
- much stronger absorption than in Stranski-Krastanov-grown QD due to close-packing of colloidal dots.

However, colloidal nanomaterials also present several difficulties:

- inferior chemical stability and electronic passivation of the nanomaterials in comparison with epitaxial materials,
- electron transfer through many barrier interfaces in a nanomaterial can be slow and lead to insulating behavior,
- problems with long-term stability due to the large density of interfaces with atoms presenting different or weaker binding,
- high level of $1/f$ noise due to disordered granular systems.

Typically, CQDs are applied to optoelectronic devices as conducting-polymer/nanocrystal blends or nanocomposites [51–53]. For IR photodetector applications, nanocomposites often feature narrow-bandgap II-VI (HgTe, HgSe) [54–56], PbSe, or PbS [57–59] CQDs. Usually, IR photodetectors that have been reported use CQDs embedded in conducting polymer matrices, such as

Infrared photon detectors

poly[2-methoxy-5-(2-ethylhexyloxy)-1,4-phenylenevinylene] (MEH-PPV), demonstrating photodetection in the near IR region (1–3 μm) corresponding to the semiconductor nanocrystal bandgap energy [53].

CQD photodetectors typically comprise a single nanocomposite layer deposited on a glass slide by spin-casting and large-area, two-terminal, vertical devices are fabricated using p- (indiumtin-oxide) and n-type (aluminum) contacts, as shown schematically in Figure 21.14a. Figure 21.14b illustrates the capture and transport mechanisms of a colloidal dot film.

The charge transport mechanisms in CQD nanocomposites exhibit subtle differences compared to epitaxial QDIPs. As shown in Figure 21.15, the intraband transitions are not exploited. Instead, bipolar, interband (or excitonic) transitions across the CQD bandgap contribute to the photoresponse of the detector. In addition, since CQDs are electron acceptors and the polymers are typically hole conductors, the photogenerated excitons are dissociated at the QD/polymer interface. Thus, photoconduction through the nanocomposite material occurs as electrons hop among QDs and holes transport through the polymer [35].

The CQD layers are amorphous and permits the fabrication of devices directly onto ROIC substrates, as illustrated in Figure 21.16, without restrictions on array or pixel size and with a fabrication cycle in the order of days. Furthermore, the monolithic integration of the CQD detectors with a ROIC means that no

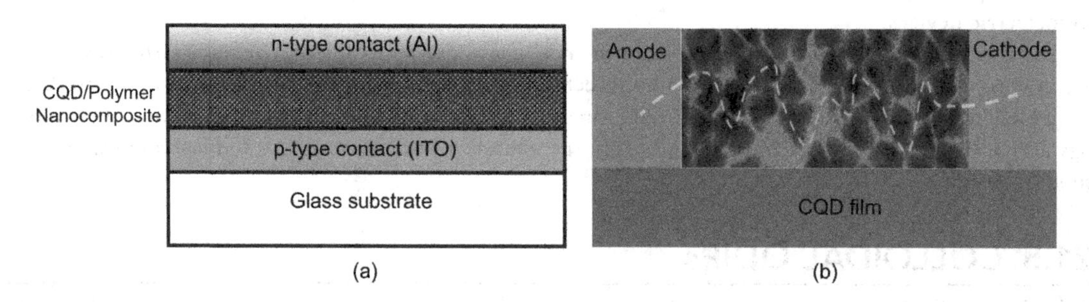

Figure 21.14 CQD photodetector: (a) schematic diagram of device heterostructure in CQD/conducting polymer nanocomposites; (b) an scanning electron microscope (SEM) image of a photovoltaic (PV) QD detector with transport illustration of photogenerated charge.

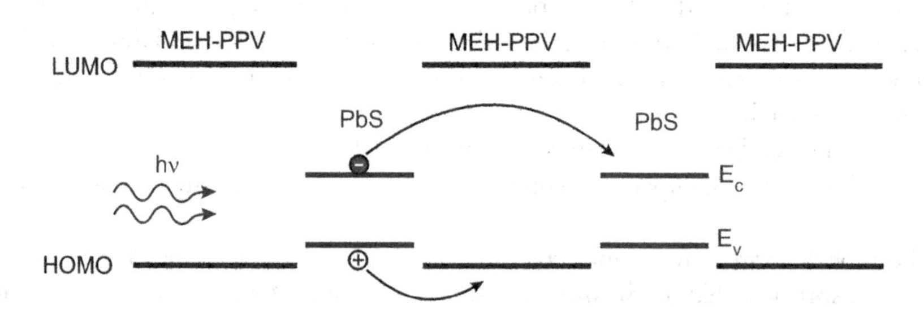

Figure 21.15 Schematic diagram of energy vs. position for interband transitions in PbS/MEH-PPV CQD-conducting polymer nanocomposites demonstrating photocurrent generation for IR photodetection (Adapted after Ref. [35])

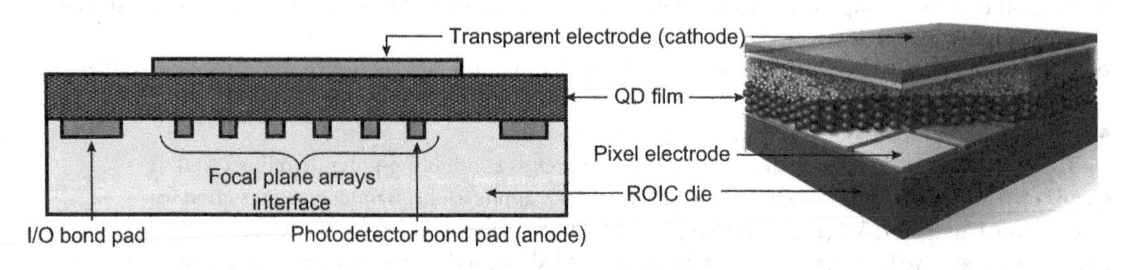

Figure 21.16 IR monolithic array structure based on CQDs.

hybridization step is required. Individual pixels are delineated by the area of the metal pads presented on the top surface of ROIC. Colloidal nanocrystals are synthesized using wet chemistry techniques. Reagents are injected into a flask and, through control of reagent concentrations, ligand selection, and temperature, the desired size and shape of nanocrystals are delivered. This so-called top-surface photodetector is compatible with postprocessing atop complementary metal oxide semiconductor (CMOS) electronics and offers a 100% fill factor.

Recently two groups [55,56] have demonstrated synthesized HgTe CQDs with IR energy gaps between 3 and 12 microns. Quantum confinement causes the dots to exhibit optical gaps larger than bulk HgTe, allowing them to be tuned through synthesis conditions to various spectral bands. A photovoltaic detector with 5.25-μm cutoff has achieved internal quantum efficiency exceeding 40% and a BLIP limit at 90 K [54]. Also, HgTe CQD FPA with noise equivalent difference temperature (NEDT) value of 102 mK at 100 K has been fabricated [55].

At present, detectivities between 10^{10} to 10^{9} Jones at 5 μm have been demonstrated while maintaining a fast response time at moderate cooling temperatures. It is unlikely that CQD IR detectors will ever achieve the performance of currently popular InGaAs, HgCdTe, or InSb photodiodes and they will likely be used in newer applications that require high-definition, low-cost imaging on smaller pixels that doesn't require extreme sensitivity. Their fundamental properties can be potentially changed depending on what IR system engineers and customers desire. It is expected that increasing dot size while maintaining good monodispersion, causes improving carrier transport and quantum efficiency while maintaining low noise levels. With continuing development of deposition and synthesis processes of this early technology, much higher performances will be reached in the future.

REFERENCES

1. Y. Arakawa and H. Sakaki, "Multidimensional quantum-well laser and temperature dependence of its threshold current," *Appl. Phys. Lett.*, 40, 939–41, 1982.
2. M. Asada, Y. Miyamoto, and Y. Suematsu, "Gain and threshold of three dimensional quantum-box lasers," *IEEE J. Quantum Electron.*, 22, 1915–21, 1986.
3. D. Bimberg, M. Grundmann, and N.N. Ledentsov, *Quantum Dot Heterostructures*, Wiley, Chichester, UK, 1999.
4. Ch. Sikorski and U. Merkt, "Spectroscopy of electronic states in InSb quantum dots," *Phys. Rev. Lett.*, 62, 2164–67, 1989.
5. T. Demel, D. Heitmann, P. Grambow, and K. Ploog, "Nonlocal dynamic response and level crossings in quantum-dot structures," *Phys. Rev. Lett.*, 64, 788–91, 1990.
6. J. Phillips, K. Kamath, and P. Bhattacharya, "Far-Infrared photoconductivity in self-organized InAs quantum wells," *Appl. Phys. Lett.*, 72, 2020–1, 1998.
7. P. Bhattacharya and Z. Mi, "Quantum-dot optoelectronic devices," *Proc. IEEE*, 95, 1723–40, 2007.
8. J.C. Campbell and A. Madhukar, "Quantum-dot infrared photodetectors," *Proc. IEEE*, 95, 1815–27, 2007.
9. P. Bhattacharya, A.D. Stiff-Roberts, and S. Chakrabarti, "Mid-Infrared Quantum Dot Photoconductors," in *Mid-Infrared Semiconductor Optoelectronics*, ed. A. Krier, 487–513, Springer Verlag, Berlin, 2007.
10. S. Krishna, S.D. Gunapala, S.V. Bandara, C. Hill, and D.Z. Ting, "Quantum dot based infrared focal plane arrays," *Proc. IEEE*, 95, 1838–52, 2007.
11. D. Leonard, M. Krishnamurthy, C.M. Reaves, S.P. Denbaars, and P.M. Petroff, "Direct formation of quantum-sized dots from uniform coherent islands of InGaAs on GaAs surface," *Appl. Phys. Lett.*, 63, 3203–5, 1993.
12. W. Seifert, N. Carlsson, J. Johansson, M-E. Pistol, and L. Samuelson, "In situ growth of nano-structures by metal-organic vapour phase epitaxy," *J. Cryst. Growth*, 170, 39–46, 1997.
13. I.N. Stranski and L. Krastanow, "Zur theorie der orientierten ausscheidung von lonenkristallen aufeinander," *Sitzungsberichte d. Akad. d. Wissenschaften in Wein. Abt. IIb*, 146, 797–810, 1937.
14. S.Y. Wang, S.D. Lin, W. Wu, and C.P. Lee, "Low dark current quantum-dot infrared photodetectors with an AlGaAs current blocking layer," *Appl. Phys. Lett.*, 78, 1023–5, 2001.
15. V. Ryzhii, "Physical model and analysis of quantum dot infrared photodetectors with blocking layer," *J. Appl. Phys.*, 89, 5117–24, 2001.
16. S.W. Lee, K. Hirakawa, and Y. Shimada, "Bound-to-continuum intersubband photoconductivity of self-assembled InAs quantum dots in modulation-doped heterostructures," *Appl. Phys. Lett.*, 75, 1428–30, 1999.
17. S. Krishna, "Quantum dots-in-a-well infrared photodetectors," *J. Phys. D*, 38, 2142–50, 2005.

18. S.D. Gunapala, S.V. Bandara, C.J. Hill, D.Z. Ting, J.K. Liu, B. Rafol, E.R. Blazejewski, et al., "640 × 512 pixels long-wavelength infrared (LWIR) quantum-dot infrared photoconductor (QDIP) imaging focal plane array," *IEEE J. Quantum Electron.*, 43, 230–7, 2007.
19. I. Vurgaftman, Y. Lam, and J. Singh, "Carrier thermalization in sub-three-dimensional electronic systems: Fundamental limits on modulation bandwidth in semiconductor lasers," *Phys. Rev. B*, 50, 14309–26, 1994.
20. J. Phillips, "Evaluation of the fundamental properties of quantum dot infrared detectors," *J. Appl. Phys.*, 91, 4590–4, 2002.
21. E. Towe and D. Pan, "Semiconductor quantum-dot nanostructures: Their application in a new class of infrared photodetectors," *IEEE J. Sel. Top. Quantum Electron.*, 6, 408–21, 2000.
22. V. Ryzhii, I. Khmyrova, V. Pipa, V. Mitin, and M. Willander, "Device model for quantum dot infrared photodetectors and their dark-current characteristics," *Semicond. Sci. Tech.*, 16, 331–8, 2001.
23. V. Ryzhii, I. Khmyrova, V. Mitin, M. Stroscio, and M. Willander, "On the detectivity of quantum-dot infrared photodetectors," *Appl. Phys. Lett.*, 78, 3523–5, 2001.
24. J. Singh, *Electronic and Optoelectronic Properties of Semiconductor Structures*, Cambridge University Press, New York, 2003.
25. H.C. Liu, "Quantum dot infrared photodetector," *Opto Electron. Rev.*, 11, 1–5, 2003.
26. J.-Y. Duboz, H.C. Liu, Z.R. Wasilewski, M. Byloss, and R. Dudek, "Tunnel current in quantum dot infrared photodetectors," *J. Appl. Phys.*, 93, 1320–2, 2003.
27. A.D. Stiff-Roberts, X.H. Su, S. Chakrabarti, and P. Bhattacharya, "Contribution of field-assisted tunneling emission to dark current in InAs-GaAs quantum dot infrared photodetectors," *IEEE Photonics Technol. Lett.*, 16, 867–9, 2004.
28. H.C. Liu, "Noise gain and operating temperature of quantum well infrared photodetectors," *Appl. Phys. Lett.*, 61, 2703–5, 1992.
29. W.A. Beck, "Photoconductive gain and generation-recombination noise in multiple-quantum-well-infrared detectors," *Appl. Phys. Lett.*, 63, 3589–1, 1993.
30. J. Phillips, P. Bhattacharya, S.W. Kennerly, D.W. Beekman, and M. Duta, "Self-assembled InAs-GaAs quantum-dot intersubband detectors," *IEEE J. Quantum Electron.*, 35, 936–43, 1999.
31. Z. Ye, J.C. Campbell, Z. Chen, E.T. Kim, and A. Madhukar, "Noise and photoconductive gain in InAs quantum dot infrared photodetectors," *Appl. Phys. Lett.*, 83, 1234–6, 2003.
32. X. Lu, J. Vaillancourt, and M.J. Meisner, "Temperature-dependent photoresponsivity and high-temperature (190 K) operation of a quantum dot infrared photodetector," *Appl. Phys. Lett.*, 91, 051115-1–3, 2007.
33. H. Lim, S. Tsao, W. Zhang, and M. Razeghi, "High-performance InAs quantum-dot infrared photoconductors grown on InP substrate operating at room temperature," *Appl. Phys. Lett.*, 90, 131112-1–3, 2007.
34. P. Martyniuk and A. Rogalski, "Quantum-dot infrared photodetectors: Status and outlook," *Prog. Quantum Electron.*, 32, 89–120, 2008.
35. A.D. Stiff-Roberts, "Quantum-dot infrared photodetectors: A review," *J. Nanophoton.*, 3, 031607-1-17, 2009.
36. A.V. Barve, S.J. Lee, S.K. Noh, and S. Krishna, "Review of current progress in quantum dot infrared photodetectors," *Laser Photon. Rev.*, 4(6), 738–750, 2010.
37. T. Chuh, "Recent developments in infrared and visible imaging for astronomy, defense and homeland security," *Proc. SPIE*, 5563, 19–34, 2004.
38. C.H. Grein, H. Cruz, M.E. Flatte, and H. Ehrenreich, "Theoretical performance of very long wavelength InAs/InxGa1–xSb superlattice based infrared detectors," *Appl. Phys. Lett.*, 65, 2530–2, 1994.
39. J. Jiang, S. Tsao, T. O'Sullivan, W. Zhang, H. Lim, T. Sills, K. Mi, M. Razeghi, G.J. Brown, and M.Z. Tidrow, "High detectivity InGaAs/InGaP quantum-dot infrared photodetectors grown by low pressure metalorganic chemical vapor deposition," *Appl. Phys. Lett.*, 84, 2166–8, 2004.
40. J. Szafraniec, S. Tsao, W. Zhang, H. Lim, M. Taguchi, A.A. Quivy, B. Movaghar, and M. Razeghi, "High-detectivity quantum-dot infrared photodetectors grown by metalorganic chemical-vapor deposition," *Appl. Phys. Lett.*, 88, 121102-1–3, 2006.
41. E.-T. Kim, A. Madhukar, Z. Ye, and J.C. Campbell, "High detectivity InAs quantum dot infrared photodetectors," *Appl. Phys. Lett.*, 84, 3277–9, 2004.
42. S. Chakrabarti, X.H. Su, P. Bhattacharya, G. Ariyawansa, and A.G.U. Perera, "Characteristics of a multicolor InGaAs-GaAs quantum-dot infrared photodetector," *IEEE Photonics Technol. Lett.*, 17, 178-180, 2005.
43. R.S. Attaluri, S. Annamalai, K.T. Posani, A. Stintz, and S. Krishna, "Influence of Si doping on the performance of quantum dots-in-well photodetectors," *J. Vac. Sci. Technol. B*, 24, 1553–5, 2006.
44. S. Chakrabarti, A.D. Stiff-Roberts, X.H. Su, P. Bhttacharya, G. Ariyawansa, and A.G.U. Perera, "High-performance mid-infrared quantum dot infrared photodetectors," *J. Phys. D*, 38, 2135–41, 2005.

45. S. Krishna, D. Forman, S. Annamalai, P. Dowd, P. Varangis, T. Tumolillo, A. Gray, et al., "Two-color focal plane arrays based on self assembled quantum dots in a well heterostructure," *Phys. Status Solidi*, 3, 439–43, 2006.
46. J. Piotrowski and A. Rogalski, *High-Operating Temperature Infrared Photodetectors*, SPIE Press, Bellingham, WA, 2007.
47. P. Martyniuk, S. Krishna, and A. Rogalski, "Assessment of quantum dot infrared photodetectors for high temperature operation," *J. Appl. Phys.*, 104, 034314-1–6, 2008.
48. X.H. Su, S. Chakrabarti, P. Bhattacharya, A. Ariyawansa, and A.G.U. Perera, "A resonant tunneling quantum-dot infrared photodetector," *IEEE J. Quantum Electron.*, 41, 974–9, 2005.
49. C.B. Murray, S. Sun, W. Gaschler, H. Doyle, T.A. Betley, and C.R. Kagan, "Colloidal synthesis of nanocrystals and nanocrystal superlattices," *IBM J. Res. Dev.*, 45, 47–56, 2001.
50. P. Guyot-Sionnest, S. Keuleyan, H. Liu, and E. Lhuillier, "Colloidal quantum dots for mid-infrared detection," *Proc. SPIE*, 8268, 82682Z-1–8, 2012.
51. N.C. Greenham, X. Peng, and A.P. Alivisatos, "Charge separation and transport in conjugated polymer/cadmium selenide nanocrystal composites studied by photoluminesence quenching and photoconductivity," *Syn. Met.*, 84, 545–6, 1997.
52. D.S. Ginger and N.C. Greenham, "Photoinduced electron transfer from conjugated polymers to CdSe nanocrystals," *Phys. Rev. B*, 59, 10622–9, 1999.
53. G. Konstantatos and E.H. Sargent, "Solution-processed quantum dot photodetectors," *Proc. IEEE*, 97(10), 1666–83, 2009.
54. P. Guyot-Sionnest and J.A. Roberts, "Background limited mid-infrared photodetection with photovoltaic HgTe colloidal quantum dots," *Appl. Phys. Lett.*, 107, 253104-1–5, 2015.
55. C. Buurma, R.E. Pimpinellaa, A.J. Ciani, J.S. Feldman, C.H. Grein, and P. Guyot-Sionnest, "MWIR imaging with low cost colloidal quantum dot films," *Proc. SPIE*, 9933, 993303-1–7, 2016.
56. C. Buurma, A.J. Ciani, R.E. Pimpinella, J.S. Feldman, C.H. Grein, and P. Guyot-Sionnes, "Advances in HgTe colloidal quantum dots for infrared detectors," *J. Electron. Mater.*, 46(11), 6685–8, 2017.
57. E.J.D. Klem, C. Gregory, D. Temple, J. Lewis, "PbS colloidal quantum dot photodiodes for low-cost SWIR sensing," *Proc. SPIE*, 9451, 945104-1–5, 2015.
58. A. De Iacovo, C. Venettacci, L. Colace, L. Scopa, and S. Foglia, "PbS colloidal quantum dot photodetectors operating in the near infrared," *Sci. Rep.*, 6, 37913, 2016.
59. M. Thambidurai, Y. Jjang, A. Shapiro, G. Yuan, H. Xiaonan, Y. Xuechao, G.J. Wang, E. Lifshitz, H.V. Demir, and C. Dang, "High performance infrared photodetectors up to 2.8 μm wavelength based on lead selenide colloidal quantum dots," *Opt. Mater. Express*, 7(7), 2336, 2017.

22 Infrared barrier photodetectors

Investigations of antimonide-based materials began at about the same time as HgCdTe—in the 1950s, and the apparent rapid success of their technology, especially low-dimensional solids, depends on the previous five decades of III-V material and device research. The sophisticated physics associated with the antimonide-based bandgap engineering concept started at the beginning of 1990s gave a new impact and interest in the development of infrared (IR) detector structures within academic and national laboratories. In addition, implementation of barriers in photoconductor structures, in so-called barrier detectors, prevents current flow in the majority carrier band of the detector's absorber, but allows unimpeded flow in the minority carrier band. As a result, this concept resurrects the performance of antimonide-based focal plane arrays (FPAs) and gives a new perspective in their applications. Significant advances have been made in the bandgap engineering of various $A^{III}B^{V}$ compound semiconductors that has led to new IR detector architectures. New emerging strategies include especially antimonide-based type-II superlattices (T2SLs), barrier structures such as nBn detector with lower generation-recombination (GR) leakage mechanisms, and multistage/cascade IR devices.

22.1 SWIR BARRIER DETECTORS

An extension of barrier detectors to short wavelength IR (SWIR) region up to 3 μm has been demonstrated using InGaAs and InGaAsSb alloy systems [1,2]. The standard grown method for making SWIR detectors is to utilize the molecular-beam epitaxy (MBE) technique.

Savich et al. [2] carried out a comparison of electrical and optical characteristics of conventional photodiodes and nBn architecture detectors with 2.8-μm cutoff wavelengths fabricated with both lattice-mismatched InGaAs and lattice-matched InGaAsSb absorbing layers. In order to minimize the number of defects in the $In_{1-x}Ga_xAs$ absorber on InP substrate, a 2-μm-thick step-graded buffer consisting of AlInAs was grown for which the lattice constant was graded from that of InP to that of $In_{0.82}Ga_{0.18}As$. Both the conventional photodiode and nBn detector include this step-graded buffer. Barrier detector includes additionally a pseudomorphic AlAsSb unipolar barrier to maintain a large conduction barrier compared to the $In_{0.82}Ga_{0.18}As$.

In case of lattice-matched solution with GaSb substrate, the quaternary composition of $In_{0.30}Ga_{0.70}As_{0.56}Sb_{0.44}$ at the edge of the miscibility gap has been used to maintain the cutoff wavelength and lattice-matching requirements. In nBn detector, also pseudomorphic AlGaSb unipolar barrier with a large conduction band offset and zero valence band offset compared to the $In_{0.30}Ga_{0.70}As_{0.56}Sb_{0.44}$ absorber was implemented.

Figure 22.1 gathers temperature-dependent dark current characteristics for both lattice-mismatched InGaAs and lattice-matched InGaAsSb detectors at an operating reverse bias of 100 mV [2]. InGaAs on InP devices suffer from low material quality, i.e., threading dislocations that occur due to the lattice mismatch, with implications for device dark current performance.

The p-n InGaAs photodiode is limited by surface leakage current below a temperature of 220 K, while the nBn detector remains diffusion limited down to 150 K. At the room-temperature background photocurrent level, the nBn detector shows dark current reduced by a factor greater than 400 as compared to the conventional photodiode.

The p-n InGaAsSb junction is limited by the depletion region current below a temperature of 250 K, while the nBn remains diffusion limited, as far as measured, down to 250 K. At the 300 K background

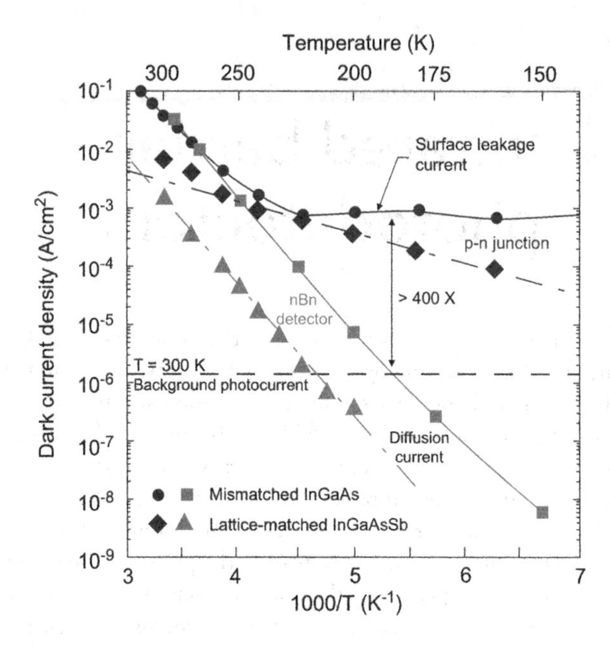

Figure 22.1 Arrhenius plots for both InGaAs and InGaAsSb p-n junctions and nBn detectors with cutoff wavelengths 2.8 μm.

photocurrent level the nBn detector shows dark current reduced by nearly three orders of magnitude as compared to the conventional photodiode.

Figure 22.2 shows that InGaAsSb nBn detector being lattice matched to a GaSb substrate, presents performance near Rule 07. Its dark current is 10–20 times lower in comparison with the lattice-mismatched InGaAs counterpart.

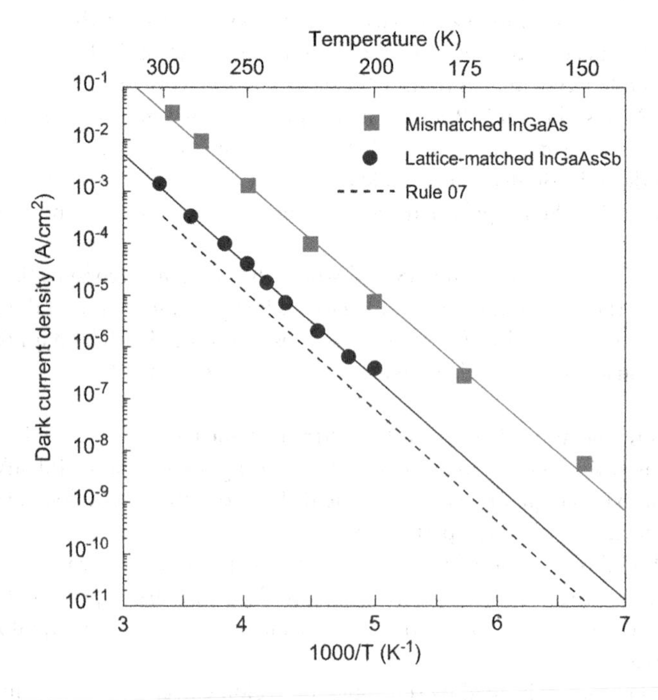

Figure 22.2 Comparison of dark current characteristics of InGaAs and InGaAsSb nBn detectors. The lattice-matched InGaAsSb barrier detector shows at least an order of magnitude reduction in the dark current compared to the mismatched InGaAs nBn detector. (Adapted after Ref. [2])

22.2 InAsSb BARRIER DETECTORS

Detailed growth procedures and device characterization of $InAs_{1-x}Sb_x/AlAs_{1-y}Sb_y$ nBn middle wavelength infrared (MWIR) detectors were the topic of several papers, for example, Refs. [3–7]. The n-type doping was usually obtained by either Si or Te elements and the InAsSb structures were grown on either GaAs(100) or GaSb(100) substrates in a Veeco Gen200 MBE system [7]. The lattice-mismatched structures that used GaAs(100) as the substrate were grown on a 4-μm-thick GaSb buffer layer, whereas the remaining structures were grown directly onto GaSb(100) substrates. The principal layers of the device structures consisted of a thick n-type InAsSb absorption layer (1.5–3 μm), a thin n-type AlSbAs barrier layer (0.2–0.35 μm), and a thin (0.2–0.3 μm) n-type InAsSb contact layer. The bottom contact layer was highly doped.

Figure 22.3 shows an example of such a nBn structure that was considered theoretically by Martyniuk and Rogalski [8] along with the *J-V* characteristics as a function of temperature that were taken from Ref. [9]. The alloy composition of $x = 0.09$ or the $InAs_{1-x}Sb_x$ absorber layer provided a cutoff wavelength of ~ 4.9 μm at 150 K. J_{dark} was 1.0×10^{-3} A cm^{-2} at 200 K and 3.0×10^{-6} A cm^{-2} at 150 K. The detectors are dominated by diffusion currents at −1.0 V bias where the quantum efficiency peaks.

An interesting difference in the dark current density versus temperature is shown in Figure 22.4 for two nominally identical devices with opposite barrier polarities, each operating at a bias of −0.1 V [10].

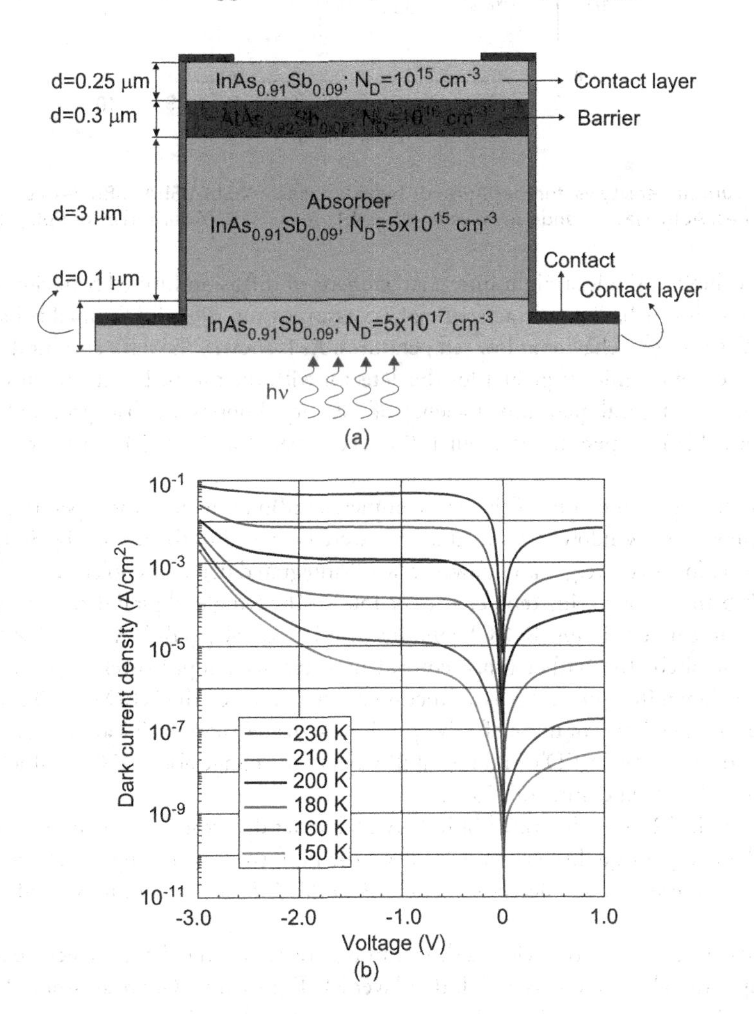

Figure 22.3 InAsSb/AlAsSb nBn MWIR detector: (a) the device structure and (b) dark current density vs. bias voltage as a function of temperature for 4,096 (18 μm pitch) detectors ($\lambda_c \approx 4.9$ μm at 150 K) tied together in parallel. (Adapted after Ref. [9])

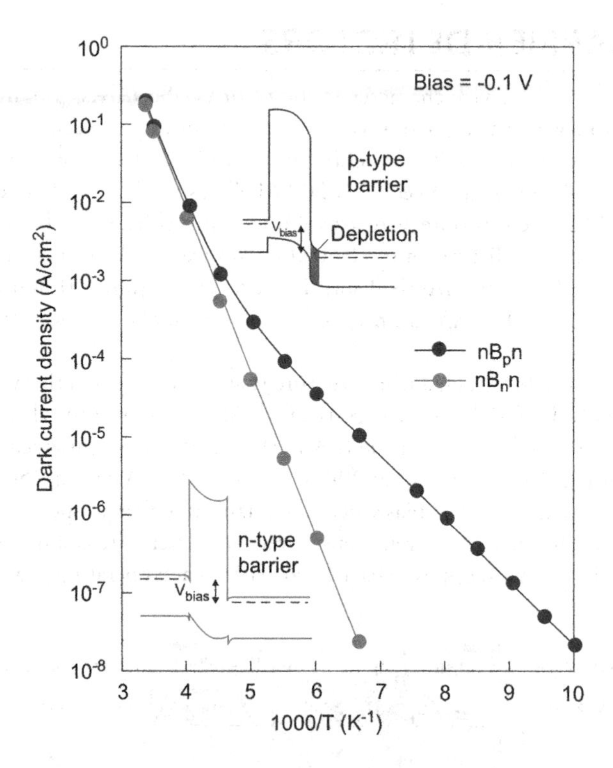

Figure 22.4 Dark current density vs. temperature for two identical InAsSb/AlSbAs nBn devices with opposite barrier doping polarities. Active layer bandgap wavelength is 4.1 μm at150 K. (Adapted after Ref. [10])

The nB_nn device exhibits a single straight line, characteristic of diffusion-limited behavior, while the nB_pn device exhibits two-slope behavior, characteristic of a crossover from diffusion-limited behavior at high temperatures to GR-limited behavior at low temperatures. As is shown, the dark current density at 150 K is more than two orders of magnitude greater for the detector with the p-type barrier because it is already GR limited. It appears, for a typical quantum efficiency of 70% at *f*/3 optics, the background limited infrared photodetector (BLIP) temperature is about 140 K as compared with ~175 K for the detector with the n-type barrier.

Klipstein et al. have presented one of the first commercial nBn array detectors operating in the blue part of the MWIR atmospheric window (3.4–4.2 μm) that were launched in the market by SCD. It is known as Kinglet and is a very low size, weight and power (SWaP)-integrated detector cooler assembly (IDCA) with an aperture of *f*/5.5 and an operating temperature of 150 K. The Kinglet digital detector based on SCD's Pelican-D readout integrated circuit (ROIC) contains nBn $InAs_{0.91}Sb_{0.09}$/B-AlAsSb 640 × 512 pixel architecture with a 15 μm pitch. The NEDT (noise equivalent difference temperature) for *f*/3.2 optics and the pixel operability is shown in Figure 22.5 as a function of temperature. The NEDT is 20 mK at 10 ms integration time and the operability of non-defective pixels was greater than 99.5% after a standard two-point nonuniformity correction. The NEDT and operability begin to change above 170 K, which is consistent with the estimated BLIP temperature of 175 K.

Recently, Lin et al. [12] have described bulk InAsSb barrier detectors with a cutoff wavelength about 10 μm at 77 K. Due to poor quality crystal structure, the performance of long wavelength infrared (LWIR) bulk InAsSb ternary alloy detectors are considerably inferior in comparison with HgCdTe photodiodes.

The schematic device cross section view is shown in Figure 22.6a. The device structure consists of a 3-μm-thick compositionally graded GaInSb buffer layer MBE-grown on GaSb substrate, 1-μm-thick $InAs_{0.60}Sb_{0.40}$ absorber, a 20-nm-thick $Al_{0.6}In_{0.4}Sb_{0.41}Sb_{0.9}$ undoped barrier, and a 20-nm-thick $InAs_{0.60}Sb_{0.40}$ top contact layer Te-doped to a level of 10^{18} cm^{-3}. The undoped AlInAsSb barrier was lattice matched to InAsSb with 40% Sb composition. As shown, the heterostructures were processed with

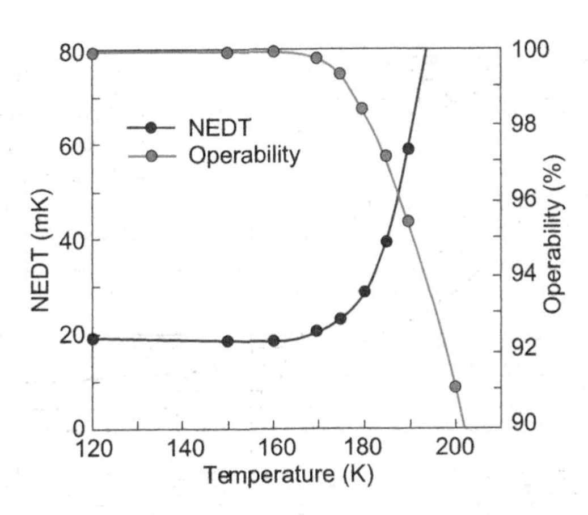

Figure 22.5 Temperature dependence of the NEDT (at optics *f*/3.2) and the pixel operability of the Kinglet detector. (Adapted after Ref. [11])

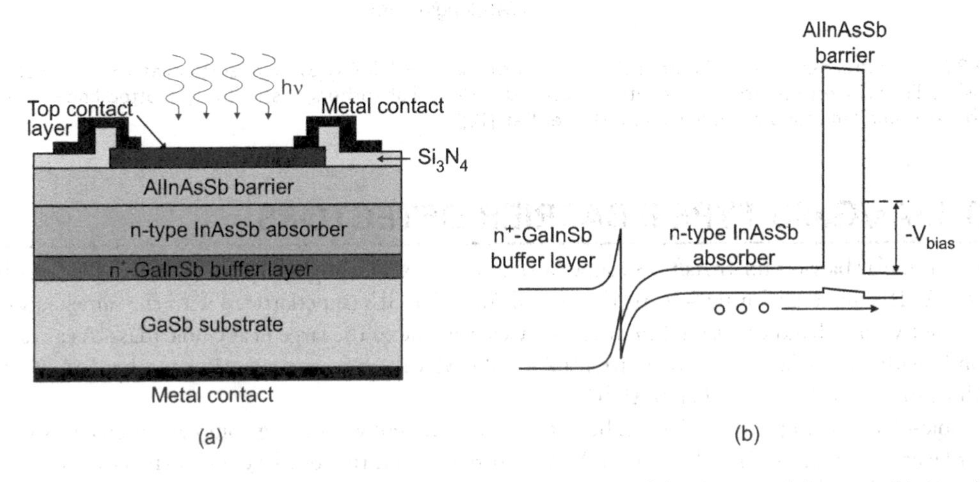

Figure 22.6 LWIR nBn InAsSb detector: (a) the schematic cross section of the processed detector, (b) the schematic bandgap diagram of the heterostructure with $InAs_{0.60}Sb_{0.40}$ bulk absorber.

a window (square with a 250 μm side) for incident radiation on top of the epilayer. The top metal contact layer was a square with a 300 μm size. The InAsSb contact layer outside the metal contact was removed down to the barrier layer by reactive ion etching. For detector passivation, Si_3N_4 was used.

A minority hole lifetime in $InAs_{0.60}Sb_{0.40}$ absorber of 185 ns and a diffusion length of 9 μm were determined at 77 K. The hole mobility of $10^3 cm^2V^{-1}s^{-1}$ was estimated with frequency response measurements.

The current-voltage characteristics have been influenced by the depletion of a part of the absorber adjacent to the barrier, which leads to domination of the GR layer and probably tunneling components. To reach diffusion-limited dark current, the valence band offset associated with the heterointerfaces must be eliminated.

The spectral detectivity curves of LWIR nBn InAsSb detector, with two absorber's doping levels, are shown in Figure 22.7. The demonstrated detectivity of 2×10^{11} cmHz$^{1/2}$ W^{-1} at 2π field of view (FOV) and wavelength of 8 μm was estimated in spite of the significant blue-shift of the absorption edge with doping. For 1-μm-thick $InAs_{0.60}Sb_{0.40}$ absorber at $\lambda = 8$ μm, an absorption coefficient was estimated at 3×10^3 cm^{-1}, which implies a quantum efficiency of 22%. The quantum efficiency was increasing with bias until it reached a constant level for a bias of −0.4 V and with increasing thickness of the absorption layer (to 40% for 3-μm-thick absorber).

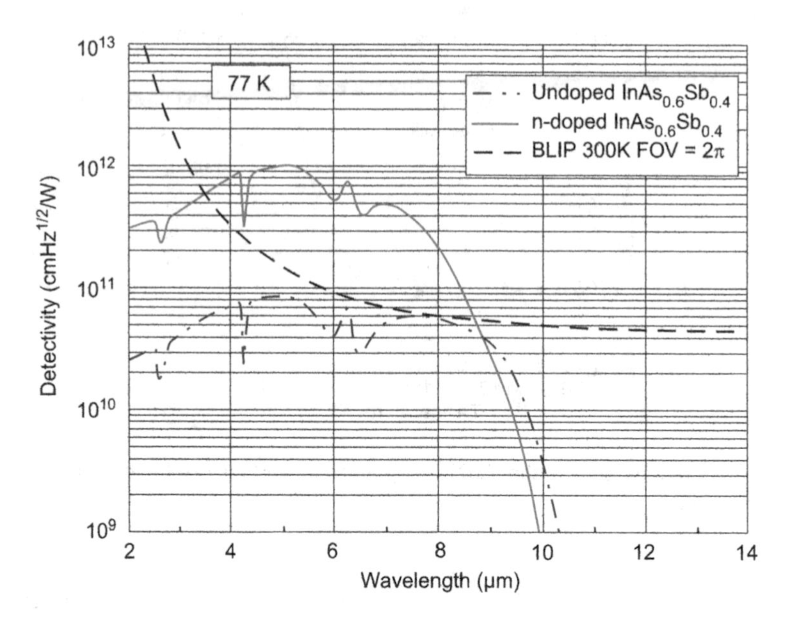

Figure 22.7 Spectral detectivity of barrier detectors with 1-μm-thick $InAs_{0.6}Sb_{0.4}$ absorbers at $T = 77$ K. Solid and dashed lines correspond to devices with doped and undoped absorbers, respectively. Dotted line shows the 300 K background limit for a 2π FOV. (Adapted after Ref. [12])

22.3 InAs/GaSb TYPE II BARRIER DETECTORS

Building unipolar barriers for InAs/GaSb superlattices is relatively straightforward because of the flexibility of the 6.1 Å III-V material family—InAs, GaSb, and AlSb. For SLs (superlattices) with the same GaSb layer widths, their valence band edges tend to lineup very closely due to the large heavy-hole mass. As a result, an electron-blocking unipolar barrier for a given InAs/GaSb SL can be formed by using another InAs/GaSb SL with thinner InAs layers or a GaSb/AlSb SL.

The hole-blocking unipolar barrier can be obtained in different ways using complex supercells, such as the four-layer InAs/GaInSb/InAs/AlGaInSb W structure [13] and the four-layer GaSb/InAs/GaSb/AlSb M structure [14]. Their designs are shown in Figure 22.8. The W structures initially developed to increase the gain in MWIR lasers are also promising as LWIR and very LWIR photodiode materials. In these structures, two InAs electron wells are located on either side of an InGaSb hole well and are bound on either side by AlGaInSb barrier layers. The barriers confine the electron wavefunctions symmetrically about the hole well, increasing the electron-hole overlap while nearly localizing the wavefunctions. The resulting quasi-dimensional densities of states give strong absorption near the band edge. Due to flexibility in adjusting the W structure, this SL has been used as a hole-blocking unipolar barrier, an absorber as well as an electron-blocking unipolar barrier.

The new design W-structured type T2SL photodiodes employ a graded-bandgap p-i-n design. The grading of the bandgap in the depletion region suppresses tunneling and GR currents in the depletion region, which have resulted in an order of magnitude improvement in dark current performance with $R_0A = 216$ Ωcm^2 at 78 K for devices with a 10.5-μm cutoff wavelength. The sidewall resistivity of ≈ 70 kΩcm for untreated mesas apparently indicates self-passivation by the graded bandgap [15].

In the M structure [16,17], the wider-energy gap AlSb layer blocks the interaction between electrons in two adjacent InAs wells, thus reducing the tunneling probability and increasing the electron effective mass. The AlSb layer also acts as a barrier for holes in the valence band and converts the GaSb hole quantum well into a double quantum well. As a result, the effective well width is reduced and the hole's energy level becomes sensitive to the well dimension. This structure significantly reduces the dark current and does not show any strong degradation of the optical properties of the devices. Moreover, it has a proven control in positioning the conduction and valence band energy levels [17]. Consequently, FPAs for imaging within

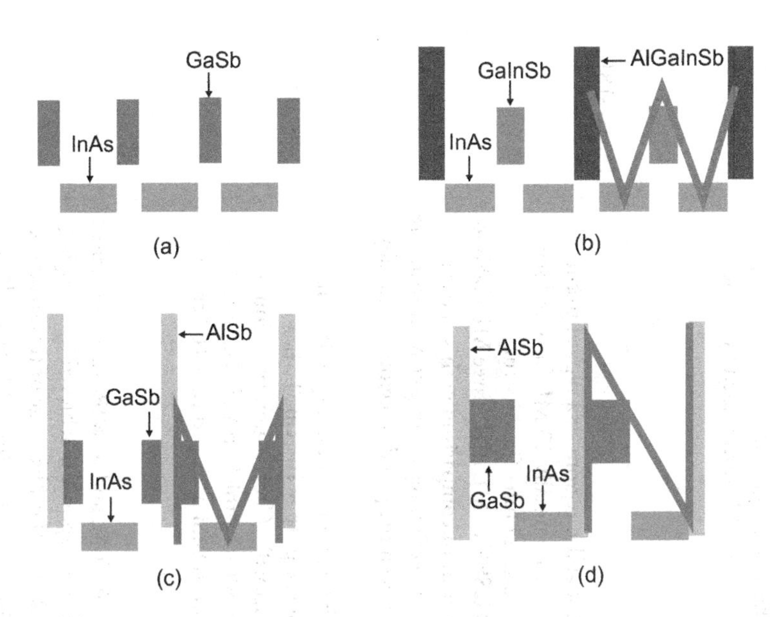

Figure 22.8 Schematic energy band diagrams of (a) InAs/GaSb SL, (b) InAs/GaInSb/InAs/AlGaInSb *W* SL, (c) GaSb/InAs/GaSb/AlSb *M* SL, and (d) GaSb/InAs/AlSb *N* SL. (Adapted after Ref. [4])

various IR regions, from SWIR to very LWIR, can be fabricated [18]. Device with a cutoff wavelength of 10.5 μm exhibits a R_0A product of 200 Ωcm^2 when a 500-nm-thick M structure was used. Using double M-structure heterojunction at the single device level, the R_0A product up to 5,300 Ωcm^2 has been obtained for a 9.3 μm cutoff at 77 K [19].

In the N structure [20], two monolayers (MLs) of AlSb are inserted asymmetrically between InAs and GaSb layers along the growth direction as an electron barrier (eB). This configuration increases significantly the electron-hole overlap under bias, and consequently increases absorption while decreasing the dark current.

Table 22.1 illustrates some flat-band energy band diagrams and describes SL-based IR detectors that make use of unipolar barriers including: double heterojunction (DH), dual band nBn structure, DH with graded gap junction and complementary barrier structure. As can be seen, these structures are based on either the nBn/pBp/XBn architecture or different double heterostructure designs.

The first LWIR InAs/InGaSb SL double heterostructure photodiode (see Table 22.1, double heterostructure) grown on GaSb substrates with photoresponse out to 10.6 μm was described by Johnson et al. in 1996 [21]. In this structure, the active SL region was surrounded by barriers made from p-GaSb and n-GaSb binary compounds. More recently, the barriers are also fabricated using different types of SLs.

The realization of dual band detection capabilities with nBn design (see Table 22.1, dual-band nBn) is schematically shown in Figure 22.9 [4,22]. Under forward bias (defined as negative voltage applied to the top contact), the photocarriers are collected from the SL absorber with λ_2 cutoff wavelength. When the device is under reverse bias (defined as positive voltage applied to the top contact), the photocarriers are collected from the SL absorber with λ_1 cutoff wavelength, while those from the absorber with λ_2 cutoff wavelength are blocked by the barrier. Thus, a two-color response is obtained under two different bias polarities.

Hood et al. [29] have modified the nBn concept to make the superior pBn LWIR device (see Figure 12.44). In this structure, the p-n junction can be located at the interface between the heavily doped p-type contact material and the lower doped barrier or within the lower-doped barrier itself. Similar to the nBn structure, the pBn structure still reduces GR currents associated with Shockley-Read-Hall (SRH) centers (the depletion region exists primarily in the barrier and does not appreciably penetrate the narrow-bandgap n-type absorber). In addition, the electric field in the barrier improves the response of the detector by sweeping from the active layers of those photogenerated carriers that reach the barrier before they can recombine.

Table 22.1 Type II superlattice barrier detectors

FLAT-BAND ENERGY DIAGRAMS	EXAMPLES	DESCRIPTION	REFS.
Double heterostructure E_c E_v	SLS: 38Å InAs/16Å $Ga_{0.64}In_{0.36}Sb$ 0.30 μm: P^+-GaSb ($2\times10^{18}cm^{-3}$), P^--GaSb ($6\times10^{16}cm^{-3}$) 0.75 μm: SLS n^- ($2\times10^{16}cm^{-3}$) 1.00 μm: N^+-GaSb ($8\times10^{17}cm^{-3}$) N-GaSb substrate ($5\times10^{17}cm^{-3}$)	**Double heterojunction (DH) photodiode** The first LWIR InAs/InGaSb SL DH photodiode grown on GaSb substrates with photoresponse out to 10.6 μm. The active region consisting of n-type 39 Å InAs/16 Å $Ga_{0.65}In_{0.35}Sb$ SL ($2\times10^{16}cm^{-3}$) is surrounded by barriers made from p-GaSb and n-GaSb.	[21]
nB_pp barrier E_c E_v	Ti/Pt/Au contact InAs/GaSb n-doped InAs/GaSb π-doped InAs/GaSb p-doped GaSb buffer Residually p-type GaSb substrate	**p-π-M-n photodiode structure** The M structure is inserted between the π and n regions of a typical p-π-n structure. The T2SL part is chosen to have nominally 13 ML InAs and seven ML GaSb for a cutoff wavelength of around 11 μm. The M structure is designed with 18 ML InAs/three ML GaSb/five ML AlSb/three ML GaSb for a cutoff wavelength of approximately 6 μm. In comparison with standard p-π-n structure, the electric field in the depletion region of p-π-M-n structure is reduced and the tunneling barrier between the p and n regions is spatially broadened. The structure consists of a 250-nm-thick GaSb:Be p^+ buffer ($p \sim 10^{18}cm^{-3}$), followed by a 500-nm-thick InAs/GaSb:Be p^+ ($p \sim 10^{18}cm^{-3}$) SL, a 2,000-nm-thick slightly p-type doped InAs:Be/GaSb region (π-region $p \sim 10^{18}cm^{-3}$), a M-structure barrier, and a 500-nm-thick InAs:Si/GaSb n^+ ($n \sim 10^{18}cm^{-3}$) region and topped with a thin InAs:Si n^+ doped ($n \sim 10^{18}cm^{-3}$) contact layer.	[14, 16–18]
Dual-band nBn E_c E_v	8 ML InAs/8 ML GaSb MWIR SL, n type, d = 97 nm $Al_{0.2}Ga_{0.8}Sb$, d = 100 nm 9 ML InAs/5 ML InGaSb LWIR SL, non intentionally doped, d = 2.5 μm 8 ML InAs/8 ML GaSb SL, n type contact layer, d = 360 nm GaSb:Te 2" wafer	**MWIR/LWIR nBn detector** In this dual-band SL nBn detector, the LWIR SL and MWIR SL are separated by AlGaSb unipolar barrier. The dual band response is achieved by changing the polarity of applied bias (see Figure 22.9). The advantage of this structure is design simplicity and compatibility with commercially available readout integrated circuits. The concerns are connected with low hole mobility and lateral diffusion.	[22]

(*Continued*)

Table 22.1 (*Continued*) Type II superlattice barrier detectors

FLAT-BAND ENERGY DIAGRAMS	EXAMPLES	DESCRIPTION	REFS.
DH with graded-gap junction E_c E_v	N-grown-junction in wider band gap SL P-graded-gap "W" SL structure p-type-II SL active layer	**Shallow-etch mesa isolation (SEMI) structure** It is a n-on-p graded-gap W photodiode structure in which the energy gap is increased in a series of steps from that of the lightly p-type absorbing region to a value typically two to three times larger. The hole-blocking unipolar barrier is typically made from a four-layer InAs/GaInSb/InAs/AlGaInSb SL. The wider gap levels is about 10 nm short of the doping defined junction, and continues for another 0.25 μm into the heavily n-doped cathode before the structure is terminated by an n^+-doped InAs top cap layer. Individual photodiodes are defined using a shallow etch that typically terminates only 10–20 nm past the junction, which is sufficient to isolate neighboring pixels while leaving the narrow-gap absorber layer buried 100–200 nm below the surface.	[23,24]
Complementary barrier E_c E_v	InAs/AlSb SL hole barrier p-type LW InAs/GaSb SL absorber MW InAs/GaSb SL electron barrier n-type InAsSb emitter GaSb buffer GaSb substrate	**Complementary barrier infrared detector (CBIRD)** This device consists of a lightly p-type InAs/GaSb SL absorber sandwiched between a n-type InAs/AlSb hB SL and wider InAs/GaSb eB. The barriers are designed in a way to have approximately zero conduction and valence band offset with respect to SL absorber. A heavily doped n-type InAsSb layer adjacent to the eB SL acts as the bottom contact layer. The n-p junction between the hB InAs/AlSb SL and the absorber SL reduces SRH-related dark current and trap-assisted tunneling. The LWIR CBIRD SL detector performance is close to the Rule 07 trend line. For a detector having a cutoff wavelength of 9.24 μm, the value of $R_0A > 10^5$ Ωcm^2 at 78 K was measured.	[4,25]

(*Continued*)

Table 22.1 (*Continued*) Type II superlattice barrier detectors

FLAT-BAND ENERGY DIAGRAMS	EXAMPLES	DESCRIPTION	REFS.
	SL p⁺-layer contact layer (5/8ML InAs/GaSb)x38; 130 nm — p SL electron blocking (eB) (7/4ML GaSb/AlSb)x45; 149 nm — B SL absorber (p-type, 10^{15} cm^{-3}) (14/7ML InAs/GaSb)x300; 1940 nm — i SL hole blocking (hB) (16/4ML InAs/AlSb)x45; 275 nm — B SL n⁺-contact layer (9/4ML InAs/GaSb)x200; 800 nm — n GaSb substrate	**pBiBn detector structure** This is another variation of the DH CBIRD structure. In this design, a pin photodiode is modified, such that the unipolar eB and hB layers are sandwiched between, respectively, the p-contact layer and the absorber, and the n-contact layer and the absorber. This design facilitates a significant reduction in the electric field drop across the narrow-gap absorber region (most of the electric field drop across the wider bandgap eB and hB layers) leading to a very small depletion region in the absorber layer, and hence reduction in the SRH, band-to-band tunneling and trap-assisted-tunneling current components.	[26,27]
pB_pp barrier E_c E_v	13 ML InAs/7 ML GaSb p-type LWIR SL active layer 15 ML InAs/4 ML AlSb SL barrier layer 13 ML InAs/7 ML GaSb p-type LWIR SL contact layer GaSb substrate	**LWIR pB_pp structure** The active (AL) and contact (CL) layers are both made from InAs/GaSb T2SLs with approximately 13 ML InAs/seven ML GaSb. The barrier layer (BL) is based on a 15 ML InAs/four ML AlSb T2SL. The interfaces are InSb-like and their presence ensures a good lattice match with the GaSb substrate.	[10,28]

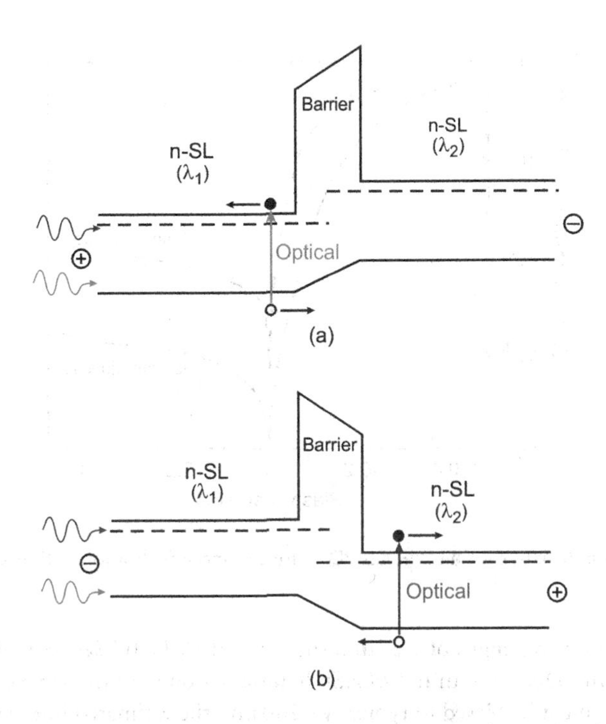

Figure 22.9 Schematic band diagram for dual band nBn detector under (a) forward and (b) reverse bias.

The p-π-M-n structure (see Table 22.1) is similar to the standard p-π-n structure, however in comparison with the latter, the electric field in the depletion region of p-π-M-n structure is reduced. As a result, the GR current is negligible in comparison with the diffusion current from the active region and the tunneling contribution is reduced since the barrier between the p and the n regions is spatially broadened.

A variation of the DH detector structure is a structure with graded bandgap in the depletion region (see Table 22.1, DH with graded gap junction). The graded gap region is inserted between the absorber and the hole barrier (hB) to reduce tunneling and GR processes. A similar structure that was developed by The Naval Research Laboratory and Teledyne enables the shallow-etch mesa isolation (SEMI) structure for surface leakage current reduction. The junction is placed in the wider gap portion of the transition-graded gap W layer. A shallow mesa etch just through the junction but not into the active layer isolates the diode, but still leaves a wide band gap surface for ease of passivation. A modest reverse bias allows efficient collection from the active layer similar to planar DH p-on-n HgCdTe photodiodes [30].

However, the culminating feature is the use of a pair of complementary barriers, namely, an electron barrier and a hole barrier (hB) formed at different depths in the growth sequence (see Table 22.1, complementary barrier). Such a structure is known as a complementary barrier infrared device (CBIRD) and was invented by Ting and others at Jet Propulsion Laboratory. An eB appears in the conduction band and a hB in the valence band. The two barriers complement one another to impede the flow of dark current. The absorber region, where the bandgap is smallest, is p-type and the top contact region is n-type, making an n-on-p polarity for the detector element. In sequence from the top, the first three regions are composed of SL material: the n-type cap, the p-type absorber, and the p-type eB. The highly doped n-InAsSb layer below the eB is an alloy. Further, at the bottom are a GaSb buffer layer and the GaSb substrate.

The introduction of device designs containing unipolar barriers has taken the LWIR CBIRD SL detector performance close to the Rule 07 trend line, which provides a heuristic predictor of the state of the-art HgCdTe photodiode performance [31]. The barriers prove to be very effective in suppressing the dark current. Figure 22.10 compares the *J-V* characteristics of a CBIRD device to a homojunction device made with the same absorber SL. Lower dark current determines a higher *RA* product.

Usually, R_0A values are plotted for devices with near zero-bias turn-on. However, since the detector is expected to operate at a higher bias, a more relevant quantity is the effective resistance-area product. In case

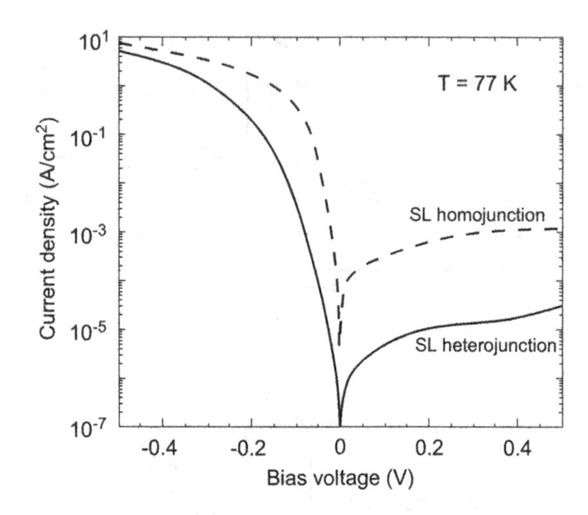

Figure 22.10 Dark *J-V* characteristics for a LWIR CBIRD detector and a SL homojunction at 77 K. (Adapted after Ref. [4])

of a detector having a cutoff wavelength of 9.24 μm, the value of $R_0A > 10^5$ Ωcm² at 78 K was measured, as compared with about 100 Ωcm² for an InAs/GaSb homojunction with the same cutoff. For a good photoresponse, the device must be biased to typically −200 mV; the estimated internal quantum efficiency is greater than 50%, while the RA_{eff} remains above 10^4 Ωcm² [4].

Recently, considerably progress has been achieved at SCD in the development of LWIR pB_pp T2SL barrier detectors. These detectors enable diffusion-limited behavior with dark currents comparable with HgCdTe Rule 07 and with high quantum efficiency. The active and contact layers of these devices are both made from InAs/GaSb T2SLs with approximately 13 ML InAs/seven ML GaSb, while the barrier layer is based on a 15 ML InAs/four ML AlSb T2SL. The individual InAs layers are terminated with In and the AlSb or GaSb layers are terminated with Sb, in order to create InSb-like interfaces with the correct amount of strain for lattice matching between the T2SL and the GaSb substrate.

As shown in Figure 22.11a, the dark current in standard LWIR n-on-p diode based solely on InAs/GaSb T2SLs is higher in the lower temperature region in comparison with pB_pp T2SL barrier detector whose schematic profile of band edges is shown in the inset of figure. The barrier device is diffusion limited down to 77 K, while the p-n diode is GR limited at this temperature with a dark current over 20 times larger. The dark current in pB_pp devices, with the thickness of active layers between 1.5 and 6.0 μm, is within one order of magnitude of HgCdTe.

Theoretical predictions of spectral response curves based on **k·p** and optical transfer matrix methods for both antireflection and no antireflection coatings (ARC and no ARC, respectively) are in good agreement with experimental data—see Figure 22.12. The detector structures include a mirror on the contact layer to reflect 80% of the light back for a second pass. The inset in Figure 22.12 shows the typical bias dependence of the quantum efficiency for a 100 × 100 μm² test device without ARC. The signal reaches its full value until a positive applied bias of ~0.6 V is applied. This bias is needed to overcome the electrostatic barrier to minority carriers caused by negative space-charge in the depleted p-type barrier. The LWIR T2SL pB_pp devices have a BLIP temperature of ~100 K at *f*/2 optics.

For the equivalent detector based on InAs/InAsSb T2SL, the predicted average quantum efficiency is only about 2/3 of the InAs/GaSb value. This estimation can be attributed to the smaller absorption coefficient of the InAs/InAsSb T2SL near the cutoff wavelength [33].

22.4 BARRIER DETECTORS VERSUS HgCdTe PHOTODIODES

There are currently two broad IR detector materials, namely, HgCdTe ternary alloys and III-V semiconductors. III-V binary InSb was among the earliest to be utilized for IR detection in the early 1950s. In the mid-1990s, U.S. Government created the research emphasis on III-V materials as an alternative technology

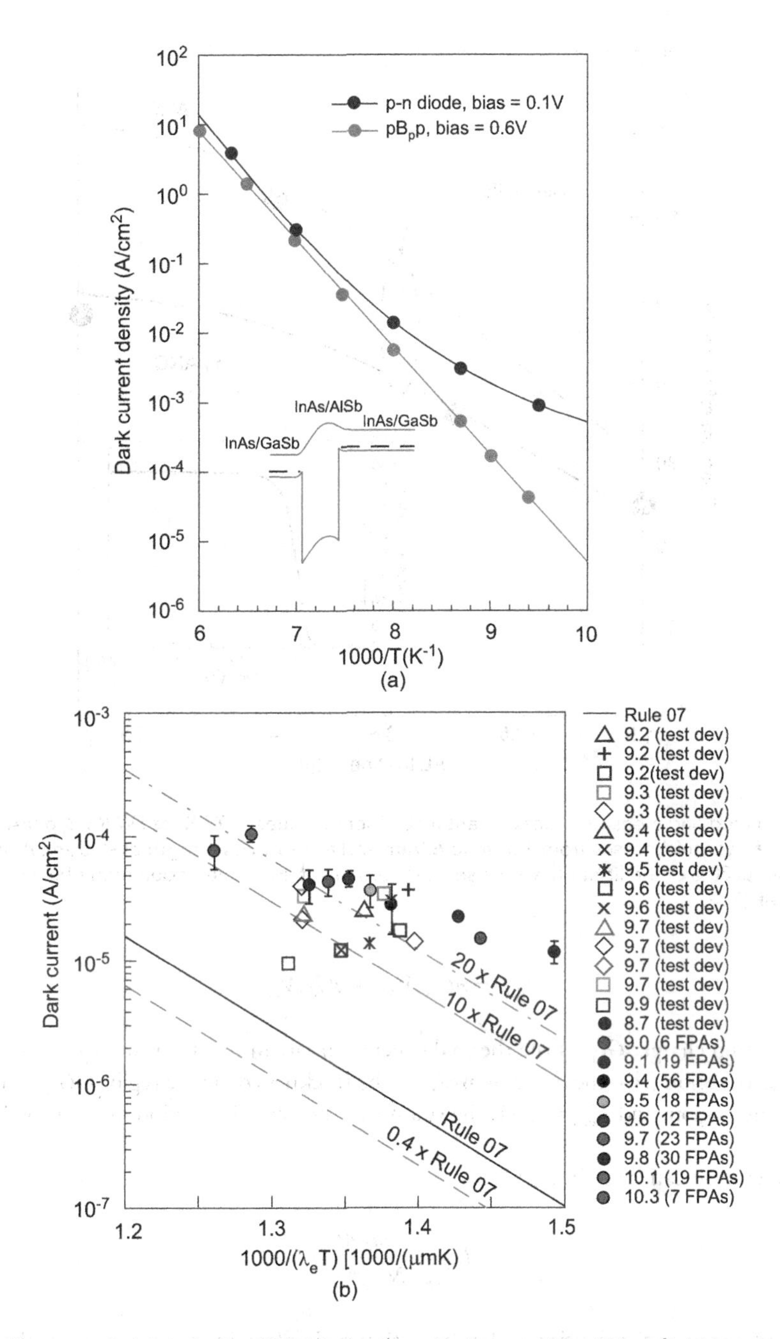

Figure 22.11 Dark current density of pB_pp T2SL barrier detector (area = 100 × 100 μm²): (a) in comparison with p-n diode (Adapted after Ref. [10]), (b) Rule 07 plot for barrier structures with different thicknesses of active layers. Range of bandgap wavelengths: $9.0 < \lambda_e < 10.3\,\mu m$. Solid line shows HgCdTe Rule 07 with uncertainty factors of 0.4, 10, and 20 (dashed lines). (Adapted after Ref. [32])

option to HgCdTe to attain its stated goal of inexpensive large-area IR FPAs. This concept typically involves the fabrication of III-V SLs to tailor the bandgap of the material to detect the desired IR radiation.

The bandgap structure and physical properties of III-V compounds are found to be remarkably similar to HgCdTe with the same bandgap. It is of interest to compare the performance of detectors composed of III-V material systems with HgCdTe photodiodes operating in both MWIR and LWIR spectral ranges.

In general, the dark current density generated in the p-n junction is the sum of the diffusion current of the active region and depletion layer region, and is given by

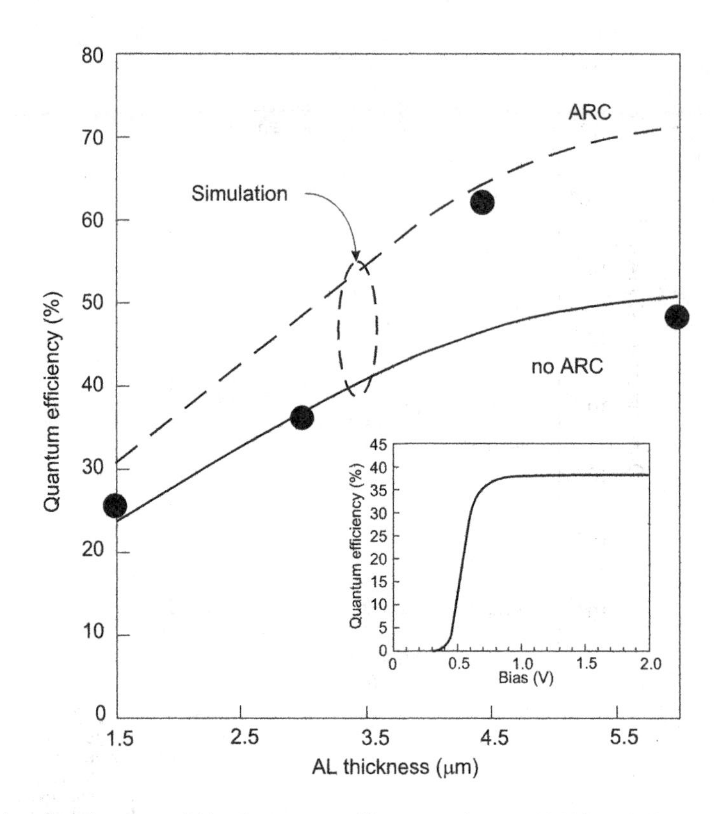

Figure 22.12 Simulated and measured (dots) quantum efficiency values at 77 K for LWIR pB$_p$p InAs/GaSb T2SL devices with active layer thicknesses from 1.5 μm to 6.0 μm and a cutoff wavelength of ~9.5 μm. A mirror on the contact layer reflects 80% of the light back for a second pass. *Inset:* Example of quantum efficiency vs. bias. (Adapted after Ref. [33])

$$I_{dark} = qG_{diff}V_{diff} + qG_{dep}V_{dep}, \tag{22.1}$$

where q is the electron charge, G_{base} is the thermal generation rate in the diffusion region, $V_{diff} = At$ is the volume of the active region (A—the detector area, t—the thickness of active region), G_{dep} is the generation rate in space-charge region, and $V_{dep} = Aw$ is the volume of the depletion region (w—the width of depletion region).

The diffusion current can be estimated as:

$$I_{diff} = \frac{qn_i^2 At}{N_{dop}\tau_{diff}}, \tag{22.2}$$

were N_{dop} is the majority carrier density in the absorption region (extrinsic regime), τ_{diff} is the diffusion carrier lifetime, and n_i is the intrinsic carrier concentration.

The depletion layer current can be given by a simplification formula:

$$I_{dep} = \frac{qn_i Aw}{2\tau_0}. \tag{22.3}$$

To obtain the last formula it is assumed that a trap is located at the intrinsic level (electron and hole concentration are equal n_i) and the SRH lifetime is equal τ_0.

From the last two equations two important conclusions result:

- for diffusion-limited photodiode, the dark current is inversely proportional to the product $N_{dop} \times \tau_{diff}$,
- the ratio between diffusion and depletion current is equal

$$\frac{I_{\text{diff}}}{I_{\text{dep}}} \approx 2\frac{n_i}{N_{\text{dop}}}\frac{\tau_0}{\tau_{\text{diff}}}\frac{t}{w}. \tag{22.4}$$

The Equation 22.4 indicates that the diffusion current dominates the depletion one for high intrinsic carrier concentration (e.g., in materials with narrow bandgap). The reverse situation is observed for semiconductors with larger bandgap. As shown, the I_{diff}/I_{dep} ratio depends also on the doping level, N_{dop}, the ratios of the respective volumes (t/w), and lifetimes. The last issue requires additional discussion of the GR mechanisms.

22.4.1 $N_{DOP} \times \tau_{DIFF}$ PRODUCT AS THE FIGURE OF MERIT OF DIFFUSION-LIMITED PHOTODETECTOR

Basic properties of the artificial T2SL materials are completely different from those of constituent layers—see Section 20.4. The effective masses of SLs are not directly dependent on the bandgap energy as is the case in a bulk semiconductor. The electron effective mass of InAs/GaSb SL is larger compared to $m_e^* = 0.009m_0$ in HgCdTe alloy with the same bandgap ($E_g \approx 0.1$ eV). Thus, diode tunneling currents in the SL can be reduced compared to the HgCdTe alloy. As a result, the doping density of T2SL is of the order of 1×10^{16} cm^{-3}, which is considerably higher than the doping level in HgCdTe (typically about 10^{15} cm^{-3}).

As seen in Equation 22.2, the figure of merit of diffusion-limited photodiode is the $N_{dop} \times \tau_{diff}$ product—so the highest lifetime is required with the highest doping. The transition between SRH lifetime (for low doping concentrations) to an Auger lifetime (for higher doping) introduces the bell function of $N_{dop} \times \tau_{diff}$ product for HgCdTe diffusion-limited photodiodes. This situation is illustrated in Figure 22.13 taking into account the lifetime data shown in Figures 17.16 and 17.17. Figure 22.13 also compares $N_{dop} \times \tau_{diff}$ product for HgCdTe photodiodes with equivalent T2SL photodiodes. As is shown, the optimum doping concentration for InAs/GaSb T2SLs is higher (about 2×10^{16} cm^{-3}) than for HgCdTe material system. As a result, the higher doping compensates for the shorter lifetime, resulting in relatively low diffusion dark current of InAs/GaSb T2SL photodiodes.

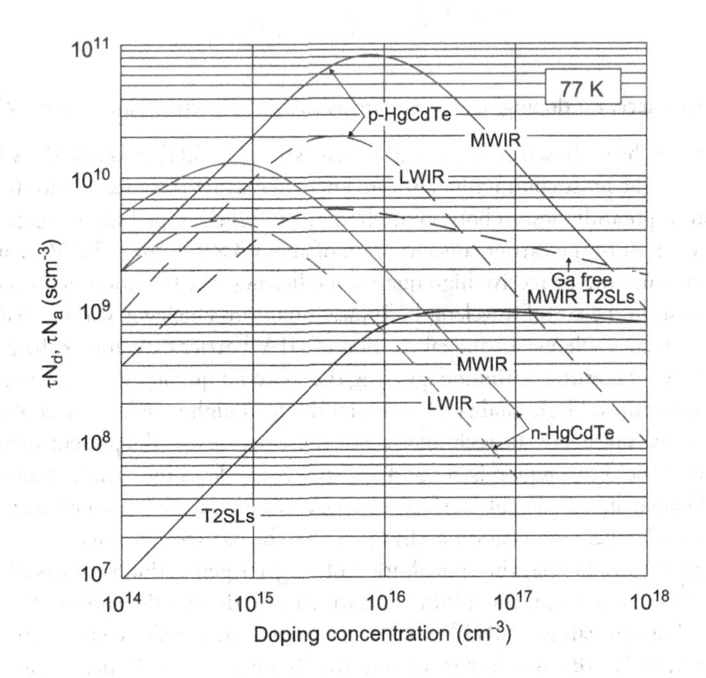

Figure 22.13 The $N_{dop} \times \tau_{diff}$ product versus doping concentration for diffusion-limited MWIR and LWIR photodiodes operated at 77 K fabricated with HgCdTe and T2SL material systems.

Another situation is observed for Ga-free T2SLs due to considerably higher carrier lifetime—the optimal doping concentration is in the middle of 10^{15} cm^{-3}.

22.4.2 DARK CURRENT DENSITY

The dark diffusion current density generated in an absorber region is given by:

$$J_{\text{dark}} = qGt, \tag{22.5}$$

where q is the electron charge, G is the thermal generation rate in the base region, and t is the thickness of active region.

Assuming that the thermal generation is a sum of Auger 1 and SRH mechanisms in the n-type absorber region,

$$G = \frac{n_i^2}{N_d \tau_{A1}} + \frac{n_i^2}{(N_d + n_i)\tau_{\text{SRH}}}, \tag{22.6}$$

and that $\tau_{A1} = 2\tau_{A1}^i / \left[1 + (N_d/n_i)^2\right]$, the dark current density is

$$J_{\text{dark}} = \frac{qN_d t}{2\tau_{A1}^i} + \frac{qn_i^2 t}{(N_d + n_i)\tau_{\text{SRH}}}. \tag{22.7}$$

The influence of generation in a depletion region ($J_{\text{dep}} = qn_i w/\tau_{SR}$) can be omitted in both III-V barrier detectors and HgCdTe photodiodes. In the present paper, the contribution of J_{dep} is taken into consideration only for a 5 µm cutoff InAsSb photodiode.

For the dark current density of p-type absorber region of detectors we can obtain a similar equation:

$$J_{\text{dark}} = \frac{qN_a t}{2\tau_{A7}^i} + \frac{qn_i^2 t}{(N_a + n_i)\tau_{\text{SRH}}}, \tag{22.8}$$

where N_a is the absorber acceptor doping, τ_{A7}^i is the intrinsic Auger 7 lifetime, and $\tau_{A7} = 2\tau_{A7}^i / \left[1 + (N_a/n_i)^2\right]$.

In the estimations, we have chosen III-V barrier detectors and HgCdTe photodiodes with cutoff wavelengths of 5 and 10 µm. The most commonly fabricated detector structures used homo- (n$^+$-on-p) and heterojunction (P-on-n) photodiodes. In both photodiodes, the lightly doped narrow-gap absorbing region (base of the photodiode: p(n)-type carrier concentration of about 3×10^{15} cm^{-3} (5×10^{14} cm^{-3})) determines the dark current and photocurrent. To receive high quantum efficiency, the thickness of an active detector layer should be equal, at least, to the cutoff wavelength. Typical quantum efficiency without ARC is about 70%.

The main technological challenge for the fabrication of III-V barrier detectors is the growth of thick active detector region SL structures without degrading the material quality. It is especially important in case of using T2SL structures. High-quality SL material thick enough to achieve acceptable quantum efficiency is crucial to the success of the technology. For these reasons, at the present stage of technology the typical thickness of the active region is about 3 µm and, as a rule, is independent on the detector's cutoff wavelength. The influence of considerable surface leakage attributed to the discontinuity in the periodic crystal structure caused by mesa delineation is eliminated in the barrier detectors.

From Equation 22.7 it results that the contribution of Auger 1 generation varies as N_d, whereas the SRH generation varies as $1/N_d$. As a result, the minimum dark current density depends on the absorber doping concentration and on the value of SRH lifetime. This dependence is shown in Figure 22.14 following Kinch et al. [34] for MWIR nBn InAsSb barrier detector. To approach BLIP performance, the detector with a 4.8 µm cutoff wavelength operating at 160 K with *f*/3 optics requires a generic IR material SRH

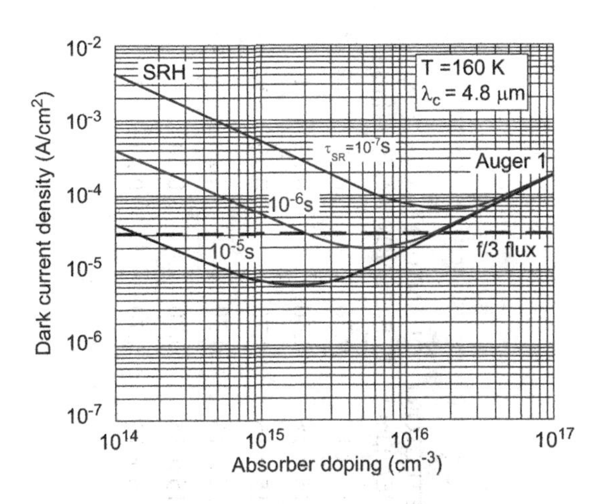

Figure 22.14 Dependence of absorber dark current density on doping concentrations for various values of SRH lifetime at an operating temperature of 160 K and a cutoff wavelength of 4.8 μm, operating at an *f*/3 background flux. (Adapted after Ref. [34])

lifetime of about 1 μs and an optimized absorber doping of ~ 10^{16}cm^{-3}. A value of 0.6 μs, relatively independent of temperature, for τ_{SR} has been suggested in Ref. [35].

Numerical estimation of dark current was performed utilizing commercial software APSYS by Crosslight Software Inc. [36]. Specific parameters are listed in Table 22.2, but other relations used in the device modeling are given in Rogalski's monograph [30] and Ting's et al. review paper [4].

It is well known that at present the performance of InAs/GaSb T2SL detectors is limited by the fairly fast SRH transition time between the conduction band and valence band states of the absorber material. To our knowledge, there are no systematic studies of the carrier lifetime and diffusion length dependence on temperature for T2SL material. Several groups have estimated the SRH carrier lifetime ranking from several tens of nanoseconds to 157 ns [37]. For these reasons, we assume that the carrier lifetime equals 200 ns in both MWIR and LWIR compositions.

Pixel reduction is mandatory to increase the detection and identification ranges of IR imaging systems. The pitch of 15 μm is in IR array production today. In our estimations we assume a detector area of 15 × 15 μm^2.

Figure 22.15 compares the predicted dependence of dark current density on operating temperature for different types of detectors with a cutoff wavelength of 5 μm. In comparison with InAsSb photodiode (with built-in depletion region), the benefits of the nBn structure is clear since it allows the operation at considerably higher temperatures. However, the HgCdTe photodiodes enable higher operating temperatures than the InAsSb nBn detector by ~20 K. The best MWIR III-V devices are heavily doped when the Auger lifetime is significantly reduced. 300 K data from Bewley et al. [38] show that Auger coefficients of SL devices are 520 times lower than those of HgCdTe. To attain their full potential, the detector developers need to realize Auger-limited devices at doping in the 10^{15}cm^{-3} range. It should be stated that III-V MWIR detector architectures fall short of the ultimate performance possible with *f*/3 optics, namely operation at about 150 K. MWIR HgCdTe systems are commercially available that operate at *f*/3 optics at 160 K.

Theoretically, LWIR T2SL materials have lower fundamental dark currents than HgCdTe. However, their performance has not achieved theoretical value. This limitation appears to be due to two main factors: relatively high background concentrations (about 10^{16}cm^{-3}) and a short minority carrier lifetime (typically tens of nanoseconds). Up until now, nonoptimized carrier lifetimes limited by SRH recombination mechanism have been observed. The minority carrier diffusion length is in the range of several micrometers. Improving these fundamental parameters is essential to realize the predicted performance of T2SL photodiodes.

Table 22.2 Material parameters of active regions

TYPE OF DETECTOR	MATERIAL	CUTOFF WAVELENGTH [μm]	DOPING OF ACTIVE REGION N_a, N_d [cm^{-3}]	THICKNESS OF ACTIVE REGION [μm]	τ_{SRH} [ms]	DETECTOR AREA [μm^2]	QUANTUM EFFICIENCY	F_1F_2	τ_{int} [ms]
nBn	InAsSb	5	1×10^{16}	3	0.40	15×15	0.70	0.28	10
nBn	InAs/GaSb	10	1×10^{16}	3	0.20	15×15	0.50	0.28	1
n-on-p	HgCdTe	5	3×10^{15}	5	—	15×15	0.70	0.20	10
n-on-p	HgCdTe	10	3×10^{15}	10	—	15×15	0.70	0.20	1
p-on-n	HgCdTe	5	5×10^{14}	5	—	15×15	0.70	0.20	10
p-on-n	HgCdTe	10	5×10^{14}	10	—	15×15	0.70	0.20	1

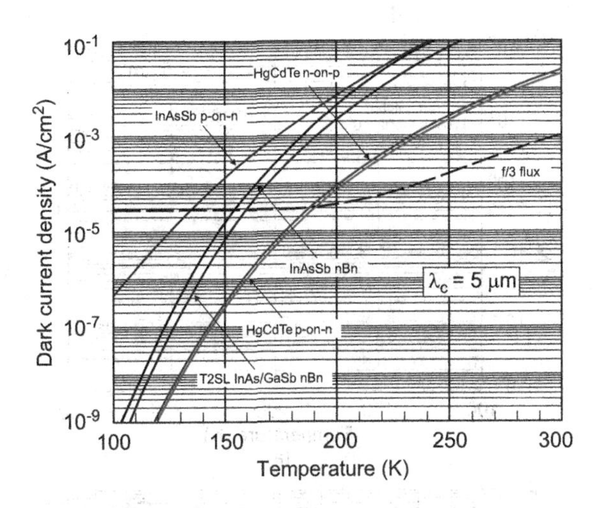

Figure 22.15 Dark current density vs. temperature for InAsSb photodiode and nBn detector, and HgCdTe photodiodes with a cutoff wavelength of 5 μm.

To date, LWIR T2SL photodiodes perform slightly worse than HgCdTe. For example, Figure 22.16 compares the predicted dependence of dark current density on operating temperature for InAs/GaSb nBn and HgCdTe photodiodes with a cutoff wavelength of 10 μm. The BLIP performance with *f*/1 optics for HgCdTe photodiode is achieved at about 130 K, about 15 K higher than for InAs/GaSb nBn detector.

22.4.3 NOISE EQUIVALENT DIFFERENCE TEMPERATURE

The detector sensitivity can be also expressed by the NEDT. This parameter is a figure of merit for thermal imagers—see Section 24.5.

NEDT can be determined knowing the dark current density, J_{dark}, the background flux (system optics), Φ_B, and the integration time, τ_{int}, according to relations (24.9) [39]. From Equation 24.9 it results that if the value of I_{dark}/I_{ϕ} ratio increases and/or the value of η decreases, then higher integration time and faster speed of optics are required. Thus, inefficient detectors can be utilized in faster optics and slower frame rate systems. Values for N are typically in the range of 1×10^6–1×10^7 electrons for a 15 μm pixel design with available node capacities for current complementary metal oxide semiconductor (CMOS) ROIC designs. In our estimation we assume 1×10^7 electrons.

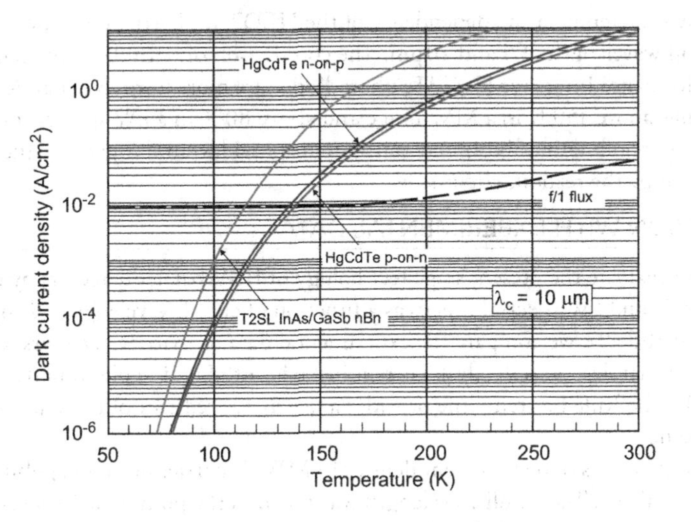

Figure 22.16 Dark current density vs. temperature for InAs/GaSb T2SL nBn detector and HgCdTe n-on-p photodiode with a cutoff wavelength of 10 μm.

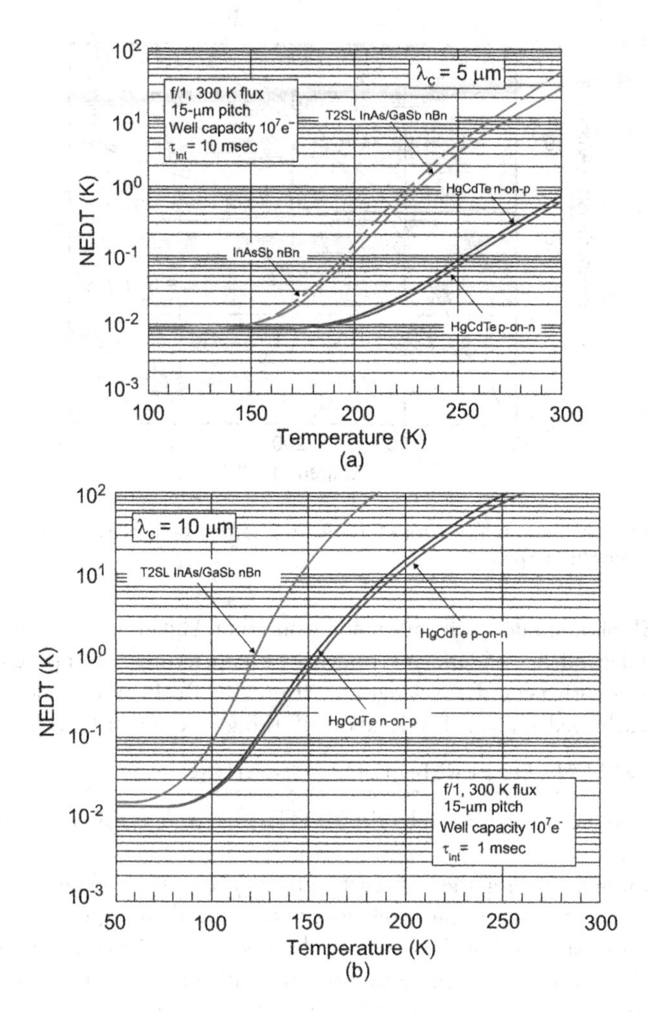

Figure 22.17 Temperature dependence of the NEDT for barrier detectors and HgCdTe photodiodes with a cutoff wavelength of 5 μm (a) and 10 μm (b).

Figure 22.17 shows the temperature dependence of the NEDT for barrier detectors and HgCdTe photodiodes with a cutoff wavelength of 5 and 10 μm. The comparison of both detector technologies indicates that theoretical performance limits for HgCdTe photodiodes are more favorable than for barrier detectors in a temperature range above 150 K in a MW range and above 80 K in a LW spectral range. In a low temperature range, the figure of merit of both material systems provides similar performance because they are predominantly limited by the readout circuits.

22.4.4 COMPARISON WITH EXPERIMENTAL DATA

At the beginning, we evaluate the present stage technology of III-V barrier detectors by examining the dark current density in MW and LW spectral ranges as a function of cutoff wavelength. Using the HgCdTe benchmark known as Rule 07, we compare the experimental data for barrier detectors with a simple empirical relationship that describes the dark current behavior of HgCdTe photodiodes with temperatures and wavelengths [31]. The Rule 07 trend line provides a heuristic predictor of the state of the art HgCdTe photodiode performance.

Figure 22.18 collects values of dark current density in MWIR barrier detectors published in literature for comparison with Rule 07. The cutoff wavelength was taken as the point of a 50% response. The empirical data are confined to 150 and 300 K. For the barrier devices, the characteristics at zero bias are not relevant. So, to extract the photogenerated minority carriers, a reverse bias about 150 mV has been chosen.

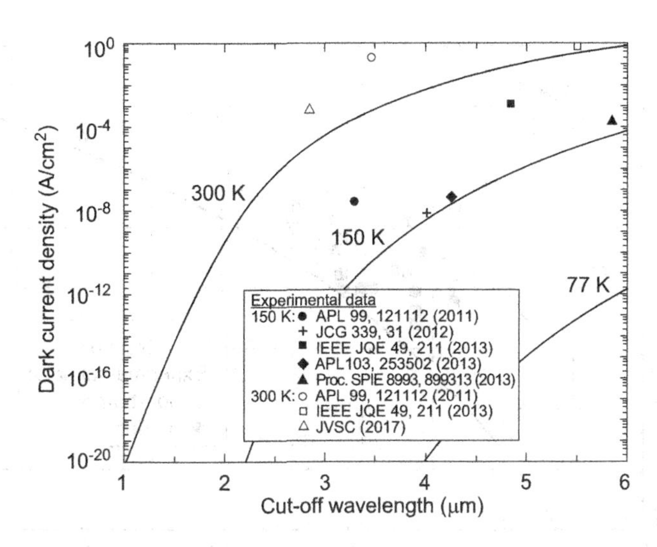

Figure 22.18 Collected values of dark current density in MWIR barrier detectors for comparison with *Rule 07* of HgCdTe photodiodes calculated for 77 K, 150 K, and 300 K. (Adapted after Ref. [36])

At liquid nitrogen temperature, the experimental data of MWIR barrier devices show considerably greater leakage current than Rule 07 by many orders of magnitude. Results close to Rule 07 are reported at a cutoff wavelength close to 4 μm at 150 K. These best quality devices are fabricated using InAsSb active region lattice matched to GaSb substrates. As we can see, the overall trend moves closer to Rule 07 as the wavelength increases. This trend is especially observed in the LWIR region—it is generally more difficult to control dark current at the shorter wavelengths.

A similar collection of experimental data for nonbarrier (homojunction) and barrier (heterojunction) LWIR T2SLs devices operating at 78 K have been gathered by Rhiger (see Figure 22.19) [40]. The nonbarrier dark currents are generally higher, with the best approaching Rule 07 to within a factor of about eight. The barrier devices clearly show lower dark currents on average, and some are close to the curve Rule 07 for a cutoff wavelength of ≥9 μm. It can be concluded that although the minority carrier lifetime in the SL detector active region is not limited by an Auger mechanism, the diffusion current does not differ so greatly from Rule 07.

Figure 22.5 shows temperature dependence of NEDT at optics *f*/3.2 for 15 μm pitch $InAs_{0.91}Sb_{0.09}$/B-AlAsSb barrier detector operated in the 3.4–4.2 μm spectral range. The NEDT is 20 mK at 10 ms integration time. A strong increase of NEDT above 170 K is consistent with the estimated BLIP temperature of 175 K [36].

To compare the performance of MWIR III-V barrier detector arrays with state of the art HgCdTe technology we choose the best quality MWIR arrays—Hawk detectors with 640 × 512 pixels, 16 μm pitch, and radiation shield *f*/4. The 50% cutoff wavelength is 5.5 μm at 80 K. These N^+-p(As) heterostructure photodiodes of $Hg_{1-x}Cd_xTe$, $x = 0.3$ and $x = 0.2867$, are optimized for high operating temperature (HOT) conditions. The devices were grown by metalorganic vapor phase epitaxy, MOCVD, on GaAs substrates. The N^+-region is doped with I at a level of 10^{16} cm^{-3} and about 3-μm-thick absorbing p-layer, of smaller energy gap, is doped with As at a level of 10^{15} cm^{-3}.

The Hawk array demonstrates good quality image at temperatures of 160 K–190 K. Although the acceptable 210 K image is slightly grainier than the 160 K image, the 210 K image is very useable. Figure 22.20 shows NEDT vs. temperature with the earlier results from 2011 and after an improvement of device technology [41]. The new results predict better detector performance together with extending the useful range of operating temperatures.

For standard production arrays (results from 2011), the NEDT remains constant up to 150 K–160 K and doubles by 185 K (see Figure 22.20). After improving the device technology, the

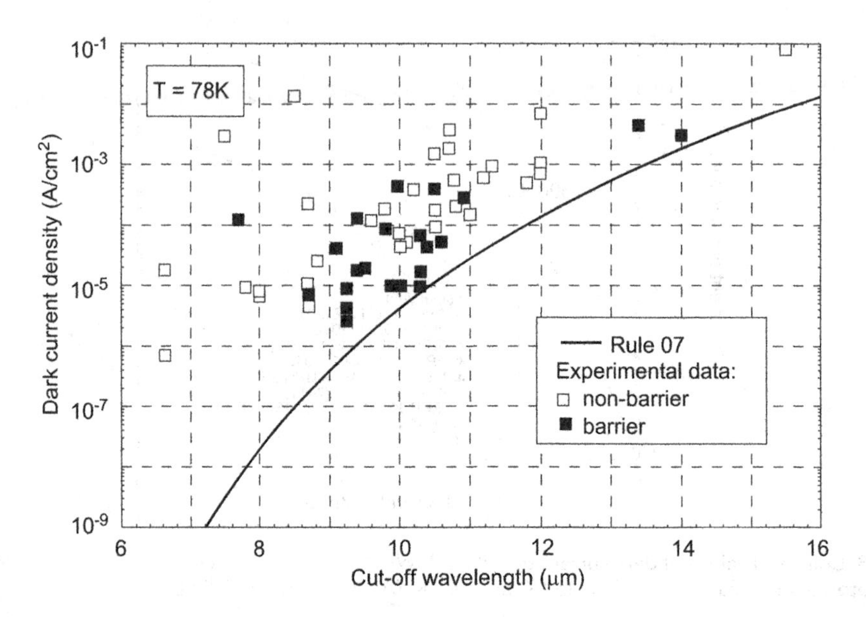

Figure 22.19 The 78 K dark current densities plotted against cutoff wavelength for T2SL nonbarrier and barrier detectors reported in the literature since late 2010. The solid line indicates the dark current density calculated using the empirical **Rule 07** model. (Adapted after Ref. [40])

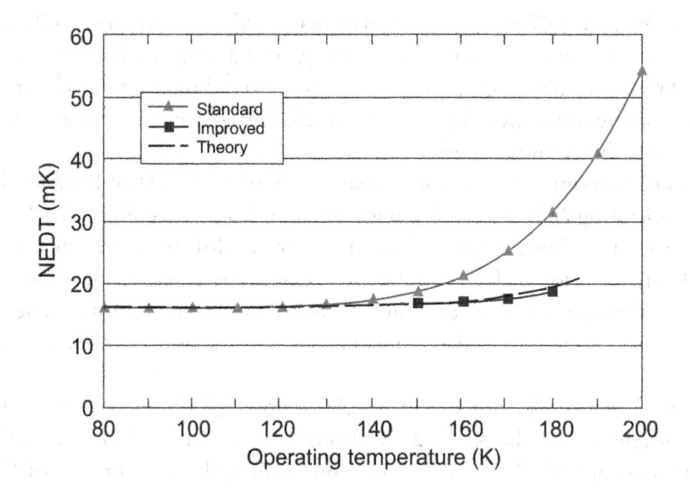

Figure 22.20 NETD performance as a function of operating temperature. (Adapted after Ref. [41])

near-background-limited performance achieved at 150 K by the standard process has been raised around 30 K to 180 K with the expectation of background-dominated performance to well above 200 K. To explain the temperature dependence of NEDT for Hawk detector we have estimated the dark current using a model given by DeWames and Pellegrino [42] where the deliberate introduction of recombination centers has emerged as a new technique in HOT detector engineering.

Usually at low temperature, the performance of MW and LWIR FPAs is limited by the readout circuits (by storage capacity of the ROIC). In this case, the NEDT is inversely proportional to the square root of the integrated charge and therefore greater the charge, higher the performance [see Equation (24.19)]. The charge handling capacity of the readout, the integration time linked to the frame time, and dark current of the sensitive material become the major issues of IR FPAs.

Figure 22.21 shows the theoretical NEDT versus charge handling capacity for different FPAs assuming that the integration capacitor is filled to half the maximum capacity (to preserve dynamic range) under

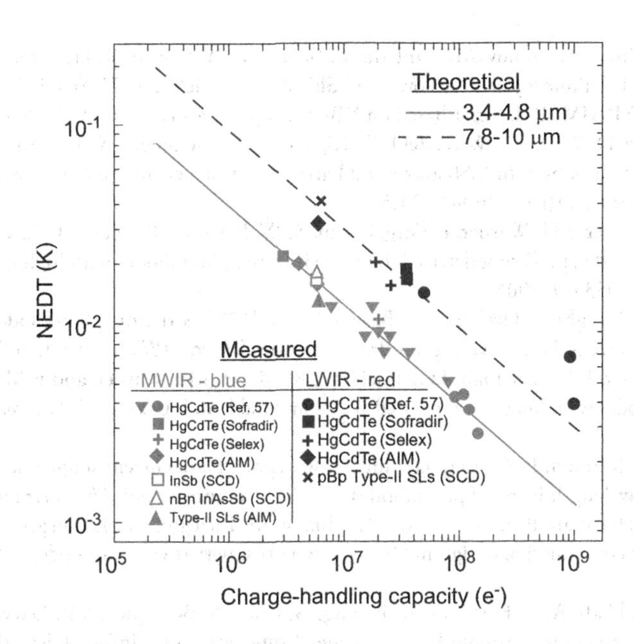

Figure 22.21 NEDT versus charge handling capacity.

nominal operating conditions in two spectral band passes: 3.4–4.8 μm and 7.8–10 μm. We can see that the measured sensitivities agree with the expected values for FPAs fabricated with different material systems, including barrier detectors and T2SLs.

REFERENCES

1. A.P. Craig, M. Jain, G. Wicks, T. Golding, K. Hossain, K. McEwan, C. Howle, B. Percy, and A.R.J. Marshall, "Short-wave infrared barriode detectors using InGaAsSb absorption material lattice matched to GaSb," *Appl. Phys. Lett.*, 106, 201103-1–4, 2015.
2. G.R. Savich, D.E. Sidor, X. Du, G.W. Wicks, M.C. Debnath, T.D. Mishima, M.B. Santos, T.D. Golding, M. Jain A.P. Craig, and A.R.J. Marshall, "III-V semiconductor extended short-wave infrared detectors," *J. Vac. Sci. Technol. B* 35(2), 02B105-1–5, 2017
3. P. Klipstein, "XBn barrier photodetectors for high sensitivity operating temperature infrared sensors," *Proc. SPIE*, 6940, 69402U–1–12, 2008.
4. D.Z. Ting, C.J. Hill, A. Soibel, J. Nguyen, S.A. Keo, M.C. Lee, J.M. Mumolo, J.K. Liu, and S.D. Gunapala, "Antimonide-based barrier infrared detectors," *Proc. SPIE*, 7660, 76601R–1–14, 2010.
5. P. Klipstein, O. Klin, S. Grossman, N. Snapi, I. Lukomsky, D. Aronov, M. Yassen, A. Glozman, T. Fishman, E. Berkowicz, O. Magen, I. Shtrichman, and E. Weiss, "XBn barrier photodetectors based on InAsSb with high operating temperatures," *Opt. Eng.*, 50, 061002–1–10, 2011.
6. A. Khoshakhlagh, S. Myers, E. Plis, M.N. Kutty, B. Klein, N. Gautam, H. Kim, E.P.G. Smith, D. Rhiger, S.M. Johnson, and S. Krishna, "Mid-wavelength InAsSb detectors based on nBn design," *Proc. SPIE*, 7660, 76602Z, 2010.
7. E. Weiss, O. Klin, S. Grossmann, N. Snapi, I. Lukomsky, D. Aronov, M. Yassen, E. Berkowicz, A. Glozman, P. Klipstein, A. Fraenkel, and I. Shtrichman, "InAsSb-based XB_nn bariodes grown by molecular beam epitaxy on GaAs," *J. Crystal Growth*, 339, 31–5, 2012.
8. P. Martyniuk and A. Rogalski, "Modeling of InAsSb/AlAsSb nBn HOT detector's performance limits," *Proc. SPIE*, 8704, 87041X-1–9, 2013.
9. A.I. D'Souza, E. Robinson, A.C. Ionescu, D. Okerlund, T.J. de Lyon, R.D. Rajavel, H. Sharifi, N.K. Dhar, P.S. Wijewarnasuriya, and C. Grein, "MWIR InAsSb barrier detector data and analysis," *Proc. SPIE*, 8704, 87041V-1-7, 2013.
10. P.C. Klipstein, "XB_nn and XB_pp infrared detectors," *J. Cryst. Growth*, 425, 351–356, 2015.

11. P.C. Klipstein, Y. Gross, A. Aronov, M. ben Ezra, E. Berkowicz, Y. Cohen, R. Fraenkel, A. Glozman, S. Grossman, O. Kin, I. Lukomsky, T. Markowitz, L. Shkedy, I. Sntrichman, N. Snapi, A. Tuito, M. Yassen, and E. Weiss, "Low SWaP MWIR detector based on XBn focal plane array," *Proc. SPIE*, 8704, 87041S–1–12, 2013.
12. Y. Lin, D. Donetsky, D. Wang, D. Westerfeld, G. Kipshidze, L. Shterengas, W.L. Sarney, S.P. Svensson, and G. Belenky, "Development of bulk InAsSb alloys and barrier heterostructures for long-wavelength infrared detectors," *J. Electron. Mater.*, 44(10), 3360–6, 2015.
13. E.H. Aifer, J.G. Tischler, J.H. Warner, I. Vurgaftman, W.W. Bewley, J.R. Meyer, C.L. Canedy, and E.M. Jackson, "W-structured type-II superlattice long-wave infrared photodiodes with high quantum efficiency," *Appl. Phys. Lett.*, 89, 053519, 2006.
14. B.-M. Nguyen, M. Razeghi, V. Nathan, G.J. Brown, "Type-II "M" structure photodiodes: An alternative material design for mid-wave to long wavelength infrared regimes," *Proc. SPIE*, 6479, 64790S, 2007.
15. C.L. Canedy, H. Aifer, I. Vurgaftman, J.G. Tischler, J.R. Meyer, J.H. Warner, and E.M. Jackson, "Antimonide type-II W photodiodes with long-wave infrared R_0A comparable to HgCdTe," *J. Electron. Mater.*, 36, 852–6, 2007.
16. B.-M. Nguyen, D. Hoffman, P.-Y. Delaunay, and M. Razeghi, "Dark current suppression in type II InAs/GaSb superlattice long wavelength infrared photodiodes with M-structure," *Appl. Phys. Lett.*, 91, 163511, 2007.
17. B.-M. Nguyen, D. Hoffman, P.-Y. Delaunay, E.K. Huang, M. Razeghi, and J. Pellegrino, "Band edge tunability of *M*-structure for heterojunction design in Sb based type II superlattice photodiodes," *Appl. Phys. Lett.*, 93, 163502, 2008.
18. M. Razeghi, H. Haddadi, A.M. Hoang, E.K. Huang, G. Chen, S. Bogdanov, S.R. Darvish, F. Callewaert, and R. McClintock, "Advances in antimonide-based Type-II superlattices for infrared detection and imaging at center for quantum devices," *Infrared Phys. Technol.*, 59, 41–52, 2013.
19. E.K. Huang, D. Hoffman, B.-M. Nguyen, P.-Y. Delaunay, and M. Razeghi, "Surface leakage reduction in narrow band gap type-II antimonide-based superlattice photodiodes," *Appl. Phys. Lett.*, 94, 053506–1–3, 2009.
20. O. Salihoglu, A. Muti, K. Kutluer, T. Tansel, R. Turan, Y. Ergun, and A. Aydinli, ""N" structure for type-II superlattice photodetectors," *Appl. Phys. Lett.*, 101, 073505, 2012.
21. J.L. Johnson, L.A. Samoska, A.C. Gossard, J.L. Merz, M.D. Jack, G.H. Chapman, B.A. Baumgratz, K. Kosai, and S.M. Johnson, "Electrical and optical properties of infrared photodiodes using the InAs/$Ga_{1-x}In_xSb$ superlattice in heterojunctions with GaSb," *J. Appl. Phys.*, 80, 1116–27, 1996.
22. A. Khoshakhlagh, J.B. Rodriguez, E. Plis, G.D. Bishop, Y.D. Sharma, H.S. Kim, L.R. Dawson, and S. Krishna, "Bias dependent dual band response from InAs/Ga(In)Sb type II strain layer superlattice detectors," *Appl. Phys. Lett.*, 91, 263504, 2007.
23. E.H. Aifer, H. Warner, C.L. Canedy, I. Vurgaftman, J.M. Jackson, J.G. Tischler, J.R. Meyer, S.P. Powell, K. Oliver, and W.E. Tennant, "Shallow-etch mesa isolation of graded-bandgap W-structured type II superlattice photodiodes," *J. Electron. Mater.*, 39, 1070–9, 2010.
24. I. Vurgaftman, E.H. Aifer, C.L. Canedy, J.G. Tischler, J.R. Meyer, and J.H. Warner, "Graded band gap for dark-current suppression in long-wave infrared W-structured type-II superlattice photodiodes," *Appl. Phys. Lett.*, 89, 121114, 2006.
25. D.Z.-Y. Ting, C.J. Hill, A. Soibel, S.A. Keo, J.M. Mumolo, J. Nguyen, and S.D. Gunapala, "A high-performance long wavelength superlattice complementary barrier infrared detector," *Appl. Phys. Lett.*, 95, 023508, 2009.
26. E.A. DeCuir, G.P. Meissner, P.S. Wijewarnasuriya, N. Gautam, S. Krishna, N.K. Dhar, R.E. Welser, and A.K. Sood, "Long-wave type-II superlattice detectors with unipolar electron and hole barriers," *Opt. Eng.*, 51(12), 124001, 2012.
27. N. Gautam, S. Myers, A.V. Barve, B. Klein, E.P. Smith, D. Rhiger, E. Plis, M.N. Kutty, N. Henry, T. Schuler-Sandyy, and S. Krishna, "Band engineering HOT midwave infrared detectors based on type-II InAs/GaSb strained layer superlattices," *Infrared Phys. Techol.*, 59, 72–7, 2013.
28. P.C. Klipstein, E. Avnon, Y. Benny, R. Fraenkel, A. Glozman, S. Grossman, O. Klin, L. Langoff, Y. Livneh, I. Lukomsky, M. Nitzani, L. Shkedy, I. Shtrichman, N. Snapi, A. Tuito, and E. Weiss, "InAs/GaSb type II superlattice barrier devices with a low dark current and a high quantum efficiency," *Proc. SPIE*, 9070, 90700U–1–10, 2014.
29. A.D. Hood, A.J. Evans, A. Ikhlassi, D.L. Lee, and W.E. Tennant, "LWIR strained-layer superlattice materials and devices at Teledyne Imaging Sensors," *J. Electron. Mater.*, 39, 1001–6, 2010.
30. A. Rogalski, *Infrared Detectors*, 2nd edition, CRC Press, Boca Raton, FL, 2010.
31. W.E. Tennant, D. Lee, M. Zandian, E. Piquette, and M. Carmody, "MBE HgCdTe Technology: A very general solution to IR detection, described by 'Rule 07', a very convenient heuristic," *J. Elect. Mat.*, 37, 1406, 2008.

32. P.C. Klipstein, E. Avnon, D. Azulai, Y. Benny, R. Fraenkel, A. Glozman, E. Hojman, O. Klin, L. Krasovitsky, L. Langof, I. Lukomsky, M. Nitzani, I. Shtrichman, N. Rappaport, N. Snapi, E. Weiss, and A. Tuito, "Type II superlattice technology for LWIR detectors," *Proc. SPIE*, 9819, 98190T, 2016.
33. P. Klipstein, "Physics and technology of antimonide heterostructure devices at SCD," *Proc. SPIE*, 9370, 937021–1–8, 2015.
34. M.A. Kinch, H.F. Schaake, R.L. Strong, P.K. Liao, M.J. Ohlson, J. Jacques, C.-F. Wan, D. Chandra, R.D. Burford, and C.A. Schaake, "High operating temperature MWIR detectors," *Proc. SPIE*, 7660, 76602V–1–13, 2010.
35. Klem J. F., Kim J. K., Cich M. J., Hawkins S. D., Fortune T. R., and Rienstra J. L., "Comparison of nBn and nBp mid-wave barrier infrared photodetectors," *Proc. SPIE*, 7608, 76081P, 2010.
36. P. Martyniuk and A. Rogalski, "Performance comparison of barier detectors and HgCdTe photodiodes," *Opt. Eng.*, 53(10), 106105–1–6, 2014.
37. D. Zuo, P. Qiao, D. Wasserman, and S.L. Chuang, "Direct observation of minority carrier lifetime improvement in InAs/GaSb type-II superlattice photodiodes via interfacial layer control," *Appl. Phys. Lett.*, 102, 141107, 2013.
38. W.W. Bewley, J.R. Lindle, C.S. Kim, M. Kim, C.L. Canedy, I. Vurgaftman, and J.R. Meyer, "Lifetime and Auger coefficients in type-II W interband cascade lasers," *Appl. Phys. Lett.*, 93, 041118, 2008.
39. M.A. Kinch, *Fundamentals of Infrared Detector Materials*, SPIE Press, Bellingham, WA, 2007.
40. D.R. Rhiger, "Performance comparison of long-wavelength infrared type II superlattice devices with HgCdTe," *J. Electron. Mater.*, 40, 1815–22, 2011.
41. R.K. McEven, D. Jeckells, S. Bains, and H. Weller, "Developments in reduced pixel geometries with MOCVD grown MCT arrays," *Proc. SPIE*, 9451, 94512D–1–9, 2015.
42. R. DeWames and J. Pellegrino, "Electrical characteristics of MOVPE grown MWIR $N^{+}p(As)$ HgCdTe heterostructure photodiodes build on GaAs substrates," *Proc. SPIE*, 8353, 83532P, 2012.

23 Cascade infrared photodetectors

In a conventional photodiode, the responsivity and diffusion length are closely coupled and an increase in the absorber thickness much beyond the diffusion length may not result in the desired improvement in the signal-to-noise (*S*/*N*) ratio. This effect is particularly pronounced at high temperatures where diffusions lengths are typically reduced. Only charge carriers that are photogenerated at distance shorter than the diffusion length from junction can be collected. In high operating temperature (HOT) detectors, the absorption depth of long wavelength infrared (LWIR) radiation is longer than the diffusion length. Therefore, only a limited fraction of the photogenerated charge contributes to the quantum efficiency.

To avoid the limitation imposed by the reduced diffusion length and to effectively increase the absorption efficiency, novel detector designs based on multistage detection and currently termed as cascade infrared detectors (CIDs) have been introduced in the last decade. CIDs contain multiple discrete absorbers where each one is shorter or narrower than the diffusion length. In this discrete CID absorber architecture, the individual absorbers are sandwiched between engineered electron (eB) and hole barriers (eR) to form a series of cascade stages. The photogenerated carriers travel only over one cascade stage before they recombine in the next stage, and every individual cascade stage can be significantly shorter than the diffusion length, while the total thickness of all the absorbers can be comparable or even longer than the diffusion length.

In this case, the *S*/*N* ratio and the detectivity will continue to increase with multiple discrete absorbers resulting in improved device performance at elevated temperatures as compared to a conventional p-n photodiode. In addition, flexibility to vary the number and thicknesses of the discrete absorbers results in the ability to tailor the CID designs for optimized performance in meeting specific applications.

23.1 MULTISTAGE INFRARED DETECTORS

Different types of multistage IR detectors have been proposed and are now grouped into two main classes: (i) so-called intersubband (IS) unipolar quantum cascade IR detectors (QCIDs), and (ii) interband (IB) ambipolar CIDs. IS QCIDs have evolved from the quantum cascade laser (QCL) research and are being built for about 10 years [1–6]. A schematic comparison between the band structure of a photoconductive quantum well infrared photodetector (QWIP) and a photovoltaic QCID is shown in Figure 23.1. The QWIP structure is polarized in order to make the electrons circulate in the external circuit and to record the variation. The active detector region consists of identical QWs (quantum wells) separated by thicker barriers. Electrons are excited from the QWs by either photoemission or by thermionic emission. In contrast, the QCIDs are usually designed to be photovoltaic detectors. They consist of several identical periods made of one active doped well and some other coupled wells. The photoexcited electrons are transported from one active well to the next one by phonon emission through cascaded levels. Figure 23.1(b) shows the conduction band of one period. Incident photon induces an electron to go from the ground state E_1 to the excited level E_2, which is next transferred to the right-hand QWs through longitudinal optical (LO)-phonon relaxations and finally to the fundamental subband of the next period. The detector period is repeated N times in order to increase the detectivity.

To describe the performance of IS QCIDs, it is convenient to use the formalism originally developed for QWIPs [7]. A theoretical model is presented in, for example, Refs. 1, 2, and 5.

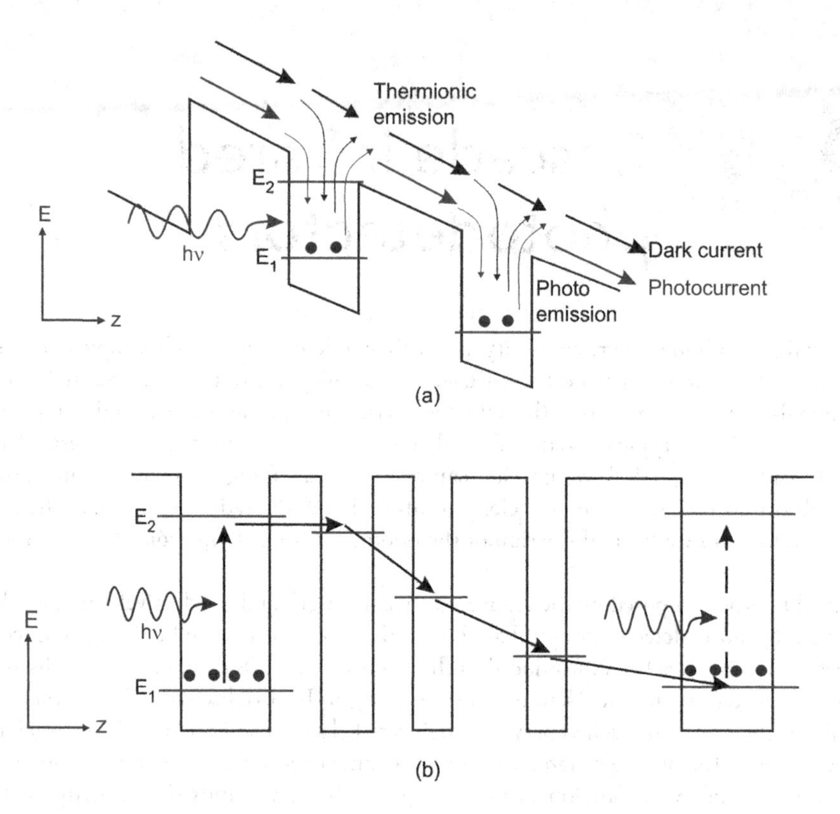

Figure 23.1 Schematic conduction band diagrams of (a) a QWIP and (b) a QCID. In the QWIP, electron transport is accomplished by an external voltage bias, whereas an internal potential ramp ensures carrier transport in QCID. (Adapted after Ref. [4])

The detectivity of QCID, including Johnson noise and electrical shot noise components, is determined by [1]

$$D^* = \frac{\eta\lambda q}{hc}\left(\frac{4kT}{NR_0A} + \frac{2qI_{\text{dark}}}{N}\right)^{-1/2}, \tag{23.1}$$

where R_0A is the resistance at zero bias times the detector area corresponding to one period of QCID, T is the detector temperature, N is the number of periods, and I_{dark} is the dark current. Equation 23.1 shows that the signal-to-noise ratio is $\propto \sqrt{N}$.

The IS QCID technology has been proven in the wavelength range from the near IR to the terahertz (THz) region and the attained detectivity is presented in Figure 23.2. At present, well-established semiconductor material systems and processing methods are available. Early QCIDs have been demonstrated in the near IR region using InGaAs/AlAsSb, in the mid-IR using InGaAs/InAlAs, and in long IR up to THz region using GaAs/AlGaAs materials. These detectors have been cryogenically cooled [3,4].

23.2 TYPE II SUPERLATTICE INTERBAND CASCADE INFRARED DETECTORS

It has been demonstrated recently that bipolar devices based on type II InAs/GaSb interband (IB) superlattice (SL) absorbers [8–21] are good candidates for detectors operating near room temperature. These IB cascade detectors combine the advantages of IB optical transitions with the excellent carrier transport properties of the IB cascade laser structures. Thermal generation rate at any specific temperature and cutoff wavelength in these devices is usually orders of magnitude smaller than that for corresponding IS QCIDs, and devices with good performance have been recently demonstrated. The operating temperatures of IB cascade detectors are considerable higher in comparison with the IS ones (see Figure 23.2).

Hinkley and Yang [14] have shown that multiple stage architecture is useful for improving the sensitivity of HOT detectors where the quantum efficiency is limited by short diffusion length. In case of HgCdTe photodiodes at room temperature, the absorption depth for LWIR radiation ($\lambda > 5$ μm) is longer than the diffusion length. Therefore, only a limited fraction of the photogenerated charge contributes to the quantum efficiency. Calculations considering the example of an uncooled 10.6 μm photodiode show that the ambipolar diffusion length is less than 2 μm, while the absorption depth is 13 μm. This reduces the quantum efficiency to ≈15% for a single pass of radiation through the detector.

A similar situation occurs in case of HOT type-II superlattices (T2SL) IB CIDs. For detector designs where the absorber lengths in each stage are equal, the multiple stage architecture offers the potential for significant detectivity improvement when $\alpha L \leq 0.2$, where α is the absorption coefficient and L is the diffusion length [14]. This theoretical prediction has been confirmed by experimental data as shown in Figure 23.2.

23.2.1 PRINCIPLE OF OPERATION

The operation idea of IB cascade photodetectors is similar to that described by Piotrowski and Rogalski (see Figure 17.54b) [22,23]. Earlier attempts to realize this type of devices using HgCdTe were carried out by utilizing tunnel junctions to electrically connect the conduction band of one absorber to the valence band of an adjacent absorber in a way similar to multi-junction solar cells. Each cell is composed of p-type doped narrow-gap absorber and heavily doped N^+ and P^+ heterojunction contacts. The incoming radiation is absorbed only in the absorber regions, while the heterojunction contacts collect the photogenerated charge carriers. Such devices are capable of achieving high quantum efficiencies, large differential resistances, and fast responses. A practical problem is associated with electrical conductivity through the adjacent N^+ and P^+ layers, however this is generally overcome by employing tunnel currents through the N^+ and P^+ interface.

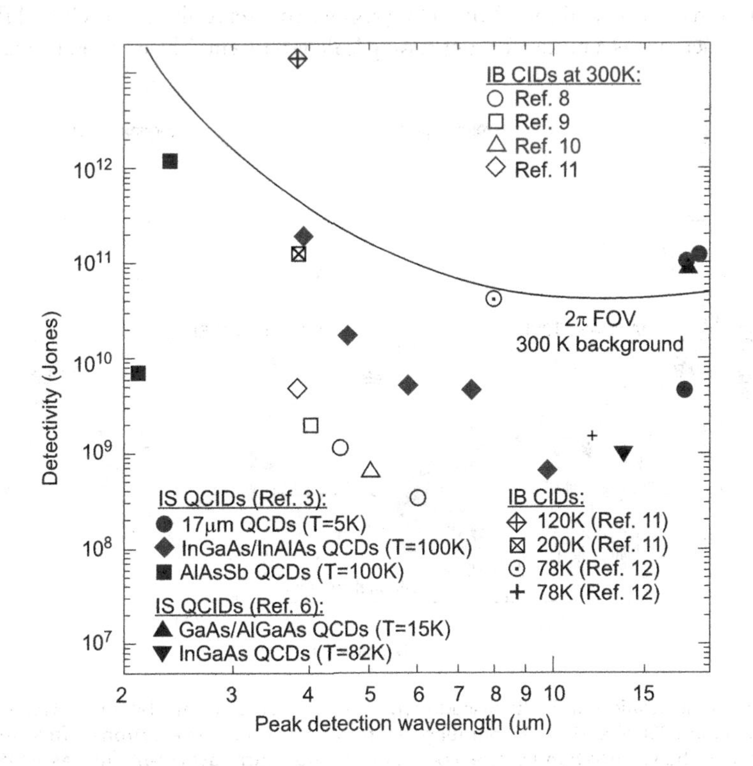

Figure 23.2 Detectivity as a function of wavelength for different types of QCIDs.

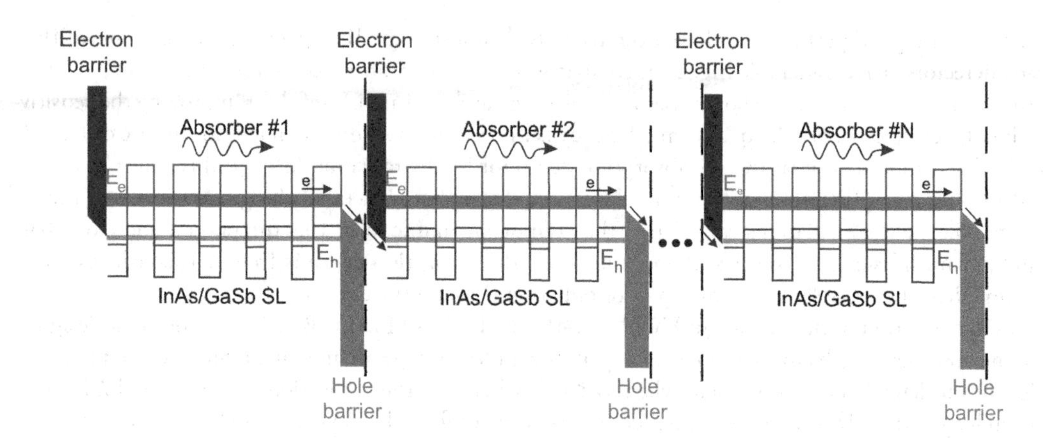

Figure 23.3 Schematic illustration of an IB QCID device with multiple stages. Each stage is composed of a SL absorber sandwiched between electron and hole barriers. E_e and E_h denote the energy for electron and hole minibands, respectively. The energy difference ($E_e - E_h$) is the bandgap, E_g, of the SL. (Adapted after Ref. [18])

The T2SL material system is a natural candidate for realizing multiple stage IB devices [24]. Figure 23.3 shows the general design structure for a T2SL cascade detector [25]. Each stage is composed of a *n*-period InAs/GaSb T2SL sandwiched between an AlSb/GaSb QW eB and an InAs/Al(In)Sb QW eR.

Because the design of IB CIDs is relatively complicated involving many interfaces and strained thin layers, their growth by molecular-beam epitaxy (MBE) is challenging. Detector designs exhibit key differences in their approach to construct the relaxation and tunneling regions as well as the contact layers. They are described in detail in Ref. [24]. Here, we focus on high quality devices presented by Tian and Krishna [11].

Tian and Krishna have proposed a cascade detector structure for which operation is shown schematically in Figure 23.4 [11]. The incoming photons are absorbed in thin InAs/GaSb T2SLs sandwiched between the electron relaxation and the IB tunneling regions that serve also as the eR and the eB, respectively. The barriers act as a means for suppressing leakage current. The electron relaxation region

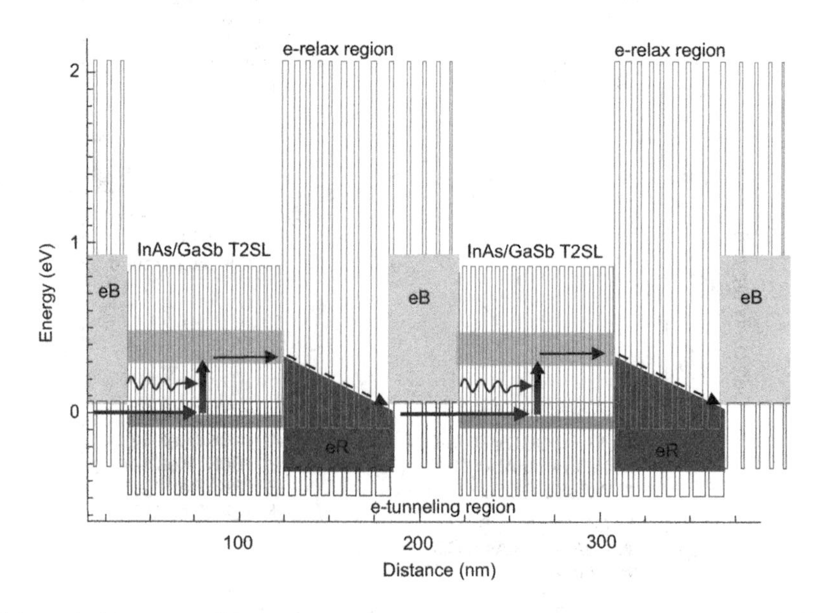

Figure 23.4 Schematic illustration of the IB cascade type II InAs/GaSb SL photodetector (after Ref. [11]). The photons are absorbed in the T2SL absorbers, generating electron-hole pairs. The electrons diffuse into the eR region, and then transport into the valence band of the next stage through ultra-fast LO-phonon assist IS relaxation and IB tunneling.

is designed to facilitate the extraction of photogenerated carriers from the conduction miniband of the absorber and transport them ideally (with little or no resistance) to the valence band of the absorber in the next stage. The energy levels of coupled InAs/AlSb multi-QWs in the conduction band form a 6-staircase, with energy-ladder separations comparable to the LO-phonon energy. The uppermost energy level of the relaxation region staircase is close to the conduction miniband in the InAs/GaSb SL and the bottom energy level is positioned below the valence-band edge of the adjacent GaSb layer, allowing the IB tunneling of extracted carriers to the next stage. The eB region consists of GaSb/AlSb QWs with estimated electron barrier thickness and height (relative to the conduction miniband minimum of the InAs/GaSb T2SL absorber) of 45 nm and 0.72 eV, respectively.

23.2.2 MWIR INTERBAND CASCADE DETECTORS

The five-stage detector structures are MBE-grown on Zn-doped 2” (001) GaSb substrates. The absorbers are composed of lightly p-doped (~5×10^{15} cm^{-3}) InAs/GaSb T2SLs with InSb interfacial layers to balance strain of the lattice-mismatched InAs. The V/III beam equivalent flux ratios for Sb/Ga and As/In are set as 4.0 and 3.2, respectively. The detectors consist of a 0.5 μm p-type GaSb buffer layer, a five-stage IB cascade absorber structure, and finally a 45-nm-thick n-type InAs as the top contact layer. The individual absorbers consist of 30, 60, and 90 periods of seven MLs InAs/eight MLs GaSb (ML stands for monolayer) T2SLs, which correspond to total absorber thicknesses of 0.73, 1.45, and 2.16 μm, respectively. Single-pixel detectors with circular mesa sizes ranging from 25 to 400 μm in diameter were fabricated. A 200-nm-thick SiN_x film is then deposited for sidewall passivation and electrical isolation. Top and bottom contacts are formed by e-beam evaporated Ti/Au. No antireflection coating is applied on top of the mesa.

Figure 23.5 shows the representative temperature-dependent dark characteristics of 90-period middle wavelength infrared (MWIR) IB cascade device. The low-temperature *J-V* curves for IB cascade detectors are relatively steep (see Figure 23.5a), suggesting contributions from tunneling components. At higher temperatures, the dark current is much less sensitive to the operation bias and is diffusion-limited.

Additional insight in dark current characteristic gives the Arrhenius plot of the dark current density at −10 mV as well as the measured zero-bias resistance-area product (R_0A) (see Figure 23.5b). The extracted activation energy (E_A) at higher operating temperatures is about 0.302 eV, which is very close to the effective bandgap in InAs/GaSb T2SL absorber, confirming that the dark current at higher temperatures is mostly due to the diffusion component. The R_0A product of the device exceeds 1.25×10^7 Ωcm^2 at 120 K, is 2,470 Ωcm^2 at 200 K, and 3.93 Ωcm^2 at room temperature, which is the highest R_0A product reported for T2SL detectors. The dark current density is as low as 1.28×10^{-7} A/cm^2 at 160 K and the extracted R_0A is 9.42×10^4 Ωcm^2, which are both slightly better than the HgCdTe Rule 07 [26].

The spectral responsivity of MWIR T2SL cascade detector is about 0.3 A/W at room temperature and wavelength 4 μm and has been observed up to 380 K. Figure 23.6 presents the Johnson-noise-limited detectivity spectra at various temperatures, extracted from the measured responsivity spectra, and R_0A product for the above described detector structure. The Johnson-limited D^* reaches 1.29×10^{13} Jones at 3.8 μm and 120 K, and 9.73×10^{11} Jones at 200 K. The background limited infrared photodetector (BLIP) performance is 180 K at 4 μm for five-stage/junction devices with 70% absorption quantum efficiency, which correspond to 14% external quantum efficiency.

In the presented design, the total thickness of the absorber is about 1 μm, and theoretically the absorption quantum efficiency could be increased by increasing the number of stages. However, the conversion quantum efficiency is lower than that of the absorption quantum efficiency by a factor of *N*.

Using the equation for spectral responsivity, $R_i = (\lambda\eta/hc)qg$, (where h, c, and g are Planck's constant, speed of light, and photoconductive gain, respectively) and the experimental data for $R_i \approx 0.3$ A/W, we can estimate room-temperature conversion quantum efficiency at a wavelength of 4 μm, as $\eta g \approx 9\%$. Since the device has five stages, one would estimate a gain of 1/5 (0.20). This leads to the absorption quantum efficiency of 45%. The absorption quantum efficiency would be increased by increasing the number of stages, provided that the absorbers are distributed closely in real space and are not very thick, ensuring equal absorption of the photon flux in each of the stages (total thickness of all the stages should be comparable to the diffusion length). However, the conversion quantum efficiency remains lower than that of the absorption quantum efficiency by the factor of number of absorbers (stages).

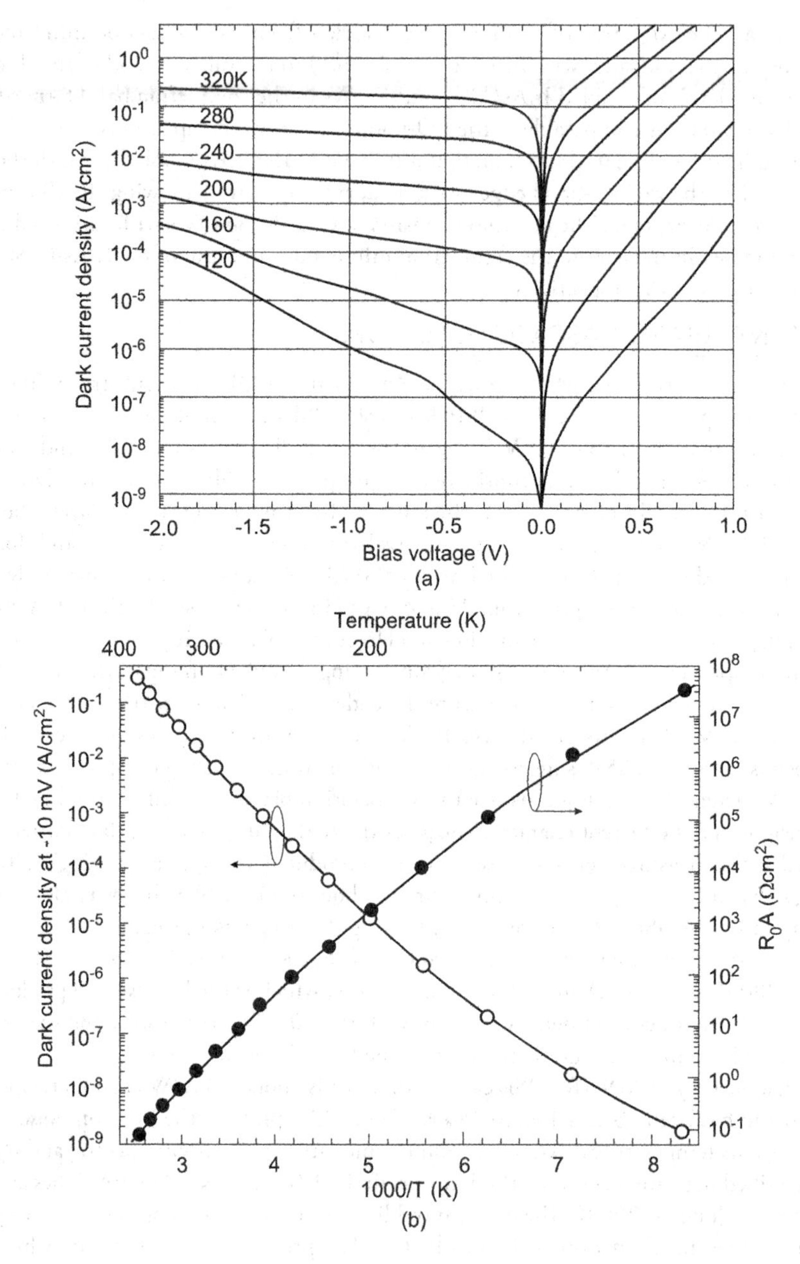

Figure 23.5 Dark characteristics of five-stage MWIR InAs/GaSb T2SL IB 90-period cascade detector: (a) current-voltage characteristics, (b) Arrhenius plots of the electrical performance. (Adapted after Ref. [11])

The transport of photoexcited carriers is very fast and occurs over a distance in each cascade stage that is much shorter than a typical diffusion length (≈50–200 nm depending on wavelength). Therefore, the lateral diffusion transport may not be significant over such a short distance and thus the deeply etched mesa structures for confining photoexcited carriers may not be necessary in QCDs in contrast to conventional photodiodes. Moreover, significant wave function overlap of energy states in the multiple QW region (relaxation region) causes the IS relaxation time (e.g., optical-phonon scattering time ≈1 ps) to be much shorter than the IB recombination time (≈1 ns or ≈0.1 ns at high temperatures with significant Auger recombination). Consequently, the photoexcited electrons in the active region are transferred to the bottom of the energy ladder with a very high efficiency. This mechanism enables the quick and efficient removal of carriers after photoexcitation.

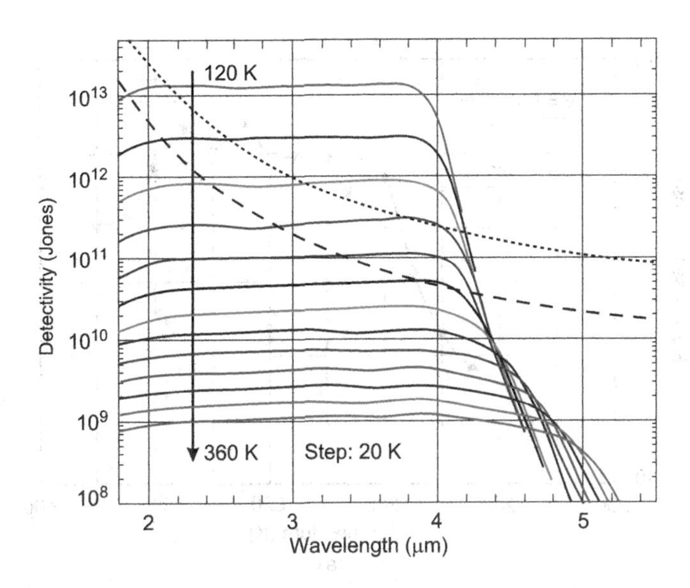

Figure 23.6 Johnson-noise-limited detectivity spectra of five-stage MWIR InAs/GaSb T2SL IB 90-period cascade detector at various temperatures (Adapted after Ref. [11]). The dashed lines represent the BLIP D^* for a photovoltaic detector with an external quantum efficiency of 70%, and the dotted lines are the BLIP D^* for five-stage devices with absorption quantum efficiency of 70%, both under 300 K background with 2π FOV.

Figure 23.7 presents the experimentally measured response time of a cascade detector versus temperature at zero bias (see Figure 23.7a) and versus bias voltage for three temperatures of operation: 225 K, 293 K, and 380 K (see Figure 23.7b). These results confirm the short response times of IB cascade detectors. At zero bias, in a temperature range of 225 K–280 K, the response time increases with increasing temperature from about 1 ns to 5 ns. It stabilizes at about 5 ns for further increase in temperature to about 360 K, while it decreases reaching a value of ~ 2 ns at 380 K.

The negative bias is beneficial for a cascade detector response time (see Figure 23.7b). The negative correlation between the response time and the applied voltage for temperatures of 225 and 293 K is probably related to the drift component decrease with increasing bias as the electric field increases across the absorber region. In this context, behavior of the response time as a function of bias above a bias of 200 mV for the detector operated at 380 K is not identified to the full extent as the time response rises with increasing voltage. The authors believe that under this condition the separation between the quantized energy levels in the GaSb QW of the tunneling region and the valence band in the transport region does not match the LO-phonon energy in AlSb, which is responsible for tunneling of holes by phonon-assisted process. In addition, ambipolar mobility is reduced, which in turn influences detector time response.

More recently, the first five-stage MWIR IB cascade detector 320×256 focal plane array has been demonstrated with pixel size of 24×24 μm and a pitch size of 30 μm [27]. This device demonstrated a BLIP performance above 150 K (300 K, 2π field of view (FOV)).

23.2.3 LWIR INTERBAND CASCADE DETECTORS

Recently, preliminary studies of the LWIR and very LWIR IB CIDs with cutoff wavelength up to 16 μm at 78 K have been presented [12,16,18–20].

Figure 23.8 shows an exemplary device structure of two-stage LWIR device grown by MBE. The absorber layers have thicknesses of 620.0 nm and 756.4 nm, with each SL period composed of 36.3 Å of InAs and 21.9 Å of GaSb. The dipper absorber is made thicker in order to achieve current matching. In each SL period, a 1.9-Å-thick InSb layer is intentionally inserted into both the InAs-on-GaSb and the GaSb-on-InAs layers as the interface strain-balancing layer. In order to make electrons the minority carriers, half of the GaSb layers in the SL absorbers are p-doped with a doping density of 3.5×10^{16} cm^{-3}. The eB and eR in each of these devices have identical designs. After the structure growth, square mesa devices with

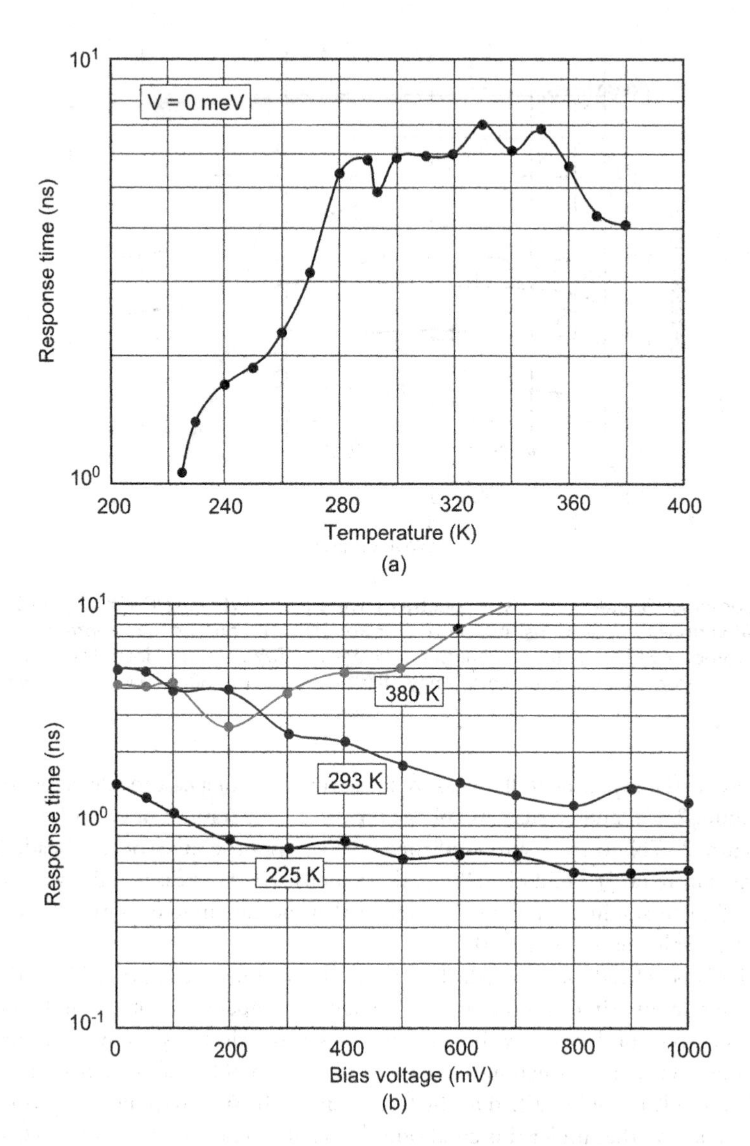

Figure 23.7 Response time of T2SL MWIR cascade detector: (a) at zero bias voltage versus temperature and (b) versus bias voltage at 225 K, 293 K, and 380 K. (after Ref. [15])

edge lengths ranging from 200 to 1,000 μm were fabricated by using conventional contact UV lithography and wet etching. For passivation, two layers consisting of 170 nm of Si_3N_4 followed by 137 nm of SiO_2 were used.

Dark current characteristics versus bias voltage at different temperatures are shown in Figure. 23.9. The activation energy is estimated to be 102 meV (see insert of the figure) in comparison with the corresponding bandgap energy at 78 K equal to 135 meV. These data imply that the detector is neither diffusion limited nor dominated by the generation-recombination process (activation energy is larger than $E_g/2$). It is suggested that the deviation from diffusion limit is probably related to the nonuniform doping that is applied to the absorber region, which creates an electric field that could affect the Shockley-Read-Hall (SRH) generation-recombination in the absorber layer. The unintentional electrostatic barrier also leads to lower collection efficiencies due to the inefficiency of hole transport in absorbers. It is suspected that the absorbers are n-type, especially at high temperatures. Then the carrier transport is less efficient compared to that of the electrons and an external bias is required to aid the collection of photocarriers.

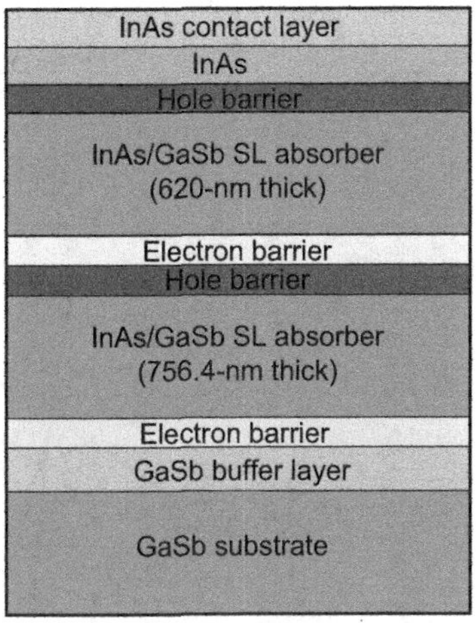

InAs contact layer
InAs
Hole barrier
InAs/GaSb SL absorber
(620-nm thick)
Electron barrier
Hole barrier
InAs/GaSb SL absorber
(756.4-nm thick)
Electron barrier
GaSb buffer layer
GaSb substrate

Figure 23.8 Device structure for two-stage LWIR IB QCID.

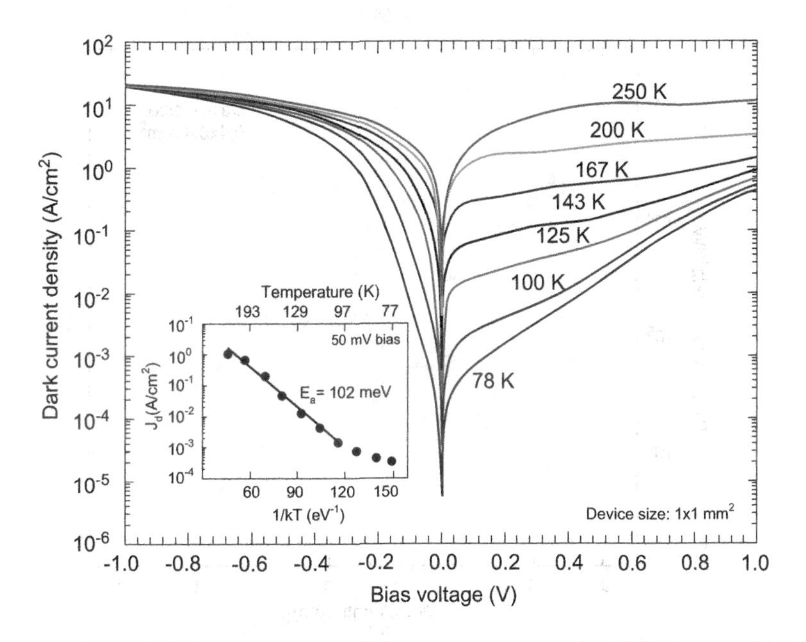

Figure 23.9 Current-voltage dark characteristics at different temperatures for LWIR detector. The inset shows the fitted activation energy for the Arrhenius plot of the dark current. (Adapted after Ref. [12])

Figure 23.10a shows spectral Johnson-limited detectivities of LWIR IB QCDs at different temperatures. At 78 K, the R_0A product value of 115 Ωcm^2 has been achieved, that corresponds to the detectivity of 3.7×10^{10} cmHz$^{1/2}$/at 8 μm [12]. Similar performance is shown in Figure 23.10b for a six-stage detector.

A comparative study of two sets of LWIR devices (current matched and non-current matched) indicates the necessity of current matching to maximize the utilization of absorbed photons for an optimal responsivity [20]. In addition, it was shown that the observed negative differential conductances in these detectors at high temperatures is identified as being related to intraband tunneling through the electron barrier [19].

Although not yet optimized, the IB QCIDs could operate at 300 K with detectivities higher than 1.0×10^8 cmHz$^{1/2}$/W, which exceeds the reported values of commercial, uncooled HgCdTe detectors [19], suggesting great potential for applications such as laser spectroscopy and free-space communication in combination with room temperature long wavelength QCLs.

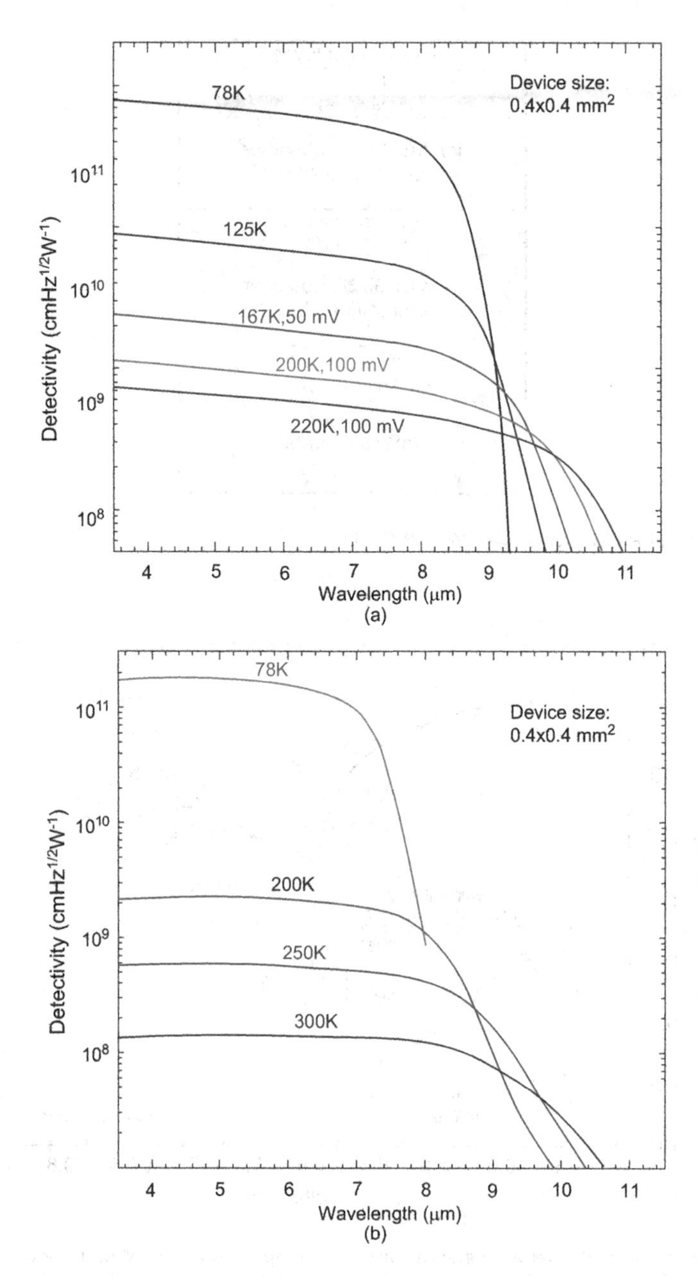

Figure 23.10 Spectral detectivities of LWIR IBQCIDs at different temperatures: (a) for two-stage detector (adapted after Ref. [12]) and (b) six-stage zero-bias detector. (Adapted after Ref. [19])

To improve the performance of LWIR T2SL QCDs, further modifications in device technology and design are required such as shorter absorbents and better bandgap alignments between the absorbers and the unipolar barriers, p-type doping of absorbers, and correction in processing.

23.3 PERFORMANCE COMPARISON WITH HgCdTe HOT PHOTODETECTORS

At present, HgCdTe is the most widely used variable gap semiconductor for IR photodetectors, including uncooled operation. However, the junction resistance of HgCdTe photodiodes operated in the LWIR region is very low due to high thermal generation. For example, small sized uncooled 10.6 μm photodiodes (50×50 μm^2) exhibit less than 1Ω zero-bias junction resistance, which is well below the series resistance of a diode. Consequently, the performance of conventional devices is very poor and they are not usable for practical applications.

Figure 23.11 compares the R_0A product of HgCdTe photodiodes with room-temperature experimental data for IB CIDs fabricated with type II InAs/GaSb SL absorbers. It is evident that at the present early stage of the CID technology, the experimentally measured R_0A values at room temperature are higher than those for the state of the art HgCdTe photodiodes. However, their quantum efficiencies are low, typically below 10%, resulting in lower detectivity for IB T2SL cascade detectors in comparison to HgCdTe photodiodes.

Figure 17.56 shows the performance of optically immersed, two-stage thermoelectrically (2TE) cooled HgCdTe devices. Without optical immersion, MWIR photovoltaic detectors are sub-BLIP devices with performance close to the GR limit. However, well-designed optically immersed devices approach the BLIP limit while thermoelectrically cooled with two-stage Peltier coolers. Situation is less favorable for > 8 μm LWIR photovoltaic detectors, which exhibit detectivities below the BLIP limit by an order of magnitude. Typically, the devices are used at zero bias. The attempts to use Auger-suppressed nonequilibrium devices were not successful due to large $1/f$ noise extending to ≈100 MHz in the extracted photodiodes.

Lenses for monolithically immersed HgCdTe detectors are formed directly from transparent GaAs substrates. Due to immersion, the apparent optical detector area increases by a factor equal to n^2 for hemispherical lenses, where n is the refraction coefficient. Using GaAs lenses, the expected increase in detectivity is about $n^2 \approx 10$. The detectivity of nonimmersed HgCdTe detectors is estimated to be about one order of magnitude below those shown in Figure 17.56. So, the performance of IB CIDs are comparable to that obtained for Peltier-cooled HgCdTe devices.

Figure 23.12 gathers the experimentally measured response times of a T2SL IB CID and HgCdTe photodetectors (mainly photodiodes) of different designs and operated in a temperature range between 220 K and room temperature. Most of the zero-bias LWIR photodiodes are characterized by response times below 10 ns. The device response time decreases under reverse bias achieving value below 1 ns. In this comparison, the response time of T2SL cascade detectors is comparable or even shorter than that of the HgCdTe photodetectors.

The IB QCIDs have complicated cascade detector architecture. Their design is particularly pronounced for high temperature operation, where diffusion lengths are considerably reduced. At the present stage, their performance is comparable with that of HgCdTe. However, due to strong covalent bonding of III-V

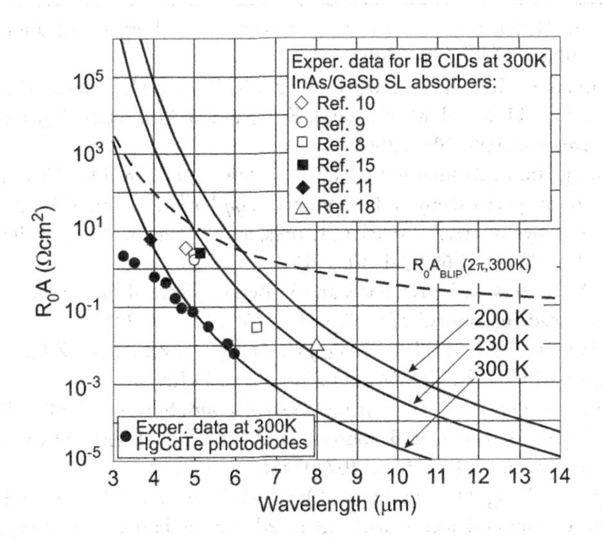

Figure 23.11 R_0A product of HgCdTe photodiodes (solid lines) in comparison with room-temperature experimental data for IB QCIDs with type II InAs/GaSb SL absorbers.

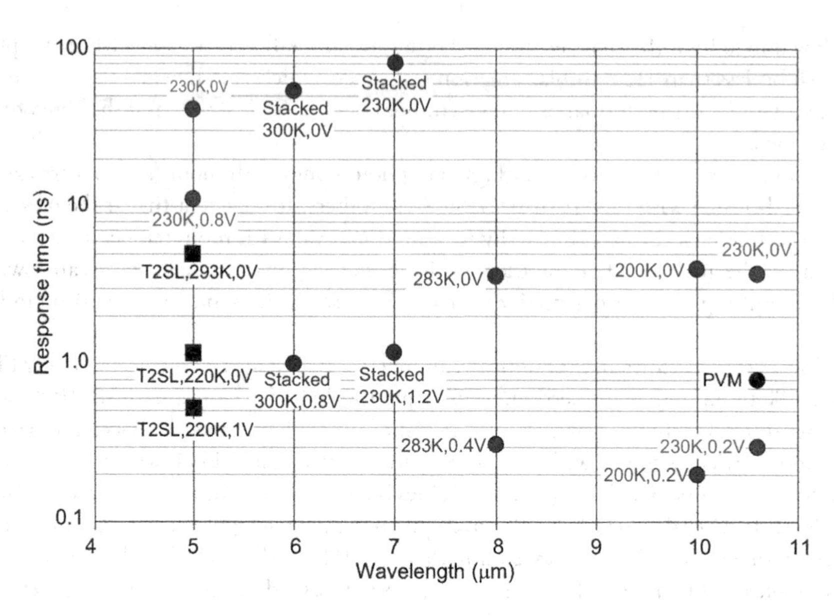

Figure 23.12 Response time versus wavelength for zero-bias and reverse bias (as indicated) HgCdTe photodiodes and a T2SL MWIR cascade detector operating in the temperature range between 220 and 300 K. *Stacked*—double-stacked photovoltaic detector; PVM—multiple heterojunction photovoltaic detector.

semiconductors, QCIDs can be operated at temperatures up to 400°C, which is not possible for the HgCdTe counterpart.

It is expected that better understanding of the QCID device physics and other aspects related to their design and material properties will enable the improvement of high performance HOT detectors. In addition, the discrete architecture of QCID provides a great deal of flexibility for manipulating carrier transports to achieve high-speed operation, which determines the maximum bandwidth. The possibility of having them monolithically integrated with active components, for instance lasers, offers entirely new avenues for telecommunication systems based on quantum devices.

REFERENCES

1. A. Gomez, M. Carras, A. Nedelcu, E. Costard, X. Marcadet, and V. Berger, "Advantages of quantum cascade detectors," *Proc. SPIE* 6900, 69000J-1–14, 2008.
2. F.R. Giorgetta, E. Baumann, M. Graf, Q. Yang, C. Manz, K. Köhler, H.E. Beere, D.A. Ritchie, E. Linfield, A.G. Davies, Y. Fedoryshyn, H. Jäckel, M. Fischer, J. Faist, and D. Hofstetter, "Quantum cascade detectors," *IEEE J. Quantum Electron* 45, 1039–1052, 2009.
3. D. Hofstetter, F.R. Giorgetta, E. Baumann, Q. Yang, C. Manz, and K. Köhler, "Mid-infrared quantum cascade detectors for applications in spectroscopy and pyrometry," *Appl. Phys. B* 100, 313–320, 2010.
4. A. Buffaz, M. Carras, L. Doyennette, A. Nedelcu, P. Bois, and V. Berger, "State of the art of quantum cascade photodetectors," *Proc. SPIE* 7660, 76603Q-1–10, 2010.
5. A. Buffaz, A. Gomez, M. Carras, L. Doyennette, and V. Berger, "Role of subband occupancy on electronic transport in quantum cascade detectors," *Phys. Rev. B* 81, 075304-1–8, 2010.
6. J.Q. Liu, S.Q. Zhai, F.Q. Liu, S.M. Liu, L.J. Wang, J.C. Zhang, N. Zhuo, and Z.G. Wang, "Quantum cascade detectors in very long wave infrared", COMMAD *2014*, pp. 127–129.
7. H. Schneider and H.C. Liu, *Quantum Well Infrared Photodetectors*, Springer, Berlin, 2007.
8. R.Q. Yang, Z. Tian, Z. Cai, J.F. Klem, M.B. Johnson, and H.C. Liu, "Interband-cascade infrared photodetectors with superlattice absorbers," *J. Appl. Phys.*, 107, 054514-1–6, 2010.
9. Z. Tian, R.T. Hinkey, R.Q. Yang, D. Lubyshev, Y. Qiu, J.M. Fastenau, W.K. Liu, and M.B. Johnson, "Interband cascade infrared photodetectors with enhanced electron barriers and p-type superlattice absorbers," *J. Appl. Phys.*, 111, 024510-1–6, 2012.

10. N. Gautam, S. Myers, A.V. Barve, B. Klein, E.P. Smith, D.R. Rhiger, L.R. Dawson, and S. Krishna, "High operating temperature interband cascade midwave infrared detector based on type-II InAs/GaSb strained layer superlattice," *Appl. Phys. Lett.*, 101, 021106-1–4, 2012.
11. Z.-B. Tian and S. Krishna, "Mid-infrared interband cascade photodetectors with different absorber designs," *IEEE J. Quant. Electron* 51(4), 4300105, 2015.
12. H. Lotfi, L. Lu, H. Ye, R.T. Hinkey, L. Lei, R.Q. Yang, J.C. Keay, T.D. Mishima, M.B. Santos, and M.B. Johnson, "Interband cascade infrared photodetectors with long and very-long cutoff wavelengths," *Infrared Phys. & Technol.*, 70, 162–167, 2015.
13. J.V. Li, R.Q. Yang, C.J. Hill, and S.L. Chung, "Interbad cascade detectors with room temperature photovoltaic operation," *Appl. Phys. Lett.*, 86, 101102-1–3, 2005.
14. R.T. Hinkey and R.Q. Yang, "Theory of multiple-stage interband photovoltaic devices and ultimate performance limit comparison of multiple-stage and single-stage interband infrared detectors," *J. Appl. Phys.*, 114, 104506-1–18, 2013.
15. W. Pusz, A. Kowalewski, P. Martyniuk, W. Gawron, E. Plis, S. Krishna, and A. Rogalski, "Mid-wavelength infrared type-II InAs/GaSb superlattice interband cascade photodetectors," *Opt. Eng.*, 53(4), 043107-1–8, 2014.
16. H. Lotfi, L. Lin, L. Lu, R.Q. Yang, J.C. Keay, M.B. Johnson, Y. Qiu, D. Lubyshev, J.M. Fastenau, and A.W.K. Liu, "High-temperature operation of interband cascade infrared photodetectors with cutoff wavelengths near 8 µm", *Opt. Eng.*, 54(6), 063103-1–9, 2015.
17. Y. Zhou, J. Chen, Z. Xu, and L. He, "High quantum efficiency mid-wavelength interband cascade infrared photodetectors with one and two stages", *Semicond. Sci. Technol.*, 31, 085005, 2016.
18. L. Lei, L. Li, H. Ye, H. Lotfi, R.Q. Yang, M.B. Johnson, J.A. Massengale, T.D. Mishima, and M.B. Santos, "Long wavelength interband cascade infrared photodetectors operating at high temperatures", *J. Appl. Phys.*, 120, 193102-1–14, 2016.
19. L. Lei, L. Li, H. Ye, H. Lotfi, R. Q. Yang, M. B. Johnson, J. A. Massengale, T. D. Mishima, and M. B. Santos, "Long wavelength interband cascade infrared photodetectors towards high temperature operations", *Proc. SPIE* 10111, 1011113, 2017.
20. W. Huang, L. Lei, L. Li, J. A. Massengale, R. Q. Yang, T. D. Mishima, and M. B. Santos, "Current-matching versus non-current-matching in long wavelength interband cascade infrared photodetectors", *J. Appl. Phys.*, 122, 083102, 2017.
21. L. Le, L. Li, H. Lotfi, H. Ye, R. Q. Yang, T. D. Mishima, M. B. Santos, and M. B. Johnson, "Midwavelength interband cascade infrared photodetectors with superlattice absorbers and gain", *Opt. Eng.*, 57(1), 011006, 2018.
22. J. Piotrowski and A. Rogalski, *High-Operating Temperature Infrared Photodetectors*, SPIE Press, Bellingham, WA, 2007.
23. J. Piotrowski, P. Brzozowski, and K. Jóźwikowski, "Stacked multijunction photodetectors of long-wavelength radiation," *J. Electron. Mater.*, 32, 672–676, 2003.
24. P. Martyniuk, J. Antoszewski, M. Martyniuk, L. Faraone, and A. Rogalski, "New concepts in infrared photodetector design", *Appl. Phys. Rev.*, 1, 041102-1–35, 2014.
25. H. Lotfi, R.T. Hinkey, L. Li, R.Q. Yang, J.F. Klem, and M.B. Johnson, "Narrow-bandgap photovoltaic devices operating at room temperature and above with high open-circuit voltage," *Appl. Phys. Lett.*, 102, 211103, 2013.
26. W.E. Tennant, D. Lee, M. Zandian, E. Piquette, and M. Carmody, "MBE HgCdTe technology: A very general solution to IR detection, described by 'Rule 07', a very convenient heuristic," *J. Electron. Mat.*, 37, 1406, 2008.
27. Z.-B. Tian, S.E. Godoy, H.S. Kim, T. Schuler-Sandy, J.A. Montoya, and S. Krishna, "High operating temperature interband cascade focal plane arrays", *Appl. Phys. Lett.*, 105, 051109-1–5, 2014.

Part IV

Infrared focal plane arrays

24 Overview of focal plane array architectures

As mentioned in Chapter 3, the development of IR (infrared) detectors was connected with thermal detectors at the beginning. The initial spectacular applications of thermal detectors in astronomy are noted in Figure 24.1 [1]. In 1856, Charles Piazzi Smyth [2,3], from the peak of Guajara on Tenerife, detected IR radiation from the Moon using a thermocouple. In the early 1900s, IR radiation was successfully detected from the planets Jupiter and Saturn and from some bright stars such as Vega and Arcturus. In 1915, William Coblentz [2] at the U. S. National Bureau of Standards developed thermopile detectors, which he used to measure the IR radiation from 110 stars. However, the low sensitivity of early IR instruments prevented the detection of other near IR sources. Work in IR astronomy remained at a low level until breakthroughs in the development of new, sensitive IR detectors were achieved in the late 1950s.

For the first time, the photoconductive effect was observed by Smith in 1873 [4] who noted that the resistance of Se decreased when exposed to light. However, the photon detectors were mainly developed during the twentieth century. In 1917, Case in the United States was able to produce thallous sulfide photoconductive detectors that were sensitive to 1.2 μm [5]. Lead sulfide cells were developed and studied intensively by the Germans before the beginning of World War II [6]. History of the development in the large family of IR photodetectors is briefly described in Chapter 3.

IR detector technology development was and continues to be primarily driven by military applications. Many of these advances were transferred to IR astronomy from the U.S. Department of Defense research. In the mid-1960s, the first IR survey of the sky was made at the Mount Wilson Observatory, California, using liquid nitrogen-cooled PbS photoconductors, which were most sensitive at 2.2 microns. The survey covered approximately 75% of the sky and found about 20,000 IR sources [2]. Many of these sources were stars that had never been seen before in visible light.

Lately, civilian applications of IR technology are being frequently called dual technology applications. One should point out the growing utilization of IR technologies in the civilian sphere at the expense of new materials and technologies and also the noticeable price decrease in these high-cost technologies. Demands to use these technologies are quickly growing due to their effective applications, for example, in global monitoring of environmental pollution and climate changes, long time prognoses of agricultural crop yield, chemical process monitoring, Fourier transform IR spectroscopy, IR astronomy, car driving, IR imaging in medical diagnostics, and others. Traditionally, IR technologies are connected with controlling functions and night vision problems with earlier applications connected simply with the detection of IR radiation, and later by forming IR images from temperature and emissivity differences (systems for recognition and surveillance, tank sight systems, anti-tank missiles, air-air missiles).

In the last five decades, different types of detectors are combined with electronic readouts to make detector arrays. The progress in integrated circuit (IC) design and fabrication techniques has resulted in a continued rapid growth in the size and performance of these solid-state arrays. In the IR technique, these devices are based on a combination of a readout array connected to an array of detectors called focal plane arrays (FPAs). The architecture of detector-readout assemblies has assumed a number of forms that are discussed below [7].

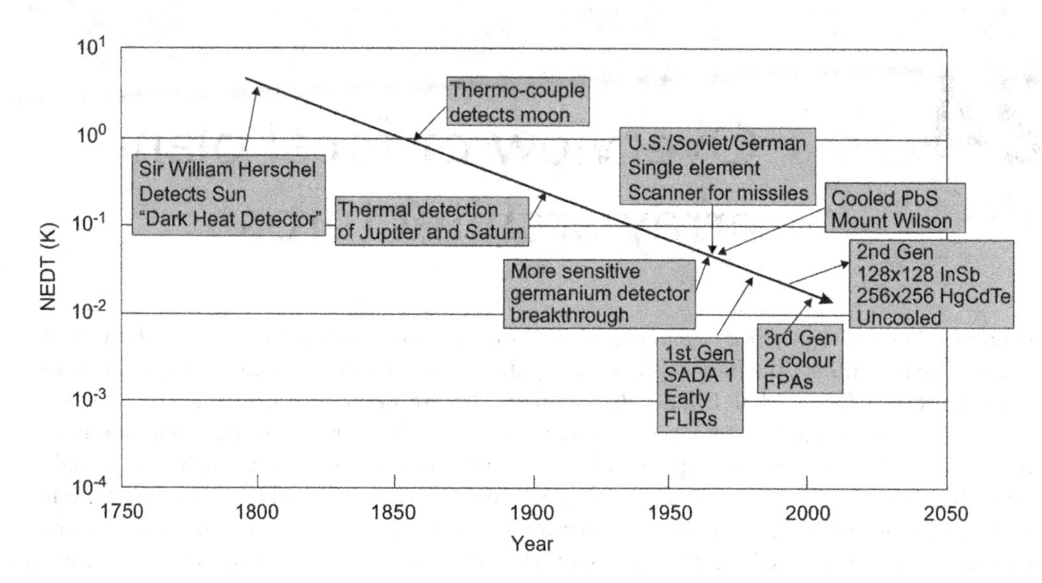

Figure 24.1 Development of IR detectors: NEDT versus era during. (From Ref. [1]. With permission)

24.1 FOCAL PLANE ARRAY OVERVIEW

Two families of multielement detectors can be considered; one used for scanning systems and the other used for staring systems. The simplest scanning linear FPA consists of a row of detectors (Figure 2.11). An image is generated by scanning the scene across the strip using, as a rule, a mechanical scanner. At standard video frame rates, at each pixel (detector), a short integration time is applied and the total charges are accommodated. A staring array is a 2D array of detector pixels that are scanned electronically (Figure 2.11b). These types of arrays can provide enhanced sensitivity and gain in camera weight.

The scanning system, which does not include multiplexing functions in the focal plane, belongs to the first-generation systems. An example of the 180 element United State common module FPA mounted on a dewar stem is shown in the upper right side of Figure 2.11 [8].

The second-generation systems (full-framing systems) typically have about 10^6 elements (pixels) on the focal plane and are configured in a 2D array. Intermediary systems are also fabricated with multiplexed, scanned linear arrays in use and with, as a rule, time delay and integration (TDI) functions. The array illustrated in Figure 24.2 is an 8 × 6 element photoconductive array elaborated in the mid-1970s and intended

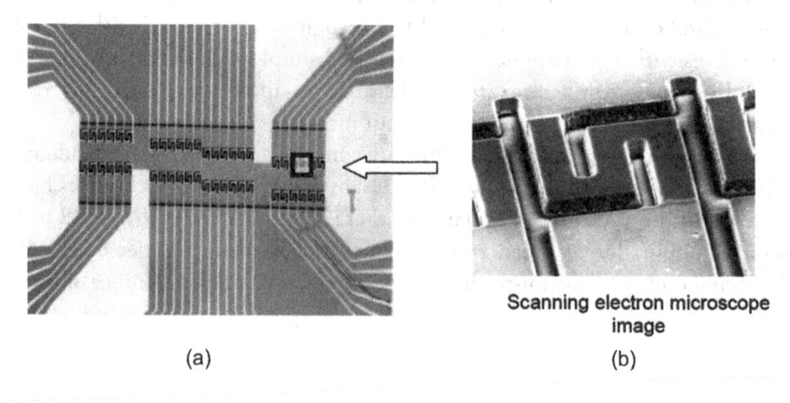

Figure 24.2 Photomicrograph of 8 × 6 element photoconductive array of 50 µm square elements using labyrinthed structure for enhanced responsivity. Staggering the elements to solve the connection problems introduces delays between image lines. (From Ref. [9]. With permission.)

for use in a serial-parallel scan image [9]. Staggering the elements to solve the connection problems introduces delays between image lines.

Development in detector FPA technology has revolutionized many kinds of imaging [7]. From γ rays to the IR and even radio waves, the rate at which images can be acquired has increased by more than a factor of a million in many cases. Figure 24.3 illustrates the trend in array size over the past 50 years. Imaging IR FPAs have been developing in-line with the ability of Si ICs technology to read and process the array signals, and also to display the resulting image. The progress in IR arrays has also been steady, mirroring the development of dense electronic structures, such as dynamic random access memories (DRAMs). FPAs have had nominally the same development rate as DRAM ICs, which have followed Moore's law with a doubling-rate period of approximately 18 months; however, with FPAs lagging DRAMs by about 5–10 years. The 18-month doubling time is evident from the slope of the graph presented in the inset of Figure 24.3 [10], which shows for middle wavelength infrared (MWIR) FPAs the log of the number of pixels per array as a function of the first year of commercial availability. Arrays exceeded 4K × 4K format—16 million pixels—in 2006, about a year later than Moore's law prediction. A subsequent expansion to 8K × 8K with 8 μm pixel array has been currently achieved [11]. Charge coupled devices (CCDs) above 3 gigapixels offer the largest formats.

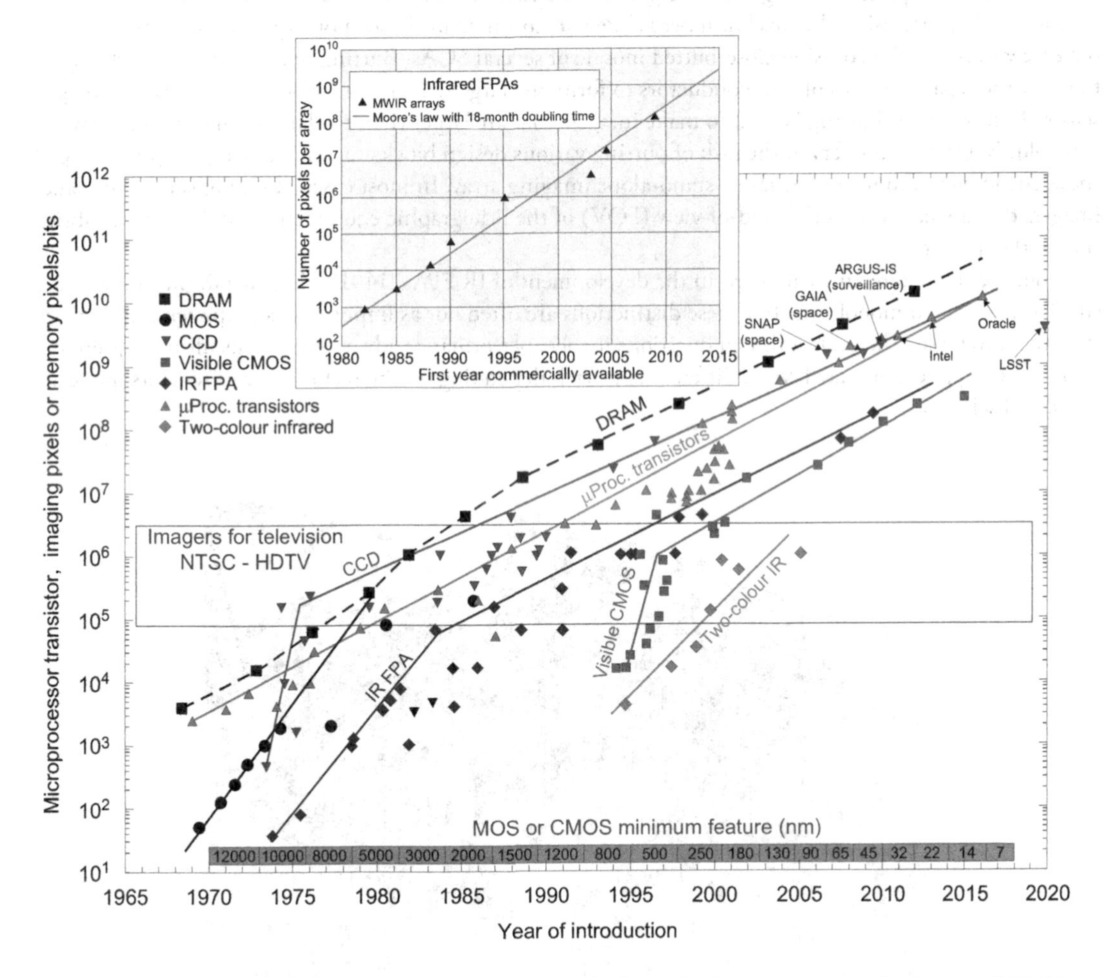

Figure 24.3 Imaging array formats compared with the complexity of Si microprocessor technology and DRAM as indicated by transistor count and memory bit capacity (adapted after Ref. [7] with completions). The timeline design rule of MOS/CMOS features is shown at the bottom. CCDs above 3 gigapixels offer the largest formats. Note the rapid rise of CMOS imagers, which are challenging CCDs in the visible spectrum. The number of pixels on an IR array has been growing exponentially, in accordance with Moore's law, for 30 years with a doubling time of approximately 18 months. IR arrays with size above 100 megapixels are now available for astronomy applications. Imaging formats of many detector types have gone beyond that required for high-definition TV.

IR array sizes will continue to increase, but perhaps at a rate that falls below Moore's law trend. An increase in array size is already technically feasible. However, the market demand for larger arrays is not as strong as it was before the megapixel milestone was achieved. In particular, astronomers were the driving force toward the day when the optoelectronic arrays could match the size of the photographic film. Since large arrays dramatically multiply the data output of a telescope system, the development of large format mosaic sensors of high sensitivity for ground-based astronomy is the goal of many astronomic observatories around the world. Raytheon manufactured a 4 × 4 mosaic of 2k × 2k HgCdTe sensor chip assemblies (SCAs) with 67 million pixels and assisted in assembling it to the final focal-plane configuration (see Figure 24.4) to survey the entire sky in the Southern Hemisphere at four IR wavelengths [12]. This is somewhat surprising given the comparative budgets of the defense market and the astronomical community.

While the size of individual arrays continues to grow very large FPAs are required for many space missions that are made by mosaicking a large number of individual arrays. An example of a large mosaic developed by Teledyne Imaging Sensors, is a 147 megapixel FPA that is comprised of 35 arrays, each with 2,048 × 2,048 pixels [12]. Although currently there are limitations on reducing the size of the gaps between active detectors on adjacent SCAs, many of these can be overcome. It is predicted that focal planes of 100 megapixels and larger will be possible, constrained only by budgets but not technology [13].

The trend of increasing the pixel number is likely to continue in the area of large format arrays. This increase will be continued using close-butted mosaic of several SCAs. Butting refers to the tiling closely together the separate pieces of semiconductors to form one large sensitive array operated as a single image sensor. In most cases, butting is used to make imagers that are larger than the largest imager a single wafer can hold. But stitching refers to the task of putting various design blocks together during the processing of the semiconductor, to make one large, stand-alone imaging array. In most cases, stitching is used to make imagers that are larger than the field-of-view (FOV) of the lithographic equipment used during the fabrication of the imagers.

A number of architectures are used in the development of IR FPAs [14–18]. In general, they may be classified as hybrid and monolithic, but these distinctions are often not as important as proponents and critics state them to be so. The central design questions involve performance advantages versus ultimate producibility. Each application may favor a different approach depending on the technical requirements, projected costs, and schedule.

Figure 24.4 16 2,048 × 2,048 HgCdTe SCAs were assembled for the VISTA (Visible and Infrared Survey Telescope for Astronomy) telescope. The SCA are attached to a precision ground plate that ensures that all pixels are within 12 μm of the desired focus. The detectors are placed in the telescope camera's vacuum chamber and cooled to 72 K. (From Ref. [10])

24.2 MONOLITHIC ARRAYS

In the monolithic approach, both detection of light and signal readout (multiplexing) are done in the detector material rather than in an external readout circuit. The integration of detector and readout onto a single monolithic piece reduces the number of processing steps, increases yields, and reduces costs. Common examples of these FPAs in the visible and near IR (0.7–1.0 μm) are found in camcorders and digital cameras. Two generic types of Si technology provide the bulk of devices in these markets: CCDs and complementary metal-oxide-semiconductor (CMOS) imagers. CCD technology has achieved the highest pixel count or largest format with numbers above 10^9 (see Figure 24.3). This approach to image acquisition was first proposed in 1970 in a paper written by Bell Lab researchers, W.S. Boyle and G.E. Smith [19]. CMOS imagers are also rapidly moving toward large formats and at present are competing with CCDs for the large format applications. Figure 24.5 shows different architectures of monolithic FPAs.

24.2.1 CCD DEVICES

The basic element of a monolithic CCD array is a metal-insulator-semiconductor (MIS) structure. Used as part of a charge transfer device, a MIS capacitor detects and integrates the generated photocurrent. Although most imaging applications tend to require high charge handling capabilities in the unit cells, a MIS capacitor fabricated in a narrow-gap semiconductor material (e.g., HgCdTe and InSb) has a limited charge capacity because of its low background potential as well as more severe problems involving

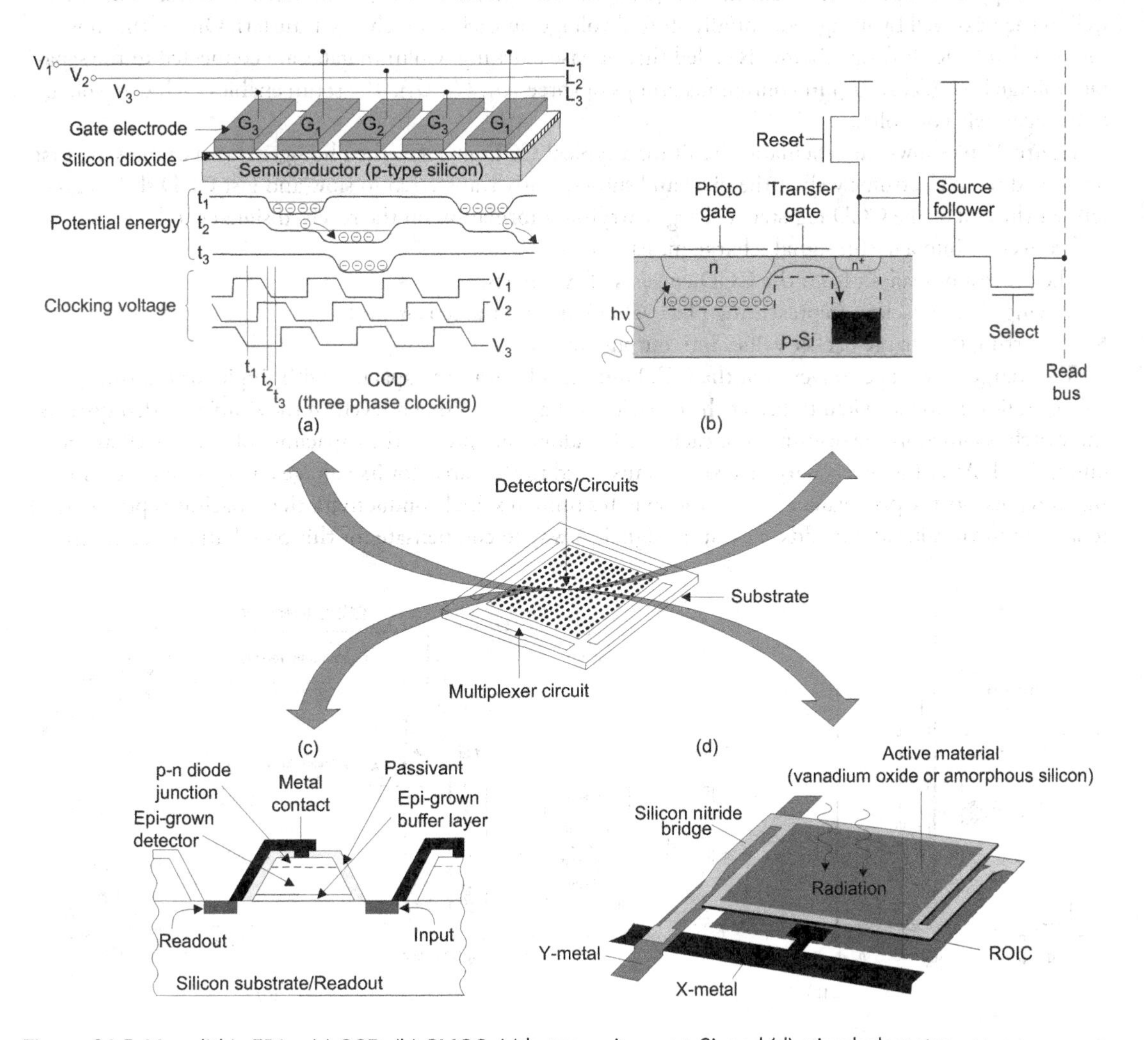

Figure 24.5 Monolithic FPAs: (a) CCD, (b) CMOS, (c) heteroepitaxy-on-Si, and (d) microbolometer.

noise, tunneling effects, and charge trapping when shifting charge through the narrow-bandgap CCD to accomplish the readout function. Because of the nonequilibrium operation of the MIS detector, much larger electric fields are set up in the depletion region than in the p-n junction, resulting in defect-related tunneling current that is orders of magnitude larger than the fundamental dark current. The MIS detector required much higher material quality than p-n junction detectors, which still has not been achieved. So, although efforts have been made to develop monolithic FPAs using narrow-gap semiconductors, Si-based FPA technology is the only mature technology with respect to fabrication yield and attainment of near-theoretical sensitivity. An example of a fully monolithic FPAs with full television (TV) resolution has been commercially available such as PtSi Schottky barrier FPAs.

The CCD technology is very mature with respect to fabrication yield and attainment of near-theoretical sensitivity. CCD technique relies on the optoelectronic properties of a well-established semiconductor architecture–the metal-oxide-semiconductor (MOS) capacitor. A MOS capacitor typically consists of an extrinsic Si substrate on which is grown an insulating layer of silicon dioxide (SiO_2). When a bias voltage is applied across p-type MOS structure, majority charge carriers (holes) are pushed away from the Si-SiO_2 interface directly below the gate, leaving a region depleted of positive charge and available as a potential energy well for any mobile minority charge carriers (electrons); see Figure 24.5a. Electrons generated in the Si through absorption (charge generation) will collect in the potential-energy well under the gate (charge collection). Linear or two-dimensional arrays of these MOS capacitors can therefore store images in the form of trapped charge carriers beneath the gates. The accumulated charges are transferred from potential well to the next well by using sequentially shifted voltage on each gate (charge transfer). One of the most successful voltage-shifting schemes is called three-phase clocking. Column gates are connected to the separate voltage lines (L_1, L_2, L_3) in contiguous groups of three (G_1, G_2, G_3). The setup enables each gate voltage to be separately controlled.

Figure 24.6a shows the schematic circuit for a typical CCD imager. The photogenerated carriers are first integrated in an electronic well at the pixel and subsequently transferred to slow and fast CCD shift registers. At the end of the CCD register, a charge carrying information on the received signal can be readout and converted into a useful signal (charge measurement).

The process of readout from the CCD consists of two parts:

- moving charge packets (representing pixel values) around the sensor, and
- converting the charge packet values into output voltages.

The charge-to-voltage converter at the CCD output is basically a capacitor with single- or multistage voltage follower and a switch to preset the capacitor voltage to a known level. In the simplest video systems, the switch is closed in the beginning of each pixel readout that presets the capacitor voltage as well as the output level. After the pixel charge packet is transferred to the capacitor its voltage changes and the output signal represents the pixel value. Due to the switch's finite residual conductivity the capacitor is pre-charged to an unknown value and it adds the output signal. A way to compensate for this pre-charge uncertainty

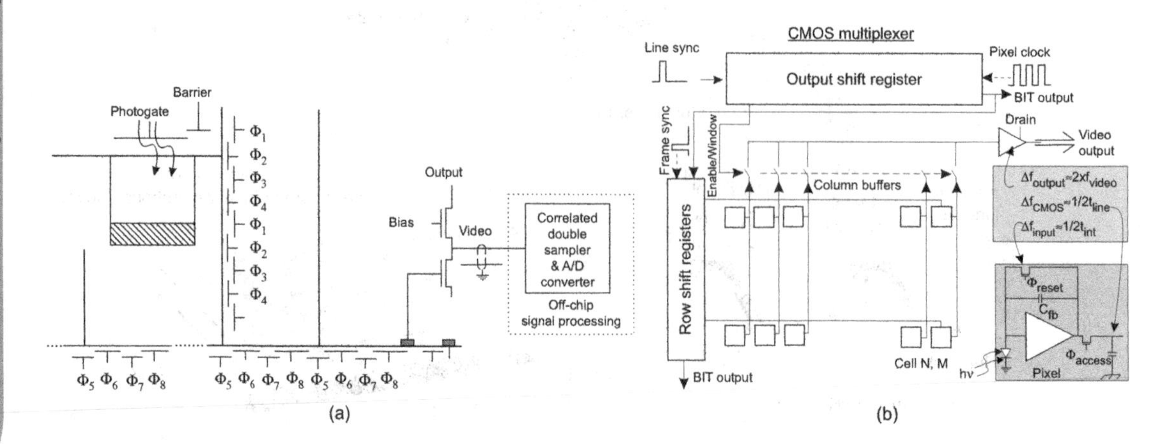

Figure 24.6 Typical readout architecture of CCD (a) and CMOS (b) images.

is the readout technique method—correlated double sampling (CDS). In this method, the output signal is sampled twice for each pixel—just after pre-charging capacitor and after the pixel charge packet is added.

Figure 24.7 shows a preamplifier, in this example—the source follower per detector (SFD—see also Figure 24.18), the output of which is connected to a clamp circuit. The output signal is initially sampled across the clamp capacitor during the onset of photon integration (after the detector is reset). The action of the clamp switch and capacitor subtracts any initial offset voltage from the output waveform. Because the initial sample is made before significant photon charge is integrated, by charging the capacitor, the final integrated photon signal swing is unaltered. However, any offset voltage or drift present at the beginning of integration is, by the action of the circuit, subtracted from the final value. This process of sampling each pixel twice, once at the beginning of the frame and again at the end, and providing the difference is called CDS. More information about readout techniques used in CCD devices (CDS, floating diffusion amplifier in each pixel, and floating gate amplifier) are described in details, for example, in Refs. [20–25].

The first CCD imager sensors were developed about 40 years ago primarily for television analogue image acquisition, transmission, and display. With increasing demand for digital image data, the traditional analogue raster scan output of image sensors has become of limited use, and there is a strong motivation to fully integrate the control, digital interface, and image sensor on a single chip.

The most popular CCD consists of Si sensor operating in visible and near infrared (NIR) wavelength ranges. These spectra could be extended into the UV using delta doping and antireflection coating. In this way, device stability and external quantum efficiency of 50%–90% at the wavelengths of 200–300 nm are obtained. CCD for scientific applications are routinely made with pixel counts exceeding 20 megapixels, and visible 50 megapixel arrays are now available with digital output shaving readout integrated circuit (ROIC) noise levels of less than ten electrons and offering a sensitivity advantage over consumer products [26].

24.2.2 CMOS DEVICES

An attractive alternative to the CCD readout is coordinative addressing with CMOS switches. In particular, Si fabrication advances now permit the implementation of CMOS transistor structures that are considerably smaller than the wavelength of visible light and have enabled the practical integration of multiple transistors within a single picture. The configuration of CCD devices requires specialized processing, unlike CMOS imagers that can be built on fabrication lines designed for commercial microprocessors. CMOS have the advantage that existing foundries, intended for application-specific integrated circuits (ASICs), can be readily used by adapting their design rules. Design rules of 14 nm are currently in production, with preproduction runs of 10 nm design rules. As a result of such fine design rules, more functionality has been designed into the unit cells of multiplexers with smaller unit cells, leading to large array sizes. Figure 24.3 shows the timelines for minimum circuit features and the resulting CCD, IR FPA, and CMOS visible imager sizes with respect to the number of imaging pixels. Along the horizontal axis is also a scale depicting the general availability of various MOS and CMOS processes. The ongoing migration to

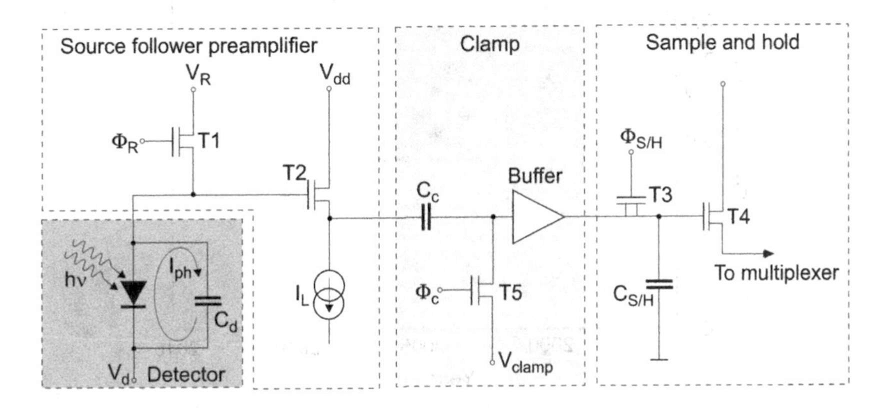

Figure 24.7 CDS circuit.

even finer lithography will thus enable the rapid development of CMOS-based imagers having even higher resolution, better image quality, higher levels of integration, and lower overall imaging system cost than CCD-based solutions. At present, CMOS with minimum features of ≤0.1 μm makes possible monolithic visible CMOS imagers because the denser photolithography allows for low-noise signal extraction and high-performance detection with high optical fill factor (FF) within each pixel. The pixel's architecture is changed to improve resolution by shrinking the pixel size. Figure 24.8 is a roadmap of where CMOS pixel pitch became smaller than CCD due to the described technological development in 2010 [27]. CMOS imagers are also rapidly moving toward large formats and at present are competing with CCDs for the large format applications. The Si wafer production infrastructure that has put high-performance personal computers into many homes makes CMOS-based imaging in consumer products, such as video and digital still cameras, widely available.

A typical CMOS multiplexer architecture (see Figure 24.6b) consists of fast (column) and slow (row) shift registers at the edges of the active area, and pixels are addressed one by one through the selection of a slow register, while the fast register scans through a column, and so on. Each image sensor is connected in parallel to a storage capacitor located in the unit cell. A column of diodes and storage capacitors is selected one at a time by a digital horizontal scan register and a row bus is selected by the vertical scan register. Therefore, each pixel can be individually addressed.

CMOS-based imagers use active and passive pixels [21,28] as shown, in simplified form, in Figure 24.5b. In comparison with passive pixel sensors (PPSs), active pixel sensors (APSs) apart from read functions exploit some form of amplification at each pixel. The PPS consists of three transistors: a reset field-effect transistor (FET), a selective switch, and a source follower (SF) for driving the signal onto the column bus. As a result, circuit overhead is low and the optical collection efficiency (FF) is high even for monolithic devices. Micro lenses, typically used in CCD and CMOS APS imagers for visible application, concentrate the incoming light into the photosensitive region when they are accurately deposited over each pixel (see Figure 24.9). In case of the per-pixel electronics, the area available in the pixel for the detector is reduced, so FF is often limited to 30%–60%. When the FF is low and micro lenses are not used, the light falling elsewhere is either lost or, in some cases, creates artifacts in the imagery by generating electrical currents in the active circuitry. Unfortunately, micro lenses are less effective when used in low *f*/# imaging systems, and they may not be appropriate for all applications.

Although micro lens improves the pixel sensitivity, there are many design challenges associated with the application of micron-scale lens arrays. One key issue is crosstalk that degrades spatial resolution, color separation, and overall sensitivity. Reducing crosstalk in small pixels has become one of the most difficult and time-consuming tasks in sensor design. At pixel dimensions below 1.45 μm, it becomes more difficult

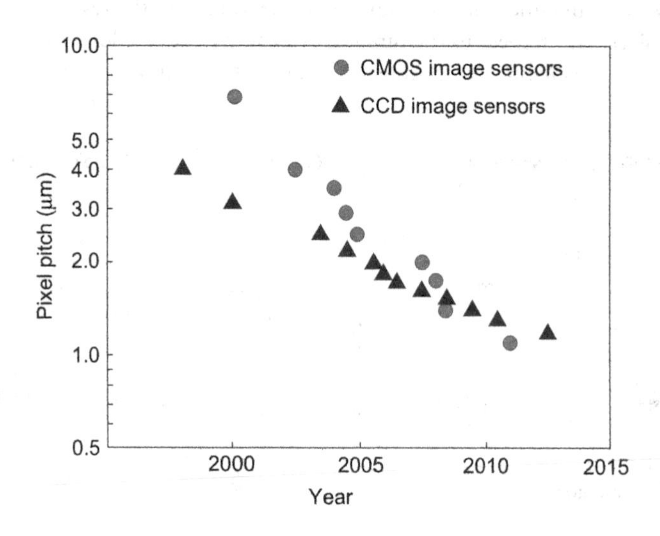

Figure 24.8 A roadmap of CMOS pixel pitch development. (Adapted after Ref. [27])

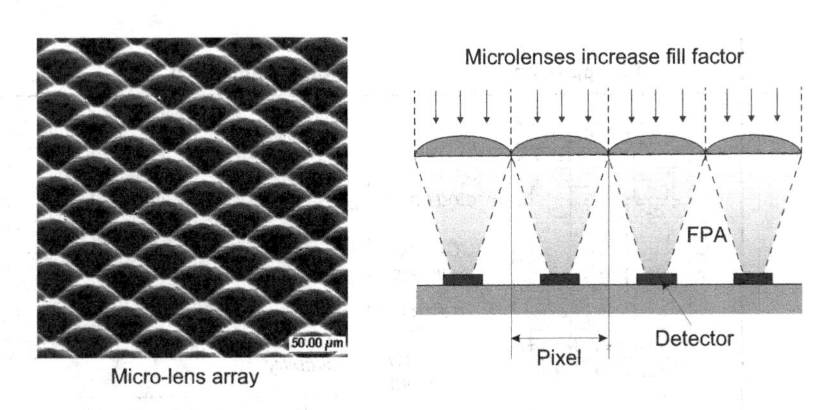

Figure 24.9 Micrograph and cross-sectional drawing of micro-lensed FPA.

in the front side metallization of a CMOS sensor and advanced light guiding structures need to be used to reduce crosstalk.

In the APS, three of the metal-oxide-semiconductor field-effect transistors (MOSFETs) have the same function as in PPS. The fourth transistor works as a transfer gate that moves charge from the photodiode to the floating diffusion. Usually, both pixels operate in rolling shutter mode. The APS is capable of performing CDS to eliminate the reset noise (*kTC* noise) and the pixel offsets. The PPS can only be used with noncorrelated double sampling, which is sufficient to reduce the pixel-to-pixel offsets but does not eliminate the temporal noise. Temporal noise can be addressed by other methods such as adding additional components (like soft reset or tapered reset), however it reduces the FF of monolithic imagers [25]. The MOSFETs incorporated in each pixel for readout are optically dead. CMOS sensors also require several metal layers to interconnect MOSFETs. The busses are stacked and interleaved above the pixel, producing an optical tunnel through which incoming photons must pass. In addition, most CMOS imagers are front side illuminated. This limits the visible sensitivity in the red because of a relatively shallow absorption material. For comparison, CCD pixels are constructed so that the entire pixel is sensitive with a 100% FF.

Figure 24.10 compares the principle of CCDs and CMOS sensors. Both detector technologies use a photosensor to generate and separate the charges in the pixel. Beyond that, however, the two sensor schemes differ significantly. During CCD readout, the collected charge is shifted from pixel to pixel all the way to the perimeter. Finally, all the charges are sequentially pushed to one common location (floating diffusion), and a single amplifier generates the corresponding output voltages. On the other hand, CMOS detectors have an independent amplifier in each pixel (APS). The amplifier converts the integrated charge into a voltage and thus eliminates the need to transfer charge from pixel to pixel. The voltages are multiplexed onto a common bus line using integrated CMOS switches. Analog and digital sensor outputs are possible by implementing either a video output amplifier or an analog-to-digital (A/D) converter on the chip.

The processing technology for CMOS is typically two to three times less complex than standard CCD technology. In comparison with CCDs, the CMOS multiplexers exhibit important advantages due to high circuit density, fewer drive voltages, fewer clocks, much lower voltages (low power consumption), and packing density compatible with many more special functions, lower cost for both digital video and still camera applications. The minimum theoretical read noise of a CCD is limited in large imagers by the output amplifier's thermal noise after CDS is applied in off-chip support circuits. The alternative CMOS paradigm offers lower temporal noise because the relevant noise bandwidth is fundamentally several orders of magnitude smaller and better matches the signal bandwidth. While CCD sensitivity is constrained by the limited design space involving the sense node and the output buffer, CMOS sensitivity is limited only by the desired dynamic range and operating voltage. CMOS-based imagers also offer practical advantages with respect to on-chip integration of camera functions including command and control electronics, digitization, and image processing. CMOS is now suitable also for TDI-type multiplexers because of the availability from foundries of design rules lower than 1.0 μm, more uniform electrical characteristics, and lower noise figures.

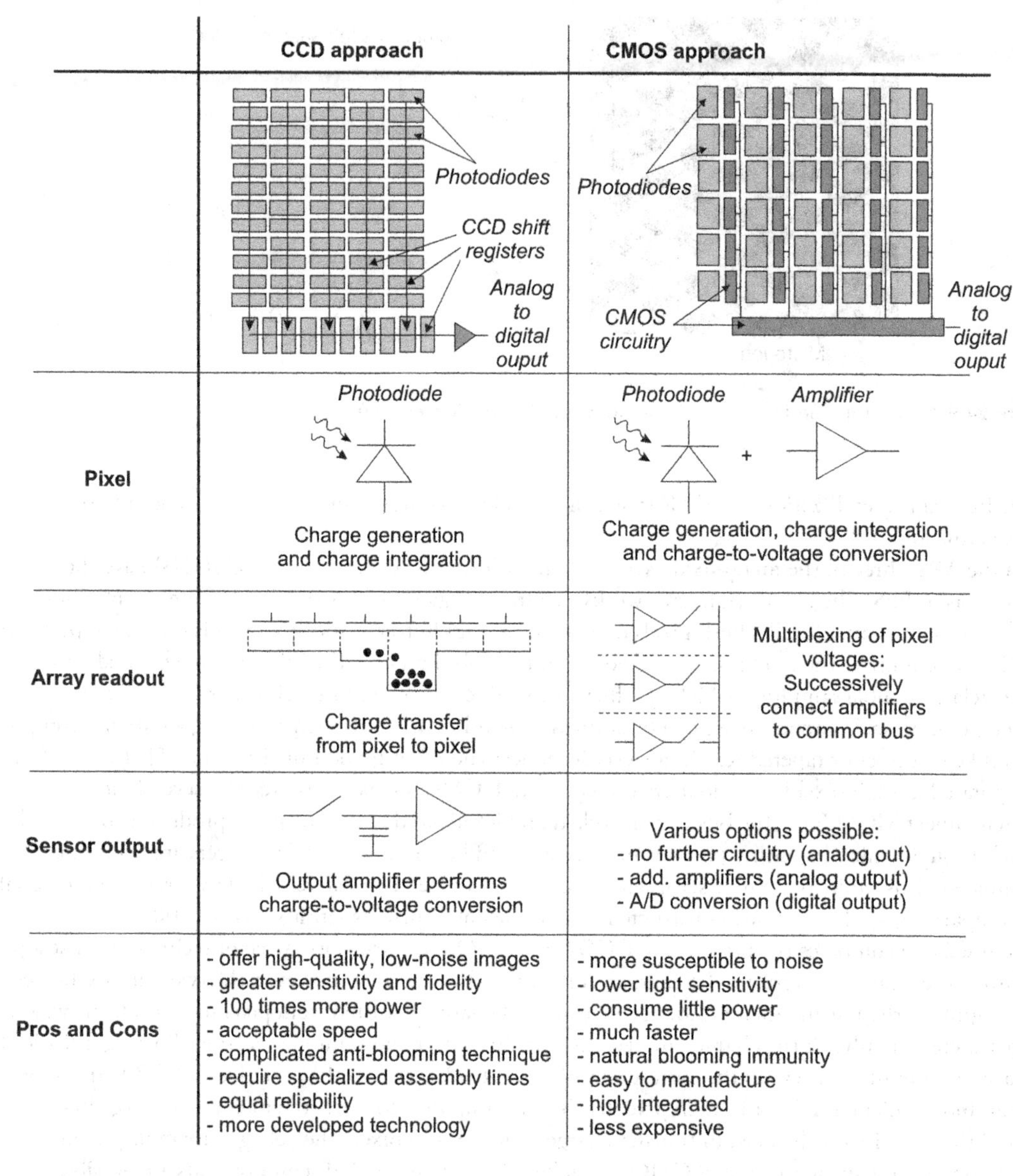

Figure 24.10 Comparison between the CCD-based and CMOS-based image sensor approaches.

24.3 HYBRID ARRAYS

In case of hybrid technology, we can optimize the detector material and multiplexer independently. Other advantages of hybrid-packaged FPAs are near-100% FFs and increased signal-processing area on the multiplexer chip. Photodiodes can be reverse-biased for even higher impedance, and can therefore better match electrically with compact low-noise Si readout preamplifier circuits. Development of hybrid packaging technology began in the late 1970s [29] and took the next decade to reach volume production. In the early 1990s, fully 2D imaging arrays provided a means for staring sensor systems to enter the production stage. In the hybrid architecture, In bump bonding with readout electronics provides for multiplexing the signals from thousands or millions of pixels onto a few output lines, greatly simplifying the interface between the vacuum-enclosed cryogenic sensor and the system electronics.

Although FPA imagers are very common in our lives, they are quite complex to fabricate. Depending on the array architecture, the process can include over 150 individual fabrication steps. The hybridization process involves flip-chip indium bonding between the top surfaces of the ROIC and detector array. The In bond must be uniform between each sensing pixel and its corresponding readout element in order to insure high-quality imaging. After hybridization, a back side thinning process is usually performed to reduce the amount of substrate absorption. The edges of the gap between the ROIC and FPA can be sealed with low viscosity epoxy before the substrate is mechanically thinned down to several microns. Some advanced FPA fabrication processes involve complete removal of the substrate material.

Innovations and progress in FPA fabrication are dependent on adjustments to the material growth parameters. Usually in-house growth has enabled manufacturers with the ability to maintain the highest material quality and to customize the layer structures for multiple applications. For example, since HgCdTe material is critical to many principal product lines, and comparable material is not available externally, most of the global manufactures continue to supply their own wafers. Figure 24.11 shows process flow for integrated IR FPA manufacturing. As shown, boule growth starts with the raw materials, polycrystalline components. In case of HgCdTe FPA process, polycrystalline ultrapure CdTe and ZnTe binary compounds are loaded into a carbon-coated quartz crucible. The crucible is mounted into an evacuated quartz ampoule, which is placed in a cylindrical furnace. Large-crystal CdZnTe boules are produced by mixing and melting the ingredients, followed by recrystallizing with the vertical gradient freeze method. Their standard diameters reach 125 millimeter (mm). The boule substrate material is then sawn into slices, diced into squares, and polished to prepare the surface for epitaxial growth. Typical substrate sizes up to $8 \times 8\ \mathrm{cm}^2$ have been produced. The HgCdTe layers are usually grown on top of the substrate by MBE or MOCVD. In case of MOCVD epitaxial technology, large-size GaAs substrates are also used. The selection of substrate depends on the specific application. The entire growth procedure is automated with each step being programmed in advance.

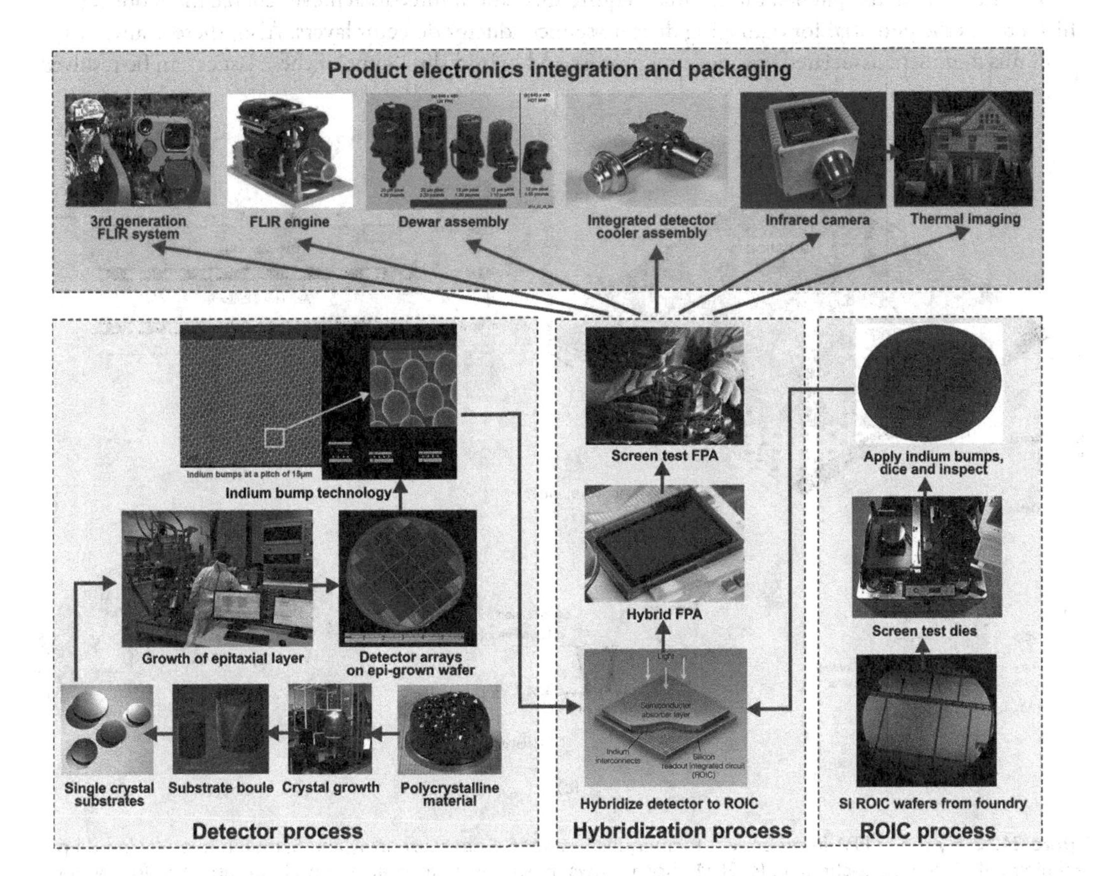

Figure 24.11 Process flow for integrated IR FPA manufacturing.

After growing the detector epitaxial structures, the wafers are nondestructively evaluated against multiple quality specifications. They are then conveyed to the array processing line where the sensing elements (pixels) are formed by photolithographic steps, including mesa etching, surface passivation, metal contact deposition, and In bump formation. After wafer dicing, the FPAs are ready for mating to the ROICs. The ROIC branch of the process is shown in the lower right of Figure 24.11. For each pixel on the detector array, there is a corresponding unit cell on the ROIC to collect the photocurrent and process the signal. Each design is delivered to a Si foundry for fabrication. Next, the ROIC wafers are diced and are ready for mating with the FPA. The most advanced flip-chip bonders, utilizing laser alignment and submicron-scale motion control, bring the two chips together (see the center of Figure 24.11). At present, FPAs with a pixel pitch size below 10 μm are aligned and hybridized with high yield. Each FPA with attached ROIC is tested according to a defined protocol and is installed in a sensor module. Finally, associated packaging and electronics are designed and assembled to complete the integrated manufacturing process.

Detector FPA has revolutionized many kinds of imaging from gamma rays to the IR and even radio waves. More general information about background, history, present stage of technology, and trends can be found, for example, in Refs. [30,31]. Information about the assemblies and applications can be found at different vendor websites.

Different hybridization approaches are in use today. The most popular is flip-chip interconnect using bump bond (see Figure 24.12a,c). In this approach, In bumps are formed on both the detector array and the ROIC chip. The array and the ROIC are aligned and force is applied to cause the In bumps to cold-weld together. In the other approach, In bumps are formed only on the ROIC; the detector array is brought into alignment and in proximity with the ROIC, the temperature is raised to cause the In to melt, and contact is made by reflow.

Large FPAs with fine-pitched bumps may require substantial force to achieve reliable interconnects, which creates the potential for damaging delicate semiconductor detector layers. Also, there is always a risk of misalignment associated with joining process and hybrid slip. Some of these issues can be resolved

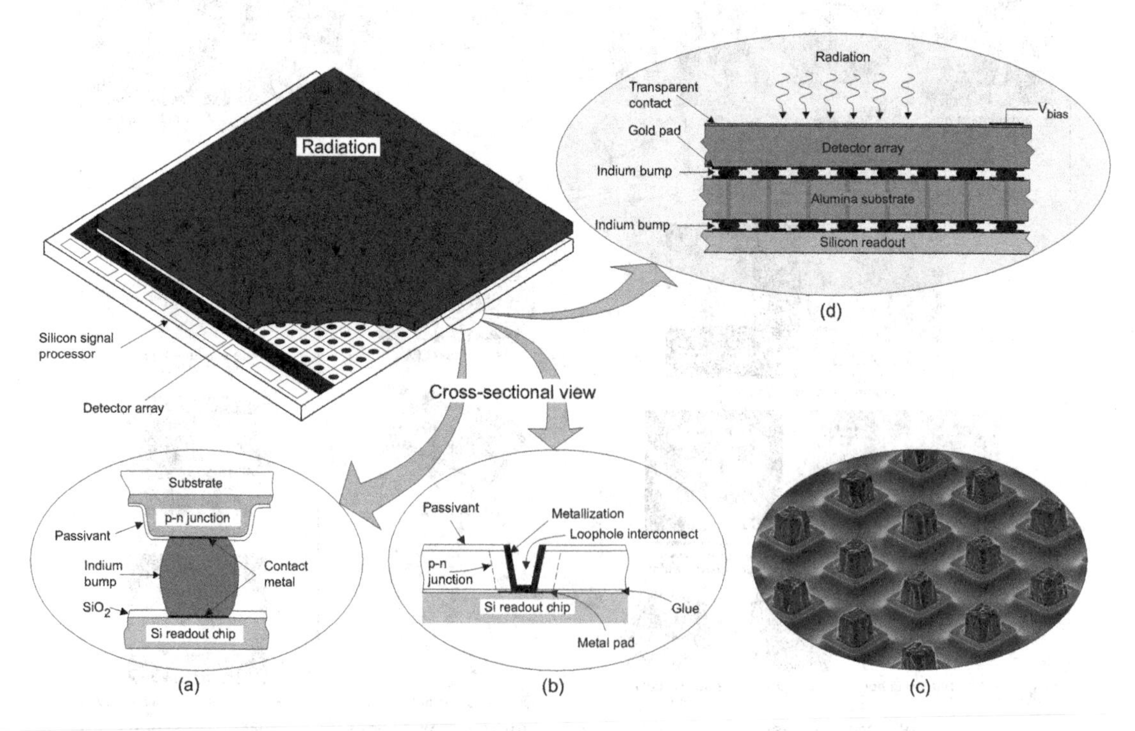

Figure 24.12 Hybrid IR FPA interconnect techniques between a detector array and Si multiplexer: (a) In bump technique, (b) loophole technique, (c) SEM photo shows mesa photodiode array with In bumps, and (d) layered-hybrid design suitable for large format far IR and sub-mm arrays.

by using a fusion bonding (or direct bond) process, which usually involves a room temperature alignment of wafers via van der Waals interactions followed by annealing that creates permanent covalent interfacial bonds between the upper and lower wafers. Direct bond process requires less than 10 pounds of force (in comparison with >1,000 pounds for In-based process) to mate the detector and ROIC. This eliminates misalignment errors and many potential problems associated with compression bonder (hybrid slip, layer damage, bump separation, etc.). Fusion bonding is especially used in the production of very large arrays up to 10 k × 12 k format and thin active layers (<10 μm). This approach is beneficial for devices with enhanced blue and UV performance and modulation transfer function (MTF). The drawback of this method is its extreme sensitivity to surface flatness, roughness, and particles.

Recently, an alternative technology based upon adhesive bonding between the active detector layer and support wafer has been introduced. It provides low temperature bonding and alignment accuracies in the 1–2 μm range. However, the adhesive has to be carefully selected and tested for mechanical rigidity, temperature stability, and outgassing and inter-via connections must be made after bonding and thinning.

IR hybrid FPA detectors and multiplexers are also fabricated using loophole interconnection—see Figure 24.12b [32,33]. In this case, the detector and the multiplexer chips are glued together to form a single chip before detector fabrication. The photovoltaic detector is formed by ion implantation and loopholes are drilled by ion milling, and electrical interconnection between each detector and its corresponding input circuit is made through a small hole formed in each detector. The junctions are connected down to the Si circuit by cutting the fine, few μm in diameter, holes through the junctions by ion milling, and then backfilling the holes with metallization. A similar type of hybrid technology called VIP™ (vertically integrated photodiode) was reported by DRS Infrared Technologies (former Texas Instruments) [34,35].

It is difficult to make small pixel pitches (below 10 μm) using bump-bonding interconnect technique, especially when high yield and 100% pixel operability are required. A new facility gives 3D integration process using wafer bonding, where such materials as Si and InP have been monolithically integrated with pixels size down to 6 μm [17,36,37]. Figure 24.13 compares three methods of vertically interconnected circuit layers: (a) bump bond, (b) insulated through-Si vias, and (c) Lincoln Laboratory's SOI-based vias. The Lincoln integration method enables the dense vertical interconnection of multiple circuit layers and is capable of achieving far smaller pixel sizes than possible with bump bonding.

The detector array can be illuminated from either the front side (with the photons passing through the transparent Si multiplexer) or back side (with photons passing through the transparent detector array substrate). In general, the latter approach is most advantageous as the multiplexer will typically have areas of metallization and other opaque regions, which can reduce the effective optical area of the structure. The epoxy is flowed into the space between the readout and the detectors to increase the bonding strength. In case of back side detector illumination, transparent substrates are required. When using opaque materials, substrates must be thinned to below 10 μm to obtain sufficient quantum efficiencies and reduce crosstalk.

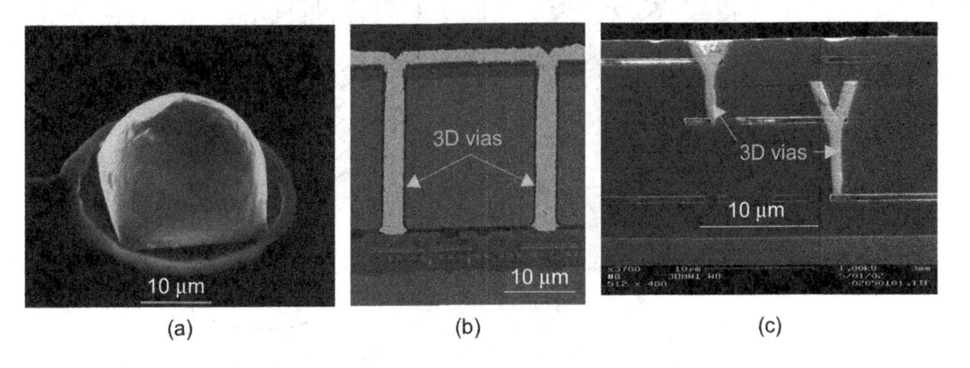

Figure 24.13 Approaches to 3D integration: (a) bump bond used to flip-chip interconnect two circuit layers, (b two-layer stack with insulated vias through thinned-bulk Si, (c) two-layer stack using Lincoln's SOI-based vias.

In some cases, the substrates are completely removed. In the direct back side illuminated configuration, both the detector array and the Si ROIC chip are bump mounted side-by-side onto a common circuit board. The indirect configuration allows the unit cell area in the Si ROIC to be larger than the detector area and is usually used for small scanning FPAs where stray capacitance is not an issue.

Readout circuit wafers are processed in standard commercial foundries and can be constrained in size by the die-size limits of the photolithography step and repeat printers. Because of field size limitations in these photography systems, CMOS imager chip sizes must currently be limited to standard lithographic field sizes of less than 32 × 26 mm for submicron lithography. To build larger sensor arrays, a new photolithographic technique called *stitching* can be used to fabricate detector arrays larger than the reticle field of photolithographic steppers. The large array is divided into smaller subblocks. Later, the complete sensor chips are stitched together from the building blocks in the reticle as shown in Figure 24.14. Each block can be photocomposed on the wafer by multiple exposures at appropriate locations. Single blocks of the detector array are exposed at one time, as the optical system allows shuttering, or there is selective exposure of only a desired section of the reticle.

It should be noted that stitching creates a seamless detector array as opposed to an assembly of closely butted sub-arrays [39,40]. The butting technique is commonly used in the fabrication of very large format sensor arrays due to the limited size of substrate wafers. For example, the 1.4 gigapixel (Gpixel) orthogonal transfer array (OTA) CCD imager (spread over an area about 40 cm^2) used in Panoramic Survey Telescope And Rapid Response System (PanSTARRS) is comprised of 64 chips, each of 22 megapixels (Mpixels). Figure 24.15 shows one of the 64 OTA devices. This above 1 gigapixel imager could not be made monolithically since it exceeds the size of the largest Si wafers used by the IC industry. Currently, Si wafer sizes do not exceed the 300 mm diameter although the Si IC industry is actively exploring a transition to 450 mm diameter wafers. Due to both cost and technological reasons, many imaging chips are made typically using process technologies that are being run on 200 mm diameter wafers. Gaps between chips are reduced to as little as a few tens of micrometers, especially for monolithic technologies. In the future, greater use of four-side buttable designs is expected.

24.4 READOUT INTEGRATED CIRCUITS

The development of FPAs using IC techniques together with development of new material growth techniques and microelectronic innovations began about 40 years ago. The combination of the last two

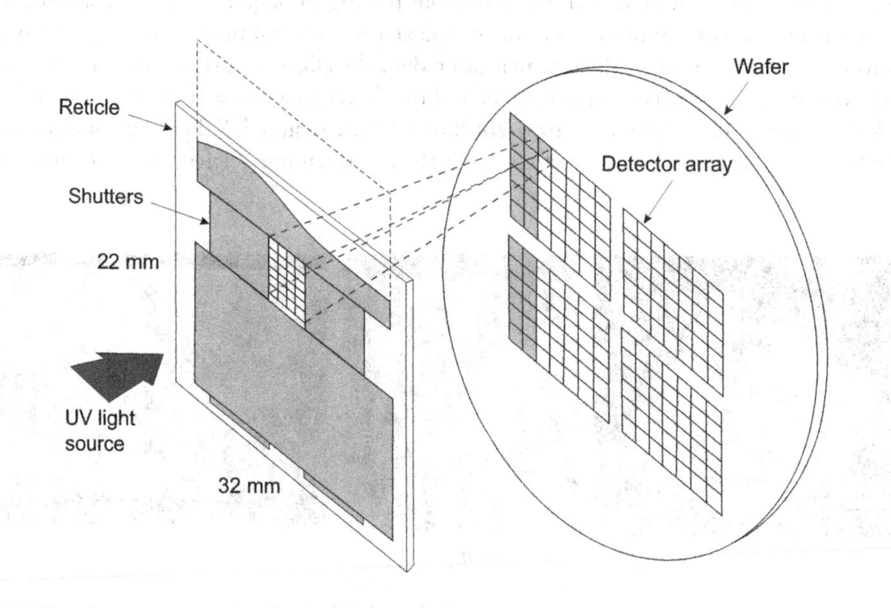

Figure 24.14 The photocomposition of a detector array die using array stitching based on photolithographic stepper. (Adapted after Ref. [38])

Figure 24.15 One of the 64 OTA devices used in PanSTARRS camera. The OTA consists of an 8 × 8 array of 600 × 600 CCD devices, each of which can be controlled and readout independently. (Adapted after Ref. [39])

techniques gives many new possibilities for imaging systems with increased sensitivity and spatial resolution. Key to the development of ROICs has been the evolution in input preamplifier technology. This evolution has been driven by increased performance requirements and Si processing technology improvements. A discussion of the various circuits is given, for example, in Refs. [20, 22, 41].

The direct injection (DI) circuit was one of the first integrated readout preamplifiers and has been used as an input to CCDs and visible imagers for many years. The direct injection is also commonly used input circuit for IR tactical applications, where the backgrounds are high and detector resistances are moderate. The goal is to fit as large a capacitor as possible into the unit cell, where signal-to-noise (SNR) ratios can be obtained through longer integration times. Photon current in DI circuits is injected, via the source of the input transistor, onto an integration capacitor (Figure 24.16a). As the photon current charges the capacitor throughout the frame, a simple charge integration takes place (Figure 24.16b). Next, a multiplexer reads out the final value and the capacitor voltage is reset prior to the beginning of the frame. To reduce detector noise, it is important that a uniform, near-zero-voltage bias be maintained across all the detectors.

The DI circuit is widely used for simplicity; however, it requires a high impedance detector interface and is not generally used for low IR backgrounds due to injection efficiency issues. Many times, the strategic applications have low backgrounds and require low noise multiplexers interfaced to high resistance detectors. A commonly used input circuit for strategic applications is the capacitive transimpedance amplifier (CTIA) input circuit.

The CTIA is a reset integrator and addresses a broad range of detector interface and performance requirements across many applications. The CTIA consists of an inverting amplifier with a gain of A, the integration capacitance C_f placed in a feedback loop, and the reset switch K (Figure 24.17). The photoelectron charge causes a slight change in voltage at the inverting input node of an amplifier. The amplifier responds with a sharp reduction in output voltage. As the detector current accumulates over the frame time, uniform illumination results in a linear ramp at the output. At the end of integration, the output

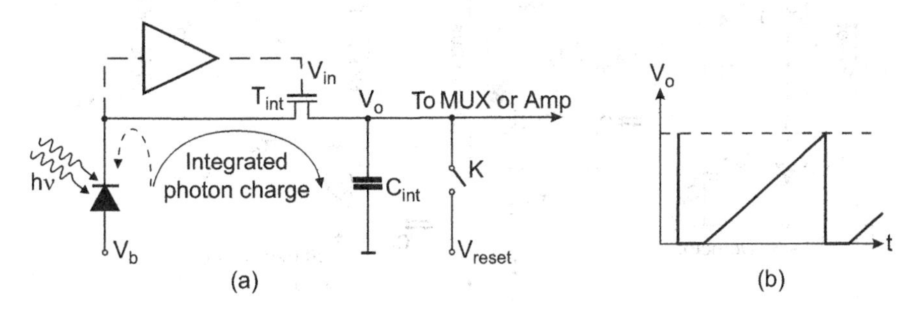

Figure 24.16 Direct injection readout circuit. (Adapted after Ref. [23])

Infrared focal plane arrays

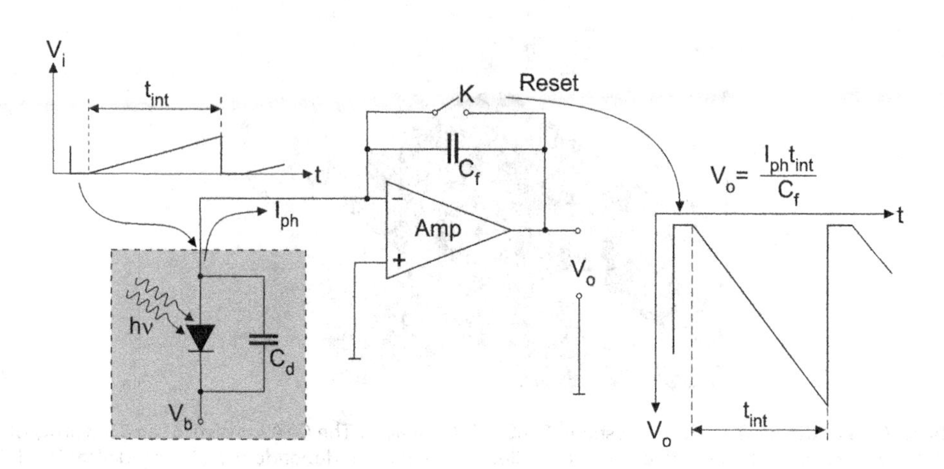

Figure 24.17 Schematic of a CTIA unit cell.

voltage is sampled and multiplexed to the output bus. Since the input impedance of the amplifier is low, the integration capacitance can be made extremely small, yielding low noise performance. The feedback, or integration, capacitor sets the gain. The switch *K* is cyclically closed to achieve reset. The CTIA provides low input impedance, stable detector bias, high gain, high frequency response, and a high photon current injection efficiency. It has very low noise from low to high backgrounds.

Besides the DI and CTIA inputs mentioned above, we can distinguish other multiplexers; the most important are: source follower per detector (SFD), buffered direct injection (BDI), and gate modulation input (GMI) circuits.

The combined SFD unit cell is shown in Figure 24.18. The unit cell consists of an integration capacitance, a reset transistor (*T*1) operated as a switch, the source-follower transistor (*T*2), and selection transistor (*T*3). The integration capacitance may just be the detector capacitance and transistor *T*2 input capacitance. The integration capacitance is reset to a reference voltage (V_R) by pulsing the reset transistor. The photocurrent is then integrated on the capacitance during the integration period. The ramping input voltage of the SFD is buffered by the source follower and then multiplexed, via the *T*3 switch, to a common bus prior to the video output buffer. After the multiplexer read cycle, the input node is reset and the integration cycle begins again. The switch must have very low current leakage characteristics when in the open state, or else this will add to the photocurrent signal. The dynamic range of the SFD is limited by the current-voltage characteristics of the detector. As the signal is integrated, the detector bias changes with time and incident light level. The SFD has low noise for low bandwidth applications such as astronomy and still has acceptable SNR at very low backgrounds (e.g., a few photons per pixel per 100 ms). It is nonlinear

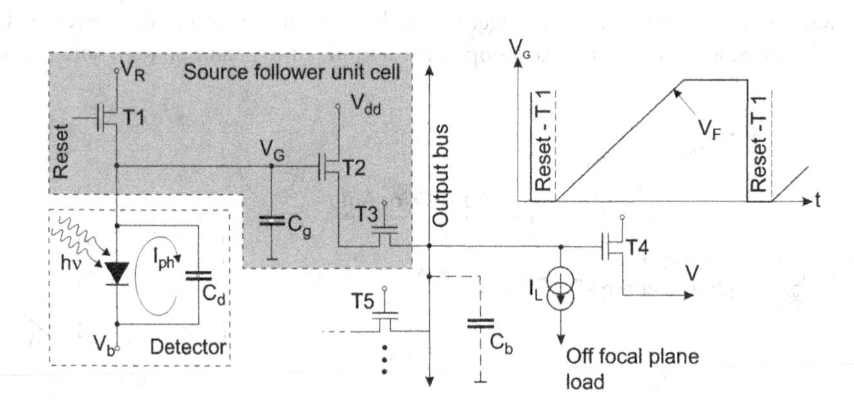

Figure 24.18 Schematic of SFD unit cell.

at medium and high backgrounds, resulting in a limited dynamic range. The gain is set by the detector responsivity and the combined detector plus source-follower-input capacitance. The major noise sources are the *kTC* noise (resulting from resetting the detector), MOSFET channel thermal, and MOSFET $1/f$ noise.

Table 24.1 provides a description of the advantages and disadvantages of DI, SFD, BDI and CTIA circuits. The DI circuit is used in higher-flux situations. The CTIA is more complex and of higher power but is extremely linear. The SFD is most commonly used in large-format hybrid astronomy arrays as well as commercial monolithic CMOS cameras.

In Table 24.2, there are gathered specifications of a family of large-format ROICs designed by FLIR and Raytheon Vision Systems (RVS). FLIR products provide an off-the-shelf solution for the most demanding

Table 24.1 Comparison of attributes of the four most common input circuits

CIRCUIT	ADVANTAGES	DISADVANTAGES	FULL WELL CAPACITY (e^-)	READOUT NOISE (e)	APPLICATIONS
Direct injection (DI)	Low impedance detector interface, stable bias at medium to high backgrounds, compact unit cell	Unstable bias at low current levels	Tens of millions	<1,000	Medium to high photon flux, terrestrial IR systems
Source follower per detector (SFD)	Compact unit cell, low power	Poor high flux performance, bias changes during integration, IPC crosstalk	100,000	<15	Low background astronomy (IR and visible detectors)
Buffered direct injection (BDI)	High injection efficiency, constant bias at medium to high backgrounds, improved frequency response, maximum flux 10 times lower than that of DI	Oscillations needs compensation, noise due to inverting amplifier, large unit cell, increased power dissipation, bias variation from cell to cell	Tens of millions	<1,000	Medium to high photon flux, space IR systems
Capacitance trans-impedance amplifier (CTIA)	Stable bias, high injection efficiency, high gain, high linearity, wide dynamic range	Complex circuit, high power and noise than SF for low flux, worse performance than DI for high flux	1–10 million	<50	Suitable for all backgrounds, space IR and visible detectors

Adapted after Ref. [18].

Table 24.2 Large-format readout integrated circuits

	FLIR						RAYTHEON VISION SYSTEMS					
	ISC9803	ISC002	ISC9901	ISC0402	ISC0403	ISC0404	ALADDIN	ORION	VIRGO	PHOENIX	AQUA-RIUS	MIRI
Format	640 × 512	640 × 512	640 × 512	640 × 512	640 × 512	1,024 × 1,024	1,024 × 1,024	1,024 × 1,024, 2048 × 2048	1,024 × 1,024, 2,048 × 2048 4096 × 4096	1,024 × 1,024, 2048 × 2048	1,024 × 1,024	1,024 × 1,024
Pixel size (μm)	25	25	20	20	15	18	27	25	20	25	30	25
ROIC type	DI	CTIA	DI	DI	DI	DI	SFD	SFD	SFD	SFD	SFD	SFD
Operating temperature (K)	80–310	80–310	80–310	80	80	80	10–30	30	77	10–30	4–10	
Integrated capacity (e^-)	1.1×10^7	2.5×10^6	7×10^6	1.1×10^7	6.5×10^6	1.2×10^7	2.0×10^5	3.0×10^5	$>3.5 \times 10^5$	3×10^5	1 or 15×10^6	2×10^5
ROIC noise (e^-)	≤550	≤360	≤350	≤1279	≤760	≤1,026	10–50	<20	<20	6–20	<1,000	10–30
Full frame rates (Hz)	30	30	30	>30	>30	>30						
Number of outputs	1, 2 or 4	1, 2 or 4	1, 2 or 4	1, 2 or 4	1, 2 or 4	4, 8 or 16	32	64	4 or 16	4	16 or 64	4

(Continued)

Table 24.2 (*Continued*) Large-format readout integrated circuits

	FLIR						RAYTHEON VISION SYSTEMS					
	ISC9803	ISC002	ISC9901	ISC0402	ISC0403	ISC0404	ALADDIN	ORION	VIRGO	PHOENIX	AQUA-RIUS	MIRI
Pack-aging	LCC*	LCC	LCC	LCC	LCC	LCC	LCC	Module—2 side buttable	Module—3 or 4 side buttable	LCC	Module—2 side buttable	Module
Detector	p-on-n	p-on-n	p-on-n	p-on-n	p-on-n	p-on-n						
Comp-atible dete-ctors	InSb or QWIP	InGaAs or HgCdTe	InSb or QWIP	InSb, InGaAs,. HgCdTe or QWIP	InSb	InSb	InSb, HgCdTe, or IBC	InSb	HgCdTe	InSb or IBC	IBC	IBC

* LCC—leadless chip carrier.

applications. The large arrays include a variety of pixel ranging from 25 to 15 microns for customers with a wide range of optical design, dewar/cooler configurations, and resolution requirements. On the contrary, RVS has a rich heritage of developing astronomy FPAs. As the demand for both finer resolution and large FOV for astronomy imagery has increased, the RVS sensor chip utilizes various detector materials and Si readouts with array size up to 4,096 × 4,096 pixels.

24.5 PERFORMANCE OF FOCAL PLANE ARRAYS

This section discusses concepts associated with the performance of FPAs. For arrays, the relevant figure of merit for determining the ultimate performance is not the detectivity, D^*, but the noise equivalent difference temperature (NEDT) and the MTF. They are considered the primary performance metrics to thermal imaging systems: thermal sensitivity and spatial resolution. Thermal sensitivity is concerned with the minimum temperature difference that can be discerned above the noise level. The MTF concerns the spatial resolution and answers one question—how small an object can be imaged by the system? The general approach of system performance is given by Lloyd in his fundamental monograph [42].

24.5.1 MODULATION TRANSFER FUNCTION

MTF is the ability of an imaging system to faithfully image a given object; it quantifies the ability of the system to resolve or transfer spatial frequencies [43]. Consider a bar pattern with a cross section of each bar being a sine wave. Since the image of a sine wave light distribution is always a sine wave, the image is always a sine wave independent of the other effects in the imaging system such as aberration.

Usually, imaging systems have no difficulty in reproducing the bar pattern when the bar pattern is closely spaced. However, an imaging system reaches its limit when the features of the bar pattern get closer and closer together. When the imaging system reaches this limit, the contrast or the modulation (M) is defined as:

$$M = \frac{E_{max} - E_{min}}{E_{max} + E_{min}}, \tag{24.1}$$

where E is the irradiance. Once the modulation of an image is measured experimentally, the MTF of the imaging system can be calculated for that spatial frequency, using,

$$\text{MTF} = \frac{M_{image}}{M_{object}}, \tag{24.2}$$

The system MTF is dominated by the optics, detector, and display MTFs and can be cascaded by simply multiplying the MTF components to obtain the MTF of the combination. In spatial frequency terms, the MTF of an imaging system at a particular operating wavelength is dominated by limits set by the size of the detector and the aperture of the optics. More details about this issue is given in section 24.6.

24.5.2 NOISE EQUIVALENT DIFFERENCE TEMPERATURE

NEDT of a detector represents the temperature change, for incident radiation, that gives an output signal equal to the rms (root mean square) noise level. While normally thought of as a system parameter, detector NEDT and system NEDT are the same except for system losses. NEDT is defined as

$$\text{NEDT} = \frac{V_n(\partial T/\partial \Phi)}{(\partial V_s/\partial \Phi)} = V_n \frac{\Delta T}{\Delta V_s}, \tag{24.3}$$

where V_n is the rms noise, Φ is the spectral photon flux density (photons cm^{-2}s^{-1}) incident on a focal plane, and ΔV_s is the signal measured for the temperature difference ΔT.

We further follow after Kinch [44] to obtain useful equations for noise equivalent irradiance (NEI) and NEDT used for estimation of detector performance.

In modern IR FPAs, the current generated in a biased photon detector is integrated onto a capacitive node with a carrier well capacity of N_w. For ideal system, in absence of excess noise, the detection limit of the node is achieved when a minimum detectable signal flux $\Delta\Phi$ creates a signal equal shot noise on the node:

$$\Delta\Phi\eta A_d\tau_{\text{int}} = \sqrt{N_w} = \sqrt{\frac{(J_d + J_\Phi)A_d\tau_{\text{int}}}{q}}, \tag{24.4}$$

where η is the detector collection efficiency, A_d is the detector area, τ_{int} is the integration time, J_d is the detector dark current, and J_Φ is the flux current.

With NEDT is connected other critical parameters such as so-called noise equivalent flux ($NE\Delta\Phi$). This parameter is defined for spectral regions in which the thermal background flux does not dominate. By equating the minimum detectable signal to the integrated current noise, we have

$$\eta\Phi_s A_d\tau_{\text{int}} = \sqrt{\frac{(J_d + J_\Phi)A_d\tau_{\text{int}}}{q}}, \tag{24.5}$$

giving

$$\text{NE}\Delta\Phi = \frac{1}{\eta}\sqrt{\frac{J_d + J_\Phi}{qA_a\tau_{\text{int}}}}. \tag{24.6}$$

This can be converted to a NEI, which is defined as the minimum observable flux power incident on the system aperture, by renormalizing the incident flux density on the detector to the system aperture area A_{opt}. The NEI is given by:

$$\text{NEI} = \text{NE}\Delta\Phi\frac{A_d h\nu}{A_{\text{apt}}}, \tag{24.7}$$

where monochromatic radiation of energy $h\nu$ is assumed.

NEI (photons/cm^{2}s) noise equivalent irradiance is the signal flux level at which the signal produces the same output as the noise present in the detector. This unit is useful because it directly gives the photon flux above which the detector will be photon noise limited.

For high-background flux conditions, the signal flux can be defined as $\Delta\Phi = \Delta T(d\Phi_B/dT)$. Thus, for shot noise, substituting in Equation 24.4, we have

$$\eta\Delta T\frac{d\Phi_B}{dT} = \sqrt{\frac{(J_d + J_\Phi)A_d\tau_{\text{int}}}{q}}. \tag{24.8}$$

Finally, after some rearrangement

$$\text{NEDT} = \frac{1+(J_d/J_\Phi)}{\sqrt{N_w}\,C}, \tag{24.9}$$

where $C = (d\Phi_B/dT)/\Phi_B$ is the scene contrast through the optics. In deriving the last equation it was assumed that the optics transmission is unity, and that the cold shield of the detector is not contributing flux. This is reasonable at low detector temperatures but not at higher operating temperatures. At higher temperatures, the scene contrast is defined in terms of the signal flux coming through the optics, whereas the flux current is defined by the total flux through the optics and the flux from the cold shield.

The above considerations are valid assuming that the temporal noise of the detector is the main source of noise. However, this assertion is not true to staring arrays, where the nonuniformity of the detectors response is a significant source of noise. This nonuniformity appears as a fixed pattern noise (spatial noise). It is defined in various ways in the literature, however, the most common definition is that it is the dark signal nonuniformity arising from electronic source (i.e., other than thermal generation of the dark current); for example, clock breakthrough or from offset variations in row, column, or pixel amplifiers/switches. So, estimation of IR sensor performance must include a treatment of spatial noise that occurs when FPA nonuniformities cannot be compensated correctly.

Mooney et al. [45] have given a comprehensive discussion of the origin of spatial noise. The total noise of a staring array is the composite of the temporal noise and the spatial noise. The spatial noise is the residual nonuniformity u after application of nonuniformity compensation, multiplied by the signal electrons N. Photon noise, equal $N^{1/2}$, is the dominant temporal noise for the high IR background signals for which spatial noise is significant. Then, the total NEDT is

$$NEDT_{total} = \frac{\left(N + u^2 N^2\right)^{1/2}}{\dfrac{\partial N}{\partial T}} = \frac{\left(\dfrac{1}{N} + u^2\right)^{1/2}}{\left(\dfrac{1}{N}\right)\left(\dfrac{\partial N}{\partial T}\right)}, \tag{24.10}$$

where $\partial N/\partial T$ is the signal change for a 1 K source temperature change. The denominator, $\left(\partial N/\partial T\right)/N$, is the fractional signal change for a 1 K source temperature change. This is the relative scene contrast.

The dependence of the total NEDT on detectivity for different residual nonuniformities is plotted in Figure 24.19 for 300 K scene temperature and a set of parameters shown in the insert of the figure. When the detectivity is approaching a value above 10^{10} cmHz$^{1/2}$ W^{-1}, the FPA performance is uniformity limited prior to correction and thus essentially it is independent of the detectivity. An improvement in nonuniformity from 0.1% to 0.01% after correction could lower the NEDT from 63 to 6.3 mK.

The NEDT characterizes also the thermal sensitivity of an IR system, that is, the amount of temperature difference required to produce a unity SNR ratio. A smaller NEDT indicates a better thermal sensitivity. In spite of its widespread use in IR literature, it is applied to different systems, in different conditions, and with different meanings [46].

To derive the equation estimating NEDT, we consider the configuration of the thermal imaging system shown in Figure 24.20. On the graph shown in this figure, A_s and A_d are, respectively, the surfaces

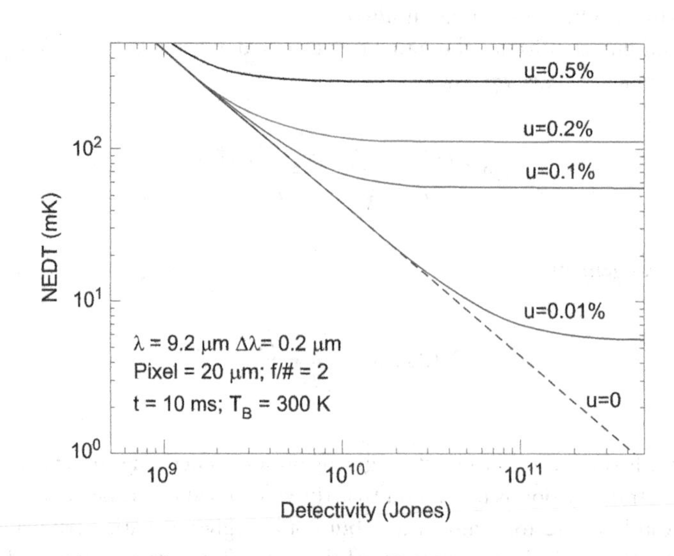

Figure 24.19 NEDT as a function of detectivity. The effects of nonuniformity are included for u = 0.01%, 0.1%, 0.2%, and 0.5%. Note that for $D^* > 10^{10}$ cmHz$^{1/2}$ W^{-1}, detectivity is not the relevant figure of merit.

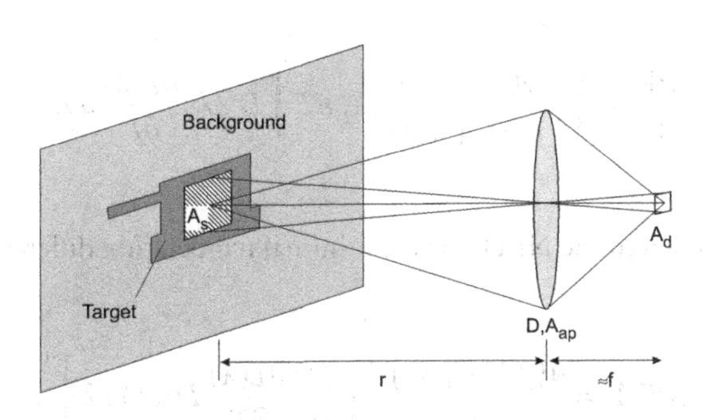

Figure 24.20 Thermal imaging system configuration.

of the object and the detector, r is the distance of the object to the lens (system optics), A_{ap} and D are the surface and the diameter of the lens (aperture, entrance-pupil). The detector is placed in the focal plane of the system in the distance $\approx f$ to the entrance pupil. The optic's system is opened to $f/\#$ (i.e., $f/\# = f/D$ with $A_{ap} = \pi D^2/4$).

Taking into consideration the methods presented in Section 1.3, we find the flux on the detector, ϕ_d, in terms of the radiance. Within the limitations of small angles ($r >> f$), flux transfer from target to detector is described in terms of the $A\Omega$ product.

$$\Phi_d = LA_{ap}\Omega_d. \tag{24.11}$$

Assuming that the solid state of the elementary field of the system is

$$\Omega_d = \frac{A_d}{f^2} = \frac{A_s}{r^2}, \tag{24.12}$$

then

$$\Phi_d = L\frac{\pi D^2}{4}\frac{A_d}{f^2} = L\frac{\pi}{4}\frac{A_d}{(f/\#)^2}. \tag{24.13}$$

For the geometry of optical system shown in Figure 24.20, we see that the irradiance on the detector, $E_d = \Phi_d/A_d$, is independent of the range r, and depends only on the source radiance and the $f/\#$ of the optics.

To obtain an expression for the signal voltage V_s produced by the detector, the expression 24.13 should be multiplied by detector responsivity and integrated over the wavelength of the product of the spectral functions, over the passband of the system:

$$V_s = \frac{\pi}{4}\frac{A_d}{(f/\#)^2}\int_{\lambda_a}^{\lambda_b} R_v(\lambda)L(\lambda)d\lambda. \tag{24.14}$$

Because we are interested in thermal sensitivity of the system,

$$\frac{dV_s}{dT} = \frac{\pi}{4}\frac{A_d}{(f/\#)^2}\int_{\lambda_a}^{\lambda_b} R_v(\lambda)\frac{\partial L(\lambda)}{\partial T}d\lambda. \tag{24.15}$$

Using Equation 3.7, we can rewrite the last equation in the form

$$\frac{dV_s}{V_n} = \frac{\pi}{4} \frac{A_d}{(f/\#)^2} \frac{1}{(A_d \Delta f)^{1/2}} dT \int_{\lambda_a}^{\lambda_b} D^*(\lambda) \frac{\partial L(\lambda)}{\partial T} d\lambda. \tag{24.16}$$

From this equation, we can determine NEDT as the minimal temperature difference ΔT required to get the SNR ratio equal to 1:1.

$$\text{NETD} = \frac{4(f/\#)^2 (\Delta f)^{1/2}}{\pi A_d} \left[\int_{\lambda_a}^{\lambda_b} \frac{\partial L(\lambda)}{\partial T} D^*(\lambda) d\lambda \right]^{-1}. \tag{24.17}$$

Taking into account that $M = \pi L$ (see Equation 1.14) results that

$$\text{NETD} = \frac{4(f/\#)^2 (\Delta f)^{1/2}}{A_d^{1/2}} \left[\int_{\lambda_a}^{\lambda_b} \frac{\partial M(\lambda)}{\partial T} D^*(\lambda) d\lambda \right]^{-1}. \tag{24.18}$$

In the above consideration, both atmosphere and optics transmissions have been assumed to be equal to 1.

The NEDT characterizes the thermal sensitivity of an IR system, that is, the amount of temperature difference required to produce a unity SNR ratio. A smaller NEDT indicates a better thermal sensitivity.

To receive best sensitivity (lowest NEDT), the spectral integral in Equations 24.17 and 24.28 should be maximized. This can be obtained when the peak of the spectral responsivity and the peak of the exitance contrast coincide. However, the thermal imaging system may not satisfy these conditions because of other constraints such as atmospheric/abscurant transmittance effects or available detector characteristics. Dependence on the square root of bandwidth is intuitive, since the rms noise is proportional to $(\Delta f)^{1/2}$. In addition, better NEDT results from lower *f*/#. A lower *f*/# number results in more flux captured by the detector that increases SNR for a given level.

The dependence of NEDT on the detector area is critical. The inverse square root dependence of NEDT on detector area results as an effect of two terms: increasing rms noise as the square root of the detector area and proportional increase of the signal voltage to the area of detector. The net result is that NEDT $\propto 1/(A_d)^{1/2}$. While the thermal sensitivity of an imager is better for larger detectors, the spatial resolution is poorer for larger detectors (pixels). Hence, a reasonable compromise between the requirement of high thermal and spatial resolution is necessary. Another parameter, the minimum resolvable difference temperature (MRDT), considers both thermal sensitivity and spatial resolution, and is more appropriate for design.

As Equation 24.18 shows, improvement of thermal resolution without detrimental effects on spatial resolution may be achieved by:

- an decrease of detector area combined with a corresponding decrease of the optics *f*-number,
- improved detector performance, and
- an increase in the number of detectors.

As mentioned before, the increase of aperture is undesirable because it increases size, mass, and price of an IR system. It is more appropriate to use a detector with higher detectivity. This can be achieved by better coupling of the detector with the incident radiation. Another possibility is the application of multielemental sensor, which reduces each element bandwidth proportionally to the number of elements for the same frame rate and other parameters.

As Table 24.3 and the last two equations indicate, the best performance of IR imaging devices can be achieved operating in the wide spectral range [47]. The spectral range limited to the atmospheric windows 8–14 μm and 3–5.5 μm will reduce the integral value NEDT (see Equations 24.17 and 24.18) to about 33% and about 6% of the 0–∞ range value, respectively. Therefore, IR systems based on unselective detectors, which are optimized for detection of ≈ 300 K objects in the atmosphere must operate in the 8–14 μm region.

Table 24.3 Calculated radiant exitance between λ_a and λ_b at different temperatures

λ (µm)		$\int_{\lambda_a}^{\lambda_b} \frac{\partial M(\lambda,T)}{\partial \lambda} d\lambda \left(\mathrm{Wcm^{-2}K^{-1}}\right)$			
λ_a	λ_b	$T = 280$ K	$T = 290$ K	$T = 300$ K	$T = 310$ K
3	5	1.1×10^{-5}	1.54×10^{-5}	2.1×10^{-5}	2.81×10^{-5}
3	5.5	2.01×10^{-5}	2.73×10^{-5}	3.62×10^{-5}	4.72×10^{-5}
3.5	5	1.06×10^{-5}	1.47×10^{-5}	2.0×10^{-5}	2.65×10^{-5}
3.5	5.5	1.97×10^{-5}	2.66×10^{-5}	3.52×10^{-5}	4.57×10^{-5}
4	5	9.18×10^{-6}	1.26×10^{-5}	1.69×10^{-5}	2.23×10^{-5}
4	5.5	1.83×10^{-5}	2.45×10^{-5}	3.22×10^{-5}	4.14×10^{-5}
8	10	8.47×10^{-5}	9.65×10^{-5}	1.09×10^{-4}	1.21×10^{-4}
8	12	1.54×10^{-4}	1.77×10^{-4}	1.97×10^{-4}	2.17×10^{-4}
8	14	2.15×10^{-4}	2.38×10^{-4}	2.62×10^{-4}	2.86×10^{-4}
10	12	7.34×10^{-5}	8.08×10^{-5}	8.81×10^{-5}	9.55×10^{-5}
10	14	1.3×10^{-4}	1.42×10^{-4}	1.53×10^{-4}	1.65×10^{-4}
12	14	5.67×10^{-5}	6.1×10^{-5}	6.52×10^{-5}	6.92×10^{-5}

Adapted after Ref. [47].

24.5.3 NEDT LIMITED BY READOUT CIRCUIT

Usually, the performance of MWIR and long wavelength infrared (LWIR) FPAs is limited by the readout circuits (by storage capacity of the ROIC). In this case [41]

$$\mathrm{NEDT} = \left(\tau C \eta_{\mathrm{BLIP}} \sqrt{N_w}\right)^{-1}, \tag{24.19}$$

where N_w is the number of photogenerated carriers integrated for one integration time, t_{int}

$$N_w = \eta A_d t_{\mathrm{int}} Q_B. \tag{24.20}$$

Percentage of background limited infrared photodetector (BLIP), η_{BLIP}, is simply the ratio of photon noise to composite FPA noise

$$\eta_{\mathrm{BLIP}} = \left(\frac{N_{\mathrm{photon}}^2}{N_{\mathrm{photon}}^2 + N_{\mathrm{FPA}}^2}\right)^{1/2}. \tag{24.21}$$

It results from the above formulas that the charge handling capacity of the readout, the integration time linked to the frame time, and dark current of the sensitive material become the major issues of IR FPAs. The NEDT is inversely proportional to the square root of the integrated charge and therefore the greater the charge, the higher the performance. The well charge capacity is the maximum amount of the charge that can be stored in the storage capacitor of each cell. The size of the unit cell is limited to the dimensions of the detector element in the array.

Figure 22.21 shows the theoretical NEDT versus charge handling capacity for different types of FPAs assuming that the integration capacitor is filled to half the maximum capacity (to preserve dynamic range) under nominal operating conditions in two spectral bandpasses: 3.4–4.8 µm and 7.8–10 µm. We can see that the measured sensitivities agree with the expected values.

It must be noted that there is distinction between integration time and FPA's frame time. At high backgrounds, it is often impossible to handle the large amounts of carriers generated over frame time compatible with standard video rates. Off-FPA frame integration can be used to attain a level of sensor sensitivity that is commensurate with the detector-limited D^* and not the charge-handling-limited D^*.

24.5.3.1 Readout-limited NEDT for HgCdTe photodiode and QWIP

The noise in HgCdTe photodiodes at 77 K is due to two sources: the shot noise from the photocurrent and the Johnson noise from the detector resistance. It can be expressed as [48]:

$$I_n = \sqrt{\left(2qI_{\text{ph}} + \frac{4kT_d}{R}\right)\Delta f}, \quad (24.22)$$

where k is the Boltzmann constant and R is the dynamic resistance of a photodiode. Assuming that the integration time τ_{int} is such that the readout node capacity is kept half full, we have

$$\Delta f = \frac{1}{2\tau_{\text{int}}}, \quad (24.23)$$

and then

$$I_n = \sqrt{\left(2qI_{\text{ph}} + \frac{4kT_d}{R}\right)\frac{1}{2\tau_{\text{int}}}}. \quad (24.24)$$

At tactical background levels, the Johnson noise is much smaller than the shot noise from the photocurrent. In case where the number of electrons collected in a frame is limited by the capacity of the ROIC charge well, which is often true, the SNR ratio is given by [48]

$$\frac{S}{N} = \frac{qN_w/2\tau_{\text{int}}}{\sqrt{2q\left(\frac{qN_w}{2\tau}\right)\frac{1}{2\tau_{\text{int}}}}} = \sqrt{\frac{N_w}{2}}. \quad (24.25)$$

Assuming that the temperature derivative of the background flux Φ can be written to a good approximation as:

$$\frac{\partial \Phi}{\partial T} = \frac{hc}{\lambda k T_B^2} Q, \quad (24.26)$$

and using Equation 24.25, the NEDT is equal to

$$\text{NEDT} = \frac{2kT_B^2\bar{\lambda}}{hc\sqrt{2N_w}}. \quad (24.27)$$

In the last two equations, $\bar{\lambda} = (\lambda_1 + \lambda_2)/2$ is the average wavelength of the spectral band between λ_1 and λ_2.

If one assumes a typical storage capacity of 2×10^7 electrons, $\bar{\lambda} = 10\,\mu\text{m}$, and $T_B = 300$ K, Equation 24.27 yields NEDT of 19.8 mK.

The same estimate can be made for QWIP. In this case, the Johnson noise is negligible compared to the generation-recombination noise, therefore

$$I_n = \sqrt{4qg\left(I_{\text{ph}} + I_d\right)\frac{1}{2\tau_{\text{int}}}}, \quad (24.28)$$

where the dark current may be approximated by

$$I_d = I_0 \exp\left(-\frac{E_a}{kT}\right). \tag{24.29}$$

In the above expressions, I_d is the dark current, I_0 is a constant that depends on the transport properties and the doping level, and E_a is the thermal activation energy, which is usually slightly less than the energy corresponding to the cutoff wavelength of the spectral response. It should be also stressed that g, I_{ph}, and I_0 are bias-dependent.

The SNR ratio for a storage-capacity-limited QWIP is given by

$$\frac{S}{N} = \frac{\dfrac{qN_w}{2\tau_{int}}}{\sqrt{4qg\left(\dfrac{qN_w}{2\tau}\right)\dfrac{1}{2\tau_{int}}}} = \frac{1}{2}\sqrt{\frac{N_w}{g}}, \tag{24.30}$$

and the NEDT is

$$\text{NEDT} = \frac{2kT_B^2\bar{\lambda}}{hc}\sqrt{\frac{g}{N_w}}. \tag{24.31}$$

Comparing Equations 24.27 and 24.31 one may notice that the value of NEDT in a charge-limited QWIP detector is better than that of HgCdTe photodiodes by a factor of $(2g)^{1/2}$ since a reasonable value of g is 0.4. Assuming the same operation conditions as for HgCdTe photodiodes, the value of NEDT is 17.7 mK. Thus, a low photoconductive gain actually increases the S/N ratio and a QWIP FPA can have a better NEDT than an HgCdTe FPA with a similar storage capacity.

The performance figures of merit of state of the art QWIP and HgCdTe FPAs are similar because the main limitations come from the readout circuits. Performance is, however, achieved with very different integration times. The very short integration time of LWIR HgCdTe devices (typically below 300 μs) is very useful to freeze a scene with rapidly moving objects. Due to excellent homogeneity and low photoelectrical gain, QWIP devices achieve an even better NEDT; the integration time, however, must be 10–100 times longer, and typically is between 5 and 20 ms. The choice of the best technology is therefore driven by the specific needs of a system.

24.6 TOWARD SMALL PIXEL FOCAL PLANE ARRAYS

It is well known that detector size, d, and F-number ($f/\#$) are primary parameters of infrared systems [49]. These two parameters have a major impact on both detection/identification range and the noise equivalent difference temperature (NEDT), since they depend on $F\lambda/d$ [44]:

$$\text{Range} = \frac{D\Delta x}{M\lambda}\left(\frac{F\lambda}{d}\right), \tag{24.32}$$

$$\text{NEDT} \approx \frac{2}{C\lambda\left(\eta\Phi_B^{2\pi}\tau_{int}\right)^{1/2}}\left(\frac{F\lambda}{d}\right), \tag{24.33}$$

where λ is the wavelength, D is the aperture, M is the number of pixels required to identify a target Δx, C is the scene contrast, η is the detector collection efficiency, $\Phi_B^{2\pi}$ is the background flux into a 2π FOV, and τ_{int} is the integration time. The above two equations indicate that the parameter space defined by $F\lambda$ and d can be utilized in the optimum design of any IR system.

Most military systems today have the classical view presented in Figure 24.21 [50], where the detector size ranges from 15 to 50 μm. For long range identification systems, high $f/\#$ optics is used (for given

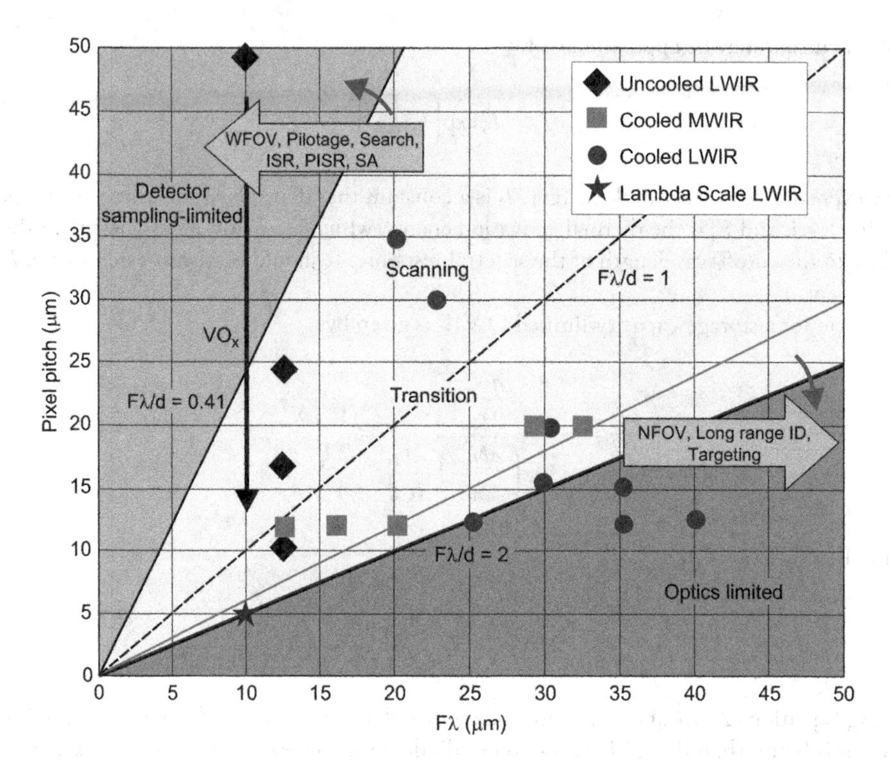

Figure 24.21 *Fλ/d* space for IR system design. Straight lines represent constant NEDT. There are an infinite number of combinations that provide the same range. (Adapted after Ref. [50])

aperture) to reduce the detector angular subtense. On the other hand, wide field-of-view (WFOV) systems are typically low *f*/# systems with short focal lengths since the focal plane had to be spread over wide angles. In recently published papers it has been shown that long-range identification does not need to be limited to high *f*/# systems and that very small detectors enable high performance with a smaller package [51,52].

The detection range of many uncooled IR imaging systems is limited by pixel resolution rather than sensitivity. Figure 24.22 presents a trade-off analysis of the detection range and sensor optics for a thermal weapon sight using the NVESD NVTherm IP model, assuming a detector sensitivity of 35 mK NEDT (*f*/1, 30 Hz) for the 25, 17, and 12 μm pitch pixel of uncooled FPAs. The advantages of small pixel pitch

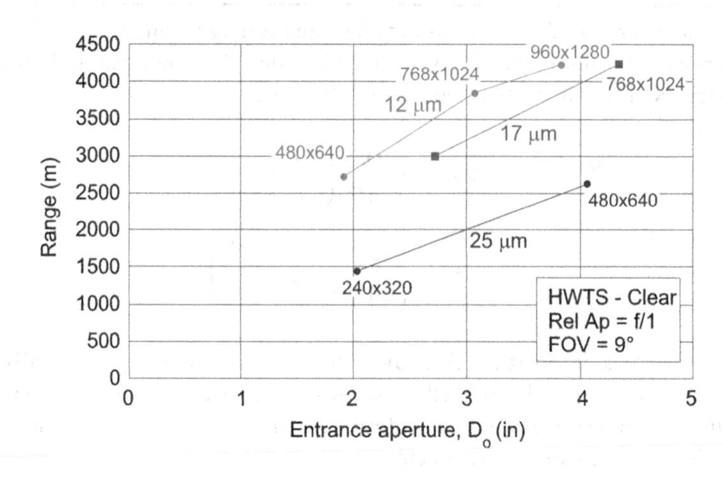

Figure 24.22 Calculated detection range as a function of sensor optics and detector pixel size and format using NVESD NVTherm IP modeling, assuming a 35 mK NETD (f/1, 30 Hz) for all detectors. (Adapted after Ref. [53])

and large format FPAs are obvious. By switching to smaller pitch and larger format detectors, the detection range of a weapon sight increases significantly with a fixed optic.

The fundamental limit of pixel size is determined by diffraction. The size of diffraction-limited optical spot or Airy disk is given by

$$d = 2.44\lambda F, \tag{24.34}$$

where d is the diameter of the spot and λ is the wavelength. The spot size for *f*/# ranging from *f*/1 to *f*/10 are shown in Figure 24.23. For typical *f*/2.0 optics at 4 μm wavelength, the spot size is 20 μm.

It is generally interesting to investigate pixel scaling beyond the diffraction limit using wavelength- and even subwavelength-scale optics that are enabled by modern nanofabrication (diffraction-limited pixel size is still relatively large compared with feature size that can be achieved with state of the art nanofabrication approaches).

FPAs of 1 cm^2 still dominate the IR market, while pixel pitch has decreased to 15 μm during the last few years, now reaching 12 μm [54], 10 μm [55,56], 8 μm [57], and even 5 μm in test devices [58,59]. This trend is expected to continue. Systems operating at shorter wavelengths are more likely to benefit from small pixel sizes because of the smaller diffraction-limited spot size. Diffraction-limited optics with low *F*-numbers (e.g., *f*/1) could benefit from pixels on the order of one wavelength across. Oversampling the diffractive spot may provide some additional resolution for smaller pixels, but this saturates quickly as the pixel size is decreased. Pixel reduction is mandatory also to cost reduction of a system (reduction of the optics diameter, dewar size and weight, together with the power, and increase of the reliability). In addition, smaller detectors provide better resolution [60]. Reduction of the focal plane proportionally to the detector size has not changed the detector FOV, so in the optics-limited region, smaller detectors have no effect on the system spatial resolution.

Figure 24.24 shows the influence of pixel shrinkage on format enlargement of Sofradir's IR arrays. A catalog of detectors with pixel pitch of 15 μm [Epsilon (384 × 288), Scorpio (640 × 512) and Jupiter (1,280 × 1,014)] is compared with Daphnis 10 μm product family.

The detector-limited region occurs when $F\lambda/d \leq 0.41$ and the optics-limited occurs when $F\lambda/d \geq 2$ (see Figure 24.21). When $F\lambda/d = 0.41$, the Airy disk is equal to the detector size. A transition in the region $0.41 \leq F\lambda/d \leq 2.0$ is large and represents a change from detector-limited to optics-limited performance. The condition $F\lambda/d = 2$ is equivalent to placing 4.88 pixels within the Rayleigh blur circle. The lines presenting a constant $F\lambda/d$ indicate a constant range and NEDT. For a given aperture D and operating wavelength λ, the detection range is given by the optimum resolution condition $F\lambda/d = 2$ and a minimum NEDT for given τ_{int}

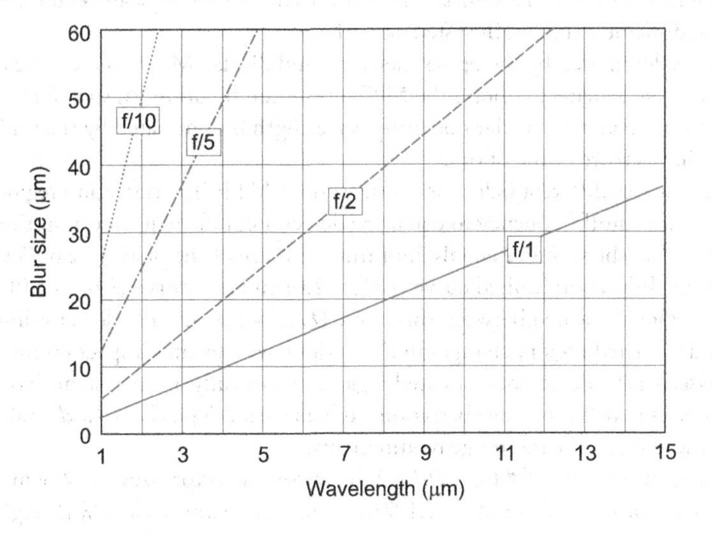

Figure 24.23 Optics diffraction limit. The spot size of a diffraction-limited optical system is the Airy disk diameter.

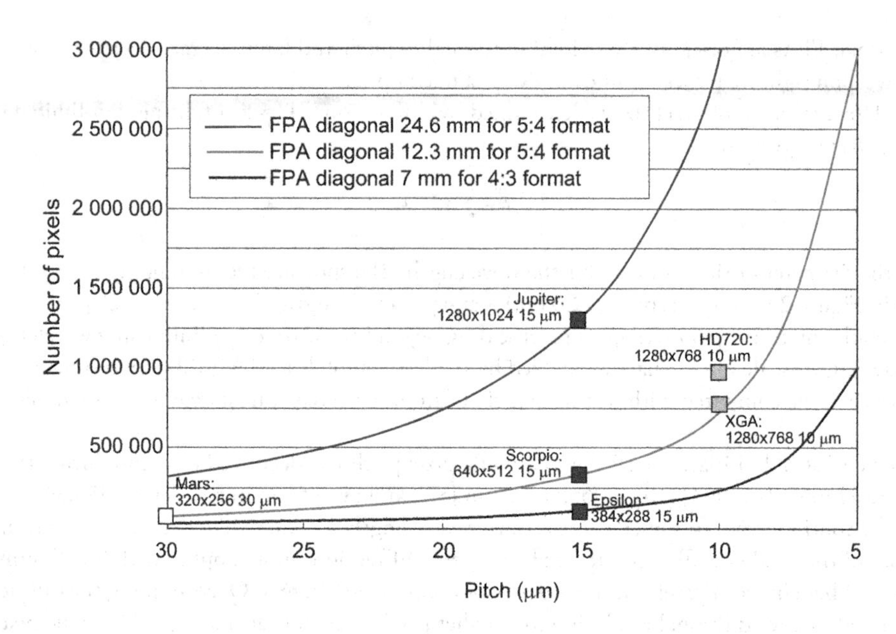

Figure 24.24 Number of pixels versus pitch size for Sofradir IR FPAs. (Adapted after Ref. [55])

(see Equation 24.33). From these considerations it results that the system *f*/# should be locked to the pixel size to predict the potential limiting performance of IR systems.

Figure 24.21 also includes experimental data points for various classes of thermal imaging systems that have been produced at DRS Technologies including both uncooled thermal imagers and cooled photon imagers. The earliest uncooled imagers fabricated at the beginning of 1990s (barium strontium titanate (BST) dielectric bolometers and VO_x microbolometers) had large pixels of approximately 50 μm pitch and fast optics to achieve useful system sensitivities. With decreasing of detector size, the relative apertures remained around *f*/1. As shown in Figure 24.21, as the pixel dimensions shrank over time, uncooled systems steadily progressed from the detector-limited regime to the optics-limited ones. However, they are still far from the ultimate range capacity for *f*/1 optics.

The cooled thermal imagers include early LWIR scanning systems and modern staring systems operating in both MWIR and LWIR bands. It is shown that LWIR imaging systems typically approach $F\lambda/d = 2$ condition, whereas for MWIR systems values of $F\lambda/d$ less than two are typically employed—lower available photon flux makes it difficult to maintain system sensitivity.

The system MTF is dominated by the optics, detector, and display MTFs and can be cascaded by simply multiplying the MTF components to obtain the MTF of the combination. In spatial frequency terms, the MTF of an imaging system at a particular operating wavelength is dominated by the limits set by the size of the detector and the aperture of the optics.

Figure 24.25 summarizes different behaviors of the system MTF. The transition region can be further split by setting the optics cutoff frequency to equal the detector cutoff frequency, resulting in an $F\lambda/d = 1.0$ [61]. When $F\lambda/d = 1.0$, the spot size equals 2.44 times the size of the pixel. The optics-dominated region lies between the diffraction-limited curve ($F\lambda/d = 2.0$) and this curve ($F\lambda/d = 1.0$), while the detector-dominated region is located between this curve ($F\lambda/d = 1.0$) and the detector-limited curve ($F\lambda/d = 0.41$). In the optics-dominated region, changes to the optics have a greater impact on the system MTF than the detector. Likewise, for the detector-dominated region. Historically, most systems have been designed to have a resulting optics blur (to include aberrations) of less than 2.5 pixels (~ $F\lambda/d < 1.0$). This is of course very dependent on the application and range requirements.

Table 24.4 provides the required *f*/# for $F\lambda/d = 2$ for various detector sizes. As shown, with *f*/1 optics, the smallest useful detector size is 2 μm in the MWIR region and 5 μm in the LWIR region. With more realistic *f*/1.2 optics, the smallest useful detector size is 3 μm in the MWIR region and 6 μm in the LWIR region.

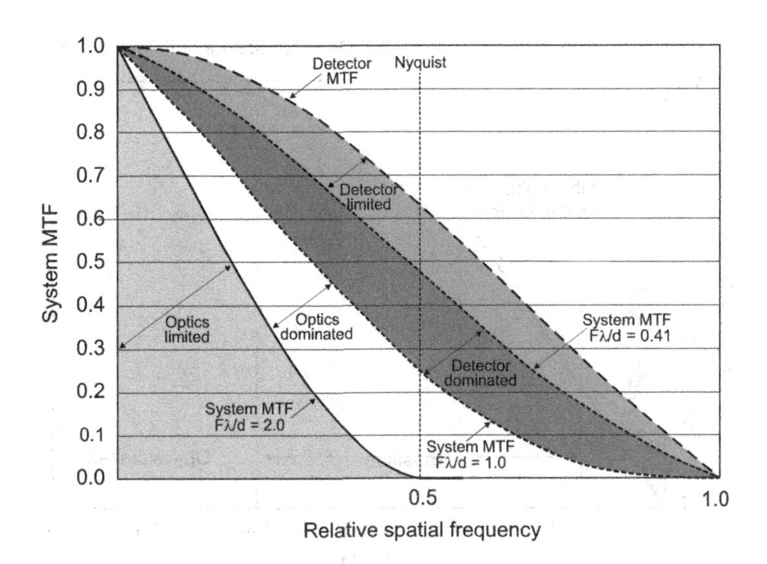

Figure 24.25 System MTF curves illustrating the different regions with the design space for various $F\lambda/d$ conditions. Spatial frequencies are normalized to the detector cutoff. (Adapted after Ref. [61])

Table 24.4 Required *f/#* for $F\lambda/d = 2.0$

d (μm)	MWIR (4 μm)	LWIR (10 μm)
2.0	1.0	—
2.5	1.25	—
3.0	1.33	—
5.0	2.5	1.0
6.0	3.0	1.3
12	6.0	2.4
15	7.5	3.0
17	8.5	3.4
20	10.0	4.5
25	12.5	5.0

Adapted after Ref. [52].
Real optics usually has *f/#* > 1.

Host and Driggers [52] have considered the influence of optics IR design on range approximation that is illustrated in Figure 24.26. When the IR system is detector limited, decreasing the detector size has a dramatic effect on range. On the other size, in the optics-limited region, decreasing the detector size has minimal effect on range performance. The acquisition range is reduced when the atmospheric transmission and the NEDT are included.

As has been indicated by Kinch [44,62], challenges that must be addressed in the fabrication of small-pixel FPAs concern:

- pixel delineation,
- pixel hybridization,
- dark current, and
- unit cell capacity.

The above topics are considered in more details in Ref. [63].

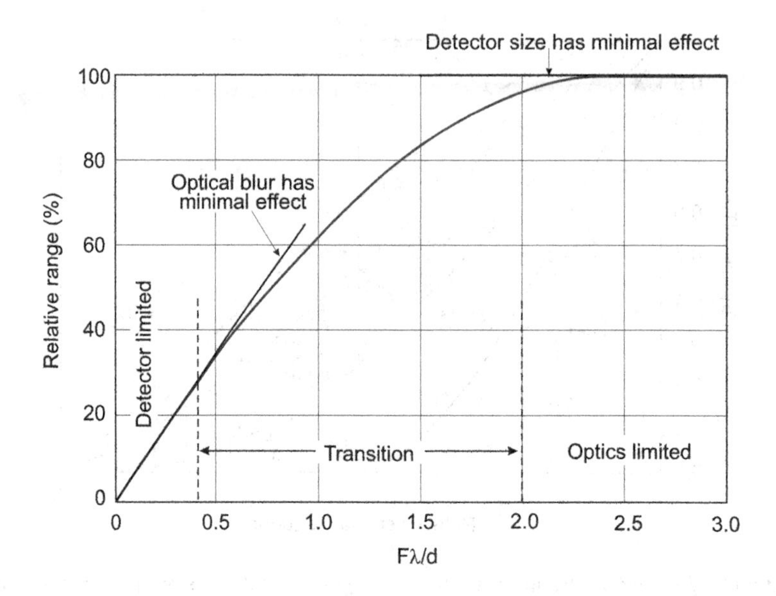

Figure 24.26 Relative range as a function of *Fλ/d*.

Unlike the visible sensors where the pixel size has been reduced to 1.4 µm, the scaling of IR pixels is much more difficult. The detector pixels are generally connected to the per-pixel electronics through In bump bonds. As the pixel size is reduced, bump bonding, ROIC, signal integrating capacitor, and signal-to-noise ratio become difficult.

While bump-bond pitches of 12 µm or larger are relatively common, pitches of less than 8 µm offer significant challenges in terms of array yield, pixel operability, and cost. It is expected that developments will continue to extend bump-bonding technology to smaller pixels as well as to improving manufacturability and reducing cost at all pixel sizes. In the past, much of the processing done on a focal plane was confined to the two-dimensional (2D) real estate directly under a given pixel. One of the novel research areas has been in the development of 3D integration technologies providing alternatives to bump bonding [64]. Recently, as many as three layers of CMOS have been stacked and vertically interconnected, offering the potential to increase the amount of processing available within a pixel footprint.

In conventional analog ROIC technology, the photocurrent generated by a detector is accumulated and stored locally in a capacitor (electron well); the maximum charge stored during an integration time is equal to the product of the total capacitance and the maximum allowable voltage across the capacitor. A simplified ROIC unit cell (pixel) circuit diagram and example ROIC layout are presented in Figure 24.27. As indicated in this figure, the integrating capacitor dominates the unit-cell area usage. Thanks to multiple gain selection at the pixel level, the charge capacity can be tuned to match the targeted applications and scenario with maximum SNR and dynamic. The gain and integration time are set depending on the application requirements (*f*/#, motion blur, IR radiation flux) mainly concerning the detection range as well as detection performance in various weather conditions.

With a small pixel pitch, the photon flux is reduced and the integration time is generally increased. As a consequence, the long-range devices involving lower instantaneous IFOV are often limited by the motion blur, a consequence of holder movement during the integration time. In this case, the impact on range is kept low with short integration time (2–5 ms). On the other hand, the integration could be increased (10 ms) with a stabilized system independent of the platform's maneuvers and vibrations.

To achieve high sensitivity (e.g., below 30 mK) LWIR FPAs with 5 µm pixels require large amounts of integrated charge to be accommodated in a very small unit cell. State of the art storage density of conventional ROIC design rules are about 2.5×10^4 $e^{-}\mu m^{-2}$. It can provide high sensitivity values (say < 30 mK) for most tactical MWIR and LWIR applications utilizing pixel dimensions of 12 µm [62]. For a 5 µm planar unit cell, the charge capacity in standard ROIC technology is less than 1 million electrons, whereas

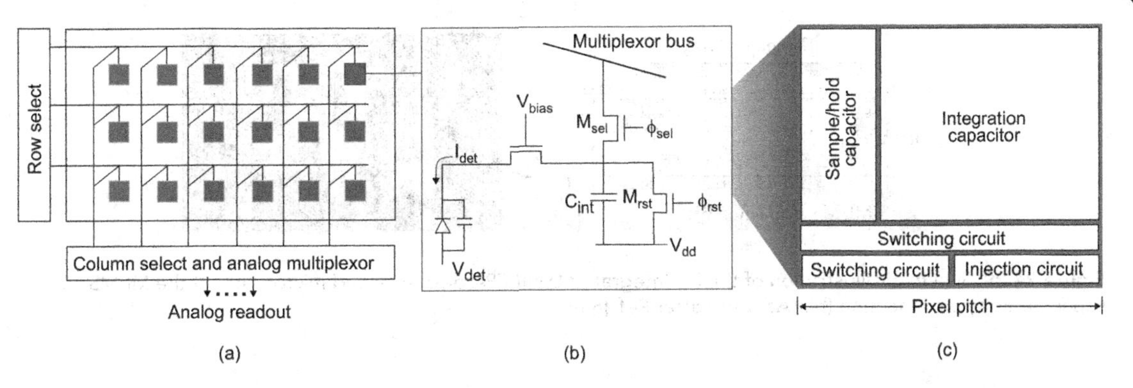

Figure 24.27 Analog ROIC architecture (a) with simplified unit cell or pixel circuit diagram (b) and unit cell layout (c). As shown in (b), photocurrent I_{det} is generated by the photodiode and subsequently integrated onto the capacitance C_{int} through the injection transistor M_i, which also provides the photodiode bias V_{bias}. The signal voltage across C_{int} can be switched onto the multiplexor bus for readout via control signal ϕ_{sel} on switch M_{sel}; the signal voltage across C_{int} can be reset by control signal ϕ_{rst} using switch M_{rst}. A maximum 2.2V process (set by V_{dd}) and C_{int} equal to 1,850 femtofarads results in a maximum stored photocharge equal to 25 million electrons. Note that circuit capacitance as shown in (c) dominates the pixel footprint. (Adapted after Ref. [65])

8–12 million electrons are required for good sensitivity. Therefore, small pitch IR detectors are not available today.

The challenge of charge storage in small pixels is being addressed by fabricating microelectromechanical system (MEMS) capacitors suited to a 3D ROIC design. In recent years, Si-foundries have developed the process technology charge handling capacities for 2 fF μm^{-2} density metal-insulator-metal capacitors (MIM) in a 0.18 μm CMOS platform [66]. Multiple stacking of MIM capacitors results in higher densities of up to ~7 fF μm^{-2}. A redesign of the AIM 640 × 515 15 μm pitch ROIC in 0.18 μm Si-CMOS technology with MIM capacitor technology causes improved NEDT values in comparison with standard ROIC (see Figure 24.28) [67].

Lincoln Laboratory has developed a digital pixel FPA with per-pixel, 16-bit full dynamic range, analog-to-digital conversion, and real-time digital image processing capability [65]. Developed and tested LWIR ROIC MEMS technology (see Figure 24.29) overcomes many performance and scaling limitations imposed by conventional ROIC technology. The MEMS capacitor array can be fabricated in a separate 8" wafer. This technology yielded 20 million electrons in a 5 μm unit cell. This breakthrough will pave the way for small pitch FPAs to operate with very high sensitivity. Figure 24.29b shows a Transmission Electron Micrograph (TEM) picture of a portion of the MEMS capacitor array. Using the HDVIP

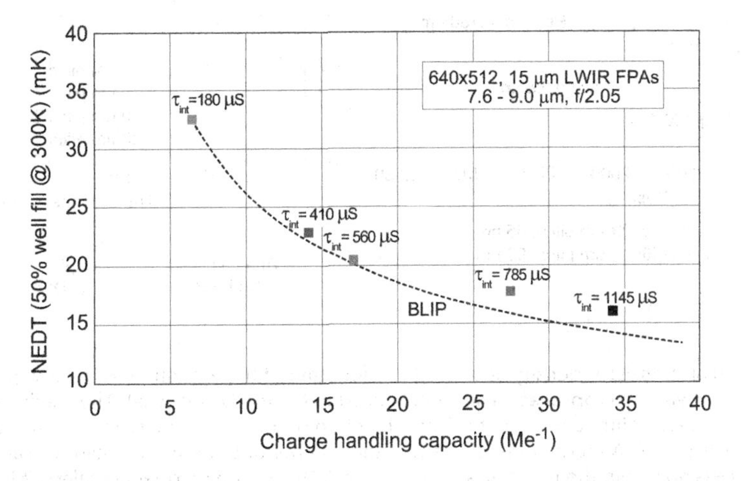

Figure 24.28 Measured NEDT versus charge handling capacity for LWIR 640 × 512 HgCdTe 15 μm pitch modules. (Adapted after Ref. [68])

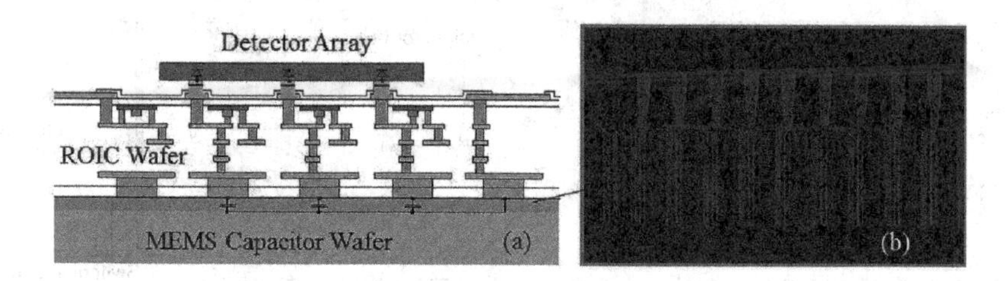

Figure 24.29 Schematic illustration of the 3D integrated LWIR FPA design (a) and micrograph of the MEMS capacitor array cross section (b). (Adapted after Ref. [68])

technology (see Figure 17.46), a fully functional 1,280 × 720, 5 μm unit cell LWIR HgCdTe FPA has been demonstrated [58,68].

To maintain or increase the unit cell's dynamic range it will require employing increasingly deeply scaled, higher-density CMOS processes. Higher-density CMOS fabrication processes can also be used to increase in-unit cell processing capacity. The extrapolated unit-cell transistor count, shown in Figure 24.30, suggests the feasibility of advanced in-pixel signal processing in smaller pixels within the next decade. As predicted in Ref. [65], sophisticated in-unit cell processing, coupled with interpixel data communication and control structures, would enable massively parallel computational imagers and resultant sensor systems with capabilities far beyond what are achievable today.

The ultimate desired ROIC's goal for photon detectors is to achieve BLIP operation at room temperature. It requires that the charge from the full 2π FOV integrated during a frame time should be effectively stored on the detector node, even as the pixel pitch is being decreased. Figure 24.31 illustrates the required well capacity for a 5 μm detector pixel as a function of cutoff wavelength for BLIP operation at room temperature in the diffraction limit. The figure indicates that a significantly larger well capacity per μm² is required to achieve BLIP performance at room temperature compared to the values available with current ROIC designs.

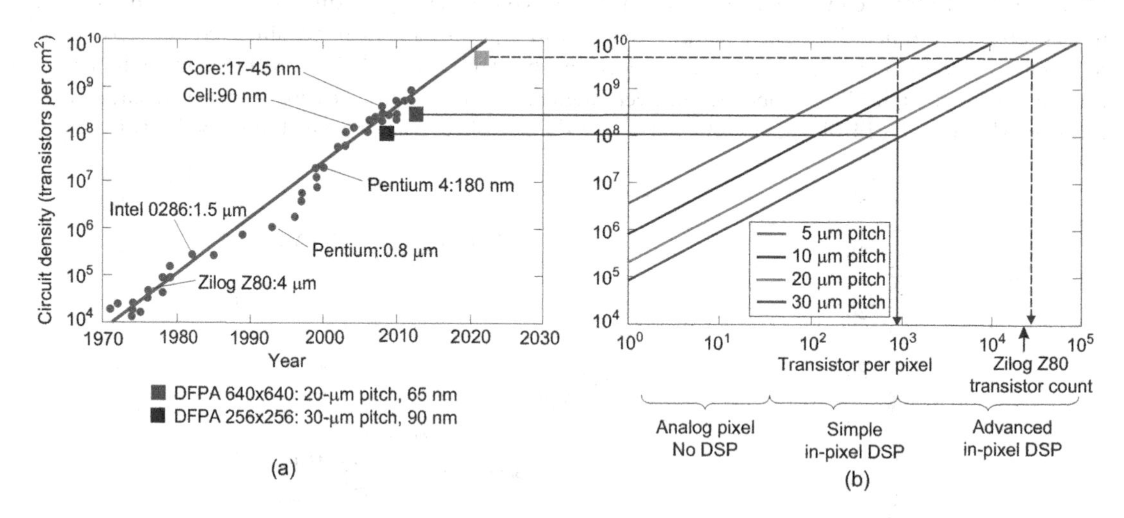

Figure 24.30 The circuit transistor density and trend line for state of the art commercial microprocessors plotted versus the year in which each microprocessor was introduced in the marketplace (a). The maximum number of transistors that can be packed into a pixel unit cell as a function of the circuit transistor density for several pixel sizes between 5 and 30 μm (b). A magnitude estimate of the number of transistors required to achieve three levels of digital processing within the unit cell is also indicated. By leveraging deeply scaled CMOS processes, digital FPA technology enables designers to miniaturize pixel pitch and/or increase on-chip processing capability depending on application-specific needs. (Adapted after Ref. [65])

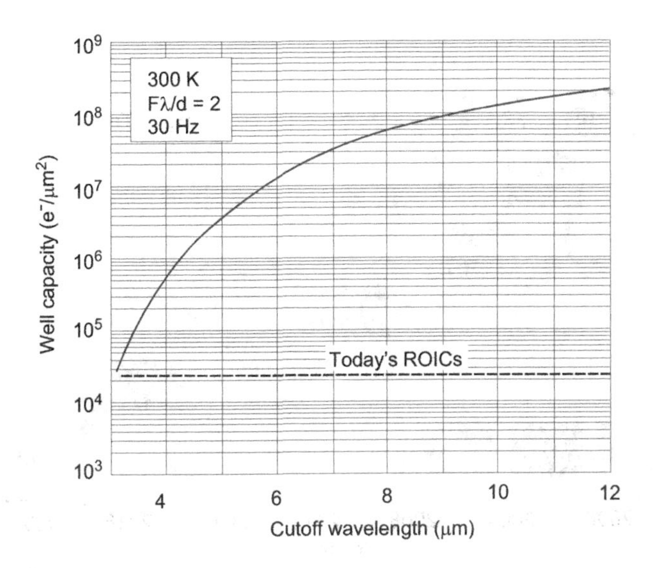

Figure 24.31 Required well capacity as a function of cutoff wavelength for BLIP operation at room temperature in the diffraction limit for a 30 Hz frame rate. (Adapted after Ref. [69])

24.6.1 SWaP CONSIDERATIONS

IR photodetectors are typically operated at cryogenic temperatures to decrease the noise of the detector arising from various mechanisms associated with the narrow bandgap. The cooling technologies are expensive and for many applications are unattractive due to their prohibitive size, weight, and power signature. There are considerable efforts to decrease system size, weight, and power consumption (SWaP)—in consequence reducing system cost—to increase the operating temperature in high operating temperature (HOT) detectors. The ultimate goal is fabrication of detector with the dark current less than the system background flux current and with $1/f$ noise insignificant relative to the shot noise of the background flux. Smaller pixels enhance the value of proposition of the imaging systems and their functionality. An example of their capability is illustrated in Figure 24.32 [50], which shows DRS's production 640 × 480 FPAs with pixel size from 20 to 12 μm. Further improvement in SWaP is achieved in HOT conditions. Smaller pitches are scheduled in the short term (see Figure 24.33a) [70]. A similar tendency has been observed in case of microbolometers (see Figure 24.33b) [71].

The cooling equipment is costly and bulky, and requires a cooling-down time, which pushes the time to first image anywhere to 10 minutes. Increasing the operating temperature of detector reduces the cooling load, allowing more compact engines with higher efficiency. For example, for an operating temperature of

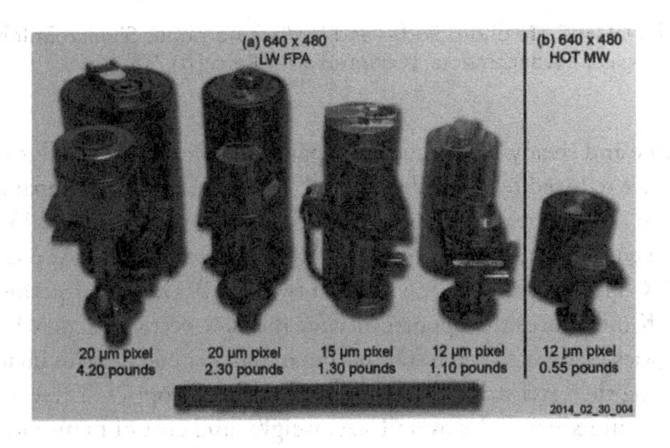

Figure 24.32 Progressive reduction of DRS's 640 × 480 LWIR package as a function of pixel pitch (a) and increasing of operating temperature (b). (Adapted after Ref. [50])

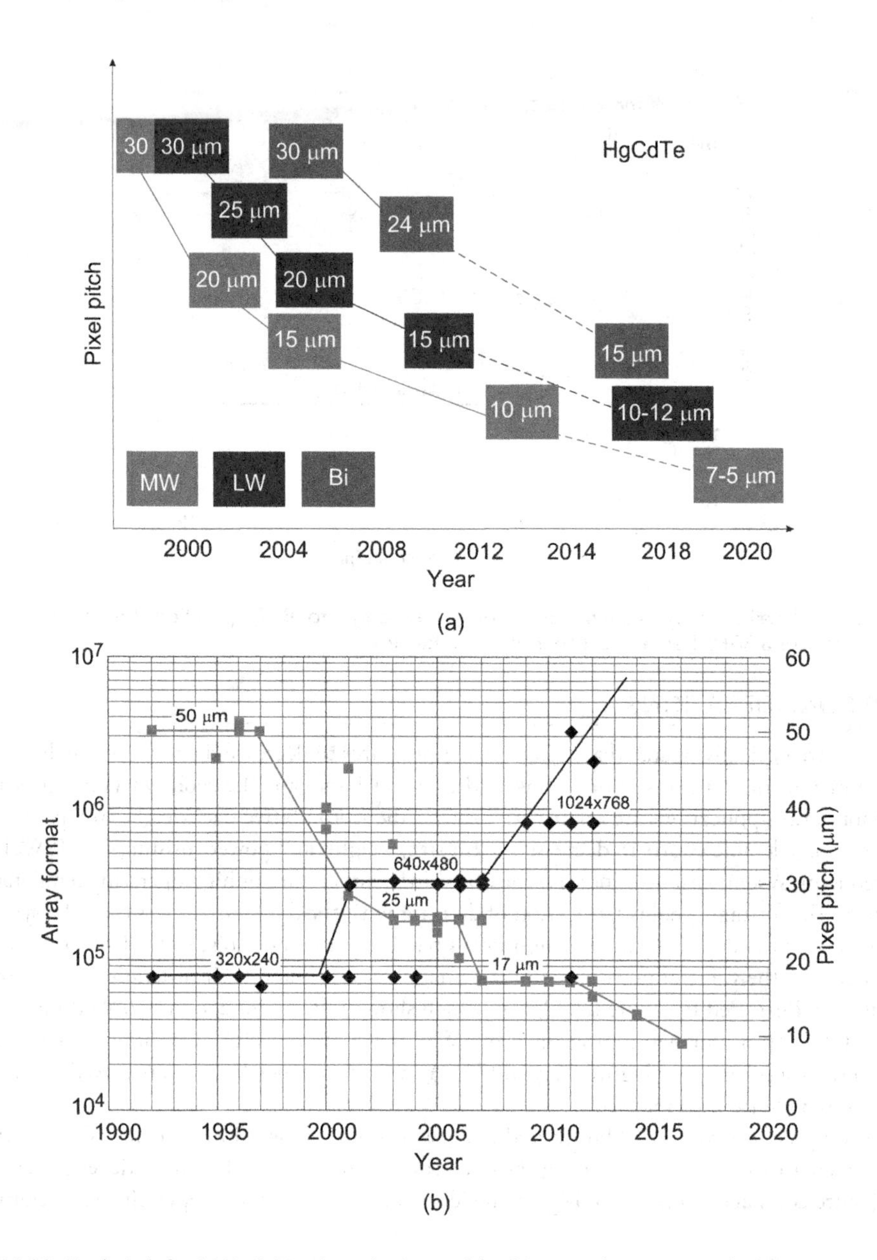

Figure 24.33 Pixel pitch for (a) HgCdTe photodiodes and for (b) amorphous Si microbolometers have continued to decrease due to technological advancements. (Adapted after Refs. [70,71])

150 K, the cool-down time and steady-state power dissipation in the standard Selex Hawk integrated dewar cooler assembly (IDCA) are reduced by around 40% and 55%, respectively, compared with 80 K operation (see Figure 24.34) [72]. To achieve near BLIP performance at temperatures above 150 K, the previously optimized, engine-cooled configuration consumed 1–2 W in the steady state. At present, after improved processing of MWIR HgCdTe photodiode arrays grown by MOCVD, similar performance at temperatures in the range 200 K–220 K introduces the opportunity for thermoelectrically cooled operations.

Because cost of the optics made of Ge, the standard material, depends approximately upon the square of the diameter, so reducing the pixel size causes reducing cost of the optics. These reductions in optics size would have a major benefit in reducing the overall size, weight, and cost of man-portable IR systems. In addition, the reduction in pixel size allows a significantly larger number of FPAs to be fabricated on each wafer.

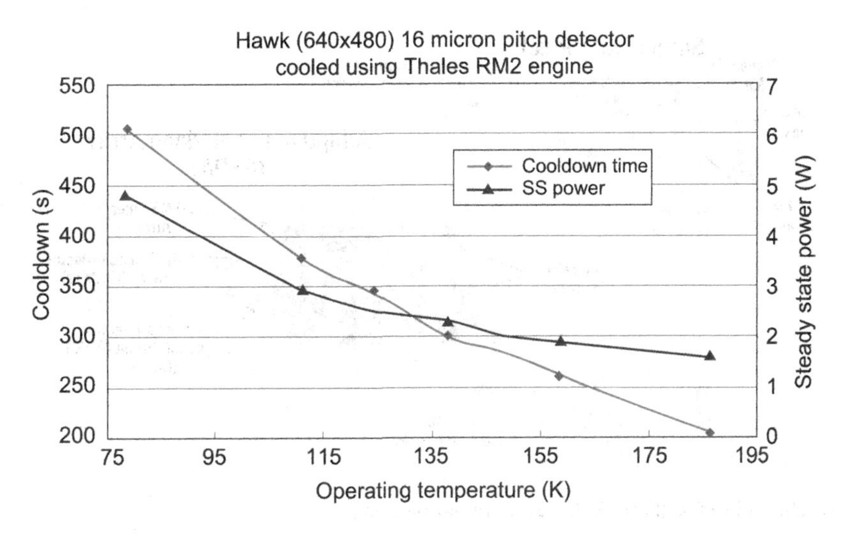

Figure 24.34 HAWK IDCA performance in room temperature ambient (power excludes cooler control electronics losses). (Adapted after Ref. [72])

24.7 ADAPTIVE FOCAL PLANE ARRAYS

A number of recent developments in the area of MEMS-based tunable IR detectors have the potential to deliver voltage-tunable, multiband IR FPAs. These technologies have been developed as part of the DARPA (Defense Advanced Research Projects Agency)-funded adaptive focal plane array (AFPA) program, and have demonstrated multispectral tunable IR HgCdTe detector structures [73–77]. The AFPAs are independently developed by other groups using HgCdTe [78–81] and IV-VI detectors [82].

Figure 24.35 presents a general concept of MEMS-based tunable IR detector. The MEMS filters are individual electrostatically actuated Fabry-Perot tunable filters. In the actual implementation, the MEMS filter array is mounted so that the filters are facing toward the detector to minimize spectral crosstalk. By the use of MEMS fabrication techniques arrays of devices, such as etalons, can be fabricated on an IR detector array that permits tuning of the incident radiation on the detector. If the etalons can be programmed to change distance from the detector surface in the order of IR wavelengths, the detector responds to all the wavelengths in a waveband sequentially.

The integration of various component technologies into an AFPA involves a complex interplay across a broad range of disciplines, involving MEMS device processing, optical coating technology, micro lenses, optical system modeling, and FPA devices. The goal of this integration is to produce an image-sensor array in which the wavelength sensitivity of each pixel can be independently tuned. In effect, the device would constitute a large-format array of electronically programmable micro-spectrometers.

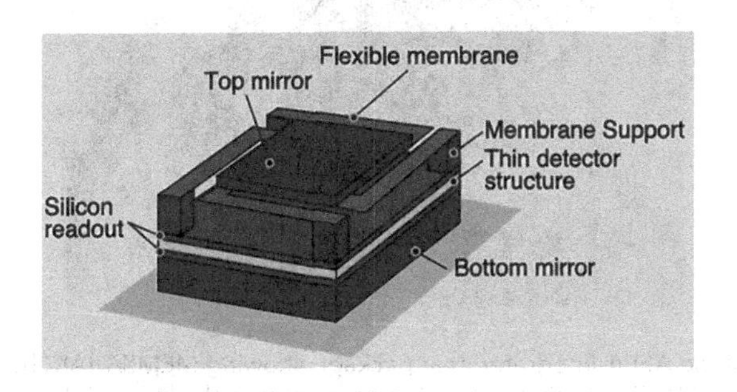

Figure 24.35 General concept of MEMS-based tunable IR detector.

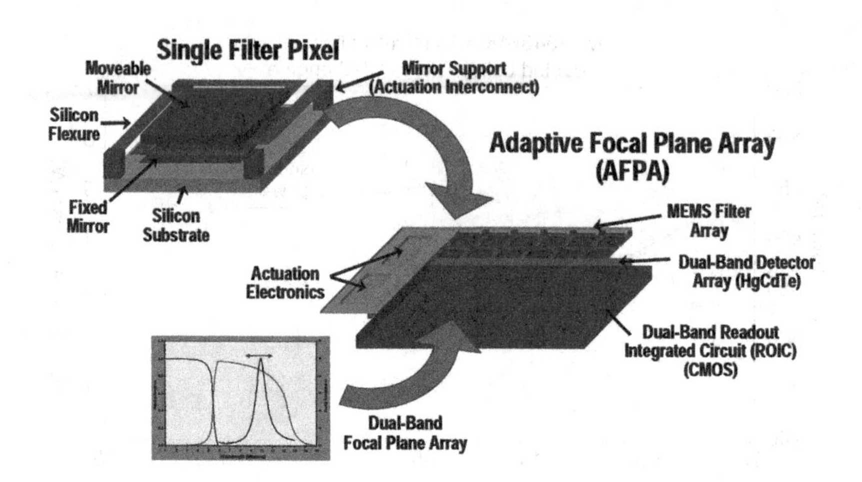

Figure 24.36 Dual-band AFPA. (From Ref. [75]. With permission.)

Teledyne Scientific & Imaging has demonstrated simultaneous spectral tuning in the LWIR region while providing broadband imagery in MWIR band using dual-band AFPA (see Figure 24.36 [75]). The filter characteristics, including LWIR passband bandwidth and tuning range, are determined by the integral thin-film reflector and antireflection coatings. The nominal dimension of each MEMS filter is between 100 and 200 μm on a side and each filter covers a small sub-array of the detector pixels. Employing dual-band FPA with 20 μm pixel pitch results in each MEMS filter covering a detector sub-array ranging from 5 × 5 to 10 × 10 pixels. The MEMS filter array will then evolve to tunable individual pixels.

The device requires a new ROIC to accommodate the additional control functions at each pixel. Therefore, MEMS filters, positioned within 100 μm of the back side of the FPA, have been designed to include a separate MEMS Actuation IC (MAIC) that is hybridized to the MEMS chip (see Figure 24.37 [76]). Connection of the MAIC is made via pin block connector on the side of the device.

Figure 24.38 shows the measured spectral response of a typical filter tuned to various wavelengths in the 8–11 μm band [70]. The LWIR passbands have measured bandwidths of about 100 nm. Data at each MEMS filter wavelength are normalized to the peak value to eliminate the dependence on FPA spectral response. The sideband oscillations were found to be caused by interference arising from residual reflectance of the back side of the FPA and the MEMS filter mirror.

The realization of the AFPA concepts offers the potential for dramatic improvements in critical military missions involving reconnaissance, battlefield surveillance, and precision targeting [73].

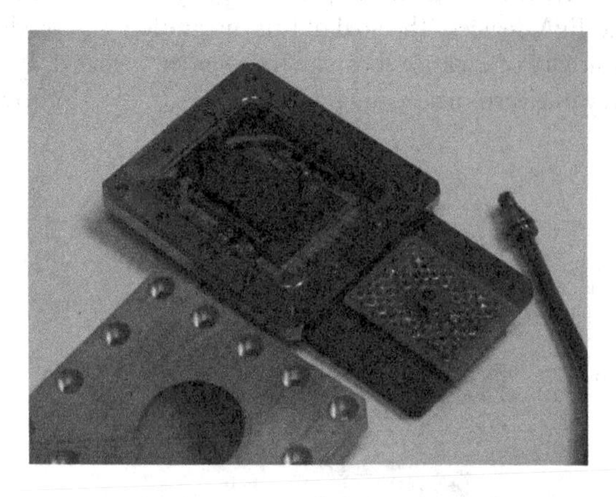

Figure 24.37 Open view of the AFPA-integrated test package showing MEMS/MAIC hybrid mounted above dual-band FPA. Connection to the MAIC is made through the connector at the lower right (with pin short connector in place). (From Ref. [76]. With permission.)

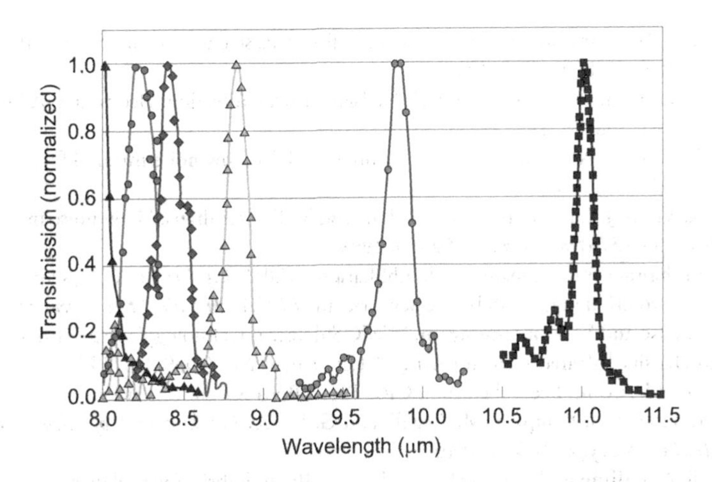

Figure 24.38 Measured normalized transmission of integrated AFPA device showing narrow-band spectral response over the 8–11 μm band. For each measurement, a MEMS filter was tuned to a fixed wavelength and the FPA output was recorded as the wavelength of the narrow-band incident illumination was scanned. (From Ref. [76]. With permission.)

REFERENCES

1. J.T. Caulfield, "Next Generation IR Focal Plane Arrays and Applications," in *Proceedings of the 32nd Applied Imagery Pattern Recognition Workshop* October, IEEE, New York, pp. 7–10, 2003.
2. http://coolcosmos.ipac.caltech.edu/cosmic_classroom/timeline/timeline_onepage.html.
3. H.J. Walker, "Brief history of infrared astronomy," *Astron. Geophys.*, 41, 10–3, 2000.
4. W. Smith, "Effect of light on selenium during the passage of an electric current," *Nature*, 7, 303, 1873.
5. T.W. Case, "Notes on the change of resistance of certain substrates in light," *Phys. Rev.*, 9, 305–10, 1917.
6. R.J. Cushman, "Film-type infrared photoconductors," *Proc. IRE*, 47, 1471–75, 1959.
7. P. Norton, "Detector Focal Plane Array Technology," in *Encyclopedia of Optical Engineering*, ed. R. Driggers, 320–48, Marcel Dekker Inc., New York, 2003.
8. M.A. Kinch, "Fifty years of HgCdTe at Texas instruments and beyond," *Proc. SPIE*, 7298, 72982T, 2009.
9. T. Elliott, "Recollections of MCT work in the UK at Malvern and Southampton," *Proc. SPIE*, 7298, 72982M, 2009.
10. A. Hoffman, "Semiconductor processing technology improves resolution of infrared arrays," *Laser Focus World*, 81–4, February, 2006.
11. H. Sharifi, M. Roebuck, S. Terterian, J. Jenkins, B. Tu, W. Strong, T.J. De Lyon, and R.D. Rajavel, J. Caulfield, and J.P. Curzan, "Advances in III-V bulk and superlattice-based high operating temperature MWIR detector technology," *Proc. SPIE*, 10177, 101770U–1–6, 2017.
12. J.W. Beletic, R. Blank, D. Gulbransen, D. Lee, M. Loose, E.C. Piquette, T. Sprafke, W.E. Tennant, M. Zandian, and J. Zino, "Teledyne imaging sensors: Infrared imaging technologies for astronomy & civil space," *Proc. SPIE*, 7021, 70210H, 2008.
13. A.W. Hoffman, P.L. Love, and J.P. Rosbeck, "Mega-pixel detector arrays: Visible to 28 μm," *Proc. SPIE*, 5167, 194–203, 2004.
14. D.A. Scribner, M.R. Kruer, and J.M. Killiany, "Infrared focal plane array technology," *Proc. IEEE*, 79, 66–85, 1991.
15. J. Janesick, "Charge coupled CMOS and hybrid detector arrays," *Proc. SPIE*, 5167, 1–18, 2003.
16. B. Burke, P. Jorden, and P. Vu, "CCD technology," *Exp. Astron.*, 19, 69–102, 2005.
17. A. Hoffman, M. Loose, and V. Suntharalingam, "CMOS detector technology," *Exp. Astron.*, 19, 111–34, 2005.
18. O. Djazovski, "Focal Plane Arrays for Optical Payloads", in *Optical Payloads for Space Missions*, Chapter 37, ed. S.E. Qian, 793–838, Wiley, Chichester, 2016.
19. W.S. Boyle and G.E. Smith, "Charge-coupled semiconductor devices," *Bell Syst. Tech. J.*, 49, 587–93, 1970.
20. J.L. Vampola, "Readout Electronics for Infrared Sensors," in *The Infrared and Electro-Optical Systems Handbook*, Vol. 3, ed. W.D. Rogatto, 285–342, SPIE Press, Bellingham, WA, 1993.
21. E.R. Fossum, "Active pixel sensors: Are CCD's dinosaurs?" *Proc. SPIE*, 1900, 2–14, 1993.

22. E.R. Fossum and B. Pain, "Infrared readout electronics for space science sensors: State of the art and future directions," *Proc. SPIE*, 2020, 262–85, 1993.
23. M.J. Hewitt, J.L. Vampola, S.H. Black, and C.J. Nielsen, "Infrared readout electronics: A historical perspective," *Proc. SPIE*, 2226, 108–19, 1994.
24. L.J. Kozlowski, J. Montroy, K. Vural, and W.E. Kleinhans, "Ultra-low noise infrared focal plane array status," *Proc. SPIE*, 3436, 162–71, 1998.
25. L.J. Kozlowski, K. Vural, J. Luo, A. Tomasini, T. Liu, and W.E. Kleinhans, "Low-noise infrared and visible focal plane arrays," *Opto Electron. Rev.*, 7, 259–69, 1999.
26. http://www.ipi.uni-hannover.de/uploads/tx_tkpublikationen/2011_GISOSTRAVA_KJ.pdf.
27. T. Hirayama, "The Evolution of CMOS Image Sensors," in *IEEE Asian Solid-State Circuits Conference* 5–8, 2013.
28. A. Hoffman, M. Loose and V. Suntharalingam, "CMOS detector technology", *Exp. Astron.*, 19, 111–134, 2005.
29. R. Thorn, "High Density Infrared Detector Arrays," U.S. Patent No. 4,039,833, 1977.
30. A. Rogalski, *Infrared Detectors*, Second edition, CRC Press, Boca Raton, FL, 2010.
31. J.D. Vincent, S.E. Hodges, J. Vampola, M. Stegall, and G. Pierce, *Fundamentals of Infrared and Visible Detector Operation and Testing*, Wiley, Hoboken, 2016.
32. I.M. Baker and R.A. Ballingall, "Photovoltaic CdHgTe-silicon hybrid focal planes," *Proc. SPIE*, 510, 121–29, 1984.
33. I.M. Baker, "Photovoltaic IR Detectors," in *Narrow-Gap II-VI Compounds for Optoelectronic and Electromagnetic Applications*, ed. P. Capper, 450–73, Chapman & Hall, London, 1997.
34. A. Turner, T. Teherani, J. Ehmke, C. Pettitt, P. Conlon, J. Beck, K. McCormack, et al., "Producibility of VIP™ scanning focal plane arrays," *Proc. SPIE*, 2228, 237–48, 1994.
35. M.A. Kinch, "HDVIP™ FPA technology at DRS," *Proc. SPIE*, 4369, 566–78, 2001.
36. *Seeing Photons: Progress and Limits of Visible and Infared Sensor Arrays*, Committee on Developments in Detector Technologies; National Research Council, 2010, http://www.nap.edu/catalog/12896.html.
37. *Handbook of 3D Integration, Technology and Applications of 3D Integrated Circuits*, Second edition, eds. P. Garrou, C. Bower, and P. Ramm, Wiley-VCH, Weinheim, 2008.
38. S.D. Gunapala, S.V. Bandara, J.K. Liu, J.M. Mumolo, C.J. Hill, D.Z. Ting, E. Kurth, J. Woolaway, P.D. LeVan, and M.Z. Tidrow, "Towards 16 megapixel focal plane arrays," *Proc. SPIE*, 6660, 66600E, 2007.
39. http://pan-starrs.ifa.hawaii.edu/public/design-features/cameras.html.
40. J.W. Beletic, R. Blank, D. Gulbransen, D. Lee, M. Loose, E.C. Piquette, T. Sprafke, W.E. Tennant, M. Zandian, and J. Zino, "Teledyne Imaging Sensors: Infrared imaging technologies for astronomy & civil space," *Proc. SPIE*, 7021, 70210H, 2008.
41. L.J. Kozlowski and W.F. Kosonocky, "Infrared Detector Arrays," in *Handbook of Optics*, Chapter 23, pp. 23.1-23.37 eds. M. Bass, E.W. Van Stryland, D.R. Williams, and W.L. Wolfe, McGraw-Hill, Inc., New York, 1995.
42. J.M. Lloyd, *Thermal Imaging Systems*, Plenum Press, New York, 1975.
43. G.C. Holst, "Infrared Imaging Testing", in *The Infrared and Electro-Optical Systems Handbook*, Vol.4, *Electro-Optical Systems Design, Analysis, and Testing*, pp. 195–243, ed. M.C. Dudzik, SPIE Press, Bellingham, WA, 1993.
44. M.A. Kinch, *State-of-the-Art. Infrared Detector Technology*, SPIE Press, Bellingham, WA, 2014.
45. J.M. Mooney, F.D. Shepherd, W.S. Ewing, and J. Silverman, "Responsivity nonuniformity limited performance of infrared staring cameras," *Opt. Eng.*, 28, 1151–61, 1989.
46. J.M. Lopez-Alonso, "Noise Equivalent Temperature Difference (NETD)," in *Encyclopedia of Optical Engineering*, ed. R. Driggers, 1466–74, Marcel Dekker Inc., New York, 2003.
47. G. Gaussorgues, *La Thermographe Infrarouge*, Lavoisier, Paris, 1984.
48. A.C. Goldberger, S.W. Kennerly, J.W. Little, H.K. Pollehn, T.A. Shafer, C.L. Mears, H.F. Schaake, M. Winn, M. Taylor, and P.N. Uppal, "Comparison of HgCdTe and QWIP dual-band focal plane arrays," *Proc. SPIE*, 4369, 532–46, 2001.
49. G.C. Holst and T.S. Lomheim, *CMOS/CCD Sensors and Camera Systems*, JCD Publishing and SPIE Press, Winter Park and Bellingham, 2007.
50. J. Robinson, M. Kinch, M. Marquis, D. Littlejohn, and K. Jeppson, "Case for small pixels: System perspective and FPA challenge," *Proc. SPIE*, 9100, 91000I–1–10, 2014.
51. R.G. Driggers, R. Vollmerhausen, J.P. Reynolds, J. Fanning, and G.C. Holst, "Infrared detector size: How low should you go?" *Opt. Eng.*, 51(6), 063202–1–6, 2012.
52. G.C. Holst and R.G. Driggers, "Small detectors in infrared system design," *Opt. Eng.*, 51(9), 096401–1–10, 2012.
53. C. Li, G. Skidmore, C. Howard, E. Clarke, and C.J. Han, "Advancement in 17 micron pixel pitch uncooled focal plane arrays," *Proc. SPIE*, 7298, 72980S–1–11, 2009.

54. R.L. Strong, M.A. Kinch, and J.M. Armstrong, "Performance of 12–15 μm-pitch MWIR and LWIR HgCdTe FPAs at elevated temperatures," *J. Electron. Mater.*, 42, 3103–7, 2013.
55. Y. Reibel, N. Pere-Laperne, T. Augey, L. Rubaldo, G. Decaens, M.L. Bourqui, S. Bisotto, O. Gravrand, and G. Destefanis, "Getting small, new 10 μm pixel pitch cooled infrared products," *Proc. SPIE*, 9070, 9070–94, 2014.
56. Y. Reibel, N. Pere-Laperne, L. Rubaldo, T. Augey, G. Decaens, V. Badet, L. Baud, J. Roumegoux, A. Kessler, P. Maillart, N. Ricard, O. Pacaud, and G. Destefanis, "Update on 10 μm pixel pitch MCT-based focal plane array with enhanced functionalities," *Proc. SPIE*, 9451, 9451–82, 2015.
57. R.K. McEven, D. Jeckells, S. Bains, and H. Weller, "Developments in reduced pixel geometries with MOCVD grown MCT arrays," *Proc. SPIE*, 9451, 94512D–1–9, 2015.
58. J.M. Armstrong, M.R. Skokan, M.A. Kinch, and J.D. Luttmer, "HDVIP five micron pitch HgCdTe focal plane arrays," *Proc. SPIE*, 9070, 907033-1–7, 2014.
59. W.E. Tennanat, D.J. Gulbransen, A. Roll, M. Carmody, D. Edwall, A. Julius, P. Dreiske, A. Chen, W. McLevige, S. Freeman, D. Lee, D.E. Cooper, and E. Piquette, "Small-pitch HgCdTe photodetectors," *J. Electron. Mater.*, 43, 3041–6, 2014.
60. R. Bates and K. Kubala, "Direct optimization of LWIR systems for maximized detection range and minimized size and weight," *Proc. SPIE*, 9100, 91000M, 2014.
61. D. Lohrmann, R. Littleton, C. Reese, D. Murphy, and J. Vizgaitis, "Uncooled long-wave infrared small pixel focal plane array and system challenges," *Opt. Eng.*, 52(6), 061305–1–6, 2013.
62. M.A. Kinch, "The rationale for ultra-small pitch IR systems," *Proc. SPIE*, 9070, 907032, 2014.
63. A. Rogalski, P. Martyniuk, and M. Kopytko, "Challenges of small-pixel infrared detectors: A review", *Rep. Prog. Phys.*, 79(4), 046501, 2016.
64. J. Fan and C.S. Tan, *Low Temperature Wafer-Level Metal Thermo-Compression Bonding Technology for 3D Integration*, InTech, Rijeka, 2012.
65. K.I. Schultz, M.W. Kelly, J.J. Baker, M.H. Blackwell, M.G. Brown, C.B. Colonero, C.L. David, B.M. Tyrrell, and J.R. Wey, "Digital-pixel focal plane array technology," *Lincoln Lab. J.*, 20(2), 36–51, 2014.
66. A. Kar-Roy, M. Racanelli, D. Howard, G. Miyagi, M. Bowler, S. Jordan, T. Zhang, W. Kreiger, "Scaling and application of commercial, feature-rich, modular mixed-signal technology platforms for large format ROICs," *Proc. SPIE*, 7660, 76603V, 2010.
67. R. Breiter, H. Figgemeier, H. Lutz, J. Wendler, S. Rutzinger, and T. Schallenberg, "Improved MCT LWIR modules for demanding imaging applications," *Proc. SPIE*, 9451, 945128–1–11, 2015.
68. N.K. Dhar and R. Dat, "Advanced imaging research and development at DARPA," *Proc. SPIE*, 8353, 835302, 2012.
69. M.A. Kinch, "The future of infrared; III–Vs or HgCdTe?" *J. Electron. Mater.* doi:10.1007/s11664-015-3717-5 (2015).
70. G. Destefanis, P. Tribolet, M. Vuillermet, and D.B. Lanfrey, "MCT IR detectors in France," *Proc. SPIE*, 8012, 801235–1–12, 2011.
71. N. Oda, private communication.
72. L. Pillans, R.M. Ash, L. Hipwood, and P. Knowles, "MWIR mercury cadmium telluride detectors for high operating temperatures," *Proc. SPIE*, 8353, 83532W, 2012.
73. J. Carrano, J. Brown, P. Perconti, and K. Barnard, "Tuning In to Detection," SPIE's OEmagazine 20–22, April 2004.
74. W.J. Gunning, J. DeNatale, P. Stupar, R. Borwick, R. Dannenberg, R. Sczupak, and P.O. Pettersson, "Adaptive focal plane array: An example of MEMS, photonics, and electronics integration," *Proc. SPIE*, 5783, 336–75, 2005.
75. W.I. Gunning, J. DeNatale, P. Stupar, R. Borwick, S. Lauxterman, P. Kobrin, and J. Auyeung, "Dual band adaptive focal plane array. An example of the challenge and potential of intelligent integrated microsystems," *Proc. SPIE*, 6232, 62320F, 2006.
76. W. Gunning, S. Lauxtermann, H. Durmas, M. Xu, P. Stupar, R. Borwick, D. Cooper, P. Kobrin, M. Kangas, J. DeNatale, and W. Tennant, "MEMS-based tunable filters for compact IR spectral imaging," *Proc. SPIE*, 7298, 729821, 2009.
77. C.A. Musca, J. Antoszewski, K.J. Winchester, A.J. Keating, T. Nguyen, K.K. M.B. D. Silva, J.M. Dell, et al., "Monolithic integration of an infrared photon detector with a MEMS-based tunable filter," *IEEE Electron Device Lett.*, 26, 888–90, 2005.
78. A.J. Keating, K.K. M.B. D. Silva, J.M. Dell, C.A. Musca, and L. Faraone, "Optical characteristics of Fabry-Perot MEMS filters integrated on tunable short-wave IR detectors," *IEEE Photonics Technol. Lett.*, 18, 1079–81, 2006.

79. J. Antoszewski, K.J. Winchester, T. Nguyen, A.J. Keating, K.K. M.B. Dilusha Silva, C.A. Musca, J.M. Dell, and L. Faraone, "Materials and processes for MEMS-based infrared microspectrometer integrated on HgCdTe detector," *IEEE J. Sel. Top. Quantum Electron.*, 14, 1031–41, 2008.
80. L.P. Schuler, J.S. Milne, J.M. Dell, and L. Faraone, "MEMS-based microspectrometer technologies for NIR and MIR wavelengths," *J. Phys. D*, 42, 13301, 2009.
81. H. Mao, D. Silva, M. Martyniuk, J. Antoszewski, J. Bumgarner, B.D. Nener, J.M. Dell, and L. Faraone, "MEMS-based tunable Fabry-Perot filters for adaptive multispectral thermal imaging," *J. Microelectromech. Syst.*, 25(1), 22735, 2016.
82. H. Zogg, M. Arnold, F. Felder, M. Rahim, C. Ebneter, I. Zasavitskiy, N. Quack, S. Blunier, and J. Dual, "Epitaxial lead chalcogenides on Si got Mid-IR detectors and emitters including cavities," *J. Electron. Mater.*, 37, 1497–503, 2008.

25 Thermal detector focal plane arrays

The use of thermal detectors for IR (infrared) imaging has been the subject of research and development for many decades. These devices achieved their fundamental limits of performance by about 1970. Thermal detectors are not useful for high-speed scanning thermal imagers. However, the speed of thermal detectors is quite adequate for nonscanned imagers with 2D detectors. Figure 25.1 shows the dependence of noise equivalent difference temperature (NEDT) on noise bandwidth for typical detectivities of thermal detectors [1]. The calculations have been carried out assuming $100 \times 100\ \mu m^2$ pixel size, 8–14 μm spectral range, *f*/1 optics, and $t_{op} = 1$ of the IR system. With large arrays of thermal detectors, the best values of NEDT below 0.1 K could be reached because effective noise bandwidths less than 100 Hz can be achieved. This compares with a bandwidth of several hundred kilohertz for conventional cooled thermal imagers with a small photon detector array and scanner. Realization of this fact caused a new revolution in thermal imaging. This is due to the development of 2D electronically scanned arrays in which moderate sensitivity can be compensated by a large number of elements. Large-scale integration (LSI) combined with micromachining has been used for manufacturing large 2D arrays of uncooled IR sensors. This enables fabrication of low-cost and high-quality thermal imagers.

The typical cost of cryogenically cooled imagers of around US$50,000 restricts their installation for critical military applications involving operations in complete darkness. The commercial systems (microbolometer imagers, radiometers, and ferroelectric imagers) are derived from military systems that are too costly for widespread use. Imaging radiometers employ linear thermoelectric (TE) arrays operating in the snapshot mode; they are less costly than the television (TV)-rate imaging radiometers that employ microbolometer arrays [2]. As the volume of production increases, the cost of commercial systems will inevitably decrease. The current market price for a low-cost thermal imager generally costs around US$1,000. Recently, the first thermal imaging smartphone has been launched [3].

The temperature fluctuation noise limit to the performance of FPAs (focal plane arrays) is determined by assuming that all other detector (pixel) and system noise sources are negligible in comparison with temperature fluctuation noise in the detector. By substituting Equation 4.23 into Equation 24.18, the temperature fluctuation noise limited NEDT (i.e., NEDT_t) is given by

$$\mathrm{NEDT}_t = \frac{8F^2 T_d \left(kG_{th}\right)^{1/2} \Delta f^{1/2}}{\varepsilon t_{op} A_d} \left[\int_{\lambda_a}^{\lambda_b} \frac{dM}{dT} d\lambda\right]^{-1}. \tag{25.1}$$

In a similar way, we can determine the background fluctuation noise limit to the NEDT. The NEDT_b is found when radiation exchange is the dominant thermal exchange mechanism. In this case, by substituting Equation 4.24 into Equation 24.18, we can obtain

$$\mathrm{NEDT}_b = \frac{8F^2 \left[2kG\sigma(T_d^5 + T_b^5)_{th}\, \Delta f\right]^{1/2}}{\left(\varepsilon A_d\right)^{1/2} t_{op}} \left[\int_{\lambda_a}^{\lambda_b} \frac{dM}{dT} d\lambda\right]^{-1}. \tag{25.2}$$

The temperature fluctuation noise and background fluctuation noise-limited NEDT of FPAs operating at 300 K and 85 K against a 300 K background, determined from Equations 25.1 and 25.2, are illustrated

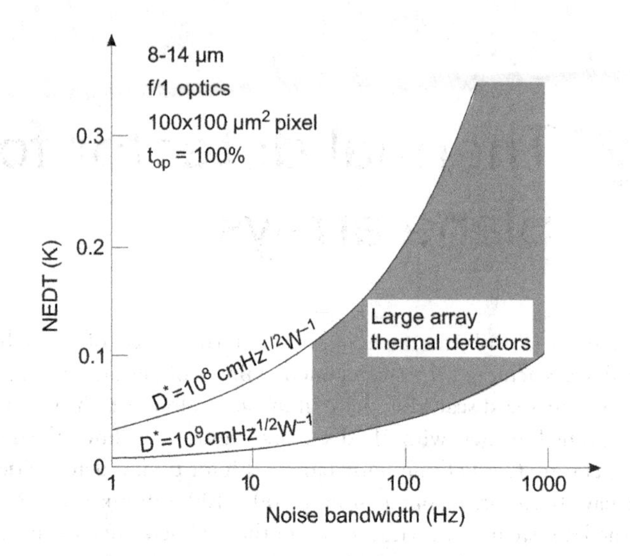

Figure 25.1 The NETD versus equivalent noise bandwidth for typical detectivities of thermal detectors. (From Ref. [1])

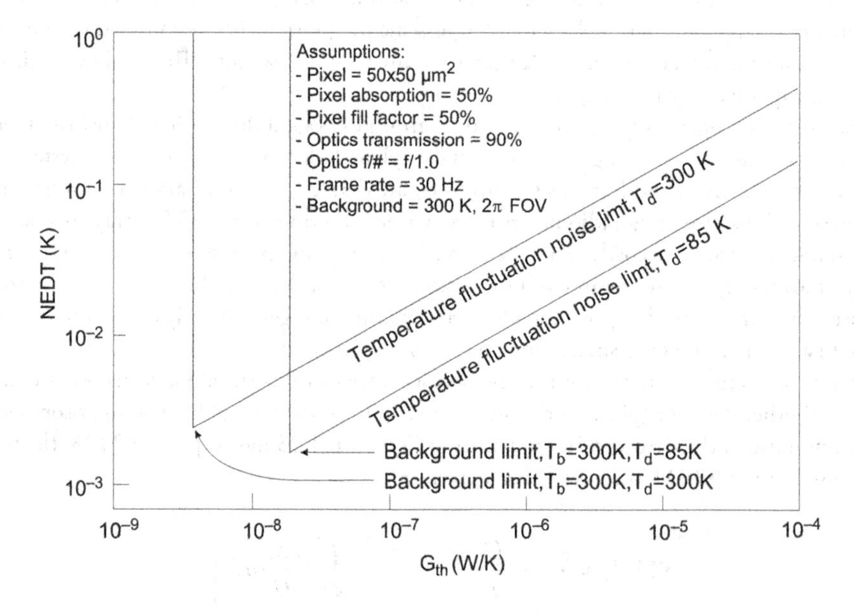

Figure 25.2 Temperature fluctuation noise limit and background fluctuation noise limit to NETD of uncooled and cryogenic thermal detector FPAs as a function of thermal conductance. Other parameters are listed on the figure. (From Ref. [4])

in Figure 25.2 [4]. Other parameters used in calculations are listed in the figure. All thermal IR detectors fall on or above the limits shown in Figure 25.2. Real detectors usually lie above the corresponding lines because of a noise greater than the temperature fluctuation noise.

The key trade-off with respect to uncooled thermal imaging systems is between sensitivity and response time. The thermal conductance is an extremely important parameter since NEDT is proportional to $G_{th}^{1/2}$, but the response time of the detector is inversely proportional to G_{th}. Therefore, a change in thermal conductance due to the improvements in material processing technique improves sensitivity at the expense of time response. Typical calculations of the trade-off between NEDT and time response carried out by Horn and colleagues [5] are shown in Figure 25.3 [6].

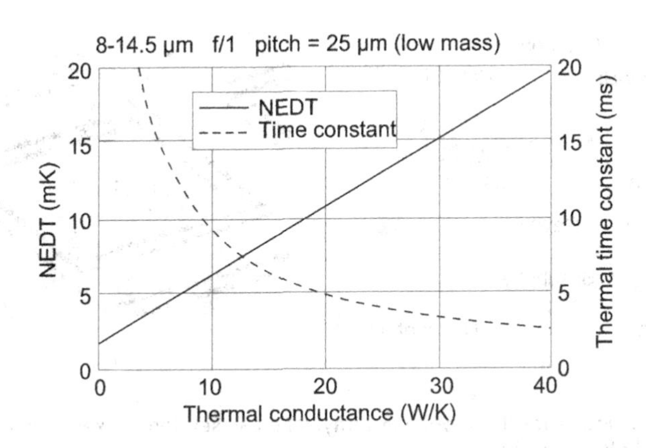

Figure 25.3 Trade-off between sensitivity and response time of uncooled thermal imaging systems. (From Ref. [6])

25.1 THERMOPILE FOCAL PLANE ARRAYS

Thermopiles are widely used, mostly as single-point detectors, in many low-power applications such as radiation temperature sensors and IR gas detectors. With recent improvements in the resolution and sensitivity of thermopile IR FPA, the thermopile detector has been regarded as among the best cost-performance ratio solution for thermal imager development. Compared with the microbolometer, the thermopile has a big advantage in price and is easy to obtain from the market. They have very useful characteristics; they are highly linear, require no optical chopper, however have D^* values lower to bolometers and pyroelectric detectors. They operate over a broad temperature range with little or no temperature stabilization. They have no electrical bias, leading to negligible $1/f$ noise and no voltage pedestal in their output signal. However, array implementations of thermopiles are limited and much less effort has been made in their development. It is mainly owing to the large pixel size required for implementing each thermopile pixel. Because of poor sensitivity the pixel pitch is limited to ~100 μm that especially limits their use for large-format detector arrays. Their responsivity (of the order of 5–15 V/W) and noise are orders of magnitude less and thus their applications in thermal imaging systems require very low-noise electronics to realize their potential performance. TE detectors found almost no use as matrix arrays in TV frame rate imagers. Instead, they are employed as linear arrays that are mechanically scanned to form an image of stationary or nearly stationary objects. The wide operating temperature range, lack of temperature stabilization owing to their inherent differential operation between hot and cold junctions, and radiometric accuracy make thermopiles well situated for same space-based scientific imaging applications [7]. However, the temperature gradient in the thermopile array may cause significant offsets. Therefore, careful array design is required to minimize spatial variation in the array temperature. These limitations prevent the use of thermopiles for IR imagers that require large FPAs, and the attention on uncooled IR detectors has shifted mainly to microbolometers.

In spite of the above limitations, there are some successful FPA implementations merged with readout electronics. As described in Chapter 7, a thermopile consists of two dissimilar conductors connected to each other, measuring the temperature difference between hot and cold contacts from TE power. They usually consist of pairs of moderately doped p-type and n-type polysilicons, which are typical materials in complementary metal oxide semiconductor (CMOS) processes. Figure 25.4 shows an example of a thermopile pixel structure and a SEM (scanning electron microscope) photo of a thermopile pixel. The cavity formed in the substrate reduces the thermal conductance and enhances the responsivity. To thermally isolate the pixels, front-access process is needed.

An interesting 128 × 128 array implementation using post-charge-coupled device (CCD) surface micromachining has been described by Kanno et al. [9]. Each thermopile pixel in the array has 32 pairs of p-polySi/n-polySi thermocouples, each 100 × 100 μm^2, with a fill-factor of 67%. Over the CCD, silicon dioxide diaphragms with thickness of 450 nm (for thermal isolation structure) are made using

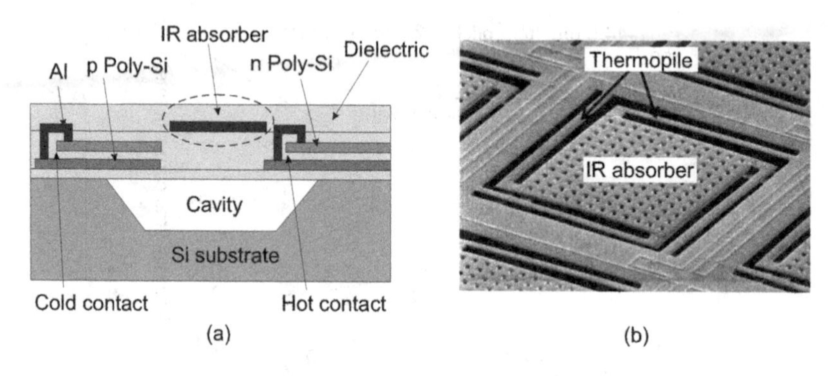

Figure 25.4 Typical pixel structure for thermopile IR array: (a) cross-sectional pixel structure and (b) SEM photo of thermopile pixel. (Adapted after Ref. [8])

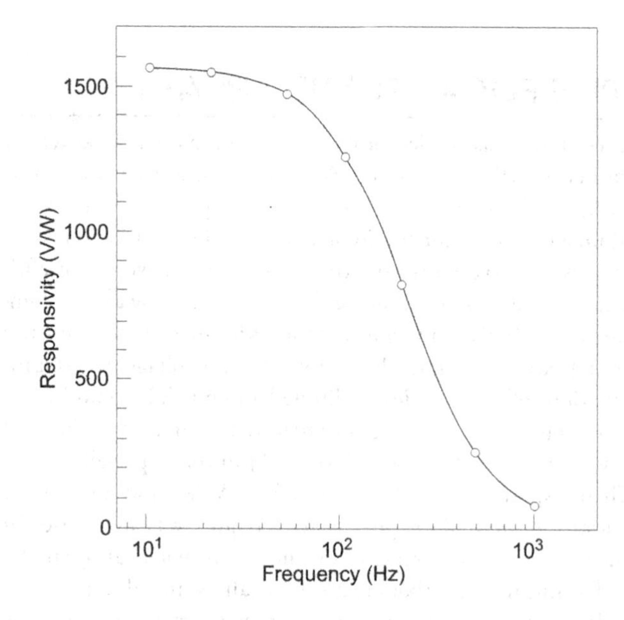

Figure 25.5 Frequency dependence of the thermopile voltage responsivity. (From Ref. [9])

micromachining technology. The polysilicon electrode is 70 nm thick and 0.6 μm wide. The hot junctions are located at the central part of the diaphragm, while the cold junctions are located on the outside edge of the diaphragm where the heat conductance is very large.

The low-frequency voltage responsivity of the 32 pair thermopile is found to be 1550 V/W (see Figure 25.5) [9]. The scanner of 128 × 128 thermopile FPA consists of a vertical and horizontal buried type CCDs. They have overlapping double-layer polysilicon electrodes. The cutoff frequency of 130 Hz was sufficiently large to take a moving object with a 30 or 60 frame/second frame rate. The reported NEDT is 0.5 K with *f*/1 optics. Although these parameters are very good for a thermopile FPA, requirements of vacuum operation packaging increase its cost. Moreover, CCD technology is today a less widespread technology in comparison with CMOS.

Another large-format thermopile FPA suitable for application to various automotive sensor systems has been demonstrated by the Nissan Research Center [10,11]. Each detector in a 120 × 90 element array consists of two pairs of p-n polysilicon thermocouples and has external dimensions of 100 × 100 μm^2 and internal electrical resistance of 90 kΩ. To thermally isolate the detectors, the front-end bulk etching is used. A precisely patterned Au-black absorbing layer and a lift-off technique utilizing a phosphorus silicate glass sacrificial layer provides high responsivity of detectors—about 3900 V/W. The thermopiles are monolithically integrated with a 0.8 μm CMOS process. The measured NEDT is 0.5 K with *f*/1 optics, which is on the same level as the TE FPA with CCD scanner [9].

Also, researchers at the University of Michigan demonstrated a 32 × 32 FPA compatible with an in-house 3 μm CMOS process. The pixel size of 375 × 375 μm^2 with an active area of 300 × 300 μm^2 (fill factor 64%) were implemented with 32 n-p polysilicon thermocouples on dielectric diaphragms [12] using mainly micromachining from the back side of the wafers, but small etch cavities were also placed on the front side of the wafers to achieve heat sink between pixels, to prevent heating of the cold junction and to achieve good thermal isolation between adjacent pixels. The device has a responsivity of 15 V/W, a thermal time constant of 1 ms, and a detectivity of 1.6 × 10^7 cmHz$^{1/2}$/W.

McManus and Mickelson [13] and Kruse [2] have described a family of imaging radiometers that employ Si microstructure linear TE arrays using chromel/constantan thermocouples. One example employs a 120 pixel linear array with a pitch of 50 μm and thermopiles consisting of three series-connected thermocouples. The linear array is mechanically scanned across the focal plane of an *f*/0.7 Ge lens in 1.44 seconds. The NEDT is 0.35 K.

Ann Arbor Sensor Systems, LLC, a small Michigan-based company specializing in noncontact temperature measurement, has teamed up with Malaysia-based MemsTech in developing the first commercial thermal imaging AXT100 camera using thermopile 32 × 32 FPA technology (see Figure 25.6a) [14]. The low production cost is achieved through conventional CMOS processing and nitrogen oxide (NO) vacuum packaging. Some of the camera features include image processing that interpolates and smoothens a 32 × 32 image to 128 × 128 resolution. Two manual focus lens options are available: a 29° (*f*/0.8) or a 22° (*f*/1.0), with a spectral range of 7–14 μm. The camera provides composite video and S-Video outputs (including National Television System Committee (NTSC) and Phase Alternation by Line (PAL) formats).

At present, various IR array sensors based on thermopile technology are adopted by some low-end markets, such as home appliances. Table 25.1 gathers their typical performances. Lapis Semiconductor and Heimann Sensor are selling relatively large-format arrays.

Heimann Sensor has fabricated fully monolithic HTPA thermopile array series with different packages, optics, and sizes up to 84 × 64 elements [15]. The TE materials are n- and p-doped polysilicon, which require a modified CMOS process. The HTPA80 × 64d modules inside a TO8 housing offer high performance dual Ge lens optics as well as low cost uncoated, single Si lenses with a varied field of view. Due to the digital software interface only six pins are needed. The speed can be set internally via the sensor clock and analog-to-digital (AD) converter resolution up to 20 Hz (highest resolution) or up to 200 Hz (lowest resolution). The thermal imager having 80 × 64-element arrays with the corresponding focusing IR optics and electronics is shown in Figure 25.6b.

Requirements for low cost and manufacturability of two-dimensional thermopile arrays led to the use of polysilicon TE material, which have relatively low TE figures of merit. Foote and coworkers [16–18] have improved the performance of thermopile linear arrays by combining Bi-Te and Bi-Sb-Te TE materials. Compared with most other TE arrays, their D^* values are highest, which is shown in Figure 7.5. The thermopile linear array technology is described in Ref. [16]. They consist of 0.5-μm-thick Si_3N_4 membranes

Figure 25.6 Thermopile imaging cameras fabricated by (a) Ann Arbor Sensor Systems (AXT100) and (b) Heimann Sensor GmbH (HTPA 80 × 64d L10/0.7 HiA Thermopile Array).

Table 25.1 Thermopile focal plane arrays

COMPANY	LAPIS	HEIMANN	EXCELITAS	MELEXIS
Array format	48 × 47	80 × 64	32 × 32	32 × 24
Pixel pitch (μm)	100	90	220	100
NRDT (K)	0.5	0.1	0.8	0.1
Field of view (deg)	N.A.	41 × 33	60	110
Frame rate/Time constant	6 Hz	30 Hz	115 ms	64 Hz
Packaging	Vacuum	Vacuum	Atmospheric pressure (N_2)	Atmospheric pressure

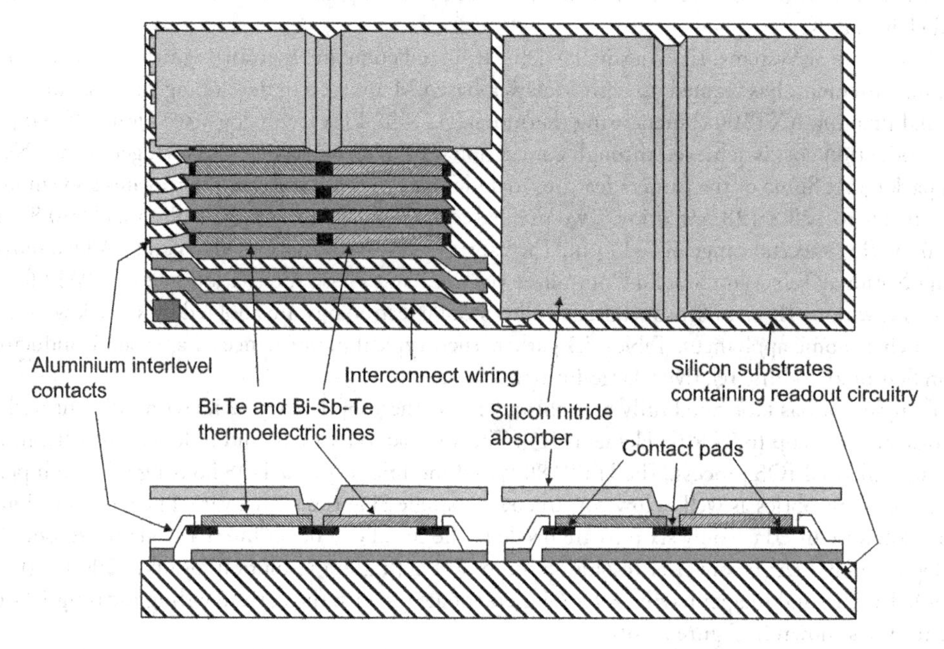

Figure 25.7 Schematic diagram of thermopile detector structure. The top diagram shows two pixels viewed from the top, with part of the left pixel cut away to show the underlying structure. The lower diagram shows a cross section side view of two pixels. (From Ref. [18])

formed by back side etching of the underlying Si substrate. On each membrane there are a number of Bi-Te and Bi-Sb-Te thermocouple running along narrow legs between the substrate and membrane. The detectors are closely spaced with slits through the membrane separating the detectors from each other and defining the detector legs.

The linear thermopile arrays have been bonded to separate CMOS readout electronic chips since Bi-Sb-Te materials are not readily available in a CMOS technology. Next, this technology has been developed to improve the performance of 2D arrays using a three-level structure with two sacrificial layers. In such a way, it is possible to improve the fill factor and incorporate a large number of thermocouple per pixel. Figure 25.7 shows the thermopile detector structure [18]. The structure allows almost 100% fill factor (FF) and the model suggests that optimized detectors will have D^* values over 10^9 cmHz$^{1/2}$/W.

25.2 BOLOMETER FOCAL PLANE ARRAYS

Today's achievements in uncooled IR technology were initiated by the U.S. Department of Defense in the 1980s, when it gave large classified contracts to both Honeywell and Texas Instruments (TI) to develop two different uncooled IR technologies [2,19,20]. TI concentrated on pyroelectric technology (such as barium strontium titinate (BST)), whereas Honeywell concentrated on microbolometer technology, and they both successfully developed uncooled IR 320 × 240 format FPAs with sensitivities less than 50 mK. These technologies were unclassified in 1992, and since then many other companies have started working on this technology. Honeywell has licensed this technology to several companies for the development and production of uncooled FPAs for commercial and military systems. The U.S. government allowed American manufactures to sell their devices to foreign countries, but not to divulge manufacturing technologies. Several countries, including the United Kingdom, France, Japan, and South Korea have picked up the ball, determined to develop their own uncooled imaging systems. In the mid-1990s, the amorphous Si (a-Si) technology was developed in other countries, especially in France. During this time, the big advantage of using a-Si was their fabrication in a Si foundry. The VO_x technology was controlled by the U.S. military, and export license was required for microbolometer cameras that were sold outside the U.S. Today, VO_x bolometers can be also produced in a Si foundry and both the above reasons disappeared. Although the U.S. has a significant lead, some of the most exciting and promising developments for low-cost uncooled IR systems may come from non-U.S. companies (e.g., microbolometer FPAs with series p-n junction elaborated by Mitsubishi Electric [21]). This approach is unique, based on an all-Si version of microbolometer.

At present, the most important manufacturers of uncooled microbolometer FPAs are: Raytheon [22–29], BAE (formerly Honeywell) [30–34], DRS (formerly Boeing) [35–38], Indigo [39,40], and L-3 Communications Infrared Products [41,42] in the United States; INO in Canada [43–46]; ULIS in France [47–52]; NEC [53–56] and Mitsubishi [21,57–61] in Japan; QuinetiQ [62] in the United Kingdom; XenICs [63] in Belgium; SCD [64–66] in Israel, DALI Technology in China [67], and Fraunhofer Institute of Microelectronic Circuits and Systems in Germany [68]. Also, many research institutions are working on uncooled microbolometer IR arrays.

At present, LSI combined with micromachining has been used for manufacturing large 2D arrays of uncooled IR sensors. This enables fabrication of low-cost and high-quality thermal imagers. The development of the microbridge detector arrays has provided a significant leap forward in sensitivity and array size for uncooled thermal imagers. The sensitivity is not as good as the cooled photon detectors, however it is sufficient for low cost, lightweight, low power IR imagers. Today, 1,024 × 768 element arrays with 17 μm pixel size are available with the predicted NEDT less than 50 mK. Although developed for military applications, low-cost IR imagers are used in nonmilitary applications such as driver's aid, aircraft aid, industrial process monitoring, community services, firefighting, portable mine detection, night vision, border surveillance, law enforcement, search and rescue, etc.

At present, VO_x microbolometer arrays are clearly the most used technology for uncooled detectors (see Figure 25.8). VO_x is winner of the battle between the technologies, and VO_x detectors are being produced at a lower cost than any of the three other technologies: a-Si, BST, and Si diodes [69,70]. However, VO_x is challenged by a-Si material and new Si-based materials introduced by new market entrants, thanks to their cost structure and easier manufacturability.

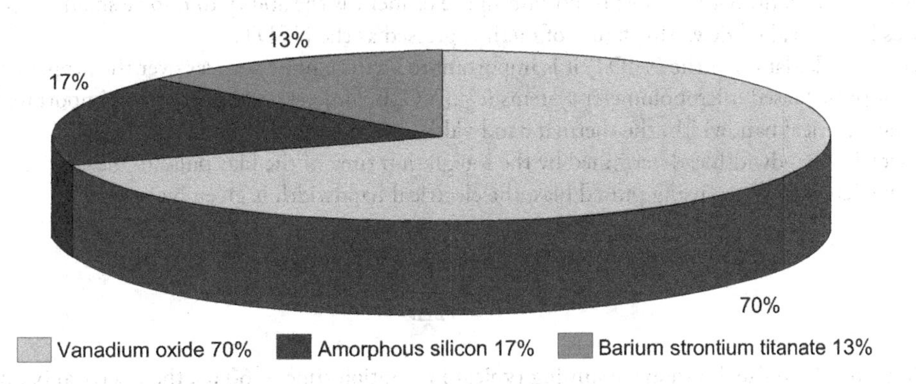

Figure 25.8 Estimated market shares for VO_x, a-Si and BST detectors. (Adapted after Ref. [69])

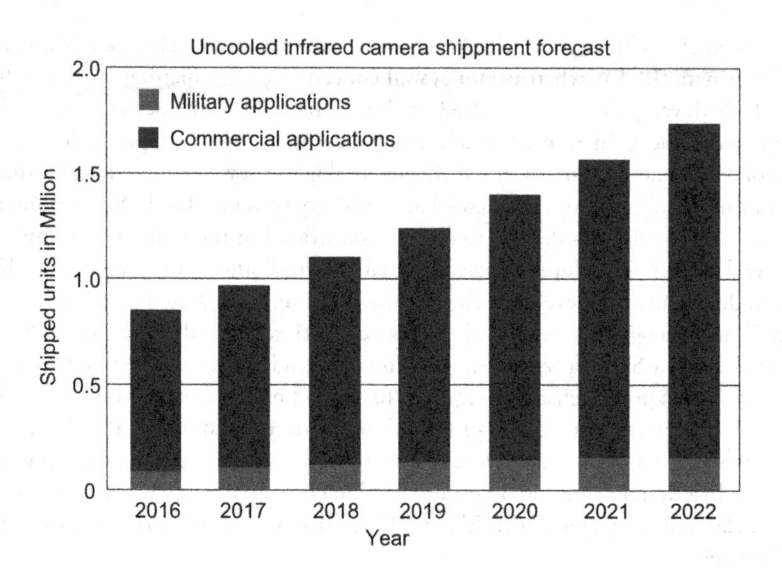

Figure 25.9 Uncooled thermal camera shipment forecast. (Adapted after Ref. [72])

Currently, the microbolometer detectors are produced in larger volumes than all the other IR array technologies together. Their cost will be drastically dropped (about 15% per year). It is expected that commercial applications in surveillance, automotive, and thermography will reach total volumes more than 1 million units in 2018 (see Figure 25.9) [71,72]. The present thermography boom is confirmed with camera prices now available for nearly $1,000 from FLIR that expand the use of IR cameras for maintenance engineers and building inspectors. Surveillance is becoming a key market with closed-circuit television (CCTV) big camera players introducing many new models of thermal cameras. In addition, it was expected that automotive will exceed 500,000 units sales in 2017.

The uncooled IR camera market is showing an 8% CAGR (compound annual growth rate) between 2016 and 2022, reaching almost US$4.4 billion at the end of the period. In 2016, two leading companies, FLIR and ULIS, owned more than 75% of the total market (in volume). Military uncooled camera markets are mainly driven by the huge U.S. Military demand for soldiers (weapon sight, portable goggles, and vehicle vision enhancement). It takes more than 85% of the world market, with a strong presence of DRS and BAE for various applications.

25.2.1 TRADE-OFF BETWEEN SENSITIVITY, RESPONSE TIME, AND DETECTOR SIZE

As mentioned in Section 24.5, when the detectors are integrated into arrays, the ability to have high detectivity is important, but the most important figure of merit is the ability to resolve small temperature differences in the field of view. This figure of merit is pressed as the NEDT.

For correct calculation of the NEDT, it is important to analyze noise sources over the proper bandwidths. For pulse biased microbolometer systems (e.g., VO_x bolometers), there are three important bandwidths: the electrical bandwidth, the thermal bandwidth, and the output bandwidth [2,73].

The electrical bandwidth is determined by the integration time of the bias pulse to measure the resistance of the detector. When using pulsed bias, the electrical bandwidth is given by [2]:

$$\Delta f = \frac{1}{2\Delta t}, \tag{25.3}$$

where Δt is the bias pulse duration. Assuming typical integration time of 60 μs, the electrical bandwidth is 8 kHz.

The electrical bandwidth is important for analyzing the system contributions of 1/*f* and Johnson noise. Often, for large FPAs that are readout in a serial manner by pulsed bias, the bandwidth can be sufficiently large so that the Johnson noise is much greater than 1/*f* noise over that bandwidth.

The thermal bandwidth is determined by the thermal time constant of the bolometer and is important for analyzing thermal fluctuation noise. Assuming the bolometer as a first order low pass filter, the thermal bandwidth is given by:

$$\Delta f = \frac{1}{4\tau_{th}}, \quad (25.4)$$

where τ_{th} is thermal time constant. For typical time constants between 5 and 20 ms, the thermal bandwidth changes between 12 and 50 Hz.

The output bandwidth is the bandwidth at which the bolometer is pulsed. Assuming the frame rate as 30 or 60 Hz, the output bandwidth given by:

$$\Delta f = \frac{\text{Frame rate}}{2}, \quad (25.5)$$

is 15–30 Hz.

From the above analysis it results that the electrical bandwidth is much greater than the thermal or output bandwidth.

It can be shown that the contributions of different types of noise to NEDT follows as [73]:

- Johnson noise

$$\text{NEDT}_{\text{Johnson}} \propto \frac{G_{th}\left(TR_B\right)^{1/2}}{V_b\alpha}, \quad (25.6)$$

- thermal fluctuation noise

$$\text{NEDT}_{tf} \propto TG_{th}^{1/2}, \quad (25.7)$$

- 1/*f* noise

$$\text{NEDT}_{1/f} \propto \frac{\beta G_{th}}{\alpha}, \quad (25.8)$$

where R_B is the bolometer resistance, β is ratio of the measured 1/*f* noise voltage, $V_{1/f}$, to the detector bias voltage, V_b.

Under the 1/*f* noise model, the noise voltage would be inversely proportional to the square root of the number of carriers, N, [74] and the volume. Therefore,

$$\beta = \frac{V_{1/f}}{V_b} \propto \frac{1}{N^{1/2}} \propto \frac{1}{\left(\text{Volume}\right)^{1/2}} \propto \frac{1}{A^{1/2}}. \quad (25.9)$$

Figure 25.10 shows the 1/*f* noise contribution to noise as a function of the surface area of a VO_x film fabricated by BAE Systems [73]. This dependence follows the relation described by Equation 25.9. Unfortunately, this dependence of 1/*f* noise predicts that smaller pixel bolometers will have higher 1/*f* noise than larger pixels.

Figure 25.11 presents the influence of different detector and electronic noise sources on VO_x BAE Systems microbolometer performance. The key conclusion that can be drawn from this figure is that current microbolometer performance is limited by 1/*f* noise in the VO_x material. Large 1/*f* noise is due to noncrystalline VO_x structure. As Equation 25.6 indicates, the Johnson noise contribution to NEDT

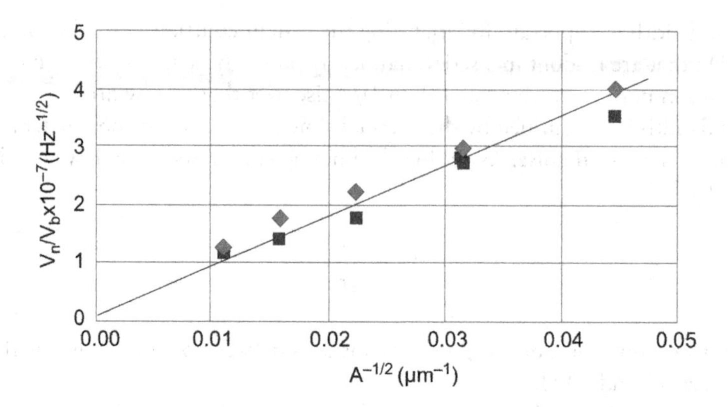

Figure 25.10 Dependence of the V_n/V_b ratio on the square root inversion of the surface area of the VO_x film. (From Ref. [73])

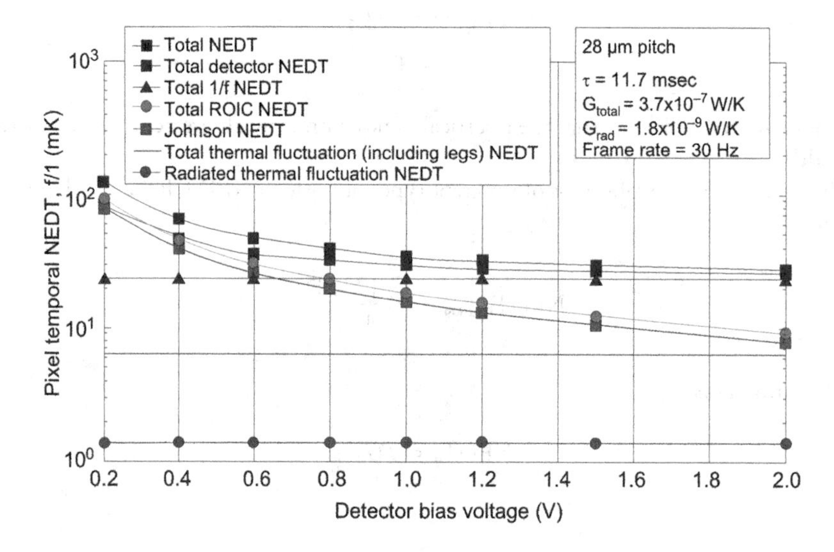

Figure 25.11 Influence of different noise sources on the performance of VO_x microbolometer. (From Ref. [73])

is inversely proportional to the bias voltage, V_b. This type of noise does not significantly degrade system performance when the bolometer is highly biased. Both ROIC (readout integrated circuit) and Johnson noise approach the thermal fluctuation noise in a high enough bias region. The thermal fluctuation noise is dominated by thermal conductance of the detector legs, not by radiated conductance.

In order to improve the performance of VO_x microbolometers, the $1/f$ noise must be reduced. From Equation 25.8 shows that it can be achieved by:

- reducing the $1/f$ noise of the detector material,
- reducing the thermal conductance of the bolometer legs, and
- increasing the α of the VO_x.

It appears that the last approach has a negative effect on the dynamic range requirements for ROIC [73].

Various methods for improving microbolometer sensitivity are pointed out in Table 25.2. As shown, many methods of improving the sensitivity negatively affect the thermal time constant.

Assuming *f*/1 optics and 100% transmission optics at direct current (DC) condition, NEDT can be estimated as [75]:

$$\text{NEDT} = \frac{4G_{th}V_n}{\text{TCR}\cdot\text{FF}\cdot\varepsilon\cdot R_B\cdot I_b\cdot A\cdot\left(\Delta P/\Delta T\right)}. \tag{25.10}$$

Table 25.2 Methods for improving sensitivity of a microbolometer system

DESIGN/PROCESS MODIFICATION	IMPACT ON NEDT	IMPACT ON THERMAL TIME CONSTANT	IMPACT ON SYSTEM SIZE	COMMENTS
Increase VO_x volume	Reduce	Increase		Pixel resistance must be high enough so VO_x resistance dominates total pixel resistance; increasing length does not negatively affect resistance
Reduce $1/f$ noise inherent in material	Reduce			How?
Increase VO_x TCR	Reduce?			Not known whether higher TCR material will have equivalent or lower $1/f$ noise
Reduce leg thermal conductance	Reduce	Increase		Pixel resistance must be high enough so VO_x resistance dominates total pixel resistance
Reduce bridge heat capacitance	Increase	Reduce		
Increase leg thermal conductance	Increase	Reduce		
Reduce *f*/#	Reduce		Increase	
Reduce pixel pitch	Increase		Reduce	Essential for smaller, cheaper systems

Source: Ref. [73].

This expression includes the detector noise, V_n, the detector electrical resistance, R_B, the detector bias current, I_b, the thermal coefficient of resistance (TCR), the detector FF, the absorption rate, ε, and the differential radiation falling on the detector, $\Delta P/\Delta T$ (also called temperature contrast).

The current sensitivity (in A/W) of bolometer is shown by the expression:

$$R_i = \frac{\mathrm{FF}\cdot\varepsilon\cdot\mathrm{TCR}\cdot V_b}{G_{th}\cdot R_B}. \tag{25.11}$$

Equation 25.10 can be manipulated to include $\tau_{th} = C_{th}/G_{th}$:

$$\mathrm{NEDT} = \frac{4C_{th}V_n}{\tau_{th}\cdot\mathrm{TCR}\cdot\mathrm{FF}\cdot\varepsilon\cdot V_b\cdot A\cdot(\Delta P/\Delta T)}. \tag{25.12}$$

If the NEDT is dominated by a noise source that is proportional to G_{th}, which takes place when Johnson and $1/f$ noises are dominant, then the figure of merit (FOM) given by:

$$\mathrm{FOM} = \mathrm{NEDT}\times\tau_{th} = \frac{4C_{th}V_n}{\mathrm{TCR}\cdot\mathrm{FF}\cdot\varepsilon\cdot V_b\cdot A\cdot(\Delta P/\Delta T)} \tag{25.13}$$

can be introduced [73]. Users are interested not only in the sensitivity but also in their thermal time constants and the *FOM* described by Equation 25.13 recognizes the trade-off between thermal time constant

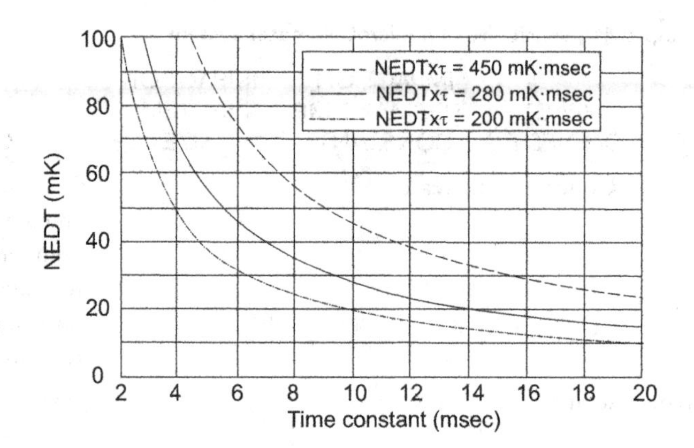

Figure 25.12 Calculated microbolometer NEDT and thermal time constant, τ_{th}, for three NEDT × τ_{th} products. (From Ref. [73])

and sensitivity. Figure 25.12 shows the dependence of NEDT on thermal time constant for three NEDT × τ_{th} products.

From Equation 25.13 shows, however, that the more-encompassing metric for tracking performance would involve the NEDT, τ_{th}, and some indicator of detector size. Such metric has been proposed recently by Skidomre, Han and Li [76] as

$$\mathrm{FOM} = \mathrm{NEDT} \times \tau_{th} \times A = \frac{4C_{th}V_n}{\mathrm{TCR} \cdot \mathrm{FF} \cdot \varepsilon \cdot V_b \cdot \left(\Delta P/\Delta T\right)}. \tag{25.14}$$

Table 25.3 shows the bolometer rank including three ways for FOM: NEDT, NEDT × τ_{th}, and NEDT × τ_{th} × A gathered from Ref. [76]. The experimental data of bolometers are taken from different manufactures. L-3 Communications and BAE report only NEDT × τ_{th} product (not independent NEDT and τ_{th}).

Considering NEDT value, at the top of the ranking is a resonant frequency bolometer from Penn State University [77]. This bolometer is a 500 μm pitch quartz bolometer with NEDT of 3.8 mK and τ_{th} of 5.9 ms built in a 3 × 3 format. The expression for NEDT for this type of bolometer is not technically accurate, but the NEDT metric is still applicable.

The demonstrated performance is getting closer to the theoretical limit with the advantages regarding weight, power consumption, and cost. Because the cost of the optics made of Ge, the standard material, depends approximately upon the square of the diameter, reducing the pixel size allows the size and cost of the optics to be reduced. However, the NEDT is inversely proportional to the pixel area, thus if the pixel size reduces from 50 × 50-μm to 12 × 12-μm, and everything else remains the same, the NEDT would increase by a factor of 16. Improvements in the readout electronics are needed to compensate for this.

To reach smaller pixels for uncooled IR FPA, further improvements in detector technology are necessary, which presents significant challenges in both fabrication process improvements and in pixel design. At the present stage of technology, the detector FF and the absorption coefficient are close to their ideal values and only a little benefit can be expected from the optimization of these two parameters. More gain can be obtained through improvement of the thermistor material, its TCR and R. A promising approach is the development of lower resistance a-Si/a-SiGe thin films [78]. The TCR of Si alloy has been increased to ~3.9%/K from a baseline of 3.2%/K without an increase in material 1/f-noise. Also, properties of the Si/SiGe single crystalline quantum well as a thermistor material are promising [79].

25.2.2 MANUFACTURING TECHNIQUES

Most modern microbolometer FPA technology derives from the pioneering efforts of a team under the direction of R. A. Wood at the Honeywell Technology Center that began in 1982 [2]. In 1985, Honeywell Technology Center received military contracts from U.S. Department of Defense, especially Defense

Table 25.3 Microbolometer rank through 2013

ORGANIZATION	NEDT (mK)	τ_{th} (ms)	PITCH (μm)	NEDT × τ_{th} (mK × ms)	NEDT × τ_{th} × A/1000
SCD	20	12	17	240	69
BAE			17	<350	<101
L-3			17	<350	<101
Ulis	45	9	17	405	117
NEC	63	14.5	12	914	132
DRS	<50	13	17	<650	<188
Mitsubishi	84	12	15	1,008	227
Toshiba	40	16	22	640	310
Fraunhofer	100	15	25	1,500	938
Univ. Elec. Sci. Tech. China	81	13.5	35	1,094	1,340
Penn State	3.8	5.9	500	22	5,605

Source: Ref. [76].

Advanced Research Projects Agency (DARPA) and U.S. Army Night Vision and Electronic Sensors Directorate (NVESD). These contracts led to the successful development of an uncooled vanadium oxide 50 μm pixel 240 × 336 arrays operating at the U.S. TV frame rate of 30 Hz.

In the period 1990–1994, Honeywell licensed this technology originally to four companies (Hughes, Amber, Rockwell, and Loral) for the development and production of uncooled FPAs for commercial and military systems. This was followed by acquisitions and merges with defense and aerospace industries there are now: British Aerospace (the original Honeywell division) and Raytheon. A great deal of activity on VO_x bolometer arrays by different manufacturers modified the Honeywell microbolometer support structure to increase the FF, decrease the size of pixels, and improve the CMOS readout.

The first 240 × 336 arrays of VO_x, 50 μm microbolometers were fabricated on industry-standard wafer (4-inch diameter) complete with monolithic readout circuits integrated into underlying Si (see Section 8.3) [80]. To obtain high thermal isolation of the microbolometer, the ambient gas pressure is typically of the order of 0.01 mbar. Thermal conduction through the bolometer legs can be as low as 3.5×10^{-8} W/K [73]. The bolometers in principle do not need to be thermally stabilized. However, to simplify the problem of pixel nonuniformity correction, the original Honeywell bolometer array incorporated a TE temperature stabilizer. Another issue was the need to readout the signal by accessing the pixels sequentially. The method chosen was by pulsing the electrical bias to the pixels sequentially. A bipolar input amplifier was normally required, and this was obtained with biCMOS technology. Horizontal and vertical pixel addressing circuitry was integrated with the array but most of the analog readout circuitry was off-chip. The dominant noise was Johnson noise in the sensitive resistor (typically 10–20 kΩ) with some additional contribution from $1/f$ noise and transistor readout noise. In operation, an array consumed about 40 mW [80,81]. An average NEDT of better than 0.05 K was demonstrated with uncooled imager fitted with as $f/1$ optics (see Figure 25.13).

Today, most of the approaches employ CMOS Si circuitry for which the power dissipation is much less than that of the bipolar one. Moreover, most of the readout electronics has been moved onto the chip where it is referred to as the ROIC. Column parallel readout architectures with integrated AD conversion are commonly used in commercial FPAs [73,82]. The basic construction of the microbolometer's ROIC is the same as those for visible CMOS image sensors (see Figure 25.14). A row of pixels is electrically activated by the row selection signal from the row multiplexer, and each column integrator (amplifier) integrates a signal from a selected pixel during the horizontal scanning period. After that, the integrated signals are transferred to the row storage capacitors. The signals at the row storage capacitors are serially read out by operating the output multiplexer.

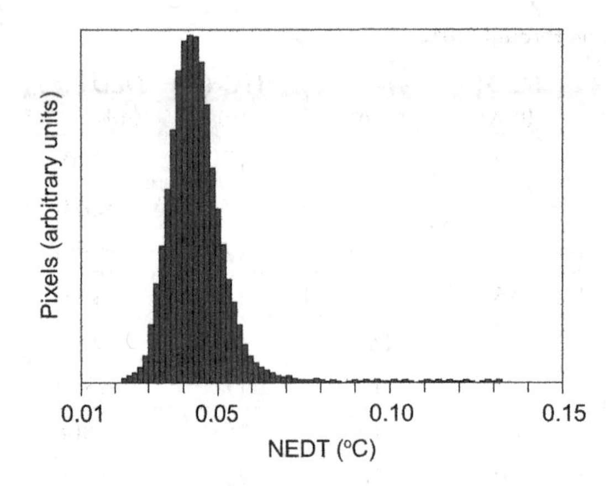

Figure 25.13 Measured pixel NEDT histogram of Honeywell uncooled imager with f/1 optics. (From Ref. [81])

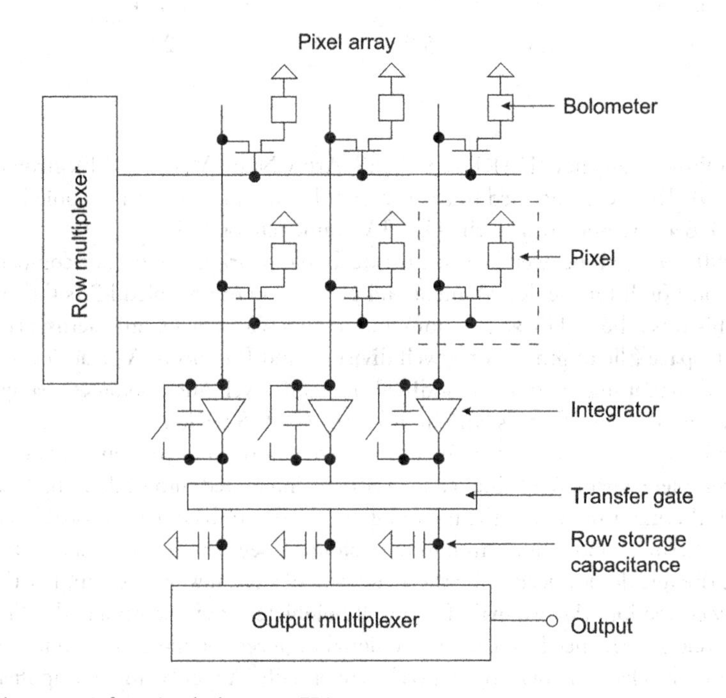

Figure 25.14 Readout circuit for microbolometer FPA.

The surface micromachined bridges on CMOS-processed wafers developed in Honeywell is one of the most widely used monolithic approaches for uncooled imaging. The simplified step processes of monolithic integration are shown in Figure 25.15 [83]. At the beginning, the ROIC is pre-manufactured and the detector materials are subsequently deposited and patterned on the ROIC wafer. For fabrication of the sacrificial layer, a high-temperature stable polyimide is used typically. Finally, the polyimide layer is removed in an oxygen plasma to obtain free-standing, thermally isolated bolometer membranes. The deposition process for the sensing bolometer material is limited to about 450°C due to damaging risk of the ROIC. Low deposition temperature precludes the receiving of monocrystalline materials, which is a potential disadvantage of monolithic integration. This disadvantage is especially serious for poly-SiGe resistive microbolometers developed by IMEC and subsequently transferred to XenICs, Belgium. Poly-SiGe needs to be deposited at high temperatures; therefore, it is not easy to integrate with CMOS [63]. This material exhibits high $1/f$ noise owing to its noncrystalline structure and requires complicated post-CMOS processing to reduce the effects of residual stress.

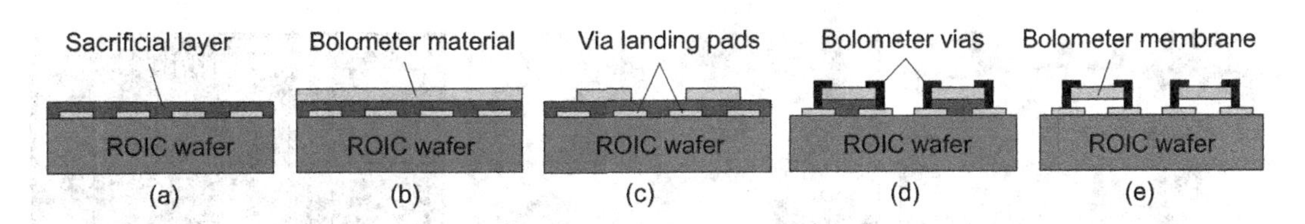

Figure 25.15 Monolithic integration for uncooled IR bolometer arrays: (a) deposition of sacrificial layer on ROIC wafer, (b) deposition of bolometer materials, (c) patterning of bolometer materials, (d) via formation, and (e) etching of sacrificial layer. (From Ref. [83])

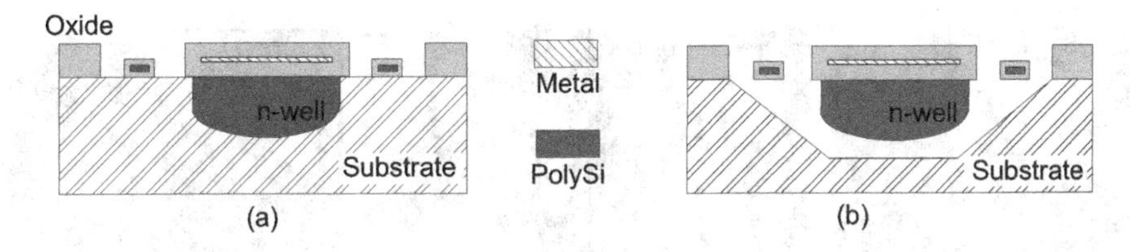

Figure 25.16 Bulk micromachining for uncooled IR bolometer arrays: (a) formation of the bolometer and the electronics for signal readout (typically side-by-side) and (b) selective etching of the bulk material underneath the bolometer membrane. (From Ref. [84])

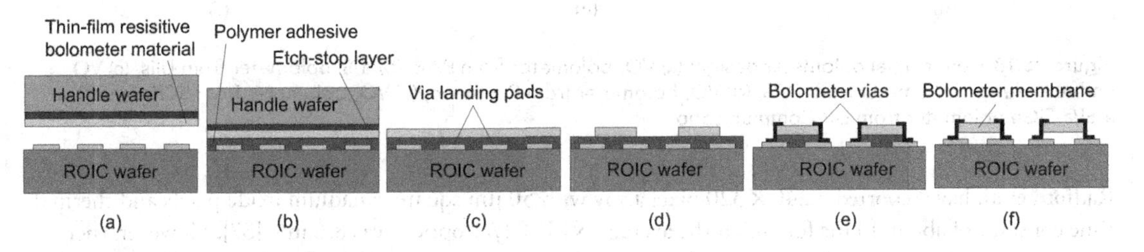

Figure 25.17 Heterogeneous 3D integration for uncooled IR bolometer arrays: (a) separate fabrication of ROIC wafer and handle wafer with resistive bolometer material, (b) adhesive wafer bonding, (c) thinning of handle wafer, (d) bolometer definition, (e) via formation, and (f) sacrificial etching of polymer adhesive. (From Ref. [83])

An alternative to the manufactured uncooled bolometers is bulk micromachining shown in Figure 25.16 [84] where the bolometers are formed in the substrate surface of a wafer. Subsequently, the substrate is selectively etched underneath the bolometers to thermally isolate them from the rest of the substrate. The micromachining process is implemented before, in between, or after processing the wafers to implement electronic components. In the bulk micromachining process, both electronics and the bolometers can typically be manufactured in a standard CMOS line, which is its advantage. However, a disadvantage of this technology is that the ROIC cannot be placed underneath the bolometer membranes, but has to be placed beside the bolometers that, in consequence, reduces the array FF. The bulk micromachining technique is successfully used in commercial fabrication of diode bolometer arrays by Mitsubishi [60].

The third technology of microbolometer array fabrication is the so-called heterogeneous three-dimensional (3D) bolometer integration shown schematically in Figure 25.17 [83]. In this case, the bolometer materials are deposited on a separate handle wafer and next, the materials are transferred from the handle wafer to the ROIC wafer using low-temperature adhesive wafer bonding. The important advantage of this technology is that it allows the use of high-performance monocrystalline sensing bolometer materials on top of standard ROICs. This 3D bolometer integration process was implemented in Acreo, Sweden [85,86].

25.2.3 FPA PERFORMANCE

The demonstrated performance of microbolometer detectors is getting closer to the theoretical limit with the advantages regarding weight, power consumption, and cost. Large microbolometer arrays are fabricated on industry-standard wafer complete with monolithic readout circuits integrated into the underlying Si.

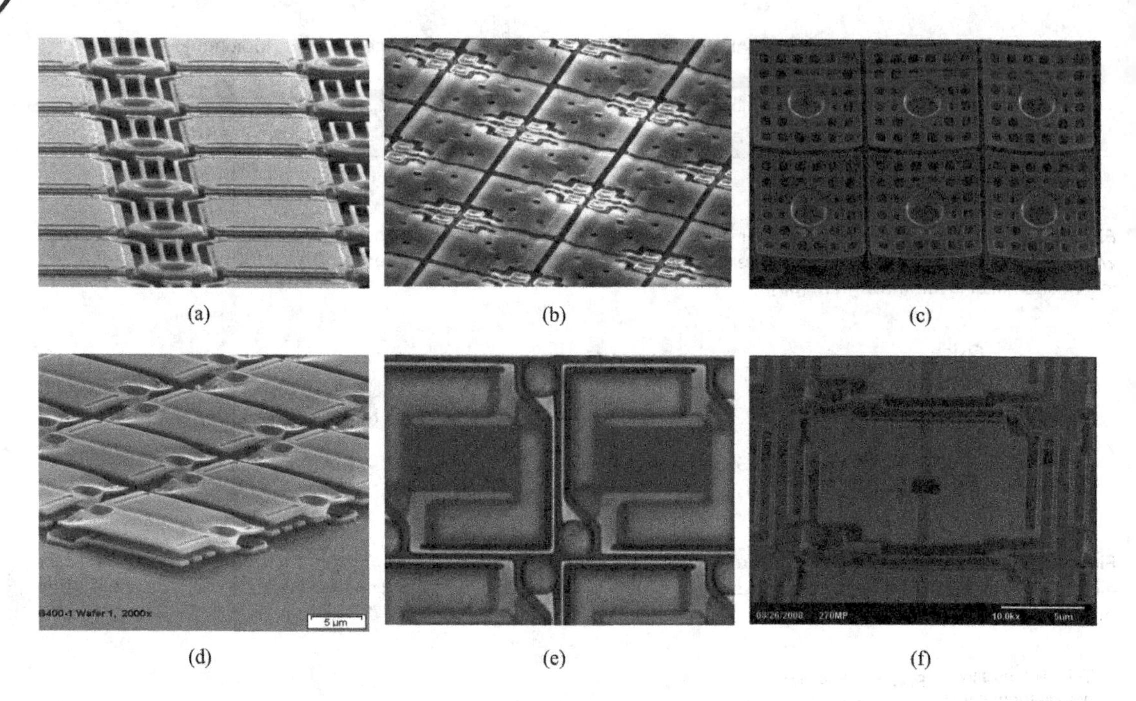

Figure 25.18 Commercial bolometer design: (a) VO_x bolometer from BAE, (b) a-Si bolometer from Ulis, (c) VO_x umbrella design bolometer from DRS, (d) VO_x bolometer from Raytheon, (e) VO_x bolometer from SCD, and (f) a-Si/a-SiGe bolometer from L-3 Communications.

Radford et al. have reported a 240 × 320 pixel array with 50 μm square vanadium oxide pixels and thermal time constant of about 40 ms for which the average NEDT (*f*/1 optics) was 8.6 mK [87]. However, there is a strong system need to reduce the pixel size to achieve several potential benefits. The detection range of many uncooled IR imaging systems is limited by pixel resolution rather than sensitivity. Figure 24.33b shows the trends in pixel reduction. While the pixel pitch of the first microbolometer reported in 1992 was 50 μm, it was reduced to 25 μm by 2002, 17 μm by 2007, and 12 μm in 2013. Today, the most advanced microbolometer IR FPA has a pixel pitch of 10 μm [38].

At present, the commercially available bolometer arrays are either made from VO_x, a-Si, or silicon diodes. Figure 25.18 shows SEM images of commercial bolometers fabricated by different manufacturers.

The micromachined microbolometers reported to date are classified in two design categories—single and double-layer. A single-layer microbolometer, shown schematically in Figure 8.8, is comprised of a transducer element and a leg structure. The pixel design shows a resonant cavity formed by an absorbing layer suspended above a reflecting metal layer. The cavity is used to amplify the absorptance of the incident IR radiation. The microbridge is supported by two beams and is thermally isolated from the ROIC to increase the sensitivity of the microbolometer.

Conventional single-level bolometer arrays typically have a FF between 60% and 70% [37,48]. To increase the FF, two-layer bolometer designs have been reported that reach FF of up to 90% [25,56]. The double deck structures were first presented by research at KAIST in 1998 [88]. A double-layer microbolometer consists of a space-filling metal/dielectric sandwich layer designed to capture the maximum amount of incoming radiant heat energy. The end result is a structure over the pixel that resembles an umbrella (see, e.g., Figure 25.18c) which is the fundamental architecture of DRS microbolometer. A similar design way has been chosen by Raytheon. The second layer can fill space over the top and ensures that approximately 95% of radiation from the 814 μm spectrum is absorbed. The umbrella designs are implemented in arrays with very small pixel sizes up to 10 × 10 μm² [38]. The punctured holes in the space-filling umbrella layer positively affect sensitivity and performance. The fact is that even though the umbrella absorber layer has less solid surface mass, it is better able to capture the incoming radiant energy. Because the thermal mass of the umbrella has been reduced by the holes, the radiant energy heats the umbrella faster for better responsivity.

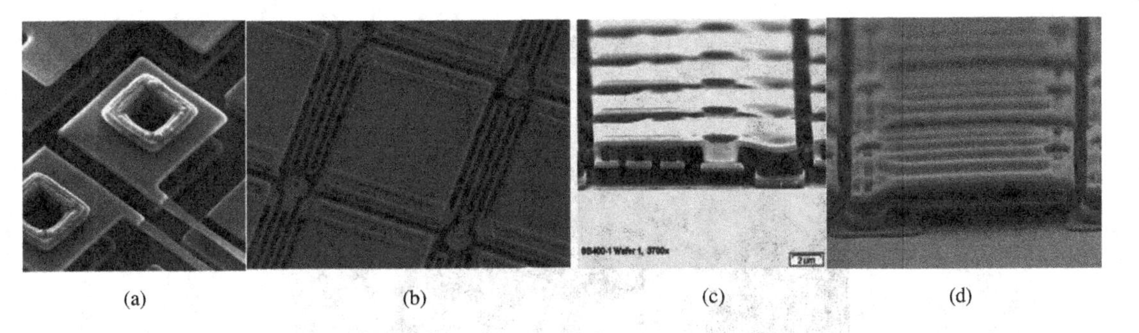

Figure 25.19 SEM of 17 µm bolometer pixels with fine line geometry for single- (a,b) and multiple-level designs (c,d).

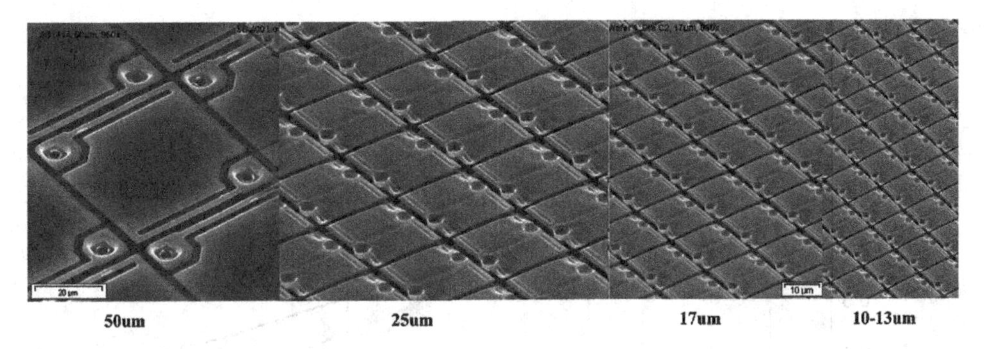

Figure 25.20 Significant design and process challenges exist when a bolometer structure is reduced from 50 µm to a notional 10 µm design. (After Ref. [89])

The development of highly sensitive microbolometer with pixel size below 10 µm presents significant challenges in both fabrication process improvements and in pixel design. Further, shrinking the pixel pitch leads to a dramatic drop in sensitivity as all the driving parameters are adversely affected. Since the responsivity is driven by the relation 25.11, one can increase the FF, the absorption factor (ε), TCR, the applied bias voltage (V_{bias}) or reduce the thermal conductance (G_{th}) or the electrical resistance (R) of the thermistor. At the present stage of technology, the detector FF and the absorption coefficient are close to their ideal values and only a little benefit can be expected from the optimization of these two parameters. In the fabrication of microbolometer arrays, deep ultraviolet photolithography with line widths <0.20 µm is used. Figure 25.19 is an example SEM of foundry-processed with leg widths of <0.02 µm for single-level and multilevel designs [89]. Decreasing the leg width and thickness or an increase in leg length maintain sensitivity, while thinning the bridge for lower mass reduces or maintains the time constant. Figure 25.20 shows SEMs for bolometer pixels of 50, 25, 17, and 10 µm with appropriate scaling. The figure indicates the importance of a fine-line photolithography capability in shrinking the pixel pitch. Figure 25.20 shows SEMs for bolometer pixels of 50, 25, 17, and 10 µm with appropriate scaling. The figure indicates the importance of a fine-line photolithography capability in shrinking the pixel pitch.

Considering influence of the applied voltage, it is limited by the ROIC CMOS technology and is down from the earlier standard (5 V) to 3.3 V and even more to 1.2 V. Thus, CMOS voltage scaling runs in the wrong way to sensitivity improvement.

In uncooled imaging systems, sensitivity usually drives the design more than resolution. Since sensitivity is inversely proportional to F^2 (higher sensitivity → lower NEDT), uncooled imaging technology drives the *f*/# to values between 1 and 1.4. Historically, most imaging systems have been designed to have a resulting optics blur of less than 2.5 pixels ($\sim F\lambda/d < 1.0$; see Figure 24.21). Recently, several companies (NEC, Ulis, Raytheon, BAE, and DRS) (29,34,90–92) have successfully demonstrated the technological integration of 12 µm pixels for imaging systems with $F\lambda/d > 1.0$. Recently, DRS has announced the successful

Figure 25.21 Image from the DRS, 10 μm pitch uncooled camera running at 30Hz and operating in 1,280 × 720 format. (After Ref. [38])

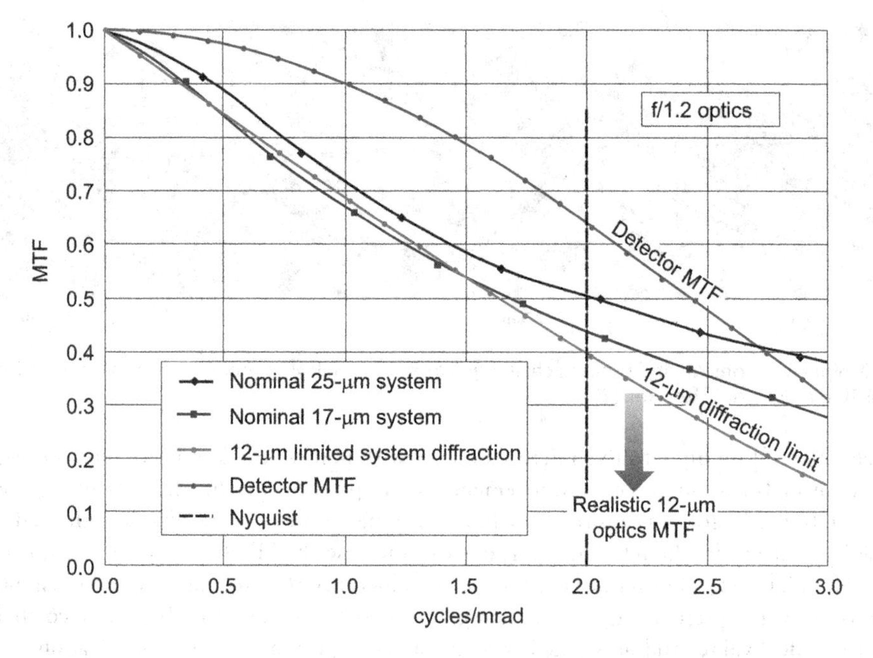

Figure 25.22 Optics and detector MTF curves for 12, 17, and 25 μm pixel systems showing the aberrated curves from the larger pixels falling slightly higher than the $F\lambda/d = 1.0$ condition for the 12 μm pixel system. (Adapted after Ref. [89])

demonstration of its 10 μm pixel pitch IR detector to select defense industry prime contractors [38]. 10 μm pitch image sensors, with NEDT capabilities of better than 50 mK, have been operated in 1,280 × 1,024 thermal camera and arrays of 640 × 512 and 320 × 256 suitable for most original equipment manufacturer integrations. Flexibility in system design suggests that pixel sizes smaller than 10 μm may be the practical limit for a focal plane.

Figure 25.21 shows an example imagery from the DRS 1,280 × 1,024, 10 μm pitch uncooled bolometer sensor. The image shows unprecedented detail as expected with such a small pixel pitch. The optics in this camera is *f*/1.1, which places this image at an $F\lambda/d = 1.1$—close to best imaging theoretically suggested by Holst and Driggers [93]. Outstanding bolometer performance results in improvements of thermal and spatial resolution and image quality of IR cameras.

Figure 25.22 shows the diffraction MTF curve of optics for 12 μm pixel at *f*/1.2 as compared to the modeled optics designs from larger pixels. The aberrated MTF curves for the 25 and 17 μm pixel designs still remain above the diffraction limit for the 12 μm optics.

Table 25.4 contains an overview of the main suppliers and specifications for existing products and for bolometer arrays that are in the R&D stage, while Table 25.5 summarizes the design and performance

Table 25.4 Representative commercial uncooled infrared bolometer array

COMPANY	BOLOMETER TYPE	ARRAY FORMAT	PIXEL PITCH (μm)	DETECTOR NEDT (mK) (f/1, 20–60 Hz)	TIME CONSTANT (ms)
L-3 (USA)	VO_x bolometer	320 × 240	37.5	50	
www.l-3com.com	a-Si bolometer	160 × 120–640 × 480	30	50	
	a-Si/a-SiGe bolometer	320 × 240–1,024 × 768	17	30–50	
BAE (USA)	VO_x bolometer	320 × 240	28	<35	<10–15
http://www.baesystems.com	VO_x bolometer (standard design)	640 × 480	12	<50	<10–15
	VO_x bolometer (standard design)	1,024 × 768	17		
DRS (USA)	VO_x bolometer (standard design)	320 × 240	25	≤40	≤18
www.drsinfrared.com	VO_x bolometer (umbrella design)	320 × 240	17	≤40	≤14
	VO_x bolometer (umbrella design)	640 × 480, 1,024 × 768	17	≤40	≤14
	VO_x bolometer (umbrella design)	1,280 × 1,024	10	<50	
Raytheon (USA)	VO_x bolometer	320 × 240, 640 × 480	25	30–40	
http://www.raytheon.com/	VO_x bolometer (umbrella design)	320 × 240, 640 × 480	17	50	
	VO_x bolometer (umbrella design)	1,024 × 480, 2,048 × 1,536	17		
ULIS (France)	a-Si bolometer	80 × 80	34	<100	
www.ulis-ir.com	a-Si bolometer	160 × 120	25	<60	<10
		320 × 240	12	<60	<10
		384 × 240	17	<55	<10
		640 × 480, 1,024 × 768	17	<50	<10
SCD (Israel)	VO_x bolometer	384 × 288	25	<20	22
www.scd.co.il	VO_x bolometer	384 × 288	25	<35	16
	VO_x bolometer	640 × 480	17	<35	16
	VO_x bolometer	1,024 × 768	17	<35	14
FLIR Systems	VO_x bolometer	640 × 512	17	<60	<12
http://www.flir.com	VO_x bolometer	336 × 256	17	<50	<15
NEC (Japan)	VO_x bolometer	320 × 240	23.5	<75	
http://www.nec.com	VO_x bolometer	640 × 480	23.5	<75	
	VO_x bolometer	640 × 480	12	60	
	VO_x bolometer	320 × 240	23.5	NEP<100pW*	

NEP–noise equivalent power.
*at 4 THz.

Table 25.5 Performance characteristics of DRS's VO_x microbolometers (www.drsinfrared.com)

PERFORMANCE PARAMETER	CAPABILITY (f/1 AND 300 K SCENE)			
Array configuration	320 × 240	320 × 240	640 × 480	1,024 × 768
Pixel size (μm²)	25 × 25	17 × 17	17 × 17	17 × 17
Spectral response (μm)	8–14	8–14	8–14	8–14
Frame rate (Hz)	60	60	30 (using 1 output) 60 (using 2 outputs)	30
NEDT @ *f*/1 (mK)	<40	<40	<50	<40
Time constant (ms)	≤18	≤14	≤14	≤14
Area FF (%)	90	90	90	90
Number of analog outputs	1	1	1/2	1/2
Output voltage range (V)	0.5–4.5	1.2–3.2	1.2–3.2	1.3–4.5
Temperature stabilization	No TEC required	No TEC required	No TEC required	No TEC required
Pixel operability (%)	>99	>99	>99	>99
On-chip NUC	6 bits parallel	7 bits parallel	7 bits parallel	7 bits parallel
Power (mW) nominal	≤300	≤120	≤220	≤450
Packaging	Ceramic—LCC	Ceramic—LCC	Ceramic—LCC	Ceramic—LCC
Dimensions L × W × H (cm)	1.83 × 1.83 × 0.37	1.83 × 1.83 × 0.37	2.40 × 2.40 × 0.37	2.92 × 2.92 × 0.37
Weight (g)	≤4	≤4	≤6	≤9
Operating temperature (°C)	−40 to +71	−40 to +85	−40 to +85	−40 to +71

TEC– thermoelectric cooler.

parameters for DRS VO_x microbolometers. As we can see, a similar performance has been described by BAE Systems, DRS, Ulis, L-3, and SCD.

The present standard microbolometer technology is based on 17 μm pixel pitch FPAs extended to both smaller arrays (320 × 240) and arrays larger than 3 megapixel. Currently, the largest microbolometer array fabricated by Raytheon is shown on a wafer in Figure 25.23a. In the fabrication of 2,048 × 1,536 staring arrays and associated ROIC circuits, a stitching technique has been used. Improvement is the transition from 150 mm in-house wafer production facilities to commercial CMOS/microelectromechanical system (MEMS) (class 1) foundries with low-defect, high-yield, and 200 mm high-volume wafer capability. Each 200 mm wafer contains nine 2,048 × 1,536 uncooled detector dies, which represent an 80% increase in yield over an equivalent 150 mm wafer [28]. Figure 25.23b shows an example of the dramatic image resolution improvement using this very large-format FPA.

In Section 8.3.1, is marked that silicon diode thermometers can be applied to develop uncooled IR FPAs. The first polysilicon p-n junction diodes appeared in the early 1990s and were characterized by large

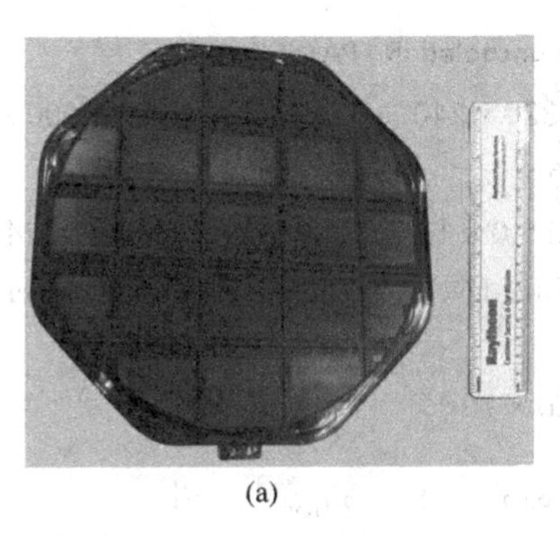

(a) (b)

Figure 25.23 2,048 × 1,536 uncooled VO_x microbolometers with 17 μm pixel pitch: (a) nine-2,048 × 1,536 uncooled detector die on a 200 mm wafer and (b) image resolution improvement using this array. (After Ref. [28])

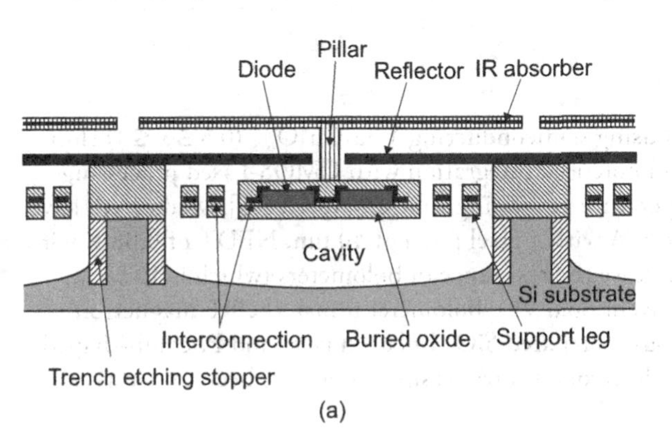

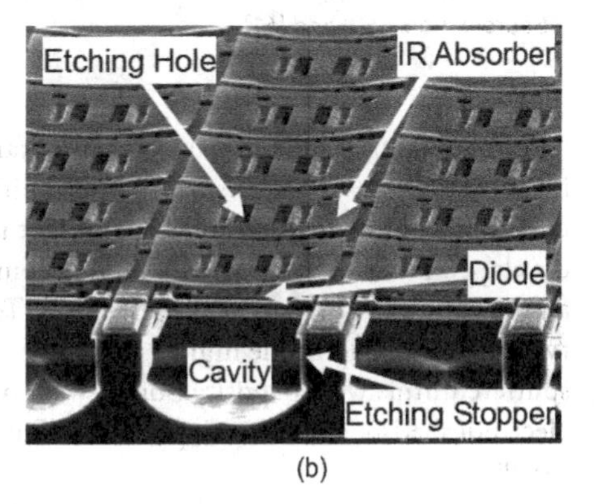

(a) (b)

Figure 25.24 SOI diode microbolometer: (a) schematic of the detector cross section and (b) SEM view of the diode pixels with 40 × 40 μm² pitch. (From Ref. [60])

noise. By devising an absorbing structure with a FF up to 90%, considerable improvement of its performance has been achieved. At present, this structure is fabricated by a MEMS process that is a combination of surface micromachining and bulk micromachining technologies. Commercial bolometer arrays manufactured by Mitsubishi using a custom silicon on insulator (SOI) CMOS technology [21,57–61] are series-connected diode microbolometers. Figure 25.24 shows a schematic of the detector cross section and an SEM view of the diode pixels. To lower the thermal conductance without degrading the efficiency of IR absorption, the most advanced pixels has a three-level structure that has an independent metal reflector for interface IR absorption between the temperature sensor (bottom level) and the IR absorbing thin metal film (top level). The MEMS process includes a XeF_2 dry bulk Si etching and a double organic sacrificial layer surface micromachining process [60]. Arrays consisting of 640 × 480 pixels are based on suspended multiple series diodes with 25 × 25 μm² pixel sizes. The reported NEDT value is 40 mK for *f*/1 optics (see Table 25.6). Although this approach provides very uniform arrays with very good potential for low-cost, high-performance uncooled detectors, its fabrication is based on a dedicated in-house SOI CMOS process. The better approach would be to implement the detector arrays together with readout circuitry fully in standard CMOS process [61].

Si diode uncooled sensors have slightly lower sensitivity than resistance bolometers. In spite of this, in 2008, Mitsubishi Electric announced a 25 μm, pixel pitch, 640 × 480 uncooled IR FPA with an NETD of 20 mK and *f*/1 optics [94]. Their noise and uniformity are much better because they are made of single crystal Si and advanced LSI technology.

Table 25.6 Specifications and performance of SOI diode uncooled IR FPAs

	320 × 240	320 × 240	320 × 240	640 × 480	2,000 × 1,000
Pixel size (μm)	40 × 40	28 × 28	25 × 25	25 × 25	15 × 15
Chip size (mm)	17.0 × 17.0	13.5 × 13.0	12.5.0 × 13.5	20.0 × 19.0	40.30 × 24.75
Pixel structure	Two-level	Three-level	Three-level	Three-level	Three-level
Number of diodes	8	6	6	6	10
Thermal conductance (W/K)	1.1×10^{-7}	4.0×10^{-8}	1.6×10^{-8}	1.6×10^{-8}	
Sensitivity (μV/K)	930	801	2,842	2,064	
Noise (μV rms)	110	70	102	83	
Nonuniformity (%)	1.46	1.25	1.45	0.90	
NEDT, *f*/1 (mK)	120	87	36	40	65(15 Hz) τ_{th}=12 ms

Source: Refs. [60] and [61].

Encouraging results have also been obtained using semiconducting $YBa_2Cu_3O_{6+x}$ ($0.5 \leq x \leq 1$) thin films on Si [95–97]. To ensure compatibility and potential integration with CMOS-based processing circuitry, Si micromachining and ambient temperature processing were employed [95]. Wada et al. have developed 320 × 240 YBaCuO microbolometer FPA with a pixel pitch of 40 μm, NEDT of 0.08 K with a prototype camera and *f*/1.0 optics [97]. To decrease the resistance of bolometers (which is 10 Ωcm, two orders of magnitude higher than that of conventional VO_x bolometer films), the RF magnetron sputtered films were deposited on Si with previously prepared SiO_2 isolation layer and Pt comb-shaped electrodes. Also, efforts to implement YBaCuO detectors in various substrates have been undertaken [98,99].

25.3 PYROELECTRIC FOCAL PLANE ARRAYS

The use of pyroelectric materials and devices in IR imaging cameras is now a well-established technology. They have been in use for many years in applications ranging from intruder sensing, people counting, through spectroscopy, and environmental monitoring to flame and fire detection [100]. In the last two decades, there has been a growth of interest in using 2D arrays of small pyroelectric elements for uncooled thermal imaging [2,101–110].

The imaging systems based on pyroelectric arrays usually need to be operated with optical modulators that chop or defocus the incoming radiation. This may be an important limitation for many applications in which a chopper-less operation is highly desirable (e.g., guided munitions). A chopper is a less reliable mechanical part, inconvenient and more bulky. However using a chopper blade, the thermal image can be produced by subtracting the field of data output by the detector while viewing the chopper from the field output when viewing the scene. This image difference processing not only removes offset variations between elements in the array but also serves as a temporal high-pass filter, eliminating 1/*f* noise components and long-term drifts (low frequency spatial noise).

In the last decade of the twentieth century, attention was turned to the solid-state readout of large pyroelectric FPAs and, in particular, to the possibility of interfacing the pyroelectric arrays directly with a Si chip. Practical arrays were demonstrated by several groups including GEC [102], RRSE [103], DERA/BAE Systems [107] in the United Kingdom, TI [105,108,110], and others [111,112]. At present, the main efforts are concentrated on linear arrays from a few tens of elements to many hundreds, but because of the absence of detector scanning along the direction of the array, bandwidths are low. In the last two decades, there has been also a growth of interest in using arrays of small pyroelectric elements for uncooled thermal imaging.

Table 25.7 Typical properties of linear arrays

NUMBER OF ELEMENTS	1 × 128	1 × 128	1 × 128	1 × 256	1 × 510
Size of elements [μm²]	90 × 100	90 × 500	90 × 1,000	42 × 100	20 × 100
Pitch [μm]	100	100	100	50	25
Responsivity R_v [V/W]	230,000	540,000	230,000	620,000	680,000
Noise voltage (mV)	0.7	0.8	1.1	0.7	0.9
NEP (nW)	3.0	1.5	4.9	1.1	1.3
MTF (R=3 lp/mm)	0.6	0.6	0.6	0.6	0.8
Uniformity of R_v (%)	5	5	5	5	10

Source: Ref. [122].

25.3.1 LINEAR ARRAYS

Linear arrays are particularly suitable for applications where there is relative motion between the sensor head and the objects being imaged (e.g., intruder alarms and pushbroom linescan) [100,102,113–121]. A thin, polished wafer of pyroelectric material, about 20 μm thick, is bonded down to a substrate. The manufacture of pyroelectric linear arrays based on $LiTaO_3$ with thicknesses of the self-supporting responsive elements of less than 5 μm has become possible by the development of special thinning techniques (ion beam etching) [118]. Table 25.7 shows the essential properties of different types of $LiTaO_3$ linear arrays.

Figure 25.25 shows the principal design of a pyroelectric linear array, which includes a lithium tantalate chip with up to 510 elements [123]. The pixel size is several tens of microns in width, *a*, and up to 1 mm in length, *b*. The chopped radiation signal, Φ_s, strikes the active surface of the pyroelectric chip where it is absorbed. The signals generated in the sensitive elements are processed in a CMOS circuit that contains both analog and digital sections. Sensitivity of the detector array is maximized by reducing the thickness of the pyroelectric chips (typical to 5 μm), in fabricated ion beam etching techniques.

The pyroelectric layers are reticulated to provide a two-dimensional array of pyroelectric islands or elements separated by grooves. A high-performance ion beam etching system is available for microstructuring.

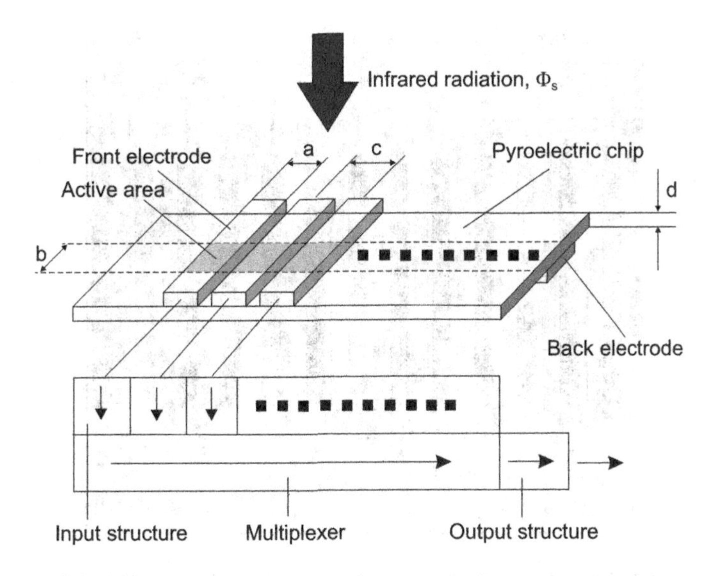

Figure 25.25 Principal design of a pyroelectric linear array. (Adapted after Ref. [123])

Ion milling enables 2D reticulation of pyroelectric material with groove widths of 10 μm or less. These structures are necessary to inhibit the lateral diffusion of heat in the plane of the arrays. Without reticulation interelement thermal diffusion becomes severe at modulating frequencies below 100 Hz. Another effect that increases the interelement coupling and therefore the crosstalk, is the capacitative coupling due to fringing effects at electrode edges. This effect is also removed by reticulation.

The pyroelectric $LiTaO_3$ sensors have the optimum signal-to-noise ratio typically between 5 and 10 Hz modulation frequency. However, because of the high thermal crosstalk in sensor chips with a pixel spacing of 5–10 μm, modulation frequencies below 80 Hz are generally not useful.

The first research pyroelectric arrays were produced using a CCD design and integrated circuit (IC) for readout. A rigorous analysis of the interface condition for direct injection mode to optimize the injection efficiency and noise has been given by Watton et al. [117]. It appears, however, that the large ferroelectric element capacitance coupled with the CCD sampling within the pixel led to dominating kTC noise that limited performance. On account of this, attention turned to the use of CMOS ROIC designs.

Readout of charge from pyroelectric detectors is by means of conventional field effect transistors (FETs). Each detector is connected to its own source follower FET that acts as an impedance buffer. Outputs of these transistors go to a multiplexer that samples the elements in turn at a rate dependent upon the particular applications.

For a long time, there has been an interest in the use of thin pyroelectric films, because of their potential for making low thermal mass elements [124]. Arrays that have been demonstrated include the linear arrays fabricated using bulk micromachining techniques with sputtered $PbTiO_3$ on (100) Si [125], La-$PbTiO_3$ on MgO [126], PVDF-TrEE on Si [127], and sputtered $Pb(Zr_{0.15}Ti_{0.85})O_3$ on (100) Si [128]. In these micromachining techniques, Si from behind the pyroelectric film has been removed, leaving a thin, low thermal mass membrane. Two major solutions have been elaborated for achieving high IR absorption coefficients of the detectors. The first one utilizes black absorption layers, usually porous metal films (black Pt or black Au). The second method uses semitransparent top electrodes (e.g., NiCr) on $\lambda/4$ thick pyroelectric layers. The first method gives larger absorption (about 90%) as compared to the second one (60%); however, the total thickness and heat capacity are increased [119]. For example, Figure 25.26 shows the linear array developed for an IR spectroscopy gas sensor. The pyroelectric material is a sol-gel deposited (111)-oriented PZT15/85 thin film [129].

More recently, the integration of a radiation collector with a pyroelectric detector has been demonstrated successfully. Both pyramidal and hemi-elliptical collector cavities can be made in Si substrates using wet and dry etching [130,131].

Figure 25.26 Top view of 50-element array with 200 μm period obtained with bulk micromachining, membrane size of 2 × 11 mm. The black Pt absorbers, the CrAu contact lines, the membrane layers between the elements, and the SiO_2 layer for reduction of parasitic capacitance are visible. (From Ref. [129])

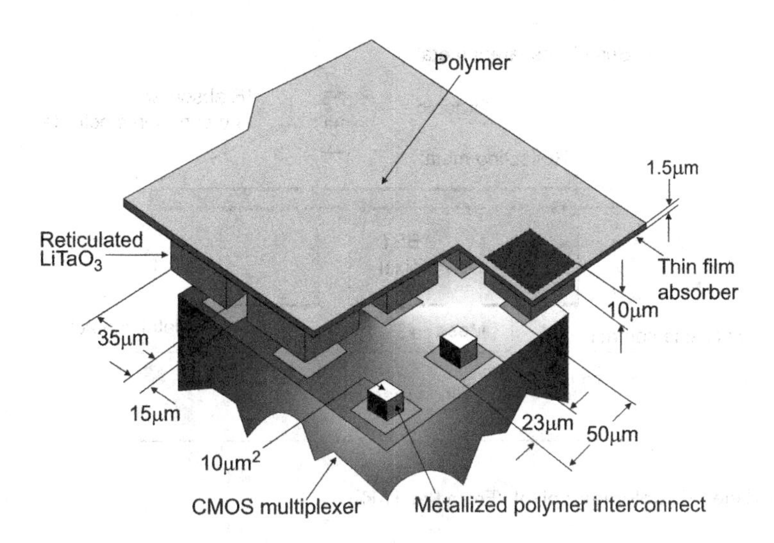

Figure 25.27 Hybrid pyroelectric array structure. (From Ref. [111])

The FPA technologies may be classified into either hybrid or monolithic varieties; hybrid fabrication and micromachining techniques compete at the moment.

25.3.2 HYBRID ARCHITECTURE

The hybrid approach is based on the reticulation of ceramic wafers that are polished down in thickness to 10–15 μm and joined with the readout chip. The interface technology must resolve the conflict between providing electrical connection for signal readout from the element and thermal isolation of the element to avoid loss of signal through thermal loading.

Figure 25.27 shows the structure of an array [111]. A $LiTaO_3$ active volume detector is bound by the common front electrode, the back electrode, and by reticulation cuts. The detector back electrode is connected to the underlying multiplexer by a metalized polymer thermal isolation link that provides connection and support for the detector with controlled thermal conductance. The nominal detector size is 35 × 35 μm² with 10 μm thickness and a 15 μm gap between detectors for an element pitch of 50 μm. For the detector with f=30 Hz, $R_v \approx 1 \times 10^6$ V/W. To obtain maximum responsivity and minimize thermal fluctuation noise, the thermal conductance from the detector to its surroundings should be minimized. The estimated total thermal conductance is about 3.3×10^{-6} W/K, and is higher than that obtained for micromachined Si bolometers. The thermal time constant is then ≈15 ms. The predicted NEDT of the prototype system with 330 × 240 elements array has been estimated as 0.07 K with f/1 optics, an optic transmission of 0.85, and a chopper efficiency of 0.85.

As discussed in Chapter 9, the performance of pyroelectric detectors can be improved with a bias voltage applied to maintain and optimize the pyroelectric effect near the phase transition. This type of pyroelectric FPAs has been developed by TI [104,105]. The TI detector array comprises 245 × 328 pixels on 48.5 μm centers. Operating near room temperature, ferroelectric BST pixels hybridized with a Si ROIC consistently yield devices with a system NEDT of 0.047°C with f/1 optics.

The fabrication process for forming theses arrays is 95% compatible with standard Si processes. Detector process commonality with a Si wafer processing format is maintained by the fabricating 100 mm diameter ceramic BST wafers with excellent dielectric properties. Highly dense, sintered ceramic BST offers cost and performance advantages not found in single crystal materials. After edging and polishing, reticulated arrays are formed by writing the array pattern on the wafer surface with a Nd-YAG laser. The laser damage is removed with an acid etchant, after which stoichiometry is restored by annealing in an oxygen ambient. The kerf is approximately 12 μm. Deposition of a parylene backfill, replanarization of the reticulated surface, and application of a common electrode and resonant-cavity IR absorber finishes the processing of the IR sensitive side of the array. The absorber layer is a transparent organic $\lambda/4$ layer, sandwiched between a semitransparent metal layer and the common electrode. Information on the composition of the absorber

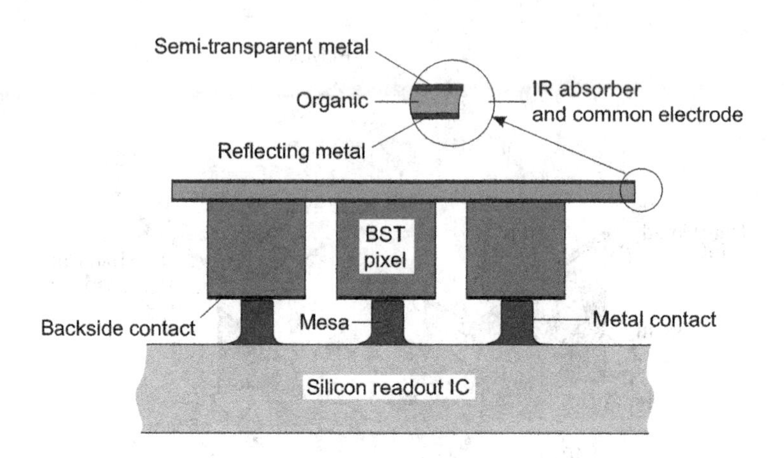

Figure 25.28 BST dielectric bolometer pixel. (From Ref. [104])

is generally scarce, this part of the process being considered as technological know-how. The IR absorber provides better than 90% average absorption over the 7.5–13.0 μm spectral band. Dicing the wafer and mounting the individual die to carriers (optical coating side down) prepare the die for thinning and polishing to final thickness of approximately 20 μm. Deposition of contact and bond metals followed by the removal of parylene by dry etching prepares the array for hybridization. ROICs are prepared for hybridization by forming matching organic mesas with over-the-edge metallization. Mating of the two parts by solder bonding prepares them for packing and final testing. Figure 25.28 shows the details of the completed pyroelectric detector device structure [104].

The CMOS readout unit cell contains a high-pass filter, a gain stage, a tunable low-pass filter, and an address switch. The array output is compatible with standard TV formatting. The array is mounted onto a single-stage TE cooler for stabilization near the ferroelectric phase transition. The ceramic device interface processor and package is completed by the attachment of the antireflection-coated Ge window that allows IR transmission in the 7.5–13 μm spectral band.

In the mid-1990s, TI was clearly leading in the development and production of uncooled, ferroelectric IR systems since NEDTs less than 0.04°C have been measured on systems with *f*/1 optics, without correcting for system-level noise and other losses. The production average was between 70 and 80 mK. A demonstrated sustained production rate in excess of 500 units per month was a small fraction of factory capacity [104]. Hybrid ferroelectric bolometer detector was the first to enter production, and was the most widely used type of thermal detector (in the United States, the Cadillac Division of General Motors pioneered this application, selling thermal imagers to customers for just under $2,000) [2].

Large hybrid arrays have also been demonstrated by BAE Systems in the United Kingdom with a pitch of 56 μm and 40 μm. In this case, the dielectric bolometers are $Pb(Sc_{0.5}Ta_{0.5})O_3$ (PST) biased at 4–5 V/μm (higher than for BST) with F_d levels between 10 to 15 × 10^{-5} $Pa^{-1/2}$. Conventional unbiased pyroelectrics give values of about 4 × 10^{-5} $Pa^{-1/2}$. The thin PST wafer is cut and polished from a hot pressed ceramic block and next reticulated by a laser-assisted etching process [132]. Pb/Sn solder bonds have been used in a liquid phase soldering process. The details of fabrication procedures of these arrays are described by Whatmore and Watton [107].

The performance of hybrid arrays that have been fabricated and tested in the United Kingdom program (BAE Systems and DERA) is listed in Table 25.8.

25.3.3 MONOLITHIC ARCHITECTURE

Although many applications for this hybrid array technology have been identified, and imagers employing these arrays are in mass production, no hybrid technology advances are foreseen. The reason is that the thermal conductance of the bump bonds is so high that the array NEDT (*f*/1 optics) is limited to about 50 mK. The best NEDT achieved with a hybrid array is about 38 mK, which is consistent with the thermal conductance of approximately 4 μW/K. Early BST products suffered from features like poor modulation

Table 25.8 Hybrid arrays demonstrated in the UK (DERA/BAE Systems) program

ARRAY ELEMENTS	PITCH (μm)	ROIC SIZE (mm^2)	PACKAGE ATMOSPHERE	NEDT (mK)	ARRAY MTF AT NYQUIST
100 × 100	100	15.3 × 13.4	N_2	87	65%
256 × 128	56	17.0 × 12.4	Xe	90	45%
384 × 288	40	19.7 × 19.0	Xe	140	35%

Source: Ref. [107].

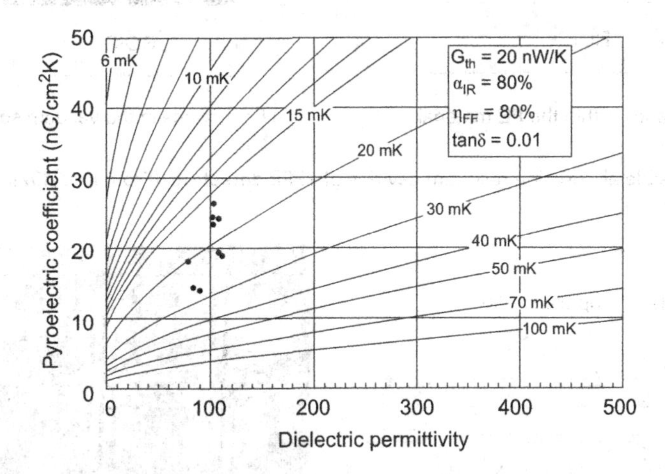

Figure 25.29 Lines of constant NEDT for 25 μm pixels as functions of dielectric permittivity and pyroelectric coefficient. Data points indicate properties of material samples in test structures. (From Ref. [108])

transfer function (MTF) resulting from thermal conductance between detector pixels and excessive noise resulting from insufficient digital resolution in the system. Pyroelectric array technology, therefore, is moving toward monolithic Si microstructure technology. The monolithic process should have fewer steps and shorter cycle time. The detector cost in high volume is limited primarily by detector packing costs, which are not significantly different for hybrid and monolithic arrays. However, the serious problem is the loss of ferroelectrics' interesting properties as the thickness is reduced.

Thin-film ferroelectric (TFFE) detectors have the performance potential of microbolometers with minimum NEDT below 20 mK [107,108,110]. The properties of the materials and the device structure are sufficient to match NEDT projections of the bolometer technologies (see Figure 25.29).

The first surface-machining 64 × 64 $PbTiO_3$ pyroelectric IR imager has been demonstrated by Polla and coworkers [133,134]. Polysilicon microbridges of 1.2 μm thickness have been formed 0.8 μm above the surface of a Si wafer. The microbridge measures 50 × 50 μm^2 and forms a low thermal mass support for a 30 × 30 μm^2 $PbTiO_3$ film with a thickness of 0.36 μm. An n-channel MOSFET (NMOS) preamplifier cell is located directly beneath each microbridge element. The measured pyroelectric coefficient for a single element is 90 nC/cm^2K. The measured blackbody voltage responsivity at 30 Hz is 1.2×10^4 V/W, whereas the detectivity is 2×10^8 $cmHz^{1/2}W^{-1}$.

The present TFFE device approach appears remarkably similar to the VO_x microbolometer structure developed by Honeywell. However, there are several key features that distinguish it from that technology [106–110,135,136]. Since the device is a capacitor rather than a resistor (as in a bolometer), the electrodes are located above and below the face of the pixel, are transparent, and do not obscure the active optical area. Usually, the electrical resistance of the leads can be quite large without degrading the signal-to-noise ratio since the detector capacitance is approximately 3 pF. This enables the use of thin, poorly conducting electrode materials to minimize the thermal conductance. A key feature of the design is that the ferroelectric film is self-supporting; there is no underlying membrane necessary to provide mechanical support. In such a way, with the use of transparent oxide electrodes, the ferroelectric material can dominate thermal conductance.

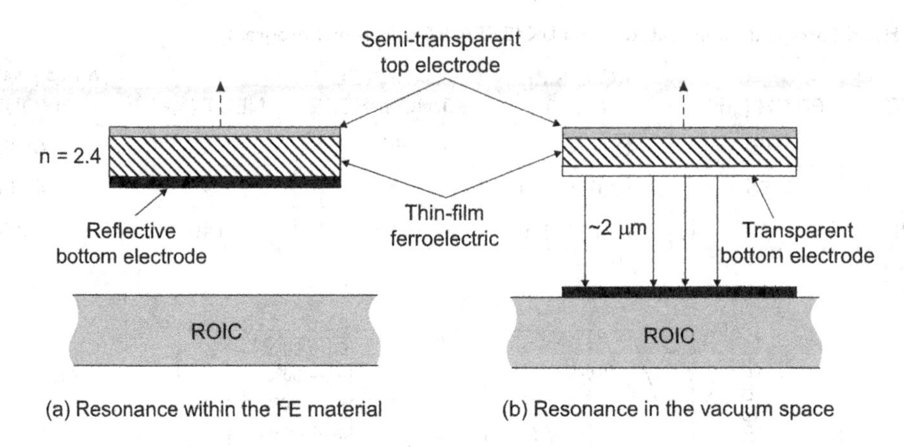

Figure 25.30 Two possible absorption resonant cavities of TFFE detectors. (From Ref. [137])

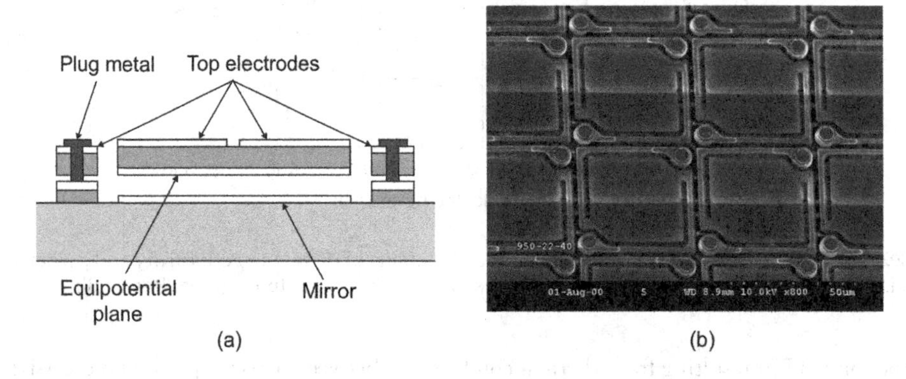

Figure 25.31 TFFE pixels: (a) schematic cross section of pixel element with split top electrode and (b) micrograph of part of a 320 × 240 array with 48.5 μm pixels. (From Ref. [108])

It is well known that absorption of IR radiation is accomplished by means of a resonant optical cavity. In monolithic bridge structure, the cavity is located within the ferroelectric itself or in the space between ferroelectric and the ROIC. This can be realized in two ways (see Figure 25.30) [136,137]:

1. The bottom electrode must be highly reflective, the top electrode must be semitransparent, and the ferroelectric must be approximately 1 μm thick for optimal tuning of the cavity for 10–12 μm radiation.
2. Both electrodes must be semitransparent, a reflective mirror must be present on the ROIC under each pixel, and the pixel must be located approximately 2 μm above the ROIC.

Figure 25.31a shows a cross section of the TFFE pixel designed according to the second way [108] where two top electrodes are connected to one of the posts, which provide electrical connection to the readout. Thus, the pixel capacitance is one-quarter the value of a similar capacitor whose connections are full-face electrodes on top and bottom. In this case, the ferroelectric was lead-calcium titanate (PCT) deposited by metal-organic decomposition of a spun-on solution. Figure 25.31b shows a micrograph of part of a 320 × 240 array with 48.5 μm pixels.

A key factor to the performance of the ceramic thin films is the high temperature processing required in achieving the correct ferroelectric crystal phase. The TFFEs of interest are refractory, and require annealing at elevated temperatures to crystallize and develop good pyroelectric properties. Thermal treatments at temperatures that exceed about 450°C may lead to adverse interaction between the Si and Al interconnects. Various techniques for the deposition of thin ferroelectric films have been investigated including spin-on metal-organic decomposition, radio frequency magnetron sputtering, dual ion beam sputtering, sol-gel processing, and laser ablation. Also, a number of surface rapid thermal annealing

Figure 25.32 Single-frame *f*/1 320 × 240 TFFE imagery extracted from a driving video. (From Ref. [108])

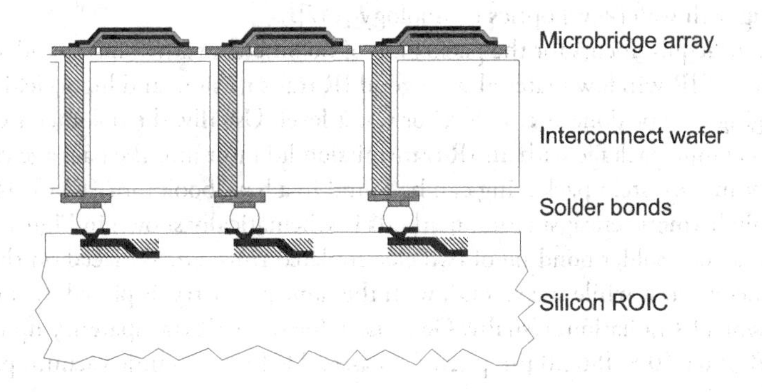

Figure 25.33 Schematic cross section of the composite detector array design. (From Ref. [106])

techniques have been investigated to obtain optimum material response while leaving the underlying Si substrate undamaged [106].

The NEDT of TFFE devices with 48.5 μm pixels is typically about 80–90 mK including all the system losses [108]. They are characterized by excellent MTF compared to the bulk BST. Figure 25.32 shows an example of the image resolution taken from a driving video using micromachined 320 × 240 TFFE pixels. The low spatial noise is apparent from the uniformity in the extended dark areas of the image. The potential for NEDT improvement can be realized by reducing the thickness and improving the thermal isolation and material modification. However, the great challenge is in reducing the pixel size. The challenge becomes even greater when we note that several microbolometer manufactures are moving toward 10 μm pixels.

It should be mentioned that the research group from DERA (United Kingdom) has developed an integrated and composite detector technology [106,138]. In the first technology, the detector material was deposited as a thin film onto free-standing microbridge structure defined on the surface of the Si ROIC. The composite technology combines elements of hybrid and integrated technologies (see Figure 25.33). Microbridge pixels are fabricated in a similar fashion to the integrated technology and next are formed onto a high-density interconnect Si wafer. The interconnect wafer uses materials that can withstand the intermediate high temperature processing stage during fabrication of thin ferroelectric films and contains a narrow conducting channel via for every pixel, permitting electrical connection to the underside. Finally, the detector wafer is solder bump bonded to the ROIC as per the established hybrid array process. It is predicted that using PST films a NEDT of 20 mK (50 Hz image rate and *f*/1 optics) is possible.

25.4 PACKAGING

Generally, the uncooled IR array packages are based on available technologies widely developed for the packaging of mass produced electronic devices. Partially, however, the packages are designed by manufacturers in-house. The lead frames enable electronic board integration like it is for standard CMOS devices, ensuring high electrical contact reliability in tough and demanding environmental applications such as military operations, firefighting, automotive applications, process control, or predictable maintenance.

For the best performance, the conventional bolometers operate with vacuum levels below 0.01 mbar [24]. Necessity of vacuum packaging is related to the fact that heat loss due to thermal conduction from bolometers through the gas gap to the substrate underneath causes an increase in the NEDT. The measurements carried out by He et al. have shown that thermal conduction through the gas starts to have an effect from a pressure in the range 0.1 mbar for a device with pixel area 50 × 50 μm^2 and an air gap of 20 μm [139].

Conventional vacuum packaging performed package by package is a time consuming and expensive process. To solve this issue, such vacuum packaging technologies like batch packaging [140,141], chip-level packaging [141,142], and wafer-level packaging [29,34,143,144] are being developed (see Figure 25.34). In addition, other techniques have been developed such as a pixel-level packaging [145,146] and extending wafer-level packaging with wafer-level optics technology [147].

The most important requirements for the packaging of bolometer arrays follow good and reliable hermetic seal, integration of IR window material with good IR transmission, and high-yield, low-cost packaging [83]. The packaging may be done at chip level or wafer level. Usually, the bolometer chips are built into a hermetic metal or ceramic package with an IR transmission lid built into the package cap. Descriptions of different methods for microsystem packaging can be found in a handbook for MEMS [148].

One of the on-chip hermetic encapsulation methods is schematically shown in Figure 25.35a [149]. The method is based on eutectic solder bonding of two electroplated rims: one is placed on the Si wafer and surrounds the bolometer array, while the second, with the same geometry, is placed on a wafer made from a material that transmits IR radiation. Usually, Ge is used due to its IR transparency up to 20 μm. Figure 25.35b shows the SB-300 640 × 480 20 μm pitch FPA assembled into ceramic vacuum packages for performance testing and imaging evolution.

At present, the wafer-level packaging technology becomes a popular vacuum packaging technology for cost reduction. In this packaging process (see Figure 25.36), a cap wafer is bonded to an array wafer and hermetically sealed under vacuum and next the bonded wafer is diced to obtain individual vacuum-packaged arrays. An example of a detailed cross section of a wafer-level vacuum-packaged arrays is shown in Figure 25.37. The back side of the cap wafer is etched to obtain a cavity for the pixel array and an opening space for bonding pads. Both sides of the cap wafer are covered by an antireflecting coating and a patterned vacuum getter is deposited on the back side of the cap wafer. After flip-chip bonding, the nanoliter volume between two substrates and two rims is pumped to the required pressure level through a

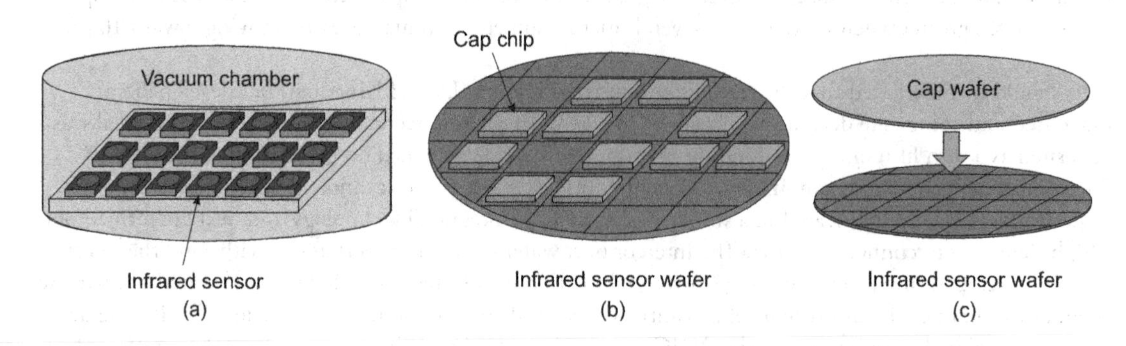

Figure 25.34 Vacuum packaging technologies: (a) batch packaging, (b) chip-level packaging, and (c) wafer-level packaging. (Adapted after Ref. [140])

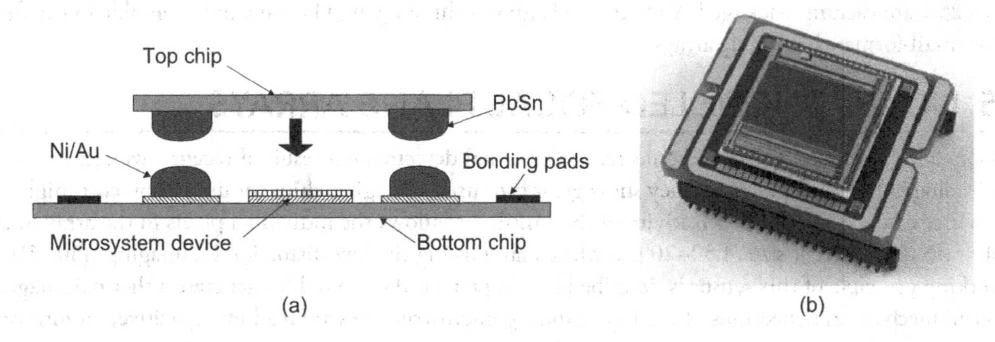

Figure 25.35 Bolometer array packaging: (a) schematic overview of the on-chip hermetic encapsulation of microbolometer array (after Ref. [149]) and (b) the SB-300 640 × 480 20-µm pitch FPA mounted in the high vacuum, low-mass and low cost ceramic package fabricated by Raytheon. (After Ref. [25])

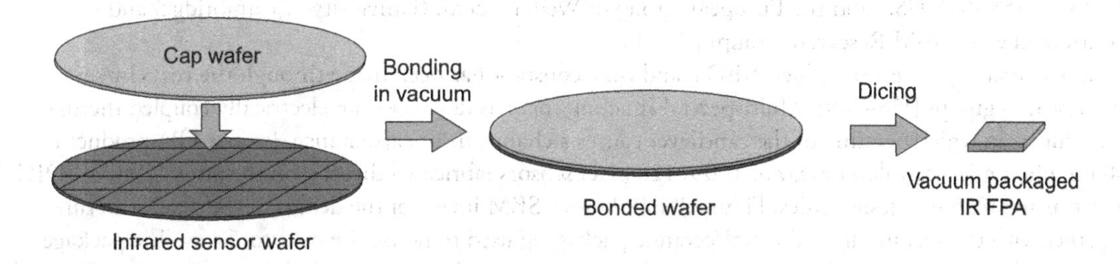

Figure 25.36 Wafer-level vacuum packaging technology.

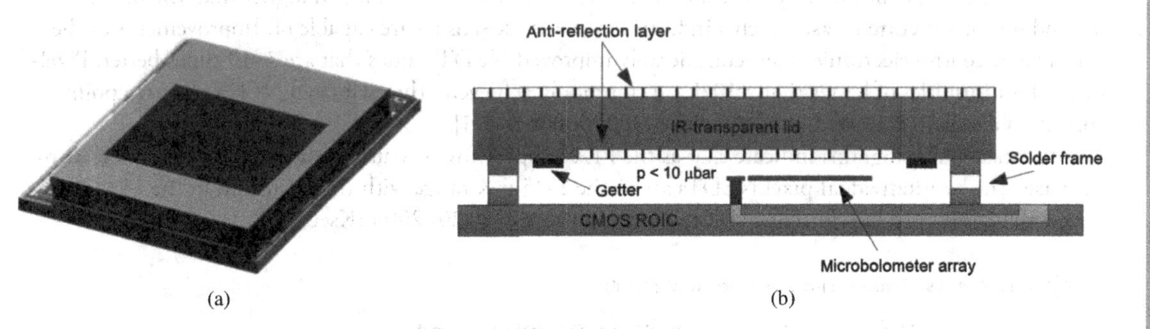

Figure 25.37 Wafer-level packaged microbolometer array: (a) stacking a cap wafer onto a CMOS ROIC wafer populated with pixels, (b) chip-scale vacuum package.

small groove in the rim substrate. Finally, heating two substrates to the eutectic temperature (for PbSn only 240°C) causes the rim material to reflow. Electrical interconnections from bonding pads to the readout circuits go through the sealing area. By stacking a cap wafer onto a CMOS ROIC, smaller and lighter devices with leading edge performance and game-changing lower costs are delivered. This packaging enables the manufacturing of compact detectors with advanced integrated features that will drive the next phase in the evolution of uncooled IR imaging.

While microbolometer arrays are encapsulated in vacuum packages, many thermopile IR arrays are operated in the atmospheric pressure. However, the low sensitivity of thermopile sensors can be partially compensated by applying vacuum packaging. The thermal conduction through the atmosphere occupies a large fraction in the total thermal conduction as the pixel size is reduced. Therefore, some thermopile arrays in Table 25.1 are vacuum packaged. Vacuum packaging technology will become indispensable in the future even for small-format thermopile arrays.

25.5 NOVEL UNCOOLED FOCAL PLANE ARRAYS

Development of novel uncooled micromechanical thermal detectors is a result of recent advances in the technology of MEMS systems. They show great promise as imaging chips for use in low-cost, high-performance cameras. The high sensitivity of the cantilevers allows the individual pixels in the array to be shrunk from their current size of 50–20 μm while maintaining the low-distortion IR imaging [150,151]. The working principle of this sensor is described in Chapter 11. As marked, to generate a thermal image, the thermomechanical deflections of the free-standing microstructures are read by capacitive, piezoresistive, or optical means.

Figure 25.38 illustrates a milestone timeline of the micro-thermomechanical IR sensor developments. Steffanson and Rangelow have reviewed the representative milestone of this technology together with the analysis of important results of notable groups during the last 20 years. [152]. Research and development activities in this field began in the early 1990s, especially by the groups of Oak Ridge National Laboratory (ORNL) [153,154], USA and the European group of Welland et al. (University of Cambridge) and Gimzewski et al. (IBM Research Group) [155–157].

Considerable progress to reduce NEDT and time constant has been made through the years by several research groups [158–169]. Multispectral Imaging Inc. has developed an electrically coupled thermal transducers in which bending of the cantilever causes a change in its capacitance. Its first FPA product is a 50 μm pitch microcantilever array of 160 × 120 pixel sensors fabricated directly on the array CMOS ROIC wafers using 0.25 μm design rules. Figure 25.39 shows a SEM image of the details of the pixel structure together with the vacuum hybrid metal/ceramic packaging used to house the detector array. The package contains a TE cooler to stabilize the array temperature along with a nonevaporable getter to remove residual gas in the package after sealing. Including a thermal compensation technique [150] enables operation of the detector arrays without the TE cooler.

The NEDT values of the early fabricated FPAs were around 1 K–2 K, much higher than the modeling and sensor structure measurements indicating what these sensors are capable of. Improvements of the camera and control electronics have considerably improved NEDT values that are 3–10 times better. Pixel-to-pixel uniformities of better than ±10% are commonly achieved. The estimated NEDTs for two point correction and gain corrected images is about 300–500 mK [151].

The measured histograms indicate that as the FPA temperature is reduced, the peak NEDT values also decrease. The best individual pixel NEDTs are in the 1–15 mK range with time constant in the 15 ms range, giving best pixel performance NEDT × τ_{th} in the range 120–200 mKsec [152].

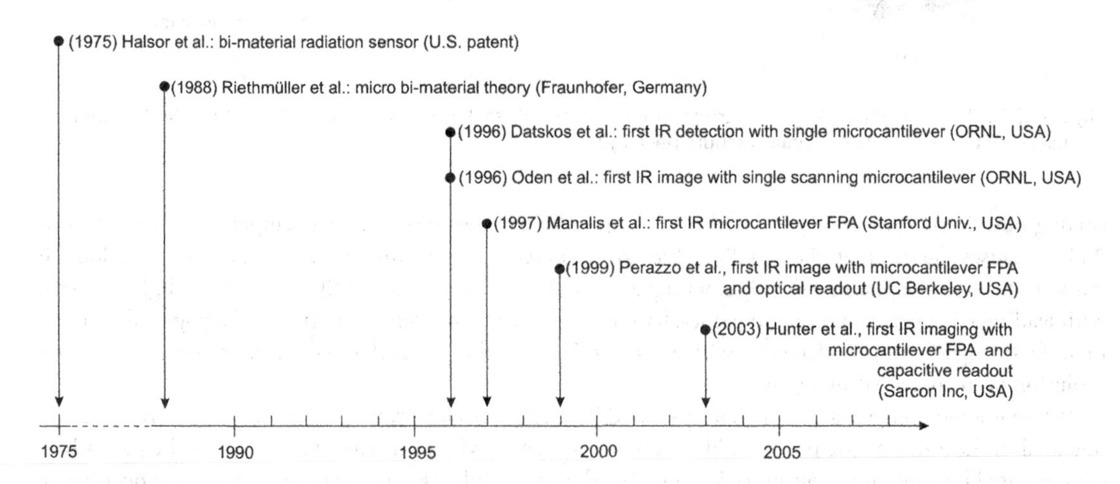

Figure 25.38 A timeline of the micro-thermomechanical IR sensor technology milestones. (Adapted after Ref. [152])

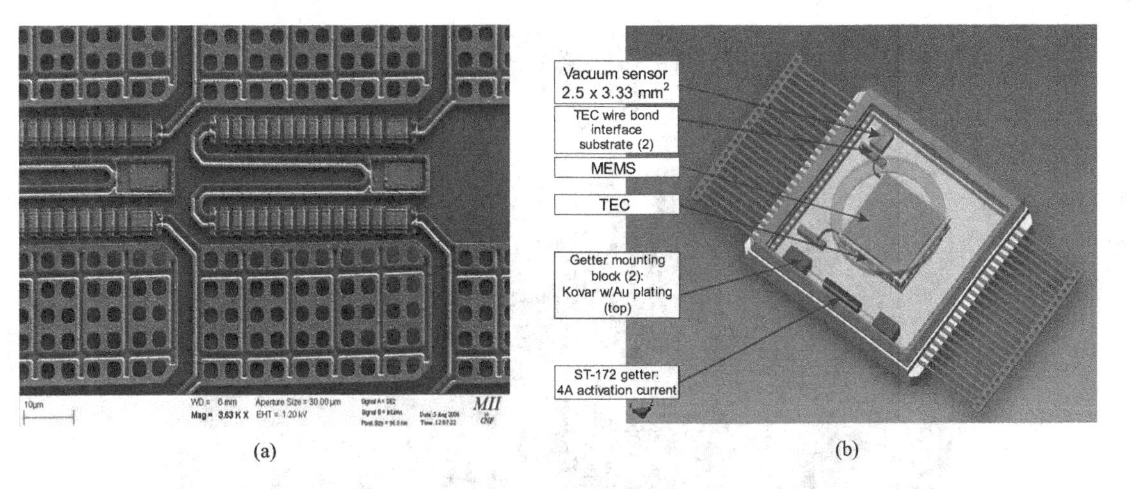

Figure 25.39 50 μm pitch microcantilever 160 × 120 FPA: (a) an SEM image of the details of the pixel structure and (b) schematic of the vacuum sensor package. (From Ref. [151])

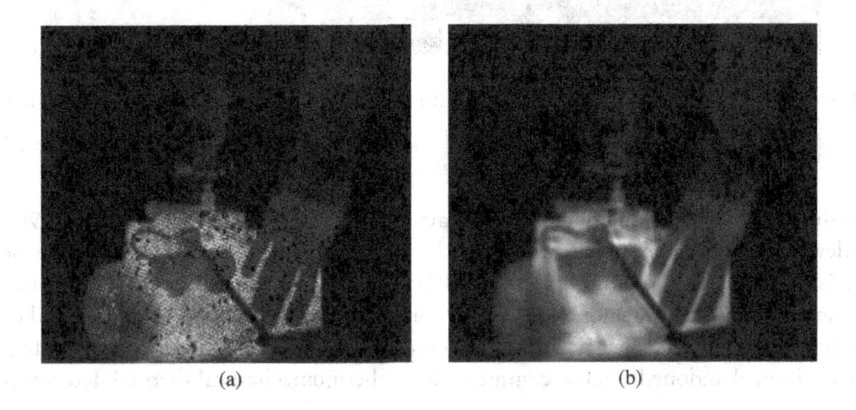

Figure 25.40 Readout of 256 × 256 MEMS IR FPA: (a) after baseline subtraction and with (b) inpainting method. (From Ref. [160])

Several research groups are involved in the development of optical-readable imaging arrays. At the SPIE Aerosense Conference in Orlando, Florida, Sarcon Microsystems of Knoxville, Tennessee, and its partner the Sarnoff Corporation of Princeton, New Jersey, presented a prototype IR imaging system that exploits a 320 × 240 pixel MEMS [170] (see Figure 11.4). In 2006, Grbovic et al. of the Oak Ridge National Laboratory (Tennessee) reported a high resolution 256 × 256 pixel array [159]. The fabricated arrays had NEDT and response time of below 500 mK and 6 ms, respectively. Image quality can be improved by automatic postprocessing of artifacts arising from noise and nonresponsive pixels, which is shown in Figure 25.40 [160]. The image quality of inpainting method is improved if an accurate mask is determined during the device calibration process. For inpainting methods, it is better to work with less true image pixels than to include corrupted pixels in the mask.

The larger micro-thermomechanical sensor array was demonstrated in 2013 by a research group from Ilmenau University of Technology, Germany. Figure 25.41 shows this 640 × 480 pixel sensor with a 50 μm pitch. It was shown that approximately 50% of the incoming IR flux is lost due to high back side illumination [152]. Removing the substrate underneath the pixel resulted in a 100% increase of responsivity.

Agiltron Inc. produced a 280 × 240 photomechanical IR sensor with an optical readout for both middle wavelength infrared (MWIR) and long wavelength infrared (LWIR) imaging at a speed of up to 1,000 frames per second [161]. Results of the detection of rapid occurrence events, such as gunfire and rocket travel, were reported. At the present stage of development, the imager has a NEDT of approximately 120 mK at *f*/1 optics.

Figure 25.41 Photography of a 640 × 480 pixel sensor (top) with its micrograph details (bottom). (After Ref. [152])

The successful creation of the microbolometer arrays by Honeywell at the beginning of 1990s has inspired the development of alternative MEMS-based technologies including the thin-film ferroelectric [108], bimaterial [171,172], and thermo-optical IR FPAs [173]. Although they challenged the microbolometer's dominance and made considerable inroads, the microbolometer seems unthreatened and continues to hold the leading position in uncooled IR array technology. For this reason, most novel uncooled technologies have already been abandoned. Today, commercially, a thermomechanical thermal detector is not yet available.

REFERENCES

1. R. Watton and M. V. Mansi, "Performance of a thermal imager employing a hybrid pyroelectric detector array with MOSFET readout," *Proc. SPIE* 865, 78–85, 1987.
2. P. W. Kruse, *Uncooled Thermal Imaging. Arrays, Systems, and Applications*, SPIE Press, Bellingham, WA, 2001.
3. http://www.flir.com/home/news/details/?ID=74197.
4. P. W. Kruse, "A comparison of the limits to the performance of thermal and photon detector imaging arrays," *Infrared Phys. Technol.*, 36, 869–82, 1995.
5. S. Horn, D. Lohrmann, P. Norton, K. McCormack, and A. Hutchinson, "Reaching for the sensitivity limits of uncooled and minimally-cooled thermal and photon infrared detectors," *Proc. SPIE* 5783, 401–11, 2005.
6. J. A. Ratches, "Current and future trends in military night vision applications," *Ferroelectrics* 342, 183–92, 2006.
7. A. W. Van Herwaarden, F. G. van Herwaarden, S. A. Molenaar, E. J. G. Goudena, M. Laros, P. M. Sarro, C. A. Schot, W. van der Vlist, L. Blarre, and J. P. Krebs, "Design and fabrication of infrared detector arrays for satellite attitude control," *Sens. Actuators* 83, 101–8, 2000.
8. M. Kimata, "Trends in small-format infrared array sensors", *Conference Sensors*, IEEE, 2013.
9. T. Kanno, M. Saga, S. Matsumoto, M. Uchida, N. Tsukamoto, A. Tanaka, S. Itoh, et al., "Uncooled infrared focal plane array having 128 × 128 thermopile detector elements," *Proc. SPIE* 2269, 450–9, 1994.
10. M. Hirota, Y. Nakajima, M. Saito, F. Satou, and M. Uchiyama, "Thermoelectric infrared imaging sensors for automotive applications," *Proc. SPIE* 5359, 111–25, 2004.
11. M. Hirota, Y. Nakajima, M. Saito, and M. Uchiyama, "120 × 90 element thermoelectric infrared focal plane array with precisely patterned au-black absorber," *Sens. Actuators* A135, 146–51, 2007.

12. A. D. Oliver and K. D. Wise, "A 1024-element bulk-micromachined thermopile infrared imaging array," *Sens. Actuators* 73, 222–31, 1999.
13. T. McManus and S. Mickelson, "Imaging radiometers employing linear thermoelectric arrays," *Proc. SPIE* 3698, 352–60, 1999.
14. AXT100 brochure, http://www.aas2.com/products/axt100/index.htm.
15. B. Forg, F. Herrmann, J. Schieferdecker, W. Leneke, M. Schulze, M. Simon, and K. Storck, "Thermopile sensor array with improved spatial resolution, sensitivity and image quality," *Sensor+Test Conference 2011 – IRS² Proceedings*, pp. 42–44.
16. M. C. Foote, E. W. Jones, and T. Caillat, "Uncooled thermopile infrared detector linear arrays with detectivity greater than 10^9 cmHz$^{1/2}$/W," *IEEE Trans. Electron Devices* 45, 1896–1902, 1998.
17. M. C. Foote and E. W. Jones, "High performance micromachined thermopile linear arrays," *Proc. SPIE* 3379, 192–7, 1998.
18. M. C. Foote and S. Gaalema, "Progress towards high-performance thermopile imaging arrays," *Proc. SPIE* 4369, 350–4, 2001.
19. R. E. Flannery and J. E. Miller, "Status of uncooled infrared imagers," *Proc. SPIE* 1689, 379–95, 1992.
20. R. A. Wood, "Micromachined bolometer arrays achieve low-cost imaging," *Laser Focus World*, 101–6, June 1993.
21. T. Ishikawa, M. Ueno, K. Endo, Y. Nakaki, H. Hata, T. Sone, and M. Kimata, "Low-cost 320 × 240 uncooled IRFPA using conventional silicon IC process," *Opto-Electron. Rev.*, 7, 297–303, 1999.
22. D. Murphy, M. Ray, R. Wyles, J. Asbrock, N. Lum, J. Wyles, C. Hewitt, A. Kennedy, and D. V. Lue, "High sensitivity 25 μm microbolometer FPAs," *Proc. SPIE* 4721, 99–110, 2002.
23. D. Murphy, A. Kennedy, M. Ray, R. Wyles, J. Wyles, J. Asbrock, C. Hewitt, D. Van Lue, and T. Sessler, "Resolution and sensitivity improvements for VO_x microbolometer FPAs," *Proc. SPIE* 5074, 402–13, 2003.
24. D. Murphy, M. Ray, J. Wyles, J. Asbrock, C. Hewitt, R. Wyles, E. Gordon, et al., "Performance improvements for VO_x microbolometer FPAs," *Proc. SPIE* 5406, 531–40, 2004.
25. D. Murphy, M. Ray, A. Kennedy, J. Wyles, C. Hewitt, R. Wyles, E. Gordon, et al., "High sensitivity 640 × 512 (20 μm pitch) microbolometer FPAs," *Proc. SPIE* 6206, 62061A, 2006.
26. D. Murphy, M. Ray, J. Wyles, C. Hewitt, R. Wyles, E. Gordon, K. Almada, et al., "640 × 512 17 μm microbolometer FPA and sensor development," *Proc. SPIE* 6542, 65421Z, 2007.
27. S. H. Black, R. Kraft, A. Medrano, T. Kocian, D. Bradstreet, R. Williams, and T. Yang, "Advances in high rate uncooled detector fabrication at Raytheon", *Proc. SPIE* 7660, 76600X-1–12, 2010.
28. S. H. Black, T. Sessler, E. Gordon, R. Kraft, T. Kocian, M. Lamb, R. Williams, and T. Yang, "Uncooled detector development at Raytheon", *Proc. SPIE* 8012, 80121A-1–12, 2011.
29. A. Kennedy, P. Masini, M. Lamb, J. Hamers, T. Kocian, E. Gordon, W. Parrish, R. Williams, and T. LeBeau, "Advanced uncooled sensor product development", *Proc. SPIE* 9451, 94511C-1–10, 2015.
30. M. N. Gurnee, M. Kohin, R. Blackwell, N. Butler, J. Whitwam, B. Backer, A. Leary, and T. Nielsen, "Developments in uncooled IR technology at BAE systems," *Proc. SPIE* 4369, 287–96, 2001.
31. R. Blackwell, S. Geldart, M. Kohin, A. Leary, and R. Murphy, "Recent technology advancements and applications of advanced uncooled imagers," *Proc. SPIE* 5406, 422–7, 2004.
32. R. J. Blackwell, T. Bach, D. O'Donnell, J. Geneczko, and M. Joswick, "17 μm pixel 640 × 480 microbolometer FPA development at BAE systems," *Proc. SPIE* 6542, 65421U, 2007.
33. R. Blackwell, D. Lacroix, T. Bach, J. Ishii, S. Hyland, J. Geneczko, S. Chan, B. Sujlana, and M. Joswick, "Uncooled VO_x systems at BAE systems," *Proc. SPIE* 6940, 694021, 2008.
34. L. Sengupta, P.-A. Auroux, D. McManus, D. A. Harris, R. J. Blackwell, J. Bryant, M. Boal, and E. Binkerd, "BAE Systems' SMART chip camera FPA development," *Proc. SPIE* 9451, 94511B-1–7, 2015.
35. P. E. Howard, J. E. Clarke, A. C. Ionescu, and C. Li, "DRS U6000 640 × 480 VO_x uncooled IR focal plane," *Proc. SPIE* 4721, 48–55, 2002.
36. P. E. Howard, J. E. Clarke, A. C. Ionescu, C. Li, and A. Frankenberger, "Advances in uncooled 1-mil pixel size focal plane products at DRS," *Proc. SPIE* 5406, 512–20, 2004.
37. C. Li, G. D. Skidmore, C. Howard, C. J. Han, L. Wood, D. Peysha, E. Williams, et al., "Recent development of ultra small pixel uncooled focal plane arrays at DRS," *Proc. SPIE* 6542, 65421Y, 2007.
38. G. D. Skidmore, "Uncooled 10 μmFPA development at DRS," *Proc. SPIE* 9819, 98191O-1–8, 2016.
39. W. Parish, J. T. Woolaway, G. Kincaid, J. L. Heath, and J. D. Frank, "Low cost 160 × 128 uncooled infrared sensor array," *Proc. SPIE* 3360, 111–9, 1998.
40. W. A. Terre, R. F. Cannata, P. Franklin, A. Gonzalez, E. Kurth, W. Parrish, K. Peters, T. Romeo, D. Salazar, and R. Van Ysseldyk, "Microbolometer production at Indigo Systems," *Proc. SPIE* 5406, 557–65, 2004.

41. T. Schimert, J. Brady, T. Fagan, M. Taylor, W. McCardel, R. Gooch, S. Ajmera, C. Hanson, and A. J. Syllaios, "Amorphous silicon based large format uncooled FPA microbolometer technology," *Proc. SPIE* 6940, 694023, 2008.
42. C. M. Hanson, S. K. Ajmera, J. Brady, T. Fagan, W. McCardel, D. Morgan, T. Schimert, A. J. Syllaios, M. F. Taylor, "Small pixel a-Si/a-SiGe bolometer focal plane array technology at L-3 Communications", *Proc. SPIE* 7660, 76600R-1–8, 2010.
43. H. Jerominek, T. D. Pope, C. Alain, R. Zhang, F. Picard, M. Lehoux, F. Cayer, S. Savard, C. Larouche, and C. Grenier, "Miniature VO_2-based bolometric detectors for high-resolution uncooled FPAs," *Proc. SPIE* 4028, 47–56, 2000.
44. T. D. Pope, H. Jeronimek, C. Alain, F. Cayer, B. Tremblay, C. Grenier, P. Topart, et al., "Commercial and custom 160 × 120, 256 × 1 and 512 × 3 pixel bolometric FPAs," *Proc. SPIE* 4721, 64–74, 2002.
45. C. Alain, H. Jerominek, P. A. Topart, T. D. Pope, F. Picard, F. Cayer, C. Larouche, S. Leclair, and B. Tremblay, "Microfabrication services at INO," *Proc. SPIE* 4979, 353–63, 2003.
46. P. Topart, C. Alain, L. LeNoc, S. Leclair, Y. Desroches, B. Tremblay, and H. Jerominek, "Hybrid micropackaging technology for uncooled FPAs," *Proc. SPIE* 5783, 544–50, 2005.
47. E. Mottin, J. Martin, J. Ouvrier-Buffet, M. Vilain, A. Bain, J. Yon, J. L. Tissot, and J. P. Chatard, "Enhanced amorphous silicon technology for 320 × 240 microbolometer arrays with a pitch of 35 μm," *Proc. SPIE* 4369, 250–6, 2001.
48. E. Mottin, A. Bain, J. Martin, J. Ouvrier-Buffet, S. Bisotto, J. J. Yon, and J. L. Tissot, "Uncooled amorphous silicon technology enhancement for 25 μm pixel pitch achievement," *Proc. SPIE* 4820, 200–07, 2003.
49. J. J. Yon, A. Astier, S. Bisotto, G. Chamming's, A. Durand, J. L. Martin, E. Mottin, J. L. Ouvrier-Buffet, and J. L. Tissot, "First demonstration of 25μm pitch uncooled amorphous silicon microbolometer IRFPA at LETI-LIR," *Proc. SPIE* 5783, 432–40, 2005.
50. J. J. Yon, E. Mottin, and J. L. Tissot, "Latest amorphous silicon microbolometer developments at LETI-LIR," *Proc. SPIE* 6940, 69401W, 2008.
51. J.L. Tissot, S. Tinnes, A. Durand, C. Minassian, P. Robert, and M. Vilain, High performance uncooled amorphous silicon VGA and XGA IRFPA with 17μm pixel-pitch," *Proc. SPIE* 7834, 78340K-1–8, 2010.
52. J.L. Tissot, P. Robert, A. Durand, S. Tinnes, E. Bercier, and A. Crastes, "Status of uncooled infrared detector technology at ULIS, France", *Def. Sci. J.*, 63(6), 545–9, 2013.
53. H. Wada, T. Sone, H. Hata, Y. Nakaki, O. Kaneda, Y. Ohta, M. Ueno, and M. Kimata, "YBaCuO uncooled microbolometer IR FPA," *Proc. SPIE* 4369, 297–304, 2001.
54. Y. Tanaka, A. Tanaka, K. Iida, T. Sasaki, S. Tohyama, A. Ajisawa, A. Kawahara, et al., "Performance of 320 × 240 uncooled bolometer-type infrared focal plane arrays," *Proc. SPIE* 5074, 414–24, 2003.
55. N. Oda, Y. Tanaka, T. Sasaki, A. Ajisawa, A. Kawahara, and S. Kurashina, "Performance of 320 × 240 bolometer-type uncooled infrared detector," *NEC Res. Dev.*, 44, 170–4, 2003.
56. S. Tohyama, M. Miyoshi, S. Kurashina, N. Ito, T. Sasaki, A. Ajisawa, and N. Oda, "New thermally isolated pixel structure for high-resolution uncooled infrared FPAs," *Proc. SPIE* 5406, 428–36, 2004.
57. T. Ishikawa, M. Ueno, Y. Nakaki, K. Endo, Y. Ohta, J. Nakanishi, Y. Kosasayama, H. Yagi, T. Sone, and M. Kimata, "Performance of 320 × 240 uncooled IRFPA with SOI diode detectors," *Proc. SPIE* 4130, 1–8, 2000.
58. Y. Kosasayama, T. Sugino, Y. Nakaki, Y. Fujii, H. Inoue, H. Yagi, H. Hata, M. Ueno, M. Takeda, and M. Kimata, "Pixel scaling for SOI-diode uncooled infrared focal plane arrays," *Proc. SPIE* 5406, 504–11, 2004.
59. T. Ishikawa, M. Ueno, K. Endo, Y. Nakaki, H. Hata, T. Sone, and M. Kimata, "640 × 480 pixel uncooled infrared with SOI diode detectors," *Proc. SPIE* 5783, 566–77, 2005.
60. M. Kimata, M. Uenob, M. Takedac, and T. Setod, "SOI diode uncooled infrared focal plane arrays," *Proc. SPIE* 6127, 61270X, 2006.
61. D. Fujisawa, T. Maeawa, Y. Ohta, Y. Kosasayama, T. Ohnakado, H. Hata, M. Ueno, H. Ohji, R. Sato, H. Katayama, T. Imai, and M. Ueno, "2-million-pixel SOI diode uncooled IRFPA with 15 μm pixel pitch," *Proc. SPIE* 8352, SPIE, 83531G-1–13, 2012.
62. P. A. Manning, J. P. Gillham, N. J. Parkinson, and T. P. Kaushal, "Silicon foundry microbolometers: The route to the mass-market thermal imager," *Proc. SPIE* 5406, 465–72, 2004.
63. V. N. Leonov, Y. Creten, P. De Moor, B. Du Bois, C. Goessens, B. Grietens, P. Merken, et al., "Small two-dimensional and linear arrays of polycrystalline SiGe microbolometers at IMEC-XenICs," *Proc. SPIE* 5074, 446–57, 2003.
64. U. Mizrahi, A. Fraenkel, L. Bykov, A. Giladi, A. Adin, E. Ilan, N. Shiloah, et al., "Uncooled detector development program at SCD," *Proc. SPIE* 5783, 551–8, 2005.

65. U. Mizrahi, L. Bikov, A. Giladi, A. Adin, N. Shiloah, E. Malkinson, T. Czyzewski, A. Amsterdam, Y. Sinai, and A. Fraenkel, "New features and development directions in SCD's μ-bolometer technology," *Proc. SPIE* 6940, 694020, 2008.
66. U. Mizrahi, N. Argaman, S. Elkind, A. Giladi, Y. Hirsh, M. Labilov, I. Pivnik, N. Shiloah, M. Singer, A. Tuito, M. Ben-Ezra, and I. Shtrichman, "Large-format 17μm high-end VO_x μ-bolometer infrared detector", *Proc. SPIE* 8704, 87041H, 2013.
67. L. Jiang, H. Liu, J. Chi, L. Qian, F. Pan, X. Liu, X. Zhu, and Z. Ma, "Uncooled infrared detector and imager development at DALI technology," *Proc. SPIE* 9451, 945119-1–6, 2015.
68. D. Weiler, F. Hochschulz, D. Würfel, R. Lerch, T. Geruschke, S. Wall, J. Heß, Q. Wang, and H. Vogt, "Uncooled digital IRFPA-family with 17μm pixel-pitch based on amorphous silicon with massively parallel Sigma-Delta-ADC readout," *Proc. SPIE* 9070, 90701M-1–6, 2014.
69. http://www.flir.com/uploadedFiles/Eurasia/Cores_and_Components/Technical_Notes/uncooled%20detectors%20BST.pdf.
70. FLIR Technical Note (2009), http://www.flir.com/uploadedFiles/Eurasia/Cores_and_Components/Technical_Notes/uncooled% 20detectors%20BST.pdf.
71. "Uncooled infrared imaging market commercial & military applications," *Market & Technology Report* – available in JULY 2011, Yole Development.
72. http://image-sensors-world.blogspot.com/2017/08/yole-uncooled-thermal-imaging-report.html.
73. M. Kohin and N. Butler, "Performance limits of uncooled VO_x microbolometer focal-plane arrays," *Proc. SPIE* 5406, 447–53, 2004.
74. A. Van Der Ziel, "Flicker noise in electronic devices," in *Adv. Electron. Electron Phys.*, 49, 225–97, 1979.
75. F. Niklaus, C. Jansson, A. Decharat, J.-E. Källhammer, H. Pettersson, and G. Stemme, "Low to medium vacuum atmosphere: performance model and tradeoffs", *Proc. SPIE* 6542, 1M-1–12, 2007.
76. G.D. Skidmore, C.J. Han, and C. Li, "Uncooled microbolometers at DRS and elsewhere through 2013", *Proc. SPIE* 9100, 910003-1–5, 2013.
77. M. Pisani, K. Ren, P. Kao, and S. Tadigadapa, "Application of micromachined Y-cut-quartz bulk acoustic wave rezonator for infrared sensing", *J. Microelectromech. Syst.*, 20, 288–296, 2011.
78. N. Roxhed, F. Niklaus, A.C. Fischer, F. Forsberg, L. Höglund, P. Ericsson, B. Samel, S. Wissmar, A. Elfving, and T.I. Simonsen, "Low-cost uncooled microbolometers for thermal imaging", *Proc. SPIE* 7726, 772611-1–10, 2010.
79. H. H. Radamson and M. Kolahdouz, "Group IV Material for Low Cost and High Performance Bolometers", in *Bolometers*, edited by U. Perera, InTech, London, 2012, Available from: http:/intechopen.com/books/bolometers/group-iv-materials-for-low-cost-and-high-performance-bolometers.
80. R. A. Wood, C. J. Han, and P. W. Kruse, "Integrated Uncooled IR Detector Imaging Arrays," in *Proceedings of IEEE Solid State Sensor and Actuator Workshop*, pp. 132–5, Hilton Head Island, SC, June 1992.
81. R. A. Wood, "Uncooled thermal imaging with monolithic silicon focal planes," *Proc. SPIE* 2020, 322–9, 1993.
82. W. J. Parrish and T. Woolaway, "Improvements in uncooled systems using bias equalization", *Proc. SPIE* 3698, 748–55, 1999.
83. F. Niklas, C. Vieider, and H. Jakobsen, "MEMS-based uncooled infrared bolometer arrays: A review," *Proc. SPIE* 6836, 68360D-1, 2007.
84. S. Eminoglu, D. Sabuncuoglu Tezcan, M. Y. Tanrikulu, and T. Akin, "Low-cost uncooled infrared detectors in CMOS process," *Sens. Actuators A* 109, 102–13, 2003.
85. C. Vieider, S. Wissmar, P. Ericsson, U. Halldin, F. Niklaus, G. Stemme, J.-E. Källhammer, et al., "Low-cost far infrared bolometer camera for automotive use," *Proc. SPIE* 6542, 65421L, 2007.
86. F. Forsberg, N. Roxhed, A. C. Fischer, B. Samel, P. Ericsson, N. Koivik, A. Lapadatu, M. Bring, G. Kittilsland, G. Stemme, and F. Niklaus, "Very large scale heterogeneous integration (VLSHI) and wafer-level vacuum packaging for infrared bolometer focal plane arrays", *Infrared Phys. Technol.*, 60, 251–9, 2013.
87. W. Radford, D. Murphy, A. Finch, K. Hay, A. Kennedy, M. Ray, A. Sayed, et al., "Sensitivity improvements in uncooled microbolometer FPAs," *Proc. SPIE* 3698, 119–30, 1999.
88. H.-K. Lee, J.-B. Yoon, E. Yoon, S.-B. Ju, Y.-J. Yong, W. Lee, and S.-G. Kim, "A high fill factor infrared bolometer using micromachined multilevel electrothermal structures," *IEEE Trans. Electron Devices* 46, 1489–91, 1999.
89. D. Lohrmann, R. Littleton, C. Reese, D. Murphy, and J. Vizgaitis, "Uncooled long-wave infrared small pixel focal plane array and system challenges," *Opt. Eng.*, 52(6), 061305-1–6, 2013.
90. S. Tohyama, T. Sasaki, T. Endoh, M. Sano, K. Katoha, S. Kurashina, M. Miyoshi, T. Yamazaki, M. Ueno, H. Katayama, and T. Imai, "Uncooled infrared detectors toward smaller pixel pitch with newly proposed pixel structure," *Proc. SPIE* 8012, 80121M-1–13, 2011.

91. S. Becker, P. Imperinetti, J.-J. Yon, J.-L. Ouvrier-Buffet, V. Goudon, A. Hamelin, C. Vialle, and A. Arnaud, "Latest pixel size reduction of uncooled IR-FPA at CEA, LETI," *Proc. SPIE* 8541, 85410C-1–7, 2012.
92. http://www.drsinfrared.com/AboutDRS/PressDetail.aspx?id=25.
93. G. Holst and R. Driggers, "Small detectors in infrared system design", *Opt. Eng.*, 51(9), 096401, 2012.
94. Y. Kosasayama, T. Sugino, Y. Nakaki, M. Ueno, and K. Kama, "High sensitive uncooled infrared FPA with SOI diode detectors," *ITE Tech. Rep.*, 32, 216, 2008.
95. A. Jahanzeb, C. M. Travers, Z. Celik-Butler, D. P. Butler, and S. G. Tan, "A semiconductor YBaCuO microbolometer for room temperature IR imaging," *IEEE Trans. Electron Devices* 44, 1795–1801, 1997.
96. M. Almasri, Z. Celik-Butler, D. P. Butler, A. Yaradanakul, and A. Yildiz, "Semiconducting YBaCuO microbolometers for uncooled broad-band IR sensing," *Proc. SPIE* 4369, 264–73, 2001.
97. H. Wada, T. Sone, H. Hata, Y. Nakaki, O. Kaneda, Y. Ohta, M. Ueno, and M. Kimata, "YBaCuO uncooled microbolometer IRFPA," *Proc. SPIE* 4369, 297–304, 2001.
98. A. Yildiz, Z. Celik-Butler, and D. P. Butler, "Microbolometers on a flexible substrate for infrared detection," *IEEE Sens. J.*, 4, 112–7, 2004.
99. S. A. Dayeh, D. P. Butler, and Z. Celik-Butler, "Micromachined infrared bolometers on flexible polyimide substrates," *Sens. Actuators* A118, 49–56, 2005.
100. R. W. Whatmore, "Pyroelectric devices and materials," *Rep. Prog. Phys.*, 49, 1335–86, 1986.
101. R. Watton, "Ferroelectric materials and design in infrared detection and imaging," *Ferroelectrics* 91, 87–108, 1989.
102. R. W. Whatmore, "Pyroelectric ceramics and devices for thermal infra-red detection and imaging," *Ferroelectrics* 118, 241–59, 1991.
103. R. Watton, "IR bolometers and thermal imaging: The role of ferroelectric materials," *Ferroelectrics* 133, 5–10, 1992.
104. C. M. Hanson, "Uncooled thermal imaging at Texas Instruments," *Proc. SPIE* 2020, 330–9, 1993.
105. H. Betatan, C. Hanson, and E. D. G. Meissner, "Low cost uncooled ferroelectric detector," *Proc. SPIE* 2274, 147–56, 1994.
106. M. A. Todd, P. A. Manning, O. D. Donohue, A. G. Brown, and R. Watton, "Thin film ferroelectric materials for microbolometer arrays," *Proc. SPIE* 4130, 128–39, 2000.
107. R. W. Whatmore and R. Watton, "Pyroelectric Materials and Devices," in *Infrared Detectors and Emitters: Materials and Devices*, eds. P. Capper and C. T. Elliott, 99–147, Kluwer Academic Publishers, Boston, MA, 2000.
108. C. M. Hanson, H. R. Beratan, and J. F. Belcher, "Uncooled infrared imaging using thin-film ferroelectrics," *Proc. SPIE* 4288, 298–303, 2001.
109. R. W. Whatmore, Q. Zhang, C. P. Shaw, R. A. Dorey, and J. R. Alock, "Pyroelectric ceramics and thin films for applications in uncooled infra-red sensor arrays," *Phys. Scr.*, *T* 129, 6–11, 2007.
110. C. M. Hanson, H. R. Beratan, and D. L. Arbuthnot, "Uncooled thermal imaging with thin-film ferroelectric detectors," *Proc. SPIE* 6940, 694025, 2008.
111. N. Butler and S. Iwasa, "Solid state pyroelectric imager," *Proc. SPIE* 1685, 146–54, 1992.
112. R. Takayama, Y. Tomita, J. Asayama, K. Nomura, and H. Ogawa, "Pyroelectric infrared array sensors made of *c*-axis-oriented La-modified $PbTiO_3$ thin films," *Sens. Actuators* A21–A23, 508–12, 1990.
113. R. Watton, F. Ainger, S. Porter, D. Pedder, and J. Gooding, "Technologies and performance for linear and two dimensional pyroelectric arrays," *Proc. SPIE* 510, 139–48, 1984.
114. D. E. Burgess, P. A. Manning, and R. Watton, "The theoretical and experimental performance of a pyroelectric array imager," *Proc. SPIE* 572, 2–6, 1985.
115. D. E. Burgess, "Pyroelectric in harsh environment," *Proc. SPIE* 930, 139–50, 1988.
116. R. Takayama, Y. Tomita, K. Iijima, and I. Ueda, "Pyroelectric properties and application to infrared sensors of $PbTiO_3$ and $PbZrTiO_3$ ferroelectric thin films," *Ferroelectrics* 118, 325–42, 1991.
117. R. Watton, P. Manning, D. Burgess, and J. Gooding, "The pyroelectric/CCD focal plane hybrid: Analysis and design for direct charge injection," *Infrared Phys.*, 22, 259–75, 1982.
118. V. Norkus, T. Sokoll, G. Gerlach, and G. Hofmann, "Pyroelectric infrared arrays and their applications," *Proc. SPIE* 3122, 409–19, 1997.
119. P. Muralt, "Micromachined infrared detectors based on pyroelectric thin films," *Rep. Prog. Phys.*, 64, 1339–88, 2001.
120. A. J. Holden, "Pyroelectric sensor arrays for detection and thermal imaging", *Proc. SPIE* 8704, 87041N-1–10, 2013.
121. R. Köhler, D. Wassilew, V. Norkus, M. Schossing, and G. Hofmann, "Enhanced pyroelectric linear arrays for infrared spectroscopy", *AMA Conferences 2017 – Sensor 2017* and *IRS² 2017*, 754–9, 2017.

122. http://www.dias-infrared.de/pdf/pyrosens_arrays_eng_mail.pdf.
123. http://www.dias-infrared.com/news/publications.
124. J. D. Zook and S. T. Liu, "Pyroelectric effects in thin films," *J. Appl. Phys.*, 49, 4604–6, 1978.
125. M. Okuyama, H. Seto, M. Kojima, Y. Matsui, and Y. Hamakawa, "Integrated pyroelectric infrared sensor using $PbTiO_3$ thin film," *Jpn. J. Appl. Phys.*, 22, 465–68, 1983.
126. R. Takayama, Y. Tomita, J. Asayama, K. Nomura, and H. Ogawa, "Pyroelectric infrared array sensors made of c-axis-oriented La-modified $PbTiO_3$ thin films," *Sens. Actuators* A21–23, 508–12, 1990.
127. N. Neumann, R. Köhler, and G. Hofmann, "Pyroelectric thin film sensors and arrays based on P(VDF/TrFE)," *Integr. Ferroelectr.*, 6, 213–30, 1995.
128. M. Kohli, C. Wuethrich, K. Brooks, B. Willing, M. Forster, P. Muralt, N. Setter, and P. Ryser, "Pyroelectric thin-film sensor array," *Sen. Actuators* A60, 147–53, 1997.
129. B. Willing, M. Kohli, P. Muralt, N. Setter, and O. Oehler, "Gas spectrometry based on pyroelectric thin film arrays integrated on silicon," *Sens. Actuators* A66, 109–13, 1998.
130. C. Shaw, S. Landi, R. Whatmore, and P. Kirby, "Development aspects of an integrated pyroelectric array incorporating and thin PZT film and radiation collectors," *Integr. Ferroelectr.*, 63, 93–97, 2004.
131. R. W. Whatmore, "Uncooled pyroelectric detector arrays using ferroelectric ceramics and thin films," *MEMS Sens. Technol.*, 10812, 1–9, 2005.
132. M. A. Todd and R. Watton, "Laser-assisted etching of ferroelectric ceramics for the reticulation of IR detector arrays," *Proc. SPIE* 1320, 95, 1990.
133. D. L. Polla, C. Ye, and T. Tamagawa, "Surface-micromachined $PbTiO_3$ pyroelectric detectors," *Appl. Phys. Lett.*, 59, 3539–41, 1991.
134. L. Pham, W. Tjhen, C. Ye, and D. L. Polla, "Surface-micromachined pyroelectric infrared imaging array with vertically integrated signal processing circuitry," *IEEE Trans. Ultrason. Ferroelectrics Freq. Control* 41, 552–5, 1994.
135. S. G. Porter, R. Watton, and R. K. McEwen, "Ferroelectric arrays: The route to low cost uncooled infrared imaging," *Proc. SPIE* 2552, 573–82, 1995.
136. J. F. Belcher, C. M. Hanson, H. R. Beratan, K. R. Udayakumar, and K. L. Soch, "Uncooled monolithic ferroelectric IRFPA technology," *Proc. SPIE* 3436, 611–22, 1998.
137. M. Z. Tidrow, W. W. Clark, W. Tipton, R. Hoffman, W. Beck, S. C. Tidrow, D. N. Robertson, et al., "Uncooled infrared detectors and focal plane arrays," *Proc. SPIE* 3553, 178–87, 1997.
138. R. K. McEwen and P. A. Manning, "European uncooled thermal imaging sensors," *Proc. SPIE* 3698, 322–37, 1999.
139. X. He, G. Karunasiri, T. Mei, W. J. Zeng, P. Neuzil, and U. Sridhar, "Performance of microbolometer focal plane arrays under varying pressure," *IEEE Electron Device Lett.*, 21, 233–5, 2000.
140. M. Kimata, "Trends in small-format infrared array sensors", *Sensors*, IEEE 2013, DOI: 10.1109/ICSENS.2013.6688495.
141. T. Ito, T. Tokuda, M. Kimata, H. Abe, and N. Tokashiki, "Vacuum packaging technology for mass production of uncooled IRFPAs", *Proc. SPIE* 7298, 72982A-1–10, 2009.
142. M. Takeda, H. Hata, Y. Nakaki, Y. Kosasayama, and M. Kimata, "Chip scale vacuum packaging for uncooled IRFPA," *IEEJ Trans. Fundam. Mater.*, 127(7), 405–410, 2007.
143. S. Ajmera, J. Brady, C. Hanson, T. Schimert, A. J. Syllaios, and M. Taylor, "Performance improvement in amorphous silicon based uncooled microbolometers through pixel design and materials development," *Proc. SPIE* 8012, 80121L-1–8, 2011.
144. C. Li, C. J. Han, G. D. Skidmore, G. Cook, K. Kubala, R. Bates, D. Temple, J. Lannon, A. Hilton, K. Glukh, and B. Hardy, "Low cost uncooled VO_x infrared camera development", *Proc. SPIE* 8704, 87041L-1–10, 2013.
145. G. Dumont, W. Rabaud, X. Baillin, J. L. Pornin, L. Carle, V. Goudon, C. Vialle, M. Pellat, and A. Amaud, "Pixel level packaging for uncooled IRFPA," *Proc. SPIE* 8012, 801211-1–7, 2011.
146. J. J. Yon, G. Dumont, V. Goudon, S. Becker, and A. Arnaud, "Latest improvements in microbolometer thin film packaging: Paving the way for low cost consumer applications," *Proc. SPIE* 9070, 90701N-1–8, 2014.
147. J. Lee, C. Rodriguez, and R. Blackwell, "BAE Systems' 17μm LWIR camera core for civil, commercial and military applications," *Proc. SPIE* 8704, 87041J-16, 2013.
148. V. K. Lindroos, M. Tilli, A. Lehto, and T. Motorka, *Handbook of Silicon Based MEMS Materials and Technologies*, William Andrew Publishing, Norwich, NY, 2008.
149. P. De Moor, J. John, S. Sedky, and C. Van Hoof, "Lineał arrays of fast uncooled poly SiGe microbolometers for IR detection," *Proc. SPIE* 4028, 27–34, 2000.
150. S. R. Hunter, G. S. Maurer, G. Simelgor, S. Radhakrishnan, and J. Gray, "High sensitivity 25μm and 50μm pitch microcantilever IR imaging arrays," *Proc. SPIE* 6542, 65421F, 2007.

151. S. R. Hunter, G. Maurer, G. Simelgor, S. Radhakrishnan, J. Gray, K. Bachir, T. Pennell, M. Bauer, and U. Jagadish, "Development and optimization of microcantilever based IR imaging arrays," *Proc. SPIE* 6940, 694013, 2008.
152. M. Steffanson and I.W. Rangelow, "Microthermomechanical infrared sensors," *Opto-Electron. Rev.*, 22, 1–15, 2014.
153. T. Thundat, S. L. Sharp, W. G. Fisher, R. J. Warmack, and E. A. Wachter, "Micromechanical radiation dosimeter", *Appl. Phys. Lett.*, 66, 1563–5, 1995.
154. T. Thundat, R. J. Warmack, G. Y. Chen, and D. P. Allison. "Thermal and ambient–induced deflections of scanning force microscope cantilevers", *Appl. Phys. Lett.*, 64, 2894–6, 1994.
155. J. R. Barnes, R. J. Stephenson, C. N. Woodburn, S. J. O'Shea, M. E. Welland, T. Rayment, J. K. Gimzevski, and C. H. Gerber, "A femtojoule colorimeter using micromechanical sensors", *Rev. Sci. Instrum.*, 65, 3793–8, 1994.
156. J. K. Gimzewski, C. H. Gerber, E. Meyer, and R. R. Schlittler, "Observation of a chemical reaction using a micromechanical sensor", *Chem. Phys. Lett.*, 217, 589–94, 1994.
157. J. R. Barnes, R. J. Stephenson, M. E. Welland, C. H. Gerber, and J. K. Gimzewski, "Photothermal spectroscopy with femtojoule sensitivity using a micromechanical device", *Nature* 372, 79–81, 1994.
158. J. Zhao, "High sensitivity photomechanical MW-LWIR imaging using an uncooled MEMS microcantilever array and optical readout," *Proc. SPIE* 5783, 506–13, 2005.
159. D. Grbovic, N. V. Lavrik, and P. G. Datskos, "Uncooled infrared imaging using bimaterial microcantilever arrays," *Appl. Phys. Lett.*, 89, 073118, 2006.
160. N. Lavrik, R. Archibald, D. Grbovic, S. Rajic, and P. Datskos, "Uncooled MEMS IR imagers with optical readout and image processing," *Proc. SPIE* 6542, 65421E, 2007.
161. J. P. Salerno, "High frame rate imaging using uncooled optical readout photomechanical IR sensor," *Proc. SPIE* 6542, 65421D, 2007.
162. M. Wagner, E. Ma, J. Heanue, and S. Wu, "Solid state optical thermal imagers," *Proc. SPIE* 6542, 65421P, 2007.
163. M. Wagner, "Solid state optical thermal imaging: Performance update," *Proc. SPIE* 6940, 694016, 2008.
164. B. Jiao, C. Li, D. Chen, T. Ye, S. Shi, Y. Qu, L. Dong, et al., "A novel opto-mechanical uncooled infrared detector," *Infrared Phys. Technol.*, 51, 66–72, 2007.
165. S. Shi, D. Chen, B. Jiao, C. Li, Y. Qu, Y. Jing, T. Ye, et al., "Design of a novel substrate-free double-layer-cantilever FPA applied for uncooled optical-readable infrared imaging system," *IEEE Sens. J.*, 7, 1703–10, 2007.
166. F. Dong, Q. Zhang, D. Chen, Z. Miao, Z. Xiong, Z. Guo, C. Li, B. Jiao, and X. Wu, "Uncooled infrared imaging device based on optimized optomechanical micro-cantilever array," *Ultramicroscopy* 108, 579–88, 2008.
167. X. Yu, Y. Yi, S. Ma, M. Liu, X. Liu, L. Dong, and Y. Zhao, "Design and fabrication of a high sensitivity focal plane array for uncooled IR imaging," *J. Micromech. Microeng.*, 18, 057001, 2008.
168. M. Steffanson, K. Gorovoy, M. Holz, T. Ivanov, R. Kampmann, R. Kleindienst, S. Sinzinger, and W. Rangelow, "Low-Cost Infrared Detector Using Thermomechanical Micro-Mirror Array with Optical Readout", in AMA Conferences 2013, *Sensor* 2013, *Opto* 2013, pp. 85–88, 2013.
169. U. Adiyan, F. Çivitçi, O. Ferhanoğlu, H. Torun, and H. Urey, "A 35-μm pitch IR thermo-mechanical MEMS sensor with AC-coupled optical readout," *IEEE J. Selec. Top. Quant. Electron.*, 21(4), 2701306, 2015.
170. http://optics.org/article/17600.
171. R. Amantea, C. M. Knoedler, F. P. Pantuso, V. K. Patel, D. J. Sauer, and J. R. Tower, "An uncooled IR imager with 5 mK NETD," *Proc. SPIE* 3061, 210–222, 1997.
172. T. Ishizuya, J. Suzuki, K. Akagawa, and T. Kazama, "Optically readable bi-materials infrared detector," *Proc. SPIE* 4369, 342–9, 2001.
173. M. Wu, J. Cook, R. D. Vito, J. Li, E. Ma, R. Murano, N. Nemchuk, M. Tabasky, and M. Wagner, "Novel low-cost uncooled infrared camera," *Proc. SPIE* 5783, 496–505, 2005.

26 Photon detector focal plane arrays

Looking back over the past several hundred years, we notice that following the invention and evolution of optical systems (telescopes, microscopes, eyeglasses, cameras, etc.) the optical image was formed on the human retina, photographic plate, or films. The birth of photodetectors can be dated back to 1873 when Smith discovered photoconductivity in selenium. Progress was slow until 1905, when Einstein explained the newly observed photoelectric effect in metals, and Planck solved the blackbody emission puzzle by introducing the quanta hypothesis. Applications and new devices soon flourished, pushed by the dawning technology of vacuum tube sensors developed in the 1920s and 1930s, culminating in the advent of television (TV). Zworykin and Morton, the celebrated fathers of videonics, on the last page of their legendary book *Television* (1939) concluded that "*when rockets will fly to the moon and to other celestial bodies, the first images we will see of them will be those taken by camera tubes, which will open to mankind new horizons.*" Their foresight became a reality with the Apollo and Explorer missions. Photolithography enabled the fabrication of silicon monolithic imaging focal planes for the visible spectrum beginning in the early 1960s. Some of these early developments were intended for a picture-phone, other efforts were for TV cameras, satellite surveillance, and digital imaging. Infrared (IR) imaging has been vigorously pursued in parallel with visible imaging because of its utility in military applications. More recently (1997), the charge-coupled device (CCD) camera aboard the Hubble space telescope delivered a deep-space picture, a result of 10 days' integration, featuring galaxies of the thirtieth magnitude—an unimaginable figure even for astronomers of our generation. Thus, photodetectors continue to open to mankind the most amazing new horizons.

IR system performance is highly scenario dependent and requires the designer to account for numerous different factors when specifying detector performance. It means that a good solution for one application may not be as suitable for a different application. In general, detector material is primarily selected based on the wavelength of interest, performance criteria, and operating temperature (see Figure 26.1). Although efforts have been made to develop monolithic structures using a variety of IR photodetector materials (including narrow-gap semiconductors) over the past 40 years, only a few have matured to a level of practical use. These include Si, PtSi, and more recently PbS, PbSe. Other IR material systems (InGaAs, InSb, HgCdTe, InAs/GaSb type-II superlattice, GaAs/AlGaAs quantum well IR photoconductor (QWIP), and extrinsic silicon) are used in hybrid configurations. Table 26.1 contains a description of representative IR focal plane arrays (FPAs) that are commercially available as standard products and/or catalogue items from the major manufacturers.

A suitable detector material for near-IR (1.0–1.7 μm) spectral range is Si and InGaAs lattice-matched to the InP. Various HgCdTe alloys, in both photovoltaic (PV) and photoconductive configurations, cover from 0.7 μm to over 20 μm. InAs/GaSb strained-layer superlattices have emerged as an alternative to the HgCdTe. Impurity-doped (Sb, As, and Ga) silicon-blocked impurity band (BIB) detectors operating at 10 K have a spectral response cutoff in the range of 16–30 μm.

This chapter is a guide over the arrays of photon detectors sensing IR radiation.

26.1 INTRINSIC SILICON ARRAYS

Silicon is the semiconductor that has dominated the electronic industry for over 50 years. While the first transistor fabricated in Ge and III-V semiconductor material compounds may have higher mobilities, higher saturation velocities, or larger bandgaps, silicon devices account for over 97% of all microelectronics [2]. The main reason is that silicon is the cheapest microelectronic technology for integrated circuits.

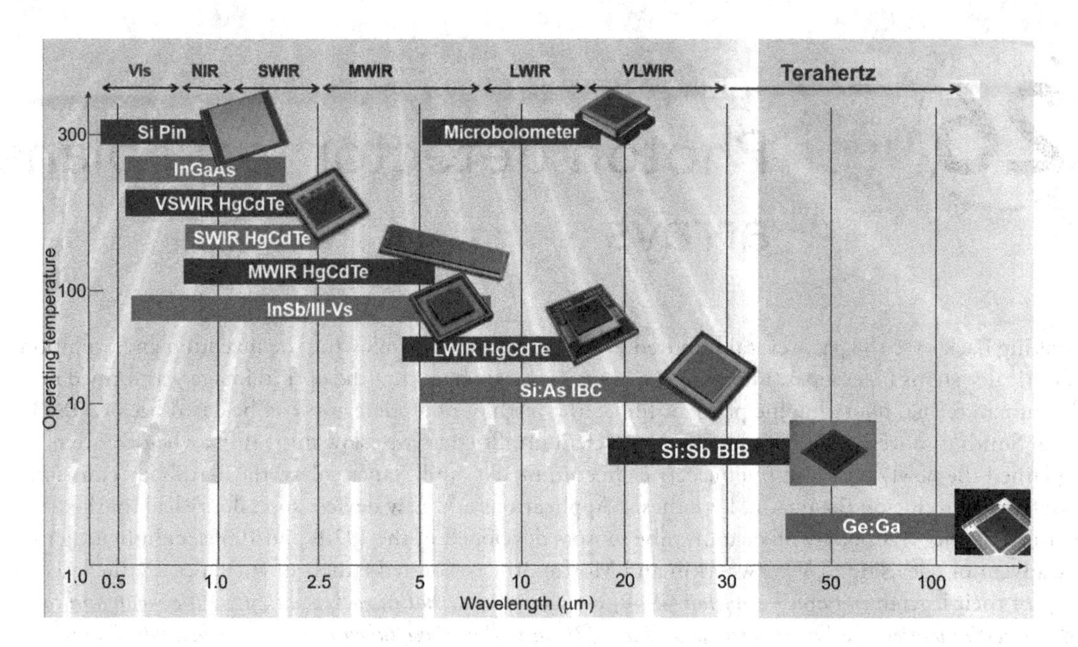

Figure 26.1 Detector materials that have the largest interest for IR detector technology. Vis, visible; NIR, near IR; Si PIN, silicon photodiodes; VSWIR, visible/short wave IR. (Adapted after Ref. [1])

The reason for the dominance of silicon can be traced to a number of natural properties of silicon but, more importantly, two insulators of silicon, SiO_2 and Si_3N_4, allow deposition and selective etching processes to be developed with exceptionally high uniformity and yield.

Monolithic imaging FPAs for the visible spectrum, both CCD and complementary metal-oxide semiconductor (CMOS), began in the late 1960s. At present, two silicon monolithic technologies provide the bulk of devices in the markets of camcorders and digital cameras: CCD and CMOS imagers. The fundamental performance parameters common to both CCD and CMOS imagers have been compared by Janesick [3,4].

Since their invention in 1969, the CCDs have become the detector of choice for high-quality imaging in a wide range of fields. They have the advantages of excellent resolution, 100% fill factor, greater than 90% peak quantum efficiency (QE), excellent charge transfer efficiency (CTE), and very low dark current with sufficient cooling. The importance of the CCD to mankind was recognized by the Nobel Prize in Physics in 2009.

Initially, devices based on CCD technology presented better image quality than the CMOS image sensor (CIS) devices. CIS devices suffered from higher noise levels and were used mainly for applications where the image quality was a less important factor than costs. CCDs were dominant in the imaging sensors market until the 1990s because they produced better image quality with the manufacturing technology available. However, the CCD technology has been surpassed in the market volume by CIS technology. The principal commercial reason is that the CCD manufacturing is costly compared to CIS. The CMOS manufacture process is largely used in most of the other larger electronic markets, such as computing and communication. This widespread utilization drives the cost per unit to much lower levels than CCD processes. Lower power consumption and on-chip functional integration are other driving factors leading to the preponderance of CIS devices. Moreover, the consistent reduction of CMOS feature size following the empirical Moore law also enables pixel size reduction which further drives their cost down. Finally, introduction of series innovations in CMOS processes including the development of buried (or pinned) photodiodes (previously used in CCD imagers) in past decades result that current CIS devices present comparable image quality to CCDs.

Digital CCD silicon image sensors were developed approximately 40 years ago, ushering in the era of digital photography. In visible imaging, CCD arrays integrate the readout and sensor in a combined

Table 26.1 Representative IR hybrid FPAs offered by some major manufactures

MANUFACTURER/WEB SITE	SIZE/ARCHITECTURE	PIXEL SIZE (μm)	DETECTOR MATERIAL	SPECTRAL RANGE (μm)	OPERATING TEMPERATURE (K)	$D^*(\lambda_p)$ (cmHz$^{1/2}$ W^{-1})/NEDT (mK)
Sensors Unlimited www.sensorsinc.com	320 × 256	12.5 × 12.5	InGaAs	0.7–1.7	300	12.9×10^{13}
	320 × 256	25 × 25	InGaAs	0.4–1.7	300	$<5 \times 10^{12}$
	640 × 512	25 × 25	InGaAs	0.7–1.7	300	4.2×10^{13}
	1280 × 1024	12.5 × 12.5	InGaAs	0.4–1.7	300	
Raytheon Vision Systems www.raytheon.com/businesses/ncs/rvs/index.html	1024 × 1024	30 × 30	InSb	0.6–5.0	50	
	2048 × 2048 (Orion II)	25 × 25	HgCdTe	0.6–5.0	32	
	2048 × 2048 (Virgo-2k)	20 × 20	HgCdTe	0.8–2.5	4–10	
	2048 × 2048	15 × 15	HgCdTe/Si	3.0–5.0	78	23
	1024 × 1024	25 × 25	Si:As	5–28	6.7	
	2048 × 1024	25 × 25	Si:As	5–28		
Teledyne Imaging Sensors http://teledynesi.com/imaging/	4096 × 4096 (H4RG)	10 × 10 or 15 × 15	HgCdTe	1.0–1.7	120	
	4096 × 4096 (H4RG)	10 × 10 or 15 × 15	HgCdTe	1.0–2.5	77	
	4096 × 4096 (H4RG)	10 × 10 or 15 × 15	HgCdTe	1.0–5.4	37	
	2048 × 2048 (H2RG)	18 × 18	HgCdTe	1.0–1.7	120	
	2048 × 2048 (H2RG)	18 × 18	HgCdTe	1.0–2.5	77	
	2048 × 2048 (H2RG)	18 × 18	HgCdTe	1.0–5.4	37	
Sofradir/www.sofradir.com/	640 × 512	15 × 15	InGaAs	0.9–1.7	300	
	640 × 512	15 × 15	InSb	3.7–4.8	80	<18
	1280 × 1024 (Jupiter)	15 × 15	HgCdTe	3.7–4.8	77–110	18
	1280 × 720 (Daphnis)	10 × 10	HgCdTe	3.4–4.9	110	<20
	640 × 512 (Scorpio)	15 × 15	HgCdTe	1.5–5.1	<90	≤16
	640 × 512 (Leo)	15 × 15	HgCdTe	3.7–4.8	110	20
	640 × 512	20 × 20	QWIP	8.0–9.0	73	31
	640 × 512	24 × 24	HgCdTe	MW (dual) MW/LW (dual)	77–80	15–20
	640 × 512	24 × 24	HgCdTe		77–80	20–25

(Continued)

Table 26.1 (*Continued*) Representative IR hybrid FPAs offered by some major manufactures

MANUFACTURER/WEB SITE	SIZE/ ARCHITECTURE	PIXEL SIZE (μm)	DETECTOR MATERIAL	SPECTRAL RANGE (μm)	OPERATING TEMPERATURE (K)	$D^*(\lambda_p)$ (cmHz$^{1/2}$ W^{-1})/NEDT (mK)
Selex www.leonardo-company.com	320 × 256 (Saphira)	24 × 24	HgCdTe APD	0.8–2.5		
	640 × 512 (Hawk)	16 × 16	HgCdTe	3–5	up to 170	17
	1280 × 720 (Horizon)	12 × 12	HgCdTe	3.7–5		
	640 × 512 (Hawk)	16 × 16	HgCdTe	8–10	up to 90	32
	640 × 512 (CondorII)	24 × 24	HgCdTe	MW/LW (dual)	80	24/26
IAM www.aim-ir.com	640 × 512	15 × 15	HgCdTe	1–5	95–120	17
	640 × 512	15 × 15	HgCdTe	8–9	67–80	30
	640 × 512	20 × 20	HgCdTe	MW/LW	80	18/25
	384 × 288	40 × 40	Type II SL	MW (dual)	80	20/25
SCD www.scd.co.il	640 × 512	15 × 15	InSb	3–5		20
	1280 × 1024	15 × 15	InSb	3–5	77	20
	1920 × 1536	10 × 10	InSb	1–5.4		<25
	1280 × 1024	15 × 15	InAsSb nBn	3.6–4.2	150	20
	640 × 512	15 × 15	InAs/GaSb T2SL	$\lambda_c = 9.5$	80	15
FLIR Systems http://www.flir.com	640 × 512	15 × 15	InGaAs	0.9–1.7	300	10^{10} ph cm^{-2}s^{-1} (NEI)
	640 × 512	15 × 15	InSb	3.4–5.2	80	<25
	640 × 512	15 × 15	InAs/GaSb T2SL	7.5–12	80	<40
DRS Technologies	1280 × 720	12 × 12	HgCdTe	3–5		20
	640 × 480	12 × 12	HgCdTe	3–5		25
	2048 × 2048	18 × 18	Si:As	5–28	7.8	
	1024 × 1024	25 × 25	Si:As	5–28	7.8	
	2048 × 2048	18 × 18	Si:Sb	5–40	7.8	

pixel and integrated red-green-blue (RGB) color filters are included. The most common checkerboard filter pattern dedicates two out of four pixels to green, and one pixel each to red and blue. As a result, the sensor gathers only 50% of the green light and 25% of the red and blue. Digital post-processing interpolates to fill in the blanks, so more than half of the image is artificially generated and innately imperfect. In this case, the fill factor is only a fraction of the pixel area (e.g., about 70%) [5]. Unfortunately, the rich warm tones and detail of color film that the world came to expect suffered over the convenience and immediacy of digital. This was due to the fact that CCD digital image sensors were only capable of recording just one color at each point in the captured image instead of the full range of colors at each location.

CCD imaging sensors use three basic architectures (see Figure 26.2) [6]:

- front-side illuminated (FSI),
- back-side illuminated (BSI), and
- deep-depletion device.

In FSI CCD, the light passes through the polysilicon gates that define each pixel and generates electric charge in the collecting well when pixels are electrically biased. However, due to reflection and absorption losses in the poly-gate structure, the QE of FSI devices is only 50% (see Figure 26.3). To improve QE, the silicon substrate material is uniformly removed to attain approximately 10–15 μm thickness. In this way, an image is focused directly onto the photosensitive area of the CCD without absorption losses in the gate structure (Figure 26.2b).

Compared to FSI CCDs, the thinned BSI devices have a higher QE with peak >90%. Further improvement of QE can be achieved by using high-resistivity silicon with a thickness ranging from 50 to 300 μm (Figure 26.2c) in order to produce a larger active photosensitive volume "deeper" in depletion region. This architecture allows fabrication of devices with longer cutoff wavelength.

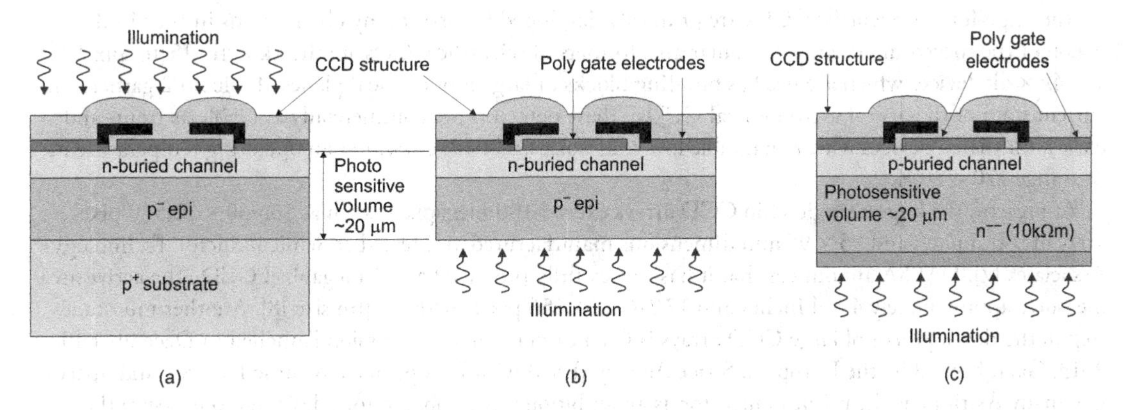

Figure 26.2 CCDs technologies: (a) FSI CCDs, n-channel on p-type, low-resistivity silicon; (b) thinned BSI CCDs n-channel on p-type, low-resistivity silicon; and (c) BSI deep-depletion CCDs, p-channel on n-type, high-resistivity silicon. (Adapted after Ref. [6])

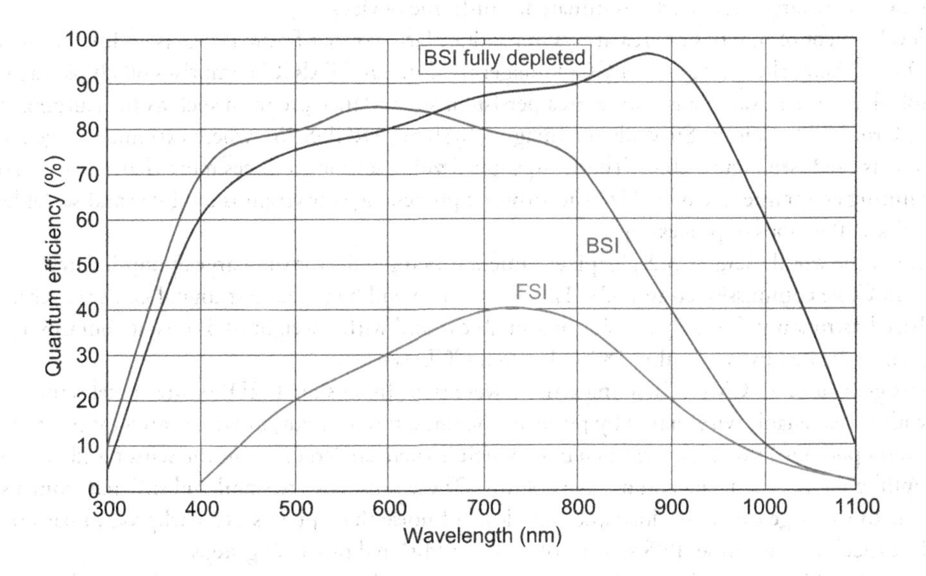

Figure 26.3 Spectral dependence of QE for FSI CCDs, n-channel on p-type, low-resistivity silicon; thinned BSI CCDs, n-channel on p-type, low resistivity silicon; and BSI deep-depletion CCDs, p-channel on n-type, high-resistivity silicon. (Adapted after Ref. [6])

Table 26.2 Characteristics of the state-of-the-art scientific CCDs

	eV2 CCD299-99	FAIRCHILD CCD6161	STA4150
SENSOR TECHNOLOGY	BACK-ILLUMINATED	FRONT-ILLUMINATED	BACK-ILLUMINATED
Format	9216 × 9232	4096 × 4096	4096 × 4096
Pixel size (μm)	10	15	15
Image area (mm)	92.2 × 92.4	61.44 × 61.44	61.44 × 61.44
Full well capacity (e^-)	90,000	50,000	200,000
Dark current (e^-/pixel/h)	4	0.6	3–5
Read noise (e^-)	2.5–4	<5	2–4
Readout rate (MHz)	3	Up to 1	0.1–1

The large-format scientific CCDs are primarily developed for astronomy observations in the visible region of the spectrum. The most popular area formats of scientific CCDs are the 2k × 4k, 15 μm pixel and 4k × 4k devices which are used as building blocks of large mosaic focal planes. Table 26.2 gathers the current state-of-the-art for astronomical CCDs. These detectors are commercially available as front- and back-illuminated devices with remarkable low read noise and dark current at an operating temperature in the range 170–220 K.

At present, the largest single-chip CCD arrays exceed 100 megapixels such as 10,560 × 10,560 pixel array of 9-μm pitch and 95 × 95 mm dimensions manufactured and tested at Semiconductor Technology Associates [7]. DALSA announces that it has successfully produced a 252-megapixel CCD. The active area measures approximately 4 × 4 inches and 17,216 × 14,656 pixels with 5.6 μm size [8]. Another milestone step in the development of large CCD arrays is Gaia camera for space mission launched in December 19, 2013. Gaia, funded by the European Space Agency (ESA) with European Aeronautic Defence and Space Company Astrium as the prime contractor, is an ambitious space observatory designed to measure the positions of around one billion stars with unprecedented accuracy. Gaia's FPA is populated with 106 back-illuminated devices, each with an active area of 45 × 59 mm^2 corresponding to 4,500 × 1,966 pixels, each 10 × 30 μm^2 in size [9]. The entire assembly represents a remarkable 937 megapixel camera system. All of the Gaia CCDs are large-area, back-illuminated, full-frame devices.

The development of mosaics of area arrays to produce large format frame image is an intriguing idea [10,11]. One of them, the 1.4-gigapixel CCD imager used in PanSTARRS comprises 60 chips, each of 22 megapixel. Another example is a wide-area persistent surveillance program such as the Autonomous Real-time Ground Ubiquitous Surveillance Imaging System (ARGUS-IS), where extremely large mosaics of visible FPAs is used (see Figure 26.4). The 1.8-gigapixel video sensor produces more than 27 gigapixels per second running at a frame rate of 15 Hz. The airborne processing subsystem is modular and scalable providing more than 10 teraops of processing.

Currently, the world's largest 3.2-gigapixel camera is being built for the Large Synoptic Survey Telescope in Chile commissioned in 2021 [12]. The camera will be the largest digital camera ever made for ground-based astronomy. Roughly the size of a small car and with a weight of 3 tons, it contains a giant imaging sensor created by combining 189 individual CCD sensors.

The first generation of CMOS imaging sensors were two-dimensional (2D) passive pixel sensor (PPS) devices with access enable wire shared by pixels in the same row and output wire shared by column. Each pixel converts photons into an electrical charge, which is then carried off from the sensor and amplified by an amplifier connected at the end of each column. These sensors were small and suffered from a slow readout due to the large capacitive load and high level of noise that appears as a background pattern in the image. To cancel out this noise, PPS sensors often use additional processing steps.

The general architecture and operation of a CMOS active pixel image sensor are described in Section 24.2.2. The monolithic devices, where the detector array and accompanying readout circuits are fabricated in the same substrate, were developed and commercialized in the mid-1990s [13]. While Sony has been

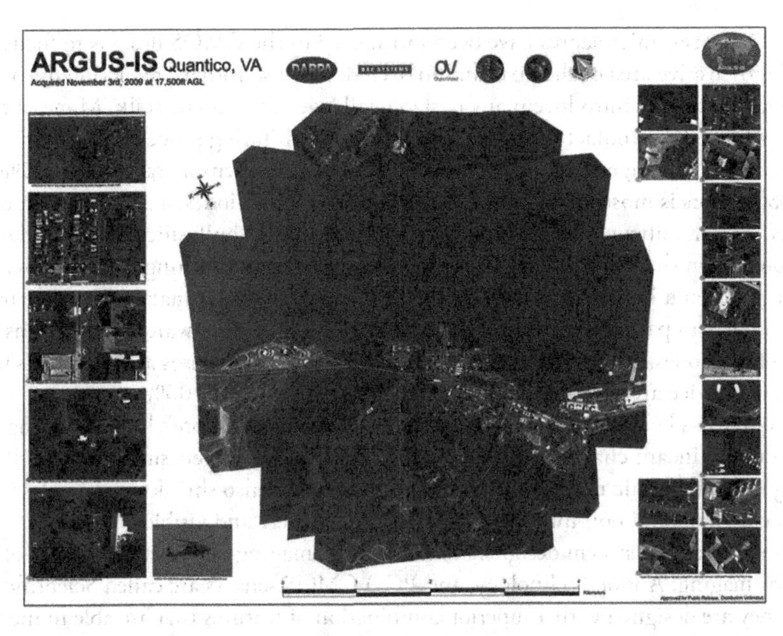

Figure 26.4 Sample of ARGUS-IS imagery. Mounted under a YEH-60B helicopter at 17,500 ft. over Quantico, VA, ARGUS-IS images an area more than 4 km wide and provides multiple 640 × 480-pixel real-time video windows. (Adapted after Ref. [11])

the dominant supplier of CCDs to the imaging and machine markets, this dominance is not evident with CMOS giving more market choice. Key players besides Sony are ON Semiconductor, CMOSIS, eV2, and Teledyne DALSA.

Visible CMOS imaging sensors use two basic architecture:

- Monolithic CMOS (both FSI and BSI), for which the photodiodes are included within the silicon readout integrated circuit (ROIC) (see Figure 26.5)
- Hybrid CMOS that uses a detector layer for detection of light and collection of photocharge into pixels and a CMOS ROIC for signal amplification and readout (see Figure 13.4)

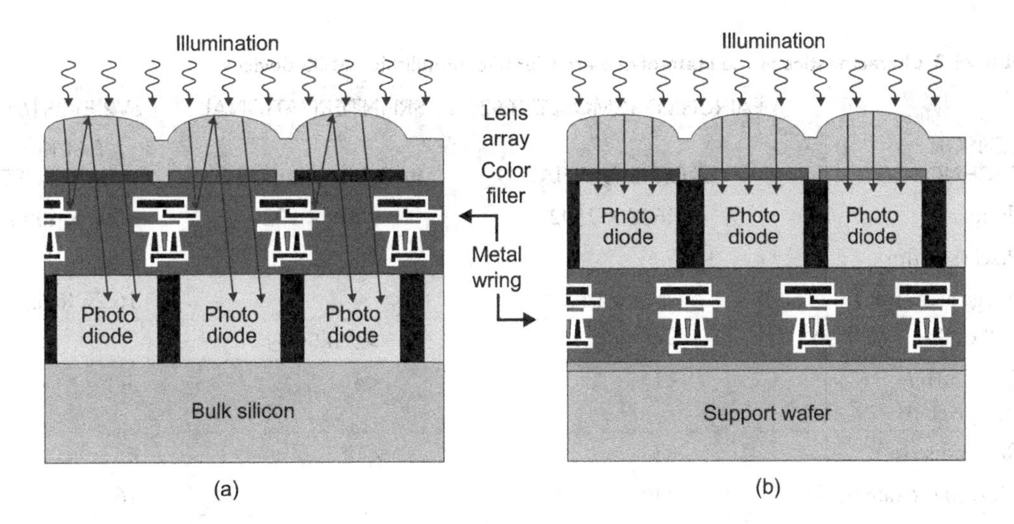

Figure 26.5 Cross-section view of FSI (a) and BSI (b) CMOS devices.

Similar to CCD devices, microlenses have been introduced to the CMOS designs to focus the incoming light onto the photosensitive area of the pixel and to overcome the sensitivity reduction due to a fill factor <100%. However, microlenses introduce many design challenges such as crosstalk. Many of these issues have been solved when CIS manufacturers introduced the back-thinning process.

The majority of CMOS imagers are FSI. Typical fill factors of FSI sensors range from 30% to 70%. Fabrication of BSI sensors is mostly the same as a FSI sensor—photodiodes are laid out first and then the metal layer. After that, the silicon wafer is flipped over and the excess bulk silicon is ground down until the wafer is only about 10 μm thick [13]. Drawbacks associated with back-thinning of CMOS sensors include lower production yields and higher costs because the thinner silicon wafer makes dies more fragile to produce and more complex to package. Grinding the silicon down allows the wafer to only be as thick enough as to expose the photodiodes. After grinding, the color filters and microlenses are laid out as with FSI sensors. In this way, the device architecture improves the fill factor to nearly 100%.

CMOS technology has been the key for meeting the challenges of Moore's law. Shrinking sensor pixel size, however, poses significant challenges to sensor design. As pixel size gets smaller, the well capacity shrinks resulting in less dynamic range. The photon collection area also shrinks, causing less sensitivity. Crosstalk, noise, coupling, and non-uniformity become more severe and visible.

CMOS technology enables system designers to tailor their imaging devices to the needs of their applications. Fairchild Imaging, Andor Technology, and PCO CMOS sensors are called Scientific CMOS (sCMOS) since they are designed with a superior combination of features that are able to meet the extreme performance requirements of many scientific applications in biomedical research, astronomy, security, and defense. Table 26.3 lists performance characteristics of monolithic CMOS devices that have been designed for scientific applications.

A new design of visible sensor has been proposed by Foveon. It combined the best of what both film and digital have to offer [5]. This is accomplished by the innovative design of Foveon's X3 direct image sensors that have three layers of pixels, just like film has three layers of chemical emulsion (see Figure 26.6) [14]. This silicon sensor is fabricated on a standard CMOS processing line. Foveon's layers are embedded in silicon to take advantage of the fact that RGB light penetrates silicon to different depths (the photodetectors sensitive to blue light are on top, the green sensitive detectors are in the middle, and the red at the bottom) forming the image sensor that captures full color at every point in the captured image. This is 100% full color with no interpolation. Figure 26.6b plots the absorption coefficient as a function of depth, which is an exponential function of depth for any wavelength. Since the higher energy photons interact more strongly, they have a smaller space constant, and thus the exponential falloff with depth is more rapid.

Currently, the most technologically advanced scientific CMOS arrays are made with a hybrid architecture using the same technology developed for IR detector arrays (see Section 24.3). In this approach, the

Table 26.3 Characteristics of the state-of-the-art scientific monolithic CMOS devices

	FAIRCHILD sCMOSLTN4625A	SRI INTERNATIONAL	eV2 EV2S16M
SENSOR TECHNOLOGY	BACK-ILLUMINATED	BACK-ILLUMINATED	BACK-ILLUMINATED
Format	4608 × 2592	2048 × 1920	4096 × 4096
Pixel size (μm)	5.5	10	2.8
Image area (mm)	25.3 × 14.3		16.22 (diagonal)
Full well capacity (e^-)	>40,000	>220,000	
Dark current (e^-/pixel/s)	<15	<4	3–5
Read noise (e^-)	<5	<14	<5
Max. frame rate (fps)	240		16
Power consumption (W)	2		1.6

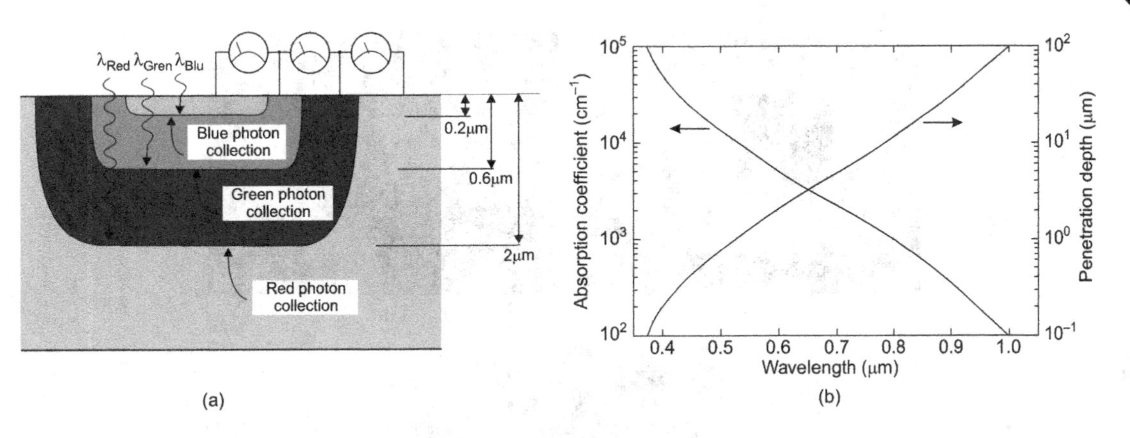

Figure 26.6 Schematic illustration of Foveon's sensor stack: (a) cross section of the sensor, and (b) absorption coefficient and penetration depth in silicon. (From Ref. [14])

CMOS ROICs are usually fabricated in the same silicon foundries that produce computer chips, though special amplifiers and circuit designs are required for electro-optical applications. With regard to that hybrid architecture combines the best quality detector material with the superior performance achieved by CMOS integrated circuit technologies, this allows to fine-tune key performance parameters such as spectral response, QE, full well capacity, noise level, modulation transfer function (MTF), etc. for each specific application. Very low power consumption of CMOS devices allows nondestructive random access to the pixels and on-chip integration of analog and digital circuitry which significantly reduces the imaging system mass, volume, power and thermal output, and overall complexity for different applications.

Until recently, CMOS imagers have been at a disadvantage relative to CCDs for readout noise, since the CMOS's readout circuit has inherently higher noise than a CCD amplifier. One reason for higher noise is the capacitance of the sense node of the three-transistor CMOS pixel. For a typical pixel size of 5 μm in a 0.25-μm CIS process, the sense node capacitance is about 5 fF, corresponding to a responsivity of 32 μV per electron, which limits the lowest readout noise to about 10 electrons. With the development of the four-transistor pixel, monolithic CMOS can achieve the lowest noise levels required by astronomy [15,16]. The state-of-the-art CMOS pixel architecture utilize up to eight transistor designs by incorporating more electronics in the pixel at the cost of increased noise and reduced fill factor [13].

For visible light detection, Teledyne Imaging Sensors (TIS) use both monolithic and hybrid CMOS detectors [15–17]. These highest performance silicon-based image arrays are addressed to the astronomical community. Low-noise silicon hybrid arrays H4RG-10 termed HyViSI™ (Hybrid Visible Silicon Imager), produced in 4k × 4k format with 10-μm pixel pitch, achieve high pixel interconnectivity (>99.99%), low readout noise (<10 e^- rms single CDS), low dark current (<0.5 e^-/pixel/s at 193 K), high QE (>90% broadband), and large dynamic range (>13 bits) [17]. The fabrication of H4RG-10 is accomplished using sub-field stitching techniques in a 0.25-μm CMOS process. The modular stitching design approach allows fabrication of ROICs ranging from 2k × 2k to 16k × 16k formats. The chip size of the H4RG-10 is 43.0 × 45.5 mm^2, slightly bigger than the 18-μm pixel H2RG (38.8 × 40.0 mm^2). There are nine H4RG-10 ROIC dies on each 200-mm CMOS wafer, while the H2RG wafer has 12 ROIC dies. Figure 26.7 shows photographs of Teledyne's hybrid silicon p-i-n CMOS sensors.

Teledyne's monolithic CMOS sensors are fully digital system-on-chip, with all bias generation, clocking, and analog-to-digital conversion included with the image array. Some examples of the arrays are listed in Table 26.4 [17]. The H4RG-10 is fully compatible with Teledyne's System for Image Digitization, Enhancement, Control And Retrieval (SIDECAR) ASIC as the control electronics. The SIDECAR is an application-specific integrated circuit that functions as focal plane electronics by providing clocks and biases to the ROIC as well as digitizing its analog outputs. Up to 36 analog inputs can be accommodated in parallel for each SIDECAR ASIC chip, with choice of 500-kHz, 16-bit analog-to-digital converter (ADC) or 10-MHz, 12-bit ADC. For a H4RG-10 ROIC operating in a 64-output mode, two SIDECAR ASIC chips are required.

Figure 26.7 Teledyne's hybrid silicon p-i-n CMOS sensors: (a) 1k × 1k H1RG-18 HyViSI, (b) 2k × 2k H2RG-18 HyViSI, and (c) 4k × 4k H4RG-10 HyViSI. (From Ref. [16]. With permission.)

Table 26.4 List of ROICs used for silicon p-i-n CMOS FPAs

	TCM6604A	CHROMA*			H1RG	H2RG	H4RG-10*
INPUT CIRCUIT	CTIA				SFD		
Array format (pixels)	640 × 480	1280 × 480			1024 × 1024	2048 × 2048	4096 × 4096
Pixel pitch (μm)	27	30			18	18	10
Number of outputs	4	8			1,2, or 16	1,4, or 32	1,4,16,32, or 64
Nominal pixel rate (MHz)	8	10			0.1 to 5	0.1 to 5	0.1 to 5
Shutter mode	Snapshot, integrate while read				Ripple read		
Window mode	Programmable in row direction				Arbitrary size and location guide window		
Peak QE (%)	>90						
Spectral band (nm)	200–1050						
Charge capacity (ke^-)	700	700	1,000	5,000	100	100	90 (40)
Readout noise** (e^- rms)	<100	<80	<110	<600	<10	<10	<10 (6)
Power dissipation*** (mW)	<70	<150			<2	<5	<10

Source: Adapted after Ref. [17].

Two noise and full well values are quoted for H4RG-10 to reflect two design versions (low gain and high gain).

* New FPAs since 2008.

** Single correlated double sample (CDS) noise, except for TCM660A.

*** Power dissipation for H1RG, H2RG, H4RG-10 is for Reading out of the maximum number of ports AT 100 kHz pixel rate per port.

State-of-the-art scientific CMOS devices with hybrid architecture are available in formats as large as 8k × 8k four-side buttable [1,15–20]. These detectors achieve the same performance as the best deep depletion CCDs. Figure 26.8 shows QE curves of monolithic and hybrid arrays demonstrating significant QE enhancement achieved with the hybrid CMOS technology, whereas Table 26.5 lists performance characteristics of CMOS FPAs fabricated by Raytheon and Teledyne.

Apart from Teledyne, for over 20 years, Raytheon has built hybrid focal plane based on silicon p-i-n photodiodes [1,17–20]. The sensor chip assemblies (SCAs) can be combined in mosaic configurations to form large composite arrays using two-, three-, and four-side buttable packages. Recent advancements in the latest generation of 8-μm pixels with formats up to 8k × 8k are presented in Refs. [1,20]. The current family of devices has very low read-noise ROICs and low detector dark current, operates with a 25 volt bias, and delivers 50% mean response operability greater than 99.995%. The detector structure is fully depleted and thickness of active region is changed from 10 to 350 μm, what allows tuning of MTF and near IR response. Raytheon's latest ROIC delivers 14-bit ADCs, windowing, >1 Gbit s^{-1} outputs, noise floor 5–7 e^-, and well capacity >200 ke^-. Figure 26.9 shows a photograph of 8k × 8k, 8-μm pixel Si p-i-n-based SCA and resulting image.

Even though silicon imagers are routinely manufactured with magapixel resolution and excellent reproducibility and uniformity, they are not sensitive to radiation longer than about 1 μm. Today, SiGe alloys are

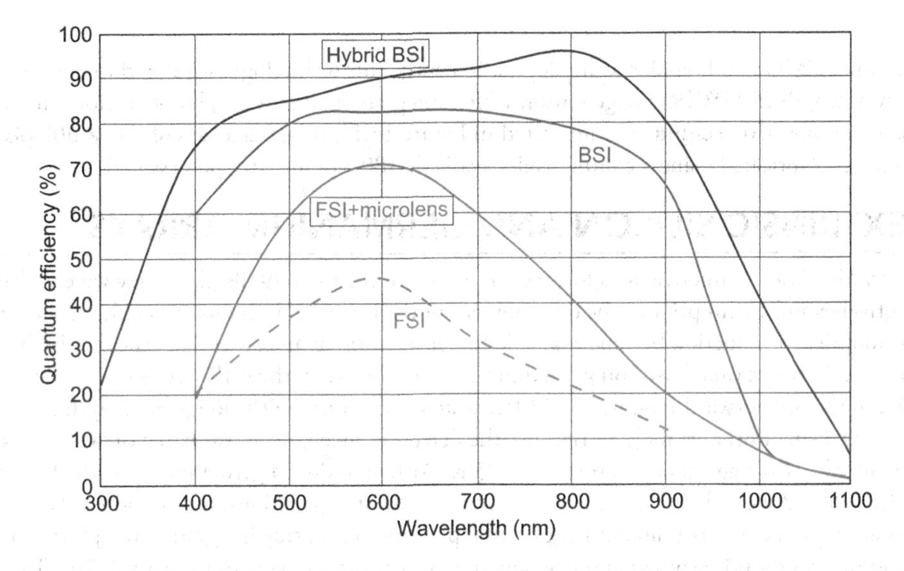

Figure 26.8 QE of CMOS imaging sensors from top to bottom: hybrid device, BSI device, FSI device with microlens, and FSI device without microlens. (Adapted after Ref. [6])

Table 26.5 Characteristics of the state-of-the-art scientific hybrid CMOS devices

PARAMETER	RAYTHEON VISION SYSTEMS	TELEDYNE IMAGING SENSORS
TECHNOLOGY	HYBRID, DIRECT BOND	HYBRID, In-BUMP
Format	8160 × 8160	4096 × 4096
	5100 × 5100	2048 × 2048
	1020 × 1020	1024 × 1024
Pixel size (μm)	8	10
Full well capacity (e^-)	200,000	100,000
Dark current (e^-/pixel/h)	<1	<0.5
Read noise (e^-)	5–7	<10

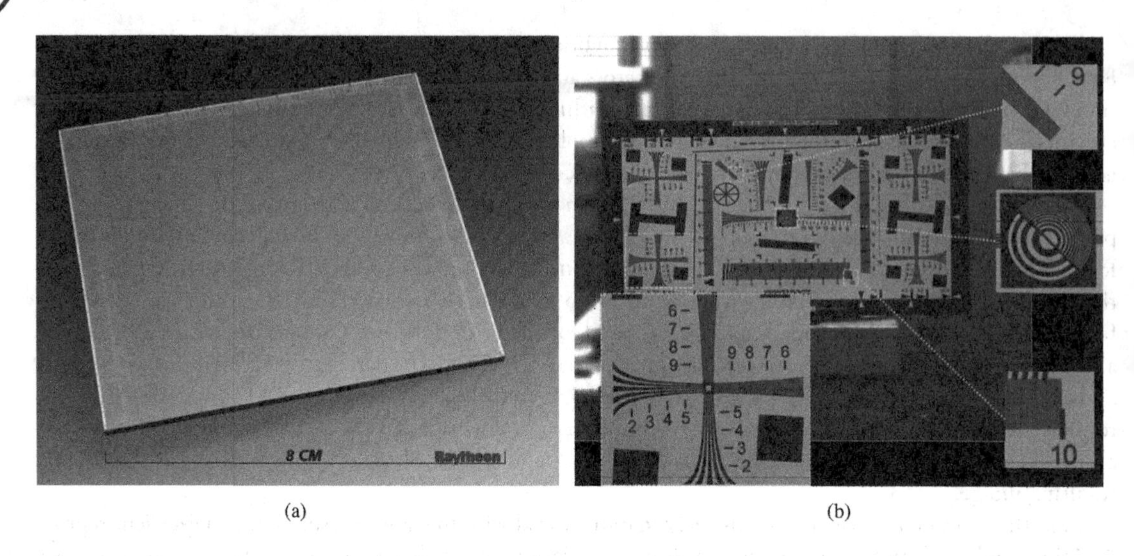

(a) (b)

Figure 26.9 Silicon p-i-n SCA: (a) photograph of 8k × 8k, 8-µm pixel array and (b) resulting high-resolution image. (Adapted after Ref. [1])

widely used in Si CMOS and bipolar technologies. Recently, this technology is adopted to develop a single-chip short-wavelength IR (SWIR) image sensor, which integrates germanium photodetectors on a standard Si process. This innovative technique is illustrated in Figure 13.14. Imaging arrays of 768 × 600 pixels at a 10-µm pitch were fabricated using a commercially available 180-nm foundry process [21].

26.2 EXTRINSIC SILICON AND GERMANIUM ARRAYS

The first extrinsic photoconductive detectors were reported in the early 1950s [22]. They were widely used at wavelengths beyond 10 µm prior to the development of the intrinsic detectors. Since the techniques for controlled impurity introduction became available for germanium at an earlier date, the first high performance extrinsic detectors were based on germanium. The discovery in the early 1960s of Hg-doped germanium led to the first forward-looking IR (FLIR) systems operating in the long-wavelength IR (LWIR) spectral window using linear arrays [23]. Because the detection mechanism was based on an extrinsic excitation, it required a two-stage cooler to operate at 25 K. Although doped germanium was the IR detector of choice in the 1960s, doped silicon replaced germanium for most applications during the 1980s.

The first attempt to develop a monolithic extrinsic photodetector array integrating the photosensitive elements and devices for primary signal processing in one crystal was made in the early 1970s. Twenty photosensitive unit cells were arranged on a Si:As plate. Each of them included a photosensitive element, load resistor based on impurity compensation, and metal-oxide-semiconductor field-effect transistor (MOSFET) connected into the circuit as a source follower [24]. This array provided useful performance but exhibited undesirable electrical and optical crosstalk between detector channels. That was corrected with its improvement.

The first report on the development of monolithic arrays with CCD multiplexing was published in 1974 [25]. Later, two more monolithic CCD versions were developed [26]. All these devices (Figure 26.10) operate at the temperature of photosensitive substrate corresponding to a low-ionized state of impurity when the concentration of free carriers is low as compared with that of impurity.

A CCD, for the accumulation mode, transfers photogenerated majority charge carriers (holes) that are accumulated at the Si-Si oxide interface. This device has not seen any application because the transfer efficiency was found to be low and the clocking frequency was limited to low frequencies [25]. The CCD approach is not compatible with low read noise because of the low operating temperature of the IR detectors. The buried channel CCDs (BCCDs) were also used to avoid trapping noise; however, there is no longer sufficient mobile charge to maintain the channel. In addition, CCDs suffer due to damage to the devices under extremely high doses of ionizing radiation, which degrades the CTE.

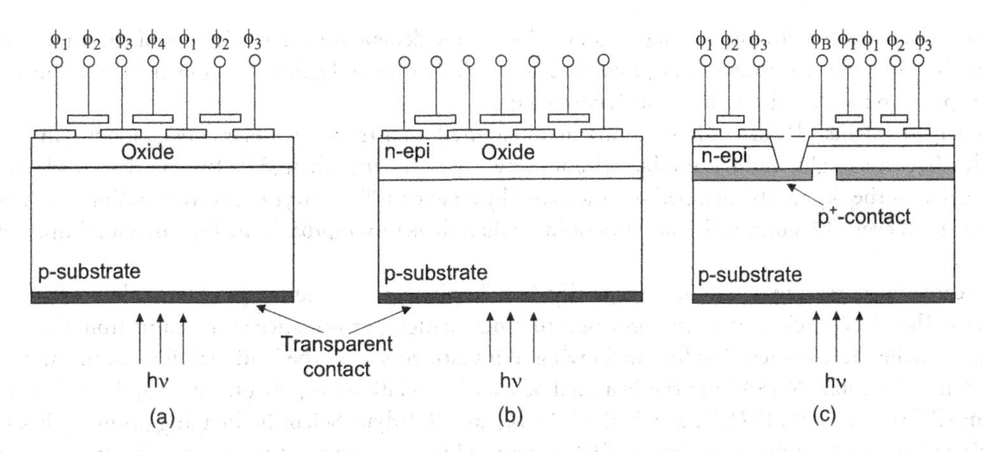

Figure 26.10 Monolithic silicon arrays with CCD multiplexing: (a) accumulation mode, (b) pseudoaccumulation mode, and (c) vias mode. (From Ref. [26]. With permission.)

In the pseudo accumulation mode (PAM), photogenerated holes are injected as minority carriers in the n-epilayer where they can then be clocked out by the CCD. The PAM has been used with Si:Ga and Si:In to make 32 × 96 element arrays [27]. It has been operated at background levels of 10^8–10^{14} photons $cm^{-2}s^{-1}$. In the vias mode, the photosensitive substrate is equipped with buried individual detector cell contacts. These contacts collect the photogenerated holes that are transferred to the n-epilayer under the action of a CCD transfer gate where they are then clocked out.

For the monolithic arrays, biasing and operational temperature were found to be critical in avoiding injection currents from the n-epitaxial, p-substrate contacts. Operating temperatures lower than those dictated by the detector characteristics were required. Further, the detector responsivity obtained for monolithic devices was found to be considerably degraded compared to that obtained on discrete detectors prepared from the same substrate material (lowering by two orders of magnitude). This degradation was supposed to be associated with thermal oxidation, p^+, n^+ diffusion, epitaxial growths, and polysilicon (gate) depositions. It was found that the loss in responsivity results from an increase in donor concentration to over $10^{14}cm^{-3}$ occurring with manufacturing a device [28].

To avoid the long shift registers inherent in CCDs, other pseudomonolithic readout mechanisms have been devised. In charge injection device (CID) photodetector arrays, a photosignal charge is assumed to be collected and stored in metal-oxide-semiconductor (MOS) capacitors of the device cells. Originally, CIDs were proposed in which the accumulation of minority charge carriers occurred [29]. In the presence of a negative bias applied to the gate on n-type Si, an electron-depleted layer emerges under this gate. Under the intrinsic irradiation of the substrate, hole-electron pairs are generated and minority charge carriers (holes) are collected and stored in this layer at the Si-Si oxide interface. With removal of the voltage, the accumulated charge carriers are injected into the substrate where they are collected to provide a signal output.

At low temperatures and irradiances when the concentration of free carriers becomes very low, such devices exhibit a capability to collect and store majority charge carriers generated via the photoionization of impurity centers. To realize this, a positive bias should be applied to the gate if the substrate is of n-type conductivity. Then, a pulse of voltage of negative polarity should be applied to the gate for the injection of accumulated charge into the substrate and its readout. A value of this voltage should be great enough to minimize the losses for recombination during the drift through the substrate. Two-dimensional 32 × 32 and 2 × 64 Si:Bi CID arrays based on this approach have been constructed [30,31]. However, it was noted that the array well capacity was significantly smaller than predicted. The cutoff frequency of photoresponse appeared to be also considerably lower than the expected value determined by the rate of the dielectric-relaxation processes in the device.

The above-mentioned shortcomings of the monolithic arrays have been overcome by the transition to a hybrid fashion of devices with CMOS detector arrays which are dominant today. This makes it possible to use lower temperatures when manufacturing it. With such a design, there also appears an additional Si space for selective input circuits and signal processing. These depend on the application of the array and can include time and add electronics to improve detector performance and gain reduction and direct current (DC) suppression electronics to increase dynamic range.

The largest extrinsic IR detector arrays are manufactured for astronomy. Their application began roughly 30 years ago [32] and has doubled about every 7 months since then [33]. The speed with which a given region of the sky can be mapped has increased by a factor 10^{18} in 40 years, corresponding to a doubling of speed every 12 months. Sensitivities of individual detectors approach the fundamental limits set by photon noise.

The early detector arrays were small (typically 32 × 32 pixels) with read noises of more than 1,000 electrons. Their basic architecture and processes to produce high-performance arrays came from the military. Further development has followed owing to investments from the National Aeronautics and Space Administration (NASA) and the National Science Foundation [34]. At present, Raytheon Vision Systems (RVS) [1,36–40], DRS Technologies [41–44], and Teledyne Scientific Imaging (formerly Rockwell Scientific Company) supply the majority of IR arrays used in astronomy between them, the most important being BIB detector arrays. The theoretical basis for impurity band conduction in germanium and silicon was established in 1950s [35], but practical devices were not developed until early 1980s. The BIB detector structure is described in detail in Section 14.5. It usually has a thin lightly doped n^--region (usually grown epitaxially) between the absorbing n-type region and the common back-side implant (p^+-region) in order to block hopping conduction currents from reaching the p^+-contact, which are substantial at the doping levels used in the active detection region. As is mentioned in Section 14.5, the BIB detector active layer should be thick and doped as heavily as possible. This limit is reached at an As concentration of about 10^{18} cm^{-3}. The layer thickness is limited by the minority impurity concentration and by incipient avalanching of charge carriers created by bias voltage. According to Love et al. [36], the minority upper limits are 1.44×10^{12} cm^{-3} for a 45-μm-thick layer and 1.85×10^{12} cm^{-3} for a 35-μm-thick layer. Therefore, the detectors are designed for arsenic doping of 7×10^{17} cm^{-3} and a thickness of 35 μm.

Both low- and high-flux versions of the BIB detector have been elaborated by significant modifications of the detector design. These adjustments included changes in doping profiles and layer thickness and tailoring buried contact resistivity to allow for the larger current densities in the higher photon flux environment. Impressive progress has been achieved especially in Si:As BIB array technology with format as large as 2,048 × 2,048 and pixels as small as 18 μm; operated in spectral bands up to 30 μm at about 10 K. The pixel size of 18 μm is smaller than the wavelength at Q band (17–24 μm); however, this does not pose a problem since an imager operating at these wavelengths will typically spread the beam out over many pixels to be fully sampled. The characteristics of the most advanced Si:As BIB arrays for astronomy are summarized in Table 26.6. Figure 26.11 shows the evolution of BIB detector arrays at RVS.

The BIB arrays applied for ground- and space-based far-IR astronomy should be operated under the most uniform possible conditions, in the most benign and constant environment possible. Array performance is strongly affected by background levels. The examples described in Table 26.6 have been optimized for low background, where the read noise is minimized by minimizing the detector plus gate integrating capacitance, C. Then, for a given charge, Q, the voltage is maximized since $V = Q/C$. If the detectors are exposed to higher background levels, the readout amplifiers saturate. Counteraction includes simply increasing the integration capacitance or using alternative amplifier architecture with handling larger signals.

To enhance array performance, Fowler and Gatley [45] proposed to reduce the read noise by performing multiple nondestructive passes through the array at the beginning and end of the integration ramp. By resetting each pixel on one pass through the array and sampling the detector node voltage on subsequent passes through the array, the *true* pedestal is removed from the data. Each time a pixel is selected, a charge is redistributed from the row and column select field-effect transistors (FETs) back onto the detector node capacitance. Fowler claims that the read noise is predominantly due to the kTC noise associated with this charge redistribution. By performing multiple nondestructive passes through the array at the beginning and end of the integration, the read noise is reduced by the square root of the number of passes.

Extrinsic silicon arrays for high background applications are less developed than that for low background applications. Detectors in conventional ground-based systems are operated in thermal backgrounds

Table 26.6 Characteristics of Si:As BIB hybrid arrays

	DRS TECHNOLOGIES	RVS	RVS	RVS
PARAMETER	WISE	JWST MIRI	AQUARIUS-1K	PHOENIX
Applications/Users	Low background	Space telescopes NASA	Ground-based telescopes	Space telescopes JAXA
Wavelength range (µm)	5–28	5–28	5–28	5–28
Format	1024 × 1024 2024 × 2024	1024 × 1024	1024 × 1024	1024 × 1024 2024 × 2024
Pixel pitch (µm)	18	25	30	25
Operating temperature (K)	7.8	6.7	7–9	8–10
Fill factor (%)	>98	>98	≥98	≥95
ROIC type	SFD	SFD	SFD	SFD
Read noise (e rms)	<40	10–30	Low gain <1,000 High gain <100	6–20
Dark current (e/s)	<5	0.1	1	
Well capacity (e)	>10^5	2×10^5	(1 or 15) × 10^6	3×10^5
QE (%)	>70	>70	>40	>70
Max frame rates (Hz)	1	0.1	150	0.1
Number of outputs	4 or 16	4	16 or 64	4
Packaging	Module	Module	Module, two-side buttable	LCC

Source: Adapted after Refs. [40,42]
LCC, leadless chip carrier; WISE, Wide-Field IR Survey Explorer.

Figure 26.11 Evolution of BIB FPAs at RVS. From left to right: SIRTF 256 × 256, CRC774 320 × 240, Aquarius-1k 1024 × 1024, and Phoenix 2048 × 2048 devices. (Adapted after Ref. [40])

up to 10^9 photons per second. The available high background Si:As BIB arrays are 256 × 256 or 240 × 320 pixels [33]. The first 1,024 × 1,024 high background Si:As BIB detector array has been developed at DRS Technologies [42].

The readouts for large BIB arrays use a circuit similar to that described in Chapter 24. However, it should be mentioned that silicon-based MOSFETs show a number of operational difficulties conditioned by the very low temperatures required for the readout circuits for these detectors [46]. They are related to freeze-out of thermally generated charge carriers, making the circuits unstable, increasing noise, and causing signal hysteresis. They are described in detail by Glidden et al. [47]. Many of them can be mitigated by growing the circuits that are heavily doped.

Silicon detectors have largely supplanted germanium extrinsic detectors for both high and low background applications where comparable spectral response can be obtained. However, for wavelengths longer than 40 μm, there are no appropriate shallow dopants for silicon; therefore, germanium devices are still of interest for very long wavelengths (LWs). Very shallow donors, such as Sb, and acceptors, such as B, In, or Ga, provide cutoff wavelengths in the region of 100 μm (see Figure 14.4). The achievement of low noise-equivalent power (*NEP*) values in the range of a few parts 10^{-17} $WHz^{-1/2}$ was made possible by advances in crystal growth development and this controls the residual minority impurities down to $10^{10}cm^{-3}$ in a doped crystal. As a result, a high lifetime and mobility value and thus a higher photoconductive gain have been obtained.

Due to small energy band gap, the germanium detectors must be operated well below the silicon "freeze-out" range, typically at liquid helium temperature. There are a number of problems with the use of germanium. For example, to control dark current, the material must be lightly doped and therefore absorption lengths become long (typically 3–5 mm). Because the diffusion lengths are also large (typically 250–300 μm), pixel dimensions of 500–700 μm are required to minimize crosstalk. In space applications, large pixels imply higher hit rates for cosmic radiation. This in turn implies very low readout noise for arrays operated in low background limit, what is difficult to achieve for large pixels with large capacitance and large noise. A solution is using the shortest possible exposure time. Moreover, germanium detectors have complicated responses that affect calibration, observing strategies, and data analysis in low background applications. The devices operate at very low bias voltages and even small changes in the operating points of amplifiers can result in unacceptable bias changes on the detectors. More details can be found by Rieke [33].

Application of uniaxial stress along the [100] axis of Ge:Ga crystals reduces the Ga acceptor binding energy, extending the cutoff wavelength to ≈ 240 μm [48]. At the same time, the operating temperature must be reduced to less than 2 K. In making practical use of this effect, it is essential to apply and maintain very uniform and controlled pressure to the detector so that the entire detector volume is placed under stress without exceeding its breaking strength at any point. A number of mechanical stress modules have been developed. The stressed Ge:Ga photoconductor systems have found a wide range of astronomical and astrophysical applications [49–63].

The Infrared Astronomical Satellite, the Infrared Space Observatory, and, for the far-IR channels, the Spitzer-Space Telescope (Spitzer) have all used bulk germanium photoconductors. In Spitzer mission, a 32 × 32-pixel Ge:Ga unstressed array was used for the 70-μm band, while the 160-μm band had a 2 × 20 array of stressed detectors [49]. The detectors are configured in the so-called Z-plane to indicate that the array has substantial size in the third dimension. The poor absorption of the Ge:Ga detector material requires that the detectors in this array are huge—2 mm long.

An innovative integral field spectrometer, called the Field Imaging Far-Infrared Line Spectrometer, which produces a 5 × 5 pixel image with 16 spectral resolution elements per pixel in each of the two bands, was constructed at the Max Planck Institut für Extraterrestrische Physik. This array, shown in Figure 26.12, was developed for the Herschel Space Observatory and Stratospheric Observatory For Infrared Astronomy [53,54]. To accomplish this, the instrument has two 16 × 25 Ge:Ga arrays, unstressed for the 45–110 μm range and stressed for the 110–210 μm range. The low-stressed blue detectors has a mechanical stress on the pixels which is reduced to about 10% of the level needed for the LW response of the red detectors. Each detector pixel is stressed in its own subassembly, and a signal wire is routed to preamplifiers housed nearby, what obviously limits this type of array to much smaller formats than are available without these constraints (Figure 26.12).

The Photodetector Array Camera and Spectrometer (PACS) is one of the three science instruments on ESA's far IR and sub-millimetre observatory—Herschel Space Laboratory [55,56]. Apart form two Ge:Ga photoconductor arrays (stressed and unstressed) with 16 × 25 pixels each, it employs two filled silicon bolometer arrays with 16 × 32 and 32 × 64 pixels, respectively, to perform integral-field spectroscopy and imaging photometry in the 60- to 210-μm wavelength regime. Figure 26.13 shows the spectral response of the filter/detector chain of the PACS photometer in its three bands. Median *NEP* values are 8.9×10^{-18} $WHz^{-1/2}$ for the stressed and 2.1×10^{-17} $WHz^{-1/2}$ for the unstressed detectors, respectively. The detectors are operated at ~1.65 K. The readout electronics is integrated into the detector modules—each linear

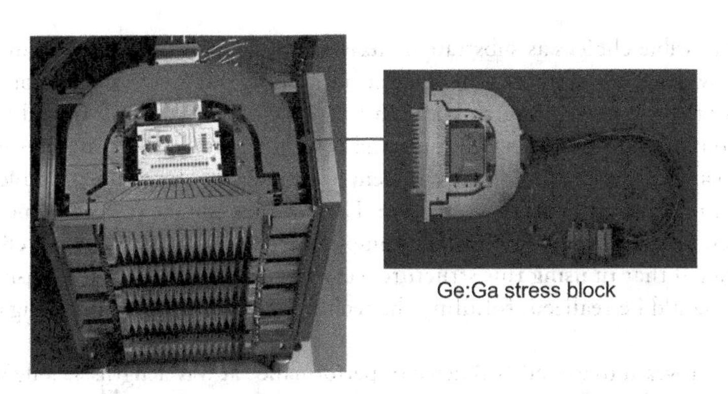

Figure 26.12 PACS photoconductor FPA. The 25 stressed and low-stress modules of PACS instrument (corresponding to 25 spatial pixels) in the red and blue arrays are integrated into their housing. (Adapted after Ref. [54])

module of 16 detectors is read out by a cryogenic amplifier/multiplexer circuit in CMOS technology but operates at temperatures 3–5 K.

The standard hybrid FPA architecture is not generally suitable for far-IR arrays (although this architecture is also used [60,61]) primarily because glow from the readout is sensed by the detector, degrading its performance. In response, a new layered-hybrid structure was introduced to alleviate these problems and make possible the construction of large format far-IR FPAs (see Figure 24.12d) [62,63]. In this design, an intermediate substrate is placed between the detector and the readout, which is pixelized on both sides in a format identical to that of the array, and the electrical contact between the corresponding pixel pads are made through embedded vias. The substrate material must be chosen to have sufficient IR-blocking property, high thermal conductivity, and an expansion coefficient that is between that of germanium and silicon. Alumina (Al_2O_3) and aluminum nitrite (AlN) have these

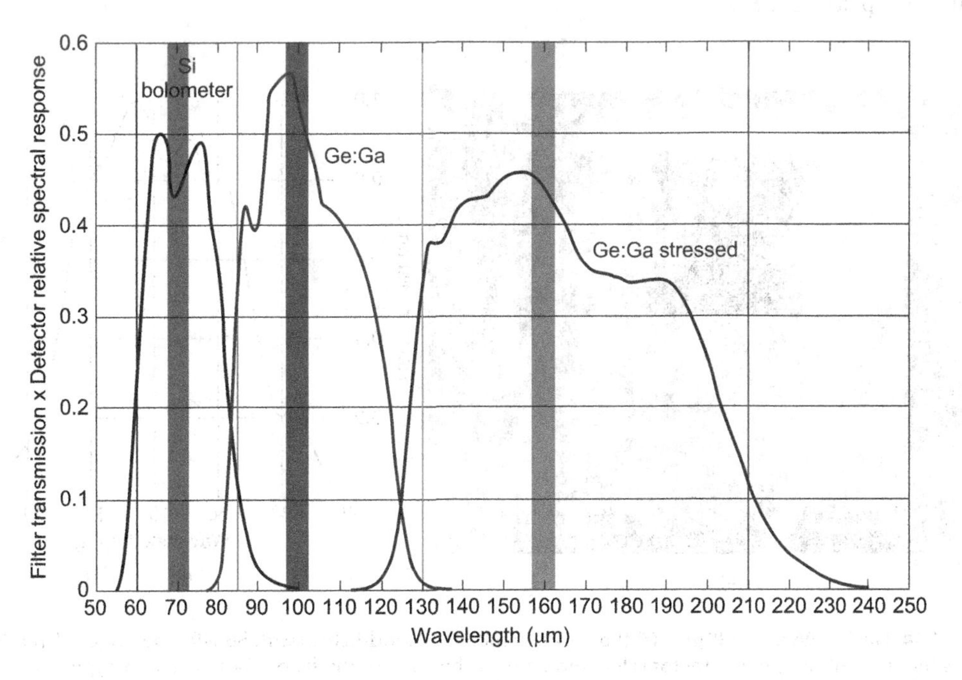

Figure 26.13 Effective spectral response of the filter/detector chain of the PACS photometer in its three bands. (Adapted after Ref. [55])

properties and are possible choices as substrate materials. Blocking of the readout glow from reaching the detector provides more efficient heat dissipation, improves temperature uniformity across the array, and mitigates the thermal mismatch between the detector and readout. In addition, the substrate serves as a fanout board providing a simple and robust way to connect the FPA to the external electronics with no additional packaging requirement. Figure 26.14 shows an assembled Ge:Sb FPA ($\lambda_c \approx 130\,\mu m$) using the layered-hybrid architecture. For low bias voltage photoconductor operated at low temperatures, a capacitive transimpedance amplifier (CTIA) design offers an effective readout solution. It is predicted that in using this structure, very large format FPAs with sensitivities better than 10^{-18} WHz$^{-1/2}$ could be realized, fulfilling the technology goals of the upcoming astronomical instruments.

A modest NASA program to extend BIB detector performance at wavelengths as long as 400 μm using gallium arsenide is rather difficult to achieve due to a problematic requirement for producing the required epitaxial low-doped blocking layer [64,65].

26.3 PHOTOEMISSIVE ARRAYS

The first Schottky-barrier FPA was the 25 × 50 element IR-CCD developed at RCA laboratories under contract to the Rome Air Development Center [66]. At the beginning of 1990s, Schottky-barrier FPAs represented the most advanced monolithic FPAs technology for medium wavelength (MW) applications [67]. Also scanning PtSi FPAs with up to 4 × 4,096 elements [68] and 2,048 × 16 time delay and integration (TDI) [69] elements were developed for space-borne remote sensing applications. Review of different configuration of staring Schottky-barrier FPAs is given by Kosonocky [70,71] and Kimata et al. [67,72–75], for example. Table 26.7 summarizes the specifications and performance of typical high-resolution PtSi Schottky-barrier FPAs that have full TV resolution.

The details of the geometry, and the method of charge transfer differ for different manufacturers. The design of a staring Schottky-barrier FPAs for a given pixel size and design rules involves a trade-off between the charge handling capacity and the fill factor. This trade-off depends also on the choice of the FPA architecture that includes:

- interline transfer CCD architecture
- charge sweep device (CSD)

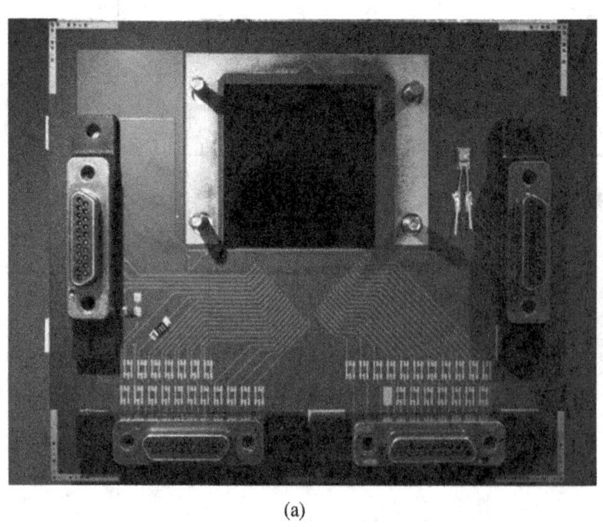

(a)

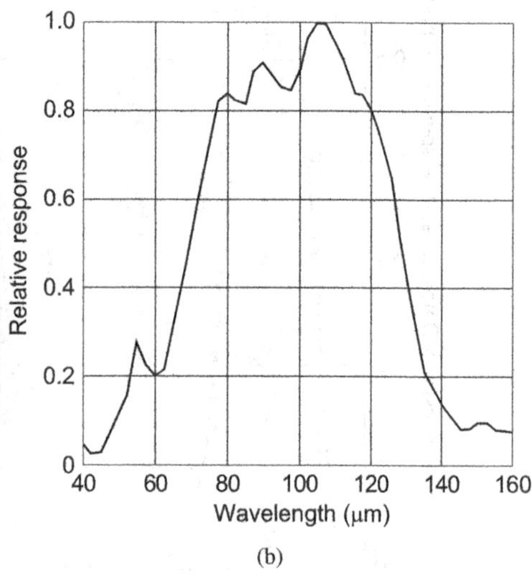

(b)

Figure 26.14 Ge:Sb FPA ($\lambda_c \approx 130\,\mu m$): (a) the fully assembled layered-hybrid with SB349 readout (CTIA readout)—shown is the readout side; the detector is located on the other side of the fanout board, and (b) typical spectral response of Ge:Sb photoconductor. (Adapted after Ref. [62])

Table 26.7 Specifications and performances of typical PtSi Schottky-barrier FPAs

ARRAY SIZE	READOUT	PIXEL SIZE (μm^2)	FILL FACTOR (%)	SATURATION (e$^-$)	NEDT/(f/#) (K)	YEAR	COMPANY
512 × 512	CSD	26 × 20	39	1.3×10^6	0.07 (1.2)	1987	Mitsubishi
512 × 488	IL-CCD	31.5 × 25	36	5.5×10^5	0.07 (1.8)	1989	Fairchild
640 × 486	LACA	30 × 30	54	4.0×10^5	0.10 (1.8)	1989	Reticon
512 × 512	IL-CCD	30 × 30	54	5.5×10^5	0.10 (2.8)	1990	Kodak
640 × 480	MOS	24 × 24	38	1.5×10^6	0.06 (1.0)	1990	Sarnoff
640 × 488	IL-CCD	21 × 21	40	5.0×10^5	0.10 (1.0)	1991	NEC
640 × 480	HB/MOS	20 × 20	80	7.5×10^5	0.10 (2.0)	1991	Hughes
1040 × 1040	CSD	17 × 17	53	1.6×10^6	0.10 (1.2)	1991	Mitsubishi
512 × 512	CSD	26 × 20	71	2.9×10^6	0.03 (1.2)	1992	Mitsubishi
656 × 492	IL-CCD	26.5 × 26.5	46	8.0×10^5	0.06 (1.8)	1993	Fairchild
640 × 480	HB/MOS	24 × 24	60	1.2×10^6	0.10 (1.4)	1996	AEG
811 × 508	IL-CCD	18 × 21	38	7.5×10^5	0.06 (1.2)	1996	Nikon
801 × 512	CSD	17 × 20	61	2.1×10^6	0.04 (1.2)	1997	Mitsubishi
1968 × 1968	IL-CCD	30 × 30	—	—	—	1998	Fairchild

Source: Adapted after Ref. [74].
IL-CCD, Interline Transfer CCD; HB: Hybrid.

- line-addressed charge-accumulation (LACA) readout
- readout by MOS switches.

Most of the reported Schottky-barrier FPAs have the interline transfer CCD architecture. Figure 26.15 shows the basic construction and operation of the most popular Schottky-barrier detector in the 3–5 μm spectral range, PtSi/p-Si, integrated with a silicon CCD readout [70]. Radiation is transmitted through the p-type silicon and is absorbed in the metal PtSi, producing hot holes that are then emitted over the potential barrier into the silicon, leaving the silicide charged negatively. Negative charge of silicide is transferred to a CCD by the direct charge injection method. The typical cross-section view of the pixel and its operation in interline transfer CCD architecture is shown in Figure 26.16 [72]. The pixel consists of a Schottky-barrier detector with an optical cavity, a transfer gate, and a stage of vertical CCD. The n-type guard ring on the periphery of the Schottky-barrier diode reduces the edge electric field and suppresses dark current. The effective detector area is determined by the inner edge of the guard ring. The transfer gate is an enhancement MOS transistor. The connection between detector and the transfer gate is made by an n^+ diffusion. A buried-channel CCD is used for the vertical transfer.

During the optical integration time, the surface-channel transfer gate is biased into accumulation. The Schottky-barrier detector is isolated from the CCD register in this condition. The IR radiation generates hot holes in the PtSi film and some of the excited hot holes are emitted into the silicon substrate leaving excess electrons in the PtSi electrode. This lowers the electrical potential of the PtSi electrode. At the end of the integration time, the transfer gate is pulsed-on to readout the signal electrons from the detector to the CCD register. At the same time, the electrical potential of the PtSi electrode is reset to the channel level of the transfer gate.

A unique feature of the Schottky-barrier IR FPAs is the built-in blooming control (blooming is a form of crosstalk in which a well saturates and the electrons spill over into neighboring pixels). A strong

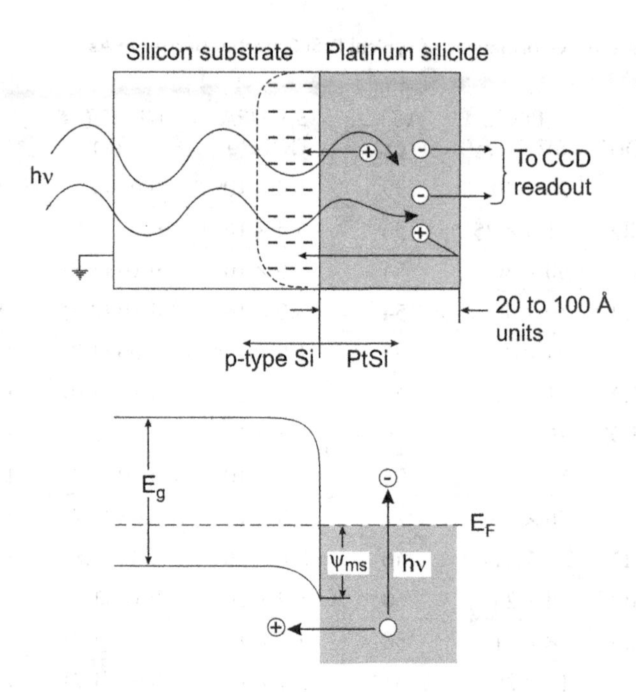

Figure 26.15 Operation of a PtSi/p-Si Schottky-barrier detector. (From Ref. [70])

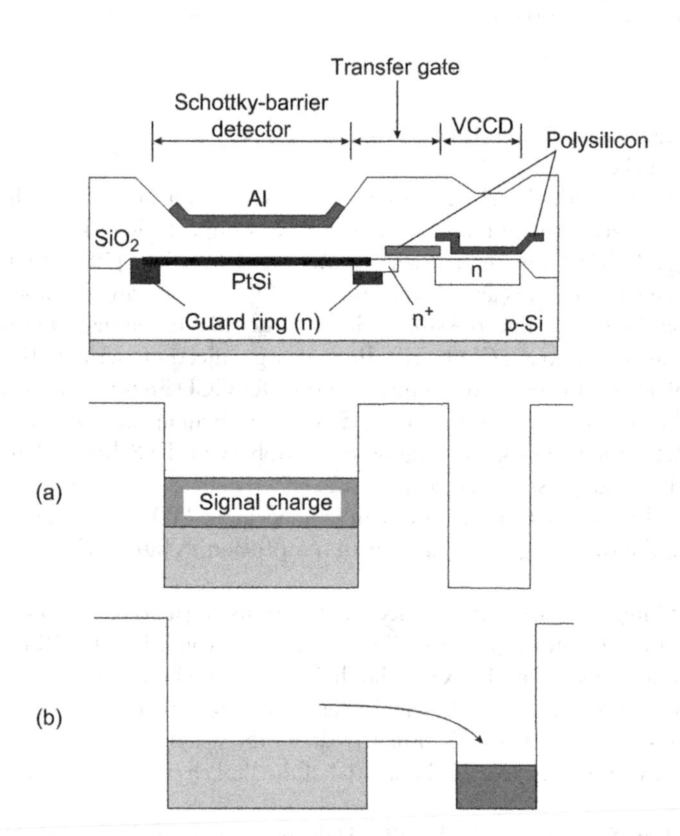

Figure 26.16 Typical construction and operation of PtSi Schottky-barrier IR FPA designed with interline transfer CCD readout architecture. (a) and (b) show the potential diagrams in the integration and readout operations, respectively. (From Ref. [72]. With permission.)

illumination forward biases the detector and no further electrons are accumulated at the detector. The small negative voltage developed at the detector is not sufficient to forward bias the guard ring to the extent that electrons are injected to the CCD register through the silicon region under the transfer gate. Therefore, unless the vertical CCD has an insufficient charge handling capacity, blooming is suppressed perfectly in the Schottky-barrier IR FPA.

The responsivity of the FPAs is proportional to their fill factor, and improvement in the fill factor has been one of the most important issues in the development of imagers. For improving the fill factor, a readout architecture called the CSD developed by Mitsubishi Corporation is also used. Kimata and coworkers have developed a series of IR image sensors with the CSD readout architecture with array sizes from 256 × 256 to 1,040 × 1,040 elements. Specifications and performance of these devices are summarized in Table 26.8 [67]. The effectiveness of this readout architecture is enhanced as the design rule becomes finer.

At the beginning of the 1990s, the 1,040 × 1,040 element CSD FPA had the smallest pixel size (17 × 17 μm²) among 2D IR FPAs [77,78]. The pixel was constructed with 1.5-μm design rules and had 53% fill factor. The array of 1,040 × 1,040 pixels was divided into four blocks of 520 × 520 pixels. Each block had a horizontal CCD and a floating diffusion amplifier. A 1 million pixel data at a 30-Hz frame rate was read by operating each horizontal CCD at a 10 MHz clock frequency. Figure 26.17 shows a photograph of the 1,040 × 1,040 array mounted in a 40-lead ceramic package and the first mega-pixel IR image with this PtSi CSD FPA [77]. The chip size of the device is 20.6 × 19.4 mm². The noise-equivalent temperature difference (NEDT) of mega-pixel array at 300 K with an *f*/1.2 cold shield and a 30-Hz frame was 0.1 K.

As Table 26.8 shows, all the 512 × 512 FPA fabricated by Mitsubishi Electric Corporation have a pixel size 26 × 20 μm². The earliest array, developed in 1987, was made using design rules of 2 μm and had a

Table 26.8 Specifications and performance of 2-D PtSi Schottky-barrier FPAs with CSD readout

Array size	256 × 256	512 × 512	512 × 512	512 × 512	801 × 512	1040 × 1040
Pixel size (μm²)	26 × 26	26 × 20	26 × 20	26 × 20	17 × 20	17 × 17
Fill factor (%)	58	39	58	71	61	53
Chip size (mm²)	9.9 × 8.3	16 × 12	16 × 12	16 × 12	16 × 12	20.6 × 19.4
Pixel capacitor	Normal	Normal	High-C	High-C	High-C	High-C
CSD	4-phase	4-phase	4-phase	4-phase	4-phase	4-phase
HCCD	4-phase	4-phase	4-phase	4-phase	4-phase	4-phase
Number of outputs	1	1	1	1	1	4
Interface	Non integration	Field integration	Frame/Field integration	Frame/Field integration	Flexible	Field integration
Number of I/O pins	30	30	30	30	25	40
Process technology	NMOS/ CCD 2 poly/2 Al	NMOS/ CCD 2 poly/2 Al	NMOS/ CCD 2 poly/2 Al	NMOS/ CCD 2 poly/2 Al	CMOS/ CCD 2 poly/2 Al	NMOS/ CCD 2 poly/2 Al
Design rule (μm)	1.5	2	1.5	1.2	1.2	1.5
Thermal response (ke/K)	—	13	—	32	22	9.6
Saturation (e)	0.7 × 10⁶	1.2 × 10⁶	—	2.9 × 10⁶	2.1 × 10⁶	1.6 × 10⁶
NEDT (K)	—	0.07	—	0.033	0.037	0.1

Source: Adapted after Ref. [67]

Figure 26.17 1040 × 1040 element PtSi/p-Si Schottky-barrier CSD FPA: (a) photograph of array mounted in a 40-lead ceramic package, and (b) the first mega-pixel IR image with this array. (From Ref. [77]. With permission.)

relatively small fill factor of 39% [79]. More recently as the design rules have been reduced, a high-performance PtSi Schottky-barrier IR image sensor has been developed with an enhanced CSD readout architecture [67,80–82]. Figure 26.18 shows a photograph of the pixel of 512 × 512 array with 71% fill factor; the array was made with 1.2 μm design rules [67]. The NEDT about 30 mK with *f*/1.2 optics at 300 K has been measured (see Table 26.8). The total power consumption of the 801 × 512 element device was less than 50 mW. Using the finest design rules makes it possible to manufacture a high-sensitivity FPA with 78% fill factor [74].

In order to improve the fill factor, other pixel designs have been proposed. One of them is the hybrid structure, which is generally used in compound semiconductor FPA. As shown in Figure 26.19, individual Schottky electrodes are fabricated so close to each other that their depletion regions merge, and diode isolation is only achieved by a 2-μm oxide gap between the silicide electrodes without the guard ring (the self-guarded detector) [83]. Using this structure, a fill factor of 80% was obtained for a 20-μm pixel array [84].

Kosonocky et al. [85] proposed a novel concept of the Schottky-barrier FPAs called the direct Schottky injection (DSI), which provides a 100% fill factor. This DSI FPA consists of a continuous silicide electrode (DSI surface) formed on one surface of a thinned (10- to 25-μm) silicon substrate with the CCD readout register on the other side as shown in Figure 26.20. During the operation, the silicon substrate is depleted between the DSI surface and charge-collecting elements of the readout structure. Injected hot holes from the DSI surface drift along the electric field line toward the collecting elements. The feasibility of the DSI

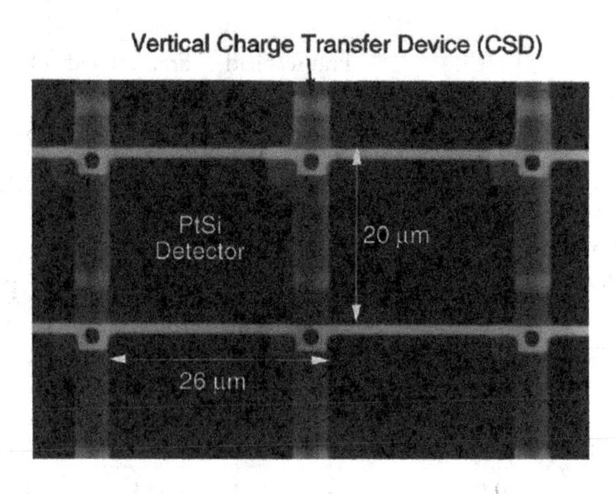

Figure 26.18 Pixel photograph of 512 × 512 PtSi Schottky-barrier CSD FPA taken just before aluminum reflector formation. (From Ref. [67]. With permission.)

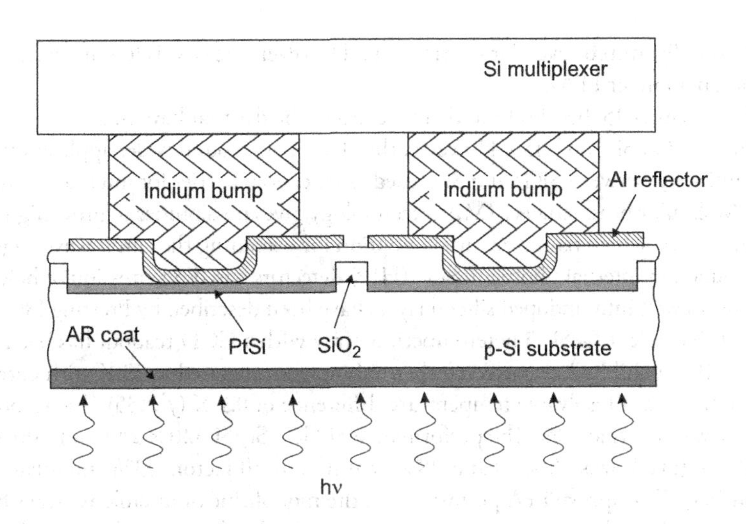

Figure 26.19 Pixel structures of hybrid Schottky-barrier FPA with self-guarded detector. (From Ref. [84]. With permission.)

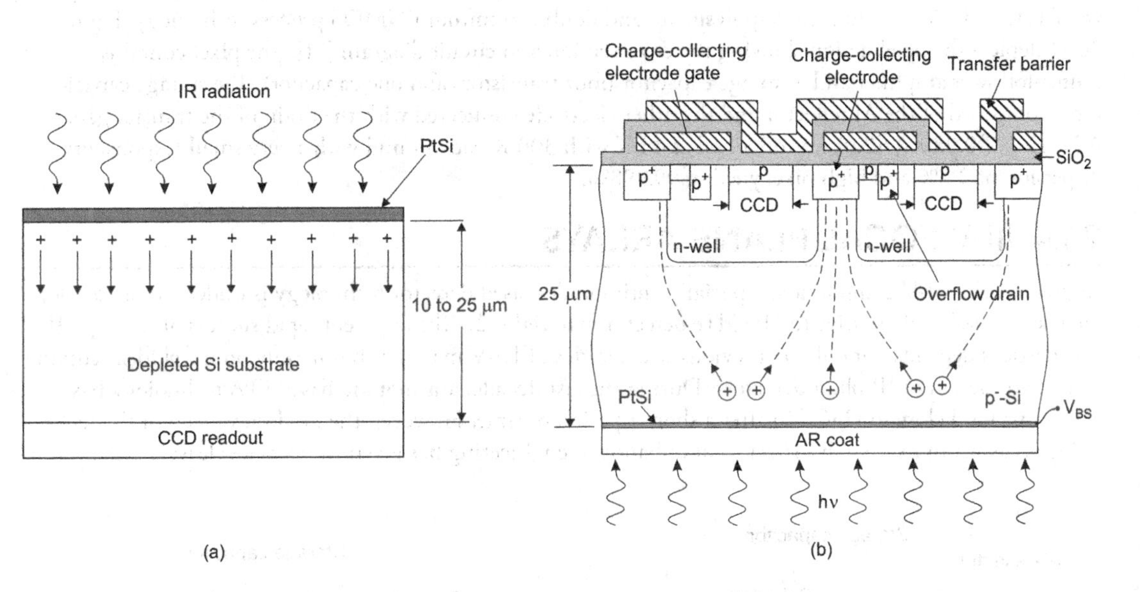

Figure 26.20 DSI FPA: (a) concept of the device—holes are injected from the continuous PtSi electrode into the readout CCD fabricated on the opposite surface, and (b) cross section of the IT-CCD DSI FPA across the BCCD channels. (From Ref. [85]. With permission.)

concept was demonstrated at the David Sarnoff Research Center with a 128 × 128 FPA with 50 × 50 μm² pixels and p-channel IT-CCD readout multiplexer [85]. The device has, however, considerable crosstalk to about 20%.

Current PtSi Schottky-barrier FPAs are mainly manufactured in 300-mm wafer process lines with around 0.15 μm lithography technologies. However, the performance of monolithic PtSi Schottky-barrier FPAs reached a plateau about 20 years ago and further slow progress is expected.

Development of PtSi Schottky-barrier FPAs was stopped in late 2010s. Both dark current and QE have reached their theoretical limits, and no further improvement is expected by refining the material and/or process technologies [75]. Si-based technology has a great cost advantage over other quantum IR FPA technologies. The state-of-the-art Si large-scale integration (LSI) technology also makes it possible to fabricate full-wafer IR FPAs on 300-mm or even larger Si wafers. However, the cost is less attractive when compared with uncooled microbolometer FPAs.

As is described in Chapter 15, besides PtSi, there are other silicides that have been used for Schottky-barrier IR detectors (Pd_2Si, IrSi, Co_2Si, and NiSi). However, they have never found wider applications [74]. Also the valence band discontinuity between SiGe and Si is used as an energy barrier for internal photoemission in the IR spectral range. Molecular beam epitaxy (MBE) technology gives possibilities to grow high-quality strained SiGe films on Si substrates, which has given another option for extending the cutoff wavelength. Different structures of heterojunction internal photoemission (HIP) detectors based on emission of holes from a highly p-doped SiGe quantum well into undoped silicon layers have been described by Presting [86].

The first 400 × 400 element GeSi/Si heterojunction array with a CCD readout has been developed by Tsaur et al. [87,88]. They exhibited uncorrected thermal imagers operated at 53 K with cutoff wavelength at 9.3 μm and with minimum resolvable temperature difference of 0.2 K (*f*/2.35). The responsivity nonuniformity of this array was less than 1%. The performance of $Ge_{1-x}Si_x$/Si 320 × 244 and 400 × 400 element arrays with $\lambda_c = 10\,\mu m$ (pixel sizes 40 × 40 and 28 × 28 μm² and fill factors 43% and 40%, respectively) has been described [89]. To improve FPA performance, the monolithic Si microlens arrays have been incorporated. Although these detectors were in an early stage of development, they already outperform IrSi detectors with the same cutoff wavelength with respect to QE.

More recently, Wada et al. have developed a high resolution 8- to 12-μm 512 × 512-element SiGe HIP MOS readout architecture FPA with pixel size 34 × 34 μm² and a fill factor of 59% [90]. The array was fabricated with 0.8-μm single-polysilicon and double-aluminum NMOS process technology. Figure 26.21 depicts the pixel design showing the cross section and circuit diagram [91]. The pixel contains a source-follower amplifier and a storage capacitor (four transistors and one capacitor). The storage capacitor is composed of the Al reflector and the other electrode connected with the drain of the transfer gate. A NEDT of 0.08 K (*f*/2.0) was obtained at 43 K with 300 K background with a very small responsivity dispersion of 2.2% and high pixel yield of 99.998%.

26.4 III-V FOCAL PLANE ARRAYS

At present, the III-V compounds, especially antimonide-based detector technology is under strong development as a possible alternative to HgCdTe detector material [92]. The apparent rapid success of the type-II superlattices depends not only on previous five decades of III-V materials but mainly on novel ideas coming recently in design of IR photodetectors. During the last decade, antimonide-based FPA technology has achieved a level close to HgCdTe after a shorter period of time. However, the modern version of the technology is as yet in its infancy. The advent of bandgap engineering has given III-Vs a new lease on life.

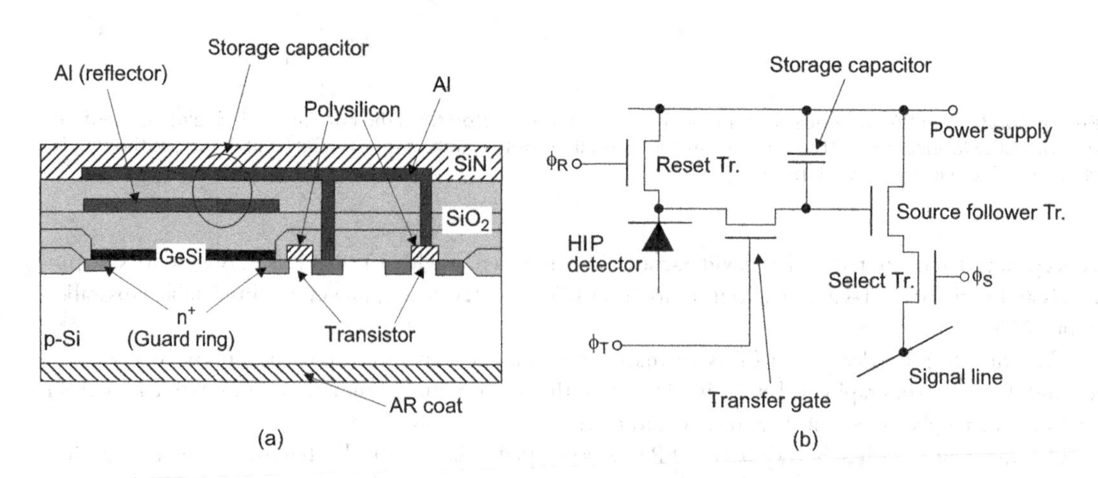

Figure 26.21 Pixel of GeSi HIP FPA: (a) the cross-section structure, and (b) the circuit diagram for a pixel. (From Ref. [91]. With permission.)

26.4.1 InGaAs ARRAYS

InGaAs ternary alloy is an optimal material choice for SWIR imaging applications due to the ability to operate at room temperature with high QE for wavelengths in the visible range to about 3 μm. These applications span both commercial and industrial opportunities including semiconductor-wafer inspection, wavefront sensing, astronomy, spectroscopy, machine vision, and military applications (surveillance, active point tracking, and laser radar). The increasing interest in the use of InGaAs detector arrays is driven primarily by potential advantages in detecting objects using target signatures which are dominated by reflection of external sources of illumination as opposed to thermal emission of radiation which occurs in the longer IR wavelengths. The InGaAs cameras produce significantly better quality images in challenging atmospheric conditions such as haze, mist, fog, and rain compared to visible imagers. They are operated at room temperature and therefore are compact, versatile, and simple to use as a commercial digital video camcorder.

Fabrication of traditional InGaAs photodiodes is described in Section 16.2. The arrays of photodiodes are hybridized to CMOS ROICs and then integrated into cameras for video-rate output to a monitor or for use with a computer for quantitative measurements and machine vision.

Linear array formats of 256, 512, 1,024, and 2,048 elements have been fabricated for operations in three ranges: 0.9–1.7 μm, 1.1–2.2 μm, or 1.1–2.6 μm with pixel pitch as small as 12.5 μm [93]. They are available in various sizes, defined by the detector height, pixel pitch, and the number of pixels and are packaged with 1-, 2-, or 3-stage thermoelectric cooling or without a cooler for externally cooled applications. Linear arrays with 2,048 square pixels are used for high-resolution imaging of fast-moving industrial processes. Arrays with tall pixels are widely used in optical spectrometers. The wide range of array pixel and readout formats enable users to make the optimal match to their applications.

The first 2D 128 × 128 $In_{0.53}Ga_{0.47}As$ hybrid FPA for the 1.0–1.7 μm spectral range was demonstrated by Olsen et al. in 1990 [94]. The 30-μm^2 pixels had 60-μm spacing and were designed to be compatible with a 2D Reticon multiplexer. Dark current below 100 pA, capacitance near 0.1 pF (−5 V, 300 K), and QE above 80% (at 1.3 μm) were measured. During the past 20 years, great strides have been made in the development of these progressing to the large 1,280 × 1,024 element arrays now readily available. A 320 × 240 array operates at room temperature, which allows development of a camera that is smaller than 25 cm^3 in volume, weighs less than 100 g and uses less than 750 mW of power [95].

At present, InGaAs FPAs are fabricated by several manufacturers including Sensors Unlimited [96–99], Indigo Systems (merged with FLIR Systems [100,101], Teledyne Judson Technologies [102], XenICs [93], SCD [103], Spectrolab [104], and Sofradir [105,106]. Table 26.9 lists the measured characteristics of near-IR cameras fabricated by Sensors Unlimited.

Detector dark current and noise are low enough that InGaAs detectors can be considered for astronomy where the bandwidth of interest is between 0.9 and 1.7 μm and a high operating temperature for the focal plane is important. In a Ref. [107] has been shown that the low temperature performance of a 1.7 μm cutoff wavelength 1k × 1k InGaAs photodiode against similar 2k × 2k HgCdTe imagers has been compared. The data indicates that InGaAs detector technology is well behaved and comparable to those obtained for state-of-the-art HgCdTe imagers. The InGaAs imagers offer potentially lower cost and higher reliability to those based on HgCdTe. In contrast to InGaAs however, HgCdTe maintains a nearly constant lattice parameter over the entire range of alloy composition without performance degradation.

The InGaAs FPAs achieves very high sensitivity in the shortwave IR bands in addition to the visible response added via substrate removal process post hybridization. The visible InGaAs detector structure (see Figure 26.22) is very similar to that shown in Figure 16.9, except that an InGaAs stop layer is added into the epi-wafer structure to allow complete removal of InP substrate [96,97]. The substrate is removed using a combination of mechanical and wet chemical etching techniques. The remaining InP contact layer thickness must be controlled to within 10 nm for consistent visible QE, typically about 40%. Thicker InP layers lead to lower QE in the visible spectrum as well as to image retention.

The largest and finest pitched imager in $In_{0.53}Ga_{0.47}As$ material system has been demonstrated recently. Goodrich has presented a high-resolution 1,280 × 1,024 InGaAs visible/SWIR imager with 15 μm pixels

Table 26.9 Specification of the near-IR InGaAs FPAs fabricated by Sensors Unlimited

	CONFIGURATION		
	640 × 512	640 × 512	1280 × 1024
Pitch (μm)	25	15	12.5
Optical fill factor (%)	100	100	100
Spectral response (μm)	0.5–1.7	0.7–1.7	0.5–1.7
QE (%)	>65 from 1.0 to 1.6 μm	>65 from 1.0 to 1.6 μm	≥65 from 1.0 to 1.6 μm
Mean detectivity ($cmHz^{1/2} W^{-1}$)	7.6×10^{12}	1.8×10^{13}	2.9×10^{13}
Noise equivalent irradiance (photons $cm^{-2}s^{-1}$)	3.6×10^{8}–1.6×10^{12} in 13 steps	1.1×10^{9}	8.5×10^{8}
Noise (rms)	225 electrons (typical)	65 electrons (typical)	35 electrons (typical)
Operability (%)	>99.2	>99	≥99
Capacity (electrons)	1×10^{6}		6×10^{6}
Exposure times	31 μs–128 ms in 13 steps	63 μs–33 ms	33 μs–33 ms
Image correction	2-point (offset and gain) pixel by pixel; user selectable	Goodrich proprietary	pixel by pixel; user selectable
Dynamic range	>3500:1	300:1(high gain), 1000:1(low gain)	1700:1
Active area (mm^{3})	16 × 12.8 × 20.5 diagonal	16 × 12.8 × 20.5 diagonal	16 × 12.8 × 20.5 diagonal

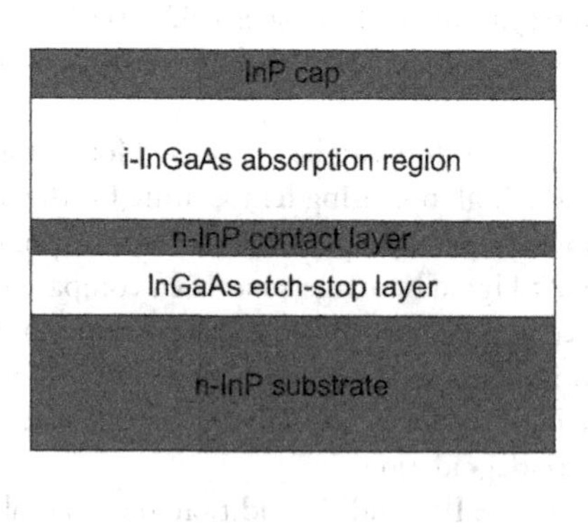

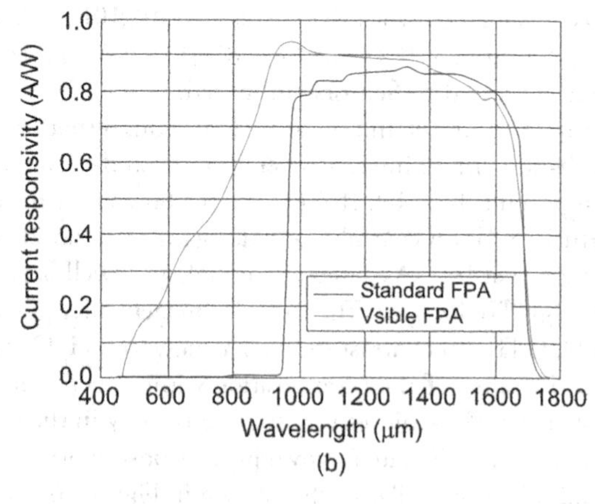

Figure 26.22 Visible InGaAs detector: (a) epitaxial wafer structure, and (b) QE in comparison with the standard detector. (From Ref. [97]. With permission.)

for day/night imaging [99]. The array with CTIA readout unit cells was designated to achieve a noise level of less than 50 electrons, due to its small integration capacitor. The ROIC was readout at 120 frames per second, and had a dynamic range of 3,000:1 using rolling, nonsnapshot integration. Total measured noise with the detector was 114 electrons using double sampling.

Figure 26.23a shows the 1,280 × 1,024 1.7 μm InGaAs sensor chip assembly (SCA) for MANTIS (Multispectral Adaptive Networked Tactical Imaging System) program [108]. The detector array is hybridized to an innovative ROIC with unit cell amplifiers designed with a capacitance transimpedance amplifier and a sample/hold circuit. Noise measurements of this SCA at a 30 Hz frame rate are shown in Figure 26.23b. A fit of noise model assuming domination of *kTC* and the detector g-r noise implies that the detector R_0A product is 8×10^6 Ωcm² at 280 K and the ROIC contributes approximately 40 electrons of noise. Above 240 K, the detector noise dominates while below 240 K ROIC noise dominates.

In recent years SCD has developed InGaAs/InP product, Cardinal 1,280, with 10-μm pitch and 1,280 × 1,024 (SXGA) format [103]. The new array is sensitive down to the visible spectrum, with a typical dark current of ~ 0.5 fA at 280 K, and a QE >80% at 1,550 nm. It also has a low noise imaging mode with 35 e readout noise with internal correlated double sampling. The FPA is integrated into a ruggedized, high vacuum integrity, metallic package, with a thermoelectric cooler for optimized performance, and a high-grade sapphire window. The operability is well above 99.5%. Figure 26.24 shows the image taken with the imager, after nonuniformity correction.

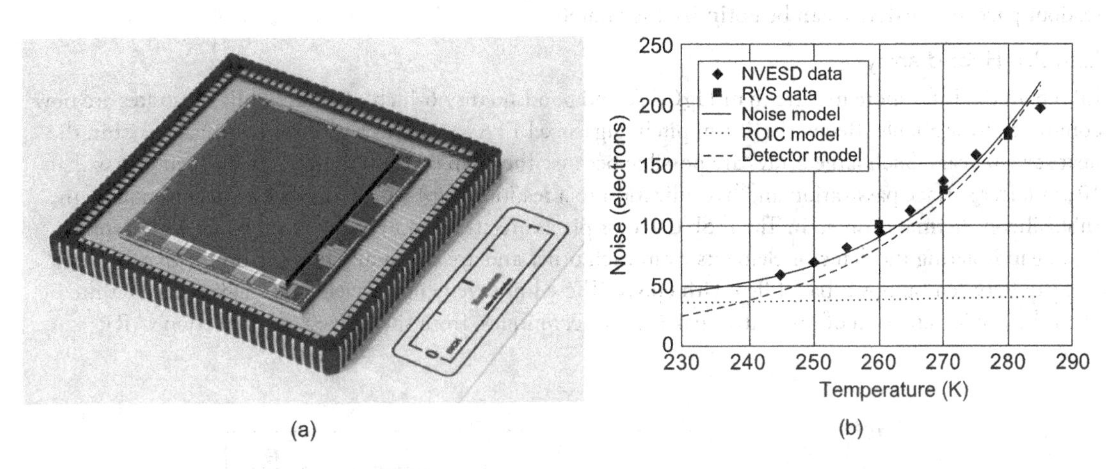

Figure 26.23 1280 × 1024 MANTIS InGaAs: (a) sensor chip assembly, and (b) noise at 30 Hz as a function of temperature. The line is a fit of the data to a noise model that includes detector g-r noise and ROIC *kTC* noise. NVESD, Night Vision and Electronic Sensors Directorate. (From Ref. [108]. With permission.)

Figure 26.24 The daylight image taken by 10-μm pitch 1280 × 1024 InGaAs visible/SWIR imager. (Adapted after Ref. [103])

26.4.2 InSb ARRAYS

InSb photodiodes have been available since the late 1950s. They are used in the 1- to 5 μm spectral region and must be cooled to approximately 77 K. InSb photodiodes can also be operated in the temperature range above 77 K. The applications include IR homing guidance, threat warning, IR astronomy, commercial thermal imaging cameras, and FLIR systems. One of the most significant recent advances in IR technology has been the development of large 2D FPAs for use in the staring arrays. Array formats are available with readouts suitable for both high-background *f*/2 operation and for low-background astronomy applications. Linear arrays are rather rarely used.

The earliest arrays fabricated in the mid-1980s were just 58 × 62 elements in size [109–112] compared to arrays that are up to 8192 × 8192 today, an increase of more than three orders of magnitude in pixel count (see Figure 26.25). The array noise has improved over this period of time from hundreds of electrons to as low as four electrons today [113]. Similarly, at the same time, detector dark current has decreased from about 10 electrons per second to as low as 0.004 electrons per second [114,115]; QE has reached the level above 90%.

The InSb FPA has been developed with the monolithic architecture as well as hybrid architecture. The best performance of an InSb FPA has been obtained using hybrid architecture where the detection and readout parts of the device can be optimized separately.

26.4.2.1 Hybrid arrays

InSb material is far more mature than HgCdTe and good-quality 6-inch diameter bulk substrates are now commercially available. Below the10-μm pitch, giga-pixel FPAs are set as a goal for realization within the next several years. Such large arrays are possible because the InSb detector material is thinned to less than 10 μm (after surface passivation and hybridization to a readout chip) which allows it to accommodate the InSb/silicon thermal mismatch. The InSb detector pixels are attached to a silicon substrate with a small 1-μm gap isolating the detector elements from each other and are essentially floating on the Si substrate. The gap between bumps is backfilled with epoxy. These improvements in yield and quality have resulted from the implementation of innovative passivation techniques, from specialized antireflection (AR)

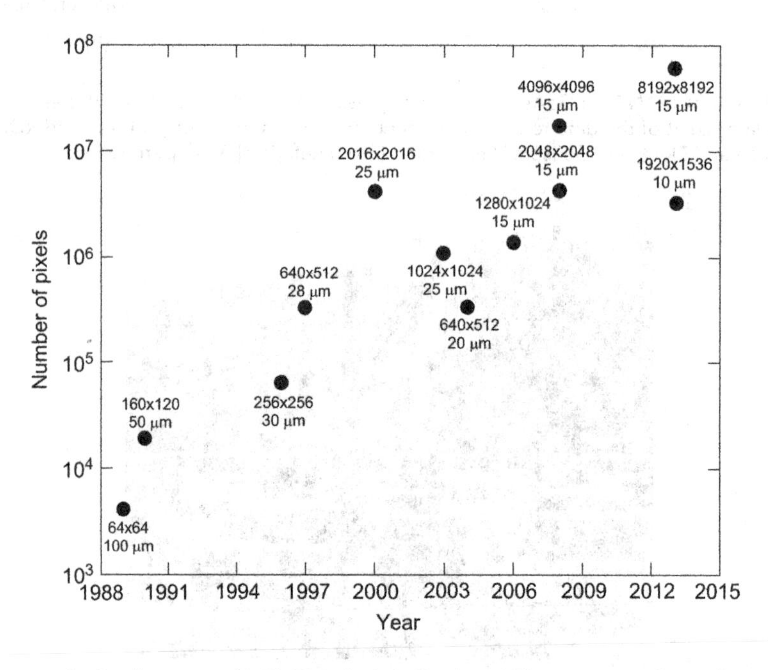

Figure 26.25 Progress in development of InSb FPAs by L-3-Cincinnati Electronics. It shows the total number of pixels with the year of first operation on the abscissa. 3 million pixel array with a pixel dimension of 10 μm has been first fabricated by SemiConductor Devices (Haifa). (Adapted after Refs. [116,117])

coatings to unique thinning processes. All of this processing is done at the wafer level. This is vitally important, since these devices are cooled from room temperature to 78 K several thousand times during their expected lifetime. The array operability (defined as the ratio of the number of nondefective pixels to total number of pixels in the array) is above 99.6% in over 15,000 cryogenic cycles [116]. At one cryogenic cycle per day, 15,000 cycles would take over 40 years.

One of the reasons for selecting InSb for IR instruments is its broad response, which is shown in Figure 26.26 [115,118]. The internal QE is nearly 100% in wide spectral band from 0.4 to 5 μm, which is an advantage of the thinned InSb arrays for this application. The limiting factor in QE is the reflection of incident light at the surface, which can be minimized with AR coatings.

InSb photodiodes demonstrate low dark current in large format arrays, which is shown in Figure 26.27 [115]. However, the dark current does not follow the predicted dark current due to generation-recombination mechanisms (see Figure 16.21a). The possibility of surface currents due to nonideal passivation can be investigated pending further funding for development.

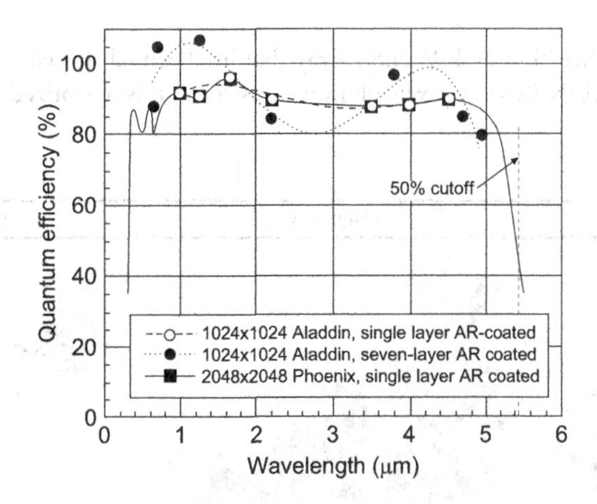

Figure 26.26 QE of InSb SCAs as a function of wavelength for a 1024 × 1024 ALADDIN SCA (single-layer and seven-layer AR-coated) and 2048 × 2048 PHOENIX SCA (single-layer AR-coated). (Adapted after Ref. [118])

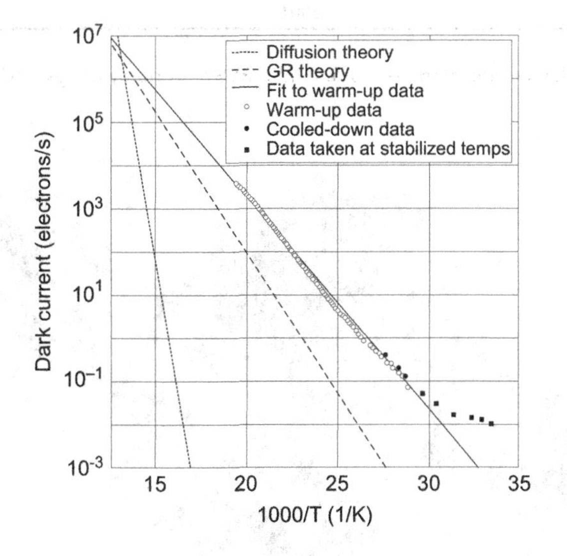

Figure 26.27 Dark current versus inverse temperature (temp) for a 2k × 2k InSb array with theoretical plots of diffusion and generation-recombination dark currents for comparison. The measured dark current follows a straight line on this semi-log plot down to 33 K and then flattens out to 0.01 electrons/second at 300 K. Data were obtained for both warming up and cooling down the detectors. (From Ref. [115]. With permission.)

Large staring InSb focal plane evolution has been driven by astronomy applications. Astronomers have funded large focal plane development to dramatically improve telescope throughput. The first InSb array to exceed one million pixels was the ALADDIN array first produced in 1993 by the Santa Barbara Research Center and demonstrated on a telescope by the National Optical Astronomy Observations (NOAO), Tucson, Arizona, in 1994 [119]. This array had 1024 × 1024 pixels spaced on 27-μm centers and was divided into four independent quadrants, each containing eight output amplifiers. This solution was chosen due to uncertain yield of large arrays at that time.

The ALLADDIN has been upgraded with the larger version, the ORION FPA family. A chronological history of the RVS astronomical FPAs is shown in Figure 26.28 [118]. The next step in the development of InSb FPAs for astronomy was the 2048 × 2048 ORION SCA (see Figure 26.29). Four ORION SCAs were deployed as a 4096 × 4096 focal plane in the NOAO near-IR camera [119], currently in operation at the Mayall 4-meter telescope on Kit Peak. This array has 64 outputs, allowing up to a 10-Hz frame rate. Many of the packaging concepts used on the ORION program are shared with the three-side buttable 2k × 2k FPA InSb modules developed by RVS for the James Webb Space Telescope (JWST) mission [120].

PHOENIX SCA is another 2k × 2k FPA InSb array that has been fabricated and tested. This detector array is identical to ORION (25-μm pixels); however, its readout is optimized for lower frame rates

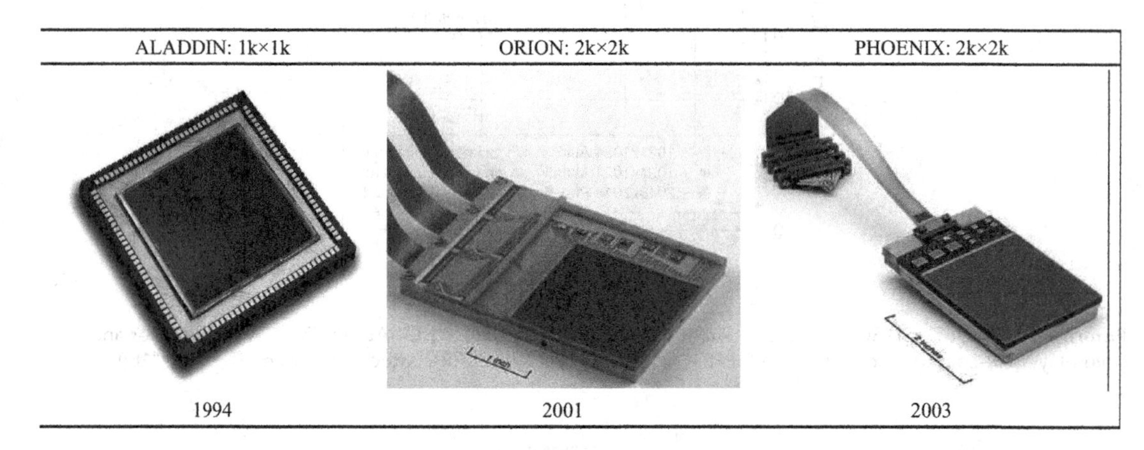

Figure 26.28 Timeline and history development of the InSb RVS astronomy arrays. (Adapted after Ref. [118])

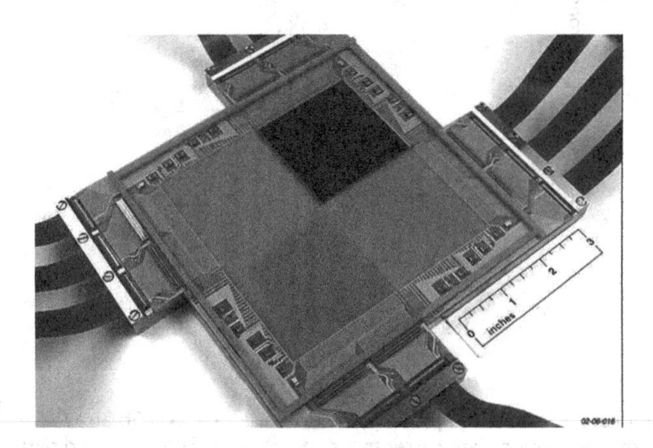

Figure 26.29 A demonstration of the two-side buttable ORION modules to create a 4k × 4k focal plane. One module contains an InSb SCA while the others have bare readouts. (Adapted after Ref. [115])

and lower power dissipation. With only four outputs, the full-frame read time is typically 10 seconds. The smaller number of outputs allows a smaller module package that is three-side buttable [115]. Three-side buttable modules allow the possibility to realize a large detection area.

Different formats of InSb FPAs have found also many high background applications including missile systems, interceptor systems, and commercial imaging camera systems. With an increasing need for higher resolution, several manufacturers have developed megapixel detectors. Table 26.10 compares performance of commercially available megapixel InSb FPAs fabricated by L-3 Communications (Cincinnati Electronics), Santa Barbara Focalplane, and SCD. Santa Barbara Focalplane has developed a new large format InSb detector with 1280 × 1024 elements and a pixel size of 12 µm [121].

L-3 Cincinnati Electronics is the manufacturer of large-format/wide-area surveillance sensors with 16.7 megapixels and an ultra-wide field of view. This imaging engine is capable of detecting and identifying features that small-format sensors would miss. It is currently in use by U.S. assets in overseas combat zones. Table 26.11 presents typical performance specifications for this InSb sensor.

SemiConductor Devices (Haifa) started in 1997 with the introduction of 320 × 256 format, 30-µm pitch detectors, and continued with the larger format arrays with pixel sizes of 25, 20, 15, and 10 µm. The MTFs of InSb FPAs with four different pitches (30, 20, 15, and 10 µm are shown in Figure 26.30. The shift to a smaller pixel dimension required the migration from a 0.5-µm CMOS process to a more advanced CMOS technology, in order to allow a higher value of capacitance per unit area, a lower operating voltage for reduced power consumption, and a denser device layout for maintaining a high level of functionality. The trend to larger format and smaller pitch continued with Hercules, an InSb detector with 1280 × 1024 pixels of 15-µm pitch. The new Blackbird detector is a natural step in this roadmap with 3 million pixels in the FPA and a pixel dimension of 10 µm.

Table 26.11 presents typical performance characteristics for Blackbird sensor module. It is packaged in a dewar which is integrated with a cryo-cooler and an electronic proximity board. This integrated detector cooler assembly (IDCA) makes a compact MWIR detector that generates 13-bit, 3-M pixel images at a frame rate of up to 120 Hz with a total power consumption of less than 30 Watt at 71°C. Figure 26.31 shows an image from the new detector with high temperature and spatial resolutions.

26.4.2.2 Monolithic InSb arrays

In the historical evolution of InSb detector technology, the monolithic FPA architecture has been also developed. An InSb metal-insulator-semiconductor (MIS) device as a PV IR detector was first proposed in 1967 by Phelan and Dimmock [123]. More comprehensive studies have been carried out by Lile and

Table 26.10 Performance of commercially available megapixel InSb FPAs

	CONFIGURATION			
	2048 × 2048	1024 × 1024	1024 × 1024	1280 × 1024
PARAMETER	(RAYTHEON ORION)	(L-3 COMMUNICATIONS)	(SANTA BARBARA FOCALPLANE)	(SCD)
Pixel pitch (µm)	25	25	19.5	15
Pixel capacity (electrons)	$>3 \times 10^5$	1.1×10^7	8.1×10^6	6×10^6
Power dissipation (mW)	<100	<100	<150	<120
NEDT (mK)	<24	<20	<20	20
Frame rate (Hz)	10	1–10	120	120
Operability (%)	>99.9	>99	>99.5	>99.5
References	www.raytheon.com	www2.l3t.com/ce/	www.sbfp.com	www.scd.co.il

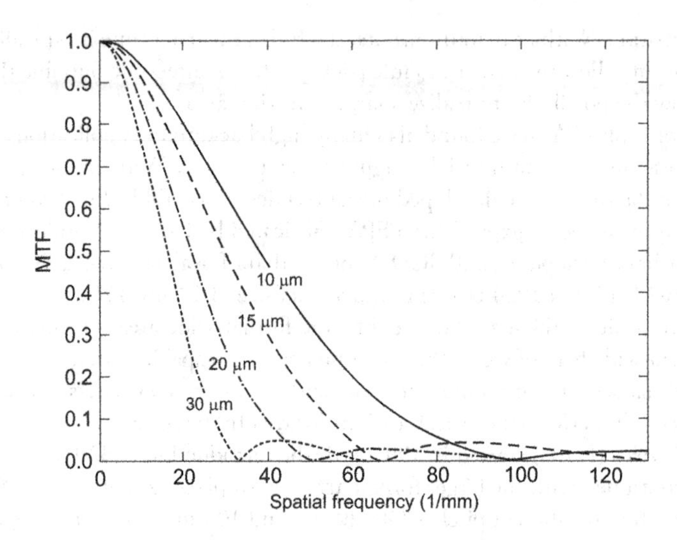

Figure 26.30 MTF curves of InSb FPAs with four different pitches: 30, 20, 15, and 10 µm, corresponding to SCDs Blue Fairy, Sebastian, Pelican, and Blackbird FPAs, respectively. (Adapted after Ref. [122])

Table 26.11 Performance characteristics of sensors with large format InSb arrays

	L-3 CINCINNATI ELECTRONICS	SCD
	(LARGE-FORMAT/WIDE-AREA SURVEILLANCE SENSORS)	(BLACKBIRD IDCA)
View of integrated detector		
Format	4096 × 4096	1920 × 1536
Pixel size (μm^2)	15 × 15	10 × 10
FPA power consumption	2/5 W	400 mW
Cooler power steady state (W)	<55	20
Weight	~15 lb.	700 gr
NEDT	Dependent upon integration time	<24 mK

Wieder [124,125]. The main motivation for developing MIS detectors was to integrate all functions needed for solid-state imaging (such as photon detection, charge storage, and multiplexed readout; see Chapter 24) together with achievements of high-performance imagers. These goals were not achieved due to fundamental limitation of narrow-gap semiconductor like InSb.

The CID was originally fabricated as a MOS silicon device to reduce the number of transfers needed for readout [126]. Shortly thereafter, CID devices were fabricated using narrow-gap material to form monolithic InSb FPAs [127]. Basic CID mechanisms and readout techniques have been described by Michon and

Figure 26.31 Image from the Blackbird detector at f/3, 2 km away. (Adapted after Ref. [122])

Burke [128]. In a CID, the detection process occurs within the unit cells that comprise two MIS structures that are readout in an x–y addressable manner. For readout mechanisms fabricated in narrow-gap semiconductors, the charge capacity is significantly less than a comparable silicon device. The bulk breakdown voltage V_{bd} of a semiconductor has been empirically related to the semiconductor's bandgap energy E_g as $V_{bd} \propto E_g^{3/2}$ [129]. Therefore, the nominal breakdown voltage of InSb is about 0.1 of silicon. Charge storage is also dependent upon the dielectric constant of the insulator thickness.

Coupling of the InSb MOS technology with that of CID has been presented by Kim [127]. Figure 26.32a shows schematically the n-type InSb MIS structure used in CID devices [130]. A fully planar (nonetch-back) process has been employed in InSb CID technology [131,132]. Figure 26.32b shows the cross-sectional and top view of a unit-cell geometry. The wafers were chemically polished and coated with a CVD SiO_2 film of about 135 nm thickness at <200°C. Then the column gates and the field plate were patterned using a thin chromium layer. A second layer of 220-nm SiO_2 was afterward deposited, and the row gates were defined with another thin chromium film. The transmission of the 7.5-nm-thick chromium layers was 60%–70% at the wavelength of 4 μm when AR coating was used. Thicker gold layers were used

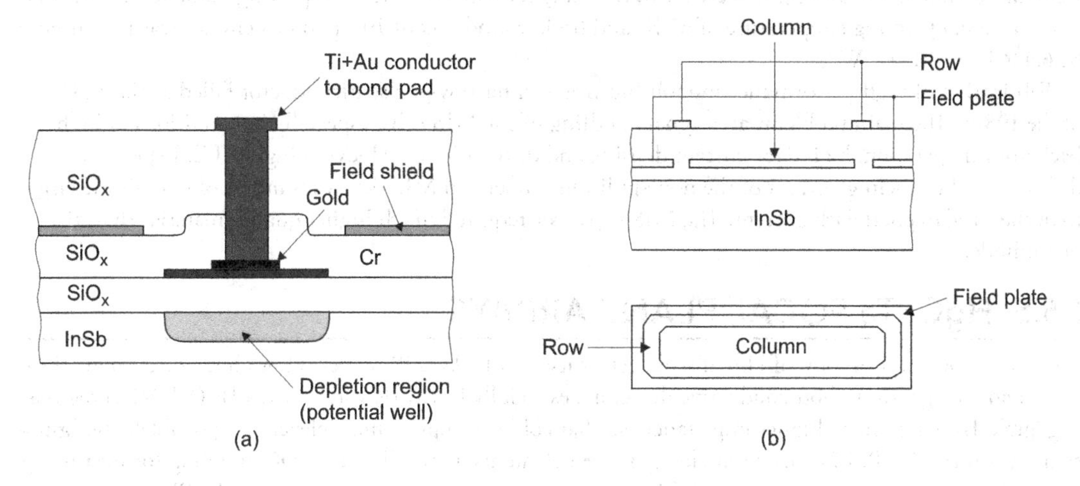

Figure 26.32 The InSb CID device: (a) construction of MIS capacitor. (From Ref. [130]. With permission.) (b) Cross-sectional and top view of a dual-gate CID cell. (From Ref. [132]. With permission.)

to form connecting runs, pads, and field shields. Instead of the conventional side-by-side capacitor layout, the improved devices had a concentric design, with one capacitor surrounding the other. The corners of the gates were rounded to minimize the electric fields in their vicinity. Arrays fabricated with that planar process exhibited nonuniformity within the gate oxide thickness of the order of 2%. Injection crosstalk was less than 1%.

The optimization of a CID design strongly depends on the readout scheme by which the array is operated. Although many different readout techniques were applied to silicon CID [128], only three (i.e., ideal mode, conventional charge-sharing mode, and sequential row injection mode) were used so far with InSb CID because of its later development [130]. A comparison of the three common readouts has been carried out by Gibbons and others [132–134]. Progress in InSb CID technology has led to the development of line arrays with 512 elements [130] and 128 × 128 staring arrays [133,135] with focal-plane Si MOS scanner/preamplifiers.

The dark current in the InSb CID is proportional to the sum of the depletion generation of carriers and minority carrier diffusion from the bulk material into the depletion layer and may be written as

$$J_{ds} = \frac{qn_i w}{2\tau} + \frac{qn_i^2 L}{N_d \tau}. \tag{26.1}$$

Assuming donor concentration $N_d = 3 \times 10^{14}\,\text{cm}^{-3}$, diffusion length $L = 25\,\mu\text{m}$, and operating temperature is 80 K, the depletion generation current is nearly three orders of magnitude higher than the diffusion current. The minority carrier lifetime, which is very sensitive to impurities and crystalline defects, has the most significant effect on the dark current. InSb material quality has been improved by growing epitaxial layers of InSb on InSb wafers by liquid phase epitaxy (LPE) [136]. The minority carrier lifetime increases more than two orders of magnitude in comparison with bulk material. Bulk InSb MIS detectors have been typically operated in the 77–90 K range, but LPE devices may allow operating temperatures above 100 K.

The first demonstration of CCDs in semiconductors other than silicon and germanium was achieved with InSb in 1975 [137]. Next, Thom et al. have fabricated a fully monolithic 20-element linear p-channel CCD array [138]. This process included planar p^+-n junction formation by Be-ion implantation to form the fat-zero input and charge output stages, and an aluminum-and-SiO_2-overlapping CCD gate structure that makes use of low-temperature CVD and plasma etching. The CTE of the device was 0.995 and was limited by lateral surface potential variations rather than surface states. The integration time was varied independently of readout rate, and operation in both the multiplexing and the time-delay and integration modes was demonstrated. The average detectivity of the array measured in the multiplexing mode for a 5-ms integration time, operating temperature of 65 K, and background flux of 10^{12} photons $\text{cm}^{-2}\text{s}^{-1}$ was measured to be $6.4 \times 10^{11}\,\text{cmHz}^{1/2}\,\text{W}^{-1}$.

Fabrication of high-performance monolithic arrays in narrow-gap semiconductor failed at the end of the 1980s. The main problems are: signal handling of the MIS cells, especially in conditions of high background operation, high dark current density, and difficulties in achieving high CTE. Especially defect-related tunneling current of the nonequilibrium operated MIS devices is orders of magnitude larger than the fundamental dark current. The MIS capacitor required much higher quality material than the photodiode.

26.5 HgCdTe FOCAL PLANE ARRAYS

The main mode of operation of HgCdTe detectors used in FPAs is PV effect. Photodiode offers many key system advantages over photoconductive detectors, especially in LWIR and very LWIR (VLWIR) regions: negligible $1/f$ noise, much higher impedance (so that cold preamps or multiplexers are possible), configuration versatility with BSI 2D arrays of closely spaced elements, better linearity, DC coupling for measuring the total incident photon flux, and a $2^{1/2}$ higher background-limited IR performance (BLIP) detectivity limit. However, photoconductive detectors will continue to be the better choice for certain instruments, such as those with relatively small numbers of detectors or with detection requirements out to extremely

LWs. Reine et al. [139,140] have presented an excellent paper that compares the performance of photoconductive and PV HgCdTe detectors for 15-μm remote sensing applications. Up to the present, PV HgCdTe FPAs have been mainly based on p-type material.

HgCdTe photodiodes are available to cover the spectral range from 1 to 20 μm. Most applications are concentrated in SWIR (1–3 μm), MWIR (3–5 μm), and LWIR (8–12 μm). Also development work on improving performance of VLWIR photodiodes in the 13- to 18-μm region for important earth-monitoring applications are undertaken.

There are a number of architectures used in the development of IR FPAs that are discussed in Chapter 23. In general, they may be classified as monolithic and hybrid. The best results have been obtained using hybrid architecture. Higher density detector configuration leads to higher image resolution as well as greater system sensitivity. HgCdTe IR FPAs have been made in linear (240, 288, 480, 960, and 1,024), 2D scanning with TDI (with common formats of 256 × 4, 288 × 4, 480 × 6), and various 2D staring formats with sizes from 64 × 64 up to 4096 × 4096 pixels (see Figure 26.33a). Efforts are also underway to develop avalanche photodiode (APD) capabilities in the 1.6 μm and at longer wavelength regions. Pixel sizes ranging from as small as 5 × 5 μm^2 have been demonstrated. The size of individual arrays continues to grow and the very large FPAs required for many space missions are fabricated by mosaicking a large number of individual arrays [141]. An example of a large mosaic developed by TIS is a 147-megapixel FPA that comprises 35 arrays, each with 2048 × 2048 pixels (see Figure 26.33b).

The fiftieth anniversary of the first publication devoted to HgCdTe ternary alloy [142] was an occasion to review the historical progress of HgCdTe material and device development in different countries. Figure 26.34 shows the timeline for HgCdTe FPA development at RVS (formerly Santa Barbara Research Center) starting from the initial bulk HgCdTe crystal wafers of 3 cm^2 and progressing through LPE on CdZnTe substrates of 60 cm^2 up to today's MBE on alternate substrates of 180 cm^2 [1,143].

The present LWIR and MWIR arrays are typically operated at liquid nitrogen temperature using Joule-Thompson or engine coolers. Some MWIR and SWIR arrays are thermoelectrically cooled to 190–240 K. Since many FPAs have very high data rates, the housings have coplanar leads to minimize parasitic impedances. Cooling of FPAs addressed for space missions are specially designed.

At higher backgrounds, it is impossible to handle the large amounts of carriers generated over frame times compatible with standard video frame rates. The FPAs are often operated at subframe rates much higher than the video update rate. Off-FPA integration of these subframes can be used to attain a level of sensor sensitivity that is commensurate with the detector-limited D^* and not the charge-handling D^*.

(a)

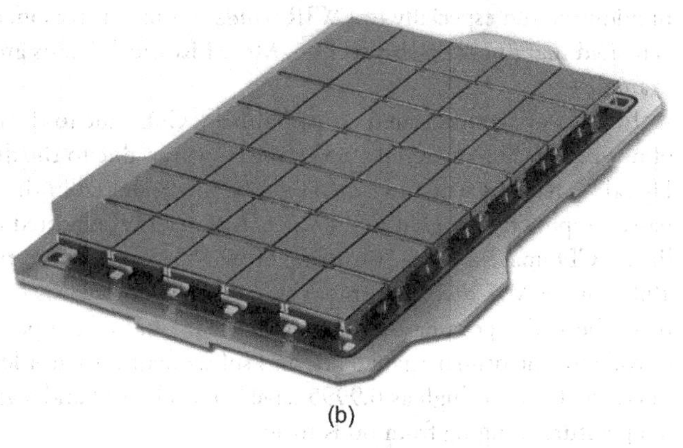

(b)

Figure 26.33 Examples of Teledyne Imaging Sensors packaging: (a) a mosaic of four Hawaii-2RGs as is being used for astronomy observations, and (b) a mechanical prototype of a mosaic of 35 Hawaii-2RG arrays as envisioned for the Microlensing Planet Finder. (From Ref. [141]. With permission.)

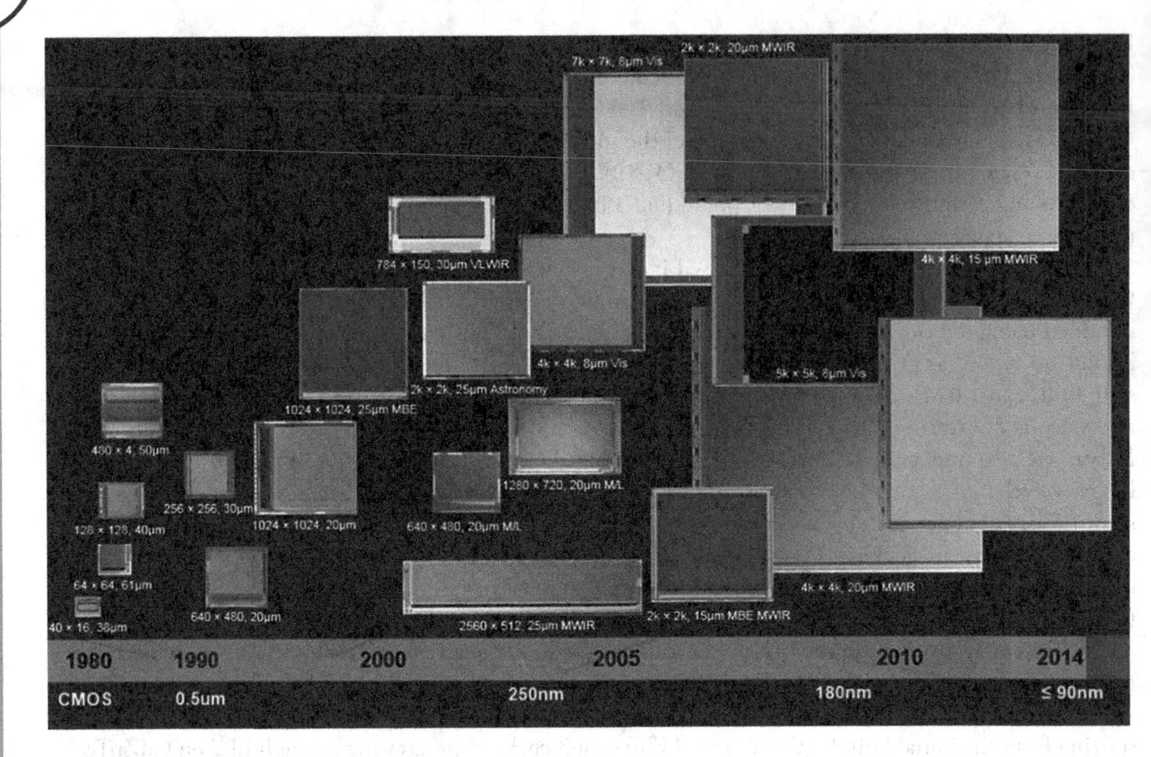

Figure 26.34 Timeline for HgCdTe development at Raytheon Vision Systems. Vis, visible. (Adapted after Ref. [1])

While the LWIR band should offer an order of magnitude better sensitivity, staring readout limitations due to charge handling limitations often constrain LWIR camera sensitivity to lower levels than competing MWIR devices due to lower LWIR contrast and similar (or lower) charge-handling capacity.

26.5.1 MONOLITHIC FOCAL PLANE ARRAYS

Monolithic HgCdTe MIS charge transfer devices (CTDs) were developed for nearly two decades between the mid-1970s to the mid-1990s. The three basic configurations of the HgCdTe CTD have been developed: CCD, CID, and charge imaging matrices. However, due to basic limitations of the monolithic charge transfer CCD FPAs associated with the use of narrow-gap HgCdTe material, high dark currents, and difficulties in achieving the high CTE, these devices did not rival state-of-the-art hybrid diode arrays in midwave and especially in LWIR bands. From this reason, an approach of monolithic HgCdTe devices is treated marginally in this section. More historical details are given in Rogalski's monograph [144], for example.

Initial work concentrated on p-channel CCDs, due to the maturity of the growth and doping control of n-type HgCdTe material [145–148]. However, due to the difficulty of forming stable p^+-n junctions in HgCdTe, readout structures could not be incorporated in the devices. After the demonstration of MISFET-based amplifiers in HgCdTe [149], Koch et al. [150] reported the development of a monolithic n-channel liner CCD imaging array consisting of two 55-bit CCD multiplexers, each addressing one-half of the 100-element MIS detector array ($x = 0.37$). The device was fabricated on an epitaxial HgCdTe layer grown by isothermal vapor phase epitaxy with the use of low-temperature photochemically vapor deposited silicon dioxide for the primary gate oxide and subsequent insulator levels in the structure. Using SiO_2/HgCdTe interface, CTEs as high as 0.9995 have been achieved and a value higher than 0.999 was measured for temperatures ranging from 60 K to 140 K.

Fully monolithic 128 × 28-element HgCdTe CCD arrays with 5-μm cutoff for low background applications have been demonstrated by Wadsworth et al. [151]. These arrays incorporate TDI detection, serial readout multiplexing, charge-to-voltage conversion, and buffer amplification in the HgCdTe detector chip.

The performance of these arrays (at 77 K the detectivity values exceed 3 × 10^{13} cmHz$^{1/2}$ W^{-1} for a background flux level of 6 × 10^{12} photon cm^{-2}s^{-1}).

The low storage capacity of 8–14 μm HgCdTe MIS structures makes them useless even for moderate background flux CCDs as the integration time (limited to ≈10 μs) becomes comparable to the transfer time, making the readout of the array impossible. LWIR system applications have been limited to scanning scenarios with relatively short integration times (e.g., 960 × 1 and 480 × 4 elements) [152]. Many device limitations were difficult to eliminate and, for these reasons, the MIS devices were superseded by the HgCdTe photodiode in hybrid FPAs.

26.5.2 HYBRID FOCAL PLANE ARRAYS

The architecture of most of the HgCdTe devices and manufacturing hybrid processes vary significantly from one manufacturer to another but have the following common features. First, they are all based on epitaxial growth of the HgCdTe material (LPE, MBE, and metal-organic chemical vapor deposition (MOCVD)); second, arrays of photodiodes are fabricated by a sequence of layering, doping, heat treatment, and etching steps; and third, all detector dies are connected to a silicon ROIC via a flip chip bonding. A BSI architecture, shown in Figure 24.12a, utilizes separately prepared detector arrays, which are then flipped over and hybridized to a silicon fanout pattern by means of indium bumps [1,129,153–155]. A high optical fill factor is easily achieved with this technique.

Baker et al. [156–159] have developed a unique interconnect technology for FSI detectors, named the loophole technique, shown in Figure 24.12b. This is a lateral collection device with a small central contact. The thermal expansion mismatch problem is approached by using a monolith of about 9-μm-thick p-type HgCdTe, bonded rigidly to the silicon so that strain is taken up elastically. This makes the devices mechanically and electrically very robust with contact obscuration typically less than 10%. Arrays up to 15 mm in length have been shown to be unaffected by multiple cycling to cryogenic temperatures [160]. The process has two simple masking stages. The first defines a photoresist film with a matrix of holes of typically 5 μm in diameter. Using ion beam milling, the HgCdTe is eroded away in the holes until the aluminum contact pads are exposed. The holes are then backfilled with a conductor to form the bridge between the HgCdTe and the underlying multiplexer pad. The junction is formed around the hole during the ion milling process. The second masking stage enables the p-type contact to be applied. The junctions are connected down to the silicon circuit by cutting the fine, few micrometers in diameter holes through the junctions by ion milling, and then backfilling the holes with metallization. The loophole technology has been applied to both LWIR and MWIR arrays yielding high-performance and reliable megapixel devices.

A modification of the lateral loophole technology is the vertically MIS approach developed at Texas Instruments [161]. More recently, vertically integrated photodiode (VIP^{TM}) technology has been developed at Texas Instruments [162]. In this case, a plasma etching stage is used to cut the via-hole and an ion implantation stage to create a stable HgCdTe junction and damage region near contact. In order to achieve higher lifetimes and lower thermal currents, Cu is introduced at the LPE growth stage. This is swept out during the diode formation and resides selectively in the p-type region, partially neutralizing the S-R centers associated with Hg vacancies. The final effect of this procedure is that dark current approaches this fully doped heterostructures. In the VIP^{TM} process, n-on-p photodiode chip is epoxy hybridized directly to the ROICs on large Si wafers by means of vias in the HgCdTe.

DRS' (formerly Texas Instruments) HgCdTe HDVIP® pixel architecture (see Figure 17.46) was scaled down up to 5-μm pitch. This technology is a major advance in the state of the art for IR FPAs. Since the via at the center of the pixel does not absorb any radiation, so with pixel size decreasing, the diode fill factor decreases. To improve fill factor of detector arrays, a new photolithography process was introduced in order to minimize the via size. A small-enough via was patterned and etched through the HgCdTe thin film down to the ROICs (see Figure 26.35). A cylindrical n-type region is formed around the via by type-converting p-type material. Thinner HgCdTe layer causes lower QE but using *f*/1 optics of typical tactical environments, LWIR diodes are not starved for flux current [163].

Highly operable 1280 × 720 5-μm pitch HDVIP® FPAs have been demonstrated for the LWIR (operability >99.6%) and MWIR (operability >99.95%) bands. Figure 26.36 presents an example of image from this array. DARPA in association with DRS Technologies has developed the 5-μm LWIR camera during

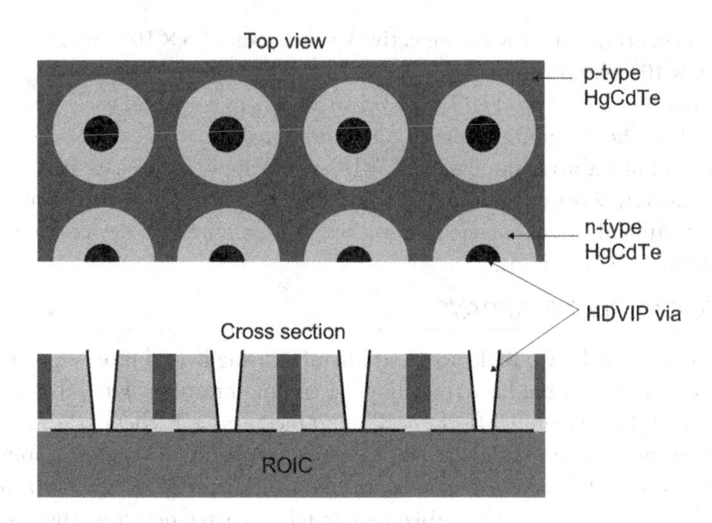

Figure 26.35 DRS' HgCdTe HDVIP® small-pitch array diagram. (Adapted after Ref. [163])

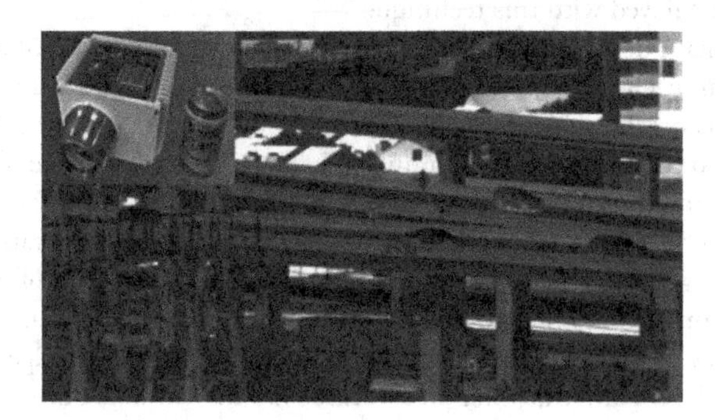

Figure 26.36 LWIR image taken with a 720 × 1280 HDVIP® FPA having 5-μm pixels. The FPA is designed for operation with *f*/1 optics, thus satisfying the criterion of $F\lambda/d = 2$. Inside is shown Lambda Scale camera with *f*/1 optics. (Adapted after Ref. [163])

the Lambda Scale program that uses a 1280 × 720 FPA (see inside of Figure 26.35). The FPA is designed for operation with *f*/1 optics, thus satisfying the criterion of $F\lambda/d = 2$. The approach is similar to that of a phone camera, which also uses smaller pixels to provide higher density in a compact package.

Arrays with pixel size as small as 5–10 μm have been fabricated in different architectures including n-on-p [164] and p-on-n [165] structures in both planar [164,165] and mesa [166] structures. Figure 26.37

Figure 26.37 Hybridization and bump pixel technology of 10-μm pitch HgCdTe pixel in comparison to a human hair. (Adapted after Ref. [167])

shows an array of 10-μm pixels in proportion to a human hair [167]. Technological challenges connected with fabrication of small pixel arrays are described in Ref. [168].

Initially, the diodes used in hybrid architecture were formed in a single p-type wafer of HgCdTe by ion implantation. After the diode arrays were hybridized, the HgCdTe wafer had to be thinned to about 10 μm to permit optimum absorption of IR radiation at the junction region and an increased R_0A product by a reduction of the diffusion volume.

Back-side illumination is readily achieved by the epitaxial growth of HgCdTe on transparent substrates. No thinning of the material after hybridization is required and the superior quality of epitaxial layers compared to bulk crystals is an additional advantage of this approach. Despite early concern over the stability of the bump interconnections, the devices have exhibited >99% interconnection yield and excellent reliability. At present, the operability is typically about 99.9%.

Advances in astronomy have spurred the need for imaging over as large a spectral range as possible, including visible to SWIR and MWIR. Recently, a process to remove the visible light blocking substrate has been developed. In addition, this allows the array to accommodate any thermal expansion by eliminating the thermal mismatch between the silicon readout and the detector array and eliminates pixel-to-pixel crosstalk. Figure 26.38 shows the typical visible and SWIR spectral response of substrate-removed HgCdTe FPAs with no degradation in detector mechanical and electrical quality and the expected improvements in visible response [15].

Sapphire buffered with CdZnTe became a standard substrate for SWIR and MWIR devices [169]. LWIR devices are typically based on CdZnTe. The progress on GaAs-based substrates has not progressed as fast as had been hoped. However, most of the MOCVD work on silicon has used a GaAs layer to buffer lattice mismatch between silicon and HgCdTe [170]. The next approach to reach production is connected with silicon-based alternative substrates, such as CdZnTe/Si [171]. Significant advantages of HgCdTe/Si are evident and available in large-area wafers because the coupling of the Si substrates with Si readout circuitry in an FPA structure allows fabrication of very large arrays exhibiting long-term thermal cycle reliability [172]. Figure 17.42b shows a schematic cross section of an MBE-grown p-on-n HgCdTe/Si double-layer heterojunction (DLHJ) design. A thin ZnTe buffer, typically 1 μm thick, is used to preserve preferred (211) orientation that can readily twin to form undesirable (552) domains depending on the grown conditions. The CdTe buffer layer is typically 6–9 μm in thickness and helps to reduce the dislocation density by annihilation.

Despite the large lattice mismatch (≈19%) between CdTe and Si, MBE has been successfully used for the heteroepitaxial growth of MWIR HgCdTe photodiodes on composite CdTe/Si substrates. However, it has been proved difficult to attain the best LW photodiode performance when grown on Si by MBE, such as is observed on lattice-matched CdZnTe substrates.

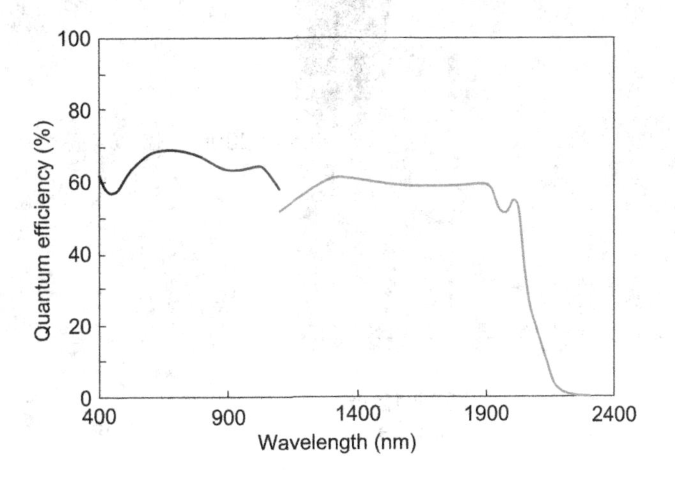

Figure 26.38 Spectral QE of substrate-removed 256 × 256 array. (From Ref. [15]. With permission.)

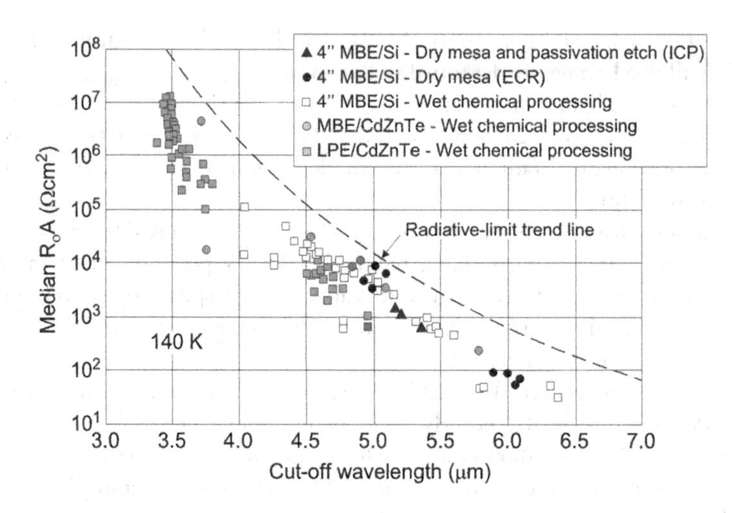

Figure 26.39 40-μm unit cell HgCdTe/Si DLHJ median R_0A detector array trend-line data as a function of measured 140 K cutoff wavelength. Trend-line data include HgCdTe material growth on Si (MBE) and CdZnTe (MBE and LPE) substrates. (From Ref. [143]. With permission.)

Figure 26.39 shows a 140-K detector median R_0A product versus cutoff trend-line that includes HgCdTe grown by MBE on both bulk CdZnTe and Si and LPE grown on bulk CdZnTe results. The diode performance with cutoff wavelength in the MWIR region for HgCdTe on Si is comparable to that on bulk CdZnTe substrates [143,173]. A series of high-performance megapixel staring SWIR and MWIR FPAs were demonstrated by RVS with operabilities exceeding 99.9%. For over a decade, RVS has fabricated HgCdTe FPAs on 6-inch silicon substrates. Utilizing fabrication processes identical to those used for HgCdTe/CdZnTe, FPAs are the largest reported for any IR detector fabrication method. Figure 26.40 shows a 6-inch-diameter HgCdTe/Si detector wafer, a 4k × 4k-format 20-μm pixel array

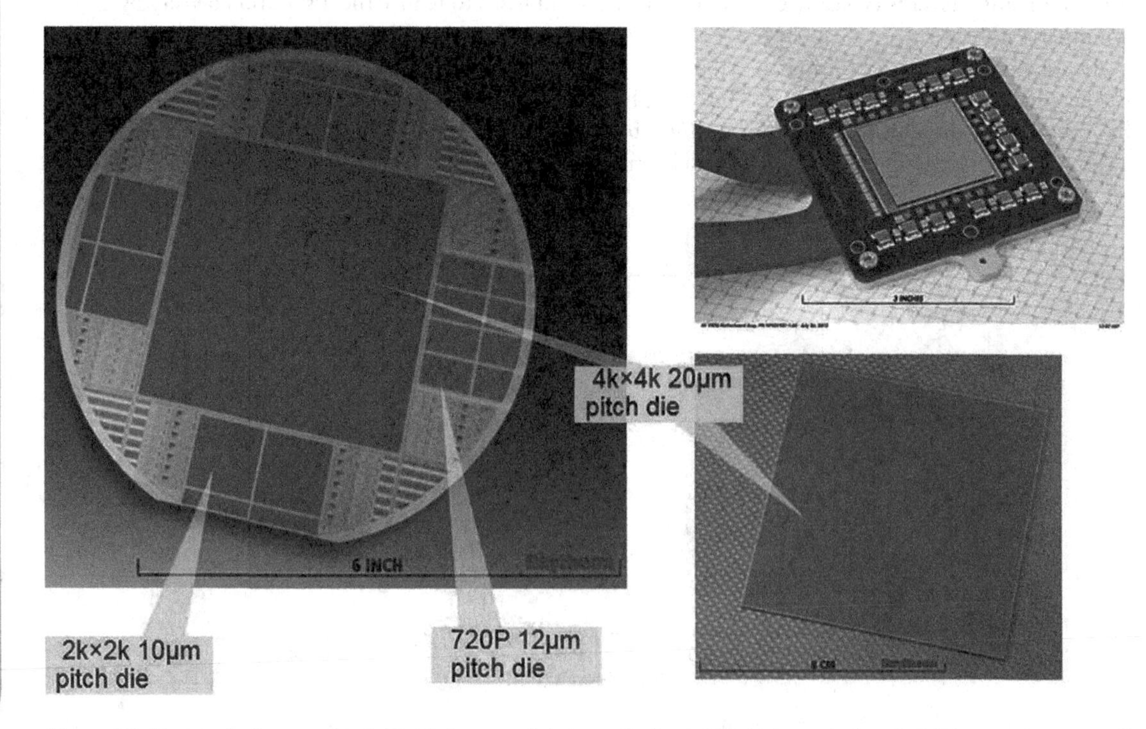

Figure 26.40 6-inch diameter HgCdTe/Si detector fabricated using MBE. (Adapted after Ref. [1])

composed of FPA operabilities greater than 99.9%. This is an equivalent-size array for an 8k × 8k 10-μm pixel format. This technology readiness provides affordable large-format arrays for current and future IR applications.

Efforts to extend the performance of HgCdTe/Si into the LWIR range has been also pursued using metalorganic vapor phase epitaxy–grown HgCdTe on GaAs and GaAs/Si substrates in the United Kingdom [174–176]. A major challenge has been achieving good I-V characteristics for material having uniformly high dislocation density values in the mid-10^6 cm^{-2} range. The median NEDT of arrays is quite good, but these arrays typically exhibit a noise tail that limits operability, particularly under background flux conditions [143].

Two generic types of silicon addressing circuits have been developed for readout of HgCdTe arrays: CCD and CMOS devices. The evolution of the indium bumps technology provided the enabling technology required for the ROIC progress achieved. In early 1980s, the more popular were CCD readouts. CMOS technology, since 1984, has improved the overall circuit and design of sensor chip assemblies to achieve ROICs with lower noise, higher yields, and higher densities [177]. The choice between CCD and CMOS depends on the application. For example, it seems much more complex to design a TDI linear array using a CMOS processor instead of a CCD processor, and on the other hand, a lot of advantages could be found for using a CMOS processor for a staring array. CMOS is now the preferred choice; it operates well at low temperatures.

At the beginning, the CCD structures have been utilized to multiplex information from the detectors in staring systems or/and to perform the TDI function in scanned systems [155,178]. The problem in the use of CCD for HgCdTe arrays is the difficulty of efficient charge injection into the source-coupled input of the CCD. To achieve 0.9 injection efficiency, the R_0A value should be about 10 times that for BLIP operation [179]. Even more stringent demands on R_0A are placed by the input gate $1/f$ noise, which is of greater importance in CCD structures. Another problem arises from a limited charge storage capacity ($\approx 10^4$ electrons μm^{-2}) and clock rate. Saturation of the storage wells by the background photocurrent in a short period makes frame rates impractically high for large arrays, unless additional charge skimming and/or partition circuitry is introduced to remove a part of the pedestrial charge. In order to achieve an adequate sensitivity, the clocking rate must be high and the background subtraction should be performed within every unit cell of the CCD structure. However, the small unit sizes desired in high-density FPAs constrain the input complexity. Despite higher ultimate performance of the IR image converters operating in LWIR, the stringent demands on R_0A value and the background limitations make SWIR and MWIR FPAs much easier to realize practically, compared to LWIR ones. In CCD technology, it is also difficult to envisage ways of disabling defective pixels with additional circuitry because of the poor packing density of mixed CCD/CMOS processes.

The actual minimum resolvable temperature difference of the system is set by fixed pattern noise. To achieve the potential value of 10 mK, the uniformity of output must be within 0.03%, while the typical standard deviation in present arrays is higher. It clearly illustrates the necessity for correction of the fixed pattern noise. The high injection efficiency of MWIR photodiodes into readout output and related linear output enables simple two-point correction for nonuniformities. It may be performed by calibrating the FPA at two different uniform background flux levels and storing the calibrated coefficient of each pixel in memory [154]. The DC offset and AC responsivity of all the pixels is then normalized by an addition and multiplication algorithm. Another source of fixed pattern noise is fluctuation of the FPA temperature. Recalibration of the DC offset with one temperature is usually sufficient.

Currently, the coordinative addressing with CMOS switches is the better attractive alternative to CCD readout in SWIR, MWIR, and especially LWIR FPAs (see Section 24.2.2). In hybrid HgCdTe FPAs, various detector interface circuits are used to appropriately condition the signal. The readouts used several detector interfaces. One of the simplest and most popular readout circuits for IR FPAs is the DI input, where dark current and photocurrent are integrated into a storage capacity. In this case, the bias varies across the array by about ±(5–10) mV due to variations in the transistor thresholds. At 80 K, HgCdTe diodes show very little dependence of leakage current with small changes in reverse-bias near zero volts. For high injection efficiency, the resistance of the FET should be small compared to the diode resistance at its operating point [180,181]. Generally, it is not a problem to fulfill this condition for MWIR HgCdTe staring designs where diode resistance is large (the R_0A product is in the range above 10^6 Ωcm²), but it can be very important for LWIR designs where diode resistance is small (the R_0A product is several hundred Ωcm²). In this case of LWIR HgCdTe photodiodes, a large bias is desirable, but it strongly depends on the material quality of the array. For very high-quality LWIR HgCdTe array, −1 V bias is possible.

Specifically, optimized input circuits are typically required for strategic and tactical applications. For tactical applications, where the backgrounds are high and detector resistances are moderate, direct injection (DI) is a commonly used input circuit. The goal is to fit as large a capacitor as possible into the unit cell, particularly for high tactical applications were signal-to-noise ratios can be obtained through longer integration times. This circuit is widely used for simplicity; however, it requires high impedance detector interface and is not generally used for low backgrounds due to injection efficiency issues. The strategic applications many times have low backgrounds and require low noise multiplexers interfaced to high resistance detectors. Commonly used input circuits for strategic applications are the CTIA input circuit. Besides the DI and CTIA inputs, we can distinguish other multiplexers; the most important are: source follower per detector (SFD; see Table 24.1), buffered DI (BDI), and MOSFET load gate modulation (BGM) input circuits [72,180–185]. Both CTIA and buffered DI give high injection efficiency and also accentuate the $1/f$ noise and the operability, but require higher power to operate.

As is marked in Section 24.5.3, the performance of MWIR and LWIR FPAs is limited by the readout circuits, and NEDT is estimated by Equation 24.19. High sensitivity can only be achieved if a large number of electrons are integrated, and this requires the integration capacitance in each pixel to be fairly high. The charge-handling capacity depends on cell pitch. For a 30 × 30-μm² pixel size, the storage capacities are limited to 1–5 × 10^7 electrons (it depends on design feature). For example, for a 5 × 10^7 electron storage capacity, the total current density of a detector with a 30 × 30-μm² pixel size has to be smaller than 27 μA cm^{-2} with a 33-ms integration time [186]. If the total current density is in the 1 mA/cm² range, the integration time has to be reduced to 1 ms. For the LWIR HgCdTe FPAs, the integration time is usually below 100 μs. Since the noise power bandwidth $\Delta f = 1/2t_{int}$, a small integration time causes extra noise in integration. Normally, the capacitance has a thin gate oxide dielectric and capacitance densities as high as 3 fFμm^{-2}. The capacitance is restricted to about 1 pF in pixels of around 25 μm², and the best NEDT that can be expected is about 10 mK per frame.

SWIR, MWIR, and LWIR electronically scanned HgCdTe arrays with CMOS multiplexer are commercially available from several manufactures. Table 26.1 presents the worldwide situation in the industry, while Tables 26.12 through 26.15 list typical performance specifications for larger SWIR, MWIR, and LWIR staring arrays fabricated by Raytheon, Sofradir, Teledyne, AIM, and Selex. Most manufactures produce their own multiplexer designs because these often have to be tailored to the applications. For example, Raytheon has a large advanced ROIC group with a portfolio of over 500 devices [1].

Reytheon's SW Virgo-2k 2,048 × 2,048/Virgo-4k 4,096 × 4,096 pixel array is fabricated for astronomy standard products. This 20-μm pitch array is characterized by high QE, low noise, low dark current, and on-chip clocking for ease of operation. Four or sixteen outputs can be selected to accommodate a wide range of input flux conditions and readout rates. Cut-on wavelength down to the visible (0.4 μm) is obtained by adjusting the optical transmission by removal of the CdZnTe or Si substrates. Very large imaging areas are realized using three-side or four-side buttable sensor chip assembles like for the Vital Infrared Sensor Technology Acceleration (VISTA) telescope with 67.1 million pixels (see Figure 24.4).

Teledyne's imaging sensors of Hawaii-2RG™ family are substrate-removed short-wavelength (SW) and medium-wavelength (MW) HgCdTe arrays with response in the visible spectrum. These arrays built with modularity in mind—four-side buttable to allow assembly of large mosaics of 2048 × 2048 H2RG modules—are dedicated for visible and IR astronomy in ground-based and space telescope applications.

Sofradir staring MW and LW snapshot arrays are dedicated to high resolution (TV format) applications (FLIR, IR search and track (IRST), reconnaissance, surveillance, airborne camera, and thermography). These FPAs can be offered in different long vacuum-time dewar and cooler configurations in order to meet the different mechanical and cooling needs of the systems. Similar snapshot arrays are offered by Selex and AIM.

Table 26.12 Raytheon's Virgo

Format	2048 × 2048
	4096 × 4096
Spectral response (μm)	0.4–2.5
Pixel size (μm^2)	20 × 20
ROIC type	SFD
Fill factor (%)	≥98
Detector materials	DLHJ HgCdTe
Integration capacity (e^-)	>3 × 10^5
Input referred noise (e^- rms)	<20
Number of outputs	4 or 16 (2k)
	8 or 32 (4k)
Max. frame rates	690 ms per frame
QE (%)	70–90
Read noise (e^-s^{-1})	<20 (Fowler 1)
Typical dark current (e^-s^{-1})	<0.05
Operating temperature (K)	70–80
Packaging	Module—three- or four-side buttable

Table 26.13 Teledyne imaging sensors—Hawaii-2RG

PARAMETER	1.7 μm	2.5 μm	5.3 μm
ROIC	Hawaii-2RG		
Number of pixels (#)	2048 × 2048		
Pixel size (μm)	18		
Outputs	Programmable 1, 4, 32		
Power dissipation (mW)	≤4/≤300		
Detector material	HgCdTe		
Detector substrate	CdZnTe - removed		
Cutoff wavelength (40–140 K) (μm)	1.65–1.80	2.45–2.65	5.3–5.5
Mean QE (%)	≥70		
Mean dark current (e^-s^{-1})		≤0.05 (goal is ≤0.01)	
Median readout noise (CDS at 100 kHz pixel readout rate (e^-)	≤30 (goal is ≤15)	≤18 (goal is ≤12)	≤15 (goal is ≤12)
Well capacity at 0.25 V bias (0.175 V bias for 5.3 μm cutoff) (e^-)	≥80,000 (goal is ≥100,000)		≥65,000 (goal is ≥85,000)
Crosstalk (%)	≤2 (goal is ≤1)		≤4 (goal is ≤2)
Operability (%)	≥95 (goal is ≥99)		
Cluster: 50 or more contiguous inoperable pixels (%)	≤1 (goal is ≤0.5) of array		
SCA flatness (μm)	≤20 (goal is 10)		

Table 26.14 MWIR HgCdTe FPAs

COMPANY	SOFRADIR (DAPHNIS)	SELEX (FALCON)	AIM
Array size	1280 × 720	1280 × 720	1280 × 1024
Pixel pitch (μm²)	10 × 10	12 × 12	15 × 15
Spectral response (μm)	3.7–4.8	3–5	3.4–5
Operating temperature (K)	up to 120	80–100	95–120
Max charge capacity (e⁻)	4.2×10^6	4×10^5	6×10^6
Pixel output rate (MHz)	up to 20	up to 10	up to 10
Frame rate (Hz)	up to 85 full frame rate		50
NEDT (mK)	<20	19	25
Operability (%)	>99.8	>99.8	>99.3

Table 26.15 LWIR HgCdTe FPAs

PARAMETER	SOFRADIR (SCORPIO)	SELEX (EAGLE)	AIM
Array size	640 × 512	1280 × 720	1280 × 1024
Pixel pitch (μm²)	15 × 15	12 × 12	15 × 15
Spectral response (μm)	7.7–9.3 at 80 K	8–10	7.6–9
Operating temperature (K)	up to 90	up to 90	70
Max charge capacity (e⁻)	1.36×10^7	1.8×10^7	6×10^6
Pixel output rate (MHz)	up to 8	up to 10	up to 10
Frame rate (Hz)	up to 210 full frame rate		50
NEDT (mK)	22	19	30
Operability (%)	>99.8	>99.8	>99.0

One of the most challenging tasks in the development of the next generation of HgCdTe FPAs is the integration of multiple functions into the detection circuit. The efforts are mainly focused on the development of multicolor detectors particularly for target recognition (see Chapter 27).

APDs are other devices with additional functionalities to the focal plane, in particular in the SW and MW ranges. The extremely low excess noise in the HgCdTe APDs is due to selective electron multiplication for wavelength $\lambda > 2\,\mu m$ and a nearly deterministic multiplication processes (see Section 17.6.4) [187]. The HgCdTe e-APDs are applied for gated-active/passive imaging [188–193]. Baker et al. at Selex were the first to demonstrate laser gated imaging in a 320 × 256, 24-μm pitch APD FPA [188]. They reported avalanche gains up to $M = 100$, low excess noise, and an input noise-equivalent photon noise $NEP_h = 15$ photons rms, for short integration times $t_{int} = 1$ μs for photodiodes with $\lambda_c = 4.2\,\mu m$. The latter is of particular interest for low-flux applications in the MW range, observing in a narrow field of view or spectral range. In addition, the amplification of the photocurrent can improve the linearity of some ROIC designs and a dynamic gain could be used to increase the dynamic range. Recently, Selex [194] and Leti [195] have made these devices available to the astronomy community.

In a partnership with the European Southern Observatory, Selex has developed a full-custom silicon ROIC as SAPHIRA (Selex Advanced Photodiode array for High speed Infrared Array). MOCVD HgCdTe epilayer is grown on low-cost GaAs substrates, which is removed after hybridization to the multiplexer. This 320 × 256, 24-μm pixel pitch array is designed for wavefront sensors and interferometry applications in astronomical telescopes. Technical specifications of SAPHIRA array is included in Table 26.16.

Recently, First Light Imaging has developed C-RED One camera based on the last version of the SAPHIRA detector developed by Selex (see Figure 26.41). This camera is capable of capturing up to 3,500

Table 26.16 Avalanche SAPHIRA array

PARAMETER	
Array	320 × 256
Spectral range (μm)	0.8–2.5
Pixel pitch (μm)	24
Active area (mm)	7.68 × 6.14
Avalanche gain	up to 80
Median sensitivity	1 photon RMS (at gain of 80)
Pixel operability (%)	>99
Power consumption (mW)	30
Modes	Snapshot or rolling
Charge capacity (electrons)	2×10^5
Number of outputs	4, 8, 16, or 32
Array operating temperature (K)	30–150

Photo—left; specification—right

Figure 26.41 C-RED One camera. The cooling system (pulse tube) can be seen on the top, whereas in the bottom are the vacuum cryostat and the readout electronics. (Adapted after Ref. [196])

full frames per second with a subelectron readout noise and very low background in spectral response over J, H, and K bands. The sensor cooled down to cryogenic temperature using an integrated pulse tube.

26.6 LEAD SALT ARRAYS

Lead salts are ones of the first polycrystalline thin-film materials sensitive to the IR radiation used for military applications. Early research works on the materials as IR detectors were carried out during the 1930s, and the first useful devices were processed by Germans, Americans, and British during and just after World War II. Since then, lead salts have been commonly used as MWIR photodetectors in multiple applications, from spectrometers for gas and flame detection to IR fuzes for artillery ammunition or passive IR cueing systems. Even though they have been extensively studied, today the mechanisms responsible of its high detectivity at room temperature are not well understood. What is widely accepted is that the material and the polycrystalline nature of the active thin film play a key role in both the reduction of the Auger mechanism and the reduction of the dark current associated with the presence of multiple intergrain depletion regions and potential barriers inside the polycrystalline thin films. Their historical prospect is given [144,197–200]. Low-cost, PbS and PbSe polycrystalline thin films remain the photoconductive detectors of choice for many applications in the MWIR spectral range [201–203].

Modern lead salt detector arrays contain more than 1,000 elements on a single substrate. Operability exceeding 99% is readily achieved for these arrays. Smaller arrays having 100 or fewer elements have been produced with operability of 100%. Arrays as large as several inches on a side use a linear configuration with a single row and either equal areas equally spaced or of variable sizes. Other configurations include dual rows either in-line or staggered (several staggered rows in staircase fashion, chevron, and double cruciform).

Northrop Grumman EOS coupled 256-pixel PbSe arrays with Si multiplexer readout chips to fabricate assemblies with scanning capabilities [204]. Table 26.17 summarizes the performance of PbS and PbSe arrays in 128- and 256-element configurations [205]. A long-lived thermoelectric element cools the detector/dewar assembly to provide lifetimes greater than 10 years. It should be noted, however, that lead salt photoconductive detectors have a significant $1/f$ noise; for example, for PbSe, a knee frequency is of the order of 300 Hz at 77 K, 750 Hz at 200 K, and 7 kHz at 300 K [206]. This generally limits the use of these materials to scanning imagers.

Figure 26.42 shows the multimode detector/multiplexer/cooler assembly manufactured by Northrop Grumman [201]. This package configuration was originally developed for multiplexed 256-element linear

Table 26.17 Typical performance of PbS and PbSe linear arrays with a CMOS multiplexed readout

	PbS		PbSe	
Configuration	128	256	128	256
Element dimensions (μm)	91 × 102 (in line)	38 × 56 (staggered)	91 × 102 (in line)	38 × 56 (staggered)
Center spacing (μm)	101.6	50.8	101.6	50.8
D^* (cmHz$^{1/2}$ W^{-1})	3×10^{11}	3×10^{11}	3×10^{10}	3×10^{10}
Responsivity (V W^{-1})	1×10^{8}	1×10^{8}	1×10^{6}	1×10^{6}
Element time constant (μs)	≤1,000	≤1,000	≤20	≤20
Nominal element temperature (K)	220	220	220	220
Operability (%)	≥98	≥98	≥98	≥98
Dynamic range	≤2000:1	≤2000:1	≤2000:1	≤2000:1
Channel uniformity (%)	±10	±10	±10	±10

Source: Adapted after Ref. [205]

Figure 26.42 PbSe FPAs mounted in Northrop's standard M2105 package. (From Ref. [201]. With permission.)

arrays. The multiplexed array is thermoelectrically cooled and closed in a long-life evacuated package with an AR-coated sapphire window and mounted on a circuit board, as shown in the photograph. Similar assemblies containing PbS elements are also fabricated.

A first attempt to realize quasimonolithic lead salt detector arrays was described by Barrett, Jhabvala, and Maldari, who elaborated on direct integration of PbS photoconductive detectors with MOS transitions [207,208]. In this process, the PbS films were chemically deposited on the overlaying SiO_2 and metallization. Detectivity at 2.0–2.5 μm of 10^{11} $cmHz^{1/2}$ W^{-1} was measured at 300 K on an integrated photoconductive PbS detector-Si MOSFET preamplifier. Elements of 25 × 25 μm^2 were easily fabricated.

In FPA fabrication, lead salt chalcogenides are deposited on Si or SiO from wet chemical baths. Such a monolithic solution avoids the use of a thick slab of these materials mated to Si, as is done with typical hybrids. The detector material is deposited from a wet chemical solution to form polycrystalline photoconductive islands on a CMOS multiplexer. Figure 24.43 shows a few of the 30-μm pixels in this detector array format. Northrop Grumman elaborated monolithic PbS FPAs in a 320 × 240 format (specified in Table 26.18) with a pixel size of 30 μm [201]. Although PbS photoconductors may be operated satisfactorily at ambient temperature, performance is enhanced by utilizing a self-contained thermoelectric cooler.

Considerably, breakthrough in developing low-cost PbSe detectors for threat-warning systems has been recently achieved by Northrop Grumman [203]. The focal planes on these detectors are PbSe-deposited on silicon readouts. The arrays are 320 × 240 pixels with 60-μm pixels and 99.6 reproducibility. Sensitivities are now 30 milliKelvin with an *f*/1 optic and a 2.5-ms integration time, operating at 230 K (see Table 28.18). Figure 26.44 exhibits broadband IR across the entire PbSe sensitivity range. The sensor was placed on a roof top 40 meters in elevation at midday under sunny conditions facing a parking lot whose furthest features are at a slant range of 1 km.

Zogg et al. [209–212] have fabricated a monolithic, staggered linear array with up to 256 PbTe and PbSnSe Schottky-barrier photodiodes (see Figure 18.29) with 30 μm diameter on a 50-μm pitch. The substrates for these arrays contain integrated transistors for each pixel as needed for the readout. The readout chips were fabricated in a combined CMOS/junction FET (JFET) technology. While CMOS designs require impedances in the MΩ range in order that the amplifier noise does not dominate, JFET input transistors can be designed with negligible noise even for low impedances (down to below 10 KΩ) without demanding high bias currents as, for example, in bipolar designs. For each channel, a charge integrator collects the photogenerated charges over a certain present time. The generated signals are then fed to a common output. Individual offset correction and multiple correlated sampling to reduce the readout noise was

Figure 26.43 Monolithic PbS pixels in a 320 × 240 format. Pixel pitch is 30 μm. (From Ref. [201]. With permission.)

Table 26.18 Specifications for 320 × 240 lead salt FPAs

FPA CONFIGURATION	MONOLITHIC 320 × 240 PbS (REF. [205])	MONOLITHIC 320 × 240 PbSe (REF. [203])
Pixel size (μm)	30 × 30	60 × 60
Detectivity (cmHz$^{1/2}$ W^{-1})/NEI*	8 × 10^{10} (ambient); 3 × 10^{11} (220 K)	NEI* = 0.07 μWcm^{-2}
NEDT (mK)		30 (*f*/1 optics) 230 K
Type of signal processor	CMOS	CMOS
Time constant (ms)	0.2 (ambient); 1 (220 K)	
Integration options	Snapshot	
Number of output lines	2	
Frame rate (Hz)	60	400
Integration period	Full frame time	
Mux dynamic range (dB)	69	
Active heat dissipation (mW)	200 max	
Operability (%)	>99	>99.6
Mux transimpedance (MΩ)	100	
Detector bias (V)	0–6 (user adjustable)	

Figure 26.44 Image frame data collected on a roof top, viewing parking lot of Northrop Grumman Rolling Meadows Campus. *f*/1 lens was broad band with no spectral filter. NEDT performance was measured to be 30 mK with 99.6% pixel operability at 230 K operating temperature. (Adapted after Ref. [203])

employed. The further processing, background subtraction, and correction of the fixed pattern noise are performed digitally.

The research group at the Swiss Federal Institute of Technology also demonstrated the first realization of monolithic PbTe FPA (96 × 128) on a Si substrate containing the active addressing electronics [213,214]. The monolithic approach overcomes the large mismatch in the thermal coefficient of expansion between group IV-VI materials and Si. Large lattice mismatches did not impede fabrication of the high-quality layers because the easy plastic deformation of the IV-VIs by dislocation glide on their main glide system without causing structural deterioration.

A schematic cross section of a PbTe pixel grown epitaxially by MBE on a Si readout structure is shown in Figure 26.45 [213]. A 2- to 3-nm-thick CaF_2 buffer layer is employed for compatibility with the Si substrate. Active layers of 2–3 μm thickness suffice to obtain only a near-reflection-loss-limited QE. The spectral response curves of PbTe photodiodes are shown in Figure 18.30. Typical quantum efficiencies are around 50% without an AR coating. Metal semiconductor Pb/PbTe detectors are employed. Each Pb-cathode contact is fed to the drain of the access transistor while the anode (sputtered Pt) is common for all pixels. Figure 26.46 shows a complete array [214].

Despite typical dislocation densities in the $10^7 cm^{-2}$ range in epitaxial lattice and thermal-expansion-mismatched IV-VI on Si(111) layers, useful photodiodes can be fabricated with R_0A products to 200 Ωcm² at 95 K (with a 5.5-μm cutoff wavelength). This is due to the high permittivities of the IV-VI materials, which shield the electric field from charged defects over short distances.

Lead salt heterostructure PV detectors were also fabricated. The heterojunctions were formed directly between the substrate and the IV-VI films [215]. However, the performance of heterostructure arrays is inferior to the Schottky-barrier and p-n junction arrays.

Experiments with development of lead salt hybrid array approach during late 1970s and early 1980s failed. A severe problem is the large thermal expansion mismatch of IV-VI materials to silicon. Therefore, the IV-VI detector/silicon hybrid technology lay in small-size photodiode arrays with 10^3 elements operating at 77 K in the 8–14 μm region. Such hybrid structures containing 32 × 32 elements of BSI PbTe [216] and PbSnTe [129] photodiodes with solder bump interconnections to silicon chips were demonstrated. Larger arrays require assembly from submodules of ≈10^3 contiguous arrays [129].

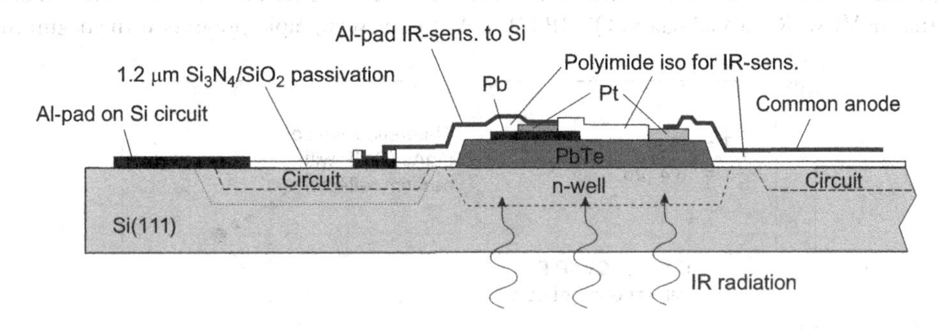

Figure 26.45 Schematic cross section of one pixel showing the PbTe island as a BSI PV IR detector, the electrical connections to the circuit (access transistor), and a common anode. IR-sens., IR sensor. (From Ref. [213]. With permission.)

Figure 26.46 Part of the completely processed monolithic 96 × 128 PbTe-on-Si IR FPA for the MWIR with the readout electronics in the Si substrate. Pixel pitch is 75 μm. (From Ref. [214]. With permission.)

Infrared focal plane arrays

Felix et al. [178] described experimental results for a hybrid island structure with FSI $Pb_{0.8}Sn_{0.2}Te$ photodiodes on a silicon CCD. The photodiodes were made by double LPE, PbTe/PbSnTe on a PbTe substrate. This technology is more complex but, because each detector is a separate physical unity, these structures avoid the problems due to thermal expansion mismatch. On the other hand, the island approach suffers from a fill factor loss because of the area contacts. The average R_0A product of photodiodes was close to 0.8 Ωcm^2, and detectivity was about 2×10^{10} $cmHz^{1/2} W^{-1}$ at 77 K for 2π field of view (FOV). Felix and colleagues [178] also gave experimental results for the readout of $Pb_{0.8}Sn_{0.2}Te$ photodiodes directly injected to a linear multiplexing silicon CCD. An injection efficiency of 65% has been obtained.

26.7 QUANTUM WELL INFRARED PHOTOCONDUCTOR ARRAYS

As is marked in Section 19.3.5, the quantum well IR photoconductors (QWIPs) is an alternative to HgCdTe, hybrid detector for the MW and LW spectral ranges. The advantages of QWIPs are linked to pixel performance uniformity and to the availability of large-size arrays; however, their drawbacks are the performance limitations for applications requiring short integration time and the requirement to operate at a lower temperature than HgCdTe of comparable wavelengths. The potential advantages of GaAs/AlGaAs quantum well devices include the use of standard manufacturing techniques based on mature GaAs growth and processing technologies, highly uniform (low frequency noise is generally negligible in QWIP arrays in contrast to HgCdTe arrays where low $1/f$ noise demands very good processing) and well-controlled MBE growth on greater than 6-inch GaAs wafers, high yield and thus low cost, more thermal stability, and extrinsic radiation hardness. Figure 26.47 shows evolution of the performance of VLWIR GaAs/AlGaAs QWIP [217]. As can be seen, rapid progress at the beginning of

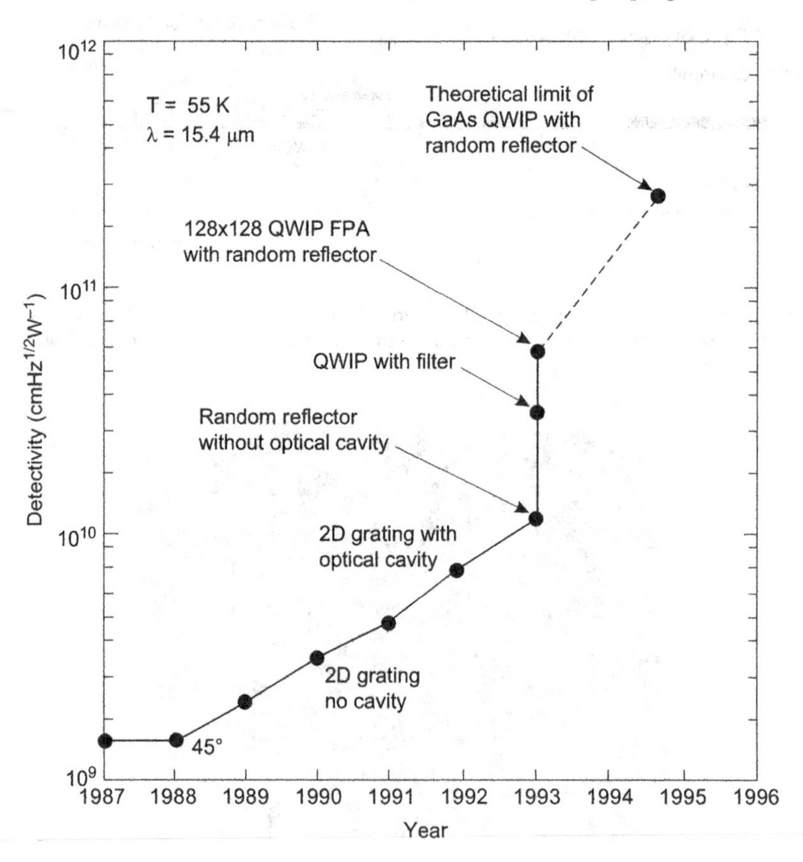

Figure 26.47 Evolution of the performance of very-LW GaAs/AlGaAs QWIP. All the data are normalized to wavelength λ = 15.4 μm at temperature T = 55 K. (From Ref. [217]. With permission.)

development has been made in detectivity, starting with bound-to-bound QWIPs, which had relatively poor sensitivity, and achieving considerable higher performance of bound-to-quasibound QWIPs with random reflectors.

QWIP detectors have relatively low quantum efficiencies, typically less than 10%. The spectral response band is also narrow for this detector, with a full-width, half-maximum of about 15%. All the QWIP data with cutoff wavelength about 9 μm is clustered between 10^{10} and 10^{11} $cmHz^{1/2}\,W^{-1}$ at about 77 K operating temperature. Investigations of the fundamental physical limitations of HgCdTe photodiodes indicate better performance with this type of detector in comparison with QWIPs operated in the range of 40–77 K (see Section 19.3.5). However, it has been shown that a low photoconductive gain actually increases the signal-to-noise ratio, and a QWIP FPA can have a better temperature resolution than the HgCdTe FPA with similar storage capacity (see Section 24.5.3.1.).

The first LWIR camera using AlGaAs/GaAs QWIPs was demonstrated by Bethea et al. in 1991 [218]. They used a commercial InSb scanning camera operating in the 3–5 μm spectral region and modified both the optics and electronics to allow operation at $\lambda = 10\,\mu m$. Also fabrication of GaAs superlattice was changed to obtain geometrical compatibility with the original 10-element linear InSb array. This resulted in a 10-pixel GaAs quantum well array consisting of 200 μm^2 pixels separated by 670 μm. The GaAs substrate was polished at 45° to allow good optical coupling to the quantum wells. Using the above-described nonoptimal IR imaging camera ($\lambda_c = 10.7\,\mu m$), NEDT < 0.1 K has been achieved. Further development of QWIP FPAs is described, for example, by Rogalski [144]. Hitherto, a variety of excellent FPAs have been developed by many groups. These include the first arrays realized at today's Lucent Technologies (Murray Hill, New Jersey) [219], and those of the Jet Propulsion Laboratory (JPL, Pasadena, California) [217,220,221], Thales Research and Technology (Palaiseau, France) [222–224], Fraunhofer IAF (Freiburg, Germany) [225,226], Acreo/IRNova (Kista, Sweden) [227,228], U.S. Army Research Laboratory (Adelphi, Maryland) [229–232], and BAE Systems (former Lockheed Martin) [233–235], among others. Also several university groups have demonstrated QWIP FPAs, between them, groups at Northwestern University (Evanston, Illinois) [236], Jerusalem College of Technology (Jerusalem, Israel), and Middle East Technical University (Ankara, Turkey) [237,238].

As is described in Section 19.4, the low noise level of PV QWIP FPAs enable longer integration time and improved thermal resolution of thermal imaging systems as compared to conventional photoconductive QWIPs. On the other side, photoconductive QWIPs are best situated if short integration times (5 ms and below) are required. In this case, thermal resolution is limited by the QE of the detector contrary to typical limitation by the storage capacity. Quantum wells of QWIPs with high QE are doped to higher electron concentrations (typically $4 \times 10^{11}\,cm^{-2}$, about four times higher than for standard photoconductive QWIPs) [226]. Photoconductive QWIPs with even higher carrier concentration ($2 \times 10^{12}\,cm^{-2}$) are exploited for MWIR with BLIP performance at about 90 K.

Figure 26.48 presents representative NEDT histograms of two types of FPAs with 640 × 512 pixels—low-noise LWIR and standard MWIR FPAs [239]. For the LW camera system with 24-μm pitch and integration time of 30 ms, a NEDT value as low as 9.6 mK has been observed, which is the best temperature resolutions ever obtained for thermal imagers operating in the 8–12 μm regime. In the case of a typical 640 × 512 MWIR QWIP FPA (Figure 26.48b), a NEDT value of 14.3 mK has been obtained at 88 K.

Properties of QWIP FPAs demonstrated by Fraunhofer IAF are summarized in Table 26.19 [226]. It is interesting to notice that a thermal resolution of 40 mK is possible for only 1.5-ms integration time, for arrays with higher doping ($4 \times 10^{11}\,cm^{-2}$ per QW) and an increased number of periods ($N = 35$).

Various types of MWIR and LWIR high-resolution hybrid QWIPs are offered for different applications (FLIR, IRST, reconnaissance, surveillance, airborne camera, etc.). The arrays can be assembled in various long-vacuum-life dewar and cooler configuration in order to meet the different mechanical and cooling needs of the systems. The examples are several hundreds of Catherine XP and Catherine MP cameras manufactured by the Thales Research Technology—the cameras contain Vega and Sirius FPAs hybridized and integrated in sensor chip assembly by Sofradir [240] (see Table 26.20 and Figure 26.49). Also a wide set of QWIP configurations is offered by Lockheed Martin Corporation (see Table 26.21).

Catherine MP's combination of functionality, long-range performance, and extended situational awareness makes it the current UK preferred in-service thermal imager with the British Army. With mexapixel

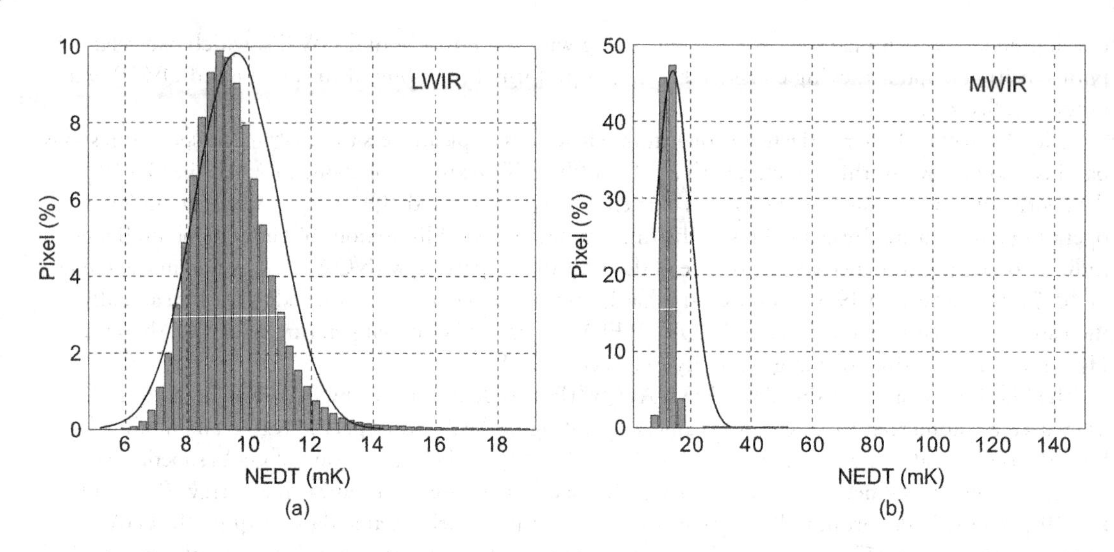

Figure 26.48 NEDT-histogram (a) of a 640 × 512 LWIR low-noise QWIP FPA for *f*/2 and 30 ms, and (b) of a 640 × 512 MWIR QWIP FPA for *f*/1.5 and 20 ms. (From Ref. [239]. With permission.)

Table 26.19 Properties of QWIP FPAs demonstrated by Fraunhofer IAF

FPA TYPE	ARRAY SIZE	PITCH (μm)	λ (μm)	f/#	τ_{int} (ms)	NEDT (mK)
256 × 256 PC	256 × 256	40	8–9.5	*f*/2	16	10
640 × 512 PC	640 × 486	24	8–9.5	*f*/2	16	20
	512 × 512					
256 × 256 LN	256 × 256	40	8–9.5	*f*/2	20	7
					40	5
384 × 288 LN	384 × 288	24	8–9.5	*f*/2	20	10
640 × 512 LN	640 × 486	24	8–9.5	*f*/2	20	10
	512 × 512					
384 × 288 PC-HQE	384 × 288	24	8–9.5	*f*/2	1.5	40
640 × 512 PC-HQE	640 × 486	24	8–9.5	*f*/2	1.5	40
	512 × 512					
640 × 512 PC-MWIR	640 × 486	24	4.3–5	*f*/1.5	20	14
	512 × 512					

Source: Adapted after Ref. [226].
PC, photoconductive; LN, low-noise; HQE, high QE; τ_{int}, integration time.

resolution (1280 × 1020 pixel) and low NEDT (below 25 mK) offers flexibility and reliability where extreme performance is demanded on land, sea, and air platforms.

One magapixel hybrid MWIR and LWIR QWIP with 18 μm pixel size has been demonstrated (see Figure 26.50) with excellent imaging performance using transitions from bound to extended states and from bound to miniband states [241–244]. Gunapala et al. [242] have demonstrated the MWIR detector arrays with a NEDT of 17 mK at 95 K operating temperature, *f*/2.5 optics, and a 300 K background and the LWIR detector array with a NEDT of 13 mK at 70 K operating temperature and the same optical and background conditions as the MWIR detector array. This technology can be readily extended to a 2K × 2K array. Figure 26.51 shows frames of video images taken with both 5.1- and 9-μm cutoff 1024 × 1024 pixel

Table 26.20 QWIP Sofradir's FPAs

PARAMETER	LW (SIRIUS)	LW (VEGA)
Array size	640 × 512	384 × 288
Pixel pitch (μm^2)	20 × 20	25 × 25
Spectral response	$\lambda_p = 8.5 \pm 0.1\ \mu m$, $\Delta\lambda = 1\ \mu m$ @ 50%	$\lambda_p = 8.5\ \mu m$, $\Delta\lambda = 1\ \mu m$ @ 50%
Operating temperature (K)	70–73	73
Integration type	Snapshot	Snapshot
Max charge capacity (e^-)	1.04×10^7	1.85×10^7
Readout noise	110 μV for gain 1	950 e^- for gain 1
Signal outputs	1, 2 or 4	1, 2 or 4
Pixel output rate (MHz)	up to 10 per output	up to 10 per output
Frame rate (Hz)	up to 120 full frame rate	up to 200 full frame rate
NEDT (mK)	31 (300 K, *f*/2, 7 ms integration time)	<35 (300 K, *f*/2, 7 ms integration time)
Operability (%)	>99.9	>99.95
Nonuniformity (%)	<5	<5

Figure 26.49 Catherine XP (left) and Catherine MP (right) LWIR QWIP cameras.

cameras. In addition to the excellent thermal resolution and contrast, both thermal images show high detail that indicates a small optical crosstalk between adjacent pixels and a good modulation transfer function.

Figure 26.52 shows the estimated NEDT as a function temperature for bias voltage of –2 V for MWIR and LWIR 1024 × 1024 QWIP FPAs. The background temperature is 300 K, and the area of the pixel is 17.5 × 17.5 μm^2. The *f*/# of the optical system is 2.5, and the frame rates are 10 and 30 Hz for MW and LW arrays, respectively.

26.8 BARRIER DETECTOR AND TYPE-II SUPERLATTICE FOCAL PLANE ARRAYS

MWIR nBn sensor arrays are manufactured by several companies. As is mentioned in Section 12.9, the nBn sensor design is self-passivating, decreasing leakage current and associated noise, while improving reliability and manufacturability. Because of its simple design (see Figure 12.41), the array technology is a major advance in the state of the art for large IR FPAs.

Table 26.21 QWIP FPAs assembled in the ImagIR camera fabricated by Lockheed Martin Corporation

PARAMETER	
Spectral range (μm)	8.5–9.1
Resolution/pixel pitch (μm)	1024 × 1024/19.5 640 × 512/24 320 × 256/30
Integration type	Snapshot
Integration time	<5 μs to full frame time
Dynamic range (bits)	14
Data rate (Mpixels/s)	32
Frame rate (Hz)	1024 × 1024 – 114 640 × 512 – 94 320 × 256 – 366
Well capacity (Me⁻)	1024 × 1024 – 8.1 640 × 512 – 8.4 320 × 256 – 20
NEDT (mK)	<35
Operability	>99.95 (typical)
Fixed focal plane (mm)	*f*/2.3 (13, 25, 50, 100)

Figure 26.50 Picture a 1024 × 1024 pixel QWIP FPA mounted on a 84-pin leadless chip carrier. (From Ref. [242]. With permission.)

The depletion dark current is limited in nBn structure and its temperature of operation is increased. For example, Figure 22.5 shows temperature dependence of NEDT for *f*/3.2 optics of the Kinglet digital detector. This sensor based on SCD's Pelican-D ROIC contains nBn $InAs_{0.91}Sb_{0.09}$/B-AlAsSb 640 × 512 pixel architecture with a 15-μm pitch. The NEDT is 20 mK at 10-ms integration time, and the operability of non-defective pixels was greater than 99.5% after a standard two-point nonuniformity correction. The NEDT and operability begin to change above 170 K, which is consistent with the estimated BLIP temperature of 175 K.

The first commercially developed nBn InAsSb arrays sensor by Lockheed Martin Santa Barbara Focalplane operates at 145–175 K temperatures. In IRCameras' implementation (see Table 26.22) of Santa Barbara Focalplane's MWIR nBn sensor, a 1280 × 1024 format, 12-μm pixel pitch detector is packaged in a 1.4-inch-diameter dewar with an overall dewar housing length of about 3.8 in., including the cooler. With a long life, 25,000-hours cryocooler consuming about 2.5 W and electronics adding another 2.5 W,

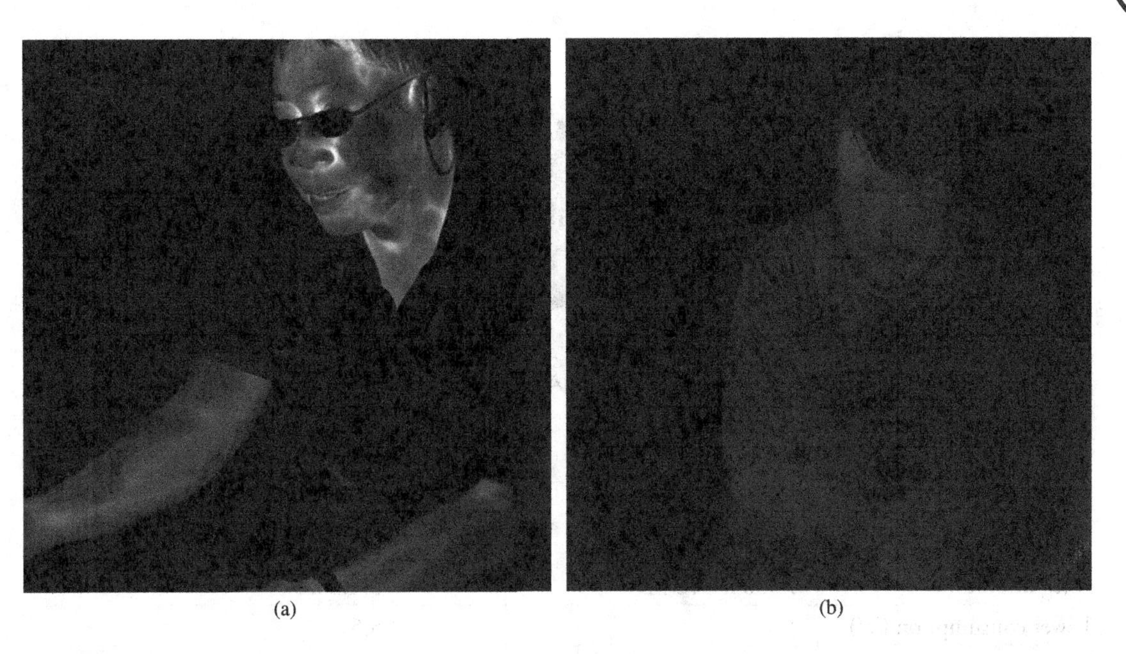

Figure 26.51 One frame of video taken with (a) the 5.1 μm, and (b) 9-μm cutoff 1024 × 1024 pixel QWIP cameras. The MW (LW) video images were taken at a frame rate of 10 Hz (30 Hz) at temperatures 90 K (72 K), using a ROIC capacitor having a charge capacity of 8×10^6 electrons. (From Ref. [242]. With permission.)

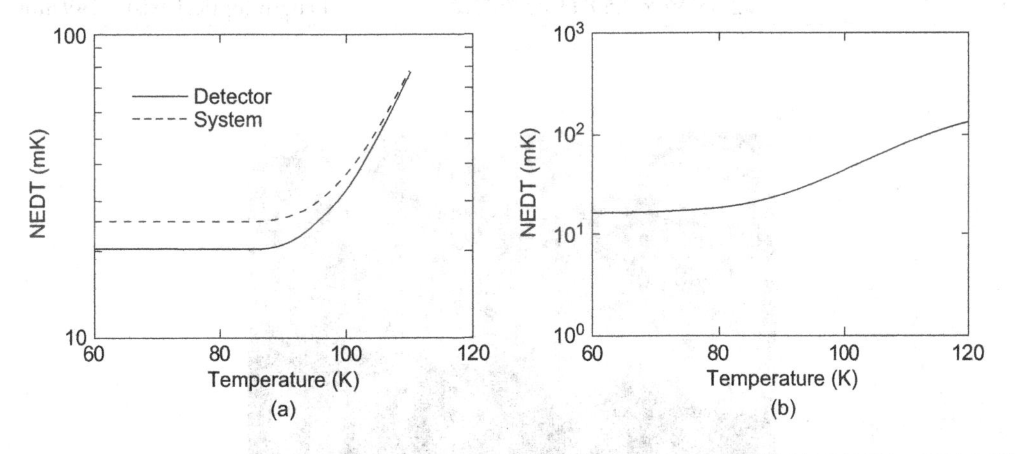

Figure 26.52 NEDT as a function temperature for bias voltage (a) of –2 V for MW, and (b) LW 1024 × 1024 QWIP FPAs. The background temperature is 300 K and the area of the pixel is 17.5 × 17.5 μm². (From Ref. [242]. With permission.)

the total camera core power draw is only about 5 W total. High spatial resolution image acquired with nBn sensor running at 160 K is shown in Figure 26.53.

At present, the VISTA U.S. government program is working on innovative approaches for IR FPA technology to enhance IR sensor capabilities. First announcement about high-definition FPA containing 5-μm pixels in a 2040 × 1156 array format has been given in Ref. [246]. Figure 26.54 shows an outdoor image from a laboratory camera with this array fabricated by Cyan Systems.

Under the VISTA program, HRL has advanced the growth and fabrication of III-V bulk MWIR detector technology on GaAs substrates for HOT applications. It has been shown that feasibility of small-pixel (5- to 10-μm pitch) technology is an attractive alternative to HgCdTe technology, mainly due to lower cost, ease of scalability to larger formats (e.g. 8k × 8k/10 μm), and better uniformity. IR FPAs with 2k × 2k/10 μm and 2k × 1k/5 μm formats have been demonstrated by developing high aspect ratio dry etching

Table 26.22 nBn InAsSb FPA characteristics

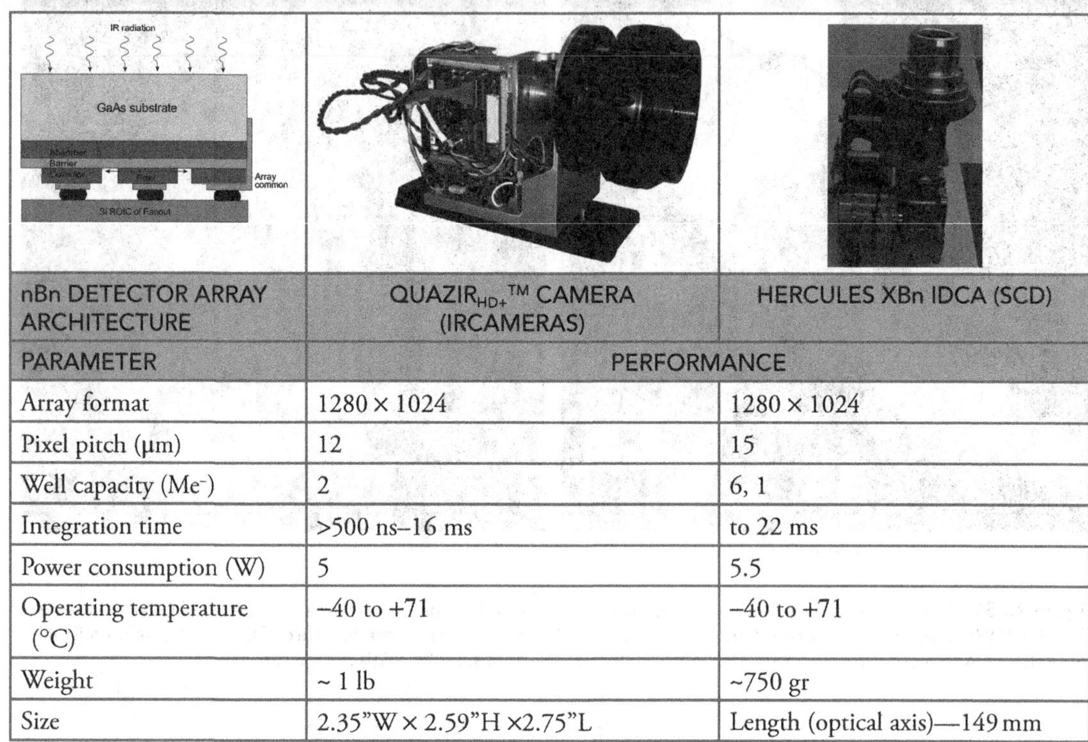

nBn DETECTOR ARRAY ARCHITECTURE	QUAZIR$_{HD+}$™ CAMERA (IRCAMERAS)	HERCULES XBn IDCA (SCD)
PARAMETER	PERFORMANCE	
Array format	1280 × 1024	1280 × 1024
Pixel pitch (μm)	12	15
Well capacity (Me$^-$)	2	6, 1
Integration time	>500 ns–16 ms	to 22 ms
Power consumption (W)	5	5.5
Operating temperature (°C)	–40 to +71	–40 to +71
Weight	~ 1 lb	~750 gr
Size	2.35”W × 2.59”H ×2.75”L	Length (optical axis)—149 mm

Figure 26.53 A MWIR nBn InAsSb FPA with 1280 × 1024 pixels images a scene from a baseball game, with a player attempting to steal second base. (Adapted after Ref. [245])

for mesa delineation (fill factor >80%) (see Figure 26.55), proper device passivation by dielectric layer and high aspect ratio indium bump schemes (operability >99.9%). Resulting hybrids operating at 150 K show low-dark current with low turn-on bias, low NEDT (<20 mK at 150 K using *f*/2.3 optics), and high operability for both 5- and 10-μm pixels.

As is marked in Section 20.3, InAs/GaSb type-II strained-layer superlattice (T2SL) can be considered as an alternative to HgCdTe and GaAs/AlGaAs IR material systems. The strained-layer superlattice structures provide high responsivity, as already reached with HgCdTe, without any need for gratings

Figure 26.54 Image from a HOT MWIR nBn InAsSb array with 5-μm pixels in a 2040 × 1156 format. (Adapted after Ref. [246])

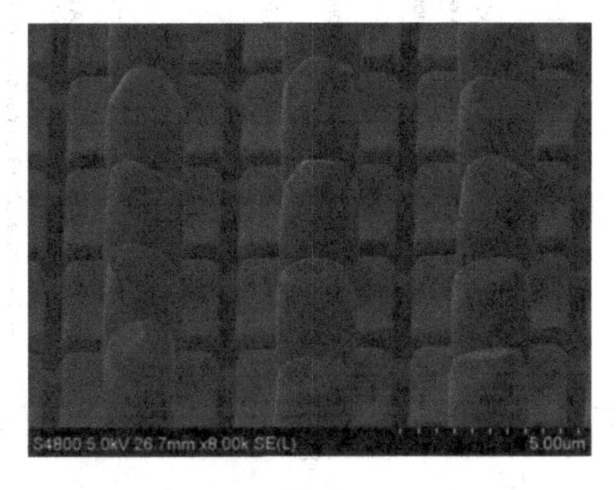

Figure 26.55 A scanning electron microscopy photo of fabricated 5-μm pixels using a high-aspect ratio dry etch process resulting >80% physical fill factor. (Adapted after Ref. [247])

necessary in QWIPs. Further advantages are a PV operation mode, operation at elevated temperatures, and well-established III-V process technology.

Type-II InAs/GaInSb-based detectors have made rapid progress over the past few years [247, 248]. The roadmap of the T2SL photodetectors is shown in Figure 26.56.

In 1987, Smith and Mailhiot proposed InAs/GaSb T2SL for IR detection applications [249]. Despite these positive theoretical predictions, high-quality growth of InAs/GaSb T2SL material was not demonstrated for almost the next two decades. Advances in the MBE superlattice material technology and device processing techniques have been gradually improved, making the fabrication of high-quality single element and FPA more routine. In the last decade, the first megapixel MWIR and LWIR type-II SL FPAs have been demonstrated with excellent imaging [92,250–259]. At about 78 K, an NEDT value of below 20 mK has been presented for MWIR arrays and just above 20 mK for LWIR arrays. Figure 26.57 shows images taken with MWIR 640 × 512 nBn array and two (MWIR and LWIR) megapixel PV arrays.

The values of NEDT for MWIR type-II SL FPAs are shown in Figure 26.58. Excellent NEDT value of approximately 10 mK measured with *f*/2 optics and integration time $\tau_{int} = 5$ ms has been presented for 256 × 256 MWIR detector of a cutoff wavelength of 5.3 μm (see Figure 26.58(a)) [250]. Tests with reduced time down to 1 ms show that the NEDT scales are inversely proportional to the square root of the integration time. It means that even for short integration time, the detectors are background limited. Similar results have been demonstrated by Nortwestern University's group [253]. The minimum NEDT stays almost constant at 11 mK with integration time of 10.02 ms up to 120 K, what suggests that the FPA is

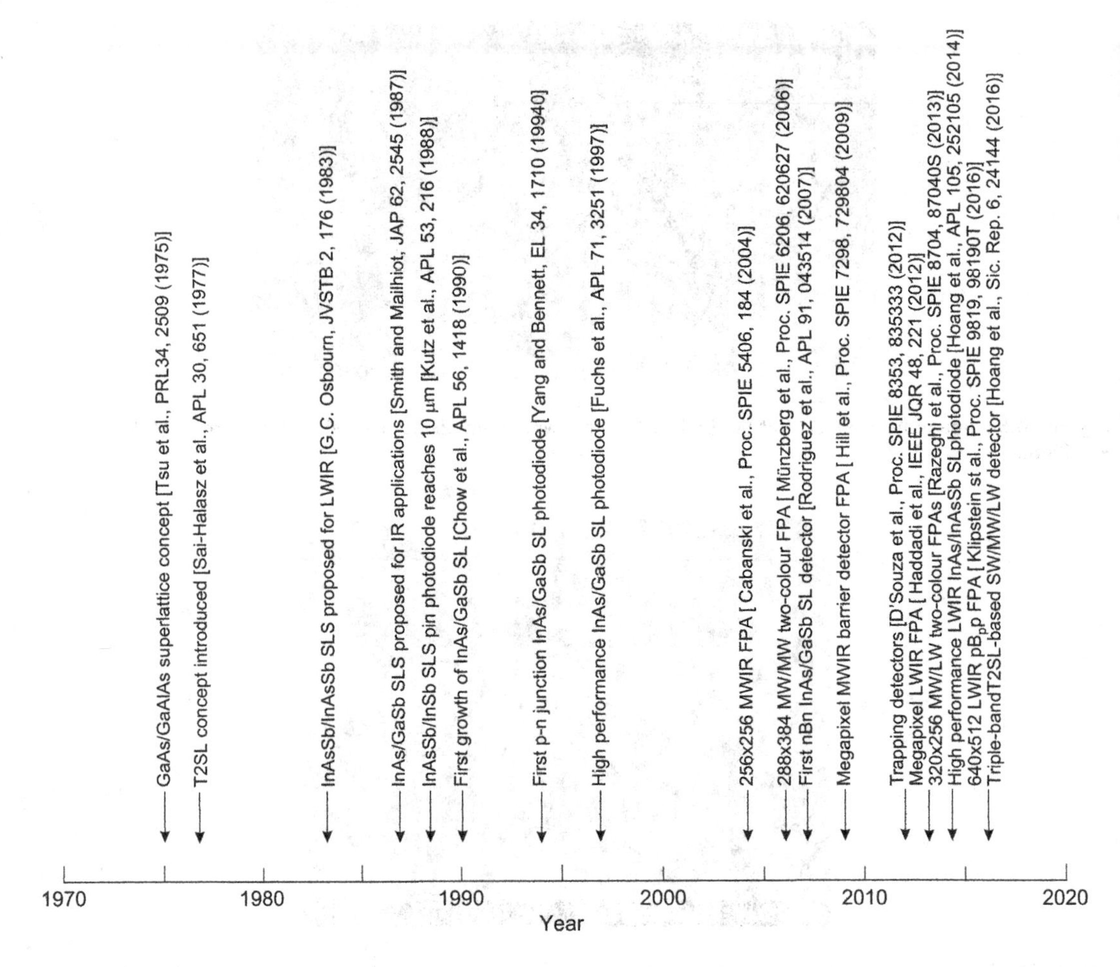

Figure 26.56 Roadmap development of type-II superlattice IR photodetectors.

mainly dominated by temperature insensitive noise such as system noise or photon noise since dark current noise should increase exponentially with temperature. From 130 K to 150 K, the measurements are performed with an integration time of 4.02 ms to avoid the saturation of the readout capacitors due to higher dark current levels. The increase in NEDT in this region might be related to the increase in the dark current. A very important feature of InAs/GaSb FPAs is their high uniformity. The responsivity spread shows a standard deviation of approximately 3%. It is estimated that the pixel outages are in the order of 1%–2%, and the pixels are statistically distributed as single pixels without large clusters [250].

Northwestern University's group has demonstrated high-quality 1024 × 1024 LWIR FPA using a p-π-M-n pixel structure shown in Figure 26.59a. This device design combines both high optical and electrical performance. The M-structure and the double heterostructure design technique help to reduce both the bulk dark current and surface leakage current. A high QE (>50%) was obtained thanks to a thick absorption region (6.5 μm). The device exhibited a dark current level below 5×10^{-5} A cm^{-2} at—50 mV applied bias voltage at 77 K. Figure 26.59b shows NEDT histogram after two-point uniformity correction measured at 81 K with an integration time of 0.13 ms and *f*/2 optics. At a reverse bias of 20 mV, the median NEDT value was 27 mK.

Recently, SCD has developed advanced pB$_p$p T2SL barrier detectors to enable diffusion-limited dark currents comparable with HgCdTe "Rule 07" and high QE, above 50% (see Section 22.3) [257]. Special attention was paid to eliminate influence of surface leakage current. A robust passivation process has been developed that allows glue underfill and substrate thinning, after bonding the sensor array with indium bumps to the custom-designed silicon readout. Polishing of the GaSb substrate down to a final thickness of

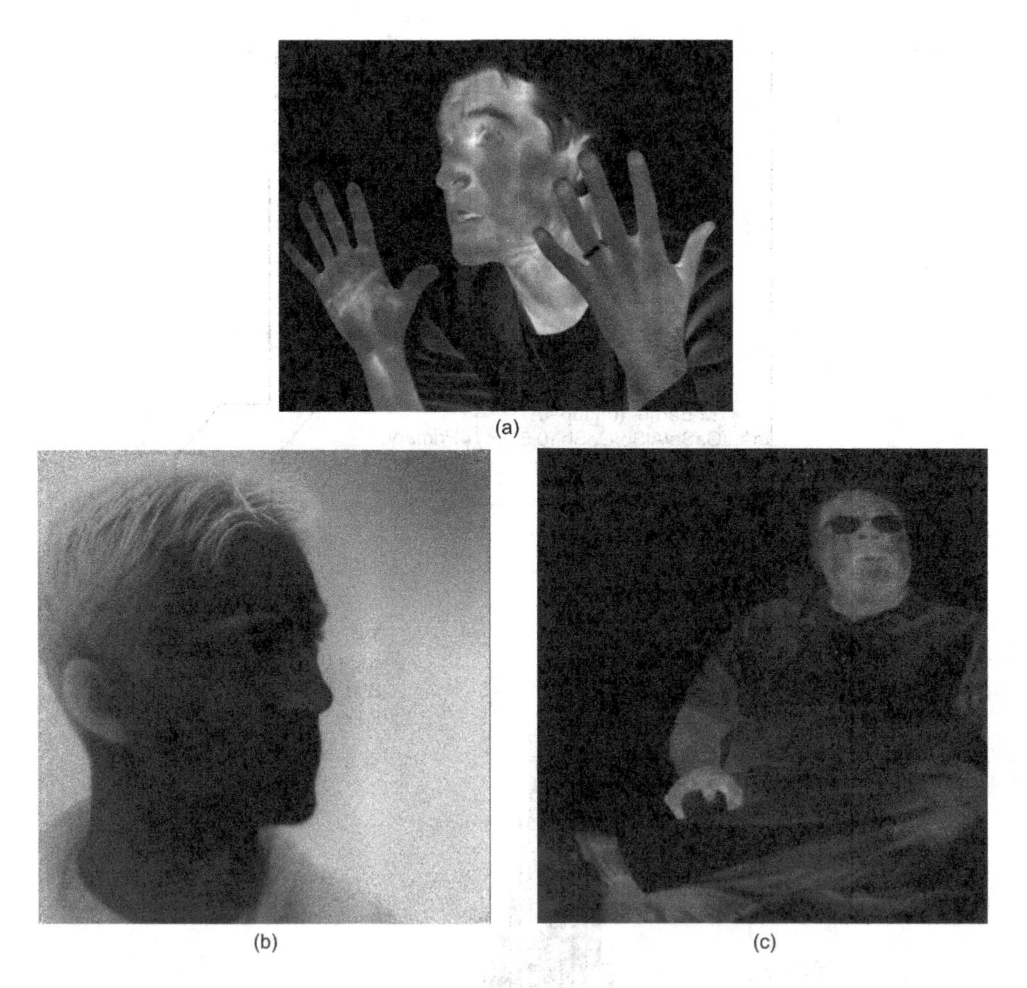

Figure 26.57 Images taken with (a) an nBn 640 × 512 MWIR FPA, as well as with megapixel (b) MWIR (p-i-n pixels) and (c) LWIR (complementary barrier infrared (CBIRD) detector—see Table 22.1) Sb-based PV FPAs. (Adapted after Refs. [251,252])

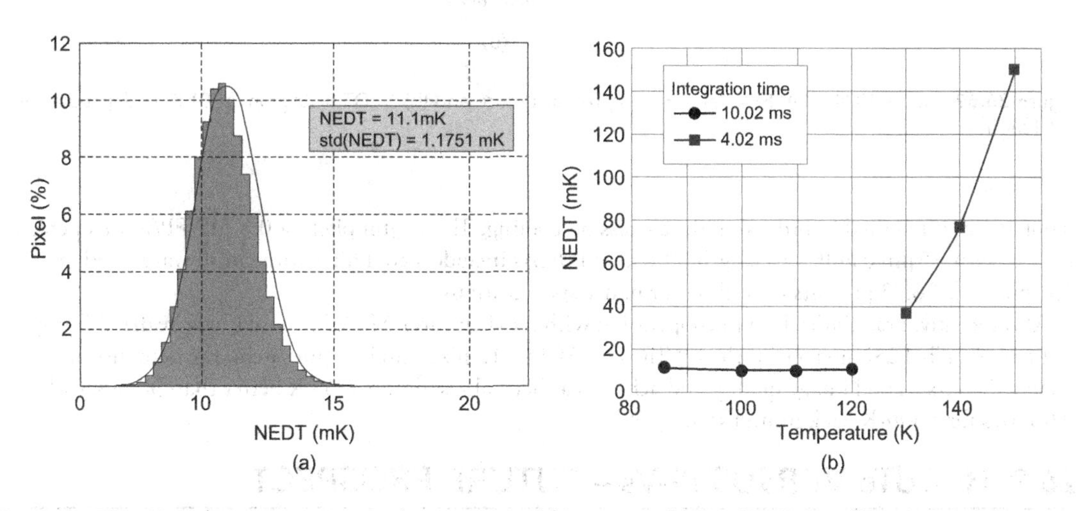

Figure 26.58 NEDT of MWIR type-II superlattice FPAs: (a) histogram of 256 × 256 FPA at *f*/2 optics, τ_{int} = 5 ms, and 77 K (Adapted after Ref. [250]) and (b) temperature dependence for 320 × 256 FPA at *f*/2.3 optics; the integration time is reduced above 120 K to avoid the saturation of the readout capacitor due to higher dark current levels. (Adapted after Ref. [253])

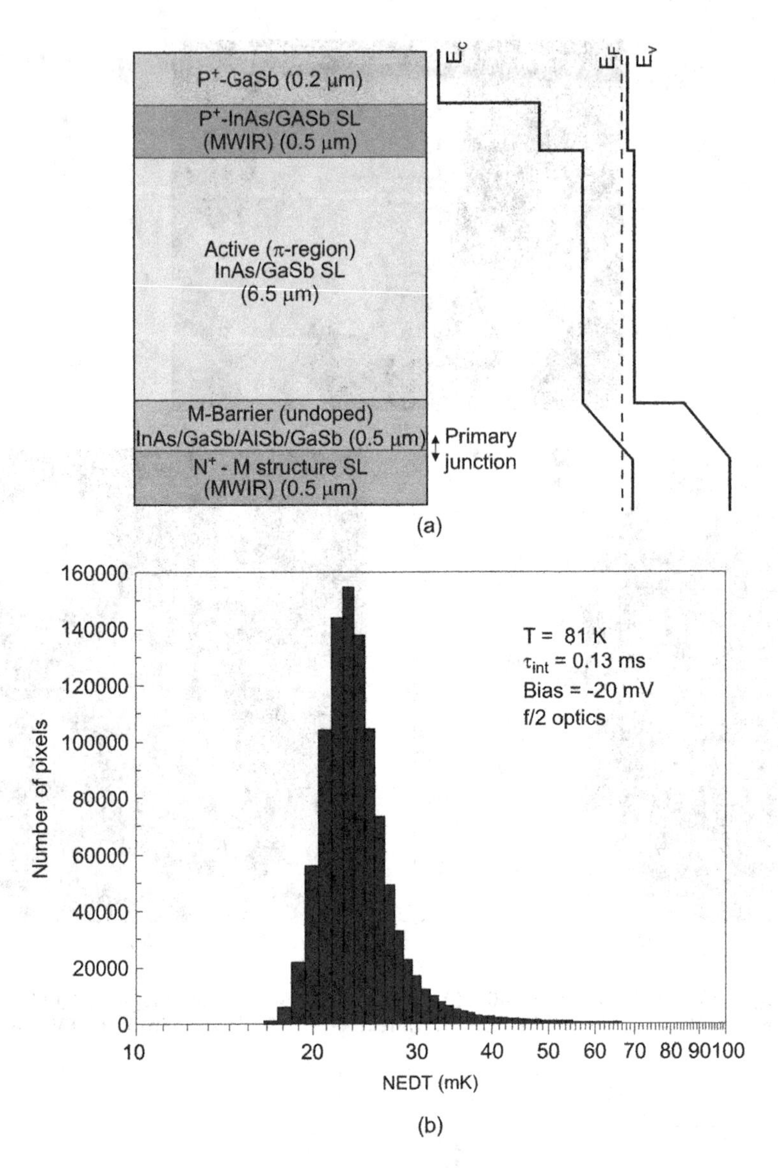

Figure 26.59 1024 × 1024 LWIR FPA: (a) p-π-M-n pixel structure and (b) NEDT histogram at 81 K. (Adapted after Ref. [254])

about 10 μm is required in order to relieve stress on cooling. The 15-μm pitch 640 × 512 FPAs are operated at 77 K with 9.5 μm cutoff wavelength. The final IDCAs include a cold filter with cutoff wavelength of 9.3 μm. Table 26.23 presents typical performance specifications.

Also recently, Fraunhofer IAF in cooperation with AIM Infrarot-Module GmbH have realized Europe's first InAs/GaSb T2SL imager for the LWIR with 640 × 512 pixels and 15-μm pitch. The demonstrator camera delivers a good image quality and achieves a thermal resolution at 55 K better than 30 mK with *f*/2-optics for a 300-K background scene [258].

26.9 HgCdTe VERSUS III-Vs—FUTURE PROSPECT

At present, the III-V antimonide-based detector technology is under strong development as a possible alternative to HgCdTe detector material. The ability to tune the positions of the conduction and valence band

Table 26.23 Specification of LW pB_pp T2SL array performance at 77 K

PARAMETER	VALUE	PELICAN-D LW IDCA
Format	640 × 512	
Pitch (μm)	15	
Cutoff wavelength (μm)	9.3 (filter)	
QE (%)	>50	
Operability (%)	>99	
RNU (%)	<0.04 STD/DR @ 10–90 well fill capacity	
NEDT	15 mK @ 65% well fill capacity, 30 Hz (by averaging 8 frames)	
Response uniformity (%)	<2.5 (STD/DR)	
Cooler	Ricor K548	
Weight (g)	750	
Environment condition (°C)	–40 to +71	
Total power at 23°C (W)	16	
Cool down time (min)	8	
MTTF (depends on mission profile) (h)	15,000	

Source: Adapted after Ref. [257].

edges independently in a broken-gap T2SL is especially helpful in the design of unipolar barriers. Unipolar barriers are used to implement the barrier detector architecture for increasing the collection efficiency of photogenerated carriers and reducing dark current originating within the depletion region without inhibiting photocurrent flow. During the last decade, antimonide-based FPA technology has achieved a level close to HgCdTe. The apparent rapid success of the T2SL depends not only on previous five decades of III-V materials, but mainly on novel ideas coming recently in design of IR photodetectors. The advent of bandgap engineering has given III-Vs a new lease on life.

For HOT-MWIR operation, T2SL materials have demonstrated higher operating temperatures (up to 150 K) compared to InSb (~ 80 K). In addition, T2SL offers both performance and manufacturability especially for large-format FPA applications. GaSb substrates with diameters up to 6 inches are commercially available.

Despite numerous advantages of III-V semiconductors (T2SLs and barrier detectors) over present-day detection technologies, including reduced tunneling and surface leakage currents, normal-incidence absorption, and suppressed Auger recombination, the promise of superior performance of these detectors has not been yet realized. The dark current density is higher than that of bulk HgCdTe photodiodes, especially in the MWIR range.

To attain their full potential, the following essential technological limitations such as short carrier lifetime, passivation, and heterostructure engineering need to be overcome. Much of the improvements can be attributed to the identification and minimization of Shockley-Read-Hall (SRH) traps. The T2SL material has the potential to outperform HgCdTe if it can overcome current SRH defect limits. Introduction of barrier designs can considerably impede the flow of dark current without photocurrent impeding when a bias voltage is applied. It should be expected that future advances in III-V barrier detector technology will push the dark current down to "Rule 07" in a wider IR spectral range.

From a performance perspective, III-V diffusion current-limited FPAs can indeed operate at levels that approach that of HgCdTe, but always with a required lower operating temperature [260,261].

Recently published Kinch's monograph [261] is largely devoted to the intense competition between HgCdTe and T2SLs for the future of IR detector technology. It is clearly shown that *the ultimate cost reduction for an IR system will only be achieved by the room temperature operation of depletion-current-limited arrays with pixel densities that are fully consistent with background- and diffraction-limited performance due to the system optics. This mandates the use of IR materials with a long S-R lifetime. Currently, the only material that meets this requirement is HgCdTe.* Kinch predicted that large-area ultra-small-pixel diffraction-limited and background-limited photon-detecting MW and LW HgCdTe FPAs operating at room temperature will be available within the next 10 years [262].

It will be rather difficult to improve SRH lifetime to overcome the disadvantage of large InAs/GaSb T2SL's depletion dark currents. InSb SRH lifetime issue since its inspection in 1950s is well known. Better situation is observed in Ga-free InAs/InAsSb T2SLs due to large values of carrier lifetimes, including SRH lifetime.

26.9.1 P-I-N HgCdTe PHOTODIODES

The p-i-n photodiodes is a popular alternative to the simple p-n photodiodes, especially for ultrafast photodetection in optical communication, measurement, and sampling systems. In p-i-n photodiode, an undoped i-region (π or ν, depending on the method of junction formation) is sandwiched between the p and n regions. Very often the absorber layer is surrounded by a wider bandgap contact layers to suppress dark current generation from these regions and to suppress tunneling current under reverse bias. Thus, the P-i-N device is essentially p-i-n diode. Figure 26.60a shows schematic representation of a p-i-n diode with an energy band diagram under reverse-bias conditions. Because of the very low density of free carriers in the i-region and its high resistivity, any applied bias drops entirely across the i-region, which is fully depleted at zero bias or very low value of reverse bias.

The P-i-N photodiode has a "controlled" depletion layer width, which can be tailored to meet the requirements of photoresponse and bandwidth. A trade-off is necessary between response speed and QE. For high response speed, the depletion layer width should be small, but for high QE (or responsivity), the width should be large. An external resonant microcavity approach has been proposed to enhance QE in such situations. In this approach, the absorption region is placed inside a cavity so that a large portion of the photons can be absorbed even with a small detection volume.

At the present stage of technology, the above requirements are fulfilled in P-on-n HgCdTe double-layer photodiode—see Figure 26.60b [263]. The absorber layer is surrounded by a wider bandgap cap and buffer region in order to suppress dark current generation from these regions. The n-doping of the absorber is sufficiently low to allow full depletion at moderate bias. To suppress tunneling current under reverse bias, the wide bandgap cap is used. The planar nature of the structure is potentially self-passivating and is analogous to the pBn geometry of III-V barrier detectors. Moreover, as is discussed in Refs. [168,261], the fully depleted structure is compatible with the small pixel pitch, fulfilling low crosstalk due to the built-in vertical electric field generated under detector reverse bias. Both fully depleted absorber and wide bandgap cap potentially reduce $1/f$ and random telegraph noise.

The general relativity (GR) current density of a fully depleted detector can be estimated by the following expression: $J_{GR} = qn_i w/\tau_{SRH}$ (see Equation 12.113), where w is the width of depletion region and τ_{SRH} is the SRH lifetime.

Experimental data at 30 K indicate a very encouraging SRH lifetime for a 10.7-μm cutoff array pixel having 18-μm pixel pitch. A lower limit to the SRH lifetime is estimated to be 100 ms [264]. Assuming this lifetime, the current density for three different absorbing doping values is shown in Figure 26.61, together with the current density from background blackbody radiation integrated over π steradians. As is shown, radiation current is observed to dominate for doping below approximately 10^{13} cm^{-3}.

An additional insight into the performance of LWIR HgCdTe photodiodes has been recently given by Rogalski *et al.* [265]. They used an enhanced computer program to estimate the performance of p-i-n HgCdTe LWIR HOT photodiodes operated at 230 K; dark current density and spectral current responsivity. It has been shown that a 5-μm-thick absorber with a low donor concentration of 5×10^{13} cm^{-3} can be fully depleted at 0.4 V reverse bias. Taking into account the realistic position of trap levels in the depletion

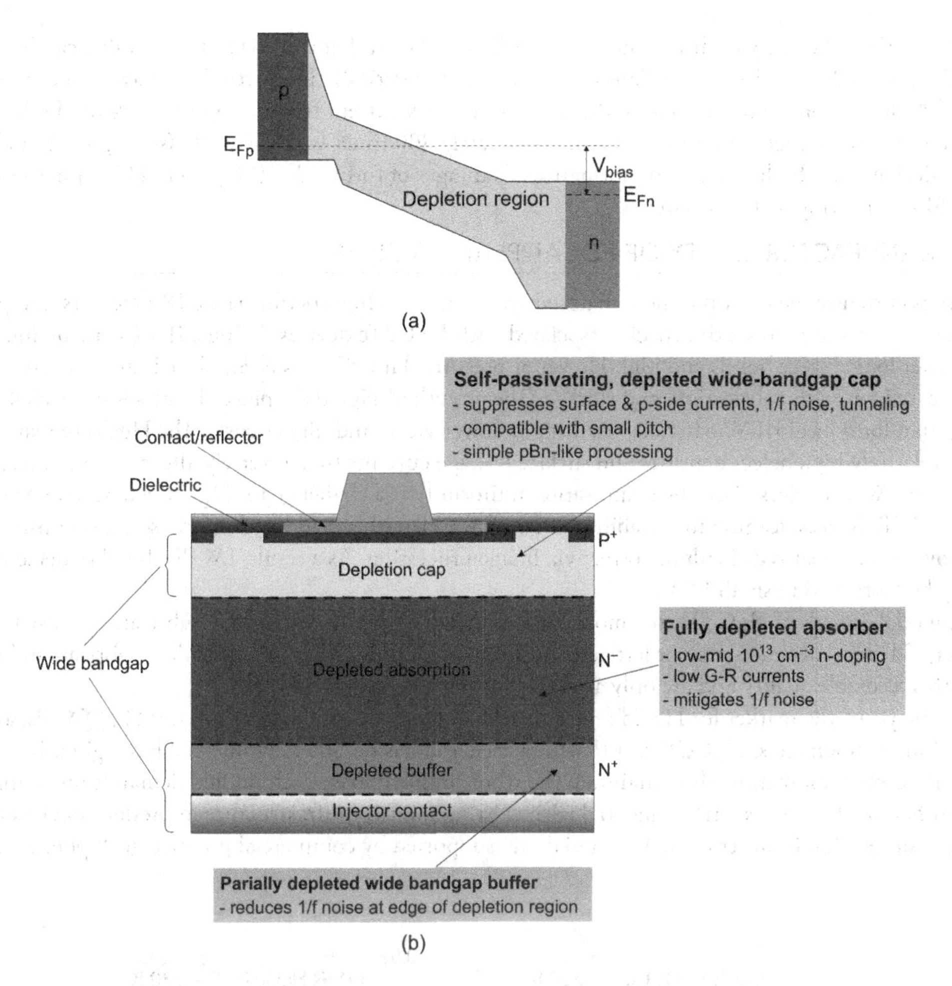

Figure 26.60 p-i-n photodiode: (a) energy band diagram under reverse bias, (b) heterojunction P-i-N photodiode architecture. (Adapted after Ref. [263])

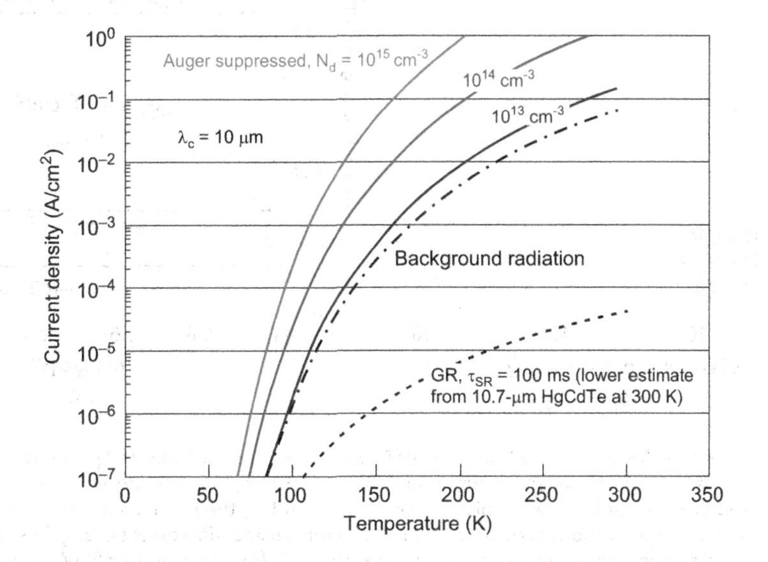

Figure 26.61 Comparison of GR current density extracted from 30-K lifetime measurements for 10-μm cutoff HgCdTe photodiodes with Auger-suppressed and background radiation current densities. (Adapted after Ref. [263])

region (0.75 E_g) and density of dislocations (10^4cm^{-2}), it is shown that the SRH lifetime value can be changed between 0.1 and 10 ms (see Figure 26.62a). The theoretically predicted dark current density of mid-10^{-3} A cm^{-2} is comparable to that of the background flux current from *f*/3 optics for photodiodes with a 9.5-μm cutoff wavelength at 230 K and is approximately 800 times below Rule 07 (see Figure 26.62b). This prediction offers both cost and performance advantages of future LWIR HgCdTe HOT photodiodes over T2SLs competing material system.

26.9.2 MANUFACTURABILITY OF FOCAL PLANE ARRAYS

HgCdTe is currently the most prevalent material system used in high-performance IR detectors. Despite this position, there are several drawbacks associated with HgCdTe devices. Being a II-VI semiconductor with weaker ionic Hg-Te bonds and high Hg vapor pressure, HgCdTe is soft and brittle and requires extreme care in growth, fabrication, and storage. The growth of HgCdTe epitaxial layers is more challenging than for typical III-V materials, resulting in lower yields and higher costs. The HgCdTe material exhibits relatively high defect densities and surface leakage currents that adversely affect performance, particularly for LWIR devices. Also the composition uniformity is a challenge for HgCdTe devices, particularly for LWIR devices, leading to variability in cutoff wavelength. In addition, 1/*f* noise causes uniformity to vary over time, what is difficult to correct via image processing. As a result, LWIR HgCdTe detectors can only be fabricated in small FPAs.

In epitaxial growth of HgCdTe, the most commonly lattice-matched CdZnTe substrates are used. However, CdZnTe substrates are not lattice-matched to Si ROICs—they are difficult to fabricate in large sizes with acceptable quality and are only available from limited sources.

Over the years, the market for HgCdTe has shrunk due to success of the InGaAs, InSb, QWIPs, and uncooled microbolometers. HgCdTe is a II-VI material and has no other commercial leveraging. In consequence, it becomes more difficult to maintain the whole industry base with limited demands on quantity. On the other hand, T2SL is a III-V material with existing industry infrastructure to produce devices at a low cost. The existing facilities of III-V materials are supported by commercial products (cell phone chips,

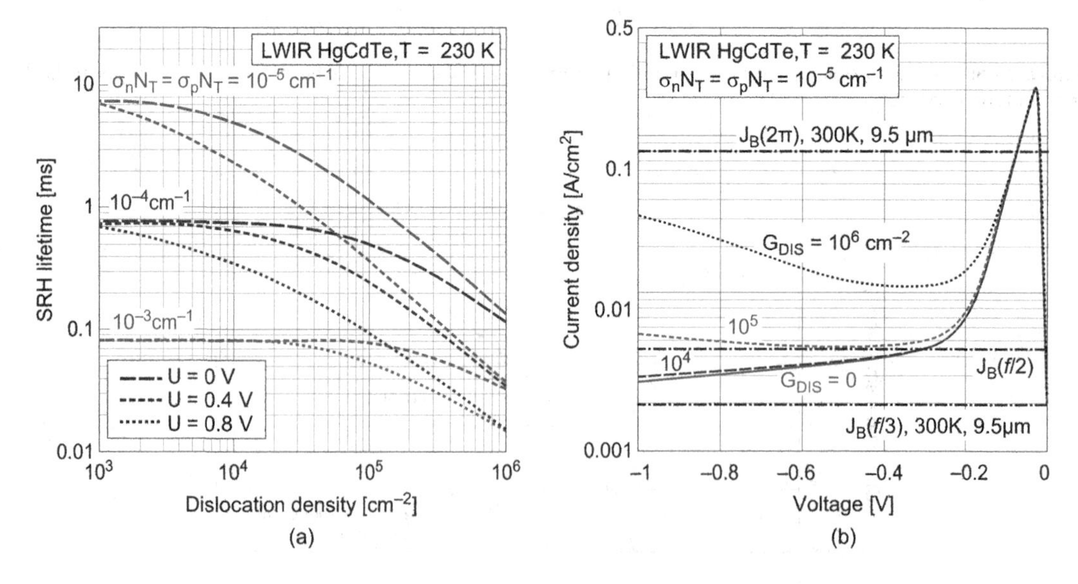

Figure 26.62 Calculated performance of 9.5-μm cut off P⁺-ν-N⁺ HgCdTe photodiode operated at 230 K with 5-μm-thick absorption layer, donor concentration of 5 × 10^{13}cm^{-3} and trap levels in depletion region of 0.75 E_g: (a) dependence of the SRH carrier lifetime on dislocation density and (b) the reverse biased current-voltage characteristics for $\sigma_n N_T$ and $\sigma_p N_T$ products of 10^5cm^{-1} and different dislocation density, G_{DIS}. The dark current densities are compared to background flux generation rates through *f*/3 optics and 2π FOV. N_T denotes the mercury vacancy concentration, σ_n and σ_p are the electron and hole captures cross-sections in SRH recombination processes.

millimeter wave integrated circuits), which present a lesser problem for the government to maintain the IR industry base.

The important advantage of T2SLs is the high quality, high uniformity, and stable nature of the material. In general, III-V semiconductors are more robust than their II-VI counterparts due to stronger, less ionic chemical bonding. As a result, III-V-based FPAs excel in operability, spatial uniformity, temporal stability, scalability, producibility, and affordability—the so-called "ibility" advantages [266]. The energy gap and electronic properties of T2SLs are determined by the layer thicknesses rather than the molar fraction as is the case for HgCdTe. The growth of T2SLs can be carried out with better control over the structure and with higher reproducibility. The spatial uniformity is also improved, since the effects of compositional change due to flux and temperature nonuniformity are not as important as they are in ternary/quaternary bulk materials.

The VISTA program implemented a horizontally integrated model that significantly departed from the traditional vertical integration used by the HgCdTe industry (see Figure 26.63). For example, HRL Laboratories served as an FPA foundry within the VISTA program. Using wafers grown by IQE and Intelligent Epitaxy Technology based on designs from JPL, the dual-band FPAs are fabricated and next hybridized to a ROIC provided by RVS [267].

Currently, very high-performance and very-large-format LWIR FPAs do not exist with the state-of-the-art technology; there is an urgency to demonstrate and produce these arrays for systems where very-large-format and high-performance LWIR FPAs are needed. VISTA's focus is in III-V superlattice epitaxial materials research for advanced IR FPAs with large formats and reduced pixel size that are capable of MWIR/LWIR detection and dual-band-sensing applications.

26.9.3 CONCLUSIONS

Summarizing above discussions:

- III-V materials have inherently short SRH lifetimes below 1 μs, and require an nBn architecture to operate at reasonable temperatures, and as such are diffusion current limited. This applies both to the simple alloy and the T2SL versions,
- HgCdTe alloys have long SRH lifetimes >200 μs–50 ms depending on the cutoff wavelength [261]. They can thus operate with either architecture and may be diffusion or depletion current limited,
- III-Vs offer similar performance to HgCdTe at an equivalent cutoff wavelength, but with a sizeable penalty in operating temperature, due to the inherent difference in SRH lifetimes.

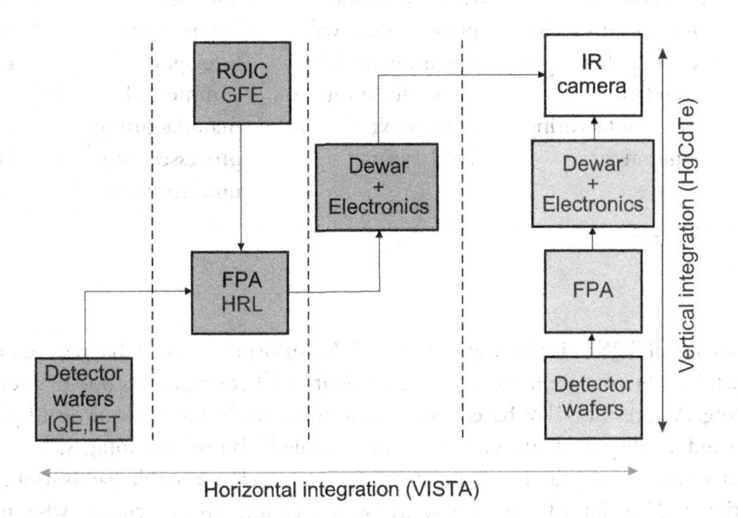

Figure 26.63 Horizontally integrated VISTA program versus existing vertically integrated HgCdTe approach within one company. (Adapted after Ref. [267])

Table 26.24 Comparison of LWIR existing state-of-the-art device systems for LWIR detectors

MATURITY LEVEL	BOLOMETER	HgCdTe	QWIP	TYPE-II SLs
	TRL 9	TRL 9	TRL 8	TRL 5–6
Status	Material of choice for applications requiring medium to low performance	Material of choice for applications requiring high performance	Commercial	Research and development
Operating temperature	Uncooled	Cooled	Cooled	Cooled
Manufacturability	Excellent	Poor	Excellent	Good
Cost	Low	High	Medium	Medium
Prospect for large format	Excellent	Very good	Excellent	Excellent
Availability of large substrate	Excellent	Poor	Excellent	Very good
Military system examples	Weapon sights, night-vision goggles, missile seekers, small UAV sensors, unattended ground sensors	Missile intercepts, tactical ground and air born imaging, hyperspectral, missile seekers, missile tracking, space-based sensing	Being evaluated for some military applications and astronomy sensing	Being developed in universities and evaluated in industry research environments
Limitations	Low sensitivity and long time constants	Performance susceptible to manufacturing variations. Difficult to extend to >14-μm cutoff	Narrow bandwidth and low sensitivity	Requires a significant investment and fundamental material breakthrough to mature
Advantages	Low cost and requires no active cooling; leverages standard Si-manufacturing equipment	Near theoretical performance; will remain material of choice for minimum of the next 10–15 years	Low-cost applications; leverages commercial manufacturing processes; very uniform material	Theoretically better then HgCdTe; leverages commercial III-V fabrication techniques

TRL, technology readiness level.

- Important advantage of T2SLs is the high quality, high uniformity, and stable nature of the material. In general, III-V semiconductors are more robust than their II-VI counterparts due to stronger, less ionic chemical bonding. As a result, III-V-based FPAs excel in operability, spatial uniformity, temporal stability, scalability, producibility, and affordability—the so-called "ibility" advantages.

In FPA fabrication, the major practical issue of T2SLs is the lack of a stable passivation. Usually, surface donor contamination and insulator-fixed positive charge are a common occurrence, what not be an issue for nBn architectures, but for p-type embodiment of the T2SL presents a highly undesirable situation. The p-type T2SL barrier photodetector will also be sensitive to donor core dislocations, instead the nBn barrier detector is relatively immune to donor core dislocations.

Table 26.24 provides a snapshot of the current state of development of LWIR detectors fabricated from different material systems including T2SLs. Note that TRL means technology readiness level. The highest level of TRL (ideal maturity) achieves a value of 10 [11]. The highest level of maturity (TRL = 9) is credited to HgCdTe photodiodes and microbolometers. A little less, TRL = 8, for QWPs. The T2SL structure has great potential for LWIR spectral range application with performance comparable to HgCdTe for the same cutoff wavelength but requires a significant investment and fundamental material breakthrough to mature.

From the economical point of view and future technology perspective, an important aspect concerns industry organization. The HgCdTe array industry is vertically integrated; there are no commercial wafer suppliers because there are no diverse commercial applications of HgCdTe FPA to be profitable. The wafers are grown within each FPA fabrication facility (or its exclusive partner). One important disadvantage of this integrity is the high cost. In the case of III-V semiconductors, horizontal integration is more profitable. This solution is especially effective for avoiding the heavy investment in capital equipment and subsequent upgrading and maintenance, as well as the cost of highly skilled engineers and technicians.

REFERENCES

1. B. Starr, L. Mears, C. Fulk, J. Getty, E. Beuville, R. Boe, C. Tracy, E. Corrales, S. Kilcoyne, J. Vampola, J. Drab, R. Peralta, C. Doyle, "RVS large format arrays for astronomy," *Proc. SPIE*, 9915, 99152X–1–14, 2016.
2. L. Pavesi, "Will silicon be the photonic material of the third millennium?" *J. Phys. Condens. Matter.*, 15, R1169–96, 2003.
3. J.R. Janesick, *Scientific Charge-Coupled Devices*, SPIE Press, Bellingham, WA, 2001.
4. J.R. Janesick, "Charge-coupled CMOS and hybrid detector arrays," *Proc. SPIE*, 5167, 1–18, 2003.
5. "X-3: new single-chip colour CCD technology," *New Technol.*, 20–24, 2002.
6. O. Djazovski, "Focal Plane Arrays for Optical Payloads," in *Optical Payloads for Space Missions*, eds. S.N. Qian, 793–837, Wiley, Chichester, UK, 2016.
7. N. Zacharias, B. Dorland, R. Bredthauer, K. Boggs, G. Bredthauer, M. Lesse, "Realization and application of a 111 million pixel backside-illuminated detector and camera," *Proc. SPIE*, 6690, 669008–1–8, 2007.
8. K. Jacobsen, "Recent Developments of Digital Cameras and Space Imagery," in *Proceedings GIS*, Ostrava 2011, 24–26 January, Ostrava, Czech Republic, pp. 8S-1–8, 2011.
9. J. Bruijne, "Science performance of Gaia, ESA's space-astrometry mission," *Astrophys. Space Sci.*, 341(1), 31–41, 2012.
10. G. Patrie, "Gigapixel frame images: Part II. Is the holy grail of airborne digital frame imaging in sight?" *GeoInformatics*, 24–29, March 2006.
11. "Seeing Photons: Progress and Limits of Visible and Infrared Sensor Arrays," Committee on Developments in Detector Technologies; National Research Council, 2010. www.nap.edu/catalog/12896.html.
12. https://petapixel.com/2015/01/15/worlds-largest-powerful-camera-gets-funding-3-2-gigapixel-sky-photos/.
13. J. Ohta, *Smart CMOS Sensors and Applications*, CRC Press, Boca Raton, FL, 2008.
14. R.F. Lyon and P. Hubel, "Eyeing the Camera: Into the Next Century," in *IS&T/SID Tenth Color Imaging Conference*, 349–55, 2001.
15. T. Chuh, "Recent developments in infrared and visible imaging for astronomy, defense and homeland security," *Proc. SPIE*, 5563, 19–34, 2004.
16. Y. Bai, J. Bajaj, J.W. Beletic, and M.C. Farris, "Teledyne imaging sensors: silicon CMOS imaging technologies for X-Ray, UV, visible and near infrared," *Proc. SPIE*, 7021, 702102, 2008.
17. Y. Bai, W. Tennant, S. Anglin, A. Wong, M. Farris, M. Xu, E. Holland, D. Cooper, J. Hosack, K. Ho, T. Sprafke, R. Kopp, B. Starr, R. Blank, and J.W. Beletic, "4K × 4K format, 10 μm pixel pitch H4RG-10 hybrid CMOS silicon visible focal plane array for space astronomy," *Proc. SPIE*, 8453, 84530M–1–18, 2012.
18. S. Kilcoyne, N. Malone, M. Harris, J. Vampola, and D. Lindsay, "Silicon p-i-n focal plane arrays at Raytheon," *Proc. SPIE*, 7082, 70820J, 2008.
19. B.W. Kean, S. Kilcoyne, N.R. Malone, G. Wilberger, R. Troup, S. Miller, K.C. Brown, J. Vampola, "Advancements in large-format SiPIN hybrid focal plane technology," *Proc. SPIE*, 8511, 851111–1–10, 2012.

20. S. Kilcoyne, N. Malone, B. Kean, J. Cantrell, J. Fierro, L. Meier, S. DeWalt, C. Hewitt, J. Wyles, J. Drab, G. Grama, G. Paloczi, J. Vampola, K. Brown, "Advancements in large-format SiPIN hybrid focal plane technology," *Proc. SPIE*, 9219, 921906–1–11, 2014.
21. B. Ackland, C. Rafferty, C. King, I. Aberg, J. O'Neill, T. Sriram, A. Lattes, C. Godek, and S. Pappas, "A monolithic Ge-on-Si CMOS imager for short wave infrared," www.imagesensors.org/Past%20Workshops/2009%20Workshop/2009%20Papers /053_paper_ackland_noblepeak_swir.pdf.
22. E. Burstein, G. Pines, and N. Sclar, "Optical and Photoconductive Properties of Silicon and Germanium," in *Photoconductivity Conference at Atlantic City*, eds. R. Breckenbridge, B. Russell, and E. Hahn, 353–413, Wiley, New York, 1956.
23. S. Borrello and H. Levinstein, "Preparation and properties of mercury doped infrared detectors," *J. Appl. Phys.*, 33, 2947–50, 1962.
24. N. Sclar, "Properties of doped silicon and germanium infrared detectors," *Prog. Quantum Electron.*, 9, 149–257, 1984.
25. R.D. Nelson, "Accumulation-mode charge-coupled device," *Appl. Phys. Lett.*, 25, 568–70, 1974.
26. D.H. Pommerrenig, "Extrinsic silicon focal plane arrays," *Proc. SPIE*, 443, 144–50, 1984.
27. R.D. Nelson, "Infrared charge transfer devices: the silicon approach," *Opt. Eng.* 16, 275–83, 1977.
28. T.T. Braggins, H.M. Hobgood, J.C. Swartz, and R.N. Thomas, "High infrared responsivity indium-doped silicon detector material compensated by neutron transmutation," *IEEE Trans. Electron Devices*, 27, 2–10, 1980.
29. C.J. Michon and H.K. Burke, "Charge Injection Imaging," in *International Solid-State Circuits Conference, Dig. Tech. Papers*, 138–39, 1973.
30. C.M. Parry, "Bismuth-doped silicon: an extrinsic detector for long-wavelength infrared (LWIR) applications," *Proc. SPIE*, 244, 2–8, 1980.
31. M.E. McKelvey, C.R. McCreight, J.H. Goebel, and A.A. Reeves, "Charge-injection-device 2 × 64 element infrared array performance," *Appl. Opt.*, 24, 2549–57, 1985.
32. W.J. Forrest, A. Moneti, C.E. Woodward, J.L. Pipher, and A. Hoffman, "The new near-infrared array camera at the University of Rochester," *Publ. Astron. Soc. Pac.*, 97, 183–98, 1985.
33. G.H. Rieke, "Infrared detector arrays for astronomy," *Annu. Rev. Astron. Astrophys.*, 45, 77–115, 2007.
34. J. Wu, W.J. Forrest, J.L. Pipher, N. Lum, and A. Hoffman, "Development of infrared focal plane arrays for space," *Rev. Sci. Instrum.*, 68, 3566–78, 1997.
35. E.M. Conwell, "Impurity band conduction in germanium and silicon," *Phys. Rev.*, 103(1), 51–61, 1956.
36. J. Love, K.J. Ando, R.E. Bornfreund, E. Corrales, R.E. Mills, J.R. Cripe, N.A. Lum, J.P. Rosbeck, and M.S. Smith, "Large-format infrared arrays for future space and ground-based astronomy applications," *Proc. SPIE*, 4486, 373–84, 2002.
37. K.J. Ando, A.W. Hoffman, P.J. Love, A. Toth, C. Anderson, G. Chapman, C.R. McCreight, K.A. Ennico, M.E. McKelvey, and R.E. McMurray, Jr., "Development of Si:As impurity band conduction (IBC) detectors for mid-infrared applications," *Proc. SPIE*, 5074, 648–57, 2003.
38. P.J. Love, A.W. Hoffman, N.A. Lum, K.J. Ando, J. Rosbeck, W.D. Ritchie, N.J. Therrien, R.S. Holcombe, and E. Corrales, "1024 × 1024 Si:As IBC detector arrays for JWST MIRI," *Proc. SPIE*, 5902, 590209, 2005.
39. M.E. Ressler, H. Cho, R.A.M. Lee, K.G. Sukhatme, J.J. Drab, G. Domingo, M.E. McKelvey, R.E. McMurray, Jr., and J.L. Dotson, "Performance of the JWST/MIRI Si:As detectors," *Proc. SPIE*, 7021, 70210O, 2008.
40. R. Mills, E. Beuville, E. Corrales, A. Hoffman, G. Finger, and D. Ives, "Evolution of large format impurity band conductor focal plane arrays for astronomy applications," *Proc. SPIE*, 8154, 81540R–1–10, 2011.
41. A.K. Mainzer, P. Eisenhardt, E.L. Wright, F.-C. Liu, W. Irace, I. Heinrichsen, R. Cutri, and V. Duval, "Preliminary design of the wide-field infrared survey explorer (WISE)," *Proc. SPIE*, 5899, 58990R, 2005.
42. P.A.K. Mainzer, H. Hogue, M. Stapelbroek, D. Molyneux, J. Hong, M. Werner, M. Ressler, and E. Young, "Characterization of a megapixel mid-infrared array for high background applications," *Proc. SPIE*, 7021, 70210T, 2008.
43. H.H. Hogue, M.G. Mlynczak, M.N. Abedin, S.A. Masterjohn, and J.E. Huffman, "Far-infrared detector development for space-based Earth observation," *Proc. SPIE*, 7082, 70820E–1–8, 2008.
44. H. Hogue, E. Atkins, D. Reynolds, M. Salcido, L Dawson, D. Molyneux, and M. Muzilla, "Update on Blocked Impurity Band detector technology from DRS," *Proc. SPIE*, 7780, 778004–1–10, 2010.
45. M. Fowler and I. Gatley, "Demonstration of an algorithm for read-noise reduction in infrared arrays," *Astrophys. J.*, 353, L33–4, 1990; "Noise reduction strategy for hybrid JR focal plane arrays," *Proc. SPIE*, 1541, 127–33, 1991.
46. E.T. Young, "Progress on readout electronics for far-infrared arrays," *Proc. SPIE*, 2226, 21–8, 1994.
47. R.M. Glidden, S.C. Lizotte, J.S. Cable, L.W. Mason, and C. Cao, "Optimization of cryogenic CMOS processes for sub-10°K applications," *Proc. SPIE*, 1684, 2–39, 1992.

48. A.G. Kazanskii, P.L. Richards, and E.E., Haller, "Far-infrared photoconductivity of uniaxially stressed germanium," *Appl. Phys. Lett.*, 31, 496–7, 1977.
49. E.T. Young, J.T. Davis, C.L. Thompson, G.H. Rieke, G. Rivlis, R. Schnurr, J. Cadien, L. Davidson, G.S. Winters, and K.A. Kormos, "Far-infrared imaging array for SIRTF," *Proc. SPIE*, 3354, 57–65, 1998.
50. E.T. Young, "Germanium detectors for the far-infrared," www.stsci.edu/stsci/meetings/space_detectors/pdf/gdfi.pdf.
51. R. Schnurr, C.L. Thompson, J.T. Davis, J.W. Beeman, J. Cadien, E.T. Young, E.E. Haller, and G.H. Rieke, "Design of the stressed Ge:Ga far-infrared array for SIRTF," *Proc. SPIE*, 3354, 322–31, 1998.
52. M. Fujiwara, T. Hirao, M. Kawada, H. Shibai, S. Matsuura, H. Kaneda, M. Patrashin, and T. Nakagawa, "Development of a gallium-doped germanium far-infrared photoconductor direct hybrid two-dimensional array," *Appl. Opt.*, 42, 2166–73, 2003.
53. A. Poglitsch, R.O. Katterloher, R. Hoenle, J.W. Beeman, E.E. Haller, H. Richter, U. Grozinger, N.M. Haegel, and A. Krabbe, "Far-infrared photoconductors for Herschel and SOFIA," *Proc. SPIE*, 4855, 115–28, 2003.
54. http://fifi-ls.mpg-garching.mpg.dr/detector.html.
55. http://pacs.mpe.mpg.de/p15n.html.
56. A. Poglitsch and B. Altieri, "The PACS Instrument," in *Astronomy in the Submillimeter and Far Infrared Domains with the Herschel Space Observatory*, eds. L. Pagani and M. Gerin, EAS Publications Series, 34, 43–62, EDP Sciences, Les Ulis, 2009.
57. M. Shirahata, S. Matsuure, S. Makiuti, M.A. Patrashin, H. Kaneda, T. Nakagawa, M. Fujiwara, et al., "Preflight performance measurements of a monolithic Ge:Ga array detector for the far-infrared surveyor onboard ASTRO-F," *Proc. SPIE*, 5487, 369–80, 2004.
58. S.M. Birkmann, K. Eberle, U. Grozinger, D. Lemke, J. Schreiber, L. Barl, R. Katterloher, A. Poglitsch, J. Schubert, and H. Richter, "Characterization of high- and low-stressed Ge:Ga array cameras for Herschel's PACS instrument," *Proc. SPIE*, 5487, 437–47, 2004.
59. N. Billot, P. Agnese, J.L. Augueres, A. Beguin, and A. Bouere, O. Boulade, C. Cara, C. Cloue, E. Doumayrou, L. Duband, B. Horeau, I. Le Mer, J.L. Pennec, J. Martignac, K. Okumura, V. Reveret, M. Sauvage, F. Simoens, and L. Vigroux, "The Herschel/PACS 2560 bolometers imaging camera," *Proc. SPIE*, 6265, 62650D, 2006.
60. M. Shirahata, S. Matsuura, T. Nakagawa, T. Wada, S. Kamiya, M. Kawada, Y. Sawayama, Y. Doi, H. Kawada, Y. Creten, B. Okcan, W. Raab, and A. Poglitsh, "Development of a far-infrared Ge:Ga monolithic array detector for a possible application to SPICA," *Proc. SPIE*, 7741, 77410B, 2010.
61. T. Wada, Y. Arai, S. Baba, M. Hanaoka, Y. Hattori, H. Ikeda, H. Kaneda, C. Kochi, A. Miyachi, K. Nagase, H. Nakaya, M. Ohno, S. Oyabu, T. Suzuki, S. Ukai, K.Watanabe, K. Yamamoto, "Development for germanium blocked impurity band far-infrared image sensors with fully-depleted silicon-on-insulator CMOS readout integrated circuit," *J. Low. Temp. Phys.*, 184, 217–24, 2016.
62. J. Farhoomand, D.L. Sisson, and J.W. Beeman, "Viability of layered-hybrid architecture for far IR focal-plane arrays," *Infrared Phys. Technol.*, 51, 152–9, 2008.
63. M. Ressler, H. Hogue, M. Muzilla, J. Blacksberg, J. Beeman, E. Haller, J. Huffman, J. Farhoomand, E. Young, "Development of large format far-infrared detectors," *Astro2010: The Astronomy and Astrophysics Decadal Survey*, Technology Development Papers, no. 18.
64. N.M. Haegel, "BIB detector development for the far infrared: from Ge to GaAs," *Proc. SPIE*, 4999, 182–94, 2003.
65. E.E. Haller and J.W. Beeman, "Far infrared photoconductors: Recent advances and future prospects," *Far-IR Sub-MM&MM Detectors Technology Workshop*, 2-06, Monterey, April 1–3, 2002.
66. E.S. Kohn, W.F. Kosonocky, and F.V. Shallcross, "Charge-Coupled Scanned IR Imaging Sensors," *Final Report RADC-TR-308*, Rome Air Development Center, 1977.
67. M. Kimata, M. Ueno, H. Yagi, T. Shiraishi, M. Kawai, K. Endo, Y. Kosasayama, T. Sone, T. Ozeki, and N. Tsubouchi, "PtSi Schottky-barrier infrared focal plane arrays," *Opto-Electron. Rev.*, 6, 1–10, 1998.
68. M. Denda, M. Kimata, S. Iwade, N. Yutani, T. Kondo, and N. Tsubouchi, "Schottky-barrier infrared linear image sensor with 4-band × 4096-element," *IEEE Trans. Electron Devices*, 38, 1145–51, 1991.
69. M.T. Daigle, D. Colvin, E.T. Nelson, S. Brickman, K. Wong, S. Yoshizumi, M. Elzinga, et al., "High resolution 2048 × 16 TDI PtSi IR imaging CCD," *Proc. SPIE*, 1308, 88–98, 1990.
70. W.F. Kosonocky, "Review of infrared image sensors with Schottky-barrier detectors," *Opto-Electron. Devices Technol.*, 6, 173–203, 1991.
71. W.F. Kosonocky, "State-of-the-art in Schottky-barrier IR image sensors," *Proc. SPIE*, 1682, 2–19, 1992.
72. M. Kimata and N. Tsubouchi, "Schottky Barrier Photoemissive Detectors," in *Infrared Photon Detectors*, ed. A. Rogalski, 299–349, SPIE Optical Engineering Press, Bellingham, WA, 1995.

73. M. Kimata, "Metal Silicide Schottky Infrared Detector Arrays," in *Infrared Detectors and Emitters: Materials and Devices*, eds. P. Capper and C.T. Elliott, 77–98, Kluwer Academic Publishers, Boston, MA, 2000.
74. M. Kimata, "Silicon Infrared Focal Plane Arrays," in *Handbook of Infrared Detection Technologies*, eds. M. Henini and M. Razeghi, 353–92, Elsevier, Oxford, UK, 2002.
75. M. Kimata, "My life in IRFPA R&D," *Proc. SPIE*, 10177, 1017727–1–11, 2017.
76. H. Yagi, N. Yutani, J. Nakanishi, M. Kimata, and M. Nunoshita, "A monolithic Schottky-barrier infrared image sensor with 71% fill factor," *Opt. Eng.*, 33, 1454–60, 1994.
77. N. Yutani, H. Yagi, M. Kimata, J. Nakanishi, S. Nagayoshi, and N. Tsubouchi, "1040 × 1040 element PtSi Schottky-barrier IR image sensor," *Tech. Dig. IEDM*, 175–78, 1991.
78. M. Kimata, N. Yutani, N. Tsubouchi, and T. Seto, "High performance 1040 × 1040 element PtSi Schottky-barrier image sensor," *Proc. SPIE*, 1762, 350–60, 1992.
79. M. Kimata, M. Denda, N. Yutani, S. Iwade, and N. Tsubouchi, "512 × 512 element PtSi Schottky-barrier infrared image sensor," *IEEE J. Solid State Circuits*, 22, 1124–29, 1987.
80. T. Shiraishi, H. Yagi, K. Endo, M. Kimata, T. Ozeki, K. Kama, and T. Seto, "PtSi FPA with improved CSD operation," *Proc. SPIE*, 2744, 33–43, 1996.
81. M. Inoue, T. Seto, S. Takahashi, S. Itoh, H. Yagi, T. Siraishi, K. Endo, and M. Kimata, "Portable high performance camera with 801 × 512 PtSi-SB IRCSD," *Proc. SPIE*, 3061, 150–58, 1997.
82. M. Kimata, T. Ozeki, M. Nunoshita, and S. Ito, "PtSi Schottky-barrier infrared FPAs with CSD readout," *Proc. SPIE*, 3179, 212–23, 1997.
83. F.D. Shepherd, "Recent advances in platinum silicide infrared focal plane arrays," *Tech. Dig. IEDM*, 370–73, 1984.
84. J.L. Gates, W.G. Connelly, T.D. Franklin, R.E. Mills, F.W. Price, and T.Y. Wittwer, "488 × 640-element hybrid platinum silicide Schottky focal plane array," *Proc. SPIE*, 1540, 262–73, 1991.
85. W.F. Kosonocky, T.S. Villani, F.V. Shallcross, G.M. Meray, and J.J. O'Neil, "A Schottky-barrier image sensor with 100% fill factor," *Proc. SPIE*, 1308, 70–80, 1990.
86. H. Presting, "Infrared Silicon/Germanium Detectors," in *Handbook of Infrared Detection Technologies*, eds. M. Henini and M. Razeghi, 393–448, Elsevier, Oxford, UK, 2002.
87. B-Y. Tsaur, C.K. Chen, and S.A. Marino, "Long-wavelength GeSi/Si heterojunction infrared detectors and 400 × 400-element imager arrays," *IEEE Electron. Device Lett.*, 12, 293–96, 1991.
88. B-Y. Tsaur, C.K. Chen, and S.A. Marino, "Long-wavelength $Ge_{1-x}Si_x$/Si heterojunction infrared detectors and focal plane arrays," *Proc. SPIE*, 1540, 580–95, 1991.
89. B-Y. Tsaur, C.K. Chen, and S.A. Marino, "Heterojunction $Ge_{1-x}Si_x$/Si infrared detectors and focal plane arrays," *Opt. Eng.*, 33, 72–78, 1994.
90. H. Wada, M. Nagashima, K. Hayashi, J. Nakanishi, M. Kimata, N. Kumada, and S. Ito, "512 × 512 element GeSi/Si heterojunction infrared focal plane array," *Opto-Electron. Rev.*, 7, 305–11, 1999.
91. M. Kimata, H. Yagi, M. Ueno, J. Nakanishi, T. Ishikawa, Y. Nakaki, M. Kawai, et al., "Silicon infrared focal plane arrays," *Proc. SPIE*, 4288, 286–97, 2001.
92. A. Rogalski, P. Martyniuk, and M. Kopytko, *Antimonide-based Infrared Detectors—A New Perspective*, SPIE Press, Bellingham, WA, 2018.
93. http://xenics.com/en/products/cameras; http://www.sensorsinc.com/products/linescan-cameras/.
94. G. Olsen, A. Joshi, M. Lange, K. Woodruff, E. Mykietyn, D. Gay, G. Erickson, D. Ackley, V. Ban, and C. Staller, "A 128 × 128 InGaAs detector array for 1.0–1.7 microns," *Proc. SPIE*, 1341, 432–37, 1990.
95. M.H. Ettenberg, M.J. Cohen, R.M. Brubaker, M.J. Lange, M.T. O'Grady, and G.H. Olsen, "Indium gallium arsenide imaging with smaller cameras, higher resolution arrays, and greater material sensitivity," *Proc. SPIE*, 4721, 26–36, 2002.
96. T.J. Martin, M.J. Cohen, J.C. Dries, and M.J. Lange, "InGaAs/InP focal plane arrays for visible light imaging," *Proc. SPIE*, 5406, 38–45, 2004.
97. T. Martin, R. Brubaker, P. Dixon, M.-A. Gagliardi, and T. Sudol, "640 × 512 InGaAs focal plane array camera for visible and SWIR imaging," *Proc. SPIE*, 5783, 12–20, 2005.
98. B.M. Onat, W. Huang, N. Masaun, M. Lange, M.H. Ettenberg, and C. Dries, "Ultra low dark current InGaAs technology for focal plane arrays for low-light level visible-shortwave infrared imaging," *Proc. SPIE*, 6542, 65420L, 2007.
99. M.D. Enriguez, M.A. Blessinger, J.V. Groppe, T.M. Sudol, J. Battaglia, J. Passe, M. Stern, and B.M. Onat, "Performance of high resolution visible-InGaAs imager for day/night vision," *Proc. SPIE*, 6940, 69400O, 2008.
100. T.R. Hoelter and J.B. Barton, "Extended short wavelength spectral response from InGaAs focal plane arrays," *Proc. SPIE*, 5074, 481–90, 2003.

101. A.D. Hood, M.H. MacDougal, J. Manzo, D. Follman, J.C. Geske, "Large-format InGaAs focal plane arrays for SWIR imaging," *Proc. SPIE*, 8353, 83530A–1–7, 2012.
102. H. Yuan, G. Apgar, J. Kim, J. Laguindanum, V. Nalavade, P. Beer, J. Kimchi, and T. Wong, "FPA development: From InGaAs, InSb, to HgCdTe," *Proc. SPIE*, 6940, 69403C, 2008.
103. R. Fraenkel, E. Berkowicz, L. Bykov, R. Dobromislin, R. Elishkov, A. Giladi, I. Grimberg, I. Hirsh, E. Ilan, C. Jacobson, I. Kogan, P. Kondrashov, I. Nevo, I. Pivnik, and S. Vasserman, "High definition 10μm pitch InGaAs detector with asynchronous laser pulse detection mode," *Proc. SPIE*, 9819, 9819–1–8, 2016.
104. P. Yuan, J. Chang, J.C. Boisvert, N. Karam, "Low-dark current 1024 × 1280 InGaAs PIN arrays," *Proc. SPIE*, 9070, 907007–1–6, 2014.
105. J. Coussement, A. Rouvié, E.H. Oubensaid, O. Huet, S. Hamard, JP. Truffer, M. Pozzi, P. Maillart, Y. Reibel, E. Costard, and D. Billon-Lanfrey, "New developments on InGaAs focal plane array," *Proc. SPIE*, 9070, 907005–1–9, 2014.
106. A. Rouvié, J. Coussement, O. Huet, JP. Truffer, M. Pozzi, E.H. Oubensaid, S. Hamard, V. Chaffraix, E. Costard, "InGaAs focal plane array developments and perspectives," *Proc. SPIE*, 9451, 945105–1–8, 2015.
107. S. Seshadri, D.M. Cole, B. Hancock, P. Ringold, C. Peay, C. Wrigley, M. Bonati, et al., "Comparison the low-temperature performance of megapixel NIR InGaAs and HgCdTe imager arrays," *Proc. SPIE*, 6690, 669006, 2007.
108. A. Hoffman, T. Sessler, J. Rosbeck, D. Acton, and M. Ettenberg, "Megapixel InGaAs arrays for low background applications," *Proc. SPIE*, 5783, 32–8, 2005.
109. G. Orias, A. Hoffman, and M. Casselman, "58 × 62 indium antimonide focal plane array for infrared astronomy," *Proc. SPIE*, 627, 408–17, 1986.
110. G.C. Baily, C.A. Niblack, and J.T. Wimmers, "Recent developments on a 128 × 128 indium antimonide/FET switch hybrid imager for low background applications," *Proc. SPIE*, 686, 76–83, 1986.
111. S. Shirouzu, T. Tsuji, N. Harada, T. Sado, S. Aihara, R. Tsunoda, and T. Kanno, "64 × 64 InSb focal plane array with improved two layer structure," *Proc. SPIE*, 661, 419–25, 1986.
112. J.T. Wimmers, R.M. Davis, C.A. Niblack, and D.S. Smith, "Indium antimonide detector technology at Cincinnati Electronics Corporation," *Proc. SPIE*, 930, 125–38, 1988.
113. C.W. McMurtry, W.J. Forrest, J.L. Pipher, and A.C. Moore, "James Webb Space Telescope characterization of flight candidate NIR InSb array," *Proc. SPIE*, 5167, 144–58, 2003.
114. G. Finger, R.J. Dorn, A.W. Hoffman, H. Mehrgan, M. Meyer, A.F.M. Moorwood, and J. Stegmeier, "Readout Techniques for Drift and Low Frequency Noise Rejection in Infrared Arrays," in *Scientific Detectors for Astronomy*, eds. P. Amico, J.W. Beletic, and J.E. Beletic, 435–44, Springer, Berlin, 2003.
115. A.W. Hoffman, E. Corrales, P.J. Love, J. Rosbeck, M. Merrill, A. Fowler, and C. McMurtry, "2K × 2K InSb for Astronomy," *Proc. SPIE*, 5499, 59–67, 2004.
116. M. Devis and M. Greiner, "Indium antimonide large-format detector arrays," *Opt. Eng.*, 50, 061016–1–6, 2011.
117. G. Gershon, A. Albo, M. Eylon, O. Cohen, Z. Calahorra, M. Brumer, M. Nitzani, E. Avnon, Y. Aghion, I. Kogan, E. Ilan, and L. Shkedy, "3 Mega-pixel InSb detector with 10 μm pitch," *Proc. SPIE*, 8704, 870438, 2013.
118. E. Beuville, D. Acton, E. Corrales, J. Drab, A. Levy, M. Merrill, R. Peralta, and W. Ritchie, "High performance large infrared and visible astronomy arrays for low background applications: instruments performance data and future developments at Raytheon," *Proc. SPIE*, 6660, 66600B, 2007.
119. A.M. Fowler, D. Bass, J. Heynssens, I. Gatley, F.J. Vrba, H.D. Ables, A. Hoffman, M. Smith, and J. Woolaway, "Next generation in InSb arrays: ALADDIN, the 1024 × 1024 InSb focal plane array readout evaluation results," *Proc. SPIE*, 2268, 340–5, 1994.
120. A.M. Fowler, K.M. Merrill, W. Ball, A. Henden, F. Vrba, and C. McCreight, "Orion: A 1-5 Micron Focal Plane for the 21st Century," in *Scientific Detectors for Astronomy: The Beginning of a New Era*, ed. P. Amico, 51–8, Kluwer, Dordrecht, The Netherlands, 2004.
121. www.sbfp.com/documents/FPA%20S019-0001-08.pdf.
122. G. Gershon, A. Albo, M. Eylon, O. Cohen, Z. Calahorra, M. Brumer, M. Nitzani, E. Avnon, Y. Aghion, I. Kogan, E. Ilan, A. Tuito, M. Ben Ezra, and L. Shkedy, "Large Format InSb Infrared Detector with 10 μm Pixels," in *Optro* 2014 *Symposium—Optoelectronics in Defence and Security,* 2830 January, 2014. www.scd.co.il/SCD/Templates/ShowPage.asp?DBID=1&LNGID=1&TMID=111&FID=1279&IID=1613.
123. R.J. Phelan and J.O. Dimmock, "InSb MOS infrared detector," *Appl. Phys. Lett.*, 10, 55–8, 1967.
124. D.L. Lile and H.H. Wieder, "The thin film MIS surface photodiode," *Thin Solid Films*, 13, 15–20, 1972.
125. D.L. Lile, "Surface photovoltage and internal photoemission at the anodized InSb surface," *Surf. Sci.* 34, 337–67, 1973.

126. H.K. Burke and G.J. Milton, "Charge-injection device imaging; operating techniques and performance characteristics," *IEEE Trans. Electron Devices*, 23, 189–95, 1976.
127. J.C. Kim, "InSb charge-injection device imaging array," *IEEE Trans. Electron Devices*, 25, 232–46, 1978.
128. G.J. Michon and H.K. Burke, "CID Image Sensing," in *Charge-Coupled Devices*, ed. D.F. Barbe, 5–24, Springer-Verlag, Berlin, 1980.
129. J.T. Longo, D.T. Cheung, A.M. Andrews, C.C. Wang, and J.M. Tracy, "Infrared focal plane in intrinsic semiconductors," *IEEE Trans. Electron Devices*, 25, 213–32, 1978.
130. M.D. Gibbons, S.C. Wang, S.R. Jost, V.F. Meikleham, T.H. Myers, and A.F. Milton, "Developments in InSb material and charge injection devices," *Proc. SPIE*, 865, 52–8, 1987.
131. C.Y. Wei and H.H. Woodbury, "Ideal mode operation of an InSb charge injection device," *IEEE Trans. Electron Devices*, 31, 1773–80, 1984.
132. S.C.H. Wang, C.Y. Wei, H.H. Woodbury, and M.D. Gibbons, "Characteristics and readout of an InSb CID two-dimensional scanning TDI array," *IEEE Trans. Electron Devices*, 32, 1599–607, 1985.
133. M.D. Gibbons and S.C. Wang, "Status of CID InSb detector technology," *Proc. SPIE*, 443, 151–66, 1984.
134. D.A. Scribner, M.R. Kruer, and J.M. Killiany, "Infrared focal plane array technology," *Proc. IEEE*, 79, 66–85, 1991.
135. A. Bahraman, C.H. Chen, J.M. Geneczko, M.H. Shelstad, R.N. Ting, and J.G. Vodicka, "Current state of the art in InSb infrared staring imaging devices," *Proc. SPIE*, 750, 27–31, 1987.
136. S.R. Jost, V.F. Meikleham, and T.H. Myers, "InSb: A key material for IR detector applications," *Mater. Res. Soc. Symp. Proc.*, 90, 429–35, 1987.
137. R.D. Thom, R.E. Eck, J.D. Philips, and J.B. Scorso, "InSb CCDs and Other MIS Devices for Infrared Applications," in *Proceedings of the 1975 International Conference on the Applications of CCDs*, 31–41, 1975.
138. R.D. Thom, T.L. Koch, J.D. Langan, and W.L. Parrish, "A fully monolithic InSb Infrared CCD array," *IEEE Trans. Electron Devices*, 27, 160–70, 1980.
139. M.B. Reine, E.E. Krueger, P. O'Dette, C.L. Terzis, B. Denley, J. Hartley, J. Rutter, and D.E. Kleinmann, "Advances in 15 μm HgCdTe photovoltaic and photoconductive detector technology for remote sensing," *Proc. SPIE*, 2816, 120–37, 1996.
140. M.B. Reine, E.E. Krueger, P. O'Dette, and C.L. Terzic, "Photovoltaic HgCdTe detectors for advanced GOES instruments," *Proc. SPIE*, 2812, 501–17, 1996.
141. J.W. Beletic, R. Blank, D. Gulbransen, D. Lee, M. Loose, E.C. Piquette, T. Sprafke, W.E. Tennant, M. Zandian, and J. Zino, "Teledyne imaging sensors: infrared imaging technologies for astronomy & civil space," *Proc. SPIE*, 7021, 70210H, 2008.
142. W.D. Lawson, S. Nielson, E.H. Putley, and A.S. Young, "Preparation and properties of HgTe and mixed crystals of HgTe-CdTe," *J. Phys. Chem. Solids*, 9, 325–9, 1959.
143. P.R. Bratt, S.M. Johnson, D.R. Rhiger, T. Tung, M.H. Kalisher, W.A. Radford, G.A. Garwood, and C.A. Cockrum, "Historical perspectives on HgCdTe material and device development at Raytheon Vision Systems," *Proc. SPIE*, 7298, 72982U, 2009.
144. A. Rogalski, *Infrared Detectors*, Gordon and Breach Science Publishers, Amsterdam, 2000.
145. R.A. Chapman, M.A. Kinch, A. Simmons, S.R. Borrello, H.B. Morris, J.S. Wrobel, and D.D. Buss, "$Hg_{0.7}Cd_{0.3}Te$ charge-coupled device shift registers," *Appl. Phys. Lett.*, 32, 434–6, 1978.
146. R.A. Chapman, S.R. Borrello, A. Simmons, J.D. Beck, A.J. Lewis, M.A. Kinch, J. Hynecek, and C.G. Roberts, "Monolithic HgCdTe charge transfer device infrared imaging arrays," *IEEE Trans. Electron Devices*, 27, 134–46, 1980.
147. A.F. Milton, "Charge Transfer Devices for Infrared Imaging," in *Optical and Infrared Detectors*, ed. R.J. Keyes, 197–228, Springer-Verlag, Berlin, 1980.
148. M.A. Kinch, "Metal-Insulator-Semiconductor Infrared Detectors," in *Semiconductors and Semimetals*, Vol. 18, eds. R.K. Willardson and A.C. Beer, 313–78, Academic Press, New York, 1981.
149. R.A. Schiebel, "Enhancement mode HgCdTe MISFETs and circuits for focal plane applications," *IEDM Tech. Dig.*, 132, 1987.
150. T.L. Koch, J.H. De Loo, M.H. Kalisher, and J.D. Phillips, "Monolithic n-Channel HgCdTe linear imaging arrays," *IEEE Trans. Electron Devices*, 32, 1592–607, 1985.
151. M.V. Wadsworth, S.R. Borrello, J. Dodge, R. Gooh, W. McCardel, G. Nado, and M.D. Shilhanek, "Monolithic CCD imagers in HgCdTe," *IEEE Trans. Electron Devices*, 42, 244–50, 1995.
152. M.A. Kinch, "MIS Devices in HgCdTe," in *Properties of Narrow Gap Cadmium-Based Compounds*, EMIS Datareviews Series No. 10, ed. P. Capper, 359–63, IEE, London, 1994.
153. R. Thorn, "High Density Infrared Detector Arrays," U.S. Patent No. 4,039,833, 1977.

154. K. Chow, J.P. Rode, D.H. Seib, and J. Blackwell, "Hybrid infrared focal-plane arrays," *IEEE Trans. Electron Devices*, 29, 3–13, 1982.
155. J.P. Rode, "HgCdTe hybrid focal plane," *Infrared Phys.*, 24, 443–53, 1984.
156. M. Baker and R.A. Ballingall, "Photovoltaic CdHgTe: silicon hybrid focal planes," *Proc. SPIE*, 510, 121–9, 1984.
157. I.M. Baker, "Infrared Radiation Imaging Devices and Methods for Their Manufacture," U. S. Patent 4,521,798, 1985.
158. I.M. Baker, G.J. Crimes, J.E. Parsons, and E.S. O'Keefe, "CdHgTe-CMOS hybrid focal plane arrays: A flexible solution for advanced infrared systems," *Proc. SPIE*, 2269, 636–47, 1994.
159. I.M. Baker, M.P. Hastings, L.G. Hipwood, C.L. Jones, and P. Knowles, "Infrared detectors for the year 2000," *III-Vs Review*, 9(2), 50–60, 1996.
160. I.M. Baker, G.J. Crimes, R.A. Lockett, M.E. Marini, and S. Alfuso, "CMOS/HgCdTe, 2D array technology for staring systems," *Proc. SPIE*, 2744, 463–72, 1996.
161. R.L. Smythe, "Monolithic HgCdTe Focal Plane Arrays," in *Government Microcircuit Applications Conference Proceedings*, 289–92, Fort Monmouth, U.S. Army ERADCOM, New Jersey, 1982.
162. A. Turner, T. Teherani, J. Ehmke, C. Pettitt, P. Conlon, J. Beck, K. McCormack, et al., "Producibility of VIP™ scanning focal plane arrays," *Proc. SPIE*, 2228, 237–48, 1994.
163. J.M. Armstrong, M.R. Skokan, M.A. Kinch, and J.D. Luttmer, "HDVIP five micron pitch HgCdTe focal plane arrays," *Proc. SPIE*, 9070, 907933, 2014.
164. Y. Reibel, N. Pere-Laperne, L. Rubaldo, T. Augey, G. Decaens, V. Badet, L. Baud, J. Roumegoux, A. Kessler, P. Maillart, N. Ricard, O. Pacaud, and G. Destefanis, "Update on 10 μm pixel pitch MCT-based focal plane array with enhanced functionalities," *Proc. SPIE*, 9451, 9451–82, 2015.
165. W.E. Tennanat, D.J. Gulbransen, A. Roll, M. Carmody, D. Edwall, A. Julius, P. Dreiske, A. Chen, W. McLevige, S. Freeman, D. Lee, D.E. Cooper, and E. Piquette, "Small-pitch HgCdTe photodetectors," *J. Electron. Mater.*, 43, 3041–6, 2014.
166. R.K. McEven, D. Jeckells, S. Bains, and H. Weller, "Developments in reduced pixel geometries with MOCVD grown MCT arrays," *Proc. SPIE*, 9451, 94512D–1–9, 2015.
167. R. Breiter, D. Eich, H. Figgemeier, H. Lutz, J. Wendler, I. Rühlich, S. Rutzinger, T. Schallenberg, "Optimized MCT IR-modules for high performance imaging applications," *Proc. SPIE*, 9070, 90702V–1–8, 2014.
168. A. Rogalski, P. Martyniuk, and M. Kopytko, "Challenges of small-pixel infrared detectors: a review," *Rep. Prog. Phys.*, 79, 046501, 2016.
169. W.E. Tennant, "Recent development in HgCdTe photovoltaic device grown on alternative substrates using heteroepitaxy," *Tech. Dig. IEDM*, 704–6, 1983.
170. C.D. Maxey, J.P. Camplin, I.T. Guilfoy, J. Gardner, R.A. Lockett, C.L. Jones, P. Capper, M. Houlton, and N.T. Gordon, "Metal-organic vapor-phase epitaxial growth of HgCdTe device heterostructures on three-inch-diameter substrates," *J. Electron. Mater.*, 32, 656–60, 2003.
171. T.J. de Lyon, S.M. Johnson, C.A. Cockrum, O.K. Wu, W.J. Hamilton, and G.S. Kamath, "CdZnTe on Si(001) and Si(112): direct MBE growth for large-area HgCdTe infrared focal-plane array applications," *J. Electrochem. Soc.*, 141, 2888–93, 1994.
172. J.M. Peterson, J.A. Franklin, M. Readdy, S.M. Johnson, E. Smith, W.A. Radford, and I. Kasai, "High-quality large-area MBE HgCdTe/Si," *J. Electron. Mater.*, 36, 1283–86, 2006.
173. R. Bornfreund, J.P. Rosbeck, Y.N. Thai, E.P. Smith, D.D. Lofgreen, M.F. Vilela, A.A. Buell, et al., "High-performance LWIR MBE-grown HgCdTe/Si focal plane arrays," *J. Electron. Mater.*, 36, 1085–91, 2007.
174. D.J. Hall, L. Buckle, N.T. Gordon, J. Giess, J.E. Hails, J.W. Cairns, R.M. Lawrence, et al., "High-performance long-wavelength HgCdTe infrared detectors grown on silicon substrates," *Appl. Phys. Lett.*, 85, 2113–15, 2004.
175. C.D. Maxey, J.C. Fitzmaurice, H.W. Lau, L.G. Hipwood, C.S. Shaw, C.L. Jones, and P. Capper. "Current status of large-area MOVPE growth of HgCdTe device structures for infrared focal plane arrays," *J. Electron. Mater.*, 35, 1275–82, 2006.
176. C.L. Jones, L.G. Hipwood, C.J. Shaw, J.P. Price, R.A. Catchpole, M. Ordish, C.D. Maxey, et al., "High performance MW and LW IRFPAs made from HgCdTe grown by MOVPE," *Proc. SPIE*, 6206, 620610, 2006.
177. M.J. Hewitt, J.L. Vampola, S.H. Black, and C.J. Nielsen, "Infrared readout electronics: a historical perspective," *Proc. SPIE*, 2226, 108–19, 1994.
178. P. Felix, M. Moulin, B. Munier, J. Portmann, and J.P. Reboul, "CCD readout of infrared hybrid focal-plane arrays," *IEEE Trans. Electron Devices*, 27, 175–88, 1980.
179. P. Knowles, "Mercury cadmium telluride detectors for thermal imaging," *GEC J. Res.*, 2, 141–56, 1984.

180. L.J. Kozlowski, K. Vural, J. Luo, A. Tomasini, T. Liu, and W.E. Kleinhans, "Low-noise infrared and visible focal plane arrays," *Opto-Electron. Rev.*, 7, 259–69, 1999.
181. J.L. Vampola, "Readout Electronics for Infrared Sensors," in *The Infrared and Electro-Optical Systems Handbook*, Vol. 3, ed. W.D. Rogatto, 285–342, SPIE Press, Bellingham, WA, 1993.
182. E.R. Fossum and B. Pain, "Infrared readout electronics for space science sensors: state of the art and future directions," *Proc. SPIE*, 2020, 262–85, 1993.
183. L.J. Kozlowski and W.F. Kosonocky, "Infrared Detector Arrays," in *Handbook of Optics*, Chapter 23, eds. M. Bass, E.W. Van Stryland, D.R. Williams, and W.L. Wolfe, 23.1–23.37, McGraw-Hill, Inc., New York, 1995.
184. J.D. Vincent, S.E. Hodges, J. Vampola, M. Stegall, and G. Pierce, *Fundamentals of Infrared and Visible Detector Operation and Testing*, Wiley, Hoboken, NJ, 2016.
185. A. Rogalski, Z. Bielecki, and J. Mikolajczak, "Detection of Optical Radiation," in *Handbook of Optoelectronics*, Second edition, Vol. 1, eds. J.P. Dakin and R.G.W. Brown, 65–123, CRC Press, New York, 2018.
186. M.Z. Tidrow, W.A. Beck, W.W. Clark, H.K. Pollehn, J.W. Little, N.K. Dhar, R.P. Leavitt, et al., "Device physics and focal plane array applications of QWIP and MCT," *Opto-Electron. Rev.*, 7, 283–96, 1999.
187. M.A. Kinch, J.D. Beck, C.-F. Wan, F. Ma, and J. Campbell, "HgCdTe electron avalanche photodiodes," *J. Electron. Mater.*, 33, 630–9, 2004.
188. I. Backer, S. Duncan, and J. Copley, "Low noise laser gated imging system for long range target identification," *Proc. SPIE*, 5406, 133–44, 2004.
189. I. Baker, P. Thorne, J. Henderson, J. Copley, D. Humphreys, and A. Millar, "Advanced multifunctional detectors for laser-gated imaging applications," *Proc. SPIE*, 6206, 620608, 2006.
190. G. Perrais, J. Rothman, G. Destefanis, J. Baylet, P. Castelein, J.-P. Chamonal, and P. Tribolet, "Demonstration of multifunctional bi-colour-avalanche gain detection in HgCdTe FPA," *Proc. SPIE*, 6395, 63950H, 2006.
191. J. Beck, M. Woodall, R. Scritchfield, M. Ohlson, L. Wood, P. Mitra, and J. Robinson, "Gated IR imaging with 128 × 128 HgCdTe electron avalanche photodiode FPA," *Proc. SPIE*, 6542, 654217, 2007.
192. I. Baker, D. Owton, K. Trundle, P. Thorne, K. Storie, P. Oakley, and J. Copley, "Advanced infrared detectors for multimode active and passive imaging applications," *Proc. SPIE*, 6940, 69402L, 2008.
193. J. Rothman, E. de Borniol, P. Ballet, L. Mollard, S. Gout, M. Fournier, J.P. Chamonal, et al., "HgCdTe APD: focal plane array performance at DEFIR," *Proc. SPIE*, 7298, 729835, 2009.
194. I. Baker, Ch. Maxey, L. Hipwood, and K. Barnes, "Leonardo (formerly Selex ES) infrared sensors for astronomy—present and future," *Proc. SPIE*, 9915, 991505, 2016.
195. J. Rothman, E. de Borniol, O. Gravrand, P. Kern, P. Feautrier, J.-B. Lebouquin, and O. Boulade, "MCT APD focal plane arrays for astronomy at CEA-LETI," *Proc. SPIE*, 9915, 99150B, 2016.
196. J.-L. Gacha, P. Feautriera, E. Stadlera, F. Clopa, S. Lemarchanda, T. Carmignania, Y. Wanwanscappela, C. Doucurea, and D. Boutolleaua, "Infrared Detectors for Wavefront Sensing," www.researchgate.net/publication/320880974_Infrared_detectors_for_wavefront_sensing.
197. R.J. Cashman, "Film-type infrared photoconductors," *Proc. IRE*, 47, 1471–75, 1959.
198. P.W. Kruse, L.D. McGlauchlin, and R.B. McQuistan, *Elements of Infrared Technology: Generation, Transmission, and Detection*, Wiley, New York, 1962.
199. A. Smith, F.E. Jones, and R.P. Chasmar, *The Detection and Measurement of Infrared Radiation*, Clarendon, Oxford, UK, 1968.
200. A. Rogalski, "IV-VI Detectors," in *Infrared Photon Detectors*, ed. E. Rogalski, 513–59, SPIE Optical Engineering Press, Bellingham, WA, 1995.
201. T. Beystrum, R. Himoto, N. Jacksen, and M. Sutton, "Low cost Pb salt FPAs," *Proc. SPIE*, 5406, 287–94, 2004.
202. G. Vergara, M.T. Montojo, M.C. Torquemada, M.T. Rodrigo, F.J. Sanchez, L.J. Gomez, R.M. Almazan, et al., "Polycrystalline lead selenide: The resurgence of an old infrared detector," *Opto-Electron. Rev.* 15, 110–17, 2007.
203. K. Green, S.-S. Yoo, and Ch. Kauffman, "Lead salt TE-cooled imaging sensor development," *Proc. SPIE*, 9070, 9070–1–7, 2014.
204. J.F. Kreider, M.K. Preis, P.C.T. Roberts, L.D. Owen, and W.M. Scott, "Multiplexed mid-wavelength IR long, linear photoconductive focal plane arrays," *Proc. SPIE*, 1488, 376–88, 1991.
205. Northrop/Grumman Electro-Optical Systems Data Sheet, 2002.
206. P.R. Norton, "Infrared image sensors," *Opt. Eng.*, 30, 1649–63, 1991.
207. M.D. Jhabvala and J.R. Barrett, "A monolithic lead sulfide-silicon MOS integrated-circuit structure," *IEEE Trans. Electron Devices*, 29, 1900–5, 1982.
208. J.R. Barrett, M.D. Jhabvala, and F.S. Maldari, "Monolithic lead salt-silicon focal plane development," *Proc. SPIE*, 409, 76–88, 1988.

209. H. Zogg, S. Blunier, T. Hoshino, C. Maissen, J. Masek, and A.N. Tiwari, "Infrared sensor arrays with 3–12 μm cutoff wavelengths in heteroepitaxial narrow-gap semiconductors on silicon substrates," *IEEE Trans. Electron Devices*, 38, 1110–17, 1991.
210. H. Zogg, A. Fach, C. Maissen, J. Masek, and S. Blunier, "Photovoltaic lead-chalcogenide on silicon infrared sensor arrays," *Opt. Eng.*, 33, 1440–9, 1994.
211. H. Zogg, A. Fach, J. John, P. Müller, C. Paglino, and A.N. Tiwari, "PbSnSe-on-Si: Material and IR-device properties," *Proc. SPIE*, 3182, 26–9, 1998.
212. H. Zogg, "Photovoltaic IV-VI on silicon infrared devices for thermal imaging applications," *Proc. SPIE*, 3629, 52–62, 1999.
213. K. Alchalabi, D. Zimin, H. Zogg, and W. Buttler, "Monolithic heteroepiraxial PbTe-on-Si infrared focal plane array with 96 × 128 pixels," *IEEE Electron Device Lett.*, 22, 110–2, 2001.
214. H. Zogg, K. Alchalabi, D. Zimin, and K. Kellermann, "Two-dimensional monolithic lead chalcogenide infrared sensor arrays on silicon read-out chips and noise mechanisms," *IEEE Trans. Electron Devices*, 50, 209–14, 2003.
215. A.J. Steckl, H. Elabd, K.Y. Tam, S.P. Sheu, and M.E. Motamedi, "The optical and detector properties of the PbS-Si heterojunction," *IEEE Trans. Electron Devices*, 27, 126–33, 1980.
216. D.R. Lamb and N.A. Foss, "The applications of charge-coupled devices to infra-red image sensing systems," *Radio Electron. Eng.*, 50, 226–36, 1980.
217. S.D. Gunapala and K.M.S.V. Bandara, "Recent Development in Quantum-Well Infrared Photodetectors,"in *Thin Films*, Vol. 21, eds. M.M. Francombe and J.L. Vossen, 113–237, Academic Press, New York, 1995.
218. C.C. Bethea, B.F. Levine, V.O. Shen, R.R. Abbott, and S.J. Hseih, "10 μm GaAs/AlGaAs multiquantum well scanned array infrared imaging camera," *IEEE Trans. Electron Devices*, 38, 1118–23, 1991.
219. B.F. Levine, "Quantum-well infrared photodetectors," *J. Appl. Phys.*, 74, R1–81, 1993.
220. S.D. Gunapala and S.V. Bandara, "Quantum Well Infrared Photodetectors (QWIP)," in *Handbook of Thin Devices*, Vol. 2, ed. M.H. Francombe, 63–99, Academic Press, San Diego, CA, 2000.
221. S.D. Gunapala and S.V. Bandara, "GaAs/AlGaAs Based Quantum Well Infrared Photodetector Focal Plane Arrays," in *Handbook of Infrared Detection Technologies*, eds. M. Henini and M. Razeghi, 83–119, Elsevier, Oxford, UK, 2002.
222. E. Costard, Ph. Bois, F. Audier, and E. Herniou, "Latest improvements in QWIP technology at Thomson-CSF/LCR," *Proc. SPIE*, 3436, 228–39, 1998.
223. J.A. Robo, E. Costard, J.P. Truffer, A. Nedelcu, X. Marcadet, and P. Bois, "QWIP focal plane arrays performances from MWIR to VLWIR," *Proc. SPIE*, 7298, 72980F, 2009.
224. M. Runtz, F. Perrier, N. Ricard, E. Costard, A. Nedelcu, V. Guériaux, "QWIP infrared detector production line results," *Proc. SPIE*, 8353, 835339-1-12, 2012.
225. H. Schneider, P. Koidl, M. Walther, J. Fleissner, R. Rehm, E. Diwo, K. Schwarz, and G. Weimann, "Ten years of QWIP development at Fraunhofer," *Infrared Phys. Technol.*, 42, 283–89, 2001.
226. H. Schneider and H.C. Liu, *Quantum Well Infrared Photodetectors*, Springer, Berlin, 2007.
227. H. Martijn, S. Smuk, C. Asplund, H. Malm, A. Gromov, J. Alverbro, and H. Bleichner, "Recent advances of QWIP development in Sweden," *Proc. SPIE*, 6542, 6542OV, 2007.
228. H. Martijn, C. Asplund, H. Malm, S. Smuk, L. Höglund, O. Gustafsson, M. Hammar, and S. Hellström, "Development of IR imaging at IRnova," *Proc. SPIE*, 7298, 72980E-1-11, 2009.
229. K.K. Choi, *The Physics of Quantum Well Infrared Photodetectors*, Word Scientific, Singapore, 1997.
230. K-K. Choi, C. Monroy, V. Swaminathan, T. Tamir, M. Leung, J. Devitt, D. Forrai, and D. Endres, "Optimization of corrugated-QWIP for large format, high quantum efficiency, and multi-color FPAs," *Infrared Phys. Technol.*, 50, 124–35, 2007.
231. K.K. Choi, M.D. Jhabvala, D.P. Forrai, A. Waczynski, J. Sun, R. Jones, "Electromagnetic design of resonator-QWIPs," *Proc. SPIE*, 8268, 82682O-1-9, 2012.
232. E.A. DeCuir Jr., K.-K. Choi, J. Sun, P.S. Wijewarnasuriya, "Progress in resonator quantum well infrared photodetector (R-QWIP) focal plane arrays," *Infrared Phys. Technol.*, 70, 138–46, 2015.
233. W.A. Beck and T.S. Faska, "Current status of quantum well focal plane arrays," *Proc. SPIE*, 2744, 193–206, 1996.
234. T. Whitaker, "Sanders' QWIPs detect two color at once," *Compd. Semicond.*, 5(7), 48–51, 1999.
235. M. Sundaram and S.C. Wang, "2-Color QWIP FPAs," *Proc. SPIE*, 4028, 311–7, 2000.
236. J. Jiang, S. Tsao, K. Mi, M. Razeghi, G.J. Brown, C. Jelen, and M.Z. Tidrow, "Advanced monolithic quantum well infrared photodetector focal plane array integrated with silicon readout integrated circuit," *Infared Phys. Technol.*, 46, 199–207, 2005.

237. S. Ozer, U. Tumkaya, and C. Besikci, "Large format AlInAs-InGaAs quantum-well infrared photodetector focal plane array for midwavelength infrared thermal imaging," *IEEE Photonics Technol. Lett.*, 19, 1371–3, 2007.
238. M. Kaldirim, Y. Arslan, S.U. Eker, and C. Besikci, "Lattice-matched AlInAs-InGaAs mid-wavelength infrared QWIPs: characteristics and focal plane array performance," *Semicond. Sci. Technol.*, 23, 085007, 2008.
239. H. Schneider, J. Fleissner, R. Rehm, M. Walther, W. Pletschen, P. Koidl, G. Weimann, J. Ziegler, R. Breiter, and W. Cabanski, "High-resolution QWIP FPAs for the 8–12 μm and 3–5 μm regimes," *Proc. SPIE*, 4820, 297–305, 2003.
240. E. Costard and Ph. Bois, "THALES long wave QWIP thermal imagers," *Infared Phys. Technol.*, 50, 260–69, 2007.
241. M. Jhabvala, K. Choi, A. Goldberg, A. La, and S. Gunapala, "Development of a 1k × 1k GaAs QWIP far IR imaging array," *Proc. SPIE*, 5167, 175–85, 2004.
242. S.D. Gunapala, S.V. Bandara, J.K. Liu, C.J. Hill, B. Rafol, J.M. Mumolo, J.T. Trinh, M.Z. Tidrow, and P.D. LeVan, "1024 × 1024 pixel mid-wavelength and long-wavelength infrared QWIP focal plane arrays for imaging applications," *Semicond. Sci. Technol.*, 20, 473–80, 2005.
243. M. Jhabvala, K.K. Choi, C. Monroy, and A. La, "Development of a 1K × 1K, 8–12 μm QWIP array," *Infared Phys. Technol.*, 50, 234–39, 2007.
244. S.D. Gunapala, S.V. Bandara, J.K. Liu, J.M. Mumolo, C.J. Hill, S.B. Rafol, D. Salazar, J. Woollaway, P.D. LeVan, and M.Z. Tidrow, "Towards dualband megapixel QWIP focal plane arrays," *Infared Phys. Technol.*, 50, 217–26, 2007.
245. A. Adams and E. Rittenberg, "HOT IR sensors improve IR camera size, weight, and power," *Laser Focus World*, January 2014, 83–87.
246. J. Caulfield, J. Curzan, J. Lewis, and N. Dhar, "Small pixel oversampled IR focal plane arrays," *Proc. SPIE*, 9451, 94512F–1–9, 2015.
247. H. Sharifi, M. Roebuck, S. Terterian, J. Jenkins, B. Tu, W. Strong, T.J. De Lyon, and R.D. Rajavel, J. Caulfield, and J.P. Curzan, "Advances in III-V bulk and superlattice-based high operating temperature MWIR detector technology," *Proc. SPIE*, 10177, 101770U–1–6, 2017.
248. M.Z. Tidrow, L. Zheng, and H. Barcikowski, "Recent success on SLS FPAs and MDA's new direction for development," *Proc. SPIE*, 7298, 72981O-1–11, 2009.
249. D.L. Smith and C. Mailhiot, "Proposal for strained type II superlattice infrared detectors," *J. Appl. Phys.*, 62, 2545–8, 1987.
250. W. Cabanski, K. Eberhardt, W. Rode, J. Wendler, J. Ziegler, J. Fleißner, F. Fuchs, R. Rehm, J. Schmitz, H. Schneider, and M. Walther, "3rd gen focal plane array IR detection modules and applications," *Proc. SPIE*, 5406, 184–92, 2005.
251. C.J. Hill, A. Soibel, S.A. Keo, J.M. Mumolo, D.Z. Ting, S.D. Gunapala, D.R. Rhiger, R.E. Kvaas, and S.F. Harris, "Demonstration of mid and long-wavelength infrared antimonide-based focal plane arrays," *Proc. SPIE*, 7298, 7298–04-1–9, 2009.
252. S.D. Gunapala, D.Z. Ting, C.J. Hill, J. Nguyen, A. Soibel, S.B. Rafol, S.A. Keo, J.M. Mumolo, M.C. Lee, J.K. Liu, and B. Yang, "Demonstration of a 1024 × 1024 pixel InAs-GaSb superlattice focal plane array," *Photonics Technol. Lett.*, 22, 1856–8, 2010.
253. M. Razeghi, H. Haddadi, A.M. Hoang, E.K. Huang, G. Chen, S. Bogdanov, S.R. Darvish, F. Callewaert, and R. McClintock, "Advances in antimonide-based type-II superlattices for infrared detection and imaging at Center for Quantum Devices," *Infrared Phys. Technol.*, 59, 41–52, 2013.
254. M. Razeghi, H. Haddadi, A.M. Hoang, E.K. Huang, G. Chen, S. Bogdanov, S.R. Darvish, F. Callewaert, P.R. Bijjam, and R. McClintock, "Antomonide-based type-II superlattices: A superior candidate for the third generation of infrared imaging systems," *J. Electron. Mater.*, 43, 2802–7, 2014.
255. P. Manurkar, S. Ramezani-Darvish, B.-M. Nguyen, M. Razeghi, and J. Hubbs, "High performance long wavelength infrared mega-pixel focal plane array based on type-II superlattices," *Appl. Phys. Lett.*, 97, 193505–1–3, 2010.
256. M. Razeghi and B.-M. Nguyen, "Advances in mid-infrared detection and imaging: a key issues review," *Rep. Prog. Phys.*, 77, 082401–1–17, 2014.
257. P.C. Klipstein, E. Avnon, D. Azulai, Y. Benny, R. Fraenkel, A. Glozman, E. Hojman, O. Klin, L. Krasovitsky, L. Langof, I. Lukomsky, M. Nitzani, I. Shtrichman, N. Rappaport, N. Snapi, E. Weiss and A. Tuito, "Type II superlattice technology for LWIR detectors," *Proc. SPIE*, 9819, 9819–20, 2016.
258. R. Rehm, V. Daumer, T. Hugger, N. Kohn, W. Luppold, R. Müller, J. Niemasz, J. Schmidt, F. Rutz, T. Stadelmann, M. Wauro, and A. Wörl, "Type-II superlattice infrared detector technology at Fraunhofer IAF," *Proc. SPIE*, 9819, 9819–24, 2016.

259. A. Rogalski, "Next decade in infrared detectors," *Proc. SPIE*, 10333, 104330L–1–25, 2017.
260. M.A. Kinch, *Fundamentals of Infrared Detector Materials*, SPIE Press, Bellingham, WA, 2007.
261. A.A. Kinch, *State-of-the-Art Infrared Detector Technology*, SPIE Press, Bellingham, WA, 2014.
262. M.A. Kinch, "An infrared journey," *Proc. SPIE*, 9451, 94512B, 2015.
263. D. Lee, M. Carmody, E. Piquette, P. Dreiske, A. Chen, A. Yulius, D. Edwall, S. Bhargava, M. Zandian, and W.E. Tennant, "High-operating temperature HgCdTe: A vision for the near future," *J. Electron. Mater.*, 45(9), 4587–95, 2016.
264. C. McMurtry, D. Lee, J. Beletic, A. Chen, R. Demers, M. Dorn, D. Edwall, C.B. Fazar, W. Forrest, F. Liu, A. Mainzer, J. Pipher, and A. Yulius, "Development of sensitive long-wave infrared detector arrays for passively cooled space missions," *Opt. Eng.*, 52(9), 091804–1–9, 2014.
265. A. Rogalski, M. Kopytko, and P. Martyniuk, "Performance prediction of p-i-n HgCdTe long wavelength infrared HOT photodiodes," *Applied Optics* 57(18), D11–D19, 2018.
266. D.Z. Ting, A. Soibel, A. Khoshakhlagh, L. Höglund, S.A. Keo, B. Rafol, C.J. Hill, A.M. Fisher, E.M. Luong, J. Nguyen, J.K. Liu, J.M. Mumolo, B.J. Pepper, and S.D. Gunapala, "Antimonide type-II superlattice barrier infrared detectors," *Proc. SPIE*, 10177, 101770N, 2017.
267. P.-Y. Delaunay, B.Z. Nosho, A.R. Gurga, S. Terterian, and R.D. Rajavel, "Advances in III-V based dual-band MWIR/LWIR FPAs at HRL," *Proc. SPIE*, 101777, 101770T–1–12, 2017.

27 Third-generation infrared detectors

Multicolor detector capabilities are highly desirable for advanced infrared (IR) imaging systems since they provide enhanced target discrimination and identification, combined with lower false-alarm rates. Systems that collect data in separate IR spectral bands can discriminate both absolute temperature as well as unique signatures of objects in the scene. By providing this new dimension of contrast, multiband detection also offers advanced color-processing algorithms to further improve sensitivity above that of single-color devices. This is extremely important for identifying temperature differences between missile targets, warheads, and decoys. Multispectral IR focal plane arrays (FPAs) are highly beneficial for a variety of applications such as missile warning and guidance, precision strike, airborne surveillance, target detection, recognition, acquisition and tracking, thermal imaging, navigational aids and night vision, and so on [1,2]. They also play an important role in Earth and planetary remote sensing, astronomy, and so forth [3]. Single-color FPAs in conjunction with spectral filters, grating spectrometers of Fourier transform spectrometers have been deployed for a variety of National Aeronautics and Space Administration (NASA) spaceborne remote-sensing applications utilizing push-broom scanning to record hyperspectral images of the earth over the visible through very-long-wavelength IR (VLWIR) spectral range.

Military surveillance, target detection, and target tracking can be undertaken using single-color FPAs if the targets are easy to identify. However, in the presence of clutter, or when the target and/or background are uncertain, or in situations where the target and/or background may change during engagement, the single-color system design involves compromises that can degrade overall capability. It is well established that in order to reduce clutter and enhance the desired features/contrast, one will require the use of multispectral FPAs. In such cases, multicolor imaging can greatly improve overall system performance.

Currently, multispectral systems rely on cumbersome imaging techniques that either disperse the optical signal across multiple IR FPAs or use a filter wheel to spectrally discriminate the image focused on a single FPA. These systems include beam-splitters, lenses, and band-pass filters in the optical path to focus the images onto separate FPAs responding to different IR bands. Also, complex optical alignment is required to map the multispectral image pixel-for-pixel. Consequently, these approaches are of relatively high cost and place additional burdens on the sensor platform because of their extensive size, complexity, and cooling requirements. The concept of a multispectral FPA with all detector arrays mounted on a single focal plane is shown in Figure 27.1. To detect broadband radiation ranging from visible to long-wavelength IR (LWIR), different detector materials are used for the various wavelengths. In the push broom imagers, linear arrays provide spatial resolution in the cross-track direction while the motion of the imaging system relative to the ground produces a scanning operation in the along-track direction producing two-dimensional (2D) images of the scene in multiple bands.

Dispersive devices based on mechanical scanning (e.g., filter wheels, monochromators) are not desirable because, in addition to their relatively large size, they are prone to vibrations and can be spectrally tuned in a relatively narrow range of a relatively slow speed [5]. Recent advances in material, electronic, and optical technologies have led to the development of novel types of electronically tunable filters, including so-called adaptive FPAs [6]. In the future, multispectral imaging systems will include very large sensors feeding an enormous amount of data to the digital mission processing subsystem. The FPAs with the number of pixels above one million are now available. As these imaging arrays grow in detector number for higher resolution, so will the computing requirements for the embedded digital image-processing system. One approach to solving this processing bottleneck problem could be to incorporate a certain amount of pixel-level processing within the detector pixel, similar to the technique implemented in biological

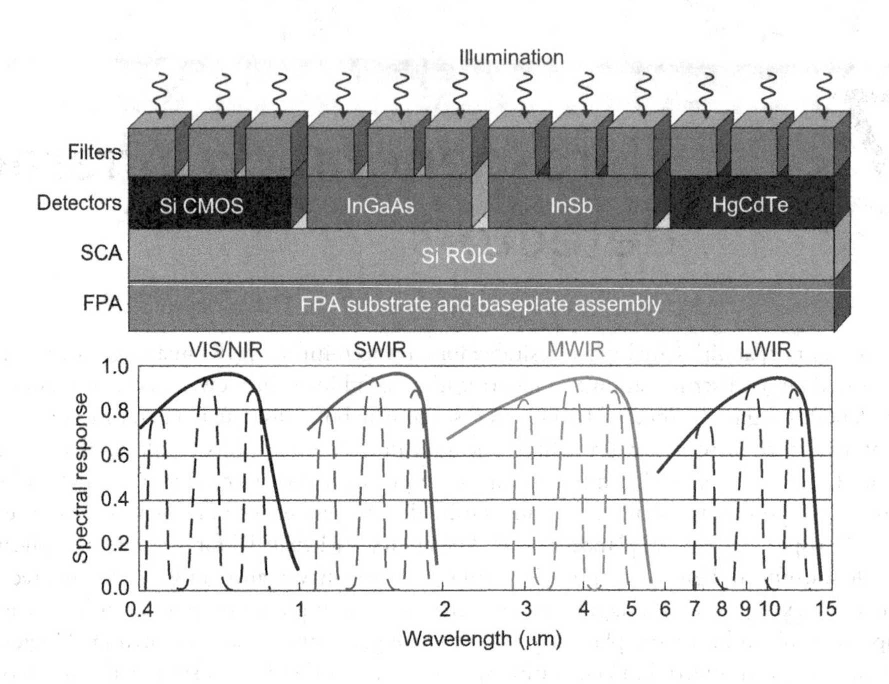

Figure 27.1 The concept of a multispectral FPA with all detector arrays mounted on a single focal plane and designed to operate at the same temperature. CMOS, complementary metal-oxide-semiconductor; SCA, sensor chip assembly; VIS, visible. (Adapted after Ref. [4])

sensor information-processing systems. Currently, several scientific groups in the world have turned to the biological retina for answers as to how to improve man-made sensors [7,8].

In this chapter, we will review the state-of-the-art multicolor detector technologies over a wide IR spectral range. In the wavelength regions of interest, such as short-wavelength IR (SWIR), medium-wavelength IR (MWIR), and LWIR, three detector technologies that are developing multicolor capability are visited here: HgCdTe, quantum well IR photodetectors (QWIPs), and antimonide-based type-II superlattices. Also quantum dot IR photodetectors (QDIPs) have been demonstrated as potential materials for multicolor detection.

Both HgCdTe photodiodes [9–22] and QWIPs [23–34] offer multicolor capability in the SWIR, MWIR, and LWIR range. The performance figures of merit of state-of-the-art QWIP and HgCdTe FPAs are similar because the main limitations are related to the readout circuits. A more detailed comparison of both technologies has been given by Tidrow et al. [2] and Rogalski [25,35].

In the last decade also, type II InAs/GaSb superlattices [36–49] and QDIPs [50–55] have emerged as possible candidates for third-generation IR detectors. Table 27.1 compares the essential properties of three types of LWIR devices at 77 K. Whether the low-dimensional solid IR photodetectors can outperform the "bulk" narrow-gap HgCdTe detectors is one of the most important questions that needs to be addressed for the future of IR photodetectors.

The subsections below describe issues associated with the development and exploitation of materials used in the fabrication of multicolor IR detectors. Finally, we discuss the on-going detector technology efforts being undertaken to realize third-generation FPAs.

27.1 REQUIREMENTS OF THIRD-GENERATION DETECTORS

The standard method to detect multiwavelength simultaneously is to use optical components such as lenses, prisms, and gratings to separate the wavelength components before they impinge on the IR detectors. Another simpler method is a stacked arrangement in which the shorter wavelength detector is placed optically ahead of the longer wavelength detector. In such a way, two-color detectors using HgCdTe [56]

Table 27.1 Essential properties of LWIR HgCdTe and type II SL photodiodes, and QWIPs at 77 K

PARAMETER	HgCdTe	QWIP (n-type)	InAs/GaSb SL
IR absorption	Normal incidence	$E_{optical} \perp$ plane of well required Normal incidence: no absorption	Normal incidence
Quantum efficiency (%)	≥70	≤10	≈50–60
Spectral sensitivity	Wide-band	Narrow-band (FWHM ≈ 1÷2 μm)	Wide-band
Optical gain	1	0.2 (30–50 wells)	1
Thermal generation lifetime	≈1 μs	≈10 ps	≈0.1 μs
R_0A product (Ωcm^2) ($\lambda_c = 10\,\mu$m)	10^3	10^4	500
Detectivity (cmHz$^{1/2}$ W^{-1}) ($\lambda_c = 10\,\mu$m, FOV = 0)	2×10^{12}	2×10^{10}	1×10^{12}

FWHM, full-width at half maximum.

and InSb/HgCdTe [57] photoconductors have been demonstrated in the early 1970s. At present, however, considerable efforts are directed to fabricating a single FPA with multicolor capability to eliminate the spatial alignment and temporal registration problems that exist whenever separate arrays are used; to simplify optical design; and to reduce size, weight, and power consumption.

In the 1990s (see Figure 3.1), third-generation IR detectors emerged after the tremendous impetus provided by detector developments. The definition of third-generation IR systems is not particularly well established. In the common understanding, third-generation IR systems provide enhanced capabilities such as larger number of pixels, higher frame rates, better thermal resolution, as well as multicolor functionality and other on-chip signal processing functions. According to Reago et al. [58], the third generation is defined by the requirement to maintain the current advantage enjoyed by the United States and allied armed forces. This class of devices includes both cooled and uncooled FPAs [1,58]:

- high-performance, high-resolution cooled imagers having multicolor bands
- medium- to high-performance uncooled imagers
- very low cost, expendable uncooled imagers.

When developing third-generation imagers, the IR community is faced with many challenges. Some of them are described in two sections of this:

- noise-equivalent temperature difference (NEDT)—see Section 24.5 and
- pixel and chip size issues—see Section 24.6,

and other

- uniformity and
- identification and detection ranges

are shortly considered here.

Current readout technology is based upon complementary metal-oxide-semiconductor (CMOS) circuitry that has benefited from dramatic and continuing progress in miniaturizing circuit dimensions. Second-generation imagers provide NEDT of about 20 mK with *f*/2 optics. A goal of third-generation imagers is to achieve sensitivity improvement corresponding to NEDT of about 1 mK. From Equation 24.19, it can be determined that in a 300-K scene in the LWIR region with thermal contrast of 0.02, the required charge storage capacity is above 10^9 electrons. This high charge-storage density cannot be obtained within the small pixel dimensions using standard CMOS capacitors [1]. Although the reduced oxide thickness of submicrometer CMOS design rules gives large capacitance per unit area, the reduced bias voltage, as illustrated in Figure 27.2, largely cancels any improvement in charge storage density. Ferroelectric capacitors may provide much greater charge storage densities than the oxide-on-silicon capacitors now used. However, such a technology is not yet incorporated into standard CMOS foundries.

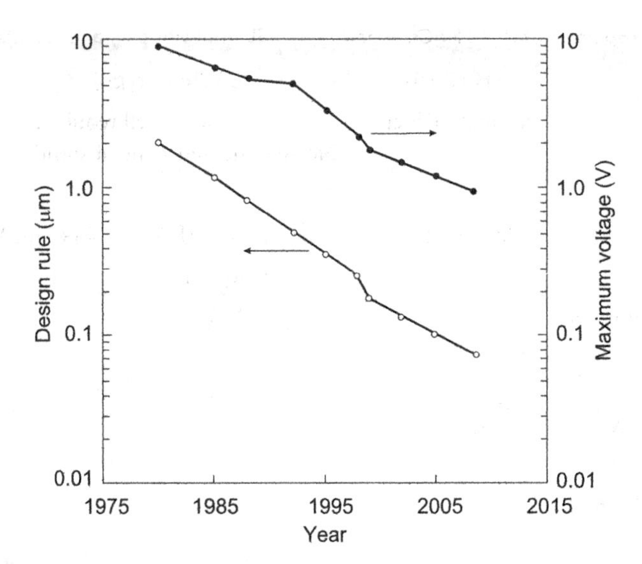

Figure 27.2 Trends for design rule minimum and maximum bias voltage of silicon foundry requirements. (From Ref. [1]. With permission.)

To provide an opportunity to significantly increase the charge storage capacity and the dynamic range, the vertically integrated sensor array (VISA) program has been sponsored by the DARPA (Defense Advanced Research Projects Agency) [59–61]. The approach being developed builds on the traditional "hybrid" structure of a detector with a 2D array of indium-bump interconnects to the silicon readout. The VISA program allows additional layers of silicon processing chips to be connected below the readout to provide more complex functionality. It will allow the use of smaller and multicolor detectors without compromising storage capacity. Signal-to-noise ratios will increase for multicolor FPAs. This will permit LWIR FPAs to improve the sensitivity by a factor of 10.

Pixel and chip sizes are important issues in association with multicolor imager formats. Small pixels reduce cost by increasing the number of readout and detector dice potentially available from processed wafers. Small pixels also allow smaller, lightweight optics to be used—see Section 24.6.

Thermal imaging systems are used first to detect an object and then to identify it. In military circles, "identification" (I) is used along with "detection" (D) and "recognition" (R) as part of the DRI criteria established by John Johnson in 1958 [62]. This standard was established to define the performance of thermal imaging cameras. According to Johnson's criteria:

- detection is defined as the ability to distinguish an object from the background,
- recognition—as the ability to classify the object class (animal, human, vehicle, boat, ...),
- identification—as the ability to describe the object in details (a man with a hat, a deer, a Jeep, ...).

The nominal range performance of IR camera is calculated for a defined task and standardized target and environmental conditions. The only standardization available to date is STANAG 4347 [63].

Figure 27.3 is a comparison of various FPAs in detecting, recognizing, and identifying a man-size target for a canonical tactical sensor (*f*/3, 454 mm focal length, 152 mm aperture operating at 60 Hz) [64]. These ranges were calculated using the NVTherm model. It should be mentioned that estimations for type-II superlattice FPAs indicate the same range performance as the HgCdTe ternary alloy.

Typically, identification ranges are between two and three times shorter than detection ranges [24]. To increase ranges, better resolution and sensitivity of the IR systems (and hence the detectors) are required. Third-generation cooled imagers are being developed to extend the range of target detection and identification and to ensure that defense forces maintain a technological advantage in night operations over any opposing force.

Identification ranges can be further increased by using multispectral detection to correlate the images at different wavelengths. It appears that in the MWIR spectral range, the IR image is washed out to the point

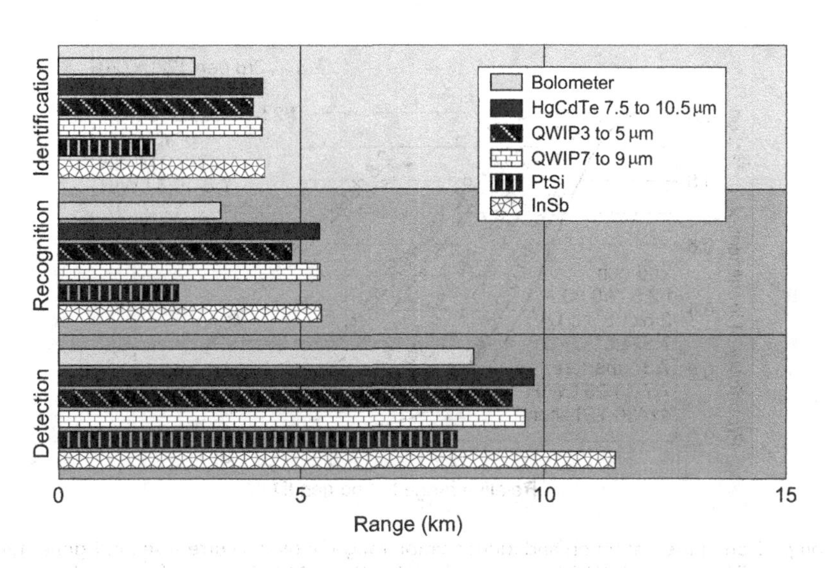

Figure 27.3 Comparison of DRI ranges for man-sized target assuming atmospheric parameters of a mid-latitude summer and rural 23 km visibility. (Adapted after Ref. [64])

that the target and the background cannot be distinguished from each other (see Figure 27.4 [24]). Detectors that cover the entire spectral range will suffer from washout because the background contrast changes from positive to negative. Alternatively, using two band detectors (up to 3.8 μm and from 3.8 up to 5 μm) and summing the inverse of the second band and the output of the first band, it will yield a contrast enhancement that is impossible to achieve if an integrated response of the entire spectral range is used.

Figure 27.5 compares the relative detection and identification ranges modeled for third-generation imagers using NVESD's (Fort Belvoir, Virginia) NVTherm program. As a range criterion, the standard 70% probability of detection or identification is assumed. Note that the identification range in the MWIR range is almost 70% of the LWIR detection range. For detection, LWIR provides superior range. In the detection mode, the optical system provides a wide field of view (WFOV, *f*/2.5) since third-generation systems will operate as an on-the-move wide area step-scanner with automated target recognition (second-generation systems relay on manual target searching) [65]. MWIR offers higher spatial resolution sensing and has an advantage for long-range identification when used with telephoto optics (narrow field of view (NFOV), *f*/6).

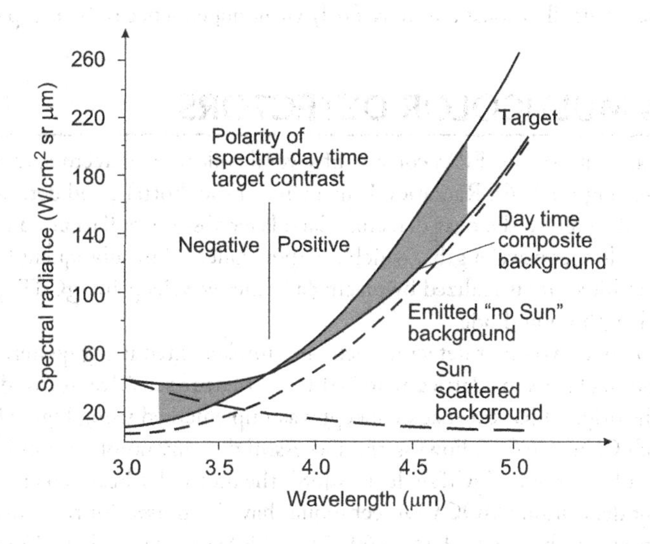

Figure 27.4 Target and background contrast reversal in the MWIR spectral range. (From Ref. [24]. With permission.)

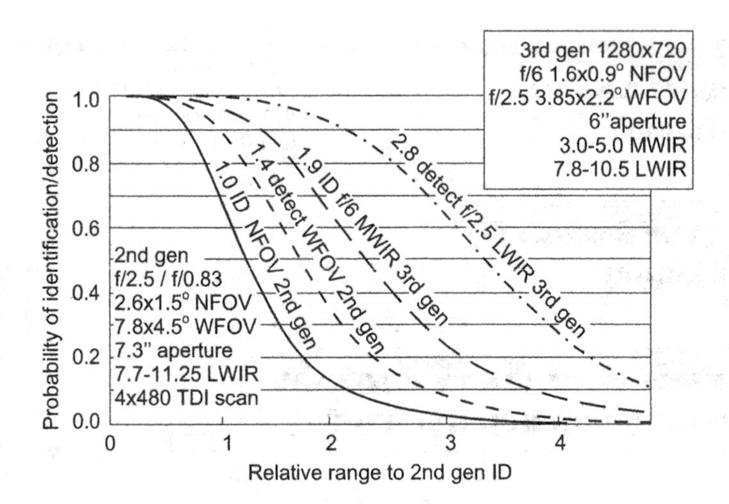

Figure 27.5 Comparison of the detection and identification range between current second-generation (gen) time delay and integration (TDI) scanned LWIR imagers and the LWIR and MWIR bands of the third-generation imager in a 1,280 × 720 format with 20-μm pixels. ID, identification. (From Ref. [65]. With permission.)

In Section 24.5, it is marked that IR images can be severely degraded by fixed pattern noise which is caused by the nonuniformity of the detector response. Figure 17.15 shows the uncertainty in cutoff wavelength of $Hg_{1-x}Cd_xTe$ for x variations of 0.1%. It is shown that the serious changes in cutoff wavelength are observed in the VLWIR region. The variation of x across the $Hg_{1-x}Cd_xTe$ wafer causes much larger spectral nonuniformity. At 77 K, a variation of $\Delta x = 0.1\%$ gives a $\Delta\lambda_c$ above 0.5 μm at $\lambda_c = 20$ μm, which cannot be corrected by either two- or three-point corrections [2]. This cutoff wavelength nonuniformity at the FPA level can be spectrally corrected by using a cold filter, but the dark current variation caused by the variation of cutoff wavelengths will still exist. For applications that require operation in the LWIR band as well as two-color LWIR/VLWIR bands, most probably, HgCdTe will not be the optimal solution.

Recently, the first megapixel format HgCdTe FPAs with pixel dimensions as small as 5 μm have been demonstrated (see Section 26.5.2). It will be an extreme challenge to deploy a two- or three-color detector structure into a small pixel such as $5 \times 5\,\mu m^2$. Current two-color simultaneous mode pixels with two indium bumps per pixel have not been built with pixels smaller than 20 μm on a side.

An alternative candidate for third-generation IR detectors are the Sb-based III-V material system. These materials are mechanically robust and have fairly weak dependence of bandgap on composition (see Figure 16.55).

27.2 HgCdTe MULTICOLOR DETECTORS

The unit cell of integrated multicolor FPAs consists of several collocated detectors, each sensitive to a different spectral band (see Figure 27.6). Radiation is incident on the shorter band detector, with the longer wave radiation passing through to the next detector. Each layer absorbs radiation up to its cutoff and hence transparent to the longer wavelengths, which are then collected in subsequent layers. In the case of HgCdTe, this device architecture is realized by placing a longer wavelength HgCdTe photodiode optically behind a shorter wavelength photodiode.

Back-to-back photodiode two-color detectors were first implemented using quaternary III-V alloy ($Ga_xIn_{1-x}As_yP_{1-y}$) absorbing layers in a lattice-matched InP structure sensitive to two different SWIR bands [66]. A variation on the original back-to-back concept was implemented using HgCdTe at Rockwell and Santa Barbara Research Center [67]. Following the successful demonstration of multispectral detectors in liquid phase epitaxial (LPE)-grown HgCdTe devices [68], the molecular-beam epitaxy (MBE) and metal-organic chemical vapor deposition (MOCVD) techniques have been used for the growth of a variety of multispectral detectors at Raytheon [10,11,14,19,69–72], BAE Systems [73], Leti [15,16,20,22,74–77],

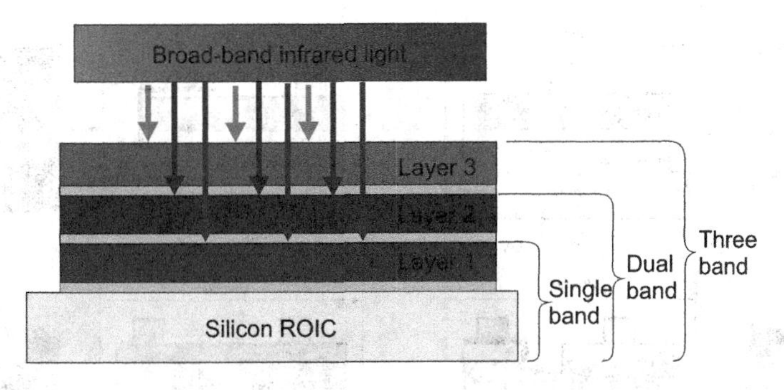

Figure 27.6 Structure of a three-color detector pixel. IR flux from the first band is absorbed in Layer 3, while LW flux is transmitted through the next layers. The thin barriers separate the absorbing bands.

Selex and QinetiQ [17,18,78–81], DRS [13,82–84], Teledyne and NVESD [85,86], and AIM [21]. For more than a decade, steady progression has been made in a wide variety of pixel sizes (to as small as 20 μm), array formats (up to 1,280 × 720), and spectral-band sensitivity (MWIR/MWIR, MWIR/LWIR, and LWIR/LWIR).

27.2.1 DUAL-BAND HgCdTe DETECTORS

Both sequential mode and simultaneous mode detectors are fabricated from multilayer materials. The simplest two-color HgCdTe detector and the first to be demonstrated was the bias selectable n-P-N triple-layer heterojunction (TLHJ), back-to-back photodiode shown in Figure 27.7a (capital letter means wider bandgap structure). The n-type base-absorbing regions are deliberately doped with indium at a level of about (1–3) × 10^{15} cm^{-3}. A critical step in device formation is ensuring that the in situ p-type As-doped layer (typically 1–2 μm thick) has good structural and electrical properties to prevent internal gain from generating spectral crosstalk. The bandgap engineering effort consists of increasing the CdTe mole fraction and the effective thickness of the p-type layer to keep out-of-band carriers from being collected at the terminal.

The sequential-mode detector has a single indium bump per unit cell that permits sequential bias selectivity of the spectral bands associated with operating back-to-back photodiodes. When the polarity of the bias voltage applied to the bump contact is positive, the top (long-wavelength (LW)) photodiode is reverse biased and the bottom (short-wavelength (SW)) photodiode is forward biased. The SW photocurrent is shunted by the low impedance of the forward-biased SW photodiode, and the only photocurrent to emerge in the external circuit is the LW photocurrent. When the bias voltage polarity is reversed, the situation reverses; only SW photocurrent is available. Switching times within the detector can be relatively short, on the order of microseconds, so detection of slowly changing targets or images can be achieved by switching rapidly between the MW and LW modes. The problems with the bias-selectable device are the following: its construction does not allow independent selection of the optimum bias voltage for each photodiode, and there can be substantial MW crosstalk in the LW detector.

Multicolor detectors require deep isolation trenches to cut completely through the relatively thick (at least 10 μm) LWIR absorbing layer. The design of small two-color TLHJ detectors of less than 20-μm pitch requires at least 15-μm deep trenches, which are no more than 5-μm wide at the top. Dry etching technology has been used for a number of years to produce two-color detectors. One of the materials technology being developed in order to meet the challenge of shrinking the pixel size to below 20 μm is advanced etching technology. Recently, Raytheon has developed an inductively coupled plasma (ICP) dry mesa etching capability to replace electron cyclotron resonance (ECR) dry mesa etching. The ICP, when compared to ECR, has shown reduced lateral mask erosion during etching, less significant etch-lag effects,

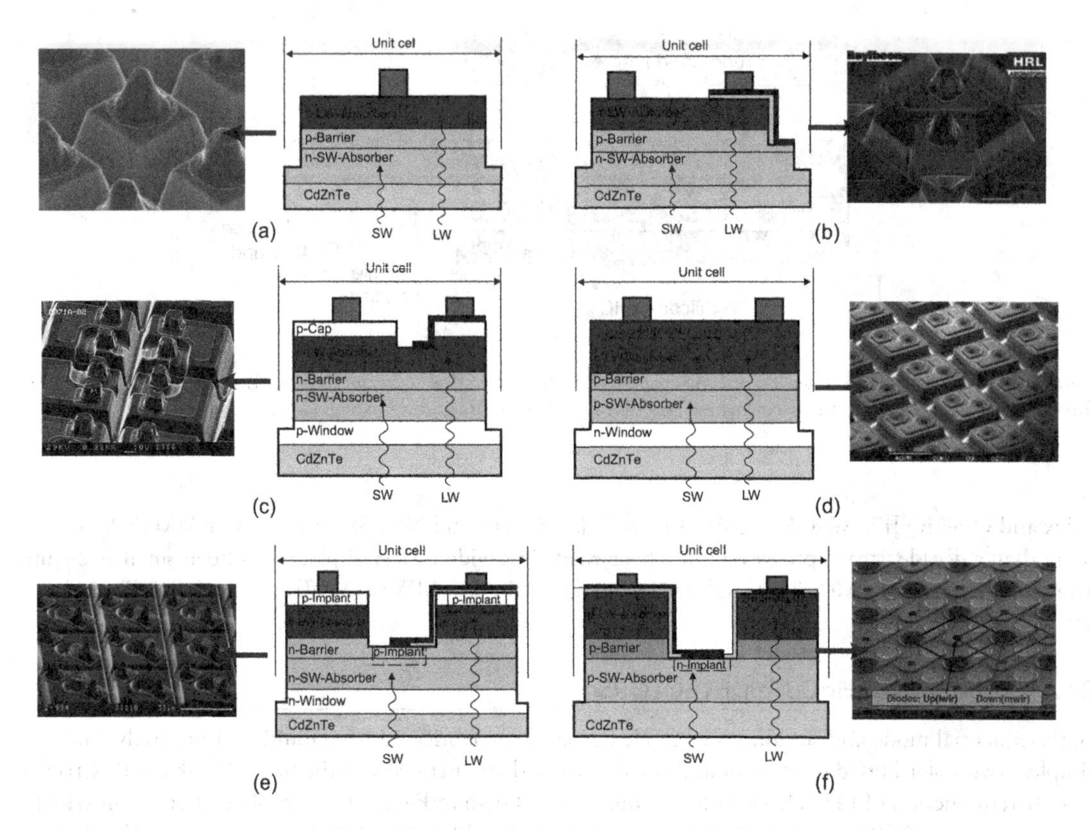

Figure 27.7 Cross-section views of unit cells for various back-illuminated dual-band HgCdTe detector approaches: (a) bias-selectable n-p-n structure reported by Raytheon (Ref. [68]. With permission.), (b) simultaneous n-p-n design reported by Raytheon (Ref. [69]. With permission.), (c) simultaneous p-n-n-p reported by BAE Systems (Ref. [73]. With permission.), (d) simultaneous n-p-p-p-n design reported by Leti (Ref. [74]. With permission.), (e) simultaneous structure based on p-on-n junctions reported by Rockwell (Ref. [85]. With permission.), and (f) simultaneous structure based on n-on-p junctions reported by Leti (Ref. [16]. With permission.)

and improved etch depth uniformity [85]. For the pseudoplanar devices, the etching step is easier to perform because of the lower aspect ratio. Moreover, there is no electrical crosstalk as the pixels are electrically independent.

Many applications require true simultaneous detection in the two spectral bands. This has been achieved in a number of ingenious architectures shown in Figure 27.7b–f. Two different architectures are shown. The first one is the classical n-P-N back-to-back photodiode structure (Figure 27.7b). In the case of the architecture developed at Leti (Figure 27.7d), the two absorption materials are p-type separated by a barrier to prevent any carrier drift between the two n-on-p diodes. Each pixel consists of two standard n-on-p photodiodes, where the p-type layers are usually doped with Hg vacancies. The shorter wavelength diode is realized during epitaxy by simply doping part of the first absorbing layer with In. The longer wavelength junction is obtained by a planar implantation process. It should be noted that the electron mobility is around 100 times greater in n-type material than in holes in p-type material and, hence, the n-on-p structures will have a much lower common resistance. This is an important consideration for large area FPAs with detection in the LW range due to the larger incident-photon flux.

The last two architectures shown in Figure 27.7e,f, called pseudoplanar, presents a totally different approach. They are close to the structure proposed by Lockwood et al. [87] in 1976 for PbTe/PbSnTe heterostructure two-color photodiodes. They are based on the concept of two p-on-n (Figure 27.7e) or n-on-p (Figure 27.7f) diodes fabricated by p-type or n-type implantation, respectively, but on two different levels of a three-layer heterostructure. The architecture developed by Rockwell is a simultaneous two-color MWIR/LWIR FPA technology based on a double-layer planar heterostructure MBE technology (Figure 27.7e). To prevent the diffusion of carriers between two bands, a wide bandgap 1-μm-thick layer separates these two absorbing layers. The diodes are formed by implanting arsenic as a p-type dopant and activating it with an anneal. This results in a unipolar operation for both bands. The implanted area of Band 2 is a concentric ring around the Band 1 dimple. Because the lateral carrier-diffusion length is larger than the pixel pitch in the MWIR material, and the Band 1 junction is shallow, each pixel is isolated by dry etching a trench around it to reduce carrier crosstalk. The entire structure is capped with a layer of material with a slightly wider bandgap to reduce surface recombination and simplify passivation.

All these simultaneous dual-band detector architectures require an additional electrical contact from an underlying layer in the multijunctional structure to both the SW and the LW photodiodes. The most important distinction is the requirement of a second readout circuit in each unit cell.

It is expected that with the TLHJ architecture, pixel size could decrease to 10 μm and array format could increase to several megapixels. With the pseudoplanar architecture, MWIR/LWIR devices should be produced more easily, with large format arrays having pixel sizes around 15 μm.

The silicon readout integrated circuits (ROICs) need to be custom designed because the flux levels in the two bands may be markedly different. The polarity of the input MOSFET and the gain within the silicon must be matched to the technology and application. Having only one bump contact per unit cell, as for single-color hybrid FPAs, is the major advantage of the bias-selectable detector. In addition, it is compatible with existing silicon readout chips. This structure achieves approximately 100% optical fill factor in each band due to total internal reflection of the incident radiation off the mesa sidewalls. In Raytheon's arrays, the ROICs share a common chip architecture and incorporate identical unit cell circuit designs and layouts. Raytheon's approach employs a ROIC with time-division multiplexed integration (TDMI) [11] (see Figure 27.8). As the detector bias is changed, the detector current is directed to separate input circuitry and integration capacitors. Bias switching is performed at times much shorter than the frame period. Fast subframe switching of less than 1 ms is typically employed. The MWIR band is integrated by summing the charge collected from the individual subframe integration periods. The LWIR band is integrated by averaging the charge collected from the individual subframe integrations.

The silicon ROICs need to be custom designed because the flux levels in the two bands may be markedly different. The polarity of the input MOSFET and the gain within the silicon must be matched to the technology and application. In Sofradir's arrays [20], two separate input stages are designed into each pixel to allow simultaneous integration and readout of both bands. In Raytheon's arrays, the ROICs share a common chip architecture and incorporate identical unit cell circuit designs and layouts.

Figure 27.9 shows the current-voltage characteristics for a single mesa, single indium bump two-color MWIR1/MWIR2 [88] and MWIR/LWIR [70] TLHJ unit-cell detector design. With appropriate polarity

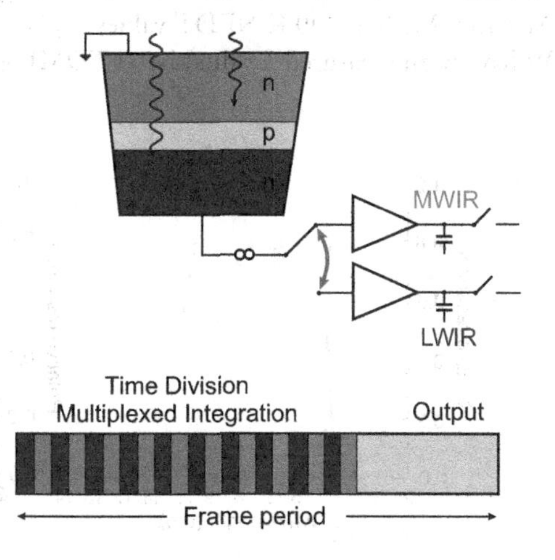

Figure 27.8 Raytheon's two-color FPAs with a time division multiplexed integration scheme in which the detector bias polarity is alternated many times within a single-frame period. (From Ref. [11]. With permission.)

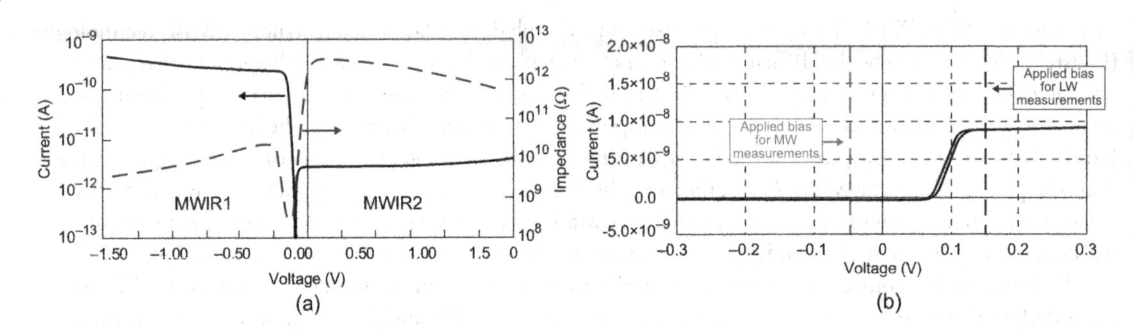

Figure 27.9 Typical I-V characteristics for a single mesa, single indium bump two-color TLHJ unit-cell detector design: (a) MWIR1/MWIR2 25-μm pixel with cutoff wavelength at 3.1 μm and 5.0 μm at 77 K and 30° FOV. (From Ref. [88]. With permission.) (b) MWIR/LWIR 20-μm pixel with cutoff wavelength at 5.5 μm and 10.5 μm. (From Ref. [70]. With permission.)

and voltage bias at the pixel contact, the junctions respond to either shorter or longer wavelength IR radiation. The I-V curves exhibit the serpentine shapes expected for a back-to-back diode structure; the "flatness" displayed in both effective reverse bias regions is a key indicator of high-quality two-color diodes. Figure 27.10 illustrates examples of the spectral response from different two-color devices [89]. Note that there is minimal crosstalk between the bands, since the SW detector absorbs nearly 100% of the shorter wavelengths. Test structures indicate that the separate photodiodes in a two-color detector perform exactly like single-color detectors in terms of achievable R_0A variation with wavelength at a given temperature (see Table 27.2) [10].

The best performing bias selectable dual-color FPAs being produced at Raytheon Vision Systems exhibit out-of-band crosstalk below 10%, 99.9% interconnect operability, and 99% response operability that is comparable to state-of-the-art, single-color technology. It is predicted that ongoing development of material growth and fabrication processes will translate to further improvements in dual-color FPA performance.

Raytheon Vision Systems has developed two-color, large-format IR FPAs to support the U.S. Army's third-generation forward-looking IR (FLIR) systems in both 640 × 480) and "high definition" 1,280 × 720 formats with 20 × 20-μm unit cells (see Figure 27.11a) [71]. The megapixel arrays are implemented in 3rd Gen eLRAS3 FLIR systems (see Figure 2.14). The ROICs share a common chip architecture and incorporate identical unit cell circuit designs and layouts; both FPAs can operate in either dual-band or single-band modes. High-quality MWIR/LWIR 1,280 × 720 FPAs with cutoffs ranging out to 11 μm at 78 K have demonstrated excellent sensitivity and pixel operabilities exceeding 99.9% in the MW band and greater than 98% in the LW band. Median 300 K NEDT values at *f*/3.5 of approximately 20 mK for the MW and 25 mK for the LW have been measured for dual-band TDMI operation at 60-Hz frame rate

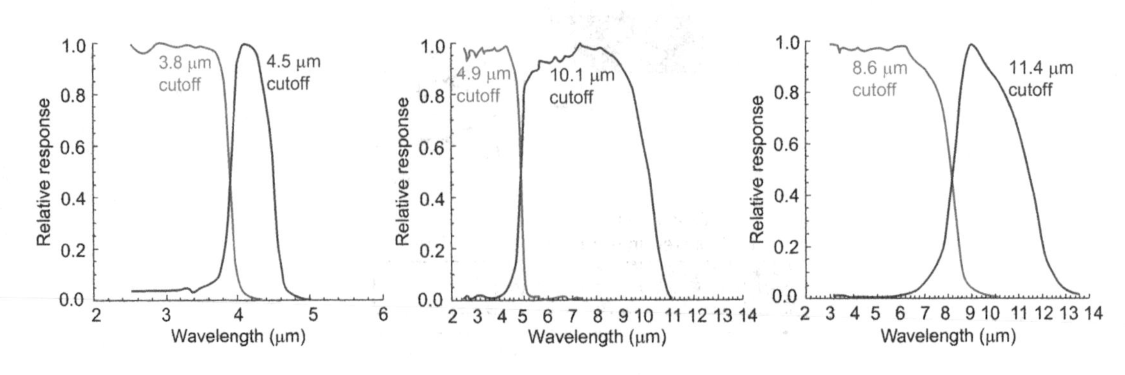

Figure 27.10 Spectral response curves for two-color HgCdTe detectors in various dual-band combinations of MWIR and LWIR spectral bands. (From Ref. [89]. With permission.)

Table 27.2 Typical measured performance parameters for single- and dual-color HgCdTe MWIR and LWIR detector configuration for 256 × 256 30-μm unit-cell FPAs

256 × 256, 30 μm UNIT-CELL PERFORMANCE PARAMETERS	DLHJ SINGLE COLOR		TLHJ SEQUENTIAL DUAL COLOR					
	MWIR	LWIR	MWIR/MWIR		MWIR/LWIR		LWIR/LWIR	
SPECTRAL BAND	MWIR	LWIR	BAND 1	BAND 2	BAND 1	BAND 2	BAND 1	BAND 2
78-K cutoff (μm)	5	10	4	5	5	10	8	10
Operating temperature (K)	78	78	120	120	70	70	70	70
Cross talk (%)	—	—	<5	<10	<5	<10	<5	<10
Quantum efficiency (%)	>70	>70	>70	>65	>70	>50	>70	>50
R_0A, zero FOV (Ωcm^2)	>1 × 10^7	>500						
RA^*, zero FOV (Ωcm^2)	—	—	6 × 10^5	2 × 10^5	1 × 10^6	2 × 10^2	5 × 10^4	5 × 10^2
Interconnect operability (%)	>99.9	>99.9	>99.9	>99.9	>99.9	>99.9	>99.9	>99.9
Response operability (%)	>99	>98	>99	>97	>99	>97	>98	>95

Source: Adapted after Ref. [10].
**RA* product at nonzero bias.
DLHJ – double layer heterojunction.
TLHJ – triple layer heterojunction.

with integration times corresponding to roughly 40% (MW) and 60% (LW) of full well charge capacities. Typical integration times were about 3 and 0.1 ms for MW and LW spectral bands, respectively. As shown in Figure 27.12 [10,91], excellent high-resolution IR camera imaging with *f*/2.8 field of view (FOV) broadband refractive optics at 60-Hz frame rate has been achieved.

Impressive results have also been demonstrated for other architectures. For example, the NEDT of 128 × 128 simultaneous MWIR1-MWIR2 FPAs (see device architecture in Figure 27.7b) for both bands (2.5–3.9 μm and 3.9–4.6 μm) was below 25 mK [89], and imagery was acquired at temperatures as high as 180 K with no visible degradation in image quality. The camera used for these measurements had a 50-mm,

(a) (b)

Figure 27.11 Dual-band megapixel MW/LW FPAs: (a) RVS 1,280 × 720 format HgCdTe FPAs mounted on dewar platforms (From Ref. [71]. With permission.), and (b) JPL 1,024 × 1,024 format QWIP FPA mounted on a 124-pin LCC. (Adapted after Ref. [90])

Figure 27.12 A still camera image taken at 78 K with *f*/2.8 FOV and 60-Hz frame rate using two-color 20-μm unit-cell MWIR/LWIR HgCdTe/CdZnTe TLHJ 1,280 × 720 FPA hybridized to a 1,280 × 720 TDMI ROIC: (a) MWIR and (b) LWIR. (From Ref. [14]. With permission.)

f/2.3 lens. Also, high-performance two-color 128 × 128 FPAs with 40-μm pitch have also been obtained using the pseudoplanar simultaneous architecture shown in Figure 27.7e. Background-limited detectivity performance has been obtained for MWIR (3–5 μm) devices at $T < 130$ K and for LWIR (8–10 μm) devices at $T \approx 80$ K (see Figure 27.13) [85]. The FPA also exhibits low NEDT values: 9.3 mK for the MW band and 13.3. mK for the LW band, similar to good-quality single-color FPAs.

Sofradir and Selex have reported on the development of TV format (640 × 512, 24-μm pixel pitch) MW/MW and MW/LW dual-band FPAs with NETD around 20 mK and operability exceeding 99.5%. Sofradir has been designed the semiplanar MBE structure (see Figure 27.7f) with a proven standard process and robust reproducibility, leading to low risk and a facilitated ramp-up to production—two separate input stages are designed into each pixel to allow simultaneous integration and readout of both bands. SELEX Galileo has used MOCVD method for growing third-generation material structures with GaAs substrates [18]. Table 27.3 specifies technical parameters of dual-band MW/MW and MW/LW arrays operated in 3-to 5-m and 8- to 10-m bands.

The HgCdTe high-density vertically integrated photodiode (HDVIP) or loophole concept (see Figure 17.46), developed at DRS and BAE Southampton, represents an alternative approach to IR FPA architecture. It differs from the more entrenched FPA architectures in both its method of diode formation

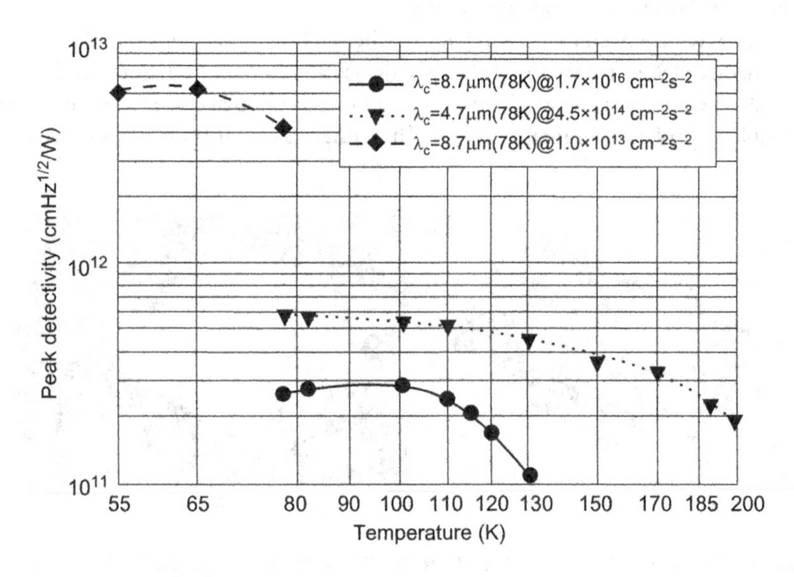

Figure 27.13 Detectivity of two-color pseudoplanar simultaneous MWIR/LWIR 128 × 128 HgCdTe FPA. (From Ref. [85]. With permission.)

Table 27.3 Specification of the dual-band HgCdTe FPA fabricated by Sofradir and Selex

PARAMETER	MW/MW (SOFRADIR)	MW/LW (SOFRADIR)	MW/LW (SELEX)
Array	640 × 512	640 × 512	640 × 512
Pixel pitch (μm)	24	24	24
Spectral range (Band 1) (μm)	3.4–4.2	3–5	3–5
Spectral range (Band 2) (μm)	4.4–4.8	8–9.5	8–10
Active area (mm)			15.36 × 12.29
NEDT (mK) (Band 1)	15–20	20–25	28 (22 dedicated)
NEDT (mK) (Band 2)	15–20	20–25	28 (11 dedicated)
Pixel operability (%)	>99.5	>99.5	>99
Charge capacity (dedicated)	3×10^6 electrons (Band 1) 1.05×10^7 electrons (Band 2)	3×10^6 electrons (Band 1) 1.05×10^7 electrons (Band 2)	8×10^6 electrons (Band 1) 8×10^6 electrons (Band 2)
Number of outputs	2 analog outputs per band	2 analog outputs per band	8
Pixel rate	Frame rate: 90 Hz	Frame rate: 90 Hz	up to 10 MHz per output
Operating temperature	80 K nominal	80 K nominal	80 K nominal

and the manner of its hybridization to the silicon ROIC [82]. The monocolor HDVIP architecture consists of a single HgCdTe epilayer grown on CdZnTe substrate by LPE or MBE [83]. After epitaxial growth, the substrate is removed and the HgCdTe layer is passivated on both surfaces with interdiffused layers of evaporated CdTe (the interdiffusion at 250°C on the Te-rich side of the phase field generates about 10^{16} cm^{-3} metal vacancies). During this process, the Cu can also be in-diffused from a doped ZnS source providing an alternative to doping during growth. This single-color architecture has been extended to two colors at DRS by gluing two monocolor layers together into a composite and forming an insulated via through the lower layer in order to readout the upper color, as illustrated in Figure 27.14. Contact to the Si ROIC is obtained by etching holes (or vias) through the HgCdTe down to contact pads on the Si (see Figure 27.14c). The ROIC used for the dual-band FPA was originally designed for a single-color 640 × 480 array with 25-μm (square) pixels. The even numbered rows of the ROIC have no detectors attached to them, so the chip is operated in a mode that only outputs the odd rows. Odd-numbered columns connect to LWIR detectors, and the MWIR detectors are on the even columns. This approach has been utilized to fabricate both MW-LW and MW-MW 240 × 320 FPAs on a 50-μm pitch. Higher densities are being investigated with dedicated two-color ROIC designs, enabling pitches of <30 μm for two-color FPAs.

Performance data for representative DRS two-color MW-LW and MW-MW 240 × 320 FPAs utilizing *f*/3 optics and a 60-Hz frame rate are described in Refs. [83,84]. Their NEDT values were ≤20 mK and operabilities in excess of 99%. However, relatively low collection efficiency (the product of quantum efficiency and unit cell fill factor) has been measured on the LWIR layer.

Two-color MWIR/LWIR HgCdTe detectors have been examined theoretically [92–94]. It has been shown that it is possible to predict, with relatively good accuracy, the performance of complex detectors by using numerical models. Furthermore, the simulation technique is also useful for understanding the effects of different material parameters and geometrical characteristics on the detector performance.

27.2.2 THREE-COLOR HgCdTe DETECTORS

Some system considerations suggest that three-color FPAs would be more generally useful than two-color ones. The successful development of three-color HgCdTe FPAs requires further improvement in the

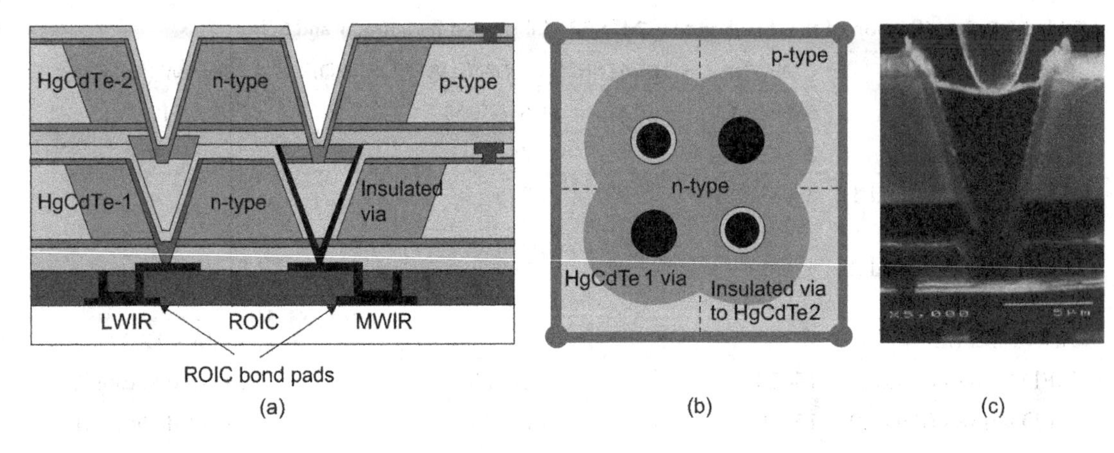

Figure 27.14 Two-color HDVIP architecture is composed of two layers of thinned HgCdTe epoxied to a silicon readout: (a) side view, (b) top view (Adapted after Ref. [82]. With permission.), and (c) small hole etched to form junction and to contact the Si readout.

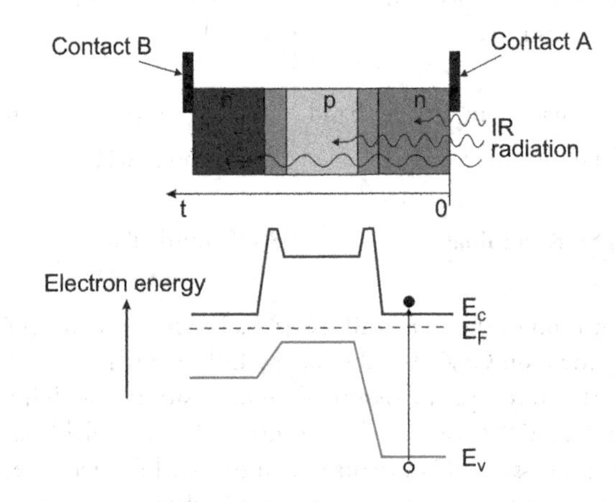

Figure 27.15 Three-color concept and associated zero-bias band diagram. (From Ref. [95]. With permission.)

material quality, adequate processing techniques, and a better understanding of imager operation both in terms of pixel performance and interaction between different pixels in the array.

The first concept for achieving three-color HgCdTe detectors has been demonstrated by British workers [95]. The concept of a back side-illuminated HgCdTe detector is shown schematically in Figure 27.15. The bias-dependent cutoff is achieved by using three absorbers in an n-p-n structure in which the first n-layer defines the SW region; the p-type layer, the intermediate wavelength region (IW); and the top layer, the LW region. Note that the terms SW and LW used here are relative and do not necessarily coincide with the SWIR and LWIR bands. The cutoff wavelengths of the SW, IW, and LW regions are, respectively, marked as λ_{c1}, λ_{c2}, and λ_{c3}. Since the barrier region is low doped, the applied bias mainly falls on this side of the junction. For the device configuration shown in Figure 27.15, the negative bias denotes higher potential of contact A in comparison with contact B.

It is expected that at low biases either the SW or LW response would dominate, depending on which junction is reverse-biased. In this case, we have the same situation as in the bias selectable two-color detector since the barriers prevent electron flow from the IW layer—both generated photoelectrons as well as direct injected carriers from the forward-biased junction. Increasing reverse bias reduces the barrier, and electrons photogenerated in the IW layer can cross the junction. As a result, the cutoff wavelength changes from the SW to the IW as the bias is increased. This situation, with negative bias, and corresponding changes in spectral response, is shown in Figure 27.16c. Changing the bias direction to positive shifts the cutoff wavelength to the LW region (see Figure 27.16d). Similarly, increasing positive bias moves the cut-on

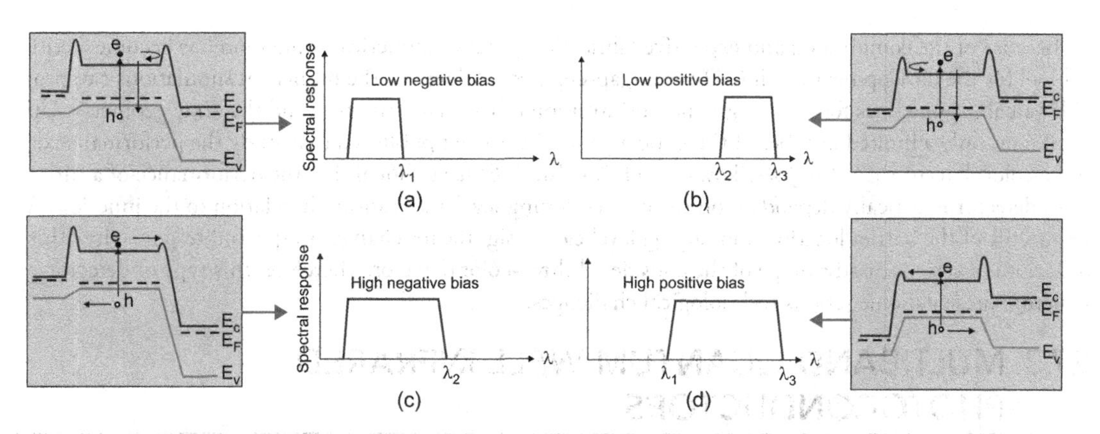

Figure 27.16 Idealized spectral responses of three-color detector. λ_1, λ_2 and λ_3 are determined by the SW, intermediate, and LW band gaps respectively. Effect of negative (a,c) and positive (b,d) bias voltages on bandgap structure is also shown. (From Ref. [95]. With permission.)

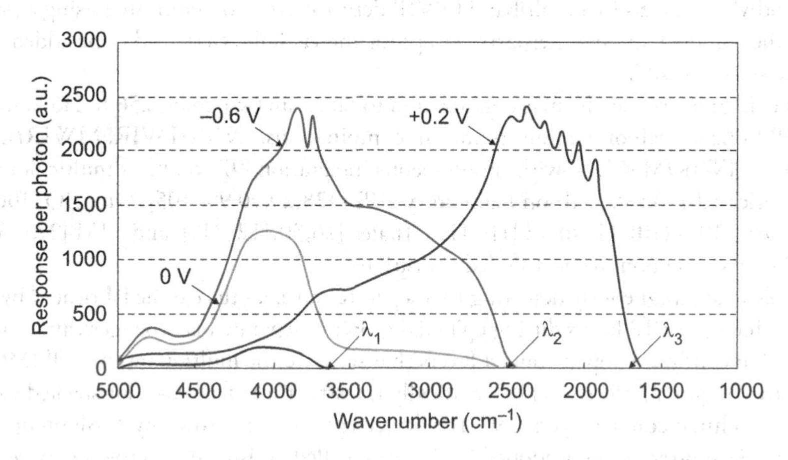

Figure 27.17 Spectral response of a three-color HgCdTe detector with cutoff wavelength 3, 4, and 6 μm at various biases. (From Ref. [95]. With permission.)

from being coincident with the IW cutoff to the SW cutoff. It should be noted that the above considerations concern the ideal case of the detector structure. The ideal LW response may not be achievable in practice due to the IW absorber being insufficiently thick to absorb all the IW radiation.

The three-color HgCdTe detectors were grown by metal-organic vapor phase epitaxy (MOVPE) on GaAs substrates oriented off the (100) direction to reduce the size of pyramidal hillock growth defects. Figure 27.17 shows the spectral response with cutoff wavelength 3 (SW), 4 (IW), and 6 μm (LW) [95]. In positive bias mode, the LW/IW junction is in reverse bias and a bias-independent LW spectrum is obtained above 0.2 V. No barrier lowering at the LW/IW junction at these applied biases is observed due to the chosen doping levels. The response below λ_2 is due to incomplete absorption in the IW layer resulting in carrier generation in the LW layer at these wavelengths (carriers generated in the IW absorber have insufficient energy to surmount the LW barrier). As the positive bias is reduced below 0.2 V, the LW response collapses and a signal from the SW layer appears with the current flowing in the opposite direction. For this bias regime, the built-in fields dominate and the largest field is at the SW/IW junction due to the larger bandgap. Further reduction in the bias voltage causes the SW response to grow. The negative voltage puts the SW/IW junction into reverse bias and results in a SW response with cutoff λ_1. Further increase of the negative bias lowers the barrier at this junction and allows a response from the IW layer, thus moving the cutoff to λ_2. The observed increase of the SW signal with increasing negative bias is caused by incomplete absorption in the SW absorption.

Because of the complicated and expensive fabrication process, numerical simulation has become a critical tool for the development of HgCdTe bandgap-engineered devices. The numerical simulations can provide valuable guidelines for the design and optimization of the pixel structure and the array geometry. Up until now, only a limited number of theoretical papers have been published that study the performance of three-color detectors [95,96]. Jóźwikowski and Rogalski [96] have shown that the performance of a three-color detector is critically dependent on the barrier doping level and position in relation to the junction. A small shift of the barrier location and doping level causes significant changes in spectral responsivity. This behavior is a serious disadvantage of the considered three-color detector. Therefore, this type of detector structure presents some serious technological challenges.

27.3 MULTIBAND QUANTUM WELL INFRARED PHOTOCONDUCTORS

QWIPs are ideal detectors for the fabrication of pixel coregistered simultaneously readable two-color IR FPAs because a QWIP absorbs IR radiation only in a narrow spectral band and is transparent outside of that absorption band. Thus it provides zero spectral crosstalk when two spectral bands are more than a few microns apart. Individual pixels in a multiband QWIP detector array are fabricated using a process similar to that used for their single-band counterparts, except for the via holes that need to be added to electrically connect with the silicon ROIC.

Lockheed Martin Sanders was the first organization to fabricate two-color, 256 × 256 bound-to-miniband QWIP FPAs in each of the four important combinations: LWIR/LWIR, MWIR/LWIR, near-IR (NIR)/LWIR, and MWIR/MWIR—with simultaneous integration [97,98]. Next multicolor QWIP detectors have been fabricated at Jet Propulsion Laboratory (JPL) [28,29,90,99–105], QmagiQ [106,107], Army Research Laboratory [108–110], Goddard [110,111], Thales [26,30,112–115], and AIM [27,116,117] with the majority being based on bound-to-extended transitions.

Devices capable of simultaneously detecting two separate wavelengths can be fabricated by vertical stacking of the different QWIP layers during epitaxial growth. Separate bias voltages can be applied to each QWIP simultaneously via doped contact layers that separate the multiquantum well (MQW) detector heterostructures. Figure 27.18a shows schematically the structure of a two-color stacked QWIP with contacts to all three ohmic-contact layers [29]. The device epilayers are grown by MBE on up to 6-inch semi-insulating GaAs substrates. An undoped GaAs layer, called an isolator, is grown between two AlGaAs etch stop layers, followed by a 0.5-μm-thick doped GaAs layer. Next, the two QWIP heterostructures are grown, separated by another ohmic contact. The long wavelength-sensitive stack (red QWIP) is grown above the SW sensitive stack (blue QWIP). Typical responsivity spectra at 77 K using a common bias of 1.5 V, recorded simultaneously for two QWIPs at the same pixel, are shown in Figure 27.18b. Each QWIP

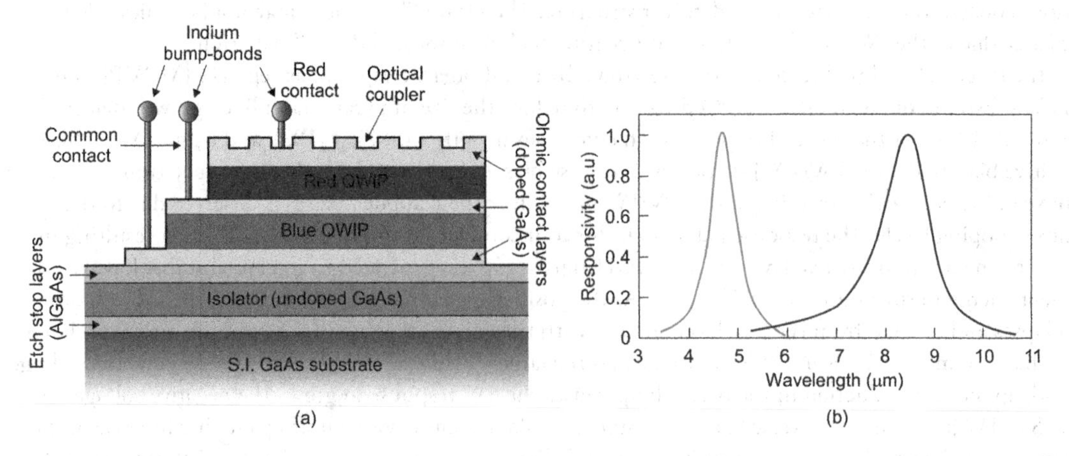

Figure 27.18 Schematic representation of the (a) dual-band QWIP detector structure and (b) typical responsivity spectra at 77 K and a common bias of 1 V, recorded simultaneously for two QWIPs at the same pixel. S.I., semi-insulating. (From Ref. [29]. With permission.)

consists of about a 20-period GaAs/$Al_xGa_{1-x}As$ MQW stack in which the thickness of the Si-doped GaAs quantum wells (QWs) (with typical electron concentration $5 \times 10^{17} cm^{-3}$) and the Al composition of the undoped $Al_xGa_{1-x}As$ barriers (≈550–600 Å thick) is adjusted to yield the desired peak responsivity position and spectral width. The gaps between FPA detectors and the readout multiplexer are backfilled with epoxy. The epoxy backfilling provides the necessary mechanical strength to the detector array and readout hybrid prior to array thinning. The initial GaAs substrate of dual-band FPAs are completely removed leaving only a 50-nm-thick GaAs membrane. This allows the array to accommodate any thermal expansion by eliminating the thermal mismatch between the silicon readout and the detector array. It also eliminates pixel-to-pixel crosstalk and, finally, significantly enhances the optical coupling of IR radiation into the QWIP pixels. Using the above-described fabrication process, significant progress has been made toward development of a megapixel dual-band QWIP FPA [28,90,104,105].

Figure 27.19 provides additional insight into dual-band QWIP processing technology developed at JPL [118], based on 4-inch wafers to fabricate 320 × 256 MWIR/LWIR dual-band QWIP devices with pixels collocated and simultaneously readable. As shown in Figure 27.19b, the carriers emitted from each MQW region are collected separately using three contacts. The middle contact layer (see Figure 27.19c) is used as the detector common. The electrical connections to the detector common and the LWIR connection are brought to the top of each pixel using via connections. Electrical connections to the common contact and the LWIR pixel connection are brought to the top of each pixel using the gold via connections visible in the figure. This elaborate processing technology could lead to 2D imaging arrays that can detect separate bands on a single pixel.

Most QWIP arrays use a 2D grating, which has the disadvantage of being very wavelength dependent, combined with an efficiency that decreases as the pixel size is reduced. Lockheed Martin has used rectangular and rotated rectangular 2D gratings for their two-color LW-LW FPAs. Although random reflectors

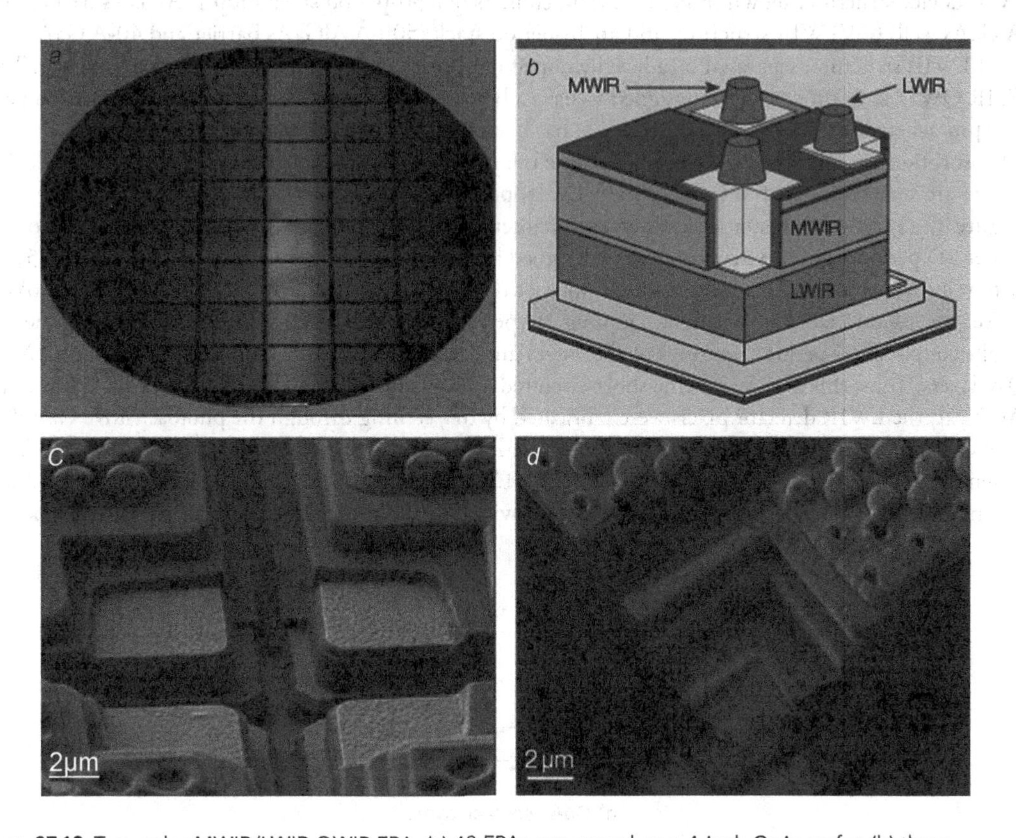

Figure 27.19 Two-color MWIR/LWIR QWIP FPA: (a) 48 FPAs processed on a 4-inch GaAs wafer, (b) three-dimensional view of pixel structure, (c) electrical connections to the common contact, and (d) the pixel connections are brought to the top of each pixel using the gold via connections. (From Ref. [118]. With permission.)

have achieved relatively high quantum efficiencies with large test device structures, it is not possible to achieve comparable quantum efficiencies with random reflectors on small FPA pixels, due to the reduced width-to-height aspect ratios [101]. In addition, it is more difficult to fabricate random reflectors for shorter wavelength detectors because feature sizes of random reflectors are linearly proportional to the peak wavelength of the detectors. Thus, quantum efficiency becomes a more difficult issue for multicolor QWIP FPAs in comparison to single-color arrays. At JPL, two different optical coupling techniques have been developed. The first technique uses a dual period Lamar grating structure and the second is based on multiple diffraction orders (see Figure 27.19c,d) [118].

Typical operating temperatures for QWIP detectors are in the range of 60–80 K. The bias across each QWIP can be adjusted separately, although it is desirable to apply the same bias to both colors. Results indicate that the complex two-color processing has not compromised the electrical and optical quality of either FPA in the two-color device, since the peak quantum efficiency for each of the 20-period QWIPs was estimated to be ≈10%. For comparison, a normal single-color QWIP with twice the number of periods has a quantum efficiency of around 20%. An accurate design methodology is needed to optimize the detector structure to meet different requirements. In the production process, the fabrication of gratings is still quite an involved process, and the detector quantum efficiency is rather uncertain in small pixels and in pixels with thick material layers.

Different design structures of dual-band QWIP FPAs has been undertaken at JPL [99–101]. One of the key issues has been the scarcity of appropriate readout multiplexers. To overcome this problem, JPL has chosen to demonstrate initial dual-band concepts with existing multiplexers developed for single-color applications and use a waveband-interlaced CMOS readout architecture (i.e., odd rows for one color and even rows for the other color). This scheme has the disadvantage that it does not provide a full fill factor for both wavelength bands, resulting in an approximate 50% fill factor for each wavelength band. The LWIR/VLWIR device structure, shown in Figure 27.20, consists of a 30-period stack (500-Å AlGaAs barrier and 60 Å GaAs well) of VLWIR structure, and an 18-period stack (500-Å AlGaAs barrier and 40-Å GaAs well) of LWIR structure, separated by a heavily doped 0.5-μm-thick intermediate GaAs contact layer. The VLWIR QWIP structure has been designed to have a bound-to-quasibound intersubband absorption peak at 14.5 μm, whereas the LWIR QWIP structure has been designed to have a bound-to-continuum intersubband absorption peak at 8.5 μm, primarily because the photocurrent and dark current of the LWIR device structure are small compared to those of the VLWIR portion.

Figure 27.21 shows a schematic side view of the interlaced dual-band GaAs/AlGaAs FPA [99]. Two different 2D periodic grating structures were designed to independently couple the 8–9 μm and 14–15 μm radiation into detector pixels in even and odd rows of the FPA, respectively. The top 0.7-μm-thick GaAs cap layer was used to fabricate the light-coupling 2D periodic gratings for 8–9 μm detector pixels, whereas the light-coupling 2D periodic gratings of the 14–15 μm detector pixels were fabricated through the LWIR MQW layers. Thus, this grating scheme short-circuited all 8–9 μm sensitive detectors in all odd rows of the FPAs. Next, the LWIR detector pixels were fabricated by dry etching through the photosensitive GaAs/AlGaAs MQW layers into the 0.5-μm-thick doped GaAs intermediate contact layer. All VLWIR pixels in the even rows of the FPAs were short circuited. The VLWIR detector pixels were fabricated by dry etching through both MQW stacks into the 0.5-μm-thick heavily doped GaAs bottom contact layer. After epoxy

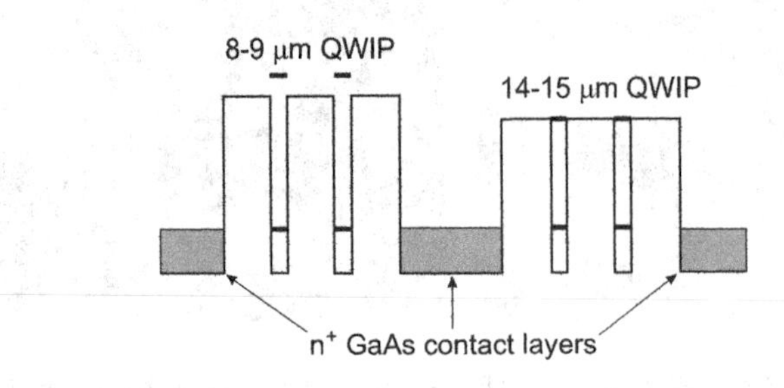

Figure 27.20 Conduction band diagram of LWIR/VLWIR two-color QWIP detector. (From Ref. [99]. With permission.)

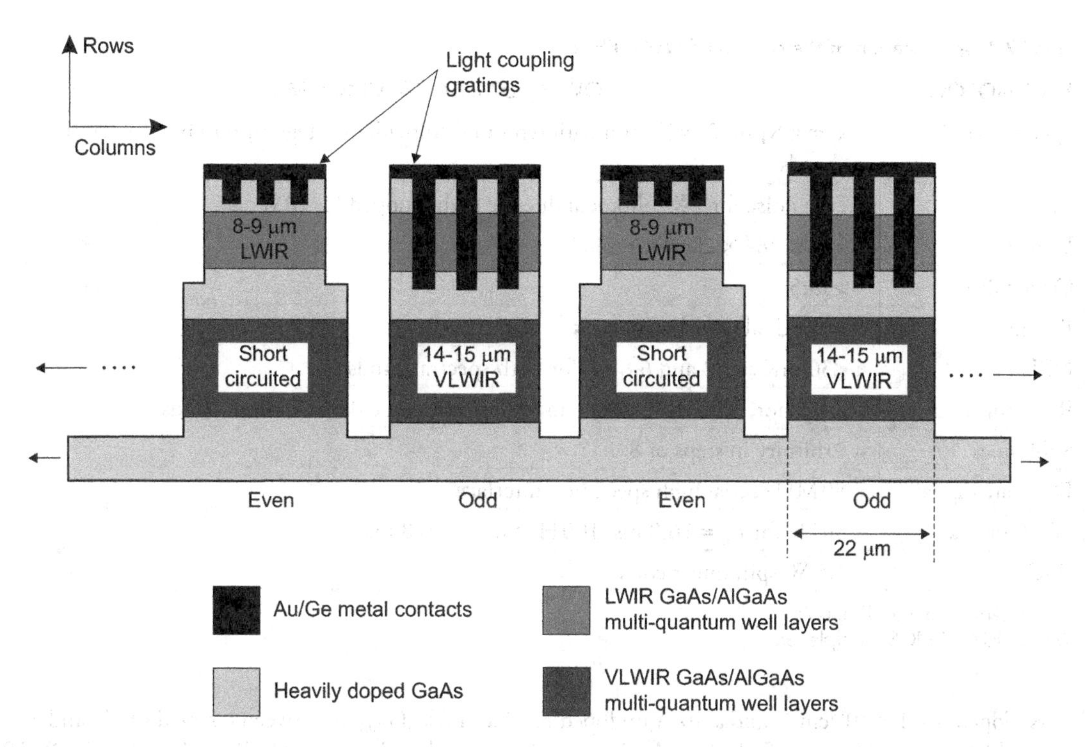

Figure 27.21 Structure cross section of the interlace dual-band FPA. (From Ref. [99]. With permission.)

backfilling of the gaps between FPA detectors and the readout multiplexer, the substrate was thinned, and finally the remaining GaAs/AlGaAs material contained only the QWIP pixels and a very thin membrane (≈1,000 Å).

The 640 × 486 GaAs/AlGaAs array provided images with 99.7% of the LWIR pixels and 98% of VLWIR pixels working, demonstrating the high yield of GaAs technology. The 8–9 μm detectors have shown background-limited performance (BLIP) at 70 K operating temperature, at 300 K background with an *f*/2 cold stop. The 14–15 μm detectors show BLIP with the same operating conditions at 45 K. The performance of these dual-band FPAs were tested at a background temperature of 300 K, with *f*/2 cold stop and a 30-Hz frame rate. The estimated NEDTs of LWIR and VLWIR detectors at 40 K were 36 and 44 mK, respectively. The experimentally measured values of the LWIR NEDT, equal to 29 mK, were lower than the estimated ones. This improvement was attributed to the light coupling efficiency of the 2D periodic grating. However, the experimental VLWIR NEDT value was higher than the estimated value. That was probably a result of inefficient light coupling in the 14–15 μm region, readout multiplexer noise, and noise of the proximity electronics. At 40 K, the performance of detector pixels in both bands was limited by photocurrent noise and readout noise.

To cover the MWIR range, a strained-layer InGaAs/AlGaAs material system is used. InGaAs in the MWIR stack produces high in-plane compressive strain, which enhances the responsivity. The MWIR/LWIR FPAs fabricated by the Sanders consisted of an 8.6-μm cutoff GaAs/AlGaAs QWIP on top of a 4.7-μm cutoff strained InGaAs/GaAs/AlGaAs heterostructure. The fabrication process allowed for fill factors of 85 and 80% for the MW and LW detectors, respectively. The first FPAs with this configuration had operability in excess of 97%, and NEDT values better than 35 mK with *f*/2 optics.

The first dual-band QWIP FPA with pixel collocation and simultaneous operation in MWIR and LWIR has been described by Goldberg et al. [108]. This 256 × 256 pixel FPA has achieved a NEDT of 30 mK in the MWIR spectral band and 34 mK in the LWIR spectral band. Also Gunapala et al. [29] have demonstrated a 320 × 256 MWIR/LWIR pixel collocated and simultaneously readable dual-band QWIP FPA. The device structures of the MWIR and LWIR devices were very similar. Each period of the MQW structure consists of coupled QWs of 40 Å containing 10-Å GaAs, 20-Å $In_{0.3}Ga_{0.7}As$, and 10-Å GaAs

Table 27.4 Specification of the dual band QWIP FPAs

TECHNOLOGY	QWIP DUAL BAND, CMOS MUX
Spectral bands	$\lambda_p = 4.8\,\mu m$; $\lambda_p = 7.8\,\mu m$ with temporal coincident integration in both spectral bands
Type	Low noise for LW; photoconductive highly doped for MW
Elements	388 × 284 × 2; 40-μm pitch
Operability	>99.5%
Biasing	Individually for both bands
NEDT	<30 mK @ *f*/2 and 6.8 ms for both spectral bands
Read put models	Snapshot, stare then scan, temporal signal coincidence in both bands
Subframes	Arbitrary in steps of 8
Data rate digital	80 MHz serial high speed link interface
Full frame rate	50 Hz for $t_{int} = 16.8$ ms; 100 Hz for $t_{int} = 6.8$ ms
IDCA	1.5 W split linear cooler

Source: Adapted after Ref. [27].
CMOC MUX – CMOS multiplexer.

layers (doped $n = 1 \times 10^{18}\,cm^{-3}$) and a 40-Å undoped barrier of $Al_{0.3}Ga_{0.7}As$ between coupled QWs, and a 400-Å-thick undoped barrier of $Al_{0.3}Ga_{0.7}As$. It is worth noting that the active MQW region of each QWIP device is transparent at other wavelengths, which is an important advantage over conventional interband detectors. The experimentally measured NEDT of MWIR and LWIR detectors at 65 K were 28 and 38 mK, respectively.

Another design structure for dual-band MWIR/LWIR QWIPs has been proposed by Schneider et al. [117]. This simultaneously integrated 384 × 288 FPA with 40-μm pitch comprising a photovoltaic and a photoconductive QWIP for the LWIR and MWIR, respectively (see Figure 19.29). Excellent NEDT (17 mK) is obtained in the MWIR band. Owing to the nonoptimized coupling for LWIR wavelengths, the observed NEDT is higher, but still shows a reasonable value of 43 mK. Due to improvements in the device design, excellent thermal resolution with NEDT <30 mK (*f*/2 optics and full frame time of 6.8 ms) for both peak wavelengths (4.8 and 8.0 μm) has been demonstrated. The features and performance of the dual-band QWIP fabricated by AIM GmbH are summarized in Table 27.4.

More recently, the research group from JPL has implemented a MWIR/LWIR pixel co-registered simultaneously readable 1,024 × 1,024 dual-band device structure that uses only two indium bumps per pixel (Figure 27.22) compared to three indium bumps per pixel with pixel collocated dual-band devices [90,105]. In this device structure, the detector common (or ground) is shorted to the bottom detector common plane via a metal bridge. Thus, this device structure reduces the number of indium bumps by 30% and has a unique advantage in large-format FPAs, since more indium bumps require additional force during the FPA hybridization process. The substrate removal process used in array processing eliminates the thermal mismatch problem between the silicon-based readout and the GaAs-based detector array, eliminates pixel-to-pixel optical cross-talk, and provides enhancement in optical coupling of IR radiation into QWIP pixels. Figure 27.11b shows a megapixel dual-band QWIP FPA mounted on a 124-pin leadless chip carrier (LCC). The pitch of the detector array is 30 μm and the actual MWIR and LWIR pixel sizes are $28 \times 28\,\mu m^2$. The estimated NEDT based on single pixel data of MWIR and LWIR detectors at 70 K are 22 and 24 mK, respectively. The experimentally measured NEDT values are 27 and 40 mK for MWIR and LWIR, respectively (see Figure 27.23). This is due to the fact that it is difficult to independently optimize the operating bias of LWIR band due to a ROIC pixel short circuit occurring at the MWIR band. An image taken with the first megapixel simultaneous pixel co-registered MWIR/LWIR dual-band QWIP camera is shown in Figure 27.24. The flame in the MWIR image (left) looks broader due to the detection of heated CO_2 (from a cigarette lighter) re-emission in a 4.1–4.3-micron band, whereas the heated CO_2 gas does not have

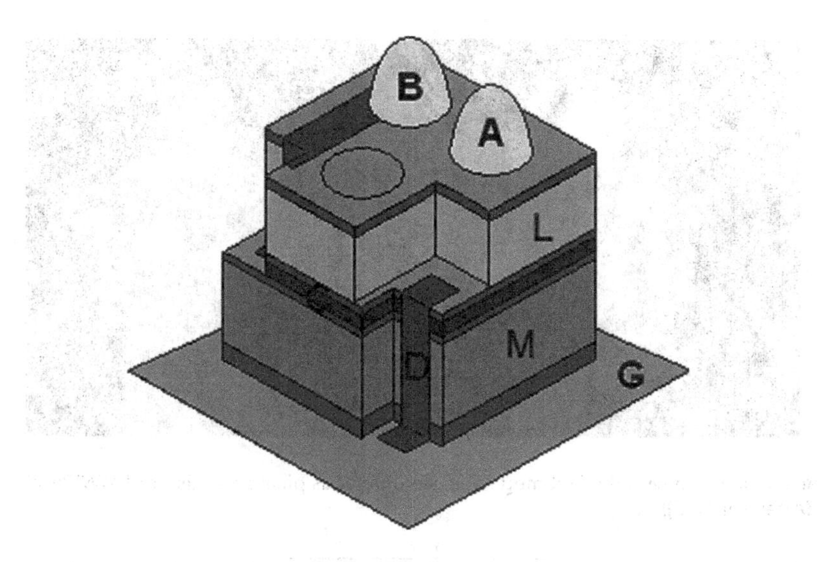

Figure 27.22 Three-dimensional view of dual-band QWIP device structure showing via connects for independent access of MWIR and LWIR devices. The color code is as follows: C, isolation layer; L, LWIR QWIP; M, MWIR QWIP; G, contact layer; D, metal bridges between MQW regions; A,B, indium bumps. (From Ref. [105]. With permission.)

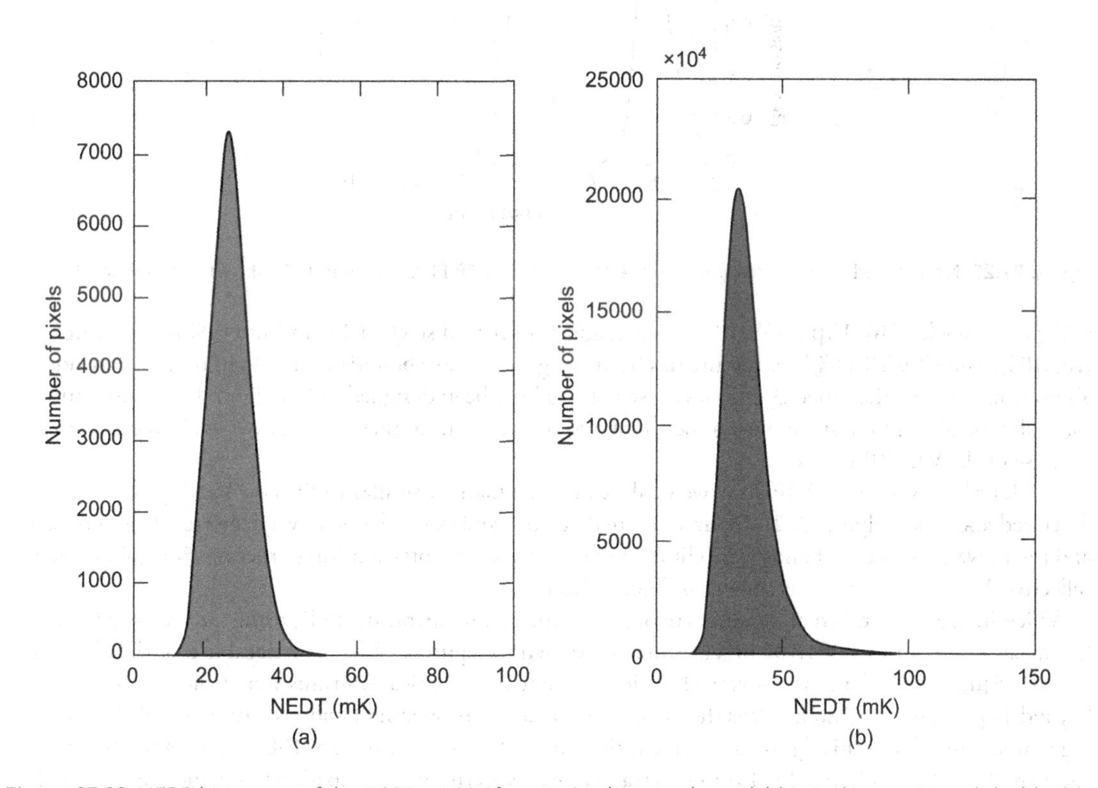

Figure 27.23 NEDT histogram of the 1,024 × 1,024 format simultaneously readable pixel co-registered dual-band QWIP FPA: (a) MWIR and (b) LWIR. (Adapted after Ref. [90])

any emission line in the LWIR band. Thus, the LWIR image shows only thermal signatures of the flame. However, the silicon wafer blocked most of the LWIR signal.

The potential of QWIP technology is connected with multicolor detection. A four-band hyper spectral 640 × 512 QWIP array was successfully developed under a joint Goddard-JPL-Army Research Laboratory project funded by the Earth Science Technology Office of NASA (see Figure 27.25). The device structure consists of a 15-period stack of 3–5 μm QWIP structure, a 25-period stack of 8.5–10 μm QWIP structure,

Figure 27.24 An image taken with the first megapixel simultaneous pixel co-registered MWIR:LWIR dual-band QWIP camera. (After Ref. [105])

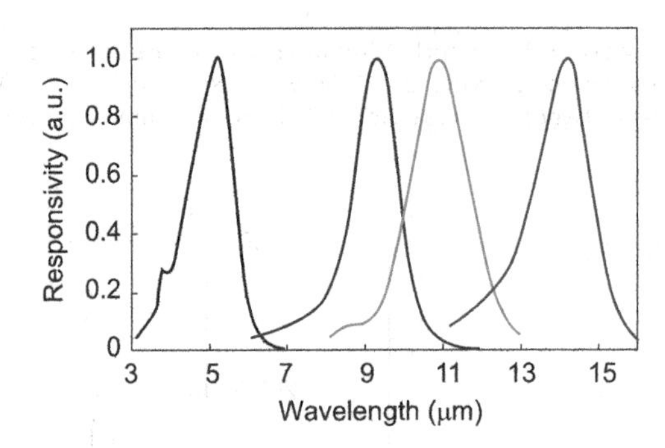

Figure 27.25 Normalized spectral response of the four-band QWIP FPA. (From Ref. [103]. With permission.)

a 25-period stack of 10–12 μm QWIP structure, and a 30-period stack of 14–15.5 μm QWIP structure [102,103]. The VLWIR QWIP structure has been designed to have bound-to-quasibound intersubband absorption, whereas the other QWIP device structures have been designed to have bound-to-continuum intersubband absorption, since the photocurrent and dark current of these devices are small in comparison to those of the VLWIR device.

The four bands of the QWIP array were fabricated in a manner similar to the two-band system described above (see Figure 27.21). Four separate detector bands were defined by a deep trench etch process and the unwanted spectral bands were eliminated by a detector short-circuiting process using gold-coated reflective 2D etched gratings as shown in Figure 27.26.

Video images were taken at a frame rate of 30 Hz and at a temperature 45 K, using a ROIC capacitor having a charge capacity of 1.1×10^7 electrons. As shown in Figure 27.27, it is noticeable that the object in the 13–15 μm spectral band is not very clear due to the reduced optical transmission of the germanium lens beyond 14 μm. Figure 27.28 displays the peak detectivities of all spectral bands as a function of the operating temperature. From this figure, it is evident that the BLIP temperatures are 100, 60, 50, and 40 K for the 4–6, 8.5–10, 10–12, and 13–15 μm spectral bands, respectively. The experimentally measured NEDT of 4–6, 8.5–10, 10–12, and 13–15 μm detectors at 40 K are 21, 45, 14, and 44 mK, respectively.

A novel four-band IR imaging system with simultaneously readable collocated pixels has been proposed in Ref. [105]. The FPAs divided into 2 × 2 subpixel areas that function as superpixels marked as Q1, Q2, Q3, and Q4 in Figure 27.29, each sensitive to one of four specific wavelength bands.

In typical QWIP arrays, an isotropic optical coupling scheme is used to eliminate polarization sensitivity. For the polarimetric QWIPs, linear instead of 2D gratings are used. The use of a microscanner makes it possible for the design of a camera that resolves the polarimetric components of the scene radiation. Such a discriminating imager, added without significant loss of sensitivity or increased cost, may be beneficial

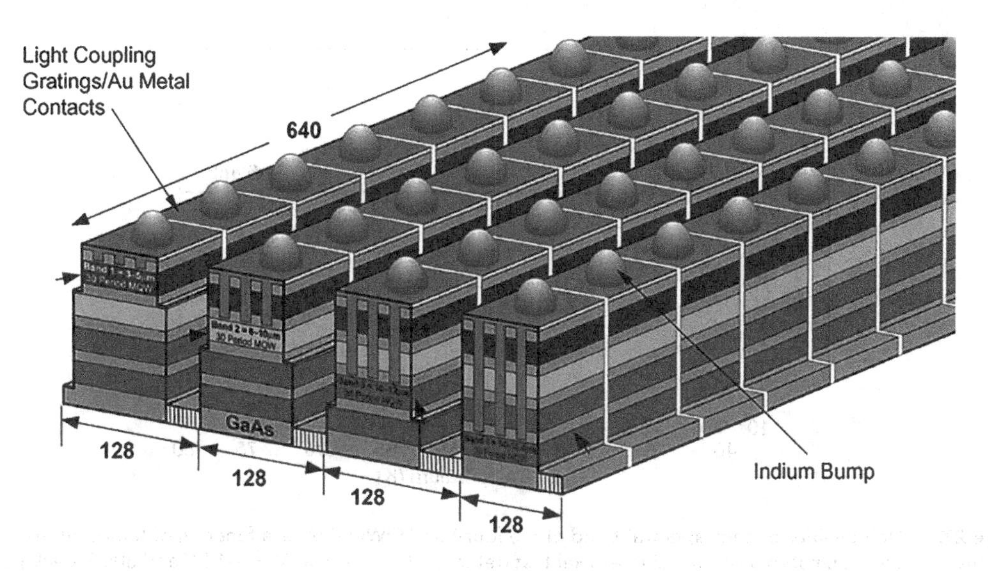

Figure 27.26 Layer diagram of the four-band QWIP device structure and the deep groove 2D periodic grating structure. Each pixel represents a 640 × 128 pixel area of the four-band FPA. (From Ref. [103]. With permission.)

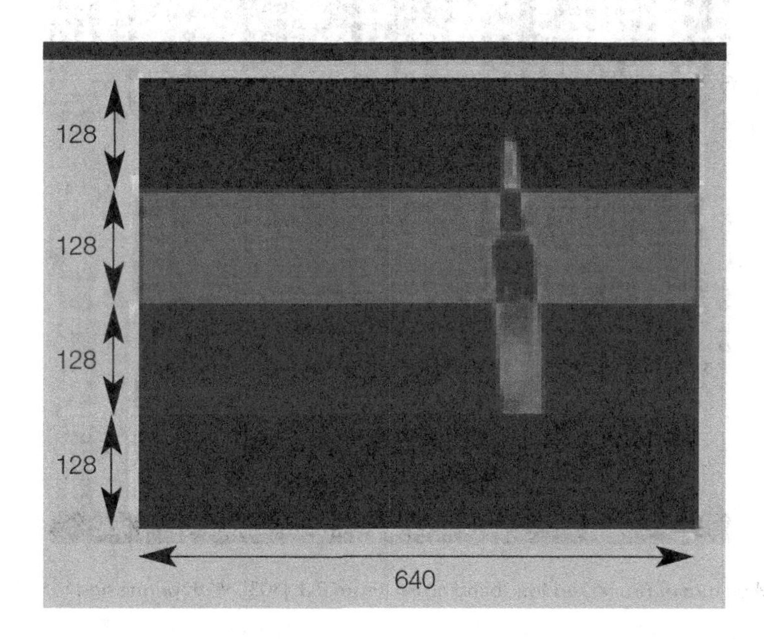

Figure 27.27 One frame of video image taken with the 4–15 μm cutoff four-band 640 × 512 pixel QWIP camera. The image is barely visible in the 13–15 μm spectral band due to the poor optical transmission of the antireflection layer coated germanium lens. (From Ref. [103]. With permission.)

in locating difficult targets. Thales formed four linear gratings rotated by 45° to each other on a set of four detector elements. This pattern is then replicated across the whole array. The layout and a scanning electron microscopy (SEM) picture from an actual array are shown in Figure 19.34 [115].

In most of the two-color QWIP FPAs, the three-bump approach is used to assure simultaneous integration of the detector output in different bands. However, this design decreases the FPA fill factor and significantly complicates the fabrication process. The research group from the Middle East Technical University in Ankara has demonstrated the voltage tunable two-color 640 × 512 MWIR/LWIR FPAs using commercially ROICs [119]. Dual-band sensors are implemented with conventional FPA fabrication process

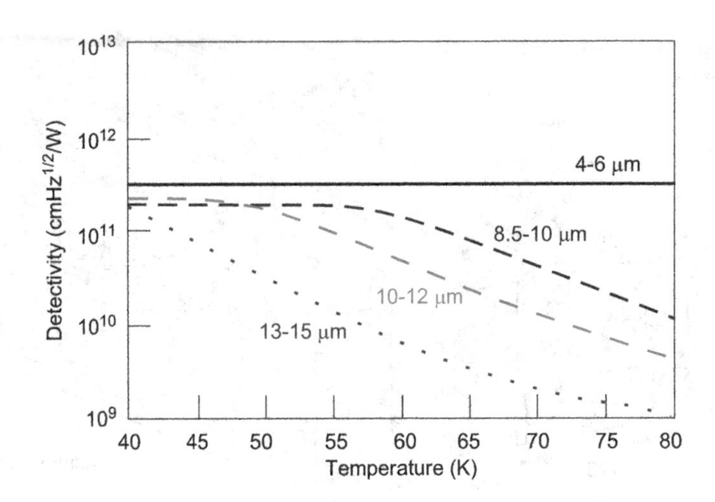

Figure 27.28 Detectivities of each spectral-band of the four-band QWIP FPA as a function of temperature. Detectivities were estimated using the single-pixel test detector data taken at $V_b = -1.5$V and 300 K background with *f*/5 optics. (From Ref. [103]. With permission.)

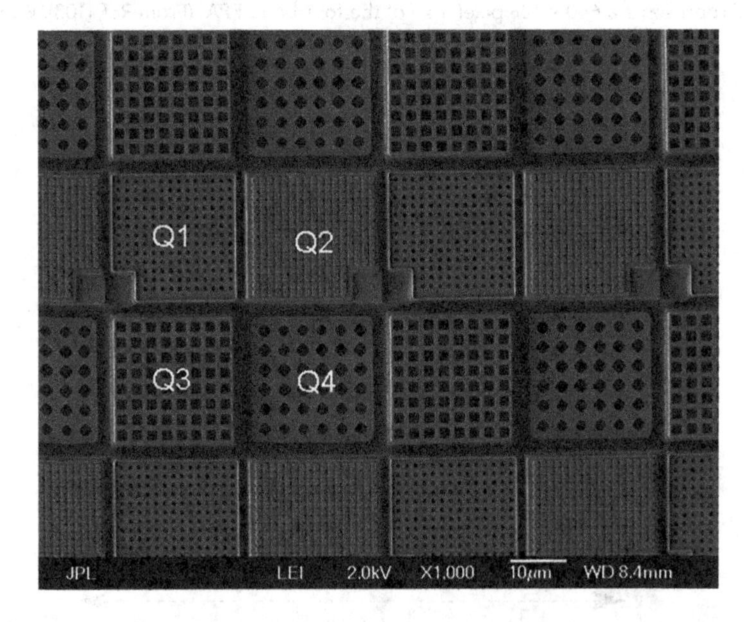

Figure 27.29 SEM picture of processed four-band array. (From Ref. [105]. With permission.)

requiring only one In bump on each pixel, making it possible to fabricate large format arrays at the cost and yield of single-band detectors.

The above results indicate that QWIPs have shown significant progress in recent years, especially in their applications to the multiband imaging problem. It is a niche in which they have an intrinsic advantage due to the comparative ease of growing multiband structures by MBE with very low defect density.

27.4 MULTIBAND TYPE-II InAs/GaSb DETECTORS

In the last decade, type II InAs/GaInSb superlattices have emerged as the third candidate for third-generation IR detectors [27,36–49,120–126].

In 2005, the worldwide first bispectral InAs/GaSb superlattice IR camera had been demonstrated by AIM Infrarot-Module GmbH, Heilbronn, Germany. Manufacture of dual-color detectors via MBE and subsequent processing of FPAs were accomplished by Fraunhofer IAF. The thickness of the entire vertical

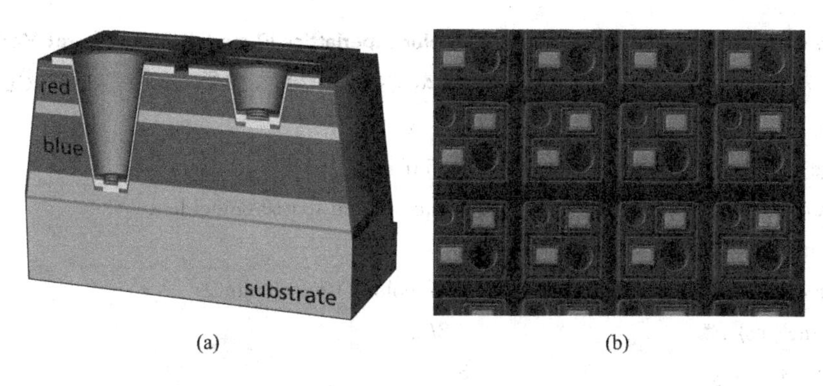

Figure 27.30 Dual-color InAs/GaSb SLS FPAs images illustrating enabling a simultaneous co-located photon detection at 3–4 μm (blue channel) and 4–5 μm (red channel): (a) schematic cross-section (Adapted after Ref. [125]), (b) at a pixel pitch of 30 μm, three contact lands per pixel permit simultaneous and spatially coincident detection of both colors. (Adapted after Ref. [43])

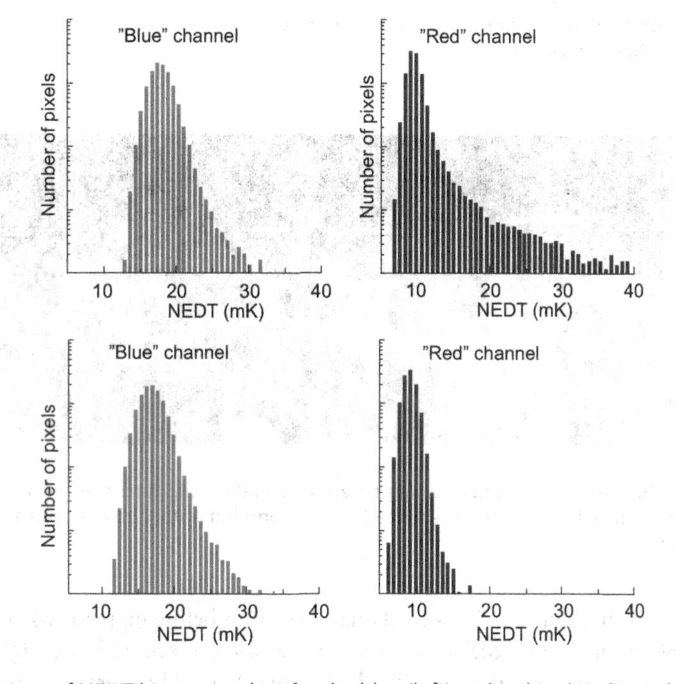

Figure 27.31 Comparison of NEDT histogram data for the blue (left) and red (right) channels of a typical dual-color 384 × 288 InAs/GaSb SL-FPA fabricated with old (upper row) and new (lower row) process technology. (Adapted after Ref. [47])

pixel structure is only 4.5 μm, which significantly reduces the technological challenge in comparison to dual-band HgCdTe FPAs with a typical total layer thickness around 15 μm. The pixel reduction to 30-μm pitch was achieved by restricting the number of contacts per pixel to two lands. A metallization grid, deposited in the trenches and connected to the ROIC outside of the active array, interconnects the common ground contact vias. Fraunhofer's dual-color MWIR superlattice detector array technology with simultaneous, co-located detection capability is ideally suited for airborne missile threat warning systems [42,43]. Figure 27.30 illustrates a fully processed dual-color 288 × 384 FPA. With 0.2-ms integration time and 78 K detector temperature, the superlattice camera achieves an NEDT of 29.5 mK for the blue channel (3.4 μm ≤ λ ≤ 4.1 μm) and 14.3 mK for the red channel (4.1 μm ≤ λ ≤ 5.1 μm).

Figure 27.31 compares the NEDT data for 384 × 288 dual-color InAs/GaSb superlattice (SL) detector arrays with a pitch size of 40 μm—each pixel with two back-to-back homojunction photodiodes detects a blue (3–4 μm) and a red (4–5 μm) wavelength band simultaneously [47]. The plots in the upper and lower

Table 27.5 Key characteristics of the 384 × 288 dual-color superlattice IR-module (AIM Infrarot-Module GmbH)

FORMAT	DUAL COLOR 384 × 288 × 2 SUPERLATTICE
Spectral range (μm) 1	3.4–x**
Spectral range (μm) 2	y–5.0**
Signal registration	Temporal and spatial coincidence
Pixel pitch (μm)	40
NEDT @ 50% well (mK)	<35 (color 1)/<25 (color 2)
Integration time (ms)/*F#*	2.8/2.0
Detector outputs analog	8
Max. pixel rate (MHz)	80
Readout mode	Snapshot, ITR
Max. full frame rate (Hz)	100 (for $\tau_{int} < 5.5$ ms)

** (3.4<x<y<5.0) details of cross over points x, y customer specific.
ITR, integrated then read; Max. maximum.

Figure 27.32 Bispectral IR image of an industrial site taken with a 384 × 288 dual-color InAs/GaSb SL camera. The two-color channels 3–4 μm and 4–5 μm are represented by the complementary colors cyan and red, respectively. (Adapted after Ref. [43])

rows represent the NEDT distribution of a typical dual-color FPA before (upper row) and after (lower row) implementation of a novel method for dielectric surface passivation. Pixels with high $1/f$ noise produce a tail in the root-mean-square noise distribution. While the blue channel histogram data are virtually unaffected by the modified process, the noisy pixels responsible for the tail of the NEDT distribution have now disappeared and the pixel operability has been increased to values well over 99%. In particular, the refined technology has dramatically reduced the number of pixels displaying burst or random telegraph noise. Table 27.5 summarizes the performance characteristics of bispectral type-II InAs/GaSb superlattice arrays fabricated by AIM.

As an example, the excellent imagery delivered by the 288 × 384 InAs/GaSb dual-color camera is presented in Figure 27.32. The image is a superposition of the images of the two channels coded in the complimentary colors cyan and red for the detection ranges of 3–4 μm and 4–5 μm, respectively. The red signatures reveal hot CO_2 emissions in the scene, whereas water vapor, for example, from steam exhausts or in clouds, appear cyan due to the frequency dependency of the Rayleigh scattering coefficient.

A research group of the Northwestern University has demonstrated different types of bias-selectable dual-band T2SL FPAs including combination of SW/MW, MW/LW, and LW1/LW2 arrays [44,45,123,124]. The MW/LW combination uses a back-to-back n-M-π-p-p-π-M-n structure, where the MW active region was achieved using 7.5 mono-layers (MLs) of InAs and 10 MLs of GaSb per period with the doping M-barrier. In the LW active region, 13 MLs of InAs and 7 MLs of GaSb were used in

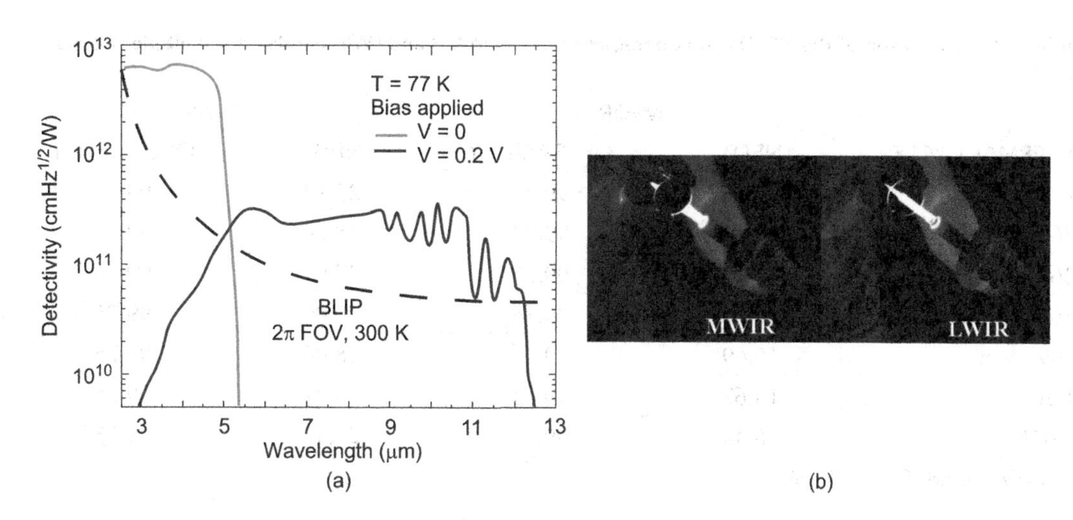

Figure 27.33 Bias selectable dual-band MW/LW T2SL array: (a) detectivity spectrum of both MWIR and LWIR channels at 77 K shown with BLIP detectivity limit (2π FOV, 300 K background), (b) MWIR and LWIR channels imaging a 11.3-μm narrow-band pass optical filter at 81 K. (Adapted after Ref. [45])

Figure 27.34 Pictures acquired with a dual-band 1,280 × 720, 12 μm pitch, T2SL MW/LW FPA. The image was captured at 80 K and *f*/4 optics. (Adapted after Ref. [126])

superlattice periods. An n-type GaSb semitransparent substrate was mechanically lapped down to a thickness of 30–40 μm and polished to a mirror-like surface. Figure 27.33a shows the detectivity spectrum of both MW and LW channels at 77 K. The resistance-area (*RA*) product of the LW channel at bias voltage of 0.2 V attained a value close to 600 Ωcm^2. The median NEDT of ~10 mK and ~30 mK were achieved using 10- and 0.18-ms integration times for MW and LW channels, respectively. The obtained images are shown in Figure 27.33b.

Excellent manufacturability and high uniformity of T2SLs superlattices have been also demonstrated by HRL Laboratories, Malibu, for HD-format (1,280 × 720, 12 μm pitch) T2SL dual-band MWIR/LWIR FPAs grown on GaSb substrates [126]. After metallization and dielectric etching, indium bumps were deposited on both the detectors and ROICs. Next, after hybridization and underfilling the hybrids, the GaSb substrates were completely etched to eliminate any transmission loss due to free carrier absorption.

Excellent imaging was achieved in both single-color and sequential modes—see Figure 27.34. The reliability test was carried out by cycling array's temperature from 70 to 290 K over 2,000 times. No degradation in either the sensitivity or the operability of the MWIR and LWIR bands (Table 27.6) was observed, confirming the stability of our hybridization and post-process. The rapid maturation of this technology makes it a strong candidate for deployment in future dual band MW/LW systems.

Recently, a novel device design using the two-terminal triple-band T2SL-based SWIR/MWIR/LWIR photodetector similar to that presented in Figure 27.35 has been demonstrated [48]. This device can

Table 27.6 Comparison of the NETDs and operabilities of the MWIR and LWIR bands after multiple thermal cycles

	MWIR		LWIR	
THERMAL CYCLE	NEDT	OPERABILITY	NEDT	OPERABILITY
4	15.13	99.71	27.12	99.76
266	15.66	99.73	28.74	99.75
368	16.17	99.75	27.91	99.76
469	16.09	99.74	27.00	99.75
669	15.69	99.74	28.06	99.75
1,269	15.67	99.73	26.87	99.75
2,031	16.44	99.74	27.47	99.75

Source: Adapted after Ref. [126]

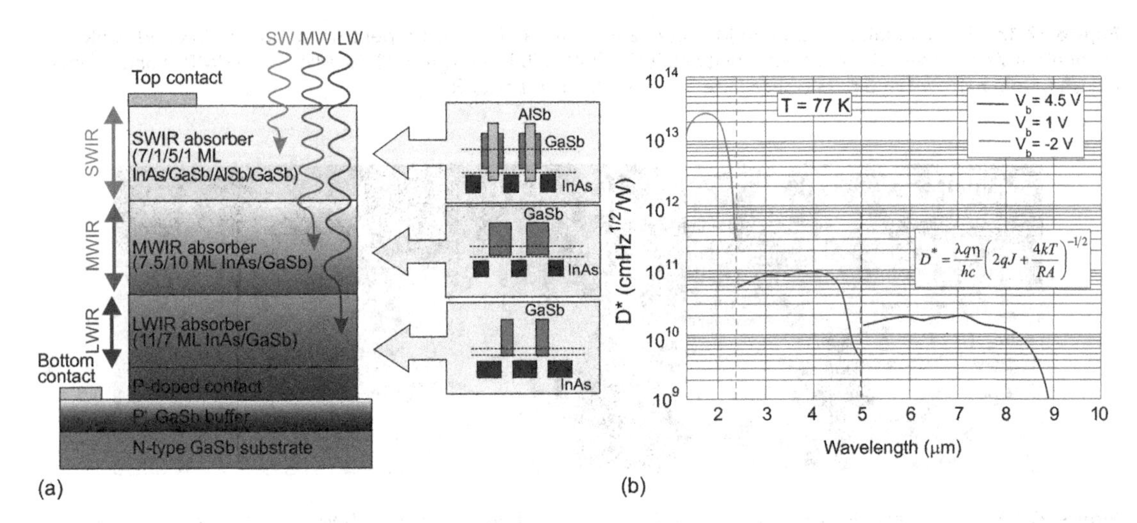

Figure 27.35 Triple-band SWIR/MWIR/LWIR T2SL photodiode: (a) structure with two terminal contacts and schematic band alignments; (b) calculated detectivity at 77 K using the equation in the inset. The SWIR detection operates at –2V. The MWIR detection operates at 1V, and LWIR detection operates at 4.5V positive bias voltages. (Adapted after Ref. [48])

perform sequentially as three individual single-color photodetectors according to the magnitude variation of applied bias (see Figure 27.35).

Triple-band SW-MW-LW photodiodes are designed to consist of 1.5-µm-thick undoped active region in SWIR, 0.5-µm-thick n-doped SWIR ($n \sim 10^{18}$ cm^{-3}), 2.0-µm-thick MWIR active region, 0.5-µm-thick undoped followed by 1.0-µm-thick p-doped LWIR active region ($p \sim 10^{16}$ cm^{-3}) and 0.5-µm-thick bottom p-contact ($p \sim 10^{18}$ cm^{-3}). The total thickness of the device structure is 6 µm.

Figure 27.35b shows the calculated shot-noise-limited detectivity of the device in its three operation modes at 77 K based on the measured quantum efficiency, the dark current, and the *RA* product. The device operates at bias of –2, 1, and 4.5 V and provides a D^* of 3.0×10^{13}, 1×10^{11}, and 2.0×10^{10} cmHz$^{1/2}$ W^{-1} at peak responsivity (λ = 1.7, 4.0, and 7.2 µm).

27.5 MULTIBAND QUANTUM DOT INFRARED PHOTODETECTORS

QDIP devices capable of detecting several separate wavelengths can be fabricated by vertical stacking of the different QWIP layers during epitaxial growth. The schematic structure is shown in Figure 27.36. In the

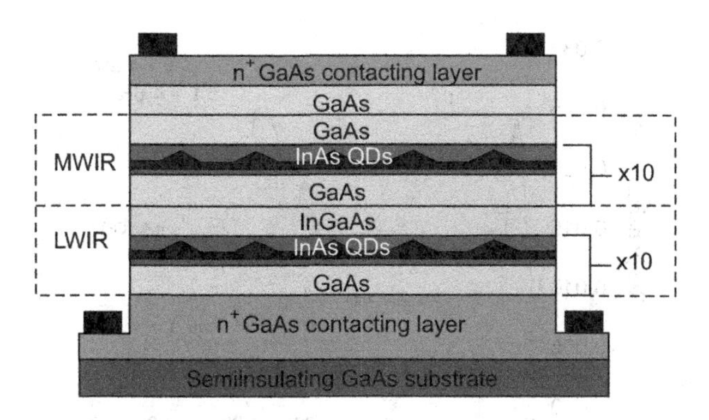

Figure 27.36 Schematic structure of the multispectral QDIP device. (From Ref. [127]. With permission.)

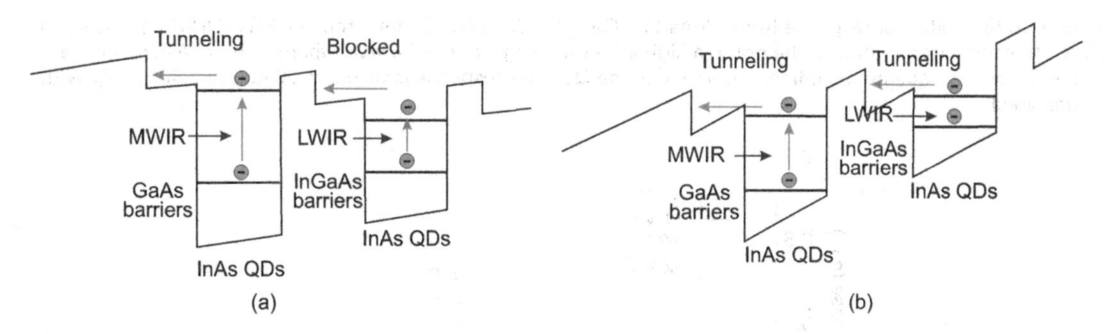

Figure 27.37 Simplified band diagram of the structure shown in Figure 27.36 at different bias levels; (a) low and (b) higher bias voltages. (From Ref. [127]. With permission.)

case of the structure described by Lu et al. [127], each of the QDIP absorption band consists of 10-period InAs/InGaAs quantum dot (QD) layers sandwiched between the top and bottom electrodes. Figure 27.37 shows the simplified band diagram of this structure at different bias levels. The bias voltage selection of detection bands originates from the asymmetric band structure. At low bias voltage, the high-energy GaAs barrier blocks the photocurrent generated by LWIR radiation and only responds to the MWIR incidence. On the contrary, as the bias voltage increases, the barrier energy decreases, allowing LWIR signals to be detected at different bias voltage levels.

The first two-color quantum dot FPA demonstration was based on a voltage-tunable InAs/InGaAs/GaAs DWELL structure [52,53]. As was described in Section 21.1, in this type of structure, InAs QDs are placed in an InGaAs well, which in turn is placed in a GaAs matrix (see Figure 21.4).

Figure 27.38 shows the multicolor response from a DWELL detector [128]. This device has demonstrated multicolor response ranging from the MWIR (3–5 μm) based on a bound-to-continuum transition to the LWIR (8–12 μm), which is based on a bound state in the dot to a bound state in the well. A VLWIR has also been observed and has been attributed to transitions between two bound states in the QDs, since the calculated energy spacing between the dot levels is about 50–60 meV. Moreover, by adjusting the voltage bias of the device, it is possible to modify the ratio of electrons promoted by MWIR, LWIR, and VLWIR absorptions. Typically, the MWIR response dominates at low to nominal voltages due to higher escape probability. With increasing voltage, the LWIR and eventually VLWIR responses are enhanced due to the increased tunneling probability of lower states in the DWELL detector (see Figure 27.39) [55]. The bias-dependent shift of the spectral response is observed due to quantum-confined Stark effect. This voltage-control of spectral response can be exploited to realize spectrally smart sensors whose wavelength and bandwidth can be tuned depending on the desired application [52,128–130].

Typically, the detector structure consists of a 15-stack asymmetric DWELL structure sandwiched between two highly doped n-GaAs contact layers. The DWELL region consists of 2.2 ML of n-doped InAs QDs in an $In_{0.15}Ga_{0.85}As$ well, itself placed within a GaAs matrix. By varying the width of the bottom

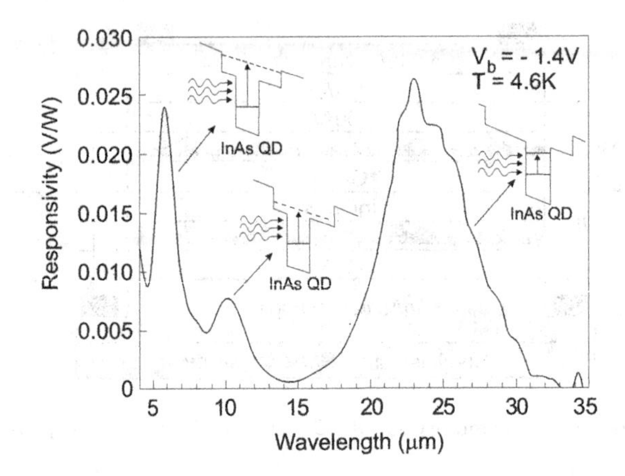

Figure 27.38 Multicolor response from a InAs/$In_{0.15}Ga_{0.85}As$/GaAs DWEL detector. The MWIR (LWIR) peak is possibly a transition from a state in the dot to a higher (lower) lying state in the well, whereas the VLWIR response is possibly from two quantum-confined levels within the QD. This response is visible to 80 K. (From Ref. [128]. With permission.)

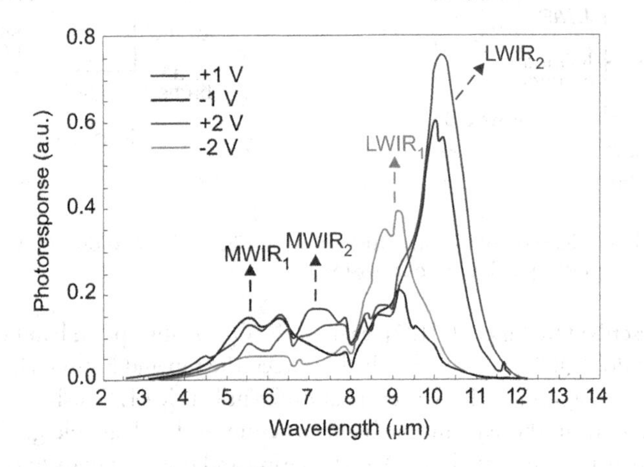

Figure 27.39 Spectral response from a DWELL detector with response at $V_b = \pm 1$ V and ± 2 V. Note the response in the two MWIR and LWIR bands can be measured using this detector. The relative intensities of the bands can be altered by the applied bias. (From Ref. [55]. With permission.)

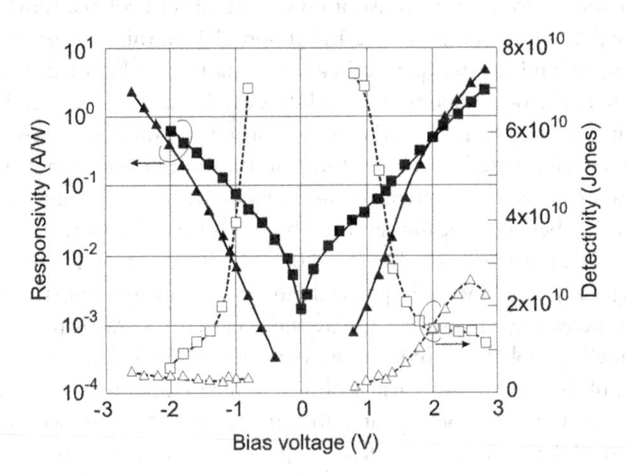

Figure 27.40 Peak responsivity for a 15-stack DWELL detector at 78 K obtained using a calibrated blackbody source. Solid squares: MWIR responsivity; solid triangles: LWIR responsivity; open square: MWIR detectivity; open triangles: LWIR detectivity. (From Ref. [53]. With permission.)

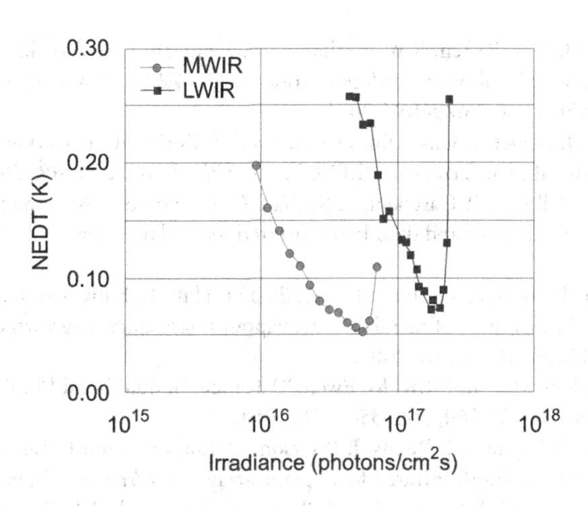

Figure 27.41 NEDT in the MWIR and LWIR bands at 77 K. Irradiance levels for MWIR and LWIR are 3–5 μm (f/2) and 8–12 μm (f/2.3), respectively. (From Ref. [54]. With permission.)

InGaAs well from 10 to 60 Å, the operating wavelength of the detector can be changed from 7.2 to 11 μm. The responsivity and detectivity obtained from the test devices at 78 K are shown in Figure 27.40 [53]. The measured detectivities were 2.6×10^{10} cmHz$^{1/2}$ W^{-1} ($V_b = 2.6$ V) for the LWIR band, and 7.1×10^{10} cmHz$^{1/2}$ W^{-1} ($V_b = 1$ V) for the MWIR band.

Varley et al. [54] have demonstrated a two-color, MWIR/LWIR, 320 × 256 FPA based on DWELL detectors. Minimum NEDT values of 55 mK (MWIR) and 70 mK (LWIR) were measured (see Figure 27.41).

REFERENCES

1. P. Norton, J. Campbell, S. Horn, and D. Reago, "Third-generation infrared imagers," *Proc. SPIE*, 4130, 226–36, 2000.
2. M.Z. Tidrow, W.A. Beck, W.W. Clark, H.K. Pollehn, J.W. Little, N.K. Dhar, P.R. Leavitt, et al., "Device physics and focal plane applications of QWIP and MCT," *Opto-Electron. Rev.*, 7, 283–96, 1999.
3. M.N. Abedin, T.F. Refaat, I. Bhat, Y. Xiao, S. Bandara and S.D. Gunapala, "Progress of multicolor single detector to detector array development for remote sensing," *Proc. SPIE*, 5543, 239–47, 2004.
4. O. Djazovski, "Focal Plane Arrays for Optical Payloads," in *Optical Payloads for Space Missions*, eds. S.N. Qian, 793–837, Wiley, Chichester, UK, 2016.
5. C.D. Tran, "Principles, instrumentation, and applications of infrared multispectral imaging, an overview," *Anal. Lett.*, 38, 735–52, 2005.
6. W.J. Gunning, J. DeNatale, P. Stupar, R. Borwick, R. Dannenberg, R. Sczupak and P.O. Pettersson, "Adaptive focal plane array: An example of MEMS, photonics, and electronics integration," *Proc. SPIE*, 5783, 336–75, 2005.
7. P. McCarley, "Recent developments in biologically inspired seeker technology," *Proc. SPIE*, 4288, 1–12, 2001.
8. J.T. Caulfield, "Next Generation IR Focal Plane Arrays and Applications," in *Proceedings of 32nd Applied Imagery Pattern Recognition Workshop*, IEEE, New York, 2003.
9. R. Keller, P. Dreiske, A. Turner, B. Seymour and H. Schaake, "Multispectral HDVIP Focal Plane Arrays," in Defense Technical Information Center, 1998. https://pdfs.semanticscholar.org/7c6b/db4cba710d59fd7ee476 2ff1721044426112.pdf
10. E.P.G. Smith, L.T. Pham, G.M. Venzor, E.M. Norton, M.D. Newton, P.M. Goetz, V.K. Randall, et al., "HgCdTe focal plane arrays for dual-color mid- and long-wavelength infrared detection," *J. Electro. Mater.*, 33, 509–16, 2004.
11. W.A. Radford, E.A. Patten, D.F. King, G.K. Pierce, J. Vodicka, P. Goetz, G. Venzor, et al., "Third generation FPA development status at Raytheon vision systems," *Proc. SPIE*, 5783, 331–9, 2005.
12. A. Rogalski, "HgCdTe infrared detector material: history, status, and outlook," *Rep. Prog. Phys.*, 68, 2267–336, 2005.
13. P.D. Dreiske, "Development of two-color focal-plane arrays based on HDVIP," *Proc. SPIE*, 5783, 325–30, 2005.

14. D.F. King, W.A. Radford, E.A. Patten, R.W. Graham, T.F. McEwan, J.G. Vodicka, R.F. Bornfreund, P.M. Goetz, G.M. Venzor, and S.M. Johnson, "3rd-generation 1280 × 720 FPA development status at Raytheon Vision Systems," *Proc. SPIE*, 6206, 62060W, 2006.
15. G. Destefanis, P. Ballet, J. Baylet, P. Castelein, O. Gravrand, J. Rothman, F. Rothan, et al., "Bi-color and dual-band HgCdTe infrared focal plane arrays at DEFIR," *Proc. SPIE*, 6206, 62060R, 2006.
16. G. Destefanis, J. Baylet, P. Ballet, P. Castelein, F. Rothan, O. Gravrand, J. Rothman, J.P. Chamonal, and A. Million, "Status of HgCdTe bicolor and dual-band infrared focal plane arrays at LETI," *J. Electron. Mater.*, 36, 1031–44, 2007.
17. N.T. Gordon, P. Abbott, J. Giess, A. Graham, J.E. Hails, D.J. Hall, L. Hipwood, C.L. Lones, C.D. Maxeh, and J. Price, "Design and assessment of metal-organic vapour phase epitaxy–grown dual wavelength infrared detectors," *J. Electron. Mater.*, 36, 931–6, 2007.
18. P. Abbott, L. Pillans, P. Knowles, and R.K. McEwen, "Advances in dual-band IRFPAs made from HgCdTe grown by MOVPE," *Proc. SPIE*, 7660, 766035–1–11, 2010.
19. E. Smith, G. Venzor, A. Gallagher, M. Reddy, J. Peterson, D. Lofgreen, and J. Randolph, "Large-format HgCdTe dual-band long-wavelength infrared focal-plane arrays," *J. Electron. Mater.*, 40(8), 1630–6, 2011.
20. Y. Reibel, F. Chabuel, C. Vaz, D. Billon-Lanfrey, J. Baylet, O. Gravrand, P. Ballet, and G. Destefanis, "Infrared dual band detectors for next generation," *Proc. SPIE*, 8012, 801238–1–13, 2011.
21. J. Ziegler, D. Eich, M. Mahlein, T. Schallenberg, R. Scheibner, J. Wendler, J. Wenischa, R. Wollrab, V. Daumer, R. Rehm, F. Rutz, and M. Walther, "The development of 3rd gen IR detectors at AIM," *Proc. SPIE*, 8012, 801237–1–13, 2011.
22. M. Vuillermet, D. Billon-Lanfrey, Y. Reibel, A. Manissadjian, L. Mollard, N. Baier, O. Gravrand, and G. Destéfanis, "Status of MCT focal plane arrays in France," *Proc. SPIE*, 8353, 83522K–1–12, 2012.
23. S.D. Gunapala and S.V. Bandara, "GaAs/AlGaAs Based Quantum Well Infrared Photodetector Focal Plane Arrays," in *Handbook of Infrared Detection Technologies*, eds. M. Henini and M. Razeghi, 83–119, Elsevier, Oxford, 2002.
24. G. Sarusi, "QWIP or other alternatives for third generation infrared systems," *Infrared Phys. Technol.*, 44, 439–44, 2003.
25. A. Rogalski, "Quantum well photoconductors in infrared detectors technology," *J. Appl. Phys.*, 93, 4355–91, 2003.
26. A. Manissadjian, D. Gohier, E. Costard, and A. Nedelcu, "Single color and dual band QWIP production results," *Proc. SPIE*, 6206, 62060E, 2006.
27. M. Münzberg, R. Breiter, W. Cabanski, H. Lutz, J. Wendler, J. Ziegler, R. Rehm, and M. Walther, "Multi spectral IR detection modules and applications," *Proc. SPIE*, 6206, 620627, 2006.
28. S.D. Gunapala, S.V. Bandara, J.K. Liu, J.M. Mumolo, C.J. Hill, D.Z. Ting, E. Kurth, J. Woolaway, P.D. LeVan, and M.Z. Tidrow, "Towards 16 megapixel focal plane arrays," *Proc. SPIE*, 6660, 66600E, 2007.
29. S.D. Gunapala, S.V. Bandara, J.K. Liu, J.M. Mumolo, C.J. Hill, S.B. Rafol, D. Salazar, J. Woollaway, P.D. LeVan, and M.Z. Tidrow, "Towards dualband megapixel QWIP focal plane arrays," *Infrared Phys. Technol.*, 50, 217–26, 2007.
30. A. Nedelcu, E. Costard, P. Bois, and X. Marcadet, "Research topics at Tales research and technology: small pixels and third generation applications," *Infrared Phys. Technol.*, 50, 227–33, 2007.
31. S.D. Gunapala, S.V. Bandara, J.K. Liu, J.M. Mumolo, D.Z. Ting, C.J. Hill, J. Nguyen, and S.B. Rafol, "Demonstration of 1024 × 1024 pixel dual-band QWIP focal plane array," *Proc. SPIE*, 7660, 76603L–1–8, 2010.
32. V. Guériaux, N. Brière de l'Isle, A. Berurier, O. Huet, A. Manissadjian, H. Facoetti, X. Marcadet, M. Carras, V. Trinite, and A. Nedelcu, "Quantum well infrared photodetectors: present and future," *Opt. Eng.*, 50(6), 061013–1–19, 2011.
33. P. Bois, V. Guériaux, N. Brière de l'Isle, A. Manissadjian, H. Facoetti, X. Marcadeta, E. Costard, and A. Nedelcu, "QWIP status and future trends at Thales," *Proc. SPIE*, 8268, 82682M–1–11, 2012.
34. M. Runtz, F. Perrier, N. Ricard, E. Costard, A. Nedelcu and V. Guériaux, "QWIP infrared detector production line results," *Proc. SPIE*, 8353, 835337–1–13, 2012.
35. A. Rogalski, "Third generation photon detectors," *Opt. Eng.*, 42, 3498–516, 2003.
36. R. Rehm, M. Walther, J. Schmitz, J. Fleißner, F. Fuchs, J. Ziegler, and W. Cabanski, "InAs/GaSb superlattice focal plane arrays for high-resolution thermal imaging," *Opto-Electron. Rev.*, 14, 283–96, 2006.
37. A. Rogalski and P. Martyniuk, "InAs/GaInSb superlattices as a promising material system for third generation infrared detectors," *Infrared Phys. Technol.*, 48, 39–52, 2006.

38. R. Rehm, M. Walther, J. Schmitz, J. Fleißner, J. Ziegler, W. Cabanski, and R. Breiter, "Bispectral thermal imaging with quantum-well infrared photodetectors and InAs/GaSb type II superlattices," *Proc. SPIE*, 6292, 629404, 2006.
39. A. Rogalski, "Competitive technologies of third generation infrared photon detectors," *Opto-Electron. Rev.*, 14, 87–101, 2006.
40. A. Rogalski, "New material systems for third generation infrared photodetectors," *Opto-Electron. Rev.*, 16, 458–82, 2008.
41. A. Rogalski, J. Antoszewski, and L. Faraone, "Third-generation infrared photodetector arrays," *J. Appl. Phys.*, 105, 091101–44, 2009.
42. F. Rutz, R. Rehm, J. Schmitz, J. Fleissner and M. Walther, "InAs/GaSb superlattice focal plane array infrared detectors: manufacturing aspects," *Proc. SPIE*, 7298, 72981R–1–10, 2009.
43. R. Rehm, M. Walther, J. Schmitz, F. Rutz, A. Wörl, R. Scheibner, and J. Ziegler, "Type-II superlattices: the Fraunhofer perspective," *Proc. SPIE*, 7660, 76601G–1–12, 2010.
44. M. Razeghi, A.M. Hoang, A. Haddadi, G. Chen, S. Ramezani-Darvish, P. Bijjam, P. Wijewarnasuriya, and E. Decuir, "High-performance bias-selectable dual-band short-/mid-wavelength infrared photodetectors and focal plane arrays based on InAs/GaSb/AlSb type-II superlattices," *Proc. SPIE*, 8704, 8704–54, 2013.
45. M. Razeghi, A. Haddadi, A.M. Hoang, G. Chen, S. Ramezani-Darvish and P. Bijjam, "High-performance bias-selectable dual-band mid-/long-wavelength infrared photodetectors and focal plane arrays based on InAs/GaSb type-II superlattices," *Proc. SPIE*, 8704, 87040S, 2013.
46. A.M. Hoang, G. Chen, A. Haddadi, and M. Razeghi, "Demonstration of high performance bias-selectable dual-band short-/mid-wavelength infrared photodetectors based on type-II InAs/GaSb/AlSb superlattices," *Appl. Phys. Lett.*, 102, 011108, 2013.
47. R. Rehm, F. Lemke, M. Masur, J. Schmitz, T. Stadelman, M. Wauro, A. Wörl, and M. Walther, "InAs/GaSb superlattice infrared detectors," *Infrared Phys. Technol.*, 70, 87–92, 2015.
48. A.M. Hoang, A. Dehzangi, S. Adhikary, and M. Razeghi, "High performance bias-selectable three-color short-wave/mid-wave/ long-wave infrared photodetectors based on type-II InAs/GaSb/AlSb superlattices," *Sci. Rep.*, 6, 24144, 2016, doi:10.1038/srep24144.
49. S. Yaoyao, H. Xi, H. Hongyue, J. Dongwei, G. Chunyan, J. Zhi, L. Yuexi, W. Guowei, X. Yingqiang, and N. Zhichuan, "320 × 256 short-/mid-wavelength dual-color infrared focal plane arrays based on type-II InAs/GaSb superlattice," *Infrared Phys. Technol.*, 82, 140–3, 2017.
50. S.M. Kim and J.S. Harris, "Multicolor InGaAs quantum-dot infrared photodetectors," *IEEE Photonics Technol. Lett.*, 16, 2538–40, 2004.
51. S. Chakrabarti, X.H. Su, P. Bhattacharya, G. Ariyawansa, and A.G.U. Perera, "Characteristics of a multicolour InGaAs-GaAs quantum-dot infrared photodetector," *IEEE Photonics Technol. Lett.*, 17, 178–80, 2005.
52. S. Krishna, D. Forman, S. Annamalai, P. Dowd, P. Varangis, T. Tumolillo, A. Gray, et al., "Demonstration of a 320 × 256 two-color focal plane array using InAs/InGaAs quantum dots in well detectors," *Appl. Phys. Lett.*, 86, 193501, 2005.
53. S. Krishna, D. Forman, S. Annamalai, P. Dowd, P. Varangis, T. Tumolillo, et al., "Two-color focal plane arrays based on self assembled quantum dots in a well heterostructure," *Phys. Status Solidi C*, 3, 439–43, 2006.
54. E. Varley, M. Lenz, S.J. Lee, J.S. Brown, D.A. Ramirez, A. Stintz, and S. Krishna, "Single bump, two-color quantum dot camera," *Appl. Phys. Lett.*, 91, 081120, 2007.
55. S. Krishna, S.D. Gunapala, S.V. Bandara, C. Hill, and D.Z. Ting, "Quantum dot based infrared focal plane arrays," *Proc. IEEE*, 95, 1838–52, 2007.
56. H. Halpert and B.I. Musicant, "N-color (Hg,Cd)Te photodetectors," *Appl. Opt.*, 11, 2157–61, 1972.
57. "InSb/HgCdTe two-color detector," www.irassociates.com/insbhgcdte.htm.
58. D. Reago, S. Horn, J. Campbell, and R. Vollmerhausen, "Third generation imaging sensor system concepts," *Proc. SPIE*, 3701, 108–17, 1999.
59. S. Horn, P. Norton, K. Carson, R. Eden, and R. Clement, "Vertically-integrated sensor arrays—VISA," *Proc. SPIE*, 5406, 332–40, 2004.
60. R. Balcerak and S. Horn, "Progress in the development of vertically-integrated sensor arrays," *Proc. SPIE*, 5783, 384–91, 2005.
61. P.R. Norton, "Third-generation sensors for night vision," *Opto-Electron. Rev.*, 14, 283–96, 2006.
62. J. Johnson, "Analysis of Image Forming Systems," in *Image Intensifier Symposium, AD 220160*, Warfare Electrical Engineering Department, U.S. Army Research and Development Laboratories, Ft. Belvoir, VA., 1958, pp. 244–73.

63. U. Adomeit, "Infrared detection, recognition and identification of handheld objects," *Proc. SPIE*, 8541, 85410O–1–9, 2012.
64. J.L. Miller, "Future sensor system needs for staring arrays," *Infrared Phys. Technol.*, 54, 164–9, 2011.
65. S. Horn, P. Norton, T. Cincotta, A. Stolz, D. Benson, P. Perconti, and J. Campbell, "Challenges for third-generation cooled imagers," *Proc. SPIE*, 5074, 44–51, 2003.
66. J.C. Campbell, A.G. Dentai, T.P. Lee, and C.A. Burrus, "Improved two-wavelength demultiplexing InGaAsP photodetector," *IEEE J. Quantum Electron.*, 16, 601, 1980.
67. E.R. Blazejewski, J.M. Arias, G.M. Williams, W. McLevige, M. Zandian, and J. Pasko, "Bias-switchable dual-band HgCdTe infrared photodetector," *J. Vac. Sci. Technol. B*, 10, 1626, 1992.
68. J.A. Wilson, E.A. Patten, G.R. Chapman, K. Kosai, B. Baumgratz, P. Goetz, S. Tighe, et al., "Integrated two-color detection for advanced FPA applications," *Proc. SPIE*, 2274, 117–25, 1994.
69. R.D. Rajavel, D.M. Jamba, J.E. Jensen, O.K. Wu, J.A. Wilson, J.L. Johnson, E.A. Patten, K. Kasai, P.M. Goetz, and S.M. Johnson, "Molecular beam epitaxial growth and performance of HgCdTe-based simultaneous-mode two-color detectors," *J. Electron. Mater.*, 27, 747–51, 1998.
70. E.P.G. Smith, E.A. Patten, P.M. Goetz, G.M. Venzor, J.A. Roth, B.Z. Nosho, J.D. Benson, et al., "Fabrication and characterization of two-color midwavelength/long wavelength HgCdTe infrared detectors," *J. Electron. Mater.*, 35, 1145–52, 2006.
71. D.F. King, J.S. Graham, A.M. Kennedy, R.N. Mullins, J.C. McQuitty, W.A. Radford, T.J. Kostrzewa, et al., "3rd-generation MW/LWIR sensor engine for advanced tactical systems," *Proc. SPIE*, 6940, 69402R, 2008.
72. E.P.G. Smith, A.M. Gallagher, T.J. Kostrzewa, M.L. Brest, R.W. Graham, C.L. Kuzen, E.T. Hughes, et al., "Large format HgCdTe focal plane arrays for dual-band long-wavelength infrared detection," *Proc. SPIE*, 7298, 72981Y, 2009.
73. M.B. Reine, A. Hairston, P. O'Dette, S.P. Tobin, F.T.J. Smith, B.L. Musicant, P. Mitra, and F.C. Case, "Simultaneous MW/LW dual-band MOCVD HgCdTe 64 × 64 FPAs," *Proc. SPIE*, 3379, 200–12, 1998.
74. J.P. Zanatta, P. Ferret, R. Loyer, G. Petroz, S. Cremer, J.P. Chamonal, P. Bouchut, A. Million, and G. Destefanis, "Single and two colour infrared focal plane arrays made by MBE in HgCdTe" *Proc. SPIE*, 4130, 441–51, 2000.
75. P. Tribolet, M. Vuillermet, and G. Destefanis, "The third generation cooled IR detector approach in France," *Proc. SPIE*, 5964, 49–60, 2005.
76. J.P. Zanatta, G. Badano, P. Ballet, C. Largeron, J. Baylet, O. Gravrand, J. Rothman, et al., "Molecular beam epitaxy of HgCdTe on Ge for third-generation infrared detectors," *J. Electron. Mater.*, 35, 1231–6, 2006.
77. P. Tribolet, G. Destefanis, P. Ballet, J. Baylet, O. Gravrand, and J. Rothman, "Advanced HgCdTe technologies and dual-band developments," *Proc. SPIE*, 6940, 69402P, 2008.
78. J. Giess, M.A. Glover, N.T. Gordon, A. Graham, M.K. Haigh, J.E. Hails, D.J. Hall, and D.J. Lees, "Dial-wavelength infrared focal plane arrays using MCT grown by MOVPE on silicon substrates," *Proc. SPIE*, 5783, 316–24, 2005.
79. N.T. Gordon, P. Abbott, J. Giess, A. Graham, J.E. Hails, D.J. Hall, L. Hipwood, C.L. Lones, C.D. Maxeh, and J. Price, "Design and assessment of metal-organic vapour phase epitaxy–grown dual wavelength infrared detectors," *J. Electron. Mater.*, 36, 931–6, 2007.
80. C.L. Jones, L.G. Hipwood, J. Price, C.J. Shaw, P. Abbott, C.D. Maxey, H.W. Lau, et al., "Multi-colour IRFPAs made from HgCdTe grown by MOVPE," *Proc. SPIE*, 6542, 654210, 2007.
81. J.P.G. Price, C.L. Jones, L.G. Hipwood, C.J. Shaw, P. Abbott, C.D. Maxey, H.W. Lau, et al., "Dual-band MW/LW IRFPAs made from HgCdTe grown by MOVPE," *Proc. SPIE*, 6940, 69402S, 2008.
82. F. Aqariden, P.D. Dreiske, M.A. Kinch, P.K. Liao, T. Murphy, H.F. Schaake, T.A. Shafer, H.D. Shih, and T.H. Teherant, "Development of molecular beam epitaxy grown $Hg_{1-x}Cd_xTe$ for high-density vertically-integrated photodiode-based focal plane arrays," *J. Electron. Mater.*, 36, 900–4, 2007.
83. M.A. Kinch, "HDVIP^{TM} FPA technology at DRS," *Proc. SPIE*, 4369, 566–78, 2001.
84. M.A. Kinch, *Fundamentals of Infrared Detector Materials*, SPIE Press, Bellingham, WA, 2007.
85. W.E. Tennant, M. Thomas, L.J. Kozlowski, W.V. McLevige, D.D. Edwall, M. Zandian, K. Spariosu, et al., "A novel simultaneous unipolar multispectral integrated technology approach for HgCdTe IR detectors and focal plane arrays," *J. Electron. Mater.*, 30, 590–4, 2001.
86. L.A. Almeida, M. Thomas, W. Larsen, K. Spariosu, D.D. Edwall, J.D. Benson, W. Mason, A.J. Stoltz, and J.H. Dinan, "Development and fabrication of two-color mid- and short-wavelength infrared simultaneous unipolar multispectral integrated technology focal-plane arrays," *J. Electron. Mater.*, 30, 669–76, 2002.
87. A.H. Lockwood, J.R. Balon, P.S. Chia, and F.J. Renda, "Two-color detector arrays by PbTe/$Pb_{0.8}Sn_{0.2}Te$ liquid phase epitaxy," *Infrared Phys.*, 16, 509–14, 1976.

88. J. Baylet, P. Ballet, P. Castelein, F. Rothan, O. Gravrand, M. Fendler, E. Laffosse, et al., "TV/4 dual-band HgCdTe infrared focal plane arrays with a 25 μm pitch and spatial coherence," *J. Electron. Mater.*, 35, 1153–8, 2006.
89. P.R. Norton, "Status of infrared detectors," *Proc. SPIE*, 3379, 102–14, 1998.
90. S. Gunapala, S.V. Bandara, J.K. Liu, J.M. Mumolo, D.Z. Ting, C.J. Hill, J. Nguyen, B. Simolon, J. Woolaway, S.C. Wang, W. Li, P.D. LeVan, and M.Z. Tidrow, "Demonstration of megapixel dual-band QWIP focal plane array," *IEEE J. Quantum Electron.*, 46, 285–93, 2010.
91. E.P.G. Smith, R.E. Bornfreund, I. Kasai, L.T. Pham, E.A. Patten, J.M. Peterson, J.A. Roth, et al., "Status of two-color and large format HgCdTe FPA technology at Reytheon Vision Systems," *Proc. SPIE*, 6127, 61271F, 2006.
92. K. Jóźwikowski and A. Rogalski, "Computer modeling of dual-band HgCdTe photovoltaic detectors," *J. Appl. Phys.*, 90, 1286–91, 2001.
93. A.K. Sood, J.E. Egerton, Y.R. Puri, E. Bellotti, D. D'Orsogna, L. Becker, R. Balcerak, K. Freyvogel, and R. Richwine, "Design and development of multicolor MWIR/LWIR and LWIR/VLWIR detector arrays," *J. Electron. Mater.*, 34, 909–12, 2005.
94. E. Bellotti and D. D'Orsogna, "Numerical analysis of HgCdTe simultaneous two-color photovoltaic infrared detectors," *IEEE J. Quantum Electron.*, 42, 418–26, 2006.
95. L.G. Hipwood, C.L. Jones, C.D. Maxey, H.W. Lau, J. Fitzmaurice, R.A. Catchpole, and M. Ordish, "Three-color MOVPE MCT diodes," *Proc. SPIE*, 6206, 620612, 2006.
96. K. Jóźwikowski and A. Rogalski, "Numerical analysis of three-colour HgCdTe detectors," *Opto-Electron. Rev.*, 15, 215–22, 2007.
97. W.A. Beck and T.S. Faska, "Current status of quantum well focal plane arrays," *Proc. SPIE*, 2744, 193–206, 1996.
98. M. Sundaram and S.C. Wang, "2-color QWIP FPAs," *Proc. SPIE*, 4028, 311–7, 2000.
99. S.D. Gunapala, S.V. Bandara, A. Sigh, J.K. Liu, S.B. Rafol, E.M. Luong, J.M. Mumolo, et al., "8–9 and 14–15 μm two-color 640 × 486 quantum well infrared photodetector (QWIP) focal plane array camera," *Proc. SPIE*, 3698, 687–97, 1999.
100. S.D. Gunapala, S.V. Bandara, A. Singh, J.K. Liu, B. Rafol, E.M. Luong, J.M. Mumolo, et al., "640 × 486 long-wavelength two-color GaAs/AlGaAs quantum well infrared photodetector (QWIP) focal plane array camera," *IEEE Trans. Electron Devices*, 47, 963–71, 2000.
101. S.D. Gunapala, S.V. Bandara, J.K. Liu, E.M. Luong, S.B. Rafol, J.M. Mumolo, D.Z. Ting, et al., "Recent developments and applications of quantum well infrared photodetector focal plane arrays," *Opto-Electron. Rev.*, 8, 150–63, 2001.
102. S.D. Gunapala, S.V. Bandara, J.K. Liu, B. Rafol, J.M. Mumolo, C.A. Shott, R. Jones, et al., "640 × 512 pixel narrow-band, four-band, and broad-band quantum well infrared photodetector focal plane arrays," *Infrared Phys. Technol.*, 44, 411–25, 2003.
103. S.D. Gunapala, S.V. Bandara, J.K. Liu, B. Rafol, and J.M. Mumolo, "640 × 512 pixel long-wavelength infrared narrowband, multiband, and broadband QWIP focal plane arrays," *IEEE Trans. Electron Devices*, 50, 2353–60, 2003.
104. S.D. Gunapala, S.V. Bandara, J.K. Liu, J.M. Mumolo, C.J. Hill, D.Z. Ting, E. Kurth, J. Woolaway, P.D. LeVan, and M.Z. Tidrow, "Development of megapixel dual-band QWIP focal plane array," *Proc. SPIE*, 6940, 69402T, 2008.
105. A. Soibel, S.D. Gunapala, S.V. Bandara, J.K. Liu, J.M. Mumolo, D.Z. Ting, C.J. Hill, and J. Nguyen, "Large format multicolor QWIP focal plane arrays," *Proc. SPIE*, 7298, 729806, 2009.
106. J. Bundas, K. Patnaude, R. Dennis, D. Burrows, R. Cook, A. Reisinger, M. Sundaram, R. Benson, J. Woolaway, J. Schlesselmann, and S. Petronio, "Two-color quantum well infrared photodetector focal plane arrays," *Proc. SPIE*, 6206, 62060G-1-11, 2006.
107. M. Sundaram, A. Reisinger, R. Dennis, K. Patnaude, D. Burrows, J. Bundas, K. Beech, and R. Faska, "Status of quantum well infrared photodetector technology at QmagiQ today," *Infrared Phys. Technol.*, 53, 194–8, 2011.
108. A. Goldberg, T. Fischer, J. Kennerly, S. Wang, M. Sundaram, P. Uppal, M. Winn, G. Milne, and M. Stevens, "Dual band QWIP MWIR/LWIRs focal plane array test results," *Proc. SPIE*, 4028, 276–87, 2000.
109. A.C. Goldberger, S.W. Kennerly, J.W. Little, H.K. Pollehn, T.A. Shafer, C.L. Mears, H.F. Schaake, M. Winn, M. Taylor, and P.N. Uppal, "Comparison of HgCdTe and QWIP dual-band focal plane arrays," *Proc. SPIE*, 4369, 532–46, 2001.
110. K.-K. Choi, M.D. Jhabvala, and R.J. Peraltas, "Voltage-tunable two-color corrugated-QWIP focal plane arrays," *IEEE Electron Device Lett.*, 29, 1011–3, 2008.

111. M. Jhabvala, "Applications of GaAs quantum well infrared photoconductors at the NASA/Goddard Space Flight Center," *Infrared Phys. Technol.*, 42, 363–76, 2001.
112. E. Costard, Ph. Bois, X. Marcadet, and A. Nedelcu, "QWIP and third generation IR imagers," *Proc. SPIE*, 5783, 728–35, 2005.
113. P. Castelein, F. Guellec, F. Rothan, S. Martin, P. Bois, E. Costard, O. Huet, X. Marcadet, and A. Nedelcu, "Demonstration of 256 × 256 dual-band QWIP infrared FPAs," *Proc. SPIE*, 5783, 804–15, 2005.
114. N. Perrin, E. Belhaire, P. Marquet, V. Besnard, E. Costard, A. Nedelcu, P. Bois, et al., "QWIP development status at Thales," *Proc. SPIE*, 6940, 694008, 2008.
115. J.A. Robo, E. Costard, J.P. Truffer, A. Nedelcu, X. Marcadet, and P. Bois, "QWIP focal plane arrays performances from MWIR to VLWIR," *Proc. SPIE*, 7298, 72980F, 2009.
116. W. Cabanski, M. Münzberg, W. Rode, J. Wendler, J. Ziegler, J. Fleißner, F. Fuchs, et al., "Third generation focal plane array IR detection modules and applications," *Proc. SPIE*, 5783, 340–9, 2005.
117. H. Schneider, T. Maier, J. Fleissner, M. Walther, P. Koidl, G. Weimann, W. Cabanski, et al., "Dual-band QWIP focal plane array for the second and third atmospheric windows," *Infrared Phys. Technol.*, 47, 53–8, 2005.
118. S. Gunapala, "Megapixel QWIPs deliver multi-color performance," *Compd. Semicond.*, 10, 25–8, 2005.
119. S.U. Eker, Y. Arslan, M. Kaldirim, and C. Besikci, "QWIP focal plane arrays on InP substrates for single and dual band thermal imagers," *Infrared Phys. Technol.*, 52, 385–90, 2009.
120. E.H. Aifer, J.G. Tischler, J.H. Warner, I. Vurgaftman, and J.R. Meyer, "Dual band LWIR/VLWIR type-II superlattice photodiodes," *Proc. SPIE*, 5783, 112–22, 2005.
121. M. Münzberg, R. Breiter, W. Cabanski, K. Hofmann, H. Lutz, J. Wendler, J. Ziegler, P. Rehm, and M. Walther, "Dual color IR detection modules, trends and applications," *Proc. SPIE*, 6542, 654207, 2007.
122. M. Razeghi, D. Hoffman, B.M. Nguyen, P.-Y. Delaunay, E.K. Huang, M.Z. Tidrow, and V. Nathan, "Recent advances in LWIR type-II InAs/GaSb superlattice photodetectors and focal plane arrays at the Center for Quantum Devices," *Proc. IEEE*, 97, 1056–66, 2009.
123. M. Razeghi, H. Haddadi, A.M. Hoang, E.K. Huang, G. Chen, S. Bogdanov, S.R. Darvish, F. Callewaert, and R. McClintock, "Advances in antimonide-based type-II superlattices for infrared detection and imaging at Center for Quantum Devices," *Infrared Phys. Technol.*, 59, 41–52, 2013.
124. M. Razeghi and B.-M. Nguyen, "Advances in mid-infrared detection and imaging: a key issues review," *Rep. Prog. Phys.*, 77, 082401–1–17, 2014.
125. R. Rehm, V. Daumer, T. Hugger, N. Kohn, W. Luppold, R. Müller, J. Niemasz, J. Schmidt, F. Rutz, T. Stadelmann, M. Wauro, and A. Wörl, "Type-II superlattice infrared detector technology at Fraunhofer IAF," *Proc. SPIE*, 9819, 9819–24, 2016.
126. Delaunay, P.-Y., Nosho, B.Z., Gurga, A.R., Terterian, S., and Rajavel, R.D., "Advances in III-V based dual-band MWIR/LWIR FPAs at HRL," *Proc. SPIE*, 101777, 101770T–1–12, 2017.
127. X. Lu, J. Vaillancourt, and M. Meisner, "A voltage-tunable multiband quantum dot infrared focal plane array with high photoconductivity," *Proc. SPIE*, 6542, 65420Q, 2007.
128. S. Krishna, "Quantum dots-in-a-well infrared photodetectors," *J. Phys. D: Appl. Phys.*, 38, 2142–50, 2005.
129. U. Sakoglu, J.S. Tyo, M.M. Hayat, S. Raghavan, and S. Krishna, "Spectrally adaptive infrared photodetectors with bias-tunable quantum dots," *J. Opt. Soc. Am. B*, 21, 7–17, 2004.
130. A.G.U. Perera, "Quantum structures for multiband photon detection," *Opto-Electron. Rev.*, 14, 99–108, 2006.

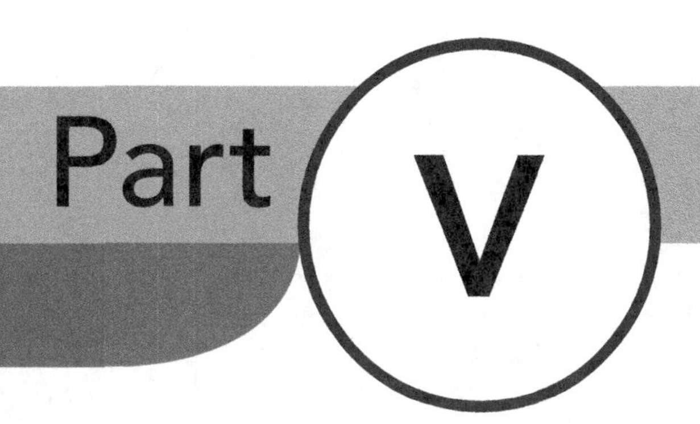

Terahertz detectors and focal plane arrays

28 Terahertz detectors and focal plane arrays

Terahertz (THz) technology is one of emerging technologies that will change our lives. A lot of attractive applications in security, medicine, biology, astronomy, and nondestructive materials testing have been demonstrated already. However, the realization of THz emitters and receivers is a challenge because the frequencies are too high for conventional electronics and the photon energies are too small for classical optics. As a result, THz radiation is resistant to the techniques commonly employed in these well-established neighboring bands.

In this chapter, issues associated with the development and exploitation of THz radiation detectors and focal plane arrays (FPA) are discussed. The historical impressive progress in THz detector sensitivity in a period more than half century is analyzed. More attention is given to the basic physical phenomena and the recent progress in both direct and heterodyne detectors. After the short description of the general classification of THz detectors, more details concern Schottky barrier diodes (SBDs), pair braking detectors, hot electron mixers, and field-effect transistor (FET) detectors, where links between THz devices and modern technologies such as micromachining are underlined. Also, the operational conditions of THz detectors and their upper performance limits are reviewed. Finally, recent advances in novel nanoelectronics materials and technologies are described. It is expected that applications of nanoscale materials and devices will open the door for further performance improvement of THz detectors.

28.1 INTRODUCTION

The THz region of the electromagnetic spectrum is often described as the final unexplored area of spectrum. First humans relied on the radiation from the Sun. Cave men used torches (approximately 500,000 years ago). Candles appeared around 1000 BC, followed by gas lighting (1772) and incandescent bulbs (Edison, 1897). Radio (1886–1895), X-rays (1895), UV radiation (1901), and radar (1936) were discovered in the end of the nineteenth and the beginning of the twentieth centuries. However, THz range of electromagnetic spectrum still presents a challenge for both electronic and photonic technologies.

Development of THz technologies began in the 1980s mainly for laboratory applications (see Figure 28.1). It was an expensive scientific device that only highly qualified staff could use. However, for the last two decades, efforts have been made toward the implementation of easy-to-use and cost-effective systems through the development of more compact and reliable components. At present, the THz techniques enter the commercial market and the first products for applications like in-line industrial process monitoring have been sold. The THz market is expected to grow from above 46M Euro in 2015 to about 100M Euro in 2020 with a compound annual growth rate (CAGR) of 16% (see Figure 28.2).

THz radiation (see Figure 28.3) is frequently treated as the spectral region within frequency range $\nu \approx$ 1–10 THz ($\lambda \approx$ 3000–30 μm) [2–4], and it is partly overlapping with loosely treated submillimeter (sub-mm) wavelength band $\nu \approx$ 0.1–3 THz ($\lambda \approx$ 3–100 μm) [5]. Even wider region $\nu \approx$ 0.1–10 THz [6,7] is treated as THz band overlapping thus with sub-mm wavelength band. As a result, frequently both these notions are used as equal ones (see, e.g. Ref. [8]). Here THz range is accepted as the range within $\nu \approx$ 0.1–10 THz. THz electronics spans the transition range from radio-electronics to photonics.

The THz region of the electromagnetic spectrum has proven to be one of the most elusive. Being situated between infrared (IR) light and microwave radiation, THz radiation is resistant to the techniques commonly employed in these well-established neighboring bands. Historically, the major use of THz spectroscopy has been by chemists and astronomers in the spectral characterization of the rotational and vibrational resonances and thermal-emission lines of simple molecules. Terahertz receivers are also used to

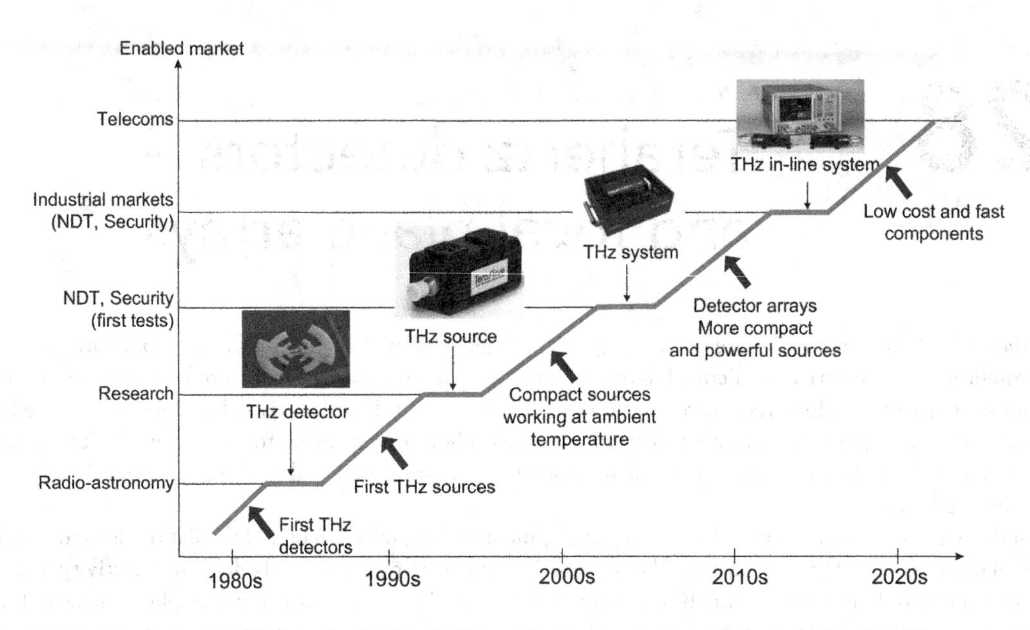

Figure 28.1 Development of THz technologies. NDT, nondestructive testing. (Adapted after Ref. [1])

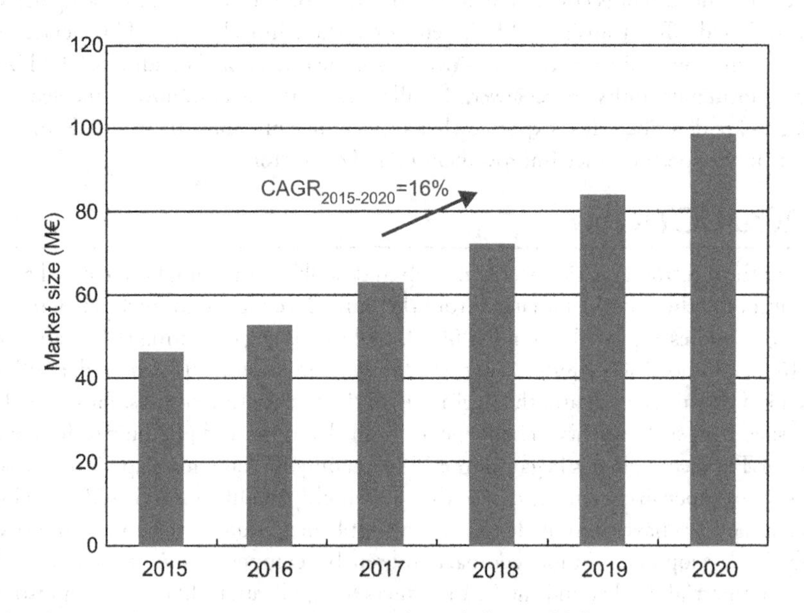

Figure 28.2 Global THz market revenue (2015–2020). (Adapted after Ref. [1])

study the trace gases in the upper atmosphere, such as ozone and the many gases involved in ozone depletion cycles, such as chloride monoxide. Air efficiently absorbs in the wide spectral THz region (except for narrow windows around $\nu \approx 35$, 96, 140 and 220 GHz, and others, see Figure 28.4 [9]). THz and millimeter (mm) waves are efficient at detecting the presence of water and thus are efficient in discriminating different objects on human bodies (water content of human body is about 60%) as the clothes are transparent. In the longer wavelength region (centimeter wavelength region), even persons hiding behind a wall (not very thick) can be visualized. It should be mentioned that about half of the luminosity of the universe and 98% of all the photons emitted since the Big Bang belong to THz radiation [10]. This relict radiation carries information about the cosmic space, galaxies, stars, and planets formation [11].

The past 20 years have seen a revolution in THz systems, as advanced materials research provided new and higher power sources, and the potential of THz for advanced physics research and commercial

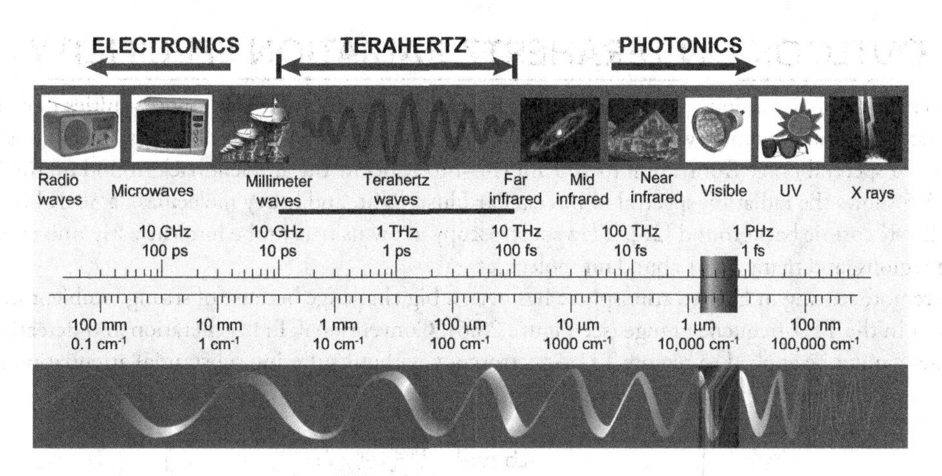

Figure 28.3 The electromagnetic spectrum.

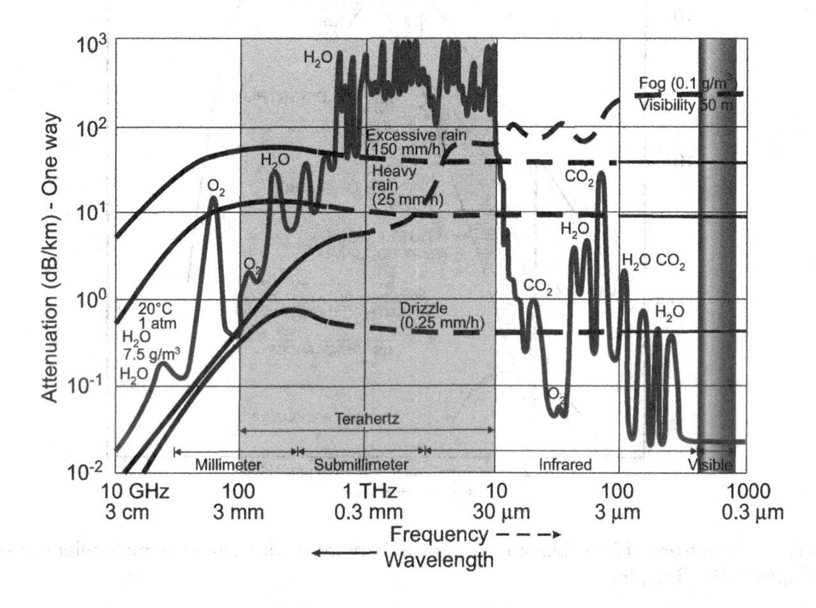

Figure 28.4 Attenuation of Earth atmosphere from visible to RF band region. (Adapted after Ref. [9])

applications was demonstrated. Numerous recent breakthroughs in the field have pushed THz research into the center stage. As examples of milestone achievements, the development of THz time-domain spectroscopy (TDS), THz imaging, and high-power THz generation by means of nonlinear effects [2–4] can be included. Researches evolved with THz technologies are now received with increasing attention, and devices exploiting this wavelength band are set to become increasingly important in diverse ranges of human activity applications (e.g., security, biological, drugs and explosion detection, gases fingerprints, imaging, etc.). The interest in the THz range is attracted by the fact that this range is the place where different physical phenomena are revealed, which frequently calls for multidisciplinary special knowledge in this research area. Nowadays, the THz technology is also of much use in fundamentals science, such as nanomaterials science and biochemistry. This is based on the fact that THz frequencies correspond to single and collective excitations in nanoelectronics devices and collective dynamics in biomolecules. In 2004, *Technology Review*'s editors selected THz technology as one of "10 emerging technologies that will change your world" [12]. Overviews on the various applications of THz technologies can be found, for example, in Refs. [2–4,7,13–25].

28.2 OUTLOOK ON TERAHERTZ RADIATION SPECIFICITY

THz radiation due to its unique properties provides a variety of applications and opportunities in different fields. Historically, astronomers were the first who focused on THz detection technology since interstellar dust covers a spectral range from 1 mm to 100 μm (14–140 K below the ambient background on the Earth). Figure 28.5 shows the radiation spectral of interstellar (dust, light, and heavy molecules), a 30-K blackbody, and the 2.7-K cosmic background [26]. THz spectroscopy allow us true probe into the early universe, star-forming regions, and many other abundant molecules.

THz remote sensing in Earth's atmosphere has been a big challenge because of strong, ambient moisture absorption in the THz frequency range (see Figure 28.6). Conventional THz generation and detection techniques cannot be applied to remote THz spectroscopy without suffering substantial attenuation during

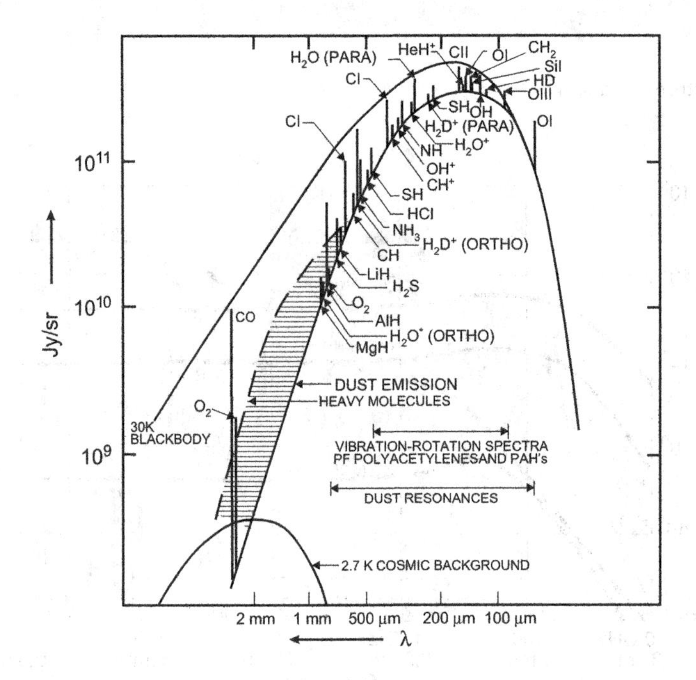

Figure 28.5 Radiation spectrum of 30-K blackbody, typical interstellar dust, and key molecular line emissions in the sub-mm. (Adapted after Ref. [26])

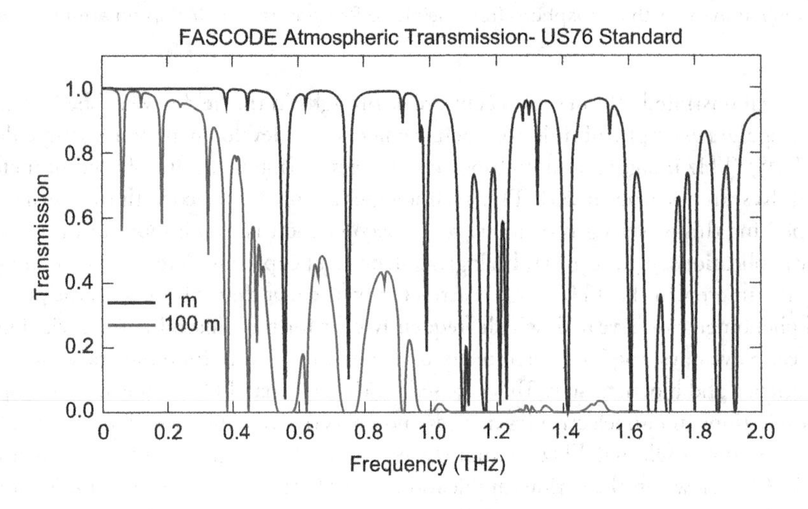

Figure 28.6 Transmission of THz radiation in air.

THz wave propagation in air. With currently available and practical THz detectors and sources, measurements over distances greater than 20 m are very difficult. Because of considerable attenuation, THz waves are not very useful for long-range communications [27]. However, due to strong absorption, the transmission spectra of a lot of materials can provide information about the physical properties of the materials investigated. In addition, an important feature of THz radiation is the ability to penetrate into and distinguish between non-metallic materials.

Unlike visible and IR detectors, the far IR and sub-mm wavelength detectors have not yet reached fundamental quantum limit characteristics. They are not limited by photon flux fluctuations (photon noise), except operation at some selected frequencies and sub-Kelvin temperature operation [28,29]. The registration of single THz photons was demonstrated using quantum dot (QD) devices [30,31].

The noise equivalent power (NEP) is one of the figures of merit for detectors and characterizes their sensitivity. It is defined as the value of root-mean-square (rms) incident power on the detector generating a signal output equal to the rms noise output (signal-to-noise ratio (SNR) = 1). Intrinsic temperature fluctuation noise of thermal detector defines its upper NEP limit as

$$\text{NEP} = \left(4k_B T G_{th}\right)^{1/2}, \tag{28.1}$$

where k_B is the Boltzmann's constant, T is the temperature of the thermistor, and G_{th} is the thermal conductance between the detector and the heat sink. For lower G_{th}, lower values of NEP can be achieved. For $T \approx 50$ mK and low phonon conductance, $G_{th} \approx 10$ fW K^{-1}; the values of electrical NEP $\approx 4 \times 10^{-20}$ WHz$^{-1/2}$ can be achieved at low background fluctuations conditions. Figure 28.7 presents the data of the background-limited detector sensitivity for THz spectroscopy in space and also the photons flux at cosmic background shot noise conditions. The presented dependences are valid both for direct detection and heterodyne detector systems, and also for coherent (mixer) FPAs, which at present are realized only in single-pixel or small number detector arrays [32].

Realization of appropriate low values of NEP ~ 4×10^{-20} WHz$^{-1/2}$ considerably depends on background temperature, spectral band, and resolution needed. The calculated NEP values for $\Delta\lambda/\lambda = 0.3$ and for diffraction limited beams, taking into account only the fluctuations of power from the background radiation,

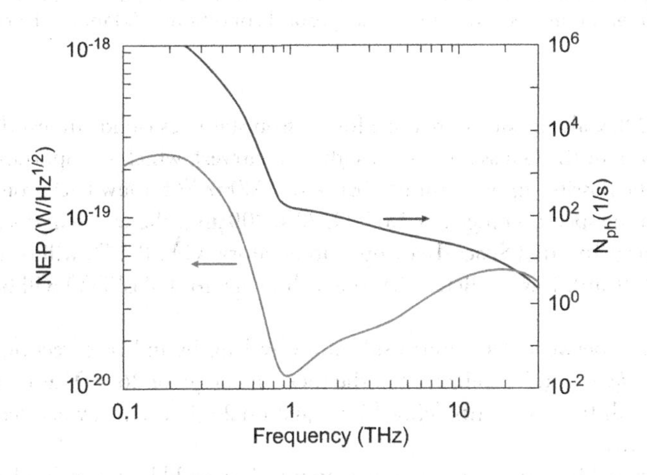

Figure 28.7 Background-limited detector sensitivity for THz spectroscopy in space. The flux of photons $N_{ph}(\nu)$ due to the background radiation from the universe was calculated from the experimental luminosity of continuum radiation within a diffraction-limited beam. This estimate assumes that the detector is sensitive to a single-photon polarization, its optical coupling efficiency is 25%, and its spectral resolution $\nu/d\nu \approx 1{,}000$ corresponds to the width of a typical extragalactic emission line ($\delta\nu = 2\nu u/c$ results from Doppler line broadening in distant galaxies with rotational velocities $u \approx 10^2$km s^{-1}). The background photon flux is very weak: $N_{ph} < 100$ photons per second at $\nu > 1$ THz. Fluctuations of this flux (the photon shot noise) determine the background-limited detector sensitivity NEP$_{ph}(\nu)$. At $\nu < 1$ THz, the detector performance is limited by the cosmic microwave background; at higher frequencies, it is limited by the radiation from the galactic core and dust clouds. (Adapted after Ref. [8])

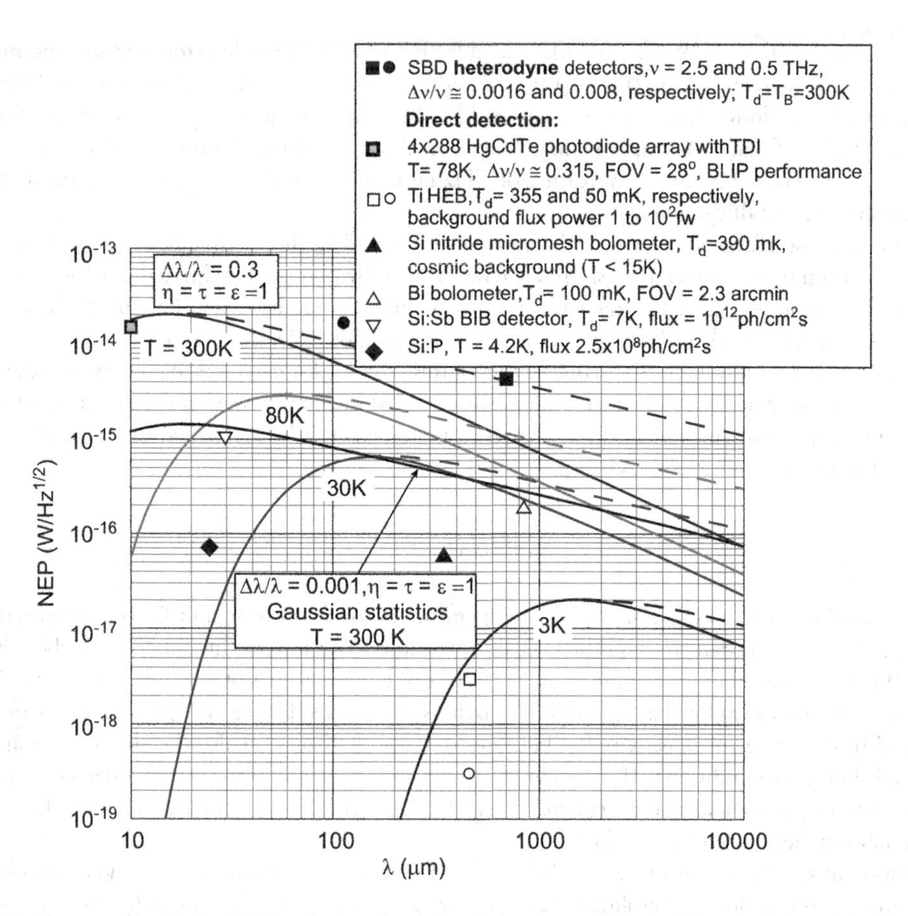

Figure 28.8 Photon noise-limited NEP calculated only for Poisson statistics (solid curves) that dominates in IR systems and with account of the Gaussian statistics (dashed curves), which is important for THz/sub-THz region. $\Delta\lambda/\lambda = 0.3$. Calculations are done for emissivity, transmittivity, and coupling efficiency equal 1 (ε, τ, $\eta = 1$). Also shown are some experimental results with known background conditions. (Adapted after Ref. [33])

are shown in Figure 28.8. Calculations were done for Poisson statistics (solid curves) that dominates in IR systems and with account of the Gaussian statistics (dashed curves), which is important for THz/sub-THz region. It is seen that the possible upper limit of NEP ~10^{-19} $WHz^{-1/2}$ for low background conditions (T = 3 K) is possible only in the spectral range $\nu > 2.6$ THz ($\lambda < 200\,\mu m$), the wavelengths region where the Cryogenic Aperture Large Infrared Space Telescope Observatory (CALISTO) will be operated (spectral range from 30 μm to 300 μm). It is predicted that the technology for CALISTO will be developed during the next decade [34].

The critical differences between detection at sub-mm wavelengths and IR detection lie in small photon energies (at $\lambda \approx 300$ μm $h\nu \approx 4$ meV, compared to the thermal energy of 26 meV at room temperature). Also the Airy Disk diameter (diffraction limit) defined by Equation 24.34 is large, what determines low spatial resolution of THz systems.

To achieve higher spatial imaging resolution, two approaches could be applied: solid immersion lenses (usually Si lenses) or near-field imaging. Compared to the visible region, the near-field imaging in the THz region has not yet been well established, because, for example, of the lack of THz fibers or other bulk media transparent in the THz region to generate near-field waves.

One of the problems, which limits an advent of heterodyne arrays in the THz spectral region needed for high-resolution spectroscopy applications ($\nu/\Delta\nu \approx 10^6$), photometry ($\nu/\Delta\nu \approx 3$–10), and imaging, lies in technology limitations of solid-state local oscillator (LO) power (see Figure 28.9). Around $\nu \approx 1$ THz, one can see that there exists the so-called "THz gap".

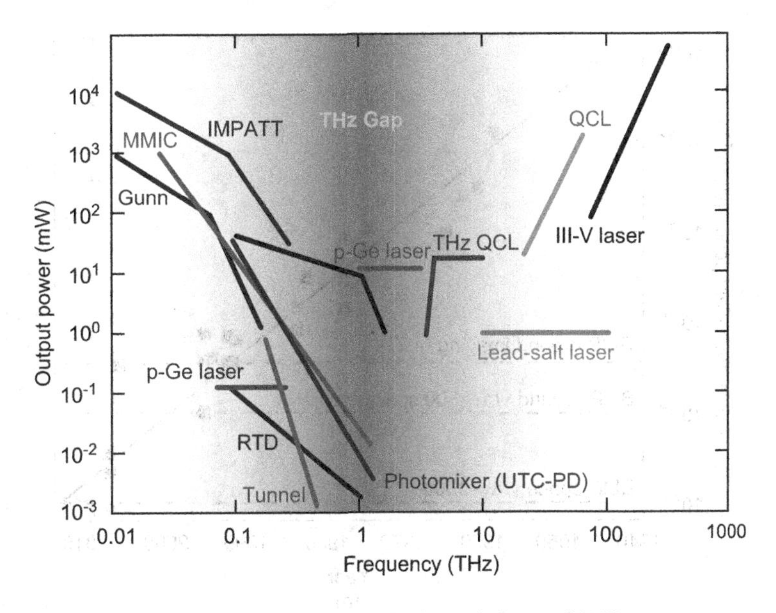

Figure 28.9 THz gap with respect to source technology. (Adapted after Ref. [20])

28.3 TRENDS IN DEVELOPMENTS OF TERAHERTZ DETECTORS

The detection of THz radiation is resistant to the commonly employed techniques in the neighboring microwave and IR frequency bands. In THz detection, the use of solid-state detectors has been hampered for the reasons of transit time of charge carriers being larger than the time of one oscillation period of THz radiation. Also the energy of radiation quanta is substantially smaller than the thermal energy at room temperature and even liquid nitrogen temperature.

Detector development is at the heart of all current plants. There exists a large variety of traditional deeply cooled mm and sub-mm wavelength detectors (mainly bolometers) as well as new propositions based on optoelectronic quantum devices [21], CNT bolometers, plasma wave detection by FETs, and hot electron room temperature bipolar semiconductor bolometers [23,24].

Progress in THz detector sensitivity has been impressive in a period more than half century what is shown in Figure 28.10a in the case of bolometers used in far-IR (FIR) and sub-mm-wave astrophysics [35,36]. The situation is analogous to the rapid development of IR detector arrays. The NEP value has decreased by a factor of 10^{11} in 70 years, corresponding to improvements by a factor of 2 every 2 years. Individual detectors achieved photon noise-limited performance for ground-based imaging in the 1990s. The photon noise from astrophysical sources, achievable in space with a cold telescope, ~10^{-18} WHz$^{-1/2}$, is now within demonstrated sensitivities. In the present decade, the studies of inflation via cosmic microwave background (CMB) polarization will be driven not by detector sensitivity but by array formats. FIR spectroscopy from a cold telescope, however, requires sensitivity ~10^{-20} WHz$^{-1/2}$ to reach the astrophysical photon noise limit. Achieving this sensitivity in working detector arrays remains a challenge for the coming decade, as the number of the detectors in the array is a key parameter that determines the information capabilities of the system [29] and the speed improvement in obtaining a complete imaging or spectrum when observing galactic objects, as the accumulation time τ_{acc} at each sensitive element is proportional to the number M_e of sensitive elements in the array and inversely proportional to the frame rate f_r, and the number of picture dots M, $\tau_{acc} = 1/f_r(M_e/M)$.

The development of pixel arrays has been comparably revolutionary [36]. Figure 28.11 shows increasing number of pixels in the period over the last three decades. Detector arrays have doubled in format every 20 months over the past 10 years producing arrays with pixels now numbering in the thousands. Steady increase in overall observing efficiency is expected in near future, which by now has reached factors in

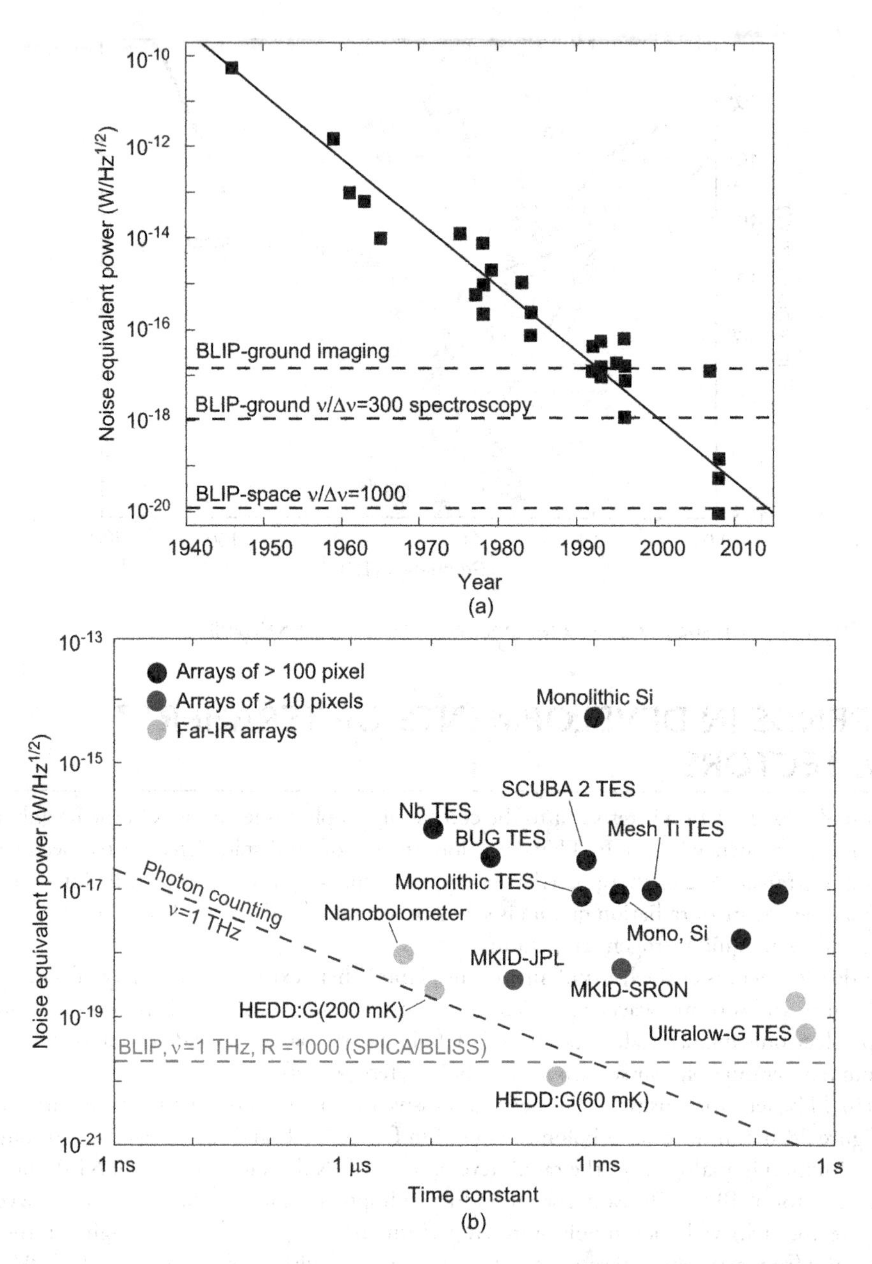

Figure 28.10 Trends in development of THz detectors: (a) improvement of the bolometer NEP value for more than half a century (sensitivity doubled every 2 years over the past 70 years), (b) NEP versus detector time constant (Adapted after Ref. [36])

the range of 10^{12} in comparison to capabilities in the early 1960s. Microwave kinetic inductance detectors (MKIDs) are currently capable of powering state-of-the-art instruments, both in number of pixels and in sensitivity, at a lower price and with significantly less focal plane and readout complexity than comparable superconducting transition edge (TES) bolometer-based instruments.

For space-board systems, the system NEP should not be limited by the thermal emission from the telescope optics or the detector noise itself. Assuming that the telescope is cooled to ~4 K, its emission should be negligible throughout the sub-mm/FIR range, and the requirement on detector performance (and ultimate system sensitivity) is set by the astronomical background and spectral resolution needed (see, e.g. Figure 28.8). At longer wavelengths, this is dominated by dust emission from the Milky Way, together with the cosmic microwave background. At shorter wavelengths ($\lambda < 100\,\mu m$), the zodiacal emission

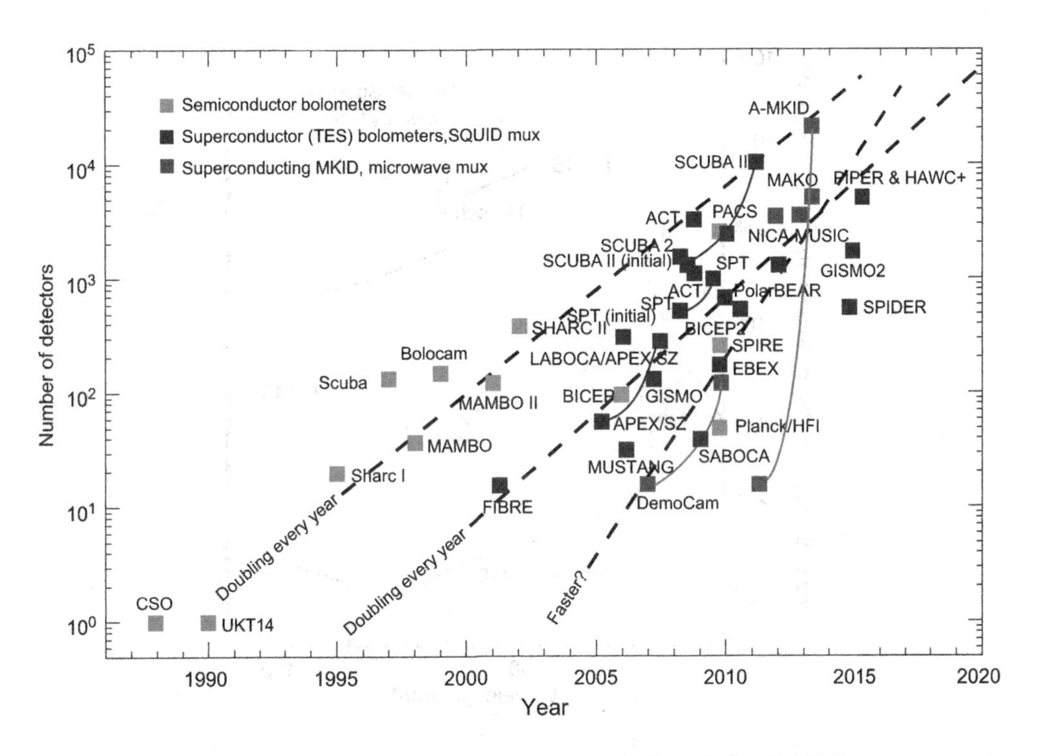

Figure 28.11 The exponential growth of detector arrays over time. (Adapted after Ref. [37])

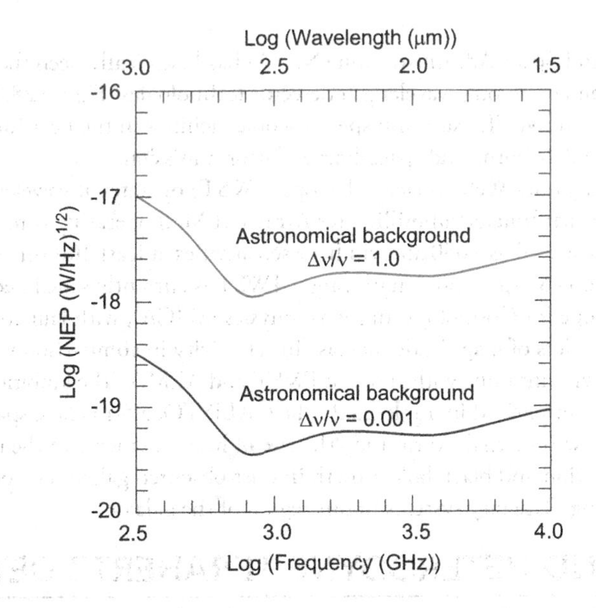

Figure 28.12 Dark sky astronomical background as a function of wavelength for two different fractional resolutions, $\nu/\Delta\nu = 1$ for photometry and $\nu/\Delta\nu = 1{,}000$ for moderate resolution extragalactic spectroscopy. The detector NEP should be well below these values in order for the overall noise to be dominated by the astronomical background. (Adapted after Ref. [34])

becomes significant. These two contributions have different angular distributions, but if one considers only the "darkest sky," the resulting detector NEP requirements should be below 3×10^{-18} WHz$^{-1/2}$ ($\nu/\Delta\nu = 1$, photometry requirements) and for $\nu/\Delta\nu = 1{,}000$ spectroscopy, below 8×10^{-20} WHz$^{-1/2}$ ($\lambda \sim 30$–$1{,}000\,\mu$m, Figure 28.12).

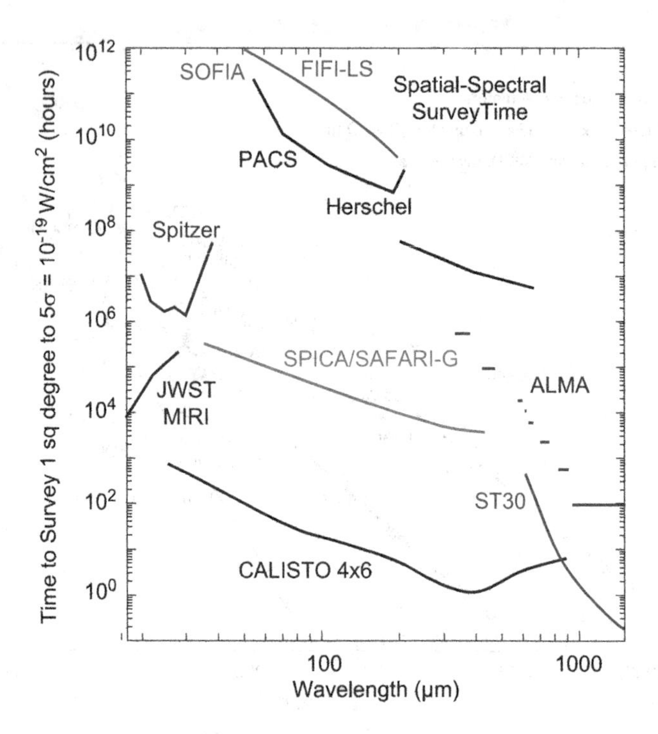

Figure 28.13 Sensitivity of FIR spectroscopy platforms. (Adapted after Ref. [35])

National Aeronautics and Space Administration (NASA) has historically been the leading U.S. agency for promoting the development of long-wavelength detector technologies. Figure 28.13 shows the sensitivities of currently planned or active FIR/sub-mm spectroscopic facilities in the near future, instead Table 28.1 describes shortly several airborne and space-borne platform missions.

As last figure shows, the James Webb Space Telescope (JWST) operates at wavelengths below about 27 μm. The Atacama Large Millimeter/submillimeter Array (ALMA) operating through a number of sub-mm atmospheric windows as well as ~650 μm will have sensitivities at least 100 times higher than Herschel spanning the intervening 60–650 μm wavelength range. JWST is currently scheduled to start operations. The Space Infrared Telescope for Cosmology and Astrophysics (SPICA), with launch envisioned in 2027, will provide two to three orders of magnitude increase in sensitivity in comparison with Herschel that will bring FIR/sub-mm sensitivity into line with those of JWST and ALMA. The ambitious requirements of future space missions are summarized in Table 28.2. The CALISTO, a 5-m class, space-borne telescope actively cooled to $T \approx 4$ K, will be enabled to study the rise of heavy elements in the universe's first billion years, chart star formation and black hole growth in dust-obscured galaxies through cosmic time, and conduct a census of forming planetary systems in our region of the galaxy.

28.4 DIRECT AND HETERODYNE TERAHERTZ DETECTION

All radiation detection systems in THz spectral ranges can be divided into two groups (see Chapter 6):

- incoherent detection systems (with direct detection sensors), which allow only signal amplitude detection and which, as a rule, are broadband detection systems, and
- coherent detection systems, which allow detecting not only the amplitude of the signal but also its phase.

Coherent signal detection systems use heterodyne circuit design since, so far, for high radiation frequency range proper amplifiers do not exist. The detected signals are transferred to much lower frequencies ($f \approx$ 1–30 GHz) where they are amplified by low-noise amplifiers. Basically, these systems are selective (narrow-band) detection systems.

Table 28.1 FIR spectroscopy platforms

Spitzer Space Telescope 2003		The Spitzer Space Telescope was launched in August 2003. It is the last of NASA's "great observatories" in space. Spitzer is much more sensitive than prior IR missions and studies the universe at a wide range of IR wavelengths. Spitzer concentrates on the study of brown dwarfs, super planets, protoplanetary and planetary debris disks, ultraluminous galaxies, active galaxies, and deep surveys of the early universe.
SOFIA 2005		SOFIA was finally completed in 2005. SOFIA, a joint project between NASA and the German Space Agency, incorporates a 2.5-meter optical/IR/sub-mm telescope mounted in a Boeing 747. Designed as a replacement for the successful Kuiper Airborne Observatory, SOFIA is the largest airborne telescope in the world.
Herschel Space Observatory 2009		The Herschel Space Observatory carried into orbit in May 2009 is a European Space Agency IR-sub-mm mission. Herschel is perform spectroscopy and photometry over a wide range of IR wavelengths and is used to study galaxy formation, interstellar matter, star formation, and the atmospheres of comets and planets. The Herschel Observatory is capable of seeing the coldest and dustiest objects in space. It is the largest space telescope ever launched carrying a single mirror of 3.5 meter in diameter.
ALMA 2011		The ALMA is an international partnership between Europe, North America, East Asia, and the Republic of Chile to build the largest astronomical project in existence. It is an astronomical interferometer, comprising an array of 66 12-m and 7-m diameter radiotelescopes observing at mm and sub-mm wavelengths. It is being built on the Chajnantor plateau at 5,000 m altitude in the Atacama desert of northern Chile. ALMA is expected to provide insight on star birth during the early universe and detailed imaging of local star and planet formation. Costing more than a billion dollars, it is the most ambitious ground-based telescope currently under construction. ALMA begun scientific observations in the second half of 2011 and was fully operational by the end of 2012.

(*Continued*)

Terahertz detectors and focal plane arrays

Table 28.1 (*Continued*) FIR spectroscopy platforms

James Webb Space Telescope 2018		The JWST is a large, IR-optimized space telescope, scheduled for launch in 2018. It is a visible/IR space mission which will have extremely good sensitivity and resolution, giving us the best views yet of the sky in the near-mid IR. JWST will be used to study the early universe and the formation of galaxies, stars, and planets. Webb will have a large mirror, 6.5 m in diameter and a sunshield the size of a tennis court. Both the mirror and sunshade won't fit onto the rocket when fully open, so both will fold up and open once Webb is in outer space. Webb will reside in an orbit about 1.5 million km from the Earth.
SPICA/BLISS 2027		The Background-Limited Infrared-Submillimeter Spectrograph (BLISS) is an FIR spectrograph concept for SPICA. The SPICA mission is a future Japanese IR astronomical satellite, with launch envisioned in 2027, to explore the universe with a cooled, large telescope. The philosophy of BLISS is to provide a rapid survey spectroscopy capability over the full FIR range. The baseline approach is a suite of broadband grating spectrometer modules with TES bolometers. SPICA will use a cooled telescope (3.5 m diameter primary, ~5 K) to achieve sensitivities currently inaccessible to existing facilities operating over this wavelength range (SOFIA, Herschel).
CALISTO		The CALISTO platform is especially compelling for wideband spectroscopy. It will obtain full-band spectra of thousands of objects ranging from the first dusty galaxies to the most heavily enshrouded young stars and proto-planetary disks in our own galaxy, as well as blind discovery of thousands more. With excellent spectral sensitivity in the 35–600 μm band, CALISTO will bring a powerful new toolkit to bear on high-redshift galaxy populations. Arrays with a total count of a few ×10^5 detector pixels installed in imaging spectrometers will achieve the photon background limit.

Table 28.2 Requirements for future space applications

SCIENCE	FUTURE OPPORTUNITIES	REQUIREMENTS		
		NEP (WHz^{-1})	τ(ms)	FORMAT
CMB polarization	Inflation probe	$(1–5) \times 10^{-18}$	1–30	10^4
Galaxy evolution Star formation Circumstellar disks	SPICA/BLISS	$(3–30) \times 10^{-20}$	100	5,000
	SAFIR/CALISTO imaging	3×10^{-19}	10	10^5
	SAFIR/CALISTO spectroscopy	3×10^{-20}	100	10^5
	SPIRIT	1×10^{-19}	0.2	256

Source: Adapted after Ref. [38]

28.4.1 DIRECT DETECTION

Detectors with direct signal detection basically are used in spectroscopic and technical vision systems of ultraviolet (UV), visible, IR, sub-mm, and mm regions. In sub-mm and mm wavelength bands, they are suitable for applications that do not require ultrahigh spectral resolution ($\nu/\Delta\nu \approx 10^6$) that is provided by heterodyne detector spectroscopic systems. But unlike heterodyne detection systems, there does not exist the problem of multielement arrays formation conditioned by LO power and fast detector response ($\tau \approx 10^{-10}–10^{-11}$ s).

In direct THz detection systems, even room temperature detectors can be used with relatively long response time ($\tau \approx 10^{-2}–10^{-3}$ s) and modest sensitivity. Among them are Golay cells, pyroelectric detectors, different kinds of thermal direct detection detectors (bolometers and microbolometers), which use antennas to couple power to small thermally absorbing regions [22]. The NEP value for uncooled detectors typically is from 10^{-10} to 10^{-9} $WHz^{-1/2}$.

Different kinds of cooled semiconductor detectors (hot electron InSb, Si, Ge bolometers, extrinsic Si, and Ge) with response time of $\tau \approx 10^{-6}–10^{-8}$ s and NEP $\approx 10^{-13}–5 \times 10^{-17}$ $WHz^{-1/2}$ and operated at $T \leq$ 4 K are also used. Direct extrinsic photoconductors (stressed Ge:Ga) can be sensitive up to a wavelength $\lambda \approx 400$ μm [39] and be assembled into arrays [40]. However, the most sensitive direct detectors in sub-mm and mm wavebands are different bolometer designs cooled to $T \approx 100–300$ mK reaching NEP limited by cosmic background radiation fluctuations [41–44]. Intrinsic and extrinsic photon detectors based on interband, intersubband, or impurity optical transitions are used only in the short range of sub-mm spectral band. The reason of that is high thermal generation rate in comparison with photoionization rate under radiation. A comprehensive review of a different kind of direct detection detectors up to 1994 is carried out in Ref. [39]. Some kind of new cooled detectors are considered in Refs. [45–47].

A schematic diagram of a direct detection is shown in Figure 28.14. Detector detects both signal radiation with signal power W_s and background radiation with power W_B. Focusing optics (lenses, mirrors, horns, etc.) is used to collect radiation over a large area to focus it to detector. Frequently, an optical filter is located before detector to remove background radiation at wavelengths other than the detected signal.

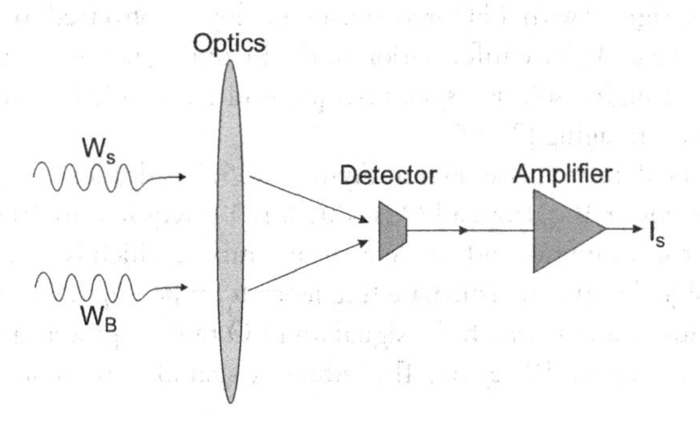

Figure 28.14 Schematic representation of a direct detection. W_s is the signal power and W_B is the background radiation power.

A relatively small electrical signal from detector is amplified by amplifier and generated signal I_s is further processed.

It can be shown that for direct detection with non-photoconductive detector, when fluctuation noise (background flux fluctuations) predominates, the minimum detectable signal in background limited performance (BLIP) conditions is equal [48,49]:

$$W_{s,\text{dir}}^{\min} = \left(\frac{2h\nu}{\eta} W_B \Delta f\right)^{1/2}, \quad (\text{W}), \tag{28.2}$$

where η is the detector quantum efficiency (coupling efficiency) and Δf is a bandwidth.

It is seen that $W_{s,\text{dir}}^{\min} \sim (\Delta f)^{1/2}$ and detectable signal can be appreciably weaker than W_B. From this expression one can also see that when comparing detectors, it is useful to normalize them to $(\Delta f)^{1/2}$ which is done comparing NEP of different detectors.

At the condition that fluctuation noise is in the signal flow itself $\left(W_{s,\text{dir}}^{\min} = W_B\right)$ for non-photoconductive direct detector for minimum detectable signal it follows:

$$W_{s,\text{dir}}^{\min} = \frac{2h\nu}{\eta} \Delta f, \quad (\text{W}). \tag{28.3}$$

It means that to detect a minimum current, at least two photons should be accepted by the detector ($\eta = 1$). Then for radiation frequency $\nu = 1$ THz, $W_{s,\text{dir}}^{\min} \approx 6.6 \times 10^{-22}$ W ($\Delta f = 1$ Hz). But the ability to detect such signals by direct detection detectors is limited by irreducible background photon noise not vanishingly small even for cosmic background. Performance of these detectors is background noise-limited compared to heterodyne detectors, the performance of which is quantum noise-limited. As a rule, the threshold power detected by direct detectors is higher, that is, caused by other noises present both in the detector itself and in circuit elements and amplifiers.

For BLIP detection, when normalizing detector NEP to $(\Delta f)^{1/2}$, from Equation 28.2, it follows:

$$\text{NEP}_{\text{dir}} = \left(\frac{2h\nu}{\eta} W_B\right)^{1/2}, \quad \left(\text{WHz}^{-1/2}\right). \tag{28.4}$$

The lower the NEP means the more sensitive the detector.

An advantage of the systems with direct detection is the relative simplicity and possibility to design large format arrays [50,51]. Most imaging systems used passive direct detection. In active systems, for operations of which the scene is illuminated, heterodyne detection can also be used in order to increase sensitivity to low radiant levels or to image through the scattering media.

28.4.2 HETERODYNE DETECTION

In heterodyne detectors, the signals with THz frequencies are down-converted to intermediate frequency (IF), preserving the amplitude and phase information of the incoming radiation. In the last several decades the detectors are of choice for high resolution spectroscopic studies, cosmic remote sensing, and relatively recently, for mm and sub-mm imaging [2–4,52].

A schematic of heterodyne detection is shown in Figure 28.15. In addition to signal W_s and background W_B radiant powers, radiant power W_{LO} from a LO is added. LO is required to drive the mixing process. The basic element of mm or sub-mm heterodyne detector is a mixer, which is needed to align W_s and W_{LO} for generating a copy of a signal at the intermediate frequency $\nu_{IF} = |\nu_s - \nu_{LO}|$, and its key component is the nonlinear mixing element (detector) at which the signal and LO radiant powers are coupled using some kind of diplexer or a beam splitter for IR region. The latter one spatially combines the signal beam and LO beam.

The mixer is the most important component of heterodyne receiver input stage that is responsible for its responsivity. Its conversion loss enhancement decreases contribution to the noise heterodyne receiver

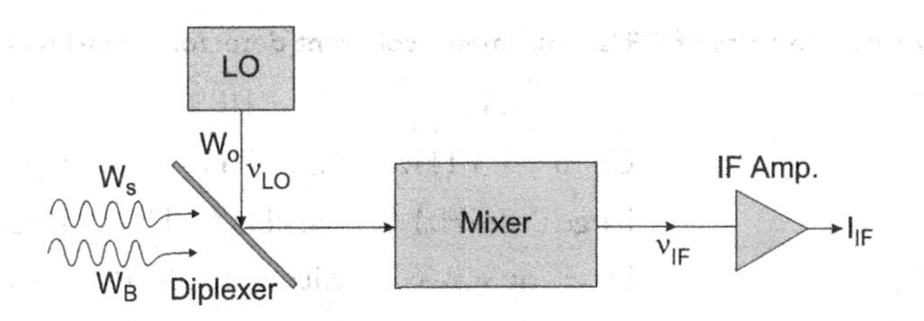

Figure 28.15 Simplified schematic representation of heterodyne receiver. W_s is the signal power with frequency ν_s, W_B is the background radiation power, W_0 is LO radiation power with frequency ν_{LO}, and ν_{IF} is intermediate frequency.

temperature and consequent intermediate frequency amplifier. Signal power losses occur at diplexer and detector, but it is a mixer and its distributing circuits, which contribute the most noise to the heterodyne receiver [53]. To be used in mm or sub-mm array, the mixer choice is dictated by the available LO power in these spectral ranges, mixer operating temperature and sensitivity needed.

Dependent on the availability of suitable LOs, two heterodyne techniques are possible. One can use a tunable LO and a fixed IF amplifier with filters. Another one uses a fixed-frequency LO in combination with IF amplifiers and filters to cover the needed frequency range. The first one is more flexible but tunable continuous wave sources (such as, e.g., backward wave oscillators or frequency multiplied mm wave sources) produce low power at wavelengths less than 500 μm, which can be not enough, for example, for detection with SBD receivers. The available narrow linewidth sources with sufficient power, as an LO for SBDs, are optically pumped sub-mm gas lasers or quantum cascade lasers (QCLs) for shorter part of sub-mm range. Although many continuous wave (CW) lines are available from the gas lasers, these are all at specific wavelengths and heterodyne detection is restricted to a relatively narrow range on either side of each available wavelength (e.g., gas lasers at $\lambda = 433$ μm (HCOOH), at $\lambda = 184$, 214, and 288 μm (CH_2F_2), $\lambda = 337$ μm (HCN), $\lambda = 118$ μm (H_2O), and others).

The serious problem, which limits an advent of heterodyne sensor arrays in sub-mm (THz) spectral region (e.g., for high-resolution spectroscopy applications ($\nu/\Delta\nu \approx 10^6$) or photometry ($\nu/\Delta\nu \approx 3$–10) and imaging), lies in technology limitations of solid-state LO power. With an exception of free electron lasers that use relativistic electrons and are capable of reaching kilowatt level THz power [54], other THz sources generate milliwatt or microwatt power levels (see Figure 28.9). Traditional electronic devices, such as transistors, do not work well much above about 150 GHz. Therefore, there are no amplifiers available throughout most of the THz band. Similarly, semiconductor lasers that have long been available in optical and IR bands are not available in most of the THz band. Although much progress is being made in the area of high-frequency transistors [55] and semiconductor lasers [56], it seems clear that the so-called "THz gap" will remain an important challenge to scientists and engineers for the foreseeable future.

The primary benefit of heterodyne detection systems is that the frequency and phase information at the signal frequency ν_s is converted to the frequency ν_{IF}, which is in a much lower frequency band ($\nu_{IF} \ll \nu_s$) appropriate to electronics time response. This transformation ($\nu_s \rightarrow \nu_{IF}$) is called heterodyne conversion. If the signal and LO frequencies are equal, then $\nu_{IF} = 0$ and the beat tone degenerates to DC, and such a detection process is called homodyne conversion.

Any nonlinear electronic device can be used as a mixer. Their choice is crucial for receiver sensitivity. To achieve efficient conversion and low noise in the mm and sub-mm wavelength bands only, several types of detectors can be used. Frequently used mixers are devices having a strong electric field quadratic nonlinearity. Examples are forward-biased SBDs, superconductor-insulator-superconductor (SIS) tunnel junctions, semiconductor and superconducting hot-electron bolometers (HEBs), and superlattices (SLs). Table 28.3 lists their key characteristics.

Schematic current-voltage characteristics of nonlinear devices are shown in Figure 28.16. Simultaneously with reasonable conversion efficiency and low noise, these nonlinear devices should possess high conversion operation speed for assurance of wide band-pass for consequent signals amplification at much lower frequencies 1–30 GHz.

Table 28.3 Key characteristics of FIR/submillimeter coherent detector (mixer) technologies

	SIS	HEB	SCHOTTKY
RF range	Up to ≈ 1.3 THz	1.3–5 THz	Up to 3 THz
IF bandwidth	Large (>8 GHz)	Small (<4 GHz)	Large (>>8 GHz)
Sensitivity[a]/T_{min}	Excellent: ≈ 2–6	Medium: ≈ 8–10	Poor: ≈ 20–40
LO power requirement	Low: ≈ 1 μW	Very low: ≤1 μW	High: ≈ 1 mW
Operating temperature	≤4 K	≤4 K	70–300 K
Astronomical space missions	Herschel-HIFI	Herschel-HIFI	SWAS, ODIN, Rosetta-MIRO

[a] In units of $T_{min} = h\nu/k_B$, the minimum achievable noise temperature for a single sideband receiver

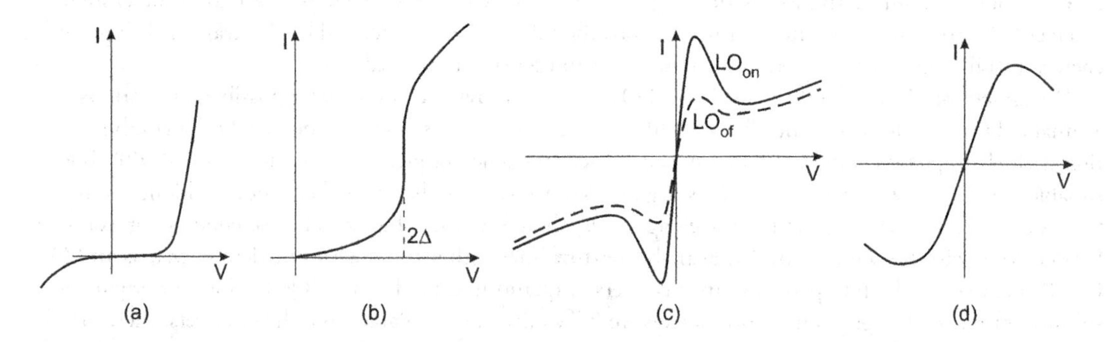

Figure 28.16 Schematic of I-V characteristics of nonlinear elements on which THz heterodyne receivers are based: (a) Schottky diode, (b) SIS, (c) HEB, and (d) SL.

At large LO power W_0, one can detect relatively small signal powers W_s. When this condition is abided by $W_0 >> W_s$, the quantum noise in the signal flux can be the dominant noise, and for internal signal gain $G = 1$, for non-photoconductive detector at SNR $S/N = 1$, it follows [48,49]:

$$W_{s,het}^{min} = \frac{h\nu}{\eta}\Delta f, \quad (W), \tag{28.5}$$

and for minimal detectable energy, one has $E_{s,het}^{min} = h\nu/\eta$. For coupling efficiency $\eta = 1$, it means the quantum limit of signal detection. Thus, the energy of one photon accepted by non-photoconductive detector is transformed into the kinetic energy of one electron, which then crosses the barrier.

For heterodyne detection in BLIP regime, it can be shown [49] that

$$NEP_{het} = \frac{W_{s,het}^{min}}{\Delta f} = \frac{h\nu}{\eta}, \quad \left(WHz^{-1}\right), \tag{28.6}$$

In heterodyne detection, the possible minimum detectable signal is twice lower in comparison with direct detection (see Equations 28.3 and 28.5). Note also that for heterodyne detection, units of NEP are WHz^{-1} instead of $WHz^{-1/2}$ (see Equation 28.4) as for direct detection. But frequently NEP still is cited in units of $WHz^{-1/2}$.

A key advantage of mm and sub-mm sensors over IR and visible ones is ultimate noise limits. Both direct and heterodyne detectors operate against fundamental noise limits that depend on the background radiation and the photon frequency. In the coherent case, the limit is simply the photon shot noise that has NEP equal to $h\nu/\eta$. This limit is plotted in Figure 28.17 for $\eta = 1$ in terms of the photon energy $h\nu$ and equivalent temperature $T = h\nu/k_B$. The same advantage is shared by radiofrequency (RF) receivers

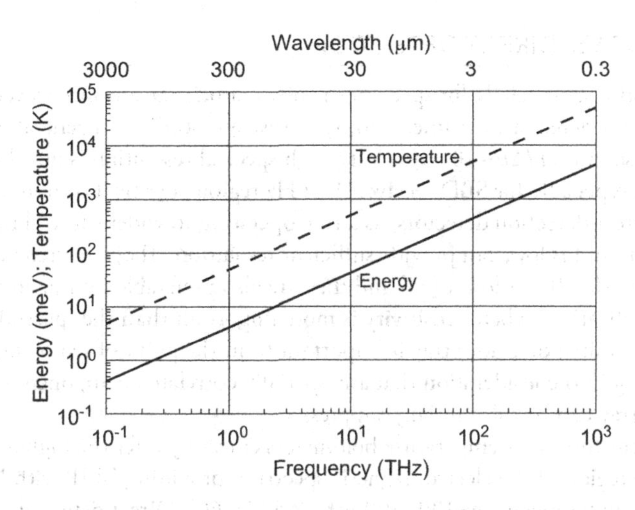

Figure 28.17 Quantum-limit defined by the minimum energy—one photon—per spatial mode from the mm-wave region through the visible region. (Adapted after Ref. [16])

operating at lower frequencies, which is one of the reasons for RF communications, be it wired or wireless, is generally superior to photonic communications in terms of sensitivity [16].

The sensitivity of heterodyne detectors is frequently given in terms of mixer noise temperature T_{mix}, which correlates with the mixer NEP

$$\mathrm{NEP_{mix}} = k_B T_{mix}. \quad (28.7)$$

For wavelength band $\lambda \approx 3\,\mathrm{mm}$ ($\nu \approx 100\,\mathrm{GHz}$), where it is atmospheric transparency window, the value $T_s^{min} = E_{s,het}^{min}/k_B = h\nu/k_B \approx 4.8$ K is the fundamental limit to the noise temperature imposed by the uncertainty principle on any simultaneous measurement of the amplitude and phase of the electromagnetic wave (see Figure 28.17). In the case of heterodyne detection with SIS tunnel junctions as the mixing elements, "the true quantum-noise-limited mixer temperature is $T_{mix}^{min} = h\nu/2k_B$" [57].

The limit values of noise temperature of heterodyne THz detectors are frequently compared using T_s^{min} values. Since heterodyne detectors measure both amplitude and phase simultaneously, they are governed by the uncertainty principle and hence they are quantum noise-limited to an absolute noise floor of 48 K THz^{-1}.

THz mixer receivers can operate in different modes, depending on the configuration of the receiver and the nature of the measurement. The signal and image frequencies may be separated in the correlator, or the image may be removed by appropriate phase switching of pairs of LOs. The function of separating or dumping the image in the receiver is to remove some of the uncorrelated noise to improve the system sensitivity.

In single-sideband (SSB) operation, the receiver is configured so that, at the image sideband, the mixer is connected to a termination within the receiver. There is no external connection to the image frequency, and the complete receiver is functionally equivalent to an amplifier followed by a frequency converter.

In double-sideband (DSB) operation, on the other hand, the mixer is connected to the same input port at both upper and lower sidebands. The DSB receivers can be operated in two modes:

- in SSB operation to measure narrow-band signals contained entirely within one sideband—for detection of such narrow-band signals, power collected in the image band of a DSB receiver degrades the measurement sensitivity
- in DSB operation to measure broadband (or continuum) sources whose spectrum covers both sidebands—for continuum radiometry, the additional signal power collected in the image band of a DSB receiver improves the measurement sensitivity.

Figure 6.10 shows DSB noise temperature of Schottky diode mixers, SIS mixers, and HEB mixers operated in terahertz spectral band.

28.4.3 HETERODYNE VS. DIRECT DETECTION

One of the critical questions, especially for space-borne observatories at sub-mm wavelengths, is whether to use heterodyne or direct detector instruments for spectroscopic studies. In general, heterodyne detection offers higher spectral resolution $\nu/\Delta\nu \sim 10^5$–10^6. Very high spectral resolution is possible, since $\nu_{IF} \ll \nu$. But for heterodyne systems, especially for SBD receivers in THz region, a critical component is the LO source.

At the same time, direct detection detectors, as a rule operating in wider spectral range, when, for example, photon background is low, can provide sufficient resolution. They are preferable for moderate spectral resolution $\nu/\Delta\nu \sim 10^3$–10^4 or lower [58] and they are also preferable for imaging. Direct detectors can be used in those applications where sensitivity is more important than the spectral resolution.

Having background limited detector array is important from the point of removing of, for example, sky background noise, taking into consideration that any spatially correlated component of this noise detected in all detectors in the array can be substantially suppressed.

Among the direct detectors, low-temperature bolometers currently offer the highest sensitivity from the FIR to millimeter-wave region of the electromagnetic spectrum providing BLIP with NEP up ~ (0.4–3) × 10^{-19} WHz$^{-1/2}$ at operation temperature ~100–300 mK [8,43,59,60]. Direct detection bolometers have been used for decades in measurements of CMB spectrum and anisotropy, including space flight COBE-FIRAS instrument [61]. In CMB experiments coherent detector systems and incoherent bolometric systems are used. For cosmic ground-based experiments, both detector types are viable [62].

Compared to direct detection, heterodyne detection exhibits both advantages and disadvantages [49]. Among the advantages of heterodyne detection are: (i) it can detect frequency modulation and phase modulation; (ii) dominant noise follows from fluctuations in W_{LO} rather than from background radiation noise, thus providing discrimination, for example, against background flux, microphonics etc.; (iii) IF conversion process provides gain so that IF detector signal output may be made large to override, for example, thermal and generation-recombination noise; (iv) conversion gain is proportional to W_{LO}/W_s and thus, much weaker radiant signal powers compared to direct detection can be detected.

Among the disadvantages of heterodyne detection are: (i) both beams should be coincident and equal of diameter, and also their Pointing vectors should be coincident; (ii) wavefronts of both beams should have the same radius of curvature and have the same transverse spatial mode structure, moreover they should be polarized in the same direction; and (iii) difficulty of producing large format arrays.

Coherent detection systems (with SIS or SBD mixers), as a rule, are limited in detection signals with frequencies above 1 THz. Heterodyne superconducting HEB mixers, TES and MKID direct detection detectors have almost no practical limitations in applications in shorter sub-mm range. In this spectral band both antenna-coupled detectors and detectors itself (except, e.g., HEB detectors) for radiation detection can be used.

28.5 PHOTOCONDUCTIVE TERAHERTZ GENERATION AND DETECTION

Before the mid-1970s the only viable source for THz radiation was thermal source. This changed in 1975, when D.H. Auston of Bell Labs demonstrated that a short laser pulse (on the order of 100 fs and with a wavelength above the bandgap) impinging on a biased semiconductor would create a picosecond current transient [63]. This time-dependent current radiates and contains frequency components in the THz frequency range. This pioneering source for a THz pulse called the "Auston switch" led to the development of photoconductive [64] and electro-optic methods [65] to generate and detect radiation in the THz frequency range.

In typical photoconductive switch (see Figure 28.18a) two biased parallel metal strip lines, called the "Grischkowsky antenna" [66], with a typical separation of several micrometers are embedded in a semiconductor substrate [15]. The THz pulse is generated when fs-laser pulse is focused near the anode of the antenna thereby creating free charge carriers, which are accelerated in the electric field between the striplines. These are radiated into the substrate (and in opposite direction) and collimated by a hyperhemispherical lens with high-resistivity silicon for better coupling of THz radiation to the free space (free space impedance $Z \approx 377\ \Omega$).

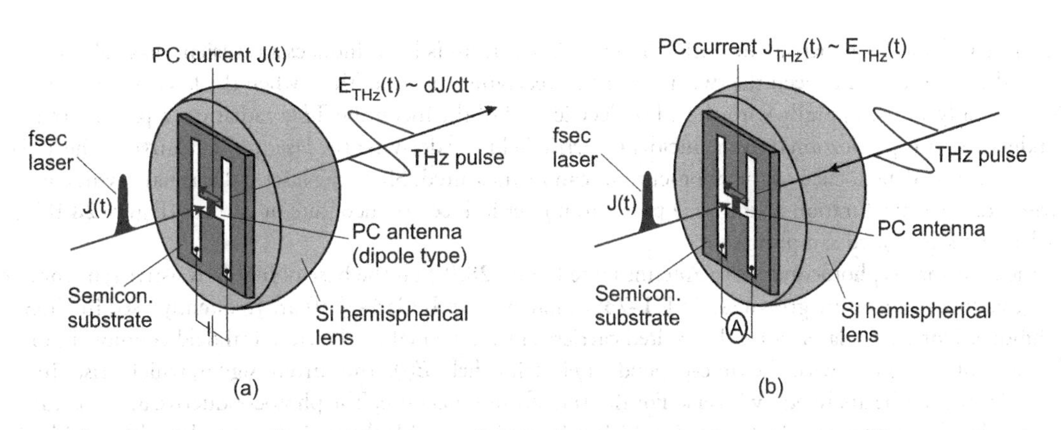

Figure 28.18 Photoconductive emitter (a) and photoconductive detector antenna (b) mounted on a hemispherical lens. (Adapted after Ref. [67])

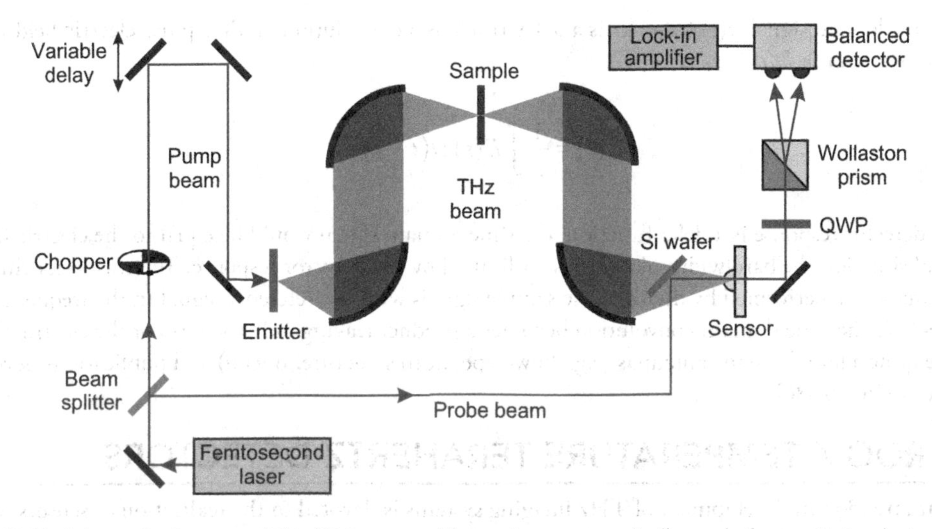

Figure 28.19 Schematic diagram of THz-TDS system. The emitters are typically made from a GaAs photoconductive switch, semiconductors (InAs, ZnTe, GaSe), and nonlinear crystals (DAST, GaP). The sensors are often a GaAs photoconductive switch or electro-optic crystals (ZnTe, GaSe, GaAs, DAST). In the second case, a quarter wave plate (QWP), Wollaston prism, and balanced detectors are used.

The photoconductive detection of broadband THz radiation is based on antenna structures similar to those used for generation of pulsed broadband THz radiation emission spectra and is used in TDS experiments or imaging—see Figure 28.19 [15,68–70]. Each pulse of a femtosecond laser producing an optical-pulse train separates into two paths. One reaches the THz emitter, such as a photoconductive antenna, semiconductor wafer or nonlinear crystal, where the optical pulses are transformed into ultrashort electromagnetic pulses. These pulses propagate in free space, and are focused onto an ultrafast detector, such as low-temperature grown (LTG) GaAs photoconductive switch or electro-optic crystal. The second part of the pulse is also delivered onto the detector after passing through a time-delay stage.

Unfortunately, LTG detector sensitivities or changes in optical polarization of electro-optical cells are small due to the weakness of the THz fields and nonlinear effect. Because of these circumstances, to obtain an image (only in the active mode) with their help, one needs to move, for example, the object mechanically, and it takes at least several minutes.

Figures 28.18b shows a schematic illustration of the photoconductive antenna sensor (detector). It consists of a H-shaped stripline structure deposited on a semiconductor substrate. In analogy to the emitter switch an incoming fs laser beam (from the same fs laser), which is focused between protruding parts of the electrodes, injects free carriers leading to a drop of the resistance across the switch below about a 100 Ω.

The electric field of the focused incoming THz radiation from fs laser, induces a transient bias voltage across the l–5 μm gap between the two arms of this receiving antenna. Thus, when the laser pulse coincides spatially and temporally with the THz electric field of the incoming THz radiation, a photocurrent is induced that is proportional to the incident electric field. By delaying the laser pulse relative to the THz pulse, the time-dependence of the photocurrent can be measured. Since the laser pulse is narrow in comparison to the time duration of the THz pulse from laser induced semiconductor emitter (Figure 28.18b), the laser acts as a gated sampling signal.

The detectors as photoconductive antennas (see Figure 28.19) on the base of highly resistive semiconductors (e.g., low temperature grown GaAs [71–73] or narrow-gap InGaAs [74]) are frequently used in TDS technique. During the laser pulse the excited carriers are accelerated by the electrical field component of the incident THz pulse with the time-dependent electrical field $E(t)$. The current signal, which arises in outer circuit, can be analysed by inverse Fourier transform procedure. The photoconductive antenna can be considered as a dipole of the length L, which is in resonance with the radiation wavelength λ_n inside the semiconductor. The resonance condition is $L = m(\lambda/2n)$, where m is the integer and n is the semiconductor refractive index.

The time domain signal, $I(\tau)$ (where τ is a delay time), is a convolution of THz pulse electric field $E(t)$ and detector response $D(t)$

$$I(t) = \frac{1}{T}\int_0^T E(t)D(t-\tau)\,dt. \tag{28.8}$$

If the detector response is a delta function, the time domain signal would be equal to the electric field. In the real situation the bandwidth of the pulse is limited by the detector response. The reason for this is that the analysis is performed by dividing the sample signals with the reference signal in the frequency domain where the time domain convolution becomes a product leaving only the ratio of the electric fields. There are quite a lot of designs antennas (e.g., bow-type, horn structure, fractal) and publications devoted to this question [75–80].

28.6 ROOM TEMPERATURE TERAHERTZ DETECTORS

Particular attention in development of THz imaging systems is devoted to the realization of sensors with a large potential for real-time imaging while maintaining a high dynamic range and room-temperature operation. Complimentary metal-oxide-semiconductor (CMOS) process technology is especially attractive due to their low price tag for industrial, surveillance, scientific, and medical applications. However, CMOS THz imagers developed thus far have mainly operated single detectors based on lock-in technique to acquire raster-scanned imagers with frame rates on the order of minutes. With this mind, much of recent developments are directed towards three types of focal plane sensors:

- SBDs compatible with CMOS process,
- FETs relay on plasmonic rectification phenomena, and
- adaptation of IR bolometers to the THz frequency range.

An important issue for a FPA is pixel uniformity. It appears however, that the production of monolithically-integrated detector arrays encounters so many technological problems that the device-to-device performance variations and even the percentage of non-functional detectors per chip tend to be unacceptably high.

The performance of monolithically integrated detector arrays with room temperature THz detectors is summarized in Table 28.4 [81]. Figure 28.20 shows representative data for a range of detector technologies at room temperature, and progress over recent years. Further insight on NEP values of state-of-the-art of direct detectors at room temperature is provided in Figure 28.21.

SBDs respond to the THz electric field and usually generate an output current or voltage through a quadratic term in their current-voltage characteristics. In general, the NEP of SBD and FET detectors is better than that of Golay cells and pyroelectric detectors around 300 GHz. Both the pyroelectric and the

Table 28.4 Parameter of some uncooled THz detectors

DEVICE TYPE	ELECTRICAL RESPONSIVITY (VW^{-1})	CONDITIONS	NEP ($WHz^{-1/2}$)
Schottky diodes			
ErAs/InGaAlAs spiral planar antenna	—	Zero bias, 639 GHz	4.0×10^{-12} NEDT = 120 mK
InGaAs log-spiral antenna	~200 for system estimate 10^3 intrinsic for the diode	0.8 THz	5.0×10^{-12}
VDI Model: WR2.8 ZBD	1,500	260–400 GHz	2.7×10^{-12}
VDI Model:WR1.5ZBD	750	500–750 GHz	5.1×10^{-12}
VDI Model: WR1.0 ZBD	200	750–1,100 GHz	20×10^{-12}
VDI Model:WR0.65 ZBD	100	1,100–1,700	40×10^{-12}
Bolometers			
$Hg_{0.8}Cd_{0.2}Te$ HEB	0.30 at 17 mV bias, 36 GHz 96 for 0.89 THz, 13 mV bias	Room temperature	2.2×10^{-9} for 17 mV bias, 35 GHz 7.4×10^{-9} for 0.89 THz, 12 mV bias
Si_xGe_y:H	170	0.934 THz, uncooled	0.2×10^{-9}
Vanadium oxide	—	Uncooled	320×10^{-12} at 4.3 THz, 9×10^{-13} @ 7.5–14 μm
Niobum film	21	3.6 mA bias, 1 kHz mod, 300 K	1.10×10^{-10}
Ti, antenna-coupled microbolometer	—	10 kHz chopper, 1.04 mA bias, 300 K	1.5×10^{-11}
Nb_5N_6	400	0.4 mA bias, >10 kHz	9.8×10^{-12}
Vanadium oxide array	1.5×10^4	1 V bias, 130 μm, uncooled	2.00×10^{-10}
Nb, polymide, antenna coupled	450	<1 THz	1.5×10^{-11}
Al/Nb; antenna coupled	85	1 kHz mod, 1.6 mA bias	2.5×10^{-11}
Free-standing Nb bridge antenna coupled	210 (average over 10 devices)	650 GHz	12.5×10^{-12}
Pyroelectrics			
Philips P5219 DLATGS	321	10 Hz mod; amplifier with gain of 4.8, 91 GHz	3.1×10^{-8}
QMC instr.	18,300 1,200	10 Hz mod; 1.89 THz, <20 Hz mod	4.4×10^{-10}
$LiTaO_3$	—	530 GHz, Melectron Model SPH-45	2.0×10^{-9}

(Continued)

Table 28.4 (*Continued*) Parameter of some uncooled THz detectors

DEVICE TYPE	ELECTRICAL RESPONSIVITY (VW^{-1})	CONDITIONS	NEP ($WHz^{-1/2}$)
Golay cells			
Tydex Golay Cell GC-1X	100,000	21 Hz chopper	1.4×10^{-10}
Microtech Instruments	10,000	12.5 Hz chopper	10×10^{-8}
Microarray, layer by layer, polimer membranes over Si	—	30 Hz mod 105 GHz	300×10^{-9}
Tydex Golay Cell, 6-mm diameter diamond window		10 Hz mod	7.0×10^{-10}
CMOS-based and plasma detectors			
BiCMOS SiGe, 0.25 μm HBT	Current R_i 1 AW^{-1} at 0.7 THz	3 × 5 array, chopper 125 kHz	50×10^{-12} at 0.7 THz
BiCMOS SiGe, 0.25 μm NMOS	Voltage R_v 80 kVW^{-1} at 0.6 THz	3 × 5 array, chopper 16 kHz	300×10^{-12} at 0.6 THz
CMOS SiGe, 65 nm NMOS	Voltage R_v 140 kVW^{-1} at 0.87 THz	32 × 32 array, chopper 5 kHz	100×10^{-12} at 0.87 THz
CMOS SiGe, 65 nm NMOS	Voltage R_v 0.8 kVW^{-1} at 1 THz	3 × 5 array, chopper 1 kHz	66×10^{-12} at 1 THz
CMOS-SBD, 130 nm	Voltage R_v 0.323 kVW^{-1} at 0.28 THz	4 × 4 array, chopper 1 kHz	29×10^{-12} at 0.28 THz
CMOS-SBD, 65 nm	—	1 element; 1 MHz mod	42×10^{-12} at 0.86 THz
CMOS, 150 nm, NMOS	Voltage R_v at 4.1 THz	1 element	133×10^{-12} at 4.1 THz
InGaAs HEMT	Voltage R_v 23 kVW^{-1} at 200 GHz	1 element	0.5×10^{-12} at 200 GHz
Asymmetric dual-grating gate InGaAs HEMT	Voltage R_v 6.4 kVW^{-1} at 1.5 THz	1 element	50×10^{-12} at 1.5 THz

bolometer FPAs with detector response times in the millisecond time range are not suited for heterodyne operation. FET detectors are clearly capable in heterodyne detection with improving sensitivity. Diffraction aspects predicts FPAs for higher frequencies (0.5 THz and above) and in conjunction with large *f*/# optics.

Today, SBDs provide NEP up to and beyond 1 THz with zero bias (zero-bias operation offers reduced 1/*f* noise), both in quasioptical configurations and on-wafer. Heterostructure backward diodes produce larger curvatures than Schottky diodes, which resulted in an NEP value of 0.18 $pWHz^{-1/2}$ at 94 GHz [84]. Microbolometers provide at frequencies beyond 2 THz at direct detection. Their drawback is required bias voltage and a millisecond response time, slower than SBD.

In the last decade, CMOS FETs have emerged as promising detectors [85]. Although the FETs are biased with a gate voltage, they draw zero bias current, enabling low noise. In the last years, an NEP value close to Schottky diode been demonstrated from a few hundred GHz to 4.3 THz, using moderate gate lengths of 150 nm [86,87]. The high performance and easy integration with other CMOS technologies opens a way for low-cost terahertz detector arrays [88].

Figure 28.22 compares the best results to date for THz room-temperature direct detectors operating above 300 GHz. These frequencies are close to 650 GHz—this is center frequency of a well-known atmospheric window—perhaps the last one useful for terrestrial sensing before the atmosphere becomes prohibitively lossy. The best popular are pyroelectric detectors and Golay cells with NEP values around

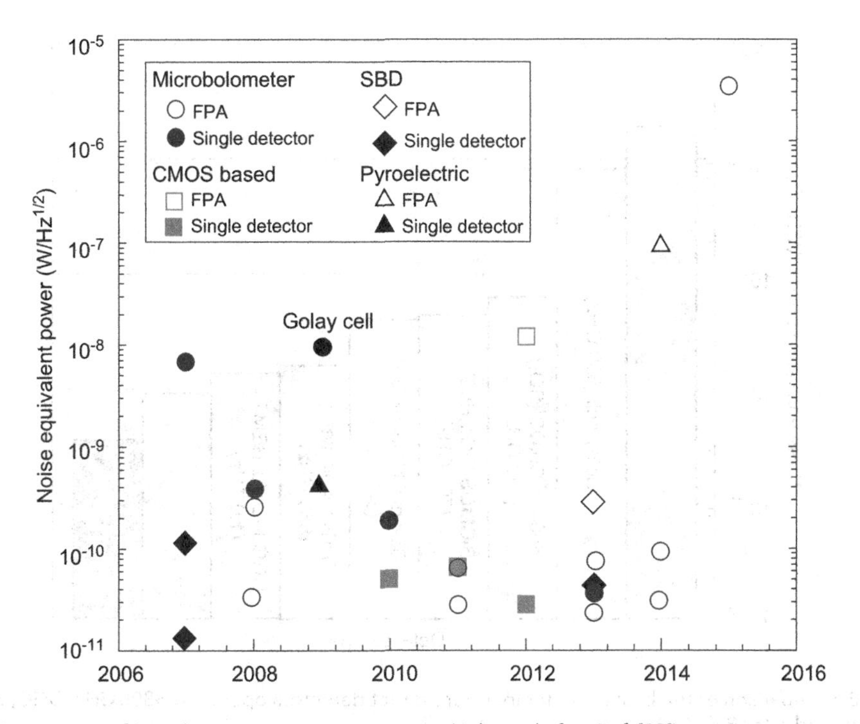

Figure 28.20 Progress of THz detectors over recent years. (Adapted after Ref. [82])

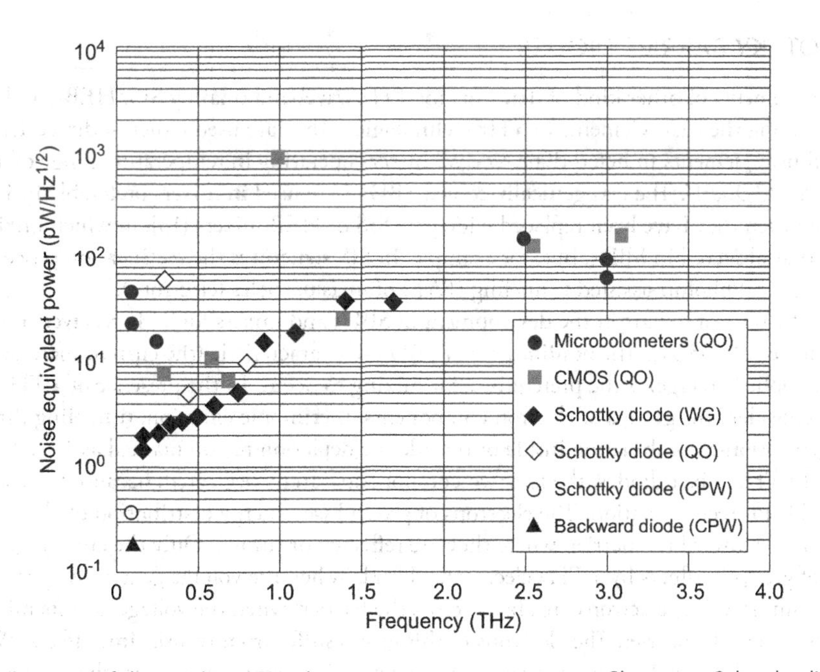

Figure 28.21 The state-of-the-art direct THz detectors at room temperature. Shown are Schottky diodes mounted in a waveguide (WG), in quasioptical configuration (QO), in on-wafer measurements (CPW), CMOS FET (QO), microbolometers (QO), and heterostructure backward diodes. (Adapted after Ref. [83])

about 1 nWHz$^{-1/2}$. Both detectors have physical wide apertures (5 or 6 nm) so it is quite easy to focus THz radiation into them. This benefit and their nanowatts sensitivity cause that they are popular devices to characterize THz sources of all types.

Below, description of different kinds of uncooled THz detectors is presented.

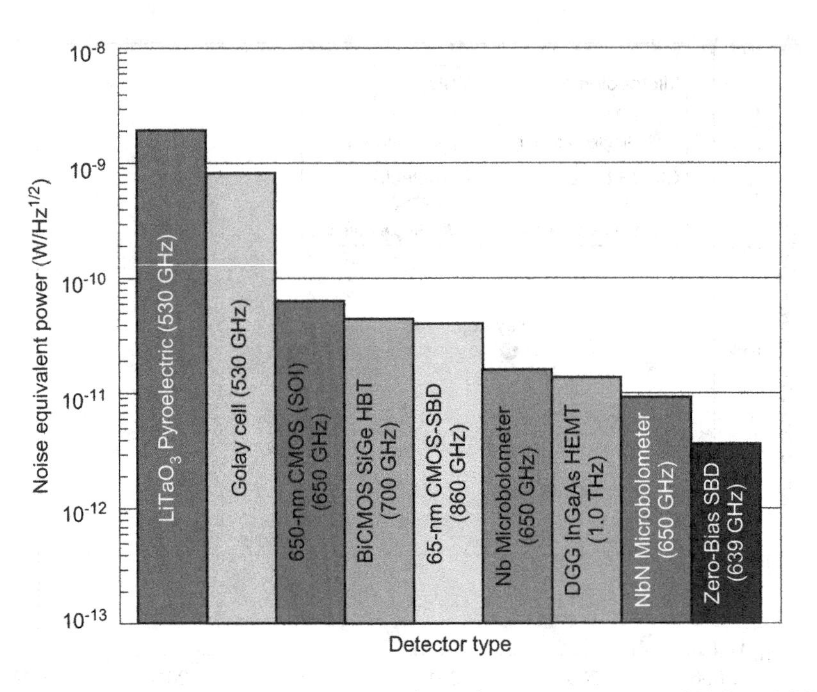

Figure 28.22 Comparison of the best room temperature direct detectors operating >300 GHz. DGG, dual grating gate. (Adapted after Ref. [89])

28.6.1 SCHOTTKY BARRIER DIODES

In spite of achievements of other kind of detectors for THz waveband (mainly SIS, HEBs, and TESs), the SBDs are among the basic elements in THz technologies. They are used either in direct detection or as nonlinear elements in heterodyne receiver mixers operating in temperature range of 4–300 K [2,6,25,28,29,44,53,90,91]. The cryogenically cooled SBDs were used in mixers preferably in 1980s and early 1990s and then they have been replaced widely by SIS or HEB mixers [13], in which mixing processes are similar to that observed in SBDs, but, for example, in SIS structures the rectification process is based on quantum-mechanical photon-assisted tunneling (PAT) of quasiparticles (electrons).

Analysis of the state of the art in the development of SBDs and mixers for THz receivers is carried out, for example, in Refs. [6,53,91]. The nonlinearity of SBD *I-V* characteristic (the current increases exponentially with the applied voltage) is the prerequisite for mixing to occur. In the presence of a THz electric field, one can consider four groups of electron components: thermionic emission, tunneling through the barrier, and generation-recombination inside or outside the depletion region marked as 1, 2, 3, and 4 in Figure 28.23. In THz mixer diodes, the last two components can be neglected, because there are almost no holes available for recombination. The electrons of group I (see energy distribution of electrons inside Figure 28.23) move toward the barrier, where they are reflected or tunnel. Only the current generated by the electrons of group II affects by a THz electric field [92]. When the voltage generated by the THz field is close to its maximum value, electrons are able to cross the barrier; when the voltage is at its minimum—the electrons cannot cross the barrier. The electrons of this group suffer from transit time effects. Within half a period of the THz field, they have to transverse the depletion layer in order to cross the barrier. Such effects as skin effect, charge inertia, dielectric relaxation, and plasma resonance, lead to degradation of the detector performance. At THz frequencies, the noise due to series resistance and hot electrons dominates over shot noise. However, cooling of the diode lowers this noise.

Historically first Schottky-barrier structures were pointed contacts of tapered metal wires (e.g., a tungsten needle) with a semiconductor surface (the so-called crystal detectors). Widely used were, for example, contacts p-Si/W. At operation temperature $T = 300$ K they have NEP $\approx 4 \times 10^{-10}$ WHz$^{-1/2}$. Also pointed tungsten or beryllium bronze contacts to n-Ge, n-GaAs, n-InSb were used (see, e.g., Refs. [93,94]). In the

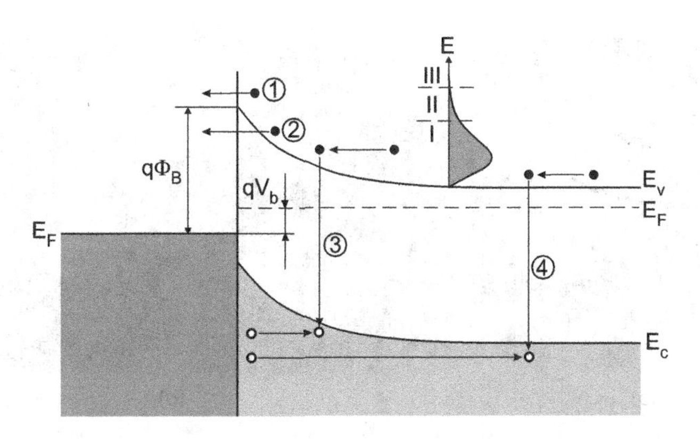

Figure 28.23 Four basic transport processes in forward-biased Schottky barrier on an n-type semiconductor: (1) thermionic emission; (2) tunneling current; and generation-recombination current (3) inside and (4) outside the depletion region.

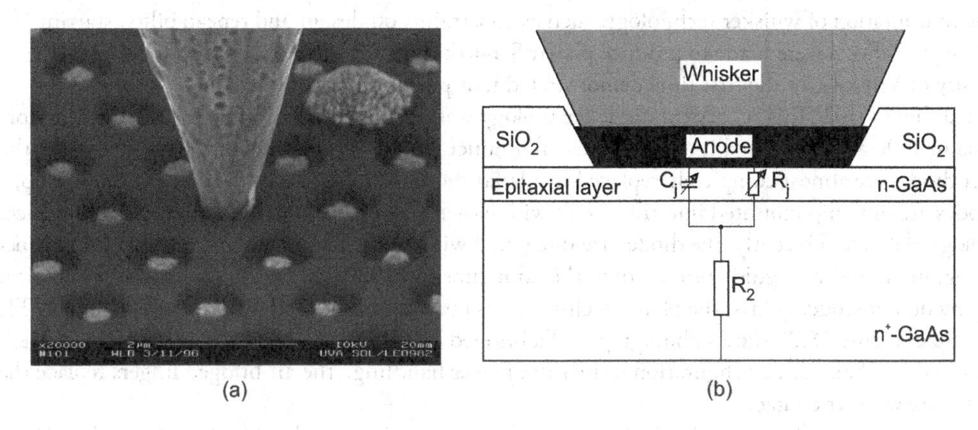

Figure 28.24 GaAs Schottky barrier whisker contacted diode: (a) scanning electron microscopy image of a contacted chip used up to 5 THz (Adapted after Ref. [96]) and (b) cross-section with equivalent circuit of the junction. (Adapted after Ref. [97])

mid 1960s Young and Irvin [95] developed the first lithographically defined GaAs Schottky diodes for high frequency applications. Their basic diode structure was next replicated by a variety of groups. The basic whiskered diode structure, depicted in Figure 28.24a, greatly improved the quality of the diode due to inherently low capacity of the wicker contact. The pointed metal wire has a tip diameter of about 0.5 μm and contacts a single anode in the array. The metal anodes are about 0.2 μm where they contact the GaAs surface which is located below the silicon dioxide passivation layer. A tipped metal whisker provides the electrical contact to the anode and serves as a long wire antenna to couple in an external radiation.

The SBD structure shown in Figure 28.24a is similar to the so-called "honeycomb" diode chip design, first produced by Young and Irvin in 1965 [95,98]. This design has been the most important steps toward a practical Schottky diode mixer for THz frequency applications, with several thousand diodes on a single chip and where parasitic losses such as the series resistance and the shunt capacitance are minimized. On highly doped GaAs substrate ($\approx 5 \times 10^{18}$ cm^{-3}) with an ohmic contact on the backside, a thin GaAs epitaxial layer, with thickness of 300 nm to 1 μm, is grown on the top of the substrate. Holes filled with a metal (Pt) in the SiO_2 insulating layer on top of the epitaxial layer define the anode area (0.25–1 μm) [99]. In order to couple the signal and the LO radiation to the mixer, a long-wire antenna in a 90° corner-cube reflector is used [100,101]. The required LO power ranges from 1 to 10 mW.

Cross section of whisker SBD with the equivalent circuit of the junction is shown in Figure 28.24b. In heterodyne operation, mixing occurs in the nonlinear junction resistance, R_j. The diode series resistance, R_s, and the voltage-dependent junction capacitance, C_j, are parasitic elements which degrade the performance.

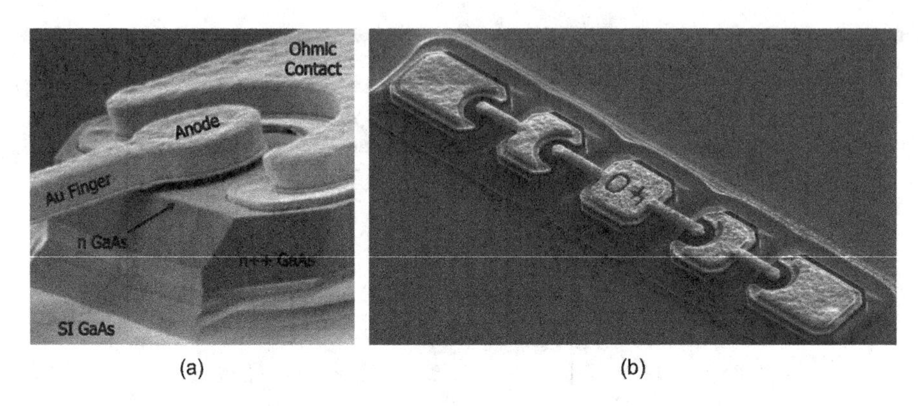

Figure 28.25 Photographs of a bridged Schottky diode (a) and a four-Schottky diodes' chip arrayed in a balanced configuration to increase power handling (b).

Due to limitation of whisker technology, such as constraints on design and repeatability, starting in the 1980s, the efforts were made to produce planar Schottky diodes. In 1993, a research group of the University of Virginia for the first time demonstrated that planar chips could compete with whisker contacted diodes [102].This successful diode technology was ultimately spun off from the University of Virginia in 1996 and developed into a commercial product line by Virginia Diodes, Inc. Nowadays, the whisker diodes are almost completely replaced by planar diodes. In the case of the discrete diode chip, the diodes are flip-chip mounted into the circuit with either solder or conductive epoxy. Using advanced technology elaborated recently, the diodes are integrated with many passive circuit elements (impedance matching, filters and waveguide probes) onto the same substrate [98,103]. By improving the mechanical arrangement and reducing loss, the planar technology is pushed well beyond 300 GHz up to several THz. For example, Figure 28.25 shows photographs of a bridged Schottky diode and a four-Schottky diodes' chip arrayed in a balanced configuration to increase power handling. The air-bridged fingers replace the now obsolete whisker contact.

At present, monolithic integration is the most common technique used to build mixers and multipliers above 1 THz. Schottky diode mixers in a waveguide configuration have an improved coupling efficiency compared to the open structure mount. To overcome a large shunt capacitance caused by the coplanar contact pads, a surface channel diode, shown in Figure 28.26a, was proposed [104]. The planar diode soldered onto a microstrip circuit (e.g., a thin quartz substrate) is mounted into a waveguide mixer block. The required LO power is about 1 mW. For mixers operating at ≈600 GHz, the DSB noise temperature is about 1,000 K. Two another diode structures are shown in Figure 28.26b,c. In the case of structure shown in Figure 28.26b, a key innovation fabrication process was the development of a robust air bridge. In the quasivertical diode structure, the Schottky contact is formed on the front side of the epitaxial layer, whereas the ohmic contact cathode is formed on its backside, directly below the corresponding Schottky contact, to ensure vertical current flow. This is done to prevent current crowding effects reducing ohmic losses.

To eliminate losses caused by support (influence of surface modes), fabrication of Schottky diodes on thin membrane was also developed [105]. In this approach, the diodes are integrated with the matching circuit and most of the GaAs substrate is removed from the chip, and the entire circuit is fabricated on the remaining GaAs membrane.

The current-voltage characteristic of the Schottky barrier junction for bias voltage values $V > 3kT/q$, can be approximated by

$$J_{\mathrm{MSt}} = J_{\mathrm{st}} \exp\left(\frac{V}{V_0}\right), \tag{28.9}$$

where the device slope parameter $V_0 = \beta kT/q$. Differentiating Equation 28.9, we can obtain the junction resistance

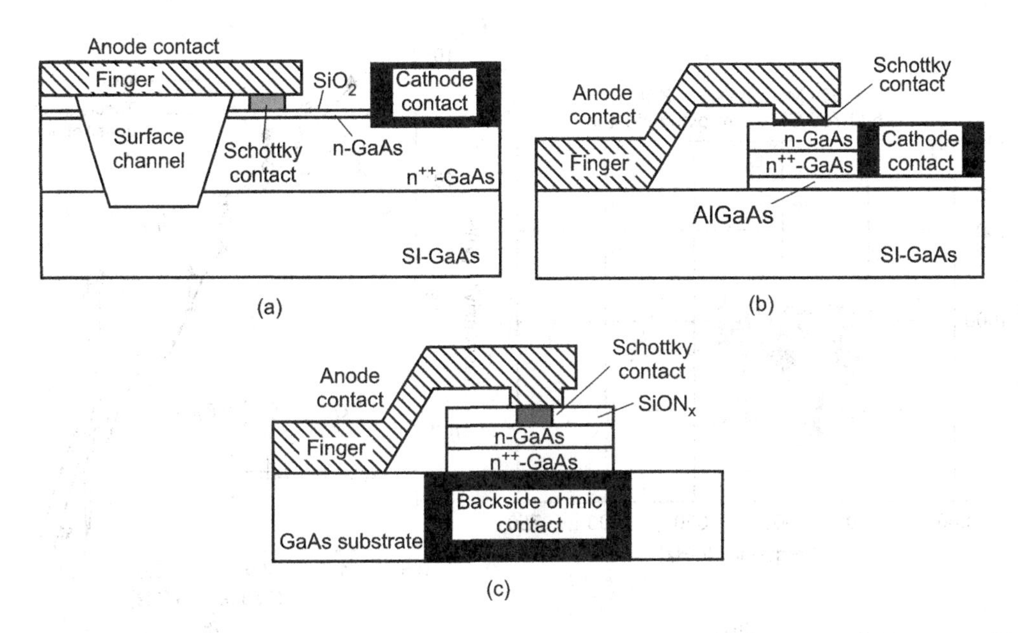

Figure 28.26 Designs of planar Schottky diodes: (a) a surface channel design for frequencies below 1 THz, (b) air-bridged diode, and (c) quasivertical diode structure.

$$R_j = \frac{V_0}{I_{\mathrm{MSt}}}. \quad (28.10)$$

Parasitic parameters, R_s and C_j, define diode critical frequency called also cutoff frequency

$$\nu_c = \left(2\pi R_s C_j\right)^{-1}, \quad (28.11)$$

which should be notably higher compared to the operation frequency.

The junction space-charge capacitance can be approximated by [106]

$$C_j(V) = \frac{\varepsilon A}{w} + 3\frac{\varepsilon A}{d}. \quad (28.12)$$

Here ε is semiconductor permittivity, A and d are the anode area and diameter, respectively, and w is the depletion region thickness dependent on the carrier density, diffusion potential, and bias. The second term is the peripheral capacitance. The junction capacitance is voltage dependent as depletion region depends on the bias applied. The specific capacity as a rule is not less than $C \approx 10^{-7}$ F cm^{-2}.

To achieve good performance at high frequencies, the diode area should be small. By reducing the junction area, one reduces junction capacities to increase the operating frequency. But at the same time, one increases the series resistance. The state-of-the-art devices have anode diameters about 0.25 μm and capacitances of 0.25 fF. For a high-frequency operation, the GaAs layers are doped up to $n \approx (5–10) \times 10^{17}$ cm^{-3} [6,97,107].

In the lower frequency range ($\nu < \approx 0.1$ THz), the operation of SBDs is well understood and could be described by mixer theories taking into account Schottky diode stray parameters (variable-capacitance and series resistance). In the THz range, however, the design and performance of the devices become increasingly complex. At higher frequencies, there appears several parasitic mechanisms credit with not only, for example, skin effect, but also high-frequency processes in semiconductor material such as carrier scattering, carrier transit time through the barrier, dielectric relaxation, etc., which become important.

Figure 28.27 presents frequency-dependent voltage sensitivity characteristic of SBDs at the room temperature [108]. The virtual desktop infrastructure (VDI) offers zero-biased detectors with full-waveguide-band operation, high sensitivity, and high responsivity for a variety of THz applications (see Figure 28.27a).

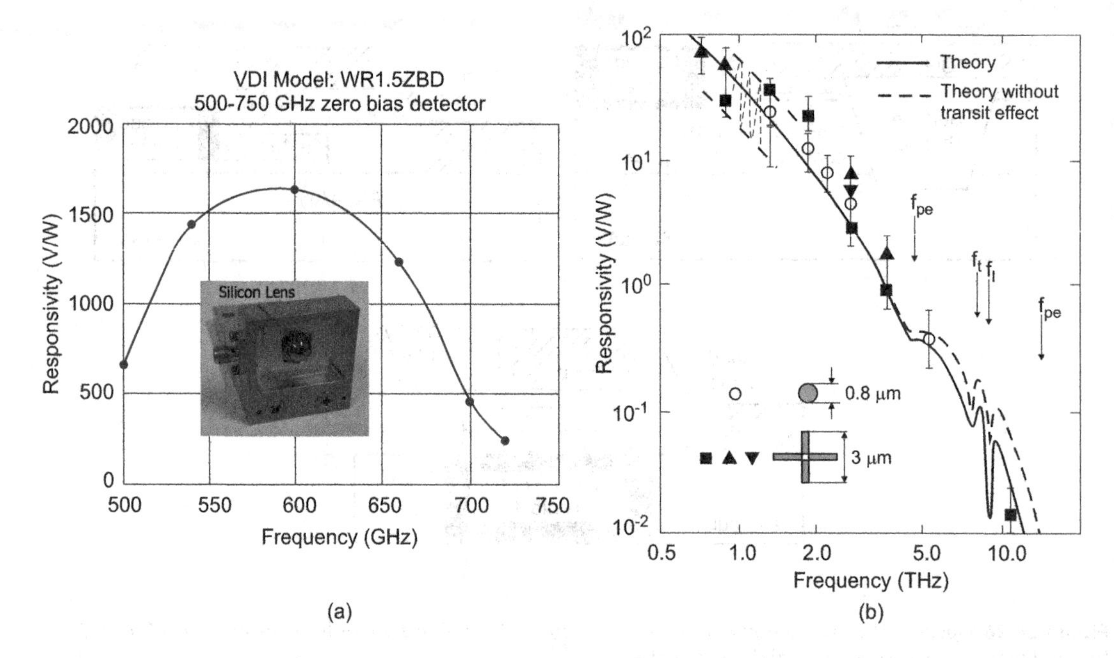

Figure 28.27 Dependence of SBD voltage sensitivity on radiation frequency dependence for diodes: (a) produced by Virginia Diode, Inc. (inside photograph of a quasioptical detector covering 100–1,000 GHz) (Adapted after Ref. [108]) and (b) with different anode shapes—experimental data are compared with theory. (Adapted after Ref. [90])

Figure 28.27b compares experimental data with theoretical prediction for diodes with different anode shapes. The solid line shows the theoretical dependence with allowance for the skin effect, the carrier inertia, the plasma resonance in the epitaxial layer (f_{pe}) and in the substrate (f_{ps}), the phonon absorption (f_t and f_l are the frequencies of the transverse and longitudinal polar optical phonons), and the transit effects. The dashed line shows the same without allowance for the transit effect. A satisfactory agreement between the experiment and the calculation takes place on the whole range of frequency. Owing to further improvement of the antenna, the detector sensitivity presented in the figure was increased by one order of magnitude to approximately 350 V W^{-1} near 1 THz. In the direct detection, SBDs reach NEP ≈ 3 × 10^{-10}–10^{-8} WHz$^{-1/2}$ at ν = 891 GHz.

The typical Schottky diodes usually have high low-frequency noise levels due to introduction of oxides, contaminants, and damage to the junction in the fabrication process. More recently, an alternative method of Schottky barrier formation has been elaborated by molecular beam epitaxy (MBE) in situ deposition of a semimetal on semiconductor to reduce the imperfections that give rise to excess low-frequency noise, particularly $1/f$ noise. The semimetal used is an ErAs film grown on Si-doped $(In_{0.53}Ga_{0.47}As)_{1-x}(In_{0.52}Al_{0.48}As)_x$ on InP substrates—see Figure 28.28 [109,110]. ErAs is a semimetal with the rock salt crystal structure and a lattice constant of 5.74 Å. This is close enough to both GaAs (5.65 Å) and InP (5.87 Å) so that high-quality epitaxial films (up to 75 Å) of ErAs can be grown by MBE on either substrate, or on $In_xAl_{1-x-y}Ga_yAs$ films lattice matched to InP. The performance of the ErAs detector can be varied by controlling the Al percentage in the InAlGaAs Schottky layer. This results in the modification of the Schottky barrier height from approximately −0.05 eV–0.45 eV in a highly controlled manner. The NEP for these detectors at 639 GHz reach 4 × 10^{-12} WHz$^{-1/2}$, which is about two orders of magnitude better than for typical GaAs Schottky diodes.

The heterodyne SBD receivers are worse compared to cooled HEB receivers and SIS mixers (see Figure 6.10). At the same time, SBD receivers operation without cooling gives opportunities for using SBD mixers in different mm and sub-mm applications. Their sensitivity is quite suitable to be used in mm wave spectrometers with moderate resolution [91,111,112]. Superconducting mixers typically require microwatt LO powers, which is roughly three to four orders of magnitude lower than their SBD predecessors. As

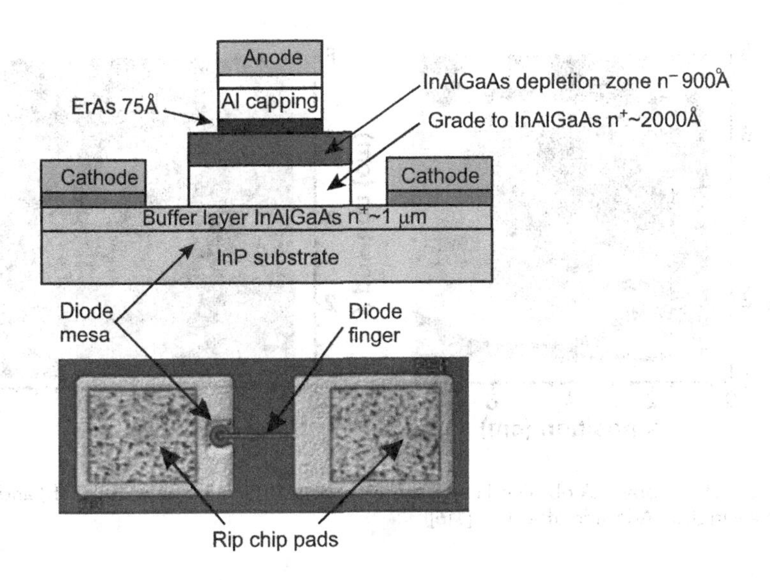

Figure 28.28 Epitaxial layer structure of the ErAs detector scaled to micron size in a planer flip chip structure. (Adapted after Ref. [109])

a result, a broader range of LO sources can be used. The technologies being used or investigated include diode multipliers, lasers and optoelectronics, and "vacuum tube" oscillators such as klystrons, including novel nanofabricated versions.

Recently, the interests for utilizing silicon integrated circuits in THz applications have increased. Silicon integrated circuits benefit from the industrially qualified CMOS process technology and can be mass-produced with significantly reduced cost. The first Schottky detector in CMOS for active imaging was reported in Refs. [113,114]. The 280-GHz detector utilized polygate separated SBDs, which have a measured cutoff frequency of ≈2 THz and are fabricated in 130-nm CMOS without any process modifications. More recently, Han et al. have demonstrated fully functional CMOS imager operating near or in the sub-mm-wave frequency range. Using a compact passive-pixel array architecture, a fully-integrated 280-GHz 4 × 4 imager has been demonstrated [115,116]. Figure 28.29 shows a die microphotograph of SBD imager mounted and wire bonded onto an FR-4 printed circuit board. The chip size is 2.4 × 2.4 mm, most

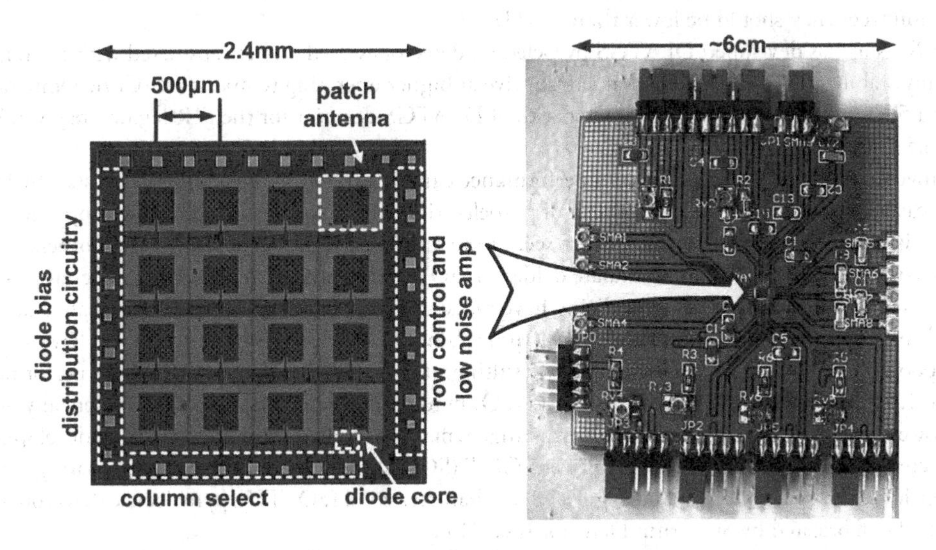

Figure 28.29 Microphotograph of the 280-GHz SBD image sensor and photograph of an imager printed circuit board. (Adapted after Ref. [116])

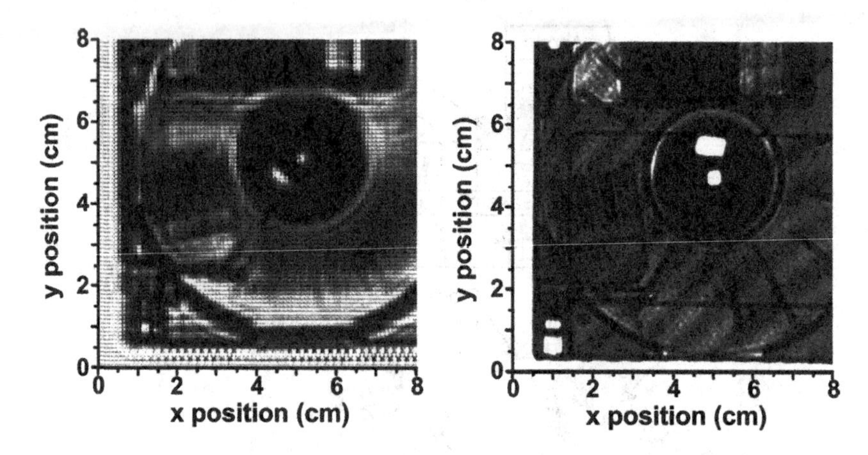

Figure 28.30 Images of a floppy disk obtained using the 16-pixel 280-GHz image sensor (left) and single 860-GHz image pixel (right). (Adapted after Ref. [116])

of which is occupied by the on-chip patch antennas. The pixel pitch is set to about a half of the wavelength in free space (≈500 μm) for lower Rayleigh diffraction.

At 1-MHz input modulation frequency, the measured peak responsivity is 5.1 kVW^{-1}. The measured minimum NEP is 29 pWHz$^{-1/2}$. Additionally, an 860-GHz SBD detector is implemented by reducing the number of unit cells in the diode and by exploiting the efficiency improvement of patch antenna with frequency. The measured NEP is 42 pWHz$^{-1/2}$ at 1 MHz modulation frequency. This is competitive to the best reported performance of metal-oxide-semiconductor FET (MOSFET)-based pixels. The 4 × 4 array increases the imaging speed by 4–8 times, due to fewer mechanical scan steps. A 280-GHz image of a floppy disk formed with 80 × 80 subimages (320 × 320 pixels) is shown in Figure 28.30. It can be clearly seen that due to a smaller wavelength, the 860-GHz scan provides a better spatial resolution.

28.6.2 PYROELECTRIC DETECTORS

Commercially available uncooled pyroelectric detectors with broadband capability in the 1- to 1,000-μm wavelength range are fabricated using such materials as $LiTaO_3$, $LiNbO_3$, and deuterated L-alanine-doped triglycene sulphate (DLATGS) (see Chapter 9). Their response time is slower than 10 ms, therefore the modulation frequency should be lower than 100 Hz.

Selex Sensors has developed DLATGS pyroelectric detectors sealed and encapsulated within a termo-electrically stabilized TO-5 package that can survive at higher operating temperature (Curie temperature is around 59°C). The performance characteristics of DLATGS detector for the FIR region are given in Table 28.5 [117]].

Improvement of pyroelectric detector's performance can be achieved by reducing the crystal thickness and increasing the coating absorption. Most of pyroelectrics tend to lose their interesting properties as the thickness is reduced. However, some of them seem to maintain their properties better than others. This seems particularly true for lithium tantalate oxide ($LiTaO_3$) and related materials. New material-processing techniques such as ion milling and ion slicing have made available $LiTaO_3$ and lithium niobate oxide ($LiNO_3$) materials with thickness of less than 10 μm. Using the new thin-film materials, we have seen current responsivity values higher than 4 μA W^{-1}, resulting in hybrid detector optical amplifier performance of less than 1.0×10^{-10} WHz$^{-1/2}$ [118]. Thin-film $LiTaO_3$ pyroelectric detectors are now commercially available. However their absorption in the terahertz range remains a challenge. Some promising developments are expected in the area of single- and multiwall-CNT (SCNT and MCNT, respectively) coatings for pyroelectric detectors. Table 28.6 and Figure 28.31 characterize $LiTaO_3$ THz pyroelectric detector—Model SPH-62 THz fabricated by Spectrum Detector Inc. [119].

Very broadband pyroelectric cameras are commercially available. For example, the Pyrocam IIIHR (see Figure 28.32), fabricated by Ophir Optics, consists of 160 × 160 $LiTaO_3$ 80-μm pixels mounted with

Table 28.5 Performance characteristics of DLATGS detector for FIR region

PARAMETERS	VALUE
IR spectral range (μm)	1–1,000
Operating temperature (K)	298
Element active area diameter (mm)	2
Frequency range of operation (Hz)	1–3,000
Thermal time constant (ms)	18
Detector window	CsI or diamond
Responsivity (VW^{-1})	2,440 (10 Hz) 300 (100 Hz) 30 (1,000 Hz)
Detectivity ($cmHz^{1/2}W^{-1}$)	6.6×10^8 (10 Hz) 6.6×10^8 (100 Hz) 3.5×10^8 (1,000 Hz)
NEP ($WHz^{-1/2}$)	2.7×10^{-10} (10 Hz) 2.7×10^{-10} (100 Hz) 5.1×10^{-10} (1,000 Hz)

Source: Adapted after Ref. [117].

Table 28.6 $LiTaO_3$ THz pyroelectric detector—Model SPH-62 THz

PARAMETERS	VALUE
Detector size (mm)	2 × 2
Electronic 3 db frequency (Hz)	15
Thermal 3 db frequency (Hz)	0.5
Voltage responsivity ($V W^{-1}$)	1.5×10^5
NEP ($WHz^{-1/2}$)	4×10^{-10} (10.6 μm, 5 Hz)
Detectivity ($cmHz^{1/2} W^{-1}$)	4×10^8 (10.6 μm, 5 Hz)

Source: Adapted after Spectrum Detector Inc.

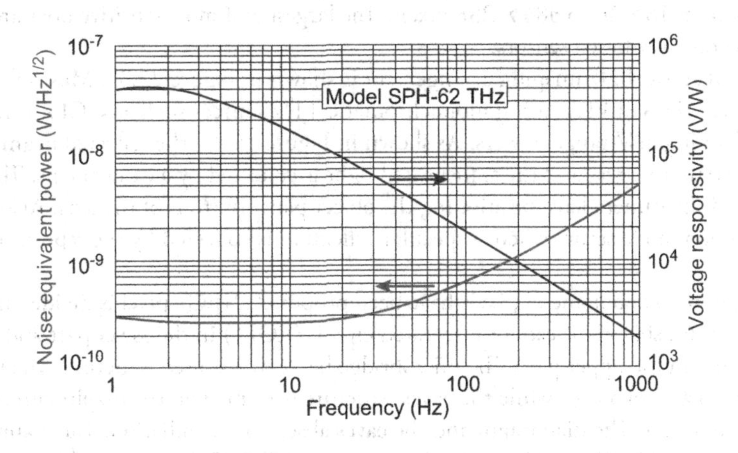

Figure 28.31 Frequency-dependent characteristics of $LiTaO_3$ THz pyroelectric detector—Model SPH-62 THz (adapted after Spectrum Detector Inc.) (Adapted after Ref. [119])

Figure 28.32 Pyrocam IIIHR beam profiling camera.

indium bumps to a solid-state readout multiplexer (MUX). However, to cover the whole wavelength range, the packaging window has to be interchanged. The camera can operate with pulsed or CW sources. For THz sources, the sensitivity is relatively low—around 300 $mWcm^{-2}$ at full output. With an SNR of 1,000, beams of 30 $mWcm^{-2}$ are detectable.

It should be mentioned that thermopiles were also demonstrated as detectors for THz imaging [120,121]. Different absorber (TiN, highly doped polySi) and thermopile materials (p-polySi/n-polySi, p-polySi/Al) have been implemented in large pixels of $1 \times 1\ cm^2$ size. 8×8 pixel array with sensitivity in the wavelength range between 50 and 500 μm gave the value of NEP ≈ 5 $nWHz^{-1/2}$ with a bandwidth of ~50 Hz.

28.6.3 MICROBOLOMETERS

An impressive promising technology is also coming from commercially available microbolometer arrays. Adaptation of IR microbolometers to the THz frequency range after the successful demonstration of active THz imaging in 2006 [122] entailed that in the period 2010–2011, three different companies/organizations announced cameras optimized for the >1 THz frequency range: NEC (Japan) [123], INO (Canada) [124], and Leti (France) [125]. The number of vendors is expected to increase soon. The array formats listed in Table 28.7 vary from 16×16 to 384×288 pixels. The largest and most sensitive ones are provided by microbolometer technology-based sensors.

The experimental active THz imaging arrangement is shown in Figure 28.33. Most of the papers presented characterizations of FPAs using monochromatic THz sources such as QCLs or far IR optically pumped lasers delivering mW-range powers. As shown in Figure 28.33, the reflected beam backlights an object with a maximum area, and the transmitted light is collected by a camera lens. The focal plane is positioned behind the camera lens, positioning the object plane in front of the lens. Also shown is the modified reflection mode setup, where a specular reflection is collected by the repositioned lens and camera.

Different designs of THz bolometer pixels have been proposed. NEC's pixel is divided into two parts (see Figure 28.34) [126], silicon Si readout integrated circuit (ROIC) in the lower part and suspended microbridge structure in the upper part. The microbridge has a two-storied structure. The first floor is composed of a diaphragm and two legs, while the eaves structure is formed on the diaphragm to increase the sensitive area and fill factor. The diaphragm and the eaves absorb THz radiation. The diaphragm is composed of VO_x bolometer thin film, SiN_x passivation layers, and TiAlV electrodes, while the eaves structure is composed of SiN_x layer and TiAlV thin film THz absorption layer. The structure composed of the thin

Table 28.7 Commercial uncooled THz imaging cameras

CAMERA—COMPANY	SENSOR	SPECTRAL RANGE	OPTICS
TZCAM—i2S http://www.i2s.fr/project/camera-terahertz-tzcam/	320 × 240 a-Si microbolometer 50-μm pitch LETI is the provider of the sensor	0.6–3 THz 25 fps	*f*/0.8, 50 mm focal plane
IR/V-T0831—NEC http://www.nec.com/en/global/prod/terahertz/	320 × 240 VO_x microbolometer 23.5-μm pitch	1–7 THz 500 fps	*f*/1, 28.2 focal plane
MICROXCAM-3841I-THz—INO http://www.ino.ca/en/products/terahertz-camera-microxcam-384i-thz/	384 × 288 microbolometer 35-μm pitch	0.94–4.25 THz 50 fps	*f*/0.9 or F/0.7 HRFZ-Si 44 mm focal plane
TicMOS-1px—Tic-Wave http://ticwave.com/products.html	100 × 100 FET	0.3–1.3 THz 500 pfs	No optics
Tera-257/1024/2056—Terasense http://terasense.com/products/sub-thz-imaging-cameras/	16 × 16/32 × 32/64 × 64 FET 1.5-mm pitch	10 GHz-0.7 THz	No optics
Pyrocam IV—Ophir Photonics http://www.ophiropt.com/laser--measurement/beam-profilers/products/Beam-Profiling/Camera-Profilingwith-BeamGage/Pyrocam-IV	320 × 240 pyroelectric 80-μm pitch	13–355 nm, 1.06–3,000 μm, 100 fps	No optics, window assembly with AR coating
Open View—Nethis http://nethis-thz.com/index.php/openview/	256 × 320, 170-μm pitch/512 × 640, 80-μm pitch based on THz to IR converter	0.1–3,000 μm 1 kfps	No optics

AR, antireflection.

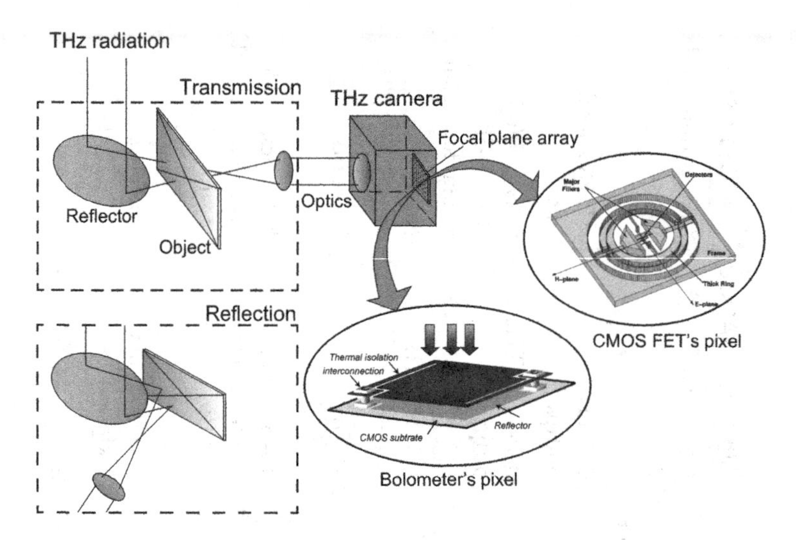

Figure 28.33 Experimental setup of THz imaging system. Cutaway depicts alternative reflection mode setup.

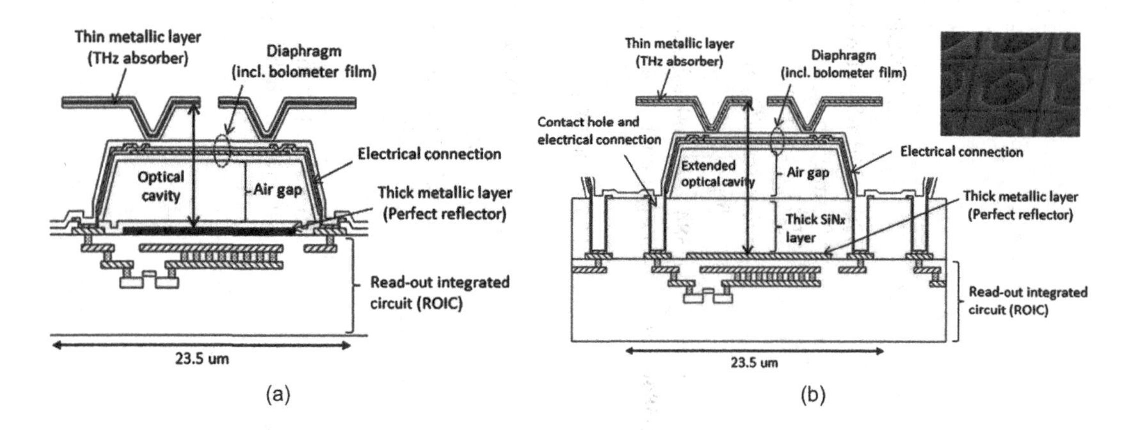

Figure 28.34 Schematic diagrams of pixel structures of uncooled bolometer-type THz-FPAs: (a) original structure and (b) modified structure. (Adapted after Ref. [126].) The picture inset in the right corner of figure (b) is the scanning electron microscopy image of the microbolometer detectors.

metallic layer and the thick metallic layer acts as an optical cavity. The estimated NEP value was 41 pW at 3.1 THz. These results have motivated the commercialization of the first real-time handy THz bolometer camera (see the picture inset in the left bottom of Figure 28.35).

However, the cavity length of the original THz detector (see Figure 28.34a) was about 3–4 µm, which is much smaller than the THz wavelength of approximately 100 µm. As a result, the sensitivities of the previously developed THz cameras became worse in the lower frequency region, especially below 1 THz [127]. This drawback is solved with the modified pixel structure, which is shown in Figure 28.34b [128]. A thick SiN layer (ca. 7-µm thick) and a layered electrical connection are inserted between the trick metallic layer and the air-gap, so that the geometrical optical-cavity length is made much longer by a factor of 3 than the previous one. Figure 28.36 illustrates the enhancement of the optical absorption versus the sheet resistance for two different optical cavity. Compared to the standard LWIR 3-µm cavity height, the NEP can be improved by another factor of 3-10 in the THz regime [129]. The modification of pixel structure is applied to fabrication of both 640 × 480 and 320 × 240 THz-FPAs to increase sensitivity in the sub-THz region.

A schematic of one Leti's pixel of amorphous silicon microbolometer array is shown in the top-center of Figure 28.35. The 50-µm pitch is associated with quasi-double-bowtie antennas to a thermometer

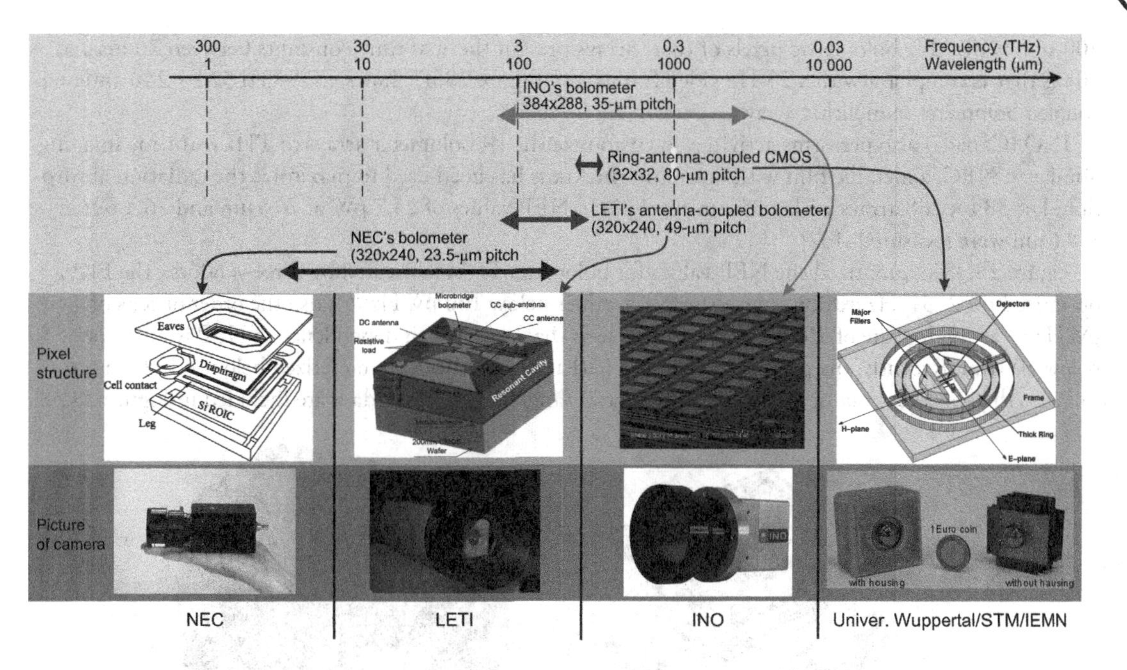

Figure 28.35 Development status of uncooled THz FPAs.

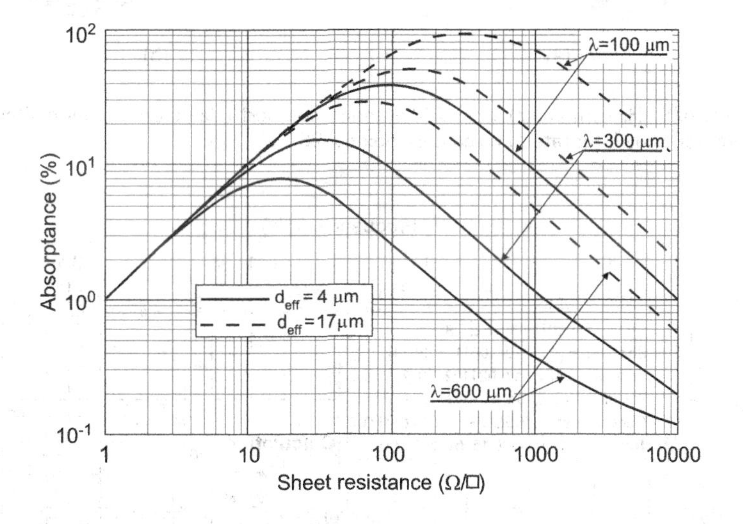

Figure 28.36 The modeled absorptance of the optical cavity versus sheet resistance of the thin metallic layer at several wavelengths (100, 300, and 600 μm) for d_{eff} equal 4 and 17 μm. The length of the effective air gap is expressed by $d_{eff} = d_{air\ gap} + n_{SiN}d_{SiN}$, where $d_{air\ gap}$ is a total length of air gaps shown in Figure 28.34; n_{SiN} is a refractive index of SiN; and d_{SiN} is a total thickness of SiN layers including the thick SiN layer (ca.7 μm).

microbridge structure derived from the standard IR bolometer. The crossed bowtie membrane is suspended over the substrate by arms and pillars. In order to enhance the antenna gain, an equivalent quarter-wavelength resonant cavity is realized under antennas with an 11-μm-thick SiO_2 layer deposited over the metallic reflector. To ensure electric contact between the bolometer pillars and CMOS metal upper contacts, the vias are etched through 11-μm cavity and then metalized. Crossed polarized antenna structures have been implemented to make the bolometer sensitive to both TE and TM polarizations. The 320 × 240 arrays are designed for an optimized sensitivity in the 2-4 THz range [130,131]. Thanks to state-of-the-art silicon microelectronic facilities and robust bolometer technology know-how, a very high yield has been achieved—more than 99.5% of the 320 × 240 pixels are functional for 56 out of 63 chips available per

200-mm wafer. The bolometric pixels of these arrays present thermal time constants between 20 ms and 40 ms that is compliant with a 25-Hz video frame rate. Figure 28.37 shows CEA-Leti 320 × 240 antenna-coupled bolometer monolithic array.

INO (Canada) also performs activities to customize the IR bolometer arrays to THz real-time imaging. Similar to NEC, a metallic film with optimum thickness has been used to maximize the radiation absorption. For 384 × 288 arrays with a 35-μm-pixel pitch, NEP values of 24.7 pW at 70.5 μm and 76.4 pW at 118.6 μm were measured [132].

Figure 28.38 summarized the NEP values for bolometer FPAs fabricated by three vendors. The FPAs optimized for 2–5 THz exhibit impressive NEP values below 100 pWHz$^{-1/2}$ (also the order of tens of pWHz$^{-1/2}$ at a video rate of 20–30 fps). It can be seen that wavelength dependence of NEP is quite flat below 200 μm. Recently, the performance of the Leti arrays has been characterized below 1 THz in order to assess the sensitivity out of the optimized spectral range [131]—these data are included in Figure 28.38

Figure 28.37 CEA-Leti/i2S THz camera: (a) 320 × 240 antenna-coupled THz bolometer monolithic array provided by Leti and (b) bolometer array integrated in a camera box by i2S.

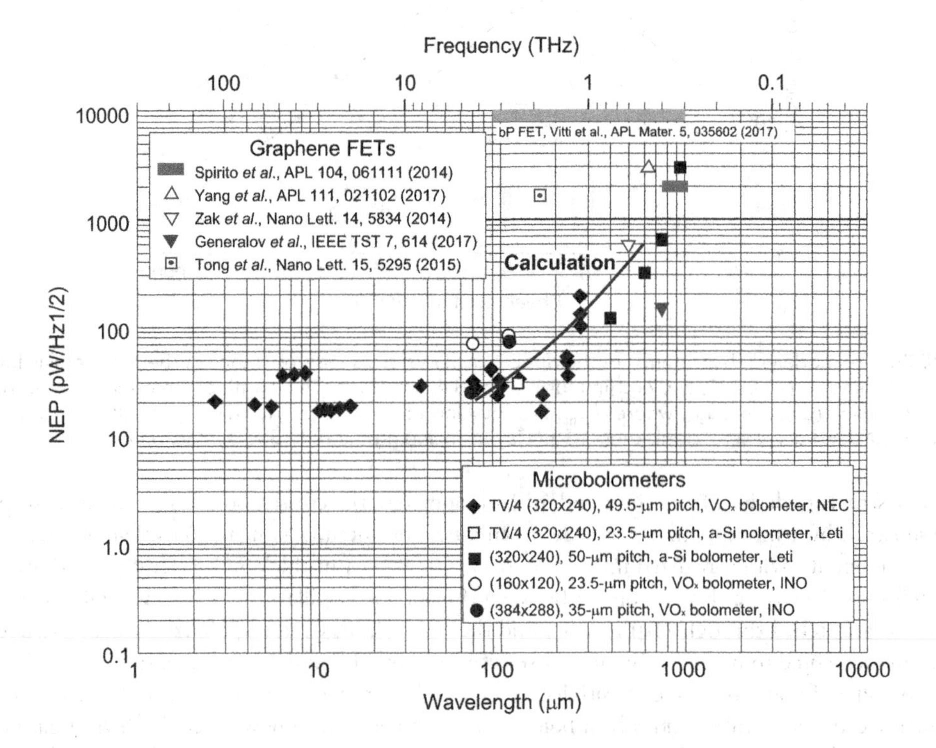

Figure 28.38 Spectral dependence of NEP for bolometer and graphene FET detector THz FPAs.

(solid square points). Further improvement of performance is possible by increasing the number of pixels, modification of antenna design while preserving pixel pitch, ROIC, and technological stack.

Terahertz room-temperature arrays are also fabricated using hybrid structure, where the pixel array is processed separately from the ROIC and then hybridized to the ROIC with microelectronic techniques like wire bonding or flip-chip assembly. A research group from the Ohio University has presented a new THz FPA sensor that incorporates ultrafast (2.5-THz intrinsic cut-off frequency), zero-biased, antimonide-based heterostructure backward diodes (Sb-HBDs), monolithically integrated with planar THz antennas for each sensor pixel [133,134]. A 31 × 31-pixel camera designer to cover the 0.6–1.2 THz range has been developed. The diode output signal is read out by a ROIC that is flip-chip-bonded to the planar FPA.

HBD THz video-like mode camera with direct detection has been commercialized by TRaycer [135]. It is specified to provide an unmodulated NEP of 200 nWHz$^{-1/2}$ with 25–100 fps for the 80 × 64 format.

A new microbolometer approach for design of array pixel has been proposed, using Si FET and exploiting the dependence of the channel's conductance on temperature [136]. Heating by radiation in the HgCdTe layers can also be used for designing of uncooled THz/sub-THz detectors with appropriate active imaging characteristics and a possibility to be assembled into arrays [137].

28.6.4 FIELD-EFFECT TRANSISTOR DETECTORS

Nonlinear properties of plasma wave excitations (the electron density waves) in nanoscale FET channels enable their response at frequencies appreciably higher than the device cutoff frequency, what is due to electron ballistic transport. In the ballistic regime of operation, the momentum relaxation time is longer than the electron transit time. The FETs can be used both for resonant (tuned to a certain wavelength) and non-resonant (broadband) THz detection [138–142] and can be directly tunable by changing the gate voltage.

The transistor receivers operate in the wide temperature range up to room temperatures [143,144]. Different material systems are used in the fabrication FET, high-electron-mobility transistor (HEMT), and MOSFET devices, including Si, GaAs/AlGaAs, InGaP/InGaAs/GaAs, and GaN/AlGaN [145–147]. Plasma oscillations can be also observed in a two-dimensional (2D) electron channel with a reverse-biased Schottky junction [148] and double quantum well (QW) FET with a periodic grating gate [149].

The use of FETs as detectors of THz radiation was first proposed by Dyakonov and Shur in 1993 [150] on the basis of a formal analogy between the equations of the electron transport in a gated 2D transistor channel and those of shallow water, or acoustic waves in music instruments. As a consequence, hydrodynamic-like phenomena should exist also in carrier dynamics in the channel. Instability of this flow in the form of plasma waves was predicted under certain boundary conditions.

The physical mechanism supporting the development of stable oscillations lies in the refection of plasma waves at the borders of the transistor with subsequent amplification of the wave's amplitude. Plasma excitations in FETs with sufficiently high electron mobility can be used for emission as well as detection of THz radiation [151,152].

The detection by FETs is due to nonlinear properties of the transistor, which lead to the rectification of the AC current induced by the coming radiation. As a result, a photoresponse appears in the form of a DC voltage between the source and drain. This voltage is proportional to the radiation intensity (photovoltaic (PV) effect). Even in the absence of an antenna, the THz radiation is coupled to the FET by contact pads and bonding wires. A big progress in sensitivity can be obtained by adding a proper antenna or a cavity coupling.

The plasma waves in FET is characterized by the linear dispersion law [150], and in the gated region:

$$\omega_p = sk = k\left[\frac{q\left(V_g - V_{th}\right)}{m^*}\right]^{1/2}. \tag{28.13}$$

where s is the plasma wave velocity in channel, V_g is the gate voltage, V_{th} is the threshold voltage, k is the wave vector, q is the electron charge, and m^* is the electron effective mass. Figure 28.39 schematically shows the resonant oscillation of plasma waves in the gated region of FET. The dispersion relations in bulk (three dimensional) and ungated regions of FET differ from Equation 28.13 and are equal

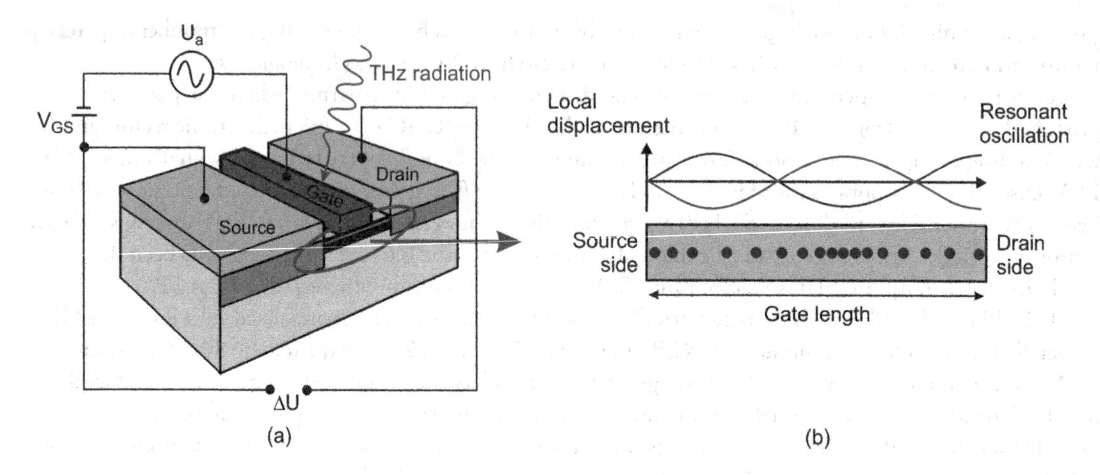

Figure 28.39 Schematic of the THz CMOS detector (a) and plasma oscillations in a transistor (b).

$$\omega_p = \left(\frac{q^2 N}{\varepsilon m^*}\right)^{1/2} \quad \text{and} \quad \omega_p = \left(\frac{q^2 n_s}{2\varepsilon m^*}k\right)^{1/2}, \tag{28.14}$$

respectively. Here N is the bulk electron concentration for alloyed regions, and n_s is the sheet electron density for channel regions.

The plasma wave velocity in the gated region is typically noticeably larger compared to the electron drift velocity. A short FET channel with length L_g acts as a resonant cavity for these waves with the eigen frequencies $\omega_n = \omega_o(1 + 2n)$ $(n = 1, 2, 3,\ldots)$. The fundamental plasma frequency is

$$\omega_0 = \frac{\pi}{2L_g}\left[\frac{q\left(V_g - V_{th}\right)}{m^*}\right]^{1/2}. \tag{28.15}$$

When $\omega_0\tau \ll 1$, where τ is the momentum relaxation time, the detector response is a smooth function of ω and V_g (broadband detector). When $\omega_0\tau \gg 1$, the FET can operate as a resonant detector, tunable by the gate voltage response frequency, and this device can operate in the THz range. The detection character (resonant or nonresonant) depends on the quality factor of the transistor resonating cavity.

Assuming $m^* \approx 0.1m_0$ (m_0 is free electron mass), $L_g \approx 100$ nm, and $V_g - V_{th} \approx 1$ V, the frequency of plasma waves is estimated as $\nu_0 = \omega_0/2\pi \approx 3$ THz. The minimum gate length can approach ≈30 nm, and thus, ν_0 can reach 12–14 THz for FETs with GaAs channels.

The plasma wave velocity s may be presented in another manner. When the thickness of the dielectric layer between the channel and the gate is small compared to the wavelength of plasma waves, it is equal to $s = (n_s q^2 d/\varepsilon m^*)^{1/2}$ [153], where n_s is the 2D electron concentration, ε is the permittivity of the dielectric layer, d is the distance from gate to channel. Then the fundamental frequency can be expressed by the relation:

$$\nu_0 = \frac{s}{4L_g} = \frac{1}{4L_g}\sqrt{\frac{n_s q^2 d}{\varepsilon m^*}}, \tag{28.16}$$

in which there are no free parameters and from which the locations of the resonant peaks can be predicted to a certain degree, though a discrepancy between the peak frequencies predicted and those experimentally found at $T \le 4$ K in GaAs/AlGaAs modulation-doped single QW were observed [154].

In a simple approximation, the electron concentration n_s is described by a plane capacitor formula: $n_s = CV_0/q$. Here, C is the capacitance between the gate and channel (per unit surface area) and $V_0 = (V_g - I_{ds}R_s - V_{th})$ is the difference between the gate voltage V_g, the voltage drop across the contact resistance R_s

(I_{ds} is the current in the transistor channel), and the threshold voltage V_{th} of the transistor. In this case, the velocity of the plasma waves is defined by the following expression:

$$s = \left[\frac{q\left(V_g - I_{ds}R_s - V_{th}\right)}{m^*}\right]^{1/2}. \tag{28.17}$$

The resonance frequency of plasma oscillations in the subgate 2D electron gas is governed by the gate length L_g and the plasma-wave velocity s, and is similar to Equation 28.15:

$$\nu_r = \frac{1}{4L_g}\left[\frac{q\left(V_g - I_{ds}R_s - V_{th}\right)}{m^*}\right]^{1/2}. \tag{28.18}$$

The resonance frequency is maximal for zero bias at the gate and tend to zero when $V_{gs} \rightarrow V_{th}$. For 2D electron gas in GaN/AlGaN with gate length $L_g = 250$ nm at $T = 4.2$ K, the resonance frequency $\nu_r = 576$ GHz (the quality factor $\omega_r\tau = 1.81$) [155].

Summarizing the above discussion for high frequencies, high mobility, and short MOS channel, the theory predicts that a standing wave can arise in the MOS channel, so the FET behaves like a resonator at the fundamental plasma frequency, delivering high induced voltage between drain and source. However, in silicon, the mobility is too low at room temperature to obtain these resonant conditions. Using low-cost CMOS process designed in a 0.13-μm technology, THz detectors meet non-resonant low-frequency case. For transistors with long channel, the THz signal that is coupled to the source does not reach the drain due to damping effect. A DC voltage variation arises, and the theory predicts that this DC output voltage between the source and the drain (see Figure 28.39) is proportional to the power of the THz arrays in the non-resonant low-frequency case:

$$\Delta U \propto U_a^2. \tag{28.19}$$

Another theory, called the self-mixing theory, was developed by Lisauskas et al. [156,157]. They presented a comprehensive review of plasma-wave-based detection of THz radiation with MOSFETs. They consider the direct detection in FETs as a distributed resistive self-mixing process in a nonbiased transistor. In this way by investigation of intrinsic carrier transport dynamics inside the transistor channel, they explained the mixing procedure and also predicted that the DC drain-to-source voltage is proportional to the squared amplitude of the THz signal present between the gate and source. Therefore, antenna gain and efficiency performances directly impact detection quality.

Figure 28.40 shows the characteristics of the 60-nm gate length InGaAs/InAlAs transistor [158]. The photoresponse of the device exposed to the radiation of 2.5 THz frequency as a function of the gate voltage measured at various temperatures is shown in Figure 28.40a. At $T > 100$ K, only non-resonant detection was observed as a broadband peak. With temperature decreasing below 80 K, the additional peak appears as a shoulder on the temperature-independent background of the non-resonant detection. This behavior can be attributed to the resonant detection of THz radiation by plasma waves. To support this assumption, additional measurements were carried out at 10 K for excitation frequencies of 1.8, 2.5, and 3.1 THz. The experimental results are displayed in Figure 28.40b. For comparison, the theoretical prediction of plasma frequency as a function of the gate voltage according to Equation 28.14 is plotted as a continuous line in Figure 28.40c. One can see that the increasing of excitation frequency from 1.8 to 3.1 THz causes moving of the plasmon resonance with the gate voltage, roughly in agreement with the theory.

Veksler et al. have predicted that by applying a drain-to-source current, it is possible to observe room temperature THz radiation resonant detection [159]. It appears that driving a transistor into the saturation region enhances the non-resonant detection and can lead to the resonant detection even if the condition $\omega_0\tau >> 1$ is not satisfied [144,155]. The physical reason is that the effective decay rate for plasma oscillations, in the condition of hot drifting electrons, becomes equal to $1/\tau_{\text{eff}} = 1/\tau - 2v/L_g$, where v is the electron drift velocity, and becomes longer when applying current. As $\omega\tau_{\text{eff}}$ becomes of the order of unity, the detection becomes resonant.

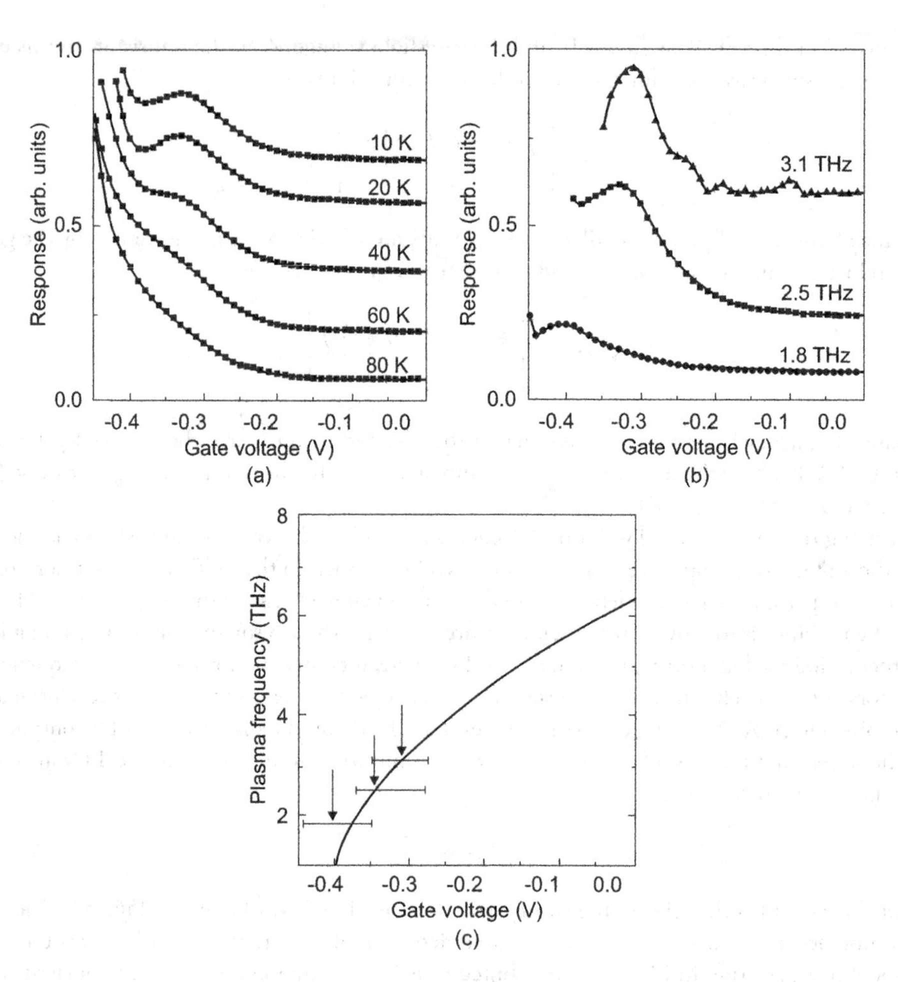

Figure 28.40 Characteristics of 60-nm gate length InGaAs/InAlAs transistor: (a) response at 2.5-THz frequency as a function of the gate voltage at different temperatures (80 K down to 10 K), (b) response at 10 K as a function of the gate voltage at different frequencies (1.8, 2.5, and 3.1 THz), (c) position (indicated by arrows) of resonance maxima as a function of gate voltage. The calculated plasmon frequency as a function of the gate voltage, using Equation 28.13 for $V_{th} = -0.41$ V, is shown by the solid line. The error bars correspond to the line width of the measured plasmon resonance peaks. (Adapted after Ref. [158])

Tauk et al. have studied Si MOSFETs with 20- to 300-nm gate lengths at room temperature and frequency $\nu = 0.7$ THz [160]. It was found that response depends on the gate length and the gate bias. The responsivity value of 200 V W^{-1} and NEP ≥ 10^{-10} WHz$^{-1/2}$ (see Figure 28.41 [141]) is comparable to the best current commercial room-temperature THz detectors. The inset of Figure 28.41 shows the detection signal for transistors with different gate lengths. One can see that the detected signal decreases when the gate length is reduced from 300 nm to 120 nm.

More recently, it has been demonstrated that by using appropriate antenna and Si FET transistor design, one can reach a responsivity of up to a few kV W^{-1} and an NEP about 10 pWHz$^{-1/2}$ for detectors operating in the atmospheric window around 300 GHz (see Figure 28.42) [161].

The large-scale interest in using FETs as THz detectors started around 2004 after the first experimental demonstration of sub-THz and THz detection in silicon-CMOS FETs [162]. Two years later, it was shown that Si-CMOS FETs can reach an NEP value competitive with the best conventional room temperature THz detectors [160]. At present, the advantages of Si-CMOS FET technology (room temperature operation, very fast response time, easy on-chip integration with read-out electronics, and high reproducibility) lead to the straightforward array fabrication. Similar to visible CMOS imagers, the readout of pixels is provided by the fully integrated analog or digital circuitry within the FPA. Readout multiplexer in FET

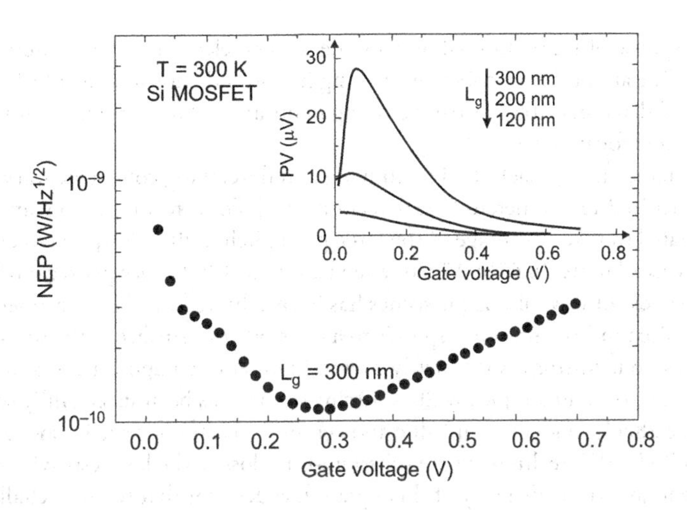

Figure 28.41 NEP as a function of the gate voltage for Si MOSFETs with a 300-nm gate length, $T = 300$ K. The inset shows the detection signal as a function of the gate length. (Adapted after Ref. [141])

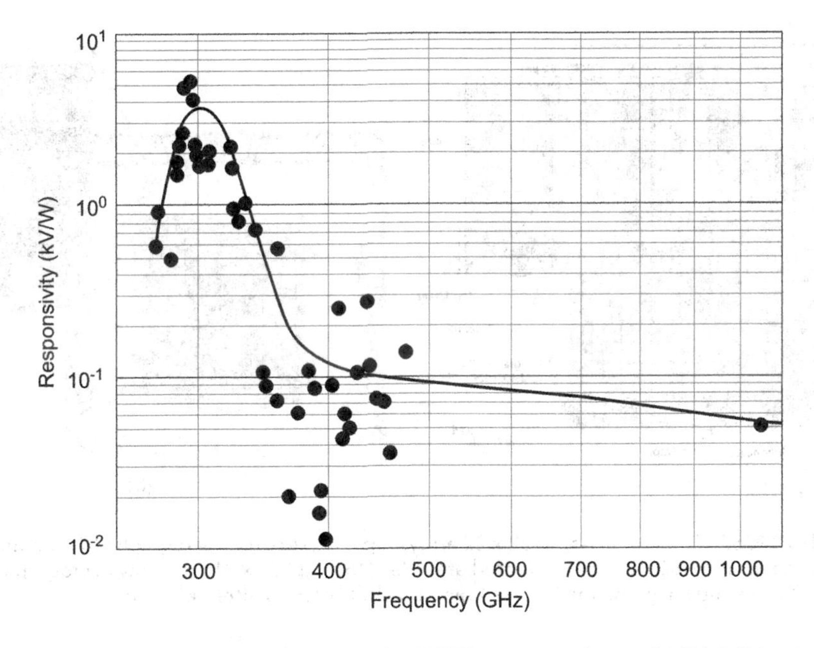

Figure 28.42 Responsivity as a function of frequency for Si FETs at gate voltage of 0.2V. Solid points: measured points, solid line: guide for the eye. (Adapted after Ref. [161])

cameras is better suited to direct detection, where the power consumption per pixel is small that enables fabrication of large-pixel arrays.

The first demonstration of a 32 × 32 CMOS FET-based 2D camera for real-time imaging operated to capture THz video streams was given by Al Hadi et al. [163]. The pixels consist of 1,024 differential on-chip ring antennas coupled to n-type MOS (NMOS) that detect directly the incoming radiation at frequencies well beyond the FET cutoff frequency (see Figure 28.43a). At 0.856 THz and 25 fps, the voltage responsivity is 115 kV W^{-1} and the camera video NEP is 12 nW integrated over a 500 kHz bandwidth. The picture of the terahertz camera module with and without its housing is shown in Figure 28.35 (bottom right). Real-time imaging has also been demonstrated at 590 GHz with a 12 × 12-pixel CMOS array operated with lock-in detection technique [164].

Due to the fast response of FETs (limited only by the readout electronics), it is interesting to use FETs in heterodyne mode. The applicable of coherent imaging has been demonstrated at 591.4 GHz with single or a few raster-scanned detectors [165]. Further developments are still to be carried out to implement a full monolithically integrated focal plane.

Apart from the bolometric approach, Leti is studying and developing complementary THz arrays based on CMOS FETs, either in direct or heterodyne detection [166]. For direct detection, an innovative readout architecture is elaborated to take advantage of the large pixel pitch (240 × 240 µm²) to enhance the flexibility and the sensitivity (see Figure 28.43b). Video sequences up to 100 frames per second have been achieved. Fast scanning of large field of view of opaque scenes has been achieved in a body scanner prototype. Each individual image acquired in real-time corresponds to a 40 × 60 mm² surface at the scene level. In order to cover the size of a chest, one mirror is successively moved in order to compose a 5 × 5 tiled array of individual images. A human trunk of a typically 20 × 30 cm² surface has been successfully scanned in less than 10 seconds (see Figure 28.44) to identify metallic and ceramic objects concealed under a shirt [167].

The key issues for THz FET technology are substrate-wave losses, the low-resistivity silicon substrate, and the low equivalent quantum efficiency of this square-law detector that requires challenging low-noise THz camera readout [168]. Parasitic capacitances tend to shunt down part of the coupled THz signals to the substrate. In addition, the antenna behavior is altered by substrate effects, resulting in considerable power carried by surface waves and multiple reflection effects at the Si-air boundary.

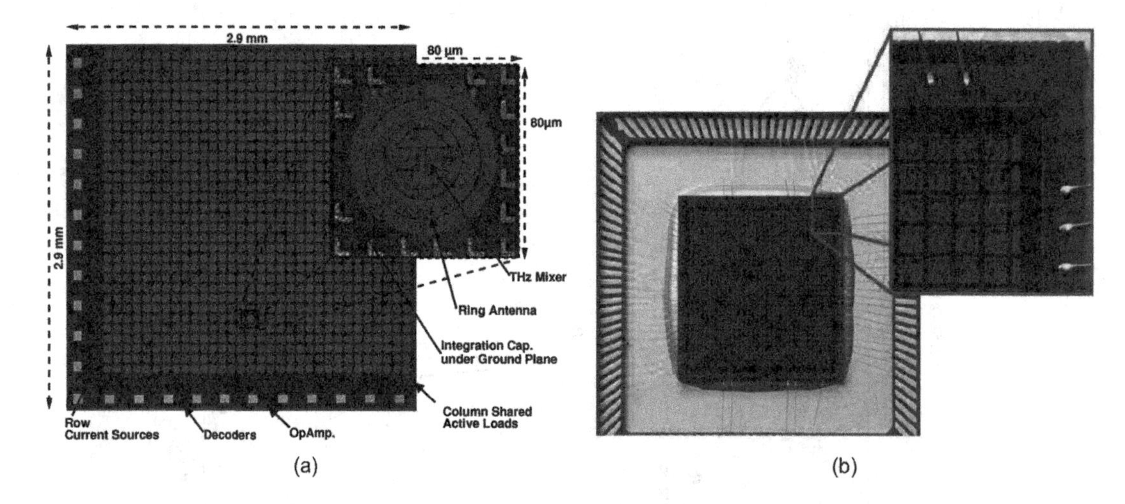

Figure 28.43 THz MOSFET imager FPAs: (a) 32 × 32 array chip complete die micrograph (2.9 × 2.9 mm²) and a single-detector topography (80 × 80 µm²) (Adapted after Ref. [163]), (b) 31 × 31 FET array micrograph (total area 8.5 × 8.5 mm²) with zoomed image on the 240 × 240 µm² pixels. (Adapted after Ref. [131])

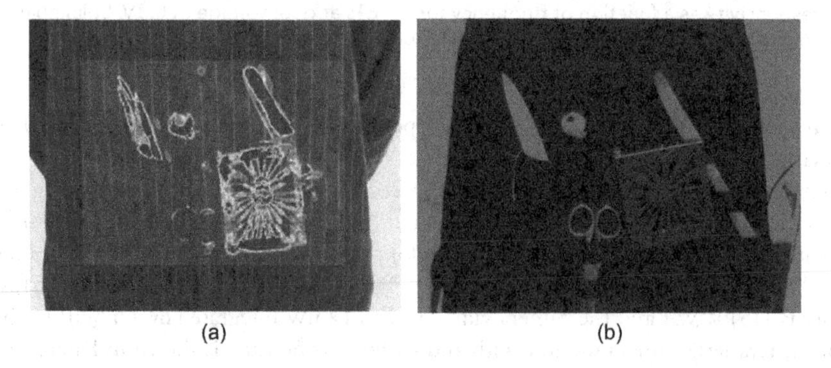

Figure 28.44 Large field of view fast scanning THz image acquired within less than 10 seconds by the scanner demonstrator (a) and visible photograph (b).

Further improvement in radiation coupling is expected by the use of silicon-on-insulator (SOI) substrates instead of bulk CMOS due to their lower device parasitics and their higher resistivity. At the same time, different packaging configurations are being investigated with front-side and backside illumination and antireflective-coated lens glued to the chip during the chip assembly process to improve the on-chip dipole antenna efficiency.

Hitherto, the signal current levels in FET channels compared to dark (noise) current levels in biased source-drain channels are still relatively small which requires the use of, for example, the lock-in amplifiers with a narrow bandwidth to suppress the noise in every channel. At present, however, the first real-time imagers are exploited with specially designed circuits with wide bandwidth.

Although currently FET-based THz detectors present poorer sensitivities than bolometer THz detectors, FET-based THz detection using standard low-cost CMOS technology appears to be an interesting way to benefit from THz imaging advantages. FET FPAs do not require vacuum packaging. The state-of-the-art CMOS FET 32 × 32-pixel camera exhibits the 12 nW at 0.856 THz and 25 fps. In the 2–4 THz range, an NEP of the order of 30 pW is detected by a 320 × 240-pixel array bolometer camera. It is expected to achieve a similar performance for bolometer arrays designed for spectral range below 1 THz [168].

28.7 EXTRINSIC DETECTORS

For several decades, THz spectral range was a rather exotic field, although there were interesting applications particularly in condensed matter research. In the 1970s, it was realized that the THz portion of the electromagnetic spectrum—at that time called the FIR or sub-mm region—offers unique opportunities in astronomical research. Subsequently, many dedicated THz telescopes were built, and among them very successful air- and space-borne observatories. Historically, an extrinsic photoconductor detector based on germanium was the first extrinsic photodetector. After it, photodetectors based on silicon and other semiconductor materials, such as GaAs or GaP, have appeared.

Extrinsic photodetectors are used in a wide range of IR spectrum extending from a few micrometers to approximately 300 μm. They are the principal detectors operating in the range $\lambda > 20$ μm. The spectral range of particular photodetectors is determined by the doping impurity and by the material into which it is introduced (see Chapter 14). Detectors based on silicon and germanium have found the widest application as compared with extrinsic photodetectors on other materials (see Chapter 14 and Section 26.2).

Between different types of extrinsic photoconductors, Ge:Sb is the first of the THz photoconductor discovered in 1959 [169]. Ge:Ga has been the most widely studied and widely used THz photoconductive detector in astronomy since its introduction in 1965 [170]. As is marked in Section 26.2, the maximum pixel numbers are 16 × 25 and 32 × 32 for stressed and unstressed arrays, respectively. Their response can be changed from 1.5 THz to 7 THz.

Nowadays, the promise of a new generation of large format FIR detector arrays seems to be attainable using another material systems. Silicon-based blocked impurity band (BIB) detectors doped with arsenic and antimony have the materials of choice for astronomical detectors at wavelengths from 5 μm to 40 μm. Conventionally designed and processed Si:As BIB detectors have a cutoff wavelength of about 28 μm, but wavelength extension to approximately 50 μm is possible [171]. Attempts have been also made to provide a similar direct detector technology for operation in the FIR by switching from silicon to a semiconductor material that would provide a shallower impurity band [172]. Both Ge-based and GaAs-based BIB systems have been attempted, with greater success achieved in germanium [173]. However, the smaller binding energy of shallow donors in GaAs compared to Ge results in response at wavelengths exceeding 300 μm without uniaxial stress.

Among THz low-temperature cooled detectors, there exist many publications about the possibility to use $Pb_{1-x}Sn_xTe$:In ($x \approx 0.25$, In content is about 2 atomic %) photoconductors as THz detectors in the range of 1 THz [174–179]. Persistent photoresponse with current responsivity about 10^3 A W^{-1} at 40 mV bias and integration time ≈ 1 s at the wavelengths of 90 μm and 116 μm has been observed in $Pb_{0.25}Sn_{0.75}Te$:In photoconductors [174]. This value is larger by a factor of about 100 than the responsivity of the Ge:Ga photoconductor in the same conditions of the experiments ($T \approx 4$ K, detectors protected from the background radiation). The photoresponse signals were observed at wavelengths up to 337 μm [176]; it is one of the highest cut-off wavelengths observed so far for extrinsic semiconductor photodetectors.

Investigation of III-group doped impurity states in narrow-gap IV-VI alloys based on the lead telluride semiconductors has begun in early 1970s. The persistent photoconductivity effects observed in these materials are analogous to the features of III-V and II-VI semiconductors with the DX-centers [180]. It was shown that in the group of III-doped IV-VI semiconductors at low temperatures, $T \ll T_c$, photoconductivity relaxation consists of two components: the "fast" one (varies between 1 ms and 1 s) and the slow one, which may exceed 10^5s [181]. The T_c value is about 25 K for indium-doped and about 80 K for the Ga-doped materials.

The persistent photoconductivity induced by THz radiation in $Pb_{1-x}Sn_xTe$:In alloys at low temperatures is clearly connected with the existence of a barrier between the localized deep and free electron states, as it is observed in other semiconductors at shorter wavelengths. The evidence of such a barrier in IV-VI semiconductors with deep impurities is also becoming apparent by current instabilities like the Gunn effect, but with much lower frequency of oscillations [182].

28.8 PAIR BRAKING PHOTON DETECTORS

One of the methods of photon detection consists in using superconductivity materials. If the temperature is far below the transition temperature, T_c, most of the electrons in them are banded into Cooper pairs. Photons with energies exceeding the binding Cooper pair energies in superconductor, 2Δ (each electron must be supplied an energy Δ), can break these pairs producing quasiparticles (electrons) (see Figure 28.45a). This process resembles the interband absorption in semiconductors, with the energy gap equal to 2Δ, when under the photons absorbed, the electron-hole pairs are created. One of the advantages of these detectors is that the fundamental noise due to the random generation and recombination of thermal quasiparticles decreases exponentially with temperature as $\exp(-\Delta/k_BT)$ [183]. The best SSB noise temperature, T_n, that can be gained is $k_BT_n \geq h\nu/\eta$. With $\eta = 1$, the quantum limit can be achieved but it never can be overcome [184]. In pair-breaking detectors, it is possible to get $\eta \rightarrow 1$ and thus one can approach nearly the quantum operation limit (see Figure 6.10).

The superconductor tunnel junction (STJ) is used to let quasiparticles pass through the junction and to separate off the Copper pairs. The first proposition of pair-breaking detector with STJ was made in early 1960s [185]. Next, several structures of pair-braking detectors which use different ways to separate

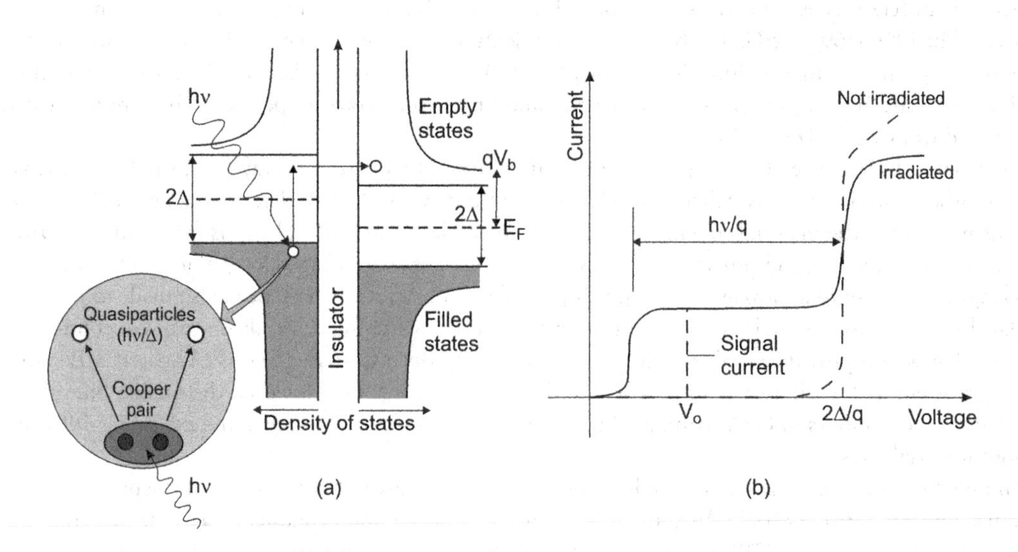

Figure 28.45 SIS junction: (a) energy diagram with applied bias voltage and illustration of photon-assisted tunnelling, and (b) current-voltage characteristic of a non-irradiated and irradiated barrier. The intensity of the incident radiation is measured as an excess of the current at a certain bias voltage V_0. Schematic creation of quasiparticle is shown inside (a).

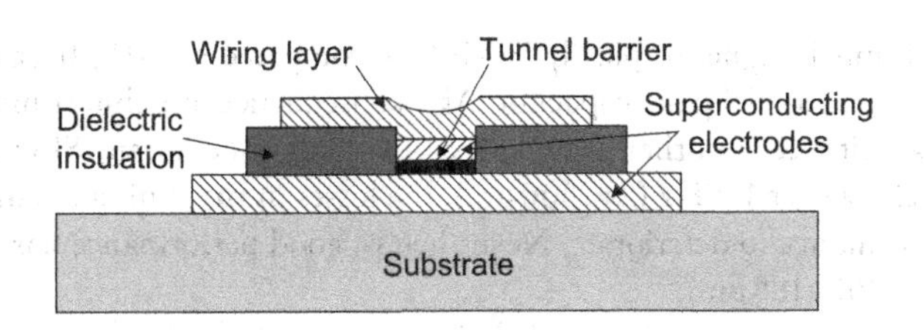

Figure 28.46 A cross section of a typical SIS junction.

quasiparticles from Cooper pairs have been proposed. Among them are SIS and superconductor-insulator-normal metal (SIN) detectors and mixers, RF kinetic inductance detectors, and superconducting quantum interference device (SQUID) kinetic inductance detectors. Superconducting detectors offer many benefits: outstanding sensitivity, lithographic fabrication, and large array sizes, especially through the development of multiplexing techniques. The basic physics of these devices and the recent progress in their developments are described in Ref. [13]. Here, we concentrate on the most important SIS detectors.

The SIS detector is a sandwich of two superconductors separated by a thin (≈ 20 Å) insulator, what is schematically shown in Figure 28.46. Nb and NbTiN are almost exclusively used as superconductors for the electrodes. For a standard junction process, the base electrode is 200-nm sputtered Nb, the tunnel barrier is made using a thin 5-nm sputtered Al layer which is either thermally oxidized (Al_2O_3) or plasma nitridized (AlN). The counter electrode is 100-nm sputtered Nb or reactively sputtered NbTiN. Typical junction areas are about 1 μm². The entire SIS structure is deposited in a single deposited run. The junction is defined by photolithography or electron-beam lithography and reactive ion etching of the counter electrode. Finally, 200-nm-thick thermally evaporated SiO or sputtered SiO_2 is deposited. A SiO_x layer insulates the junction and serves as the dielectric for the wiring and RF tuning circuit on top of the junction.

The SIS operation is based on photoassisted tunneling of quasiparticles through the insulating layer. Although the physics of this effect was demonstrated and theoretically explained already in the 1960s [186,187], it took almost two decades to make use of the effect in a mixer [188,189]. Nowadays, SIS tunnel junctions are mainly used as mixers in heterodyne type mm and sub-mm receivers, because of their strong non-linear I-V characteristic. They can be also used as direct detection detectors [189,190]. The operating temperature of SIS junctions is below 1 K; typically $T \le 300$ mK.

The SIS operation can be described by using the energy band representation known for semiconductors. The states below the energy gap are considered to be occupied, and those above the gap are empty (Figure 28.45a). The curves indicate the electron density of states. When a bias voltage, V_b, is applied to the junction, there is a relative energy shift of qV_b between the Fermi levels of the two superconductors. If qV_b is lower than the energy gap, 2Δ, no current flows (electrons may only tunnel into unoccupied states at the same energy). However, if the junction is irradiated, photons of energy $h\nu$ may assist the tunneling, which now may occur for $qV_b > 2\Delta - h\nu$.

The current-voltage characteristic of an SIS device is shown in Figure 28.45b. When the bias voltage reaches the gap voltage, a steep increase of the current occurs. At this particular voltage, the divergent densities of states of both superconductor layers cross, and the Cooper pairs on one side of the insulating layer break up into two electrons (quasiparticles). Next, these quasiparticles tunnel from one side of the insulator to the other, where they recombine. A sharp onset of normal tunneling current arises beyond a DC threshold voltage equal to a superconductor energy gap 2Δ, and this abrupt nonlinearity in the single-particle tunneling is used for mixing. The *I-V* nonlinearity is small, about few tenths of a millivolt, and might be compared with the nonlinearity of a Schottky diode (in the order of 1 mV). However, the SIS nonlinearity at the gap voltage is substantially less than the energy of THz photons and the classical mixer theory can not be applied. To explain the possibility of using SIS in THz detection, a detailed description of mixing theory including quantum effects is necessary [190].

Although the IF bandwidth of and SIS junction itself is very large due to the inherently fast tunneling process, it is limited in practical applications by the electric circuitry. A typical IF band is 4–8 GHz. One of the main challenges for high-frequency SIS mixer design is dealing with the large parallel-plate capacitance of the SIS junction. Although the area of an SIS junction is similar to the area of a Schottky diode, its parasitic capacitance is much higher (typically 50–100 fF compared to ≈1fF) because the two superconducting electrodes form a parallel plate capacitor. As a consequence, on-chip tuning circuits are needed to compensate for the capacitance, and their proper design is a key issue for any SIS mixer [13,44]. At higher frequencies, especially over 1 THz ($\lambda = 300\,\mu m$), the losses in the tuning circuit become important and cause the mixer performance to deteriorate. Nevertheless, good performance has been obtained up to around 1.5–1.6 THz ($\lambda = 200–188\,\mu m$).

There are two major ways to accomplish proper design issues for SIS mixers: waveguide coupling and quasioptical coupling [13]. Figure 28.47 shows the examples of both configurations. The more traditional approach is waveguide coupling, in which the radiation is first collected by a horn into a single-mode waveguide and then coupled onto a lithographed thin-film transmission line on the SIS chip itself. The chip shown in Figure 28.47a is approximately 2-mm long and 0.24-mm wide and is fabricated on 25-µm-thick silicon utilizing silicon-on-insulator-bonded wafers. The 1-µm-thick gold beam leads extend beyond the edges of the substrate and are in electrical contact with the metal waveguide probe. A serious complication of waveguide coupling is that the mixer chip must be very narrow and must be fabricated on ultrathin substrate.

In the quasioptical coupling, the intermediate step of radiation collection into a waveguide is omitted and instead uses a lithographed antenna on the SIS chip itself. Such mixers are substantially simpler to fabricate and may be produced using thick substrates. In the design structure shown in Figure 28.47b, the radial stubs serve as RF short circuits and couple the radiation received by the slots into a Nb/SiO/Nb superconducting microstrip. This chip uses two $Nb/Al_2O_3/Nb$ SIS junctions and the short strip section in between the junctions provide the tuning inductance needed to compensate the junction capacitance.

SIS mixers are among the most sensitive and low intrinsic noise structures at $\nu \approx 0.3–0.7$ THz. Nb-based SIS mixers with Nb wiring yield almost quantum limited, that is, the noise temperature is below $3h\nu/kT$ (see Figure 6.10) [190,191]. At larger frequencies, $\nu \approx 1.0–1.3$ THz, SISs intrinsic noise quickly increases due to high-frequency losses. Further gain in sensitivity is possible using multielement or matrix arrays. However, up to now SIS detectors are difficult to integrate into large arrays. There is only a success in fabrication of small number element arrays because of appreciable difficulties in their creation [192]. SIS

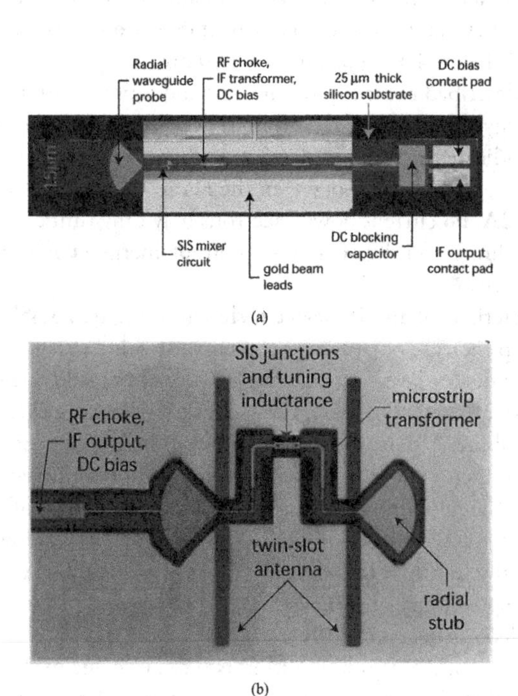

Figure 28.47 Images of a waveguide SIS mixer operated in the 200–300GHz band (a) and a quasioptical SIS mixer (b). (Adapted after Ref. [13])

mixers seem to be the best solution for ground-based radio-astronomy at mm and sub-mm wavelengths region in the frequency range $\nu < 1$ THz [193].

The signal bandwidth of an SIS mixer is 10–30% of its centre frequency with the large fractional bandwidth at low frequency band. Up to 1 THz, the mixers are in a waveguide mount, while the 1.2–1.25 THz mixers employ a quasioptical coupling.

Single-pixel SIS mixers typically require approximately 40–100 μW of LO pump power, which is appreciably lower compared to LO pump power for single-pixel SBD mixers ($P > 1$ mW) [194]. Much lower LO powers require superconductor hot electron bolometer mixers (<100 nW^{-1} μW) [195] though they also operate at very low temperature. Unlike Schottky diodes or SIS detectors, the hot electron bolometers are thermal detectors.

The presence of the capacity in SIS structures is a reason of current shorting in them because of Josephson's effect. To exclude this effect, the SIN structure was proposed to change one of the superconductors by normal metal contact [196]. Though in SIN structures, *I-V* characteristics are not so nonlinear as in SIS structures, which is a reason of sensitivity decrease, the influence of Josephson's effect is excluded. To force the electron tunneling from the normal metal into the superconductor, the energy of electrons above the Fermi level should be not less than $\Delta - qV_b$, where V_b is the junction bias voltage [13]. Thus, the junction current probes the tail of the Fermi distribution of electrons in the normal metal and is exponentially sensitive to the electron temperature, T_e, as $\exp[-(\Delta - qV_b)/kT_e]$. Thus, SIN junction is a thermometer for measuring the electron temperature in a normal metal.

Figure 28.48 shows schematic configuration of SIN radiation coupling and temperature readout [197,198]. A thin strip of metal (black strip) with micron dimensions serves as a resistive load to thermalize the RF currents from the superconducting antenna. The resulting temperature rise of the electrons in the strip are measured as a change in the voltage across the junction which is biased at a constant current *I*. Variations of the electron temperature in the absorber strip are detected from smearing of *I-V* curves of the SIN junction (see Figure 28.48b). The contact to the superconducting electrode is made of a superconductor whose T_c is much higher than that of the electrode. A 300-Å-thick Cu layer can be used as a normal metal absorber [197]. In order to avoid energy losses through diffusion of the electrons into the antenna, the absorber layer is contacted via superconducting electrodes (e.g., Al), since the Andreev effect prohibits energy transport from the normal metal to the superconductor at an NS-interface [200].

The principles of SIN structures operation are discussed in Ref. [201]. The NEP of these detectors can achieve values close to 10^{-17} WHz^{-1} at $T \approx 300$ mK. Schmidt et al. [202] have measured the NEP value of 7×10^{-17} WHz$^{-1/2}$ for a normal metal volume of 4.5 μm^3 at an operating temperature $T = 270$ mK with the time constant $\tau = 1.2$ μs.

SIS tunnel junctions are mainly used as mixers in heterodyne type mm and sub-mm receivers because of their strong non-linear I-V characteristic. They can be also used as direct detection detectors. However, up to now SIS detectors are difficult to integrate into large arrays.

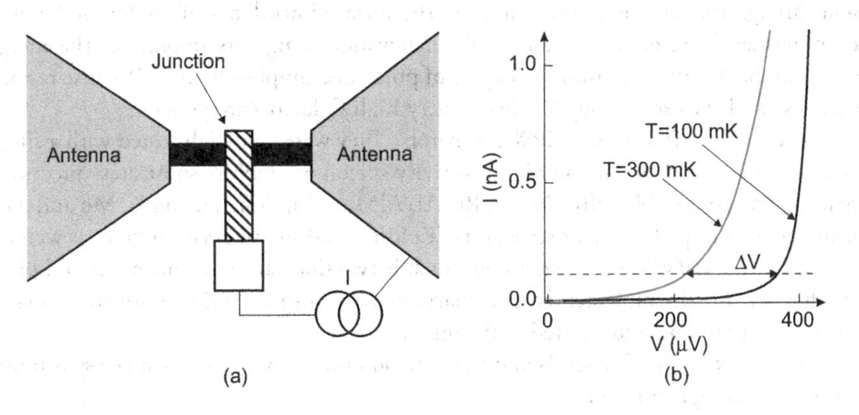

Figure 28.48 SIN detector: (a) schematic configuration of radiation coupling and temperature readout (after Ref. [195]) and (b) principle of operation. (Adapted after Ref. [199])

High-temperature superconductor SIS devices are still under investigation. In spite of possible high-frequency operation, the noise level of such devices is appreciably higher.

28.9 MICROWAVE KINETIC INDUCTANCE DETECTORS

Cryogenic detectors, with operating temperatures on the order of ~100 mK, are currently the preferred technology for astronomical observations over most of the electromagnetic spectrum. Progress in development of large-format, high-sensitivity FPAs is especially promising with two detector technologies: TES bolometers (see Section 28.12) and MKIDs based on different principles of superconductivity. Multiple instruments are currently in development based on arrays up to 10,000 detectors (see Figure 28.11) using both time-domain multiplexing and frequency-domain multiplexing with SQUIDs [38]. Both sensors show potential to realize the very low ~10^{-20} $WHz^{-1/2}$ sensitivity needed for space-borne spectroscopy.

MKIDs is much easier to fabricate and multiplex than TES or STJ detectors, and

- megapixel arrays are on the horizon, and they can be significantly easier to fabricate and readout out than any competing technology,
- can work from UV (~ 0.1 μm) to IR (~ 10 μm) to mm wavelengths,
- has to be kept on a superconducting temperature, but the readout electronics does not—their readouts can leverage the tremendous advances in room temperature microwave integrated circuits developed for wireless communication industry.

MKIDs were first developed by scientists at the California Institute of Technology and the Jet Propulsion Laboratory in 2003 [203]. They are a type of non-equilibrium superconducting photon detector in which the energy absorbed from an incoming photon breaks Cooper pairs. The incident photons change the surface impedance of a superconductor through the kinetic inductance effect, which occurs because energy can be stored in the supercurrent of a superconductor. An extra inductance can be measured using a thin-film superconducting resonant circuit, resulting in a measurement of the energy of the incident photons. The difference between arrays designed for ultraviolet, optical, and IR astrophysics is the method used to couple the photon energy into MKID [204]. The physics of MKIDs has been extensively discussed elsewhere, and only a short description will be present here [204,205].

An MKID is essentially a high-Q resonant circuit made out of either superconducting microwave transmission lines or a lumped element *LC* resonator (fabricated from thin aluminum and niobium films). In the first case, a meandered quarter-wavelength strip of superconducting material is coupled by means of a coupling capacitance to a coplanar waveguide (CPW) through line used for excitation and readout. Lumped element is instead created from an *LC* series resonant circuit inductively coupled to a microstrip feed line placed in a high-frequency resonant circuit (see Figure 28.49). Photons hitting an MKID break the Cooper pairs, which changes the surface impedance of the transmission line or inductive element producing a number of quasiparticles. This causes the resonant frequency and quality factor to shift an amount proportional to the energy deposited by the photon. The amplitude (c) and phase (d) of a microwave excitation signal sent through the resonator. The change in the surface impedance of the film following a photon absorption event pushes the resonance to a lower frequency and changes its amplitude. The energy of the absorber photon can be determined from the degree of phase and amplitude shift. Because resonators made of superconductors can have exceedingly low losses, very high Q factors are possible.

The first studied type of MKIDs was CPW resonators. They were easily fabricated with a single metal layer on a crystalline dielectric, mainly on high-resistivity silicon or sapphire substrates. Successful CPW MKIDs have been made out of Nb, NbTiN, Ta, Re, Al, AlMn, Mo, PtSi, Ti, and Ir. Mo and Ti films are very difficult to make high-quality factor resonators. Ta films tend to not perform quite as well as Al and Nb. Figure 28.50 shows a CPW transmission line which is two slots cut into a metal ground plane to form a center strip. They are usually implemented as a quarter-wave transmission line resonator, capacitively coupled to a feedline at one end and shorted at the other.

The primary drawback of CPW MKIDs and transmission line resonators is that the sensitivity to quasi-particle peaks in areas of high current.

Figure 28.51 shows examples of two other MKIDs. A lumped element MKIDs (see Figure 28.51a) uses a separate inductor and capacitor to form a resonator. In this design, an interdigitated capacitor is attached to

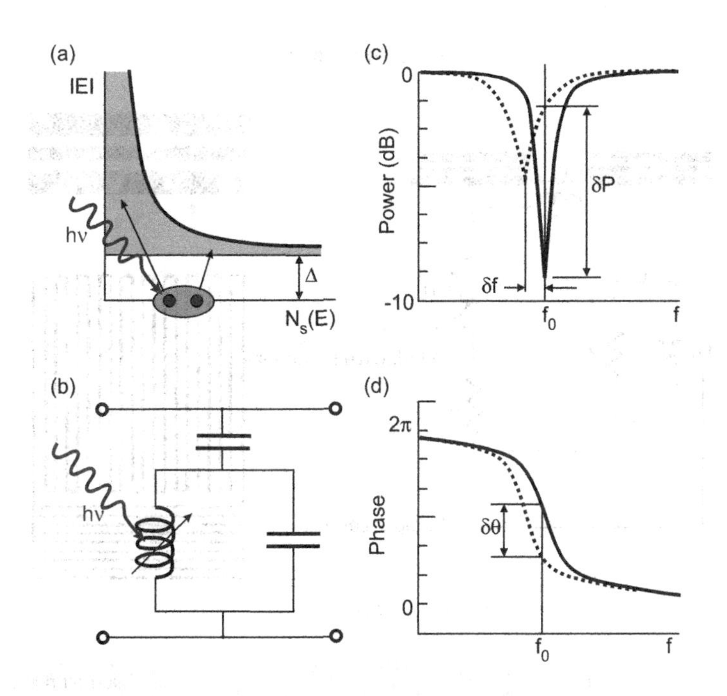

Figure 28.49 An illustration of the operational principle behind an MKID: (a) a photon with energy, $h\nu > 2\Delta$, breaks Cooper pairs and creates quasiparticles in a superconducting strip cooled to $T < T_c$, (b) the superconducting strip is used as an inductive element in a microwave resonant circuit. The increase in the quasiparticle density changes the impedance, (c) the transmission through the resonant circuit has a narrow dip at the resonance frequency f_0 which moves when inductance changes, and (d) the microwave probe signal acquires a phase shift when f_0 changes.

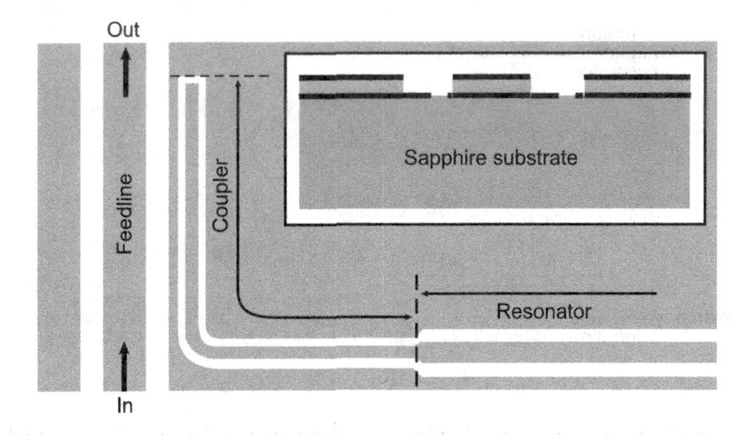

Figure 28.50 An illustration of the CPW coupler and resonator. The inset shows a cross-sectional view of the CPW. The contour of the metal surface and the contour of the exposed surface of the substrate are indicated by the solid line and the dashed line, respectively. (Adapted after Ref. [206])

an inductive meander. The advantage of lumped element MKIDs is that they are very simple to make since they do not require quasiparticle trapping. The second MKIDs design uses aspects of both lumped element and transmission line resonators.

The readout is almost entirely at room temperature and can be highly multiplexed; in principle, hundreds or even thousands of resonators could be read out on a single feedline [22,23].

In the case of MKIDs, it is possible to frequency multiplex many detectors on a single output line by having each pixel tuned to a slightly different microwave frequency. This tuning is accomplished geometrically in the fabrication using standard microlithographic techniques. During detector operation, each pixel

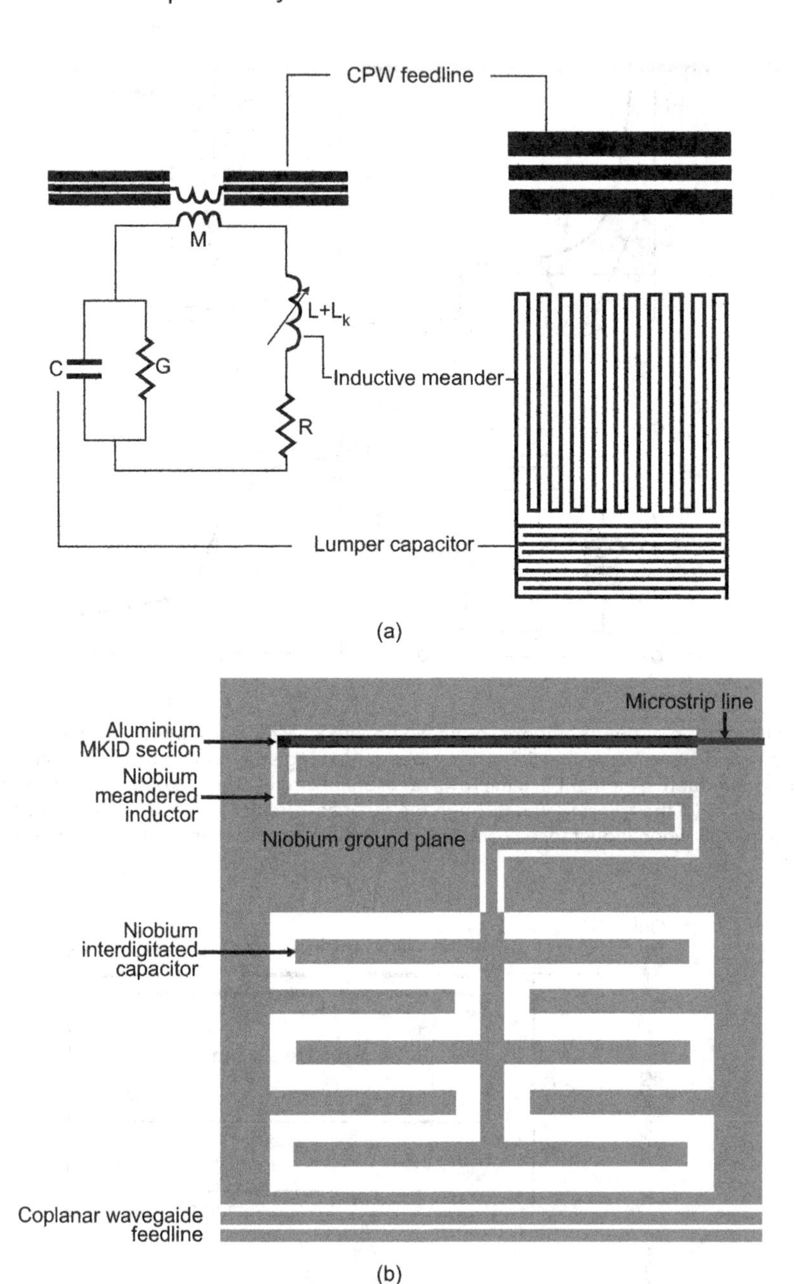

Figure 28.51 Two types of MKIDs: (a) an example of a lumped element MKID and (b) a mixed MKID with an interdigitated capacitor and CPW inductor. (Adapted after Ref. [206])

is excited by a tone matched to its resonant frequency as illustrated in Figure 28.52. In this way, each pixel is associated with a single microwave frequency that can be analyzed by a microwave spectrometer. The individual oscillators for each pixel shown in Figure 28.52 can be replaced by a single high-speed digital-to-analog converter that puts out a waveform with the necessary Fourier components to replicate the desired comb spectrum.

MKID arrays are an active area of detector work, and a number of cameras for ground-based telescopes have been developed including the MUSIC (Multiwavelength Submillimeter kinetic Inductance Camera) for Caltech Submillimeter Observatory (CSO) [206], APEX [207], and IRAM [208]. Recently also, demonstrating their suitability for space applications at FIR and sub-mm wavelengths for low-background (typical for space-borne applications) has been presented [209].

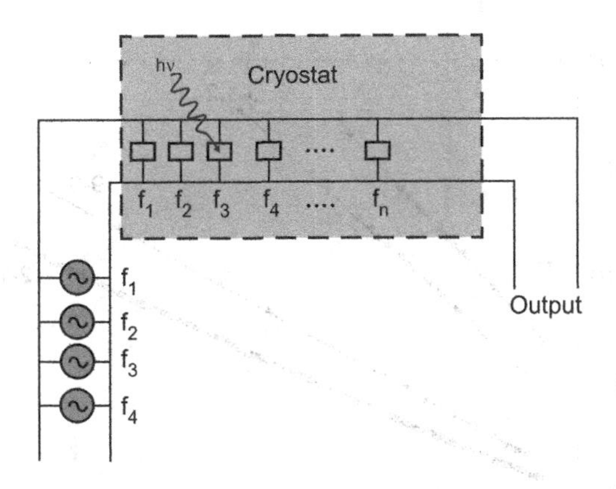

Figure 28.52 Frequency multiplexed MKID array.

Figure 28.53 Inside the MKID camera cryostat. This shows the 4-K radiation shield and the split section of the 50-K shield (at bottom). Cooling to 4 K is provided by a pulse-tube cooler; refrigeration of the focal plane to temperatures below the transition temperature of Al/Nb ($T_c = 1.2$ K) is accomplished with a helium sorption fridge. (Adapted after Ref. [206])

Figure 28.53 shows the MKID camera constructed for the CSO with 576 antenna-coupled spatial pixels each simultaneously sensitive in four bands at 750, 850, 1,100 and 1,300 μm and with a total 2,304 detectors.

28.10 SEMICONDUCTOR BOLOMETERS

The classic bolometers contain a heavily doped and compensated semiconductor which conducts by a hopping process that yields a resistance $R(T) = R_0 \exp(T/T_0)^p$, where R is the resistance at temperature T, and T_0 and R_0 are constants which depend on the doping, and for R_0, on the thermistor dimensions [39]. The exponent p is a constant, and it is often assumed that $p = 0.5$. The thermistors are made by ion implantation in Si or by neutron transmutations doping (NTD) in Ge [50]. Figure 28.54 shows the experimental results at the temperature range from 70 mK to 1 K for some of the NTD Ge samples together with fitting curves,

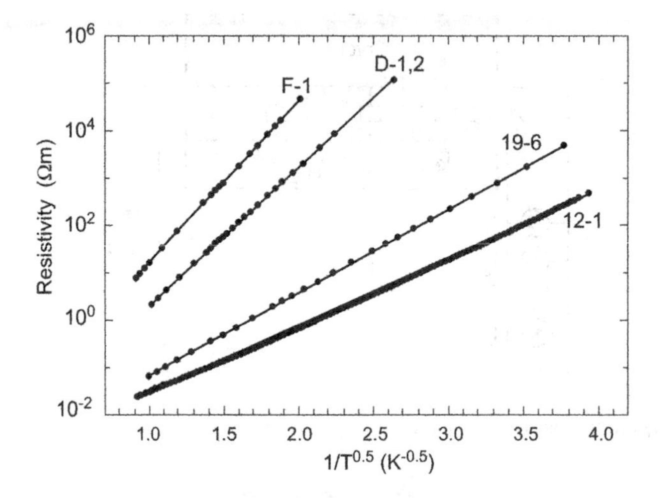

Figure 28.54 Zero-bias resistivity as a function of temperature for several different NDT Ge samples. (Adapted after Ref. [210])

where the p-value for all samples is about 0.5 [210]. NTD converts ^{70}Ge to ^{71}Ga (acceptor) and ^{74}Ge to ^{75}As (donor). The doping level depends on the neutron flux, instead the compensation ratio can be changed by altering the isotope ratios.

The thermistors are typically fabricated by lithography on membranes of Si or SiN. The impedance is selected to a few MΩ to minimize the noise in junction FET (JFET) amplifiers operated at about 100 K. Limitation of this technology is assertion of thermal mechanical and electrical interface between the bolometers at 100–300 mK and the amplifiers at ≈ 100 K. Usually, JFET amplifiers are sited on membranes, which isolate them so effectively that the environment remains at a much low temperature (about 10 K) (see Figure 28.55). In addition, the equipment at 10 K is itself thermally isolated from the nearby components at 0.1–0.3 K. There are not practical approaches to multiplexing many such bolometers to one JFET amplifier. Current arrays require one amplifier per pixel and are limited to a few hundred pixels.

In bolometer metal films that can be continuous or patterned in a mesh absorb the photons. The patterning is designed to select the spectral band, to provide polarization sensitivity, or to control the

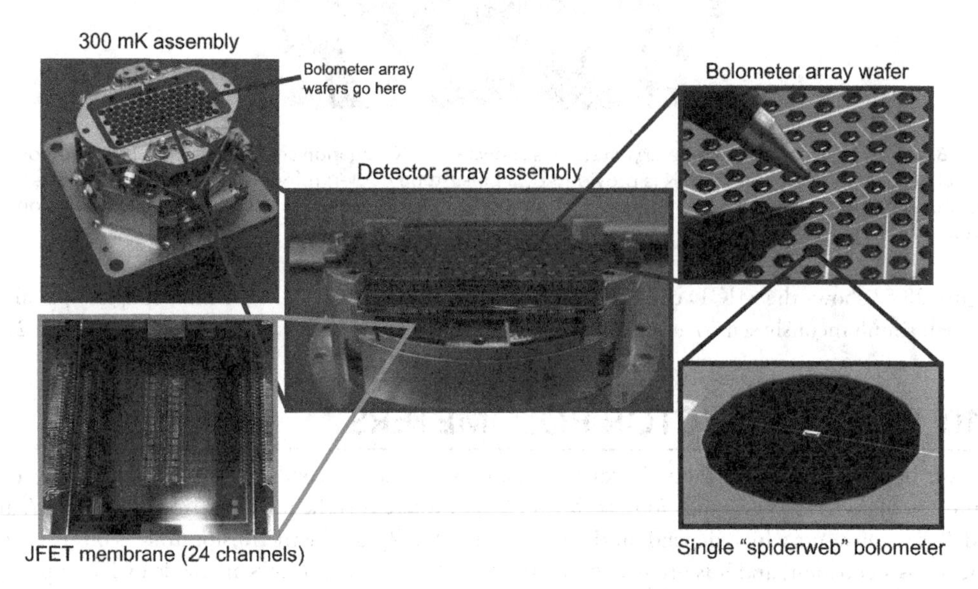

Figure 28.55 Bolometer array of the Spectral and Photometric Imaging Receiver (SPIRE). (Adapted after Ref. [211])

throughput. Different bolometer architectures are used. In close-packed arrays and spiderwebs, the pop-up structures or two-layer bump bonded structures are fabricated. Agnese et al. have described a different array architecture, which is assembled from two wafers by indium bump bonds [212]. Other types of bolometers are integrated in horn-coupled arrays. To minimize low-frequency noise an AC bias is used.

The present-day technology exists to produce arrays of hundreds of pixels that are operated in the spectral range from 40 to 3,000 μm in many experiments including NASA Pathfinder ground-based instruments, and balloon experiments such as BOOMERANG, MAXIMA, and BAM. In Table 28.8, requirements for typical astronomic instruments containing bolometers are gathered. To fulfil these requirements, the compromise between the response time and NEP is needful. A spiderweb bolometer from the BOOMERANG experiment, which is being used on space experiments (Planck and Herschel missions), is shown in Figure 28.56.

An alternative approach is array for the SHARCII instrument on Stratospheric Observatory For Infrared Astronomy (SOFIA) shown in Figure 28.56a [213]. This array construction, with 12 × 32 pixels, involves a pop-up configuration, where the absorber is deposited on a dielectric film that is subsequently folded. The 12 × 32 bolometer array, load resistors, and thermally isolated JFETs are housed in a structure approximately 18 × 17 × 18 cm^3 volume, having a total mass of 5 kg, and heat sink to 4 K. Each bolometer is fabricated on a 1-μm silicon membrane and has a collecting area of 1 × 1 mm^2. The full area is ion implanted with phosphorus and boron, to a depth of ≈0.4 μm, to form a thermistor. Electrical contact between the thermistor and aluminum traces on the silicon frame is accomplished with degenerately doped leads on

Table 28.8 Requirements for typical astronomic instruments

INSTRUMENT	WAVELENGTH RANGE (μm)	NEP (WHz$^{-1/2}$)	τ (ms)	NEP$\tau^{1/2}$ (×10^{-19}J)	COMMENTS
SCUBA	350–850	1.5×10^{-16}	6	9	High background, needs reasonable τ
SCUBA-2	450–850	7×10^{-17}	1–2	1	Lower background, need faster τ
BoloCAM	1,100–2,000	3×10^{-17}	10	3	Lower background, slower device okay
SPIRE	250–500	3×10^{-17}	8	2.4	Space background, slower device okay
Planck-HFI	350–3,000	1×10^{-17}	5	0.5	Lowest background, need quite fast τ

SPIRE, Spectral and Photometric Imaging Receiver.

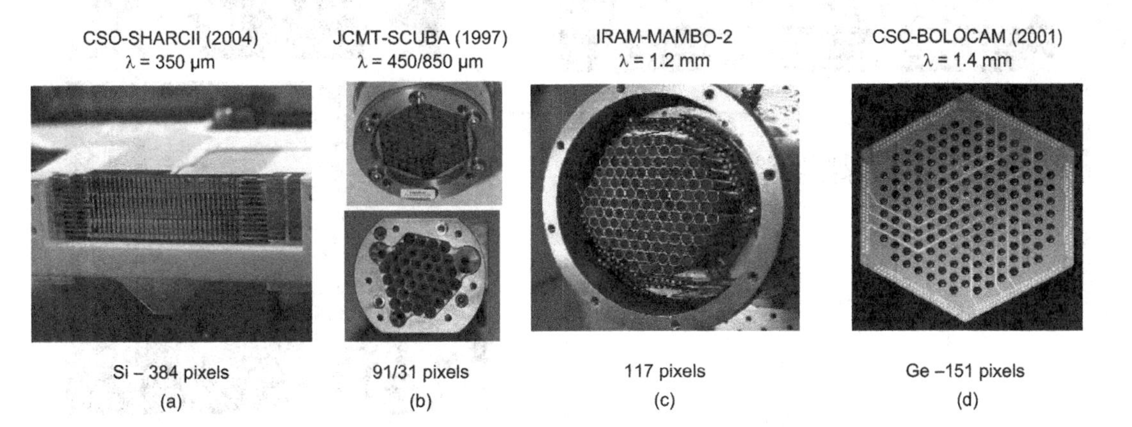

Figure 28.56 Arrays installed in ground-based telescopes.

the edge of the thermistor and running down the bolometer legs. Each of the four thermally isolating legs is 16-μm wide and 420-μm long. Prior to folding, each bolometer is coated with an absorbing ≈200 Å bismuth film and a protective ≈160 Å SiO film. At a base temperature of 0.36 K and in the dark, the bolometers have a peak responsivity of approximately 4×10^{8} V W^{-1} and a minimum NEP of approximately 6×10^{-17} $WHz^{-1/2}$ at 10 Hz. Phonon noise is the dominant contributor, followed by bolometer Johnson noise.

With the development of low-noise readouts that can operate near the bolometer temperature, the first true high-performance bolometer arrays for the far IR and sub-mm spectral ranges are just becoming available. For example, the Herschel/photodetector array camera and spectrometer (PACS) instrument uses a 2048-pixel array of bolometers [214] and is an alternative to JFET amplifiers. The architecture of this array is vaguely similar to the direct hybrid mid-IR arrays, where one silicon wafer is patterned with bolometers, each in the form of a silicon mesh, as shown in Figure 28.57. The development of silicon micromachining has enabled substantial advances in bolometer construction generally and is central to making large-scale arrays. To achieve appropriate response and time constant characteristics, the rods and mesh are designed carefully. The mesh is blackened with a thin layer of titanium nitride with sheet resistance matched to the impedance of free space (377 Ω on square section of film) to provide an efficiency of 50% over a broad spectral band. Each bolometer located at the center of the mesh, containing a silicon-based thermometer doped by ion implantation, is characterized by appropriate temperature-sensitive resistance. Their large resistances ($>10^{10}$ Ω) are well adjusted to MOSFET readout amplifiers. In the final step of hybrid array fabrication, the MOSFET-based readouts and silicon bolometer wafer are joined by indium bump bonding. Performance is currently limited by the noise in the MOSFET amplifiers to the NEP ≈ 10^{-16} $WHz^{-1/2}$ regime, but this technology allows the construction of very large arrays suitable for higher background applications. Further details are given in Billot et al. [214].

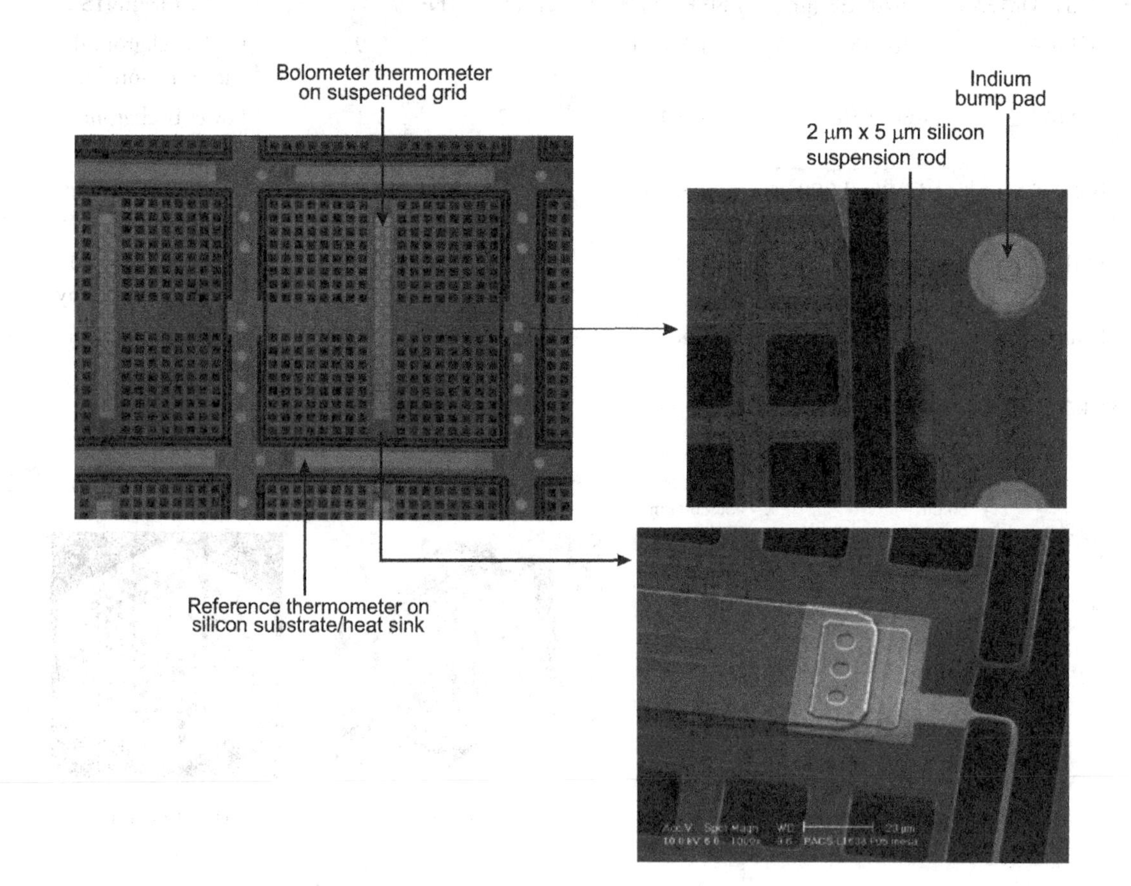

Figure 28.57 The Herschel/PACS bolometer array with pixel size 750 μm. (Adapted after Ref. [215])

28.10.1 SEMICONDUCTOR HOT ELECTRON BOLOMETERS

If the bolometer is to be used as a THz mixer, it has to be fast enough to follow the IF, i.e., the overall time constant of the processes involved in the mixing has to be a few tens of picoseconds at the maximum. In other words, high heat conductivity and small heat capacity are required [41]. These requirements can be fulfilled by such a subsystem as electrons in the semiconductor or superconductor interacting with the lattice (phonons). Electron heat capacity is many orders lower compared to the lattice one.

The term "hot electrons" was introduced to describe nonequilibrium electrons in semiconductors [216]. In this case, the electron distributions could be formally described by the Fermi function distribution, but with an effective elevated temperature. This concept fruitfully was applicable to semiconductors, where the carrier mobility depends on the effective temperature. In metals, mobility changes are much less pronounced and electron heating does not affect the metal resistance unless the change in the effective temperature is comparable with the temperature at Fermi level.

In the normal bolometer, the crystal lattice absorbs energy and transfers it to the free carriers via collisions. However, in hot electron bolometer, the incident radiation power is absorbed directly by free carriers, the crystal lattice temperature remaining essentially constant. Note that this mechanism differs from photoconductivity in that free-electron mobility rather than electron number is created by incident light. At a low temperature, the mobility of the electrons varies as $T_e^{3/2}$, where T_e is the electron temperature, and the conductivity of the material is modulated by mobility. This mechanism offers submicrosecond response and broad FIR coverage out to mm wavelengths but requires liquid-helium cooling.

The first bolometer with "hot electrons" (HEB) was low temperature bulk n-InSb [217,218]. Currently, this detector uses a specially shaped high purity n-type InSb crystal which may be coupled directly to a very low noise preamplifier. Parameters of InSb HEB manufactured by Infrared Laboratories are given in Table 28.9 [219]. Figure 28.58 shows a liquid-helium-cooled InSb HRB bolometer and its spectral dependence of detectivity [220].

Presently, also other semiconductor materials have been proposed in HEBs fabrication [221–223]. In spite of the fact that the rate of electron heating is extremely high because of a high rate of photon-electron interaction, the maximum transformation frequency is restricted by the thermal relaxation rate, which in semiconductors is governed by electron-phonon interaction time $\tau \approx 10^{-7}$s at low temperatures [224]. This response time is relatively short compared to convention thermal detectors with lattice heating, but long compared to τ in superconducting HEBs. Thus, for direct detection semiconductor systems, the speed of response is quite suitable but not for the mixers. Their NEP can reach 5×10^{-13} WHz$^{-1/2}$ at an operation temperature of about 4 K and below.

Nonlinearity of semiconductor HEB current-voltage characteristic, needed for heterodyne detector operation, is conditioned by the dependence of conductivity on electron mobility, which is a function of applied electric field, thereby the function of electron temperature. Higher IFs and broader Δf can be obtained in semiconductor HEB frequency converters increasing their temperature to about 80 K (where electron-phonon interaction is much stronger and $\tau \approx 10^{-11}$ s), but in this case, the noise level of such frequency converters increases appreciably and conversion losses increase fast too.

Table 28.9 Parameters of InSb HEB manufactured by Infrared Laboratories

PARAMETER	VALUE
Detector area (mm)	5 × 5
Detector mounting	Sapphire substrate set into an integrating cavity
Operating temperature (K)	1.5–4.2
Spectral response (mm)	0.2–6
Frequency response, 3 db (kHz)	600
NEP (WHz$^{-1/2}$)	<8 × 10^{-13}

Source: Adapted after Ref. [219]

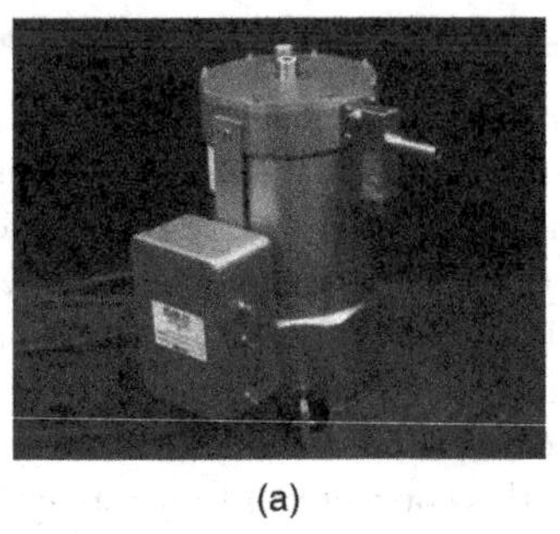

(a)

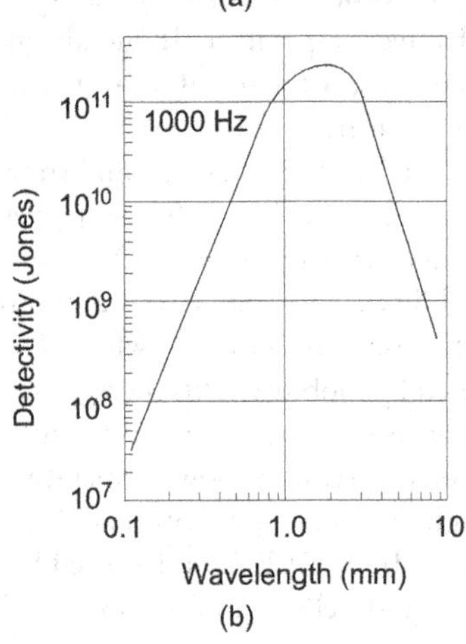

(b)

Figure 28.58 InSb HEB detector: (a) photo of the bolometer, (b) detectivity versus cutoff wavelength.

In low-dimensional semiconductor structures, the electron-phonon interaction can be substantially increased (τ decreased) and, thus, such kind of structures can be considered as frequency converters with higher IFs and wider bandwidth up to 10^9 Hz [225–227]. Direct measurements of photoresponse relaxation time have shown that τ is about 0.5 ns in the temperature range 4.2–20 K [228]. Thus, the IF can be increased by about three orders compared to bulk semiconductor HEBs.

Historically, HEB mixers using semiconductors were invented in the early 1970s [215] and played an important role in early sub-mm astronomy [229] but were superseded by SIS mixers by early 1990s. However, the development of superconducting HEB versions led to the most sensitive THz mixers at frequencies beyond the reach of SIS mixers. The main difference between HEB mixers and ordinary bolometers is the speed of their response. HEB mixers are fast enough to allow GHz output IF bandwidths.

28.11 SUPERCONDUCTING BOLOMETERS

The next generation of resistive bolometers to reach very high sensitivities requires detectors that are able to detect photons over a wide frequency range (100 GHz to a few THz), with the NEP value around 10^{-18} $WHz^{-1/2}$ or below. More recently, based on the same detectors developed for Herschel-PACS, the sub-mm wide-field camera ArTéMiS has been elaborated [230]. This camera is specifically designed to interface with the 12-meter aperture APEX sub-mm telescope located at the Llano de Chajnantor in the Atacama Desert in Chile. ArTeMiS detectors consist of Si:P:B bolometers (thermometers are made of silicon doped with phosphorous and boron) and include 20 subarrays of 16 × 18 pixels, operating simultaneously at three wavelengths (4 arrays at 200 μm, 8 arrays at 350 μm, and 8 arrays at 450 μm, 5,760 pixels in total) and at 300 mK or a bit below [231].

In comparison with PACS bolometers, the new detectors (see Figure 28.59) are not hybridized by indium bumps—each layer that composes the sensor is grown above the ROIC. The sub-mm radiation is efficiently absorbed thanks to a superconducting TiN layer deposited on a silicon grid above a backshort, what is shown in Figure 28.60 [232,233]. The absorption is achieved through the combination of vertical

Figure 28.59 The ArTéMiS 350-μm FPA successfully commissioned at 350 μm on APEX in the 2014 run. The 300-mK ribbon cables visible on one side of each 16 × 18 array connect the arrays to the 4 K electronic stage. (Adapted after Ref. [231])

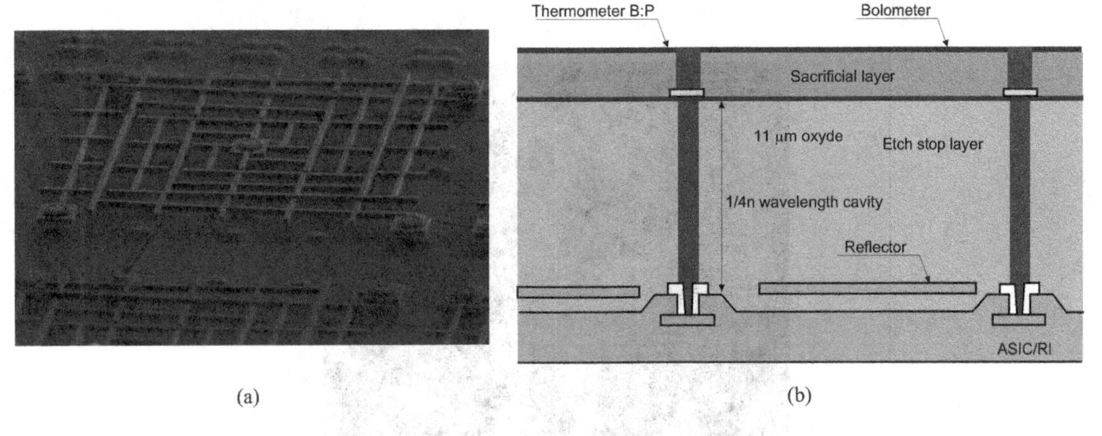

Figure 28.60 ArTéMiS 150-μm pixel size: (a) scanning electron microscopy image of polarization sensitive pixel. Two interlaced silicon meanders are maintained at 2 μm above the SiO_2 layer thanks to copper through-silicon-vias. (b) Design of pixel structure.

resonance (quarter wave cavity) and horizontal resonance (metallic pattern deposited on the meander, forming two networks of planar antennas). The designed pixel is adapted to the relatively short waves (80–150 μm). The absorption in longer wavelengths is achieved by developing a system based on an antireflecting layer that can be tuned to enhance the absorption in a particular band. The spectral response measured in the laboratory is above 92% in the range 200–550 μm. Table 28.10 sums up the measured performance of the ArTeMiS instrument in comparison with previously installed resistive bolometer arrays.

The ArTeMiS camera has already delivered a spectacularly detailed view of the Cat's Paw Nebula (NGC6334), a star located in the constellation Scorpius (see Figure 28.61). The background image comes from VISTA observations, and the ArTéMiS data at 350 μm are in orange. The filaments are clearly visible around the most active areas that correspond to pre-stellar cores

28.11.1 SUPERCONDUCTING HOT-ELECTRON BOLOMETERS

General information about HEB is included in Section 8.6. As is mentioned there, the HEBs operation depends on interaction mechanism electrons and phonons. The ratio of phonon to electron specific heats, C_p/C_e, controls the energy flow from electrons to phonons and the energy backflow due to reabsorption of nonequilibrium phonons by electrons. This ratio is 0.85 for Nb, 6.5 for NbN, and 38 for YBaCuO layers [234]. In very thin films, the phonons can escape into the substrate before being reabsorbed by the electrons. In thin, below 10-nm-thick, Nb films deposited on a substrate, $\tau_{phe} > \tau_{eph}$, and the effective escape of phonons to the substrate prevails in energy backflow to electrons. As a result, τ_{eph} alone controls the response time, which is approximately equal to about 5 ns. Thus, Nb devices are sensitive in a wide range of

Table 28.10 Performance of ground-based bolometer instruments

INSTRUMENT	CENTRAL WAVELENGTH (μm)	NUMBER OF PIXELS	FIELD OF VIEW	BEAM FWHM	PIXEL NEFD (mJy.s$^{1/2}$,1σ,1s)	IMAGING SPEED
Artemis (on APEX)	350	4 subarrays 16 × 18	4.7′ × 2.3′	8.5″	600 (best values at 300)	5 times quicker than SABOCA[a]
P-Artemis (on APEX)	450	8 sub-arrays 16 × 18	1.0′ × 1.0′	9.4″	2,000	0.3
SHARC-2	350	12 × 32	0.9′ × 2.5′	8.5″	1,000	1
SCUBA-2	450	80 × 80	8’ × 8′	7.5″	600	40

NEFD, noise equivalent flux density.
[a] Sixteen times quicker when considering the full instrument.

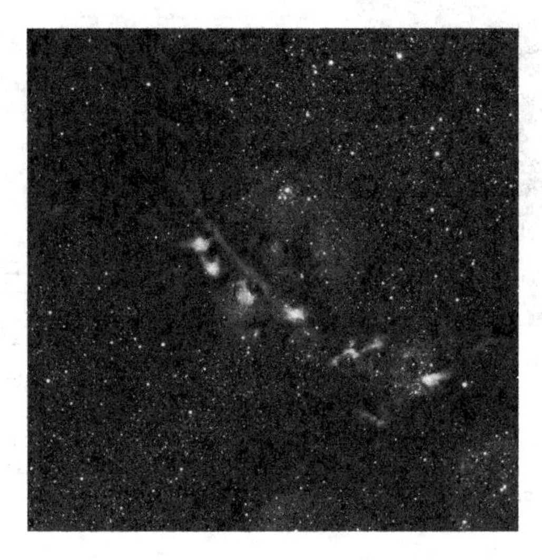

Figure 28.61 The star forming region NGC6334 (Cat's Paw Nebula) observed at 350 μm with ArTéMiS. (Adapted after Ref. [233])

spectra, are much faster compared to bulk semiconductor bolometers (operating at $T \approx 4$ K), and can reach NEP ≈ 3 × 10^{-13} WHz$^{-1/2}$ [235].

Karasik et al. have demonstrated hot-electron superconducting direct detection Ti nanobolometers fabricated on Si planar bulk substrates with Nb contacts and electrical NEP value of 3 × 10^{-19} WHz$^{-1/2}$ at 300 mK [43]. The time constant of cooled HEBs can reach value in a wide range from 10^{-5} to 10^{-10}s. The thermal time constant τ_{eph} = 25 μs at T = 190 mK for larger devices has been demonstrated. For the first time, the record optical NEP = 3 × 10^{-19} WHz$^{-1/2}$ at λ = 460 μm and T = 50 mK has been achieved [60]. Such high sensitivity meets the requirements for SAFARI instrument on the SPICA telescope.

NbN films in comparison to Nb ones have much shorter τ_{eph} and τ_{phe} because of stronger electron-phonon interaction. In ultrathin 3-nm-thick NbN films, both τ_{eph} and τ_{phe} determine the response time of the detector, which can be about 30 ps near T_c ($\tau_{eph} \approx$ 10 ps) [236]. NEP can reach values of 10^{-12} WHz$^{-1/2}$ [237].

Since for YBaCuO detector layers, the ratio $C_p/C_e \approx 38$, the layers are mainly the phonon-cooled type; the energy backflow from phonons to electrons can be neglected and the thermalization time is about an order faster ($\tau_{eph} \approx$ 1 ps) compared to NbN layers. In YBaCuO films excited by fs pulses, the non-thermal (hot-electron) and thermal bolometric (phonon) processes are practically decoupled, with the former one dominating the early stage of electron relaxation. To decouple electrons from phonons, non-equilibrium phonons in the film should escape from it (into the substrate) in short time compared to phonon-electron time, τ_{phe}.

As is shown in Figure 6.10, HEBs did not demonstrate any frequency limitation of the detection mechanism in contrast to the competitor such as SIS tunnel junctions. The HEB mixer is the most attractive candidate for heterodyne observations at frequencies above 1 THz. The successful operation of practical instruments (the Heinrich Hertz Telescope, the Receiver Lab Telescope, APEX, SOFIA, Hershel) ensures the importance of the HEB technology despite the lack of rigorous theoretical routine for predicting the performance [238,239].

The standard setup for a heterodyne mixer is shown in Figure 28.62. The QCL is mounted in a liquid helium cryostat and operated at about 10 K. The beam entered a blackbody hot/cold vacuum setup attached to the HEB cryostat via a window and then is reflected by a 3-μm mylar beam splitter. This cryostat is cooled to 4.2 K. The HEB is mounted to the backside of a 10-mm Si lens with an anti-reflection coating. The first-stage low-noise amplifier is attached to the cold plate and operated at 4.2 K. Outside the dewar, room temperature amplifiers and wide band-pass filter is used to further condition the IF signal before the total power is read using a power meter.

A major issue for HEB mixers is achieving a thermal time constant that is fast enough to yield a useful IF output bandwidth of a few GHz. For a fixed thermal relaxation time, the heat capacity sets the required LO power and can be minimized by using a very small volume ($< 10^{-2}\,\mu m^3$) of a superconducting film. The required LO power is about one order of magnitude lower (typically between 100 and 500 nW) compared to SIS mixers and much lower (roughly 3–4 orders) in comparison with SBDs. The LO power scales with the volume of the microbridge and decreases with increasing critical temperature [240].

A large number of antennas were proposed (see, e.g. Refs. [241,242]), including dish and horn antennas, log-periodic, spiral, slots/apertures, bow-tie, etc. (see Figure 8.20). For printed antennas for radiation effective transmission and reception, they should be about half of the wavelength λ. Detector's sensitivity is proportional to the antenna effective area S_{eff}. Antenna effective area S_{eff} and antenna gain G are linked by [241]

$$S_{eff} = \frac{\lambda^2}{4\pi} G. \tag{28.20}$$

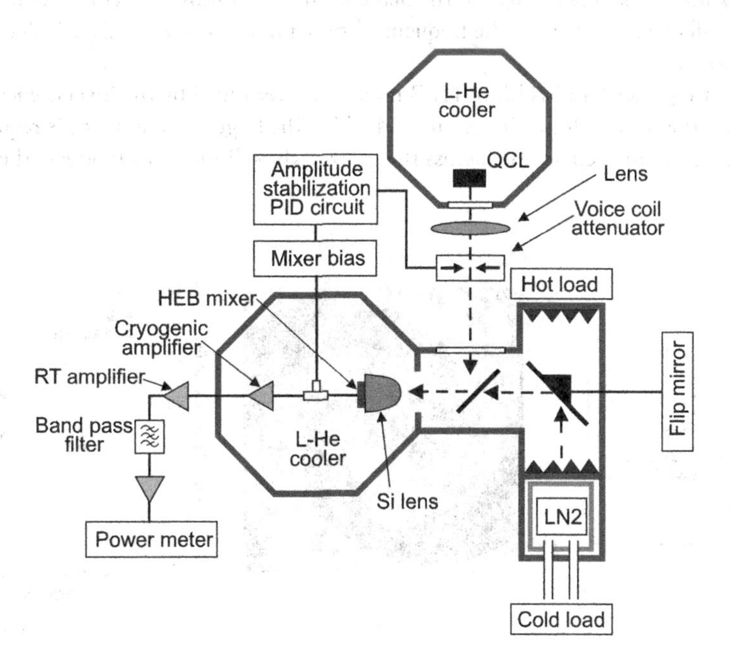

Figure 28.62 Standard setup for a heterodyne mixer. The HEB mixer is mounted to a Si lens and indicated in red with HEB chip. An LO is applied consisting of a QCL. As a signal, a black body source is applied as a 77-K load from liquid nitrogen and a room temperature load at 295 K.

Thus, with frequency $\nu = c/\lambda$ increase, the effective antenna area reduces, as the gain G depends on antenna design, dielectric substrate properties, etc. and increases with a decrease in wavelength.

Two methods are used to obtain useful IF output bandwidth:

- phonon cooling, using ultrathin NbN or NbTiN films characterized by large electron-phonon interaction,
- diffusion cooling, using submicron Nb, Ta, NbAu, or Al devices coupled to normal-metal cooling "pads" or electrodes.

Competitive sensitivities have been demonstrated for both types of devices.

Typically, phonon-cooled HEBs are made from ultrathin films of NbN, whereas diffusion-cooled devices use Nb or Al. Current state-of-the-art NbN technology is capable of routinely delivering 3-nm-thick devices that are 500 nm^2 in size. NbN films are deposited on a dielectric (typically high resistivity >10 kΩcm silicon). The superconducting bridge is defined by means of electron beam lithography. Its length varies between 0.1 and 0.4 μm and the width between 1 and 4 μm.

Figure 28.63a presents an example cross-sectional view of the NbN mixer chip. About 150-nm-thick Au spiral structure is connected to the contacts pads. The supeconducting NbN film extends underneath the contact layer/antenna. The central area of a mixer chip shown in Figure 28.63b is manufactured from a 3.5-nm-thick superconducting NbN film on a high-resistive Si substrate with an e-beam evaporated MgO buffer layer [243]. Ultrathin NbN films are deposited by reactive magnetron sputtering in the Ar+N_2 gas mixture. The active NbN film area is determined by the dimensions of the 0.2-μm gap between the gold contact pads. The NbN critical temperature depends on the thickness of film deposited on the substrate. An improvement of superconducting properties of NbN films due to the presence of MgO buffer layer on silicon substrates is evident (see Figure 28.63c). The superconducting transition temperature is about 9 K and the transition widths are ≈ 0.5 K.

The NbN superconductive HEB mixers are characterized by strong electron-phonon interaction. The response time can achieve a value of 10^{-11} s [236] and because of no principal restrictions for operation at $\nu > 1$ THz (the absence of their noticeable capacities), these devices can be effectively used for heterodyne detection in the wide spectral range up to the visible one, where operation, for example, of SIS mixers, is hampered. For example, Figure 28.64 shows intermediate frequently dependent output power for the 3.5-nm-thick NbN film devices deposited on the plane of the MgO substrate (curve A) and on the Si substrate with MgO buffer layer (curve B). The frequency limited range to several GHz is due to influence of free carrier relaxation rate.

An intermediate frequency bandwidth in HEB mixers is determined by the inverse energy relaxation time. In NbN HEB, the bandwidth is limited to 3–4 GHz. The large IF bandwidth is required for high-resolution spectroscopy of molecular lines across the galaxy. These limitations motivate the search for new

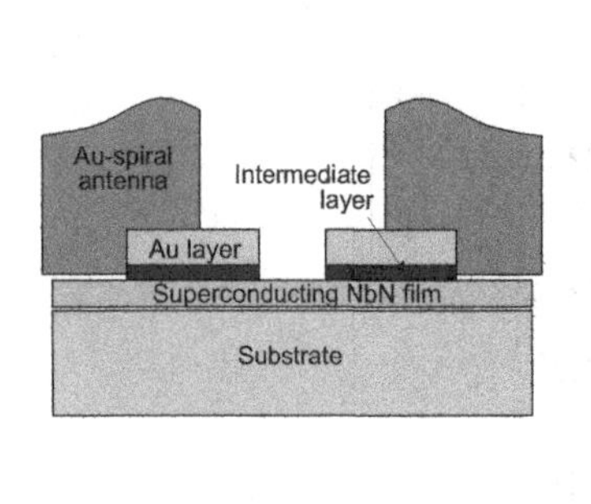

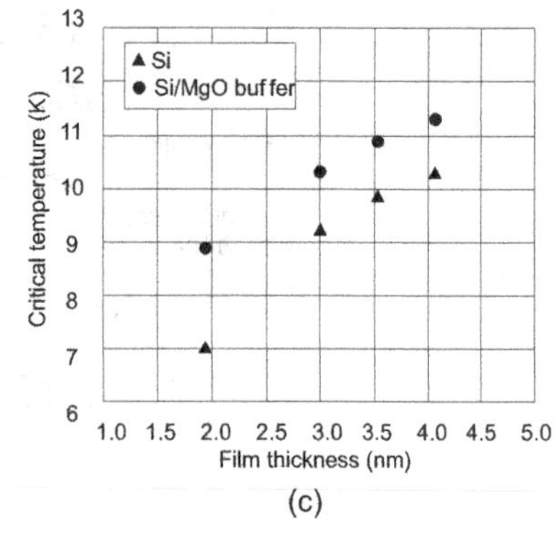

Figure 28.63 NbN HEB mixer chip: (a) cross section view, (b) SEM micrograph of the central area of mixer, (c) critical temperature versus thickness for NbN films deposited on Si substrates (triangles) and on Si with MgO buffer layer (circles). (Adapted after Ref. [243].)

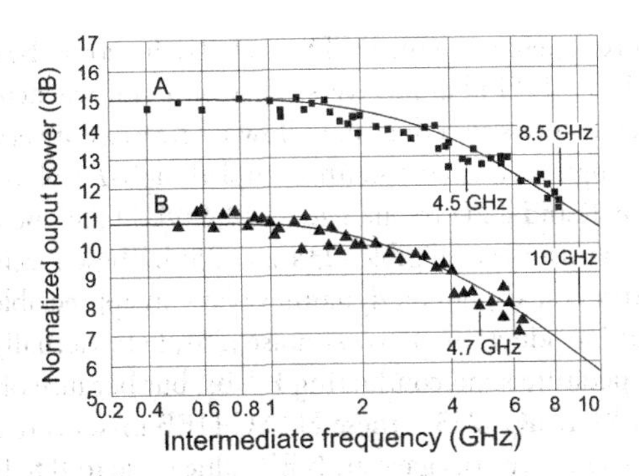

Figure 28.64 Output power as a function of intermediate frequency for the 3.5-nm-thick NbN film devices deposited on plain of MgO substrate (curve A) and on Si substrate with MgO buffer layer (curve B). (Adapted after Ref. [243].)

materials with higher critical temperatures where the electron thermal relaxation may go faster. Also HEB mixers with high operating temperature are sought in view of the reduced cryocooling requirement that is very important for applications in space [244]. As a promising HEB material, Cherednichenko et al. [245] have proposed MgB_2 with substantially shorter phonon escape time due to the very high sound velocity $\approx 8\,km\,s^{-1}$, which is a factor of 2.5 greater than that in NbN. More recently [246], an MgB_2 mixer fabricated using the hybrid physical-chemical vapor deposition process has demonstrated a large IF bandwidth of 8–9 GHz in a 15-nm-thick film with $T_C = 36–38$ K (see Figure 28.65). Even a larger IF bandwidth is expected since the bandwidth scales inversely proportional to the film thickness [247].

HEBs are significantly more sensitive than SBDs but somewhat less sensitive than SIS mixers (see Figure 6.10). This figure shows that DSB noise temperature achieved with HEB mixers range from 400 K at 600 GHz up to 6,800 K at 5.2 THz. In the lower frequency range up to 2.5 THz, the noise temperature follows closely the 10 $h\nu/k$ line. Above this frequency, the correspondence becomes somewhat worse, what is caused by increasing losses in the optical components, lower efficiency of the antenna, and skin effect contributions in the supeconducting bridge.

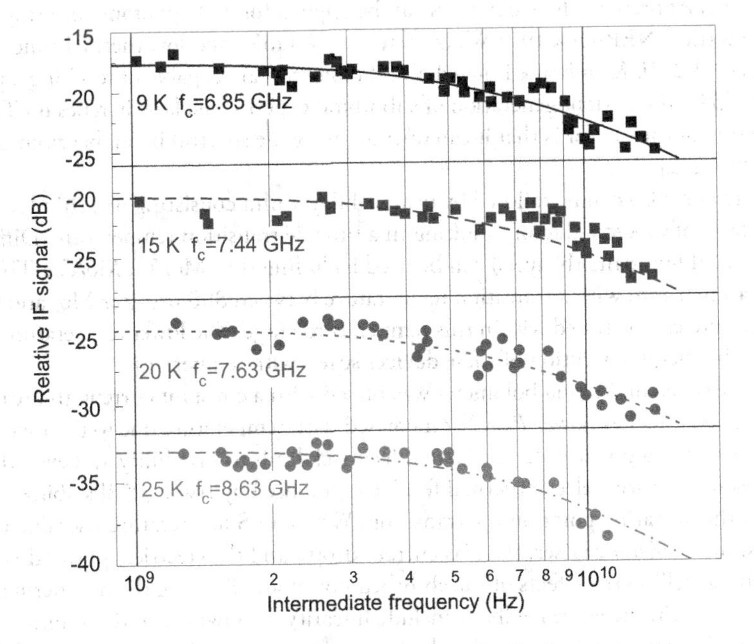

Figure 28.65 IF spectra in a 15-nm-thick MgB_2 mixer device on SiC substrate coupled to a broadband planar microantenna. The results obtained using two monochromatic 600-GHz sources. f_c denotes the −3 dB cut-off frequency. (Adapted after Ref. [246])

Concerning high-temperature superconductor (HTSC) HEBs, it should be noted that there do not exist numerous publications devoted to this kind of receivers. They have not reached high state of technological maturity since their complicated composition does not allow fabrication of very thin layers with high critical temperature. Kreisler and Gaugue reviewed antenna-coupled high-T_c bolometers for both homodyne and heterodyne applications in the FIR and THz frequencies [248,249]. HTSC belongs to the phonon cooled type and electron diffusion mechanism is negligible [234,249,250]. These receivers are noticeably noisier compared to low-temperature devices, as phonon dynamics plays an appreciable role due to the relatively high operating temperature and introduction of excess noise [251,252]. Actually, HTSC HEB mixers do not reach the sensitivity of low-temperature superconducting HEBs, but because of very short electron-phonon relaxation time ($\tau_{eph} \approx 1.1$ ps in YBaCuO [253]), these HTSC HEB mixers are the wide bandwidth devices.

Experimental data for Josephson detectors gave an NEP value close to 8×10^{-15} WHz$^{-1/2}$ at $T = 80$ K ($\nu = 86$ GHz) and NEP $\approx 3 \times 10^{-13}$ WHz$^{-1/2}$ for $T = 55$ K ($\nu = 692$ GHz) [254]. Lyatti et al. have concluded [255] that HTSC HEB mixers can reach an NEP value of 5×10^{-15} WHz$^{-1/2}$.

28.12 TRANSITION EDGE SENSOR BOLOMETERS

The name of the TES bolometer is derived from its thermometer, which is based on thin superconducting films held within the transition region, where it change from the superconducting to normal state over a temperature range of a few millikelvin (see Figure 8.18) [256]. The film has stable but very steep dependence of resistance on temperature in the transition region. Changes in temperature transition can be set by using a bilayer film consisting of a normal material and a layer of superconductor. Such design enables diffusion of the Cooper pairs from the superconductor into the normal metal and makes it weakly superconducting—this process is called the proximity effect. As a result, the transition temperature is lowered relative to that for the pure superconducting film. Thus in principle, the TES bolometers are quite similar to the HEBs. In the case of HEB, high speed is achieved by allowing the radiation power to be directly absorbed by the electrons in the superconductor. In TES bolometers, however, rather a separate radiation absorber is used that allows the energy to flow to the superconducting TES via phonons, as ordinary bolometers do.

TES bolometers are superior to current-biased particle detectors in terms of linearity, resolution, and maximum count rate. At present, these detectors can be applied for THz photons counting because of their high sensitivity (electrical NEP ~ 3×10^{-19} WHz$^{-1/2}$ at $T = 300$ mK) and low thermal time constant ($\tau = 25$ μs at $T = 190$ mK) [43,257]. Membrane isolated TES bolometers are capable of reaching a phonon NEP $\approx$ 4×10^{-20} WHz$^{-1/2}$ [39]. The current generation of suborbital experiments largely relies on TES bolometers. An important feature of this sensor is that it can operate in a wide spectral band, between the radio and gamma rays [13,258–262].

The temperature of a TES can be tailored by using a bilayer film consisting of a thin layer of normal metal and a thin layer of superconductor, resulting in a tunable transition temperature. Different types of superconducting metal film pairs (bilayers) can be used including thin Mo/Au, Mo/Cu, Ti/Au, etc.). Two metals behave as a single film with a transition temperature between 800 mK (for Mo) and 0 K (for Au). Transition temperature can be tuned within this temperature range. The lower temperature ($T < 200$ mK) is needed because the energy resolution of these devices scales with temperature.

Traditionally, the superconducting bolometer was biased with a constant current and read out with a voltage amplifier. Then, the bias power $P_b = I^2R$ increased with temperature due to the increase of resistance R near T_c. In consequence, a positive electrothermal feedback leads to instability and even thermal runaway. The new idea of a negative electrothermal feedback proposed by Irwin [258] stabilizes the temperature of the TES at the operating point on the transition. When TES temperature rises due to power from absorbed photons, their resistance rises, the bias current drops, and the electrical power dissipation in them decreases, partially cancelling the effects of the absorbed power and limiting the net thermal excursion. The advantages of TES with negative feedback include linearity, bandwidth, and immunity of the response to changes in external parameters (e.g., the absorbed optical power and the temperature of the heat sink). Consequently, these devices are suitable for fabrication of large format horn-coupled and filled arrays required for many new missions [263–265] offering advantages over semiconducting bolometers.

In practice, the bias voltage V_b is chosen so that for small optical power P, the TES will be heated to a steep point on the temperature transition. For intermediate values of P, the electrothermal feedback keeps the total power input $P + V^2/R$ (and, thus, the temperature) constant. The current responsivity is defined as the response of the bolometer current I to a change in the optical power. Then, for a thermal circuit with a single pole response is equal [13,266]

$$R_i = \frac{dI}{dP} = -\frac{1}{V_b}\frac{L}{(L+1)}\frac{1}{(1+i\omega\tau)}, \tag{28.21}$$

where $L = \alpha P/GT$ is the loop gain, $\alpha = (T/R)dR/dT$ is a measure of the steepness of the superconducting transition and is a bolometer figure of merit, $G = dP/dT$ is the differential thermal conductance, and τ is the effective time constant. For a typical loop gain $L \approx 10^2$ [258,261,265], the low-frequency responsivity becomes $R_i \approx -1/V_b$ and depends only on the bias voltage and is independent of the signal power and the heat sink temperature. The effective time constant $\tau = \tau_0/(1 + L)$ is much shorter than the time constant without feedback $\tau_0 = C/G$ for thermal detectors. Negative electrothermal feedback can make the bolometers operate tens or even hundreds of times faster. The values of α for Mo/Cu proximity-effect layers (T_c = 190 mK) consistent with thermal fluctuation noise [267] are within 100–250 [268].

The resistance of a TES is low, so it can deliver significant power only to low-input impedance amplifiers, which rules out JFETs and MOSFETs. Instead, the signals are fed into SQUIDs, which are the basis for a growing family of electronic devices that operate by superconductivity. In this case, the TES is transformer coupled to the SQUID by an input coil. A current-biased shunt resistor is used to provide a constant voltage bias to the TES. When the shunt resistor is operated close to the detector temperature, a negligible Johnson noise from the bias network is given. The SQUID readout has a number of advantages, including it operates near the bolometer temperature, has very low power dissipation and large noise margin, and low sensitivity to microphonic pickup. In addition, the fabrication and lithographic processes used in both SQUID readouts and TES bolometers are similar, what helps in their integration on the same chip.

Figure 28.66 shows the typical TES circuit wired in series with the coil of a SQUID amplifier. The voltage bias is achieved by current bias of a cold shunt resistor R_{sh}, whose about 10 mΩ resistance is much smaller than the $R \approx 1$ Ω resistance of the TES. The current through the TES is measured with a SQUID ammeter, and the in-band reactance of the SQUID input is much less than R. When the bias to a SQUID is turned off, the whole device goes into a superconducting state where it adds no noise. Thus by switching on or off rows or columns of SQUIDs in an array (one for each pixel), a cold multiplexer may be realized. The biases across the SQUIDs are controlled by the address lines. Each SQUID can be switched from an operational state to a superconducting one if it is biased to carry about 100 μA. The address lines are set so all the SQUIDs in series are superconducting except one, and then only that one contributes to the output voltage. By a suitable series of bias settings, each SQUID amplifier can be read out in turn. To avoid very large numbers of leads leaving the cryostat, lines of 30–50 detectors can be multiplexed before amplification.

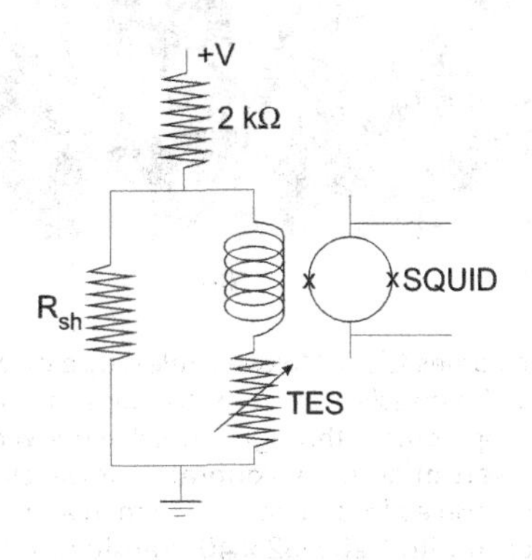

Figure 28.66 Typical TES bias circuit and low noise, low power SQUID.

In general, SQUID-base multiplexers developed for TES bolometers and microcalorimeters used both time-division [269] and frequency-division [270] approaches. We have described the time-division approach, where multiplexer uses a SQUID for each bolometer to switch the outputs sequentially through a single SQUID amplifier. In the case of frequency domain, each TES is biased with a sinusoidally varying voltage and the signals from a number of TESs are encoded in amplitude-modulated carrier signals by summing them. The signals are then amplified by a single SQUID and recovered with ambient temperature lock-in amplifiers. For more SQUID multiplexer's details, see, for example, Refs. [271–276].

The most ambitious example of TES bolometer array is that used in the sub-mm camera submillimeter common-user bolometer array-2 (SCUBA-2) with 10,240 pixels [264,277,278]. It is the world's largest sub-mm camera. The camera operated at wavelengths of 450 and 850 μm has been mounted on the James Clerk Maxwell Telescope in Hawaii (see Figure 28.67). Each SCUBA-2 array is made of four-side-buttable subarrays, each with 1,280 (32 × 40) transition-edge sensors. The design is illustrated in Figure 28.68

(a)

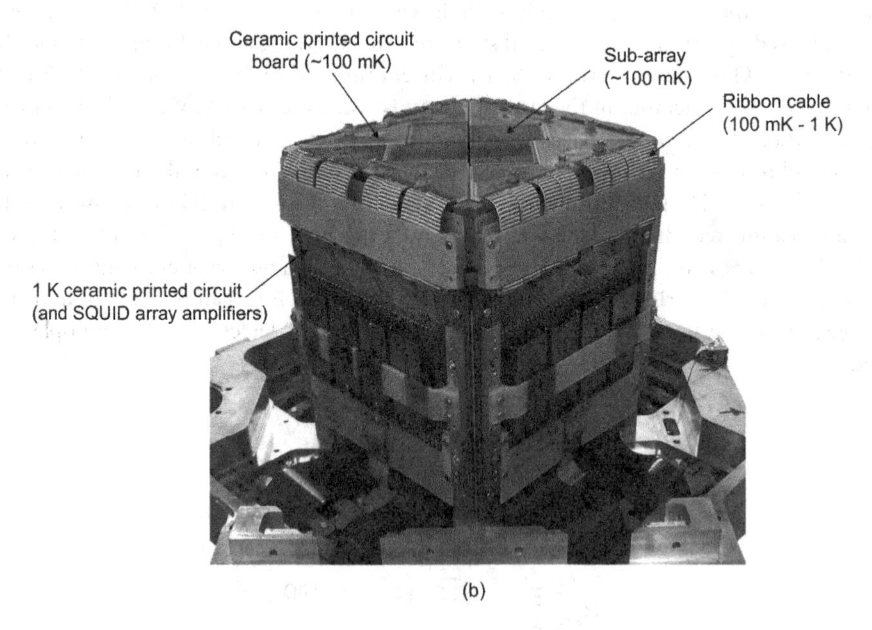

(b)

Figure 28.67 SCUBA-2 mounted on the James Clerk Maxwell Telescope on Mauna Kea, Hawaii (a). The instrument weighs 4.5 tonnes and is 3 m high. The massive blue box contains the camera and keeps it cold at about 0.1 K. Submillimeter light from the telescope enters through a small window on the left-hand side (behind the white bars) and is directed onto the two sets of detectors operated at wavelengths of 450 and 850 μm. Figure (b) shows the photograph of four subarray modules folded into position in a focal plane unit. Each SCUBA-2 array is made of four-side-buttable subarrays, each with 1,280 (32 × 40) transition-edge sensors. The principal components are also highlighted.

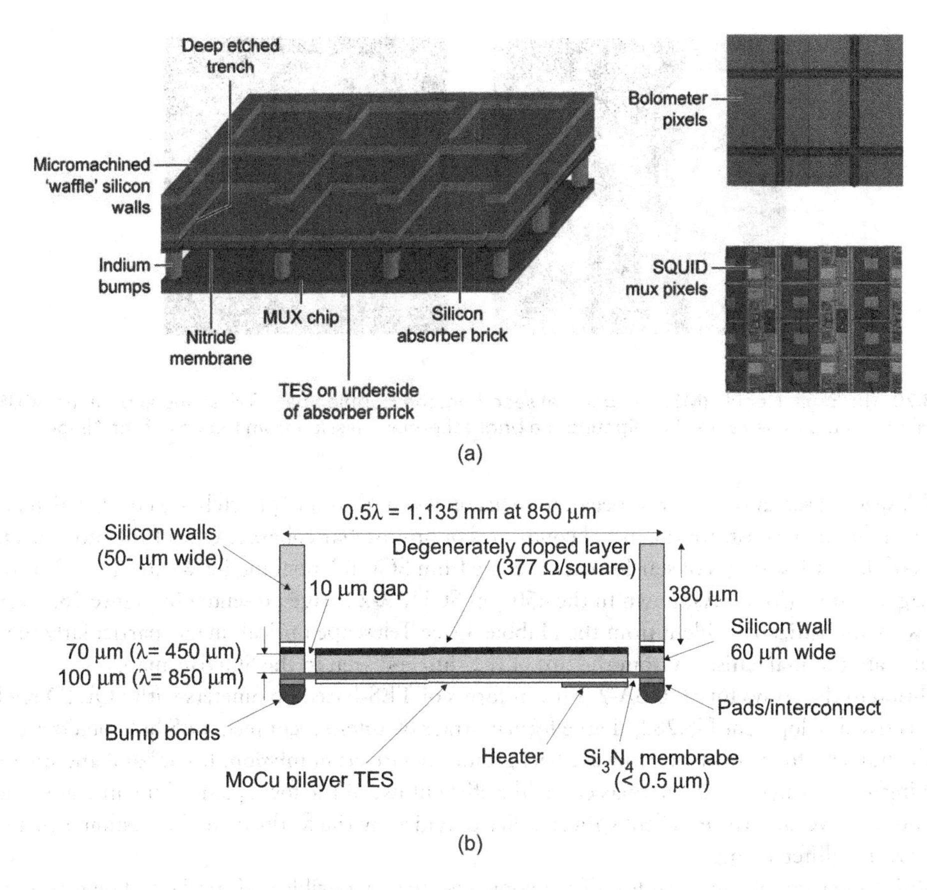

Figure 28.68 SCUBA-2 bolometer array: (a) design features, (b) cross-section of single pixel. (Adapted after Ref. [264])

together with the cross-section of the detector architecture. The detector technology is based on silicon micromachining. Each pixel consists of two silicon wafers bonded together. The upper wafer with its square wells supports the silicon nitride membrane on which Mo-Cu bilayer TES detectors and the absorbing silicon brick are suspended. The lower wafer is thinned to 1/4 wavelength in silicon (70–850 μm). The upper surface of this wafer has been previously implanted with phosphorus to match the impedance of free space (377 Ω/square). The detector elements are separated from their heat sinks by a deep-etched trench that is bridged by only a thin silicon nitride membrane. The superconducting electronics that read out the bolometers are fabricated on separate wafers (see Figure 28.69). The two components are assembled into an array using indium bump bonding. Further details are in Walton et al. [264] and Woodcraft et al. [279].

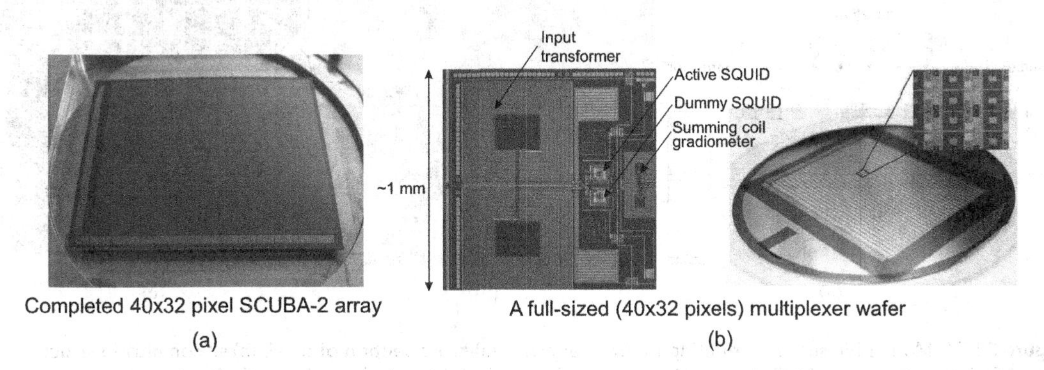

Figure 28.69 SCUBA-2: (a) bolometer array and (b) SQUID multiplexer pixel. (Adapted after Ref. [280])

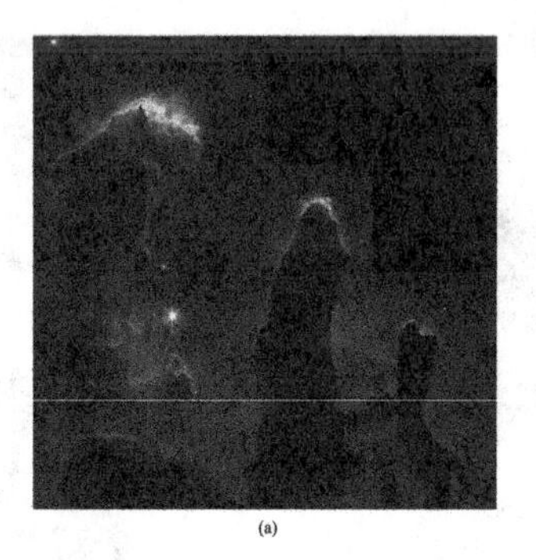

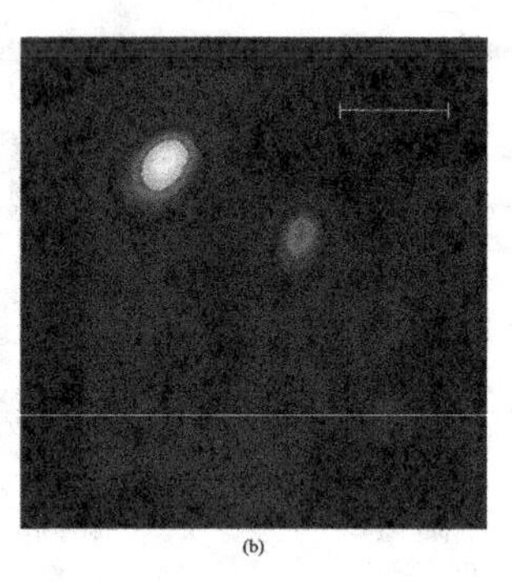

Figure 28.70 The Eagle Nebula (M16) imagers: (a) seen from the Hubble Space Telescope and (b) the SCUBA (450-μm image) of the same region highlighting the bright thermal emission from the tips of the "fingers".

SCUBA space observations interest began to focus on the starless (or "pre-stellar") cores, which are significant in that they constrain the initial conditions of protostellar collapse. One of the most spectacular images of the earliest stages of star formation came from SCUBA imaging by White et al. [281] of the famous Eagle Nebula (M16). As shown in the 450-μm SCUBA-2 image presented in Figure 28.70, some differences are immediately evident from the Hubble Space Telescope optical image, particularly in terms of the dominant thermal emission from the tips of the "fingers" seen in the SCUBA map.

In addition to the arrays for SCUBA-2, various forms of TES-based bolometers with SQUID readouts are under active development [45,282]. Large format arrays of antenna-coupled TES bolometers are very attractive candidates for a cosmic microwave background polarization mission. Broadband antennas with RF diplexing and/or interleaved antennas can make efficient use of the focal plane. Antennas are inherently polarization sensitive, and the excellent gain stability provided by the feedback in TES bolometers facilitates polarization differencing.

The fabrication technologies used for TES bolometers are very flexible and specialized detectors are being developed to meet the needs of specific observations. A bolometer design allows the production of large monolithic detector arrays with a very high fill factor by standard planar lithography. Figure 28.71 shows the 1024-pixel array structure together with a single pixel [283]. The absorbing element is a square mesh of 1-μm-thick low-stress (non-stoichiometric) silicon nitride (LSN), which is metallized with gold to produce an average sheet resistance of 377 Ω per square. A conducting backshort is located at a distance $\lambda/4$ behind the mesh. This mesh absorber is supported at four points (shown by arrows) by low thermal conductivity beams of LSN. To produce $T_c \approx 400$ mK, a proximity-effect sandwich of Al and Ti is made at the center of the mesh. To connect the thermistors to the edges of the array, fully superconducting Nb leads are

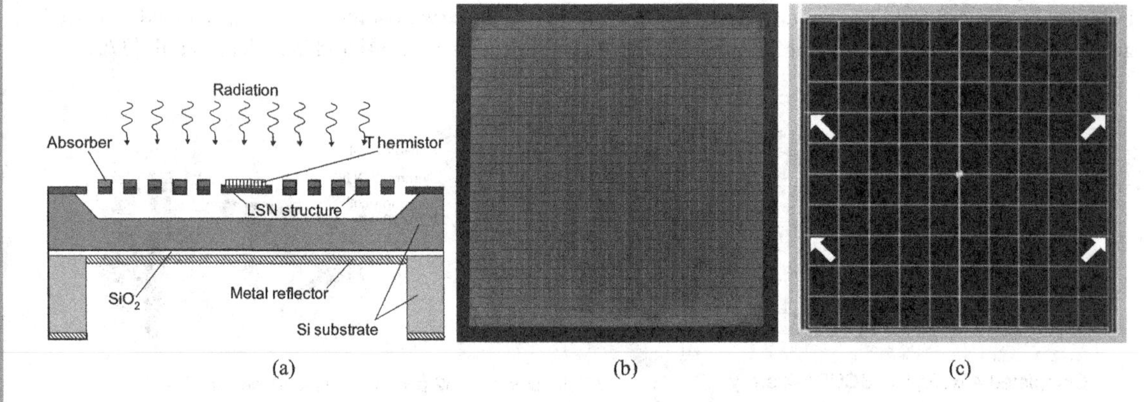

Figure 28.71 Monolithic superconducting bolometer array: (a) cross section of pixel, (b) silicon nitride structure for a 1,024 element array of bolometers (1.5 × 1.5 mm² pixel size), (c) complete pixel including sensor and metallization. (Adapted after Ref. [285])

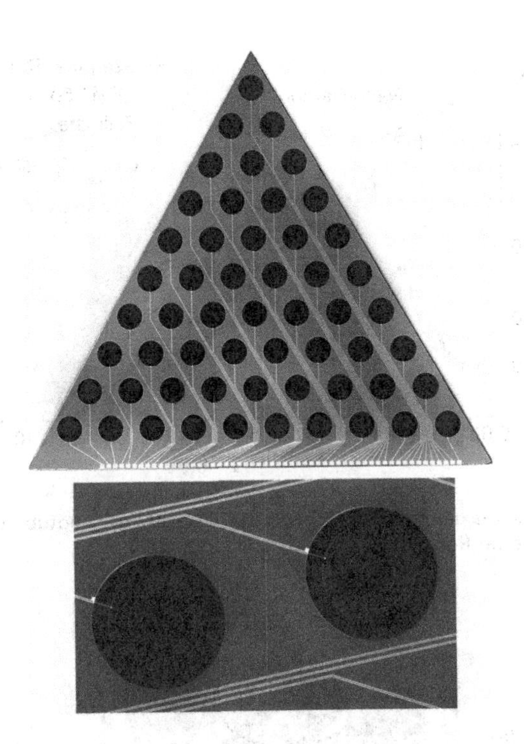

Figure 28.72 Array of 55 TES spiderweb bolometers and closeup of bolometers. Six wedges of the type shown are assembled to form a 330-element hexagonal horn-coupled array. (Adapted after Ref. [13])

used by way of the beams and the dividing strips between pixels. Other designs use micromachining and folding to bring the leads out of the third dimension [284].

Figure 28.72 shows a close-packed horn-coupled array with radial support legs. The bolometers which are located at the small ends of the horns are then separated sufficiently for easy support and wiring. The array is complete fully lithographed [285].

28.13 NOVEL TERAHERTZ DETECTORS

In comparison to the technical nature of other frequency regions, in the THz region, detectors have not been fully established due to mainly two reasons: (i) the frequency of the THz wave is too high to be handled with existing high-frequency semiconductor technology and (ii) the photon energy of the THz wave is much lower than the bandgap energy of semiconductors. It is expected that applications of nanoscale materials and devices will open the door to overcome such difficulties. Today, research activities in the field of THz radiation detectors focus also on the development of novel nanoelectronic materials and technologies. Here we give several examples of new solutions.

28.13.1 NOVEL NANOELECTRONIC DETECTORS

Carbon nanotubes (CNTs) are the hottest topic in physics and microelectronics. From the material point of view, CNTs offer an excellent alternative to their solid-state counterparts because of their small junction areas due to their physical dimensions (<1–2 nm diameter), high electron mobilities (up to 200 000 $cm^2 Vs^{-1}$), and low estimated capacitances (tens of aF μm^{-1}), leading to predicted cutoff frequencies in the THz range. Figure 28.73 shows predictions of the maximum frequency for nanotube transistors vs. gate length and comparison to other technologies. In the estimations, the largest transconductance, $g_m = 20\ \mu S$, has been assumed.

The CNT with unique one-dimensional structure has attracted much attention and has been extensively studied for future nanoelectronics, nanophotonics, and nanomechanics [288]. A sensitive and frequency tunable THz detector based on a CNT QD transistor has been also demonstrated (see Figure 28.74) [289]. Photon-assisted tunneling (PAT) is a process that can be used in THz photodetectors. Figure 28.74c

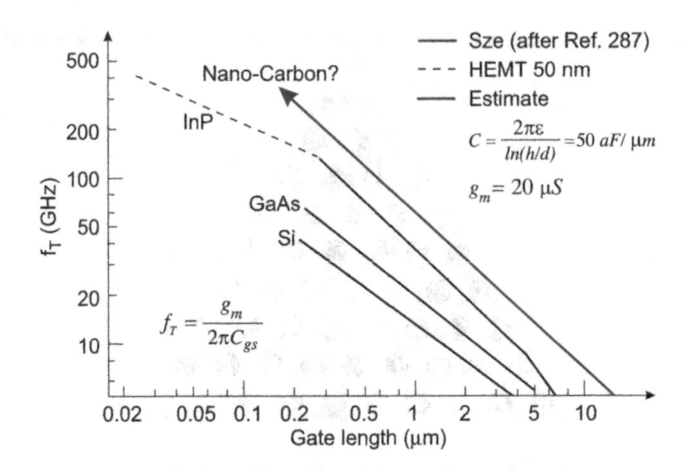

Figure 28.73 Predictions of the maximum frequencies vs. gate length for nanotube transistors as compared with other technologies. (Adapted after Ref. [286])

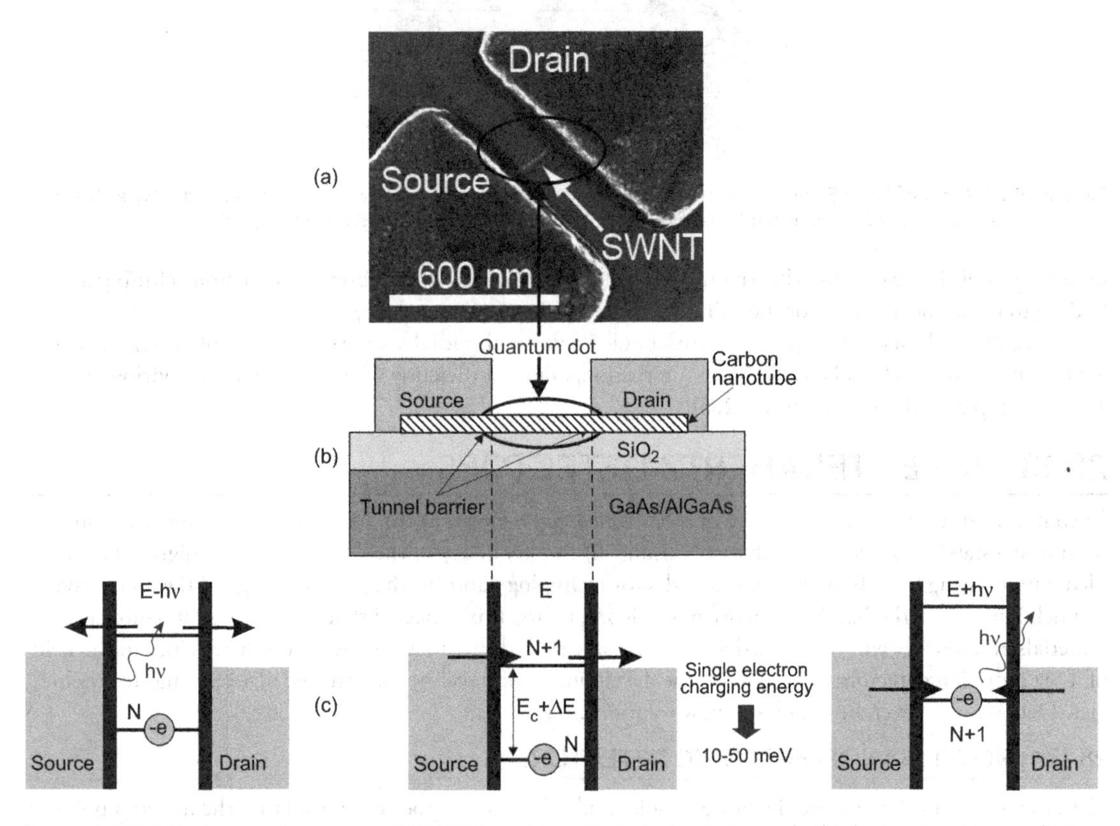

Figure 28.74 THz detector based on a CNT QD transistor: (a) device's photograph, (b) device structure, and (c) schematic diagram of electron tunneling processes in a QD in the presence of an electromagnetic wave. When the Fermi level in the source lead aligns with a level in the QD (see middle panel of (c)), a source-drain current flows via elastic tunneling. When electrons exchange photons, a new current is generated via inelastic tunneling, that is, PAT (see left and right panels). (Adapted after Ref. [289])

presents a schematic diagram of electron tunneling processes in a QD in the presence of an electromagnetic wave. Energy states in the QD can be tuned by changing the electrostatic potential with application of a gate voltage. When the Fermi level in the source lead aligns with a level in the QD (see middle panel of c), a source-drain current flows via elastic tunneling. When electrons exchange photons, a new current is generated via inelastic tunneling, that is, PAT (see left and right panels).

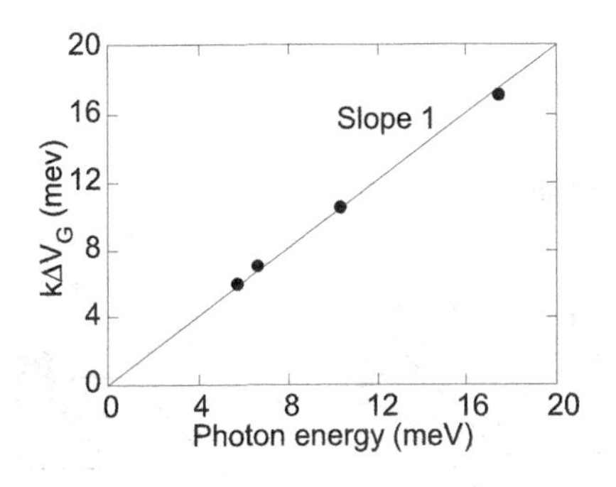

Figure 28.75 CNT QD device: the energy spacing between the original and satellite peaks as a function of the photon energy of the THz wave measured at temperature 1.5 K. (Adapted after Ref. [289])

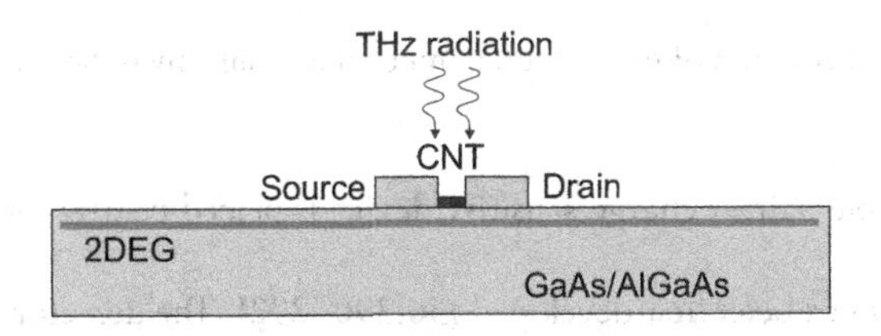

Figure 28.76 Schematic representation of the CNT/2DEG terahertz detector. (Adapted after Ref. [290])

These experimental results provide evidence for PAT in the THz region and the peak position of the satellite currents varies linearly with the THz photon energy (see Figure 28.75). It was thus demonstrated that the CNT QD works as a THz detector with frequency tunability by changing the gate voltage.

Although the frequency-selective THz detection has been achieved, the detection sensitivity was not high. It is due to detection mechanism, where one-photon absorption generates only one electron even if quantum efficiency of 100% is assumed. To resolve this problem, a CNT single-electron transistor (SET) integrated with a GaAs/AlGaAs heterostructure chip having a 2D electron gas (2DEG) [290]. In this hybrid structure, THz absorption takes place in the 2DEG but signal readout in the CNT-SET (see Figure 28.76). The operation principle of this device is that the CNT transistor senses electrical polarization induced by terahertz-excited electron-hole pairs in the 2DEG. The CNT-SET has the source drain electrodes with an interval of about 600 nm and the side-gate electrode, and was operated at 2.5 K. The NEP of this detector is estimated to be 10^{-18}–10^{-19} $WHz^{-1/2}$.

It was also shown that nanotubes can be used as bolometers up to frequencies of ≈2.5 THz [291] as well as the nanoantenna [292,293].

Among a variety of novel THz detection schemes, only semiconductor quantum devices have demonstrated to produce clear single-photon signal against incident THz photons [30,31,294–298]. Komiyama have described and analyzed two types of detectors [31]:

- QD detector, where a QD is electrically polarized by photoexcitation and the induced polarization is sensed by a nearby SET, and
- charge-sensitive IR phototransistor (CSIP) in which an isolated QW island is charged up by photoexcitation and the induced charge is detected by a conducting channel of 2DEG.

28.13.1.1 Quantum dot detectors

In conventional photoconductor (see Figure 28.77a), an electron is excited by one photon, next is carried to the drain, and usually the photoconductivity gain is below 1. Several of the THz single-photon detectors described by Komiyama adopt a different detection scheme. As is shown in Figure 28.77b, a photon absorbed by an isolated small semiconductor island creates an electron which next tunnels out of the island. Losing one electron, the island is positively charged—referred as a "hole". The excited electron outside the island is separated from the hole by a potential barrier and, as a result, yields a long recombination time,

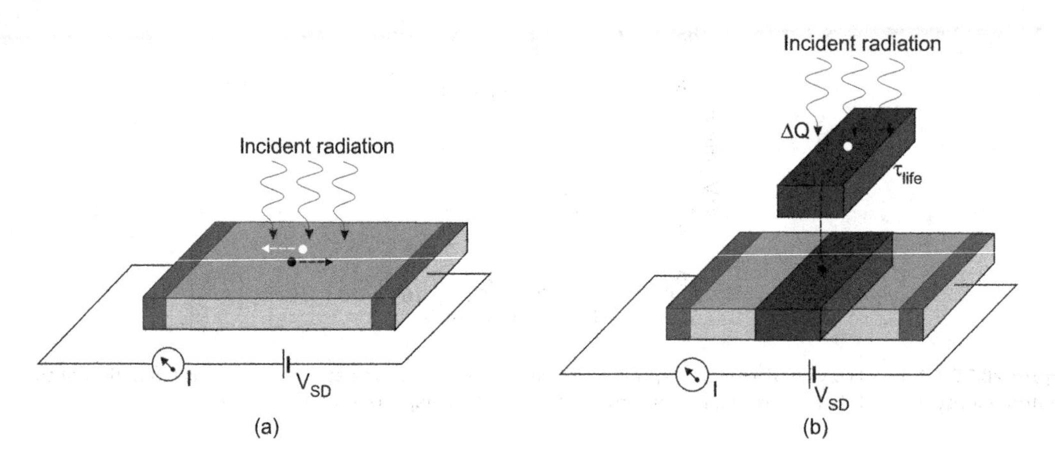

Figure 28.77 Schematic representation of the detection mechanism: (a) conventional scheme, (b) novel scheme. (Adapted after Ref. [31])

τ_{life}, of the exited electron-hole pair. A charge-sensitive device is placed nearby the island to detect the charging-up of the island.

Historically, a single QD has been first developed [30,290–292]. The detector shown in Figure 28.78 is fabricated in GaAs/AlGaAs single heterostructure crystals and utilizes cyclotron resonance excitation in magnetic field. The metal gates forming a SET (Figure 28.78a) extend to the opposite directions. A bowtie antenna (Figure 28.78b) couples incident radiation to the QD.

The most straightforward realization of the QD detector is a double QD (DQD) single-photon detector implemented in a GaAs/AlGaAs heterostructure crystal [296]. Figure 28.79 shows an example of DQD detector, where metal gates (cross gate (CG), barrier gate (BG), floating gate (FG), and antenna) are deposited on top of the crystal surface. About 100 nm below the crystal surface, a heterointerface with a high-mobility 10-nm-thick 2DEG layer, with electron sheet density of $3 \times 10^{11}\,cm^{-2}$, is located. By negatively biasing the gates, the 2DEG is depleted from the regions below the gates, forming an SET (QD1) coupled with QD2. Two electrodes define QD2 and form a planar dipole antenna to couple incident radiation with QD2. The radiation is absorbed by plasma resonance mechanism in which collective oscillation of electrons in the confining potential of QD2 is excited.

The outstanding sensitivity is the distinct advantage of QD detectors. However, the application of these detectors is restricted because of ultralow temperature operation (below 1 K; see Figure 28.80) and sophisticated fabrication technique. It appears also that the spectral range of detection is limited to relatively long wavelengths due to excitation mechanism relying on plasma resonance and magnetoplasma resonance. An attractive solution alternative is the photosensitive field-effect transition fabricated in a double-QW structure, called a CISP [294,295,299].

28.13.1.2 Charge-sensitive IR phototransistors

Figure 28.81 demonstrates operation of the CISP detector. The photoexcited electrons generated via intersubband transition in the upper QW escape out of the upper QW through the tunnel barrier and relax into the lower QW. The upper QW is electrically isolated from the lower QW by negatively biasing the surface

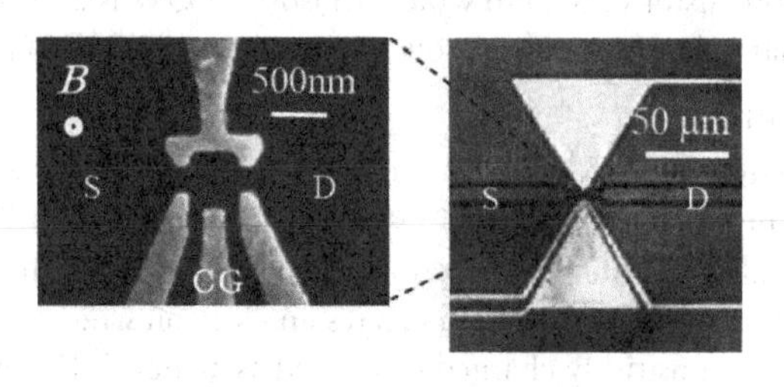

Figure 28.78 Single QD THz detector: (a) closed-up top view, (b) bird view. (Adapted after Ref. [31])

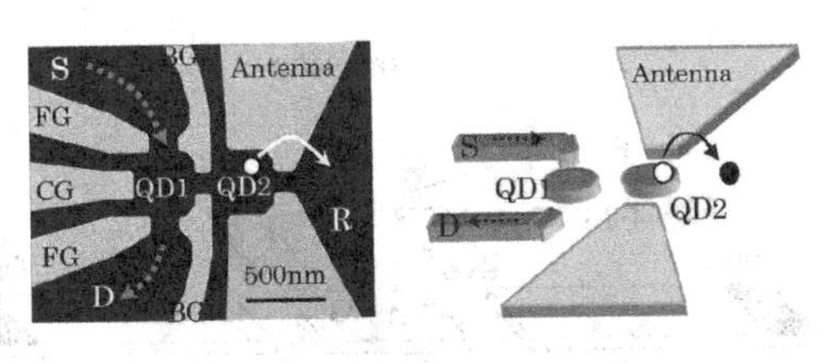

Figure 28.79 Double QD single-photon detector: (a) top view, (b) schematic representation including antennas. (Adapted after Ref. [31])

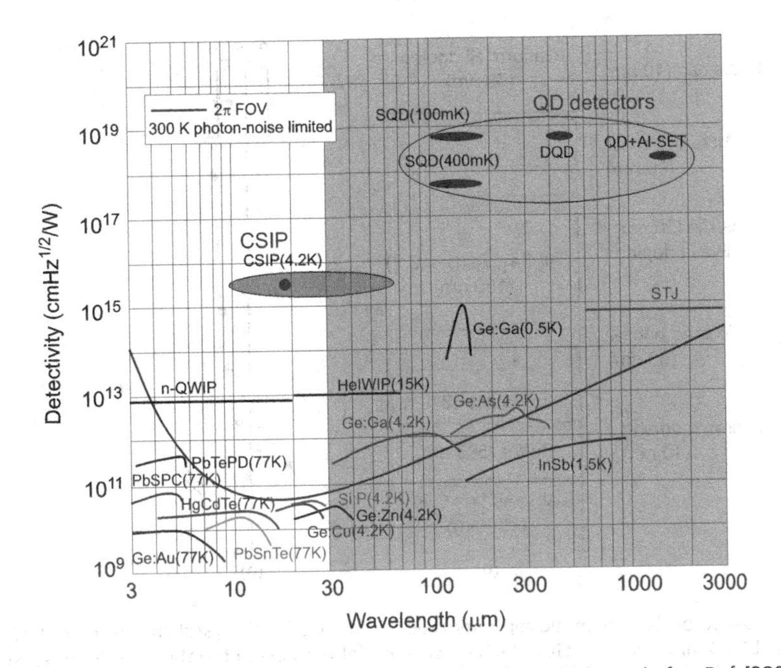

Figure 28.80 Spectral dependence of detectivity of various IR detectors. (Adapted after Ref. [299])

metal gates and then the isolated upper QW is positively charged up due to photexcitation. The gathered positive charge in the isolated upper QW is detected by an increase in conductance through the lower QW, what is shown in Figure 28.81b. Summarizing, the CISP detector operates as a photosensitive FET with a photoactive floating gate served by the upper QW.

The detection sensitivity demonstrated by the QD and CISP detectors is more orders of magnitude superior to any other detectors at low background conditions (see Figure 28.80). In addition, remarkable signal amplitude is a consequence of large current responsivity or the large photoconductive gain. Their low output impedance (about 200 kΩ for QD detectors and 0.1–10 kΩ for CISPs) makes them unique detectors [299]. Small submicrometer CSIPs are promising for integration to large-scale arrays.

A new approach demonstrating possibilities of modern semiconductor nanotechnology to fabricate compact THz detector has been presented by Seliuta et al. [300,301]. This device is an example of a combination of nanostructure physics (GaAs/AlGaAs modulation doped structure) and antennae approach and relies on the non-uniform 2D electron gas heating in external high-frequency fields. Figure 28.82 shows the voltage sensitivity of this structure as a function of frequency. The detector exhibits broad detection bandwidth ranging from microwaves up to THz frequencies (between 10 GHz and 0.8 THz) with nearly constant voltage sensitivity of around 0.3 V W^{-1}.

Detection in the THz region is also demonstrated using quantum rings (QRs) [302,303]. QRs are derived from epitaxially grown self-organized QDs by post growth annealing. Confinement in these nanostructures is stronger than that in dots because of the altered shape. It is found that the QR intersublevel

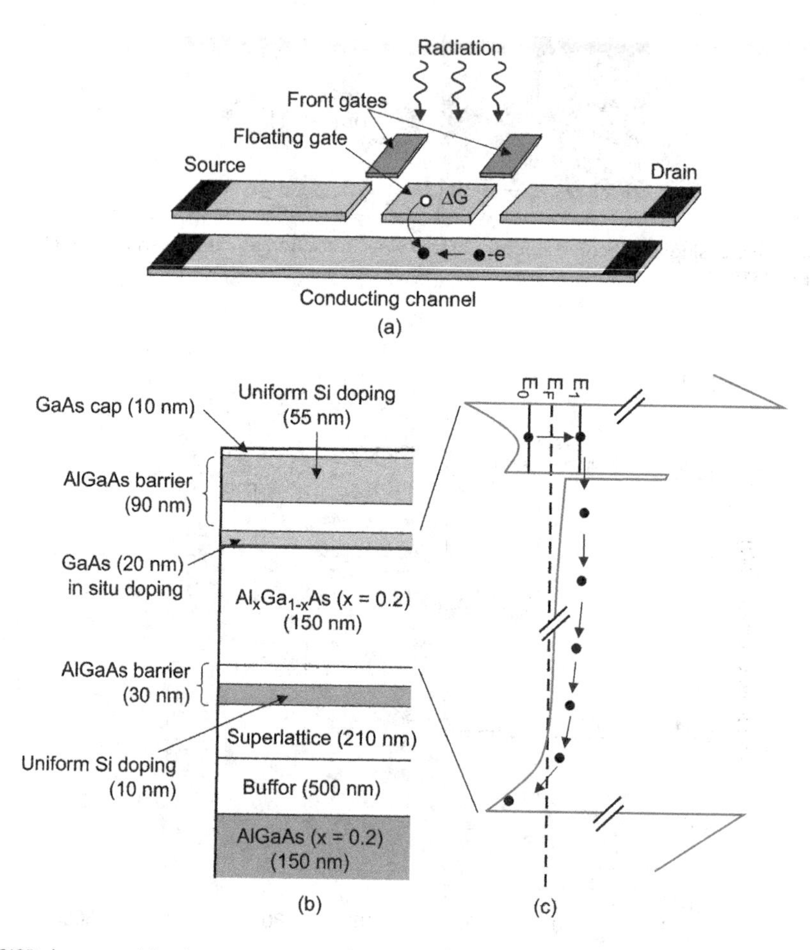

Figure 28.81 CISP detector: (a) schematic representation of the CSIP, (b) crystal structure, and (c) energy diagram. In figure (b), a 2-nm-thick $Al_{0.1}Ga_{0.9}$ As barrier layer (labeled using the thick red line) is just under the 20-nm-thick GaAs QW. (Adapted after Ref. [294])

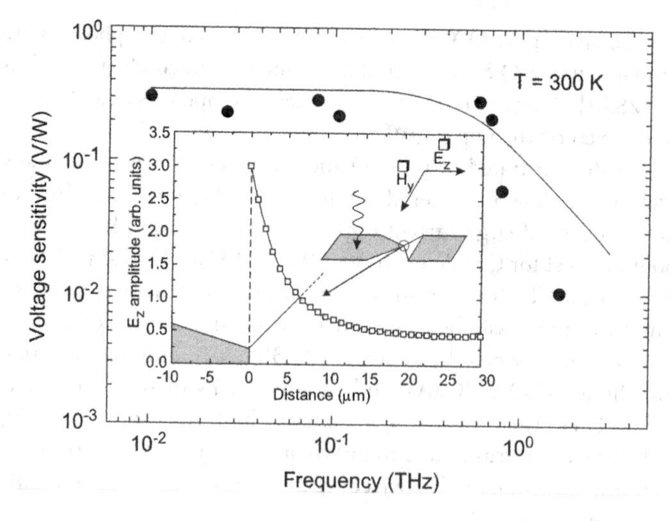

Figure 28.82 Voltage sensitivity as function of frequency for the 2D electron gas bow-tie diode. Dark circles denote experimental data, solid line shows the fit using a phenomenological approach. Inset: distribution of the amplitude of electric field in the active part of the diode exposed to radiation of 0.75 THz. The background of the inset is decorated schematically as the shape of the device. (Adapted after Ref. [301])

detectors exhibit very low dark current and strong response in the 1–3 THz range, with the peak response measured at 1.82 THz (165 μm) in the temperature range of 5–10 K. This detection peak is characterized by a peak responsivity of 25 A W^{-1} and specific detectivity of 1 × 10^{16} Jones [303].

28.13.2 GRAPHENE DETECTORS

Graphene has been extensively studied since 2004 due to its unique electronic and optical properties [304–306]. The most intriguing electronic property of graphene is its linear dispersion relation between the energy and the wave vector. This relativistic-like energy dispersion is accompanied by electrons travelling at a Fermi velocity only 100 times smaller than the speed of light.

28.13.2.1 Relevant graphene properties

Graphene is made out of sp^2-hybridized carbon atoms arranged on a honeycomb lattice with lattice constant $a = 1.42$ Å. The formed valence and conduction bands touch at the Brillouin zone corners (so-called Dirac points) making graphene a zero-band-gap semiconductor, as shown in Figure 28.83. Due to the zero density of states at the Dirac points, electronic conductivity is actually quite low. However, the Fermi level can be changed by doping (with electrons or holes) to create a material that is potentially better at conducting electricity than, for example, copper at room temperature. Carbon atoms have a total of six electrons: two in the inner shell and four in the outer shell. The four outer shell electrons in an individual carbon atom are available for chemical bonding; but in graphene, each atom is connected to three other carbon atoms on the two-dimensional (2D) plane, leaving one electron freely available in the third dimension for electronic conduction. These highly mobile electrons are called π-electrons and are located above and below the graphene sheet. These π-orbitals overlap and help to enhance the carbon-to-carbon bonds in graphene. Fundamentally, the electronic properties of graphene are dictated by the bonding and antibonding (the valance and conduction bands) of these π-orbitals.

Graphene boasts the potential for ballistic carrier transport with an inferred mean free path >2 μm at room temperature. Its carriers propagate via diffraction (like optical light in a waveguide), rather than by carrier diffusion as is common with carriers in conventional semiconductors.

The electrical engineers are interested in the high carrier mobility and saturation velocity in graphene what offer the promise of graphene-based high-speed photonic devices [307]. Graphene layer structures with long momentum relaxation time of electrons and holes promise a significant enhancement of the performance of future optoelectronic devices. Theoretically, graphene possesses a room temperature electron mobility of 250,000 cm^2 Vs^{-1}; however, the transport behavior is extremely dependent on the local environment and processing done to the material. Graphene achieved via exfoliation is characterized by extremely high carrier mobilities >200,000 cm^2 Vs^{-1} at room temperature. Unfortunately, these films have a very

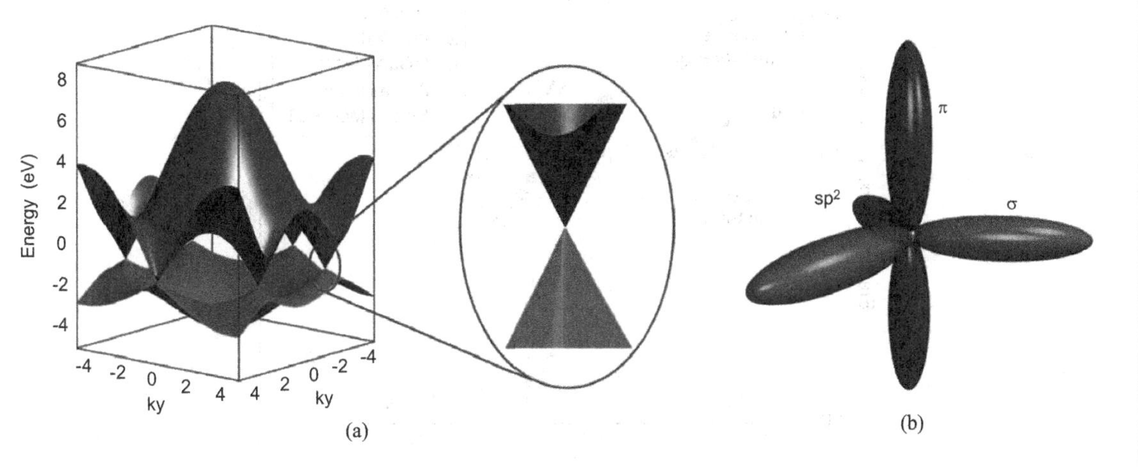

Figure 28.83 (a) The band structure of graphene in the honeycomb lattice. The enlarged picture shows the energy bands close to one of the Dirac points. (b) Schematic of electron σ- and π-orbitals of one carbon atom in graphene.

small area (100 μm^2), and this makes it expensive for industrial applications. When placed on a substrate, charged impurity scattering and remote interfacial phonon scattering reduce mobility (see Figure 28.84). On SiO_2, interfacial phonon scattering limits the graphene mobility to 40,000 cm^2 Vs^{-1} [308]. Atmospheric exposure and processing contaminants such as resist residue, water, and metallic impurities act as scattering sources and degrade the mobility.

Other features which make graphene of interest is its high thermal conductivity (about 10× cooper and 2× diamond) and high conductivity (about 100× copper). It is also characterized by high tensile strength (130 GPa, compared to 400 MPa for A36 structural steel).

Unlike metals with abundance of free charges, graphene is a semimetal, where carriers can be induced through chemical doping or electrical gating with great ease due to its 2D nature. In such way, the doping concentration from 10^{12} to 10^{13}cm^{-2} can be obtained, which is significantly smaller than that of 1 per noble metals. Therefore, semimetallic nature of graphene allows for electrical tenability not possible with conventional metals.

Also optical properties of graphene are fascinating [309]. Its optical conductivity is universal conductance, $\pi\beta$, where β is equal $(1/4\pi\epsilon_o)(e^2/\hbar c)s$, e is the electron charge, $\hbar$ is Planck's constant, and c is the speed of light. This provides graphene with broadband (visible and infrared) linear absorption of 2.3% per monolayer. For its 0.33-nm continuous monolayer thickness, it absorbs roughly 2.3% of the incident light what makes it 10–1,000× more absorbing than semiconductors like Si and GaAs, while covering a much broader spectral bandwidth.

Band gap structure of graphene can be modified in different ways: by the addition of multiple layers as shown in Figure 28.85a, by substitutional doping (Figure 28.85b), by the addition of two layers (Figure 28.85c), and doping bilayer (Figure 28.85d). The doping of graphene layer can move the Fermi level either up or down what results in a decrease in the carrier mobility (both electrons and holes). The thickness restriction for graphene creates large resistance and chemical inertness, making its use for pure conductive applications less attractive.

Graphene has the highest specific interaction strength (absorption per atom of material) known. Silicon has typically a 10-μm absorption depth which causes 2.3% of the light to be absorbed in a 200-nm thickness as opposed to the same optical absorption in a much smaller 0.3-nm thickness (interplane spacing) of graphene.

Figure 28.86 illustrates a typical absorption spectrum of doped graphene [310]. In terahertz region, in energy range below $2E_F$, the absorption is mainly imparted to a Drude peak response. In doped graphene

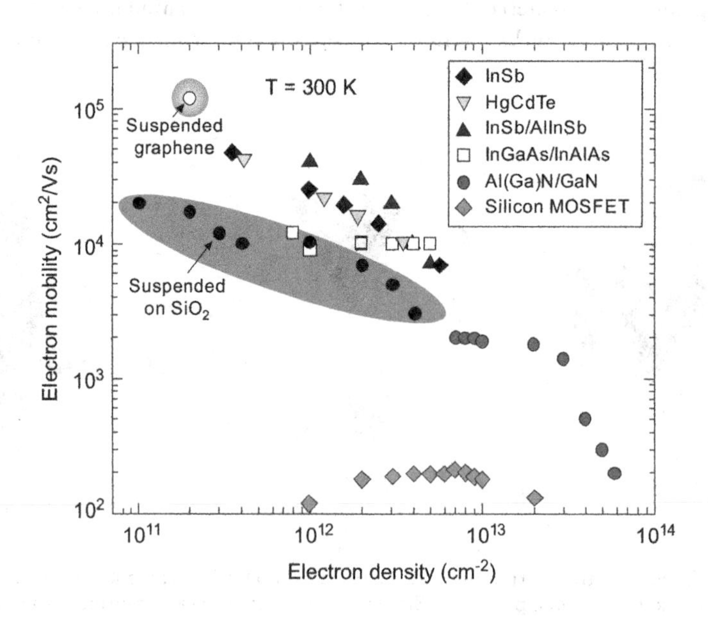

Figure 28.84 Electron mobility in graphene at room temperature in comparison with other material systems.

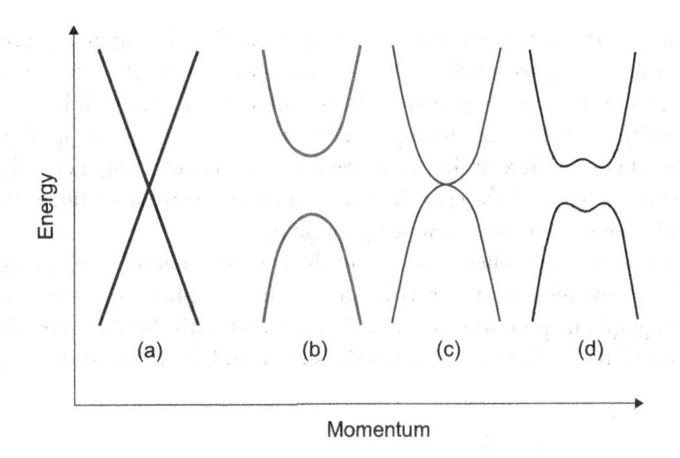

Figure 28.85 Modification of graphene's band gap structure: (a) the Dirac Fermi cone, (b) substitutional doping, (c) bilayer graphene, and (d) doped bilayers.

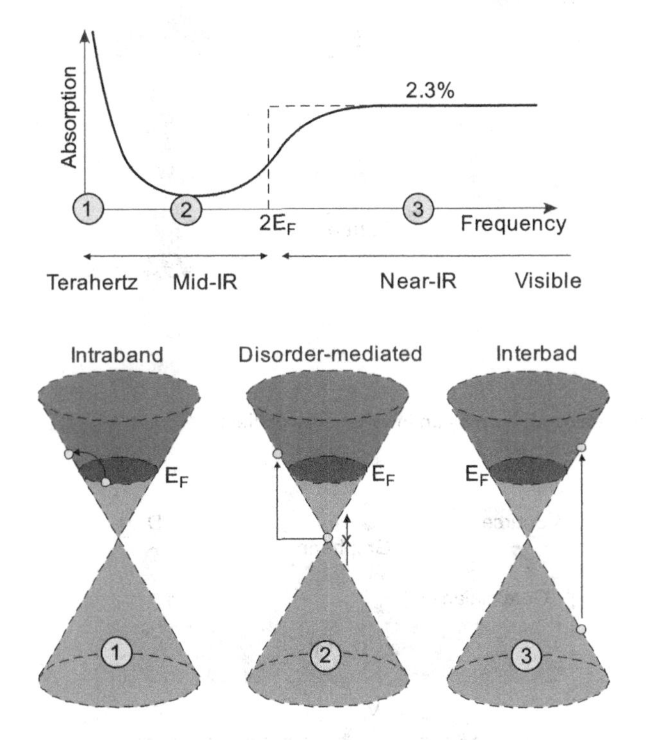

Figure 28.86 Typical absorption spectrum of doped graphene. (Adapted after Ref. [310])

in the mid-infrared region, the optical absorption is minimal, and the residual absorption is generally attributed to disorder in imparting the momentum for the optical transition. A transition occurs around $2E_F$, where direct interband processes lead to a universal 2.3% absorption.

28.13.2.2 Photodetection mechanisms in graphene detectors

Graphene detectors can be separated into two separate categories: thermal (the bolometer effect) and photon detectors (the photovoltaic effect). Another more recent study utilized the photo-thermoelectric effect (Seebeck effect) to create a net electric field due to electron diffusion into dissimilar metal contacts. In the following texts, we describe these mechanisms.

Photovoltaic (PV) photocurrent generation is based on the separation of photogenerated electron-hole pairs by built-in electric fields at junctions between positively (p-type) and negatively (n-type) doped regions of graphene (see Figure 28.87). The same effect can be achieved by applying a source-drain bias voltage, producing an external electric field. In the last case however, this effect is generally avoided, since graphene is a semimetal and therefore it generates a large dark current. The built-in field can be introduced in different ways: either by local chemical doping, electrostatically by the use of (split) gates, or by taking advantage of the work-function difference between graphene and a contacting metal. Typically, p-type doping is achieved for metals with a work function higher than that of intrinsic graphene (4.45 eV), whereas the graphene channel can be adjusted to p- or n-state by the gate.

Electron-electron scattering in graphene can lead to the conversion of one high-energy e-h pair into multiple e-h pairs of lower energy, what potentially can enhance the photodetection efficiency [311].

Figure 28.88 shows graphene phototransistor design together with the short-circuit current induced by light. In the absence of an applied bias between the source and the drain, minimal photocurrent is

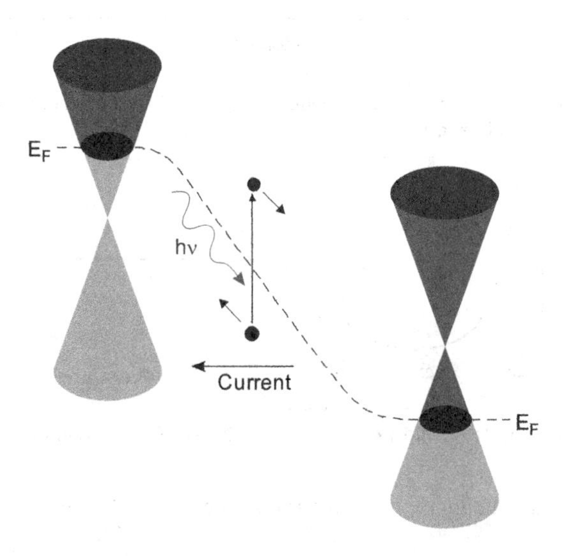

Figure 28.87 Separation of electron hole by an internal electric field.

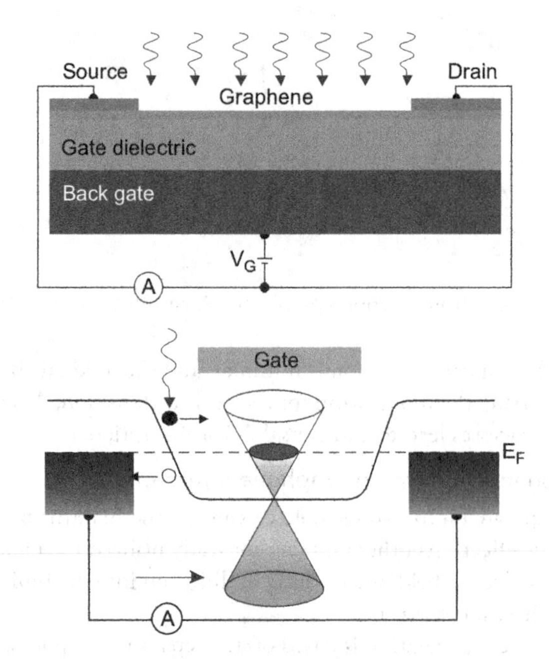

Figure 28.88 Graphene phototransistor: (a) structure of transistor, (b) a schematic view of photocurrent generation.

recorded when the light spot is focused on the middle of the graphene channel. Significant photocurrent is observed when the light is incident on the metal–graphene interface area, what is attributed to the conventional PV effect. The built-in electric field in graphene (due to different work functions of the graphene and metal) separates electron-hole pairs, and hence photocurrent is created in the external circuit. In the middle of the channel, there is no built-in electric field, and as a result, no photocurrent is observed. The built-in electric field can be further adjusted by a gate bias, what influences on the value the photocurrent.

The photothermal electric effect (PTE—Seebeck effect) also plays an important in photocurrent generation [310,311]. Because the optical phonon energy in graphene is large (~200 meV), hot carriers created by the radiation field remain at a temperature higher than that of the lattice for many picoseconds. Equilibration of the hot electrons and the lattice occurs via the slower scattering between charge carriers and acoustic phonons (nanosecond timescale), although they are a substantial speedup attributed to disorder-assisted collisions. Incident radiation of the light spot induces carrier temperature variations, and hot carriers generated by photons diffuse due to the temperature gradient, leading to photocurrent generation, as shown in Figure 28.89—the carrier and lattice can have different temperatures. The polarities of photocurrent due to both PV and PTE effects are the same, making the experimental determination of their relative contributions difficult.

Ryzhii et al. have proposed to utilize multiple graphene layer structures with lateral p-i-n junctions for THz detection [312,313]. The p- and n-regions of the structure are formed due to the voltages, V_p and V_n, and i-region consists of several graphene layers (see Figure 28.90a). As in customary p-i-n photodiodes, the electrons and holes photogenerated in the i-region induce the terminal current constitutes for output electric signal. It was predicted that these structures can exhibit high responsivity and detectivity in the terahertz region at room temperatures. Due to relatively high quantum efficiency and low thermal generation rate, the photodetectors can substantially surpass other THz detectors.

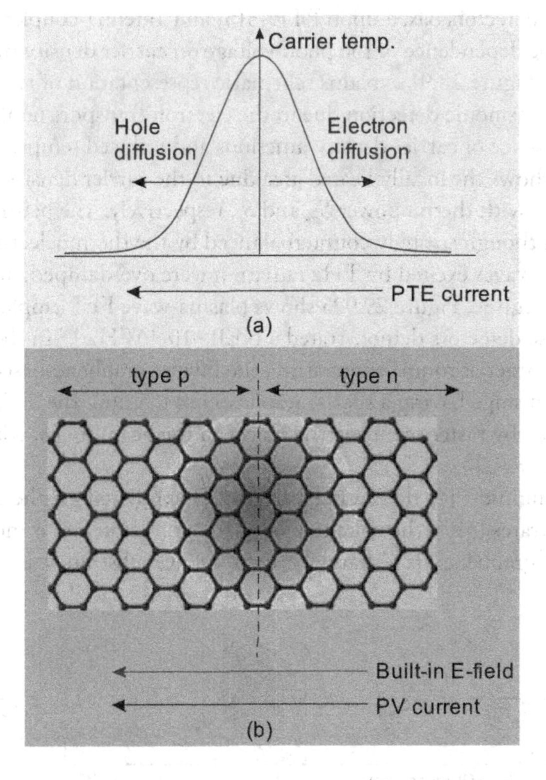

Figure 28.89 Photocurrent generation in a graphene p-n junction: (a) profile of carrier concentrations due to light intensity distribution, (b) built-in electric field of p-n junction as well as photothermal electric effect lead to PV current flowing from n-type region to p-type region. (Adapted after Ref. [307])

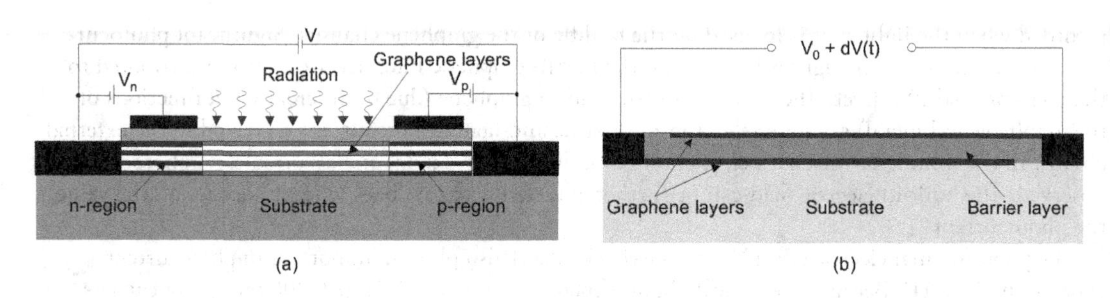

Figure 28.90 Structure of the graphene-based photodetectors: (a) with electrically induced p-i-n junction, and (b) resonance-based detector.

Another design of graphene detector is a resonant structure of two graphene sheets separated by a dielectric to tune the photon wavelength of absorption as shown in Figure 28.90b. The responsivity of the detector exhibits the resonant peaks when the frequency of incoming terahertz radiation approaches the resonant plasma frequencies. These frequencies are tuned by the bias voltage [314]. The pronounced resonant response of the detector requires the frequency of electron and hole collisions with impurities and acoustic phonons be sufficiently low.

Graphene can be also used for the detection of terahertz radiation as a field effect transistor (FET)—see Section 28.6.4. In this detection scheme, a DC voltage is generated in response to an oscillating radiation field. In the so-called resonant regime, the plasma waves are weakly damped (when a plasma wave launched at the source can reach the drain in a time shorter than the momentum relaxation time) the detection mechanism exploits interference of the plasma waves in the cavity, which results in a resonantly enhanced response [151]. Broadband detection occurs when plasma waves are overdamped, when plasma waves launched at the source decay before reaching the drain.

Room-temperature THz detectors based upon PTE [315] and antenna-coupled graphene FETs [316–318] have been demonstrated. The dependence of the photovoltage on carrier density in the FET channel also displays PTE contributions. Figure 28.91 explains schematic representation of two competitive independent detection mechanisms: the plasmonic detection due to the electron transport nonlinearity and the thermoelectric effect due to the presence of carrier density junctions and induced temperature gradient across the FET channel. Shaded area shows the locally heated area due to the carrier density junction at the interface of ungated and gated regions with thermopower S_{ug} and S_g, respectively. The plasma wave detection is the dominant mechanism, even though strongly counterbalanced by the thermoelectric response.

In Ref. [316], the plasma waves excited by THz radiation were overdamped, and thus the detectors did not operate in the resonant regime. Figure 29.92a shows plasma-wave FET employing a top-gate antenna-coupled configuration. These detectors demonstrated a NEP ~10^{-9} WHz$^{-1/2}$ in the range 0.29–0.38 THz. During measurements of a target at room temperature, the bilayer-graphene-based FET at $V_g = 3$ V was mounted on a x-y translation stage having a spatial resolution of 0.5 μm. The THz image consists of 200 × 550 scanned points, collected by raster scanning the object in the beam focus, with integration time of 20 ms per point (see Figure 28.92b).

In comparison with plasmonic THz detectors fabricated by exfoliated graphene [317] or CVD graphene transferred on Si/SiO_2 substrates [319], the epitaxial graphene grown on SiC is more promising [318]. The photoresponsivity of bilayer graphene FET channel grown on SiC substrate was estimated in ≈ 0.25 V W^{-1} and NEP ≈ 80 nW Hz^{-2}.

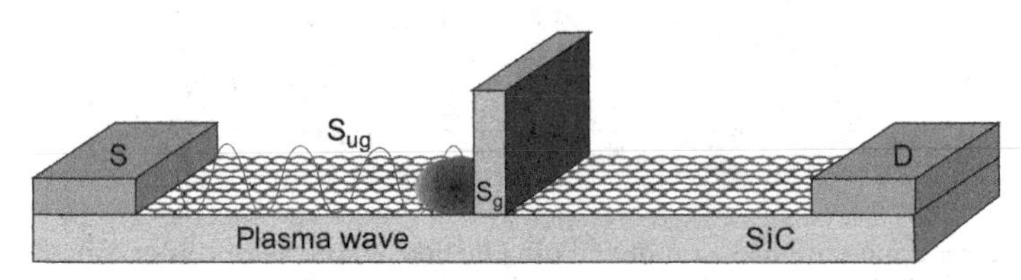

Figure 28.91 Schematic representation of the detection mechanism in graphene FET THz photodetector.

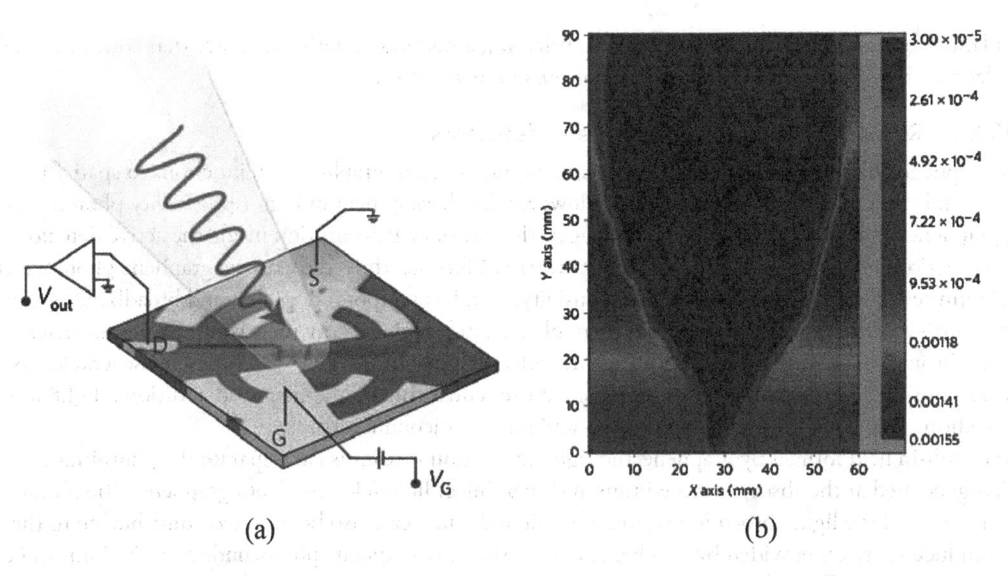

Figure 28.92 Plasma-wave FET terahertz detector: (a) schematics of the THz detection configuration in a FET embedding the optical image of the central area of a bilayer graphene-based FET, and (b) 0.3 GHz transmission mode image of a leaf. (Adapted after Ref. [316])

The key parameters of a bolometer are the thermal resistance and the heat capacity (see Section 4.1). Graphene has small volume for a given area and low density of states, which results in low heat capacity, thus a fast device response. The cooling of electrons by acoustic phonons is inefficient (owing to the small Fermi surface), and cooling by optical phonons requires high temperature ($kT > 200$ meV). Thus, thermal resistance is relatively high, giving rise to high bolometric sensitivity [311,320].

Two types of graphene-based bolometer are shown in Figure 28.93. Yan et al. have considered graphene as a hot-electron bolometer [321]. The device structure is shown in Figure 28.93a. Due to weak electron–phonon interaction, they used bilayer graphene, which has a tunable bandgap. Application of perpendicular electric field gives rise to electron-temperature-dependent resistance at low temperature, making the device suitable for thermometry. The extrapolated NEP value for a 1 μm² sample at 100 mK is about 5×10^{-21} WHz$^{-1/2}$, similar to state-of-the-art transition edge sensor (TES) bolometer. The schematic view of graphene-based detector with temperature coefficient of resistance above 4%/K is shown in Figure 28.93b, where the pyroelectric response of a $LiNbO_3$ crystal is transduced with high gain (up to 200) into

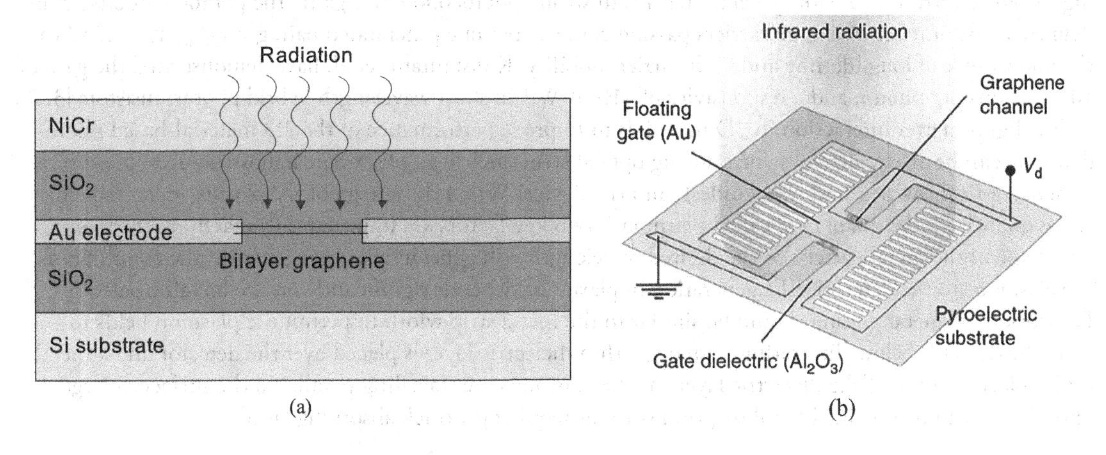

Figure 28.93 Schematic view of graphene bolometers: (a) side view of the bilayer graphene hot-electron bolometer (semitransparent NiCr top gate covers the graphene device and silicon oxide surrounds the graphene), (b) pyroelectric bolometer (conductance of graphene channel is modulated by the pyroelectric substrate and by a floating gate).

modulation for graphene. This is achieved by fabricating a floating metallic structure that concentrates the pyroelectric charge on the top-gate capacitor of the graphene channel.

28.13.2.3 Responsivity enhanced graphene detectors

Most graphene photodetectors utilize graphene-metal junctions or graphene p-n junctions to spatially separate and extract photogenerated carriers. However, for development of high-responsivity photodetectors, graphene represents the two major challenges: the low optical absorption inside the active detector's region junctions (~100–200 nm) and short photocarrier lifetime. Thus, the existing graphene photodetectors remain limited by tradeoff between high responsivity, ultrafast temporal response, and broadband operation.

Responsivity enhanced photodetection in graphene can be achieved by increasing the photocarrier lifetime through band-structure engineering and defect engineering. Table 28.11 presents several kinds of novel responsivity enhanced photoconductor structures consisting of graphene and additional light absorption mediums (e.g., quantum dots, nanowires, and bulk semiconductors).

The built-in field formed by graphene and light absorption mediums can separate the photoinduced carriers generated at the absorption mediums and then inject holes/electrons into graphene. The photoresponse beyond the light absorption region of semiconductors can also be detected contributing to the photo-induced carriers provided by graphene. Unlike the pure graphene photoconductor, the built-in field at the interface can efficiently separate the photo-induced carriers and prolong their lifetime, resulting in the relatively high responsivity.

Figure 29.94 explains schematically the differences between pure graphene photoconductor and hybrid photoconductor. This comparison concerns also differences between ultrafast and ultrasensitive graphane photodetectors. In the early reports, photocurrent was generated by local illumination of one of the metal/graphene interfaces of a back-gated graphene FET. An asymmetric metallization scheme was used to break mirror symmetry of the built-in potential profile within channel, allowing giving the overall photocurrent. Interdigitated metal fingers were used, leading to the creation of greatly enlarged, high electric field, light-detection region (see Figure 28.94a). The high carrier mobility and short carrier lifetime in graphene (see Figure 28.94c) allow metal-graphene-metal photodetectors to operate at high data rates.

The main feature of the hybrid photodetector (see Figure 29.94d) is its ultrahigh gain, which originates from the high carrier mobility of the graphene sheet and the recirculation of charge carriers during the lifetime of the carriers that remain trapped in the quantum dots (Figure 28.94e); also other light-absorbing media (e.g., carbon nanotubes and nanoplates) can be used. Photoexcited holes in the quantum dots are transferred to the graphene layer and drift by means of a voltage bias V_{DS} to the drain, with a typical timescale of transit, $\tau_{transit}$, which is inversely proportional to the carrier mobility. Electrons remain trapped (with a typical lifetime, $\tau_{lifetme}$) in the quantum dots. Multiple circulation of holes in the graphene channel following a single electron-hole photogeneration leads to strong photoconductive gain. The photoconductive gain defined as the number of charge carriers passing contacts per one generatated pair, $g = \tau_{lifetme}/\tau_{transit}$, indicate the importance of long lifetime and high carrier mobility. Konstantatos et al. have demonstrated the gain of 10^8 electrons per photon and a responsivity of ~10^7 A W^{-1} in short wavelength hybrid phototransistors [322].

The light–matter interaction in 2D materials to improve performance of the 2D material-based photodetectors can be also realized by introducing optical structures (e.g., plasmonic nanostructures, photonic crystals, optical cavities, and waveguides) onto the device. When the energy of the plasma-wave frequencies is quantized, the quanta are called plasmons. Two key factors are important: (i) matching the size and shape of the metal pattern so the desired wavelength will generate plasmons, and (ii) the coupling of the plasmons to the detector. The generation of plasmons depends significantly on the metallic pattern. Dimensions of metal grating should be similar to the metal strip width to permit the plasmon fields to enter the detector below the grating. Usually, a thin dielectric layer is placed over the detector and a metal grid is placed on top of the dielectric layer. As the plasmons are travelling parallel to the surface, a large optical path can be obtained for absorption without requiring a thick absorbing layer.

28.13.2.4 Related 2D material detectors

The performance of photodetectors is mainly dependent on the inherent characteristics of the photodetector's active materials, such as absorption coefficient and lifetime of electron-hole pair and charge mobility.

Table 28.11 Responsivity enhanced graphene detectors

	ADVANTAGES	DISADVANTAGES	REFERENCES
Hybrid graphene–quantum dots detector 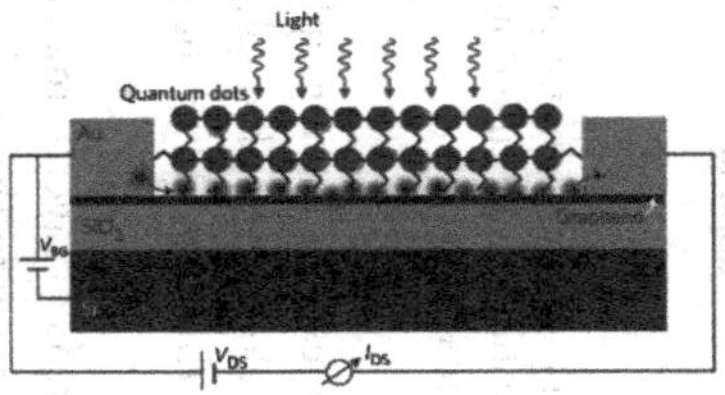	Increasing absorption and introducing large carrier multiplication factors	Bandwidth and response time are restricted by the narrow spectral bandwidth and long carrier trapping times of the quantum dots	[322–325]
Two graphene layers separated by a thin tunnel barrier 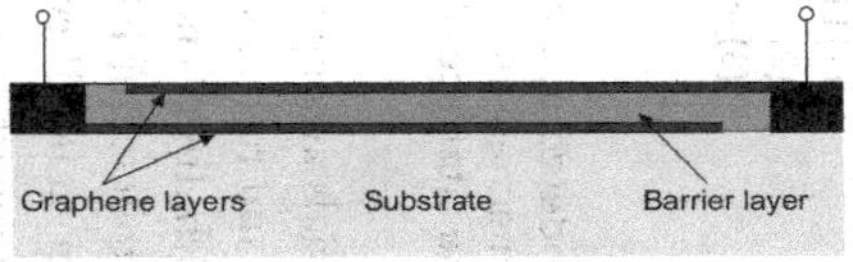	Broadband responsivity via separation of the photogenerated electrons and holes through quantum tunneling and minimization of their recombination	Response times limited by the long carrier trapping times in the tunneling barriers utilized	[314,326]
Waveguide-integrated graphene detector 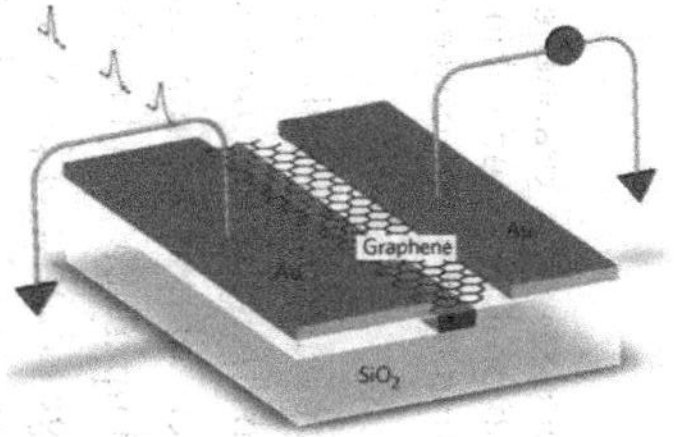	Ultrafast responsivity by increasing the interaction length of light within graphene and process compatibility with standard photonic integrated circuits	Spectral bandwidth restricted by the bandwidth limitations of the waveguides utilized	[327–330]
Microcavities, plasmonic structures, and optical antennae integrated with graphene 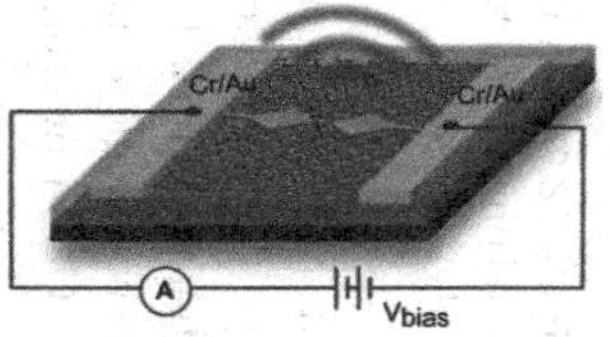	High responsivities by increasing the interaction length of light within graphene	Bandwidth is limited by the resonant nature of the structures utilized	[310,331–345]

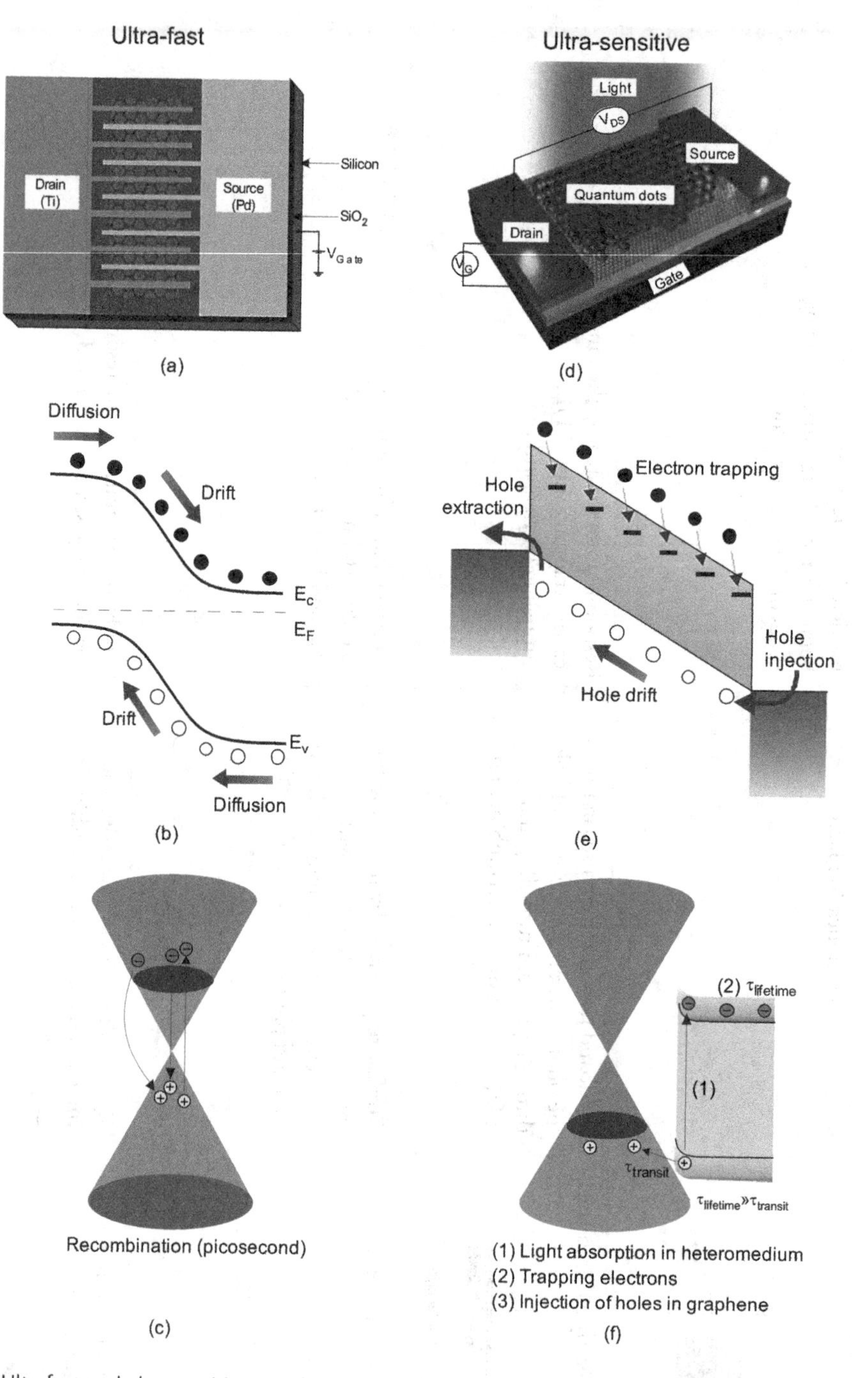

Figure 28.94 Ultrafast and ultrasensitive graphene photodetectors: (a) schematic structure of metal-graphene-metal photodetector, (b) band profile, (c) recombination mechanism, (d) hybrid graphene/quantum dots photodetector, (e) trapping process, and (f) dynamic process at the interface of graphene/quantum dots.

The high dark current of conventional graphene materials arising from the gapless nature of graphene significantly reduces the sensitivity of photodetection and restricts further developments of graphene-based photodetectors. The discovery of new 2D materials with direct energy gaps in the infrared to the visible spectral regions has opened up a new window for photodetector fabrication.

Graphene is one of a large number of possible 2D crystals. There are hundreds of layered materials that retain their stability down to monolayers, and whose properties are complementary to those of graphene

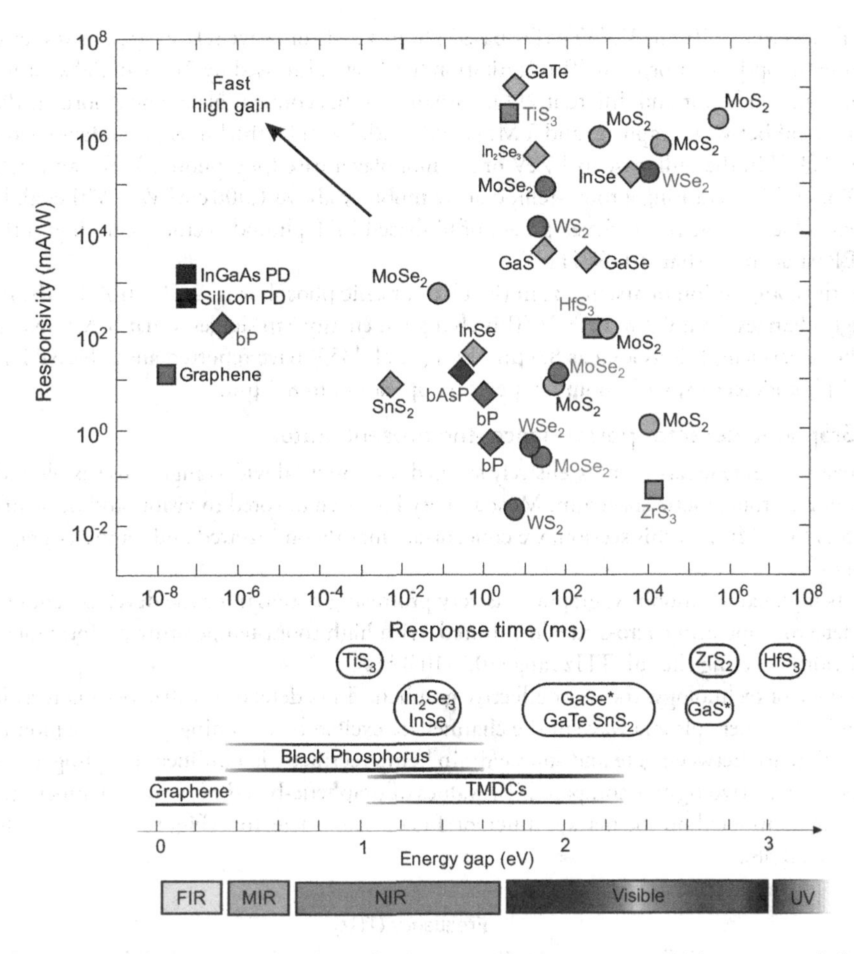

Figure 28.95 Responsivity against response time for 2D materials in comparison with commercial silicon and InGaAs photodiodes. At the bottom, bandgap of the different layered semiconductors and electromagnetic spectrum are shown. The exact bandgap value depends on the number of layers, strain level, and chemical doping. The asterisk indicates that the material's fundamental bandgap is indirect. FIR, far infrared; MIR, mid infrared; NIR, near-infrared; UV, ultraviolet. (Adapted after Ref. [348])

[346–350]. Even though the technology readiness levels are still low and device manufacturability and reproducibility remain a challenge, 2D materials technology can be found in research labs around the globe including materials like silicene, germanene, stanene, phosphorene, transition metal dichalcogenides (TMDs), black phosphorus, and recently discovered all-inorganic perovscites. Compared to graphene, TMDs like molybdenum disulfide (MoS_2), tungsten disulfide (WS_2), and molybdenum diselenide ($MoSe_2$) exhibit even higher absorption in the visible and the near-infrared range, i.e., above their respective energy band gaps, which makes this 2D material class a candidate to act as the thinnest photo-active materials, 2D semiconductors cover a very broad portion of the spectrum from the infrared to the ultraviolet. Figure 28.95 shows the responsivity against response time for 2D materials (graphene-based devices and TMDs). It can be seen that photodetectors based on semiconducting layered materials display a large (about 10 orders of magnitude) variation in their responsivity. Their response time is considerably longer than commercial silicon and InGaAs photodiodes and is longer than ~1×10^{-2} ms. The slow response is attributed to traps and enhanced capacitance. However, there are multiple ways to reduce trap times and densities.

TMDs show a direct, finite band gap (0.4–2.3 eV) which endows them with a very high carrier density tunability (on/off current ratios in a transistor up to 10^{10}), but limits the carrier mobility to relatively low values, typically less than 250 cm^2 Vs^{-1}. TMDs-based phototransistors have relatively poor responsivity owing to their weak optical absorption. This limitation has been addressed by 2D material synthesized with other material (semiconductor and different 2D materials). In this context, black phosphorus (bP), appears as a natural trade-off between graphene and TMDs [351,352]. With its thickness dependent energy gap, spanning from 0.3 eV in the bulk case to 1.7 eV of the monolayer case (phosphorene), bP can reach on/off ratio up to 10^5 in a FET, retaining a room-temperature mobility above 1,000 cm^2 Vs^{-1}. Viti et al. [352] have reviewed recent achievements in the development of bP-based FET photodetectors operating in the 0.3–3.4 THz with NEP value lower than 10 nWHz$^{-1/2}$.

By varying the composition of arsenic, x, in the black arsenic phosphorus As_xP_{1x} (b-AsP), the band gap correspondingly changes from 0.3 to 0.15 eV. This change in energy gap suggests that b-AsP may interact with light, whose wavelength is as long as 8.5 μm. Long et al. [353] have reported about b-AsP long-wavelength IR photodetectors, with room temperature operation to 8.2 μm.

28.13.2.5 Graphene detector performance–the present status

Since its discovery, graphene has been extensively studied for potential wide-ranging use as photodetector in wide range of electromagnetic spectrum. Most activity has been devoted to visible and near-infrared photon detectors [311,346]. In this section, we concentrate mainly on infrared and terahertz graphene photodetectors.

Owing to its high carrier mobility, graphene is very promising material for the development of room-temperature detectors operating across the far infrared, with high room-temperature performance for high spectral bandwidth covering the full THz range (0.1–10 THz).

At present stage of technology, the most effective graphene THz detectors utilize plasma rectification phenomena in FETs, where plasma wave in the channel are excited by incoming THz wave modulating the potential difference between gate and source/drain being rectified via nonlinear coupling and transfer characteristics in FET. Two figures compare NEP values of graphene-based FET room-temperature detectors with existing dominated on the market different THz photon detectors (Figure 28.96) and thermal detectors (Figure 28.38).

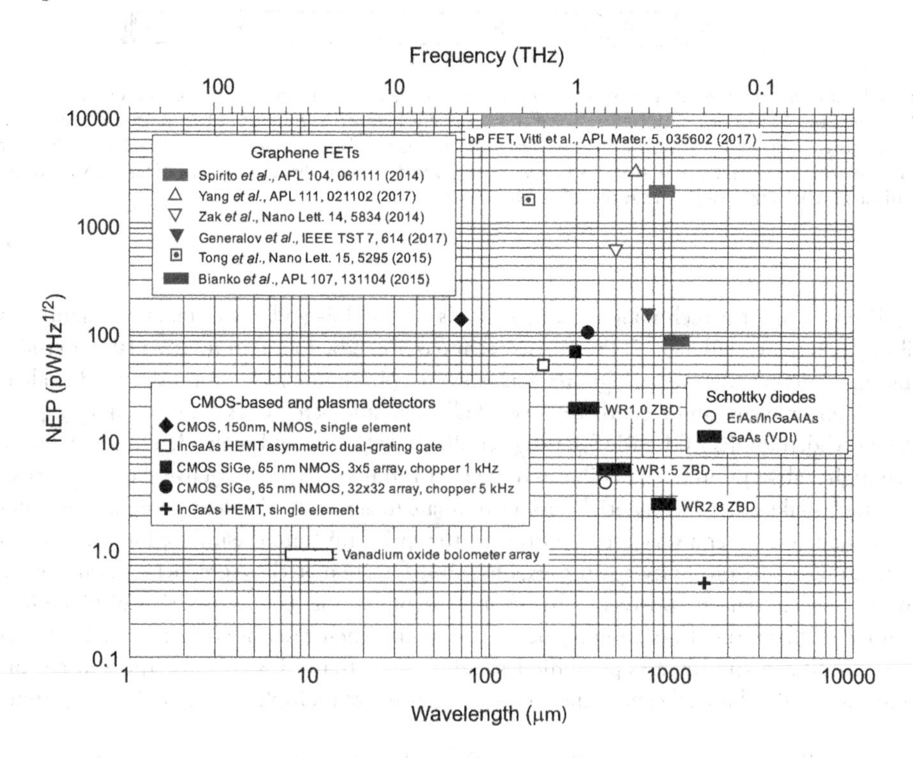

Figure 28.96 Spectral dependence of NEP for graphene FET detectors and different photon THz (CMOS-based, Schottky diodes).

Most experimental data gathered in literature is given for graphene detectors operated above wavelength of 100 μm (frequency below 3 THz). Generally, the performance of graphene FET detectors is poorer w comparison with CMOS-based and plasma detectors fabricated using Si-, SiGe-, and InGaAs-based materials. However, in comparison with vanadium oxide and amorphous silicon microbolometers, the performance of graphene detectors is close to trend line estimated for microbolometers (see Figure 28.38). It should be marked here, that microbolometer data are addressed to monolithic arrays. In this case, an important issue for a focal plane arrays (FPA) is pixel uniformity. It appears that the production of monolithically integrated THz detector arrays encounters so many technological problems that the device-to-device performance variations and even the percentage of nonfunctional detectors per chip tend to be unacceptably high.

As is shown in Chapter 4, the ratio of the absorption coefficient to the thermal generation rates, α/G, is the main figure of merit of any materials for infrared photon detectors. The thermal generation rate is inversely proportional to the recombination lifetime. Graphene is an attractive material for optical detection due to its broad absorption spectrum and ultrashort response time. However, it remains a great challenge to achieve high responsivity in graphene detectors because of weak optical absorption (only 2.3% in the monolayer graphene sheet) and short carrier lifetime (<1 ps). In other words, the applications of graphene-based photodetectors are limited in comparison to traditionally detectors. Various approaches have been proposed to enhance sensitivity by introduction of a bandgap, electron trap layers (quantum dot structures), or nanoribbons; however, these methods degraded the electronic performance of graphene such as the high mobility. Despite the large amount of funding and research invested on 2D materials, there is currently very limited set of 2D material covering the infrared and terahertz regions of spectrum.

Advanced imaging systems require arrays of detectors. Development of a scalable graphene detector array concept with appropriate fabrication tools is a noteworthy indicator that the graphene will enable new detector systems. The combination of scalability, the prospects for integration with Si-platforms, and the potential for implementing flexible devices can make graphene competitive for future generation of detection systems.

REFERENCES

1. https://tematys.fr/Publications/en/terahertz/39-terahertz-components-systems-technology-and-market-trends-update-2016.html.
2. P.H. Siegel, "Terahertz technology," *IEEE Trans. Microw. Theory Tech.*, 50, 910–28, 2002.
3. P.H. Siegel and R.J. Dengler, "Terahertz heterodyne imaging Part I: Introduction and techniques," *Int. J. Infrared Millim. Waves*, 27, 465–80, 2006.
4. P.H. Siegel and R.J. Dengler, "Terahertz heterodyne imaging Part II: Instruments," *Int. J. Infrared Millim. Waves*, 27, 631–55, 2006.
5. G. Chattopadhyay, "Submillimeter-Wave Coherent and Incoherent Sensors for Space Applications," in *Sensors. Advancements in Modeling, Design Issues, Fabrication and Practical Applications*, eds. S.C. Mukhopadhyay and R.Y.M. Huang, 387–14, Springer, New York, 2008.
6. T.W. Crowe, W.L. Bishop, D.W. Porterfield, J.L. Hesler, and R.M. Weikle, "Opening the terahertz window with integrated diode circuits," *IEEE J. Solid-State Circuits*, 40, 2104–10, 2005.
7. D. Dragoman and M. Dragoman, "Terahertz fields and applications," *Prog. Quantum Electron.*, 28, 1–66, 2004.
8. J. Wei, D. Olaya, B.S. Karasik, S.V. Pereverzev, A.V. Sergeev, and M.E. Gershenzon, "Ultrasensitive hot-electron nanobolometers for terahertz astrophysics," *Nat. Nanotechnol.*, 3, 496–500, 2008.
9. A.H. Lettington, I.M. Blankson, M. Attia, and D. Dunn, "Review of imaging architecture," *Proc. SPIE*, 4719, 327–40, 2002.
10. A.W. Blain, I. Smail, R.J. Ivison, J.-P. Kneib, and D.T. Frayer, "Submillimetre galaxies," *Phys. Rep.*, 369, 111–76, 2002.
11. D. Leisawitz, W.C. Danchi, M.J. DiPirro, L.D. Feinberg, D.Y. Gezari, M. Hagopian, W.D. Langer, J.C. Mather, S.H. Moseley, M. Shao, R.F. Silverberg, J.G. Staguhn, M.R. Swain, H.W. Yorke, and X. Zhang, "Scientific motivation and technology requirements for the SPIRIT and SPECS far-infrared/submillimeter space interferometers," *Proc. SPIE*, 4013, 36–46, 2000.
12. D. Arnone, "10 emerging technologies that will change your world," *Technol. Rev.*, 107(1), 32–50, 2004.
13. J. Zmuidzinas and P.L. Richards, "Superconducting detectors and mixers for millimeter and submillimeter astrophysics," *Proc. IEEE*, 92, 1597–616, 2004.

14. B. Ferguson and X.-C. Zhang, "Materials for terahertz science and technology," *Nature Mater.*, 1, 26–33, 2002.
15. D. Mittleman, *Sensing with Terahertz Radiation*, Springer-Verlag, Berlin, 2003.
16. E.R. Brown, "Fundamentals of terrestrial millimetre-wave and THz remote sensing," *Int. J. High Speed Electron. Syst.*, 13, 99–1097, 2003.
17. R.M. Woodward, "Terahertz technology in global homeland security," *Proc. SPIE*, 5781, 22–31, 2005.
18. D.L. Woolard, R. Brown, M. Pepper, and M. Kemp, "Terahertz frequency sensing and imaging: A time of reckoning future applications?," *Proc. IEEE*, 93, 1722–43, 2005.
19. H. Zhong, A. Redo-Sanchez, and X.-C. Zhang, "Identification and classification of chemicals using terahertz reflective spectroscopic focal-plane imaging system," *Opt. Express*, 14, 9130–41, 2006.
20. M. Tonouchi, "Cutting-edge terahertz technology," *Nat. Photonics*, 1, 97–105, 2007.
21. A. Rostami, H. Rasooli, and H. Baghban, *Terahertz Technology. Fundamentals and Applications*, Springer, Berlin, 2011.
22. E. Bründermann, H.-W. Hübers, and M.F. Kimmitt, *Terahertz Techniques*, Springer, Dordrecht, The Netherlands, 2012.
23. D. Saeedkia, *Handbook of Terahertz Technology for Imaging, Sensing and Communications*, Woodhead Publishing, Oxford, UK, 2013.
24. M. Perenzoni and D.J. Paul, *Physics and Applications of Terahertz Radiation*, Springer, Dordrecht, The Netherlands, 2014.
25. H.-J. Song and T. Nagatsuma, *Handbook of Terahertz Technologies: Devices and Applications*, CRC Press, Boca Raton, FL, 2015.
26. T.G. Phillips and J. Keene, "Submillimeter astronomy," *Proc. IEEE*, 80, 1662–78, 1992.
27. R. Piesiewicz, T. Kleine-Ostmann, N. Krumbholz, D. Mittleman, M. Koch, J. Schoebel, and T. Kuerner, "Short-range ultra-broadband terahertz communications: Concept and perspectives," *IEEE Antenn. Propag. Mag.*, 49, 24–35, 2007.
28. F. Sizov, "THz radiation sensors," *Opto-Electron. Rev.*, 18, 10–36, 2010.
29. F. Sizov and A. Rogalski, "THz detectors," *Prog. Quantum Electron.*, 34, 278–347, 2010.
30. S. Komiyama, O. Astafiev, V. Antonov, T. Kutsuwa, and H. Hirai, "A single-photon detector in the far-infrared range," *Nature*, 403, 405–7, 2000.
31. S. Komiyama, "Single-photon detectors in terahertz region," *IEEE J. Sel. Top. Quantum Electron.*, 17, 54–66, 2011.
32. G. Chattopadhyay, "Heterodyne Arrays at Submillimeter Wavelengths," in *38-th General Assembly of International Union of Radio Science*, New Delhi, October, 2005.
33. F.F. Sizov, V.V. Zabudsky, A.G. Golenkov, and A. Shevchik-Shekera, "Millimeter-wave narrow-gap uncooled hot-carrier detectors for active imaging," *Opt. Eng.*, 53(3), 033203, 2013.
34. P.F. Goldsmith, Ph. Appleton, L. Armus, J. Bauer, D. Benford, A. Blaind, M. Bradford, G. Bryden, M. Dragovan, M. Harwit, G. Helou, W.D. Langer, D. Leisawitz, C. Paineb, and H. Yorke, "CALISTO: The Cryogenic Aperture Large Infrared Space Telescope Observatory." www.ipac.caltech.edu/DecadalSurvey/farir.html.
35. M. Harwit, G. Helou, L. Armus, C.M. Bradford, P.F. Goldsmith, M. Hauser, D. Leisawitz, D.F. Lester, G. Rieke, and S.A. Rinehart, "Far-Infrared/Submillimeter Astronomy from Space Tracking an Evolving Universe and the Emergence of Life." www.ipac.caltech.edu/DecadalSurvey/farir.html.
36. S.H. Moseley, J. Zmuidzinas, V. Sarohia, and E. Smith, "Mm/Submm/Far-IR Detector Arrays," A Coordinated GSFC/JPL Program, 2008.
37. D. Benford, "On the Application of MKIDs and TES Bolometers for FIR Astrophysics,"in *Community Workshop on Far-Infrared Space Astrophysics*, 2014. http://asd.gsfc.nasa.gov/conferences/FIR/posters/poster-Benford_FIR2014_MKID_vs_TES_rev_ small.pdf.
38. J.J. Bock, "Superconducting Detector Arrays for Far-Infrared to mm-Wave Astrophysics." http://cmbpol.uchicago.edu/depot/pdf/white-paper_j-bock.pdf.
39. P.L. Richards, "Bolometers for infrared and millimeter waves," *J. Appl. Phys.*, 76, 1–24, 1994.
40. A. Poglish, R.O. Katterloher, R. Hoenle, J.W. Beeman, E.E. Haller, H. Richter, U. Groezinger, N.M. Haegel, and A. Krabbe, "Far-infrared photoconductors for Herschel and SOFIA," *Proc. SPIE*, 4855, 115–28, 2003.
41. M. Kenyon, P.K. Day, C.M. Bradford, J.J. Bock, and H.G. Leduc, "Progress on background-limited membrane-isolated TES bolometers for far-IR/submillimeter spectroscopy," *Proc. SPIE*, 6275, 627508, 2006.
42. A.D. Turner, J.J. Bock, J.W. Beeman, J. Glenn, P.C. Hargrave, V.V. Hristov, H.T. Nguyen, F. Rahman, S. Sethuraman, and A.L. Woodcraft, "Silicon nitride micromesh bolometer array for submillimeter astrophysics," *Appl. Opt.*, 40, 4921–32, 2001.
43. B.S. Karasik, D. Olaya, J. Wei, S. Pereverzev, M.E. Gershenson, J.H. Kawamura, W.R. McGrath, and A.V. Sergeev, "Record-low NEP in hot-electron titanium nanobolometers," *IEEE Trans. Appl. Supercond.*, 17, 293–7, 2007.

44. H.-W. Hübers, "Terahertz heterodyne receivers," *IEEE J. Sel. Top. Quantum. Electron.*, 14, 378–91, 2008
45. D.J. Benford, "Transition Edge Sensor Bolometers for CMB Polarimetry." http://cmbpol.uchicago.edu/workshops/technology2008/depot/cmbpol_technologies_benford_jcps_4.pdf.
46. P.L. Richards, "Cosmic Microwave Background Experiments—Past, Present and Future." http://sciencestage.com/d/5334058/.
47. T. Otsuji, "Trends in the research of modern terahertz detectors: Plasmon detectors," *IEEE Trans. Terahertz Sci. Technol.*, 5(4), 1110–20, 2015.
48. F. Sizov, *Photoelectronics for Vision Systems in Invisible Spectral Ranges*, Akademperiodika, Kiev, 2008. (In Russian).
49. N. Kopeika, *A System Engineering Approach to Imaging*, SPIE Optical Eng. Press, Bellingham, WA, 1998.
50. "Detectors Needs for Long Wavelength Astrophysics," *A Report by the Infrared, Submillimeter, and Millimeter Detector Working Group*, June, 2002. http://safir.gsfc.nasa.gov/docs/ISMDWG_final.pdf.
51. D. Farrah, K.E. Smith, D. Ardila, C.M. Bradford, M. Dipirro, C. Ferkinhoff, J. Glenn, P. Goldsmith, D. Leisawitz, T. Nikola, N. Rangwala, S.A. Rinehart, J. Staguhn, M. Zemcov, J. Zmuidzinas, J. Bartlett, S. Carey, W.J. Fischer, J. Kamenetzky, J. Kartaltepe, M. Lacy, D.C. Lis, E. Lopez-Rodriguez, M. MacGregor, S.H. Moseley, E.J. Murphy, A. Rhodes, M. Richter, D. Rigopoulou, D. Sanders, R. Sankrit, G. Savini, J.-D. Smith, and S. Stierwalt, "Review: Far-Infrared Instrumentation and Technology Development for the Next Decade," arXiv:1709.02389v1 [astro-ph.IM], 7 September, 2017.
52. W. Wild, "Coherent Far-Infrared/Submillimetre Detectors," in *Detectors in Observing Photons in Space: A Guide to Experimental Space Astronomy*, eds. M.C.E. Huber, A. Pauluhn, J.L. Culhane, J.G. Timothy, K. Wilhelm, and A. Zehnder, 503–23, Springer, New York, 2013.
53. T.W. Crowe, R.J. Mattauch, H.-P. Roser, W.L. Bishop, W.C.B. Peatman, and X. Liu, "GaAs Schottky diodes for THz mixing applications," *Proc. IEEE*, 80, 1827–41, 1992.
54. G.L. Carr, M.C. Martin, W.R. McKinney, G.R. Neil, K. Jordan, and G.P. Williams, "High power terahertz radiation from relativistic electrons," *Nature*, 420, 153, 2002.
55. M. Rodwell, E. Lobisser, M. Wistey, V. Jain, A. Baraskar, E. Lind, J. Koo, B. Thibeault, A.C. Gossard, Z. Griffith, J. Hacker, M. Urteaga, D. Mensa, R. Pierson, and B. Brar, "Development of THz Transistors and (300–3000 GHz) Sub-mm-Wave Integrated Circuits," in *The 11th International Symposium on Wireless Personal Multimedia Communications (WPMC 2008)*. www.ece.ucsb.edu/Faculty/Rodwell/publications/2008_9_sept_wpmc_rodwell_digest.pdf.
56. B.S. Williams, "Terahertz quantum-cascade lasers," *Nat. Photonics*, 1, 517–25, 2007.
57. J.R. Tucker and M.J. Feldman, "Quantum detection at millimeter wavelength," *Rev. Modern Phys.*, 57, 1055–113, 1985.
58. C.M. Bradford, B.J. Naylor, J. Zmuidzinas, J.J. Bock, J. Gromke, H. Nguyen, M. Dragovan, M. Yun, L. Earle, J. Glenn, H. Matsuhara, P.A.R. Ade, and L. Duband, "WaFIRS: A waveguide far-IR spectrometer: Enabling spectroscopy of high-z galaxies in the far-IR and submillimeter," *Proc. SPIE*, 4850, 1137–48, 2003.
59. M. Kenyon, P.K. Day, C.M. Bradford, J.J. Bock, and H.G. Leduc, "Progress on background-limited membrane-isolated TES bolometers for far-IR/submillimeter spectroscopy," *Proc. SPIE*, 6275, 627508, 2006.
60. B.S. Karasik and R. Cantor, "Optical NEP in Hot-Electron Nanobolometers," in *21st International Symposium on Space Terahertz Technology*, Oxford, UK, March 23–25, 2010.
61. J.C. Mather, E.S. Cheng, D.A. Cottingham, R.E. Eplee, D.J. Fixsen, T. Hewagama, R.B. Isaacman, K.A. Jensen, S.S. Meyer, P.D. Noerdlinger, S.M. Read, L.P. Rosen, R.A. Shafer, E.L. Wright, C.L. Bennett, N.W. Boggess, M.G. Hauser, T. Kelsall, S.H. Moseley, R.F. Silverberg, G.F. Smoot, R. Weiss, and D.T. Wilkinson, "Measurement of the cosmic microwave background spectrum by the COBE FIRAS instrument," *Astrophys. J.*, 420, 439–44, 1994.
62. J. Dunkley, A. Amblard, C. Baccigalupi, M. Betoule, D. Chuss, A. Cooray, J. Delabrouille, C. Dickinson, G. Dobler, J. Dotson, H.K. Eriksen, D. Finkbeiner, D. Fixsen, P. Fosalba, A. Fraisse, C. Hirata, A. Kogut, J. Kristiansen, C. Lawrence, A.M. Magalhaes, M.A. Miville-Deschenes, S. Meyer, A. Miller, S.K. Naess, L. Page, H.V. Peiris, N. Phillips, E. Pierpaoli, G. Rocha, J.E. Vaillancourt, and L. Verde, "A Program of Technology Development and of Sub-Orbital Observations of the Cosmic Microwave Background Polarization Leading to and Including a Satellite Mission," *in A Report for the Astro-2010 Decadal Committee on Astrophysics*, April, 2009.
63. D.H. Auston, "Picosecond optoelectronic switching and gating in silicon," *Appl. Phys. Lett.*, 26, 101–3, 1975.
64. P. LeFur and D.H. Auston, "A kilovolt picosecond optoelectronic switch and Pockels cell," *Appl. Phys. Lett.*, 28, 21–33, 1976.
65. J.A. Valdmani, G. Mourou, and C.W. Gabel, "Picosecond electrooptic sampling system," *Appl. Phys. Lett.*, 41, 211–2, 1982.

66. D. Grischkowsky, S. Keiding, M. van Exter, and C. Fattinger, "Far-infrared time-domain spectroscopy with terahertz beams of dielectrics and semiconductors," *J. Opt. Soc.*, B7, 2006–15, 1990.
67. M. Tani, Y. Hirota, C. Que, S. Tanaka, R. Hattori, M. Yamaguchi, S. Nishizawa, and M. Hangyo, "Novel terahertz photoconductive antennas," *Int. J. Infrared Millim. Waves*, 27, 531–46, 2006.
68. D.M. Mittleman, M. Gupta, R. Neelamani, R.G. Baraniuk, J.V. Rudd, M. Koch, "Recent advances in terahertz imaging," *Appl. Phys. B*, 68, 1085–94, 1999.
69. W.L. Chan, J. Deibel, and D.M. Mittleman, "Imaging with terahertz radiation," *Rep. Prog. Phys.*, 70, 1325–79, 2007.
70. M. Jarrahi, "Advanced photoconductive terahertz optoelectronics based on nano-antennas and nano-plasmonic light concentrators," *IEEE Trans. Terahertz Sci. Technol.*, 5(3), 391–7, 2015.
71. L. Xu, X.-C. Zhang, and D.H. Auston, "Terahertz beam generation by femtosecond optical pulses in electro-optic materials," *Appl. Phys. Lett.*, 61, 1784–6, 1992.
72. E.R. Brown, K.A. McIntosh, F.W. Smith, K.B. Nichols, M.J. Manfra, C.L. Dennis, and J.P. Mattia, "Milliwatt output levels and superquadratic bias dependence in a low-temperature-grown GaAs photomixer," *Appl. Phys. Lett.*, 64, 3311–3, 1994.
73. M. Tani, K.-S. Lee, and X.-C. Zhang, "Detection of terahertz radiation with low-temperature-grown GaAs based photoconductive antenna using 1.55 μm probe," *Appl. Phys. Lett.*, 77, 1396–8, 2000.
74. M. Suzukia and M. Tonouchi, "Fe-implanted InGaAs photoconductive terahertz detectors triggered by 1.56 μm femtosecond optical pulses," *Appl. Phys. Lett.*, 86, 163504, 2005.
75. H. Page, S. Malik, M. Evans, I. Gregory, I. Farrer, and D. Ritchie, "Waveguide coupled terahertz photoconductive antennas: Toward integrated photonic terahertz devices," *Appl. Phys. Lett.*, 92, 163502, 2008.
76. D.P. Neikirk, D.B. Rutledge, and M.S. Mucha, "Far-infrared imaging antenna arrays," *Appl. Phys. Lett.*, 40, 203–5, 1982.
77. D.B. Rutledge, D.P. Neikirk, and D.P. Kasilingam, "Integrated-circuit antennas," in *Infrared and Millimeter Waves*, Vol. 10, ed. K.J. Button, 1–90, Academic Press, New York, 1983.
78. J. Zhang, Y. Hong, S.L. Braunstein, and K.A. Shore, "Terahertz pulse generation and detection with LT-GaAs photoconductive antenna," *IEE Proc. Optoelectron.*, 151, 98–101, 2004.
79. E.R. Brown, A.W.M. Lee, B.S. Navi, and J.E. Bjarnason, "Characterization of a planar self-complementary squarespiral antenna in the THz region," *Microw. Opt. Technol. Lett.*, 48, 524–9, 2006.
80. J. Grade, P. Haydon, and D. van der Weide, "Electronic terahertz antennas and probes for spectroscopic detection and diagnostics," *Proc. IEEE*, 95, 1583–91, 2007.
81. E.R. Brown and D. Segovia-Vargas, "Principles of THz Direct Detection," in *Semiconductor Terahertz Technology: Devices and Systems at Room Temperature Operation*, eds. G. Carpintero, L.E. Garcia Muñoz, L. Hartnagel, S. Preu, and A.V. Räisäinem, 212–53, Wiley, Chichester, UK, 2015.
82. D.R.S. Cumming, F. Simoens, I. Escorcia-Carranza, and J. Grant, "Components for terahertz imaging," *J. Phys. D*, 50, 043001–14–17, 2017.
83. A. Westlund, "Self-Switching Diodes for Zero-Bias Terahertz Detection," Thesis, Chalmers University of Technology, Gothenburg, Sweden, March 2015. http://publications.lib.chalmers.se/records/fulltext/214629/214629.pdf.
84. Z. Zhang, R. Rajavel, P. Deelman, and P. Fay, "Sub-micron area heterojunction backward diode millimeter-wave detectors with 0.18 pWHz$^{-1/2}$ noise equivalent power," *IEEE Microw. Wirel. Compon. Lett.*, 21(5), 267–9, 2011.
85. R. Tauk, F. Teppe, S. Boubanga, D. Coquillat, W. Knap, Y.M. Meziani, C. Gallon, F. Boeuf, T. Skotnicki, C. Fenouillet-Beranger, D.K. Maude, S. Rumyantsev, and M.S. Shur, "Plasma wave detection of terahertz radiation by silicon field effects transistors: Responsivity and noise equivalent power," *Appl. Phys. Lett.*, 89, 253511, 2006.
86. F. Schuster, D. Coquillat, H. Videlier, M. Sakowicz, F. Teppe, L. Dussopt, B. Giffard, T. Skotnicki, and W. Knap, "Broadband terahertz imaging with highly sensitive silicon CMOS detectors," *Opt. Express*, 19, 7827–32, 2011.
87. S. Boppel, A. Lisauskas, M. Mundt, D. Seliuta, L. Minkeviius, I. Kašalynas, G. Valušis, M. Mittendorff, S. Winnerl, V. Krozer, and H.G. Roskos, "CMOS integrated antenna-coupled field-effect transistors for the detection of radiation from 0.2 to 4.3 THz," *IEEE Trans. Microw. Theory Tech.*, 60, 3834–43, 2012.
88. R. Al Hadi, H. Sherry, J. Grzyb, Y. Zhao, W. Förster, H.M. Keller, A. Cathelin, A. Kaiser, and U.R. Pfeiffer, "A 1 k-pixel video camera for 0.7-1.1 terahertz imaging applications in 65-nm CMOS," *IEEE J. Solid-State Circuits*, 47, 2999–3012, 2012.
89. E.R. Brown and D. Segovia-Vargas, "Principles of THz Direct Detection," in *Semiconductor Terahertz Technology*, eds. Dr. G. Carpintero, L.E. García Muńoz, H. Hartnagel, S. Preu, and A.V. Räisänen, 212–53, Wiley, Chichester, UK, 2015.

90. V.G. Bozhkov, "Semiconductor detectors, mixers, and frequency multipliers for the terahertz band," *Radiophys. Quantum Electron.*, 46, 631–56, 2003.
91. I. Mehdi, J.V. Siles, C. Lee, and E. Schlecht, "THz diode technology: Status, prospects, and applications," *Proc. IEEE*, 105(6), 990–1007, 2017.
92. A. Van Der Ziel, "Infrared detection and mixing in heavily doped Schottky barrier diodes," *J. Appl. Phys.*, 47, 2059–68, 1976.
93. H.A. Watson, *Microwave Semiconductor Devices and Their Circuit Applications*, McGraw-Hill, New York, 1969.
94. E.J. Becklake, C.D. Payne, and B.E. Pruer, "Submillimetre performance of diode detectors using Ge, Si and GaAs," *J. Phys. D*, 3, 473–81, 1970.
95. D.T. Young and J.C. Irvin, "Millimeter frequency conversion using Au-n-type GaAs Schottky barrier epitaxial diodes with a novel contacting technique," *Proc. IEEE*, 53, 2130–2, 1965.
96. T.W. Crowe, D.P. Porterfield, J.L. Hesler, W.L. Bishop, D.S. Kurtz, and K. Hui, "Terahertz sources and detectors," *Proc. SPIE*, 5790, 271–80, 2005.
97. H.P. Röser, H.-W. Hübers, E. Bründermann, and M.F. Kimmitt, "Observation of mesoscopic effects in Schottky diodes at 300 K when used as mixers at THz frequencies," *Semicond. Sci. Technol.*, 11, 1328–32, 1996.
98. T.W. Crowe and W.C.B. Peatman, "GaAs Schottky Diodes for Mixing Applications Beyond 1 THz,"in *Second International Symposium on Space Terahertz Technology*, pp. 323–39, Pasadena, CA, February 26–28, 1991. http://www.nrao.edu/meetings/isstt/papers/ 1991/1991323339.pdf.
99. T.W. Crowe, "GaAs Schottky barrier mixer diodes for the frequency range 1–10 THz," *Int. J. Infrared Milim. Waves*, 11, 765–77, 1990.
100. H. Kräutle, E. Sauter, and G.V. Schultz, "Antenna characteristics of whisker diodes used at submillimeter receivers," *Infrared Phys.*, 17, 477–83, 1977.
101. R. Titz, B. Auel, W. Esch, H.P. Röser, and G.W. Schwaab, "Antenna measurements of open-structure Schottky mixers and determination of optical elements for a heterodyne system at 184, 214 and 287 μm," *Infrared Phys.*, 30, 435–41, 1990.
102. B. J. Rizzi, T.W. Crowe, and N.R. Erickson, "A high-power millimeter-wave frequency doubler using a planar diode array," *IEEE Microw. Guided Wave Lett.*, 3(6), 188–90, 1993.
103. B. I. Mehdi, G. Chattopadhyay, E. Schlecht, J. Ward, J. Gill, F. Maiwald, and A. Maestrini, "THz Multiplier Circuits," in *IEEE MTT-S International Microwave Symposium Digest*, pp. 341–44, San Francisco, 2006.
104. S.M. Marazita, W.L. Bishop, J.L. Hesler, K. Hui, W.E. Bowen, and T.W. Crowe, "Integrated GaAs Schottky mixers by spin-on-dielectric wafer bonding," *IEEE Trans. Electron Devices*, 47, 1152–6, 2000.
105. P. Siegel, R.P. Smith, M.C. Gaidis, and S. Martin, "2.5-THz GaAs monolithic membrane-diode mixer," *IEEE Trans. Microw. Theory Tech.*, 47, 596–604, 1999.
106. J.A. Copeland, "Diode edge effects on doping profile measurements," *IEEE Trans. Electron Devices*, 17, 404–7, 1970.
107. V.I. Piddyachiy, V.M. Shulga, A.M. Korolev, and V.V. Myshenko, "High doping density Schottky diodes in the 3 mm wavelength cryogenic heterodyne receiver," *Int. J. Infrared Millim. Waves*, 26, 1307–15, 2005.
108. J.L. Hesler and T.W. Crowe "Responsivity and noise measurements of zero-bias Schottky diode detectors." http://www.virginiadiodes.com/VDI/pdf/VDI%20Detector%20Char%20ISSTT2007.pdf.
109. H. Kazemi, G. Nagy, L Tran, E. Grossman, E.R. Brown, A.C. Gossard, G.D. Boreman, B. Lail, A.C. Young, and J.D. Zimmerman, "Ultra Sensitive ErAs/InAlGaAs Direct Detectors for Millimeter Wave and THz Imaging Applications," in *IEEE/MTT International Microwave Symposium*, pp. 1367–70, 2007.
110. E.R. Brown, A.C. Young, J.E. Bjarnason, J.D. Zimmerman, A.C. Gossard, and H. Kazemi, "Millimeter and sub-millimeter wave performance of an ErAs:InAlGaAs Schottky diode coupled to a single-turn square spiral," *Int. J. High Speed Electron. Syst.*, 17, 383–94, 2007.
111. F. Maiwald, F. Lewen, B. Vowinkel, W. Jabs, D.G. Paveljev, M. Winnerwisser, and G. Winnerwisser, "Planar Schottky diode frequency multiplier for molecular spectroscopy up to 1.3 THz," *IEEE Microw. Guided Wave Lett.*, 9, 198–200, 1999.
112. D.H. Martin, *Spectroscopic Techniques for Far-infrared, Submillimeter and Millimeter Waves*, North-Holland, Amsterdam, 1967.
113. R. Han, Y. Zhang, D. Coquillat, J. Hoy, H. Videlier, W. Knap, E. Brown, and K.K. O, "280-GHz Schottky Diode Detector in 130-nm Digital CMOS," in *Proceedings of 2010 IEEE Custom Integrated Circuits Conference (CICC)*, September 19–22, pp. 1–4, 2010.
114. R. Han, Y. Zhang, D. Coquillat, H. Videlier, W. Knap, E. Brown, and K.K. O, "A 280-GHz Schottky diode detector in 130-nm digital CMOS," *IEEE J. Solid-State Circuits*, 46(11), 2602–12, 2011.

115. R. Han, Y. Zhang, Y. Kim, D.Y. Kim, H. Shichijo, E. Afshari, and K. O, "280 GHz and 860 GHz Image Sensors Using Schottky-Barrier Diodes in 0.13μm Digital CMOS," in *IEEE International Solid-State Circuits Conference*, p. 253, 2012.
116. R. Han, Y. Zhang, Y. Kim, D.Y. Kim, H. Shichijo, E. Afshari, and K.K. O, "Active terahertz imaging using Schottky diodes in CMOS: Array and 860-GHz pixel," *IEEE J. Solid-State Circuits*, 48(10), 2296–308, 2013
117. SELEX GALILEO. www.selex-sas.com/EN/Common/files/SELEX_Galileo/Products/DLATGS_dsh.pdf.
118. D. Dooley, "Sensitivity of broadband pyroelectric terahertz detectors continues to improve," *Laser Focus World*, May, 2010.
119. www.spectrumdetector.com/pdf/datasheets/THZ.pdf.
120. I. Kašalynas, A.J.L. Adam, T.O. Klaassen, N.J. Hovenier, G. Pandraud, V.P. Iordanov, and P.M. Sarro, "Some properties of a room temperature THz detection array," *Proc. SPIE*, 6596, 65960J, 2007.
121. J.A. Cox, R. Higashi, F. Nusseibeh, K. Newstrom-Peitso, and C. Zins, "Uncooled MEMS-based detector arrays for THz imaging applications," *Proc. SPIE*, 7311, 73110R–1, 2009.
122. A.W.M. Lee, B.S. Williams, S. Kumar, Q. Hu, and J.L. Reno, "Real-time imaging using a 4.3-THz quantum cascade laser and a 320 × 240 microbolometer focal-plane array," *IEEE Photonics Technol. Lett.*, 18, 1415–7, 2006.
123. N. Oda, "Uncooled bolometer-type terahertz focal-plane array and camera for real-time imaging," *C. R. Phys.*, 11, 496–509, 2010.
124. M. Bolduc, M. Terroux, B. Tremblay, L. Marchese, E. Savard, M. Doucet, H. Oulachgar, C. Alain, H. Jerominek, and A. Bergeron, "Noise-equivalent power characterization of an uncooled microbolometer-based THz imaging camera," *Proc. SPIE*, 8023, 80230C–1–10, 2011.
125. D.-T. Nguyen, F. Simoens, J.-L. Ouvrier-Buffet, J. Meilhan, and J.-L. Coutaz, "Broadband THz uncooled antenna-coupled microbolometer array—electromagnetic design, simulations and measurements," *IEEE Trans. Terahertz Sci. Technol.*, 2, 299–305, 2012.
126. N. Oda, T. Ishi, S. Kurashina, T. Sudou, T. Morimoto, M. Miyoshi, K. Doi, H. Goto, T. Sasaki, G. Isoyamab, R. Kato, A. Irizawa, and K. Kawase, "Performance of THz cameras with enhanced sensitivity in sub-terahertz region," *Proc. SPIE*, 9483, 94830S–1–9, 2015.
127. N. Oda, "Uncooled bolometer-type terahertz focal plane array and camera for real-time imaging," *C. R. Phys.*, 11(7), 496–509, 2010.
128. N. Nemoto, N. Kanda, R. Imai, K. Konishi, M. Miyoshi, S. Kurashina, T. Sasaki, N. Oda, and M. Kuwata-Gonokami, "High-sensitivity and broadband, real-time terahertz camera incorporating a micro-bolometer array with resonant cavity structure," *IEEE Trans. Terahertz Sci. Technol.*, 6(2), 175–182, 2016.
129. N. Oda, S. Kurashina, M. Miyoshi, K. Doi, T. Ishi, T. Sudou, T. Morimoto, H. Goto, T. Sasaki, "Microbolometer terahertz focal plane array and camera with improved sensitivity in the sub-terahertz region," *J. Infrared Millim. Terahertz Waves*, 36, 947–960, 2015.
130. F. Simoens and J. Meilhan, "Terahertz real-time imaging uncooled array based on antenna- and cavity-coupled bolometers," *Phil. Trans. R. Soc. A*, 372, 20130111, 2014.
131. F. Simoens, J. Meilhan, L. Dussopt, J.-A. Nicolas, N. Monnier, G. Sicarda, A. Siligaris, and B. Hiberty, "Uncooled terahertz real-time imaging 2D arrays developed at LETI: Present status and perspectives," *Proc. SPIE*, 10194, 101942N–1–13, 2017.
132. D. Dufour, L. Marchese, M. Terroux, H. Oulachgar, F. Généreux, M. Doucet, L. Mercier, B. Tremblay, C. Alain, P. Beaupré, N. Blanchard, M. Bolduc, C. Chevalier, D. D'Amato, Y. Desroches, F. Duchesne, L. Gagnon, S. Ilias, H. Jerominek, F. Lagacé, J. Lambert, F. Lamontagne, L. Le Noc, A. Martel, O. Pancrati, J.-E. Paultre, T. Pope, F. Provençal, P. Topart, C. Vachon, S. Verreault, and A. Bergeron, "Review of terahertz technology development at INO," *J. Infrared Millim. Terahertz Waves*, 36, 922–46, 2015.
133. D.J. Burdette, J. Alverbro, Z. Zhang, P. Fay, Y. Ni, P. Potet, K. Sertel, G. Trichopoulos, K. Topalli, J. Volakis, and H.L. Mosbacker, "Development of an 80 × 64 pixel, broadband, real-time THz imager," *Proc. SPIE*, 8023, 80230F–1–12, 2011.
134. G.C. Trichopoulos, H.L. Mosbaker, D. Burdette, and K. Sertel, "A broadband focal plane array camera for real-time THz imaging applications," *IEEE Trans. Antenn. Propag.*, 61(4), 1733–40, 2013.
135. www.traycer.com.
136. D. Corcos, I. Brouk, M. Malits, A. Svetlitza, S. Stolyarova, A. Abramovich, E. Farber, N. Bachar, D. Elad, and Y. Nemirovsky, "The TeraMOS Sensor for Monolithic Passive THz Imagers," in *2011 IEEE International Conference Microwaves, Communications, Antennas and Electronic Systems (COMCAS)*, 2011.
137. V. Dobrovolsky and F.F. Sizov, "THz/sub-THz bolometer based on the electron heating in a semiconductor waveguide," *Opto-Electron. Rev.*, 18, 250–8, 2010.

138. W. Knap, V. Kachorowskii, Y. Deng, S. Rumyantsev, J.-Q. Lu, R. Gaska, M.S. Shur, G. Simin, X. Hu, M.A. Khan, C.A. Saylor, and L.C. Brunal, "Nonresonant detection of terahertz radiation in field effect transistors," *J. Appl. Phys.*, 91, 9346–53, 2002.
139. A. El Fatimy, F. Teppe, N. Dyakonova, W. Knap, D. Seliuta, G. Valusis, A. Shchepetov, Y. Roelens, S. Bollaert, A. Cappy, and S. Rumyantsev, "Resonant and voltage-tunable terahertz detection in InGaAs/InP nanometer transistors," *Appl. Phys. Lett.*, 89, 131926, 2006.
140. Y.M. Meziani, J. Lusakowski, N. Dyakonova, W. Knap, D. Seliuta, E. Sirmulis, J. Deverson, G. Valusis, F. Boeuf, and T. Skotnicki, "Non resonant response to terahertz radiation by submicron CMOS transistors," *IEICE Trans. Electron.*, E89-C, 993–8, 2006.
141. W. Knap, M. Dyakonov, D. Coquillat, F. Teppe, N. Dyakonova, J. Łusakowski, K. Karpierz, M. Sakowicz, G. Valusis, D. Seliuta, I. Kasalynas, A. El Fatimy, Y.M. Meziani, T. Otsuji, "Field effect transistors for terahertz detection: Physics and first imaging applications," *J. Infrared Millim Terahertz Waves*, 30, 1319–37, 2009.
142. W. Knap, D. Coquillat, N. Dyakonova, F. Teppe, O. Klimenko, H. Videlier, S. Nadar, J. Łusakowski, G. Valusis, F. Schuster, B. Giffardd, T. Skotnicki, C. Gaquiere, and A. El Fatimy, "Plasma excitations in field effect transistors for terahertz detection and emission," *C. R. Phys.*, 11, 433–43, 2010.
143. W. Knap, F. Teppe, Y. Meziani, N. Dyakonova, J. Lusakowski, F. Boeuf, T. Skotnicki, D. Maude, S. Rumyantsev, and M.S. Shur, "Plasma wave detection of sub-terahertz and terahertz radiation by silicon field-effect transistors," *Appl. Phys. Lett.*, 85, 675–7, 2002.
144. F. Teppe, M. Orlov, A. El Fatimy, A. Tiberj, W. Knap, J. Torres, V. Gavrilenko, A. Shchepetov, Y. Roelens, and S. Bollaert, "Room temperature tunable detection of subterahertz radiation by plasma waves in nanometer InGaAs transistors," *Appl. Phys. Lett.*, 89, 222109, 2006.
145. W. Knap, D. Coquillat, N. Dyakonova, D. But, T. Otsuji, and F. Teppe, "Terahertz Plasma Field Effect Transistors," in *Physics and Applications of Terahertz Radiation*, eds. M. Perenzoni and D.J. Paul, 77–100, Springer, Dordrecht, The Netherlands, 2014.
146. W. Knap and M.I. Dyakonov, "Field Effect Transistors for Terahertz Applications," in *Handbook of Terahertz Technology for Imaging, Sensing and Communications*, eds. D. Saeedkia, 121–55, Woodhead Publishing, Oxford, UK, 2013.
147. J. Sun, *Field-effect Self-mixing Terahertz Detectors*, Springer-Verlag, Berlin, 2016.
148. V. Ryzhii, A. Satou, I. Khmyrova, M. Ryzhii, T. Otsuji, V. Mitin, and M.S. Shur, "Plasma effects in lateral Schottky junction tunneling transit-time terahertz oscillator," *J. Phys. Conf. Ser.*, 38, 228–33, 2006.
149. X.G. Peralta, S.J. Allen, M.C. Wanke, N.E. Harff, J.A. Simmons, M.P. Lilly, J.L. Reno, P.J. Burke, and J.P. Eisenstein, "Terahertz photoconductivity and plasmon modes in double-quantum-well field-effect transistors," *Appl. Phys. Lett.*, 81, 1627–30, 2002.
150. M. Dyakonov and M.S. Shur, "Shallow water analogy for a ballistic field effect transistor: New mechanism of plasma wave generation by the dc current," *Phys. Rev. Lett.*, 71, 2465–8, 1993.
151. M. Dyakonov and M. Shur, "Plasma wave electronics: Novel terahertz devices using two dimensional electron fluid," *IEEE Trans. Electron. Devices*, 43, 1640–6, 1996.
152. M. Shur and V. Ryzhii, "Plasma wave electronics," *Int. J. High Speed Electron. Syst.*, 13, 575–600, 2003.
153. A. Eguiluz, T.K. Lee, J.J. Quinn, and K.W. Chiu, "Interface excitations in metal-insulator-semiconductor structures," *Phys. Rev. B*, 11, 4989–93, 1975.
154. S. Kang, P.J. Burke, L.N. Pfeifer, and K.W. West, "Resonant frequency response of plasma wave detector," *Appl. Phys. Lett.*, 89, 213512, 2006.
155. V.I. Gavrilenko, E.V. Demidov, K.V. Marem'yanin, S.V. Morozov, W. Knap, and J. Lusakowski, "Electron transport and detection of terahertz radiation in a GaN/AlGaN submicrometer field-effect transistor," *Semiconductors*, 41, 232–4, 2007.
156. S. Boppel, A. Lisauskas, and H.G. Roskos, "Terahertz array imagers: towards the implementation of terahertz camera with plasma-wave-based silicon MOSFET detectors," in *Handbook of Terahertz Technology for Imaging, Sensing and Communications*, ed. D. Saeedkia, 231–71, Woodhead Publishing, Oxford, UK, 2013.
157. A. Lisauskas, U. Pfeiffer, E. Öjefors, P. Bolivar, D. Glaab, and H.G. Roskos, "Rational design of high-responsivity detectors of terahertz radiation based on distributed self-mixing in silicon field-effect transistors," *J. Appl. Phys.*, 105(11), 114511, 2009.
158. F. Teppe, A. El Fatimy, S. Boubanga, D. Seliuta, G. Valusis, B. Chenaud, and W. Knap, "Terahertz resonant detection by plasma waves in nanometric transistors," *Acta Phys. Polon. A*, 113, 815–20, 2008.
159. D. Veksler, F. Teppe, A.P. Dmitriev, V.Yu. Kachorovskii, W. Knap, and M.S. Shur, "Detection of terahertz radiation in gated two-dimensional structures governed by dc current," *Phys. Rev. B*, 73, 125328, 2006.

160. R. Tauk, F. Teppe, S. Boubanga, D. Coquillat, W. Knap, Y.M. Meziani, C. Gallon, F. Boeuf, T. Skotnicki, and C. Fenouillet-Beranger, "Plasma wave detection of terahertz radiation by silicon field effects transistors: Responsivity and noise equivalent power," *Appl. Phys. Lett.*, 89, 253511, 2006.
161. F. Schuster, D. Coquillat, H. Videlier, M. Sakowicz, F. Teppe, L. Dussopt, B. Giffard, T. Skotnicki, W. Knap, "Broadband terahertz imaging with highly sensitive silicon CMOS detectors," *Opt. Express*, 19(8), 7827–32, 2011.
162. W. Knap, F. Teppe, Y. Meziani, N. Dyakonova, J. Lusakowski, F. Boeuf, T. Skotnicki, D. Maude, S. Rumyantsev, and M.S. Shur, "Plasma wave detection of sub-terahertz and terahertz radiation by silicon field effect transistors," *Appl. Phys. Lett.*, 85(4), 675–7, 2004.
163. R. Al Hadi, H. Sherry, J. Grzyb, Z. Yan, W. Forster, H.M. Keller, A. Cathelin, A. Kaiser, and U.R. Pfeiffer, "A 1 k-pixel video camera for 0.7–1.1 terahertz imaging applications in 65-nm CMOS," *IEEE J. Solid-State Circuits*, 47(12), 2999–3012, 2012.
164. A. Lisauskas, S. Boppel, M. Saphar, V. Krozer, L. Minkevičius, R. Venckevičius, D. Seliuta, I. Kašalynas, V. Tamošiūnas, G. Valušis, and H.G. Roskos, "Detectors for terahertz multi-pixel coherent imaging and demonstration of real-time imaging with a 12 × 12-pixel CMOS array," *Proc. SPIE*, 8496, 84960J–1–9, 2012.
165. S. Boppel, A. Lisauskas, A. Max, V. Krozer, and H.G. Roskos, "CMOS detector arrays in a virtual 10-kilopixel camera for coherent terahertz real-time imaging," *Opt. Lett.*, 37(4), 536–8, 2012.
166. F. Simoens, J. Meilhan, J.-A. Nicolas, "Terahertz real-time imaging uncooled arrays based on antenna-coupled bolometers or FET developed at CEA-LETI," *J. Infrared Millim. Terahertz Waves*, 36, 961–85, 2015.
167. www.leti-cea.fr/cea-tech/leti/Documents/Rapport%20scientifique/DOPT_annual%20research%20report%20 2015.pdf.
168. F. Simoens, "Terahertz cameras," in *Handbook of Terahertz Technologies: Devices and Applications*, eds. H.-J. Song and T. Nagatsuma, 395–428, CRC Press, Boca Raton, FL, 2015.
169. S.J. Fray and J.F.C. Oliver, "Photoconductive detector of radiation of wavelength greater than 50 μm," *J. Sci. Instrum.*, 36, 195, 1959.
170. W.J. Moore and H. Shenker, "A high-detectivity gallium-doped germanium detector for the 40–120 μm region," *Infrared Phys.*, 5, 99, 1965.
171. H.H. Houge, M.G. Mlynczak, M.N. Abedin, S.A. Masterjohn, and J.E. Huffman, "Far-infrared detector development for space-based Earth observation," *Proc. SPIE*, 7082, 70820E–1–8, 2008.
172. J. Bandaru, J.W. Beeman, and E.E. Haller, "Growth and performance of Ge:Sb blocked impurity band (BIB) detectors," *Proc. SPIE*, 4486, 193–9, 2002.
173. L.A. Reichertz, J.W. Beeman, B.L. Cardozo, G. Jakob, R. Katterloher, N.M. Haegel, and E.E. Haller, "Development of a GaAs-based BIB detector for sub-mm wavelengths," *Proc. SPIE*, 6275, 62751S, 2006.
174. D.R. Khokhlov, I.I. Ivanchik, S.N. Raines, D.M. Watson, and J.L. Pipher, "Performance and spectral response of $Pb_{1-x}Sn_xTe(In)$ far-infrared photodetectors," *Appl. Phys. Lett.*, 76, 2835–7, 2000.
175. K.G. Kristovskii, A.E. Kozhanov, D.E. Dolzhenko, I.I. Ivanchik, D. Watson, and D.R. Khokhlov, "Photoconductivity of lead telluride-based doped alloys in the submillimeter wavelength range," *Phys. Solid State*, 46, 122–4, 2004.
176. A.N. Akimov, V.G. Erkov, V.V. Kubarev, E.L. Molodtsova, A.E. Klimov, and V.N. Shumskyi, "Photosensitivity of $Pb_{1-x}Sn_xTe$:In films in the terahertz region of the spectrum," *Semiconductors*, 40, 164–8, 2006.
177. A. Artamkin, A. Nikorici, L. Ryabova, V. Shklover, and D. Khokhlov, "Continuous focal plane array for detection of terahertz radiation," *Proc. SPIE*, 6297, 62970B, 2006.
178. A.N. Akimov, A.E. Klimov, I.G. Neizvestny, V.N. Shumsky, V.V. Kubarev, O.V. Smolin, and E.V. Susov, "Sensitivity of $Pb_{1-x}Sn_xTe$ films in submillimeter spectral range," *Prikladnaya Fizika*, 6, 12–17, 2007 (in Russian).
179. A. Klimov, V. Shumsky, and V. Kubarev, "Terahertz sensitivity of $Pb_{1-x}Sn_xTe$:In," *Ferroelectrics*, 347, 111–9, 2007.
180. A.G. Milnes, *Deep Impurities in Semiconductors*, Wiley Interscience, New York, 1973.
181. B.A. Volkov, L.I. Ryabova, and D.R. Khokhlov, "Mixed-valence impurities in lead telluride-based solid solutions," *Physics-Uspekhi*, 45, 819–46, 2002.
182. Yu.G. Troyan, F.F. Sizov, and V.M. Lakeenkov, "Relaxation time and current instabilities in highly resistive PbTe:Ga single crystals," *Ukr. J. Phys.*, 32, 467–71, 1987.
183. C. Wilson, L. Frunzio, and D. Prober, "Time-resolved measurements of thermodynamic fluctuations of the particle number in a nondegenerate Fermi gas," *Phys. Rev. Lett.*, 87, 067004, 2001.
184. C.A. Mears, Q. Hu, P.L. Richards, A.H. Worsham, D.E. Prober, and A.V. Raisanen, "Quantum limited heterodyne detection of millimeter waves using super conducting tantalum tunnel junctions," *Appl. Phys. Lett.*, 57, 2487–9, 1990.

185. E. Burstein, D.N. Langenberg, and B.N. Taylor, "Superconductors as quantum detectors for microwave and sub-millimeter radiation," *Phys. Rev. Lett.*, 6, 92–4, 1961.
186. A.H. Dayem and R.J. Martin, "Quantum interaction of microwave radiation with tunnelling between superconductors," *Phys. Rev. Lett.*, 8, 246–8, 1962.
187. P.K. Tien and J.P. Gordon, "Multiphoton process observed in the interaction of microwave fields with the tunnelling between superconductor films," *Phys. Rev.*, 129, 647–51, 1963.
188. P.L. Richards, T.M. Shen, R.E. Harris, and F.L. Lloyd, "Quasiparticle heterodyne mixing in SIS tunnel junctions," *Appl. Phys. Lett.*, 34, 345–7, 1979.
189. G.J. Dolan, T.G. Phillips, and D.P. Woody, "Low-noise 115-GHz mixing in superconducting oxide-barrier tunnel junctions," *Appl. Phys. Lett.*, 34, 347–9, 1979.
190. J.R. Tucker and M.J. Feldman, "Quantum detection at millimeter wavelength," *Rev. Mod. Phys.*, 57, 1055–113, 1985.
191. C.A. Mears, Q. Hu, P.L. Richards, A.H. Worsham, D.E. Prober, and A.V. Raisanen, "Quantum limited heterodyne detection of millimeter waves using super conducting tantalum tunnel junctions," *Appl. Phys. Lett.*, 57, 2487–9, 1990.
192. V.P. Koshelets, S.V. Shitov, L.V. Filippenko, P.N. Dmitriev, A.N. Ermakov, A.S. Sobolev, and M.Yu. Torgashin, "Integrated superconducting sub-mm wave receivers," *Radiophys. Quantum Electron.*, 46, 618–30, 2003.
193. A. Karpov, D. Miller, F. Rice, J.A. Stern, B. Bumble, H.G. LeDuc, and J. Zmuidzinas, "Low noise SIS mixer for far infrared radio astronomy," *Proc. SPIE*, 5498, 616–21, 2004.
194. G. Chattopadhyay, "Future of Heterodyne Receivers at Submillimeter Wavelengths," in *Digest IRMMW-THz-2005 Conference*, pp. 461–2, 2005.
195. G.N. Gol'tsman, "Hot electron bolometric mixers: New terahertz technology," *Infrared Phys. Technol.*, 40, 199–206, 1999.
196. R. Blundell and K.H. Gundlach, "A quasioptical SIN mixer for 230 GHz frequency range," *Int. J. Infrared Millim. Waves*, 8, 1573–9, 1987.
197. M. Nahum, P.L. Richards, and C.A. Mears, "Design analysis of a novel hot-electron microbolometer," *IEEE Trans. Appl. Supercond.*, 3, 2124–7, 1993.
198. M. Nahum and J. Martinis, "Ultrasensitive hot-electron microbolometer," *Appl. Phys. Lett.*, 63, 3075–7, 1993.
199. D. Chouvaev, D. Sandgren, M. Tarasov, and L. Kuzmin, "Optical Qualification of the Normal Metal Hot-Electron Microbolometer (NHEB)," in *12th International Symposium Space THz Technology*, pp. 446–56, San Diego, 2001.
200. D. Sandgren, D. Chouvaev, M. Tarasov, and L. Kuzmin, "Fabrication and optical characterization of the normal metal hot-electron microbolometer with Andreev mirrors," *Physica C*, 372, 444–7, 2002.
201. D. Golubev and L. Kuzmin, "Nonequilibrium theory of a hot-electron bolometer with normal metal-insulator-superconductor tunnel junction," *J. Appl. Phys.*, 89, 6464–72, 2001.
202. D.R. Schmidt, K.W. Lehnert, A.M. Clark, W.D. Duncan, K.D. Irwin, N. Miller, and J.N. Ullom, "A superconductor-insulator-normal metal bolometer with microwave readout suitable for large-format arrays," *Appl. Phys. Lett.*, 86, 053505, 2005.
203. P.K. Day, H.G. Leduc, B.A. Mazin, A. Vayonakis, and J. Zmuidzinas, "A broadband superconducting detector suitable for use in large arrays," *Nature*, 425, 817–21, 2003.
204. B.A. Mazin, B. Bumble, S.R. Meeker, K. O'Brien, S. McHugh, and E. Langman, "A superconducting focal plane array for ultraviolet, optical, and near-infrared astrophysics," *Opt. Express*, 20, 1503–11, 2012.
205. B.A. Mazin, "Microwave Kinetic Inductance Detectors," *Phd Thesis*, California Institute of Technology, Pasadena, California, 2005.
206. P.R. Maloney, N.G. Czakon, P.K. Day, T.P. Downes, R. Duan, J. Gao, J. Glenn, S.R. Golwala, M.I. Hollister, H.G. LeDuc, B.A. Mazin, S.G. McHugh, O. Noroozian, H.T. Nguyen, J. Sayers, J.A. Schlaerth, S. Siegel, J.E. Vaillancourt, A. Vayonakis, P. Wilson, and J. Zmuidzinas, "MUSIC for sub/millimeter astrophysics," *Proc. SPIE*, 7741, 77410F–1–11, 2010.
207. S. Heyminck, B. Klein, R. Güsten, C. Kasemann, A. Baryshev, J. Baselmans, S. Yates, T.M. Klapwijk, "Development of a MKID Camera for APEX," in *Twenty-First International Symposium on Space Terahertz Technology*, March 23–25, Oxford, UK, 262, 2010.
208. A. Monfardini, A. Benoit, A. Bideaud, L. Swenson, A. Cruciani, P. Camus, C. Hoffmann, F.X. Désert, S. Doyle, P. Ade, P. Mauskopf, C. Tucker, M. Roesch, S. Leclercq, K.F. Schuster, A. Endo, A. Baryshev, J.J.A. Baselmans, L. Ferrari, S.J.C Yates, O. Bourrion, J. Macias-Perez, C. Vescovi, M. Calvo, and C. Giordano, "A dual-band millimeter-wave kinetic inductance camera for the IRAM 30 m telescope," *Astrophys. J. Suppl. Ser.*, 194, 24, 11, 2011.

209. M. Griffin, J. Baselmans, A. Baryshev, S. Doyle, M. Grim, P. Hargrave, T. Klapwijk, J. Martin-Pintado, A. Monfardini, A. Neto, H. Steenbeck, I. Walker, K. Wood, A. D'Addabbo, P. Barry, A. Bideaud, B. Blazquez, J. Bueno, M. Calvo, J-L Costa-Kramer, L. Ferrari, A. Gomez-Gutierrez, J. Goupy, N. Llombart, and S. Yates, "SPACEKIDS: Kinetic inductance detectors for space applications," *Proc. SPIE*, 9914, 991407–1–11, 2016.
210. A.L. Woodcraft, R.V. Sudiwal, E. Wakui, and C. Paine, "Hopping conduction in NTD germanium: Comparison between measurement and theory," *J. Low Temp. Phys.*, 134, 925–44, 2004.
211. Herschel Space Observatory. http://herschel.jpl.nasa.gov/spireInstrument.shtml.
212. P. Agnese, C. Buzzi, P. Rey, L. Rodriguez, and J.L. Tissot, "New technological development for far-infrared bolometer arrays," *Proc. SPIE*, 3698, 284–90, 1999.
213. C. Dowell, C.A. Allen, S. Babu, M.M. Freund, M.B. Gardner, J. Groseth, M. Jhabvala, A. Kovacs, D.C. Lis, S.H. Moseley, T.G. Phillips, R. Silverberg, G. Voellmer, and H. Yoshida, "SHARC II: A Caltech Submillimeter Observatory facility camera with 384 pixels," *Proc. SPIE*, 4855, 73–87, 2003.
214. N. Billot, P. Agnese, J.L. Augueres, A. Beguin, A. Bouere, O. Boulade, C. Cara, C. Cloue, E. Doumayrou, L. Duband, B. Horeau, I. Le Mer, J.L. Pennec, J. Martignac, K. Okumura, V. Reveret, M. Sauvage, F. Simoens, and L. Vigroux, "The Herschel/PACS 2560 bolometers imaging camera," *Proc. SPIE*, 6265, 62650D, 2006.
215. G.H. Rieke, "Infrared detector arrays for astronomy," *Annu. Rev. Astrophys.*, 45, 77–115, 2007.
216. E.M. Conwell, "High Field Transport in Semiconductors," *Solid State Physics*, Supplement 9, Academic Press, New York, 1967.
217. T.G. Phillips and K.B. Jefferts, "A low temperature bolometer heterodyne receiver for millimeter wave astronomy," *Rev. Sci. Instrum.*, 44, 1009–14, 1973.
218. E.H. Putley, "InSb Submilimeter Photoconductive Detectors," in *Semiconductors and Semimetals*, Vol. 12, eds. R.K. Willardson and A.C. Beer, 143–67, Academic Press, New York, 1977.
219. www.infraredlaboratories.com/InSb_Hot_e_Bolometers.html.
220. P.R. Norton, "Photodetectors," in *Handbook of Optics*, third edition, Vol. 1, Chapter 24, pp. 24.3–24.102, ed. M. Bass, McGraw Hill, New York, 2010.
221. K.S. Yngvesson, J.-X. Yang, F. Agahi, D. Dai, C. Musante, W. Grammer, and K.M. Lau, "AlGaAs/GaAs Quasi-Bulk Effect Mixers: Analysis and Experiments,"in *Third International Symposium Space THz Technology*, pp. 688–705, 1992.
222. Yu.B. Vasilyev, A.A. Usikova, N.D. Il'inskaya, P.V. Petrov, and Yu.L. Ivanov, "Highly sensitive submillimeter InSb photodetectors," *Semiconductors*, 42, 1234–6, 2008.
223. H. Moseley and D. McCammon, "High Performance Silicon Hot Electron Bolometers," in *Ninth International Workshop on Low Temperature Detectors, AIP Proceedings*, 605, pp. 103–6, 2002.
224. K. Seeger, *Semiconductor Physics*, Springer, Berlin, 1991.
225. S.M. Smith, M.J. Cronin, R.J. Nicholas, M.A. Brummell, J.J. Harris, and C.T. Foxon, "Millimeter and submillimeter detection using $Ga_{1-x}Al_xAs$/GaAs heterosructures," *Int. J. Infrared Millim. Waves*, 8, 793–802, 1987.
226. J.-X. Yang, F. Agahi, D. Dai, C.F. Musante, W. Grammer, K.M. Lau, and K.S. Yngvesson, "Wide-bandwidth electron bolometric mixers: A 2DEG prototype and potential for low-noise THz receivers," *IEEE Trans. Microw. Theory Tech.*, 41, 581–9, 1993.
227. G.N. Gol'tsman and K.V. Smirnov, "Electron-phonon interaction in a two-dimensional electron gas of semiconductor heterostructures at low temperatures," *JETP Lett.*, 74, 474–9, 2001.
228. A.A. Verevkin, N.G. Ptitsina, K.V. Smirnov, G.N. Gol'tsman, E.M. Gershenzon, and K.S. Ingvesson, "Direct measurements of energy relaxation times on an AlGaAs/GaAs heterointerface in the range 4.2–50 K," *JETP Lett.*, 64, 404–9, 1996.
229. T. Phillips and D. Woody, "Millimeter-wave and submillimeter-wave receivers," *Annu. Rev. Astron. Astrophys.*, 20, 285–321, 1982.
230. M. Talvard, P. André, L. Rodriguez, V. Minier, A. Benoit, B. Leriche, F. Pajot, L. Vigroux, P. Agnèse, O. Boulade, E. Doumayrou, D. Dubreuil, G. Durand, P. Gallais, B. Horeau, P. Lagage, Y. Le-Pennec, M. Lortholary, J. Martignac, V. Revéret, N. Schneider, J. Stutzki, C. Veyssière, C. Walter, "ArTeMiS: Filled bolometer arrays for next generation submm telescopes," *Proc. SPIE*, 6275, 627503–1–12, 2006.
231. V. Revéret, P. André, J. Le Pennec, M. Talvard, P. Agnčse, A. Arnaud, L. Clerc, C. de Breuck, J.-C. Cigna, C. Delisle, E. Doumayrou, L. Duband, D. Dubreuil, L. Dumaye, E. Ercolani, P. Gallais, E. Groult, T. Jourdanc, B. Leriche, B. Maffei, M. Lortholary, J. Martignac, W. Rabaudb, J. Relland, L. Rodriguez, A. Vandeneynde, F. Visticot, "The ArTéMiS wide-field submillimeter camera: Preliminary on-sky performances at 350 microns," *Proc. SPIE*, 9153, 915305–1–11, 2014.
232. V. Revéret, L. Rodriguez, and P. Agnèse, "Enhancing the spectral response of filled bolometer arrays for submillimeter astronomy," *Appl. Opt.*, 49(35), 6726–36, 2010.

233. V. Revéret and L. Rodriguez, "Progress on Silicon Bolometers for (sub)-Millimeter Astronomy: From ArTéMiS to Future B-Mode Detection Space Missions," in *26th International Symposium on Space Terahertz Technology*, T36, Cambridge, UK, March 16–18, 2015.
234. A.D. Semenov, G.N. Gol'tsman, and R. Sobolewski, "Hot-electron effect in semiconductors and its applications for radiation sensors," *Semicond. Sci. Technol.*, 15, R1–16, 2002.
235. E.M. Gershenson, M.E. Gershenson, G.N. Goltsman, B.S. Karasik, A.M. Lyul'kin, and A.D. Semenov, "Ultrafast superconducting electron bolometer," *J. Tech. Phys. Lett.*, 15, 118–9, 1989.
236. K.S. Il'in, M. Lindgren, M. Currie, A.D. Semenov, G.N. Gol'tsman, R. Sobolewski, S.I. Cherednichenko, and E.M. Gershenzon, "Picosecond hot-electron energy relaxation in NbN superconducting photodetectors," *Appl. Phys. Lett.*, 76, 2752–4, 2000.
237. Y. Gousev, G. Gol'tsman, A. Semenov, E. Gershenzon, R. Nebosis, M. Heusinger, and K. Renk, "Broad-band ultrafast superconducting NbN detector for electromagnetic-radiation," *J. Appl. Phys.*, 75, 3695–7, 1994.
238. H. Maezawa, "Application of superconducting hot-electron bolometer mixers for terahertz-band astronomy," *IEICE Trans. Electron.*, E98-C, 196–206, 2015.
239. A. Shurakov, Y. Lobanov, and G. Goltsman, "Superconducting hot-electron bolometer: From the discovery of hot-electron phenomena to practical applications," *Supercond. Sci. Technol.*, 29, 023001–1–27, 2016.
240. J.J.A. Baselmans, A. Baryshev, S.F. Reker, M. Hajenius, J. Gao, T. Klapwijk, B. Voronov, and G. Gol'tsman, "Influence of the direct response on the heterodyne sensitivity of hot electron bolometer mixers," *J. Appl. Phys.*, 100, 184103, 2006.
241. A. Balanis, *Antenna Theory: Analysis and Design*, Third edition, Wiley & Sons, New York, 2005.
242. J. Volakis, *Antenna Engineering Handbook*, Fourth edition, McGraw-Hill, New York, 2007.
243. G.N. Gol'tsman, Yu.B. Vachtomin, S.V. Antipov, M.I. Finkel, S.N. Maslennikov, K.V. Smirnov, S.L. Poluakov, S.I. Svechnikov, N.S. Kaurova, E.V. Grishina, and B.M. Voronov, "NbN phonon-cooled hot-electron bolometer mixer for terahertz heterodyne receivers," *Proc. SPIE*, 5727, 95–106, 2005.
244. B.S. Karasik, D.P. Cunnane, and A.V. Sergeev, "High-TC THz HEB Mixers: Progress and Prospects,"in *40th International Conference on Infrared, Millimeter, and Terahertz Waves (IRMMW-THz)*, August 23–28, 2015.
245. S. Cherednichenko, V. Drakinskiy, K. Ueda, and M. Naito, "Terahertz mixing in MgB_2 microbolometers," *App. Phys. Lett.*, 90, 023507–1–3, 2007.
246. D.P. Cunnane, J.H. Kawamura, M.A. Wolak, N. Acharya, T. Tan, X.X. Xi, and B.S. Karasik "Characterization of MgB_2 superconducting hot electron bolometers," *IEEE Trans. Appl. Supercond.*, 25(3), 2300206, 2015.
247. M.A. Wolak, N. Acharya, T. Tan, D. Cunnane, B.S. Karasik, and X. Xi "Fabrication and characterization of ultrathin MgB_2 films for hot electron bolometer applications," *IEEE Trans. Appl. Supercond.*, 25(3), 7500905, 2015.
248. A.J. Kreisler and A. Gaugue, "Recent progress in HTSC bolometric detectors at terahertz frequencies," *Proc. SPIE*, 3481, 457–68, 1998.
249. A.J. Kreisler and A. Gaugue, "Recent progress in high-temperature superconductor bolometric detectors: From the mid-infrared to the far-infrared (THz) range," *Supercond. Sci. Technol.*, 13, 1235–45, 2000.
250. O. Harnack, B. Karasik, W. McGrath, A. Kleinsasser, and J. Barner, "Submicron-long HTS hot-electron mixers," *Supercond. Sci. Technol.*, 12, 850–2, 1999.
251. B. Karasik, W. McGrath, and M. Gaidis, "Analysis of a high-T_c hot-electron mixer for terahertz applications," *J. Appl. Phys.*, 81, 1581–9, 1997.
252. F. Ronnung, S. Cherednichenko, G. Gol'tsman, E. Gershenzon, and D. Winkler, "A nanoscale YBCO mixer optically coupled with a bow tie antenna," *Supercond. Sci. Technol.*, 12, 853–5, 1999.
253. M. Lindgren, M. Currie, C. Williams, T.Y. Hsiang, P.M. Fauchet, R. Sobolewsky, S.H. Moffat, R.A. Hughes, J.S. Preston, and F.A. Hegmann, "Intrinsic picosecond response times of Y-Ba-Cu-O superconducting photoresponse," *Appl. Phys. Lett.*, 74, 853–5, 1999.
254. V.V. Shirotov and Yu.Ya. Divin, "Frequency-selective Josephson detector: Power dynamic range at subterahertz frequencies," *Tech. Phys. Lett.*, 30, 522–4, 2004.
255. M.V. Lyatti, D.A. Tkachev, and Yu.Ya. Divin, "Signal and noise characteristics of a terahertz frequency-selective $YBa_2Cu_3O_{7-\delta}$ Josephson detector," *Tech. Phys. Lett.*, 32, 860–2, 2006.
256. D.J. Benford and S.H. Moseley, "Superconducting Transition Edge Sensor Bolometer Arrays for Submillimeter Astronomy," in *Proceedings of the International Sympsium on Space and THz Technology*. www.eecs.umich.edu/~jeast/benford_2000_4_1.pdf.
257. D. Olaya, J. Wei, S. Pereverzev, B.S. Karasik, J.H. Kawamura, W.R. McGrath, A.V. Sergeev, and M.E. Gershenson, "An untrasensitive hot-electron bolometer for low-background SMM applications," *Proc. SPIE*, 6275, 627506, 2006.

258. K. Irwin, "An application of electrothermal feedback for high-resolution cryogenic particle-detection," *Appl. Phys. Lett.*, 66, 1998–2000, 1995.
259. K. Irwin, G. Hilton, D. Wollman, and J. Martinis, "X-ray detection using a superconducting transition-edge sensor microcalorimeter with electrothermal feedback," *Appl. Phys. Lett.*, 69, 1945–7, 1996.
260. A.T. Lee. P.L. Richards, S.W. Nam, B. Cabrera, and K.D. Irwin, "A superconducting bolometer with strong electrothermal feedback," *Appl. Phys. Lett.*, 69, 1801–3, 1996.
261. B. Cabrera, R. Clarke, P. Colling, A. Miller, S. Nam, and R. Romani, "Detection of single infrared, optical, and ultraviolet photons using superconducting transition edge sensors," *Appl. Phys. Lett.*, 73, 735–7, 1998.
262. G.C. Hilton, J.M. Martinis, K.D. Irwin, N.F. Bergren, D.A. Wollman, M.E. Huber, S. Deiker, and S.W. Nam, "Microfabricated transition-edge X-ray detectors," *IEEE Trans. Appl. Supercond.*, 11, 739–42, 2001.
263. W. Duncan, W.S. Holland, M.D. Audley, M. Cliffe, T. Hodson, B.D. Kelly, X. Gao, D.C. Gostick, M. MacIntosh, H. McGregor, T. Peacocke, K.D. Irwin, G.C. Hilton, S.W. Deiker, J. Beier, C.D. Reintsema, A.J. Walton, W. Parkes, T. Stevenson, A.M. Gundlach, C. Dunare, and P.A.R. Ade, "SCUBA-2: Developing the detectors," *Proc. SPIE*, 4855, 19–29, 2003.
264. A.J. Walton, W. Parkes, J.G. Terry, C. Dunare, J.T.M. Stevenson, A.M. Gundlach, G.C. Hilton, K.D. Irwin, J.N. Ullom, W.S. Holland, W. Duncan, M.D. Audley, P.A.R. Ade, R.V. Sudiwala, and E. Schulte, "Design and fabrication of the detector technology for SCUBA-2," *IEE Proc. Sci. Meas. Technol.*, 151, 110–20, 2004.
265. A.-D. Brown, D. Chuss, V. Mikula, R. Henry, E. Wollack, Y. Zhao, G.C. Hilton, and J.A. Chervenak, "Auxiliary components for kilopixel transition edge sensor arrays," *Solid State Electron.*, 52, 1619–24, 2008.
266. S. Lee, J. Gildemeister, W. Holmes, A. Lee, and P. Richards, "Voltage-biased superconducting transition-edge bolometer with strong electrothermal feedback operated at 370 mK," *Appl. Opt.*, 37, 3391–7, 1998.
267. H.F.C. Hoevers, A.C. Bento, M.P. Bruijn, L. Gottardi, M.A.N. Korevaar, W.A. Mels, and P.A.J. de Korte, "Thermal fluctuation noise in a voltage biased superconducting transition edge thermometer," *Appl. Phys. Lett.*, 77, 4422–4, 2000.
268. M.D. Audley, D.M. Glowacka, D.J. Goldie, A.N. Lasenby, V.N. Tsaneva, S. Withington, P.K. Grimes, C.E. North, G. Yassin, L. Piccirillo, G. Pisano, P.A.R. Ade, G. Teleberg, K.D. Irwin, W.D. Duncan, C.D. Reintsema, M. Halpern, and E.S. Battistellik, "Tests of Finline-Coupled TES Bolometers for COVER," in *Digest IRMMW-THz-2007 Conference*, pp. 180, 181, Cardiff, Wales, 2007.
269. J.A. Chervenak, K.D. Irwin, E.N. Grossman, J.M. Martinis, C.D. Reintsema, M.E. Huber, "Superconducting multiplexer for arrays of transition edge sensors," *Appl. Phys. Lett.*, 74, 4043–5, 1999.
270. P.J. Yoon, J. Clarke, J.M. Gildemeister, A.T. Lee, M.J. Myers, P.L. Richards, J.T. Skidmore, "Single superconducting quantum interference device multiplexer for arrays of low-temperature sensors," *Appl. Phys. Lett.*, 78, 371–3, 2001.
271. *The SQUID Handbook*, Vol. II: Applications, eds. J. Clarke and A.I. Braginski, Wiley-VCH, Weinheim, Germany, 2006.
272. K.D. Irvin, "SQUID multiplexers for transition-edge sensors," *Physica C*, 368, 203–10, 2002.
273. K.D. Irwin, M.D. Audley, J.A. Beall, J. Beyer, S. Deiker, W. Doriese, W.D. Duncan, G.C. Hilton, W.S. Holland, C.D. Reintsema, J.N. Ullom, L.R. Vale, and Y. Xu, "In-focal-plane SQUID multiplexer," *Nucl. Instrum. Methods Phys. Res.*, A520, 544–7, 2004.
274. K.D. Irvin and G.C. Hilton, "Transition-edge Sensors," in *Cryogenic Particle Detection*, ed. C. Enss, 63–149, Springer-Verlag, Berlin, 2005.
275. T.M. Lanting, H.M. Cho, J. Clarke, W.L. Holzapfel, A.T. Lee, M. Lueker, P.L. Richards, M.A. Dobbs, H. Spieler, and A. Smith, "Frequency-domain multiplexed readout of transition-edge sensor arrays with a superconducting quantum interference device," *Appl. Phys. Lett.*, 86, 112511, 2005.
276. R.L. Fagaly, "Superconducting quantum interference device instruments and applications," *Rev. Sci. Instrum.*, 77, 101101–45, 2006.
277. W.S. Holland, W. Duncan, B.D. Kelly, K.D. Irwin, A.J. Walton, P.A.R. Ade, and E.I. Robson, "SCUBA-2: A new generation submillimeter imager for the James Clerk Maxwell Telescope," *Proc. SPIE*, 4855, 1–18, 2003.
278. W.S. Holland, D. Bintley, E.L. Chapin, A. Chrysostomou, G.R. Davis, J.T. Dempsey, W.D. Duncan, M. Fich, P. Friberg, M. Halpern, K.D. Irwin, T. Jenness, B.D. Kelly, M.J. MacIntosh, E.I. Robson, D. Scott, P.A.R. Ade, E. Atad-Ettedgui, D.S. Berry, S.C. Craig, X. Gao, A.G. Gibb, G.C. Hilton, M.I. Hollister, J.B. Kycia, D.W. Lunney, H. McGregor, D. Montgomery, W. Parkes, R.P.J. Tilanus, J.N. Ullom, C.A. Walther, A.J. Walton, A.L. Woodcraft, M. Amiri, D. Atkinson, B. Burger, T. Chuter, I.M. Coulson, W.B. Doriese, C. Dunare, F. Economou, M.D. Niemack, H.A.L. Parsons, C.D. Reintsema, B. Sibthorpe, I. Smail, R. Sudiwala, and H.S. Thomas, "SCUBA-2: The 10 000 pixel bolometer camera on the James Clerk Maxwell Telescope," *MNRAS*, 430, 2513–33, 2013.

279. A.L. Woodcraft, M.I. Hollister, D. Bintley, M.A. Ellis, X. Gao, W.S. Holland, M.J. MacIntosh, P.A.R. Ade, J.S. House, C.L. Hunt, and R.V. Sudiwala, "Characterization of a prototype SCUBA-2 1280-pixel submillimetre superconducting bolometer array," *Proc. SPIE*, 6275, 62751F, 2006.
280. SCUBA-2. www.roe.ac.uk/ukatc/projects/scubatwo/.
281. G.J. White, R.P. Nelson, W.S. Holland, E.I. Robson, J.S. Greaves, M.J. McCaughrean, G.L. Pilbratt, D.S. Balser, T. Oka, S. Sakamoto, T. Hasegawa, W.H. McCutcheon, H.E. Matthews, C.V.M. Fridlund, N.F.H. Tothill, M. Huldtgren, and J.R. Deane, "The Eagle Nebula's fingers—pointers to the earliest stages of star formation?" *Astron. Astrophys.*, 342, 233–56, 1999.
282. D.J. Benford, J.G. Steguhn, T.J. Ames, C.A. Allen, J.A. Chervenak, C.R. Kennedy, S. Lefranc, S.F. Maher, S.H. Moseley, F. Pajot, C. Rioux, R.A. Shafer, and G.M. Voellmer, "First astronomical images with a multiplexed superconducting bolometer array," *Proc. SPIE*, 6275, 62751C, 2006.
283. J. Gildemeister, A. Lee, and P. Richards, "Monolithic arrays of absorber-coupled voltage- biased superconducting bolometers," *Appl. Phys. Lett.*, 77, 4040–2, 2000.
284. D.J. Benford, G.M. Voellmer, J.A. Chervenak, K.D. Irwin, S.H. Moseley, R.A. Shafer, G.J. Stacey, and J.G. Staguhn, "Thousand-Element Multiplexed Superconducting Bolometer Arrays," in *Proceedings of Far-IR, Sub-MM, and MM Detector Workshop*, vol. NASA/CP-2003–211 408, pp. 272–5, eds. J. Wolf, J. Farhoomand, and C.R. McCreight, 2003.
285. J. Gildemeister, A. Lee, and P. Richards, "A fully lithographed voltage-biased superconducting spiderweb bolometer," *Appl. Phys. Lett.*, 74, 868–70, 1999.
286. P.J. Burke, "Carbon nanotube devices for GHz to THz applications," *Proc. SPIE*, 5593, 52–61, 2004.
287. C.M. Sze. *Physics of Semiconductor Devices*, Wiley, New York, 1981.
288. S. Reich, C. Thomsen, and J. Maultzsch, *Carbon Nanotubes: Basic Concepts and Physical Properties*, Wiley, Berlin, 2004.
289. Y. Kawano, T. Fuse, S. Toyokawa, T. Uchida, and K. Ishibashi, "Terahertz photon-assisted tunneling in carbon nanotube quantum dots," *J. Appl. Phys.*, 103, 034307, 2008.
290. Y. Kawano, T. Uchida, and K. Ishibashi, "Terahertz sensing with a carbon nanotube/two-dimensional electron gas hybrid transistor," *Appl. Phys. Lett.*, 95, 083123–1–3, 2009.
291. K.S. Yngvesson, K. Fu, B. Fu, R. Zannoni, J. Nicholson, S.H. Adams, A. Ouarraoui, J. Donovan and E. Polizzi, "Experimental Detection of Terahertz Radiation in Bundles of Single Wall Carbon Nanotubes," in *19th International Symposium Space THz Technology*, 304–13, Groningen, The Netherlands, 2008.
292. Y. Wang, K. Kempa, B. Kimball, J.B. Carlson, G. Benham, W.Z. Li T. Kempa, J. Rybczynski, A. Herczynski, and Z.F. Ren, "Receiving and transmitting light-like radio waves: Antenna effect in arrays of aligned carbon nanotubes," *Appl. Phys. Lett.*, 85, 2607–9, 2004.
293. Y. Wang, Q. Wu, X. He, X. Sun, and T. Gui, "Radiation properties of carbon nanotubes antenna at terahertz/infrared range," *Int. J. Infrared Millim. Waves*, 29, 35–42, 2008.
294. Z. Wang, K. Ishibashi, S. Komiyama, M. Patrashin, and I. Hosako, "Charge-sensitive infrared phototransistors with integrated plasmonic photocouplers," *J. Phys. D*, 46, 165107 (6pp), 2013.
295. O. Astavief, S. Komiyama, T. Kutsuwa, V. Antonov, Y. Kawaguchi, and K. Hirakawa, "Single-photon detector in the microwave range," *Appl. Phys. Lett.* 80, 4250–2, 2002.
296. H. Hashiba, V. Antonov, L. Kulik, A. Tzalenchuk, P. Kleindschmid, S. Giblin, and S. Komiyama, "Isolated quantum dot in application to terahertz photon counting," *Phys. Rev. B*, 73, 081310:1–4, 2006.
297. X.H. Su, J. Yang, P. Bhattacharya, G. Ariyawansa, and A.G.U. Perera, "Terahertz detection with tunneling quantum dot intersublevel photodetector," *Appl. Phys. Lett.*, 89, 031117–1–3, 2006.
298. T. Ueda and S. Komiyama, "Novel ultra-sensitive detectors in the 10–50 μm wavelength range," *Sensors*, 10, 8411–23, 2010.
299. T. Ueda, Z. An, and S. Komiyama, "Temperature dependence of novel single-photon detectors in the long-wavelength infrared range," *J. Infrared Millim. Terahertz Waves*, 2010. doi:10.1007/s10762-010-9659–3.
300. D. Seliuta, I. Kašalynas, V. Tamošiūnas, S. Balakauskas, Z. Martūnas, S. Ašmontas, G. Valušis, A. Lisauskas, H.G. Roskos, and K. Köhler, "Silicon lens-coupled bow-tie InGaAs-based broadband terahertz sensor operating at room temperature," *Electron. Lett.*, 44, 825–7, 2006.
301. G. Valušis, D. Seliuta, V. Tamošiūnas, R. Simniškis, S. Balakauskas, and I. Kašalynas, "Selective and Broadband Terahertz Sensors Based on GaAs Nanostructures," in *Workshop THz Wave Technology*, Bucharest, 19–20 May, 2008.
302. J.-H. Dai, J.-H. Lee, Y.-L. Lin, and S.-C. Lee, "In(Ga)As quantum rings for terahertz detectors," *J. Appl. Phys.*, 47, 2924–6, 2008.
303. S. Bhowmick, G. Huang, W. Guo, C.S. Lee, P. Bhattacharya, G. Ariyawansa, and A.G.U. Perera, "High-performance quantum ring detector for the 1–3 terahertz range," *Appl. Phys. Lett.*, 96, 231103–1–3, 2010.

304. K.S. Novoselov, A.K. Geim, S.V. Morozov, D. Jiang, Y. Zhang, S.V. Dubonos, I.V. Grigorieva, and A.A. Firsov, "Electric field effect in atomically thin carbon films," *Science*, 306, 666–9, 2004.
305. K.S. Novoselov, A.K. Geim, S.V. Morozov, D. Jiang, M.I. Katsnelson, I.V. Grigorieva, S.V. Dubonos, and A.A. Firsov, "Two-dimensional gas of massless Dirac fermions in graphene," *Nature*, 438, 197–200, 2005.
306. A.K. Geim and K.S. Novoselov, "The rise of graphene," *Nat. Mater.*, 6, 183–91, 2007.
307. F. Xia, H. Yan, and P. Avouris, "The interaction of light and graphene: Basic, devices, and applications," *Proc. IEEE*, 101(7), 1717–31, 2013.
308. J.-H. Chen, C. Jang, S. Xiao, M. Ishigami, and M.S. Fuhrer, "Intrinsic and extrinsic performance limits of graphene devices on SiO_2," *Nat. Nanotechnol.*, 3, 206–9, 2008.
309. R.R. Nair, P. Blake, A.N. Grigorenko, K.S. Novoselov, T.J. Booth, T. Stauber, N.M.R. Peres, and A.K. Geim, "Fine structure constant defines visual transparency of graphene," *Science*, 320, 1308, 2008.
310. T. Low and P. Avouris, "Graphene plasmonic for terhertz to mid-infrared applications," *ACS Nano*, 8(2), 1086–101, 2014.
311. F.H.L. Koppens, T. Mueller, Ph. Avouris, A.C. Ferrari, M.S. Vitiello, and M. Polini, "Photodetectors based on graphene, other two-dimensional materials and hybrid systems," *Nat. Nanotechnol.*, 9, 780–93, 2014.
312. V. Ryzhii, M. Ryzhii, V. Mitin, and T. Otsuji, "Terahertz and infrared photodetection using p-i-n multiple-graphene-layer structures," *J. Appl. Phys.*, 107, 054512–1–7, 2010.
313. V. Ryzhii, N. Ryabova, M. Ryzhii, N.V. Baryshnikov, V.R. Karasik, V. Mitin, and T. Otsuji, "Terahertz and infrared photodetectors based on multiple graphene layer and nanoribbon structures," *Opto-Electron. Rev.*, 20(1), 15–25, 2012.
314. V. Ryzhii, T. Otsuji, M. Ryzhii, and M.S. Shur, "Double graphene-layer plasma resonances terahertz detector," *J. Phys. D Appl. Phys.*, 45, 302001 (6pp), 2012.
315. X. Cai, A.B. Sushkov, R.J. Suess, M.M. Jadidi, G.S. Jenkins, L.O. Nyakiti, R.L. Myers-Ward, J. Yan, D.K. Gaskill, T.E. Murphy, H.D. Drewl, M.S. Fuhrer, "Sensitive room-temperature terahertz detection via the photothermoelectric effect in graphene," *Nat. Nanotechnol.*, 9(10), 814–20, 2013.
316. L. Vicarelli, M.S. Vitiello, D. Coquillat, A. Lombardo, A.C. Ferrari, W. Knap, M. Polini, V. Pellegrini, and A. Tredicucci, "Graphene effect as room-temperature terahertz detectors," *Nat. Mater.*, 11, 865–71, 2012.
317. D. Spirito, D. Coquillat, S.L. De Bonis, A. Lombardo, M. Bruna, A.C. Ferrari, V. Pellegrini, A. Tredicucci, W. Knap, and M.S. Vitiello, "High performance bilayer-graphene terahertz detectors," *Appl. Phys. Lett.*, 104, 061111–1–5, 2014.
318. F. Bianco, D. Perenzoni, D. Convertino, S.L. De Bonis, D. Spirito, M. Perenzoni, C. Coletti, M.S. Vitiello, and A. Tredicucci, "Terahertz detection by epitaxial-graphene field-effect-transistors on silicon carbide," *Appl. Phys. Lett.*, 107, 131104–1–5, 2015.
319. A. Zak, M.A. Andersson, M. Bauer, J. Matukas, A. Lisauskas, H.G. Roskos, and J. Stake, "Antenna-integrated 0.6 THz FET direct detectors based on CVD graphene," *Nano Lett.*, 14, 5834, 2014.
320. X. Du, D.E. Prober, H. Vora, and Ch.B. Mckitterick, "Graphene-based bolometers," *Graphene 2D Mater.*, 1, 1–22, 2014.
321. J. Yan, M.-H. Kim, J.A. Elle, A.B. Sushkov, G.S. Jenkins, H.M. Milchberg, M.S. Fuhrer, and H.D. Drew, "Dual-gated bilayer graphene hot electron bolometer," *Nat. Nanotechnol.*, 7(7), 472–8, 2012.
322. G. Konstantatos, M. Badioli, L. Gaudreau, J. Osmond, M. Bernechea, F.P.G. de Arquer, F. Gatti, and F.H.L. Koppens, "Hybrid graphene–quantum dot phototransistors with ultrahigh gain," *Nat. Nanotechnol.*, 7, 363–8, 2012.
323. Z. Sun, Z. Liu, J. Li, G.-an Tai, S.-P. Lau, and F. Yan, "Infrared photodetectors based on CVD-grown graphene and PbS quantum dots with ultrahigh responsivity," *Adv. Mater.*, 24, 5878–83, 2012.
324. W. Guo, S. Xu, Z. Wu, N. Wang, M.M.T. Loy, and S. Du, "Oxygen-assisted charge transfer between ZnO quantum dots and graphene," *Small*, 9, 3031–6, 2013.
325. I. Nikitskiy, S. Goossens, D. Kufer, T. Lasanta, G. Navickaite, F.H. Koppens, G. Konstantatos, "Integrating an electrically active colloidal quantum dot photodiode with a graphene phototransistor," *Nat. Commun.*, 7, 11954, 2016.
326. C.H. Liu, Y.C. Chang, T.B. Norris, and Z.H. Zhong, "Graphene photodetectors with ultra-broadband and high responsivity at room temperature," *Nat. Nanotechnol.*, 9, 273–8, 2014.
327. X. Gan, R.-J. Shiue, G. Yuanda, I. Meric, T.F. Heinz, K. Shepard, J. Hone, S. Assefa, and D. Englund, "Chip-integrated ultrafast graphene photodetector with high responsivity," *Nat. Photonics*, 7, 883–7, 2013.
328. A. Pospischil, M. Humer, M.M. Furchi, D. Bachmann, R. Guider, T. Fromherz, and T. Mueller, "CMOS-compatible graphene photodetector covering all optical communication bands," *Nat. Photonics*, 7, 892–6, 2013.

329. X.M. Wang, Z.Z. Cheng, K. Xu, H.K. Tsang, and J.B. Xu, "High-responsivity graphene/silicon-heterostructure waveguide photodetectors," *Nat. Photonics*, 7, 888–91, 2013.
330. D. Schall, D. Neumaier, M. Mohsin, B. Chmielak, J. Bolten, C. Porschatis, A. Prinzen, C. Matheisen, W. Kuebart, B. Junginger, W. Templ, A.L. Giesecke, and H. Kurz, "50 GBit/s photodetectors based on wafer-scale graphene for integrated silicon photonic communication systems," *ACS Photonics*, 1, 781–4, 2014.
331. T.J. Echtermeyer, L. Britnell, P.K. Jasnos, A. Lombardo, R.V. Gorbachev, A.N. Grigorenko, A.K. Geim, A.C. Ferrari, and K.S. Novoselov, "Strong plasmonic enhancement of photovoltage in graphene," *Nat. Commun.*, 2, 458, 2011.
332. Y. Liu, R. Cheng, L. Liao, H. Zhou, J. Bai, G. Liu, L. Liu, Y. Huang, and X. Duan, "Plasmon resonance enhanced multicolour photodetection by graphene," *Nat. Commun.*, 2, 579, 2011.
333. A.N. Grigorenko, M. Polini, and K.S. Novoselov, "Graphene plasmonics," *Nat. Photonics*, 6, 749–58, 2012.
334. M. Engel, M. Steiner, A. Lombardo, A.C. Ferrari, H.V. Löhneysen, P. Avouris, and R. Krupke, "Light-matter interaction in a microcavity-controlled graphene transistor," *Nat. Commun.*, 3, 906, 2012.
335. X. Gan, K.F. Mak, Y. Gao, Y. You, F. Hatami, J. Hone, T.F. Heinz, and D. Englund, "Strong enhancement of light-matter interaction in graphene coupled to a photonic crystal nanocavity," *Nano Lett.*, 12, 5626–31, 2012.
336. M. Furchi, A. Urich, A. Pospischil, G. Lilley, K. Unterrainer, H. Detz, P. Klang, A.M. Andrews, W. Schrenk, G. Strasser, and T. Mueller, "Microcavity-integrated graphene photodetector," *Nano Lett.*, 12, 2773–7, 2012.
337. M. Freitag, T. Low, W. Zhu, H. Yan, F. Xia, and P. Avouris, "Photocurrent in graphene harnessed by tunable intrinsic plasmons," *Nat. Commun.*, 4, 1951, 2013.
338. Y. Liu, R. Cheng, L. Liao, H. Zhou, J. Bai, G. Liu, L. Liu, Y. Huang, and X. Duan, "Plasmon resonance enhanced multicolour photodetection by graphene," *Nat. Commun.*, 2, 579, 2011.
339. R.-J. Shiue, X. Gan, Y. Gao, L. Li, X. Yao, A. Szep, D. Walker Jr., J. Hone, and D. Eglund, "Enhanced photodetection in graphene-integrated photonic crystal cavity," *Appl. Phys. Lett.*, 103, 241109, 2013.
340. Z. Fang, Z. Liu, Y. Wang, P.M. Ajayan, P. Nordlander, and N.J. Halas, "Graphene–antenna sandwich photodetector," *Nano Lett.*, 12, 3808–13, 2012.
341. Y. Yao, R. Shankar, P. Rauter, Y. Song, J. Kong, M. Loncar, and F. Capasso, "High-responsivity mid-infrared graphene detectors with antenna enhanced photocarrier generation and collection," *Nano Lett.*, 14, 3749–54, 2014.
342. C. Chakraborty, R. Beams, K.M. Goodfellow, G.W. Wicks, L. Novotny, and A.N. Vamivakas, "Optical antenna enhanced graphene photodetector," *Appl. Phys. Lett.*, 105, 241114, 2014.
343. T.J. Echtermeyer, S. Milana, U. Sassi, A. Eiden, M. Wu, E. Lidorikis, and A.C. Ferrari, "Surface plasmon polariton graphene photodetectors," *Nano Lett.*, 16, 8–20, 2015.
344. J. Fang, D. Wang, C.T. DeVault, T.-F. Chung, Chen, A. Baltasseva, V.M. Shalaev, and A.V. Kildishev, "Enhanced graphene photodetector with fractal metasurface," *Nano Lett.*, 17, 57–62, 2016.
345. Z. Chen, X. Li, J. Wang, L. Tao, M. Long, S.-J. Liang, L.K. Ang, C. Shu, H.K. Tsang, and J.-B. Xu, "Synergistic effects of plasmonics and electron trapping in graphene short-wave infrared photodetectors with ultrahigh responsivity," *ACS Nano*, 11, 430–7, 2017.
346. G. Wang, Y. Zhang, C. You, B. Liu, Y. Yang, H. Li, A. Cui, D. Liu, and H. Yan, "Two dimensional materials based photodetectors," *Infrared Phys. Technol.*, 88, 149–73, 2018.
347. X. Li, L. Tao, Z. Chen, H. Fang, X. Li, X. Wang, J.-B. Xu, and H. Zhu, "Graphene and related two-dimensional materials: Structure-property relationships for electronics and optoelectronics," *Appl. Phys. Rev.*, 4, 021306, 2017.
348. M. Buscema, J.O. Island, D.J. Groenendijk, S.I. Blanter, G.A. Steele, H.S.J. van der Zant, and A. Castellanos-Gomez, "Photocurrent generation with two-dimensional van der Waals semiconductor," *Chem. Soc. Rev.*, 44, 3691–718, 2015.
349. C.L. Tan and H. Mohseni, "Emerging technologies for high performance infrared detectors," *Nanophotonics*, 7(1), 169–97, 2018.
350. Q. Cui, Y. Yang, J. Li, F. Teng, and X. Wang, "Material and device architecture engineering toward high performance two-dimensional (2D) photodetectors," *Crystals*, 7, 149, 2017.
351. L. Viti, J. Hu, D. Coquillat, W. Knap, A. Tredicucci, A. Politano, and M.S. Vitiello, "Black phosphorus terahertz photodetectors," *Adv. Mater.*, 27, 5567–72, 2015.
352. L. Viti, A. Politano, and M.S. Vitiello, "Black phosphorus nanodevices at terahertz frequencies: Photodetectors and future challenges," *APL Mater.*, 5, 035602, 2017.
353. M. Long, A. Gao, P. Wang, H. Xia, C. Ott, C. Pan, Y. Fu, E. Liu, X. Chen, W. Lu, T. Nilges, J. Xu, X. Wang, W. Hu, and F. Miao, "Room temperature high-detectivity mid-infrared photodetectors based on black arsenic phosphorus," *Sci. Adv.*, 3, e1700589, 2017.

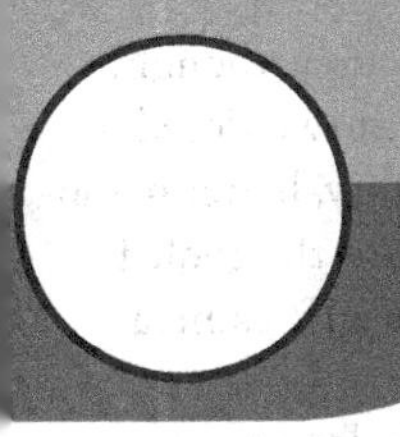

Final Remarks

There are many critical challenges for future civilian and military infrared (IR) and terahertz (THz) detector applications. For many systems, such as night-vision goggles, the IR image is viewed by the human eye, which can discern resolution improvements only up to about 1 megapixel, roughly the same resolution as high-definition television. Most high-volume applications can be completely satisfied with a format of 1280 × 1024. Although wide-area surveillance and astronomy applications could make use of larger formats, funding limits may prevent the exponential growth that was seen in past decades.

The future applications imaging IR systems require the following:

- Higher pixel sensitivity
- Further increase in pixel count to above 10^8 pixels (mosaicing may be used) with pixel size decreasing to about 5 μm for both cooled and uncooled LWIR applications
- Cost reduction in imaging array systems through the use of integration of detectors and signal processing functions (with much more on-chip signal processing) and less cooling sensor technology
- Improvements in the functionality of imaging arrays through development of multispectral sensors

Small-pitch IR FPAs will require the development of larger effective ROIC well capacities per unit area, possibly faster optics than *f*/1, and improved hybridization technologies. Leveraging deeply scaled CMOS process technology enables designers to miniaturize pixel pitch and/or increase on-chip processing capability depending on application-specific needs. Array sizes will continue to increase but perhaps at a rate that falls below the Moore's Law curve. An increase in array size is already technically feasible. However, the market forces that have demanded larger arrays are not as strong now that the megapixel barrier has been broken.

Currently, HgCdTe ternary alloy is the most prevalent material system used in high-performance IR detectors. However, in last two decades, the III–V antimonide-based detector technology is under strong development as a possible alternative to HgCdTe detector material. The apparent rapid success of the T2SL depends mainly on novel ideas coming recently in design of IR photodetectors. The III–Vs offer similar performance to HgCdTe at an equivalent cutoff wavelength but with a sizeable penalty in operating temperature, due to the inherent difference in SRH lifetimes. Important advantage of T2SLs is the high-quality, high-uniformity, and stable nature of the material. In general, III–V semiconductors are more robust than their II–VI counterparts due to stronger, less ionic chemical bonding. As a result, III–V-based FPAs excel in operability, spatial uniformity, temporal stability, scalability, producibility, and affordability—the so-called ibility advantages.

From an economical point of view and a future technology perspective, an important aspect concerns industry organization. The HgCdTe array industry is vertically integrated; there are no commercial wafer suppliers because there are no diverse commercial applications of HgCdTe FPA to be profitable. The wafers are grown within each FPA fabrication facility (or its exclusive partner). One important disadvantage of this integrity is the high cost. In the case of III–V semiconductors, horizontal integration is more profitable. This solution is especially effective for avoiding the heavy investment in capital equipment and subsequent upgrading and maintenance, as well as the cost of highly skilled engineers and technicians.

Although in an early stage of development, the potential to deliver FPAs that can adapt their spectral response to match the sensor requirements in real time presents a compelling case for future multispectral IR imaging systems. Such systems have the potential to deliver a much-improved threat and target recognition capabilities for future defense combat systems.

Until only a few decades ago, IR technology was mainly the domain of military technology, though, it has invaded an increasing number of new applications in our everyday lives. In the recent two decades, thermal detectors have realized many significant technological advances, improved reliability, better manufacturability, and lower costs. Uncooled IR detectors have become an excellent alternative to the cooled detectors and are much more commonly used in many commercial, industrial, and military IR camera products.

Near-future high-performance uncooled thermal imaging will be dominated by VO_x bolometers. However, their sensitivity limitations and the still significant prices will encourage many research teams to explore other IR sensing techniques with the potential for improved performance with reduced detector costs. In the near future both VO_x and amorphous silicon will be challenged by silicon derivatives (a-SiGe, poly-SiGe, and a-$Ge_xSi_{1-x}O_y$). Recent advances in MEMS systems have led to the development of uncooled IR detectors operating as micromechanical thermal detectors.

The THz detectors are actively spreading in different areas of human activity and will receive an increasing importance in a very diverse range of applications including detection of biological and chemical hazardous agents, explosive detection, building and airport security, radio astronomy and space research, biology, and medicine. The future sensitivity improvement of THz instruments will come with the use of large format arrays with readouts in the focal plane to provide the vision demands of high-resolution spectroscopy. Progress in THz detector technology achieves the level that performance many discrete and low pixel arrays operated at low or sub-kelvin temperature are close to ultimate performance at low background in the whole THz range. There exists a large variety of traditional deeply cooled millimeter- and sub-millimeter-wavelength detectors as well as new propositions based on novel optoelectronic quantum devices, for example, carbon nanotube sensors and plasma wave detection by field-effect transistors.

Low-temperature operating heterodyne detectors based on SIS structures operated at low or sub-kelvin temperature are now the most sensitive devices, giving high spectral resolution ($\nu/\Delta\nu \approx 10^6$) and operation close to quantum limit at $\nu < 0.7$ THz. In the frequency range above 1 THz, HEB mixers on the base of superconducting ultrathin NbN layers have the best performance and are promising to be used in large format arrays, due to low LO powers needed for their operation.

Direct detection detectors such as superconducting HEBs have high sensitivity and are fast. Also, direct detection TESs (bolometers) with small volumes are extremely sensitive at sub-kelvin operation conditions and are relatively fast allowing high data rate transfer and counting of THz phonons. Their performance can be close to BLIP regime in the case of very low cold background. They can be much easily assembled into large format arrays compared to heterodyne detectors, as there is no need of LOs.

Serious obstacles in the THz technology are the relatively low-detection sensitivity of uncooled detectors used that operate far away of BLIP conditions and the low imaging resolution. Recently, several types of photon detectors have been created with nanostructured semiconductors, superconductors, carbon nanotube, and graphene-based devices. At the present stage, the applications of graphene-based photodetectors are limited in comparison to traditionally detectors. Despite the large amount of funding and research invested on two-dimensional materials, there is currently very limited set of materials covering the IR and THz regions of the spectrum with performance comparable to detectors existing on global market.

Index

9781032338668